Tetrahydrofolate (transfer of C_1 units)

S-Adenosylmethionine (methyl transfer)

Lipoic acid (acyl transfer)

**Pyridoxal phosphate
(amino acid metabolism)**

**Thiamin diphosphate
(decarboxylation)**

Biotin (carboxylation)

Department of Chemistry and Chemical Biology

John McMurry
Professor of Chemistry and
Chemical Biology

Baker Laboratory
Ithaca, New York 14853-4203

Dear Colleague:

All of us who teach organic chemistry know that most of the students in our courses, even the chemistry majors, are interested primarily in the life sciences rather than in pure chemistry. Because we are teaching so many future biologists, biochemists, and doctors rather than younger versions of ourselves, more and more of us are questioning why we continue to teach the way we do. Why do we spend so much time discussing the details of reactions that are of interest to research chemists but have little connection to biology? Why don't we instead spend more time discussing the organic chemistry of living organisms?

There is still much to be said for teaching organic chemistry in the traditional way, but it is also true that until now there has been no real alternative for those instructors who want to teach somewhat differently. And that is why I wrote **Organic Chemistry with Biological Applications**. As chemical biology continues to gain in prominence, I suspect that more and more faculty will be changing their teaching accordingly.

Make no mistake: this is still a textbook on organic chemistry. But my guiding principle in deciding what to include and what to leave out has been to focus almost exclusively on those reactions that have a direct counterpart in biological chemistry. The space saved by leaving out nonbiological reactions has been put to good use, for almost every reaction discussed is followed by a biological example and approximately 25% of the book is devoted entirely to biomolecules and the organic chemistry of their biotransformations. In addition, **Organic Chemistry with Biological Applications** is nearly 200 pages shorter than standard texts, making it possible for faculty to cover the entire book in a typical two-semester course.

Organic Chemistry with Biological Applications is different from any other text; I believe that it is ideal for today's students.

Sincerely,

John McMurry

All royalties from *Organic Chemistry with Biological Applications* will be donated to the Cystic Fibrosis (CF) Foundation. This book and donation are dedicated to the author's eldest son and to the thousands of others who daily fight this disease. To learn more about CF and the programs and services provided by the CF Foundation, please visit http://www.cff.org.

Organic Chemistry

with Biological Applications 2e

John McMurry
Cornell University

BROOKS/COLE
CENGAGE Learning™

Australia · Brazil · Japan · Korea · Mexico · Singapore · Spain · United Kingdom · United States

BROOKS/COLE
CENGAGE Learning™

Organic Chemistry with Biological Applications 2e
John McMurry

Publisher: Mary Finch

Senior Acquisitions Editor: Lisa Lockwood

Senior Development Editor: Sandra Kiselica

Assistant Editor: Elizabeth Woods

Senior Media Editor: Lisa Weber

Marketing Manager: Amee Mosley

Marketing Assistant: Kevin Carroll

Marketing Communications Manager:
Linda Yip

Project Manager, Editorial Production:
Teresa L. Trego

Creative Director: Rob Hugel

Art Director: John Walker

Print Buyer: Judy Inouye

Permissions Editor: Robert Kauser

Production Service: Graphic World Inc.

Text Designer: Jeanne Calabrese

Photo Researcher: John Hill

Copy Editor: Graphic World Inc.

Illustrator: Graphic World Inc., 2064design

OWL Producers: Stephen Battisti, Cindy Stein, and David Hart in the Center for Educational Software Development at the University of Massachusetts, Amherst, and Cow Town Productions

Cover Designer: Jeanne Calabrese

Cover Image: Neil Fletcher and Matthew Ward/Getty Images

Compositor: Graphic World Inc.

We gratefully acknowledge SDBS for providing data for the following figures: 10.12, 10.14, 10.16, 10.17, 13.9, 13.10, 13.7, 14.15, 18.5; and data for the spectra in Problems 10.31, 10.45, 10.46, 13.72, 15.54, and 16.62 (http://riodb01 .ibase.aist.go.jp/sdbs/, National Institute of Advanced Industrial Science and Technology, 8/26/05, 2/7/09, 2/13/09, 3/10/09).

For product information and technology assistance, contact us at
Cengage Learning Customer & Sales Support, 1-800-354-9706.

For permission to use material from this text or product, submit all requests online at www.cengage.com/permissions. Further permissions questions can be e-mailed to **permissionrequest@cengage.com.**

Library of Congress Control Number: 2009928764

Student Edition:

ISBN-13: 978-0-495-39144-9

ISBN-10: 0-495-39144-1

Brooks/Cole
20 Davis Drive
Belmont, CA 94002-3098
USA

Cengage Learning is a leading provider of customized learning solutions with office locations around the globe, including Singapore, the United Kingdom, Australia, Mexico, Brazil, and Japan. Locate your local office at **www.cengage .com/global.**

Cengage Learning products are represented in Canada by Nelson Education, Ltd.

To learn more about Brooks/Cole, visit **www.cengage.com/brookscole**

Purchase any of our products at your local college store or at our preferred online store **www.ichapters.com.**

Printed in Canada
2 3 4 5 6 7 13 12 11 10

Brief Contents

Key to Sequence of Topics (chapter numbers are color coded as follows):

- **Traditional foundations of organic chemistry**
- Organic reactions and their biological counterparts
- **The organic chemistry of biological molecules and pathways**

Detailed Contents

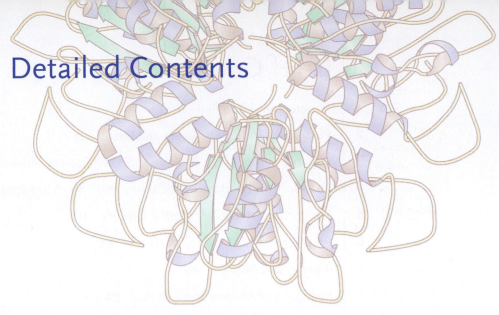

Organic Compounds: Alkanes and Their Stereochemistry 70

3

Organic Compounds: Cycloalkanes and Their Stereochemistry 105

4

10   Structure Determination: Mass Spectrometry, Infrared Spectroscopy, and Ultraviolet Spectroscopy 367

11   Structure Determination: Nuclear Magnetic Resonance Spectroscopy 404

Organohalides: Nucleophilic Substitutions and Eliminations   444

12

Alcohols, Phenols, and Thiols; Ethers and Sulfides   501

13

Carboxylic Acids and Nitriles 610

Carboxylic Acid Derivatives: Nucleophilic Acyl Substitution Reactions 643

Carbonyl Alpha-Substitution and Condensation Reactions 695

18   Amines and Heterocycles   749

19   Biomolecules: Amino Acids, Peptides, and Proteins   791

Amino Acid Metabolism   832

20

Biomolecules: Carbohydrates   862

21

Secondary Metabolites: An Introduction to Natural Products Chemistry 1015

Appendices

Preface

I've taught organic chemistry many times for many years, and it has often struck me what a disconnect there is between the interests and expectations of me—the teacher—and the interests and expectations of those being taught—my students. I love the logic and beauty of organic chemistry, and I want to pass that feeling on to others. My students, however, seem to worry primarily about getting into medical school. That may be an exaggeration, but there is also a lot of truth in it. All of us who teach organic chemistry know that the large majority of our students—90% or more, including many chemistry majors—are interested primarily in medicine, biology, and other life sciences rather than in pure chemistry.

But if we are primarily teaching future physicians, biologists, biochemists, and others in the life sciences (not to mention the occasional lawyer and businessperson), why do we continue to teach the way we do? Why do we spend so much time discussing details of topics that interest research chemists but have no connection to biology? Wouldn't the limited amount of time we have be better spent paying more attention to the organic chemistry of living organisms and less to the organic chemistry of the research laboratory? I believe so, and I have written this book, *Organic Chemistry with Biological Applications,* to encourage others who might also be thinking that the time has come to try doing things a bit differently.

This is, first and foremost, a textbook on organic chemistry, and you will find that almost all of the standard topics are here. Nevertheless, my guiding principle in writing this text has been to emphasize organic reactions and topics that are relevant to biological chemistry.

Organization of the Text

When looking through the text, three distinct groups of chapters are apparent. The first group (Chapters 1–6 and 10–11) covers the traditional principles of organic chemistry that are essential for building the background necessary to further understanding. The second group (Chapters 7–9 and 12–18) covers the common organic reactions found in all texts. As each laboratory reaction is discussed, however, a biological example is also shown to make the material more interesting to students. As an example, trans fatty acids are described at the same time that catalytic hydrogenation is discussed (see Section 8.5, page 261). The third group of chapters (19–25) is unique to this text in their depth of coverage. These chapters deal exclusively with the main classes of biomolecules—amino acids and proteins, carbohydrates, lipids, and nucleic acids—and show how thoroughly organic chemistry permeates biological chemistry. Following an introduction to each class, major metabolic pathways for that class are discussed from the perspective of mechanistic organic chemistry. Finally, the book ends with a chapter devoted to natural products and their biosynthesis.

Content Changes in the Second Edition

Text content has been revised substantially for this second edition as a result of user feedback. Consequently, the text covers most of the standard topics found in typical organic courses yet still retains an emphasis on biological reactions and molecules. Perhaps the most noticeable change is that the book is now titled *Organic Chemistry with Biological Applications* to emphasize that it is, above all, written for the standard organic chemistry course found in colleges and universities everywhere.

Within the text itself, a particularly important change is that the chapter on chirality and stereochemistry at tetrahedral centers, a topic crucial to understanding biological chemistry, has been moved forward to Chapter 4 from its previous placement in Chapter 9. In addition, the chapter on organohalides has been moved from Chapter 10 to Chapter 12, thereby placing spectroscopy earlier (Chapters 10 and 11).

Other Changes and Newly Added Content

- Alkene ozonolysis and diol cleavage—added in Section 8.8
- Addition of carbenes to alkenes—added in Section 8.9
- The Diels–Alder cycloaddition reaction—added in Section 8.14
- Acetylide alkylations—added in Section 8.15
- Aromatic ions—added in Section 9.4
- Nucleophilic aromatic substitution—added in Section 9.9
- Aromatic hydrogenation—added in Section 9.10
- Allylic bromination of alkenes—added in Section 12.2
- Dess–Martin oxidation of alcohols—added in Section 13.5
- Protection of alcohols as silyl ethers—added in Section 13.6
- Claisen rearrangement—added in Section 13.10
- Protection of ketones and aldehydes as acetals—added in Section 14.8
- Conjugate addition of diorganocuprates to enones—added in Section 14.11
- Grignard reaction of nitriles—added in Section 15.7
- Reaction of diorganocuprates with acid halides—added in Section 16.4
- Alpha bromination of carboxylic acids—added in Section 17.3
- Amino acid metabolism—simplified coverage, Section 20.4
- Amino acid biosynthesis—simplified coverage, Section 20.5
- Final comments on metabolism—added in Section 23.10
- Nucleotide metabolism—simplified coverage, Section 24.9

- Nucleotide biosynthesis—simplified coverage, Section 24.10
- "Secondary Metabolites: An Introduction to Natural Products Chemistry"—new Chapter 25

There is more than enough organic chemistry in this book, along with a coverage of biological chemistry that far surpasses what is found in any other text. My hope is that all the students we teach, including those who worry about medical school, will come to agree that there is also logic and beauty here.

Features of the Second Edition

Reaction Mechanisms

The innovative vertical presentation of reaction mechanisms that has become a hallmark of all my texts in retained in *Organic Chemistry with Biological Applications.* Mechanisms in this format have the reaction steps printed vertically, while the changes taking place in each step are explained next to the reaction arrows. With this format, students can see what is occurring at each step in a reaction without having to jump back and forth between structures and text. See Figure 14.10 on page 581 for a chemical example and Figure 22.7 on page 912 for a biochemical example.

Visualization of Biological Reactions

One of the most important goals of this book is to demystify biological chemistry—to show students how the mechanisms of biological reactions are the same as those of laboratory organic reactions. Toward this end, and to let students more easily visualize the changes that occur during reactions of large biomolecules, I use an innovative method for focusing attention on the reacting parts in large molecules by "ghosting" the nonreacting parts. See Figure 13.6 on page 522, for example.

Other Features

- "Why do we have to learn this?" I've been asked this question by students so many times that I thought I should answer it upfront. Thus, the introduction to every chapter now includes "Why This Chapter?"—a brief paragraph that tells students why the material about to be covered is important and explains how the organic chemistry in each chapter relates to biological chemistry.
- The Worked Examples in each chapter are titled to give students a frame of reference. Each Worked Example includes a Strategy and worked-out Solution, followed by Problems for students to try on their own.
- A Lagniappe—a Louisiana Creole word meaning "something extra"—is provided at the end of each chapter to relate real-world concepts to students' lives. New Lagniappes in this edition include essays on Green Chemistry and Ionic Liquids as green reaction solvents.
- Visualizing Chemistry problems at the end of each chapter offer students an opportunity to see chemistry in a different way by visualizing molecules rather than simply interpreting structural formulas.
- Summaries and Key Word lists at the ends of chapters help students focus on the key concepts in that chapter.

- Reaction Summaries at the ends of chapters bring together the key reactions from that chapter into one complete list.

- An overview titled "A Preview of Carbonyl Chemistry," following Chapter 13, highlights the idea that studying organic chemistry works by both summarizing past ideas and looking ahead to new ones.

- The latest IUPAC nomenclature rules, as updated in 1993, are used in this text.

- Thorough media integration with OWL for Organic Chemistry, an online homework assessment program, is provided to help students practice and test their knowledge of important concepts. For this second edition, OWL includes parameterized end-of-chapter questions from the text (marked in the text with ▪). An access code is required. Visit www .cengage.com/owl to register.

- Students can work through animated versions of the text's Active Figures at the Student Companion site, which is accessible from www.cengage .com/chemistry/mcmurry.

Acknowledgments

I thank all the people who helped to shape this book and its message. At Brooks/Cole Cengage Learning they include: Lisa Lockwood, executive editor; Sandra Kiselica, senior development editor; Amee Mosley, executive marketing manager; Teresa Trego, senior production manager; Lisa Weber, senior media editor; Elizabeth Woods, assistant editor, and Suzanne Kastner at Graphic World.

I am grateful to colleagues who reviewed the manuscript for this book. They include:

REVIEWERS OF THE SECOND EDITION

Peter Alaimo, Seattle University

Sheila Browne, Mount Holyoke College

Gordon Gribble, Dartmouth College

John Grunwell, Miami University

Eric Kantorowski, California Polytechnic State University

Kevin Kittredge, Siena College

Rizalia Klausmeyer, Baylor University

Bette Kreuz, University of Michigan–Dearborn

Manfred Reinecke, Texas Christian University

Frank Rossi, State University of New York, Cortland

Miriam Rossi, Vassar College

Paul Sampson, Kent State University

Martin Semmelhack, Princeton University

Megan Tichy, Texas A&M University

Bernhard Vogler, University of Alabama, Huntsville

REVIEWERS OF FIRST EDITION

Helen E. Blackwell, University of Wisconsin

Joseph Chihade, Carleton College

Robert S. Coleman, Ohio State University

John Hoberg, University of Wyoming

Eric Kantorowski, California Polytechnic State University

Thomas Lectka, Johns Hopkins University

Paul Martino, Flathead Valley Community College

Eugene Mash, University of Arizona

Pshemak Maslak, Pennsylvania State University

Kevin Minbiole, James Madison University

Andrew Morehead, East Carolina University

K. Barbara Schowen, University of Kansas

Ancillaries to Accompany This Book

For Students

STUDY GUIDE AND SOLUTIONS MANUAL Written by Susan McMurry, this manual provides complete answers and explanations to all in-text and end-of-chapter exercises. The PowerLecture Instructor's CD contains a three-chapter preview. ISBN: 0-495-39145-X

 OWL FOR ORGANIC CHEMISTRY (ONLINE WEB LEARNING)
Instant Access to OWL for Organic Chemistry (four semesters): ISBN-10: 0-495-05102-0; ISBN-13: 978-0-495-05102-2
Instant Access to OWL with e-Book for McMurry's Second Edition (four semesters): ISBN-10: 0-495-39150-6; ISBN-13: 978-0-495-39150-0

Authored by Steve Hixson and Peter Lillya of the University of Massachusetts, Amherst, and William Vining of the State University of New York at Oneonta. Developed at the University of Massachusetts, Amherst, used by thousands of chemistry students, and featuring an updated and more intuitive instructor interface, **OWL for Organic Chemistry** is a customizable online learning system and assessment tool that reduces faculty workload and facilitates instruction. You can select from various types of assignments—tutors, simulations, and short answer questions that are numerically, chemically, and contextually parameterized—and **OWL** can accept superscript and subscript as well as structure drawings. With parameterization, **OWL for Organic Chemistry** offers more than 6000 questions and includes an upgrade to the latest version of MarvinSketch, an advanced molecular drawing program for drawing gradable structures. For this second edition, OWL includes parameterized end-of-chapter questions from the text (marked in the text with ■). New questions are authored by David W. Brown, Florida Gulf Coast University.

When you become an **OWL** user, you can expect service that goes far beyond the ordinary. **OWL** is continually enhanced with online learning tools to address the various learning styles of today's students such as:

- **e-Books**, which offer a fully integrated electronic textbook linked to OWL questions

- **Quick Prep** review courses that help students learn essential skills to succeed in General and Organic Chemistry

- **Jmol** molecular visualization program for rotating molecules and measuring bond distances and angles

To view an OWL demo and for more information, visit www.cengage.com/owl or contact your Brooks/Cole Cengage Learning representative.

STUDENT COMPANION WEBSITE Students can work through animated versions of the text's Active Figures at the Student Companion site, which is accessible from www.cengage.com/chemistry/mcmurry.

PUSHING ELECTRONS: A GUIDE FOR STUDENTS OF ORGANIC CHEMISTRY, THIRD EDITION Written by Daniel P. Weeks, this workbook is designed to help students learn techniques of electron pushing. Its programmed approach emphasizes repetition and active participation. ISBN: 0-03-020693-6

SPARTANMODEL ELECTRONIC MODELING KIT A set of easy-to-use builders allow for the construction and 3-D manipulation of molecules of any size or complexity—from a hydrogen atom to DNA and everything in between. This kit includes the SpartanModel software on CD-ROM, an extensive molecular database, 3-D glasses, and a *Tutorial and Users Guide* that includes a wealth of activities to help you get the most out of your course. ISBN: 0-495-01793-0

For Instructors

POWERLECTURE WITH EXAMVIEW® AND JOININ™ INSTRUCTOR'S CD/DVD PACKAGE ISBN-10: 0-495-39146-8; ISBN-13: 978-0-495-39146-3

PowerLecture is a dual-platform, one-stop digital library and presentation tool that includes:

- Prepared Microsoft® PowerPoint® Lecture Slides by Richard Morrison of the University of Georgia that cover all key points from the text in a convenient format that you can enhance with your own materials or with additional interactive video and animations from the CD-ROM for personalized, media-enhanced lectures.

- Image Libraries in PowerPoint and in JPEG format that provide electronic files for all text art, most photographs, and all numbered tables in the text. These files can be used to print transparencies or to create your own PowerPoint lectures.

- Electronic files for the Test Bank.

- Sample chapters from the *Student Solutions Manual and Study Guide.*

- ExamView testing software, with all test items from the printed Test Bank in electronic format, which enables you to create customized tests of up to 250 items in print or online.

- JoinIn clicker questions authored for this text, for use with the classroom response system of your choice. Assess student progress with instant quizzes and polls, and display student answers seamlessly within the Microsoft PowerPoint slides of your own lecture. Consult your Brooks/Cole Cengage Learning representative for more details.

FACULTY COMPANION WEBSITE Accessible from www.cengage.com/chemistry/mcmurry, this website provides downloadable files for the WebCT and Blackboard versions of ExamView Computerized Testing.

TEST BANK Revised by Bette Kreuz of the University of Michigan–Dearborn, this Test Bank includes more than 1000 multiple-choice and matching questions, with detailed answers, in preprinted test forms corresponding to the main text organization. The *Test Bank* is available on the instructor's PowerLecture CD as electronic files and in ExamView format. Instructors can customize tests using the Test Bank files on the PowerLecture CD-ROM. ISBN: 0-495-39149-2

ORGANIC CHEMISTRY LABORATORY MANUALS Brooks/Cole, Cengage Learning is pleased to offer you a choice of organic chemistry laboratory manuals catered to fit your needs. Visit www.cengage.com/chemistry. Customizable laboratory manuals also can be assembled. Go to www.signature-labs.com/specializations/chemistry.html for more information.

Author royalties from this book are being
donated to the Cystic Fibrosis Foundation.

1 Structure and Bonding

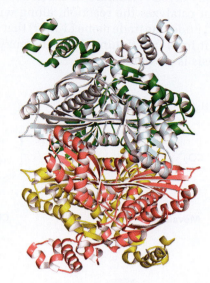

A model of the enzyme HMG-CoA reductase, which catalyzes a crucial step in the body's synthesis of cholesterol.

A scientific revolution is now taking place—a revolution that will give us safer and more effective medicines, cure our genetic diseases, increase our life spans, and improve the quality of our lives. The revolution is based in understanding the structure and function of the approximately 21,000 genes in the human body, but it relies on organic chemistry as the enabling science. It is our fundamental chemical understanding of biological processes at the molecular level that has made the revolution possible and that continues to drive it. Anyone who wants to understand or be a part of the remarkable advances now occurring in medicine and the biological sciences must first understand organic chemistry.

As an example of how organic and biological chemistry together are affecting modern medicine, look at coronary heart disease—the buildup of cholesterol-containing plaques on the walls of arteries in the heart, leading to restricted blood flow and eventual heart attack. Coronary heart disease is the leading cause of death for both men and women older than age 20, and it's estimated that up to one-third of women and one-half of men will develop the disease at some point in their lives.

The onset of coronary heart disease is directly correlated with blood cholesterol levels, and the first step in disease prevention is to lower those levels. It turns out that only about 25% of our blood cholesterol comes from what we eat; the remaining 75% (about 1000 mg each day) is made, or *biosynthesized,* by our bodies from dietary fats and carbohydrates. Thus, any effective plan for lowering our cholesterol level means limiting the amount that our bodies biosynthesize, which in turn means understanding and controlling the chemical reactions that make up the metabolic pathway for cholesterol biosynthesis.

Now look at Figure 1.1. Although the figure may seem unintelligible at this point, don't worry; before long it will make perfectly good sense. What's shown in Figure 1.1 is the biological conversion of a compound called 3-hydroxy-3-methylglutaryl coenzyme A (HMG-CoA) to mevalonate, a crucial

OWL Online homework for this chapter can be assigned in Organic OWL, an online homework assessment tool.

step in the pathway by which our bodies synthesize cholesterol. Also shown in the figure is an X-ray crystal structure of the active site in the HMG-CoA reductase enzyme that catalyzes the reaction, along with a molecule of the drug atorvastatin (sold under the trade name Lipitor) that binds to the enzyme's active site and stops it from functioning. With the enzyme thus inactivated, cholesterol biosynthesis is prevented.

FIGURE 1.1 The metabolic conversion of 3-hydroxy-3-methylglutaryl coenzyme A (HMG-CoA) to mevalonate is a crucial step in the body's pathway for biosynthesizing cholesterol. An X-ray crystal structure of the active site in the HMG-CoA reductase enzyme that catalyzes the reaction is shown, along with a molecule of atorvastatin (Lipitor) that is bound in the active site and stops the enzyme from functioning. With the enzyme thus inactivated, cholesterol biosynthesis is prevented.

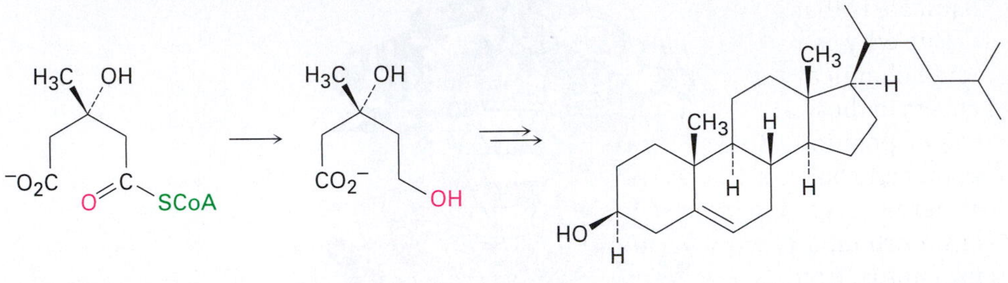

3-Hydroxy-3-methyl-glutaryl coenzyme A (HMG-CoA)   **Mevalonate**   **Cholesterol**

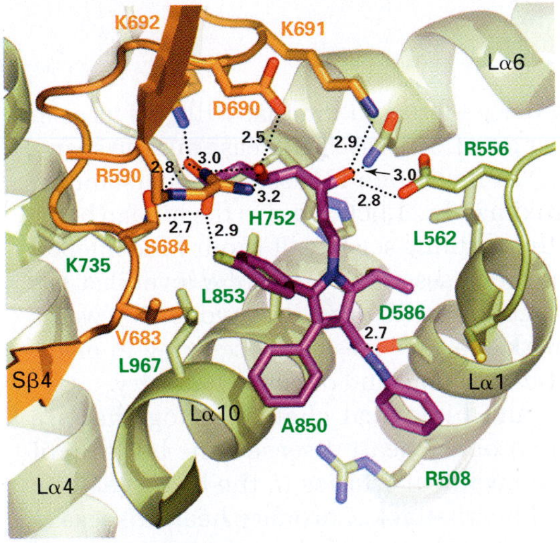

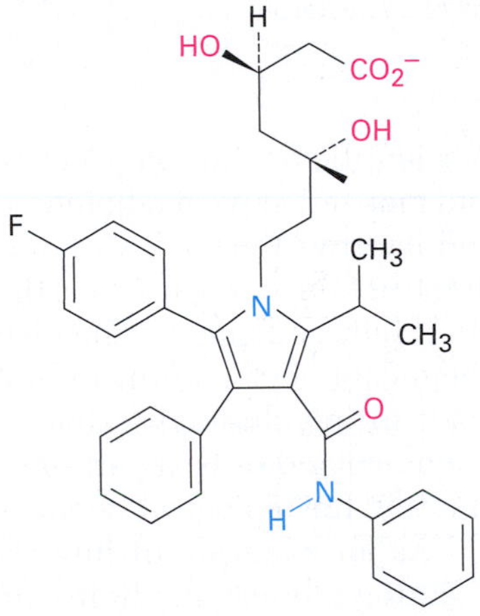

Atorvastatin (Lipitor)

Atorvastatin is one of a widely prescribed class of drugs called *statins,* which reduce a person's risk of coronary heart disease by lowering the level of cholesterol in their blood. Taken together, the statins—atorvastatin (Lipitor), simvastatin (Zocor), rosuvastatin (Crestor), pravastatin (Pravachol), lovastatin (Mevacor), and several others—are the most widely prescribed drugs in the world, with an estimated $14.6 billion in annual sales.

The statins function by blocking the HMG-CoA reductase enzyme and preventing it from converting HMG-CoA to mevalonate, thereby limiting the body's biosynthesis of cholesterol. As a result, blood cholesterol levels drop and coronary heart disease becomes less likely. It sounds simple, but it would be impossible without a detailed knowledge of the steps in the pathway for cholesterol biosynthesis, the enzymes that catalyze those steps, and how precisely shaped organic molecules can be designed to block those steps. Organic chemistry is what makes it all happen.

Historically, the term **organic chemistry** was used to mean the chemistry of compounds found in living organisms. At that time, in the late 1700s, little was known about chemistry, and the behavior of the "organic" substances isolated from plants and animals seemed different from that of the "inorganic" substances found in minerals. Organic compounds were generally low-melting solids and were usually more difficult to isolate, purify, and work with than high-melting inorganic compounds. By the mid-1800s, however, it was clear that there was no fundamental difference between organic and inorganic compounds. The same principles explain the behaviors of all substances, regardless of origin or complexity. The only distinguishing characteristic of organic chemicals is that *all contain the element carbon.*

But why is carbon special? Why, of the more than 37 million presently known chemical compounds, do more than 99% of them contain carbon? The answers to these questions come from carbon's electronic structure and its consequent position in the periodic table (Figure 1.2). As a group 4A element, carbon can share four valence electrons and form four strong covalent bonds. Furthermore, carbon atoms can bond to one another, forming long chains and rings. Carbon, alone of all elements, is able to form an immense diversity of compounds, from the simple to the staggeringly complex—from methane, with one carbon atom, to DNA, which can have more than *100 million* carbons.

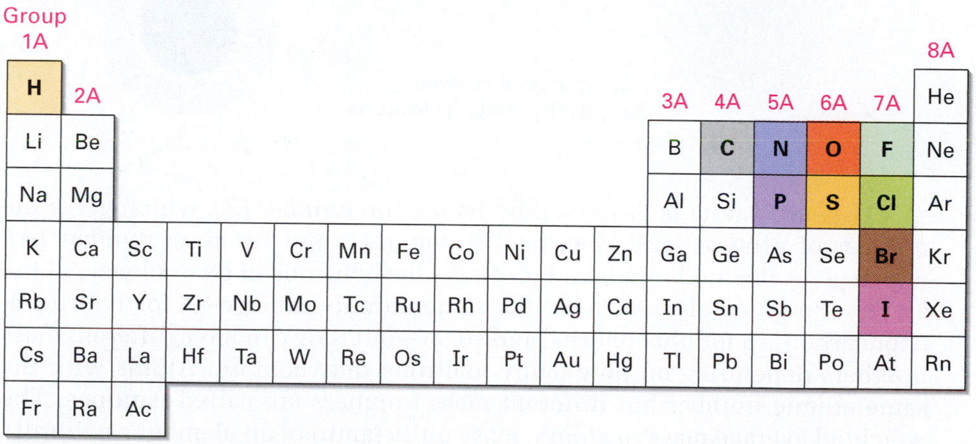

FIGURE 1.2 Carbon, hydrogen, and other elements commonly found in organic compounds are shown in the colors typically used to represent them.

Not all carbon compounds are derived from living organisms of course, and over the years chemists have developed a remarkably sophisticated ability to design and synthesize new organic compounds in the laboratory—medicines, dyes, polymers, and a host of other substances. Organic chemistry touches the lives of everyone; its study can be a fascinating undertaking.

WHY THIS CHAPTER?

We'll ease into the study of organic chemistry by first reviewing some ideas about atoms, bonds, and molecular geometry that you may recall from your general chemistry course. Much of the material in this chapter and the next is likely to be familiar to you, but it's nevertheless a good idea to make sure you understand it before going on.

1.1 Atomic Structure: The Nucleus

As you probably know from your general chemistry course, an atom consists of a dense, positively charged *nucleus* surrounded at a relatively large distance by negatively charged *electrons* (Figure 1.3). The nucleus consists of subatomic particles called *neutrons,* which are electrically neutral, and *protons,* which are positively charged. Because an atom is neutral overall, the number of positive protons in the nucleus and the number of negative electrons surrounding the nucleus are the same.

Although extremely small—about 10^{-14} to 10^{-15} meter (m) in diameter—the nucleus nevertheless contains essentially all the mass of the atom. Electrons have negligible mass and circulate around the nucleus at a distance of approximately 10^{-10} m. Thus, the diameter of a typical atom is about 2×10^{-10} m, or 200 picometers (pm), where 1 pm = 10^{-12} m. To give you an idea of how small this is, a thin pencil line is about 3 million carbon atoms wide. Many organic chemists and biochemists still use the unit *angstrom* (Å) to express atomic distances, where 1 Å = 100 pm = 10^{-10} m, but we'll stay with the SI unit picometer in this book.

FIGURE 1.3 A schematic view of an atom. The dense, positively charged nucleus contains most of the atom's mass and is surrounded by negatively charged electrons. The three-dimensional view on the right shows calculated electron-density surfaces. Electron density increases steadily toward the nucleus and is 40 times greater at the blue solid surface than at the gray mesh surface.

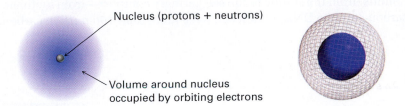

Nucleus (protons + neutrons)

Volume around nucleus occupied by orbiting electrons

A specific atom is described by its *atomic number (Z),* which gives the number of protons (and electrons) it contains, and its *mass number (A),* which gives the total number of protons plus neutrons in its nucleus. All the atoms of a given element have the same atomic number—1 for hydrogen, 6 for carbon, 15 for phosphorus, and so on—but they can have different mass numbers depending on how many neutrons they contain. Atoms with the same atomic number but different mass numbers are called **isotopes**. The weighted average mass in atomic mass units (amu) of an element's naturally occurring isotopes is called the element's *atomic mass (or atomic weight)*—1.008 amu for hydrogen, 12.011 amu for carbon, 30.974 amu for phosphorus, and so on.

1.2 Atomic Structure: Orbitals

How are the electrons distributed in an atom? According to the quantum mechanical model, the behavior of a specific electron in an atom can be described by a mathematical expression called a *wave equation*—the same sort of expression used to describe the motion of waves in a fluid. The solution to a wave equation is called a *wave function,* or **orbital**, and is denoted by the Greek letter psi, ψ.

By plotting the square of the wave function, ψ^2, in three-dimensional space, the orbital describes the volume of space around a nucleus that an electron is most likely to occupy. You might therefore think of an orbital as looking like a photograph of the electron taken at a slow shutter speed. In such a photo, the orbital would appear as a blurry cloud indicating the region of space around the nucleus where the electron has been. This electron cloud doesn't have a sharp boundary, but for practical purposes we can set the limits by saying that an orbital represents the space where an electron spends most (90%–95%) of its time.

What do orbitals look like? There are four different kinds of orbitals, denoted *s*, *p*, *d*, and *f*, each with a different shape. Of the four, we'll be concerned primarily with *s* and *p* orbitals because these are the most common in organic and biological chemistry. An *s* orbital is spherical, with the nucleus at its center; a *p* orbital is dumbbell-shaped; and four of the five *d* orbitals are cloverleaf-shaped, as shown in Figure 1.4. The fifth *d* orbital is shaped like an elongated dumbbell with a doughnut around its middle.

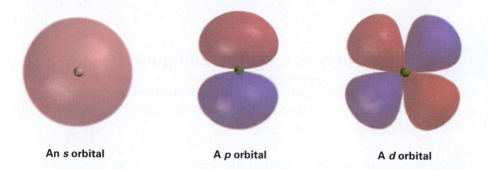

An *s* orbital   **A *p* orbital**   **A *d* orbital**

FIGURE 1.4 Representations of *s*, *p*, and *d* orbitals. An *s* orbital is spherical, a *p* orbital is dumbbell-shaped, and four of the five *d* orbitals are cloverleaf-shaped. Different lobes of *p* orbitals are often drawn for convenience as teardrops, but their true shape is more like that of a doorknob, as indicated.

The orbitals in an atom are organized into different layers, or **electron shells**, of successively larger size and energy. Different shells contain different numbers and kinds of orbitals, and each orbital within a shell can be occupied by two electrons. The first shell contains only a single *s* orbital, denoted 1*s*, and thus holds only 2 electrons. The second shell contains one 2*s* orbital and three 2*p* orbitals and thus holds a total of 8 electrons. The third shell contains a 3*s* orbital, three 3*p* orbitals, and five 3*d* orbitals, for a total capacity of 18 electrons. These orbital groupings and their energy levels are shown in Figure 1.5.

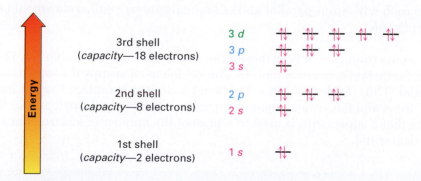

3rd shell
(*capacity*—18 electrons)

2nd shell
(*capacity*—8 electrons)

1st shell
(*capacity*—2 electrons)

Energy

FIGURE 1.5 The energy levels of electrons in an atom. The first shell holds a maximum of 2 electrons in one 1*s* orbital; the second shell holds a maximum of 8 electrons in one 2*s* and three 2*p* orbitals; the third shell holds a maximum of 18 electrons in one 3*s*, three 3*p*, and five 3*d* orbitals; and so on. The two electrons in each orbital are represented by up and down arrows, ↑↓. Although not shown, the energy level of the 4*s* orbital falls between 3*p* and 3*d*.

The three different *p* orbitals within a given shell are oriented in space along mutually perpendicular directions, denoted p_x, p_y, and p_z. As shown in Figure 1.6, the two lobes of each *p* orbital are separated by a region of zero electron density called a **node**. Furthermore, the two orbital regions separated by the node have different algebraic signs, + and −, in the wave function, as represented by the different colors in Figure 1.6. As we'll see in Section 1.11, the algebraic signs of the different orbital lobes have important consequences with respect to chemical bonding and chemical reactivity.

FIGURE 1.6 Shapes of the $2p$ orbitals. Each of the three mutually perpendicular, dumbbell-shaped orbitals has two lobes separated by a node. The two lobes have different algebraic signs in the corresponding wave function, as indicated by the different colors.

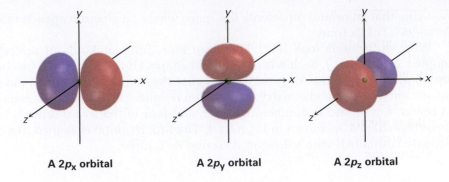

A $2p_x$ orbital   A $2p_y$ orbital   A $2p_z$ orbital

1.3 Atomic Structure: Electron Configurations

The lowest-energy arrangement, or **ground-state electron configuration**, of an atom is a listing of the orbitals occupied by its electrons. We can predict this arrangement by following three rules:

Rule 1

The lowest-energy orbitals fill up first, according to the order $1s \rightarrow 2s \rightarrow 2p \rightarrow 3s \rightarrow 3p \rightarrow 4s \rightarrow 3d$, a statement called the *aufbau principle*. Note that the $4s$ orbital lies between the $3p$ and $3d$ orbitals in energy.

Rule 2

Electrons act in some ways as if they were spinning around an axis, in much the same way that the earth spins. This spin can have two orientations, denoted as up ↑ and down ↓. Only two electrons can occupy an orbital, and they must be of opposite spin, a statement called the *Pauli exclusion principle*.

Rule 3

If two or more empty orbitals of equal energy are available, one electron occupies each with spins parallel until all orbitals are half-full, a statement called *Hund's rule*.

Some examples of how these rules apply are shown in Table 1.1. Hydrogen, for instance, has only one electron, which must occupy the lowest-energy orbital. Thus, hydrogen has a $1s$ ground-state configuration. Carbon has six electrons and the ground-state configuration $1s^2 \, 2s^2 \, 2p_x^1 \, 2p_y^1$, and so forth. Note that a superscript is used to represent the number of electrons in a particular orbital.

TABLE 1.1
Ground-State Electron Configurations of Some Elements

Element	Atomic number	Configuration		Element	Atomic number	Configuration
Hydrogen	1	$1s$ ⤊		Phosphorus	15	$3p$ ↿ ↿ ↿
						$3s$ ⇅
Carbon	6	$2p$ ↿ ↿ —				$2p$ ⇅ ⇅ ⇅
		$2s$ ⇅				$2s$ ⇅
		$1s$ ⇅				$1s$ ⇅

Problem 1.1

Give the ground-state electron configuration for each of the following elements:

(a) Oxygen   **(b)** Phosphorus   **(c)** Sulfur

Problem 1.2

How many electrons does each of the following biological trace elements have in its outermost electron shell?

(a) Magnesium   **(b)** Cobalt   **(c)** Selenium

1.4 Development of Chemical Bonding Theory

By the mid-1800s, the new science of chemistry was developing rapidly and chemists had begun to probe the forces holding compounds together. In 1858, August Kekulé and Archibald Couper independently proposed that, in all its compounds, carbon is *tetravalent*—it always forms four bonds when it joins other elements to form stable compounds. Furthermore, said Kekulé, carbon atoms can bond to one another to form extended chains of linked atoms.

Shortly after the tetravalent nature of carbon was proposed, extensions to the Kekulé–Couper theory were made when the possibility of *multiple* bonding between atoms was suggested. Emil Erlenmeyer proposed a carbon–carbon triple bond for acetylene, and Alexander Crum Brown proposed a carbon–carbon double bond for ethylene. In 1865, Kekulé provided another major advance when he suggested that carbon chains can double back on themselves to form *rings* of atoms.

Although Kekulé and Couper were correct in describing the tetravalent nature of carbon, chemistry was still viewed in a two-dimensional way until 1874. In that year, Jacobus van't Hoff and Joseph Le Bel added a third dimension to our ideas about organic compounds. They proposed that the four bonds of carbon are not oriented randomly but have specific spatial directions. Van't Hoff went even further and suggested that the four atoms to which carbon is bonded sit at the corners of a regular tetrahedron, with carbon in the center.

A representation of a tetrahedral carbon atom is shown in Figure 1.7. Note the conventions used to show three-dimensionality: solid lines represent bonds in the plane of the page, the heavy wedged line represents a bond coming out of the page toward the viewer, and the dashed line represents a bond receding back behind the page away from the viewer. These representations will be used throughout this text.

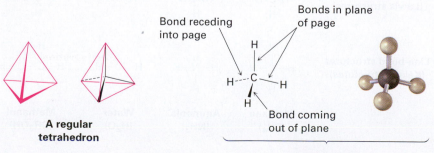

A regular tetrahedron

Bond receding into page

Bonds in plane of page

Bond coming out of plane

A tetrahedral carbon atom

FIGURE 1.7 A representation of van't Hoff's tetrahedral carbon atom. The solid lines represent bonds in the plane of the paper, the heavy wedged line represents a bond coming out of the plane of the page, and the dashed line represents a bond going back behind the plane of the page.

Why, though, do atoms bond together, and how can bonds be described electronically? The *why* question is relatively easy to answer: atoms bond together because the compound that results is more stable and lower in energy than the separate atoms. Energy (usually as heat) is always released and flows *out of* the chemical system when a chemical bond forms. Conversely, energy must be put *into* the system to break a chemical bond. Making bonds always releases energy, and breaking bonds always absorbs energy. The *how* question is more difficult. To answer it, we need to know more about the electronic properties of atoms.

We know through observation that eight electrons (an electron *octet*) in an atom's outermost shell, or **valence shell**, impart special stability to the noble-gas elements in group 8A of the periodic table: Ne (2 + 8); Ar (2 + 8 + 8); Kr (2 + 8 + 18 + 8). We also know that the chemistry of main-group elements is governed by their tendency to take on the electron configuration of the nearest noble gas. The alkali metals in group 1A, for example, achieve a noble-gas configuration by losing the single *s* electron from their valence shell to form a cation, while the halogens in group 7A achieve a noble-gas configuration by gaining a *p* electron to fill their valence shell and form an anion. The resultant ions are held together in compounds like $Na^+ Cl^-$ by an electrostatic attraction that we call an *ionic bond.*

But how do elements closer to the middle of the periodic table form bonds? Look at methane, CH_4, the main constituent of natural gas, for example. The bonding in methane is not ionic because it would take too much energy for carbon ($1s^2 2s^2 2p^2$) to either gain or lose *four* electrons to achieve a noble-gas configuration. As a result, carbon bonds to other atoms, not by gaining or losing electrons, but by *sharing* them. Such a shared-electron bond, first proposed in 1916 by G. N. Lewis, is called a **covalent bond**. The neutral collection of atoms held together by covalent bonds is called a **molecule**.

A simple way of indicating the covalent bonds in molecules is to use what are called *Lewis structures,* or **electron-dot structures**, in which the valence-shell electrons of an atom are represented as dots. Thus, hydrogen has one dot representing its 1s electron, carbon has four dots ($2s^2 2p^2$), oxygen has six dots ($2s^2 2p^4$), and so on. A stable molecule results whenever a noble-gas configuration is achieved for all the atoms—eight dots (an octet) for main-group atoms or two dots for hydrogen. Simpler still is the use of *Kekulé structures,* or **line-bond structures**, in which a two-electron covalent bond is indicated as a line drawn between atoms.

Electron-dot structures
(Lewis structures)

Line-bond structures
(Kekulé structures)

| Methane | Ammonia | Water | Methanol |
| (CH4) | (NH3) | (H2O) | (CH3OH) |

The number of covalent bonds an atom forms depends on how many additional valence electrons it needs to reach a noble-gas configuration. Hydrogen has one valence electron ($1s$) and needs one more to reach the helium configuration ($1s^2$), so it forms one bond. Carbon has four valence electrons ($2s^2\ 2p^2$) and needs four more to reach the neon configuration ($2s^2\ 2p^6$), so it forms four bonds. Nitrogen has five valence electrons ($2s^2\ 2p^3$), needs three more, and forms three bonds; oxygen has six valence electrons ($2s^2\ 2p^4$), needs two more, and forms two bonds; and the halogens have seven valence electrons, need one more, and form one bond.

H—	—C—	—N̈—	—Ö—	:F̈— :C̈l—
				:B̈r— :Ï—
One bond	**Four bonds**	**Three bonds**	**Two bonds**	**One bond**

Valence electrons that are not used for bonding are called **lone-pair electrons**, or *nonbonding electrons*. The nitrogen atom in ammonia (NH_3), for instance, shares six valence electrons in three covalent bonds and has its remaining two valence electrons in a nonbonding lone pair. As a time-saving shorthand, nonbonding electrons are often omitted when drawing line-bond structures, but you still have to keep them in mind since they're often crucial in chemical reactions.

Nonbonding,
lone-pair electrons

H:N̈:H   or   H—N̈—H   [or   H—N—H]
  H | |
                        H   H

Ammonia

WORKED EXAMPLE 1.1	Predicting the Number of Bonds Formed by Atoms in a Molecule

How many hydrogen atoms does phosphorus bond to in phosphine, $PH_?$?

Strategy

Identify the periodic group of phosphorus, and tell from that how many electrons (bonds) are needed to make an octet.

Solution

Phosphorus, like nitrogen, is in group 5A of the periodic table and has five valence electrons. It thus needs to share three more electrons to make an octet and therefore bonds to three hydrogen atoms, giving PH_3.

Problem 1.3

Draw a molecule of chloroform, $CHCl_3$, using solid, wedged, and dashed lines to show its tetrahedral geometry.

Problem 1.4

Convert the following representation of ethane, C_2H_6, into a conventional drawing that uses solid, wedged, and dashed lines to indicate tetrahedral geometry around each carbon (gray = C, ivory = H).

Ethane

Problem 1.5

What are likely formulas for the following substances?
(a) $CH_?Cl_2$ **(b)** $CH_3SH_?$ **(c)** $CH_3NH_?$

Problem 1.6

Draw line-bond structures for the following substances, showing all nonbonding electrons:
(a) CH_3CH_2OH, ethanol   **(b)** H_2S, hydrogen sulfide
(c) CH_3NH_2, methylamine   **(d)** $N(CH_3)_3$, trimethylamine

Problem 1.7

Why can't an organic molecule have the formula C_2H_7?

1.5 The Nature of Chemical Bonds: Valence Bond Theory

How does electron sharing lead to bonding between atoms? Two models have been developed to describe covalent bonding: *valence bond theory* and *molecular orbital theory*. Each model has its strengths and weaknesses, and chemists tend to use them interchangeably depending on the circumstances. Valence bond theory is the more easily visualized of the two, so most of the descriptions we'll use in this book derive from that approach.

According to **valence bond theory**, a covalent bond forms when two atoms approach each other closely and a singly occupied orbital on one atom *overlaps* a singly occupied orbital on the other atom. The electrons are now paired in the overlapping orbitals and are attracted to the nuclei of both atoms, thus bonding the atoms together. In the H_2 molecule, for example, the H–H bond results from the overlap of two singly occupied hydrogen $1s$ orbitals:

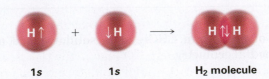

$1s$ $1s$ H_2 molecule

The overlapping orbitals in the H_2 molecule have the elongated egg shape we might get by pressing two spheres together. If a plane were to pass through the middle of the bond, the intersection of the plane and the overlapping orbitals would be a circle. In other words, the H–H bond is *cylindrically symmetrical,* as shown in Figure 1.8. Such bonds, which are formed by the head-on overlap of two atomic orbitals along a line drawn between the nuclei, are called **sigma (σ) bonds**.

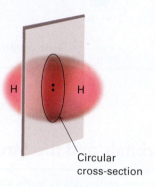

Circular
cross-section

FIGURE 1.8 The cylindrical symmetry of the H–H σ bond in an H_2 molecule. The intersection of a plane cutting through the σ bond is a circle.

During the bond-forming reaction $2\,H\cdot \rightarrow H_2$, 436 kJ/mol (104 kcal/mol) of energy is released. Because the product H_2 molecule has 436 kJ/mol less energy than the starting $2\,H\cdot$ atoms, the product is more stable than the reactant and we say that the H–H bond has a **bond strength** of 436 kJ/mol. In other words, we would have to put 436 kJ/mol of energy *into* the H–H bond to break the H_2 molecule apart into H atoms (Figure 1.9.) [For convenience, we'll generally give energies in both kilocalories (kcal) and the SI unit kilojoules (kJ): 1 kJ = 0.2390 kcal; 1 kcal = 4.184 kJ.]

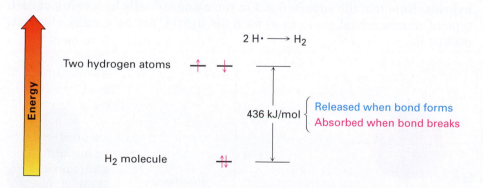

Energy

Two hydrogen atoms

$2\,H\cdot \longrightarrow H_2$

436 kJ/mol

Released when bond forms
Absorbed when bond breaks

H_2 molecule

FIGURE 1.9 Relative energy levels of H atoms and the H_2 molecule. The H_2 molecule has 436 kJ/mol (104 kcal/mol) less energy than the two H atoms, so 436 kJ/mol of energy is released when the H–H bond forms. Conversely, 436 kJ/mol must be added to the H_2 molecule to break the H–H bond.

How close are the two nuclei in the H_2 molecule? If they are too close, they will repel each other because both are positively charged, yet if they are too far apart, they won't be able to share the bonding electrons. Thus, there is an optimum distance between nuclei that leads to maximum stability (Figure 1.10). Called the **bond length**, this distance is 74 pm in the H_2 molecule. Every covalent bond has both a characteristic bond strength and bond length.

FIGURE 1.10 A plot of energy versus internuclear distance for two hydrogen atoms. The distance between nuclei at the minimum energy point is the bond length.

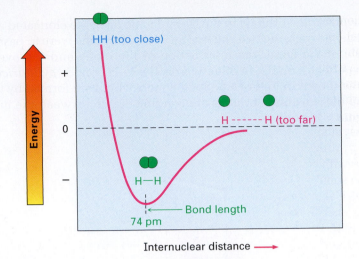

1.6 *sp*³ Hybrid Orbitals and the Structure of Methane

The bonding in the hydrogen molecule is fairly straightforward, but the situation is more complicated in organic molecules with tetravalent carbon atoms. Take methane, CH_4, for instance. As we've seen, carbon has four valence electrons ($2s^2\ 2p^2$) and forms four bonds. Because carbon uses two kinds of orbitals for bonding, $2s$ and $2p$, we might expect methane to have two kinds of C–H bonds. In fact, though, all four C–H bonds in methane are identical and are spatially oriented toward the corners of a regular tetrahedron (Figure 1.7). How can we explain this?

An answer was provided in 1931 by Linus Pauling, who showed mathematically how an *s* orbital and three *p* orbitals on an atom can combine, or *hybridize,* to form four equivalent atomic orbitals with tetrahedral orientation. Shown in Figure 1.11, these tetrahedrally oriented orbitals are called **sp^3 hybrids**. Note that the superscript 3 in the name sp^3 tells how many of each type of atomic orbital combine to form the hybrid, not how many electrons occupy it.

ACTIVE FIGURE 1.11 Four *sp*³ hybrid orbitals (green), oriented to the corners of a regular tetrahedron, are formed by combination of an *s* orbital (red) and three *p* orbitals (red/blue). The *sp*³ hybrids have two lobes and are unsymmetrical about the nucleus, giving them a directionality and allowing them to form strong bonds to other atoms. **Go to this book's student companion site at** www.cengage.com/chemistry/ mcmurry **to explore an interactive version of this figure.**

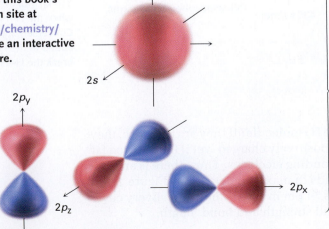

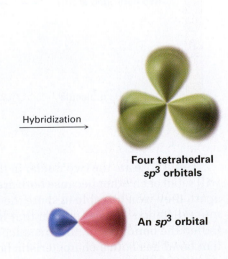

The concept of hybridization explains *how* carbon forms four equivalent tetrahedral bonds but not *why* it does so. The shape of the hybrid orbital suggests the answer. When an *s* orbital hybridizes with three *p* orbitals, the resultant *sp*^{3} hybrid orbitals are unsymmetrical about the nucleus. One of the two lobes is much larger than the other and can therefore overlap more effectively with an orbital from another atom when it forms a bond. As a result, *sp*^{3} hybrid orbitals form stronger bonds than do unhybridized *s* or *p* orbitals.

The asymmetry of *sp*^{3} orbitals arises because, as noted previously, the two lobes of a *p* orbital have different algebraic signs, $+$ and $-$. Thus, when a *p* orbital hybridizes with an *s* orbital, the positive *p* lobe adds to the *s* orbital but the negative *p* lobe subtracts from the *s* orbital. The resultant hybrid orbital is therefore unsymmetrical about the nucleus and is strongly oriented in one direction.

When each of the four identical *sp*^{3} hybrid orbitals of a carbon atom overlaps with the 1*s* orbital of a hydrogen atom, four identical C–H bonds are formed and methane results. Each C–H bond in methane has a strength of 439 kJ/mol (105 kcal/mol) and a length of 109 pm. Because the four bonds have a specific geometry, we also can define a property called the **bond angle**. The angle formed by each H–C–H is 109.5°, the so-called tetrahedral angle. Methane thus has the structure shown in Figure 1.12.

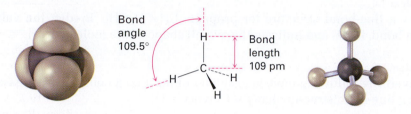

Bond
angle
109.5°

Bond
length
109 pm

FIGURE 1.12 The structure of methane, showing its 109.5° bond angles.

1.7 *sp*^{3} Hybrid Orbitals and the Structure of Ethane

The same kind of orbital hybridization that accounts for the methane structure also accounts for the bonding together of carbon atoms into chains and rings to make possible many millions of organic compounds. Ethane, C_2H_6, is the simplest molecule containing a carbon–carbon bond:

$$H\!:\!\overset{\displaystyle H}{\underset{\displaystyle H}{\overset{..}{C}}}\!:\!\overset{\displaystyle H}{\underset{\displaystyle H}{\overset{..}{C}}}\!:\!H \qquad H\!-\!\overset{\displaystyle H}{\underset{\displaystyle H}{C}}\!-\!\overset{\displaystyle H}{\underset{\displaystyle H}{C}}\!-\!H \qquad CH_3CH_3$$

Some representations of ethane

We can picture the ethane molecule by imagining that the two carbon atoms bond to each other by σ overlap of an *sp*^{3} hybrid orbital from each (Figure 1.13). The remaining three *sp*^{3} hybrid orbitals of each carbon overlap with the 1*s* orbitals of three hydrogens to form the six C–H bonds. The C–H bonds in ethane are similar to those in methane, although a bit weaker— 421 kJ/mol (101 kcal/mol) for ethane versus 439 kJ/mol for methane. The C–C bond is 154 pm long and has a strength of 377 kJ/mol (90 kcal/mol). All the bond angles of ethane are near, although not exactly at, the tetrahedral value of 109.5°.

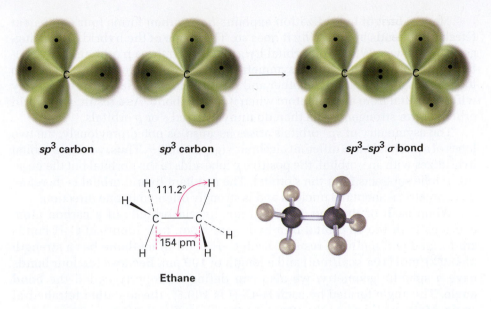

sp^3 **carbon** sp^3 **carbon** sp^3–sp^3 σ **bond**

Ethane

Problem 1.8

Draw a line-bond structure for propane, $CH_3CH_2CH_3$. Predict the value of each bond angle, and indicate the overall shape of the molecule.

Problem 1.9

Convert the following molecular model of hexane, a component of gasoline, into a line-bond structure (gray = C, ivory = H).

Hexane

1.8 sp^2 Hybrid Orbitals and the Structure of Ethylene

Although sp^3 hybridization is the most common electronic state of carbon, it's not the only possibility. Look at ethylene, C_2H_4, for example. It was recognized more than 100 years ago that ethylene carbons can be tetravalent only if they share *four* electrons and are linked by a *double* bond. Furthermore, ethylene is planar (flat) and has bond angles of approximately 120° rather than 109.5°.

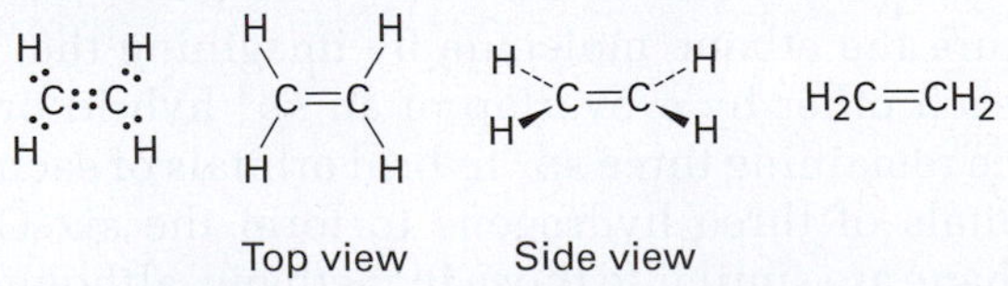

Top view   Side view

Some representations of ethylene

When we discussed sp^3 hybrid orbitals in Section 1.6, we said that the four valence-shell atomic orbitals of carbon combine to form four equivalent *sp³* hybrids. Imagine instead that the 2*s* orbital combines with only *two* of the three available 2*p* orbitals. Three ***sp²* hybrid orbitals** result, and one 2*p* orbital remains unchanged. The three *sp²* orbitals lie in a plane at angles of 120° to one another, with the remaining *p* orbital perpendicular to the *sp²* plane, as shown in Figure 1.14.

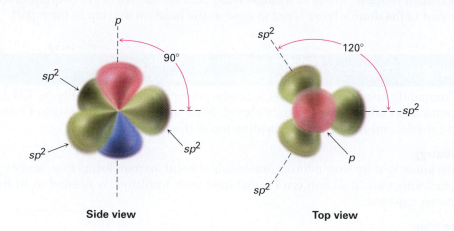

Side view   **Top view**

FIGURE 1.14 An *sp²*-hybridized carbon. The three equivalent *sp²* hybrid orbitals (green) lie in a plane at angles of 120° to one another, and a single unhybridized *p* orbital (red/blue) is perpendicular to the *sp²* plane.

When two *sp²*-hybridized carbons approach each other, they form a σ bond by *sp²*–*sp²* overlap. At the same time, the unhybridized *p* orbitals approach with the correct geometry for *sideways* overlap, leading to the formation of what is called a **pi (π) bond**. The combination of an *sp²*–*sp²* σ bond and a 2*p*–2*p* π bond results in the sharing of four electrons and the formation of a carbon–carbon double bond (Figure 1.15). Note that the electrons in the σ bond occupy the region centered between nuclei, while the electrons in the π bond occupy regions above and below a line drawn between nuclei.

To complete the structure of ethylene, four hydrogen atoms form σ bonds with the remaining four *sp²* orbitals. Ethylene thus has a planar structure, with H–C–H and H–C–C bond angles of approximately 120°. (The actual values are 117.4° for the H–C–H bond angle and 121.3° for the H–C–C bond angle.) Each C–H bond has a length of 108.7 pm and a strength of 464 kJ/mol (111 kcal/mol).

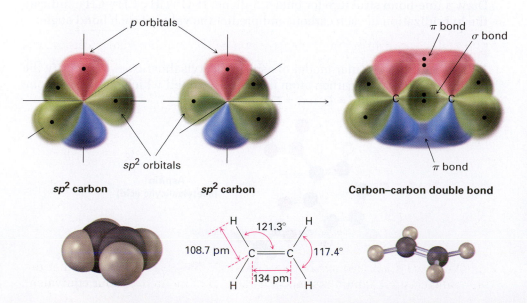

FIGURE 1.15 The structure of ethylene. Orbital overlap of two *sp²*-hybridized carbons forms a carbon–carbon double bond. One part of the double bond results from σ (head-on) overlap of *sp²* orbitals (green), and the other part results from π (sideways) overlap of unhybridized *p* orbitals (red/blue). The π bond has regions of electron density above and below a line drawn between nuclei.

As you might expect, the carbon–carbon double bond in ethylene is both shorter and stronger than the single bond in ethane because it has four electrons bonding the nuclei together rather than two. Ethylene has a C=C bond length of 134 pm and a strength of 728 kJ/mol (174 kcal/mol) versus a C–C length of 154 pm and a strength of 377 kJ/mol for ethane. The carbon–carbon double bond is less than twice as strong as a single bond because the sideways overlap in the π part of the double bond is not as great as the head-on overlap in the σ part.

WORKED EXAMPLE 1.2 — Predicting the Structures of Simple Molecules from Their Formulas

Commonly used in biology as a tissue preservative, formaldehyde, CH_2O, contains a carbon–*oxygen* double bond. Draw the line-bond structure of formaldehyde, and indicate the hybridization of the carbon atom.

Strategy

We know that hydrogen forms one covalent bond, carbon forms four, and oxygen forms two. Trial and error, combined with intuition, is needed to fit the atoms together.

Solution

There is only one way that two hydrogens, one carbon, and one oxygen can combine:

$$\begin{array}{c} :\ddot{O}: \\ \| \\ C \\ H \quad H \end{array} \qquad \textbf{Formaldehyde}$$

Like the carbon atoms in ethylene, the carbon atom in formaldehyde is in a double bond and is therefore sp^2-hybridized.

Problem 1.10

Draw a line-bond structure for propene, $CH_3CH{=}CH_2$; indicate the hybridization of each carbon; and predict the value of each bond angle.

Problem 1.11

Draw a line-bond structure for buta-1,3-diene, $H_2C{=}CH{-}CH{=}CH_2$; indicate the hybridization of each carbon; and predict the value of each bond angle.

Problem 1.12

Following is a molecular model of aspirin (acetylsalicylic acid). Identify the hybridization of each carbon atom in aspirin, and tell which atoms have lone pairs of electrons (gray = C, red = O, ivory = H).

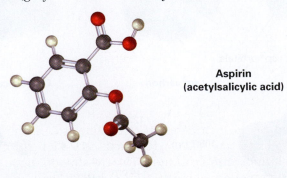

**Aspirin
(acetylsalicylic acid)**

1.9 *sp* Hybrid Orbitals and the Structure of Acetylene

In addition to forming single and double bonds by sharing two and four electrons, respectively, carbon also can form a *triple* bond by sharing six electrons. To account for the triple bond in a molecule such as acetylene, H—C≡C—H, we need a third kind of hybrid orbital, an **sp hybrid**. Imagine that, instead of combining with two or three *p* orbitals, a carbon 2*s* orbital hybridizes with only a single *p* orbital. Two *sp* hybrid orbitals result, and two *p* orbitals remain unchanged. The two *sp* orbitals are oriented 180° apart on the *x*-axis, while the remaining two *p* orbitals are perpendicular on the *y*-axis and the *z*-axis, as shown in Figure 1.16.

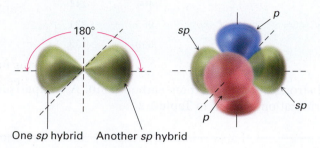

One *sp* hybrid   Another *sp* hybrid

FIGURE 1.16 An *sp*-hybridized carbon atom. The two *sp* hybrid orbitals (green) are oriented 180° away from each other, perpendicular to the two remaining *p* orbitals (red/blue).

When two *sp*-hybridized carbon atoms approach each other, *sp* hybrid orbitals on each carbon overlap head-on to form a strong *sp*–*sp* σ bond. In addition, the p_z orbitals from each carbon form a p_z–p_z π bond by sideways overlap, and the p_y orbitals overlap similarly to form a p_y–p_y π bond. The net effect is the sharing of six electrons and formation of a carbon–carbon triple bond. The two remaining *sp* hybrid orbitals each form a σ bond with hydrogen to complete the acetylene molecule (Figure 1.17).

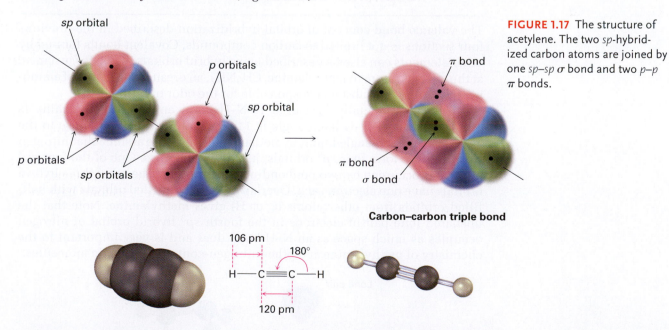

sp orbital

p orbitals

sp orbital

p orbitals

sp orbitals

π bond

π bond

σ bond

Carbon–carbon triple bond

106 pm

180°

H — C ≡ C — H

120 pm

FIGURE 1.17 The structure of acetylene. The two *sp*-hybridized carbon atoms are joined by one *sp*–*sp* σ bond and two *p*–*p* π bonds.

As suggested by *sp* hybridization, acetylene is a linear molecule with H–C–C bond angles of 180°. The C–H bonds have a length of 106 pm and a strength of 558 kJ/mol (133 kcal/mol). The C–C bond length in acetylene is 120 pm, and its strength is about 965 kJ/mol (231 kcal/mol), making it the

TABLE 1.2
Comparison of C–C and C–H Bonds in Methane, Ethane, Ethylene, and Acetylene

Molecule	Bond	Bond strength (kJ/mol)	(kcal/mol)	Bond length (pm)
Methane, CH_4	(sp^3) C—H	439	105	109
Ethane, CH_3CH_3	(sp^3) C—C (sp^3)	377	90	154
	(sp^3) C—H	420	100	109
Ethylene, $H_2C=CH_2$	(sp^2) C=C (sp^2)	728	174	134
	(sp^2) C—H	464	111	109
Acetylene, HC≡CH	(sp) C≡C (sp)	965	231	120
	(sp) C—H	558	133	106

shortest and strongest of any carbon–carbon bond. A comparison of sp, sp^2, and sp^3 hybridization is given in Table 1.2.

Problem 1.13
Draw a line-bond structure for propyne, $CH_3C≡CH$; indicate the hybridization of each carbon; and predict a value for each bond angle.

1.10 Hybridization of Nitrogen, Oxygen, Phosphorus, and Sulfur

The valence-bond concept of orbital hybridization described in the previous four sections is not limited to carbon compounds. Covalent bonds formed by other elements can also be described using hybrid orbitals. Look, for instance, at the nitrogen atom in methylamine, CH_3NH_2, an organic derivative of ammonia (NH_3) and the substance responsible for the odor of rotting fish.

The experimentally measured H–N–H bond angle in methylamine is 107.1° and the C–N–H bond angle is 110.3°, both of which are close to the 109.5° tetrahedral angle found in methane. We therefore assume that nitrogen hybridizes to form four sp^3 orbitals, just as carbon does. One of the four sp^3 orbitals is occupied by two nonbonding electrons, and the other three hybrid orbitals have one electron each. Overlap of these half-filled orbitals with half-filled orbitals from other atoms (C or H) gives methylamine. Note that the unshared lone pair of electrons in the fourth sp^3 hybrid orbital of nitrogen occupies as much space as an N–H bond does and is very important to the chemistry of methylamine and other nitrogen-containing organic molecules.

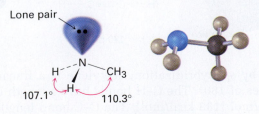

Methylamine

Like the carbon atom in methane and the nitrogen atom in methylamine, the oxygen atom in methanol (methyl alcohol) and many other organic molecules can be described as sp^3-hybridized. The C–O–H bond angle in methanol is 108.5°, very close to the 109.5° tetrahedral angle. Two of the four sp^3 hybrid orbitals on oxygen are occupied by nonbonding electron lone pairs, and two are used to form bonds.

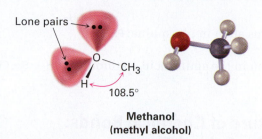

**Methanol
(methyl alcohol)**

Phosphorus and sulfur are the third-row analogs of nitrogen and oxygen, and the bonding in both can be described using hybrid orbitals. Because of their positions in the third row, however, both phosphorus and sulfur can expand their outer-shell octets and form more than the typical number of covalent bonds. Phosphorus, for instance, often forms five covalent bonds, and sulfur occasionally forms four.

Phosphorus is most commonly encountered in biological molecules in *organophosphates,* compounds that contain a phosphorus atom bonded to four oxygens, with one of the oxygens also bonded to carbon. Methyl phosphate, $CH_3OPO_3{}^{2-}$, is the simplest example. The O–P–O bond angle in such compounds is typically in the range 110° to 112°, implying sp^3 hybridization for the phosphorus.

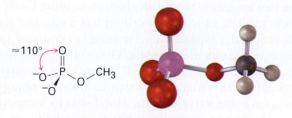

**Methyl phosphate
(an organophosphate)**

Sulfur is most commonly encountered in biological molecules either in compounds called *thiols,* which have a sulfur atom bonded to one hydrogen and one carbon, or in *sulfides,* which have a sulfur atom bonded to two carbons. Produced by some bacteria, methanethiol (CH_3SH) is the simplest example of a thiol, and dimethyl sulfide [$(CH_3)_2S$] is the simplest example of a sulfide. Both can be described by approximate sp^3 hybridization around sulfur, although both have significant deviation from the 109.5° tetrahedral angle.

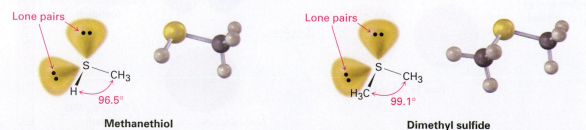

Methanethiol   **Dimethyl sulfide**

Problem 1.14

Identify all nonbonding lone pairs of electrons in the following molecules, tell what geometry you expect for each of the indicated atoms, and tell the kind of hybridized orbital occupied by the lone pairs.

(a) The oxygen atom in dimethyl ether: CH_3—O—CH_3

(b) The nitrogen atom in trimethylamine:
$$H_3C-N-CH_3$$
$$\overset{\mid}{\underset{CH_3}{}}$$

(c) The phosphorus atom in phosphine: PH_3

(d) The sulfur atom in the amino acid methionine:
$$CH_3-S-CH_2CH_2\overset{\overset{\displaystyle O}{\parallel}}{\underset{\underset{\displaystyle NH_2}{\mid}}{C}HCOH}$$

1.11 The Nature of Chemical Bonds: Molecular Orbital Theory

We said in Section 1.5 that chemists use two models for describing covalent bonds: valence bond theory and molecular orbital theory. Having now seen the valence bond approach, which uses hybrid atomic orbitals to account for geometry and assumes the overlap of atomic orbitals to account for electron sharing, let's look briefly at the molecular orbital approach to bonding.

Molecular orbital (MO) theory describes covalent bond formation as arising from a mathematical combination of atomic orbitals (wave functions) on different atoms to form *molecular orbitals,* so called because they belong to the entire *molecule* rather than to an individual atom. Just as an *atomic* orbital, whether unhybridized or hybridized, describes a region of space around an *atom* where an electron is likely to be found, so a *molecular* orbital describes a region of space in a *molecule* where an electron is most likely to be found.

Like an atomic orbital, a molecular orbital has a specific size, shape, and energy. In the H_2 molecule, for example, two singly occupied $1s$ atomic orbitals combine to form two molecular orbitals. The orbital combination can occur in two ways—an additive way or a subtractive way. The additive combination leads to formation of a molecular orbital that is lower in energy and roughly egg-shaped, while the subtractive combination leads to formation of a molecular orbital that is higher in energy and has a node between nuclei (Figure 1.18). Note that the additive combination is a *single* egg-shaped molecular orbital; it is not the same as the two overlapping $1s$ atomic orbitals of the valence bond description. Similarly, the subtractive combination is a single molecular orbital with the shape of an elongated dumbbell.

FIGURE 1.18 Molecular orbitals of H_2. Combination of two hydrogen $1s$ atomic orbitals leads to two H_2 molecular orbitals. The lower-energy, bonding MO is filled, and the higher-energy, antibonding MO is unfilled.

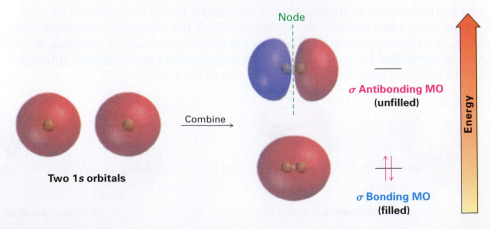

The additive combination is lower in energy than the two hydrogen 1s atomic orbitals and is called a **bonding MO** because electrons in this MO spend part of their time in the region between the two nuclei, thereby bonding the atoms together. The subtractive combination is higher in energy than the two hydrogen 1s orbitals and is called an **antibonding MO** because any electrons it contains *can't* occupy the central region between the nuclei, where there is a node, and can't contribute to bonding. The two nuclei therefore repel each other.

Just as bonding and antibonding σ molecular orbitals result from the combination of two s atomic orbitals in H_2, so bonding and antibonding π molecular orbitals result from the combination of two p atomic orbitals in ethylene. As shown in Figure 1.19, the lower-energy π bonding MO has no node between nuclei and results from combination of p orbital lobes with the same algebraic sign. The higher-energy π antibonding MO has a node between nuclei and results from combination of lobes with opposite algebraic signs. Only the bonding MO is occupied; the higher-energy, antibonding MO is vacant. We'll see in Sections 8.12 and 9.2 that molecular orbital theory is particularly useful for describing π bonds in compounds that have more than one double bond.

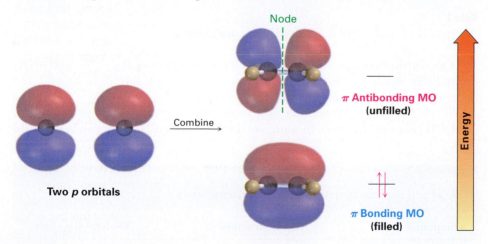

FIGURE 1.19 A molecular orbital description of the C–C π bond in ethylene. The lower-energy π bonding MO results from an additive combination of atomic orbitals and is filled. The higher-energy π antibonding MO results from a subtractive combination of atomic orbitals and is unfilled.

1.12 Drawing Chemical Structures

Let's cover one more point before ending this introductory chapter. In the structures we've been drawing until now, a line between atoms has represented the two electrons in a covalent bond. Drawing every bond and every atom is tedious, however, so chemists have devised several shorthand ways for writing structures. In **condensed structures**, carbon–hydrogen and carbon–carbon single bonds aren't shown; instead, they're understood. If a carbon has three hydrogens bonded to it, we write CH_3; if a carbon has two hydrogens bonded to it, we write CH_2; and so on. The compound called 2-methylbutane, for example, is written as follows:

Condensed structures

$$CH_3CH_2CHCH_3 \quad \text{or} \quad CH_3CH_2CH(CH_3)_2$$

2-Methylbutane

Notice that the horizontal bonds between carbons aren't shown in condensed structures—the CH_3, CH_2, and CH units are simply placed next to each other—but the vertical carbon–carbon bond in the first of the condensed structures just drawn is shown for clarity. Notice also in the second of the condensed structures that the two CH_3 units attached to the CH carbon are grouped together as $(CH_3)_2$.

Even simpler than condensed structures is the use of **skeletal structures**, such as those shown in Table 1.3. The rules for drawing skeletal structures are straightforward:

Rule 1

Carbon atoms aren't usually shown. Instead, a carbon atom is assumed to be at each intersection of two lines (bonds) and at the end of each line. Occasionally, a carbon atom might be indicated for emphasis or clarity.

Rule 2

Hydrogen atoms bonded to carbon aren't shown. Since carbon always has a valence of 4, we mentally supply the correct number of hydrogen atoms for each carbon.

Rule 3

Atoms other than carbon and hydrogen *are* shown.

One further comment: although such groupings as $-CH_3$, $-OH$, and $-NH_2$ are usually written with the C, O, or N atom first and the H atom second, the order of writing is sometimes inverted to H_3C-, $HO-$, and H_2N- if needed to make the bonding connections in a molecule clearer. Larger units such as $-CH_2CH_3$ are not inverted, though; we don't write H_3CH_2C- because it would

TABLE 1.3

Kekulé and Skeletal Structures for Some Compounds

Compound	Kekulé structure	Skeletal structure
Isoprene, C_5H_8	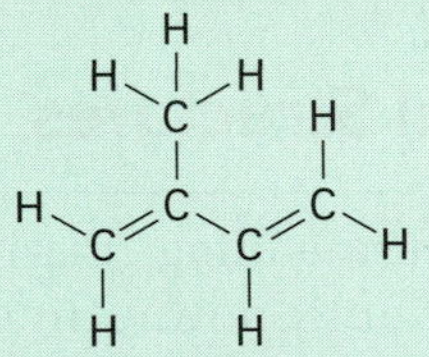	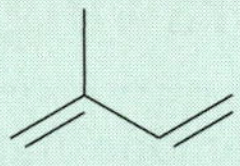
Methylcyclohexane, C_7H_{14}	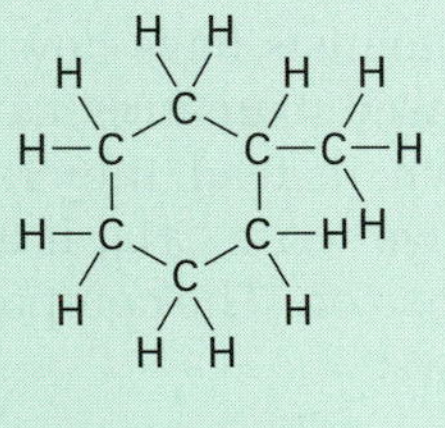	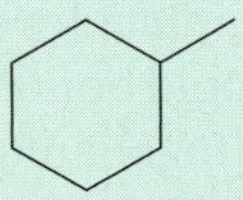
Phenol, C_6H_6O	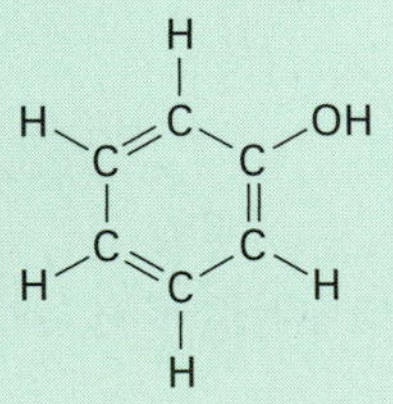	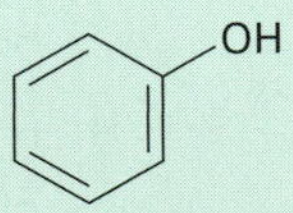

be confusing. There are, however, no well-defined rules that cover all cases; it's largely a matter of preference.

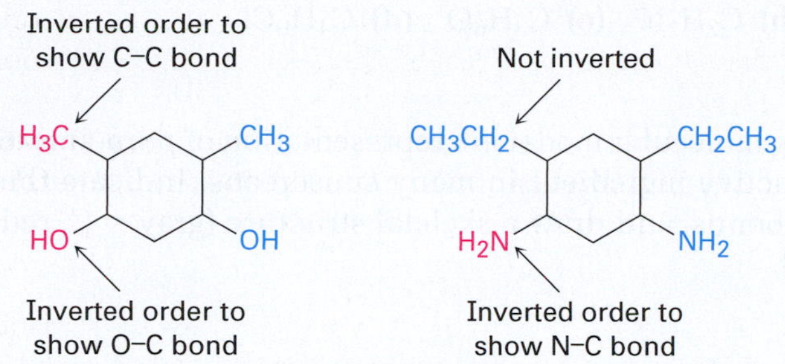
Inverted order to show C–C bond
Not inverted
H3C
CH3
CH3CH2
CH2CH3
HO
OH
H2N
NH2
Inverted order to show O–C bond
Inverted order to show N–C bond

WORKED EXAMPLE 1.3 Interpreting Line-Bond Structures

Carvone, a compound responsible for the odor of spearmint, has the following structure. Tell how many hydrogens are bonded to each carbon, and give the molecular formula of carvone.

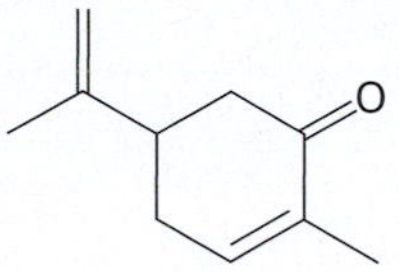

Carvone

Strategy

The end of a line represents a carbon atom with 3 hydrogens, CH_3; a two-way intersection is a carbon atom with 2 hydrogens, CH_2; a three-way intersection is a carbon atom with 1 hydrogen, CH; and a four-way intersection is a carbon atom with no attached hydrogens.

Solution

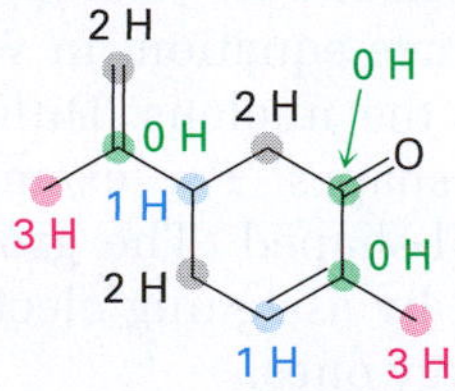
2 H
0 H
2 H
0 H
O
1 H
3 H
0 H
2 H
1 H
3 H

Carvone ($C_{10}H_{14}O$)

Problem 1.15

Tell how many hydrogens are bonded to each carbon in the following compounds, and give the molecular formula of each substance:

(a)

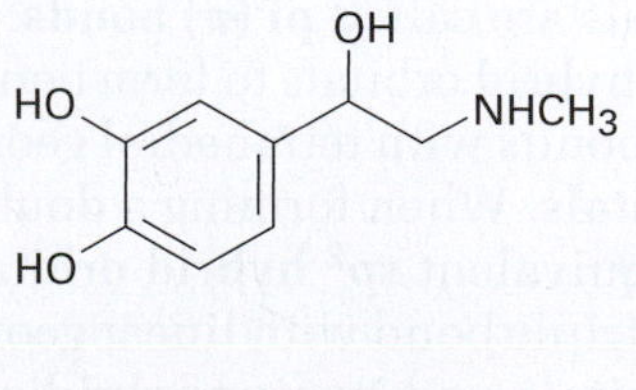
OH
HO
NHCH3
HO

Adrenaline

(b)

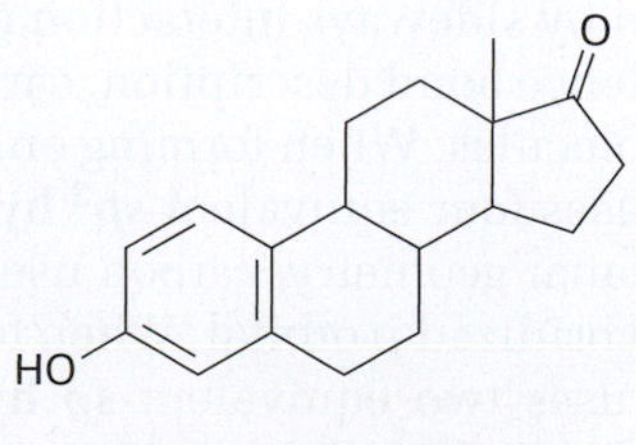
O
HO

Estrone (a hormone)

Problem 1.16

Propose skeletal structures for compounds that satisfy the following molecular formulas. There is more than one possibility in each case.

(a) C_5H_{12} **(b)** C_2H_7N **(c)** C_3H_6O **(d)** C_4H_9Cl

Problem 1.17

The following molecular model is a representation of *para*-aminobenzoic acid (PABA), the active ingredient in many sunscreens. Indicate the positions of the multiple bonds, and draw a skeletal structure (gray = C, red = O, blue = N, ivory = H).

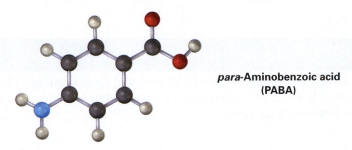

para-**Aminobenzoic acid
(PABA)**

Summary

Key Words

The purpose of this chapter has been to get you up to speed—to review some ideas about atoms, bonds, and molecular geometry.

As we've seen, **organic chemistry** is the study of carbon compounds. Although a division into organic and inorganic chemistry occurred historically, there is no scientific reason for the division.

An atom consists of a positively charged nucleus surrounded by one or more negatively charged electrons. The electronic structure of an atom can be described by a quantum mechanical wave equation, in which electrons are considered to occupy **orbitals** around the nucleus. Different orbitals have different energy levels and different shapes. For example, *s* orbitals are spherical, and *p* orbitals are dumbbell-shaped. The **ground-state electron configuration** of an atom can be found by assigning electrons to the proper orbitals, beginning with the lowest-energy ones.

A **covalent bond** is formed when an electron pair is shared between atoms. According to **valence bond theory**, electron sharing occurs by overlap of two atomic orbitals. According to **molecular orbital (MO) theory**, bonds result from the mathematical combination of atomic orbitals to give molecular orbitals, which belong to the entire molecule. Bonds that have a circular cross-section and are formed by head-on interaction are called **sigma (σ) bonds**; bonds formed by sideways interaction of *p* orbitals are called **pi (π) bonds**.

In the valence bond description, carbon uses hybrid orbitals to form bonds in organic molecules. When forming only single bonds with tetrahedral geometry, carbon uses four equivalent **sp^3 hybrid orbitals**. When forming a double bond with planar geometry, carbon uses three equivalent **sp^2 hybrid orbitals** and one unhybridized *p* orbital. When forming a triple bond with linear geometry, carbon uses two equivalent **sp hybrid orbitals** and two unhybridized *p* orbitals. Other atoms such as nitrogen, phosphorus, oxygen, and sulfur also use hybrid orbitals to form strong, oriented bonds.

Organic molecules are usually drawn using either condensed structures or skeletal structures. In **condensed structures**, carbon–carbon and carbon–hydrogen bonds aren't shown. In **skeletal structures**, only the bonds and not the atoms are shown. A carbon atom is assumed to be at the ends and at the junctions of lines (bonds), and the correct number of hydrogens is mentally supplied.

Lagniappe

Chemicals, Toxicity, and Risk

Lagniappe, pronounced lan-yap, is a word in the Creole dialect of southern Louisiana meaning an extra benefit, or a little something extra. Which is just what these small pieces at the ends of chapters are intended to be. You might find them interesting to read when you need a short break from studying.

KEITH LARRETT/AP Photo

We all take many risks each day, some more dangerous than others.

We hear and read a lot these days about the dangers of "chemicals"—about pesticide residues on our food, toxic wastes on our land, unsafe medicines, and so forth. What's a person to believe?

Life is not risk-free; we all take many risks each day. We decide to ride a bike rather than drive, even though there is a ten times greater likelihood per mile of dying in a bicycling accident than in a car accident. We decide to walk down stairs rather than take an elevator, even though 7000 people die from falls each year in the United States. Some of us decide to smoke cigarettes, even though it increases our chance of getting cancer by 50%. Making decisions that affect our health is something we do routinely without even thinking about it.

What about risks from chemicals? Risk evaluation of chemicals is carried out by exposing test animals, usually mice or rats, to the chemical and then monitoring for signs of harm. To limit the expense and time needed, the amounts administered are hundreds or thousands of times greater than those a person might normally encounter. The *acute chemical toxicity* (as opposed to chronic toxicity) observed in animal tests is reported as a single number called an LD_{50}, the amount of a substance per kilogram body weight that is lethal to 50% of the test animals. The LD_{50} values of some common substances are shown in Table 1.4. The lower the value, the more toxic the substance.

Even with an LD_{50} value established in test animals, the risk of human exposure is still hard to assess. If a substance is harmful to animals, is it necessarily harmful to humans? How can a large dose for a small animal be translated into a small dose for a large human? All substances are toxic to some organisms to some extent, and the difference between help and harm is often a matter of degree. Vitamin A, for example, is necessary for vision, yet it can promote cancer at high dosages. Arsenic trioxide is the most classic of poisons, yet it induces remissions in some types of leukemia and is sold for drug use under the name Trisenox. Even water is toxic if drunk in large amounts because it dilutes the salt in body fluids and causes a potentially life-threatening condition called *hyponatremia,* which has resulted in the death of several marathon runners. Furthermore, how we evaluate risk is strongly influenced by familiarity. Many foods contain small amounts of natural ingredients that are far more toxic than synthetic additives or pesticide residues, but the ingredients are ignored because the foods are familiar.

All decisions involve tradeoffs. Does the benefit of increased food production outweigh possible health risks of a pesticide? Do the beneficial effects of a new drug outweigh a potentially dangerous side effect in a small fraction of users? Different people will have different opinions, but an honest evaluation of the facts is surely a good way to start.

TABLE 1.4
Some LD$_{50}$ Values

Substance	LD$_{50}$ (mg/kg)	Substance	LD$_{50}$ (mg/kg)
Strychnine	5	Chloroform	1,200
Arsenic trioxide	15	Iron(II) sulfate	1,500
DDT	115	Ethyl alcohol	7,100
Aspirin	1,100	Sodium cyclamate	12,800

WORKING PROBLEMS

There is no surer way to learn organic chemistry than by working problems. Although careful reading and rereading of this text are important, reading alone isn't enough. You must also be able to use the information you've read and be able to apply your knowledge in new situations. Working problems gives you practice at doing this.

Each chapter in this book provides many problems of different sorts. The in-chapter problems are placed for immediate reinforcement of ideas just learned; the end-of-chapter problems provide additional practice and are of several types. They begin with a short section called "Visualizing Chemistry," which helps you "see" the microscopic world of molecules and provides practice for working in three dimensions. After the visualizations are many "Additional Problems." Early problems are primarily of the drill type, providing an opportunity for you to practice your command of the fundamentals. Later problems tend to be more thought-provoking, and some are real challenges.

As you study organic chemistry, take the time to work the problems. Do the ones you can, and ask for help on the ones you can't. If you're stumped by a particular problem, check the accompanying *Study Guide and Solutions Manual* for an explanation that will help clarify the difficulty. Working problems takes effort, but the payoff in knowledge and understanding is immense.

Exercises

VISUALIZING CHEMISTRY

(Problems 1.1–1.17 appear within the chapter.)

1.18 ■ Convert each of the following molecular models into a skeletal structure, and give the formula of each. Only the connections between atoms are shown; multiple bonds are not indicated (gray = C, red = O, blue = N, ivory = H).

(a)

Coniine (the toxic substance
in poison hemlock)

(b)

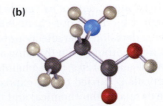

Alanine (an amino acid)

1.19 ▪ The following model is a representation of citric acid, a compound in the so-called citric acid cycle by which food molecules are metabolized in the body. Only the connections between atoms are shown; multiple bonds are not indicated. Complete the structure by indicating the positions of multiple bonds and lone-pair electrons (gray = C, red = O, ivory = H).

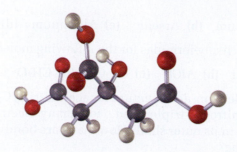

1.20 ▪ The following model is a representation of acetaminophen, a pain reliever sold in drugstores under a variety of names, including Tylenol. Identify the hybridization of each carbon atom in acetaminophen, and tell which atoms have lone pairs of electrons (gray = C, red = O, blue = N, ivory = H).

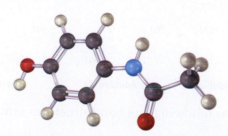

1.21 The following model is a representation of aspartame, $C_{14}H_{18}N_2O_5$, known commercially under many names, including NutraSweet. Only the connections between atoms are shown; multiple bonds are not indicated. Draw a skeletal structure for aspartame, and indicate the positions of multiple bonds (gray = C, red = O, blue = N, ivory = H).

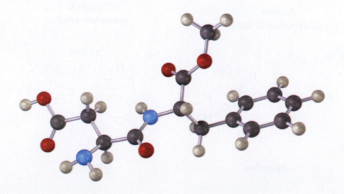

▪ Problems assignable in Organic OWL.

ADDITIONAL PROBLEMS

1.22 ▪ How many valence electrons does each of the following dietary trace elements have?

(a) Zinc **(b)** Iodine **(c)** Silicon **(d)** Iron

1.23 ▪ Give the ground-state electron configuration for each of the following elements:

(a) Potassium **(b)** Arsenic **(c)** Aluminum **(d)** Germanium

1.24 ▪ What are likely formulas for the following molecules?

(a) $NH_?OH$ **(b)** $AlCl_?$ **(c)** $CF_2Cl_?$ **(d)** $CH_?O$

1.25 Draw an electron-dot structure for acetonitrile, C_2H_3N, which contains a carbon–nitrogen triple bond. How many electrons does the nitrogen atom have in its outer shell? How many are bonding, and how many are nonbonding?

1.26 What is the hybridization of each carbon atom in acetonitrile (Problem 1.25)?

1.27 ▪ Draw a line-bond structure for vinyl chloride, C_2H_3Cl, the starting material from which PVC [poly(vinyl chloride)] plastic is made.

1.28 ▪ Fill in any nonbonding valence electrons that are missing from the following structures:

(a)

H3C
S
S
CH3

Dimethyl disulfide

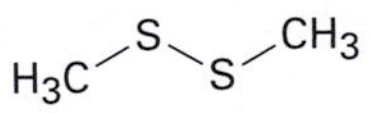

(b)

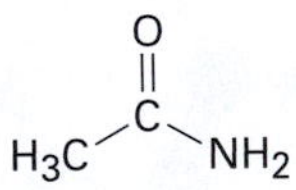

O
C
H3C
NH2

Acetamide

(c)

O
C
H3C
O−

Acetate ion

1.29 ▪ Convert the following line-bond structures into molecular formulas:

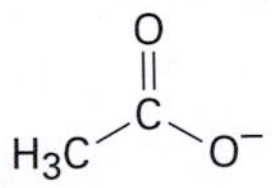

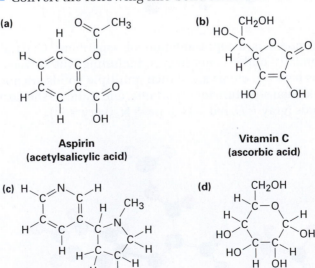

(a)

Aspirin (acetylsalicylic acid)

(b)

Vitamin C (ascorbic acid)

(c)

Nicotine

(d)

Glucose

▪ Problems assignable in Organic OWL.

1.30 ■ Convert the following molecular formulas into structures that are consistent with the usual bonding patterns:

(a) C_3H_8 (b) CH_5N

(c) C_2H_6O (2 possibilities)   (d) C_3H_7Br (2 possibilities)

(e) C_2H_4O (3 possibilities)   (f) C_3H_9N (4 possibilities)

1.31 ■ What kind of hybridization do you expect for each carbon atom in the following molecules?

(a) Propane, $CH_3CH_2CH_3$

(b) 2-Methylpropene, CH_3
                          |
              $CH_3C{=}CH_2$

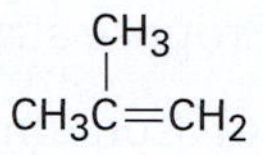

(c) But-1-en-3-yne, $H_2C{=}CH{-}C{\equiv}CH$

(d) Acetic acid,   O
                    ||
            CH_3COH

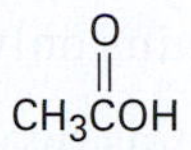

1.32 What is the overall shape of benzene, and what hybridization do you expect for each carbon?

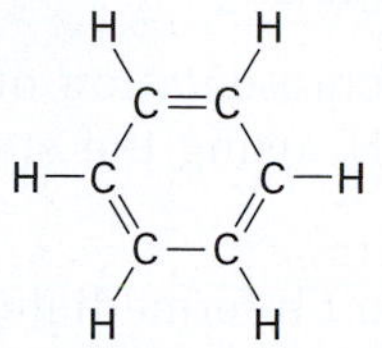

Benzene

1.33 ■ What values do you expect for the indicated bond angles in each of the following molecules, and what kind of hybridization do you expect for the central atom in each?

(a)   O
       ||
$H_2N{-}CH_2{-}C{-}OH$

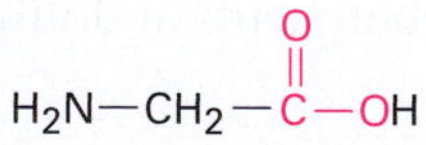

Glycine
(an amino acid)

(b) Pyridine

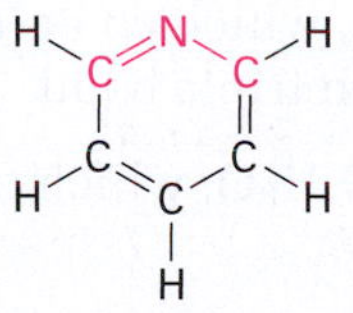

(c)   OH O
       | ||
$CH_3{-}CH{-}C{-}OH$

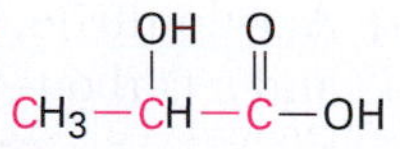

Lactic acid
(in sour milk)

1.34 ■ Convert the following structures into skeletal drawings:

(a) Indole

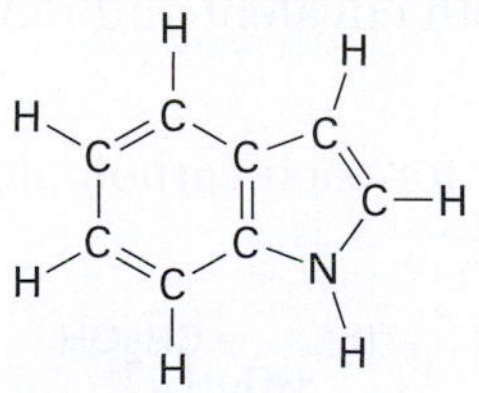

(b) Penta-1,3-diene

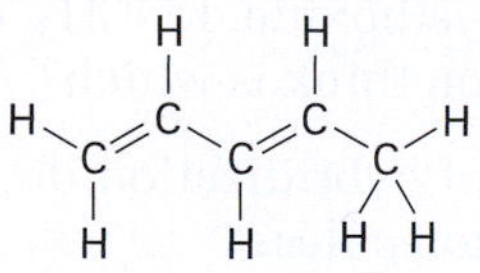

(c) 1,2-Dichlorocyclopentane

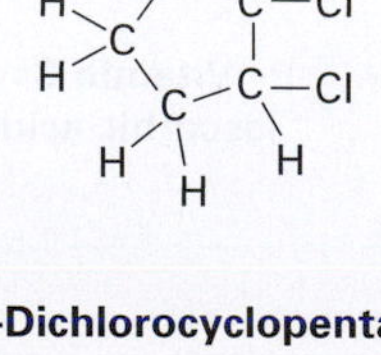

(d)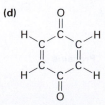

Benzoquinone

■ Problems assignable in Organic OWL.

1.35 ■ Tell the number of hydrogens bonded to each carbon atom in the following substances, and give the molecular formula of each:

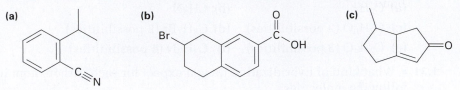

(a)   **(b)**   **(c)**

1.36 Propose structures for molecules that meet the following descriptions:

(a) Contains two sp^2-hybridized carbons and two sp^3-hybridized carbons

(b) Contains only four carbons, all of which are sp^2-hybridized

(c) Contains two sp-hybridized carbons and two sp^2-hybridized carbons

1.37 ■ Why can't molecules with the following formulas exist?

(a) CH_5 **(b)** C_2H_6N **(c)** $C_3H_5Br_2$

1.38 Draw a three-dimensional representation of the oxygen-bearing carbon atom in ethanol, CH_3CH_2OH, using the standard convention of solid, wedged, and dashed lines.

1.39 Oxaloacetic acid, an important intermediate in food metabolism, has the formula $C_4H_4O_5$ and contains three C=O bonds and two O–H bonds. Propose two possible structures.

1.40 ■ Draw structures for the following molecules, showing lone pairs:

(a) Acrylonitrile, C_3H_3N, which contains a carbon–carbon double bond and a carbon–nitrogen triple bond

(b) Ethyl methyl ether, C_3H_8O, which contains an oxygen atom bonded to two carbons

(c) Butane, C_4H_{10}, which contains a chain of four carbon atoms

(d) Cyclohexene, C_6H_{10}, which contains a ring of six carbon atoms and one carbon–carbon double bond

1.41 Potassium methoxide, $KOCH_3$, contains both covalent and ionic bonds. Which do you think is which?

1.42 What kind of hybridization do you expect for each carbon atom in the following molecules?

(a)

Procaine

(b)

Vitamin C
(ascorbic acid)

1.43 Pyridoxal phosphate, a close relative of vitamin B_6, is involved in a large number of metabolic reactions. Tell the hybridization, and predict the bond angles for each nonterminal atom.

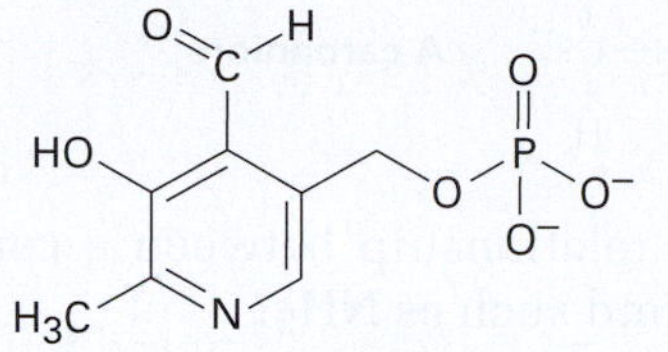

Pyridoxal phosphate

1.44 Why do you suppose no one has ever been able to make cyclopentyne as a stable molecule?

Cyclopentyne

1.45 Allene, $H_2C=C=CH_2$, is somewhat unusual in that it has two adjacent double bonds. Draw a picture showing the orbitals involved in the σ and π bonds of allene. Is the central carbon atom sp^2- or sp-hybridized? What about the hybridization of the terminal carbons? What shape do you predict for allene?

1.46 Allene (see Problem 1.45) is related structurally to carbon dioxide, CO_2. Draw a picture showing the orbitals involved in the σ and π bonds of CO_2, and identify the likely hybridization of carbon.

1.47 Complete the electron-dot structure of caffeine, showing all lone-pair electrons, and identify the hybridization of the indicated atoms.

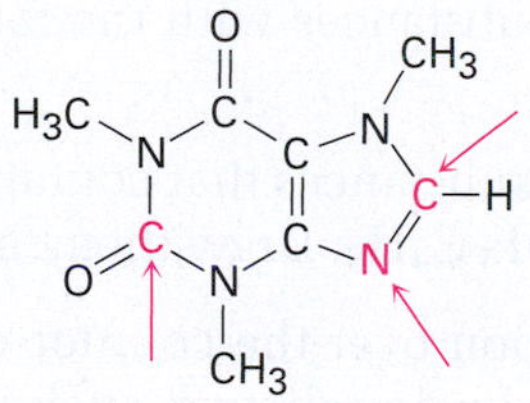

Caffeine

1.48 Almost all stable organic species have tetravalent carbon atoms, but species with trivalent carbon atoms also exist. *Carbocations* are one such class of compounds.

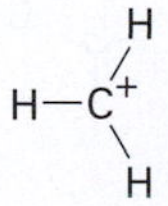

A carbocation

(a) How many valence electrons does the positively charged carbon atom have?

(b) What hybridization do you expect this carbon atom to have?

(c) What geometry is the carbocation likely to have?

1.49 A *carbanion* is a species that contains a negatively charged, trivalent carbon.

$$\underset{\overset{\displaystyle |}{H}}{\overset{\overset{\displaystyle H}{|}}{H-C{:}^{-}}} \qquad \textbf{A carbanion}$$

(a) What is the electronic relationship between a carbanion and a trivalent nitrogen compound such as NH_3?

(b) How many valence electrons does the negatively charged carbon atom have?

(c) What hybridization do you expect this carbon atom to have?

(d) What geometry is the carbanion likely to have?

1.50 Divalent carbon species called *carbenes* are capable of fleeting existence. For example, methylene, $:CH_2$, is the simplest carbene. The two unshared electrons in methylene can be either spin-paired in a single orbital or unpaired in different orbitals. Predict the type of hybridization you expect carbon to adopt in singlet (spin-paired) methylene and triplet (spin-unpaired) methylene. Draw a picture of each, and identify the valence orbitals on carbon.

1.51 There are two different substances with the formula C_4H_{10}. Draw both, and tell how they differ.

1.52 There are two different substances with the formula C_3H_6. Draw both, and tell how they differ.

1.53 There are two different substances with the formula C_2H_6O. Draw both, and tell how they differ.

1.54 There are three different substances that contain a carbon–carbon double bond and have the formula C_4H_8. Draw them, and tell how they differ.

1.55 ■ Among the most common over-the-counter drugs you might find in a medicine cabinet are mild pain relievers such ibuprofen (Advil, Motrin), naproxen (Aleve), and acetaminophen (Tylenol).

Ibuprofen   **Naproxen**   **Acetaminophen**

(a) How many sp^3-hybridized carbons does each molecule have?

(b) How many sp^2-hybridized carbons does each molecule have?

(c) What similarities do you see in their structures?

2 Polar Covalent Bonds; Acids and Bases

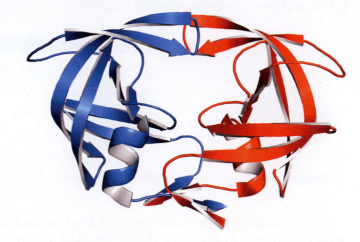

HIV protease processes proteins during the life cycle of the AIDS virus.

We saw in the last chapter how covalent bonds between atoms are described, and we looked at the valence bond model, which uses hybrid orbitals to account for the observed shapes of organic molecules. Before going on to a systematic study of organic chemistry, however, we still need to review a few fundamental topics. In particular, we need to look more closely at how electrons are distributed in covalent bonds and at some of the consequences that arise when the electrons in a bond are not shared equally between atoms.

WHY THIS CHAPTER?

Understanding biological organic chemistry means knowing not just what happens but also why and how it happens at the molecular level. This chapter reviews some of the ways that chemists describe and account for chemical reactivity, thereby providing a foundation for understanding the specific reactions discussed in subsequent chapters. Topics such as bond polarity, the acid–base behavior of molecules, and hydrogen-bonding are a particularly important part of that foundation.

2.1 Polar Covalent Bonds: Electronegativity

Up to this point, we've treated chemical bonds as either ionic or covalent. The bond in sodium chloride, for instance, is ionic. Sodium transfers an electron to chlorine to give Na^+ and Cl^- ions, which are held together in the solid by electrostatic attractions between the unlike charges. The C–C bond in ethane, however, is covalent. The two bonding electrons are shared equally by the two equivalent carbon atoms, resulting in a symmetrical electron

OWL Online homework for this chapter can be assigned in Organic OWL.

distribution in the bond. Most bonds, however, are neither fully ionic nor fully covalent but are somewhere between the two extremes. Such bonds are called **polar covalent bonds**, meaning that the bonding electrons are attracted more strongly by one atom than the other so that the electron distribution between atoms is not symmetrical (Figure 2.1).

FIGURE 2.1 The continuum in bonding from covalent to ionic is a result of an unequal distribution of bonding electrons between atoms. The symbol δ (lowercase Greek delta) means *partial* charge, either partial positive (δ+) for the electron-poor atom or partial negative (δ−) for the electron-rich atom.

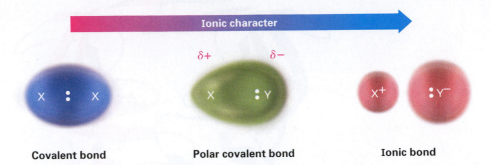

Bond polarity is due to differences in **electronegativity (EN)**, the intrinsic ability of an atom to attract the shared electrons in a covalent bond. As shown in Figure 2.2, electronegativities are based on an arbitrary scale, with fluorine being the most electronegative (EN = 4.0) and cesium, the least (EN = 0.7). Metals on the left side of the periodic table attract electrons weakly and have lower electronegativities, while oxygen, nitrogen, and halogens on the right side of the periodic table attract electrons strongly and have higher electronegativities. Carbon, the most important element in organic compounds, has an electronegativity value of 2.5.

FIGURE 2.2 Electronegativity values and trends. Electronegativity generally increases from left to right across the periodic table and decreases from top to bottom. The values are on an arbitrary scale, with F = 4.0 and Cs = 0.7. Elements in red-orange are the most electronegative, those in yellow are medium, and those in green are the least electronegative.

H 2.1																		He
Li 1.0	Be 1.6											B 2.0	C 2.5	N 3.0	O 3.5	F 4.0	Ne
Na 0.9	Mg 1.2											Al 1.5	Si 1.8	P 2.1	S 2.5	Cl 3.0	Ar
K 0.8	Ca 1.0	Sc 1.3	Ti 1.5	V 1.6	Cr 1.6	Mn 1.5	Fe 1.8	Co 1.9	Ni 1.9	Cu 1.9	Zn 1.6	Ga 1.6	Ge 1.8	As 2.0	Se 2.4	Br 2.8	Kr
Rb 0.8	Sr 1.0	Y 1.2	Zr 1.4	Nb 1.6	Mo 1.8	Tc 1.9	Ru 2.2	Rh 2.2	Pd 2.2	Ag 1.9	Cd 1.7	In 1.7	Sn 1.8	Sb 1.9	Te 2.1	I 2.5	Xe
Cs 0.7	Ba 0.9	La 1.0	Hf 1.3	Ta 1.5	W 1.7	Re 1.9	Os 2.2	Ir 2.2	Pt 2.2	Au 2.4	Hg 1.9	Tl 1.8	Pb 1.9	Bi 1.9	Po 2.0	At 2.1	Rn

As a rough guide, bonds between atoms whose electronegativities differ by less than 0.5 are nonpolar covalent, bonds between atoms whose electronegativities differ by 0.5 to 2 are polar covalent, and bonds between atoms whose electronegativities differ by more than 2 are largely ionic. Carbon–hydrogen bonds, for example, are relatively nonpolar because carbon (EN = 2.5) and hydrogen (EN = 2.1) have similar electronegativities. Bonds between carbon and *more* electronegative elements, such as oxygen (EN = 3.5) and nitrogen (EN = 3.0), by contrast, are polarized so that the bonding electrons are drawn away from carbon toward the electronegative atom. This leaves carbon with a partial positive charge, denoted by δ+, and the electronegative atom with a partial negative charge, δ− (δ is the lowercase Greek letter delta). An example is the C–O bond in methanol, CH_3OH (Figure 2.3a). Bonds between carbon and *less* electronegative elements are polarized so that carbon bears a partial negative charge and the other atom bears a partial positive charge. An example is the C–Li bond in methyllithium, CH_3Li (Figure 2.3b).

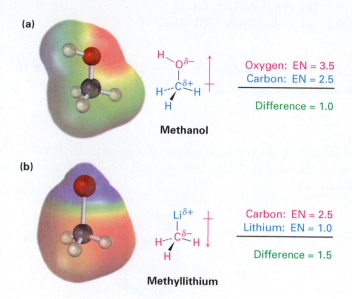

(a)

$$H\diagdown O^{\delta-}$$
$$H-\underset{H}{\overset{|}{C}}{}^{\delta+}$$

Methanol

Oxygen: EN = 3.5
Carbon: EN = 2.5

Difference = 1.0

(b)

$$Li^{\delta+}$$
$$H-\underset{H}{\overset{|}{C}}{}^{\delta-}H$$

Methyllithium

Carbon: EN = 2.5
Lithium: EN = 1.0

Difference = 1.5

FIGURE 2.3 **(a)** Methanol, CH_3OH, has a polar covalent C–O bond, and **(b)** methyllithium, CH_3Li, has a polar covalent C–Li bond. The computer-generated representations, called electrostatic potential maps, use color to show calculated charge distributions, ranging from red (electron-rich; $\delta-$) to blue (electron-poor; $\delta+$).

Note in the representations of methanol and methyllithium in Figure 2.3 that a crossed arrow $\longmapsto$ is used to indicate the direction of bond polarity. By convention, *electrons are displaced in the direction of the arrow.* The tail of the arrow (which looks like a plus sign) is electron-poor ($\delta+$), and the head of the arrow is electron-rich ($\delta-$).

Note also in Figure 2.3 that calculated charge distributions in molecules can be displayed visually using so-called electrostatic potential maps, which use color to indicate electron-rich (red; $\delta-$) and electron-poor (blue; $\delta+$) regions. In methanol, oxygen carries a partial negative charge and is colored red, while the carbon and hydrogen atoms carry partial positive charges and are colored blue-green. In methyllithium, lithium carries a partial positive charge (blue), while carbon and the hydrogen atoms carry partial negative charges (red). Electrostatic potential maps are useful because they show at a glance the electron-rich and electron-poor atoms in molecules. We'll make frequent use of these maps throughout the text and will see how electronic structure often correlates with chemical reactivity.

When speaking of an atom's ability to polarize a bond, we often use the term *inductive effect.* An **inductive effect** is simply the shifting of electrons in a σ bond in response to the electronegativity of nearby atoms. Metals, such as lithium and magnesium, inductively donate electrons, whereas reactive nonmetals, such as oxygen and nitrogen, inductively withdraw electrons. Inductive effects play a major role in understanding chemical reactivity, and we'll use them many times throughout this text to explain a variety of chemical phenomena.

Problem 2.1
Which element in each of the following pairs is more electronegative?
(a) Li or H   **(b)** B or Br   **(c)** Cl or I   **(d)** C or H

Problem 2.2

Use the $\delta+/\delta-$ convention to show the direction of expected polarity for each of the bonds indicated.

(a) $H_3C{-}Cl$ **(b)** $H_3C{-}NH_2$ **(c)** $H_2N{-}H$
(d) $H_3C{-}SH$ **(e)** $H_3C{-}MgBr$ **(f)** $H_3C{-}F$

Problem 2.3

Use the electronegativity values shown in Figure 2.2 to rank the following bonds from least polar to most polar: $H_3C{-}Li$, $H_3C{-}K$, $H_3C{-}F$, $H_3C{-}MgBr$, $H_3C{-}OH$

Problem 2.4

Look at the following electrostatic potential map of methylamine, a substance responsible for the odor of rotting fish, and tell the direction of polarization of the C–N bond:

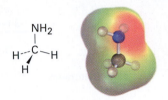

Methylamine

2.2 Polar Covalent Bonds: Dipole Moments

Just as individual bonds are often polar, molecules as a whole are often polar also. Molecular polarity results from the vector summation of all individual bond polarities and lone-pair contributions in the molecule. As a practical matter, strongly polar substances are often soluble in polar solvents like water, whereas nonpolar substances are insoluble in water.

Net molecular polarity is measured by a quantity called the *dipole moment* and can be thought of in the following way: assume that there is a center of mass of all positive charges (nuclei) in a molecule and a center of mass of all negative charges (electrons). If these two centers don't coincide, then the molecule has a net polarity.

The **dipole moment**, μ (Greek mu), is defined as the magnitude of the charge Q at either end of the molecular dipole times the distance r between the charges, $\mu = Q \times r$. Dipole moments are expressed in *debyes* (D), where 1 D = 3.336×10^{-30} coulomb meter (C · m) in SI units. For example, the unit charge on an electron is 1.60×10^{-19} C. Thus, if one positive charge and one negative charge were separated by 100 pm (a bit less than the length of a typical covalent bond), the dipole moment would be 1.60×10^{-29} C · m, or 4.80 D.

$$\mu = Q \times r$$

$$\mu = (1.60 \times 10^{-19}\ C)(100 \times 10^{-12}\ m)\left(\frac{1\ D}{3.336 \times 10^{-30}\ C \cdot m}\right) = 4.80\ D$$

It's relatively easy to measure dipole moments in the laboratory, and values for some common substances are given in Table 2.1. Of the compounds shown in the table, sodium chloride has the largest dipole moment (9.00 D) because it is ionic. Even small molecules like water (μ = 1.85 D), methanol (CH_3OH; μ = 1.70 D), and ammonia (μ = 1.47 D), have substantial dipole moments, however, both because they contain strongly electronegative atoms (oxygen and nitrogen) and because all three molecules have lone-pair electrons. The lone-pair electrons on oxygen and nitrogen stick out into space away from the positively charged nuclei, giving rise to a considerable charge separation and making a large contribution to the dipole moment.

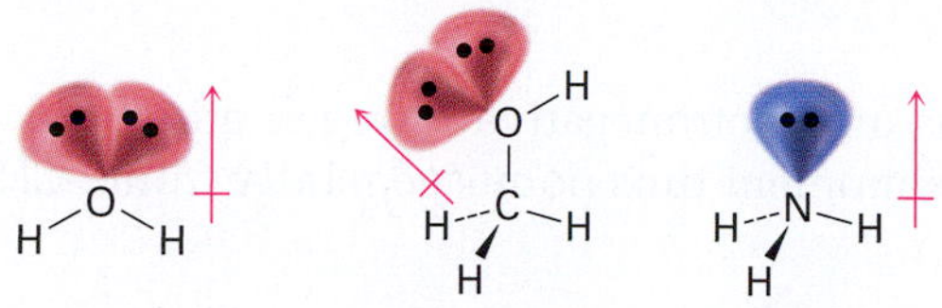

Water
(μ = **1.85 D**)

Methanol
(μ = **1.70 D**)

Ammonia
(μ = **1.47 D**)

In contrast with water, methanol, ammonia, and other substances in Table 2.1, carbon dioxide, methane, ethane, and benzene have zero dipole moments. Because of the symmetrical structures of these molecules, the individual bond polarities and lone-pair contributions exactly cancel.

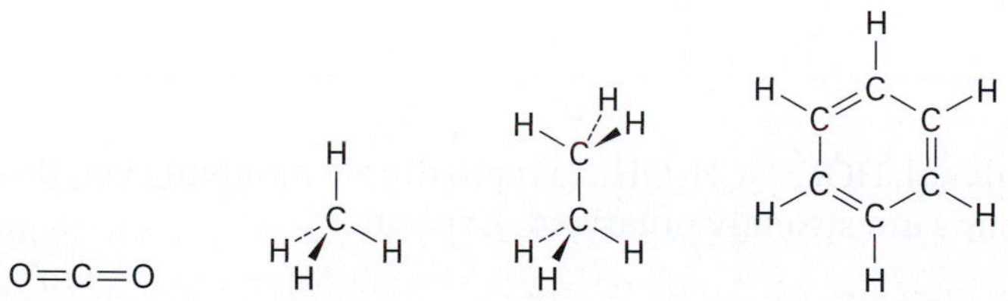

Carbon dioxide
($\mu = 0$)

Methane
($\mu = 0$)

Ethane
($\mu = 0$)

Benzene
($\mu = 0$)

TABLE 2.1
Dipole Moments of Some Compounds

Compound	Dipole moment (D)	Compound	Dipole moment (D)
NaCl	9.00	NH_3	1.47
CH_2O	2.33	CH_3NH_2	1.31
CH_3Cl	1.87	CO_2	0
H_2O	1.85	CH_4	0
CH_3OH	1.70	CH_3CH_3	0
CH_3CO_2H	1.70	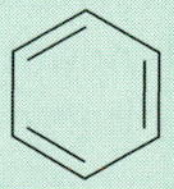	0
CH_3SH	1.52		
		Benzene	

WORKED EXAMPLE 2.1 Predicting the Direction of a Dipole Moment

Make a three-dimensional drawing of methylamine, CH_3NH_2, and show the direction of its dipole moment ($\mu = 1.31$).

Strategy

Look for any lone-pair electrons, and identify any atom with an electronegativity substantially different from that of carbon. (Usually, this means O, N, F, Cl, or Br.) Electron density will be displaced in the general direction of the electronegative atoms and the lone pairs.

Solution

Methylamine has an electronegative nitrogen atom and a lone pair of electrons. The dipole moment thus points generally from –CH_3 toward nitrogen.

Methylamine
($\mu = 1.31$)

Problem 2.5

Ethylene glycol, $HOCH_2CH_2OH$, has zero dipole moment even though carbon–oxygen bonds are strongly polarized. Explain.

Problem 2.6

Make three-dimensional drawings of the following molecules, and predict whether each has a dipole moment. If you expect a dipole moment, show its direction.
(a) $H_2C{=}CH_2$ **(b)** $CHCl_3$ **(c)** CH_2Cl_2 **(d)** $H_2C{=}CCl_2$

2.3 Formal Charges

Closely related to the ideas of bond polarity and dipole moment is the concept of assigning *formal charges* to specific atoms within a molecule, particularly atoms that have an apparently "abnormal" number of bonds. Look at dimethyl sulfoxide (CH_3SOCH_3), for instance, a solvent commonly used for preserving biological cell lines at low temperatures. The sulfur atom in dimethyl sulfoxide has three bonds rather than the usual two and has a formal positive charge. The oxygen atom, by contrast, has one bond rather than the usual two and has a formal negative charge. Note that an electrostatic potential map of dimethyl sulfoxide shows the oxygen as negative (red) and the sulfur as relatively positive (blue), just as the formal charges suggest.

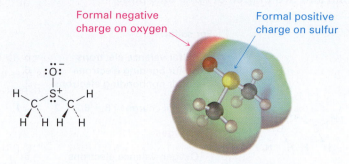

Dimethyl sulfoxide

Formal charges, as the name suggests, are a formalism and don't imply the presence of actual ionic charges in a molecule. Instead, they're a device for electron "bookkeeping" and can be thought of in the following way: a typical covalent bond is formed when each atom donates one electron. Although the bonding electrons are shared by both atoms, each atom can still be considered to own one electron for bookkeeping purposes. In methane, for instance, the carbon atom owns one electron in each of the four C–H bonds, for a total of four. Because a neutral, isolated carbon atom has four valence electrons, and because the carbon atom in methane still owns four, the methane carbon atom is neutral and has no formal charge.

An isolated carbon atom owns 4 valence electrons.

This carbon atom also owns $\frac{8}{2}$ = 4 valence electrons.

The same is true for the nitrogen atom in ammonia, which has three covalent N–H bonds and two nonbonding electrons (a lone pair). Atomic nitrogen has five valence electrons, and the ammonia nitrogen also has five—one in each of three shared N–H bonds plus two in the lone pair. Thus, the nitrogen atom in ammonia has no formal charge.

An isolated nitrogen atom owns 5 valence electrons.

This nitrogen atom also owns $\frac{6}{2}$ + 2 = 5 valence electrons.

The situation is different in dimethyl sulfoxide. Atomic sulfur has six valence electrons, but the dimethyl sulfoxide sulfur owns only *five*—one in each of the two S–C single bonds, one in the S–O single bond, and two in a lone pair. Thus, the sulfur atom has formally lost an electron and therefore has a positive charge. A similar calculation for the oxygen atom shows that it has formally gained an electron and has a negative charge: atomic oxygen has six valence electrons, but the oxygen in dimethyl sulfoxide has seven—one in the O–S bond and two in each of three lone pairs.

For sulfur:

Sulfur valence electrons	= 6
Sulfur bonding electrons	= 6
Sulfur nonbonding electrons	= 2

Formal charge = 6 − 6/2 − 2 = +1

For oxygen:

Oxygen valence electrons	= 6
Oxygen bonding electrons	= 2
Oxygen nonbonding electrons	= 6

Formal charge = 6 − 2/2 − 6 = −1

To express the calculations in a general way, the **formal charge** on an atom is equal to the number of valence electrons in a neutral, isolated atom minus the number of electrons owned by that atom in a molecule. The number of electrons in the bonded atom, in turn, is equal to half the number of bonding electrons plus the nonbonding, lone-pair electrons.

$$\text{Formal charge} = \left(\begin{array}{c}\text{Number of}\\\text{valence electrons}\\\text{in free atom}\end{array}\right) - \left(\begin{array}{c}\text{Number of}\\\text{valence electrons}\\\text{in bonded atom}\end{array}\right)$$

$$= \left(\begin{array}{c}\text{Number of}\\\text{valence electrons}\\\text{in free atom}\end{array}\right) - \left(\frac{\begin{array}{c}\text{Number of}\\\text{bonding electrons}\end{array}}{2}\right) - \left(\begin{array}{c}\text{Number of}\\\text{nonbonding}\\\text{electrons}\end{array}\right)$$

A summary of commonly encountered formal charges and the bonding situations in which they occur is given in Table 2.2. Although only a bookkeeping

TABLE 2.2
A Summary of Common Formal Charges

Atom	C		N		O		S		P
Structure	$-\overset{+}{\underset{\vert}{C}}-$	$-\overset{\cdot\cdot}{\underset{\vert}{C}}-$	$-\overset{\vert}{\underset{\vert}{N}}{}^{+}-$	$-\overset{\cdot\cdot}{\underset{\cdot\cdot}{N}}-$	$-\overset{\cdot\cdot}{\underset{\vert}{O}}{}^{+}-$	$-\overset{\cdot\cdot}{\underset{\cdot\cdot}{O}}:^{-}$	$-\overset{\cdot\cdot}{\underset{\vert}{S}}-$	$-\overset{\cdot\cdot}{\underset{\cdot\cdot}{S}}:^{-}$	$-\overset{\vert}{\underset{\vert}{P}}{}^{+}-$
Valence electrons	4	4	5	5	6	6	6	6	5
Number of bonds	3	3	4	2	3	1	3	1	4
Number of lone pairs	0	1	0	2	1	3	1	3	0
Formal charge	+1	−1	+1	−1	+1	−1	+1	−1	+1

device, formal charges often give clues about chemical reactivity, so it's helpful to be able to identify and calculate them correctly.

Problem 2.7

Calculate formal charges for the nonhydrogen atoms in the following molecules:

(a) Diazomethane, $H_2C{=}N{=}\ddot{N}:$

(b) Acetonitrile oxide, $H_3C{-}C{\equiv}N{-}\ddot{\underset{\cdot\cdot}{O}}:$

(c) Methyl isocyanide, $H_3C{-}N{\equiv}C:$

Problem 2.8

Organic phosphate groups occur commonly in biological molecules. Calculate formal charges on the four O atoms in the methyl phosphate ion.

$$\left[\begin{array}{c} \text{H} \qquad :\!\overset{\cdot\cdot}{O}\!: \\ | \qquad\; \| \\ \text{H}-\text{C}-\overset{\cdot\cdot}{\underset{\cdot\cdot}{O}}-\text{P}-\overset{\cdot\cdot}{\underset{\cdot\cdot}{O}}: \\ | \qquad\; | \\ \text{H} \qquad :\!\overset{\cdot\cdot}{\underset{\cdot\cdot}{O}}\!: \end{array} \right]^{2-}$$

Methyl phosphate ion

2.4 Resonance

Most substances can be represented by the Kekulé line-bond structures we've been using up to this point, but an interesting problem sometimes arises. Look at the acetate ion, for instance. When we draw a line-bond structure for acetate, we need to show a double bond to one oxygen and a single bond to the other. But which oxygen is which? Should we draw a double bond to the "top" oxygen and a single bond to the "bottom" oxygen or vice versa?

Double bond to this oxygen?

Acetate ion   **Or to this oxygen?**

Although the two oxygen atoms in the acetate ion appear different in line-bond structures, they are in fact equivalent. Both carbon–oxygen bonds, for example, are 127 pm in length, midway between the length of a typical C–O single bond (135 pm) and a typical C=O double bond (120 pm). In other words, *neither* of the two structures for acetate is correct by itself. The true structure is intermediate between the two, and an electrostatic potential map shows that both oxygen atoms share the negative charge and have equal electron densities (red).

Acetate ion—two resonance forms

The two individual line-bond structures for acetate are called **resonance forms**, and their special resonance relationship is indicated by the double-headed arrow between them. *The only difference between resonance forms is the placement of the π and nonbonding valence electrons.* The atoms themselves occupy exactly the same place in both resonance forms, the connections between atoms are the same, and the three-dimensional shapes of the resonance forms are the same.

A good way to think about resonance forms is to realize that a substance like the acetate ion is no different from any other. Acetate doesn't jump back and forth between two resonance forms, spending part of the time looking like one and part of the time looking like the other. Rather, acetate has a single unchanging structure that we say is a **resonance hybrid** of the two individual forms and has characteristics of both. The only "problem" with acetate is that we can't draw it accurately using a familiar line-bond structure—line-bond structures just don't work well for resonance hybrids. The difficulty, however, lies with the *representation* of acetate on paper, not with acetate itself.

Resonance is a very useful concept that we'll return to on numerous occasions throughout the rest of this book. We'll see in Section 9.2, for instance, that the six carbon–carbon bonds in so-called aromatic compounds such as benzene are equivalent and that benzene is best represented as a hybrid of two resonance forms. Although an individual resonance form seems to imply that benzene has alternating single and double bonds, neither form is correct by itself. The true benzene structure is a hybrid of the two individual forms, and all six carbon–carbon bonds are equivalent. This symmetrical distribution of electrons around the molecule is evident in an electrostatic potential map.

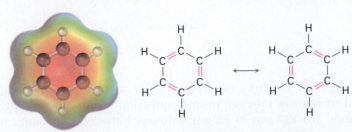

Benzene (two resonance forms)

2.5 Rules for Resonance Forms

When first dealing with resonance forms, it's useful to have a set of guidelines that describe how to draw and interpret them. The following rules should be helpful:

Rule 1

Individual resonance forms are imaginary, not real. The real structure is a composite, or resonance hybrid, of the different forms. Species such as the acetate ion and benzene are no different from any other. They have single, unchanging structures, and they do not switch back and forth between resonance forms. The only difference between these and other substances is in the way they must be represented in drawings on paper.

Rule 2

Resonance forms differ only in the placement of their π or nonbonding electrons. Neither the position nor the hybridization of any atom changes from one resonance form to another. In the acetate ion, for example, the carbon atom is sp^2-hybridized and the oxygen atoms remain in exactly the same place in both resonance forms. Only the positions of the π electrons in the C=O double bond and the lone-pair electrons on oxygen differ from one form to another. This movement of electrons from one resonance structure to another can be indicated by using curved arrows. *A curved arrow always indicates the movement of electrons, not the movement of atoms.* An arrow shows that a pair of electrons moves *from* the atom or bond at the tail of the arrow *to* the atom or bond at the head of the arrow.

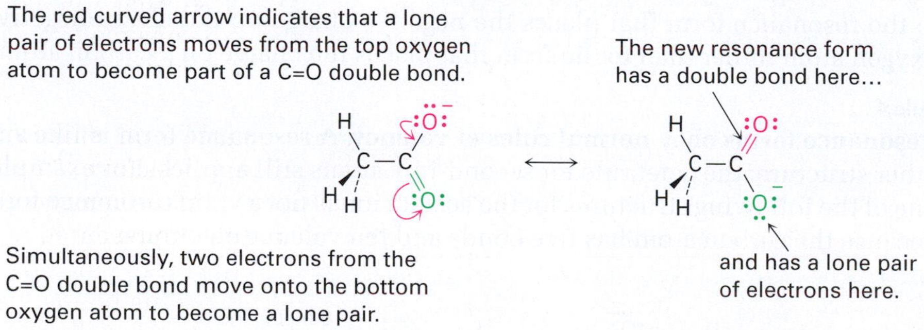

The red curved arrow indicates that a lone pair of electrons moves from the top oxygen atom to become part of a C=O double bond.

The new resonance form has a double bond here...

Simultaneously, two electrons from the C=O double bond move onto the bottom oxygen atom to become a lone pair.

and has a lone pair of electrons here.

The situation with benzene is similar to that with acetate: the π electrons in the double bonds move, as shown with curved arrows, but the carbon and hydrogen atoms remain in place.

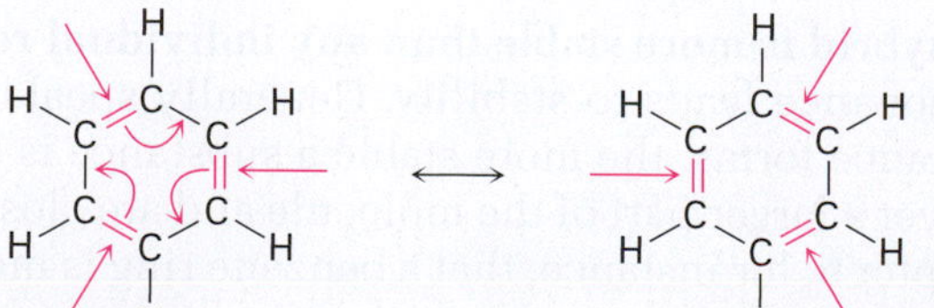

Rule 3

Different resonance forms of a substance don't have to be equivalent. For example, we'll see in Chapter 17 that compounds containing a C=O double bond, such as acetyl coenzyme A, an intermediate in carbohydrate and fat metabolism, can be converted into an anion by reaction with a base. (For now, we'll abbreviate the coenzyme A part of the structure as "CoA.") The resultant anion has two resonance forms. One form contains a carbon–*oxygen* double bond and has a negative charge on the adjacent *carbon*, while the other contains a carbon–*carbon* double bond and has a negative charge on *oxygen*. Even though the two resonance forms aren't equivalent, both contribute to the overall resonance hybrid.

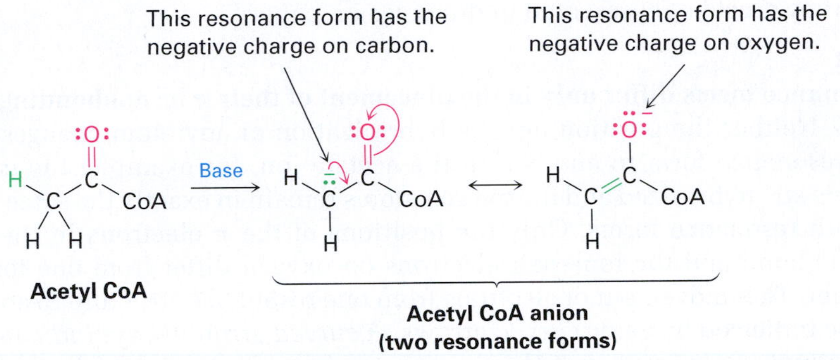

This resonance form has the negative charge on carbon.

This resonance form has the negative charge on oxygen.

Acetyl CoA

Acetyl CoA anion (two resonance forms)

When two resonance forms are not equivalent, the actual structure of the resonance hybrid is closer to the more stable form than to the less stable form. Thus, we might expect the true structure of the acetyl CoA anion to be closer to the resonance form that places the negative charge on the electronegative oxygen atom rather than to the form that places the charge on a carbon atom.

Rule 4

Resonance forms obey normal rules of valency. A resonance form is like any other structure: the octet rule for second-row atoms still applies. For example, one of the following structures for the acetate ion is not a valid resonance form because the carbon atom has five bonds and ten valence electrons:

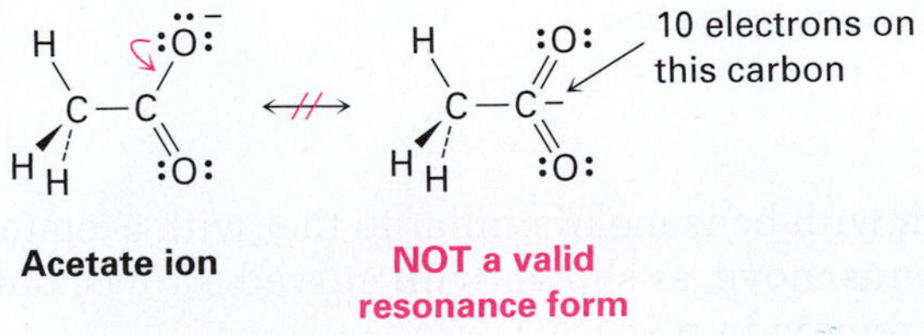

10 electrons on this carbon

Acetate ion

NOT a valid resonance form

Rule 5

The resonance hybrid is more stable than any individual resonance form. In other words, resonance leads to stability. Generally speaking, the larger the number of resonance forms, the more stable a substance is because electrons are spread out over a larger part of the molecule and are closer to more nuclei. We'll see in Chapter 9, for instance, that a benzene ring is more stable because of resonance than might otherwise be expected.

2.6 Drawing Resonance Forms

Look back at the resonance forms of the acetate ion and acetyl CoA anion shown in the previous section. The pattern seen there is a common one that leads to a useful technique for drawing resonance forms. In general, *any three-atom grouping with a p orbital on each atom has two resonance forms:*

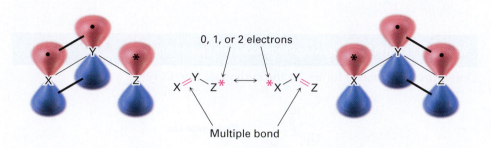

The atoms X, Y, and Z in the general structure might be C, N, O, P, or S, and the asterisk (*) might mean that the *p* orbital on atom Z is vacant, that it contains a single electron, or that it contains a lone pair of electrons. The two resonance forms differ simply by an exchange in position of the multiple bond and the asterisk from one end of the three-atom grouping to the other.

By learning to recognize such three-atom groupings within larger structures, resonance forms can be systematically generated. Look, for instance, at the anion produced when H$^+$ is removed from pentane-2,4-dione by reaction with a base. How many resonance structures does the resultant anion have?

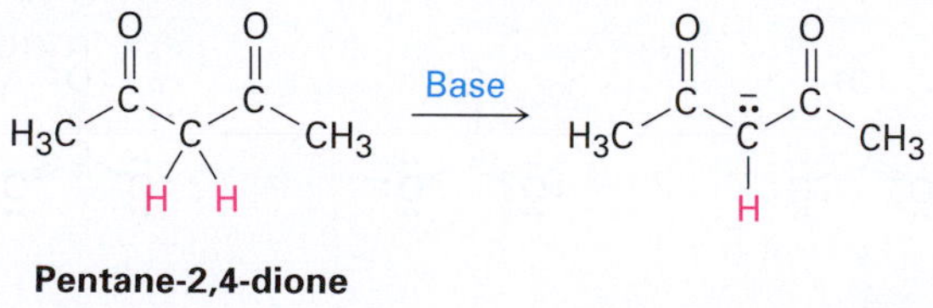

Pentane-2,4-dione

The pentane-2,4-dione anion has a lone pair of electrons and a formal negative charge on the central carbon atom, next to a C=O bond on the left. The O=C–C:$^-$ grouping is a typical one for which two resonance structures can be drawn:

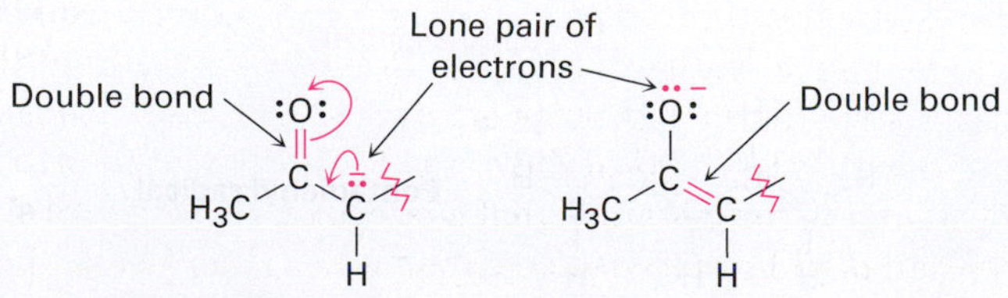

Just as there is a C=O bond to the left of the lone pair, there is a second C=O bond to the right. Thus, we can draw a total of three resonance structures for the pentane-2,4-dione anion:

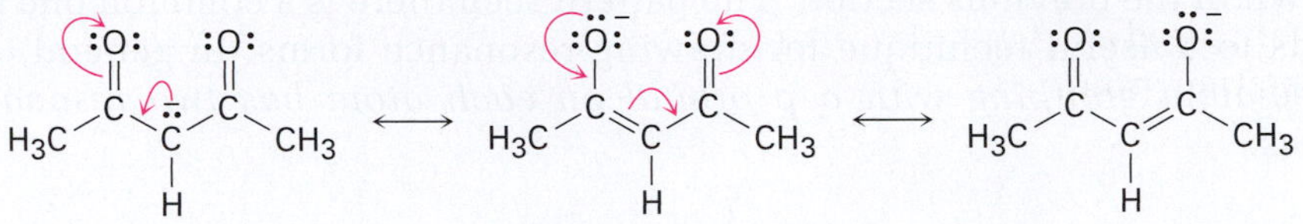

WORKED EXAMPLE 2.2 Drawing Resonance Forms for an Anion

Draw three resonance forms for the carbonate ion, CO_3^{2-}.

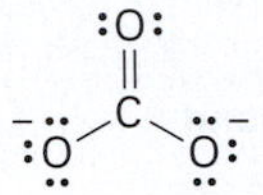

Carbonate ion

Strategy

Look for three-atom groupings that contain a multiple bond next to an atom with a p orbital. Then exchange the positions of the multiple bond and the electrons in the p orbital. In the carbonate ion, each of the singly bonded oxygen atoms with its lone pairs and negative charge is next to the C=O double bond, giving the grouping O=C–O:$^-$.

Solution

Exchanging the position of the double bond and an electron lone pair in each grouping generates three resonance structures:

Three-atom groupings

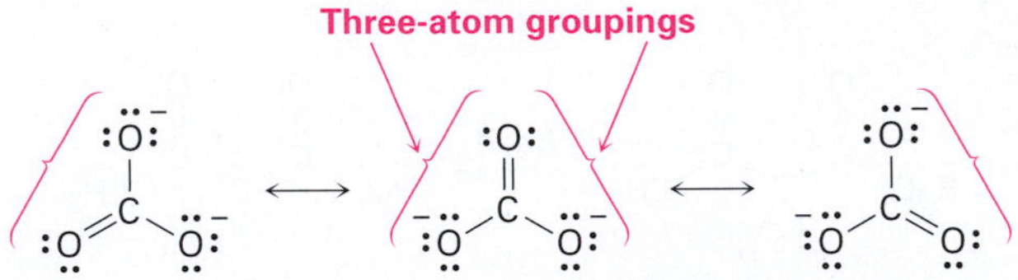

WORKED EXAMPLE 2.3 Drawing Resonance Forms for a Radical

Draw three resonance forms for the pentadienyl radical, where a *radical* is a substance that contains a single, unpaired electron in one of its orbitals, denoted by a dot (·).

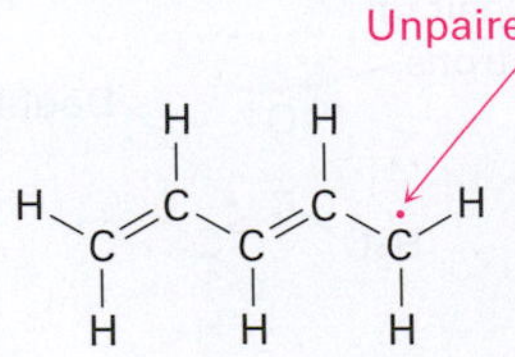

Unpaired electron

Pentadienyl radical

Strategy

Find the three-atom groupings that contain a multiple bond next to a *p* orbital.

Solution

The unpaired electron is on a carbon atom next to a C=C bond, giving a typical three-atom grouping that has two resonance forms:

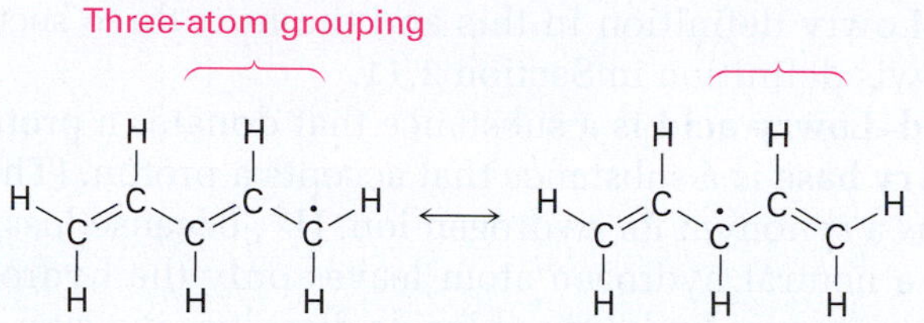

Three-atom grouping

In the second resonance form, the unpaired electron is next to another double bond, giving another three-atom grouping and leading to another resonance form:

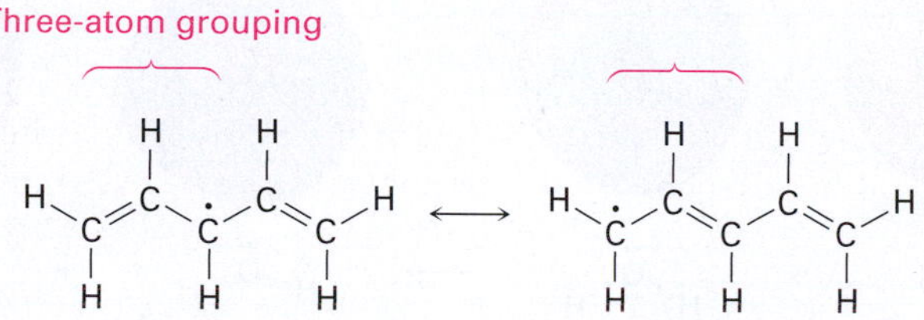

Three-atom grouping

Thus, the three resonance forms for the pentadienyl radical are:

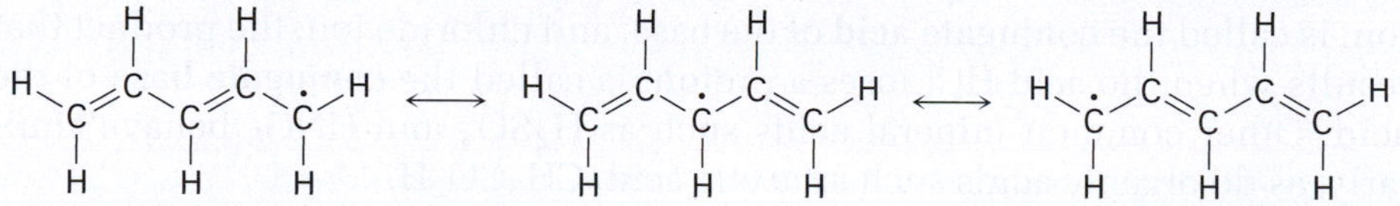

Problem 2.9

Draw the indicated number of resonance structures for each of the following species:

(a) The methyl phosphate anion, $CH_3OPO_3{}^{2-}$ (3)

(b) The nitrate anion, $NO_3{}^-$ (3)

(c) The allyl cation, $H_2C=CH-CH_2{}^+$ (2)

(d) The benzoate anion (4)

2.7 Acids and Bases: The Brønsted–Lowry Definition

A further important concept related to electronegativity and polarity is that of *acidity* and *basicity*. We'll see, in fact, that much of the chemistry of organic molecules can be explained by their acid–base behavior. You may recall from a course in general chemistry that there are two frequently used definitions of acidity: the *Brønsted–Lowry definition* and the *Lewis definition*. We'll look at the Brønsted–Lowry definition in this and the next three sections and then discuss the Lewis definition in Section 2.11.

A **Brønsted–Lowry acid** is a substance that donates a proton (H^+), and a **Brønsted–Lowry base** is a substance that accepts a proton. (The name *proton* is often used as a synonym for hydrogen ion, H^+, because loss of the valence electron from a neutral hydrogen atom leaves only the hydrogen nucleus— a proton.) When gaseous hydrogen chloride dissolves in water, for example, a polar HCl molecule acts as an acid and donates a proton, while a water molecule acts as a base and accepts the proton, yielding hydronium ion (H_3O^+) and chloride ion (Cl^-).

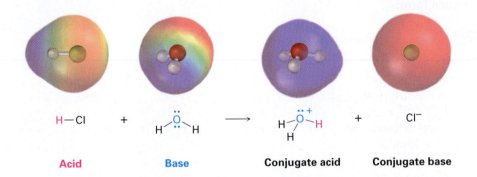

| Acid | Base | Conjugate acid | Conjugate base |

Hydronium ion, the product that results when the base H_2O gains a proton, is called the **conjugate acid** of the base, and chloride ion, the product that results when the acid HCl loses a proton, is called the **conjugate base** of the acid. Other common mineral acids such as H_2SO_4 and HNO_3 behave similarly, as do organic acids such as acetic acid, CH_3CO_2H.

In a general sense,

$$H-A \ + \ :B \ \rightleftharpoons \ :A^- \ + \ H-B^+$$

| Acid | Base | Conjugate base | Conjugate acid |

For example:

| | Acid | Base | | Conjugate base | Conjugate acid |

Notice that water can act either as an acid or as a base, depending on the circumstances. In its reaction with HCl, water is a base that accepts a proton to give the hydronium ion, H_3O^+. In its reaction with ammonia, NH_3, however, water is an acid that donates a proton to give ammonium ion, NH_4^+, and hydroxide ion, HO^-.

Problem 2.10

Nitric acid (HNO_3) reacts with ammonia (NH_3) to yield ammonium nitrate. Write the reaction, and identify the acid, the base, the conjugate acid product, and the conjugate base product.

2.8 Acid and Base Strength

Acids differ in their ability to donate H^+. Stronger acids, such as HCl, react almost completely with water, whereas weaker acids, such as acetic acid (CH_3CO_2H), react only slightly. The exact strength of a given acid HA in water solution is described using the **acidity constant (K_a)** for the acid-dissociation equilibrium. Remember from general chemistry that the concentration of solvent is ignored in the equilibrium expression and that brackets [ ] around a substance refer to the concentration of the enclosed species in moles per liter.

$$HA + H_2O \rightleftharpoons A^- + H_3O^+$$

$$K_a = \frac{[H_3O^+][A^-]}{[HA]}$$

Stronger acids have their equilibria toward the right and thus have larger acidity constants, whereas weaker acids have their equilibria toward the left and have smaller acidity constants. The range of K_a values for different acids is enormous, running from about 10^{15} for the strongest acids to about 10^{-60} for the weakest. The common inorganic acids such as H_2SO_4, HNO_3, and HCl have K_a's in the range of 10^2 to 10^9, while organic acids generally have K_a's in the range of 10^{-5} to 10^{-15}. As you gain more experience, you'll develop a rough feeling for which acids are "strong" and which are "weak" (always remembering that the terms are relative).

Acid strengths are normally expressed using pK_a values rather than K_a values, where the **pK_a** is the negative common logarithm of the K_a:

$$pK_a = -\log K_a$$

A *stronger* acid (larger K_a) has a *smaller* pK_a, and a *weaker* acid (smaller K_a) has a *larger* pK_a. Table 2.3 lists the pK_a's of some common acids in order of their strength, and a more comprehensive table is given in Appendix B.

TABLE 2.3
Relative Strengths of Some Common Acids and Their Conjugate Bases

	Acid	Name	pK_a	Conjugate base	Name	
Weaker acid	CH_3CH_2OH	Ethanol	16.00	$CH_3CH_2O^-$	Ethoxide ion	**Stronger base**
	H_2O	Water	15.74	HO^-	Hydroxide ion	
	HCN	Hydrocyanic acid	9.31	CN^-	Cyanide ion	
	$H_2PO_4^-$	Dihydrogen phosphate ion	7.21	HPO_4^{2-}	Hydrogen phosphate ion	
	CH_3CO_2H	Acetic acid	4.76	$CH_3CO_2^-$	Acetate ion	
	H_3PO_4	Phosphoric acid	2.16	$H_2PO_4^-$	Dihydrogen phosphate ion	
	HNO_3	Nitric acid	−1.3	NO_3^-	Nitrate ion	
Stronger acid	HCl	Hydrochloric acid	−7.0	Cl^-	Chloride ion	**Weaker base**

Notice that the pK_a value shown in Table 2.3 for water is 15.74, which results from the following calculation. Because water is both the acid and the solvent, the equilibrium expression is

$$H_2O + H_2O \rightleftharpoons OH^- + H_3O^+$$
$$\text{(acid)} \quad \text{(solvent)}$$

$$K_a = \frac{[H_3O^+][A^-]}{[HA]} = \frac{[H_3O^+][OH^-]}{[H_2O]}$$

$$= \frac{[1.0 \times 10^{-7}][1.0 \times 10^{-7}]}{[55.4]} = [1.8 \times 10^{-16}]$$

$$pK_a = 15.74$$

The numerator in this expression is the so-called ion-product constant for water, $K_w = [H_3O^+][OH^-] = 1.00 \times 10^{-14}$, and the denominator is the molar concentration of pure water, $[H_2O] = 55.4$ M at 25 °C. The calculation is artificial in that the concentration of "solvent" water is ignored while the concentration of "acid" water is not, but it is nevertheless useful in allowing us to make a comparison of water with other weak acids on a similar footing.

Notice also in Table 2.3 that there is an inverse relationship between the acid strength of an acid and the base strength of its conjugate base. That is, a *strong* acid has a *weak* conjugate base, and a *weak* acid has a *strong* conjugate base. To understand this relationship, think about what happens to the acidic hydrogen in an acid–base reaction: a strong acid is one that loses an H^+ easily, meaning that its conjugate base holds on to the H^+ weakly and is therefore a weak base. A weak acid is one that loses an H^+ with difficulty, meaning that its conjugate base holds on to the H^+ strongly and is therefore a strong base. HCl, for instance, is a strong acid, meaning that Cl^- holds on to the H^+ weakly and is thus a weak base. Water, on the other hand, is a weak acid, meaning that OH^- holds on to the H^+ strongly and is a strong base.

Problem 2.11

The amino acid phenylalanine has pK_a = 1.83, and tryptophan has pK_a = 2.83. Which is the stronger acid?

Phenylalanine
(pK_a = 1.83)

Tryptophan
(pK_a = 2.83)

Problem 2.12

Amide ion, H_2N^-, is a much stronger base than hydroxide ion, HO^-. Which is the stronger acid, NH_3 or H_2O? Explain.

2.9 Predicting Acid–Base Reactions from pK_a Values

Compilations of pK_a values like those in Table 2.3 and Appendix B are useful for predicting whether a given acid–base reaction will take place because H^+ will always go *from* the stronger acid *to* the stronger base. That is, an acid will donate a proton to the conjugate base of a weaker acid, and the conjugate base of a weaker acid will remove the proton from a stronger acid. For example, since water (pK_a = 15.74) is a weaker acid than acetic acid (pK_a = 4.76), hydroxide ion holds a proton more tightly than acetate ion does. Hydroxide ion will therefore react with acetic acid, CH_3CO_2H, to yield acetate ion and H_2O.

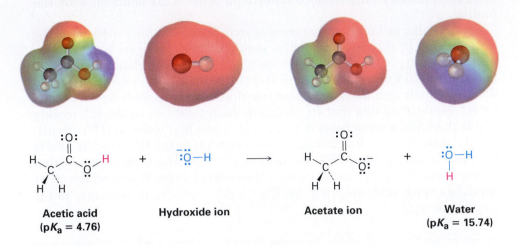

Another way to predict acid–base reactivity is to remember that the product conjugate acid in an acid–base reaction must be weaker and less reactive than the starting acid and that the product conjugate base must be weaker and less reactive than the starting base. In the reaction of acetic acid with hydroxide ion, for example, the product conjugate acid (H_2O) is weaker than the starting acid (CH_3CO_2H) and the product conjugate base ($CH_3CO_2^-$) is weaker than the starting base (OH^-).

$$CH_3\overset{O}{\overset{\|}{C}}H \;+\; HO^- \rightleftharpoons HOH \;+\; CH_3\overset{O}{\overset{\|}{C}}O^-$$

| Stronger acid | Stronger base | Weaker acid | Weaker base |

WORKED EXAMPLE 2.4 Predicting Acid Strengths from pK_a Values

Water has pK_a = 15.74, and acetylene has pK_a = 25. Which is the stronger acid? Does hydroxide ion react with acetylene?

$$H-C\equiv C-H \;+\; OH^- \overset{?}{\longrightarrow} H-C\equiv C\overset{..}{\underset{}{}}^- \;+\; H_2O$$

Acetylene

Strategy

In comparing two acids, the one with the lower pK_a is stronger. Thus, water is a stronger acid than acetylene and gives up H^+ more easily.

Solution

Because water is a stronger acid and gives up H^+ more easily than acetylene does, the HO^- ion must have less affinity for H^+ than the $HC\equiv C:^-$ ion has. In other words, the anion of acetylene is a stronger base than hydroxide ion, and the reaction will not proceed as written.

WORKED EXAMPLE 2.5 Calculating K_a from pK_a

According to the data in Table 2.3, acetic acid has pK_a = 4.76. What is its K_a?

Strategy

Since pK_a is the negative logarithm of K_a, it's necessary to use a calculator with an ANTILOG or INV LOG function. Enter the value of the pK_a (4.76), change the sign (−4.76), and then find the antilog (1.74 × 10^{-5}).

Solution

K_a = 1.74 × 10^{-5}.

Problem 2.13

Will either of the following reactions take place as written, according to the pK_a data in Table 2.3?

(a) HCN + $CH_3CO_2^-$ Na^+ $\xrightarrow{?}$ Na^+ $^-$CN + CH_3CO_2H

(b) CH_3CH_2OH + Na^+ $^-$CN $\xrightarrow{?}$ $CH_3CH_2O^-$ Na^+ + HCN

Problem 2.14

Ammonia, NH_3, has pK_a ≈ 36, and acetone has pK_a ≈ 19. Will the following reaction take place?

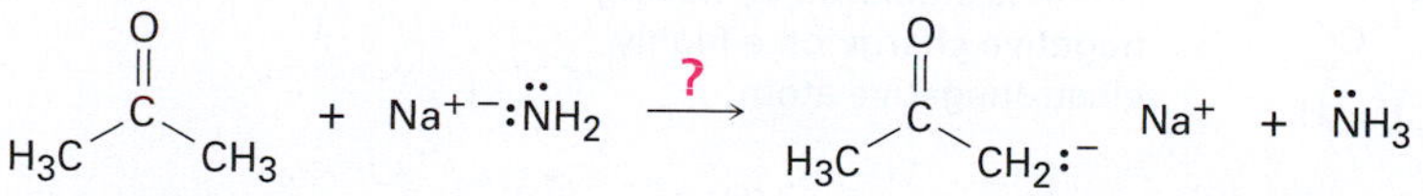

Acetone

Problem 2.15

What is the K_a of HCN if its pK_a = 9.31?

2.10 Organic Acids and Organic Bases

Almost all biological reactions involve organic acids and organic bases. Although it's too early to go into the details of these processes now, you might keep the following generalities in mind as your study progresses.

Organic Acids

Organic acids are characterized by the presence of a positively polarized hydrogen atom (blue in electrostatic potential maps) and are of two main kinds: those acids such as methanol and acetic acid that contain a hydrogen atom bonded to an electronegative oxygen atom (O–H) and those such as acetone and acetyl CoA (Section 2.5) that contain a hydrogen atom bonded to a carbon atom next to a C=O double bond (O=C–C–H).

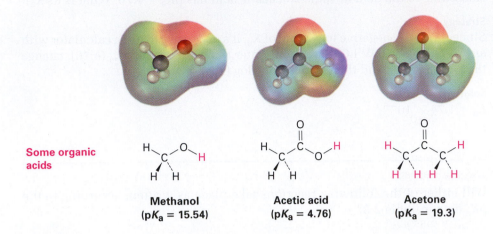

Some organic acids

Methanol
(pK_a = 15.54)

Acetic acid
(pK_a = 4.76)

Acetone
(pK_a = 19.3)

Methanol contains an O–H bond and is a weak acid, while acetic acid also contains an O–H bond and is a somewhat stronger acid. In both cases, acidity is due to the fact that the conjugate base resulting from loss of H⁺ is stabilized by having its negative charge on a strongly electronegative oxygen atom. In addition, the conjugate base of acetic acid is stabilized by resonance (Sections 2.4 and 2.5).

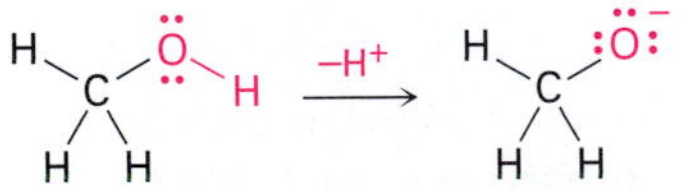

Anion is stabilized by having negative charge on a highly electronegative atom.

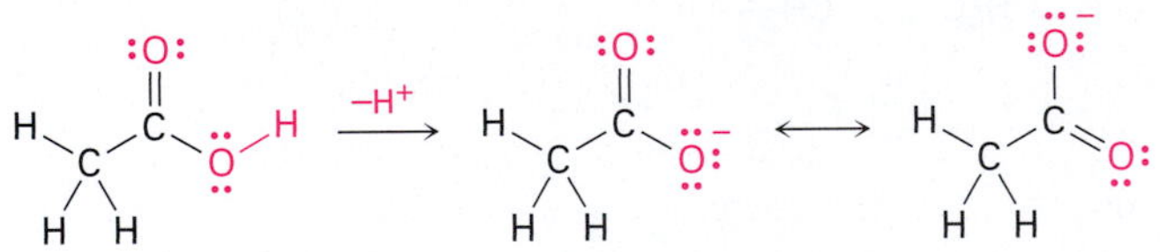

Anion is stabilized by having negative charge on a highly electronegative atom and by resonance.

The acidity of acetone, acetyl CoA, and other compounds with C=O double bonds is due to the fact that the conjugate base resulting from loss of H⁺ is stabilized by resonance. In addition, one of the resonance forms stabilizes the negative charge by placing it on an electronegative oxygen atom.

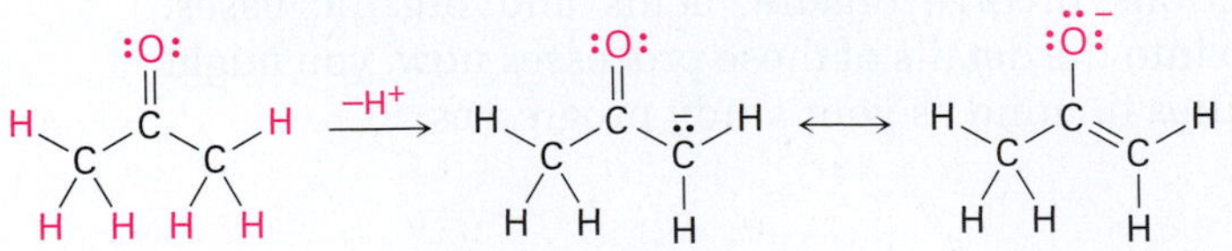

Anion is stabilized by resonance and by having negative charge on a highly electronegative atom.

Electrostatic potential maps of the conjugate bases from methanol, acetic acid, and acetone are shown in Figure 2.4. As you might expect, all three show a substantial amount of negative charge (red) on oxygen.

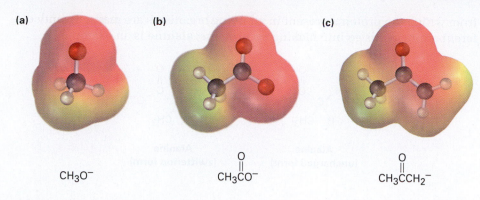

(a)   (b)   (c)

CH₃O⁻   O   O
                       ‖   ‖
                       CH₃CO⁻   CH₃CCH₂⁻

FIGURE 2.4 Electrostatic potential maps of the conjugate bases of **(a)** methanol, **(b)** acetic acid, and **(c)** acetone. The electronegative oxygen atoms stabilize the negative charge in all three.

Compounds called *carboxylic acids,* which contain the –CO₂H grouping, occur abundantly in all living organisms and are involved in almost all metabolic pathways. Acetic acid, pyruvic acid, and citric acid are examples. You might note that at the typical pH of 7.3 found within cells, carboxylic acids are usually dissociated and exist as their carboxylate anions, –CO₂⁻.

Acetic acid   **Pyruvic acid**   **Citric acid**

Organic Bases

Organic bases are characterized by the presence of an atom (reddish in electrostatic potential maps) with a lone pair of electrons that can bond to H⁺. Nitrogen-containing compounds such as methylamine are the most common organic bases and are involved in almost all metabolic pathways, but oxygen-containing compounds can also act as bases when reacting with a sufficiently strong acid. Note that some oxygen-containing compounds can act both as acids and as bases depending on the circumstances, just as water can. Methanol and acetone, for instance, act as *acids* when they donate a proton but as *bases* when their oxygen atom accepts a proton.

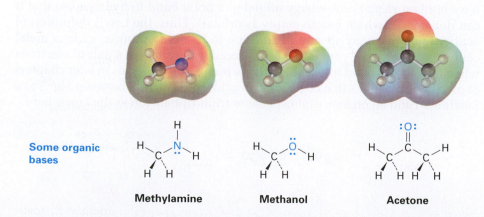

Some organic bases

Methylamine   **Methanol**   **Acetone**

We'll soon see that substances called *amino acids,* so named because they are both amines (–NH₂) and carboxylic acids (–CO₂H), are the building blocks from which the proteins present in all living organisms are made. Twenty different amino acids go into making up proteins; alanine is an example.

Alanine
(uncharged form)

Alanine
(zwitterion form)

Interestingly, alanine and other amino acids exist primarily in a doubly charged form called a *zwitterion* rather than in the uncharged form. The zwitterion form arises because amino acids have both acidic and basic sites within the same molecule and therefore undergo an *internal* acid–base reaction.

2.11 Acids and Bases: The Lewis Definition

The *Lewis definition* of acids and bases is broader and more encompassing than the Brønsted–Lowry definition because it's not limited to substances that donate or accept protons. A **Lewis acid** is a substance that *accepts an electron pair*, and a **Lewis base** is a substance that *donates an electron pair*. The donated electron pair is shared between the acid and the base in a covalent bond.

Filled orbital

Vacant orbital

B + A $\longrightarrow$ B—A

Lewis base   **Lewis acid**

Lewis Acids and the Curved Arrow Formalism

The fact that a Lewis acid is able to accept an electron pair means that it must have either a vacant, low-energy orbital or a polar bond to hydrogen so that it can donate H$^+$ (which has an empty 1s orbital). Thus, the Lewis definition of acidity includes many species in addition to H$^+$. For example, various metal cations, such as Mg^{2+}, are Lewis acids because they accept a pair of electrons when they form a bond to a base. We'll see numerous instances in later chapters of metabolic reactions that begin with an acid–base reaction between Mg^{2+} as a Lewis acid and an organic diphosphate or triphosphate ion as the Lewis base.

Lewis acid

Lewis base
(an organic diphosphate ion)

Acid–base complex

In the same way, compounds of group 3A elements, such as BF_3 and $AlCl_3$, are Lewis acids because they have unfilled valence orbitals and can accept electron pairs from Lewis bases, as shown in Figure 2.5. Similarly, many transition-metal compounds, such as $TiCl_4$, $FeCl_3$, $ZnCl_2$, and $SnCl_4$, are Lewis acids.

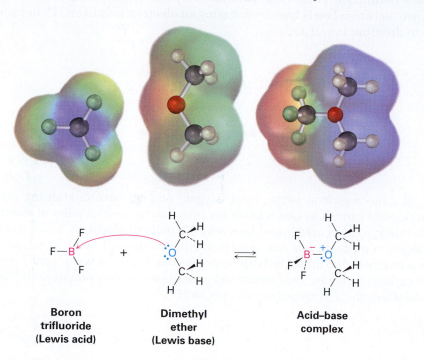

Boron trifluoride (Lewis acid) **Dimethyl ether (Lewis base)** **Acid–base complex**

ACTIVE FIGURE 2.5 The reaction of boron trifluoride, a Lewis acid, with dimethyl ether, a Lewis base. The Lewis acid accepts a pair of electrons, and the Lewis base donates a pair of nonbonding electrons. Note how the movement of electrons *from* the Lewis base *to* the Lewis acid is indicated by a curved arrow. Note also how, in electrostatic potential maps, the boron becomes more negative (red) after reaction because it has gained electrons and the oxygen atom becomes more positive (blue) because it has donated electrons. **Go to this book's student companion site at www.cengage .com/chemistry/mcmurry to explore an interactive version of this figure.**

Look closely at the acid–base reaction in Figure 2.5, and note how it is shown. Dimethyl ether, the Lewis base, donates an electron pair to a vacant valence orbital of the boron atom in BF_3, a Lewis acid. The direction of electron-pair flow from the base to the acid is shown using curved arrows, just as the direction of electron flow in going from one resonance structure to another was shown using curved arrows in Section 2.5. *A curved arrow always means that a pair of electrons moves* from *the atom at the tail of the arrow* to *the atom at the head of the arrow.* We'll use this curved-arrow notation throughout the remainder of this text to indicate electron flow during reactions.

Some further examples of Lewis acids follow:

Some Lewis acids

Some neutral proton donors:

H_2O HCl   HBr HNO_3 H_2SO_4

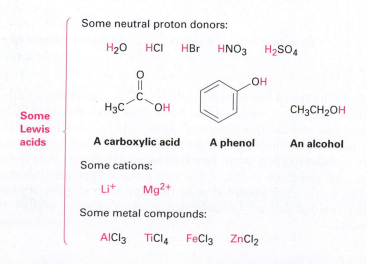

A carboxylic acid   **A phenol**   **An alcohol**

Some cations:

Li^+ Mg^{2+}

Some metal compounds:

$AlCl_3$ $TiCl_4$ $FeCl_3$ $ZnCl_2$

Lewis Bases

The Lewis definition of a base—a compound with a pair of nonbonding electrons that it can use in bonding to a Lewis acid—is similar to the Brønsted–Lowry definition. Thus, H_2O, with its two pairs of nonbonding electrons on oxygen, acts as a Lewis base by donating an electron pair to an H^+ in forming the hydronium ion, H_3O^+.

| Acid | Base | Hydronium ion |

In a more general sense, most oxygen- and nitrogen-containing organic compounds can act as Lewis bases because they have lone pairs of electrons. A divalent oxygen compound has two lone pairs of electrons, and a trivalent nitrogen compound has one lone pair. Note in the following examples that some compounds can act as both acids and bases, just as water can. Alcohols and carboxylic acids, for instance, act as acids when they donate an H^+ but as bases when their oxygen atom accepts an H^+.

Some Lewis bases

| $CH_3CH_2\ddot{O}H$ | $CH_3\ddot{O}CH_3$ | CH_3CH with $:O:$ double bond | CH_3CCH_3 with $:O:$ double bond |
| An alcohol | An ether | An aldehyde | A ketone |

| CH_3CCl with $:O:$ double bond | $CH_3C\ddot{O}H$ with $:O:$ double bond | $CH_3C\ddot{O}CH_3$ with $:O:$ double bond | $CH_3C\ddot{N}H_2$ with $:O:$ double bond |
| An acid chloride | A carboxylic acid | An ester | An amide |

| $CH_3\ddot{N}CH_3$ / CH_3 | $CH_3\ddot{S}CH_3$ | An organic triphosphate ion: $CH_3O-P-O-P-O-P-\ddot{O}:^-$ with three O double bonds and three $:\ddot{O}:^-$ groups |
| An amine | A sulfide | An organic triphosphate ion |

Notice in the list of Lewis bases just given that some compounds, such as carboxylic acids, esters, and amides, have more than one atom with a lone pair of electrons and can therefore react at more than one site. Acetic acid, for example, can be protonated either on the doubly bonded oxygen atom or on the singly bonded oxygen atom. Reaction normally occurs only once in such instances, and the more stable of the two possible protonation products is formed. For acetic acid, protonation by reaction with sulfuric acid occurs on the doubly bonded oxygen because that product is stabilized by two resonance forms.

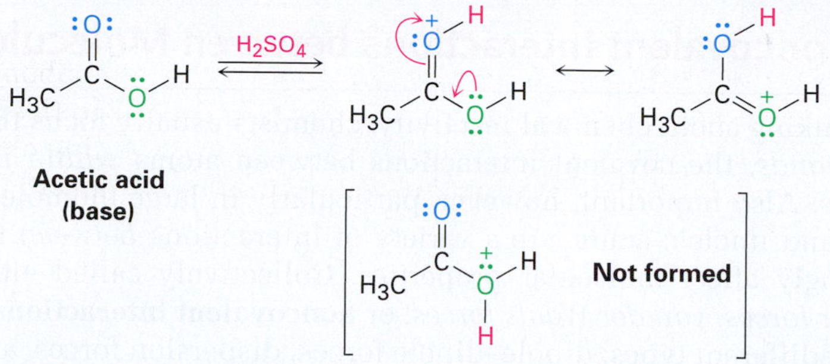

:O:
H_3C — C — O — H
(Acetic acid structure) $\xrightarrow{H_2SO_4}$ (protonated structure) $\longleftrightarrow$ (resonance structure)

**Acetic acid
(base)**

:O:
H_3C — C — O — H   **Not formed**

WORKED EXAMPLE 2.6 Using Curved Arrows to Show Electron Flow

Using curved arrows, show how acetaldehyde, CH_3CHO, can act as a Lewis base.

Strategy

A Lewis base donates an electron pair to a Lewis acid. We therefore need to locate the electron lone pairs on acetaldehyde and use a curved arrow to show the movement of a pair toward the H atom of the acid.

Solution

:O:
H_3C — C — H $+$ H — A $\rightleftharpoons$ (protonated product) $+$:A^-

Acetaldehyde

Problem 2.16

Using curved arrows, show how the species in part (a) can act as Lewis bases in their reactions with HCl, and show how the species in part (b) can act as Lewis acids in their reaction with OH^-.

(a) CH_3CH_2OH, $HN(CH_3)_2$, $P(CH_3)_3$ **(b)** H_3C^+, $B(CH_3)_3$, $MgBr_2$

Problem 2.17

Imidazole, which forms part of the structure of the amino acid histidine, can act as both an acid and a base.

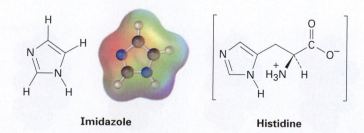

Imidazole   **Histidine**

(a) Look at the electrostatic potential map of imidazole, and identify the most acidic hydrogen atom and the more basic nitrogen atom.

(b) Draw resonance structures for the products that result when imidazole is protonated by an acid and deprotonated by a base.

2.12 Noncovalent Interactions between Molecules

When thinking about chemical reactivity, chemists usually focus their attention on *bonds,* the covalent interactions between atoms *within* individual molecules. Also important, however, particularly in large biomolecules like proteins and nucleic acids, are a variety of interactions *between* molecules that strongly affect molecular properties. Collectively called either *intermolecular forces, van der Waals forces,* or **noncovalent interactions**, they are of several different types: dipole–dipole forces, dispersion forces, and hydrogen bonds.

Dipole–dipole forces occur between polar molecules as a result of electrostatic interactions among dipoles. The forces can be either attractive or repulsive depending on the orientation of the molecules—attractive when unlike charges are together and repulsive when like charges are together. The attractive geometry is lower in energy and therefore predominates (Figure 2.6).

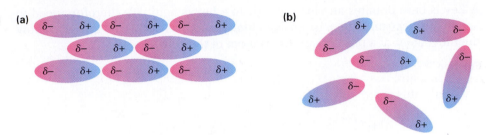

FIGURE 2.6 Dipole–dipole forces cause polar molecules **(a)** to attract one another when they orient with unlike charges together but **(b)** to repel one another when they orient with like charges together.

Dispersion forces occur between all neighboring molecules and arise because the electron distribution within molecules is constantly changing. Although uniform on a time-averaged basis, the electron distribution even in nonpolar molecules is likely to be nonuniform at any given instant. One side of a molecule may, by chance, have a slight excess of electrons relative to the opposite side, giving the molecule a temporary dipole. This temporary dipole in one molecule causes a nearby molecule to adopt a temporarily opposite dipole, with the result that a tiny attraction is induced between the two (Figure 2.7). Temporary molecular dipoles have only a fleeting existence and are constantly changing, but their cumulative effect is often strong enough to hold molecules close together so that a substance is a liquid or solid rather than a gas.

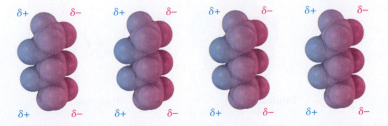

FIGURE 2.7 Attractive dispersion forces in nonpolar molecules are caused by temporary dipoles, as shown in these models of pentane, C_5H_{12}.

Perhaps the most important noncovalent interaction in biological molecules is the **hydrogen bond,** an attractive interaction between a hydrogen bonded to an electronegative O or N atom and an unshared electron pair on another O or N atom. In essence, a hydrogen bond is a very strong dipole–dipole interaction involving polarized O–H or N–H bonds. Electrostatic potential maps of water and ammonia clearly show the positively polarized hydrogens (blue) and the negatively polarized oxygens and nitrogens (red).

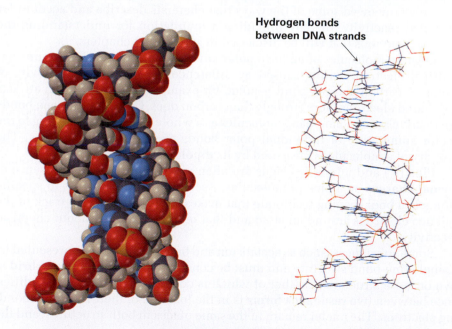

Hydrogen-bonding has enormous consequences for living organisms. Hydrogen bonds cause water to be a liquid rather than a gas at ordinary temperatures, they hold enzymes in the shapes necessary for catalyzing biological reactions, and they cause strands of deoxyribonucleic acid (DNA) to pair up and coil into the double helix that stores genetic information.

Hydrogen bonds between DNA strands

A deoxyribonucleic acid segment

One further point before leaving the subject of noncovalent interactions: biochemists frequently use the term **hydrophilic**, meaning "water-loving," to describe a substance that is strongly attracted to water and the term **hydrophobic**, meaning "water-fearing," to describe a substance that is not strongly attracted to water. Hydrophilic substances, such as table sugar, usually have a number of ionic charges or polar –OH groups in their structure so they can form hydrogen bonds, whereas hydrophobic substances, such as vegetable oil, do not have groups that form hydrogen bonds, so their attraction to water is limited to weak dispersion forces.

Problem 2.18

Of the two vitamins A and C, one is hydrophilic and water-soluble while the other is hydrophobic and fat-soluble. Which do you think is which?

Vitamin A
(retinol)

Vitamin C
(ascorbic acid)

Summary

Key Words

acidity constant (K_a), 49

Brønsted–Lowry acid, 48

Brønsted–Lowry base, 48

conjugate acid, 48

conjugate base, 48

dipole moment (μ),36

electronegativity (EN), 34

formal charge, 40

hydrogen bond, 60

hydrophilic, 61

hydrophobic, 61

inductive effect, 35

Lewis acid, 56

Lewis base, 56

noncovalent interaction, 60

pK_a, 49

polar covalent bond, 34

resonance form, 42

resonance hybrid, 42

Understanding biological organic chemistry means knowing not just what happens but also why and how it happens at the molecular level. In this chapter, we've reviewed some of the ways that chemists describe and account for chemical reactivity, thereby providing a foundation for understanding the specific reactions that will be discussed in subsequent chapters.

Organic molecules often have **polar covalent bonds** as a result of unsymmetrical electron sharing caused by differences in the **electronegativity** of atoms. A carbon–oxygen bond is polar, for example, because oxygen attracts the shared electrons more strongly than carbon does. Carbon–hydrogen bonds are relatively nonpolar. Many molecules as a whole are also polar owing to the vector summation of individual polar bonds and electron lone pairs. The polarity of a molecule is measured by its **dipole moment**, μ.

Plus (+) and minus (−) signs are often used to indicate the presence of **formal charges** on atoms in molecules. Assigning formal charges to specific atoms is a bookkeeping technique that makes it possible to keep track of the valence electrons around an atom and that offers some clues about chemical reactivity.

Some substances, such as acetate ion and benzene, can't be represented by a single line-bond structure and must be considered as a **resonance hybrid** of two or more structures, neither of which is correct by itself. The only difference between two **resonance forms** is in the location of their π and nonbonding electrons. The nuclei remain in the same places in both structures, and the hybridization of the atoms remains the same.

Acidity and basicity are closely related to the ideas of polarity and electronegativity. A **Brønsted–Lowry acid** is a compound that can donate a proton (hydrogen ion, H^+), and a **Brønsted–Lowry base** is a compound that can accept a proton. The strength of a Brønsted–Lowry acid or base is expressed by its **acidity constant**, K_a, or by the negative logarithm of the acidity constant, pK_a. The larger the pK_a, the weaker the acid. More useful is the Lewis definition of acids and bases. A **Lewis acid** is a compound that has a low-energy empty orbital that can accept an electron pair; Mg^{2+}, BF_3, $AlCl_3$, and H^+ are examples. A **Lewis base** is a compound that can donate an unshared electron pair;

NH$_3$ and H$_2$O are examples. Most organic molecules that contain oxygen or nitrogen can act as Lewis bases toward sufficiently strong acids.

A variety of **noncovalent interactions** have a significant effect on the properties of large biomolecules. **Hydrogen-bonding**—the attractive interaction between a positively polarized hydrogen atom bonded to an O or N atom with an unshared electron pair on another O or N atom, is particularly important in giving proteins and nucleic acids their shapes.

Lagniappe

Alkaloids: Naturally Occurring Bases

The coca bush *Erythroxylon coca*, native to upland rain forest areas of Colombia, Ecuador, Peru, Bolivia, and western Brazil, is the source of the alkaloid cocaine.

Just as ammonia, NH$_3$, is a weak base, there are a large number of nitrogen-containing organic compounds called *amines* that are also weak bases. In the early days of organic chemistry, basic amines derived from natural sources were known as *vegetable alkali,* but they are now called *alkaloids.* The study of alkaloids provided much of the impetus for the growth of organic chemistry in the 19th century and remains today an active and fascinating area of research.

Alkaloids vary widely in structure, from the simple to the enormously complex. The odor of rotting fish, for example, is caused largely by methylamine, CH$_3$NH$_2$, a simple relative of ammonia in which one of the NH$_3$ hydrogens has been replaced by an organic CH$_3$ group. In fact, the use of lemon juice to mask fish odors is simply an acid–base reaction of the citric acid in lemons with methylamine base in the fish.

Many alkaloids have pronounced biological properties, and a substantial number of the pharmaceutical agents used today are derived from naturally occurring amines. As a few examples, morphine, an analgesic agent, is obtained from the opium poppy *Papaver somniferum.* Cocaine, both an anesthetic and a central nervous system stimulant, is obtained from the coca bush *Erythroxylon coca,* endemic to upland rain forest areas of Colombia, Ecuador, Peru, Bolivia, and western Brazil. Reserpine, an antianxiety agent and antihypertensive, comes from powdered roots of the semitropical plant *Rauwolfia serpentina.* Ephedrine, a bronchodilator and decongestant, is obtained from the Chinese plant *Ephedra sinica.*

Morphine

Cocaine

Reserpine

Ephedrine

A recent report from the U.S. National Academy of Sciences estimates than less than 1% of all living species have been characterized. Thus, alkaloid chemistry today remains an active area of research, and innumerable substances with potentially useful properties remain to be discovered.

Exercises

■ indicates problems that are assignable in Organic OWL.

Go to this book's companion website at **www.cengage.com/chemistry/mcmurry** to explore interactive versions of the Active Figures from this text.

VISUALIZING CHEMISTRY

(Problems 2.1–2.18 appear within the chapter.)

2.19 Fill in the multiple bonds in the following molecular model of naphthalene, $C_{10}H_8$, the main ingredient in mothballs (gray = C, ivory = H). How many resonance structures does naphthalene have?

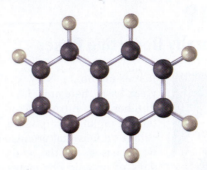

2.20 *cis*-1,2-Dichloroethylene and *trans*-1,2-dichloroethylene are *isomers,* compounds with the same formula but different chemical structures. Look at the following electrostatic potential maps, and tell whether either compound has a dipole moment:

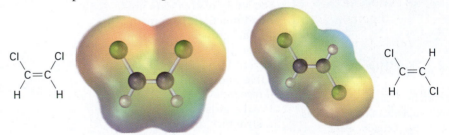

cis-1,2-Dichloroethylene   **trans-1,2-Dichloroethylene**

2.21 ■ The following molecular models are representations of **(a)** adenine and **(b)** cytosine, constituents of DNA. Indicate the positions of multiple bonds and lone pairs for both, and draw skeletal structures (gray = C, red = O, blue = N, ivory = H).

(a)   **(b)**

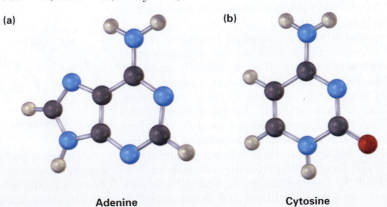

Adenine   **Cytosine**

■ Problems assignable in Organic OWL.

2.22 Electrostatic potential maps of **(a)** acetamide and **(b)** methylamine are shown. Which has the more basic nitrogen atom? Which has the more acidic hydrogen atoms?

(a)

(b)

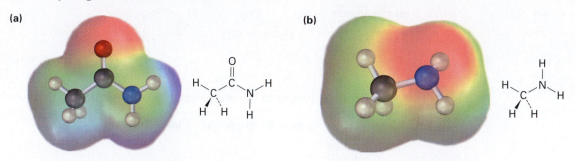

Acetamide   **Methylamine**

ADDITIONAL PROBLEMS

2.23 ▪ Identify the most electronegative element in each of the following molecules:

(a) CH_2FCl **(b)** $FCH_2CH_2CH_2Br$

(c) $HOCH_2CH_2NH_2$ **(d)** CH_3OCH_2Li

2.24 ▪ Use the electronegativity table (Figure 2.2) to predict which bond in each of the following sets is more polar, and indicate the direction of bond polarity for each compound:

(a) $H_3C—Cl$ or $Cl—Cl$ **(b)** $H_3C—H$ or $H—Cl$

(c) $HO—CH_3$ or $(CH_3)_3Si—CH_3$ **(d)** $H_3C—Li$ or $Li—OH$

2.25 ▪ Which of the following molecules has a dipole moment? Indicate the expected direction of each.

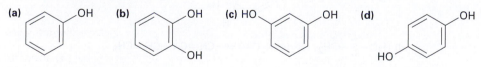

2.26 Phosgene, $Cl_2C{=}O$, has a smaller dipole moment than formaldehyde, $H_2C{=}O$, even though it contains electronegative chlorine atoms in place of hydrogen. Explain.

2.27 (a) The H–Cl bond length is 136 pm. What would the dipole moment of HCl be if the molecule were 100% ionic, $H^+ Cl^-$?

(b) The actual dipole moment of HCl is 1.08 D. What is the percent ionic character of the H–Cl bond?

2.28 Fluoromethane (CH_3F, $\mu = 1.81$ D) has a smaller dipole moment than chloromethane (CH_3Cl, $\mu = 1.87$ D), even though fluorine is more electronegative than chlorine. Explain.

2.29 Methanethiol, CH_3SH, has a substantial dipole moment ($\mu = 1.52$), even though carbon and sulfur have identical electronegativities. Explain.

▪ Problems assignable in Organic OWL.

2.30 ▪ Calculate the formal charges on the atoms shown in red:

(a) $(CH_3)_2\overset{\cdot\cdot}{\underset{}{O}}BF_3$ (b) $H_2\overset{\cdot\cdot}{C}-N\equiv N:$ (c) $H_2C=N=\overset{\cdot\cdot}{\underset{\cdot\cdot}{N}}:$

(d) $:\overset{\cdot\cdot}{O}=\overset{\cdot\cdot}{O}-\overset{\cdot\cdot}{\underset{\cdot\cdot}{O}}:$ (e)

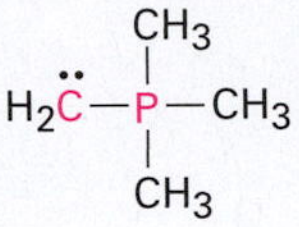

$$H_2\overset{\cdot\cdot}{C}-\underset{\underset{CH_3}{|}}{\overset{\overset{CH_3}{|}}{P}}-CH_3$$

(f)

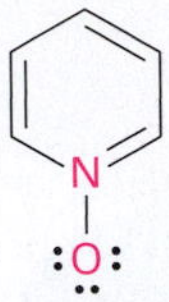

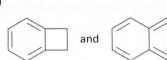

2.31 Assign formal charges to the atoms in each of the following molecules:

(a)

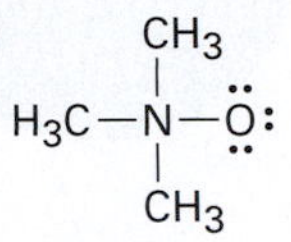

$$H_3C-\underset{\underset{CH_3}{|}}{\overset{\overset{CH_3}{|}}{N}}-\overset{\cdot\cdot}{\underset{\cdot\cdot}{O}}:$$

(b) $H_3C-\overset{\cdot\cdot}{\underset{\cdot\cdot}{N}}-N\equiv N:$ (c) $H_3C-\overset{\cdot\cdot}{\underset{\cdot\cdot}{N}}=N=\overset{\cdot\cdot}{\underset{\cdot\cdot}{N}}:$

2.32 ▪ Which of the following pairs of structures represent resonance forms?

(a)

[structures] and [structures]

(b) $:\overset{\cdot\cdot}{\overset{\cdot\cdot}{O}}\overset{-}{}$ [structure] and $:O:$ [structure] $\overset{\cdot\,-}{}$

(c) $:\overset{\cdot\cdot}{\overset{\cdot\cdot}{O}}\overset{-}{}$ [structure] and $:O:$ [structure] $\overset{\cdot\,-}{}$

(d) $:\overset{\cdot\cdot}{\overset{\cdot\cdot}{O}}\overset{-}{}$ [structure] and $:O:$ [structure]

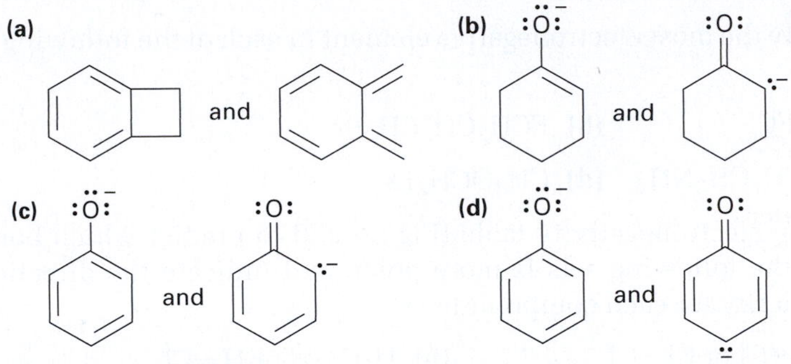

2.33 ▪ Draw as many resonance structures as you can for the following species:

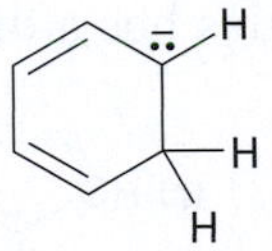

(a)

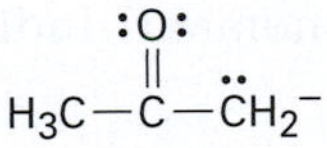

$$H_3C-\overset{\overset{\displaystyle :O:}{\|}}{C}-\overset{\cdot\cdot}{C}H_2^-$$

(b) [structure with H atoms]

(c)

$$H_2\overset{\cdot\cdot}{N}-\underset{}{\overset{\overset{\displaystyle :NH_2}{|}}{C}}=\overset{+}{N}H_2$$

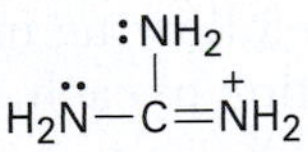

(d) $H_3C-\overset{\cdot\cdot}{\underset{\cdot\cdot}{S}}\overset{+}{}-\overset{\cdot\cdot}{C}H_2$ (e) $H_2C=CH-CH=CH-\overset{+}{C}H-CH_3$

2.34 Cyclobutadiene is a rectangular molecule with two shorter double bonds and two longer single bonds. Why do the following structures *not* represent resonance forms?

2.35 ▪ Alcohols can act either as weak acids or as weak bases, just as water can. Show the reaction of methanol, CH_3OH, with a strong acid such as HCl and with a strong base such as Na^+ $^-NH_2$.

2.36 The O–H hydrogen in acetic acid is much more acidic than any of the C–H hydrogens. Explain.

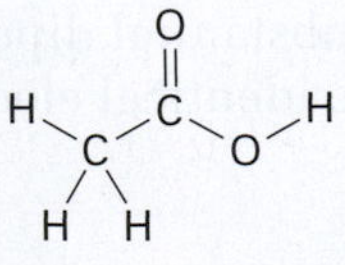

$$\underset{\underset{\displaystyle H\ \ H}{\underset{\displaystyle |\ \ |}{C}}}{\overset{\overset{\displaystyle O}{\|}}{\underset{}{H\diagdown C\diagup}}}\overset{C}{\diagup}\overset{\diagdown}{O}\diagdown H \qquad \textbf{Acetic acid}$$

▪ Problems assignable in Organic OWL.

2.37 ■ Which of the following are likely to act as Lewis acids and which as Lewis bases? Explain.

(a) $AlBr_3$ (b) $CH_3CH_2NH_2$ (c) BH_3

(d) HF   (e) CH_3SCH_3 (f) $TiCl_4$

2.38 Maleic acid has a dipole moment, but the closely related fumaric acid, a substance involved in the citric acid cycle by which food molecules are metabolized, does not. Explain.

Maleic acid   **Fumaric acid**

2.39 ■ Rank the following substances in order of increasing acidity:

| **Acetone** | **Pentane-2,4-dione** | **Phenol** | **Acetic acid** |
| **(pK_a = 19.3)** | **(pK_a = 9)** | **(pK_a = 9.9)** | **(pK_a = 4.76)** |

2.40 Which, if any, of the four substances in Problem 2.39 is a strong enough acid to react almost completely with NaOH? (The pK_a of H_2O is 15.74.)

2.41 The ammonium ion (NH_4^+, pK_a = 9.25) has a lower pK_a than the methylammonium ion ($CH_3NH_3^+$, pK_a = 10.66). Which is the stronger base, ammonia (NH_3) or methylamine (CH_3NH_2)? Explain.

2.42 Is *tert*-butoxide anion a strong enough base to react with water? In other words, can a solution of potassium *tert*-butoxide be prepared in water? The pK_a of *tert*-butyl alcohol is approximately 18.

$$K^+ \ ^-O-\overset{\overset{CH_3}{|}}{\underset{\underset{CH_3}{|}}{C}}-CH_3$$ **Potassium *tert*-butoxide**

2.43 Predict the structure of the product formed in the reaction of the organic base pyridine with the organic acid acetic acid, and use curved arrows to indicate the direction of electron flow.

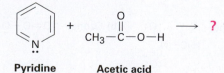

Pyridine   **Acetic acid**

2.44 ■ Calculate K_a values from the following pK_a's:

(a) Acetone, pK_a = 19.3   (b) Formic acid, pK_a = 3.75

■ Problems assignable in Organic OWL.

2.45 ■ Calculate pK_a values from the following K_a's:

(a) Nitromethane, $K_a = 5.0 \times 10^{-11}$ (b) Acrylic acid, $K_a = 5.6 \times 10^{-5}$

2.46 ■ What is the pH of a 0.050 M solution of formic acid (see Problem 2.44)?

2.47 Sodium bicarbonate, $NaHCO_3$, is the sodium salt of carbonic acid (H_2CO_3), $pK_a = 6.37$. Which of the substances shown in Problem 2.39 will react with sodium bicarbonate?

2.48 Assume that you have two unlabeled bottles, one of which contains phenol ($pK_a = 9.9$) and one of which contains acetic acid ($pK_a = 4.76$). In light of your answer to Problem 2.47, suggest a simple way to determine what is in each bottle.

2.49 ■ Identify the acids and bases in the following reactions:

(a)

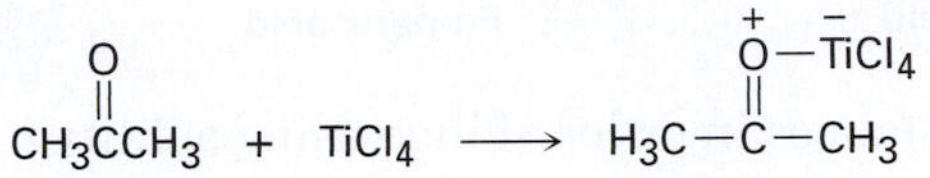

CH₃CCH₃ + TiCl₄ ⟶ H₃C—C—CH₃

(b)

+ NaH ⟶ + H₂

(c)

+ BH₃ ⟶

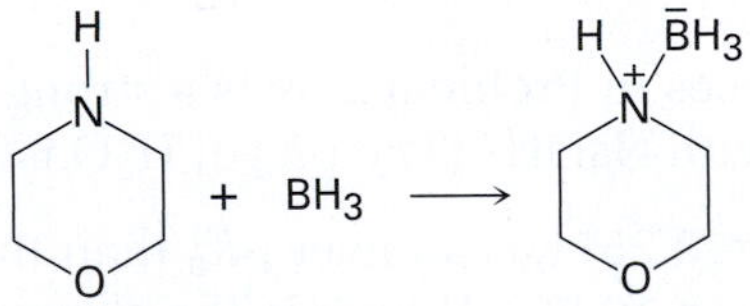

2.50 ■ Which of the following pairs represent resonance structures?

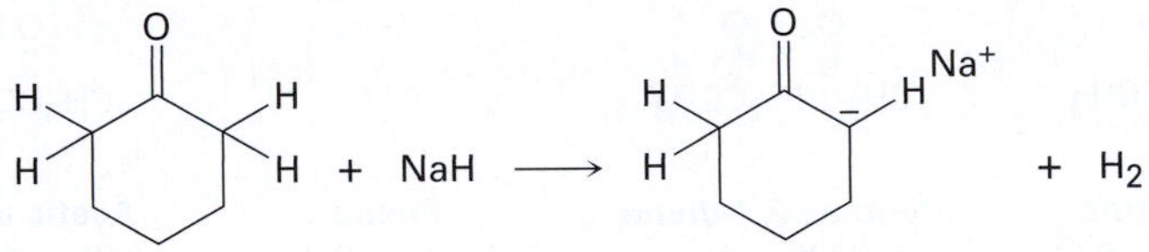

(a)

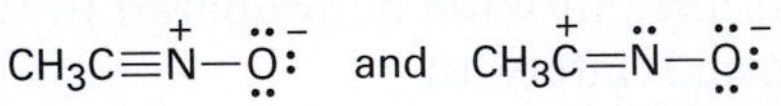

$CH_3C≡N—O$ and $CH_3C=N—O$

(b)

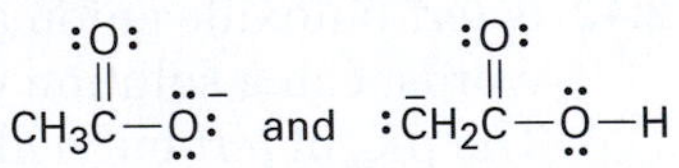

$CH_3C—O$ and $CH_2C—O—H$

(c)

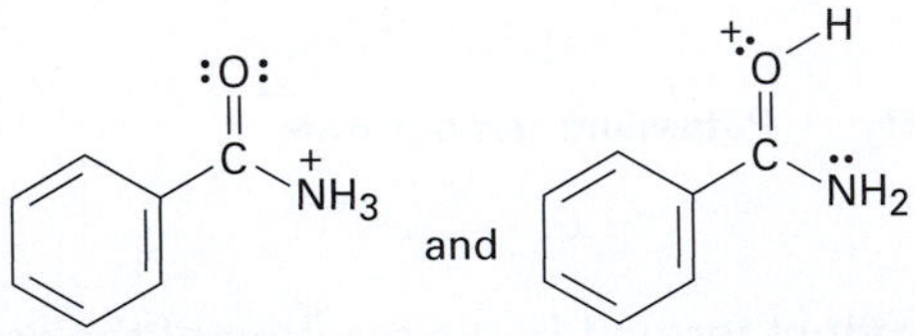

and

(d)

$CH_2=N$ and $CH_2—N$

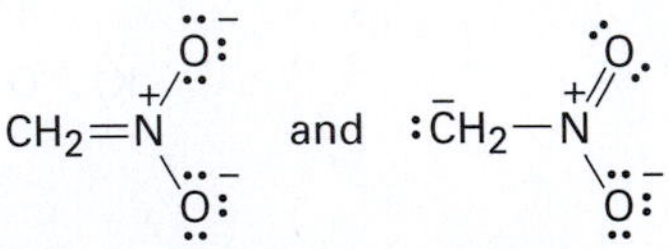

2.51 Draw as many resonance structures as you can for the following species, adding appropriate formal charges in each case:

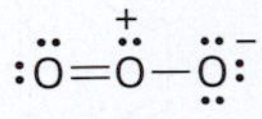

(a) Nitromethane,

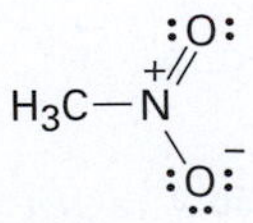

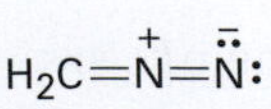

H₃C—N

(b) Ozone, $O=O—O$

(c) Diazomethane, $H_2C=N=N$

2.52 We'll see at the beginning of the next chapter that organic molecules can be classified according to the *functional groups* they contain, where a functional group is a collection of atoms with a characteristic chemical reactivity. Use the electronegativity values given in Figure 2.2 to predict the polarity of the following functional groups:

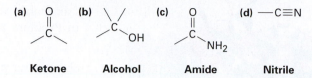

Ketone   **Alcohol**   **Amide**   **Nitrile**

2.53 Phenol, C_6H_5OH, is a stronger acid than methanol, CH_3OH, even though both contain an O–H bond. Draw the structures of the anions resulting from loss of H^+ from phenol and methanol, and use resonance structures to explain the difference in acidity.

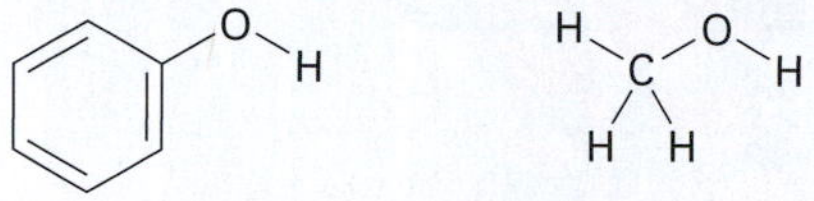

Phenol (pK_a = 9.89)   Methanol (pK_a = 15.54)

2.54 Carbocations, ions that contain a trivalent, positively charged carbon atom, react with water to give alcohols:

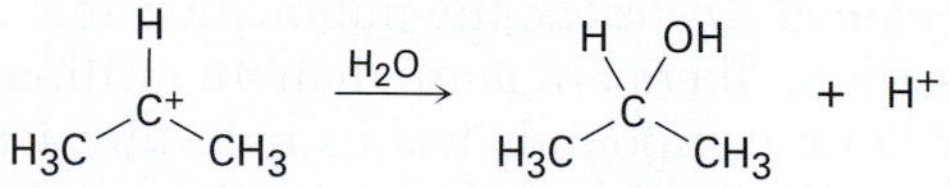

A carbocation   An alcohol

How can you account for the fact that the following carbocation gives a mixture of *two* alcohols on reaction with water?

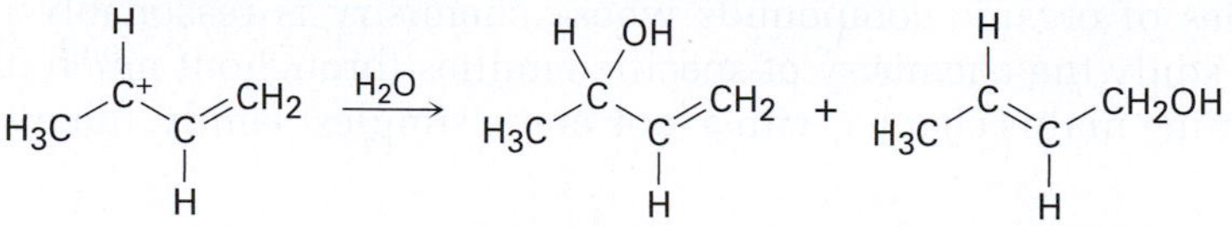

3 Organic Compounds: Alkanes and Their Stereochemistry

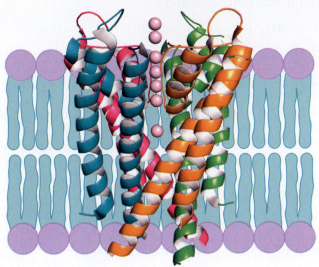

A membrane channel protein that conducts K$^+$ ions across cell membranes.

According to *Chemical Abstracts,* the publication that abstracts and indexes the chemical literature, there are more than 40 million known organic compounds. Each of these compounds has its own physical properties, such as melting point and boiling point, and each has its own chemical reactivity.

Chemists have learned through many years of experience that organic compounds can be classified into families according to their structural features and that the members of a given family often have similar chemical behavior. Instead of 40 million compounds with random reactivity, there are a few dozen families of organic compounds whose chemistry is reasonably predictable. We'll study the chemistry of specific families throughout much of this book, beginning in this chapter with a look at the simplest family, the *alkanes.*

WHY THIS CHAPTER?

Alkanes are relatively unreactive and are rarely involved in chemical reactions, but they nevertheless provide a useful vehicle for introducing some important general ideas. In this chapter, we'll use alkanes to introduce the basic approach to naming organic compounds and to take an initial look at some of the three-dimensional aspects of molecules, a topic of particular importance in understanding biological organic chemistry.

3.1 Functional Groups

The structural features that make it possible to classify compounds into families are called *functional groups.* A **functional group** is a group of atoms within a molecule that has a characteristic chemical behavior. Chemically, a given

OWL Online homework for this chapter can be assigned in Organic OWL.

functional group behaves in nearly the same way in every molecule it's a part of. For example, compare ethylene, a plant hormone that causes fruit to ripen, with menthene, a much more complicated molecule. Both substances contain a carbon–carbon double-bond functional group, and both therefore react with Br_2 in the same way to give products in which a Br atom has added to each of the double-bond carbons (Figure 3.1). This example is typical: *the chemistry of every organic molecule, regardless of size and complexity, is determined by the functional groups it contains.*

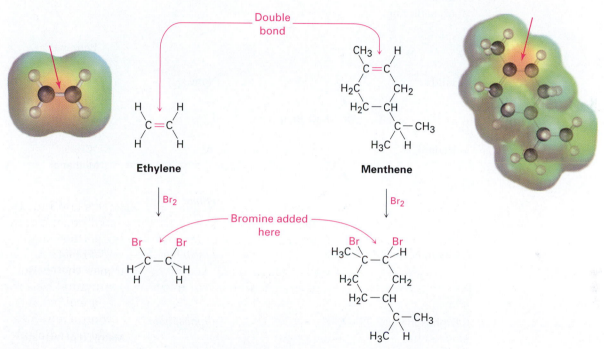

FIGURE 3.1 The reactions of ethylene and menthene with bromine. In both molecules, the carbon–carbon double-bond functional group has a similar polarity pattern, so both molecules react with Br_2 in the same way. The size and complexity of the remainders of the molecules are not important.

Look carefully at Table 3.1, which lists many of the common functional groups and gives simple examples of their occurrence. Some functional groups have only carbon–carbon double or triple bonds; others have halogen atoms; and still others contain oxygen, nitrogen, sulfur, or phosphorus. Much of the chemistry you'll be studying in subsequent chapters is the chemistry of these functional groups.

Functional Groups with Carbon–Carbon Multiple Bonds

Alkenes, alkynes, and arenes (aromatic compounds) all contain carbon–carbon multiple bonds. *Alkenes* have a double bond, *alkynes* have a triple bond, and *arenes* have alternating double and single bonds in a six-membered ring of

TABLE 3.1
Structures of Some Common Functional Groups

Name	Structure[a]	Name ending	Example
Alkene (double bond)	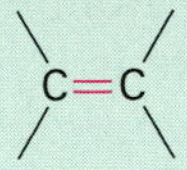 C=C	-ene	$H_2C=CH_2$ Ethene
Alkyne (triple bond)	—C≡C—	-yne	HC≡CH Ethyne
Arene (aromatic ring)	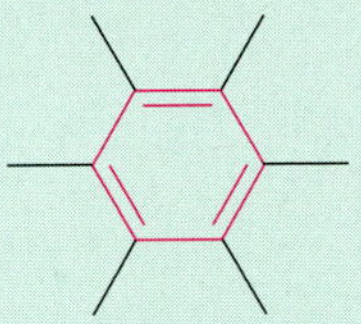	None	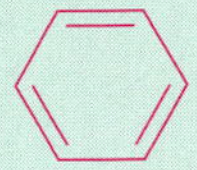 Benzene
Halide	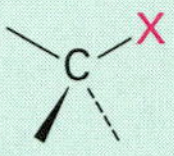 C—X (X = F, Cl, Br, I)	None	CH_3Cl Chloromethane
Alcohol	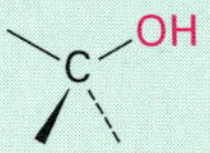 C—OH	-ol	CH_3OH Methanol
Ether	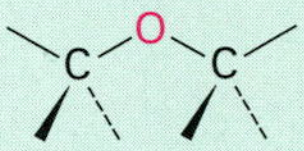 C—O—C	ether	CH_3OCH_3 Dimethyl ether
Monophosphate	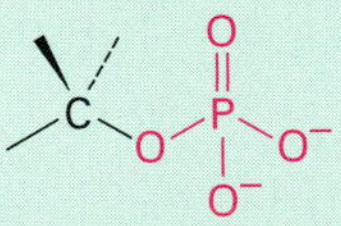 C—O—PO	phosphate	$CH_3OPO_3^{2-}$ Methyl phosphate
Diphosphate	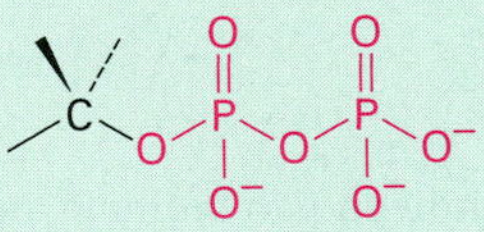 C—O—P—O—P	diphosphate	$CH_3OP_2O_6^{3-}$ Methyl diphosphate
Amine	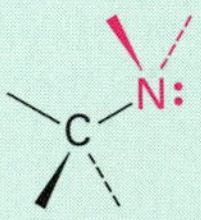 C—N:	-amine	CH_3NH_2 Methylamine
Imine (Schiff base)	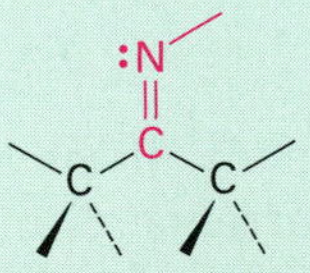 C—C—C	None	CH_3CCH_3 Acetone imine
Nitrile	—C≡N	-nitrile	$CH_3C≡N$ Ethanenitrile
Thiol	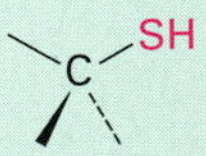 C—SH	-thiol	CH_3SH Methanethiol

[a]The bonds whose connections aren't specified are assumed to be attached to carbon or hydrogen atoms in the rest of the molecule.

(Continued)

TABLE 3.1
Structures of Some Common Functional Groups *continued*

Name	Structure[a]	Name ending	Example
Sulfide	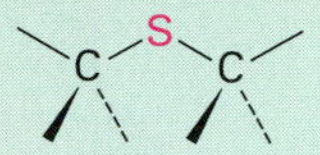	*sulfide*	CH_3SCH_3 Dimethyl sulfide
Disulfide	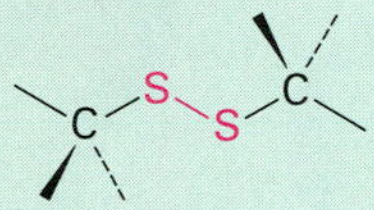	*disulfide*	CH_3SSCH_3 Dimethyl disulfide
Sulfoxide	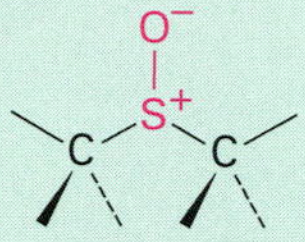	*sulfoxide*	CH_3SCH_3 Dimethyl sulfoxide
Aldehyde	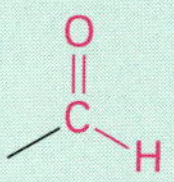	*-al*	CH_3CH Ethanal
Ketone	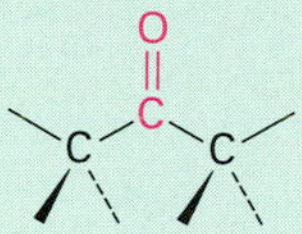	*-one*	CH_3CCH_3 Propanone
Carboxylic acid	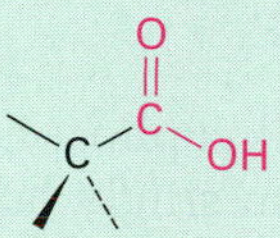	*-oic acid*	CH_3COH Ethanoic acid
Ester	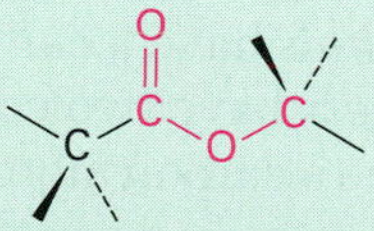	*-oate*	CH_3COCH_3 Methyl ethanoate
Thioester	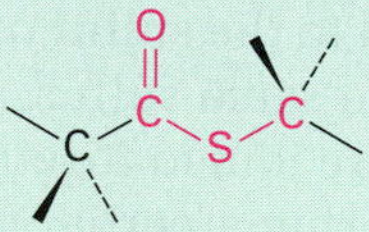	*-thioate*	CH_3CSCH_3 Methyl ethanethioate
Amide	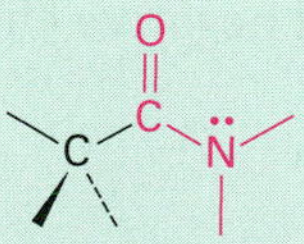	*-amide*	CH_3CNH_2 Ethanamide
Acid chloride	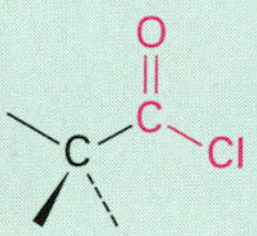	*-oyl chloride*	CH_3CCl Ethanoyl chloride
Carboxylic acid anhydride	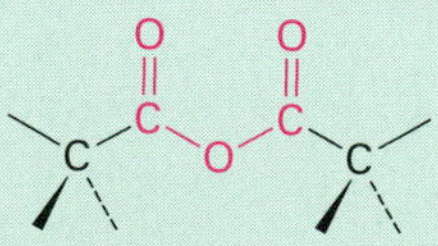	*-oic anhydride*	CH_3COCCH_3 Ethanoic anhydride

[a]The bonds whose connections aren't specified are assumed to be attached to carbon or hydrogen atoms in the rest of the molecules.

carbon atoms. Because of their structural similarities, these compounds also have chemical similarities.

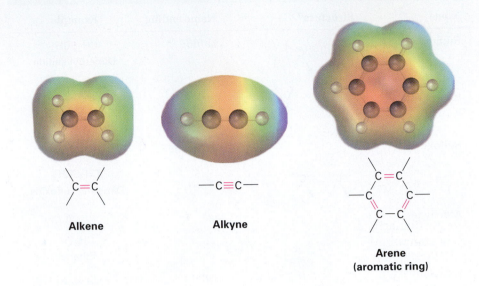

Alkene

Alkyne

**Arene
(aromatic ring)**

Functional Groups with Carbon Singly Bonded to an Electronegative Atom

Alkyl halides (haloalkanes), alcohols, ethers, alkyl phosphates, amines, thiols, sulfides, and disulfides all have a carbon atom singly bonded to an electronegative atom—halogen, oxygen, nitrogen, or sulfur. *Alkyl halides* have a carbon atom bonded to halogen (–X), *alcohols* have a carbon atom bonded to the oxygen of a hydroxyl group (–OH), *ethers* have two carbon atoms bonded to the same oxygen, *organophosphates* have a carbon atom bonded to the oxygen of a phosphate group (–OPO$_3{}^{2-}$), *amines* have a carbon atom bonded to a nitrogen, *thiols* have a carbon atom bonded to the sulfur of an –SH group, *sulfides* have two carbon atoms bonded to the same sulfur, and *disulfides* have carbon atoms bonded to two sulfurs that are joined together. In all cases, the bonds are polar, with the carbon atom bearing a partial positive charge ($\delta+$) and the electronegative atom bearing a partial negative charge ($\delta-$).

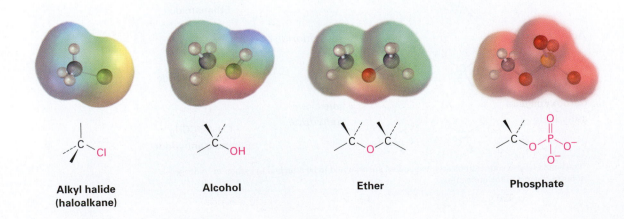

**Alkyl halide
(haloalkane)**

Alcohol

Ether

Phosphate

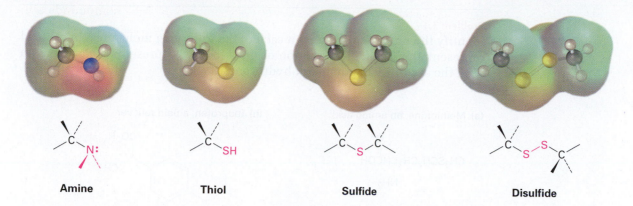

Amine   **Thiol**   **Sulfide**   **Disulfide**

Functional Groups with a Carbon–Oxygen Double Bond (Carbonyl Groups)

Note particularly the last eight entries in Table 3.1, which give different families of compounds that contain the *carbonyl group,* C=O (pronounced car-bo-**neel**). Functional groups with a carbon–oxygen double bond are present in the great majority of organic compounds and in practically all biological molecules. These compounds behave similarly in many respects but differ depending on the identity of the atoms bonded to the carbonyl-group carbon. *Aldehydes* have at least one hydrogen bonded to the C=O, *ketones* have two carbons bonded to the C=O, *carboxylic acids* have an –OH group bonded to the C=O, *esters* have an ether-like oxygen bonded to the C=O, *thioesters* have a sulfide-like sulfur bonded to the C=O, *amides* have an amine-like nitrogen bonded to the C=O, *acid chlorides* have a chlorine bonded to the C=O, and so on. The carbonyl carbon atom bears a partial positive charge ($\delta+$), and the oxygen bears a partial negative charge ($\delta-$).

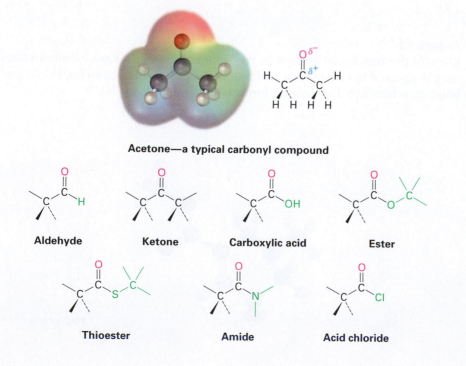

Acetone—a typical carbonyl compound

Aldehyde   **Ketone**   **Carboxylic acid**   **Ester**

Thioester   **Amide**   **Acid chloride**

Problem 3.1

Identify the functional groups in each of the following molecules. In skeletal representations, each intersection of lines (bonds) represents a carbon atom with the appropriate number of hydrogens attached.

(a) Methionine, an amino acid:

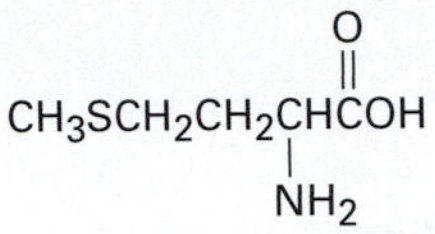

$$CH_3SCH_2CH_2\overset{\overset{\displaystyle O}{\|}}{C}HCOH$$
$$\underset{\displaystyle NH_2}{|}$$

(b) Ibuprofen, a pain reliever:

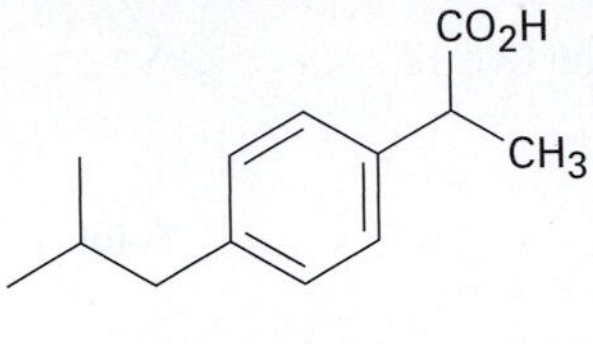

CO$_2$H
CH$_3$

(c) Capsaicin, the pungent substance in chili peppers:

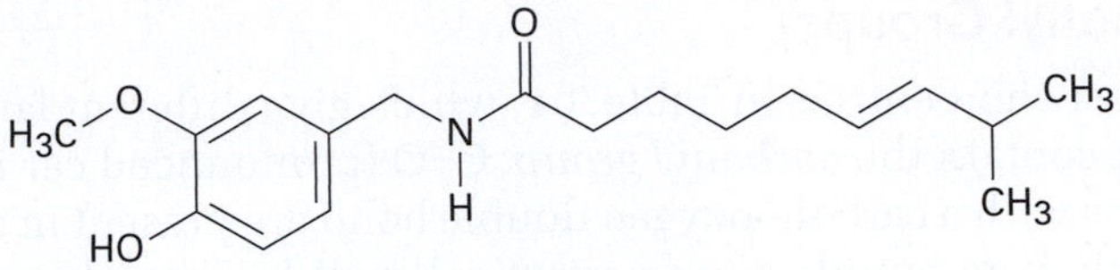

H$_3$C—O
HO
N
H
O
CH$_3$
CH$_3$

Problem 3.2

Propose structures for simple molecules that contain the following functional groups:

(a) Alcohol   **(b)** Aromatic ring
(c) Carboxylic acid   **(d)** Amine
(e) Both ketone and amine   **(f)** Two double bonds

Problem 3.3

Identify the functional groups in the following model of arecoline, a veterinary drug used to control worms in animals. Convert the drawing into a line-bond structure and a molecular formula (red = O, blue = N).

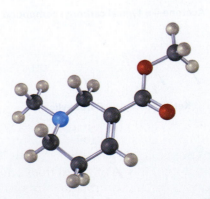

3.2 Alkanes and Alkane Isomers

Before beginning a systematic study of the different functional groups, let's look first at the simplest family of molecules—the *alkanes*—to develop some general ideas that apply to all families. We saw in Section 1.7 that the carbon–carbon single bond in ethane results from σ (head-on) overlap of carbon sp^3 orbitals. If we imagine joining three, four, five, or even more carbon atoms by C–C single bonds, we can generate the large family of molecules called **alkanes**.

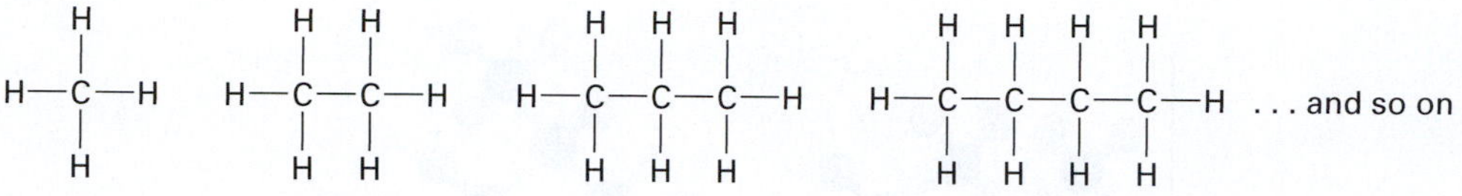

Methane　　　**Ethane**　　　**Propane**　　　**Butane**

Alkanes are often described as *saturated hydrocarbons*—**hydrocarbons** because they contain only carbon and hydrogen; **saturated** because they have only C–C and C–H single bonds and thus contain the maximum possible number of hydrogens per carbon. They have the general formula C_nH_{2n+2}, where n is an integer. Alkanes are also occasionally referred to as **aliphatic** compounds, a name derived from the Greek *aleiphas,* meaning "fat." We'll see in Section 23.1 that many animal fats contain long carbon chains similar to alkanes.

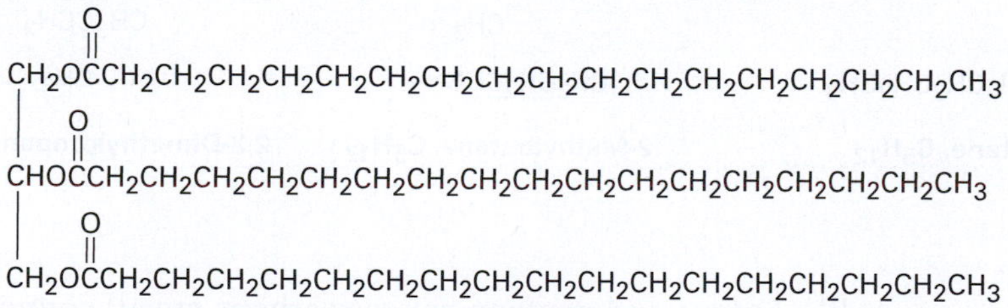

A typical animal fat

Think about the ways that carbon and hydrogen might combine to make alkanes. With one carbon and four hydrogens, only one structure is possible: methane, CH_4. Similarly, there is only one combination of two carbons with six hydrogens (ethane, CH_3CH_3) and only one combination of three carbons with eight hydrogens (propane, $CH_3CH_2CH_3$). When larger numbers of carbons and hydrogens combine, however, more than one structure is possible. For example, there are *two* substances with the formula C_4H_{10}: the four carbons can all be in a row (butane), or they can branch (isobutane). Similarly, there are three C_5H_{12} molecules, and so on for larger alkanes.

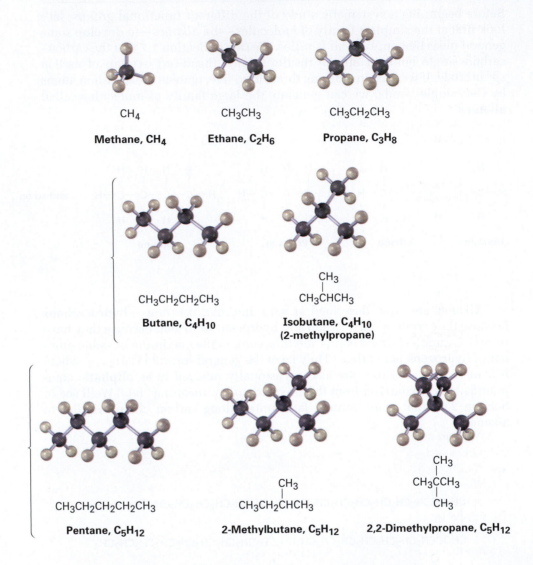

CH$_4$

Methane, CH$_4$

CH$_3$CH$_3$

Ethane, C$_2$H$_6$

CH$_3$CH$_2$CH$_3$

Propane, C$_3$H$_8$

CH$_3$CH$_2$CH$_2$CH$_3$

Butane, C$_4$H$_{10}$

CH$_3$
|
CH$_3$CHCH$_3$

**Isobutane, C$_4$H$_{10}$
(2-methylpropane)**

CH$_3$CH$_2$CH$_2$CH$_2$CH$_3$

Pentane, C$_5$H$_{12}$

CH$_3$
|
CH$_3$CH$_2$CHCH$_3$

2-Methylbutane, C$_5$H$_{12}$

CH$_3$
|
CH$_3$CCH$_3$
|
CH$_3$

2,2-Dimethylpropane, C$_5$H$_{12}$

Compounds like butane and pentane, whose carbons are all connected in a row, are called **straight-chain alkanes**, or *normal alkanes.* Compounds like 2-methylpropane (isobutane), 2-methylbutane, and 2,2-dimethylpropane, whose carbon chains branch, are called **branched-chain alkanes**. The difference between the two is that you can draw a line connecting all the carbons of a straight-chain alkane without retracing your path or lifting your pencil from the paper. For a branched-chain alkane, however, you either have to retrace your path or lift your pencil from the paper to draw a line connecting all the carbons.

Compounds like the two C_4H_{10} molecules and the three C_5H_{12} molecules, which have the same formula but different structures, are called *isomers,* from the Greek *isos + meros,* meaning "made of the same parts." **Isomers** are compounds that have the same numbers and kinds of atoms but differ in the way the atoms are arranged. Compounds like butane and isobutane, whose atoms are connected differently, are called **constitutional isomers**. We'll see shortly that other kinds of isomers are also possible, even among compounds whose atoms are connected in the same order. As Table 3.2 shows, the number of possible alkane isomers increases dramatically as the number of carbon atoms increases.

Constitutional isomerism is not limited to alkanes—it occurs widely throughout organic chemistry. Constitutional isomers may have different carbon skeletons (as in isobutane and butane), different functional groups (as in ethanol and dimethyl ether), or different locations of a functional group along the chain (as in isopropylamine and propylamine). Regardless of the reason for the isomerism, constitutional isomers are always different compounds with different properties but with the same formula.

TABLE 3.2
Number of Alkane Isomers

Formula	Number of isomers
C_6H_{14}	5
C_7H_{16}	9
C_8H_{18}	18
C_9H_{20}	35
$C_{10}H_{22}$	75
$C_{15}H_{32}$	4,347
$C_{20}H_{42}$	366,319
$C_{30}H_{62}$	4,111,846,763

Different carbon skeletons
C_4H_{10}

$$CH_3CHCH_3 \quad \text{and} \quad CH_3CH_2CH_2CH_3$$
(with CH_3 on the central carbon)

2-Methylpropane (isobutane)   **Butane**

Different functional groups
C_2H_6O

$$CH_3CH_2OH \quad \text{and} \quad CH_3OCH_3$$

Ethanol   **Dimethyl ether**

Different position of functional groups
C_3H_9N

$$CH_3CHCH_3 \quad \text{and} \quad CH_3CH_2CH_2NH_2$$
(with NH_2 on the central carbon)

Isopropylamine   **Propylamine**

A given alkane can be drawn arbitrarily in many ways. For example, the straight-chain, four-carbon alkane called butane can be represented by any of the structures shown in Figure 3.2. These structures don't imply any particular three-dimensional geometry for butane; they indicate only the connections among atoms. In practice, as noted in Section 1.12, chemists rarely draw all the bonds in a molecule and usually refer to butane by the condensed structure, $CH_3CH_2CH_2CH_3$ or $CH_3(CH_2)_2CH_3$. Still more simply, butane can be represented as $n\text{-}C_4H_{10}$, where n denotes *normal* (straight-chain) butane.

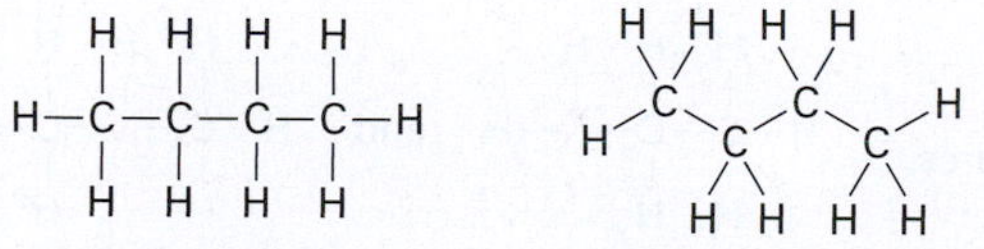

$$CH_3-CH_2-CH_2-CH_3 \qquad CH_3CH_2CH_2CH_3 \qquad CH_3(CH_2)_2CH_3$$

FIGURE 3.2 Some representations of butane, C_4H_{10}. The molecule is the same regardless of how it's drawn. These structures imply only the connections between atoms; they don't imply any specific geometry.

Straight-chain alkanes are named according to the number of carbon atoms they contain, as shown in Table 3.3. With the exception of the first four compounds—methane, ethane, propane, and butane—whose names have historical roots, the alkanes are named based on Greek numbers. The suffix -*ane* is added to the end of each name to indicate that the molecule identified is an

TABLE 3.3
Names of Straight-Chain Alkanes

Number of carbons (n)	Name	Formula (C_nH_{2n+2})	Number of carbons (n)	Name	Formula (C_nH_{2n+2})
1	Methane	CH_4	9	Nonane	C_9H_{20}
2	Ethane	C_2H_6	10	Decane	$C_{10}H_{22}$
3	Propane	C_3H_8	11	Undecane	$C_{11}H_{24}$
4	Butane	C_4H_{10}	12	Dodecane	$C_{12}H_{26}$
5	Pentane	C_5H_{12}	13	Tridecane	$C_{13}H_{28}$
6	Hexane	C_6H_{14}	20	Icosane	$C_{20}H_{42}$
7	Heptane	C_7H_{16}	30	Triacontane	$C_{30}H_{62}$
8	Octane	C_8H_{18}			

alkane. Thus, pent*ane* is the five-carbon alkane, hex*ane* is the six-carbon alkane, and so on. We'll soon see that these alkane names form the basis for naming all other organic compounds, so at least the first ten should be memorized.

WORKED EXAMPLE 3.1 Drawing the Structures of Isomers

Propose structures for two isomers with the formula C_2H_7N.

Strategy
We know that carbon forms four bonds, nitrogen forms three, and hydrogen forms one. Write down the carbon atoms first, and then use a combination of trial and error plus intuition to put the pieces together.

Solution
There are two isomeric structures. One has the connection C–C–N, and the other has the connection C–N–C.

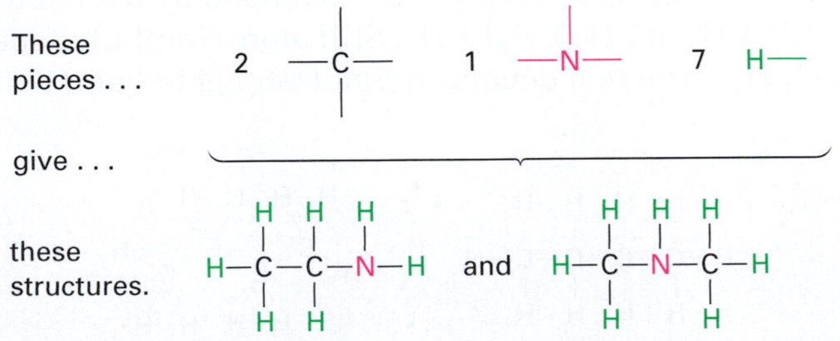

Problem 3.4
Draw structures of the five isomers of C_6H_{14}.

Problem 3.5
Propose structures that meet the following descriptions:
(a) Two isomeric esters with the formula $C_5H_{10}O_2$
(b) Two isomeric disulfides with the formula $C_4H_{10}S_2$

Problem 3.6

How many isomers are there that meet the following descriptions?
(a) Alcohols with the formula C_3H_8O
(b) Bromoalkanes with the formula C_4H_9Br
(c) Thioesters with the formula C_4H_8OS

3.3 Alkyl Groups

If you imagine removing a hydrogen atom from an alkane, the partial structure that remains is called an **alkyl group**. Alkyl groups are not stable compounds themselves; they are simply parts of larger compounds. Alkyl groups are named by replacing the -*ane* ending of the parent alkane with an -*yl* ending. For example, removal of a hydrogen from methane, CH_4, generates a *methyl* group, $-CH_3$, and removal of a hydrogen from ethane, CH_3CH_3, generates an *ethyl* group, $-CH_2CH_3$. Similarly, removal of a hydrogen atom from the end carbon of any straight-chain alkane gives the series of straight-chain alkyl groups shown in Table 3.4. Combining an alkyl group with any of the functional groups listed earlier makes it possible to generate and name many thousands of compounds. For example:

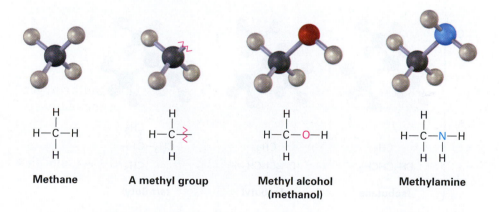

| Methane | A methyl group | Methyl alcohol (methanol) | Methylamine |

TABLE 3.4
Some Straight-Chain Alkyl Groups

Alkane	Name	Alkyl group	Name (abbreviation)
CH_4	Methane	$-CH_3$	Methyl (Me)
CH_3CH_3	Ethane	$-CH_2CH_3$	Ethyl (Et)
$CH_3CH_2CH_3$	Propane	$-CH_2CH_2CH_3$	Propyl (Pr)
$CH_3CH_2CH_2CH_3$	Butane	$-CH_2CH_2CH_2CH_3$	Butyl (Bu)
$CH_3CH_2CH_2CH_2CH_3$	Pentane	$-CH_2CH_2CH_2CH_2CH_3$	Pentyl, or amyl
:			
:			

Just as straight-chain alkyl groups are generated by removing a hydrogen from an *end* carbon, branched alkyl groups are generated by removing a hydrogen atom from an *internal* carbon. Two 3-carbon alkyl groups and four 4-carbon alkyl groups are possible (Figure 3.3).

FIGURE 3.3 Alkyl groups generated from straight-chain alkanes.

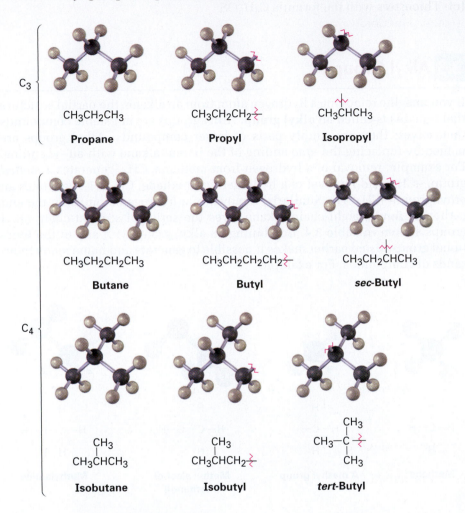

C_3

CH$_3$CH$_2$CH$_3$
Propane

CH$_3$CH$_2$CH$_2\!\!\!\lambda$
Propyl

CH$_3$CHCH$_3$
Isopropyl

C_4

CH$_3$CH$_2$CH$_2$CH$_3$
Butane

CH$_3$CH$_2$CH$_2$CH$_2\!\!\!\lambda$
Butyl

CH$_3$CH$_2$CHCH$_3$
***sec*-Butyl**

CH$_3$
CH$_3$CHCH$_3$
Isobutane

CH$_3$
CH$_3$CHCH$_2\!\!\!\lambda$
Isobutyl

CH$_3$
CH$_3$—C$\!\!\!\lambda$
CH$_3$
***tert*-Butyl**

One further word about naming alkyl groups: the prefixes *sec-* (for secondary) and *tert-* (for tertiary) used for the C$_4$ alkyl groups in Figure 3.3 refer to *the number of other carbon atoms attached to the branching carbon atom*. There are four possibilities: primary (1°), secondary (2°), tertiary (3°), and quaternary (4°):

R   H
 \ /
  C
 / \
H   H

Primary carbon (1°)
is bonded to one
other carbon.

R   R
 \ /
  C
 / \
H   H

Secondary carbon (2°)
is bonded to two
other carbons.

R   R
 \ /
  C
 / \
R   H

Tertiary carbon (3°)
is bonded to three
other carbons.

R   R
 \ /
  C
 / \
R   R

Quaternary carbon (4°)
is bonded to four
other carbons.

The symbol **R** is used in organic chemistry to represent a *generalized* organic group. The R group can be methyl, ethyl, propyl, or any of a multitude of others. You might think of **R** as representing the **R**est of the molecule, which we aren't bothering to specify.

The terms *primary, secondary, tertiary,* and *quaternary* are routinely used in organic chemistry, and their meanings need to become second nature. For example, if we were to say, "Citric acid is a tertiary alcohol," we would mean that it has an alcohol functional group (–OH) bonded to a carbon atom that is itself bonded to three other carbons. (These other carbons may in turn connect to other functional groups.)

$$R-\underset{\underset{R}{|}}{\overset{\overset{OH}{|}}{C}}-R$$

General class of tertiary alcohols, R$_3$COH

$$HO_2CCH_2-\underset{\underset{CO_2H}{|}}{\overset{\overset{OH}{|}}{C}}-CH_2CO_2H$$

Citric acid—a specific tertiary alcohol

In addition, we also speak about hydrogen atoms as being primary, secondary, or tertiary. Primary hydrogen atoms are attached to primary carbons (RCH$_3$), secondary hydrogens are attached to secondary carbons (R$_2$CH$_2$), and tertiary hydrogens are attached to tertiary carbons (R$_3$CH). There is, of course, no such thing as a quaternary hydrogen. (Why not?)

Primary hydrogens (CH$_3$)

Secondary hydrogens (CH$_2$)

A tertiary hydrogen (CH)

$$CH_3CH_2CHCH_3 = $$

Problem 3.7
Draw the eight 5-carbon alkyl groups (pentyl isomers).

Problem 3.8
Identify the carbon atoms in the following molecules as primary, secondary, tertiary, or quaternary:

(a)
$$\underset{CH_3CHCH_2CH_2CH_3}{\overset{\overset{CH_3}{|}}{}}$$

(b)
$$\underset{CH_3CH_2CHCH_2CH_3}{\overset{\overset{CH_3CHCH_3}{|}}{}}$$

(c)
$$CH_3CHCH_2CCH_3$$ with CH_3 groups

Problem 3.9
Identify the hydrogen atoms on the compounds shown in Problem 3.8 as primary, secondary, or tertiary.

Problem 3.10
Draw structures of alkanes that meet the following descriptions:
(a) An alkane with two tertiary carbons
(b) An alkane that contains an isopropyl group
(c) An alkane that has one quaternary and one secondary carbon

3.4 Naming Alkanes

In earlier times, when relatively few pure organic chemicals were known, new compounds were named at the whim of their discoverer. Thus, urea (CH_4N_2O) is a crystalline substance isolated from urine; morphine ($C_{17}H_{19}NO_3$) is an analgesic (painkiller) named after Morpheus, the Greek god of dreams; and acetic acid, the primary organic constituent of vinegar, is named from the Latin word for vinegar, *acetum.*

As the science of organic chemistry slowly grew in the 19th century, so too did the number of known compounds and the need for a systematic method of naming them. The system of nomenclature we'll use in this book is that devised by the International Union of Pure and Applied Chemistry (IUPAC, usually spoken as **eye**-you-pac).

A chemical name typically has four parts in the IUPAC system of nomenclature: prefix, parent, locant, and suffix. The prefix specifies the location and identity of various substituent groups in the molecule, the parent selects a main part of the molecule and tells how many carbon atoms are in that part, the locant gives the location of the primary functional group, and the suffix identifies the primary functional group.

Prefix—Parent—Locant—Suffix

Where and what are   How many   Where is the primary   What is the primary
the substituents?   carbons?   functional group?   functional group?

As we cover new functional groups in later chapters, the applicable IUPAC rules of nomenclature will be given. In addition, Appendix A at the back of this book gives an overall view of organic nomenclature and shows how compounds that contain more than one functional group are named. For the present, let's see how to name branched-chain alkanes and learn some general naming rules that are applicable to all compounds.

All but the most complex branched-chain alkanes can be named by following four steps. For a very few compounds, a fifth step is needed.

Step 1
Find the parent hydrocarbon.

(a) Find the longest continuous chain of carbon atoms in the molecule, and use the name of that chain as the parent name. The longest chain may not always be apparent from the manner of writing; you may have to "turn corners."

CH₃CH₂CH₂CH—CH₃ with CH₂CH₃ branch — Named as a substituted hexane

CH₃—CHCH—CH₂CH₃ with CH₃/CH₂ branch and CH₂CH₂CH₃ branch — Named as a substituted heptane

(b) If two different chains of equal length are present, choose the one with the larger number of branch points as the parent:

$$CH_3CHCHCH_2CH_2CH_3 \atop \displaystyle{CH_3 \atop |} \ \displaystyle{| \atop CH_2CH_3}$$

NOT

$$CH_3CH-CHCH_2CH_2CH_3 \atop \displaystyle{CH_3 \atop |} \ \displaystyle{| \atop CH_2CH_3}$$

Named as a hexane with *two* substituents

as a hexane with *one* substituent

Step 2

Number the atoms in the longest chain.

(a) Beginning at the end nearer the first branch point, number each carbon atom in the parent chain:

$$\underset{3}{CH_3}-\underset{}{\overset{\overset{\displaystyle 2\ \ 1}{CH_2CH_3}}{\underset{\underset{5\quad6\quad7}{CH_2CH_2CH_3}}{|}}}\overset{}{CH}CH-CH_2CH_3$$

NOT

$$\underset{}{CH_3}-\underset{}{\overset{\overset{\displaystyle 6\ \ 7}{CH_2CH_3}}{\underset{\underset{3\quad2\quad1}{CH_2CH_2CH_3}}{|}}}\overset{}{CH}CH-CH_2CH_3$$

The first branch occurs at C3 in the proper system of numbering, not at C4.

(b) If there is branching an equal distance from both ends of the parent chain, begin numbering at the end nearer the second branch point:

$$\underset{7\ \ 6}{CH_3-CHCH_2CH_2CH}\underset{5\quad\quad4}{-}\underset{3\ \ 2\ \ 1}{CHCH_2CH_3} \atop \displaystyle{\overset{8\ \ 9}{CH_2CH_3} \quad CH_3 \ CH_2CH_3}$$

NOT

$$\underset{3\ \ 4}{CH_3-CHCH_2CH_2CH}\underset{5\quad\quad6}{-}\underset{7\ \ 8\ \ 9}{CHCH_2CH_3} \atop \displaystyle{\overset{2\ \ 1}{CH_2CH_3} \quad CH_3 \ CH_2CH_3}$$

Step 3

Identify and number the substituents.

(a) Assign a number (called a *locant*) to each substituent to locate its point of attachment to the parent chain:

$$\underset{7\quad6\quad\ 5\quad4\ \ 3\ \ 2\ \ 1}{CH_3-CHCH_2CH_2CHCHCH_2CH_3} \atop \displaystyle{\overset{9\ \ 8}{CH_3CH_2} \quad H_3C \ CH_2CH_3}$$

Named as a nonane

Substituents: On C3, CH_2CH_3   (3-ethyl)
On C4, CH_3   (4-methyl)
On C7, CH_3   (7-methyl)

(b) If there are two substituents on the same carbon, give them both the same number. There must be as many numbers in the name as there are substituents.

$$\underset{6\quad5\quad\ \ 3\quad2\ \ 1}{CH_3CH_2CCH_2CHCH_3} \atop \displaystyle{\overset{CH_3 \ CH_3}{|\quad\ |}} \atop \displaystyle{\underset{CH_2CH_3}{|}}$$

Named as a hexane

Substituents: On C2, CH_3   (2-methyl)
On C4, CH_3   (4-methyl)
On C4, CH_2CH_3   (4-ethyl)

Step 4

Write the name as a single word.

Use hyphens to separate the different prefixes, and use commas to separate numbers. If two or more different substituents are present, cite them in alphabetical order. If two or more identical substituents are present on the parent chain, use one of the multiplier prefixes *di-*, *tri-*, *tetra-*, and so forth, but don't use these prefixes for alphabetizing. Full names for some of the examples we have been using follow:

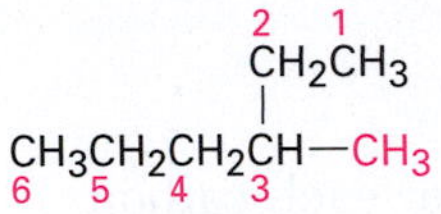

3-Methylhexane 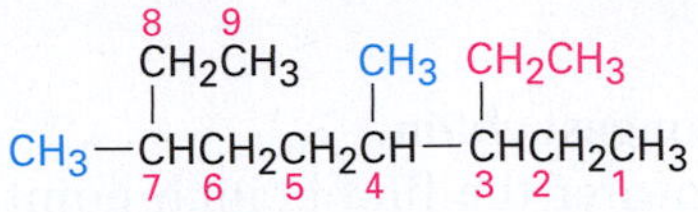   **3-Ethyl-4,7-dimethylnonane** 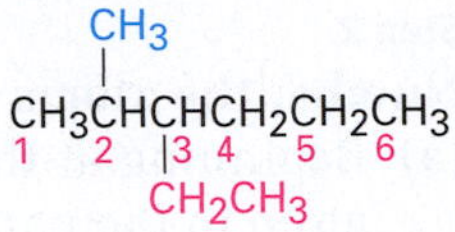   **3-Ethyl-2-methylhexane**

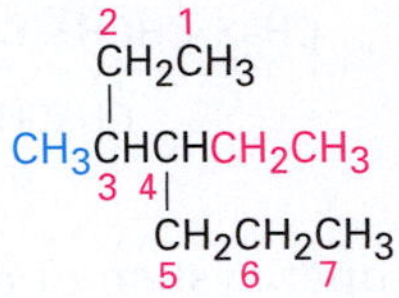

4-Ethyl-3-methylheptane 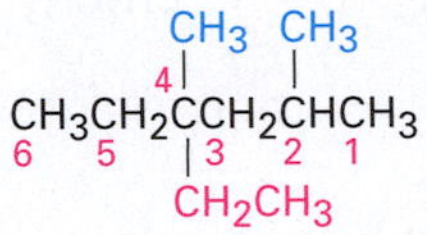   **4-Ethyl-2,4-dimethylhexane**

Step 5

Name a branched substituent as though it were itself a compound.

In some particularly complex cases, a fifth step is necessary. It occasionally happens that a substituent on the main chain is itself branched. In the following case, for instance, the substituent at C6 is a three-carbon chain with a methyl sub-branch. To name the compound fully, the branched substituent must first be named.

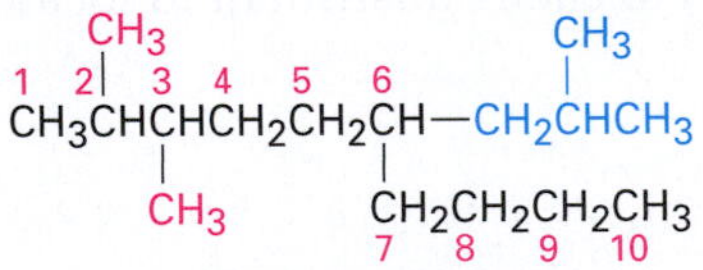

Named as a 2,3,6-   **A 2-methylpropyl group**
trisubstituted decane

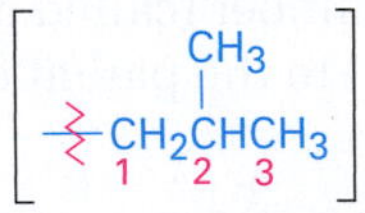

Begin numbering the branched substituent at its point of its attachment to the main chain, and identify it as a 2-methylpropyl group. The substituent is alphabetized according to the first letter of its complete name, including any numerical prefix, and is set off in parentheses when naming the entire molecule:

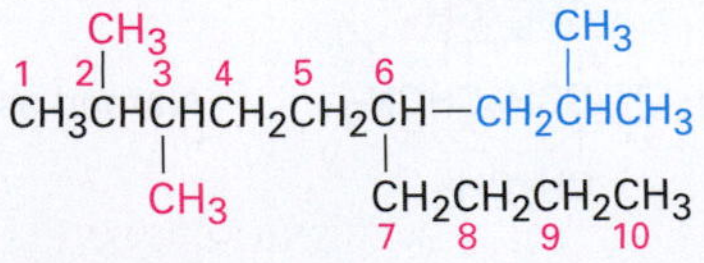

2,3-Dimethyl-6-(2-methylpropyl)decane

As a further example:

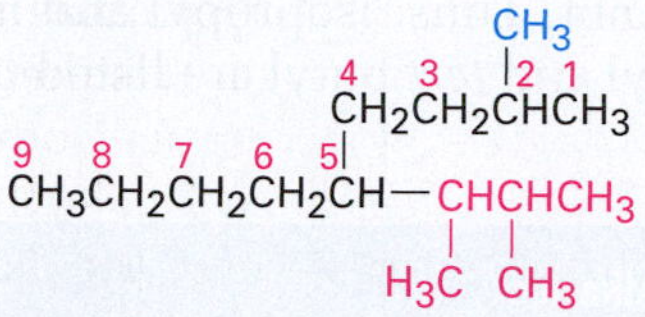

$$\overset{4}{C}H_2\overset{3}{C}H_2\overset{2}{C}H\overset{1}{C}H_3$$

$$\underset{9}{C}H_3\underset{8}{C}H_2\underset{7}{C}H_2\underset{6}{C}H_2\underset{5}{C}H{-}CHCHCH_3$$

CH₃ (blue, attached at position 2)

H₃C CH₃

$$\left[-\overset{1}{C}H\overset{2}{C}H\overset{3}{C}H_3 \atop H_3C \quad CH_3 \right]$$

5-(1,2-Dimethylpropyl)-2-methylnonane 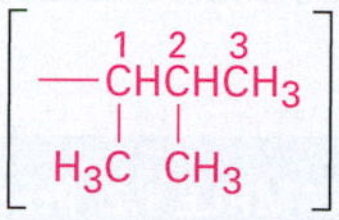   **A 1,2-dimethylpropyl group**

For historical reasons, some of the simpler branched-chain alkyl groups also have nonsystematic, common names, as noted earlier.

1. Three-carbon alkyl group:

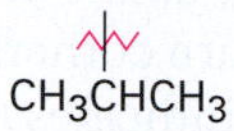

CH₃CHCH₃

Isopropyl (*i*-Pr)

2. Four-carbon alkyl groups:

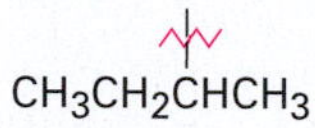

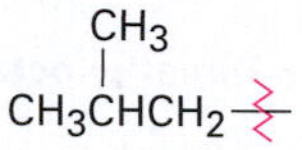

CH₃CH₂CHCH₃   CH₃ \
                        CH₃CHCH₂—   CH₃ \
                                            CH₃—C— \
                                            CH₃

sec-Butyl   **Isobutyl** 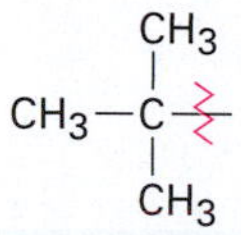   **tert-Butyl**
(sec-Bu)   **(t-butyl or t-Bu)**

3. Five-carbon alkyl groups:

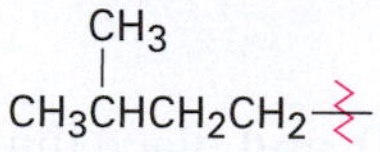

CH₃ \
CH₃CHCH₂CH₂—   CH₃ \
                        CH₃—C—CH₂—   CH₃ \
                        CH₃   CH₃CH₂—C— \
                                            CH₃

Isopentyl, also called   **Neopentyl** 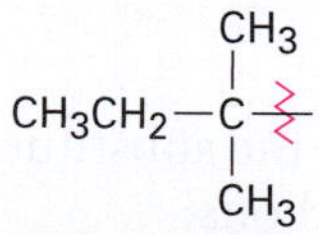   **tert-Pentyl, also called**
isoamyl (*i*-amyl)   **tert-amyl (*t*-amyl)**

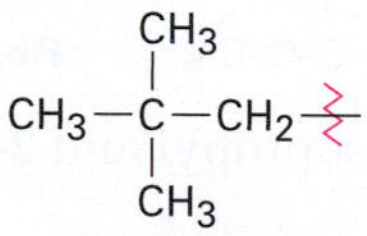

The common names of these simple alkyl groups are so well entrenched in the chemical literature that IUPAC rules make allowance for them. Thus, the following compound is properly named either 4-(1-methylethyl)heptane or 4-isopropylheptane. There's no choice but to memorize these common names; fortunately, there are only a few of them.

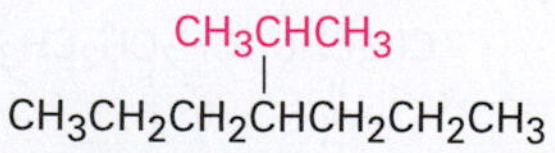

CH₃CHCH₃ \
CH₃CH₂CH₂CHCH₂CH₂CH₃

4-(1-Methylethyl)heptane   or   **4-Isopropylheptane**

When writing an alkane name, the nonhyphenated prefix iso- is considered part of the alkyl-group name for alphabetizing purposes, but the hyphenated and italicized prefixes *sec-* and *tert-* are not. Thus, isopropyl and isobutyl are listed alphabetically under *i*, but *sec*-butyl and *tert*-butyl are listed under *b*.

WORKED EXAMPLE 3.2 Practice in Naming Alkanes

What is the IUPAC name of the following alkane?

$$\begin{array}{cc} CH_2CH_3 & CH_3 \\ | & | \\ CH_3CHCH_2CH_2CH_2CHCH_3 \end{array}$$

Strategy

Find the longest continuous carbon chain in the molecule, and use that as the parent name. This molecule has a chain of eight carbons—octane—with two methyl substituents. (You have to turn corners to see it.) Numbering from the end nearer the first methyl substituent indicates that the methyls are at C2 and C6.

Solution

$$\begin{array}{cc} \overset{7\quad 8}{CH_2CH_3} & \overset{}{CH_3} \\ | & | \\ \underset{6\ \ 5\ \ 4\ \ 3\ \ 2\ \ 1}{CH_3CHCH_2CH_2CH_2CHCH_3} \end{array}$$

2,6-Dimethyloctane

WORKED EXAMPLE 3.3 Converting a Chemical Name into a Structure

Draw the structure of 3-isopropyl-2-methylhexane.

Strategy

This is the reverse of Worked Example 3.2 and uses a reverse strategy. Look at the parent name (hexane), and draw its carbon structure.

C–C–C–C–C–C   **Hexane**

Next, find the substituents (3-isopropyl and 2-methyl), and place them on the proper carbons:

$$\begin{array}{l} CH_3CHCH_3 \longleftarrow \text{\textbf{An isopropyl group at C3}}\\ | \\ \underset{1\ \ 2\ \ \ 3\ \ \ 4\ \ \ 5\ \ \ 6}{C-C-C-C-C-C}\\ \quad\ |\\ \quad\ CH_3 \longleftarrow \text{\textbf{A methyl group at C2}} \end{array}$$

Finally, add hydrogens to complete the structure.

Solution

$$\begin{array}{l} CH_3CHCH_3 \\ | \\ CH_3CHCHCH_2CH_2CH_3 \\ \quad\ | \\ \quad\ CH_3 \end{array}$$

3-Isopropyl-2-methylhexane

Problem 3.11

Give IUPAC names for the following compounds:

(a) The three isomers of C_5H_{12}

(b)
$$CH_3CH_2CHCHCH_3$$
with CH_3 above the third carbon and CH_3 below the fourth carbon

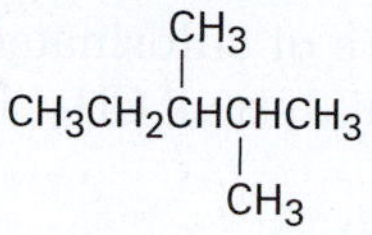

(c)
$$(CH_3)_2CHCH_2CHCH_3$$
with CH_3 above the fourth carbon

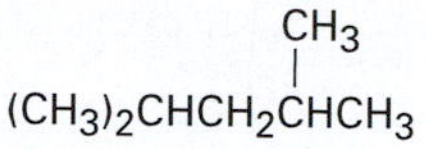

(d)
$$(CH_3)_3CCH_2CH_2CH$$
with CH_3 above and CH_3 below the last carbon

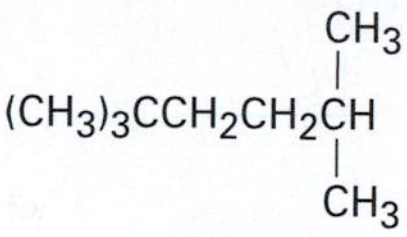

Problem 3.12

Draw structures corresponding to the following IUPAC names:
(a) 3,4-Dimethylnonane   **(b)** 3-Ethyl-4,4-dimethylheptane
(c) 2,2-Dimethyl-4-propyloctane   **(d)** 2,2,4-Trimethylpentane

Problem 3.13

Name the eight 5-carbon alkyl groups you drew in Problem 3.7.

Problem 3.14

Give the IUPAC name for the following hydrocarbon, and convert the drawing into a skeletal structure:

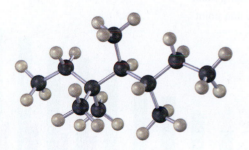

3.5 Properties of Alkanes

Alkanes are sometimes referred to as *paraffins,* a word derived from the Latin *parum affinis,* meaning "little affinity." This term aptly describes their behavior, for alkanes show little chemical affinity for other substances and are chemically inert to most laboratory reagents. They are also relatively inert biologically and are not often involved in the chemistry of living organisms. Alkanes do, however, react with oxygen, halogens, and a few other substances under the appropriate conditions.

Reaction with oxygen occurs during combustion in an engine or furnace when the alkane is used as a fuel. Carbon dioxide and water are formed as products, and a large amount of heat is released. For example, methane (natural gas) reacts with oxygen according to the equation

$$CH_4 + 2\,O_2 \rightarrow CO_2 + 2\,H_2O + 890\ \text{kJ/mol (213 kcal/mol)}$$

The reaction of an alkane with Cl_2 occurs when a mixture of the two is irradiated with ultraviolet light (denoted $h\nu$, where ν is the Greek letter nu). Depending on the relative amounts of the two reactants and on the time allowed, a sequential substitution of the alkane hydrogen atoms by chlorine occurs, leading to a mixture of chlorinated products. Methane, for example, reacts with Cl_2 to yield a mixture of CH_3Cl, CH_2Cl_2, $CHCl_3$, and CCl_4.

$$CH_4 \; + \; Cl_2 \; \xrightarrow{h\nu} \; CH_3Cl \; + \; HCl$$
$$\xrightarrow{Cl_2} \; CH_2Cl_2 \; + \; HCl$$
$$\xrightarrow{Cl_2} \; CHCl_3 \; + \; HCl$$
$$\xrightarrow{Cl_2} \; CCl_4 \; + \; HCl$$

Alkanes show regular increases in both boiling point and melting point as molecular weight increases (Figure 3.4), an effect due to the presence of weak dispersion forces between molecules (Section 2.12). Only when sufficient thermal energy is applied to overcome these forces does the solid melt or liquid boil. As you might expect, dispersion forces increase as molecular size increases, accounting for the higher melting and boiling points of larger alkanes.

FIGURE 3.4 A plot of melting and boiling points versus number of carbon atoms for the C_1–C_{14} straight-chain alkanes. There is a regular increase with molecular size.

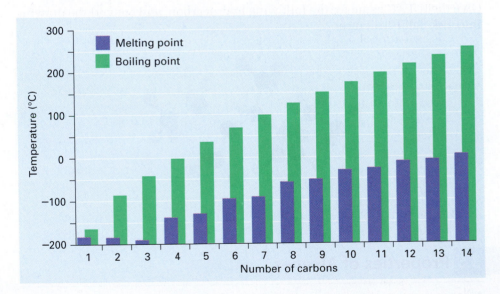

3.6 Conformations of Ethane

Up to this point, we've viewed molecules primarily in a two-dimensional way and have given little thought to any consequences that might arise from the spatial arrangement of atoms in molecules. Now it's time to add a third dimension to our study. **Stereochemistry** is the branch of chemistry concerned with the three-dimensional aspects of molecules. We'll see on many occasions in future chapters that the exact three-dimensional structure of a molecule is often crucial to determining its properties and biological behavior.

We know from Section 1.5 that σ bonds are cylindrically symmetrical. In other words, the intersection of a plane cutting through a carbon–carbon single-bond orbital looks like a circle. Because of this cylindrical symmetry, *rotation* is possible around carbon–carbon bonds in open-chain molecules. In ethane, for instance, rotation around the C–C bond occurs freely, constantly changing the geometric relationships between the hydrogens on one carbon and those on the other (Figure 3.5).

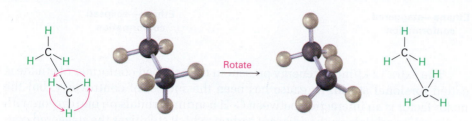

FIGURE 3.5 Rotation occurs around the carbon–carbon single bond in ethane because of σ bond cylindrical symmetry.

The different arrangements of atoms that result from bond rotation are called **conformations**, and molecules that have different arrangements are called conformational isomers, or **conformers**. Unlike constitutional isomers, however, different conformers can't usually be isolated because they intercovert too rapidly.

Conformational isomers are represented in two ways, as shown in Figure 3.6. A *sawhorse representation* views the carbon–carbon bond from an oblique angle and indicates spatial orientation by showing all C–H bonds. A **Newman projection** views the carbon–carbon bond directly end-on and represents the two carbon atoms by a circle. Bonds attached to the front carbon are represented by lines to the center of the circle, and bonds attached to the rear carbon are represented by lines to the edge of the circle.

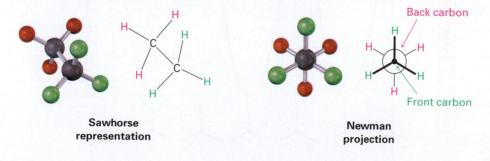

Sawhorse representation

Newman projection

FIGURE 3.6 A sawhorse representation and a Newman projection of ethane. The sawhorse representation views the molecule from an oblique angle, while the Newman projection views the molecule end-on. Note that the molecular model of the Newman projection appears at first to have six atoms attached to a single carbon. Actually, the front carbon, with three attached green atoms, is directly in front of the rear carbon, with three attached red atoms.

Despite what we've just said, we actually don't observe *perfectly* free rotation in ethane. Experiments show that there is a small (12 kJ/mol; 2.9 kcal/mol) barrier to rotation and that some conformers are more stable than others. The lowest-energy, most stable conformer is the one in which all six C–H bonds are as far away from one another as possible—**staggered** when viewed end-on in a Newman projection. The highest-energy, least stable conformer is the one in which the six C–H bonds are as close as possible—**eclipsed** in a Newman projection. At any given instant, about 99% of ethane molecules have an approximately staggered conformation and only about 1% are near the eclipsed conformation.

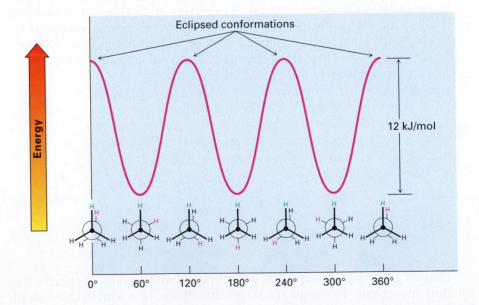

Ethane—staggered conformation

Rotate rear carbon 60°

4.0 kJ/mol

4.0 kJ/mol   4.0 kJ/mol

Ethane—eclipsed conformation

The extra 12 kJ/mol of energy present in the eclipsed conformer of ethane is called **torsional strain**. Its cause has been the subject of controversy, but the major factor is an interaction between C–H bonding orbitals on one carbon with antibonding orbitals on the adjacent carbon, which stabilizes the staggered conformer relative to the eclipsed conformer. Because the total strain of 12 kJ/mol arises from three equal hydrogen–hydrogen eclipsing interactions, we can assign a value of approximately 4.0 kJ/mol (1.0 kcal/mol) to each single interaction. The barrier to rotation that results can be represented on a graph of potential energy versus degree of rotation in which the angle between C–H bonds on front and back carbons as viewed end-on (the *dihedral angle*) goes full circle from 0° to 360°. Energy minima occur at staggered conformations, and energy maxima occur at eclipsed conformations, as shown in Figure 3.7.

FIGURE 3.7 A graph of potential energy versus bond rotation in ethane. The staggered conformers are 12 kJ/mol lower in energy than the eclipsed conformers.

Eclipsed conformations

Energy

12 kJ/mol

0°   60°   120°   180°   240°   300°   360°

3.7 Conformations of Other Alkanes

Propane, the next higher member in the alkane series, also has a torsional barrier that results in hindered rotation around the carbon–carbon bonds. The barrier is slightly higher in propane than in ethane—a total of 14 kJ/mol (3.4 kcal/mol) versus 12 kJ/mol.

The eclipsed conformer of propane has three interactions—two ethane-type hydrogen–hydrogen interactions and one additional hydrogen–methyl interaction. Since each eclipsing $H \leftrightarrow H$ interaction is the same as that in ethane and thus has an energy "cost" of 4.0 kJ/mol, we can assign a value of $14 - (2 \times 4.0) = 6.0$ kJ/mol (1.4 kcal/mol) to the eclipsing $H \leftrightarrow CH_3$ interaction (Figure 3.8).

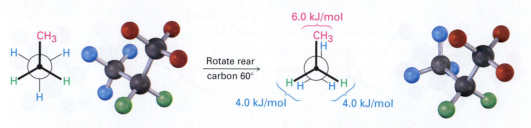

Staggered propane

Eclipsed propane

FIGURE 3.8 Newman projections of propane showing staggered and eclipsed conformations. The staggered conformer is lower in energy by 14 kJ/mol.

The conformational situation becomes more complex for larger alkanes because not all staggered conformations have the same energy and not all eclipsed conformations have the same energy. In butane, for instance, the lowest-energy arrangement, called the **anti conformation**, is the one in which the two methyl groups are as far apart as possible—180° away from each other. As rotation around the C2–C3 bond occurs, an eclipsed conformation is reached in which there are two $CH_3 \leftrightarrow H$ interactions and one $H \leftrightarrow H$ interaction. Using the energy values derived previously from ethane and propane, this eclipsed conformation is more strained than the anti conformation by 2×6.0 kJ/mol + 4.0 kJ/mol (two $CH_3 \leftrightarrow H$ interactions plus one $H \leftrightarrow H$ interaction), for a total of 16 kJ/mol (3.8 kcal/mol).

Butane—anti conformation (0 kJ/mol)

Butane—eclipsed conformation (16 kJ/mol)

As bond rotation continues, an energy minimum is reached at the staggered conformation where the methyl groups are 60° apart. Called the **gauche conformation**, it lies 3.8 kJ/mol (0.9 kcal/mol) higher in energy than the anti conformation *even though it has no eclipsing interactions.* This energy difference occurs because the hydrogen atoms of the methyl groups are near one another in the gauche conformation, resulting in what is called *steric strain*. **Steric strain** is the repulsive interaction that occurs when atoms are forced closer together than their atomic radii allow. It's the result of trying to force two atoms to occupy the same space.

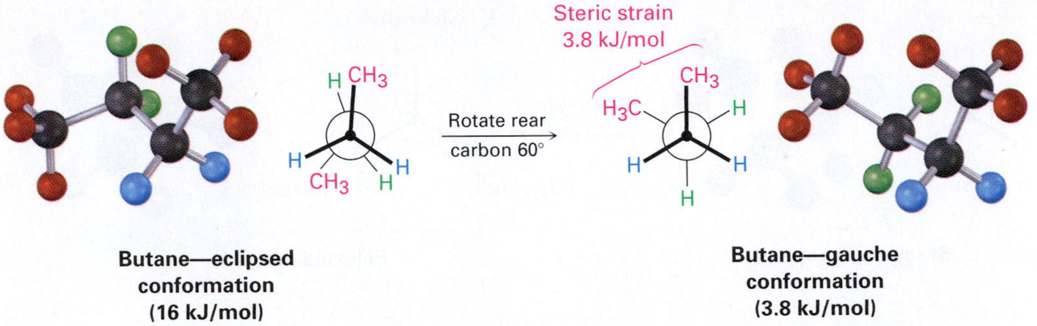

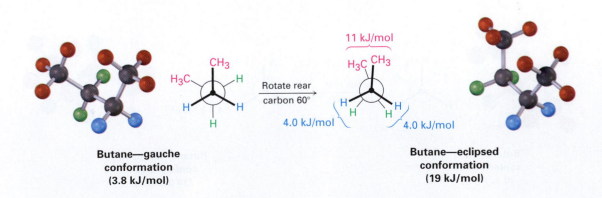

Butane—eclipsed conformation (16 kJ/mol)

Steric strain 3.8 kJ/mol

Rotate rear carbon 60°

Butane—gauche conformation (3.8 kJ/mol)

As the dihedral angle between the methyl groups approaches 0°, an energy maximum is reached at a second eclipsed conformation. Because the methyl groups are forced even closer together than in the gauche conformation, both torsional strain and steric strain are present. A total strain energy of 19 kJ/mol (4.5 kcal/mol) has been estimated for this conformation, making it possible to calculate a value of 11 kJ/mol (2.6 kcal/mol) for the $CH_3 \leftrightarrow CH_3$ eclipsing interaction: total strain of 19 kJ/mol less the strain of two $H \leftrightarrow H$ eclipsing interactions (2 × 4.0 kcal/mol) equals 11 kJ/mol.

Butane—gauche conformation (3.8 kJ/mol)

CH3

Rotate rear carbon 60°

4.0 kJ/mol

11 kJ/mol

4.0 kJ/mol

Butane—eclipsed conformation (19 kJ/mol)

After 0°, the rotation becomes a mirror image of what we've already seen: another gauche conformation is reached, another eclipsed conformation, and finally a return to the anti conformation. A plot of potential energy versus rotation about the C2–C3 bond is shown in Figure 3.9.

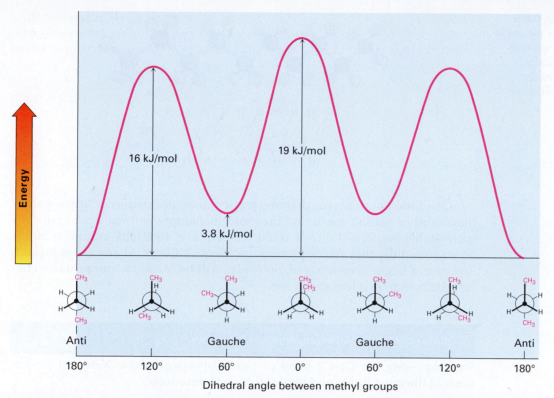

ACTIVE FIGURE 3.9 A plot of potential energy versus rotation for the C2–C3 bond in butane. The energy maximum occurs when the two methyl groups eclipse each other, and the energy minimum occurs when the two methyl groups are 180° apart (anti). **Go to this book's student companion site at www.cengage.com/chemistry/mcmurry to explore an interactive version of this figure.**

The notion of assigning definite energy values to specific interactions within a molecule is a very useful one that we'll return to in the next chapter. A summary of what we've seen thus far is given in Table 3.5.

TABLE 3.5
Energy Costs for Interactions in Alkane Conformations

Interaction	Cause	Energy cost (kJ/mol)	(kcal/mol)
H ↔ H eclipsed	Torsional strain	4.0	1.0
H ↔ CH₃ eclipsed	Mostly torsional strain	6.0	1.4
CH₃ ↔ CH₃ eclipsed	Torsional and steric strain	11	2.6
CH₃ ↔ CH₃ gauche	Steric strain	3.8	0.9

The same principles just developed for butane apply to pentane, hexane, and all higher alkanes. The most favorable conformation for any alkane has the carbon–carbon bonds in staggered arrangements, with large substituents arranged anti to one another. A generalized alkane structure is shown in Figure 3.10.

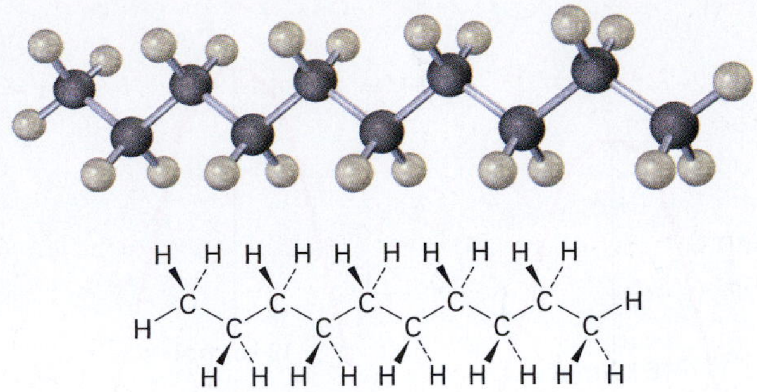

FIGURE 3.10 The most stable alkane conformation is the one in which all substituents are staggered and the carbon–carbon bonds are arranged anti, as shown in this model of decane.

One final point: saying that one particular conformation is "more stable" than another doesn't mean that the molecule adopts and maintains only the more stable conformation. At room temperature, rotations around σ bonds occur so rapidly that all conformers are in equilibrium. At any given instant, however, a larger percentage of molecules will be found in a more stable conformation than in a less stable one.

WORKED EXAMPLE 3.4 Drawing Newman Projections

Sighting along the C1–C2 bond of 1-chloropropane, draw Newman projections of the most stable and least stable conformations.

Strategy

The most stable conformation of a substituted alkane is generally a staggered one in which large groups have an anti relationship. The least stable conformation is generally an eclipsed one in which large groups are as close as possible.

Solution

Most stable (staggered)   **Least stable (eclipsed)**

Problem 3.15

Make a graph of potential energy versus angle of bond rotation for propane, and assign values to the energy maxima.

Problem 3.16

Consider 2-methylpropane (isobutane). Sighting along the C2–C1 bond:
(a) Draw a Newman projection of the most stable conformation.
(b) Draw a Newman projection of the least stable conformation.
(c) Make a graph of energy versus angle of rotation around the C2–C1 bond.
(d) Since an H $\leftrightarrow$ H eclipsing interaction costs 4.0 kJ/mol and an H $\leftrightarrow$ CH$_3$ eclipsing interaction costs 6.0 kJ/mol, assign relative values to the maxima and minima in your graph.

Problem 3.17

Sight along the C2–C3 bond of 2,3-dimethylbutane, and draw a Newman projection of the most stable conformation.

Problem 3.18

Draw a Newman projection along the C2–C3 bond of the following conformation of 2,3-dimethylbutane, and calculate a total strain energy:

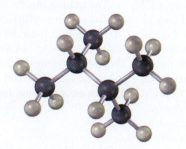

Summary

Even though alkanes are relatively unreactive and rarely involved in chemical reactions, they nevertheless provide a useful vehicle for introducing some important general ideas. In this chapter, we've used alkanes to introduce the basic approach to naming organic compounds and to take an initial look at some of the three-dimensional aspects of molecules.

A **functional group** is a group of atoms within a larger molecule that has a characteristic chemical reactivity. Because functional groups behave approximately the same way in all molecules where they occur, the chemical reactions of an organic molecule are largely determined by its functional groups.

Alkanes are a class of **saturated hydrocarbons** with the general formula C_nH_{2n+2}. They contain no functional groups, are relatively inert, and can be either **straight-chain** (*normal*) or **branched**. Alkanes are named by a series of IUPAC rules of nomenclature. Compounds that have the same chemical formula but different structures are called **isomers**. More specifically, compounds such as butane and isobutane, which differ in their connections between atoms, are called **constitutional isomers**.

Carbon–carbon single bonds in alkanes are formed by σ overlap of carbon sp^3 hybrid orbitals. Rotation is possible around σ bonds because of their cylindrical symmetry, and alkanes therefore exist in a large number of rapidly interconverting **conformations**. **Newman projections** make it possible to visualize the spatial consequences of bond rotation by sighting directly along a carbon–carbon bond axis. Not all alkane conformations are equally stable. The **staggered** conformation of ethane is 12 kJ/mol (2.9 kcal/mol) more stable than the **eclipsed** conformation because of **torsional strain**. In general, any alkane is most stable when all its bonds are staggered.

Key Words

aliphatic, 77
alkane, 77
alkyl group, 81
anti conformation, 93
branched-chain alkane, 78
conformation, 91
conformers, 91
constitutional isomers, 79
eclipsed conformation, 91
functional group, 70
gauche conformation, 93
hydrocarbon, 77
isomers, 78
Newman projection, 91
R group, 82
saturated, 77
staggered conformation, 91
stereochemistry, 90
steric strain, 94
straight-chain alkane, 78
torsional strain, 92

Lagniappe

Gasoline

Gasoline is a finite resource. It won't be around forever.

British Foreign Minister Ernest Bevin once said that "The Kingdom of Heaven runs on righteousness, but the Kingdom of Earth runs on alkanes." Well, actually he said "runs on oil" not "runs on alkanes," but they're essentially the same. By far the major sources of alkanes are the world's natural gas and petroleum deposits. Laid down eons ago, these deposits are thought to be derived from the decomposition of plant and animal matter, primarily of marine origin. *Natural gas* consists chiefly of methane but also contains ethane, propane, and butane. *Petroleum* is a complex mixture of hydrocarbons that must be separated into fractions and then further refined before it can be used.

The petroleum era began in August 1859, when the world's first oil well was drilled near Titusville, Pennsylvania. The petroleum was distilled into fractions according to boiling point, but it was high-boiling kerosene, or lamp oil, rather than gasoline that was primarily sought. Literacy was becoming widespread at the time, and people wanted better light for reading than was available from candles. Gasoline was too volatile for use in lamps and was initially considered a waste by-product. The world has changed greatly since those early days, however, and it is now gasoline rather than lamp oil that is prized.

Petroleum refining begins by fractional distillation of crude oil into three principal cuts according to boiling point (bp): straight-run gasoline (bp 30–200 °C), kerosene (bp 175–300 °C), and heating oil or diesel fuel (bp 275–400 °C). Further distillation under reduced pressure then yields lubricating oils and waxes and leaves a tarry residue of asphalt. The distillation of crude oil is only the first step in gasoline production, however. Straight-run gasoline turns out to be a poor fuel in automobiles because of *engine knock,* an uncontrolled combustion that can occur in a hot engine.

The *octane number* of a fuel is the measure by which its antiknock properties are judged. It was recognized long ago that straight-chain hydrocarbons are far more prone to induce engine knock than are highly branched compounds. Heptane, a particularly bad fuel, is assigned a base value of 0 octane number, and 2,2,4-trimethylpentane, commonly known as isooctane, has a rating of 100.

$$CH_3CH_2CH_2CH_2CH_2CH_2CH_3$$

Heptane
(octane number = 0)

$$CH_3CCH_2CHCH_3$$

2,2,4-Trimethylpentane
(octane number = 100)

Because straight-run gasoline burns so poorly in engines, petroleum chemists have devised numerous methods for producing higher-quality fuels. One of these methods, *catalytic cracking,* involves taking the high-boiling kerosene cut (C_{11}–C_{14}) and "cracking" it into smaller branched molecules suitable for use in gasoline. Another process, called *reforming,* is used to convert C_6–C_8 alkanes to aromatic compounds such as benzene and toluene, which have substantially higher octane numbers than alkanes. The final product that goes in your tank has an approximate composition of 15% C_4–C_8 straight-chain alkanes, 25% to 40% C_4–C_{10} branched-chain alkanes, 10% cyclic alkanes, 10% straight-chain and cyclic alkenes, and 25% arenes (aromatics).

Exercises

VISUALIZING CHEMISTRY

(Problems 3.1–3.18 appear within the chapter.)

3.19 ■ Identify the functional groups in the following substances, and convert each drawing into a molecular formula (red = O, blue = N):

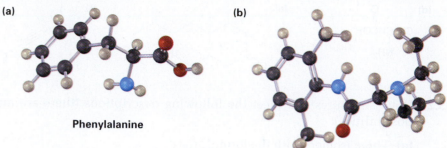

(a)

Phenylalanine

(b)

Lidocaine

3.20 ■ Give IUPAC names for the following alkanes, and convert each drawing into a skeletal structure:

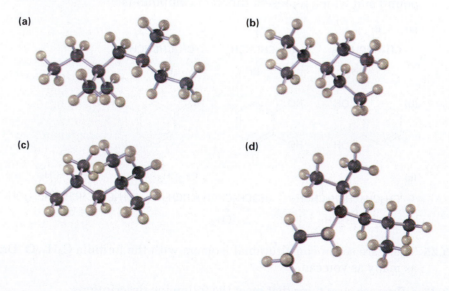

(a)

(b)

(c)

(d)

■ indicates problems that are assignable in Organic OWL.

Go to this book's companion website at **www.cengage.com/chemistry/mcmurry** to explore interactive versions of the Active Figures from this text.

ADDITIONAL PROBLEMS

3.21 ▪ Locate and identify the functional groups in the following molecules:

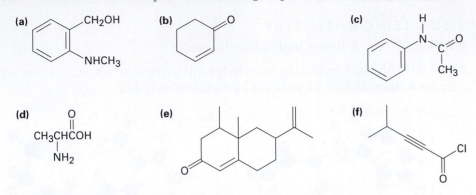

3.22 Draw structures that meet the following descriptions (there are many possibilities):

(a) Three isomers with the formula C_8H_{18}

(b) Two isomers with the formula $C_4H_8O_2$

3.23 Draw structures of the nine isomers of C_7H_{16}.

3.24 ▪ In each of the following sets, which structures represent the same compound and which represent different compounds?

(a)

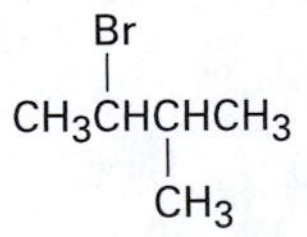

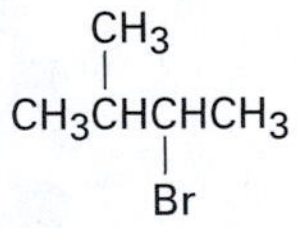

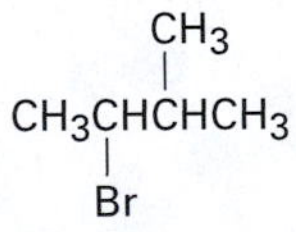

$$Br$$
$$CH_3CHCHCH_3$$
$$CH_3$$

$$CH_3$$
$$CH_3CHCHCH_3$$
$$Br$$

$$CH_3$$
$$CH_3CHCHCH_3$$
$$Br$$

(b)

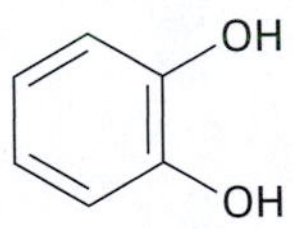

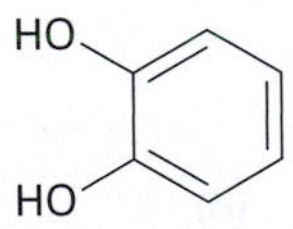

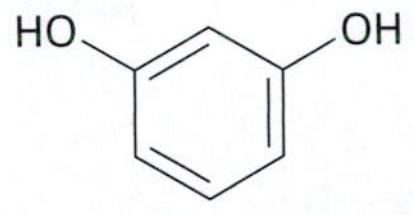

(c)

$$CH_3$$
$$CH_3CH_2CHCH_2CHCH_3$$
$$CH_2OH$$

$$CH_2CH_3$$
$$HOCH_2CHCH_2CHCH_3$$
$$CH_3$$

$$CH_3 \quad CH_3$$
$$CH_3CH_2CHCH_2CHCH_2OH$$

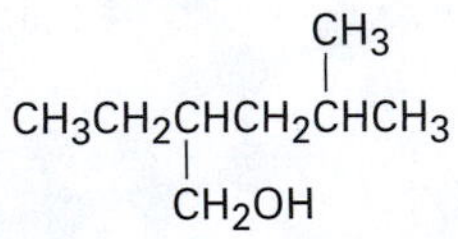

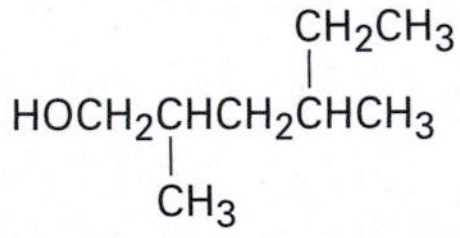

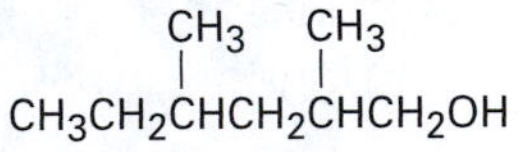

3.25 There are seven constitutional isomers with the formula $C_4H_{10}O$. Draw as many as you can.

3.26 ▪ Propose structures that meet the following descriptions:

(a) A ketone with five carbons   **(b)** A four-carbon amide

(c) A five-carbon ester   **(d)** An aromatic aldehyde

(e) A keto ester   **(f)** An amino alcohol

3.27 ▪ Propose structures for the following:

    (a) A ketone, C_4H_8O **(b)** A nitrile, C_5H_9N

    (c) A dialdehyde, $C_4H_6O_2$ **(d)** A bromoalkene, $C_6H_{11}Br$

    (e) An alkane, C_6H_{14} **(f)** A *cyclic* saturated hydrocarbon, C_6H_{12}

    (g) A diene (dialkene), C_5H_8 **(h)** A keto alkene, C_5H_8O

3.28 Draw as many compounds as you can that fit the following descriptions:

    (a) Alcohols with formula $C_4H_{10}O$

    (b) Amines with formula $C_5H_{13}N$

    (c) Ketones with formula $C_5H_{10}O$

    (d) Aldehydes with formula $C_5H_{10}O$

    (e) Esters with formula $C_4H_8O_2$

    (f) Ethers with formula $C_4H_{10}O$

3.29 ▪ Draw compounds that contain the following:

    (a) A primary alcohol   **(b)** A tertiary nitrile

    (c) A secondary thiol   **(d)** Both primary and secondary alcohols

    (e) An isopropyl group   **(f)** A quaternary carbon

3.30 Draw and name all monobromo derivatives of pentane, $C_5H_{11}Br$.

3.31 Draw and name all monochloro derivatives of 2,5-dimethylhexane, $C_8H_{17}Cl$.

3.32 Predict the hybridization of the carbon atom in each of the following functional groups:

    (a) Ketone   **(b)** Nitrile   **(c)** Carboxylic acid   **(d)** Thioester

3.33 ▪ Draw the structures of the following molecules:

    (a) *Biacetyl,* $C_4H_6O_2$, a substance with the aroma of butter; it contains no rings or carbon–carbon multiple bonds.

    (b) *Ethylenimine,* C_2H_5N, a substance used in the synthesis of melamine polymers; it contains no multiple bonds.

    (c) *Glycerol,* $C_3H_8O_3$, a substance isolated from fat and used in cosmetics; it has an –OH group on each carbon.

3.34 ▪ Draw structures for the following:

    (a) 2-Methylheptane   **(b)** 4-Ethyl-2,2-dimethylhexane

    (c) 4-Ethyl-3,4-dimethyloctane   **(d)** 2,4,4-Trimethylheptane

    (e) 3,3-Diethyl-2,5-dimethylnonane   **(f)** 4-Isopropyl-3-methylheptane

3.35 Draw a compound that:

(a) Has only primary and tertiary carbons

(b) Has no secondary or tertiary carbons

(c) Has four secondary carbons

3.36 Draw a compound that:

(a) Has nine primary hydrogens

(b) Has only primary hydrogens

3.37 For each of the following compounds, draw an isomer that has the same functional groups:

(a)
$$CH_3$$
$$CH_3CHCH_2CH_2Br$$

(b) [cyclopentane with OCH₃ substituent]

(c) $CH_3CH_2CH_2C{\equiv}N$

(d) [cyclohexane with OH substituent]

(e) CH_3CH_2CHO

(f) [benzene ring with CH_2CO_2H substituent]

3.38 ■ Give IUPAC names for the following compounds:

(a)
$$CH_3$$
$$CH_3CHCH_2CH_2CH_3$$

(b)
$$CH_3$$
$$CH_3CH_2CCH_3$$
$$CH_3$$

(c)
$$H_3C \quad CH_3$$
$$CH_3CHCHCH_2CH_2CH_3$$
$$CH_3$$

(d)
$$CH_2CH_3 \quad CH_3$$
$$CH_3CH_2CHCH_2CH_2CHCH_3$$

(e)
$$CH_3 \quad CH_2CH_3$$
$$CH_3CH_2CH_2CHCH_2CCH_3$$
$$CH_3$$

(f)
$$H_3C \quad CH_3$$
$$CH_3C{-}CCH_2CH_2CH_3$$
$$H_3C \quad CH_3$$

3.39 ■ Name the five isomers of C_6H_{14}.

3.40 ■ Explain why each of the following names is incorrect:

(a) 2,2-Dimethyl-6-ethylheptane   (b) 4-Ethyl-5,5-dimethylpentane

(c) 3-Ethyl-4,4-dimethylhexane   (d) 5,5,6-Trimethyloctane

(e) 2-Isopropyl-4-methylheptane

3.41 Propose structures and give IUPAC names for the following:

(a) A diethyldimethylhexane   (b) A (3-methylbutyl)-substituted alkane

3.42 ■ Consider 2-methylbutane (isopentane). Sighting along the C2–C3 bond:

(a) Draw a Newman projection of the most stable conformation.

(b) Draw a Newman projection of the least stable conformation.

(c) Since a $CH_3 \leftrightarrow CH_3$ eclipsing interaction costs 11 kJ/mol (2.5 kcal/mol) and a $CH_3 \leftrightarrow CH_3$ gauche interaction costs 3.8 kJ/mol (0.9 kcal/mol), make a quantitative plot of energy versus rotation about the C2–C3 bond.

■ Problems assignable in Organic OWL.

3.43 ▪ What are the relative energies of the three possible staggered conformations around the C2–C3 bond in 2,3-dimethylbutane? (See Problem 3.42.)

3.44 Construct a qualitative potential-energy diagram for rotation about the C–C bond of 1,2-dibromoethane. Which conformation would you expect to be more stable? Label the anti and gauche conformations of 1,2-dibromoethane.

3.45 Which conformation of 1,2-dibromoethane (Problem 3.44) would you expect to have the larger dipole moment? The observed dipole moment of 1,2-dibromoethane is $\mu = 1.0$ D. What does this tell you about the actual conformation of the molecule?

3.46 ▪ The barrier to rotation about the C–C bond in bromoethane is 15 kJ/mol (3.6 kcal/mol).

(a) What energy value can you assign to an H ↔ Br eclipsing interaction?

(b) Construct a quantitative diagram of potential energy versus bond rotation for bromoethane.

3.47 Draw the most stable conformation of pentane, using wedges and dashes to represent bonds coming out of the paper and going behind the paper, respectively.

3.48 Draw the most stable conformation of 1,4-dichlorobutane, using wedges and dashes to represent bonds coming out of the paper and going behind the paper, respectively.

3.49 Malic acid, $C_4H_6O_5$, has been isolated from apples. Because this compound reacts with 2 molar equivalents of base, it is a dicarboxylic acid.

(a) Draw at least five possible structures.

(b) If malic acid is a secondary alcohol, what is its structure?

3.50 ▪ Formaldehyde, $H_2C=O$, is known to all biologists because of its usefulness as a tissue preservative. When pure, formaldehyde *trimerizes* to give trioxane, $C_3H_6O_3$, which, surprisingly enough, has no carbonyl groups. Only one monobromo derivative ($C_3H_5BrO_3$) of trioxane is possible. Propose a structure for trioxane.

3.51 ▪ Increased substitution around a bond leads to increased strain. Take the four substituted butanes listed here, for example. For each compound, sight along the C2–C3 bond and draw Newman projections of the most stable and least stable conformations. Use the data in Table 3.5 to assign strain energy values to each conformation. Which of the eight conformations is most strained? Which is least strained?

(a) 2-Methylbutane   **(b)** 2,2-Dimethylbutane

(c) 2,3-Dimethylbutane   **(d)** 2,2,3-Trimethylbutane

3.52 ■ The cholesterol-lowering agents called *statins,* such as simvastatin (Zocor) and pravastatin (Pravachol), are among the most widely prescribed drugs in the world (see the Chapter 1 Introduction). Identify the functional groups in both, and tell how the two substances differ.

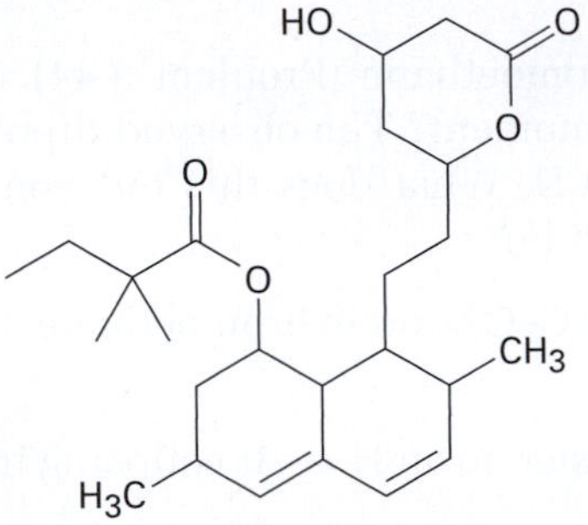

Simvastatin (Zocor)

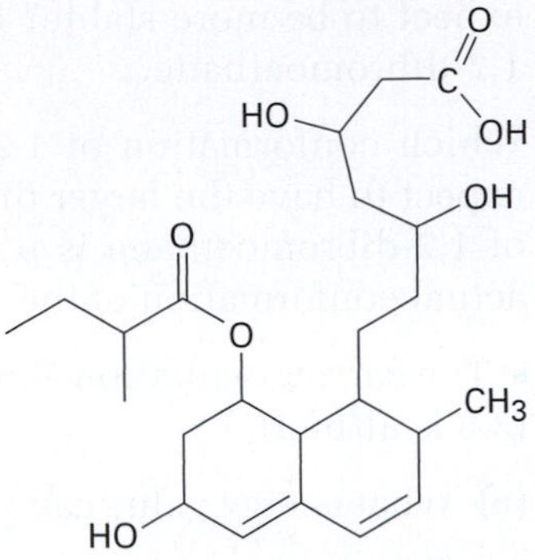

Pravastatin (Pravachol)

3.53 We'll look in the next chapter at *cycloalkanes*—saturated cyclic hydrocarbons—and we'll see that the molecules generally adopt puckered, nonplanar conformations. Cyclohexane, for instance, has a puckered shape like a lounge chair rather than a flat shape. Why?

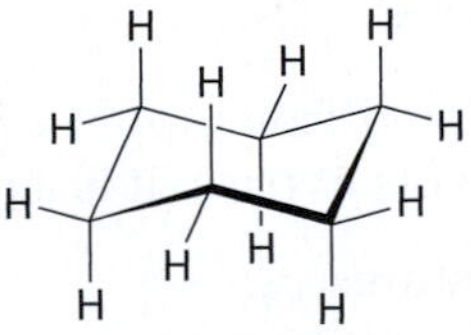

Nonplanar cyclohexane

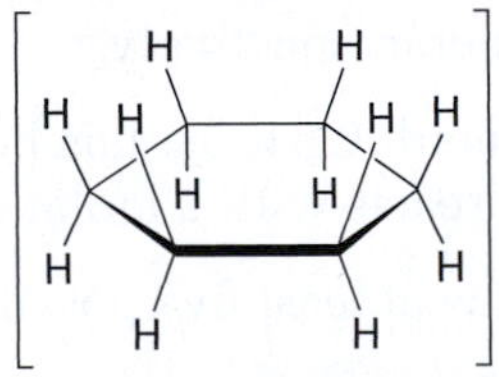

Planar cyclohexane

3.54 We'll see in the next chapter that there are two isomeric substances both named 1,2-dimethylcyclohexane. See if you can figure out why.

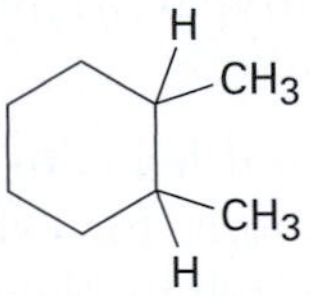

1,2-Dimethylcyclohexane

■ Problems assignable in Organic OWL.

4 Organic Compounds: Cycloalkanes and Their Stereochemistry

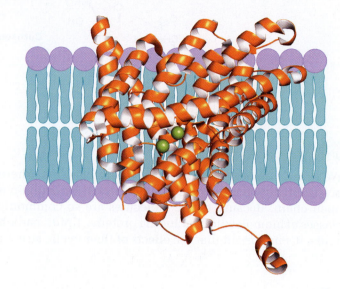

A membrane channel protein that conducts Cl⁻ ions across cell membranes.

Although we've discussed only open-chain compounds up to this point, most organic compounds contain *rings* of carbon atoms. Chrysanthemic acid, for instance, whose esters occur naturally as the active insecticidal constituents of chrysanthemum flowers, contains a three-membered (cyclopropane) ring.

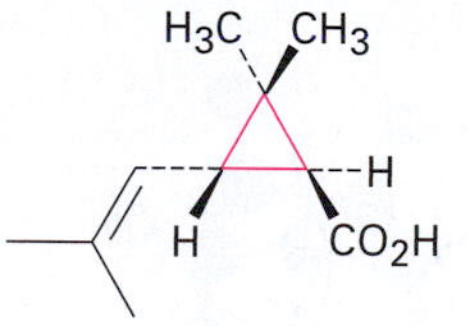

Chrysanthemic acid

Prostaglandins, potent hormones that control an extraordinary variety of physiological functions in humans, contain a five-membered (cyclopentane) ring.

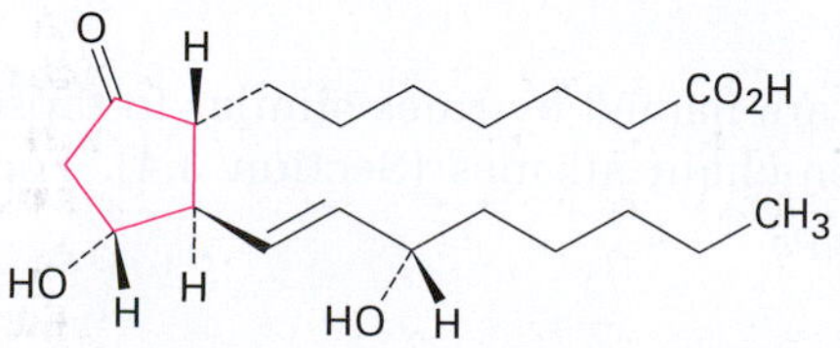

Prostaglandin E₁

ŎWL Online homework for this chapter can be assigned in Organic OWL.

Steroids, such as cortisone, contain four rings joined together—3 six-membered (cyclohexane) and 1 five-membered. We'll discuss steroids and their properties in more detail in Sections 23.8 and 23.9.

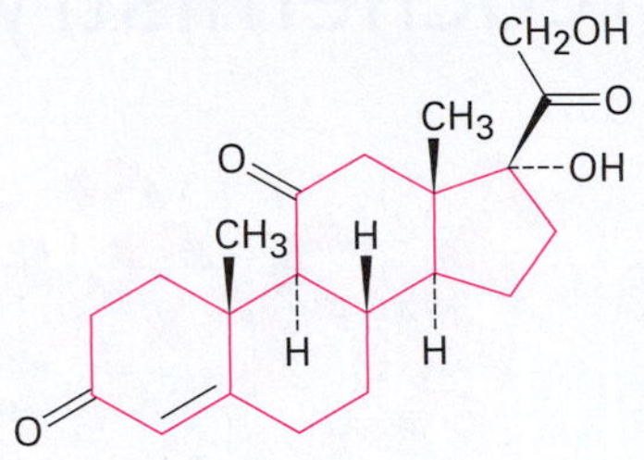

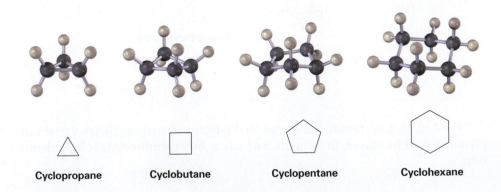

Cortisone

WHY THIS CHAPTER?

We'll see numerous instances in future chapters where the chemistry of a given functional group is strongly affected by being in a ring rather than an open chain. Because cyclic molecules are so commonly encountered in all classes of biomolecules, including proteins, lipids, carbohydrates, and nucleic acids, it's important that the effects of their cyclic structures be understood.

4.1 Naming Cycloalkanes

Saturated cyclic hydrocarbons are called **cycloalkanes**, or **alicyclic** compounds (**ali**phatic **cyclic**). Because cycloalkanes consist of rings of $-CH_2-$ units, they have the general formula $(CH_2)_n$, or C_nH_{2n}, and can be represented by polygons in skeletal drawings:

Cyclopropane   **Cyclobutane**   **Cyclopentane**   **Cyclohexane**

Substituted cycloalkanes are named by rules similar to those we saw in the previous chapter for open-chain alkanes (Section 3.4). For most compounds, there are only two steps:

Rule 1
Find the parent.
Count the number of carbon atoms in the ring and the number in the largest substituent chain. If the number of carbon atoms in the ring is equal to or

greater than the number in the substituent, the compound is named as an alkyl-substituted cycloalkane. If the number of carbon atoms in the largest substituent is greater than the number in the ring, the compound is named as a cycloalkyl-substituted alkane. For example:

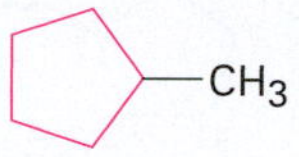

3 carbons   4 carbons

Methylcyclopentane 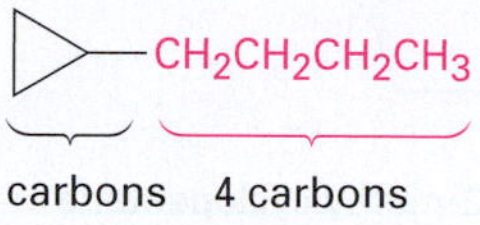   **1-Cyclopropylbutane**

Rule 2
Number the substituents, and write the name.

For an alkyl- or halo-substituted cycloalkane, choose a point of attachment as carbon 1 and number the substituents on the ring so that the *second* substituent has as low a number as possible. If ambiguity still exists, number so that the third or fourth substituent has as low a number as possible, until a point of difference is found.

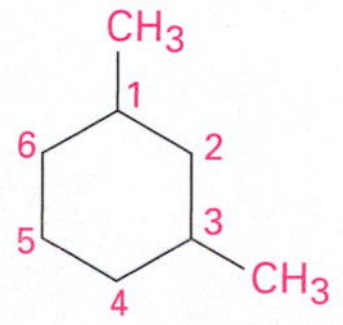

NOT

1,3-Dimethylcyclohexane   **1,5-Dimethylcyclohexane**

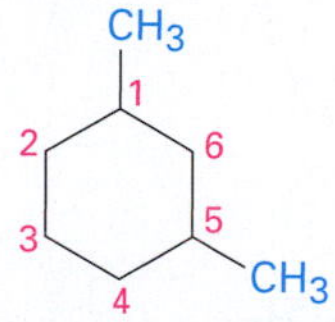

Lower   Higher

NOT

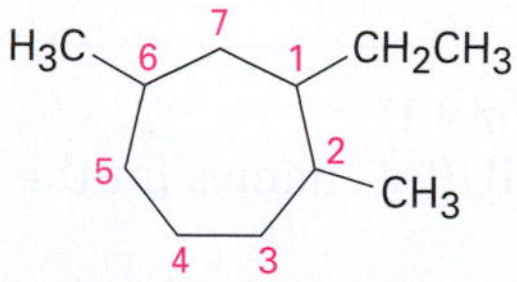

1-Ethyl-2,6-dimethylcycloheptane

Higher

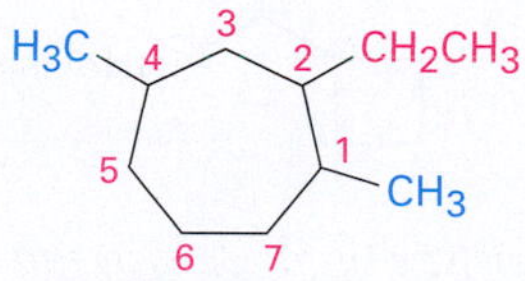

2-Ethyl-1,4-dimethylcycloheptane

Lower   Lower

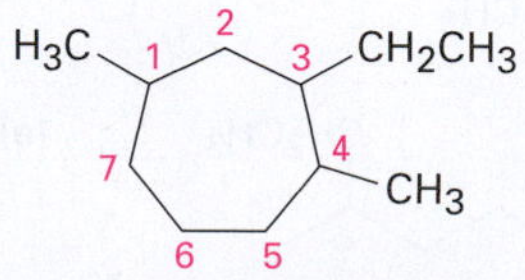

3-Ethyl-1,4-dimethylcycloheptane

Higher

(a) When two or more different alkyl groups that could potentially receive the same numbers are present, number them by alphabetical priority, ignoring numerical prefixes such as di- and tri-.

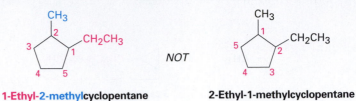

NOT

1-Ethyl-2-methylcyclopentane   **2-Ethyl-1-methylcyclopentane**

(b) If halogens are present, treat them just like alkyl groups:

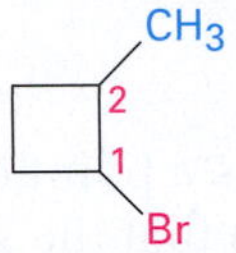

NOT

1-Bromo-2-methylcyclobutane   **2-Bromo-1-methylcyclobutane**

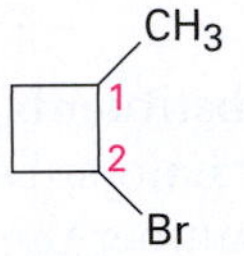

Some additional examples follow:

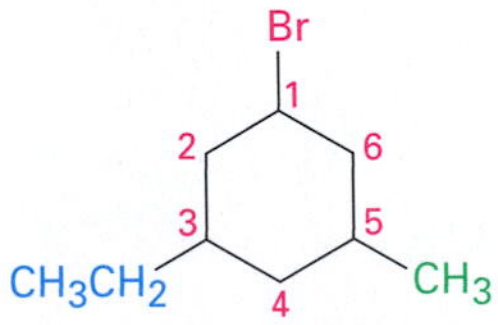

1-Bromo-3-ethyl-5-methyl-   **(1-Methylpropyl)cyclobutane**   **1-Chloro-3-ethyl-2-methyl-**
cyclohexane   **or *sec*-butylcyclobutane**   **cyclopentane**

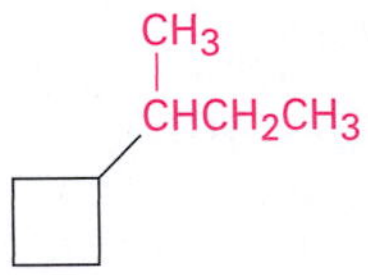

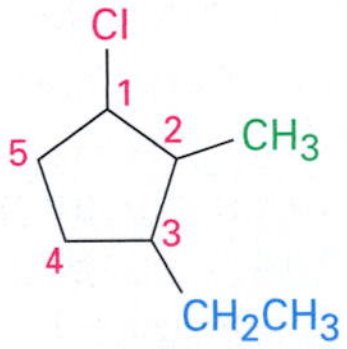

Problem 4.1

Give IUPAC names for the following cycloalkanes:

(a) CH₃

(b) CH₂CH₂CH₃

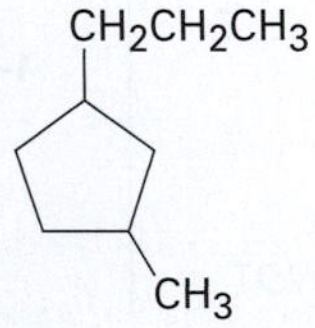

(c)

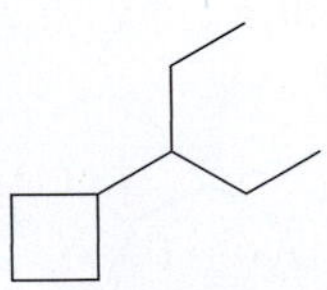

(d) CH₂CH₃

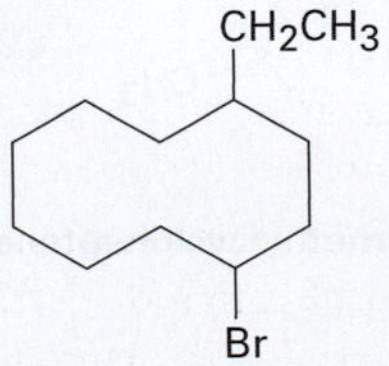

(e)

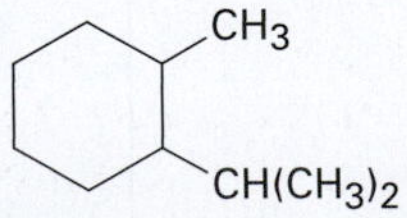

(f) Br

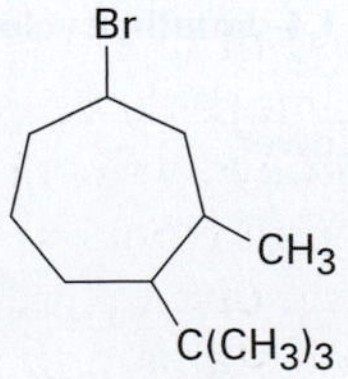

Problem 4.2

Draw structures corresponding to the following IUPAC names:

(a) 1,1-Dimethylcyclooctane   **(b)** 3-Cyclobutylhexane

(c) 1,2-Dichlorocyclopentane   **(d)** 1,3-Dibromo-5-methylcyclohexane

Problem 4.3

Name the following cycloalkane:

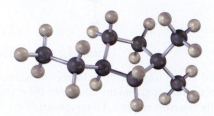

4.2 Cis–Trans Isomerism in Cycloalkanes

In many respects, the chemistry of cycloalkanes is like that of open-chain alkanes: both are nonpolar and fairly inert. There are, however, some important differences. One difference is that cycloalkanes are less flexible than open-chain alkanes. In contrast with the rotational freedom around single bonds seen in open-chain alkanes (Sections 3.6 and 3.7), there is much less freedom in cycloalkanes. Cyclopropane, for example, must be a rigid, planar molecule because three points (the carbon atoms) define a plane. No bond rotation can take place around a cyclopropane carbon–carbon bond without breaking open the ring (Figure 4.1).

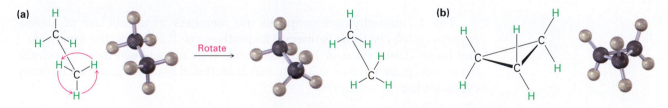

FIGURE 4.1 **(a)** Rotation occurs around the carbon–carbon bond in ethane, but **(b)** no rotation is possible around the carbon–carbon bonds in cyclopropane without breaking open the ring.

Larger cycloalkanes have increasing rotational freedom, and the very large rings (C_{25} and up) are so floppy that they are nearly indistinguishable from open-chain alkanes. The common ring sizes (C_3–C_7), however, are severely restricted in their molecular motions.

Because of their cyclic structures, cycloalkanes have two faces as viewed edge-on, a "top" face and a "bottom" face. As a result, isomerism is possible in substituted cycloalkanes. For example, there are two different 1,2-dimethyl-cyclopropane isomers, one with the two methyl groups on the same face of the ring and one with the methyl groups on opposite faces (Figure 4.2). Both isomers are stable compounds, and neither can be converted into the other without breaking and reforming chemical bonds. Make molecular models to prove this to yourself.

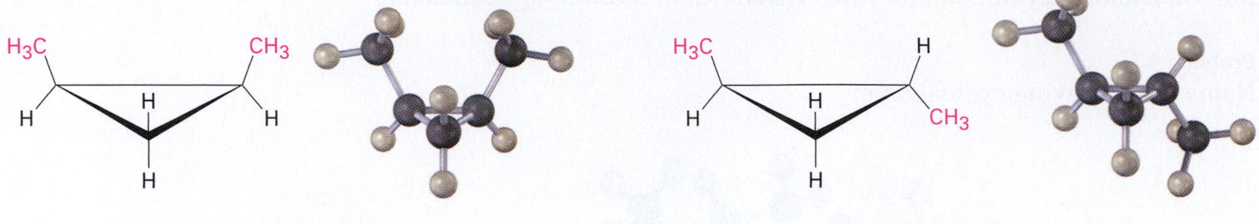

cis-1,2-Dimethylcyclopropane

trans-1,2-Dimethylcyclopropane

FIGURE 4.2 There are two different 1,2-dimethylcyclopropane isomers, one with the methyl groups on the same face of the ring (cis) and the other with the methyl groups on opposite faces of the ring (trans). The two isomers do not interconvert.

Unlike the constitutional isomers butane and isobutane (Section 3.2), which have their atoms connected in a different order, the two 1,2-dimethyl-cyclopropanes have the same order of connections but differ in the spatial orientation of the atoms. Such compounds, which have their atoms connected in the same order but differ in three-dimensional orientation, are called stereochemical isomers, or **stereoisomers**.

Constitutional isomers (different connections between atoms)

$$CH_3-\overset{\overset{\displaystyle CH_3}{|}}{CH}-CH_3 \quad \text{and} \quad CH_3-CH_2-CH_2-CH_3$$

Stereoisomers (same connections but different three-dimensional geometry)

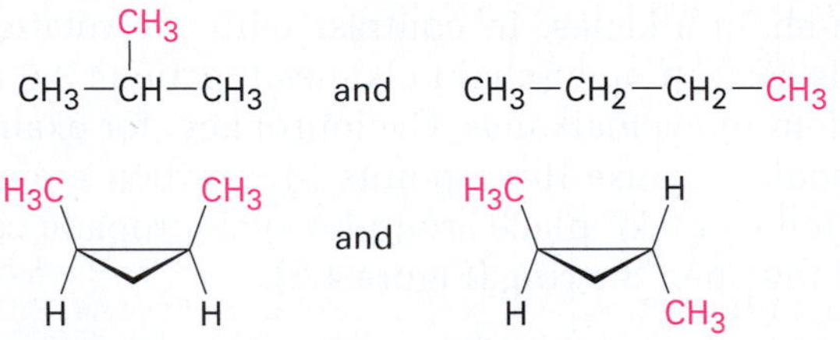

and

The 1,2-dimethylcyclopropanes are members of a subclass of stereo-isomers called **cis–trans isomers**. The prefixes *cis-* (Latin, "on the same side") and *trans-* (Latin, "across") are used to distinguish between them. Cis–trans isomerism is a common occurrence in substituted cycloalkanes and in many cyclic biological molecules.

cis-1,3-Dimethylcyclobutane

trans-1-Bromo-3-ethylcyclopentane

WORKED EXAMPLE 4.1 Naming Cycloalkanes

Name the following substances, including the *cis-* or *trans-* prefix:

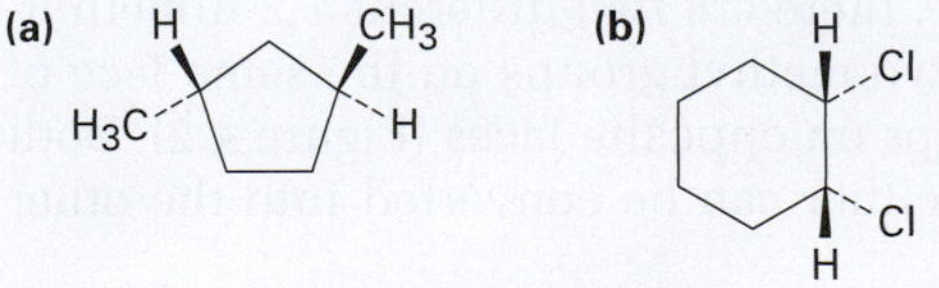

(a)   **(b)**

Strategy

In these views, the ring is roughly in the plane of the page, a wedged bond protrudes out of the page, and a dashed bond recedes into the page. Two substituents are cis if they are both out of or both into the page, and they are trans if one is out of and one is into the page.

Solution

(a) *trans*-1,3-Dimethylcyclopentane   **(b)** *cis*-1,2-Dichlorocyclohexane

---

Problem 4.4

Name the following substances, including the *cis*- or *trans*- prefix:

(a)

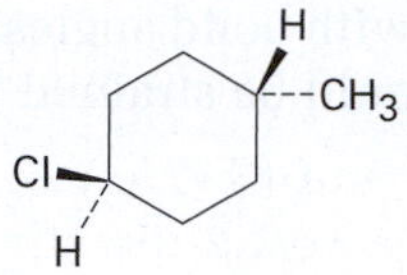

(b) H3C   CH2CH3

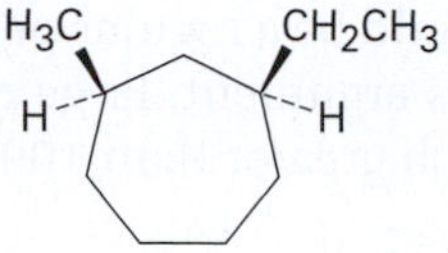

Problem 4.5

Draw the structures of the following molecules:
(a) *trans*-1-Bromo-3-methylcyclohexane   **(b)** *cis*-1,2-Dimethylcyclobutane
(c) *trans*-1-*tert*-Butyl-2-ethylcyclohexane

Problem 4.6

Prostaglandin F$_{2\alpha}$, a hormone that causes uterine contraction during childbirth, has the following structure. Are the two hydroxyl groups (–OH) on the cyclopentane ring cis or trans to each other? What about the two carbon chains attached to the ring?

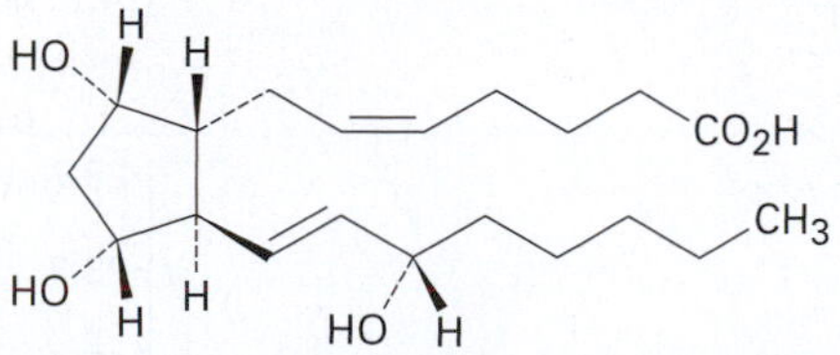

Prostaglandin F$_{2\alpha}$

Problem 4.7

Name the following substances, including the *cis*- or *trans*- prefix (redbrown = Br):

(a)   **(b)**

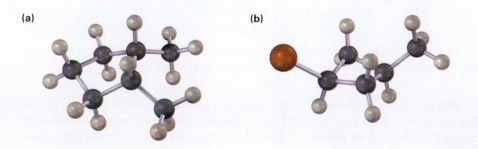

4.3 Stability of Cycloalkanes: Ring Strain

Chemists in the late 1800s knew that cyclic molecules existed, but the limitations on ring size were unclear. Although numerous compounds containing five-membered and six-membered rings were known, smaller and larger ring sizes had not been prepared despite many efforts.

A theoretical interpretation of this observation was proposed in 1885 by Adolf von Baeyer, who suggested that small and large rings might be unstable due to **angle strain**—the strain induced in a molecule when bond angles are forced to deviate from the ideal 109° tetrahedral value. Baeyer based his suggestion on the simple geometric notion that a three-membered ring (cyclopropane) should be an equilateral triangle with bond angles of 60° rather than 109°, a four-membered ring (cyclobutane) should be a square with bond angles of 90°, a five-membered ring should be a regular pentagon with bond angles of 108°, and so on. Continuing this argument, large rings should be strained by having bond angles that are much greater than 109°.

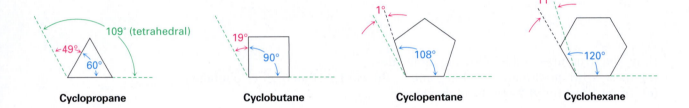

Cyclopropane Cyclobutane Cyclopentane Cyclohexane

Experimental data on strain energy in cycloalkanes show that Baeyer's theory is only partially correct (Figure 4.3). Cyclopropane and cyclobutane are indeed strained, just as predicted, but cyclopentane is more strained than predicted and cyclohexane is strain-free. Cycloalkanes of intermediate size have only modest strain, and rings of more than 14 carbons are strain-free. Why is Baeyer's theory wrong?

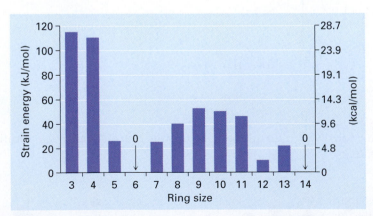

FIGURE 4.3 Cycloalkane strain energies, calculated from thermodynamic heats of formation. Small and medium rings are strained, but cyclohexane rings are strain-free.

Baeyer's theory is wrong for the simple reason that he assumed all cycloalkanes to be flat. In fact, as we'll see in the next section, most cycloalkanes are *not* flat; they adopt puckered three-dimensional conformations that allow bond angles to be nearly tetrahedral. As a result, angle strain occurs only in small rings that have little flexibility. For most ring sizes, torsional strain caused by H ↔ H eclipsing interactions on adjacent carbons (Section 3.6) and steric strain caused by the repulsion between nonbonded atoms that approach too closely (Section 3.7) are the most important factors. Thus, three kinds of strain contribute to the overall energy of a cycloalkane:

- **Angle strain**—the strain due to expansion or compression of bond angles
- **Torsional strain**—the strain due to eclipsing of bonds on neighboring atoms
- **Steric strain**—the strain due to repulsive interactions when atoms approach each other too closely

Problem 4.8

Each H ↔ H eclipsing interaction in ethane costs about 4.0 kJ/mol. How many such interactions are present in cyclopropane? What fraction of the overall 115 kJ/mol (27.5 kcal/mol) strain energy of cyclopropane is due to torsional strain?

Problem 4.9

cis-1,2-Dimethylcyclopropane has more strain than *trans*-1,2-dimethylcyclopropane. How can you account for this difference? Which of the two compounds is more stable?

4.4 Conformations of Cycloalkanes

Cyclopropane

Cyclopropane is the most strained of all rings, primarily because of the angle strain caused by its 60° C–C–C bond angles. In addition, cyclopropane has considerable torsional strain because the C–H bonds on neighboring carbon atoms are eclipsed (Figure 4.4).

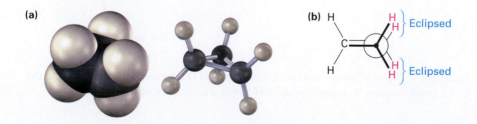

(a) **(b)**

H ⟩ Eclipsed
H ⟩ Eclipsed

FIGURE 4.4 The structure of cyclopropane, showing the eclipsing of neighboring C–H bonds that gives rise to torsional strain. Part (b) is a Newman projection along a C–C bond.

How can the hybrid-orbital model of bonding account for the large distortion of bond angles from the normal 109° tetrahedral value to 60° in cyclopropane? The answer is that cyclopropane has *bent bonds*. In an unstrained alkane, maximum bonding is achieved when two atoms have their overlapping orbitals pointing directly toward each other. In cyclopropane, though, the orbitals can't point directly toward each other; rather, they overlap at a slight angle. The result is that cyclopropane bonds are weaker and more reactive than typical alkane bonds—255 kJ/mol (61 kcal/mol) for a C–C bond in cyclopropane versus 370 kJ/mol (88 kcal/mol) for a C–C bond in open-chain propane.

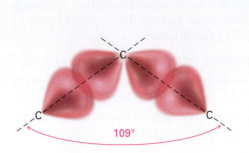

Typical alkane C–C bonds   **Typical bent cyclopropane C–C bonds**

Cyclobutane

Cyclobutane has less angle strain than cyclopropane but has more torsional strain because of its larger number of ring hydrogens. As a result, the total strain for the two compounds is nearly the same—110 kJ/mol (26.4 kcal/mol) for cyclobutane versus 115 kJ/mol (27.5 kcal/mol) for cyclopropane. Experiments show that cyclobutane is not quite flat but is slightly bent so that one carbon atom lies about 25° above the plane of the other three (Figure 4.5). The effect of this slight bend is to *increase* angle strain but to *decrease* torsional strain until a minimum-energy balance between the two opposing effects is achieved.

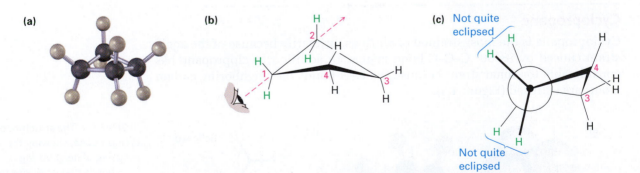

FIGURE 4.5 The conformation of cyclobutane. Part (c) is a Newman projection along the C1–C2 bond showing that neighboring C–H bonds are not quite eclipsed.

Cyclopentane

Cyclopentane was predicted by Baeyer to be nearly strain-free, but it actually has a total strain energy of 26 kJ/mol (6.2 kcal/mol). Although planar cyclopentane has practically no angle strain, it has a large amount of torsional strain. Cyclopentane therefore twists to adopt a puckered, nonplanar conformation that strikes a balance between increased angle strain and decreased torsional strain. Four of the cyclopentane carbon atoms are in approximately the same plane, with the fifth carbon atom bent out of the plane. Most of the hydrogens are nearly staggered with respect to their neighbors (Figure 4.6).

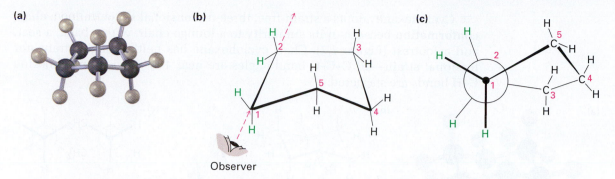

ACTIVE FIGURE 4.6 The conformation of cyclopentane. Carbons 1, 2, 3, and 4 are nearly planar, but carbon 5 is out of the plane. Part (c) is a Newman projection along the C1–C2 bond showing that neighboring C–H bonds are nearly staggered. **Go to this book's student companion site at www.cengage.com/chemistry/mcmurry to explore an interactive version of this figure.**

Problem 4.10

How many H ↔ H eclipsing interactions would be present if cyclopentane were planar? Assuming an energy cost of 4.0 kJ/mol for each eclipsing interaction, how much torsional strain would planar cyclopentane have? Since the measured total strain of cyclopentane is 26 kJ/mol, how much of the torsional strain is relieved by puckering?

Problem 4.11

Two conformations of *cis*-1,3-dimethylcyclobutane are shown. What is the difference between them, and which do you think is likely to be more stable?

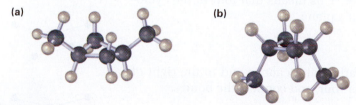

(a)   (b)

4.5 Conformations of Cyclohexane

Substituted cyclohexanes are the most common cycloalkanes and occur widely in nature. A large number of compounds, including steroids and many pharmaceutical agents, have cyclohexane rings. The flavoring agent menthol, for instance, has three substituents on a six-membered ring.

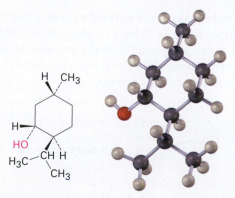

Menthol

Cyclohexane adopts a strain-free, three-dimensional shape called a **chair conformation** because of its similarity to a lounge chair, with a back, a seat, and a footrest (Figure 4.7). Chair cyclohexane has neither angle strain nor torsional strain—all C–C–C bond angles are near 109°, and all neighboring C–H bonds are staggered.

(a) (b) Observer (c)

FIGURE 4.7 The strain-free chair conformation of cyclohexane. All C–C–C bond angles are 111.5°, close to the ideal 109.5° tetrahedral angle, and all neighboring C–H bonds are staggered.

The easiest way to visualize chair cyclohexane is to build a molecular model. (In fact, do it now.) Two-dimensional drawings like that in Figure 4.7 are useful, but there's no substitute for holding, twisting, and turning a three-dimensional model in your own hands.

The chair conformation of cyclohexane can be drawn in three steps:

Step 1
Draw two parallel lines, slanted downward and slightly offset from each other. This means that four of the cyclohexane carbons lie in a plane.

Step 2
Place the topmost carbon atom above and to the right of the plane of the other four, and connect the bonds.

Step 3
Place the bottommost carbon atom below and to the left of the plane of the middle four, and connect the bonds. Note that the bonds to the bottommost carbon atom are parallel to the bonds to the topmost carbon.

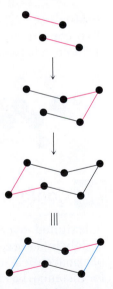

When viewing cyclohexane, it's helpful to remember that the lower bond is in front and the upper bond is in back. If this convention is not defined, an optical illusion can make it appear that the reverse is true. For clarity, all cyclohexane rings drawn in this book will have the front (lower) bond heavily shaded to indicate nearness to the viewer.

This bond is in back.

This bond is in front.

In addition to the chair conformation of cyclohexane, an alternative called the *twist-boat conformation* is also nearly free of angle strain. It does, however, have both steric strain and torsional strain and is about 23 kJ/mol (5.5 kcal/mol) higher in energy than the chair conformation. As a result, molecules adopt the twist-boat geometry only under special circumstances.

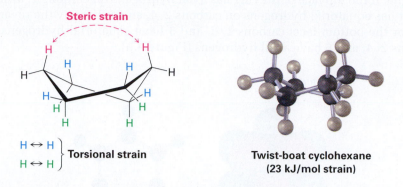

Twist-boat cyclohexane
(23 kJ/mol strain)

4.6 Axial and Equatorial Bonds in Cyclohexane

The chair conformation of cyclohexane has many consequences. We'll see in Section 12.12, for instance, that the chemical behavior of many substituted cyclohexanes is influenced by their conformation. In addition, we'll see in Section 21.5 that simple carbohydrates, such as glucose, adopt a conformation based on the cyclohexane chair and that their chemistry is directly affected as a result.

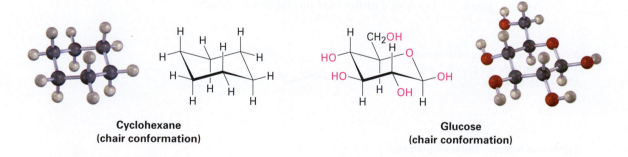

Cyclohexane
(chair conformation)

Glucose
(chair conformation)

Another consequence of the chair conformation is that there are two kinds of positions for substituents on the cyclohexane ring: *axial* positions and *equatorial* positions (Figure 4.8). The six **axial** positions are perpendicular to the ring, parallel to the ring axis, and the six **equatorial** positions are in the rough plane of the ring, around the ring equator.

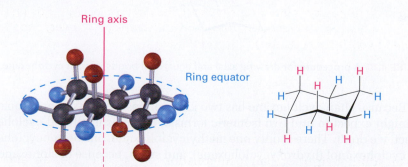

FIGURE 4.8 Axial (red) and equatorial (blue) positions in chair cyclohexane. The six axial hydrogens are parallel to the ring axis, and the six equatorial hydrogens are in a band around the ring equator.

As shown in Figure 4.8, each carbon atom in chair cyclohexane has one axial and one equatorial hydrogen. Furthermore, each face of the ring has three axial and three equatorial hydrogens in an alternating arrangement. For example, if the top face of the ring has axial hydrogens on carbons 1, 3, and 5, then it has equatorial hydrogens on carbons 2, 4, and 6. Exactly the reverse is true for the bottom face: carbons 1, 3, and 5 have equatorial hydrogens, but carbons 2, 4, and 6 have axial hydrogens (Figure 4.9).

FIGURE 4.9 Alternating axial and equatorial positions in chair cyclohexane, as shown in a view looking directly down the ring axis. Each carbon atom has one axial and one equatorial position, and each face has alternating axial and equatorial positions.

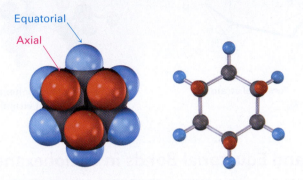

Note that we haven't used the words *cis* and *trans* in this discussion of cyclohexane conformation. Two hydrogens on the same face of the ring are always cis, regardless of whether they're axial or equatorial and regardless of whether they're adjacent. Similarly, two hydrogens on opposite faces of the ring are always trans.

Axial and equatorial bonds can be drawn following the procedure in Figure 4.10. Look at a molecular model as you practice.

Axial bonds: The six axial bonds, one on each carbon, are parallel and alternate up–down.

Equatorial bonds: The six equatorial bonds, one on each carbon, come in three sets of two parallel lines. Each set is also parallel to two ring bonds. Equatorial bonds alternate between sides around the ring.

Completed cyclohexane

FIGURE 4.10 A procedure for drawing axial and equatorial bonds in chair cyclohexane.

Because chair cyclohexane has two kinds of positions, axial and equatorial, we might expect to find two isomeric forms of a monosubstituted cyclohexane. In fact, we don't. There is only *one* methylcyclohexane, *one* bromocyclohexane, *one* cyclohexanol (hydroxycyclohexane), and so on, because cyclohexane rings are *conformationally mobile* at room temperature. Different chair conformations readily interconvert, exchanging axial and equatorial positions. This interconversion, usually called a **ring-flip**, is shown in Figure 4.11.

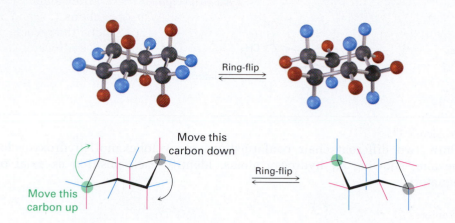

FIGURE 4.11 A ring-flip in chair cyclohexane interconverts axial and equatorial positions. What is axial (red) in the starting structure becomes equatorial in the ring-flipped structure, and what is equatorial (blue) in the starting structure is axial after ring-flip.

As shown in Figure 4.11, a chair cyclohexane can be ring-flipped by keeping the middle four carbon atoms in place while folding the two end carbons in opposite directions. In so doing, an axial substituent in one chair form becomes an equatorial substituent in the ring-flipped chair form and vice versa. For example, axial bromocyclohexane becomes equatorial bromocyclohexane after ring-flip. Since the energy barrier to chair–chair interconversion is only about 45 kJ/mol (10.8 kcal/mol), the process is rapid at room temperature and we see what appears to be a single structure rather than distinct axial and equatorial isomers.

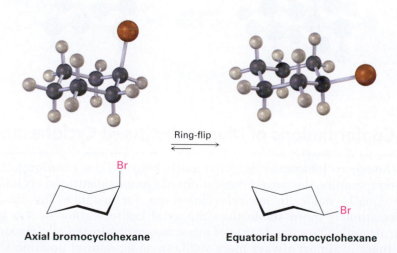

Axial bromocyclohexane   **Equatorial bromocyclohexane**

WORKED EXAMPLE 4.2 Drawing the Chair Conformation of a Substituted Cyclohexane

Draw 1,1-dimethylcyclohexane in a chair conformation, indicating which methyl group in your drawing is axial and which is equatorial.

Strategy
Draw a chair cyclohexane ring using the procedure in Figure 4.10, and then put two methyl groups on the same carbon. The methyl group in the rough plane of the ring is equatorial, and the one directly above or below the ring is axial.

Solution

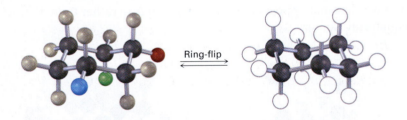

Axial methyl group

CH₃

CH₃

Equatorial methyl group

Problem 4.12

Draw two different chair conformations of cyclohexanol (hydroxycyclo-hexane) showing all hydrogen atoms. Identify each position as axial or equatorial.

Problem 4.13

Draw two different chair conformations of *trans*-1,4-dimethylcyclohexane, and label all positions as axial or equatorial.

Problem 4.14

Identify each of the colored positions—red, blue, and green—as axial or equatorial. Then carry out a ring-flip, and show the new positions occupied by each color.

Ring-flip

4.7 Conformations of Monosubstituted Cyclohexanes

Even though cyclohexane rings flip rapidly between chair conformations at room temperature, the two conformations of a monosubstituted cyclohexane aren't equally stable. In methylcyclohexane, for instance, the equatorial conformation is more stable than the axial conformation by 7.6 kJ/mol (1.8 kcal/mol). The same is true of other monosubstituted cyclohexanes: a substituent is almost always more stable in an equatorial position than in an axial position.

You might recall from your general chemistry course that it's possible to calculate the percentages of two isomers at equilibrium using the equation $\Delta E = -RT \ln K$, where ΔE is the energy difference between isomers, R is the gas constant [8.315 J/(K · mol)], T is the Kelvin temperature, and K is the equilibrium constant between isomers. For example, an energy difference of 7.6 kJ/mol means that about 95% of methylcyclohexane molecules have the methyl group equatorial at any given instant and only 5% have the methyl group axial. Figure 4.12 plots the relationship between energy and isomer percentages.

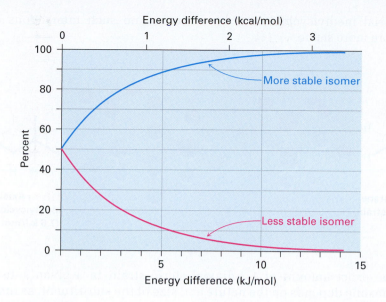

FIGURE 4.12 A plot of the percentages of two isomers at equilibrium versus the energy difference between them. The curves are calculated using the equation $\Delta E = -RT \ln K$.

The energy difference between axial and equatorial conformations is due to steric strain caused by **1,3-diaxial interactions**. The axial methyl group on C1 is too close to the axial hydrogens three carbons away on C3 and C5, resulting in 7.6 kJ/mol of steric strain (Figure 4.13).

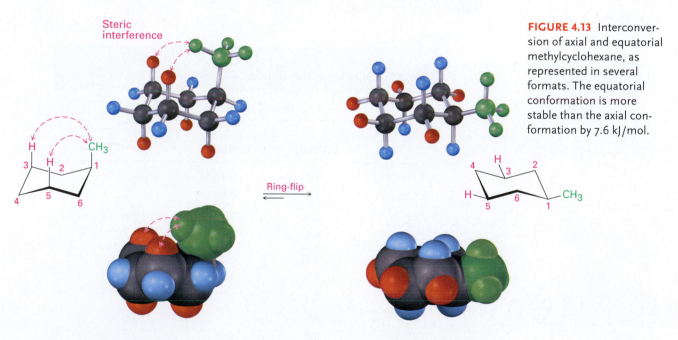

FIGURE 4.13 Interconversion of axial and equatorial methylcyclohexane, as represented in several formats. The equatorial conformation is more stable than the axial conformation by 7.6 kJ/mol.

The 1,3-diaxial steric strain in substituted methylcyclohexane is already familiar—we saw it previously as the steric strain between methyl groups in gauche butane. Recall from Section 3.7 that gauche butane is less stable than anti butane by 3.8 kJ/mol (0.9 kcal/mol) because of steric interference between hydrogen atoms on the two methyl groups. Comparing a four-carbon fragment of axial methylcyclohexane with gauche butane shows that the steric interaction is the same in both cases (Figure 4.14). Because axial methylcyclohexane has two such interactions, though, it has 2 × 3.8 = 7.6 kJ/mol of steric strain.

Equatorial methylcyclohexane, however, has no such interactions and is therefore more stable.

FIGURE 4.14 The origin of 1,3-diaxial interactions in methylcyclohexane. The steric strain between an axial methyl group and an axial hydrogen atom three carbons away is identical to the steric strain in gauche butane. Note that the –CH$_3$ group in methylcyclohexane moves slightly away from a true axial position to minimize the strain.

Gauche butane
(3.8 kJ/mol strain)

Axial
methylcyclohexane
(7.6 kJ/mol strain)

The exact amount of 1,3-diaxial steric strain in a given substituted cyclohexane depends on the nature and size of the substituent, as indicated in Table 4.1. Not surprisingly, the amount of steric strain increases through the series H$_3$C– < CH$_3$CH$_2$– < (CH$_3$)$_2$CH– << (CH$_3$)$_3$C–, paralleling the increasing bulk of the alkyl groups. Note that the values in Table 4.1 refer to 1,3-diaxial interactions of the substituent with a *single* hydrogen atom. These values must be doubled to arrive at the amount of strain in a monosubstituted cyclohexane.

TABLE 4.1
Steric Strain in Monosubstituted Cyclohexanes

Y	1,3-Diaxial strain	
	(kJ/mol)	(kcal/mol)
F	0.5	0.12
Cl, Br	1.0	0.25
OH	2.1	0.5
CH$_3$	3.8	0.9
CH$_2$CH$_3$	4.0	0.95
CH(CH$_3$)$_2$	4.6	1.1
C(CH$_3$)$_3$	11.4	2.7
C$_6$H$_5$	6.3	1.5
CO$_2$H	2.9	0.7
CN	0.4	0.1

Problem 4.15
What is the energy difference between the axial and equatorial conformations of cyclohexanol (hydroxycyclohexane)?

Problem 4.16
Why do you suppose an axial cyano (–CN) substituent causes practically no 1,3-diaxial steric strain (0.4 kJ/mol)? Use molecular models to help with your answer.

Problem 4.17
Look at Figure 4.12, and estimate the percentages of axial and equatorial conformers present at equilibrium in bromocyclohexane.

4.8 Conformations of Disubstituted Cyclohexanes

Monosubstituted cyclohexanes are always more stable with their substituent in an equatorial position, but the situation in disubstituted cyclohexanes is more complex because the steric effects of both substituents must be taken into account. All steric interactions in both possible chair conformations must be analyzed before deciding which conformation is favored.

Let's look at 1,2-dimethylcyclohexane as an example. There are two isomers, *cis*-1,2-dimethylcyclohexane and *trans*-1,2-dimethylcyclohexane, which must be considered separately. In the cis isomer, both methyl groups are on the same face of the ring, and the compound can exist in either of the two chair conformations shown in Figure 4.15. (It may be easier for you to see whether a compound is cis- or trans-disubstituted by first drawing the ring as a flat representation and then converting to a chair conformation.)

Both chair conformations of *cis*-1,2-dimethylcyclohexane have one axial methyl group and one equatorial methyl group. The top conformation in Figure 4.15 has an axial methyl group at C2, which has 1,3-diaxial interactions with hydrogens on C4 and C6. The ring-flipped conformation has an axial methyl group at C1, which has 1,3-diaxial interactions with hydrogens on C3 and C5. In addition, both conformations have gauche butane interactions between the two methyl groups. *The two conformations are equal in energy,* with a total steric strain of 3 × 3.8 kJ/mol = 11.4 kJ/mol (2.7 kcal/mol).

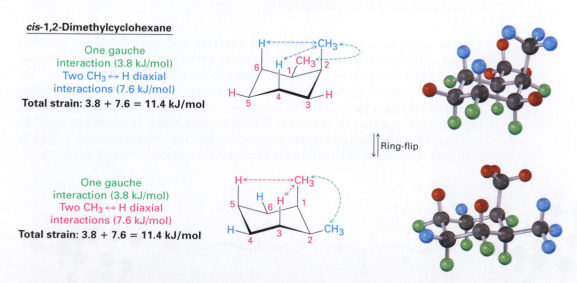

FIGURE 4.15 Conformations of *cis*-1,2-dimethylcyclohexane. The two chair conformations are equal in energy because each has one axial methyl group and one equatorial methyl group.

In *trans*-1,2-dimethylcyclohexane, the two methyl groups are on opposite faces of the ring and the compound can exist in either of the two chair conformations shown in Figure 4.16. The situation here is quite different from that of the cis isomer. The top trans conformation in Figure 4.16 has both methyl groups equatorial and therefore has only a gauche butane interaction between methyls (3.8 kJ/mol) but no 1,3-diaxial interactions. The ring-flipped conformation, however, has both methyl groups axial. The axial methyl group at C1 interacts with axial hydrogens at C3 and C5, and the axial methyl group at C2 interacts with axial hydrogens at C4 and C6. These four 1,3-diaxial interactions produce a steric strain of 4×3.8 kJ/mol $= 15.2$ kJ/mol and make the diaxial conformation $15.2 - 3.8 = 11.4$ kJ/mol less favorable than the diequatorial conformation. We therefore predict that *trans*-1,2-dimethylcyclohexane will exist almost exclusively in the diequatorial conformation.

trans-1,2-Dimethylcyclohexane

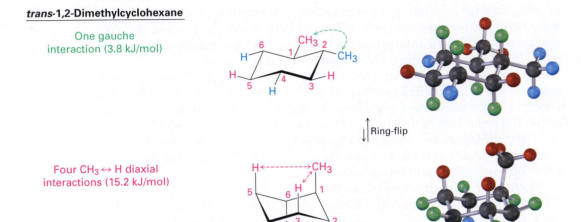

One gauche interaction (3.8 kJ/mol)

Ring-flip

Four CH₃ ↔ H diaxial interactions (15.2 kJ/mol)

FIGURE 4.16 Conformations of *trans*-1,2-dimethylcyclohexane. The conformation with both methyl groups equatorial (top) is favored by 11.4 kJ/mol (2.7 kcal/mol) over the conformation with both methyl groups axial (bottom).

The same kind of **conformational analysis** just carried out for *cis*- and *trans*-1,2-dimethylcyclohexane can be done for any substituted cyclohexane, such as *cis*-1-*tert*-butyl-4-chlorocyclohexane (see Worked Example 4.3). As you might imagine, though, the situation becomes more complex as the number of substituents increases. For instance, compare glucose with mannose, a carbohydrate present in seaweed. Which do you think is more strained? In glucose, all substituents on the six-membered ring are equatorial, while in mannose, one of the –OH groups is axial, making mannose more strained.

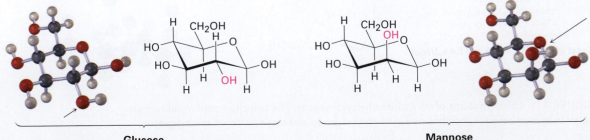

Glucose

Mannose

| WORKED EXAMPLE 4.3 | Drawing the Most Stable Conformation of a Substituted Cyclohexane |

Draw the most stable conformation of *cis*-1-*tert*-butyl-4-chlorocyclohexane. By how much is it favored?

Strategy

Draw the possible conformations, and calculate the strain energy in each. Remember that equatorial substituents cause less strain than axial substituents.

Solution

First draw the two chair conformations of the molecule:

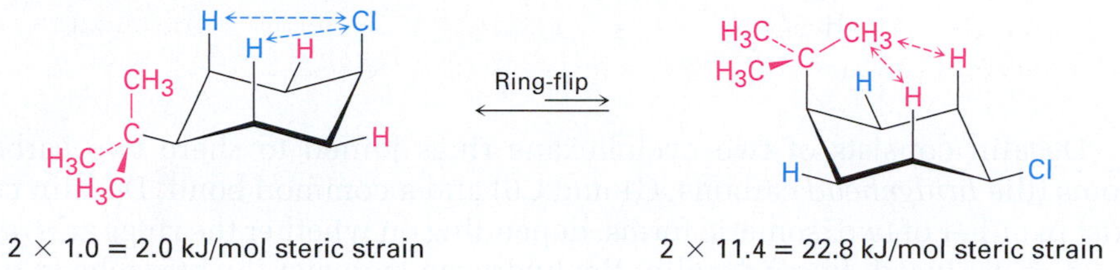

2 × 1.0 = 2.0 kJ/mol steric strain   2 × 11.4 = 22.8 kJ/mol steric strain

In the left-hand conformation, the *tert*-butyl group is equatorial and the chlorine is axial. In the right-hand conformation, the *tert*-butyl group is axial and the chlorine is equatorial. These conformations aren't of equal energy because an axial *tert*-butyl substituent and an axial chloro substituent produce different amounts of steric strain. Table 4.1 shows that the 1,3-diaxial interaction between a hydrogen and a *tert*-butyl group costs 11.4 kJ/mol (2.7 kcal/mol), whereas the interaction between a hydrogen and a chlorine costs only 1.0 kJ/mol (0.25 kcal/mol). An axial *tert*-butyl group therefore produces (2 × 11.4 kJ/mol) − (2 × 1.0 kJ/mol) = 20.8 kJ/mol (4.9 kcal/mol) more steric strain than does an axial chlorine, and the compound preferentially adopts the conformation with the chlorine axial and the *tert*-butyl equatorial.

Problem 4.18

Draw the most stable chair conformation of the following molecules, and estimate the amount of strain in each:

(a) *trans*-1-Chloro-3-methylcyclohexane
(b) *cis*-1-Ethyl-2-methylcyclohexane
(c) *cis*-1-Bromo-4-ethylcyclohexane
(d) *cis*-1-*tert*-Butyl-4-ethylcyclohexane

Problem 4.19

Identify each substituent in the following compound as axial or equatorial, and tell whether the conformation shown is the more stable or less stable chair form (yellow-green = Cl):

4.9 Conformations of Polycyclic Molecules

The final point we'll consider about cycloalkane stereochemistry is to see what happens when two or more cycloalkane rings are fused together along a common bond to construct a **polycyclic** compound—for example, decalin.

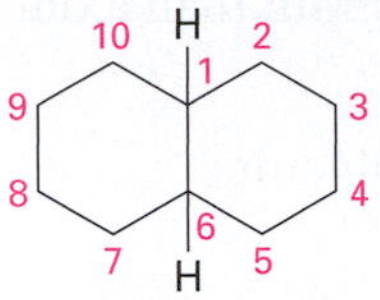

Decalin—two fused cyclohexane rings

Decalin consists of two cyclohexane rings joined to share two carbon atoms (the *bridgehead* carbons, C1 and C6) and a common bond. Decalin can exist in either of two isomeric forms, depending on whether the rings are trans fused or cis fused. In *cis*-decalin, the hydrogen atoms at the bridgehead carbons are on the same face of the rings; in *trans*-decalin, the bridgehead hydrogens are on opposite faces. Figure 4.17 shows how both compounds can be represented using chair cyclohexane conformations. Note that *cis*- and *trans*-decalin are not interconvertible by ring-flips or other rotations. They are cis–trans stereoisomers and have the same relationship to each other that *cis*- and *trans*-1,2-dimethylcyclohexane have.

FIGURE 4.17 Representations of *cis*- and *trans*-decalin. The red hydrogen atoms at the bridgehead carbons are on the same face of the rings in the cis isomer but on opposite faces in the trans isomer.

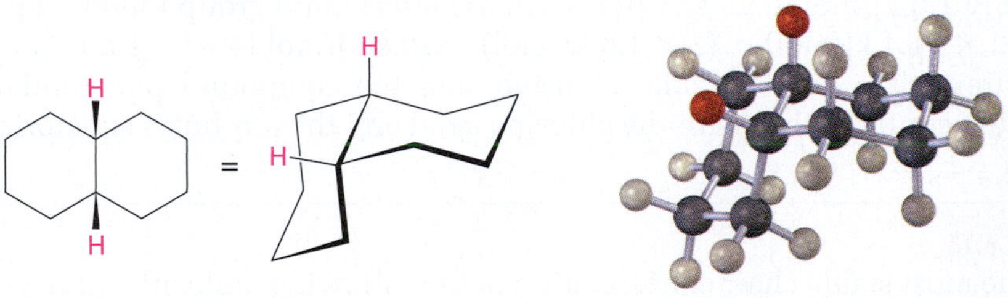

cis-Decalin

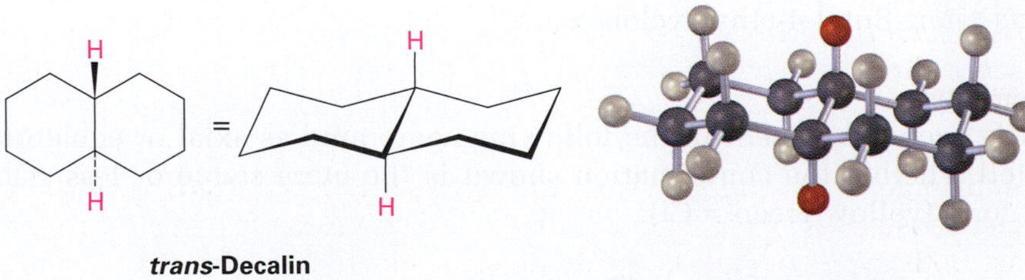

trans-Decalin

Polycyclic compounds are common in nature, and many valuable substances have fused-ring structures. For example, steroids, such as the male hormone testosterone, have 3 six-membered rings and 1 five-membered ring fused together. Although steroids look complicated compared with cyclohexane or decalin, the same principles that apply to the conformational analysis of simple cyclohexane rings apply equally well (and often better) to steroids.

Testosterone (a steroid)

Problem 4.20

Which isomer is more stable, *cis*-decalin or *trans*-decalin? Explain.

Summary

Cyclic molecules are so commonly encountered in all classes of biomolecules, including proteins, lipids, carbohydrates, and nucleic acids, that it's important to understand the effects of their cyclic structures. Thus, we've taken a close look at some of those effects in this chapter.

A **cycloalkane** is a saturated cyclic hydrocarbon with the general formula C_nH_{2n}. In contrast to open-chain alkanes, where nearly free rotation occurs around C–C bonds, rotation is greatly reduced in cycloalkanes. Disubstituted cycloalkanes can therefore exist as **cis–trans isomers**. The cis isomer has both substituents on the same face of the ring; the trans isomer has substituents on opposite faces. Cis–trans isomers are just one kind of **stereoisomers**—isomers that have the same connections between atoms but different three-dimensional arrangements.

Not all cycloalkanes are equally stable. Three kinds of strain contribute to the overall energy of a cycloalkane: (1) **angle strain** is the resistance of a bond angle to compression or expansion from the normal 109° tetrahedral value, (2) *torsional strain* is the energy cost of having neighboring C–H bonds eclipsed rather than staggered, and (3) *steric strain* is the repulsive interaction that arises when two groups attempt to occupy the same space.

Cyclopropane (115 kJ/mol strain) and cyclobutane (110.4 kJ/mol strain) have both angle strain and torsional strain. Cyclopentane is free of angle strain but has a substantial torsional strain due to its large number of eclipsing interactions. Both cyclobutane and cyclopentane pucker slightly away from planarity to relieve torsional strain.

Cyclohexane is strain-free because it adopts a puckered **chair conformation**, in which all bond angles are near 109° and all neighboring C–H bonds are staggered. Chair cyclohexane has two kinds of positions: **axial** and **equatorial**. Axial positions are oriented up and down, parallel to the ring axis, whereas equatorial positions lie in a belt around the equator of the ring. Each carbon atom has one axial and one equatorial position.

Key Words

alicyclic, 106
angle strain, 112
axial position, 117
chair conformation, 116
cis–trans isomers, 110
conformational analysis, 124
cycloalkane, 106
1,3-diaxial interaction, 121
equatorial position, 117
polycyclic compound, 126
ring-flip (cyclohexane), 119
stereoisomers, 110

Chair cyclohexanes are conformationally mobile and can undergo a **ring-flip**, which interconverts axial and equatorial positions. Substituents on the ring are more stable in the equatorial position because axial substituents cause **1,3-diaxial interactions**. The amount of 1,3-diaxial steric strain caused by an axial substituent depends on its bulk.

Lagniappe

Molecular Mechanics

Computer programs make it possible to portray accurate representations of molecular geometry.

All the structural models in this book are computer-drawn. To make sure they accurately portray bond angles, bond lengths, torsional interactions, and steric interactions, the most stable geometry of each molecule has been calculated on a desktop computer using a commercially available *molecular mechanics* program based on work by N. L. Allinger of the University of Georgia.

The idea behind molecular mechanics is to begin with a rough geometry for a molecule and then calculate a total strain energy for that starting geometry, using mathematical equations that assign values to specific kinds of molecular interactions. Bond angles that are too large or too small cause angle strain; bond lengths that are too short or too long cause stretching or compressing strain; unfavorable eclipsing interactions around single bonds cause torsional strain; and nonbonded atoms that approach each other too closely cause steric, or *van der Waals,* strain.

$$E_{\text{total}} = E_{\text{bond stretching}} + E_{\text{angle strain}} + E_{\text{torsional strain}} + E_{\text{van der Waals}}$$

After calculating a total strain energy for the starting geometry, the program automatically changes the geometry slightly in an attempt to lower strain—perhaps by lengthening a bond that is too short or decreasing an angle that is too large. Strain is recalculated for the new geometry, more changes are made, and more calculations are done. After dozens or hundreds of iterations, the calculation ultimately converges on a minimum energy that corresponds to the most favorable, least strained conformation of the molecule.

Molecular mechanics calculations have proved to be enormously useful in pharmaceutical research, where the complementary fit between a drug molecule and a receptor molecule in the body is often a key to designing new pharmaceutical agents (Figure 4.18).

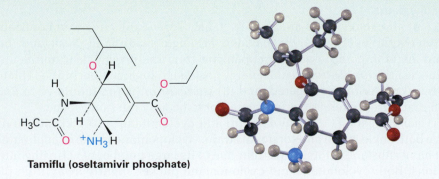

Tamiflu (oseltamivir phosphate)

FIGURE 4.18 The structure of Tamiflu (oseltamivir phosphate), an antiviral agent active against type A influenza, and a molecular model of its minimum-energy conformation as calculated by molecular mechanics.

Exercises

VISUALIZING CHEMISTRY

(Problems 4.1–4.20 appear within the chapter.)

4.21 ■ Name the following cycloalkanes:

(a)   **(b)**

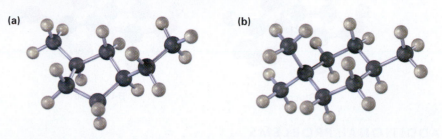

■ indicates problems that are assignable in Organic OWL.

Go to this book's companion website at **www.cengage.com/ chemistry/mcmurry** to explore interactive versions of the Active Figures from this text.

4.22 ■ Name the following compound, identify each substituent as axial or equatorial, and tell whether the conformation shown is the more stable or less stable chair form (yellow-green = Cl):

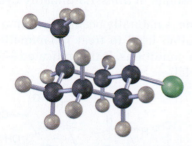

4.23 A trisubstituted cyclohexane with three substituents—red, green, and blue—undergoes a ring-flip to its alternative chair conformation. Identify each substituent as axial or equatorial, and show the positions occupied by the three substituents in the ring-flipped form.

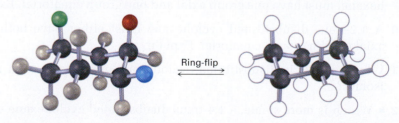

Ring-flip

4.24 Glucose exists in two forms having a 36:64 ratio at equilibrium. Draw a skeletal structure of each, describe the difference between them, and tell which of the two you think is more stable (red = O).

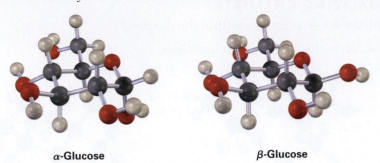

α-Glucose   **β-Glucose**

ADDITIONAL PROBLEMS

4.25 Draw the five cycloalkanes with the formula C_5H_{10}.

4.26 ▪ Draw two constitutional isomers of *cis*-1,2-dibromocyclopentane.

4.27 ▪ Draw a stereoisomer of *trans*-1,3-dimethylcyclobutane.

4.28 ▪ Hydrocortisone, a naturally occurring hormone produced in the adrenal glands, is often used to treat inflammation, severe allergies, and numerous other conditions. Is the indicated —OH group in the molecule axial or equatorial?

Hydrocortisone

4.29 A 1,2-cis disubstituted cyclohexane, such as *cis*-1,2-dichlorocyclohexane, must have one group axial and one group equatorial. Explain.

4.30 A 1,2-trans disubstituted cyclohexane must either have both groups axial or both groups equatorial. Explain.

4.31 Why is a 1,3-cis disubstituted cyclohexane more stable than its trans isomer?

4.32 ▪ Which is more stable, a 1,4-trans disubstituted cyclohexane or its cis isomer?

4.33 *cis*-1,2-Dimethylcyclobutane is less stable than its trans isomer, but *cis*-1,3-dimethylcyclobutane is more stable than its trans isomer. Draw the most stable conformations of both, and explain.

4.34 ▪ Draw the two chair conformations of *cis*-1-chloro-2-methylcyclohexane. Which is more stable, and by how much?

4.35 ▪ Draw the two chair conformations of *trans*-1-chloro-2-methylcyclohexane. Which is more stable?

▪ Problems assignable in Organic OWL.

4.36 ▪ Galactose, a sugar related to glucose, contains a six-membered ring in which all the substituents except the –OH group indicated below in red are equatorial. Draw galactose in its more stable chair conformation.

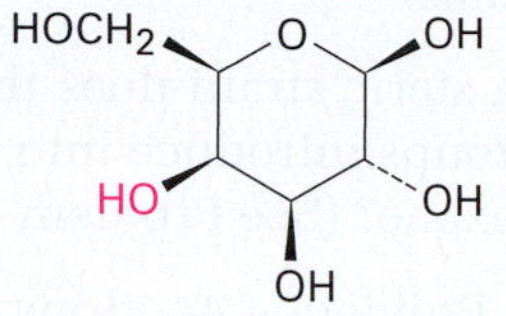

HOCH$_2$ O   OH

HO   OH

OH

Galactose

4.37 ▪ Draw the two chair conformations of menthol, and tell which is more stable.

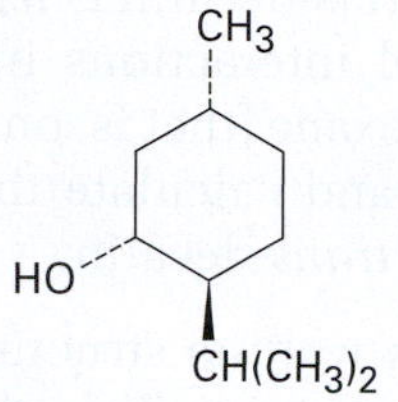

CH$_3$

HO

CH(CH$_3$)$_2$

Menthol

4.38 ▪ There are four cis–trans isomers of menthol (Problem 4.37), including the one shown. Draw the other three.

4.39 ▪ Identify each pair of relationships among the –OH groups in glucose (red–blue, red–green, red–black, blue–green, blue–black, green–black) as cis or trans.

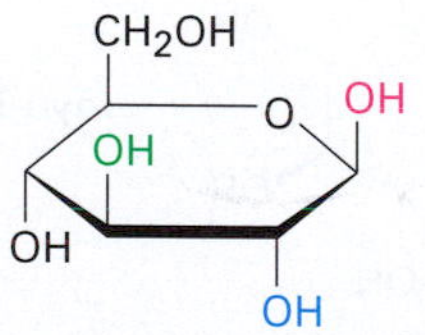

CH$_2$OH

O   OH

OH

OH

OH

Glucose

4.40 Draw 1,3,5-trimethylcyclohexane using a hexagon to represent the ring. How many cis–trans stereoisomers are possible?

4.41 ▪ From the data in Figure 4.12 and Table 4.1, estimate the percentages of molecules that have their substituents in an axial orientation for the following compounds:

(a) Isopropylcyclohexane **(b)** Fluorocyclohexane

(c) Cyclohexanecarbonitrile, C$_6$H$_{11}$CN

4.42 ▪ Assume that you have a variety of cyclohexanes substituted in the positions indicated. Identify the substituents as either axial or equatorial. For example, a 1,2-cis relationship means that one substituent must be axial and one equatorial, whereas a 1,2-trans relationship means that both substituents are axial or both are equatorial.

(a) 1,3-Trans disubstituted   **(b)** 1,4-Cis disubstituted

(c) 1,3-Cis disubstituted   **(d)** 1,5-Trans disubstituted

(e) 1,5-Cis disubstituted   **(f)** 1,6-Trans disubstituted

▪ Problems assignable in Organic OWL.

4.43 The diaxial conformation of *cis*-1,3-dimethylcyclohexane is approximately 23 kJ/mol (5.4 kcal/mol) less stable than the diequatorial conformation. Draw the two possible chair conformations, and suggest a reason for the large energy difference.

4.44 Approximately how much steric strain does the 1,3-diaxial interaction between the two methyl groups introduce into the diaxial conformation of *cis*-1,3-dimethylcyclohexane? (See Problem 4.43.)

4.45 In light of your answer to Problem 4.44, draw the two chair conformations of 1,1,3-trimethylcyclohexane, and estimate the amount of strain energy in each. Which conformation is favored?

4.46 We saw in Problem 4.20 that *cis*-decalin is less stable than *trans*-decalin. Assume that the 1,3-diaxial interactions in *cis*-decalin are similar to those in axial methylcyclohexane [that is, one $CH_2 \leftrightarrow H$ interaction costs 3.8 kJ/mol (0.9 kcal/mol)], and calculate the magnitude of the energy difference between *cis*- and *trans*-decalin.

4.47 Using molecular models as well as structural drawings, explain why *trans*-decalin is rigid and cannot ring-flip, whereas *cis*-decalin can easily ring-flip.

4.48 *myo*-Inositol, one of the isomers of 1,2,3,4,5,6-hexahydroxycyclohexane, acts as a growth factor in both animals and microorganisms. Draw the most stable chair conformation of *myo*-inositol.

myo-**Inositol**

4.49 How many cis–trans stereoisomers of *myo*-inositol (Problem 4.48) are there? Draw the structure of the most stable isomer.

4.50 ■ One of the two chair structures of *cis*-1-chloro-3-methylcyclohexane is more stable than the other by 15.5 kJ/mol (3.7 kcal/mol). Which is it? What is the energy cost of a 1,3-diaxial interaction between a chlorine and a methyl group?

4.51 ■ Tell whether each of the following substituents on a steroid is axial or equatorial. (A substituent that is "up" is on the top face of the molecule as drawn, and a substituent that is "down" is on the bottom face.)

(a) Substituent up at C3

(b) Substituent down at C7

(c) Substituent down at C11

■ Problems assignable in Organic OWL.

4.52 Amantadine is an antiviral agent that is active against influenza type A infection. Draw a three-dimensional representation of amantadine showing the chair cyclohexane rings.

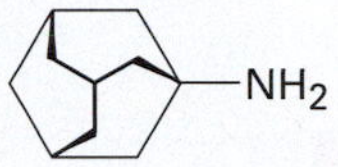

—NH₂   **Amantadine**

4.53 Alcohols undergo an *oxidation* reaction to yield carbonyl compounds on treatment with CrO_3. For example, 2-*tert*-butylcyclohexanol gives 2-*tert*-butylcyclohexanone. If axial –OH groups are generally more reactive than their equatorial isomers, which do you think would react faster, the cis isomer of 2-*tert*-butylcyclohexanol or the trans isomer? Explain.

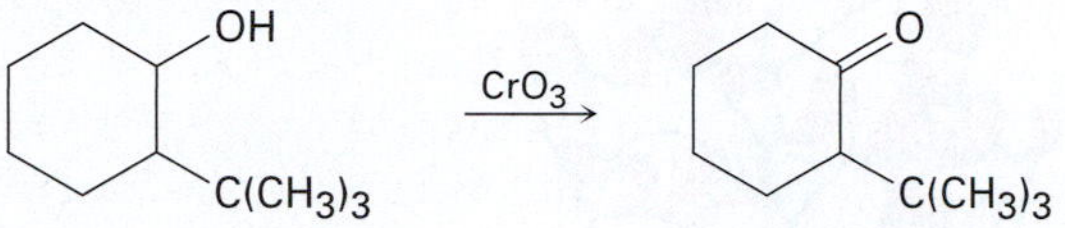

OH

$\xrightarrow{\text{CrO}_3}$

C(CH₃)₃

C(CH₃)₃

2-*tert*-Butylcyclohexanol   **2-*tert*-Butylcyclohexanone**

4.54 Ketones react with alcohols to yield products called *acetals*. Why is it that the all-cis isomer of 4-*tert*-butylcyclohexane-1,3-diol reacts readily with acetone and an acid catalyst to form an acetal but other stereoisomers do not react? In formulating your answer, draw the more stable chair conformations of all four stereoisomers and the product acetal from each. Use molecular models for help.

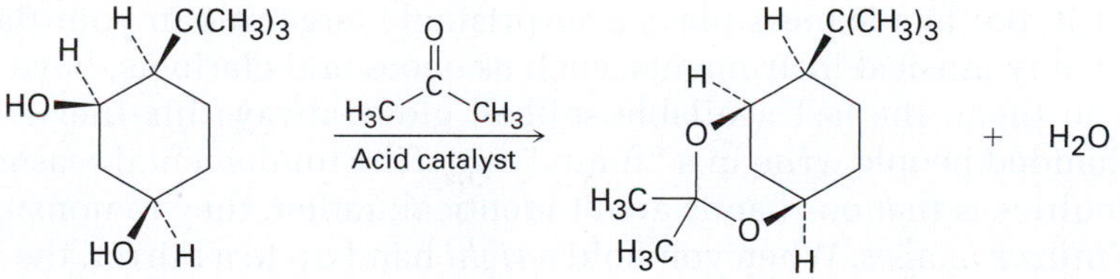

An acetal

5 Stereochemistry at Tetrahedral Centers

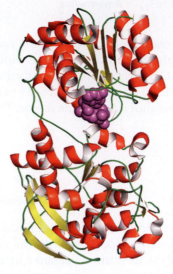

Glycogen synthase catalyzes the conversion of glucose to glycogen for energy storage.

Are you right-handed or left-handed? You may not spend much time thinking about it, but handedness plays a surprisingly large role in your daily activities: many musical instruments, such as oboes and clarinets, have a handedness to them; the last available softball glove always fits the wrong hand; left-handed people write in a "funny" way. The fundamental reason for these difficulties is that our hands aren't identical; rather, they're nonsuperimposable *mirror images*. When you hold a *right* hand up to a mirror, the image you see looks like a *left* hand. Try it.

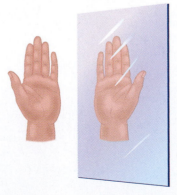

Left hand   **Right hand**

Handedness is also important in organic and biological chemistry, where it primarily arises as a consequence of the tetrahedral stereochemistry of

OWL Online homework for this chapter can be assigned in Organic OWL.

sp^3-hybridized carbon atoms. Many drugs and almost all the molecules in our bodies, for instance, are handed. Furthermore, it is molecular handedness that makes possible the specific interactions between enzymes and their substrates that are necessary for enzyme function.

WHY THIS CHAPTER?

Understanding the causes and consequences of molecular handedness is crucial to understanding biological chemistry. The subject can be a bit complex, but the material covered in this chapter nevertheless forms the basis for much of the remainder of the book.

5.1 Enantiomers and the Tetrahedral Carbon

What causes molecular handedness? Look at generalized molecules of the type CH_3X, CH_2XY, and CHXYZ shown in Figure 5.1. On the left are three molecules, and on the right are their images reflected in a mirror. The CH_3X and CH_2XY molecules are identical to their mirror images and thus are not handed. If you make molecular models of each molecule and its mirror image, you find that you can superimpose one on the other. The CHXYZ molecule, by contrast, is *not* identical to its mirror image. You can't superimpose a model of the molecule on a model of its mirror image for the same reason that you can't superimpose a left hand on a right hand: they simply aren't the same.

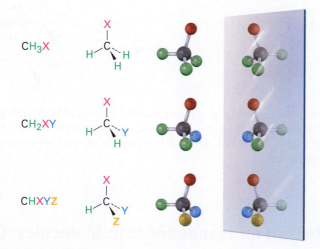

FIGURE 5.1 Tetrahedral carbon atoms and their mirror images. Molecules of the type CH_3X and CH_2XY are identical to their mirror images, but a molecule of the type CHXYZ is not. A CHXYZ molecule is related to its mirror image in the same way that a right hand is related to a left hand.

Molecules that are not identical to their mirror images are kinds of stereoisomers called **enantiomers** (Greek *enantio,* meaning "opposite"). Enantiomers are related to each other as a right hand is related to a left hand and result whenever a tetrahedral carbon is bonded to four different substituents (one need not be H). For example, lactic acid (2-hydroxypropanoic acid) exists as a pair of enantiomers because there are four different groups (–H, –OH, –CH$_3$, and –CO$_2$H) bonded to the central carbon atom. The enantiomers are called

(+)-lactic acid and (−)-lactic acid. Both are found in sour milk, but only the (+) enantiomer occurs in muscle tissue.

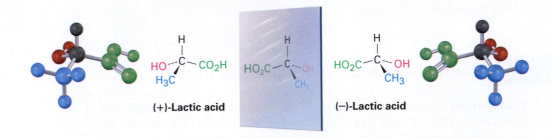

Lactic acid: a molecule of general formula CHXYZ

(+)-Lactic acid   **(−)-Lactic acid**

No matter how hard you try, you can't superimpose a molecule of (+)-lactic acid on a molecule of (−)-lactic acid; they simply aren't identical. If any two groups match up, say −H and −CO$_2$H, the remaining two groups don't match (Figure 5.2).

(a)

Mismatch

Mismatch

(b)

Mismatch

Mismatch

FIGURE 5.2 Attempts at superimposing the mirror-image forms of lactic acid. **(a)** When the −H and −OH substituents match up, the −CO$_2$H and −CH$_3$ substituents don't; **(b)** when −CO$_2$H and −CH$_3$ match up, −H and −OH don't. Regardless of how the molecules are oriented, they aren't identical.

5.2 The Reason for Handedness in Molecules: Chirality

A molecule that is not identical to its mirror image is said to be **chiral** (**ky**-ral, from the Greek *cheir*, meaning "hand"). You can't take a chiral molecule and its enantiomer and place one on the other so that all atoms coincide.

How can you predict whether a given molecule is or is not chiral? *A molecule is not chiral if it has a plane of symmetry.* A plane of symmetry is a plane that cuts through the middle of a molecule (or any object) in such a way that one half of the molecule or object is a mirror image of the other half.

A laboratory flask, for example, has a plane of symmetry. If you were to cut the flask in half, one half would be a mirror image of the other half. A hand, however, does not have a plane of symmetry. One "half" of a hand is not a mirror image of the other half (Figure 5.3).

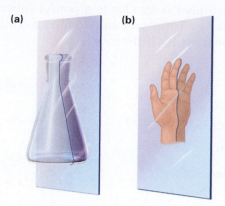

FIGURE 5.3 The meaning of *symmetry plane.* **(a)** An object like the flask has a symmetry plane cutting through it, making right and left halves mirror images. **(b)** An object like a hand does not have a symmetry plane; the right half of a hand is not a mirror image of the left half.

A molecule that has a plane of symmetry in any of its possible conformations must be identical to its mirror image and hence must be nonchiral, or **achiral**. Thus, propanoic acid, $CH_3CH_2CO_2H$, has a plane of symmetry when lined up as shown in Figure 5.4 and is achiral, while lactic acid, $CH_3CH(OH)CO_2H$, has no plane of symmetry in any conformation and is chiral.

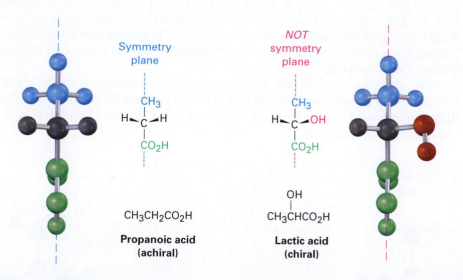

FIGURE 5.4 The achiral propanoic acid molecule versus the chiral lactic acid molecule. Propanoic acid has a plane of symmetry that makes one side of the molecule a mirror image of the other side. Lactic acid has no such symmetry plane.

The most common, although not the only, cause of chirality in an organic molecule is the presence of a carbon atom bonded to four different groups—for example, the central carbon atom in lactic acid. Such carbons are referred to as **chirality centers**, although other terms, such as *stereocenter, asymmetric center,* and *stereogenic center,* have also been used. Note that *chirality* is a property of an entire molecule, whereas a chirality *center* is the *cause* of chirality.

Detecting chirality centers in a complex molecule takes practice because it's not always immediately apparent whether four different groups are bonded to a given carbon. The differences don't necessarily appear right next to the chirality center. For example, 5-bromodecane is a chiral molecule because four different groups are bonded to C5, the chirality center (marked with an asterisk). A butyl substituent is *similar* to a pentyl substituent, but it isn't identical. The difference isn't apparent until four carbon atoms away from the chirality center, but there's still a difference.

Br
|
$CH_3CH_2CH_2CH_2CH_2C CH_2CH_2CH_2CH_3$
|*
H

5-Bromodecane (chiral)

Substituents on carbon 5

—H

—Br

—$CH_2CH_2CH_2CH_3$ (butyl)

—$CH_2CH_2CH_2CH_2CH_3$ (pentyl)

As other possible examples, look at methylcyclohexane and 2-methylcyclohexanone. Methylcyclohexane is achiral because no carbon atom in the molecule is bonded to four different groups. You can immediately eliminate all –CH_2– carbons and the –CH_3 carbon from consideration, but what about C1 on the ring? The C1 carbon atom is bonded to a –CH_3 group, to an –H atom, and to C2 and C6 of the ring. Carbons 2 and 6 are equivalent, however, as are carbons 3 and 5. Thus, the C6–C5–C4 "substituent" is equivalent to the C2–C3–C4 substituent, and methylcyclohexane is achiral. Another way of reaching the same conclusion is to realize that methylcyclohexane has a symmetry plane passing through the methyl group and through C1 and C4 of the ring.

The situation is different for 2-methylcyclohexanone. 2-Methylcyclohexanone has no symmetry plane and is chiral because C2 is bonded to four different groups: a –CH_3 group, an –H atom, a –$COCH_2$– ring bond (C1), and a –CH_2CH_2– ring bond (C3).

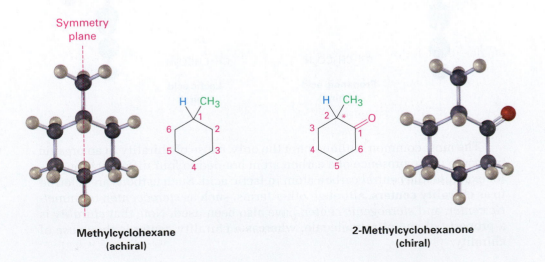

Symmetry plane

Methylcyclohexane (achiral)

2-Methylcyclohexanone (chiral)

Several more examples of chiral molecules follow. Check for yourself that the labeled carbons are chirality centers. You might note that carbons in $-CH_2-$, $-CH_3$, C=O, C=C, and C≡C groups *can't* be chirality centers. (Why not?)

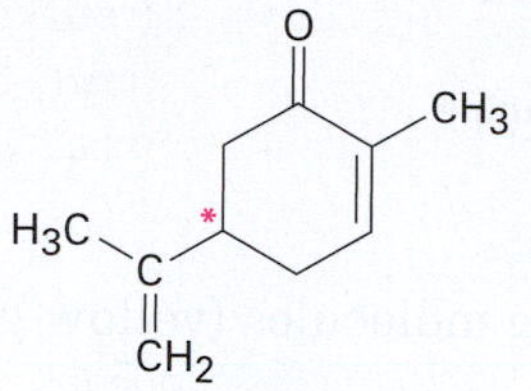

Carvone (spearmint oil)

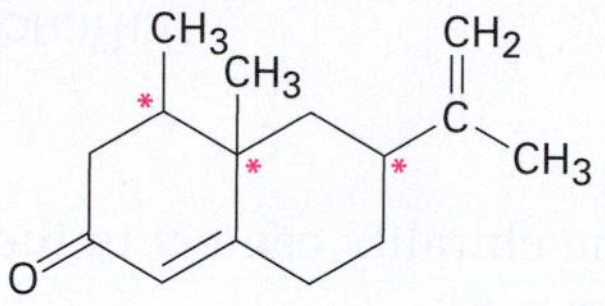

Nootkatone (grapefruit oil)

WORKED EXAMPLE 5.1   Drawing the Three-Dimensional Structure of a Chiral Molecule

Draw the structure of a chiral alcohol.

Strategy

An alcohol is a compound that contains the $-OH$ functional group. To make an alcohol chiral, we need to have four different groups bonded to a single carbon atom, say $-H$, $-OH$, $-CH_3$, and $-CH_2CH_3$.

Solution

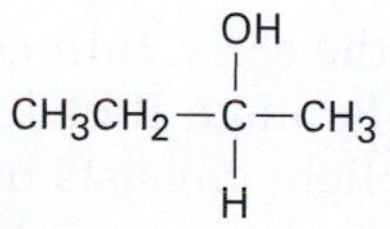

$$CH_3CH_2-\underset{\underset{H}{|}}{\overset{\overset{OH}{|}}{C}}-CH_3$$

Butan-2-ol (chiral)

Problem 5.1

Which of the following objects are chiral?
(a) Screwdriver   **(b)** Screw   **(c)** Shoe   **(d)** Beanstalk

Problem 5.2

Which of the following molecules are chiral? Identify the chirality center(s) in each.

(a)

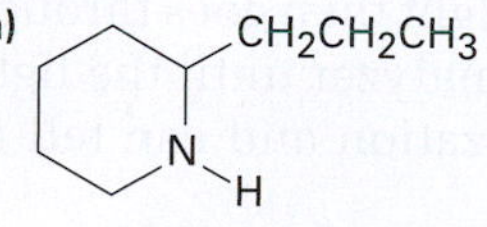

Coniine (poison hemlock)

(b)

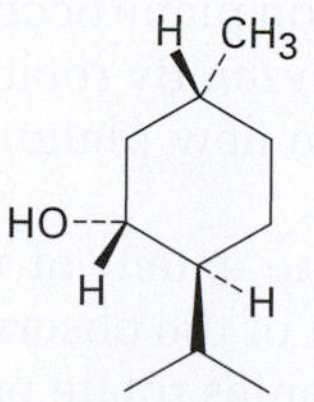

Menthol (flavoring agent)

(c)

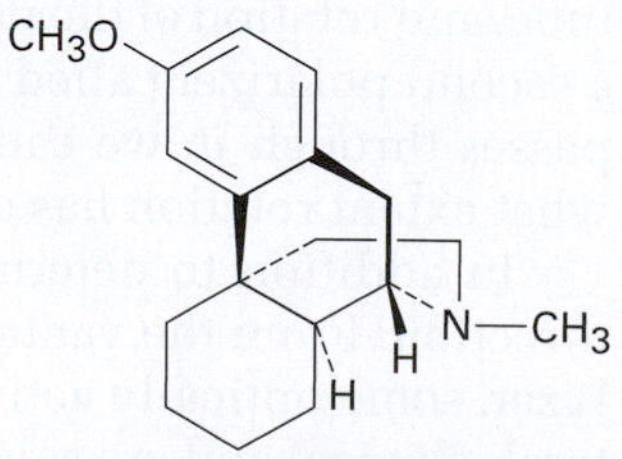

Dextromethorphan (cough suppressant)

Problem 5.3

Alanine, an amino acid found in proteins, is chiral. Draw the two enantiomers of alanine using the standard convention of solid, wedged, and dashed lines.

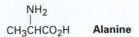

$$NH_2$$
$$CH_3CHCO_2H \quad \textbf{Alanine}$$

Problem 5.4

Identify the chirality centers in the following molecules (yellow-green = Cl, pale yellow = F):

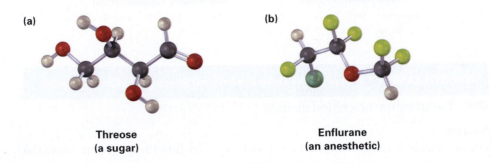

(a)

**Threose
(a sugar)**

(b)

**Enflurane
(an anesthetic)**

5.3 Optical Activity

The study of chirality originated in the early 19th century during investigations by the French physicist Jean-Baptiste Biot into the nature of *plane-polarized light.* A beam of ordinary light consists of electromagnetic waves that oscillate in an infinite number of planes at right angles to the direction of light travel. When a beam of ordinary light is passed through a device called a *polarizer,* however, only the light waves oscillating in a single plane pass through and the light is said to be plane-polarized. Light waves in all other planes are blocked out.

Biot made the remarkable observation that when a beam of plane-polarized light passes through a solution of certain organic molecules such as sugar or camphor, the plane of polarization is *rotated* through an angle, α. Not all organic substances exhibit this property, but those that do are said to be **optically active**.

The angle of rotation can be measured with an instrument called a *polarimeter,* represented in Figure 5.5. A solution of optically active organic molecules is placed in a sample tube, plane-polarized light is passed through the tube, and rotation of the polarization plane occurs. The light then goes through a second polarizer called the *analyzer.* By rotating the analyzer until the light passes through *it,* we can find the new plane of polarization and can tell to what extent rotation has occurred.

In addition to determining the extent of rotation, we can also find the direction. From the vantage point of the observer looking directly at the analyzer, some optically active molecules rotate polarized light to the left (counterclockwise) and are said to be **levorotatory**, whereas others rotate polarized light to the right (clockwise) and are said to be **dextrorotatory**. By convention, rotation to the left is given a minus sign ($-$), and rotation to the right is given a plus sign ($+$). ($-$)-Morphine, for example, is levorotatory, and ($+$)-sucrose is dextrorotatory.

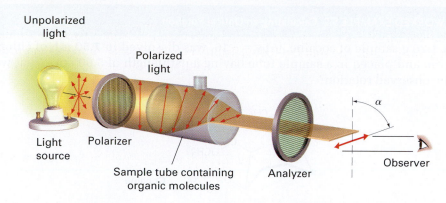

FIGURE 5.5 Schematic representation of a polarimeter. Plane-polarized light passes through a solution of optically active molecules, which rotate the plane of polarization.

The extent of rotation observed in a polarimetry experiment depends on the number of optically active molecules encountered by the light beam. This number, in turn, depends on sample concentration and sample pathlength. If the concentration of sample is doubled, the observed rotation doubles. If the concentration is kept constant but the length of the sample tube is doubled, the observed rotation is doubled. It also happens that the angle of rotation depends on the wavelength of the light used.

To express optical rotations in a meaningful way so that comparisons can be made, we have to choose standard conditions. The **specific rotation, $[\alpha]_D$,** of a compound is defined as the observed rotation when light of 589.6 nanometer (nm; 1 nm = 10^{-9} m) wavelength is used with a sample pathlength l of 1 decimeter (dm; 1 dm = 10 cm) and a sample concentration c of 1 g/cm^3. (Light of 589.6 nm, the so-called sodium D line, is the yellow light emitted from common sodium lamps.)

$$[\alpha]_D = \frac{\text{Observed rotation (degrees)}}{\text{Pathlength, } l \text{ (dm)} \times \text{Concentration, } c \text{ (g/cm}^3)} = \frac{\alpha}{l \times c}$$

When optical rotation data are expressed in this standard way, the specific rotation, $[\alpha]_D$, is a physical constant characteristic of a given optically active compound. For example, (+)-lactic acid has $[\alpha]_D = +3.82$, and (−)-lactic acid has $[\alpha]_D = -3.82$. That is, the two enantiomers rotate plane-polarized light to exactly the same extent but in opposite directions. Note that the units of specific rotation are [(deg · cm^2)/g] but that values are usually expressed without the units. Some additional examples are listed in Table 5.1.

TABLE 5.1
Specific Rotation of Some Organic Molecules

Compound	$[\alpha]_D$	Compound	$[\alpha]_D$
Penicillin V	+233	Cholesterol	−31.5
Sucrose	+66.47	Morphine	−132
Camphor	+44.26	Cocaine	−16
Chloroform	0	Acetic acid	0

WORKED EXAMPLE 5.2 Calculating an Optical Rotation

A 1.20 g sample of cocaine, $[\alpha]_D = -16$, was dissolved in 7.50 mL of chloroform and placed in a sample tube having a pathlength of 5.00 cm. What was the observed rotation?

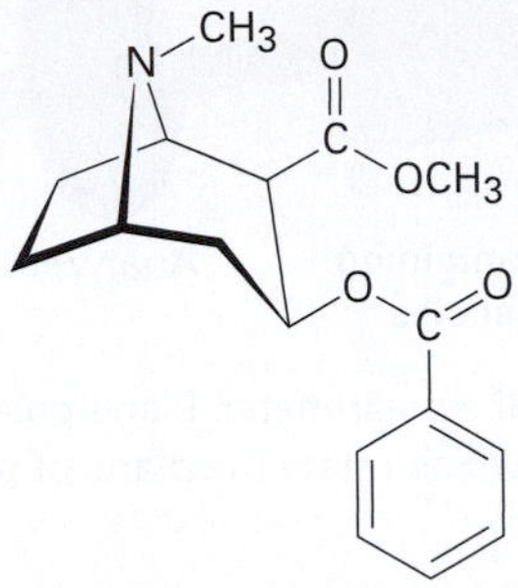

Cocaine

Strategy

$$\text{Since } [\alpha]_D = \frac{\alpha}{l \times c}$$

$$\text{Then } \alpha = l \times c \times [\alpha]_D$$

where $[\alpha]_D = -16$, $l = 5.00$ cm $= 0.500$ dm, and $c = 1.20$ g$/7.50$ cm$^3 = 0.160$ g/cm^3.

Solution

$\alpha = (-16)(0.500)(0.160) = -1.3°$.

Problem 5.5

Is cocaine (Worked Example 5.2) dextrorotatory or levorotatory?

Problem 5.6

A 1.50 g sample of coniine, the toxic extract of poison hemlock, was dissolved in 10.0 mL of ethanol and placed in a sample cell with a 5.00 cm pathlength. The observed rotation at the sodium D line was +1.21°. Calculate $[\alpha]_D$ for coniine.

5.4 Pasteur's Discovery of Enantiomers

Little was done after Biot's discovery of optical activity until 1848, when Louis Pasteur began work on a study of crystalline tartaric acid salts derived from wine. On crystallizing a concentrated solution of sodium ammonium tartrate below 28 °C, Pasteur made the surprising observation that two distinct kinds of crystals precipitated. Furthermore, the two kinds of crystals were nonsuperimposable mirror images and were related in the same way that a right hand is related to a left hand.

Working carefully with tweezers, Pasteur was able to separate the crystals into two piles, one of "right-handed" crystals and one of "left-handed" crystals, like those shown in Figure 5.6. Although the original sample, a 50:50 mixture of right and left, was optically inactive, *solutions of the crystals from each of the sorted piles were optically active,* and their specific rotations were equal in amount but opposite in sign.

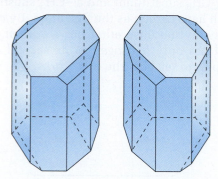

$$\begin{array}{c} CO_2^- \ Na^+ \\ | \\ H-C-OH \\ | \\ HO-C-H \\ | \\ CO_2^- \ NH_4^+ \end{array}$$

Sodium ammonium tartrate

FIGURE 5.6 Drawings of sodium ammonium tartrate crystals taken from Pasteur's original sketches. One of the crystals is "right-handed" and one is "left-handed."

Pasteur was far ahead of his time. Although the structural theory of Kekulé had not yet been proposed, Pasteur explained his results by speaking of the molecules themselves, saying, "There is no doubt that [in the *dextro* tartaric acid] there exists an asymmetric arrangement having a nonsuperimposable image. It is no less certain that the atoms of the *levo* acid possess precisely the inverse asymmetric arrangement." Pasteur's vision was extraordinary, for it was not until 25 years later that his ideas regarding the asymmetric carbon atom were confirmed.

Today, we would describe Pasteur's work by saying that he had discovered enantiomers. Enantiomers, also called *optical isomers,* have identical physical properties, such as melting point and boiling point, but differ in the direction in which their solutions rotate plane-polarized light.

5.5 Sequence Rules for Specifying Configuration

Structural drawings provide a visual representation of stereochemistry, but a verbal method for indicating the three-dimensional arrangement, or **configuration**, of substituents at a chirality center is also needed. The method used employs a set of *sequence rules* to rank the four groups attached to the chirality center and then looks at the handedness with which those groups are attached. Called the **Cahn–Ingold–Prelog rules** after the chemists who proposed them, the sequence rules are as follows:

Rule 1

Look at the four atoms directly attached to the chirality center, and rank them according to atomic number. The atom with the highest atomic number has the highest ranking (first), and the atom with the lowest atomic number (usually hydrogen) has the lowest ranking (fourth). When different isotopes of the same element are compared, such as deuterium (^2H) and protium (^1H), the heavier isotope ranks higher than the lighter isotope. Thus, atoms commonly found in organic compounds have the following order.

Atomic number	35		17		16		15		8		7		6		(2)		(1)	
Higher ranking	Br	>	Cl	>	S	>	P	>	O	>	N	>	C	>	^2H	>	^1H	**Lower ranking**

Rule 2

If a decision can't be reached by ranking the first atoms in the substituent, look at the second, third, or fourth atoms away from the chirality center until the first difference is found. A $-CH_2CH_3$ substituent and a $-CH_3$ substituent are equivalent by rule 1 because both have carbon as the first atom. By rule 2, however, ethyl ranks higher than methyl because ethyl has a *carbon* as its highest second atom, while methyl has only *hydrogen* as its second atom. Look at the following pairs of examples to see how the rule works:

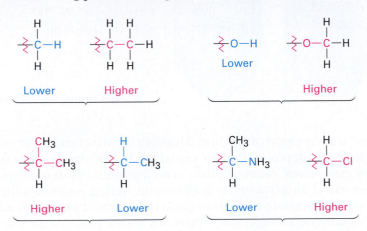

Rule 3

Multiple-bonded atoms are equivalent to the same number of single-bonded atoms. For example, an aldehyde substituent ($-CH=O$), which has a carbon atom *doubly* bonded to *one* oxygen, is equivalent to a substituent having a carbon atom *singly* bonded to *two* oxygens:

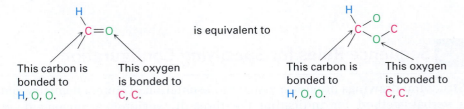

As further examples, the following pairs are equivalent:

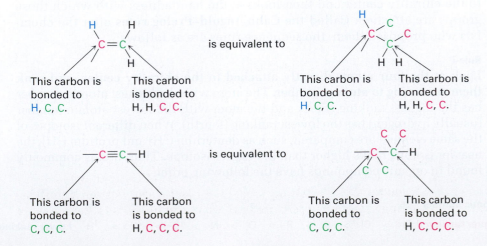

Having ranked the four groups attached to a chiral carbon, we describe the stereochemical configuration around the carbon by orienting the molecule so that the group with the lowest ranking (4) points directly back, away from us. We then look at the three remaining substituents, which now appear to radiate toward us like the spokes on a steering wheel (Figure 5.7). If a curved arrow drawn from the highest to second-highest to third-highest ranked substituent ($1 \rightarrow 2 \rightarrow 3$) is clockwise, we say that the chirality center has the **R configuration** (Latin *rectus*, meaning "right"). If an arrow from $1 \rightarrow 2 \rightarrow 3$ is counterclockwise, the chirality center has the **S configuration** (Latin *sinister*, meaning "left"). To remember these assignments, think of a car's steering wheel when making a *R*ight (clockwise) turn.

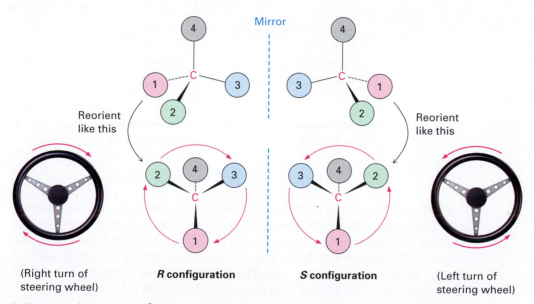

FIGURE 5.7 Assigning configuration to a chirality center. When the molecule is oriented so that the lowest-ranked group (4) is toward the rear, the remaining three groups radiate toward the viewer like the spokes of a steering wheel. If the direction of travel $1 \rightarrow 2 \rightarrow 3$ is clockwise (right turn), the center has the R configuration. If the direction of travel $1 \rightarrow 2 \rightarrow 3$ is counterclockwise (left turn), the center is S.

Look at (−)-lactic acid in Figure 5.8 for an example of how to assign configuration. Sequence rule 1 says that –OH is ranked 1 and –H is ranked 4, but it doesn't allow us to distinguish between –CH_3 and –CO_2H because both groups have carbon as their first atom. Sequence rule 2, however, says that –CO_2H ranks higher than –CH_3 because O (the highest second atom in –CO_2H) outranks H (the highest second atom in –CH_3). Now, turn the molecule so that the fourth-ranked group (–H) is oriented toward the rear, away from the observer. Since a curved arrow from 1 (–OH) to 2 (–CO_2H) to 3 (–CH_3) is clockwise (right turn of the steering wheel), (−)-lactic acid has the R configuration. Applying the same procedure to (+)-lactic acid leads to the opposite assignment.

FIGURE 5.8 Assigning configuration to **(a)** (*R*)-(−)-lactic acid and **(b)** (*S*)-(+)-lactic acid.

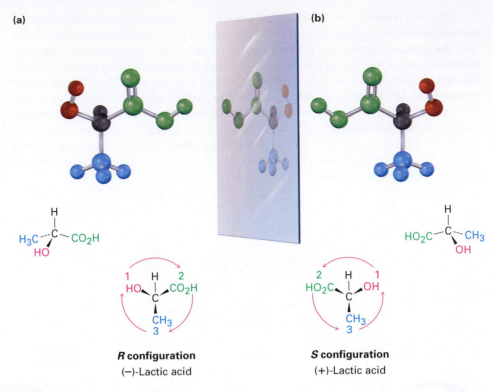

(a)

H₃C—C—CO₂H
HO
(H on top)

1 HO ⟋ C ⟍ CO₂H 2
CH₃
3

R configuration
(−)-Lactic acid

(b)

HO₂C—C—CH₃
OH
(H on top)

2 HO₂C ⟋ C ⟍ OH 1
CH₃
3

S configuration
(+)-Lactic acid

Further examples are provided by naturally occurring (−)-glyceraldehyde and (+)-alanine, which both have the *S* configuration, as shown in Figure 5.9. Note that the sign of optical rotation, (+) or (−), is not related to the *R,S* designation. (*S*)-Glyceraldehyde happens to be levorotatory (−), and (*S*)-alanine happens to be dextrorotatory (+). There is no simple correlation between *R,S* configuration and direction or magnitude of optical rotation.

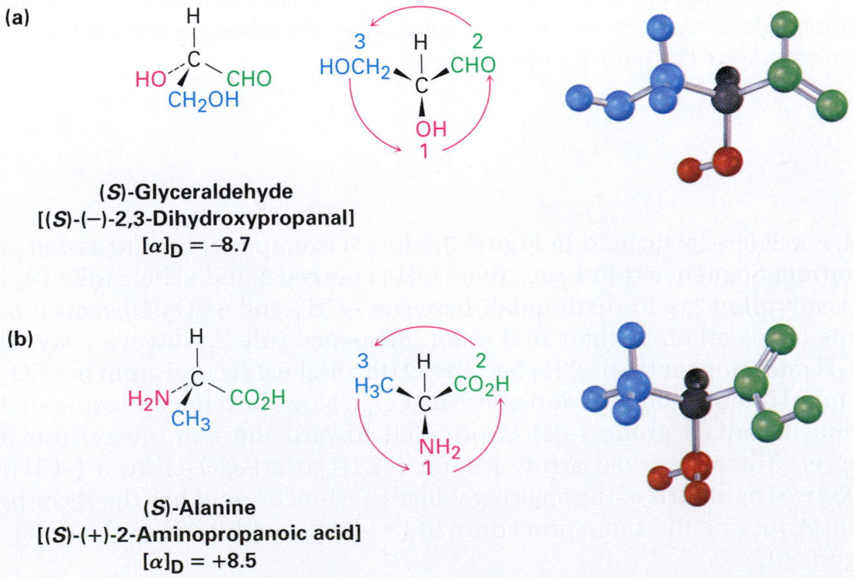

FIGURE 5.9 Assigning configuration to **(a)** (−)-glyceraldehyde and **(b)** (+)-alanine. Both happen to have the *S* configuration, although one is levorotatory and the other is dextrorotatory.

(a)

HO—C—CHO
CH₂OH
(H on top)

3 HOCH₂ ⟋ C ⟍ CHO 2
OH
1

(S)-Glyceraldehyde
[(S)-(−)-2,3-Dihydroxypropanal]
$[\alpha]_D = -8.7$

(b)

H₂N—C—CO₂H
CH₃
(H on top)

3 H₃C ⟋ C ⟍ CO₂H 2
NH₂
1

(S)-Alanine
[(S)-(+)-2-Aminopropanoic acid]
$[\alpha]_D = +8.5$

One additional point needs to be mentioned—the matter of **absolute configuration**. How do we know that the assignments of *R* and *S* configuration are correct in an *absolute*, rather than a relative, sense? Since we can't see the molecules themselves, how do we know that the *R* configuration belongs to the levorotatory enantiomer of lactic acid? This difficult question was solved in 1951, when an X-ray diffraction method for determining the absolute spatial arrangement of atoms in a molecule was found. Based on those results, we can say with certainty that the *R,S* conventions are correct.

WORKED EXAMPLE 5.3 Assigning Configuration to Chirality Centers

Orient each of the following drawings so that the lowest-ranked group is toward the rear, and then assign *R* or *S* configuration:

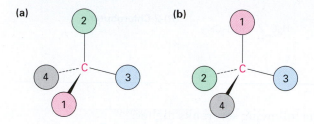

Strategy

It takes practice to be able to visualize and orient a chirality center in three dimensions. You might start by indicating where the observer must be located—180° opposite the lowest-ranked group. Then imagine yourself in the position of the observer, and redraw what you would see.

Solution

In **(a)**, you would be located in front of the page toward the top right of the molecule, and you would see group 2 to your left, group 3 to your right, and group 1 below you. This corresponds to an *R* configuration.

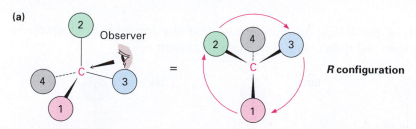

In **(b)**, you would be located behind the page toward the top left of the molecule from your point of view, and you would see group 3 to your left, group 1 to your right, and group 2 below you. This also corresponds to an *R* configuration.

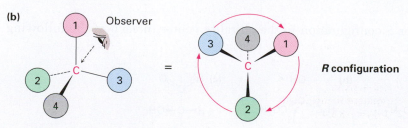

WORKED EXAMPLE 5.4   Drawing the Three-Dimensional Structure of a Specific Enantiomer

Draw a tetrahedral representation of (*R*)-2-chlorobutane.

Strategy

Begin by ranking the four substituents bonded to the chirality center: (1) –Cl, (2) –CH$_2$CH$_3$, (3) –CH$_3$, (4) –H. To draw a tetrahedral representation of the molecule, orient the lowest-ranked group (–H) away from you and imagine that the other three groups are coming out of the page toward you. Then place the remaining three substituents such that the direction of travel 1 → 2 → 3 is clockwise (right turn), and tilt the molecule toward you to bring the rear hydrogen into view. Using molecular models is a great help in working problems of this sort.

Solution

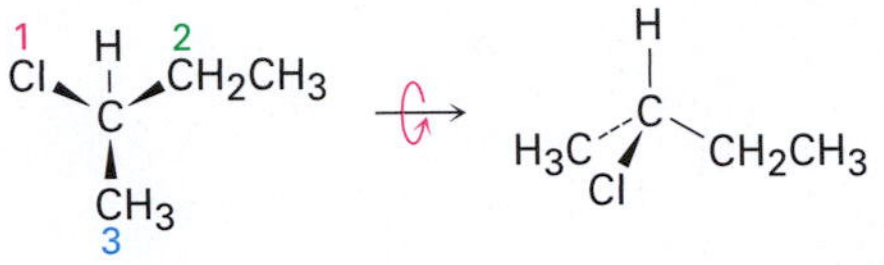

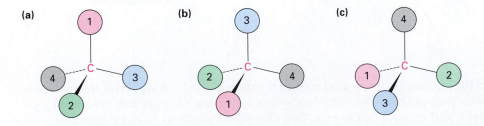

Problem 5.7

Which member in each of the following sets ranks higher?
(a) –H or –Br   **(b)** –Cl or –Br   **(c)** –CH$_3$ or –CH$_2$CH$_3$
(d) –NH$_2$ or –OH   **(e)** –CH$_2$OH or –CH$_3$ **(f)** –CH$_2$OH or –CH=O

Problem 5.8

Rank the substituents in each of the following sets according to the Cahn–Ingold–Prelog rules:
(a) –H, –OH, –CH$_2$CH$_3$, –CH$_2$CH$_2$OH
(b) –CO$_2$H, –CO$_2$CH$_3$, –CH$_2$OH, –OH
(c) –CN, –CH$_2$NH$_2$, –CH$_2$NHCH$_3$, –NH$_2$
(d) –SH, –CH$_2$SCH$_3$, –CH$_3$, –SSCH$_3$

Problem 5.9

Orient each of the following drawings so that the lowest-ranked group is toward the rear, and then assign *R* or *S* configuration:

(a)

(b)

(c)

Problem 5.10

Assign *R* or *S* configuration to the chirality center in each of the following molecules:

(a)   CH$_3$
H–C–CO$_2$H
HS

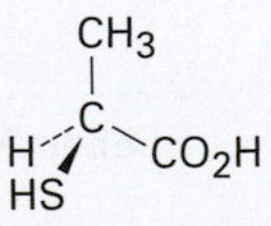

(b)   OH
H$_3$C–C–CO$_2$H
H

(c) H–C=O
H–C–OH
CH$_2$OH

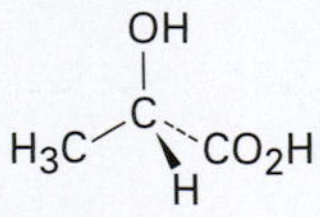

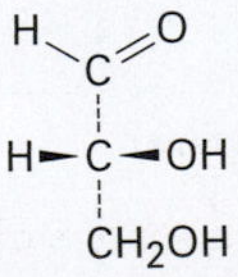

Problem 5.11
Draw a tetrahedral representation of (*S*)-pentan-2-ol (2-hydroxypentane).

Problem 5.12
Assign *R* or *S* configuration to the chirality center in the following molecular model of the amino acid methionine (yellow = S):

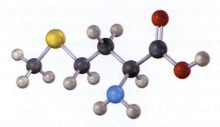

5.6 Diastereomers

Molecules like lactic acid, alanine, and glyceraldehyde are relatively simple because each has only one chirality center and only two stereoisomers. The situation becomes more complex, however, with molecules that have more than one chirality center. As a general rule, a molecule with *n* chirality centers can have up to 2^n stereoisomers (although it may have fewer, as we'll see shortly). Take the amino acid threonine (2-amino-3-hydroxybutanoic acid), for example. Since threonine has two chirality centers (C2 and C3), there are four possible stereoisomers, as shown in Figure 5.10. Check for yourself that the *R,S* configurations are correct.

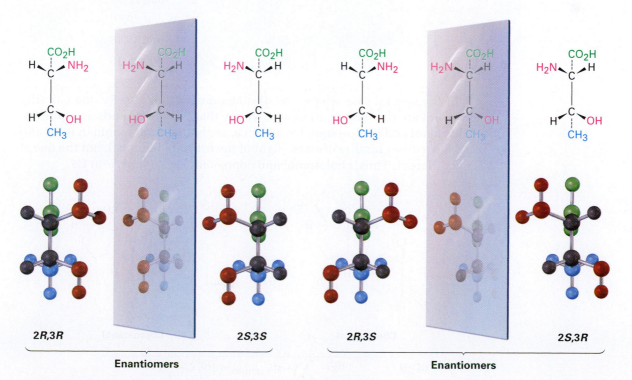

FIGURE 5.10 The four stereoisomers of 2-amino-3-hydroxybutanoic acid.

The four stereoisomers of 2-amino-3-hydroxybutanoic acid can be grouped into two pairs of enantiomers. The 2*R*,3*R* stereoisomer is the mirror image of 2*S*,3*S*, and the 2*R*,3*S* stereoisomer is the mirror image of 2*S*,3*R*. But what is the relationship between any two stereoisomers that are not mirror images? What, for instance, is the relationship between the 2*R*,3*R* isomer and the 2*R*,3*S* isomer? They are stereoisomers, yet they aren't enantiomers. To describe such a relationship, we need a new term—*diastereomer.*

Diastereomers are stereoisomers that are not mirror images. Since we used the right hand/left hand analogy to describe the relationship between two enantiomers, we might extend the analogy by saying that the relationship between diastereomers is like that of hands from different people. Your hand and your friend's hand look *similar,* but they aren't identical and they aren't mirror images. The same is true of diastereomers: they're similar, but they aren't identical and they aren't mirror images.

Note carefully the difference between enantiomers and diastereomers: enantiomers have opposite configurations at *all* chirality centers, whereas diastereomers have opposite configurations at *some* (one or more) chirality centers but the same configuration at others. A full description of the four stereoisomers of threonine is given in Table 5.2. Of the four, only the 2*S*,3*R* isomer, $[\alpha]_D = -28.3$, occurs naturally in plants and animals and is an essential human nutrient. This result is typical: most biological molecules are chiral, and usually only one stereoisomer is found in nature.

TABLE 5.2
Relationships among the Four Stereoisomers of Threonine

Stereoisomer	Enantiomer	Diastereomer
2*R*,3*R*	2*S*,3*S*	2*R*,3*S* and 2*S*,3*R*
2*S*,3*S*	2*R*,3*R*	2*R*,3*S* and 2*S*,3*R*
2*R*,3*S*	2*S*,3*R*	2*R*,3*R* and 2*S*,3*S*
2*S*,3*R*	2*R*,3*S*	2*R*,3*R* and 2*S*,3*S*

In the special case where two diastereomers differ at only one chirality center but are the same at all others, we say that the compounds are **epimers**. Cholestanol and coprostanol, for instance, are both found in human feces and both have nine chirality centers. Eight of the nine are identical, but the one at C5 is different. Thus, cholestanol and coprostanol are *epimeric* at C5.

Cholestanol   **Coprostanol**

Epimers

Problem 5.13

One of the following molecules **(a)–(d)** is D-erythrose 4-phosphate, an intermediate in the Calvin photosynthetic cycle by which plants incorporate CO_2 into carbohydrates. If D-erythrose 4-phosphate has R stereochemistry at both chirality centers, which of the structures is it? Which of the remaining three structures is the enantiomer of D-erythrose 4-phosphate, and which are diastereomers?

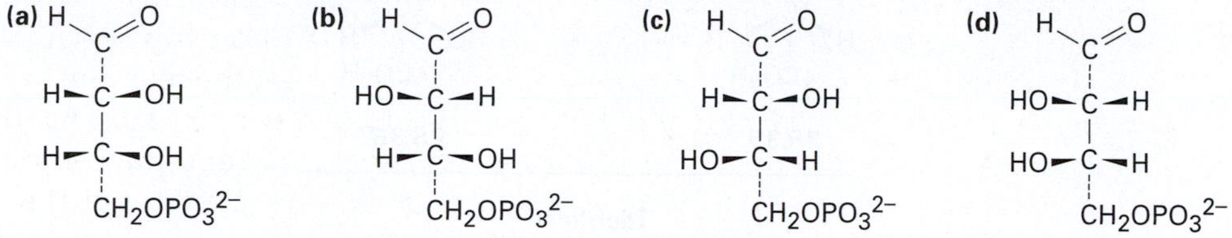

(a)

H—C=O
H—C—OH
H—C—OH
$CH_2OPO_3^{2-}$

(b)

H—C=O
HO—C—H
H—C—OH
$CH_2OPO_3^{2-}$

(c)

H—C=O
H—C—OH
HO—C—H
$CH_2OPO_3^{2-}$

(d)

H—C=O
HO—C—H
HO—C—H
$CH_2OPO_3^{2-}$

Problem 5.14

Assign R,S configuration to each chirality center in the following molecular model of the amino acid isoleucine:

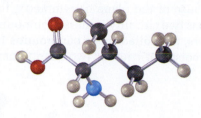

Problem 5.15

How many chirality centers does morphine have? How many stereoisomers of morphine are possible in principle?

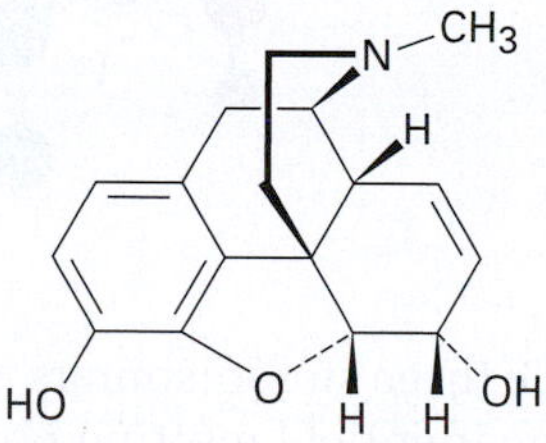

Morphine

5.7 Meso Compounds

Let's look at one more example of a compound with more than one chirality center: the tartaric acid used by Pasteur. The four stereoisomers can be drawn as follows:

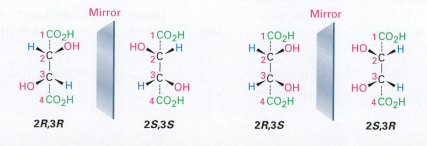

2R,3R   2S,3S   2R,3S   2S,3R

The mirror-image $2R,3R$ and $2S,3S$ structures are not identical and therefore represent a pair of enantiomers. A close look, however, shows that the $2R,3S$ and $2S,3R$ structures *are* identical, as can be seen by rotating one structure 180°:

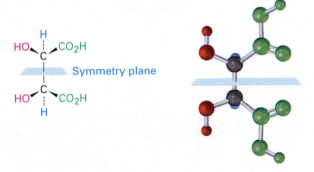

Rotate 180°

2R,3S   **2S,3R**

Identical

The $2R,3S$ and $2S,3R$ structures are identical because the molecule has a plane of symmetry and is therefore achiral. The symmetry plane cuts through the C2–C3 bond, making one half of the molecule a mirror image of the other half (Figure 5.11). Because of the plane of symmetry, the molecule is achiral despite the fact that it has two chirality centers. Compounds that are achiral, yet contain chirality centers, are called **meso compounds** (**me**-zo). Thus, tartaric acid exists in three stereoisomeric forms: two enantiomers and one meso form.

FIGURE 5.11 A symmetry plane through the C2–C3 bond of *meso*-tartaric acid makes the molecule achiral.

Some physical properties of the three stereoisomers are listed in Table 5.3. The (+)- and (−)-tartaric acids have identical melting points, solubilities, and densities but differ in the sign of their rotation of plane-polarized light. The meso isomer, by contrast, is diastereomeric with the (+) and (−) forms. As such, it has no mirror-image relationship to (+)- and (−)-tartaric acids, is a different compound altogether, and has different physical properties.

TABLE 5.3
Some Properties of the Stereoisomers of Tartaric Acid

Stereoisomer	Melting point (°C)	$[\alpha]_D$	Density (g/cm³)	Solubility at 20 °C (g/100 mL H₂O)
(+)	168–170	+12	1.7598	139.0
(−)	168–170	−12	1.7598	139.0
Meso	146–148	0	1.6660	125.0

WORKED EXAMPLE 5.5 Distinguishing Chiral Compounds from Meso Compounds

Does *cis*-1,2-dimethylcyclobutane have any chirality centers? Is it chiral?

Strategy

To see whether a chirality center is present, look for a carbon atom bonded to four different groups. To see whether the molecule is chiral, look for the presence or absence of a symmetry plane. Not all molecules with chirality centers are chiral overall—meso compounds are an exception.

Solution

A look at the structure of *cis*-1,2-dimethylcyclobutane shows that both methyl-bearing ring carbons (C1 and C2) are chirality centers. Overall, though, the compound is achiral because there is a symmetry plane bisecting the ring between C1 and C2. Thus, *cis*-1,2-dimethylcyclobutane is a meso compound.

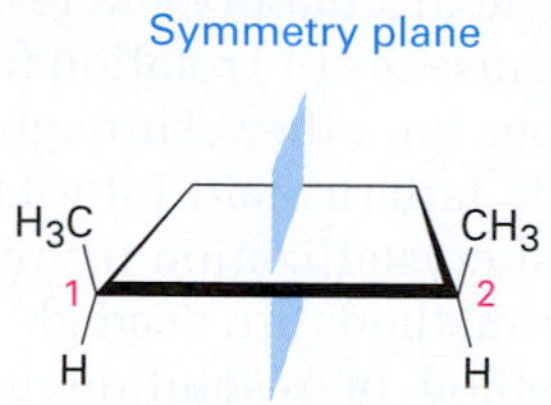

Symmetry plane

H₃C   CH₃

1   2

H   H

Problem 5.16

Which of the following structures represent meso compounds?

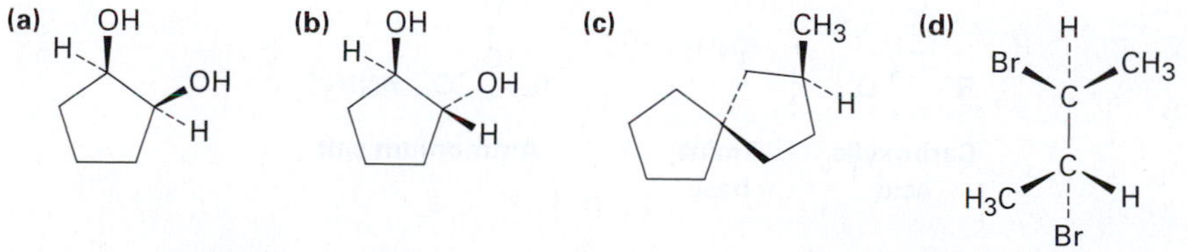

(a)   OH
H.
          OH
       H

(b)   OH
H.
          OH
         H

(c)   CH₃
                H

(d)   H
Br   CH₃
        C
H₃C   H
        C
        Br

Problem 5.17

Which of the following have a meso form? (Recall that the *-ol* suffix refers to an alcohol, ROH.)

(a) Butane-2,3-diol   **(b)** Pentane-2,3-diol   **(c)** Pentane-2,4-diol

Problem 5.18

Does the following structure represent a meso compound? If so, indicate the symmetry plane.

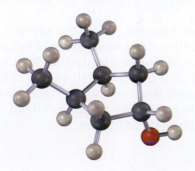

5.8 Racemic Mixtures and the Resolution of Enantiomers

Let's return for a last look at Pasteur's pioneering work described in Section 5.4. Pasteur took an optically inactive tartaric acid salt and found that he could crystallize from it two optically active forms having what we would now call the 2R,3R and 2S,3S configurations. But what was the optically inactive form he started with? It couldn't have been *meso*-tartaric acid, because *meso*-tartaric acid is a different chemical compound and can't interconvert with the two chiral enantiomers without breaking and re-forming chemical bonds.

The answer is that Pasteur started with a 50:50 *mixture* of the two chiral tartaric acid enantiomers. Such a mixture is called a **racemate** (**ra**-suh-mate) or *racemic mixture,* and is denoted by either the symbol (±) or the prefix *d,l* to indicate an equal mixture of dextrorotatory and levorotatory forms. Racemates show no optical rotation because the (+) rotation from one enantiomer exactly cancels the (−) rotation from the other. Through luck, Pasteur was able to separate, or **resolve**, racemic tartaric acid into its (+) and (−) enantiomers. Unfortunately, the fractional crystallization technique he used doesn't work for most racemates, so other methods are needed.

The most common method of resolution uses an acid–base reaction between the racemate of a chiral carboxylic acid (RCO_2H) and an amine base (RNH_2) to yield an ammonium salt:

| Carboxylic acid | Amine base | Ammonium salt |

To understand how this method of resolution works, let's see what happens when a racemic mixture of chiral acids, such as (+)- and (−)-lactic acids, reacts with an achiral amine base, such as methylamine, CH_3NH_2. Stereochemically, the situation is analogous to what happens when left and right hands (chiral) pick up a ball (achiral). Both left and right hands pick up the ball equally well, and the products—ball in right hand versus ball in left hand—are mirror images. In the same way, both (+)- and (−)-lactic acid react with methylamine equally well, and the product is a racemic mixture of methylammonium (+)-lactate and methylammonium (−)-lactate (Figure 5.12).

Now let's see what happens when the racemic mixture of (+)- and (−)-lactic acids reacts with a single enantiomer of a chiral amine base, such as (R)-1-phenylethylamine. Stereochemically, the situation is analogous to what happens when left and right hands (chiral) put on a right-handed glove *(also chiral).* Left and right hands don't put on the same glove in the same way. The products—right hand in right glove versus left hand in right glove—are not mirror images; they're altogether different.

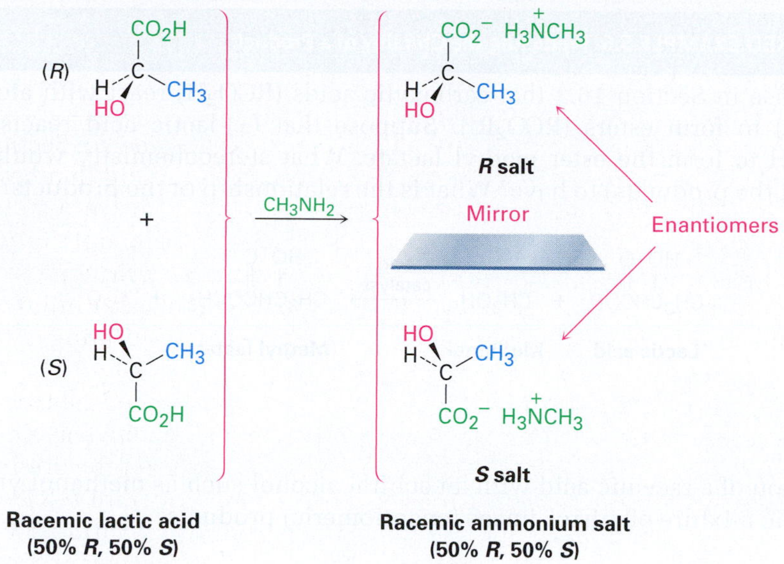

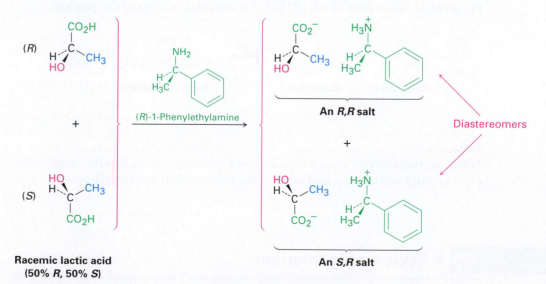

FIGURE 5.12 Reaction of racemic lactic acid with achiral methylamine leads to a racemic mixture of ammonium salts.

In the same way, (+)- and (−)-lactic acids react with (*R*)-1-phenylethyl-amine to give two different products (Figure 5.13). (*R*)-Lactic acid reacts with (*R*)-1-phenylethylamine to give the *R,R* salt, and (*S*)-lactic acid reacts with the *R* amine to give the *S,R* salt. *The two salts are diastereomers;* they are different compounds, with different chemical and physical properties. It may therefore be possible to separate them by crystallization or some other means. Once separated, acidification of the two diastereomeric salts with a strong acid then allows us to isolate the two pure enantiomers of lactic acid and to recover the chiral amine for reuse.

FIGURE 5.13 Reaction of racemic lactic acid with (*R*)-1-phenylethylamine yields a mixture of diastereomeric ammonium salts, which have different properties and can be separated.

WORKED EXAMPLE 5.6 Predicting the Chirality of a Product

We'll see in Section 16.3 that carboxylic acids (RCO_2H) react with alcohols ($R'OH$) to form esters (RCO_2R'). Suppose that (±)-lactic acid reacts with CH_3OH to form the ester methyl lactate. What stereochemistry would you expect the product(s) to have? What is the relationship of the products?

$$CH_3CHCOH + CH_3OH \xrightarrow[\text{catalyst}]{\text{Acid}} CH_3CHCOCH_3 + H_2O$$

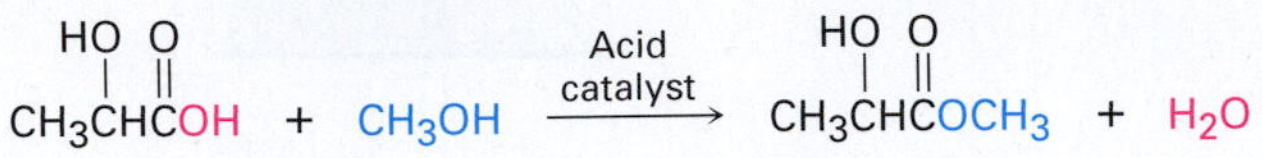

Lactic acid   Methanol   Methyl lactate

Solution

Reaction of a racemic acid with an achiral alcohol such as methanol yields a racemic mixture of mirror-image (enantiomeric) products:

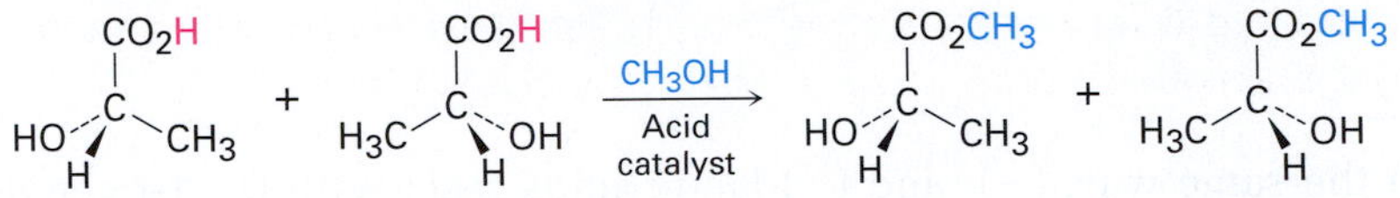

(S)-Lactic acid   (R)-Lactic acid   Methyl   Methyl
                                                    (S)-lactate   (R)-lactate

---

Problem 5.19

Suppose that acetic acid (CH_3CO_2H) reacts with (S)-butan-2-ol to form an ester (see Worked Example 5.6). What stereochemistry would you expect the product(s) to have, assuming that the singly bonded oxygen atom comes from the alcohol rather than the acid? What is the relationship of the products?

$$CH_3COH + CH_3CHCH_2CH_3 \xrightarrow[\text{catalyst}]{\text{Acid}} CH_3COCHCH_2CH_3 + H_2O$$

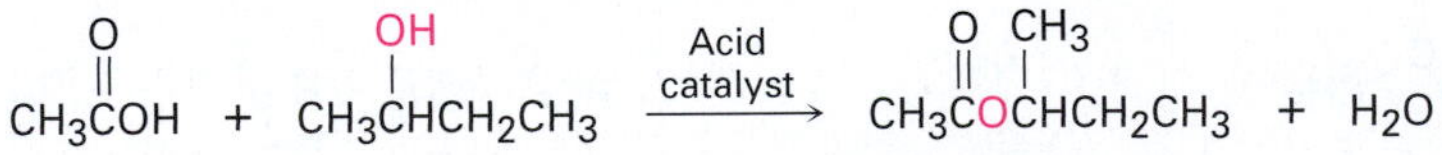

Acetic acid   Butan-2-ol   *sec*-Butyl acetate

Problem 5.20

What stereoisomers would result from reaction of (±)-lactic acid with (S)-1-phenylethylamine, and what is the relationship between them?

5.9 A Review of Isomerism

As noted on several previous occasions, isomers are compounds that have the same chemical formula but different structures. We've seen several kinds of isomers in the past few chapters, and it's a good idea at this point to see how they relate to one another (Figure 5.14).

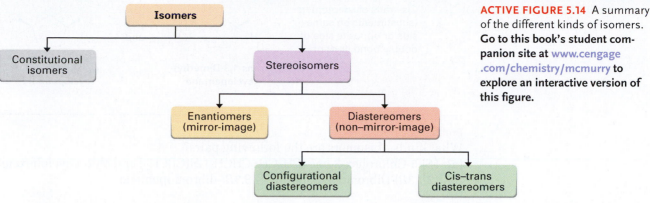

ACTIVE FIGURE 5.14 A summary of the different kinds of isomers. **Go to this book's student companion site at** www.cengage .com/chemistry/mcmurry **to explore an interactive version of this figure.**

There are two fundamental types of isomers, both of which we've now encountered: constitutional isomers and stereoisomers.

- **Constitutional isomers** (Section 3.2) are compounds whose atoms are connected differently. Among the kinds of constitutional isomers we've seen are skeletal, functional, and positional isomers.

Different carbon skeletons	CH$_3$ $\mid$ CH$_3$CHCH$_3$ **2-Methylpropane**	and	CH$_3$CH$_2$CH$_2$CH$_3$ **Butane**
Different functional groups	CH$_3$CH$_2$OH **Ethyl alcohol**	and	CH$_3$OCH$_3$ **Dimethyl ether**
Different position of functional groups	NH$_2$ $\mid$ CH$_3$CHCH$_3$ **Isopropylamine**	and	CH$_3$CH$_2$CH$_2$NH$_2$ **Propylamine**

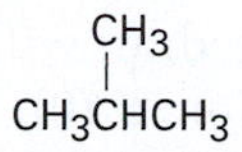

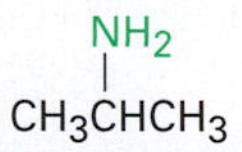

- **Stereoisomers** (Section 4.2) are compounds whose atoms are connected in the same order but with a different arrangement in space. Among the kinds of stereoisomers we've seen are enantiomers, diastereomers, and cis–trans isomers of cycloalkanes. Actually, cis–trans isomers are just one class of diastereomers because they are non–mirror-image stereoisomers:

Enantiomers (nonsuperimposable mirror-image stereoisomers)

Diastereomers (nonsuperimposable non–mirror-image stereoisomers)

Configurational diastereomers

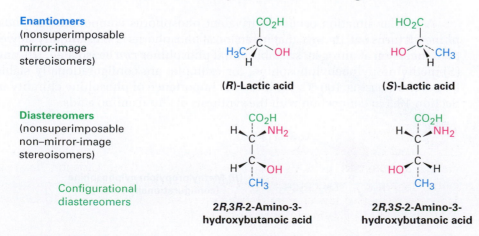

(**R**)-Lactic acid   (**S**)-Lactic acid

2R,3R-2-Amino-3-hydroxybutanoic acid   **2R,3S-2-Amino-3-hydroxybutanoic acid**

Cis–trans diastereomers
(substituents on same
side or opposite side of
double bond or ring)

**trans-1,3-Dimethyl-
cyclopentane**   and   **cis-1,3-Dimethyl-
cyclopentane**

Problem 5.21

What kinds of isomers are the following pairs?

(a) (S)-5-Chlorohex-2-ene [$CH_3CH=CHCH_2CH(Cl)CH_3$] and chlorocyclohexane

(b) (2R,3R)-Dibromopentane and (2S,3R)-dibromopentane

5.10 Chirality at Nitrogen, Phosphorus, and Sulfur

Although the most common cause of chirality is the presence of four different substituents bonded to a tetrahedral atom, that atom doesn't necessarily have to be carbon. Nitrogen, phosphorus, and sulfur are all commonly encountered in organic molecules, and all can be chirality centers. We know, for instance, that trivalent nitrogen is tetrahedral, with its lone pair of electrons acting as the fourth "substituent" (Section 1.10). Is trivalent nitrogen chiral? Does a compound such as ethylmethylamine exist as a pair of enantiomers?

The answer is both yes and no. Yes in principle, but no in practice. Trivalent nitrogen compounds undergo a rapid umbrella-like inversion that interconverts enantiomers. We therefore can't isolate individual enantiomers except in special cases.

A similar situation occurs in trivalent phosphorus compounds, or *phosphines*. It turns out, though, that inversion at phosphorus is substantially slower than inversion at nitrogen, so stable chiral phosphines *can* be isolated. (R)- and (S)-methylpropylphenylphosphine, for example, are configurationally stable for several hours at 100 °C. We'll see the importance of phosphine chirality in Section 19.3 in connection with the synthesis of chiral amino acids.

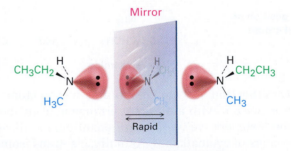

Lowest ranked

**(R)-Methylpropylphenylphosphine
(configurationally stable)**

Divalent sulfur compounds are achiral, but trivalent sulfur compounds called *sulfonium salts* (R_3S^+) can be chiral. Like phosphines, sulfonium salts undergo relatively slow inversion, so chiral sulfonium salts are configurationally stable and can be isolated. Perhaps the best known example is the coenzyme *S*-adenosylmethionine, the so-called biological methyl donor, which is involved in many metabolic pathways as a source of CH_3 groups. (The "*S*" in the name *S*-adenosylmethionine stands for *sulfur* and means that the adenosyl group is attached to the sulfur atom of methionine.) The molecule has *S* stereochemistry at sulfur and is configurationally stable for several days at room temperature. Its *R* enantiomer is also known but has no biological activity.

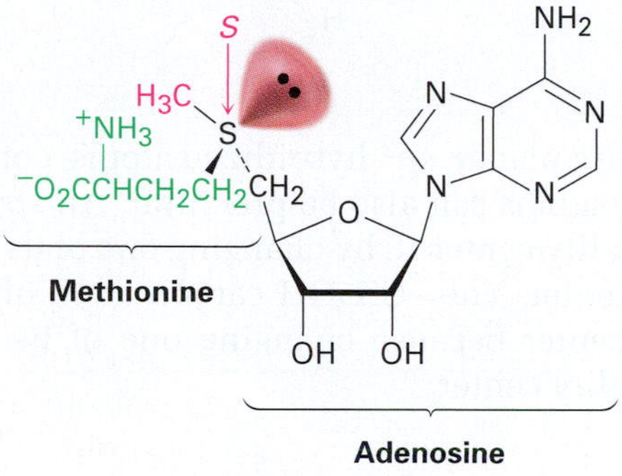

(*S*)-*S*-Adenosylmethionine

5.11 Prochirality

Closely related to the concept of chirality, and particularly important in biological chemistry, is the notion of *prochirality*. A molecule is said to be **prochiral** if it can be converted from achiral to chiral in a single chemical step. For instance, an unsymmetrical ketone like butan-2-one is prochiral because it can be converted to the chiral alcohol butan-2-ol by addition of hydrogen, as we'll see in Section 13.3.

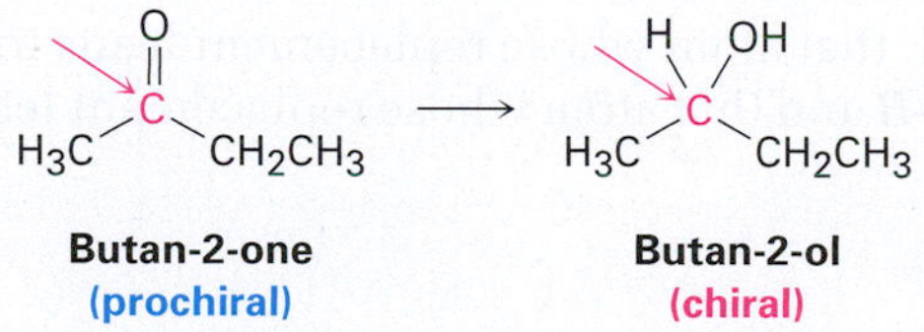

Which enantiomer of butan-2-ol is produced depends on which face of the planar carbonyl group undergoes reaction. To distinguish between the possibilities, we use the stereochemical descriptors *Re* and *Si*. Rank the three groups attached to the trigonal, sp^2-hybridized carbon, and imagine curved arrows from the highest to second-highest to third-highest ranked substituents. The face on which the arrows curve clockwise is designated **Re** (similar to *R*), and the face on which the arrows curve counterclockwise is designated **Si** (similar to *S*). In this particular example, addition of hydrogen from the *Re* faces gives (*S*)-butan-2-ol, and addition from the *Si* face gives (*R*)-butan-2-ol.

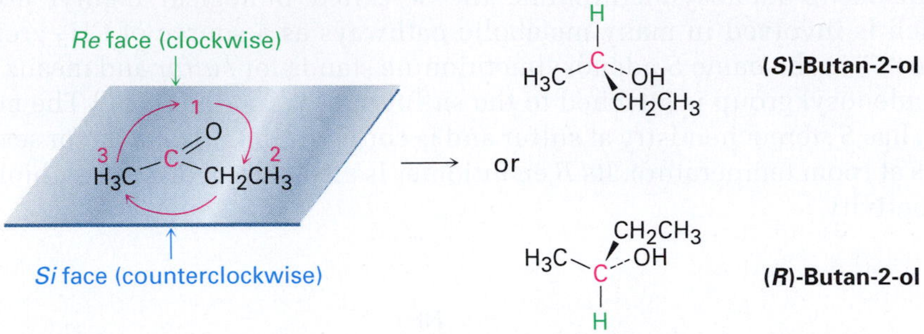

In addition to compounds with planar, sp^2-hybridized atoms, compounds with tetrahedral, sp^3-hybridized atoms can also be prochiral. An sp^3-hybridized atom is said to be a **prochirality center** if, by changing one of its attached groups, it becomes a chirality center. The $-CH_2OH$ carbon atom of ethanol, for instance, is a prochirality center because changing one of its attached $-H$ atoms converts it into a chirality center.

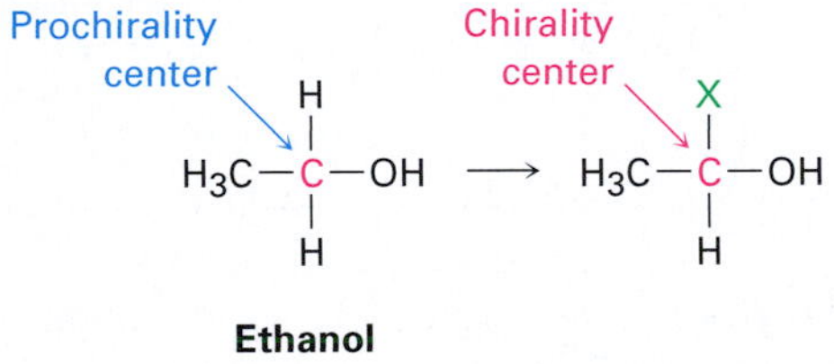

Ethanol

To distinguish between the two identical atoms (or groups of atoms) on a prochirality center, we imagine a change that will raise the ranking of one atom over the other without affecting its rank with respect to other attached groups. On the $-CH_2OH$ carbon of ethanol, for instance, we might imagine replacing one of the ^1H atoms (protium) by ^2H (deuterium). The newly introduced ^2H atom ranks higher than the remaining ^1H atom, but it remains lower than other groups attached to the carbon. Of the two identical atoms in the original compound, that atom whose replacement leads to an *R* chirality center is said to be **pro-R** and that atom whose replacement leads to an *S* chirality center is **pro-S**.

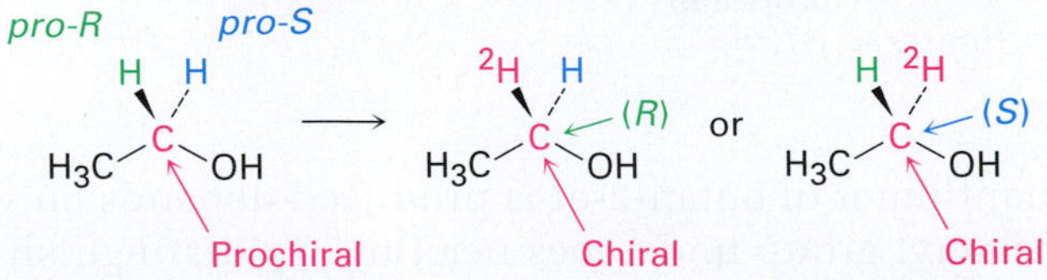

A large number of biological reactions involve prochiral compounds. One of the steps in the citric acid cycle by which food is metabolized, for instance, is the addition of H_2O to fumarate to give malate. Addition of $-OH$ occurs on the *Si* face of a fumarate carbon and gives (*S*)-malate as product.

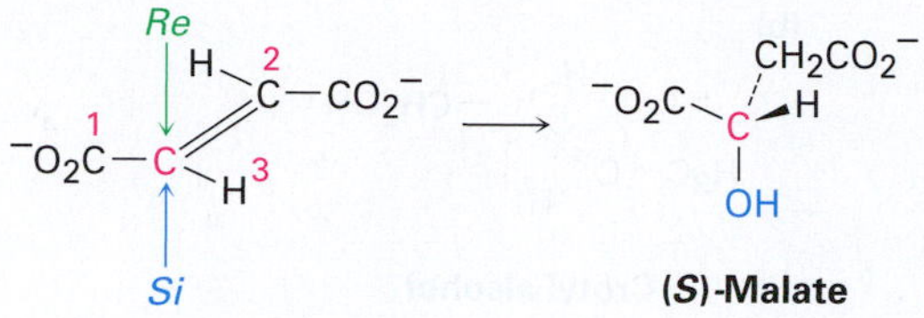

(*S*)-**Malate**

As another example, studies with deuterium-labeled substrates have shown that the reaction of ethanol with the coenzyme nicotinamide adenine dinucleotide (NAD$^+$) catalyzed by yeast alcohol dehydrogenase occurs with exclusive removal of the *pro-R* hydrogen from ethanol and with addition only to the *Re* face of NAD$^+$.

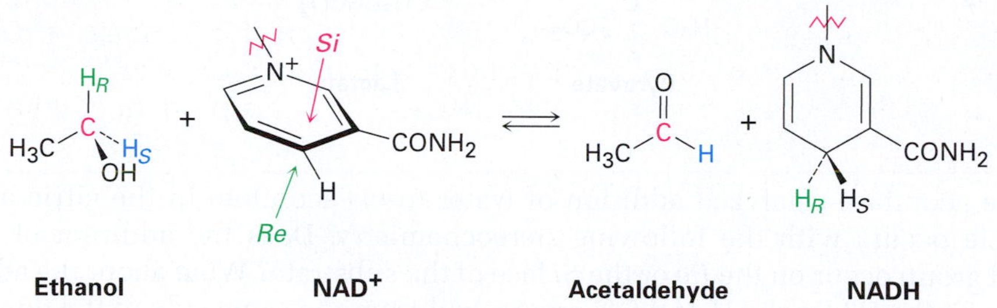

Ethanol   **NAD$^+$**   **Acetaldehyde**   **NADH**

Elucidating the stereochemistry of reactions at prochirality centers is a powerful method for studying detailed mechanisms in biochemical reactions. As just one example, the conversion of citrate to (*cis*)-aconitate in the citric acid cycle has been shown to occur with loss of a *pro-R* hydrogen, implying that the OH and H groups leave from opposite sides of the molecule.

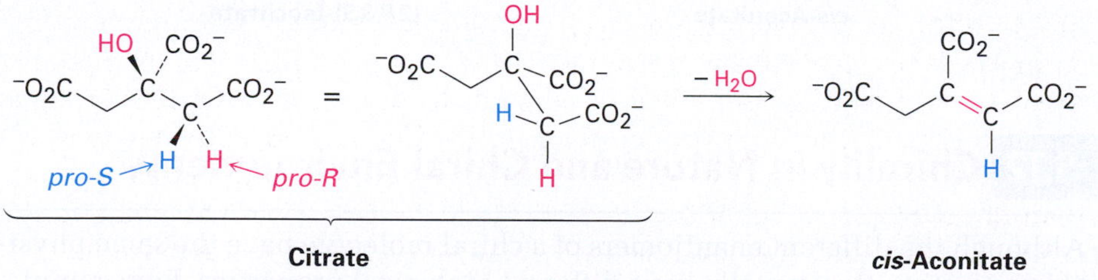

Citrate   *cis*-**Aconitate**

Problem 5.22

Identify the indicated hydrogens in the following molecules as *pro-R* or *pro-S*:

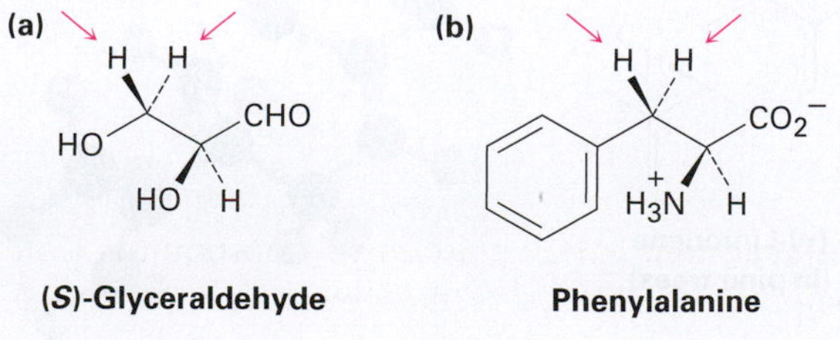

(a)   (b)

(*S*)-**Glyceraldehyde**   **Phenylalanine**

Problem 5.23

Identify the indicated faces in the following molecules as *Re* or *Si:*

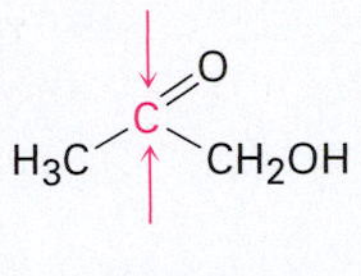

(a)

H3C—C(=O)—CH2OH

Hydroxyacetone

(b)

H3C—C(H)=C(H)—CH2OH

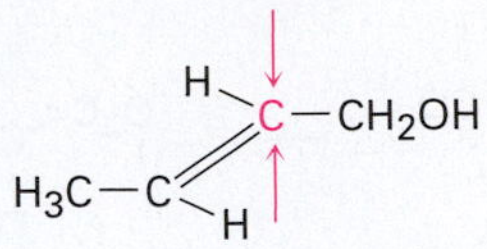

Crotyl alcohol

Problem 5.24

The lactic acid that builds up in tired muscles is formed from pyruvate. If the reaction occurs with addition of hydrogen to the *Re* face of pyruvate, what is the stereochemistry of the product?

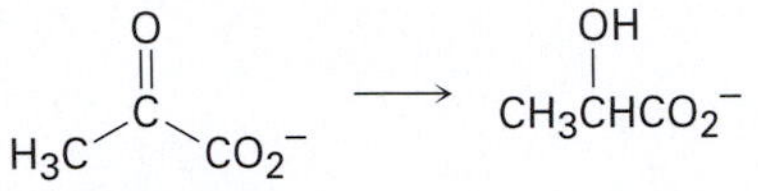

$$H_3C-C(=O)-CO_2^- \longrightarrow CH_3CH(OH)CO_2^-$$

Pyruvate   **Lactate**

Problem 5.25

The aconitase-catalyzed addition of water to *cis*-aconitate in the citric acid cycle occurs with the following stereochemistry. Does the addition of the OH group occur on the *Re* or the *Si* face of the substrate? What about the addition of the H? Do the H and OH groups add from the same side of the double bond or from opposite sides?

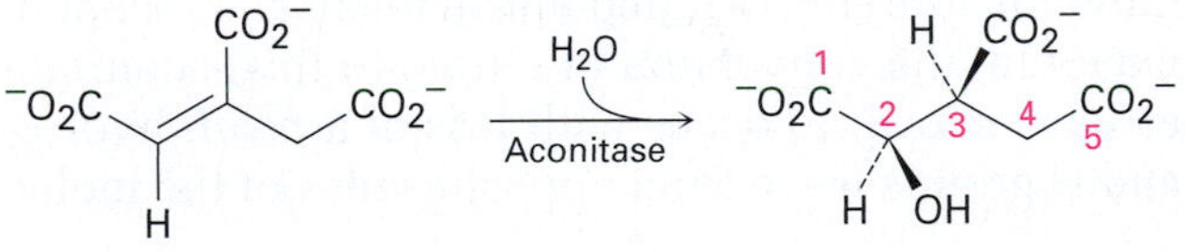

$$^-O_2C-C(CO_2^-)=CH-CO_2^- \xrightarrow[\text{Aconitase}]{H_2O} {}^-O_2C-CH(OH)-CH(CO_2^-)-CO_2^-$$

cis-Aconitate   **(2R,3S)-Isocitrate**

5.12 Chirality in Nature and Chiral Environments

Although the different enantiomers of a chiral molecule have the same physical properties, they usually have different biological properties. For example, the (+) enantiomer of limonene has the odor of oranges and lemons, but the (−) enantiomer has the odor of pine trees.

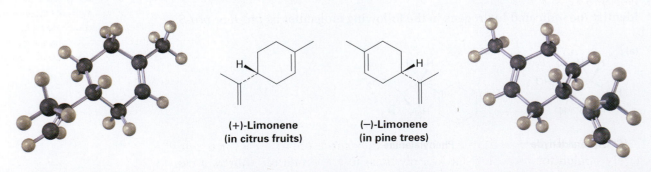

(+)-Limonene
(in citrus fruits)

(−)-Limonene
(in pine trees)

More dramatic examples of how a change in chirality can affect the biological properties of a molecule are found in many drugs, such as fluoxetine, a heavily prescribed medication sold under the trade name Prozac. Racemic fluoxetine is an extraordinarily effective antidepressant but has no activity against migraine. The pure *S* enantiomer, however, works remarkably well in preventing migraine. Other examples of how chirality affects biological properties are given in the *Lagniappe* at the end of this chapter.

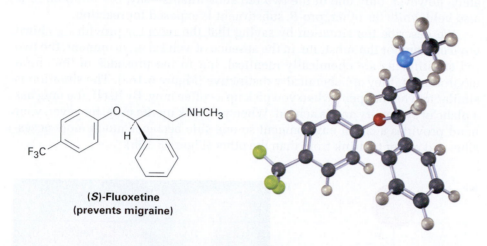

**(S)-Fluoxetine
(prevents migraine)**

Why do different enantiomers have different biological properties? To have a biological effect, a substance typically must fit into an appropriate receptor that has an exactly complementary shape. But because biological receptors are chiral, only one enantiomer of a chiral substrate can fit in, just as only a right hand will fit into a right-handed glove. The mirror-image enantiomer will be a misfit, like a left hand in a right-handed glove. A representation of the interaction between a chiral molecule and a chiral biological receptor is shown in Figure 5.15. One enantiomer fits the receptor perfectly, but the other does not.

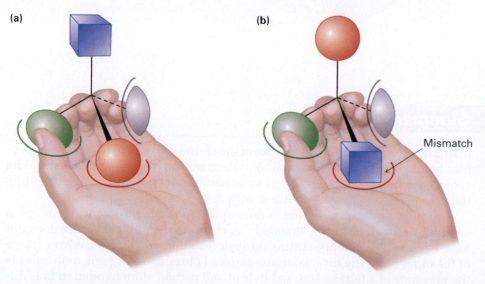

FIGURE 5.15 Imagine that a left hand interacts with a chiral object, much as a biological receptor interacts with a chiral molecule. **(a)** One enantiomer fits into the hand perfectly: green thumb, red palm, and gray pinkie finger, with the blue substituent exposed. **(b)** The other enantiomer, however, can't fit into the hand. When the green thumb and gray pinkie finger interact appropriately, the palm holds a blue substituent rather than a red one, with the red substituent exposed.

The hand-in-glove fit of a chiral substrate into a chiral receptor is relatively straightforward, but it's less obvious how a prochiral substrate can undergo a selective reaction. Take the reaction of ethanol with NAD$^+$ catalyzed by yeast alcohol dehydrogenase. As we saw at the end of Section 5.11, the reaction occurs with exclusive removal of the *pro-R* hydrogen from ethanol and with addition only to the *Re* face of the NAD$^+$ carbon.

We can understand this result by imagining that the chiral enzyme receptor again has three binding sites, as was previously the case in Figure 5.15. When green and gray substituents of a prochiral substrate are held appropriately, however, only one of the two red substituents—say, the *pro-S* one—is also held while the other, *pro-R*, substituent is exposed for reaction.

We describe the situation by saying that the receptor provides a **chiral environment** for the substrate. In the absence of a chiral environment, the two red substituents are chemically identical, but in the presence of the chiral environment, they are chemically distinctive (Figure 5.16a). The situation is similar to what happens when you pick up a coffee mug. By itself, the mug has a plane of symmetry and is achiral. When you pick up the mug, however, your hand provides a chiral environment so one side becomes much more accessible and easier to drink from than the other (Figure 5.16b).

FIGURE 5.16 (a) When a prochiral molecule is held in a chiral environment, the two seemingly identical substituents (red) are distinguishable. **(b)** Similarly, when an achiral coffee mug is held in the chiral environment of your hand, it's much easier to drink from one side than the other because the two sides of the mug are now distinguishable.

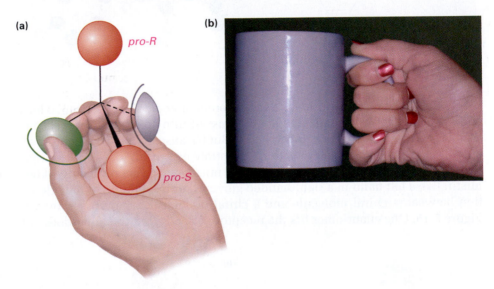

(a)   pro-R

pro-S

(b)

Summary

In this chapter, we've looked at some of the causes and consequences of molecular handedness—a topic crucial to understanding biological chemistry. The subject can be a bit complex, but is so important that it's worthwhile spending the time needed to become familiar with it.

An object or molecule that is not superimposable on its mirror image is said to be **chiral**, meaning "handed." A chiral molecule is one that does not have a plane of symmetry cutting through it so that one half is a mirror image of the other half. The most common cause of chirality in organic molecules is the presence of a tetrahedral, sp^3-hybridized carbon atom bonded to four different groups—a so-called **chirality center**. Chiral compounds can exist as a pair of nonsuperimposable mirror-image stereoisomers called **enantiomers**.

Key Words

absolute configuration, 147
achiral, 137
Cahn–Ingold–Prelog rules, 143
chiral, 136
chiral environment, 164
chirality center, 137
configuration, 143
dextrorotatory, 140
diastereomers, 150
enantiomers, 135

Enantiomers are identical in all physical properties except for their **optical activity**, or direction in which they rotate plane-polarized light.

The stereochemical **configuration** of a carbon atom can be specified as either **R** *(rectus)* or **S** *(sinister)* by using the **Cahn–Ingold–Prelog rules**. First rank the four substituents on the chiral carbon atom, and then orient the molecule so that the lowest-ranked group points directly back. If a curved arrow drawn in the direction of decreasing rank ($1 \rightarrow 2 \rightarrow 3$) for the remaining three groups is clockwise, the chirality center has the *R* configuration. If the direction is counterclockwise, the chirality center has the *S* configuration.

Some molecules have more than one chirality center. Enantiomers have opposite configuration at all chirality centers, whereas **diastereomers** have the same configuration in at least one center but opposite configurations at the others. **Epimers** are diastereomers that differ in configuration at only one chirality center. A compound with n chirality centers can have a maximum of 2^n stereoisomers.

Meso compounds contain chirality centers but are achiral overall because they have a plane of symmetry. Racemic mixtures, or **racemates**, are 50:50 mixtures of $(+)$ and $(-)$ enantiomers. Racemates and individual diastereomers differ in their physical properties, such as solubility, melting point, and boiling point.

A molecule is **prochiral** if can be converted from achiral to chiral in a single chemical step. A prochiral sp^2-hybridized atom has two faces, described as either **Re** or **Si**. An sp^3-hybridized atom is a **prochirality center** if, by changing one of its attached atoms, a chirality center results. The atom whose replacement leads to an *R* chirality center is **pro-R**, and the atom whose replacement leads to an *S* chirality center is **pro-S**.

Lagniappe

Chiral Drugs

The *S* enantiomer of ibuprofen soothes the aches and pains of athletic injuries much more effectively than the *R* enantiomer.

The hundreds of different pharmaceutical agents approved for use by the U.S. Food and Drug Administration come from many sources (see the *Lagniappe* in the next chapter). Many drugs are isolated directly from plants or bacteria, and others are made by chemical modification of naturally occurring compounds, but an estimated 33% are made entirely in the laboratory and have no relatives in nature.

Those drugs that come from natural sources, either directly or after chemical modification, are usually chiral and are generally found only as a single enantiomer rather than as a racemate. Penicillin V, for example, an antibiotic isolated from the *Penicillium* mold, has the 2*S*,5*R*,6*R* configuration. Its enantiomer, which does not occur naturally but can be made in the laboratory, has no antibiotic activity.

Penicillin V (2*S*,5*R*,6*R* configuration)

In contrast to drugs from natural sources, those drugs that are made entirely in the laboratory either are achiral or, if chiral, are often produced and sold as racemic mixtures. Ibuprofen, for example, has one chirality center and is sold commercially under such trade names as Advil, Nuprin, and Motrin as a 50:50 mixture of *R* and *S*. It turns out, however, that only the *S* enantiomer is active as an

continued

© Heath Robbins/Photanica/Getty Images

Lagniappe *continued*

analgesic and anti-inflammatory agent. The *R* enantiomer of ibuprofen is inactive, although it is slowly converted in the body to the active *S* form.

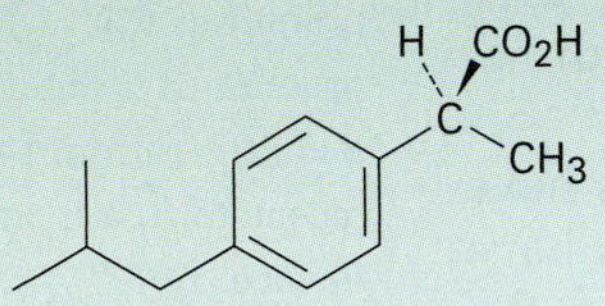

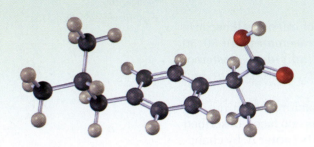

**(S)-Ibuprofen
(an active analgesic agent)**

Not only is it chemically wasteful to synthesize and administer an enantiomer that does not serve the intended purpose, many examples are now known where the presence of the "wrong" enantiomer in a racemic mixture either affects the body's ability to utilize the "right" enantiomer or has unintended pharmacological effects of its own. The presence of (*R*)-ibuprofen in the racemic mixture, for instance, slows substantially the rate at which the *S* enantiomer takes effect in the body, from 12 minutes to 38 minutes.

To get around this problem, pharmaceutical companies attempt to devise methods of *enantioselective synthesis,* which allow them to prepare only a single enantiomer rather than a racemic mixture. Viable methods have been developed for the preparation of (*S*)-ibuprofen, which is now being marketed in Europe. We'll look further into enantioselective synthesis in the Chapter 14 *Lagniappe.*

Exercises

indicates problems that are assignable in Organic OWL.

Go to this book's companion website at **www.cengage.com/chemistry/mcmurry** to explore interactive versions of the Active Figures from this text.

VISUALIZING CHEMISTRY

(Problems 5.1–5.25 appear within the chapter.)

5.26 ■ Which of the following structures are identical? (Yellow-green = Cl.)

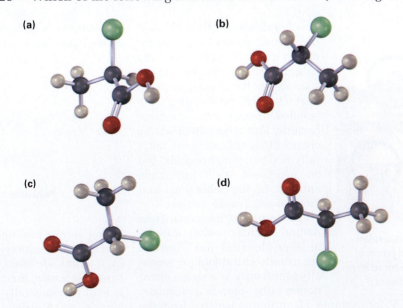

(a)

(b)

(c)

(d)

5.27 ■ Assign *R* or *S* configuration to the chirality centers in the following molecules:

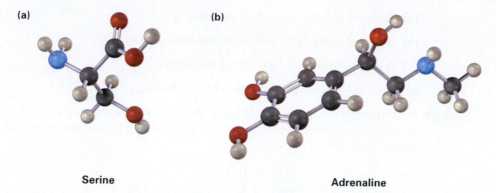

(a)

(b)

Serine

Adrenaline

5.28 ■ Which, if any, of the following structures represent meso compounds? (Yellow-green = Cl.)

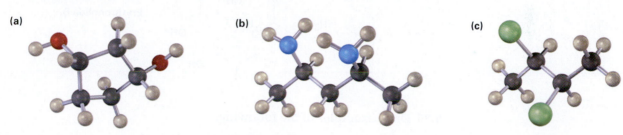

(a)

(b)

(c)

5.29 ■ Assign *R* or *S* configuration to each chirality center in pseudo-ephedrine, an over-the-counter decongestant found in cold remedies.

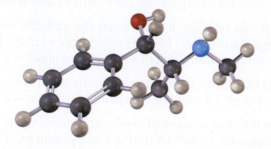

ADDITIONAL PROBLEMS

5.30 ■ Which of the following compounds are chiral? Draw them, and label the chirality centers.

    (a) 2,4-Dimethylheptane   **(b)** 5-Ethyl-3,3-dimethylheptane

    (c) *cis*-1,4-Dichlorocyclohexane   **(d)** *trans*-1,4-Dimethylcyclohexane

5.31 Draw chiral molecules that meet the following descriptions:

    (a) A chloroalkane, $C_5H_{11}Cl$ **(b)** An alcohol, $C_6H_{14}O$

    (c) An alkene, C_6H_{12} **(d)** An alkane, C_8H_{18}

■ Problems assignable in Organic OWL.

5.32 Eight alcohols have the formula $C_5H_{12}O$. Draw them. Which are chiral?

5.33 ▪ Draw compounds that fit the following descriptions:

(a) A chiral alcohol with four carbons

(b) A chiral carboxylic acid with the formula $C_5H_{10}O_2$

(c) A compound with two chirality centers

(d) A chiral hydroxy aldehyde with the formula $C_3H_6O_2$

5.34 ▪ Erythronolide B is the biological precursor of erythromycin, a broad-spectrum antibiotic. How many chirality centers does erythronolide B have?

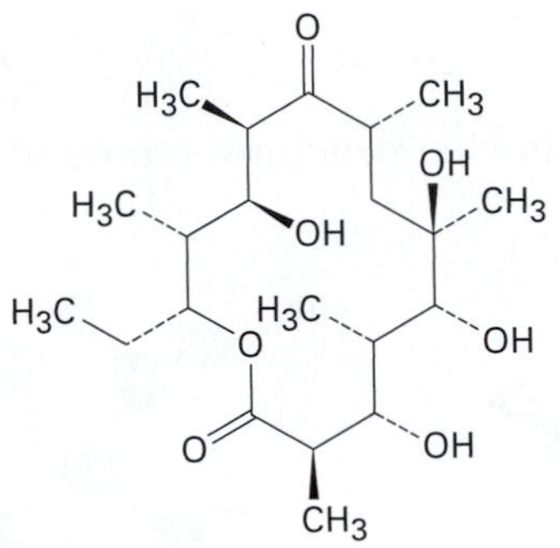

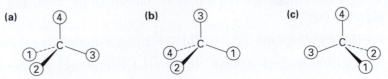

Erythronolide B

5.35 Draw examples of the following:

(a) A meso compound with the formula C_8H_{18}

(b) A meso compound with the formula C_9H_{20}

(c) A compound with two chirality centers, one R and the other S

5.36 ▪ What is the relationship between the specific rotations of $(2R,3R)$-dichloropentane and $(2S,3S)$-dichloropentane? Between $(2R,3S)$-dichloropentane and $(2R,3R)$-dichloropentane?

5.37 ▪ What is the stereochemical configuration of the enantiomer of $(2S,4R)$-octane-2,4-diol? (A diol is a compound with two –OH groups.)

5.38 What are the stereochemical configurations of the two diastereomers of $(2S,4R)$-octane-2,4-diol? (A diol is a compound with two –OH groups.)

5.39 Orient each of the following drawings so that the lowest-ranked group is toward the rear, and then assign R or S configuration:

(a)   (b)   (c)

5.40 ■ Assign Cahn–Ingold–Prelog rankings to the following sets of substituents:

(a) —CH=CH$_2$, —CH(CH$_3$)$_2$, —C(CH$_3$)$_3$, —CH$_2$CH$_3$

(b) —C≡CH, —CH=CH$_2$, —C(CH$_3$)$_3$,

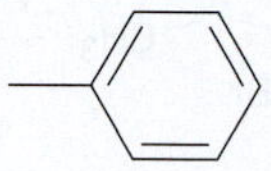

(c) —CO$_2$CH$_3$, —COCH$_3$, —CH$_2$OCH$_3$, —CH$_2$CH$_3$

(d) —C≡N, —CH$_2$Br, —CH$_2$CH$_2$Br, —Br

5.41 ■ Assign R or S configurations to the chirality centers in the following molecules:

(a) 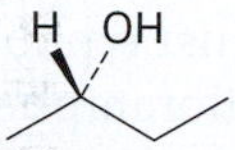 H OH

(b) 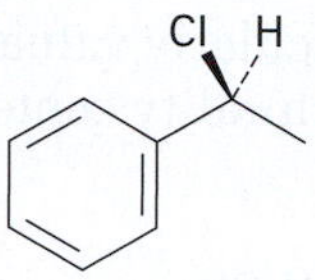 Cl H

(c) 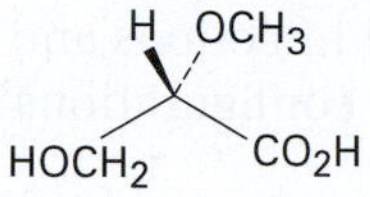 H OCH$_3$ / HOCH$_2$ CO$_2$H

5.42 ■ Assign R or S configuration to each chirality center in the following molecules:

(a) 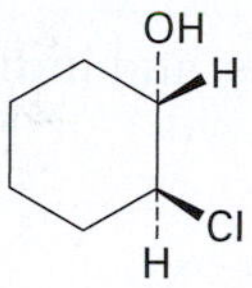 OH H / Cl H

(b) 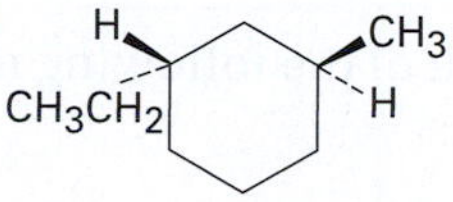 H CH$_3$ / CH$_3$CH$_2$ H

(c) 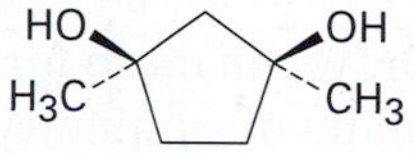 HO OH / H$_3$C CH$_3$

5.43 ■ Assign R or S configuration to each chirality center in the following biological molecules:

(a)

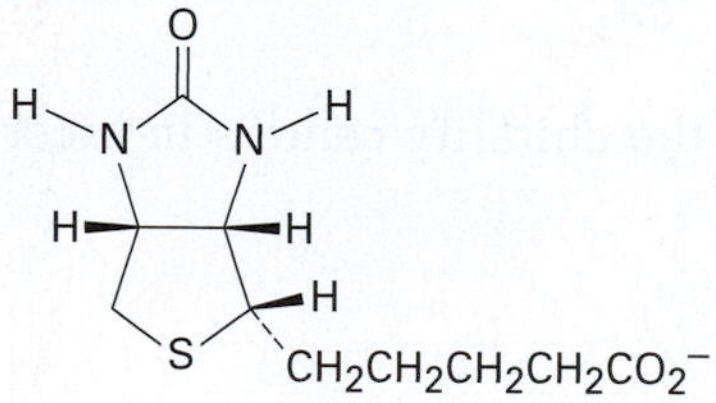

Biotin

(b)

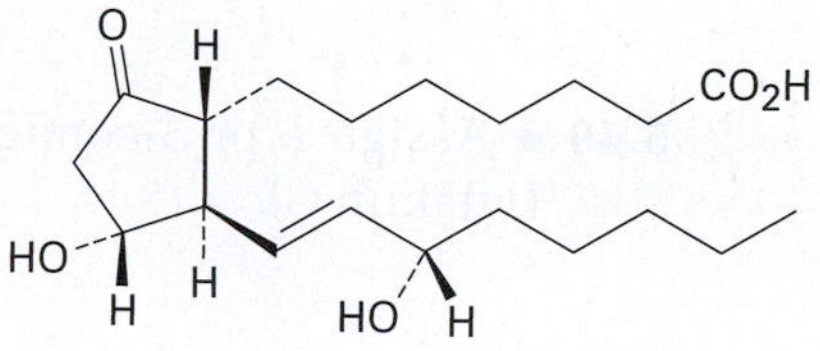

Prostaglandin E$_1$

5.44 ■ Draw tetrahedral representations of the two enantiomers of the amino acid cysteine, HSCH$_2$CH(NH$_2$)CO$_2$H, and identify each as R or S.

5.45 The naturally occurring form of the amino acid cysteine (Problem 5.44) has the S configuration at its chirality center. On treatment with a mild oxidizing agent, two cysteines join to give cystine, a disulfide. Assuming that the chirality center is not affected by the reaction, is cystine optically active?

$$2 \text{ HSCH}_2\overset{\overset{\displaystyle NH_2}{|}}{\text{C}}\text{HCO}_2\text{H} \longrightarrow \text{HO}_2\text{C}\overset{\overset{\displaystyle NH_2}{|}}{\text{C}}\text{HCH}_2\text{S}-\text{SCH}_2\overset{\overset{\displaystyle NH_2}{|}}{\text{C}}\text{HCO}_2\text{H}$$

Cysteine   **Cystine**

■ Problems assignable in Organic OWL.

5.46 ▪ Which of the following pairs of structures represent the same enantiomer, and which represent different enantiomers?

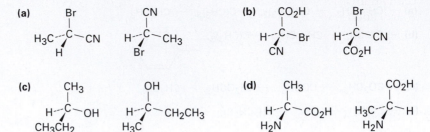

5.47 ▪ Chloramphenicol, a powerful antibiotic isolated in 1949 from the *Streptomyces venezuelae* bacterium, is active against a broad spectrum of bacterial infections and is particularly valuable against typhoid fever. Assign *R,S* configurations to the chirality centers in chloramphenicol.

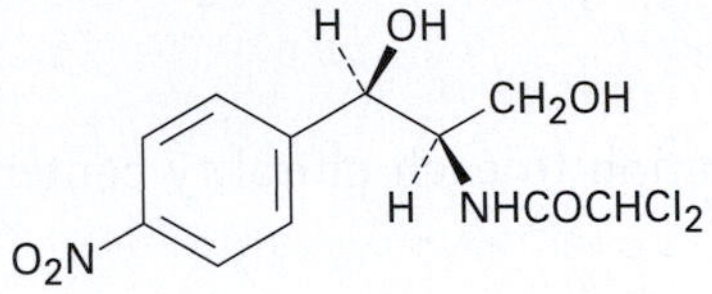

Chloramphenicol

5.48 Draw the meso form of each of the following molecules, and indicate the plane of symmetry in each:

(a)

$$CH_3CHCH_2CH_2CHCH_3$$
with OH, OH above

(b) CH₃ ... CH₃

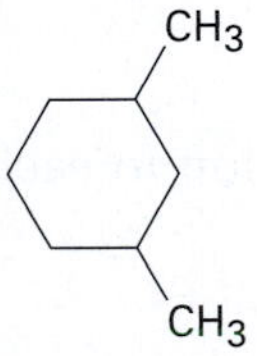

(c) H₃C ... H₃C ... OH

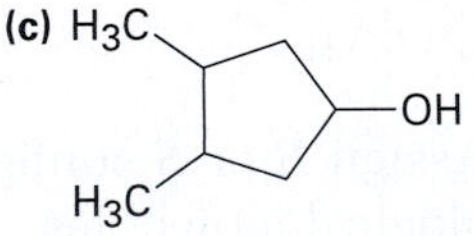

5.49 ▪ Assign *R* or *S* configurations to the chirality centers in ascorbic acid (vitamin C).

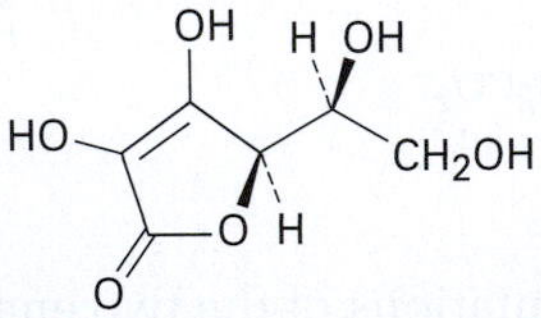

Ascorbic acid

5.50 ▪ Assign *R* or *S* stereochemistry to the chirality centers in the following Newman projections:

(a)

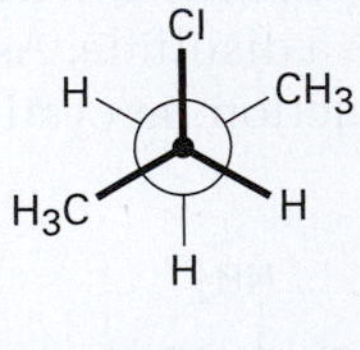

(b)

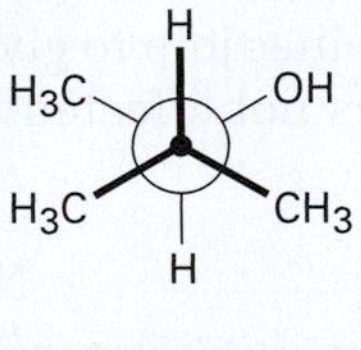

▪ Problems assignable in Organic OWL.

5.51 ▪ Xylose is a common sugar found in many types of wood, including maple and cherry. Because it is much less prone to cause tooth decay than sucrose, xylose has been used in candy and chewing gum. Assign *R* or *S* configurations to the chirality centers in xylose.

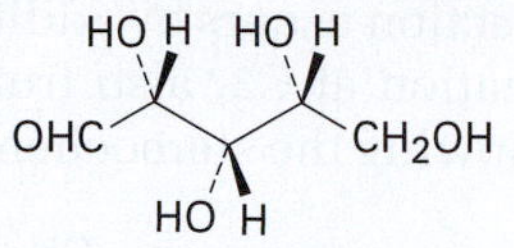

OHC — CH₂OH   **(+)-Xylose**

5.52 ▪ Ribose, an essential part of ribonucleic acid (RNA), has the following structure:

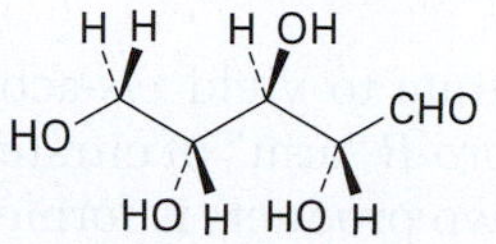

HO — CHO   **Ribose**

(a) How many chirality centers does ribose have? Identify them.

(b) How many stereoisomers of ribose are there?

(c) Draw the structure of the enantiomer of ribose.

(d) Draw the structure of a diastereomer of ribose.

5.53 On catalytic hydrogenation over a platinum catalyst, ribose (Problem 5.52) is converted into ribitol. Is ribitol optically active or inactive? Explain.

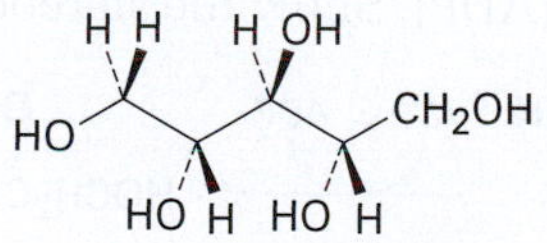

HO — CH₂OH   **Ribitol**

5.54 ▪ Identify the indicated hydrogens in the following molecules as *pro-R* or *pro-S*:

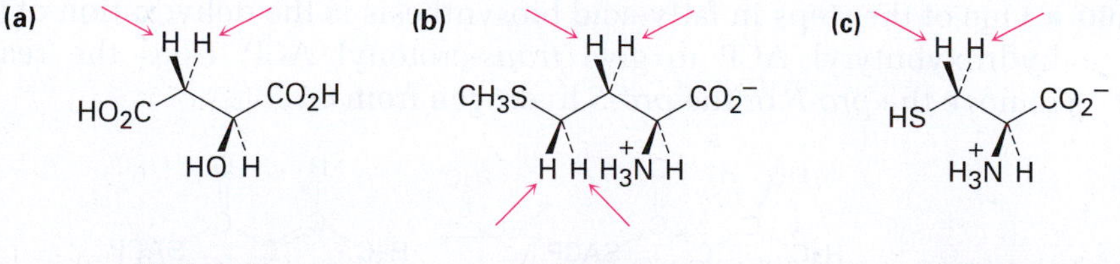

(a)

HO_2C — CO_2H

Malic acid

(b)

CH_3S — CO_2^-

Methionine

(c)

HS — CO_2^-

Cysteine

5.55 ▪ Identify the indicated faces in the following molecules as *Re* or *Si*:

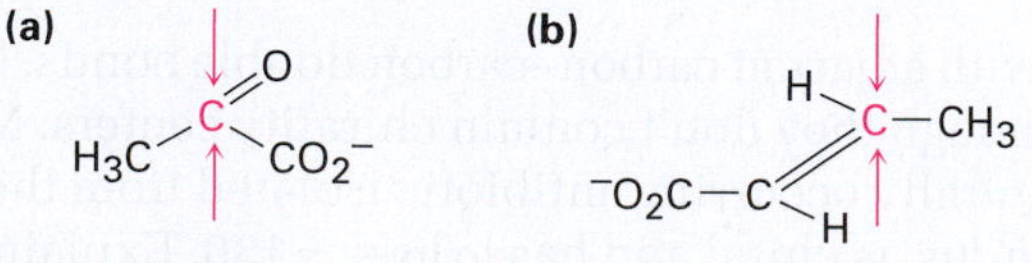

(a)

H_3C — CO_2^-

Pyruvate

(b)

^-O_2C — C — CH_3

Crotonate

▪ Problems assignable in Organic OWL.

5.56 Draw all possible stereoisomers of cyclobutane-1,2-dicarboxylic acid, and indicate the interrelationships. Which, if any, are optically active? Do the same for cyclobutane-1,3-dicarboxylic acid.

5.57 ■ One of the steps in fat metabolism is the hydration of crotonate to yield 3-hydroxybutyrate. The reaction occurs by addition of −OH to the *Si* face at C3, followed by protonation at C2, also from the *Si* face. Draw the product of the reaction, showing the stereochemistry of each step.

Crotonate   **3-Hydroxybutyrate**

5.58 ■ The dehydration of citrate to yield *cis*-aconitate, a step in the citric acid cycle, involves the *pro-R* "arm" of citrate rather than the *pro-S* arm. Which of the following two products is formed?

Citrate   **cis-Aconitate**

5.59 ■ The first step in the metabolism of glycerol, formed by digestion of fats, is phosphorylation of the *pro-R* −CH₂OH group by reaction with adenosine triphosphate (ATP) to give the corresponding glycerol phosphate plus adenosine diphosphate (ADP). Show the stereochemistry of the product.

Glycerol   **Glycerol phosphate**

5.60 ■ One of the steps in fatty-acid biosynthesis is the dehydration of (*R*)-3-hydroxybutyryl ACP to give *trans*-crotonyl ACP. Does the reaction remove the *pro-R* or the *pro-S* hydrogen from C2?

(R)-3-Hydroxybutyryl ACP   **trans-Crotonyl ACP**

5.61 *Allenes* are compounds with adjacent carbon–carbon double bonds. Many allenes are chiral, even though they don't contain chirality centers. Myco-mycin, for example, a naturally occurring antibiotic isolated from the bacterium *Nocardia acidophilus,* is chiral and has $[\alpha]_D = -130$. Explain why mycomycin is chiral. Making a molecular model should be helpful.

$$HC\equiv C-C\equiv C-CH=C=CH-CH=CH-CH=CH-CH_2CO_2H$$

Mycomycin

■ Problems assignable in Organic OWL.

5.62 Long before chiral allenes were known (Problem 5.61), the resolution of 4-methylcyclohexylideneacetic acid into two enantiomers had been carried out. Why is it chiral? What geometric similarity does it have to allenes?

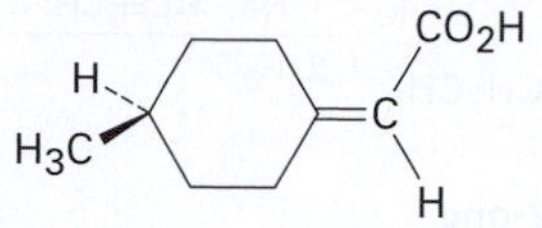

4-Methylcyclohexylideneacetic acid

5.63 (*S*)-1-Chloro-2-methylbutane undergoes reaction with Cl_2 to yield a mixture of products, among which are 1,4-dichloro-2-methylbutane and 1,2-dichloro-2-methylbutane.

  (a) Write the reaction, showing the correct stereochemistry of the reactant.

  (b) One of the two products is optically active, but the other is optically inactive. Which is which?

5.64 Draw the structure of a meso compound that has five carbons and three chirality centers.

5.65 Draw both *cis*- and *trans*-1,4-dimethylcyclohexane in their most stable chair conformations.

  (a) How many stereoisomers are there of *cis*-1,4-dimethylcyclohexane and how many of *trans*-1,4-dimethylcyclohexane?

  (b) Are any of the structures chiral?

  (c) What are the stereochemical relationships among the various stereoisomers of 1,4-dimethylcyclohexane?

5.66 Draw both *cis*- and *trans*-1,3-dimethylcyclohexane in their most stable chair conformations.

  (a) How many stereoisomers are there of *cis*-1,3-dimethylcyclohexane and how many of *trans*-1,3-dimethylcyclohexane?

  (b) Are any of the structures chiral?

  (c) What are the stereochemical relationships among the various stereoisomers of 1,3-dimethylcyclohexane?

5.67 We'll see in Chapter 12 that alkyl halides react with hydrosulfide ion (HS^-) to give a product whose stereochemistry at carbon is *inverted* from that of the reactant:

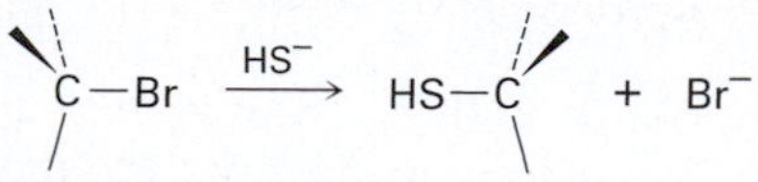

An alkyl bromide

Draw the reaction of (*S*)-2-bromobutane with HS^- ion to yield butane-2-thiol, $CH_3CH_2CH(SH)CH_3$. What is the stereochemistry of the product, *R* or *S*?

■ Problems assignable in Organic OWL.

5.68 ▪ Ketones react with acetylide ion (HC≡C:⁻) to give alcohols. For example, the reaction of sodium acetylide with butan-2-one yields 3-methylpent-1-yn-3-ol:

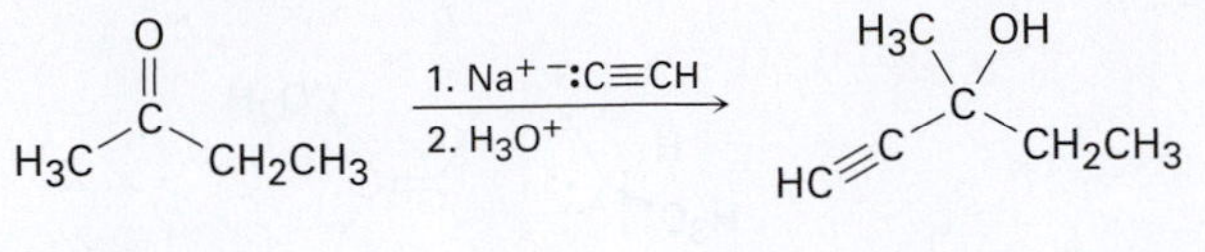

Butan-2-one   **3-Methylpent-1-yn-3-ol**

(a) Is the product chiral? Is it optically active?

(b) How many stereoisomers of the product are likely to be formed, and what are their stereochemical relationships?

5.69 Imagine that a reaction similar to that in Problem 5.68 is carried out between sodium acetylide and (R)-2-phenylpropanal to yield 4-phenylpent-1-yn-3-ol:

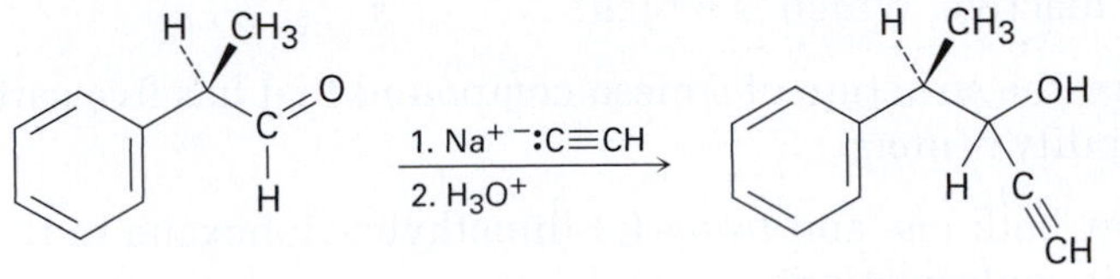

(R)-2-Phenylpropanal   **4-Phenylpent-1-yn-3-ol**

(a) Is the product chiral? Is it optically active?

(b) How many stereoisomers of the product are likely to be formed, and what are their stereochemical relationships?

6   An Overview of Organic Reactions

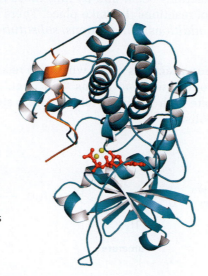

Protein kinase A catalyzes the phosphorylation of various amino acids in proteins.

When first approached, organic chemistry might seem overwhelming. It's not so much that any one part is difficult to understand; it's that there are so many parts: tens of millions of compounds, dozens of functional groups, and an apparently endless number of reactions. With study, though, it becomes evident that there are only a few fundamental ideas that underlie all organic reactions. Far from being a collection of isolated facts, organic chemistry is a beautifully logical subject that is unified by a few broad themes. When these themes are understood, learning organic chemistry becomes much easier and memorization is minimized. The aim of this book is to describe the themes and clarify the patterns that unify organic chemistry.

WHY THIS CHAPTER?

All chemical reactions, whether they take place in the laboratory or in living organisms, follow the same "rules." Reactions in living organisms often look more complex than laboratory reactions because of the size of the biomolecules and the involvement of biological catalysts called *enzymes,* but the principles governing all reactions are the same.

To understand both organic and biological chemistry, it's necessary to know not just *what* occurs but also *why* and *how* chemical reactions take place. In this chapter, we'll start with an overview of the fundamental kinds of organic reactions, we'll see why reactions occur, and we'll see how reactions can be described. Once this background is out of the way, we'll then be ready to begin studying the details of organic and biological chemistry.

OWL Online homework for this chapter can be assigned in Organic OWL.

6.1 Kinds of Organic Reactions

Organic chemical reactions can be organized broadly in two ways—by *what kinds* of reactions occur and by *how* those reactions occur. Let's look first at the kinds of reactions that take place. There are four general types of organic reactions: *additions*, *eliminations*, *substitutions*, and *rearrangements*.

- **Addition reactions** occur when two reactants add together to form a single product with no atoms "left over." An example is the reaction of fumarate with water to yield malate, a step in the citric acid cycle of food metabolism.

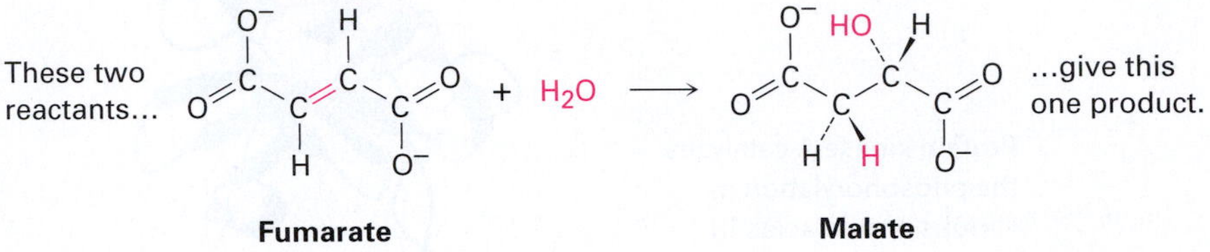

These two reactants...

Fumarate　　　　　　**Malate**

...give this one product.

- **Elimination reactions** are, in a sense, the opposite of addition reactions. They occur when a single reactant splits into two products, often with formation of a small molecule such as water. An example is the reaction of hydroxybutyryl ACP to yield *trans*-crotonyl ACP plus water, a step in the biosynthesis of fat molecules. (The abbreviation ACP stands for "acyl carrier protein.")

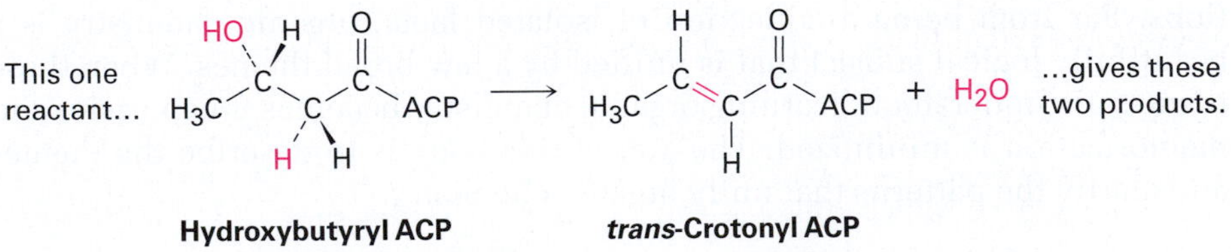

This one reactant...

Hydroxybutyryl ACP　　　　**trans-Crotonyl ACP**

...gives these two products.

- **Substitution reactions** occur when two reactants exchange parts to give two new products. An example is the reaction of an ester such as methyl acetate with water to yield a carboxylic acid plus an alcohol. Similar reactions occur in many biological pathways, including the metabolism of dietary fats.

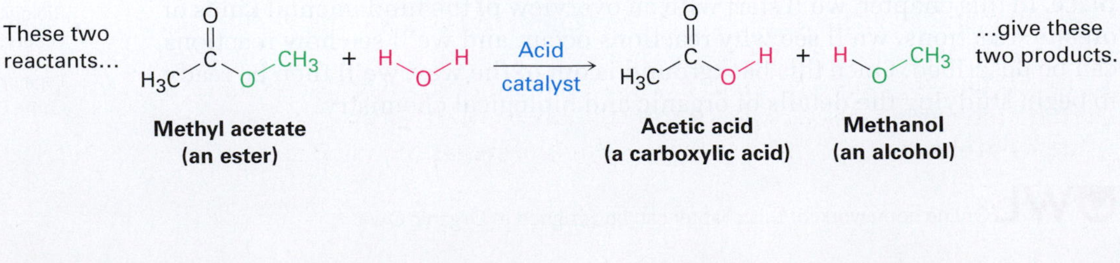

These two reactants...

Methyl acetate (an ester)　　　**Acetic acid (a carboxylic acid)**　　**Methanol (an alcohol)**

...give these two products.

- **Rearrangement reactions** occur when a single reactant undergoes a reorganization of bonds and atoms to yield an isomeric product. An example is the conversion of dihydroxyacetone phosphate into its constitutional isomer glyceraldehyde 3-phosphate, a step in the glycolysis pathway by which carbohydrates are metabolized.

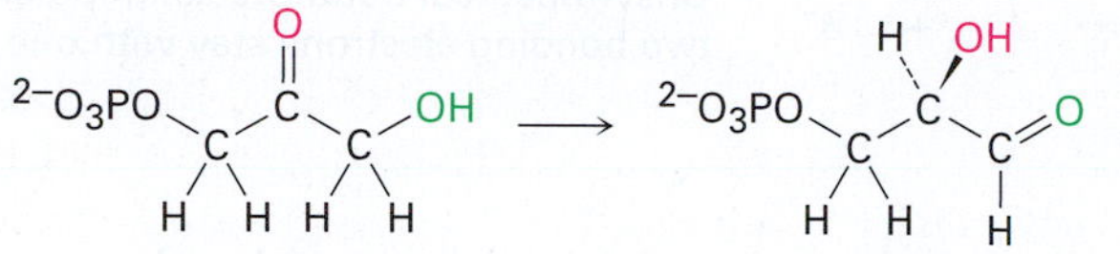

This reactant... $^{2-}O_3PO$ ... → ... $^{2-}O_3PO$ ...gives this isomeric product.

Dihydroxyacetone phosphate   **Glyceraldehyde 3-phosphate**

Problem 6.1

Classify each of the following reactions as an addition, elimination, substitution, or rearrangement:

(a) $CH_3Br + KOH \rightarrow CH_3OH + KBr$

(b) $CH_3CH_2OH \rightarrow H_2C{=}CH_2 + H_2O$

(c) $H_2C{=}CH_2 + H_2 \rightarrow CH_3CH_3$

6.2 How Organic Reactions Occur: Mechanisms

Having looked at the kinds of reactions that take place, let's now see how reactions occur. An overall description of how a reaction occurs is called a **reaction mechanism**. A mechanism describes in detail exactly what takes place at each stage of a chemical transformation—which bonds are broken and in what order, which bonds are formed and in what order, and what the relative rates of the steps are. A complete mechanism must also account for all reactants used and all products formed.

All chemical reactions involve bond-breaking and bond-making. When two molecules come together, react, and yield products, specific bonds in the reactant molecules are broken and specific bonds in the product molecules are formed. Fundamentally, there are two ways in which a covalent two-electron bond can break: a bond can break in an electronically *symmetrical* way so that one electron remains with each product fragment, or a bond can break in an electronically *unsymmetrical* way so that both bonding electrons remain with one product fragment, leaving the other with a vacant orbital. The symmetrical cleavage is said to be *homolytic*, and the unsymmetrical cleavage is said to be *heterolytic*.

We'll develop the point in more detail later, but you might note for now that the movement of *one* electron in the symmetrical process is indicated using a half-headed, or "fishhook," arrow ($\curvearrowright$), whereas the movement of *two* electrons in the unsymmetrical process is indicated using a full-headed curved arrow ($\frown$).

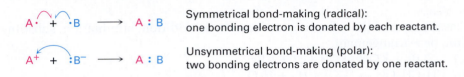

A : B $\longrightarrow$ A· + ·B   Symmetrical bond-breaking (radical): one bonding electron stays with each product.

A : B $\longrightarrow$ A$^+$ + :B$^-$ Unsymmetrical bond-breaking (polar): two bonding electrons stay with one product.

Just as there are two ways in which a bond can break, there are two ways in which a covalent two-electron bond can form. A bond can form in an electronically symmetrical way if one electron is donated to the new bond by each reactant or in an unsymmetrical way if both bonding electrons are donated by one reactant.

A· + ·B $\longrightarrow$ A : B   Symmetrical bond-making (radical): one bonding electron is donated by each reactant.

A$^+$ + :B$^-$ $\longrightarrow$ A : B   Unsymmetrical bond-making (polar): two bonding electrons are donated by one reactant.

Processes that involve symmetrical bond-breaking and bond-making are called **radical reactions**. A **radical**, often called a *free radical,* is a neutral chemical species that contains an odd number of electrons and thus has a single, unpaired electron in one of its orbitals. Processes that involve unsymmetrical bond-breaking and bond-making are called **polar reactions**. Polar reactions involve species that have an even number of electrons and thus have only electron pairs in their orbitals. Polar processes are by far the more common reaction type in both organic and biological chemistry, and a large part of this book is devoted to their description.

In addition to polar and radical reactions, there is a third, less commonly encountered process called a *pericyclic reaction.* Rather than explain pericyclic reactions now, though, we'll look at them more carefully in Section 8.14.

6.3 Radical Reactions

Radical reactions are not as common as polar reactions but are nevertheless important in some industrial processes and in numerous biological pathways. Let's see briefly how they occur.

A radical is highly reactive because it contains an atom with an odd number of electrons (usually seven) in its valence shell rather than a stable, noble-gas octet. A radical can achieve a valence-shell octet in several ways. For example, the radical might abstract an atom and one bonding electron from another reactant, leaving behind a new radical. The net result is a radical substitution reaction:

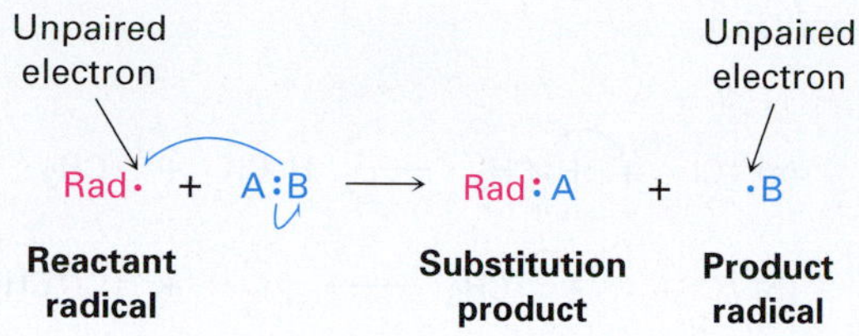

Alternatively, a reactant radical might add to a double bond, taking one electron from the double bond and leaving one behind to form a new radical. The net result is a radical addition reaction:

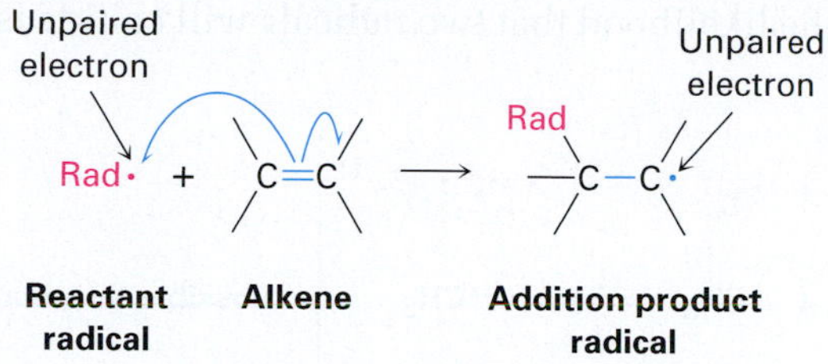

As an example of an industrially useful radical reaction, look at the chlorination of methane to yield chloromethane. This substitution reaction is the first step in the preparation of the solvents dichloromethane (CH_2Cl_2) and chloroform ($CHCl_3$).

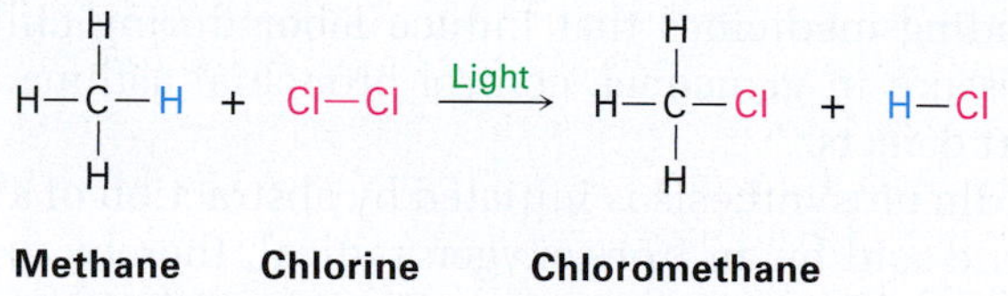

Like many radical reactions in the laboratory, methane chlorination requires three kinds of steps: *initiation, propagation,* and *termination.*

Initiation   Irradiation with ultraviolet light begins the reaction by breaking the relatively weak Cl–Cl bond of a small number of Cl_2 molecules to give a few reactive chlorine radicals.

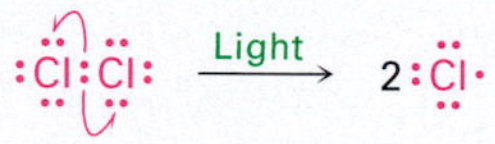

Propagation   Once produced, a reactive chlorine radical collides with a methane molecule in a propagation step, abstracting a hydrogen atom to give HCl and a methyl radical ($\cdot CH_3$). This methyl radical reacts further with Cl_2 in a second propagation step to give the product chloromethane plus a new chlorine radical, which cycles back and repeats the first propagation step. Thus, once the sequence has been initiated, it becomes a self-sustaining cycle of repeating steps (a) and (b), making the overall process a *chain reaction.*

(a) $:\ddot{C}l\cdot \ + \ H:CH_3 \longrightarrow H:\ddot{C}l: \ + \ \cdot CH_3$

(b) $:\ddot{C}l:\ddot{C}l: \ + \ \cdot CH_3 \longrightarrow :\ddot{C}l\cdot \ + \ :\ddot{C}l:CH_3$

Termination   Occasionally, two radicals might collide and combine to form a stable product. When that happens, the reaction cycle is broken and the chain is ended. Such termination steps occur infrequently, however, because the concentration of radicals in the reaction at any given moment is very small. Thus, the likelihood that two radicals will collide is also small.

$:\ddot{C}l\cdot \ + \ \cdot\ddot{C}l: \longrightarrow :\ddot{C}l:\ddot{C}l:$

$:\ddot{C}l\cdot \ + \ \cdot CH_3 \longrightarrow :\ddot{C}l:CH_3$ **Possible termination steps**

$H_3C\cdot \ + \ \cdot CH_3 \longrightarrow H_3C:CH_3$

As a biological example of radical reactions, look at the synthesis of *prostaglandins,* a large class of molecules found in virtually all body tissues and fluids. A number of pharmaceuticals are based on or derived from prostaglandins, including medicines that induce labor during childbirth, reduce intraocular pressure in glaucoma, control bronchial asthma, and help treat congenital heart defects.

Prostaglandin biosynthesis is initiated by abstraction of a hydrogen atom from arachidonic acid by an iron–oxygen radical, thereby generating a new carbon radical in a substitution reaction. Don't be intimidated by the size of the molecules; focus only on the changes occurring in each step. To help you do that, the unchanged part of the molecule is "ghosted," with only the reactive part clearly visible.

Radical substitution

Arachidonic acid   Carbon radical

Following the initial abstraction of a hydrogen atom, the carbon radical then reacts with O_2 to give an oxygen radical, which reacts with a C=C bond within the same molecule in an addition reaction. Several further transformations ultimately yield prostaglandin H_2.

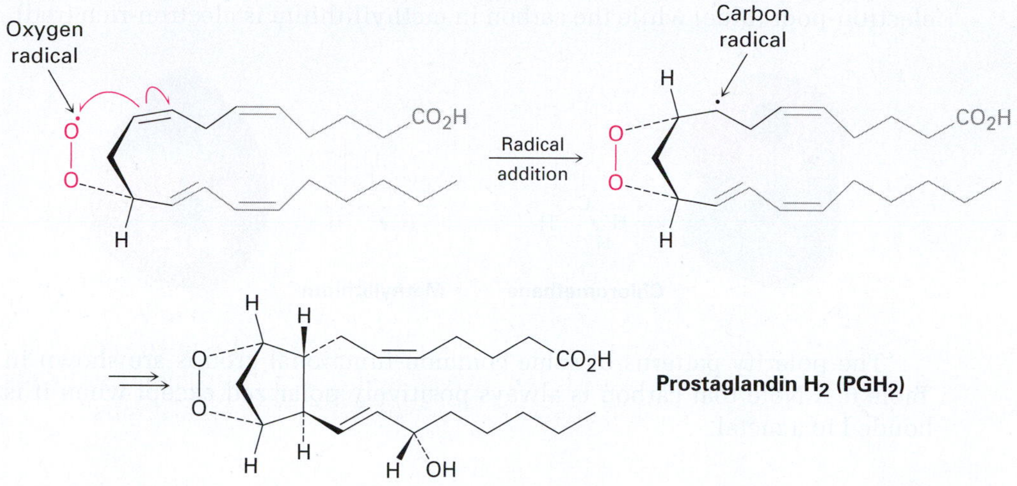

Oxygen radical

Carbon radical

Radical addition

Prostaglandin H_2 (PGH$_2$)

Problem 6.2

Radical chlorination of alkanes is not generally useful because mixtures of products often result when more than one kind of C–H bond is present in the substrate. Draw and name all monochloro substitution products $C_6H_{13}Cl$ you might obtain by reaction of 2-methylpentane with Cl_2.

Problem 6.3

Using a curved arrow, propose a mechanism for formation of the cyclopentane ring of prostaglandin H_2. What kind of reaction is occurring?

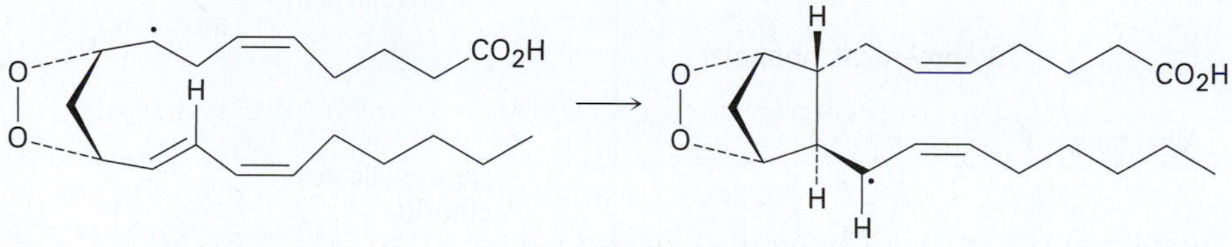

6.4 Polar Reactions

Polar reactions occur because of the electrical attraction between positively polarized and negatively polarized centers on functional groups in molecules. To see how these reactions take place, let's first recall the discussion of polar covalent bonds in Section 2.1 and then look more deeply into the effects of bond polarity on organic molecules.

Most organic compounds are electrically neutral; they have no net charge, either positive or negative. We saw in Section 2.1, however, that certain bonds within a molecule, particularly the bonds in functional groups, are polar. Bond polarity is a consequence of an unsymmetrical electron distribution in a bond and is due to the difference in electronegativity of the bonded atoms.

Elements such as oxygen, nitrogen, fluorine, and chlorine are more electronegative than carbon, so a carbon atom bonded to one of these atoms has a partial positive charge ($\delta+$). Conversely, metals are less electronegative than carbon, so a carbon atom bonded to a metal has a partial negative charge ($\delta-$). Electrostatic potential maps of chloromethane and methyllithium illustrate these charge distributions, showing that the carbon atom in chloromethane is electron-poor (blue) while the carbon in methyllithium is electron-rich (red).

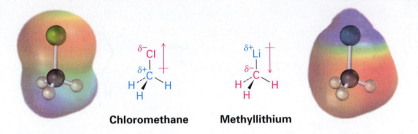

Chloromethane   **Methyllithium**

The polarity patterns of some common functional groups are shown in Table 6.1. Note that carbon is always positively polarized except when it is bonded to a metal.

TABLE 6.1
Polarity Patterns in Some Common Functional Groups

Compound type	Functional group structure	Compound type	Functional group structure
Alcohol	$\overset{\delta+}{C}-\overset{\delta-}{O}H$	Carbonyl	$\overset{\delta+}{C}=\overset{\delta-}{O}$
Alkene	C=C Symmetrical, nonpolar	Carboxylic acid	$-\overset{\delta+}{C}\overset{\delta-}{\underset{\delta-}{O}}OH$
Alkyl halide	$\overset{\delta+}{C}-\overset{\delta-}{X}$	Carboxylic acid chloride	$-\overset{\delta+}{C}\overset{\delta-}{\underset{\delta-}{O}}Cl$
Amine	$\overset{\delta+}{C}-\overset{\delta-}{N}H_2$	Thioester	$-\overset{\delta+}{C}\overset{\delta-}{\underset{\delta-}{O}}S-C$
Ether	$\overset{\delta+}{C}-\overset{\delta-}{O}-\overset{\delta+}{C}$	Aldehyde	$-\overset{\delta+}{C}\overset{\delta-}{\underset{}{O}}H$
Thiol	$\overset{\delta+}{C}-\overset{\delta-}{S}H$	Ester	$-\overset{\delta+}{C}\overset{\delta-}{\underset{\delta-}{O}}O-C$
Nitrile	$\overset{\delta+}{C}\equiv\overset{\delta-}{N}$	Ketone	$-\overset{\delta+}{C}\overset{\delta-}{\underset{}{O}}C$
Grignard reagent	$\overset{\delta-}{C}-\overset{\delta+}{Mg}Br$		
Alkyllithium	$\overset{\delta-}{C}-\overset{\delta+}{Li}$		

Polar bonds can also result from the interaction of functional groups with acids or bases. Take an alcohol such as methanol, for example. In neutral methanol, the carbon atom is somewhat electron-poor because the electronegative oxygen attracts the electrons in the C–O bond. On protonation of the methanol oxygen by an acid, however, a full positive charge on oxygen attracts the electrons in the C–O bond much more strongly and makes the carbon much more electron-poor. We'll see numerous examples throughout this book of reactions that are catalyzed by acids because of the resultant increase in bond polarity on protonation.

Methanol—weakly electron-poor carbon

Protonated methanol—strongly electron-poor carbon

Yet a further consideration is the *polarizability* (as opposed to polarity) of atoms in a molecule. As the electric field around a given atom changes because of changing interactions with solvent or other polar molecules nearby, the electron distribution around that atom also changes. The measure of this response to an external electrical influence is called the polarizability of the atom. Larger atoms with more, loosely held electrons are more polarizable, and smaller atoms with fewer, tightly held electrons are less polarizable. Thus, sulfur is more polarizable than oxygen, and iodine is more polarizable than chlorine. The effect of this higher polarizability for sulfur and iodine is that carbon–sulfur and carbon–iodine bonds, although nonpolar according to electronegativity values (Figure 2.2), nevertheless usually react as if they were polar.

What does functional-group polarity mean with respect to chemical reactivity? Because unlike charges attract, the fundamental characteristic of all polar organic reactions is that electron-rich sites react with electron-poor sites. Bonds are made when an electron-rich atom donates a pair of electrons to an electron-poor atom, and bonds are broken when one atom leaves with both electrons from the former bond.

As we saw in Section 2.11, chemists indicate the movement of an electron pair during a polar reaction by using a curved, full-headed arrow. A curved arrow shows where electrons move when reactant bonds are broken and product bonds are formed. It means that an electron pair moves *from* the atom (or bond) at the tail of the arrow *to* the atom at the head of the arrow during the reaction.

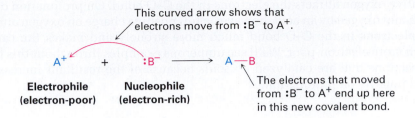

This curved arrow shows that electrons move from :B⁻ to A⁺.

A^+ + $:B^-$ ⟶ $A{-}B$

Electrophile   **Nucleophile**
(electron-poor)   **(electron-rich)**

The electrons that moved from :B⁻ to A⁺ end up here in this new covalent bond.

In referring to the electron-rich and electron-poor species involved in polar reactions, chemists use the words *nucleophile* and *electrophile*. A **nucleophile** is a substance that is "nucleus-loving." (Remember that a nucleus is positively charged.) A nucleophile has a negatively polarized, electron-rich atom and can form a bond by donating a pair of electrons to a positively polarized, electron-poor atom. Nucleophiles can be either neutral or negatively charged; ammonia, water, hydroxide ion, and chloride ion are examples. An **electrophile**, by contrast, is "electron-loving." An electrophile has a positively polarized, electron-poor atom and can form a bond by accepting a pair of electrons from a nucleophile. Electrophiles can be either neutral or positively charged. Acids (H⁺ donors), alkyl halides, and carbonyl compounds are examples (Figure 6.1).

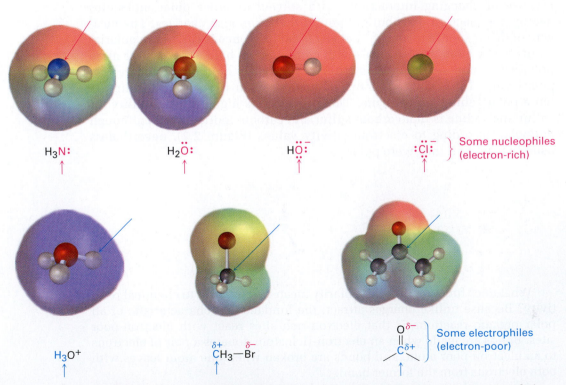

$H_3N\colon$ $H_2\ddot{O}\colon$ $H\ddot{O}\colon^-$ $\colon\ddot{C}\ddot{l}\colon^-$ } Some nucleophiles (electron-rich)

H_3O^+ $\overset{\delta+}{C}H_3{-}\overset{\delta-}{Br}$ $\overset{O^{\delta-}}{\underset{}{\overset{\|}{C^{\delta+}}}}$ } Some electrophiles (electron-poor)

FIGURE 6.1 Some nucleophiles and electrophiles. Electrostatic potential maps identify the nucleophilic (red; negative) and electrophilic (blue; positive) atoms.

Note that neutral compounds can often react either as nucleophiles or as electrophiles, depending on the circumstances. After all, if a compound is neutral yet has an electron-*rich* nucleophilic site, it must also have a corresponding electron-*poor* electrophilic site. Water, for instance, acts as an electrophile when it donates H^+ but acts as a nucleophile when it donates a nonbonding pair of electrons. Similarly, a carbonyl compound acts as an electrophile when it reacts at its positively polarized carbon atom, yet acts as a nucleophile when it reacts at its negatively polarized oxygen atom.

If the definitions of nucleophiles and electrophiles sound similar to those given in Section 2.11 for Lewis acids and Lewis bases, that's because there is indeed a correlation. Lewis bases are electron donors and behave as nucleophiles, whereas Lewis acids are electron acceptors and behave as electrophiles. Thus, much of organic chemistry is explainable in terms of acid–base reactions. The main difference is that the words *acid* and *base* are used broadly in all fields of chemistry, while the words *nucleophile* and *electrophile* are used primarily in organic chemistry when bonds to carbon are involved.

WORKED EXAMPLE 6.1 Identifying Electrophiles and Nucleophiles

Which of the following species is likely to behave as a nucleophile and which as an electrophile?
(a) $(CH_3)_3S^+$ **(b)** ^-CN **(c)** CH_3NH_2

Strategy

Nucleophiles have an electron-rich site, either because they are negatively charged or because they have a functional group containing an atom that has a lone pair of electrons. Electrophiles have an electron-poor site, either because they are positively charged or because they have a functional group containing an atom that is positively polarized.

Solution

(a) $(CH_3)_3S^+$ (trimethylsulfonium ion) is likely to be an electrophile because it is positively charged.

(b) $^-:C{\equiv}N$ (cyanide ion) is likely to be a nucleophile because it is negatively charged.

(c) CH_3NH_2 (methylamine) might be either a nucleophile or an electrophile depending on the circumstances. The lone pair of electrons on the nitrogen atom makes methylamine a potential nucleophile, while positively polarized N–H hydrogens make methylamine a potential acid (electrophile).

Problem 6.4

Which of the following species are likely to be nucleophiles and which electrophiles?

(a) CH_3Cl **(b)** CH_3S^- **(c)**   **(d)**

Problem 6.5

An electrostatic potential map of boron trifluoride is shown. Is BF_3 likely to be a nucleophile or an electrophile? Draw a Lewis structure for BF_3, and explain your answer.

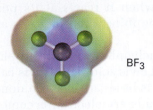

BF_3

6.5 An Example of a Polar Reaction: Addition of H_2O to Ethylene

Let's look at a typical polar process—the acid-catalyzed addition reaction of an alkene, such as ethylene, with water. When ethylene is heated to 250 °C with water and a strong acid catalyst such as H_2SO_4, ethanol is produced. Related processes that add water to a double bond and give an alcohol occur throughout biochemistry.

$$\underset{\textbf{Ethylene}}{\text{H}_2\text{C}=\text{CH}_2} + \text{H}-\overset{}{\underset{}{\text{O}}}-\text{H} \xrightarrow[\text{250 °C}]{\text{H}_2\text{SO}_4 \text{ catalyst}} \underset{\textbf{Ethanol}}{\text{CH}_3\text{CH}_2\text{OH}}$$

The reaction is an example of a polar reaction type known as an *electrophilic addition reaction* and can be understood using the general ideas discussed in the previous section. Let's begin by looking at the two reactants.

What do we know about ethylene? We know from Section 1.8 that a carbon–carbon double bond results from orbital overlap of two sp^2-hybridized carbon atoms. The σ part of the double bond results from sp^2–sp^2 overlap, and the π part results from p–p overlap.

What kind of chemical reactivity might we expect of a C=C bond? We know that *alkanes,* such as ethane, are relatively inert because the valence electrons are tied up in strong, nonpolar C–C and C–H bonds. Furthermore, the bonding electrons in alkanes are relatively inaccessible to approaching reactants because they are sheltered in σ bonds between nuclei. The electronic situation in *alkenes* is quite different, however. For one thing, double bonds have a greater electron density than single bonds—four electrons in a double bond versus only two in a single bond. Furthermore, the electrons in the π bond are accessible to approaching reactants because they are located above and below the plane of the double bond rather than being sheltered between nuclei (Figure 6.2). As a result, the double bond is nucleophilic and the chemistry of alkenes is dominated by reactions with electrophiles.

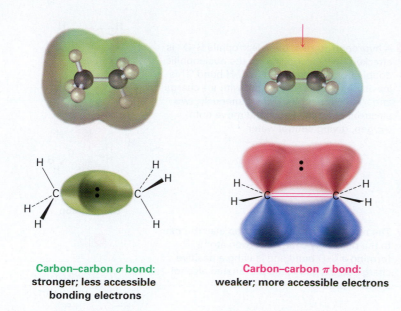

Carbon–carbon σ bond:
stronger; less accessible
bonding electrons

Carbon–carbon π bond:
weaker; more accessible electrons

FIGURE 6.2 A comparison of carbon–carbon single and double bonds. A double bond is both more accessible to approaching reactants than a single bond and more electron-rich (more nucleophilic). An electrostatic potential map of ethylene indicates that the double bond is the region of highest negative charge (red).

What about the second reactant, H_2O? In the presence of a strong acid such as H_2SO_4, water is protonated to give the hydronium ion H_3O^+, itself a powerful proton (H^+) donor and electrophile. Thus, the reaction between H_3O^+ and ethylene is a typical electrophile–nucleophile combination, characteristic of all polar reactions.

We'll see more details about alkene electrophilic addition reactions shortly, but for the present we can imagine the reaction as taking place by the pathway shown in Figure 6.3. The reaction begins when the alkene nucleophile donates a pair of electrons from its C=C bond to H_3O^+ to form a new C–H bond plus H_2O, as indicated by the path of the curved arrows in the first step of Figure 6.3. One curved arrow begins at the middle of the double bond (the source of the electron pair) and points to a hydrogen atom in H_3O^+ (the atom to which a bond will form). This arrow indicates that a new C–H bond forms using electrons from the former C=C bond. Simultaneously, a second curved arrow begins in the middle of the H–O bond and points to the O, indicating that the H–O bond breaks and the electrons remain with the O atom, giving neutral H_2O.

When one of the alkene carbon atoms bonds to the incoming hydrogen, the other carbon atom, having lost its share of the double-bond electrons, now has only six valence electrons and is left with a formal positive charge. This positively charged species—a carbon-cation, or **carbocation**—is itself an electrophile that can accept an electron pair from nucleophilic H_2O in a second step, forming a C–O bond and yielding a protonated alcohol addition product. Once again, a curved arrow in Figure 6.3 shows the electron-pair movement, in this case from O to the positively charged carbon. Finally, a second water molecule acts as a base to remove H^+ from the protonated addition product, regenerating H_3O^+ catalyst and giving the neutral alcohol.

ACTIVE FIGURE 6.3

MECHANISM: The acid-catalyzed electrophilic addition reaction of ethylene and H_2O. The reaction takes place in three steps, all of which involve electrophile–nucleophile interactions. **Go to this book's student companion site at www.cengage.com/chemistry/mcmurry to explore an interactive version of this figure.**

1 A hydrogen atom on the electrophile H_3O^+ is attacked by π electrons from the nucleophilic double bond, forming a new C–H bond. This leaves the other carbon atom with a + charge and a vacant p orbital. Simultaneously, two electrons from the H–O bond move onto oxygen, giving neutral water.

Ethylene

Carbocation

2 The nucleophile H_2O donates an electron pair to the positively charged carbon atom, forming a C–O bond and leaving a positive charge on oxygen in the protonated alcohol addition product.

Protonated ethanol

3 Water acts as a base to remove H^+, regenerating H_3O^+ and yielding the neutral alcohol addition product.

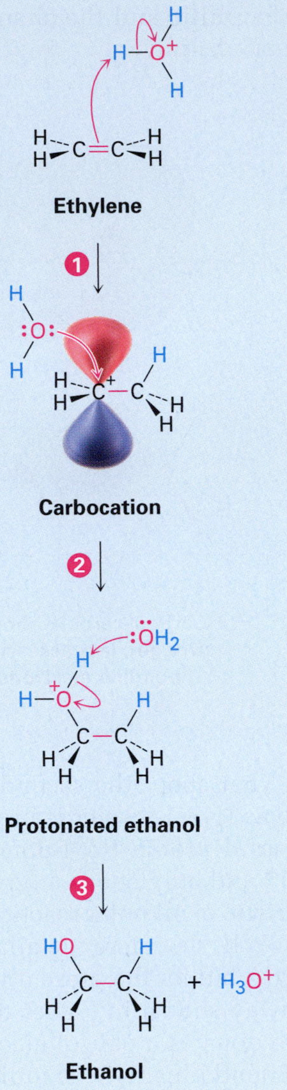

Ethanol

The electrophilic addition of H_2O to ethylene is only one example of a polar process; we'll study many others in detail in later chapters. But regardless of the details of individual reactions, all polar reactions take place between an electron-poor site and an electron-rich site and involve the donation of an electron pair from a nucleophile to an electrophile.

Problem 6.6

What product would you expect from acid-catalyzed reaction of cyclohexene with H_2O?

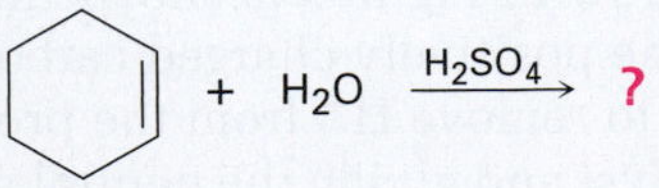

$$+ \; H_2O \; \xrightarrow{H_2SO_4} \; ?$$

Cyclohexene

Problem 6.7

Acid-catalyzed reaction of H_2O with 2-methylpropene yields 2-methyl-propan-2-ol. What is the structure of the carbocation formed during the reaction? Show the mechanism of the reaction.

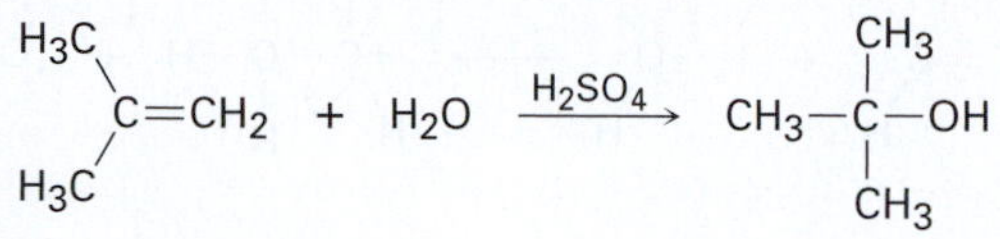

$$
\begin{array}{ccc}
\underset{H_3C}{\overset{H_3C}{\diagdown}}C{=}CH_2 \ + \ H_2O & \xrightarrow{H_2SO_4} & CH_3-\underset{CH_3}{\overset{CH_3}{\overset{|}{\underset{|}{C}}}}-OH
\end{array}
$$

2-Methylpropene   **2-Methylpropan-2-ol**

6.6 Using Curved Arrows in Polar Reaction Mechanisms

It takes practice to use curved arrows properly in reaction mechanisms, but there are a few rules and a few common patterns you should look for that will help you become more proficient:

Rule 1

Electrons move *from* a nucleophilic source (Nu: or Nu:⁻) *to* an electrophilic sink (E or E⁺). The nucleophilic source must have an electron pair available, usually either as a lone pair or in a multiple bond. For example:

Electrons usually flow *from* one of these nucleophiles.

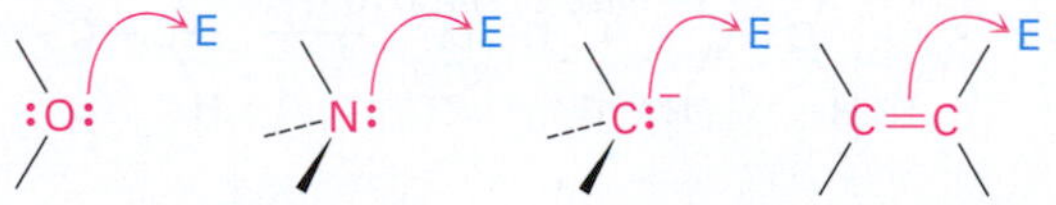

The electrophilic sink must be able to accept an electron pair, usually because it has either a positively charged atom or a positively polarized atom in a functional group. For example:

Electrons usually flow *to* one of these electrophiles.

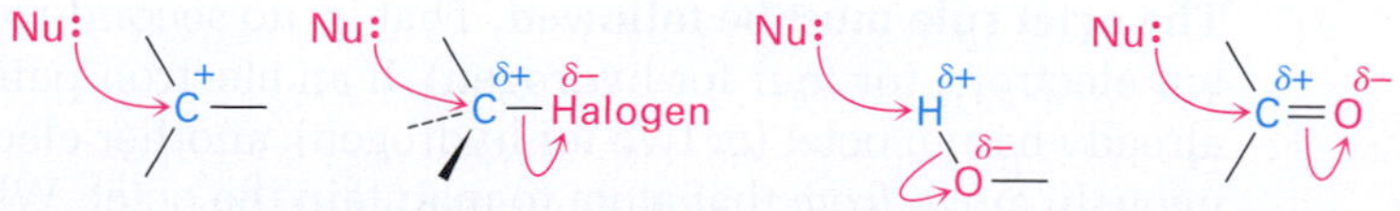

Rule 2

The nucleophile can be either negatively charged or neutral. If the nucleophile is negatively charged, the atom that donates an electron pair becomes neutral. For example:

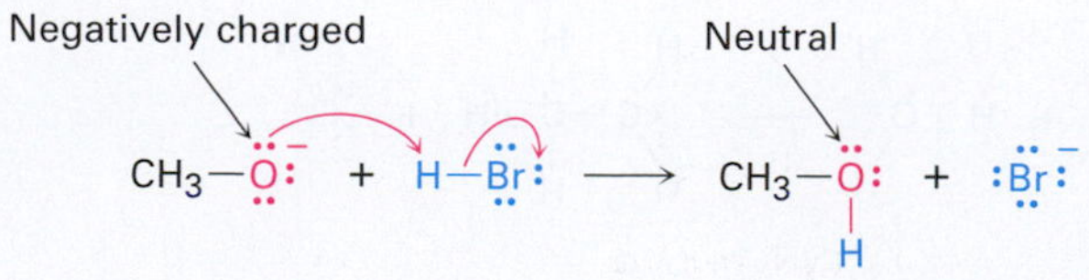

If the nucleophile is neutral, the atom that donates an electron pair acquires a positive charge. For example:

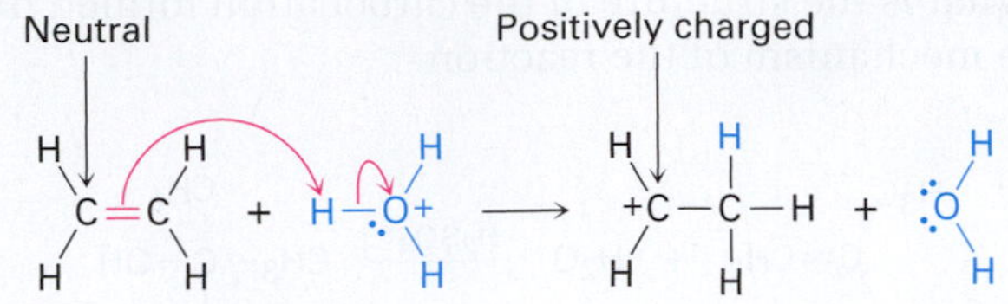

Rule 3

The electrophile can be either positively charged or neutral. If the electrophile is positively charged, the atom bearing that charge becomes neutral after accepting an electron pair. For example:

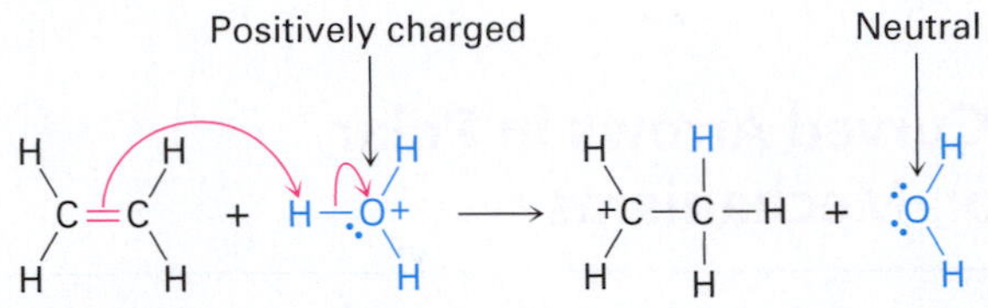

If the electrophile is neutral, the atom that ultimately accepts the electron pair acquires a negative charge. For this to happen, however, the negative charge must be stabilized by being on an electronegative atom such as oxygen, nitrogen, or a halogen. Carbon and hydrogen do not typically stabilize a negative charge. For example:

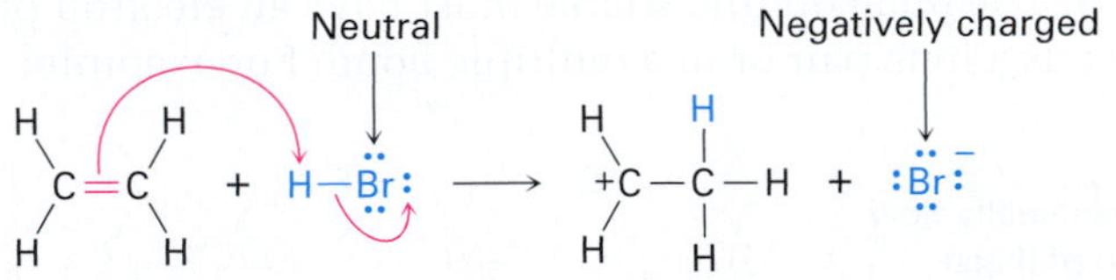

The result of Rules 2 and 3 together is that charge is conserved during the reaction. A negative charge in one of the reactants gives a negative charge in one of the products, and a positive charge in one of the reactants gives a positive charge in one of the products.

Rule 4

The octet rule must be followed. That is, no second-row atom can be left with ten electrons (or four for hydrogen). If an electron pair moves *to* an atom that already has an octet (or two for hydrogen), another electron pair must simultaneously move *from* that atom to maintain the octet. When two electrons move from the C=C bond of ethylene to the hydrogen atom of H_3O^+, for instance, two electrons must leave that hydrogen. This means that the H–O bond must break and the electrons must stay with the oxygen, giving neutral water.

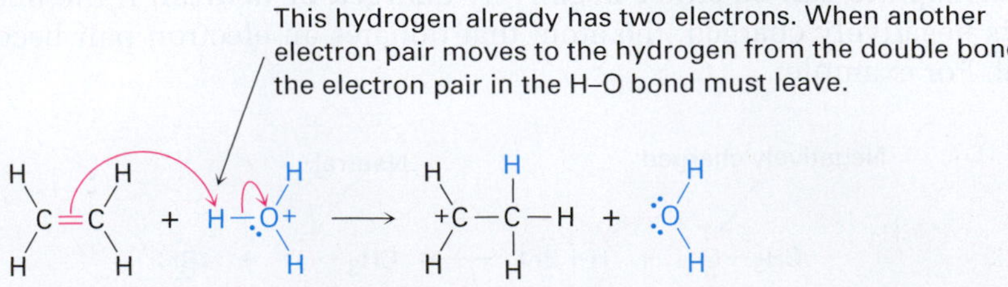

Worked Example 6.2 gives another example of drawing curved arrows.

| **WORKED EXAMPLE 6.2 Using Curved Arrows in Reaction Mechanisms** |

Add curved arrows to the following polar reaction to show the flow of electrons:

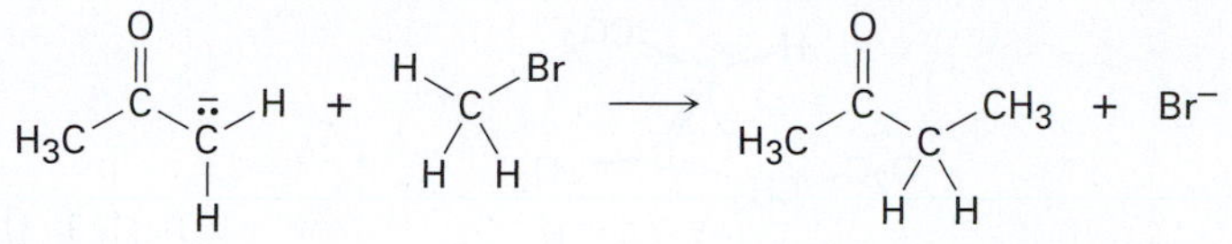

Strategy

Look at the reaction, and identify the bonding changes that have occurred. In this case, a C–Br bond has broken and a C–C bond has formed. The formation of the C–C bond involves donation of an electron pair from the nucleophilic carbon atom of the reactant on the left to the electrophilic carbon atom of CH_3Br, so we draw a curved arrow originating from the lone pair on the negatively charged C atom and pointing to the C atom of CH_3Br. At the same time the C–C bond forms, the C–Br bond must break so that the octet rule is not violated. We therefore draw a second curved arrow from the C–Br bond to Br. The bromine is now a stable Br^- ion.

Solution

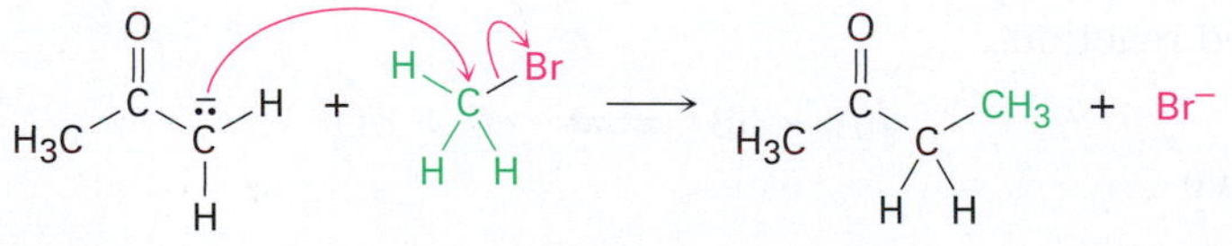

Problem 6.8

Add curved arrows to the following polar reactions to indicate the flow of electrons in each:

(a)

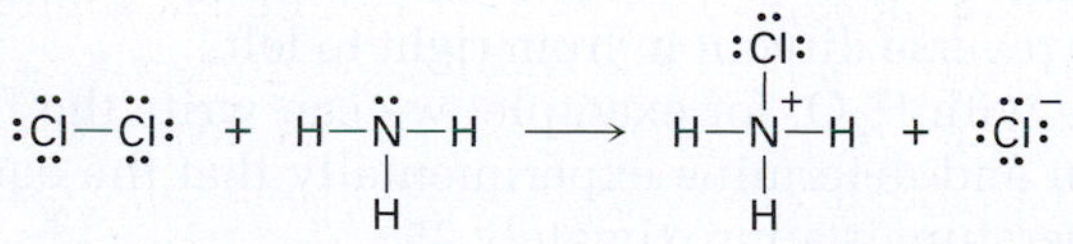

(b)

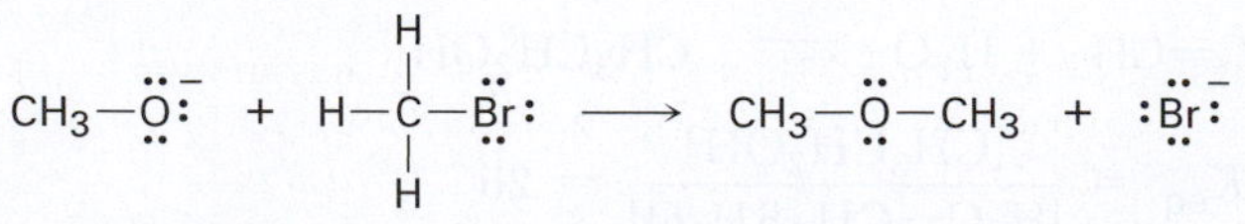

(c)

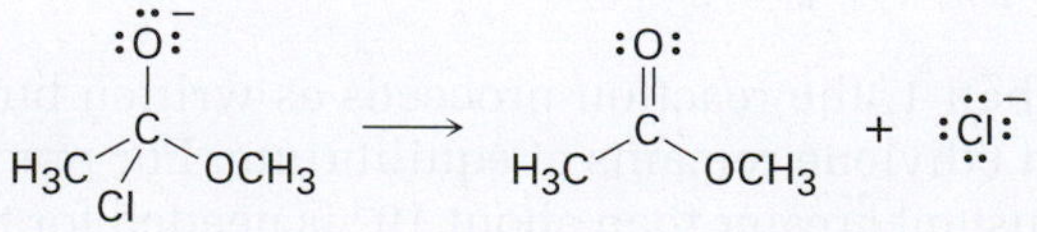

Problem 6.9

Predict the products of the following polar reaction, a step in the citric acid cycle for food metabolism, by interpreting the flow of electrons indicated by the curved arrows.

6.7 Describing a Reaction: Equilibria, Rates, and Energy Changes

Every chemical reaction can go in either forward or reverse direction. Reactants can go forward to products, and products can revert to reactants. As you may remember from your general chemistry course, the position of the resulting chemical equilibrium is expressed by an equation in which K_{eq}, the equilibrium constant, is equal to the product concentrations multiplied together, divided by the reactant concentrations multiplied together, with each concentration raised to the power of its coefficient in the balanced equation. For the generalized reaction

$$aA + bB \;\rightleftharpoons\; cC + dD$$

we have

$$K_{eq} = \frac{[C]^c [D]^d}{[A]^a [B]^b}$$

The value of the equilibrium constant tells which side of the reaction arrow is energetically favored. If K_{eq} is much larger than 1, then the product concentration term $[C]^c [D]^d$ is much larger than the reactant concentration term $[A]^a [B]^b$ and the reaction proceeds as written from left to right. If K_{eq} is near 1, appreciable amounts of both reactant and product are present at equilibrium. And if K_{eq} is much smaller than 1, the reaction does not take place as written but instead goes in the reverse direction, from right to left.

In the reaction of ethylene with H_2O, for example, we can write the following equilibrium expression and determine experimentally that the equilibrium constant at room temperature is approximately 25.

$$H_2C=CH_2 + H_2O \;\rightleftharpoons\; CH_3CH_2OH$$

$$K_{eq} = \frac{[CH_3CH_2OH]}{[H_2C=CH_2][H_2O]} = 25$$

Because K_{eq} is a bit larger than 1, the reaction proceeds as written but a substantial amount of unreacted ethylene remains at equilibrium. For practical purposes, an equilibrium constant greater than about 10^3 is needed for the amount of reactant left over to be barely detectable (less than 0.1%).

What determines the magnitude of the equilibrium constant? For a reaction to have a favorable equilibrium constant and proceed as written, the energy of the products must be lower than the energy of the reactants. In other words, energy must be *released*. The situation is analogous to that of a rock poised precariously in a high-energy position near the top of a hill. When it rolls downhill, the rock releases energy until it reaches a more stable low-energy position at the bottom.

The energy change that occurs during a chemical reaction is called the **Gibbs free-energy change (ΔG)**, which is equal to the free energy of the products minus the free energy of the reactants: $\Delta G = G_{\text{products}} - G_{\text{reactants}}$. For a favorable reaction, ΔG has a negative value, meaning that energy is lost by the chemical system and released to the surroundings. Such reactions are said to be **exergonic**. For an unfavorable reaction, ΔG has a positive value, meaning that energy is absorbed by the chemical system *from* the surroundings. Such reactions are said to be **endergonic**.

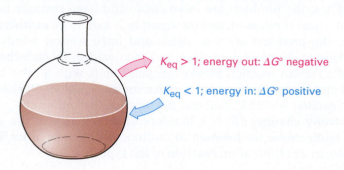

$K_{\text{eq}} > 1$; energy out: $\Delta G°$ negative

$K_{\text{eq}} < 1$; energy in: $\Delta G°$ positive

You might also recall from general chemistry that the *standard* free-energy change for a reaction is denoted $\Delta G°$, where the superscript ° means that the reaction is carried out under standard conditions, with pure substances in their most stable form at 1 atm pressure and a specified temperature, usually 298 K. For biological reactions, the standard free-energy change is symbolized $\Delta G°'$ and refers to a reaction carried out at pH = 7.0 with solute concentrations of 1.0 M.

Because the equilibrium constant, K_{eq}, and the standard free-energy change, $\Delta G°$, both measure whether a reaction is favored, they are mathematically related:

$$\Delta G° = -RT \ln K_{\text{eq}} \quad \text{or} \quad K_{\text{eq}} = e^{-\Delta G°/RT}$$

where $R = 8.314 \text{ J/(K} \cdot \text{mol)} = 1.987 \text{ cal/(K} \cdot \text{mol)}$

$T = $ Kelvin temperature

$e = 2.718$

$\ln K_{\text{eq}} = $ natural logarithm of K_{eq}

For example, the reaction of ethylene with H_2O has $K_{\text{eq}} = 25$, so $\Delta G° = -7.9$ kJ/mol (-1.9 kcal/mol) at 298 K:

$$K_{\text{eq}} = 25 \quad \text{and} \quad \ln K_{\text{eq}} = 3.2$$

$$\Delta G° = -RT \ln K_{\text{eq}} = -[8.314 \text{ J/(K} \cdot \text{mol)}] (298 \text{ K}) (3.2)$$

$$= -7900 \text{ J/mol} = -7.9 \text{ kJ/mol}$$

The free-energy change ΔG is made up of two terms, an *enthalpy* term, ΔH, and a temperature-dependent *entropy* term, $T\Delta S$. Of the two terms, the enthalpy term is often larger and more dominant.

$$\Delta G° = \Delta H° - T\Delta S°$$

For the reaction of ethylene with H_2O at room temperature (298 K), the approximate values are

$$H_2C=CH_2 \ + \ H_2O \ \rightleftharpoons \ CH_3CH_2OH \quad \begin{cases} \Delta G° = -7.9 \text{ kJ/mol} \\ \Delta H° = -44 \text{ kJ/mol} \\ \Delta S° = -0.12 \text{ kJ/(K} \cdot \text{mol)} \end{cases}$$

The **enthalpy change**, ΔH, also called the **heat of reaction**, is a measure of the change in total bonding energy during a reaction. If ΔH is negative, as in the reaction of H_2O with ethylene, the products have less energy than the reactants. Thus, the products are more stable and have stronger bonds than the reactants, heat is released, and the reaction is said to be **exothermic**. If ΔH is positive, the products are less stable and have weaker bonds than the reactants, heat is absorbed, and the reaction is said to be **endothermic**. For example, if a certain reaction breaks reactant bonds with a total strength of 380 kJ/mol and forms product bonds with a total strength of 400 kJ/mol, then ΔH for the reaction is -20 kJ/mol and the reaction is exothermic.

The **entropy change**, ΔS, is a measure of the change in the amount of molecular randomness, or freedom of motion, that accompanies a reaction. For example, in an elimination reaction of the type

A → B + C

there is more freedom of movement and molecular randomness in the products than in the reactant because one molecule has split into two. Thus, there is a net increase in entropy during the reaction and ΔS has a positive value.

On the other hand, for an addition reaction of the type

A + B → C

the opposite is true. Because such reactions restrict the freedom of movement of two molecules by joining them together, the product has less randomness than the reactants and ΔS has a negative value. The reaction of ethylene and H_2O to yield ethanol, which has $\Delta S° = -120$ J/(K · mol), is an example. Table 6.2 describes the thermodynamic terms more fully.

Knowing the value of K_{eq} for a reaction is useful, but it's important to realize the limitations. An equilibrium constant tells only the *position* of the equilibrium, or how much product is theoretically possible. It doesn't tell the *rate* of reaction, or how fast the equilibrium is established. Some reactions are extremely slow even though they have favorable equilibrium constants. Gasoline is stable at room temperature, for instance, because the rate of its reaction with oxygen is slow at 298 K. At higher temperatures, however, such as contact with a lighted match, gasoline reacts rapidly with oxygen and undergoes complete conversion to the equilibrium products water and carbon dioxide. Rates (*how fast* a reaction occurs) and equilibria (*how much* a reaction occurs) are entirely different.

Rate → Is the reaction fast or slow?

Equilibrium → In what direction does the reaction proceed?

TABLE 6.2
Explanation of Thermodynamic Quantities: $\Delta G° = \Delta H° - T\Delta S°$

Term	Name	Explanation
$\Delta G°$	Gibbs free-energy change	The energy difference between reactants and products. When $\Delta G°$ is negative, the reaction is **exergonic**, has a favorable equilibrium constant, and can occur spontaneously. When $\Delta G°$ is positive, the reaction is **endergonic**, has an unfavorable equilibrium constant, and cannot occur spontaneously.
$\Delta H°$	Enthalpy change	The heat of reaction, or difference in strength between the bonds broken in a reaction and the bonds formed. When $\Delta H°$ is negative, the reaction releases heat and is **exothermic**. When $\Delta H°$ is positive, the reaction absorbs heat and is **endothermic**.
$\Delta S°$	Entropy change	The change in molecular randomness during a reaction. When $\Delta S°$ is negative, randomness decreases; when $\Delta S°$ is positive, randomness increases.

Problem 6.10
Which reaction is more energetically favored, one with $\Delta G° = -44$ kJ/mol or one with $\Delta G° = +44$ kJ/mol?

Problem 6.11
Which reaction is likely to be more exergonic, one with $K_{eq} = 1000$ or one with $K_{eq} = 0.001$?

6.8 Describing a Reaction: Bond Dissociation Energies

We've just seen that heat is released (negative ΔH) when a bond is formed because the products are more stable and have stronger bonds than the reactants. Conversely, heat is absorbed (positive ΔH) when a bond is broken because the products are less stable and have weaker bonds than the reactants. The measure of the heat change that occurs on breaking a bond is called the *bond strength*, or **bond dissociation energy (D)**, defined as the amount of energy required to break a given bond to produce two radical fragments when the molecule is in the gas phase at 25 °C.

$$A : B \xrightarrow[\text{energy}]{\text{Bond dissociation}} A\cdot \ + \ \cdot B$$

Each specific bond has its own characteristic strength, and extensive tables of data are available. For example, a C–H bond in methane has a bond dissociation energy $D = 439.3$ kJ/mol (105.0 kcal/mol), meaning that 439.3 kJ/mol must be added to break a C–H bond of methane to give the two radical fragments $\cdot CH_3$ and $\cdot H$. Conversely, 439.3 kJ/mol of energy is released when a methyl radical and a hydrogen atom combine to form methane. Table 6.3 lists some other bond strengths.

TABLE 6.3
Some Bond Dissociation Energies, D

Bond	D (kJ/mol)	Bond	D (kJ/mol)	Bond	D (kJ/mol)
H—H	436	$(CH_3)_3C$—I	227	C_2H_5—CH_3	370
H—F	570	H_2C=CH—H	464	$(CH_3)_2CH$—CH_3	369
H—Cl	431	H_2C=CH—Cl	396	$(CH_3)_3C$—CH_3	363
H—Br	366	H_2C=CHCH$_2$—H	369	H_2C=CH—CH_3	426
H—I	298	H_2C=CHCH$_2$—Cl	298	H_2C=CHCH$_2$—CH_3	318
Cl—Cl	242	C$_6$H$_5$—H	472	H_2C=CH_2	728
Br—Br	194			C$_6$H$_5$—CH_3	427
I—I	152	C$_6$H$_5$—Cl	400		
CH_3—H	439			C$_6$H$_5$—CH_2—CH_3	325
CH_3—Cl	350	C$_6$H$_5$—CH_2—H	375		
CH_3—Br	294			$CH_3\overset{O}{\underset{\parallel}{C}}$—H	374
CH_3—I	239	C$_6$H$_5$—CH_2—Cl	300		
CH_3—OH	385			HO—H	497
CH_3—NH_2	386	C$_6$H$_5$—Br	336	HO—OH	211
C_2H_5—H	421			CH_3O—H	440
C_2H_5—Cl	352	C$_6$H$_5$—OH	464	CH_3S—H	366
C_2H_5—Br	293			C_2H_5O—H	441
C_2H_5—I	233	HC≡C—H	558	$CH_3\overset{O}{\underset{\parallel}{C}}$—$CH_3$	352
C_2H_5—OH	391	CH_3—CH_3	377		
$(CH_3)_2CH$—H	410			CH_3CH_2O—CH_3	355
$(CH_3)_2CH$—Cl	354			NH_2—H	450
$(CH_3)_2CH$—Br	299			H—CN	528
$(CH_3)_3C$—H	400				
$(CH_3)_3C$—Cl	352				
$(CH_3)_3C$—Br	293				

Think again about the connection between bond strengths and chemical reactivity. In an exothermic reaction, more heat is released than is absorbed. But because making bonds in the products releases heat and breaking bonds in the reactants absorbs heat, the bonds in the products must be stronger than the bonds in the reactants. In other words, exothermic reactions are favored by products with strong bonds and by reactants with weak, easily broken bonds.

Sometimes, particularly in biochemistry, reactive substances that undergo highly exothermic reactions, such as ATP (adenosine triphosphate), are referred to as "energy-rich" or "high-energy" compounds. Such a label doesn't mean that ATP is special or different from other compounds; it means only that ATP has relatively weak bonds that require a relatively small amount of heat to break, thus leading to a larger release of heat when a strong new bond forms in a reaction. When a typical organic phosphate such as glycerol 3-phosphate reacts with water, for instance, only 9 kJ/mol of heat is released ($\Delta H = -9$ kJ/mol), but when ATP reacts with water, 30 kJ/mol of heat is released ($\Delta H = -30$ kJ/mol). The difference between the two reactions is due to the fact that the bond broken in ATP is substantially weaker than the bond broken in glycerol 3-phosphate. We'll see the metabolic importance of this reaction in future chapters.

$\Delta H°' = -9$ kJ/mol

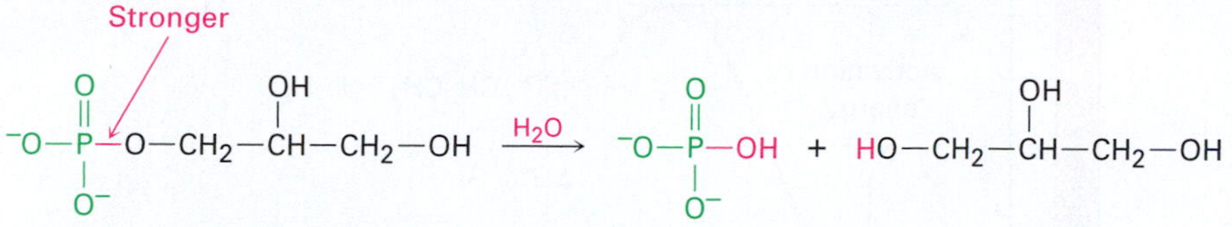

Glycerol 3-phosphate   **Glycerol**

$\Delta H°' = -30$ kJ/mol

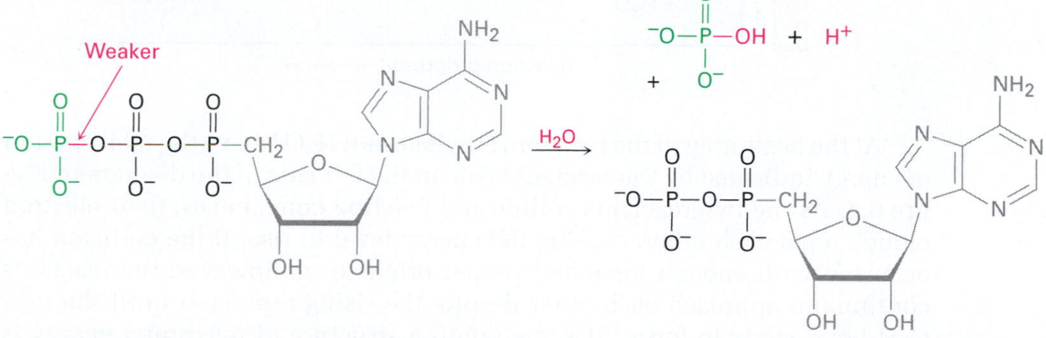

Adenosine triphosphate (ATP)   **Adenosine diphosphate (ADP)**

6.9 Describing a Reaction: Energy Diagrams and Transition States

For a reaction to take place, reactant molecules must collide and reorganization of atoms and bonds must occur. Let's again look at the three-step addition reaction of H_2O and ethylene:

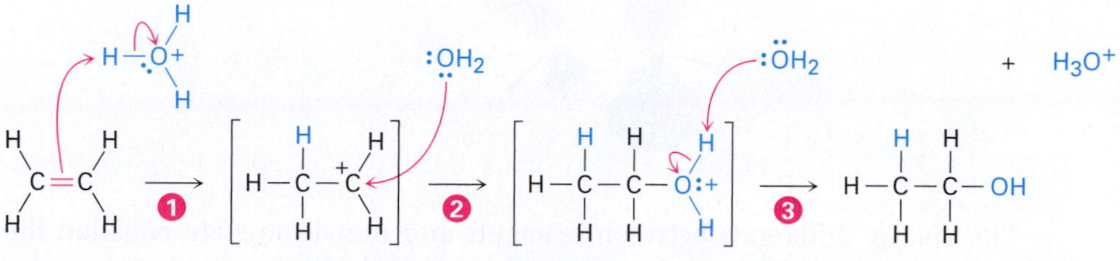

Carbocation   **Protonated alcohol**

As the reaction proceeds, ethylene and H_3O^+ must approach each other, the ethylene π bond and an H–O bond must break, a new C–H bond must form in the first step, and a new C–O bond must form in the second step.

To depict graphically the energy changes that occur during a reaction, chemists use energy diagrams, such as that shown in Figure 6.4. The vertical axis of the diagram represents the total energy of all reactants, and the horizontal axis, called the *reaction coordinate*, represents the progress of the reaction from beginning to end. Let's see how the addition of H_2O to ethylene can be described in an energy diagram.

FIGURE 6.4 An energy diagram for the first step in the reaction of ethylene with H_2O. The energy difference between reactants and transition state, $\Delta G^{\ddagger}$, defines the reaction rate. The energy difference between reactants and carbocation product, $\Delta G°$, defines the position of the equilibrium.

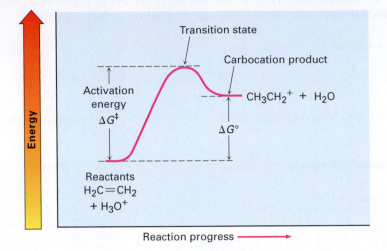

At the beginning of the reaction, ethylene and H_3O^+ have the total amount of energy indicated by the reactant level on the left side of the diagram in Figure 6.4. As the two reactants collide and reaction commences, their electron clouds repel each other, causing the energy level to rise. If the collision has occurred with enough force and proper orientation, however, the reactants continue to approach each other despite the rising repulsion until the new C–H bond starts to form. At some point, a structure of maximum energy is reached, a structure we call the *transition state*.

The **transition state** represents the highest-energy structure involved in this step of the reaction. It is unstable and can't be isolated, but we can nevertheless imagine it to be an activated complex of the two reactants in which both the C–C π bond and H–O bond are partially broken and the new C–H bond is partially formed (Figure 6.5).

FIGURE 6.5 A hypothetical transition-state structure for the first step of the reaction of ethylene with H_3O^+. The C=C π bond and O–H bond are just beginning to break, and the C–H bond is just beginning to form.

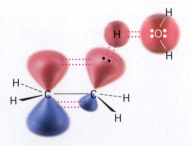

The energy difference between reactants and transition state is called the **activation energy**, $\Delta G^{\ddagger}$, and determines how rapidly the reaction occurs at a given temperature. (The double-dagger superscript, $\ddagger$, always refers to the transition state.) A large activation energy results in a slow reaction because few collisions occur with enough energy for the reactants to reach the transition state. A small activation energy results in a rapid reaction because almost all collisions occur with enough energy for the reactants to reach the transition state.

As an analogy, you might think of reactants that need enough energy to climb the activation barrier from reactant to transition state as similar to hikers who need enough energy to climb a mountain pass. If the pass is a high one, the hikers need a lot of energy and surmount the barrier with difficulty. If the pass is low, however, the hikers need less energy and reach the top easily.

As a rough generalization, many organic reactions have activation energies in the range 40 to 150 kJ/mol (10–35 kcal/mol). Reactions with activation energies less than 80 kJ/mol take place at or below room temperature, whereas reactions with higher activation energies normally require a higher temperature to give the reactants enough energy to climb the activation barrier.

Once the transition state is reached, the reaction can either continue on to give the carbocation product or revert back to reactants. When reversion to reactants occurs, the transition-state structure comes apart and an amount of free energy corresponding to $-\Delta G^{\ddagger}$ is released. When the reaction continues on to give the carbocation, the new C–H bond forms fully and an amount of energy corresponding to the difference between transition state and carbocation product is released. The net change in energy for the step, ΔG°, is represented in the diagram as the difference in level between reactant and product. Since the carbocation is higher in energy than the starting alkene, the step is endergonic, has a positive value of ΔG°, and absorbs energy.

Not all energy diagrams are like that shown for the reaction of ethylene and H_3O^+. Each reaction has its own energy profile. Some reactions are fast (small $\Delta G^{\ddagger}$) and some are slow (large $\Delta G^{\ddagger}$); some have a negative ΔG°, and some have a positive ΔG°. Figure 6.6 illustrates some different possibilities.

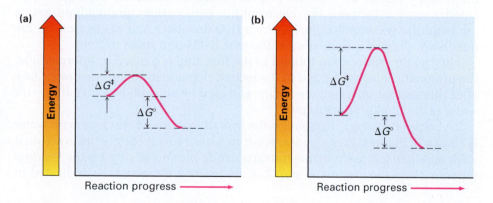

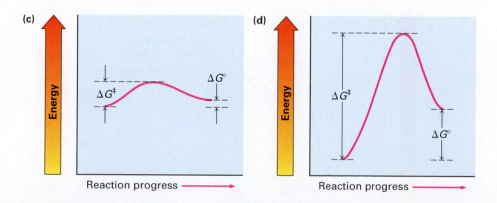

FIGURE 6.6 Some hypothetical energy diagrams: **(a)** a fast exergonic reaction (small $\Delta G^{\ddagger}$, negative ΔG°); **(b)** a slow exergonic reaction (large $\Delta G^{\ddagger}$, negative ΔG°); **(c)** a fast endergonic reaction (small $\Delta G^{\ddagger}$, small positive ΔG°); **(d)** a slow endergonic reaction (large $\Delta G^{\ddagger}$, positive ΔG°).

Problem 6.12
Which reaction is faster, one with $\Delta G^{\ddagger} = +45$ kJ/mol or one with $\Delta G^{\ddagger} = +70$ kJ/mol?

6.10 Describing a Reaction: Intermediates

How can we describe the carbocation formed in the first step of the reaction of ethylene with water? The carbocation is clearly different from the reactants, yet it isn't a transition state and it isn't a final product.

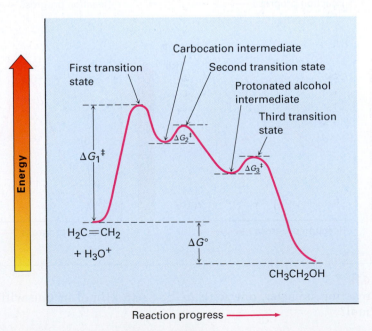

We call the carbocation, which exists only transiently during the course of the multistep reaction, a **reaction intermediate**. As soon as the intermediate is formed in the first step by reaction of ethylene with H_3O^+, it reacts further with H_2O in the second step to give the protonated alcohol product. This second step has its own activation energy $\Delta G^{\ddagger}$, its own transition state, and its own energy change $\Delta G°$. We can picture the second transition state as an activated complex between the electrophilic carbocation intermediate and a nucleophilic water molecule, in which H_2O donates a pair of electrons to the positively charged carbon atom as the new C–O bond starts to form.

Just as the carbocation formed in the first step is a reaction intermediate, the protonated alcohol formed in the second step is also an intermediate. Only after this second intermediate is deprotonated by an acid–base reaction with water is the final product formed.

A complete energy diagram for the overall reaction of ethylene with water is shown in Figure 6.7. In essence, we draw a diagram for each of the individual steps and then join them so that the carbocation *product* of step 1 is the *reactant* for step 2 and the product of step 2 is the reactant for step 3. As indicated in

FIGURE 6.7 An overall energy diagram for the reaction of ethylene with water. Three steps are involved, each with its own transition state. The energy minimum between steps 1 and 2 represents the carbocation reaction intermediate, and the minimum between steps 2 and 3 represents the protonated alcohol intermediate.

Figure 6.7, the reaction intermediates lie at energy minima between steps. Because the energy level of each intermediate is higher than the level of either the reactant that formed it or the product it yields, intermediates can't normally be isolated. They are, however, more stable than the two transition states that neighbor them.

Each step in a multistep process can always be considered separately. Each step has its own $\Delta G^{\ddagger}$ and its own $\Delta G°$. The overall $\Delta G°$ of the reaction, however, is the energy difference between initial reactants and final products.

The biological reactions that take place in living organisms have the same energy requirements as reactions that take place in the laboratory and can be described in similar ways. They are, however, constrained by the fact that they must have low enough activation energies to occur at moderate temperatures, and they must release energy in relatively small amounts to avoid overheating the organism. These constraints are generally met through the use of large, structurally complex, enzyme catalysts that change the mechanism of a reaction to an alternative pathway that proceeds through a series of small steps rather than one or two large steps. Thus, a typical energy diagram for a biological reaction might look like that in Figure 6.8.

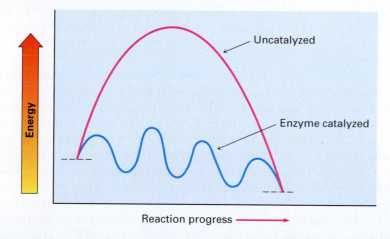

FIGURE 6.8 An energy diagram for a typical, enzyme-catalyzed biological reaction (blue curve) versus an uncatalyzed laboratory reaction (red curve). The biological reaction involves many steps, each of which has a relatively small activation energy and small energy change. The end result is the same, however.

WORKED EXAMPLE 6.3 Drawing Energy Diagrams for Reactions

Sketch an energy diagram for a one-step reaction that is fast and highly exergonic.

Strategy

A fast reaction has a small $\Delta G^{\ddagger}$, and a highly exergonic reaction has a large negative $\Delta G°$.

Solution

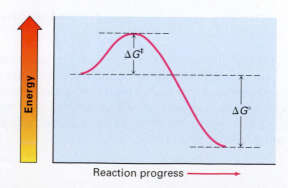

Problem 6.13
Sketch an energy diagram for a two-step reaction with an endergonic first step and an exergonic second step. Label the parts of the diagram corresponding to reactant, product, and intermediate.

6.11 A Comparison between Biological Reactions and Laboratory Reactions

Beginning in Chapter 7, we'll be seeing a lot of reactions. Although we'll keep the focus largely on those processes that have counterparts in biological pathways, we'll also discuss some reactions that are particularly important in laboratory chemistry yet do not occur in nature.

In comparing laboratory reactions with biological reactions, several differences are apparent. For one thing, laboratory reactions are usually carried out in an organic solvent such as diethyl ether or dichloromethane to dissolve the reactants and bring them into contact, whereas biological reactions occur in the aqueous medium inside cells. For another thing, laboratory reactions often take place over a wide range of temperatures without catalysts, while biological reactions take place at the temperature of the organism and are catalyzed by enzymes.

We'll be mentioning specific enzymes frequently throughout this book (all enzyme names end with the suffix -ase) and will look at them in more detail in Chapter 19. You may already be aware, however, that an **enzyme** is a large, globular, protein molecule that contains in its structure a protected pocket called its **active site**. The active site is lined by acidic or basic groups as needed for catalysis and has precisely the right shape to bind and hold a substrate molecule in the orientation necessary for reaction. Figure 6.9 shows a molecular model of hexokinase, along with an X-ray crystal structure of the glucose substrate and adenosine diphosphate (ADP) bound in the active site. Hexokinase is an enzyme that catalyzes the initial step of glucose metabolism—the transfer of a phosphate group from ATP to glucose, giving glucose 6-phosphate and ADP. The structures of ATP and ADP were shown at the end of Section 6.8.

Glucose   Glucose 6-phosphate

Note how the hexokinase-catalyzed phosphorylation reaction of glucose is shown. It's common when writing biological equations to show only the structure of the primary reactant and product, while abbreviating the structures of various biological "reagents" and by-products such as ATP and ADP. A curved arrow intersecting the straight reaction arrow indicates that ATP is also a reactant and ADP also a product.

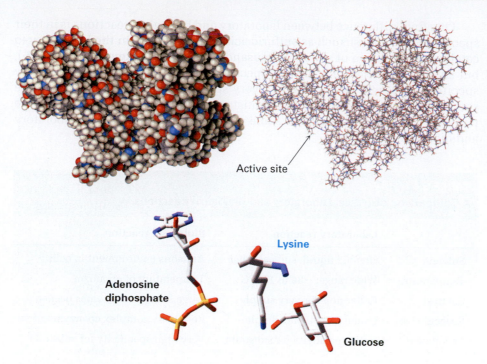

FIGURE 6.9 Models of hexo-kinase in space-filling and wire-frame formats, showing the cleft that contains the active site where substrate binding and reaction catalysis occur. At the bottom is an X-ray crystal structure of the enzyme active site, showing the positions of both glucose and ADP as well as a lysine amino acid that acts as a base to deprotonate glucose.

Active site

Lysine

**Adenosine
diphosphate**

Glucose

Yet another difference is that laboratory reactions are often done using relatively small, simple reagents such as Br_2, HCl, $NaBH_4$, CrO_3, and so forth, while biological reactions usually involve relatively complex "reagents" called *coenzymes*. In the hexokinase-catalyzed phosphorylation of glucose just shown, for instance, ATP is the coenzyme. As another example, compare the H_2 molecule, a laboratory reagent that adds to a carbon–carbon double bond to yield an alkane, with the reduced nicotinamide adenine dinucleotide (NADH) molecule, a coenzyme that effects an analogous addition of hydrogen to a double bond in many biological pathways. Of all the atoms in the entire coenzyme, only the one hydrogen atom shown in red is transferred to the double-bond substrate.

**Reduced nicotinamide adenine dinucleotide, NADH
(a coenzyme)**

Don't be intimidated by the size of the NADH molecule; most of the structure is there to provide an overall shape for binding to the enzyme and to provide appropriate solubility behavior. When looking at biological molecules, focus on the small part of the molecule where the chemical change takes place.

One final difference between laboratory and biological reactions is in their specificity. A catalyst such as sulfuric acid might be used in the laboratory to catalyze the addition of water to thousands of different alkenes (Section 6.5), but an enzyme, because it binds a specific substrate molecule having a very specific shape, will catalyze only a very specific reaction. It's this exquisite specificity that makes biological chemistry so remarkable and that makes life possible. Table 6.4 summarizes some of the differences between laboratory and biological reactions.

TABLE 6.4
A Comparison of Typical Laboratory and Biological Reactions

	Laboratory reaction	Biological reaction
Solvent	Organic liquid, such as ether	Aqueous environment in cells
Temperature	Wide range; −80 to 150 °C	Temperature of organism
Catalyst	Either none or very simple	Large, complex enzymes needed
Reagent size	Usually small and simple	Relatively complex coenzymes
Specificity	Little specificity for substrate	Very high specificity for substrate

Summary

Key Words

activation energy ($\Delta G^{\ddagger}$), 198

active site, 202

addition reaction, 176

bond dissociation energy (D), 195

carbocation, 187

electrophile, 184

elimination reaction, 176

endergonic, 193

endothermic, 194

enthalpy change (ΔH), 194

entropy change (ΔS), 194

enzyme, 202

exergonic, 193

exothermic, 194

Gibbs free-energy change (ΔG), 193

heat of reaction, 194

nucleophile, 184

polar reaction, 178

radical, 178

radical reaction, 178

reaction intermediate, 200

reaction mechanism, 177

rearrangement reaction, 177

substitution reaction, 176

transition state, 198

All chemical reactions, whether in the laboratory or in living organisms, follow the same "rules." To understand both organic and biological chemistry, it's necessary to know not just *what* occurs but also *why* and *how* chemical reactions take place. In this chapter, we've taken a brief look at the fundamental kinds of organic reactions, we've seen why reactions occur, and we've seen how reactions can be described.

There are four common kinds of reactions: **addition reactions** take place when two reactants add together to give a single product; **elimination reactions** take place when one reactant splits apart to give two products; **substitution reactions** take place when two reactants exchange parts to give two new products; and **rearrangement reactions** take place when one reactant undergoes a reorganization of bonds and atoms to give an isomeric product.

A full description of how a reaction occurs is called its **mechanism**. There are two general kinds of mechanisms by which reactions take place: **radical** mechanisms and **polar** mechanisms. Polar reactions, the most common type, occur because of an attractive interaction between a **nucleophilic** (electron-rich) site in one molecule and an **electrophilic** (electron-poor) site in another molecule. A bond is formed in a polar reaction when the nucleophile donates an electron pair to the electrophile. This movement of electrons is indicated by a curved arrow showing the direction of electron travel from the nucleophile to the electrophile. Radical reactions involve species that have an odd number of electrons. A bond is formed when each reactant donates one electron.

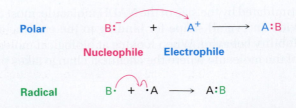

The energy changes that take place during reactions can be described by considering both rates (how fast the reactions occur) and equilibria (how much the reactions occur). The position of a chemical equilibrium is determined by the value of the **free-energy change (ΔG)** for the reaction, where $\Delta G = \Delta H - T\Delta S$. The **enthalpy** term (ΔH) corresponds to the net change in strength of chemical bonds broken and formed during reaction; the **entropy** term (ΔS) corresponds to the change in the amount of disorder during the reaction. Reactions that have negative values of ΔG release energy, are said to be **exergonic**, and have favorable equilibria. Reactions that have positive values of ΔG absorb energy, are said to be **endergonic**, and have unfavorable equilibria.

A reaction can be described pictorially using an energy diagram that follows the reaction course from reactant through transition state to product. The **transition state** is an activated complex occurring at the highest-energy point of a reaction. The amount of energy needed by reactants to reach this high point is the **activation energy, $\Delta G^{\ddagger}$**. The higher the activation energy, the slower the reaction.

Many reactions take place in more than one step and involve the formation of a **reaction intermediate**. An intermediate is a species that lies at an energy minimum between steps on the reaction curve and is formed briefly during the course of a reaction.

Lagniappe

Where Do Drugs Come From?

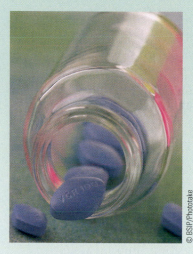

Approved for sale in March 1998 to treat male impotency, Viagra has been used by more than 16 million men. It is also used to treat pulmonary hypertension and is currently undergoing study as a treatment for preeclampsia, a complication of pregnancy that is responsible for as many as 70,000 deaths each year. Where do new drugs like this come from?

It has been estimated that major pharmaceutical companies in the United States spend some $33 billion per year on drug research and development, while government agencies and private foundations spend another $28 billion. What does this money buy? For the period 1981–2004, the money resulted in a total of 912 new molecular entities (NMEs)—new biologically active chemical substances approved for sale as drugs by the U.S. Food and Drug Administration (FDA). That's an average of only 38 new drugs each year spread over all diseases and conditions, and the number has been steadily falling: in 2004, only 23 NMEs were approved.

Where do the new drugs come from? According to a study carried out at the U.S. National Cancer Institute, only 33% of new drugs are entirely synthetic and completely unrelated to any naturally occurring substance. The remaining 67% take their lead, to a greater or lesser extent, from nature. Vaccines and genetically engineered proteins of biological origin account for 15% of NMEs, but most new drugs come from *natural products,* a catchall term generally taken to mean small molecules found in bacteria, plants, and other living organisms. Unmodified natural products isolated directly from the producing organism account for 28% of NMEs, while natural products that have been chemically modified in the laboratory account for the remaining 24%.

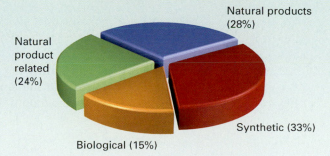

Origin of New Drugs 1981–2002

Natural products (28%)

Natural product related (24%)

Synthetic (33%)

Biological (15%)

Many years of work go into screening many thousands of substances to identify a single compound that might ultimately gain approval as an NME. But after that single compound has been identified, the work has just begun

continued

Lagniappe *continued*

because it takes an average of 9 to 10 years for a drug to make it through the approval process. First, the safety of the drug in animals must be demonstrated and an economical method of manufacture must be devised. With these preliminaries out of the way, an Investigational New Drug (IND) application is submitted to the FDA for permission to begin testing in humans.

Human testing takes 5 to 7 years and is divided into three phases. Phase I clinical trials are carried out on a small group of healthy volunteers to establish safety and look for side effects. Several months to a year are needed, and only about 70% of drugs pass at this point. Phase II clinical trials next test the drug for 1 to 2 years in several hundred patients with the target disease or condition, looking both for safety and for efficacy, and only about 33% of the original group pass. Finally, phase III trials are undertaken on a large sample of patients to document definitively the drug's safety, dosage, and efficacy. If the drug is one of the 25% of the original group that make it to the end of phase III, all the data are then gathered into a New Drug Application (NDA) and sent to the FDA for review and approval, which can take another 2 years. Ten years have elapsed and at least $500 million has been spent, with only a 20% success rate for the drugs that began testing. Finally, though, the drug will begin to appear in medicine cabinets. The following timeline shows the process.

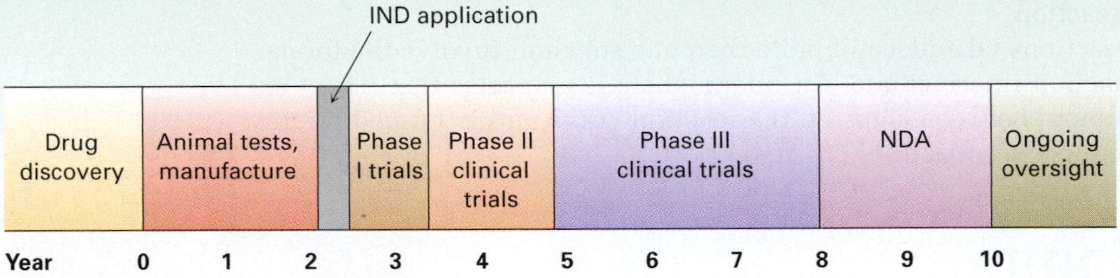

IND application

Drug discovery	Animal tests, manufacture	Phase I trials	Phase II clinical trials	Phase III clinical trials	NDA	Ongoing oversight

| Year | 0 | 1 | 2 | 3 | 4 | 5 | 6 | 7 | 8 | 9 | 10 |

Exercises

■ indicates problems that are assignable in Organic OWL.

Go to this book's companion website at **www.cengage.com/ chemistry/mcmurry** to explore interactive versions of the Active Figures from this text.

VISUALIZING CHEMISTRY

(Problems 6.1–6.13 appear within the chapter.)

6.14 ■ The following alcohol can be prepared by addition of H_2O to two different alkenes. Draw the structures of both (red = O).

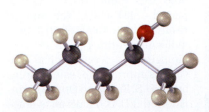

6.15 ■ The following structure represents the carbocation intermediate formed in the acid-catalyzed addition reaction of H_2O to an alkene to yield an alcohol. Draw the structure of the alkene.

■ Problems assignable in Organic OWL.

6.16 Electrostatic potential maps of **(a)** formaldehyde (CH₂O) and **(b)** methane-thiol (CH₃SH) are shown. Is the formaldehyde carbon atom likely to be electrophilic or nucleophilic? What about the methanethiol sulfur atom? Explain.

(a)

(b)

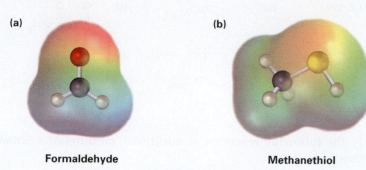

Formaldehyde   Methanethiol

6.17 ■ Look at the following energy diagram:

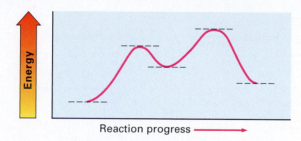

(a) Is $\Delta G°$ for the reaction positive or negative? Label it on the diagram.

(b) How many steps are involved in the reaction?

(c) How many transition states are there? Label them on the diagram.

6.18 Look at the following energy diagram for an enzyme-catalyzed reaction:

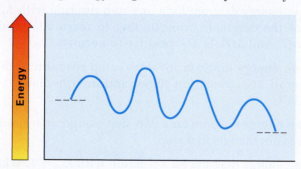

(a) How many steps are involved?

(b) Which step is most exergonic?

(c) Which step is the slowest?

■ Problems assignable in Organic OWL.

ADDITIONAL PROBLEMS

6.19 ▪ Identify the functional groups in the following molecules, and show the polarity of each:

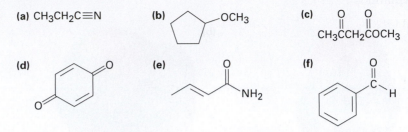

(a) $CH_3CH_2C\equiv N$

(b) [cyclopentane]—OCH_3

(c) $CH_3\overset{O}{\overset{\|}{C}}CH_2\overset{O}{\overset{\|}{C}}OCH_3$

(d) [cyclohexadienedione]

(e) [with NH_2]

(f) [benzene ring]—$\overset{O}{\overset{\|}{C}}$—H

6.20 ▪ Identify the following reactions as additions, eliminations, substitutions, or rearrangements:

(a) $CH_3CH_2Br + NaCN \longrightarrow CH_3CH_2CN (+ NaBr)$

(b) 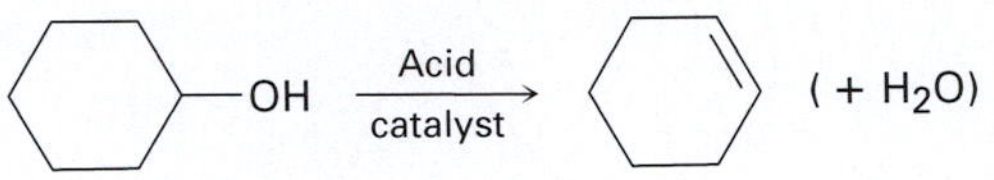 [cyclohexane]—OH $\xrightarrow[\text{catalyst}]{\text{Acid}}$ [cyclohexene] $(+ H_2O)$

(c) 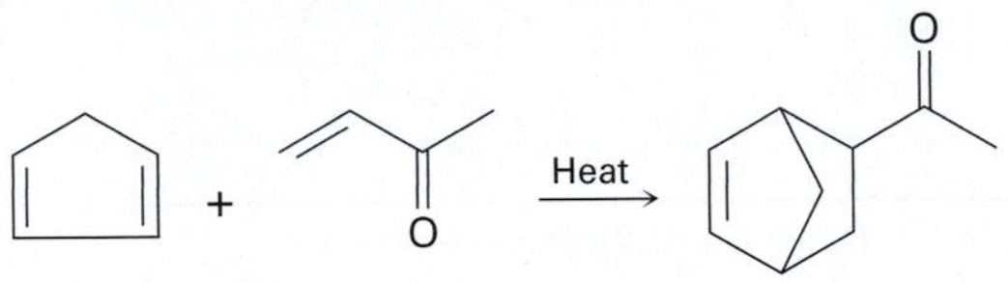 [cyclopentadiene] $+$ [but-3-en-2-one] $\xrightarrow{\text{Heat}}$ [bicyclic ketone product]

(d) 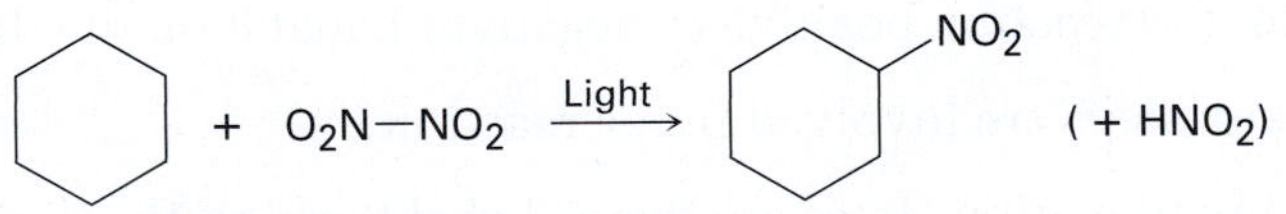 [cyclohexane] $+ O_2N-NO_2 \xrightarrow{\text{Light}}$ [cyclohexane with NO_2] $(+ HNO_2)$

6.21 What is the difference between a transition state and an intermediate?

6.22 Draw an energy diagram for a one-step reaction with $K_{eq} < 1$. Label the parts of the diagram corresponding to reactants, products, transition state, $\Delta G°$, and $\Delta G^{\ddagger}$. Is $\Delta G°$ positive or negative?

6.23 Draw an energy diagram for a two-step reaction with $K_{eq} > 1$. Label the overall $\Delta G°$, transition states, and intermediate. Is $\Delta G°$ positive or negative?

6.24 Draw an energy diagram for a two-step exergonic reaction whose second step is faster than its first step.

6.25 Draw an energy diagram for a reaction with $K_{eq} = 1$. What is the value of $\Delta G°$ in this reaction?

6.26 When a mixture of methane and chlorine is irradiated, reaction commences immediately. When irradiation is stopped, the reaction gradually slows down but does not stop immediately. Explain.

6.27 Radical chlorination of pentane is a poor way to prepare 1-chloropentane, but radical chlorination of neopentane, $(CH_3)_4C$, is a good way to prepare neopentyl chloride, $(CH_3)_3CCH_2Cl$. Explain.

▪ Problems assignable in Organic OWL.

6.28 ▪ Despite the limitations of radical chlorination of alkanes, the reaction is still useful for synthesizing certain halogenated compounds. For which of the following compounds does radical chlorination give a single monochloro product?

(a) C_2H_6 (b) $CH_3CH_2CH_3$ (c)

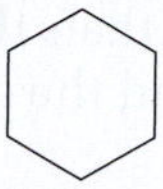

(d) $(CH_3)_3CCH_2CH_3$ (e) 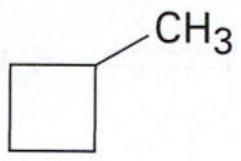   (f) $CH_3C\equiv CCH_3$

6.29 ▪ Add curved arrows to the following reactions to indicate the flow of electrons in each:

(a)

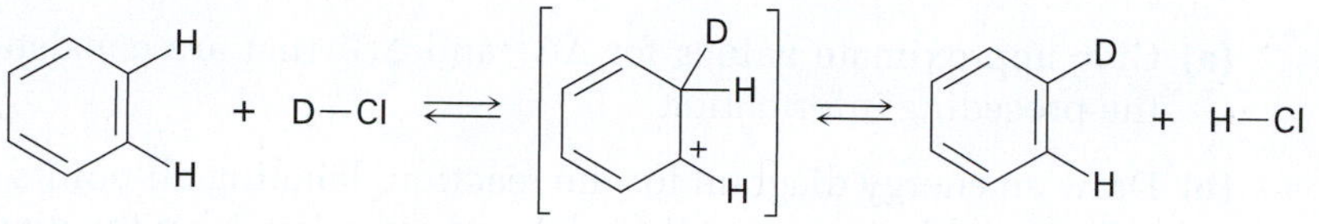

(b)

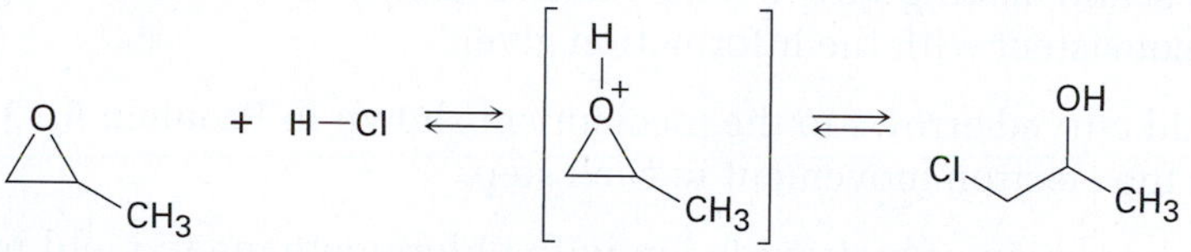

6.30 ▪ Follow the flow of electrons indicated by the curved arrows in each of the following reactions, and predict the products that result:

(a)   (b)

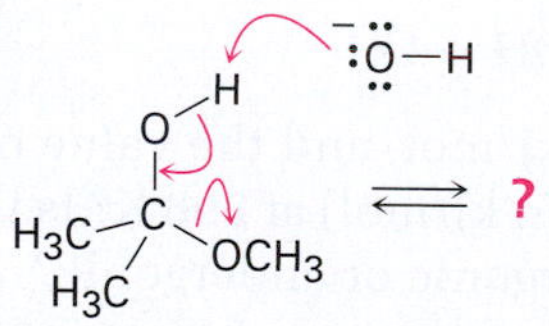

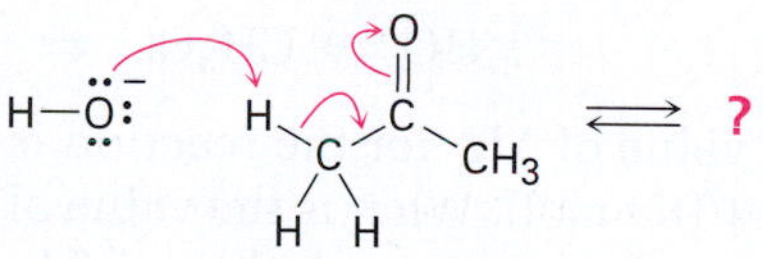

6.31 When isopropylidenecyclohexane is treated with strong acid at room temperature, isomerization occurs by the mechanism shown below to yield 1-isopropylcyclohexene:

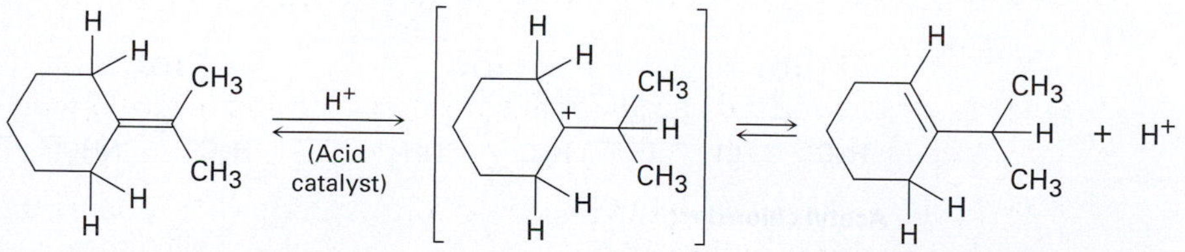

Isopropylidenecyclohexane   **1-Isopropylcyclohexene**

At equilibrium, the product mixture contains about 30% isopropylidene-cyclohexane and about 70% 1-isopropylcyclohexene.

(a) What is an approximate value of K_{eq} for the reaction?

(b) Since the reaction occurs slowly at room temperature, what is its approximate $\Delta G^{\ddagger}$?

(c) Draw an energy diagram for the reaction.

▪ Problems assignable in Organic OWL.

6.32 ■ Add curved arrows to the mechanism shown in Problem 6.31 to indicate the electron movement in each step.

6.33 2-Chloro-2-methylpropane reacts with water in three steps to yield 2-methylpropan-2-ol. The first step is slower than the second, which in turn is much slower than the third. The reaction takes place slowly at room temperature, and the equilibrium constant is near 1.

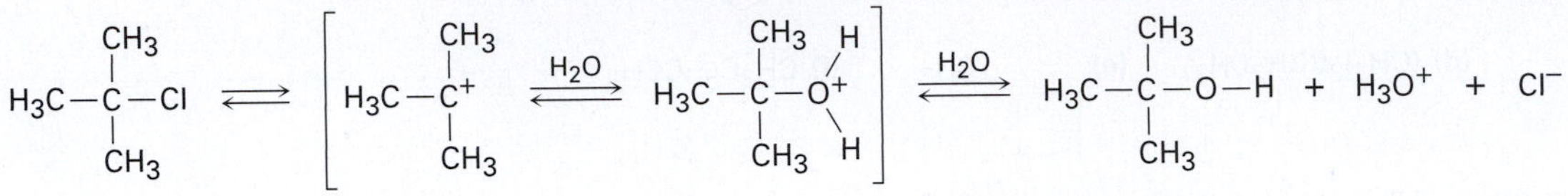

2-Chloro-2-methylpropane

2-Methylpropan-2-ol

(a) Give approximate values for $\Delta G^{\ddagger}$ and ΔG° that are consistent with the preceding information.

(b) Draw an energy diagram for the reaction, labeling all points of interest and making sure that the relative energy levels on the diagram are consistent with the information given.

6.34 ■ Add curved arrows to the mechanism shown in Problem 6.33 to indicate the electron movement in each step.

6.35 ■ The reaction of hydroxide ion with chloromethane to yield methanol and chloride ion is an example of a general reaction type called a *nucleophilic substitution reaction*:

$$HO^- + CH_3Cl \rightleftarrows CH_3OH + Cl^-$$

The value of ΔH° for the reaction is -75 kJ/mol, and the value of ΔS° is $+54$ J/(K · mol). What is the value of ΔG° (in kJ/mol) at 298 K? Is the reaction exothermic or endothermic? Is it exergonic or endergonic?

6.36 ■ Ammonia reacts with acetyl chloride (CH_3COCl) to give acetamide (CH_3CONH_2). Identify the bonds broken and formed in each step of the reaction, and draw curved arrows to represent the flow of electrons in each step.

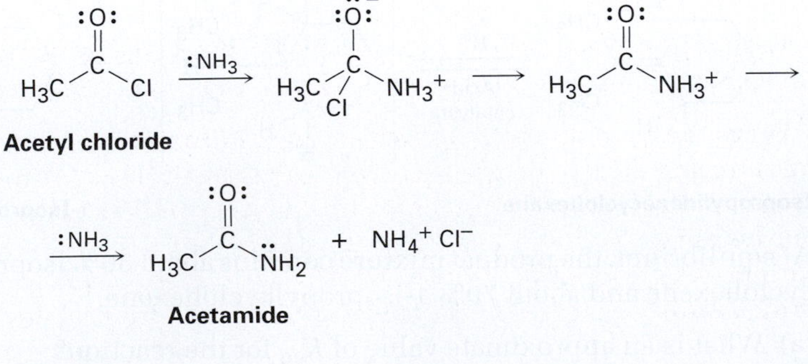

Acetyl chloride

Acetamide

6.37 ■ The naturally occurring molecule α-terpineol is biosynthesized by a route that includes the following step:

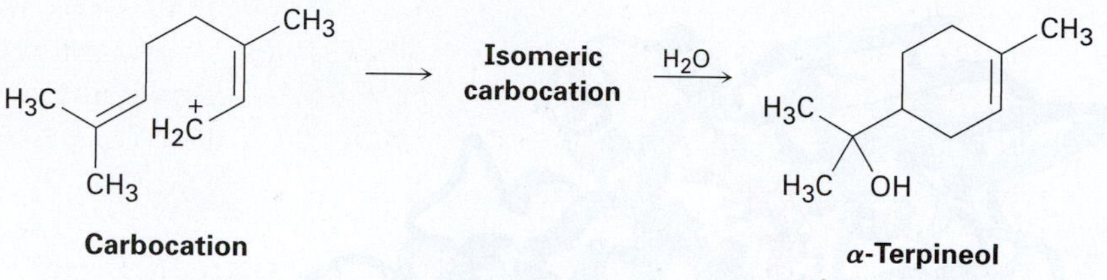

Carbocation   **α-Terpineol**

(a) Propose a likely structure for the isomeric carbocation intermediate.

(b) Show the mechanism of each step in the biosynthetic pathway, using curved arrows to indicate electron flow.

6.38 ■ Predict the product(s) of each of the following biological reactions by interpreting the flow of electrons as indicated by the curved arrows:

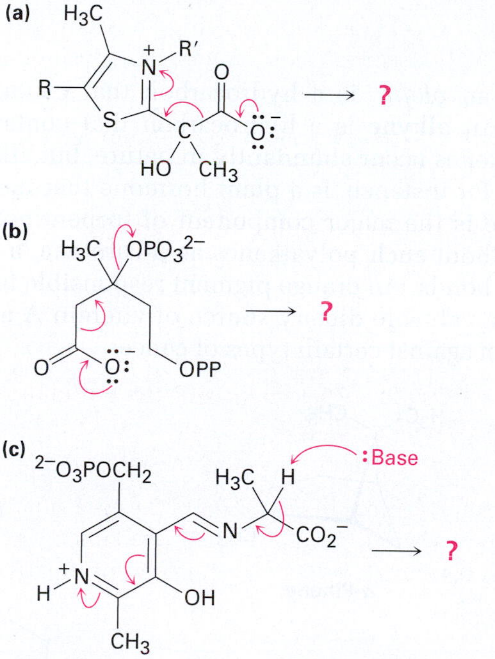

(a)

(b)

(c)

6.39 ■ Reaction of 2-methylpropene with H_3O^+ might, in principle, lead to a mixture of two alcohol addition products. Draw their structures.

6.40 ■ Draw the structures of the two carbocation intermediates that might form during the reaction of 2-methylpropene with H_3O^+ (Problem 6.39). We'll see in the next chapter that the stability of carbocations depends on the number of alkyl substituents attached to the positively charged carbon—the more alkyl substituents there are, the more stable the cation. Which of the two carbocation intermediates you drew is more stable?

■ Problems assignable in Organic OWL.

7   Alkenes and Alkynes

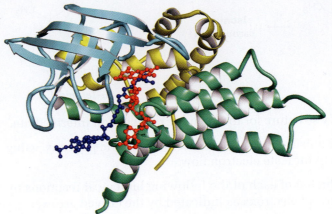

Acyl CoA dehydrogenase catalyzes the introduction of a C=C double bond into fatty acids during their metabolism.

An **alkene**, sometimes called an *olefin,* is a hydrocarbon that contains a carbon–carbon double bond. An **alkyne** is a hydrocarbon that contains a carbon–carbon triple bond. Alkenes occur abundantly in nature, but alkynes are much more rare. Ethylene, for instance, is a plant hormone that induces ripening in fruit, and α-pinene is the major component of turpentine. Life itself would be impossible without such polyalkenes as β-carotene, a compound that contains 11 double bonds. An orange pigment responsible for the color of carrots, β-carotene is a valuable dietary source of vitamin A and is thought to offer some protection against certain types of cancer.

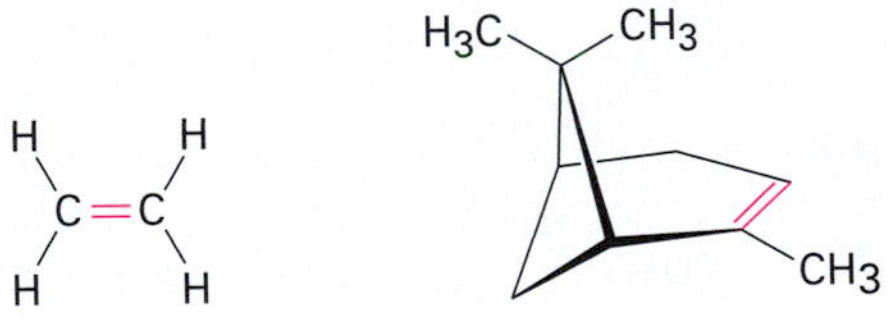

Ethylene

α-Pinene

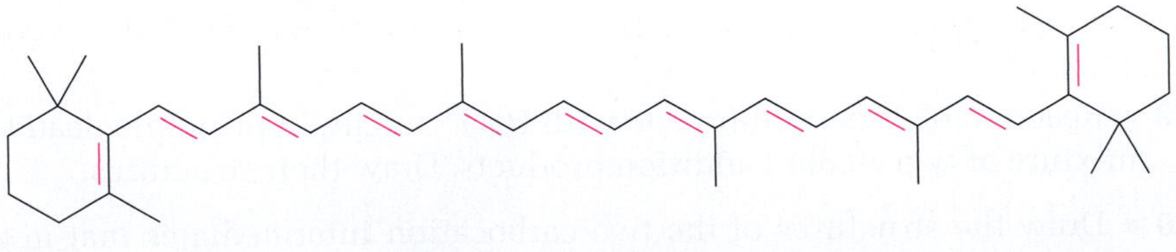

β-Carotene
(orange pigment and vitamin A precursor)

Ethylene and propylene, the simplest alkenes, are the two most important organic chemicals produced industrially. Approximately 28 million tons of ethylene and 17 million tons of propylene are produced each year in the United States for use in the synthesis of polyethylene, polypropylene, ethylene

OWL Online homework for this chapter can be assigned in Organic OWL.

glycol, acetic acid, acetaldehyde, and a host of other substances. Both are synthesized industrially by the "cracking" of C_2–C_8 alkanes on heating to temperatures up to 900 °C.

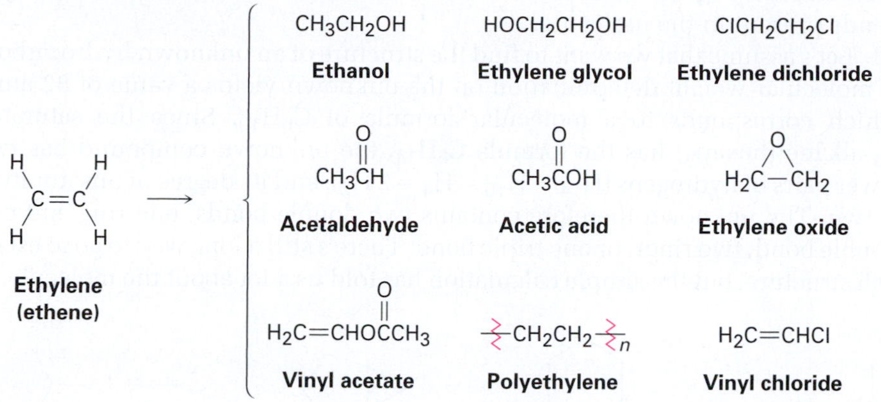

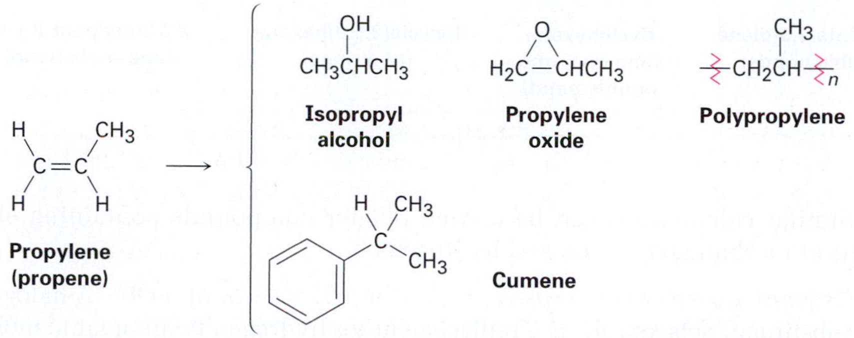

WHY THIS CHAPTER?

Carbon–carbon double bonds are present in most organic and biological molecules, so a good understanding of their behavior is needed. In this chapter, we'll look at some consequences of alkene stereoisomerism and then focus in detail on the broadest and most general class of alkene reactions, the electrophilic addition reaction. Carbon–carbon *triple* bonds, by contrast, occur only rarely in biological molecules and pathways, so we'll not spend much time on their chemistry.

7.1 Calculating a Degree of Unsaturation

Because of its double bond, an alkene has fewer hydrogens than an alkane with the same number of carbons—C_nH_{2n} for an alkene versus C_nH_{2n+2} for an alkane—and is therefore referred to as **unsaturated**. Ethylene, for example, has the formula C_2H_4, whereas ethane has the formula C_2H_6.

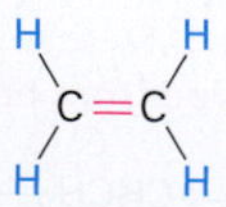

Ethylene: C_2H_4
(fewer hydrogens—*unsaturated*)

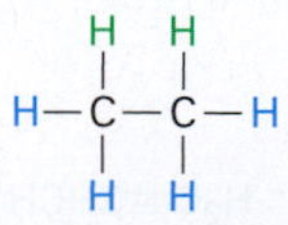

Ethane: C_2H_6
(more hydrogens—*saturated*)

In general, each ring or double bond in a molecule corresponds to a loss of two hydrogens from the related alkane formula C_nH_{2n+2}. Knowing this relationship, it's possible to work backward from a molecular formula to calculate a molecule's **degree of unsaturation**—the number of rings and/or multiple bonds present in the molecule.

Let's assume that we want to find the structure of an unknown hydrocarbon. A molecular weight determination on the unknown yields a value of 82 amu, which corresponds to a molecular formula of C_6H_{10}. Since the saturated C_6 alkane (hexane) has the formula C_6H_{14}, the unknown compound has two fewer pairs of hydrogens ($H_{14} - H_{10} = H_4 = 2\ H_2$), and its degree of unsaturation is two. The unknown therefore contains two double bonds, one ring and one double bond, two rings, or one triple bond. There's still a long way to go to establish structure, but the simple calculation has told us a lot about the molecule.

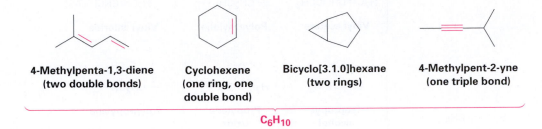

| **4-Methylpenta-1,3-diene**
(two double bonds) | **Cyclohexene**
(one ring, one
double bond) | **Bicyclo[3.1.0]hexane**
(two rings) | **4-Methylpent-2-yne**
(one triple bond) |

C_6H_{10}

Similar calculations can be carried out for compounds containing elements other than just carbon and hydrogen.

- **Organohalogen compounds (C, H, X, where X = F, Cl, Br, or I)** A halogen substituent acts simply as a replacement for hydrogen in an organic molecule, so we can add the number of halogens and hydrogens to arrive at an equivalent hydrocarbon formula from which the degree of unsaturation can be found. For example, the organohalogen formula $C_4H_6Br_2$ is equivalent to the hydrocarbon formula C_4H_8 and thus has one degree of unsaturation.

Replace 2 Br by 2 H

$$BrCH_2CH=CHCH_2Br\ =\ HCH_2CH=CHCH_2H$$

$$C_4H_6Br_2\ =\ \text{``}C_4H_8\text{''}\quad \begin{array}{l}\text{One unsaturation:}\\\text{one double bond}\end{array}$$

Add

- **Organooxygen compounds (C, H, O)** Oxygen doesn't affect the formula of an equivalent hydrocarbon and can be ignored when calculating the degree of unsaturation. You can convince yourself of this by seeing what happens when an oxygen atom is inserted into an alkane bond: C–C becomes C–O–C or C–H becomes C–O–H, and there is no change in the number of hydrogen atoms. For example, the formula C_5H_8O is equivalent to the hydrocarbon formula C_5H_8 and thus has two degrees of unsaturation:

O removed from here

$$H_2C=CHCH=CHCH_2OH\ =\ H_2C=CHCH=CHCH_2{-}H$$

$$C_5H_8O\ =\ \text{``}C_5H_8\text{''}\quad \begin{array}{l}\text{Two unsaturations:}\\\text{two double bonds}\end{array}$$

- **Organonitrogen compounds (C, H, N)** An organonitrogen compound has one more hydrogen than a related hydrocarbon, so you have to subtract the number of nitrogens from the number of hydrogens to arrive at the equivalent hydrocarbon formula. Again, you can convince yourself of this by seeing what happens when a nitrogen atom is inserted into an alkane bond: C–C becomes C–NH–C or C–H becomes C–NH_2, meaning that one additional hydrogen atom has been added. We must therefore subtract this extra hydrogen atom to arrive at the equivalent hydrocarbon formula. For example, the formula C_5H_9N is equivalent to C_5H_8 and thus has two degrees of unsaturation:

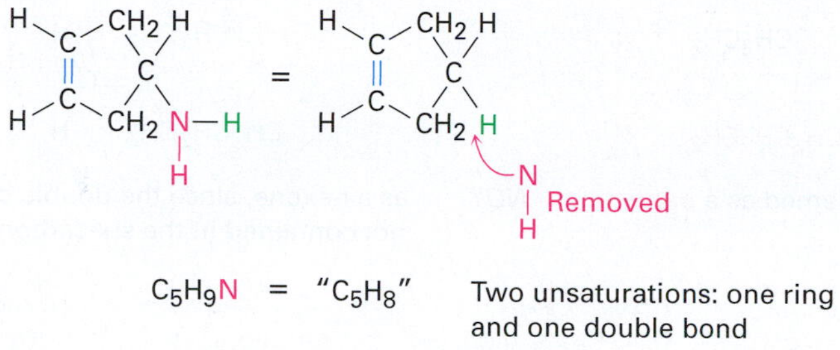

$$C_5H_9N \;=\; \text{``}C_5H_8\text{''} \qquad \text{Two unsaturations: one ring and one double bond}$$

To summarize:

- **Add** the number of halogens to the number of hydrogens.
- **Ignore** the number of oxygens.
- **Subtract** the number of nitrogens from the number of hydrogens.

Problem 7.1

Calculate the degree of unsaturation in the following formulas:
(a) C_8H_{14} **(b)** C_5H_6 **(c)** $C_{12}H_{20}$
(d) C_6H_5N **(e)** $C_6H_5NO_2$ **(f)** $C_8H_9Cl_3$

Problem 7.2

Calculate the degree of unsaturation in the following formulas, and then draw as many structures as you can for each:
(a) C_4H_8 **(b)** C_4H_6 **(c)** C_3H_4

Problem 7.3

Diazepam, marketed as an antianxiety medication under the name Valium, has three rings, eight double bonds, and the formula $C_{16}H_?ClN_2O$. How many hydrogens does diazepam have? (Calculate the answer; don't count hydrogens in the structure.)

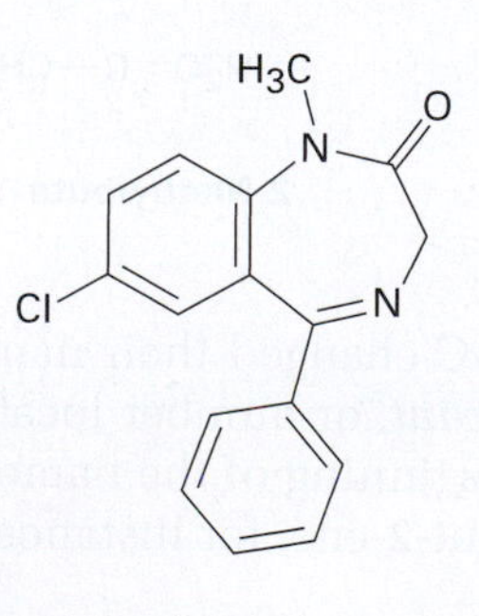

Diazepam

7.2 Naming Alkenes and Alkynes

Alkenes are named using a series of rules similar to those for alkanes (Section 3.4), with the suffix *-ene* used instead of *-ane* to identify the family. There are three steps:

Step 1

Name the parent hydrocarbon. Find the longest carbon chain containing the double bond, and name the compound accordingly, using the suffix *-ene* in place of *-ane*.

Named as a *pentene*   NOT   as a hexene, since the double bond is not contained in the six-carbon chain

Step 2

Number the carbon atoms in the chain. Begin at the end nearer the double bond, or if the double bond is equidistant from the two ends, begin at the end nearer the first branch point. This rule ensures that the double-bond carbons receive the lowest possible numbers:

$$CH_3CH_2CH_2CH\!=\!CHCH_3$$
$$6 \quad 5 \quad 4 \quad 3 \quad\; 2 \quad 1$$

$$\underset{1 \quad 2 \quad 3 \quad\; 4 \quad 5 \quad 6}{CH_3CHCH\!=\!CHCH_2CH_3}$$
with CH_3 on carbon 2

Step 3

Write the full name. Number the substituents according to their positions in the chain, and list them alphabetically. Indicate the position of the double bond by giving the number of the first alkene carbon and placing that number directly before the *-ene* suffix. If more than one double bond is present, indicate the position of each and use one of the suffixes *-diene*, *-triene*, and so on.

Hex-2-ene

2-Methylhex-3-ene

2-Ethylpent-1-ene

2-Methylbuta-1,3-diene

We might also note that IUPAC changed their naming recommendations in 1993. Prior to that time, the *locant*, or number locating the position of the double bond, was placed at the beginning of the name rather than before the *-ene* suffix: 2-butene rather than but-2-ene, for instance. Changes always need time to be fully accepted, so the new rules have not yet been adopted universally, and some texts have not yet been updated. We'll use the new naming system in this book, although you will probably encounter the old system elsewhere. Fortunately, the difference between old and new is minor and rarely causes problems.

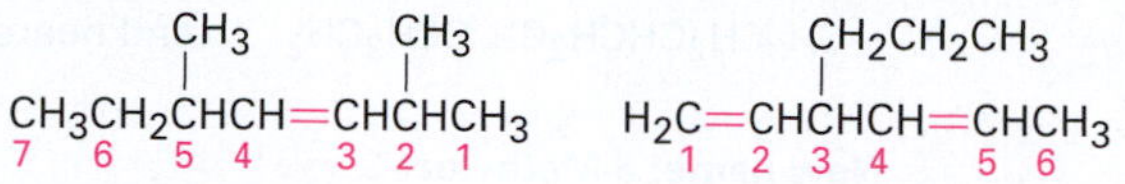

$$CH_3 \qquad CH_3 \qquad\qquad CH_2CH_2CH_3$$
$$| \qquad\quad | \qquad\qquad\qquad |$$
$$CH_3CH_2CHCH{=}CHCHCH_3 \qquad H_2C{=}CHCHCH{=}CHCH_3$$
$$\ \ 7 \ \ \ 6 \ \ \ 5 \ \ \ 4 \quad\ \ 3\ \ \ 2\ \ 1 \qquad\quad 1\ \ \ 2\ \ \ 3\ \ \ 4 \quad\ \ 5\ \ 6$$

Newer naming system: **2,5-Dimethylhept-3-ene**   **3-Propylhexa-1,4-diene**

(Older naming system: **2,5-Dimethyl-3-heptene**   **3-Propyl-1,4-hexadiene)**

Cycloalkenes are named similarly, but because there is no chain end to begin from, we number the cycloalkene so that the double bond is between C1 and C2 and the first substituent has as low a number as possible. Note that it's not necessary to indicate the position of the double bond in the name because it's always between C1 and C2.

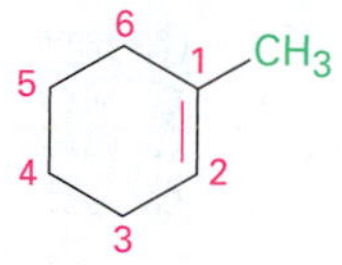

1-Methylcyclohexene 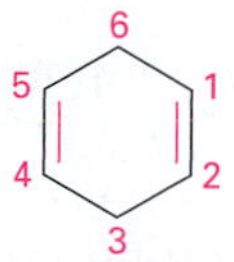   **New name: Cyclohexa-1,4-diene**   **1,5-Dimethylcyclopentene**
                             (Old name: 1,4-Cyclohexadiene)

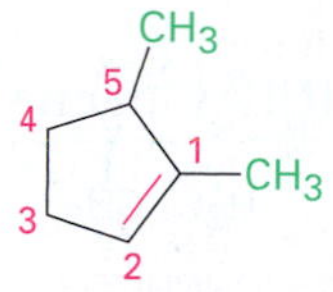

For historical reasons, there are a few alkenes whose names are firmly entrenched in common usage but don't conform to the rules. For example, the alkene derived from ethane should be called *ethene,* but the name *ethylene* has been used so long that it is accepted by IUPAC. Table 7.1 lists several other common names that are often used and are recognized by IUPAC. Note also that a =CH$_2$ substituent is called a *methylene group,* a H$_2$C=CH– substituent is called a *vinyl group,* and a H$_2$C=CHCH$_2$– substituent is called an *allyl group:*

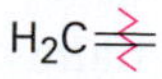

     A methylene group 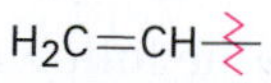   **A vinyl group** 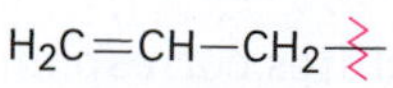   **An allyl group**

TABLE 7.1
Common Names of Some Alkenes

Compound	Systematic name	Common name
$H_2C{=}CH_2$	Ethene	Ethylene
$CH_3CH{=}CH_2$	Propene	Propylene
$CH_3\overset{\displaystyle CH_3}{\underset{\displaystyle |}{C}}{=}CH_2$	2-Methylpropene	Isobutylene
$H_2C{=}\overset{\displaystyle CH_3}{\underset{\displaystyle |}{C}}{-}CH{=}CH_2$	2-Methylbuta-1,3-diene	Isoprene

Alkynes are named just like alkenes, with the suffix -*yne* used in place of -*ene*. Numbering the main chain begins at the end nearer the triple bond so that the triple bond receives as low a number as possible, and the locant is again placed immediately before the -*yne* suffix in the post-1993 naming system.

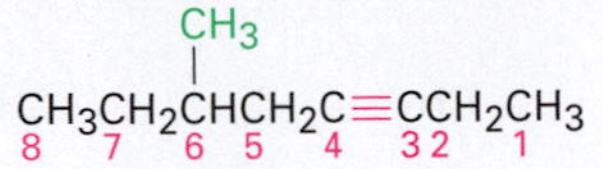

$$CH_3CH_2CHCH_2C\equiv CCH_2CH_3$$
$$\underset{8}{}\quad\underset{7}{}\quad\underset{6}{}\quad\underset{5}{}\quad\underset{4}{}\quad\underset{3\,2}{}\quad\underset{1}{}$$

CH₃ at position 6

Begin numbering at the end nearer the triple bond.

New name: 6-Methyloct-3-yne
(Old name: 6-Methyl-3-octyne)

As with alkyl groups derived from alkanes, *alkenyl* and *alkynyl* groups are also possible:

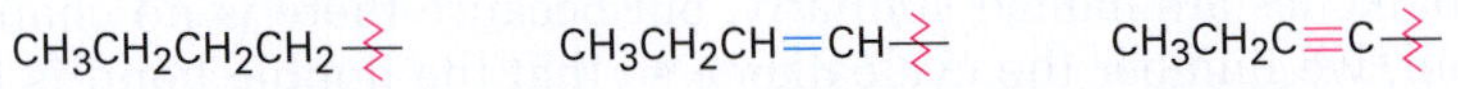

$CH_3CH_2CH_2CH_2$— $CH_3CH_2CH=CH$— $CH_3CH_2C\equiv C$—

Butyl   **But-1-enyl**   **But-1-ynyl**
(an alkyl group)   **(a vinylic group)**   **(an alkynyl group)**

Problem 7.4

Give IUPAC names for the following compounds:

(a) 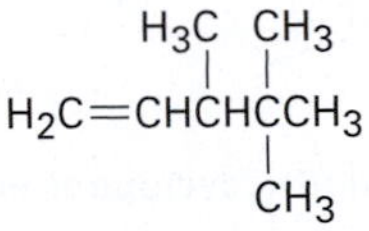
$$H_2C=CHCHCCH_3$$
with H₃C, CH₃ groups and CH₃ below

(b) 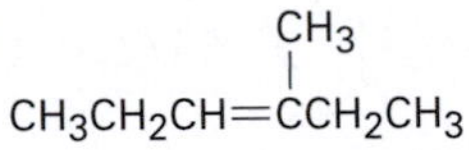
$$CH_3CH_2CH=CCH_2CH_3$$
with CH₃ above

(c) 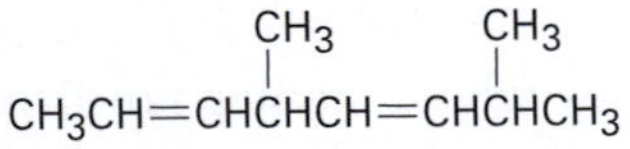
$$CH_3CH=CHCHCH=CHCHCH_3$$
with CH₃ and CH₃ above

(d) 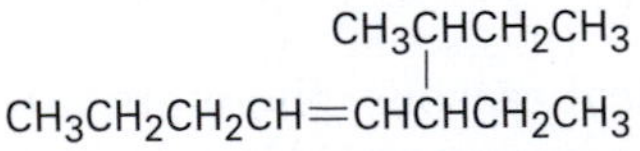
$$CH_3CH_2CH_2CH=CHCHCH_2CH_3$$
with CH₃CHCH₂CH₃ above

(e) 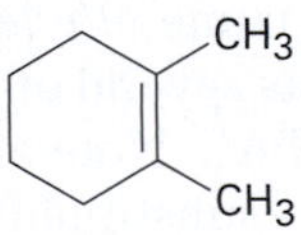 cyclohexene with two CH₃ groups

(f) 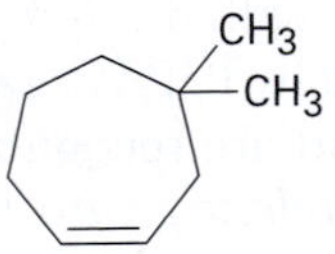 cycloheptene with two CH₃ groups

(g) 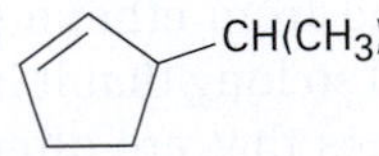 cyclopentene with CH(CH₃)₂ group

Problem 7.5

Draw structures corresponding to the following IUPAC names:
(a) 2-Methylhexa-1,5-diene
(b) 3-Ethyl-2,2-dimethylhept-3-ene
(c) 2,3,3-Trimethylocta-1,4,6-triene
(d) 3,4-Diisopropyl-2,5-dimethylhex-3-ene

Problem 7.6

Name the following alkynes:

(a) 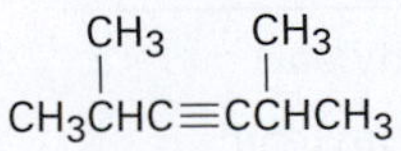
$$CH_3CHC\equiv CCHCH_3$$
with CH₃ and CH₃ above

(b) 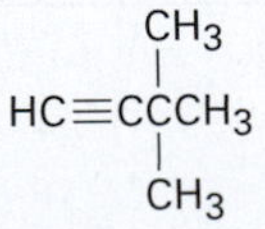
$$HC\equiv CCCH_3$$
with CH₃ above and CH₃ below

(c) 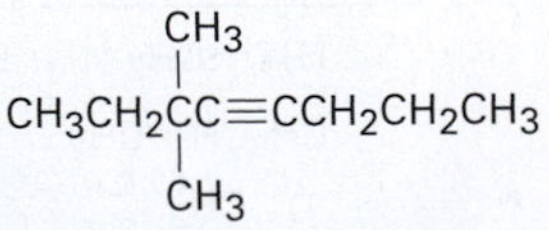
$$CH_3CH_2CC\equiv CCH_2CH_2CH_3$$
with CH₃ above and CH₃ below

(d) 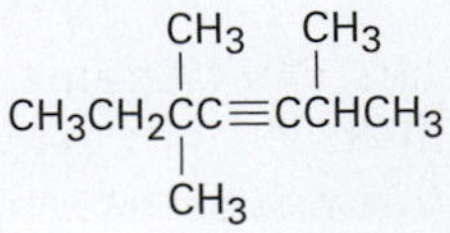
$$CH_3CH_2CC\equiv CCHCH_3$$
with CH₃, CH₃ above and CH₃ below

(e) 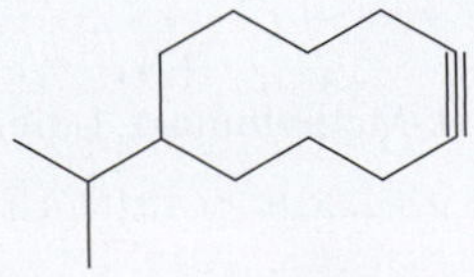 a cyclic structure with an isopropyl group and a triple bond

Problem 7.7

Change the following old names to new, post-1993 names, and draw the structure of each compound:

(a) 2,5,5-Trimethyl-2-hexene   **(b)** 2,2-Dimethyl-3-hexyne

7.3 Cis–Trans Isomerism in Alkenes

We saw in Chapter 1 that the carbon–carbon double bond can be described in two ways. In valence bond language (Section 1.8), the carbons are sp^2-hybridized and have three equivalent hybrid orbitals that lie in a plane at angles of 120° to one another. The carbons form a σ bond by head-on overlap of sp^2 orbitals and a π bond by sideways overlap of unhybridized p orbitals oriented perpendicular to the sp^2 plane, as shown in Figure 1.15 on page 15. In molecular orbital language (Section 1.11), interaction between the p orbitals leads to one bonding and one antibonding π molecular orbital. The π bonding MO has no node between nuclei and results from a combination of p orbital lobes with the same algebraic sign. The π antibonding MO has a node between nuclei and results from a combination of lobes with different algebraic signs, as shown in Figure 1.19 on page 21.

Although essentially free rotation is possible around single bonds (Section 3.6), the same is not true of double bonds. For rotation to occur around a double bond, the π bond must break and re-form (Figure 7.1). Thus, the barrier to double-bond rotation must be at least as great as the strength of the π bond itself, an estimated 350 kJ/mol (84 kcal/mol). Recall that the barrier to bond rotation in ethane is only 12 kJ/mol.

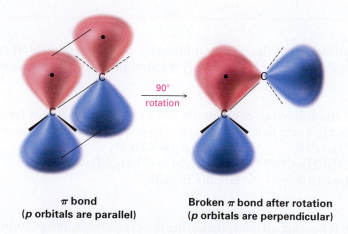

FIGURE 7.1 The π bond must break for rotation to take place around a carbon–carbon double bond.

90°
rotation

π bond
(*p* orbitals are parallel)

Broken π bond after rotation
(*p* orbitals are perpendicular)

The lack of rotation around carbon–carbon double bonds is of more than just theoretical interest; it also has chemical consequences. Imagine the situation for a disubstituted alkene such as but-2-ene. (*Disubstituted* means that two substituents other than hydrogen are attached to the double-bond carbons.) The two methyl groups in but-2-ene can be either on the same side of the double bond or on opposite sides, a situation reminiscent of disubstituted cycloalkanes (Section 4.2).

Since bond rotation can't occur, the two but-2-enes can't spontaneously interconvert; they are different, isolable compounds. As with disubstituted cycloalkanes, we call such compounds *cis–trans stereoisomers*.

The compound with substituents on the same side of the double bond is called *cis*-but-2-ene, and the isomer with substituents on opposite sides is *trans*-but-2-ene (Figure 7.2).

FIGURE 7.2 Cis and trans isomers of but-2-ene. The cis isomer has the two methyl groups on the same side of the double bond, and the trans isomer has the methyl groups on opposite sides.

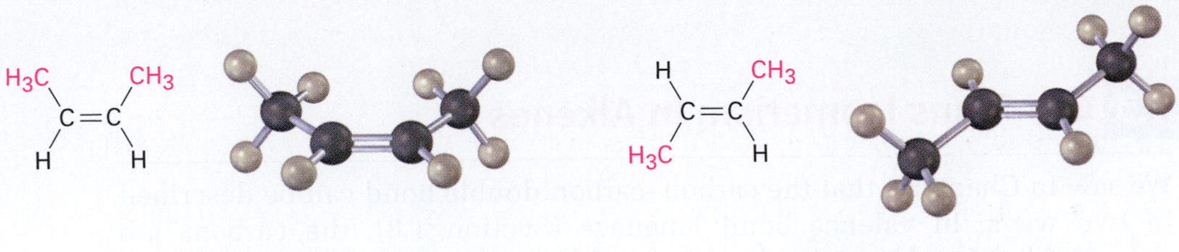

cis-But-2-ene

trans-But-2-ene

Cis–trans isomerism is not limited to *di*substituted alkenes. It can occur whenever both double-bond carbons are attached to two different groups. If one of the double-bond carbons is attached to two identical groups, however, cis–trans isomerism is not possible (Figure 7.3).

FIGURE 7.3 The requirement for cis–trans isomerism in alkenes. Compounds that have one of their carbons bonded to two identical groups can't exist as cis–trans isomers. Only when both carbons are bonded to two different groups are cis–trans isomers possible.

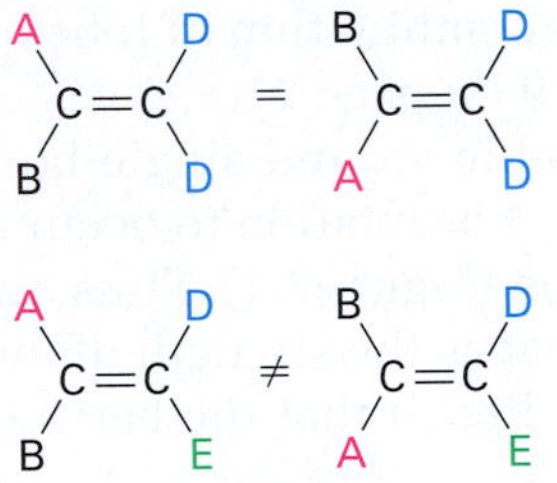

These two compounds are identical; they are not cis–trans isomers.

These two compounds are not identical; they are cis–trans isomers.

Problem 7.8

The sex attractant of the common housefly is an alkene named *cis*-tricos-9-ene. Draw its structure. (Tricosane is the straight-chain alkane $C_{23}H_{48}$.)

Problem 7.9

Which of the following compounds can exist as pairs of cis–trans isomers? Draw each cis–trans pair, and indicate the geometry of each isomer.
(a) $CH_3CH{=}CH_2$ **(b)** $(CH_3)_2C{=}CHCH_3$
(c) $CH_3CH_2CH{=}CHCH_3$ **(d)** $(CH_3)_2C{=}C(CH_3)CH_2CH_3$
(e) $ClCH{=}CHCl$ **(f)** $BrCH{=}CHCl$

Problem 7.10

Name the following alkenes, including the cis or trans designation:

(a)   **(b)**

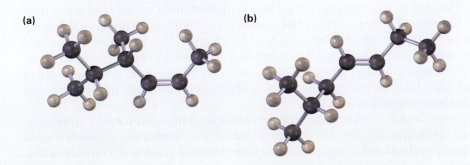

7.4 Alkene Stereochemistry and the *E,Z* Designation

The cis–trans naming system used in the previous section works only with disubstituted alkenes—compounds that have two substituents other than hydrogen on the double bond. With trisubstituted and tetrasubstituted double bonds, however, a more general method is needed for describing double-bond geometry. (*Trisubstituted* means three substituents other than hydrogen on the double bond; *tetrasubstituted* means four substituents other than hydrogen.)

The method used for describing alkene stereochemistry is called the **E,Z system** and employs the same Cahn–Ingold–Prelog sequence rules given in Section 5.5 for specifying the configuration of a chirality center. Let's briefly review the sequence rules and then see how they're used to specify double-bond geometry. For a more thorough review, you should reread Section 5.5.

Rule 1

Considering each of the double-bond carbons separately, look at the two substituents attached and rank them according to the atomic number of the first atom in each. An atom with higher atomic number ranks higher than an atom with lower atomic number.

Rule 2

If a decision can't be reached by ranking the first atoms in the two substituents, look at the second, third, or fourth atoms away from the double-bond until the first difference is found.

Rule 3

Multiple-bonded atoms are equivalent to the same number of single-bonded atoms.

Once the two groups attached to each doubly bonded carbon atom have been ranked as either higher or lower, look at the entire molecule. If the higher-ranked groups on each carbon are on the same side of the double bond, the alkene is designated **Z**, for the German *zusammen,* meaning "together." If the higher-ranked groups are on opposite sides, the alkene is designated **E**, for the German *entgegen,* meaning "opposite." (A simple way to remember which is which is to note that the groups are on "ze zame zide" in the *Z* isomer.)

As an example, look at the following two isomers of 2-chlorobut-2-ene. Because chlorine has a higher atomic number than carbon, a –Cl substituent is ranked higher than a –CH$_3$ group. Methyl is ranked higher than hydrogen, however, so isomer (a) is assigned *E* geometry because the higher-ranked groups are on opposite sides of the double bond. Isomer (b) has *Z* geometry because its higher-ranked groups are on ze zame zide of the double bond.

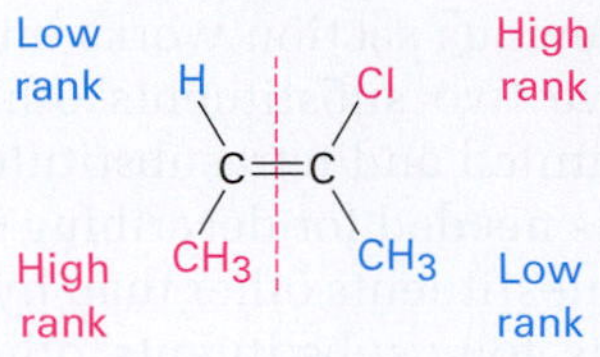

(a) (*E*)-2-Chlorobut-2-ene 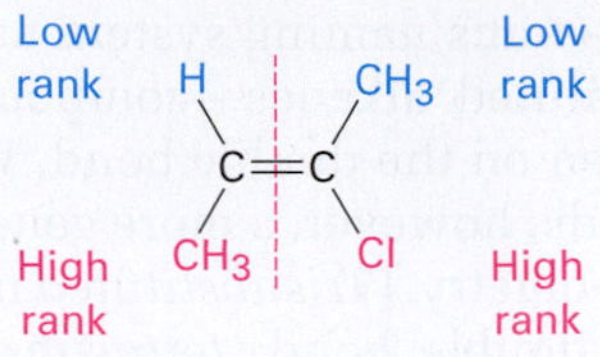   **(b) (*Z*)-2-Chlorobut-2-ene**

For further practice, work through each of the following examples to convince yourself that the assignments are correct:

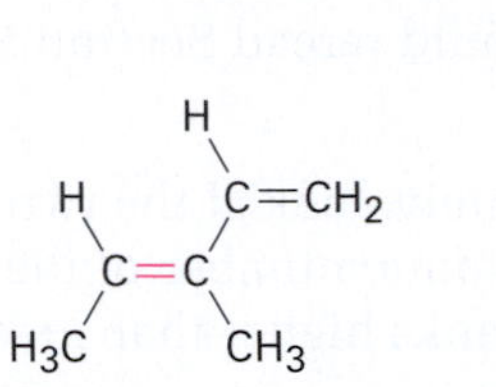

(*E*)-3-Methylpenta-1,3-diene 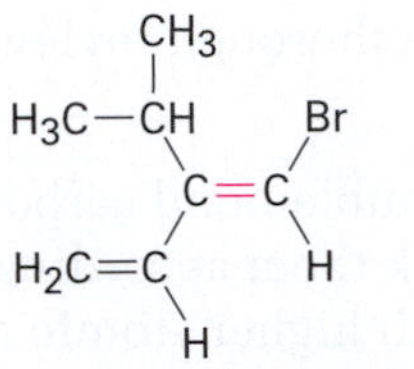   **(*E*)-1-Bromo-2-isopropyl-   **(*Z*)-2-Hydroxymethyl-
                                       buta-1,3-diene**   but-2-enoic acid**

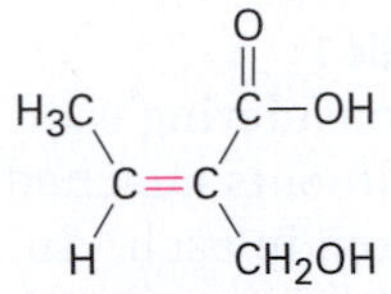

WORKED EXAMPLE 7.1 Assigning *E* and *Z* Configurations to Substituted Alkenes

Assign *E* or *Z* configuration to the double bond in the following compound:

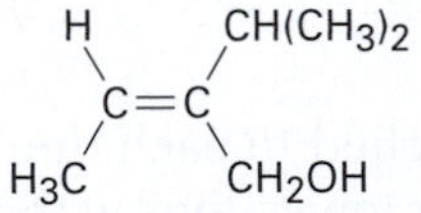

Strategy

Look at the two substituents connected to each double-bond carbon, and determine their ranking using the Cahn–Ingold–Prelog rules. Then see whether the two higher-ranked groups are on the same or opposite sides of the double bond.

Solution

The left-hand carbon has –H and –CH$_3$ substituents, of which –CH$_3$ ranks higher by sequence rule 1. The right-hand carbon has –CH(CH$_3$)$_2$ and –CH$_2$OH substituents, which are equivalent by rule 1. By rule 2, however, –CH$_2$OH ranks higher than –CH(CH$_3$)$_2$. The substituent –CH$_2$OH has an *oxygen* as its highest second atom, but –CH(CH$_3$)$_2$ has a *carbon* as its highest second atom. The two higher-ranked groups are on the same side of the double bond, so we assign *Z* configuration.

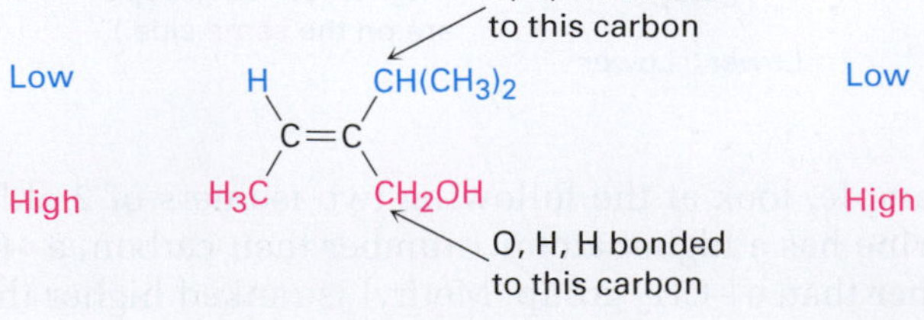

***Z* configuration**

Problem 7.11

Which member in each of the following sets ranks higher?

(a) –H or –CH₃   (b) –Cl or –CH₂Cl
(c) –CH₂CH₂Br or –CH=CH₂   (d) –NHCH₃ or –OCH₃
(e) –CH₂OH or –CH=O   (f) –CH₂OCH₃ or –CH=O

Problem 7.12

Rank the substituents in each of the following sets according to the sequence rules:

(a) –CH₃, –OH, –H, –Cl
(b) –CH₃, –CH₂CH₃, –CH=CH₂, –CH₂OH
(c) –CO₂H, –CH₂OH, –C≡N, –CH₂NH₂
(d) –CH₂CH₃, –C≡CH, –C≡N, –CH₂OCH₃

Problem 7.13

Assign *E* or *Z* configuration to the double bonds in the following compounds:

(a)   H₃C   CH₂OH
              C=C
       CH₃CH₂   Cl

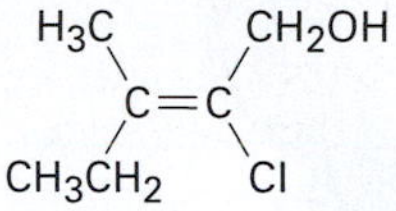

(b)   Cl   CH₂CH₃
             C=C
       CH₃O   CH₂CH₂CH₃

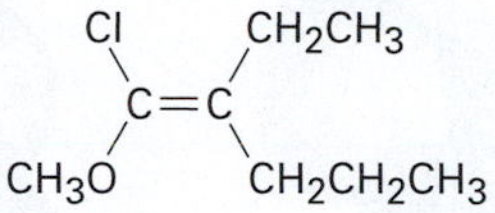

(c) CH₃
                 CO₂H
            C=C
              CH₂OH

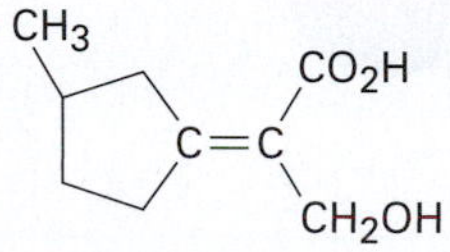

(d)   H   CN
            C=C
       H₃C   CH₂NH₂

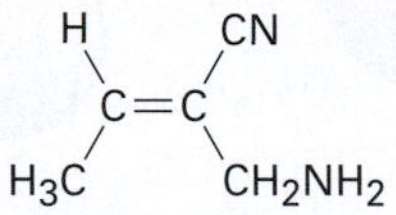

Problem 7.14

Assign stereochemistry (*E* or *Z*) to the double bond in the following compound, and convert the drawing into a skeletal structure (red = O):

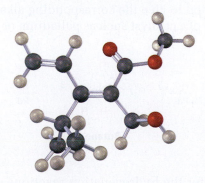

7.5 Stability of Alkenes

Although the cis–trans interconversion of alkene isomers does not occur spontaneously, it can often be brought about by treating the alkene with a strong acid catalyst. If we interconvert *cis*-but-2-ene with *trans*-but-2-ene and allow them to reach equilibrium, we find that they aren't of equal stability. The trans isomer is more stable than the cis isomer by 2.8 kJ/mol (0.66 kcal/mol) at room temperature, corresponding to a 76 : 24 ratio:

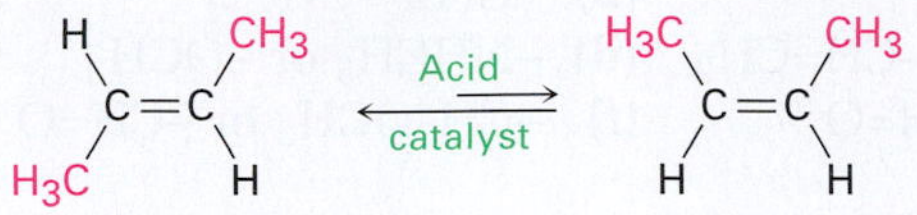

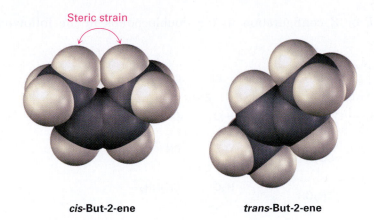

Cis alkenes are less stable than their trans isomers because of steric strain between the two larger substituents on the same side of the double bond. This is the same kind of steric interference that we saw previously in the axial conformation of methylcyclohexane (Section 4.7).

Steric strain

cis-But-2-ene   *trans*-But-2-ene

Although it's sometimes possible to find relative stabilities of alkene isomers by establishing a cis–trans equilibrium through treatment with strong acid, a more general method is to take advantage of the fact that alkenes undergo a *hydrogenation* reaction to give the corresponding alkane on treatment with H_2 gas in the presence of a catalyst such as palladium or platinum:

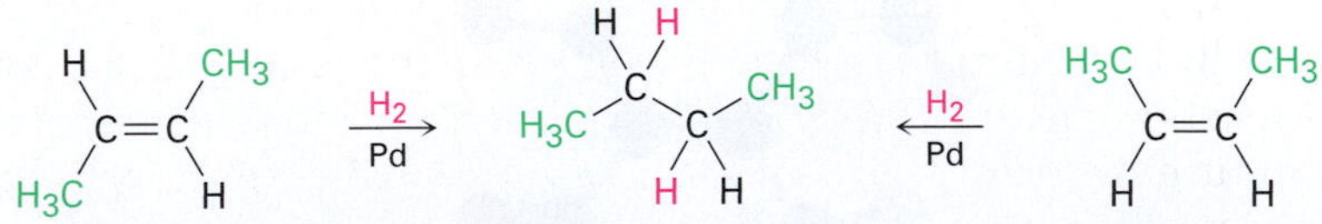

trans-But-2-ene   Butane   *cis*-But-2-ene

Energy diagrams for the hydrogenation reactions of *cis*- and *trans*-but-2-ene are shown in Figure 7.4. Because *cis*-but-2-ene is less stable than *trans*-but-2-ene by 2.8 kJ/mol, the energy diagram shows the cis alkene at a higher energy level. After reaction, however, both curves are at the same energy level (butane). It therefore follows that $\Delta G°$ for reaction of the cis isomer must be larger than $\Delta G°$ for reaction of the trans isomer by 2.8 kJ/mol. In other words, more energy is released in the hydrogenation of the cis isomer than the trans isomer because the cis isomer is higher in energy to begin with.

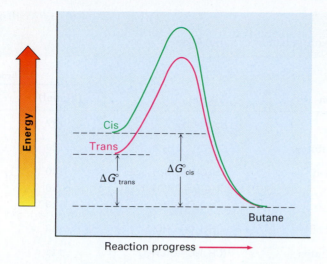

FIGURE 7.4 Energy diagrams for hydrogenation of *cis*- and *trans*-but-2-ene. The cis isomer is higher in energy than the trans isomer by about 2.8 kJ/mol and therefore releases more energy in the reaction.

If we were to measure the so-called heats of hydrogenation ($\Delta H°_{hydrog}$) for two double-bond isomers and find their difference, we could determine the relative stabilities of cis and trans isomers without having to measure an equilibrium position. *cis*-But-2-ene, for instance, has $\Delta H°_{hydrog} = -120$ kJ/mol (-28.6 kcal/mol), while *trans*-but-2-ene has $\Delta H°_{hydrog} = -116$ kJ/mol (-27.6 kcal/mol)—a difference of 4 kJ/mol.

Cis isomer
$\Delta H°_{hydrog} = -120$ kJ/mol

Trans isomer
$\Delta H°_{hydrog} = -116$ kJ/mol

The 4 kJ/mol energy difference between the but-2-ene isomers calculated from heats of hydrogenation agrees reasonably well with the 2.8 kJ/mol energy difference calculated from equilibrium data, but the numbers aren't exactly the same for two reasons. First, there is probably some experimental error, since heats of hydrogenation are difficult to measure accurately. Second, heats of reaction and equilibrium constants don't measure exactly the same thing. Heats of reaction measure enthalpy changes, $\Delta H°$, whereas equilibrium constants measure free-energy changes, $\Delta G°$, so we might expect a slight difference between the two.

Table 7.2 lists some representative data for the hydrogenation of different alkenes and shows that alkenes become more stable with increasing substitution. That is, alkenes follow the stability order:

Tetrasubstituted > Trisubstituted > Disubstituted > Monosubstituted

We'll see some consequences of this stability order in later chapters.

TABLE 7.2
Heats of Hydrogenation of Some Alkenes

Substitution	Alkene	$\Delta H°_{hydrog}$ (kJ/mol)	(kcal/mol)
Ethylene	$H_2C{=}CH_2$	−137	−32.8
Monosubstituted	$CH_3CH{=}CH_2$	−126	−30.1
Disubstituted	$CH_3CH{=}CHCH_3$ (cis)	−120	−28.6
	$CH_3CH{=}CHCH_3$ (trans)	−116	−27.6
	$(CH_3)_2C{=}CH_2$	−119	−28.4
Trisubstituted	$(CH_3)_2C{=}CHCH_3$	−113	−26.9
Tetrasubstituted	$(CH_3)_2C{=}C(CH_3)_2$	−111	−26.6

The stability order of substituted alkenes is due to a combination of two factors. One is a stabilizing interaction between the C=C π bond and adjacent C–H σ bonds on substituents. In valence-bond language, the interaction is called **hyperconjugation**. In a molecular orbital description, there is a bonding MO that extends over the four-atom C=C–C–H grouping, as shown in Figure 7.5. The more substituents that are present on the double bond, the more hyperconjugation there is and the more stable the alkene.

FIGURE 7.5 Hyperconjugation is a stabilizing interaction between the C=C π bond and adjacent C–H σ bonds on substituents, as indicated by this molecular orbital. The more substituents there are, the greater the stabilization of the alkene.

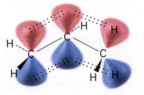

A second factor that contributes to alkene stability involves bond strengths. A bond between an sp^2 carbon and an sp^3 carbon is somewhat stronger than a bond between two sp^3 carbons. Thus, in comparing but-1-ene and but-2-ene, the monosubstituted isomer has one $sp^3–sp^3$ bond and one $sp^3–sp^2$ bond, while the disubstituted isomer has two $sp^3–sp^2$ bonds. More highly substituted alkenes always have a higher ratio of $sp^3–sp^2$ bonds to $sp^3–sp^3$ bonds than less highly substituted alkenes and are therefore more stable.

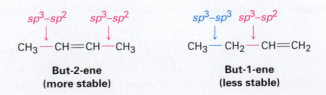

But-2-ene
(more stable)   But-1-ene
(less stable)

Problem 7.15

Name the following alkenes, and tell which compound in each pair is more stable:

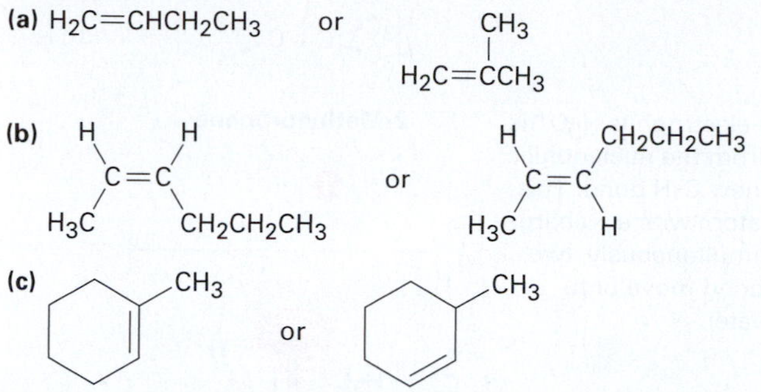

(a) $H_2C=CHCH_2CH_3$ or

$$H_2C=CCH_3$$
with CH_3 substituent

(b)

$$\underset{H_3C}{\overset{H}{>}}C=C\underset{CH_2CH_2CH_3}{\overset{H}{<}}$$ or $$\underset{H_3C}{\overset{H}{>}}C=C\underset{H}{\overset{CH_2CH_2CH_3}{<}}$$

(c) (cyclohexene with CH3 on double-bond carbon)   or   (cyclohexene with CH3 allylic)

7.6 Electrophilic Addition Reactions of Alkenes

Before beginning a detailed discussion of alkene reactions, let's review briefly some conclusions from the previous chapter. We said in Section 6.5 that alkenes behave as nucleophiles (Lewis bases) in polar reactions, donating a pair of electrons from their electron-rich C=C bond to an electrophile (Lewis acid). For example, acid-catalyzed reaction of 2-methylpropene with H_2O yields 2-methylpropan-2-ol, where the *-ol* name ending on the product indicates an alcohol. A careful study of this and similar reactions has led to the generally accepted mechanism shown in Figure 7.6 for the **electrophilic addition reaction**.

The reaction begins with an attack on a hydrogen of the electrophile, H_3O^+, by the electrons of the nucleophilic π bond. Two electrons from the π bond form a new σ bond between the entering hydrogen and an alkene carbon, as shown by the curved arrow at the top of Figure 7.6. The carbocation intermediate that results is itself an electrophile, which can accept an electron pair from nucleophilic H_2O to form a C–O bond and yield a protonated alcohol addition product. Removal of H^+ by acid–base reaction with water then gives the alcohol product and regenerates the acid catalyst.

Electrophilic addition to alkenes is successful not only with H_2O but with HBr, HCl, and HI as well, although the addition of halogen acids is not common in living organisms.

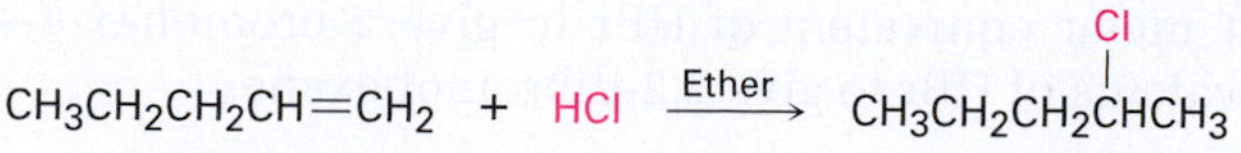

$$CH_3CH_2CH_2CH=CH_2 + \text{HCl} \xrightarrow{\text{Ether}} CH_3CH_2CH_2\overset{Cl}{\underset{|}{C}}HCH_3$$

Pent-1-ene   **2-Chloropentane**

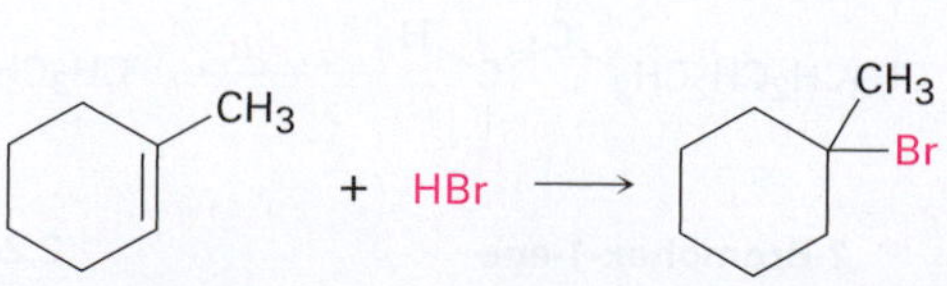

(1-methylcyclohexene structure) + HBr ⟶ (1-bromo-1-methylcyclohexane structure)

1-Methylcyclohexene   **1-Bromo-1-methylcyclohexane**

ACTIVE FIGURE 7.6

MECHANISM: Mechanism of the acid-catalyzed electrophilic addition of H_2O to 2-methylpropene to give the alcohol 2-methylpropan-2-ol. The reaction involves a carbocation intermediate. **Go to this book's student companion site at** www.cengage.com/chemistry/mcmurry **to explore an interactive version of this figure.**

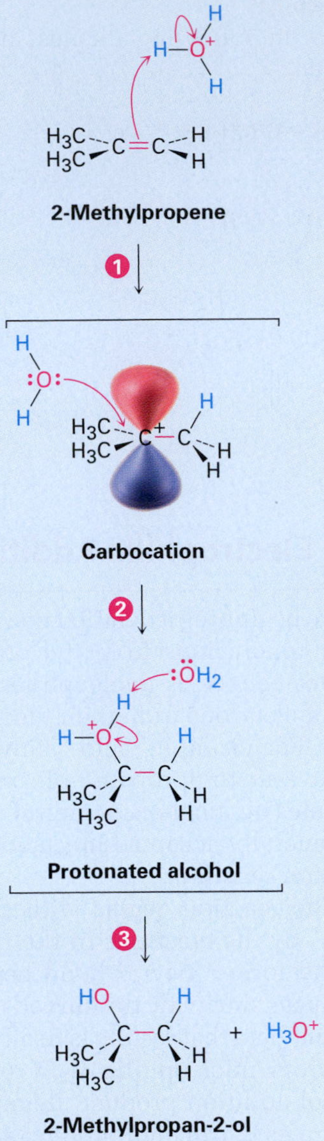

1 A hydrogen atom on the electrophile H_3O^+ is attacked by π electrons from the nucleophilic double bond, forming a new C–H bond. This leaves the other carbon atom with a + charge and a vacant p orbital. Simultaneously, two electrons from the H–O bond move onto oxygen, giving neutral water.

2 The nucleophile H_2O donates an electron pair to the positively charged carbon atom, forming a C–O bond and leaving a positive charge on oxygen in the protonated alcohol addition product.

3 Water acts as a base to remove H^+, regenerating H_3O^+ and yielding the neutral alcohol addition product.

© John McMurry

Alkynes, too, undergo electrophilic addition reactions, although their reactivity is substantially less than that of alkenes. Hex-1-yne, for instance, reacts with 1 molar equivalent of HBr to give 2-bromohex-1-ene and with 2 molar equivalents of HBr to give 2,2-dibromohexane.

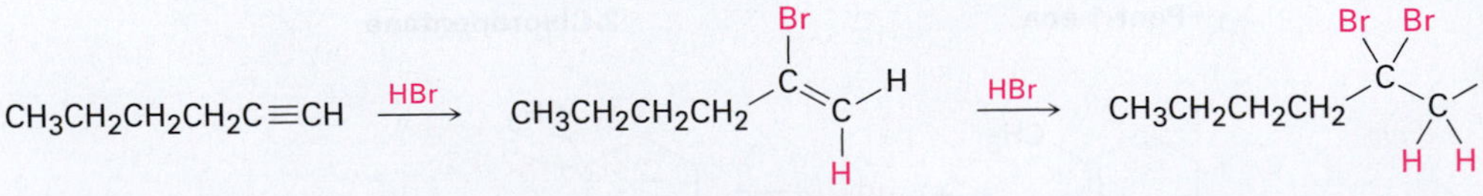

WRITING ORGANIC REACTIONS

This is a good time to mention that organic reaction equations are sometimes written in different ways to emphasize different points. In describing a laboratory process, for example, the reaction of 2-methylpropene with HCl might be written in the format A + B → C to emphasize that both reactants are equally important for the purposes of the discussion. The solvent and notes about other reaction conditions, such as temperature, are written either above or below the reaction arrow:

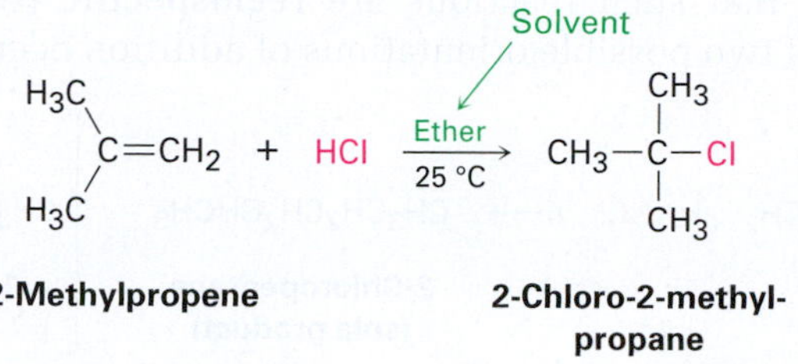

2-Methylpropene   **2-Chloro-2-methyl-
                                      propane**

Alternatively, we might write the same reaction in a format to emphasize that 2-methylpropene is the reactant whose chemistry is of greater interest. The second reactant, HCl, is placed above the reaction arrow together with notes about solvent and reaction conditions:

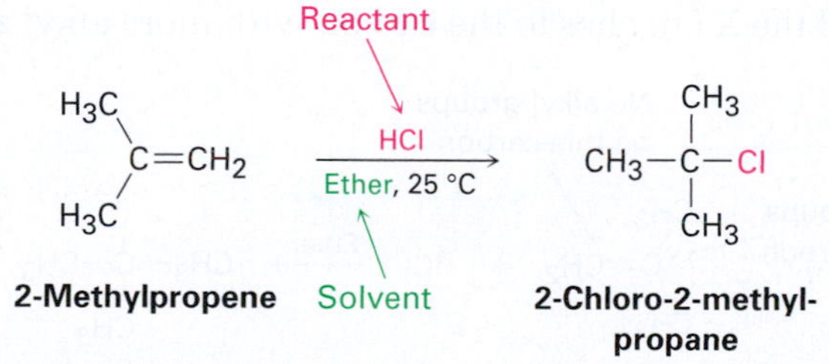

2-Methylpropene   Solvent   **2-Chloro-2-methyl-
                                    propane**

In describing a biological process, the reaction is usually written to show only the structure of the primary reactant and product, while abbreviating the structures of various biological "reagents" and by-products by using a curved arrow that intersects the straight reaction arrow. The reaction of glucose with ATP (Section 6.11) to give glucose 6-phosphate plus ADP would be written as follows:

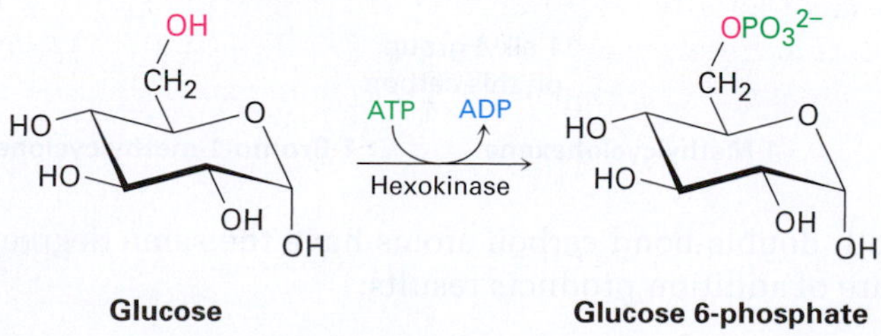

Glucose   **Glucose 6-phosphate**

7.7 Orientation of Electrophilic Addition: Markovnikov's Rule

Look carefully at the electrophilic addition reactions shown in the previous section. In each case, an unsymmetrically substituted alkene gives a single addition product, rather than the mixture that might be expected. For example, pent-1-ene *might* react with HCl to give both 1-chloropentane and 2-chloropentane, but it doesn't. It gives only 2-chloropentane. Similarly, it's invariably the case in biological alkene addition reactions that only a single product is formed. We say that such reactions are **regiospecific** (**ree**-jee-oh-specific) when only one of two possible orientations of addition occurs.

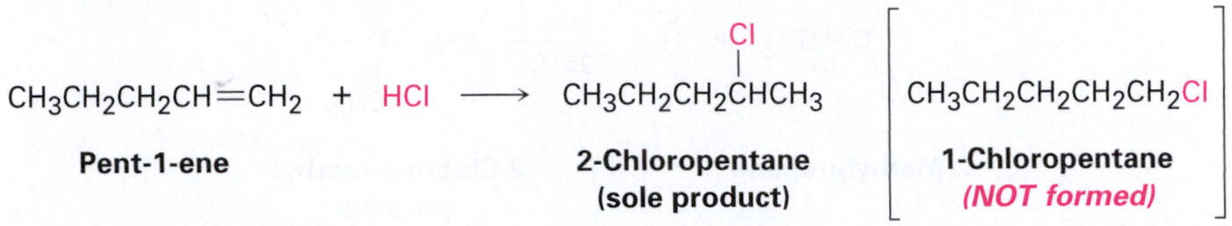

$$CH_3CH_2CH_2CH=CH_2 \ + \ HCl \ \longrightarrow \ CH_3CH_2CH_2\overset{\overset{\displaystyle Cl}{|}}{C}HCH_3 \qquad \left[\ CH_3CH_2CH_2CH_2CH_2Cl\ \right]$$

Pent-1-ene   **2-Chloropentane**   **1-Chloropentane**
                                                         (sole product)   *(NOT formed)*

After looking at the results of many such reactions, the Russian chemist Vladimir Markovnikov proposed in 1869 what has become known as **Markovnikov's rule**:

Markovnikov's rule
In the addition of HX to an alkene, the H attaches to the carbon with fewer alkyl substituents and the X attaches to the carbon with more alkyl substituents.

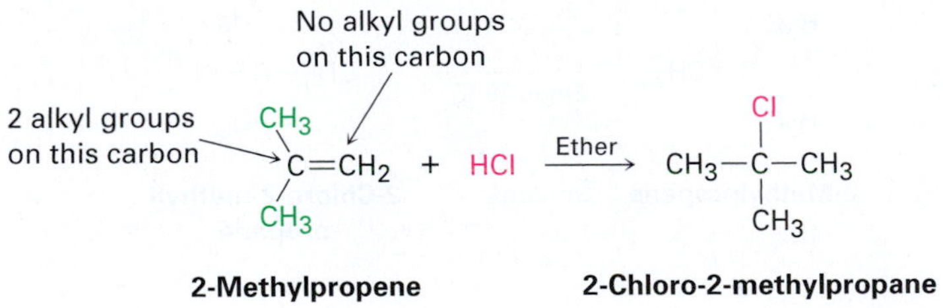

2-Methylpropene   **2-Chloro-2-methylpropane**

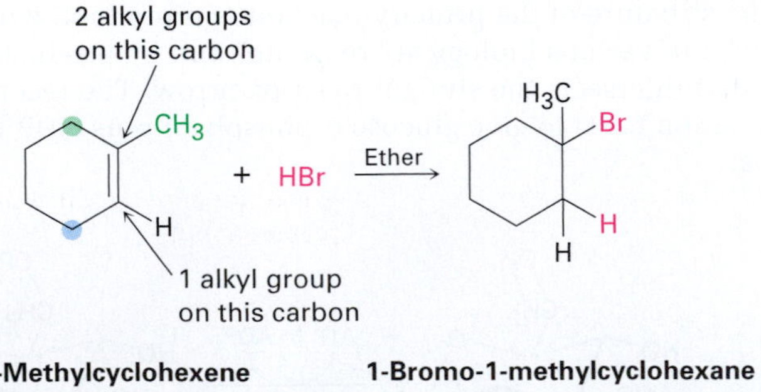

1-Methylcyclohexene   **1-Bromo-1-methylcyclohexane**

When both double-bond carbon atoms have the same degree of substitution, a mixture of addition products results:

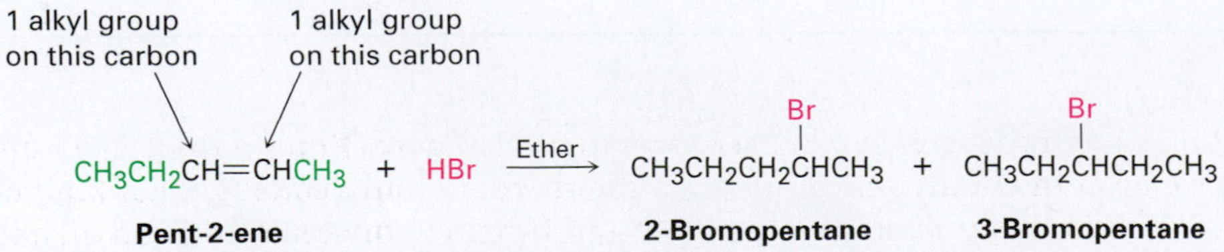

Pent-2-ene   **2-Bromopentane**   **3-Bromopentane**

Because carbocations are involved as intermediates in these electrophilic addition reactions, Markovnikov's rule can be restated in the following way:

Markovnikov's rule restated
In the addition of HX to an alkene, the more highly substituted carbocation is formed as the intermediate rather than the less highly substituted one.

For example, addition of H^+ to 2-methylpropene yields the intermediate *tertiary* carbocation rather than the alternative primary carbocation, and addition to 1-methylcyclohexene yields a tertiary cation rather than a secondary one. Why should this be?

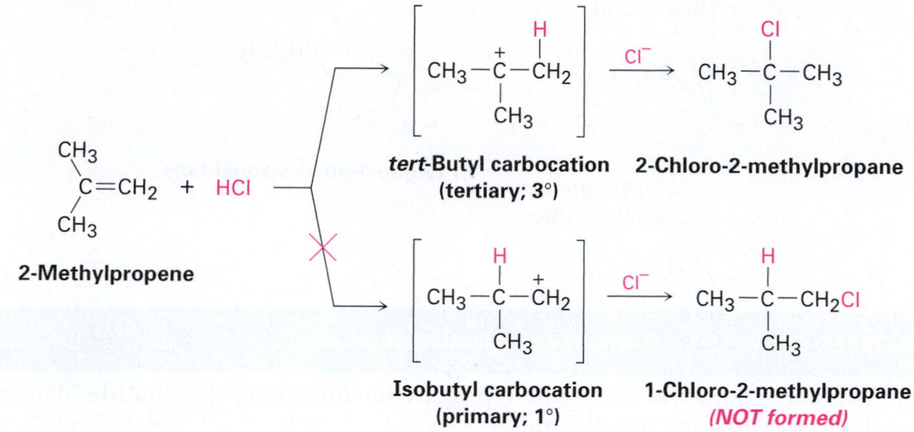

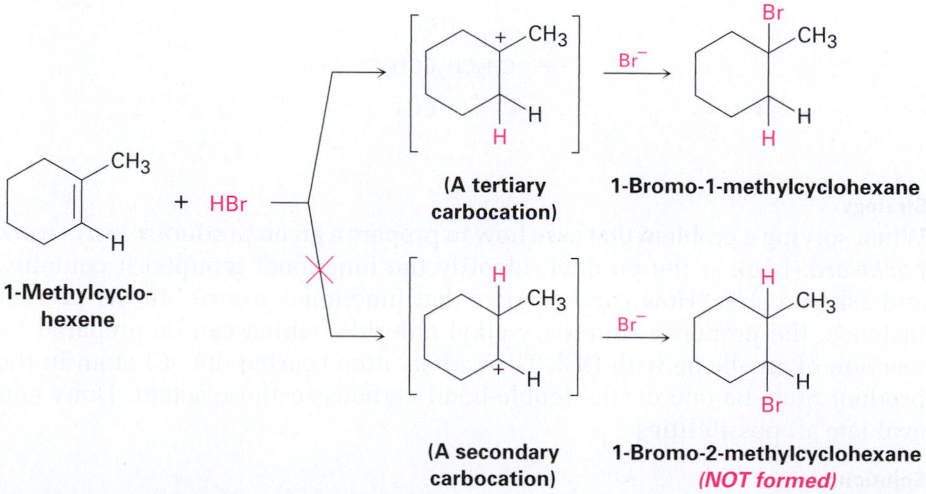

WORKED EXAMPLE 7.2 **Predicting the Product of an Electrophilic Addition Reaction**

What product would you expect from reaction of HCl with 1-ethylcyclopentene?

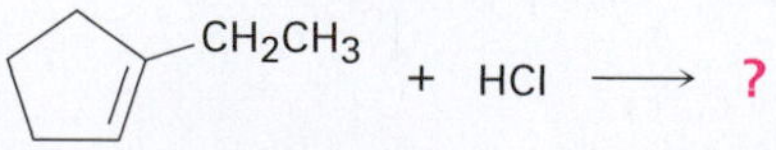

Strategy
When solving a problem that asks you to predict a reaction product, begin by looking at the functional group(s) in the reactants and deciding what kind of reaction is likely to occur. In the present instance, the reactant is an alkene

that will probably undergo an electrophilic addition reaction with HCl. Next, recall what you know about electrophilic addition reactions and use your knowledge to predict the product. You know that electrophilic addition reactions follow Markovnikov's rule, so H$^+$ will add to the double-bond carbon that has one alkyl group (C2 on the ring) and the Cl will add to the double-bond carbon that has two alkyl groups (C1 on the ring).

Solution
The expected product is 1-chloro-1-ethylcyclopentane.

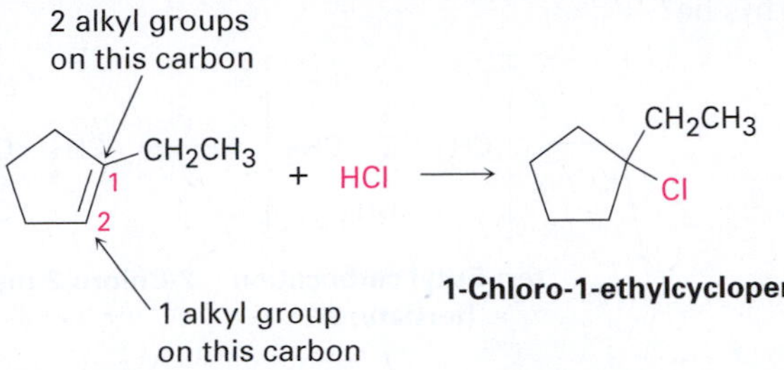

1-Chloro-1-ethylcyclopentane

WORKED EXAMPLE 7.3 Synthesizing a Specific Compound

What alkene would you start with to prepare the following alkyl halide? There may be more than one possibility.

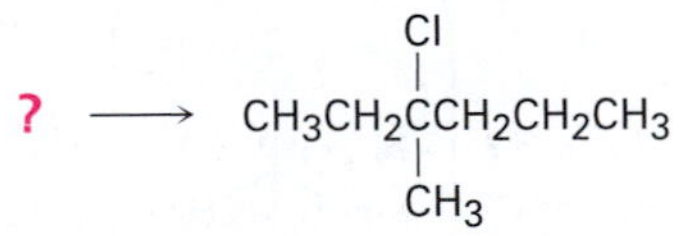

Strategy
When solving a problem that asks how to prepare a given product, *always work backward.* Look at the product, identify the functional group(s) it contains, and ask yourself, "How can I prepare that functional group?" In the present instance, the product is a tertiary alkyl chloride, which can be prepared by reaction of an alkene with HCl. The carbon atom bearing the –Cl atom in the product must be one of the double-bond carbons in the reactant. Draw and evaluate all possibilities.

Solution
There are three possibilities, any one of which could give the desired product.

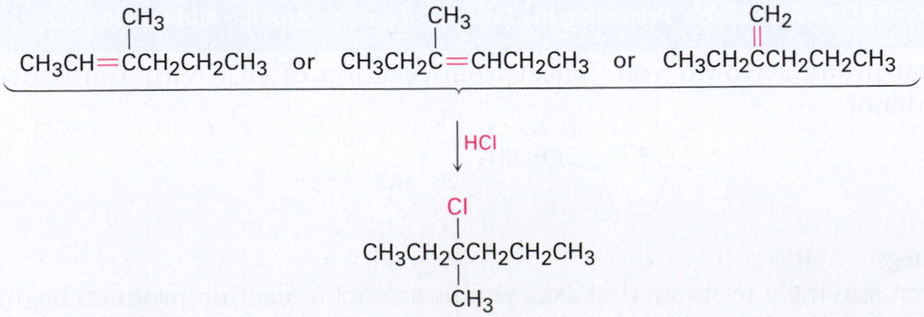

Problem 7.16
Predict the products of the following reactions:

(a) [cyclohexene] $\xrightarrow{\text{HCl}}$?

(b) $\underset{\underset{}{\overset{\displaystyle CH_3}{|}}}{CH_3C}{=}CHCH_2CH_3 \xrightarrow{\text{HBr}}$?

(c) $\underset{\underset{}{\overset{\displaystyle CH_3}{|}}}{CH_3CHCH_2CH}{=}CH_2 \xrightarrow[\text{H}_2\text{SO}_4]{\text{H}_2\text{O}}$?

(d) [cyclohexane with =CH$_2$] $\xrightarrow{\text{HBr}}$?

Problem 7.17
What alkenes would you start with to prepare the following products?

(a) [cyclopentane]—Br

(b) [cyclohexane with CH$_2$CH$_3$ and —OH]

(c) $\underset{\underset{}{\overset{\displaystyle Br}{|}}}{CH_3CH_2CHCH_2CH_2CH_3}$

(d) [cyclohexane with CHCH$_3$ and Cl]

7.8 Carbocation Structure and Stability

To understand why Markovnikov's rule works, we need to learn more about the structure and stability of carbocations and about the general nature of reactions and transition states. The first point to explore involves structure.

A great deal of experimental evidence has shown that carbocations are planar. The trivalent carbon is sp^2-hybridized, and the three substituents are oriented toward the corners of an equilateral triangle, as indicated in Figure 7.7. Because there are only six valence electrons on carbon and all six are used in the three σ bonds, the p orbital extending above and below the plane is unoccupied.

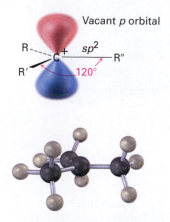

Vacant p orbital

sp^2

$120°$

FIGURE 7.7 The structure of a carbocation. The trivalent carbon is sp^2-hybridized and has a vacant p orbital perpendicular to the plane of the carbon and three attached groups.

The second point to explore involves carbocation stability. 2-Methyl-propene might react with H^+ to form a carbocation having three alkyl substituents (a tertiary ion, 3°), or it might react to form a carbocation having one alkyl substituent (a primary ion, 1°). Since the tertiary alkyl chloride,

2-chloro-2-methylpropane, is the only product observed, formation of the tertiary cation is evidently favored over formation of the primary cation. Thermodynamic measurements show that, indeed, the stability of carbocations increases with increasing substitution so that the stability order is tertiary > secondary > primary > methyl.

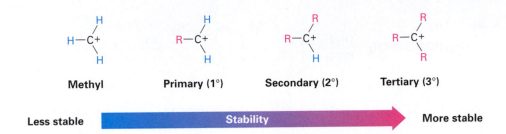

Why are more highly substituted carbocations more stable than less highly substituted ones? There are at least two reasons. Part of the answer has to do with inductive effects, and part has to do with hyperconjugation. Inductive effects, discussed in Section 2.1 in connection with polar covalent bonds, result from the shifting of electrons in a σ bond in response to the electronegativity of nearby atoms. In the present instance, electrons from a relatively larger and more polarizable alkyl group can shift toward a neighboring positive charge more easily than the electron from a hydrogen. Thus, the more alkyl groups there are attached to the positively charged carbon, the more electron density shifts toward the charge and the more inductive stabilization of the cation occurs (Figure 7.8).

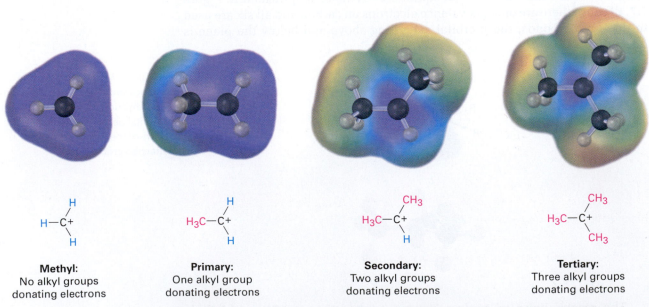

FIGURE 7.8 A comparison of inductive stabilization for methyl, primary, secondary, and tertiary carbocations. The more alkyl groups there are bonded to the positively charged carbon, the more electron density shifts toward the charge making the charged carbon less electron poor (blue in electrostatic potential maps).

Hyperconjugation, discussed in Section 7.5 in connection with the stability of substituted alkenes, is the stabilizing interaction between a *p* orbital and C–H σ bonds on neighboring carbons that are roughly parallel to the *p* orbital (Figure 7.9). The more alkyl groups there are on the carbocation, the more stable the carbocation.

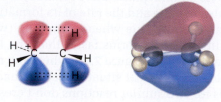

FIGURE 7.9 In the ethyl carbocation, $CH_3CH_2^+$, there is a stabilizing interaction between the carbocation *p* orbital and adjacent C–H σ bonds on the methyl substituent, as indicated by this molecular orbital. The more substituents there are, the greater the stabilization of the cation. Only the C–H bonds that are roughly parallel to the neighboring *p* orbital are oriented properly to take part in hyperconjugation.

Problem 7.18

Show the structures of the carbocation intermediates you would expect in the following reactions:

(a)
$$CH_3CH_2\underset{\underset{CH_3}{|}}{C}=\underset{\underset{CH_3}{|}}{C}HCHCH_3 \xrightarrow{HBr} \ ?$$

(b)
 =CHCH₃ $\xrightarrow[H_2SO_4]{H_2O}$?

Problem 7.19

Draw a skeletal structure of the following carbocation. Identify it as primary, secondary, or tertiary, and identify the hydrogen atoms that have the proper orientation for hyperconjugation in the conformation shown.

7.9 The Hammond Postulate

Let's summarize our knowledge of electrophilic addition reactions to this point:

- **Electrophilic addition to an unsymmetrically substituted alkene gives the more highly substituted carbocation intermediate.** A more highly substituted carbocation forms faster than a less highly substituted one and, once formed, rapidly goes on to give the final product.

- **A more highly substituted carbocation is more stable than a less highly substituted one.** That is, the stability order of carbocations is tertiary > secondary > primary > methyl.

What we have not yet seen is how these two points are related. Why does the *stability* of the carbocation intermediate affect the *rate* at which it's formed and thereby determine the structure of the final product? After all, carbocation stability is determined by the free-energy change $\Delta G°$, but reaction rate is determined by the activation energy $\Delta G^{\ddagger}$. The two quantities aren't directly related.

Although there is no simple quantitative relationship between the stability of a carbocation intermediate and the rate of its formation, there *is* an intuitive relationship. It's generally true when comparing two similar reactions that the more stable intermediate forms faster than the less stable one. The situation is shown graphically in Figure 7.10, where the reaction energy profile in part (a) represents the typical situation rather than the profile in part (b). That is, the curves for two similar reactions don't cross one another.

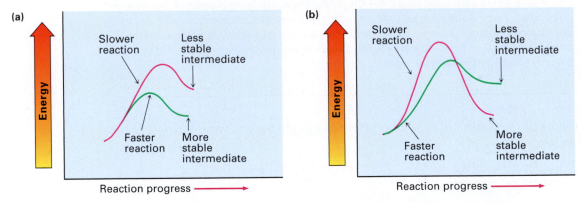

FIGURE 7.10 Energy diagrams for two similar competing reactions. In **(a)**, the faster reaction yields the more stable intermediate. In **(b)**, the slower reaction yields the more stable intermediate. The curves shown in **(a)** represent the typical situation.

Called the **Hammond postulate**, the explanation of the relationship between reaction rate and intermediate stability goes like this: Transition states represent energy maxima. They are high-energy activated complexes that occur transiently during the course of a reaction and immediately go on to a more stable species. Although we can't actually *observe* transition states, because they have no finite lifetime, the Hammond postulate says that we can get an idea of a particular transition state's structure by looking at the structure of the nearest stable species. Imagine the two cases shown in Figure 7.11, for example. The reaction profile in part (a) shows the energy curve for an endergonic reaction step, and the profile in part (b) shows the curve for an exergonic step.

FIGURE 7.11 Energy diagrams for endergonic and exergonic steps. **(a)** In an endergonic step, the energy levels of transition state and *product* are closer. **(b)** In an exergonic step, the energy levels of transition state and *reactant* are closer.

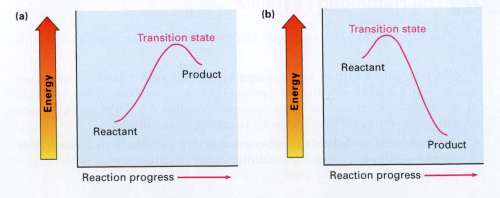

In an endergonic reaction (Figure 7.11a), the energy level of the transition state is closer to that of the product than to that of the reactant. Since the transition state is closer energetically to the product, we make the natural assumption that it's also closer structurally. In other words, *the transition state for an endergonic reaction step structurally resembles the product of that step.* Conversely, the transition state for an exergonic reaction (Figure 7.11b) is closer energetically, and thus structurally, to the reactant than to the product. We therefore say that *the transition state for an exergonic reaction step structurally resembles the reactant for that step.*

Hammond postulate
The structure of a transition state resembles the structure of the nearest stable species. Transition states for endergonic steps structurally resemble products, and transition states for exergonic steps structurally resemble reactants.

How does the Hammond postulate apply to electrophilic addition reactions? The formation of a carbocation by protonation of an alkene is an endergonic step. Thus, the transition state for alkene protonation structurally resembles the carbocation intermediate, and any factor that stabilizes the carbocation will also stabilize the nearby transition state. Since increasing alkyl substitution stabilizes carbocations, it also stabilizes the transition states leading to those ions, thus resulting in faster reaction. More stable carbocations form faster because their greater stability is reflected in the lower-energy transition state leading to them (Figure 7.12).

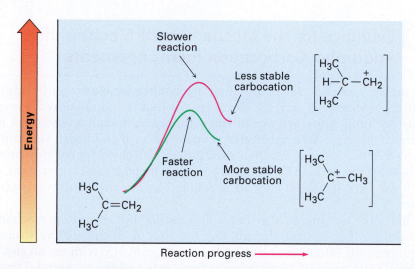

FIGURE 7.12 Energy diagrams for carbocation formation. The more stable tertiary carbocation is formed faster (green curve) because its increased stability lowers the energy of the transition state leading to it.

We can imagine the transition state for alkene protonation to be a structure in which one of the alkene carbon atoms has almost completely rehybridized from sp^2 to sp^3 and in which the remaining alkene carbon bears much of the positive charge (Figure 7.13). This transition state is stabilized by hyperconjugation and inductive effects in the same way as the product carbocation. The more alkyl groups that are present, the greater the extent of stabilization and the faster the transition state forms.

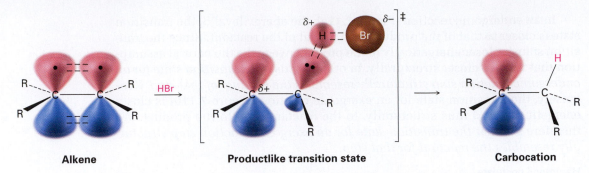

FIGURE 7.13 The hypothetical structure of a transition state for alkene protonation. The transition state is closer in both energy and structure to the carbocation than to the alkene. Thus, an increase in carbocation stability (lower $\Delta G°$) also causes an increase in transition-state stability (lower $\Delta G^{\ddagger}$), thereby increasing the rate of its formation.

Problem 7.20

What about the second step in the electrophilic addition of HCl to an alkene—the reaction of chloride ion with the carbocation intermediate? Is this step exergonic or endergonic? Does the transition state for this second step resemble the reactant (carbocation) or product (alkyl chloride)? Make a rough drawing of what the transition-state structure might look like.

7.10 Evidence for the Mechanism of Electrophilic Additions: Carbocation Rearrangements

How do we know that the carbocation mechanism for electrophilic addition reactions of alkenes is correct? The answer is that we *don't* know it's correct; at least we don't know with complete certainty. Although an incorrect reaction mechanism can be disproved by demonstrating that it doesn't account for observed data, a correct reaction mechanism can never be entirely proved. The best we can do is to show that a proposed mechanism is consistent with all known facts. If enough facts are accounted for, the mechanism is probably correct.

One of the best pieces of evidence supporting the carbocation mechanism proposed for the electrophilic addition reaction of alkenes is that structural *rearrangements* often occur during the reaction of HX with an alkene. For example, reaction of HCl with 3-methylbut-1-ene yields a substantial amount of 2-chloro-2-methylbutane in addition to the "expected" product, 2-chloro-3-methylbutane:

3-Methylbut-1-ene + HCl → **2-Chloro-3-methylbutane** (approx. 50%) + **2-Chloro-2-methylbutane** (approx. 50%)

If the reaction takes place in a single step, it would be difficult to account for rearrangement, but if the reaction takes place in several steps through a carbocation intermediate, rearrangement is more easily explained. The secondary carbocation intermediate formed by protonation of 3-methylbut-1-ene evidently rearranges to a more stable tertiary carbocation by a **hydride shift**—the shift of a hydrogen atom and its electron pair (a hydride ion, :H⁻) between neighboring carbons:

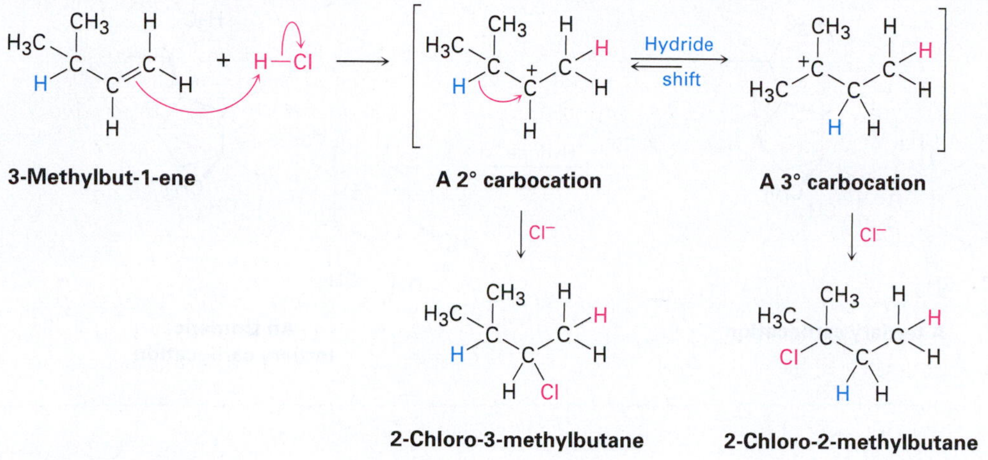

3-Methylbut-1-ene   A 2° carbocation   A 3° carbocation

2-Chloro-3-methylbutane   2-Chloro-2-methylbutane

Carbocation rearrangements can also occur by the shift of an alkyl group with its electron pair. For example, reaction of 3,3-dimethylbut-1-ene with HCl leads to an equal mixture of unrearranged 2-chloro-3,3-dimethylbutane and rearranged 2-chloro-2,3-dimethylbutane. In this instance, a secondary carbocation rearranges to a more stable tertiary carbocation by the shift of a methyl group:

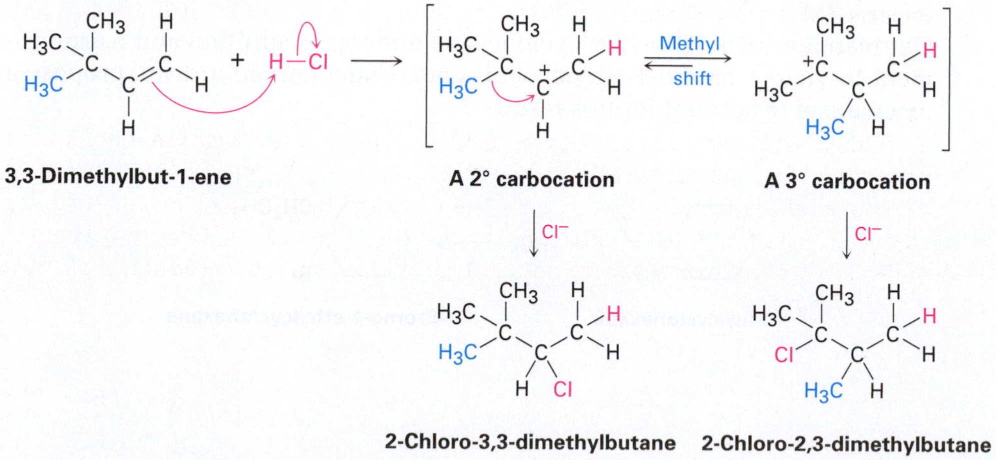

3,3-Dimethylbut-1-ene   A 2° carbocation   A 3° carbocation

2-Chloro-3,3-dimethylbutane   2-Chloro-2,3-dimethylbutane

Note the similarities between the two carbocation rearrangements: in both cases, a group (:H⁻ or :CH₃⁻) moves to an adjacent positively charged carbon, taking its bonding electron pair with it. Also in both cases, a less stable carbocation rearranges to a more stable ion. Rearrangements of this kind are a common feature of carbocation chemistry and are particularly important in the biological pathways by which steroids and related substances are synthesized. An example is the following hydride shift that occurs during the biosynthesis of cholesterol; Sections 23.8 and 23.9 show many others.

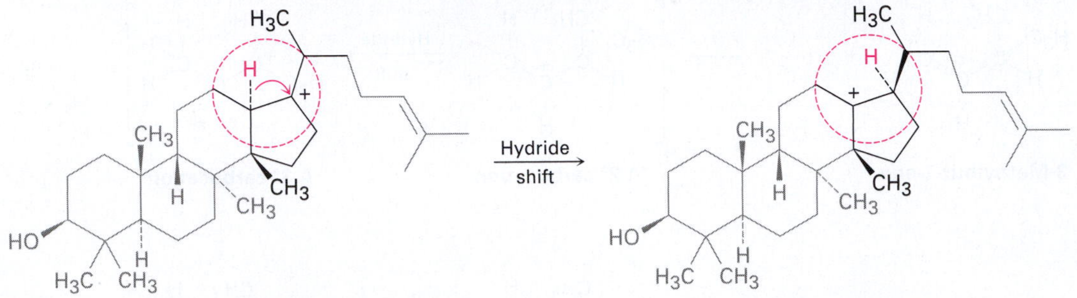

A tertiary carbocation

**An isomeric
tertiary carbocation**

A word of advice that we've noted before and will repeat on occasion: biological molecules are often larger and more complex in appearance than the molecules chemists work with in the laboratory, but don't be intimidated. When looking at *any* chemical transformation, whether biochemical or not, focus on the part of the molecule where the change is occurring and don't worry about the rest. The tertiary carbocation just pictured looks complicated, but all the chemistry is taking place in the small part of the molecule inside the red circle.

Problem 7.21

On treatment with HBr, vinylcyclohexane undergoes addition and rearrangement to yield 1-bromo-1-ethylcyclohexane. Using curved arrows, propose a mechanism to account for this result.

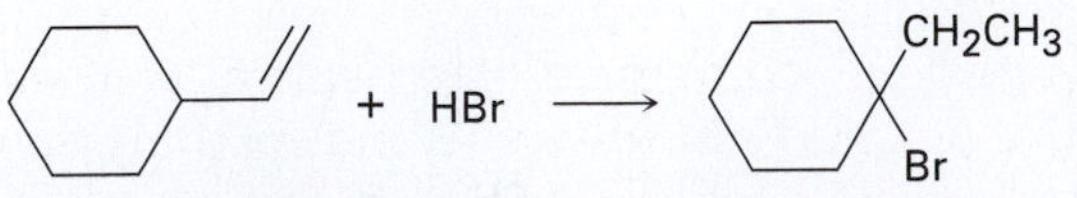

Vinylcyclohexane

1-Bromo-1-ethylcyclohexane

Summary

Carbon–carbon double bonds are present in most organic and biological molecules, so a good understanding of their behavior is needed. In this chapter, we've looked at some consequences of alkene stereoisomerism and at the details of the broadest and most general class of alkene reactions—the electrophilic addition reaction.

An **alkene** is a hydrocarbon that contains a carbon–carbon double bond, and an **alkyne** is a hydrocarbon that contains a triple bond. Because they contain fewer hydrogens than alkanes with the same number of carbons, alkenes and alkynes are said to be **unsaturated**.

Because rotation around the double bond can't occur, substituted alkenes can exist as cis–trans stereoisomers. The geometry of a double bond can be specified by application of the Cahn–Ingold–Prelog rules, which rank the substituents on each double-bond carbon. If the higher-ranking groups on each carbon are on the same side of the double bond, the geometry is **Z** (*zusammen*, "together"); if the higher-ranking groups on each carbon are on opposite sides of the double bond, the geometry is **E** (*entgegen,* "apart").

Alkene chemistry is dominated by **electrophilic addition reactions**. When HX reacts with an unsymmetrically substituted alkene, **Markovnikov's rule** predicts that the H will add to the carbon having fewer alkyl substituents and the X group will add to the carbon having more alkyl substituents. Electrophilic additions to alkenes take place through carbocation intermediates formed by reaction of the nucleophilic alkene π bond with electrophilic H^+. Carbocation stability follows the order

Tertiary (3°) > Secondary (2°) > Primary (1°) > Methyl

R_3C^+ > R_2CH^+ > RCH_2^+ > CH_3^+

Markovnikov's rule can be restated by saying that, in the addition of HX to an alkene, the more stable carbocation intermediate is formed. This result is explained by the **Hammond postulate**, which says that the transition state of an exergonic reaction step structurally resembles the reactant, whereas the transition state of an endergonic reaction step structurally resembles the product. Since an alkene protonation step is endergonic, the stability of the more highly substituted carbocation is reflected in the stability of the transition state leading to its formation.

Evidence in support of a carbocation mechanism for electrophilic additions comes from the observation that structural rearrangements often take place during reaction. Rearrangements occur by shift of either a hydride ion, :H^- (a **hydride shift**), or an alkyl anion, :R^-, from a carbon atom to the adjacent positively charged carbon. The result is isomerization of a less stable carbocation to a more stable one.

Key Words

alkene ($R_2C=CR_2$), 212

alkyne ($RC\equiv CR$), 212

degree of unsaturation, 214

E geometry, 221

electrophilic addition reaction, 227

Hammond postulate, 236

hydride shift, 239

hyperconjugation, 226

Markovnikov's rule, 230

regiospecific, 230

unsaturated, 213

Z geometry, 221

Lagniappe

Terpenes: Naturally Occurring Alkenes

© Photodisc Green/Getty Images

The wonderful fragrance of leaves from the California bay laurel is due primarily to myrcene, a simple terpene.

It has been known for centuries that codistillation of many plant materials with steam produces a fragrant mixture of liquids called *essential oils*. For hundreds of years, such plant extracts have been used as medicines, spices, and perfumes. The investigation of essential oils also played a major role in the emergence of organic chemistry as a science during the 19th century.

Chemically, plant essential oils consist largely of mixtures of compounds known as *terpenoids*—small organic molecules with an immense diversity of structure. More than 35,000 different terpenoids are known. Some are open-chain molecules, and others contain rings; some are hydrocarbons, and others contain oxygen. Hydrocarbon terpenoids, in particular, are known as *terpenes,* and all contain double bonds. For example:

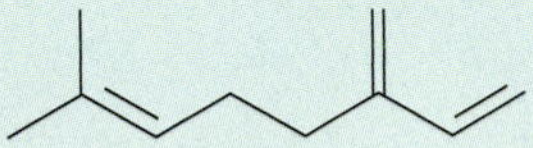

**Myrcene
(oil of bay)**

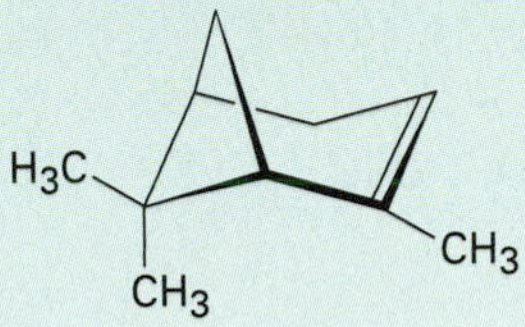

**α-Pinene
(turpentine)**

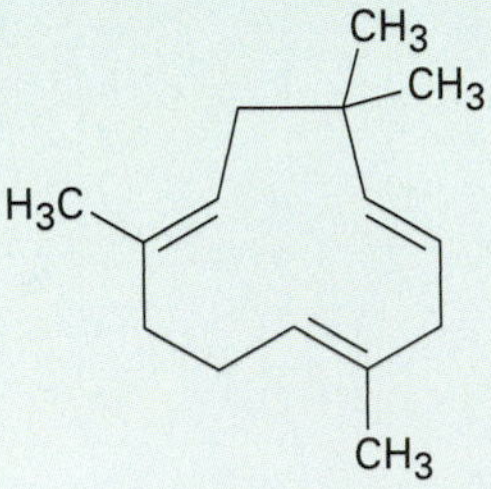

**Humulene
(oil of hops)**

Regardless of their apparent structural differences, all terpenoids are related. According to a formalism called the *isoprene rule,* they can be thought of as arising from head-to-tail joining of 5-carbon isoprene units (2-methyl-buta-1,3-diene). Carbon 1 is the head of the isoprene unit, and carbon 4 is the tail. For example, myrcene contains two isoprene units joined head to tail, forming an 8-carbon chain with two 1-carbon branches. α-Pinene similarly contains two isoprene units assembled into a more complex cyclic structure, and humulene contains three isoprene units. See if you can identify the isoprene units in α-pinene and humulene.

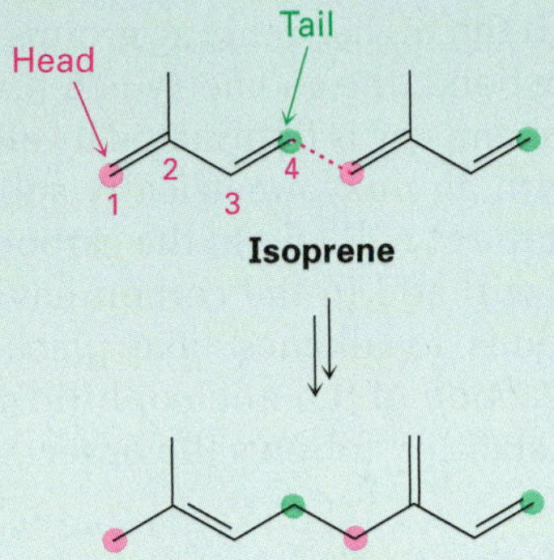

Head   Tail

1   2   3   4

Isoprene

Myrcene

Terpenes (and terpenoids) are further classified according to the number of 5-carbon units they contain. Thus, *monoterpenes* are 10-carbon substances derived from two isoprene units, *sesquiterpenes* are 15-carbon molecules derived from three isoprene units, *diterpenes* are 20-carbon substances derived from four isoprene units, and so on. Monoterpenes and sesquiterpenes are found primarily in plants, but the higher terpenoids occur in both plants and animals, and many have important biological roles. The triterpenoid lanosterol, for example, is the precursor from which all steroid hormones are made.

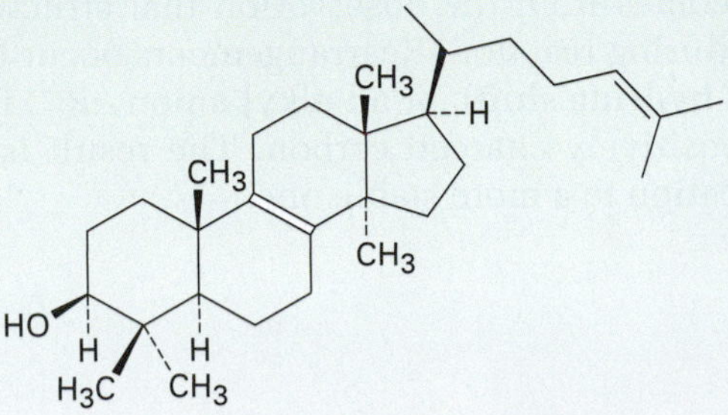

Lanosterol, a triterpene (C_{30})

continued

Lagniappe *continued*

Isoprene itself is not the true biological precursor of terpenoids. As we'll see in Section 23.7, nature instead uses two "isoprene equivalents"—isopentenyl diphosphate and dimethylallyl diphosphate—which are themselves made by two different routes depending on the organism. Lanosterol, in particular, is biosynthesized from acetic acid by a complex pathway that has been worked out in great detail.

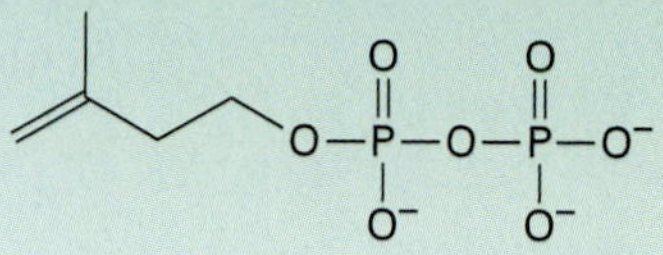

Isopentenyl diphosphate

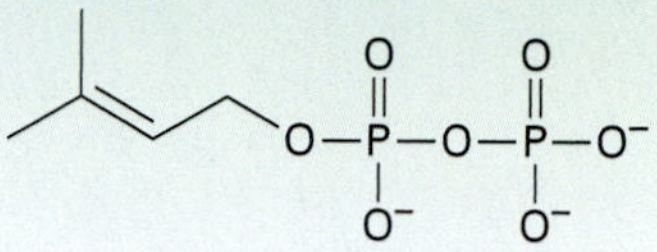

Dimethylallyl diphosphate

Exercises

VISUALIZING CHEMISTRY

(Problems 7.1–7.21 appear within the chapter.)

7.22 ▪ Name the following alkenes, and convert each drawing into a skeletal structure:

(a)   (b)

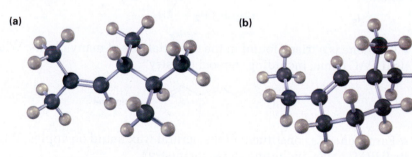

▪ indicates problems that are assignable in Organic OWL.

Go to this book's companion website at **www.cengage.com/ chemistry/mcmurry** to explore interactive versions of the Active Figures from this text.

7.23 ▪ Assign stereochemistry (*E* or *Z*) to the double bonds in each of the following alkenes, and convert each drawing into a skeletal structure (red = O, yellow-green = Cl):

(a)   (b)

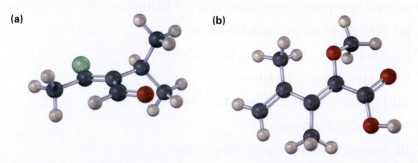

▪ Problems assignable in Organic OWL.

7.24 ■ The following carbocation is an intermediate in the electrophilic addition reaction of HCl with two different alkenes. Identify both, and tell which C–H bonds in the carbocation are aligned for hyperconjugation with the vacant *p* orbital on the positively charged carbon.

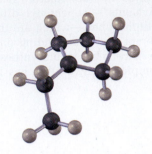

ADDITIONAL PROBLEMS

7.25 ■ Name the following alkenes:

(a)

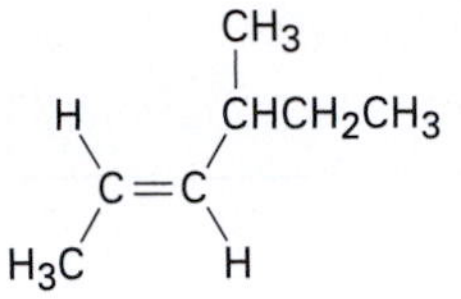

$$CH_3$$
$$H \quad CHCH_2CH_3$$
$$C=C$$
$$H_3C \quad H$$

(b)

$$CH_3 \quad CH_2CH_3$$
$$CH_3CHCH_2CH_2CH \quad CH_3$$
$$C=C$$
$$H_3C \quad H$$

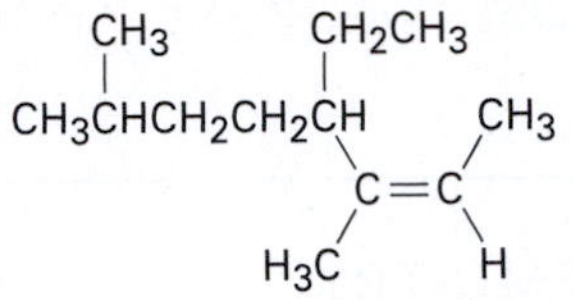

(c)

$$CH_2CH_3$$
$$H_2C=CCH_2CH_3$$

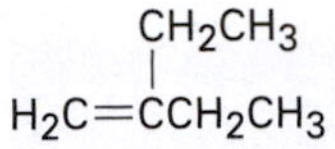

(d)

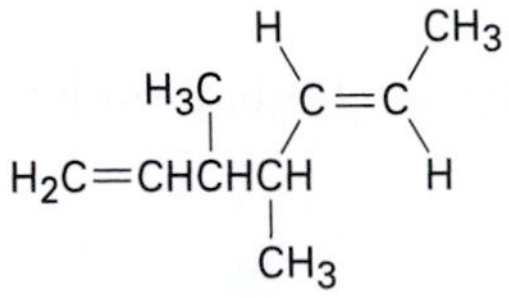

$$H \quad CH_3$$
$$H_3C \quad C=C$$
$$H_2C=CHCHCH \quad H$$
$$CH_3$$

(e)

$$H \quad H$$
$$H_3C \quad C=C$$
$$C=C \quad CH_3$$
$$CH_3CH_2CH_2 \quad CH_3$$

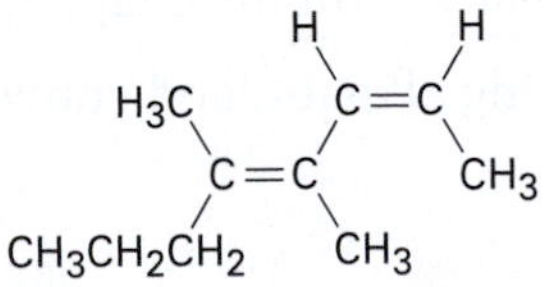

(f) $H_2C=C=CHCH_3$

7.26 ■ Ocimene is a triene found in the essential oils of many plants. What is its IUPAC name, including stereochemistry?

 **Ocimene**

7.27 α-Farnesene is a constituent of the natural wax found on apples. What is its IUPAC name, including stereochemistry?

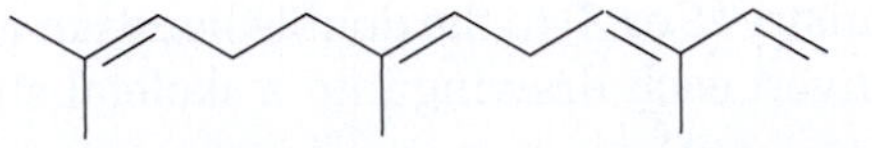

α-Farnesene

7.28 ■ Draw structures corresponding to the following systematic names:

(a) (4*E*)-2,4-Dimethylhexa-1,4-diene

(b) *cis*-3,3-Dimethyl-4-propylocta-1,5-diene

(c) 4-Methylpenta-1,2-diene

(d) (3*E*,5*Z*)-2,6-Dimethylocta-1,3,5,7-tetraene

(e) 3-Butylhept-2-ene

(f) *trans*-2,2,5,5-Tetramethylhex-3-ene

7.29 There are seven isomeric alkynes with the formula C_6H_{10}. Draw and name them.

7.30 ■ Tridec-1-ene-3,5,7,9,11-pentayne is a hydrocarbon isolated from sunflowers. Draw its structure. (Tridecane is the straight-chain alkane $C_{13}H_{28}$.)

7.31 ■ Menthene, a hydrocarbon found in mint plants, has the systematic name 1-isopropyl-4-methylcyclohexene. Draw its structure.

7.32 ■ Calculate the degree of unsaturation in the following formulas:

(a) $C_{20}H_{32}$ (b) $C_9H_{16}Br_2$ (c) $C_{10}H_{12}N_2O_3$

(d) $C_{20}H_{32}ClN$ (e) $C_{40}H_{56}$ (β-carotene)

7.33 ■ How many hydrogens does each of the following compounds have?

(a) $C_8H_?O_2$, has two rings and one double bond

(b) $C_7H_?N$, has two double bonds

(c) $C_9H_?NO$, has one ring and three double bonds

7.34 ■ Loratadine, marketed as an antiallergy medication under the trade name Claritin, has four rings, eight double bonds, and the formula $C_{22}H_?ClN_2O_2$. How many hydrogens does loratadine have? (Calculate the answer; don't count hydrogens in the structure.)

Loratadine

7.35 Draw and name the 6 alkene isomers, C_5H_{10}, including *E,Z* isomers.

7.36 Draw and name the 17 alkene isomers, C_6H_{12}, including *E,Z* isomers.

7.37 *trans*-But-2-ene is more stable than *cis*-but-2-ene by only 4 kJ/mol, but *trans*-2,2,5,5-tetramethylhex-3-ene is more stable than its cis isomer by 39 kJ/mol. Explain.

7.38 Cyclodecene can exist in both cis and trans forms, but cyclohexene cannot. Explain. (Making molecular models is helpful.)

7.39 Normally, a trans alkene is *more* stable than its cis isomer, but *trans*-cyclooctene is *less* stable than its cis isomer by 38.5 kJ/mol. Explain.

7.40 *trans*-Cyclooctene is less stable than *cis*-cyclooctene by 38.5 kJ/mol, but *trans*-cyclononene is less stable than *cis*-cyclononene by only 12.2 kJ/mol. Explain.

7.41 Allene (propa-1,2-diene), $H_2C=C=CH_2$, has two adjacent double bonds. What kind of hybridization must the central carbon have? Sketch the bonding π orbitals in allene. What shape do you predict for allene?

■ Problems assignable in Organic OWL.

7.42 The heat of hydrogenation for allene (Problem 7.41) to yield propane is −295 kJ/mol, and the heat of hydrogenation for a typical monosubstituted alkene such as propene is −126 kJ/mol. Is allene more stable or less stable than you might expect for a diene? Explain.

7.43 ▪ Predict the major product of each of the following reactions:

(a) 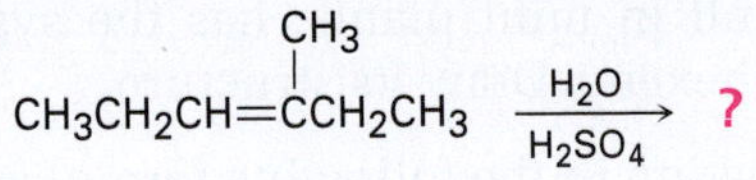
$$CH_3CH_2CH=\overset{\overset{\displaystyle CH_3}{|}}{C}CH_2CH_3 \xrightarrow[H_2SO_4]{H_2O} \ ?$$

(b)
CH$_2$CH$_3$ $\xrightarrow{HBr}$ **?**

(c)
CH$_3$ $\xrightarrow{HBr}$ **?**

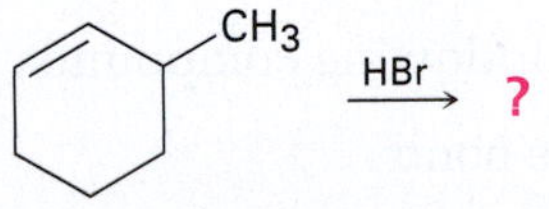

(d) $H_2C=CHCH_2CH_2CH_2CH=CH_2 \xrightarrow{2\ HCl} \ ?$

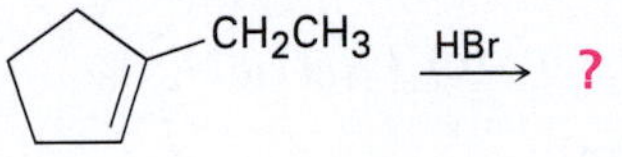

7.44 ▪ Predict the major product from addition of HBr to each of the following alkenes:

(a) 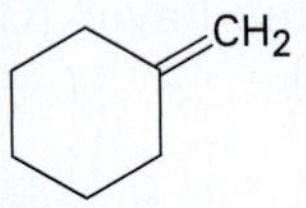 CH$_2$

(b)

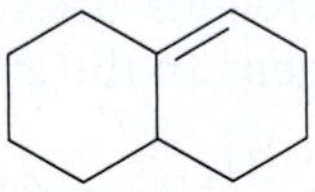

(c) 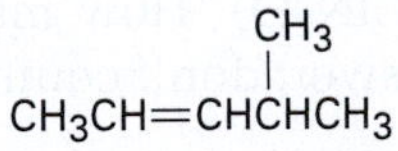
$$CH_3CH=\overset{\overset{\displaystyle CH_3}{|}}{C}HCHCH_3$$

7.45 ▪ Rank the substituents in each of the following sets according to the Cahn–Ingold–Prelog rules:

(a) –CH$_3$, –Br, –H, –I

(b) –OH, –OCH$_3$, –H, –CO$_2$H

(c) –CO$_2$H, –CO$_2$CH$_3$, –CH$_2$OH, –CH$_3$

(d) –CH$_3$, –CH$_2$CH$_3$, –CH$_2$CH$_2$OH, $-\overset{\overset{\displaystyle O}{\|}}{C}CH_3$

(e) –CH=CH$_2$, –CN, –CH$_2$NH$_2$, –CH$_2$Br

(f) –CH=CH$_2$, –CH$_2$CH$_3$, –CH$_2$OCH$_3$, –CH$_2$OH

7.46 ▪ Assign *E* or *Z* configuration to the double bonds in each of the following compounds:

(a) 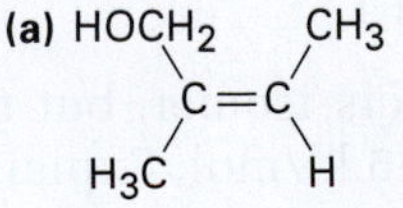
$$\underset{H_3C}{\overset{HOCH_2}{}}C=C\underset{H}{\overset{CH_3}{}}$$

(b)
$$\underset{Cl}{\overset{HO_2C}{}}C=C\underset{OCH_3}{\overset{H}{}}$$

(c)
$$\underset{CH_3CH_2}{\overset{NC}{}}C=C\underset{CH_2OH}{\overset{CH_3}{}}$$

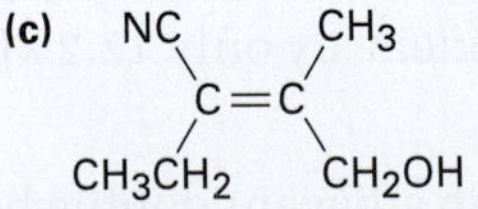

(d) 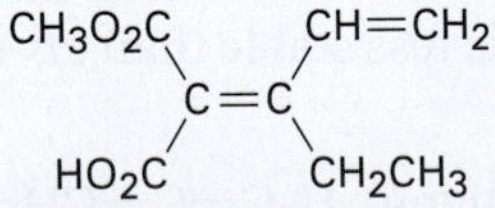
$$\underset{HO_2C}{\overset{CH_3O_2C}{}}C=C\underset{CH_2CH_3}{\overset{CH=CH_2}{}}$$

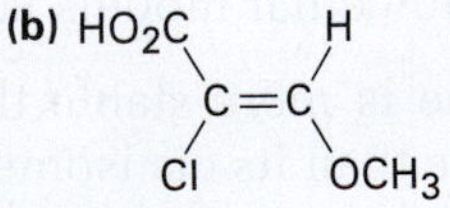

▪ Problems assignable in Organic OWL.

7.47 ■ Name the following cycloalkenes:

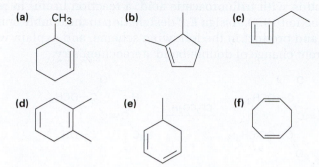

7.48 Fucoserraten, ectocarpen, and multifidene are sex pheromones produced by marine brown algae. What are their systematic names? (The latter two are very difficult, but give them a try. Make your best guess, and then check your answer in the *Study Guide and Solutions Manual*.)

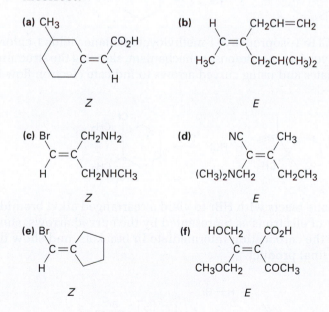

Fucoserraten   **Ectocarpen**   **Multifidene**

7.49 ■ Which of the following *E,Z* designations are correct, and which are incorrect?

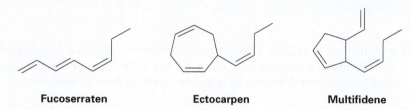

7.50 *tert*-Butyl esters [RCO$_2$C(CH$_3$)$_3$] are converted into carboxylic acids (RCO$_2$H) by reaction with trifluoroacetic acid, a reaction useful in protein synthesis (Section 19.7). Assign *E,Z* designation to the double bonds of both reactant and product in the following scheme, and explain why there is an apparent change of double-bond stereochemistry:

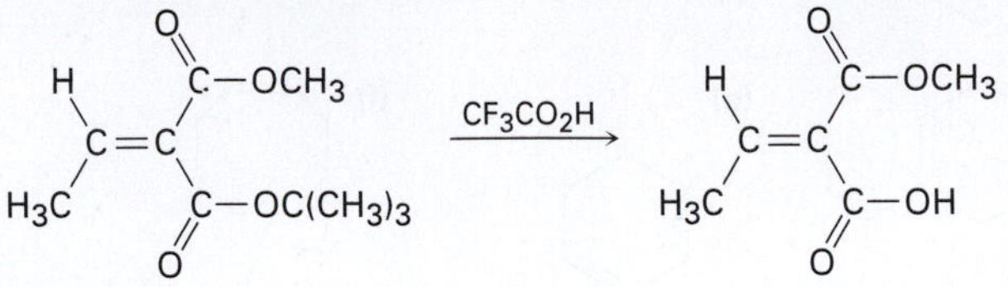

7.51 ■ Each of the following carbocations can rearrange to a more stable ion. Propose structures for the likely rearrangement products.

(a) CH$_3$CH$_2$CH$_2$CH$_2$$^+$ **(b)** CH$_3$CHCHCH$_3$ **(c)**

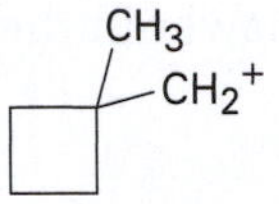

7.52 ■ Addition of HCl to 1-isopropylcyclohexene yields a rearranged product. Propose a mechanism, showing the structures of the intermediates and using curved arrows to indicate electron flow in each step.

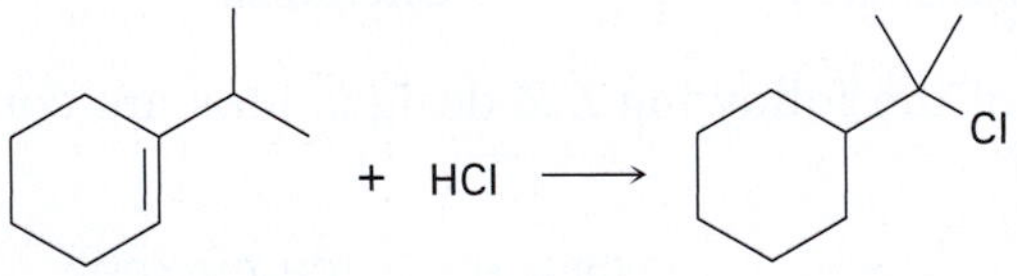

7.53 ■ Addition of HCl to 1-isopropenyl-1-methylcyclopentane yields 1-chloro-1,2,2-trimethylcyclohexane. Propose a mechanism, showing the structures of the intermediates and using curved arrows to indicate electron flow in each step.

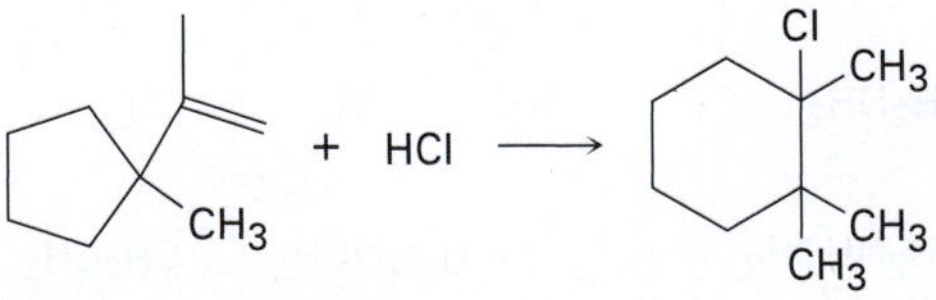

7.54 Vinylcyclopropane reacts with HBr to yield a rearranged alkyl bromide. Follow the flow of electrons as represented by the curved arrows, show the structure of the carbocation intermediate in brackets, and show the structure of the final product.

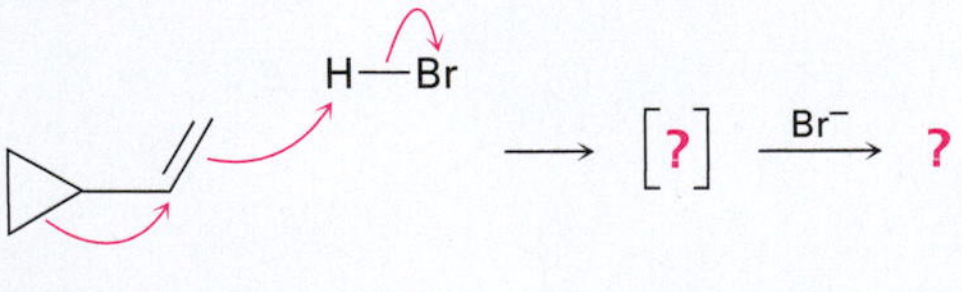

Vinylcyclopropane

■ Problems assignable in Organic OWL.

7.55 The isobutyl cation spontaneously rearranges to the *tert*-butyl cation by a hydride shift. Is the rearrangement exergonic or endergonic? Draw what you think the transition state for the hydride shift might look like according to the Hammond postulate.

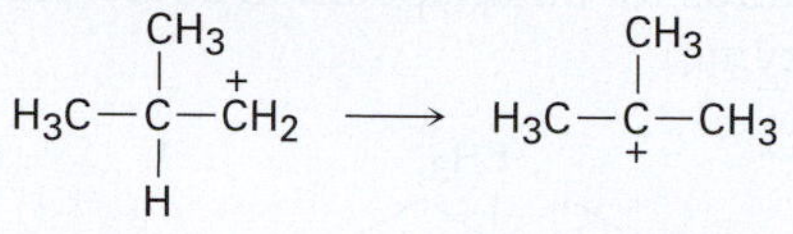

Isobutyl cation   **tert-Butyl cation**

7.56 Draw an energy diagram for the addition of HBr to pent-1-ene. Let one curve on your diagram show the formation of 1-bromopentane product and another curve on the same diagram show the formation of 2-bromopentane product. Label the positions for all reactants, intermediates, and products. Which curve has the higher-energy carbocation intermediate? Which curve has the higher-energy first transition state?

7.57 Make sketches of the transition-state structures involved in the reaction of HBr with pent-1-ene (Problem 7.56). Tell whether each structure resembles reactant or product.

7.58 ■ Limonene, a fragrant hydrocarbon found in lemons and oranges, is biosynthesized from geranyl diphosphate by the following pathway. Add curved arrows to show the mechanism of each step. Which step involves an alkene electrophilic addition? (The ion $OP_2O_6^{4-}$ is the diphosphate ion, and "Base" is an unspecified base in the enzyme that catalyzes the reaction.)

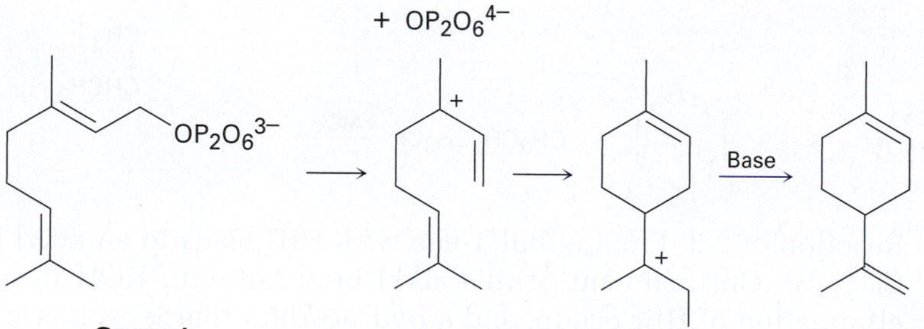

Geranyl diphosphate   **Limonene**

7.59 *epi*-Aristolochene, a hydrocarbon found in both pepper and tobacco, is biosynthesized by the following pathway. Add curved arrows to show the mechanism of each step. Which steps involve alkene electrophilic addition(s), and which involve carbocation rearrangement(s)? The abbreviation H—A stands for an unspecified acid, and "Base" is an unspecified base in the enzyme.

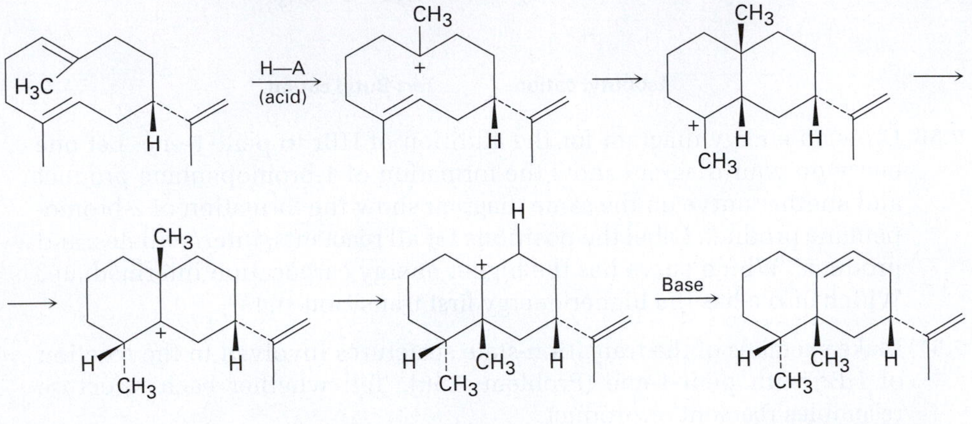

epi-Aristolochene

7.60 Aromatic compounds such as benzene react with alkyl chlorides in the presence of $AlCl_3$ catalyst to yield alkylbenzenes. The reaction occurs through a carbocation intermediate, formed by reaction of the alkyl chloride with $AlCl_3$ ($R–Cl + AlCl_3 \rightarrow R^+ + AlCl_4^-$). How can you explain the observation that reaction of benzene with 1-chloropropane yields isopropylbenzene as the major product?

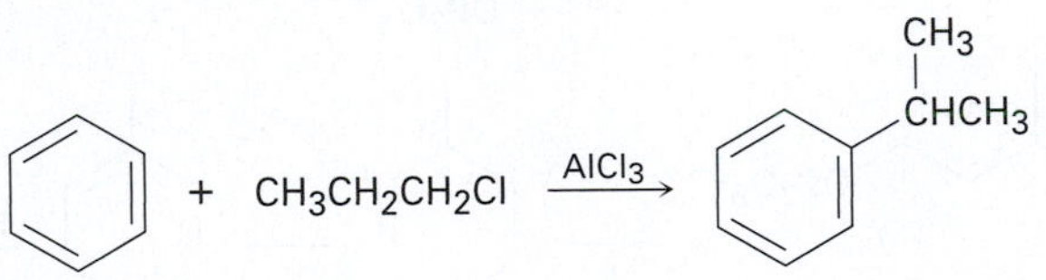

7.61 Reaction of 2,3-dimethylbut-1-ene with HBr leads to an alkyl bromide, $C_6H_{13}Br$. On treatment of this alkyl bromide with KOH in methanol, elimination of HBr occurs and a hydrocarbon that is isomeric with the starting alkene is formed. What is the structure of this hydrocarbon, and how do you think it is formed from the alkyl bromide?

8 Reactions of Alkenes and Alkynes

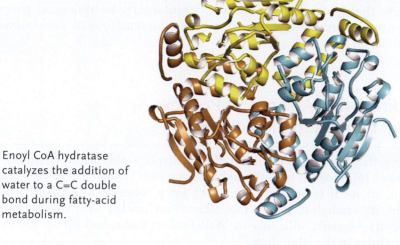

Enoyl CoA hydratase catalyzes the addition of water to a C=C double bond during fatty-acid metabolism.

Alkene addition reactions occur widely, both in the laboratory and in living organisms. Although we've studied only the addition of H_2O and HX thus far, many closely related reactions also take place. In this chapter, we'll see briefly how alkenes are prepared and we'll discuss further examples of alkene addition reactions. Particularly important are the addition of a halogen to give a 1,2-dihalide, addition of a hypohalous acid to give a halohydrin, addition of water to give an alcohol, addition of hydrogen to give an alkane, addition of a single oxygen to give a three-membered cyclic ether called an *epoxide,* and addition of two hydroxyl groups to give a 1,2-diol.

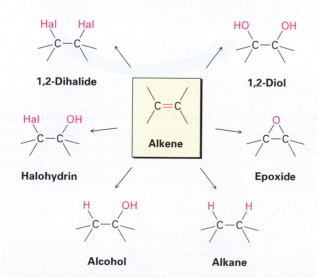

CONTENTS

OWL Online homework for this chapter can be assigned in Organic OWL.

Reactions of Alkene

WHY THIS CHAPTER?

Much of the background needed to understand organic reactions has now been covered, and it's time to begin a systematic description of the major functional groups. Both in this chapter on alkenes and in future chapters on other functional groups, we'll discuss a variety of reactions but try to focus on those that have direct or indirect counterparts in biological pathways. There are no shortcuts: you have to know the reactions to understand biological chemistry.

8.1 Preparing Alkenes: A Preview of Elimination Reactions

Before getting to the main subject of this chapter—the reactions of alkenes—let's take a brief look at how alkenes are prepared. The subject is a bit complex, though, so we'll return in Chapter 12 for a more detailed study. For the present, it's enough to realize that alkenes are readily available from simple precursors—usually alcohols in biological systems and either alcohols or alkyl halides in the laboratory.

Just as the chemistry of alkenes is dominated by addition reactions, the preparation of alkenes is dominated by elimination reactions. Additions and eliminations are, in many respects, two sides of the same coin. That is, an addition reaction might involve the addition of H_2O to an alkene to form an alcohol, whereas an elimination reaction might involve the loss of H_2O from an alcohol to form an alkene.

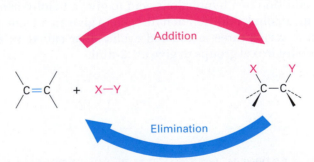

The two most common elimination reactions are **dehydrohalogenation**—the loss of HX from an alkyl halide—and **dehydration**—the loss of water from an alcohol. Dehydrohalogenation usually occurs by reaction of an alkyl halide with a strong base such as potassium hydroxide. For example, bromocyclohexane yields cyclohexene when treated with KOH in ethanol solution:

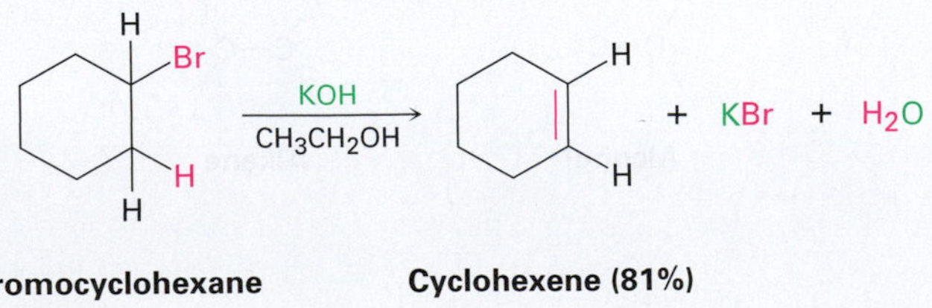

Bromocyclohexane   **Cyclohexene (81%)**

Dehydration is often carried out in the laboratory by treatment of an alcohol with a strong acid. For example, when 1-methylcyclohexanol is warmed with aqueous sulfuric acid in tetrahydrofuran (THF) solvent, loss of water occurs and 1-methylcyclohexene is formed.

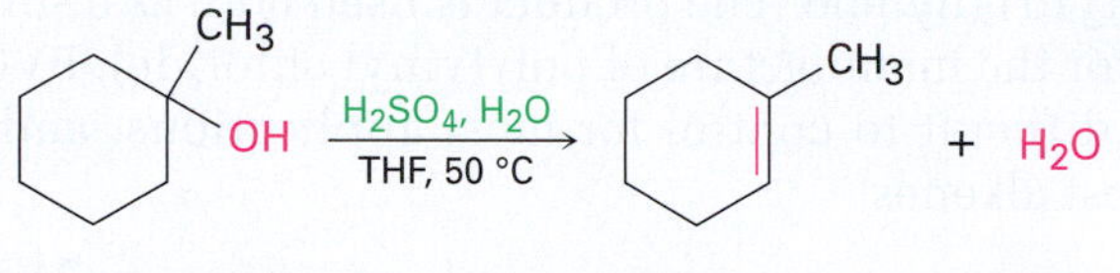
CH3
OH
H2SO4, H2O
THF, 50 °C
CH3
+ H2O

1-Methylcyclohexanol **1-Methylcyclohexene (91%)**

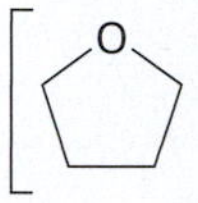
O

Tetrahydrofuran (THF)—a common solvent

In biological pathways, dehydrations rarely occur with isolated alcohols. Instead, they normally take place on substrates in which the –OH is positioned two carbons away from a carbonyl group. In the biosynthesis of fats, for instance, β-hydroxybutyryl ACP is converted by dehydration to *trans*-crotonyl ACP, where ACP is an abbreviation for *acyl carrier protein*. We'll see the reason for this requirement in Section 12.13.

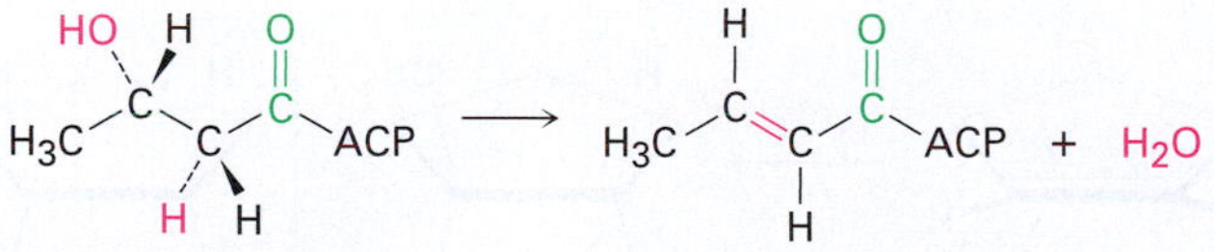
HO H O
H3C C C C ACP
H H
H O
H3C C C C ACP + H2O
H

β-Hydroxybutyryl ACP ***trans*-Crotonyl ACP**

Problem 8.1

One problem with elimination reactions is that mixtures of products are often formed. For example, treatment of 2-bromo-2-methylbutane with KOH in ethanol yields a mixture of two alkene products. What are their likely structures?

Problem 8.2

How many alkene products, including *E,Z* isomers, might be obtained by dehydration of 3-methylhexan-3-ol with aqueous sulfuric acid?

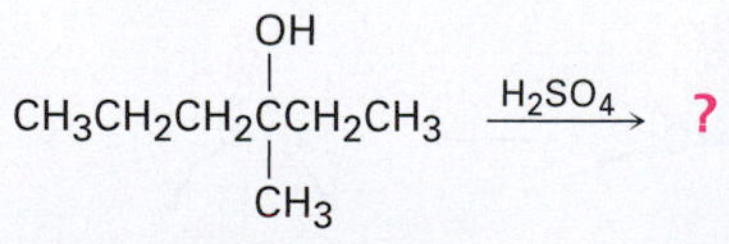
OH
CH3CH2CH2CCH2CH3
CH3
H2SO4
?

3-Methylhexan-3-ol

8.2 Halogenation of Alkenes

Bromine and chlorine add to alkenes to yield 1,2-dihalides, a process called **halogenation**. For example, each year approximately 6 million tons of 1,2-dichloroethane (ethylene dichloride) are synthesized industrially by the addition of Cl_2 to ethylene. The product is used both as a solvent and as starting material for the manufacture of poly(vinyl chloride), PVC. Fluorine is too reactive and difficult to control for most applications, and iodine does not react with most alkenes.

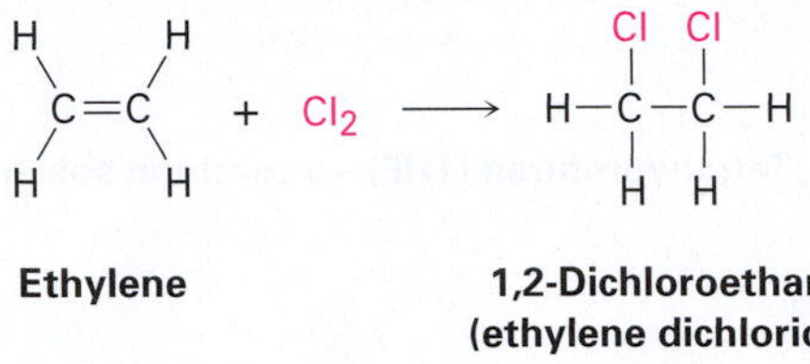

Ethylene   **1,2-Dichloroethane
                              (ethylene dichloride)**

Interestingly, when the halogenation reaction is carried out on a cycloalkene, such as cyclopentene, only the *trans* stereoisomer of the dihalide product is formed rather than a mixture of cis and trans isomers. We therefore say that the reaction occurs with **anti stereochemistry**, meaning that the two halogen atoms come from opposite faces of the double bond—one from the top face and one from the bottom face.

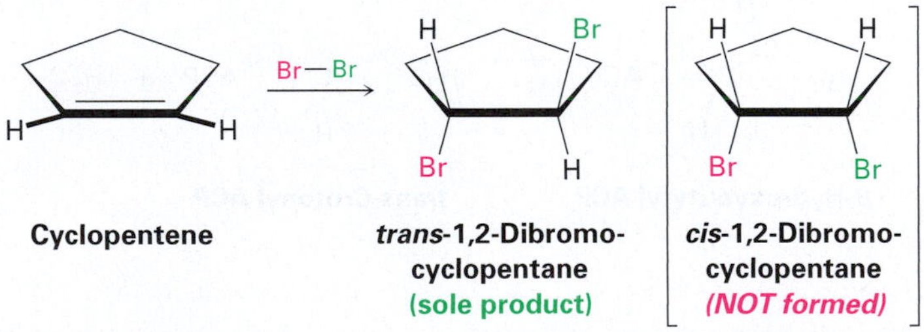

Cyclopentene   ***trans*-1,2-Dibromo-   ***cis*-1,2-Dibromo-
                          cyclopentane**   cyclopentane**
                          (sole product)   **(NOT formed)**

An explanation for the observed stereochemistry of alkene addition came in 1937 with the suggestion that the reaction occurs through an intermediate **bromonium ion (R_2Br^+)**, formed by electrophilic addition of Br^+ to the alkene. (Similarly, a *chloronium ion* contains a positively charged, divalent chlorine, R_2Cl^+.) The bromonium ion is formed in a single step by interaction of the alkene with Br_2 and simultaneous loss of Br^- (Figure 8.1).

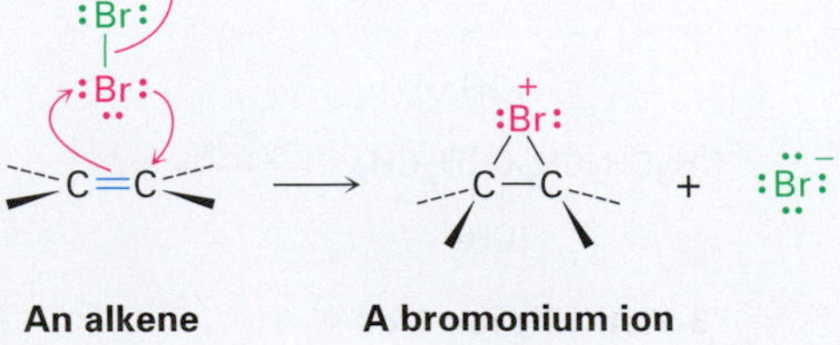

FIGURE 8.1 Formation of a bromonium ion intermediate by reaction of Br_2 with an alkene. The reaction occurs in a single step and results in overall electrophilic addition of Br^+ to the alkene.

An alkene   **A bromonium ion**

How does the formation of a bromonium ion account for the observed anti stereochemistry of addition to cyclopentene? If a bromonium ion is formed as an intermediate, we can imagine that the large bromine atom might "shield" one side of the molecule. Reaction with Br⁻ ion in the second step can then occur only from the opposite, unshielded side to give the trans product.

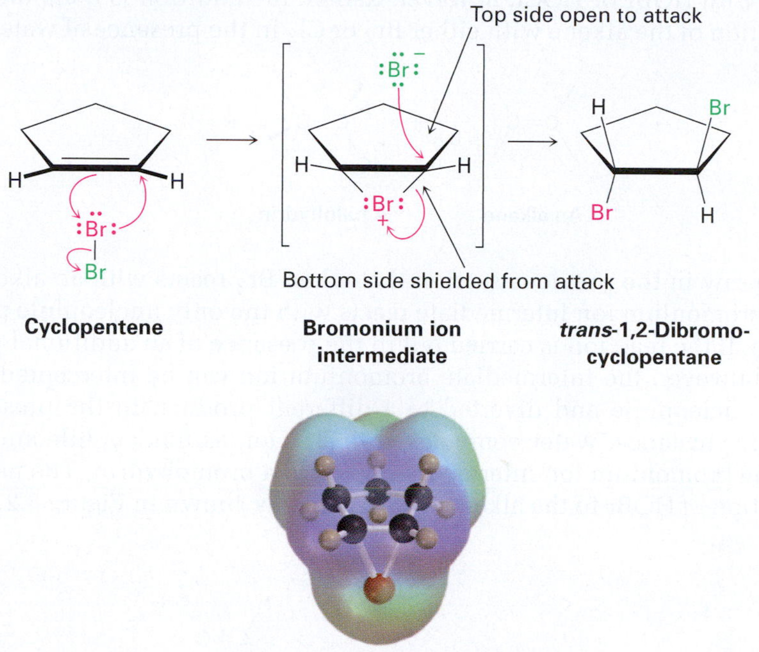

Cyclopentene   **Bromonium ion**   ***trans*-1,2-Dibromo-**
                              intermediate   **cyclopentane**

Alkene halogenation reactions occur in nature just as they do in the laboratory but are limited primarily to marine organisms, which live in a halide-rich environment. The reactions are carried out by enzymes called *haloperoxidases,* which use H_2O_2 to oxidize Br^- or Cl^- ions to a biological equivalent of Br^+ or Cl^+. Electrophilic addition to the double bond of a substrate molecule then yields a bromonium or chloronium ion intermediate just as in the laboratory, and reaction with another halide ion completes the process. For example, the following tetrahalide isolated from the red alga *Plocamium cartilagineum* is thought to arise from β-ocimene by twofold addition of BrCl through the corresponding bromonium ions.

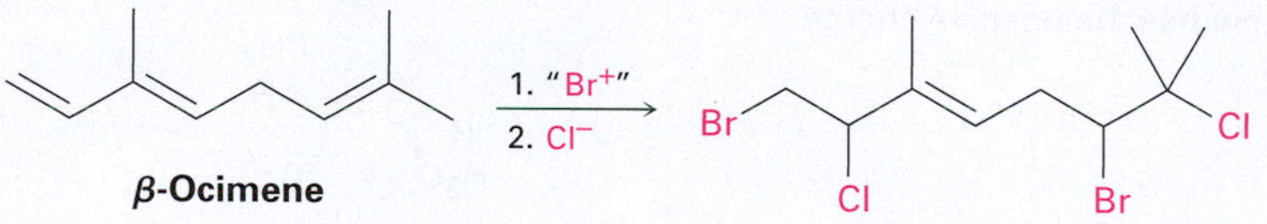

β-Ocimene

Problem 8.3
What product would you expect to obtain from addition of Cl_2 to 1,2-dimethylcyclohexene? Show the stereochemistry of the product.

Problem 8.4
Addition of HCl to 1,2-dimethylcyclohexene yields a mixture of two products. Show the stereochemistry of each, and explain why a mixture is formed.

8.3 Halohydrins from Alkenes

Another example of an electrophilic addition is the reaction of alkenes with the hypohalous acids HO–Cl or HO–Br to yield 1,2-halo alcohols, called **halohydrins**. Halohydrin formation doesn't take place by direct reaction of an alkene with HOBr or HOCl, however. Rather, the addition is done indirectly by reaction of the alkene with either Br_2 or Cl_2 in the presence of water.

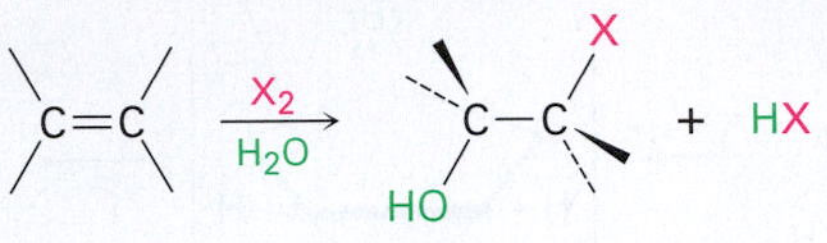

An alkene   **A halohydrin**

We saw in the previous section that when Br_2 reacts with an alkene, the cyclic bromonium ion intermediate reacts with the only nucleophile present, Br^- ion. If the reaction is carried out in the presence of an additional nucleophile, however, the intermediate bromonium ion can be intercepted by the added nucleophile and diverted to a different product. In the presence of water, for instance, water competes with Br^- ion as nucleophile and reacts with the bromonium ion intermediate to yield a bromohydrin. The net effect is addition of HOBr to the alkene by the pathway shown in Figure 8.2.

FIGURE 8.2 MECHANISM: Bromohydrin formation by reaction of an alkene with Br_2 in the presence of water. Water acts as a nucleophile to react with the intermediate bromonium ion.

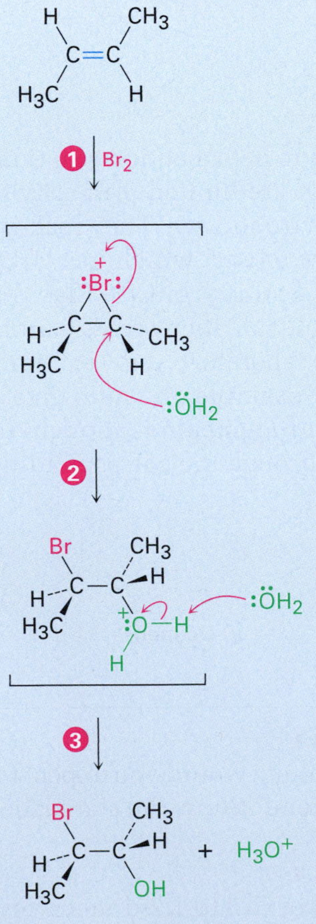

1 Reaction of the alkene with Br_2 yields a bromonium ion intermediate, as previously discussed.

2 Water acts as a nucleophile, using a lone pair of electrons to open the bromonium ion ring and form a bond to carbon. Since oxygen donates its electrons in this step, it now has the positive charge.

3 Loss of a proton (H^+) from oxygen then gives H_3O^+ and the neutral bromohydrin addition product.

3-Bromobutan-2-ol

There are a number of biological examples of halohydrin formation, particularly in marine organisms. As with halogenation (Section 8.2), halohydrin formation is carried out by haloperoxidases, which function by oxidizing Br⁻ or Cl⁻ ions to the corresponding HOBr or HOCl bonded to a metal atom in the enzyme. Electrophilic addition to the double bond of a substrate molecule then yields a bromonium or chloronium ion intermediate, and reaction with water gives the halohydrin. For example:

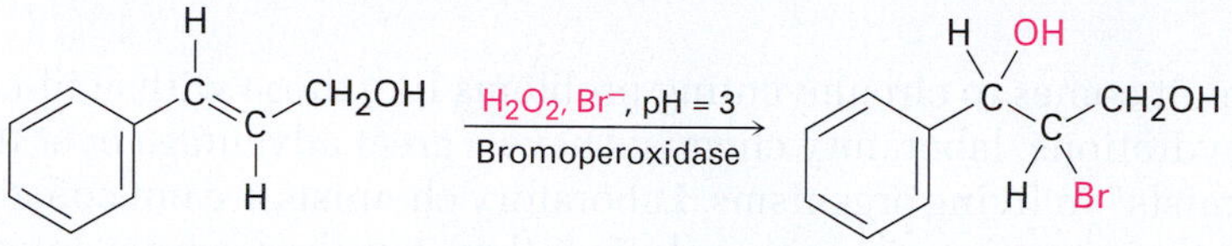

Problem 8.5

When an unsymmetrically substituted alkene such as propene is treated with Br_2 and water, the major product has the bromine atom bonded to the less highly substituted carbon atom. Is this Markovnikov or non-Markovnikov orientation? Explain.

$$CH_3CH=CH_2 \xrightarrow{Br_2,\ H_2O} CH_3CHCH_2Br$$
$$\overset{\displaystyle OH}{|}$$

8.4 Hydration of Alkenes

We saw in Section 7.6 that alkenes undergo an acid-catalyzed addition reaction with water to yield alcohols. The process is particularly suited to large-scale industrial procedures, and approximately 300,000 tons of ethanol are manufactured each year in the United States by hydration of ethylene. Unfortunately, the reaction is not of much use in the laboratory because of the high temperatures needed—250 °C in the case of ethylene.

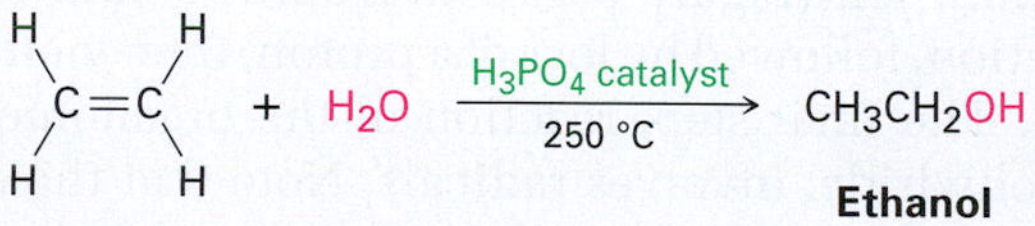

Acid-catalyzed hydration of isolated double bonds, although known, is also uncommon in biological pathways. More frequently, biological hydrations require that the double bond be adjacent to a carbonyl group for reaction to proceed. Fumarate, for instance, is hydrated to give malate as one step in the citric acid cycle of food metabolism. Note that the requirement for an adjacent carbonyl group in the addition of water is the same as that we saw in Section 8.1 for the elimination of water. We'll see the reason for the requirement in Section 14.11, but might note for now that the reaction is not an electrophilic addition but instead occurs through a mechanism that involves formation of an anion intermediate followed by protonation by an acid HA.

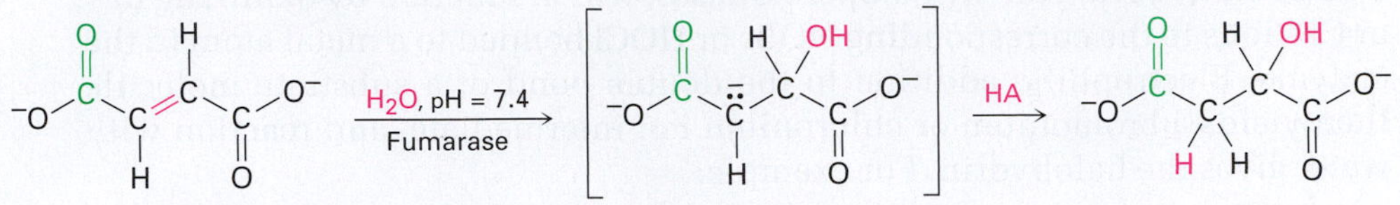

Fumarate   **Anion intermediate**   **Malate**

When it comes to circumventing problems like those with acid-catalyzed alkene hydrations, laboratory chemists have a great advantage over the cellular "chemists" in living organisms. Laboratory chemists are not constrained to carry out their reactions in water solution; they can choose from any of a large number of solvents. Laboratory reactions don't need to be carried out at a fixed temperature; they can take place over a wide range of temperatures. And laboratory reagents aren't limited to containing carbon, oxygen, nitrogen, and a few other elements; they can contain any element in the periodic table.

The general theme of this text is to focus on reactions that have a direct relevance to the chemistry of living organisms. Every so often, though, we'll discuss a particularly useful laboratory reaction that has no biological counterpart. In the present case, alkenes are often hydrated in the laboratory by two nonbiological procedures, *oxymercuration* and *hydroboration/oxidation*, which give complementary results.

Oxymercuration involves electrophilic addition of Hg^{2+} to the alkene on treatment with mercury(II) acetate [$(CH_3CO_2)_2Hg$, or $Hg(OAc)_2$] in aqueous tetrahydrofuran (THF) solvent. The intermediate *organomercury* compound is then treated with sodium borohydride, $NaBH_4$, and an alcohol is produced. For example:

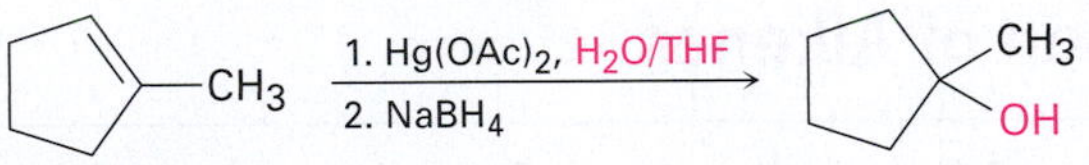

1-Methylcyclopentene   **1-Methylcyclopentanol**
                                   (92%)

Alkene oxymercuration is closely analogous to halohydrin formation. The reaction is initiated by electrophilic addition of Hg^{2+} (mercuric ion) to the alkene to give an intermediate *mercurinium ion,* whose structure resembles that of a bromonium ion (Figure 8.3). Nucleophilic addition of water as in halohydrin formation, followed by loss of a proton, then yields a stable organomercury product. The final step, reaction of the organomercury compound with sodium borohydride, involves radicals. Note that the regiochemistry of the reaction corresponds to Markovnikov addition of water; that is, the $-OH$ group attaches to the more highly substituted carbon atom, and the $-H$ attaches to the less highly substituted carbon.

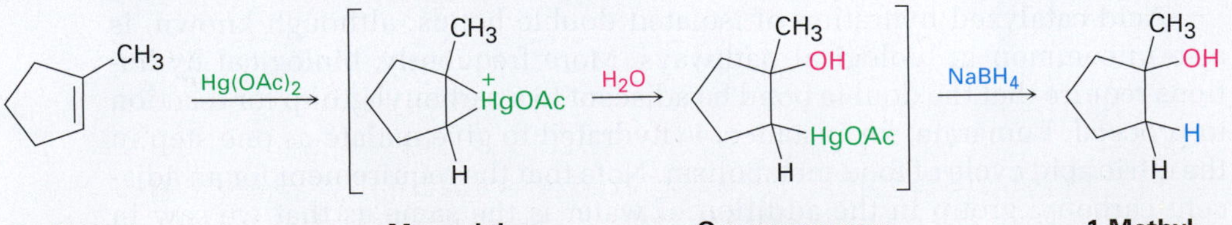

FIGURE 8.3 Mechanism of the oxymercuration of an alkene to yield an alcohol. The reaction involves a mercurinium ion intermediate and proceeds by a mechanism similar to that of halohydrin formation. The product of the reaction is the more highly substituted alcohol, corresponding to Markovnikov regiochemistry.

1-Methyl-   **Mercurinium**   **Organomercury**   **1-Methyl-**
cyclopentene   **ion**   **compound**   **cyclopentanol**
                                                              (92% yield)

In addition to the oxymercuration method, which yields the Markovnikov product, a complementary hydroboration/oxidation method that yields the non-Markovnikov product is also used in the laboratory. Hydroboration/oxidation involves addition of a B–H bond of borane, BH_3, to an alkene to yield an organoborane intermediate, RBH_2. Oxidation of the organoborane by reaction with basic hydrogen peroxide, H_2O_2, then gives the alcohol. For example:

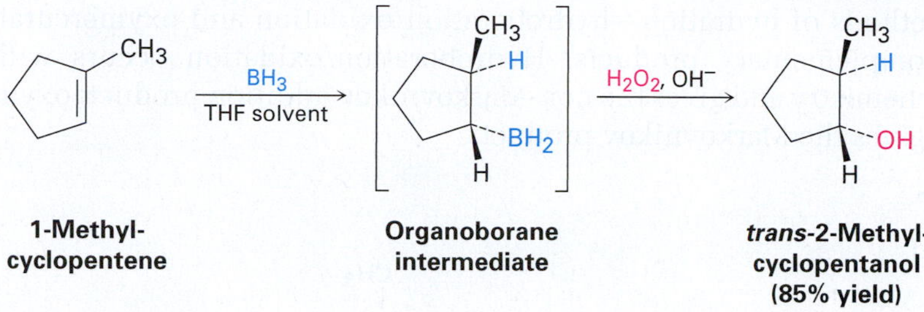
CH3
BH3
THF solvent
CH3
H
BH2
H
H2O2, OH−
CH3
H
OH
H
1-Methyl-cyclopentene
Organoborane intermediate
trans-2-Methyl-cyclopentanol (85% yield)

Note that during the initial addition step, both boron and hydrogen add to the double bond from the same face of the double bond—that is, with **syn stereochemistry**, the opposite of anti. In this step, boron attaches to the less highly substituted carbon. During the subsequent oxidation, the boron is replaced by an –OH with the same stereochemistry, resulting in an overall syn, non-Markovnikov addition of water.

Why does hydroboration/oxidation take place with syn, non-Markovnikov regiochemistry to yield the less highly substituted alcohol? Hydroboration differs from many other alkene addition reactions in that it occurs in a single step without a carbocation intermediate. Because both C–H and C–B bonds form at the same time and from the same face of the alkene, syn stereochemistry results. Non-Markovnikov regiochemistry is found because attachment of boron is favored at the less sterically crowded carbon atom of the alkene rather than at the more crowded carbon (Figure 8.4).

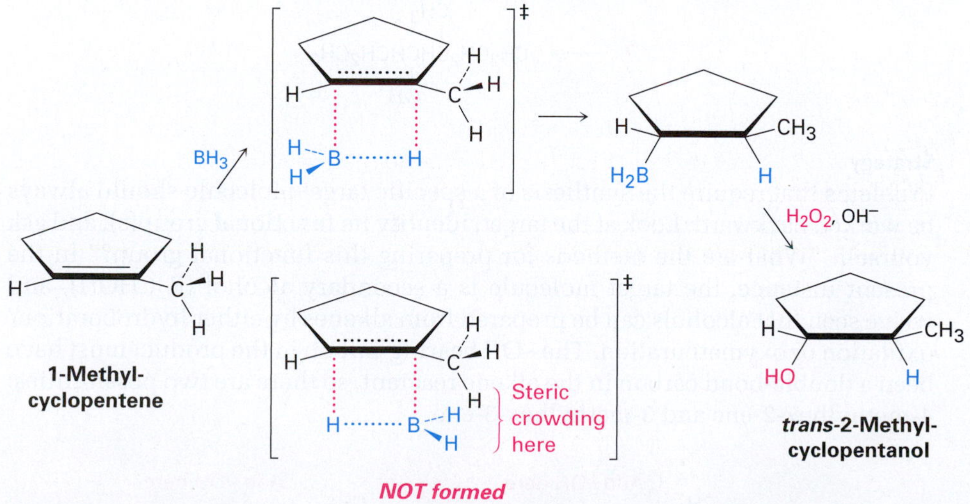
BH3
1-Methyl-cyclopentene
‡
H2B
H2O2, OH−
HO
trans-2-Methyl-cyclopentanol
Steric crowding here
NOT formed

FIGURE 8.4 Alkene hydroboration. The reaction occurs in a single step in which both C–H and C–B bonds form at the same time and on the same face of the double bond. The lower energy, more rapidly formed transition state is the one with less steric crowding, leading to non-Markovnikov regiochemistry.

WORKED EXAMPLE 8.1 Predicting the Products of a Hydration Reaction

What products would you obtain from reaction of 2-methylpent-2-ene with:
(a) BH_3, followed by H_2O_2, OH^- **(b)** $Hg(OAc)_2$, followed by $NaBH_4$

Strategy

When predicting the product of a reaction, you have to recall what you know about the kind of reaction being carried out and then apply that knowledge to the specific case you're dealing with. In the present instance, recall that the two methods of hydration—hydroboration/oxidation and oxymercuration—give complementary products. Hydroboration/oxidation occurs with syn stereochemistry and gives the non-Markovnikov addition product; oxymercuration gives the Markovnikov product.

Solution

$$CH_3CH_2CH=\overset{\overset{\displaystyle CH_3}{|}}{C}CH_3$$

(a)   **2-Methylpent-2-ene**   (b)

1. BH_3
2. H_2O_2, OH^-

1. $Hg(OAc)_2$, H_2O
2. $NaBH_4$

$$CH_3CH_2\overset{\overset{\displaystyle H}{|}}{\underset{\underset{\displaystyle HO}{|}}{C}}-\overset{\overset{\displaystyle CH_3}{|}}{\underset{\underset{\displaystyle H}{|}}{C}}CH_3$$

2-Methylpentan-3-ol

$$CH_3CH_2\overset{\overset{\displaystyle H}{|}}{\underset{\underset{\displaystyle H}{|}}{C}}-\overset{\overset{\displaystyle CH_3}{|}}{\underset{\underset{\displaystyle OH}{|}}{C}}CH_3$$

2-Methylpentan-2-ol

WORKED EXAMPLE 8.2 Synthesizing an Alcohol

How might you prepare the following alcohol?

$$?\ \longrightarrow\ CH_3CH_2\overset{\overset{\displaystyle CH_3}{|}}{\underset{\underset{\displaystyle OH}{|}}{CH}}CHCH_2CH_3$$

Strategy

Problems that require the synthesis of a specific target molecule should always be worked backward. Look at the target, identify its functional group(s), and ask yourself, "What are the methods for preparing this functional group?" In the present instance, the target molecule is a secondary alcohol (R_2CHOH), and we've seen that alcohols can be prepared from alkenes by either hydroboration/oxidation or oxymercuration. The –OH bearing carbon in the product must have been a double-bond carbon in the alkene reactant, so there are two possibilities: 4-methylhex-2-ene and 3-methylhex-3-ene.

Add –OH here

$$CH_3CH_2\overset{\overset{\displaystyle CH_3}{|}}{CH}CH=CHCH_3$$

4-Methylhex-2-ene

Add –OH here

$$CH_3CH_2\overset{\overset{\displaystyle CH_3}{|}}{C}=CHCH_2CH_3$$

3-Methylhex-3-ene

4-Methylhex-2-ene has a disubstituted double bond, RCH=CHR', and would probably give a mixture of two alcohols with either hydration method since Markovnikov's rule does not apply to symmetrically substituted alkenes. 3-Methylhex-3-ene, however, has a trisubstituted double bond and would give only the desired product on non-Markovnikov hydration using the hydroboration/oxidation method.

Solution

$$CH_3CH_2\overset{\overset{\displaystyle CH_3}{|}}{C}=CHCH_2CH_3 \quad \xrightarrow[\text{2. } H_2O_2,\ OH^-]{\text{1. } BH_3,\ THF} \quad CH_3CH_2\overset{\overset{\displaystyle CH_3}{|}}{CH}CHCH_2CH_3$$
$$\underset{\text{OH}}{}$$

3-Methylhex-3-ene

Problem 8.6

What products would you expect from oxymercuration of the following alkenes? From hydroboration/oxidation?

(a)
$$CH_3\overset{\overset{\displaystyle CH_3}{|}}{C}=CHCH_2CH_3$$

(b)

Problem 8.7

What alkenes might the following alcohols have been prepared from?

(a)
$$CH_3\overset{\overset{\displaystyle CH_3}{|}}{CH}CH_2CH_2OH$$

(b)
$$\overset{H_3C}{\underset{}{}}\ \overset{OH}{\underset{}{}}$$
$$CH_3CHCHCH_3$$

(c) (cyclohexane with CH₂OH group)

Problem 8.8

The following cycloalkene gives a mixture of two alcohols on hydroboration/oxidation. Draw the structures of both, and explain the result.

8.5 Reduction of Alkenes: Hydrogenation

Alkenes are converted to alkanes by addition of two hydrogen atoms. In the laboratory, the reaction is usually carried out by reaction of the alkene with gaseous H_2 in the presence of a metal catalyst such as palladium or platinum. We describe the result by saying that the double bond has been **hydrogenated**, or *reduced*. Note that the words *oxidation* and *reduction* are used somewhat differently in organic chemistry than what you might have learned previously.

In general chemistry, a reduction is defined as the gain of one or more electrons by an atom. In organic chemistry, however, a **reduction** is a reaction that results in a gain of electron density by carbon, caused either by bond-making between carbon and a less electronegative atom or by bond-breaking between carbon and a more electronegative atom.

Reduction   Increases electron density on carbon by:

– forming this: C–H

– or breaking one of these: C–O   C–N   C–X

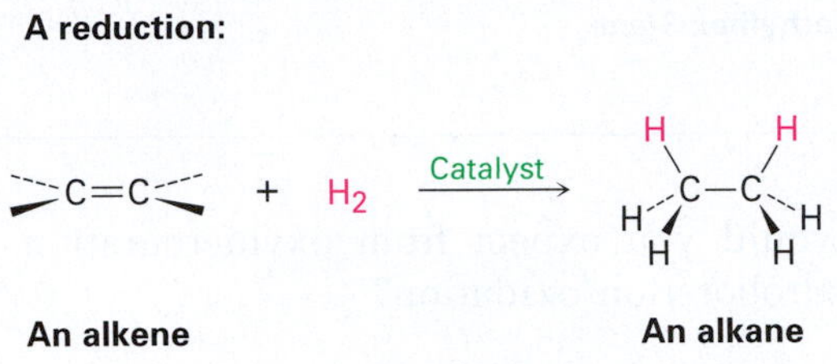

A reduction:

An alkene   An alkane

Platinum and palladium are the most common catalysts for alkene hydrogenations. Palladium is normally used as a very fine powder "supported" on an inert material such as charcoal (Pd/C) to maximize surface area. Platinum is normally used as PtO_2, a reagent known as *Adams' catalyst* after its discoverer, Roger Adams.

Catalytic hydrogenation, unlike most other organic reactions, is a *heterogeneous* process rather than a homogeneous one. That is, the hydrogenation reaction does not occur in a homogeneous solution but instead takes place on the surface of insoluble catalyst particles. Hydrogenation usually occurs with syn stereochemistry—both hydrogens add to the double bond from the same face.

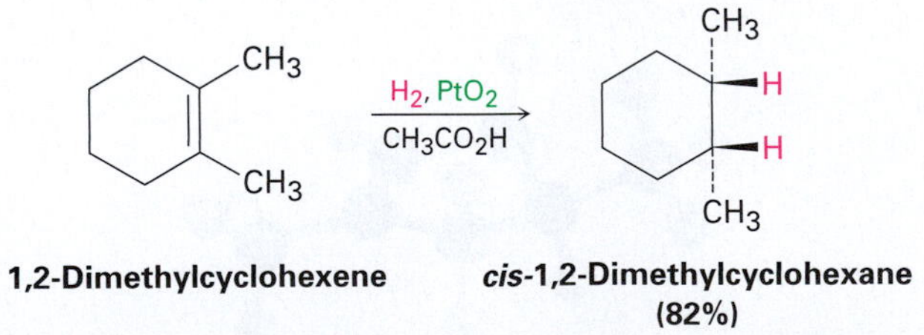

1,2-Dimethylcyclohexene   *cis*-**1,2-Dimethylcyclohexane**
                                                        (82%)

The first step in the reaction is adsorption of H_2 onto the catalyst surface. Complexation between catalyst and alkene then occurs as a vacant orbital on the metal interacts with the filled alkene π orbital. In the final steps, hydrogen is inserted into the double bond and the saturated product diffuses away from the catalyst (Figure 8.5). The stereochemistry of hydrogenation is syn because both hydrogens add to the double bond from the same catalyst surface.

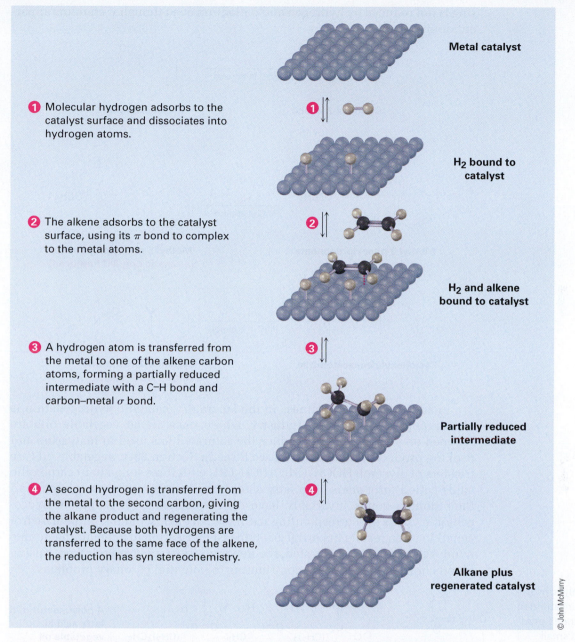

1 Molecular hydrogen adsorbs to the catalyst surface and dissociates into hydrogen atoms.

Metal catalyst

H$_2$ bound to catalyst

2 The alkene adsorbs to the catalyst surface, using its π bond to complex to the metal atoms.

H$_2$ and alkene bound to catalyst

3 A hydrogen atom is transferred from the metal to one of the alkene carbon atoms, forming a partially reduced intermediate with a C–H bond and carbon–metal σ bond.

Partially reduced intermediate

4 A second hydrogen is transferred from the metal to the second carbon, giving the alkane product and regenerating the catalyst. Because both hydrogens are transferred to the same face of the alkene, the reduction has syn stereochemistry.

Alkane plus regenerated catalyst

© John McMurry

ACTIVE FIGURE 8.5 MECHANISM: Mechanism of alkene hydrogenation. The reaction takes place with syn stereochemistry on the surface of insoluble catalyst particles. **Go to this book's student companion site at** www.cengage.com/chemistry/mcmurry **to explore an interactive version of this figure.**

Alkenes are much more reactive than most other unsaturated functional groups toward catalytic hydrogenation, and the reaction is therefore quite selective. Other functional groups such as aldehydes, ketones, and esters survive normal alkene hydrogenation conditions unchanged, although reaction with these groups does occur under more vigorous conditions. Note particularly in the hydrogenation of methyl 3-phenylpropenoate that the aromatic ring is not reduced by hydrogen and palladium even though it contains apparent double bonds.

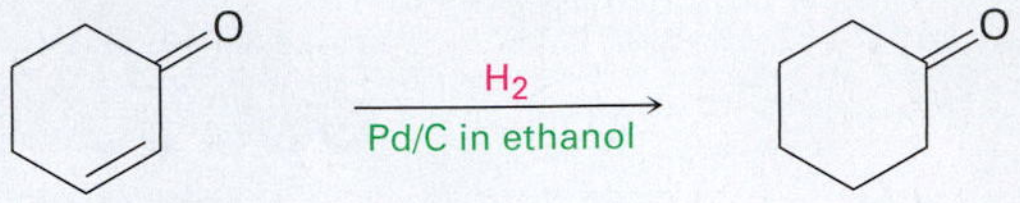

Cyclohex-2-enone → $\xrightarrow[\text{Pd/C in ethanol}]{H_2}$ → **Cyclohexanone**
(ketone *NOT* reduced)

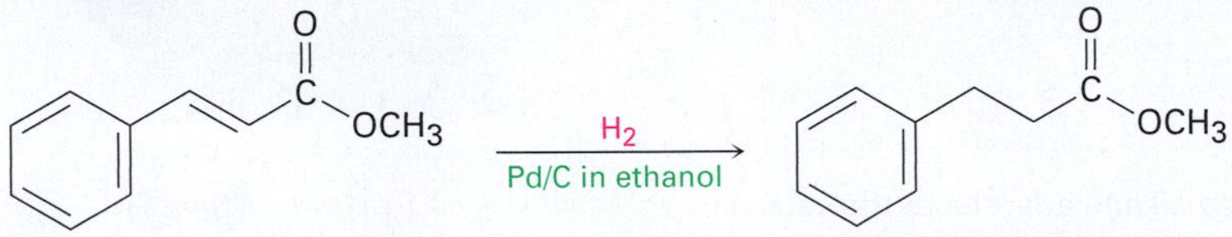

Methyl 3-phenylpropenoate → $\xrightarrow[\text{Pd/C in ethanol}]{H_2}$ → **Methyl 3-phenylpropanoate**
(aromatic ring *NOT* reduced)

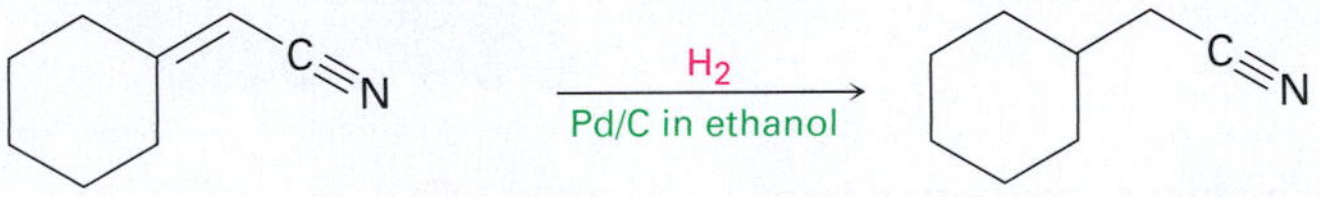

Cyclohexylideneacetonitrile → $\xrightarrow[\text{Pd/C in ethanol}]{H_2}$ → **Cyclohexylacetonitrile**
(nitrile *NOT* reduced)

In addition to its usefulness in the laboratory, catalytic hydrogenation is also important in the food industry, where unsaturated vegetable oils are reduced on a large scale to produce the saturated fats used in margarine and cooking products (Figure 8.6). As we'll see in Section 23.1, vegetable oils are triesters of glycerol, $HOCH_2CH(OH)CH_2OH$, with three long-chain carboxylic acids called *fatty acids.* The fatty acids are generally polyunsaturated, and their double bonds invariably have cis stereochemistry. Complete hydrogenation yields the corresponding saturated fatty acids, but incomplete hydrogenation often results in partial cis–trans isomerization of a remaining double bond. When eaten and digested, the free trans fatty acids are released, raising blood cholesterol levels and contributing to potential coronary problems.

FIGURE 8.6 Catalytic hydrogenation of polyunsaturated fats leads to saturated products, along with a small amount of isomerized trans fats.

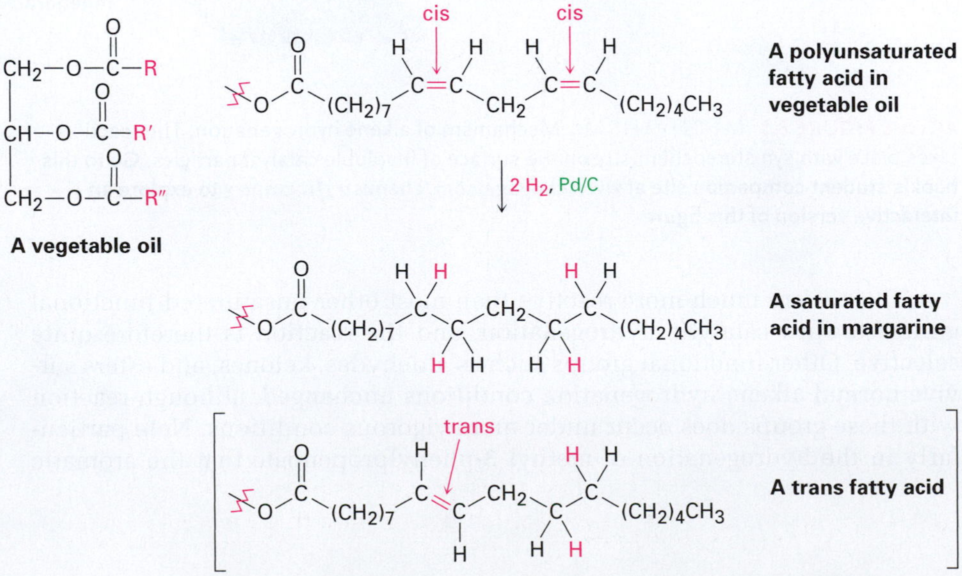

Double-bond reductions are extremely common in biological pathways, although the mechanism of the process is of course different from that of laboratory catalytic hydrogenation over palladium. As with hydrations (Section 8.4), the reduction of isolated double bonds is rare in biological pathways. Instead, biological reductions usually occur in two steps and require that the double bond be adjacent to a carbonyl group. In the first step, the coenzyme reduced nicotinamide adenine dinucleotide phosphate, abbreviated NADPH, adds a hydride ion ($H:^-$) to the double bond to give an anion. In the second, the anion is protonated by acid HA, leading to overall addition of H_2.

As an example of a biological hydrogenation, one of the steps in fatty-acid biosynthesis involves the reduction of *trans*-crotonyl ACP to yield butyryl ACP (Figure 8.7). Note the similarity of this mechanism with the mechanism for biological hydrations that we saw at the beginning of Section 8.4. In both, a carbonyl group next to the double bond is needed and an anion intermediate is involved.

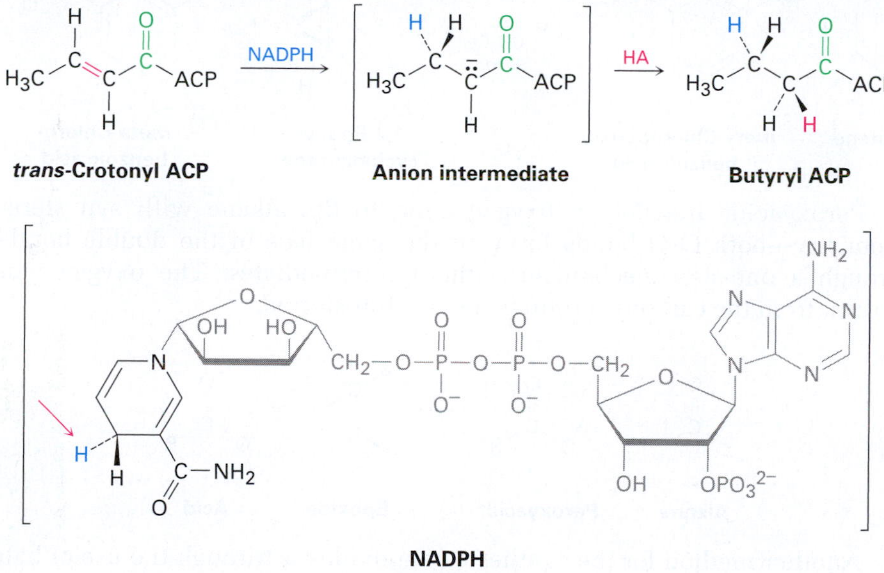

trans-Crotonyl ACP   **Anion intermediate**   **Butyryl ACP**

NADPH

FIGURE 8.7 Reduction of the carbon–carbon double bond in *trans*-crotonyl ACP, a step in the biosynthesis of fatty acids. One hydrogen (blue) is delivered from NADPH as a hydride ion, $H:^-$; the other hydrogen (red) is delivered by protonation of the anion intermediate with an acid, HA. As is often the case in biological reactions, the structure of the biochemical reagent, NADPH in this case, is relatively complex considering the apparent simplicity of the transformation itself.

Problem 8.9

What product would you obtain from catalytic hydrogenation of the following alkenes?

(a)

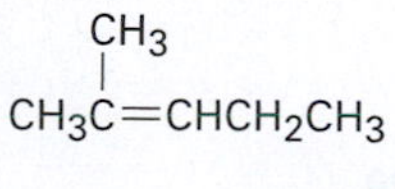

$$CH_3C{=}CHCH_2CH_3$$
with CH_3 substituent

(b)

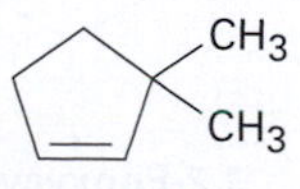

8.6 Oxidation of Alkenes: Epoxidation

Like the word *reduction* used in the previous section for addition of hydrogen to a double bond, the word *oxidation* has a slightly different meaning in organic chemistry than what you might have previously learned. In general chemistry, an oxidation is defined as the loss of one or more electrons by an atom. In organic chemistry, however, an **oxidation** is a reaction that results in a loss of electron density by carbon, caused either by bond-making between carbon and a more electronegative atom—usually oxygen, nitrogen, or a halogen—or by bond-breaking between carbon and a less electronegative atom—usually hydrogen. Note that an *oxidation* often adds oxygen, while a *reduction* often adds hydrogen.

Oxidation Decreases electron density on carbon by:

  – forming one of these: C–O   C–N   C–X

  – or breaking this: C–H

In the laboratory, alkenes are oxidized to give *epoxides* on treatment with a peroxyacid, RCO_3H, such as *meta*-chloroperoxybenzoic acid. An **epoxide**, also called an *oxirane,* is a cyclic ether with an oxygen atom in a three-membered ring. For example:

| Cycloheptene | *meta*-Chloroperoxy- benzoic acid | 1,2-Epoxy- cycloheptane | *meta*-Chloro- benzoic acid |

Peroxyacids transfer an oxygen atom to the alkene with syn stereochemistry—both C–O bonds form on the same face of the double bond—through a one-step mechanism without intermediates. The oxygen atom farthest from the carbonyl group is the one transferred.

| Alkene | Peroxyacid | Epoxide | Acid |

Another method for the synthesis of epoxides is through the use of halohydrins, prepared by electrophilic addition of HO–X to alkenes (Section 8.3). When halohydrins are treated with base, HX is eliminated and an epoxide is produced.

| Cyclohexene | *trans*-2-Chloro- cyclohexanol | 1,2-Epoxycyclohexane (73%) |

Epoxides are also produced from alkenes as intermediates in various biological pathways, although peroxyacids are not involved. An example is the conversion of squalene into 2,3-oxidosqualene, a key step in the biosynthesis of steroids. The reaction is carried out by a flavin hydroperoxide, which is formed by reaction of O_2 with the coenzyme reduced flavin adenine dinucleotide, abbreviated $FADH_2$. Note the specificity of the reaction in which only one double bond out of six in the substrate molecule undergoes reaction. Note also that, once again, the structure of the biochemical reagent, flavin hydroperoxide, is relatively complex given the apparent simplicity of the transformation (Figure 8.8).

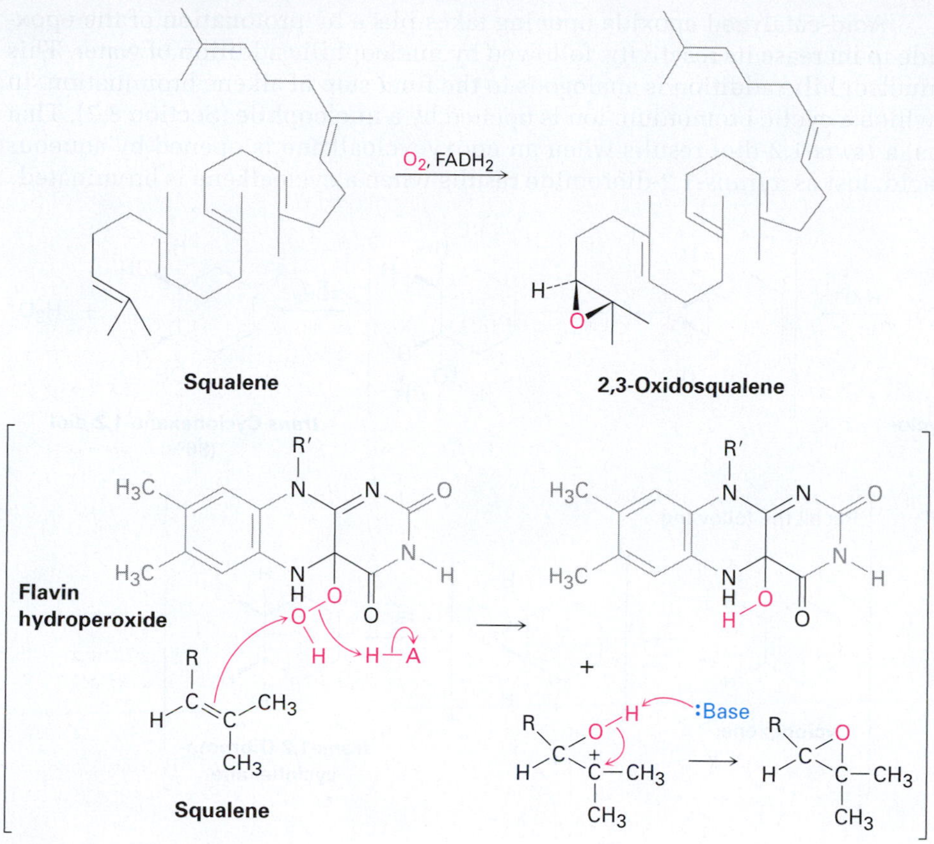

FIGURE 8.8 A biological epoxidation reaction of the alkene squalene, a step in steroid biosynthesis. The reaction is effected by a flavin hydroperoxide formed by reaction of O_2 with the coenzyme reduced flavin adenine dinucleotide, $FADH_2$.

Problem 8.10

What product would you expect from reaction of *cis*-but-2-ene with *meta*-chloroperoxybenzoic acid? Show the stereochemistry.

8.7 Oxidation of Alkenes: Hydroxylation

Both in the laboratory and in living organisms, epoxides undergo an acid-catalyzed ring-opening reaction with water (a *hydrolysis*) to give the corresponding 1,2-dialcohol, or *diol*, also called a *glycol*. Thus, the net result of the two-step alkene epoxidation/hydrolysis is **hydroxylation**—the addition of an —OH group to each of the two double-bond carbons. In fact, more than 3 million tons of ethylene glycol, $HOCH_2CH_2OH$, most of it used for automobile antifreeze, are produced each year in the United States by epoxidation of ethylene followed by hydrolysis.

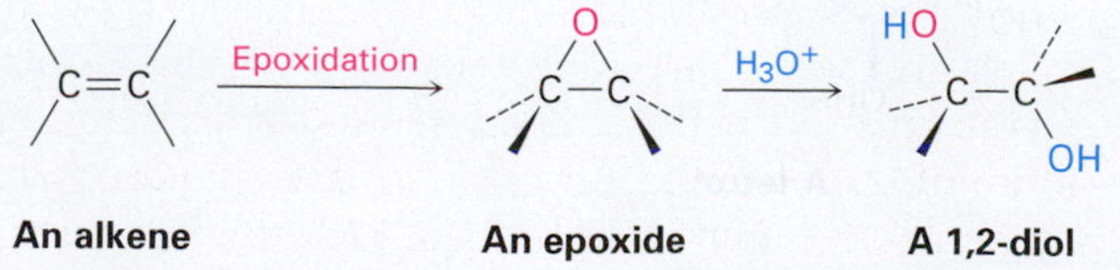

Acid-catalyzed epoxide opening takes place by protonation of the epoxide to increase its reactivity, followed by nucleophilic addition of water. This nucleophilic addition is analogous to the final step of alkene bromination, in which a cyclic bromonium ion is opened by a nucleophile (Section 8.2). That is, a *trans*-1,2-diol results when an epoxycycloalkane is opened by aqueous acid, just as a *trans*-1,2-dibromide results when a cycloalkene is brominated.

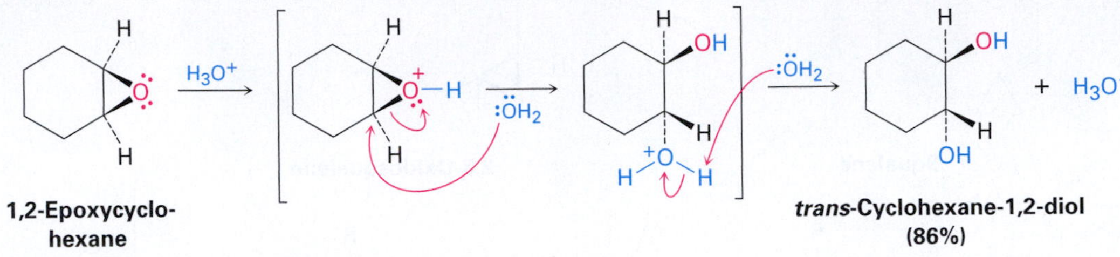

1,2-Epoxycyclo-hexane

trans-Cyclohexane-1,2-diol
(86%)

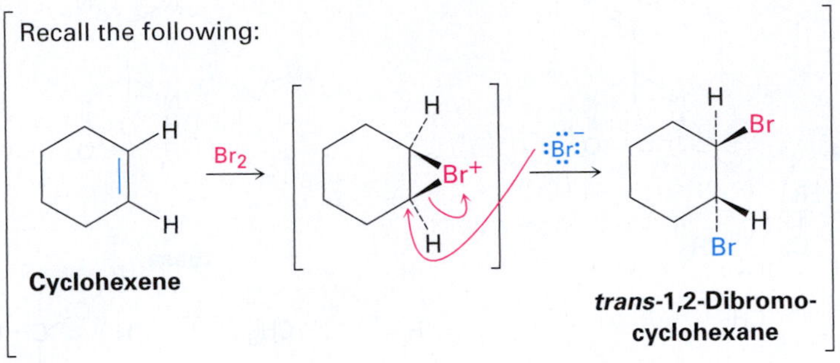

Recall the following:

Cyclohexene

trans-1,2-Dibromo-cyclohexane

Biological examples of epoxide hydrolysis are common, particularly in the pathways that animals use to detoxify harmful substances. The cancer-causing *(carcinogenic)* substance benzo[a]pyrene, for instance, is found in cigarette smoke, chimney soot, and barbecued meat. In the human liver, benzo[a]pyrene is detoxified by conversion to a diol epoxide, which then undergoes enzyme-catalyzed hydrolysis to give a soluble tetrol.

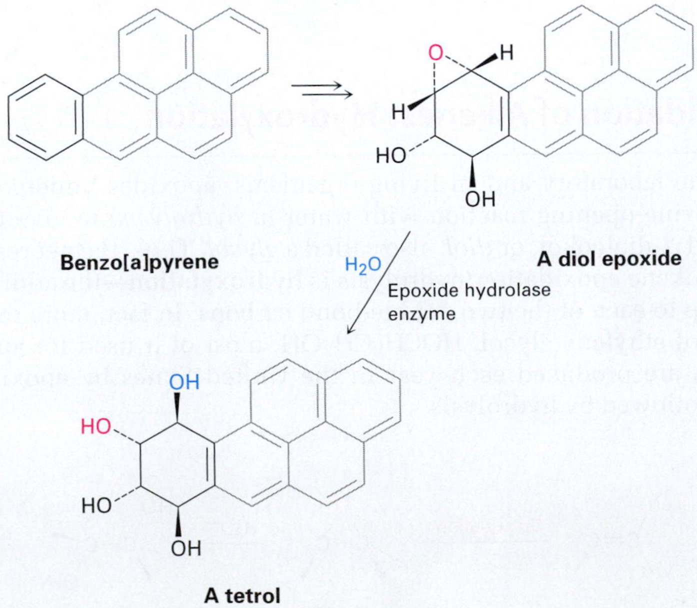

Benzo[a]pyrene H_2O **A diol epoxide**

Epoxide hydrolase
enzyme

A tetrol

In the laboratory, hydroxylation can also be carried out without going through an intermediate epoxide by treating an alkene with osmium tetroxide, OsO_4. The reaction occurs with syn stereochemistry and does not involve a carbocation intermediate. Instead, it takes place through an intermediate cyclic *osmate*, which is formed in a single step by addition of OsO_4 to the alkene. This cyclic osmate is then cleaved using aqueous sodium bisulfite, $NaHSO_3$.

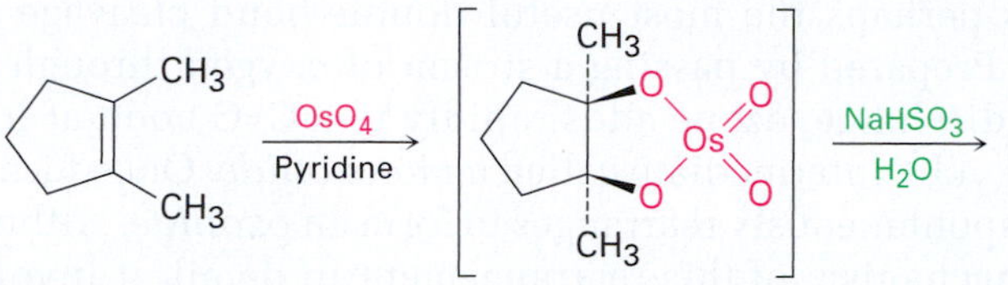

1,2-Dimethylcyclopentene   **A cyclic osmate**   ***cis*-1,2-Dimethylcyclo-**
                                       intermediate   **pentane-1,2-diol (87%)**

Because OsO_4 is both very expensive and *very* toxic, the reaction is usually carried out using only a small, catalytic amount of OsO_4 in the presence of a stoichiometric amount of a safe and inexpensive co-oxidant such as *N*-methylmorpholine *N*-oxide, abbreviated NMO. The initially formed osmate intermediate reacts rapidly with NMO to yield the product diol plus *N*-methylmorpholine and reoxidized OsO_4. The OsO_4 then reacts with more alkene in a catalytic cycle.

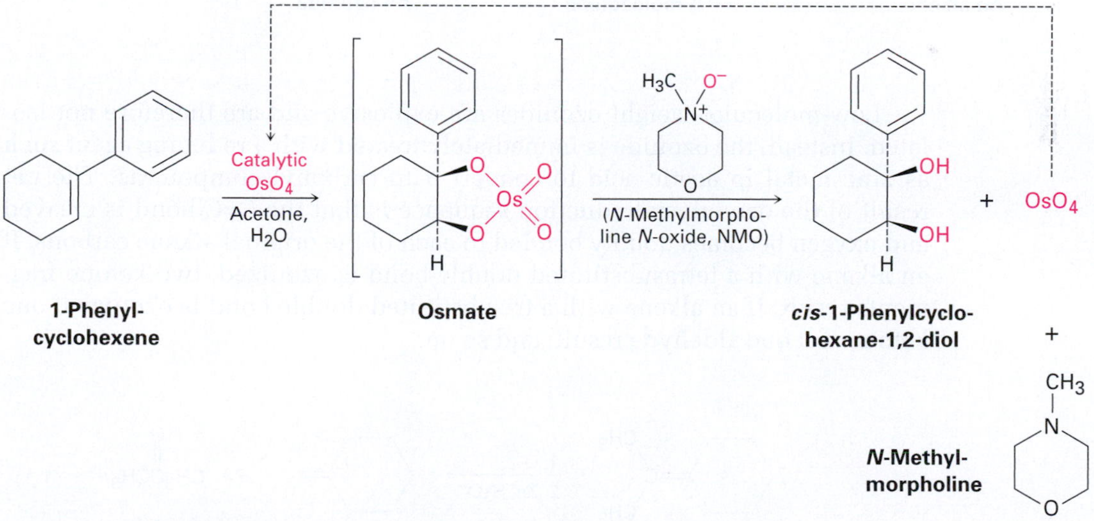

1-Phenyl-   **Osmate**   ***cis*-1-Phenylcyclo-**
cyclohexene   **hexane-1,2-diol**

N-Methyl-
morpholine

Problem 8.11

How would you prepare each of the following compounds starting with an alkene?

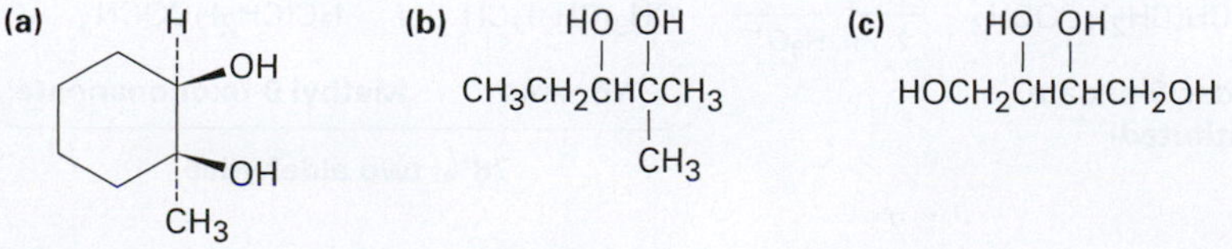

8.8 Oxidation of Alkenes: Cleavage to Carbonyl Compounds

In the alkene addition reactions we've seen thus far, the carbon–carbon double bond has been converted into a single bond but the carbon skeleton has been left intact. There are, however, powerful oxidizing reagents that will cleave C=C bonds and produce two carbonyl-containing fragments.

Ozone (O_3) is perhaps the most useful double-bond cleavage reagent in the laboratory. Prepared by passing a stream of oxygen through a high-voltage electrical discharge, ozone adds rapidly to a C=C bond at low temperature to give a cyclic intermediate called a *molozonide*. Once formed, the molozonide then spontaneously rearranges to form an *ozonide*. Although we won't study the mechanism of this rearrangement in detail, it involves the molozonide coming apart into two fragments, which then recombine in a different way.

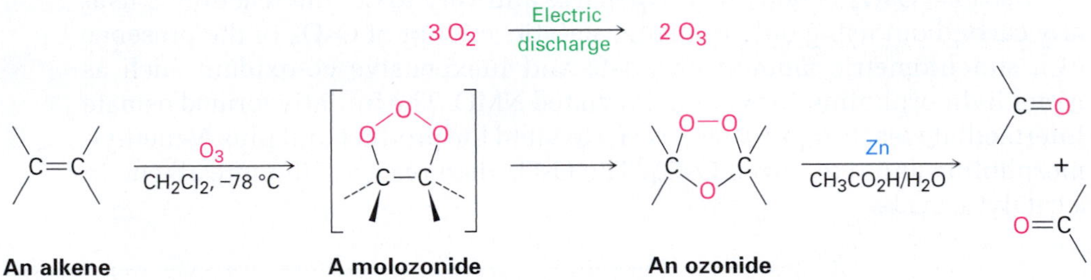

An alkene   **A molozonide**   **An ozonide**

Low-molecular-weight ozonides are explosive and are therefore not isolated. Instead, the ozonide is immediately treated with a reducing agent such as zinc metal in acetic acid to convert it to carbonyl compounds. The net result of the ozonolysis/reduction sequence is that the C=C bond is cleaved and oxygen becomes doubly bonded to each of the original alkene carbons. If an alkene with a tetrasubstituted double bond is ozonized, two ketone fragments result; if an alkene with a trisubstituted double bond is ozonized, one ketone and one aldehyde result; and so on.

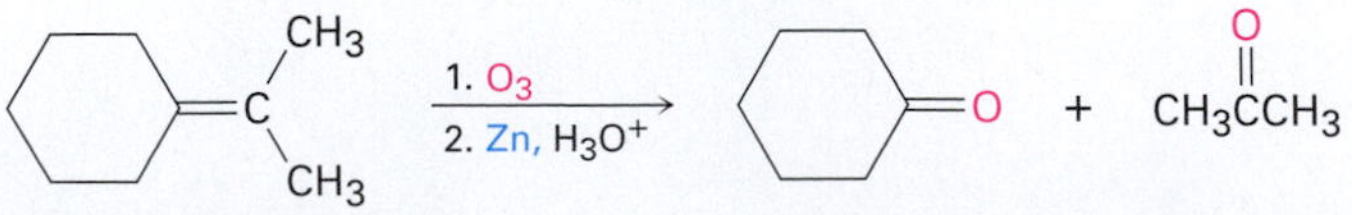

Isopropylidenecyclohexane   **Cyclohexanone**   **Acetone**
(tetrasubstituted)

84%; two ketones

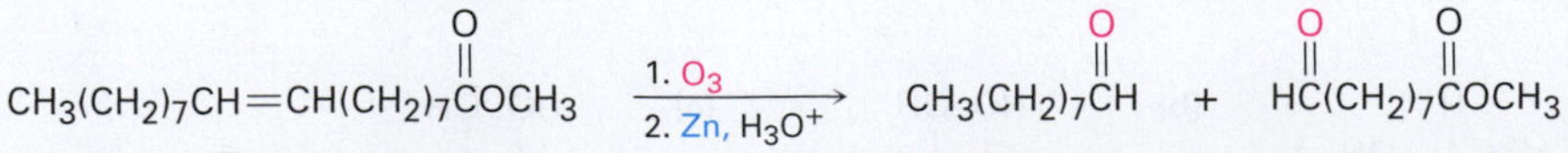

Methyl octadec-9-enoate   **Nonanal**   **Methyl 9-oxononanoate**
(disubstituted)

78%; two aldehydes

Several oxidizing reagents other than ozone also cause double-bond cleavage, although the reaction is not often used. For example, potassium permanganate ($KMnO_4$) in neutral or acidic solution cleaves alkenes to give carbonyl-containing products. If hydrogens are present on the double bond, carboxylic acids are produced; if two hydrogens are present on one carbon, CO_2 is formed.

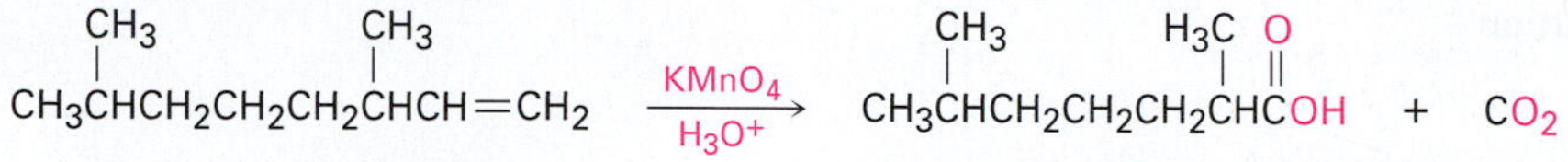

3,7-Dimethyloct-1-ene   **2,6-Dimethylheptanoic acid (45%)**

In addition to direct cleavage with ozone or $KMnO_4$, an alkene can also be cleaved by hydroxylation to a 1,2-diol, as discussed in the previous section, followed by treatment of the diol with periodic acid, HIO_4. If the two —OH groups are in an open chain, two carbonyl compounds result. If the two —OH groups are on a ring, a single, open-chain dicarbonyl compound is formed. As indicated in the following examples, the cleavage reaction takes place through a cyclic periodate intermediate.

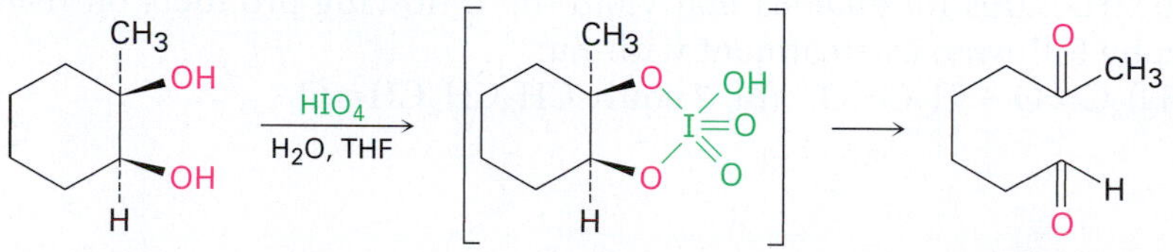

A 1,2-diol   **Cyclic periodate**   **6-Oxoheptanal (86%)**
                        intermediate

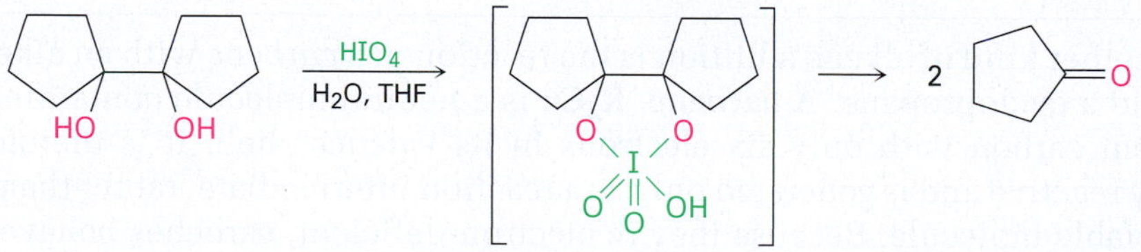

A 1,2-diol   **Cyclic periodate**   **Cyclopentanone (81%)**
                        intermediate

WORKED EXAMPLE 8.3 Predicting the Reactant in an Ozonolysis Reaction

What alkene would yield a mixture of cyclopentanone and propanal on treatment with ozone followed by reduction with zinc?

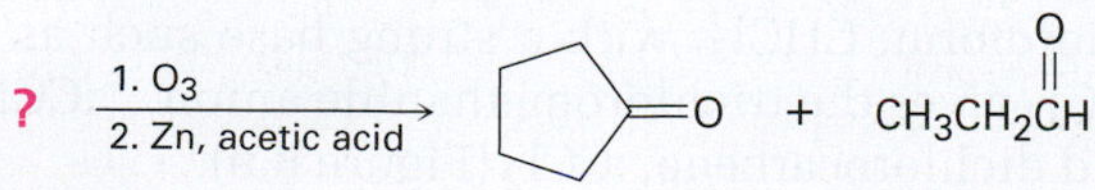

Strategy

Reaction of an alkene with ozone, followed by reduction with zinc, cleaves the carbon–carbon double bond and gives two carbonyl-containing fragments. That is, the C=C bond becomes two C=O bonds. Working backward from the carbonyl-containing products, the alkene precursor can be found by removing the oxygen from each product and joining the two carbon atoms to form a double bond.

Solution

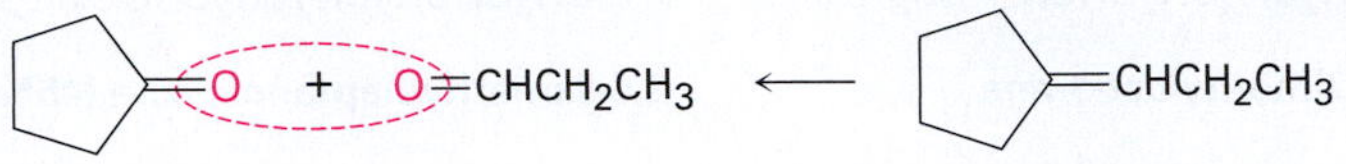

Problem 8.12

What products would you expect from reaction of 1-methylcyclohexene with the following reagents?
(a) Aqueous acidic $KMnO_4$ **(b)** O_3, followed by Zn, CH_3CO_2H

Problem 8.13

Propose structures for alkenes that yield the following products on reaction with ozone followed by treatment with Zn:
(a) $(CH_3)_2C=O + H_2C=O$ **(b)** 2 equiv $CH_3CH_2CH=O$

8.9 Addition of Carbenes to Alkenes: Cyclopropane Synthesis

Yet another kind of alkene addition is the reaction of a *carbene* with an alkene to yield a cyclopropane. A **carbene**, **R_2C:**, is a neutral molecule containing a divalent carbon with only six electrons in its valence shell. It is therefore highly reactive and is generated only as a reaction intermediate, rather than as an isolable molecule. Because they're electron-deficient, carbenes behave as electrophiles and react with nucleophilic C=C bonds. The reaction occurs in a single step without intermediates.

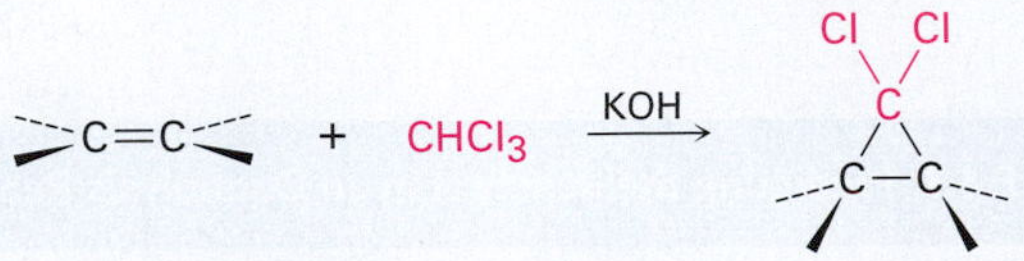

One of the simplest methods for generating a substituted carbene is by treatment of chloroform, $CHCl_3$, with a strong base such as KOH. Loss of a proton from $CHCl_3$ gives the trichloromethanide anion, $^-$:CCl_3, which expels a Cl^- ion to yield dichlorocarbene, :CCl_2 (Figure 8.9).

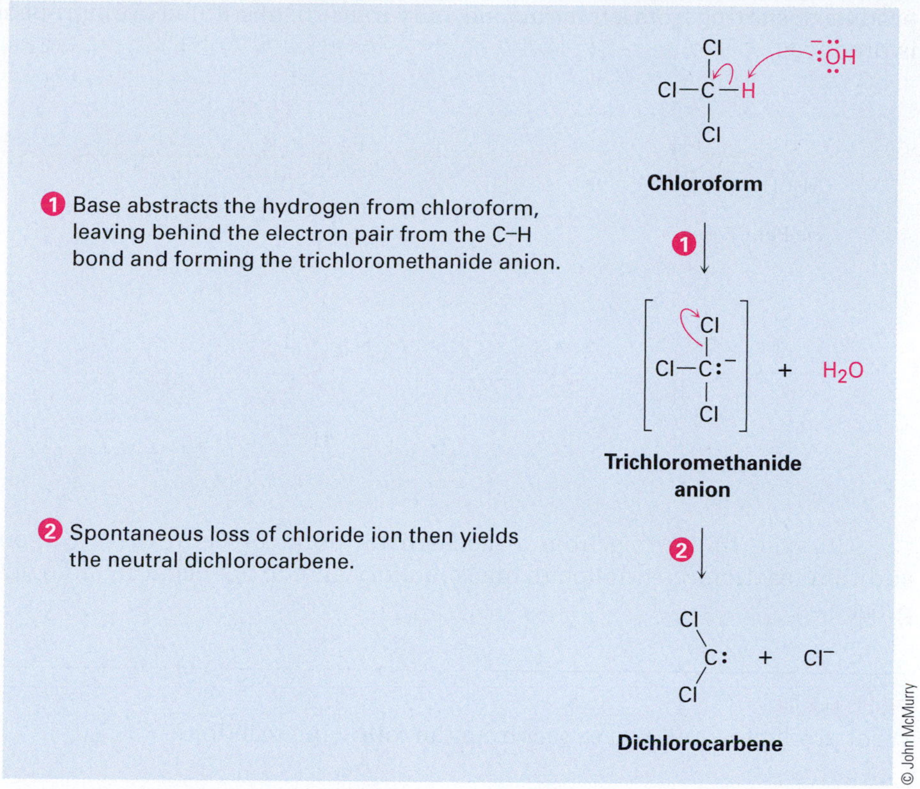

FIGURE 8.9 MECHANISM: Mechanism of the formation of dichlorocarbene by reaction of chloroform with strong base.

① Base abstracts the hydrogen from chloroform, leaving behind the electron pair from the C–H bond and forming the trichloromethanide anion.

Chloroform

② Spontaneous loss of chloride ion then yields the neutral dichlorocarbene.

Trichloromethanide anion

Dichlorocarbene

© John McMurry

The dichlorocarbene carbon atom is sp^2-hybridized, with a vacant p orbital extending above and below the plane of the three atoms and with an unshared pair of electrons occupying the third sp^2 lobe. Note that this electronic description of dichlorocarbene is similar to that of a carbocation (Section 7.8) with respect to both the sp^2 hybridization of carbon and the vacant p orbital. Electrostatic potential maps further show this similarity (Figure 8.10).

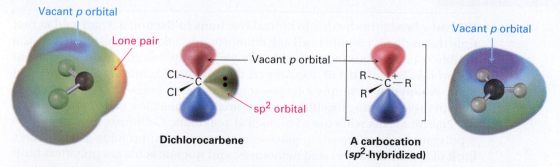

Vacant p orbital

Lone pair

Vacant p orbital — — —

Vacant p orbital

sp^2 orbital

Dichlorocarbene

A carbocation (sp^2-hybridized)

FIGURE 8.10 The structure of dichlorocarbene. Electrostatic potential maps show how the positive region (blue) coincides with the empty p orbital in both dichlorocarbene and a carbocation (CH_3^+). The negative region (red) in the dichlorocarbene map coincides with the lone-pair electrons.

If dichlorocarbene is generated in the presence of an alkene, addition to the double bond occurs and a dichlorocyclopropane is formed. As the reaction of dichlorocarbene with *cis*-pent-2-ene demonstrates, the addition is **stereospecific**, meaning that only a single stereoisomer is formed as product. Starting from a cis alkene, for instance, only cis-disubstituted cyclopropane is produced; starting from a trans alkene, only trans-disubstituted cyclopropane is produced.

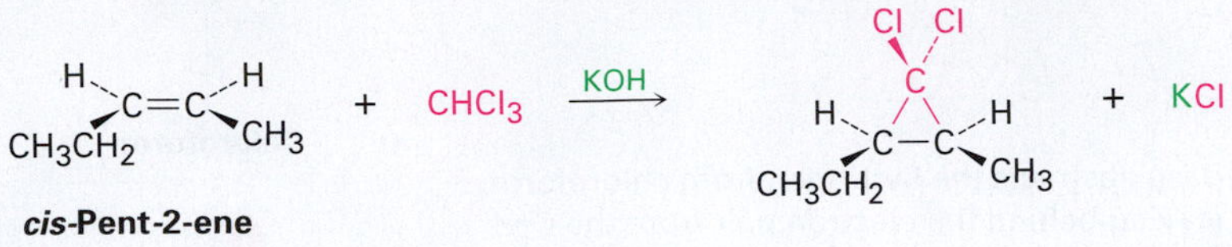

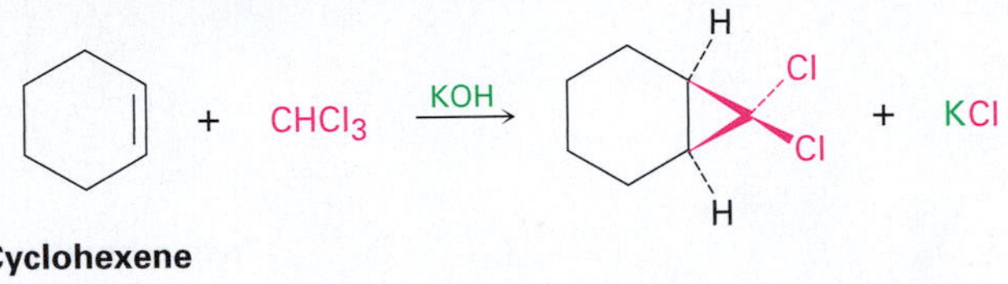

cis-Pent-2-ene

Cyclohexene

Although interesting from a mechanistic point of view, these carbene addition reactions are limited to the laboratory and do not occur in biological processes.

Problem 8.14
What product would you expect from the following reaction?

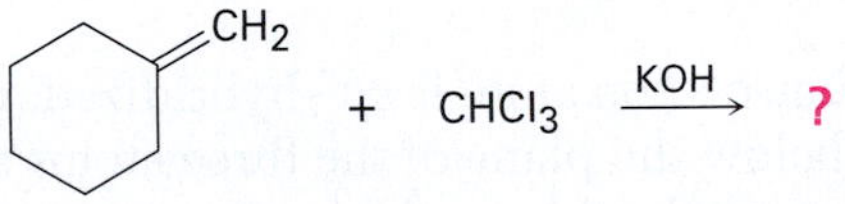

8.10 Radical Additions to Alkenes: Alkene Polymers

We had a brief introduction to radical reactions in Section 6.3 and said at that time that radicals can add to alkene double bonds, taking one electron from the double bond and leaving one behind to yield a new radical. Let's now look at the process in more detail, focusing on the industrial synthesis of alkene polymers. A **polymer** is simply a large—sometimes *very* large—molecule built up by repetitive bonding together of many smaller molecules, called **monomers**.

Nature makes wide use of biological polymers. Cellulose, for instance, is a polymer built of repeating glucose monomer units; proteins are polymers built of repeating amino acid monomers; and nucleic acids are polymers built of repeating nucleotide monomers.

Cellulose—a glucose polymer

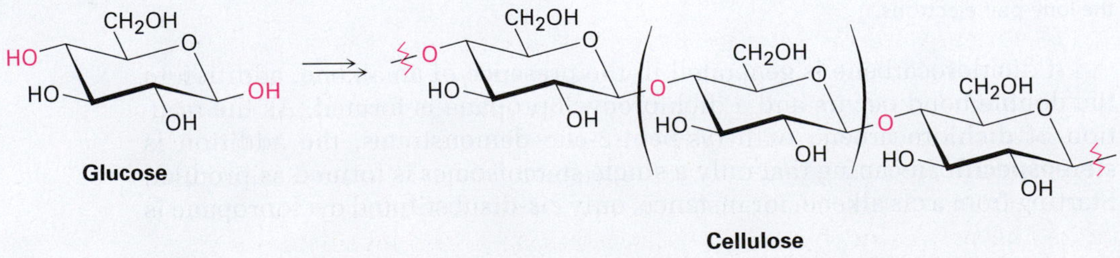

Glucose

Cellulose

Protein—an amino acid polymer

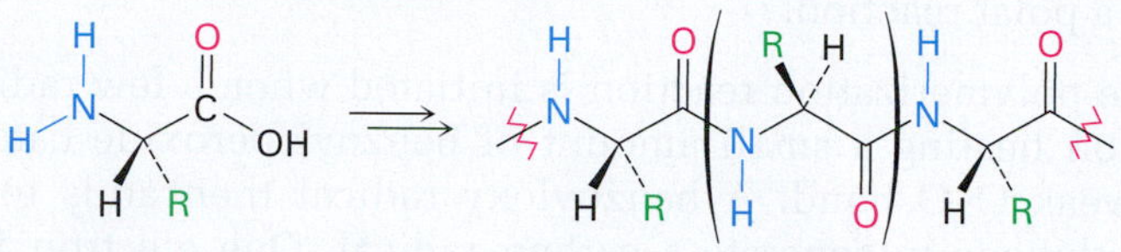

An amino acid   **A protein**

Nucleic acid—a nucleotide polymer

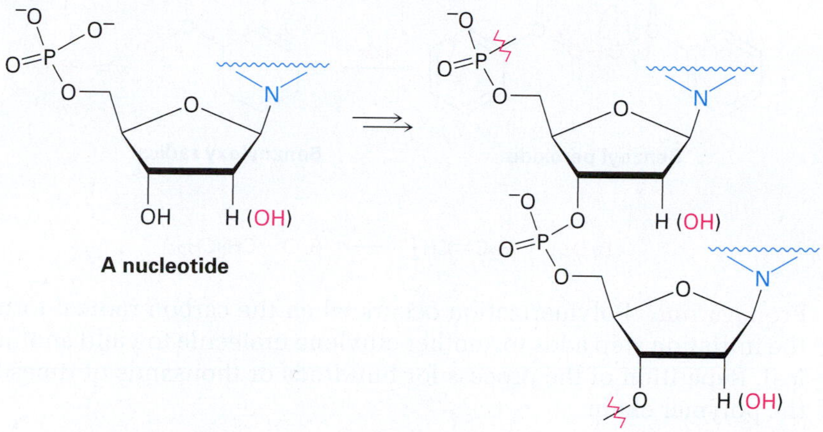

A nucleotide

A nucleic acid

Synthetic polymers, such as polyethylene, are chemically much simpler than biopolymers, but there is still a great diversity to their structures and properties, depending on the identity of the monomers and on the reaction conditions used for polymerization. The simplest synthetic polymers are those that result when an alkene is treated with a small amount of a radical as catalyst. Ethylene, for example, yields polyethylene, an enormous alkane that may have up to *200,000* monomer units incorporated into a gigantic hydrocarbon chain. Approximately 19 million tons per year of polyethylene are manufactured in the United States alone.

Polyethylene—a synthetic alkene polymer

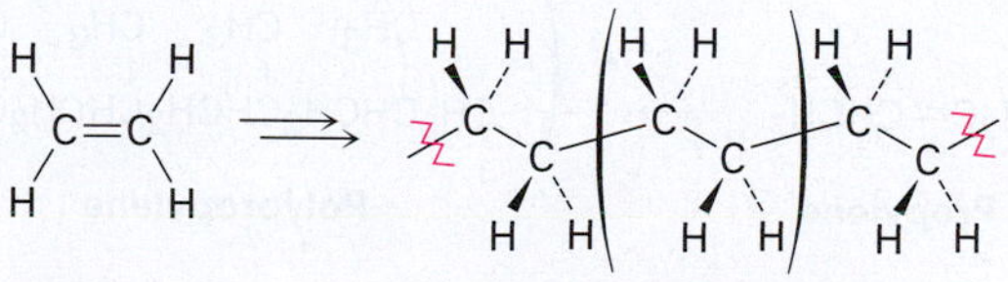

Ethylene   **Polyethylene**

Historically, ethylene polymerization was carried out at high pressure (1000–3000 atm) and high temperature (100–250 °C) in the presence of a catalyst such as benzoyl peroxide, although other catalysts and reaction conditions are now more often used. The key step is the addition of a radical to the ethylene double bond, a reaction similar in many respects to what takes place in the addition of an electrophile. In writing the mechanism, recall that a curved half-arrow, or "fishhook" ⌢, is used to show the movement of a single electron, as opposed to the full curved arrow used to show the movement of an electron pair in a polar reaction.

- **Initiation**   The polymerization reaction is initiated when a few radicals are generated on heating a small amount of benzoyl peroxide catalyst to break the weak O–O bond. A benzoyloxy radical then adds to the C=C bond of ethylene to generate a carbon radical. One electron from the C=C bond pairs up with the odd electron on the benzoyloxy radical to form a C–O bond, and the other electron remains on carbon.

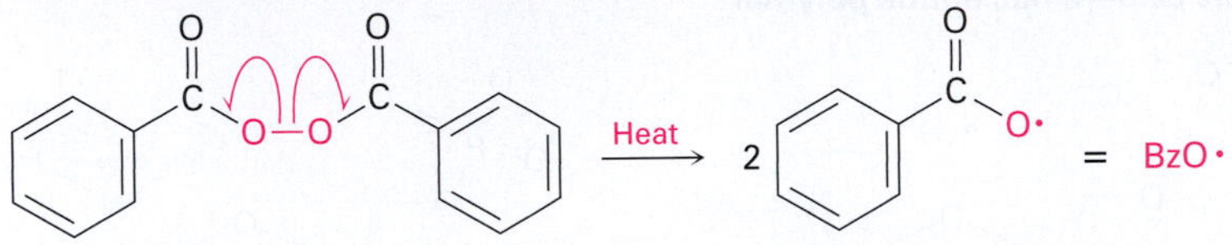

Benzoyl peroxide   **Benzoyloxy radical**

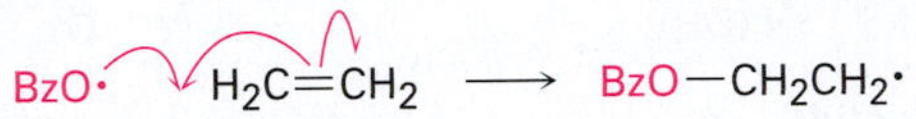

$$BzO \cdot \quad H_2C=CH_2 \longrightarrow BzO-CH_2CH_2\cdot$$

- **Propagation**   Polymerization occurs when the carbon radical formed in the initiation step adds to another ethylene molecule to yield another radical. Repetition of the process for hundreds or thousands of times builds the polymer chain.

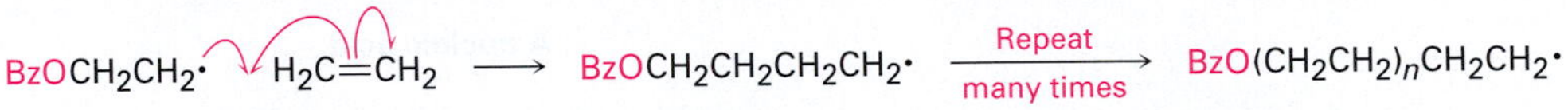

$$BzOCH_2CH_2\cdot \quad H_2C=CH_2 \longrightarrow BzOCH_2CH_2CH_2CH_2\cdot \xrightarrow[\text{many times}]{\text{Repeat}} BzO(CH_2CH_2)_nCH_2CH_2\cdot$$

- **Termination**   The chain process is eventually ended by a reaction that consumes the radical. Combination of two growing chains is one possible chain-terminating reaction:

$$2\,R-CH_2CH_2\cdot \longrightarrow R-CH_2CH_2CH_2CH_2-R$$

Ethylene is not unique in its ability to form a polymer. Many substituted ethylenes, called *vinyl monomers,* also undergo polymerization to yield polymers with substituent groups regularly spaced on alternating carbon atoms along the chain. Propylene, for example, yields polypropylene, and styrene yields polystyrene.

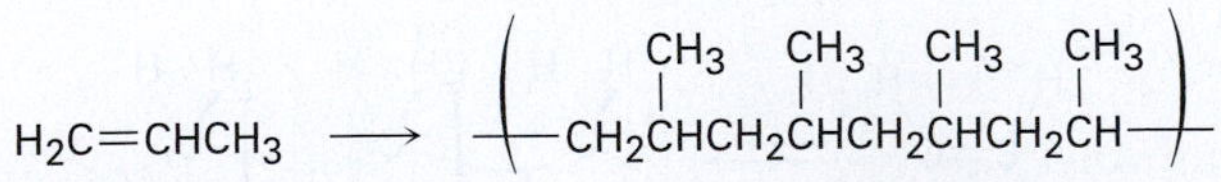

$$H_2C=CHCH_3 \longrightarrow$$

Propylene   **Polypropylene**

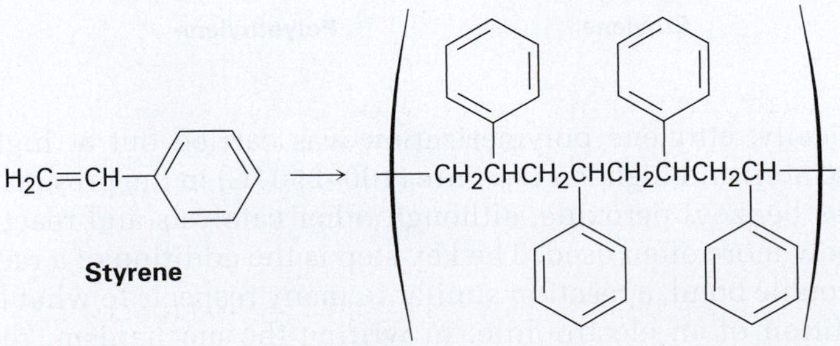

$$H_2C=CH-$$

Styrene   **Polystyrene**

When an unsymmetrically substituted vinyl monomer, such as propylene or styrene is polymerized, the radical addition steps can take place at either end of the double bond to yield either a primary radical intermediate ($RCH_2\cdot$) or a secondary radical ($R_2CH\cdot$). Just as in electrophilic addition reactions, however, we find that only the more highly substituted, secondary radical is formed.

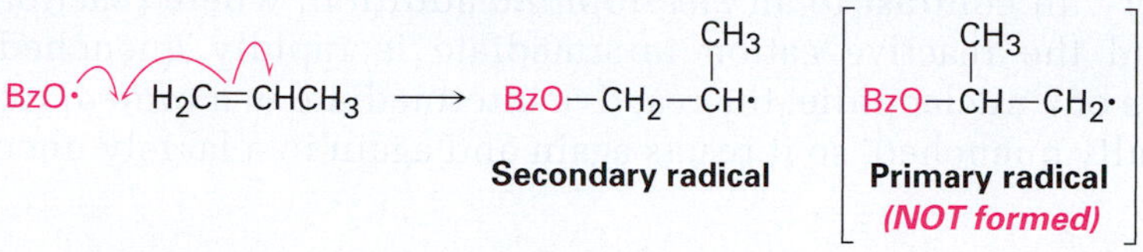

$$BzO\cdot \quad H_2C=CHCH_3 \longrightarrow BzO-CH_2-\overset{\overset{\textstyle CH_3}{|}}{CH}\cdot \qquad \left[BzO-\overset{\overset{\textstyle CH_3}{|}}{CH}-CH_2\cdot \right]$$

Secondary radical   **Primary radical**
                               (NOT formed)

WORKED EXAMPLE 8.4 Predicting the Structure of a Polymer

Show the structure of poly(vinyl chloride), a polymer made from $H_2C=CHCl$, by drawing several repeating units.

Strategy

Mentally break the carbon–carbon double bond in the monomer unit, and form single bonds by connecting numerous units together.

Solution

The general structure of poly(vinyl chloride) is

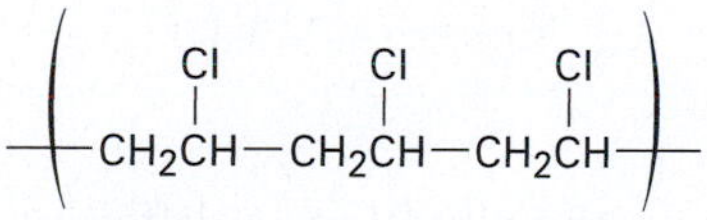

$$\left(CH_2\overset{\overset{\textstyle Cl}{|}}{CH}-CH_2\overset{\overset{\textstyle Cl}{|}}{CH}-CH_2\overset{\overset{\textstyle Cl}{|}}{CH} \right)$$

Problem 8.15

What monomer units would you would use to prepare the following polymers?

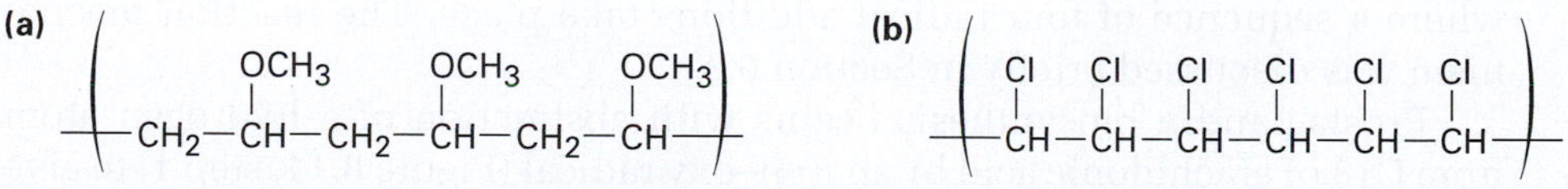

(a)

$$\left(CH_2-\overset{\overset{\textstyle OCH_3}{|}}{CH}-CH_2-\overset{\overset{\textstyle OCH_3}{|}}{CH}-CH_2-\overset{\overset{\textstyle OCH_3}{|}}{CH} \right)$$

(b)

$$\left(\overset{\overset{\textstyle Cl}{|}}{CH}-\overset{\overset{\textstyle Cl}{|}}{CH}-\overset{\overset{\textstyle Cl}{|}}{CH}-\overset{\overset{\textstyle Cl}{|}}{CH}-\overset{\overset{\textstyle Cl}{|}}{CH}-\overset{\overset{\textstyle Cl}{|}}{CH} \right)$$

Problem 8.16

One of the chain-termination steps that sometimes occurs to interrupt polymerization is the following reaction between two radicals. Propose a mechanism for the reaction, using fishhook arrows to indicate electron flow.

$$2 \ \ \sim CH_2\overset{\cdot}{C}H_2 \longrightarrow \sim CH_2CH_3 + \sim CH=CH_2$$

8.11 Biological Additions of Radicals to Alkenes

The same high reactivity of radicals that makes possible the alkene polymerization we saw in the previous section also makes it difficult to carry out controlled radical reactions on complex molecules. As a result, there are severe limitations on the usefulness of radical addition reactions in the laboratory. In contrast to an *electrophilic* addition, where reaction occurs once and the reactive cation intermediate is rapidly quenched in the presence of a nucleophile, the reactive intermediate in a *radical* reaction is not usually quenched, so it reacts again and again in a largely uncontrollable way.

Electrophilic addition
(Intermediate is quenched,
so reaction stops.)

Radical addition
(Intermediate is not quenched,
so reaction does not stop.)

In biological reactions, the situation is different from that in the laboratory. Only one substrate molecule at a time is present in the active site of the enzyme where reaction takes place, and that molecule is held in a precise position, with coenzymes and other necessary reacting groups nearby. As a result, biological radical reactions are both more controlled and more common than laboratory or industrial radical reactions. A particularly impressive example occurs in the biosynthesis of prostaglandins from arachidonic acid, where a sequence of four radical additions take place. The reaction mechanism was discussed briefly in Section 6.3.

Prostaglandin biosynthesis begins with abstraction of a hydrogen atom from C13 of arachidonic acid by an iron–oxy radical (Figure 8.11, step 1) to give a carbon radical that reacts with O_2 at C11 through a resonance form (step 2). The oxygen radical that results adds to the C8–C9 double bond (step 3) to give a carbon radical at C8, which then adds to the C12–C13 double bond and gives a carbon radical at C13 (step 4). A resonance form of this carbon radical adds at C15 to a second O_2 molecule (step 5), completing the prostaglandin skeleton, and reduction of the O–O bond then gives prostaglandin H_2 (step 6). The pathway looks complicated, but the entire process is catalyzed with exquisite control by a single enzyme.

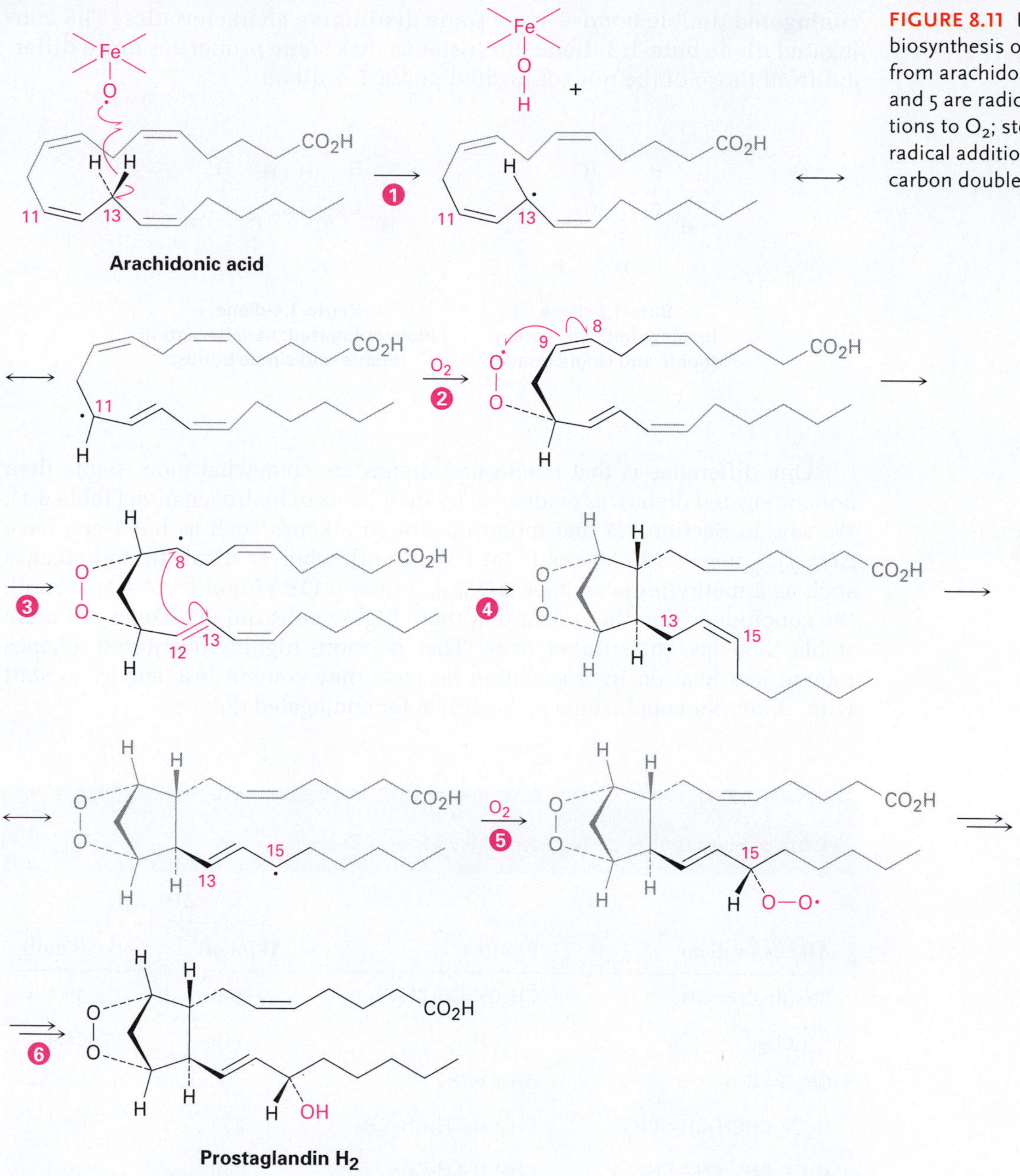

FIGURE 8.11 Pathway for the biosynthesis of prostaglandins from arachidonic acid. Steps 2 and 5 are radical addition reactions to O_2; steps 3 and 4 are radical additions to carbon–carbon double bonds.

Arachidonic acid

Prostaglandin H$_2$

8.12 Conjugated Dienes

Thus far, we've looked primarily at compounds with just one double bond, but many compounds have numerous sites of unsaturation. If the different unsaturations are well separated in a molecule, they often react independently, but if they're close together, they may interact with one another. In particular, double bonds that alternate with single bonds—so-called

conjugated double bonds—have some distinctive characteristics. The conjugated diene buta-1,3-diene, for instance, has some properties quite different from those of the nonconjugated penta-1,4-diene.

Buta-1,3-diene
(conjugated; alternating
double and single bonds)

Penta-1,4-diene
(nonconjugated; nonalternating
double and single bonds)

One difference is that conjugated dienes are somewhat more stable than nonconjugated dienes, as evidenced by their heats of hydrogenation (Table 8.1). We saw in Section 7.5 that monosubstituted alkenes such as but-1-ene have $\Delta H°_{\text{hydrog}}$ near -126 kJ/mol (-30.1 kcal/mol), whereas disubstituted alkenes such as 2-methylpropene have $\Delta H°_{\text{hydrog}}$ near -119 kJ/mol (-28.4 kcal/mol). We concluded from these data that more highly substituted alkenes are more stable than less substituted ones. That is, more highly substituted alkenes release less heat on hydrogenation because they contain less energy to start with. A similar conclusion can be drawn for conjugated dienes.

TABLE 8.1
Heats of Hydrogenation for Some Alkenes and Dienes

Alkene or diene	Product	$\Delta H°_{\text{hydrog}}$ (kJ/mol)	(kcal/mol)
$CH_3CH_2CH{=}CH_2$	$CH_3CH_2CH_2CH_3$	-126	-30.1
$CH_3\overset{\displaystyle CH_3}{\underset{}{C}}{=}CH_2$	CH_3CHCH_3 (CH_3)	-119	-28.4
$H_2C{=}CHCH_2CH{=}CH_2$	$CH_3CH_2CH_2CH_2CH_3$	-253	-60.5
$H_2C{=}CH{-}CH{=}CH_2$	$CH_3CH_2CH_2CH_3$	-236	-56.4
$H_2C{=}CH{-}C{=}CH_2$ (CH_3)	$CH_3CH_2CHCH_3$ (CH_3)	-229	-54.7

Because a monosubstituted alkene has a $\Delta H°_{\text{hydrog}}$ of approximately -126 kJ/mol, we might expect that a compound with two monosubstituted double bonds would have a $\Delta H°_{\text{hydrog}}$ approximately twice that value, or -252 kJ/mol. Nonconjugated dienes, such as penta-1,4-diene ($\Delta H°_{\text{hydrog}} = -253$ kJ/mol), meet this expectation, but the conjugated diene buta-1,3-diene

($\Delta H°_{hydrog} = -236$ kJ/mol) does not. Buta-1,3-diene is approximately 16 kJ/mol (3.8 kcal/mol) more stable than expected.

$\Delta H°_{hydrog}$ (kJ/mol)

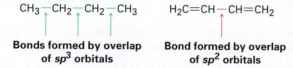

$H_2C\!\!=\!\!CHCH_2CH\!\!=\!\!CH_2$

Penta-1,4-diene

$-126 + (-126) = -252$	Expected	
-253	Observed	
1	Difference	

$H_2C\!\!=\!\!CHCH\!\!=\!\!CH_2$

Buta-1,3-diene

$-126 + (-126) = -252$	Expected	
-236	Observed	
-16	Difference	

What accounts for the stability of conjugated dienes? According to valence bond theory (Sections 1.5 and 1.8), the stability is due to orbital hybridization. Typical C–C single bonds like those in alkanes result from σ overlap of sp^3 orbitals on both carbons, but in a conjugated diene, the central C–C single bond results from σ overlap of sp^2 orbitals on both carbons. Since sp^2 orbitals have more s character (33% s) than sp^3 orbitals (25% s), the electrons in sp^2 orbitals are closer to the nucleus and the bonds they form are somewhat shorter and stronger. Thus, the "extra" stability of a conjugated diene results in part from the greater amount of s character in the orbitals forming the C–C single bond.

$$CH_3\text{—}CH_2\text{—}CH_2\text{—}CH_3 \qquad H_2C\!\!=\!\!CH\text{—}CH\!\!=\!\!CH_2$$

Bonds formed by overlap
of sp^3 orbitals

Bond formed by overlap
of sp^2 orbitals

According to molecular orbital theory (Section 1.11), the stability of a conjugated diene arises because of an interaction between the π orbitals of the two double bonds. To review briefly, when two p atomic orbitals combine to form a π bond, two π molecular orbitals (MOs) result. One is lower in energy than the starting p orbitals and is therefore bonding; the other is higher in energy, has a node between nuclei, and is antibonding. The two π electrons occupy the low-energy, bonding orbital, resulting in formation of a stable bond between atoms (Figure 8.12).

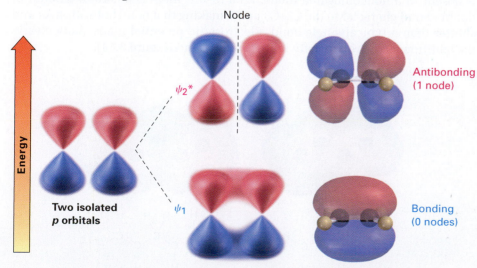

FIGURE 8.12 Two p orbitals combine to form two π molecular orbitals. Both electrons occupy the low-energy, bonding orbital, leading to a net lowering of energy and formation of a stable bond. The asterisk on $\psi_2{}^*$ indicates an antibonding orbital.

Now let's combine four adjacent *p* atomic orbitals, as occurs in a conjugated diene. In so doing, we generate a set of four π molecular orbitals, two of which are bonding and two of which are antibonding (Figure 8.13). The four π electrons occupy the two bonding orbitals, leaving the antibonding orbitals vacant.

FIGURE 8.13 Four π molecular orbitals in buta-1,3-diene. Note that the number of nodes between nuclei increases as the energy level of the orbital increases.

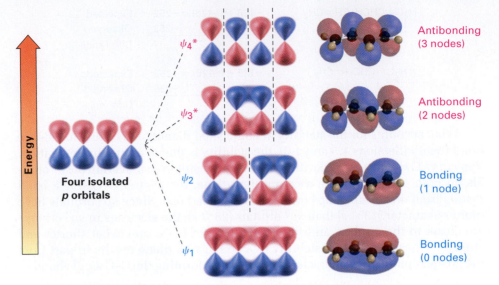

The lowest-energy π molecular orbital (denoted ψ_1, Greek psi) has no nodes between the nuclei and is therefore bonding. The π MO of next lowest energy, ψ_2, has one node between nuclei and is also bonding. Above ψ_1 and ψ_2 in energy are the two antibonding π MOs, ψ_3^* and ψ_4^*. (The asterisks indicate antibonding orbitals.) Note that the number of nodes between nuclei increases as the energy level of the orbital increases. The ψ_3^* orbital has two nodes between nuclei, and ψ_4^*, the highest-energy MO, has three nodes between nuclei.

Comparing the π molecular orbitals of buta-1,3-diene (two conjugated double bonds) with those of penta-1,4-diene (two isolated double bonds) shows why the conjugated diene is more stable. In a conjugated diene, the lowest-energy π MO (ψ_1) has a favorable bonding interaction between C2 and C3 that is absent in a nonconjugated diene. As a result, there is a certain amount of double-bond character to the C2–C3 bond, making that bond both stronger and shorter than a typical single bond. Electrostatic potential maps show clearly the additional electron density in the central bond (Figure 8.14).

FIGURE 8.14 Electrostatic potential maps of buta-1,3-diene (conjugated) and penta-1,4-diene (nonconjugated) show additional electron density (red) in the central C–C bond of buta-1,3-diene, corresponding to partial double-bond character.

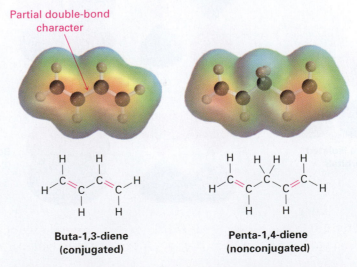

Partial double-bond character

Buta-1,3-diene (conjugated)   **Penta-1,4-diene (nonconjugated)**

In describing buta-1,3-diene, we say that the π electrons are spread out, or *delocalized*, over the entire π framework rather than localized between two specific nuclei. Electron delocalization and consequent dispersal of charge always leads to lower energy and greater stability.

8.13 Reactions of Conjugated Dienes

One of the most striking differences between conjugated and isolated double bonds is in their electrophilic addition reactions. Conjugated dienes undergo electrophilic addition reactions readily, but mixtures of products are invariably obtained. Addition of HBr to buta-1,3-diene, for instance, yields a mixture of two products (not counting cis–trans isomers). 3-Bromobut-1-ene is the typical Markovnikov product of *1,2-addition* to a double bond, but 1-bromobut-2-ene appears unusual. The double bond in this product has moved to a position between carbons 2 and 3, and HBr has added to carbons 1 and 4, a result described as *1,4-addition*.

3-Bromobut-1-ene
(71%; 1,2-addition)

Buta-1,3-diene

1-Bromobut-2-ene
(29%; 1,4-addition)

[handwritten: 1,2-addition means H + x are #1 & #2 substituents]

How can we account for the formation of the 1,4-addition product? The answer is that an *allylic carbocation* is involved as an intermediate, where the word **allylic** means "next to a double bond." When buta-1,3-diene reacts with an electrophile such as H^+, two carbocation intermediates are possible—a primary carbocation and a secondary allylic cation. Because an allylic cation is stabilized by resonance between two forms (Section 2.4), it is more stable and forms faster than a nonallylic carbocation.

Buta-1,3-diene

Secondary, allylic

Primary, nonallylic
(NOT formed)

When the allylic cation reacts with Br^- to complete the electrophilic addition, reaction can occur either at C1 or at C3 because both carbons share the positive charge (Figure 8.15). Thus, a mixture of 1,2- and 1,4-addition products results.

FIGURE 8.15 An electrostatic potential map of the carbocation produced by protonation of buta-1,3-diene shows that the positive charge is shared by carbons 1 and 3. Reaction of Br^- with the more positive carbon (C3; blue) gives predominantly the 1,2-addition product.

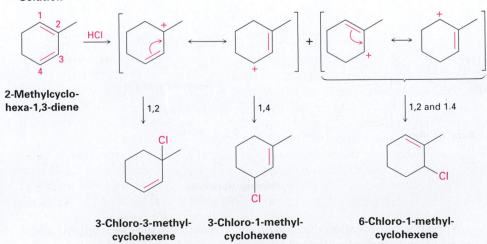

1,4-Addition (29%)   **1,2-Addition (71%)**

WORKED EXAMPLE 8.5   **Predicting the Product of Electrophilic Addition to a Conjugated Diene**

Give the structures of the likely products from reaction of 1 equivalent of HCl with 2-methylcyclohexa-1,3-diene. Show both 1,2- and 1,4-adducts.

Strategy

Electrophilic addition of HCl to a conjugated diene involves the formation of allylic carbocation intermediates. Thus, the first step is to protonate the two ends of the diene and draw the resonance forms of the two allylic carbocations that result. Then allow each resonance form to react with Cl^-, generating a maximum of four possible products.

In the present instance, protonation of the C1–C2 double bond gives a carbocation that can react further to give the 1,2-adduct 3-chloro-3-methylcyclohexene and the 1,4-adduct 3-chloro-1-methylcyclohexene. Protonation of the C3–C4 double bond gives a symmetrical carbocation, whose two resonance forms are equivalent. Thus, the 1,2-adduct and the 1,4-adduct have the same structure: 6-chloro-1-methylcyclohexene. Of the two possible modes of protonation, the first is more likely because it yields a tertiary allylic cation rather than a secondary allylic cation.

Solution

2-Methylcyclohexa-1,3-diene

3-Chloro-3-methylcyclohexene   **3-Chloro-1-methylcyclohexene**   **6-Chloro-1-methylcyclohexene**

Problem 8.17
Give the structures of both 1,2- and 1,4-adducts resulting from reaction of 1 equivalent of HCl with penta-1,3-diene.

Problem 8.18
Give the structures of both 1,2- and 1,4-adducts resulting from reaction of 1 equivalent of HBr with the following compound:

8.14 | The Diels–Alder Cycloaddition Reaction

Perhaps the most striking difference between conjugated and nonconjugated dienes is that conjugated dienes undergo a reaction with alkenes to yield substituted cyclohexene products. For example, buta-1,3-diene and but-3-en-2-one give cyclohex-3-enyl methyl ketone.

Buta-1,3-diene   **But-3-en-2-one**   Toluene, Heat → **Cyclohex-3-enyl methyl ketone (96%)**

This process, named the **Diels–Alder cycloaddition reaction** after its discoverers, is extremely useful in the laboratory because it forms two carbon–carbon bonds in a single step and is one of the few general methods available for making cyclic molecules. (As the name implies, a *cycloaddition* reaction is one in which two reactants add together to give a cyclic product.) The 1950 Nobel Prize in Chemistry was awarded to Diels and Alder in recognition of the importance of their discovery.

The mechanism of the Diels–Alder cycloaddition is different from that of other reactions we've studied because it is neither polar nor radical. Rather, the Diels–Alder reaction is a so-called pericyclic process. **Pericyclic reactions**, which are considerably less common than either polar or radical reactions, take place in a single step by a cyclic redistribution of bonding electrons. The two reactants simply join together through a cyclic transition state in which the two new carbon–carbon bonds form at the same time.

We can picture a Diels–Alder addition as occurring by head-on (σ) overlap of the two alkene p orbitals with the two p orbitals on carbons 1 and 4 of the diene (Figure 8.16). This is, of course, a *cyclic* orientation of the reactants.

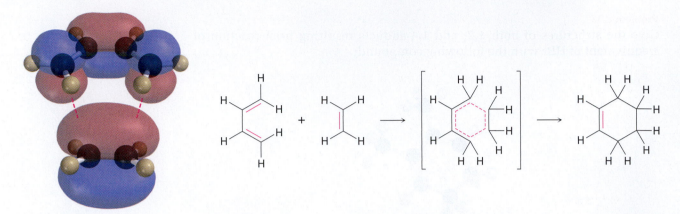

FIGURE 8.16 Mechanism of the Diels–Alder cycloaddition reaction. The reaction occurs in a single step through a cyclic transition state in which the two new carbon–carbon bonds form simultaneously.

In the Diels–Alder transition state, the two alkene carbons and carbons 1 and 4 of the diene rehybridize from sp^2 to sp^3 to form two new single bonds, while carbons 2 and 3 of the diene remain sp^2 hybridized to form the new double bond in the cyclohexene product.

The Diels–Alder cycloaddition reaction occurs most rapidly if the alkene component, or **dienophile** ("diene lover"), has an electron-withdrawing substituent group. Thus, ethylene itself reacts sluggishly, but propenal, ethyl propenoate, maleic anhydride, benzoquinone, propenenitrile, and similar compounds are highly reactive. Note also that alkynes, such as methyl propynoate, can act as Diels–Alder dienophiles.

Some Diels–Alder dienophiles

Ethylene: unreactive

Propenal (acrolein)

Ethyl propenoate (ethyl acrylate)

Maleic anhydride

Benzoquinone

Propenenitrile (acrylonitrile)

Methyl propynoate

In all the preceding cases, the double or triple bond of the dienophile is adjacent to the positively polarized carbon of an electron-withdrawing substituent. As a result, the double-bond carbons in these substances are substantially less electron-rich than the carbons in ethylene (Figure 8.17).

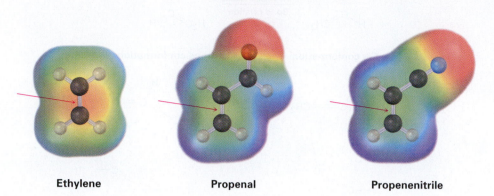

Ethylene   **Propenal**   **Propenenitrile**

FIGURE 8.17 Electrostatic potential maps of ethylene, propenal, and propenenitrile show that electron-withdrawing groups make the double-bond carbons less electron-rich (less red).

One of the most useful features of the Diels–Alder reaction is that it is *stereospecific,* meaning that a single product stereoisomer is formed (Section 8.9). Furthermore, the stereochemistry of the reactant is maintained. If we carry out the cycloaddition with a cis dienophile, such as methyl *cis*-but-2-enoate, only the cis-substituted cyclohexene product is formed. With methyl *trans*-but-2-enoate, only the trans-substituted cyclohexene product is formed.

Buta-1,3-diene   **Methyl (Z)-but-2-enoate**   **Cis product**

Buta-1,3-diene   **Methyl (E)-but-2-enoate**   **Trans product**

Just as the dienophile component has certain constraints that affect its reactivity, so too with the conjugated diene component. The diene must adopt what is called an *s-cis conformation,* meaning "cis-like" about the *single bond,* to undergo a Diels–Alder reaction. Only in the *s-cis* conformation are carbons 1 and 4 of the diene close enough to react through a cyclic transition state. In the alternative *s*-trans conformation, the ends of the diene partner are too far apart to overlap with the dienophile *p* orbitals.

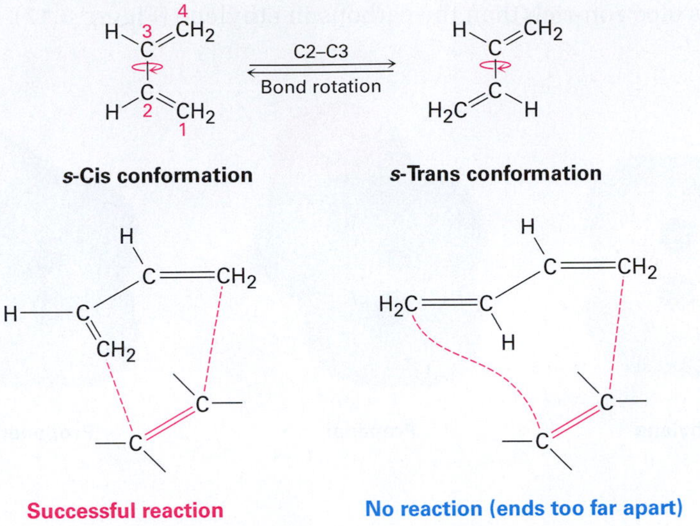

s-Cis conformation   **s-Trans conformation**

Successful reaction   **No reaction (ends too far apart)**

Two examples of dienes that can't adopt an *s*-cis conformation, and thus don't undergo Diels–Alder reactions, are shown in Figure 8.18. In the bicyclic (two-ring) diene, the double bonds are rigidly fixed in an *s*-trans arrangement by geometric constraints of the rings. In (2*Z*,4*Z*)-hexa-2,4-diene, steric strain between the two methyl groups prevents the molecule from adopting *s*-cis geometry.

FIGURE 8.18 Two dienes that can't achieve an *s*-cis conformation and thus can't undergo Diels–Alder reactions.

A bicyclic diene (rigid s-trans diene)   **Severe steric strain in s-cis form**   **(2Z,4Z)-Hexa-2,4-diene (s-trans, more stable)**

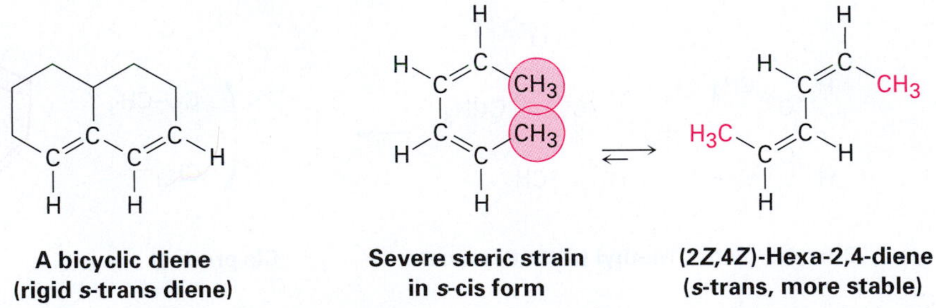

In contrast to those unreactive dienes that can't achieve an *s*-cis conformation, other dienes are fixed only in the correct *s*-cis geometry and are therefore highly reactive in the Diels–Alder cycloaddition reaction. Cyclopenta-1,3-diene, for example, is so reactive that it reacts with itself. At room temperature, cyclopenta-1,3-diene *dimerizes*. One molecule acts as diene, and a second molecule acts as dienophile in a self Diels–Alder reaction.

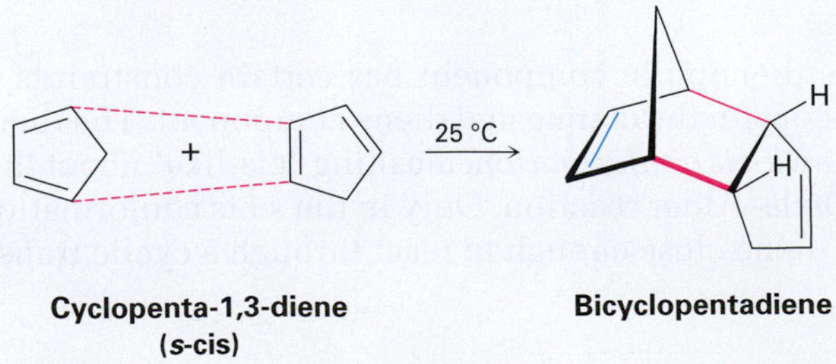

Cyclopenta-1,3-diene (s-cis)   **Bicyclopentadiene**

25 °C

Biological Diels–Alder reactions are known but uncommon. One example occurs in the biosynthesis of the cholesterol-lowering drug lovastatin (Chapter 1 Introduction) isolated from the bacterium *Aspergillus terreus*. The key step is the Diels–Alder reaction of a triene in which the diene and dienophile components are within the same molecule. Following this *intramolecular* Diels–Alder reaction, several subsequent transformations yield lovastatin.

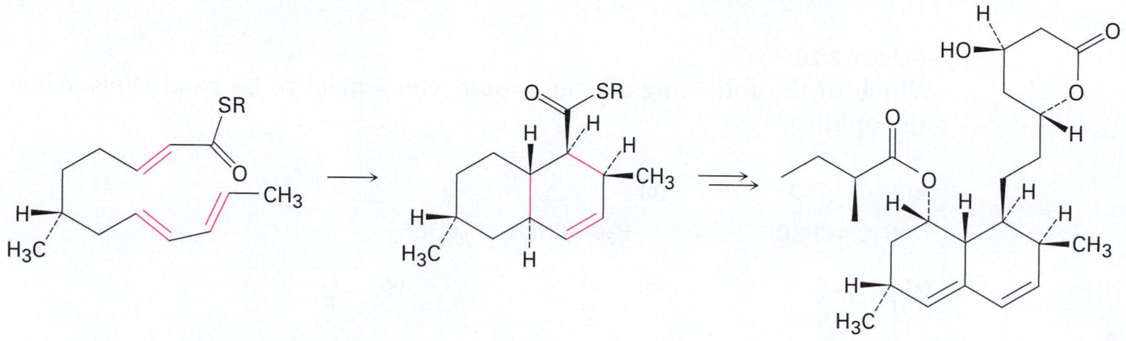

Lovastatin

WORKED EXAMPLE 8.6 Predicting the Product of a Diels–Alder Reaction

Predict the product of the following Diels–Alder reaction:

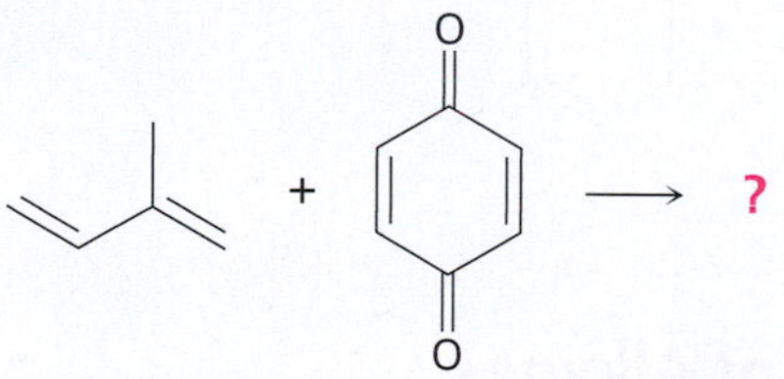

Strategy

Draw the diene so that the ends of the two double bonds are near the dienophile double bond. Then form two single bonds between the partners, convert the three double bonds into single bonds, and convert the former single bond of the diene into a double bond. Because the dienophile double bond is cis to begin with, the two attached hydrogens must remain cis in the product.

Solution

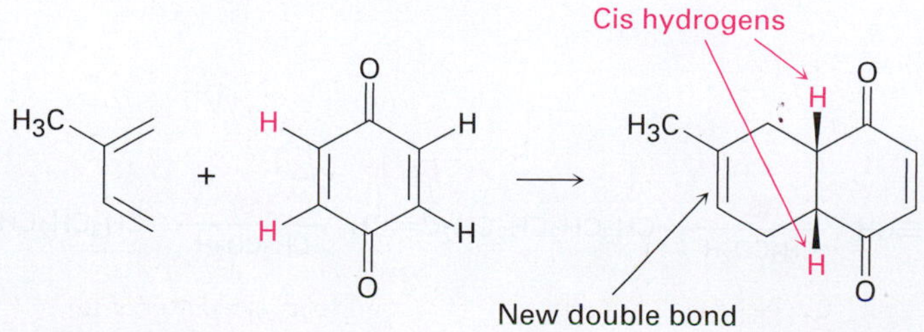

Cis hydrogens

New double bond

Problem 8.19

Predict the product of the following Diels–Alder reaction:

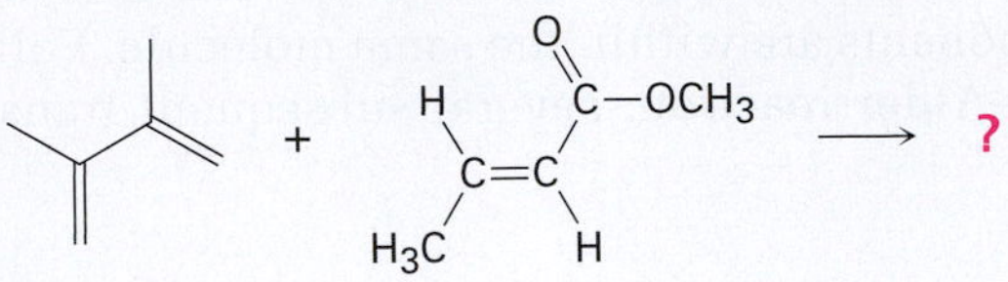

Problem 8.20

Which of the following alkenes would you expect to be good Diels–Alder dienophiles?

(a)

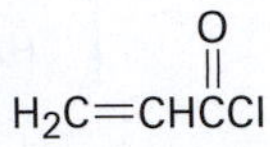

$H_2C=CHCCl$ with O

(b)

$H_2C=CHCH_2CH_2COCH_3$ with O

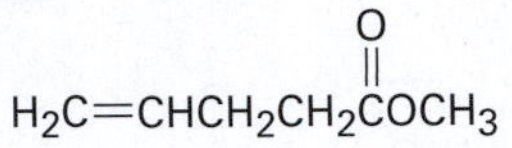

(c)

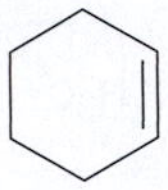

(d)

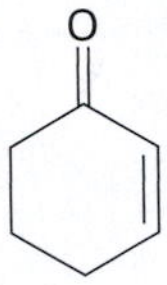

(e)

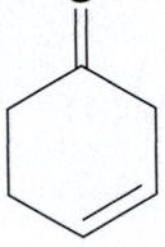

Problem 8.21

Which of the following dienes have an *s*-cis conformation, and which have an *s*-trans conformation? Of the *s*-trans dienes, which can readily rotate to *s*-cis?

(a)

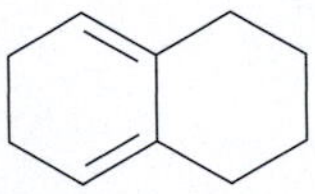

(b)

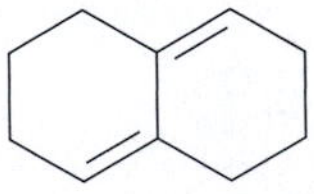

(c)

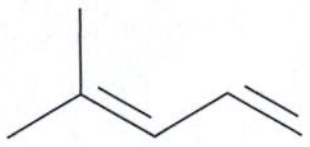

8.15 Reactions of Alkynes

Alkyne Addition Reactions

We mentioned briefly in Section 7.6 that alkynes behave similarly to alkenes in much of their chemistry. Thus, they undergo many addition reactions just as alkenes do. As a general rule, however, alkynes are somewhat less reactive than alkenes, so the various reactions can often be stopped at the monoaddition stage if only one molar equivalent of reagent is used. The additions typically show Markovnikov regiochemistry. Note that for the addition of 1 molar equivalent of H_2 to an alkyne to give an alkene, a special hydrogenation catalyst called the *Lindlar catalyst* is needed. The alkene that results has cis stereochemistry.

HBr addition

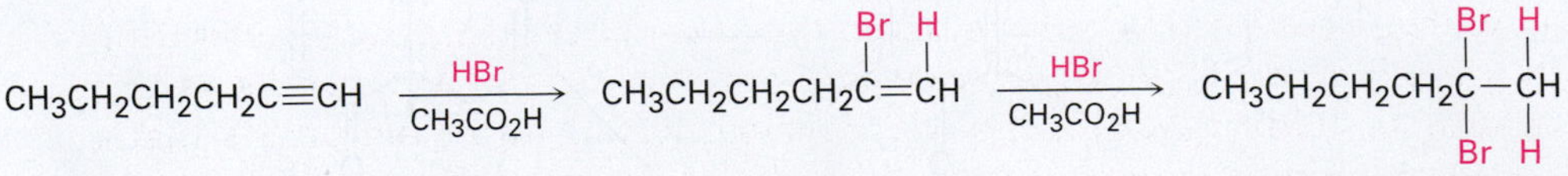

$$CH_3CH_2CH_2CH_2C\equiv CH \xrightarrow[CH_3CO_2H]{HBr} CH_3CH_2CH_2CH_2C=CH \xrightarrow[CH_3CO_2H]{HBr} CH_3CH_2CH_2CH_2C-CH$$

Hex-1-yne   **2-Bromohex-1-ene**   **2,2-Dibromohexane**

HCl addition

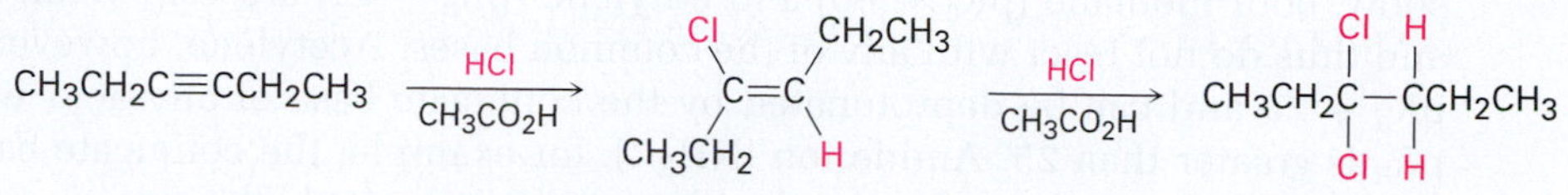

CH₃CH₂C≡CCH₂CH₃ $\xrightarrow[\text{CH}_3\text{CO}_2\text{H}]{\text{HCl}}$ (structure) $\xrightarrow[\text{CH}_3\text{CO}_2\text{H}]{\text{HCl}}$ CH₃CH₂C—CCH₂CH₃

Hex-3-yne   **(Z)-3-Chlorohex-3-ene**   **3,3-Dichlorohexane**

Br₂ addition

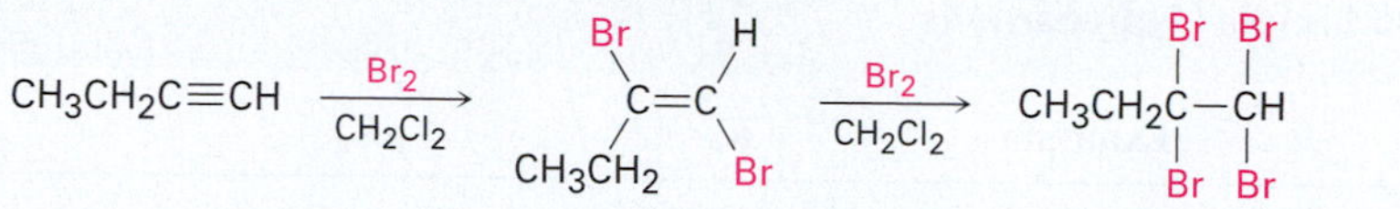

CH₃CH₂C≡CH $\xrightarrow[\text{CH}_2\text{Cl}_2]{\text{Br}_2}$ (structure) $\xrightarrow[\text{CH}_2\text{Cl}_2]{\text{Br}_2}$ CH₃CH₂C—CH

But-1-yne   **(E)-1,2-Dibromobut-1-ene**   **1,1,2,2-Tetrabromobutane**

H₂ addition

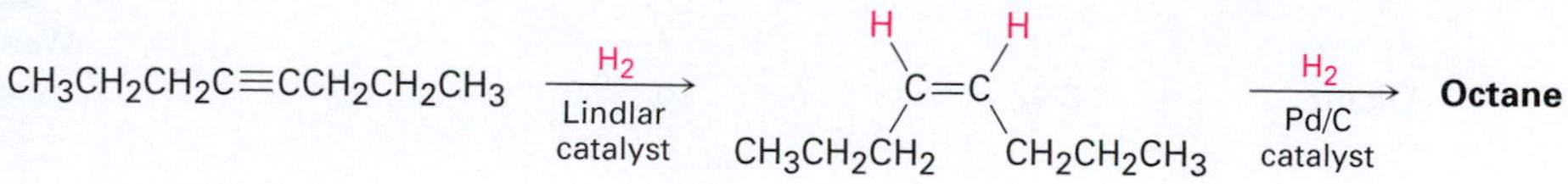

CH₃CH₂CH₂C≡CCH₂CH₂CH₃ $\xrightarrow[\substack{\text{Lindlar}\\\text{catalyst}}]{\text{H}_2}$ (structure) $\xrightarrow[\text{Pd/C catalyst}]{\text{H}_2}$ **Octane**

Oct-4-yne   **cis-Oct-4-ene**

Problem 8.22

What products would you expect from the following reactions?

(a) CH₃CH₂CH₂C≡CH + 2 Cl₂ ⟶ **?**   **(b)** (cyclopentyl)—C≡CH + 1 HBr ⟶ **?**

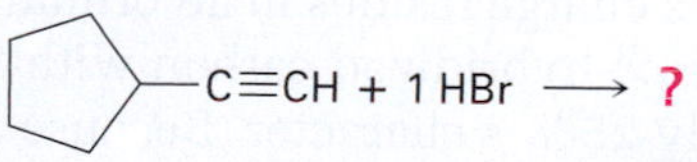

(c) CH₃CH₂CH₂CH₂C≡CCH₃ + 1 HBr ⟶ **?**

Alkyne Acidity

The most striking difference in properties between alkenes and alkynes is that terminal alkynes (RC≡CH) are relatively acidic. When a terminal alkyne is treated with a strong base, such as sodium amide, Na⁺ ⁻NH₂, the terminal hydrogen is removed and the corresponding **acetylide anion** is formed:

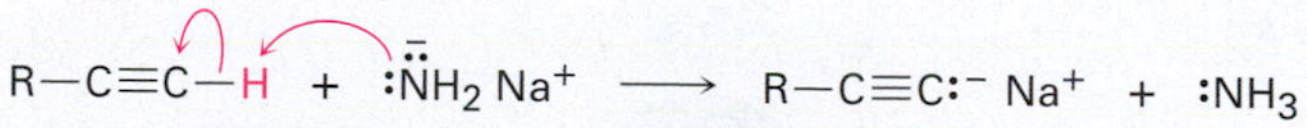

R—C≡C—H + :N̈H₂ Na⁺ ⟶ R—C≡C:⁻ Na⁺ + :NH₃

Acetylide anion

According to the Brønsted–Lowry definition (Section 2.7), an acid is a substance that donates H⁺. Although we usually think of oxyacids (H_2SO_4, HNO_3) or halogen acids (HCl, HBr) in this context, any compound containing a hydrogen atom can be an acid under the right circumstances. By measuring dissociation constants of different acids and expressing the results as pK_a values, an acidity order can be established. Recall from Section 2.8 that a lower pK_a corresponds to a stronger acid and a higher pK_a corresponds to a weaker acid.

Where do hydrocarbons lie on the acidity scale? As the data in Table 8.2 show, both methane ($pK_a \approx 60$) and ethylene ($pK_a = 44$) are very weak acids and thus do not react with any of the common bases. Acetylene, however, has $pK_a = 25$ and can be deprotonated by the conjugate base of any acid whose pK_a is greater than 25. Amide ion (NH_2^-), for example, the conjugate base of ammonia ($pK_a = 35$), is often used to deprotonate terminal alkynes.

TABLE 8.2
Acidity of Simple Hydrocarbons

Family	Example	K_a	pK_a	
Alkyne	HC≡CH	10^{-25}	25	**Stronger acid**
Alkene	$H_2C=CH_2$	10^{-44}	44	
Alkane	CH_4	10^{-60}	60	**Weaker acid**

Why are terminal alkynes more acidic than alkenes or alkanes? In other words, why are acetylide anions more stable than vinylic (alkenyl) or alkyl anions? The simplest explanation involves the hybridization of the negatively charged carbon atom. An acetylide anion has an sp-hybridized carbon, so the negative charge resides in an orbital that has 50% s character. A vinylic anion has an sp^2-hybridized carbon with 33% s character, and an alkyl anion (sp^3) has only 25% s character. Because s orbitals are nearer the positive nucleus and lower in energy than p orbitals, the negative charge is stabilized to a greater extent in an orbital with higher s character (Figure 8.19).

FIGURE 8.19 A comparison of alkyl, vinylic, and acetylide anions. The acetylide anion, with sp hybridization, has more s character and is more stable. Electrostatic potential maps show that placing the negative charge closer to the carbon nucleus makes carbon appear less negative (red).

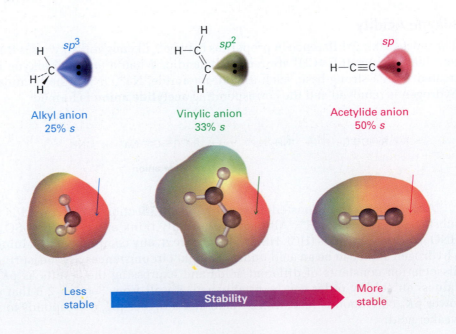

The presence of a negative charge and an unshared electron pair on carbon makes acetylide anions strongly nucleophilic. As a result, they react with many different kinds of electrophiles, such as alkyl halides, in a process that replaces the halide and yields a new alkyne product.

$$H-C\equiv C:^- \ Na^+ \ + \ H-\underset{\underset{H}{|}}{\overset{\overset{H}{|}}{C}}-Br \longrightarrow H-C\equiv C-\underset{\underset{H}{|}}{\overset{\overset{H}{|}}{C}}-H \ + \ NaBr$$

Acetylide anion   **Propyne**

We'll study the details of this substitution reaction in Section 12.6 but might note for now that the reaction is not limited to acetylene itself. *Any* terminal alkyne can be converted by base into its corresponding anion and then allowed to react with an alkyl halide to give an internal alkyne product. Hex-1-yne, for instance, gives dec-5-yne when treated first with $NaNH_2$ and then with 1-bromobutane.

$$CH_3CH_2CH_2CH_2C\equiv CH \quad \xrightarrow[\text{2. } CH_3CH_2CH_2CH_2Br]{\text{1. } NaNH_2, \ NH_3} \quad CH_3CH_2CH_2CH_2C\equiv CCH_2CH_2CH_2CH_3$$

Hex-1-yne   **Dec-5-yne (76%)**

Summary

With the background needed to understand organic reactions now covered, this chapter has begun the systematic description of major functional groups. A large variety of reactions have been covered, but we've focused on those reactions that have direct or indirect counterparts in biological pathways.

Methods for the preparation of alkenes generally involve *elimination reactions,* such as **dehydrohalogenation**—the elimination of HX from an alkyl halide—and **dehydration**—the elimination of water from an alcohol. The flip side of that elimination reaction to prepare alkenes is the addition of various substances to the alkene double bond to give saturated products.

The hydrohalic acids HCl and HBr add to alkenes by a two-step electrophilic addition mechanism. Initial reaction of the nucleophilic double bond with H^+ gives a carbocation intermediate, which then reacts with halide ion. Bromine and chlorine add to alkenes via three-membered-ring **bromonium ion** or **chloronium ion** intermediates to give addition products having **anti stereochemistry**. If water is present during halogen addition reactions, a **halohydrin** is formed.

Hydration of an alkene—the addition of water—is carried out in the laboratory by either of two complementary procedures, depending on the product desired. Oxymercuration gives the product of Markovnikov addition, whereas hydroboration/oxidation gives the product with non-Markovnikov **syn stereochemistry**.

Alkenes are reduced by addition of H_2 in the presence of a catalyst such as platinum or palladium to yield alkanes, a process called catalytic **hydrogenation**. Alkenes are also converted into **epoxides** by reaction with a

Key Words

acetylide ion, 291
allylic, 283
anti stereochemistry, 254
bromonium ion (R_2Br^+), 254
carbene ($R_2C:$), 272
conjugation, 280
dehydration, 252
dehydrohalogenation, 252
Diels–Alder cycloaddition
    reaction, 285
dienophile, 286
epoxide, 266
halogenation, 254
halohydrin, 256
hydrogenation, 261
hydroxylation, 267
monomer, 274
oxidation, 265
pericyclic reaction, 285
polymer, 274
reduction, 262
stereospecific, 273
syn stereochemistry, 259

peroxyacid and thence into trans-1,2-diols by acid-catalyzed epoxide hydro-lysis. The corresponding cis-1,2-diols can be made directly from alkenes by **hydroxylation** with OsO_4, and the diol can be cleaved to produce two carbonyl compounds by treatment with HIO_4. Alkenes can also be cleaved to produce carbonyl compounds directly by reaction with ozone followed by treatment with zinc metal. In addition, alkenes react with divalent substances called **carbenes** to yield cyclopropanes.

Alkene **polymers**—large molecules resulting from repetitive bonding together of many hundreds or thousands of small **monomer** units—are formed by reaction of simple alkenes with a radical initiator at high temperature and pressure. Polyethylene, polypropylene, and polystyrene are examples. As a general rule, radical addition reactions are not common in the laboratory but occur much more frequently in biological pathways.

A **conjugated** diene is one that contains alternating double and single bonds. One characteristic of conjugated dienes is that they are more stable than their nonconjugated counterparts. This unexpected stability can be explained by a molecular orbital description in which four p atomic orbitals combine to form four π molecular orbitals. A π bonding interaction in the lowest-energy MO introduces some partial double-bond character between carbons 2 and 3, thereby strengthening the C2–C3 bond and stabilizing the molecule. When a conjugated diene is treated with an electrophile such as HCl, a resonance-stabilized **allylic** carbocation intermediate is formed, from which both 1,2-addition and 1,4-addition products result.

Another reaction unique to conjugated dienes is the **Diels–Alder cyclo-addition**. Conjugated dienes react with electron-poor alkenes (**dienophiles**) in a single step through a cyclic transition state to yield a cyclohexene product. The reaction is **stereospecific**, meaning that only a single product stereo-isomer is formed, and can occur only if the diene is able to adopt an *s*-cis conformation.

Alkynes undergo addition reactions in much the same way that alkenes do, although their reactivity is typically less than that of alkenes. In addition, terminal alkynes (RC≡CH) are weakly acidic and can be converted into their corresponding **acetylide anions** on treatment with a sufficiently strong base.

LEARNING REACTIONS

What's seven times nine? Sixty-three, of course. You didn't have to stop and fig-ure it out; you knew the answer immediately because you long ago learned the multiplication tables. Learning the reactions of organic chemistry requires the same approach: reactions have to be learned for immediate recall if they are to be useful.

Different people take different approaches to learning reactions. Some peo-ple make flashcards; others find studying with friends to be helpful. To help guide your study, most chapters in this book end with a summary of the reac-tions just presented. In addition, the accompanying *Study Guide and Solutions Manual* has several appendixes that organize organic reactions from other viewpoints. Fundamentally, though, there are no shortcuts. Learning organic chemistry takes effort.

Summary of Reactions

Note: No stereochemistry is implied unless specifically indicated with wedged, solid, and dashed lines.

1. Addition reactions of alkenes

    (a) Addition of HCl and HBr (Sections 7.6 and 7.7)

    Markovnikov regiochemistry occurs, with H adding to the less highly substituted alkene carbon and halogen adding to the more highly substituted carbon.

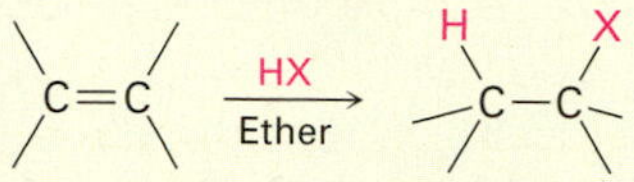

    (b) Addition of halogens Cl_2 and Br_2 (Section 8.2)

    Anti addition is observed through a halonium ion intermediate.

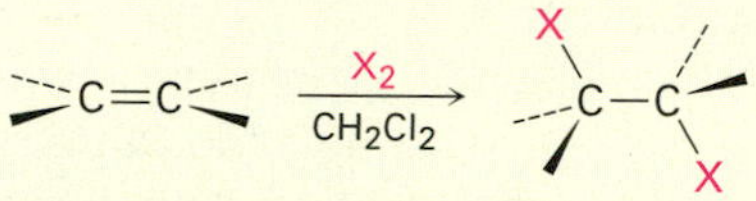

    (c) Halohydrin formation (Section 8.3)

    Markovnikov regiochemistry and anti stereochemistry occur.

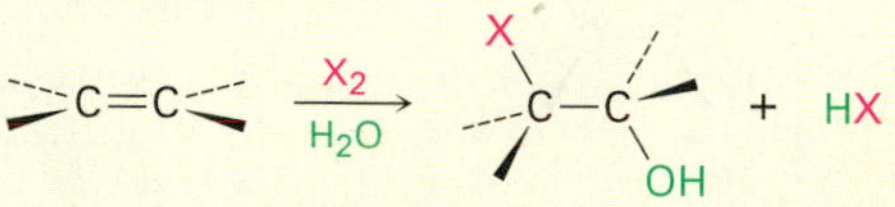

    (d) Addition of water by acid catalyzed reaction (Sections 7.6 and 7.7)

    Markovnikov regiochemistry occurs.

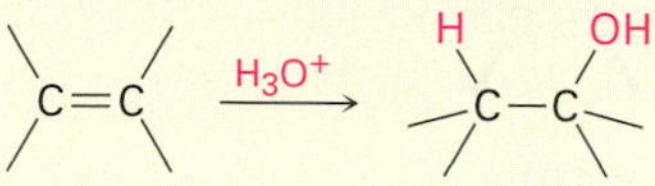

    (e) Addition of water by oxymercuration (Section 8.4)

    Markovnikov regiochemistry occurs.

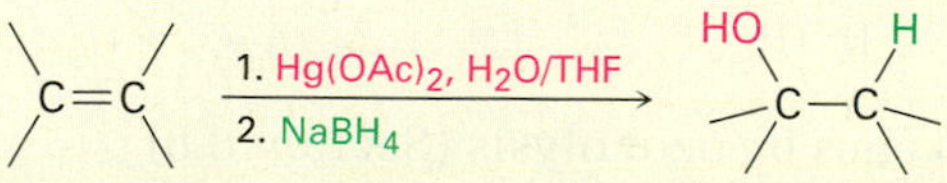

    (f) Addition of water by hydroboration/oxidation (Section 8.4)

    Non-Markovnikov syn addition occurs.

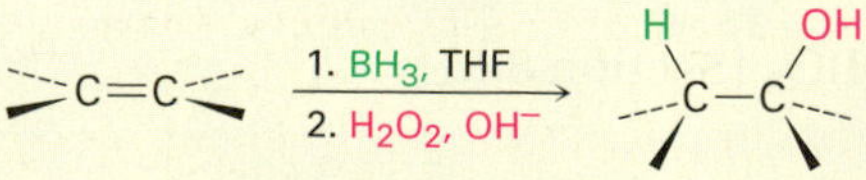

(g) Catalytic hydrogenation (Section 8.5)

Syn addition occurs.

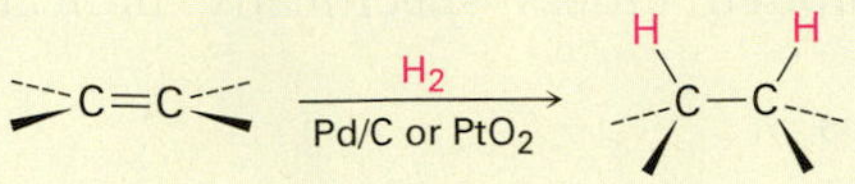

$$\text{C}=\text{C} \xrightarrow[\text{Pd/C or PtO}_2]{\text{H}_2} \overset{\text{H}}{\text{C}}-\overset{\text{H}}{\text{C}}$$

(h) Epoxidation with a peroxyacid (Section 8.6)

Syn addition occurs.

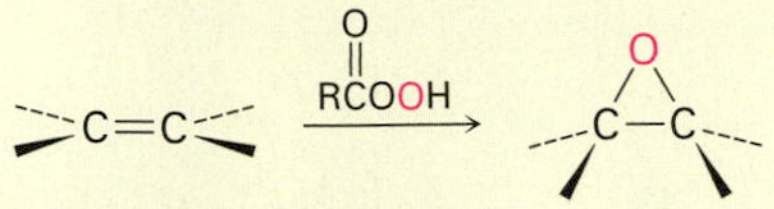

$$\text{C}=\text{C} \xrightarrow{\text{RCOOH}} \overset{\text{O}}{\text{C}-\text{C}}$$

(i) Hydroxylation by acid-catalyzed epoxide hydrolysis (Section 8.7)

Anti stereochemistry occurs.

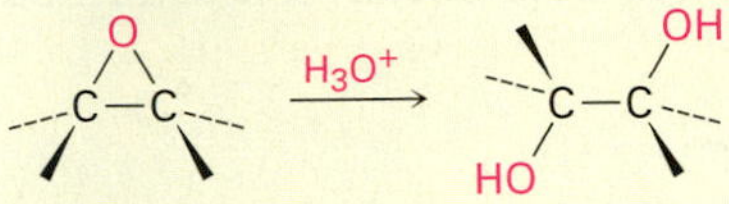

$$\overset{\text{O}}{\text{C}-\text{C}} \xrightarrow{\text{H}_3\text{O}^+} \text{HO}-\text{C}-\text{C}-\text{OH}$$

(j) Hydroxylation with OsO$_4$ (Section 8.7)

Syn addition occurs.

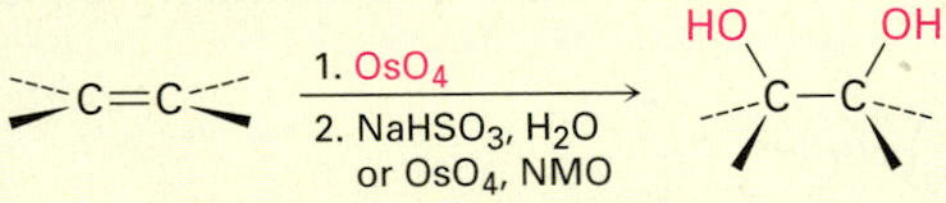

$$\text{C}=\text{C} \xrightarrow[\substack{\text{2. NaHSO}_3,\text{ H}_2\text{O} \\ \text{or OsO}_4,\text{ NMO}}]{\text{1. OsO}_4} \text{HO}-\text{C}-\text{C}-\text{OH}$$

(k) Addition of carbenes to give cyclopropanes (Section 8.9)

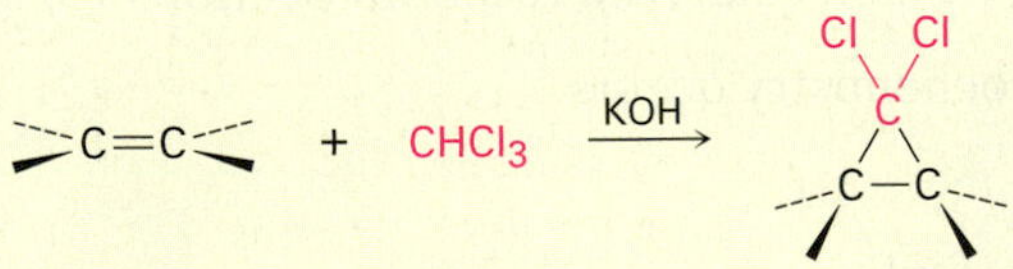

$$\text{C}=\text{C} + \text{CHCl}_3 \xrightarrow{\text{KOH}} \overset{\text{Cl}\quad\text{Cl}}{\underset{\text{C}-\text{C}}{\text{C}}}$$

(l) Radical polymerization (Section 8.10)

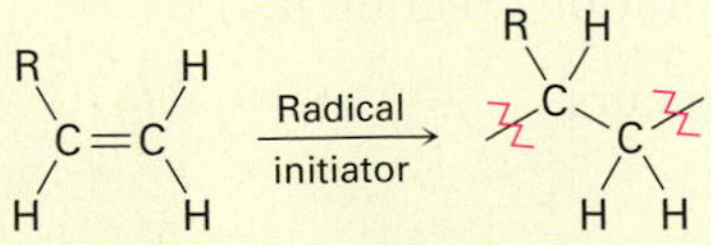

$$\overset{R}{\underset{H}{\text{C}}}=\overset{H}{\underset{H}{\text{C}}} \xrightarrow[\text{initiator}]{\text{Radical}} \cdots\overset{R\ \ H}{\underset{H\ \ H}{\text{C}-\text{C}}}\cdots$$

2.   Oxidative cleavage of alkenes by ozonolysis (Section 8.8)

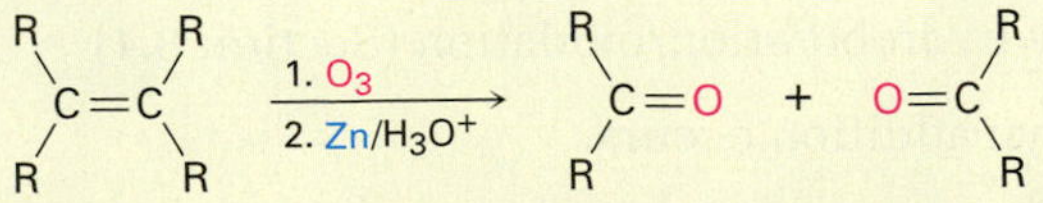

$$\overset{R}{\underset{R}{\text{C}}}=\overset{R}{\underset{R}{\text{C}}} \xrightarrow[\text{2. Zn/H}_3\text{O}^+]{\text{1. O}_3} \overset{R}{\underset{R}{\text{C}}}=\text{O} + \text{O}=\overset{R}{\underset{R}{\text{C}}}$$

3.   Cleavage of 1,2-diols with HIO$_4$ (Section 8.8)

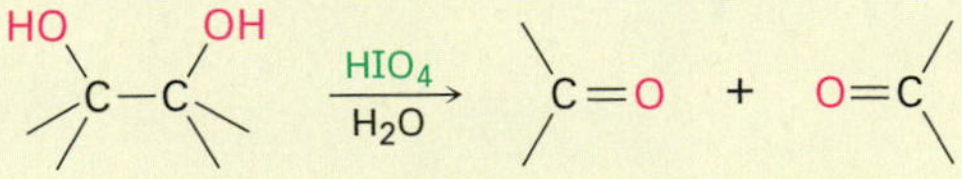

$$\overset{\text{HO}\quad\text{OH}}{\text{C}-\text{C}} \xrightarrow[\text{H}_2\text{O}]{\text{HIO}_4} \text{C}=\text{O} + \text{O}=\text{C}$$

4. Addition reactions of conjugated dienes (Section 8.13)

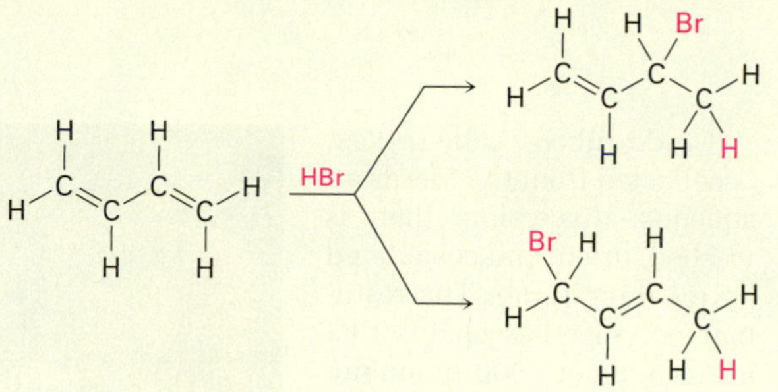

5. Diels–Alder cycloaddition reaction (Section 8.14)

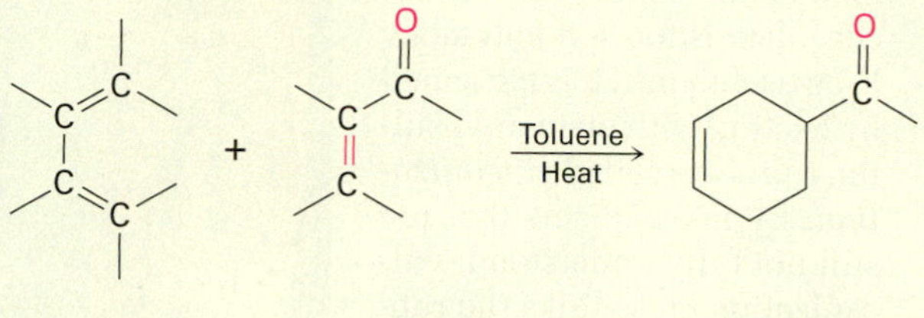

A diene   **A dienophile**   **A cyclohexene**

6. Reactions of alkynes (Section 8.15)

(a) Catalytic hydrogenation

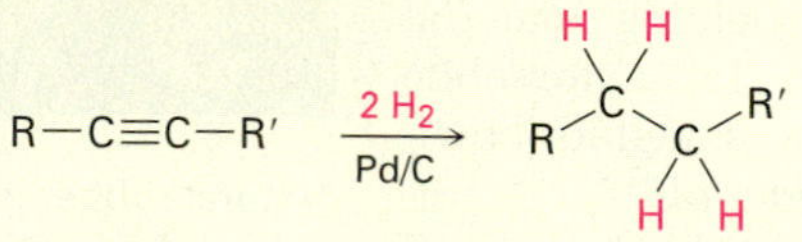

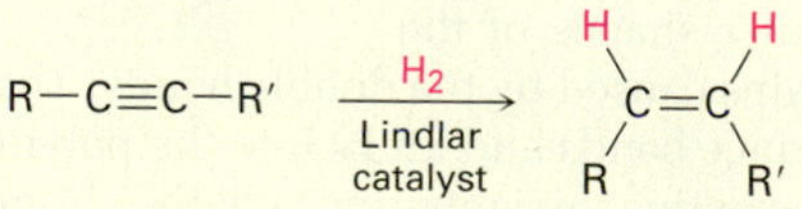

$$R—C≡C—R' \xrightarrow[\text{Pd/C}]{2\ H_2}$$

$$R—C≡C—R' \xrightarrow[\substack{\text{Lindlar}\\\text{catalyst}}]{H_2}$$

A cis alkene

(b) Conversion into acetylide anions

$$R—C≡C—H \xrightarrow[\text{NH}_3]{\text{NaNH}_2} R—C≡C:^- \ Na^+ \ + \ NH_3$$

(c) Reaction of acetylide anions with alkyl halides

$$HC≡CH \xrightarrow{\text{NaNH}_2} HC≡C^-\ Na^+ \xrightarrow{RCH_2Br} HC≡CCH_2R$$

Acetylene   **A terminal alkyne**

$$RC≡CH \xrightarrow{\text{NaNH}_2} RC≡C^-\ Na^+ \xrightarrow{R'CH_2Br} RC≡CCH_2R'$$

A terminal alkyne   **An internal alkyne**

Lagniappe

Natural Rubber

Rubber—an unusual name for an unusual substance—is a naturally occurring alkene polymer produced by more than 400 different plants. The major source is the so-called rubber tree, *Hevea brasiliensis,* from which the crude material is harvested as it drips from a slice made through the bark. The name *rubber* was coined by Joseph Priestley, the discoverer of oxygen and early researcher of rubber chemistry, for the simple reason that one of its early uses was to rub out pencil marks on paper.

Unlike polyethylene and other simple alkene polymers, natural rubber is a polymer of a conjugated diene, isoprene (2-methylbuta-1,3-diene). The polymerization takes place by 1,4-addition of isoprene monomer units to the growing chain, leading to formation of a polymer that still contains double bonds spaced regularly at four-carbon intervals. As the following structure shows, these double bonds have *Z* stereochemistry:

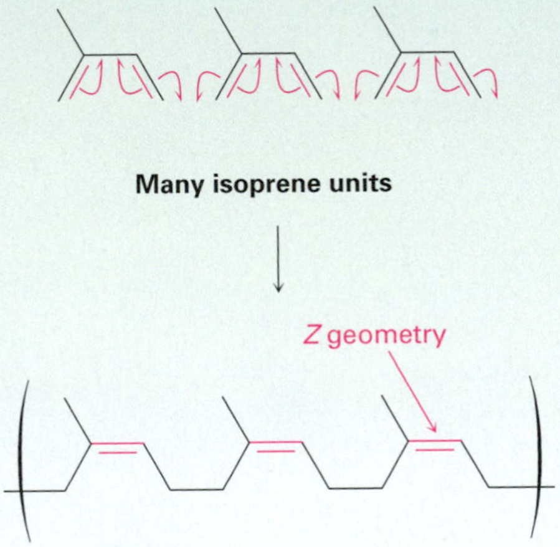

Many isoprene units

A segment of natural rubber

Crude rubber, called *latex,* is collected from the tree as an aqueous dispersion that is washed, dried, and coagulated by warming in air. The resultant polymer has chains that average about 5000 monomer units in length and have molecular weights of 200,000 to 500,000 amu. This crude coagulate is too soft and tacky to be useful until it is hardened by heating with elemental sulfur, a process called *vulcanization.* By mechanisms that are still not fully understood, vulcanization cross-links the rubber chains together by forming carbon–sulfur bonds between them, thereby hardening and stiffening the polymer. The exact degree of hardening can be varied, yielding material soft enough for automobile tires or hard enough for bowling balls *(ebonite).*

Natural rubber is obtained from the bark of the rubber tree, *Hevea brasiliensis,* grown on enormous plantations in Southeast Asia.

The remarkable ability of rubber to stretch and then contract to its original shape is due to the irregular shapes of the polymer chains caused by the double bonds. These double bonds introduce bends and kinks into the polymer chains, thereby preventing neighboring chains from nestling together. When stretched, the randomly coiled chains straighten out and orient along the direction of the pull but are kept from sliding over one another by the cross-links. When the stretch is released, the polymer reverts to its original random state.

Exercises

VISUALIZING CHEMISTRY
(Problems 8.1–8.22 appear within the chapter.)

8.23 ▪ Name the following alkenes, and predict the products of their reaction with (i) *meta*-chloroperoxybenzoic acid followed by (ii) acid-catalyzed hydrolysis:

▪ indicates problems that are assignable in Organic OWL.

Go to this book's companion website at **www.cengage.com/ chemistry/mcmurry** to explore interactive versions of the Active Figures from this text.

(a)   **(b)**

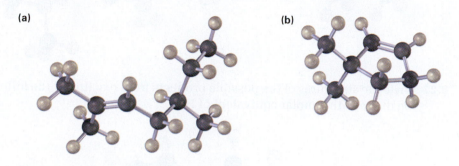

8.24 ▪ Draw the structures of alkenes that would yield the following alcohols on hydration (red = O). Tell in each case whether you would use hydroboration/oxidation or oxymercuration.

(a)   **(b)**

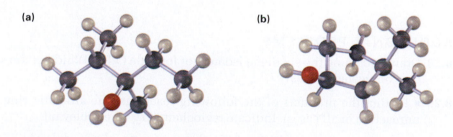

8.25 The following alkene undergoes hydroboration/oxidation to yield a single product rather than a mixture. Explain the result, and draw the product showing its stereochemistry.

▪ Problems assignable in Organic OWL.

8.26 Name the following alkynes, and predict the product of their reactions with (i) 1 molar equivalent of H_2 in the presence of Lindlar catalyst and (ii) 1 molar equivalent of Br_2:

(a)   **(b)**

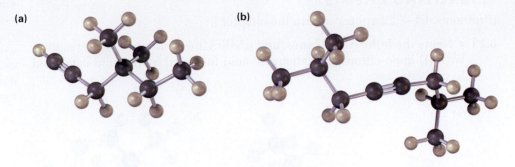

8.27 Write the structures of the possible products from reaction of the following diene with 1 molar equivalent of HCl:

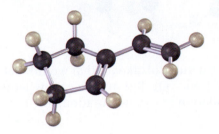

ADDITIONAL PROBLEMS

8.28 Draw and name the six diene isomers of formula C_5H_8. Which of the six are conjugated dienes?

8.29 ■ Predict the products of the following reactions (the aromatic ring is unreactive in all cases). Indicate regiochemistry when relevant.

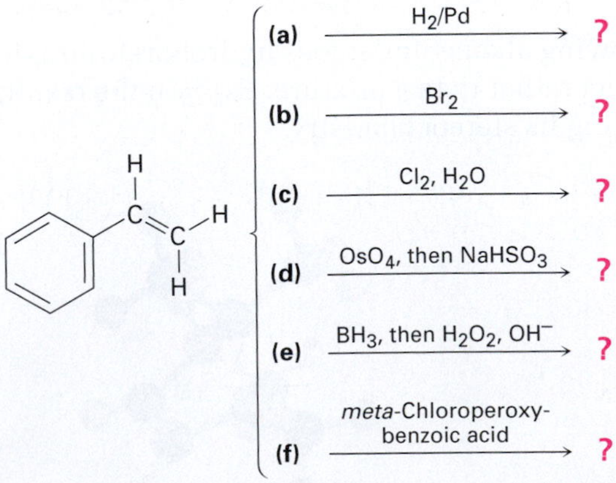

(a) $\xrightarrow{H_2/Pd}$ **?**

(b) $\xrightarrow{Br_2}$ **?**

(c) $\xrightarrow{Cl_2,\ H_2O}$ **?**

(d) $\xrightarrow{OsO_4,\ then\ NaHSO_3}$ **?**

(e) $\xrightarrow{BH_3,\ then\ H_2O_2,\ OH^-}$ **?**

(f) $\xrightarrow[\text{benzoic acid}]{meta\text{-Chloroperoxy-}}$ **?**

8.30 ■ Suggest structures for alkenes that give the following reaction products. There may be more than one answer for some cases.

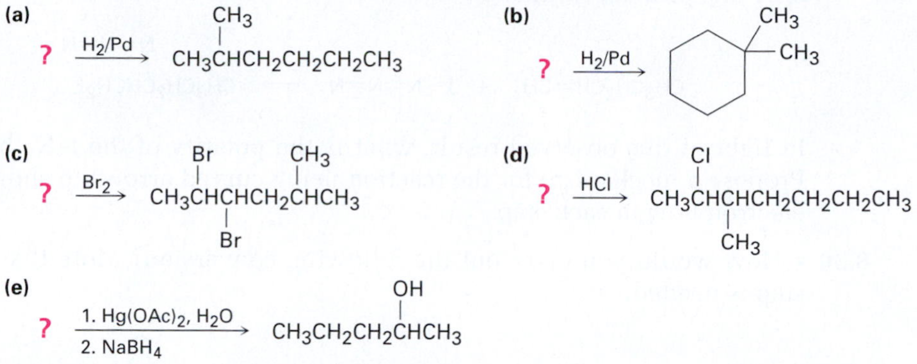

(a)

$? \xrightarrow{\text{H}_2/\text{Pd}}$ CH₃CHCH₂CH₂CH₂CH₃ (with CH₃ on second carbon)

(b)

$? \xrightarrow{\text{H}_2/\text{Pd}}$ (cyclohexane with CH₃ and CH₃ substituents)

(c)

$? \xrightarrow{\text{Br}_2}$ CH₃CHCHCH₂CHCH₃ (with Br, CH₃, and Br substituents)

(d)

$? \xrightarrow{\text{HCl}}$ CH₃CHCHCH₂CH₂CH₃ (with Cl and CH₃ substituents)

(e)

$? \xrightarrow[\text{2. NaBH}_4]{\text{1. Hg(OAc)}_2, \text{H}_2\text{O}}$ CH₃CH₂CH₂CHCH₃ (with OH substituent)

8.31 ■ How would you carry out the following transformations? Indicate the reagents you would use in each case.

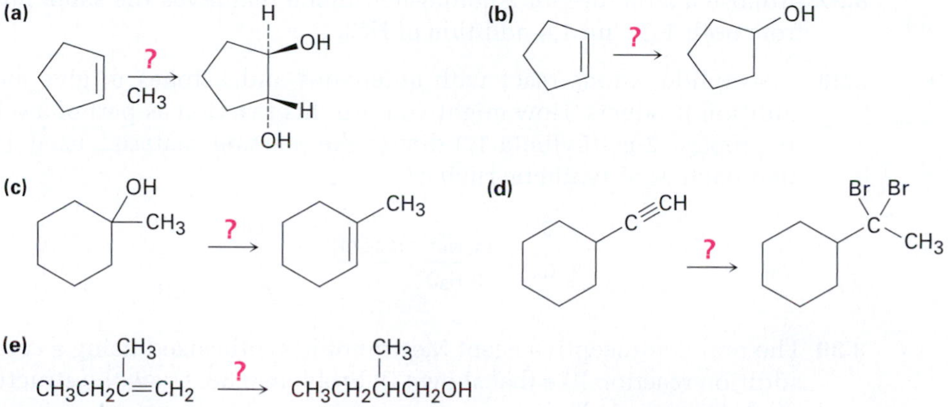

(a) cyclopentene $\xrightarrow[\text{CH}_3]{?}$ (cyclopentane with H, OH, H, OH substituents)

(b) cyclopentene $\xrightarrow{?}$ (cyclopentane with OH)

(c) (cyclohexane with OH and CH₃) $\xrightarrow{?}$ (cyclohexene with CH₃)

(d) (cyclohexane with C≡CH) $\xrightarrow{?}$ (cyclohexane with C(Br)(Br)CH₃)

(e) CH₃CH₂C=CH₂ (with CH₃) $\xrightarrow{?}$ CH₃CH₂CHCH₂OH (with CH₃)

8.32 What product will result from hydroboration/oxidation of 1-methylcyclopentene with deuterated borane, BD₃? Show both the stereochemistry (spatial arrangement) and the regiochemistry (orientation) of the product.

8.33 Which reaction would you expect to be faster, addition of HBr to cyclohexene or to 1-methylcyclohexene? Explain.

8.34 ■ Predict the products of the following reactions, and indicate regiochemistry if relevant:

(a) CH₃CH=CHCH₃ $\xrightarrow{\text{HBr}}$?

(b) CH₃CH=CHCH₃ $\xrightarrow{\text{BH}_3}$ A? $\xrightarrow[\text{⁻OH}]{\text{H}_2\text{O}_2}$ B?

8.35 Iodine azide, IN_3, adds to alkenes by an electrophilic mechanism similar to that of bromine. If a monosubstituted alkene such as but-1-ene is used, only one product results:

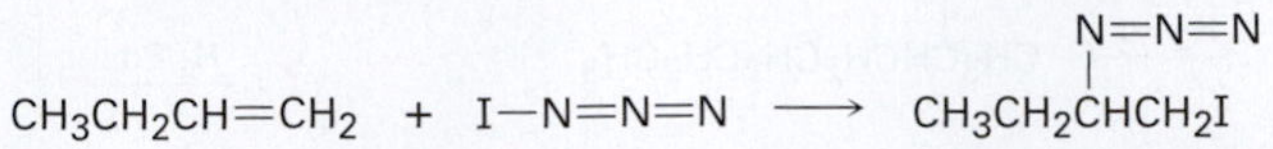

N=N=N
CH3CH2CH=CH2 + I—N=N=N ⟶ CH3CH2CHCH2I

In light of this observed result, what is the polarity of the I–N_3 bond? Propose a mechanism for the reaction using curved arrows to show the electron flow in each step.

8.36 ▪ How would you carry out the following conversion? More than one step is needed.

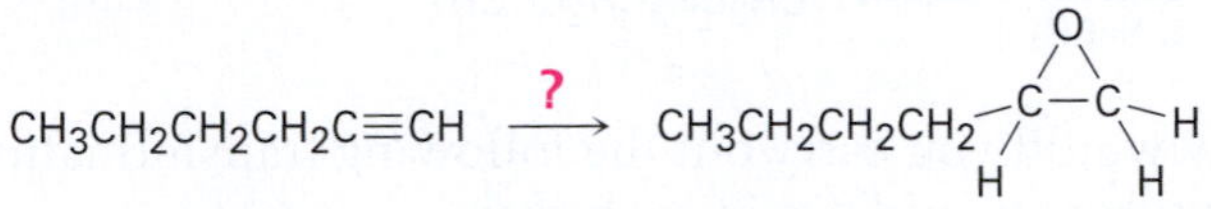

CH3CH2CH2CH2C≡CH
?
CH3CH2CH2CH2
O
C—C
H
H
H

8.37 Propose a structure for a conjugated diene that gives the same product from both 1,2- and 1,4-addition of HCl.

8.38 ▪ Acetylide anions react with aldehydes and ketones to give alcohol addition products. How might you use this reaction as part of a scheme to prepare 2-methylbuta-1,3-diene, the starting material used in the manufacture of synthetic rubber?

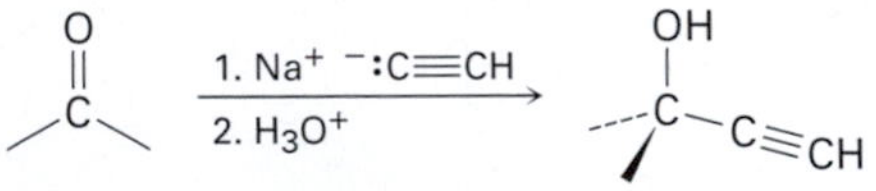

O
C
1. Na+ −:C≡CH
2. H3O+
OH
C
C≡CH

8.39 The oral contraceptive agent Mestranol is synthesized using a carbonyl addition reaction like that shown in Problem 8.38. Draw the structure of the ketone needed.

CH3
OH
C≡CH
H
H
H
CH3O

Mestranol

▪ Problems assignable in Organic OWL.

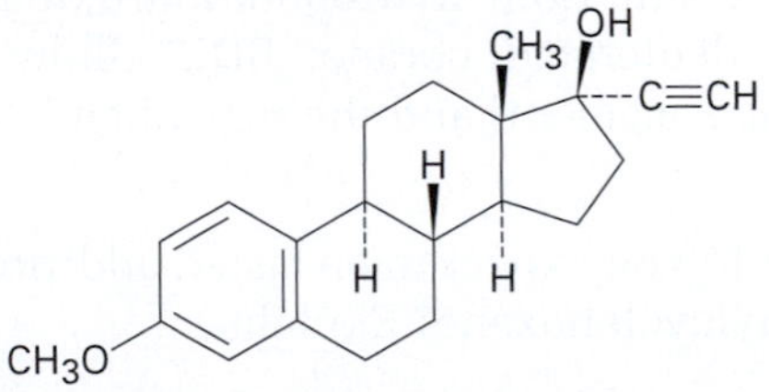

8.40 ▪ In planning the synthesis of one compound from another, it's just as important to know what *not* to do as to know what to do. The following reactions all have serious drawbacks to them. Explain the potential problems of each.

(a)

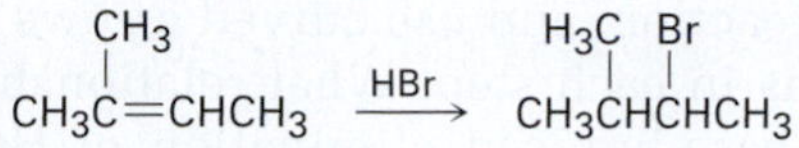

(b)

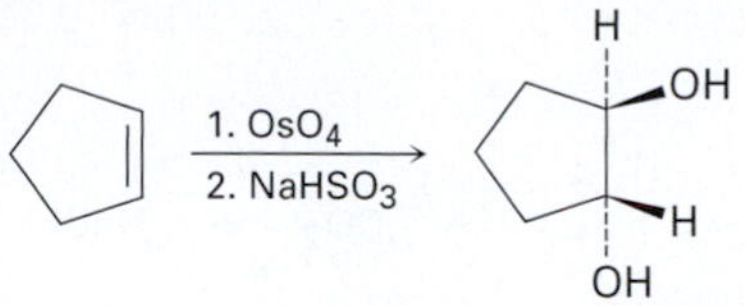

(c)

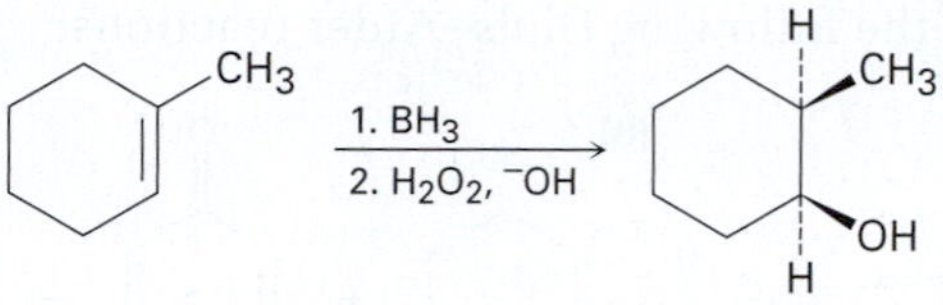

8.41 Which of the following alcohols could *not* be made selectively by hydroboration/oxidation of an alkene? Explain.

(a)

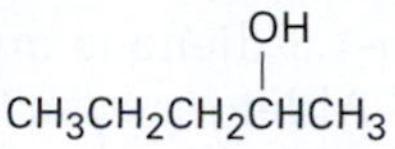

$$CH_3CH_2CH_2\overset{\overset{\displaystyle OH}{|}}{C}HCH_3$$

(b)

$$(CH_3)_2CH\overset{\overset{\displaystyle OH}{|}}{C}(CH_3)_2$$

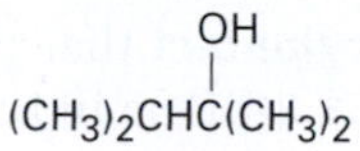

(c)

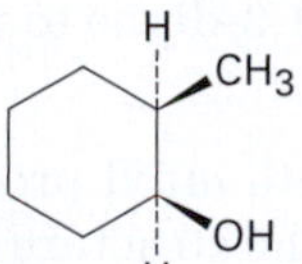

(d)

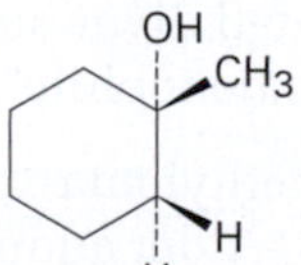

8.42 ▪ Predict the products of the following reactions. Don't worry about the size of the molecule; concentrate on the functional groups.

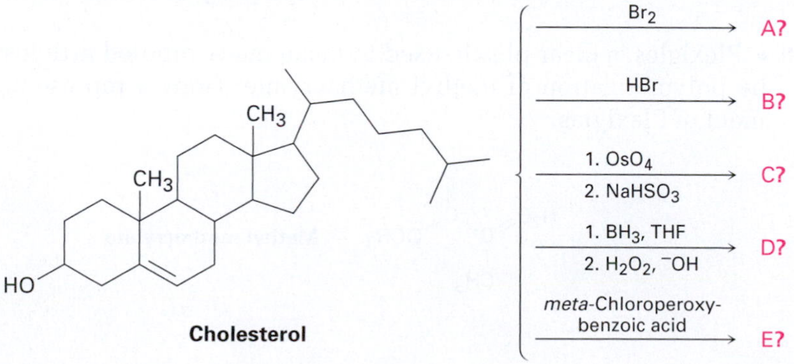

Cholesterol

$\xrightarrow{\text{Br}_2}$ A?

$\xrightarrow{\text{HBr}}$ B?

$\xrightarrow[\text{2. NaHSO}_3]{\text{1. OsO}_4}$ C?

$\xrightarrow[\text{2. H}_2\text{O}_2,\ ^-\text{OH}]{\text{1. BH}_3,\ \text{THF}}$ D?

$\xrightarrow[\text{benzoic acid}]{\textit{meta}\text{-Chloroperoxy-}}$ E?

8.43 ▪ The cis and trans isomers of but-2-ene give different dichlorocyclo-propane products when treated with $CHCl_3$ and KOH. Show the structure of each, and explain the difference.

8.44 Dichlorocarbene can be generated by heating sodium trichloroacetate. Propose a mechanism for the reaction, and use curved arrows to indicate the movement of electrons in each step. What relationship does your mechanism bear to the base-induced elimination of HCl from chloroform?

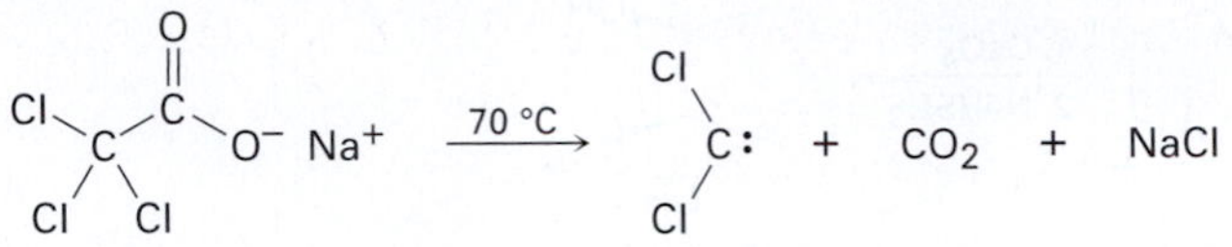

8.45 ▪ Predict the products of the following Diels–Alder reactions:

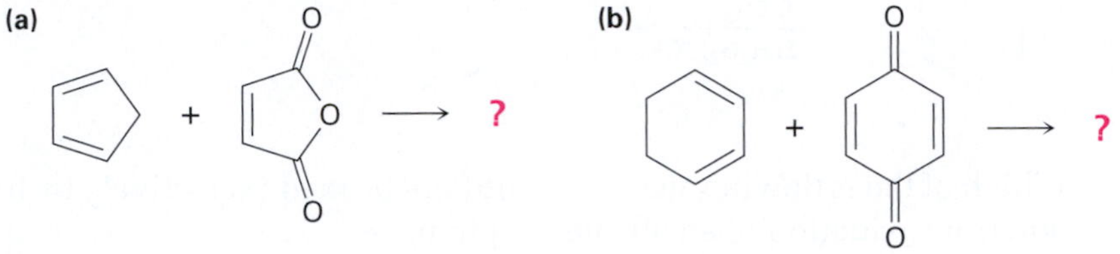

(a)

(b)

8.46 How can you account for the fact that *cis*-penta-1,3-diene is much less reactive than *trans*-penta-1,3-diene in the Diels–Alder reaction?

8.47 Would you expect a conjugated diyne such as buta-1,3-diyne to undergo Diels–Alder reaction with a dienophile? Explain.

8.48 ▪ Reaction of isoprene (2-methylbuta-1,3-diene) with ethyl propenoate gives a mixture of two Diels–Alder adducts. Show the structure of each, and explain why a mixture is formed.

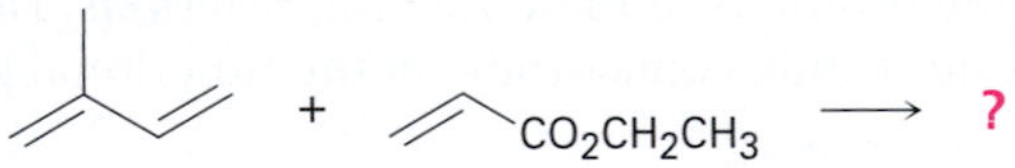

8.49 ▪ Plexiglas, a clear plastic used to make many molded articles, is made by polymerization of methyl methacrylate. Draw a representative segment of Plexiglas.

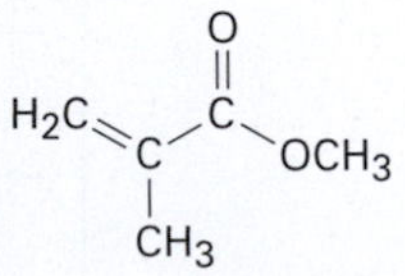

Methyl methacrylate

8.50 ■ Poly(vinyl pyrrolidone), prepared from *N*-vinylpyrrolidone, is used both in cosmetics and as a synthetic blood substitute. Draw a representative segment of the polymer.

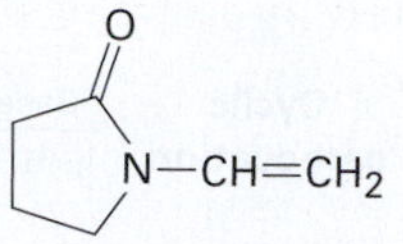

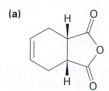

N-**Vinylpyrrolidone**

8.51 Reaction of 2-methylpropene with CH_3OH in the presence of H_2SO_4 catalyst yields methyl *tert*-butyl ether, $CH_3OC(CH_3)_3$, by a mechanism analogous to that of acid-catalyzed alkene hydration. Write the mechanism, using curved arrows for each step.

8.52 When pent-4-en-1-ol is treated with aqueous Br_2, a cyclic bromo ether is formed, rather than the expected bromohydrin. Propose a mechanism, using curved arrows to show electron movement.

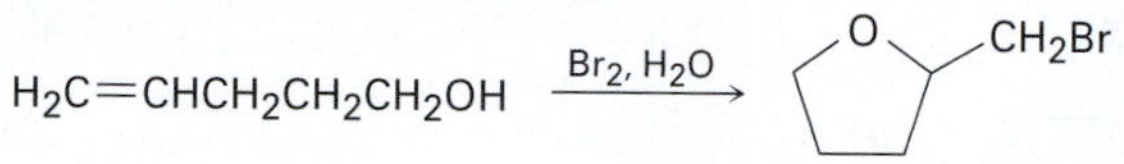

$$H_2C=CHCH_2CH_2CH_2OH \xrightarrow{Br_2,\ H_2O}$$

Pent-4-en-1-ol   **2-(Bromomethyl)tetrahydrofuran**

8.53 ■ How could you use Diels–Alder reactions to prepare the following products? Show the starting diene and dienophile in each case.

(a)

(b)

(c)

(d)

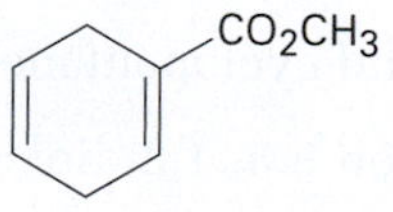

8.54 The Diels–Alder reaction is reversible and can go either forward, from diene plus dienophile to a cyclohexene, or backward, from a cyclohexene to diene plus dienophile. In light of that information, propose a mechanism for the following reaction:

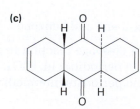

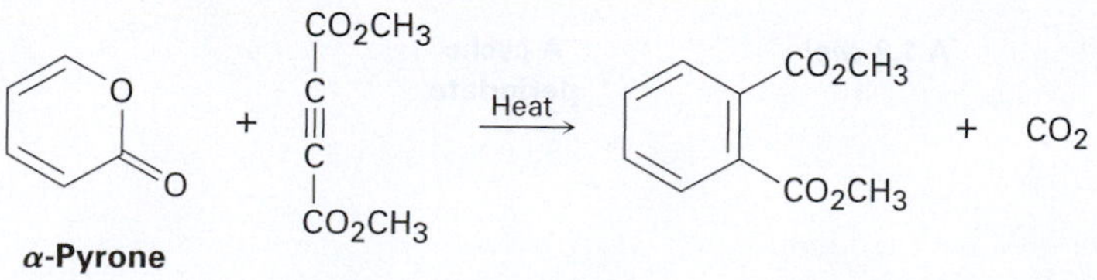

α-**Pyrone**

$+$

$\xrightarrow{\text{Heat}}$

$+$ CO_2

8.55 ■ 10-Bromo-α-chamigrene, a compound isolated from marine algae, is thought to be biosynthesized from γ-bisabolene by the following route:

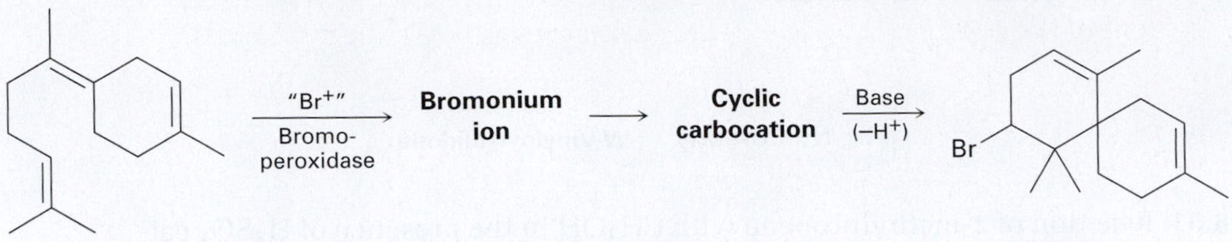

γ-Bisabolene

"Br+" / Bromo-peroxidase → **Bromonium ion** → **Cyclic carbocation** → Base (−H+) →

Br

10-Bromo-α-chamigrene

Draw the structures of the intermediate bromonium ion and cyclic carbocation, and propose mechanisms for all three steps.

8.56 ■ Isolated from marine algae, prelaureatin is thought to be biosynthesized from laurediol by the following route. Propose a mechanism. (See Problem 8.55.)

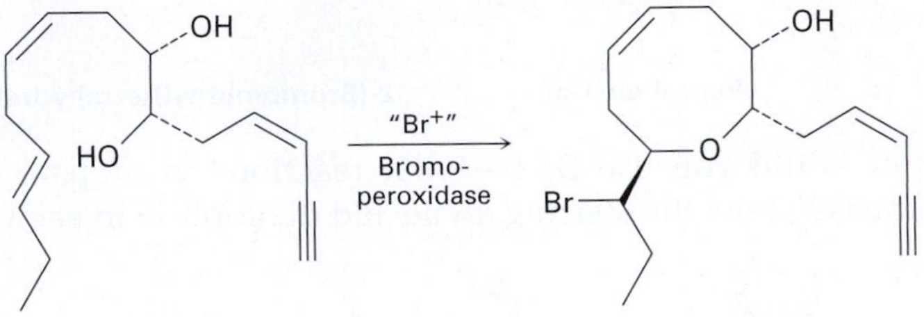

Laurediol

"Br+" / Bromo-peroxidase →

Prelaureatin

8.57 How would you distinguish between the following pairs of compounds using simple chemical tests? Tell what you would do and what you would see.

(a) Cyclopentene and cyclopentane   **(b)** Hex-2-ene and benzene

8.58 As we saw in Section 8.8, 1,2-diols undergo a cleavage reaction to give carbonyl-containing products on treatment with periodic acid, HIO_4. The reaction occurs through a five-membered cyclic periodate intermediate:

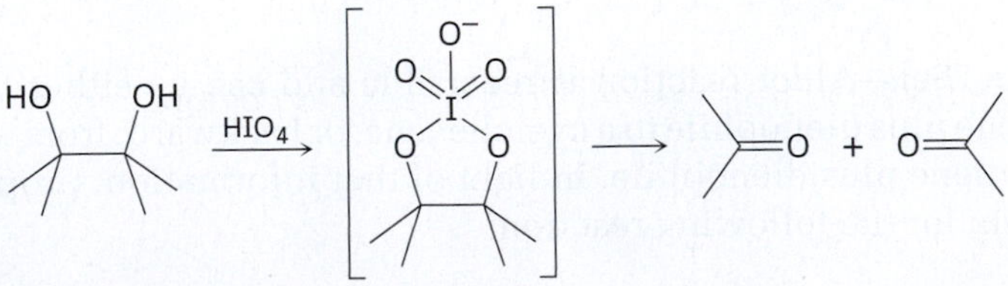

A 1,2-diol   **A cyclic periodate**

■ Problems assignable in Organic OWL.

When diols A and B were prepared and the rates of their reaction with HIO_4 were measured, it was found that diol A cleaved approximately 1 million times faster than diol B. Make molecular models of A and B and of the potential periodate intermediates, and explain the results.

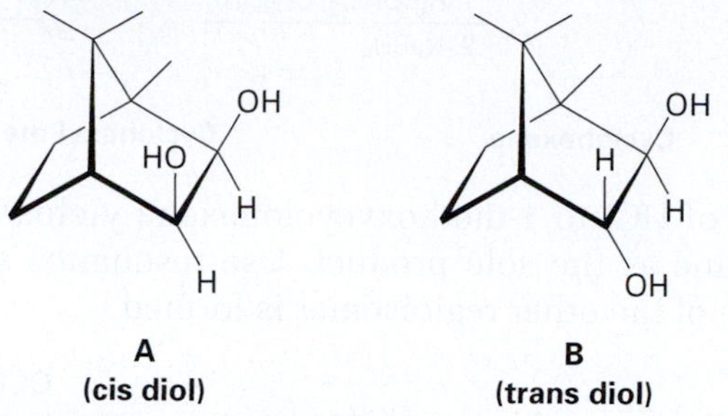

A
(cis diol)

B
(trans diol)

8.59 ■ Reaction of HBr with 3-methylcyclohexene yields a mixture of four products: *cis-* and *trans*-1-bromo-3-methylcyclohexane and *cis-* and *trans*-1-bromo-2-methylcyclohexane. The analogous reaction of HBr with 3-bromocyclohexene yields *trans*-1,2-dibromocyclohexane as the sole product. Draw structures of the possible intermediates, and explain why only a single product is formed in the reaction of HBr with 3-bromo-cyclohexene.

cis, trans   cis, trans

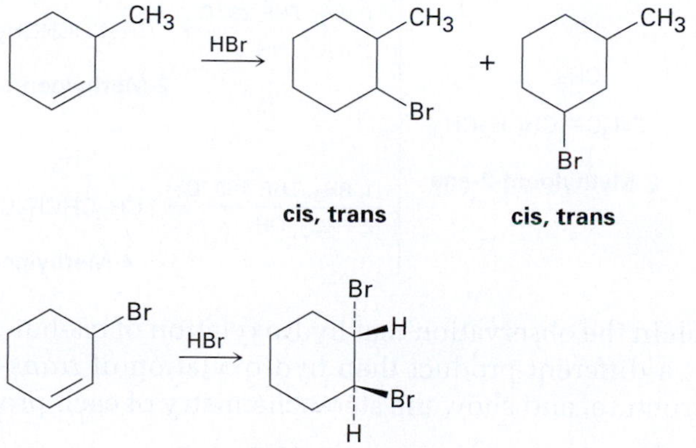

8.60 The following reaction takes place in high yield. Use your general knowledge of alkene electrophilic additions to propose a mechanism, even though you've never seen the exact reaction before.

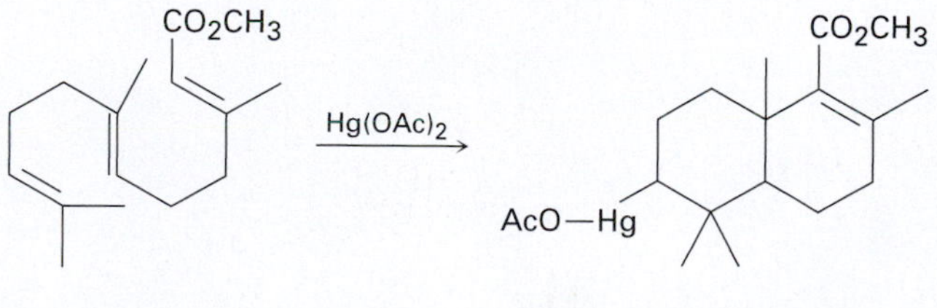

8.61 Reaction of cyclohexene with mercury(II) acetate in CH_3OH rather than H_2O, followed by treatment with $NaBH_4$, yields cyclohexyl methyl ether rather than cyclohexanol. Suggest a mechanism.

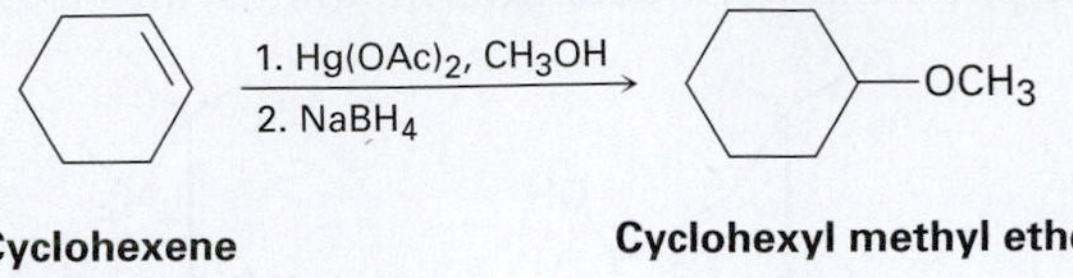

$$\text{Cyclohexene} \xrightarrow[\text{2. NaBH}_4]{\text{1. Hg(OAc)}_2,\ \text{CH}_3\text{OH}} \text{Cyclohexyl methyl ether} \ (\text{—OCH}_3)$$

Cyclohexene   **Cyclohexyl methyl ether**

8.62 Addition of HCl to 1-methoxycyclohexene yields 1-chloro-1-methoxy-cyclohexane as the sole product. Use resonance structures to explain why none of the other regioisomer is formed.

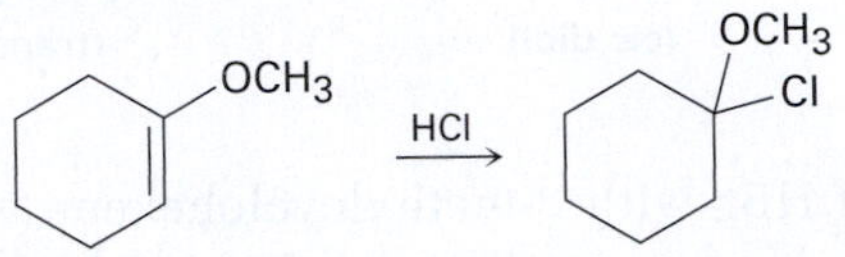

$$\text{(—OCH}_3) \xrightarrow{\text{HCl}} \text{(OCH}_3,\ \text{Cl)}$$

8.63 Addition of BH_3 to a double bond is reversible under some conditions. Explain why hydroboration of 2-methylpent-2-ene at 25 °C followed by oxidation with alkaline H_2O_2 yields 2-methylpentan-3-ol, but hydroboration at 160 °C followed by oxidation yields 4-methylpentan-1-ol.

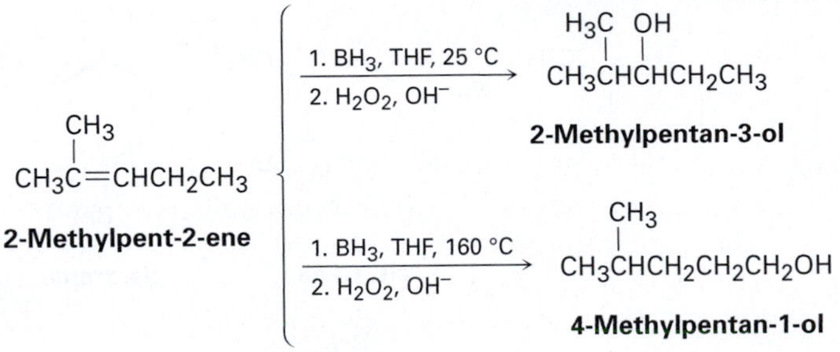

$$\underset{\textbf{2-Methylpent-2-ene}}{\overset{\overset{\displaystyle CH_3}{|}}{CH_3C=CHCH_2CH_3}}$$

1. BH_3, THF, 25 °C
2. H_2O_2, OH^-

$$\underset{\textbf{2-Methylpentan-3-ol}}{\overset{\overset{\displaystyle H_3C\ \ \ OH}{|\ \ \ \ |}}{CH_3CHCHCH_2CH_3}}$$

1. BH_3, THF, 160 °C
2. H_2O_2, OH^-

$$\underset{\textbf{4-Methylpentan-1-ol}}{\overset{\overset{\displaystyle CH_3}{|}}{CH_3CHCH_2CH_2CH_2OH}}$$

8.64 ■ Explain the observation that hydroxylation of *cis*-but-2-ene with OsO_4 yields a different product than hydroxylation of *trans*-but-2-ene. Draw the structure, and show the stereochemistry of each product.

9 Aromatic Compounds

Hemoglobin, the oxygen-carrying protein in blood, contains a large aromatic cofactor called heme.

In the early days of organic chemistry, the word *aromatic* was used to describe such fragrant substances as benzene (from coal distillate), benzaldehyde (from cherries, peaches, and almonds), and toluene (from Tolu balsam). It was soon realized, however, that substances classed as aromatic differed from most other organic compounds in their chemical behavior.

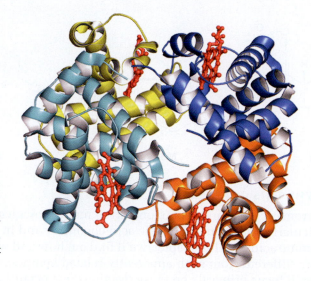

Benzene   **Benzaldehyde**   **Toluene**

Today, the association of aromaticity with fragrance has long been lost, and we now use the word **aromatic** to refer to the class of compounds that contain six-membered benzene-like rings with three double bonds. Many naturally occurring compounds are aromatic in part, such as the steroidal hormone estrone and the analgesic morphine. In addition, many synthetic drugs are aromatic in part, such as the antidepressant fluoxetine (Prozac). Benzene itself has been found to cause bone marrow depression and consequent leukopenia, or lowered white blood cell count, on prolonged

OWL Online homework for this chapter can be assigned in Organic OWL.

exposure. Benzene should therefore be handled cautiously if used as a laboratory solvent.

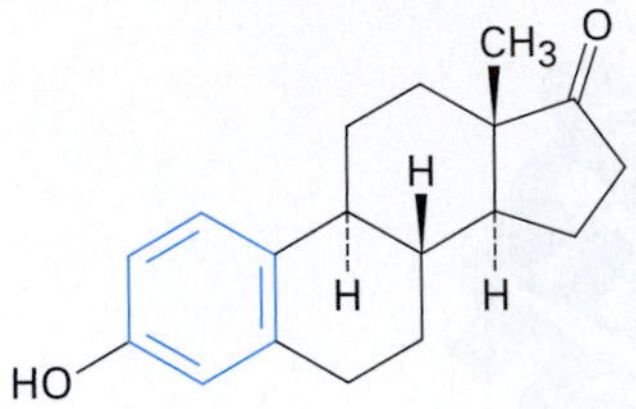

Estrone

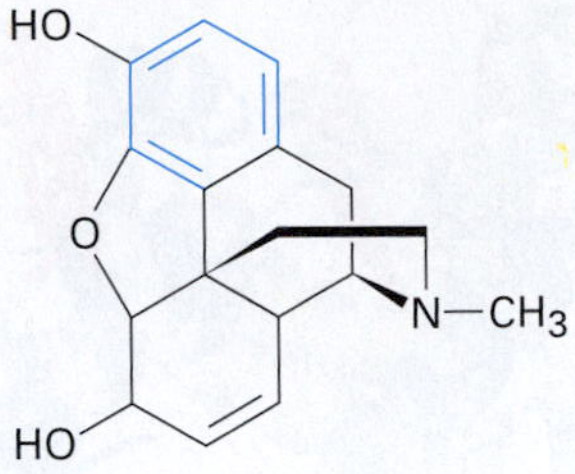

Morphine

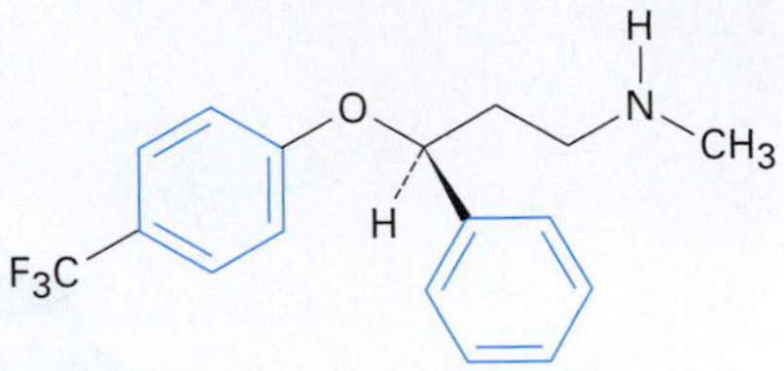

Fluoxetine (Prozac)

WHY THIS CHAPTER?

Aromatic rings are a common part of many biological structures and are particularly important in nucleic acid chemistry and in the chemistry of several amino acids. In this chapter, we'll find out how and why aromatic compounds are different from such apparently related compounds as alkenes. As usual, we'll focus primarily on those reactions that occur in both the laboratory and living organisms.

9.1 Naming Aromatic Compounds

Aromatic substances, more than any other class of organic compounds, have acquired a large number of nonsystematic names. IUPAC rules discourage the use of most such names but do allow for some of the more widely used ones to be retained (Table 9.1). Thus, methylbenzene is known commonly as *toluene,* hydroxybenzene as *phenol,* aminobenzene as *aniline,* and so on.

TABLE 9.1
Common Names of Some Aromatic Compounds

Structure	Name	Structure	Name
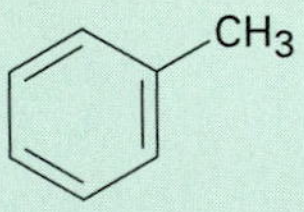 CH₃	Toluene (bp 111 °C)	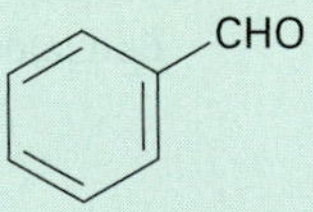 CHO	Benzaldehyde (bp 178 °C)
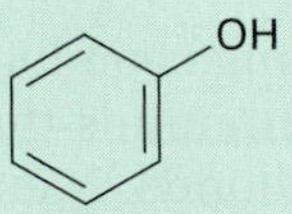 OH	Phenol (mp 43 °C)	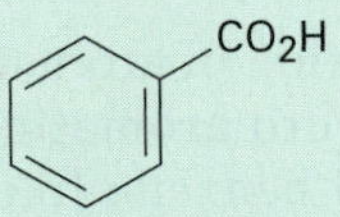 CO₂H	Benzoic acid (mp 122 °C)
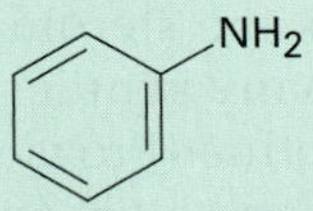 NH₂	Aniline (bp 184 °C)	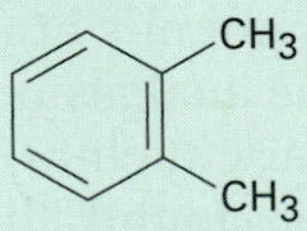 CH₃ CH₃	*ortho*-Xylene (bp 144 °C)
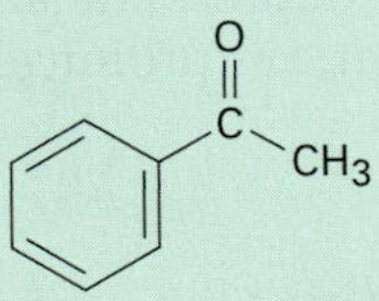 O C CH₃	Acetophenone (mp 21 °C)	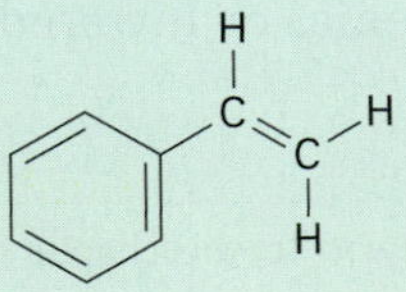 H C C H H	Styrene (bp 145 °C)

Monosubstituted benzenes are systematically named in the same manner as other hydrocarbons, with *-benzene* as the parent name. Thus, C_6H_5Br is bromobenzene, $C_6H_5NO_2$ is nitrobenzene, and $C_6H_5CH_2CH_2CH_3$ is propylbenzene.

Bromobenzene   **Nitro**benzene   **Propyl**benzene

Alkyl-substituted benzenes are sometimes referred to as **arenes** and are named in different ways depending on the size of the alkyl group. If the alkyl substituent is smaller than the ring (six or fewer carbons), the arene is named as an alkyl-substituted benzene. If the alkyl substituent is larger than the ring (seven or more carbons), the compound is named as a phenyl-substituted alkane. The name **phenyl**, pronounced **fen**-nil and sometimes abbreviated as Ph or Φ (Greek phi), is used for the $-C_6H_5$ unit when the benzene ring is considered as a substituent. The word is derived from the Greek *pheno* ("I bear light"), commemorating the discovery of benzene by Michael Faraday in 1825 from the oily residue left by the illuminating gas used in London street lamps. In addition, the name **benzyl** is used for the $C_6H_5CH_2-$ group.

A phenyl group   **2-Phenyl**heptane   **A benzyl group**

Disubstituted benzenes are named using one of the prefixes **ortho (o)**, **meta (m)**, or **para (p)**. An ortho-disubstituted benzene has its two substituents in a 1,2 relationship on the ring, a meta-disubstituted benzene has its substituents in a 1,3 relationship, and a para-disubstituted benzene has its substituents in a 1,4 relationship.

ortho-**Dichlorobenzene**
1,2 disubstituted

meta-**Dimethylbenzene**
(*meta*-xylene)
1,3 disubstituted

para-**Chlorobenzaldehyde**
1,4 disubstituted

Benzenes with more than two substituents are named by choosing a point of attachment as carbon 1 and numbering the substituents on the ring so that the second substituent has as low a number as possible. If ambiguity still exists, number so that the third or fourth substituent has as low a number as possible, until a point of difference is found. The substituents are listed alphabetically when writing the name.

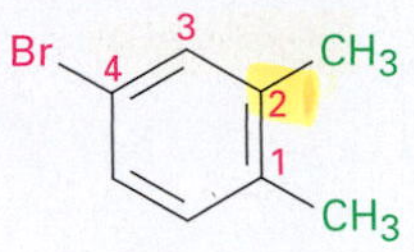

Trisubstituted

4-Bromo-1,2-dimethylbenzene   **2,5-Dimethyl**phenol   **2,4,6-Trinitro**toluene (TNT)

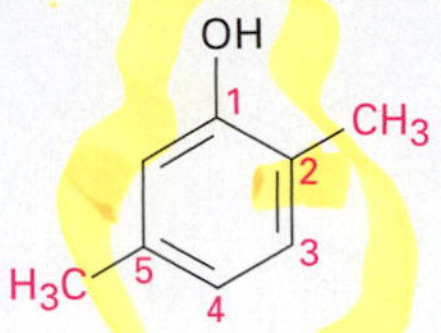

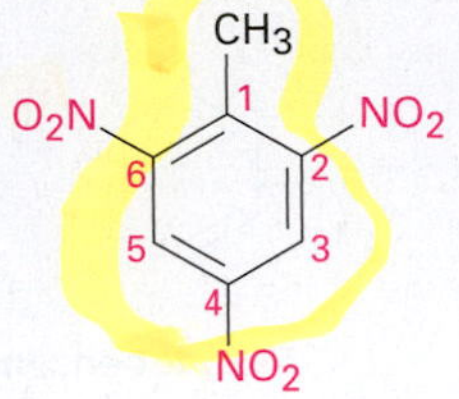

Note in the second and third examples shown that *-phenol* and *-toluene* are used as the parent names rather than *-benzene*. Any of the monosubstituted aromatic compounds shown in Table 9.1 can serve as a parent name, with the principal substituent (–OH in phenol or –CH$_3$ in toluene) attached to C1 on the ring.

Problem 9.1
Tell whether the following compounds are ortho-, meta-, or para-disubstituted:

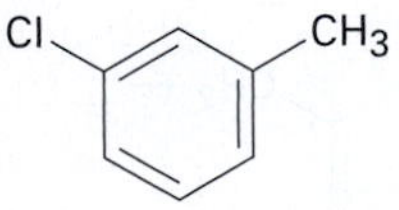

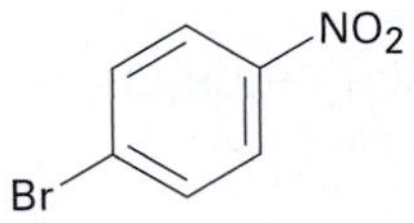

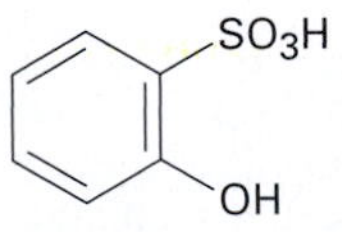

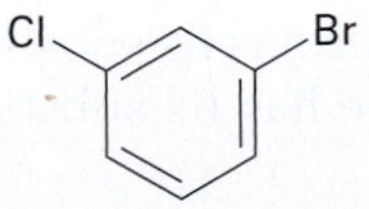

Problem 9.2
Give IUPAC names for the following compounds:

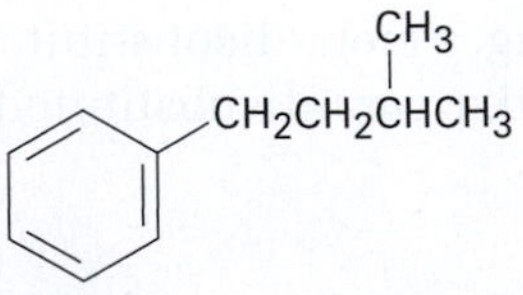

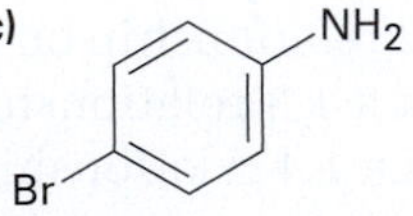

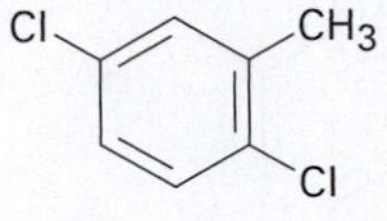

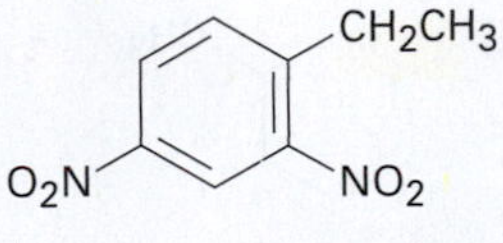

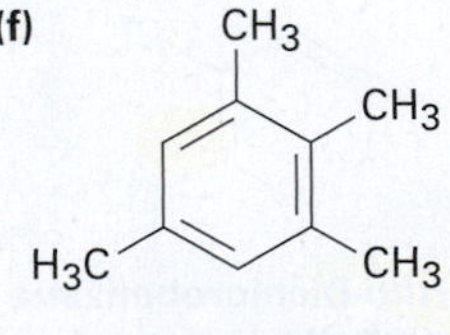

Problem 9.3
Draw structures corresponding to the following IUPAC names:
(a) *p*-Bromochlorobenzene   **(b)** *p*-Bromotoluene
(c) *m*-Chloroaniline   **(d)** 1-Chloro-3,5-dimethylbenzene

9.2 Structure and Stability of Benzene

Although benzene is clearly unsaturated, it is much less reactive than typical alkenes and fails to undergo the usual alkene addition reactions. Cyclohexene, for instance, reacts rapidly with Br_2 and gives the addition product 1,2-dibromocyclohexane, but benzene reacts only slowly with Br_2 and gives the *substitution* product C_6H_5Br.

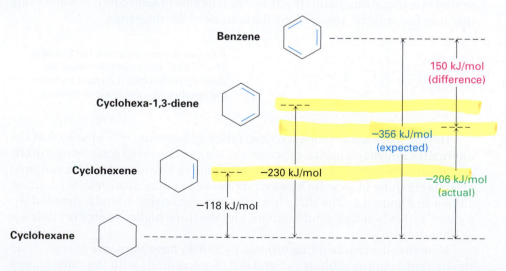

Benzene   **Bromobenzene**   **(Addition product)**
                       (substitution product)   ***NOT* formed**

We can get a quantitative idea of benzene's stability by measuring heats of hydrogenation (Section 7.5). Cyclohexene, an isolated alkene, has $\Delta H°_{hydrog} = -118$ kJ/mol (-28.2 kcal/mol), and cyclohexa-1,3-diene, a conjugated diene, has $\Delta H°_{hydrog} = -230$ kJ/mol (-55.0 kcal/mol). As noted in Section 8.12, this value for cyclohexa-1,3-diene is a bit less than twice that for cyclohexene because conjugated dienes are more stable than isolated dienes.

Carrying the process one step further, we might expect $\Delta H°_{hydrog}$ for "cyclohexatriene" (benzene) to be a bit less than -356 kJ/mol, or three times the cyclohexene value. The actual value, however, is -206 kJ/mol, some 150 kJ/mol (36 kcal/mol) less than expected. Since 150 kJ/mol less heat than expected is released during hydrogenation of benzene, benzene must have 150 kJ/mol less energy to begin with. In other words, benzene is more stable than expected by 150 kJ/mol (Figure 9.1).

Benzene

150 kJ/mol
(difference)

Cyclohexa-1,3-diene

-356 kJ/mol
(expected)

Cyclohexene -230 kJ/mol

-118 kJ/mol -206 kJ/mol
                                                   (actual)

Cyclohexane

FIGURE 9.1 A comparison of the heats of hydrogenation for cyclohexene, cyclohexa-1,3-diene, and benzene. Benzene is 150 kJ/mol (36 kcal/mol) more stable than might be expected for "cyclohexatriene."

Further evidence for the unusual nature of benzene is that all its carbon–carbon bonds have the same length—139 pm—intermediate between typical single (154 pm) and double (134 pm) bonds. In addition, an electrostatic potential map shows that the electron density in all six carbon–carbon bonds is identical. Thus, benzene is a planar molecule with the shape of a regular hexagon. All C–C–C bond angles are 120°, all six carbon atoms are sp^2-hybridized, and each carbon has a p orbital perpendicular to the plane of the six-membered ring.

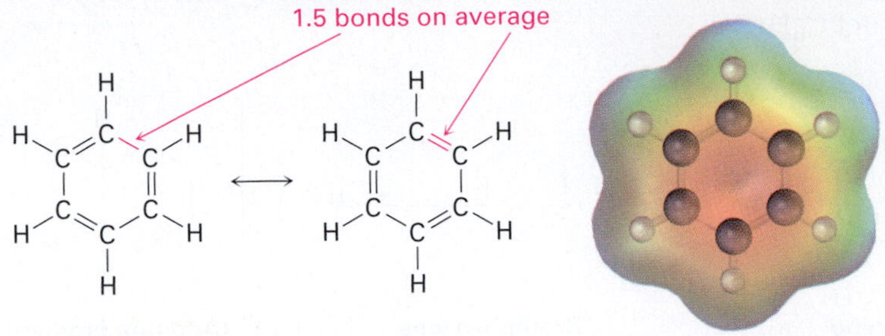

1.5 bonds on average

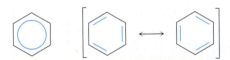

Because all six carbon atoms and all six p orbitals in benzene are equivalent, it's impossible to define three localized π bonds in which a given p orbital overlaps only one neighboring p orbital. Rather, each p orbital overlaps equally well with both neighboring p orbitals, leading to a picture of benzene in which the six π electrons are completely delocalized around the ring. In resonance terms (Sections 2.4 and 2.5), benzene is a hybrid of two equivalent forms. Neither form is correct by itself; the true structure of benzene is somewhere in between the two resonance forms but is impossible to draw with our usual conventions.

Chemists sometimes represent the two benzene resonance forms by using a circle to indicate the equivalence of the carbon–carbon bonds. This kind of representation has to be used carefully, however, because it doesn't indicate the number of π electrons in the ring. (How many electrons does a circle represent?) In this book, benzene and other aromatic compounds will be represented by a single line-bond structure. We'll be able to keep count of π electrons this way but must be aware of the limitations of the drawings.

Alternative representations of benzene. The "circle" representation must be used carefully since it doesn't indicate the number of π electrons in the ring.

Having just seen a resonance description of benzene, let's now look at the alternative molecular orbital description. We can construct π molecular orbitals for benzene just as we did for buta-1,3-diene in Section 8.12. If six p atomic orbitals combine in a cyclic manner, six benzene molecular orbitals result, as shown in Figure 9.2. The three lower-energy molecular orbitals, denoted ψ_1, ψ_2, and ψ_3, are bonding combinations, and the three higher-energy orbitals are antibonding.

Note that the two bonding orbitals ψ_2 and ψ_3 have the same energy, as do the two antibonding orbitals ψ_4^* and ψ_5^*. Such orbitals with the same energy are said to be *degenerate*. Note also that the two orbitals ψ_3 and ψ_4^* have nodes passing through ring carbon atoms, thereby leaving no π electron density on these carbons. The six p electrons of benzene occupy the three bonding molecular orbitals and are delocalized over the entire conjugated system, leading to the observed 150 kJ/mol stabilization of benzene.

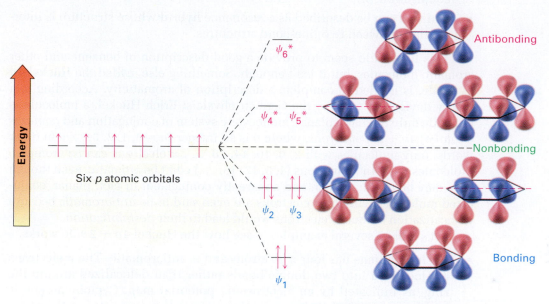

Six benzene molecular orbitals

ACTIVE FIGURE 9.2 The six benzene π molecular orbitals. The bonding orbitals ψ_2 and ψ_3 have the same energy and are said to be degenerate, as are the antibonding orbitals ψ_4^* and ψ_5^*. The orbitals ψ_3 and ψ_4^* have no π electron density on two carbons because of a node passing through the atoms. **Go to this book's student companion site at** www .cengage.com/chemistry/mcmurry **to explore an interactive version of this figure.**

Problem 9.4

Pyridine is a flat, hexagonal molecule with bond angles of 120°. It undergoes substitution rather than addition and generally behaves like benzene. Draw a picture of the π orbitals of pyridine to explain its properties. Check your answer by looking ahead to Section 9.4.

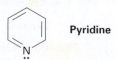

Pyridine

9.3 Aromaticity and the Hückel 4*n* + 2 Rule

Let's list what we've said thus far about benzene and, by extension, about other benzene-like aromatic molecules:

- Benzene is cyclic and conjugated.
- Benzene is unusually stable, having a heat of hydrogenation 150 kJ/mol less negative than we might expect for a conjugated cyclic triene.
- Benzene is planar and has the shape of a regular hexagon. All bond angles are 120°, all carbon atoms are sp^2-hybridized, and all carbon–carbon bond lengths are 139 pm.

- Benzene undergoes substitution reactions that retain the cyclic conjugation rather than electrophilic addition reactions that would destroy the conjugation.

- Benzene can be described as a resonance hybrid whose structure is intermediate between two line-bond structures.

This list would seem to provide a good description of benzene and other aromatic molecules, but it isn't enough. Something else, called the **Hückel 4n + 2 rule**, is needed to complete a description of aromaticity. According to a theory devised in 1931 by the German physicist Erich Hückel, a molecule is aromatic only if it has a planar, monocyclic system of conjugation and *contains a total of 4n + 2 π electrons*, where n is an integer ($n = 0, 1, 2, 3, \ldots$). In other words, only molecules with 2, 6, 10, 14, 18, . . . π electrons can be aromatic. Molecules with $4n$ π electrons (4, 8, 12, 16, . . .) *can't* be aromatic, even though they may be cyclic, planar, and apparently conjugated. In fact, planar, conjugated molecules with $4n$ π electrons are even said to be *antiaromatic* because delocalization of their π electrons would lead to their *destabilization*.

Let's look at several examples to see how the Hückel $4n + 2$ rule works.

- **Cyclobutadiene** has four π electrons and is antiaromatic. The π electrons are localized into two double bonds rather than delocalized around the ring, as indicated by an electrostatic potential map. Cyclobutadiene is highly reactive and shows none of the properties associated with aromaticity. In fact, it was not even prepared until 1965.

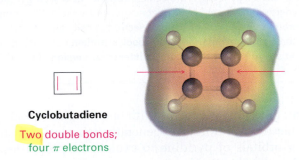

Cyclobutadiene

Two double bonds;
four π electrons

- **Benzene** has six π electrons ($4n + 2 = 6$ when $n = 1$) and is aromatic:

Benzene

Three double bonds;
six π electrons

- **Cyclooctatetraene** has eight π electrons and is not aromatic. The π electrons are localized into four double bonds rather than delocalized around the ring, and the molecule is tub-shaped rather than planar. It has no cyclic conjugation because neighboring p orbitals don't have the necessary parallel alignment for overlap, and it resembles an open-chain polyene in its reactivity.

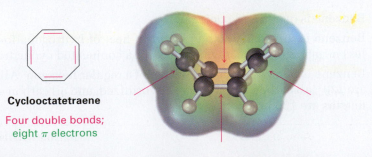

Cyclooctatetraene

Four double bonds;
eight π electrons

What's so special about $4n + 2$ π electrons? The answer comes from molecular orbital theory. When the energy levels of molecular orbitals for cyclic conjugated molecules are calculated, it turns out that there is always a single lowest-lying MO, above which the MOs come in degenerate pairs. Thus, when electrons fill the various molecular orbitals, it takes two electrons, or one pair, to fill the lowest-lying orbital and four electrons, or two pairs, to fill each of n successive energy levels—a total of $4n + 2$. Any other number would leave a bonding energy level partially unfilled.

As shown previously in Figure 9.2 for benzene, the lowest-energy MO, ψ_1, occurs singly and contains two electrons. The next two lowest-energy orbitals, ψ_2 and ψ_3, are degenerate, and it therefore takes four electrons to fill them. The result is a stable six-π-electron aromatic molecule with filled bonding orbitals.

Problem 9.5

To be aromatic, a molecule must have $4n + 2$ π electrons and must be planar for cyclic conjugation. Cyclodecapentaene fulfills one of these criteria but not the other and has resisted all attempts at synthesis. Explain.

9.4 Aromatic Ions and Aromatic Heterocycles

Look back again at the definition of aromaticity in the previous section: "... a cyclic, conjugated molecule containing $4n + 2$ π electrons." Nothing in this definition says that the number of π electrons must be the same as the number of atoms in the ring or that all the atoms in the ring must be carbon. In fact, both *ions* and *heterocyclic* compounds, which contain atoms of different elements in their ring, can also be aromatic. The cyclopentadienyl anion and the cycloheptatrienyl cation are perhaps the best known aromatic ions, while pyridine and pyrrole are common aromatic heterocycles.

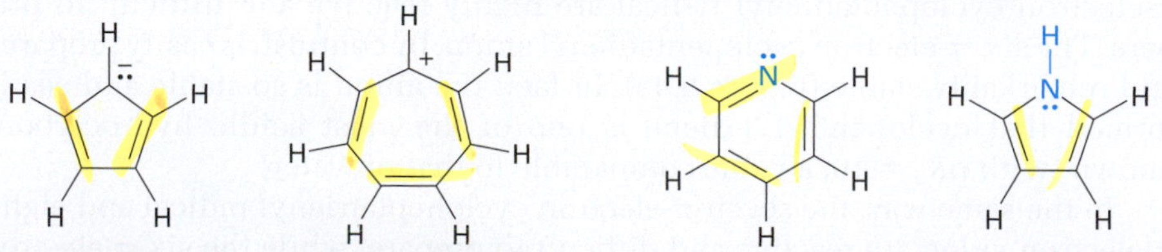

Cyclopentadienyl   Cycloheptatrienyl   Pyridine   Pyrrole
anion   cation

Aromatic ions   Aromatic heterocycles

Aromatic Ions

To see why the cyclopentadienyl anion and the cycloheptatrienyl cation are aromatic, imagine starting from the related neutral hydrocarbons, cyclopenta-1,3-diene and cyclohepta-1,3,5-triene, and removing one hydrogen from the saturated CH_2 carbon in each. If that carbon then rehybridizes from sp^3 to sp^2, the products are fully conjugated, with a p orbital on every carbon. There are three ways in which the hydrogen might be removed.

- The hydrogen can be removed with *both* electrons (H:$^-$) from the C–H bond, leaving a carbocation as product.
- The hydrogen can be removed with *one* electron (H·) from the C–H bond, leaving a carbon radical as product.
- The hydrogen can be removed with *no* electrons (H$^+$) from the C–H bond, leaving a carbon anion, or *carbanion,* as product.

All the potential products formed by removing a hydrogen from cyclopenta-1,3-diene and from cyclohepta-1,3,5-triene can be drawn with numerous resonance structures, but only the six-π-electron cyclopentadienyl anion and cycloheptatrienyl cation are predicted by the $4n + 2$ rule to be aromatic (Figure 9.3).

FIGURE 9.3 The aromatic six-π-electron cyclopentadienyl anion can be formed by removing a hydrogen ion (H$^+$) from the CH_2 group of cyclopenta-1,3-diene. Similarly, the aromatic six-π-electron cycloheptatrienyl cation can be generated by removing a hydride ion (H:$^-$) from the CH_2 group of cyclohepta-1,3,5-triene.

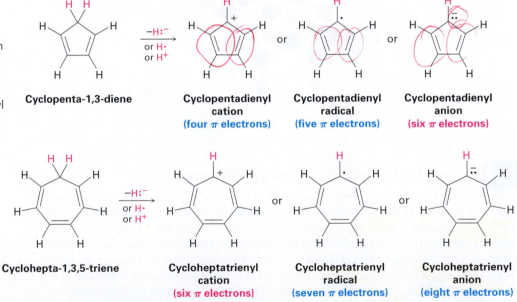

In practice, both the four-π-electron cyclopentadienyl cation and the five-π-electron cyclopentadienyl radical are highly reactive and difficult to prepare. The six-π-electron cyclopentadienyl anion, by contrast, is easily prepared and remarkably stable (Figure 9.4a). In fact, the anion is so stable and easily formed that cyclopenta-1,3-diene is one of the most acidic hydrocarbons known, with pK_a = 16, a value comparable to that of water!

In the same way, the seven-π-electron cycloheptatrienyl radical and eight-π-electron anion are reactive and difficult to prepare, while the six-π-electron cycloheptatrienyl cation is extraordinarily stable (Figure 9.4b). In fact, the cycloheptatrienyl cation was first prepared more than a century ago by reaction of cyclohepta-1,3,5-triene with Br_2, although its structure was not recognized at the time.

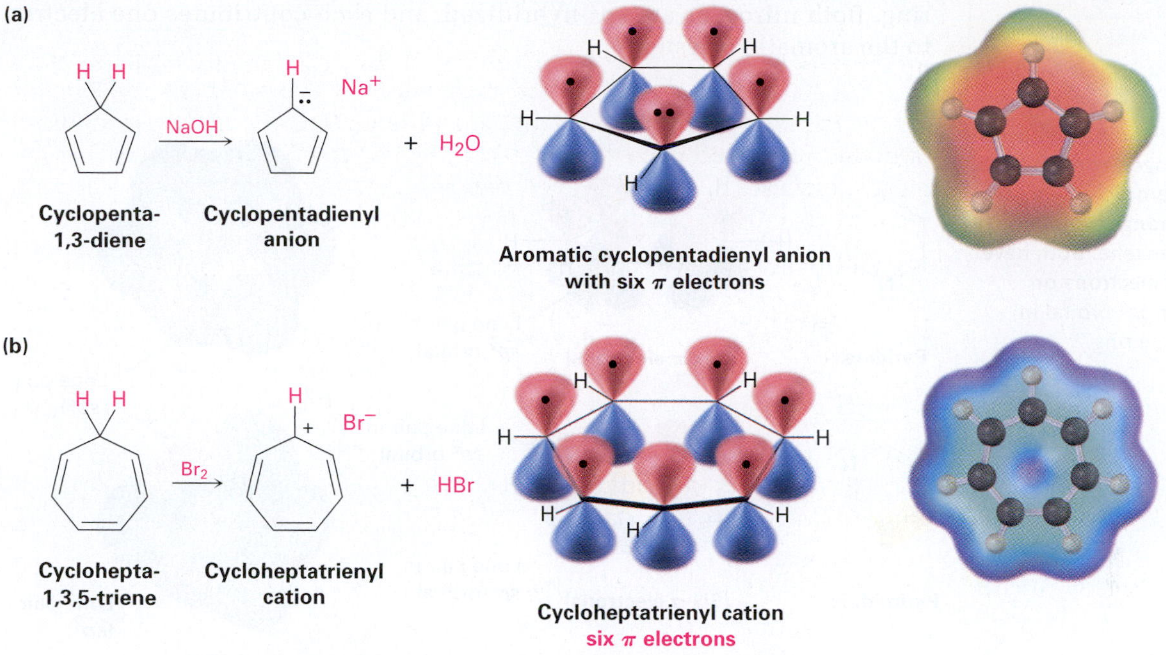

(a)

Cyclopenta-   Cyclopentadienyl
1,3-diene   anion

Aromatic cyclopentadienyl anion
with six π electrons

(b)

Cyclohepta-   Cycloheptatrienyl
1,3,5-triene   cation

Cycloheptatrienyl cation
six π electrons

FIGURE 9.4 **(a)** The aromatic cyclopentadienyl anion, showing the cyclic conjugation and six π electrons in five p orbitals, and **(b)** the aromatic cycloheptatrienyl cation, showing the cyclic conjugation and six π electrons in seven p orbitals. Electrostatic potential maps indicate that both ions are symmetrical, with the charge equally shared among all atoms in each ring.

Problem 9.6

Cycloocta-1,3,5,7-tetraene readily reacts with potassium metal to form the stable cyclooctatetraene dianion, $C_8H_8^{2-}$. Why do you suppose this reaction occurs so easily? What geometry do you think the cyclooctatetraene dianion might have?

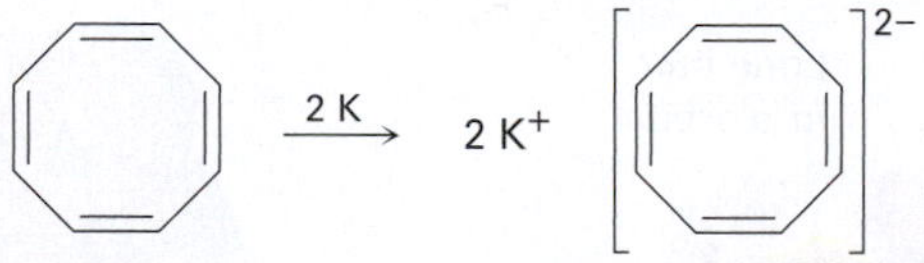

Aromatic Heterocycles

A **heterocycle**, as noted earlier in this section, is a cyclic compound that contains atoms of two or more elements in its ring, usually carbon along with nitrogen, oxygen, or sulfur. Pyridine and pyrimidine, for example, are six-membered heterocycles with nitrogen in their rings.

Pyridine is much like benzene in its π electron structure. Each of the five sp^2-hybridized carbons has a p orbital perpendicular to the plane of the ring, and each p orbital contains one π electron. The nitrogen atom is also sp^2-hybridized and has one electron in a p orbital, bringing the total to six π electrons. The nitrogen lone-pair electrons (red in an electrostatic potential map) are in an sp^2 orbital in the plane of the ring and are not part of the aromatic π system (Figure 9.5). Pyrimidine, also shown in Figure 9.5, is a benzene analog that has two nitrogen atoms in a six-membered, unsaturated ring. Both nitrogens are sp^2-hybridized, and each contributes one electron to the aromatic π system.

FIGURE 9.5 Pyridine and pyrimidine are nitrogen-containing aromatic heterocycles with π electron arrangements much like that of benzene. Both have a lone pair of electrons on nitrogen in an sp^2 orbital in the plane of the ring.

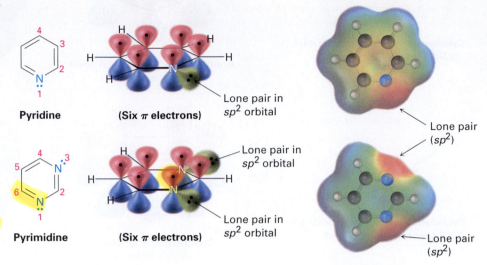

Pyrrole (spelled with two *r*'s and one *l*) and imidazole are *five*-membered heterocycles, yet both have *six* π electrons and are aromatic. In pyrrole, each of the four sp^2-hybridized carbons contributes one π electron, and the sp^2-hybridized nitrogen atom contributes the two from its lone pair, which occupies a *p* orbital (Figure 9.6). Imidazole, also shown in Figure 9.6, is an analog of pyrrole that has two nitrogen atoms in a five-membered, unsaturated ring. Both nitrogens are sp^2-hybridized, but one is in a double bond and contributes only one electron to the aromatic π system, while the other is not in a double bond and contributes two from its lone pair.

FIGURE 9.6 Pyrrole and imidazole are five-membered, nitrogen-containing heterocycles but have six π electrons and are aromatic. Both have a lone pair of electrons on nitrogen in a *p* orbital perpendicular to the ring.

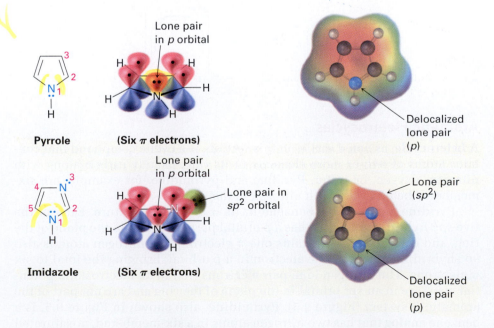

Note that nitrogen atoms have different roles depending on the structure of the molecule. The nitrogen atoms in pyridine and pyrimidine are both in double bonds and contribute only *one* π electron to the aromatic sextet, just as a carbon atom in benzene does. The nitrogen atom in pyrrole, however, is not in a double bond and contributes *two* π electrons (its lone pair) to the aromatic sextet. In imidazole, both kinds of nitrogen are present in the same molecule—a double-bonded "pyridine-like" nitrogen that contributes one π electron and a "pyrrole-like" nitrogen that contributes two.

Pyrimidine and imidazole rings are particularly important in biological chemistry. Pyrimidine, for instance, is the parent ring system in cytosine, thymine, and uracil, three of the five heterocyclic amine bases found in nucleic acids. An aromatic imidazole ring is present in histidine, 1 of the 20 amino acids found in proteins.

Cytosine
(in DNA and RNA)

Thymine
(in DNA)

Uracil
(in RNA)

Histidine
(an amino acid)

WORKED EXAMPLE 9.1 Accounting for the Aromaticity of a Heterocycle

Thiophene, a sulfur-containing heterocycle, undergoes typical aromatic substitution reactions rather than addition reactions. Why is thiophene aromatic?

Thiophene

Strategy
Recall the requirements for aromaticity—a planar, cyclic, conjugated molecule with $4n + 2$ π electrons—and see how these requirements apply to thiophene.

Solution
Thiophene is the sulfur analog of pyrrole. The sulfur atom is sp^2-hybridized and has a lone pair of electrons in a *p* orbital perpendicular to the plane of the ring. Sulfur also has a second lone pair of electrons in the ring plane.

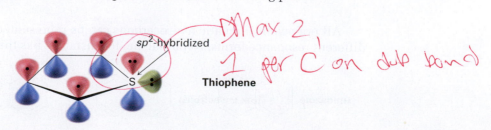

sp^2-hybridized

Thiophene

Problem 9.7

Draw an orbital picture of furan to show how the molecule is aromatic.

Furan

(handwritten: 2, 2, 2, 4n+2=6, n=1, i.A.)

Problem 9.8

Thiamin, or vitamin B$_1$, contains a positively charged five-membered nitrogen–sulfur heterocycle called a *thiazolium* ring. Explain why the thiazolium ring is aromatic.

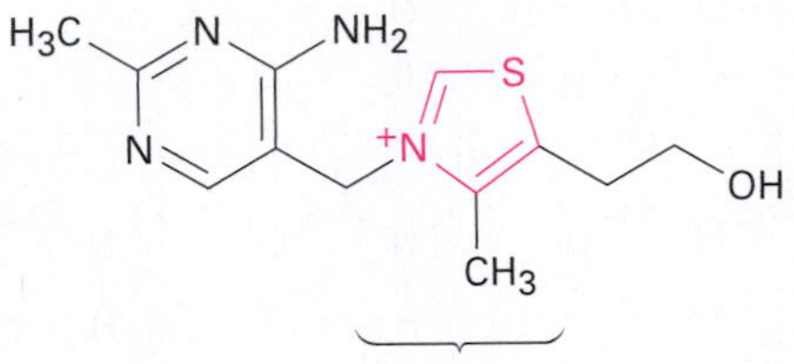

H_3C N NH_2 S

$+N$

OH **Thiamin**

CH_3

Thiazolium ring

9.5 Polycyclic Aromatic Compounds

The Hückel rule is strictly applicable only to monocyclic compounds, but the general concept of aromaticity can be extended to include *polycyclic* aromatic compounds. Naphthalene, with two benzene-like rings fused together; anthracene, with three rings; benzo[a]pyrene, with five rings; and coronene, with six rings, are all well-known aromatic hydrocarbons. Benzo[a]pyrene is particularly interesting because it is one of the cancer-causing substances found in tobacco smoke.

Naphthalene   **Anthracene**   **Benzo[a]pyrene**   **Coronene**

All polycyclic aromatic hydrocarbons can be represented by a number of different resonance forms. Naphthalene, for instance, has three:

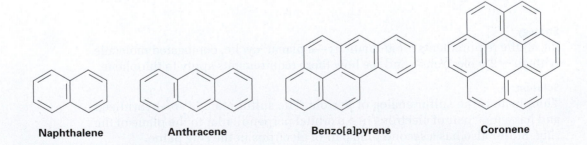

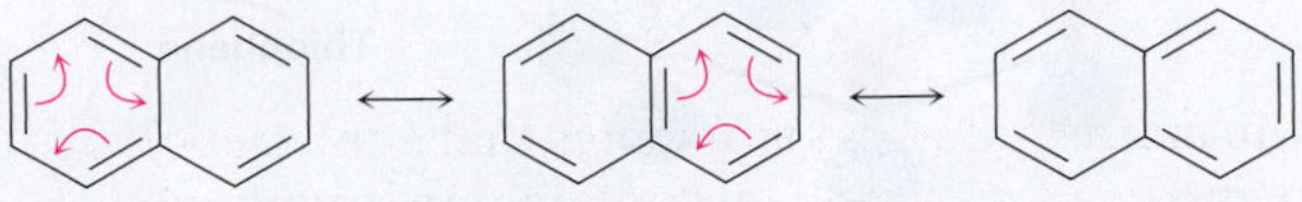

Naphthalene

Naphthalene and other polycyclic aromatic hydrocarbons show many of the chemical properties associated with aromaticity. Thus, heat of hydrogenation measurements show an aromatic stabilization energy of approximately 250 kJ/mol (60 kcal/mol). Furthermore, naphthalene reacts slowly with electrophiles such as Br_2 to give substitution products rather than double-bond addition products.

Naphthalene   → $\xrightarrow[\text{Heat}]{\text{Br}_2, \text{Fe}}$ → Br **1-Bromonaphthalene (75%)** + HBr

The aromaticity of naphthalene is explained by the orbital picture in Figure 9.7. Naphthalene has a cyclic, conjugated π electron system, with *p* orbital overlap both around the ten-carbon periphery of the molecule and across the central bond. Since 10 is a Hückel number, there is π electron delocalization and consequent aromaticity in naphthalene.

Naphthalene

FIGURE 9.7 An orbital picture and electrostatic potential map of naphthalene, showing that the ten π electrons are fully delocalized throughout both rings.

Just as there are heterocyclic analogs of benzene, there are also many heterocyclic analogs of naphthalene. Among the most common are quinoline, isoquinoline, indole, and purine. Quinoline, isoquinoline, and purine all contain pyridine-like nitrogens that are part of a double bond and contribute one electron to the aromatic π system. Indole and purine both contain pyrrole-like nitrogens that contribute two π electrons.

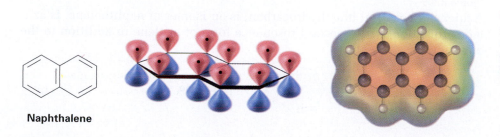

Quinoline   **Isoquinoline**   **Indole**   **Purine**

Among the many biological molecules that contain polycyclic aromatic rings, the amino acid tryptophan contains an indole ring and the antimalarial drug quinine contains a quinoline ring. Adenine and guanine, two of the five heterocyclic amine bases found in nucleic acids, have rings based on purine.

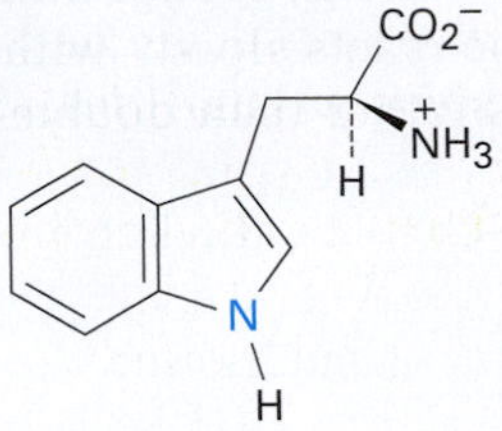

Tryptophan
(an amino acid)

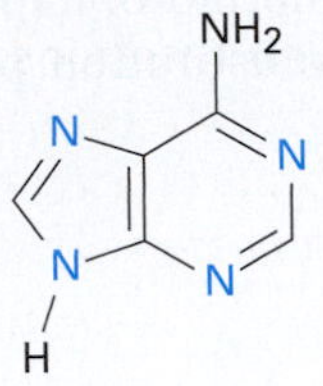

Adenine
(in DNA and RNA)

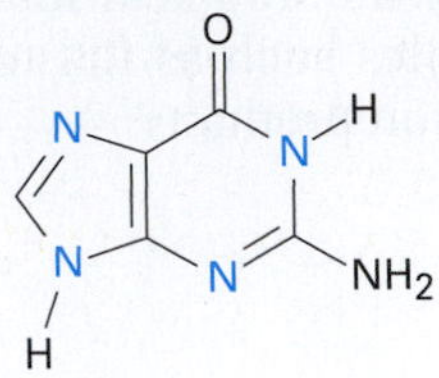

Guanine
(in DNA and RNA)

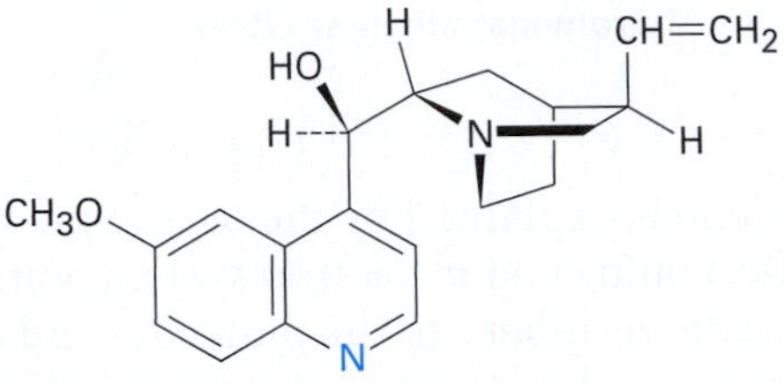

Quinine
(an antimalarial agent)

Problem 9.9

Azulene, a beautiful blue hydrocarbon, is an isomer of naphthalene. Is azulene aromatic? Draw a second resonance form of azulene in addition to the one shown.

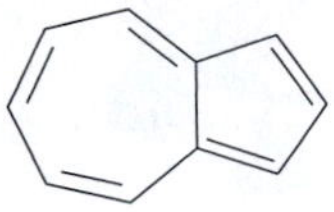

Azulene

Problem 9.10

How many electrons does each of the four nitrogen atoms in purine contribute to the aromatic π system?

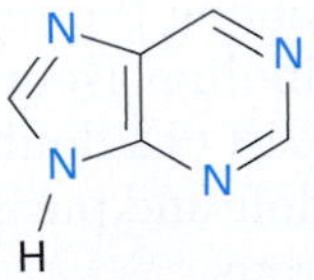

Purine

9.6 Reactions of Aromatic Compounds: Electrophilic Substitution

The most common reaction of aromatic compounds is **electrophilic aromatic substitution**, a process in which an electrophile (E^+) reacts with an aromatic ring and substitutes for one of the hydrogens:

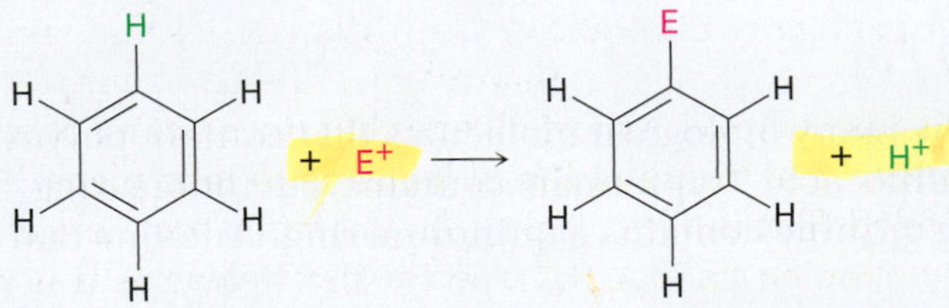

$$+ \ E^+ \longrightarrow \ + \ H^+$$

The reaction is characteristic of all aromatic rings, not just benzene and substituted benzenes. In fact, the ability of a compound to undergo electrophilic substitution is a good test of aromaticity.

Many different substituents can be introduced onto an aromatic ring through electrophilic substitution reactions. To list some possibilities, an aromatic ring can be substituted by a halogen (–Cl, –Br, –I), a nitro group (–NO$_2$), a sulfonic acid group (–SO$_3$H), a hydroxyl group (–OH), an alkyl group (–R), or an acyl group (–COR). Starting from only a few simple materials, it's possible to prepare many thousands of substituted aromatic compounds.

Halogenation   **Acylation**

Nitration   **Alkylation**

Aromatic ring

Sulfonation   **Hydroxylation**

Before seeing how electrophilic aromatic substitutions occur, let's briefly recall what we said in Chapter 6 about electrophilic alkene additions. When a reagent such as HCl adds to an alkene, the electrophilic hydrogen approaches the π electrons of the double bond and forms a bond to one carbon, leaving a positive charge at the other carbon. This carbocation intermediate then reacts with the nucleophilic Cl$^-$ ion to yield the addition product.

Alkene   **Carbocation intermediate**   **Addition product**

An electrophilic aromatic substitution reaction begins in a similar way, but there are a number of differences. One difference is that aromatic rings are less reactive toward electrophiles than alkenes are. For example, Br$_2$ in CH$_2$Cl$_2$ solution reacts instantly with most alkenes but does not react with benzene at room temperature. For bromination of benzene to take place, a catalyst such as FeBr$_3$ is needed. The catalyst makes the Br$_2$ molecule more electrophilic by polarizing it to give an FeBr$_4^-$ Br$^+$ species that reacts as if it were Br$^+$. The polarized Br_2 molecule then reacts with the nucleophilic benzene ring to yield a nonaromatic carbocation intermediate. This carbocation is doubly allylic (Section 8.12) and has three resonance forms:

$$Br—Br \;+\; FeBr_3 \;\longrightarrow\; Br^+ \; ^-FeBr_4$$

Although more stable than a typical alkyl carbocation because of resonance, the intermediate in electrophilic aromatic substitution is nevertheless much less stable than the starting benzene ring itself, with its 150 kJ/mol of aromatic stability. Thus, the reaction of an electrophile with a benzene ring is endergonic, has a substantial activation energy, and is rather slow.

Another difference between alkene addition and aromatic substitution occurs after the carbocation intermediate has formed. Instead of adding Br^- to give an addition product, the carbocation intermediate loses H^+ from the bromine-bearing carbon to give a substitution product. The net effect is the substitution of H^+ by Br^+ by the overall mechanism shown in Figure 9.8.

FIGURE 9.8 MECHANISM: The mechanism of the electrophilic bromination of benzene. The reaction occurs in two steps and involves a resonance-stabilized carbocation intermediate.

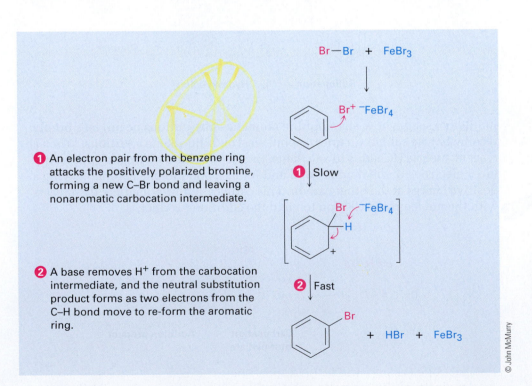

① An electron pair from the benzene ring attacks the positively polarized bromine, forming a new C–Br bond and leaving a nonaromatic carbocation intermediate.

② A base removes H^+ from the carbocation intermediate, and the neutral substitution product forms as two electrons from the C–H bond move to re-form the aromatic ring.

© John McMurry

Why does the reaction of Br_2 with benzene take a different course than its reaction with an alkene? The answer is straightforward. If addition occurred, the 150 kJ/mol stabilization energy of the aromatic ring would be lost and the overall reaction would be endergonic. When substitution occurs, though, the stability of the aromatic ring is retained and the reaction is exergonic.

There are many other kinds of electrophilic aromatic substitutions besides bromination, and all are thought to occur by the same general mechanism. Let's look at some of these other reactions briefly.

Aromatic Halogenation

Chlorine, bromine, and iodine can be introduced into aromatic rings by electrophilic substitution reactions, but fluorine is too reactive and only poor yields of monofluoroaromatic products are obtained by direct fluorination. Aromatic rings react with Cl_2 in the presence of $FeCl_3$ catalyst to yield chlorobenzenes, just as they react with Br_2 and $FeBr_3$. This kind of reaction is used in the synthesis of numerous pharmaceutical agents, including the anti-anxiety agent diazepam, marketed as Valium.

Benzene   + Cl_2 $\xrightarrow{\text{FeCl}_3\text{ catalyst}}$ Chlorobenzene (86%)   + HCl

Diazepam

Iodine itself is unreactive toward aromatic rings, so an oxidizing agent such as hydrogen peroxide or a copper salt such as $CuCl_2$ must be added to the reaction. These substances accelerate the iodination reaction by oxidizing I_2 to a more powerful electrophilic species that reacts as if it were I^+. The aromatic ring then reacts with I^+ in the typical way, yielding a substitution product.

$$I_2 + 2\,Cu^{2+} \longrightarrow 2\,I^+ + 2\,Cu^+$$

Benzene $\xrightarrow{I_2 + CuCl_2}$ Iodobenzene (65%)

Electrophilic aromatic halogenations occur in the biosynthesis of numerous naturally occurring molecules, particularly those produced by marine organisms. In humans, the best-known example occurs in the thyroid gland during the biosynthesis of thyroxine, a thyroid hormone involved in regulating growth and metabolism. The amino acid tyrosine is first iodinated by thyroid peroxidase, and two of the iodinated tyrosine molecules then couple. The electrophilic iodinating agent is an I^+ species, perhaps hypoiodous acid (HIO), that is formed from iodide ion by oxidation with H_2O_2.

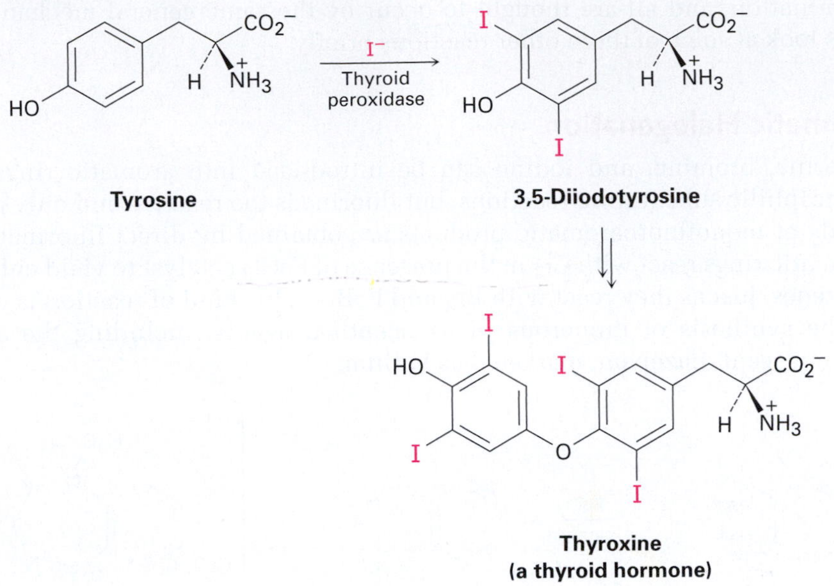

Tyrosine

3,5-Diiodotyrosine

Thyroxine
(a thyroid hormone)

Aromatic Nitration

Aromatic rings can be nitrated by reaction with a mixture of concentrated nitric and sulfuric acids. The electrophile is the nitronium ion, NO_2^+, which is generated from HNO_3 by protonation and loss of water. The nitronium ion reacts with benzene to yield a carbocation intermediate, and loss of H^+ from this intermediate gives the neutral substitution product, nitrobenzene (Figure 9.9).

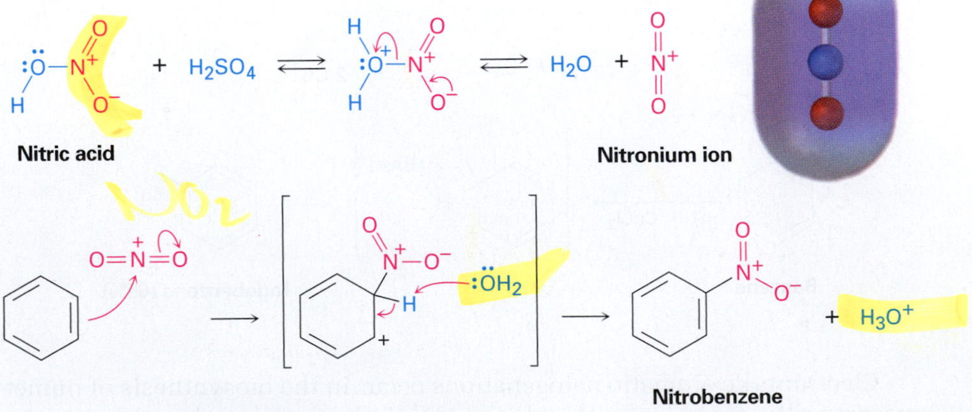

FIGURE 9.9 The mechanism of electrophilic nitration of an aromatic ring. An electrostatic potential map of the reactive electrophile NO_2^+ shows that the nitrogen atom is most positive (blue).

Nitric acid

Nitronium ion

Nitrobenzene

Nitration of an aromatic ring does not occur in nature but is particularly important in the laboratory because the nitro-substituted product can be reduced by reagents such as iron or tin metal or to yield an *arylamine*, $ArNH_2$, such as aniline. Attachment of an amino group to an aromatic ring by the two-step nitration–reduction sequence is a key part of the industrial synthesis of many dyes and pharmaceutical agents.

Nitrobenzene

1. Fe, H_3O^+
2. HO^-

Aniline (95%)

Aromatic Sulfonation

Aromatic rings can be sulfonated in the laboratory by reaction with fuming sulfuric acid, a mixture of H_2SO_4 and SO_3. The reactive electrophile is either HSO_3^+ or neutral SO_3, depending on reaction conditions, and substitution occurs by the same two-step mechanism seen previously for bromination and nitration (Figure 9.10).

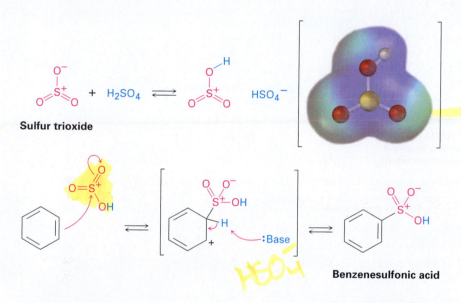

FIGURE 9.10 The mechanism of electrophilic sulfonation of an aromatic ring. An electrostatic potential map of the reactive electrophile $HOSO_2^+$ shows that sulfur and hydrogen are the most positive atoms (blue).

Sulfur trioxide   + H_2SO_4 ⇌ HSO_4^-

Benzenesulfonic acid

Like nitration, aromatic sulfonation does not occur naturally but is widely used in the preparation of dyes and pharmaceutical agents. For example, the sulfa drugs, such as sulfanilamide, were among the first clinically useful antibiotics. Although largely replaced today by more effective agents, sulfa drugs are still used in the treatment of meningitis and urinary tract infections. These drugs are prepared commercially by a process that involves aromatic sulfonation as the key step.

Sulfanilamide (an antibiotic)

Aromatic Hydroxylation

Direct hydroxylation of an aromatic ring to yield a hydroxybenzene (a *phenol*) is difficult and rarely done in the laboratory, but it occurs much more frequently in biological pathways. An example is the hydroxylation of *p*-hydroxyphenylacetate to give 3,4-dihydroxyphenylacetate. The reaction is catalyzed by *p*-hydroxyphenylacetate-3-hydroxylase and requires molecular oxygen plus the coenzyme reduced flavin adenine dinucleotide, abbreviated $FADH_2$.

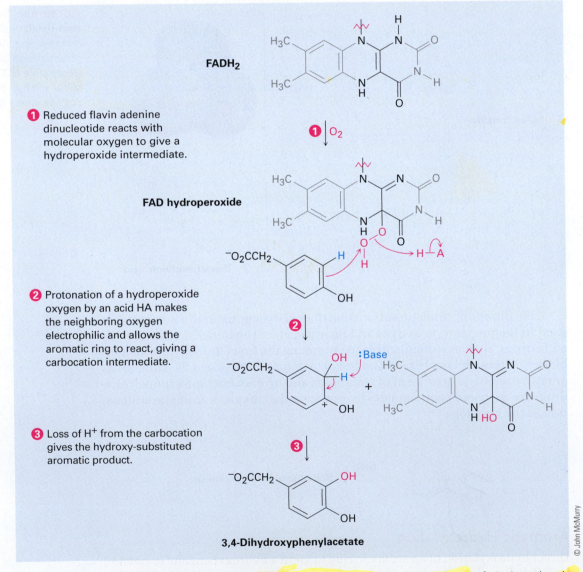

p-Hydroxyphenylacetate   **3,4-Dihydroxyphenylacetate**

By analogy with other electrophilic aromatic substitutions, you might expect that an electrophilic oxygen species acting as an "OH^+ equivalent" is needed for the hydroxylation reaction. That is exactly what happens, with the electrophilic oxygen arising by protonation of FAD hydroperoxide, RO–OH (Figure 9.11); that is, $RO–OH + H^+ \rightarrow ROH + OH^+$. The FAD hydroperoxide is itself formed by reaction of $FADH_2$ with O_2.

① Reduced flavin adenine dinucleotide reacts with molecular oxygen to give a hydroperoxide intermediate.

② Protonation of a hydroperoxide oxygen by an acid HA makes the neighboring oxygen electrophilic and allows the aromatic ring to react, giving a carbocation intermediate.

③ Loss of H^+ from the carbocation gives the hydroxy-substituted aromatic product.

FIGURE 9.11 MECHANISM: Mechanism of the electrophilic hydroxylation of *p*-hydroxyphenyl-acetate, by reaction with FAD hydroperoxide. The hydroxylating species is an "OH^+ equivalent" that arises by protonation of FAD hydroperoxide, $RO–OH + H^+ \rightarrow ROH + OH^+$.

Problem 9.11

Monobromination of toluene gives a mixture of three bromotoluene products. Draw and name them.

Problem 9.12

How many products might be formed on chlorination of *o*-xylene (*o*-dimethyl-benzene), *m*-xylene, and *p*-xylene?

Problem 9.13

When benzene is treated with D_2SO_4, deuterium slowly replaces all six hydrogens in the aromatic ring. Explain.

9.7 Alkylation and Acylation of Aromatic Rings: The Friedel–Crafts Reaction

Among the most useful electrophilic aromatic substitution reactions in the laboratory is **alkylation**—the introduction of an alkyl group onto the benzene ring. Called the **Friedel–Crafts reaction** after its discoverers, the reaction is carried out by treating the aromatic compound with an alkyl chloride, RCl, in the presence of $AlCl_3$ to generate a carbocation electrophile, R^+. Aluminum chloride catalyzes the reaction by helping the alkyl halide to dissociate in much the same way that $FeBr_3$ catalyzes aromatic brominations by polarizing Br_2 (Section 9.6). Loss of H^+ then completes the reaction (Figure 9.12).

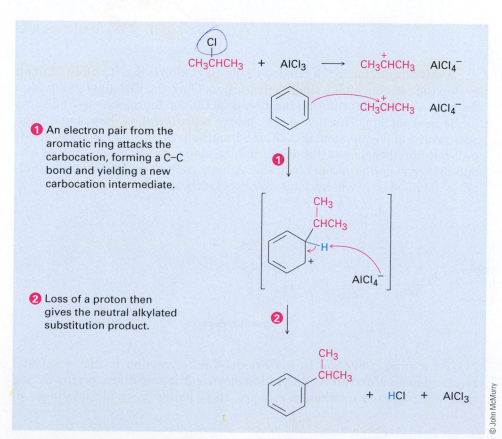

① An electron pair from the aromatic ring attacks the carbocation, forming a C–C bond and yielding a new carbocation intermediate.

② Loss of a proton then gives the neutral alkylated substitution product.

FIGURE 9.12 MECHANISM: Mechanism of the Friedel–Crafts alkylation reaction. The electrophile is a carbocation, generated by $AlCl_3$-assisted dissociation of an alkyl halide.

© John McMurry

Despite its utility, the Friedel–Crafts alkylation has several limitations. For one thing, only *alkyl* halides can be used. Aromatic (aryl) halides and vinylic halides do not react because aryl and vinylic carbocations are too high in energy to form under Friedel–Crafts conditions. (The word *vinylic* means that a substituent is attached directly to a double bond, C=C–Cl.)

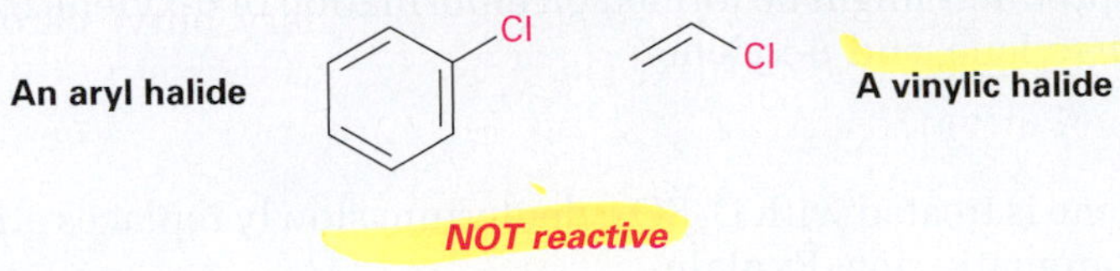

An aryl halide　　　　　　　　　**A vinylic halide**

NOT reactive

Another limitation is that Friedel–Crafts reactions don't succeed on aromatic rings that are substituted either by a strongly electron-withdrawing group such as carbonyl (C=O) or by an amino group (–NH₂, –NHR, –NR₂). We'll see in the next section that the presence of a substituent group already on a ring can have a dramatic effect on that ring's subsequent reactivity toward further electrophilic substitution. Rings that contain any of the substituents listed in Figure 9.13 do not undergo Friedel–Crafts alkylation.

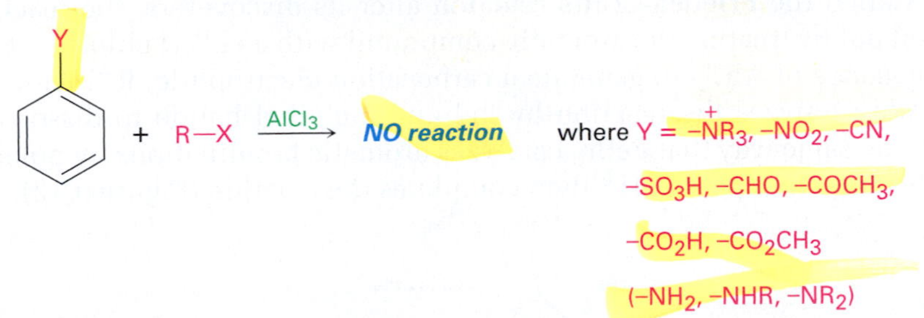

FIGURE 9.13 Limitations on the aromatic substrate in Friedel–Crafts reactions. No reaction occurs if the substrate has either an electron-withdrawing substituent or an amino group.

$$Y\text{-}C_6H_5 + R-X \xrightarrow{\text{AlCl}_3} \textbf{NO reaction}$$

where $Y = -\overset{+}{N}R_3, -NO_2, -CN,$
$-SO_3H, -CHO, -COCH_3,$
$-CO_2H, -CO_2CH_3$
$(-NH_2, -NHR, -NR_2)$

A third limitation to the Friedel–Crafts alkylation is that it's often difficult to stop the reaction after a single substitution. Once the first alkyl group is on the ring, a second substitution reaction is facilitated for reasons we'll discuss in the next section. Thus, we often observe *polyalkylation.* Reaction of benzene with 1 mol equivalent of 2-chloro-2-methylpropane, for example, yields *p*-di-*tert*-butylbenzene as the major product, along with small amounts of *tert*-butylbenzene and unreacted benzene. A high yield of monoalkylation product is obtained only when a large excess of benzene is used.

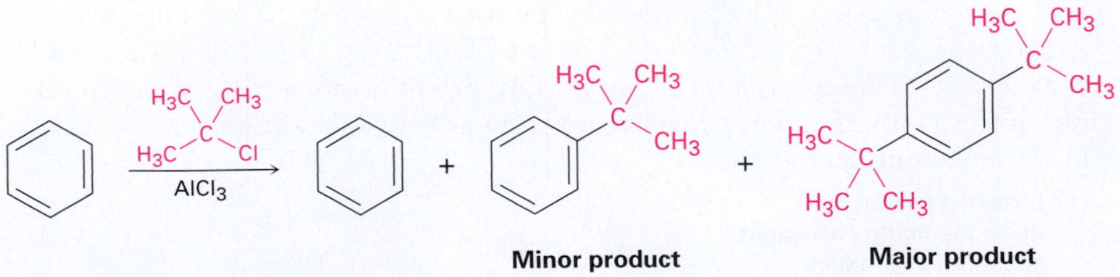

Minor product　　　　　　　**Major product**

Yet a final limitation to the Friedel–Crafts reaction is that a skeletal rearrangement of the alkyl carbocation electrophile sometimes occurs during reaction, particularly when a primary alkyl halide is used. Treatment of benzene with 1-chlorobutane at 0 °C, for instance, gives an approximately 2:1 ratio of rearranged (*sec*-butyl) to unrearranged (butyl) products.

The carbocation rearrangements that accompany Friedel–Crafts reactions are like those that accompany electrophilic additions to alkenes (Section 7.10) and occur either by hydride shift or alkyl shift. For example, the relatively unstable primary butyl carbocation produced by reaction of 1-chlorobutane with AlCl$_3$ rearranges to the more stable secondary butyl carbocation by shift of a hydrogen atom and its electron pair (a hydride ion, H:$^-$) from C2 to C1. Similarly, alkylation of benzene with 1-chloro-2,2-dimethylpropane yields (1,1-dimethylpropyl)benzene. The initially formed primary carbocation rearranges to a tertiary carbocation by shift of a methyl group and its electron pair from C2 to C1.

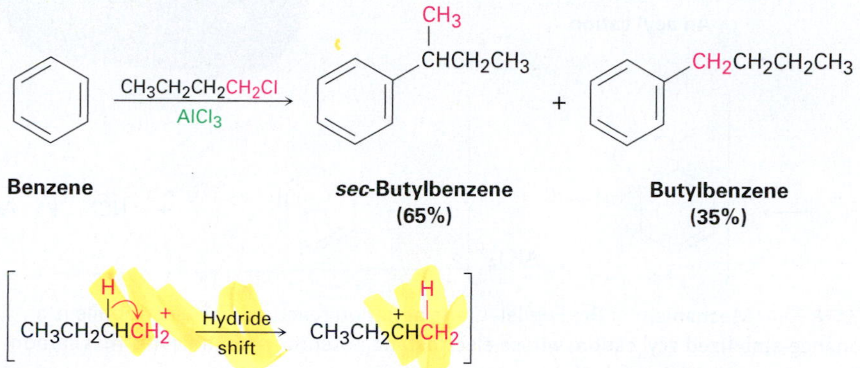

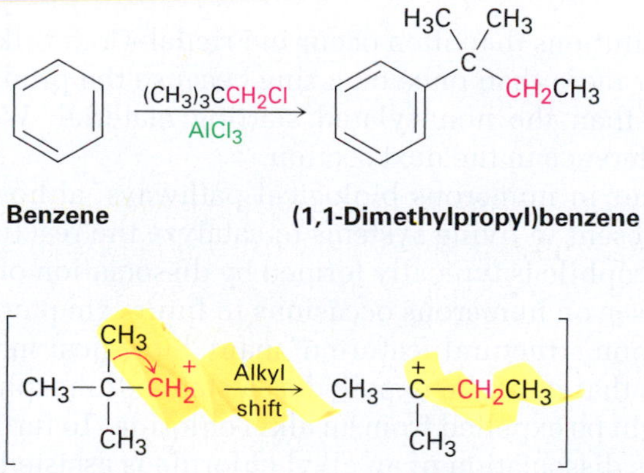

Benzene

sec-Butylbenzene
(65%)

Butylbenzene
(35%)

Benzene

(1,1-Dimethylpropyl)benzene

Just as an aromatic ring is alkylated by reaction with an alkyl chloride, it is **acylated** by reaction with a carboxylic acid chloride, RCOCl, in the presence of AlCl$_3$. That is, an **acyl group** (–COR; pronounced **a**-sil) is substituted onto the aromatic ring. For example, reaction of benzene with acetyl chloride yields the ketone acetophenone.

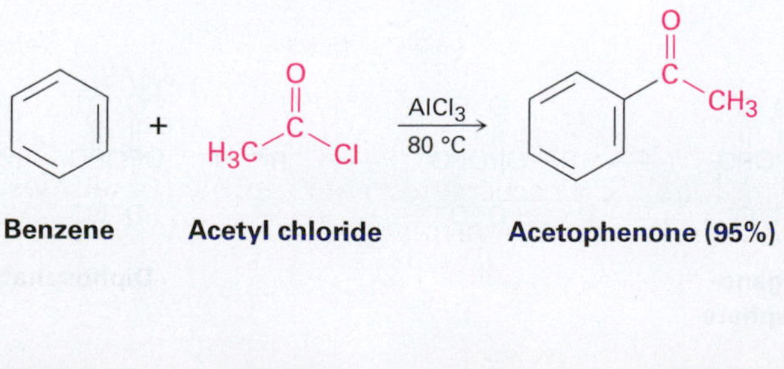

Benzene   **Acetyl chloride**   **Acetophenone (95%)**

The mechanism of the Friedel–Crafts acylation reaction is similar to that of Friedel–Crafts alkylation, and the same limitations on the aromatic substrate noted previously in Figure 9.13 for alkylation also apply to acylation. The reactive electrophile is a resonance-stabilized acyl cation, generated by reaction between the acid chloride and $AlCl_3$ (Figure 9.14). As the resonance structures in the figure indicate, an acyl cation is stabilized by interaction of the vacant orbital on carbon with lone-pair electrons on the neighboring oxygen. Because of this stabilization, no carbocation rearrangement occurs during acylation.

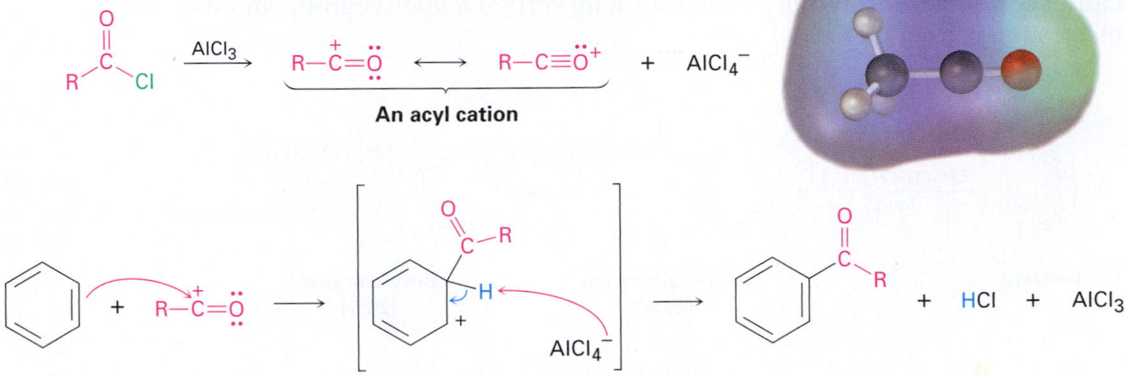

FIGURE 9.14 Mechanism of the Friedel–Crafts acylation reaction. The electrophile is a resonance-stabilized acyl cation, whose electrostatic potential map indicates that carbon is the most positive atom (blue).

Unlike the multiple substitutions that often occur in Friedel–Crafts alkylations, acylations never occur more than once on a ring because the product acylbenzene is less reactive than the nonacylated starting material. We'll account for this reactivity difference in the next section.

Aromatic alkylations occur in numerous biological pathways, although there is of course no $AlCl_3$ present in living systems to catalyze the reaction. Instead, the carbocation electrophile is typically formed by dissociation of an organodiphosphate. As we'll see on numerous occasions in future chapters, a diphosphate group is a common structural feature of many biological molecules. Among its functions is that it can be expelled as a stable diphosphate ion, much as chloride ion might be expelled from an alkyl chloride. To further strengthen the analogy, just as dissociation of an alkyl chloride is assisted by $AlCl_3$ in the Friedel–Crafts reaction, the dissociation of an organodiphosphate in a biological reaction is typically assisted by complexation to a divalent metal cation such as Mg^{2+} to help neutralize charge.

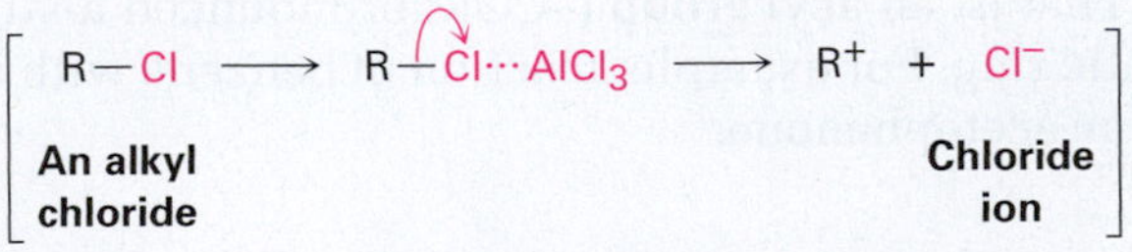

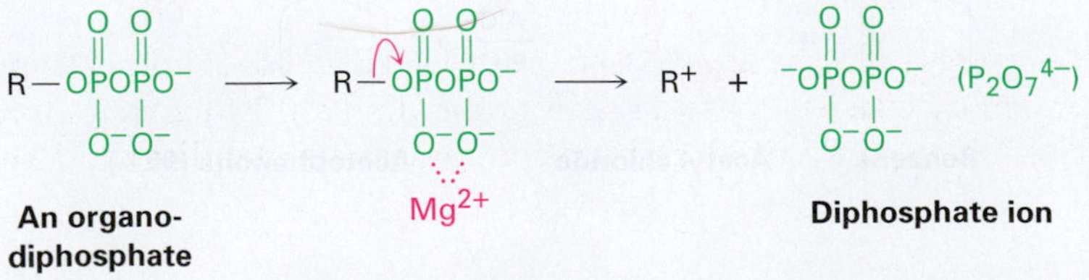

An example of a biological electrophilic aromatic substitution occurs during the biosynthesis of phylloquinone, or vitamin K_1, the human blood-clotting factor. Phylloquinone is formed by reaction of 1,4-dihydroxy-naphthoic acid with phytyl diphosphate. Phytyl diphosphate first dissociates to a resonance-stabilized allylic carbocation, which then substitutes onto the aromatic ring in the typical way. Several further transformations lead to phylloquinone (Figure 9.15).

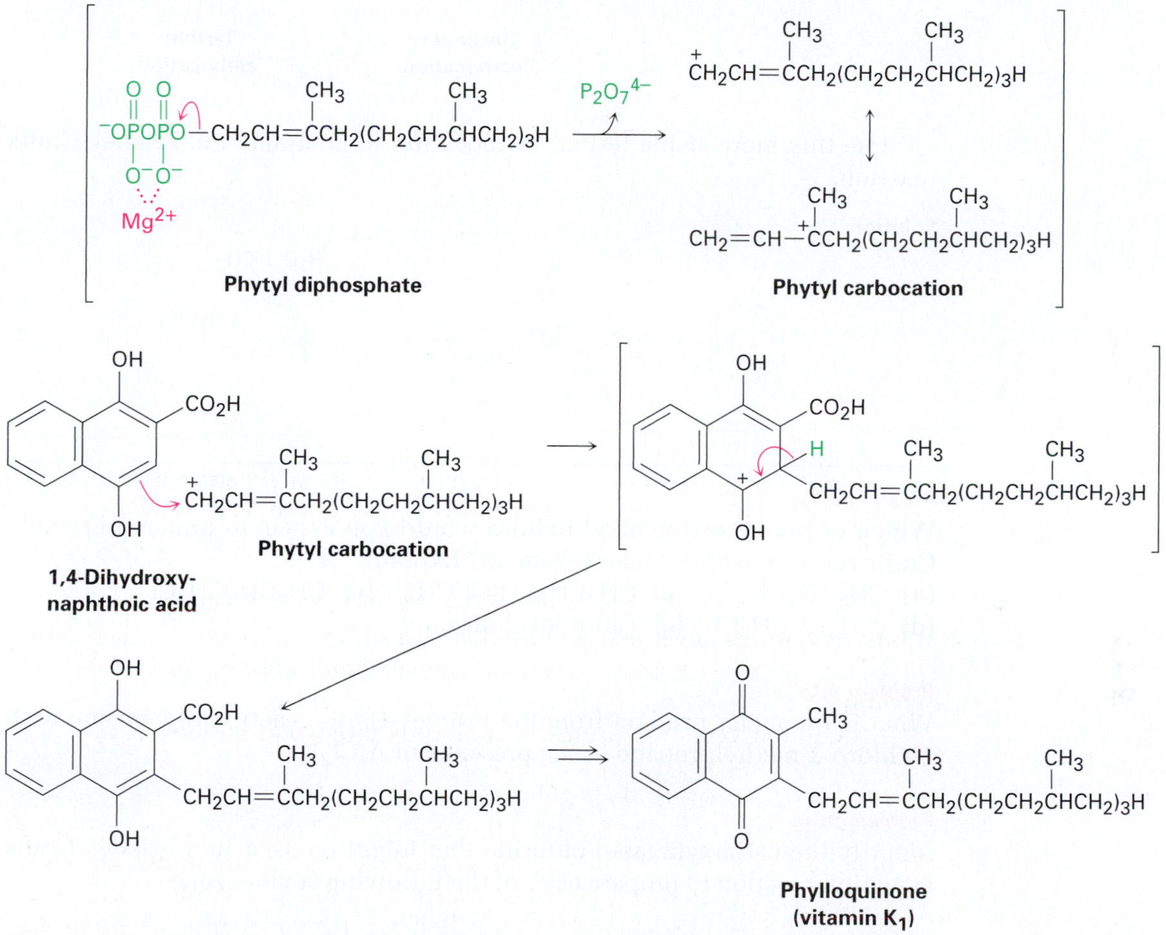

FIGURE 9.15 Biosynthesis of phylloquinone (vitamin K_1) from 1,4-dihydroxynaphthoic acid. The key step that joins the 20-carbon phytyl side chain to the aromatic ring is an electrophilic substitution reaction.

WORKED EXAMPLE 9.2 Predicting the Product of a Carbocation Rearrangement

The Friedel–Crafts reaction of benzene with 2-chloro-3-methylbutane in the presence of $AlCl_3$ occurs with a carbocation rearrangement. What is the structure of the product?

Strategy

A Friedel–Crafts reaction involves initial formation of a carbocation, which can rearrange by either a hydride shift or an alkyl shift to give a more stable carbocation. Draw the initial carbocation, assess its stability, and see if the shift of a hydride ion or an alkyl group from a neighboring carbon will result in increased stability. In the present instance, the initial carbocation is a secondary one that can rearrange to a more stable tertiary one by a hydride shift:

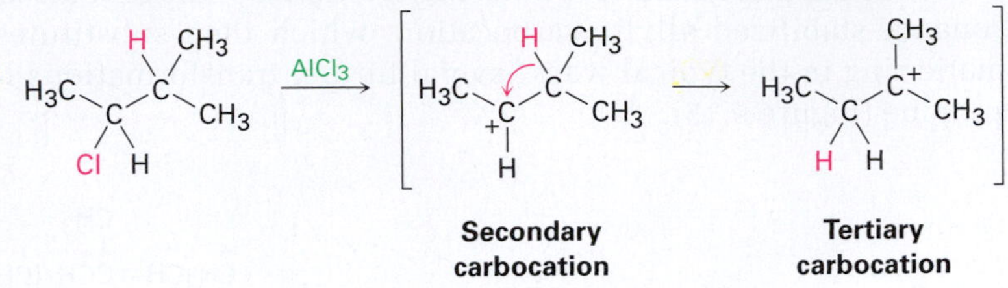

Secondary carbocation   **Tertiary carbocation**

Use this more stable tertiary carbocation to complete the Friedel–Crafts reaction.

Solution

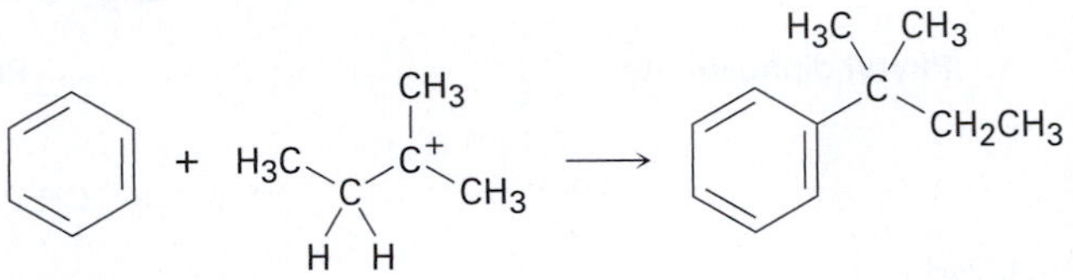

Problem 9.14

Which of the following alkyl halides would you expect to undergo Friedel–Crafts reaction *without* rearrangement? Explain.
(a) CH_3CH_2Cl　　(b) $CH_3CH_2CH(Cl)CH_3$　(c) $CH_3CH_2CH_2Cl$
(d) $(CH_3)_3CCH_2Cl$　(e) Chlorocyclohexane

Problem 9.15

What is the major product from the Friedel–Crafts reaction of benzene with 1-chloro-2-methylpropane in the presence of $AlCl_3$?

Problem 9.16

Identify the carboxylic acid chloride that might be used in a Friedel–Crafts acylation reaction to prepare each of the following acylbenzenes:

(a)   (b)

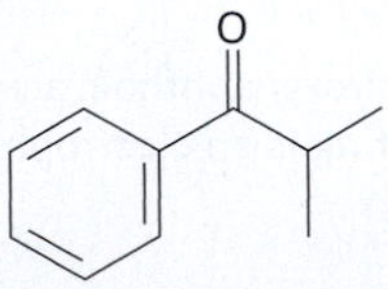

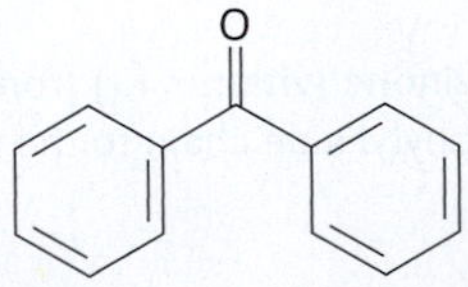

9.8 Substituent Effects in Electrophilic Substitutions

Only one product can form when an electrophilic substitution occurs on benzene, but what would happen if we were to carry out a reaction on an aromatic ring that already has a substituent? A substituent already present on the ring has two effects:

- **Substituents affect the *reactivity* of the aromatic ring.** Some substituents activate the ring, making it more reactive than benzene, and some deactivate the ring, making it less reactive than benzene. In aromatic nitration, for instance, the presence of an –OH substituent makes the ring 1000 times more reactive than benzene, while an –NO$_2$ substituent makes the ring more than 10 million times less reactive.

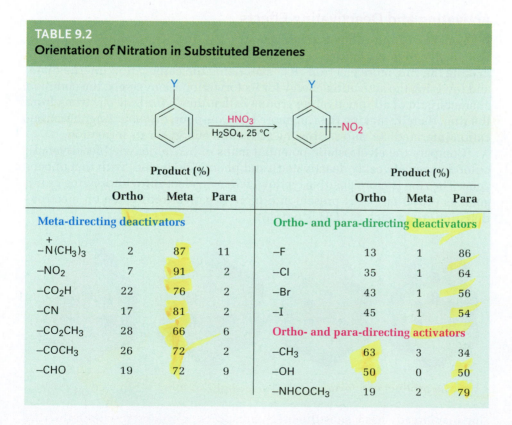

Relative rate of nitration	6×10^{-8}	0.033	1	1000

Reactivity

- **Substituents affect the *orientation* of the reaction.** The three possible disubstituted products—ortho, meta, and para—are usually not formed in equal amounts. Instead, the nature of the substituent already present on the benzene ring determines the position of the second substitution. An –OH group directs substitution toward the ortho and para positions, for instance, while a carbonyl group such as –CHO directs substitution primarily toward the meta position. Table 9.2 lists experimental results for the nitration of substituted benzenes.

TABLE 9.2
Orientation of Nitration in Substituted Benzenes

	Product (%)				Product (%)		
	Ortho	Meta	Para		Ortho	Meta	Para
Meta-directing deactivators				**Ortho- and para-directing deactivators**			
$-\overset{+}{N}(CH_3)_3$	2	87	11	–F	13	1	86
–NO$_2$	7	91	2	–Cl	35	1	64
–CO$_2$H	22	76	2	–Br	43	1	56
–CN	17	81	2	–I	45	1	54
–CO$_2$CH$_3$	28	66	6	**Ortho- and para-directing activators**			
–COCH$_3$	26	72	2	–CH$_3$	63	3	34
–CHO	19	72	9	–OH	50	0	50
				–NHCOCH$_3$	19	2	79

Substituents can be classified into three groups, as shown in Figure 9.16: *meta-directing deactivators*, *ortho- and para-directing deactivators*, and *ortho- and para-directing activators*. There are no meta-directing activators.

Notice how the directing effect of a group correlates with its reactivity. All meta-directing groups are strongly deactivating, and most ortho- and para-directing groups are activating. The halogens are unique in being ortho- and para-directing but weakly deactivating.

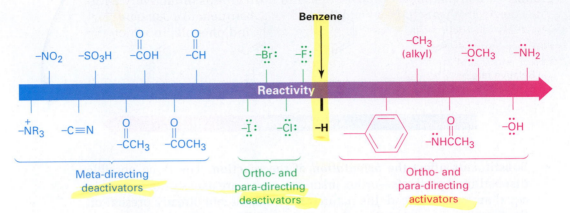

FIGURE 9.16 Classification of substituent effects in electrophilic aromatic substitution. All activating groups are ortho- and para-directing, and all deactivating groups other than halogen are meta-directing. The halogens are unique in being deactivating but ortho- and para-directing.

Activating and Deactivating Effects

What makes a group either activating or deactivating? The common characteristic of all activating groups is that they *donate* electrons to the ring, thereby making the ring more electron-rich, stabilizing the carbocation intermediate, and lowering the activation energy for its formation. Conversely, the common characteristic of all deactivating groups is that they *withdraw* electrons from the ring, thereby making the ring more electron-poor, destabilizing the carbocation intermediate, and raising the activation energy for its formation.

Compare the electrostatic potential maps of benzaldehyde (deactivated), chlorobenzene (weakly deactivated), and phenol (activated) with that of benzene. The ring is more positive (yellow) when an electron-withdrawing group such as –CHO or –Cl is present and more negative (red) when an electron-donating group such as –OH is present.

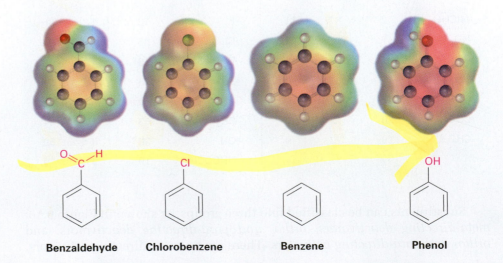

Benzaldehyde   Chlorobenzene   Benzene   Phenol

The withdrawal or donation of electrons by a substituent group is controlled by an interplay of inductive effects and resonance effects. Recall from Section 2.1 that an *inductive effect* is the withdrawal or donation of electrons through a σ bond due to electronegativity. Halogens, hydroxyl groups, carbonyl groups, cyano groups, and nitro groups inductively withdraw electrons through the σ bond linking the substituent to a benzene ring. The effect is most pronounced in halobenzenes and phenols, in which the electronegative atom is directly attached to the ring, but is also significant in carbonyl compounds, nitriles, and nitro compounds, in which the electronegative atom is farther removed. Alkyl groups, on the other hand, inductively donate electrons. This is the same hyperconjugative donating effect that causes alkyl substituents to stabilize alkenes (Section 7.5) and carbocations (Section 7.8).

Inductive electron withdrawal

Inductive electron donation

A *resonance effect* is the withdrawal or donation of electrons through a π bond due to the overlap of a p orbital on the substituent with a p orbital on the aromatic ring. Carbonyl, cyano, and nitro substituents, for example, withdraw electrons from the aromatic ring by resonance. The π electrons flow from the rings to the substituents, leaving a positive charge in the ring. Note that substituents with an electron-withdrawing resonance effect have the general structure –Y=Z, where the Z atom is more electronegative than Y.

Conversely, halogen, hydroxyl, alkoxyl (–OR), and amino substituents donate electrons to the aromatic ring by resonance. Lone-pair electrons flow from the substituents to the ring, placing a negative charge in the ring. Substituents with an electron-donating resonance effect have the general structure –Ÿ, where the Y atom has a lone pair of electrons available for donation to the ring.

Resonance electron-withdrawing groups

Resonance electron-donating groups

One further point: inductive effects and resonance effects don't necessarily act in the same direction. Halogen, hydroxyl, alkoxyl, and amino substituents, for instance, have electron-*withdrawing* inductive effects because of the electronegativity of the –X, –O, or –N atom bonded to the aromatic ring but have electron-*donating* resonance effects because of the lone-pair electrons on those –X, –O, or –N atoms. When the two effects act in opposite directions, the stronger effect dominates. Thus, hydroxyl, alkoxyl, and amino substituents are activators because their stronger electron-donating resonance effect outweighs their weaker electron-withdrawing inductive effect. Halogens, however, are deactivators because their stronger electron-withdrawing inductive effect outweighs their weaker electron-donating resonance effect.

Problem 9.17

Rank the compounds in each group in order of their reactivity to electrophilic substitution:

(a) Nitrobenzene, phenol, toluene, benzene
(b) Phenol, benzene, chlorobenzene, benzoic acid
(c) Benzene, bromobenzene, benzaldehyde, aniline

Problem 9.18

Use Figure 9.16 to explain why Friedel–Crafts alkylations often give polysubstitution products but Friedel–Crafts acylations do not.

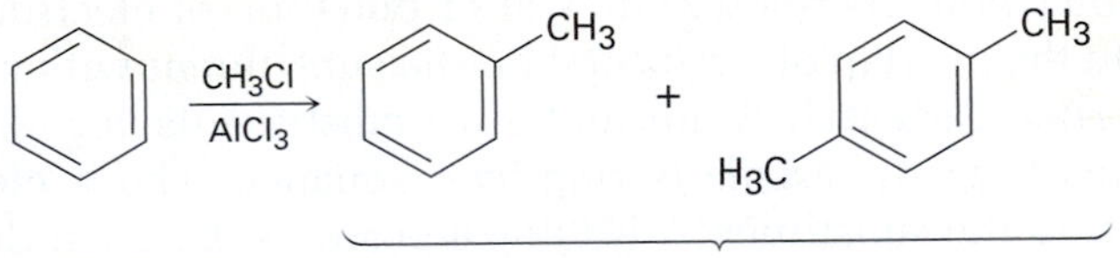

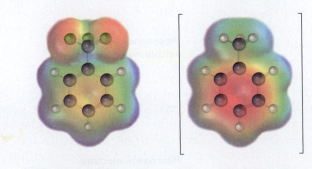

(Product mixture)

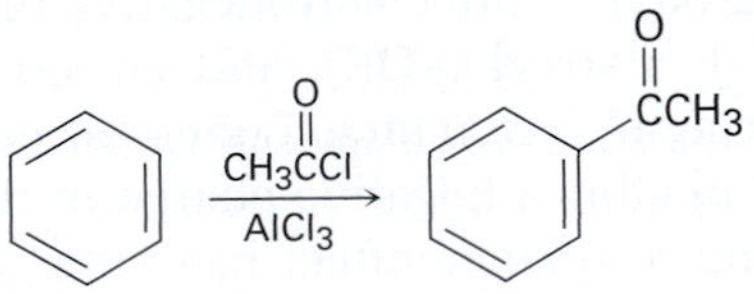

(Sole product)

Problem 9.19

An electrostatic potential map of (trifluoromethyl)benzene, $C_6H_5CF_3$, is shown. Would you expect (trifluoromethyl)benzene to be more reactive or less reactive than toluene toward electrophilic substitution? Explain.

(Trifluoromethyl)benzene   **Toluene**

Orienting Effects: Ortho and Para Directors

Inductive and resonance effects account not only for reactivity but also for the orientation of electrophilic aromatic substitutions. Take alkyl groups, for instance, which have an electron-donating inductive effect and are ortho and para directors. The results of toluene nitration are shown in Figure 9.17.

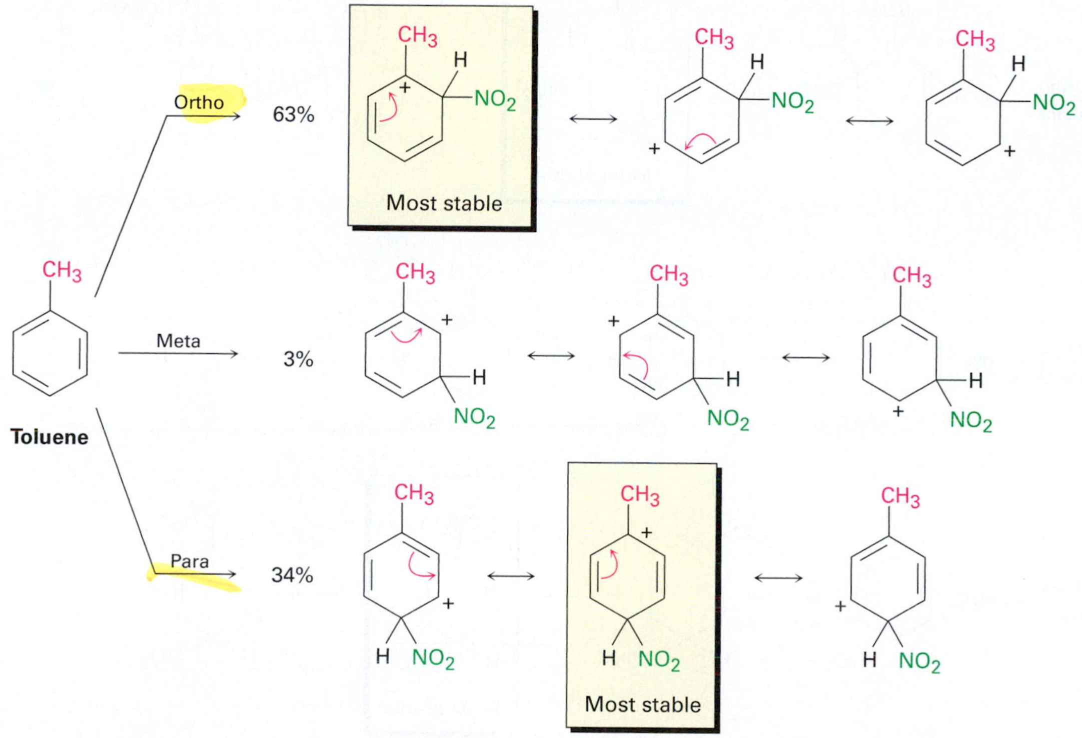

FIGURE 9.17 Carbocation intermediates in the nitration of toluene. Ortho and para intermediates are more stable than the meta intermediate because the positive charge is on a tertiary carbon rather than a secondary carbon.

Nitration of toluene might occur either ortho, meta, or para to the methyl group, giving the three carbocation intermediates shown in Figure 9.17. Although all three intermediates are resonance-stabilized, the ortho and para intermediates are more stabilized than the meta intermediate. For both the ortho and para reactions, but not for the meta reaction, a resonance form places the positive charge directly on the methyl-substituted carbon, where it can be stabilized by the electron-donating inductive effect of the methyl group. The ortho and para intermediates are thus lower in energy than the meta intermediate and form faster.

Halogen, hydroxyl, alkoxyl, and amino groups are also ortho–para directors, but for a different reason than for alkyl groups. As described earlier in this section, halogen, hydroxyl, alkoxyl, and amino groups have an electron-donating resonance effect because the atom attached to the ring—halogen, O, or N—has a lone pair of electrons. In the nitration of phenol, for instance, reaction with the electrophile NO_2^+ can occur either ortho, meta, or para to the −OH group, giving the carbocation intermediates shown in

Figure 9.18. The ortho and para intermediates are more stable than the meta intermediate because they have more resonance forms, including one that allows the positive charge to be stabilized by electron donation from the substituent oxygen atom. The intermediate from meta reaction has no such stabilization.

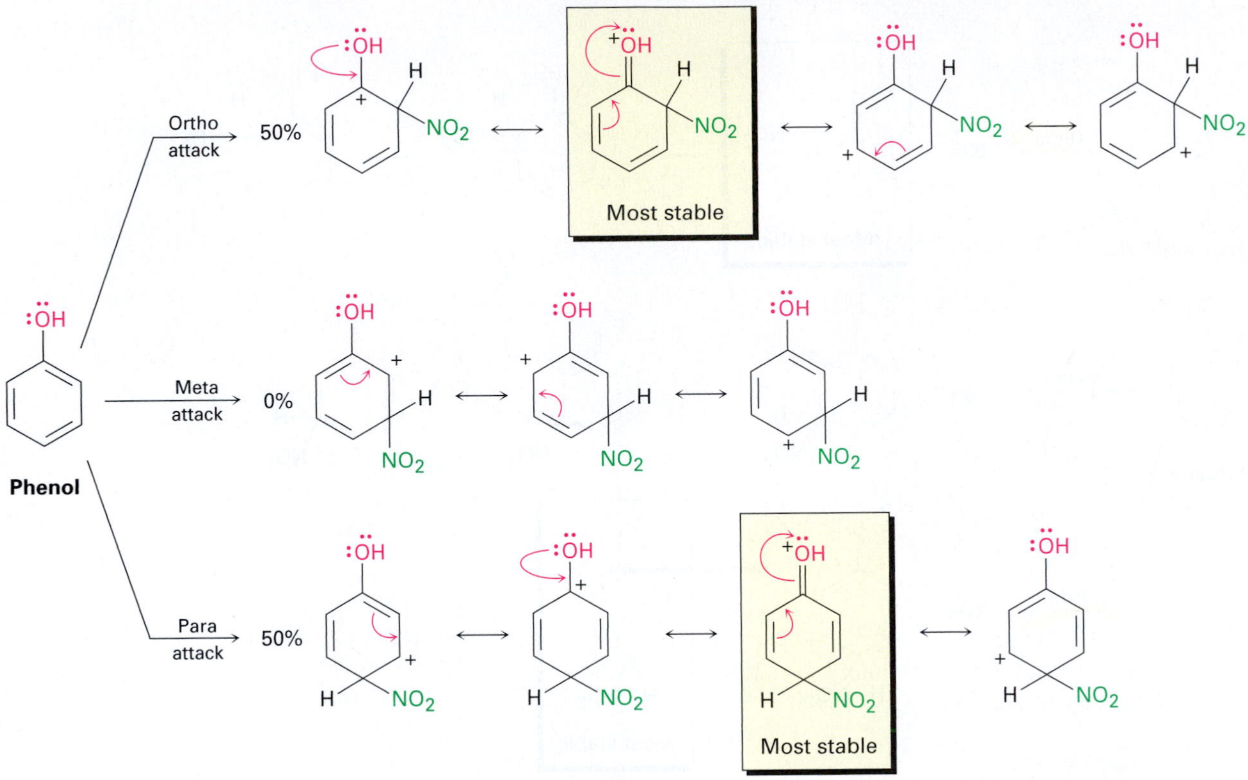

FIGURE 9.18 Cation intermediates in the nitration of phenol. The ortho and para intermediates are more stable than the meta intermediate because they have more resonance forms, including one that involves electron donation from the oxygen atom.

Orienting Effects: Meta Directors

The influence of meta-directing substituents can be explained using the same kinds of arguments used for ortho and para directors. Look at the nitration of benzaldehyde, for instance (Figure 9.19). Of the three possible carbocation intermediates, the meta intermediate has three favorable resonance forms, while the ortho and para intermediates have only two. In both ortho and para intermediates, the third resonance form is unfavorable because it places the positive charge directly on the carbon that bears the aldehyde group, where it is disfavored by a repulsive interaction with the positively polarized carbon atom of the C=O group. Hence, the meta intermediate is more favored and is formed faster than the ortho and para intermediates.

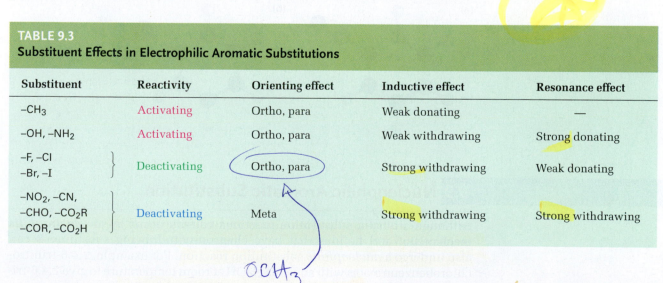

FIGURE 9.19 Intermediates in the nitration of benzaldehyde. The meta intermediate is more favorable than ortho and para intermediates because it has three favorable resonance forms rather than two.

In general, any substituent that has a positively polarized atom ($\delta+$) directly attached to the ring makes one of the resonance forms of the ortho and para intermediates unfavorable, and thus acts as a meta director.

A Summary of Substituent Effects in Electrophilic Substitutions

A summary of the activating and directing effects of substituents in electrophilic aromatic substitution is shown in Table 9.3.

TABLE 9.3
Substituent Effects in Electrophilic Aromatic Substitutions

Substituent	Reactivity	Orienting effect	Inductive effect	Resonance effect
$-CH_3$	Activating	Ortho, para	Weak donating	—
$-OH, -NH_2$	Activating	Ortho, para	Weak withdrawing	Strong donating
$-F, -Cl$ $-Br, -I$	Deactivating	Ortho, para	Strong withdrawing	Weak donating
$-NO_2, -CN,$ $-CHO, -CO_2R$ $-COR, -CO_2H$	Deactivating	Meta	Strong withdrawing	Strong withdrawing

OCH₃

Predict the major product of the sulfonation of toluene.

Strategy

Identify the substituent present on the ring, and decide whether it is ortho- and para-directing or meta-directing. According to Figure 9.16, an alkyl substituent is ortho- and para-directing, so sulfonation of toluene will give primarily a mixture of *o*-toluenesulfonic acid and *p*-toluenesulfonic acid.

Solution

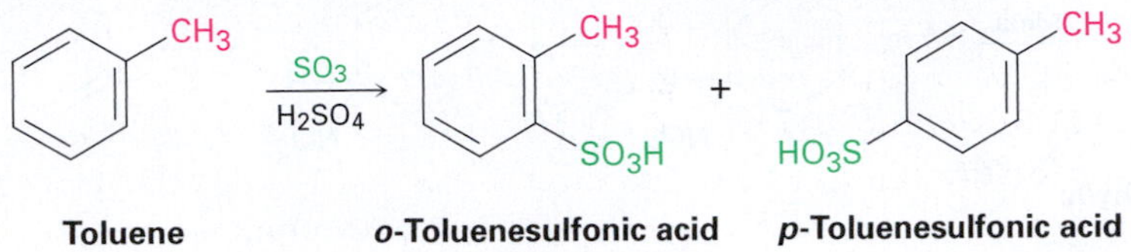

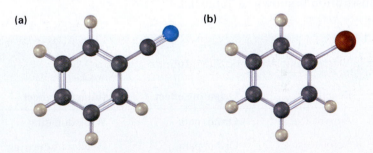

Toluene   *o*-Toluenesulfonic acid   *p*-Toluenesulfonic acid

---

Problem 9.20

Predict the major products of the following reactions:
(a) Nitration of methyl benzoate, $C_6H_5CO_2CH_3$
(b) Bromination of nitrobenzene
(c) Chlorination of phenol
(d) Bromination of aniline

Problem 9.21

Write resonance structures for *o*-, *m*-, and *p*- intermediates in the nitration of chlorobenzene to show the electron-donating resonance effect of the chloro group.

Problem 9.22

Predict the major product you would expect from reaction of each of the following substances with Cl_2 and $FeCl_3$ (blue = N, reddish brown = Br):

(a)   (b)

9.9 Nucleophilic Aromatic Substitution

Although aromatic substitution reactions usually occur by an *electrophilic* mechanism, aryl halides that have electron-withdrawing substituents can also undergo a *nucleophilic* substitution reaction. For example, 2,4,6-trinitrochlorobenzene reacts with aqueous NaOH at room temperature to give 2,4,6-trinitrophenol. The nucleophile OH^- has substituted for Cl^-.

2,4,6-Trinitrochlorobenzene   2,4,6-Trinitrophenol (100%)

Nucleophilic aromatic substitution is much less common than electrophilic substitution but nevertheless does have certain uses. One such use is the reaction of proteins with 2,4-dinitrofluorobenzene, known as *Sanger's reagent,* to attach a "label" to the terminal NH$_2$ group of the amino acid at one end of the protein chain.

2,4-Dinitro-fluorobenzene   A protein   A labeled protein

Nucleophilic substitutions on an aromatic ring proceed by the mechanism shown in Figure 9.20. The nucleophile first adds to the electron-deficient aryl halide, forming a resonance-stabilized negatively charged intermediate called a *Meisenheimer complex.* Halide ion is then eliminated in the second step.

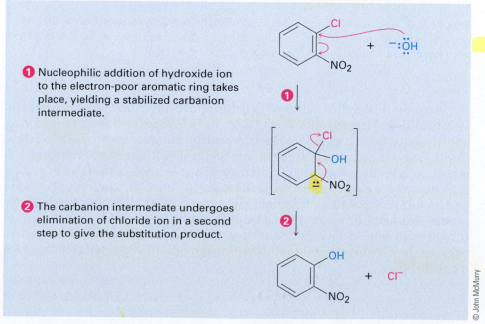

1 Nucleophilic addition of hydroxide ion to the electron-poor aromatic ring takes place, yielding a stabilized carbanion intermediate.

2 The carbanion intermediate undergoes elimination of chloride ion in a second step to give the substitution product.

FIGURE 9.20 MECHANISM: Mechanism of nucleophilic aromatic substitution. The reaction occurs in two steps and involves a resonance-stabilized carbanion intermediate.

© John McMurry

Nucleophilic aromatic substitution occurs only if the aromatic ring has an electron-withdrawing substituent in a position ortho or para to the leaving group to stabilize the anion intermediate through resonance (Figure 9.21). Thus, *p*-chloronitrobenzene and *o*-chloronitrobenzene react with hydroxide ion to yield substitution products, but *m*-chloronitrobenzene is inert to OH⁻.

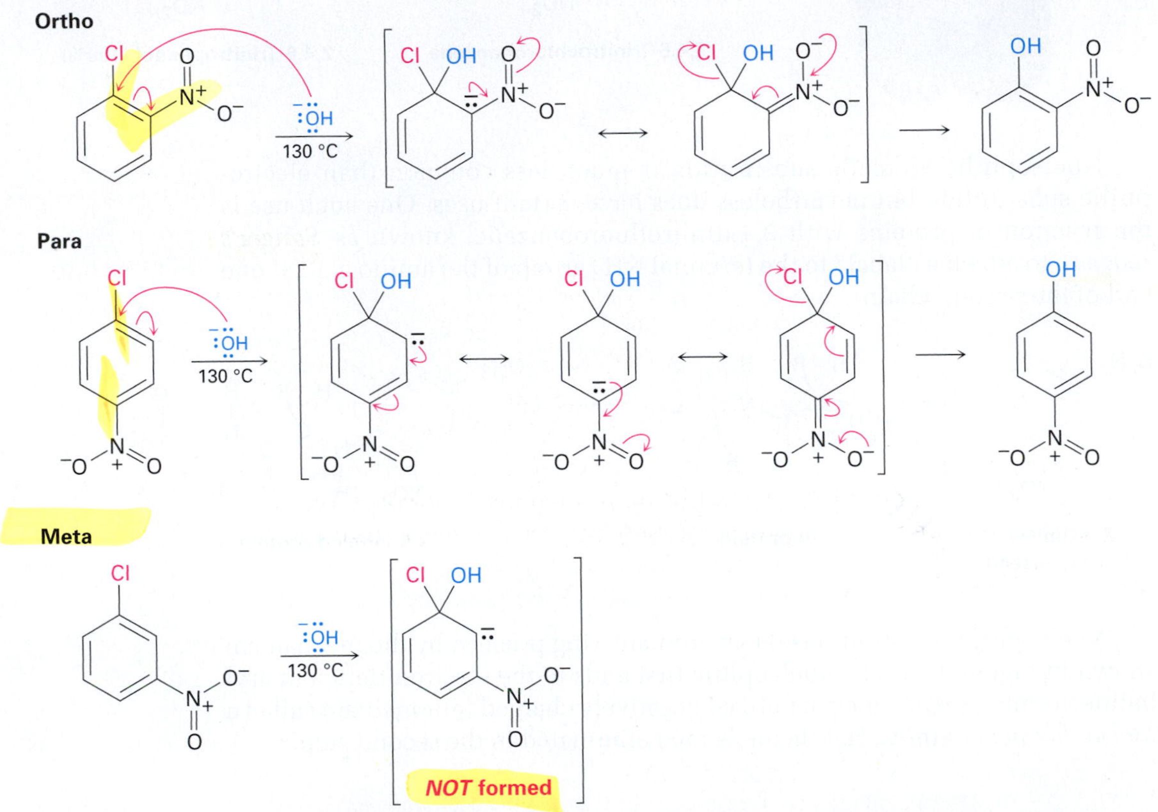

FIGURE 9.21 Nucleophilic aromatic substitution on nitrochlorobenzenes. Only in the ortho and para intermediates is the negative charge stabilized by a resonance interaction with the nitro group, so only the ortho and para isomers undergo reaction.

Note the differences between electrophilic and nucleophilic aromatic substitutions. Electrophilic substitutions are favored by electron-*donating* substituents, which stabilize the carbocation intermediate, while nucleophilic substitutions are favored by electron-*withdrawing* substituents, which stabilize a carbanion intermediate. The electron-withdrawing groups that *deactivate* rings for electrophilic substitution (nitro, carbonyl, cyano, and so on) *activate* them for nucleophilic substitution. What's more, these groups are meta directors in electrophilic substitution but are ortho–para directors in nucleophilic substitution. In addition, electrophilic substitutions replace hydrogen on the ring, while nucleophilic substitutions replace a halide ion.

Problem 9.23

The herbicide oxyfluorfen can be prepared by reaction between a phenol and an aryl fluoride. Propose a mechanism.

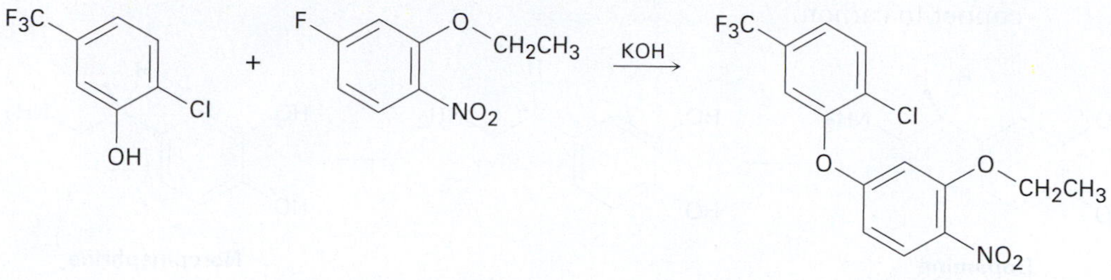

Oxyfluorfen

9.10 Oxidation and Reduction of Aromatic Compounds

Oxidation of Alkylbenzenes

Despite its unsaturation, the benzene ring is inert to oxidizing agents such as *m*-chloroperoxybenzoic acid and OsO_4, reagents that react readily with alkene double bonds (Sections 8.6 and 8.7). It turns out, however, that the presence of the aromatic ring has a dramatic effect on alkyl substituents. Alkyl substituents on the aromatic ring react readily with common laboratory oxidizing agents such as aqueous $KMnO_4$ or $Na_2Cr_2O_7$ and are converted into carboxyl groups, $-CO_2H$. The net effect is conversion of an alkylbenzene into a benzoic acid, $Ar-R \rightarrow Ar-CO_2H$. Butylbenzene is oxidized by aqueous $KMnO_4$ to give benzoic acid, for instance.

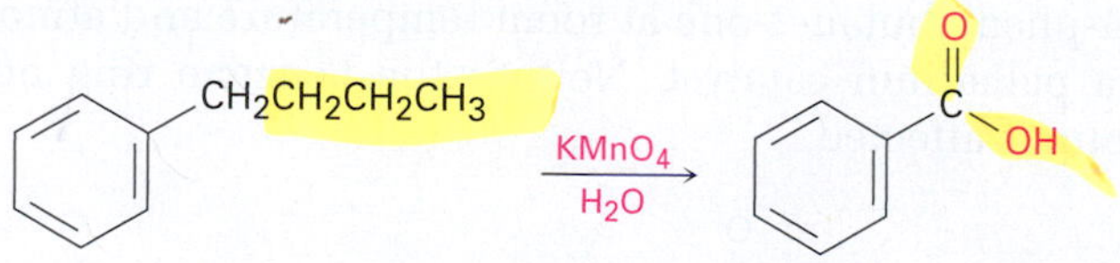

Butylbenzene $\quad\quad$ **Benzoic acid (85%)**

The mechanism of side-chain oxidation is complex and involves reaction of a C–H bond at the position next to the aromatic ring (the **benzylic** position) to form an intermediate radical. Benzylic radicals are stabilized by resonance and thus form more readily than typical alkyl radicals. If the alkylbenzene has no benzylic C–H bonds, however, as in *tert*-butylbenzene, it is inert to oxidation.

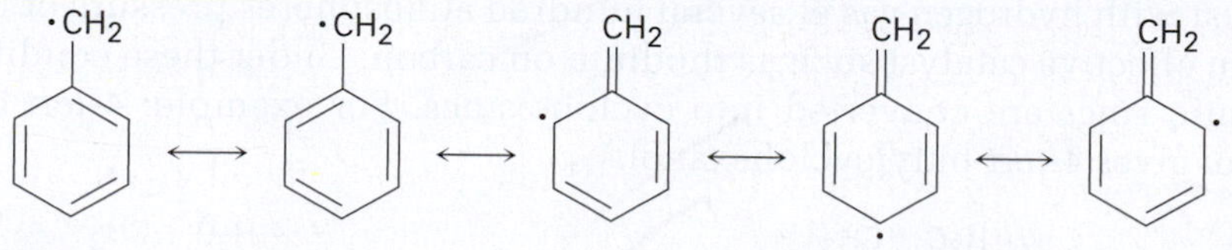

Analogous oxidations occur in various biosynthetic pathways. The neurotransmitter norepinephrine, for instance, is biosynthesized from dopamine by a benzylic hydroxylation reaction. The process is catalyzed by the copper-containing enzyme dopamine β-monooxygenase and occurs by a radical mechanism. A copper–oxygen species in the enzyme first abstracts the *pro-R* benzylic hydrogen to give a radical, and a hydroxyl is then transferred from copper to carbon.

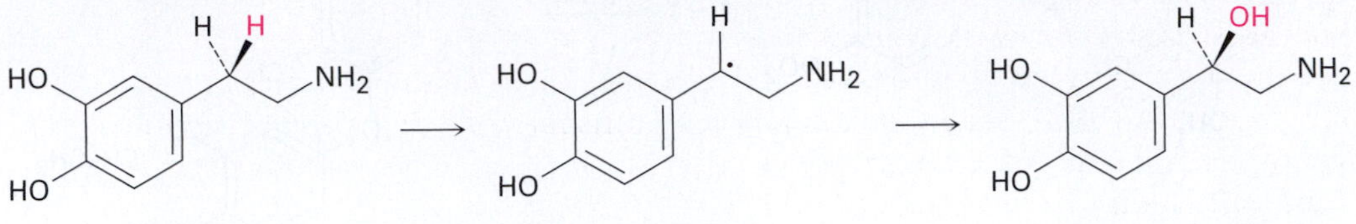

Dopamine   **Norepinephrine**

Problem 9.24

What aromatic products would you obtain from the KMnO$_4$ oxidation of the following substances?

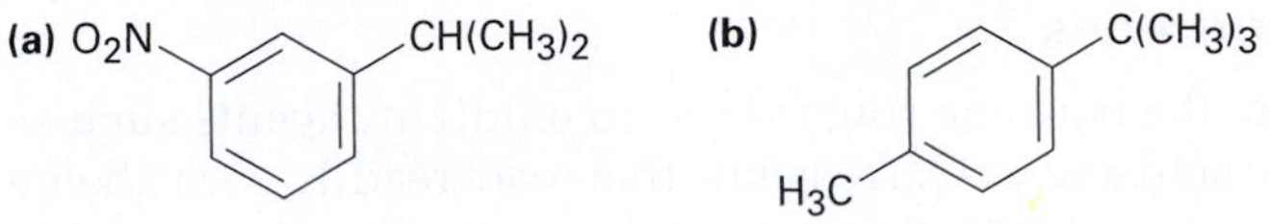

(a)   **(b)**

Hydrogenation of Aromatic Rings

Just as aromatic rings are generally inert to oxidation, they're also inert to catalytic hydrogenation under conditions that reduce typical alkene double bonds. As a result, it's possible to reduce an alkene double bond selectively in the presence of an aromatic ring. For example, 4-phenylbut-3-en-2-one is reduced to 4-phenylbutan-2-one at room temperature and atmospheric pressure using a palladium catalyst. Neither the benzene ring nor the ketone carbonyl group is affected.

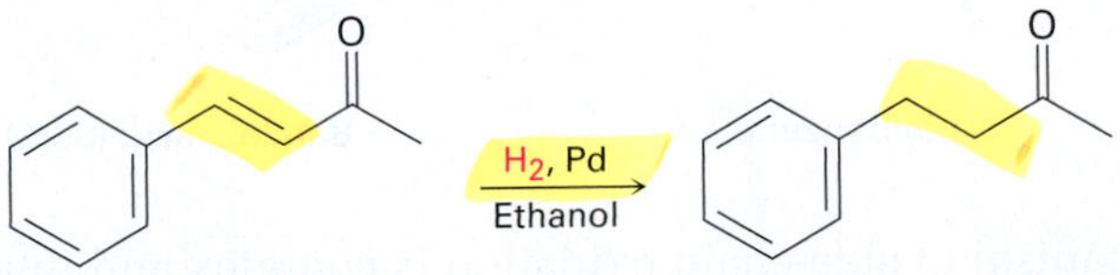

4-Phenylbut-3-en-2-one   **4-Phenylbutan-2-one**
(100%)

To hydrogenate an aromatic ring, it's necessary either to use a platinum catalyst with hydrogen gas at several hundred atmospheres pressure or to use a more effective catalyst such as rhodium on carbon. Under these conditions, aromatic rings are converted into cyclohexanes. For example, 4-*tert*-butylphenol gives 4-*tert*-butylcyclohexanol.

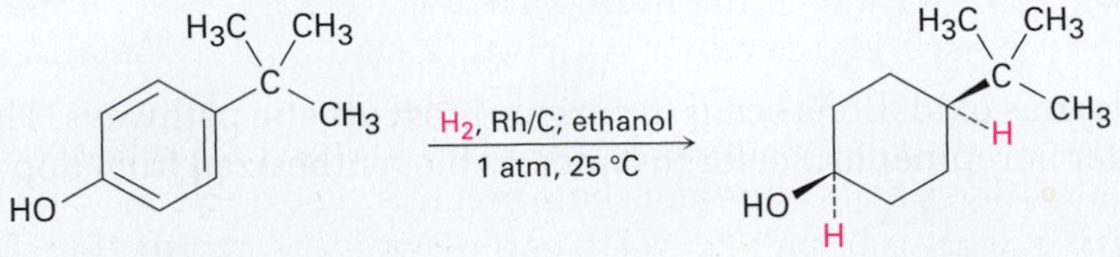

4-*tert*-Butylphenol   **cis-4-*tert*-Butyl-cyclohexane**

Reduction of Aryl Alkyl Ketones

In the same way that an aromatic ring activates a neighboring (benzylic) C–H position toward oxidation, it also activates a neighboring carbonyl group toward reduction. Thus, an aryl alkyl ketone prepared by Friedel–Crafts acylation of an aromatic ring can be converted into an alkylbenzene by catalytic hydrogenation over a palladium catalyst. Propiophenone, for instance, is reduced to propylbenzene by catalytic hydrogenation. Since the net effect of Friedel–Crafts acylation followed by reduction is the preparation of a primary alkylbenzene, this two-step sequence of reactions makes it possible to circumvent the carbocation rearrangement problems associated with direct Friedel–Crafts alkylation using a primary alkyl halide (Section 9.7).

Propiophenone (95%)   **Propylbenzene (100%)**

Propylbenzene   **Isopropylbenzene**

Mixture of two products

Problem 9.25

How might you prepare diphenylmethane, $(Ph)_2CH_2$, from benzene and an appropriate acid chloride? More than one step is needed.

9.11 An Introduction to Organic Synthesis: Polysubstituted Benzenes

There are many reasons for carrying out the laboratory synthesis of an organic molecule. In the pharmaceutical industry, new molecules are designed and synthesized in the hope that some might be useful new drugs. In the chemical industry, syntheses are done to devise more economical routes to known compounds. In biochemistry laboratories, the synthesis of molecules designed to probe enzyme mechanisms is often undertaken.

The ability to plan a successful multistep synthesis of a complex molecule requires a working knowledge of the uses and limitations of numerous organic reactions. Not only must you know which reactions to use, you must also know when to use them, because the order in which reactions are carried out is often critical to the success of the overall scheme.

There's no secret to planning an organic synthesis: all it takes is a knowledge of the different reactions and some practice. The only real trick is to *work backward,* in what is often referred to as a *retrosynthetic* direction. Don't look at a potential starting material and ask yourself what reactions it might undergo. Instead, look at the final product and ask, "What was the immediate precursor of that product?" For example, if the final product is an alkyl halide, the immediate precursor might be an alkene, to which you could add HX. If the final product is a substituted benzoic acid, the immediate precursor might be a substituted alkylbenzene, which could be oxidized. Having found an immediate precursor, work backward again, one step at a time, until you get back to the starting material. You have to keep the starting material in mind, of course, so that you can work back to it, but you don't want that starting material to be your main focus.

Let's look at some examples of synthetic planning using polysubstituted aromatic compounds as the targets. First, however, it's necessary to point out that electrophilic substitution on a disubstituted benzene ring is governed by the same resonance and inductive effects that affect monosubstituted rings. The only difference is that it's necessary to consider the additive effects of two groups. In practice, this isn't as difficult as it sounds; three rules are usually sufficient:

1. If the directing effects of the two groups reinforce each other, the situation is straightforward. In *p*-nitrotoluene, for instance, both the methyl and the nitro group direct further substitution to the same position (ortho to the methyl = meta to the nitro). A single product is thus formed on electrophilic substitution.

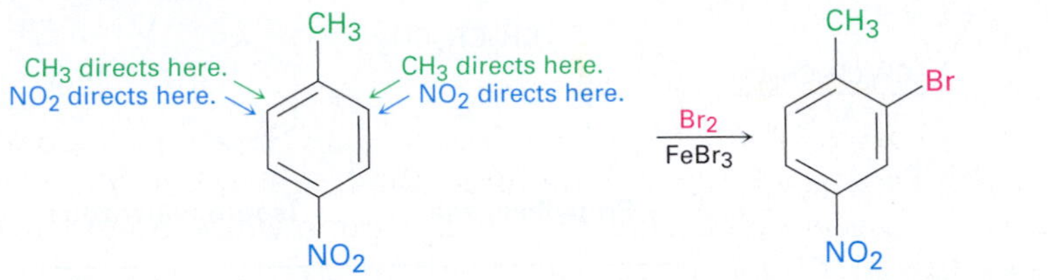
CH3
CH3 directs here.
NO2 directs here.
CH3 directs here.
NO2 directs here.
NO2
Br2
FeBr3
CH3
Br
NO2

***p*-Nitrotoluene** **2-Bromo-4-nitrotoluene**

2. If the directing effects of the two groups oppose each other, the more powerful activating group has the dominant influence. For example, nitration of *p*-methylphenol yields primarily 4-methyl-2-nitrophenol because –OH is a more powerful activator than $-CH_3$.

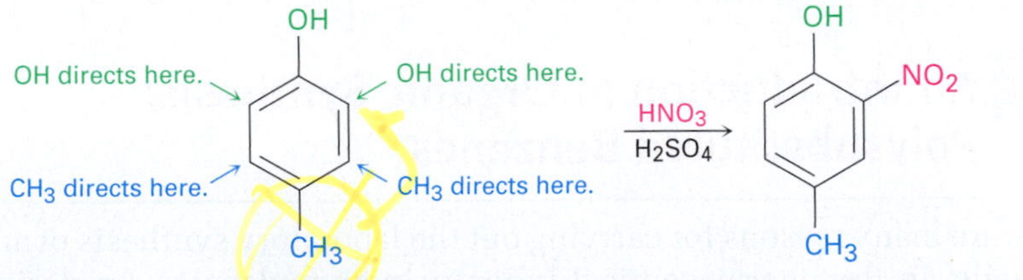
OH
OH directs here.
OH directs here.
CH3 directs here.
CH3 directs here.
CH3
HNO3
H2SO4
OH
NO2
CH3

***p*-Methylphenol** **4-Methyl-2-nitrophenol**

3. Further substitution rarely occurs between the two groups in a meta-disubstituted compound because this site is too hindered. Aromatic rings with three adjacent substituents must therefore be prepared by some other route, such as the substitution of an ortho-disubstituted compound.

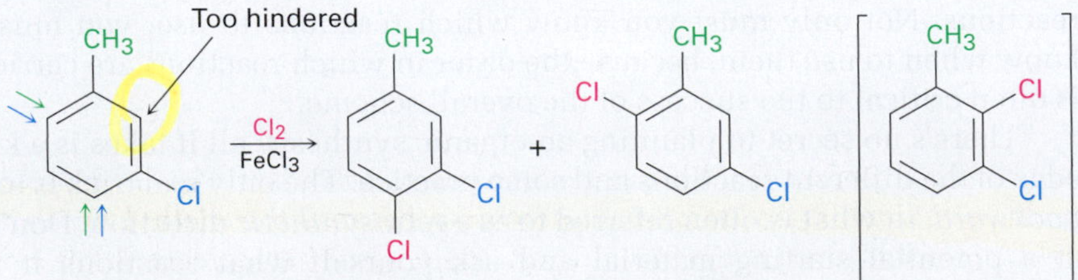
Too hindered
CH3
Cl
Cl2
FeCl3
CH3
Cl
Cl
+
CH3
Cl
Cl
CH3
Cl
Cl

***m*-Chlorotoluene** **3,4-Dichlorotoluene** **2,5-Dichlorotoluene** ***NOT* formed**

Now let's work several examples:

Synthesize 4-bromo-2-nitrotoluene from benzene.

Strategy
Draw the target molecule, identify the substituents, and recall how each group can be introduced separately. Then plan retrosynthetically.

4-Bromo-2-nitrotoluene

The three substituents on the ring are a bromine, a methyl group, and a nitro group. A bromine can be introduced by bromination with $Br_2/FeBr_3$, a methyl group can be introduced by Friedel–Crafts alkylation with $CH_3Cl/AlCl_3$, and a nitro group can be introduced by nitration with HNO_3/H_2SO_4.

Solution
"What is an immediate precursor of the target?" The final step will involve introduction of one of three groups—bromine, methyl, or nitro—so we have to consider three possibilities. Of the three, the bromination of *o*-nitrotoluene could be used because the activating methyl group would dominate the deactivating nitro group and direct bromination to the right position. Unfortunately, a mixture of product isomers would be formed. A Friedel–Crafts reaction can't be used as the final step because this reaction doesn't work on a nitro-substituted (strongly deactivated) benzene. The best precursor of the desired product is probably *p*-bromotoluene, which can be nitrated ortho to the activating methyl group to give a single product.

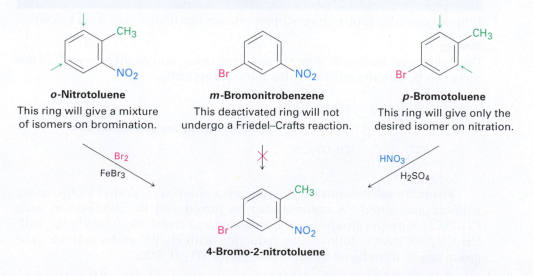

o-Nitrotoluene
This ring will give a mixture of isomers on bromination.

m-Bromonitrobenzene
This deactivated ring will not undergo a Friedel–Crafts reaction.

p-Bromotoluene
This ring will give only the desired isomer on nitration.

4-Bromo-2-nitrotoluene

Next ask yourself, "What is an immediate precursor of *p*-bromotoluene?" Perhaps toluene is an immediate precursor because the methyl group would direct bromination to the ortho and para positions. Alternatively, bromobenzene might be an immediate precursor because we could carry out a Friedel–Crafts methylation and obtain a mixture of ortho and para products. Both answers are satisfactory, although both would also lead unavoidably to a product mixture that would have to be separated.

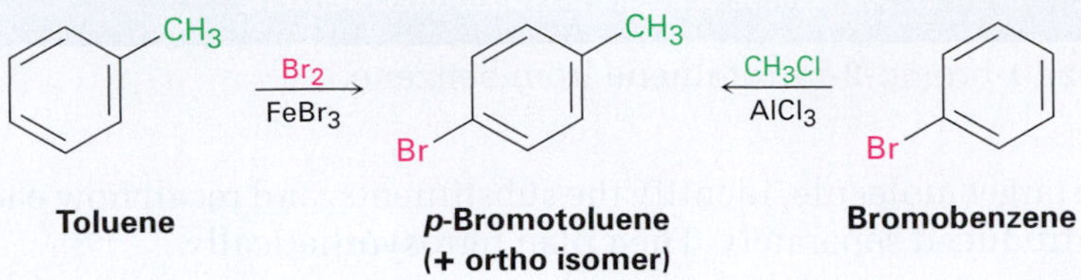

Toluene   **p-Bromotoluene**   **Bromobenzene**
                (+ ortho isomer)

"What is an immediate precursor of toluene?" Benzene, which could be methylated in a Friedel–Crafts reaction. Alternatively, "What is an immediate precursor of bromobenzene?" Benzene, which could be brominated.

The retrosynthetic analysis has provided two valid routes from benzene to 4-bromo-2-nitrotoluene.

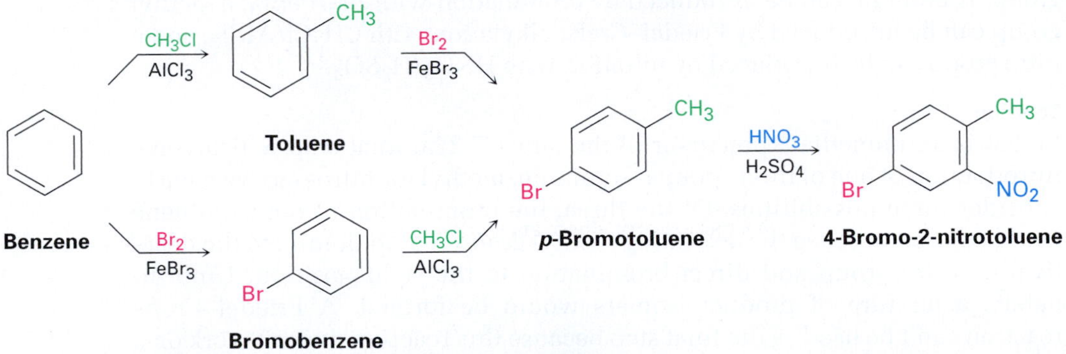

Benzene   **Toluene**   **p-Bromotoluene**   **4-Bromo-2-nitrotoluene**

                Bromobenzene

WORKED EXAMPLE 9.5 Synthesizing a Polysubstituted Benzene

Propose a synthesis of 4-chloro-2-propylbenzenesulfonic acid from benzene.

Strategy

Draw the target molecule, identify its substituents, and recall how each of the three can be introduced. Then plan retrosynthetically.

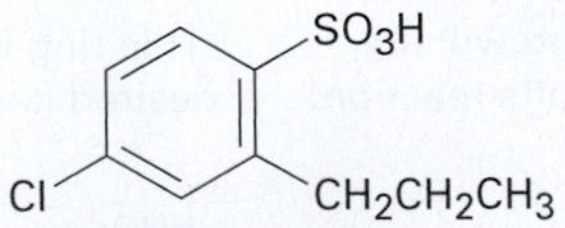

4-Chloro-2-propylbenzenesulfonic acid

The three substituents on the ring are a chlorine, a propyl group, and a sulfonic acid group. A chlorine can be introduced by chlorination with $Cl_2/FeCl_3$, a propyl group can be introduced by Friedel–Crafts acylation with $CH_3CH_2COCl/AlCl_3$ followed by reduction with H_2/Pd, and a sulfonic acid group can be introduced by sulfonation with SO_3/H_2SO_4.

Solution

"What is an immediate precursor of the target?" The final step will involve introduction of one of three groups—chlorine, propyl, or sulfonic acid—so we have to consider three possibilities. Of the three, the chlorination of o-propylbenzenesulfonic acid can't be used because the reaction would occur at the wrong position. Similarly, a Friedel–Crafts reaction can't be used as the final step because this reaction doesn't work on sulfonic acid–substituted (strongly deactivated) benzenes. Thus, the immediate precursor of the desired product is probably m-chloropropylbenzene, which can be sulfonated to give a mixture of product isomers that must then be separated.

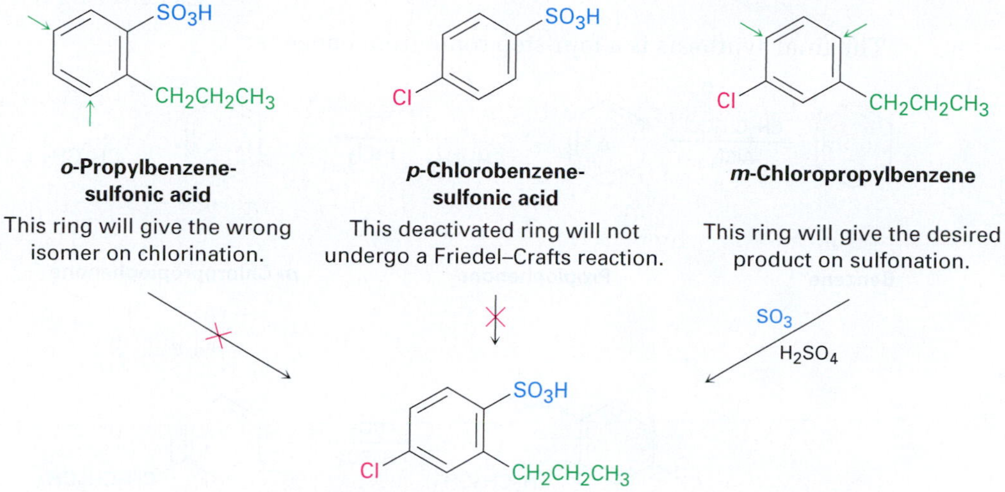

**o-Propylbenzene-
sulfonic acid**

This ring will give the wrong isomer on chlorination.

**p-Chlorobenzene-
sulfonic acid**

This deactivated ring will not undergo a Friedel–Crafts reaction.

m-Chloropropylbenzene

This ring will give the desired product on sulfonation.

4-Chloro-2-propylbenzenesulfonic acid

"What is an immediate precursor of m-chloropropylbenzene?" Because the two substituents have a meta relationship, the first substituent placed on the ring must be a meta director so that the second substitution will take place at the proper position. Furthermore, because primary alkyl groups such as propyl can't be introduced directly by Friedel–Crafts alkylation, the precursor of m-chloropropylbenzene is probably m-chloropropiophenone, which could be catalytically reduced.

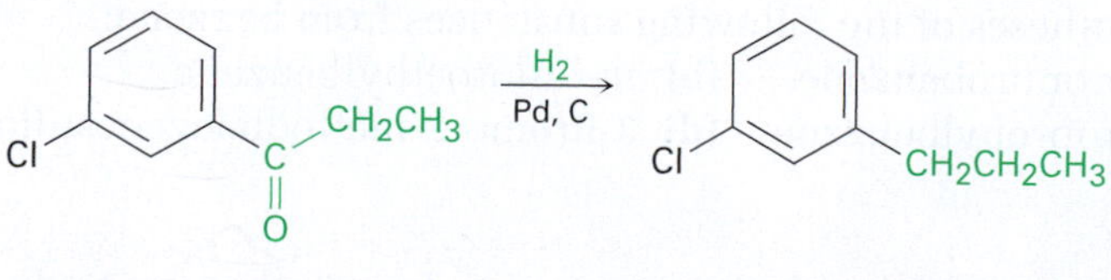

m-Chloropropiophenone

m-Chloropropylbenzene

"What is an immediate precursor of m-chloropropiophenone?" Propiophenone, which could be chlorinated in the meta position.

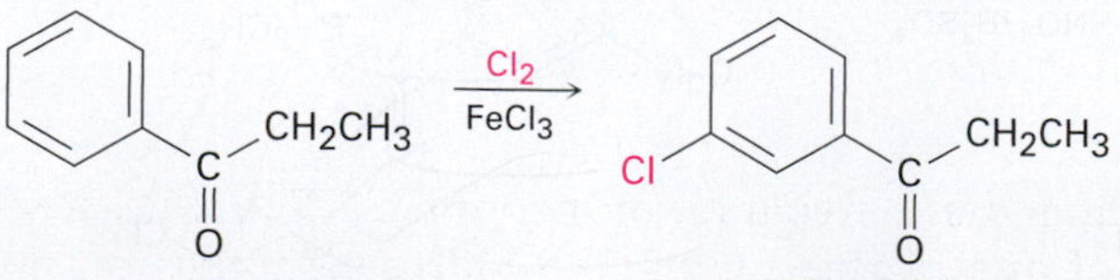

Propiophenone

m-Chloropropiophenone

"What is an immediate precursor of propiophenone?" Benzene, which could undergo Friedel–Crafts acylation with propanoyl chloride and $AlCl_3$.

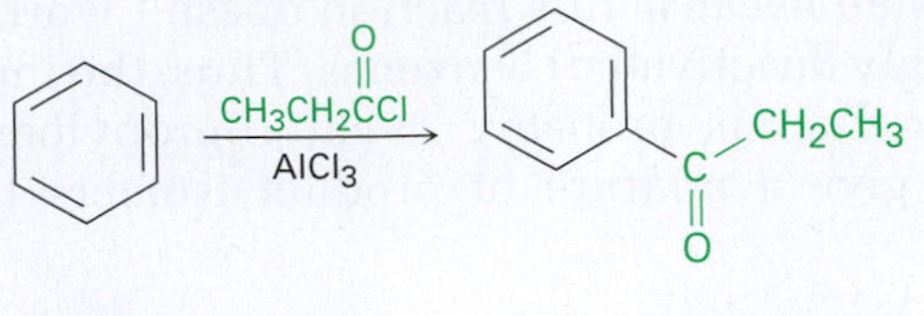

Benzene   **Propiophenone**

The final synthesis is a four-step route from benzene:

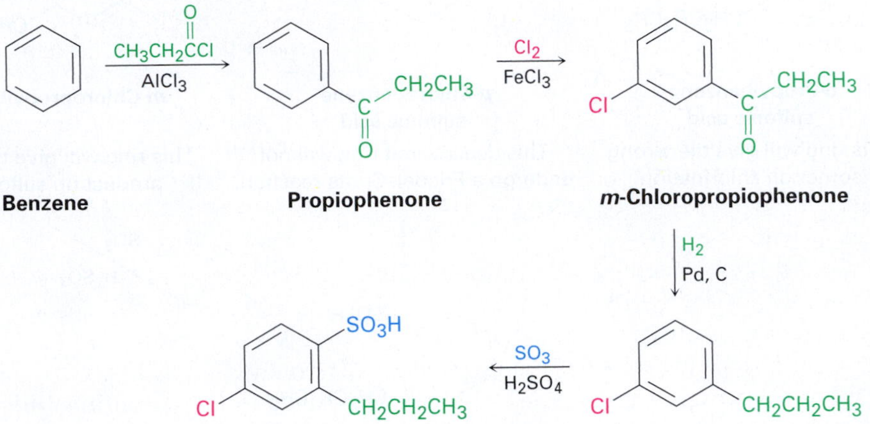

Benzene   **Propiophenone**   *m*-**Chloropropiophenone**

4-Chloro-2-propyl-   *m*-**Chloropropylbenzene**
benzenesulfonic acid

Planning organic syntheses has been compared with playing chess. There are no tricks; all that's required is a knowledge of the allowable moves (the organic reactions) and the discipline to plan ahead, carefully evaluating the consequences of each move. Practicing is not always easy, but it's a great way to learn organic chemistry.

Problem 9.26
Propose syntheses of the following substances from benzene:
(a) *m*-Chloronitrobenzene   **(b)** *m*-Chloroethylbenzene
(c) *p*-Chloropropylbenzene   **(d)** 3-Bromo-2-methylbenzenesulfonic acid

Problem 9.27
In planning a synthesis, it's as important to know what not to do as to know what to do. As written, the following reaction schemes have flaws in them. What is wrong with each?

(a)

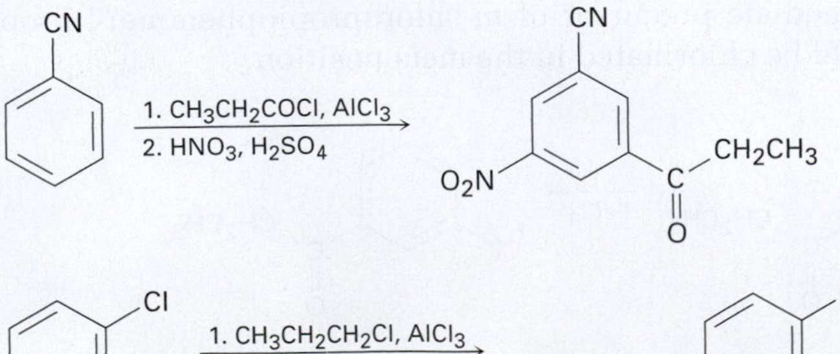

(b)

Summary

Aromatic rings are a common part of many biological structures and are particularly important in nucleic acid chemistry and in the chemistry of several amino acids. In this chapter, we've seen how and why aromatic compounds are different from such apparently related compounds as alkenes, and we've seen some of their most common reactions.

The word **aromatic** is used for historical reasons to refer to the class of compounds related structurally to benzene. Aromatic compounds are systematically named according to IUPAC rules, but many common names are also used. Disubstituted benzenes are named as *ortho* (1,2 disubstituted), *meta* (1,3 disubstituted), or *para* (1,4 disubstituted) derivatives. The C_6H_5- unit itself is referred to as a **phenyl** group.

Benzene is described by resonance theory as a resonance hybrid of two equivalent structures and is described by molecular orbital theory as a planar, cyclic, conjugated molecule with six π electrons. According to the **Hückel rule**, a molecule must have **$4n + 2$ π electrons**, where $n = 0, 1, 2, 3$, and so on, to be aromatic.

Other kinds of molecules besides benzene-like compounds can also be aromatic. The cyclopentadienyl anion and cycloheptatrienyl cation, for instance, are aromatic ions. Pyridine and pyrimidine are six-membered, nitrogen-containing, aromatic **heterocycles**. Pyrrole and imidazole are five-membered, nitrogen-containing heterocycles. Naphthalene, quinoline, indole, and many others are polycyclic aromatic compounds.

The chemistry of aromatic compounds is dominated by **electrophilic aromatic substitution reactions**, both in the laboratory and in biological pathways. Many variations of the reaction can be carried out, including halogenation, nitration, sulfonation, and hydroxylation. Friedel–Crafts alkylation and acylation, which involve reaction of an aromatic ring with carbocation electrophiles, are particularly useful.

Substituents on the benzene ring affect both the reactivity of the ring toward further substitution and the orientation of that substitution. Groups can be classified as ortho- and para-directing activators, ortho- and para-directing deactivators, or meta-directing deactivators. Substituents influence aromatic rings by a combination of electron-donating and electron-withdrawing effects.

Halobenzenes with a strongly electron-withdrawing substituent in the ortho or para position undergo a *nucleophilic* substitution, which occurs by addition of a nucleophile to the ring, followed by elimination of halide from the intermediate anion.

The entire side chain of an alkylbenzene can be degraded to a carboxyl group by oxidation with aqueous $KMnO_4$. Although aromatic rings are less reactive than isolated alkene double bonds, they can be reduced to cyclohexanes by hydrogenation over a platinum or rhodium catalyst. In addition, aryl alkyl ketones are reduced to alkylbenzenes by hydrogenation over a palladium catalyst.

Key Words

acyl group, 333
acylation, 333
alkylation, 331
arene, 311
aromatic, 309
benzyl group, 311
benzylic, 347
electrophilic aromatic
   substitution reaction, 324
Friedel–Crafts reaction, 331
heterocycle, 319
Hückel $4n + 2$ rule, 316
meta (*m*), 311
ortho (*o*), 311
para (*p*), 311
phenyl group, 311

Summary of Reactions

1. Electrophilic aromatic substitution (Section 9.6)

   (a) Bromination

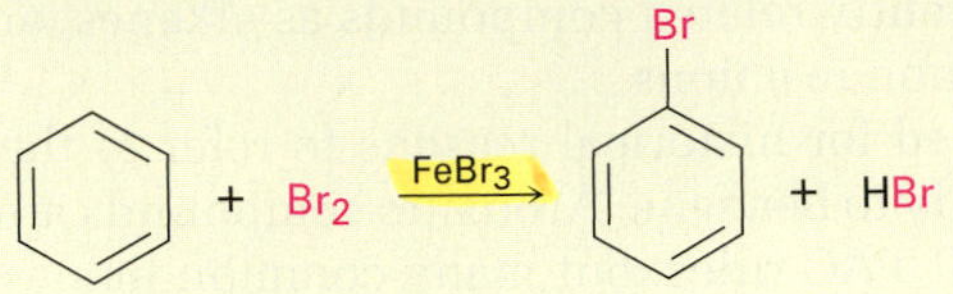

   Benzene + Br_2 $\xrightarrow{FeBr_3}$ Bromobenzene + HBr

   (b) Chlorination

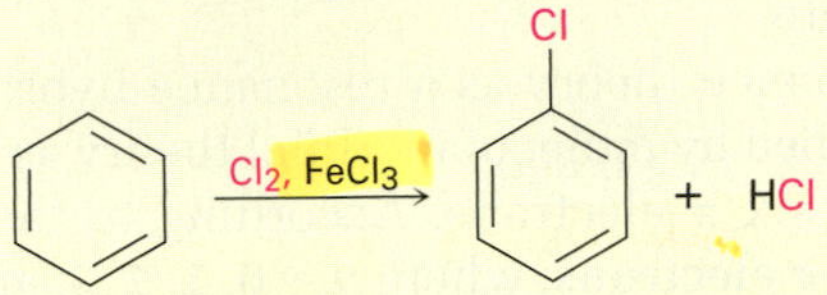

   Benzene $\xrightarrow{Cl_2,\ FeCl_3}$ Chlorobenzene + HCl

   (c) Iodination

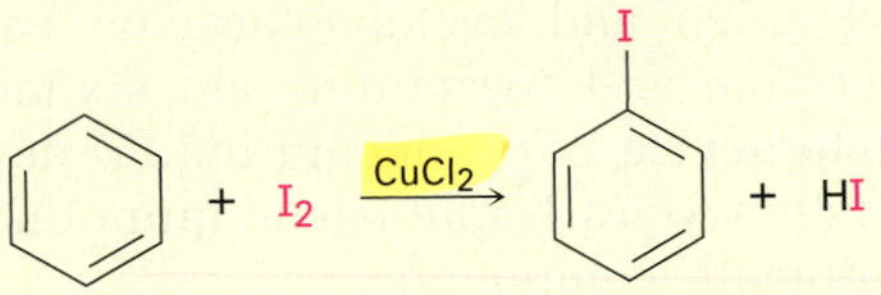

   Benzene + I_2 $\xrightarrow{CuCl_2}$ Iodobenzene + HI

   (d) Nitration

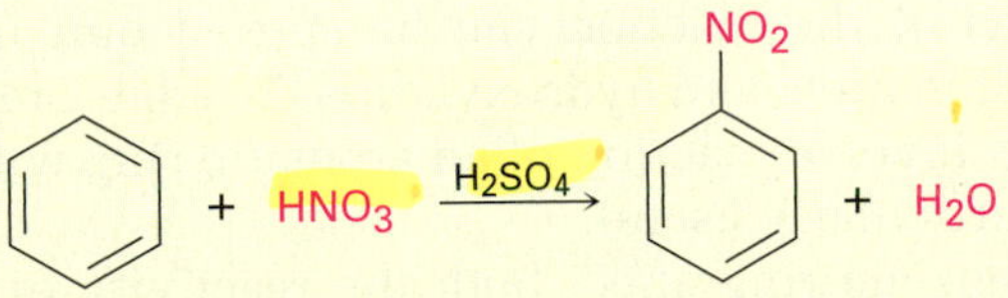

   Benzene + HNO_3 $\xrightarrow{H_2SO_4}$ Nitrobenzene + H_2O

   (e) Sulfonation

   Benzene + SO_3 $\xrightarrow{H_2SO_4}$ Benzenesulfonic acid

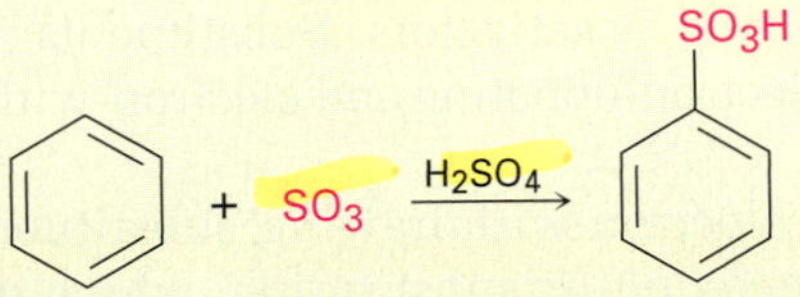

   (f) Friedel–Crafts alkylation (Section 9.7)

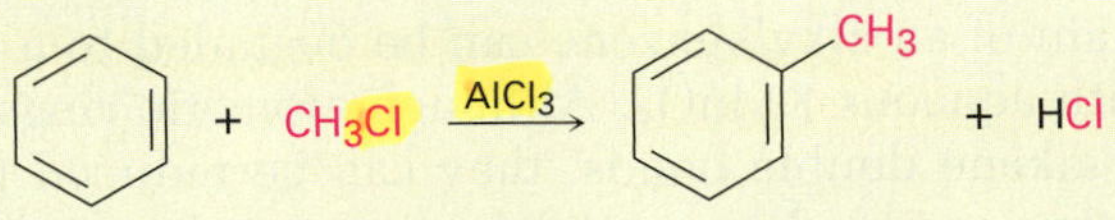

   Benzene + CH_3Cl $\xrightarrow{AlCl_3}$ Toluene + HCl

   (g) Friedel–Crafts acylation (Section 9.7)

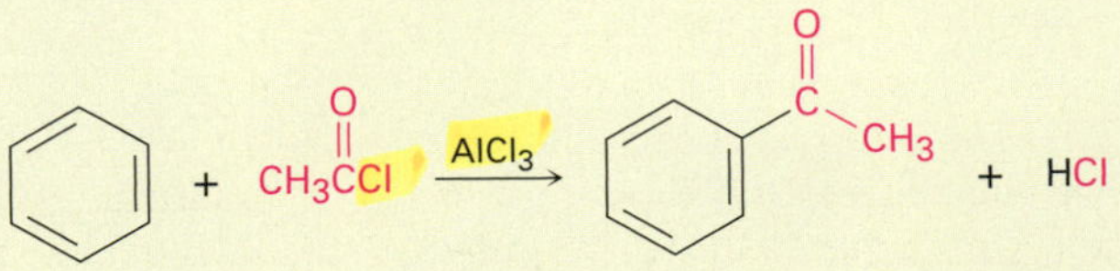

   Benzene + CH_3CCl (O) $\xrightarrow{AlCl_3}$ Acetophenone + HCl

2. Reduction of aromatic nitro groups (Section 9.6)

$$\text{(NO}_2\text{)} \xrightarrow[\text{2. HO}^-]{\text{1. Fe, H}_3\text{O}^+} \text{(NH}_2\text{)}$$

3. Nucleophilic aromatic substitution (Section 9.9)

$$\xrightarrow[\text{H}_2\text{O}]{\text{Na}^+ \ ^-\text{OH}} \quad + \quad \text{NaCl}$$

4. Oxidation of alkylbenzene side chain (Section 9.10)

$$\text{(CH}_3\text{)} \xrightarrow[\text{H}_2\text{O}]{\text{KMnO}_4} \text{(CO}_2\text{H)}$$

5. Catalytic hydrogenation of aromatic rings (Section 9.10)

$$\xrightarrow[\text{Rh/C catalyst}]{\text{H}_2}$$

6. Reduction of aryl alkyl ketones (Section 9.10)

$$\xrightarrow{\text{H}_2/\text{Pd}}$$

Lagniappe

Aspirin, NSAIDs, and COX-2 Inhibitors

Whatever the cause—tennis elbow, a sprained ankle, or a wrenched knee—pain and inflammation seem to go together. They are, however, different in their origin, and powerful drugs are available for treating each separately. Codeine, for example, is a powerful *analgesic,* or pain reliever, used in the management of debilitating pain, while cortisone and related steroids are potent *anti-inflammatory* agents, used for treating arthritis and other crippling inflammations. For minor pains and inflammation, both problems are often treated at the same time by using a common, over-the-counter medication called an *NSAID,* or *nonsteroidal anti-inflammatory drug.*

The most common NSAID is aspirin, or acetylsalicylic acid, whose use goes back to the late 1800s. It had been known from before the time of Hippocrates in 400 BC that fevers could be lowered by chewing the bark of willow trees. The active agent in willow bark was found in 1827 to be an aromatic compound called *salicin,* which could be converted by reaction with water into salicyl alcohol and then oxidized to give salicylic acid. Salicylic acid turned out to be even more effective than salicin for reducing fever and to have analgesic and anti-inflammatory action as well. Unfortunately, it also turned out to be too corrosive to the walls of the stomach for everyday use. Conversion of the

continued

Lagniappe *continued*

phenol –OH group into an acetate ester, however, yielded acetylsalicylic acid, which proved just as potent as salicylic acid but less corrosive to the stomach.

Salicyl alcohol   **Salicylic acid**

**Acetylsalicylic acid
(aspirin)**

Although extraordinary in its powers, aspirin is also more dangerous than commonly believed. A dose of only about 15 g can be fatal to a small child, and aspirin can cause stomach bleeding and allergic reactions in long-term users. Even more serious is a condition called *Reye's syndrome,* a potentially fatal reaction to aspirin sometimes seen in children recovering from the flu. As a result of these problems, numerous other NSAIDs have been developed in the last several decades, most notably ibuprofen and naproxen.

Like aspirin, both ibuprofen and naproxen are relatively simple aromatic compounds containing a side-chain carboxylic acid group. Ibuprofen, sold under the names Advil, Nuprin, Motrin, and others, has roughly the same potency as aspirin but is less prone to cause stomach upset. Naproxen, sold under the names Aleve and Naprosyn, also has about the same potency as aspirin but remains active in the body six times longer.

Aspirin and other NSAIDs function by blocking the cyclooxygenase (COX) enzymes that carry out the body's synthesis of prostaglandins (Section 6.3). There are two forms of the enzyme, COX-1, which carries out the normal physiological production of prostaglandins, and COX-2, which mediates the body's response to arthritis and other inflammatory conditions. Unfortunately, both COX-1 and COX-2 enzymes are blocked by aspirin, ibuprofen, and other NSAIDs, thereby shutting down not only the response to inflammation but also various protective functions, including the control mechanism for production of acid in the stomach.

Medicinal chemists have devised a number of drugs that act as selective inhibitors of the COX-2 enzyme. Inflammation is thereby controlled without blocking protective functions. Originally heralded as a breakthrough in arthritis treatment, the first generation of COX-2 inhibitors, including Vioxx, Celebrex, and Bextra, turned out to cause potentially serious heart problems, particularly in elderly or compromised patients. The second generation of COX-2 inhibitors now under development promises to be safer but will be closely scrutinized for side effects before gaining approval.

© Doug Berry/Corbis

Many athletes rely on NSAIDs to help with pain and soreness.

**Ibuprofen
(Advil, Nuprin, Motrin)**

**Naproxen
(Aleve, Naprosyn)**

**Celecoxib
(Celebrex)**

**Rofecoxib
(Vioxx)**

Exercises

■ indicates problems that are assignable in Organic OWL.

Go to this book's companion website at www.cengage.com/chemistry/mcmurry to explore interactive versions of the Active Figures from this text.

VISUALIZING CHEMISTRY

(Problems 9.1–9.27 appear within the chapter.)

9.28 ■ Give IUPAC names for the following substances (red = O, blue = N):

(a)   (b)

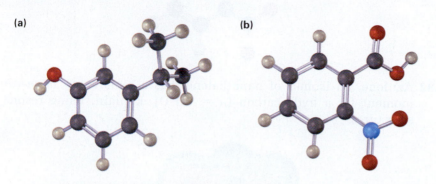

9.29 ■ The following molecular model is that of a carbocation. Draw two resonance structures for the carbocation, indicating the positions of the double bonds.

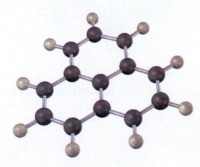

9.30 ■ Draw the product from reaction of each of the following substances with (a) Br₂, FeBr₃ and (b) CH₃COCl, AlCl₃. (Red = O.)

(a)   (b)

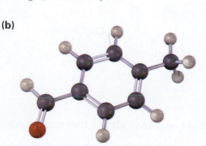

9.31 ■ How would you synthesize the following compound starting from benzene? More than one step is needed (red = O, blue = N).

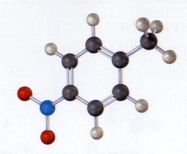

9.32 Azulene, an isomer of naphthalene, has a remarkably large dipole moment for a hydrocarbon (μ = 1.0 D). Explain, using resonance structures.

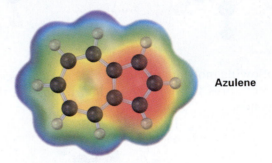

Azulene

ADDITIONAL PROBLEMS

9.33 ■ Give IUPAC names for the following compounds:

(a) CH₃   CH₃
  CHCH₂CH₂CHCH₃

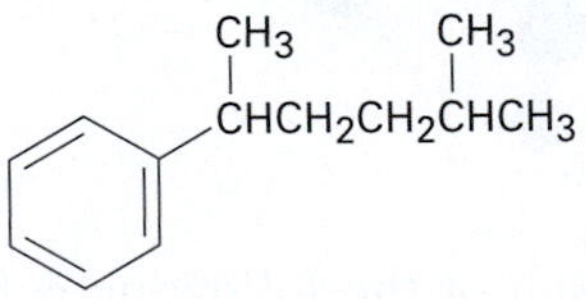

(b) CO₂H
  Br

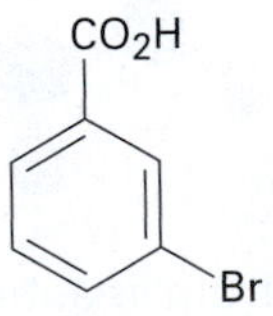

(c) Br
  H₃C   CH₃

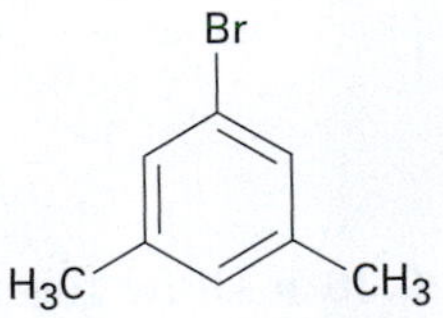

(d) Br
  CH₂CH₂CH₃

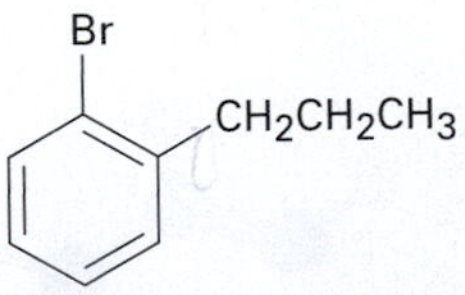

(e) 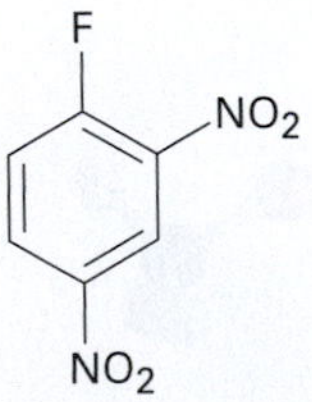 F
  NO₂
  NO₂

(f) NH₂
  Cl

9.34 ■ Draw structures corresponding to the following names:

(a) 3-Methyl-2-nitrobenzoic acid   (b) Benzene-1,3,5-triol

(c) 3-Methyl-2-phenylhexane   (d) *o*-Aminobenzoic acid

(e) *m*-Bromophenol   (f) 2,4,6-Trinitrophenol (picric acid)

(g) *p*-Iodonitrobenzene

■ Problems assignable in Organic OWL.

9.35 ▪ Draw and name all aromatic compounds with the formula C_7H_7Cl.

9.36 Rank the following aromatic compounds in their expected order of reactivity toward Friedel–Crafts alkylation. Which compounds are unreactive?

   (a) Bromobenzene   **(b)** Toluene   **(c)** Phenol

   (d) Benzoic acid   **(e)** Nitrobenzene   **(f)** p-Bromotoluene

9.37 ▪ Rank the compounds in each group according to their reactivity toward electrophilic substitution.

   (a) Chlorobenzene, o-dichlorobenzene, benzene

   (b) p-Bromonitrobenzene, nitrobenzene, phenol

   (c) Fluorobenzene, benzaldehyde, o-xylene

   (d) Benzonitrile, p-methylbenzonitrile, p-methoxybenzonitrile

9.38 ▪ Propose structures for aromatic hydrocarbons that meet the following descriptions:

   (a) C_9H_{12}; gives only one $C_9H_{11}Br$ product on substitution with bromine

   (b) $C_{10}H_{14}$; gives only one $C_{10}H_{13}Cl$ product on substitution with chlorine

   (c) C_8H_{10}; gives three C_8H_9Br products on substitution with bromine

   (d) $C_{10}H_{14}$; gives two $C_{10}H_{13}Cl$ products on substitution with chlorine

9.39 ▪ Predict the major product(s) of the following reactions:

(a) [benzene ring with Cl substituent] $\xrightarrow[\text{AlCl}_3]{\text{CH}_3\text{CH}_2\text{Cl}}$ **?**

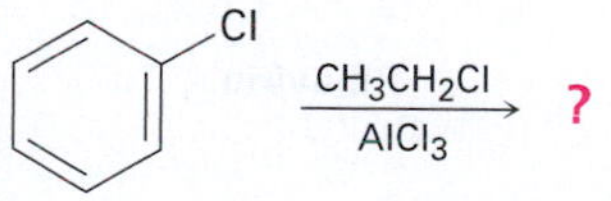

(b) [diphenyl ether] $\xrightarrow[\text{AlCl}_3]{\text{CH}_3\text{CH}_2\text{COCl}}$ **?**

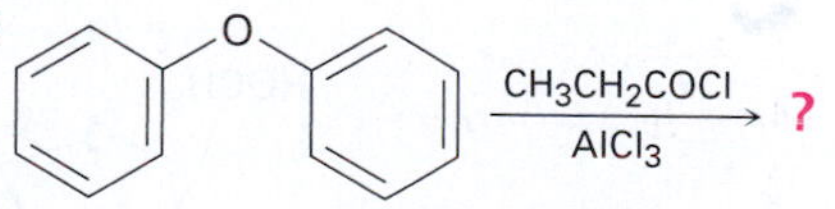

(c) [benzene ring with CO$_2$H substituent] $\xrightarrow[\text{H}_2\text{SO}_4]{\text{HNO}_3}$ **?**

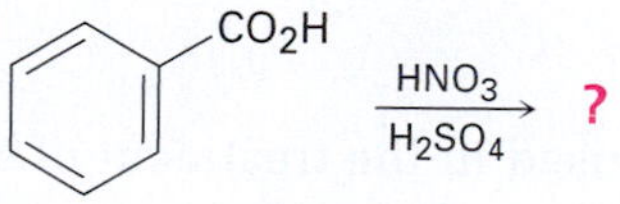

(d) [benzene ring with N(CH$_2$CH$_3$)$_2$ substituent] $\xrightarrow[\text{H}_2\text{SO}_4]{\text{SO}_3}$ **?**

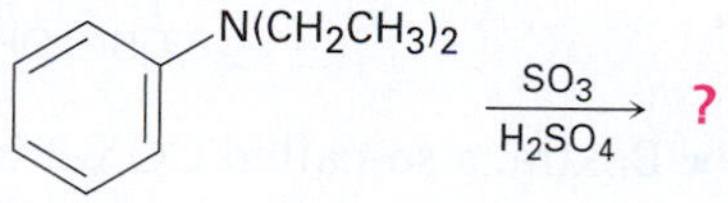

9.40 Identify each of the following groups as an activator or deactivator and as an o,p-director or m-director:

(a) $\ce{-N(CH3)2}$ **(b)** [cyclopentyl]   **(c)** $\ce{-OCH2CH3}$ **(d)** [acyl group attached to cyclohexane]

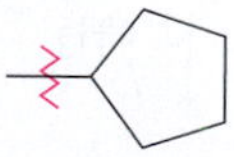

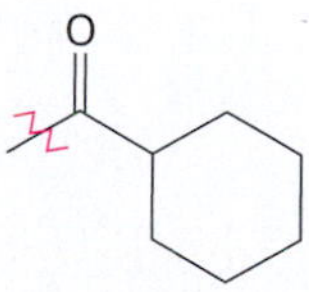

9.41 ▪ Predict the major product(s) of mononitration of the following substances. Which react faster than benzene, and which slower?

   (a) Bromobenzene   **(b)** Benzonitrile   **(c)** Benzoic acid

   (d) Nitrobenzene   **(e)** Benzenesulfonic acid   **(f)** Methoxybenzene

▪ Problems assignable in Organic OWL.

9.42 ■ Predict the major monoalkylation products you would expect to obtain from reaction of the following substances with chloromethane and AlCl$_3$:

(a) *p*-Chloroaniline   (b) *m*-Bromophenol

(c) 2,4-Dichlorophenol   (d) 2,4-Dichloronitrobenzene

(e) *p*-Methylbenzenesulfonic acid   (f) 2,5-Dibromotoluene

9.43 ■ Name and draw the major product(s) of electrophilic chlorination of the following compounds:

(a) *m*-Nitrophenol   (b) *o*-Xylene (dimethylbenzene)

(c) *p*-Nitrobenzoic acid   (d) *p*-Bromobenzenesulfonic acid

9.44 Aromatic iodination can be carried out with a number of reagents, including iodine monochloride, ICl. What is the direction of polarization of ICl? Propose a mechanism for the iodination of an aromatic ring with ICl.

9.45 ■ The carbocation electrophile in a Friedel–Crafts reaction can be generated in ways other than by reaction of an alkyl chloride with AlCl$_3$. For example, reaction of benzene with 2-methylpropene in the presence of H$_3$PO$_4$ yields *tert*-butylbenzene. Propose a mechanism for this reaction.

9.46 ■ Ribavirin, an antiviral agent used against hepatitis C and viral pneumonia, contains a 1,2,4-triazole ring. Why is the ring aromatic?

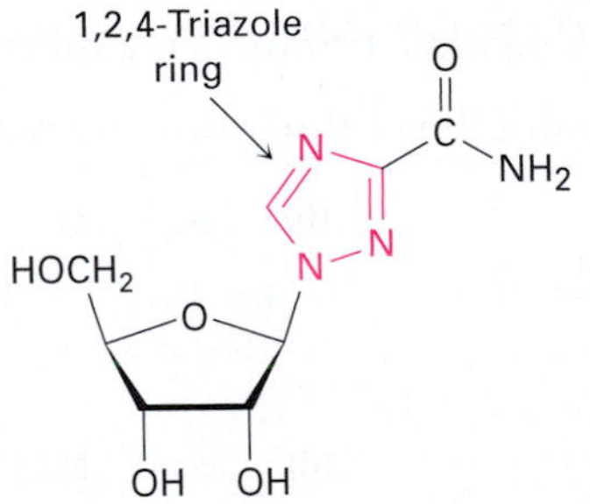

9.47 ■ Bextra, a so-called COX-2 inhibitor used in the treatment of arthritis, contains an isoxazole ring. Why is the ring aromatic?

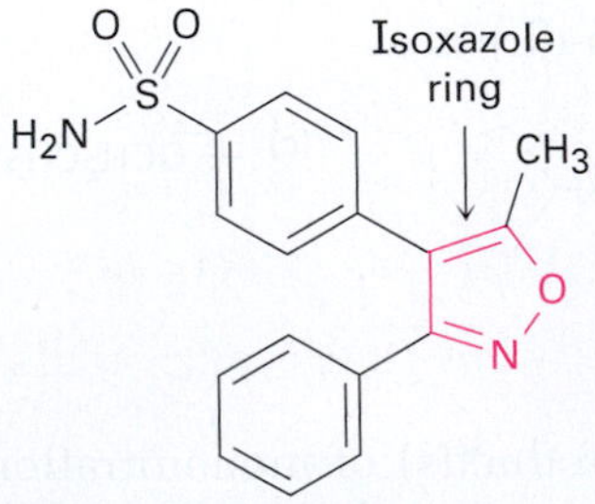

9.48 Look at the three resonance structures of naphthalene shown in Section 9.5, and account for the fact that not all carbon–carbon bonds have the same length. The C1–C2 bond is 136 pm long, whereas the C2–C3 bond is 139 pm long.

9.49 ▪ There are four resonance structures for anthracene, one of which is shown. Draw the other three.

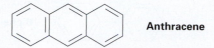

Anthracene

9.50 ▪ There are five resonance structures of phenanthrene, one of which is shown. Draw the other four.

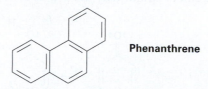

Phenanthrene

9.51 Look at the five resonance structures for phenanthrene (Problem 9.50) and predict which of its carbon–carbon bonds is shortest.

9.52 Which would you expect to be most stable, cyclononatetraenyl radical, cation, or anion?

9.53 How might you convert cyclonona-1,3,5,7-tetraene to an aromatic substance?

9.54 Calicene, like azulene (Problem 9.32), has an unusually large dipole moment for a hydrocarbon. Explain, using resonance structures.

Calicene

9.55 Pentalene is a most elusive molecule that has been isolated only at liquid-nitrogen temperature. The pentalene dianion, however, is well known and quite stable. Explain.

Pentalene   **Pentalene dianion**

9.56 Indole is an aromatic heterocycle that has a benzene ring fused to a pyrrole ring. Draw an orbital picture of indole.

(a) How many π electrons does indole have?

(b) What is the electronic relationship of indole to naphthalene?

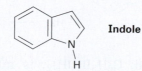

Indole

9.57 The nitroso group, $-N=O$, is one of the few nonhalogens that is an ortho- and para-directing deactivator. Explain by drawing resonance structures of the carbocation intermediates in ortho, meta, and para electrophilic reaction on nitrosobenzene, $C_6H_5N=O$.

▪ Problems assignable in Organic OWL.

9.58 Using resonance structures of the intermediates, explain why bromination of biphenyl occurs at ortho and para positions rather than at meta.

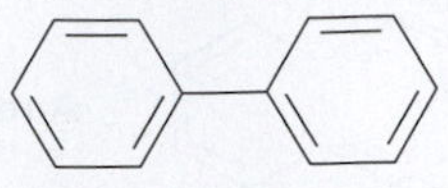

Biphenyl

9.59 On reaction with acid, 4-pyrone is protonated on the carbonyl-group oxygen to give a stable cationic product. Using resonance structures and the Hückel $4n + 2$ rule, explain why the protonated product is so stable.

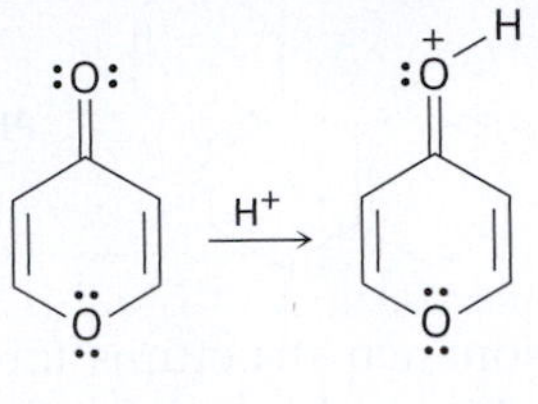

4-Pyrone

9.60 *N*-Phenylsydnone, so named because it was first studied at the University of Sydney, Australia, behaves like a typical aromatic molecule. Explain, using the Hückel $4n + 2$ rule.

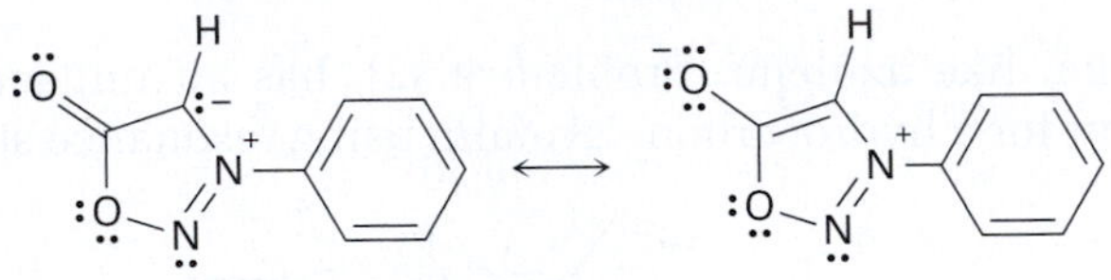

N-Phenylsydnone

9.61 Electrophilic substitution on 3-phenylpropanenitrile occurs at the ortho and para positions, but reaction with 3-phenylpropenenitrile occurs at the meta position. Explain, using resonance structures of the intermediates.

CH₂CH₂CN

CN

3-Phenylpropanenitrile   **3-Phenylpropenenitrile**

9.62 Addition of HBr to 1-phenylpropene yields only (1-bromopropyl)-benzene. Propose a mechanism for the reaction, and explain using resonance structures why none of the other regioisomer is produced.

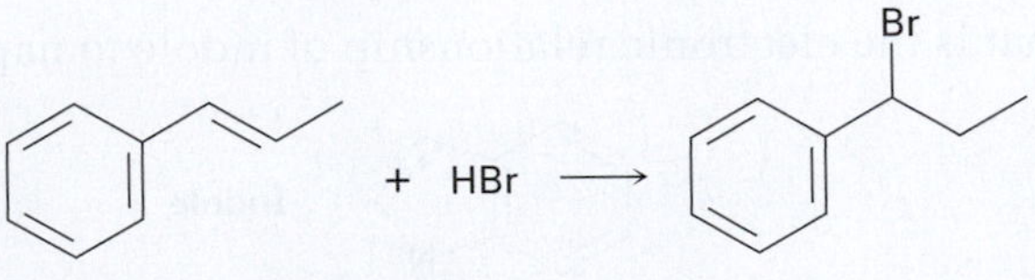

+ HBr ⟶

9.63 Phenylboronic acid, $C_6H_5B(OH)_2$, is nitrated to give 15% ortho-substitution product and 85% meta. Explain the meta-directing effect of the $-B(OH)_2$ group.

9.64 Draw resonance structures of the intermediate carbocations in the bromination of naphthalene, and account for the fact that naphthalene undergoes electrophilic substitution at C1 rather than C2.

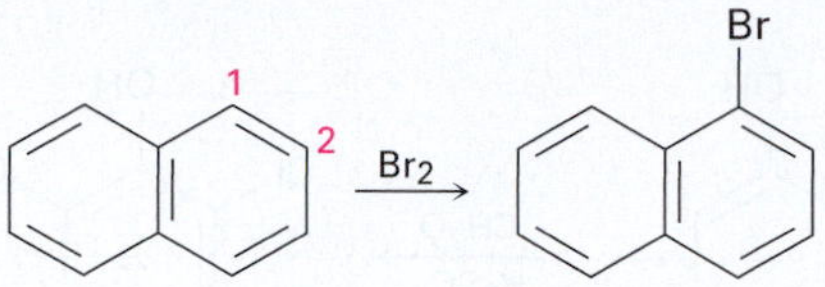

9.65 ■ How would you synthesize the following substances starting from benzene? Assume that ortho- and para-substitution products can be separated.

(a) *p*-Bromoaniline **(b)** *m*-Bromoaniline

(c) 2,4,6-Trinitrobenzoic acid **(d)** 3,5-Dinitrobenzoic acid

9.66 ■ Starting with either benzene or toluene, how would you synthesize the following substances? Assume that ortho and para isomers can be separated.

(a) 2-Bromo-4-nitrotoluene **(b)** 2,4,6-Tribromoaniline

(c) 3-Bromo-4-*tert*-butylbenzoic acid **(d)** 1,3-Dichloro-5-ethylbenzene

9.67 Benzene and alkyl-substituted benzenes can be hydroxylated by reaction with H_2O_2 in the presence of a strong acid catalyst. What is the likely structure of the reactive electrophile? Review Figure 9.11 on page 330, and then propose a mechanism for the reaction.

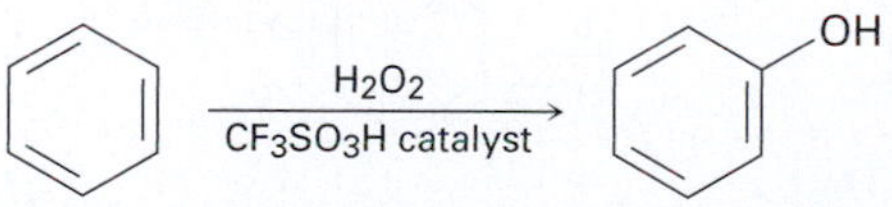

9.68 Propose a mechanism to account for the following reaction:

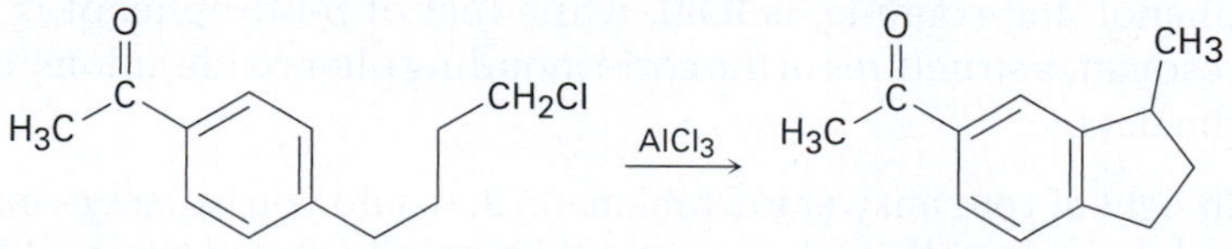

9.69 In the *Gatterman–Koch reaction,* a formyl group (–CHO) is introduced directly onto a benzene ring. For example, reaction of toluene with CO and HCl in the presence of $AlCl_3$ gives *p*-methylbenzaldehyde. Propose a mechanism.

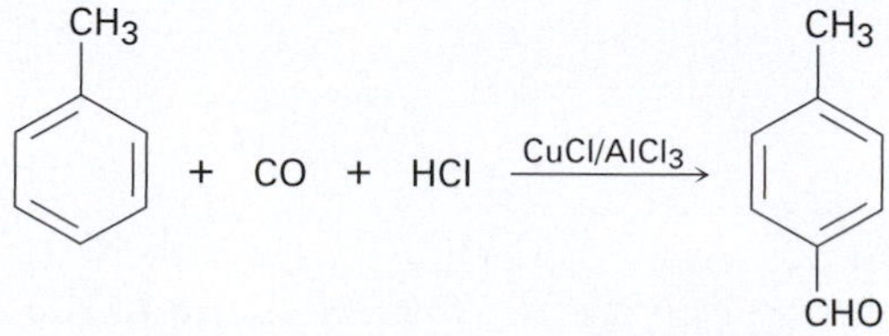

■ Problems assignable in Organic OWL.

9.70 ▪ Hexachlorophene, a substance used in the manufacture of germicidal soaps, is prepared by reaction of 2,4,5-trichlorophenol with formaldehyde in the presence of concentrated sulfuric acid. Propose a mechanism for the reaction.

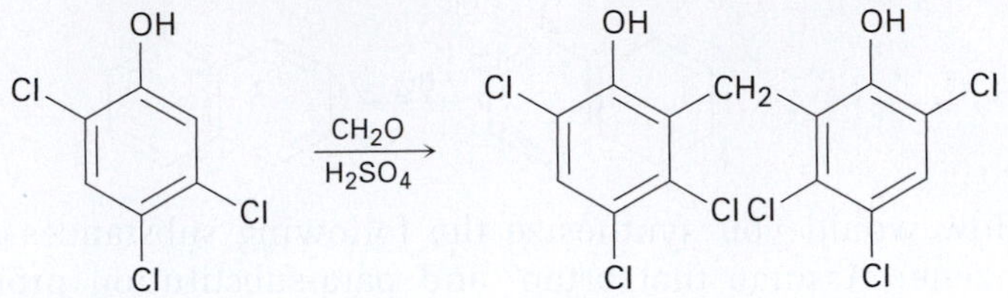

Hexachlorophene

9.71 Use your knowledge of directing effects, along with the following data, to deduce the directions of the dipole moments in aniline and bromobenzene.

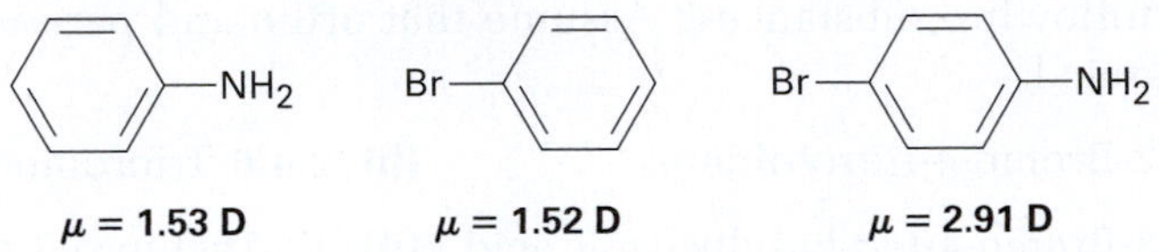

$\mu = 1.53\ D$ $\mu = 1.52\ D$ $\mu = 2.91\ D$

9.72 Identify the reagents represented by the letters **a–c** in the following scheme:

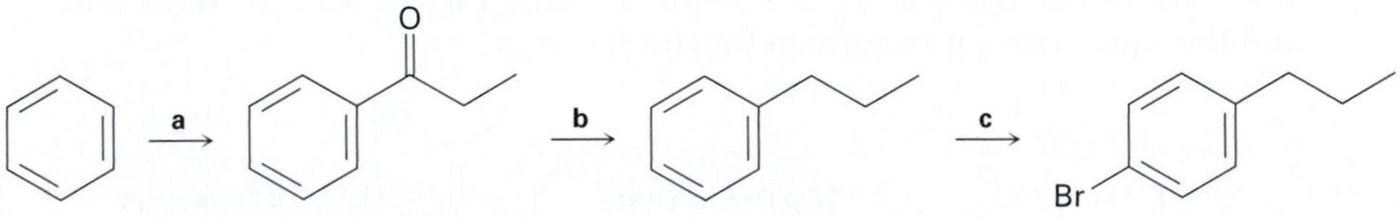

9.73 Phenols (ArOH) are relatively acidic, and the presence of a substituent group on the aromatic ring has a large effect. The pK_a of unsubstituted phenol, for example, is 9.89, while that of p-nitrophenol is 7.15. Draw resonance structures of the corresponding phenoxide anions, and explain the data.

9.74 In light of your answer to Problem 9.73, would you expect p-methylphenol to be more acidic or less acidic than unsubstituted phenol? What about p-bromophenol? Explain.

▪ Problems assignable in Organic OWL.

10 Structure Determination: Mass Spectrometry, Infrared Spectroscopy, and Ultraviolet Spectroscopy

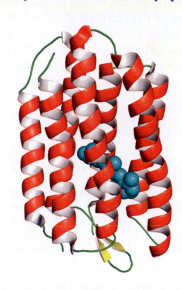

Bacteriorhodopsin is a membrane protein involved in the chemistry of vision.

Determining the structure of an organic compound was a difficult and time-consuming process until the mid-20th century, but powerful techniques are now available that greatly simplify the problem. In this and the next chapter, we'll look at four such techniques—mass spectrometry (MS), infrared (IR) spectroscopy, ultraviolet spectroscopy (UV), and nuclear magnetic resonance spectroscopy (NMR)—and we'll see the kind of information that can be obtained from each.

Mass spectrometry   What is the size and formula?

Infrared spectroscopy   What functional groups are present?

Ultraviolet spectroscopy   Is a conjugated π electron system present?

Nuclear magnetic resonance spectroscopy   What is the carbon–hydrogen framework?

WHY THIS CHAPTER?

Finding the structures of new molecules, whether small ones synthesized in the laboratory or large proteins and nucleic acids found in living organisms, is central to progress in chemistry and biochemistry. We can only scratch the surface of structure determination in this book, but after reading this and the following chapter, you should have a good idea of the range of structural techniques available and of how and when each is used.

 Online homework for this chapter can be assigned in Organic OWL.

367

10.1 Mass Spectrometry of Small Molecules: Magnetic-Sector Instruments

At its simplest, **mass spectrometry (MS)** is a technique for measuring the mass, and therefore the molecular weight (MW), of a molecule. In addition, it's often possible to gain structural information about a molecule by measuring the masses of the fragments produced when molecules are broken apart.

More than 20 different kinds of commercial mass spectrometers are available depending on the intended application, but all have three basic parts: an *ionization source* in which sample molecules are given an electrical charge, a *mass analyzer* in which ions are separated by their mass-to-charge ratio, and a *detector* in which the separated ions are observed and counted.

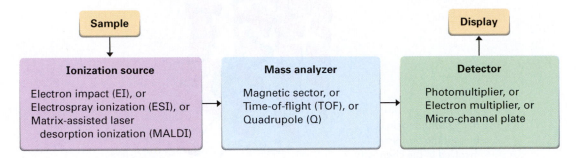

Sample		Display
Ionization source	**Mass analyzer**	**Detector**
Electron impact (EI), or Electrospray ionization (ESI), or Matrix-assisted laser desorption ionization (MALDI)	Magnetic sector, or Time-of-flight (TOF), or Quadrupole (Q)	Photomultiplier, or Electron multiplier, or Micro-channel plate

Perhaps the most common mass spectrometer used for routine purposes in the laboratory is the electron-impact, magnetic-sector instrument shown schematically in Figure 10.1. A small amount of sample is vaporized into the ionization source, where it is bombarded by a stream of high-energy electrons. The energy of the electron beam can be varied but is commonly around 70 electron volts (eV), or 6700 kJ/mol. When a high-energy electron strikes an organic molecule, it dislodges a valence electron from the molecule, producing a *cation radical—cation* because the molecule has lost an electron and now has a positive charge; *radical* because the molecule now has an odd number of electrons.

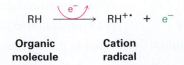

$$RH \xrightarrow{\ e^- \ } RH^{+\cdot} + e^-$$

Organic molecule   **Cation radical**

Electron bombardment transfers so much energy that most of the cation radicals *fragment* after formation. They fly apart into smaller pieces, some of which retain the positive charge, and some of which are neutral. The fragments then flow through a curved pipe in a strong magnetic field, which deflects them into different paths according to their mass-to-charge ratio (m/z). Neutral fragments are not deflected by the magnetic field and are lost on the walls of the pipe, but positively charged fragments are sorted by the mass spectrometer onto a detector, which records them as peaks at the various m/z ratios. Since the number of charges z on each ion is usually 1, the value of m/z for each ion is simply its mass, m. Masses up to approximately 2500 atomic mass units (amu) can be analyzed.

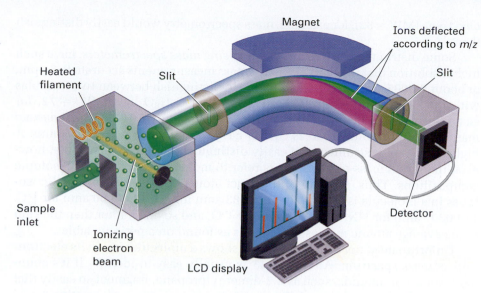

Magnet

Ions deflected
according to m/z

Heated
filament

Slit

Slit

Sample
inlet

Ionizing
electron
beam

LCD display

Detector

ACTIVE FIGURE 10.1 A representation of an electron-ionization, magnetic-sector mass spectrometer. Molecules are ionized by collision with high-energy electrons, causing some of the molecules to fragment. Passage of the charged fragments through a magnetic field then sorts them according to their mass. **Go to this book's student companion site at www.cengage.com/chemistry/mcmurry to explore an interactive version of this figure.**

The **mass spectrum** of a compound is typically presented as a bar graph with masses (m/z values) on the x-axis and intensity, or relative abundance of ions of a given m/z striking the detector, on the y-axis. The tallest peak, assigned an intensity of 100%, is called the **base peak**, and the peak that corresponds to the unfragmented cation radical is called the **parent peak**, or the *molecular ion (M$^+$)*. Figure 10.2 shows the mass spectrum of propane.

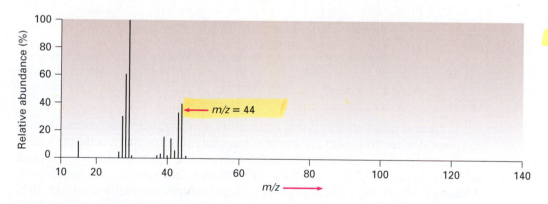

m/z = 44

FIGURE 10.2 Mass spectrum of propane (C$_3$H$_8$; MW = 44).

Mass spectral fragmentation patterns are usually complex, and the molecular ion is often not the base peak. The mass spectrum of propane in Figure 10.2, for instance, shows a molecular ion at m/z = 44 that is only about 30% as high as the base peak at m/z = 29. In addition, many other fragment ions are present.

10.2 Interpreting Mass Spectra

What kinds of information can we get from a mass spectrum? The most obvious information is the molecular weight of the sample, which in itself can be invaluable. If we were given samples of hexane (MW = 86), hex-1-ene (MW = 84), and hex-1-yne (MW = 82), for example, mass spectrometry would easily distinguish them.

Some instruments, called *double-focusing mass spectrometers*, have such high resolution that they provide exact mass measurements accurate to 5 ppm, or about 0.0005 amu, making it possible to distinguish between two formulas with the same nominal mass. Both C_5H_{12} and C_4H_8O have MW = 72, for example, but they differ slightly beyond the decimal point: C_5H_{12} has an exact mass of 72.0939 amu, whereas C_4H_8O has an exact mass of 72.0575 amu. A high-resolution instrument can easily distinguish between them. Note, however, that exact mass measurements refer to molecules with specific isotopic compositions. Thus, the sum of the exact atomic masses of the specific isotopes in a molecule is measured—1.00783 amu for ^1H, 12.00000 amu for ^{12}C, 14.00307 amu for ^{14}N, 15.99491 amu for ^{16}O, and so on—rather than the sum of the average atomic masses of elements as found on a periodic table.

Unfortunately, not every compound shows a molecular ion in its electron-impact mass spectrum. Although M^+ is usually easy to identify if it's abundant, some compounds, such as 2,2-dimethylpropane, fragment so easily that no molecular ion is observed (Figure 10.3). In such cases, alternative "soft" ionization methods that do not use electron bombardment can prevent or minimize fragmentation.

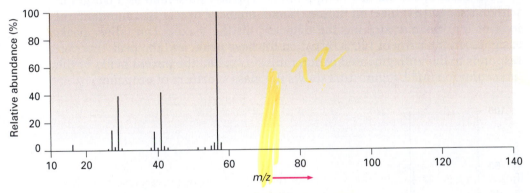

FIGURE 10.3 Mass spectrum of 2,2-dimethyl-propane (C_5H_{12}; MW = 72). No molecular ion is observed when electron-impact ionization is used. (What do you think is the structure of the M^+ peak at m/z = 57?)

A further point about mass spectrometry, noticeable in the spectra of both propane (Figure 10.2) and 2,2-dimethylpropane (Figure 10.3), is that the peak for the molecular ion is not at the highest m/z value. There is also a small peak at M+1 because of the presence of different isotopes in the molecules. Although ^{12}C is the most abundant carbon isotope, a small amount (1.10% natural abundance) of ^{13}C is also present. Thus, a certain percentage of the molecules analyzed in the mass spectrometer are likely to contain a ^{13}C atom, giving rise to the observed M+1 peak. In addition, a small amount of ^2H (deuterium; 0.015% natural abundance) is present, making a further contribution to the M+1 peak.

In addition to obtaining molecular weight, it's also possible to derive structural information about a molecule by interpreting its fragmentation pattern. Fragmentation occurs when the high-energy cation radical flies apart by spontaneous cleavage of a chemical bond. One of the two fragments retains the positive charge and is a carbocation, while the other fragment is a neutral radical.

Not surprisingly, the positive charge often remains with the fragment that is best able to stabilize it. In other words, a relatively stable carbocation is often formed during fragmentation. For example, 2,2-dimethylpropane tends to fragment in such a way that the positive charge remains with the *tert*-butyl group. 2,2-Dimethylpropane therefore has a base peak at $m/z = 57$, corresponding to $C_4H_9^+$ (Figure 10.3).

$$\left[\begin{array}{c} CH_3 \\ | \\ H_3C-C-CH_3 \\ | \\ CH_3 \end{array}\right]^{+\cdot} \longrightarrow \begin{array}{c} CH_3 \\ | \\ H_3C-C^+ \\ | \\ CH_3 \end{array} + \ \cdot CH_3$$

$$m/z = 57$$

Because mass-spectral fragmentation patterns are usually complex, it's often difficult to assign structures to fragment ions. Most hydrocarbons fragment in many ways, as the mass spectrum of hexane shown in Figure 10.4 demonstrates. The hexane spectrum shows a moderately abundant molecular ion at $m/z = 86$ and fragment ions at $m/z = 71, 57, 43,$ and 29. Since all the carbon–carbon bonds of hexane are electronically similar, all break to a similar extent, giving rise to the observed mixture of ions.

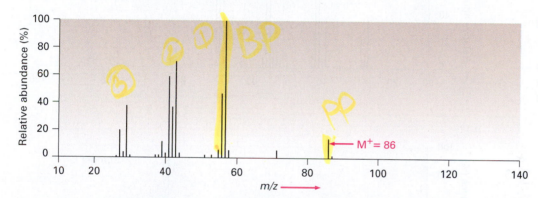

FIGURE 10.4 Mass spectrum of hexane (C_6H_{14}; MW = 86). The base peak is at $m/z = 57$, and numerous other ions are present.

Figure 10.5 shows how the hexane fragments might arise. The loss of a methyl radical from the hexane cation radical ($M^+ = 86$) gives rise to a fragment of mass 71, the loss of an ethyl radical accounts for a fragment of mass 57, the loss of a propyl radical accounts for a fragment of mass 43, and the loss of a butyl radical accounts for a fragment of mass 29. With skill and practice, chemists can learn to analyze the fragmentation patterns of unknown compounds and work backward to a structure that is compatible with the data.

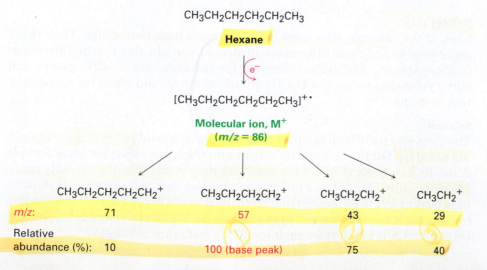

FIGURE 10.5 Fragmentation of hexane in a mass spectrometer.

We'll see in the next section and in later chapters that specific functional groups, such as alcohols, ketones, aldehydes, and amines, show specific kinds of mass spectral fragmentations that can be interpreted to provide structural information.

WORKED EXAMPLE 10.1 Using Mass Spectra to Identify Compounds

Assume that you have two unlabeled samples, one of methylcyclohexane and the other of ethylcyclopentane. How could you use mass spectrometry to identify them? The mass spectra of both are shown in Figure 10.6.

FIGURE 10.6 Mass spectra of unlabeled samples A and B for Worked Example 10.1.

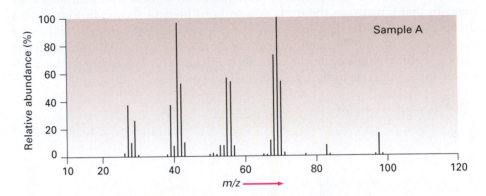

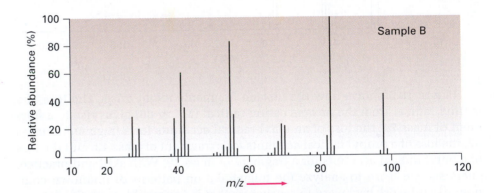

Strategy

Look at the possible structures and determine how they differ. Then think about how any of these differences in structure might give rise to differences in mass spectra. Methylcyclohexane, for instance, has a $-CH_3$ group, and ethylcyclopentane has a $-CH_2CH_3$ group, which should affect the fragmentation patterns.

Solution

The mass spectra of both samples show molecular ions at $M^+ = 98$, corresponding to C_7H_{14}, but the two spectra differ in their fragmentation patterns. Sample A has its base peak at $m/z = 69$, corresponding to the loss of a CH_2CH_3 group (29 mass units), but B has a rather small peak at $m/z = 69$. Sample B shows a base peak at $m/z = 83$, corresponding to the loss of a CH_3 group (15 mass units), but sample A has only a small peak at $m/z = 83$. We can therefore be reasonably certain that A is ethylcyclopentane and B is methylcyclohexane.

Problem 10.1

The male sex hormone testosterone contains only C, H, and O and has a mass of 288.2089 amu, as determined by high-resolution mass spectrometry. What is the likely molecular formula of testosterone?

Problem 10.2

Two mass spectra are shown in Figure 10.7. One spectrum corresponds to 2-methylpent-2-ene; the other, to hex-2-ene. Which is which? Explain.

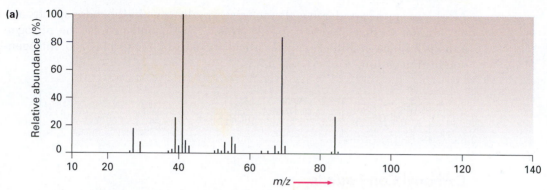

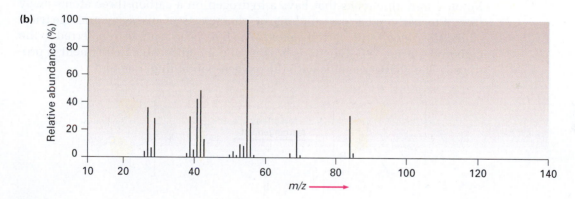

FIGURE 10.7
Mass spectra for Problem 10.2.

10.3 Mass Spectrometry of Some Common Functional Groups

As each functional group is discussed in future chapters, mass-spectral fragmentations characteristic of that group will be described. As a preview, though, we'll point out some distinguishing features of several common functional groups.

Alcohols

Alcohols undergo fragmentation in the mass spectrometer by two pathways: *alpha cleavage* and *dehydration*. In the α–cleavage pathway, a C–C bond nearest the hydroxyl group is broken, yielding a neutral radical plus a resonance-stabilized, oxygen-containing cation.

In the dehydration pathway, water is eliminated, yielding an alkene radical cation with a mass 18 units less than M⁺:

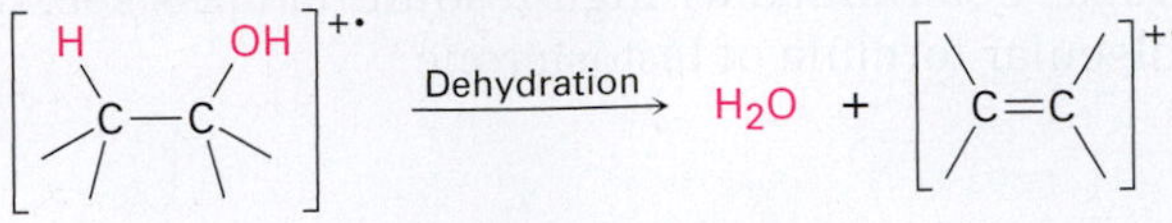

$$\left[\begin{array}{c} \text{H} \quad\quad \text{OH} \\ \diagdown \text{C} - \text{C} \diagup \end{array} \right]^{+\cdot} \xrightarrow{\text{Dehydration}} \text{H}_2\text{O} \ + \ \left[\begin{array}{c} \diagdown \text{C} = \text{C} \diagup \end{array} \right]^{+\cdot}$$

Amines

Aliphatic amines undergo a characteristic α cleavage in the mass spectrometer, similar to that observed for alcohols. A C–C bond nearest the nitrogen atom is broken, yielding an alkyl radical and a resonance-stabilized, nitrogen-containing cation.

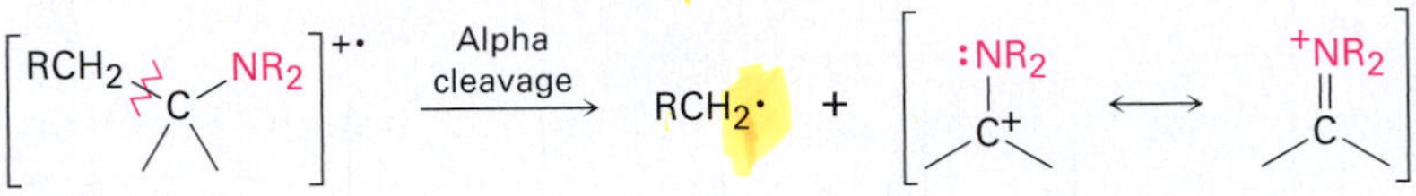

Radical

$$\left[\text{RCH}_2 \diagdown \text{C} \diagdown \text{NR}_2 \right]^{+\cdot} \xrightarrow[\text{cleavage}]{\text{Alpha}} \text{RCH}_2\cdot \ + \ \left[\begin{array}{c} :\text{NR}_2 \\ \diagup \text{C}^+ \diagdown \end{array} \longleftrightarrow \begin{array}{c} ^+\text{NR}_2 \\ \diagup \text{C} \diagdown \end{array} \right]$$

Carbonyl Compounds

Ketones and aldehydes that have a hydrogen on a carbon three atoms away from the carbonyl group undergo a characteristic mass-spectral cleavage called the *McLafferty rearrangement*. The hydrogen atom is transferred to the carbonyl oxygen, a C–C bond is broken, and a neutral alkene fragment is produced. The charge remains with the oxygen-containing fragment.

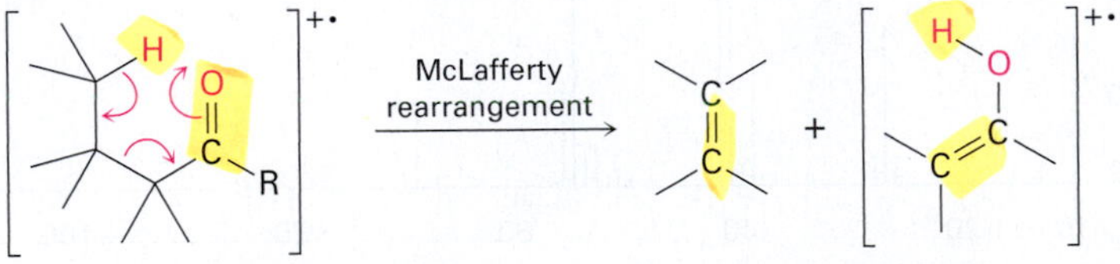

$$\left[\begin{array}{c} \text{H} \\ \text{O} \\ \| \\ \text{C} \diagdown \text{R} \end{array} \right]^{+\cdot} \xrightarrow[\text{rearrangement}]{\text{McLafferty}} \begin{array}{c} \diagup \text{C} \\ \| \\ \text{C} \diagdown \end{array} \ + \ \left[\begin{array}{c} \text{H} \quad \text{O} \\ \diagdown \text{C} = \text{C} \diagdown \end{array} \right]^{+\cdot}$$

In addition, ketones and aldehydes frequently undergo α cleavage of the bond between the carbonyl group and the neighboring carbon. Alpha cleavage yields a neutral radical and a resonance-stabilized acyl cation.

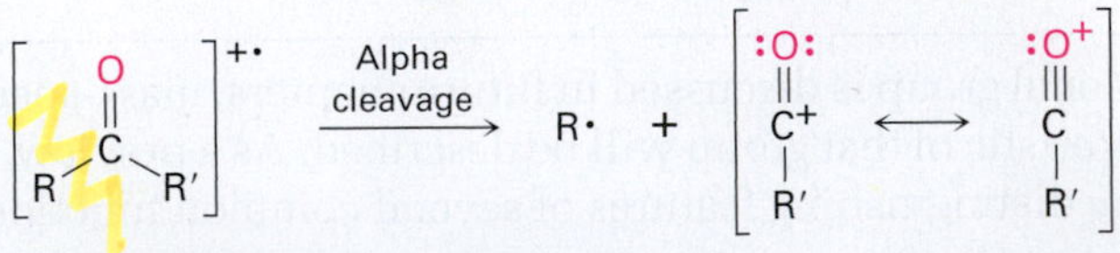

$$\left[\begin{array}{c} \text{O} \\ \| \\ \text{R} \diagup \text{C} \diagdown \text{R}' \end{array} \right]^{+\cdot} \xrightarrow[\text{cleavage}]{\text{Alpha}} \text{R}\cdot \ + \ \left[\begin{array}{c} :\text{O}: \\ \| \\ \text{C}^+ \\ | \\ \text{R}' \end{array} \longleftrightarrow \begin{array}{c} :\text{O}^+ \\ \| \\ \text{C} \\ | \\ \text{R}' \end{array} \right]$$

WORKED EXAMPLE 10.2 Identifying Fragmentation Patterns in a Mass Spectrum

The mass spectrum of 2-methylpentan-3-ol is shown in Figure 10.8. What fragments can you identify?

Strategy

Calculate the mass of the molecular ion, and identify the functional groups in the molecule. Then write the fragmentation processes you might expect, and compare the masses of the resultant fragments with the peaks present in the spectrum.

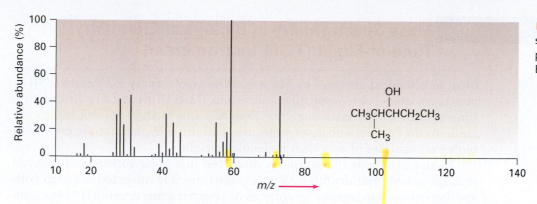

Solution

2-Methylpentan-3-ol, an open-chain alcohol, has $M^+ = 102$ and might be expected to fragment by α cleavage and by dehydration. These processes would lead to fragment ions of $m/z = 84$, 73, and 59. Of the three expected fragments, dehydration is not observed (no $m/z = 84$ peak), but both α cleavages take place ($m/z = 73, 59$).

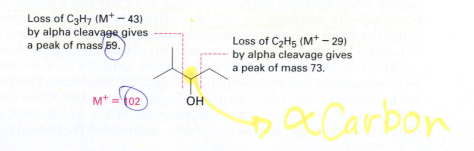

Problem 10.3

What are the masses of the charged fragments produced in the following cleavage pathways?

(a) Alpha cleavage of pentan-2-one ($CH_3COCH_2CH_2CH_3$)
(b) Dehydration of cyclohexanol
(c) McLafferty rearrangement of 4-methylpentan-2-one [$CH_3COCH_2CH(CH_3)_2$]
(d) Alpha cleavage of triethylamine [$(CH_3CH_2)_3N$]

Problem 10.4

List the masses of the parent ion and of several fragments you might expect to find in the mass spectrum of the following molecule:

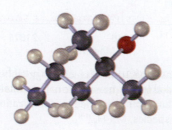

10.4   Mass Spectrometry in Biological Chemistry: Time-of-Flight (TOF) Instruments

Most biochemical analyses by MS use either electrospray ionization (ESI) or matrix-assisted laser desorption ionization (MALDI), typically linked to a time-of-flight (TOF) mass analyzer. Both ESI and MALDI are soft ionization methods that produce charged molecules with little fragmentation, even with biological samples of very high molecular weight.

In an ESI source, the sample is dissolved in a polar solvent and sprayed through a steel capillary tube. As it exits the tube, it is subjected to a high voltage that causes it to become protonated by removing one or more H^+ ions from the solvent. The volatile solvent is then evaporated, giving variably protonated sample molecules ($M+H_n^{n+}$). In a MALDI source, the sample is adsorbed onto a suitable matrix compound, such as 2,5-dihydroxybenzoic acid, which is ionized by a short burst of laser light. The matrix compound then transfers the energy to the sample and protonates it, forming $M+H_n^{n+}$ ions.

Following ion formation, the variably protonated sample molecules are electrically focused into a small packet with a narrow spatial distribution, and the packet is given a sudden kick of energy by an accelerator electrode. Since each molecule in the packet is given the same energy, $E = mv^2/2$, it begins moving with a velocity that depends on the square root of its mass, $v = \sqrt{2E/m}$. Lighter molecules move faster, and heavier molecules move slower. The analyzer itself—the *drift tube*—is simply an electrically grounded metal tube inside which the different charged molecules become separated as they move along at different velocities and take different amounts of time to complete their flight.

The TOF technique is considerably more sensitive than the magnetic-sector alternative, and protein samples of up to 100 kilodaltons (100,000 amu) can be separated with a mass accuracy of 3 ppm. Figure 10.9 shows a MALDI–TOF spectrum of chicken egg-white lysozyme, MW = 14,306.7578 daltons. (Biochemists generally use the unit *dalton,* abbreviated Da, instead of amu.)

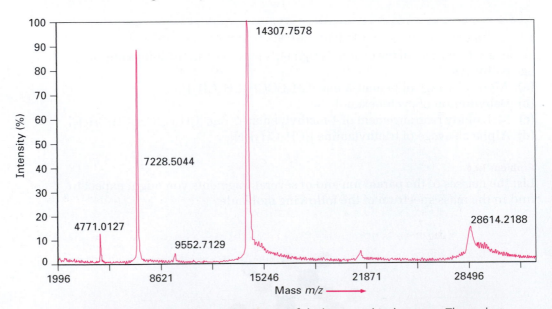

FIGURE 10.9 MALDI–TOF mass spectrum of chicken egg-white lysozyme. The peak at 14,307.7578 daltons (amu) is due to the monoprotonated protein, $M+H^+$, and that at 28,614.2188 daltons is due to an impurity formed by dimerization of the protein. Other peaks are various protonated species, $M+H_n^{n+}$.

10.5 Spectroscopy and the Electromagnetic Spectrum

Infrared, ultraviolet, and nuclear magnetic resonance spectroscopies differ from mass spectrometry in that they are nondestructive and involve the interaction of molecules with electromagnetic energy rather than with an ionizing source. Before beginning a study of these techniques, however, let's briefly review the nature of radiant energy and the electromagnetic spectrum.

Visible light, X rays, microwaves, radio waves, and so forth, are all different kinds of *electromagnetic radiation*. Collectively, they make up the **electromagnetic spectrum**, shown in Figure 10.10. The spectrum is arbitrarily divided into regions, with the familiar visible region accounting for only a small portion, from 3.8×10^{-7} m to 7.8×10^{-7} m in wavelength. The visible region is flanked by the infrared and ultraviolet regions.

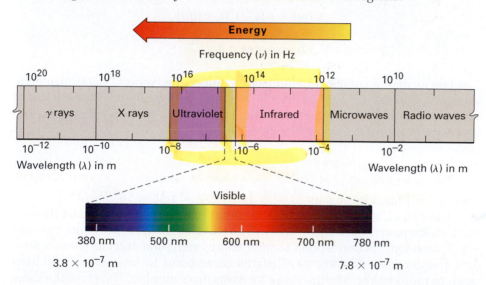

FIGURE 10.10 The electromagnetic spectrum covers a continuous range of wavelengths and frequencies, from radio waves at the low-frequency end to gamma (γ) rays at the high-frequency end. The familiar visible region accounts for only a small portion near the middle of the spectrum.

Electromagnetic radiation is often said to have dual behavior. In some respects, it has the properties of a particle (called a *photon*), yet in other respects it behaves as an energy wave. Like all waves, electromagnetic radiation is characterized by a *wavelength*, a *frequency*, and an *amplitude* (Figure 10.11). The

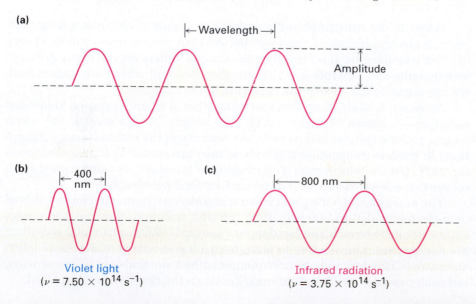

FIGURE 10.11 Electromagnetic waves are characterized by a wavelength, a frequency, and an amplitude. **(a)** Wavelength (λ) is the distance between two successive wave maxima. Amplitude is the height of the wave measured from the center. **(b)**, **(c)** What we perceive as different kinds of electromagnetic radiation are simply waves with different wavelengths and frequencies.

wavelength, λ (Greek lambda), is the distance from one wave maximum to the next. The **frequency**, ν (Greek nu), is the number of waves that pass by a fixed point per unit time, usually given in reciprocal seconds (s^{-1}), or **hertz**, **Hz** ($1 \text{ Hz} = 1 \text{ s}^{-1}$). The **amplitude** is the height of a wave, measured from midpoint to peak. The intensity of radiant energy, whether a feeble glow or a blinding glare, is proportional to the square of the wave's amplitude.

Multiplying the wavelength of a wave in meters (m) by its frequency in reciprocal seconds (s^{-1}) gives the speed of the wave in meters per second (m/s). The rate of travel of all electromagnetic radiation in a vacuum is a constant value, commonly called the "speed of light" and abbreviated c. Its numerical value is defined as exactly $2.997\ 924\ 58 \times 10^8$ m/s, often rounded off to 3.00×10^8 m/s.

$$\text{Wavelength} \times \text{Frequency} = \text{Speed}$$

$$\lambda \text{ (m)} \times \nu \text{ (s}^{-1}) = c \text{ (m/s)}$$

$$\lambda = \frac{c}{\nu} \quad \text{or} \quad \nu = \frac{c}{\lambda}$$

Just as matter comes only in discrete units called atoms, electromagnetic energy is transmitted only in discrete amounts called *quanta*. The amount of energy, ϵ, corresponding to 1 quantum of energy (1 photon) of a given frequency ν is expressed by the Planck equation

$$\epsilon = h\nu = \frac{hc}{\lambda}$$

where h = Planck's constant (6.62×10^{-34} J·s = 1.58×10^{-34} cal·s).

The Planck equation says that the energy of a given photon varies *directly* with its frequency ν but *inversely* with its wavelength λ. High frequencies and short wavelengths correspond to high-energy radiation such as gamma rays; low frequencies and long wavelengths correspond to low-energy radiation such as radio waves. Multiplying ϵ by Avogadro's number, N_A, gives the same equation in more familiar units, where E represents the energy of Avogadro's number (one "mole") of photons of wavelength λ:

$$E = \frac{N_A hc}{\lambda} = \frac{1.20 \times 10^{-4} \text{ kJ/mol}}{\lambda \text{ (m)}} \quad \text{or} \quad \frac{2.86 \times 10^{-5} \text{ kcal/mol}}{\lambda \text{ (m)}}$$

When an organic compound is exposed to a beam of electromagnetic radiation, it absorbs energy of some wavelengths but passes, or transmits, energy of other wavelengths. If we irradiate the sample with energy of many different wavelengths and determine which are absorbed and which are transmitted, we can measure the **absorption spectrum** of the compound.

An example of an absorption spectrum—that of ethanol exposed to infrared radiation—is shown in Figure 10.12. The horizontal axis records the wavelength, and the vertical axis records the intensity of the various energy absorptions in percent transmittance. The baseline corresponding to 0% absorption (or 100% transmittance) runs along the top of the chart, so a downward spike means that energy absorption has occurred at that wavelength.

The energy a molecule gains when it absorbs radiation must be distributed over the molecule in some way. With infrared radiation, the absorbed energy causes bonds to stretch and bend more vigorously. With ultraviolet radiation, the energy causes an electron to jump from a lower-energy orbital to a higher-energy one. Different radiation frequencies affect molecules in different ways, but each provides structural information when the results are interpreted.

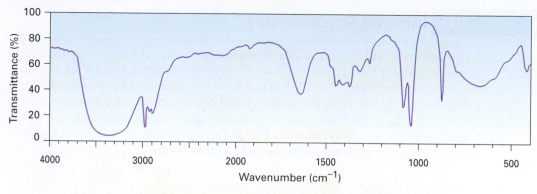

FIGURE 10.12 An infrared absorption spectrum of ethanol, CH_3CH_2OH. A transmittance of 100% means that all the energy is passing through the sample, whereas a lower transmittance means that some energy is being absorbed. Thus, each downward spike corresponds to an energy absorption.

There are many kinds of spectroscopies, which differ according to the region of the electromagnetic spectrum that is used. We'll look at three—infrared spectroscopy, ultraviolet spectroscopy, and nuclear magnetic resonance spectroscopy. Let's begin by seeing what happens when an organic sample absorbs infrared energy.

WORKED EXAMPLE 10.3 Correlating Energy and Frequency of Radiation

Which is higher in energy, FM radio waves with a frequency of 1.015×10^8 Hz (101.5 MHz) or visible green light with a frequency of 5×10^{14} Hz?

Strategy

Remember the equations $\epsilon = h\nu$ and $\epsilon = hc/\lambda$, which say that energy increases as frequency increases and as wavelength decreases.

Solution

Since visible light has a higher frequency than radio waves, it is higher in energy.

Problem 10.5

Which has higher energy, infrared radiation with $\lambda = 1.0 \times 10^{-6}$ m or an X ray with $\lambda = 3.0 \times 10^{-9}$ m? Radiation with $\nu = 4.0 \times 10^9$ Hz or with $\lambda = 9.0 \times 10^{-6}$ m?

Problem 10.6

It's useful to develop a feeling for the amounts of energy that correspond to different parts of the electromagnetic spectrum. Calculate the energies in kJ/mol of each of the following kinds of radiation:
(a) A gamma ray with $\lambda = 5.0 \times 10^{-11}$ m
(b) An X ray with $\lambda = 3.0 \times 10^{-9}$ m
(c) Ultraviolet light with $\nu = 6.0 \times 10^{15}$ Hz
(d) Visible light with $\nu = 7.0 \times 10^{14}$ Hz
(e) Infrared radiation with $\lambda = 2.0 \times 10^{-5}$ m
(f) Microwave radiation with $\nu = 1.0 \times 10^{11}$ Hz

10.6 | Infrared Spectroscopy

The **infrared (IR)** region of the electromagnetic spectrum covers the range from just above the visible (7.8×10^{-7} m) to approximately 10^{-4} m, but only the midportion from 2.5×10^{-6} m to 2.5×10^{-5} m is used by organic chemists (Figure 10.13). Wavelengths within the IR region are usually given in micrometers (1 μm = 10^{-6} m), and frequencies are given in *wavenumbers* rather than in hertz. The **wavenumber ($\tilde{\nu}$)** is the reciprocal of the wavelength in centimeters and is therefore expressed in units of cm^{-1}:

$$\text{Wavenumber: } \tilde{\nu}(\text{cm}^{-1}) = \frac{1}{\lambda \text{ (cm)}}$$

Thus, the useful IR region is from 4000 to 400 cm^{-1}, corresponding to energies of 48.0 kJ/mol to 4.80 kJ/mol (11.5–1.15 kcal/mol).

FIGURE 10.13 The infrared region of the electromagnetic spectrum.

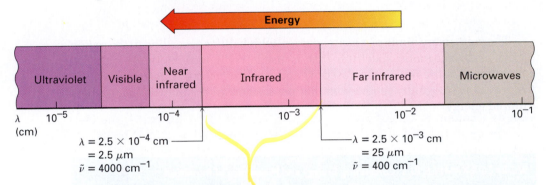

λ (cm)

$\lambda = 2.5 \times 10^{-4}$ cm
$= 2.5\ \mu$m
$\tilde{\nu} = 4000$ cm^{-1}

$\lambda = 2.5 \times 10^{-3}$ cm
$= 25\ \mu$m
$\tilde{\nu} = 400$ cm^{-1}

Why does an organic molecule absorb some wavelengths of IR radiation but not others? All molecules have a certain amount of energy and are in constant motion. Their bonds stretch and contract, atoms wag back and forth, and other molecular vibrations occur. Some of the kinds of allowed vibrations are shown:

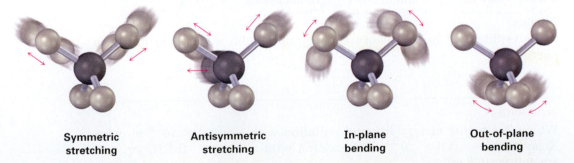

Symmetric stretching   **Antisymmetric stretching**   **In-plane bending**   **Out-of-plane bending**

The amount of energy a molecule contains is not continuously variable but is *quantized*. That is, a molecule can stretch or bend only at specific frequencies corresponding to specific energy levels. Take bond stretching, for example. Although we usually speak of bond lengths as if they were fixed, the numbers given are really averages. In fact, a typical C–H bond with an average bond length of 110 pm is actually vibrating at a specific frequency, alternately stretching and contracting as if there were a spring connecting the two atoms.

When a molecule is irradiated with electromagnetic radiation, energy is absorbed if the frequency of the radiation matches the frequency of the vibration. The result of this energy absorption is an increased amplitude for the vibration; in other words, the "spring" connecting the two atoms stretches and compresses a bit further. Since each frequency absorbed by a molecule corresponds to a specific molecular motion, we can find what kinds of motions a molecule has by measuring its IR spectrum. By then interpreting those motions, we can find out what kinds of bonds (functional groups) are present in the molecule.

IR spectrum → What molecular motions? → What functional groups?

10.7 Interpreting Infrared Spectra

Complete interpretation of an IR spectrum is difficult because most organic molecules have dozens of different bond stretching and bending motions and thus have dozens of absorptions. On the one hand, this complexity is a problem because it generally limits the laboratory use of IR spectroscopy to pure samples of fairly small molecules—little can be learned from IR spectroscopy of large, complex biomolecules. On the other hand, the complexity is useful because an IR spectrum serves as a unique fingerprint of a compound. In fact, the complex region of the IR spectrum from 1500 cm^{-1} to around 400 cm^{-1} is called the *fingerprint region*. If two samples have identical IR spectra, they are almost certainly identical compounds.

Fortunately, we don't need to interpret an IR spectrum fully to get useful structural information. Most functional groups have characteristic IR absorption bands that don't change from one compound to another. The C=O absorption of a ketone is almost always in the range 1670 to 1750 cm^{-1}, the O–H absorption of an alcohol is almost always in the range 3400 to 3650 cm^{-1}, the C=C absorption of an alkene is almost always in the range 1640 to 1680 cm^{-1}, and so forth. By learning where characteristic functional-group absorptions occur, it's possible to get structural information from IR spectra. Table 10.1 lists the characteristic IR bands of some common functional groups.

TABLE 10.1
Characteristic IR Absorptions of Some Functional Groups

Functional group	Absorption (cm^{-1})	Intensity	Functional group	Absorption (cm^{-1})	Intensity
Alkane			Amine		
C–H	2850–2960	Medium	N–H	3300–3500	Medium
Alkene			C–N	1030–1230	Medium
=C–H	3020–3100	Medium	Carbonyl		
C=C	1640–1680	Medium	compound		
Alkyne			C=O	1670–1780	Strong
≡C–H	3300	Strong	Carboxylic acid		
C≡C	2100–2260	Medium	O–H	2500–3100	Strong, broad
Alkyl halide			Nitrile		
C–Cl	600–800	Strong	C≡N	2210–2260	Medium
C–Br	500–600	Strong	Nitro		
Alcohol			NO$_2$	1540	Strong
O–H	3400–3650	Strong, broad			
C–O	1050–1150	Strong			
Arene					
C–H	3030	Weak			
Aromatic ring	1660–2000	Weak			
	1450–1600	Medium			

Broad = OH
Strong = OH, Cl, Br NO$_2$

Look at the IR spectra of hexane, hex-1-ene, and hex-1-yne in Figure 10.14 to see an example of how IR spectroscopy can be used. Although all three IR spectra contain many peaks, there are characteristic absorptions of the C=C and C≡C functional groups that allow the three compounds to be distinguished. Thus, hex-1-ene shows a characteristic C=C absorption at 1660 cm⁻¹ and a vinylic =C–H absorption at 3100 cm⁻¹, whereas hex-1-yne has a C≡C absorption at 2100 cm⁻¹ and a terminal alkyne ≡C–H absorption at 3300 cm⁻¹.

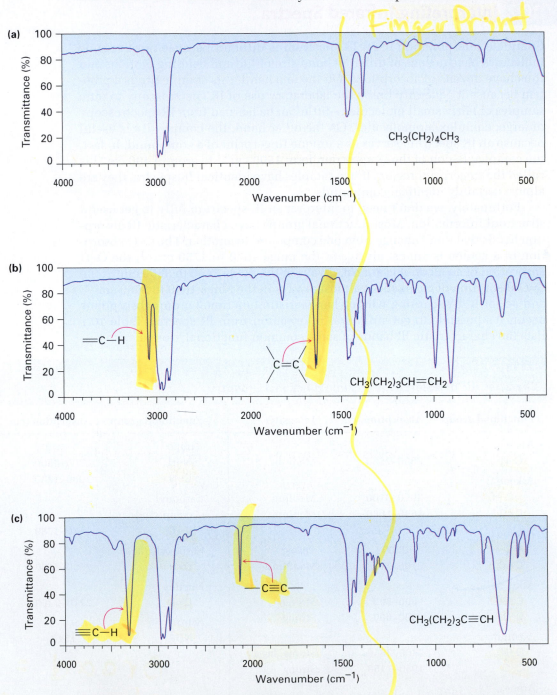

FIGURE 10.14 IR spectra of **(a)** hexane, **(b)** hex-1-ene, and **(c)** hex-1-yne. Spectra like these are easily obtained on submilligram amounts of material in a few minutes using commercially available instruments.

It helps in remembering the position of specific IR absorptions to divide the IR region from 4000 cm^{-1} to 400 cm^{-1} into four parts, as shown in Figure 10.15:

- The region from 4000 to 2500 cm^{-1} corresponds to absorptions caused by N–H, C–H, and O–H single-bond stretching motions. N–H and O–H bonds absorb in the 3300 to 3600 cm^{-1} range; C–H bond stretching occurs near 3000 cm^{-1}.
- The region from 2500 to 2000 cm^{-1} is where triple-bond stretching occurs. Both C≡N and C≡C bonds absorb here.
- The region from 2000 to 1500 cm^{-1} is where double bonds (C=O, C=N, and C=C) absorb. Carbonyl groups generally absorb in the range 1670 to 1780 cm^{-1}, and alkene stretching normally occurs in the narrow range 1640 to 1680 cm^{-1}.
- The region below 1500 cm^{-1} is the fingerprint portion of the IR spectrum. A large number of absorptions due to a variety of C–C, C–O, C–N, and C–X single-bond vibrations occur here.

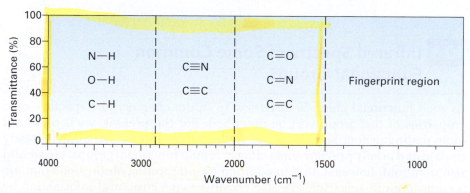

FIGURE 10.15 The four regions of the infrared spectrum: single bonds to hydrogen, triple bonds, double bonds, and fingerprint.

Why do different functional groups absorb where they do? As noted previously, a good analogy is that of two weights (atoms) connected by a spring (a bond). Short, strong bonds vibrate at a higher energy and higher frequency than do long, weak bonds, just as a short, strong spring vibrates faster than a long, weak spring. Thus, triple bonds absorb at a higher frequency than double bonds, which in turn absorb at a higher frequency than single bonds. In addition, springs connecting small weights vibrate faster than springs connecting large weights. Thus, C–H, O–H, and N–H bonds vibrate at a higher frequency than bonds between heavier C, O, and N atoms.

WORKED EXAMPLE 10.4 Distinguishing Isomeric Compounds by IR Spectroscopy

Acetone (CH$_3$COCH$_3$) and prop-2-en-1-ol (H$_2$C=CHCH$_2$OH) are isomers. How could you distinguish them by IR spectroscopy?

Strategy
Identify the functional groups in each molecule, and refer to Table 10.1.

Solution

Acetone has a strong C=O absorption at 1715 cm^{-1}, while prop-2-en-1-ol has an –OH absorption at 3500 cm^{-1} and a C=C absorption at 1660 cm^{-1}.

Problem 10.7

What functional groups might the following molecules contain?

(a) A compound with a strong absorption at 1710 cm^{-1}
(b) A compound with a strong absorption at 1540 cm^{-1}
(c) A compound with strong absorptions at 1720 cm^{-1} and at 2500 to 3100 cm^{-1}

Problem 10.8

How might you use IR spectroscopy to distinguish between the following pairs of isomers?

(a) CH_3CH_2OH and CH_3OCH_3
(b) Cyclohexane and hex-1-ene
(c) $CH_3CH_2CO_2H$ and $HOCH_2CH_2CHO$

10.8 Infrared Spectra of Some Common Functional Groups

As each functional group is discussed in future chapters, the spectroscopic properties of that group will be described. For the present, we'll point out some distinguishing features of the hydrocarbon functional groups already studied and briefly preview some other common functional groups. We should also point out, however, that in addition to interpreting absorptions that *are* present in an IR spectrum, it's also possible to get structural information by noticing which absorptions are *not* present. If the spectrum of a compound has no absorptions at 3300 and 2150 cm^{-1}, the compound is not a terminal alkyne; if the spectrum has no absorption near 3400 cm^{-1}, the compound is not an alcohol; and so on.

Alkanes

The IR spectrum of an alkane is fairly uninformative because no functional groups are present and all absorptions are due to C–H and C–C bonds. Alkane C–H bonds show a strong absorption from 2850 to 2960 cm^{-1}, and saturated C–C bonds show a number of bands in the 800 to 1300 cm^{-1} range. Since most organic compounds contain saturated alkane-like portions, most organic compounds have these characteristic IR absorptions. The C–H and C–C bands are clearly visible in the three spectra shown in Figure 10.14.

Alkanes   —C—H   2850–2960 cm^{-1}

—C—C—   800–1300 cm^{-1}

Alkenes

Alkenes show several characteristic stretching absorptions. Vinylic =C–H bonds absorb from 3020 to 3100 cm^{-1}, and alkene C=C bonds usually absorb near 1650 cm^{-1}, although in some cases the peaks can be rather small and difficult to see clearly. Both absorptions are visible in the hex-1-ene spectrum in Figure 10.14b.

Monosubstituted and disubstituted alkenes have characteristic =C–H out-of-plane bending absorptions in the 700 to 1000 cm^{-1} range, thereby allowing the substitution pattern on a double bond to be determined. Monosubstituted alkenes such as hex-1-ene show strong characteristic bands at 910 and 990 cm^{-1}, and 2,2-disubstituted alkenes (R$_2$C=CH$_2$) have an intense band at 890 cm^{-1}.

Alkenes	=C–H	3020–3100 cm^{-1}
	C=C	1640–1680 cm^{-1}
	RCH=CH$_2$	910 and 990 cm^{-1}
	R$_2$C=CH$_2$	890 cm^{-1}

[handwritten: Higher A when connected to –H]

Alkynes

Alkynes show a C≡C stretching absorption at 2100 to 2260 cm^{-1}, an absorption that is much more intense for terminal alkynes than for internal alkynes. In fact, symmetrically substituted triple bonds like that in hex-3-yne show no absorption at all, for reasons we won't go into. Terminal alkynes such as hex-1-yne also have a characteristic ≡C–H stretch at 3300 cm^{-1} (Figure 10.14c). This band is diagnostic for terminal alkynes because it is fairly intense and quite sharp.

| Alkynes | —C≡C— | 2100–2260 cm^{-1} |
| | ≡C–H | 3300 cm^{-1} |

Aromatic Compounds

Aromatic compounds, such as benzene, have a weak C–H stretching absorption at 3030 cm^{-1}, just to the left of a typical saturated C–H band. In addition, up to four absorptions are observed in the 1450 to 1600 cm^{-1} region because of complex molecular motions of the ring itself. Two bands, one at 1500 cm^{-1} and one at 1600 cm^{-1}, are usually the most intense. In addition, aromatic compounds show weak absorptions in the 1660 to 2000 cm^{-1} region and strong absorptions in the 690 to 900 cm^{-1} range due to C–H out-of-plane bending. The exact position of both sets of absorptions is diagnostic of the substitution pattern of the aromatic ring.

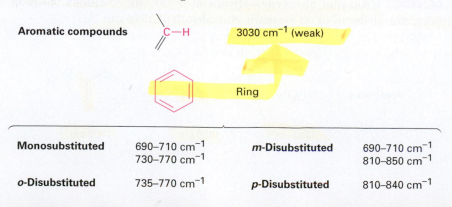

Aromatic compounds   C–H   3030 cm^{-1} (weak)

Ring

| Monosubstituted | 690–710 cm^{-1}
730–770 cm^{-1} | *m*-Disubstituted | 690–710 cm^{-1}
810–850 cm^{-1} |
| *o*-Disubstituted | 735–770 cm^{-1} | *p*-Disubstituted | 810–840 cm^{-1} |

The IR spectrum of toluene in Figure 10.16 shows these characteristic absorptions.

FIGURE 10.16 The infrared spectrum of toluene.

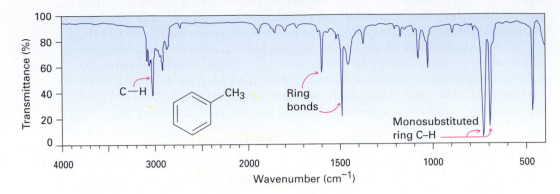

Alcohols

The O–H functional group of alcohols is easy to spot. Alcohols have a characteristic band in the range 3400 to 3650 cm^{-1} that is usually broad and intense. If present, it's hard to miss this band or to confuse it with anything else.

Alcohols   —O—H   3400–3650 cm^{-1} (broad, intense)

Amines

The N–H functional group of amines is also easy to spot in the IR, with a characteristic absorption in the 3300 to 3500 cm^{-1} range. Although alcohols absorb in the same range, an N–H absorption is sharper and less intense than an O–H band.

Amines   —N—H   3300–3500 cm^{-1} (sharp, medium intensity)

Carbonyl Compounds

Carbonyl functional groups are the easiest to identify of all IR absorptions because of their sharp, intense peak in the range 1670 to 1780 cm^{-1}. Most important, the exact position of absorption within the range can often identify the exact kind of carbonyl functional group—aldehyde, ketone, ester, and so forth.

ALDEHYDES Saturated aldehydes absorb at 1730 cm^{-1}; aldehydes next to either a double bond or an aromatic ring absorb at 1705 cm^{-1}.

Aldehydes   CH$_3$CH$_2$CH(=O)   CH$_3$CH=CHCH(=O)

1730 cm^{-1} 1705 cm^{-1} 1705 cm^{-1}

KETONES Saturated open-chain ketones and six-membered cyclic ketones absorb at 1715 cm^{-1}, five-membered cyclic ketones absorb at 1750 cm^{-1}, and ketones next to a double bond or an aromatic ring absorb at 1685 cm^{-1}.

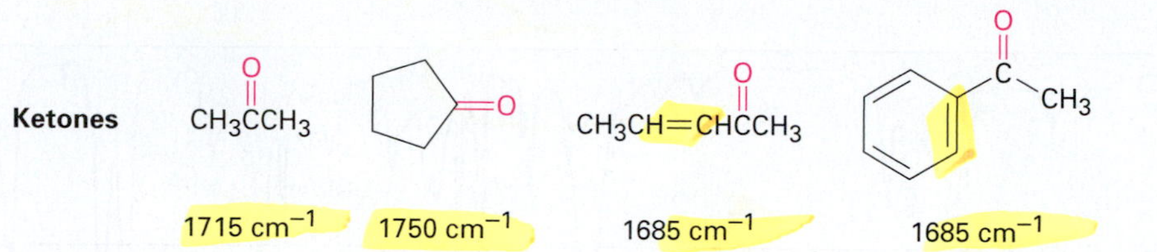

Ketones   CH$_3$CCH$_3$ CH$_3$CH=CHCCH$_3$

1715 cm^{-1} 1750 cm^{-1} 1685 cm^{-1} 1685 cm^{-1}

Beside = bond? · lower A.

ESTERS Saturated esters absorb at 1735 cm^{-1}; esters next to either a double bond or an aromatic ring absorb at 1715 cm^{-1}.

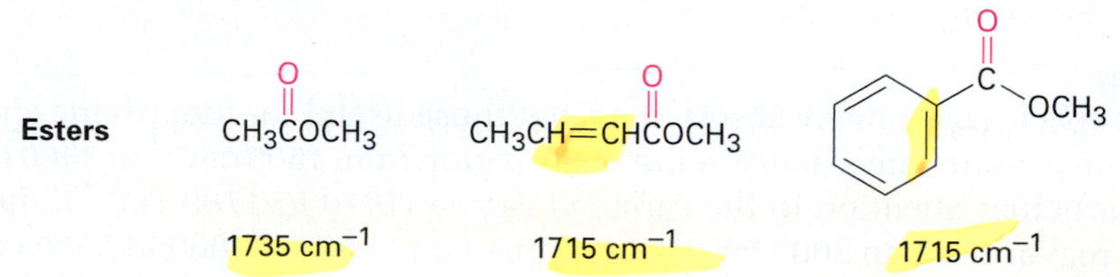

Esters   CH$_3$COCH$_3$ CH$_3$CH=CHCOCH$_3$

1735 cm^{-1} 1715 cm^{-1} 1715 cm^{-1}

WORKED EXAMPLE 10.5 Predicting IR Absorptions of Compounds

Where might the following compounds have IR absorptions?

(a) 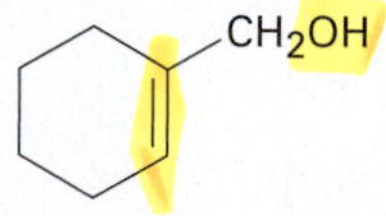   CH$_2$OH

(b) 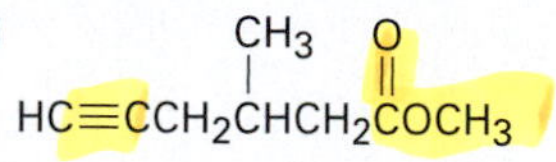   CH$_3$ O
       HC≡CCH$_2$CHCH$_2$COCH$_3$

Strategy

Identify the functional groups in each molecule, and check Table 10.1 to see where those groups absorb.

Solution

(a) This molecule has an alcohol O–H group and an alkene double bond. *Absorptions:* 3400–3650 cm^{-1} (O–H), 3020–3100 cm^{-1} (=C–H), 1640–1680 cm^{-1} (C=C).

(b) This molecule has a terminal alkyne triple bond and a saturated ester carbonyl group. *Absorptions:* 3300 cm^{-1} (≡C–H), 2100–2260 cm^{-1} (C≡C), 1735 cm^{-1} (C=O).

WORKED EXAMPLE 10.6 Identifying Functional Groups from an IR Spectrum

The IR spectrum of an unknown compound is shown in Figure 10.17. What functional groups does the compound contain?

FIGURE 10.17 The IR spectrum for Worked Example 10.6.

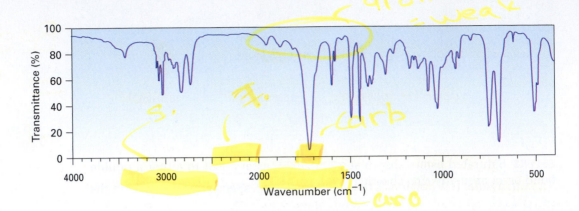

Strategy

All IR spectra have many absorptions, but those useful for identifying specific functional groups are usually found in the region from 1500 cm^{-1} to 3300 cm^{-1}. Pay particular attention to the carbonyl region (1670 to 1780 cm^{-1}), the aromatic region (1660 to 2000 cm^{-1}), the triple-bond region (2000 to 2500 cm^{-1}), and the C–H region (2500 to 3500 cm^{-1}).

Solution

The spectrum shows an intense absorption at 1725 cm^{-1} due to a carbonyl group (perhaps an aldehyde, –CHO), a series of weak absorptions from 1800 to 2000 cm^{-1} characteristic of aromatic compounds, and a C–H absorption near 3030 cm^{-1}, also characteristic of aromatic compounds. In fact, the compound is phenylacetaldehyde.

Phenylacetaldehyde

Problem 10.9

Where might the following compounds have IR absorptions?

(a)   (b)   (c)

Problem 10.10

Where might the following compound have IR absorptions?

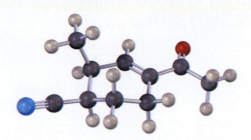

10.9 Ultraviolet Spectroscopy

The **ultraviolet (UV)** region of the electromagnetic spectrum extends from the low-wavelength end of the visible region (4×10^{-7} m) to the long-wavelength end of the X-ray region (10^{-8} m), but the narrow range from 2×10^{-7} m to 4×10^{-7} m is the portion of greatest interest to organic chemists. Absorptions in this region are usually measured in nanometers (nm), where 1 nm = 10^{-9} m. Thus, the ultraviolet range of interest is from 200 to 400 nm (Figure 10.18).

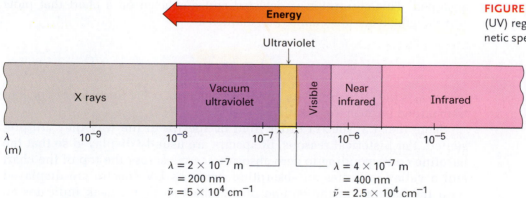

FIGURE 10.18 The ultraviolet (UV) region of the electromagnetic spectrum.

We've just seen that when a molecule is subjected to IR irradiation, the energy absorbed corresponds to the amount necessary to increase molecular vibrations. With UV radiation, the energy absorbed corresponds to the amount necessary to promote an electron from one orbital to another in a conjugated molecule. The conjugated diene buta-1,3-diene, for example, has four π molecular orbitals, as shown previously in Figure 8.13 on page 382. The two lower-energy, bonding MOs are occupied in the ground state, and the two higher-energy, antibonding MOs are unoccupied.

On irradiation with ultraviolet light ($h\nu$), buta-1,3-diene absorbs energy and a π electron is promoted from the **highest occupied molecular orbital**, or **HOMO**, to the **lowest unoccupied molecular orbital**, or **LUMO**. Since the electron is promoted from a bonding π molecular orbital to an antibonding

π^* molecular orbital, we call this a $\pi \rightarrow \pi^*$ excitation (read as "pi to pi star"). The energy gap between the HOMO and the LUMO of buta-1,3-diene is such that UV light of 217 nm wavelength is required to accomplish the $\pi \rightarrow \pi^*$ electronic transition (Figure 10.19).

FIGURE 10.19 Ultraviolet irradiation of buta-1,3-diene results in promotion of an electron from ψ_2, the highest occupied molecular orbital (HOMO), to ψ_3^*, the lowest unoccupied molecular orbital (LUMO).

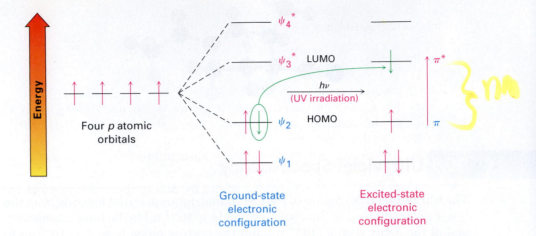

An ultraviolet spectrum is recorded by irradiating the sample with UV light of continuously changing wavelength. When the wavelength corresponds to the energy level required to excite an electron to a higher level, energy is absorbed. This absorption is detected and displayed on a chart that plots wavelength versus *absorbance (A)*, defined as

$$A = \log \frac{I_0}{I}$$

where I_0 is the intensity of the incident light and I is the intensity of the light transmitted through the sample.

Note that UV spectra differ from IR spectra in the way they are presented. For historical reasons, IR spectra are usually displayed so that the baseline corresponding to zero absorption runs across the top of the chart and a valley indicates an absorption, whereas UV spectra are displayed with the baseline at the bottom of the chart so that a peak indicates an absorption (Figure 10.20).

FIGURE 10.20 The ultraviolet spectrum of buta-1,3-diene, $\lambda_{max} = 217$ nm.

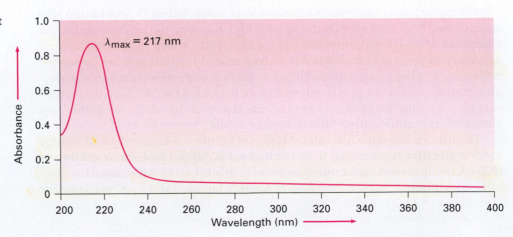

The amount of UV light absorbed is expressed as the sample's *molar absorptivity* (ε), defined by the equation

$$\epsilon = \frac{A}{c \times l}$$

where A = Absorbance

c = Concentration in mol/L

l = Sample pathlength in cm

Molar absorptivity is a physical constant, characteristic of the particular substance being observed and thus characteristic of the particular π electron system in the molecule. Typical values for conjugated dienes are in the range ϵ = 10,000 to 25,000. Note that the units are usually dropped.

Unlike IR spectra, which show many absorptions for a given molecule, UV spectra are usually quite simple—often only a single peak. The peak is usually broad, and we identify its position by noting the wavelength at the very top of the peak—λ_{max}, read as "lambda max."

Problem 10.11

Calculate the energy range of radiation in the UV region of the spectrum from 200 to 400 nm. How does this value compare with the value calculated previously for IR radiation in Section 10.6?

Problem 10.12

A knowledge of molar absorptivities is particularly useful in biochemistry, where UV spectroscopy can provide an extremely sensitive method of detection. Imagine, for instance, that you wanted to determine the concentration of vitamin A in a sample. If pure vitamin A has λ_{max} = 325 (ϵ = 50,100), what is the vitamin A concentration in a sample whose absorbance at 325 nm is A = 0.735 in a cell with a pathlength of 1.00 cm?

10.10 Interpreting Ultraviolet Spectra: The Effect of Conjugation

The wavelength necessary to effect the $\pi \rightarrow \pi^*$ transition in a conjugated molecule depends on the energy gap between HOMO and LUMO, which in turn depends on the nature of the conjugated system. Thus, by measuring the UV spectrum of an unknown, we can derive structural information about the nature of any conjugated π electron system present in a molecule.

One of the most important factors affecting the wavelength of UV absorption by a molecule is the extent of conjugation. Experiments show that the energy difference between HOMO and LUMO decreases as the extent of conjugation increases. Thus, buta-1,3-diene absorbs at λ_{max} = 217 nm, hexa-1,3,5-triene absorbs at λ_{max} = 258 nm, and octa-1,3,5,7-tetraene absorbs at λ_{max} = 290 nm. (Remember: longer wavelength means lower energy.)

Other kinds of conjugated systems, such as conjugated enones and aromatic rings, also have characteristic UV absorptions that are useful in structure determination. The UV absorption maxima of some representative conjugated molecules are given in Table 10.2.

TABLE 10.2
Ultraviolet Absorptions of Some Conjugated Molecules

Name	Structure	λ_{max} (nm)
2-Methybuta-1,3-diene	$H_2C{=}\overset{\underset{\displaystyle CH_3}{|}}{C}{-}CH{=}CH_2$	220
Cyclohexa-1,3-diene	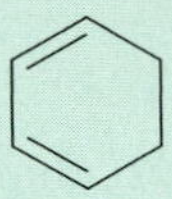	256
Hexa-1,3,5-triene	$H_2C{=}CH{-}CH{=}CH{-}CH{=}CH_2$	258
Octa-1,3,5,7-tetraene	$H_2C{=}CH{-}CH{=}CH{-}CH{=}CH{-}CH{=}CH_2$	290
But-3-en-2-one	$H_2C{=}CH{-}\overset{\displaystyle O}{\overset{\|}{C}}{-}CH_3$	219
Benzene	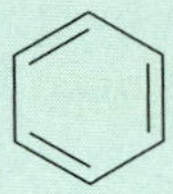	203

Problem 10.13

Which of the following compounds would you expect to show ultraviolet absorptions in the 200 to 400 nm range?

(a)

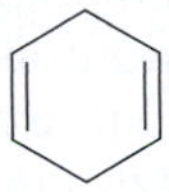

(b)

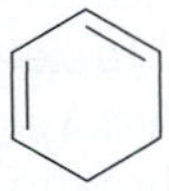

(c) 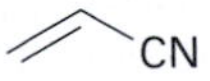　CN

(d) 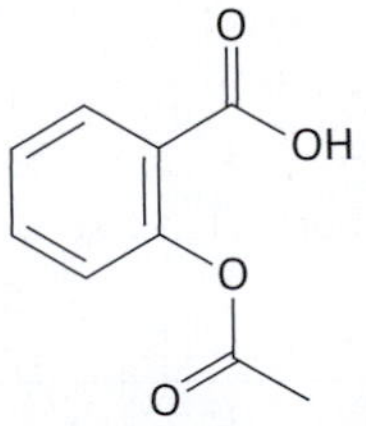　OH

Aspirin

(e) 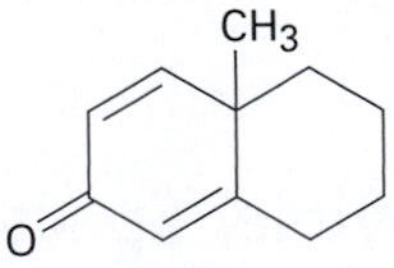　CH₃

(f)

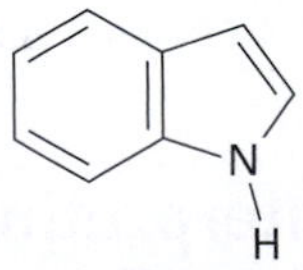

Indole

10.11 Conjugation, Color, and the Chemistry of Vision

Why are some organic compounds colored while others aren't? β-Carotene, the pigment in carrots, is purple-orange, for instance, while cholesterol is colorless. The answer involves both the chemical structure of colored molecules and the way we perceive light.

The visible region of the electromagnetic spectrum is adjacent to the ultraviolet region, extending from approximately 400 to 800 nm. Colored compounds have such extended systems of conjugation that their "UV" absorptions extend into the visible region. β-Carotene, for example, has 11 double bonds in conjugation, and its absorption occurs at λ_{max} = 455 nm (Figure 10.21).

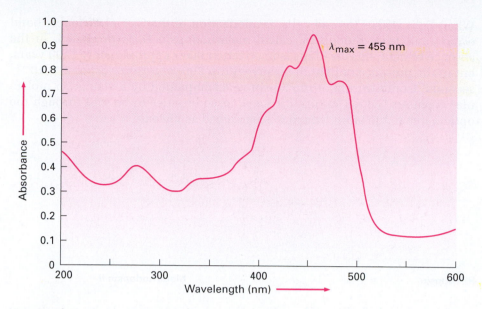

FIGURE 10.21 Ultraviolet spectrum of β-carotene, a conjugated molecule with 11 double bonds. The absorption occurs in the visible region.

$\lambda_{max} = 455$ nm

"White" light from the sun or from a lamp consists of all wavelengths in the visible region. When white light strikes β-carotene, the wavelengths from 400 to 500 nm (blue) are absorbed while all other wavelengths are transmitted and reach our eyes. We therefore see the white light with the blue removed, and we perceive a yellow-orange color for β-carotene.

Conjugation is crucial not only for the colors we see in organic molecules but also for the light-sensitive molecules on which our visual system is based. The key substance for vision is dietary β-carotene, which is converted to vitamin A by enzymes in the liver, oxidized to an aldehyde called 11-*trans*-retinal, and then isomerized by a change in geometry of the C11–C12 double bond to produce 11-*cis*-retinal.

β-Carotene

Vitamin A

11-*cis*-Retinal

There are two main types of light-sensitive receptor cells in the retina of the human eye, *rod* cells and *cone* cells. The 3 million or so rod cells are primarily responsible for seeing in dim light, whereas the 100 million cone cells are responsible for seeing in bright light and for the perception of bright colors. In the rod cells of the eye, 11-*cis*-retinal is converted into rhodopsin, a light-sensitive substance formed from the protein opsin and 11-*cis*-retinal.

When light strikes the rod cells, isomerization of the C11–C12 double bond occurs and *trans*-rhodopsin, called metarhodopsin II, is produced. In the absence of light, this cis–trans isomerization takes approximately 1100 years, but in the presence of light, it occurs within 200 *femtoseconds,* or 2×10^{-13} seconds! Isomerization of rhodopsin is accompanied by a change in molecular geometry, which in turn causes a nerve impulse to be sent through the optic nerve to the brain, where it is perceived as vision.

Rhodopsin   **Metarhodopsin II**

Metarhodopsin II is then recycled back into rhodopsin by a multistep sequence involving cleavage to all-*trans*-retinal and cis–trans isomerization back to 11-*cis*-retinal.

Summary

Key Words

absorption spectrum, 378

amplitude, 378

base peak, 369

electromagnetic spectrum, 377

frequency (ν), 378

hertz (Hz), 378

highest occupied molecular
  orbital (HOMO), 389

infrared (IR) spectroscopy, 380

lowest unoccupied molecular
  orbital (LUMO), 389

mass spectrometry (MS), 368

mass spectrum, 369

parent peak, 369

ultraviolet (UV) spectroscopy,
  389

wavelength (λ), 378

wavenumber ($\tilde{\nu}$), 380

Finding the structure of a new molecule, whether a small one synthesized in the laboratory or a large protein found in living organisms, is central to progress in chemistry and biochemistry. As we saw in this chapter, the structure of an organic molecule is usually determined using spectroscopic methods, including mass spectrometry, infrared spectroscopy, and ultraviolet spectroscopy. **Mass spectrometry (MS)** tells the molecular weight and formula of a molecule, **infrared (IR) spectroscopy** identifies the functional groups present in the molecule, and **ultraviolet (UV) spectroscopy** tells whether the molecule has a conjugated π electron system.

In small-molecule mass spectrometry, molecules are first ionized by collision with a high-energy electron beam. The ions then fragment into smaller pieces, which are magnetically sorted according to their mass-to-charge ratio (*m/z*). The ionized sample molecule is called the *molecular ion, M⁺*, and measurement of its mass gives the molecular weight of the sample. Structural clues about unknown samples can be obtained by interpreting the fragmentation pattern of the molecular ion. Mass-spectral fragmentations are usually complex, however, and interpretation is often difficult. In biological mass spectrometry, molecules are protonated using either electrospray ionization (ESI) or matrix-assisted laser desorption ionization (MALDI), and the protonated molecules are separated by time-of-flight (TOF).

Infrared spectroscopy involves the interaction of a molecule with **electromagnetic radiation**. When an organic molecule is irradiated with infrared energy, certain **frequencies** are absorbed by the molecule. The frequencies absorbed correspond to the amounts of energy needed to increase the amplitude of specific molecular vibrations, such as bond stretchings and bendings. Since every functional group has a characteristic combination of bonds, every functional group has a characteristic set of infrared absorptions. By observing which frequencies of infrared radiation are absorbed by a molecule and which are not, it's possible to determine the functional groups a molecule contains.

Ultraviolet spectroscopy is applicable only to conjugated systems. When a conjugated molecule is irradiated with ultraviolet light, energy absorption occurs and a π electron is promoted from the **highest occupied molecular orbital (HOMO)** to the **lowest unoccupied molecular orbital (LUMO)**. The greater the extent of conjugation, the less the energy needed and the longer the wavelength of radiation required.

Lagniappe

Chromatography: Purifying Organic Compounds

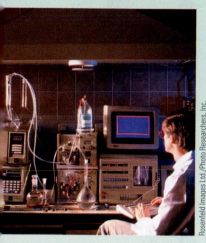

High-pressure liquid chromatography (HPLC) is used to separate and purify the products of laboratory reactions.

Even before a new organic substance has its structure determined, it must be purified by separating it from solvents and all contaminants. Purification was an enormously time-consuming, hit-or-miss proposition in the 19th and early 20th centuries, but powerful instruments developed in the past few decades now simplify the problem.

Most organic purification is done by *chromatography* (literally, "color writing"), a separation technique that dates from the work of the Russian chemist Mikhail Tswett in 1903. Tswett accomplished the separation of the pigments in green leaves by dissolving the leaf extract in an organic solvent and allowing the solution to run down through a vertical glass tube packed with chalk powder. Different pigments passed down the column at different rates, leaving a series of colored bands on the white chalk column.

A variety of chromatographic techniques are now in common use, all of which work on a similar principle. The mixture to be separated is dissolved in a solvent, called the *mobile phase,* and passed over an adsorbent material, called the *stationary phase.* Because different compounds adsorb to the stationary phase to different extents, they migrate along the phase at different rates and are separated as they emerge *(elute)* from the end of the chromatography column.

Liquid chromatography, or *column chromatography,* is perhaps the most often used chromatographic method. As in Tswett's original experiments, a mixture of organic compounds is dissolved in a suitable solvent and adsorbed onto a stationary phase such as alumina (Al_2O_3) or silica gel (hydrated SiO_2) packed into a glass column. More solvent is then passed down the column, and different compounds elute at different times.

The time at which a compound is eluted is strongly influenced by its polarity. Molecules with polar functional groups are generally adsorbed more strongly and therefore migrate through the stationary phase more slowly than nonpolar molecules. A mixture of an alcohol and an alkene, for example, can be easily separated with liquid chromatography because the nonpolar alkene passes through the column much faster than the more polar alcohol.

High-pressure (or *high-performance*) *liquid chromatography* (HPLC) is a variant of the simple column technique, based on the discovery that chromatographic separations are vastly improved if the stationary phase is made up of very small, uniformly sized spherical particles. Small particle size ensures a large surface area for better adsorption, and a uniform spherical shape allows a tight, uniform packing of particles. In practice, coated SiO_2 microspheres 2 to 5 μm diameter are often used.

High-pressure pumps operating at up to 15,000 psi are required to force solvent through a tightly packed HPLC column, and electronic detectors are used to monitor the appearance of material eluting from the column. Alternatively, the column can be interfaced to a mass spectrometer to record the mass spectrum of every substance as it elutes. Figure 10.22 shows the results of HPLC analysis of a mixture of ten fat-soluble vitamins on 5 μm silica spheres with acetonitrile as solvent.

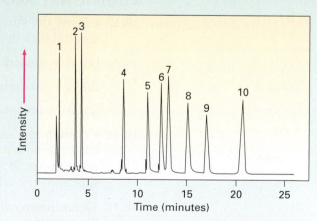

1. Menadione (vitamin K₃)
2. Retinol (vitamin A)
3. Retinol acetate
4. Menaquinone (vitamin K₂)
5. δ-Tocopherol
6. Ergocalciferol (vitamin D₂)
7. Cholecalciferol (vitamin D₃)
8. α-Tocopherol (vitamin E)
9. α-Tocopherol acetate
10. Phylloquinone (vitamin K₁)

FIGURE 10.22 Results of an HPLC analysis of a mixture of ten fat-soluble vitamins.

Exercises

■ indicates problems that are assignable in Organic OWL.

Go to this book's companion website at www.cengage.com/chemistry/mcmurry to explore interactive versions of the Active Figures from this text.

VISUALIZING CHEMISTRY

(Problems 10.1–10.13 appear within the chapter.)

10.14 ■ Show the structures of the likely fragments you would expect in the mass spectra of the following molecules:

(a)   **(b)**

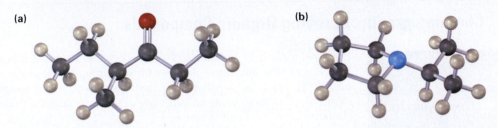

10.15 ■ Where in the IR spectrum would you expect each of the following molecules to absorb?

(a)   **(b)**   **(c)**

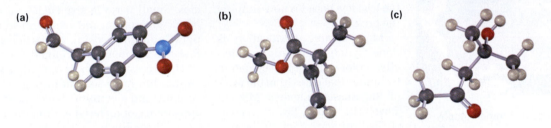

10.16 Which, if any, of the compounds shown in Problems 10.14 and 10.15 have UV absorptions?

ADDITIONAL PROBLEMS

10.17 ■ Draw the structure of a molecule that is consistent with the mass-spectral data in each of the following molecules:

(a) A hydrocarbon with $M^+ = 132$

(b) A hydrocarbon with $M^+ = 166$

(c) A hydrocarbon with $M^+ = 84$

10.18 ■ Camphor, a saturated monoketone from the Asian camphor tree, is used as a moth repellent and as a constituent of embalming fluid, among other things. If camphor has $M^+ = 152.1201$ by high-resolution mass spectrometry, what is its molecular formula?

10.19 The *nitrogen rule* of mass spectrometry says that a compound containing an odd number of nitrogens has an odd-numbered molecular ion. Conversely, a compound containing an even number of nitrogens has an even-numbered M^+ peak. Explain.

10.20 ■ In light of the nitrogen rule mentioned in Problem 10.19, what is the molecular formula of pyridine, $M^+ = 79$?

■ Problems assignable in Organic OWL.

10.21 ▪ Halogenated compounds are particularly easy to identify by their mass spectra because both chlorine and bromine occur naturally as mixtures of two abundant isotopes. Chlorine occurs as ^{35}Cl (75.8%) and ^{37}Cl (24.2%); bromine occurs as ^{79}Br (50.7%) and ^{81}Br (49.3%). At what masses do the molecular ions occur for the following formulas? What are the relative percentages of each molecular ion?

(a) Bromomethane, CH_3Br **(b)** 1-Chlorohexane, $C_6H_{13}Cl$

10.22 ▪ Propose structures for compounds that fit the following data:

(a) A ketone with $M^+ = 86$ and fragments at $m/z = 71$ and $m/z = 43$

(b) An alcohol with $M^+ = 88$ and fragments at $m/z = 73$, $m/z = 70$, and $m/z = 59$

10.23 ▪ 2-Methylpentane (C_6H_{14}) has the mass spectrum shown. Which peak represents M^+? Which is the base peak? Propose structures for fragment ions of $m/z = 71, 57, 43$, and 29. Why does the base peak have the mass it does?

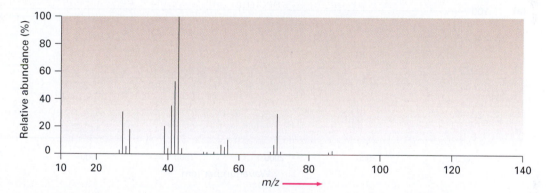

10.24 Assume that you are in a laboratory carrying out the catalytic hydrogenation of cyclohexene to cyclohexane. How could you use mass spectrometry to determine when the reaction is finished?

10.25 What fragments might you expect in the mass spectra of the following compounds?

(a) **(b)** **(c)**

10.26 ▪ How might you use IR spectroscopy to distinguish among the three isomers but-1-yne, buta-1,3-diene, and but-2-yne?

10.27 Would you expect two enantiomers such as (*R*)-2-bromobutane and (*S*)-2-bromobutane to have identical or different IR spectra? Explain.

10.28 Would you expect two diastereomers such as *meso*-2,3-dibromobutane and (2*R*,3*R*)-dibromobutane to have identical or different IR spectra? Explain.

10.29 ▪ Propose structures for compounds that meet the following descriptions:

(a) C_5H_8, with IR absorptions at 3300 and 2150 cm^{-1}

(b) C_4H_8O, with a strong IR absorption at 3400 cm^{-1}

(c) C_4H_8O, with a strong IR absorption at 1715 cm^{-1}

(d) C_8H_{10}, with IR absorptions at 1600 and 1500 cm^{-1}

10.30 ▪ How could you use infrared spectroscopy to distinguish between the following pairs of isomers?

(a) $HC{\equiv}CCH_2NH_2$ and $CH_3CH_2C{\equiv}N$

(b) CH_3COCH_3 and CH_3CH_2CHO

10.31 ▪ Two infrared spectra are shown. One is the spectrum of cyclohexane, and the other is the spectrum of cyclohexene. Identify them, and explain your answer.

(a)

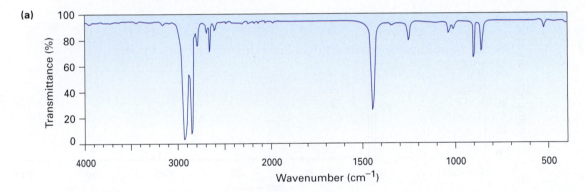

(b)

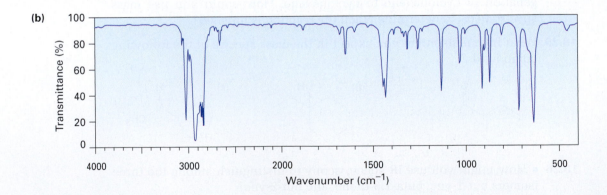

10.32 ▪ At what approximate positions might the following compounds show IR absorptions?

(a) 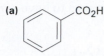CO₂H   (b) 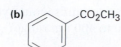CO₂CH₃   (c) 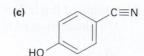C≡N

(d) O

(e)

$$CH_3\overset{O}{\underset{\|}{C}}CH_2CH_2\overset{O}{\underset{\|}{C}}OCH_3$$

10.33 ▪ How would you use infrared spectroscopy to distinguish between the following pairs of constitutional isomers?

(a) CH₃C≡CCH₃   and   CH₃CH₂C≡CH

(b)

$$CH_3\overset{O}{\underset{\|}{C}}CH=CHCH_3$$ and $$CH_3\overset{O}{\underset{\|}{C}}CH_2CH=CH_2$$

(c) H₂C=CHOCH₃   and   CH₃CH₂CHO

10.34 ▪ At what approximate positions might the following compounds show IR absorptions?

(a)

$$CH_3CH_2\overset{O}{\underset{\|}{C}}CH_3$$

(b)

$$CH_3\overset{CH_3}{\underset{|}{C}H}CH_2C≡CH$$

(c)

$$CH_3\overset{CH_3}{\underset{|}{C}H}CH_2CH=CH_2$$

(d)

$$CH_3CH_2CH_2\overset{O}{\underset{\|}{C}}OCH_3$$

(e)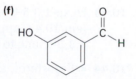

(f)

10.35 Assume you are carrying out the dehydration of 1-methylcyclohexanol to yield 1-methylcyclohexene. How could you use infrared spectroscopy to determine when the reaction is complete?

10.36 Assume that you are carrying out an elimination reaction on 3-bromo-3-methylpentane to yield an alkene. How could you use IR spectroscopy to tell which of two possible elimination products is formed, 3-methylpent-2-ene or 2-ethylbut-1-ene?

10.37 Which is stronger, the C=O bond in an ester (1735 cm⁻¹) or the C=O bond in a saturated ketone (1715 cm⁻¹)? Explain.

10.38 ▪ Carvone is an unsaturated ketone responsible for the odor of spearmint. If carvone has M⁺ = 150 in its mass spectrum and contains three double bonds and one ring, what is its molecular formula?

10.39 Carvone (Problem 10.38) has an intense infrared absorption at 1690 cm⁻¹. What kind of ketone does carvone contain?

▪ Problems assignable in Organic OWL.

10.40 Would you expect allene, $H_2C{=}C{=}CH_2$, to show a UV absorption in the 200 to 400 nm range? Explain.

10.41 Which of the following compounds would you expect to have a $\pi \to \pi^*$ UV absorption in the 200 to 400 nm range?

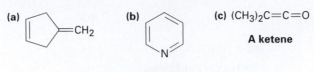

(a) $=CH_2$

(b) **Pyridine**

(c) $(CH_3)_2C{=}C{=}O$

A ketene

10.42 The following ultraviolet absorption maxima have been measured:

Buta-1,3-diene	217 nm
2-Methylbuta-1,3-diene	220 nm
Penta-1,3-diene	223 nm
2,3-Dimethylbuta-1,3-diene	226 nm
Hexa-2,4-diene	227 nm
2,4-Dimethylpenta-1,3-diene	232 nm
2,5-Dimethylhexa-2,4-diene	240 nm

What conclusion can you draw about the effect of alkyl substitution on UV absorption maxima? Approximately what effect does each added alkyl group have?

10.43 Hexa-1,3,5-triene has $\lambda_{max} = 258$ nm. In light of your answer to Problem 10.42, approximately where would you expect 2,3-dimethylhexa-1,3,5-triene to absorb? Explain.

10.44 ■ Ergosterol, a precursor of vitamin D, has $\lambda_{max} = 282$ nm and molar absorptivity $\epsilon = 11,900$. What is the concentration of ergosterol in a solution whose absorbance $A = 0.065$ with a sample pathlength $l = 1.00$ cm?

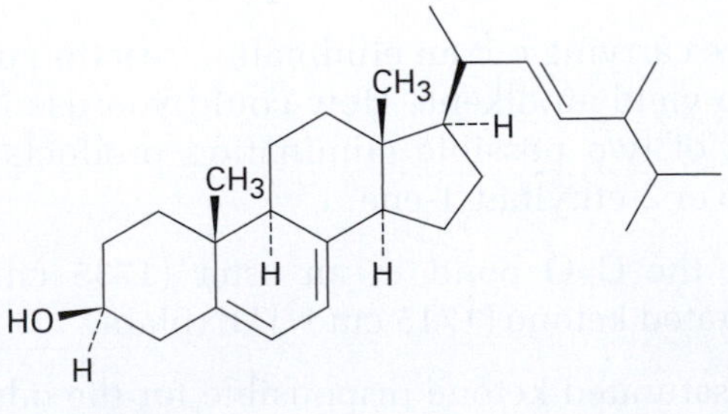

Ergosterol ($C_{28}H_{44}O$)

10.45 The mass spectrum **(a)** and the infrared spectrum **(b)** of an unknown hydrocarbon are shown. Propose as many structures as you can.

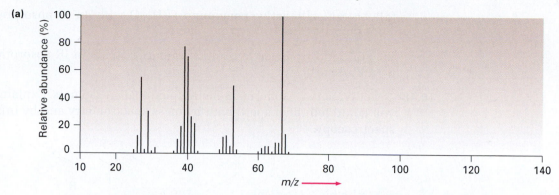

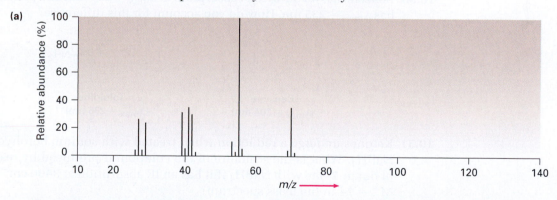

10.46 The mass spectrum **(a)** and the infrared spectrum **(b)** of another unknown hydrocarbon are shown. Propose as many structures as you can.

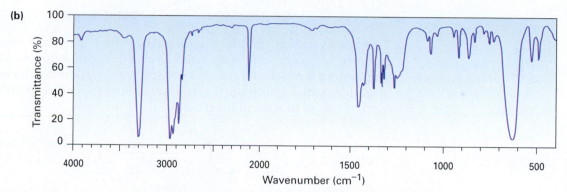

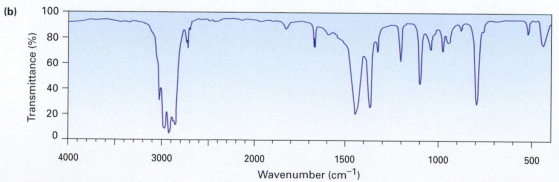

■ Problems assignable in Organic OWL.

10.47 ▪ Propose structures for compounds that meet the following descriptions:

(a) An optically active compound $C_5H_{10}O$ with an IR absorption at 1730 cm^{-1}

(b) An optically inactive compound C_5H_9N with an IR absorption at 2215 cm^{-1}

10.48 4-Methylpentan-2-one and 3-methylpentanal are isomers. Explain how you could tell them apart, both by mass spectrometry and by infrared spectroscopy.

4-Methylpentan-2-one   **3-Methylpentanal**

10.49 Organomagnesium halides (R–Mg–X), called *Grignard reagents*, undergo a general and very useful reaction with ketones. Methylmagnesium bromide, for example, reacts with cyclohexanone to yield a product with the formula $C_7H_{14}O$. What is the structure of this product if it has an IR absorption at 3400 cm^{-1}?

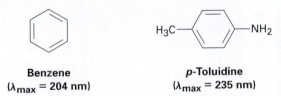

Cyclohexanone

10.50 Benzene has an ultraviolet absorption at $\lambda_{max} = 204$ nm, and *p*-toluidine has $\lambda_{max} = 235$ nm. How do you account for this difference?

H$_3$C———NH$_2$

Benzene   **p-Toluidine**
(λ_{max} = 204 nm)   (λ_{max} = 235 nm)

10.51 Ketones undergo a reduction when treated with sodium borohydride, NaBH$_4$. What is the structure of the compound produced by reaction of butan-2-one with NaBH$_4$ if it has an IR absorption at 3400 cm^{-1} and M$^+$ = 74 in the mass spectrum?

$$CH_3CH_2\overset{\displaystyle O}{\overset{\displaystyle \|}{C}}CH_3 \quad \xrightarrow[\text{2. H}_3\text{O}^+]{\text{1. NaBH}_4} \quad ?$$

Butan-2-one

10.52 Nitriles, R–C≡N, undergo a *hydrolysis* reaction when heated with aqueous acid. What is the structure of the compound produced by hydrolysis of propanenitrile, $CH_3CH_2C≡N$, if it has IR absorptions at 2500 to 3100 cm^{-1} and 1710 cm^{-1} and has $M^+ = 74$?

10.53 Enamines (C=C–N; alkene + amine) typically have a UV absorption near $\lambda_{max} = 230$ nm and are much more nucleophilic than alkenes. Assuming the nitrogen atom is *sp*^{2}-hybridized, explain both the UV absorption and the nucleophilicity of enamines.

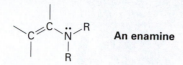 **An enamine**

11 Structure Determination: Nuclear Magnetic Resonance Spectroscopy

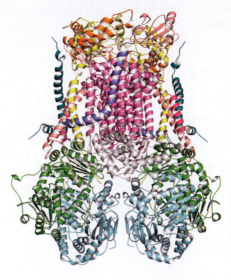

Ubiquinone-cytochrome *c* reductase catalyzes a redox pathway called the Q cycle, a crucial step in biological energy production.

Nuclear magnetic resonance (NMR) spectroscopy is the most valuable spectroscopic technique available to laboratory organic chemists. It's the method of structure determination that organic chemists turn to first.

We saw in Chapter 10 that mass spectrometry gives a molecule's formula, infrared spectroscopy identifies a molecule's functional groups, and ultraviolet spectroscopy identifies a molecule's conjugated π electron system. Nuclear magnetic resonance spectroscopy complements these other techniques by "mapping" a molecule's carbon–hydrogen framework. Taken together, mass spectrometry, IR, UV, and NMR make it possible to determine the structures of even very complex molecules.

Mass spectrometry	Molecular formula
Infrared spectroscopy	Functional groups
Ultraviolet spectroscopy	Extent of conjugation
NMR spectroscopy	Map of carbon–hydrogen framework

WHY THIS CHAPTER?

The opening sentence above says it all: NMR is by far the most valuable spectroscopic technique for structure determination. Although we'll just give an overview of the subject in this chapter, focusing on NMR applications to small molecules, more advanced NMR techniques are also used in biological chemistry to study protein structure and folding.

OWL Online homework for this chapter can be assigned in Organic OWL.

11.1 Nuclear Magnetic Resonance Spectroscopy

Many kinds of atomic nuclei behave as if they were spinning about an axis, much as the earth spins daily. Since they're positively charged, these spinning nuclei act like tiny bar magnets and interact with an external magnetic field, denoted $\boldsymbol{B}_0$. Not all nuclei act this way, but fortunately for organic chemists, both the proton (^{1}H) and the ^{13}C nucleus do have spins. (In speaking about NMR, the words *proton* and *hydrogen* are often used interchangeably.) Let's see what the consequences of nuclear spin are and how we can use the results.

In the absence of an external magnetic field, the spins of magnetic nuclei are oriented randomly. When a sample containing these nuclei is placed between the poles of a strong magnet, however, the nuclei adopt specific orientations, much as a compass needle orients in the earth's magnetic field. A spinning ^{1}H or ^{13}C nucleus can orient so that its own tiny magnetic field is aligned either with (parallel to) or against (antiparallel to) the external field. The two orientations don't have the same energy, however, and aren't equally likely. The parallel orientation is slightly lower in energy by an amount that depends on the strength of the external field, making this spin state slightly favored over the antiparallel orientation (Figure 11.1).

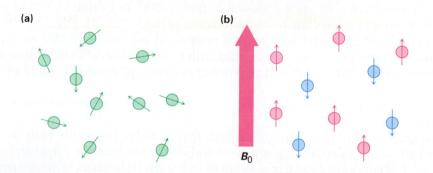

(a) (b)

B_0

FIGURE 11.1 (a) Nuclear spins are oriented randomly in the absence of an external magnetic field but (b) have a specific orientation in the presence of an external field, $\boldsymbol{B}_0$. Some of the spins (red) are aligned parallel to the external field while others (blue) are antiparallel. The parallel spin state is slightly lower in energy and therefore favored.

If the oriented nuclei are now irradiated with electromagnetic radiation of the proper frequency, energy absorption occurs and the lower-energy state "spin-flips" to the higher-energy state. When this spin-flip occurs, the magnetic nuclei are said to be in resonance with the applied radiation—hence the name *nuclear magnetic resonance*.

The exact frequency necessary for resonance depends both on the strength of the external magnetic field and on the identity of the nuclei. If a very strong magnetic field is applied, the energy difference between the two spin states is larger and higher-frequency (higher-energy) radiation is required for a spin-flip. If a weaker magnetic field is applied, less energy is required to effect the transition between nuclear spin states (Figure 11.2).

In practice, superconducting magnets that produce enormously powerful fields up to 21.2 tesla (T) are sometimes used, but field strengths in the range of 4.7 to 7.0 T are more common. At a magnetic field strength of 4.7 T, so-called radiofrequency (rf) energy in the 200 MHz range (1 MHz = 10^6 Hz) brings a ^{1}H nucleus into resonance, and rf energy of 50 MHz brings a ^{13}C nucleus into resonance. At the highest field strength currently available in commercial instruments (21.2 T), 900 MHz energy is required for ^{1}H spectroscopy. These energies needed for NMR are much smaller than those required for IR spectroscopy; 200 MHz rf energy corresponds to only 8.0×10^{-5} kJ/mol versus the 4.8 to 48 kJ/mol needed for IR spectroscopy.

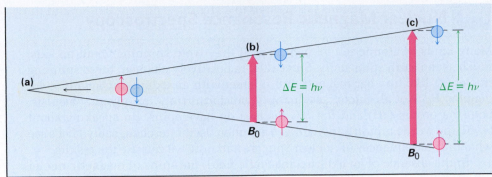

Strength of applied field, B_0 →

FIGURE 11.2 The energy difference ΔE between nuclear spin states depends on the strength of the applied magnetic field. Absorption of energy with frequency ν converts a nucleus from a lower spin state to a higher spin state. **(a)** Spin states have equal energies in the absence of an applied magnetic field but **(b)** have unequal energies in the presence of a magnetic field. At $\nu = 200$ MHz, $\Delta E = 8.0 \times 10^{-5}$ kJ/mol (1.9×10^{-5} kcal/mol). **(c)** The energy difference between spin states is greater at larger applied fields. At $\nu = 500$ MHz, $\Delta E = 2.0 \times 10^{-4}$ kJ/mol.

TABLE 11.1
The NMR Behavior of Some Common Nuclei

Magnetic nuclei	Nonmagnetic nuclei
^{1}H	^{12}C
^{13}C	^{16}O
^{2}H	^{32}S
^{14}N	
^{19}F	
^{31}P	

H and ^{13}C nuclei are not unique in their ability to exhibit the NMR phenomenon. All nuclei with an odd number of protons (^{1}H, ^{2}H, ^{14}N, ^{19}F, ^{31}P, for example) and all nuclei with an odd number of neutrons (^{13}C, for example) show magnetic properties. Only nuclei with even numbers of both protons and neutrons (^{12}C, ^{16}O, ^{32}S) do not give rise to magnetic phenomena (Table 11.1).

Problem 11.1

The amount of energy required to spin-flip a nucleus depends both on the strength of the external magnetic field and on the nucleus. At a field strength of 4.7 T, rf energy of 200 MHz is required to bring a ^{1}H nucleus into resonance, but energy of only 187 MHz will bring a ^{19}F nucleus into resonance. Calculate the amount of energy required to spin-flip a ^{19}F nucleus. Is this amount greater or less than that required to spin-flip a ^{1}H nucleus?

11.2 The Nature of NMR Absorptions

From the description thus far, you might expect all ^{1}H nuclei in a molecule to absorb energy at the same frequency and all ^{13}C nuclei to absorb at the same frequency. If so, we would observe only a single NMR absorption band in the ^{1}H or ^{13}C spectrum of a molecule, a situation that would be of little use. In fact, the absorption frequency is not the same for all ^{1}H or all ^{13}C nuclei.

All nuclei in molecules are surrounded by electrons. When an external magnetic field is applied to a molecule, the electrons moving around nuclei set up tiny local magnetic fields of their own. These local magnetic fields act in opposition to the applied field so that the effective field actually felt by the nucleus is a bit weaker than the applied field.

$$B_{\text{effective}} = B_{\text{applied}} - B_{\text{local}}$$

In describing the effect of local fields, we say that nuclei are **shielded** from the full effect of the applied field by the surrounding electrons. Because each specific nucleus in a molecule is in a slightly different electronic environment, each nucleus is shielded to a slightly different extent and the effective magnetic field felt by each is slightly different. These tiny differences in the effective magnetic fields experienced by different nuclei can be detected, and we thus see a distinct NMR signal for each chemically distinct ^{13}C or ^{1}H nucleus in a molecule. As a result, an NMR spectrum effectively maps the carbon–hydrogen framework of an organic molecule. With practice, it's possible to read the map and derive structural information.

Figure 11.3 shows both the ^{1}H and the ^{13}C NMR spectra of methyl acetate, $CH_3CO_2CH_3$. The horizontal axis shows the effective field strength felt by the nuclei, and the vertical axis indicates the intensity of absorption of rf energy. Each peak in the NMR spectrum corresponds to a chemically distinct ^{1}H or ^{13}C nucleus in the molecule. (Note that NMR spectra are formatted with the zero absorption line at the bottom, whereas IR spectra are formatted with the zero absorption line at the top; Section 10.5.) Note also that ^{1}H and ^{13}C spectra can't be observed simultaneously on the same spectrometer because different amounts of energy are required to spin-flip the different kinds of nuclei. The two spectra must be recorded separately.

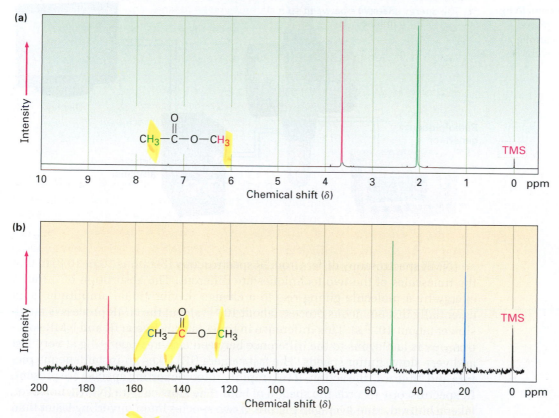

FIGURE 11.3 **(a)** The ^{1}H NMR spectrum and **(b)** the ^{13}C NMR spectrum of methyl acetate, $CH_3CO_2CH_3$. The small peak labeled "TMS" at the far right of each spectrum is a calibration peak, as explained in the next section.

The ^{13}C spectrum of methyl acetate in Figure 11.3b shows three peaks, one for each of the three chemically distinct carbon atoms in the molecule. The ^{1}H NMR spectrum in Figure 11.3a shows only two peaks, however, even though methyl acetate has six hydrogens. One peak is due to the $CH_3C=O$ hydrogens, and the other, to the $-OCH_3$ hydrogens. Because the three hydrogens in each methyl group have the same electronic environment, they are shielded to the same extent and are said to be *equivalent*. Chemically equivalent nuclei always show a single absorption. The two methyl groups themselves, however, are nonequivalent, so the two sets of hydrogens absorb at different positions.

The operation of a basic NMR spectrometer is illustrated in Figure 11.4. An organic sample is dissolved in a suitable solvent (usually deuteriochloroform, $CDCl_3$, which has no hydrogens) and placed in a thin glass tube between the poles of a magnet. The strong magnetic field causes the ^1H and ^{13}C nuclei in the molecule to align in one of the two possible orientations, and the sample is irradiated with rf energy. If the frequency of the rf irradiation is held constant and the strength of the applied magnetic field is varied, each nucleus comes into resonance at a slightly different field strength. A sensitive detector monitors the absorption of rf energy, and the electronic signal is then amplified and displayed as a peak.

FIGURE 11.4 Schematic operation of a basic NMR spectrometer. A thin glass tube containing the sample solution is placed between the poles of a strong magnet and irradiated with rf energy.

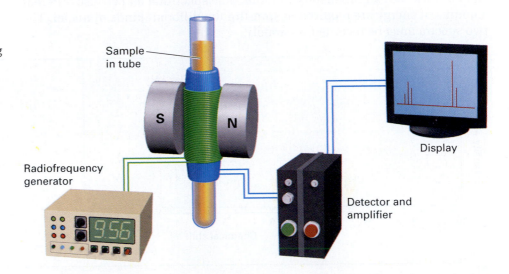

NMR spectroscopy differs from IR spectroscopy (Sections 10.6–10.8) in that the timescales of the two techniques are different. The absorption of infrared energy by a molecule giving rise to a change in vibrational amplitude is an essentially instantaneous process (about 10^{-13} s), but the NMR process is much slower (about 10^{-3} s). This difference in timescales between IR and NMR spectroscopy is analogous to the difference between cameras operating at very fast and very slow shutter speeds. The fast camera (IR) takes an instantaneous picture and "freezes" the action. If two rapidly interconverting species are present, IR spectroscopy records the spectra of both. The slow camera (NMR), however, takes a blurred, time-averaged picture. If two species interconverting faster than 10^3 times per second are present in a sample, NMR records only a single, averaged spectrum, rather than separate spectra of the two discrete species.

Because of this blurring effect, NMR spectroscopy can be used to measure the rates and activation energies of very fast processes. In cyclohexane, for example, a ring-flip (Section 4.6) occurs so rapidly at room temperature that axial and equatorial hydrogens can't be distinguished by NMR; only a single, averaged ^1H NMR absorption is seen for cyclohexane at 25 °C. At −90 °C, however, the ring-flip is slowed down enough that two absorption peaks are seen, one for the six axial hydrogens and one for the six equatorial hydrogens. Knowing the temperature and the rate at which signal blurring begins to occur, it's possible to calculate that the activation energy for the cyclohexane ring-flip is 45 kJ/mol (10.8 kcal/mol).

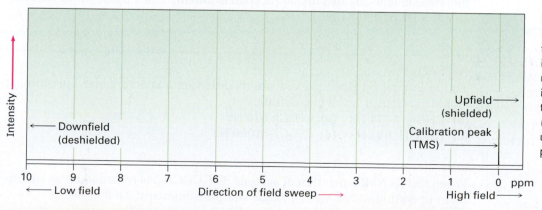

^{1}H NMR: 1 peak at 25 °C
2 peaks at −90 °C

Problem 11.2

2-Chloropropene shows signals for three kinds of protons in its ^{1}H NMR spectrum. Explain.

11.3 Chemical Shifts

NMR spectra are displayed on charts that show the applied field strength increasing from left to right (Figure 11.5). Thus, the left part of the chart is the low-field, or **downfield**, side, and the right part is the high-field, or **upfield**, side. Nuclei that absorb on the downfield side of the chart require a lower field strength for resonance, implying that they have relatively less shielding. Nuclei that absorb on the upfield side require a higher field strength for resonance, implying that they have relatively more shielding.

To define the position of an absorption, the NMR chart is calibrated and a reference point is used. In practice, a small amount of tetramethylsilane [TMS; (CH$_3$)$_4$Si] is added to the sample so that a reference absorption peak is produced when the spectrum is run. TMS is used as reference for ^{1}H and ^{13}C measurements because it produces in both a single peak that occurs upfield of other absorptions normally found in organic compounds. The ^{1}H and ^{13}C spectra of methyl acetate in Figure 11.3 have the TMS reference peak indicated.

FIGURE 11.5 The NMR chart. The downfield, deshielded, side is on the left, and the upfield, shielded, side is on the right. The tetramethylsilane (TMS) absorption is used as reference point.

The position on the chart at which a nucleus absorbs is called its **chemical shift**. The chemical shift of TMS is set as the zero point, and other absorptions normally occur downfield, to the left on the chart. NMR charts are calibrated using an arbitrary scale called the **delta (δ) scale**, where 1 δ equals 1 part per million (1 ppm) of the spectrometer operating frequency. For example, if we were measuring the ^{1}H NMR spectrum of a sample using an instrument operating at 200 MHz, 1 δ would be 1 millionth of 200,000,000 Hz, or 200 Hz. If we were measuring the spectrum using a 500 MHz instrument, 1 δ would be 500 Hz. The following equation can be used for any absorption:

$$\delta = \frac{\text{Chemical shift (number of Hz downfield from TMS)}}{\text{Spectrometer frequency in MHz}}$$

Although this method of calibrating NMR charts may seem complex, there's a good reason for it. As we saw earlier, the rf frequency required to bring a given nucleus into resonance depends on the spectrometer's magnetic field strength. But because there are many different kinds of spectrometers with many different magnetic field strengths available, chemical shifts given in frequency units (Hz) vary from one instrument to another. Thus, a resonance that occurs at 120 Hz downfield from TMS on one spectrometer might occur at 600 Hz downfield from TMS on another spectrometer with a more powerful magnet.

By using a system of measurement in which NMR absorptions are expressed in relative terms (parts per million relative to spectrometer frequency) rather than absolute terms (Hz), it's possible to compare spectra obtained on different instruments. *The chemical shift of an NMR absorption in δ units is constant, regardless of the operating frequency of the spectrometer.* A ^{1}H nucleus that absorbs at 2.0 δ on a 200 MHz instrument also absorbs at 2.0 δ on a 500 MHz instrument.

The range in which most NMR absorptions occur is quite narrow. Almost all ^{1}H NMR absorptions occur from 0 to 10 δ downfield from the proton absorption of TMS, and almost all ^{13}C absorptions occur from 1 to 220 δ downfield from the carbon absorption of TMS. Thus, there is a likelihood that accidental overlap of nonequivalent signals will occur. The advantage of using an instrument with higher field strength (say, 500 MHz) rather than lower field strength (200 MHz) is that different NMR absorptions are more widely separated at the higher field strength. The chances that two signals will accidentally overlap are therefore lessened, and interpretation of spectra becomes easier. For example, two signals that are only 20 Hz apart at 200 MHz (0.1 ppm) are 50 Hz apart at 500 MHz (still 0.1 ppm).

Problem 11.3
The following ^{1}H NMR peaks were recorded on a spectrometer operating at 200 MHz. Convert each into δ units.
(a) $CHCl_3$; 1454 Hz   **(b)** CH_3Cl; 610 Hz
(c) CH_3OH; 693 Hz   **(d)** CH_2Cl_2; 1060 Hz

Problem 11.4
When the ^{1}H NMR spectrum of acetone, CH_3COCH_3, is recorded on an instrument operating at 200 MHz, a single sharp resonance at 2.1 δ is seen.
(a) How many hertz downfield from TMS does the acetone resonance correspond to?
(b) If the ^{1}H NMR spectrum of acetone were recorded at 500 MHz, what would the position of the absorption be in δ units?
(c) How many hertz downfield from TMS does this 500 MHz resonance correspond to?

11.4 ¹³C NMR Spectroscopy: Signal Averaging and FT–NMR

Everything we've said thus far about NMR spectroscopy applies to both ¹H and ¹³C spectra. Now, though, let's focus only on ¹³C spectroscopy because it's much easier to interpret. What we learn now about interpreting ¹³C spectra will simplify the subsequent discussion of ¹H spectra.

In some ways, it's surprising that carbon NMR is even possible. After all, ¹²C, the most abundant carbon isotope, has no nuclear spin and can't be seen by NMR. Carbon-13 is the only naturally occurring carbon isotope with a nuclear spin, but its natural abundance is only 1.1%. Thus, only about 1 of every 100 carbons in an organic sample is observable by NMR. The problem of low abundance has been overcome, however, by the use of *signal averaging* and *Fourier-transform NMR* (**FT–NMR**). Signal averaging increases instrument sensitivity, and FT–NMR increases instrument speed.

The low natural abundance of ¹³C means that any individual NMR spectrum is extremely "noisy." That is, the signals are so weak that they are cluttered with random background electronic noise, as shown in Figure 11.6a. If, however, hundreds or thousands of individual runs are added together by a computer and then averaged, a greatly improved spectrum results (Figure 11.6b). Background noise, because of its random nature, averages to zero, while the nonzero NMR signals stand out clearly. Unfortunately, the value of signal averaging is limited when using the method of NMR spectrometer operation described in Section 11.2, because it takes about 5 to 10 minutes to obtain a single spectrum. Thus, a faster way to obtain spectra is needed if signal averaging is to be used.

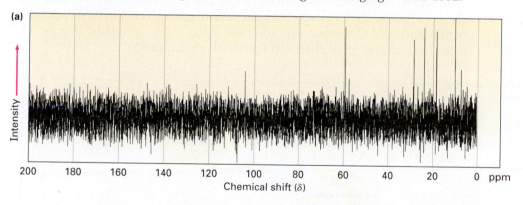

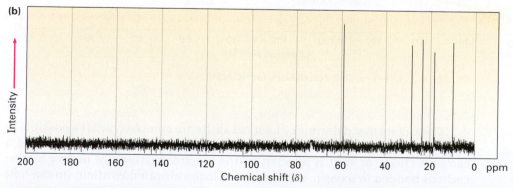

FIGURE 11.6 Carbon-13 NMR spectra of pentan-1-ol, CH₃CH₂CH₂CH₂CH₂OH. Spectrum **(a)** is a single run, showing the large amount of background noise. Spectrum **(b)** is an average of 200 runs.

In the method of NMR spectrometer operation described in Section 11.2, the rf frequency is held constant while the strength of the magnetic field is varied so that all signals in the spectrum are recorded sequentially. In the FT-NMR technique used by modern spectrometers, however, all the signals are recorded simultaneously. A sample is placed in a magnetic field of constant strength and is irradiated with a short pulse of rf energy that covers the entire range of useful frequencies. All ^1H or ^{13}C nuclei in the sample resonate at once, giving a complex, composite signal that is mathematically manipulated using so-called Fourier transforms and then displayed in the usual way. Because all resonance signals are collected at once, it takes only a few seconds rather than a few minutes to record an entire spectrum.

Combining the speed of FT-NMR with the sensitivity enhancement of signal averaging is what gives modern NMR spectrometers their power. Literally thousands of spectra can be taken and averaged in a few hours, resulting in sensitivity so high that a ^{13}C NMR spectrum can be obtained on less than 0.1 mg of sample, and a ^1H spectrum can be recorded on only a few *micro*grams.

11.5   Characteristics of ^{13}C NMR Spectroscopy

At its simplest, ^{13}C NMR makes it possible to count the number of different carbon atoms in a molecule. Look at the ^{13}C NMR spectra of methyl acetate and pentan-1-ol shown previously in Figures 11.3b and 11.6b. In each case, a single sharp resonance line is observed for each different carbon atom.

Most ^{13}C resonances are between 0 and 220 ppm downfield from the TMS reference line, with the exact chemical shift of each ^{13}C resonance dependent on that carbon's electronic environment within the molecule. Figure 11.7 shows the correlation of chemical shift with environment.

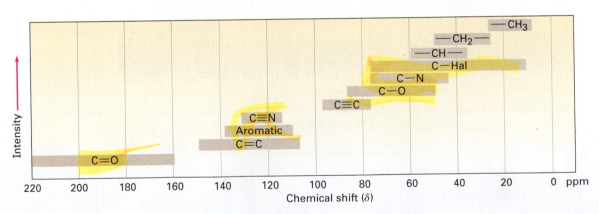

FIGURE 11.7 Chemical shift correlations for ^{13}C NMR.

The factors that determine chemical shifts are complex, but it's possible to make some generalizations from the data in Figure 11.7. One trend is that a carbon's chemical shift is affected by the electronegativity of nearby atoms: carbons bonded to oxygen, nitrogen, or halogen absorb downfield (to the left) of typical alkane carbons. Because electronegative atoms attract electrons, they pull electrons away from neighboring carbon atoms, causing those carbons to be deshielded and to come into resonance at a lower field.

Another trend is that sp^3-hybridized carbons generally absorb from 0 to 90 δ, while sp^2 carbons absorb from 110 to 220 δ. Carbonyl carbons (C=O) are particularly distinct in ^{13}C NMR and are always found at the low-field end of the spectrum, from 160 to 220 δ. Figure 11.8 shows the ^{13}C NMR spectra of butan-2-one and p-bromoacetophenone and indicates the peak assignments. Note that the C=O carbons are at the left edge of the spectrum in each case.

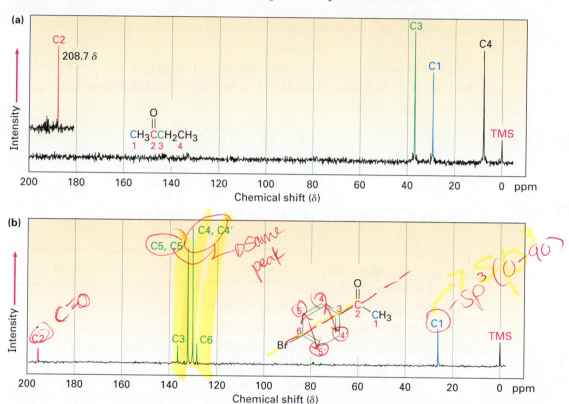

FIGURE 11.8 Carbon-13 NMR spectra of **(a)** butan-2-one and **(b)** p-bromoacetophenone.

The ^{13}C NMR spectrum of p-bromoacetophenone is interesting in several ways. Note particularly that only six carbon absorptions are observed, even though the molecule contains eight carbons. p-Bromoacetophenone has a symmetry plane that makes ring carbons 4 and 4', and ring carbons 5 and 5' equivalent. Thus, the six ring carbons show only four absorptions in the 128 to 137 δ range.

para-Bromoacetophenone

A second interesting point about both spectra in Figure 11.8 is that the peaks aren't uniform in size. Some peaks are larger than others even though they are one-carbon resonances (except for the two 2-carbon peaks of p-bromoacetophenone). This difference in peak size is a general feature of ^{13}C NMR spectra.

WORKED EXAMPLE 11.1   Predicting Chemical Shifts in ^{13}C NMR Spectra

At what approximate positions would you expect ethyl acrylate, $H_2C\!=\!CHCO_2CH_2CH_3$, to show ^{13}C NMR absorptions?

Strategy

Identify the distinct carbons in the molecule, and note whether each is alkyl, vinylic, aromatic, or in a carbonyl group. Then predict where each absorbs, using Figure 11.7 as necessary.

Solution

Ethyl acrylate has five distinct carbons: two different C=C, one C=O, one O–C, and one alkyl C. From Figure 11.7, the likely absorptions are

$\sim130\,\delta \qquad \sim180\,\delta \quad \sim60\,\delta \qquad \sim15\,\delta$

The actual absorptions are at 14.1, 60.5, 128.5, 130.3, and 166.0 δ.

Problem 11.5

How many carbon resonance lines would you expect in the ^{13}C NMR spectra of the following compounds?

(a) Methylcyclopentane   **(b)** 1-Methylcyclohexene

(c) 1,2-Dimethylbenzene   **(d)** 2-Methylbut-2-ene

(e)

(f) H_3C CH_2CH_3
$\quad\quad C\!=\!C$
H_3C CH_3

Problem 11.6

Propose structures for compounds that fit the following descriptions:

(a) A hydrocarbon with seven lines in its ^{13}C NMR spectrum
(b) A six-carbon compound with only five lines in its ^{13}C NMR spectrum
(c) A four-carbon compound with three lines in its ^{13}C NMR spectrum

Problem 11.7

Assign the resonances in the ^{13}C NMR spectrum of methyl propanoate, $CH_3CH_2CO_2CH_3$ (Figure 11.9).

FIGURE 11.9
^{13}C NMR spectrum
of methyl propanoate
for Problem 11.7.

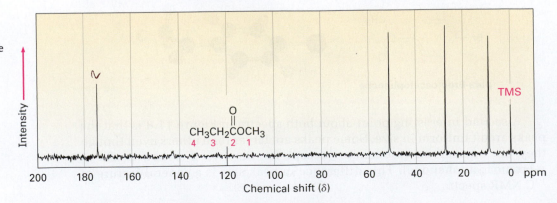

11.6 DEPT ^{13}C NMR Spectroscopy

Numerous techniques developed in recent years have made it possible to obtain enormous amounts of information from ^{13}C NMR spectra. Among the more useful of these techniques is one called *DEPT-NMR,* for *distortionless enhancement by polarization transfer,* which makes it possible to distinguish among signals due to CH_3, CH_2, CH, and quaternary carbons. That is, the number of hydrogens attached to each carbon in a molecule can be determined.

A DEPT experiment is usually done in three stages, as shown in Figure 11.10 for 6-methylhept-5-en-2-ol. The first stage is to run an ordinary spectrum (called a *broadband-decoupled spectrum*) to locate the chemical shifts of all carbons. Next, a second spectrum called a DEPT-90 is run, using special conditions under

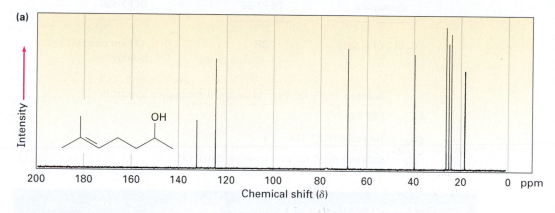

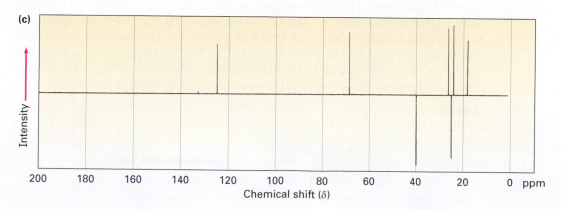

FIGURE 11.10

DEPT-NMR spectra for 6-methylhept-5-en-2-ol. Part **(a)** is an ordinary broadband-decoupled spectrum, which shows signals for all eight carbons. Part **(b)** is a DEPT-90 spectrum, which shows signals only for the two CH carbons. Part **(c)** is a DEPT-135 spectrum, which shows positive signals for the two CH and three CH_3 carbons and negative signals for the two CH_2 carbons.

which only signals due to CH carbons appear. Signals due to CH_3, CH_2, and quaternary carbons are absent. Finally, a third spectrum called a DEPT-135 is run, using conditions under which CH$_3$ and CH resonances appear as positive signals, CH$_2$ resonances appear as *negative* signals—that is, as peaks below the baseline—and quaternary carbons are again absent.

Putting together the information from all three spectra makes it possible to tell the number of hydrogens attached to each carbon. The CH carbons are identified in the DEPT-90 spectrum, the CH_2 carbons are identified as the negative peaks in the DEPT-135 spectrum, the CH_3 carbons are identified by subtracting the CH peaks from the positive peaks in the DEPT-135 spectrum, and quaternary carbons are identified by subtracting all peaks in the DEPT-135 spectrum from the peaks in the broadband-decoupled spectrum.

Broadband-decoupled	DEPT-90	DEPT-135
C, CH, CH$_2$, CH$_3$	CH	CH$_3$, CH are positive CH$_2$ is negative

C	Subtract DEPT-135 from broadband-decoupled spectrum
CH	DEPT-90
CH$_2$	Negative DEPT-135
CH$_3$	Subtract DEPT-90 from positive DEPT-135

WORKED EXAMPLE 11.2 Assigning a Chemical Structure from a ^{13}C NMR Spectrum

Propose a structure for an alcohol, $C_4H_{10}O$, that has the following ^{13}C NMR spectral data:

> Broadband-decoupled ^{13}C NMR: 19.0, 31.7, 69.5 δ
>
> DEPT-90: 31.7 δ
>
> DEPT-135: positive peak at 19.0 δ, negative peak at 69.5 δ

Strategy

Let's begin by noting that the unknown alcohol has *four* carbon atoms, yet has only *three* NMR absorptions, which implies that two carbons must be equivalent. Looking at chemical shifts, two of the absorptions are in the typical alkane region (19.0 and 31.7 δ) while one is in the region of a carbon bonded to an electronegative atom (69.5 δ)—oxygen in this instance. The DEPT-90 spectrum tells us that the alkyl carbon at 31.7 δ is tertiary (CH); the DEPT-135 spectrum tells us that the alkyl carbon at 19.0 δ is a methyl (CH_3) and that the carbon bonded to oxygen (69.5 δ) is secondary (CH_2). The two equivalent carbons are probably both methyls bonded to the same tertiary carbon, $(CH_3)_2CH-$. We can now put the pieces together to propose a structure: 2-methylpropan-1-ol.

Solution

2-Methylpropan-1-ol

Problem 11.8

Assign a chemical shift to each carbon in 6-methylhept-5-en-2-ol (Figure 11.10).

Problem 11.9

Estimate the chemical shift of each carbon in the following molecule. Predict which carbons will appear in the DEPT-90 spectrum, which will give positive peaks in the DEPT-135 spectrum, and which will give negative peaks in the DEPT-135 spectrum.

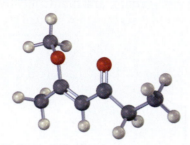

Problem 11.10

Propose a structure for an aromatic hydrocarbon, $C_{11}H_{16}$, that has the following ^{13}C NMR spectrum:

Broadband-decoupled ^{13}C NMR: 29.5, 31.8, 50.2, 125.5, 127.5, 130.3, 139.8 δ

DEPT-90: 125.5, 127.5, 130.3 δ

DEPT-135: positive peaks at 29.5, 125.5, 127.5, 130.3 δ; negative peak at 50.2 δ

11.7 Uses of ^{13}C NMR Spectroscopy

The information derived from ^{13}C NMR spectroscopy is extraordinarily useful for structure determination. Not only can we count the number of nonequivalent carbon atoms in a molecule, we can also get information about the electronic environment of each carbon and can even find how many protons each is attached to. As a result, we can answer many structural questions that go unanswered by IR spectroscopy or mass spectrometry.

Here's an example: does the elimination reaction of 1-chloro-1-methylcyclohexane on treatment with a strong base give predominantly the trisubstituted alkene 1-methylcyclohexene or the disubstituted alkene methylenecyclohexane?

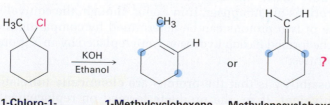

1-Chloro-1-
methylcyclohexane

1-Methylcyclohexene
(trisubstituted)

Methylenecyclohexane
(disubstituted)

1-Methylcyclohexene will have five sp^3-carbon resonances in the 20 to 50 δ range and two sp^2-carbon resonances in the 100 to 150 δ range. Methylene-cyclohexane, however, because of its symmetry, will have only three sp^3-carbon resonance peaks and two sp^2-carbon peaks. The spectrum of the actual reaction product, shown in Figure 11.11, clearly identifies 1-methylcyclohexene as the product of this elimination reaction. In fact, we'll see in the next chapter (Section 12.11) that this result is general. Elimination reactions usually give the more highly substituted alkene product rather than the less highly substituted alkene.

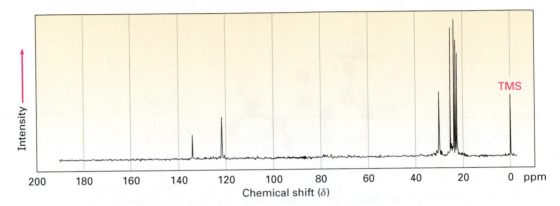

FIGURE 11.11 The ^{13}C NMR spectrum of 1-methylcyclohexene, the elimination reaction product from treatment of 1-chloro-1-methylcyclohexane with a strong base.

Problem 11.11
We saw in Section 8.15 that addition of HBr to a terminal alkyne leads to the Markovnikov addition product, with the Br bonding to the more highly substituted carbon. How could you use ^{13}C NMR to identify the product of the addition of 1 equivalent of HBr to hex-1-yne?

11.8 ^{1}H NMR Spectroscopy and Proton Equivalence

Having looked at ^{13}C spectra, let's now focus on ^{1}H NMR spectroscopy. Because each electronically distinct hydrogen in a molecule has its own unique absorption, one use of ^{1}H NMR is to find out how many kinds of electronically nonequivalent hydrogens are present. In the ^{1}H NMR spectrum of methyl acetate shown previously in Figure 11.3a, for instance, there are two signals, corresponding to the two kinds of nonequivalent protons present, $CH_3C=O$ protons and $-OCH_3$ protons.

For relatively small molecules, a quick look at a structure is often enough to decide how many kinds of protons are present and thus how many NMR absorptions might appear. If in doubt, though, the equivalence or non-equivalence of two protons can be determined by comparing the structures that would be formed if each hydrogen were replaced by an X group. There are four possibilities:

- One possibility is that the protons are chemically unrelated and thus nonequivalent. If so, the products formed on replacement of H by X would be different constitutional isomers. In butane, for instance, the –CH$_3$ protons are different from the –CH$_2$– protons, would give different products on replacement by X, and would likely show different NMR absorptions.

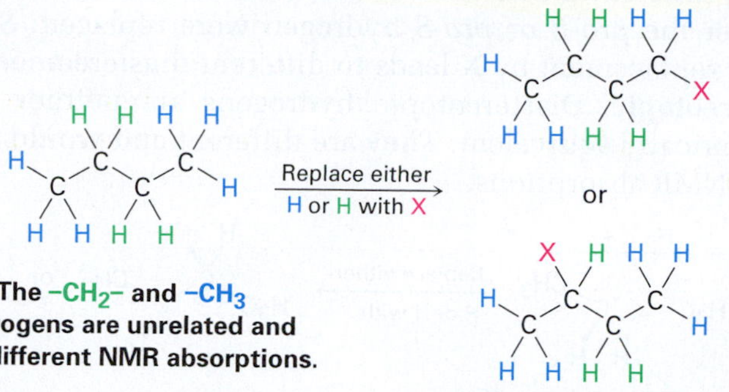

The –CH$_2$– and –CH$_3$ hydrogens are unrelated and have different NMR absorptions.

Replace either H or H with X

or

The two replacement products are constitutional isomers.

- A second possibility is that the protons are chemically identical and thus electronically equivalent. If so, the same product would be formed regardless of which H is replaced by X. In butane, for instance, the six –CH$_3$ hydrogens on C1 and C4 are identical, would give the identical structure on replacement by X, and would show the identical NMR absorption. Such protons are said to be **homotopic**.

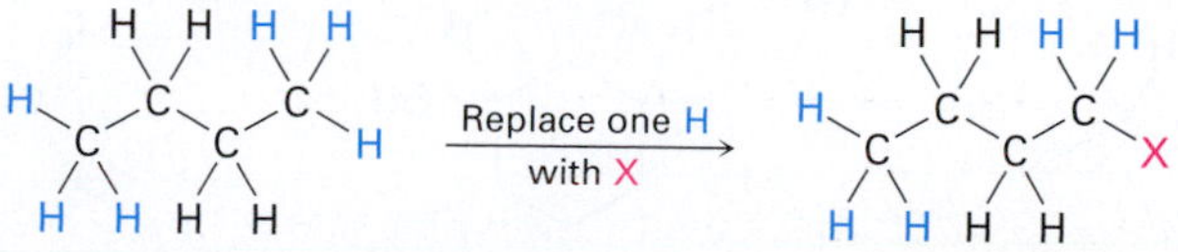

The 6 –CH$_3$ hydrogens are *homotopic* and have the same NMR absorptions.

Replace one H with X

Only one replacement product is possible.

- The third possibility is a bit more subtle. Although they might at first seem homotopic, the two –CH$_2$– hydrogens on C2 in butane (and the two –CH$_2$– hydrogens on C3) are in fact not identical. Replacement of a hydrogen at C2 (or C3) would form a new chirality center, so different enantiomers (Section 5.1) would result depending on whether the *pro-R* or *pro-S* hydrogen were replaced (Section 5.11). Such hydrogens, whose replacement by X would lead to different enantiomers, are said to be **enantiotopic**. Enantiotopic hydrogens, even though not identical, are nevertheless electronically equivalent and thus have the same NMR absorption.

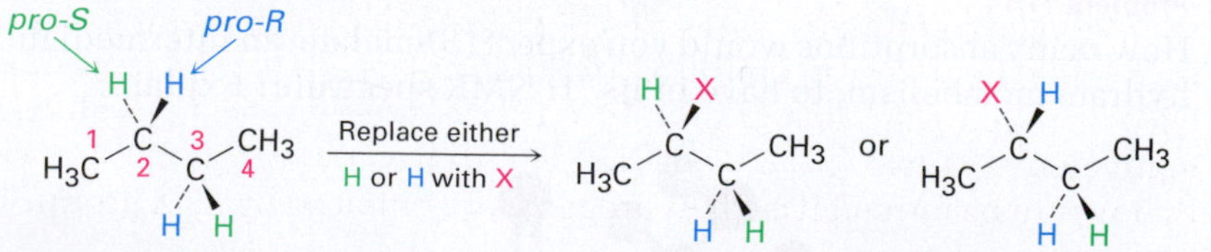

The two hydrogens on C2 (and the two hydrogens on C3) are *enantiotopic* and have the same NMR absorption.

Replace either H or H with X

The two possible replacement products are enantiomers.

- The fourth possibility arises in chiral molecules, such as *R*-butan-2-ol. The two –CH_2– hydrogens at C3 are neither homotopic nor enantiotopic. Since replacement of a hydrogen at C3 would form a *second* chirality center, different *diastereomers* (Section 5.6) would result depending on whether the *pro-R* or *pro-S* hydrogen were replaced. Such hydrogens, whose replacement by X leads to different diastereomers, are said to be **diastereotopic**. Diastereotopic hydrogens are neither chemically nor electronically equivalent. They are different and would likely show different NMR absorptions.

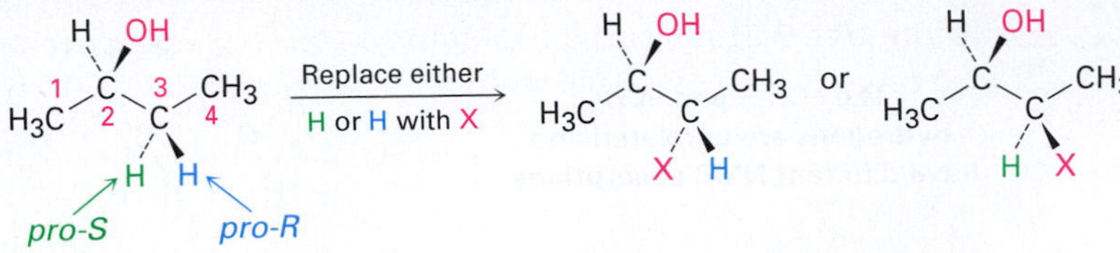

The two hydrogens on C3 are *diastereotopic* and have different NMR absorptions.

The two possible replacement products are diastereomers.

Problem 11.12

Identify the indicated sets of protons as unrelated, homotopic, enantiotopic, or diastereotopic:

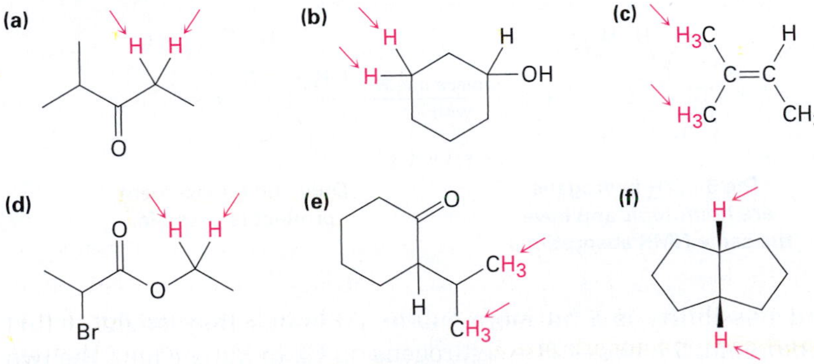

(a)

(b)

(c)

(d)

(e)

(f)

Problem 11.13

How many kinds of electronically nonequivalent protons are present in each of the following compounds, and thus how many NMR absorptions might you expect in each?

(a) CH_3CH_2Br (b) $CH_3OCH_2CH(CH_3)_2$ (c) $CH_3CH_2CH_2NO_2$
(d) Toluene (e) 2-Methylbut-1-ene (f) *cis*-Hex-3-ene

Problem 11.14

How many absorptions would you expect (*S*)-malate, an intermediate in carbohydrate metabolism, to have in its ^1H NMR spectrum? Explain.

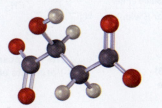

(*S*)-Malate

11.9 | Chemical Shifts in ¹H NMR Spectroscopy

We said previously that differences in chemical shifts are caused by the small local magnetic fields of electrons surrounding the different nuclei. Nuclei that are more strongly shielded by electrons require a higher applied field to bring them into resonance and therefore absorb on the right side of the NMR chart. Nuclei that are less strongly shielded need a lower applied field for resonance and therefore absorb on the left of the NMR chart.

Most ¹H chemical shifts fall within the 0 to 10 δ range, which can be divided into the five regions shown in Table 11.2. By remembering the positions of these regions, it's often possible to tell at a glance what kinds of protons a molecule contains.

TABLE 11.2
Regions of the ¹H NMR Spectrum

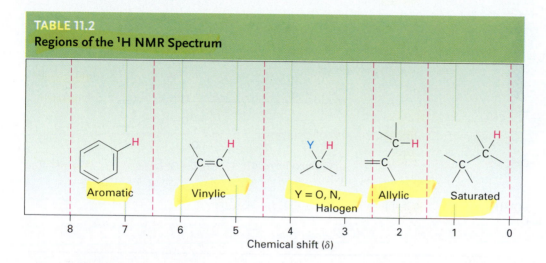

Table 11.3 shows the correlation of ¹H chemical shift with electronic environment in more detail. In general, protons bonded to saturated, sp^3-hybridized carbons absorb at higher fields, whereas protons bonded to sp^2-hybridized carbons absorb at lower fields. Protons on carbons that are bonded to electronegative atoms, such as N, O, or halogen, also absorb at lower fields.

WORKED EXAMPLE 11.3 Predicting Chemical Shifts in ¹H NMR Spectra

Methyl 2,2-dimethylpropanoate $(CH_3)_3CCO_2CH_3$ has two peaks in its ¹H NMR spectrum. What are their approximate chemical shifts?

Strategy
Identify the types of hydrogens in the molecule, and note whether each is alkyl, vinylic, or next to an electronegative atom. Then predict where each absorbs, using Table 11.3 if necessary.

Solution
The $-OCH_3$ protons absorb around 3.5 to 4.0 δ because they are on carbon bonded to oxygen. The $(CH_3)_3C-$ protons absorb near 1.0 δ because they are typical alkane-like protons.

TABLE 11.3
Correlation of ¹H Chemical Shift with Environment

Type of hydrogen		Chemical shift (δ)	Type of hydrogen		Chemical shift (δ)
Reference	Si(CH₃)₄	0	Alcohol	—C—O—H	2.5–5.0
Alkyl (primary)	—CH₃	0.7–1.3			
Alkyl (secondary)	—CH₂—	1.2–1.6	Alcohol, ether	H / —C—O—	3.3–4.5
Alkyl (tertiary)	—CH—	1.4–1.8			
Allylic	C=C—C— with H	1.6–2.2	Vinylic	C=C with H	4.5–6.5
			Aryl	Ar—H	6.5–8.0
Methyl ketone	O‖ —C—CH₃	2.0–2.4	Aldehyde	O‖ —C—H	9.7–10.0
Aromatic methyl	Ar—CH₃	2.4–2.7			
Alkynyl	—C≡C—H	2.5–3.0	Carboxylic acid	O‖ —C—O—H	11.0–12.0
Alkyl halide	H —C—Hal	2.5–4.0			

Problem 11.15
Each of the following compounds has a single ¹H NMR peak. Approximately where would you expect each compound to absorb?

(a)  [hexane ring structure]

(b) 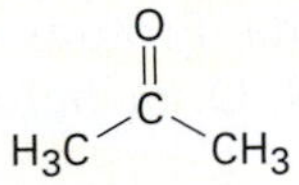 O‖ H₃C—C—CH₃

(c) [benzene ring structure]

(d) CH₂Cl₂

(e) 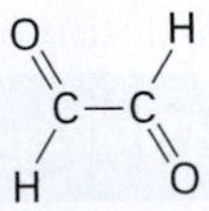 O=C(H)—C(H)=O

(f) 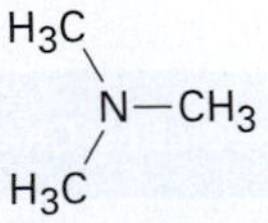 H₃C / N—CH₃ with H₃C

Problem 11.16
Identify the different kinds of nonequivalent protons in the following molecule, and tell where you would expect each to absorb:

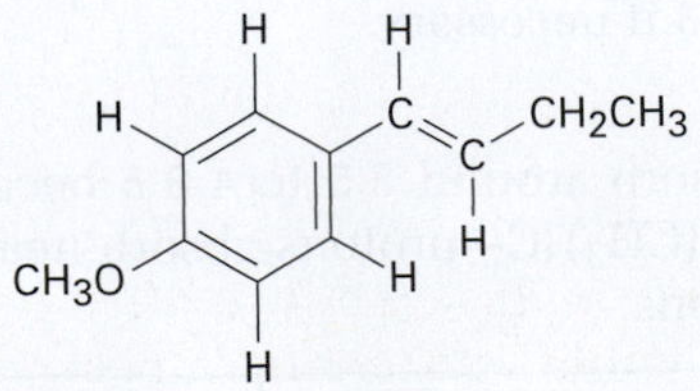

[structure of a substituted benzene with CH₃O— group, CH₂CH₃ group, and vinyl group]

11.10 Integration of ^{1}H NMR Absorptions: Proton Counting

Look at the ^{1}H NMR spectrum of methyl 2,2-dimethylpropanoate in Figure 11.12. There are two peaks, corresponding to the two kinds of protons, but the peaks aren't the same size. The peak at 1.20 δ, due to the $(CH_3)_3C-$ protons, is larger than the peak at 3.65 δ, due to the $-OCH_3$ protons.

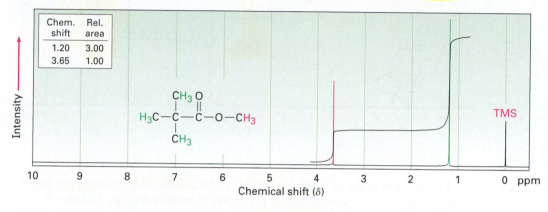

Chem. shift	Rel. area
1.20	3.00
3.65	1.00

FIGURE 11.12 The ^{1}H NMR spectrum of methyl 2,2-dimethyl-propanoate. Integrating the two peaks in a stair-step manner shows that they have a 1:3 ratio, corresponding to the 3:9 ratio of protons responsible. Modern instruments give a direct digital readout of relative peak areas.

The area under each peak is proportional to the number of protons causing that peak. By electronically measuring, or **integrating,** the area under each peak, it's possible to measure the relative numbers of the different kinds of protons in a molecule. Modern NMR instruments provide a digital readout of relative peak areas, but an older, more visual method displays the integrated peak areas as a "stair-step" line, with the height of each step proportional to the area under the peak, and therefore proportional to the relative number of protons causing the peak. To compare the size of one peak against another, simply take a ruler and measure the heights of the various steps. For example, the two steps for the peaks in methyl 2,2-dimethylpropanoate are found to have a 1:3 (or 3:9) height ratio when integrated—exactly what we expect since the three $-OCH_3$ protons are equivalent and the nine $(CH_3)_3C-$ protons are equivalent.

Problem 11.17

How many peaks would you expect in the ^{1}H NMR spectrum of 1,4-dimethyl-benzene (*p*-xylene)? What ratio of peak areas would you expect on integration of the spectrum? Refer to Table 11.3 for approximate chemical shifts, and sketch what the spectrum would look like.

p-Xylene

11.11 Spin–Spin Splitting in ^{1}H NMR Spectra

In the ^{1}H NMR spectra we've seen thus far, each different kind of proton in a molecule has given rise to a single peak. It often happens, though, that the absorption of a proton splits into *multiple* peaks, called a **multiplet.** For example, in the ^{1}H NMR spectrum of bromoethane shown in Figure 11.13, the –CH$_2$Br protons appear as four peaks (a *quartet*) centered at 3.42 δ and the –CH$_3$ protons appear as three peaks (a *triplet*) centered at 1.68 δ.

ACTIVE FIGURE 11.13

The ^{1}H NMR spectrum of bromoethane, CH$_3$CH$_2$Br. The –CH$_2$Br protons appear as a quartet at 3.42 δ, and the –CH$_3$ protons appear as a triplet at 1.68 δ. **Go to this book's student companion site at www.cengage .com/chemistry/ mcmurry to explore an interactive version of this figure.**

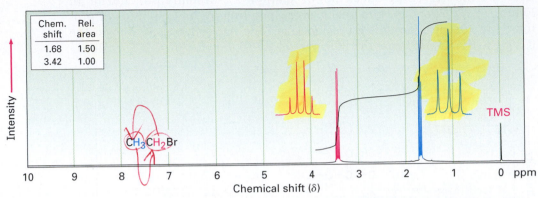

Chem. shift	Rel. area
1.68	1.50
3.42	1.00

Called **spin–spin splitting**, multiple absorptions of a nucleus are caused by the interaction, or **coupling**, of the spins of nearby nuclei. In other words, the tiny magnetic field produced by one nucleus affects the magnetic field felt by neighboring nuclei. Look at the –CH$_3$ protons in bromoethane, for example. The three equivalent –CH$_3$ protons are neighbored by two other magnetic nuclei—the two protons on the adjacent –CH$_2$Br group. Each of the neighboring –CH$_2$Br protons has its own nuclear spin, which can align either with or against the applied field, producing a tiny effect that is felt by the –CH$_3$ protons.

There are three ways in which the spins of the two –CH$_2$Br protons can align, as shown in Figure 11.14. If both proton spins align with the applied field, the total effective field felt by the neighboring –CH$_3$ protons is slightly larger than it would otherwise be. Consequently, the applied field necessary to cause resonance is slightly reduced. Alternatively, if one of the –CH$_2$Br proton spins aligns with the field and one aligns against the field, there is no effect on the neighboring –CH$_3$ protons. (There are two ways this arrangement can occur, depending on which of the two proton spins aligns which way.) Finally, if both –CH$_2$Br proton spins align against the applied field, the effective field felt by the –CH$_3$ protons is slightly smaller than it would otherwise be and the applied field needed for resonance is slightly increased.

Any given molecule has only one of the three possible alignments of –CH$_2$Br spins, but in a large collection of molecules, all three spin states are represented in a 1 : 2 : 1 statistical ratio. We therefore find that the neighboring –CH$_3$ protons come into resonance at three slightly different values of the applied field, and we see a 1 : 2 : 1 triplet in the NMR spectrum. One resonance is a little above where it would be without coupling, one is at the same place it would be without coupling, and the third resonance is a little below where it would be without coupling.

In the same way that the –CH$_3$ absorption of bromoethane is split into a triplet, the –CH$_2$Br absorption is split into a quartet. The three spins of the neighboring –CH$_3$ protons can align in four possible combinations: all three with the applied field, two with and one against (three ways), one with and two against (three ways), or all three against. Thus, four peaks are produced for the –CH$_2$Br protons in a 1 : 3 : 3 : 1 ratio.

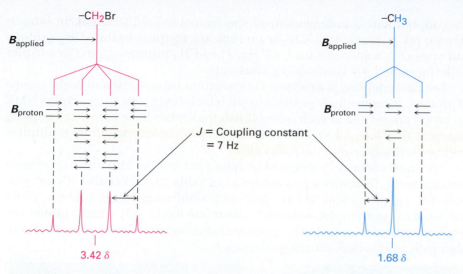

Quartet due to coupling with –CH$_3$ Triplet due to coupling with –CH$_2$Br

FIGURE 11.14 The origin of spin–spin splitting in bromoethane. The nuclear spins of neighboring protons, indicated by horizontal arrows, align either with or against the applied field, causing the splitting of absorptions into multiplets.

As a general rule, called the **n + 1 rule**, protons that have *n* equivalent neighboring protons show *n* + 1 peaks in their NMR spectrum. For example, the spectrum of 2-bromopropane in Figure 11.15 shows a doublet at 1.71 δ and a seven-line multiplet, or *septet,* at 4.28 δ. The septet is caused by splitting of the –CHBr– proton signal by six equivalent neighboring protons on the two methyl groups (*n* = 6 leads to 6 + 1 = 7 peaks). The doublet is due to signal splitting of the six equivalent methyl protons by the single –CHBr– proton (*n* = 1 leads to 2 peaks). Integration confirms the expected 6:1 ratio.

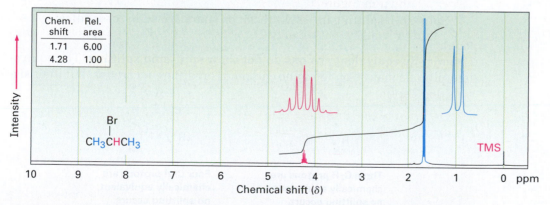

Chem. shift	Rel. area
1.71	6.00
4.28	1.00

FIGURE 11.15 The ^{1}H NMR spectrum of 2-bromopropane. The –CH$_3$ proton signal at 1.71 δ is split into a doublet, and the –CHBr– proton signal at 4.28 δ is split into a septet. Note that the distance between peaks—the *coupling constant*—is the same in both multiplets. Note also that the outer two peaks of the septet are so small as to be nearly lost.

The distance between peaks in a multiplet is called the **coupling constant** and is denoted *J*. Coupling constants are measured in hertz and generally fall in the range 0 to 18 Hz. The exact value of the coupling constant between two neighboring protons depends on the geometry of the molecule, but a typical value for an open-chain alkane is *J* = 6 to 8 Hz. The same coupling constant is shared by both groups of hydrogens whose spins are coupled and is independent of spectrometer field strength. In bromo-ethane, for instance, the $-CH_2Br$ protons are coupled to the $-CH_3$ protons and appear as a quartet with $J = 7$ Hz. The $-CH_3$ protons appear as a triplet with the same $J = 7$ Hz coupling constant.

Because coupling is a reciprocal interaction between two adjacent groups of protons, it's sometimes possible to tell which multiplets in a complex NMR spectrum are related to each other. If two multiplets have the same coupling constant, they are probably related, and the protons causing those multiplets are therefore adjacent in the molecule.

The most commonly observed coupling patterns and the relative intensities of lines in their multiplets are listed in Table 11.4. Note that it's not possible for a given proton to have *five* equivalent neighboring protons. (Why not?) A six-line multiplet, or sextet, is therefore found only when a proton has five *nonequivalent* neighboring protons that coincidentally happen to be coupled with an identical coupling constant J.

TABLE 11.4
Some Common Spin Multiplicities

Number of equivalent adjacent protons	Multiplet	Ratio of intensities
0	Singlet	1
1	Doublet	1:1
2	Triplet	1:2:1
3	Quartet	1:3:3:1
4	Quintet	1:4:6:4:1
6	Septet	1:6:15:20:15:6:1

Spin–spin splitting in ^1H NMR can be summarized in three rules:

Rule 1

Chemically equivalent protons do not show spin–spin splitting. The equivalent protons may be on the same carbon or on different carbons, but their signals don't split.

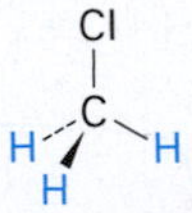

Three C–H protons are chemically equivalent; no splitting occurs.

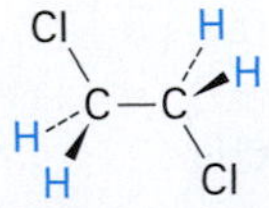

Four C–H protons are chemically equivalent; no splitting occurs.

Rule 2

The signal of a proton that has *n* equivalent neighboring protons is split into a multiplet of *n* + 1 peaks with coupling constant *J*. Protons that are farther than two carbon atoms apart don't usually couple, although they sometimes show small coupling when they are separated by a π bond.

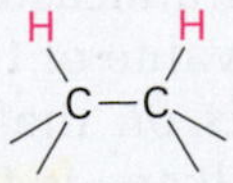

Splitting observed

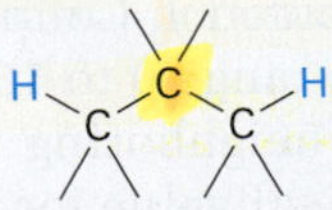

Splitting not usually observed

Rule 3

Two groups of protons coupled to each other have the same coupling constant, J.

The spectrum of *p*-methoxypropiophenone in Figure 11.16 further illustrates the three rules. The downfield absorptions at 6.91 and 7.93 δ are due to the four aromatic ring protons. There are two kinds of aromatic protons, each of which gives a signal that is split into a doublet by its neighbor. The —OCH$_3$ signal is unsplit and appears as a sharp singlet at 3.84 δ. The —CH$_2$— protons next to the carbonyl group appear at 2.93 δ in the region expected for protons on carbon next to an unsaturated center, and their signal is split into a quartet by coupling with the protons of the neighboring methyl group. The methyl protons appear as a triplet at 1.20 δ in the usual upfield region.

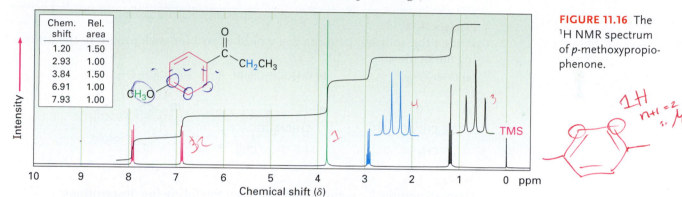

FIGURE 11.16 The ^{1}H NMR spectrum of *p*-methoxypropiophenone.

One further question needs to be answered before leaving the topic of spin–spin splitting: why is spin–spin splitting seen only for ^{1}H NMR? That is, why is there no splitting of *carbon* signals into multiplets in ^{13}C NMR? After all, you might expect that the spin of a given ^{13}C nucleus would couple with the spin of an adjacent magnetic nucleus, either ^{13}C or ^{1}H.

No coupling of a ^{13}C nucleus with nearby *carbons* is seen because the low natural abundance makes it unlikely that two ^{13}C nuclei will be adjacent. No coupling of a ^{13}C nucleus with nearby *hydrogens* is seen because ^{13}C spectra, as previously noted (Section 11.6), are normally recorded using broadband decoupling. At the same time that the sample is irradiated with a pulse of rf energy to cover the *carbon* resonance frequencies, it is also irradiated by a second band of rf energy covering all the *hydrogen* resonance frequencies. This second irradiation makes the hydrogens spin-flip so rapidly that their local magnetic fields average to zero and no coupling with carbon spins occurs.

WORKED EXAMPLE 11.4 Assigning a Chemical Structure from a ^{1}H NMR Spectrum

Propose a structure for a compound, $C_5H_{12}O$, that fits the following ^{1}H NMR data: 0.92 δ (3 H, triplet, $J = 7$ Hz), 1.20 δ (6 H, singlet), 1.50 δ (2 H, quartet, $J = 7$ Hz), 1.64 δ (1 H, broad singlet).

Strategy

Let's look at each absorption individually. The three-proton absorption at 0.92 δ is due to a methyl group in an alkane-like environment, and the triplet splitting pattern implies that the CH$_3$ is next to a CH$_2$. Thus, our molecule contains an ethyl group, CH$_3$CH$_2$—. The six-proton singlet at 1.20 δ is due to two equivalent alkane-like methyl groups attached to a carbon with no hydrogens, (CH$_3$)$_2$C, and the two-proton quartet at 1.50 δ is due to the CH_2 of the ethyl group. All 5 carbons and 11 of the 12 hydrogens in the molecule are now accounted for. The remaining hydrogen, which appears as a broad one-proton singlet at 1.64 δ, is probably due to an OH group, since there is no other way to account for it. Putting the pieces together gives the structure: 2-methylbutan-2-ol.

Solution

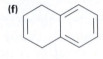

2-Methylbutan-2-ol

Problem 11.18

Predict the splitting patterns you would expect for each proton in the following molecules:

(a) $CHBr_2CH_3$

(b) $CH_3OCH_2CH_2Br$

(c) $ClCH_2CH_2CH_2Cl$

(d)

$$CH_3CHCOCH_2CH_3$$
with a C=O (O above the C) and CH_3 below

(e)

$$CH_3CH_2COCHCH_3$$
with a C=O (O above the C) and CH_3 below

(f)

(bicyclic structure shown)

Problem 11.19

Draw structures for compounds that meet the following descriptions:

(a) C_2H_6O; one singlet

(b) C_3H_7Cl; one doublet and one septet

(c) $C_4H_8Cl_2O$; two triplets

(d) $C_4H_8O_2$; one singlet, one triplet, and one quartet

Problem 11.20

The integrated ^1H NMR spectrum of a compound of formula $C_4H_{10}O$ is shown in Figure 11.17. Propose a structure.

FIGURE 11.17 An integrated ^1H NMR spectrum for Problem 11.20.

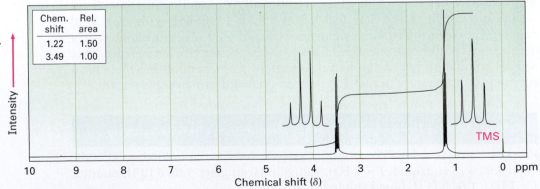

Chem. shift	Rel. area
1.22	1.50
3.49	1.00

11.12 More Complex Spin–Spin Splitting Patterns

In the ^1H NMR spectra we've seen thus far, the chemical shifts of different protons have been distinct and the spin–spin splitting patterns have been straightforward. It often happens, however, that different kinds of hydrogens in a molecule have accidentally *overlapping* signals. The spectrum of toluene in Figure 11.18, for example, shows that the five aromatic ring protons give a complex, overlapping pattern centered at 7.19 δ, even though they aren't all equivalent.

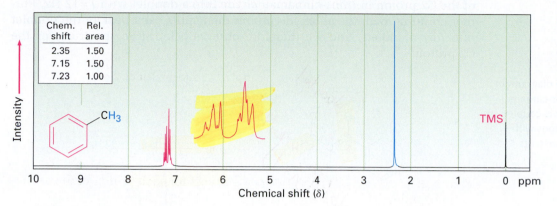

FIGURE 11.18 The ¹H NMR spectrum of toluene, showing the accidental overlap of the five nonequivalent aromatic ring protons.

Chem. shift	Rel. area
2.35	1.50
7.15	1.50
7.23	1.00

Yet another complication in ¹H NMR spectroscopy arises when a signal is split by two or more *nonequivalent* kinds of protons, as is the case with *trans*-cinnamaldehyde, isolated from oil of cinnamon (Figure 11.19). Although the n + 1 rule predicts splitting caused by *equivalent* protons, splittings caused by nonequivalent protons are more complex.

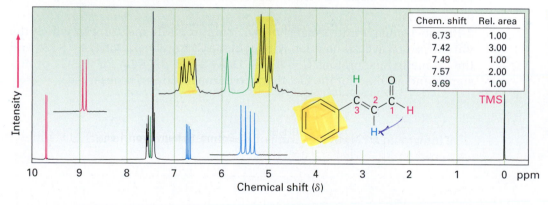

FIGURE 11.19 The ¹H NMR spectrum of *trans*-cinnamaldehyde. The signal of the proton at C2 (blue) is split into four peaks—a doublet of doublets—by the two nonequivalent neighboring protons.

Chem. shift	Rel. area
6.73	1.00
7.42	3.00
7.49	1.00
7.57	2.00
9.69	1.00

To understand the ¹H NMR spectrum of *trans*-cinnamaldehyde, we have to isolate the different parts and look at the signal of each proton individually:

- The five aromatic proton signals (black in Figure 11.19) overlap into a complex pattern with a large peak at 7.42 δ and a broad absorption at 7.57 δ.

- The aldehyde proton signal at C1 (red) appears in the normal downfield position at 9.69 δ and is split into a doublet with $J = 6$ Hz by the adjacent proton at C2.

- The vinylic proton at C3 (green) is next to the aromatic ring and is therefore shifted downfield from the normal vinylic region. This C3 proton signal appears as a doublet centered at 7.49 δ. Because it has one neighbor proton at C2, its signal is split into a doublet, with $J = 12$ Hz.

- The C2 vinylic proton signal (blue) appears at 6.73 δ and shows an interesting, four-line absorption pattern. It is coupled to the two nonequivalent protons at C1 and C3 with two different coupling constants: $J_{1-2} = 6$ Hz and $J_{2-3} = 12$ Hz.

A good way to understand the effect of multiple coupling such as occurs for the C2 proton of *trans*-cinnamaldehyde is to draw a *tree diagram,* like that in Figure 11.20. The diagram shows the individual effect of each coupling constant on the overall pattern. Coupling with the C3 proton splits the signal of the C2 proton in *trans*-cinnamaldehyde into a doublet with $J = 12$ Hz. Further coupling with the aldehyde proton then splits each peak of the doublet into new doublets, and we therefore observe a four-line spectrum for the C2 proton.

FIGURE 11.20 A tree diagram for the C2 proton of *trans*-cinnamaldehyde shows how it is coupled to the C1 and C3 protons with different coupling constants.

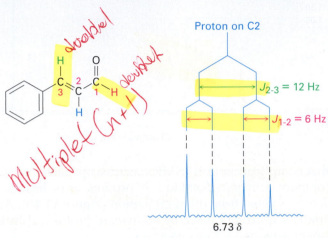

Problem 11.21

3-Bromo-1-phenylprop-1-ene shows a complex NMR spectrum in which the vinylic proton at C2 is coupled with both the C1 vinylic proton ($J = 16$ Hz) and the C3 methylene protons ($J = 8$ Hz). Draw a tree diagram for the C2 proton signal, and account for the fact that a five-line multiplet is observed.

3-Bromo-1-phenylprop-1-ene

11.13 Uses of ^{1}H NMR Spectroscopy

NMR can be used to help identify the product of nearly every reaction run in the laboratory. For example, we said in Section 8.4 that hydroboration/oxidation of alkenes occurs with non-Markovnikov regiochemistry to yield the less highly substituted alcohol. With the help of NMR, we can prove this statement.

Does hydroboration/oxidation of methylenecyclohexane yield cyclohexylmethanol or 1-methylcyclohexanol?

Methylenecyclohexane   1. BH$_3$, THF   2. H$_2$O$_2$, OH$^-$ **Cyclohexylmethanol**   or   **1-Methylcyclohexanol**   ?

The ^{1}H NMR spectrum of the reaction product is shown in Figure 11.21a. The spectrum shows a two-proton peak at 3.40 δ, indicating that the product has a –CH$_2$– group bonded to an electronegative oxygen atom (–CH$_2$OH). Furthermore, the spectrum shows *no* large three-proton singlet absorption near 1 δ, where we would expect the signal of a quaternary –CH$_3$ group to appear. (Figure 11.21b gives the spectrum of 1-methylcyclohexanol, the alternative product.) Thus, it's clear that cyclohexylmethanol is the reaction product.

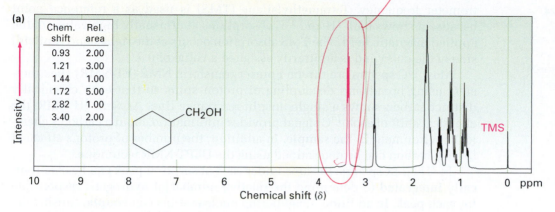

Chem. shift	Rel. area
0.93	2.00
1.21	3.00
1.44	1.00
1.72	5.00
2.82	1.00
3.40	2.00

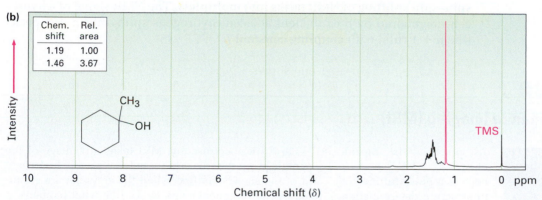

Chem. shift	Rel. area
1.19	1.00
1.46	3.67

FIGURE 11.21
(a) The ^{1}H NMR spectrum of cyclo-hexylmethanol, the product from hydroboration/oxidation of methylenecyclohexane, and (b) the ^{1}H NMR spectrum of 1-methyl-cyclohexanol, the possible alternative reaction product.

Problem 11.22

How could you use ^{1}H NMR to determine the regiochemistry of electrophilic addition to alkenes? For example, does addition of HCl to 1-methylcyclohexene yield 1-chloro-1-methylcyclohexane or 1-chloro-2-methylcyclohexane?

Summary

Nuclear magnetic resonance spectroscopy, or **NMR**, is the most valuable of the numerous spectroscopic techniques used for structure determination. Although we focused in this chapter on NMR applications to small molecules, more advanced NMR techniques are also used in biological chemistry to study protein structure and folding.

When magnetic nuclei such as ^{1}H and ^{13}C are placed in a strong magnetic field, their spins orient either with or against the field. On irradiation with radiofrequency (rf) waves, energy is absorbed and the nuclei "spin-flip" from

Key Words

chemical shift, 409
coupling, 424
coupling constant (*J*), 425
delta (δ) scale, 409
diastereotopic, 420
downfield, 409
enantiotopic, 419

the lower energy state to the higher energy state. This absorption of rf energy is detected, amplified, and displayed as an NMR spectrum.

Each electronically distinct ^1H or ^{13}C nucleus in a molecule comes into resonance at a slightly different value of the applied field, thereby producing a unique absorption signal. The exact position of each peak is called the **chemical shift**. Chemical shifts are caused by electrons setting up tiny local magnetic fields that **shield** nearby nuclei from the applied field.

The NMR chart is calibrated in **delta (δ) units**, where 1 δ = 1 ppm of spectrometer frequency. Tetramethylsilane (TMS) is used as a reference point because it shows both ^1H and ^{13}C absorptions at unusually high values of the applied magnetic field. The TMS absorption occurs at the right-hand (**upfield**) side of the chart and is arbitrarily assigned a value of 0 δ.

Most ^{13}C spectra are run on Fourier-transform NMR (**FT-NMR**) spectrometers using broadband decoupling of proton spins so that each chemically distinct carbon shows a single unsplit resonance line. As with ^1H NMR, the chemical shift of each ^{13}C signal provides information about a carbon's chemical environment in the sample. In addition, the number of protons attached to each carbon can be determined using the DEPT-NMR technique.

In ^1H NMR spectra, the area under each absorption peak can be electronically **integrated** to determine the relative number of hydrogens responsible for each peak. In addition, neighboring nuclear spins can **couple**, causing the **spin–spin splitting** of NMR peaks into **multiplets**. The NMR signal of a hydrogen neighbored by *n* equivalent adjacent hydrogens splits into *n* + 1 peaks (the **n + 1 rule**) with **coupling constant** *J*.

Lagniappe

Magnetic Resonance Imaging (MRI)

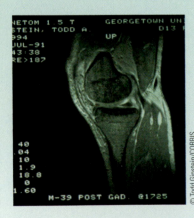

If you're a runner, you really don't want this to happen to you. The MRI of this left knee shows the presence of a ganglion cyst.

As practiced by organic chemists, NMR spectroscopy is a powerful method of structure determination. A small amount of sample, typically a few milligrams or less, is dissolved in a small amount of solvent, the solution is placed in a thin glass tube, and the tube is placed into the narrow (1–2 cm) gap between the poles of a strong magnet. Imagine, though, that a much larger NMR instrument were available. Instead of a few milligrams, the sample size could be tens of kilograms; instead of a narrow gap between magnet poles, the gap could be large enough for a whole person to climb into so that an NMR spectrum of body parts could be obtained. That large instrument is exactly what's used for *magnetic resonance imaging (MRI)*, a diagnostic technique of enormous value to the medical community.

Like NMR spectroscopy, MRI takes advantage of the magnetic properties of certain nuclei, typically hydrogen, and of the signals emitted when those nuclei are stimulated by radiofrequency energy. Unlike what happens in NMR spectroscopy, though, MRI instruments use data manipulation techniques to look at the three-dimensional *location* of magnetic nuclei in the body rather than at the chemical nature of the nuclei. As noted, most MRI instruments currently look at hydrogen, present in abundance wherever there is water or fat in the body.

The signals detected by MRI vary with the density of hydrogen atoms and with the nature of their surroundings, allowing identification of different types of tissue and even allowing the visualization of motion. For example, the volume of blood leaving the heart in a single stroke can be measured, and heart motion can be observed. Soft tissues that don't show up well on X-ray films can be seen clearly, allowing diagnosis of brain tumors, strokes, and other conditions. The technique is also valuable in diagnosing damage to knees or other joints and is a noninvasive alternative to surgical explorations.

Several types of atoms in addition to hydrogen can be detected by MRI, and the applications of images based on ^{31}P atoms are being explored. The technique holds great promise for studies of metabolism.

Exercises

VISUALIZING CHEMISTRY

(Problems 11.1–11.22 appear within the chapter.)

11.23 ▪ Into how many peaks would you expect the ^{1}H NMR signals of the indicated protons to be split? (Yellow-green = Cl.)

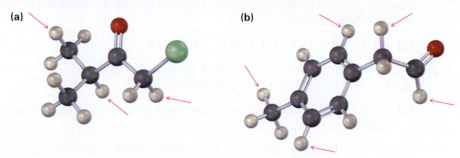

(a)   **(b)**

▪ indicates problems that are assignable in Organic OWL.

Go to this book's companion website at **www.cengage.com/ chemistry/mcmurry** to explore interactive versions of the Active Figures from this text.

11.24 ▪ Sketch what you might expect the ^{1}H and ^{13}C NMR spectra of the following compound to look like (yellow-green = Cl):

11.25 ▪ How many electronically nonequivalent kinds of protons and how many kinds of carbons are present in the following compound?

11.26 ▪ Identify the indicated protons in the following molecules as unrelated, homotopic, enantiotopic, or diastereotopic:

(a)   **(b)**

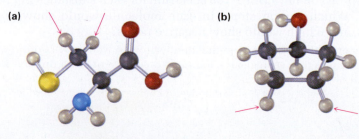

Cysteine

▪ Problems assignable in Organic OWL.

ADDITIONAL PROBLEMS

11.27 ▪ The following ^1H NMR absorptions were obtained on a spectrometer operating at 200 MHz and are given in hertz downfield from the TMS standard. Convert the absorptions to δ units.

(a) 436 Hz **(b)** 956 Hz **(c)** 1504 Hz

11.28 ▪ The following ^1H NMR absorptions were obtained on a spectrometer operating at 300 MHz. Convert the chemical shifts from δ units to hertz downfield from TMS.

(a) 2.1 δ **(b)** 3.45 δ **(c)** 6.30 δ **(d)** 7.70 δ

11.29 When measured on a spectrometer operating at 200 MHz, chloroform ($CHCl_3$) shows a single sharp absorption at 7.3 δ.

(a) How many parts per million downfield from TMS does chloroform absorb?

(b) How many hertz downfield from TMS would chloroform absorb if the measurement were carried out on a spectrometer operating at 360 MHz?

(c) What would be the position of the chloroform absorption in δ units when measured on a 360 MHz spectrometer?

11.30 How many signals would you expect each of the following molecules to have in its ^1H and ^{13}C spectra?

(a)

(b)

(c)

(d)

(e)

(f)

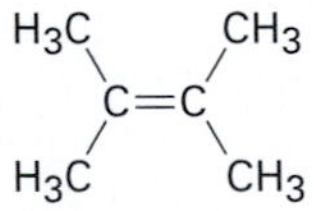

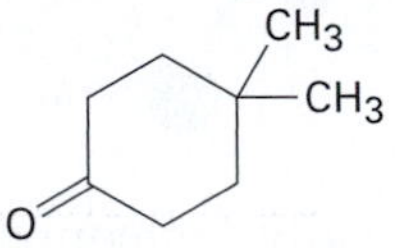

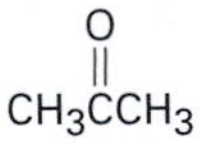

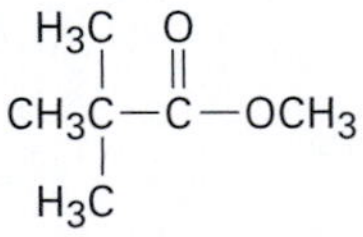

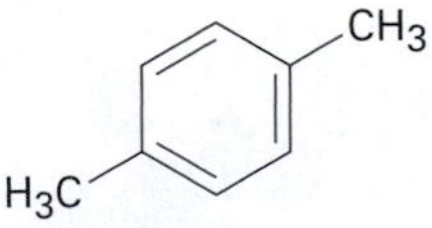

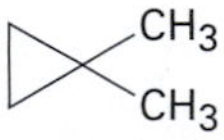

11.31 ▪ How many absorptions would you expect to observe in the ^{13}C NMR spectra of the following compounds?

(a) 1,1-Dimethylcyclohexane **(b)** $CH_3CH_2OCH_3$

(c) *tert*-Butylcyclohexane **(d)** 3-Methylpent-1-yne

(e) *trans*-1,2-Dimethylcyclohexane **(f)** Cyclohexanone

11.32 ▪ Suppose you ran a DEPT-135 spectrum for each substance in Problem 11.31. Which carbon atoms in each molecule would show positive peaks, and which would show negative peaks?

▪ Problems assignable in Organic OWL.

11.33 ▪ Is a nucleus that absorbs at 6.50 δ more shielded or less shielded than a nucleus that absorbs at 3.20 δ? Does the nucleus that absorbs at 6.50 δ require a stronger applied field or a weaker applied field to come into resonance than the nucleus that absorbs at 3.20 δ?

11.34 ▪ Identify the indicated sets of protons as unrelated, homotopic, enantiotopic, or diastereotopic:

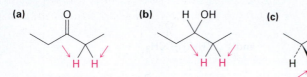

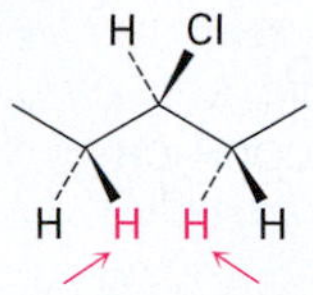

11.35 How many types of nonequivalent protons are present in each of the following molecules?

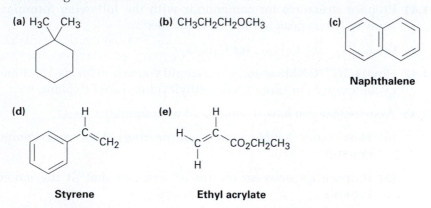

(a) H₃C CH₃

(b) CH₃CH₂CH₂OCH₃

(c)

Naphthalene

(d)

Styrene

(e)

Ethyl acrylate

11.36 ▪ Identify the indicated sets of protons as unrelated, homotopic, enantiotopic, or diastereotopic:

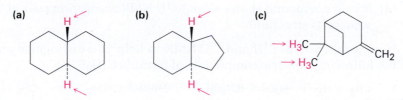

11.37 ▪ The following compounds all show a single line in their ¹H NMR spectra. List them in expected order of increasing chemical shift:

CH₄, CH₂Cl₂, cyclohexane, CH₃COCH₃, H₂C=CH₂, benzene

11.38 ▪ Predict the splitting pattern for each kind of hydrogen in the following molecules:

(a) (CH₃)₃CH   **(b)** CH₃CH₂CO₂CH₃   **(c)** *trans*-But-2-ene

11.39 Predict the splitting pattern for each kind of hydrogen in isopropyl propanoate, CH₃CH₂CO₂CH(CH₃)₂.

11.40 How could you use ^{1}H NMR to distinguish between the following pairs of compounds?

(a) $CH_3CH=CHCH_2CH_3$ and

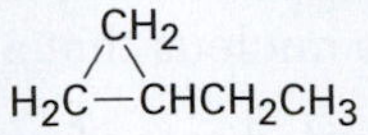

$$\overset{CH_2}{\overset{\diagup\diagdown}{H_2C-CHCH_2CH_3}}$$

(b) $CH_3CH_2OCH_2CH_3$ and $CH_3OCH_2CH_2CH_3$

(c)

$$\overset{O}{\overset{\parallel}{CH_3COCH_2CH_3}}$$ and $\overset{O}{\overset{\parallel}{CH_3CH_2CCH_3}}$

(d)

$$\overset{O}{\overset{\parallel}{H_2C=C(CH_3)CCH_3}}$$ and $\overset{O}{\overset{\parallel}{CH_3CH=CHCCH_3}}$

11.41 Propose structures for compounds with the following formulas that show only one peak in their ^{1}H NMR spectra:

(a) C_5H_{12} (b) C_5H_{10} (c) $C_4H_8O_2$

11.42 How many ^{13}C NMR absorptions would you expect for *cis*-1,3-dimethylcyclohexane? For *trans*-1,3-dimethylcyclohexane? Explain.

11.43 Assume that you have a compound with formula C_3H_6O.

(a) How many double bonds and/or rings does your compound contain?

(b) Propose as many structures as you can that fit the molecular formula.

(c) If your compound shows an infrared absorption peak at 1715 cm^{-1}, what functional group does it have?

(d) If your compound shows a single ^{1}H NMR absorption peak at 2.1 δ, what is its structure?

11.44 How would you use ^{1}H and ^{13}C NMR to help you distinguish among the following isomeric compounds of formula C_4H_8?

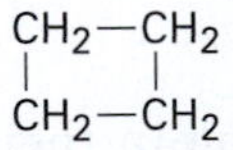

$$\begin{array}{l}CH_2-CH_2\\ |\quad\quad |\\ CH_2-CH_2\end{array}\qquad H_2C=CHCH_2CH_3\qquad CH_3CH=CHCH_3\qquad \begin{array}{l}CH_3\\ |\\ CH_3C=CH_2\end{array}$$

11.45 How could you use ^{1}H NMR, ^{13}C NMR, IR, and UV spectroscopy to help you distinguish between the following structures?

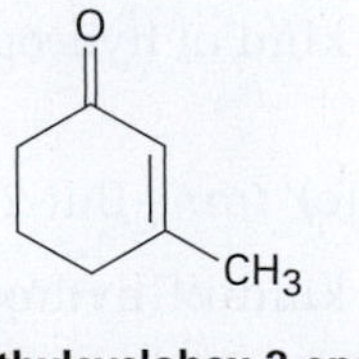

3-Methylcyclohex-2-enone 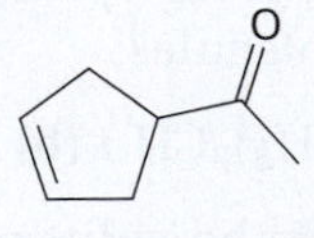   Cyclopent-3-enyl methyl ketone

11.46 ■ The compound whose ^1H NMR spectrum is shown has the molecular formula $C_3H_6Br_2$. Propose a structure.

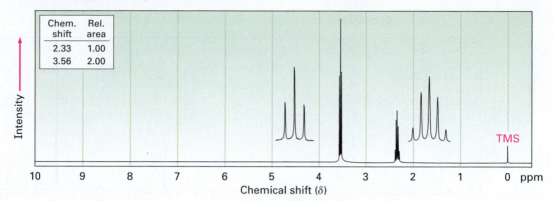

Chem. shift	Rel. area
2.33	1.00
3.56	2.00

11.47 ■ Propose structures for compounds that fit the following ^1H NMR data:

(a) $C_5H_{10}O$
0.95 δ (6 H, doublet, $J = 7$ Hz)
2.10 δ (3 H, singlet)
2.43 δ (1 H, multiplet)

(b) C_3H_5Br
2.32 δ (3 H, singlet)
5.35 δ (1 H, broad singlet)
5.54 δ (1 H, broad singlet)

11.48 ■ The compound whose ^1H NMR spectrum is shown has the molecular formula $C_4H_7O_2Cl$ and has an infrared absorption peak at 1740 cm^{-1}. Propose a structure.

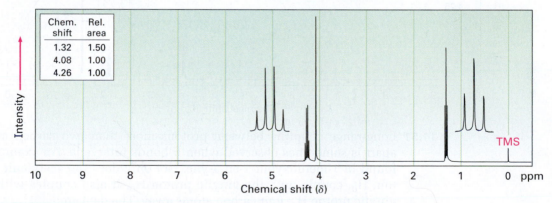

Chem. shift	Rel. area
1.32	1.50
4.08	1.00
4.26	1.00

11.49 ■ Propose structures for compounds that fit the following ^1H NMR data:

(a) $C_4H_6Cl_2$
2.18 δ (3 H, singlet)
4.16 δ (2 H, doublet, $J = 7$ Hz)
5.71 δ (1 H, triplet, $J = 7$ Hz)

(b) $C_{10}H_{14}$
1.30 δ (9 H, singlet)
7.30 δ (5 H, singlet)

(c) C_4H_7BrO
2.11 δ (3 H, singlet)
3.52 δ (2 H, triplet, $J = 6$ Hz)
4.40 δ (2 H, triplet, $J = 6$ Hz)

(d) $C_9H_{11}Br$
2.15 δ (2 H, quintet, $J = 7$ Hz)
2.75 δ (2 H, triplet, $J = 7$ Hz)
3.38 δ (2 H, triplet, $J = 7$ Hz)
7.22 δ (5 H, singlet)

■ Problems assignable in Organic OWL.

11.50 ■ Propose structures for the two compounds whose ^1H NMR spectra are shown.

(a) C_4H_9Br

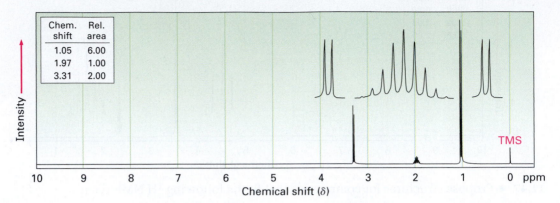

Chem. shift
Rel. area
1.05 6.00
1.97 1.00
3.31 2.00
Intensity
TMS
10 9 8 7 6 5 4 3 2 1 0 ppm
Chemical shift (δ)

(b) $C_4H_8Cl_2$

Chem. shift
Rel. area
1.56 3.00
2.13 2.00
3.72 2.00
4.25 1.00
Intensity
TMS
10 9 8 7 6 5 4 3 2 1 0 ppm
Chemical shift (δ)

11.51 Long-range coupling between protons more than two carbon atoms apart is sometimes observed when π bonds intervene. An example is found in 1-methoxybut-1-en-3-yne. Not only does the acetylenic proton, H_a, couple with the vinylic proton H_b, it also couples with the vinylic proton H_c, *four* carbon atoms away. The data are:

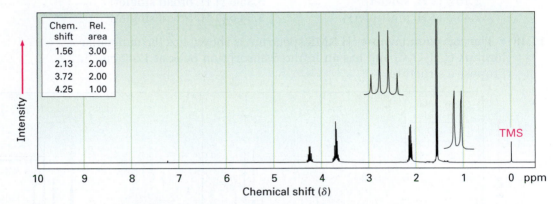

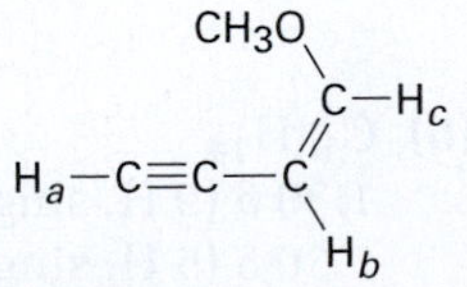

CH3O
C—Hc
Ha—C≡C—C
Hb

H_a (3.08 δ) H_b (4.52 δ) H_c (6.35 δ)

$J_{a\text{-}b}$ = 3 Hz $J_{a\text{-}c}$ = 1 Hz $J_{b\text{-}c}$ = 7 Hz

1-Methoxybut-1-en-3-yne

Construct tree diagrams that account for the observed splitting patterns of H_a, H_b, and H_c.

■ Problems assignable in Organic OWL.

11.52 Assign as many of the resonances as you can to specific carbon atoms in the ^{13}C NMR spectrum of ethyl benzoate.

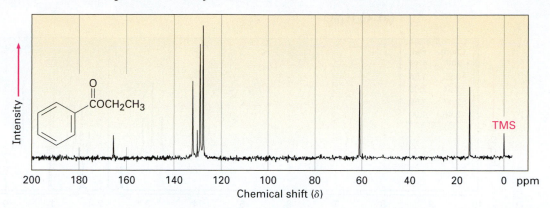

11.53 ■ The ^{1}H and ^{13}C NMR spectra of compound A, C_8H_9Br, are shown. Propose a structure for A, and assign peaks in the spectra to your structure.

Chem. shift	Rel. area
1.20	3.00
2.58	2.00
7.07	2.00
7.39	2.00

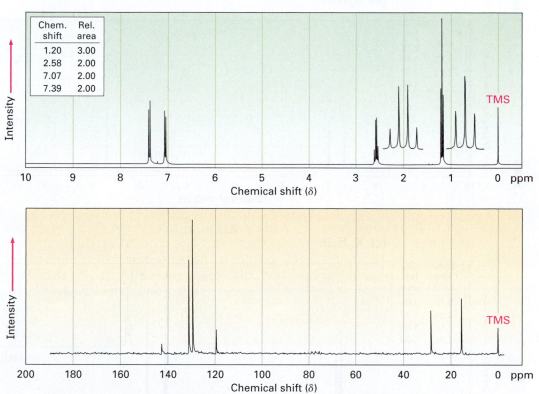

■ Problems assignable in Organic OWL.

11.54 ■ Propose structures for the three compounds whose ^{1}H NMR spectra are shown.

(a) $C_5H_{10}O$

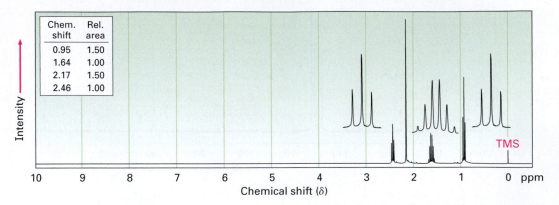

Chem. shift | Rel. area
0.95 | 1.50
1.64 | 1.00
2.17 | 1.50
2.46 | 1.00
Intensity
TMS
10 9 8 7 6 5 4 3 2 1 0 ppm
Chemical shift (δ)

(b) C_7H_7Br

Chem. shift | Rel. area
2.31 | 1.50
7.01 | 1.00
7.35 | 1.00
Intensity
TMS
10 9 8 7 6 5 4 3 2 1 0 ppm
Chemical shift (δ)

(c) C_8H_9Br

Chem. shift | Rel. area
3.16 | 1.00
3.56 | 1.00
7.18 | 1.00
7.29 | 1.50
Intensity
TMS
10 9 8 7 6 5 4 3 2 1 0 ppm
Chemical shift (δ)

■ Problems assignable in Organic OWL.

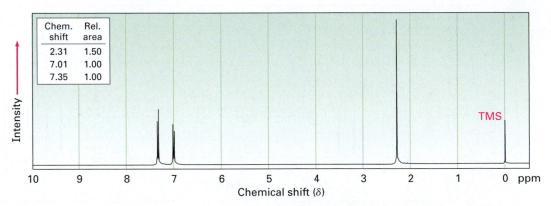

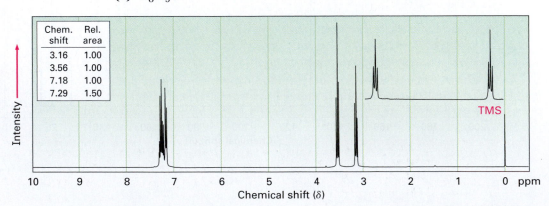

11.55 The mass spectrum and ^{13}C NMR spectrum of a hydrocarbon are shown. Propose a structure, and explain the spectral data.

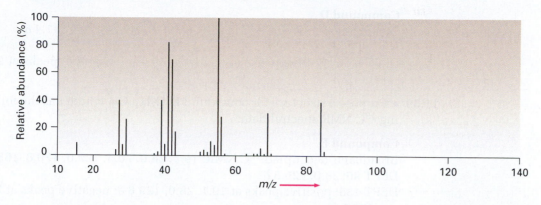

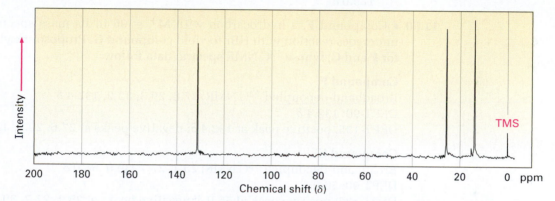

11.56 ■ Compound **A**, a hydrocarbon with $M^+ = 96$ in its mass spectrum, has the ^{13}C spectral data that follow. On reaction with BH_3 followed by treatment with basic H_2O_2, **A** is converted into **B**, whose ^{13}C spectral data are also given. Propose structures for **A** and **B**.

Compound A
Broadband-decoupled ^{13}C NMR: 26.8, 28.7, 35.7, 106.9, 149.7 δ
DEPT-90: no peaks
DEPT-135: no positive peaks; negative peaks at 26.8, 28.7, 35.7, 106.9 δ

Compound B
Broadband-decoupled ^{13}C NMR: 26.1, 26.9, 29.9, 40.5, 68.2 δ
DEPT-90: 40.5 δ
DEPT-135: positive peak at 40.5 δ; negative peaks at 26.1, 26.9, 29.9, 68.2 δ

11.57 ■ Propose a structure for compound **C**, which has $M^+ = 86$ in its mass spectrum, an IR absorption at 3400 cm^{-1}, and the following ^{13}C NMR spectral data:

Compound C
Broadband-decoupled ^{13}C NMR: 30.2, 31.9, 61.8, 114.7, 138.4 δ
DEPT-90: 138.4 δ
DEPT-135: positive peak at 138.4 δ; negative peaks at 30.2, 31.9, 61.8, 114.7 δ

■ Problems assignable in Organic OWL.

11.58 ■ Compound **D** is isomeric with compound **C** (Problem 11.57) and has the following ^{13}C NMR spectral data. Propose a structure.

Compound D
Broadband-decoupled ^{13}C NMR: 9.7, 29.9, 74.4, 114.4, 141.4 δ
DEPT-90: 74.4, 141.4 δ
DEPT-135: positive peaks at 9.7, 74.4, 141.4 δ; negative peaks at 29.9, 114.4 δ

11.59 ■ Propose a structure for compound **E**, $C_7H_{12}O_2$, which has the following ^{13}C NMR spectral data:

Compound E
Broadband-decoupled ^{13}C NMR: 19.1, 28.0, 70.5, 129.0, 129.8, 165.8 δ
DEPT-90: 28.0, 129.8 δ
DEPT-135: positive peaks at 19.1, 28.0, 129.8 δ; negative peaks at 70.5, 129.0 δ

11.60 ■ Compound **F**, a hydrocarbon with $M^+ = 96$ in its mass spectrum, undergoes reaction with HBr to yield compound **G**. Propose structures for **F** and **G**, whose ^{13}C NMR spectral data follow.

Compound F
Broadband-decoupled ^{13}C NMR: 27.6, 29.3, 32.2, 132.4 δ
DEPT-90: 132.4 δ
DEPT-135: positive peak at 132.4 δ; negative peaks at 27.6, 29.3, 32.2 δ

Compound G
Broadband-decoupled ^{13}C NMR: 25.1, 27.7, 39.9, 56.0 δ
DEPT-90: 56.0 δ
DEPT-135: positive peak at 56.0 δ; negative peaks at 25.1, 27.7, 39.9 δ

11.61 3-Methylbutan-2-ol has five signals in its ^{13}C NMR spectrum at 17.90, 18.15, 20.00, 35.05, and 72.75 δ. Why are the two methyl groups attached to C3 nonequivalent?

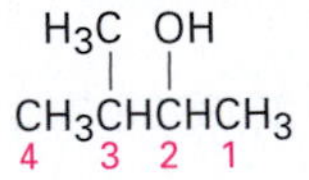

$$\begin{array}{cc} H_3C & OH \\ | & | \\ CH_3CHCHCH_3 \\ 4\quad3\quad2\quad1 \end{array}$$ **3-Methylbutan-2-ol**

11.62 A ^{13}C NMR spectrum of commercially available pentane-2,4-diol shows *five* peaks at 23.3, 23.9, 46.5, 64.8, and 68.1 δ. Explain.

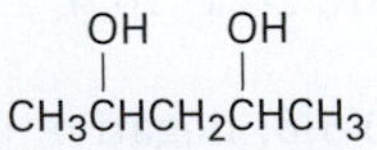

$$\begin{array}{cc} OH & OH \\ | & | \\ CH_3CHCH_2CHCH_3 \end{array}$$ **Pentane-2,4-diol**

11.63 Carboxylic acids (RCO$_2$H) react with alcohols (R'OH) in the presence of an acid catalyst. The reaction product of propanoic acid with methanol has the following spectroscopic properties. Propose a structure.

$$CH_3CH_2\overset{\overset{\displaystyle O}{\|}}{C}OH \xrightarrow[\text{H}^+\text{ catalyst}]{CH_3OH} \text{?}$$

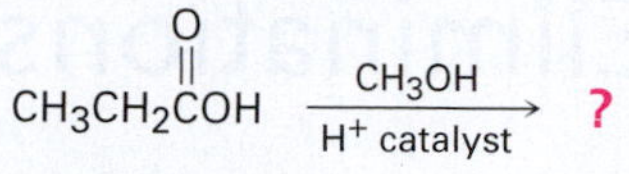

Propanoic acid

MS: M$^+$ = 88
IR: 1735 cm^{-1}
^{1}H NMR: 1.11 δ (3 H, triplet, J = 7 Hz); 2.32 δ (2 H, quartet, J = 7 Hz); 3.65 δ (3 H, singlet)
^{13}C NMR: 9.3, 27.6, 51.4, 174.6 δ

11.64 Nitriles (RC≡N) react with Grignard reagents (R'MgBr). The reaction product from 2-methylpropanenitrile with methylmagnesium bromide has the following spectroscopic properties. Propose a structure.

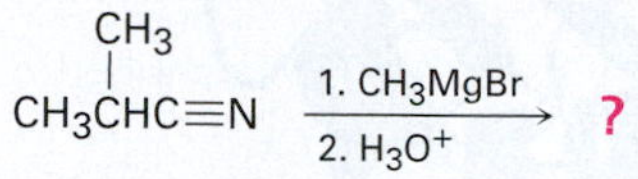

$$\overset{\overset{\displaystyle CH_3}{|}}{CH_3CHC}\equiv N \xrightarrow[\text{2. H}_3\text{O}^+]{\text{1. CH}_3\text{MgBr}} \text{?}$$

2-Methylpropanenitrile

MS: M$^+$ = 86
IR: 1715 cm^{-1}
^{1}H NMR: 1.05 δ (6 H, doublet, J = 7 Hz); 2.12 δ (3 H, singlet); 2.67 δ (1 H, septet, J = 7 Hz)
^{13}C NMR: 18.2, 27.2, 41.6, 211.2 δ

12 Organohalides: Nucleophilic Substitutions and Eliminations

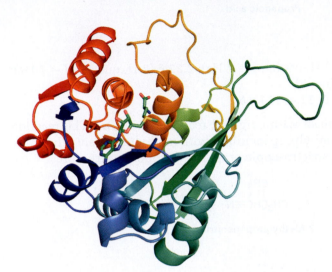

N6-Adenine methyltransferase catalyzes the methylation of pyrimidine nucleotides in DNA.

Now that we've covered the chemistry of hydrocarbons, it's time to start looking at more complex substances that contain elements in addition to just C and H. We'll begin by discussing the chemistry of **organohalides**, compounds that contain one or more halogen atoms.

Halogen-substituted organic compounds are widespread in nature, and more than 5000 organohalides have been found in algae and various other marine organisms. Chloromethane, for instance, is released in large amounts by ocean kelp, as well as by forest fires and volcanoes. Halogen-containing compounds also have a vast array of industrial applications, including their use as solvents, inhaled anesthetics, refrigerants, and pesticides.

Trichloroethylene
(a solvent)

Halothane
(an inhaled anesthetic)

Dichlorodifluoromethane
(a refrigerant)

Bromomethane
(a fumigant)

Still other halo-substituted compounds are providing important leads to new medicines. The pentahalogenated alkene halomon, for instance, has been

OWL Online homework for this chapter can be assigned in Organic OWL.

isolated from the red alga *Portieria hornemannii* and found to have anticancer activity against several human tumor cell lines.

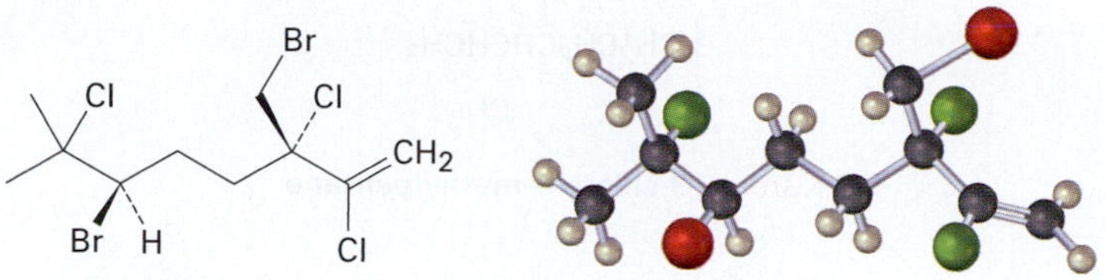

Halomon

A large variety of organohalides are known. The halogen might be bonded to an alkynyl group (C≡C–X), a vinylic group (C=C–X), an aromatic ring (Ar–X), or an alkyl group. We'll be concerned in this chapter, however, primarily with **alkyl halides**, compounds with a halogen atom bonded to a saturated, sp^3-hybridized carbon atom.

WHY THIS CHAPTER?

Alkyl halides themselves are not often involved in the biochemical pathways of terrestrial organisms, but the *kinds* of reactions they undergo—nucleophilic substitutions and eliminations—*are* frequently involved. Thus, alkyl halide chemistry acts as a relatively simple model for many mechanistically similar but structurally more complex reactions found in biomolecules. We'll begin with a look at how to name and prepare alkyl halides, and we'll then make a detailed study of their substitution and elimination reactions—two of the most important and well-studied reaction types in organic chemistry.

12.1 Names and Structures of Alkyl Halides

Although members of the class are commonly called *alkyl halides,* they are named systematically as *haloalkanes* (Section 3.4), treating the halogen as a substituent on a parent alkane chain. There are three steps:

Step 1
Find the longest chain, and name it as the parent. If a double or triple bond is present, the parent chain must contain it.

Step 2
Number the carbons of the parent chain beginning at the end nearer the first substituent, whether alkyl or halo. Assign each substituent a number according to its position on the chain.

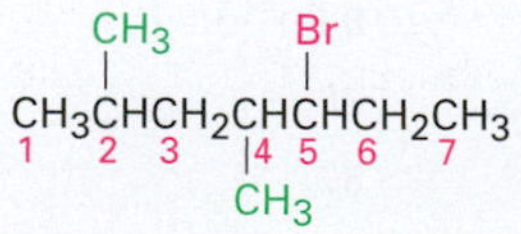

5-Bromo-2,4-dimethylheptane 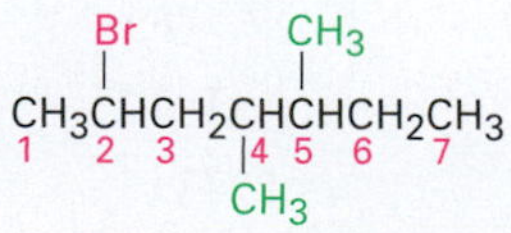   **2-Bromo-4,5-dimethylheptane**

If different halogens are present, number all and list them in alphabetical order when writing the name.

$$\underset{1}{Br}\underset{2}{CH_2}\underset{}{CH_2}\underset{3}{\overset{\overset{Cl}{|}}{C}}H\underset{4}{\overset{\overset{}{|}}{C}}H\underset{5}{CH_3}$$
$$CH_3$$

1-Bromo-3-chloro-4-methylpentane

Step 3
If the parent chain can be properly numbered from either end by step 2, begin at the end nearer the substituent that has alphabetical precedence.

$$\underset{6}{CH_3}\underset{5}{\overset{\overset{CH_3}{|}}{C}}H\underset{4}{CH_2}\underset{3}{CH_2}\underset{2}{\overset{\overset{Br}{|}}{C}}H\underset{1}{CH_3}$$

2-Bromo-5-methylhexane
(**NOT** 5-bromo-2-methylhexane)

In addition to their systematic names, many simple alkyl halides can also be named by identifying first the alkyl group and then the halogen. For example, CH_3I can be called either iodomethane or methyl iodide. Such names are well entrenched in the chemical literature and in daily usage, but they won't be used in this book.

$$CH_3I$$

Iodomethane
(or methyl iodide)

$$CH_3\overset{\overset{Cl}{|}}{C}HCH_3$$

2-Chloropropane
(or isopropyl chloride)

Bromocyclohexane
(or cyclohexyl bromide)

Halogens increase in size going down the periodic table, so the lengths of the corresponding carbon–halogen bonds increase accordingly (Table 12.1). In addition, C–X bond strengths decrease going down the periodic table. As we've been doing thus far, we'll continue to use the abbreviation X to represent any of the halogens F, Cl, Br, or I.

TABLE 12.1
A Comparison of the Halomethanes

Halomethane	Bond length (pm)	Bond strength (kJ/mol)	(kcal/mol)	Dipole moment (D)
CH_3F	139	460	110	1.85
CH_3Cl	178	350	84	1.87
CH_3Br	193	294	70	1.81
CH_3I	214	239	57	1.62

In our discussion of bond polarity in functional groups in Section 6.4, we noted that halogens are more electronegative than carbon. The C–X bond is therefore polar, with the carbon atom bearing a slight positive charge ($\delta+$) and the halogen a slight negative charge ($\delta-$). This polarity results in a substantial dipole moment for halomethanes (Table 12.1) and implies that the alkyl halide C–X carbon atom should behave as an electrophile in polar reactions. We'll soon see that this is indeed the case.

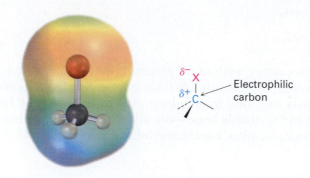

$\delta-$ X

$\delta+$ C —— Electrophilic carbon

Problem 12.1

Give IUPAC names for the following alkyl halides:

(a)

$CH_3CH_2CH_2CH_2I$

(b)

$\begin{array}{c} CH_3 \\ | \\ CH_3CHCH_2CH_2Cl \end{array}$

(c)

$\begin{array}{c} CH_3 \\ | \\ BrCH_2CH_2CH_2CCH_2Br \\ | \\ CH_3 \end{array}$

(d)

$\begin{array}{c} CH_3 \\ | \\ CH_3CCH_2CH_2Cl \\ | \\ Cl \end{array}$

(e)

$\begin{array}{c} I \;\; CH_2CH_2Cl \\ | \;\;\; | \\ CH_3CHCHCH_2CH_3 \end{array}$

(f)

$\begin{array}{c} Br \qquad\quad Cl \\ | \qquad\quad | \\ CH_3CHCH_2CH_2CHCH_3 \end{array}$

Problem 12.2

Draw structures corresponding to the following IUPAC names:

(a) 2-Chloro-3,3-dimethylhexane (b) 3,3-Dichloro-2-methylhexane
(c) 3-Bromo-3-ethylpentane (d) 1,1-Dibromo-4-isopropylcyclohexane
(e) 4-*sec*-Butyl-2-chlorononane (f) 1,1-Dibromo-4-*tert*-butylcyclohexane

---

12.2 Preparing Alkyl Halides from Alkenes: Allylic Bromination

We've already seen several methods for preparing alkyl halides, including the reactions of HX and X_2 with alkenes in electrophilic addition reactions (Sections 7.6 and 8.2). The hydrogen halides HCl, HBr, and HI react with alkenes by a polar mechanism to give the product of Markovnikov addition. Bromine and chlorine undergo anti addition through a halonium ion intermediate to give 1,2-dihalogenated products.

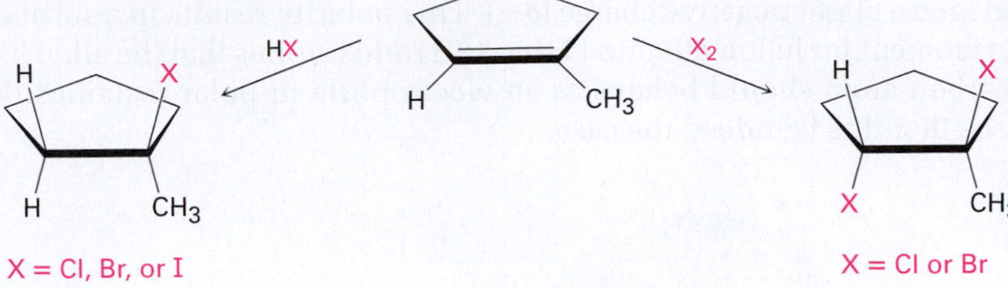

X = Cl, Br, or I

X = Cl or Br

Another laboratory method for preparing alkyl halides from alkenes is by reaction with *N*-bromosuccinimide (abbreviated NBS) in the presence of light to give products resulting from substitution of hydrogen by bromine at the position next to the double bond—the allylic position (Section 8.13). Cyclohexene, for example, gives 3-bromocyclohexene.

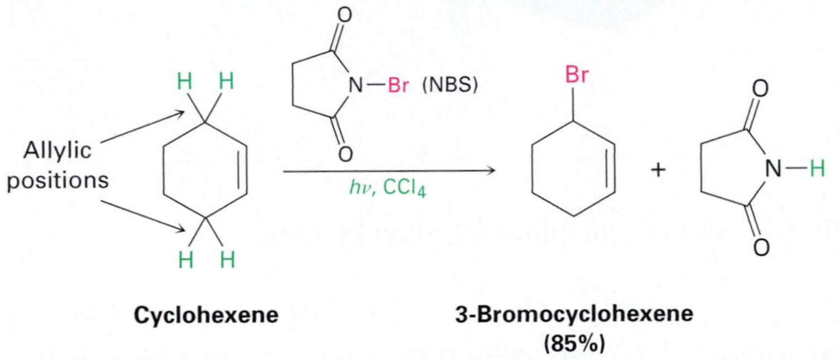

Cyclohexene

3-Bromocyclohexene
(85%)

This allylic bromination with NBS is analogous to the methane chlorination reaction discussed in Section 6.3 and occurs by a similar radical chain reaction mechanism. As in methane halogenation, Br· radical abstracts an allylic hydrogen atom of the alkene, thereby forming an allylic radical plus HBr. This allylic radical then reacts with Br_2 to yield the product and a Br· radical, which cycles back into the first step and carries on the chain. The Br_2 results from reaction of NBS with the HBr formed in the first step.

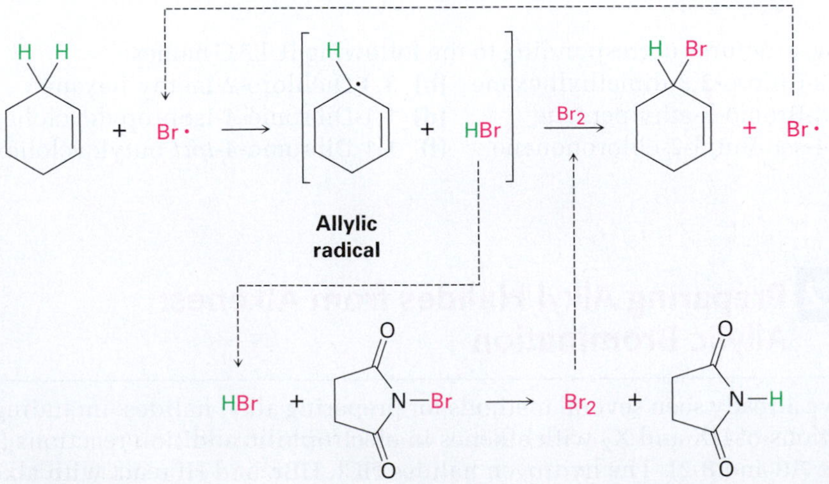

Allylic
radical

Why does bromination with NBS occur exclusively at an allylic position rather than elsewhere in the molecule? The answer has to do with the relative stabilities of various kinds of radicals. There are three sorts of C–H bonds in cyclohexene, and Table 6.3 on page 196 gives an estimate of their relative strengths. Although a typical secondary alkyl C–H bond has a strength of about 410 kJ/mol (98 kcal/mol) and a typical vinylic C–H bond has a strength of 465 kJ/mol (111 kcal/mol), an allylic C–H bond has a strength of only about 370 kJ/mol (88 kcal/mol). An allylic radical is therefore more stable than a related alkyl radical by about 40 kJ/mol (9 kcal/mol) and, according to the Hammond postulate (Section 7.9), should form faster.

Allylic
370 kJ/mol (88 kcal/mol)

Alkyl
410 kJ/mol (98 kcal/mol)

Vinylic
465 kJ/mol (111 kcal/mol)

Allylic *radicals* are stable for the same reason that allylic *carbocations* are stable (Section 8.13). Like an allylic carbocation, an allylic radical has two resonance forms. One form has the unpaired electron on the left and the double bond on the right, and one form has the unpaired electron on the right and the double bond on the left (Figure 12.1). Neither structure is correct by itself; the true structure of the allyl radical is a resonance hybrid of the two. In molecular orbital terms, the unpaired electron is delocalized, or spread out, over an extended π orbital network rather than localized at only one site. Thus, the two terminal carbons share the unpaired electron.

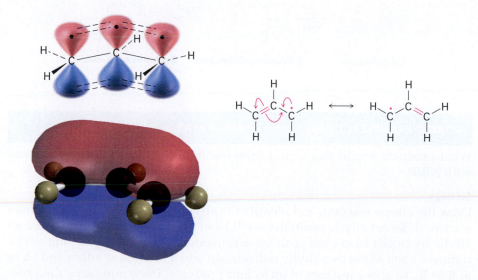

FIGURE 12.1 An orbital view of the allyl radical. The *p* orbital on the central carbon can overlap equally well with a *p* orbital on either neighboring carbon, giving rise to two resonance structures.

Because the unpaired electron in an allylic radical is delocalized over both ends of the π orbital system, reaction with Br_2 can occur at either end.

As a result, allylic bromination of an unsymmetrical alkene often leads to a mixture of products. For example, bromination of oct-1-ene gives a mixture of 3-bromooct-1-ene and 1-bromooct-2-ene. The two products are not formed in equal amounts, however, because the intermediate allylic radical is not symmetrical and reaction at the two ends is not equally likely. Reaction at the less hindered, primary end is favored.

$$CH_3CH_2CH_2CH_2CH_2CH_2CH=CH_2$$

Oct-1-ene

NBS, CCl_4

$$\left[CH_3CH_2CH_2CH_2CH_2\overset{\centerdot}{C}HCH=CH_2 \longleftrightarrow CH_3CH_2CH_2CH_2CH_2CH=CH\overset{\centerdot}{C}H_2 \right]$$

Br
|
$$CH_3CH_2CH_2CH_2CH_2CHCH=CH_2 \quad + \quad CH_3CH_2CH_2CH_2CH_2CH=CHCH_2Br$$

3-Bromooct-1-ene (17%)   **1-Bromooct-2-ene (83%)**
                                    (53 : 47 trans : cis)

The products of allylic bromination reactions are useful for conversion into conjugated dienes by dehydrohalogenation with base. Cyclohexene can be converted into cyclohexa-1,3-diene, for example.

Cyclohexene   **3-Bromocyclohexene**   **Cyclohexa-1,3-diene**

WORKED EXAMPLE 12.1   Predicting the Product of an Allylic Bromination Reaction

What products would you expect from reaction of 4,4-dimethylcyclohexene with NBS?

Strategy

Draw the alkene reactant, and identify the allylic positions. In this case, there are two different allylic positions; we'll label them **A** and **B**. Now abstract an allylic hydrogen from each position to generate the two corresponding allylic radicals. Each of the two allylic radicals can add a Br atom at either end (**A** or **a**; **B** or **b**), to give a mixture of up to four products. Draw and name the products. In the present instance, the "two" products from reaction at positions **B** and **b** are identical, so a total of only three products are formed in this reaction.

Solution

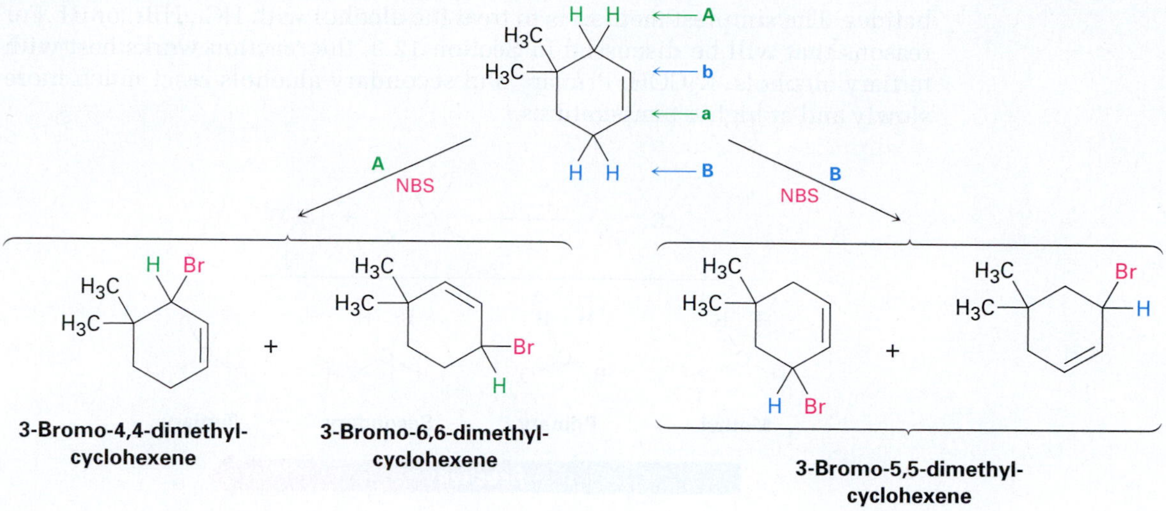

3-Bromo-4,4-dimethyl-cyclohexene

3-Bromo-6,6-dimethyl-cyclohexene

3-Bromo-5,5-dimethyl-cyclohexene

Problem 12.3

Draw three resonance forms for the cyclohexadienyl radical.

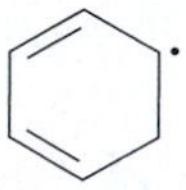

Cyclohexadienyl radical

Problem 12.4

The major product of the reaction of methylenecyclohexane with *N*-bromo-succinimide is 1-(bromomethyl)cyclohexene. Explain.

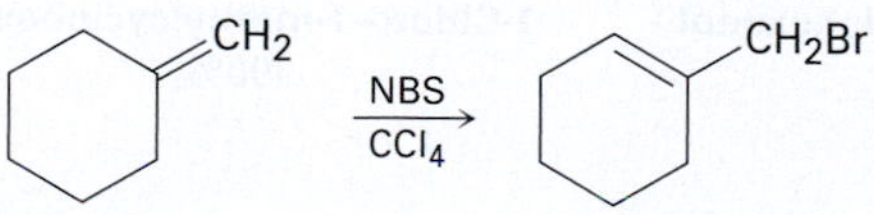

Major product

Problem 12.5

What products would you expect from reaction of the following alkenes with NBS? If more than one product is formed, show the structures of all.

(a)

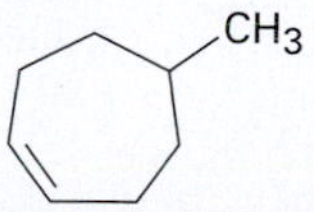

(b)

$$CH_3CHCH=CHCH_2CH_3$$
with a CH_3 group on the second carbon

12.3 Preparing Alkyl Halides from Alcohols

The most generally useful method for preparing alkyl halides is to make them from alcohols, which themselves can be obtained from carbonyl compounds as we'll see in Section 13.3. Because of the importance of the process, many different methods have been developed to transform alcohols into alkyl halides. The simplest method is to treat the alcohol with HCl, HBr, or HI. For reasons that will be discussed in Section 12.9, the reaction works best with tertiary alcohols, R₃COH. Primary and secondary alcohols react much more slowly and at higher temperatures.

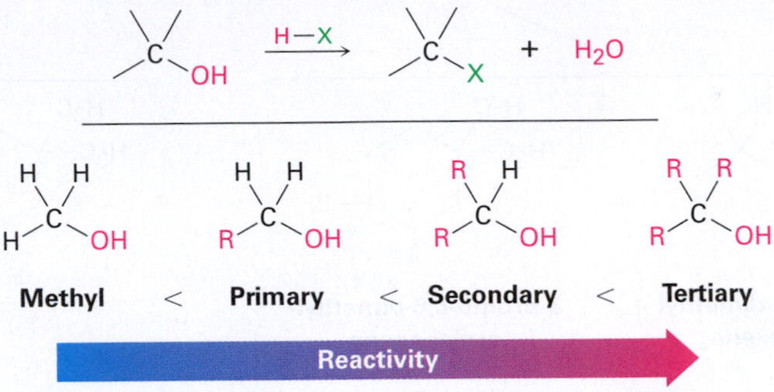

Methyl < Primary < Secondary < Tertiary

Reactivity

The reaction of HX with a tertiary alcohol is so rapid that it's often carried out simply by bubbling the pure HCl or HBr gas into a cold ether solution of the alcohol. 1-Methylcyclohexanol, for example, is converted into 1-chloro-1-methylcyclohexane by treating with HCl:

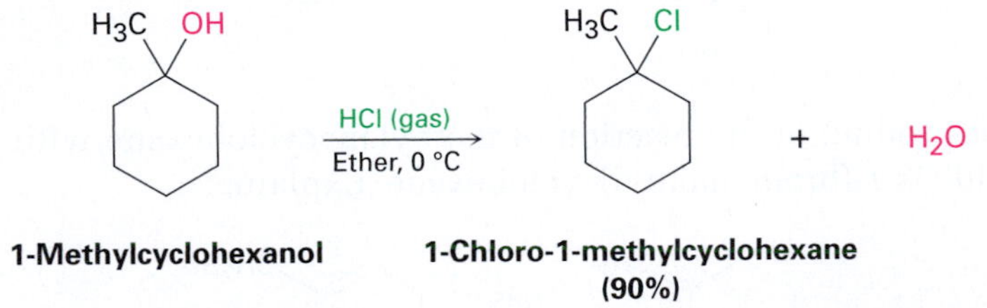

1-Methylcyclohexanol   **1-Chloro-1-methylcyclohexane**
                                          (90%)

Primary and secondary alcohols are best converted into alkyl halides by treatment with either thionyl chloride (SOCl₂) or phosphorus tribromide (PBr₃). These reactions, which normally take place readily under mild conditions, are less acidic and less likely to cause acid-catalyzed rearrangements than the HX method. We'll look at the mechanisms of these substitution reactions in Section 12.7.

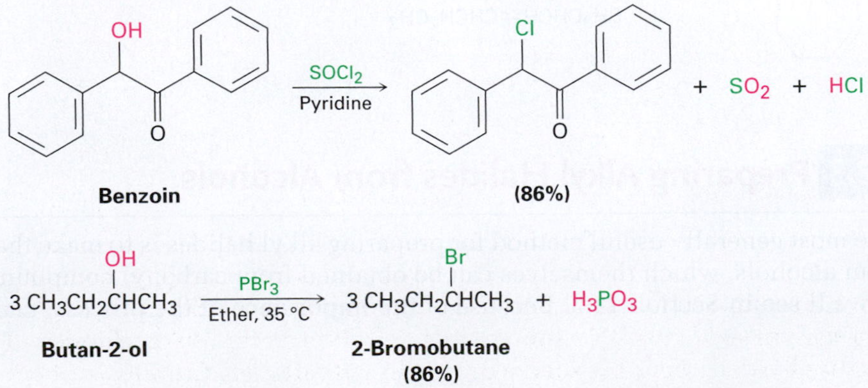

Benzoin   **(86%)**

$$3\ CH_3CH_2CHCH_3 \xrightarrow[\text{Ether, 35 °C}]{PBr_3} 3\ CH_3CH_2CHCH_3\ +\ H_3PO_3$$

Butan-2-ol   **2-Bromobutane**
                                   (86%)

How would you prepare the following alkyl halides from the corresponding alcohols?

(a)
$$CH_3\overset{\underset{\displaystyle |}{Cl}}{\underset{\underset{\displaystyle CH_3}{|}}{C}}CH_3$$

(b)
$$CH_3\overset{\underset{\displaystyle |}{Br}}{C}HCH_2\overset{\underset{\displaystyle |}{CH_3}}{C}HCH_3$$

(c)
$$BrCH_2CH_2CH_2CH_2\overset{\underset{\displaystyle |}{CH_3}}{C}HCH_3$$

(d)
$$CH_3CH_2\overset{\underset{\underset{\displaystyle CH_3}{|}}{CH_3}}{C}HCH_2\overset{\underset{\displaystyle |}{Cl}}{C}CH_3$$

12.4 Reactions of Alkyl Halides: Grignard Reagents

Alkyl halides, RX, react with magnesium metal in ether or tetrahydrofuran (THF; Section 8.1) solvent to yield alkylmagnesium halides, RMgX. The products, called **Grignard reagents** after their discoverer, Victor Grignard, are examples of *organometallic* compounds because they contain a carbon–metal bond. In addition to alkyl halides, Grignard reagents can also be made from alkenyl (vinylic) and aryl (aromatic) halides. The halogen can be Cl, Br, or I, although chlorides are less reactive than bromides and iodides. Organofluorides rarely react with magnesium.

$$
\left.\begin{array}{l}1°\text{ alkyl}\\2°\text{ alkyl}\\3°\text{ alkyl}\\ \text{alkenyl}\\ \text{aryl}\end{array}\right\}\;\longrightarrow\; R{-}X \;\longleftarrow\; \left\{\begin{array}{l}\text{Cl}\\\text{Br}\\\text{I}\end{array}\right.
$$

$$Mg \Big|\; \text{Ether or THF}$$

$$R{-}Mg{-}X$$

As you might expect from the discussion of electronegativity and bond polarity in Section 6.4, the carbon–magnesium bond is polarized, making the carbon atom of Grignard reagents both nucleophilic and basic. An electrostatic potential map of methylmagnesium iodide, for instance, indicates the electron-rich (red) character of the carbon bonded to magnesium.

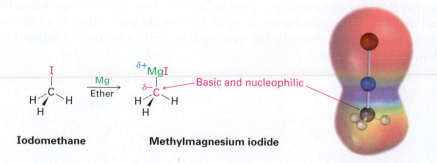

Iodomethane   **Methylmagnesium iodide**

In a formal sense, a Grignard reagent is the magnesium salt, $R_3C^- {}^+MgX$, of a carbon acid, $R_3C{-}H$, and is thus a carbon anion, or **carbanion**. But because hydrocarbons are such weak acids, with pK_a's in the range of 44 to 60

(Section 8.15), carbon anions are very strong bases. Grignard reagents therefore react with such weak acids as H_2O, ROH, RCO_2H, and RNH_2 to abstract a proton and yield hydrocarbons. Thus, an organic halide can be converted to a hydrocarbon by formation of a Grignard reagent followed by protonation, R–X $\rightarrow$ R–MgX $\rightarrow$ R–H.

$$CH_3CH_2CH_2CH_2CH_2CH_2Br \xrightarrow[\text{Ether}]{\text{Mg}} CH_3CH_2CH_2CH_2CH_2CH_2MgBr \xrightarrow{H_2O} CH_3CH_2CH_2CH_2CH_2CH_3$$

1-Bromohexane **1-Hexylmagnesium bromide** **Hexane (85%)**

Although Grignard reagents themselves have no role in biochemistry, they are useful carbon-based nucleophiles in many laboratory reactions and act as a simple model for other, more complex carbon-based nucleophiles that *are* important in biological chemistry. We'll see numerous examples in Chapter 17.

Problem 12.7

How strong a base would you expect a Grignard reagent to be? Look at Table 8.2 on page 292, and then predict whether the following reactions will occur as written. (The pK_a of NH_3 is 35.)
(a) $CH_3MgBr + H-C\equiv C-H \rightarrow CH_4 + H-C\equiv C-MgBr$
(b) $CH_3MgBr + NH_3 \rightarrow CH_4 + H_2N-MgBr$

Problem 12.8

How might you replace a halogen substituent by a deuterium atom if you wanted to prepare a deuterated compound?

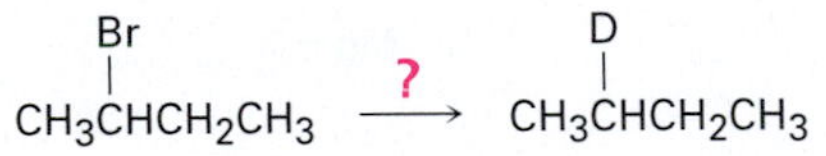

$$\underset{CH_3\overset{\mid}{C}HCH_2CH_3}{\overset{Br}{}} \xrightarrow{?} \underset{CH_3\overset{\mid}{C}HCH_2CH_3}{\overset{D}{}}$$

12.5 Discovery of the Nucleophilic Substitution Reaction

Because they are electrophiles, alkyl halides do one of two things when they react with nucleophiles/bases, such as hydroxide ion: either they undergo *substitution* of the X group by the nucleophile, or they undergo *elimination* of HX to yield an alkene.

Substitution

Elimination

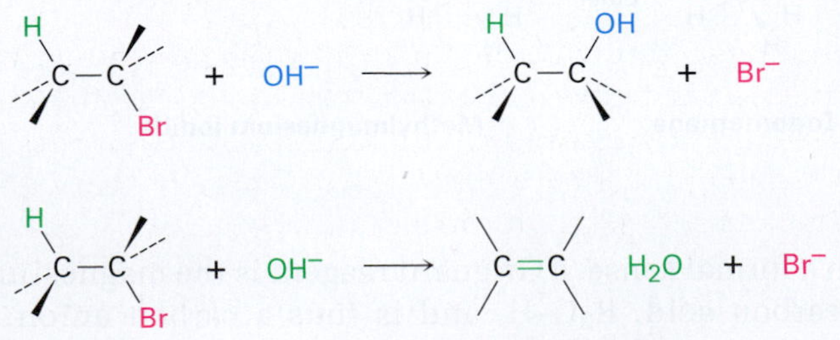

Let's look first at substitution reactions. The discovery of the nucleophilic substitution reaction of alkyl halides dates back to work carried out in 1896 by the German chemist Paul Walden. Walden found that the pure enantiomeric (+)- and (−)-malic acids could be interconverted through a series of simple substitution reactions. When Walden treated (−)-malic acid with PCl$_5$, he isolated (+)-chlorosuccinic acid. This, on treatment with wet Ag$_2$O, gave (+)-malic acid. Similarly, reaction of (+)-malic acid with PCl$_5$ gave (−)-chlorosuccinic acid, which was converted into (−)-malic acid when treated with wet Ag$_2$O. The full cycle of reactions is shown in Figure 12.2.

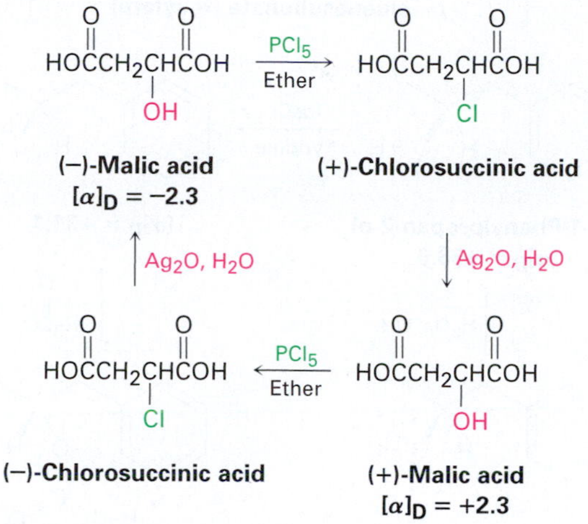

FIGURE 12.2 Walden's cycle of reactions interconverting (+)- and (−)-malic acids.

At the time, the results were astonishing. The eminent chemist Emil Fischer called Walden's discovery "the most remarkable observation made in the field of optical activity since the fundamental observations of Pasteur." Because (−)-malic acid was converted into (+)-malic acid, *some reactions in the cycle must have occurred with a change, or inversion, in configuration at the chirality center.* But which ones, and how? (Remember from Section 5.5 that the direction of light rotation and the configuration of a chirality center aren't directly related. You can't tell by looking at the sign of rotation whether a change in configuration has occurred during a reaction.)

Today, we refer to the transformations taking place in Walden's cycle as **nucleophilic substitution reactions** because each step involves the substitution of one nucleophile (chloride ion, Cl$^-$, or hydroxide ion, HO$^-$) by another. Nucleophilic substitution reactions are one of the most common and versatile reaction types in organic chemistry.

$$R—X \ + \ Nu{:}^- \ \longrightarrow \ R—Nu \ + \ X{:}^-$$

Following the work of Walden, a further series of investigations was undertaken during the 1920s and 1930s to clarify the mechanism of nucleophilic substitution reactions and to find out how inversions of configuration occur. Among the first series studied was one that interconverted the two enantiomers of 1-phenylpropan-2-ol (Figure 12.3). Although this particular series of reactions involves nucleophilic substitution of an alkyl *p*-toluenesulfonate (called a *tosylate*) rather than an alkyl halide, exactly the same type of reaction is involved as that studied by Walden. For all practical purposes, the entire tosylate group acts as if it were simply a halogen substituent. In fact, when you see a tosylate substituent in a molecule, do a mental substitution and tell yourself you're dealing with an alkyl halide.

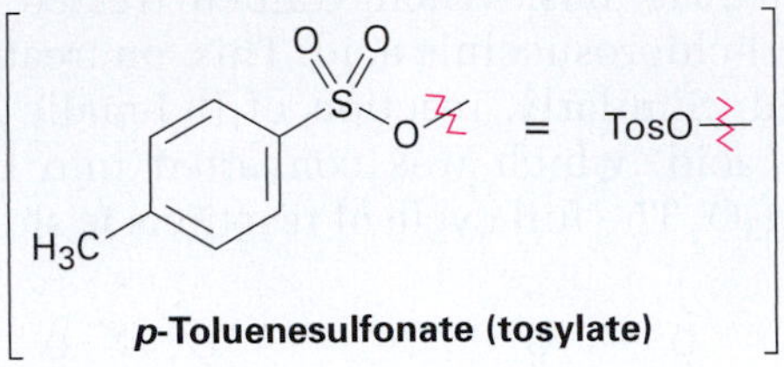

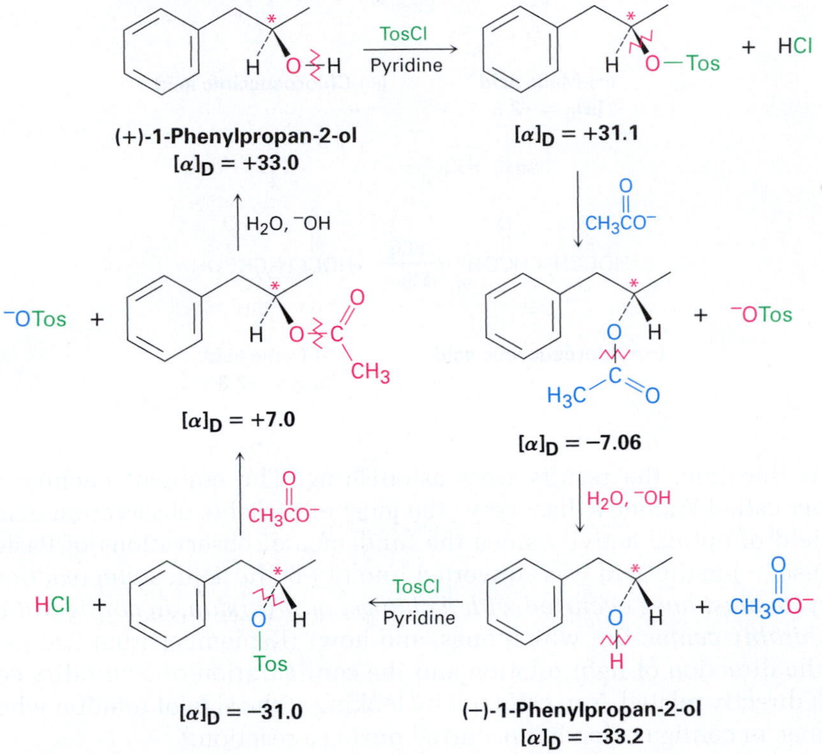

FIGURE 12.3 A Walden cycle interconverting (+) and (−) enantiomers of 1-phenyl-propan-2-ol. Chirality centers are marked by asterisks, and the bonds broken in each reaction are indicated by red wavy lines.

In the three-step reaction sequence shown in Figure 12.3, (+)-1-phenyl-propan-2-ol is interconverted with its (−) enantiomer, so at least one of the three steps must involve an inversion of configuration at the chirality center. The first step, formation of a toluenesulfonate, occurs by breaking the O–H bond of the alcohol rather than the C–O bond to the chiral carbon, so the configuration around carbon is unchanged. Similarly, the third step, hydroxide-ion cleavage of the acetate, takes place without breaking the C–O bond at the chirality center. *The inversion of stereochemical configuration must therefore take place in the second step, the nucleophilic substitution of tosylate ion by acetate ion.*

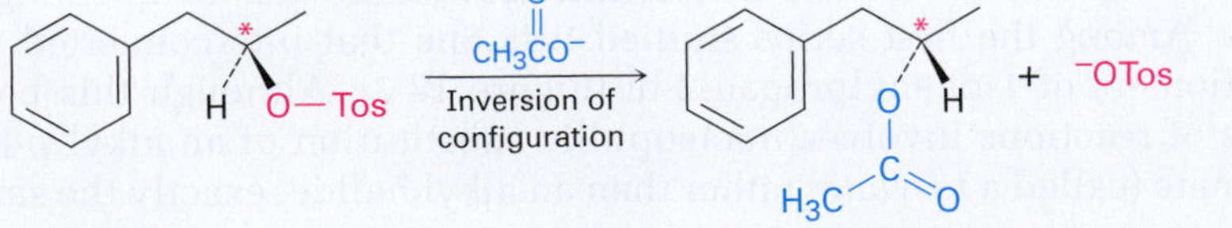

From this and nearly a dozen other series of similar reactions, workers concluded that the nucleophilic substitution reaction of a primary or secondary alkyl halide or tosylate always proceeds with inversion of configuration. (Tertiary alkyl halides and tosylates, as we'll see shortly, give different stereochemical results and react by a different mechanism.)

WORKED EXAMPLE 12.2   **Predicting the Stereochemistry of a Nucleophilic Substitution Reaction**

What product would you expect from a nucleophilic substitution reaction of (R)-1-bromo-1-phenylethane with cyanide ion, ⁻C≡N, as nucleophile? Show the stereochemistry of both reactant and product, assuming that inversion of configuration occurs.

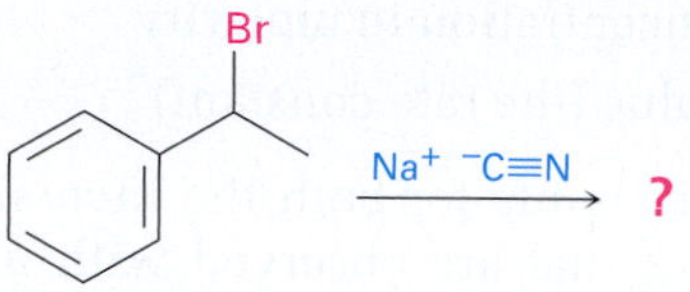

Strategy

Draw the R enantiomer of the reactant, and then change the configuration of the chirality center while replacing the –Br with a –CN.

Solution

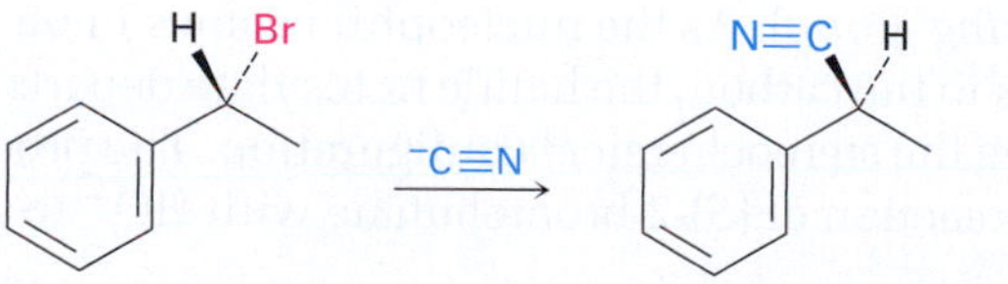

(R)-1-Bromo-1-phenylethane   (S)-2-Phenylpropanenitrile

Problem 12.9

What product would you expect to obtain from a nucleophilic substitution reaction of (S)-2-bromohexane with acetate ion, $CH_3CO_2^-$? Assume that inversion of configuration occurs, and show the stereochemistry of both reactant and product.

12.6 The S$_N$2 Reaction

In every chemical reaction, there is a direct relationship between the rate at which the reaction occurs and the concentrations of the reactants. When we measure this relationship, we measure the **kinetics** of the reaction. For example, let's look at the kinetics of a simple nucleophilic substitution—the reaction of CH_3Br with OH^- to yield CH_3OH plus Br^-.

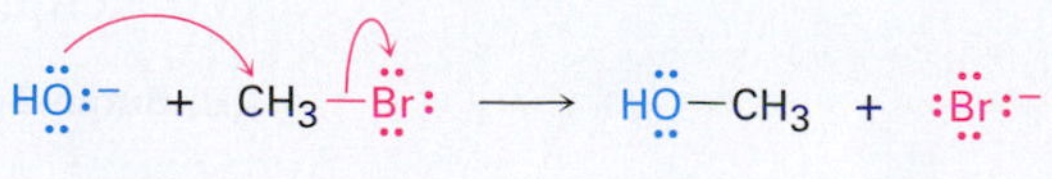

At a given temperature, solvent, and concentration of reactants, the substitution occurs at a certain rate. If we double the concentration of OH^-, the frequency of encounter between the reaction partners doubles, and we find that the reaction rate also doubles. Similarly, if we double the concentration of CH_3Br, the reaction rate again doubles. We call such a reaction, in which the rate is linearly dependent on the concentrations of two species, a **second-order reaction**. Mathematically, we can express this second-order dependence of the nucleophilic substitution reaction by setting up a *rate equation*. As either [RX] or [^-OH] changes, the rate of the reaction changes proportionately.

$$\text{Reaction rate} = \text{Rate of disappearance of reactant}$$

$$= k \times [RX] \times [^-OH]$$

where $[RX] = CH_3Br$ concentration in molarity

$[^-OH] = {}^-OH$ concentration in molarity

$k = $ A constant value (the rate constant)

A mechanism that accounts for both the inversion of configuration and the second-order kinetics that are observed with nucleophilic substitution reactions was suggested in 1937 by E. D. Hughes and Christopher Ingold, who formulated what they called the **S_N2 reaction**—short for *substitution, nucleophilic, bimolecular*. (*Bimolecular* means that two molecules, nucleophile and alkyl halide, take part in the step whose kinetics are measured.)

The essential feature of the S_N2 mechanism is that it takes place in a single step without intermediates when the incoming nucleophile reacts with the alkyl halide or tosylate (the *substrate*) from a direction opposite the group that is displaced (the *leaving group*). As the nucleophile comes in on one side of the substrate and bonds to the carbon, the halide or tosylate departs from the other side, thereby inverting the stereochemical configuration. The process is shown in Figure 12.4 for the reaction of (*S*)-2-bromobutane with HO^- to give (*R*)-butan-2-ol.

FIGURE 12.4 MECHANISM: The mechanism of the S_N2 reaction. The reaction takes place in a single step when the incoming nucleophile approaches from a direction 180° away from the leaving halide ion, thereby inverting the stereochemistry at carbon.

① The nucleophile ^-OH uses its lone-pair electrons to attack the alkyl halide carbon 180° away from the departing halogen. This leads to a transition state with a partially formed C–OH bond and a partially broken C–Br bond.

② The stereochemistry at carbon is inverted as the C–OH bond forms fully and the bromide ion departs with the electron pair from the former C–Br bond.

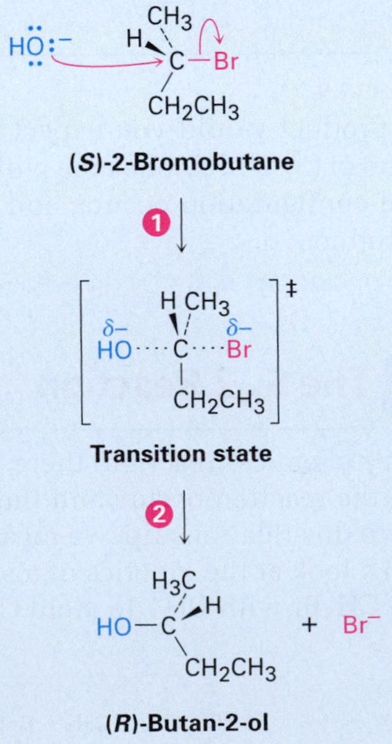

(*S*)-2-Bromobutane

Transition state

(*R*)-Butan-2-ol

© John McMurry

As shown in Figure 12.4, the S$_N$2 reaction occurs when an electron pair on the nucleophile Nu:⁻ forces out the group X:⁻, which takes with it the electron pair from the former C–X bond. This occurs through a transition state in which the new Nu–C bond is partially forming at the same time that the old C–X bond is partially breaking, and in which the negative charge is shared by both the incoming nucleophile and the outgoing halide ion. The transition state for this inversion has the remaining three bonds to carbon in a planar arrangement (Figure 12.5).

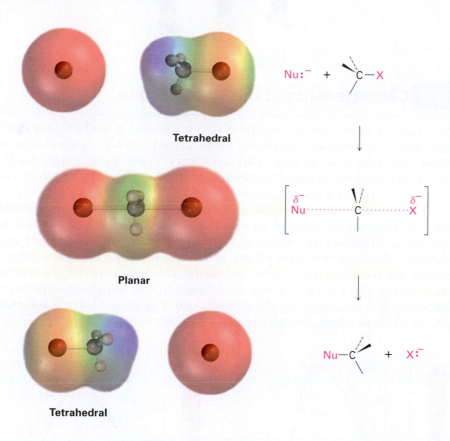

Tetrahedral

Planar

Tetrahedral

FIGURE 12.5 The transition state of an S$_N$2 reaction has a planar arrangement of the carbon atom and the remaining three groups. Electrostatic potential maps show that negative charge (red) is delocalized in the transition state.

The mechanism proposed by Hughes and Ingold is fully consistent with experimental results, explaining both stereochemical and kinetic data. Thus, the requirement for backside approach of the entering nucleophile from a direction 180° away from the leaving group causes the stereochemistry of the substrate to invert, much like an umbrella turning inside out in the wind. The Hughes–Ingold mechanism also explains why second-order kinetics are found: the S$_N$2 reaction occurs in a single step that involves both alkyl halide and nucleophile. Two molecules are involved in the step whose rate is measured.

Problem 12.10
What product would you expect to obtain from S$_N$2 reaction of OH⁻ with (*R*)-2-bromobutane? Show the stereochemistry of both reactant and product.

Problem 12.11

Assign configuration to the following substance, and draw the structure of the product that would result on nucleophilic substitution reaction with HS⁻ (reddish brown = Br):

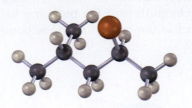

12.7 Characteristics of the S$_N$2 Reaction

Now that we know how S$_N$2 reactions occur, we need to see how they can be used and what variables affect them. Some S$_N$2 reactions are fast and some are slow; some take place in high yield and others, in low yield. Understanding the factors involved can be of tremendous value. Let's begin by recalling a few things about reaction rates in general.

The rate of a chemical reaction is determined by $\Delta G^{\ddagger}$, the energy difference between reactant ground state and transition state. A change in reaction conditions can affect $\Delta G^{\ddagger}$ either by changing the reactant energy level or by changing the transition-state energy level. Lowering the reactant energy or raising the transition-state energy increases $\Delta G^{\ddagger}$ and decreases the reaction rate; raising the reactant energy or decreasing the transition-state energy decreases $\Delta G^{\ddagger}$ and increases the reaction rate (Figure 12.6). We'll see examples of all these effects as we look at S$_N$2 reaction variables.

FIGURE 12.6 The effects of changes in reactant and transition-state energy levels on reaction rate. **(a)** A higher reactant energy level (red curve) corresponds to a faster reaction (smaller $\Delta G^{\ddagger}$). **(b)** A higher transition-state energy level (red curve) corresponds to a slower reaction (larger $\Delta G^{\ddagger}$).

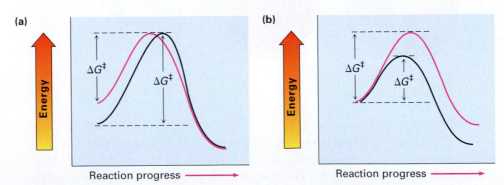

The Substrate: Steric Effects in the S$_N$2 Reaction

The first S$_N$2 reaction variable to look at is the structure of the substrate. Because the S$_N$2 transition state involves partial bond formation between the incoming nucleophile and the alkyl halide carbon atom, it seems reasonable that a hindered, bulky substrate should prevent easy approach of the nucleophile, making bond formation difficult. In other words, the transition state for reaction of a sterically hindered substrate, whose carbon atom is shielded from approach of the incoming nucleophile, is higher in energy and forms more slowly than the corresponding transition state for a less hindered substrate (Figure 12.7).

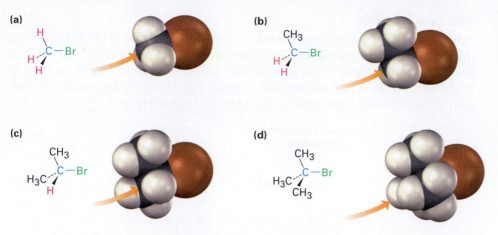

FIGURE 12.7 Steric hindrance to the S$_{N}$2 reaction. As the models indicate, the carbon atom in **(a)** bromomethane is readily accessible, resulting in a fast S$_{N}$2 reaction. The carbon atoms in **(b)** bromoethane (primary), **(c)** 2-bromopropane (secondary), and **(d)** 2-bromo-2-methylpropane (tertiary) are successively more hindered, resulting in successively slower S$_{N}$2 reactions.

As Figure 12.7 shows, the difficulty of nucleophilic approach increases as the three substituents bonded to the halo-substituted carbon atom increase in size. Methyl halides are by far the most reactive substrates in S$_{N}$2 reactions, followed by primary alkyl halides such as ethyl and propyl. Alkyl branching at the reacting center, as in isopropyl halides (2°), slows the reaction greatly, and further branching, as in *tert*-butyl halides (3°), effectively halts the reaction. Even branching one carbon removed from the reacting center, as in 2,2-dimethylpropyl (*neopentyl*) halides, greatly slows nucleophilic displacement. As a result, *S$_{N}$2 reactions occur only at relatively unhindered sites* and are normally useful only with methyl halides, primary halides, and a few simple secondary halides. Relative reactivities for some different substrates are as follows:

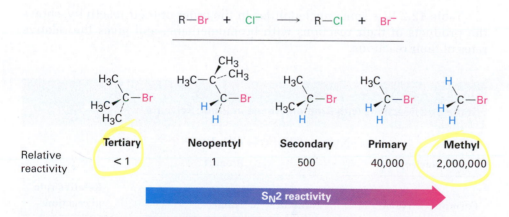

Although not shown in the preceding reactivity order, vinylic halides (R$_2$C=CRX) and aryl halides are unreactive toward S$_{N}$2 reaction. This lack of reactivity is due to steric factors, because the incoming nucleophile would have to approach in the plane of the carbon–carbon double bond to carry out a backside displacement.

The Nucleophile

Another variable that has a major effect on the S_N2 reaction is the nature of the nucleophile. Any species, either neutral or negatively charged, can act as a nucleophile as long as it has an unshared pair of electrons, that is, as long as it is a Lewis base. If the nucleophile is negatively charged, the product is neutral; if the nucleophile is neutral, the product is positively charged.

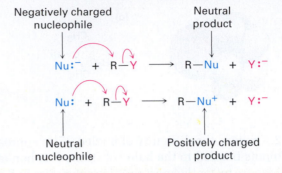

A wide array of substances can be prepared using nucleophilic substitution reactions. In fact, we've already seen an example: the reaction of an acetylide anion with an alkyl halide discussed in Section 8.15 is an S_N2 reaction in which the acetylide nucleophile replaces a halide leaving group.

$$R-C{\equiv}C{:}^- \;+\; CH_3Br \xrightarrow[\text{reaction}]{S_N2} \; R-C{\equiv}C-CH_3 \;+\; Br^-$$

An acetylide anion

Table 12.2 lists some nucleophiles in the order of their reactivity, shows the products of their reactions with bromomethane, and gives the relative rates of their reactions.

TABLE 12.2
Some S_N2 Reactions with Bromomethane in Protic Solvents

$$Nu{:}^- + CH_3Br \rightarrow CH_3Nu + Br^-$$

Nucleophile		Product		Relative rate
Formula	Name	Formula	Name	of reaction
H_2O	Water	$CH_3OH_2{}^+$	Methylhydronium ion	1
$CH_3CO_2{}^-$	Acetate	$CH_3CO_2CH_3$	Methyl acetate	500
NH_3	Ammonia	$CH_3NH_3{}^+$	Methylammonium ion	700
Cl^-	Chloride	CH_3Cl	Chloromethane	1,000
HO^-	Hydroxide	CH_3OH	Methanol	10,000
CH_3O^-	Methoxide	CH_3OCH_3	Dimethyl ether	25,000
I^-	Iodide	CH_3I	Iodomethane	100,000
^-CN	Cyanide	CH_3CN	Acetonitrile	125,000
HS^-	Hydrosulfide	CH_3SH	Methanethiol	125,000

The data in Table 12.2 show that there are large differences in the rates at which various nucleophiles react. Detailed explanations for the observed reactivities aren't always simple, but some trends can be detected in the data of Table 12.2.

- **Nucleophilicity roughly parallels basicity** when comparing nucleophiles that have the same reacting atom. Thus, OH^- is more basic and more nucleophilic than acetate ion, $CH_3CO_2^-$, which in turn is more basic and more nucleophilic than H_2O. Since "nucleophilicity" is usually taken as the affinity of a Lewis base for a carbon atom in the S_N2 reaction and "basicity" is the affinity of a base for a proton, it's easy to see why there might be a correlation between the two kinds of behavior.

- **Nucleophilicity usually increases going down a column of the periodic table.** Thus, HS^- is more nucleophilic than HO^-, and the halide reactivity order is $I^- > Br^- > Cl^-$. Going down the periodic table, elements have their valence electrons in successively larger shells, where they are successively farther from the nucleus, less tightly held, and consequently more reactive. The matter is complex, though, and the nucleophilicity order can change depending on the solvent.

- **Negatively charged nucleophiles are usually more reactive than neutral ones.** As a result, S_N2 reactions are often carried out under basic conditions rather than neutral or acidic conditions.

Aprotic

$Cl^- > Br^- > I^-$

Problem 12.12

What product would you expect from S_N2 reaction of 1-bromobutane with each of the following?
(a) NaI **(b)** KOH **(c)** H—C≡C—Li **(d)** NH_3

Problem 12.13

Which substance in each of the following pairs is more reactive as a nucleophile? Explain.
(a) $(CH_3)_2N^-$ or $(CH_3)_2NH$ **(b)** $(CH_3)_3B$ or $(CH_3)_3N$ **(c)** H_2O or H_2S

The Leaving Group

Still another variable that can affect the S_N2 reaction is the nature of the group displaced by the incoming nucleophile. Because the leaving group is expelled with a negative charge in most S_N2 reactions, the best leaving groups are those that best stabilize the negative charge in the transition state. The greater the extent of charge stabilization by the leaving group, the lower the energy of the transition state and the more rapid the reaction. But as we saw in Section 2.8, those groups that best stabilize a negative charge are also the weakest bases. Thus, weak bases such as Cl^- and tosylate ion make good leaving groups, while strong bases such as OH^- and NH_2^- make poor leaving groups.

	OH^-, NH_2^-, OR^-	F^-	Cl^-	Br^-	I^-	$TosO^-$
Relative reactivity	<<1	1	200	10,000	30,000	60,000

Leaving group reactivity →

It's just as important to know which are poor leaving groups as to know which are good, and the preceding data clearly indicate that F^-, HO^-, RO^-, and H_2N^- are not displaced by nucleophiles. In other words, alkyl fluorides, alcohols, ethers, and amines do not typically undergo S_N2 reactions. To carry out an S_N2 reaction with an alcohol, it's necessary to convert the –OH into a better leaving group. This, in fact, is just what happens when a primary or secondary alcohol is converted into either an alkyl chloride by reaction with $SOCl_2$ or an alkyl bromide by reaction with PBr_3 (Section 12.3).

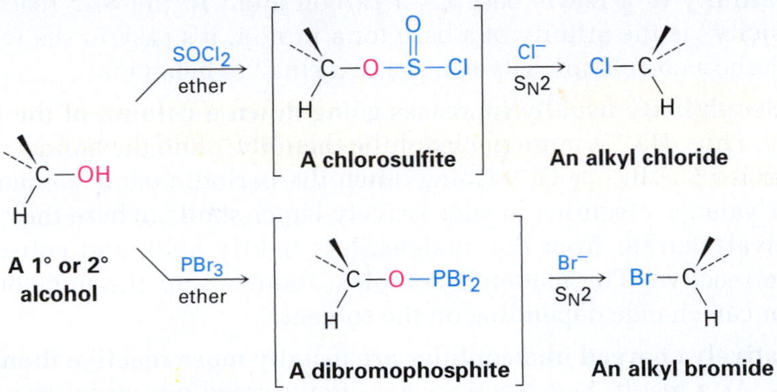

Alternatively, an alcohol can be made more reactive toward nucleophilic substitution by treating it with *p*-toluenesulfonyl chloride to form a tosylate. As noted on several previous occasions, tosylates are even more reactive than halides in nucleophilic substitutions. Note that tosylate formation does not change the configuration of the oxygen-bearing carbon because the C–O bond is not broken.

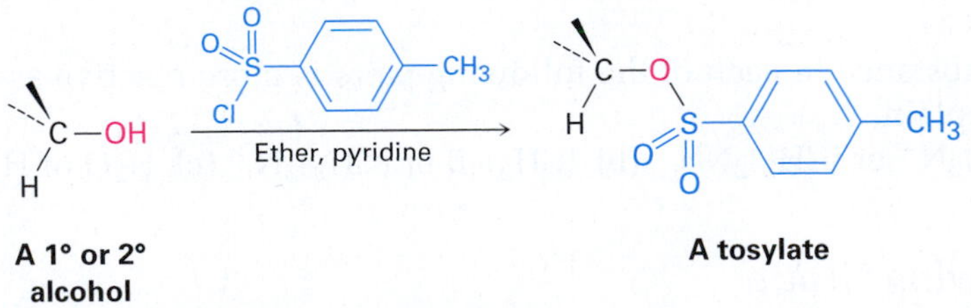

The one general exception to the rule that ethers don't typically undergo S_N2 reactions occurs with epoxides, the three-membered cyclic ethers that we saw in Section 8.6. Epoxides, because of the angle strain in the three-membered ring, are much more reactive than other ethers. They react with aqueous acid to give 1,2-diols and they react readily with many other nucleophiles as well. Propene oxide, for instance, reacts with HCl to give 1-chloropropan-2-ol by S_N2 backside attack on the less hindered primary carbon atom.

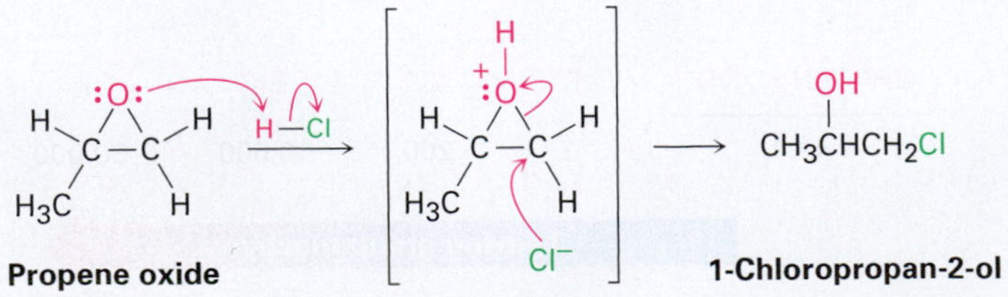

Problem 12.14

Rank the following compounds in order of their expected reactivity toward S$_N$2 reaction:

$$CH_3Cl, \; CH_3OTos, \; CH_3NH_2, \; (CH_3)_2CHCl$$

The Solvent

The rates of S$_N$2 reactions are strongly affected by the solvent. *Protic solvents*—those that contain an –OH or –NH group—are generally the worst for S$_N$2 reactions, while *polar aprotic solvents*, which are polar but don't have an –OH or –NH group, are the best.

Protic solvents, such as methanol and ethanol, slow down S$_N$2 reactions by **solvation** of the reactant nucleophile. The solvent molecules hydrogen bond to the nucleophile and form a "cage" around it, thereby lowering its energy and reactivity.

A solvated anion
(reduced nucleophilicity due to enhanced ground-state stability)

In contrast with protic solvents, which *decrease* the rates of S$_N$2 reactions by *lowering* the ground-state energy of the nucleophile, polar aprotic solvents *increase* the rates of S$_N$2 reactions by *raising* the ground-state energy of the nucleophile. Acetonitrile (CH$_3$CN), dimethylformamide [(CH$_3$)$_2$NCHO, abbreviated DMF], dimethyl sulfoxide [(CH$_3$)$_2$SO, abbreviated DMSO], and hexamethylphosphoramide {[(CH$_3$)$_2$N]$_3$PO, abbreviated HMPA} are particularly useful. These solvents can dissolve many salts because of their high polarity, but they solvate metal cations rather than nucleophilic anions. As a result, the relatively unsolvated anions have a greater nucleophilicity, and S$_N$2 reactions take place at correspondingly faster rates. For instance, a rate increase of 200,000 has been observed on changing from methanol to HMPA for the reaction of azide ion with 1-bromobutane.

$$CH_3CH_2CH_2CH_2-Br \; + \; N_3^- \; \longrightarrow \; CH_3CH_2CH_2CH_2-N_3 \; + \; Br^-$$

Solvent	CH$_3$OH	H$_2$O	DMSO	DMF	CH$_3$CN	HMPA
Relative reactivity	1	7	1300	2800	5000	200,000

Solvent reactivity →

Problem 12.15

Organic solvents such as ether and chloroform are neither protic nor strongly polar. What effect would you expect these solvents to have on the reactivity of a nucleophile in S$_N$2 reactions?

A Summary of S$_N$2 Reaction Characteristics

The effects on S$_N$2 reactions of the four variables—substrate structure, nucleophile, leaving group, and solvent—are summarized in the following statements and in the energy diagrams of Figure 12.8:

Substrate

Steric hindrance raises the energy of the S$_N$2 transition state, increasing $\Delta G^{\ddagger}$ and decreasing the reaction rate (Figure 12.8a). As a result, S$_N$2 reactions are best for methyl and primary substrates. Secondary substrates react slowly, and tertiary substrates do not react by an S$_N$2 mechanism.

Nucleophile

Basic, negatively charged nucleophiles are less stable and have a higher ground-state energy than neutral ones, decreasing $\Delta G^{\ddagger}$ and increasing the S$_N$2 reaction rate (Figure 12.8b).

Leaving group

Good leaving groups (more stable anions) lower the energy of the transition state, decreasing $\Delta G^{\ddagger}$ and increasing the S$_N$2 reaction rate (Figure 12.8c).

Solvent

Protic solvents solvate the nucleophile, thereby lowering its ground-state energy, increasing $\Delta G^{\ddagger}$, and decreasing the S$_N$2 reaction rate. Polar aprotic solvents surround the accompanying cation but not the nucleophilic anion, thereby raising the ground-state energy of the nucleophile, decreasing $\Delta G^{\ddagger}$, and increasing the reaction rate (Figure 12.8d).

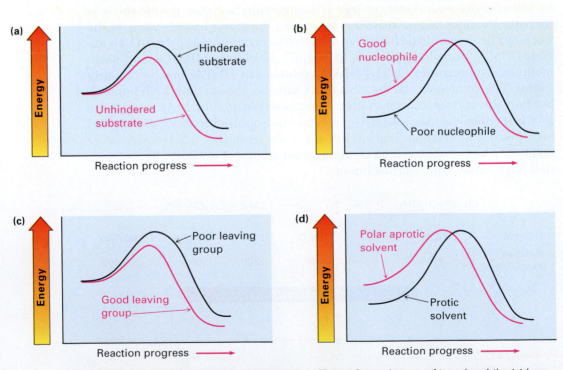

FIGURE 12.8 Energy diagrams showing the effects of **(a)** substrate, **(b)** nucleophile, **(c)** leaving group, and **(d)** solvent on S$_N$2 reaction rates. Substrate and leaving group effects are felt primarily in the transition state. Nucleophile and solvent effects are felt primarily in the reactant ground state.

12.8 The S$_N$1 Reaction

As we've seen, the <mark>S$_N$2 reaction is best when carried out with an unhindered substrate and a negatively charged nucleophile in a polar aprotic solvent but is worst when carried out with a hindered substrate and a neutral nucleophile in a protic solvent.</mark> You might therefore expect the reaction of a tertiary substrate (hindered) with water (neutral, protic) to be among the slowest of substitution reactions. Remarkably, however, the opposite is true. Reaction of the tertiary halide $(CH_3)_3CBr$ with H_2O to give the alcohol 2-methylpropan-2-ol is more than *1 million times* as fast as the corresponding reaction of CH_3Br to give methanol.

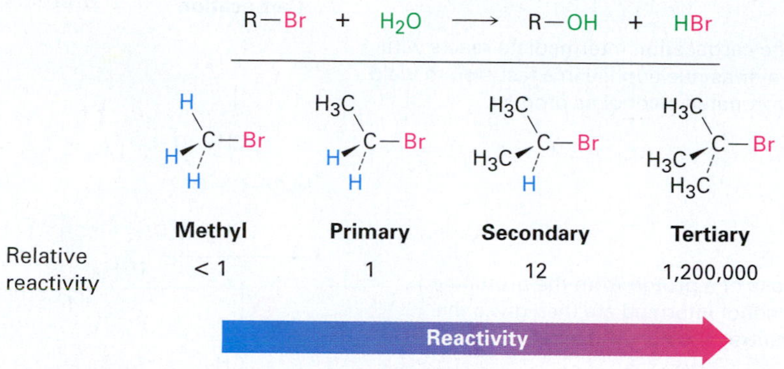

$$R-Br \ + \ H_2O \ \longrightarrow \ R-OH \ + \ HBr$$

	Methyl	Primary	Secondary	Tertiary
Relative reactivity	< 1	1	12	1,200,000

Reactivity →

What's going on here? Clearly, a nucleophilic substitution reaction is occurring, yet the reactivity order seems backward. These reactions can't be taking place by the S$_N$2 mechanism we've been discussing, and we must therefore conclude that they are occurring by *an alternative substitution mechanism.* This alternative mechanism is called the **S$_N$1 reaction**, for *substitution, nucleophilic, unimolecular.*

In contrast to the S$_N$2 reaction of CH_3Br with OH^-, the S$_N$1 reaction of $(CH_3)_3CBr$ with H_2O has a rate that depends only on the alkyl halide concentration and is independent of the H_2O concentration. In other words, the reaction is a **first-order process**; the concentration of the nucleophile does not appear in the rate equation.

Reaction rate = Rate of disappearance of alkyl halide

$$= k \times [RX]$$

To explain this result, we need to know more about kinetics measurements. Many organic reactions occur in several steps, one of which is usually slower than the others. We call this slow step the *rate-limiting step,* or *rate-determining step.* No reaction can proceed faster than its rate-limiting step, which acts as a kind of traffic jam, or bottleneck. In the S$_N$1 reaction of $(CH_3)_3CBr$ with H_2O, the fact that the nucleophile does not appear in the first-order rate equation means that the alkyl halide is involved in a *unimolecular* rate-limiting step. But if the nucleophile is not involved in the rate-limiting step, then it must be involved in some other, non–rate-limiting step. The mechanism shown in Figure 12.9 accounts for these observations.

Unlike what happens in an S$_N$2 reaction, where the leaving group is displaced at the same time the incoming nucleophile approaches, an

ACTIVE FIGURE 12.9

MECHANISM: The mechanism of the S_N1 reaction of 2-bromo-2-methylpropane with H_2O involves three steps. The first step—spontaneous, unimolecular dissociation of the alkyl bromide to yield a carbocation—is rate-limiting. **Go to this book's student companion site at www.cengage.com/chemistry/mcmurry to explore an interactive version of this figure.**

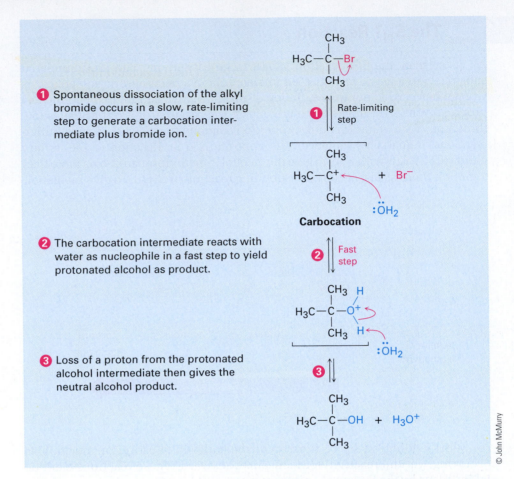

① Spontaneous dissociation of the alkyl bromide occurs in a slow, rate-limiting step to generate a carbocation intermediate plus bromide ion.

② The carbocation intermediate reacts with water as nucleophile in a fast step to yield protonated alcohol as product.

③ Loss of a proton from the protonated alcohol intermediate then gives the neutral alcohol product.

© John McMurry

S_N1 reaction takes place by loss of the leaving group *before* the nucleophile approaches. 2-Bromo-2-methylpropane spontaneously dissociates to the *tert*-butyl carbocation plus Br^- in a slow, rate-limiting step, and the intermediate carbocation is then immediately trapped by the nucleophile water in a faster second step. *Water is not a reactant in the step whose rate is measured.* The energy diagram is shown in Figure 12.10.

FIGURE 12.10 An energy diagram for an S_N1 reaction. The slower, rate-limiting step is the spontaneous dissociation of the alkyl halide to give a carbocation intermediate. Reaction of the carbocation with a nucleophile then occurs in a second, faster step.

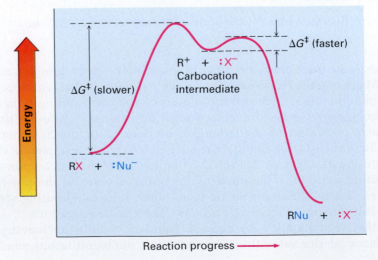

Because an S$_N$1 reaction occurs through a carbocation intermediate, its stereochemical outcome is different from that of an S$_N$2 reaction. Carbocations, as we've seen, are planar, sp^2-hybridized, and achiral. Thus, if we carry out an S$_N$1 reaction on one enantiomer of a chiral reactant and go through an achiral carbocation intermediate, the molecule loses its chirality so the product must be optically inactive. That is, the symmetrical intermediate carbocation can react with a nucleophile equally well from either side, leading to a racemic, 50:50 mixture of enantiomers (Figure 12.11).

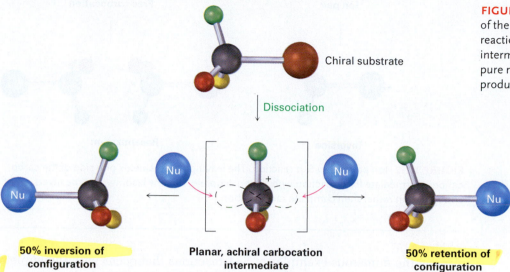

Chiral substrate

Dissociation

50% inversion of configuration

Planar, achiral carbocation intermediate

50% retention of configuration

FIGURE 12.11 Stereochemistry of the S$_N$1 reaction. Because the reaction goes through an achiral intermediate, an enantiomerically pure reactant gives a racemic product.

The conclusion that S$_N$1 reactions on enantiomerically pure substrates should give racemic products is nearly, but not exactly, what is found. In fact, few S$_N$1 displacements occur with complete racemization. Most give a minor (0%–20%) excess of inversion. The reaction of (R)-6-chloro-2,6-dimethyloctane with H$_2$O, for example, leads to an alcohol product that is approximately 80% racemized and 20% inverted (80% R,S + 20% S is equivalent to 40% R + 60% S):

H$_3$C
CH$_2$CH$_3$ CH$_3$
Cl—C
CH$_2$CH$_2$CH$_2$CHCH$_3$

$\xrightarrow[\text{Ethanol}]{\text{H}_2\ddot{\text{O}}:}$

CH$_3$CH$_2$ CH$_3$
CH$_3$ C—OH
CH$_3$CHCH$_2$CH$_2$CH$_2$

+

H$_3$C
CH$_2$CH$_3$ CH$_3$
HO—C
CH$_2$CH$_2$CH$_2$CHCH$_3$

(R)-6-Chloro-2,6-dimethyloctane

60% S (inversion)

40% R (retention)

This lack of complete racemization in most S$_N$1 reactions is due to the fact that *ion pairs* are involved. Dissociation of the substrate occurs to give a structure in which the two ions are still loosely associated and in which the carbocation is effectively shielded from reaction on one side by the departing anion. If a certain amount of substitution occurs before the two ions fully diffuse apart, then a net inversion of configuration will be observed (Figure 12.12).

Problem 12.16

What product(s) would you expect from reaction of (S)-3-chloro-3-methyloctane with acetic acid? Show the stereochemistry of both reactant and product.

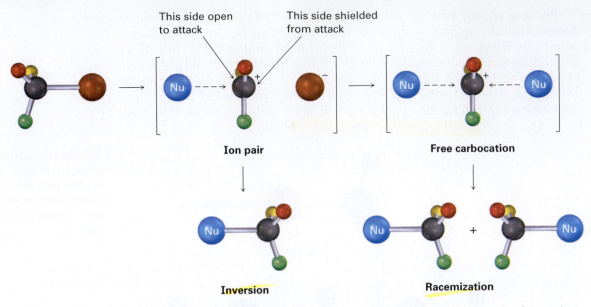

FIGURE 12.12 Ion pairs in an S_N1 reaction. The leaving group shields one side of the carbocation intermediate from reaction with the nucleophile, thereby leading to some inversion of configuration rather than complete racemization.

Problem 12.17

Among the numerous examples of S_N1 reactions that occur with incomplete racemization, the optically pure tosylate of 2,2-dimethyl-1-phenylpropan-1-ol ($[\alpha]_D = -30.3$) gives the corresponding acetate ($[\alpha]_D = +5.3$) when heated in acetic acid. If complete inversion had occurred, the optically pure acetate would have had $[\alpha]_D = +53.6$. What percentage racemization and what percentage inversion occurred in this reaction?

Problem 12.18

Assign configuration to the following substrate, and show the stereochemistry and identity of the product you would obtain by S_N1 reaction with water (reddish brown = Br):

12.9 Characteristics of the S$_N$1 Reaction

Just as the S$_N$2 reaction is strongly influenced by the structure of the substrate, the leaving group, the nucleophile, and the solvent, the S$_N$1 reaction is similarly influenced. Factors that lower $\Delta G^{\ddagger}$, either by lowering the energy level of the transition state or by raising the energy level of the ground state, favor faster S$_N$1 reactions. Conversely, factors that raise $\Delta G^{\ddagger}$, either by raising the energy level of the transition state or by lowering the energy level of the reactant, slow down the S$_N$1 reaction.

The Substrate

According to the Hammond postulate (Section 7.9), any factor that stabilizes a high-energy intermediate also stabilizes the transition state leading to that intermediate. Since the rate-limiting step in an S$_N$1 reaction is the spontaneous, unimolecular dissociation of the substrate to yield a carbocation, the reaction is favored whenever a stabilized carbocation intermediate is formed. The more stable the carbocation intermediate, the faster the S$_N$1 reaction.

We saw in Section 7.8 that the stability order of alkyl carbocations is $3° > 2° > 1° > -CH_3$. To this list we must also add the resonance-stabilized allylic and benzylic cations. As Figure 12.13 indicates, an allylic cation has two resonance forms. In one form the double bond is on the "left"; in the other form it's on the "right." A benzylic cation has five resonance forms, all of which contribute to the overall resonance hybrid.

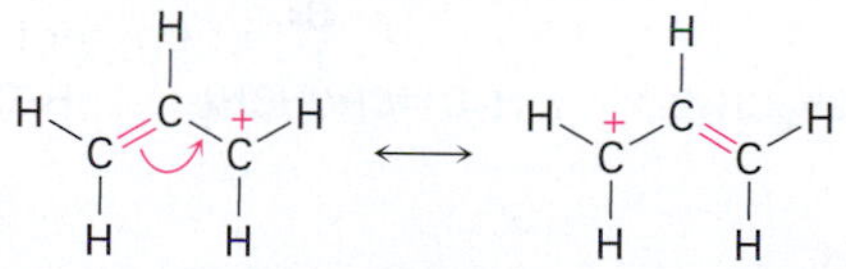

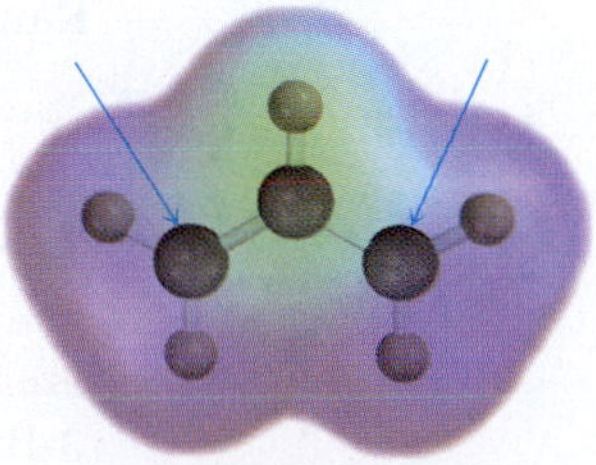

Allyl carbocation

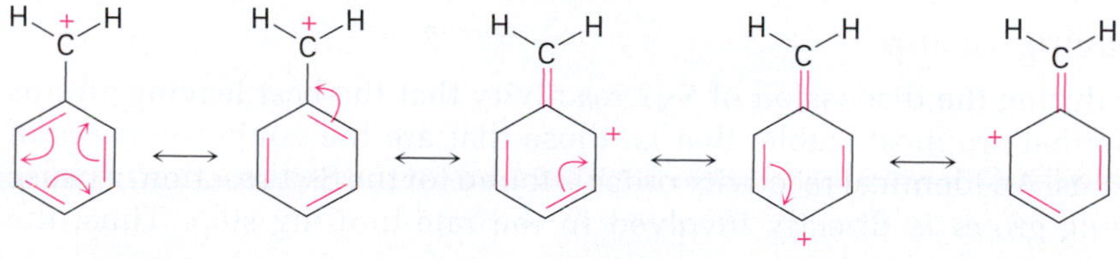

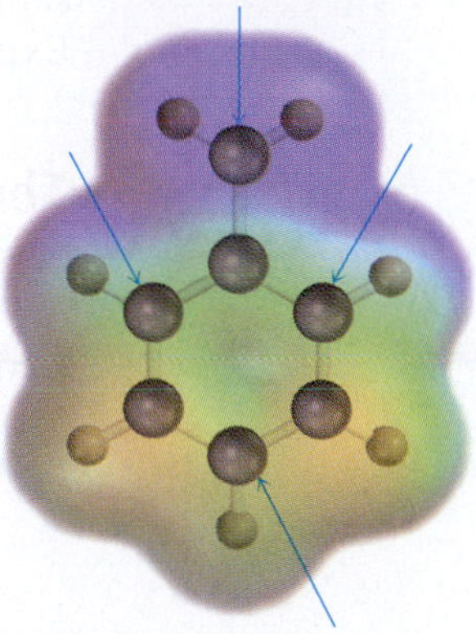

Benzyl carbocation

FIGURE 12.13 Resonance forms of allylic and benzylic carbocations. The positive charge (blue) is delocalized over the π system in both. Electron-poor atoms are indicated by blue arrows.

Because of resonance stabilization, a *primary* allylic or benzylic carbocation is about as stable as a *secondary* alkyl carbocation and a *secondary* allylic or benzylic carbocation is about as stable as a *tertiary* alkyl carbocation.

This stability order of carbocations is the same as the order of S_N1 reactivity for alkyl halides and tosylates.

H—C+   H3C—C+   H—C=C—C+   [benzene ring]—C+   H3C—C+   H3C—C+

Methyl < Primary < Allylic ≈ Benzylic ≈ Secondary < Tertiary

Carbocation stability →

Parenthetically, we might also note that primary allylic and benzylic substrates are particularly reactive in S_N2 reactions as well as in S_N1 reactions. Allylic and benzylic C–X bonds are about 50 kJ/mol (12 kcal/mol) weaker than the corresponding saturated bonds and are therefore more easily broken.

CH_3CH_2—Cl H_2C=$CHCH_2$—Cl   [benzene ring]CH_2—Cl

352 kJ/mol   298 kJ/mol   300 kJ/mol
(84 kcal/mol)   (71 kcal/mol)   (72 kcal/mol)

Problem 12.19
Rank the following substances in order of their expected S_N1 reactivity:

                                  Br   Br
                                  | |
CH_3CH_2Br H_2C=$CHCHCH_3$ H_2C=$CHBr$ CH_3CHCH_3

Problem 12.20
3-Bromobut-1-ene and 1-bromobut-2-ene undergo S_N1 reaction at nearly the same rate even though one is a secondary halide and the other is primary. Explain.

The Leaving Group

We said during the discussion of S_N2 reactivity that the best leaving groups are those that are most stable, that is, those that are the conjugate bases of strong acids. An identical reactivity order is found for the S_N1 reaction because the leaving group is directly involved in the rate-limiting step. Thus, the S_N1 reactivity order is

$HO^- < Cl^- < Br^- < I^- ≈ TosO^- ≈ H_2O$

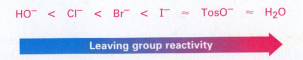

Leaving group reactivity →

Note that in the S_N1 reaction, which is often carried out under acidic conditions, neutral water is sometimes the leaving group. This occurs, for example, when an alkyl halide is prepared from a tertiary alcohol by reaction with HBr or HCl (Section 12.3). The alcohol is first protonated and then spontaneously loses H_2O to generate a carbocation, which reacts with halide ion to give the alkyl halide (Figure 12.14). Knowing that an S$_N$1 reaction is involved in the conversion of alcohols to alkyl halides explains why the reaction works well only for tertiary alcohols: tertiary alcohols react fastest because they give the most stable carbocation intermediates.

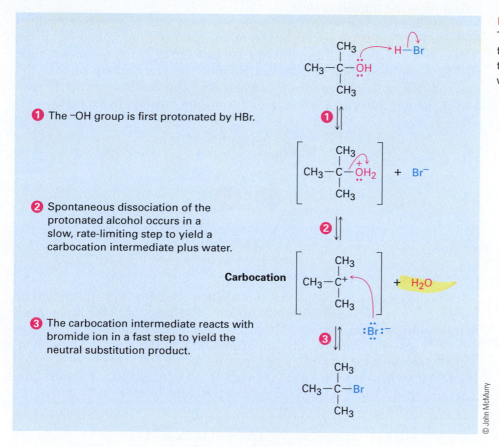

1 The −OH group is first protonated by HBr.

2 Spontaneous dissociation of the protonated alcohol occurs in a slow, rate-limiting step to yield a carbocation intermediate plus water.

3 The carbocation intermediate reacts with bromide ion in a fast step to yield the neutral substitution product.

FIGURE 12.14 MECHANISM: The mechanism of the S$_N$1 reaction of a tertiary alcohol with HBr to yield an alkyl halide. Neutral water is the leaving group.

© John McMurry

The Nucleophile

The nature of the nucleophile plays a major role in the S$_N$2 reaction but does not affect an S$_N$1 reaction. Because the S$_N$1 reaction occurs through a rate-limiting step in which the added nucleophile has no part, the nucleophile can't affect the reaction rate. The reaction of 2-methylpropan-2-ol with HX, for instance, occurs at the same rate regardless of whether X is Cl, Br, or I. Furthermore, neutral nucleophiles are just as effective as negatively charged ones, so S$_N$1 reactions frequently occur under neutral or acidic conditions.

2-Methylpropan-2-ol   (Same rate for X = Cl, Br, I)

The Solvent

What about solvent? Do solvents have the same effect in S_N1 reactions that they have in S_N2 reactions? The answer is both yes and no. Yes, solvents have a large effect on S_N1 reactions, but no, the reasons for the effects on S_N1 and S_N2 reactions are not the same. Solvent effects in the S_N2 reaction are due largely to stabilization or destabilization of the nucleophile *reactant*. Solvent effects in the S_N1 reaction, however, are due largely to stabilization or destabilization of the *transition state*.

The Hammond postulate says that any factor stabilizing the intermediate carbocation should increase the rate of an S_N1 reaction. Solvation of the carbocation—the interaction of the ion with solvent molecules—has just such an effect. Solvent molecules orient around the carbocation so that the electron-rich ends of the solvent dipoles face the positive charge (Figure 12.15), thereby lowering the energy of the ion and favoring its formation.

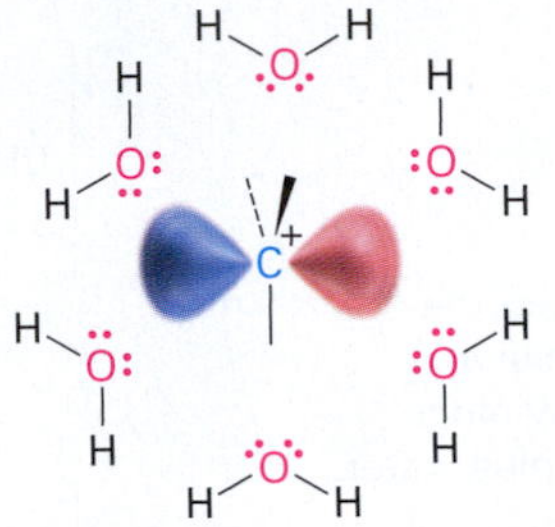

FIGURE 12.15 Solvation of a carbocation by water. The electron-rich oxygen atoms of solvent molecules orient around the positively charged carbo-cation and thereby stabilize it.

The properties of a solvent that contribute to its ability to stabilize ions by solvation are related to the solvent's polarity. S_N1 reactions take place much more rapidly in strongly polar solvents, such as water and methanol, than in less polar solvents, such as ether and chloroform. In the reaction of 2-chloro-2-methylpropane, for example, a rate increase of 100,000 is observed on going from ethanol (less polar) to water (more polar). The rate increases on going from a hydrocarbon solvent to water are so large they can't be measured accurately.

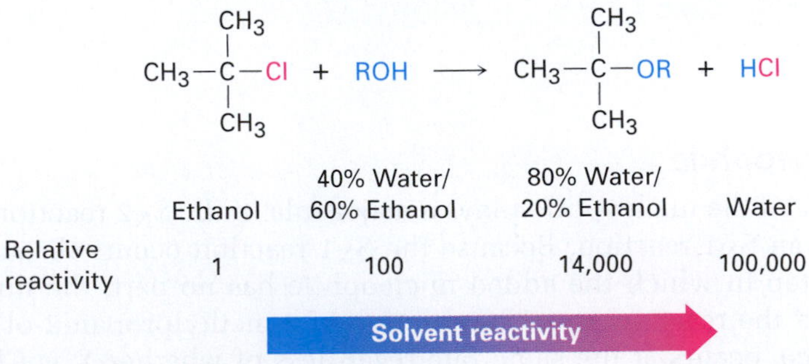

	Ethanol	40% Water/ 60% Ethanol	80% Water/ 20% Ethanol	Water
Relative reactivity	1	100	14,000	100,000

Solvent reactivity

A Summary of S_N1 Reaction Characteristics

The effects on S_N1 reactions of the four variables—substrate, leaving group, nucleophile, and solvent—are summarized in the following statements:

Substrate
The best substrates yield the most stable carbocations. As a result, S_N1 reactions are best for tertiary, allylic, and benzylic halides.

Leaving group

Good leaving groups increase the reaction rate by lowering the energy level of the transition state for carbocation formation.

Nucleophile

The nucleophile must be nonbasic to prevent a competitive elimination of HX (Section 12.12), but otherwise does not affect the reaction rate. Neutral nucleophiles work well.

Solvent

Polar solvents stabilize the carbocation intermediate by solvation, thereby increasing the reaction rate.

WORKED EXAMPLE 12.3	Predicting the Mechanism of a Nucleophilic Substitution Reaction

Predict whether each of the following substitution reactions is likely to be S$_N$1 or S$_N$2:

(a)

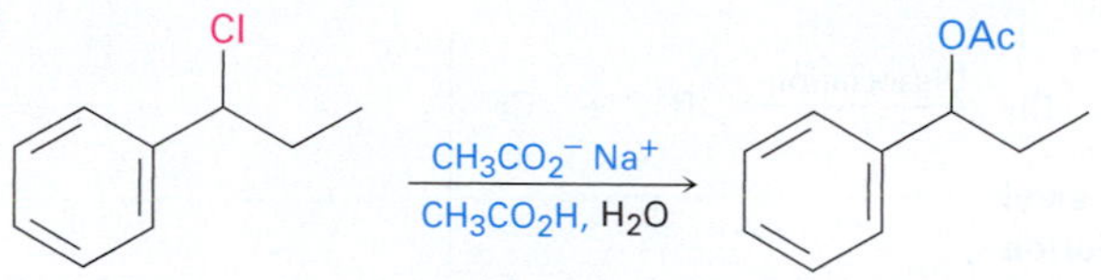

$$\xrightarrow[\text{CH}_3\text{CO}_2\text{H, H}_2\text{O}]{\text{CH}_3\text{CO}_2^- \text{ Na}^+}$$

(b)

$$\xrightarrow[\text{DMF}]{\text{CH}_3\text{CO}_2^- \text{ Na}^+}$$

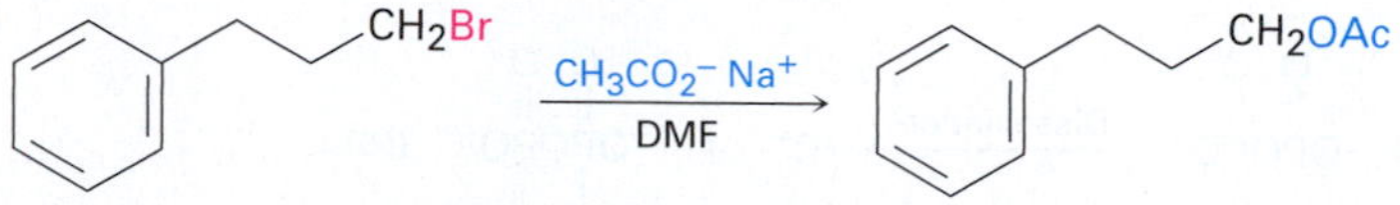

Strategy

Look at the substrate, leaving group, nucleophile, and solvent. Then decide from the summaries at the ends of Sections 12.7 and 12.9 whether an S$_N$1 or an S$_N$2 reaction is favored. S$_N$1 reactions are favored by tertiary, allylic, or benzylic substrates; by good leaving groups; by nonbasic nucleophiles; and by protic solvents. S$_N$2 reactions are favored by primary substrates, by good leaving groups, by good nucleophiles, and by polar aprotic solvents.

Solution

(a) This is likely to be an S$_N$1 reaction because the substrate is secondary and benzylic, the nucleophile is weakly basic, and the solvent is protic.

(b) This is likely to be an S$_N$2 reaction because the substrate is primary, the nucleophile is a reasonably good one, and the solvent is polar aprotic.

Problem 12.21

Predict whether each of the following substitution reactions is likely to be S$_N$1 or S$_N$2:

(a)

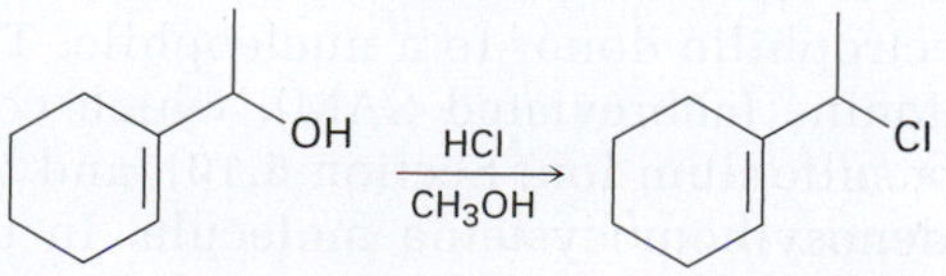

$$\xrightarrow[\text{CH}_3\text{OH}]{\text{HCl}}$$

(b)

$$\text{H}_2\text{C}=\overset{\overset{\displaystyle \text{CH}_3}{|}}{\text{C}}\text{CH}_2\text{Br} \xrightarrow[\text{CH}_3\text{CN}]{\text{Na}^+ \ ^-\text{SCH}_3} \text{H}_2\text{C}=\overset{\overset{\displaystyle \text{CH}_3}{|}}{\text{C}}\text{CH}_2\text{SCH}_3$$

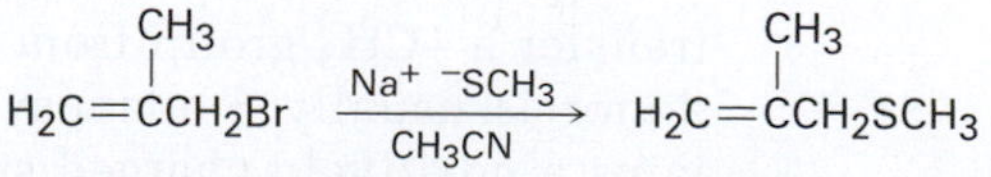

12.10 Biological Substitution Reactions

Both S_N1 and S_N2 reactions are well known in biological chemistry, particularly in the pathways for biosynthesis of the many thousands of terpenes (Chapter 7 *Lagniappe*). Unlike what typically happens in the laboratory, however, the substrate in a biological substitution reaction is often an organodiphosphate rather than an alkyl halide. Thus, the leaving group is the diphosphate ion, abbreviated PP$_i$, rather than a halide ion. In fact, it's useful to think of the diphosphate group as the "biological equivalent" of a halogen. You might recall from Section 9.7, for instance, that organodiphosphates can react with aromatic compounds in Friedel–Crafts-like alkylation reactions much as alkyl halides do. You might also recall from Section 9.7 that the dissociation of an organodiphosphate in a biological reaction is typically assisted by complexation to a divalent metal cation such as Mg^{2+} to help neutralize charge.

Two S_N1 reactions occur during the biosynthesis of geraniol, a fragrant alcohol found in roses and used in perfumery. Geraniol biosynthesis begins with dissociation of dimethylallyl diphosphate to give an allylic carbocation, which reacts with isopentenyl diphosphate (Figure 12.16). From the viewpoint of isopentenyl diphosphate, the reaction is an electrophilic alkene addition, but from the viewpoint of dimethylallyl diphosphate, the process is an S_N1 reaction in which the carbocation intermediate reacts with a double bond as the nucleophile.

Following this initial S_N1 reaction, loss of the *pro-R* hydrogen gives geranyl diphosphate, itself an allylic diphosphate that dissociates a second time. Reaction of the geranyl carbocation with water in a second S_N1 reaction, followed by loss of a proton, then yields geraniol.

S_N2 reactions are involved in almost all biological methylations, which transfer a –CH$_3$ group from an electrophilic donor to a nucleophile. The donor is usually *S*-adenosylmethionine (abbreviated SAM), which contains a positively charged sulfur (a sulfonium ion; Section 5.10), and the leaving group is the neutral *S*-adenosylhomocysteine molecule. In the

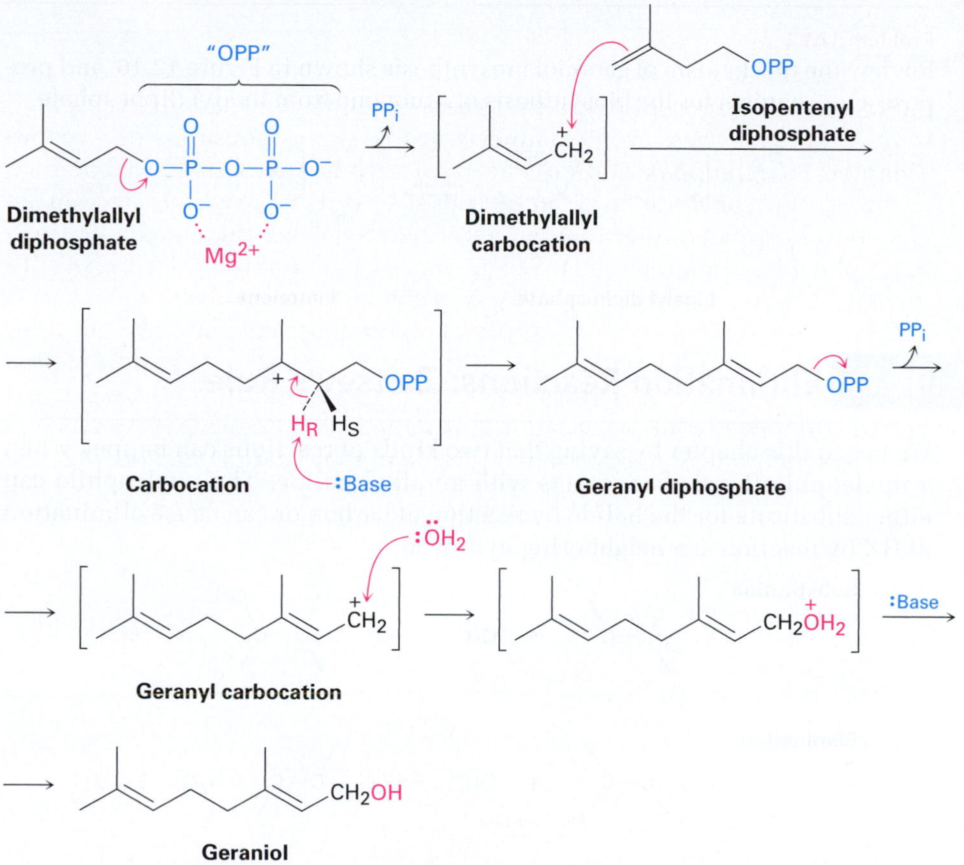

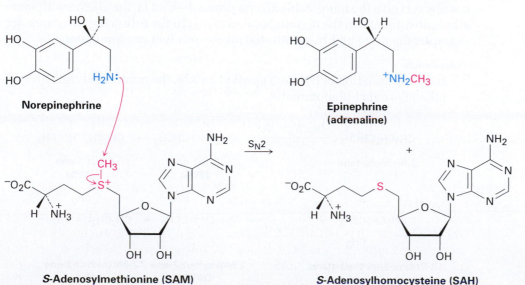

FIGURE 12.16 Biosynthesis of geraniol from dimethylallyl diphosphate. Two $S_{N}1$ reactions occur, both with diphosphate ion as the leaving group.

biosynthesis of epinephrine (adrenaline) from norepinephrine, for instance, the nucleophilic nitrogen atom of norepinephrine attacks the electrophilic methyl carbon atom of *S*-adenosylmethionine in an $S_{N}2$ reaction, displacing *S*-adenosylhomocysteine (Figure 12.17). In effect, *S*-adenosylmethionine is simply a biological equivalent of CH_3Cl.

FIGURE 12.17 The biosynthesis of epinephrine from norepinephrine occurs by an $S_{N}2$ reaction with *S*-adenosylmethionine.

Problem 12.22

Review the mechanism of geraniol biosynthesis shown in Figure 12.16, and propose a mechanism for the biosynthesis of limonene from linalyl diphosphate.

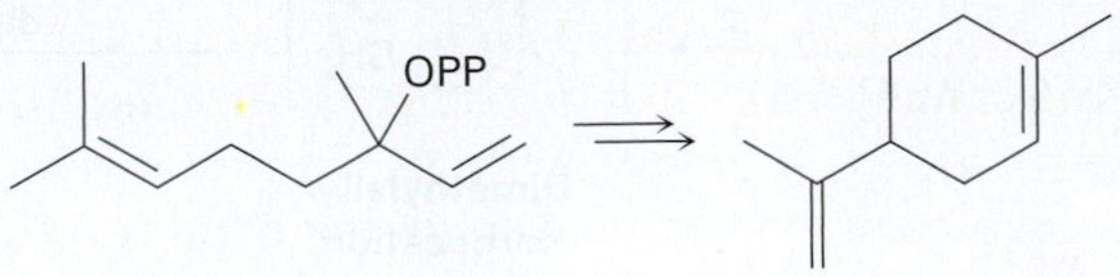

Linalyl diphosphate   **Limonene**

12.11   Elimination Reactions: Zaitsev's Rule

We began this chapter by saying that two kinds of reactions can happen when a nucleophile/Lewis base reacts with an alkyl halide. The nucleophile can either substitute for the halide by reaction at carbon or can cause elimination of HX by reaction at a neighboring hydrogen:

Substitution

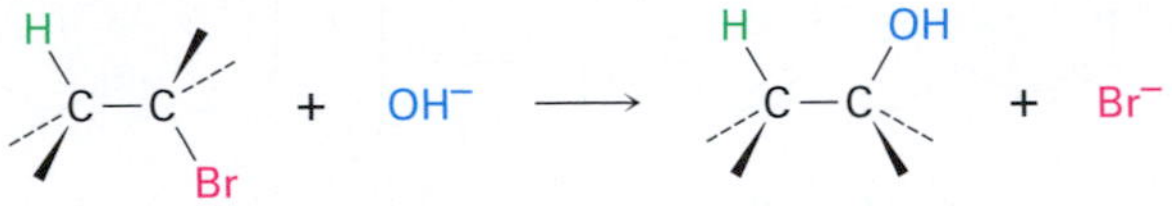

Elimination

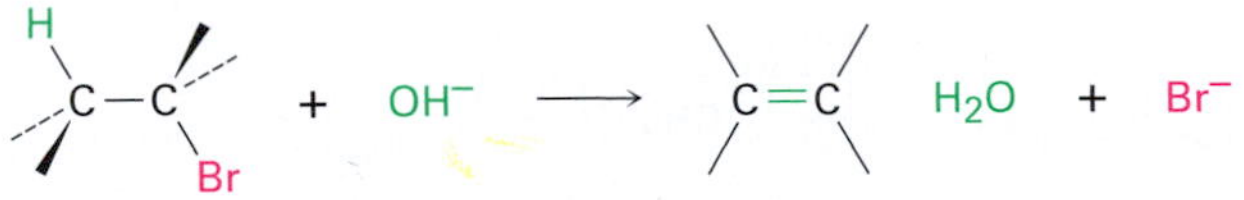

Elimination reactions are more complex than substitution reactions for several reasons. There is, for instance, the problem of regiochemistry: what products result by loss of HX from an unsymmetrical halide? In fact, elimination reactions almost always give mixtures of alkene products, and the best we can usually do is to predict which will be the major product.

According to **Zaitsev's rule**, formulated in 1875 by the Russian chemist Alexander Zaitsev, base-induced elimination reactions generally (although not always) give the more stable alkene product—that is, the alkene with more alkyl substituents on the double-bond carbons. In the following two cases, for example, the more highly substituted alkene product predominates.

Zaitsev's rule

In the elimination of HX from an alkyl halide, the more highly substituted alkene product predominates.

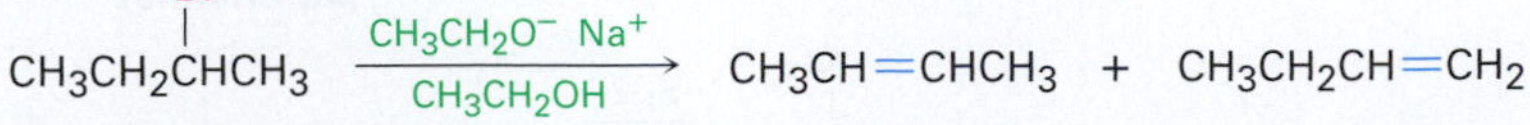

$$CH_3CH_2CHCH_3 \xrightarrow[\text{CH}_3\text{CH}_2\text{OH}]{\text{CH}_3\text{CH}_2\text{O}^-\ \text{Na}^+} CH_3CH=CHCH_3 + CH_3CH_2CH=CH_2$$

Br (on the second carbon)

2-Bromobutane   **But-2-ene**   **But-1-ene**
                                      (81%)   **(19%)**

$$CH_3CH_2CCH_3 \xrightarrow[\text{CH}_3\text{CH}_2\text{OH}]{\text{CH}_3\text{CH}_2\text{O}^-\ \text{Na}^+} CH_3CH=CCH_3 + CH_3CH_2C=CH_2$$

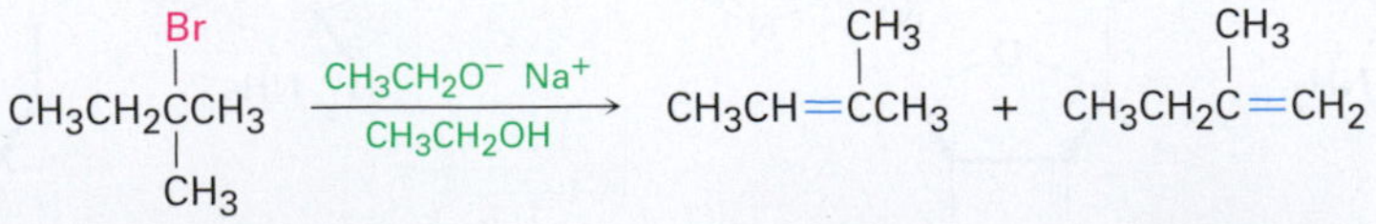

2-Bromo-2-methylbutane   **2-Methylbut-2-ene**   **2-Methylbut-1-ene**
                                    (70%)   **(30%)**

A second factor that complicates a study of elimination reactions is that they can take place by different mechanisms, just as substitutions can. We'll consider three of the most common mechanisms—the E1, E2, and E1cB reactions—which differ in the timing of C–H and C–X bond-breaking. In the E1 reaction, the C–X bond breaks first to give a carbocation intermediate that undergoes subsequent base abstraction of H^+ to yield the alkene. In the E2 reaction, base-induced C–H bond cleavage is simultaneous with C–X bond cleavage, giving the alkene in a single step. In the E1cB reaction (cB for "conjugate base"), base abstraction of the proton occurs first, giving a carbanion ($R:^-$) intermediate. This anion, the conjugate base of the reactant "acid," then undergoes loss of X^- in a subsequent step to give the alkene. All three mechanisms occur frequently in the laboratory, but the E1cB mechanism predominates in biological pathways.

E1 Reaction: C–X bond breaks first to give a carbocation intermediate, followed by base removal of a proton to yield the alkene.

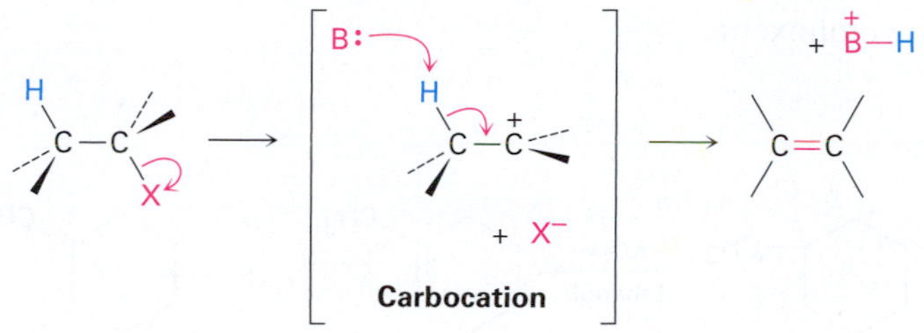

E2 Reaction: C–H and C–X bonds break simultaneously, giving the alkene in a single step without intermediates.

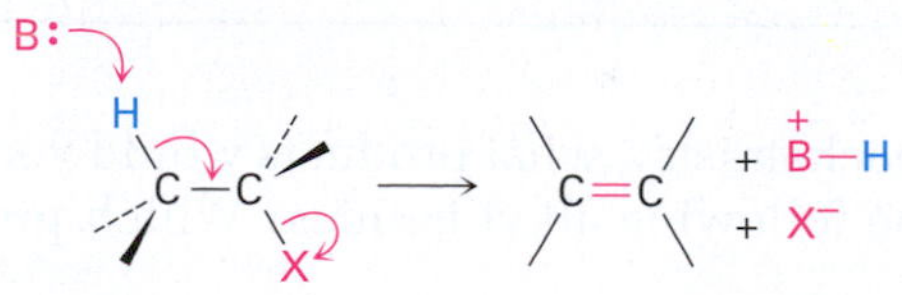

E1cB Reaction: C–H bond breaks first, giving a carbanion intermediate that loses X^- to form the alkene.

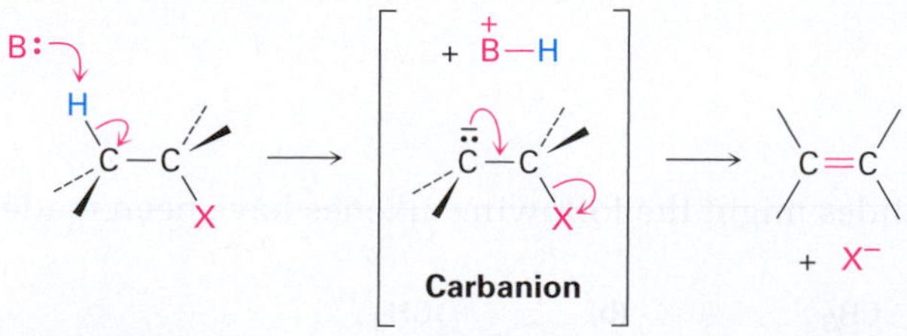

WORKED EXAMPLE 12.4 Predicting the Product of an Elimination Reaction

What product would you expect from reaction of 1-chloro-1-methylcyclohexane with KOH in ethanol?

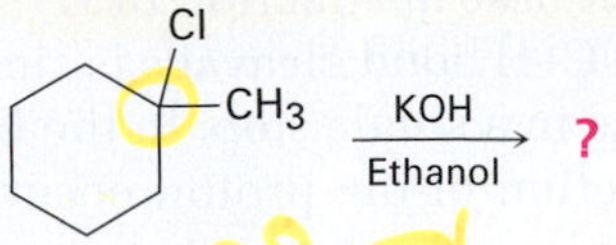
Cl
CH3
KOH
Ethanol
?

Strategy

Treatment of an alkyl halide with a strong base such as KOH yields an alkene. To find the products in a specific case, locate the hydrogen atoms on each carbon next to the leaving group and then generate the potential alkene products by removing HX in as many ways as possible. The major product will be the one that has the most highly substituted double bond—in this case, 1-methylcyclohexene.

Solution

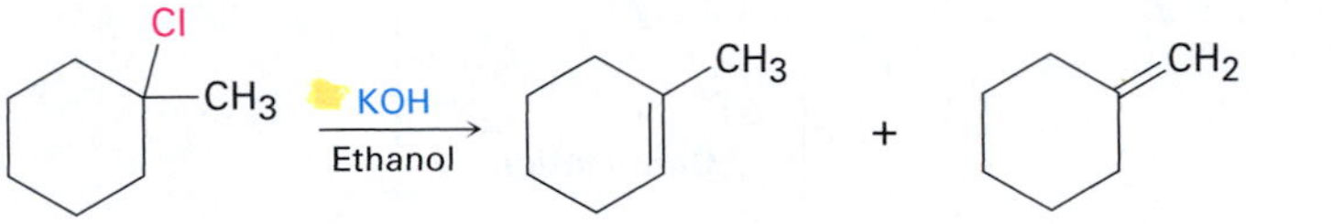
Cl
CH3
KOH
Ethanol
CH3
+
CH2

1-Chloro-1-methylcyclohexane | **1-Methylcyclohexene (major)** | **Methylenecyclohexane (minor)**

Problem 12.23

Ignoring double-bond stereochemistry, what products would you expect from elimination reactions of the following alkyl halides? Which product will be major in each case?

(a)

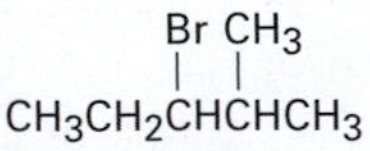
Br CH3
CH3CH2CHCHCH3

(b)

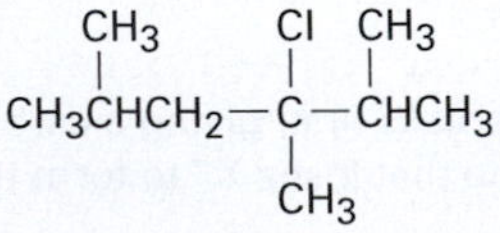
CH3
Cl CH3
CH3CHCH2—C—CHCH3
CH3

(c)

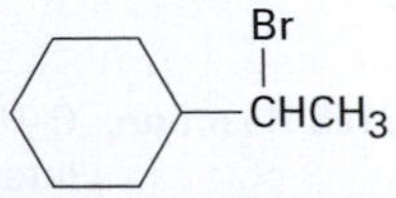
Br
CHCH3

Problem 12.24

What alkyl halides might the following alkenes have been made from?

(a)

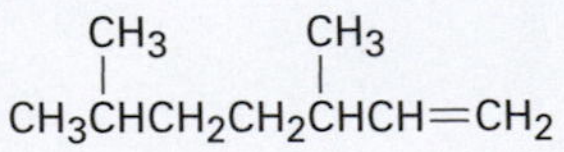
CH3
CH3
CH3CHCH2CH2CHCH=CH2

(b)

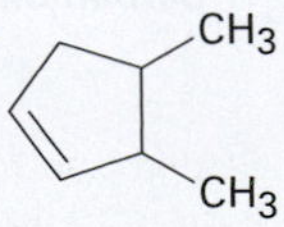
CH3
CH3

12.12 The E2 Reaction

The **E2 reaction** (for *elimination, bimolecular*) occurs when an alkyl halide is treated with a strong base, such as hydroxide ion or alkoxide ion (RO^-). It is the most commonly occurring pathway for elimination in the laboratory and can be formulated as shown in Figure 12.18.

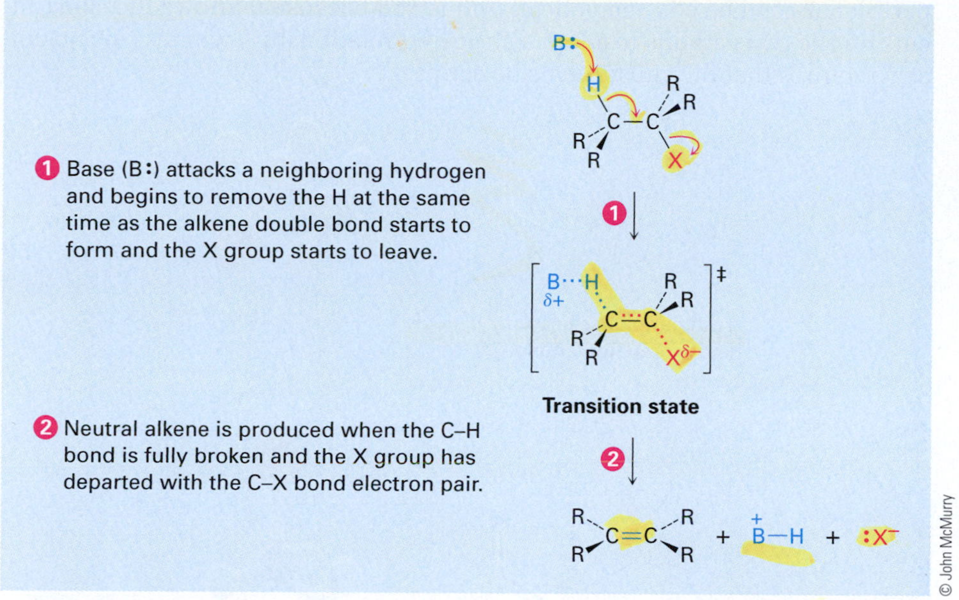

1 Base (B:) attacks a neighboring hydrogen and begins to remove the H at the same time as the alkene double bond starts to form and the X group starts to leave.

Transition state

2 Neutral alkene is produced when the C–H bond is fully broken and the X group has departed with the C–X bond electron pair.

FIGURE 12.18 MECHANISM: Mechanism of the E2 reaction of an alkyl halide. The reaction takes place in a single step through a transition state in which the double bond begins to form at the same time the H and X groups are leaving.

© John McMurry

Like the S_N2 reaction, the E2 reaction takes place in one step without intermediates. As the base begins to abstract H^+ from a carbon next to the leaving group, the C–H bond begins to break, a C=C bond begins to form, and the leaving group begins to depart, taking with it the electron pair from the C–X bond. Among the pieces of evidence supporting this mechanism is that E2 reactions show second-order kinetics and follow the rate law: rate = $k \times [RX] \times [Base]$. That is, both base and alkyl halide take part in the rate-limiting step.

A second piece of evidence in support of the E2 mechanism is provided by a phenomenon known as the *deuterium isotope effect*. For reasons that we won't go into, a carbon–*hydrogen* bond is weaker by about 5 kJ/mol (1.2 kcal/mol) than the corresponding carbon–*deuterium* bond. Thus, a C–H bond is more easily broken than an equivalent C–D bond, and the rate of C–H bond cleavage is faster. For instance, the base-induced elimination of HBr from 1-bromo-2-phenylethane proceeds 7.11 times as fast as the corresponding elimination of DBr from 1-bromo-2,2-dideuterio-2-phenylethane. This result tells us that the C–H (or C–D) bond is broken *in the rate-limiting step*, consistent with our picture of the E2 reaction as a one-step process. If it were otherwise, we couldn't measure a rate difference.

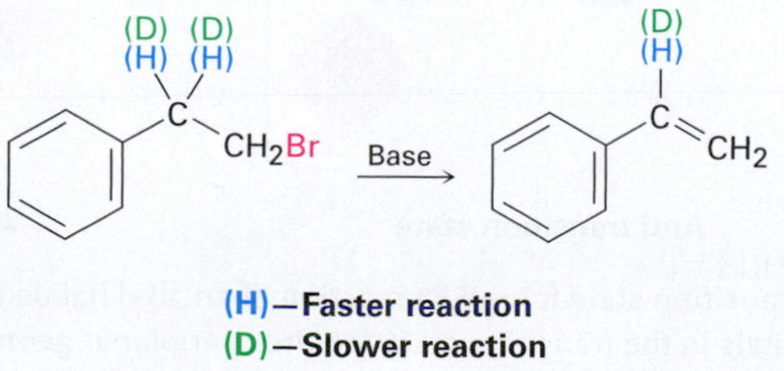

(H)—Faster reaction
(D)—Slower reaction

Yet a third piece of mechanistic evidence involves the stereochemistry of E2 eliminations. As shown by a large number of experiments, E2 reactions occur with *periplanar* geometry, meaning that all four reacting atoms—the hydrogen, the two carbons, and the leaving group—lie in the same plane. Two such geometries are possible: **syn periplanar** geometry, in which the H and the X are on the same side of the molecule, and **anti periplanar** geometry, in which the H and the X are on opposite sides of the molecule. Of the two, anti periplanar geometry is energetically preferred because it allows the substituents on the two carbons to adopt a staggered relationship, whereas syn geometry requires that the substituents be eclipsed.

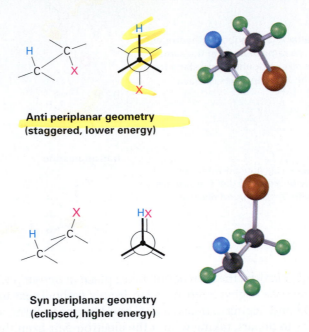

**Anti periplanar geometry
(staggered, lower energy)**

**Syn periplanar geometry
(eclipsed, higher energy)**

What's so special about periplanar geometry? Because the sp^3 σ orbitals in the reactant C–H and C–X bonds must overlap and become p π orbitals in the alkene product, there must also be some overlap in the transition state. This can occur most easily if all the orbitals are in the same plane to begin with— that is, if they're periplanar (Figure 12.19).

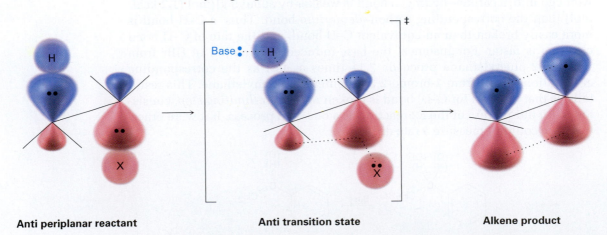

Anti periplanar reactant   Anti transition state   Alkene product

FIGURE 12.19 The transition state for the E2 reaction of an alkyl halide with base. Overlap of the developing *p* orbitals in the transition state requires periplanar geometry of the reactant.

Anti periplanar geometry for E2 eliminations has specific stereochemical consequences. To take just one example, *meso*-1,2-dibromo-1,2-diphenyl-ethane undergoes E2 elimination on treatment with base to give only the *E* alkene. None of the isomeric *Z* alkene is formed because the transition state leading to the *Z* alkene would have to have syn periplanar geometry and thus be higher in energy.

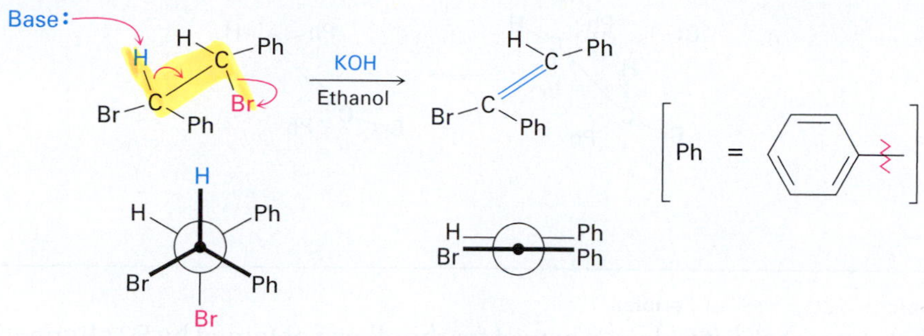

meso-1,2-Dibromo-1,2-diphenylethane (anti periplanar geometry)

(E)-1-Bromo-1,2-diphenylethylene

Anti periplanar geometry is particularly important in cyclohexane rings, where chair geometry forces a rigid relationship between substituents on adjacent carbon atoms (Section 4.6). Only if the hydrogen and the leaving group are trans diaxial can an E2 reaction occur (Figure 12.20). If either the leaving group or the hydrogen is equatorial, E2 elimination can't occur.

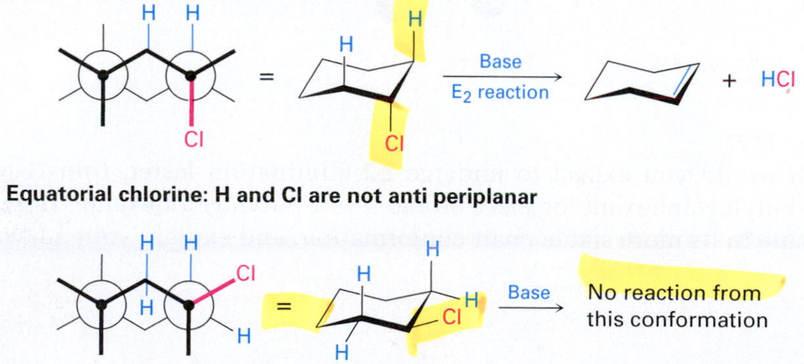

Axial chlorine: H and Cl are anti periplanar

Equatorial chlorine: H and Cl are not anti periplanar

No reaction from this conformation

FIGURE 12.20 The geometric requirement for E2 reaction in a substituted cyclohexane. The hydrogen and the leaving group must both be axial for anti periplanar elimination to occur.

WORKED EXAMPLE 12.5   **Predicting the Stereochemistry of an E2 Reaction**

What stereochemistry do you expect for the alkene obtained by E2 elimination of (1*S*,2*S*)-1,2-dibromo-1,2-diphenylethane?

Strategy

Draw (1*S*,2*S*)-1,2-dibromo-1,2-diphenylethane so that you can see its stereochemistry and so that the –H and –Br groups to be eliminated are anti periplanar. Then carry out the elimination while keeping all substituents in approximately their same positions, and see what alkene results.

Solution

Anti periplanar elimination of HBr gives (*Z*)-1-bromo-1,2-diphenylethylene.

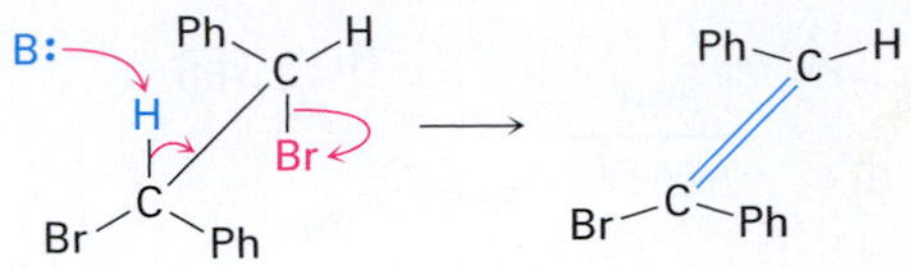

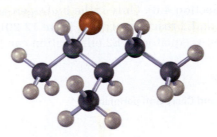

Problem 12.25

What stereochemistry do you expect for the alkene obtained by E2 elimination of (1*R*,2*R*)-1,2-dibromo-1,2-diphenylethane? Draw a Newman projection of the reacting conformation.

Problem 12.26

What stereochemistry do you expect for the trisubstituted alkene obtained by E2 elimination of the following alkyl halide on treatment with KOH? (Reddish brown = Br.)

Problem 12.27

Which would you expect to undergo E2 elimination faster, *trans*-1-bromo-4-*tert*-butylcyclohexane or *cis*-1-bromo-4-*tert*-butylcyclohexane? Draw each molecule in its more stable chair conformation, and explain your answer.

12.13 The E1 and E1cB Reactions

The E1 Reaction

Just as the E2 reaction is analogous to the S$_N$2 reaction, the S$_N$1 reaction has a close analog called the **E1 reaction** (for *elimination, unimolecular*). The E1 reaction can be formulated as shown in Figure 12.21 for the elimination of HCl from 2-chloro-2-methylpropane.

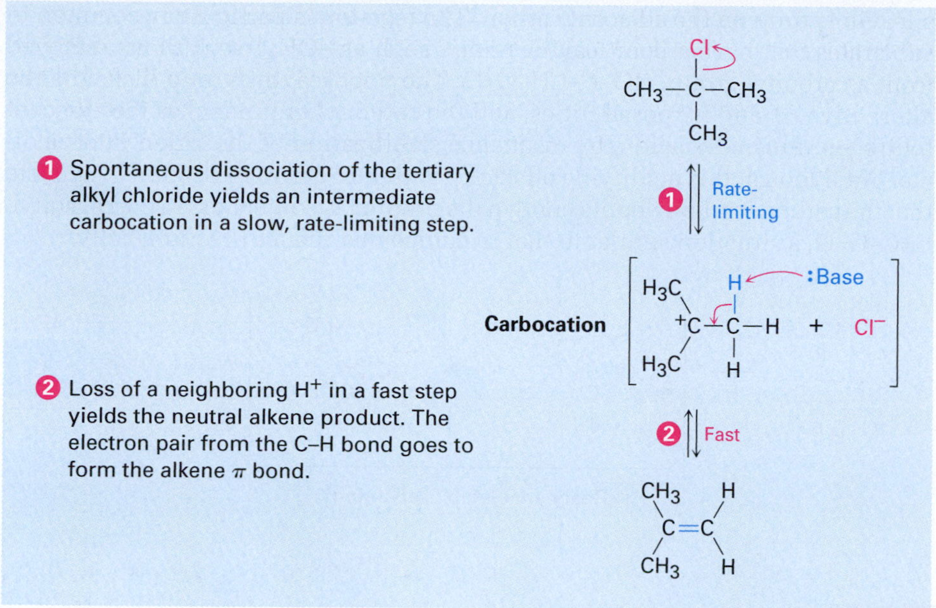

FIGURE 12.21 MECHANISM: Mechanism of the E1 reaction. Two steps are involved, the first of which is rate-limiting, and a carbocation intermediate is present.

① Spontaneous dissociation of the tertiary alkyl chloride yields an intermediate carbocation in a slow, rate-limiting step.

Carbocation

② Loss of a neighboring H+ in a fast step yields the neutral alkene product. The electron pair from the C–H bond goes to form the alkene π bond.

© John McMurry

E1 eliminations begin with the same unimolecular dissociation we saw in the S_N1 reaction, but the dissociation is followed by loss of H+ from the adjacent carbon rather than by substitution. In fact, the E1 and S_N1 reactions normally occur together whenever an alkyl halide is treated in a protic solvent with a nonbasic nucleophile. Thus, the best E1 substrates are also the best S_N1 substrates, and mixtures of substitution and elimination products are usually obtained. For example, when 2-chloro-2-methylpropane is warmed to 65 °C in 80% aqueous ethanol, a 64:36 mixture of 2-methylpropan-2-ol (S_N1) and 2-methylpropene (E1) results:

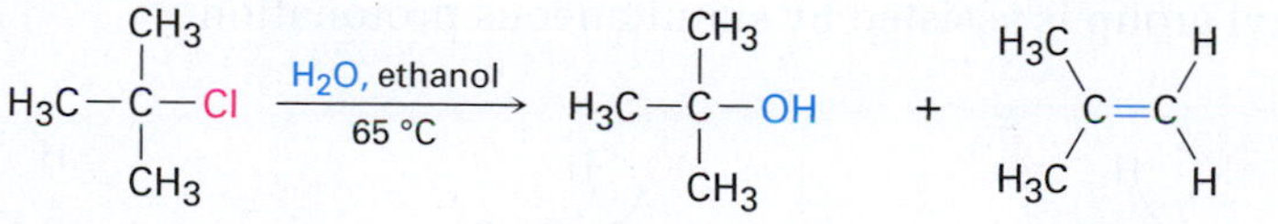

2-Chloro-2-methylpropane   **2-Methylpropan-2-ol** (64%)   **2-Methylpropene** (36%)

Much evidence has been obtained in support of the E1 mechanism. For example, E1 reactions show first-order kinetics, consistent with a rate-limiting, unimolecular dissociation process. Furthermore, E1 reactions show no deuterium isotope effect. Because rupture of the C–H (or C–D) bond occurs *after* the rate-limiting step rather than during it, we can't measure a rate difference between a deuterated and nondeuterated substrate. A final piece of evidence is that there is no geometric requirement on the E1 reaction because the halide and the hydrogen are lost in separate steps, unlike the E2 reaction, where anti periplanar geometry is required.

The E1cB Reaction

In contrast to the E1 reaction, which involves a carbocation intermediate, the **E1cB reaction** takes place through a *carbanion* intermediate. Base-induced abstraction of a proton in a slow, rate-limiting step gives an anion, which expels a leaving group on the adjacent carbon. The reaction is particularly common in substrates that have a poor leaving group, such as —OH, two carbons removed from a carbonyl group, HO–C–CH–C=O. The poor leaving group disfavors the alternative E1 and E2 possibilities, and the carbonyl group makes the adjacent hydrogen unusually acidic by resonance stabilization of the anion intermediate. We'll look at this acidifying effect of a carbonyl group in Section 17.4. Note that the carbon–carbon double bond in the product is conjugated to the carbonyl, C=C–C=O, a situation similar to that in conjugated dienes (Section 8.12).

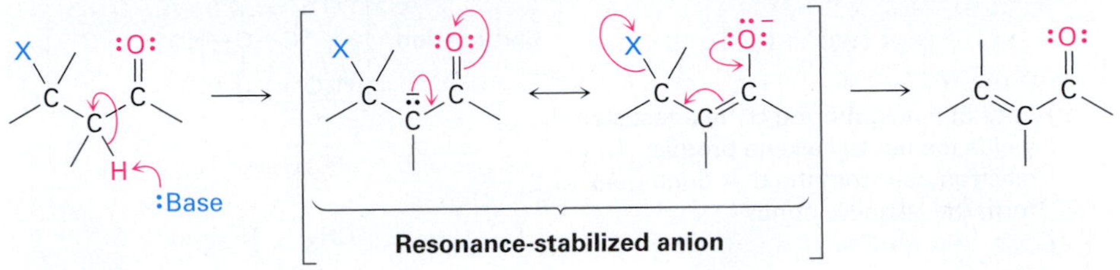

Resonance-stabilized anion

12.14 Biological Elimination Reactions

All three elimination reactions—E2, E1, and E1cB—occur in biological pathways, but the E1cB mechanism is particularly common. The substrate is usually an alcohol, and the H atom removed is usually adjacent to a carbonyl group, just as in laboratory reactions. Thus, 3-hydroxy carbonyl compounds are frequently converted to conjugated unsaturated carbonyl compounds by elimination reactions. A typical example occurs during the biosynthesis of fats when a 3-hydroxybutyryl thioester is dehydrated to the corresponding unsaturated (crotonyl) thioester. The base in this reaction is the imidazole ring (Section 9.4) of a histidine amino acid in the enzyme, and loss of the hydroxyl group is assisted by simultaneous protonation.

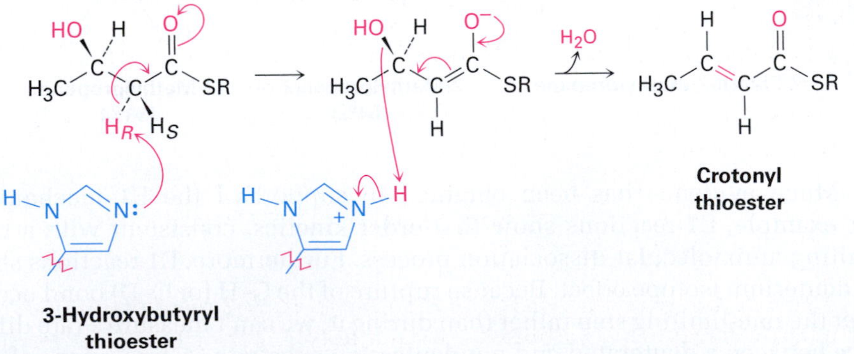

Crotonyl thioester

3-Hydroxybutyryl thioester

12.15 A Summary of Reactivity: S_N1, S_N2, E1, E1cB, and E2

S_N1, S_N2, E1, E1cB, E2—how can you keep it all straight and predict what will happen in any given case? Will substitution or elimination occur? Will the reaction be bimolecular or unimolecular? There are no rigid answers to these questions, but it's possible to recognize some trends and make some generalizations.

- **Primary alkyl halides** S$_N$2 substitution occurs if a good nucleophile is used, E2 elimination occurs if a strong, sterically hindered base is used, and E1cB elimination occurs if the leaving group is two carbons away from a carbonyl group.

- **Secondary alkyl halides** S$_N$2 substitution occurs if a weakly basic nucleophile is used in a polar aprotic solvent, E2 elimination predominates if a strong base is used, and E1cB elimination takes place if the leaving group is two carbons away from a carbonyl group. Secondary allylic and benzylic alkyl halides can also undergo S$_N$1 and E1 reactions if a weakly basic nucleophile is used in a protic solvent.

- **Tertiary alkyl halides** E2 elimination occurs when a base is used, but S$_N$1 substitution and E1 elimination occur together under neutral conditions, such as in pure ethanol or water. E1cB elimination takes place if the leaving group is two carbons away from a carbonyl group.

WORKED EXAMPLE 12.6 Predicting the Product and Mechanism of a Reaction

Tell whether each of the following reactions is likely to be S$_N$1, S$_N$2, E1, E1cB, or E2, and predict the product of each:

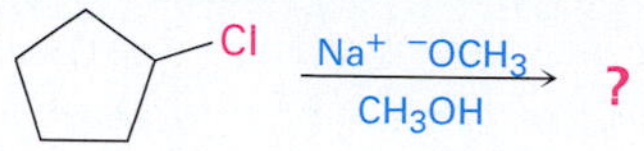

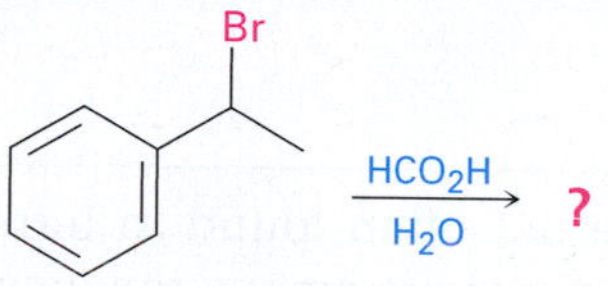

Strategy

Look carefully in each reaction at the substrate, leaving group, nucleophile, and solvent. Then decide from the preceding summary which kind of reaction is likely to be favored.

Solution

(a) A secondary, nonallylic substrate can undergo an S$_N$2 reaction with a good nucleophile in a polar aprotic solvent but will undergo an E2 reaction on treatment with a strong base in a protic solvent. In this case, E2 reaction is likely to predominate.

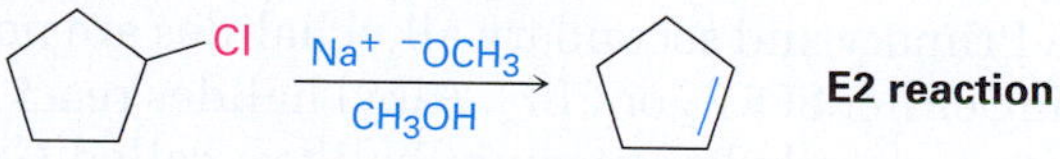

(b) A secondary benzylic substrate can undergo an S$_N$2 reaction on treatment with a nonbasic nucleophile in a polar aprotic solvent and will undergo an E2 reaction on treatment with a base. Under protic conditions, such as aqueous formic acid (HCO$_2$H), an S$_N$1 reaction is likely, along with some E1 reaction.

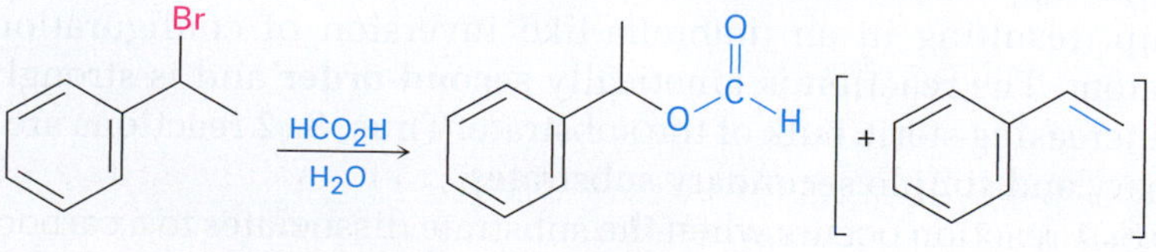

Problem 12.28

Tell whether each of the following reactions is likely to be S_N1, S_N2, E1, E1cB, or E2:

(a) $CH_3CH_2CH_2CH_2Br \xrightarrow[\text{THF}]{\text{NaN}_3} CH_3CH_2CH_2CH_2N{=}N{=}N$

(b)

$CH_3CH_2CHCH_2CH_3 \xrightarrow[\text{Ethanol}]{\text{KOH}} CH_3CH_2CH{=}CHCH_3$

with Cl substituent

(c)

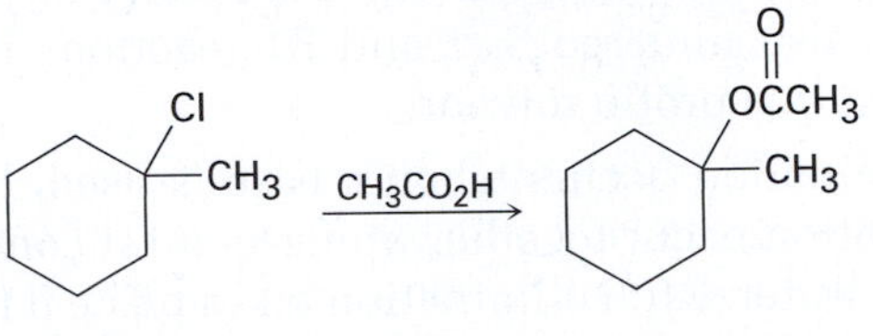

$\xrightarrow{CH_3CO_2H}$

(d)

$\xrightarrow[\text{Ethanol}]{\text{NaOH}}$

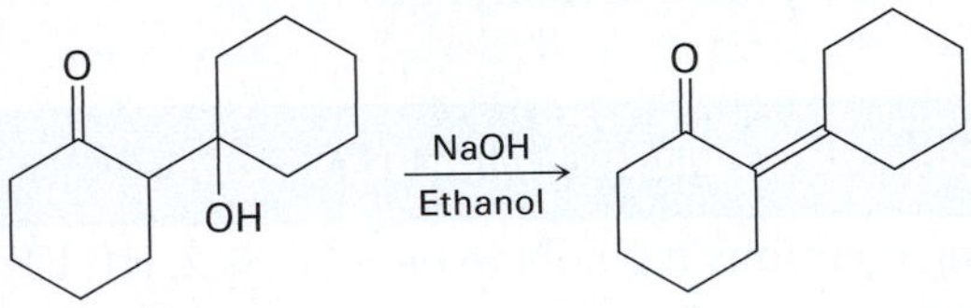

Summary

Key Words

alkyl halide, 445

anti periplanar, 482

carbanion, 453

E1 reaction, 484

E1cB reaction, 485

E2 reaction, 481

first-order reaction, 467

Grignard reagent (RMgX), 453

kinetics, 457

nucleophilic substitution reaction, 455

organohalide, 444

second-order reaction, 458

S_N1 reaction, 467

S_N2 reaction, 458

solvation, 465

syn periplanar, 482

Zaitsev's rule, 478

Alkyl halides are not often found in terrestrial organisms, but the kinds of reactions they undergo are among the most important and well-studied reaction types in organic chemistry. In this chapter, we saw how to name and prepare alkyl halides, and we made a detailed study of their substitution and elimination reactions.

Alkyl halides contain a halogen bonded to a saturated, sp^3-hybridized carbon atom. The C–X bond is polar, and alkyl halides can therefore behave as electrophiles. Alkyl halides can be prepared from alkenes by reaction with *N*-bromosuccinimide (NBS) to give the product of allylic bromination. The NBS bromination of an alkene takes place through an intermediate allylic radical, which is stabilized by resonance. Alkyl halides are also prepared from alcohols by reaction with HX, but the process works well only for tertiary alcohols, R_3COH. Primary and secondary alkyl halides are normally prepared from alcohols using either $SOCl_2$ or PBr_3. Alkyl halides react with magnesium in ether solution to form alkylmagnesium halides, called **Grignard reagents (RMgX)**. Because Grignard reagents are both nucleophilic and basic, they react with acids to yield hydrocarbons.

The reaction of an alkyl halide or tosylate with a nucleophile/base results either in *substitution* or in *elimination*. Nucleophilic substitutions are of two types: **S_N2 reactions** and **S_N1 reactions**. In the S_N2 reaction, the entering nucleophile approaches the halide from a direction 180° away from the leaving group, resulting in an umbrella-like inversion of configuration at the carbon atom. The reaction is kinetically **second-order** and is strongly inhibited by increasing steric bulk of the substrate. Thus, S_N2 reactions are favored for primary and simple secondary substrates.

The S_N1 reaction occurs when the substrate dissociates to a carbocation in a slow rate-limiting step, followed by a rapid reaction with the nucleophile.

As a result, S_N1 reactions are kinetically **first-order** and take place with substantial racemization of configuration at the carbon atom. They are most favored for tertiary substrates. Both S_N1 and S_N2 reactions occur in biological pathways, although the leaving group is typically a diphosphate ion rather than a halide.

Eliminations of alkyl halides commonly occur by three mechanisms, **E2 reactions**, **E1 reactions**, and **E1cB reactions**, which differ in the timing of C–H and C–X bond-breaking. In the E2 reaction, C–H and C–X bond-breaking occur simultaneously when a base abstracts H^+ from one carbon at the same time the leaving group departs from the neighboring carbon. The reaction takes place preferentially through an **anti periplanar** transition state in which the four reacting atoms—hydrogen, two carbons, and leaving group—are in the same plane. The reaction shows second-order kinetics and a deuterium isotope effect and occurs when a secondary or tertiary substrate is treated with a strong base. These elimination reactions usually give a mixture of alkene products in which the more highly substituted alkene predominates (**Zaitsev's rule**).

In the E1 reaction, C–X bond-breaking occurs first. The substrate dissociates to yield a carbocation in the slow rate-limiting step before losing H^+ from an adjacent carbon in a second step. The reaction shows first-order kinetics and occurs with tertiary substrates in polar, nonbasic solution.

In the E1cB reaction, C–H bond-breaking occurs first. A base abstracts a proton to give an anion, followed by loss of the leaving group from the adjacent carbon in a second step. The reaction is favored when the leaving group is two carbons removed from a carbonyl, which stabilizes the intermediate anion by resonance. Biological elimination reactions typically occur by this E1cB mechanism.

Summary of Reactions

1.   Preparation of allylic bromides from alkenes (Section 12.2)

2.   Preparation of alkyl halides from alcohols (Section 12.3)

   (a) Reaction with HCl and HBr

   Reactivity order: $3° > 2° > 1°$

   (b) Reaction of 1° and 2° alcohols with $SOCl_2$

(c) Reaction of 1° and 2° alcohols with PBr_3

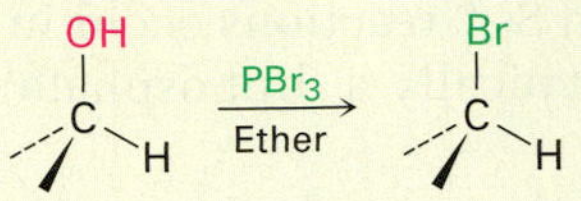

$$\text{C(OH)(H)} \xrightarrow[\text{Ether}]{PBr_3} \text{C(Br)(H)}$$

3.   Formation of Grignard reagents from alkyl halides (Section 12.4)

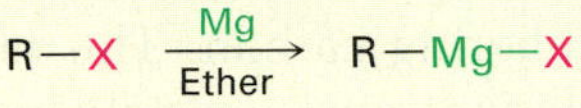

$$R-X \xrightarrow[\text{Ether}]{Mg} R-Mg-X$$

4.   Nucleophilic substitutions

(a) S_N1 reaction of 3°, allylic, and benzylic halides (Sections 12.8 and 12.9)

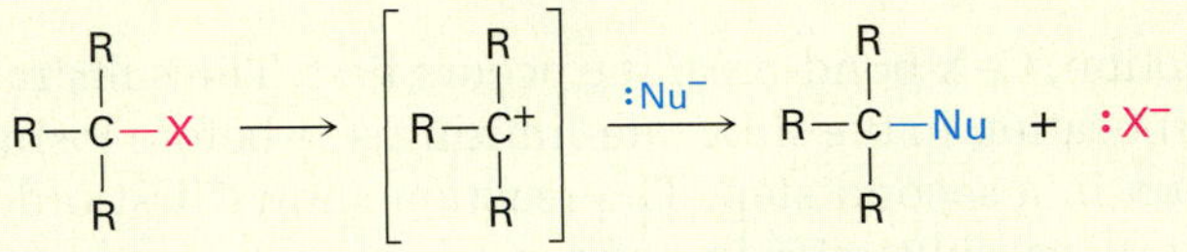

$$R-\underset{R}{\overset{R}{C}}-X \longrightarrow \left[R-\underset{R}{\overset{R}{C^+}} \right] \xrightarrow{:Nu^-} R-\underset{R}{\overset{R}{C}}-Nu \ + \ :X^-$$

(b) S_N2 reaction of 1° and simple 2° halides (Sections 12.6 and 12.7)

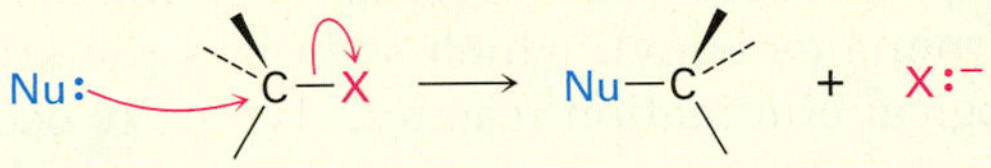

$$Nu: \quad C-X \longrightarrow Nu-C \ + \ X:^-$$

5.   Eliminations

(a) E1 reaction (Section 12.13)

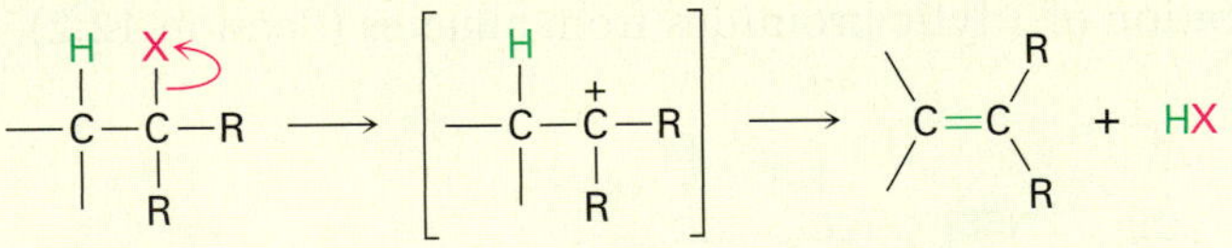

$$-\underset{R}{\overset{H \ X}{\underset{|}{C}-\underset{|}{C}}}-R \longrightarrow \left[-\underset{R}{\overset{H}{\underset{|}{C}-\overset{+}{\underset{|}{C}}}}-R \right] \longrightarrow \overset{R}{\underset{R}{C=C}} \ + \ HX$$

(b) E1cB reaction (Section 12.13)

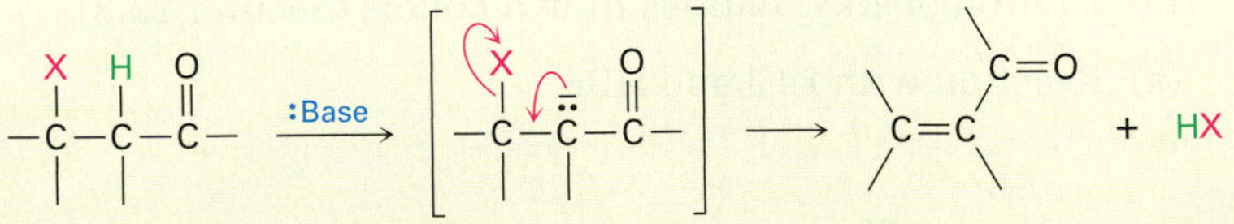

$$-\overset{X \ H \ O}{\underset{|}{C}-\underset{|}{C}-\overset{\|}{C}-} \xrightarrow{:Base} \left[-\overset{X \ \ O}{\underset{|}{C}-\ddot{C}-\overset{\|}{C}-} \right] \longrightarrow \overset{C=O}{C=C} \ + \ HX$$

(c) E2 reaction (Section 12.12)

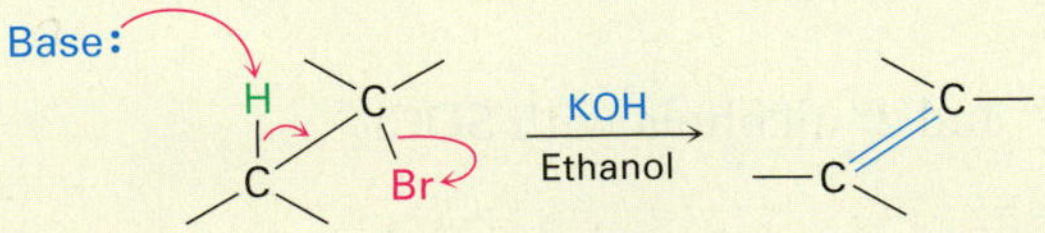

$$\text{Base:} \quad \overset{H}{\underset{C}{\ }} C \quad \overset{KOH}{\underset{\text{Ethanol}}{\longrightarrow}} \quad C=C$$

Lagniappe

Green Chemistry

Organic chemistry in the 20th century gave us new medicines, insecticides, adhesives, textiles, dyes, building materials, composites, and all manner of polymers. But these advances did not come without a cost: every chemical process produces wastes that must be dealt with, including reaction solvents and toxic by-products that might evaporate into the air or be leached into groundwater if not disposed of properly. Even apparently harmless by-products must be safely buried or otherwise sequestered. As always, there's no such thing as a free lunch; with the good also comes the bad.

Let's hope disasters like this are never repeated.

It may never be possible to make organic chemistry completely benign, but awareness of the environmental problems caused by many chemical processes has grown dramatically in recent years, giving rise to a movement called *green chemistry*. Green chemistry is the design and implementation of chemical products and processes that reduce waste and attempt to eliminate the generation of hazardous substances. There are 12 principles of green chemistry:

Prevent waste. Waste should be prevented rather than treated or cleaned up after it has been created.

Maximize atom economy. Synthetic methods should maximize the incorporation of all materials used in a process into the final product so that waste is minimized.

Use less hazardous processes. Synthetic methods should use reactants and generate wastes with minimal toxicity to health and the environment.

Design safer chemicals. Chemical products should be designed to have minimal toxicity.

Use safer solvents. Minimal use should be made of solvents, separation agents, and other auxiliary substances in a reaction.

Design for energy efficiency. Energy requirements for chemical processes should be minimized, with reactions carried out at room temperature if possible.

Use renewable feedstocks. Raw materials should come from renewable sources when feasible.

Minimize derivatives. Syntheses should be designed with minimal use of protecting groups to avoid extra steps and reduce waste.

Use catalysis. Reactions should be catalytic rather than stoichiometric.

Design for degradation. Products should be designed to be biodegradable at the end of their useful lifetimes.

Monitor pollution in real time. Processes should be monitored in real time for the formation of hazardous substances.

Prevent accidents. Chemical substances and processes should minimize the potential for fires, explosions, or other accidents.

The foregoing 12 principles won't all be met in most real-world applications, but they provide a worthy goal to aim for and they can make chemists think more carefully about the environmental implications of their work. Real success stories are already occurring, and more are in progress. Approximately 7 million pounds per year of ibuprofen (6 billion tablets!) is now made by a "green" process that produces approximately 99% less waste than the process it replaces. Only three steps are needed, the anhydrous HF solvent used in the first step is recovered and reused, and the second and third steps are catalytic.

Isobutylbenzene

Ibuprofen

Exercises

■ indicates problems that are assignable in Organic OWL.

Go to this book's companion website at **www.cengage.com/ chemistry/mcmurry** to explore interactive versions of the Active Figures from this text.

VISUALIZING CHEMISTRY

(Problems 12.1–12.28 appear within the chapter.)

12.29 ■ The following alkyl bromide can be prepared by reaction of the alcohol (*S*)-pentan-2-ol with PBr₃. Name the compound, assign (*R*) or (*S*) stereochemistry, and tell whether the reaction of the alcohol occurs with retention or inversion of configuration (reddish brown = Br).

12.30 ■ Write the product you would expect from reaction of each of the following alkyl chlorides with (i) Na⁺ ⁻SCH₃ and (ii) Na⁺ ⁻OH:

(a)  **(b)**  **(c)**

12.31 ■ From what alkyl bromide was the following alkyl acetate made by S_N2 reaction? Write the reaction, showing the stereochemistry.

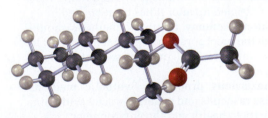

12.32 ■ Assign *R* or *S* configuration to the following chloro alcohol, write the product from S_N2 reaction with NaCN, and assign *R* or *S* configuration to the product:

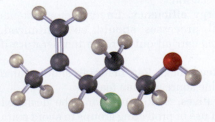

12.33 ■ Draw the structure and assign *Z* or *E* stereochemistry to the product you expect from E2 reaction of the following molecule with NaOH:

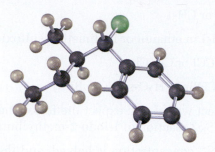

ADDITIONAL PROBLEMS

12.34 ■ Name the following alkyl halides according to IUPAC rules:

(a) 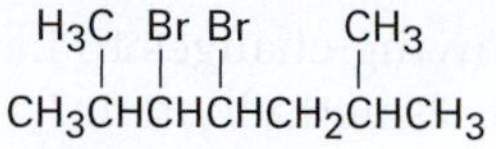   H₃C Br Br   CH₃
      | | | |
   CH₃CHCHCHCH₂CHCH₃

(b) 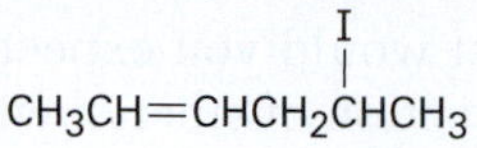   I
                      |
      CH₃CH=CHCH₂CHCH₃

(c) 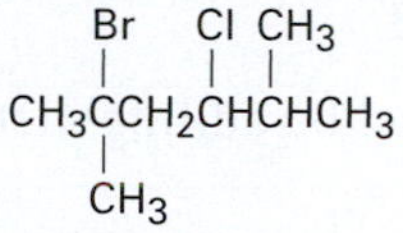   Br   Cl CH₃
         | | |
   CH₃CCH₂CHCHCH₃
         |
         CH₃

(d) 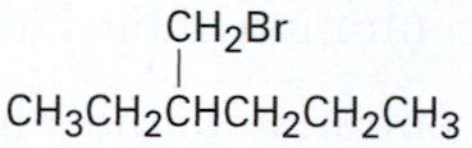   CH₂Br
               |
   CH₃CH₂CHCH₂CH₂CH₃

(e) ClCH₂CH₂CH₂C≡CCH₂Br

12.35 ■ Draw structures corresponding to the following IUPAC names:

(a) 2,3-Dichloro-4-methylhexane

(b) 4-Bromo-4-ethyl-2-methylhexane

(c) 3-Iodo-2,2,4,4-tetramethylpentane

(d) *cis*-1-Bromo-2-ethylcyclopentane

12.36 Why do you suppose it's not possible to prepare a Grignard reagent from a bromo alcohol such as 4-bromopentan-1-ol?

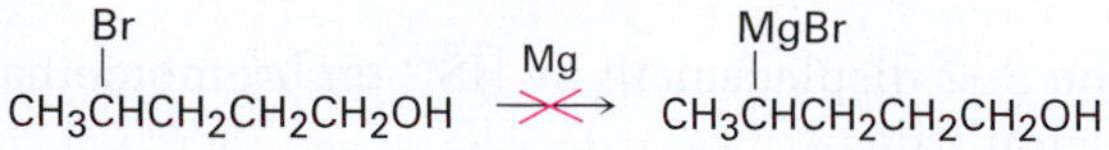

```
      Br                            MgBr
      |                 Mg          |
CH₃CHCH₂CH₂CH₂OH  ──✗──→  CH₃CHCH₂CH₂CH₂OH
```

12.37 What product(s) would you expect from the reaction of 1-methylcyclohexene with NBS?

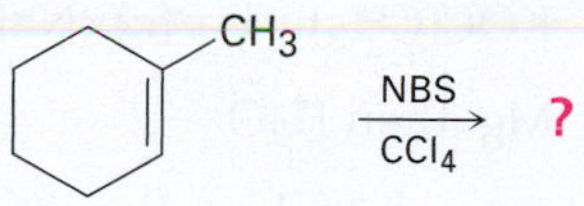

$$\text{1-methylcyclohexene (with CH₃)} \quad \xrightarrow[\text{CCl}_4]{\text{NBS}} \quad ?$$

12.38 ▪ Which compound in each of the following pairs will react faster in an S_N2 reaction with OH^-?

    (a) CH_3Br or CH_3I

    (b) CH_3CH_2I in ethanol or in dimethyl sulfoxide

    (c) $(CH_3)_3CCl$ or CH_3Cl

    (d) $H_2C=CHBr$ or $H_2C=CHCH_2Br$

12.39 ▪ What effect would you expect the following changes to have on the rate of the S_N2 reaction of 1-iodo-2-methylbutane with cyanide ion?

    (a) The CN^- concentration is halved, and the 1-iodo-2-methylbutane concentration is doubled.

    (b) Both the CN^- and the 1-iodo-2-methylbutane concentrations are tripled.

12.40 ▪ What effect would you expect the following changes to have on the rate of the reaction of ethanol with 2-iodo-2-methylbutane?

    (a) The concentration of the halide is tripled.

    (b) The concentration of the ethanol is halved by adding diethyl ether as an inert solvent.

12.41 ▪ How might you prepare each of the following molecules using a nucleophilic substitution reaction at some step?

    (a) $CH_3CH_2CH_2CH_2CN$ **(b)** $H_3C-O-C(CH_3)_3$ **(c)** $CH_3CH_2CH_2NH_2$

12.42 ▪ Which reaction in each of the following pairs would you expect to be faster?

    (a) The S_N2 displacement by I^- on CH_3Cl or on CH_3OTos

    (b) The S_N2 displacement by $CH_3CO_2^-$ on bromoethane or on bromocyclohexane

    (c) The S_N2 displacement on 2-bromopropane by $CH_3CH_2O^-$ or by CN^-

    (d) The S_N2 displacement by HS^- on bromomethane in toluene or in acetonitrile

12.43 ▪ What products would you expect from the reaction of 1-bromopropane with each of the following?

    (a) $NaNH_2$ **(b)** $KOC(CH_3)_3$ **(c)** NaI

    (d) $NaCN$ **(e)** Mg, then H_2O

12.44 ▪ Which reactant in each of the following pairs is more nucleophilic in protic solvents? Explain.

    (a) $^-NH_2$ or NH_3 **(b)** H_2O or $CH_3CO_2^-$ **(c)** BF_3 or F^-

    (d) $(CH_3)_3P$ or $(CH_3)_3N$ **(e)** I^- or Cl^- **(f)** $^-C\equiv N$ or $^-OCH_3$

12.45 The following Walden cycle has been carried out. Explain the results, and indicate where Walden inversion is occurring.

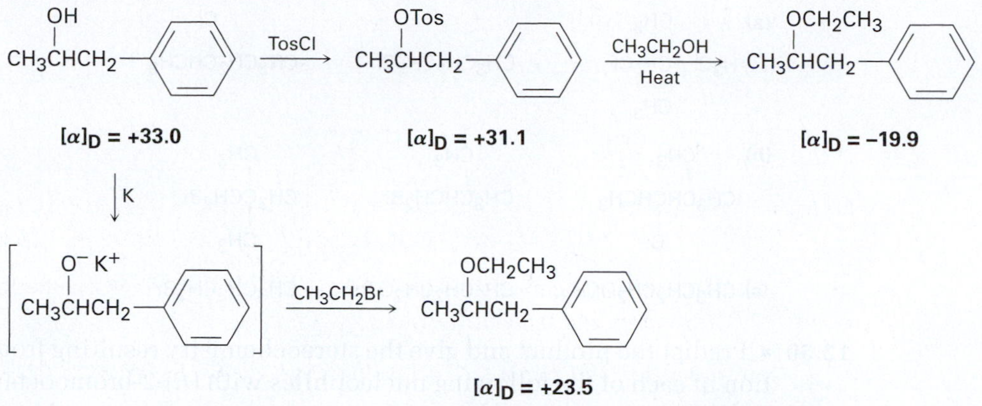

$[\alpha]_D = +33.0$ $[\alpha]_D = +31.1$ $[\alpha]_D = -19.9$

$[\alpha]_D = +23.5$

12.46 Propose structures for compounds that fit the following descriptions:

(a) An alkyl halide that gives a mixture of three alkenes on E2 reaction

(b) An organic halide that will not undergo nucleophilic substitution

(c) An alkyl halide that gives the non-Zaitsev product on E2 reaction

(d) An alcohol that reacts rapidly with HCl at 0 °C

12.47 ▪ The following reactions are unlikely to occur as written. Tell what is wrong with each, and predict the actual product.

(a)

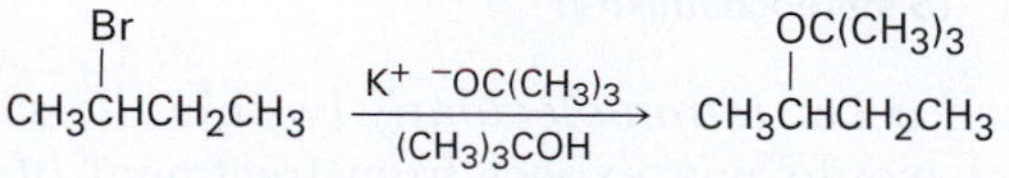

(b)

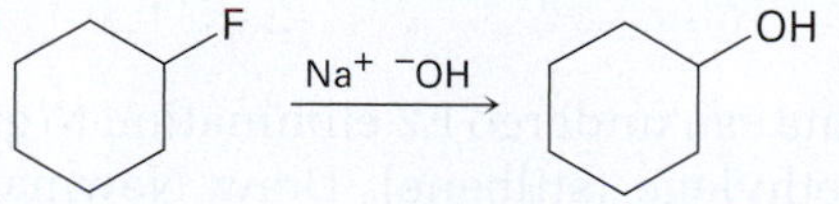

(c)

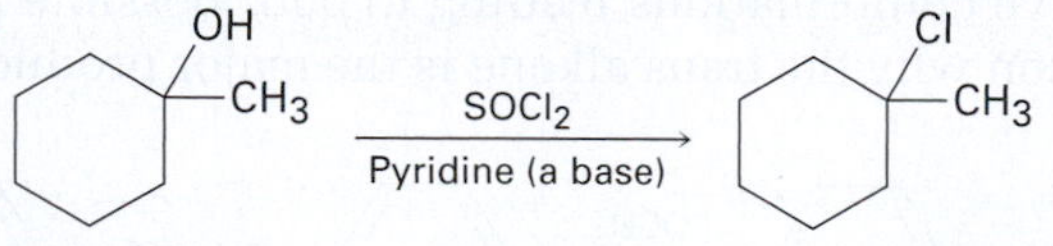

12.48 ▪ Order each of the following sets of compounds with respect to S_N1 reactivity:

(a)

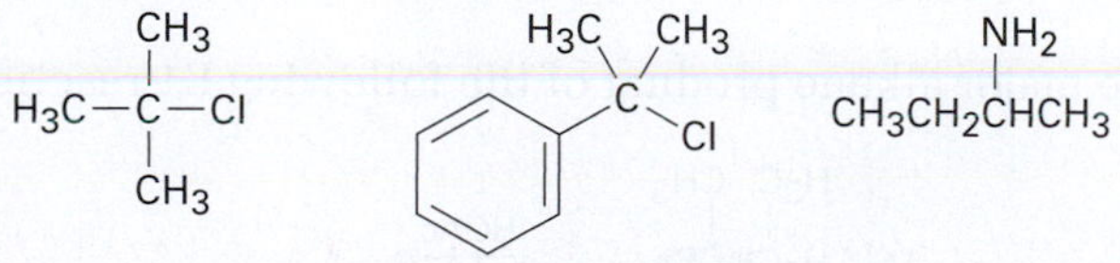

(b) $(CH_3)_3CCl$ $(CH_3)_3CBr$ $(CH_3)_3COH$

(c)

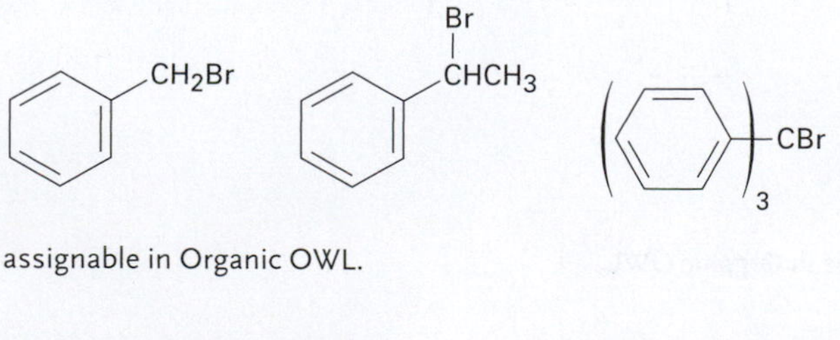

▪ Problems assignable in Organic OWL.

12.49 ■ Order each of the following sets of compounds with respect to S_N2 reactivity:

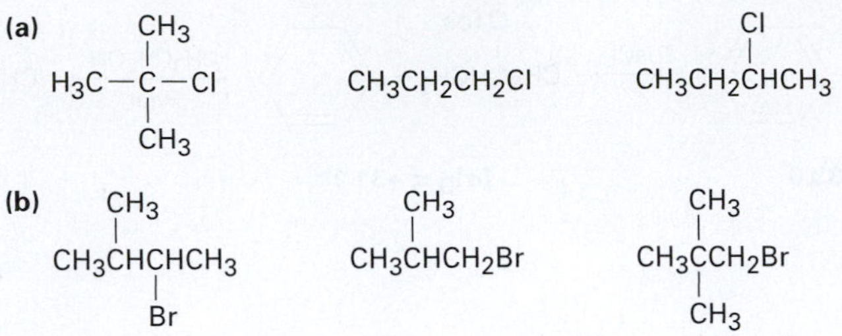

(a)

$$H_3C-\underset{\underset{CH_3}{|}}{\overset{\overset{CH_3}{|}}{C}}-Cl \qquad CH_3CH_2CH_2Cl \qquad CH_3CH_2\underset{\underset{}{}}{\overset{\overset{Cl}{|}}{C}}HCH_3$$

(b)

$$CH_3\underset{\underset{Br}{|}}{\overset{\overset{CH_3}{|}}{C}}HCHCH_3 \qquad CH_3\overset{\overset{CH_3}{|}}{C}HCH_2Br \qquad CH_3\underset{\underset{CH_3}{|}}{\overset{\overset{CH_3}{|}}{C}}CH_2Br$$

(c) $CH_3CH_2CH_2OCH_3 \qquad CH_3CH_2CH_2OTos \qquad CH_3CH_2CH_2Br$

12.50 ■ Predict the product and give the stereochemistry resulting from reaction of each of the following nucleophiles with (*R*)-2-bromooctane:

(a) ⁻CN **(b)** $CH_3CO_2^-$ **(c)** CH_3S^-

12.51 (*R*)-2-Bromooctane undergoes racemization to give (±)-2-bromooctane when treated with NaBr in dimethyl sulfoxide. Explain.

12.52 Reaction of the following *S* tosylate with cyanide ion yields a nitrile product that also has *S* stereochemistry. Explain.

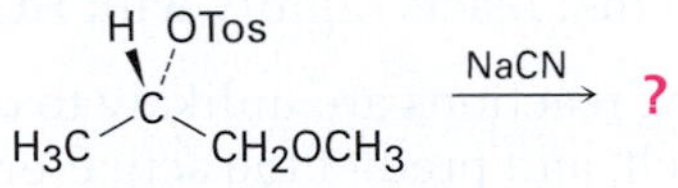

$$\underset{H_3C}{\overset{H\quad OTos}{\diagdown}}\overset{|}{\underset{\diagup}{C}}\underset{CH_2OCH_3}{} \quad\xrightarrow{\text{NaCN}}\quad \textcolor{red}{?}$$

(*S* stereochemistry)

12.53 The S_N2 reaction can occur *intramolecularly* (within the same molecule). What product would you expect from treatment of 4-bromobutan-1-ol with base?

12.54 1-Chloro-1,2-diphenylethane can undergo E2 elimination to give either *cis*- or *trans*-1,2-diphenylethylene (stilbene). Draw Newman projections of the reactive conformations leading to both possible products, and suggest a reason why the trans alkene is the major product.

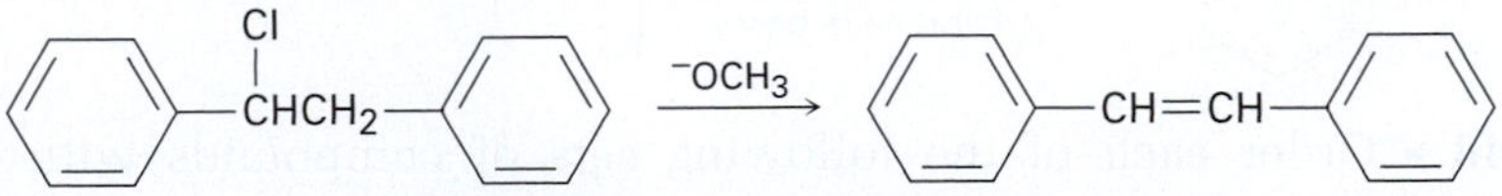

1-Chloro-1,2-diphenylethane $\xrightarrow{\text{⁻OCH}_3}$ **trans-1,2-Diphenylethylene**

12.55 ■ Predict the major alkene product of the following E1 reaction:

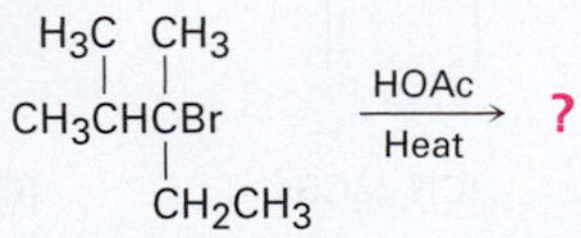

$$CH_3\overset{\overset{H_3C\;\;CH_3}{|\quad\;\;|}}{\underset{\underset{CH_2CH_3}{|}}{C}}HCBr \quad\xrightarrow[\text{Heat}]{\text{HOAc}}\quad \textcolor{red}{?}$$

12.56 The tosylate of (2*R*,3*S*)-3-phenylbutan-2-ol undergoes E2 elimination on treatment with sodium ethoxide to yield (*Z*)-2-phenylbut-2-ene. Explain, using Newman projections.

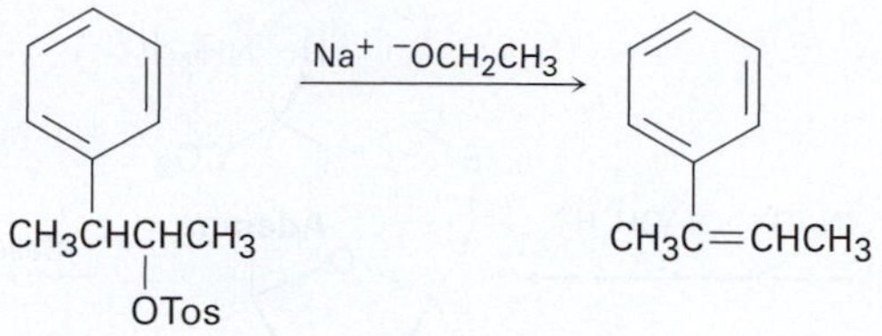
Na+ −OCH2CH3
CH3CHCHCH3
OTos
CH3C=CHCH3

12.57 In light of your answer to Problem 12.56, which alkene, *E* or *Z*, would you expect from an E2 reaction on the tosylate of (2*R*,3*R*)-3-phenylbutan-2-ol? Which alkene would result from E2 reaction on the (2*S*,3*R*) and (2*S*,3*S*) tosylates? Explain.

12.58 Arenes such as ethylbenzene react with NBS to give products in which bromine substitution has occurred at the benzylic position. Explain the result, and propose a mechanism.

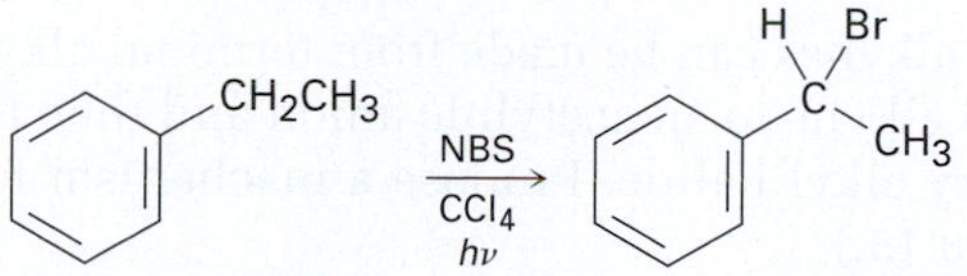
CH2CH3
NBS
CCl4
hν
H
Br
C
CH3

12.59 How can you explain the fact that *trans*-1-bromo-2-methylcyclohexane yields the non-Zaitsev elimination product 3-methylcyclohexene on treatment with base?

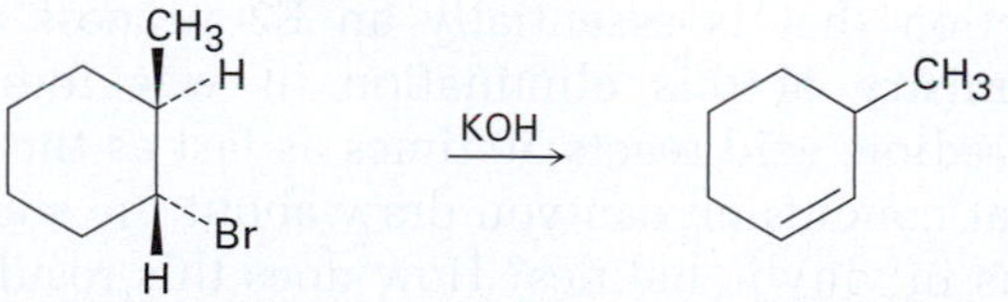
CH3
H
Br
H
KOH
CH3

***trans*-1-Bromo-2-methylcyclohexane** **3-Methylcyclohexene**

12.60 ▪ Predict the product(s) of the following reaction, indicating stereochemistry:

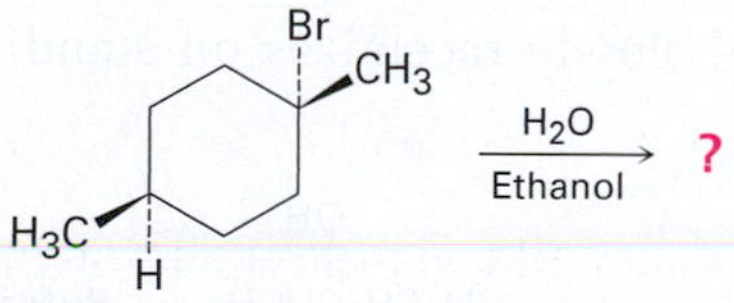
Br
CH3
H3C
H
H2O
Ethanol
?

▪ Problems assignable in Organic OWL.

12.61 Metabolism of *S*-adenosylhomocysteine (Section 12.10) involves the following steps. Propose mechanisms for both.

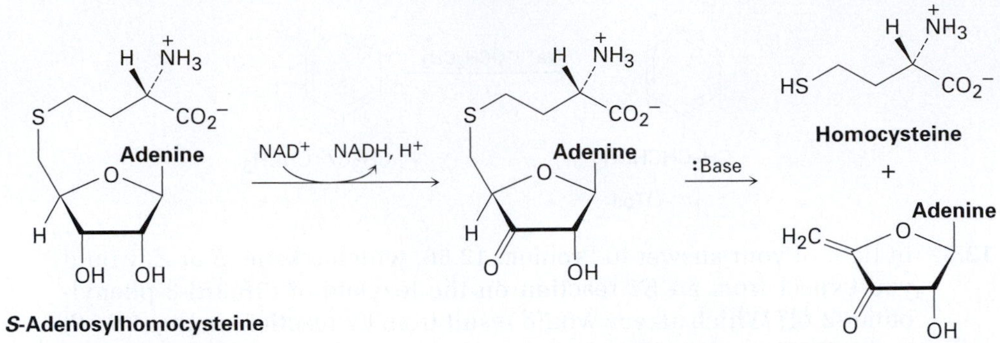

12.62 Reaction of iodoethane with CN^- yields a small amount of *isonitrile*, $CH_3CH_2N{\equiv}C$, along with the nitrile $CH_3CH_2C{\equiv}N$ as the major product. Write electron-dot structures for both products, and propose mechanisms to account for their formation.

12.63 Internal alkynes can be made from terminal alkynes by converting the terminal alkyne to an acetylide anion and then treating the anion with a primary alkyl halide. Propose a mechanism for the alkylation. (See Section 8.15.)

$$RC{\equiv}CH \xrightarrow[\text{2. R'CH}_2\text{Br}]{\text{1. NaNH}_2} RC{\equiv}CCH_2R'$$

12.64 Alkynes can be made by dehydrohalogenation of vinylic halides in a reaction that is essentially an E2 process. In studying the stereochemistry of this elimination, it was found that (*Z*)-2-chlorobut-2-enedioic acid reacts 50 times as fast as the corresponding *E* isomer. What conclusion can you draw about the stereochemistry of eliminations in vinylic halides? How does this result compare with eliminations of alkyl halides?

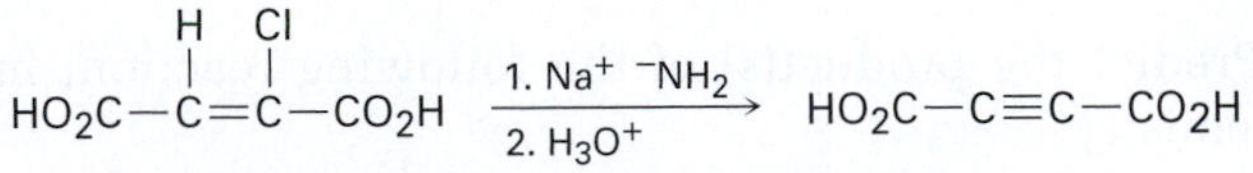

$$HO_2C-\underset{H}{\overset{}{C}}=\underset{Cl}{\overset{}{C}}-CO_2H \xrightarrow[\text{2. H}_3\text{O}^+]{\text{1. Na}^+ \ ^-\text{NH}_2} HO_2C-C{\equiv}C-CO_2H$$

12.65 (*S*)-Butan-2-ol slowly racemizes on standing in dilute sulfuric acid. Explain.

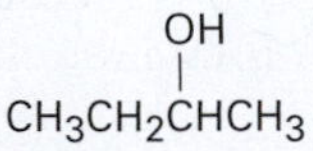

$$\underset{\textbf{Butan-2-ol}}{CH_3CH_2\overset{OH}{\underset{|}{C}}HCH_3}$$

12.66 Reaction of HBr with (*R*)-3-methylhexan-3-ol leads to racemic 3-bromo-3-methylhexane. Explain.

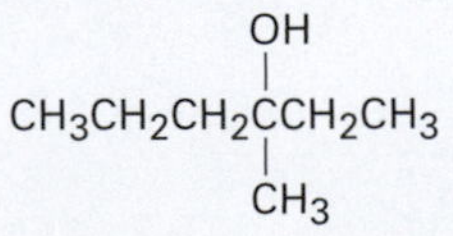

$$\underset{\textbf{3-Methylhexan-3-ol}}{CH_3CH_2CH_2\overset{OH}{\underset{\underset{CH_3}{|}}{\overset{|}{C}}}CH_2CH_3}$$

■ Problems assignable in Organic OWL.

12.67 Treatment of 1-bromo-2-deuterio-2-phenylethane with strong base leads to a mixture of deuterated and nondeuterated phenylethylenes in a ratio of approximately 7 : 1. Explain.

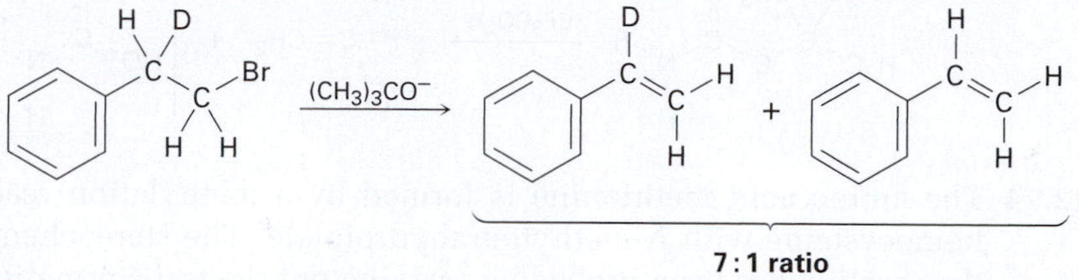

7 : 1 ratio

12.68 One step in the urea cycle for ridding the body of ammonia is the conversion of argininosuccinate to the amino acid arginine plus fumarate. Propose a mechanism for the reaction, and show the structure of arginine.

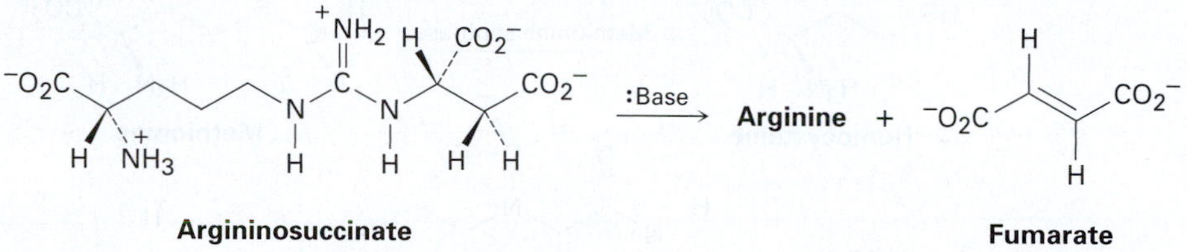

Argininosuccinate   **Fumarate**

12.69 There are eight diastereomers of 1,2,3,4,5,6-hexachlorocyclohexane. Draw each in its more stable chair conformation. One isomer loses HCl in an E2 reaction nearly 1000 times more slowly than the others. Which isomer reacts so slowly, and why?

12.70 Propose a structure, including stereochemistry, for an alkyl halide that gives only (*E*)-3-methyl-2-phenylpent-2-ene on E2 elimination.

12.71 When a primary alcohol is treated with *p*-toluenesulfonyl chloride at room temperature in the presence of an organic base such as pyridine, a tosylate is formed. When the same reaction is carried out at higher temperature, an alkyl chloride is often formed. Explain.

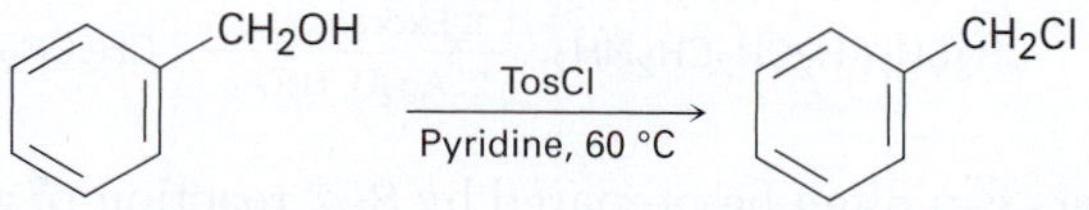

12.72 S_N2 reactions take place with inversion of configuration, and S_N1 reactions take place with racemization. The following substitution reaction, however, occurs with complete *retention* of configuration. Propose a mechanism.

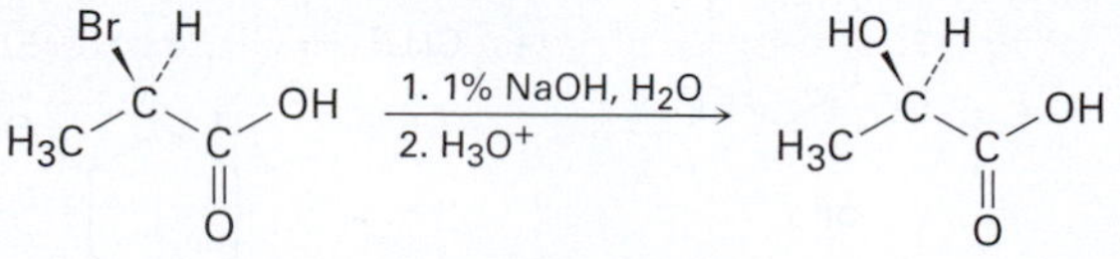

12.73 Propose a mechanism for the following reaction, an important step in the laboratory synthesis of proteins:

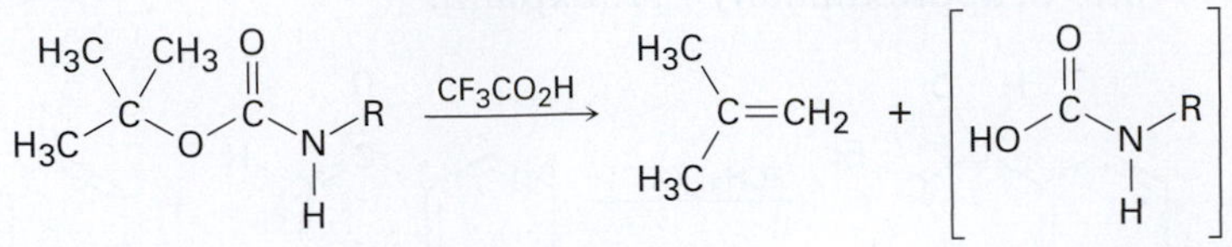

12.74 The amino acid methionine is formed by a methylation reaction of homocysteine with *N*-methyltetrahydrofolate. The stereochemistry of the reaction has been probed by carrying out the transformation using a donor with a "chiral methyl group" that contains protium (H), deuterium (D), and tritium (T) isotopes of hydrogen. Does the methylation reaction occur with inversion or retention of configuration? How might you explain this result?

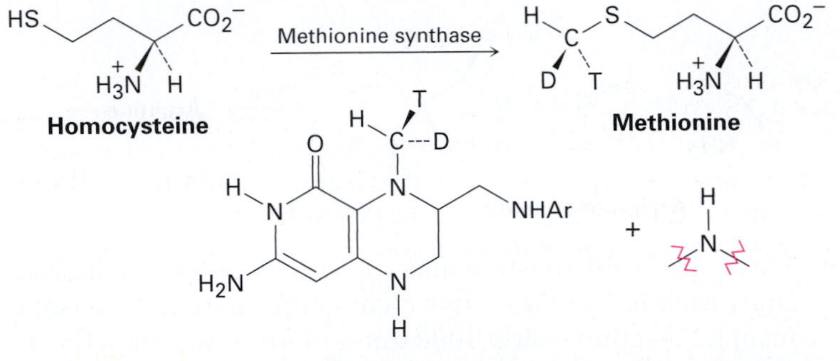

Homocysteine

Methionine

N-Methyltetrahydrofolate   **Tetrahydrofolate**

12.75 Amines are converted into alkenes by a two-step process called the *Hofmann elimination*. Reaction of the amine with excess CH_3I in the first step yields an intermediate that undergoes E2 reaction when treated with basic silver oxide. Pentylamine, for example, yields pent-1-ene. Propose a structure for the intermediate, and explain why it undergoes ready elimination.

$$CH_3CH_2CH_2CH_2CH_2NH_2 \xrightarrow[\text{2. Ag}_2\text{O, H}_2\text{O}]{\text{1. Excess CH}_3\text{I}} CH_3CH_2CH_2CH=CH_2$$

12.76 ■ Ethers can often be prepared by S_N2 reaction of alkoxide ions, RO^-, with alkyl halides. Suppose you wanted to prepare cyclohexyl methyl ether. Which of the two possible routes shown below would you choose?

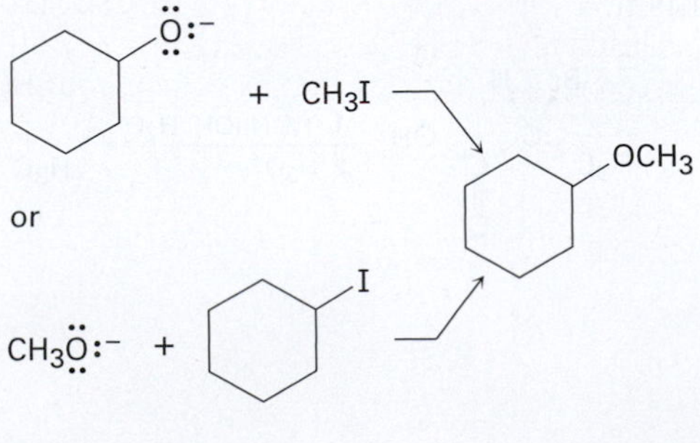

or

■ Problems assignable in Organic OWL.

13 Alcohols, Phenols, and Thiols; Ethers and Sulfides

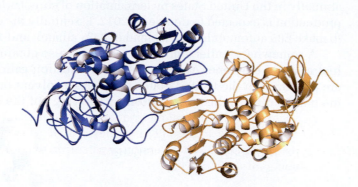

Liver alcohol dehydrogenase catalyzes the oxidation of ethanol to acetaldehyde.

Alcohols, **phenols**, and **ethers** can be thought of as organic derivatives of water in which one or both of the water hydrogens is replaced by an organic group: H–O–H versus R–O–H, Ar–O–H, and R–O–R′. **Thiols** and **sulfides** are the corresponding sulfur analogs, R–S–H and R–S–R′, respectively.

In practice, the names *alcohol* and *thiol* are restricted to compounds that have their –OH or –SH group bonded to a saturated, sp^3-hybridized carbon atom. Compounds with their –OH or –SH bonded to an aromatic ring are called *phenols* (Section 9.1) and *thiophenols*, while compounds with the –OH or –SH group bonded to a vinylic, sp^2-hybridized carbon are called *enols* and *enethiols*. We'll look in detail at the chemistry of enols in Chapter 17.

An alcohol   **A thiol**   **A phenol**   **A thiophenol**   [**An enol**   **An enethiol**]

Alcohols occur widely and have many industrial and pharmaceutical applications. Methanol, for instance, is one of the most important of all industrial chemicals. Historically, methanol was prepared by heating wood in the absence of air and thus came to be called *wood alcohol*. Today, approximately 1.3 billion gallons of methanol are manufactured each year in the United States by catalytic reduction of carbon monoxide with hydrogen. Methanol is toxic to humans, causing blindness in small doses (15 mL) and death in larger amounts (100–250 mL). Industrially, it is used both as a solvent and as a starting material for production of formaldehyde (CH_2O) and acetic acid (CH_3CO_2H).

$$CO + 2 H_2 \xrightarrow[\text{Zinc oxide/chromia catalyst}]{400\ ^\circ C} CH_3OH$$

OWL Online homework for this chapter can be assigned in Organic OWL.

Ethanol was one of the first organic chemicals to be prepared and purified. Its production by fermentation of grains and sugars has been carried out for perhaps 9000 years, and its purification by distillation goes back at least as far as the 12th century. Today, approximately 4 billion gallons of ethanol are produced annually in the United States by fermentation of corn, barley, and sorghum, and production is expected to double by 2012. Essentially the entire amount is used to make E85 automobile fuel, a blend of 85% ethanol and 15% gasoline.

A somewhat smaller amount of ethanol is also obtained by acid-catalyzed hydration of ethylene. Approximately 110 million gallons of ethanol a year are produced in the United States for use as a solvent or as a chemical intermediate in other industrial reactions.

$$H_2C=CH_2 \quad \xrightarrow[\substack{H_3PO_4 \\ 250\ °C}]{H_2O} \quad CH_3CH_2OH$$

Phenols occur widely in living organisms and are intermediates in the industrial synthesis of products as diverse as adhesives and antiseptics. Phenol itself is a general disinfectant found in coal tar, methyl salicylate is a flavoring agent found in oil of wintergreen; and the urushiols are the allergenic constituents of poison oak and poison ivy. Note that the word *phenol* is the name both of the specific compound (hydroxybenzene) and of a class of compounds.

Phenol
(also known as carbolic acid)

Methyl salicylate

Urushiols
(R = different C$_{15}$ alkyl and alkenyl chains)

Perhaps the most well-known ether is diethyl ether, which has a long history of medicinal use as an anesthetic and industrial use as a solvent. Other useful ethers include anisole, a pleasant-smelling aromatic ether used in perfumery, and tetrahydrofuran (THF), a cyclic ether often used as a solvent. Thiols and sulfides are found in various biomolecules, although not as commonly as their oxygen-containing relatives.

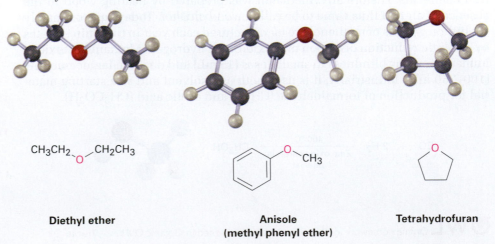

CH$_3$CH$_2$—O—CH$_2$CH$_3$

Diethyl ether

Anisole
(methyl phenyl ether)

Tetrahydrofuran

Up to this point, we've focused on developing some general ideas of organic reactivity, on looking at the chemistry of hydrocarbons and alkyl halides, and on seeing some of the tools used in structural studies. With that background, it's now time to begin a study of the oxygen-containing functional groups that lie at the heart of organic and biological chemistry. To understand the chemistry of living organisms, it's necessary to understand oxygen-containing functional groups. We'll look at compounds with C–O and C–S single bonds in this chapter and then move on to carbonyl compounds in Chapters 14 through 17.

13.1 Naming Alcohols, Phenols, and Thiols

Alcohols are classified as primary (1°), secondary (2°), or tertiary (3°), depending on the number of organic groups bonded to the hydroxyl-bearing carbon.

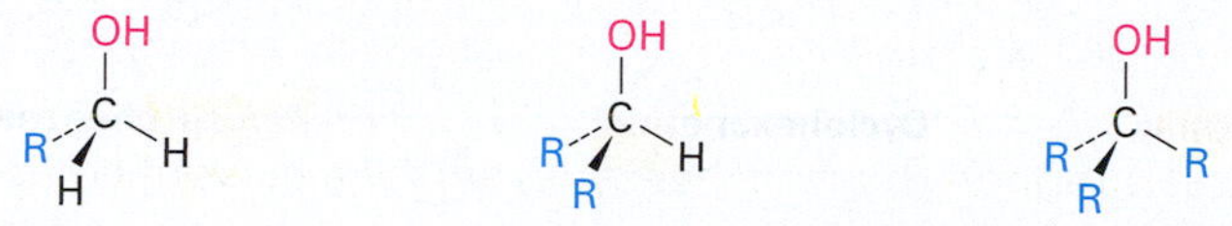

A primary (1°) alcohol   A secondary (2°) alcohol   A tertiary (3°) alcohol

Simple alcohols are named in the IUPAC system as derivatives of the parent alkane, using the suffix -ol:

Rule 1

Select the longest carbon chain containing the hydroxyl group, and derive the parent name by replacing the -e ending of the corresponding alkane with -ol. The -e is deleted to prevent the occurrence of two adjacent vowels: propanol rather than propaneol, for example.

Rule 2

Number the alkane chain beginning at the end nearer the hydroxyl group.

Rule 3

Number the substituents according to their position on the chain, and write the name, listing the substituents in alphabetical order and identifying the position to which the —OH is bonded. Note that in naming cis-cyclohexane-1,4-diol, the final -e of cyclohexane is not deleted because the next letter (d) is not a vowel; that is, cyclohexanediol is used rather than cyclohexandiol.

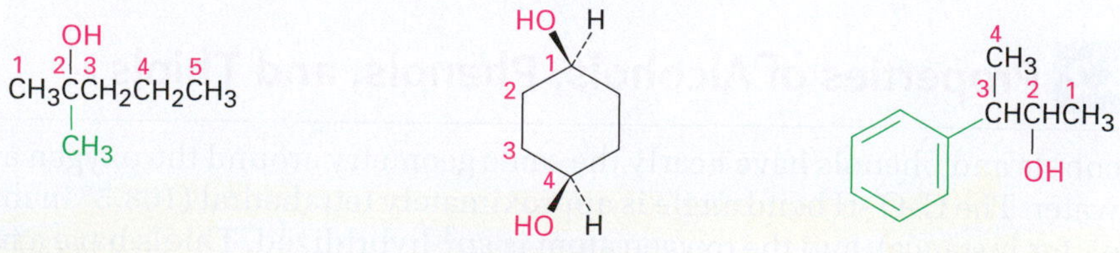

2-Methylpentan-2-ol   cis-Cyclohexane-1,4-diol   3-Phenylbutan-2-ol

Some simple and widely occurring alcohols also have common names that are accepted by IUPAC. For example:

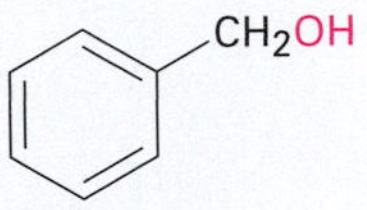

Benzyl alcohol (phenylmethanol)

$H_2C{=}CHCH_2OH$

Allyl alcohol (prop-2-en-1-ol)

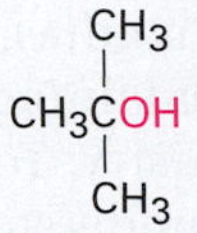

***tert*-Butyl alcohol (2-methylpropan-2-ol)**

$HOCH_2CH_2OH$

Ethylene glycol (ethane-1,2-diol)

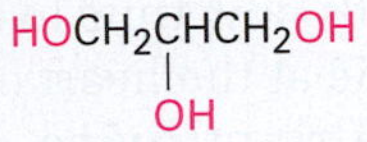

Glycerol (propane-1,2,3-triol)

Phenols are named as described previously in Section 9.1 for aromatic compounds. Thiols are named by the same system used for alcohols, with the suffix *-thiol* in place of *-ol.* The –SH group itself is sometimes referred to as a **mercapto group**, and thiols are sometimes called *mercaptans.*

CH_3CH_2SH

Ethanethiol

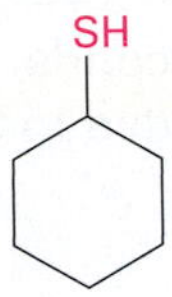

Cyclohexanethiol

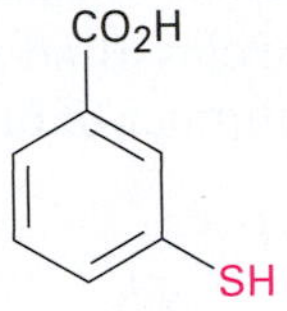

***m*-Mercaptobenzoic acid**

Problem 13.1

Give IUPAC names for the following compounds:

(a)

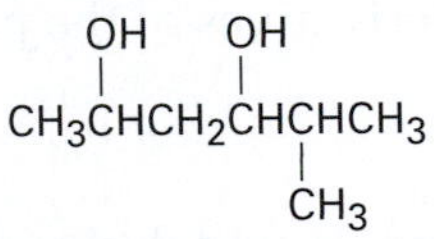

(b)

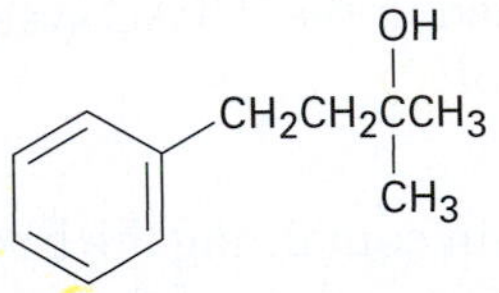

(c)

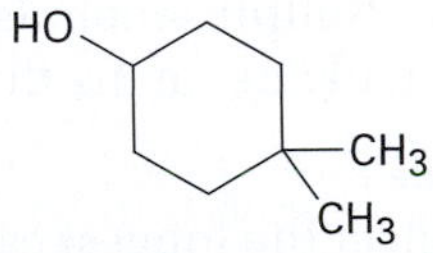

(d)

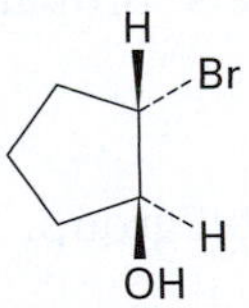

(e)

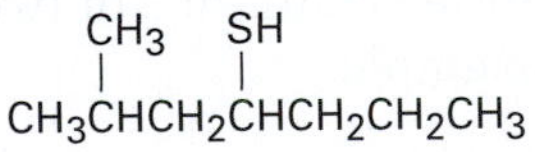

(f)

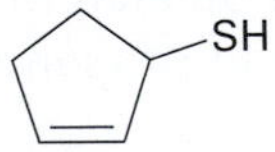

Problem 13.2

Draw structures corresponding to the following IUPAC names:

(a) 2-Ethylbut-2-en-1-ol **(b)** Cyclohex-3-en-1-ol
(c) *trans*-3-Chlorocycloheptanol **(d)** Pentane-1,4-dithiol
(e) 2,4-Dimethylphenol **(f)** *o*-(2-Hydroxyethyl)phenol

13.2 Properties of Alcohols, Phenols, and Thiols

Alcohols and phenols have nearly the same geometry around the oxygen atom as water. The C–O–H bond angle is approximately tetrahedral (108.5° in methanol, for instance), and the oxygen atom is sp^3-hybridized. Thiols have a more compressed C–S–H bond angle (96.5° in methanethiol).

Also like water, alcohols and phenols have higher boiling points than might be expected because of hydrogen-bonding (Section 2.12). A positively polarized —OH hydrogen atom from one molecule is attracted to a lone pair of electrons on the electronegative oxygen atom of another molecule, resulting in a weak force that holds the molecules together (Figure 13.1). These intermolecular attractions must be overcome for a molecule to break free from the liquid and enter the vapor state, so the boiling temperature is raised. Thiols do not typically form hydrogen bonds because sulfur is not sufficiently electronegative.

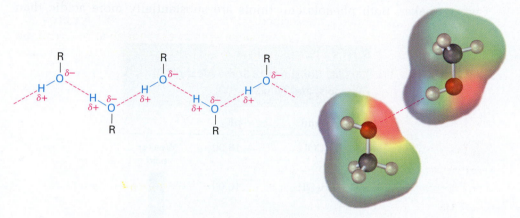

FIGURE 13.1 Hydrogen-bonding in alcohols and phenols. Attraction between a positively polarized OH hydrogen and a negatively polarized oxygen holds molecules together.

Yet another similarity with water is that alcohols and phenols are both weakly basic and weakly acidic. As weak bases, they are reversibly protonated by strong acids to yield oxonium ions, ROH_2^+:

$$R\overset{..}{\underset{..}{O}}H \ + \ H{-}X \ \rightleftharpoons \ R\overset{\overset{H}{|}}{\underset{..}{O^+}}H \ \ :X^-$$

An alcohol   **An oxonium ion**

$$\left[\ \text{or} \quad ArOH \ + \ HX \ \rightleftharpoons \ Ar\overset{+}{O}H_2 \ \ X^- \ \right]$$

As weak acids, they dissociate slightly in dilute aqueous solution by donating a proton to water, generating H_3O^+ and an **alkoxide ion (RO⁻)** or a **phenoxide ion (ArO⁻).**

$$R{-}\overset{..}{\underset{..}{O}}: \ + \ H\overset{..}{\underset{..}{O}}H \ \rightleftharpoons \ R{-}\overset{..}{\underset{..}{O}}:^- \ + \ H\overset{\overset{H}{|}}{\underset{..}{O^+}}H$$

An alcohol   **An alkoxide ion**

$$\left[\ \text{or} \quad Y{-}\bigcirc{-}\overset{..}{\underset{..}{O}}H \ + \ H_2\overset{..}{\underset{..}{O}}: \ \rightleftharpoons \ Y{-}\bigcirc{-}O^- \ + \ H_3O^+ \ \right]$$

A phenol   **A phenoxide ion**

Recall from the earlier discussion of acidity in Sections 2.8 through 2.10 that the strength of any acid HA in water can be expressed by an acidity constant, K_a:

$$K_a = \frac{[A^-][H_3O^+]}{[HA]} \qquad pK_a = -\log K_a$$

[handwritten: $K_a = 10^{-pH}$]

Compounds with a smaller K_a and larger pK_a are less acidic, whereas compounds with a larger K_a and smaller pK_a are more acidic. As shown by the data in Table 13.1, simple alcohols like methanol and ethanol are very close to water in acidity, and the more highly substituted *tert*-butyl alcohol is somewhat weaker. Both phenols and thiols are substantially more acidic than water.

TABLE 13.1
Acidity Constants of Some Alcohols, Phenols, and Thiols

Compound	pK_a	
$(CH_3)_3COH$	18.00	Weaker acid
CH_3CH_2OH	16.00	
H_2O	15.74	
CH_3OH	15.54	
CH_3SH	10.3	
p-Methylphenol	10.17	
Phenol	9.89	
p-Nitrophenol	7.15	Stronger acid

Because alcohols are weak acids, they don't react with weak bases such as amines or bicarbonate ion, and they react to only a limited extent with metal hydroxides such as NaOH. Alcohols do, however, react with alkali metals and with strong bases such as sodium hydride (NaH) and sodium amide ($NaNH_2$). Alkoxides are themselves bases that are frequently used as reagents in organic chemistry. They are named systematically by adding the *-ate* suffix to the name of the alcohol. Methanol becomes methanolate, for instance.

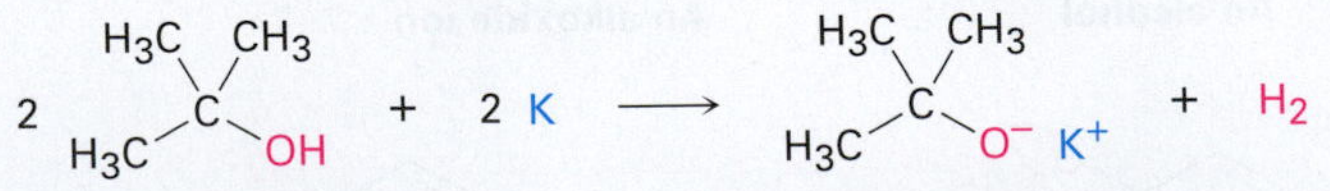

tert-Butyl alcohol
(2-methylpropan-2-ol)

Potassium *tert*-butoxide
(potassium 2-methyl-propan-2-olate)

$$CH_3OH \ + \ NaH \longrightarrow CH_3O^- \ Na^+ \ + \ H_2$$

Methanol **Sodium methoxide**
(sodium methanolate)

$$CH_3CH_2OH \ + \ NaNH_2 \longrightarrow CH_3CH_2O^- \ Na^+ \ + \ NH_3$$

Ethanol **Sodium ethoxide**
(sodium ethanolate)

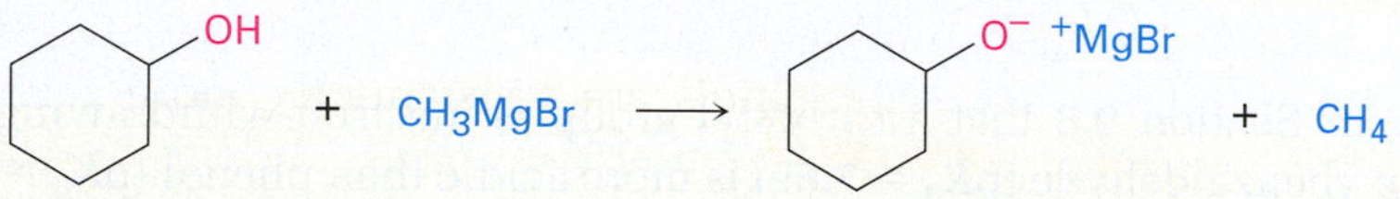

Cyclohexanol   **Bromomagnesium**
cyclohexanolate

==Phenols and thiols are about a million times more acidic than alcohols.==
Both are therefore soluble in dilute aqueous NaOH and can often be separated
from a mixture by basic extraction into aqueous solution, followed by reacidi-
fication. ==Phenols are more acidic than alcohols because the phenoxide anion
is resonance-stabilized.== Delocalization of the negative charge over the ortho
and para positions of the aromatic ring results in increased stability of the
phenoxide anion relative to undissociated phenol and in a consequently lower
$\Delta G°$ for dissociation.

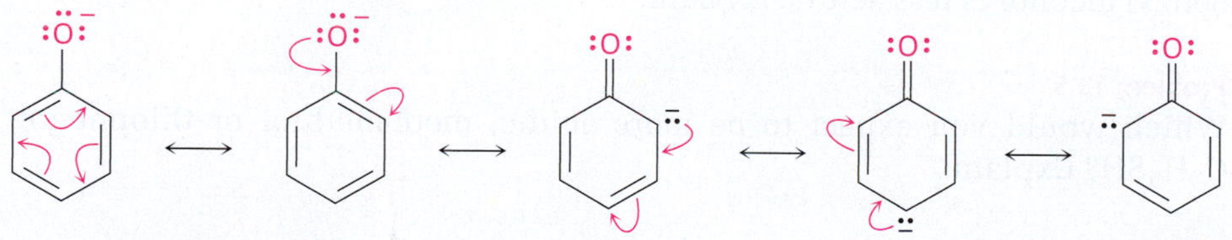

Substituted phenols can be either more acidic or less acidic than phenol
itself, depending on whether the substituent is electron-withdrawing or
electron-donating (Section 9.8). ==Phenols with an electron-withdrawing
substituent are more acidic because these substituents delocalize the negative charge; phenols with an electron-donating substituent== are less acidic
because these substituents concentrate the charge. The acidifying effect of
an electron-withdrawing substituent is particularly noticeable in phenols
with a nitro group at the ortho or para position.

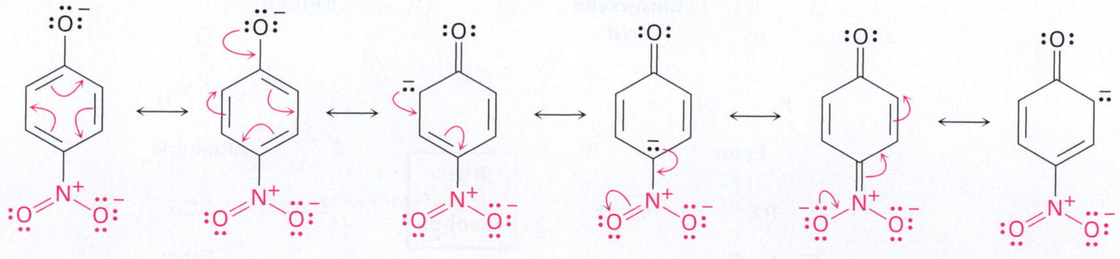

WORKED EXAMPLE 13.1 Predicting the Relative Acidity of a Substituted Phenol

Is *p*-hydroxybenzaldehyde more acidic or less acidic than phenol?

Strategy

Identify the substituent on the aromatic ring, and decide whether it is electron-donating or electron-withdrawing. Electron-withdrawing substituents make the phenol more acidic by stabilizing the phenoxide anion, and electron-donating substituents make the phenol less acidic by destabilizing the anion.

Solution

We saw in Section 9.8 that a carbonyl group is electron-withdrawing. Thus, *p*-hydroxybenzaldehyde ($pK_a = 7.89$) is more acidic than phenol ($pK_a = 9.89$).

p-Hydroxybenzaldehyde
($pK_a = 7.89$)

Problem 13.3

Rank the following substances in order of increasing acidity:
(a) Phenol, *p*-methylphenol, *p*-(trifluoromethyl)phenol
(b) Benzyl alcohol, phenol, *p*-hydroxybenzoic acid

Problem 13.4

p-Nitrobenzyl alcohol is more acidic than benzyl alcohol, but *p*-methoxybenzyl alcohol is less acidic. Explain.

Problem 13.5

Which would you expect to be more acidic, methanethiol or thiophenol, C_6H_5SH? Explain.

13.3 Preparing Alcohols from Carbonyl Compounds

Alcohols occupy a central position in organic chemistry. They can be prepared from many other kinds of compounds (alkenes, alkyl halides, ketones, esters, and aldehydes, among others), and they can be transformed into an equally wide assortment of products (Figure 13.2).

FIGURE 13.2 The central position of alcohols in organic chemistry. Alcohols can be prepared from, and converted into, many other kinds of compounds.

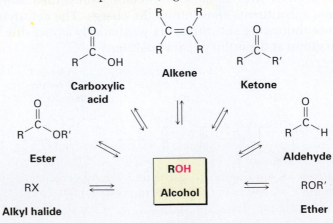

We've already seen several methods of alcohol synthesis:

- Alcohols can be prepared by hydration of alkenes. Because the direct hydration of alkenes with aqueous acid is generally a poor reaction in the laboratory, two indirect methods are commonly used. Hydroboration/oxidation yields the product of syn, non-Markovnikov hydration, whereas oxymercuration/reduction yields the product of Markovnikov hydration (Section 8.4).

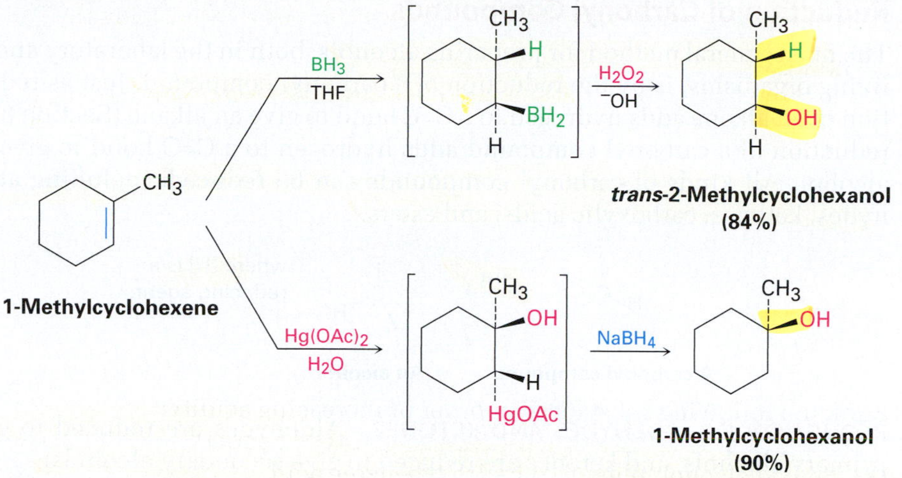

**trans-2-Methylcyclohexanol
(84%)**

1-Methylcyclohexene

**1-Methylcyclohexanol
(90%)**

- 1,2-Diols can be prepared either by direct hydroxylation of an alkene with OsO_4 followed by reduction with $NaHSO_3$ or by acid-catalyzed hydrolysis of an epoxide (Section 8.7). The OsO_4 reaction occurs with syn stereochemistry to give a cis diol, and epoxide opening occurs with anti stereochemistry to give a trans diol.

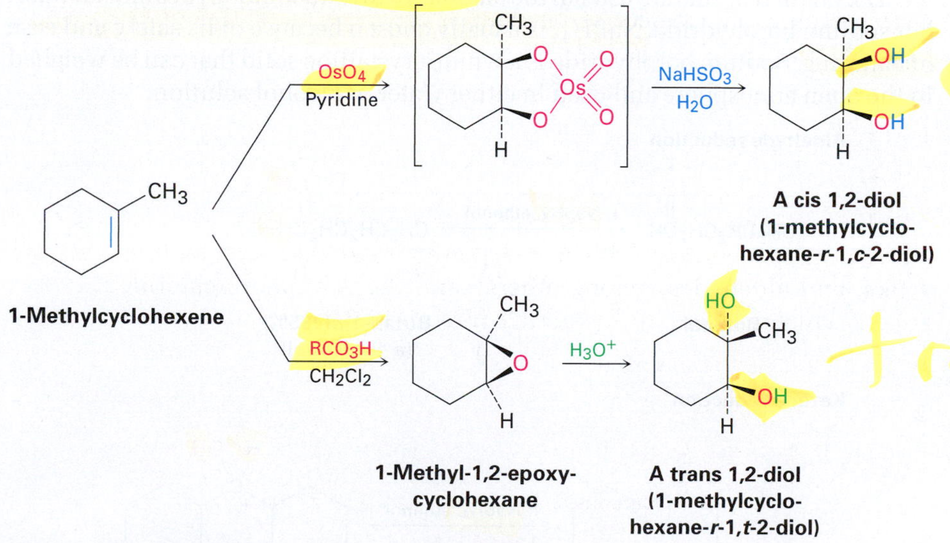

**A cis 1,2-diol
(1-methylcyclo-
hexane-r-1,c-2-diol)**

1-Methylcyclohexene

**1-Methyl-1,2-epoxy-
cyclohexane**

**A trans 1,2-diol
(1-methylcyclo-
hexane-r-1,t-2-diol)**

[Parenthetically, we might note that the prefixes *cis*- and *trans*- are ambiguous when naming the diols derived from 1-methylcyclohexene because the ring has three substituents. In such cases, the substituent with the lowest number is taken as the reference substituent, denoted *r*, and the other substituents are identified as being cis (*c*) or trans (*t*) to that reference substituent. When two substituents share the same lowest number, the substituent with the highest priority by the Cahn–Ingold–Prelog sequence rules is taken as the reference. In the case of the two 1-methylcyclohexane-1,2-diol isomers, the –OH group at C1 is the reference (r-1), and the –OH at C2 is either cis (c-2) or trans (t-2) to that reference. Thus, the diol isomer derived by cis hydroxylation is named 1-methylcyclohexane-r-1,c-2-diol, and the isomer derived by trans hydroxylation is named 1-methylcyclohexane-r-1,t-2-diol.]

Reduction of Carbonyl Compounds

The most general method for preparing alcohols, both in the laboratory and in living organisms, is by the reduction of a carbonyl compound. Just as reduction of an alkene adds hydrogen to a C=C bond to give an alkane (Section 8.5), reduction of a carbonyl compound adds hydrogen to a C=O bond to give an alcohol. All kinds of carbonyl compounds can be reduced, including aldehydes, ketones, carboxylic acids, and esters.

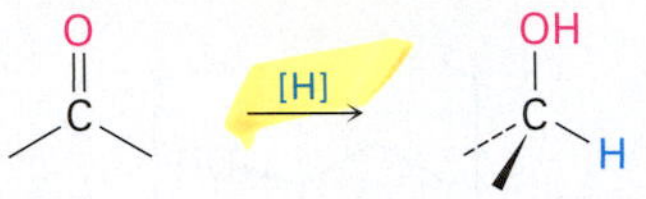

A carbonyl compound   **An alcohol**

where [H] is a reducing agent

REDUCTION OF ALDEHYDES AND KETONES   Aldehydes are reduced to give primary alcohols, and ketones are reduced to give secondary alcohols:

An aldehyde   **A primary alcohol**   **A ketone**   **A secondary alcohol**

Dozens of reagents are used in the laboratory to reduce aldehydes and ketones, but sodium borohydride, $NaBH_4$, is usually chosen because of its safety and ease of handling. Sodium borohydride is a white, crystalline solid that can be weighed in the open atmosphere and used in either water or alcohol solution.

Aldehyde reduction

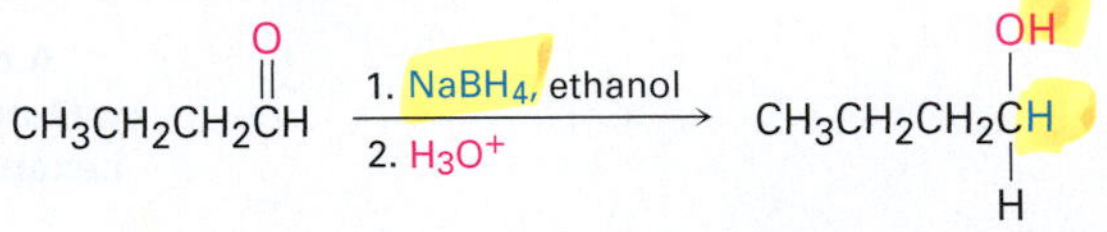

Butanal   **Butan-1-ol (85%)**
                  (a 1° alcohol)

Ketone reduction

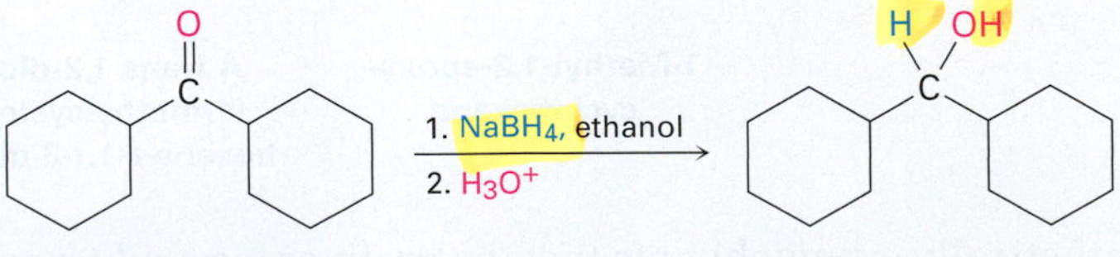

Dicyclohexyl ketone   **Dicyclohexylmethanol (88%)**
                              (a 2° alcohol)

Lithium aluminum hydride, $LiAlH_4$, is another reducing agent often used for reduction of aldehydes and ketones. A grayish powder that is soluble in ether and tetrahydrofuran, $LiAlH_4$ is much more reactive than $NaBH_4$ but also more dangerous. It reacts violently with water and decomposes explosively when heated above 120 °C.

Cyclohex-2-enone   **Cyclohex-2-enol (94%)**

We'll defer a detailed discussion of the mechanisms of these reductions until Section 14.6. For the moment, we'll simply note that they involve the ==addition of a nucleophilic hydride ion (:H⁻) to the positively polarized, electrophilic carbon atom of the carbonyl group.== The initial product is an alkoxide ion, which is protonated by addition of H_3O^+ in a second step to yield the alcohol product.

A carbonyl   **An alkoxide ion**   **An alcohol**
compound   **intermediate**

In living organisms, aldehyde and ketone reductions are carried out by either of the coenzymes NADH (reduced nicotinamide adenine dinucleotide) or NADPH (reduced nicotinamide adenine dinucleotide phosphate). Although these biological "reagents" are substantially more complex than $NaBH_4$ or $LiAlH_4$, the mechanisms of laboratory and biological reactions are similar. The coenzyme acts as a hydride-ion donor to give an alkoxide anion, and the intermediate anion is then protonated by acid. An example is the reduction of acetoacetyl ACP to β-hydroxybutyryl ACP, a step in the biological synthesis of fats (Figure 13.3). Note that the *pro-R* hydrogen of NADPH is the one transferred in this example. Enzyme-catalyzed reactions usually occur with high specificity, although it's not usually possible to predict the stereochemical result before the fact.

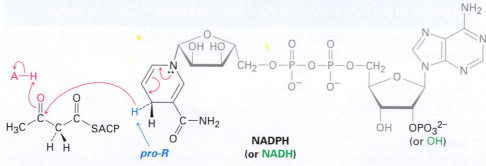

FIGURE 13.3 The biological reduction of a ketone (aceto-acetyl ACP) to an alcohol (β-hydroxybutyryl ACP) by NADPH.

REDUCTION OF CARBOXYLIC ACIDS AND ESTERS Carboxylic acids and esters are reduced to give primary alcohols.

A carboxylic acid   An ester   A primary alcohol

These reactions aren't as rapid as the reductions of aldehydes and ketones and are usually carried out with the more reactive reducing agent $LiAlH_4$. All carbonyl groups, including acids, esters, ketones, and aldehydes, are reduced by $LiAlH_4$. Note that one hydrogen atom is delivered to the carbonyl carbon atom during aldehyde and ketone reductions but that two hydrogens become bonded to the former carbonyl carbon during carboxylic acid and ester reductions. We'll defer a discussion of the mechanisms of these reactions until Chapter 16.

Carboxylic acid reduction

$$CH_3(CH_2)_7CH{=}CH(CH_2)_7COH \xrightarrow[\text{2. } H_3O^+]{\text{1. } LiAlH_4, \text{ ether}} CH_3(CH_2)_7CH{=}CH(CH_2)_7CH_2OH$$

Octadec-9-enoic acid
(oleic acid)   **Octadec-9-en-1-ol (87%)**

Ester reduction

$$CH_3CH_2CH{=}CHCOCH_3 \xrightarrow[\text{2. } H_3O^+]{\text{1. } LiAlH_4, \text{ ether}} CH_3CH_2CH{=}CHCH_2OH \; + \; CH_3OH$$

Methyl pent-2-enoate   **Pent-2-en-1-ol (91%)**

WORKED EXAMPLE 13.2 Predicting the Structure of a Reactant, Given the Product

What carbonyl compounds would you reduce to obtain the following alcohols?

(a)
$$\underset{CH_3\ \ \ \ \ OH}{CH_3CH_2CHCH_2CHCH_3}$$

(b)

Strategy
Identify the target alcohol as primary, secondary, or tertiary. A primary alcohol can be prepared by reduction of an aldehyde, an ester, or a carboxylic acid; a secondary alcohol can be prepared by reduction of a ketone; and a tertiary alcohol can't be prepared by reduction.

Solution
(a) The target molecule is a secondary alcohol, which can be prepared only by reduction of a ketone. Either $NaBH_4$ or $LiAlH_4$ can be used.

$$\underset{CH_3\ \ \ \ \ O}{CH_3CH_2CHCH_2CCH_3} \xrightarrow[\text{2. } H_3O^+]{\text{1. } NaBH_4 \text{ or } LiAlH_4} \underset{CH_3\ \ \ \ OH}{CH_3CH_2CHCH_2CHCH_3}$$

(b) The target molecule is a primary alcohol, which can be prepared by reduction of an aldehyde, an ester, or a carboxylic acid. $LiAlH_4$ rather than $NaBH_4$ is needed for the ester and carboxylic acid reductions.

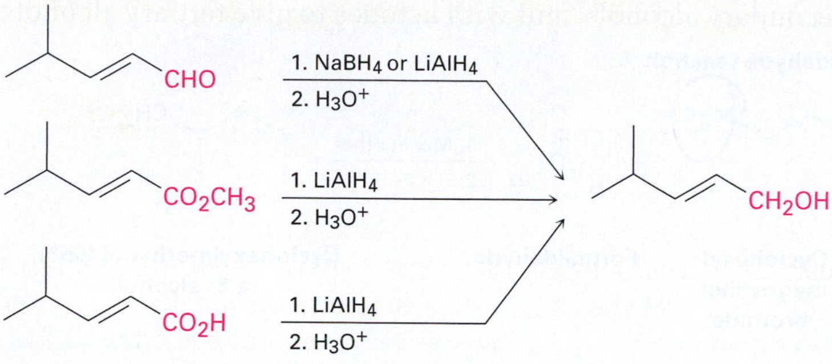

Problem 13.6

What carbonyl compounds give the following alcohols on reduction? Show all possibilities.

(a)

(b)

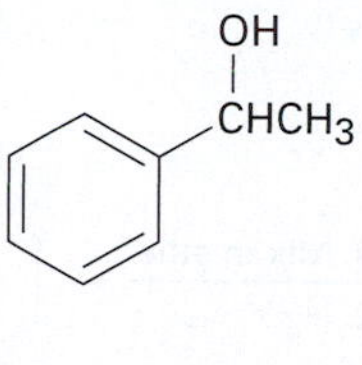

(c)

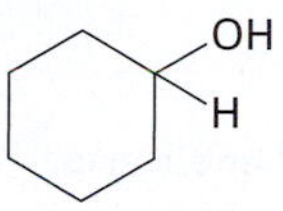

(d) $(CH_3)_2CHCH_2OH$

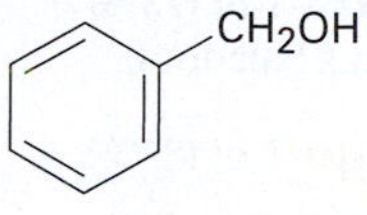

Grignard Reactions of Carbonyl Compounds

We saw in Section 12.4 that alkyl, aryl, and vinylic halides react with magnesium in ether or tetrahydrofuran to generate Grignard reagents, RMgX, which act as carbon-based nucleophiles. These Grignard reagents react with carbonyl compounds to yield alcohols in much the same way that hydride reducing agents do.

$$\left[\; R{-}X \;+\; Mg \;\longrightarrow\; \overset{\delta-}{R}{-}\overset{\delta+}{MgX} \qquad \begin{cases} R = 1°,\ 2°,\ \text{or } 3° \text{ alkyl, aryl, or vinylic} \\ X = Cl,\ Br,\ I \end{cases} \;\right]$$

A Grignard reagent

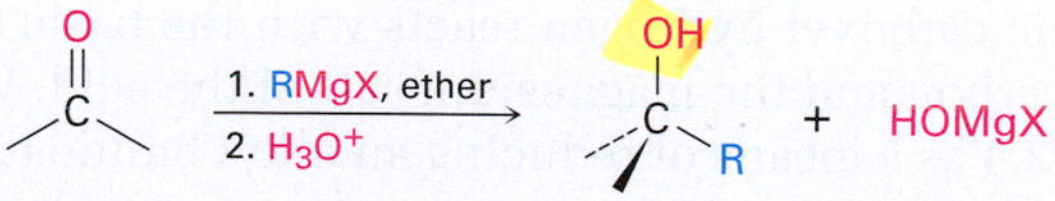

The reaction of Grignard reagents with carbonyl compounds has no direct biological counterpart because organomagnesium compounds are too strongly basic to exist in an aqueous medium. Nevertheless, the reaction is worth knowing about for two reasons. First, the reaction is an unusually broad and useful method of alcohol synthesis and demonstrates again the relative freedom with which chemists can operate in the laboratory. Second, the reaction *does* have an *indirect* biological counterpart, for we'll see in Chapter 17 that the addition of stabilized carbon nucleophiles to carbonyl compounds is used in almost all metabolic pathways as the major process for forming carbon–carbon bonds.

As examples of their addition to carbonyl compounds, Grignard reagents react with formaldehyde, $H_2C{=}O$, to give primary alcohols, with aldehydes to give secondary alcohols, and with ketones to give tertiary alcohols:

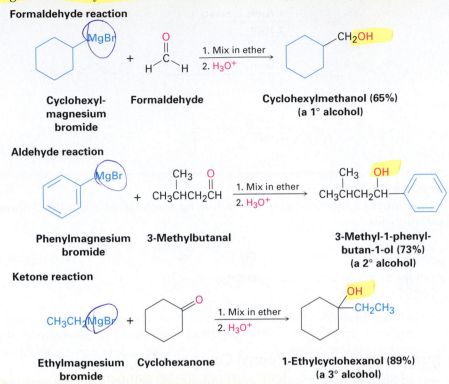

Formaldehyde reaction

Cyclohexyl-
magnesium
bromide

Formaldehyde

Cyclohexylmethanol (65%)
(a 1° alcohol)

Aldehyde reaction

Phenylmagnesium
bromide

3-Methylbutanal

3-Methyl-1-phenyl-
butan-1-ol (73%)
(a 2° alcohol)

Ketone reaction

Ethylmagnesium
bromide

Cyclohexanone

1-Ethylcyclohexanol (89%)
(a 3° alcohol)

Esters react with Grignard reagents to yield tertiary alcohols in which *two* of the substituents bonded to the hydroxyl-bearing carbon have come from the Grignard reagent, just as $LiAlH_4$ reduction of an ester adds two hydrogens.

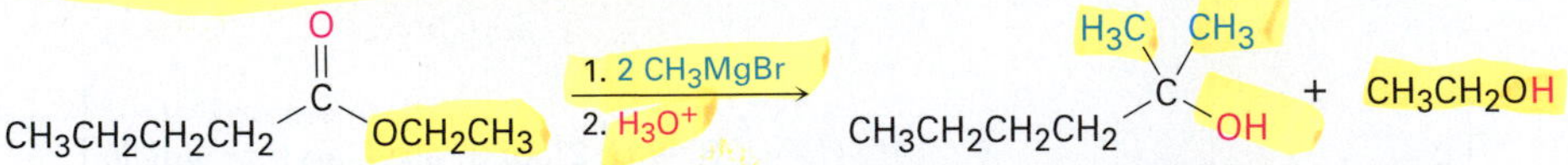

Ethyl pentanoate

2-Methylhexan-2-ol (85%)
(a 3° alcohol)

Carboxylic acids don't give addition products with Grignard reagents because the acidic carboxyl hydrogen reacts with the basic Grignard reagent to yield a hydrocarbon and the magnesium salt of the acid. We saw this reaction in Section 12.4 as a means of reducing an alkyl halide to an alkane.

$$RBr \ + \ Mg \ \longrightarrow \ RMgBr$$

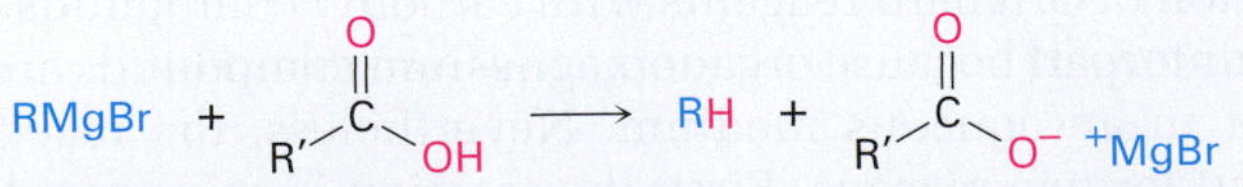

A carboxylic acid   A carboxylic acid salt

As with the reduction of carbonyl compounds, we'll defer a discussion of the mechanism of Grignard reactions until Section 14.6. For the moment, it's sufficient to note that Grignard reagents act as nucleophilic carbanions (R:⁻) and that the addition of a Grignard reagent to a carbonyl compound is analogous to the addition of hydride ion. The intermediate is an alkoxide ion, which is protonated by addition of H_3O^+ in a second step.

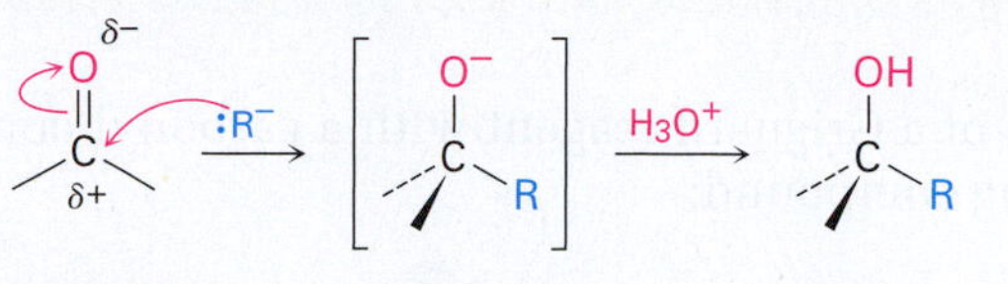

A carbonyl compound — An alkoxide ion intermediate — An alcohol

WORKED EXAMPLE 13.3 Using Grignard Reactions to Synthesize Alcohols

How could you use the reaction of a Grignard reagent with a carbonyl compound to synthesize 2-methylpentan-2-ol?

Strategy

Draw the product, and identify the three groups bonded to the alcohol carbon atom. If the three groups are all different, the starting carbonyl compound must be a ketone. If two of the three groups are identical, the starting carbonyl compound might be either a ketone or an ester.

Solution

In the present instance, the product is a tertiary alcohol with two methyl groups and one propyl group. Starting from a ketone, the possibilities are addition of methylmagnesium bromide to pentan-2-one and addition of propylmagnesium bromide to acetone:

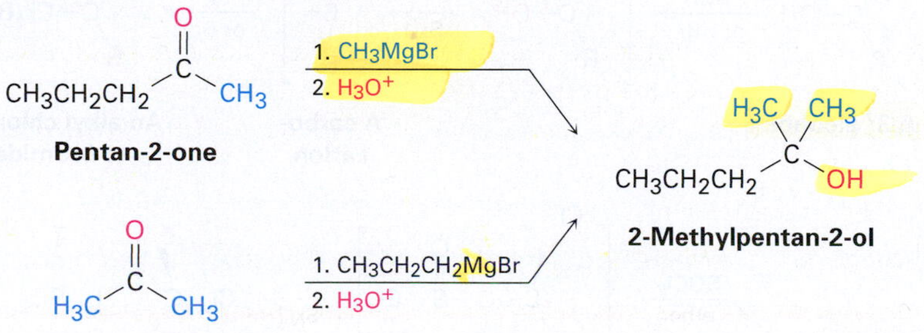

Pentan-2-one — Acetone — 2-Methylpentan-2-ol

Starting from an ester, the only possibility is addition of methylmagnesium bromide to an ester of butanoic acid, such as methyl butanoate:

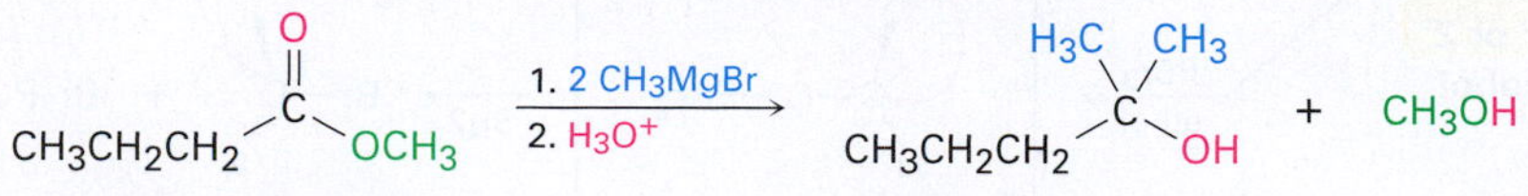

Methyl butanoate — 2-Methylpentan-2-ol

Problem 13.7

Show the products obtained from addition of methylmagnesium bromide to the following compounds:

(a) Cyclopentanone **(b)** Hexan-3-one

Problem 13.8
Use a Grignard reaction to prepare the following alcohols:
(a) 2-Methylpropan-2-ol   **(b)** 1-Methylcyclohexanol
(c) 2-Phenylbutan-2-ol   **(d)** Benzyl alcohol

Problem 13.9
Use the reaction of a Grignard reagent with a carbonyl compound to synthesize the following compound:

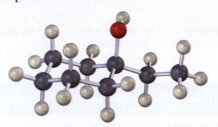

13.4 Reactions of Alcohols

We've already seen one general reaction of alcohols—their conversion to alkyl halides (Section 12.3). Tertiary alcohols react with HCl and HBr by an S_N1 mechanism through a carbocation intermediate. Primary and secondary alcohols react with $SOCl_2$ and PBr_3 by an S_N2 mechanism through backside attack on a chlorosulfite or dibromophosphite intermediate.

Dehydration of Alcohols

A second important reaction of alcohols, both in the laboratory and in biological pathways, is their dehydration to give alkenes. One method that works particularly well for tertiary alcohols is the acid-catalyzed reaction, which usually follows Zaitsev's rule (Section 12.11) and yields the more stable alkene as the major product. Thus, 2-methylbutan-2-ol gives primarily 2-methylbut-2-ene (trisubstituted double bond) rather than 2-methylbut-1-ene (disubstituted double bond).

$$H_3C-\underset{\underset{OH}{|}}{\overset{\overset{CH_3}{|}}{C}}-CH_2CH_3 \xrightarrow[25\ °C]{H_3O^+,\ THF} \underset{\underset{CH_3}{|}}{\overset{\overset{CH_3}{|}}{C}}=CHCH_3 + \underset{\underset{CH_3}{|}}{\overset{\overset{CH_2}{\|}}{C}}-CH_2CH_3$$

2-Methylbutan-2-ol   **2-Methylbut-2-ene**   **2-Methylbut-1-ene**
                                 (trisubstituted)   **(disubstituted)**

Major product   **Minor product**

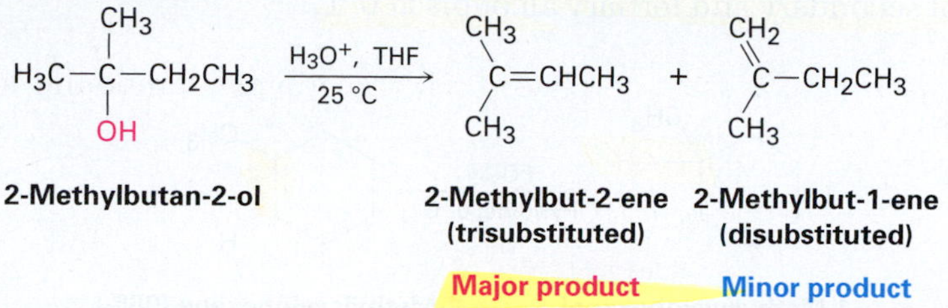

The reaction is an E1 process (Section 12.13) and occurs by a three-step mechanism involving protonation of the alcohol oxygen, unimolecular loss of water to generate a carbocation intermediate, and final loss of a proton from the neighboring carbon atom (Figure 13.4). As usual for E1 reactions, tertiary alcohols react fastest because they lead to stabilized, tertiary carbocation intermediates. Primary and secondary alcohols are much less reactive and require much higher temperatures for reaction.

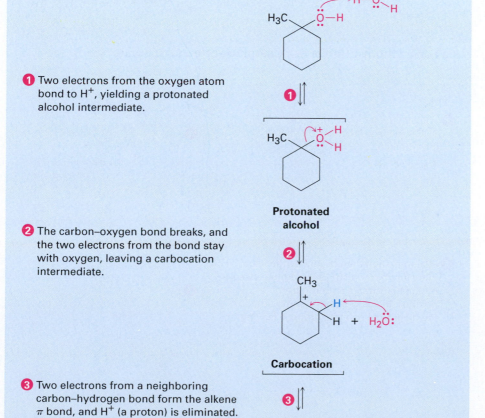

① Two electrons from the oxygen atom bond to H⁺, yielding a protonated alcohol intermediate.

② The carbon–oxygen bond breaks, and the two electrons from the bond stay with oxygen, leaving a carbocation intermediate.

③ Two electrons from a neighboring carbon–hydrogen bond form the alkene π bond, and H⁺ (a proton) is eliminated.

ACTIVE FIGURE 13.4

MECHANISM: Mechanism of the acid-catalyzed dehydration of a tertiary alcohol to yield an alkene. The process is an E1 reaction and involves a carbocation intermediate. **Go to this book's student companion site at** www.cengage.com/chemistry/mcmurry **to explore an interactive version of this figure.**

© John McMurry

To circumvent the need for strong acid and allow the dehydration of secondary alcohols in a gentler way, reagents have been developed that are effective under mild, basic conditions. One such reagent, phosphorus oxychloride (POCl$_3$) in the basic amine solvent pyridine, is often able to effect the dehydration of secondary and tertiary alcohols at 0 °C.

1-Methylcyclohexanol   **1-Methylcyclohexene (96%)**

Alcohol dehydrations carried out with POCl$_3$ in pyridine take place by an E2 mechanism, as shown in Figure 13.5. Because hydroxide ion is a poor leaving group (Section 12.7), direct E2 elimination of water from an alcohol does not occur. On reaction with POCl$_3$, however, the −OH group is converted into a dichlorophosphate group (−OPOCl$_2$), which is a good leaving group and is readily eliminated. Pyridine is both the reaction solvent and the base that removes a neighboring proton in the E2 elimination step.

FIGURE 13.5 MECHANISM: Mechanism of the dehydration of secondary and tertiary alcohols by reaction with POCl$_3$ in pyridine. The reaction is an E2 process.

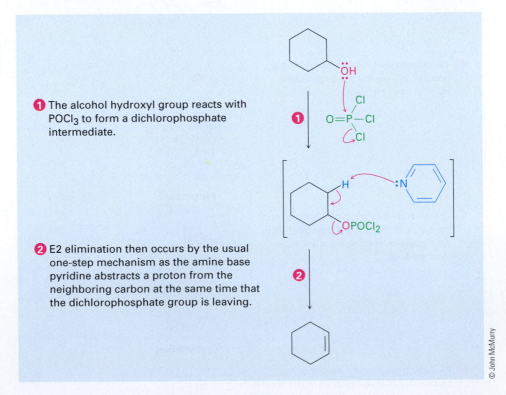

1 The alcohol hydroxyl group reacts with POCl$_3$ to form a dichlorophosphate intermediate.

2 E2 elimination then occurs by the usual one-step mechanism as the amine base pyridine abstracts a proton from the neighboring carbon at the same time that the dichlorophosphate group is leaving.

© John McMurry

As noted previously in Section 12.14, biological dehydrations are also common and usually occur by an E1cB mechanism on a substrate in which the −OH group is two carbons away from a carbonyl group. An example occurs in the biosynthesis of the aromatic amino acid tyrosine. A base (:B) first abstracts a proton from the carbon adjacent to the carbonyl group, and the anion intermediate then expels the –OH group with simultaneous protonation by an acid (HA) to form water.

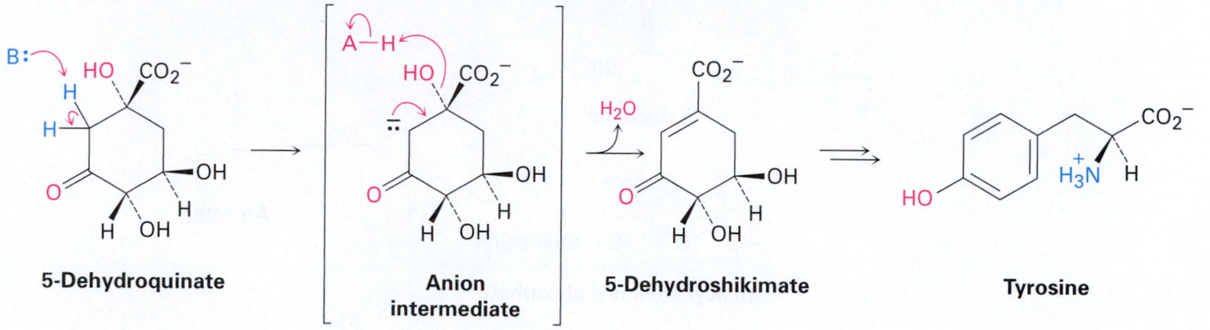

5-Dehydroquinate   Anion intermediate   5-Dehydroshikimate   Tyrosine

---

Problem 13.10

What product(s) would you expect from dehydration of the following alcohols with POCl$_3$ in pyridine? Indicate the major product in each case.

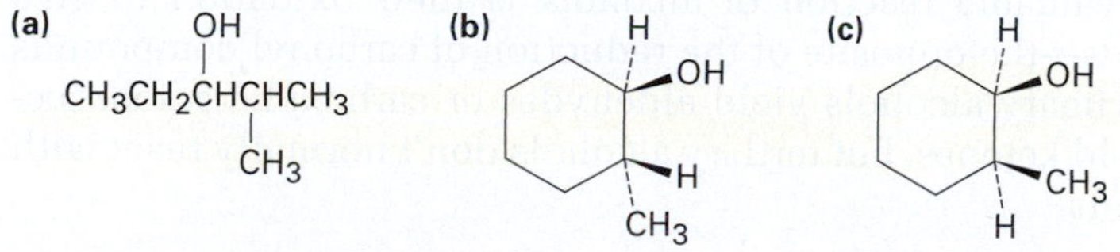

(a)   **(b)**   **(c)**

Conversion of Alcohols into Esters

Alcohols react with carboxylic acids to give esters, a reaction that is common in both the laboratory and living organisms. In the laboratory, the reaction can be carried out in a single step if a strong acid is used as catalyst. More frequently, though, the reactivity of the carboxylic acid is enhanced by first converting it into a carboxylic acid chloride, which then reacts with the alcohol. For example:

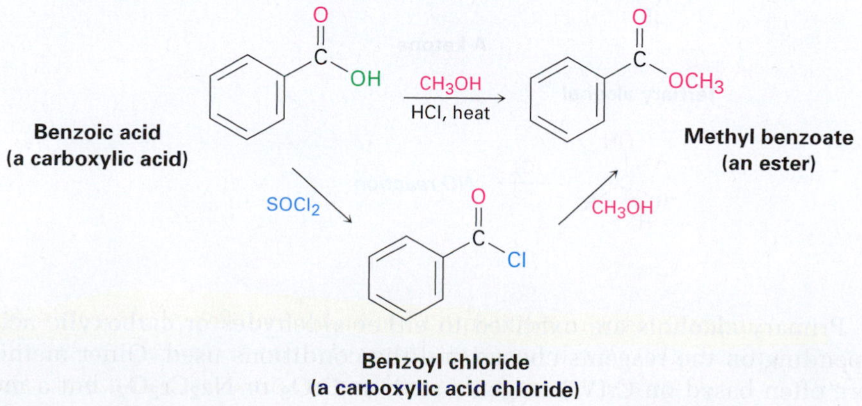

Benzoic acid
(a carboxylic acid)

CH_3OH
HCl, heat

Methyl benzoate
(an ester)

$SOCl_2$ CH_3OH

Benzoyl chloride
(a carboxylic acid chloride)

In living organisms, a similar process occurs, although a thioester or acyl adenosyl phosphate is the substrate rather than a carboxylic acid chloride. We'll look in detail at the mechanisms of these reactions in Chapter 16.

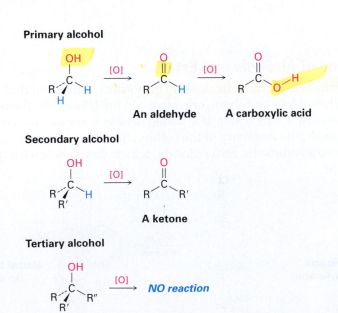

13.5 Oxidation of Alcohols and Phenols

Oxidation of Alcohols

Perhaps the most valuable reaction of alcohols is their oxidation to give carbonyl compounds—the opposite of the reduction of carbonyl compounds to give alcohols. Primary alcohols yield aldehydes or carboxylic acids, secondary alcohols yield ketones, but tertiary alcohols don't normally react with most oxidizing agents.

Primary alcohols are oxidized to either aldehydes or carboxylic acids, depending on the reagents chosen and the conditions used. Older methods were often based on Cr(VI) reagents such as CrO_3 or $Na_2Cr_2O_7$, but a more common current choice for preparing an aldehyde from a primary alcohol in the laboratory is to use the I(V)-containing *Dess–Martin periodinane* in dichloromethane solvent.

Geraniol → **Geranial (84%)**
Dess–Martin periodinane, CH_2Cl_2

−OAc = acetate

Dess–Martin periodinane

Most other commonly used oxidizing agents, such as chromium trioxide (CrO_3) in aqueous acid, oxidize primary alcohols directly to carboxylic acids. An aldehyde is involved as an intermediate in this reaction but can't usually be isolated because it is further oxidized too rapidly.

$$CH_3(CH_2)_8CH_2OH \xrightarrow[H_3O^+, \text{ acetone}]{CrO_3} CH_3(CH_2)_8\overset{O}{\overset{\|}{C}}OH$$

Decan-1-ol → **Decanoic acid (93%)**

Secondary alcohols are easily oxidized to give ketones. For a sensitive or costly alcohol, the Dess–Martin procedure is often used because the reaction is nonacidic and occurs at lower temperatures. For a large-scale oxidation, however, an inexpensive reagent such as $Na_2Cr_2O_7$ in aqueous acetic acid might be used.

4-*tert*-Butylcyclohexanol → **4-*tert*-Butylcyclohexanone (91%)**
$Na_2Cr_2O_7$, H_2O, CH_3CO_2H, heat

All these oxidations occur by a mechanism that is closely related to the E2 reaction (Section 12.12). In the Dess–Martin oxidation, for instance, the first step involves a substitution reaction between the alcohol and the I(V) reagent to form a new periodinane intermediate, followed by expulsion of reduced I(III) as the leaving group. Similarly, when a Cr(VI) reagent, such as CrO_3, is the oxidant, reaction with the alcohol gives a chromate intermediate followed by expulsion of a reduced Cr(IV) species. Although we usually think of the E2 reaction as a means of generating a carbon–*carbon* double bond by elimination of a halide leaving group, the reaction is also useful for generating a carbon–*oxygen* double bond by elimination of a reduced iodine or metal as the leaving group.

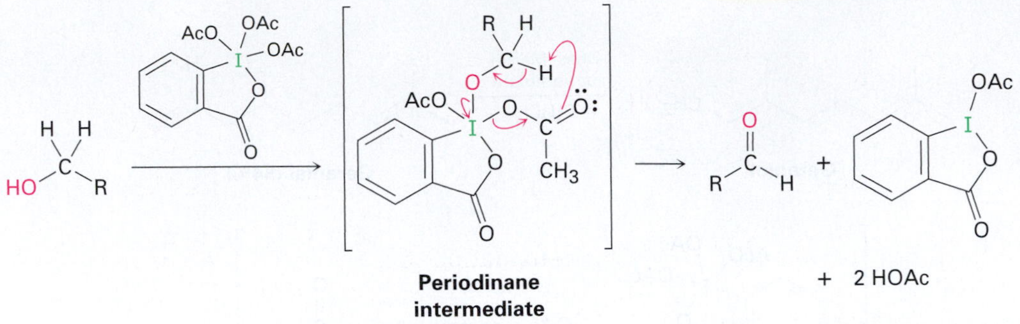

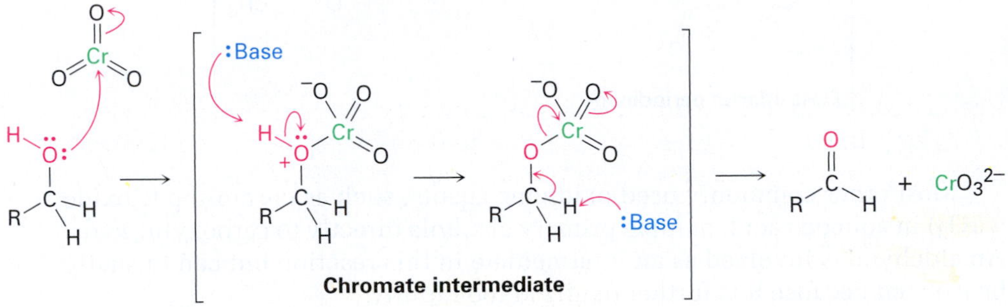

Periodinane intermediate

Chromate intermediate

Biological alcohol oxidations are the opposite of biological carbonyl reductions and are carried by the coenzymes NAD^+ and $NADP^+$. A base removes the –OH proton, and the alkoxide ion transfers a hydride ion to the coenzyme. An example is the oxidation of *sn*-glycerol 3-phosphate to dihydroxyacetone phosphate, a step in the biological metabolism of fats (Figure 13.6). Note that addition occurs exclusively on the *Re* face of the NAD^+ ring (Section 5.11), adding a hydrogen with *pro-R* stereochemistry.

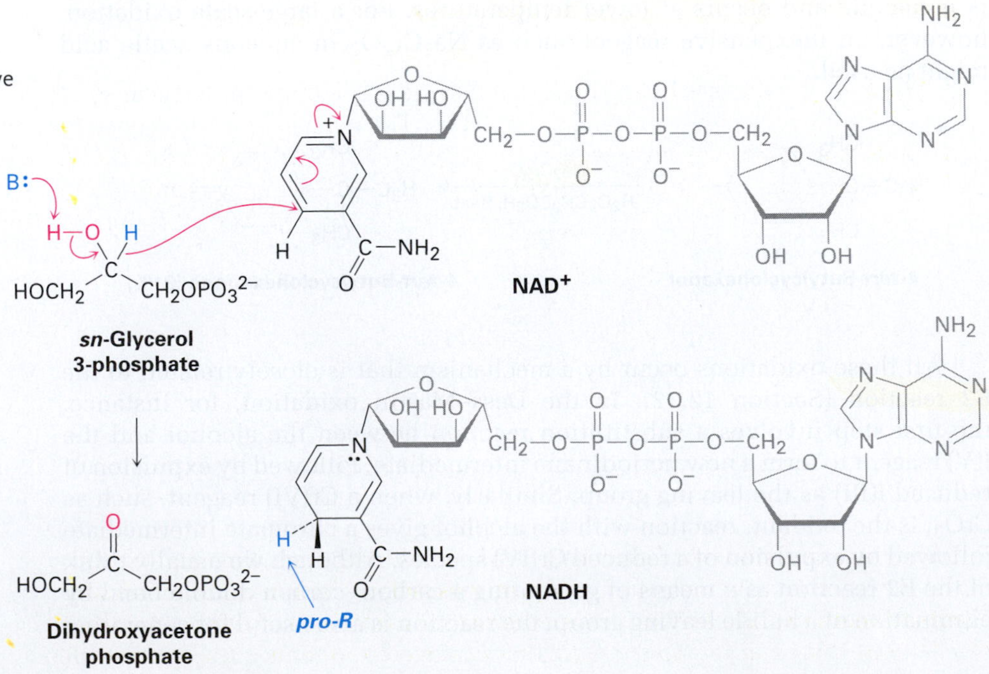

FIGURE 13.6 The biological oxidation of an alcohol (*sn*-glycerol 3-phosphate) to give a ketone (dihydroxyacetone phosphate). This mechanism is the exact opposite of the ketone reduction shown previously in Figure 13.3.

sn-Glycerol 3-phosphate

Dihydroxyacetone phosphate

Problem 13.11

What alcohols would give the following products on oxidation?

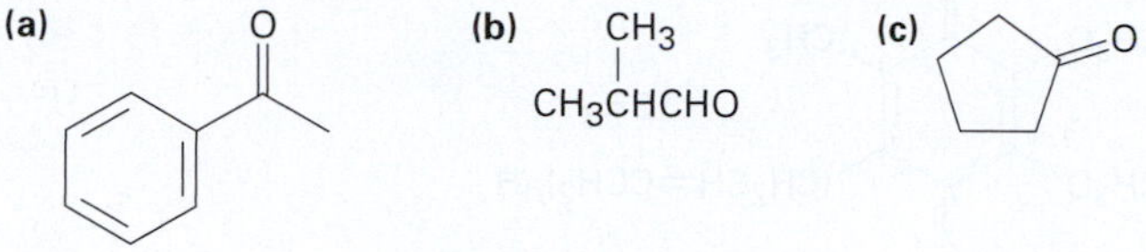

(a) **(b)** CH$_3$
CH$_3$CHCHO **(c)**

Problem 13.12

What products would you expect from oxidation of the following compounds with CrO$_3$ in aqueous acid? With the Dess–Martin periodinane?
(a) Hexan-1-ol **(b)** Hexan-2-ol **(c)** Hexanal

Oxidation of Phenols: Quinones

Phenols don't undergo oxidation in the same way that alcohols do because they don't have a hydrogen atom on the hydroxyl-bearing carbon. Instead, oxidation of a phenol yields a cyclohexa-2,5-diene-1,4-dione, or **quinone**. The most commonly used oxidant is Fremy's salt [potassium nitrosodisulfonate, $(KSO_3)_2NO$], and the reaction takes place through a radical mechanism.

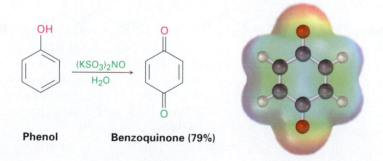

Phenol   **Benzoquinone (79%)**

Quinones are an interesting and valuable class of compounds because of their oxidation–reduction, or redox, properties. They can be easily reduced to **hydroquinones** (*p*-dihydroxybenzenes) by reagents such as NaBH$_4$ and SnCl$_2$, and hydroquinones can be easily reoxidized back to quinones by Fremy's salt.

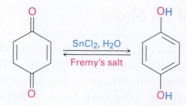

Benzoquinone   **Hydroquinone**

The redox properties of quinones are crucial to the functioning of living cells, where compounds called *ubiquinones* act as biochemical oxidizing agents to mediate the electron-transfer processes involved in energy production. Ubiquinones, also called *coenzymes Q*, are components of the cells of all aerobic organisms, from the simplest bacterium to humans. They are so named because of their ubiquitous occurrence in nature.

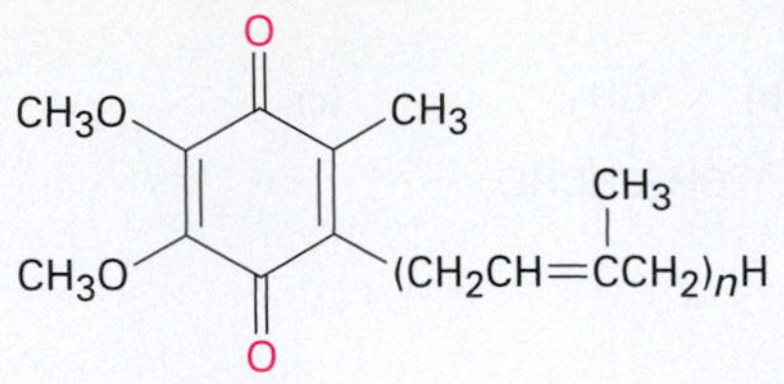

Ubiquinones (n = 1–10)

Ubiquinones function within the mitochondria of cells to mediate the respiration process in which electrons are transported from the biological reducing agent NADH to molecular oxygen. Through a complex series of steps, the ultimate result is a cycle whereby NADH is oxidized to NAD$^+$, O$_2$ is reduced to water, and energy is produced. Ubiquinone acts only as an intermediary and is itself unchanged.

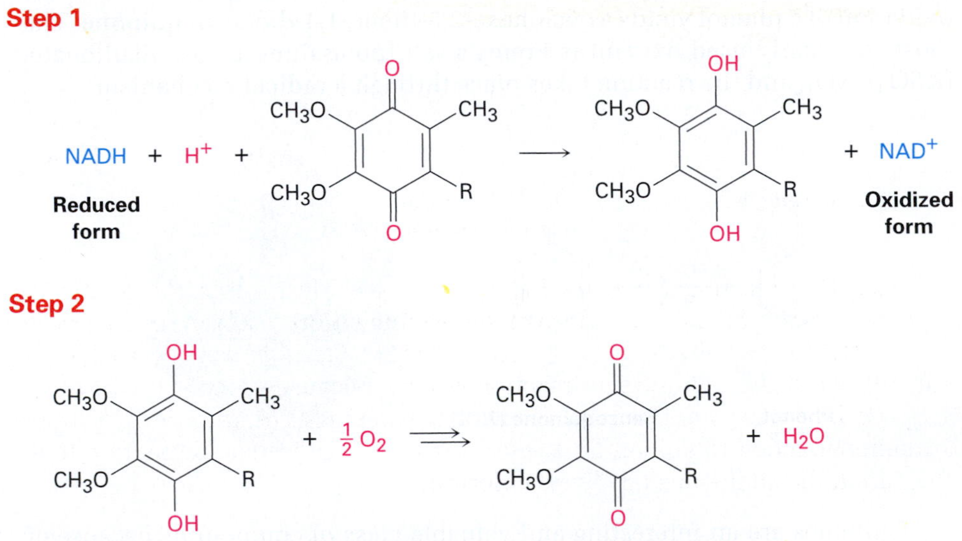

Step 1

Step 2

Net change: NADH   + $\frac{1}{2}$O$_2$ +   H$^+$ $\longrightarrow$ NAD$^+$ +   H$_2$O

13.6 Protection of Alcohols

It often happens, particularly during the preparation of complex molecules, that one functional group in a molecule interferes with an intended reaction on another functional group elsewhere in the same molecule. For example, a Grignard reagent can't be prepared from a halo alcohol because the C–Mg bond is not compatible with the presence of an acidic –OH group in the same molecule.

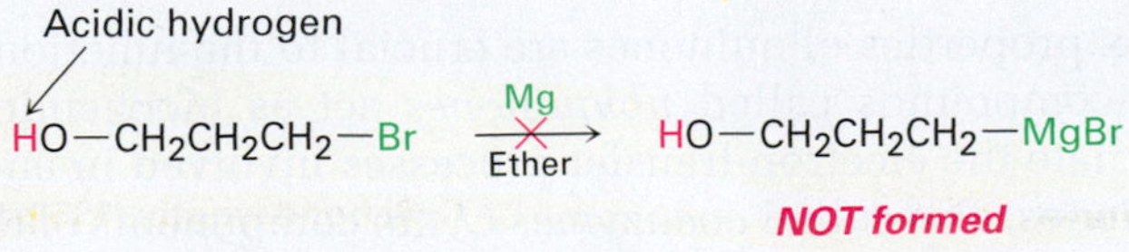

NOT formed

When this kind of incompatibility arises, it's sometimes possible to circumvent the problem by *protecting* the interfering functional group. Protection involves three steps: (1) introducing a **protecting group** to block the interfering function, (2) carrying out the desired reaction, and (3) removing the protecting group.

One of the more common methods of alcohol protection is by reaction with a chlorotrialkylsilane, Cl–SiR$_3$, to yield a trialkylsilyl ether, R'–O–SiR$_3$. Chlorotrimethylsilane is often used, and the reaction is carried out in the presence of a base, such as triethylamine, to help form the alkoxide anion from the alcohol and to remove the HCl by-product from the reaction.

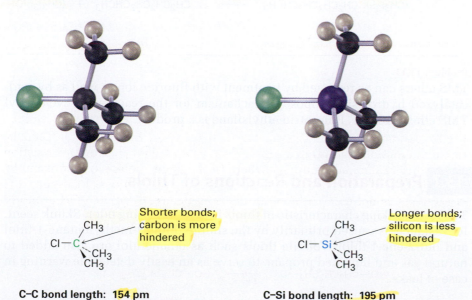

| An alcohol | Chlorotrimethyl-silane | A trimethylsilyl (TMS) ether |

For example:

Cyclohexanol → Cyclohexyl trimethylsilyl ether (94%)

The ether-forming step is an S$_N$2-like reaction of the alkoxide ion on the silicon atom, with concurrent loss of the leaving chloride anion. Unlike most S$_N$2 reactions, though, this reaction takes place at a *tertiary* center—a trialkyl-substituted silicon atom. The reaction occurs because silicon, a third-row atom, is larger than carbon and forms longer bonds. The three methyl substituents attached to silicon thus offer less steric hindrance to reaction than they do in the analogous *tert*-butyl chloride.

Shorter bonds; carbon is more hindered

Longer bonds; silicon is less hindered

C–C bond length: **154 pm**   C–Si bond length: **195 pm**

Like most other ethers, which we'll study later in this chapter, TMS ethers are relatively unreactive. They have no acidic hydrogens and don't react with oxidizing agents, reducing agents, or Grignard reagents. They do, however, react with aqueous acid or with fluoride ion to regenerate the alcohol.

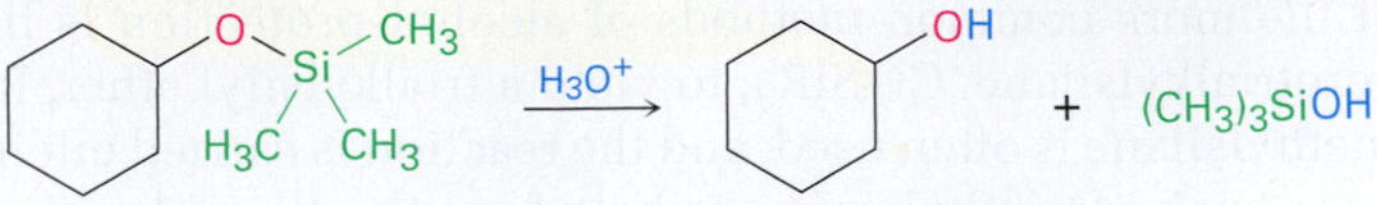

Cyclohexyl trimethylsilyl ether   **Cyclohexanol**   + (CH$_3$)$_3$SiOH

To now solve the problem posed at the beginning of this section, it's possible to use a halo alcohol in a Grignard reaction by employing a protection sequence. For example, we can add 3-bromopropan-1-ol to acetaldehyde by the route shown in Figure 13.7.

FIGURE 13.7 Use of a TMS-protected alcohol during a Grignard reaction.

Step 1   Protect alcohol:

HOCH$_2$CH$_2$CH$_2$Br + (CH$_3$)$_3$SiCl $\xrightarrow{\text{(CH}_3\text{CH}_2\text{)}_3\text{N}}$ (CH$_3$)$_3$SiOCH$_2$CH$_2$CH$_2$Br

Step 2a   Form Grignard reagent:

(CH$_3$)$_3$SiOCH$_2$CH$_2$CH$_2$Br $\xrightarrow[\text{Ether}]{\text{Mg}}$ (CH$_3$)$_3$SiOCH$_2$CH$_2$CH$_2$MgBr

Step 2b   Do Grignard reaction:

Step 3   Remove protecting group:

Problem 13.13

TMS ethers can be removed by treatment with fluoride ion as well as by acid-catalyzed hydrolysis. Propose a mechanism for the reaction of cyclohexyl TMS ether with LiF. Fluorotrimethylsilane is a product.

13.7 Preparation and Reactions of Thiols

The most striking characteristic of thiols is their appalling odor. Skunk scent, for instance, is caused primarily by the simple thiols 3-methylbutane-1-thiol and but-2-ene-1-thiol. Volatile thiols such as ethanethiol are also added to natural gas and liquefied propane to serve as an easily detectable warning in case of leaks.

Thiols are usually prepared from alkyl halides by S_N2 displacement with a sulfur nucleophile such as hydrosulfide anion, ^-SH.

CH$_3$CH$_2$CH$_2$CH$_2$CH$_2$CH$_2$CH$_2$CH$_2$—Br + $^-\!:\!SH$ $\longrightarrow$ CH$_3$CH$_2$CH$_2$CH$_2$CH$_2$CH$_2$CH$_2$CH$_2$—SH + Br$^-$

1-Bromooctane   **Octane-1-thiol (83%)**

The reaction often works poorly unless an excess of the nucleophile is used because the product thiol can undergo a second S_N2 reaction with alkyl halide to give a sulfide as a by-product. To circumvent this problem, thiourea, $(NH_2)_2C\!=\!S$, is often used as the nucleophile in the preparation of a thiol from an alkyl halide. The reaction occurs by displacement of the halide ion to yield an intermediate alkylisothiourea salt, which is hydrolyzed by subsequent reaction with aqueous base.

CH$_3$CH$_2$CH$_2$CH$_2$CH$_2$CH$_2$CH$_2$CH$_2$—Br +

Thiourea (H$_2$N—C(=S)—NH$_2$)

$\longrightarrow$

$\left[\text{CH}_3\text{CH}_2\text{CH}_2\text{CH}_2\text{CH}_2\text{CH}_2\text{CH}_2\text{CH}_2 \overset{+}{S}{=}C(\text{NH}_2){-}\text{NH}_2 \right]$ Br$^-$

1-Bromooctane   **Thiourea**

$\downarrow$ H$_2$O, NaOH

CH$_3$CH$_2$CH$_2$CH$_2$CH$_2$CH$_2$CH$_2$CH$_2$—SH + H$_2$N—C(=O)—NH$_2$

Octane-1-thiol (83%)   **Urea**

Thiols can be oxidized by Br$_2$ or I$_2$ to yield **disulfides (RSSR′)**. The reaction is easily reversed, and a disulfide can be reduced back to a thiol by treatment with zinc and acid:

$$2\,\text{R—SH} \underset{\text{Zn, H}^+}{\overset{\text{I}_2}{\rightleftharpoons}} \text{R—S—S—R} + 2\,\text{HI}$$

A thiol   **A disulfide**

This thiol–disulfide interconversion is a key part of numerous biological processes. We'll see in Section 19.8, for instance, that disulfide formation is involved in defining the structure and three-dimensional conformations of proteins, where disulfide "bridges" often form cross-links between cysteine amino acid units in the protein chains. Disulfide formation is also involved in the process by which cells protect themselves from oxidative degradation. A cellular component called *glutathione* removes potentially harmful oxidants and is itself oxidized to glutathione disulfide in the process.

H$_2$O$_2$ / FADH$_2$

Glutathione (GSH)   **Glutathione disulfide (GSSG)**

Problem 13.14

But-2-ene-1-thiol is one component of skunk spray. How would you synthesize this substance from methyl but-2-enoate?

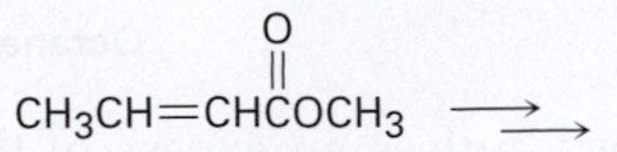

$$CH_3CH=CHCOCH_3 \longrightarrow CH_3CH=CHCH_2SH$$

Methyl but-2-enoate   **But-2-ene-1-thiol**

13.8 Ethers and Sulfides

Simple ethers with no other functional groups are named by identifying the two organic substituents and adding the word *ether*:

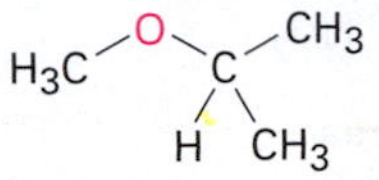

Isopropyl methyl ether 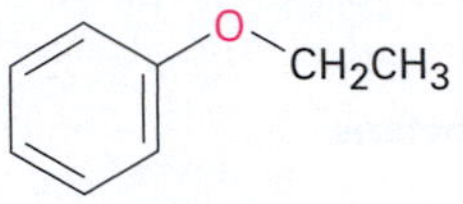   **Ethyl phenyl ether**

If other functional groups are present, the ether part is considered an *alkoxy* substituent. For example:

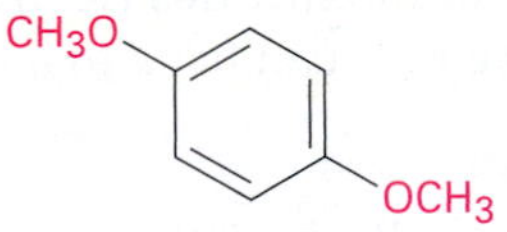

p-**Dimethoxy**benzene   4-*tert*-**Butoxy**cyclohexene

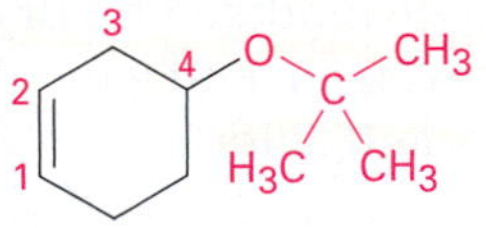

Sulfides are named by following the same rules used for ethers, with *sulfide* used in place of *ether* for simple compounds and *alkylthio* used in place of *alkoxy* for more complex substances.

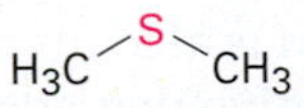

Dimethyl **sulfide** 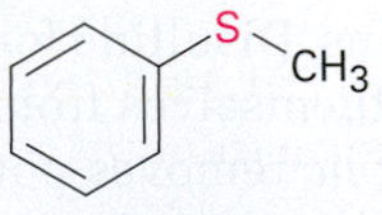   Methyl phenyl **sulfide** 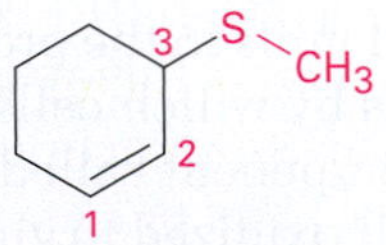   3-(**Methylthio**)cyclohexene

Like alcohols, ethers have nearly the same geometry as water. The R–O–R bonds have an approximately tetrahedral bond angle (112° in dimethyl ether), and the oxygen atom is sp^3-hybridized.

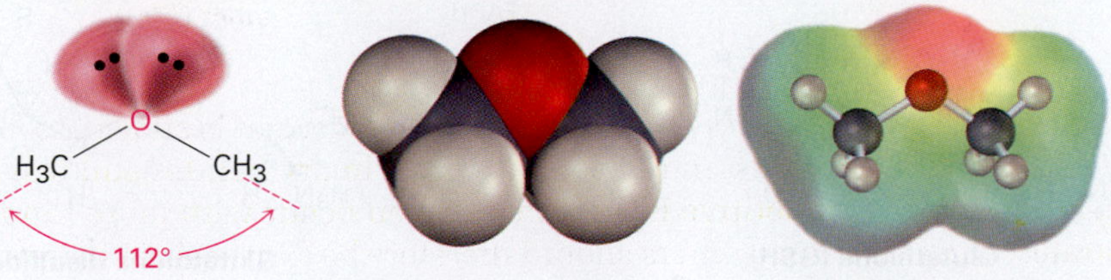

H_3C CH_3

112°

Problem 13.15

Name the following ethers and sulfides according to IUPAC rules:

(a) 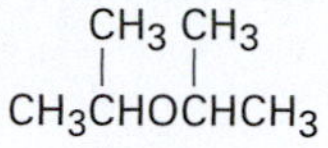
$$CH_3CHOCHCH_3$$
with CH₃ groups: CH₃ CH₃

(b) 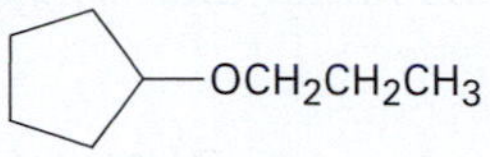 cyclopentyl—OCH₂CH₂CH₃

(c) 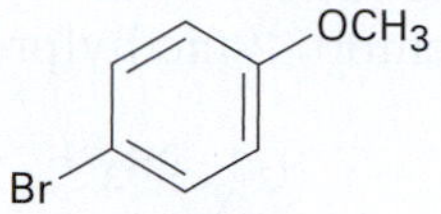 structure with Br on benzene ring and OCH₃

(d) 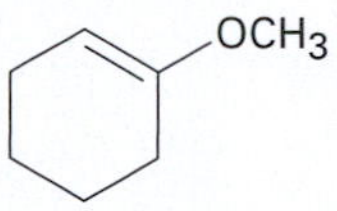 cyclohexene with OCH₃

(e) 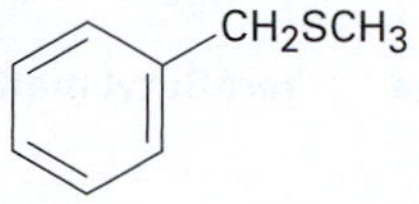 benzyl group with CH₂SCH₃

(f) $CH_3SCH_2CH=CH_2$

13.9 Preparing Ethers

Diethyl ether and other simple symmetrical ethers are prepared industrially by the sulfuric acid–catalyzed reaction of alcohols. The process occurs by S_N2 displacement of water from a protonated ethanol molecule by the oxygen atom of a second ethanol. Unfortunately, the method is limited to use with primary alcohols because secondary and tertiary alcohols dehydrate by an E1 mechanism to yield alkenes.

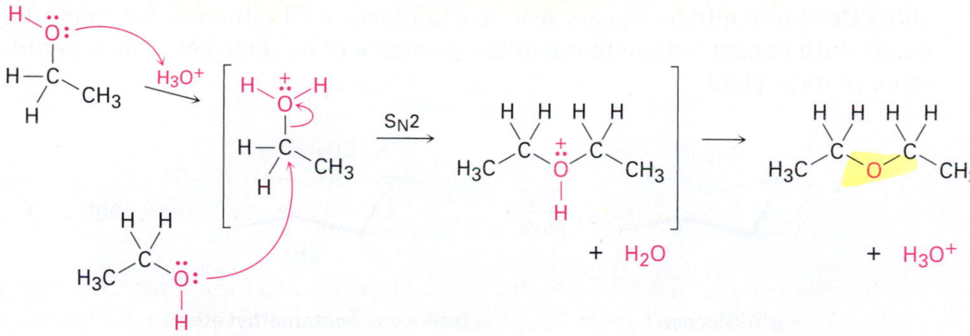

The most generally useful method of preparing ethers is by the *Williamson ether synthesis,* in which an alkoxide ion reacts with a primary alkyl halide or tosylate in an S_N2 reaction. As we saw earlier in Section 13.2, the alkoxide ion is normally prepared by reaction of an alcohol with a strong base such as sodium hydride, NaH.

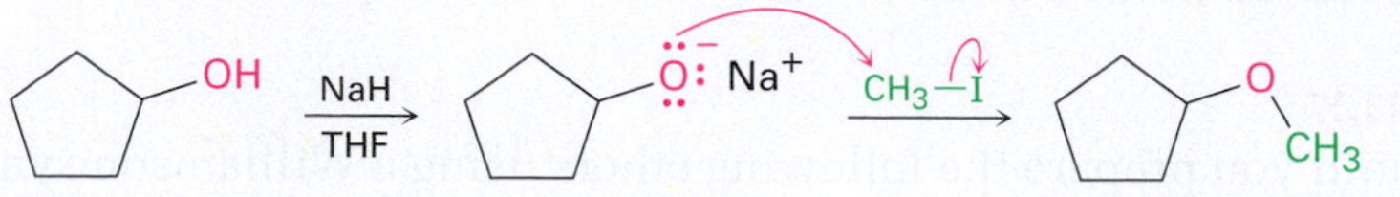

Cyclopentanol   **Alkoxide ion**   **Cyclopentyl methyl ether (74%)**

Because the Williamson synthesis is an S_N2 reaction, it is subject to all the usual constraints, as discussed in Section 12.7. Primary halides and tosylates work best because competitive E2 elimination can occur with more hindered substrates. Unsymmetrical ethers should therefore be synthesized by reaction between the more hindered alkoxide partner and less hindered halide partner rather than vice versa. For example, *tert*-butyl methyl ether, a substance used in the 1990s as an octane booster in gasoline, is best prepared by reaction of *tert*-butoxide ion with iodomethane rather than by reaction of methoxide ion with 2-chloro-2-methylpropane.

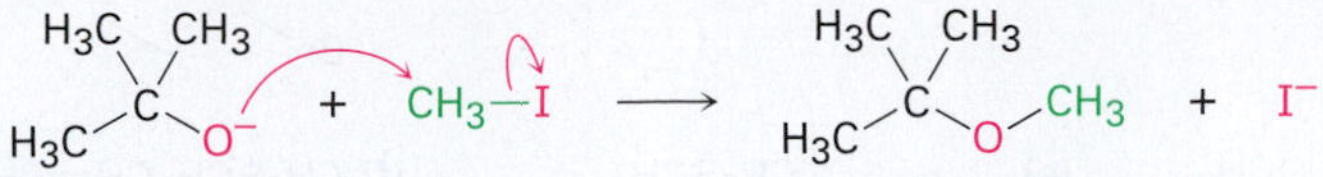

tert-Butoxide   **Iodomethane**   **tert-Butyl methyl ether**

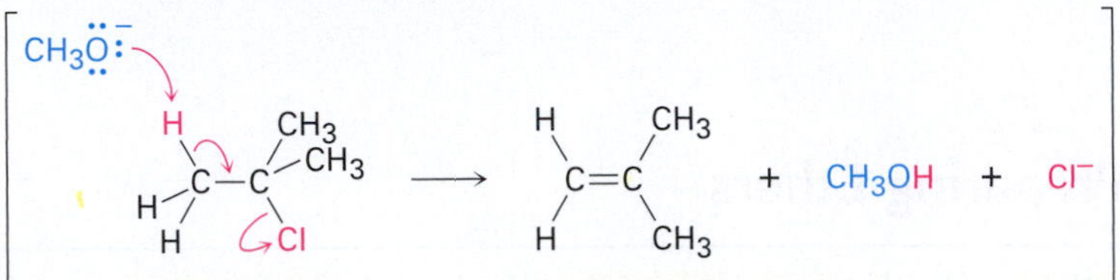

2-Chloro-2-   **2-Methylpropene**
methylpropane

A variation of the Williamson synthesis involves using silver oxide, Ag_2O, as a mild base rather than NaH. Under these conditions, the free alcohol reacts directly with alkyl halide, so there is no need to preform the metal alkoxide intermediate. Sugars react particularly well; glucose, for example, reacts with excess iodomethane in the presence of Ag_2O to generate a *penta-ether* in 85% yield.

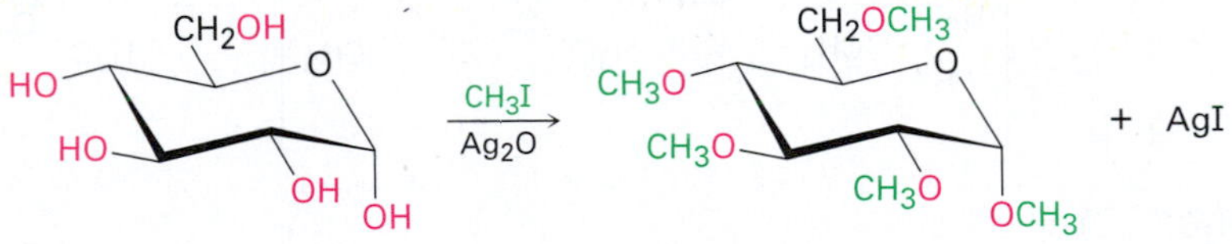

α-D-Glucose   **α-D-Glucose pentamethyl ether**
                              (85%)

Problem 13.16

Why do you suppose only symmetrical ethers are prepared by the sulfuric acid–catalyzed dehydration procedure? What product(s) would you expect if ethanol and propan-1-ol were allowed to react together? In what ratio would the products be formed if the two alcohols were of equal reactivity?

Problem 13.17

How would you prepare the following ethers using a Williamson synthesis?
(a) Methyl propyl ether   **(b)** Anisole (methyl phenyl ether)
(c) Benzyl isopropyl ether   **(d)** Ethyl 2,2-dimethylpropyl ether

Problem 13.18

Rank the following halides in order of their reactivity in the Williamson synthesis:
(a) Bromoethane, 2-bromopropane, bromobenzene
(b) Chloroethane, bromoethane, 1-iodopropene

13.10 Reactions of Ethers

Ethers are unreactive to many reagents used in organic chemistry, a property that accounts for their wide use as reaction solvents. Halogens, dilute acids, bases, and nucleophiles have no effect on most ethers.

Cleavage of Ethers

Ethers undergo only one reaction of truly general use—they are cleaved by strong acids. Aqueous HBr and HI both work well, but HCl does not cleave ethers.

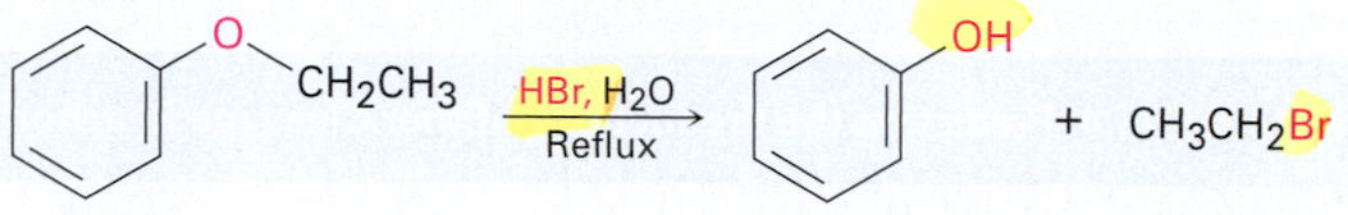

Ethyl phenyl ether   **Phenol**   **Bromoethane**

Acidic ether cleavages are typical nucleophilic substitution reactions and take place by either S_N1 or S_N2 mechanisms depending on the structure of the substrate. Ethers with only primary and secondary alkyl groups react by an S_N2 mechanism, in which I^- or Br^- attacks the protonated ether at the less hindered site. This usually results in a selective cleavage into a single alcohol and a single alkyl halide. For example, ethyl isopropyl ether yields exclusively isopropyl alcohol and iodoethane on cleavage by HI because nucleophilic attack by iodide ion occurs at the less hindered primary site rather than at the more hindered secondary site.

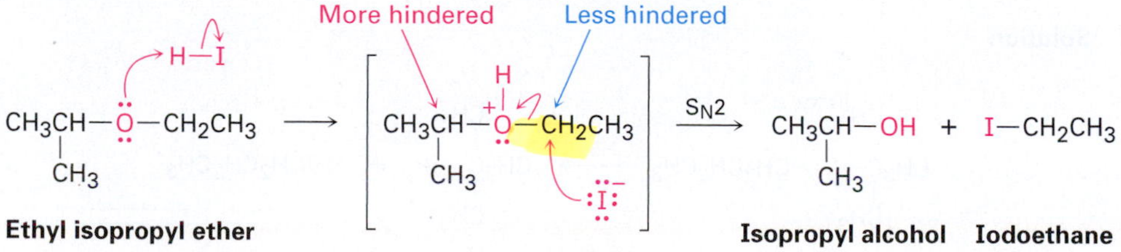

Ethyl isopropyl ether   **Isopropyl alcohol   Iodoethane**

Ethers with a tertiary, benzylic, or allylic group cleave by either an S_N1 or E1 mechanism because these substrates can produce stable intermediate carbocations. These reactions are often fast and take place at moderate temperatures. *tert*-Butyl ethers, for example, react by an E1 mechanism on treatment with trifluoroacetic acid at 0 °C. We'll see in Section 19.7 that the reaction is often used in the laboratory synthesis of peptides.

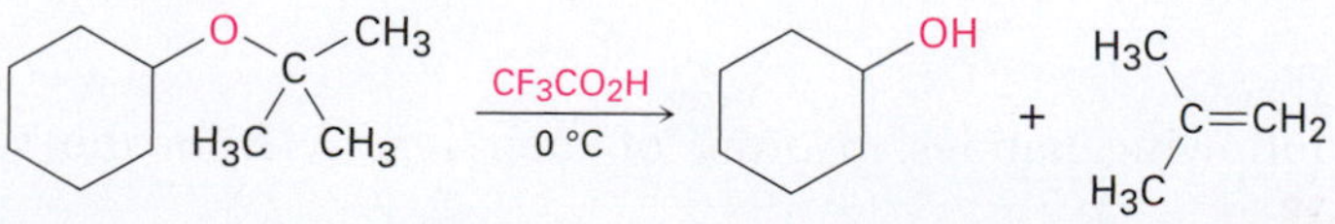

tert-Butyl cyclohexyl ether   **Cyclohexanol   2-Methylpropene**
                                          (90%)

You might also recall from Section 12.7 that epoxides, cyclic ethers with a three-membered ring, are unusually reactive in S_N2 processes because of angle strain. Methylenecyclohexane oxide, for instance, undergoes a *base-induced* S_N2 ring-opening on treatment with hydroxide ion at 100 °C.

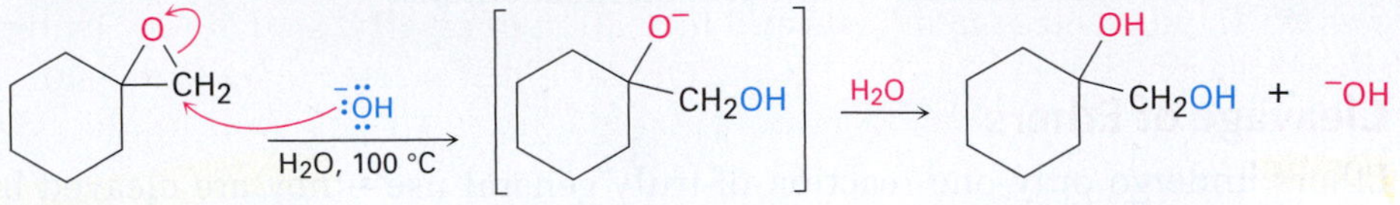

Methylenecyclohexane oxide

1-Hydroxymethyl-cyclohexanol (70%)

WORKED EXAMPLE 13.4 Predicting the Product of an Ether Cleavage Reaction

Predict the products of the following reaction:

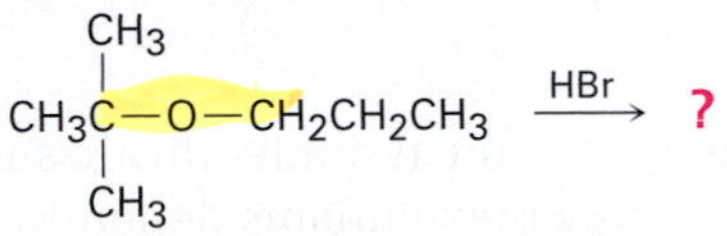

Strategy

Identify the substitution pattern of the two groups attached to oxygen—in this case a tertiary alkyl group and a primary alkyl group. Then recall the guidelines for ether cleavages. An ether with only primary and secondary alkyl groups usually undergoes cleavage by S_N2 attack of a nucleophile on the less hindered alkyl group, but an ether with a tertiary alkyl group usually undergoes cleavage by an S_N1 mechanism. In this case, an S_N1 cleavage of the tertiary C–O bond will occur, giving propan-1-ol and a tertiary alkyl bromide.

Solution

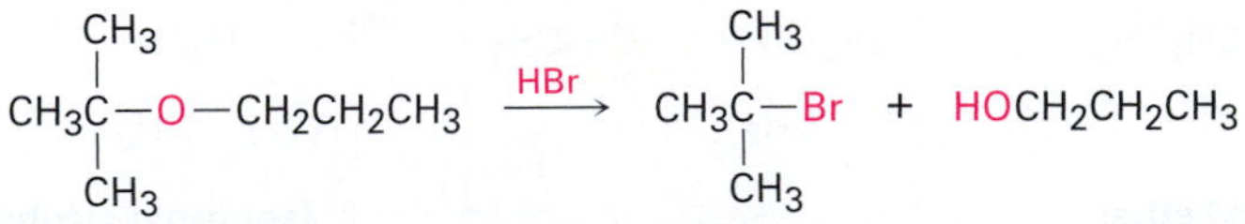

tert-Butyl propyl ether

2-Bromo-2-methylpropane

Propan-1-ol

Problem 13.19

Predict the products of the following reactions:

(a)

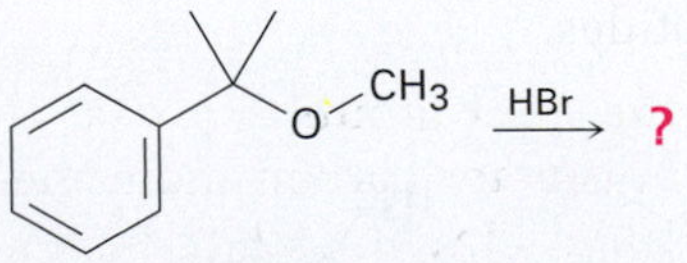

(b)

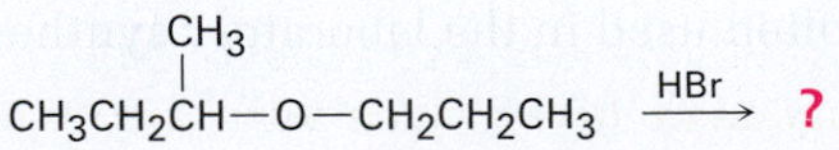

Problem 13.20

Write a mechanism for the acid-catalyzed cleavage of *tert*-butyl cyclohexyl ether with trifluoroacetic acid to yield cyclohexanol and 2-methylpropene.

Claisen Rearrangement of Allyl Aryl Ethers

Unlike the acid-catalyzed ether cleavage reaction just discussed, which is general for all ethers, the **Claisen rearrangement** is specific to allyl aryl ethers (Ar—O—$CH_2CH{=}CH_2$) and allyl vinyl ethers ($H_2C{=}CH$—O—$CH_2CH{=}CH_2$). Treatment of a phenoxide ion with 3-bromopropene (allyl bromide) results in a Williamson ether synthesis and formation of an allyl aryl ether. Heating the allyl aryl ether to 200–250 °C then effects Claisen rearrangement, leading to an *o*-allylphenol. The net result is alkylation of the phenol in an ortho position.

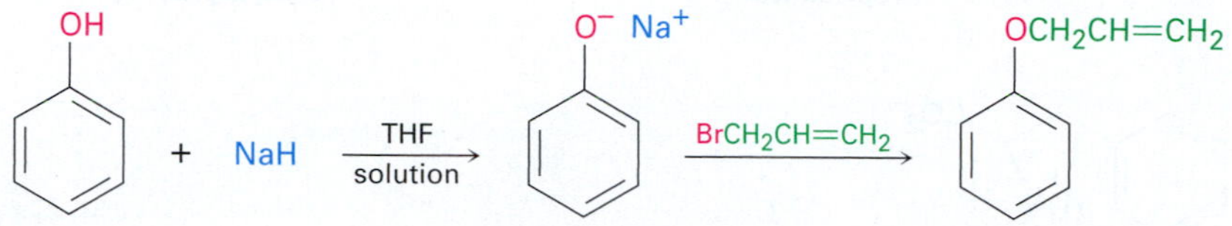

Phenol **Sodium phenoxide** **Allyl phenyl ether**

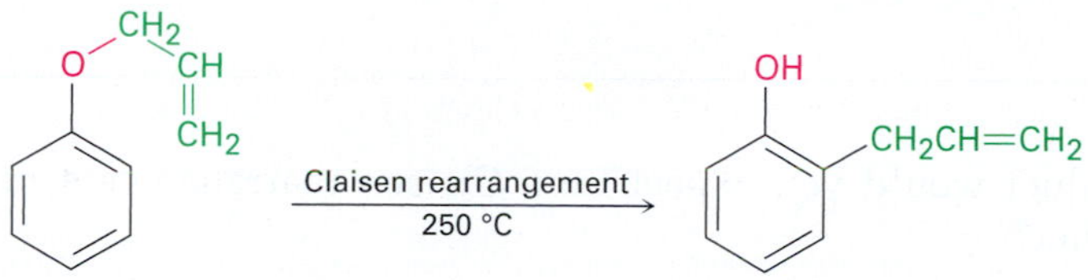

Allyl phenyl ether ***o*-Allylphenol**

Like the Diels–Alder cycloaddition reaction discussed in Section 8.14, the Claisen rearrangement reaction takes place in a single step through a pericyclic mechanism in which a reorganization of bonding electrons occurs through a six-membered, cyclic transition state. The 6-allylcyclohexa-2,4-dienone intermediate then isomerizes to *o*-allylphenol (Figure 13.8).

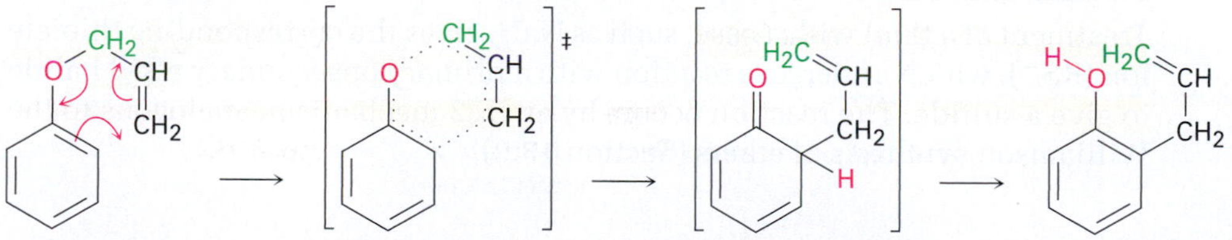

Allyl phenyl ether **Transition state** **Intermediate (6-allylcyclohexa-2,4-dienone)** ***o*-Allylphenol**

FIGURE 13.8 The mechanism of the Claisen rearrangement. The C–O bond-breaking and C–C bond-making occur simultaneously.

Evidence for this mechanism comes from the observation that the rearrangement takes place with an inversion of the allyl group. That is, allyl phenyl ether containing a ^{14}C label on the allyl *ether* carbon atom yields *o*-allylphenol in which the label is on the *terminal* vinylic carbon (green in Figure 13.8).

Claisen rearrangements are uncommon in biological pathways, but a well-studied example does occur during biosynthesis of the amino acids phenylalanine and tyrosine. Both phenylalanine and tyrosine arise from a precursor called prephenate, which is itself formed by a biological Claisen rearrangement of the allylic vinyl ether chorismate.

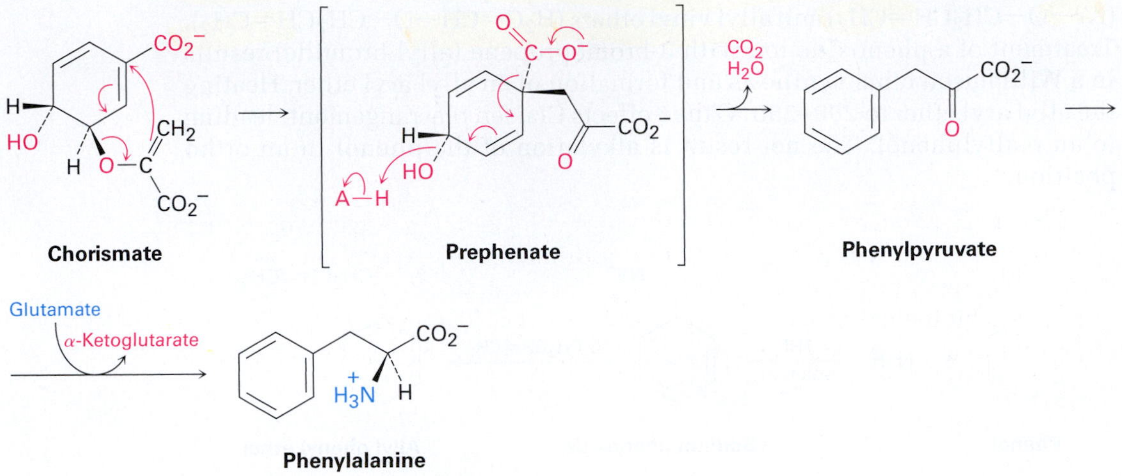

Chorismate   **Prephenate**   **Phenylpyruvate**

Phenylalanine

Problem 13.21

What product would you expect from Claisen rearrangement of but-2-enyl phenyl ether?

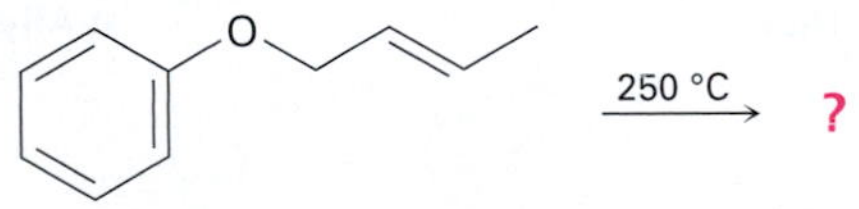

$\xrightarrow{250\ °C}$ **?**

But-2-enyl phenyl ether

13.11 Preparation and Reactions of Sulfides

Treatment of a thiol with a base, such as NaH, gives the corresponding thiolate ion (RS^-), which undergoes reaction with a primary or secondary alkyl halide to give a sulfide. The reaction occurs by an S_N2 mechanism, analogous to the Williamson synthesis of ethers (Section 13.9).

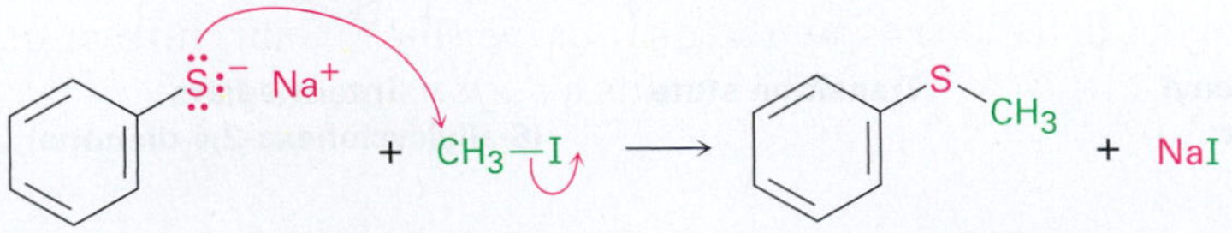

Sodium benzenethiolate   **Methyl phenyl sulfide (96%)**

Perhaps surprisingly in light of their close structural similarity, sulfides and ethers differ substantially in their chemistry. Because the valence electrons on sulfur are farther from the nucleus and are less tightly held than those on oxygen ($3p$ electrons versus $2p$ electrons), sulfur compounds are more nucleophilic than their oxygen analogs. Unlike dialkyl ethers, dialkyl sulfides react rapidly with primary alkyl halides by an S_N2 mechanism to give *sulfonium ions* (R_3S^+).

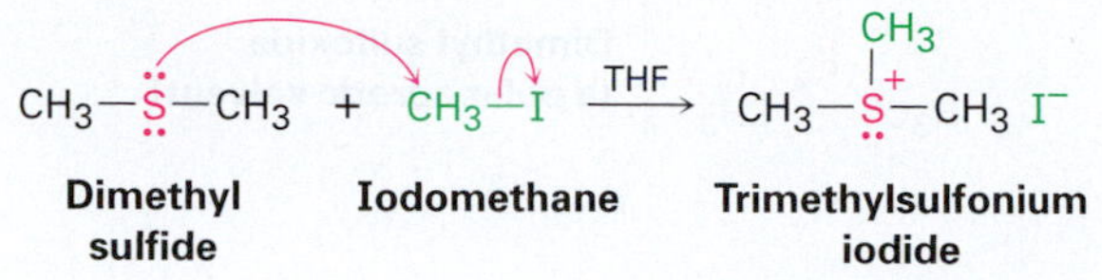

| Dimethyl sulfide | Iodomethane | Trimethylsulfonium iodide |

The most common example of this process in living organisms is the reaction of the amino acid methionine with adenosine triphosphate (ATP; Section 6.8) to give *S*-adenosylmethionine. The reaction is somewhat unusual in that the biological leaving group in this S_N2 process is the *triphosphate ion* rather than the more frequently seen *diphosphate* ion (Section 12.10).

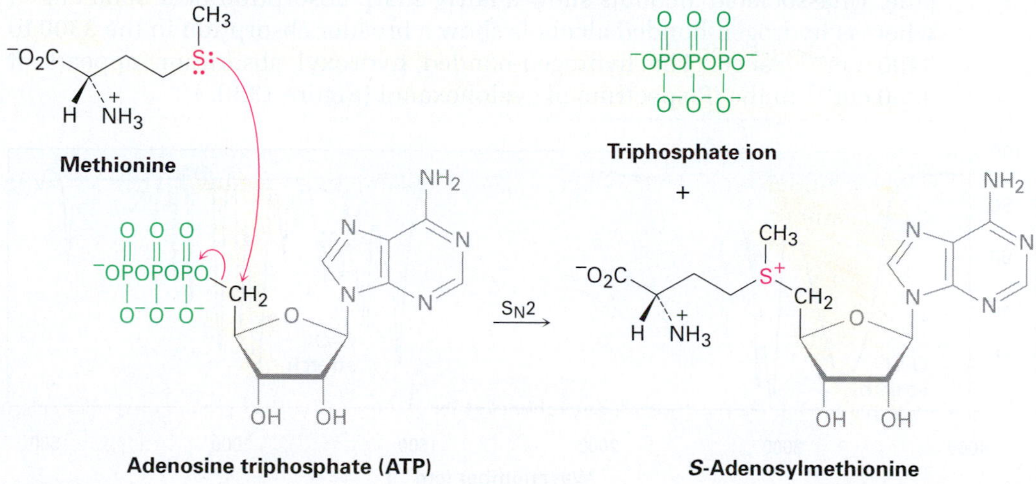

Methionine   **Triphosphate ion**

+

Adenosine triphosphate (ATP)   ***S*-Adenosylmethionine**

Sulfonium ions are themselves useful alkylating agents because a nucleophile can attack one of the groups bonded to the positively charged sulfur, displacing a neutral sulfide as leaving group. We saw an example in Section 12.10 (Figure 12.17) in which *S*-adenosylmethionine transferred a methyl group to norepinephrine to give adrenaline.

Another difference between sulfides and ethers is that sulfides are easily oxidized. Treatment of a sulfide with hydrogen peroxide, H_2O_2, at room temperature yields the corresponding *sulfoxide (R_2SO)*, and further oxidation of the sulfoxide with a peroxyacid yields a *sulfone (R_2SO_2)*.

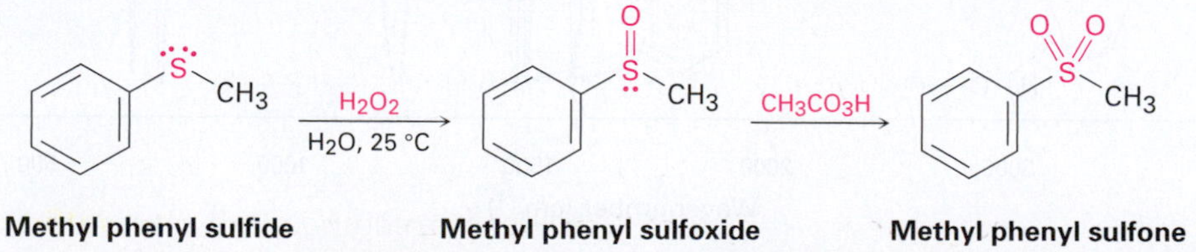

Methyl phenyl sulfide   **Methyl phenyl sulfoxide**   **Methyl phenyl sulfone**

Dimethyl sulfoxide (DMSO) is a particularly well-known sulfoxide that is often used as a polar aprotic solvent. It must be handled with care, however, because it has a remarkable ability to penetrate the skin, carrying along whatever is dissolved in it.

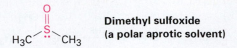

Dimethyl sulfoxide (a polar aprotic solvent)

13.12 Spectroscopy of Alcohols, Phenols, and Ethers

Infrared Spectroscopy

Alcohols have a strong C–O stretching absorption near 1050 cm^{-1} and a characteristic O–H stretching absorption at 3300 to 3600 cm^{-1}. The exact position of the O–H stretch depends on the extent of hydrogen-bonding in the molecule. Unassociated alcohols show a fairly sharp absorption near 3600 cm^{-1}, whereas hydrogen-bonded alcohols show a broader absorption in the 3300 to 3400 cm^{-1} range. The hydrogen-bonded hydroxyl absorption appears at 3350 cm^{-1} in the IR spectrum of cyclohexanol (Figure 13.9).

FIGURE 13.9
Infrared spectrum of cyclohexanol. Characteristic O–H and C–O stretching absorptions are indicated.

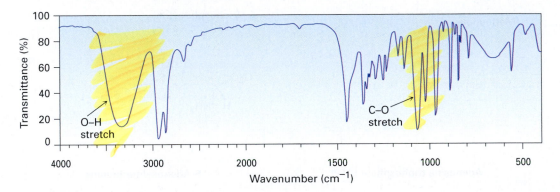

Phenols also show a characteristic broad absorption at 3500 cm^{-1} due to the –OH group, as well as the usual 1500 and 1600 cm^{-1} aromatic bands (Figure 13.10). In phenol itself, the monosubstituted aromatic-ring peaks at 690 and 760 cm^{-1} are visible.

FIGURE 13.10
Infrared spectrum of phenol.

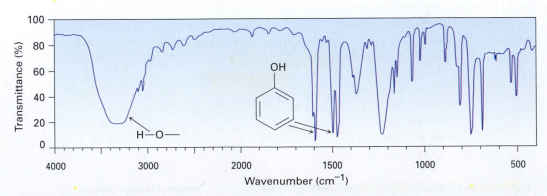

Ethers are difficult to identify by IR spectroscopy. Although they show an absorption due to C–O single-bond stretching in the range 1050 to 1150 cm^{-1}, many other kinds of absorptions occur in the same range.

Nuclear Magnetic Resonance Spectroscopy

Carbon atoms bonded to electron-withdrawing oxygen atoms are deshielded and absorb at a lower field in the ^{13}C NMR spectrum than do typical alkane carbons. Most alcohol and ether carbon absorptions fall in the range 50 to 80 δ.

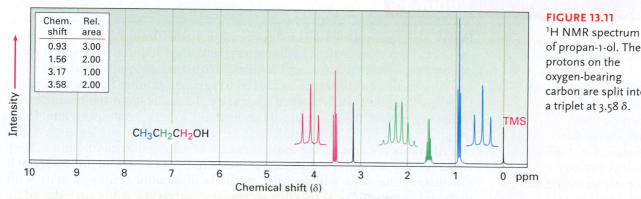

Alcohols also show characteristic absorptions in the ^{1}H NMR spectrum. Hydrogens on the oxygen-bearing carbon atom are deshielded by the electron-withdrawing effect of the nearby oxygen, and their absorptions occur in the range 3.4 to 4.5 δ. Spin–spin splitting, however, is not usually observed between the O–H proton of an alcohol and the neighboring protons on carbon. Most samples contain small amounts of acidic impurities, which catalyze an exchange of the O–H proton on a timescale so rapid that the effect of spin–spin splitting is removed. It's often possible to take advantage of this rapid proton exchange to identify the position of the O–H absorption. If a small amount of deuterated water, D_2O, is added to the NMR sample tube, the O–H proton is rapidly exchanged for deuterium, and the hydroxyl absorption disappears from the spectrum.

Typical spin–spin splitting *is* observed between protons on the oxygen-bearing carbon and other neighbors in both alcohols and ethers. Figure 13.11 shows the ^{1}H NMR spectrum of propan-1-ol.

Chem. shift	Rel. area
0.93	3.00
1.56	2.00
3.17	1.00
3.58	2.00

$CH_3CH_2CH_2OH$

FIGURE 13.11
^{1}H NMR spectrum of propan-1-ol. The protons on the oxygen-bearing carbon are split into a triplet at 3.58 δ.

Phenols, like all aromatic compounds, show ^{1}H NMR absorptions near 7 to 8 δ, the expected position for aromatic-ring protons (Section 11.9). In addition, phenol O–H protons absorb at 3 to 8 δ. In neither case are these absorptions uniquely diagnostic for phenols, since other kinds of protons absorb in the same range.

Mass Spectrometry

As noted previously in Section 10.3, alcohols undergo fragmentation in the mass spectrometer by two characteristic pathways, *alpha cleavage* and *dehydration*. In the alpha-cleavage pathway, a C–C bond nearest the hydroxyl group is broken, yielding a neutral radical plus a charged oxygen-containing fragment. In the dehydration pathway, water is eliminated, yielding an alkene radical cation. Both fragmentation modes are apparent in the mass spectrum of butan-1-ol (Figure 13.12). The peak at $m/z = 56$ is due to loss of water from the molecular ion, and the peak at $m/z = 31$ is due to an alpha cleavage.

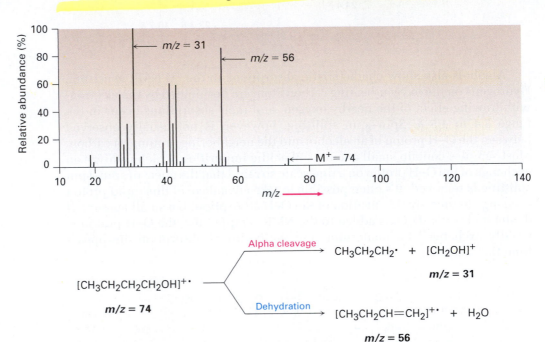

FIGURE 13.12 Mass spectrum of butan-1-ol ($M^+ = 74$). Dehydration gives a peak at $m/z = 56$, and fragmentation by alpha cleavage gives a peak at $m/z = 31$.

Alpha cleavage → $CH_3CH_2CH_2\cdot$ + $[CH_2OH]^+$ $m/z = 31$

$[CH_3CH_2CH_2CH_2OH]^{+\cdot}$ $m/z = 74$

Dehydration → $[CH_3CH_2CH=CH_2]^{+\cdot}$ + H_2O $m/z = 56$

Summary

In past chapters, we focused on developing general ideas of organic reactivity, looking at the chemistry of hydrocarbons and alkyl halides, and seeing some of the tools used in structural studies. With that accomplished, we have now begun in this chapter to study the oxygen-containing functional groups that lie at the heart of biological chemistry. To understand the chemistry of living organisms, it's necessary to understand oxygen-containing functional groups.

Alcohols are among the most versatile of all organic compounds. They occur widely in nature, are important industrially, and have an unusually rich chemistry. The most widely used methods of alcohol synthesis start with carbonyl compounds. Aldehydes, esters, and carboxylic acids are reduced by reaction with $LiAlH_4$ to give primary alcohols (RCH_2OH); ketones are reduced to yield secondary alcohols (R_2CHOH). Alcohols are also prepared by reaction of carbonyl compounds with Grignard reagents, RMgX. Addition of a Grignard reagent to formaldehyde yields a primary alcohol, addition to an aldehyde yields a secondary alcohol, and addition to a ketone or an ester yields a tertiary alcohol.

Alcohols undergo many reactions and can be converted into many other functional groups. They can be dehydrated to give alkenes by treatment with $POCl_3$ and can be transformed into alkyl halides by treatment with PBr_3 or $SOCl_2$. Furthermore, alcohols are weakly acidic and react with strong bases to form **alkoxide anions**, which are used frequently in organic synthesis. Perhaps the most important reaction of alcohols is their oxidation to carbonyl compounds. Primary alcohols yield either aldehydes or carboxylic acids, secondary alcohols yield ketones, but tertiary alcohols are not normally oxidized. An alcohol can be **protected** by formation of a trimethylsilyl (TMS) ether when the presence of the –OH group might interfere with a reaction elsewhere in the molecule.

Phenols are aromatic counterparts of alcohols but are much more acidic ($pK_a \approx 10$) because their anions are resonance stabilized by delocalization of the negative charge into the aromatic ring. Phenols can be oxidized to **quinones** by reaction with Fremy's salt (potassium nitrosodisulfonate), and quinones can be reduced to **hydroquinones** by reaction with $NaBH_4$.

Ethers are compounds that have two organic groups bonded to the same oxygen atom, ROR'. They are often prepared by the Williamson ether synthesis, which involves S_N2 reaction of an alkoxide ion with a primary alkyl halide. Ethers are inert to most reagents but react with HI and HBr to give cleavage products. The cleavage reaction takes place by an S_N2 mechanism at the less highly substituted site if only primary and secondary alkyl groups are bonded to the ether oxygen but by an S_N1 or E1 mechanism if one of the alkyl groups bonded to oxygen is tertiary. Aryl allyl ethers undergo **Claisen rearrangement** to give o-allylphenols.

Thiols, the sulfur analogs of alcohols, are usually prepared by S_N2 reaction of an alkyl halide with thiourea. Mild oxidation of a thiol yields a **disulfide**, and mild reduction of a disulfide gives back the thiol. **Sulfides**, the sulfur analogs of ethers, are prepared by S_N2 reaction between a thiolate anion and a primary or secondary alkyl halide. Sulfides are much more nucleophilic than ethers and can be oxidized to sulfoxides and to sulfones. Sulfides can also be alkylated by reaction with a primary alkyl halide to yield sulfonium ions.

Summary of Reactions

1.   Synthesis of alcohols (Section 13.3)

   (a) Reduction of carbonyl compounds

      (1) Aldehydes

Primary alcohol

      (2) Ketones

Secondary alcohol

(3) Esters

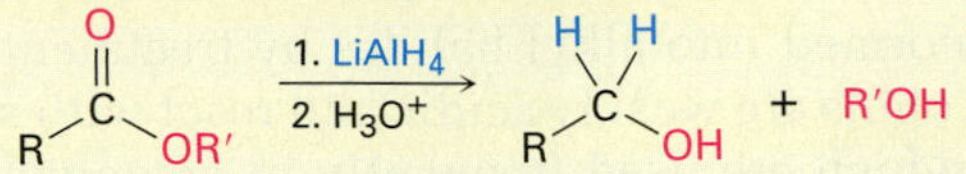

Primary alcohol

(4) Carboxylic acids

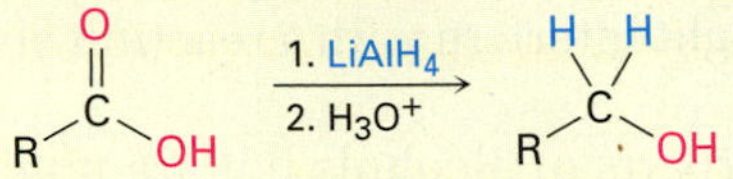

Primary alcohol

(b) Grignard addition to carbonyl compounds

(1) Formaldehyde

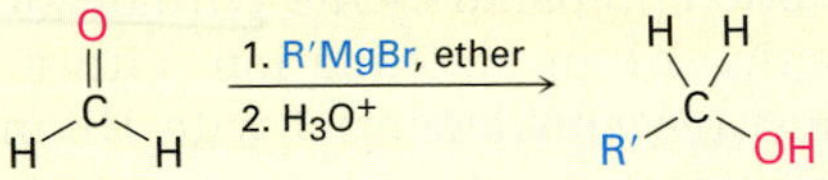

Primary alcohol

(2) Aldehydes

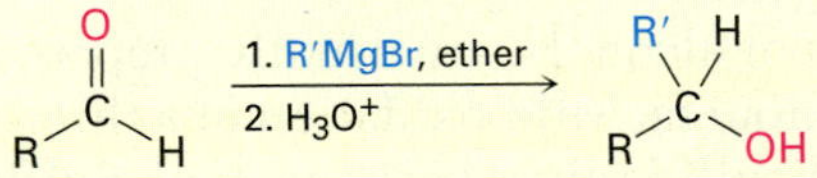

Secondary alcohol

(3) Ketones

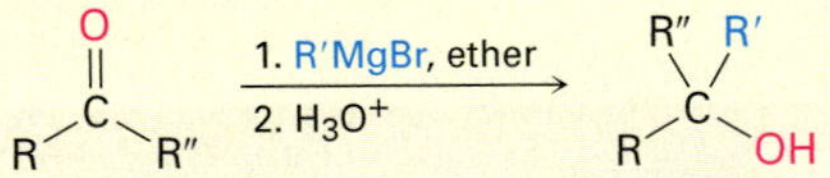

Tertiary alcohol

(4) Esters

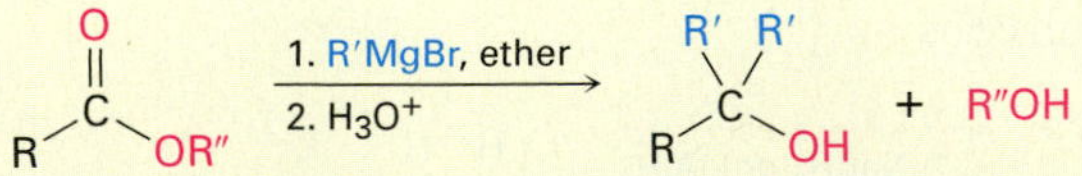

Tertiary alcohol

2.   Reactions of alcohols

(a) Dehydration (Section 13.4)

(1) Tertiary alcohols

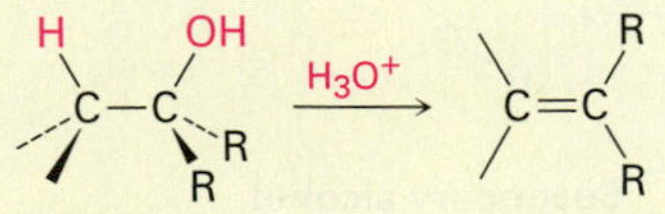

(2) Secondary and tertiary alcohols

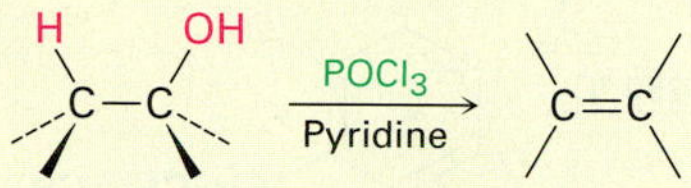

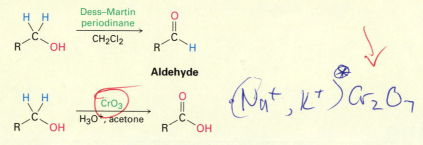

(b) Oxidation (Section 13.5)

(1) Primary alcohols

Aldehyde

Carboxylic acid

$(Na^+, K^+) Cr_2O_7$

(2) Secondary alcohols

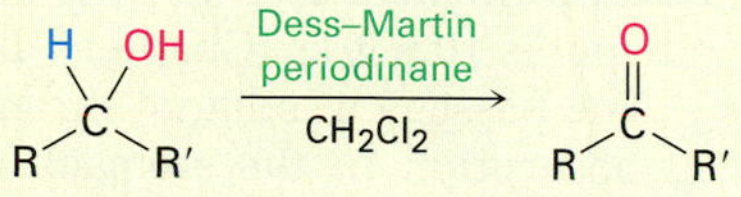

Ketone

3. Oxidation of phenols to quinones (Section 13.5)

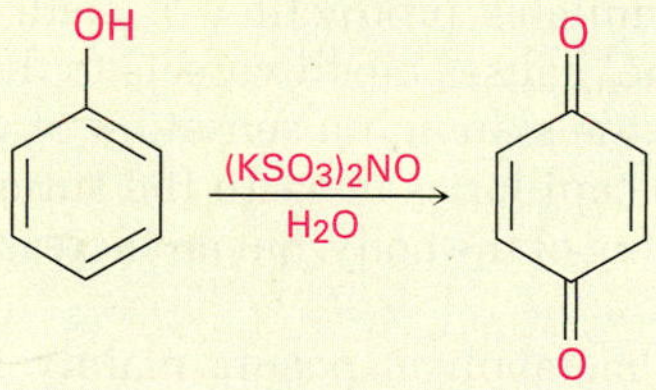

4. Synthesis of thiols (Section 13.7)

$$RCH_2Br \xrightarrow[\text{2. } H_2O, \text{ NaOH}]{\text{1. } (H_2N)_2C=S} RCH_2SH$$

5. Oxidation of thiols to disulfides (Section 13.7)

$$2\,RSH \xrightarrow{I_2, H_2O} RS-SR$$

6. Synthesis of ethers (Section 13.9)

$$RO^- + R'CH_2X \longrightarrow ROCH_2R' + X^-$$

7. Reactions of ethers (Section 13.10)

(a) Cleavage by HBr or HI

$$R-O-R' \xrightarrow[H_2O]{HX} RX + R'OH$$

(b) Claisen rearrangement of allyl aryl ethers

$$\text{OCH}_2\text{CH}=\text{CH}_2 \xrightarrow{250\,°\text{C}} \text{OH} \quad \text{CH}_2\text{CH}=\text{CH}_2$$

8. Synthesis of sulfides (Section 13.11)

$$RS^- + R'CH_2Br \longrightarrow RSCH_2R' + Br^-$$

Lagniappe

Ethanol: Chemical, Drug, and Poison

The Harger Drunkometer was introduced in 1938 to help keep drunk drivers off the road.

The production of ethanol by fermentation of grains and sugars is one of the oldest known organic reactions, going back at least 8000 years in the Middle East and perhaps as many as 9000 years in China. Fermentation is carried out by adding yeast to an aqueous sugar solution, where enzymes break down carbohydrates into ethanol and CO_2. As noted in the chapter introduction, approximately 4 billion gallons of ethanol are produced each year in the United States by fermentation, with essentially the entire amount used to make E85 automobile fuel.

$$\text{C}_6\text{H}_{12}\text{O}_6 \xrightarrow{\text{Yeast}} 2\,\text{CH}_3\text{CH}_2\text{OH} + 2\,\text{CO}_2$$

A carbohydrate

Ethanol is classified medically as a central nervous system (CNS) depressant. Its effects—that is, being drunk—resemble the human response to anesthetics. There is an initial excitability and increase in sociable behavior, but this results from depression of inhibition rather than from stimulation. At a blood alcohol concentration of 0.1% to 0.3%, motor coordination is affected, accompanied by loss of balance, slurred speech, and amnesia. When blood alcohol concentration rises to 0.3% to 0.4%, nausea and loss of consciousness occur. Above 0.6%, spontaneous respiration and cardiovascular regula-

tion are affected, ultimately leading to death. The LD_{50} of ethanol is 10.6 g/kg (Chapter 1 *Lagniappe*).

The passage of ethanol through the body begins with its absorption in the stomach and small intestine, followed by rapid distribution to all body fluids and organs. In the pituitary gland, ethanol inhibits the production of a hormone that regulates urine flow, causing increased urine production and dehydration. In the stomach, ethanol stimulates production of acid. Throughout the body, ethanol causes blood vessels to dilate, resulting in flushing of the skin and a sensation of warmth as blood moves into capillaries beneath the surface. The result is not a warming of the body, but an increased loss of heat at the surface.

Ethanol metabolism occurs mainly in the liver and proceeds by oxidation in two steps, first to acetaldehyde (CH_3CHO) and then to acetic acid (CH_3CO_2H). When continuously present in the body, ethanol and acetaldehyde are toxic, leading to the devastating physical and metabolic deterioration seen in chronic alcoholics. The liver usually suffers the worst damage since it is the major site of alcohol metabolism.

Approximately 17,000 people are killed each year in the United States in alcohol-related automobile accidents. Thus, all 50 states—Massachusetts was the final holdout—have made it illegal to drive with a blood alcohol concentration (BAC) above 0.08%. Fortunately, simple tests have been devised for measuring blood alcohol concentration. The *Breathalyzer test* measures alcohol concentration in expired air by the color change that occurs when the bright orange oxidizing agent potassium dichromate ($K_2Cr_2O_7$) is reduced to blue-green chromium(III). The *Intoxilyzer* test uses IR spectroscopy to measure blood alcohol levels in expired air. Just breathe into the machine, and let the spectrum tell the tale.

Exercises

VISUALIZING CHEMISTRY

(Problems 13.1–13.21 appear within the chapter.)

13.22 ■ Give IUPAC names for the following compounds:

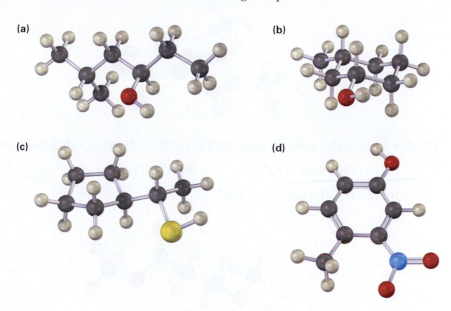

(a)

(b)

(c)

(d)

13.23 ■ Draw the structure of the carbonyl compound(s) from which each of the following alcohols might have been prepared, and show the products you would obtain by treatment of each alcohol with (i) NaH, (ii) $SOCl_2$, and (iii) the Dess–Martin periodinane reagent:

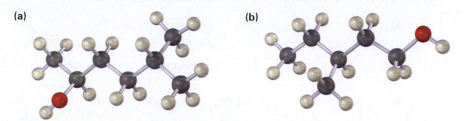

(a)

(b)

■ indicates problems that are assignable in Organic OWL.

Go to this book's companion website at **www.cengage.com/ chemistry/mcmurry** to explore interactive versions of the Active Figures from this text.

13.24 ▪ Predict the product from reaction of the following substance (reddish brown = Br) with:

(a) PBr$_3$ **(b)** Aqueous H$_2$SO$_4$ **(c)** SOCl$_2$

(d) Dess–Martin periodinane   **(e)** Br$_2$, FeBr$_3$

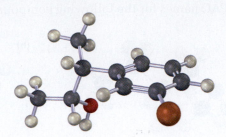

13.25 ▪ Predict the product from reaction of the following substance with:

(a) NaBH$_4$; then H$_3$O$^+$ **(b)** LiAlH$_4$; then H$_3$O$^+$

(c) 2 CH$_3$CH$_2$MgBr; then H$_3$O$^+$

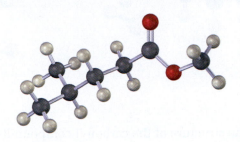

ADDITIONAL PROBLEMS

13.26 ▪ Give IUPAC names for the following compounds:

(a) 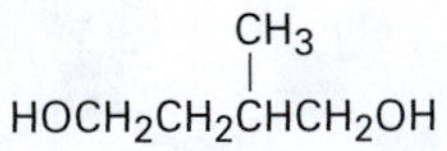
$$\underset{\text{HOCH}_2\text{CH}_2\text{CHCH}_2\text{OH}}{\overset{\text{CH}_3}{|}}$$

(b) 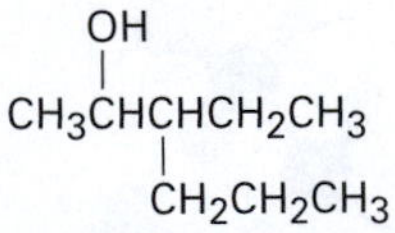
$$\underset{\text{CH}_3\text{CHCHCH}_2\text{CH}_3}{\overset{\text{OH}}{\underset{\underset{\text{CH}_2\text{CH}_2\text{CH}_3}{|}}{}}}$$

(c)

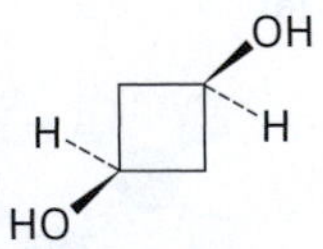

(d)

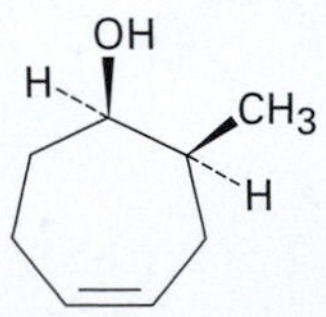

(e) 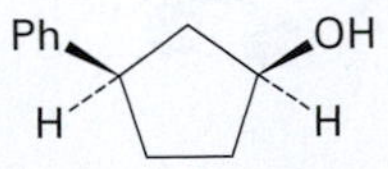 Ph

(f) 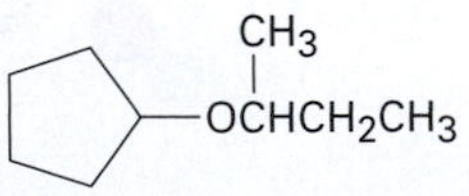
$$\overset{\text{CH}_3}{\underset{\text{OCHCH}_2\text{CH}_3}{|}}$$

(g) 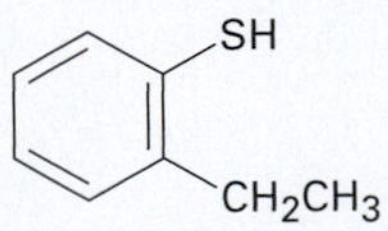 SH
CH$_2$CH$_3$

(h) 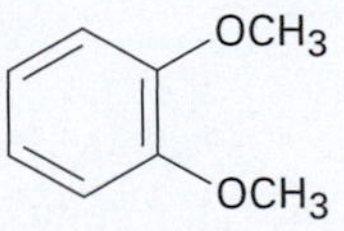 OCH$_3$
OCH$_3$

(i) 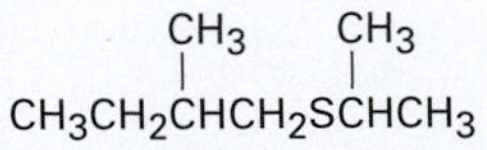
$$\underset{\text{CH}_3\text{CH}_2\text{CHCH}_2\text{SCHCH}_3}{\overset{\text{CH}_3 \quad\quad \text{CH}_3}{|\quad\quad\quad\;|}}$$

▪ Problems assignable in Organic OWL.

13.27 Draw and name the eight isomeric alcohols with formula $C_5H_{12}O$. Which are chiral?

13.28 Which of the eight alcohols you identified in Problem 13.27 react with CrO_3 in aqueous acid? Show the products you would expect from each reaction.

13.29 Named *bombykol,* the sex pheromone secreted by the female silkworm moth has the formula $C_{16}H_{30}O$ and the systematic name (10*E*,12*Z*)-hexadeca-10,12-dien-1-ol. Draw bombykol showing correct geometry for the two double bonds.

13.30 *Carvacrol* is a naturally occurring substance isolated from oregano, thyme, and marjoram. What is its IUPAC name?

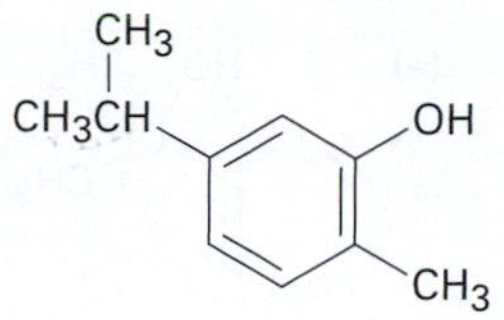

Carvacrol

13.31 Predict the products of the following ether cleavage reactions:

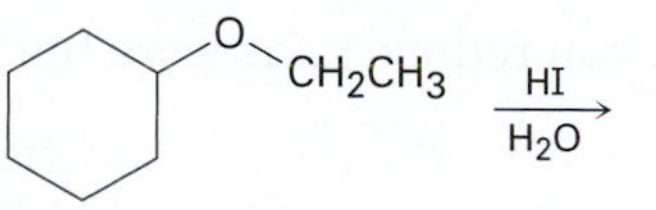

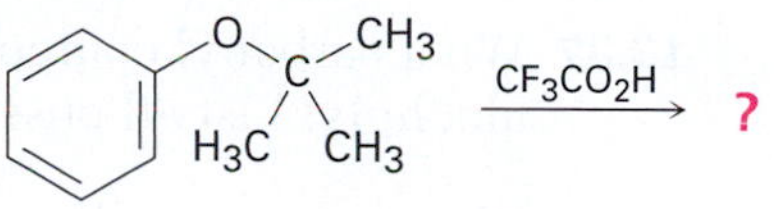

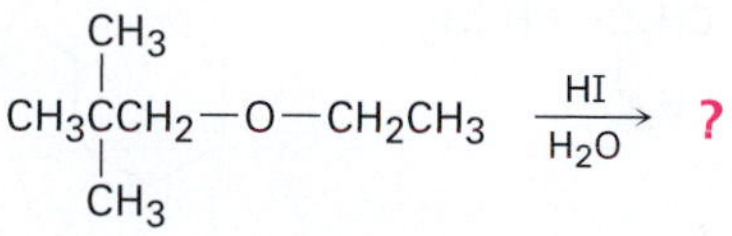

13.32 How would you prepare the following ethers?

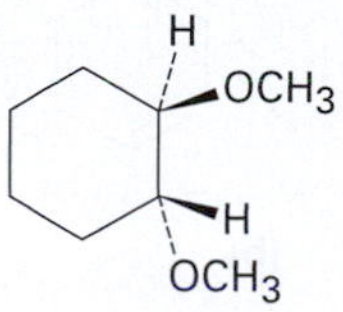

13.33 ▪ What products would you obtain from reaction of pentan-1-ol with the following reagents?

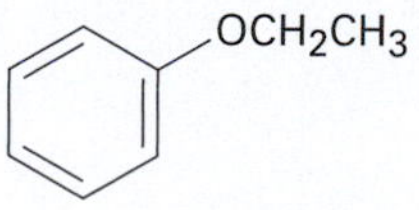

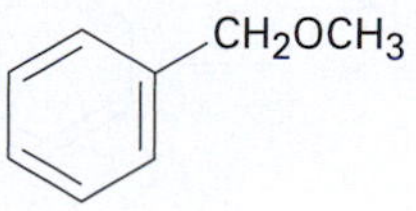

(a) PBr_3 (b) $SOCl_2$ (c) CrO_3, H_2O, H_2SO_4 (d) Dess–Martin reagent

13.34 ▪ How would you prepare the following compounds from 2-phenyl-ethanol? More than one step may be required.

(a) Styrene ($PhCH=CH_2$)

(b) Phenylacetaldehyde ($PhCH_2CHO$)

(c) Phenylacetic acid ($PhCH_2CO_2H$)

(d) Benzoic acid

(e) Ethylbenzene

(f) 1-Phenylethanol

▪ Problems assignable in Organic OWL.

13.35 ▪ How would you prepare the following compounds from 1-phenylethanol? More than one step may be required.

(a) Acetophenone ($PhCOCH_3$) **(b)** *m*-Bromobenzoic acid

(c) *p*-Chloroethylbenzene **(d)** 2-Phenylpropan-2-ol

(e) Methyl 1-phenylethyl ether **(f)** 1-Phenylethanethiol

13.36 ▪ What Grignard reagent and what carbonyl compound might you start with to prepare the following alcohols?

(a)
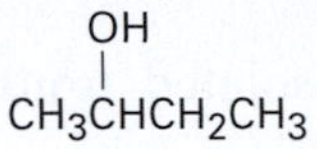
OH
CH3CHCH2CH3

(b)
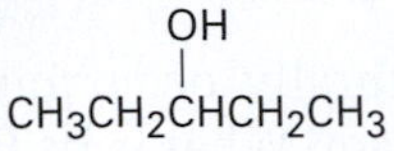
OH
CH3CH2CHCH2CH3

(c)
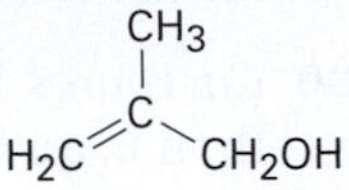
CH3
C
H2C
CH2OH

(d)
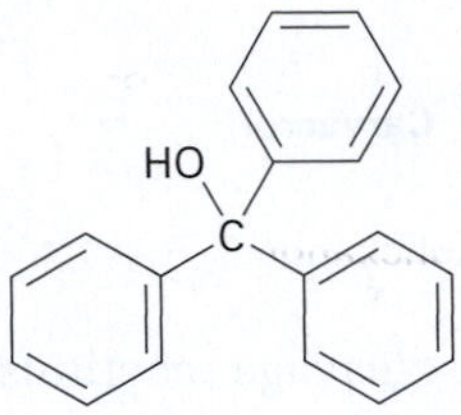
HO
C

(e)
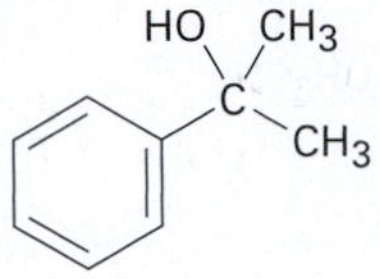
HO
CH3
C
CH3

(f)
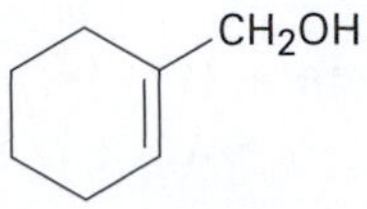
CH2OH

13.37 What carbonyl compounds would you reduce to prepare the following alcohols? List all possibilities.

(a)
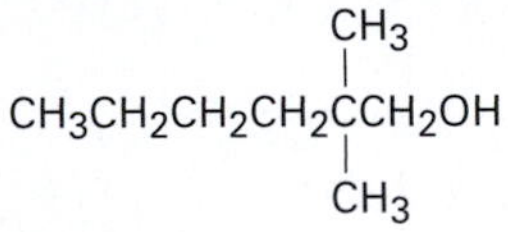
CH3
CH3CH2CH2CH2CCH2OH
CH3

(b)
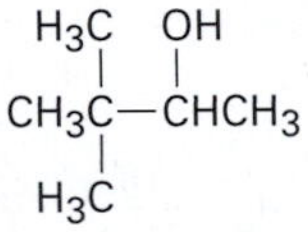
H3C
OH
CH3C—CHCH3
H3C

(c)
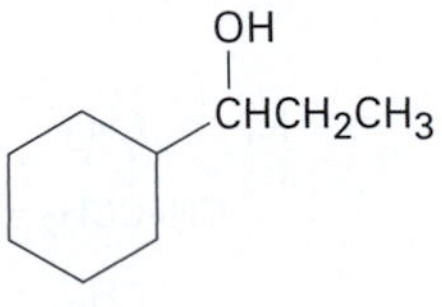
OH
CHCH2CH3

13.38 ▪ How would you carry out the following transformations?

(a)
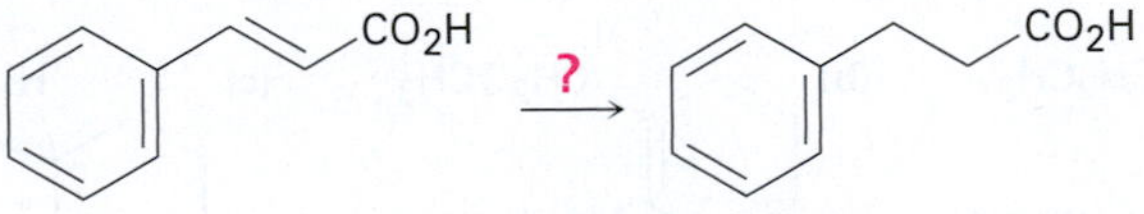
CO2H
?
CO2H

(b)
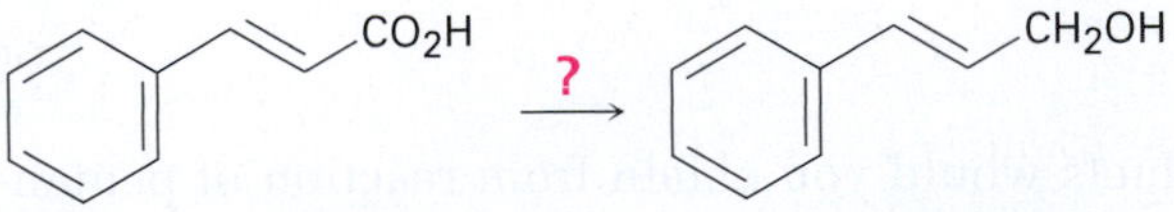
CO2H
?
CH2OH

(c)
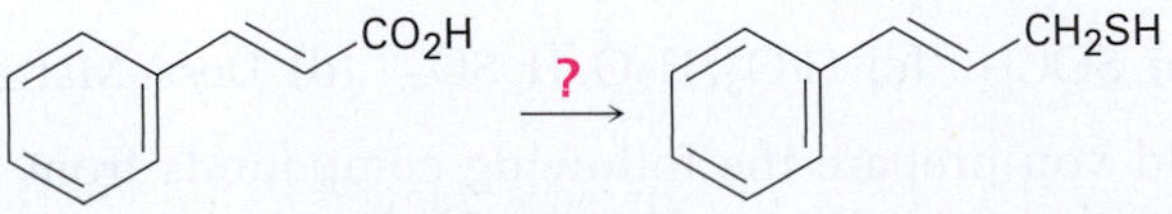
CO2H
?
CH2SH

13.39 How would you carry out the following transformations? More than one step may be required.

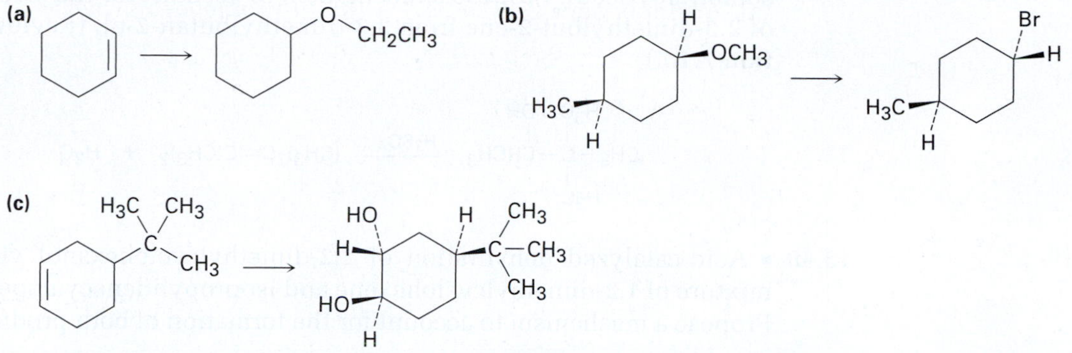

(a)

(b)

(c)

13.40 ■ What carbonyl compounds might you start with to prepare the following compounds by Grignard reaction? List all possibilities.

(a) 2-Methylpropan-2-ol

(b) 1-Ethylcyclohexanol

(c) 3-Phenylpentan-3-ol

(d) 2-Phenylpentan-2-ol

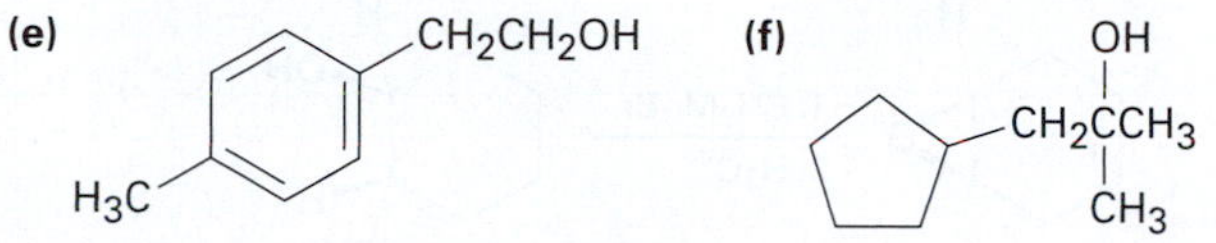

(e)

(f)

13.41 When 4-chlorobutan-1-ol is treated with a strong base such as sodium hydride, NaH, tetrahydrofuran is produced. Suggest a mechanism.

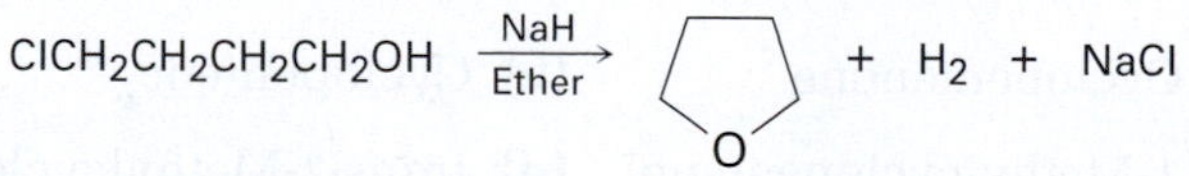

$$ClCH_2CH_2CH_2CH_2OH \xrightarrow[\text{Ether}]{\text{NaH}} \quad + \quad H_2 \quad + \quad NaCl$$

13.42 What product would you expect from cleavage of tetrahydrofuran with HI?

13.43 When 2-methylpentane-2,5-diol is treated with sulfuric acid, dehydration occurs and 2,2-dimethyltetrahydrofuran is formed. Suggest a mechanism for this reaction. Which of the two oxygen atoms is most likely to be eliminated, and why?

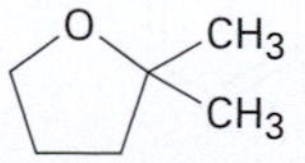

2,2-Dimethyltetrahydrofuran

13.44 Methyl aryl ethers, such as anisole, are cleaved to iodomethane and a phenoxide ion by treatment with LiI in hot DMF. Propose a mechanism for this reaction.

13.45 *tert*-Butyl ethers can be prepared by the reaction of an alcohol with 2-methylpropene in the presence of an acid catalyst. Propose a mechanism.

13.46 Why do you suppose HI and HBr are more effective than HCl in cleaving ethers? (See Section 12.7.)

■ Problems assignable in Organic OWL.

13.47 ▪ Evidence for carbocation intermediates in the acid-catalyzed dehydration of alcohols comes from the observation that rearrangements sometimes occur. Propose a mechanism to account for the formation of 2,3-dimethylbut-2-ene from 3,3-dimethylbutan-2-ol. (Review Section 7.10.)

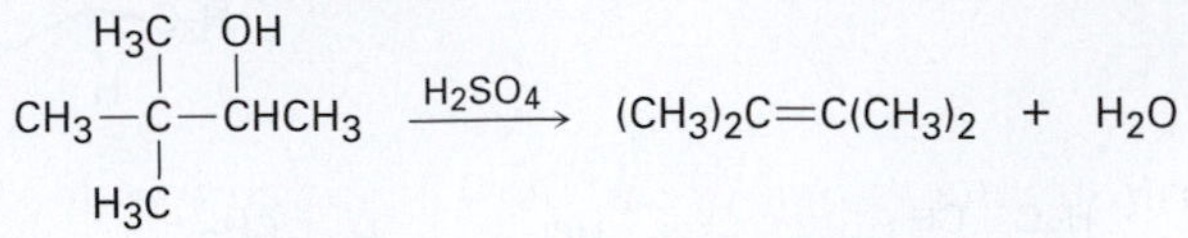

$$\underset{\underset{\underset{H_3C}{|}}{\overset{\overset{H_3C\ \ \ OH}{|\ \ \ \ |}}{CH_3-C-CHCH_3}}}{} \quad \xrightarrow{H_2SO_4} \quad (CH_3)_2C{=}C(CH_3)_2 \ + \ H_2O$$

13.48 ▪ Acid-catalyzed dehydration of 2,2-dimethylcyclohexanol yields a mixture of 1,2-dimethylcyclohexene and isopropylidenecyclopentane. Propose a mechanism to account for the formation of both products.

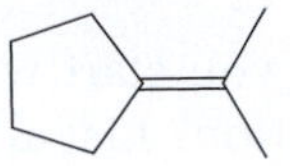

Isopropylidenecyclopentane

13.49 Epoxides react with Grignard reagents to yield alcohols. Propose a mechanism.

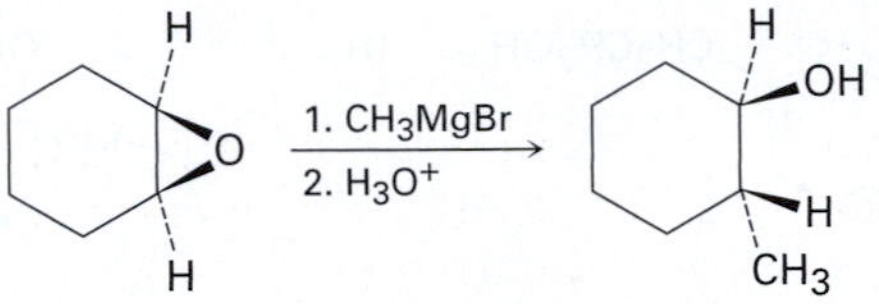

$$\xrightarrow[\text{2. }H_3O^+]{\text{1. }CH_3MgBr}$$

13.50 ▪ How would you prepare the following substances from cyclopentanol? More than one step may be required.

(a) Cyclopentanone   (b) Cyclopentene

(c) 1-Methylcyclopentanol   (d) trans-2-Methylcyclopentanol

13.51 ▪ What products would you expect to obtain from reaction of 1-methylcyclohexanol with the following reagents?

(a) HBr   (b) NaH   (c) H_2SO_4 (d) $Na_2Cr_2O_7$

13.52 ▪ Rank the following substituted phenols in order of increasing acidity:

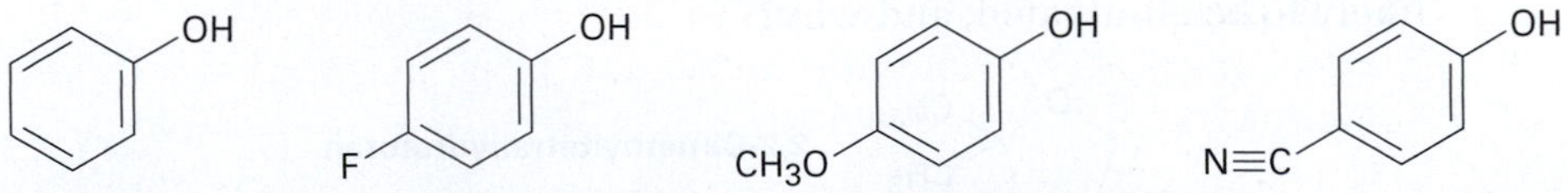

13.53 Reduction of butan-2-one with $NaBH_4$ yields butan-2-ol. Is the product chiral? Is it optically active? Explain.

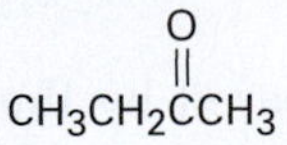

$$\underset{CH_3CH_2CCH_3}{\overset{\overset{O}{\|}}{}} \quad \textbf{Butan-2-one}$$

▪ Problems assignable in Organic OWL.

13.54 Reaction of (*S*)-3-methylpentan-2-one with methylmagnesium bromide followed by acidification yields 2,3-dimethylpentan-2-ol. What is the stereochemistry of the product? Is the product optically active?

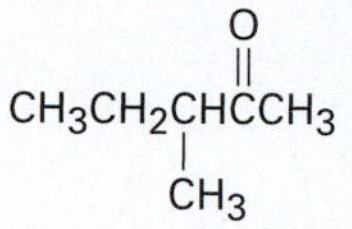

$$CH_3CH_2\overset{\overset{\displaystyle O}{\|}}{C}HCCH_3 \qquad \text{3-Methylpentan-2-one}$$
$$\underset{|}{\overset{|}{C}H_3}$$

13.55 ■ Testosterone is one of the most important male steroid hormones. When testosterone is dehydrated by treatment with acid, rearrangement occurs to yield the product shown. Propose a mechanism to account for this reaction.

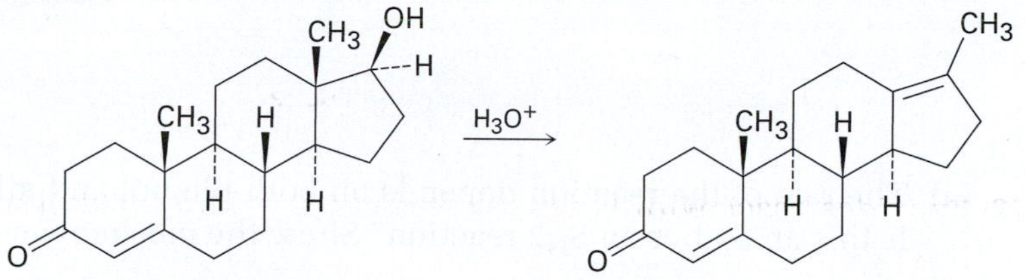

Testosterone

13.56 Dehydration of *trans*-2-methylcyclopentanol with POCl$_3$ in pyridine yields predominantly 3-methylcyclopentene. Is the stereochemistry of this dehydration syn or anti? Can you suggest a reason for formation of the observed product?

13.57 ■ 2,3-Dimethylbutane-2,3-diol has the common name *pinacol*. On heating with aqueous acid, pinacol rearranges to *pinacolone,* 3,3-dimethyl-butan-2-one. Suggest a mechanism.

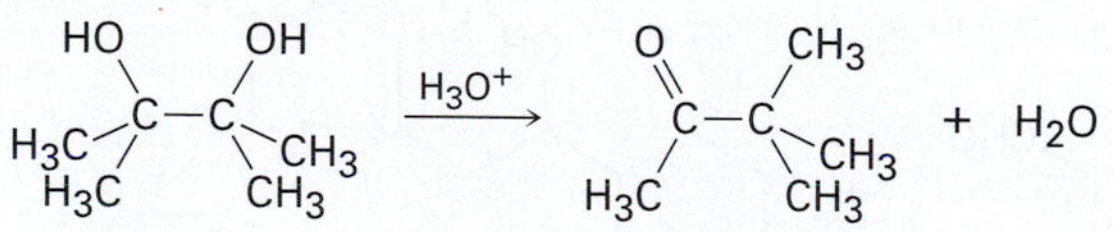

Pinacol   **Pinacolone**

13.58 As a rule, axial alcohols oxidize somewhat faster than equatorial alcohols. Which would you expect to oxidize faster, *cis*-4-*tert*-butylcyclohexanol or *trans*-4-*tert*-butylcyclohexanol? Draw the more stable chair conformation of each molecule.

13.59 Propose a synthesis of bicyclohexylidene, starting from cyclohexanone as the only source of carbon.

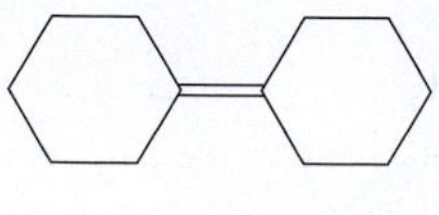

Bicyclohexylidene

13.60 Fluoxetine, a heavily prescribed antidepressant marketed under the name Prozac, can be prepared by a route that begins with reaction between a phenol and an alkyl chloride.

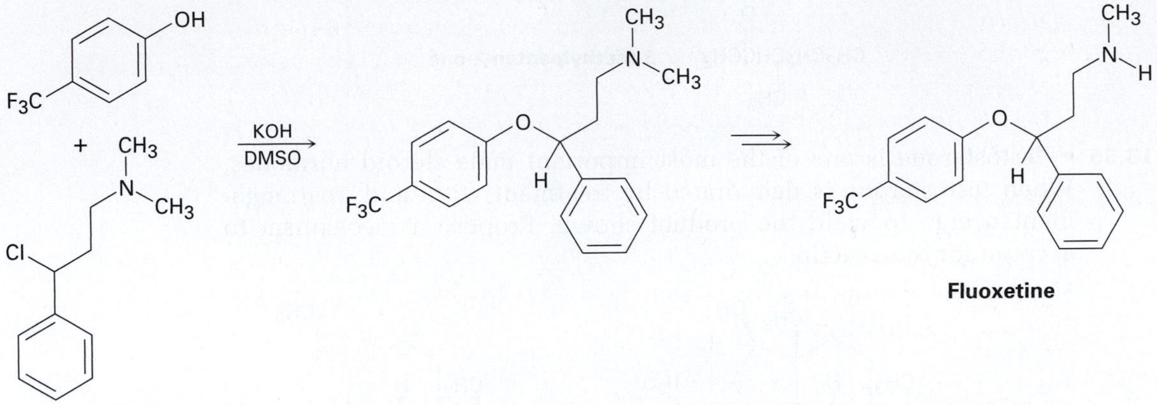

Fluoxetine

(a) The rate of the reaction depends on both phenol and alkyl halide. Is this an S_N1 or an S_N2 reaction? Show the mechanism.

(b) The physiologically active enantiomer of fluoxetine has (S) stereochemistry. Based on your answer in part (a), draw the structure of the alkyl chloride you would need, showing the correct stereochemistry.

13.61 ■ Identify the reagents **a–f** in the following scheme:

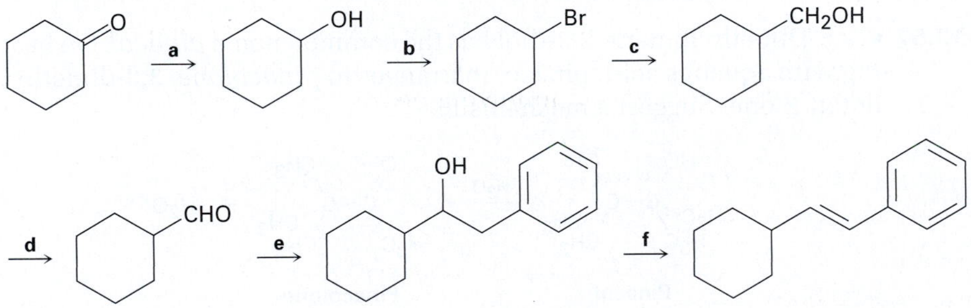

13.62 ■ Identify the reagents **a–e** in the following scheme:

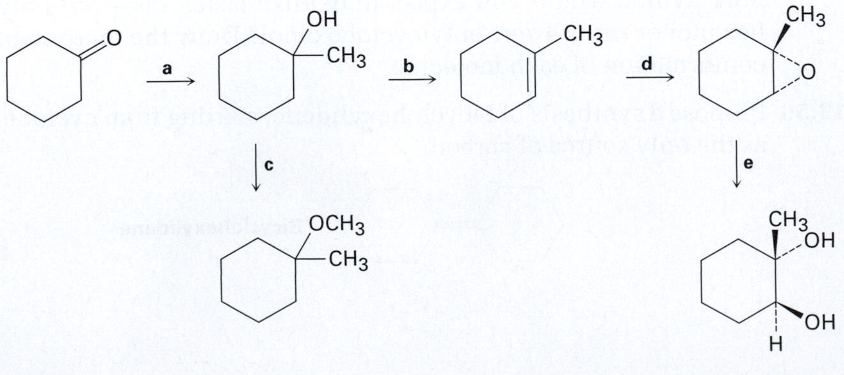

■ Problems assignable in Organic OWL.

13.63 *Disparlure,* $C_{19}H_{38}O$, is a sex attractant released by the female gypsy moth, *Lymantria dispar.* The ^1H NMR spectrum of disparlure shows a large absorption in the alkane region, 1 to 2 δ, and a triplet at 2.8 δ. Reaction of disparlure, first with aqueous acid and then with $KMnO_4$, yields two carboxylic acids identified as undecanoic acid and 6-methyl-heptanoic acid. ($KMnO_4$ cleaves 1,2-diols to yield carboxylic acids.) Neglecting stereochemistry, propose a structure for disparlure. The actual compound is a chiral molecule with 7R,8S stereochemistry. Draw disparlure, showing the correct stereochemistry.

13.64 Galactose, a constituent of the disaccharide lactose found in dairy products, is metabolized by a pathway that includes the isomerization of UDP-galactose to UDP-glucose, where UDP = uridylyl diphosphate. The enzyme responsible for the transformation uses NAD^+ as cofactor. Propose a mechanism.

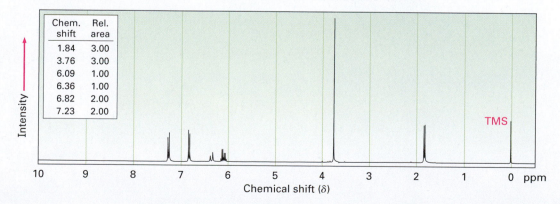

UDP-galactose　　　　　　　　　　　　**UDP-glucose**

13.65 ■ The red fox *(Vulpes vulpes)* uses a chemical communication system based on scent marks in urine. One component of fox urine is a sulfide. Mass spectral analysis of the pure scent-mark component shows M^+ = 116, IR spectroscopy shows an intense band at 890 cm^{-1}, and ^1H NMR spectroscopy reveals the following peaks. Propose a structure for the molecule. [*Note:* $(CH_3)_2S$ absorbs at 2.1 δ.]

1.74 δ (3 H, singlet); 2.11 δ (3 H, singlet); 2.27 δ (2 H, triplet, J = 4.2 Hz); 2.57 δ (2 H, triplet, J = 4.2 Hz); 4.73 δ (2 H, broad)

13.66 ■ *Anethole,* $C_{10}H_{12}O$, a major constituent of the oil of anise, has the ^1H NMR spectrum shown. On oxidation with $Na_2Cr_2O_7$, anethole yields *p*-methoxybenzoic acid. What is the structure of anethole? Assign all peaks in the NMR spectrum, and account for the observed splitting patterns.

Chem. shift	Rel. area
1.84	3.00
3.76	3.00
6.09	1.00
6.36	1.00
6.82	2.00
7.23	2.00

TMS

10   9   8   7   6   5   4   3   2   1   0 ppm

Chemical shift (δ)

13.67 ■ Propose a structure consistent with the following spectral data for a compound $C_8H_{18}O_2$:

IR: 3350 cm^{-1}

1H NMR: $1.24 \, \delta$ (12 H, singlet); $1.56 \, \delta$ (4 H, singlet); $1.95 \, \delta$ (2 H, singlet)

13.68 ■ The ^{1}H NMR spectrum shown is that of 3-methylbut-3-en-1-ol. Assign all the observed resonance peaks to specific protons, and account for the splitting patterns.

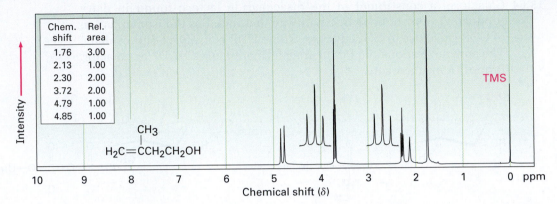

Chem. shift	Rel. area
1.76	3.00
2.13	1.00
2.30	2.00
3.72	2.00
4.79	1.00
4.85	1.00

13.69 ■ Compound **A**, $C_5H_{10}O$, is one of the basic building blocks of nature. All steroids and many other naturally occurring compounds are built from compound **A**. Spectroscopic analysis of **A** yields the following information:

IR: 3400 cm^{-1}; 1640 cm^{-1}

^{1}H NMR: $1.63 \, \delta$ (3 H, singlet); $1.70 \, \delta$ (3 H, singlet); $3.83 \, \delta$ (1 H, broad singlet); $4.15 \, \delta$ (2 H, doublet, $J = 7$ Hz); $5.70 \, \delta$ (1 H, triplet, $J = 7$ Hz)

(a) From the IR spectrum, what is the nature of the oxygen-containing functional group?

(b) What kinds of protons are responsible for the NMR absorptions listed?

(c) Propose a structure for **A**.

13.70 ■ A compound of unknown structure gave the following spectroscopic data:

Mass spectrum:	$M^+ = 88.1$
IR:	$3600\ cm^{-1}$
^1H NMR:	1.4 δ (2 H, quartet, $J = 7$ Hz); 1.2 δ (6 H, singlet); 1.0 δ (1 H, singlet); 0.9 δ (3 H, triplet, $J = 7$ Hz)
^{13}C NMR:	74, 35, 27, 25 δ

(a) Assuming that the compound contains C and H but may or may not contain O, give three possible molecular formulas.

(b) How many protons (H) does the compound contain?

(c) What functional group(s) does the compound contain?

(d) How many carbons does the compound contain?

(e) What is the molecular formula of the compound?

(f) What is the structure of the compound?

(g) Assign the peaks in the ^1H NMR spectrum of the molecule to specific protons.

13.71 ■ The following ^1H NMR spectrum is that of an alcohol, $C_8H_{10}O$. Propose a structure.

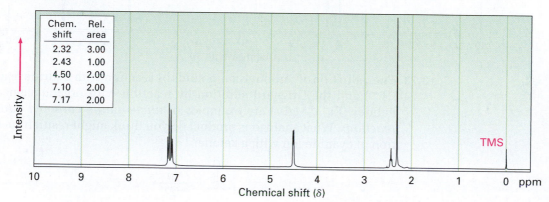

Chem. shift Rel. area
2.32 3.00
2.43 1.00
4.50 2.00
7.10 2.00
7.17 2.00
Intensity
TMS
10 9 8 7 6 5 4 3 2 1 0 ppm
Chemical shift (δ)

13.72 ■ Compound **A**, $C_8H_{10}O$, has the IR and ^1H NMR spectra shown. Propose a structure consistent with the observed spectra, and assign each peak in the NMR spectrum. Note that the absorption at 5.5 δ disappears when D_2O is added.

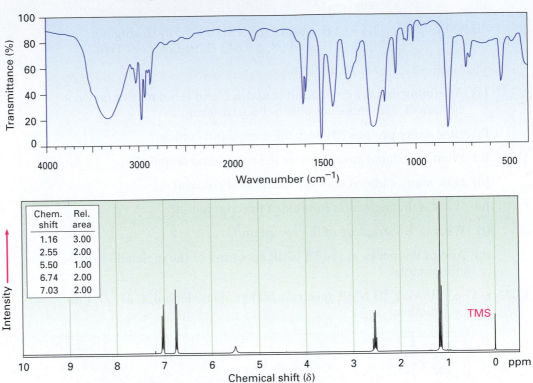

Chem. shift	Rel. area
1.16	3.00
2.55	2.00
5.50	1.00
6.74	2.00
7.03	2.00

13.73 The reduction of carbonyl compounds by reaction with hydride reagents ($H{:}^-$) and the Grignard addition by reaction with organomagnesium halides ($R{:}^-$ ^+MgBr) are examples of *nucleophilic carbonyl addition reactions*. What analogous product do you think might result from reaction of cyanide ion with a ketone?

$$\underset{\text{C}}{\overset{\text{O}}{\|}} \quad \xrightarrow[\text{H}_3\text{O}^+]{\text{CN}^-} \quad \textbf{?}$$

13.74 Aldehydes and ketones undergo acid-catalyzed reaction with alcohols to yield *hemiacetals*, compounds that have one alcohol-like oxygen and one ether-like oxygen bonded to the same carbon. Further reaction of a hemiacetal with alcohol then yields an *acetal*, a compound that has two ether-like oxygens bonded to the same carbon.

$$\underset{\text{C}}{\overset{\text{O}}{\|}} \;+\; \text{ROH} \;\xrightarrow[\text{catalyst}]{\text{H}^+}\; \underset{\substack{\text{OR}\\ \\ \text{OH}}}{\text{C}} \;\xrightarrow[\text{H}^+]{\text{ROH}}\; \underset{\substack{\text{OR}\\ \\ \text{OR}}}{\text{C}} \;+\; \text{H}_2\text{O}$$

A hemiacetal   **An acetal**

(a) Show the structures of the hemiacetal and acetal you would obtain by reaction of cyclohexanone with ethanol.

(b) Propose a mechanism for the conversion of a hemiacetal into an acetal.

■ Problems assignable in Organic OWL.

Preview of Carbonyl Chemistry

Carbonyl compounds are everywhere. Most biological molecules contain carbonyl groups, as do most pharmaceutical agents and many of the synthetic chemicals that touch our everyday lives. Citric acid, found in lemons and oranges; acetaminophen, the active ingredient in many over-the-counter headache remedies; and Dacron, the polyester material used in clothing, all contain different kinds of carbonyl groups.

Citric acid
(a carboxylic acid)

Acetaminophen
(an amide)

Dacron
(a polyester)

To a great extent, the chemistry of living organisms is the chemistry of carbonyl compounds. Thus, we'll spend the next four chapters discussing the **carbonyl group** (pronounced car-bo-**neel**). There are many different kinds of carbonyl compounds and many different reactions, but there are only a few fundamental principles that tie the entire field together. The purpose of this brief preview is not to show details of specific reactions but rather to provide a framework for learning carbonyl-group chemistry. Read through this preview now, and return to it on occasion to remind yourself of the larger picture.

I. Kinds of Carbonyl Compounds

Table 1 shows some of the many different kinds of carbonyl compounds. All contain an **acyl group** (R–C=O) bonded to another substituent. The R of the acyl group can be any practically organic part-structure, and the other substituent to which the acyl group is bonded can be a carbon, hydrogen, oxygen, halogen, nitrogen, or sulfur.

It's useful to classify carbonyl compounds into two categories based on the kinds of chemistry they undergo. In one category are aldehydes and ketones; in the other are carboxylic acids and their derivatives. The acyl group in an aldehyde or ketone is bonded to an atom (H or C, respectively) that can't stabilize a negative charge and therefore can't act as a leaving group in a nucleophilic substitution reaction. The acyl group in a carboxylic acid or its derivative, however, is bonded to an atom (oxygen, halogen, sulfur, or

TABLE 1
Types of Carbonyl Compounds

Name	General formula	Name ending	Name	General formula	Name ending
Aldehyde		-al	Ester		-oate
Ketone		-one	Lactone (cyclic ester)		None
Carboxylic acid		-oic acid	Thioester		-thioate
Acid halide		-yl or -oyl halide	Amide		-amide
Acid anhydride		-oic anhydride	Lactam (cyclic amide)		None
Acyl phosphate		-yl phosphate			

nitrogen) that *can* stabilize a negative charge and therefore *can* act as a leaving group in a nucleophilic substitution reaction.

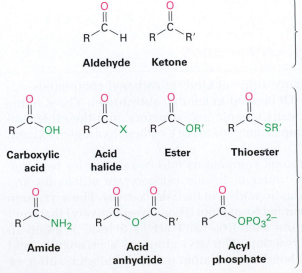

The –R′ and –H in these compounds *can't* act as leaving groups in nucleophilic substitution reactions.

Aldehyde   Ketone

Carboxylic acid   Acid halide   Ester   Thioester

Amide   Acid anhydride   Acyl phosphate

The –OH, –X, –OR′, –SR, –NH$_2$, –OCOR′, and –OPO$_3^{2-}$ in these compounds *can* act as leaving groups in nucleophilic substitution reactions.

II. Nature of the Carbonyl Group

The carbon–oxygen double bond of a carbonyl group is similar in many respects to the carbon–carbon double bond of an alkene. The carbonyl carbon atom is sp^2-hybridized and forms three σ bonds. The fourth valence electron remains in a carbon p orbital and forms a π bond to oxygen by overlap with an oxygen p orbital. The oxygen atom also has two nonbonding pairs of electrons, which occupy its remaining two orbitals.

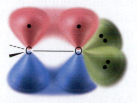
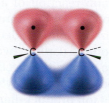

Carbonyl group   **Alkene**

Like alkenes, carbonyl compounds are planar about the double bond and have bond angles of approximately 120°. Figure 1 shows the structure of acetaldehyde and indicates its bond lengths and angles. As you might expect, the carbon–oxygen double bond is both shorter (122 pm versus 143 pm) and stronger [732 kJ/mol (175 kcal/mol) versus 385 kJ/mol (92 kcal/mol)] than a C–O single bond.

Bond angle	(°)	Bond length	(pm)
H—C—C	118	C=O	122
C—C=O	121	C—C	150
H—C=O	121	OC—H	109

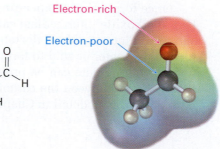

FIGURE 1 Structure of acetaldehyde.

As indicated by the electrostatic potential map in Figure 1, the carbon–oxygen double bond is strongly polarized because of the high electronegativity of oxygen relative to carbon. Thus, the carbonyl carbon atom carries a partial positive charge, is an electrophilic (Lewis acidic) site, and reacts with nucleophiles. Conversely, the carbonyl oxygen atom carries a partial negative charge, is a nucleophilic (Lewis basic) site, and reacts with electrophiles. We'll see in the next four chapters that the majority of carbonyl-group reactions can be rationalized by simple polarity arguments.

III. General Reactions of Carbonyl Compounds

Both in the laboratory and in living organisms, the reactions of carbonyl compounds take place by one of four general mechanisms: *nucleophilic addition, nucleophilic acyl substitution, alpha substitution,* and *carbonyl condensation.* These mechanisms have many variations, just as alkene electrophilic addition reactions and S$_N$2 reactions do, but the variations are much easier to learn when the fundamental features of the mechanisms are made clear. Let's see what the four mechanisms are and what kinds of chemistry carbonyl compounds undergo.

Nucleophilic Addition Reactions of Aldehydes and Ketones (Chapter 14)

The most common reaction of aldehydes and ketones is the **nucleophilic addition reaction**, in which a nucleophile, :Nu$^-$, adds to the electrophilic carbon of the carbonyl group. Because the nucleophile uses an electron pair to form a new bond to carbon, two electrons from the carbon–oxygen double bond must move toward the electronegative oxygen atom to give an alkoxide anion. The carbonyl carbon rehybridizes from sp^2 to sp^3 during the reaction, and the alkoxide ion product therefore has tetrahedral geometry.

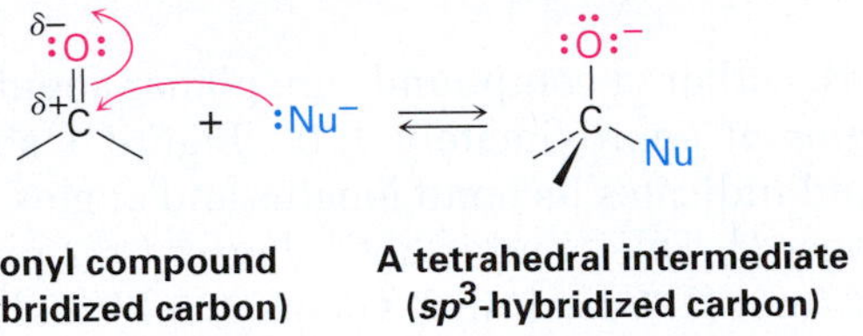

A carbonyl compound
(**sp^2-hybridized carbon**)

A tetrahedral intermediate
(**sp^3-hybridized carbon**)

Once formed, and depending on the nature of the nucleophile, the tetrahedral alkoxide intermediate can undergo either of two further reactions, as shown in Figure 2. Often, the tetrahedral alkoxide intermediate is simply protonated by water or acid to form an alcohol product. Alternatively, the tetrahedral intermediate can be protonated and expel the oxygen to form a new double bond between the carbonyl carbon and the nucleophile. We'll study both processes in detail in Chapter 14.

FIGURE 2 The addition reaction of an aldehyde or a ketone with a nucleophile. Depending on the nucleophile, either an alcohol or a compound with a C=Nu double bond is formed.

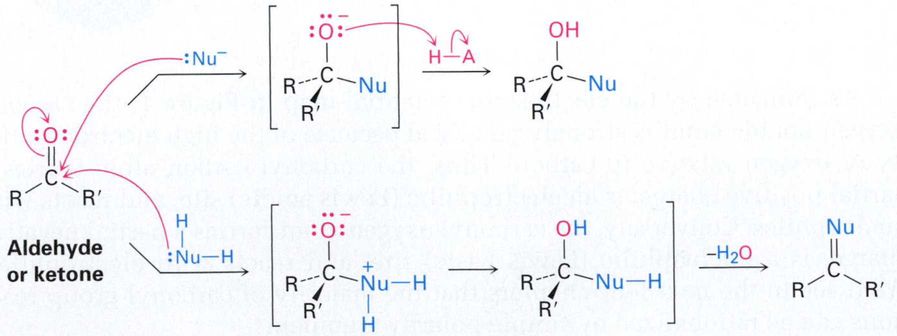

FORMATION OF AN ALCOHOL The simplest reaction of a tetrahedral alkoxide intermediate is protonation to yield an alcohol. We've already seen two examples of this kind of process during reduction of aldehydes and ketones with hydride reagents, such as NaBH$_4$ or LiAlH$_4$, and during Grignard reactions (Section 13.3). During a reduction, the nucleophile that adds to the carbonyl group is a hydride ion, H:⁻, while during a Grignard reaction, the nucleophile is a carbanion, $R_3C:^-$.

Reduction

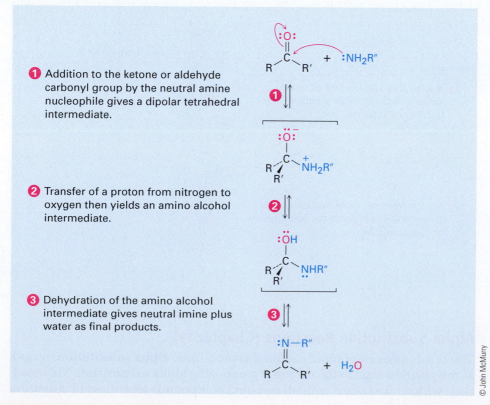

FORMATION OF C=NU The second mode of nucleophilic addition, which often occurs with amine nucleophiles, involves elimination of oxygen and formation of a C=Nu double bond. For example, aldehydes and ketones react with primary amines, RNH_2, to form *imines*, R_2C=NR′. These reactions proceed through exactly the same kind of tetrahedral intermediate as that formed during hydride reduction and Grignard reaction, but the initially formed alkoxide ion is not isolated. Instead, it is protonated and then loses water to form an imine, as shown in Figure 3.

❶ Addition to the ketone or aldehyde carbonyl group by the neutral amine nucleophile gives a dipolar tetrahedral intermediate.

❷ Transfer of a proton from nitrogen to oxygen then yields an amino alcohol intermediate.

❸ Dehydration of the amino alcohol intermediate gives neutral imine plus water as final products.

FIGURE 3 MECHANISM: Formation of an imine, R_2C=NR′, by reaction of an amine with an aldehyde or a ketone.

© John McMurry

Nucleophilic Acyl Substitution Reactions of Carboxylic Acid Derivatives (Chapter 16)

The second fundamental reaction of carbonyl compounds, **nucleophilic acyl substitution**, is related to the nucleophilic addition reaction just discussed but occurs only with carboxylic acid derivatives rather than with aldehydes and ketones. When the carbonyl group of a carboxylic acid derivative reacts with a nucleophile, addition occurs in the usual way, but the initially formed tetrahedral alkoxide intermediate is not isolated. Because carboxylic acid derivatives have a leaving group bonded to the carbonyl-group carbon, the tetrahedral intermediate can react further by expelling the leaving group and forming a new carbonyl compound:

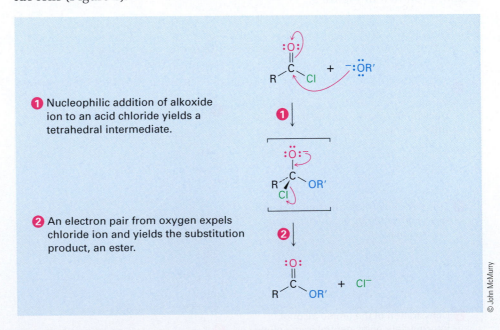

The net effect of nucleophilic acyl substitution is the replacement of the leaving group by the entering nucleophile. We'll see in Chapter 16, for instance, that acid chlorides are rapidly converted into esters by treatment with alkoxide ions (Figure 4).

FIGURE 4 MECHANISM: The nucleophilic acyl substitution reaction of an acid chloride with an alkoxide ion yields an ester.

1 Nucleophilic addition of alkoxide ion to an acid chloride yields a tetrahedral intermediate.

2 An electron pair from oxygen expels chloride ion and yields the substitution product, an ester.

© John McMurry

Alpha-Substitution Reactions (Chapter 17)

The third major reaction of carbonyl compounds, **alpha substitution**, occurs at the position *next to* the carbonyl group—the alpha (α) position. This reaction, which takes place with all carbonyl compounds regardless of structure, results in the substitution of an α hydrogen by an electrophile through the formation of an intermediate *enol* or *enolate ion:*

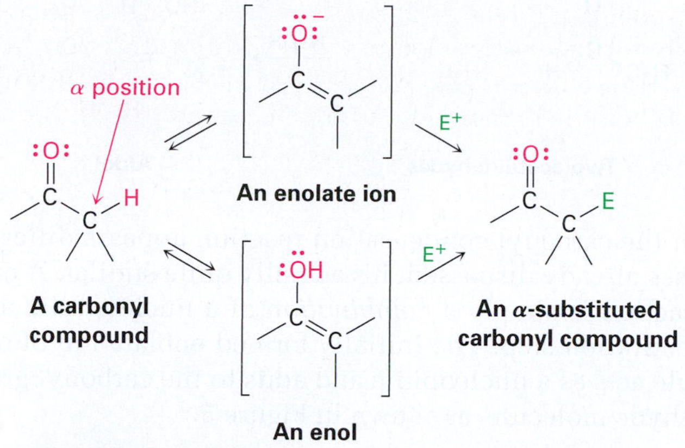

An enolate ion

An α-substituted carbonyl compound

A carbonyl compound

An enol

For reasons that we'll explore in Chapter 17, the presence of a carbonyl group renders the hydrogens on the α-carbon acidic. Carbonyl compounds therefore react with strong base to yield enolate ions.

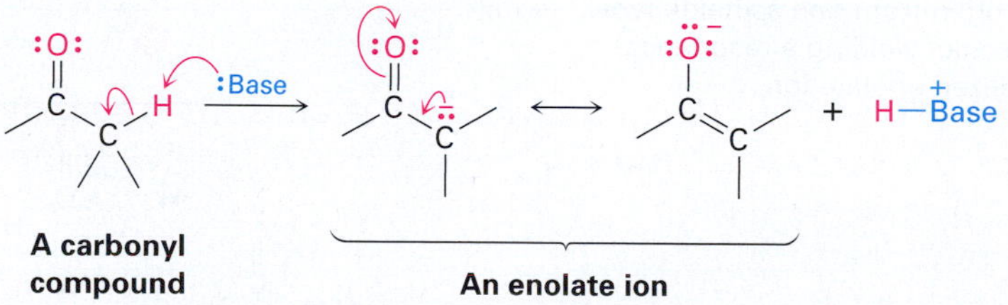

A carbonyl compound

An enolate ion

Because they're negatively charged, enolate ions act as nucleophiles and undergo many of the reactions we've already studied. For example, enolates react with primary alkyl halides in the S_N2 reaction. The nucleophilic enolate ion displaces halide ion, and a new C–C bond forms:

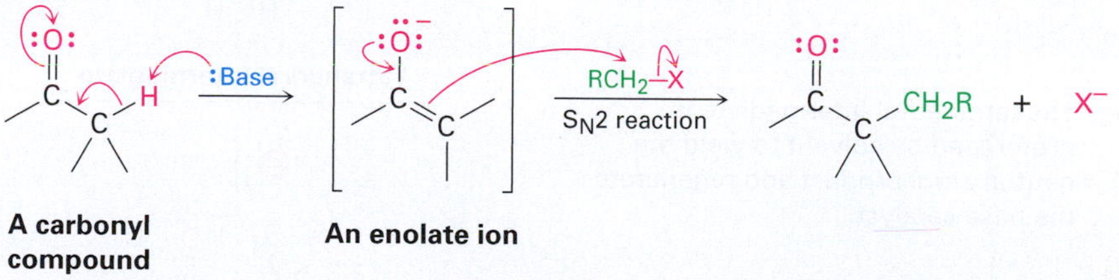

A carbonyl compound

An enolate ion

The S_N2 alkylation reaction between an enolate ion and an alkyl halide is a powerful method for making C–C bonds, thereby building up larger molecules from smaller precursors. We'll study the alkylation of many kinds of carbonyl compounds in Chapter 17.

Carbonyl Condensation Reactions (Chapter 17)

The fourth and last fundamental reaction of carbonyl groups, **carbonyl condensation**, takes place when two carbonyl compounds react with each other. When acetaldehyde is treated with base, for instance, two molecules combine to yield the hydroxy aldehyde product known as *aldol* (*ald*ehyde + alcoh*ol*):

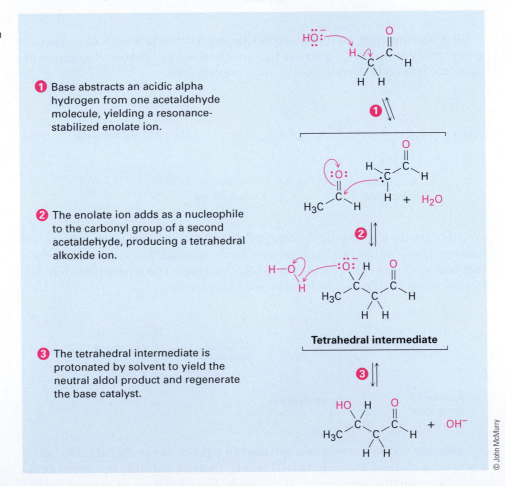

Two acetaldehydes   **Aldol**

Although the carbonyl condensation reaction appears different from the three processes already discussed, it's actually quite similar. A carbonyl condensation reaction is simply a *combination* of a nucleophilic addition step and an α-substitution step. The initially formed enolate ion of one acetaldehyde molecule acts as a nucleophile and adds to the carbonyl group of a second acetaldehyde molecule, as shown in Figure 5.

FIGURE 5 MECHANISM:
A carbonyl condensation reaction between two molecules of acetaldehyde yields a hydroxy aldehyde product.

1 Base abstracts an acidic alpha hydrogen from one acetaldehyde molecule, yielding a resonance-stabilized enolate ion.

2 The enolate ion adds as a nucleophile to the carbonyl group of a second acetaldehyde, producing a tetrahedral alkoxide ion.

3 The tetrahedral intermediate is protonated by solvent to yield the neutral aldol product and regenerate the base catalyst.

© John McMurry

IV. Summary

To a great extent, the chemistry of living organisms is the chemistry of carbonyl compounds. We have not looked at the details of specific carbonyl reactions in this short preview but rather have laid the groundwork for the next four chapters. All the carbonyl-group reactions we'll be studying in Chapters 14 through 17 fall into one of the four fundamental categories discussed in this preview. Knowing where we'll be heading should help you keep matters straight in understanding this most important of all functional groups.

Exercises

1. Judging from the following electrostatic potential maps, which kind of carbonyl compound has the more electrophilic carbonyl carbon atom, a ketone or an acid chloride? Which has the more nucleophilic carbonyl oxygen atom? Explain.

Acetone   **Acetyl chloride**
(ketone)   **(acid chloride)**

2. Predict the product formed by nucleophilic addition of cyanide ion (CN⁻) to the carbonyl group of acetone, followed by protonation to give an alcohol:

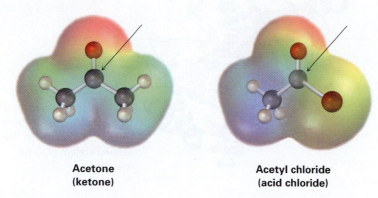

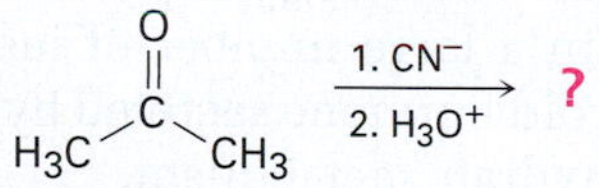

Acetone

3. Identify each of the following reactions as a nucleophilic addition, nucleophilic acyl substitution, an α substitution, or a carbonyl condensation:

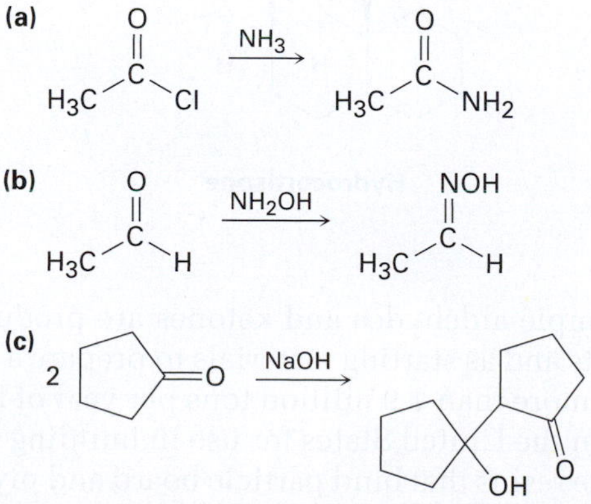

14 Aldehydes and Ketones: Nucleophilic Addition Reactions

Phosphoglucoisomerase catalyzes the isomerization of glucose 6-phosphate to fructose 6-phosphate, the second step in glucose metabolism.

Aldehydes (RCHO) and **ketones (R_2CO)** are among the most widely occurring of all compounds. In nature, many substances required by living organisms are aldehydes or ketones. The aldehyde pyridoxal phosphate, for instance, is a coenzyme involved in a large number of metabolic reactions; the ketone hydrocortisone is a steroid hormone secreted by the adrenal glands to regulate fat, protein, and carbohydrate metabolism.

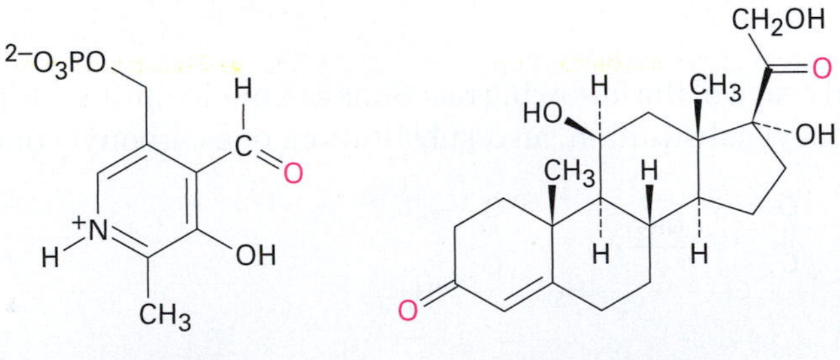

Pyridoxal phosphate (PLP)

Hydrocortisone

In the chemical industry, simple aldehydes and ketones are produced in large quantities for use as solvents and as starting materials to prepare a host of other compounds. For example, more than 1.9 million tons per year of formaldehyde, $H_2C{=}O$, are produced in the United States for use in building insulation materials and in the adhesive resins that bind particle board and plywood.

⭕**WL** Online homework for this chapter can be assigned in Organic OWL.

Acetone, $(CH_3)_2C=O$, is widely used as an industrial solvent; approximately 1.2 million tons per year are produced in the United States.

WHY THIS CHAPTER?

The chemistry of living organisms is, in many ways, the chemistry of carbonyl compounds. Aldehydes and ketones, in particular, are intermediates in almost all biological pathways, so an understanding of their properties and reactions is essential. We'll look in this chapter at some of their most important reactions.

14.1 Naming Aldehydes and Ketones

Aldehydes are named by replacing the terminal *-e* of the corresponding alkane name with *-al*. The parent chain must contain the –CHO group, and the –CHO carbon is numbered as C1. In the following examples, note that the longest chain in 2-ethyl-4-methylpentanal is actually a hexane, but this chain does not include the –CHO group and thus is not considered the parent.

Ethanal
(acetaldehyde)

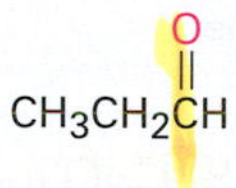

Propanal
(propionaldehyde)

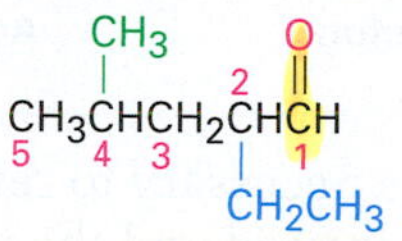

2-Ethyl-4-methylpentanal

For cyclic aldehydes in which the –CHO group is directly attached to a ring, the suffix *-carbaldehyde* is used:

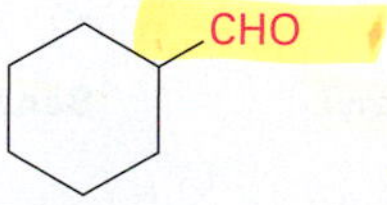

Cyclohexanecarbaldehyde 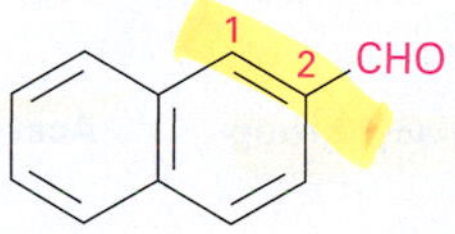   **Naphthalene-2-carbaldehyde**

A few simple and well-known aldehydes have common names that are recognized by IUPAC. Several that you might encounter are listed in Table 14.1.

TABLE 14.1
Common Names of Some Simple Aldehydes

Formula	Common name	Systematic name
HCHO	Formaldehyde	Methanal
CH_3CHO	Acetaldehyde	Ethanal
$H_2C=CHCHO$	Acrolein	Propenal
$CH_3CH=CHCHO$	Crotonaldehyde	But-2-enal
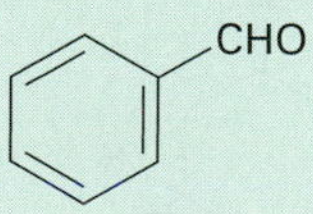 CHO	Benzaldehyde	Benzenecarbaldehyde

Ketones are named by replacing the terminal *-e* of the corresponding alkane name with *-one*. The parent chain is the longest one that contains the ketone group, and the numbering begins at the end nearer the carbonyl carbon. As with alkenes (Section 7.2) and alcohols (Section 13.1), the numerical locant is placed before the parent name in older rules but before the suffix in newer IUPAC recommendations. For example:

$$CH_3CH_2CCH_2CH_2CH_3$$
1 2 3 4 5 6

Hexan-3-one

$$CH_3CH=CHCH_2CCH_3$$
6 5   4 3   2 1

Hex-4-en-2-one

$$CH_3CH_2CCH_2CCH_3$$
6 5   4 3   2 1

Hexane-2,4-dione

A few ketones are allowed by IUPAC to retain their common names:

$$CH_3CCH_3$$

Acetone   **Acetophenone**   **Benzophenone**

When it's necessary to refer to the R—C=O as a substituent, the name **acyl** (**a**-sil) **group** is used and the name ending *-yl* is attached. Thus, CH_3CO- is an *acetyl* group, $-CHO$ is a *formyl* group, and C_6H_5CO- is a *benzoyl* group.

An acyl group   **Acetyl**   **Formyl**   **Benzoyl**

If other functional groups are present and the doubly bonded oxygen is considered a substituent on a parent chain, the prefix *oxo-* is used. For example:

$$CH_3CH_2CH_2CCH_2COCH_3$$
6 5 4   3 2 1

Methyl 3-oxohexanoate

Problem 14.1

Name the following aldehydes and ketones:

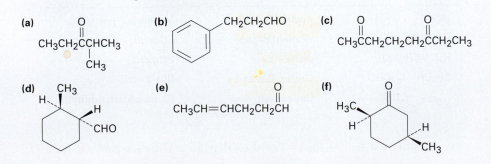

(a)
$$CH_3CH_2CCHCH_3$$
                |
               CH_3

(b) CH_2CH_2CHO

(c)
$$CH_3CCH_2CH_2CH_2CCH_2CH_3$$

(d)

(e) $CH_3CH=CHCH_2CH_2CH$

(f)

Problem 14.2
Draw structures corresponding to the following names:

(a) 3-Methylbutanal **(b)** 4-Chloropentan-2-one
(c) Phenylacetaldehyde **(d)** *cis*-3-*tert*-Butylcyclohexanecarbaldehyde
(e) 3-Methylbut-3-enal **(f)** 2-(1-Chloroethyl)-5-methylheptanal

14.2 Preparing Aldehydes and Ketones

One of the best methods of aldehyde synthesis is by oxidation of primary alcohols, as we saw in Section 13.5. The reaction is often carried out using the Dess–Martin periodinane reagent in dichloromethane solvent at room temperature:

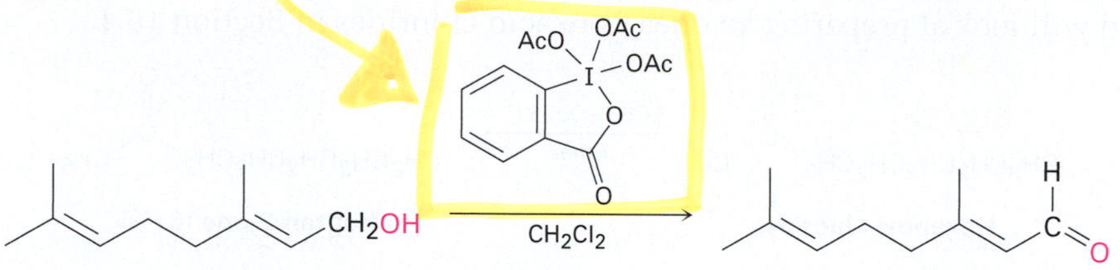
AcO
OAc
OAc
I
O
O
CH2OH
CH2Cl2
H
C
O

Geraniol **Geranial (84%)**

A second method of aldehyde synthesis is one that we'll mention here just briefly and then return to in Section 16.6. Certain carboxylic acid derivatives can be *partially* reduced to yield aldehydes. The partial reduction of an ester by diisobutylaluminum hydride (DIBAH), for instance, is an important laboratory-scale method of aldehyde synthesis, and mechanistically related processes also occur in biological pathways.

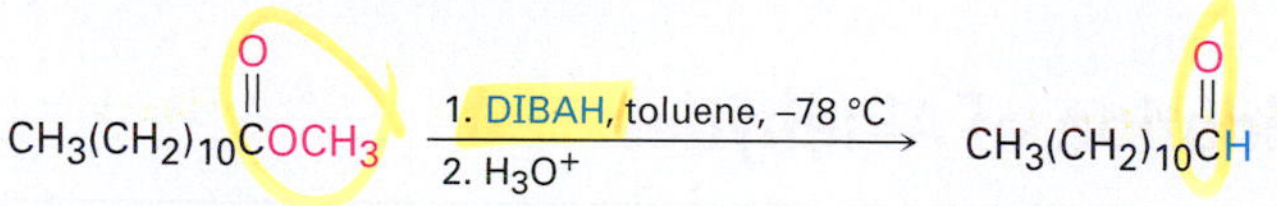
CH3(CH2)10COCH3
1. DIBAH, toluene, –78 °C
2. H3O+
CH3(CH2)10CH

Methyl dodecanoate **Dodecanal (88%)**

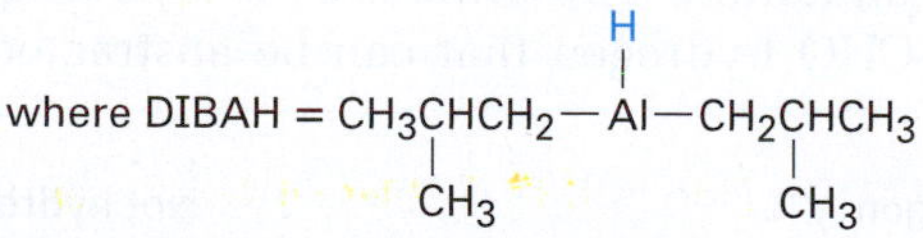
where DIBAH = CH3CHCH2—Al—CH2CHCH3
H
CH3
CH3

For the most part, methods of ketone synthesis are similar to those for aldehydes. Secondary alcohols are oxidized by a variety of reagents to give ketones (Section 13.5). The choice of oxidant depends on such factors as reaction scale, cost, and acid or base sensitivity of the alcohol, with either the Dess–Martin periodinane or a Cr(VI) regent such as CrO_3 being a common choice.

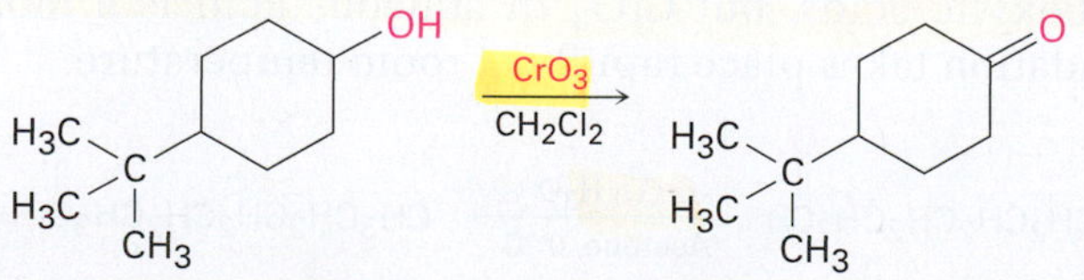
OH
H3C
H3C
C
CH3
CrO3
CH2Cl2
O

4-*tert*-Butylcyclohexanol **4-*tert*-Butylcyclohexanone (90%)**

Aryl ketones can be prepared by Friedel–Crafts acylation of an aromatic ring with an acid chloride in the presence of $AlCl_3$ catalyst (Section 9.7):

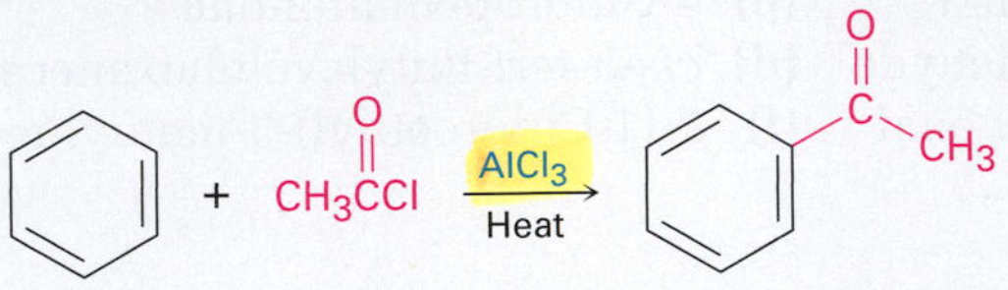

Benzene   **Acetyl chloride**   **Acetophenone (95%)**

In addition, ketones can be prepared from certain carboxylic acid derivatives, just as aldehydes can. Among the most useful reactions of this type is that between an acid chloride and a lithium diorganocopper reagent, R_2CuLi. We'll discuss lithium diorganocopper reagents later in this chapter (Section 14.11) and will look at preparing ketones from acid chlorides in Section 16.4.

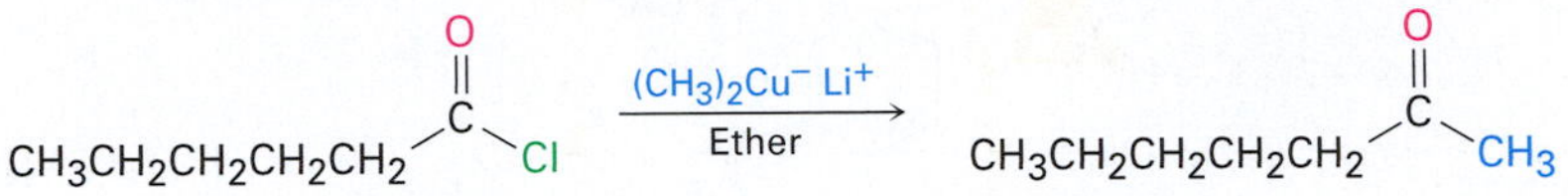

Hexanoyl chloride   **Heptan-2-one (81%)**

Problem 14.3

How would you carry out the following reactions? More than one step may be required.

(a) Benzene → *m*-Bromoacetophenone
(b) Bromobenzene → Acetophenone
(c) 1-Methylcyclohexene → 2-Methylcyclohexanone

14.3 Oxidation of Aldehydes

Aldehydes are easily oxidized to yield carboxylic acids, but ketones are generally inert toward oxidation. The difference is a consequence of structure: aldehydes have a –CHO hydrogen that can be abstracted during oxidation, but ketones do not.

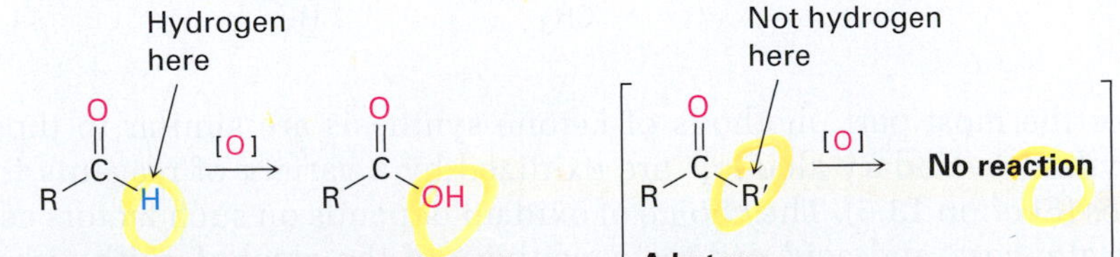

An aldehyde   **A carboxylic acid**   **A ketone**

Many oxidizing agents, including $KMnO_4$ and hot HNO_3, convert aldehydes into carboxylic acids, but CrO_3 in aqueous acid is a more common choice. The oxidation takes place rapidly at room temperature.

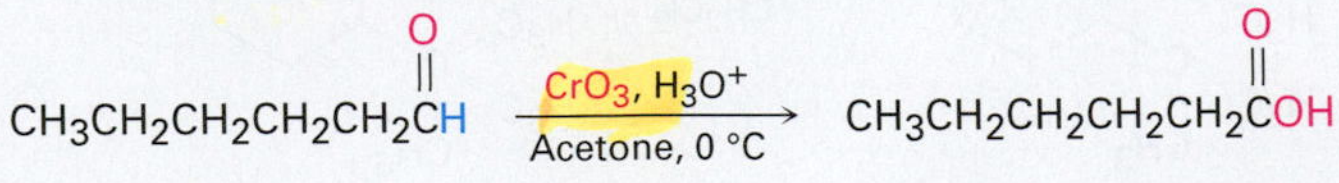

Hexanal   **Hexanoic acid (85%)**

Aldehyde oxidations occur through intermediate 1,1-diols, or *hydrates,* which are formed by a reversible nucleophilic addition of water to the carbonyl group. Even though formed to only a small extent at equilibrium, the hydrate reacts like any typical primary or secondary alcohol and is rapidly oxidized to a carbonyl compound.

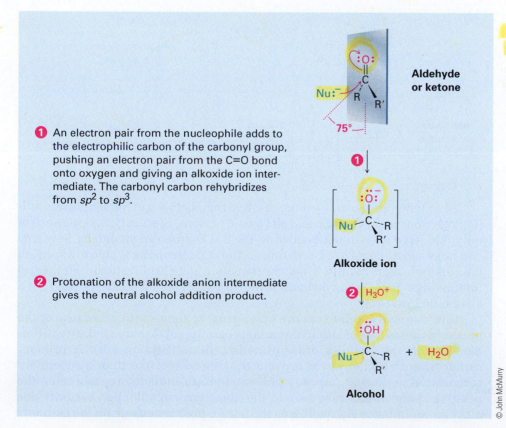

14.4 Nucleophilic Addition Reactions of Aldehydes and Ketones

As we saw in the *Preview of Carbonyl Chemistry,* the most general reaction of aldehydes and ketones is the **nucleophilic addition reaction**. A nucleophile, :Nu⁻, approaches along the C=O bond from an angle of about 75° to the plane of the carbonyl group and adds to the electrophilic C=O carbon atom. At the same time, rehybridization of the carbonyl carbon from sp^2 to sp^3 occurs, an electron pair from the C=O bond moves toward the electronegative oxygen atom, and a tetrahedral alkoxide ion intermediate is produced (Figure 14.1).

FIGURE 14.1 MECHANISM: A nucleophilic addition reaction to an aldehyde or ketone. The nucleophile approaches the carbonyl group from an angle of approximately 75° to the plane of the sp^2 orbitals, the carbonyl carbon rehybridizes from sp^2 to sp^3, and an alkoxide ion is formed.

1 An electron pair from the nucleophile adds to the electrophilic carbon of the carbonyl group, pushing an electron pair from the C=O bond onto oxygen and giving an alkoxide ion intermediate. The carbonyl carbon rehybridizes from sp^2 to sp^3.

2 Protonation of the alkoxide anion intermediate gives the neutral alcohol addition product.

© John McMurry

The nucleophile can be either negatively charged (:Nu⁻) or neutral (:Nu). If it's neutral, however, it usually carries a hydrogen atom that can subsequently be eliminated, :Nu–H. For example:

Some negatively charged nucleophiles

HO:⁻ (hydroxide ion)

H:⁻ (hydride ion)

R₃C:⁻ (a carbanion)

RO:⁻ (an alkoxide ion)

N≡C:⁻ (cyanide ion)

Some neutral nucleophiles

HOH (water)

ROH (an alcohol)

H₃N: (ammonia)

RNH₂ (an amine)

Nucleophilic additions to aldehydes and ketones have two general variations, as shown in Figure 14.2. In one variation, the tetrahedral intermediate is protonated by water or acid to give an alcohol as the final product; in the second variation, the carbonyl oxygen atom is protonated and then eliminated as HO⁻ or H₂O to give a product with a C=Nu bond.

FIGURE 14.2 Two general reaction pathways following addition of a nucleophile to an aldehyde or ketone. The top pathway leads to an alcohol product; the bottom pathway leads to a product with a C=Nu bond.

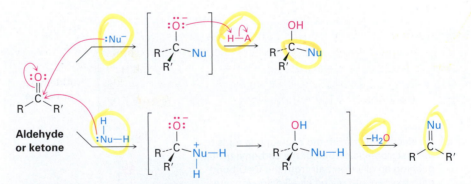

Aldehydes are generally more reactive than ketones in nucleophilic addition reactions for both steric and electronic reasons. Sterically, the presence of only one large substituent bonded to the C=O carbon in an aldehyde versus two large substituents in a ketone means that a nucleophile is able to approach the aldehyde more readily. Thus, the transition state leading to the tetrahedral intermediate is less crowded and lower in energy for an aldehyde than for a ketone (Figure 14.3).

Electronically, aldehydes are more reactive than ketones because of the greater polarization of aldehyde carbonyl groups. To see this polarity difference, recall the stability order of carbocations (Section 7.8). A primary carbocation is higher in energy and thus more reactive than a secondary carbocation because it has only one alkyl group inductively stabilizing the positive charge rather than two. In the same way, an aldehyde has only one alkyl group inductively stabilizing the partial positive charge on the carbonyl

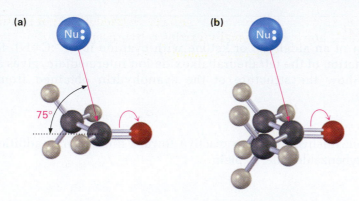

(a)   (b)

FIGURE 14.3 **(a)** Nucleophilic addition to an aldehyde is sterically less hindered because only one relatively large substituent is attached to the carbonyl-group carbon. **(b)** A ketone, however, has two large substituents and is more hindered. The approach of the nucleophile is along the C=O bond at an angle of about 75° to the plane of the carbon sp^2 orbitals.

carbon rather than two, is a bit more electrophilic, and is therefore more reactive than a ketone.

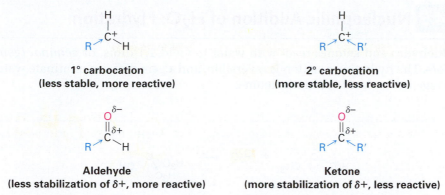

1° carbocation
(less stable, more reactive)

2° carbocation
(more stable, less reactive)

Aldehyde
(less stabilization of δ+, more reactive)

Ketone
(more stabilization of δ+, less reactive)

One further comparison: aromatic aldehydes, such as benzaldehyde, are less reactive in nucleophilic addition reactions than aliphatic aldehydes because the electron-donating resonance effect of the aromatic ring makes the carbonyl group less electrophilic. Comparing electrostatic potential maps of formaldehyde and benzaldehyde, for example, shows that the carbonyl carbon atom in the aromatic aldehyde is less positive (less blue).

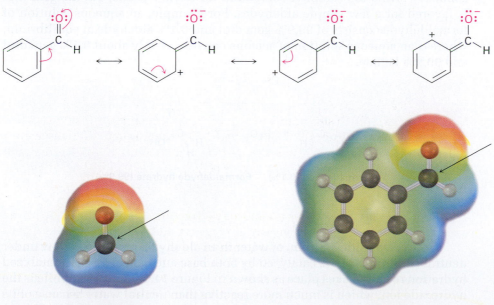

Formaldehyde   **Benzaldehyde**

Problem 14.4

Treatment of an aldehyde or ketone with cyanide ion ($^-$:C≡N), followed by protonation of the tetrahedral alkoxide ion intermediate, gives a *cyanohydrin*. Show the structure of the cyanohydrin obtained from cyclohexanone.

Problem 14.5

p-Nitrobenzaldehyde is more reactive toward nucleophilic additions than *p*-methoxybenzaldehyde. Explain.

14.5 Nucleophilic Addition of H₂O: Hydration

Aldehydes and ketones react with water to yield 1,1-diols, or *geminal (gem) diols*. The hydration reaction is reversible, and a gem diol can eliminate water to regenerate the aldehyde or ketone.

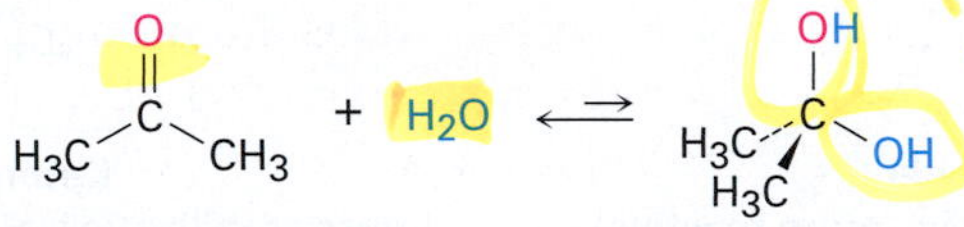

Acetone (99.9%)   Acetone hydrate (0.1%)

The position of the equilibrium between a gem diol and an aldehyde or ketone depends on the structure of the carbonyl compound. The equilibrium generally favors the carbonyl compound for steric reasons, but the gem diol is favored for a few simple aldehydes. For example, an aqueous solution of formaldehyde consists of 99.9% gem diol and 0.1% aldehyde at equilibrium, whereas an aqueous solution of acetone consists of only about 0.1% gem diol and 99.9% ketone.

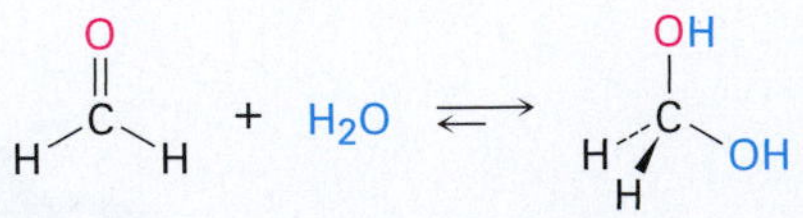

Formaldehyde (0.1%)   Formaldehyde hydrate (99.9%)

The nucleophilic addition of water to an aldehyde or ketone is slow under neutral conditions but is catalyzed by both base and acid. The base-catalyzed hydration reaction takes place as shown in Figure 14.4. The nucleophile is the hydroxide ion, which is much more reactive than neutral water because of its negative charge.

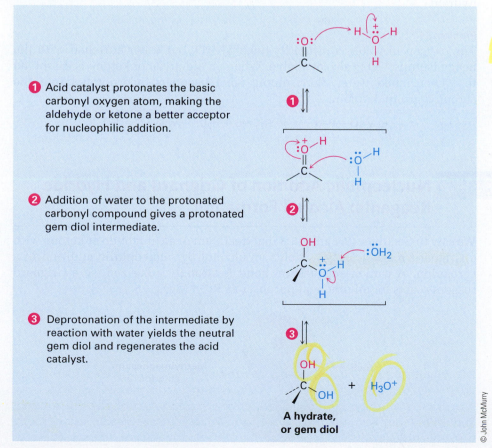

FIGURE 14.4 MECHANISM: The mechanism of base-catalyzed hydration of an aldehyde or ketone. Hydroxide ion is a more reactive nucleophile than neutral water.

① The nucleophilic hydroxide ion adds to the aldehyde or ketone and yields a tetrahedral alkoxide ion intermediate.

② The alkoxide ion is protonated by water to give the gem diol product and regenerate the hydroxide ion catalyst.

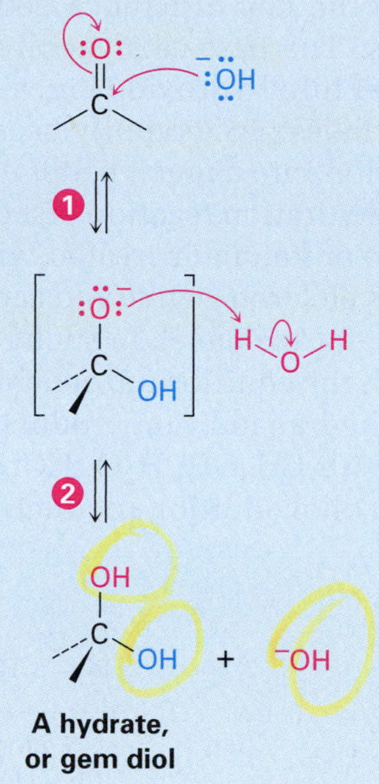

A hydrate, or gem diol

The acid-catalyzed hydration reaction begins with protonation of the carbonyl oxygen atom, which places a positive charge on oxygen and makes the carbonyl group more electrophilic. Subsequent nucleophilic addition of water to the protonated aldehyde or ketone then yields a protonated gem diol, which loses H⁺ to give the neutral product (Figure 14.5).

FIGURE 14.5 MECHANISM: The mechanism of acid-catalyzed hydration of an aldehyde or ketone. Acid protonates the carbonyl group, making it more electrophilic and more reactive.

① Acid catalyst protonates the basic carbonyl oxygen atom, making the aldehyde or ketone a better acceptor for nucleophilic addition.

② Addition of water to the protonated carbonyl compound gives a protonated gem diol intermediate.

③ Deprotonation of the intermediate by reaction with water yields the neutral gem diol and regenerates the acid catalyst.

A hydrate, or gem diol

Note the key difference between the base-catalyzed and acid-catalyzed reactions. The base-catalyzed reaction takes place rapidly because water is converted into hydroxide ion, a much better *nucleophile*. The acid-catalyzed reaction takes place rapidly because the carbonyl compound is converted by protonation into a much better *electrophile*.

The hydration reaction just described is typical of what happens when an aldehyde or ketone is treated with a nucleophile of the type H–Y, where the Y atom is electronegative and can stabilize a negative charge (oxygen, halogen, or sulfur, for instance). In such reactions, the nucleophilic addition is reversible, with the equilibrium generally favoring the carbonyl reactant rather than the tetrahedral addition product. In other words, treatment of an aldehyde or ketone with CH_3OH, H_2O, HCl, HBr, or H_2SO_4 does not normally lead to a stable alcohol addition product.

Favored when
$$Y = -OCH_3, -OH, -Br, -Cl, HSO_4^-$$

Problem 14.6

When dissolved in water, trichloroacetaldehyde (chloral, CCl_3CHO) exists primarily as chloral hydrate, $CCl_3CH(OH)_2$. Show the structure of chloral hydrate.

Problem 14.7

The oxygen in water is primarily (99.8%) ^{16}O, but water enriched with the heavy isotope ^{18}O is also available. When an aldehyde or ketone is dissolved in ^{18}O-enriched water, the isotopic label becomes incorporated into the carbonyl group. Explain.

$$R_2C{=}O + H_2O \longrightarrow R_2C{=}O + H_2O \qquad \text{where } O = {}^{18}O$$

14.6 Nucleophilic Addition of Grignard and Hydride Reagents: Alcohol Formation

We saw in Sections 12.4 and 13.3 that treatment of an aldehyde or ketone with a Grignard reagent, RMgX, yields an alcohol by nucleophilic addition of a carbanion. A carbon–magnesium bond is strongly polarized, so a Grignard reagent reacts for all practical purposes as $R:^- \ ^+MgX$.

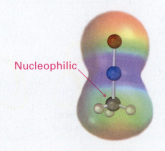

Nucleophilic

Methylmagnesium chloride

A Grignard reaction begins with an acid–base complexation of Mg^{2+} to the carbonyl oxygen atom of the aldehyde or ketone, thereby making the carbonyl group a better electrophile. Nucleophilic addition of $R:^-$ then produces a tetrahedral magnesium alkoxide intermediate, and protonation by addition of water or dilute aqueous acid in a separate step yields the neutral alcohol (Figure 14.6). Unlike the nucleophilic addition of water, Grignard additions are effectively irreversible because a carbanion is too poor a leaving group to be expelled in a reversal step.

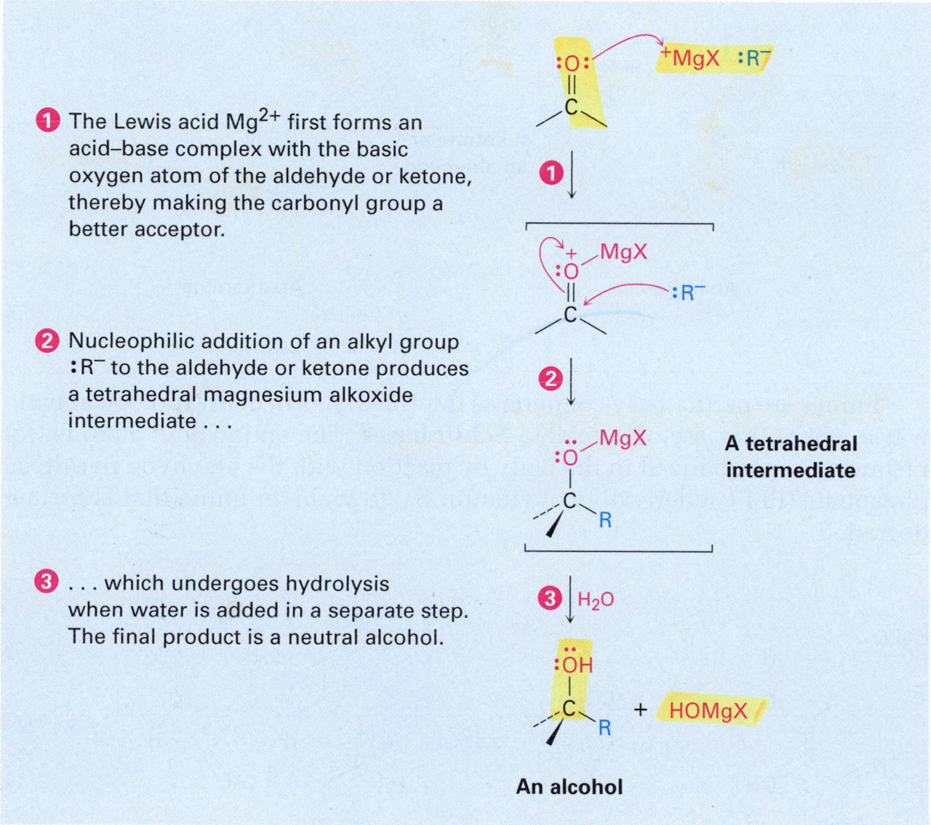

1. The Lewis acid Mg^{2+} first forms an acid–base complex with the basic oxygen atom of the aldehyde or ketone, thereby making the carbonyl group a better acceptor.

2. Nucleophilic addition of an alkyl group :R^- to the aldehyde or ketone produces a tetrahedral magnesium alkoxide intermediate . . .

A tetrahedral intermediate

3. . . . which undergoes hydrolysis when water is added in a separate step. The final product is a neutral alcohol.

An alcohol

© John McMurry

FIGURE 14.6 MECHANISM: Mechanism of the Grignard reaction. Nucleophilic addition of a carbanion to an aldehyde or ketone, followed by protonation of the alkoxide intermediate, yields an alcohol.

Just as addition of a Grignard reagent to an aldehyde or ketone yields an alcohol, so does addition of hydride ion, :H^- (Section 13.3). Although the details of carbonyl-group reductions are complex, $LiAlH_4$ and $NaBH_4$ act as if they were donors of hydride ion in a nucleophilic addition reaction (Figure 14.7). Addition of water or aqueous acid after the hydride addition step protonates the tetrahedral alkoxide intermediate and gives the alcohol product.

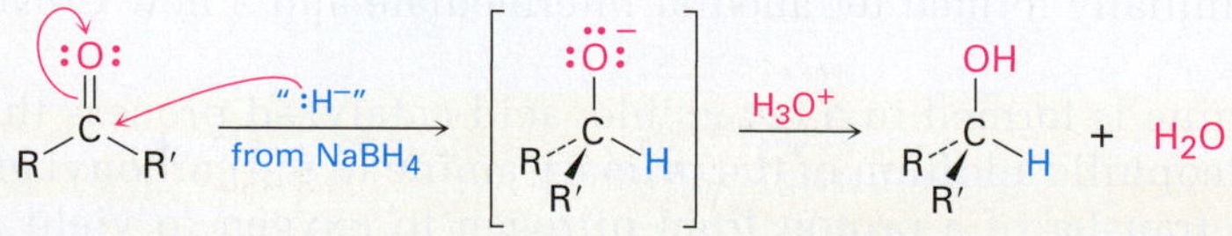

ACTIVE FIGURE 14.7 Mechanism of carbonyl-group reduction by nucleophilic addition of "hydride ion" from $NaBH_4$ or $LiAlH_4$. **Go to this book's student companion site at www.cengage.com/chemistry/mcmurry to explore an interactive version of this figure.**

14.7 Nucleophilic Addition of Amines: Imine and Enamine Formation

Primary amines, RNH_2, add to aldehydes and ketones to yield **imines**, $R_2C=NR$. Secondary amines, R_2NH, add similarly to yield **enamines**, $R_2N-CR=CR_2$ (*ene* + *amine* = unsaturated amine).

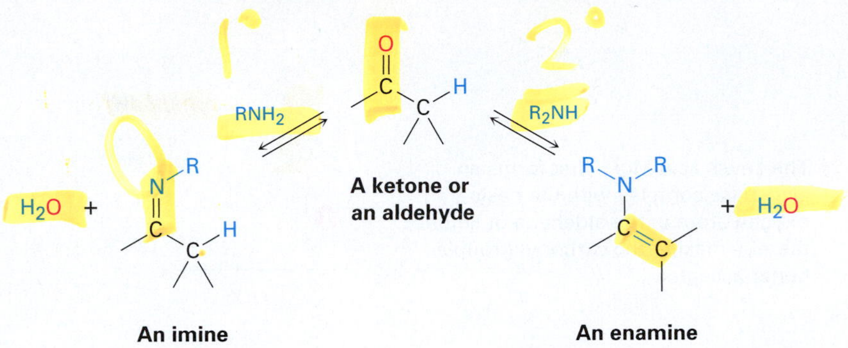

An imine   **A ketone or an aldehyde**   **An enamine**

Imines are particularly common as intermediates in many biological pathways, where they are often called **Schiff bases**. The amino acid alanine, for instance, is metabolized in the body by reaction with the aldehyde pyridoxal phosphate (PLP), a derivative of vitamin B_6, to yield an imine that is further degraded.

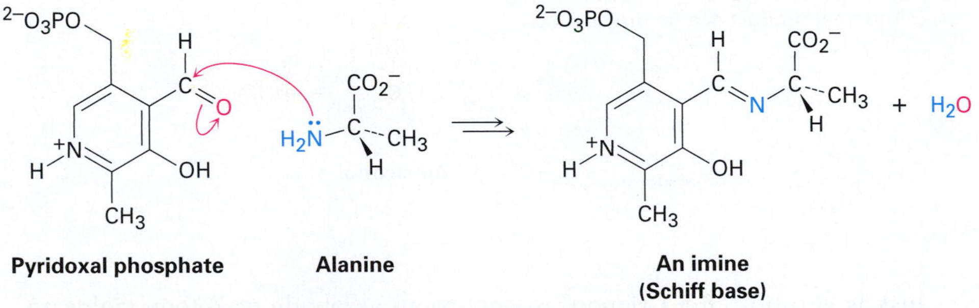

Pyridoxal phosphate   **Alanine**   **An imine (Schiff base)**

Imine formation and enamine formation appear different because one leads to a product with a C=N bond and the other leads to a product with a C=C bond. Actually, though, the reactions are quite similar. Both are typical examples of nucleophilic addition reactions in which water is eliminated from the initially formed tetrahedral intermediate and a new C=Nu bond is formed.

An imine is formed in a reversible, acid-catalyzed process that begins with nucleophilic addition of the primary amine to the carbonyl group, followed by transfer of a proton from nitrogen to oxygen to yield a neutral amino alcohol, or *carbinolamine*. Protonation of the carbinolamine oxygen by an acid catalyst then converts the –OH into a better leaving group ($-OH_2^+$), and E1-like loss of water produces an iminium ion. Loss of a proton from nitrogen gives the final product and regenerates the acid catalyst (Figure 14.8).

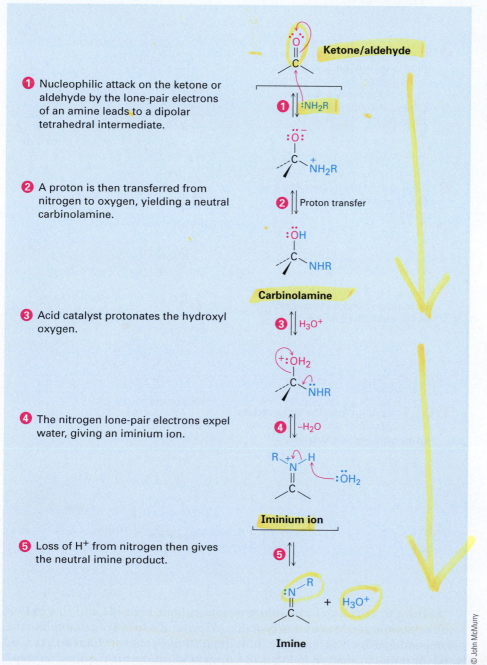

1 Nucleophilic attack on the ketone or aldehyde by the lone-pair electrons of an amine leads to a dipolar tetrahedral intermediate.

2 A proton is then transferred from nitrogen to oxygen, yielding a neutral carbinolamine.

3 Acid catalyst protonates the hydroxyl oxygen.

4 The nitrogen lone-pair electrons expel water, giving an iminium ion.

5 Loss of H^+ from nitrogen then gives the neutral imine product.

FIGURE 14.8 MECHANISM: Mechanism of imine formation by reaction of an aldehyde or ketone with a primary amine. The key step is nucleophilic addition to yield a carbinolamine intermediate, which loses water to give the imine.

© John McMurry

Reaction of an aldehyde or ketone with a secondary amine, R_2NH, rather than a primary amine yields an enamine. The process is identical to imine formation up to the iminium ion stage, but at this point there is no proton on nitrogen that can be lost to form a neutral imine product. Instead, a proton is lost from the *neighboring* carbon (the α carbon), yielding an enamine (Figure 14.9).

FIGURE 14.9 MECHANISM: Mechanism of enamine formation by reaction of an aldehyde or ketone with a secondary amine, R_2NH. The iminium ion intermediate produced in step 3 has no hydrogen attached to N and so must lose H$^+$ from the carbon two atoms away.

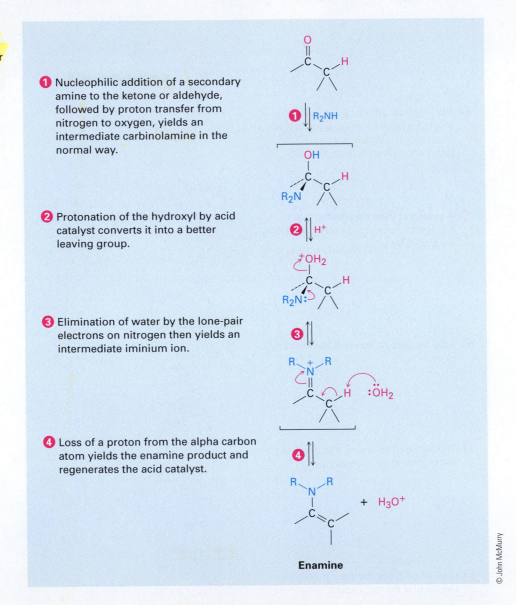

❶ Nucleophilic addition of a secondary amine to the ketone or aldehyde, followed by proton transfer from nitrogen to oxygen, yields an intermediate carbinolamine in the normal way.

❷ Protonation of the hydroxyl by acid catalyst converts it into a better leaving group.

❸ Elimination of water by the lone-pair electrons on nitrogen then yields an intermediate iminium ion.

❹ Loss of a proton from the alpha carbon atom yields the enamine product and regenerates the acid catalyst.

Enamine

© John McMurry

Imine and enamine formation are slow at both high pH and low pH but reach a maximum rate at a weakly acidic pH around 4 to 5. We can explain this pH dependence by looking at the individual steps in the mechanism. As indicated for imine formation in Figure 14.8, an acid catalyst is required in step 3 to protonate the intermediate carbinolamine, thereby converting the –OH into a better leaving group. Thus, reaction will be slow if not enough acid is present (that is, at high pH). On the other hand, if too much acid is present (low pH), the basic amine nucleophile is completely protonated, so the initial nucleophilic addition step can't occur.

Evidently, a pH of 4.5 represents a compromise between the need for *some* acid to catalyze the rate-limiting dehydration step but *not too much* acid so as to avoid complete protonation of the amine. Each individual nucleophilic addition reaction has its own requirements, and reaction conditions must be optimized to obtain maximum reaction rates.

| WORKED EXAMPLE 14.1 | Predicting the Product of Reaction between a Ketone and an Amine |

Show the products you would obtain by acid-catalyzed reaction of pentan-3-one with methylamine, CH_3NH_2, and with dimethylamine, $(CH_3)_2NH$.

Strategy

An aldehyde or ketone reacts with a primary amine, RNH_2, to yield an imine, in which the carbonyl oxygen atom has been replaced by the $=N-R$ group of the amine. Reaction of the same aldehyde or ketone with a secondary amine, R_2NH, yields an enamine, in which the oxygen atom has been replaced by the $-NR_2$ group of the amine and the double bond has moved to a position between the former carbonyl carbon and the neighboring carbon.

Solution

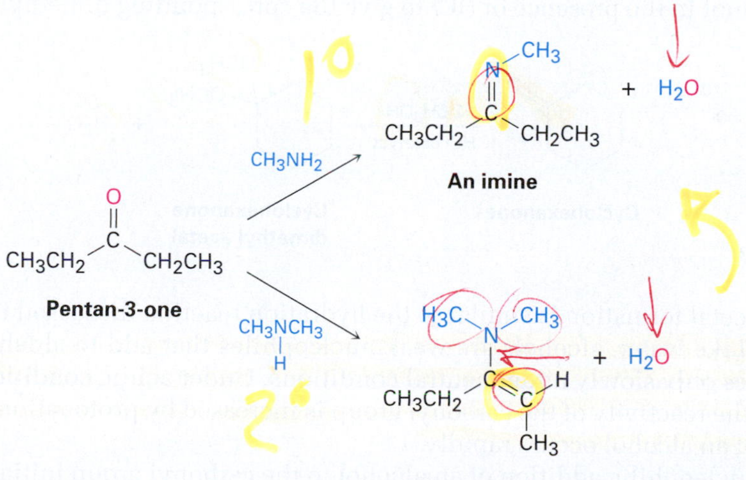

Pentan-3-one

An imine

An enamine

Problem 14.8

Show the products you would obtain by acid-catalyzed reaction of cyclohexanone with ethylamine, $CH_3CH_2NH_2$, and with diethylamine, $(CH_3CH_2)_2NH$.

Problem 14.9

Imine formation is reversible. Show all the steps involved in the acid-catalyzed reaction of an imine with water (hydrolysis) to yield an aldehyde or ketone plus primary amine.

Problem 14.10

Draw the following molecule as a skeletal structure, and show how it can be prepared from a ketone and an amine.

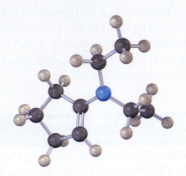

14.8 Nucleophilic Addition of Alcohols: Acetal Formation

Aldehydes and ketones react reversibly with two equivalents of an alcohol in the presence of an acid catalyst to yield **acetals**, $R_2C(OR')_2$, frequently called *ketals* if derived from a ketone. Cyclohexanone, for instance, reacts with methanol in the presence of HCl to give the corresponding dimethyl acetal.

O + 2 CH₃OH → Cyclohexanone dimethyl acetal (OCH₃, OCH₃) + H₂O

HCl catalyst

Cyclohexanone   **Cyclohexanone dimethyl acetal**

Acetal formation is similar to the hydration reaction discussed in Section 14.5. Like water, alcohols are weak nucleophiles that add to aldehydes and ketones only slowly under neutral conditions. Under acidic conditions, however, the reactivity of the carbonyl group is increased by protonation, so addition of an alcohol occurs rapidly.

Nucleophilic addition of an alcohol to the carbonyl group initially yields a hydroxy ether called a **hemiacetal**, analogous to the gem diol formed by addition of water. Hemiacetals are formed reversibly, with the equilibrium normally favoring the carbonyl compound. In the presence of acid, however, a further reaction occurs. Protonation of the –OH group, followed by an E1-like loss of water, leads to an oxonium ion, $R_2C=OR^+$, which undergoes a second nucleophilic addition of alcohol to yield the acetal. The mechanism is shown in Figure 14.10.

Because all the steps in acetal formation are reversible, the reaction can be driven either forward (from carbonyl compound to acetal) or backward (from acetal to carbonyl compound), depending on the conditions. The forward reaction is favored by conditions that remove water from the medium and thus drive the equilibrium to the right. In practice, this is often done by distilling off water as it forms. The reverse reaction is favored by treating the acetal with a large excess of aqueous acid to drive the equilibrium to the left.

FIGURE 14.10 MECHANISM:
Mechanism of acid-catalyzed
acetal formation by reaction of
an aldehyde or ketone with an
alcohol.

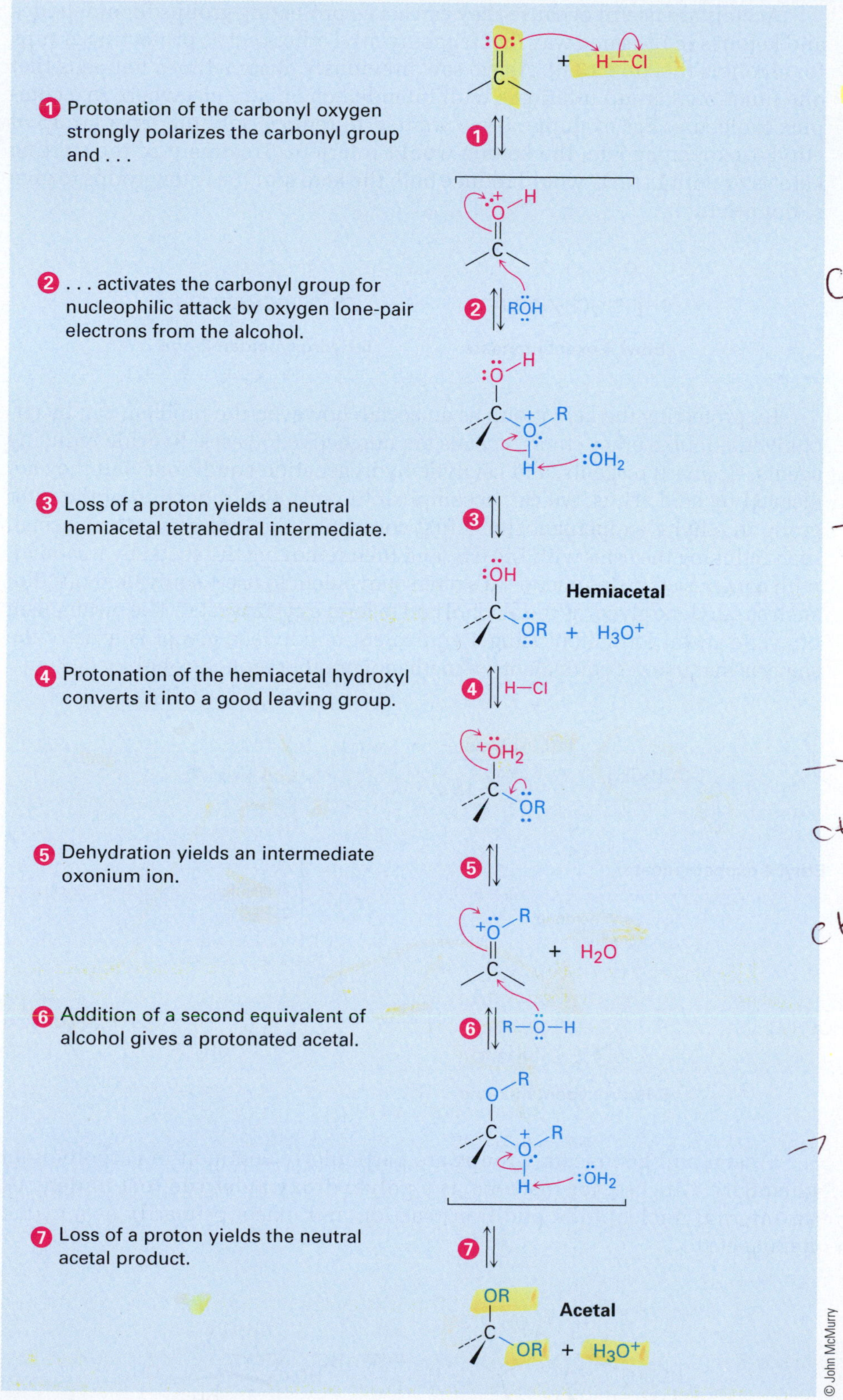

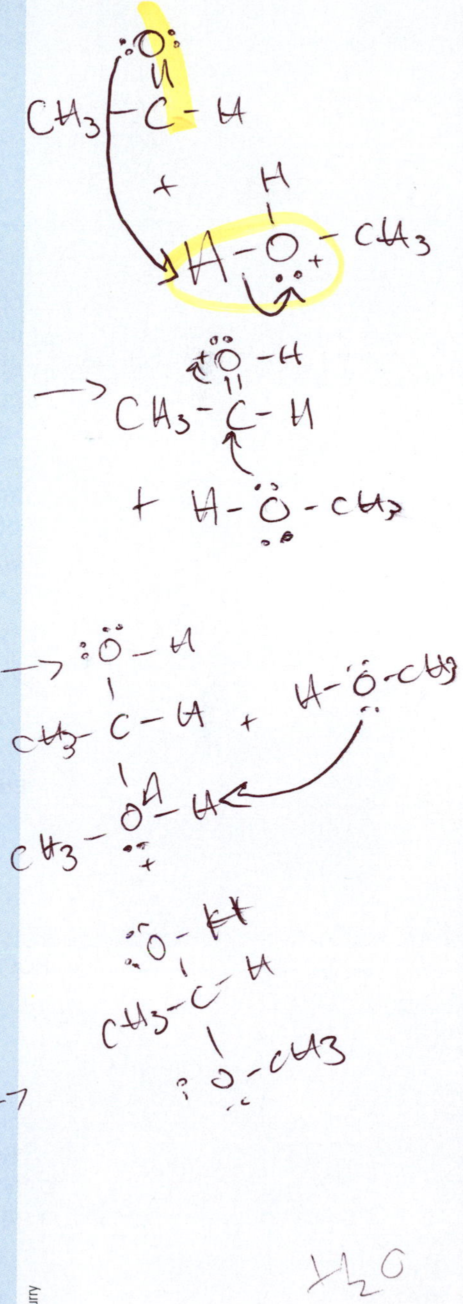

1 Protonation of the carbonyl oxygen
strongly polarizes the carbonyl group
and . . .

2 . . . activates the carbonyl group for
nucleophilic attack by oxygen lone-pair
electrons from the alcohol.

3 Loss of a proton yields a neutral
hemiacetal tetrahedral intermediate.

4 Protonation of the hemiacetal hydroxyl
converts it into a good leaving group.

5 Dehydration yields an intermediate
oxonium ion.

6 Addition of a second equivalent of
alcohol gives a protonated acetal.

7 Loss of a proton yields the neutral
acetal product.

Hemiacetal

Acetal

© John McMurry

Acetals are useful because they can act as protecting groups for aldehydes and ketones in the same way that trimethylsilyl ethers act as protecting groups for alcohols (Section 13.6). As we saw previously, it sometimes happens that one functional group interferes with intended chemistry elsewhere in a complex molecule. For example, if we wanted to reduce only the ester group of ethyl 4-oxopentanoate, the ketone would interfere. Treatment of the starting keto ester with LiAlH$_4$ would reduce both the keto and the ester groups to give a diol product.

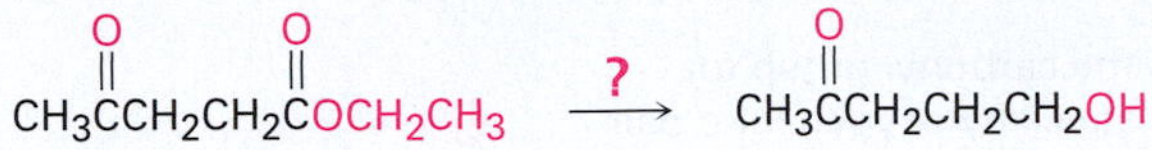

Ethyl 4-oxopentanoate   **5-Hydroxypentan-2-one**

By protecting the keto group as an acetal, however, the problem can be circumvented. Like other ethers, acetals are unreactive to bases, hydride reducing agents, Grignard reagents, and catalytic hydrogenation conditions, but they are cleaved by acid. Thus, we can accomplish the selective reduction of the ester group in ethyl 4-oxopentanoate by first converting the keto group to an acetal, then reducing the ester with LiAlH$_4$, and then removing the acetal by treatment with aqueous acid. (In practice, it's often convenient to use 1 equivalent of a diol such as ethylene glycol as the alcohol and to form a *cyclic* acetal. The mechanism of cyclic acetal formation using 1 equivalent of ethylene glycol is exactly the same as that using 2 equivalents of methanol or other monoalcohol.)

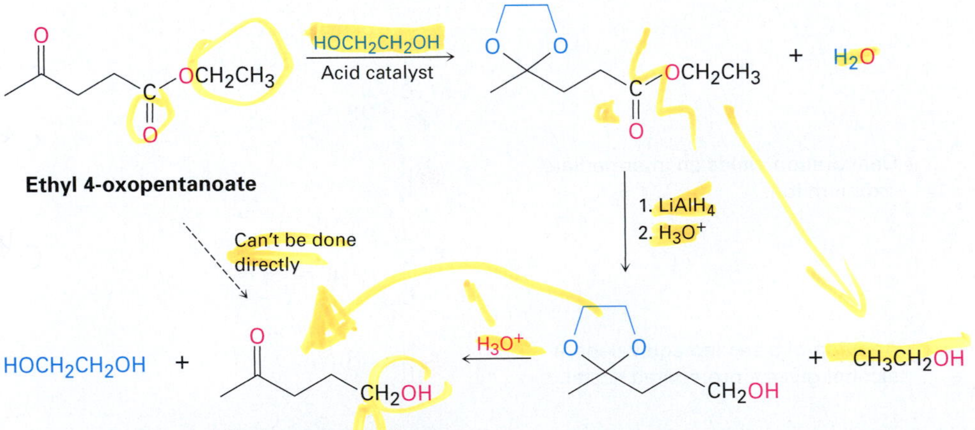

Ethyl 4-oxopentanoate

Can't be done directly

5-Hydroxypentan-2-one

Acetal and hemiacetal groups are particularly common in carbohydrate chemistry. Glucose, for instance, is a polyhydroxy aldehyde that undergoes an *internal* nucleophilic addition reaction and exists primarily as a cyclic hemiacetal.

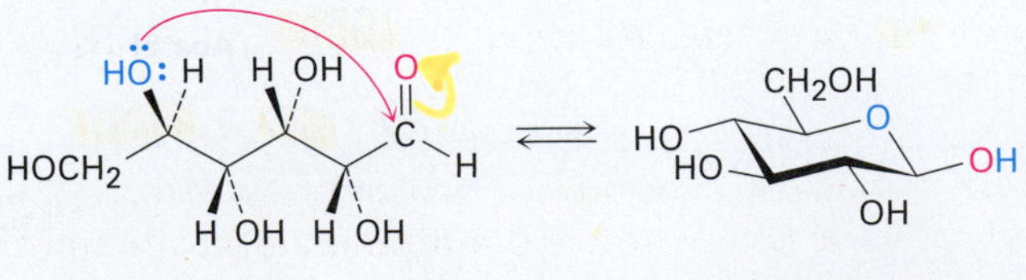

Glucose—open chain   **Glucose—cyclic hemiacetal**

WORKED EXAMPLE 14.2	Predicting the Product of Reaction between a Ketone and an Alcohol

Show the structure of the acetal you would obtain by acid-catalyzed reaction of pentan-2-one with propane-1,3-diol.

Strategy

Acid-catalyzed reaction of an aldehyde or ketone with 2 equivalents of a monoalcohol or 1 equivalent of a diol yields an acetal, in which the carbonyl oxygen atom is replaced by two —OR groups from the alcohol.

Solution

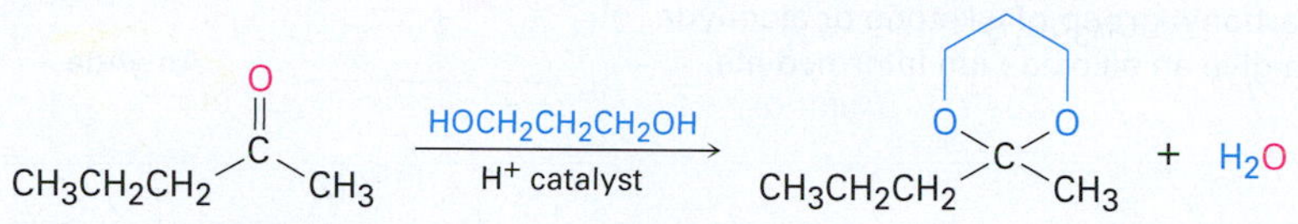

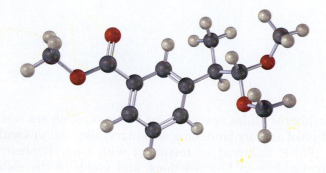

Problem 14.11

Show all the steps in the acid-catalyzed formation of a cyclic acetal from ethylene glycol and an aldehyde or ketone.

Problem 14.12

Identify the carbonyl compound and the alcohol that were used to prepare the following acetal:

14.9 Nucleophilic Addition of Phosphorus Ylides: The Wittig Reaction

Aldehydes and ketones are converted into alkenes by means of a nucleophilic addition called the **Wittig reaction**. The reaction has no direct biological counterpart but is worth knowing about both because of its wide use in the laboratory and drug manufacture and because of its mechanistic similarity to reactions of the coenzyme thiamin diphosphate, which we'll see in Section 22.3.

In the Wittig reaction, a phosphorus *ylide*, $R_2\overset{-}{C}-\overset{+}{P}(C_6H_5)_3$, also called a *phosphorane* and sometimes written in the resonance form $R_2C=P(C_6H_5)_3$, adds to an aldehyde or ketone to yield a dipolar, alkoxide ion intermediate. (An **ylide**—pronounced **ill**-id—is a neutral, dipolar compound with adjacent plus and minus charges.)

The dipolar intermediate is not isolated; rather, it spontaneously decomposes through a four-membered ring to yield alkene plus triphenylphosphine oxide, $(Ph)_3P{=}O$. The net result is replacement of the carbonyl oxygen atom by the $R_2C{=}$ group originally bonded to phosphorus (Figure 14.11).

FIGURE 14.11 MECHANISM:
The mechanism of the Wittig reaction between a phosphorus ylide and an aldehyde or ketone to yield an alkene.

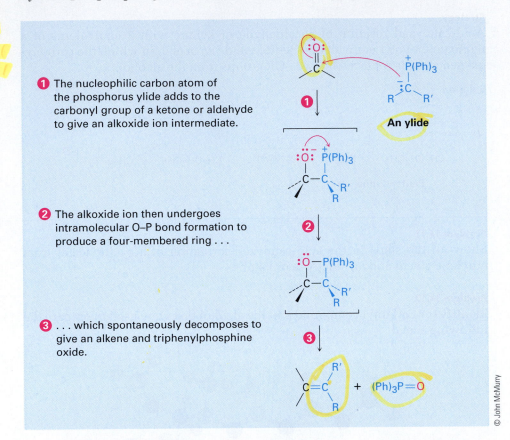

1 The nucleophilic carbon atom of the phosphorus ylide adds to the carbonyl group of a ketone or aldehyde to give an alkoxide ion intermediate.

2 The alkoxide ion then undergoes intramolecular O–P bond formation to produce a four-membered ring . . .

3 . . . which spontaneously decomposes to give an alkene and triphenylphosphine oxide.

© John McMurry

The phosphorus ylides necessary for Wittig reaction are easily prepared by S_N2 reaction of primary (and some secondary) alkyl halides with triphenylphosphine, $(Ph)_3P$, followed by treatment with base. Triphenylphosphine is a good nucleophile in S_N2 reactions, and yields of the resultant alkyltriphenylphosphonium salts are high. Because of the positive charge on phosphorus, the hydrogen on the neighboring carbon is weakly acidic and can be removed by a strong base such as butyllithium (BuLi) to generate the neutral ylide. For example:

Triphenylphosphine + CH$_3$—Br **Bromomethane** $\xrightarrow{S_N2}$ **Methyltriphenylphosphonium bromide** $\xrightarrow[\text{THF}]{\text{BuLi}}$ **Methylenetriphenylphosphorane**

The Wittig reaction is extremely general, and a great many monosubstituted, disubstituted, and trisubstituted alkenes can be prepared from the appropriate combination of phosphorane and aldehyde or ketone. Tetrasubstituted alkenes can't be prepared, however, because of steric hindrance during the reaction.

The real value of the Wittig reaction is that it yields a pure alkene of defined structure. The C=C bond in the product is always exactly where the C=O group was in the reactant, and no alkene isomers (except E,Z isomers) are formed. For example, Wittig reaction of cyclohexanone with methylenetriphenylphosphorane yields only the single alkene product methylenecyclohexane. By contrast, addition of methylmagnesium bromide to cyclohexanone, followed by dehydration with $POCl_3$, yields a roughly 9:1 mixture of two alkenes:

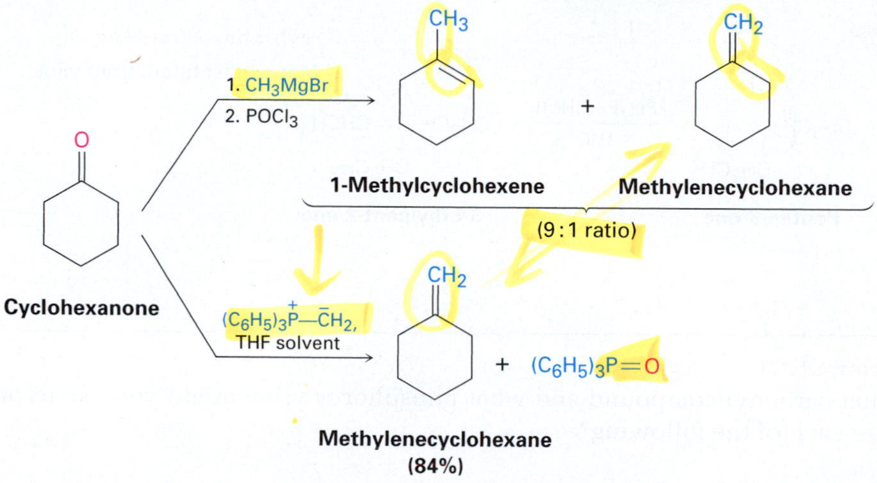

Wittig reactions are used commercially in the synthesis of numerous pharmaceuticals. For example, the German chemical company BASF prepares vitamin A by using a Wittig reaction between a 15-carbon ylide and a 5-carbon aldehyde.

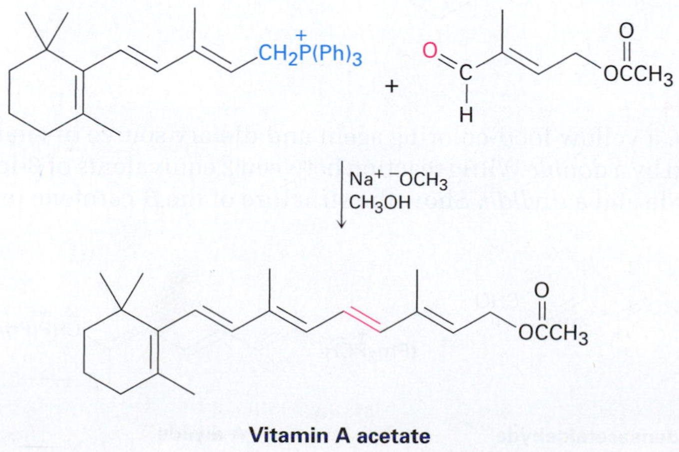

WORKED EXAMPLE 14.3 Synthesizing an Alkene Using a Wittig Reaction

What carbonyl compound and what phosphorus ylide might you use to prepare 3-ethylpent-2-ene?

Strategy

An aldehyde or ketone reacts with a phosphorus ylide to yield an alkene in which the oxygen atom of the carbonyl reactant is replaced by the $=CR_2$ of the ylide. Preparation of the phosphorus ylide itself usually involves S_N2 reaction of a primary alkyl halide with triphenylphosphine, so the ylide is typically primary, $RCH=P(Ph)_3$. This means that the disubstituted alkene carbon in the product comes from the carbonyl reactant, while the monosubstituted alkene carbon comes from the ylide.

Solution

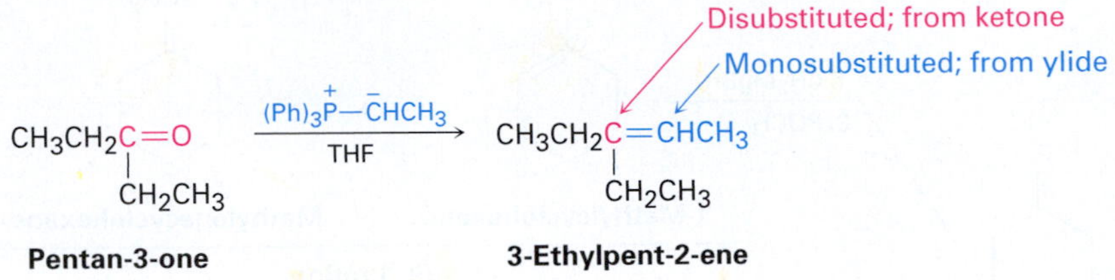

Pentan-3-one   3-Ethylpent-2-ene

Problem 14.13

What carbonyl compound and what phosphorus ylide might you use to prepare each of the following?

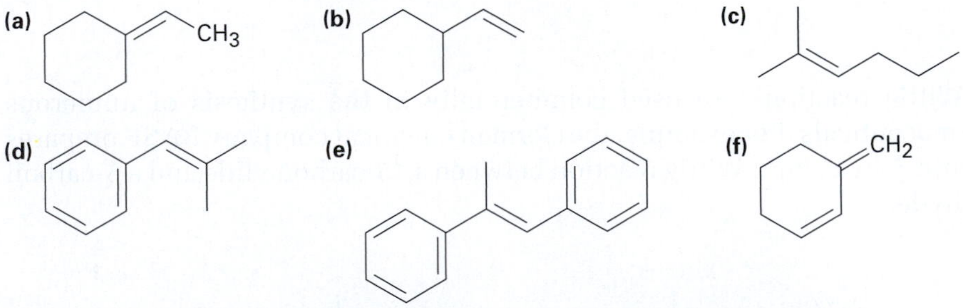

(a)   (b)   (c)

(d)   (e)   (f)

Problem 14.14

β-Carotene, a yellow food-coloring agent and dietary source of vitamin A can be prepared by a *double* Wittig reaction between 2 equivalents of β-ionylidene-acetaldehyde and a *diylide*. Show the structure of the β-carotene product.

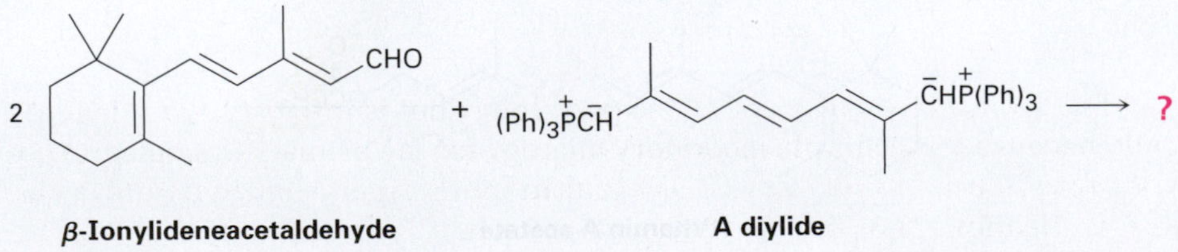

β-Ionylideneacetaldehyde   A diylide

14.10 Biological Reductions

As a general rule, nucleophilic addition reactions are characteristic only of aldehydes and ketones, not of carboxylic acid derivatives. The reason for the difference is structural. As discussed previously in the *Preview of Carbonyl Chemistry* and shown in Figure 14.12, the tetrahedral intermediate produced by addition of a nucleophile to a carboxylic acid derivative can eliminate a leaving group, leading to a net nucleophilic acyl substitution reaction. The tetrahedral intermediate produced by addition of a nucleophile to an aldehyde or ketone, however, has only alkyl or hydrogen substituents and thus can't usually expel a stable leaving group.

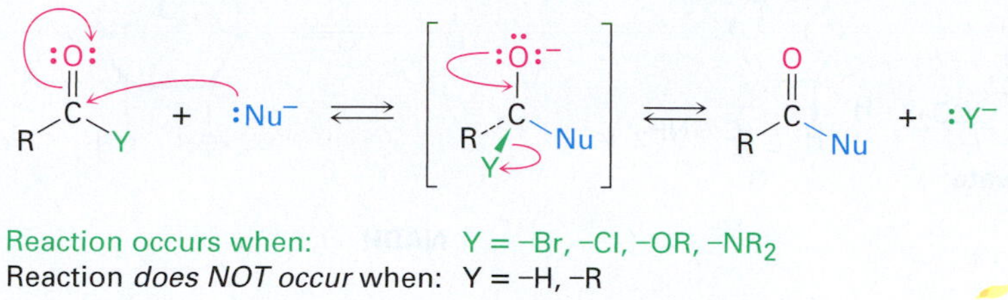

Reaction occurs when: $Y = -Br, -Cl, -OR, -NR_2$
Reaction *does NOT occur* when: $Y = -H, -R$

FIGURE 14.12 Carboxylic acid derivatives have an electronegative substituent $Y = -Br$, $-Cl, -OR, -NR_2$ that can be expelled as a leaving group from the tetrahedral intermediate formed by nucleophilic addition. Aldehydes and ketones have no such leaving group and thus do not usually undergo this reaction.

One exception to the rule that nucleophilic acyl substitutions don't occur with aldehydes and ketones is the *Cannizzaro reaction*, discovered in 1853. The Cannizzaro reaction takes place by nucleophilic addition of OH⁻ to an aldehyde to give a tetrahedral intermediate, *which expels hydride ion as a leaving group* and is thereby oxidized. A second aldehyde molecule accepts the hydride ion in another nucleophilic addition step and is thereby reduced. Benzaldehyde, for instance, yields benzyl alcohol plus benzoic acid when heated with aqueous NaOH.

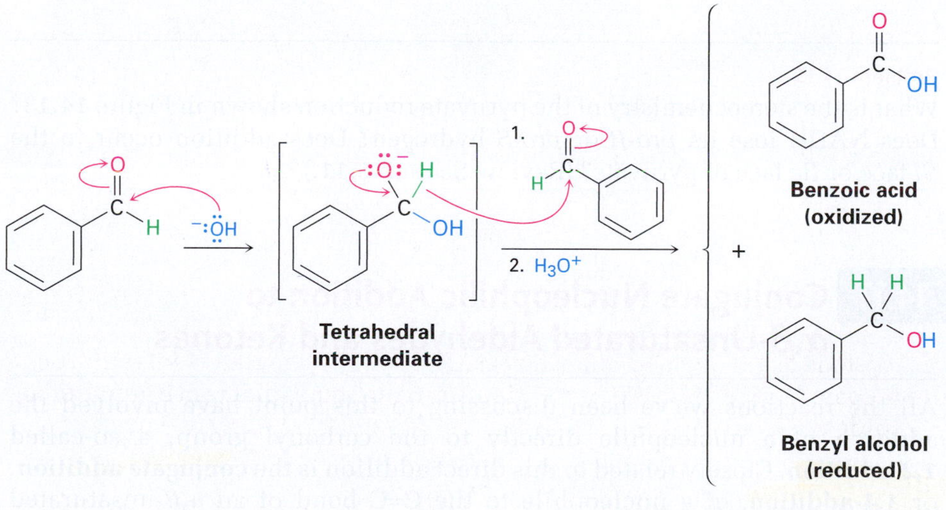

**Benzoic acid
(oxidized)**

**Tetrahedral
intermediate**

**Benzyl alcohol
(reduced)**

The Cannizzaro reaction is little used today but is interesting mechanistically because it is a simple laboratory analogy for the primary biological pathway by which carbonyl reductions occur in living organisms. In nature, as we saw in Section 13.3, one of the most important reducing agents is NADH, reduced nicotinamide adenine dinucleotide. NADH donates H⁻ to aldehydes and ketones, thereby reducing them, in much the same way that the tetrahedral alkoxide intermediate in a Cannizzaro reaction does. The electron lone pair on a nitrogen atom of NADH expels H⁻ as leaving group, which adds to a carbonyl group in another molecule (Figure 14.13). As an example, pyruvate is converted during intense muscle activity to (*S*)-lactate, a reaction catalyzed by lactate dehydrogenase.

FIGURE 14.13 Mechanism of biological aldehyde and ketone reductions by the coenzyme NADH.

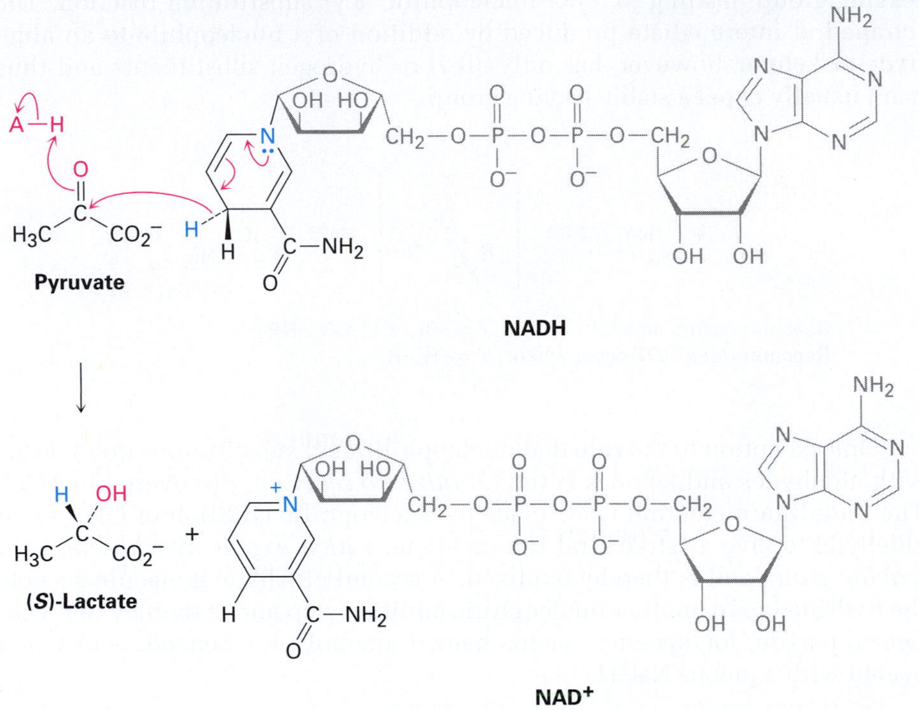

Problem 14.15

What is the stereochemistry of the pyruvate reduction shown in Figure 14.13? Does NADH lose its *pro-R* or *pro-S* hydrogen? Does addition occur to the *Si* face or *Re* face of pyruvate? (Review Section 5.11.)

14.11 Conjugate Nucleophilic Addition to α,β-Unsaturated Aldehydes and Ketones

All the reactions we've been discussing to this point have involved the addition of a nucleophile directly to the carbonyl group, a so-called **1,2-addition**. Closely related to this direct addition is the **conjugate addition**, or **1,4-addition**, of a nucleophile to the C=C bond of an α,β-unsaturated aldehyde or ketone. (The carbon atom next to a carbonyl group is often called the α *carbon,* the next one is the β *carbon,* and so on. Thus, an α,β-*unsaturated* aldehyde or ketone has a double bond conjugated with the carbonyl group.) The initial product of conjugate addition is a resonance-stabilized *enolate ion*, which typically undergoes protonation on the α carbon to give a saturated aldehyde or ketone product (Figure 14.14).

Direct (1,2) addition

Conjugate (1,4) addition

FIGURE 14.14 A comparison of direct (1,2) and conjugate (1,4) nucleophilic addition reactions.

α,β-Unsaturated aldehyde/ketone

Enolate ion

Saturated aldehyde/ketone

The conjugate addition of a nucleophile to an *α,β-unsaturated aldehyde* or ketone is caused by the same electronic factors that are responsible for direct addition: the electronegative oxygen atom of the α,β-unsaturated carbonyl compound withdraws electrons from the β carbon, thereby making it electron-poor and more electrophilic than a typical alkene carbon.

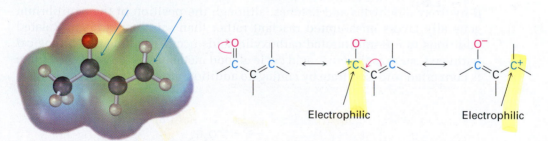

Electrophilic   Electrophilic

As noted previously, conjugate addition of a nucleophile to the β carbon of an α,β-unsaturated aldehyde or ketone leads to an enolate ion intermediate, which is protonated on the α carbon to give the saturated product (Figure 14.14). The net effect is addition of the nucleophile to the C=C bond, with the carbonyl group itself unchanged. In fact, of course, the carbonyl group is crucial to the success of the reaction. The C=C bond would not be activated for addition and no reaction would occur without the carbonyl group.

Activated double bond

1. :Nu⁻
2. H₃O⁺

Unactivated double bond

:Nu⁻ ⟶ *NO reaction*

Conjugate Addition of Amines

Both primary and secondary amines add to α,β-unsaturated aldehydes and ketones to yield β-amino aldehydes and ketones rather than the alternative imines. Under typical reaction conditions, both modes of addition occur rapidly. But because the reactions are reversible, the more stable conjugate addition product accumulates and is often obtained to the complete exclusion of the less stable direct addition product.

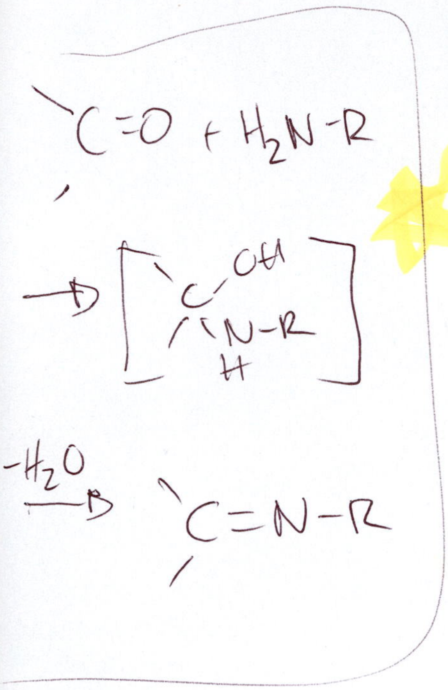

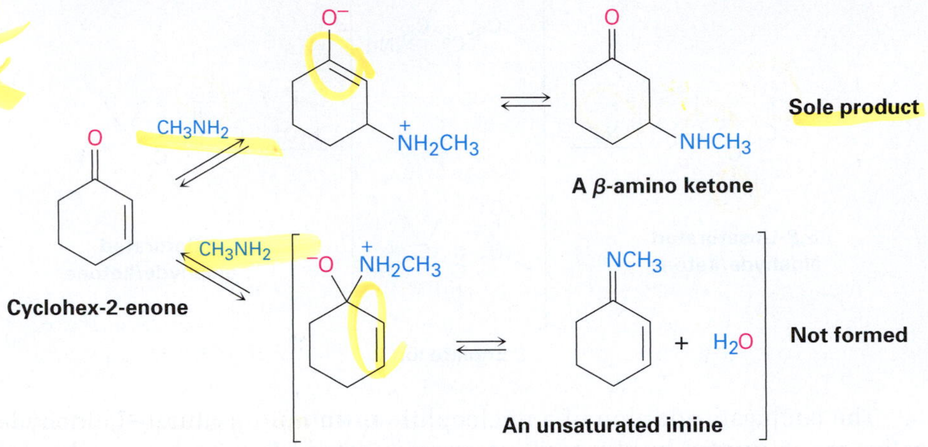

Cyclohex-2-enone

A β-amino ketone

Sole product

An unsaturated imine

Not formed

Conjugate Addition of Water

Water can add reversibly to α,β-unsaturated aldehydes and ketones to yield β-hydroxy aldehydes and ketones, although the position of the equilibrium generally favors unsaturated reactant rather than saturated adduct. Related additions to α,β-unsaturated carboxylic acids occur in numerous biological pathways, such as the citric acid cycle of food metabolism where *cis*-aconitate is converted into isocitrate by conjugate addition of water to a double bond.

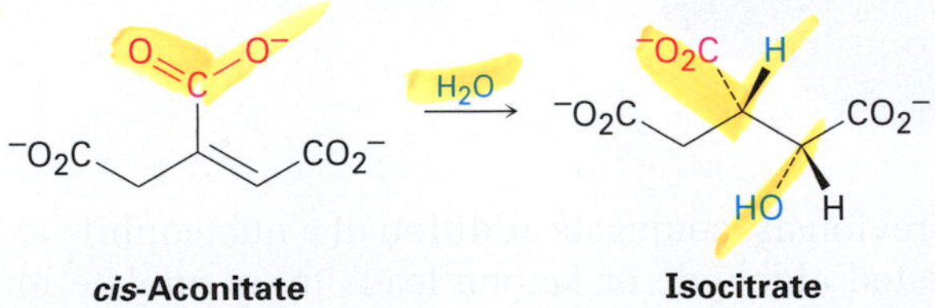

cis-Aconitate

Isocitrate

Conjugate Addition of Alkyl Groups

The conjugate addition of an alkyl or other organic group to an α,β-unsaturated ketone (but not aldehyde) is one of the more useful 1,4-addition reactions, just as direct addition of a Grignard reagent is one of the more useful 1,2-additions.

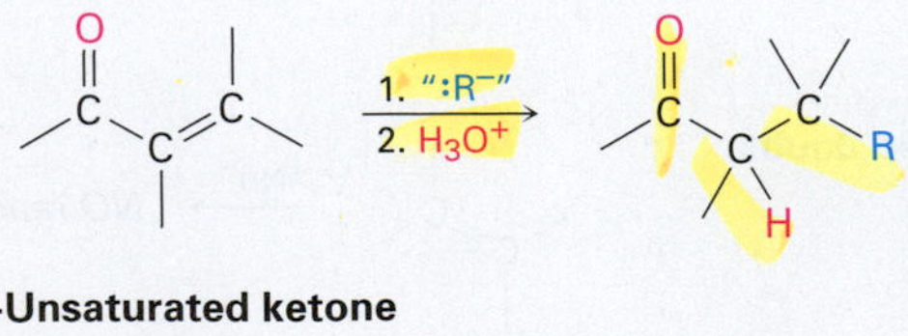

α,β-**Unsaturated ketone**

Conjugate addition of an organic group is carried out by treating the α,β-unsaturated ketone with a lithium diorganocopper reagent, R_2CuLi, which is prepared by reaction between 1 equivalent of copper(I) iodide and 2 equivalents of an organolithium regent, RLi. The organolithium reagent, in turn, is formed by reaction of lithium metal with an organohalide in the same way that a Grignard reagent is prepared by reaction of magnesium metal with an organohalide.

$$RX \xrightarrow[\text{Pentane}]{2\ Li} RLi + Li^+\ X^-$$

$$2\ RLi \xrightarrow[\text{Ether}]{CuI} Li^+(R\overset{-}{Cu}R) + Li^+\ I^-$$

A lithium diorganocopper (Gilman reagent)

Primary, secondary, and even tertiary alkyl groups undergo the conjugate addition reaction, as do aryl and alkenyl groups. Alkynyl groups, however, react poorly in the conjugate addition process. Diorganocopper reagents are unique in their ability to give conjugate addition products. Related compounds, such as Grignard reagents and organolithiums, normally give direct carbonyl addition on reaction with α,β-unsaturated ketones.

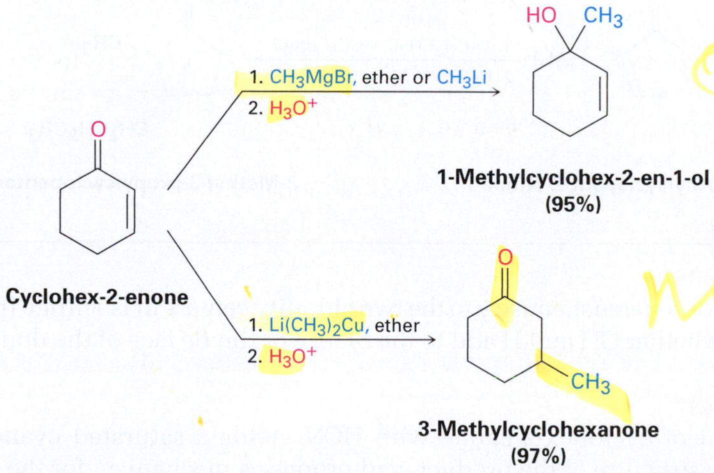

HO CH3
1. CH3MgBr, ether or CH3Li
2. H3O+
O
1-Methylcyclohex-2-en-1-ol
(95%)
Cyclohex-2-enone
1. Li(CH3)2Cu, ether
2. H3O+
CH3
3-Methylcyclohexanone
(97%)

The mechanism of the reaction is thought to involve conjugate nucleophilic addition of the diorganocopper anion, R_2Cu^-, to the unsaturated ketone to give a copper-containing intermediate. Transfer of an R group from copper to carbon, followed by elimination of a neutral organocopper species, RCu, gives the final product.

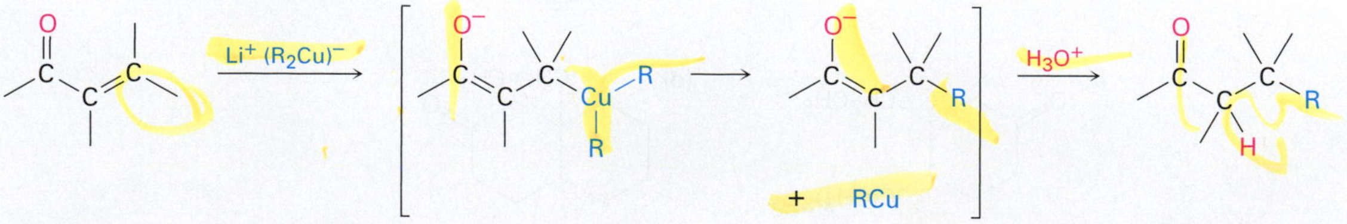

Li+ (R2Cu)−
Cu
R
+ RCu
H3O+
H

The lithium diorganocopper reaction has no direct counterpart in biological chemistry, although we'll see in Section 17.12 that conjugate addition of various other carbon-based nucleophiles to α,β-unsaturated carbonyl compounds does occur frequently in many biological pathways.

WORKED EXAMPLE 14.4 Using a Conjugate Addition Reaction

How might you use a conjugate addition reaction to prepare 2-methyl-3-propyl-cyclopentanone?

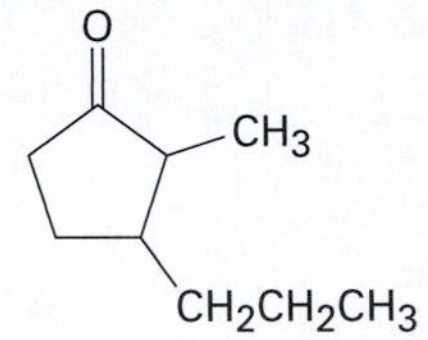

2-Methyl-3-propylcyclopentanone

Strategy

A ketone with a substituent group in its β position might be prepared by a conjugate addition of that group to an α,β-unsaturated ketone. In the present instance, the target molecule has a propyl substituent on the β carbon and might therefore be prepared from 2-methylcyclopenten-2-one by reaction with lithium dipropylcopper.

Solution

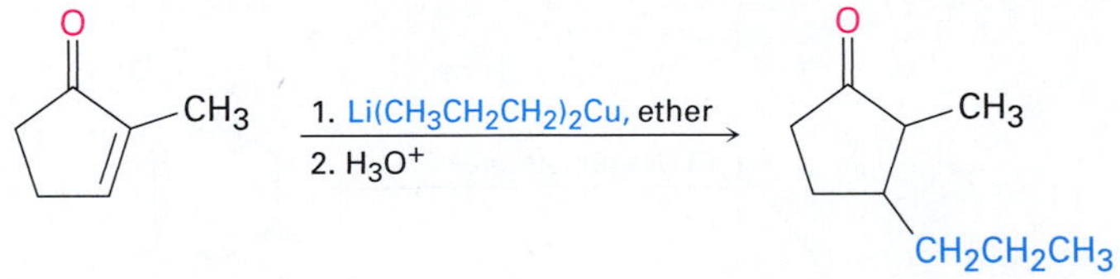

2-Methylcyclopent-2-enone   **2-Methyl-3-propylcyclopentanone**

Problem 14.16
Assign *R* or *S* stereochemistry to the two chirality centers in isocitrate (page 590), and tell whether OH and H add to the *Si* face or the *Re* face of the double bond.

Problem 14.17
Treatment of cyclohex-2-enone with HCN yields a saturated cyano ketone. Show the structure of the product, and propose a mechanism for the reaction.

Problem 14.18
How might conjugate addition reactions of lithium diorganocopper reagents be used to synthesize the following compounds?

(a)
$$CH_3CH_2CH_2CH_2CH_2CCH_3$$

(b)

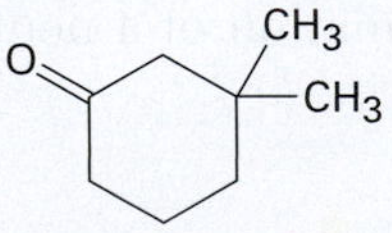

(c)

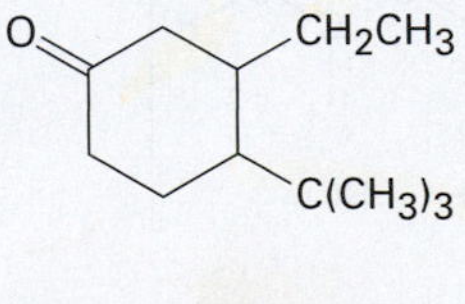

(d)

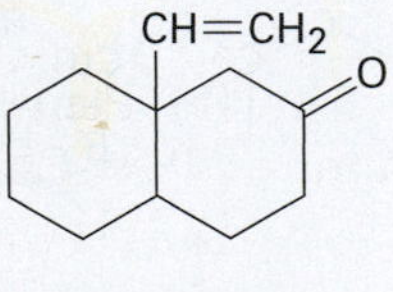

14.12 Spectroscopy of Aldehydes and Ketones

Infrared Spectroscopy

Aldehydes and ketones show a strong C=O bond absorption in the IR region from 1660 to 1770 cm^{-1}, as the spectra of benzaldehyde and cyclohexanone demonstrate (Figure 14.15). In addition, aldehydes show two characteristic C–H absorptions in the range 2720 to 2820 cm^{-1}.

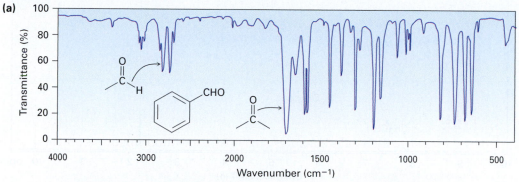

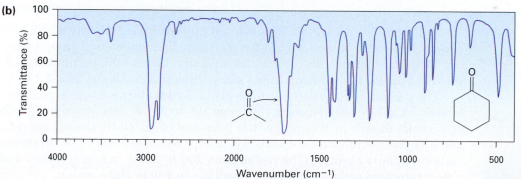

FIGURE 14.15 Infrared spectra of **(a)** benzaldehyde and **(b)** cyclohexanone.

The exact position of the C=O absorption is diagnostic of the nature of the carbonyl group. As the data in Table 14.2 indicate, saturated aldehydes usually show carbonyl absorptions near 1730 cm^{-1} in the IR spectrum, but conjugation of the aldehyde to an aromatic ring or a double bond lowers the absorption by 25 cm^{-1} to near 1705 cm^{-1}. Saturated aliphatic ketones and cyclohexanones both absorb near 1715 cm^{-1}, and conjugation with a double bond or an aromatic ring again lowers the absorption by 30 cm^{-1} to 1685 to 1690 cm^{-1}. Angle strain in the carbonyl group caused by reducing the ring size of cyclic ketones to four or five raises the absorption position.

TABLE 14.2

Infrared Absorptions of Some Aldehydes and Ketones

Carbonyl type	Absorption (cm^{-1})
Saturated aldehyde	1730
Aromatic aldehyde	1705
α,β-Unsaturated aldehyde	1705
Saturated ketone	1715
Cyclohexanone	1715
Cyclopentanone	1750
Cyclobutanone	1785
Aromatic ketone	1690
α,β-Unsaturated ketone	1685

Problem 14.19

How might you use IR spectroscopy to determine whether reaction between cyclohex-2-enone and dimethylamine gives the direct addition product or the conjugate addition product?

Problem 14.20

Where would you expect each of the following compounds to absorb in its IR spectrum?
(a) Pent-4-en-2-one
(b) Pent-3-en-2-one
(c) 2,2-Dimethylcyclopentanone
(d) *m*-Chlorobenzaldehyde
(e) Cyclohex-3-enone
(f) Hex-2-enal

Nuclear Magnetic Resonance Spectroscopy

Aldehyde protons (RCHO) absorb near 10 δ in the ^{1}H NMR spectrum and are very distinctive because no other absorptions occur in this region. The aldehyde proton shows spin–spin coupling with protons on the neighboring carbon, with coupling constant $J \approx 3$ Hz. Acetaldehyde, for example, shows a quartet at 9.8 δ for the aldehyde proton, indicating that there are three protons neighboring the –CHO group (Figure 14.16).

FIGURE 14.16
^{1}H NMR spectrum of acetaldehyde. The absorption of the aldehyde proton appears at 9.8 δ and is split into a quartet.

Hydrogens on the carbon next to a carbonyl group are slightly deshielded and normally absorb near 2.0 to 2.3 δ. The acetaldehyde methyl group in Figure 14.16, for instance, absorbs at 2.20 δ. Methyl ketones are particularly distinctive because they always show a sharp three-proton singlet near 2.1 δ.

The carbonyl-group carbon atoms of aldehydes and ketones have characteristic ^{13}C NMR resonances in the range 190 to 215 δ. Since no other kinds of carbons absorb in this range, the presence of an NMR absorption near 200 δ is clear evidence for a carbonyl group. Saturated aldehyde or ketone carbons usually absorb in the region from 200 to 215 δ, while aromatic and α,β-unsaturated carbonyl carbons absorb in the 190 to 200 δ region.

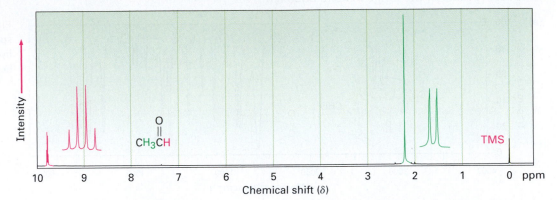

Mass Spectrometry

Aliphatic aldehydes and ketones that have hydrogens on their gamma (γ) carbon atoms undergo a characteristic mass spectral cleavage called the *McLafferty rearrangement*. A hydrogen atom is transferred from the γ carbon to the carbonyl oxygen, the bond between the α and β carbons is broken, and a neutral alkene fragment is produced. The charge remains with the oxygen-containing fragment.

Aldehydes and ketones also undergo fragmentation by cleavage of the bond between the carbonyl group and the α carbon, a so-called α *cleavage*. Alpha cleavage yields a neutral radical and an oxygen-containing cation.

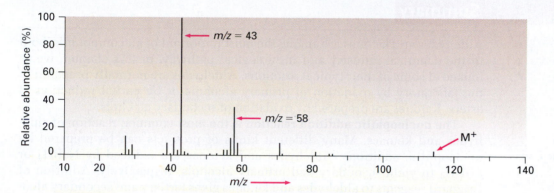

Fragment ions from both McLafferty rearrangement and α cleavage are visible in the mass spectrum of 5-methylhexan-2-one shown in Figure 14.17. McLafferty rearrangement with loss of 2-methylpropene yields a fragment with $m/z = 58$. Alpha cleavage occurs primarily at the more substituted side of the carbonyl group, leading to a $[CH_3CO]^+$ fragment with $m/z = 43$.

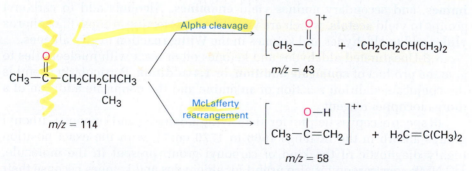

FIGURE 14.17 Mass spectrum of 5-methylhexan-2-one. The peak at $m/z = 58$ is due to McLafferty rearrangement. The abundant peak at $m/z = 43$ is due to α cleavage at the more highly substituted side of the carbonyl group. Note that the peak due to the molecular ion is very small.

Problem 14.21

How might you use mass spectrometry to distinguish between the following pairs of isomers?

(a) 3-Methylhexan-2-one and 4-methylhexan-2-one

(b) Heptan-3-one and heptan-4-one

(c) 2-Methylpentanal and 3-methylpentanal

Problem 14.21

Tell the prominent IR absorptions and mass spectral peaks you would expect for the following compound:

Summary

Key Words

acetal [R₂C(OR′)₂], 580

acyl group, 566

1,2-addition, 588

1,4-addition, 588

aldehyde (RCHO), 564

conjugate addition, 588

enamine (R₂N—CR═CR₂), 576

hemiacetal, 580

imine (R₂C═NR), 576

ketone (R₂C═O), 564

nucleophilic addition
  reaction, 569

Schiff base, 576

Wittig reaction, 583

ylide, 583

Aldehydes and ketones are among the most important of all compounds, both in the chemical industry and in biological pathway. In this chapter, we've looked at some of their typical reactions. Aldehydes are normally prepared in the laboratory by oxidation of primary alcohols or by partial reduction of esters. Ketones are prepared by oxidation of secondary alcohols.

The **nucleophilic addition reaction** is the most common reaction of aldehydes and ketones. Many different kinds of products can be prepared by nucleophilic additions. Aldehydes and ketones are reduced by NaBH₄ or LiAlH₄ to yield secondary and primary alcohols, respectively. Addition of Grignard reagents to aldehydes and ketones gives tertiary and secondary alcohols, respectively. Primary amines add to carbonyl compounds yielding **imines**, and secondary amines yield **enamines**. Alcohols add to carbonyl groups to yield **acetals**, which are valuable as protecting groups. Phosphorus **ylides** add to aldehydes and ketones in the **Wittig reaction** to give alkenes.

α,β-Unsaturated aldehydes and ketones often react with nucleophiles to give the product of **conjugate addition**, or **1,4-addition**. Particularly useful is the conjugate addition reaction of an amine and the conjugate addition of a diorganocopper reagent.

IR spectroscopy is helpful for identifying aldehydes and ketones. Carbonyl groups absorb in the IR range 1660 to 1770 cm⁻¹, with the exact position highly diagnostic of the kind of carbonyl group present in the molecule. ¹³C NMR spectroscopy is also useful for aldehydes and ketones because their carbonyl carbons show resonances in the 190 to 215 δ range. Aldehydes and ketones undergo two characteristic kinds of fragmentation in the mass spectrometer: α cleavage and McLafferty rearrangement.

Summary of Reactions

1. Preparation of aldehydes

    (a) Oxidation of primary alcohols (Section 13.5)

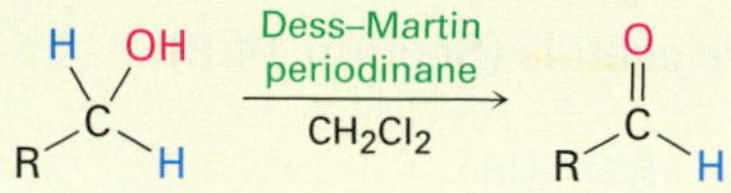

    (b) Partial reduction of esters (Section 14.2)

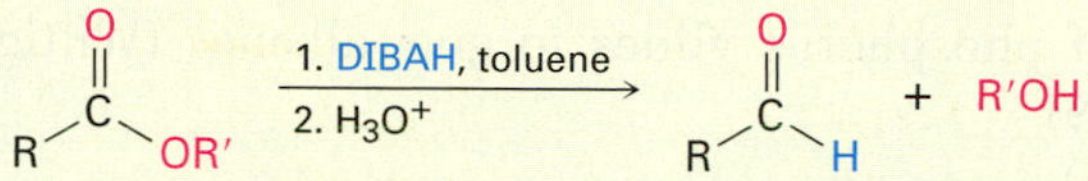

2. Preparation of ketones

    (a) Oxidation of secondary alcohols (Section 13.5)

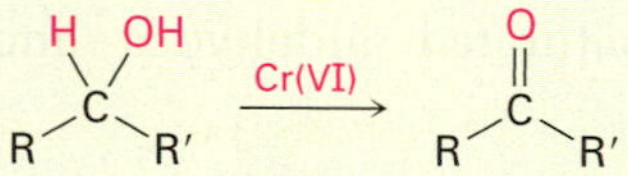

    (b) Friedel–Crafts acylation (Section 9.7)

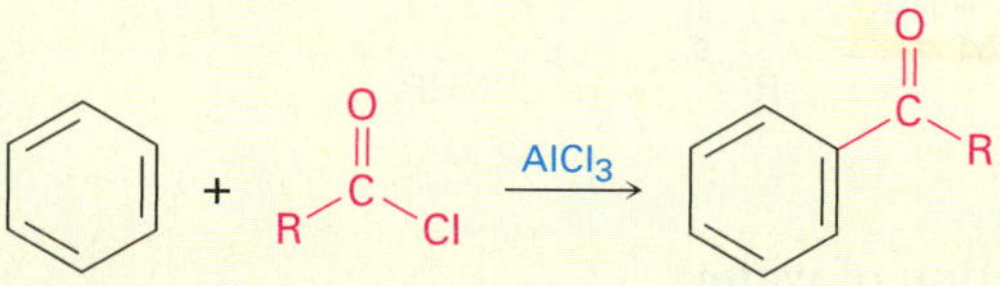

3. Oxidation of aldehydes (Section 14.3)

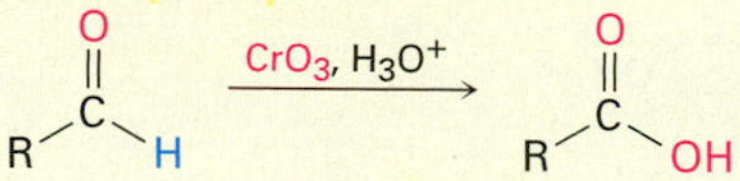

4. Nucleophilic addition reactions of aldehydes and ketones

    (a) Addition of hydride: reduction (Sections 13.3 and 14.6)

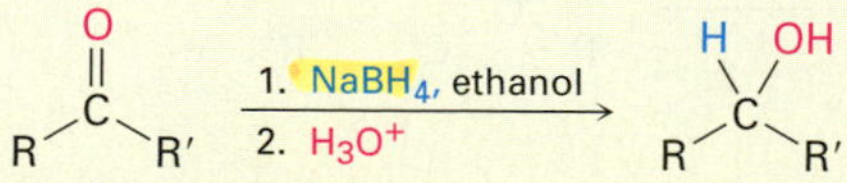

    (b) Addition of Grignard reagents (Sections 13.3 and 14.6)

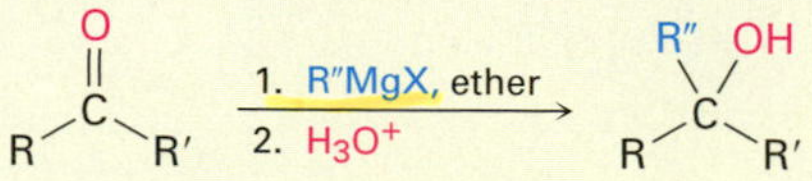

    (c) Addition of primary amines to give imines (Section 14.7)

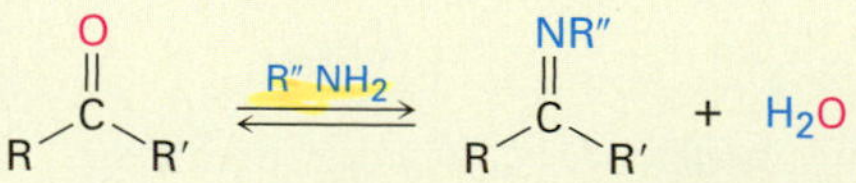

(d) Addition of secondary amines to give enamines (Section 14.7)

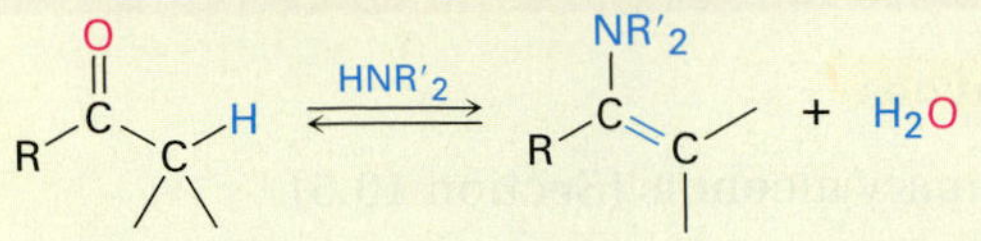

(e) Addition of alcohols to give acetals (Section 14.8)

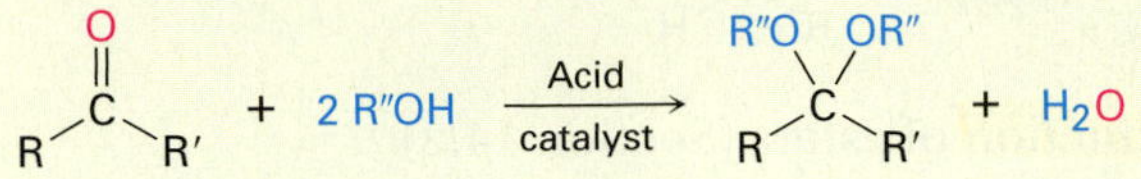

(f) Addition of phosphorus ylides to give alkenes (Wittig reaction; Section 14.9)

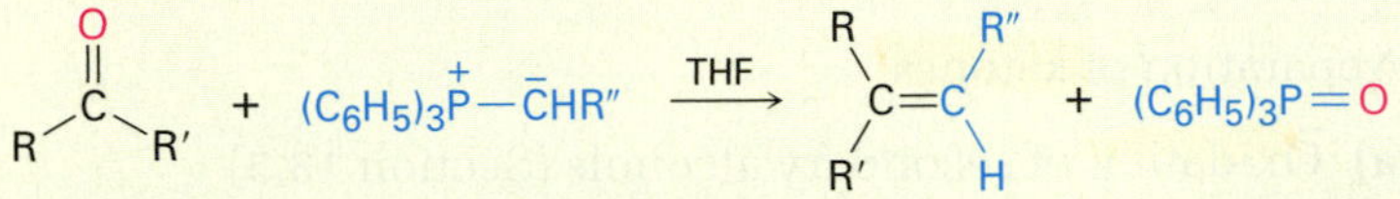

5. Conjugate additions to α,β-unsaturated aldehydes and ketones (Section 14.11)

(a) Conjugate addition of amines

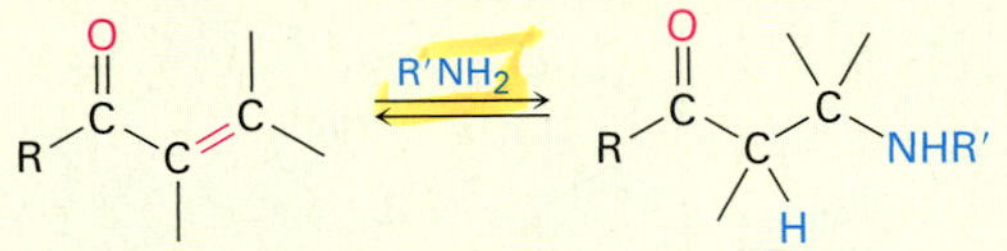

(b) Conjugate addition of water

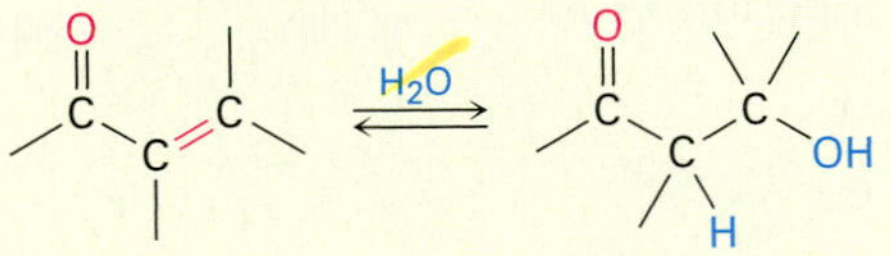

(c) Conjugate addition of alkyl groups

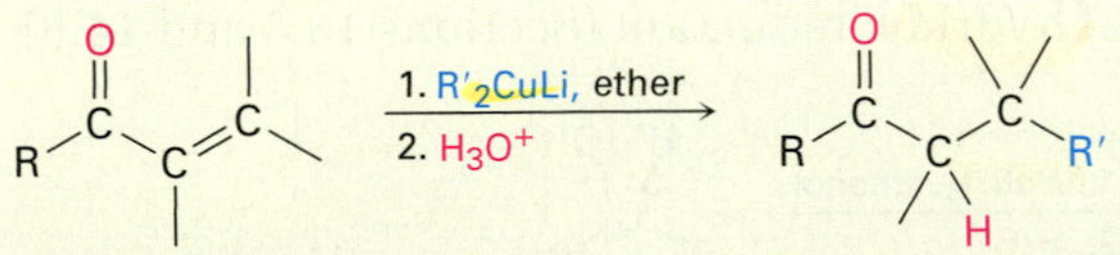

Lagniappe

Enantioselective Synthesis

Whenever a chiral product is formed by reaction between achiral reagents, the product is racemic; that is, both enantiomers of the product are formed in equal amounts. The epoxidation reaction of geraniol with *m*-chloroperoxybenzoic acid, for instance, gives a racemic mixture of (2*R*,3*S*) and (2*S*,3*R*) epoxides.

50%

+

50%

A substance made from the tartaric acid found at the bottom of this wine vat catalyzes enantioselective reactions.

© Charles O'Rear/CORBIS

Unfortunately, it's usually the case that only a *single* enantiomer of a given drug or other important substance has the desired biological properties. The other enantiomer might be inactive or even dangerous. Thus, much work is currently being done on developing *enantioselective* methods of synthesis, which yield only one of two possible enantiomers. So important has enantioselective synthesis become that the 2001 Nobel Prize in Chemistry was awarded to three pioneers in the field: William S. Knowles, K. Barry Sharpless, and Ryoji Noyori.

Several approaches to enantioselective synthesis have been taken, but the most efficient are those that use chiral catalysts to temporarily hold a substrate molecule in an unsymmetrical environment—the same strategy that nature uses when catalyzing reactions with chiral enzymes. While in that unsymmetrical environment, the substrate may be more open to reaction on one side than on another, leading to an excess of one enantiomeric product over another. As an analogy, think about picking up a coffee mug in your right hand to take a drink. The mug by itself is achiral, but as soon as you pick it up by the handle, it becomes chiral. One side of the mug now faces toward you so you can drink from it, but the other side faces away. The two sides are different, with one side much more accessible to you than the other.

Among the thousands of enantioselective reactions now known, one of the most general is the so-called Sharpless epoxidation, in which an allylic alcohol, such as geraniol, is treated with *tert*-butyl hydroperoxide, $(CH_3)_3C—OOH$, in the presence of titanium tetraisopropoxide and diethyl tartrate (DET) as a chiral auxiliary reagent. When the (*R*,*R*) tartrate is used, geraniol is converted into its 2*R*,3*S* epoxide with 98% selectivity,

whereas use of the (*S*,*S*) tartrate gives the 2*S*,3*R* epoxide enantiomer. We say that the major product in each case is formed with an *enantiomeric excess* of 96%, meaning that 4% of the product is racemic (2% 2*R*,3*S* plus 2% 2*S*,3*R*) and an extra 96% of a single enantiomer is formed. The mechanistic details by which the chiral catalyst works are a bit complex, although it appears that a chiral complex of two tartrate molecules with one titanium is involved.

(*R*,*R*)-Diethyl tartrate (*S*,*S*)-Diethyl tartrate

2*R*,3*S* isomer—98%

Geraniol

2*S*,3*R* isomer—98%

Exercises

■ indicates problems that are assignable in Organic OWL.

Go to this book's companion website at www.cengage.com/chemistry/mcmurry to explore interactive versions of the Active Figures from this text.

VISUALIZING CHEMISTRY

(Problems 14.1–14.22 appear within the chapter.)

14.23 ■ Each of the following substances can be prepared by a nucleophilic addition reaction between an aldehyde or ketone and a nucleophile. Identify the reactants from which each was prepared. If the substance is an acetal, identify the carbonyl compound and the alcohol; if it is an imine, identify the carbonyl compound and the amine; and so forth.

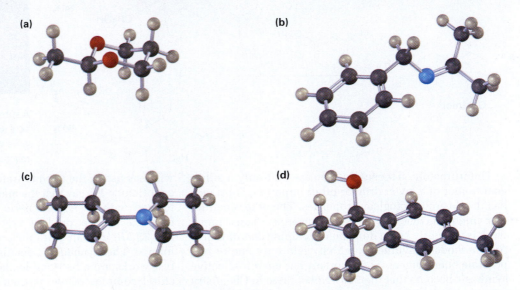

(a)   (b)

(c)   (d)

14.24 ■ The following molecular model represents a tetrahedral intermediate resulting from addition of a nucleophile to an aldehyde or ketone. Identify the reactants, and write the structure of the final product when the nucleophilic addition reaction is complete.

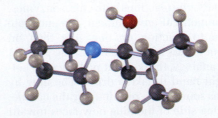

14.25 The enamine prepared from acetone and dimethylamine is shown in its lowest-energy form.

(a) What is the geometry and hybridization of the nitrogen atom?

(b) What orbital on nitrogen holds the lone pair of electrons?

(c) What is the geometric relationship between the *p* orbitals of the double bond and the nitrogen orbital that holds the lone pair? Why do you think this geometry represents the minimum energy?

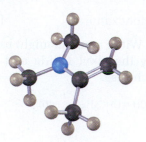

ADDITIONAL PROBLEMS

14.26 ■ Draw structures corresponding to the following names:

(a) Bromoacetone
(b) (*S*)-2-Hydroxypropanal
(c) 2-Methylheptan-3-one
(d) (2*S*,3*R*)-2,3,4-Trihydroxybutanal
(e) 2,2,4,4-Tetramethylpentan-3-one
(f) 4-Methylpent-3-en-2-one
(g) Butanedial
(h) 3-Phenylprop-2-enal
(i) 6,6-Dimethylcyclohexa-2,4-dienone
(j) *p*-Nitroacetophenone

14.27 ■ Draw and name the seven aldehydes and ketones with the formula $C_5H_{10}O$. Which are chiral?

14.28 ■ Give IUPAC names for the following structures:

(a)

(b)

(c)

(d)

(e)

(f)

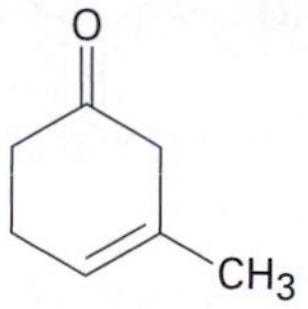

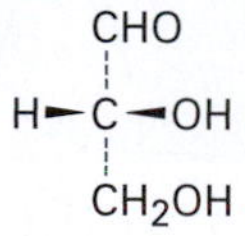

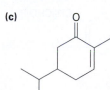

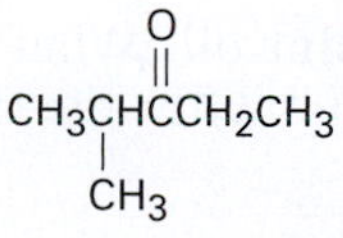

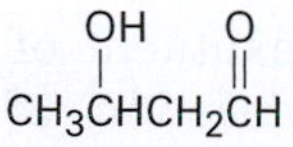

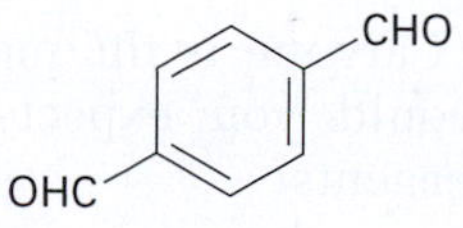

14.29 ■ Give structures that fit the following descriptions:

(a) An α,β-unsaturated ketone, C_6H_8O
(b) An α-diketone
(c) An aromatic ketone, $C_9H_{10}O$
(d) A diene aldehyde, C_7H_8O

■ Problems assignable in Organic OWL.

14.30 ▪ Predict the products of the reaction of (i) phenylacetaldehyde and (ii) acetophenone with the following reagents:

(a) $NaBH_4$, then H_3O^+ **(b)** 2 CH_3OH, HCl catalyst

(c) $NH_2CH(CH_3)_2$, HCl catalyst   **(d)** CH_3MgBr, then H_3O^+

14.31 ▪ How would you use a Grignard reaction on an aldehyde or ketone to synthesize the following compounds?

(a) Pentan-2-ol   **(b)** Butan-1-ol

(c) 1-Phenylcyclohexanol   **(d)** Diphenylmethanol

14.32 ▪ Show how the Wittig reaction might be used to prepare the following alkenes. Identify the alkyl halide and the carbonyl compound you would use in each case.

(a) $C_6H_5CH=CH-CH=CHC_6H_5$ **(b)**

(c) CH_2 **(d)** $CH=CH_2$

14.33 When 4-hydroxybutanal is treated with methanol in the presence of an acid catalyst, 2-methoxytetrahydrofuran is formed. Propose a mechanism.

$$HOCH_2CH_2CH_2CHO \xrightarrow[\text{HCl}]{CH_3OH}$$

14.34 ▪ How would you synthesize the following substances from benzaldehyde and any other reagents needed?

(a) CH_2CHO **(b)**   **(c)**

14.35 ▪ Carvone is the major constituent of spearmint oil. What products would you expect from reaction of carvone with the following reagents?

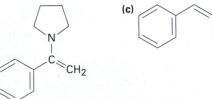

Carvone

(a) $HOCH_2CH_2OH$, HCl   **(b)** $LiAlH_4$, then H_3O^+ **(c)** CH_3NH_2

(d) C_6H_5MgBr, then H_3O^+ **(e)** 2 equiv. H_2/Pd **(f)** CrO_3, H_3O^+

▪ Problems assignable in Organic OWL.

14.36 One of the steps in the metabolism of fats is the reaction of an unsaturated acyl CoA with water to give a β-hydroxyacyl CoA. Propose a mechanism.

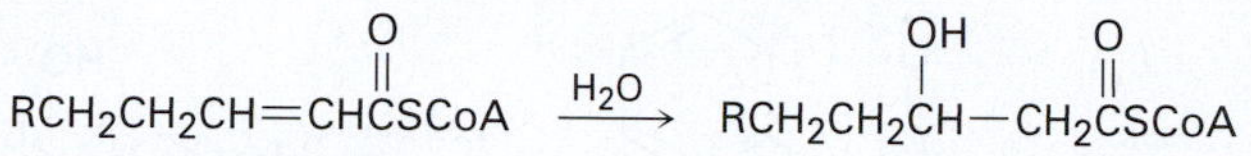
RCH2CH2CH=CHCSCoA
H2O
RCH2CH2CH—CH2CSCoA

Unsaturated acyl CoA **β-Hydroxyacyl CoA**

14.37 The amino acid methionine is biosynthesized by a multistep route that includes reaction of an imine of pyridoxal phosphate (PLP; Section 14.7) to give an unsaturated imine, which then reacts with cysteine. What kinds of reactions are occurring in the two steps?

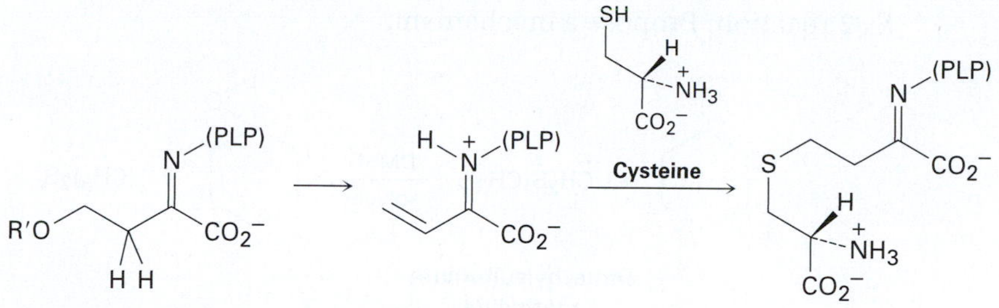
(PLP)
Cysteine

***O*-Succinylhomoserine–PLP imine** **Unsaturated imine**

14.38 The S_N2 reaction of (dibromomethyl)benzene, $C_6H_5CHBr_2$, with NaOH yields benzaldehyde rather than (dihydroxymethyl)benzene, $C_6H_5CH(OH)_2$. Explain.

14.39 Reaction of butan-2-one with phenylmagnesium bromide yields a chiral product. What stereochemistry does the product have? Is it optically active?

14.40 ■ How would you synthesize the following compounds from cyclohexanone?

(a) 1-Methylcyclohexene **(b)** 2-Phenylcyclohexanone

(c) *cis*-Cyclohexane-1,2-diol **(d)** 1-Cyclohexylcyclohexanol

14.41 Aldehydes and ketones react with thiols to yield *thioacetals* just as they react with alcohols to yield acetals. Predict the product of the following reaction, and propose a mechanism:

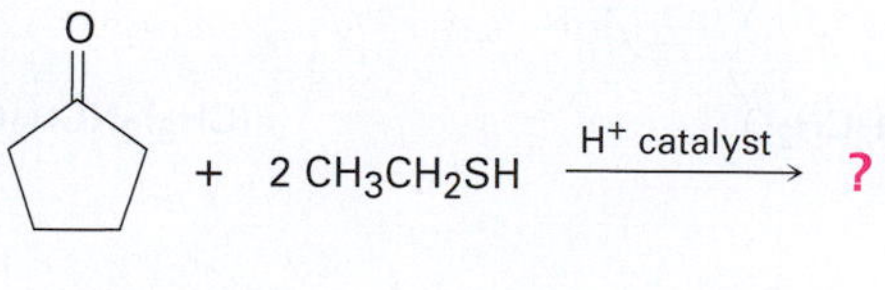
+ 2 CH3CH2SH
H+ catalyst
?

■ Problems assignable in Organic OWL.

14.42 In the benzilic acid rearrangement, an α-diketone reacts with base to yield a rearranged hydroxy acid by a process similar to the Cannizzaro reaction. Propose a mechanism.

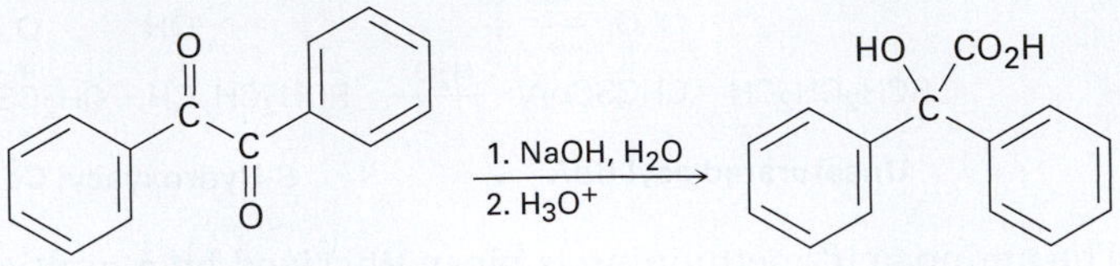

Benzil 1. NaOH, H_2O 2. H_3O^+ **Benzilic acid**

14.43 ■ Ketones react with dimethylsulfonium methylide to yield epoxides. The reaction occurs by an initial nucleophilic addition, followed by an S_N2 reaction. Propose a mechanism.

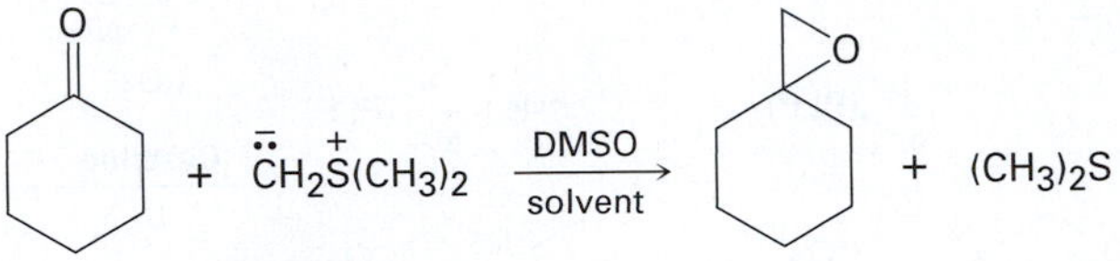

$+$ $\overset{..}{C}H_2\overset{+}{S}(CH_3)_2$ $\xrightarrow{\text{DMSO} \atop \text{solvent}}$ $+$ $(CH_3)_2S$

Dimethylsulfonium methylide

14.44 Treatment of an alcohol with dihydropyran yields an acetal called a tetrahydropyranyl ether. Propose a mechanism.

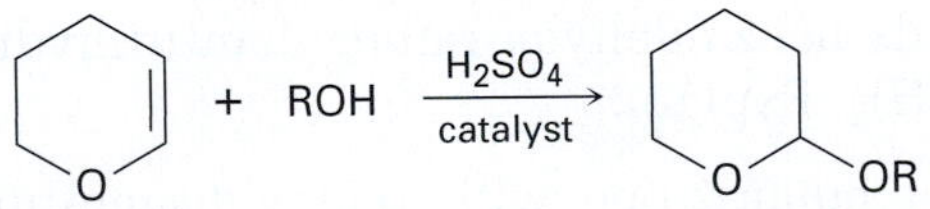

$+$ ROH $\xrightarrow{H_2SO_4 \atop \text{catalyst}}$

Dihydropyran   **A tetrahydropyranyl ether**

14.45 Tamoxifen is a drug used in the treatment of breast cancer. How would you prepare tamoxifen from benzene, the following ketone, and any other reagents needed?

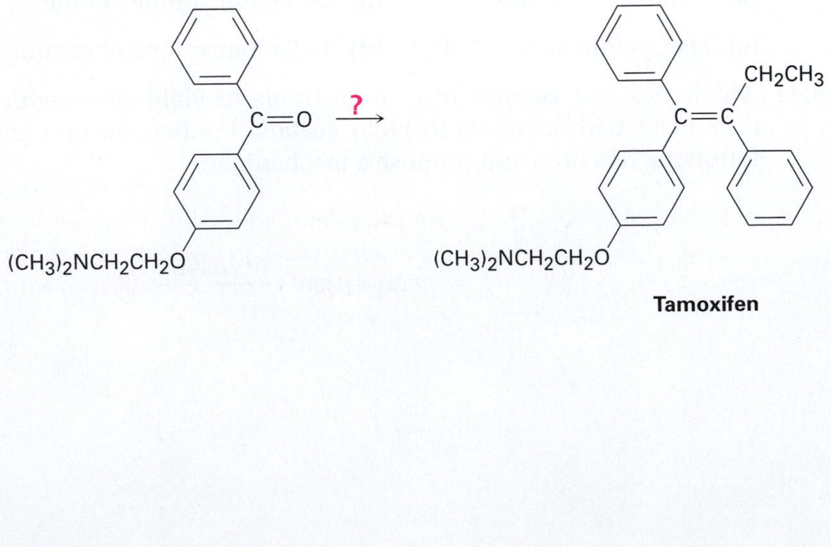

$C=O$ $\xrightarrow{?}$

$(CH_3)_2NCH_2CH_2O$

CH_2CH_3

$C=C$

$(CH_3)_2NCH_2CH_2O$

Tamoxifen

■ Problems assignable in Organic OWL.

14.46 Paraldehyde, a sedative and hypnotic agent, is prepared by treatment of acetaldehyde with an acidic catalyst. Propose a mechanism.

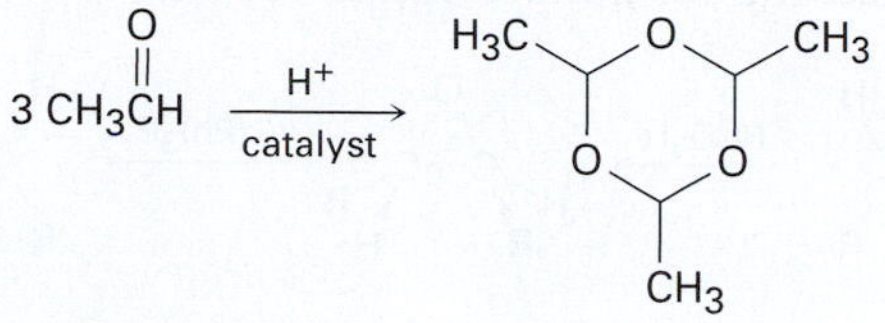

$$3 \text{ CH}_3\text{CH} \xrightarrow[\text{catalyst}]{\text{H}^+}$$

Paraldehyde

14.47 The Meerwein–Ponndorf–Verley reaction involves reduction of a ketone by treatment with an excess of aluminum triisopropoxide. The mechanism of the process is closely related to the Cannizzaro reaction in that a hydride ion acts as a leaving group. Propose a mechanism.

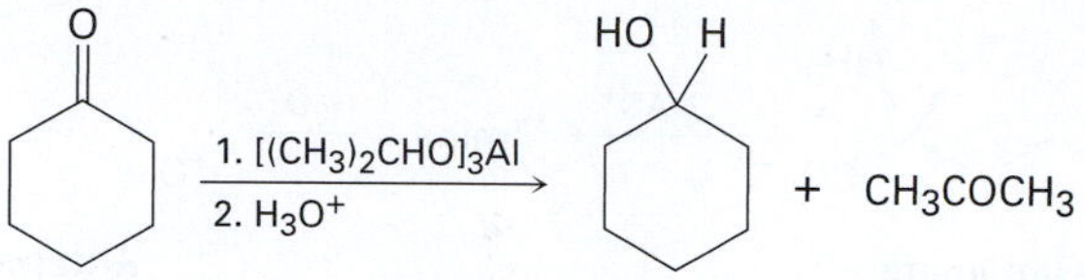

$$\xrightarrow[\text{2. H}_3\text{O}^+]{\text{1. [(CH}_3)_2\text{CHO]}_3\text{Al}} \quad + \quad \text{CH}_3\text{COCH}_3$$

14.48 Propose a mechanism to account for the formation of 3,5-dimethyl-pyrazole from hydrazine and pentane-2,4-dione. What has happened to each carbonyl carbon in going from starting material to product?

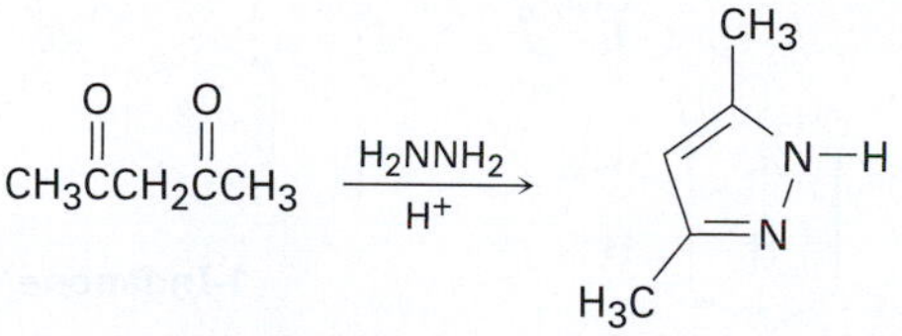

$$\text{CH}_3\text{CCH}_2\text{CCH}_3 \xrightarrow[\text{H}^+]{\text{H}_2\text{NNH}_2}$$

Pentane-2,4-dione   **3,5-Dimethylpyrazole**

14.49 In light of your answer to Problem 14.48, propose a mechanism for the formation of 3,5-dimethylisoxazole from hydroxylamine (NH$_2$OH) and pentane-2,4-dione.

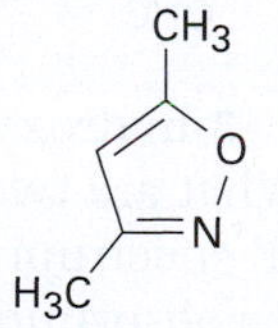

3,5-Dimethylisoxazole

14.50 ■ Treatment of an α,β-unsaturated ketone with basic aqueous hydrogen peroxide yields an epoxy ketone. The reaction is specific to unsaturated ketones and occurs by an initial conjugate addition followed by an S$_\text{N}$2 reaction. Propose a mechanism.

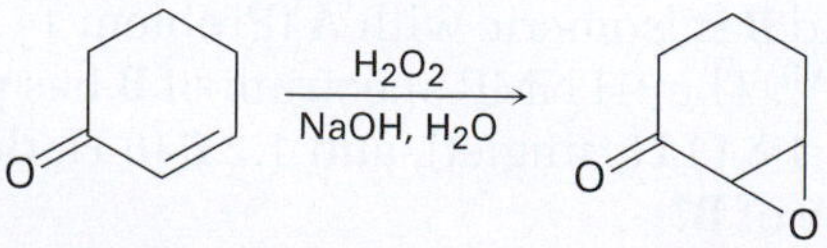

$$\xrightarrow[\text{NaOH, H}_2\text{O}]{\text{H}_2\text{O}_2}$$

■ Problems assignable in Organic OWL.

14.51 ■ Trans alkenes can be converted into their cis isomers and vice versa by epoxidation followed by reaction of the epoxide with triphenylphosphine. Propose a mechanism for the epoxide → alkene reaction.

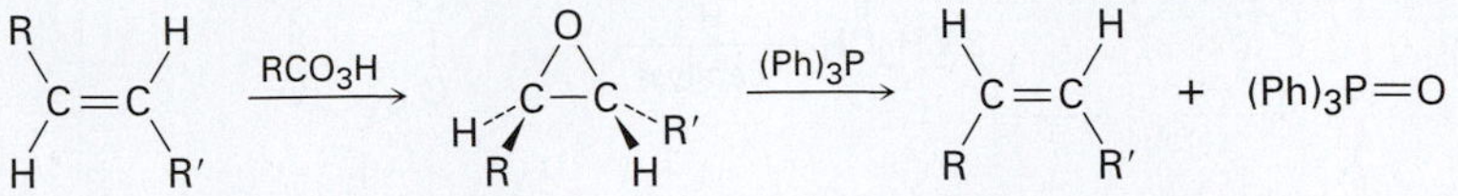

14.52 ■ One of the biological pathways by which an amine is converted to a ketone involves two steps: (1) oxidation of the amine by NAD^+ to give an imine, and (2) hydrolysis of the imine to give a ketone plus ammonia. Glutamate, for instance, is converted by this process into α-ketoglutarate. Show the structure of the imine intermediate, and propose mechanisms for both steps.

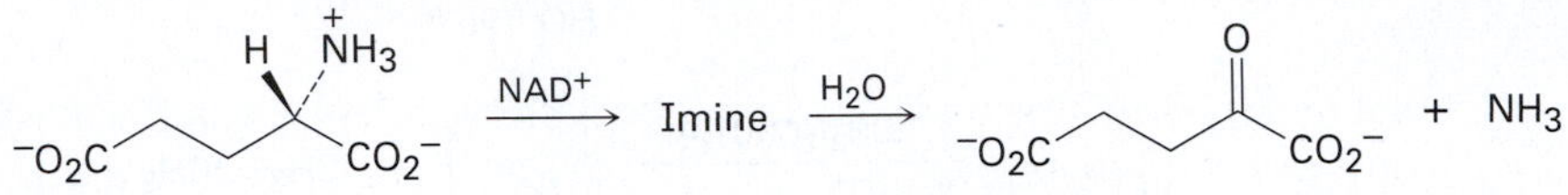

Glutamate   **α-Ketoglutarate**

14.53 At what position would you expect to observe IR absorptions for the following molecules?

(a)

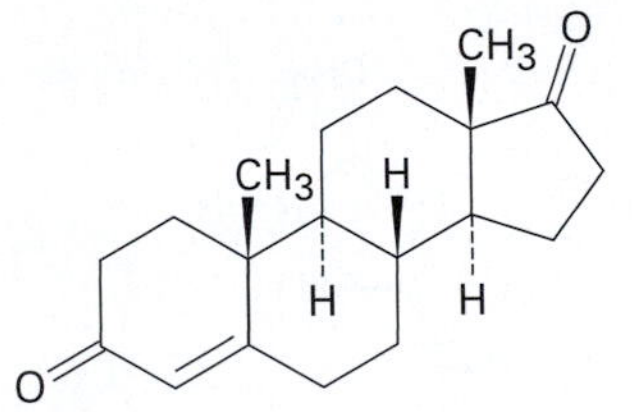

Androst-4-ene-3,17-dione

(b)

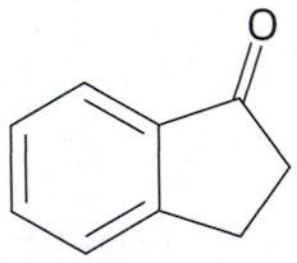

1-Indanone

(c)

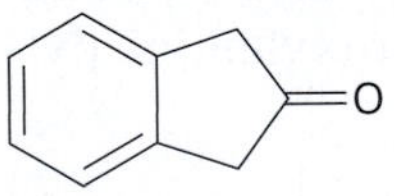

(d)

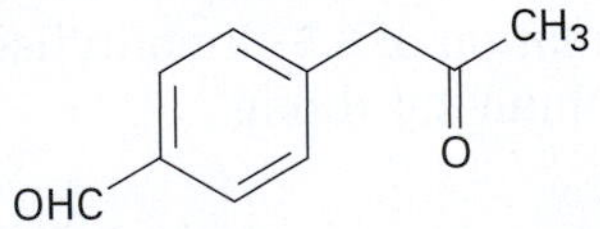

14.54 Acid-catalyzed dehydration of 3-hydroxy-3-phenylcyclohexanone leads to an unsaturated ketone. What are two likely structures for the product? At what position in the IR spectrum would you expect each to absorb? If the actual product has an absorption at 1670 cm^{-1}, which of your structures is correct?

14.55 ■ Compound **A**, MW = 86, shows an IR absorption at 1730 cm^{-1} and a simple ^{1}H NMR spectrum with peaks at 9.7 δ (1 H, singlet) and 1.2 δ (9 H, singlet). Propose a structure for **A**.

14.56 ■ Compound **B** is isomeric with **A** (Problem 14.55) and shows an IR peak at 1715 cm^{-1}. The ^{1}H NMR spectrum of **B** has peaks at 2.4 δ (1 H, septet, J = 7 Hz), 2.1 δ (3 H, singlet), and 1.2 δ (6 H, doublet, J = 7 Hz). What is the structure of **B**?

■ Problems assignable in Organic OWL.

14.57 ▪ The ^{1}H NMR spectrum shown is that of a compound with formula $C_9H_{10}O$. If the unknown has an IR absorption at 1690 cm^{-1}, what is a likely structure?

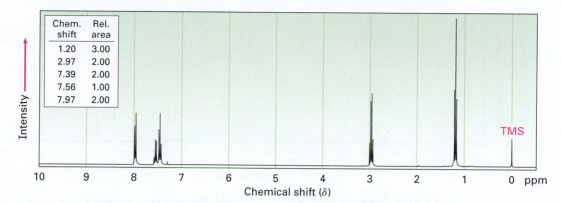

Chem. shift	Rel. area
1.20	3.00
2.97	2.00
7.39	2.00
7.56	1.00
7.97	2.00

14.58 ▪ The ^{1}H NMR spectrum shown is that of a compound isomeric with the one in Problem 14.57. This isomer has an IR absorption at 1730 cm^{-1}. Propose a structure. [*Note:* Aldehyde protons (CHO) often show very small coupling constants to adjacent hydrogens, so the splitting of aldehyde signals is not always apparent.]

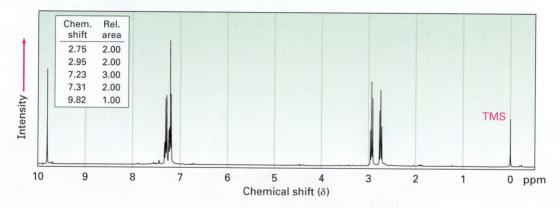

Chem. shift	Rel. area
2.75	2.00
2.95	2.00
7.23	3.00
7.31	2.00
9.82	1.00

14.59 ▪ Propose structures for molecules that meet the following descriptions. Assume that the kinds of carbons (1°, 2°, 3°, or 4°) have been assigned by DEPT-NMR.

(a) $C_6H_{12}O$; IR: 1715 cm^{-1}; ^{13}C NMR: 8.0 δ (1°), 18.5 δ (1°), 33.5 δ (2°), 40.6 δ (3°), 214.0 δ (4°)

(b) $C_5H_{10}O$; IR: 1730 cm^{-1}; ^{13}C NMR: 22.6 δ (1°), 23.6 δ (3°), 52.8 δ (2°), 202.4 δ (3°)

(c) C_6H_8O; IR: 1680 cm^{-1}; ^{13}C NMR: 22.9 δ (2°), 25.8 δ (2°), 38.2 δ (2°), 129.8 δ (3°), 150.6 δ (3°), 198.7 δ (4°)

▪ Problems assignable in Organic OWL.

14.60 ■ Compound **A**, $C_8H_{10}O_2$, has an intense IR absorption at 1750 cm^{-1} and gives the ^{13}C NMR spectrum shown. Propose a structure for **A**.

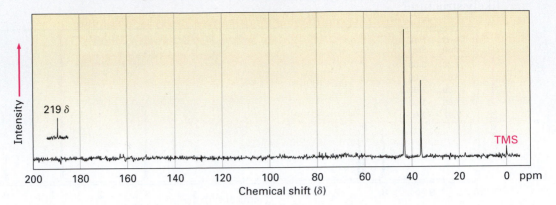

219 δ
TMS
Intensity
200
180
160
140
120
100
80
60
40
20
0 ppm
Chemical shift (δ)

14.61 ■ Propose structures for aldehydes or ketones that have the following ^1H NMR spectra:

(a) C_4H_7ClO

IR: 1715 cm^{-1}

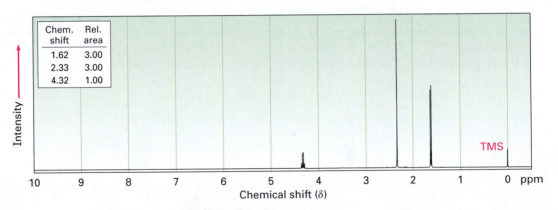

Chem. shift
Rel. area
1.62 3.00
2.33 3.00
4.32 1.00
TMS
Intensity
10
9
8
7
6
5
4
3
2
1
0 ppm
Chemical shift (δ)

(b) $C_7H_{14}O$

IR: 1710 cm^{-1}

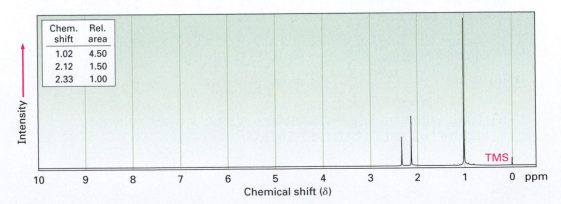

Chem. shift
Rel. area
1.02 4.50
2.12 1.50
2.33 1.00
TMS
Intensity
10
9
8
7
6
5
4
3
2
1
0 ppm
Chemical shift (δ)

■ Problems assignable in Organic OWL.

(c) $C_9H_{10}O_2$

IR: 1695 cm^{-1}

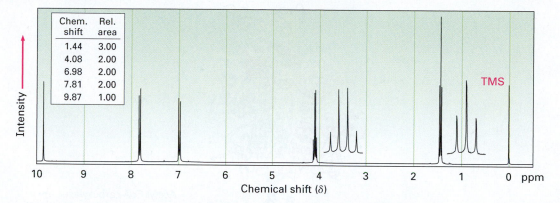

Chem. shift	Rel. area
1.44	3.00
4.08	2.00
6.98	2.00
7.81	2.00
9.87	1.00

14.62 Primary amines react with esters to yield amides: $RCO_2R' + R''NH_2 \rightarrow$ $RCONHR'' + R'OH$. Propose a mechanism for the following reaction of an α,β-unsaturated ester.

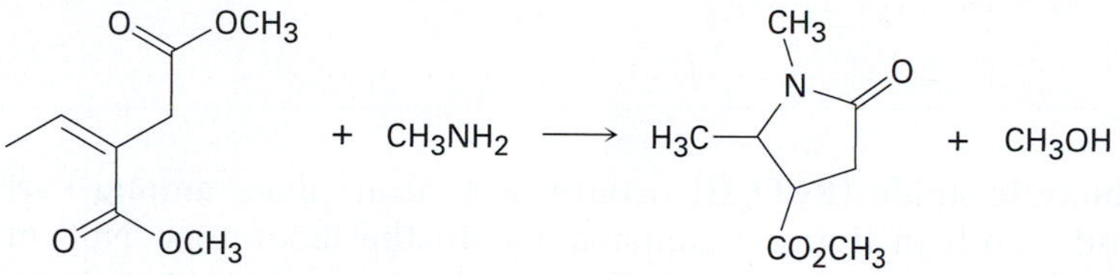

14.63 When crystals of pure α-glucose are dissolved in water, isomerization slowly occurs to produce β-glucose. Propose a mechanism for the isomerization.

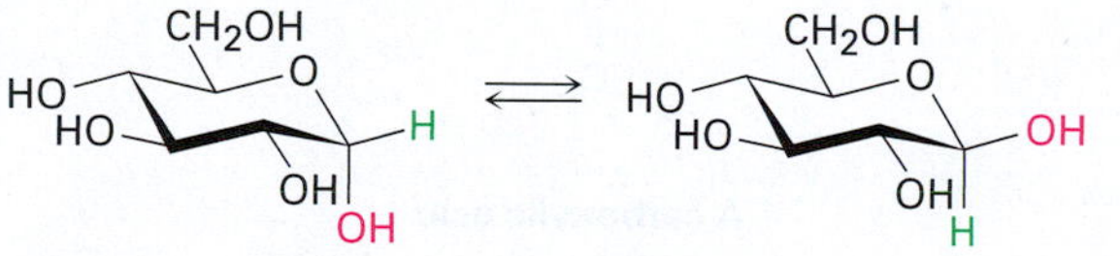

α-**Glucose**   *β*-**Glucose**

14.64 When glucose (Problem 14.63) is treated with NaBH$_4$, reaction occurs to yield *sorbitol,* a polyalcohol commonly used as a food additive. Show how this reduction occurs.

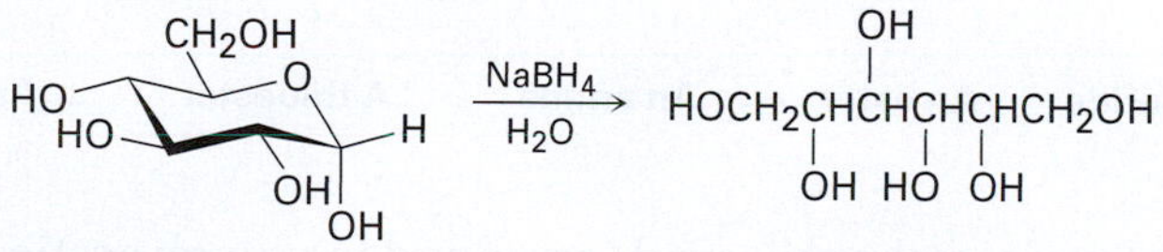

Glucose   **Sorbitol**

■ Problems assignable in Organic OWL.

15   Carboxylic Acids and Nitriles

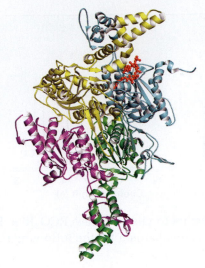

Acetyl CoA carboxylase catalyzes the carboxylation of acetyl CoA to give malonyl CoA, the first step in fatty-acid biosynthesis.

Carboxylic acids (RCO$_2$H) occupy a central place among carbonyl compounds, both in living organisms and in the laboratory. Not only are they valuable in themselves, they also serve as starting materials for preparing numerous *acyl derivatives* such as acid chlorides, esters, amides, thioesters, and acyl phosphates.

A carboxylic acid

An acid chloride   **An ester**   **An amide**   **A thioester**   **An acyl phosphate**

A great many carboxylic acids are found in nature: acetic acid, CH$_3$CO$_2$H, is the chief organic component of vinegar; butanoic acid, CH$_3$CH$_2$CH$_2$CO$_2$H, is responsible for the rancid odor of sour butter; and hexanoic acid (caproic acid), CH$_3$(CH$_2$)$_4$CO$_2$H, is responsible for the unmistakable aroma of goats and dirty gym socks (the name comes from the Latin *caper,* meaning "goat"). Other examples are cholic acid, a major component of human bile, and long-chain

OWL Online homework for this chapter can be assigned in Organic OWL.

aliphatic acids such as palmitic acid, $CH_3(CH_2)_{14}CO_2H$, a biological precursor of fats and vegetable oils.

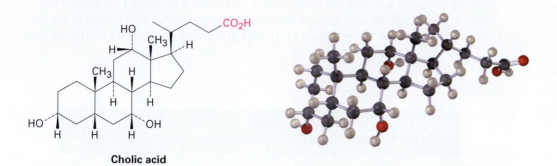

Cholic acid

WHY THIS CHAPTER?

Carboxylic acids are present in many industrial processes and most biological pathways and are the starting materials from which other acyl derivatives are made. Thus, an understanding of their properties and reactions is fundamental to understanding biological chemistry. In this chapter, we'll look both at acids and at their close relatives, *nitriles (RC≡N)*. In the next chapter, we'll look at acyl derivatives.

15.1 Naming Carboxylic Acids and Nitriles

Carboxylic Acids, RCO$_2$H

Simple carboxylic acids derived from open-chain alkanes are systematically named by replacing the terminal *-e* of the corresponding alkane name with *-oic acid*. The –CO_2H carbon atom is numbered C1.

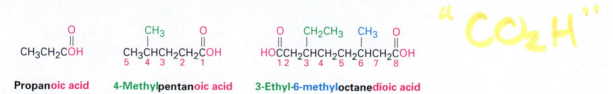

Propanoic acid   **4-Methyl**pentan**oic acid   **3-Ethyl-6-methyl**octane**dioic acid**

Compounds that have a –CO_2H group bonded to a ring are named using the suffix *-carboxylic acid*. The CO_2H carbon is attached to C1 in this system and is not itself numbered. As a substituent, the CO_2H group is called a **carboxyl group**.

trans-4-**Hydroxy**cyclohexane**carboxylic acid**   Cyclopent-1-ene**carboxylic acid**

Because many carboxylic acids were among the first organic compounds to be isolated and purified, a large number of common names exist (Table 15.1). Biological chemists make frequent use of these names, so you may find yourself referring back to this list on occasion. Also listed in Table 15.1 are the common names used for acyl groups derived from the parent acids. Except for the eight entries at the top of Table 15.1, acyl groups are named by changing the *-ic acid* or *-oic acid* ending to *-oyl*.

TABLE 15.1
Common Names of Some Carboxylic Acids and Acyl Groups

Structure	Name	Acyl group
HCO_2H	Formic	Formyl
CH_3CO_2H	Acetic	Acetyl
$CH_3CH_2CO_2H$	Propionic	Propionyl
$CH_3CH_2CH_2CO_2H$	Butyric	Butyryl
HO_2CCO_2H	Oxalic	Oxalyl
$HO_2CCH_2CO_2H$	Malonic	Malonyl
$HO_2CCH_2CH_2CO_2H$	Succinic	Succinyl
$HO_2CCH_2CH_2CH_2CO_2H$	Glutaric	Glutaryl
$HO_2CCH_2CH_2CH_2CH_2CO_2H$	Adipic	Adipoyl
$H_2C{=}CHCO_2H$	Acrylic	Acryloyl
$HO_2CCH{=}CHCO_2H$	Maleic (cis)	Maleoyl
	Fumaric (trans)	Fumaroyl
$HOCH_2CO_2H$	Glycolic	Glycoloyl
$CH_3\overset{\text{OH}}{\underset{}{C}}HCO_2H$	Lactic	Lactoyl
$CH_3\overset{\text{O}}{\underset{}{C}}CO_2H$	Pyruvic	Pyruvoyl
$HOCH_2\overset{\text{OH}}{\underset{}{C}}HCO_2H$	Glyceric	Glyceroyl
$HO_2C\overset{\text{OH}}{\underset{}{C}}HCH_2CO_2H$	Malic	Maloyl
$HO_2C\overset{\text{O}}{\underset{}{C}}CH_2CO_2H$	Oxaloacetic	Oxaloacetyl
(benzene ring)—CO_2H	Benzoic	Benzoyl
(benzene ring with two CO_2H)	Phthalic	Phthaloyl

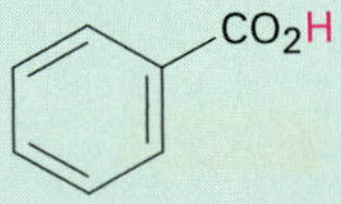

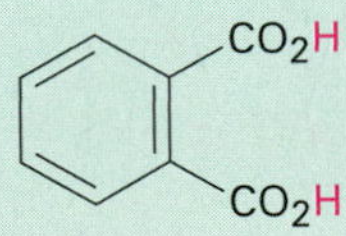

Nitriles, RC≡N

Compounds containing the –C≡N functional group are called **nitriles** and undergo some chemistry similar to that of carboxylic acids. Simple open-chain nitriles are named by adding -*nitrile* as a suffix to the alkane name, with the nitrile carbon numbered C1:

CH₃
|
CH₃CHCH₂CH₂CN   **4-Methylpentanenitrile**
5   4 3   2 1

Nitriles can also be named as derivatives of carboxylic acids by replacing the -*ic acid* or -*oic acid* ending with -*onitrile* or by replacing the -*carboxylic acid* ending with -*carbonitrile*. The nitrile carbon atom is attached to C1 but is not itself numbered.

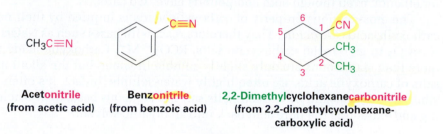

CH₃C≡N

Acetonitrile
(from acetic acid)

Benzonitrile
(from benzoic acid)

2,2-Dimethylcyclohexanecarbonitrile
(from 2,2-dimethylcyclohexane-carboxylic acid)

Problem 15.1
Give IUPAC names for the following compounds:

(a)
CH₃ O
| ||
CH₃CHCH₂COH

(b)
Br   O
| ||
CH₃CHCH₂CH₂COH

(c)
CO₂H
|
CH₃CH₂CHCH₂CH₂CH₃

(d)
H   H
 \   /
  C=C
 /   \
H₃C   CH₂CH₂COH
                ‖
                O

(e)
CH₃ CN
| |
CH₃CHCH₂CHCH₃

(f)
H   H
  HO₂C⸺⟨   ⟩⸺CO₂H

Problem 15.2
Draw structures corresponding to the following IUPAC names:
(a) 2,3-Dimethylhexanoic acid   **(b)** 4-Methylpentanoic acid
(c) *trans*-Cyclobutane-1,2-dicarboxylic acid   **(d)** *o*-Hydroxybenzoic acid
(e) (9Z,12Z)-Octadeca-9,12-dienoic acid
(f) Pent-2-enenitrile

15.2 Structure and Properties of Carboxylic Acids

Carboxylic acids are similar in some respects to both ketones and alcohols. Like ketones, the carboxyl carbon is *sp²*-hybridized, and carboxylic acid groups are therefore planar with C–C=O and O=C–O bond angles of approximately 120°. Like alcohols, carboxylic acids are strongly associated because of hydrogen-bonding. Most carboxylic acids exist as cyclic dimers held together by two hydrogen bonds.

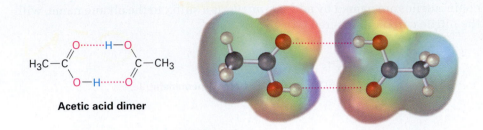

Acetic acid dimer

This strong hydrogen-bonding has a noticeable effect on boiling points, making carboxylic acids much higher boiling than the corresponding alcohols. Acetic acid, for instance, has a boiling point of 117.9 °C, versus 78.3 °C for ethanol, even though both compounds have two carbons.

The most obvious property of carboxylic acids is implied by their name: carboxylic acids are *acidic*. They therefore react with bases such as NaOH and NaHCO$_3$ to give metal carboxylate salts, RCO$_2^-$ M$^+$. Carboxylic acids with more than six carbons are only slightly soluble in water, but the alkali metal salts of carboxylic acids are often highly water-soluble. In fact, it's often possible to purify an acid by extracting its salt into aqueous base, then reacidifying and extracting the pure acid back into an organic solvent.

A carboxylic acid
(water-insoluble)   + NaOH $\xrightarrow{H_2O}$ **A carboxylic acid salt**
(water-soluble)   + H$_2$O

Like other Brønsted–Lowry acids discussed in Section 2.7, carboxylic acids dissociate slightly in dilute aqueous solution to give H$_3$O$^+$ and the corresponding carboxylate anion, RCO$_2^-$. The extent of dissociation is given by an acidity constant, K_a:

$$K_a = \frac{[RCO_2^-][H_3O^+]}{[RCO_2H]} \quad \text{and} \quad pK_a = -\log K_a$$

A list of K_a values for various carboxylic acids is given in Table 15.2. For most, K_a is approximately 10^{-4} to 10^{-5}. Acetic acid, for instance, has $K_a = 1.75 \times 10^{-5}$ at 25 °C, which corresponds to a pK_a of 4.76. In practical terms, a K_a value near 10^{-5} means that only about 0.1% of the molecules in a 0.1 M solution are dissociated, as opposed to the 100% dissociation found with strong mineral acids like HCl.

TABLE 15.2
Acidity of Some Carboxylic Acids

Structure	K_a	pK_a	
CF_3CO_2H	0.59	0.23	**Stronger acid**
HCO_2H	1.77×10^{-4}	3.75	
$HOCH_2CO_2H$	1.5×10^{-4}	3.83	
$C_6H_5CO_2H$	6.46×10^{-5}	4.19	
$H_2C{=}CHCO_2H$	5.6×10^{-5}	4.25	
CH_3CO_2H	1.75×10^{-5}	4.76	
$CH_3CH_2CO_2H$	1.34×10^{-5}	4.87	
CH_3CH_2OH (ethanol)	(1.00×10^{-16})	(16.00)	**Weaker acid**

[handwritten note: Small Ka, Low dissociation]

Although much weaker than mineral acids, carboxylic acids are neverthe-less much stronger acids than alcohols and phenols. The K_a of ethanol, for example, is approximately 10^{-16}, making ethanol a weaker acid than acetic acid by a factor of 10^{11}.

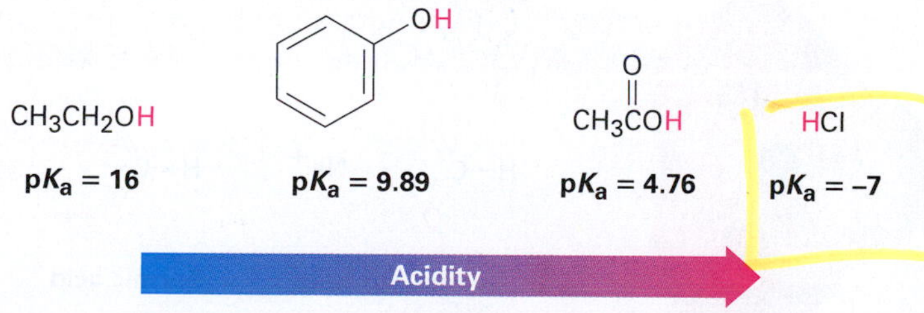

CH_3CH_2OH (phenol, OH) CH_3COH (O)   HCl   *[handwritten: Acidic]*

$pK_a = 16$ $pK_a = 9.89$ $pK_a = 4.76$ $pK_a = -7$

Acidity →

Why are carboxylic acids so much more acidic than alcohols, even though both contain –OH groups? An alcohol dissociates to give an alkoxide ion, in which the negative charge is localized on a single electronegative atom. A carboxylic acid, however, gives a carboxylate ion, in which the neg-ative charge is delocalized over *two* equivalent oxygen atoms (Figure 15.1). In resonance terms (Section 2.4), a carboxylate ion is a stabilized resonance hybrid of two equivalent structures. Since a carboxylate ion is more stable than an alkoxide ion, it is lower in energy and more favored in the dissocia-tion equilibrium.

FIGURE 15.1 An alkoxide ion has its charge localized on one oxygen atom and is less stable, while a carboxylate ion has the charge spread equally over both oxygens and is therefore more stable.

Ethanol $\xrightarrow{H_2O}$ H_3O^+ + Ethoxide ion (localized charge)

Acetic acid $\xrightarrow{H_2O}$ H_3O^+ + Acetate ion (delocalized charge)

Resonance stabilized

Experimental evidence for the equivalence of the two carboxylate oxygens comes from X-ray crystallographic studies on sodium formate. Both carbon–oxygen bonds are 127 pm in length, midway between the C=O double bond (120 pm) and C–O single bond (134 pm) of formic acid. An electrostatic potential map of the formate ion also shows how the negative charge (red) is spread equally over both oxygens.

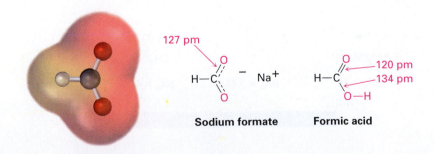

127 pm

Sodium formate

120 pm
134 pm

Formic acid

Problem 15.3
Assume you have a mixture of naphthalene and benzoic acid that you want to separate. How might you take advantage of the acidity of one component in the mixture to effect a separation?

Problem 15.4
The K_a for dichloroacetic acid is 3.32×10^{-2}. Approximately what percentage of the acid is dissociated in a 0.10 M aqueous solution?

15.3 Biological Acids and the Henderson–Hasselbalch Equation

In acidic solution at low pH, a carboxylic acid is completely undissociated and exists entirely as RCO_2H. In basic solution at high pH, a carboxylic acid is completely dissociated and exists entirely as RCO_2^-. Inside living cells, however, the pH is neither acidic nor basic but is instead buffered to nearly neutral pH— in humans, to pH = 7.3, a value often referred to as *physiological pH*. In what form, then, do carboxylic acids exist inside cells? The question is an important one for understanding the acid catalysts so often found in biological reactions.

If the pK_a value of a given acid and the pH of the medium are known, the percentages of dissociated and undissociated forms can be calculated using the **Henderson–Hasselbalch equation**.

For any acid HA, we have

$$pK_a = -\log\frac{[H_3O^+][A^-]}{[HA]} = -\log[H_3O^+] - \log\frac{[A^-]}{[HA]}$$

$$= pH - \log\frac{[A^-]}{[HA]}$$

which can be rearranged to give

$$pH = pK_a + \log\frac{[A^-]}{[HA]} \qquad \text{\textcolor{red}{\textbf{Henderson–Hasselbalch equation}}}$$

so $\log\dfrac{[A^-]}{[HA]} = pH - pK_a$

This equation says that the logarithm of the concentration of dissociated acid $[A^-]$ divided by the concentration of undissociated acid [HA] is equal to the pH of the solution minus the pK_a of the acid. Thus, if we know both the pH of the solution and the pK_a of the acid, we can calculate the ratio of $[A^-]$ to [HA]. Furthermore, when pH = pK_a, the two forms HA and A^- are present in equal amounts because log 1 = 0.

As an example of how to use the Henderson–Hasselbalch equation, let's find out what species are present in a 0.0010 M solution of acetic acid at pH = 7.3. According to Table 15.2, the pK_a of acetic acid is 4.76. From the Henderson–Hasselbalch equation, we have

$$\log\frac{[A^-]}{[HA]} = pH - pK_a = 7.3 - 4.76 = 2.54$$

$$\frac{[A^-]}{[HA]} = \text{antilog}(2.54) = 3.5 \times 10^2 \qquad \text{so } [A^-] = (3.5 \times 10^2)[HA]$$

In addition, we know that

$$[A^-] + [HA] = 0.0010 \text{ M}$$

Solving the two simultaneous equations gives $[A^-] = 0.0010$ M and [HA] = 3×10^{-6} M. In other words, at a physiological pH of 7.3, essentially 100% of acetic acid molecules in a 0.0010 M solution are dissociated to the acetate ion.

What is true for acetic acid is also true for other carboxylic acids: at the physiological pH that exists inside cells, carboxylic acids are almost entirely dissociated. To reflect this fact, we always refer to cellular carboxylic acids by the name of their anion—acetate, lactate, citrate, and so forth, rather than acetic acid, lactic acid, and citric acid.

Problem 15.5

Calculate the percentages of dissociated and undissociated forms present in the following solutions:

(a) 0.0010 M glycolic acid ($HOCH_2CO_2H$; pK_a = 3.83) at pH = 4.50

(b) 0.0020 M propanoic acid (pK_a = 4.87) at pH = 5.30

15.4 Substituent Effects on Acidity

The listing of pK_a values shown previously in Table 15.2 indicates that there are substantial differences in acidity from one carboxylic acid to another. For example, trifluoroacetic acid (pK_a = −0.23) is 33,000 times as strong as acetic acid (pK_a = 4.76). How can we account for such differences?

Because the dissociation of a carboxylic acid is an equilibrium process, any factor that stabilizes the carboxylate anion relative to undissociated carboxylic acid will drive the equilibrium toward increased dissociation and result in increased acidity. For instance, three electron-withdrawing fluorine atoms delocalize the negative charge in the trifluoroacetate anion, thereby stabilizing the ion and increasing the acidity of CF_3CO_2H. In the same way, glycolic acid ($HOCH_2CO_2H$; pK_a = 3.83) is stronger than acetic acid because of the electron-withdrawing effect of the electronegative oxygen atom.

pK_a = 4.76 pK_a = 3.83 pK_a = −0.23

Acidity →

Substituent effects on acidity are also found in substituted benzoic acids. We said during the discussion of electrophilic aromatic substitution in Section 9.8 that substituents on the aromatic ring strongly affect reactivity. Aromatic rings with electron-donating groups are activated toward further electrophilic substitution, and aromatic rings with electron-withdrawing groups are deactivated. Exactly the same effects are noticed on the acidity of substituted benzoic acids. Thus, an electron-withdrawing (deactivating) group such as nitro increases acidity by stabilizing the carboxylate anion, and an electron-donating (activating) group such as methoxy decreases acidity by destabilizing the carboxylate anion.

p-Methoxybenzoic acid   Benzoic acid   *p*-Nitrobenzoic acid
(pK_a = 4.46)   (pK_a = 4.19)   (pK_a = 3.41)

Acidity →

Because it's much easier to measure the acidity of a substituted benzoic acid than it is to determine the relative reactivity of an aromatic ring toward electrophilic substitution, the correlation between the two effects is useful for predicting reactivity. If we want to know the effect of a certain substituent on electrophilic reactivity, we can simply find the acidity of the corresponding benzoic acid. Worked Example 15.1 gives an illustration.

Finding the K_a of this acid . . .

. . . lets us predict the reactivity of this substituted benzene to electrophilic attack.

WORKED EXAMPLE 15.1	Predicting the Effect of a Substituent on the Reactivity of an Aromatic Ring toward Electrophilic Substitution

The pK_a of *p*-(trifluoromethyl)benzoic acid is 3.6. Is the trifluoromethyl substituent an activating or deactivating group in electrophilic aromatic substitution?

Strategy

Decide whether *p*-(trifluoromethyl)benzoic acid is stronger or weaker than benzoic acid. A substituent that strengthens the acid is a deactivating group because it withdraws electrons, and a substituent that weakens the acid is an activating group because it donates electrons.

Solution

A pK_a of 3.6 means that *p*-(trifluoromethyl)benzoic acid is stronger than benzoic acid, whose pK_a is 4.19. Thus, the trifluoromethyl substituent favors dissociation by helping stabilize the negative charge. Trifluoromethyl must therefore be an electron-withdrawing, deactivating group.

Problem 15.6

Which would you expect to be a stronger acid, the lactic acid found in tired muscles or acetic acid? Explain.

$$\underset{CH_3CHCOH}{\overset{HO\ \ O}{| \ \ ||}}$$ **Lactic acid**

Problem 15.7

Dicarboxylic acids have two dissociation constants, one for the initial dissociation into a monoanion and one for the second dissociation into a dianion. For oxalic acid, HO_2C-CO_2H, the first ionization constant is $pK_{a1} = 1.2$ and the second ionization constant is $pK_{a2} = 4.2$. Why is the second carboxyl group so much less acidic than the first?

Problem 15.8

The pK_a of *p*-cyclopropylbenzoic acid is 4.45. Is cyclopropylbenzene likely to be more reactive or less reactive than benzene toward electrophilic bromination? Explain.

15.5 Preparing Carboxylic Acids

Let's review briefly the methods for preparing carboxylic acids that we've seen in past chapters:

- Oxidation of a substituted alkylbenzene with $KMnO_4$ or $Na_2Cr_2O_7$ gives a substituted benzoic acid (Section 9.10). Both primary and secondary alkyl groups can be oxidized, but tertiary groups are not affected.

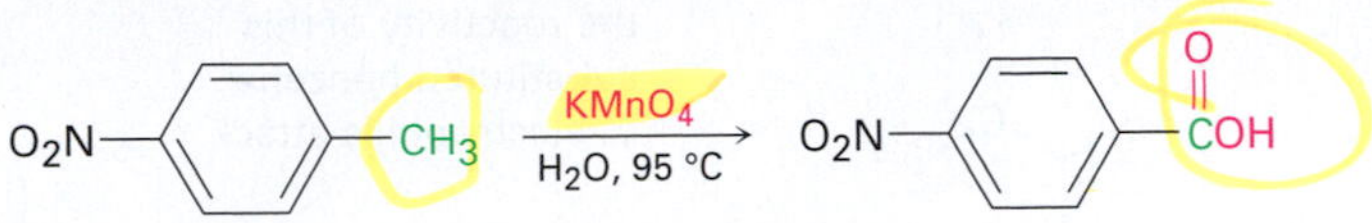

p-Nitrotoluene   **p-Nitrobenzoic acid (88%)**

- Oxidation of a primary alcohol or an aldehyde yields a carboxylic acid (Sections 13.5 and 14.3). Both oxidations are often carried out with CrO_3 in aqueous acid.

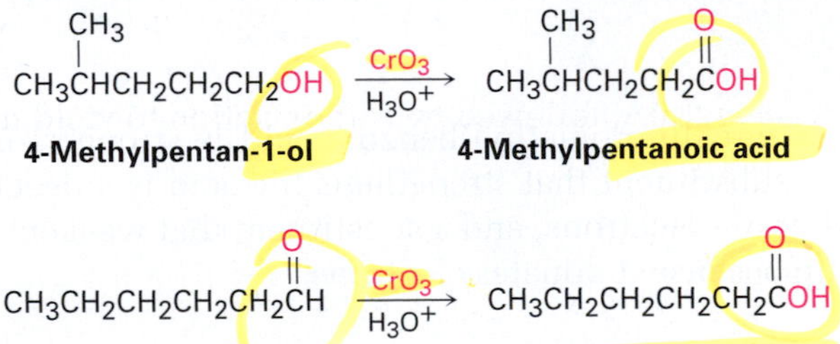

4-Methylpentan-1-ol   **4-Methylpentanoic acid**

Hexanal   **Hexanoic acid**

Hydrolysis of Nitriles

Carboxylic acids can be prepared from nitriles by heating with aqueous acid or base by a mechanism that we'll see in Section 15.7. Since nitriles themselves are easily made by S_N2 reaction of a primary or secondary alkyl halide with CN^-, the two-step sequence of cyanide displacement followed by nitrile hydrolysis is a good way to make a carboxylic acid from an alkyl halide ($RBr \rightarrow RC{\equiv}N \rightarrow RCO_2H$). Note that the product acid has one more carbon than the starting alkyl halide. An example occurs in one commercial route for the synthesis of the nonsteroidal anti-inflammatory drug ibuprofen.

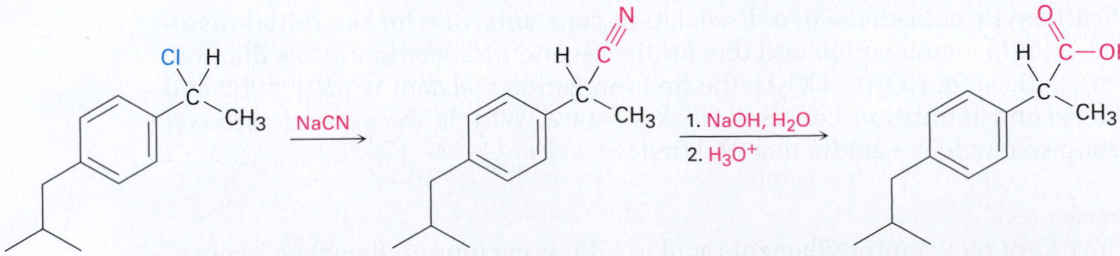

Ibuprofen

Carboxylation of Grignard Reagents

Another method for preparing carboxylic acids is by reaction of a Grignard reagent with CO_2 to yield a metal carboxylate, followed by protonation to give the carboxylic acid. This **carboxylation** reaction is usually carried out by bubbling a stream of dry CO_2 gas through a solution of the Grignard reagent. The organomagnesium halide adds to a C=O bond of carbon dioxide in a typical nucleophilic carbonyl addition reaction, and protonation of the carboxylate by addition of aqueous HCl in a separate step then gives the free carboxylic acid. For example

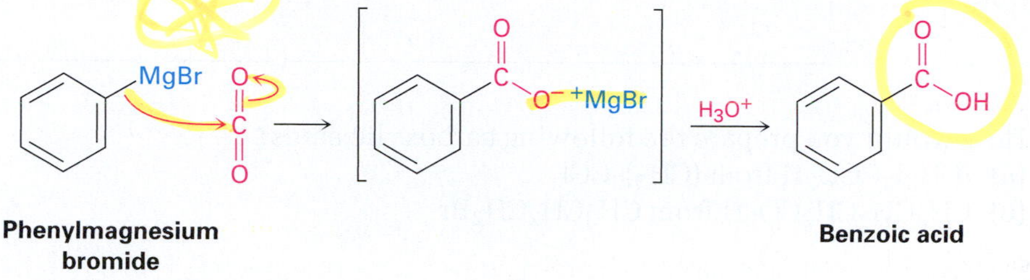

Phenylmagnesium bromide

Benzoic acid

As noted previously, there are no Grignard reagents inside living cells, but there are other types of stabilized carbanions that are often carboxylated. One of the initial steps in fatty-acid biosynthesis, for instance, involves formation of a carbanion from acetyl CoA, followed by carboxylation to yield malonyl CoA.

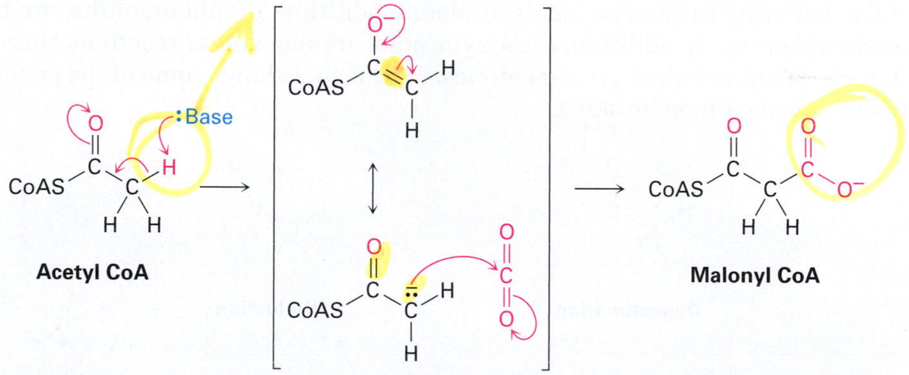

Acetyl CoA

Malonyl CoA

WORKED EXAMPLE 15.2 Devising a Synthesis Route for a Carboxylic Acid

How would you prepare phenylacetic acid ($PhCH_2CO_2H$) from benzyl bromide ($PhCH_2Br$)?

Strategy

We've seen two methods for preparing carboxylic acids from alkyl halides: (1) cyanide ion displacement followed by hydrolysis and (2) formation of a Grignard reagent followed by carboxylation. The first method involves an S_N2 reaction and is therefore limited to use with primary and some secondary alkyl halides. The second method involves formation of a Grignard reagent and is therefore limited to use with organic halides that have no acidic hydrogens or reactive functional groups elsewhere in the molecule. In the present instance, either method would work well.

Solution

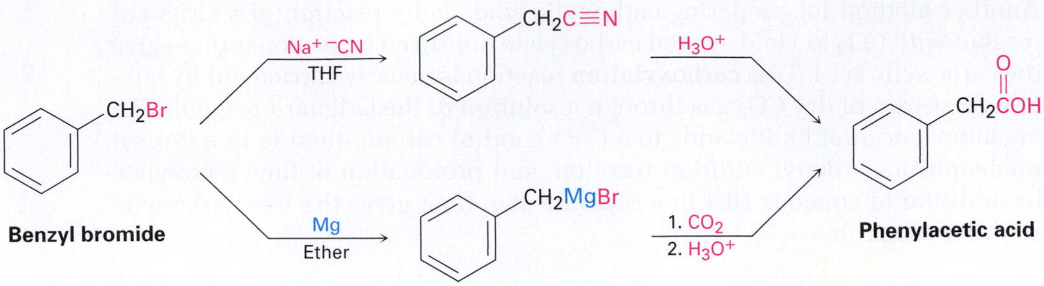

Benzyl bromide → (Na+ −CN / THF) → CH₂C≡N → (H₃O+) → Phenylacetic acid (CH₂COH)

Benzyl bromide → (Mg / Ether) → CH₂MgBr → (1. CO₂ 2. H₃O+) → Phenylacetic acid

Problem 15.9

How would you prepare the following carboxylic acids?

(a) $(CH_3)_3CCO_2H$ from $(CH_3)_3CCl$

(b) $CH_3CH_2CH_2CO_2H$ from $CH_3CH_2CH_2Br$

15.6 Reactions of Carboxylic Acids: An Overview

We commented earlier in this chapter that carboxylic acids are similar in some respects to both alcohols and ketones. Like alcohols, carboxylic acids can be deprotonated to give anions, which are good nucleophiles in S_N2 reactions. Like ketones, carboxylic acids undergo addition of nucleophiles to the carbonyl group. In addition, carboxylic acids undergo other reactions characteristic of neither alcohols nor ketones. Figure 15.2 shows some of the general reactions of carboxylic acids.

ACTIVE FIGURE 15.2 Some general reactions of carboxylic acids. **Go to this book's student companion site at www.cengage .com/chemistry/mcmurry to** explore an interactive version of this figure.

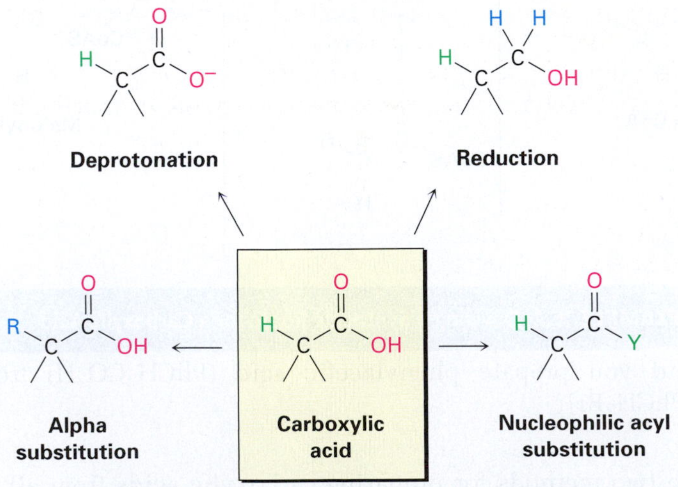

Deprotonation   Reduction

Alpha substitution   Carboxylic acid   Nucleophilic acyl substitution

Reactions of carboxylic acids can be grouped into the four categories indicated in Figure 15.2. Of the four, we've already discussed the acidic behavior of carboxylic acids in Sections 15.2 and 15.3, and we mentioned reduction by reaction of the acid with $LiAlH_4$ in Section 13.3. The remaining two categories are examples of fundamental carbonyl-group reaction mechanisms— nucleophilic acyl substitution and α substitution—that will be discussed in detail in Chapters 16 and 17.

Problem 15.10

How might you prepare 2-phenylethanol from benzyl bromide? More than one step is needed.

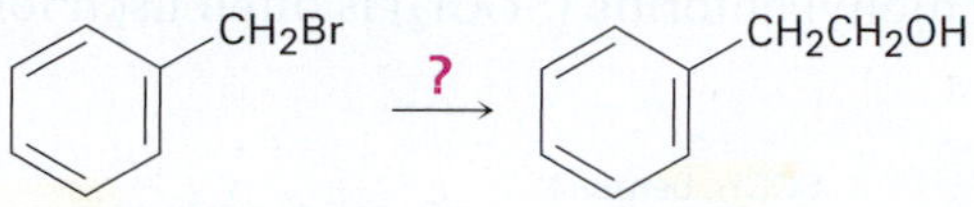

15.7 Chemistry of Nitriles

Nitriles are analogous to carboxylic acids in that both have a carbon atom with three bonds to an electronegative atom and both contain a π bond. Thus, some reactions of nitriles and carboxylic acids are similar. Both kinds of compounds are electrophiles, for instance, and both undergo nucleophilic addition reactions.

A nitrile—three bonds to nitrogen

An acid—three bonds to two oxygens

Nitriles occur infrequently in living organisms, although several hundred examples are known. Cyanocycline A, for instance, has been isolated from the bacterium *Streptomyces lavendulae* and found to have both antimicrobial and antitumor activity. In addition, more than 1000 compounds called *cyanogenic glycosides* are known. Derived primarily from plants, cyanogenic glycosides contain a sugar with an acetal carbon, one oxygen of which is bonded to a nitrile-bearing carbon (Sugar–O–C–CN). On hydrolysis with aqueous acid, the acetal is cleaved (Section 14.8), generating a cyanohydrin (HO–C–CN), which releases hydrogen cyanide. It's thought that the primary function of cyanogenic glycosides is to protect the plant by poisoning any animal foolish enough to eat it. Lotaustralin from the cassava plant is an example.

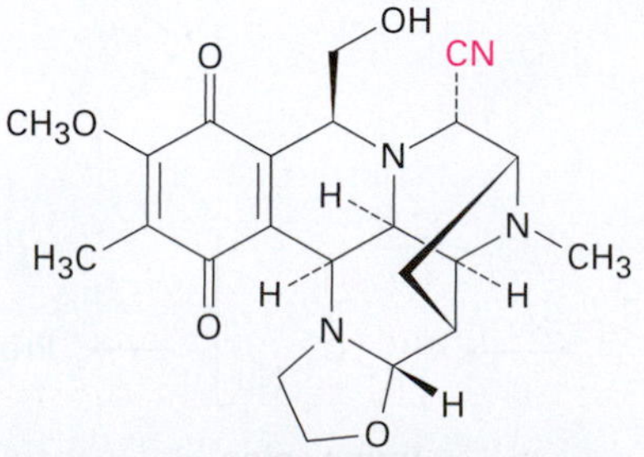

Cyanocycline A

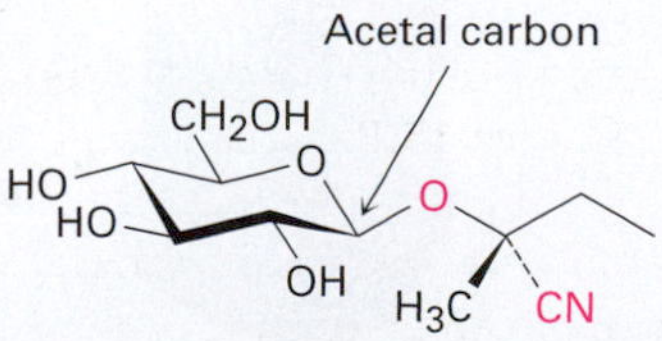

Lotaustralin (a cyanogenic glycoside)

Preparation of Nitriles

The simplest method of nitrile preparation in the laboratory is the S_N2 reaction of CN^- with a primary or secondary alkyl halide, as mentioned in Section 15.5. Another method for preparing nitriles is by dehydration of a primary amide, $RCONH_2$. Thionyl chloride ($SOCl_2$) is often used for the reaction.

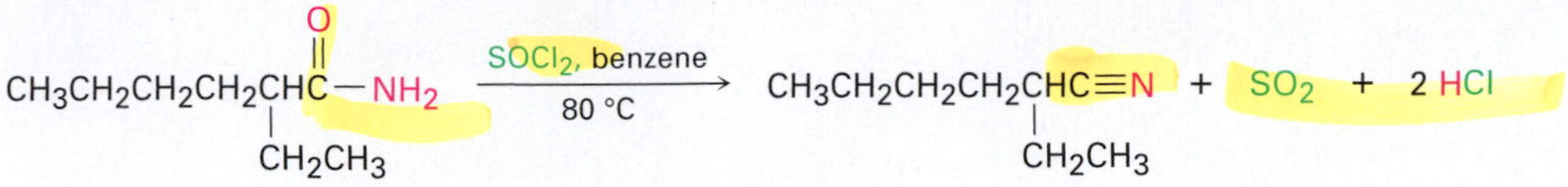

$$CH_3CH_2CH_2CH_2\underset{\underset{CH_2CH_3}{|}}{CH}\overset{\overset{O}{\|}}{C}-NH_2 \xrightarrow[80\ °C]{SOCl_2,\ benzene} CH_3CH_2CH_2CH_2\underset{\underset{CH_2CH_3}{|}}{CH}C\equiv N\ +\ SO_2\ +\ 2\ HCl$$

2-Ethylhexanamide   **2-Ethylhexanenitrile (94%)**

The dehydration occurs by initial reaction of $SOCl_2$ on the nucleophilic amide oxygen atom, followed by deprotonation and a subsequent E2-like elimination reaction.

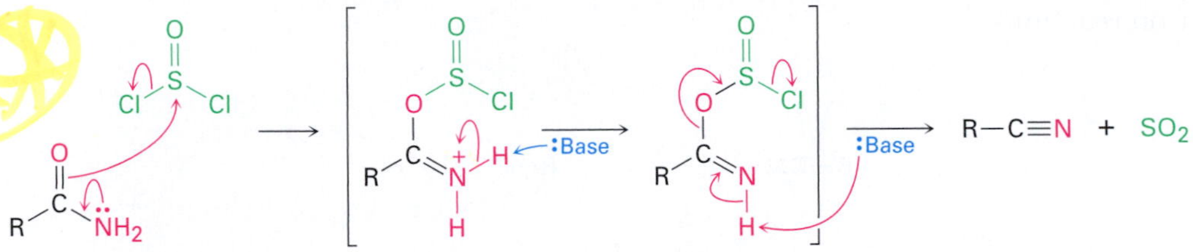

Both methods of nitrile synthesis—S_N2 displacement by CN^- on an alkyl halide and amide dehydration—are useful, but the synthesis from amides is more general because it is not limited by steric hindrance.

Reactions of Nitriles

Like a carbonyl group, a nitrile group is strongly polarized and has an electrophilic carbon atom. Nitriles therefore react with nucleophiles to yield sp^2-hybridized imine anions in a reaction analogous to the formation of an sp^3-hybridized alkoxide ion by nucleophilic addition to a carbonyl group.

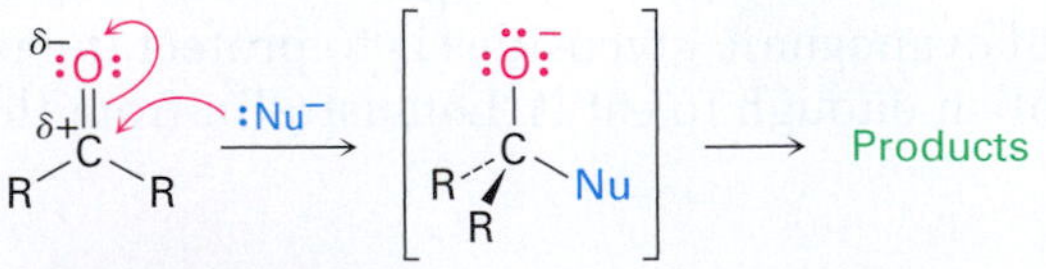

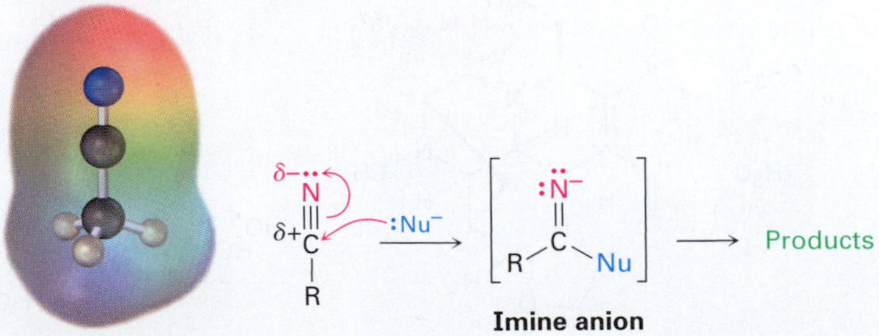

Imine anion

HYDROLYSIS: CONVERSION OF NITRILES TO CARBOXYLIC ACIDS Among the most useful reactions of nitriles is their hydrolysis to yield first an amide and then a carboxylic acid plus ammonia. The reaction occurs in either basic or acidic aqueous solution:

R—C≡N $\xrightarrow{\text{H}_3\text{O}^+ \text{ or NaOH, H}_2\text{O}}$ amide $\xrightarrow{\text{H}_3\text{O}^+ \text{ or NaOH, H}_2\text{O}}$ carboxylic acid + NH_3

A nitrile   **An amide**   **A carboxylic acid**

Base catalyzed nitrile hydrolysis involves nucleophilic addition of hydroxide ion to the polar C≡N bond to give an imine anion. Protonation then gives a hydroxy imine, which isomerizes to an amide. The mechanism is shown in Figure 15.3.

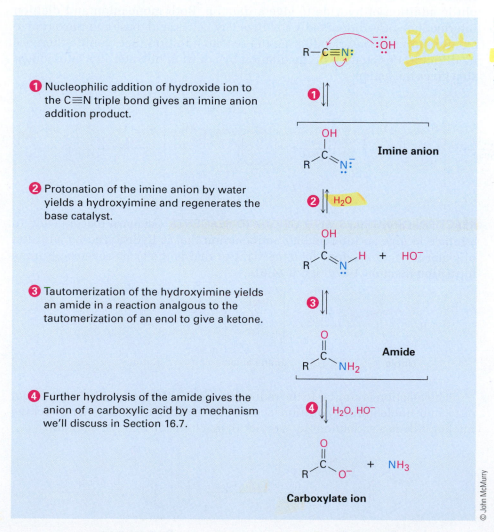

❶ Nucleophilic addition of hydroxide ion to the C≡N triple bond gives an imine anion addition product.

Imine anion

❷ Protonation of the imine anion by water yields a hydroxyimine and regenerates the base catalyst.

❸ Tautomerization of the hydroxyimine yields an amide in a reaction analgous to the tautomerization of an enol to give a ketone.

Amide

❹ Further hydrolysis of the amide gives the anion of a carboxylic acid by a mechanism we'll discuss in Section 16.7.

Carboxylate ion

FIGURE 15.3 MECHANISM: Mechanism of the basic hydrolysis of a nitrile to yield an amide, which is subsequently hydrolyzed further to a carboxylic acid anion.

© John McMurry

At higher reaction temperatures, the amide undergoes further hydrolysis to give a carboxylic acid by a nucleophilic acyl substitution mechanism that is the subject of the next chapter. In essence, a second nucleophilic addition of hydroxide ion to the amide carbonyl group yields a tetrahedral alkoxide ion, which expels amide ion, NH_2^-, as leaving group and gives the carboxylate ion, thereby driving the reaction toward products. Subsequent acidification in a separate step yields the carboxylic acid.

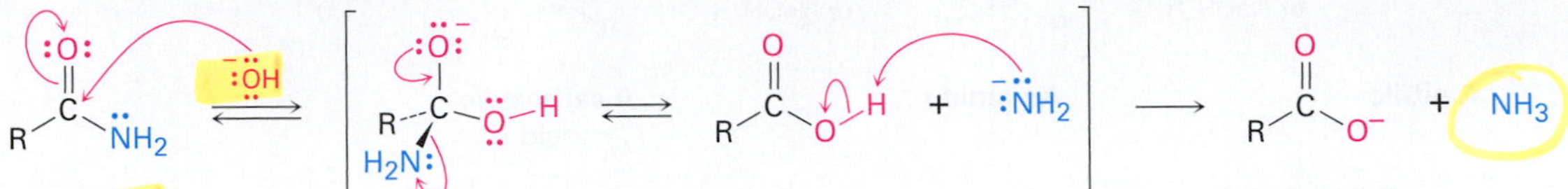

An amide — **A carboxylate ion** + NH_3

REDUCTION: CONVERSION OF NITRILES TO AMINES

Reduction of a nitrile with $LiAlH_4$ gives a primary amine, RNH_2. The reaction occurs by nucleophilic addition of hydride ion to the polar $C\equiv N$ bond, yielding an imine anion, which still contains a $C=N$ bond and therefore undergoes a second nucleophilic addition of hydride to give a *dianion*. Both monoanion and dianion intermediates are undoubtedly stabilized by Lewis acid–base complexation to an aluminum species, facilitating the second addition that would otherwise be difficult. Protonation of the dianion by addition of water in a subsequent step gives the amine.

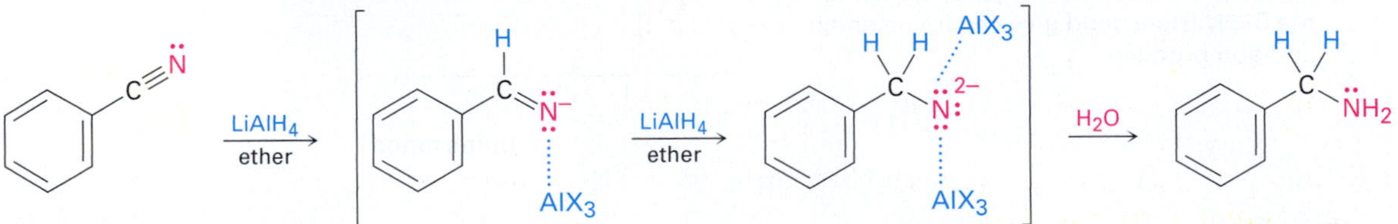

Benzonitrile — $LiAlH_4$, ether — $LiAlH_4$, ether — H_2O — **Benzylamine**

REACTION OF NITRILES WITH GRIGNARD REAGENTS

Grignard reagents add to a nitrile to give an intermediate imine anion that is hydrolyzed by addition of water to yield a ketone by a mechanism that is the exact reverse of imine formation (Section 14.7; Figure 14.8).

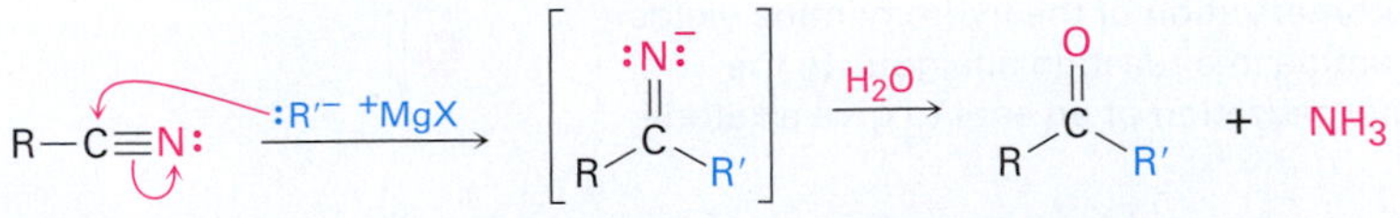

Nitrile — $:R'^- {}^+MgX$ — **Imine anion** — H_2O — **Ketone** + NH_3

The reaction is similar to the reduction of a nitrile to an amine, except that only one nucleophilic addition occurs rather than two, and the attacking nucleophile is a carbanion ($R:^-$) rather than a hydride ion. For example:

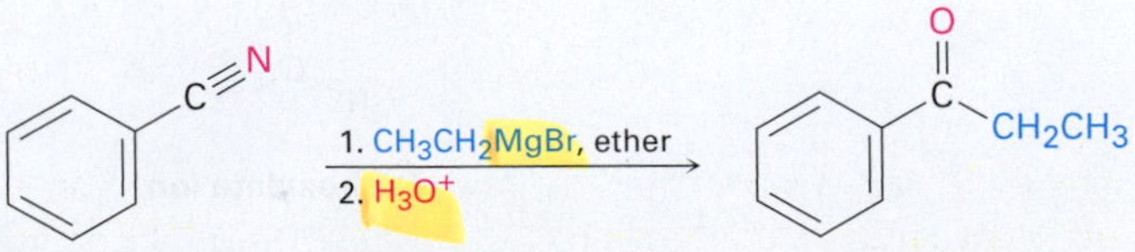

Benzonitrile — 1. CH_3CH_2MgBr, ether 2. H_3O^+ — **Propiophenone (89%)**

Problem 15.11

How would you prepare the following carbonyl compounds from a nitrile?

(a)

$$CH_3CH_2\overset{\overset{\displaystyle O}{\|}}{C}CH_2CH_3$$

(b)

(structure of a benzene ring with a C(=O)CH₃ group and an O₂N group)

Problem 15.12

How would you prepare 4-methylpentan-1-ol starting with a nitrile? More than one step is needed.

$$R-C\equiv N \overset{?}{\Longrightarrow} CH_3\overset{\overset{\displaystyle CH_3}{|}}{CH}CH_2CH_2CH_2OH$$

15.8 Spectroscopy of Carboxylic Acids and Nitriles

Infrared Spectroscopy

Carboxylic acids have two characteristic IR absorptions that make the $-CO_2H$ group easily identifiable. The O–H bond of the carboxyl group gives rise to a very broad absorption over the range 2500 to 3300 cm^{-1}, and the C=O bond shows an absorption between 1710 and 1760 cm^{-1}. The exact position of C=O absorption depends both on the structure of the molecule and on whether the acid is free (monomeric) or hydrogen-bonded (dimeric). Free carboxyl groups absorb at 1760 cm^{-1}, but the more commonly encountered dimeric carboxyl groups absorb in a broad band centered around 1710 cm^{-1}. Both the broad O–H absorption and the C=O absorption at 1710 cm^{-1} (dimeric) are visible in the IR spectrum of butanoic acid shown in Figure 15.4.

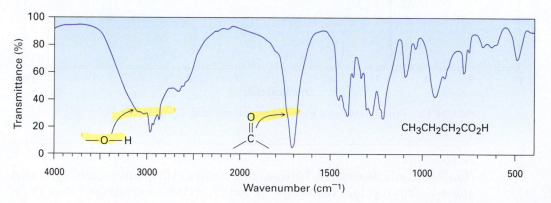

FIGURE 15.4
IR spectrum of butanoic acid, $CH_3CH_2CH_2CO_2H$.

Nitriles show an intense and easily recognizable C≡N bond absorption near 2250 cm^{-1} for saturated compounds and 2230 cm^{-1} for aromatic and conjugated molecules. Few other functional groups absorb in this region, so IR spectroscopy is highly diagnostic for nitriles.

Problem 15.13
Cyclopentanecarboxylic acid and 4-hydroxycyclohexanone have the same formula ($C_6H_{10}O_2$), and both contain an –OH and a C=O group. How could you distinguish between them by IR spectroscopy?

Nuclear Magnetic Resonance Spectroscopy

In the ^{13}C NMR spectrum, carboxyl carbon atoms absorb in the range 165 to 185 δ, with aromatic and α,β-unsaturated acids near the upfield end of the range (~165 δ) and saturated aliphatic acids near the downfield end (~185 δ). Nitrile carbons absorb in the range 115 to 130 δ.

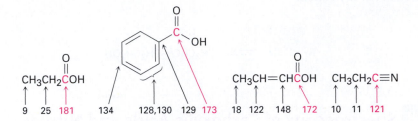

In the ^1H NMR spectrum, the acidic –CO_2**H** proton normally absorbs as a singlet near 12 δ. As with alcohols (Section 13.12), the –CO_2H proton can be replaced by deuterium when D_2O is added to the sample tube, causing the absorption to disappear from the NMR spectrum. Figure 15.5 shows the ^1H NMR spectrum of phenylacetic acid. Note that the carboxyl proton absorption occurs at 12.0 δ.

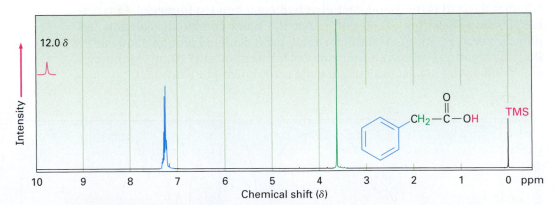

FIGURE 15.5 Proton NMR spectrum of phenylacetic acid, $C_6H_5CH_2CO_2H$.

Problem 15.14
How could you distinguish between the isomers cyclopentanecarboxylic acid and 4-hydroxycyclohexanone by ^1H and ^{13}C NMR spectroscopy? (See Problem 15.13.)

Summary

Carboxylic acids are among the most useful building blocks for synthesizing other molecules, both in nature and in the chemical laboratory. Thus, an understanding of their properties and reactions is fundamental to understanding biological chemistry. In this chapter, we've looked both at acids and at their close relatives, **nitriles (RC≡N)**.

Carboxylic acids are named systematically by replacing the terminal *-e* of the corresponding alkane name with *-oic acid.* Like aldehydes and ketones, the carbonyl carbon atom is sp^2-hybridized; like alcohols, carboxylic acids are associated through hydrogen-bonding and therefore have high boiling points.

The distinguishing characteristic of carboxylic acids is their acidity. Although weaker than mineral acids such as HCl, carboxylic acids dissociate much more readily than alcohols because the resultant carboxylate ions are stabilized by resonance between two equivalent forms.

Most carboxylic acids have pK_a values near 5, but the exact pK_a of a given acid depends on structure. Carboxylic acids substituted by electron-withdrawing groups are more acidic (have a lower pK_a) because their carboxylate ions are stabilized. Carboxylic acids substituted by electron-donating groups are less acidic (have a higher pK_a) because their carboxylate ions are destabilized. The extent of dissociation of a carboxylic acid in a buffered solution of a given pH can be calculated with the **Henderson–Hasselbalch equation**. Inside living cells, where the physiological pH = 7.3, carboxylic acids are entirely dissociated and exist as their carboxylate anions.

Methods of synthesis for carboxylic acids include (1) oxidation of alkylbenzenes, (2) oxidation of primary alcohols or aldehydes, (3) reaction of Grignard reagents with CO_2 **(carboxylation)**, and (4) hydrolysis of nitriles. General reactions of carboxylic acids include (1) loss of the acidic proton, (2) nucleophilic acyl substitution at the carbonyl group, (3) substitution on the α carbon, and (4) reduction.

Nitriles are similar in some respects to carboxylic acids and are prepared either by S_N2 reaction of an alkyl halide with cyanide ion or by dehydration of an amide. Nitriles undergo nucleophilic addition to the polar C≡N bond in the same way that carbonyl compounds do. The most important reactions of nitriles are their hydrolysis to carboxylic acids, reduction to primary amines, and reaction with Grignard reagents to yield ketones.

Carboxylic acids and nitriles are easily distinguished spectroscopically. Acids show a characteristic IR absorption at 2500 to 3300 cm^{-1} due to the O–H bond and another at 1710 to 1760 cm^{-1} due to the C=O bond; nitriles have an absorption at 2250 cm^{-1}. Acids also show ^{13}C NMR absorptions at 165 to 185 δ and ^1H NMR absorptions near 12 δ; nitriles have a ^{13}C NMR absorption in the range 115 to 130 δ.

Key Words

carboxyl group (CO_2H), 611
carboxylation, 621
carboxylic acid (RCO_2H), 610
Henderson–Hasselbalch
  equation, 617
nitrile (RC≡N), 613

Summary of Reactions

1.   Preparation of carboxylic acids

   (a) Oxidation of alkylbenzenes (Section 9.10)

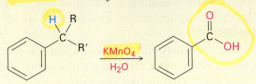

   (b) Oxidation of primary alcohols (Section 13.5)

$$R-\underset{\underset{OH}{H}}{\overset{H}{C}}-H \xrightarrow{CrO_3,\ H_3O^+} R-\overset{O}{\underset{}{C}}-OH$$

   (c) Oxidation of aldehydes (Section 14.3)

$$R-\overset{O}{\underset{}{C}}-H \xrightarrow{CrO_3,\ H_3O^+} R-\overset{O}{\underset{}{C}}-OH$$

   (d) Carboxylation of Grignard reagents (Section 15.5)

$$R-MgX \xrightarrow[2.\ H_3O^+]{1.\ CO_2} R-\overset{O}{\underset{}{C}}-OH$$

   (e) Hydrolysis of nitriles (Section 15.7)

$$R-C{\equiv}N \xrightarrow[NaOH,\ H_2O]{H_3O^+} R-\overset{O}{\underset{}{C}}-OH$$

2.   Reactions of carboxylic acids (Section 15.6)

   Reduction with LiAlH$_4$ to give alcohols

$$R-\overset{O}{\underset{}{C}}-OH \xrightarrow[2.\ H_3O^+]{1.\ LiAlH_4} R-\underset{\underset{OH}{}}{\overset{H\ \ H}{C}}-OH$$

3.   Preparation of nitriles (Section 15.7)

   (a) S$_N$2 reaction of alkyl halides

$$RCH_2Br \xrightarrow{NaCN} RCH_2C{\equiv}N$$

   (b) Dehydration of amides

$$R-\overset{O}{\underset{}{C}}-NH_2 \xrightarrow{SOCl_2} R-C{\equiv}N\ +\ SO_2\ +\ 2\ HCl$$

4.   Reactions of nitriles (Section 15.7)

(a) Hydrolysis to give amides

$$R-C\equiv N \xrightarrow[\text{or NaOH, H}_2\text{O}]{\text{H}_3\text{O}^+} R-\overset{\overset{\displaystyle O}{\parallel}}{C}-NH_2$$

(b) Reduction to give amines

$$R-C\equiv N \xrightarrow[\text{2. H}_2\text{O}]{\text{1. LiAlH}_4} R-\overset{\overset{\displaystyle H\ \ H}{\diagup}}{C}-NH_2$$

(c) Reaction with Grignard reagents to yield ketones

$$R-C\equiv N \xrightarrow[\text{2. H}_3\text{O}^+]{\text{1. R'MgX, ether}} R-\overset{\overset{\displaystyle O}{\parallel}}{C}-R' \ + \ NH_3$$

Lagniappe

Vitamin C

Vitamin C, or ascorbic acid, is surely the best known of all vitamins. It was the first vitamin to be discovered (1928), the first to be structurally characterized (1933), and the first to be synthesized in the laboratory (1933). Over 200 million pounds of vitamin C are synthesized worldwide each year—more than the total amount of all other vitamins combined. In addition to its use as a vitamin supplement, vitamin C is used as a food preservative, a "flour improver" in bakeries, and an animal food additive.

Vitamin C is perhaps most famous for its antiscorbutic properties, meaning that it prevents the onset of scurvy, a bleeding disease affecting those with a deficiency of fresh vegetables and citrus fruits in their diet. Sailors in the Age of Exploration were particularly susceptible to scurvy, and the death toll was high. The Portuguese explorer Vasco da Gama lost more than half his crew to scurvy during his 2-year voyage around the Cape of Good Hope in 1497–1499.

In more recent times, large doses of vitamin C have been claimed to prevent the common cold, cure infertility, delay the onset of symptoms in acquired immunodeficiency syndrome (AIDS), and inhibit the development of gastric and cervical cancers. None of these claims have been backed by medical evidence, however. In the largest study yet done of the effect of vitamin C on the common cold, a meta-analysis of more than 100 separate trials covering 40,000 people found no difference in the incidence of colds between those who took supplemental vitamin C regularly and those who did not. When taken *during* a cold, however, vitamin C does appear to decrease the cold's duration by 8%.

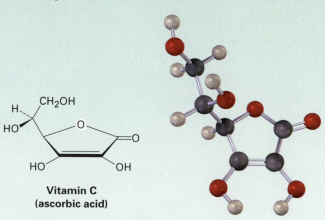

Vitamin C
(ascorbic acid)

continued

Lagniappe *continued*

The industrial preparation of vitamin C involves an unusual blend of biological and laboratory organic chemistry. The Hoffmann-La Roche company synthesizes ascorbic acid from glucose through the five-step route shown in Figure 15.6. Glucose, a pentahydroxy aldehyde, is first reduced to sorbitol, which is then oxidized by the microorganism *Acetobacter suboxydans*. No chemical reagent is known that is selective enough to oxidize only one of the six alcohol groups in sorbitol, so an enzymatic reaction is used. Treatment with acetone and an acid catalyst then converts four of the other hydroxyl groups into acetal linkages, and the remaining hydroxyl group is chemically oxidized to a carboxylic acid by reaction with aqueous NaOCl (household bleach). Hydrolysis with acid then removes the two acetal groups and causes an internal ester-forming reaction to take place to give ascorbic acid. Each of the five steps takes place in better than 90% yield.

In addition to the hazards of weather, participants in early polar expeditions often suffered from scurvy, caused by a vitamin C deficiency.

Underwood & Underwood/CORBIS

FIGURE 15.6 The industrial synthesis of ascorbic acid from glucose.

Exercises

VISUALIZING CHEMISTRY

(Problems 15.1–15.14 appear within the chapter.)

15.15 ■ Give IUPAC names for the following carboxylic acids (reddish brown = Br):

■ indicates problems that are assignable in Organic OWL.

Go to this book's companion website at **www.cengage.com/ chemistry/mcmurry** to explore interactive versions of the Active Figures from this text.

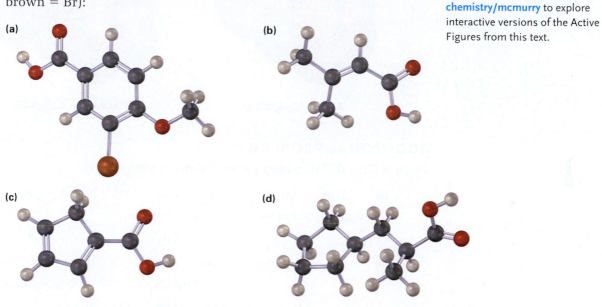

(a)   **(b)**

(c)   **(d)**

15.16 Would you expect the following carboxylic acids to be more acidic or less acidic than benzoic acid? Explain. (Reddish brown = Br.)

(a)   **(b)**

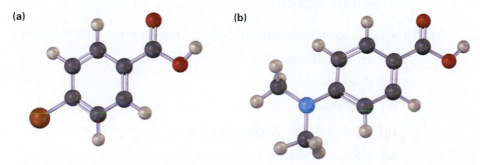

15.17 The following carboxylic acid can't be prepared from an alkyl halide by either the nitrile hydrolysis route or the Grignard carboxylation route. Explain.

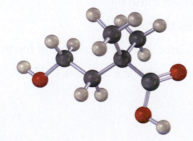

15.18 Electrostatic potential maps of anisole and thioanisole are shown. Which do you think is the stronger acid, *p*-methoxybenzoic acid or *p*-(methylthio)benzoic acid? Explain.

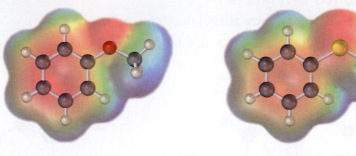

Anisole (C₆H₅OCH₃)   **Thioanisole (C₆H₅SCH₃)**

ADDITIONAL PROBLEMS

15.19 ■ Give IUPAC names for the following compounds:

(a) 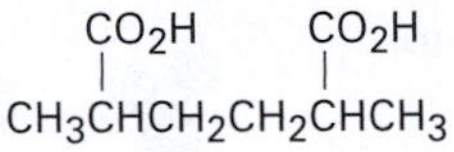 CO_2H CO_2H
    $CH_3CHCH_2CH_2CHCH_3$

(b) 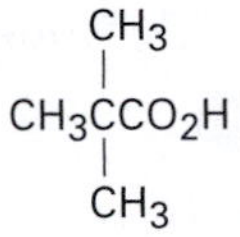 CH_3
    CH_3CCO_2H
       CH_3

(c) 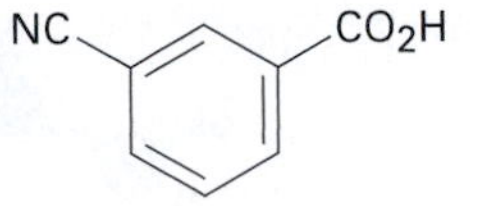 NC ⬡ CO_2H

(d) 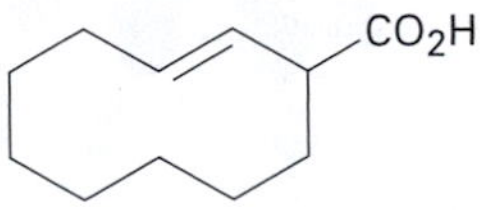 ⬡ CO_2H

(e) 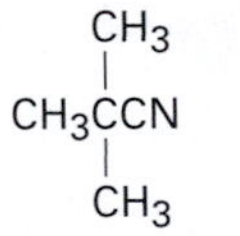 CH_3
    CH_3CCN
      CH_3

(f) 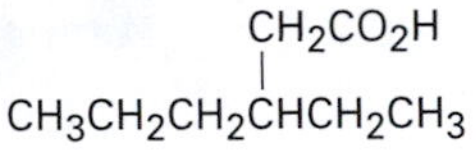 CH_2CO_2H
   $CH_3CH_2CH_2CHCH_2CH_3$

(g) 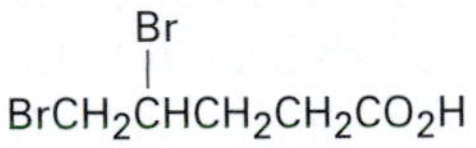 Br
  $BrCH_2CHCH_2CH_2CO_2H$

(h) 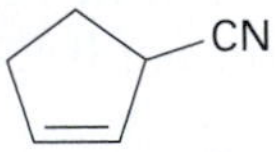 ⬠ CN

15.20 ■ Draw structures corresponding to the following IUPAC names:

(a) *cis*-Cyclohexane-1,2-dicarboxylic acid

(b) Heptanedioic acid

(c) Hex-2-en-4-ynoic acid

(d) 4-Ethyl-2-propyloctanoic acid

(e) 3-Chlorophthalic acid

(f) Triphenylacetic acid

(g) Cyclobut-2-enecarbonitrile

(h) *m*-Benzoylbenzonitrile

15.21 Draw and name the following:

(a) The eight carboxylic acids with the formula $C_6H_{12}O_2$

(b) Three nitriles with the formula C_5H_7N

15.22 ▪ Order the compounds in each set in order of increasing acidity:

(a) Acetic acid, oxalic acid, formic acid

(b) *p*-Bromobenzoic acid, *p*-nitrobenzoic acid, 2,4-dinitrobenzoic acid

(c) Fluoroacetic acid, 3-fluoropropanoic acid, 4-fluorobutanoic acid

15.23 ▪ Arrange the compounds in each set in order of increasing basicity:

(a) Magnesium acetate, magnesium hydroxide, methylmagnesium bromide

(b) Sodium benzoate, sodium *p*-nitrobenzoate, sodium acetylide

(c) Lithium hydroxide, lithium ethoxide, lithium formate

15.24 Isocitric acid, an intermediate in the citric acid cycle of food metabolism, has the systematic name (2*R*,3*S*)-3-carboxy-2-hydroxypentanedioic acid. Draw the structure.

15.25 ▪ How could you convert butanoic acid into the following compounds? More than one step may be required.

(a) Butan-1-ol (b) 1-Bromobutane

(c) Pentanoic acid (d) But-1-ene

15.26 ▪ How could you convert each of the following compounds into butanoic acid? More than one step may be required.

(a) Butan-1-ol (b) 1-Bromobutane

(c) But-1-ene (d) 1-Bromopropane

15.27 ▪ How could you convert butanenitrile into the following compounds? More than one step may be required.

(a) Butan-1-ol (b) Butanal (c) Pentanoic acid

15.28 ▪ How would you prepare the following compounds from benzene? More than one step is required in each case.

(a) *m*-Chlorobenzoic acid (b) *p*-Bromobenzoic acid

(c) Phenylacetic acid, $C_6H_5CH_2CO_2H$

15.29 ▪ Calculate pK_a's for the following acids:

(a) Lactic acid, $K_a = 8.4 \times 10^{-4}$ (b) Acrylic acid, $K_a = 5.6 \times 10^{-6}$

15.30 ▪ Calculate K_a's for the following acids:

(a) Citric acid, $pK_a = 3.14$ (b) Tartaric acid, $pK_a = 2.98$

15.31 ▪ Thioglycolic acid, $HSCH_2CO_2H$, a substance used in depilatory agents (hair removers) has $pK_a = 3.42$. What is the percent dissociation of thioglycolic acid in a buffer solution at pH = 3.00?

▪ Problems assignable in Organic OWL.

15.32 ■ In humans, the final product of purine degradation from DNA is uric acid, $pK_a = 5.61$, which is excreted in the urine. What is the percent dissociation of uric acid in urine at a typical pH = 6.0? Why do you think uric acid is acidic even though it does not have a CO_2H group?

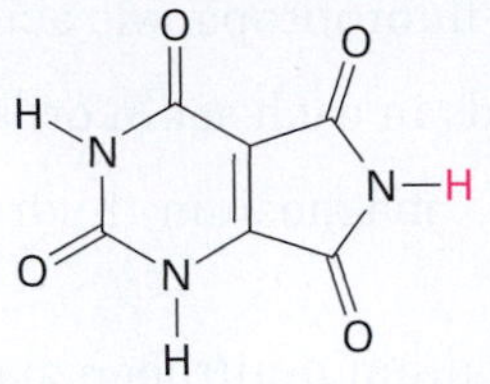

Uric acid

15.33 Following are some pK_a data for simple dibasic acids. How can you account for the fact that the difference between the first and second ionization constants decreases with increasing distance between the carboxyl groups?

Name	Structure	pK_{a1}	pK_{a2}
Oxalic	HO_2CCO_2H	1.2	4.2
Succinic	$HO_2C(CH_2)_2CO_2H$	4.2	5.6
Adipic	$HO_2C(CH_2)_4CO_2H$	4.4	5.4

15.34 ■ Predict the product of the reaction of *p*-methylbenzoic acid with each of the following:

(a) CH_3MgBr in ether, then H_3O^+ **(b)** $KMnO_4$, H_3O^+

(c) $LiAlH_4$, then H_3O^+

15.35 Using $^{13}CO_2$ as your only source of labeled carbon, along with any other compounds needed, how would you synthesize the following compounds?

(a) $CH_3CH_2{}^{13}CO_2H$ **(b)** $CH_3{}^{13}CH_2CO_2H$

15.36 Nitriles can be prepared from aldehydes by a two-step procedure that involves formation of an imine with NH_2OH, followed by dehydration with $SOCl_2$. Show the mechanism of the reaction.

■ Problems assignable in Organic OWL.

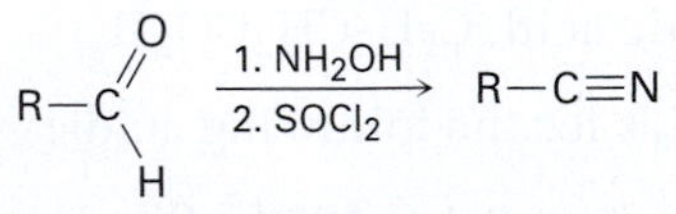

15.37 In plants, terpenes (see Chapter 7 *Lagniappe*) are biosynthesized by a pathway that involves loss of CO_2 from 3-phosphomevalonate 5-diphosphate to yield isopentenyl diphosphate. Use curved arrows to show the mechanism of this reaction.

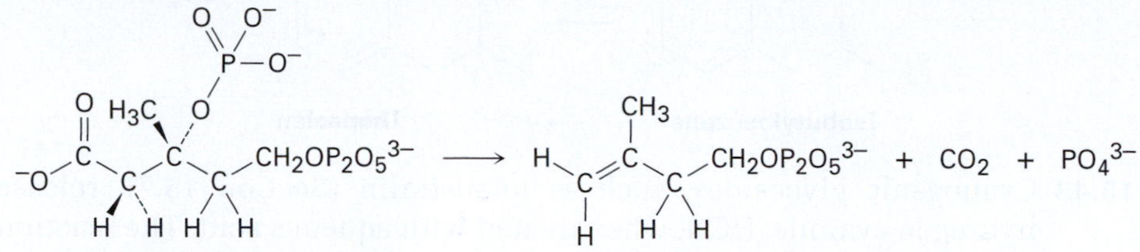

**3-Phosphomevalonate
5-diphosphate**   **Isopentenyl
diphosphate**

15.38 How would you carry out the following transformations?

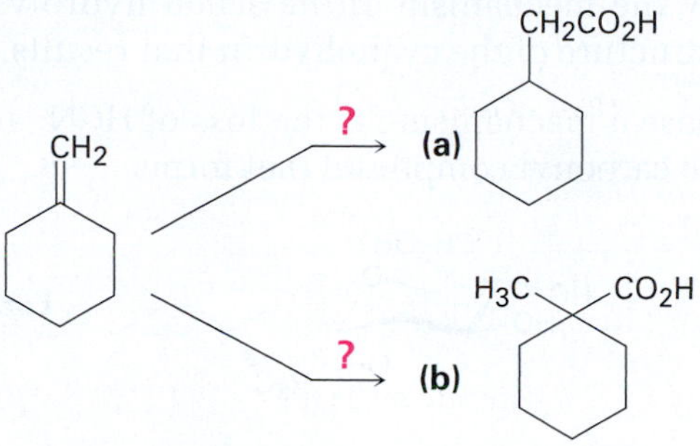

15.39 Which method—Grignard carboxylation or nitrile hydrolysis—would you use for each of the following reactions? Explain.

(a)

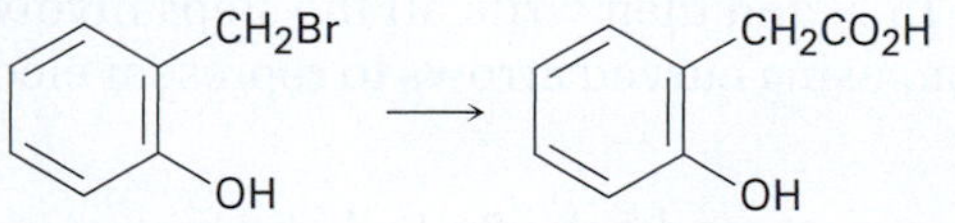

(b)

$$CH_3CH_2\overset{\overset{\displaystyle Br}{|}}{C}HCH_3 \longrightarrow CH_3CH_2\overset{\overset{\displaystyle CH_3}{|}}{C}HCO_2H$$

(c)

$$CH_3\overset{\overset{\displaystyle O}{\|}}{C}CH_2CH_2CH_2I \longrightarrow CH_3\overset{\overset{\displaystyle O}{\|}}{C}CH_2CH_2CH_2CO_2H$$

(d) $HOCH_2CH_2CH_2Br \longrightarrow HOCH_2CH_2CH_2CO_2H$

15.40 Hexane-1,6-diamine, a starting material needed for making nylon, can be made from buta-1,3-diene. How would you accomplish this synthesis?

$$H_2C{=}CHCH{=}CH_2 \overset{?}{\longrightarrow} H_2NCH_2CH_2CH_2CH_2CH_2CH_2NH_2$$

15.41 A chemist in need of 2,2-dimethylpentanoic acid decided to synthesize some by reaction of 2-chloro-2-methylpentane with NaCN, followed by hydrolysis of the product. After the reaction sequence was carried out, however, none of the desired product could be found. What do you suppose went wrong?

■ Problems assignable in Organic OWL.

15.42 Show how you might prepare the anti-inflammatory agent ibuprofen starting from isobutylbenzene. More than one step is needed.

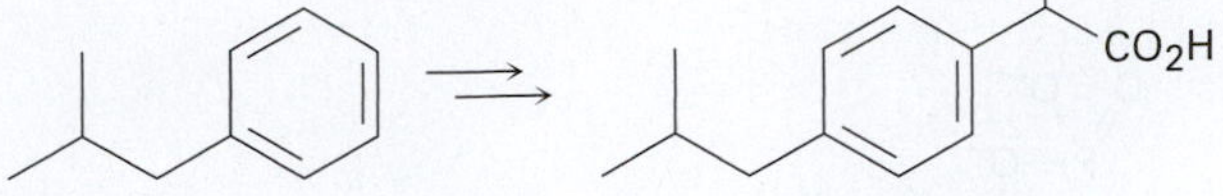

Isobutylbenzene   **Ibuprofen**

15.43 Cyanogenic glycosides, such as lotaustralin (Section 15.7), release hydrogen cyanide, HCN, when treated with aqueous acid. The reaction occurs by hydrolysis of the acetal linkage to form a cyanohydrin—a compound with a hydroxyl group and a cyano group bonded to the same carbon. The cyanohydrin then expels HCN and gives a carbonyl compound.

   (a) Show the mechanism of the acetal hydrolysis (Section 14.8) and the structure of the cyanohydrin that results.

   (b) Propose a mechanism for the loss of HCN, and show the structure of the carbonyl compound that forms.

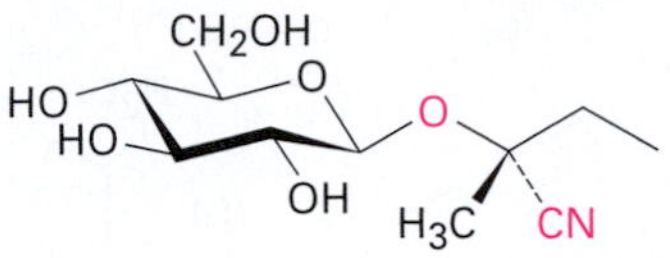

Lotaustralin

15.44 Acid-catalyzed hydrolysis of a nitrile to give a carboxylic acid occurs by initial protonation of the nitrogen atom, followed by nucleophilic addition of water. Review the mechanism of base-catalyzed nitrile hydrolysis in Section 15.7, and then write all the steps involved in the acid-catalyzed reaction, using curved arrows to represent electron flow in each step.

15.45 Predict the product of reaction of lithocholic acid, a steroid found in human bile, with each of the following reagents. Don't worry about the size of the molecule; concentrate on the functional groups.

   (a) CrO_3, H_3O^+ **(b)** CH_3MgBr, then H_3O^+ **(c)** $LiAlH_4$, then H_3O^+

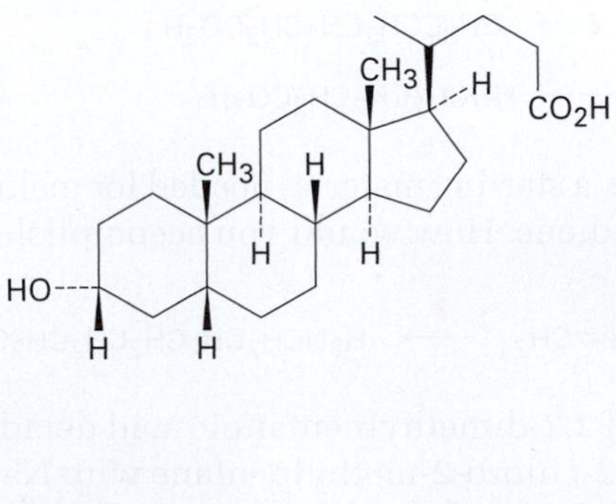

Lithocholic acid

■ Problems assignable in Organic OWL.

15.46 The pK_a's of five *p*-substituted benzoic acids (YC$_6$H$_4$CO$_2$H) follow. Rank the corresponding substituted benzenes (YC$_6$H$_5$) in order of their increasing reactivity toward electrophilic aromatic substitution. If benzoic acid has pK_a = 4.19, which of the substituents are activators and which are deactivators?

Substituent Y	pK_a of Y—⬡—CO$_2$H 
—Si(CH$_3$)$_3$	4.27
—CH=CHC≡N	4.03
—HgCH$_3$	4.10
—OSO$_2$CH$_3$	3.84
—PCl$_2$	3.59

15.47 The following pK_a values have been measured. Explain why a hydroxyl group in the para position decreases the acidity while a hydroxyl group in the meta position increases the acidity.

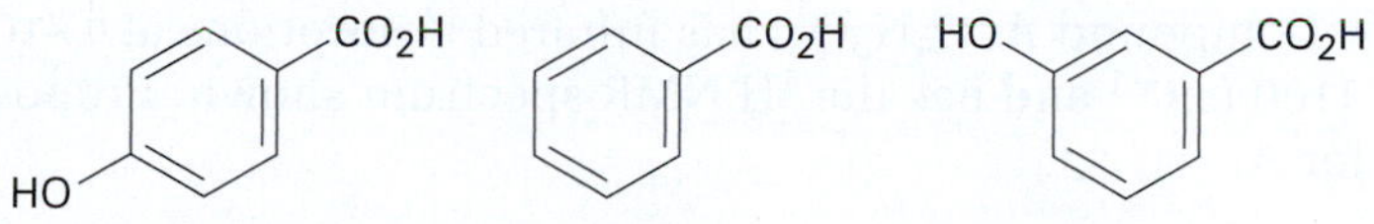

pK_a = 4.48   pK_a = 4.19   pK_a = 4.07

15.48 Identify the missing reagents **a**–**e** in the following scheme:

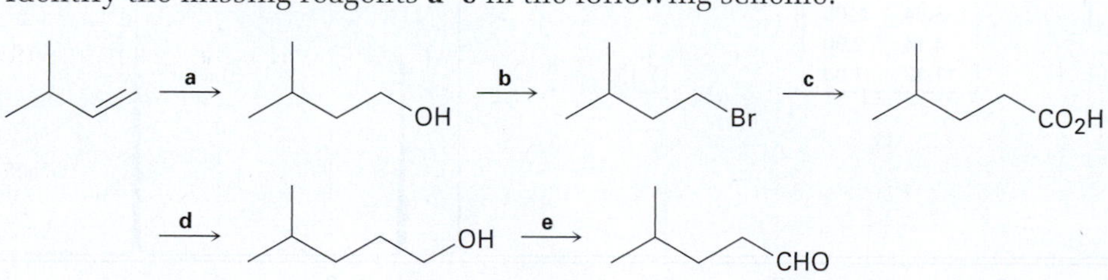

15.49 ■ 2-Bromo-6,6-dimethylcyclohexanone gives 2,2-dimethylcyclopentanecarboxylic acid on treatment with aqueous NaOH followed by acidification, a process called the *Favorskii reaction*. The reaction takes place by initial nucleophilic addition to the carbonyl group, followed by a rearrangement with loss of bromide ion. Propose a mechanism.

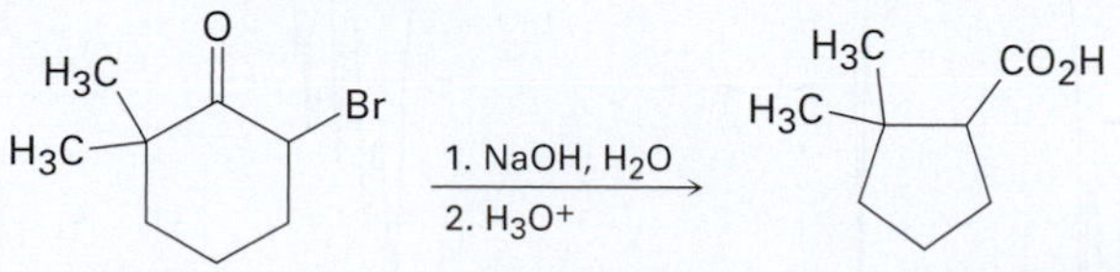

1. NaOH, H$_2$O
2. H$_3$O$^+$

15.50 ■ Propose a structure for a compound C$_6$H$_{12}$O$_2$ that dissolves in dilute NaOH and shows the following ^{1}H NMR spectrum: 1.08 δ (9 H, singlet), 2.2 δ (2 H, singlet), and 11.2 δ (1 H, singlet).

■ Problems assignable in Organic OWL.

15.51 What spectroscopic method could you use to distinguish among the following three isomeric acids? Tell what characteristic features you would expect for each acid.

$CH_3(CH_2)_3CO_2H$ $(CH_3)_2CHCH_2CO_2H$ $(CH_3)_3CCO_2H$

Pentanoic acid   **3-Methylbutanoic acid**   **2,2-Dimethylpropanoic acid**

15.52 How could you use NMR (either ^{13}C or ^{1}H) to distinguish between the following isomeric pairs?

(a)

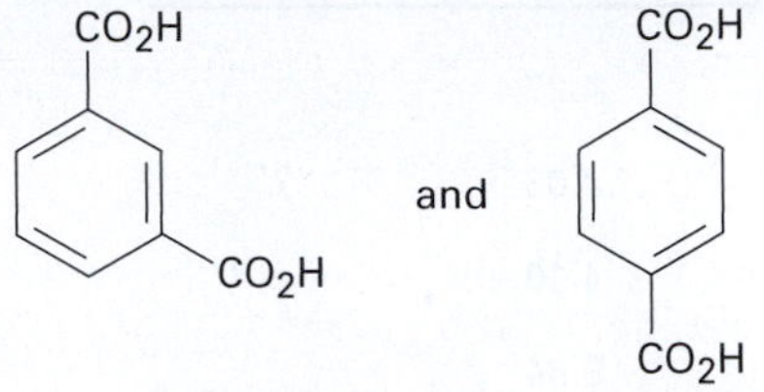

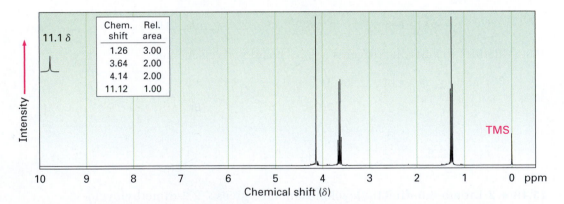

(b) $HO_2CCH_2CH_2CO_2H$ and $CH_3CH(CO_2H)_2$

(c) $CH_3CH_2CH_2CO_2H$ and $HOCH_2CH_2CH_2CHO$

(d) $(CH_3)_2C{=}CHCH_2CO_2H$ and $\bigcirc\!\!-CO_2H$

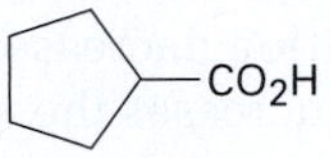

15.53 ■ Compound **A**, $C_4H_8O_3$, has infrared absorptions at 1710 and 2500 to 3100 cm^{-1} and has the ^{1}H NMR spectrum shown. Propose a structure for **A**.

Chem. shift	Rel. area
1.26	3.00
3.64	2.00
4.14	2.00
11.12	1.00

11.1 δ

15.54 ■ Propose a structure for a compound, C_4H_7N, that has the following IR and ^{1}H NMR spectra:

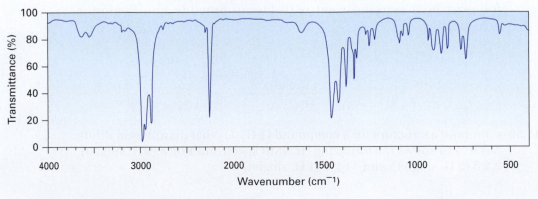

■ Problems assignable in Organic OWL.

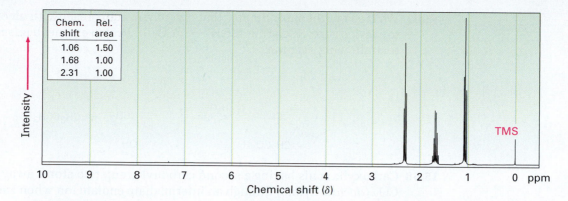

15.55 ■ The two ^{1}H NMR spectra shown here belong to crotonic acid (*trans*-CH$_3$CH=CHCO$_2$H) and methacrylic acid [H$_2$C=C(CH$_3$)CO$_2$H]. Which spectrum corresponds to which acid? Explain.

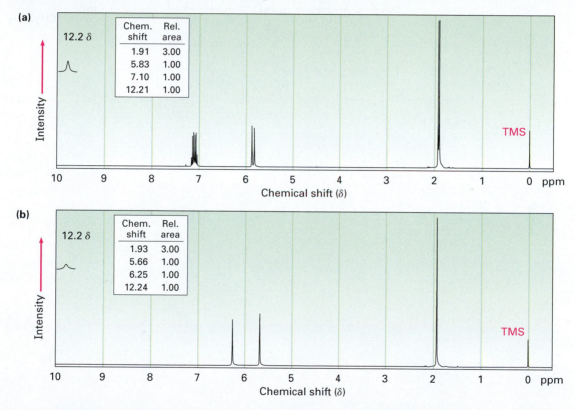

15.56 ■ Propose structures for carboxylic acids that show the following peaks in their ^{13}C NMR spectra. Assume that the kinds of carbons (1°, 2°, 3°, or 4°) have been assigned by DEPT-NMR.

(a) C$_7$H$_{12}$O$_2$: 25.5 δ (2°), 25.9 δ (2°), 29.0 δ (2°), 43.1 δ (3°), 183.0 δ (4°)

(b) C$_8$H$_8$O$_2$: 21.4 δ (1°), 128.3 δ (4°), 129.0 δ (3°), 129.7 δ (3°), 143.1 δ (4°), 168.2 δ (4°)

■ Problems assignable in Organic OWL.

15.57 We'll see in the next chapter that carboxylic acids react with alcohols to yield esters: RCO$_2$H + R'OH → RCO$_2$R'. Propose a mechanism for the following reaction.

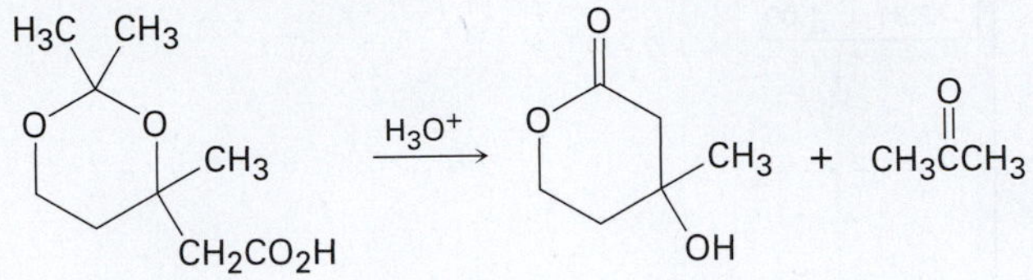

15.58 Carboxylic acids having a second carbonyl group two atoms away lose CO$_2$ *(decarboxylate)* through an intermediate enolate ion when treated with base. Write the mechanism of this decarboxylation reaction using curved arrows to show the electron flow in each step.

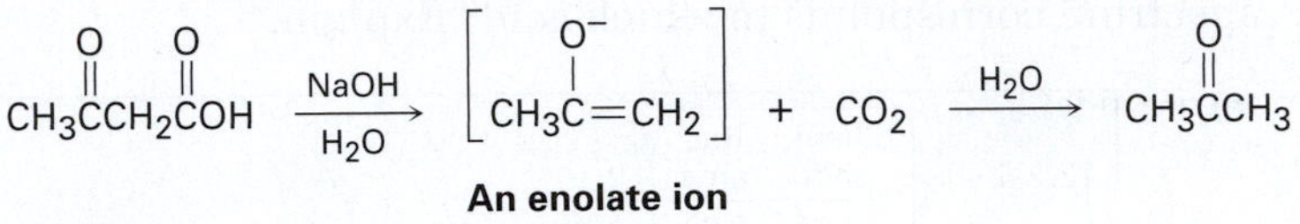

An enolate ion

16 Carboxylic Acid Derivatives: Nucleophilic Acyl Substitution Reactions

Pancreatic lipase catalyzes the hydrolysis of ester bonds in dietary fats, the initial step in fat metabolism.

Closely related to the carboxylic acids and nitriles discussed in the previous chapter are the **carboxylic acid derivatives**, compounds in which an acyl group is bonded to an electronegative atom or substituent that can act as a leaving group in the nucleophilic acyl substitution reaction that we saw briefly in the *Preview of Carbonyl Chemistry*:

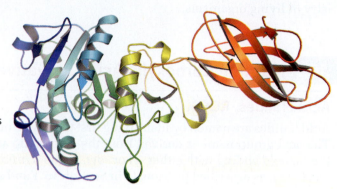

Many kinds of acid derivatives are known, but we'll be concerned primarily with four of the more common ones: **acid halides**, **acid anhydrides**, **esters**, and **amides**. Acid halides and acid anhydrides are used only in the laboratory, while esters and amides are common in both laboratory and biological chemistry. In addition, carboxylic acid derivatives called **thioesters** and **acyl phosphates** are encountered primarily in biological chemistry. Note the structural similarity between acid anhydrides and acyl phosphates.

Carboxylic acid   Acid halide (X = Cl, Br)   Acid anhydride   Ester

Amide   Thioester   Acyl phosphate

OWL Online homework for this chapter can be assigned in Organic OWL.

643

WHY THIS CHAPTER?

Carboxylic acid derivatives are among the most widely occurring of all molecules, and their primary reaction—nucleophilic acyl substitution—is one of the four fundamental carbonyl-group processes. Nucleophilic acyl substitutions are found in one variation or another in almost all biological pathways, so a detailed understanding of them is necessary for understanding the chemistry of living organisms.

16.1 Naming Carboxylic Acid Derivatives

Acid Halides, RCOX

Acid halides are named by identifying first the acyl group and then the halide. The acyl group name is derived from the carboxylic acid name by replacing the *-ic acid* ending with either *-yl* or *-oyl*, or the *-carboxylic acid* ending with *-carbonyl*, as described previously in Section 15.1 and shown in Table 15.1 on page 612.

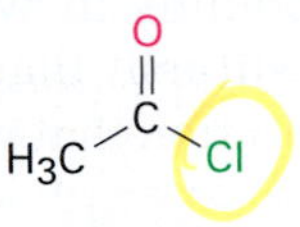

Acetyl chloride 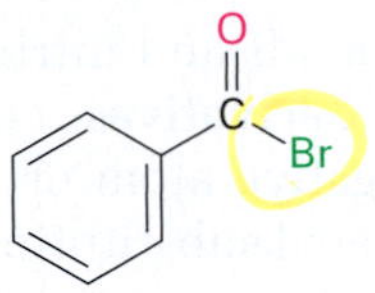   **Benzoyl bromide** 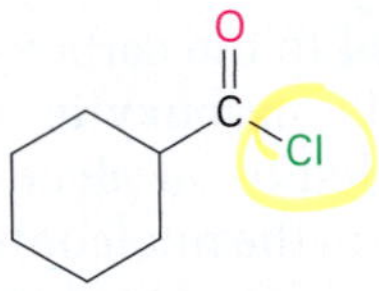   **Cyclohexanecarbonyl chloride**

Acid Anhydrides, RCO$_2$COR′

Symmetrical anhydrides of unsubstituted monocarboxylic acids and cyclic anhydrides of dicarboxylic acids are named by replacing the word *acid* with *anhydride*:

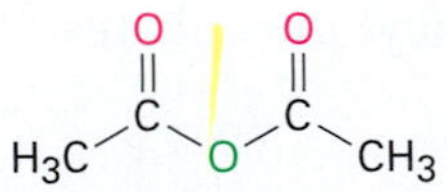

Acetic anhydride 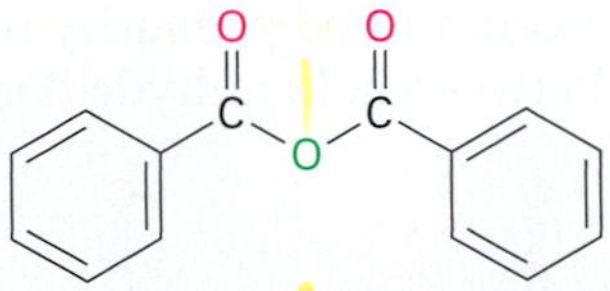   **Benzoic anhydride** 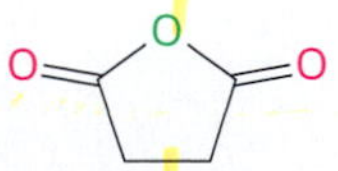   **Succinic anhydride**

Unsymmetrical anhydrides—those prepared from two different carboxylic acids—are named by citing the two acids alphabetically and then adding *anhydride*:

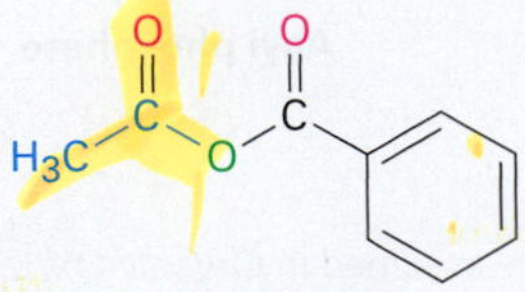

Acetic benzoic anhydride

Esters, RCO₂R′

Esters are named by first identifying the alkyl group attached to oxygen and then the carboxylic acid, with the *-ic acid* ending replaced by *-ate*:

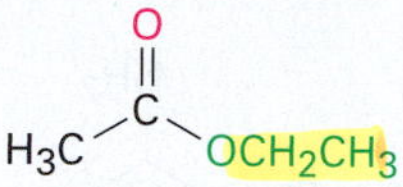

Ethyl acetate 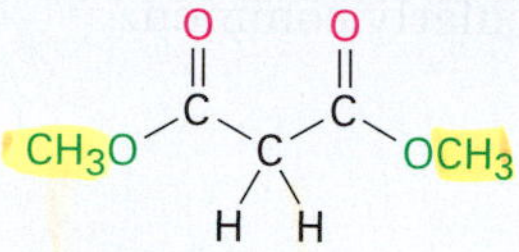   **Dimethyl malonate** 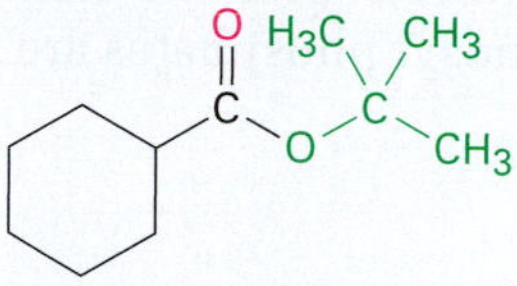   *tert*-**Butyl cyclohexane-carboxylate**

Amides, RCONH₂

Amides with an unsubstituted –NH₂ group are named by replacing the *-oic acid* or *-ic acid* ending with *-amide*, or by replacing the *-carboxylic acid* ending with *-carboxamide.*

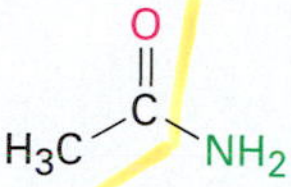

Acetamide 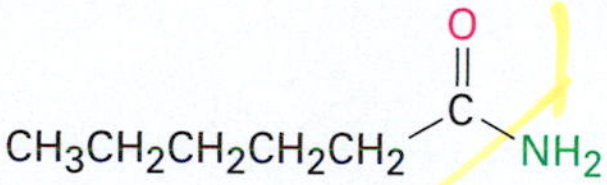   **Hexanamide** 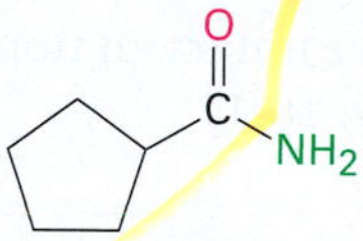   **Cyclopentane-carboxamide**

If the nitrogen atom is further substituted, the compound is named by first identifying the substituent groups and then the parent amide. The substituents are preceded by the letter *N* to identify them as being directly attached to nitrogen.

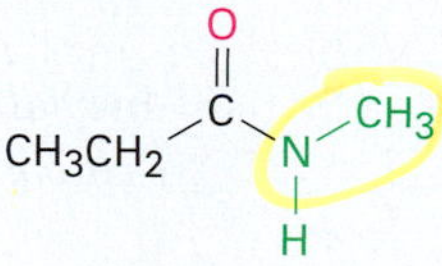

N-**Methylpropanamide** 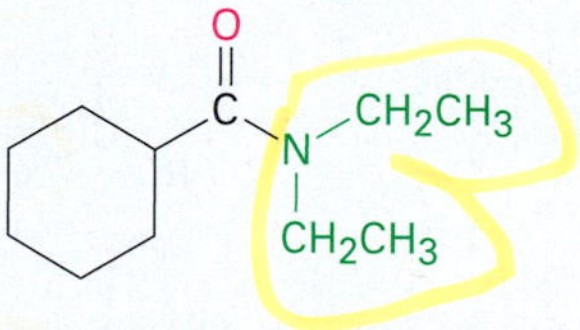   *N,N*-**Diethylcyclohexanecarboxamide**

Thioesters, RCOSR′

Thioesters are named like the corresponding esters. If the related ester has a common name, the prefix *thio-* is added to the name of the carboxylate: acetate becomes thioacetate, for instance. If the related ester has a systematic name, the *-oate* or *-carboxylate* ending is replaced by *-thioate* or *-carbothioate:* butanoate becomes butanethioate and cyclohexanecarboxylate becomes cyclohexanecarbothioate, for instance.

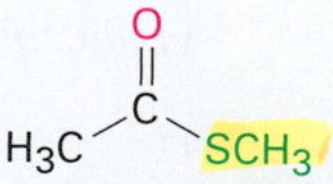

Methyl thioacetate 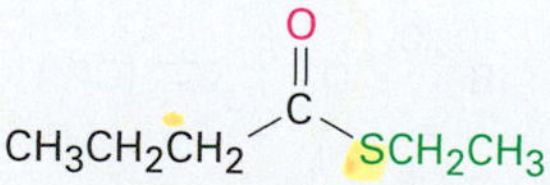   **Ethyl butanethioate** 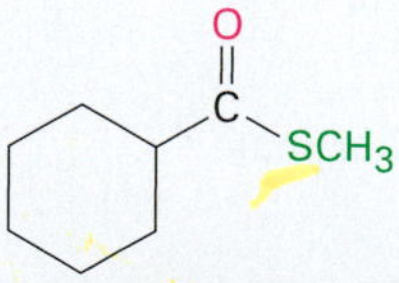   **Methyl cyclohexane-carbothioate**

Acyl Phosphates, RCO$_2$PO$_3$$^{2-}$ and RCO$_2$PO$_3$R'$^-$

Acyl phosphates are named by citing the acyl group and adding the word *phosphate.* If an alkyl group is attached to one of the phosphate oxygens, it is identified after the name of the acyl group. In biological chemistry, acyl adenosyl phosphates are particularly common.

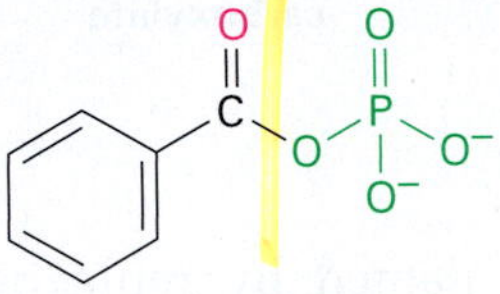

Benzoyl phosphate

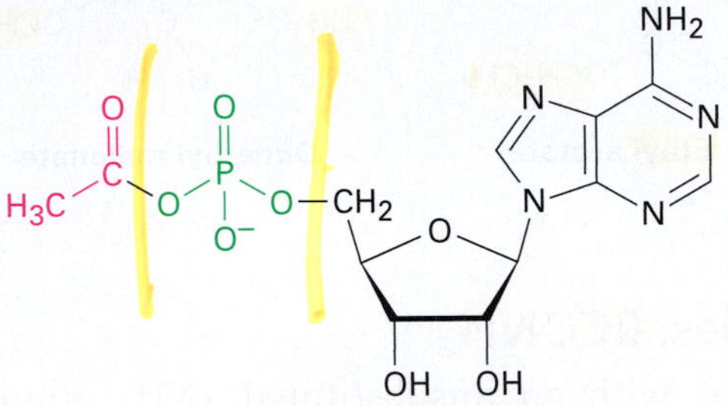

Acetyl adenosyl phosphate

A summary of nomenclature rules for carboxylic acid derivatives is given in Table 16.1.

TABLE 16.1
Nomenclature of Carboxylic Acid Derivatives

Functional group	Structure	Name ending
Carboxylic acid		-ic acid (-carboxylic acid)
Acid halide		-oyl halide (-carbonyl halide)
Acid anhydride		anhydride
Ester		-ate (-carboxylate)
Amide		-amide (-carboxamide)
Thioester		-thioate (-carbothioate)
Acyl phosphate		-yl phosphate

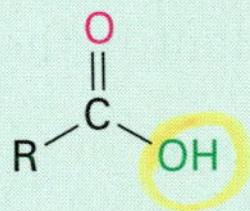

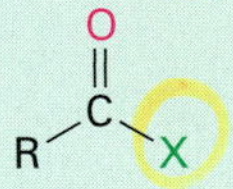

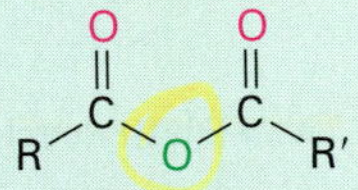

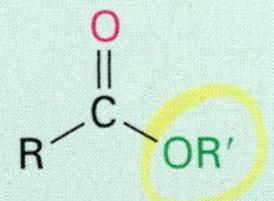

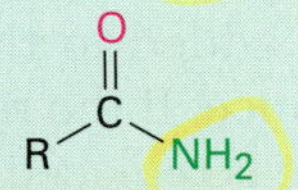

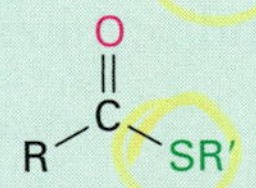

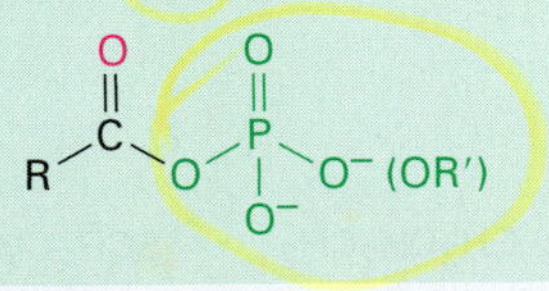

Problem 16.1

Give IUPAC names for the following substances:

(a)

$$CH_3CHCH_2CH_2CCl$$
with CH_3 above first carbon and O (double bond) above the carbonyl carbon

(b) cyclohexane with CH_2CNH_2 substituent, O double-bonded to the C

(c)

$$CH_3CHCOCHCH_3$$
with O (double bond) above the carbonyl C and CH_3 CH_3 below the two CH carbons

(d) benzene ring with
$$\left(\begin{array}{c} O \\ \| \\ C \end{array} \right)_2 O$$

(e) cyclopentane with
$$\begin{array}{c} O \\ \| \\ C-OCHCH_3 \\ CH_3 \end{array}$$

(f) cyclopentane with
$$\begin{array}{c} CH_3 \\ | \\ O \quad CHCH_3 \\ \| \\ C \end{array}$$

(g)

$$H_2C{=}CHCH_2CH_2CNHCH_3$$
with O (double bond) above the carbonyl C

(h)

$$\begin{array}{c} O \\ H_3C \quad \| \\ \diagdown C \diagup C \diagdown OPO_3{}^{2-} \\ HO \quad H \end{array}$$

(i)

$$\begin{array}{c} O \\ H_3C \quad \| \\ \diagdown C{-}SCH_2CH_3 \\ C{=}C \\ H_3C \quad CH_3 \end{array}$$

Problem 16.2

Draw structures corresponding to the following names:
(a) Phenyl benzoate
(b) *N*-Ethyl-*N*-methylbutanamide
(c) 2,4-Dimethylpentanoyl chloride
(d) Methyl 1-methylcyclohexanecarboxylate
(e) Ethyl 3-oxopentanoate
(f) Methyl *p*-bromothiobenzoate
(g) Formic propanoic anhydride
(h) *cis*-2-Methylcyclopentanecarbonyl bromide

16.2 Nucleophilic Acyl Substitution Reactions

The addition of a nucleophile to a polar C=O bond is the key step in three of the four major carbonyl-group reactions. We saw in Chapter 14 that when a nucleophile adds to an aldehyde or ketone, the initially formed tetrahedral intermediate either can be protonated to yield an alcohol or can eliminate the carbonyl oxygen, leading to a new C=Nu bond. When a nucleophile adds to a carboxylic acid derivative, however, a different reaction course is followed. The initially formed tetrahedral intermediate eliminates one of the two substituents originally bonded to the carbonyl carbon, leading to a net **nucleophilic acyl substitution reaction** (Figure 16.1).

The difference in behavior between aldehydes/ketones and carboxylic acid derivatives is a consequence of structure. Carboxylic acid derivatives have an acyl carbon bonded to a group −Y that can leave as a stable anion. As soon as the tetrahedral intermediate is formed, the leaving group is expelled

FIGURE 16.1 MECHANISM:
General mechanism of a nucleophilic acyl substitution reaction.

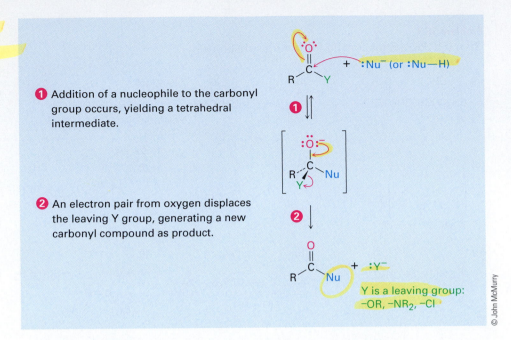

❶ Addition of a nucleophile to the carbonyl group occurs, yielding a tetrahedral intermediate.

❷ An electron pair from oxygen displaces the leaving Y group, generating a new carbonyl compound as product.

Y is a leaving group: −OR, −NR₂, −Cl

© John McMurry

to generate a new carbonyl compound. Aldehydes and ketones have no such leaving group, however, and therefore don't undergo substitution.

A leaving group

A carboxylic acid derivative

NOT a leaving group

An aldehyde   A ketone

The net effect of the addition/elimination sequence is a substitution of the nucleophile for the −Y group originally bonded to the acyl carbon. Thus, the overall reaction is superficially similar to the kind of nucleophilic substitution that occurs during an S_N2 reaction (Section 12.6), but the mechanisms of the two processes are completely different. An S_N2 reaction occurs in a single step by backside displacement of the leaving group, whereas a nucleophilic acyl substitution takes place in two steps and involves a tetrahedral intermediate.

Problem 16.3

Show the mechanism of the following nucleophilic acyl substitution reaction, using curved arrows to indicate the electron flow in each step:

Relative Reactivity of Carboxylic Acid Derivatives

Both the initial addition step and the subsequent elimination step can affect the overall rate of a nucleophilic acyl substitution reaction, but the addition step is generally the rate-limiting one. Thus, any factor that makes the carbonyl group more reactive toward nucleophiles favors the substitution process.

Steric and electronic factors are both important in determining reactivity. Sterically, we find within a series of similar acid derivatives that unhindered, accessible carbonyl groups react with nucleophiles more readily than do sterically hindered groups. The reactivity order is:

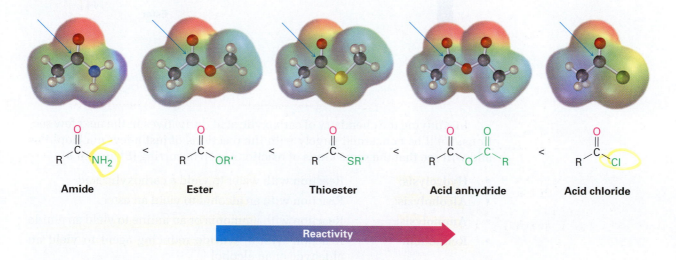

Electronically, we find that strongly polarized acyl compounds react more readily than less polar ones. Thus, acid chlorides are the most reactive because the electronegative chlorine atom withdraws electrons from the carbonyl carbon, whereas amides are the least reactive. Although subtle, electrostatic potential maps of various carboxylic acid derivatives indicate the differences by the relative blueness on the C=O carbons. Acyl phosphates are hard to place on this scale because they are not often used in the laboratory, but in biological systems they appear to be somewhat more reactive than thioesters.

The way in which various substituents affect the polarization of a carbonyl group is similar to the way they affect the reactivity of an aromatic ring toward electrophilic substitution (Section 9.8). A chlorine substituent, for example, inductively withdraws electrons from an acyl group in the same way that it withdraws electrons from and thus deactivates an aromatic ring. Similarly, amino, methoxyl, and methylthio substituents donate electrons to acyl groups by resonance in the same way that they donate electrons to and thus activate aromatic rings.

As a consequence of these reactivity differences, it's usually possible to convert a more reactive acid derivative into a less reactive one. Acid chlorides, for instance, can be directly converted into anhydrides, thioesters, esters, and amides, but amides can't be directly converted into esters, thioesters, anhydrides, or acid chlorides. Remembering the reactivity order is therefore a way to keep track of a large number of reactions (Figure 16.2). Another consequence, as noted previously, is that only acyl phosphates, thioesters, esters, and amides are commonly found in nature. Acid halides and acid anhydrides react with water so rapidly that they can't exist for long in living organisms.

FIGURE 16.2 Interconversions of carboxylic acid derivatives. A more reactive acid derivative can be converted into a less reactive one, but not vice versa.

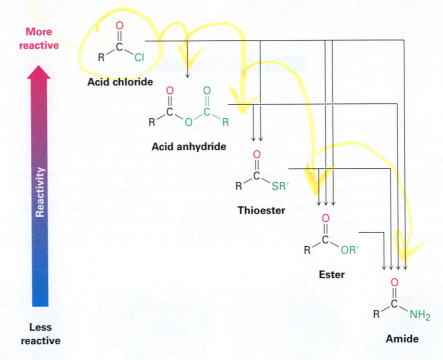

In studying the chemistry of carboxylic acid derivatives in the next few sections, we'll be concerned largely with the reactions of just a few nucleophiles and will see that the same kinds of reactions keep occurring (Figure 16.3).

- **Hydrolysis:** Reaction with water to yield a carboxylic acid
- **Alcoholysis:** Reaction with an alcohol to yield an ester
- **Aminolysis:** Reaction with ammonia or an amine to yield an amide
- **Reduction:** Reaction with a hydride reducing agent to yield an aldehyde or an alcohol
- **Grignard reaction:** Reaction with an organomagnesium reagent to yield an alcohol

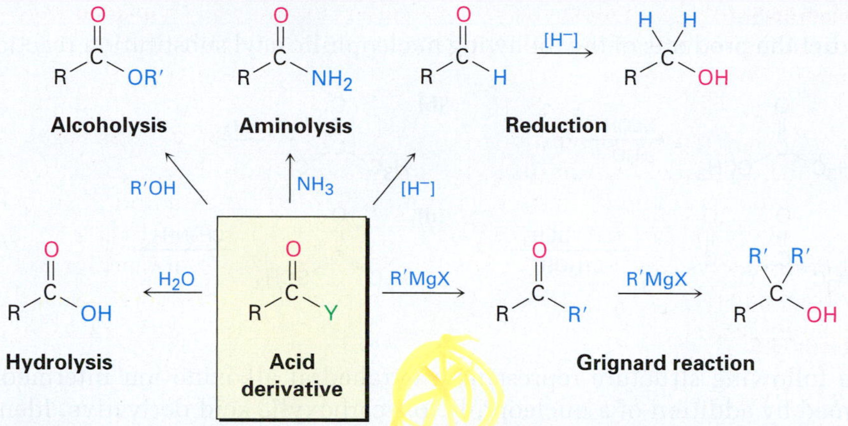

FIGURE 16.3 Some general reactions of carboxylic acid derivatives.

Predicting the Product of a Nucleophilic Acyl Substitution Reaction

Predict the product of the following nucleophilic acyl substitution reaction of benzoyl chloride with propan-2-ol:

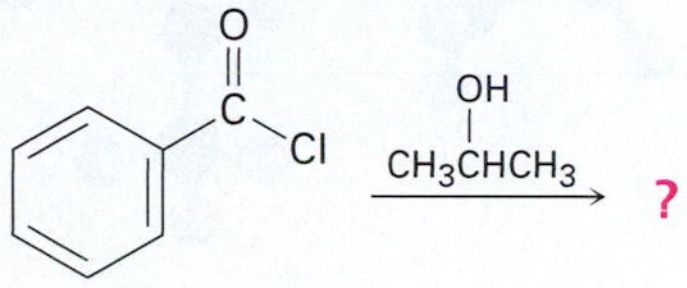

Benzoyl chloride

Strategy

A nucleophilic acyl substitution reaction involves the substitution of a nucleophile for a leaving group in a carboxylic acid derivative. Identify the leaving group (Cl⁻ in the case of an acid chloride) and the nucleophile (an alcohol in this case), and replace one by the other. The product is isopropyl benzoate.

Solution

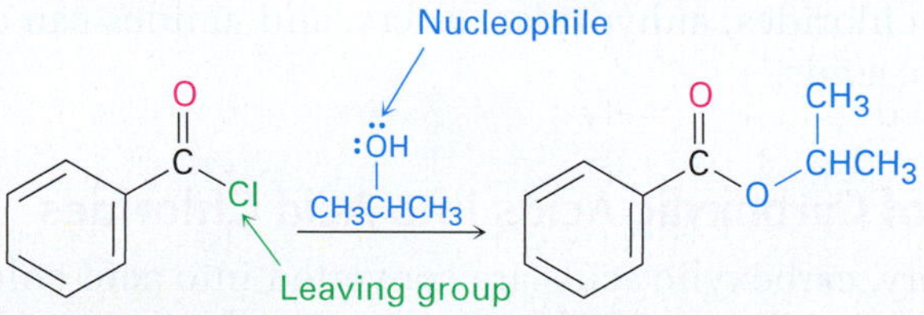

Benzoyl chloride   **Isopropyl benzoate**

Problem 16.4

Methyl trifluoroacetate, $CF_3CO_2CH_3$, is more reactive than methyl acetate, $CH_3CO_2CH_3$, in nucleophilic acyl substitution reactions. Explain.

Problem 16.5

Predict the products of the following nucleophilic acyl substitution reactions:

(a)

$$\underset{H_3C}{}\overset{O}{\underset{}{\parallel}}\underset{OCH_3}{C} \xrightarrow[H_2O]{NaOH} \textbf{?}$$

(b)

$$\underset{H_3C}{}\overset{O}{\underset{}{\parallel}}\underset{Cl}{C} \xrightarrow{NH_3} \textbf{?}$$

(c)

$$\underset{H_3C}{}\overset{O}{\underset{}{\parallel}}\underset{O}{C}\overset{O}{\underset{}{\parallel}}\underset{CH_3}{C} \xrightarrow[CH_3OH]{Na^+ \, ^-OCH_3} \textbf{?}$$

(d)

$$\underset{H_3C}{}\overset{O}{\underset{}{\parallel}}\underset{SCH_3}{C} \xrightarrow{CH_3NH_2} \textbf{?}$$

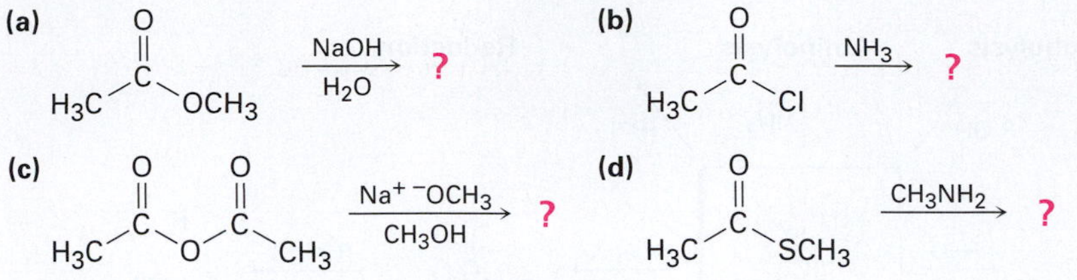

Problem 16.6

The following structure represents a tetrahedral alkoxide ion intermediate formed by addition of a nucleophile to a carboxylic acid derivative. Identify the nucleophile, the leaving group, the starting acid derivative, and the ultimate product.

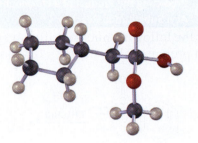

16.3 Nucleophilic Acyl Substitution Reactions of Carboxylic Acids

The direct nucleophilic acyl substitution of a carboxylic acid is difficult in the laboratory because –OH is a poor leaving group (Section 12.7). Thus, it's usually necessary to enhance the reactivity of the acid, either by using a strong acid catalyst to protonate the carboxyl and make it a better acceptor or by converting the –OH into a better leaving group. Under the right circumstances, however, acid chlorides, anhydrides, esters, and amides can all be prepared from carboxylic acids.

Conversion of Carboxylic Acids into Acid Chlorides

In the laboratory, carboxylic acids are converted into acid chlorides by treatment with thionyl chloride, $SOCl_2$:

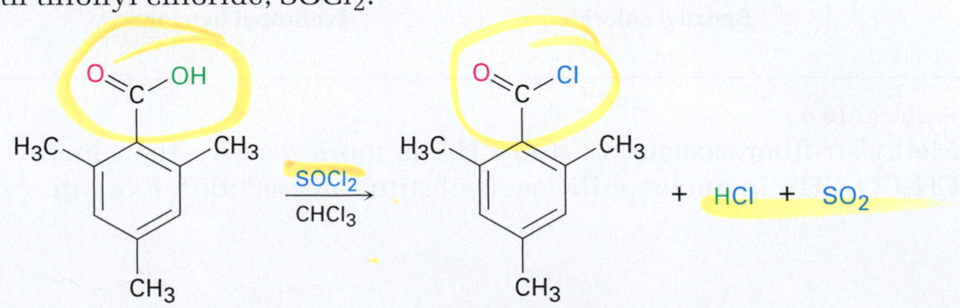

$$\xrightarrow[CHCl_3]{SOCl_2} \qquad + \; HCl \; + \; SO_2$$

2,4,6-Trimethylbenzoic acid   **2,4,6-Trimethylbenzoyl chloride (90%)**

The reaction occurs by a nucleophilic acyl substitution pathway in which the carboxylic acid is first converted into an acyl chlorosulfite intermediate, thereby replacing the –OH of the acid with a much better leaving group. The chlorosulfite then reacts with a nucleophilic chloride ion. You might recall from Section 13.4 that an analogous chlorosulfite is involved in reaction of an alcohol with $SOCl_2$ to yield an alkyl chloride.

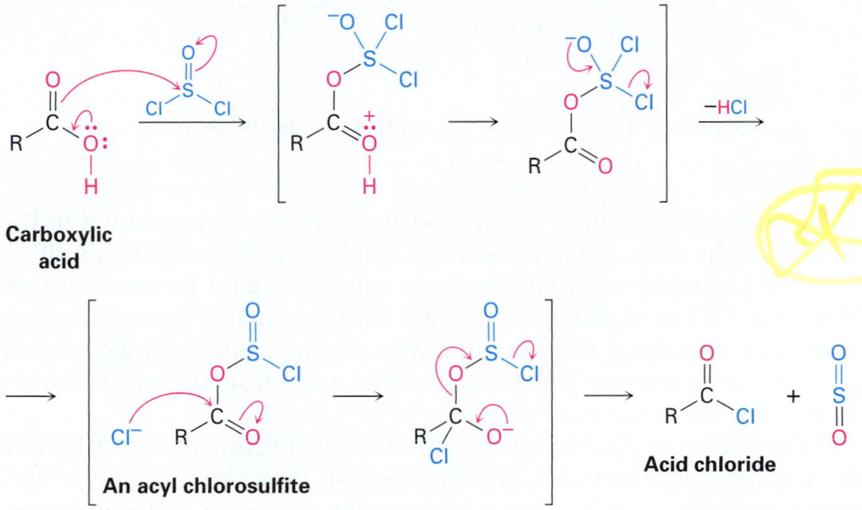

Carboxylic acid

An acyl chlorosulfite

Acid chloride

Conversion of Carboxylic Acids into Acid Anhydrides

Acid anhydrides can be derived from two molecules of carboxylic acid by strong heating to remove 1 equivalent of water. Because of the high temperatures needed, however, only acetic anhydride is commonly prepared this way.

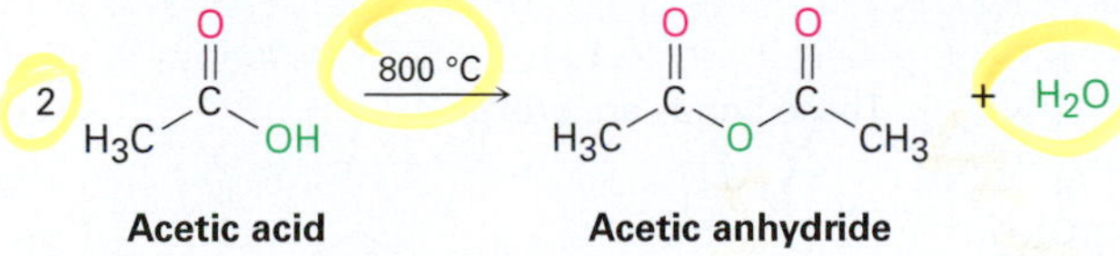

Acetic acid   Acetic anhydride

Conversion of Carboxylic Acids into Esters

Perhaps the most useful reaction of carboxylic acids is their conversion into esters. There are many methods for accomplishing the transformation, including the S_N2 reaction of a carboxylate anion with a primary alkyl halide that we saw in Section 12.7.

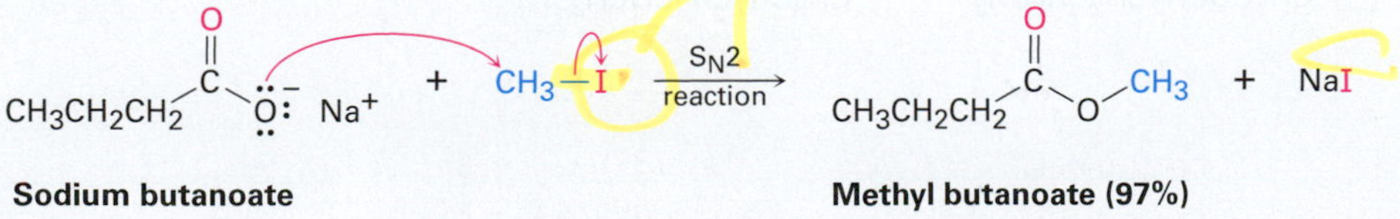

Sodium butanoate   Methyl butanoate (97%)

Esters can also be synthesized by an acid-catalyzed nucleophilic acyl substitution reaction of a carboxylic acid with an alcohol, a process called the *Fischer esterification reaction.* Unfortunately, the need to use an excess of a liquid alcohol as solvent effectively limits the method to the synthesis of methyl, ethyl, propyl, and butyl esters.

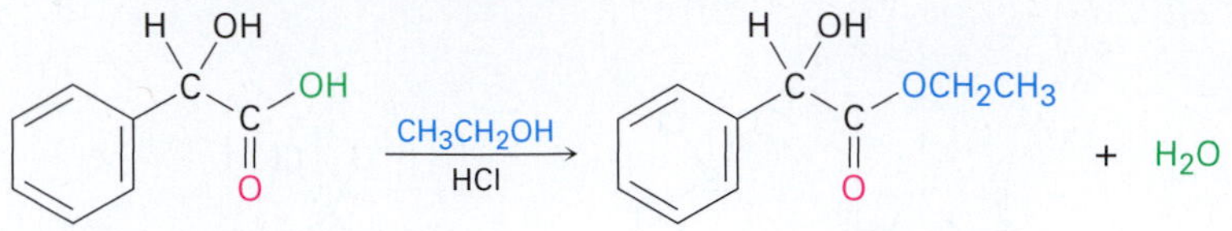
H OH
OH
CH_3CH_2OH
HCl
H OH
OCH_2CH_3
+ H_2O

Mandelic acid **Ethyl mandelate (86%)**

The mechanism of the Fischer esterification reaction is shown in Figure 16.4. Carboxylic acids are not reactive enough to undergo nucleophilic addition directly, but their reactivity is greatly enhanced in the presence of a strong acid such as HCl or H_2SO_4. The mineral acid protonates the carbonyl-group oxygen atom, thereby giving the carboxylic acid a positive charge and rendering it much more reactive. Subsequent loss of water from the tetrahedral intermediate then yields the ester product.

The net effect of Fischer esterification is substitution of an –OH group by –OR′. All steps are reversible, and the reaction typically has an equilibrium constant close to 1. Thus, the reaction can be driven in either direction by choice of reaction conditions. Ester formation is favored when a large excess of alcohol is used as solvent, but carboxylic acid formation is favored when a large excess of water is present.

Evidence in support of the mechanism shown in Figure 16.4 comes from isotope-labeling experiments. When ^{18}O-labeled methanol reacts with benzoic acid, the methyl benzoate produced is found to be ^{18}O-labeled but the water produced is unlabeled. Thus, it is the C–OH bond of the carboxylic acid that is broken during the reaction rather than the CO–H bond and the RO–H bond of the alcohol that is broken rather than the R–OH bond.

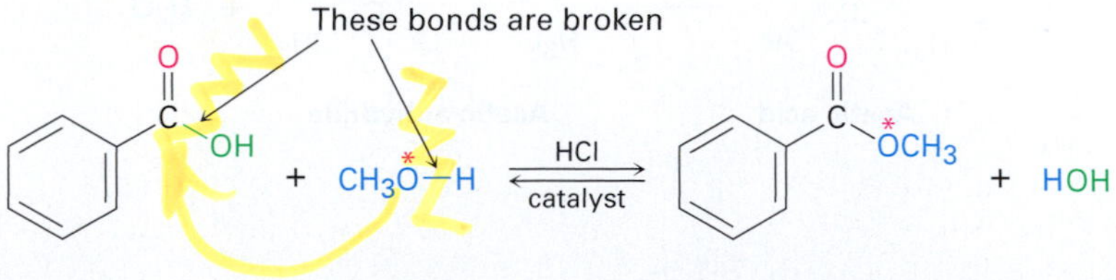
These bonds are broken
OH
+ CH_3O–H
HCl
catalyst
OCH_3
+ HOH

Problem 16.7

How might you prepare the following esters from the corresponding acids?

(a) 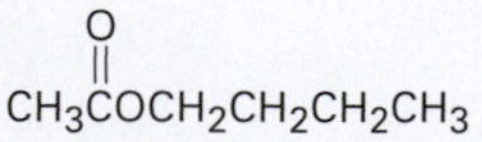
$CH_3COCH_2CH_2CH_2CH_3$

(b) 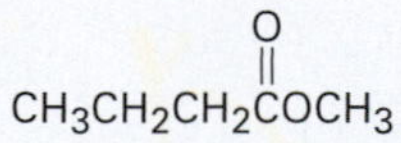
$CH_3CH_2CH_2COCH_3$

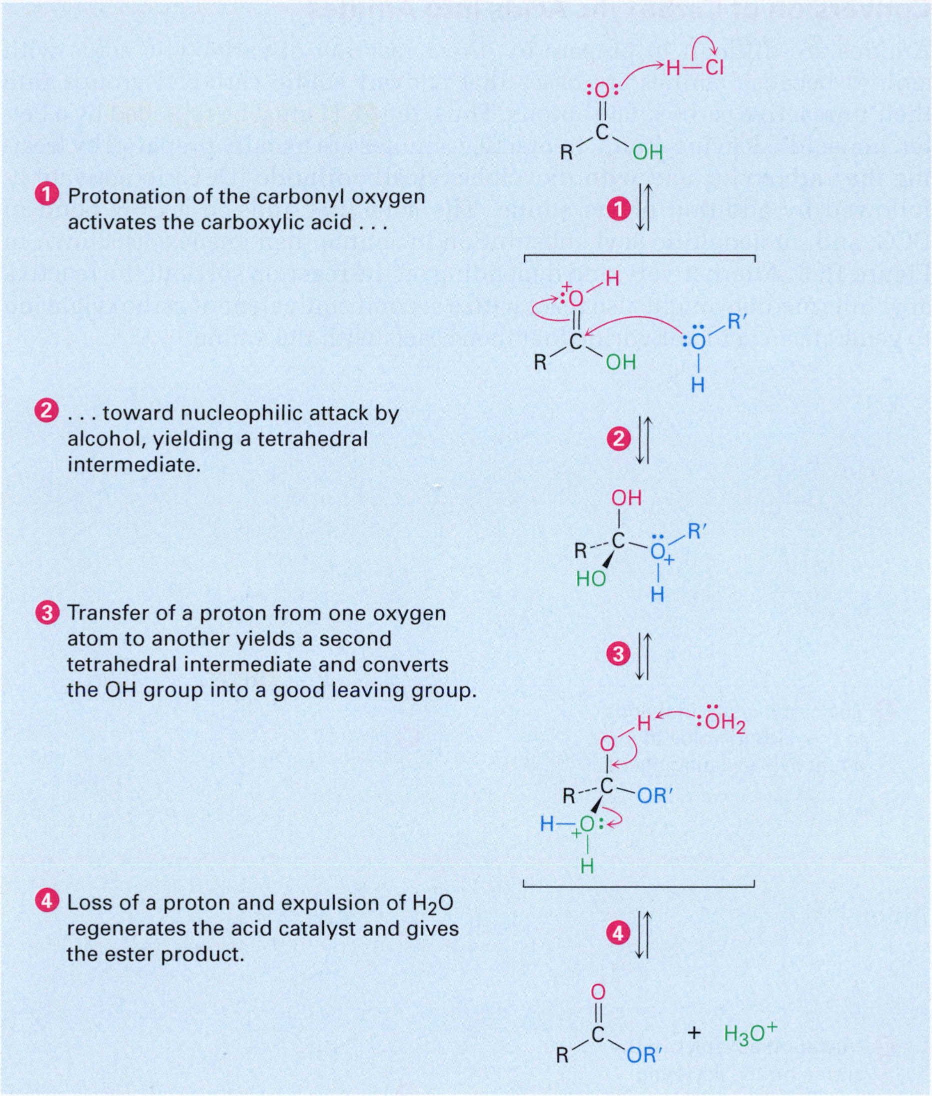

ACTIVE FIGURE 16.4

MECHANISM: Mechanism of Fischer esterification. The reaction is an acid-catalyzed, nucleophilic acyl substitution of a carboxylic acid. **Go to this book's student companion site at** www.cengage.com/chemistry/ mcmurry **to explore an interactive version of this figure.**

1 Protonation of the carbonyl oxygen activates the carboxylic acid . . .

2 . . . toward nucleophilic attack by alcohol, yielding a tetrahedral intermediate.

3 Transfer of a proton from one oxygen atom to another yields a second tetrahedral intermediate and converts the OH group into a good leaving group.

4 Loss of a proton and expulsion of H_2O regenerates the acid catalyst and gives the ester product.

© John McMurry

Problem 16.8

If the following molecule is treated with acid catalyst, an intramolecular esterification reaction occurs. What is the structure of the product? (*Intramolecular* means within the same molecule.)

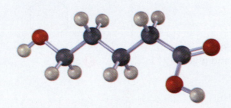

Conversion of Carboxylic Acids into Amides

Amides are difficult to prepare by direct reaction of carboxylic acids with amines because amines are bases that convert acidic carboxyl groups into their unreactive carboxylate anions. Thus, the —OH must be replaced by a better, nonacidic leaving group. In practice, amides are usually prepared by treating the carboxylic acid with dicyclohexylcarbodiimide (DCC) to activate it, followed by addition of the amine. The acid first adds to a C=N bond of DCC, and nucleophilic acyl substitution by amine then ensues, as shown in Figure 16.5. Alternatively, and depending on the reaction solvent, the reactive acyl intermediate might also react with a second equivalent of carboxylate ion to generate an acid anhydride that then reacts with the amine.

FIGURE 16.5 MECHANISM: Mechanism of amide formation by reaction of a carboxylic acid and an amine with dicyclohexyl-carbodiimide (DCC).

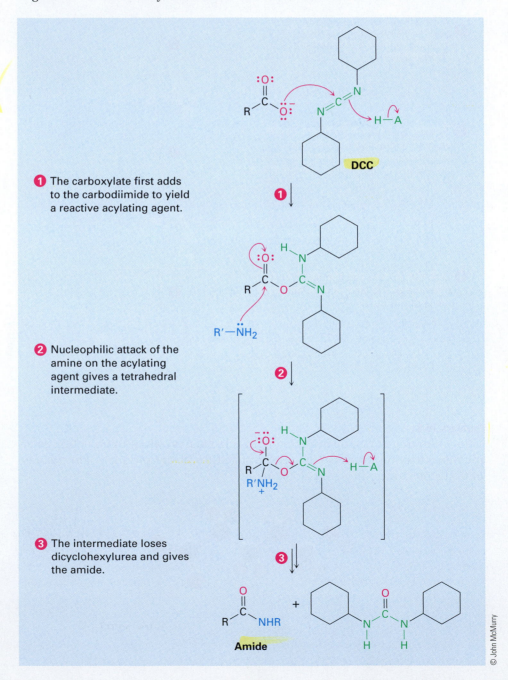

1. The carboxylate first adds to the carbodiimide to yield a reactive acylating agent.

2. Nucleophilic attack of the amine on the acylating agent gives a tetrahedral intermediate.

3. The intermediate loses dicyclohexylurea and gives the amide.

Amide

We'll see in Section 19.7 that this DCC-induced method of amide formation is the key step in the laboratory synthesis of small proteins, or *peptides.* For instance, when one amino acid with its NH_2 rendered unreactive and a second amino acid with its $-CO_2H$ rendered unreactive are treated with DCC, a dipeptide is formed:

Amino acid 1   Amino acid 2   A dipeptide

Conversion of Carboxylic Acids into Alcohols

We said in Section 13.3 that carboxylic acids are reduced by $LiAlH_4$ to give primary alcohols, but we deferred a discussion of the reaction mechanism at that time. In fact, the reduction is a nucleophilic acyl substitution reaction in which –H replaces –OH to give an aldehyde, which is further reduced to a primary alcohol by nucleophilic addition. The aldehyde intermediate is much more reactive than the starting acid, so it reacts immediately and is not isolated.

A carboxylic acid   An aldehyde (not isolated)   An alkoxide ion   A 1° alcohol

Because hydride ion is a base as well as a nucleophile, the actual nucleophilic acyl substitution step takes place on the carboxylate ion rather than on the free carboxylic acid and gives a high-energy *dianion* intermediate. In this intermediate, the two oxygens are undoubtedly complexed to a Lewis acidic aluminum species. Thus, the reaction is relatively difficult, and acid reductions require higher temperatures and extended reaction times.

A carboxylic acid   A carboxylate   A dianion   An aldehyde

Biological Conversions of Carboxylic Acids

The direct conversion of a carboxylic acid to an acyl derivative by nucleophilic acyl substitution does not occur in biological chemistry. As in the laboratory, the acid must first be activated. This activation is often accomplished in living organisms by reaction of the acid with ATP to give an acyl adenosyl phosphate, or *acyl adenylate,* a mixed anhydride between a carboxylic acid and adenosine monophosphate (AMP, also known as adenylic acid). In the biosynthesis of fats, for example, a long-chain carboxylic acid reacts with ATP to give an acyl adenylate, followed by subsequent nucleophilic acyl substitution of a thiol group in coenzyme A to give the corresponding acyl CoA (Figure 16.6).

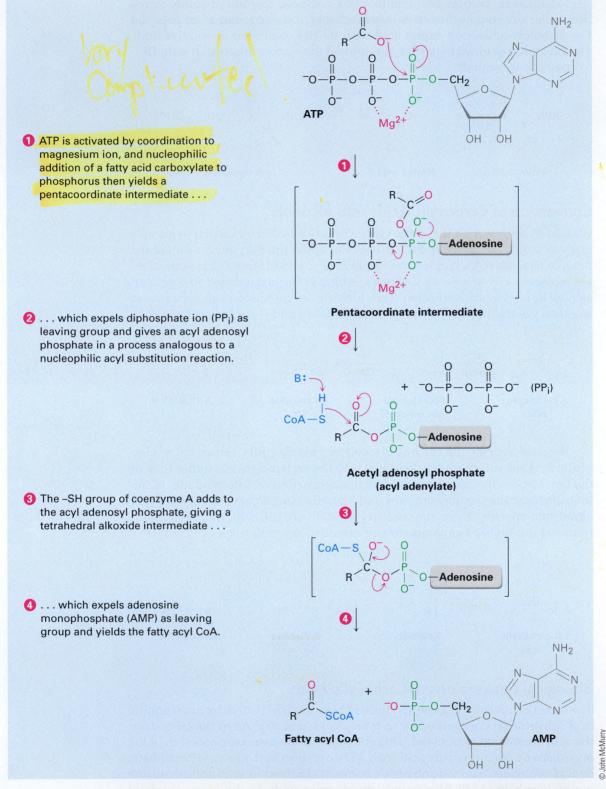

① ATP is activated by coordination to magnesium ion, and nucleophilic addition of a fatty acid carboxylate to phosphorus then yields a pentacoordinate intermediate . . .

② . . . which expels diphosphate ion (PPᵢ) as leaving group and gives an acyl adenosyl phosphate in a process analogous to a nucleophilic acyl substitution reaction.

③ The –SH group of coenzyme A adds to the acyl adenosyl phosphate, giving a tetrahedral alkoxide intermediate . . .

④ . . . which expels adenosine monophosphate (AMP) as leaving group and yields the fatty acyl CoA.

© John McMurry

FIGURE 16.6 MECHANISM: In fatty-acid biosynthesis, a carboxylic acid is activated by reaction with ATP to give an acyl adenylate, which undergoes nucleophilic acyl substitution with the –SH group on coenzyme A. (ATP= adenosine triphosphate; AMP = adenosine monophosphate.)

Note that the first step in Figure 16.6—reaction of the carboxylate with ATP to give an acyl adenylate—is itself a nucleophilic acyl substitution on *phosphorus*. The carboxylate first adds to a P=O double bond, giving a five-coordinate phosphorus intermediate that expels diphosphate ion as leaving group.

16.4 Chemistry of Acid Halides

Preparation of Acid Halides

Acid chlorides are prepared from carboxylic acids by reaction with thionyl chloride ($SOCl_2$), as we saw in the previous section. Similar reaction of a carboxylic acid with phosphorus tribromide (PBr_3) yields the acid bromide.

Reactions of Acid Halides

Acid halides are among the most reactive of carboxylic acid derivatives and can be converted into many other kinds of compounds by nucleophilic acyl substitution mechanisms. The halogen can be replaced by –OH to yield an acid, by –OCOR to yield an anhydride, by –OR to yield an ester, by –NH$_2$ to yield an amide, or by R′ to yield a ketone. Although the reactions we'll be discussing in this section are illustrated only for acid chlorides, similar processes take place with other acid halides.

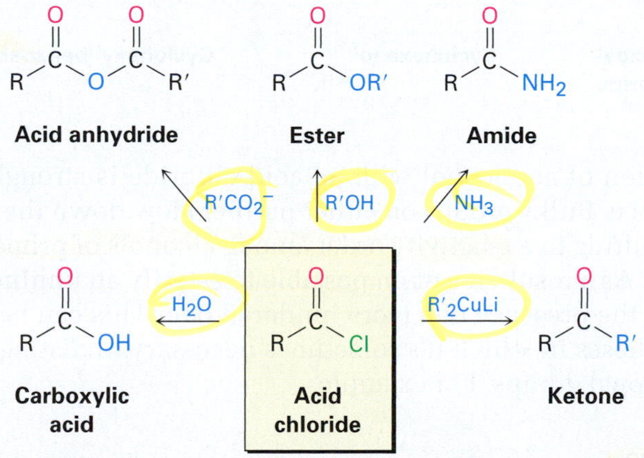

CONVERSION OF ACID HALIDES INTO ACIDS: HYDROLYSIS Acid chlorides react with water to yield carboxylic acids. This hydrolysis reaction is a typical nucleophilic acyl substitution process and is initiated by attack of water on the acid chloride carbonyl group. The tetrahedral intermediate undergoes elimination of Cl⁻ and loss of H⁺ to give the product carboxylic acid plus HCl.

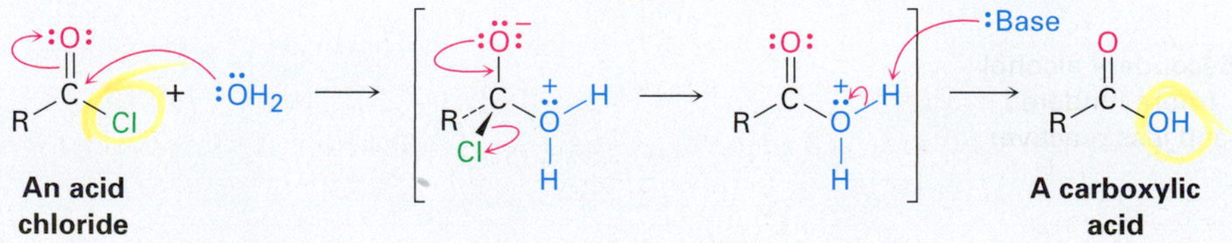

Because HCl is generated during the hydrolysis, the reaction is often carried out in the presence of a base such as pyridine or NaOH to remove the HCl and prevent it from causing side reactions.

CONVERSION OF ACID HALIDES INTO ANHYDRIDES Nucleophilic acyl substitution reaction of an acid chloride with a carboxylate anion gives an acid anhydride. Both symmetrical and unsymmetrical acid anhydrides can be prepared in this way.

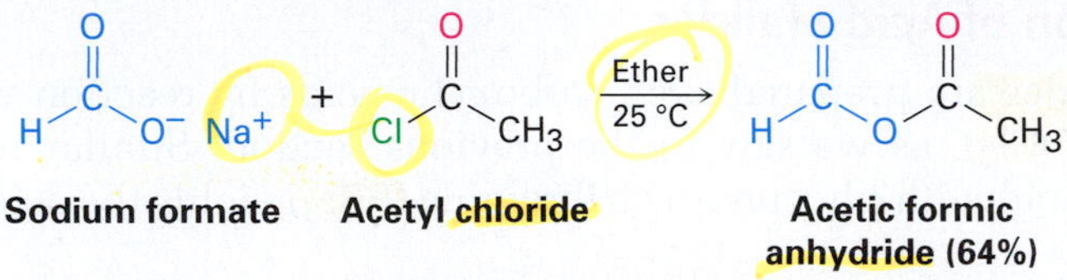

Sodium formate   **Acetyl chloride**   **Acetic formic anhydride (64%)**

CONVERSION OF ACID HALIDES INTO ESTERS: ALCOHOLYSIS Acid chlorides react with alcohols to yield esters in a process analogous to their reaction with water to yield acids. In fact, this reaction is probably the most common method for preparing esters in the laboratory. As with hydrolysis, alcoholysis reactions are usually carried out in the presence of pyridine or NaOH to react with the HCl formed.

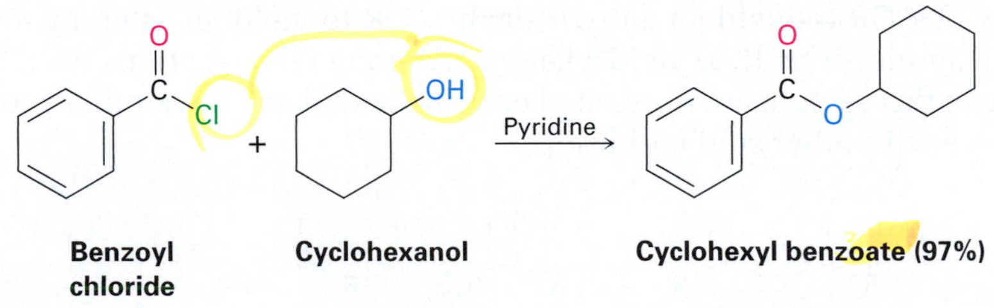

Benzoyl chloride   **Cyclohexanol**   **Cyclohexyl benzoate (97%)**

The reaction of an alcohol with an acid chloride is strongly affected by steric hindrance. Bulky groups on either partner slow down the reaction considerably, resulting in a reactivity order among alcohols of primary > secondary > tertiary. As a result, it's often possible to esterify an unhindered alcohol selectively in the presence of a more hindered one. This can be important in complex syntheses in which it's sometimes necessary to distinguish between similar functional groups. For example,

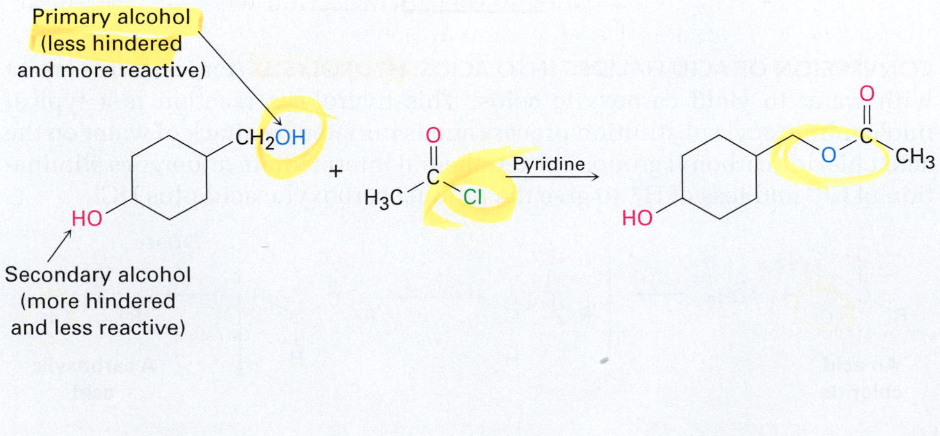

Primary alcohol (less hindered and more reactive)

Secondary alcohol (more hindered and less reactive)

Problem 16.9

How might you prepare the following esters using a nucleophilic acyl substitution reaction of an acid chloride?

(a) $CH_3CH_2CO_2CH_3$ **(b)** $CH_3CO_2CH_2CH_3$ **(c)** Ethyl benzoate

Problem 16.10

Which method would you choose if you wanted to prepare cyclohexyl benzoate—Fischer esterification of the carboxylic acid or reaction of the acid chloride with an alcohol? Explain.

CONVERSION OF ACID HALIDES INTO AMIDES: AMINOLYSIS Acid chlorides react rapidly with ammonia and amines to give amides. As with the acid chloride–plus–alcohol method for preparing esters, this reaction of acid chlorides with amines is the most commonly used laboratory method for preparing amides. Both monosubstituted and disubstituted amines can be used, but not trisubstituted amines (R_3N).

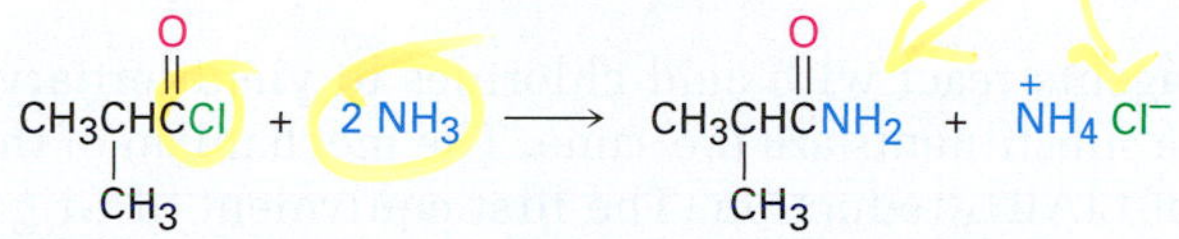

2-Methylpropanoyl chloride

2-Methylpropanamide (83%)

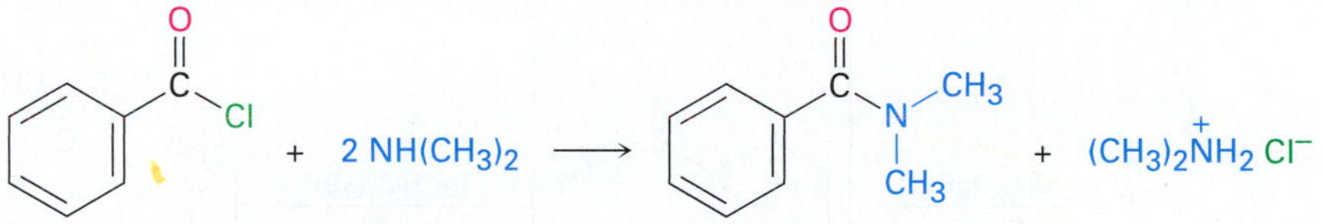

Benzoyl chloride

N,N-Dimethylbenzamide (92%)

Because HCl is formed during the reaction, 2 equivalents of the amine must be used. One equivalent reacts with the acid chloride, and one equivalent reacts with the HCl by-product to form an ammonium chloride salt. If, however, the amine component is valuable, amide synthesis is often carried out using 1 equivalent of the amine plus 1 equivalent of an inexpensive base such as NaOH. For example, the sedative trimetozine is prepared commercially by reaction of 3,4,5-trimethoxybenzoyl chloride with the amine morpholine in the presence of 1 equivalent of NaOH.

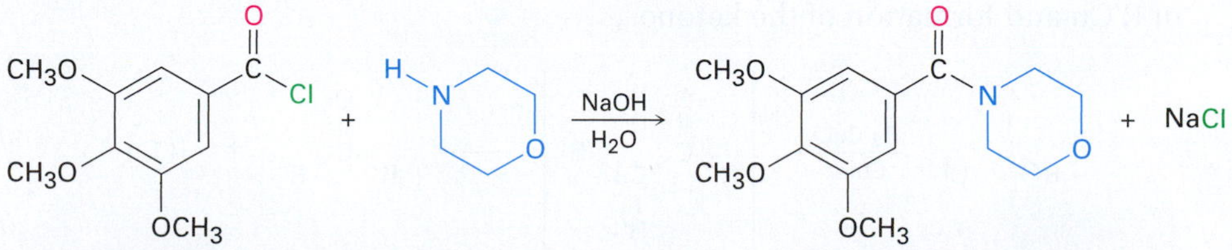

3,4,5-Trimethoxy-benzoyl chloride   **Morpholine**   **Trimetozine (an amide)**

CONVERSION OF ACID CHLORIDES INTO ALCOHOLS: REDUCTION AND GRIGNARD REACTION Acid chlorides are reduced by $LiAlH_4$ to yield primary alcohols. The reaction is of little practical value, however, because the parent carboxylic acids are generally more readily available and can themselves be reduced by $LiAlH_4$ to yield alcohols.

Reduction occurs via a typical nucleophilic acyl substitution mechanism in which a hydride ion ($H:^-$) adds to the carbonyl group, yielding a tetrahedral intermediate that expels Cl^-. The net effect is a substitution of $-Cl$ by $-H$ to yield an aldehyde, which is then immediately reduced by $LiAlH_4$ in a second step to yield the primary alcohol.

Benzoyl chloride   **Benzaldehyde** *(NOT isolated)*   **Benzyl alcohol**

Grignard reagents react with acid chlorides to yield tertiary alcohols in which two of the substituents are the same. The mechanism of the reaction is similar to that of $LiAlH_4$ reduction. The first equivalent of Grignard reagent adds to the acid chloride, loss of Cl^- from the tetrahedral intermediate yields a ketone, and a second equivalent of Grignard reagent immediately adds to the ketone to produce an alcohol.

Benzoyl chloride   **Acetophenone** *(NOT isolated)*   **2-Phenylpropan-2-ol** **(92%)**

The ketone intermediate formed during the reaction of an acid chloride with a Grignard reagent can't usually be isolated because addition of the second equivalent of organomagnesium reagent occurs too rapidly. A ketone *can*, however, be isolated from the reaction of an acid chloride with a lithium diorganocopper reagent, $Li^+ R_2Cu^-$, which is itself formed by reaction of CuI with 2 equivalents of an organolithium (Section 14.11). The reaction occurs by initial nucleophilic acyl substitution on the acid chloride by the diorganocopper anion to yield an acyl diorganocopper intermediate, followed by loss of $R'Cu$ and formation of the ketone.

An acid chloride   **An acyl diorganocopper**   **A ketone**

As an example of the process, manicone, a substance secreted by male ants to coordinate ant pairing and mating, has been synthesized by reaction of lithium diethylcopper with (*E*)-2,4-dimethylhex-2-enoyl chloride.

CH₃CH₂CH(CH₃)—CH=C(CH₃)—C(=O)Cl
2,4-Dimethylhex-2-enoyl chloride

(CH₃CH₂)₂CuLi
Ether, −78 °C

CH₃CH₂CH(CH₃)—CH=C(CH₃)—C(=O)CH₂CH₃
4,6-Dimethyloct-4-en-3-one (manicone, 92%)

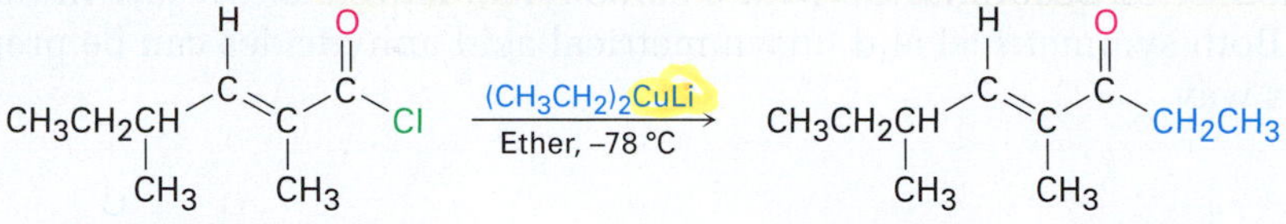

Note that the diorganocopper reaction occurs only with acid chlorides. Carboxylic acids, esters, acid anhydrides, and amides do not react with lithium diorganocopper reagents.

Problem 16.11

Write the full mechanism of the reaction just shown between 3,4,5-trimethoxybenzoyl chloride and morpholine to form trimetozine. Use curved arrows to show the electron flow in each step.

Problem 16.12

Write the full mechanism of the reaction between LiAlH₄ and benzoyl chloride to yield benzyl alcohol.

Problem 16.13

How could you prepare the following amides using an acid chloride and an amine or ammonia?
(a) CH₃CH₂CONHCH₃
(b) *N,N*-Diethylbenzamide
(c) Propanamide

Problem 16.14

How could you prepare the following ketones by reaction of an acid chloride with a lithium diorganocopper reagent?

(a)   **(b)**

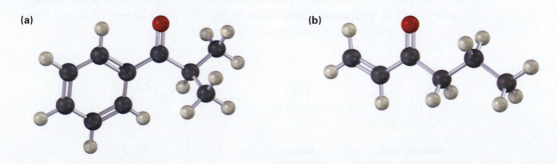

16.5 Chemistry of Acid Anhydrides

Preparation of Acid Anhydrides

Acid anhydrides are typically prepared by nucleophilic acyl substitution reaction of an acid chloride with a carboxylate anion, as we saw in Section 16.4. Both symmetrical and unsymmetrical acid anhydrides can be prepared in this way.

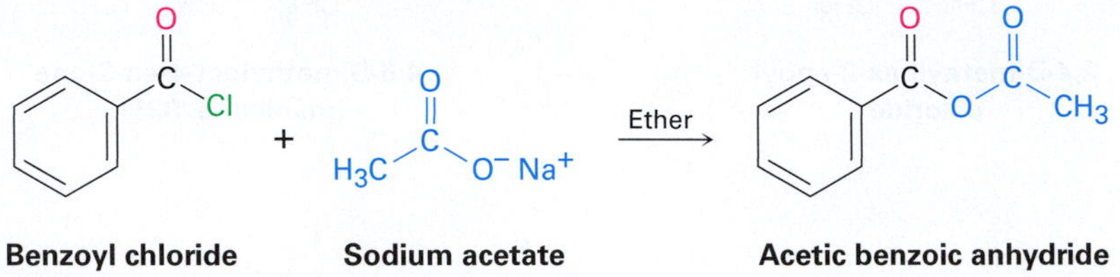

Benzoyl chloride + **Sodium acetate** → (Ether) **Acetic benzoic anhydride**

Reactions of Acid Anhydrides

The chemistry of acid anhydrides is similar to that of acid chlorides, although anhydrides react more slowly. Thus, acid anhydrides react with water to form acids, with alcohols to form esters, with amines to form amides, and with LiAlH$_4$ to form primary alcohols (Figure 16.7). Only the ester and amide forming reactions are commonly used, however.

FIGURE 16.7 Some reactions of acid anhydrides.

Alcoholysis **Aminolysis**

R'OH NH$_3$

Hydrolysis **Acid anhydride** [H$^-$] [H$^-$] **Reduction**

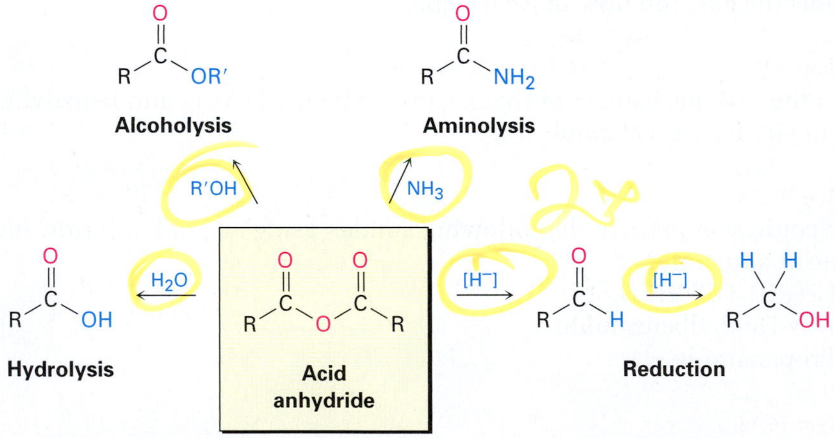

CONVERSION OF ACID ANHYDRIDES INTO ESTERS Acetic anhydride is often used to prepare acetate esters from alcohols. For example, aspirin (acetylsalicylic acid) is prepared commercially by the acetylation of *o*-hydroxybenzoic acid (salicylic acid) with acetic anhydride.

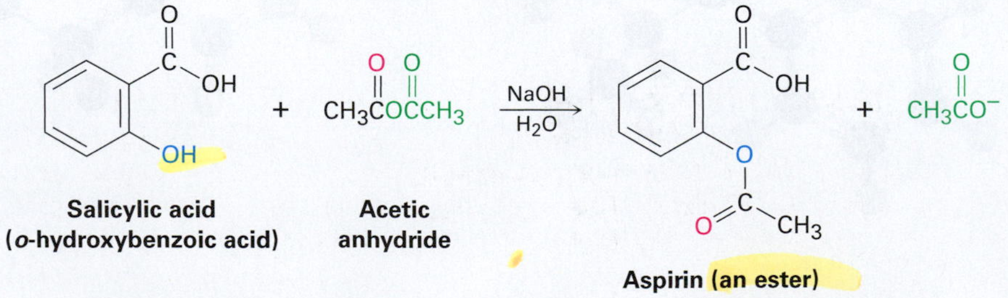

Salicylic acid (*o*-hydroxybenzoic acid) + **Acetic anhydride** → (NaOH / H$_2$O) **Aspirin (an ester)** + CH$_3$CO$^-$

CONVERSION OF ACID ANHYDRIDES INTO AMIDES Acetic anhydride is also commonly used to prepare *N*-substituted acetamides from amines. For example, acetaminophen, a drug used in over-the-counter analgesics such as Tylenol, is prepared by reaction of *p*-hydroxyaniline with acetic anhydride. Only the more nucleophilic –NH$_2$ group reacts rather than the less nucleophilic –OH group.

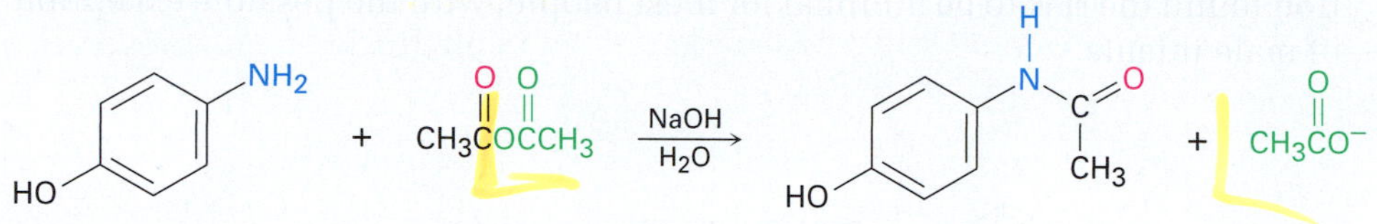

p-Hydroxyaniline   **Acetic anhydride**   **Acetaminophen**

Notice in both of the previous reactions that only "half" of the anhydride molecule is used, while the other half acts as the leaving group during the nucleophilic acyl substitution step and produces acetate ion as a by-product. Thus, anhydrides are inefficient to use, and acid chlorides are normally preferred for introducing acyl substituents other than acetyl groups.

Problem 16.15

Write the mechanism of the reaction just shown between *p*-hydroxyaniline and acetic anhydride to prepare acetaminophen.

Problem 16.16

What product would you expect from reaction of 1 equivalent of methanol with a cyclic anhydride, such as phthalic anhydride (benzene-1,2-dicarboxylic anhydride)? What is the fate of the second "half" of the anhydride?

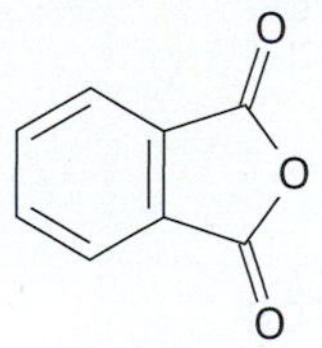

Phthalic anhydride

16.6 Chemistry of Esters

Esters are among the most widespread of all naturally occurring compounds. Many simple esters are pleasant-smelling liquids that are responsible for the fragrant odors of fruits and flowers. For example, methyl butanoate is found in pineapple oil, and isopentyl acetate is a constituent of banana oil. The ester linkage is also present in animal fats and in many biologically important molecules.

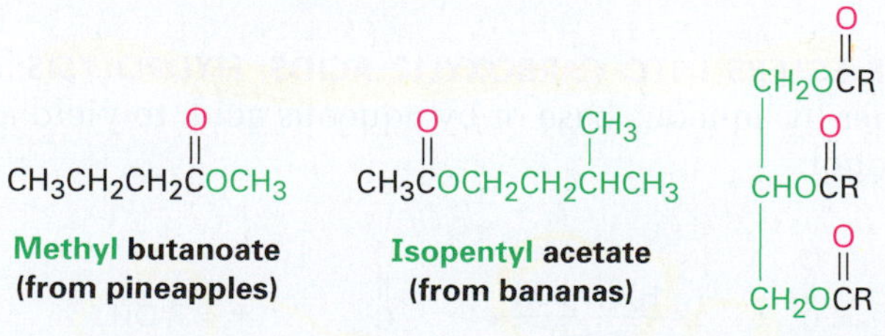

Methyl butanoate (from pineapples)   **Isopentyl acetate** (from bananas)   **A fat** (R = C$_{11-17}$ chains)

The chemical industry uses esters for a variety of purposes. Ethyl acetate, for instance, is a commonly used solvent, and dialkyl phthalates are used as plasticizers to keep polymers from becoming brittle. You may be aware that there is current concern about possible toxicity of phthalates at high concentrations, although a recent assessment by the U.S. Food and Drug Administration found the risk to be minimal for most people, with the possible exception of male infants.

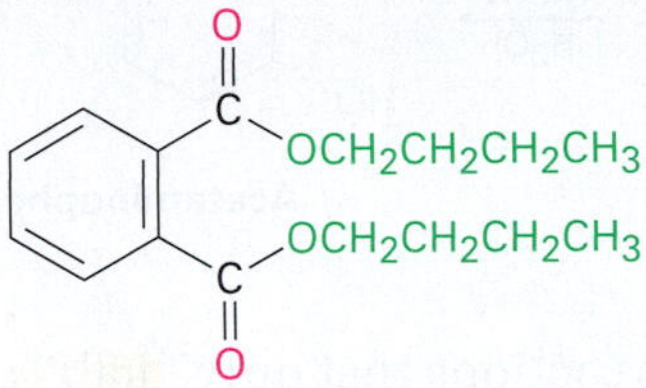

Dibutyl phthalate
(a plasticizer)

Preparation of Esters

Esters are usually prepared from carboxylic acids by the methods already discussed. Thus, carboxylic acids are converted directly into esters by S$_N$2 reaction of a carboxylate ion with a primary alkyl halide or by Fischer esterification of a carboxylic acid with an alcohol in the presence of a mineral acid catalyst. In addition, acid chlorides are converted into esters by treatment with an alcohol in the presence of base (Section 16.4).

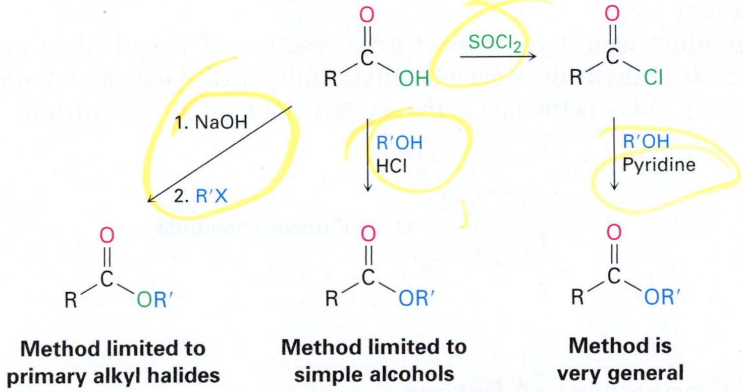

| Method limited to primary alkyl halides | Method limited to simple alcohols | Method is very general |

Reactions of Esters

Esters undergo the same kinds of reactions that we've seen for other carboxylic acid derivatives, but they are less reactive toward nucleophiles than either acid chlorides or anhydrides. All their reactions are equally applicable to both acyclic and cyclic esters, called **lactones**.

CONVERSION OF ESTERS INTO CARBOXYLIC ACIDS: HYDROLYSIS An ester is hydrolyzed, either by aqueous base or by aqueous acid, to yield a carboxylic acid plus an alcohol:

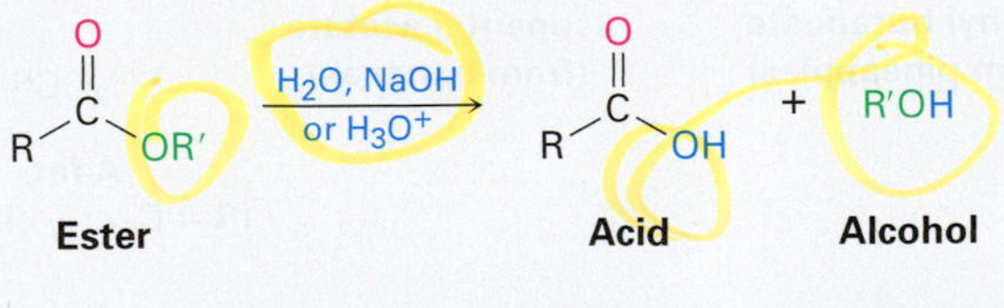

Ester   **Acid**   **Alcohol**

Ester hydrolysis in basic solution is called **saponification**, after the Latin word *sapo*, meaning "soap." As we'll see in Section 23.2, soap is in fact made by boiling animal fat with aqueous base to hydrolyze the ester linkages.

Ester hydrolysis occurs through a typical nucleophilic acyl substitution pathway in which hydroxide ion is the nucleophile that adds to the ester carbonyl group to give a tetrahedral intermediate. Loss of alkoxide ion then gives a carboxylic acid, which is deprotonated to give the carboxylate ion. Addition of aqueous HCl in a separate step after the saponification is complete protonates the carboxylate ion and gives the carboxylic acid (Figure 16.8).

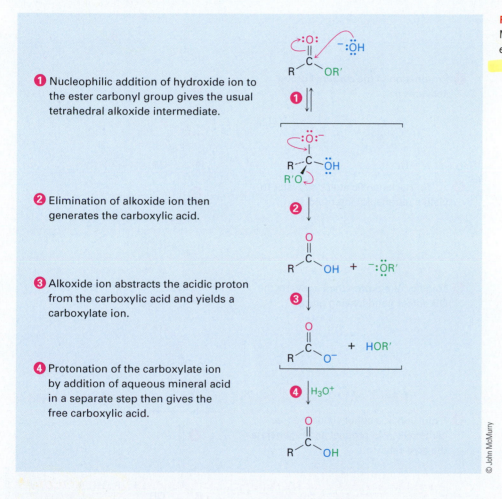

① Nucleophilic addition of hydroxide ion to the ester carbonyl group gives the usual tetrahedral alkoxide intermediate.

② Elimination of alkoxide ion then generates the carboxylic acid.

③ Alkoxide ion abstracts the acidic proton from the carboxylic acid and yields a carboxylate ion.

④ Protonation of the carboxylate ion by addition of aqueous mineral acid in a separate step then gives the free carboxylic acid.

FIGURE 16.8 MECHANISM: Mechanism of base-induced ester hydrolysis (saponification).

© John McMurry

The mechanism shown in Figure 16.8 is supported by isotope-labeling studies. When ethyl propanoate labeled with ^{18}O in the ether-like oxygen is hydrolyzed in aqueous NaOH, the ^{18}O label shows up exclusively in the ethanol product. None of the label remains with the propanoic acid, indicating that saponification occurs by cleavage of the C–OR′ bond rather than the CO–R′ bond.

Acid-catalyzed ester hydrolysis can occur by more than one mechanism, depending on the structure of the ester. The usual pathway, however, is just the reverse of a Fischer esterification reaction (Section 16.3). The ester is first activated toward nucleophilic attack by protonation of the carboxyl oxygen atom, and nucleophilic addition of water then occurs. Transfer of a proton and elimination of alcohol yields the carboxylic acid (Figure 16.9). Because this hydrolysis reaction is the reverse of a Fischer esterification reaction, Figure 16.9 is the reverse of Figure 16.4.

FIGURE 16.9 MECHANISM: Mechanism of acid-catalyzed ester hydrolysis. The forward reaction is a hydrolysis; the back-reaction is a Fischer esterification and is thus the reverse of Figure 16.4.

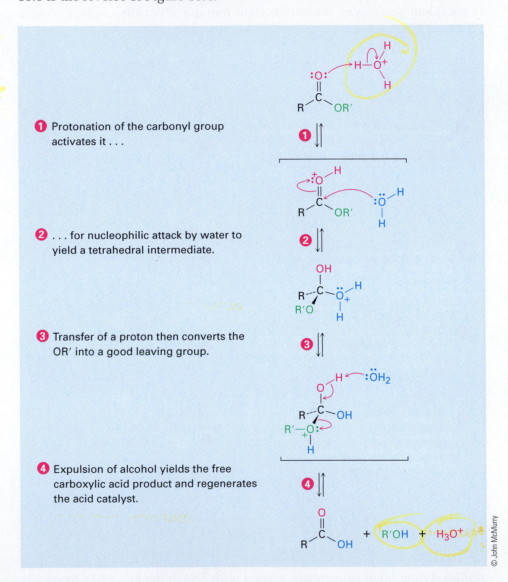

1. Protonation of the carbonyl group activates it . . .

2. . . . for nucleophilic attack by water to yield a tetrahedral intermediate.

3. Transfer of a proton then converts the OR' into a good leaving group.

4. Expulsion of alcohol yields the free carboxylic acid product and regenerates the acid catalyst.

© John McMurry

Ester hydrolysis is common in biological chemistry, particularly in the digestion of dietary fats and oils. We'll save a complete discussion of the mechanistic details of fat hydrolysis until Section 23.4 but will note for now that the reaction is catalyzed by various lipase enzymes and involves two sequential nucleophilic acyl substitution reactions. The first is a *transesterification* reaction in which an alcohol group on the lipase adds to an ester linkage in the fat molecule to give a tetrahedral intermediate that expels alcohol and forms an acyl enzyme intermediate. The second is an addition of water to the acyl enzyme, followed by expulsion of the enzyme to give a hydrolyzed acid.

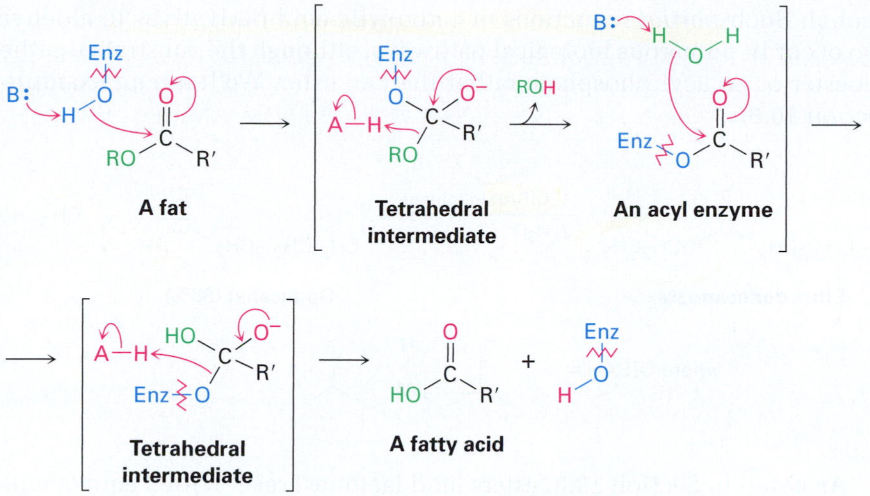
A fat
Tetrahedral intermediate
An acyl enzyme
ROH
Tetrahedral intermediate
A fatty acid

Problem 16.17

Why is the saponification of an ester irreversible? In other words, why doesn't treatment of a carboxylic acid with an alkoxide ion yield an ester?

CONVERSION OF ESTERS INTO AMIDES: AMINOLYSIS Esters react with ammonia and amines to yield amides. The reaction is not often used, however, because it's usually easier to prepare an amide by starting with an acid chloride (Section 16.4).

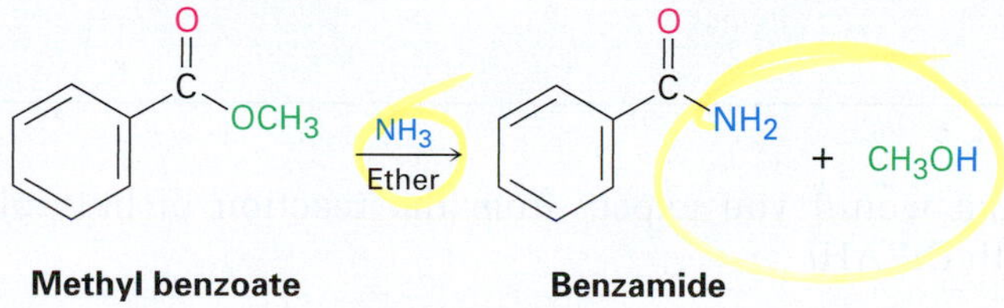
NH3
Ether
+ CH3OH
Methyl benzoate
Benzamide

CONVERSION OF ESTERS INTO ALCOHOLS: REDUCTION AND GRIGNARD REACTION Esters are reduced by treatment with $LiAlH_4$ to yield primary alcohols, as we saw in Section 13.3. The mechanism is similar to that of acid chloride reduction in that a hydride ion first adds to the carbonyl group, followed by elimination of alkoxide ion to yield an aldehyde. Further reduction of the aldehyde gives the primary alcohol.

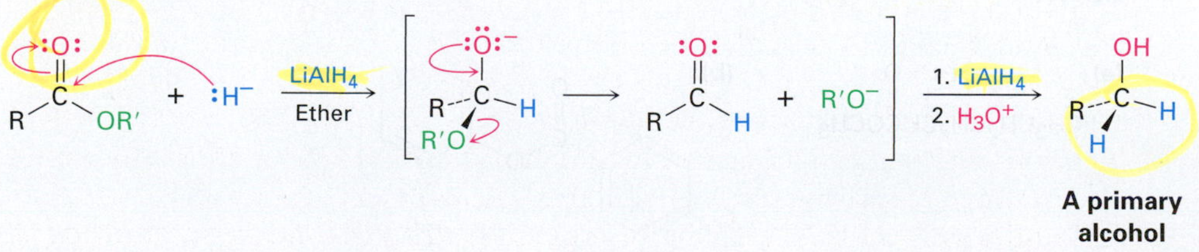
LiAlH4
Ether
1. LiAlH4
2. H3O+
A primary alcohol

The aldehyde intermediate can be isolated if 1 equivalent of the less reactive reducing agent diisobutylaluminum hydride (DIBAH) is used instead of $LiAlH_4$. The reaction is carried out at −78 °C to avoid further reduction to the alcohol. Such *partial* reductions of carboxylic acid derivatives to aldehydes also occur in numerous biological pathways, although the substrate is either a thioester or an acyl phosphate rather than an ester. We'll see an example in Section 16.8.

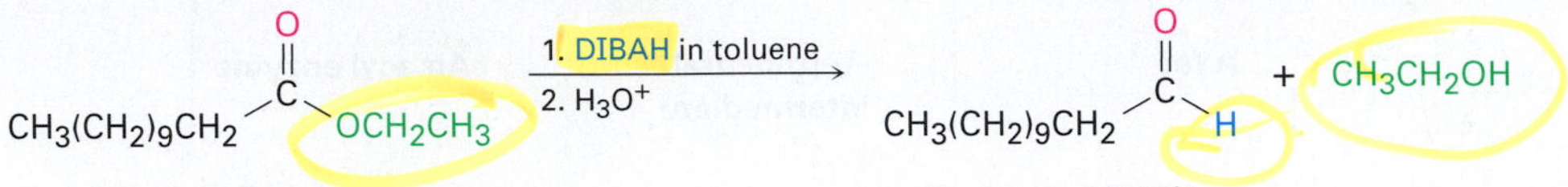
O
C
CH3(CH2)9CH2
OCH2CH3
1. DIBAH in toluene
2. H3O+
O
C
CH3(CH2)9CH2
H
+ CH3CH2OH

Ethyl dodecanoate **Dodecanal (88%)**

where DIBAH =

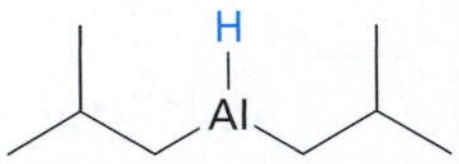
H
Al

As noted in Section 13.3, esters (and lactones) react with 2 equivalents of a Grignard reagent to yield a tertiary alcohol in which two of the substituents are identical. The reaction occurs by the usual nucleophilic substitution mechanism to give an intermediate ketone, which reacts further with the Grignard reagent to yield a tertiary alcohol.

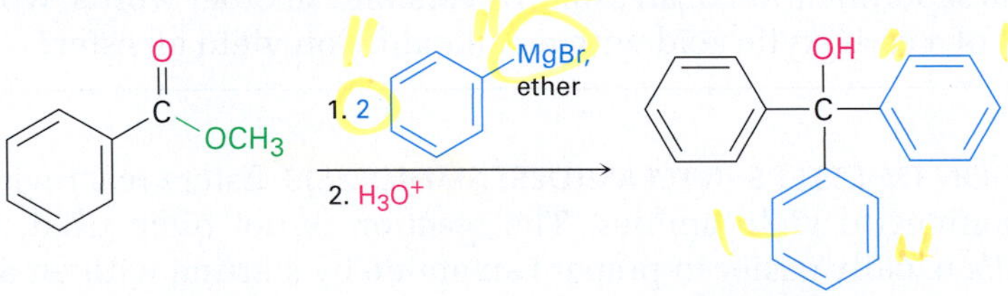
O
C
OCH3
1. 2
MgBr,
ether
2. H3O+
OH
C

Methyl benzoate **Triphenylmethanol (96%)**

Problem 16.18

What product would you expect from the reaction of butyrolactone with $LiAlH_4$? With DIBAH?

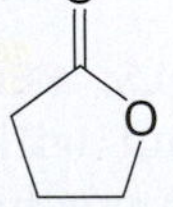
O
O

Butyrolactone

Problem 16.19

Show the products you would obtain by reduction of the following esters with $LiAlH_4$:

(a)

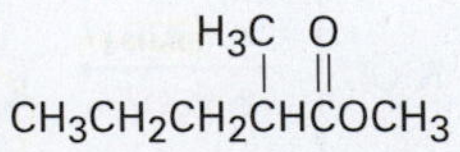
H3C O
CH3CH2CH2CHCOCH3

(b)

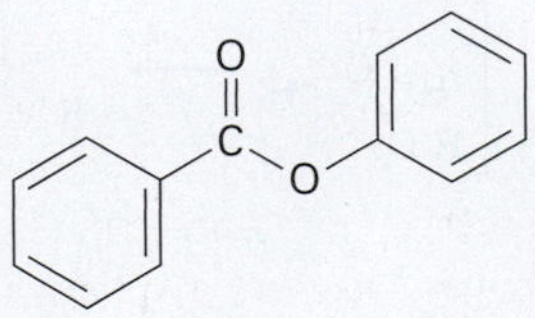
O
C
O

16.7 Chemistry of Amides

Amides, like esters, are abundant in all living organisms. Proteins, nucleic acids, and many pharmaceutical agents have amide functional groups. The reason for this abundance of amides is that they are stable to the aqueous conditions found in living organisms. Amides are the least reactive of the common acid derivatives and undergo relatively few nucleophilic acyl substitution reactions.

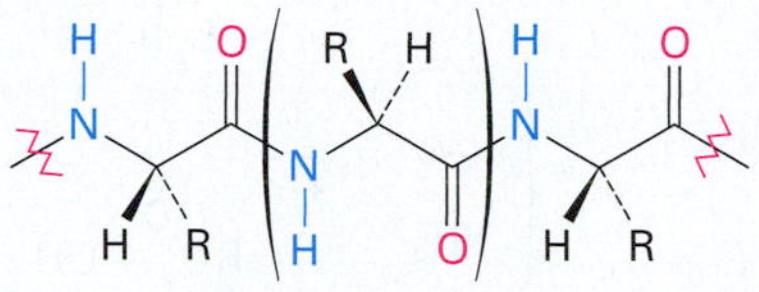

A protein segment

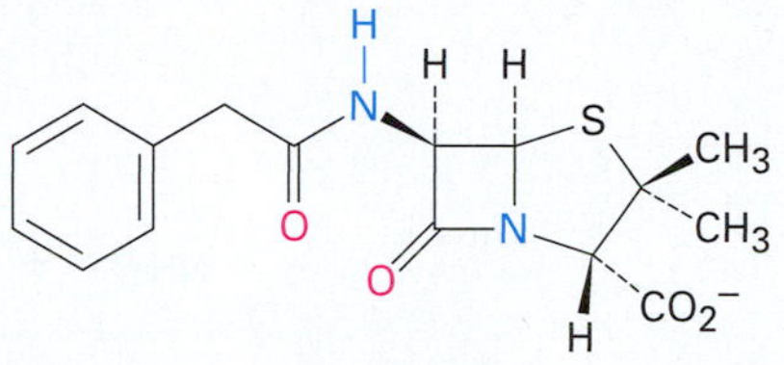

Benzylpenicillin (penicillin G)

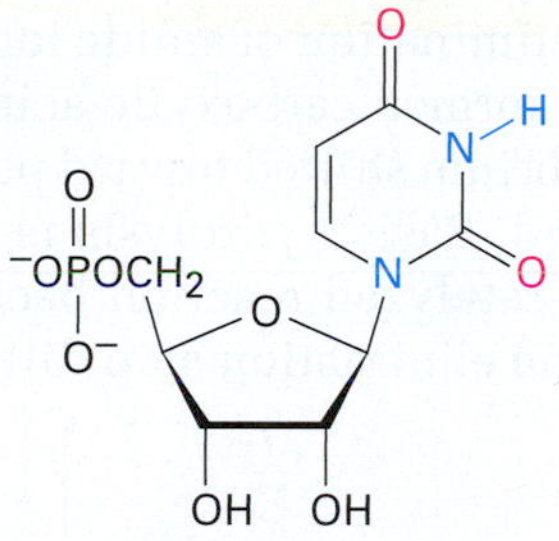

Uridine 5'-phosphate (a ribonucleotide)

Preparation of Amides

Amides are usually prepared by reaction of an acid chloride with an amine (Section 16.4). Ammonia, monosubstituted amines, and disubstituted amines all undergo the reaction.

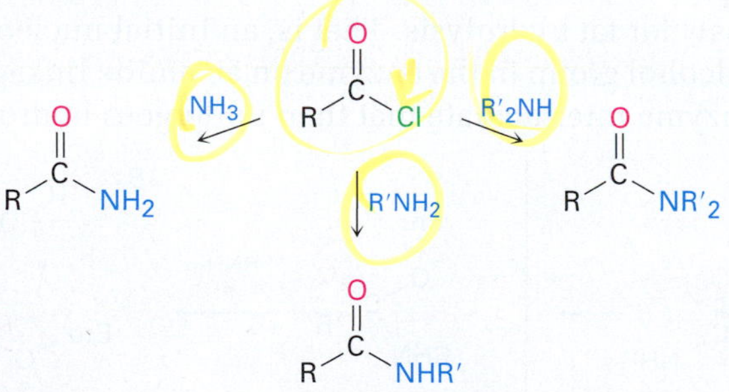

Reactions of Amides

CONVERSION OF AMIDES INTO CARBOXYLIC ACIDS: HYDROLYSIS Amides undergo hydrolysis to yield carboxylic acids plus ammonia or an amine on heating in either aqueous acid or aqueous base. The conditions required for amide hydrolysis are more severe than those required for the hydrolysis of acid chlorides or esters, but the mechanisms are similar. Acidic hydrolysis reaction occurs by nucleophilic addition of water to the protonated amide, followed by transfer of a proton from oxygen to nitrogen to make the nitrogen a better leaving group and subsequent elimination. The steps are reversible, with the equilibrium shifted toward product by protonation of NH_3 in the final step.

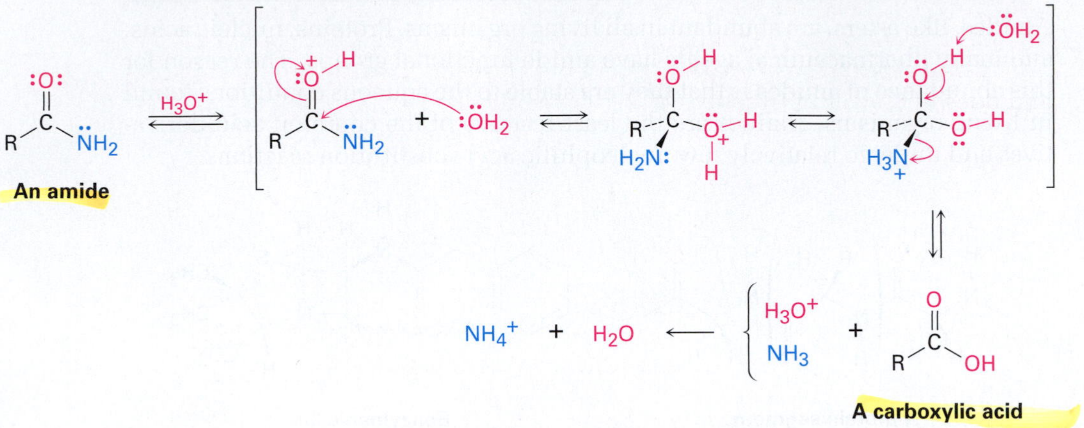

An amide

A carboxylic acid

Basic hydrolysis occurs by nucleophilic addition of OH^- to the amide carbonyl group, followed by elimination of amide ion ($^-NH_2$) and subsequent deprotonation of the initially formed carboxylic acid by ammonia. The steps are reversible, with the equilibrium shifted toward product by the final deprotonation of the carboxylic acid. Basic hydrolysis is substantially more difficult than the analogous acid-catalyzed reaction because amide ion is a very poor leaving group, making the elimination step difficult.

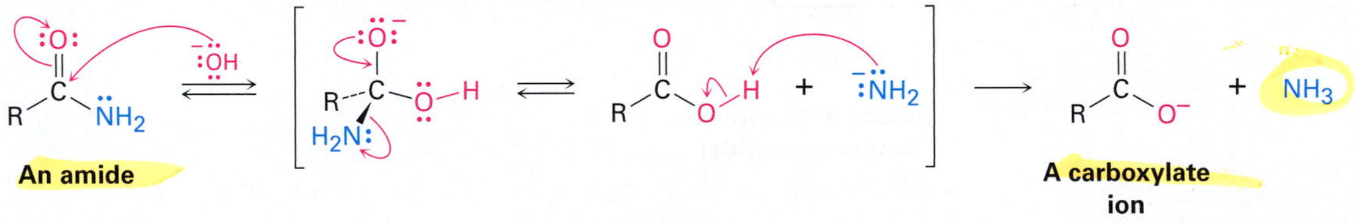

An amide

A carboxylate ion

Amide hydrolysis is common in biological chemistry. Just as the hydrolysis of esters is the initial step in the digestion of dietary fats, the hydrolysis of amides is the initial step in the digestion of dietary proteins. The reaction is catalyzed by protease enzymes and occurs by a mechanism almost identical to that we just saw for fat hydrolysis. That is, an initial nucleophilic acyl substitution of an alcohol group in the enzyme on an amide linkage in the protein gives an acyl enzyme intermediate that then undergoes hydrolysis.

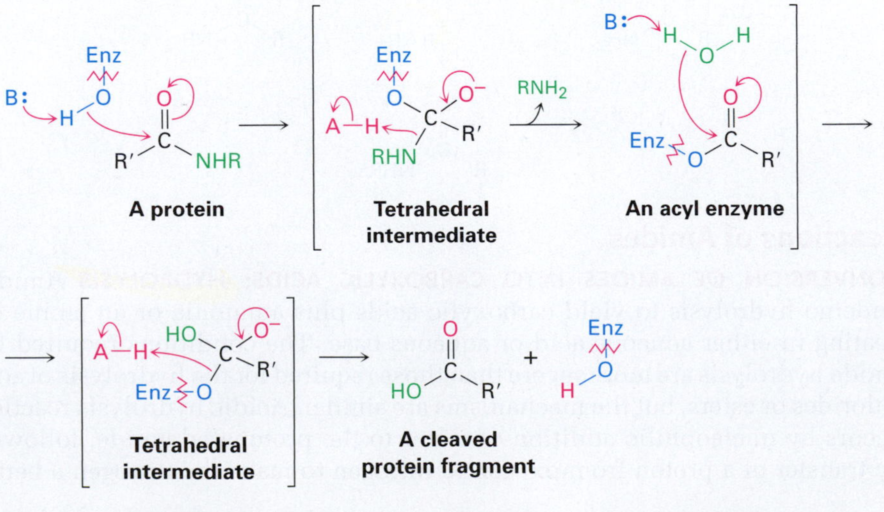

A protein

Tetrahedral intermediate

An acyl enzyme

Tetrahedral intermediate

A cleaved protein fragment

CONVERSION OF AMIDES INTO AMINES: REDUCTION Like other carboxylic acid derivatives, amides can be reduced by LiAlH$_4$. The product of the reduction, however, is an *amine* rather than an alcohol. The net effect of an amide reduction reaction is thus the conversion of the amide carbonyl group into a methylene group (C=O → CH$_2$). This kind of reaction is specific for amides and does not occur with other carboxylic acid derivatives.

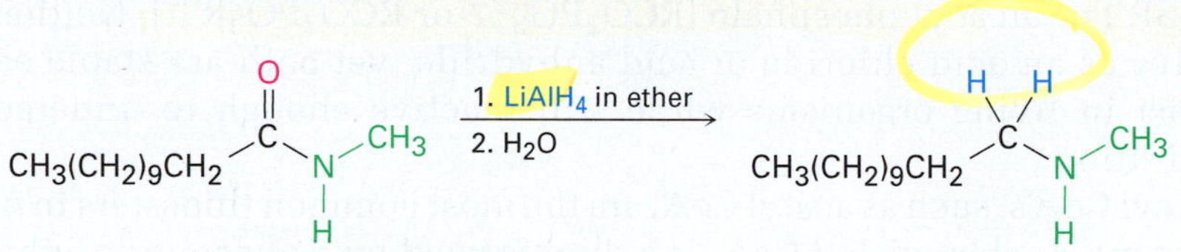

N-Methyldodecanamide

Dodecylmethylamine (95%)

Amide reduction occurs by nucleophilic addition of hydride ion to the amide carbonyl group, followed by expulsion of the *oxygen* atom as an aluminate anion leaving group to give an iminium ion intermediate. The intermediate iminium ion is then further reduced by LiAlH$_4$ to yield the amine.

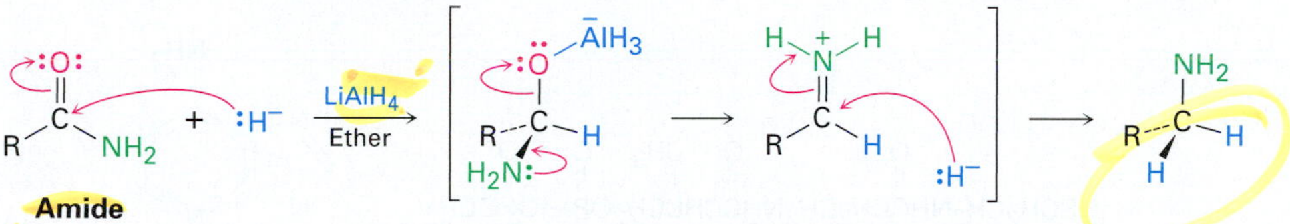

Amide

Iminium ion

The reaction is effective with both acyclic and cyclic amides, or **lactams**, and is a good method for preparing cyclic amines.

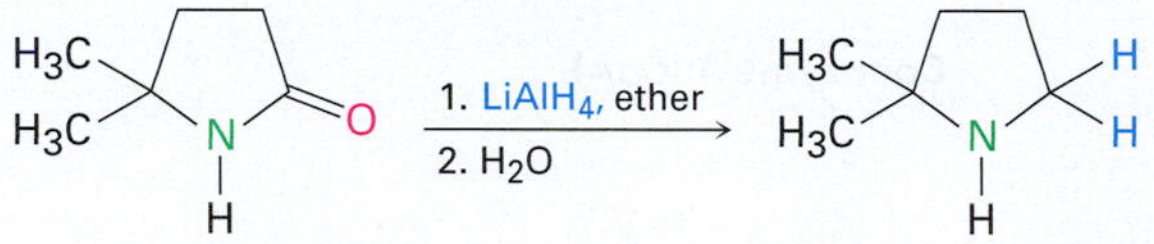

A lactam

A cyclic amine (80%)

Problem 16.20

How would you convert *N*-ethylbenzamide to each of the following products?
(a) Benzoic acid **(b)** Benzyl alcohol **(c)** C$_6$H$_5$CH$_2$NHCH$_2$CH$_3$

Problem 16.21

How would you use the reaction of an amide with LiAlH$_4$ as the key step in going from bromocyclohexane to (*N,N*-dimethylaminomethyl)cyclohexane? Write all the steps in the reaction sequence.

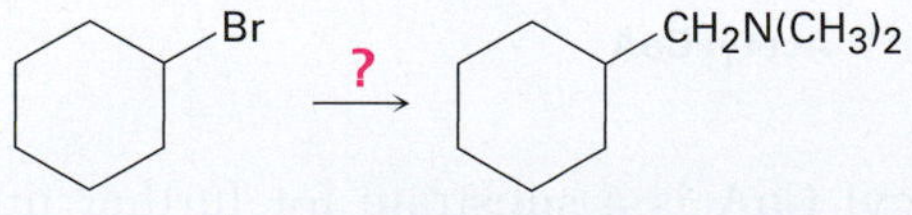

(N,N-Dimethylaminomethyl)cyclohexane

16.8 Chemistry of Thioesters and Acyl Phosphates: Biological Carboxylic Acid Derivatives

As mentioned in the chapter introduction, the substrate for a nucleophilic acyl substitution reaction in living organisms is generally either a thioester ($RCOSR'$) or an acyl phosphate [$RCO_2PO_3^{2-}$ or $RCO_2PO_3R'^-$]. Neither is as reactive as an acid chloride or acid anhydride, yet both are stable enough to exist in living organisms while still reactive enough to undergo acyl substitution.

Acyl CoA's, such as acetyl CoA, are the most common thioesters in nature. Coenzyme A, abbreviated CoA, is a thiol formed by a phosphoric anhydride linkage ($O=P-O-P=O$) between phosphopantetheine and adenosine 3',5'-bisphosphate. (The prefix *bis*- means "two" and indicates that adenosine 3',5'-bisphosphate has two phosphate groups, one on C3' and one on C5'.) Reaction of coenzyme A with an acyl phosphate or acyl adenylate gives the acyl CoA (Figure 16.10). As we saw in Section 16.5 (Figure 16.6), formation of the acyl adenylate occurs by reaction of a carboxylic acid with ATP and is itself a nucleophilic acyl substitution reaction that takes place on phosphorus.

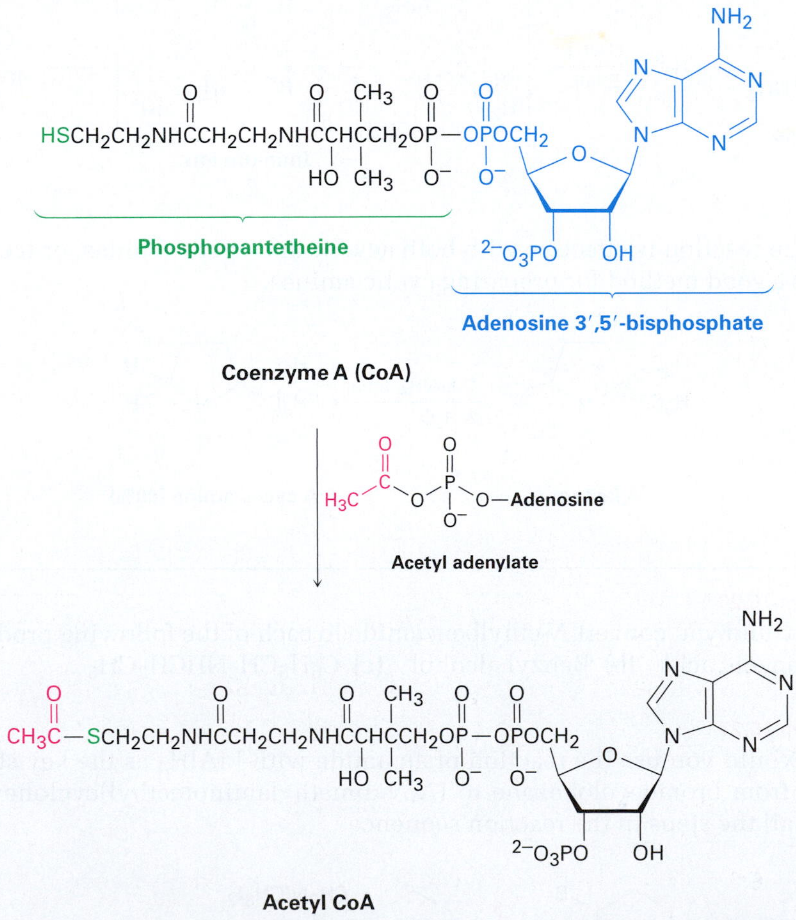

FIGURE 16.10 Formation of the thioester acetyl CoA by nucleophilic acyl substitution reaction of coenzyme A (CoA) with acetyl adenylate.

Once formed, an acyl CoA is a substrate for further nucleophilic acyl substitution reactions. For example, *N*-acetylglucosamine, a component of cartilage and other connective tissues, is synthesized by an aminolysis reaction between glucosamine and acetyl CoA.

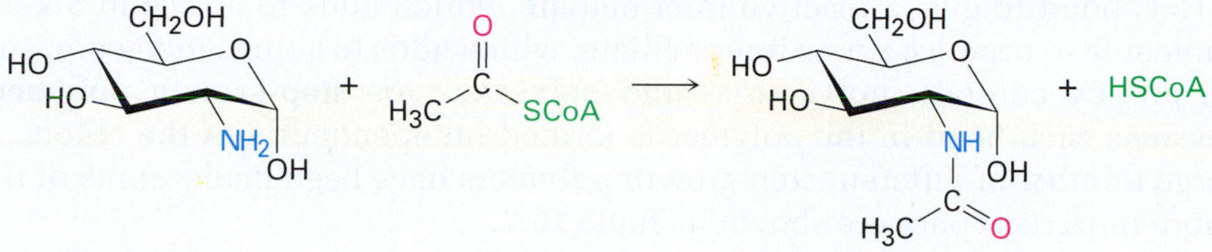

Glucosamine
(an amine)

N-Acetylglucosamine
(an amide)

Another example of a nucleophilic acyl substitution reaction on a thioester—this one a substitution by hydride ion to effect partial reduction of a thioester to an aldehyde—occurs in the biosynthesis of mevaldehyde, an intermediate in terpenoid synthesis (Chapter 7 *Lagniappe* and Section 23.7). In this reaction, (3*S*)-3-hydroxy-3-methylglutaryl CoA is reduced by hydride donation from NADPH.

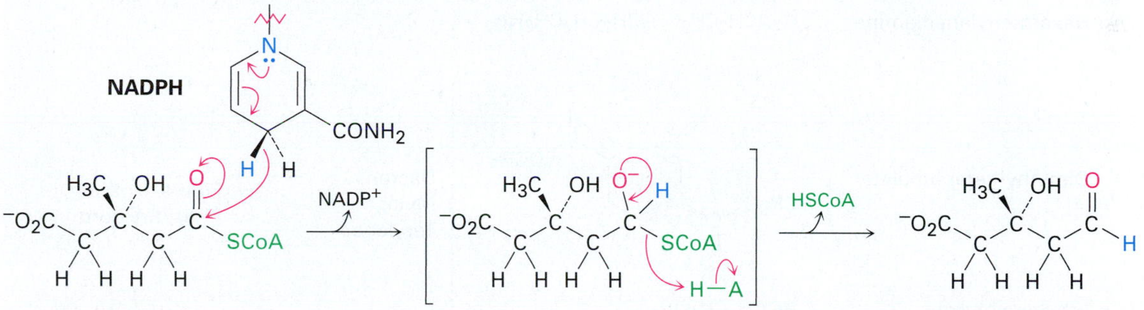

(3*S*)-3-Hydroxy-3-methylglutaryl CoA

(R)-Mevaldehyde

Problem 16.22
Write the mechanism of the reaction shown in Figure 16.10 between coenzyme A and acetyl adenylate to give acetyl CoA.

16.9 Polyamides and Polyesters: Step-Growth Polymers

When an amine reacts with an acid chloride, an amide is formed. What would happen, though, if a *diamine* and a *diacid chloride* were allowed to react? Each partner could form *two* amide bonds, linking more and more molecules together until a giant polyamide resulted. In the same way, reaction of a diol with a diacid would lead to a polyester.

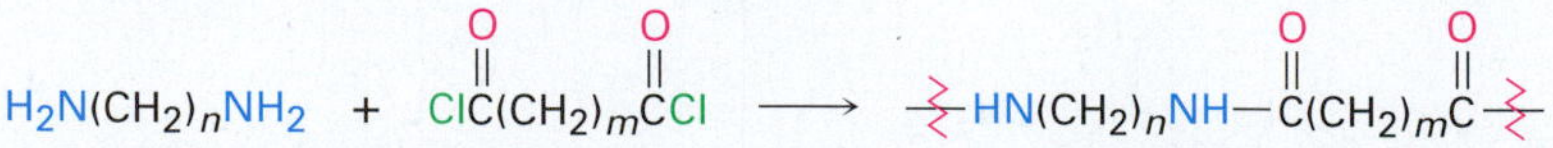

A diamine   **A diacid chloride**   **A polyamide (nylon)**

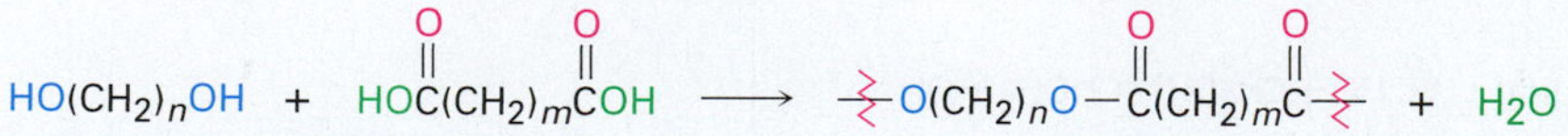

A diol   **A diacid**   **A polyester**

The alkene polymers discussed in Section 8.10 are called **chain-growth polymers** because they are produced by chain reactions. An initiator adds to a C=C bond to give a reactive intermediate, which adds to a second alkene molecule to produce a new intermediate, which adds to a third molecule, and so on. By contrast, polyamides and polyesters are **step-growth polymers** because each bond in the polymer is formed independently of the others. A large number of different step-growth polymers have been made; some of the more important ones are shown in Table 16.2.

TABLE 16.2
Some Common Step-Growth Polymers and Their Uses

Monomers	Structure	Polymer	Uses
Adipic acid + Hexamethylenediamine		Nylon 66	Fibers, clothing, tire cord
Dimethyl terephthalate + Ethylene glycol	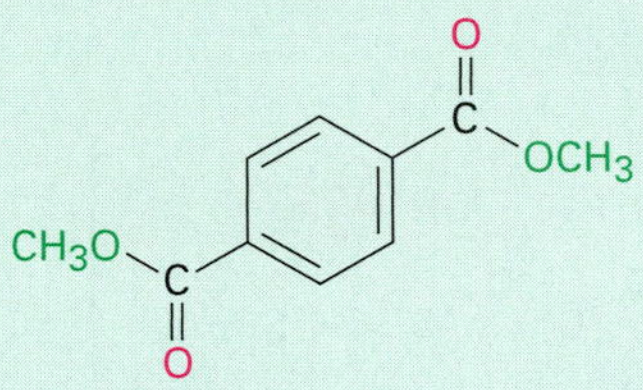	Dacron, Mylar, Terylene	Fibers, clothing, films, tire cord
Caprolactam	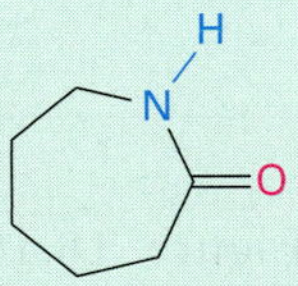	Nylon 6, Perlon	Fibers, castings
Diphenyl carbonate + Bisphenol A	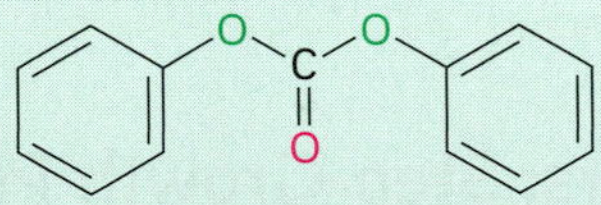	Lexan, polycarbonate	Equipment housing, molded articles
Toluene-2,6-diisocyanate + Poly(but-2-ene-1,4-diol)		Polyurethane, Spandex	Fibers, coatings, foams

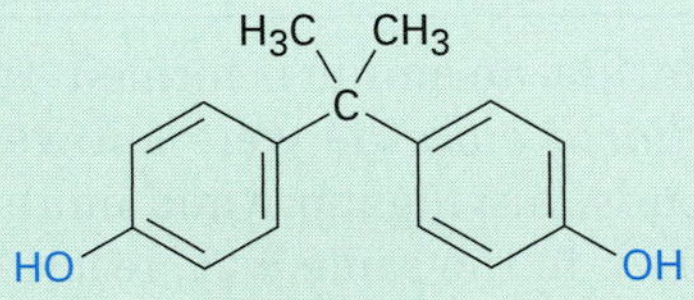

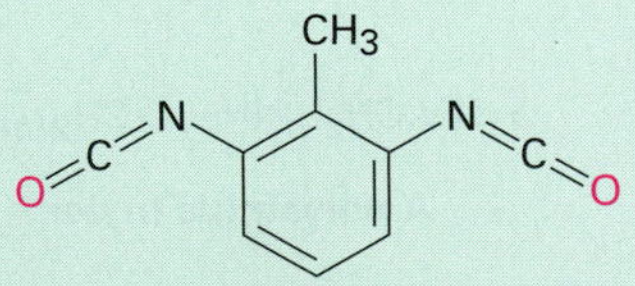

Polyamides (Nylons)

The best-known step-growth polymers are the polyamides, or *nylons*, first prepared by heating a diamine with a diacid. For example, nylon 66 is prepared by reaction of adipic acid (hexanedioic acid) with hexamethylene-diamine (hexane-1,6-diamine) at 280 °C. The designation "66" tells the number of carbon atoms in the diamine (the first 6) and the diacid (the second 6).

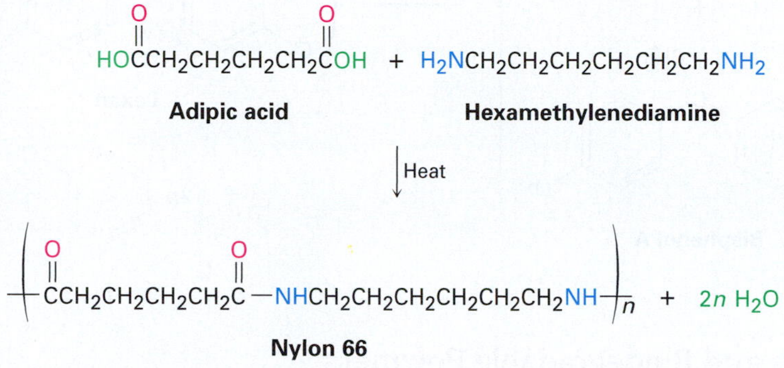

$$HOCCH_2CH_2CH_2CH_2COH \; + \; H_2NCH_2CH_2CH_2CH_2CH_2CH_2NH_2$$

Adipic acid   **Hexamethylenediamine**

Heat

$$\left(CCH_2CH_2CH_2CH_2C-NHCH_2CH_2CH_2CH_2CH_2CH_2NH\right)_n \; + \; 2n\,H_2O$$

Nylon 66

Nylons are used both in engineering applications and in making fibers. A combination of high impact strength and abrasion resistance makes nylon an excellent metal substitute for bearings and gears. As fiber, nylon is used in a variety of applications, from clothing to tire cord to Perlon mountaineering ropes.

Polyesters

The most generally useful polyester is that made by reaction between dimethyl terephthalate (dimethyl benzene-1,4-dicarboxylate) and ethylene glycol (ethane-1,2-diol). The product is used under the trade name Dacron to make clothing fiber and tire cord, and under the name Mylar to make recording tape. The tensile strength of poly(ethylene terephthalate) film is nearly equal to that of steel.

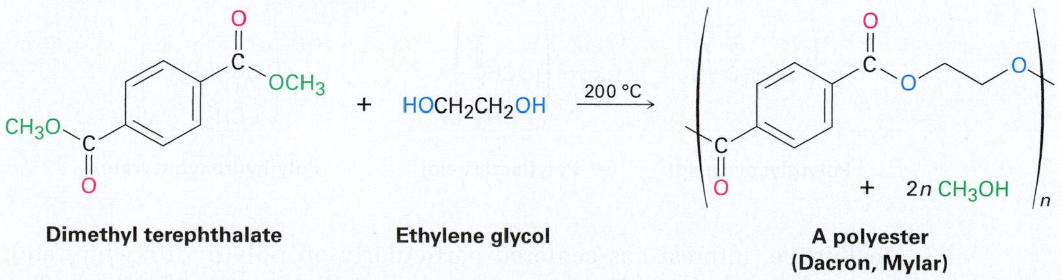

Dimethyl terephthalate   **Ethylene glycol**   **A polyester (Dacron, Mylar)**

$+ \; HOCH_2CH_2OH \xrightarrow{200\,°C} \qquad + \; 2n\,CH_3OH$

Lexan, a polycarbonate prepared from diphenyl carbonate and bisphenol A, is another commercially valuable polyester. Lexan has an unusually high impact strength, making it valuable for use in telephones, bicycle safety helmets, and laptop computer cases.

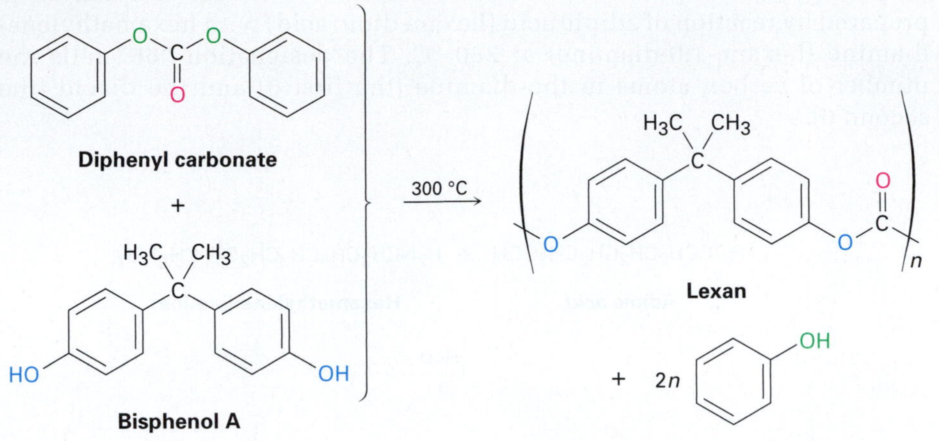

Diphenyl carbonate

+

Bisphenol A

$\xrightarrow{300\ °C}$

Lexan

+ 2n

Sutures and Biodegradable Polymers

Because plastics are too often thrown away rather than recycled, much work has been carried out on developing *biodegradable* polymers, which can be broken down rapidly in landfills by soil microorganisms. Among the most common biodegradable polymers are poly(glycolic acid) (PGA), poly(lactic acid) (PLA), and poly(hydroxybutyrate) (PHB). All are polyesters and are therefore susceptible to hydrolysis of their ester links. Copolymers of PGA with PLA have found a particularly wide range of uses. A 90/10 copolymer of poly(glycolic acid) with poly(lactic acid) is used to make absorbable sutures, for instance. The sutures are entirely hydrolyzed and absorbed by the body within 90 days after surgery.

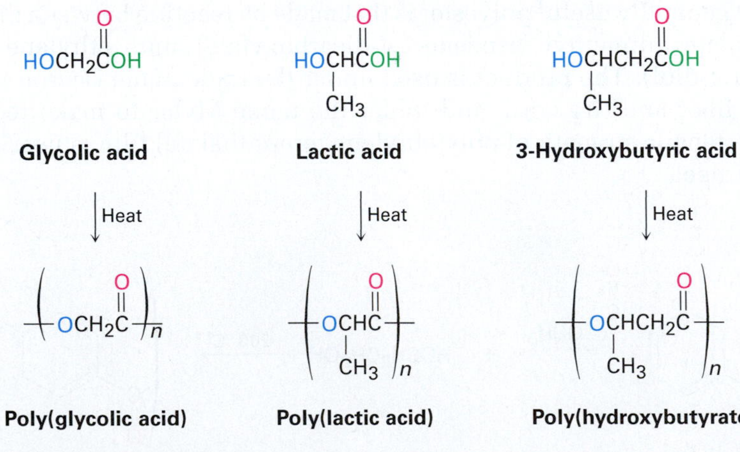

Glycolic acid

Lactic acid

3-Hydroxybutyric acid

Poly(glycolic acid)

Poly(lactic acid)

Poly(hydroxybutyrate)

In Europe, interest has centered particularly on poly(hydroxybutyrate), which can be made into films for packaging as well as into molded items. The polymer degrades within 4 weeks in landfills, both by ester hydrolysis and by an E1cB elimination reaction of the oxygen atom β to the carbonyl group. The use of poly(hydroxybutyrate) is limited at present by its cost—about four times that of polypropylene.

Problem 16.23

Draw structures of the step-growth polymers you would expect to obtain from the following reactions:

(a) $BrCH_2CH_2CH_2Br$ + $HOCH_2CH_2CH_2OH$ $\xrightarrow{\text{Base}}$ **?**

(b) $HOCH_2CH_2OH$ + $HO_2C(CH_2)_6CO_2H$ $\xrightarrow{\text{H}_2\text{SO}_4 \text{ catalyst}}$ **?**

(c)

$$H_2N(CH_2)_6NH_2 \;+\; \underset{\text{O}}{\overset{\text{O}}{\underset{\|}{\text{ClC}}}}(CH_2)_4\underset{\text{O}}{\overset{\text{O}}{\underset{\|}{\text{CCl}}}} \longrightarrow \; \textbf{?}$$

Problem 16.24

Kevlar, a nylon polymer prepared by reaction of benzene-1,4-dicarboxylic acid (terephthalic acid) with benzene-1,4-diamine (*p*-phenylenediamine), is so strong that it's used to make bulletproof vests. Draw the structure of a segment of Kevlar.

16.10 Spectroscopy of Carboxylic Acid Derivatives

Infrared Spectroscopy

All carbonyl-containing compounds have intense IR absorptions in the range 1650 to 1850 cm^{-1}. As shown in Table 16.3, the exact position of the absorption provides information about the specific kind of carbonyl group. For comparison, the IR absorptions of aldehydes, ketones, and carboxylic acids are included in the table, along with values for carboxylic acid derivatives.

Acid chlorides are easily detected by their characteristic absorption near 1800 cm^{-1}. Acid anhydrides can be identified by the fact that they show two absorptions in the carbonyl region, one at 1820 cm^{-1} and another at 1760 cm^{-1}. Esters are detected by their absorption at 1735 cm^{-1}, a position somewhat higher than that for either aldehydes or ketones. Amides, by contrast, absorb near the low wavenumber end of the carbonyl region, with the degree of substitution on nitrogen affecting the exact position of the IR band.

TABLE 16.3

Infrared Absorptions of Some Carbonyl Compounds

Carbonyl type	Absorption (cm^{-1})
Saturated acid chloride	1810
Aromatic acid chloride	1770
Saturated acid anhydride	1820, 1760
Saturated ester	1735
Aromatic ester	1720
Saturated amide	1690
N-Substituted amide	1680
N,N-Disubstituted amide	1650
Aromatic amide	1675

Problem 16.25

What kinds of functional groups might compounds have if they show the following IR absorptions?
(a) Absorption at 1735 cm^{-1}
(b) Absorption at 1810 cm^{-1}
(c) Absorptions at 2500–3300 cm^{-1} and 1710 cm^{-1}
(d) Absorption at 1715 cm^{-1}

Problem 16.26

Propose structures for compounds that have the following formulas and IR absorptions:
(a) $C_6H_{12}O_2$, 1735 cm^{-1}
(b) C_4H_9NO, 1650 cm^{-1}
(c) C_4H_5ClO, 1780 cm^{-1}

Nuclear Magnetic Resonance Spectroscopy

Hydrogens on the carbon next to a carbonyl group are slightly deshielded and absorb near 2 δ in the ^{1}H NMR spectrum. The identity of the carbonyl group can't be determined by ^{1}H NMR, however, because the α hydrogens of all acid derivatives absorb in the same range. Figure 16.11 shows the ^{1}H NMR spectrum of ethyl acetate.

FIGURE 16.11
^{1}H NMR spectrum of ethyl acetate.

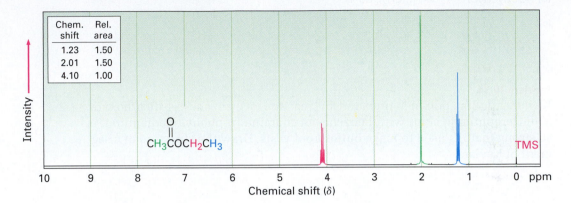

Chem. shift	Rel. area
1.23	1.50
2.01	1.50
4.10	1.00

Although ^{13}C NMR is useful for determining the presence or absence of a carbonyl group in a molecule, the identity of the carbonyl group is difficult to determine. Aldehydes and ketones absorb near 200 δ, while the carbonyl carbon atoms of various acid derivatives absorb in the range 160 to 180 δ (Table 16.4).

TABLE 16.4
^{13}C NMR Absorptions in Some Carbonyl Compounds

Compound	Absorption (δ)	Compound	Absorption (δ)
Acetic acid	177.3	Acetic anhydride	166.9
Ethyl acetate	170.7	Acetone	205.6
Acetyl chloride	170.3	Acetaldehyde	201.0
Acetamide	172.6		

Summary

Key Words

acid anhydride (RCO$_2$COR'), 643

acid halide (RCOX), 643

acyl phosphate (RCOPO$_3$$^{2-}$), 643

amide (RCONH$_2$), 643

carboxylic acid derivative, 643

chain-growth polymer, 676

ester (RCO$_2$R'), 643

lactam, 673

lactone, 666

Carboxylic acid derivatives—compounds in which the –OH group of a carboxylic acid has been replaced by another substituent—are among the most widely occurring of all molecules and are involved in almost all biological pathways. In this chapter, we covered the chemistry necessary for understanding them and thus also necessary for understanding the chemistry of living organisms. **Acid halides**, **acid anhydrides**, **esters**, and **amides** are the most common such derivatives in the laboratory; **thioesters** and **acyl phosphates** are common in biological molecules.

The chemistry of carboxylic acid derivatives is dominated by the **nucleophilic acyl substitution reaction**. Mechanistically, these substitutions take place by addition of a nucleophile to the polar carbonyl group of the acid derivative to give a tetrahedral intermediate, followed by expulsion of a leaving group.

The reactivity of an acid derivative toward substitution depends both on the steric environment near the carbonyl group and on the electronic nature of the substituent, Y. The reactivity order is acid halide > acid anhydride > thioester > ester > amide.

The most common reactions of carboxylic acid derivatives are substitution by water *(hydrolysis)* to yield an acid, by an alcohol *(alcoholysis)* to yield an ester, by an amine *(aminolysis)* to yield an amide, by hydride ion to yield an alcohol *(reduction)*, and by an organometallic reagent to yield an alcohol *(Grignard reaction)*.

Step-growth polymers, such as polyamides and polyesters, are prepared by reactions between difunctional molecules. Polyamides (nylons) are formed by reaction between a diacid and a diamine; polyesters are formed from a diacid and a diol.

IR spectroscopy is a valuable tool for the structural analysis of acid derivatives. Acid chlorides, anhydrides, esters, and amides all show characteristic IR absorptions that can be used to identify these functional groups.

Summary of Reactions

1. Reactions of carboxylic acids (Section 16.3)

    (a) Conversion into acid chlorides

    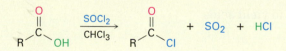

    (b) Conversion into esters

    (c) Conversion into amides

    (d) Reduction to yield primary alcohols

2. Reactions of acid chlorides (Section 16.4)

(a) Hydrolysis to yield acids

$$
\underset{\substack{R}}{\overset{\substack{O\\\|}}{C}}{}_{Cl} + H_2O \longrightarrow \underset{\substack{R}}{\overset{\substack{O\\\|}}{C}}{}_{OH} + HCl
$$

(b) Reaction with carboxylates to yield anhydrides

$$
\underset{\substack{R}}{\overset{\substack{O\\\|}}{C}}{}_{Cl} + RCO_2^- \longrightarrow \underset{\substack{R}}{\overset{\substack{O\\\|}}{C}}{}_O\underset{\substack{R}}{\overset{\substack{O\\\|}}{C}} + Cl^-
$$

(c) Alcoholysis to yield esters

$$
\underset{\substack{R}}{\overset{\substack{O\\\|}}{C}}{}_{Cl} + R'OH \xrightarrow{\text{Pyridine}} \underset{\substack{R}}{\overset{\substack{O\\\|}}{C}}{}_{OR'} + HCl
$$

(d) Aminolysis to yield amides

$$
\underset{\substack{R}}{\overset{\substack{O\\\|}}{C}}{}_{Cl} + R'OH \xrightarrow{\text{Pyridine}} \underset{\substack{R}}{\overset{\substack{O\\\|}}{C}}{}_{OR'} + HCl
$$

(e) Lithium diorganocopper reaction to yield ketones

$$
\underset{\substack{R}}{\overset{\substack{O\\\|}}{C}}{}_{Cl} \xrightarrow[\text{Ether}]{R'_2CuLi} \underset{\substack{R}}{\overset{\substack{O\\\|}}{C}}{}_{R'}
$$

3. Reactions of acid anhydrides (Section 16.5)

(a) Hydrolysis to yield acids

$$
\underset{\substack{R}}{\overset{\substack{O\\\|}}{C}}{}_O\underset{\substack{R}}{\overset{\substack{O\\\|}}{C}} + H_2O \longrightarrow 2\ \underset{\substack{R}}{\overset{\substack{O\\\|}}{C}}{}_{OH}
$$

(b) Alcoholysis to yield esters

$$
\underset{\substack{R}}{\overset{\substack{O\\\|}}{C}}{}_O\underset{\substack{R}}{\overset{\substack{O\\\|}}{C}} + R'OH \longrightarrow \underset{\substack{R}}{\overset{\substack{O\\\|}}{C}}{}_{OR'} + \underset{\substack{R}}{\overset{\substack{O\\\|}}{C}}{}_{OH}
$$

(c) Aminolysis to yield amides

$$
\underset{\substack{R}}{\overset{\substack{O\\\|}}{C}}{}_O\underset{\substack{R}}{\overset{\substack{O\\\|}}{C}} + 2\ NH_3 \longrightarrow \underset{\substack{R}}{\overset{\substack{O\\\|}}{C}}{}_{NH_2} + \underset{\substack{R}}{\overset{\substack{O\\\|}}{C}}{}_{O^-\ {}^+NH_4}
$$

4. Reactions of esters and lactones (Section 16.6)

(a) Hydrolysis to yield acids

$$
\underset{\substack{R}}{\overset{\substack{O\\\|}}{C}}{}_{OR'} \xrightarrow[\text{or NaOH, } H_2O]{H_3O^+} \underset{\substack{R}}{\overset{\substack{O\\\|}}{C}}{}_{OH} + R'OH
$$

(b) Reduction to yield primary alcohols

$$\underset{\substack{R \quad OR'}}{C}\overset{O}{\parallel} \xrightarrow[\text{2. } H_3O^+]{\text{1. LiAlH}_4\text{, ether}} \underset{\substack{R \quad OH}}{C}\overset{H \quad H}{} + R'OH$$

(c) Partial reduction to yield aldehydes

$$\underset{\substack{R \quad OR'}}{C}\overset{O}{\parallel} \xrightarrow[\text{2. } H_3O^+]{\text{1. DIBAH, toluene}} \underset{\substack{R \quad H}}{C}\overset{O}{\parallel} + R'OH$$

(d) Grignard reaction to yield tertiary alcohols

$$\underset{\substack{R \quad OR'}}{C}\overset{O}{\parallel} \xrightarrow[\text{2. } H_3O^+]{\text{1. 2 R''MgX, ether}} \underset{\substack{R \quad OH}}{C}\overset{R'' \quad R''}{} + R'OH$$

5. Reactions of amides (Section 16.7)

(a) Hydrolysis to yield acids

$$\underset{\substack{R \quad NH_2}}{C}\overset{O}{\parallel} \xrightarrow[\text{or NaOH, } H_2O]{H_3O^+} \underset{\substack{R \quad OH}}{C}\overset{O}{\parallel} + NH_3$$

(b) Reduction to yield amines

$$\underset{\substack{R \quad NH_2}}{C}\overset{O}{\parallel} \xrightarrow[\text{2. } H_3O^+]{\text{1. LiAlH}_4\text{, ether}} \underset{\substack{R \quad NH_2}}{C}\overset{H \quad H}{}$$

Lagniappe

β-Lactam Antibiotics

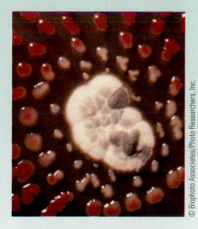

Penicillium mold growing in a petri dish.

The value of hard work and logical thinking shouldn't be underestimated, but pure luck also plays a role in most real scientific breakthroughs. What has been called "the supreme example [of luck] in all scientific history" occurred in the late summer of 1928, when the Scottish bacteriologist Alexander Fleming went on vacation, leaving in his lab a culture plate recently inoculated with the bacterium *Staphylococcus aureus*.

While Fleming was away, an extraordinary chain of events occurred. First, a 9-day cold spell lowered the laboratory temperature to a point where the *Staphylococcus* on the plate could not grow. During this time, spores from a colony of the mold *Penicillium notatum* being grown on the floor below wafted up into Fleming's lab and landed in the culture plate. The temperature then rose, and both *Staphylococcus* and *Penicillium* began to grow. On returning from vacation, Fleming discarded the plate into a tray of antiseptic, intending to sterilize it. Evidently, though, the plate did not sink deeply enough into the antiseptic, because when Fleming happened to glance at it a few days later, what he saw changed the course of human history. He noticed that the growing *Penicillium* mold appeared to dissolve the colonies of staphylococci.

Fleming realized that the *Penicillium* mold must be producing a chemical that killed the *Staphylococcus* bacteria, and he spent several years trying to isolate the substance. Finally, in 1939, the Australian pathologist Howard Florey and the German refugee Ernst Chain managed to isolate the active substance, called *penicillin*. The dramatic ability of penicillin to cure infections in mice was soon demonstrated, and successful tests in humans followed shortly

continued

Lagniappe *continued*

thereafter. By 1943, penicillin was being produced on a large scale for military use in World War II, and by 1944 it was being used on civilians. Fleming, Florey, and Chain shared the 1945 Nobel Prize in Medicine.

Now called benzylpenicillin, or penicillin G, the substance first discovered by Fleming is but one member of a large class of so-called β-lactam antibiotics, compounds with a four-membered lactam (cyclic amide) ring. The four-membered lactam ring is fused to a five-membered, sulfur-containing ring, and the carbon atom next to the lactam carbonyl group is bonded to an acylamino substituent, RCONH–. This acylamino side chain can be varied in the laboratory to provide many hundreds of penicillin analogs with different biological activity profiles. Ampicillin, for instance, has an α-aminophenylacetamido substituent [PhCH(NH$_2$)CONH–].

Closely related to the penicillins are the *cephalosporins,* a group of β-lactam antibiotics that contain an unsaturated six-membered, sulfur-containing ring. Cephalexin, marketed under the trade name Keflex, is an example. Cephalosporins generally have much greater antibacterial activity than penicillins, particularly against resistant strains of bacteria.

Cephalexin (a cephalosporin)

The biological activity of penicillins and cephalosporins is due to the presence of the strained β-lactam ring, which reacts with and deactivates the transpeptidase enzyme needed to synthesize and repair bacterial cell walls. With the wall either incomplete or weakened, the bacterial cell ruptures and dies.

Benzylpenicillin (penicillin G)

Exercises

■ indicates problems that are assignable in Organic OWL.

Go to this book's companion website at **www.cengage.com/ chemistry/mcmurry** to explore interactive versions of the Active Figures from this text.

VISUALIZING CHEMISTRY

(Problems 16.1–16.26 appear within the chapter.)

16.27 ■ Name the following compounds:

(a)

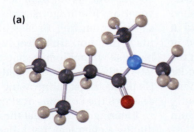

(b)

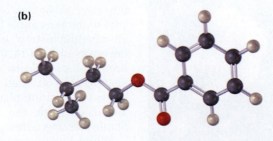

■ Problems assignable in Organic OWL.

16.28 How would you prepare the following compounds starting with an appropriate carboxylic acid and any other reagents needed? (Reddish brown = Br.)

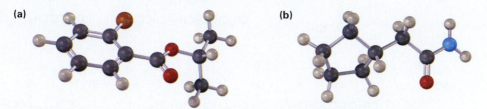

(a) **(b)**

16.29 ■ The following structure represents a tetrahedral alkoxide-ion intermediate formed by addition of a nucleophile to a carboxylic acid derivative. Identify the nucleophile, the leaving group, the starting acid derivative, and the ultimate product (yellow-green = Cl).

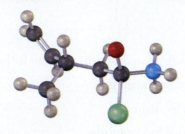

16.30 Electrostatic potential maps of a typical amide (acetamide) and an acyl azide (acetyl azide) are shown. Which of the two do you think is more reactive in nucleophilic acyl substitution reactions? Explain.

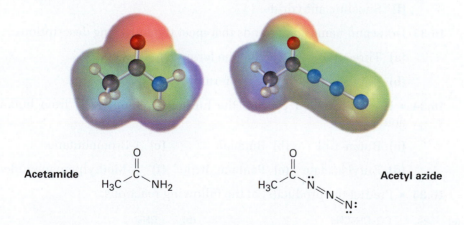

Acetamide

H_3C—C(=O)—NH_2

Acetyl azide

H_3C—C(=O)—$\ddot{N}=N=\ddot{N}:$

ADDITIONAL PROBLEMS

16.31 ■ Give IUPAC names for the following compounds:

(a)

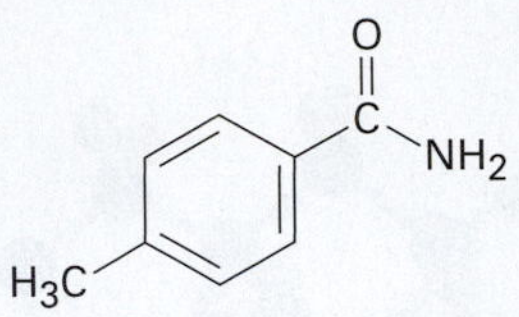

(b) 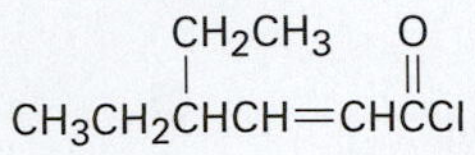 $CH_3CH_2CHCH=CHCCl$ with CH_2CH_3 and O

(c) 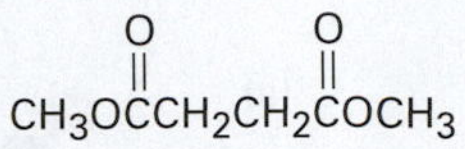 $CH_3OCCH_2CH_2COCH_3$ with two O

(d) 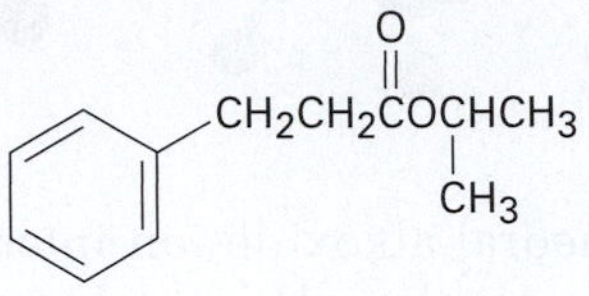 $CH_2CH_2COCHCH_3$ with O and CH_3

(e) 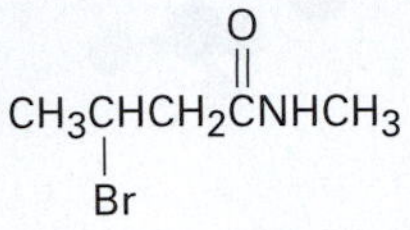 $CH_3CHCH_2CNHCH_3$ with O and Br

(f) 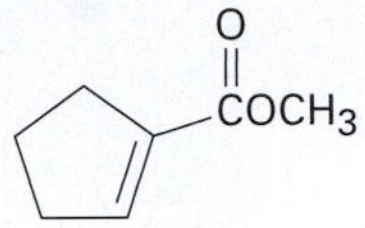 $COCH_3$ with O (cyclopentene ring)

(g)

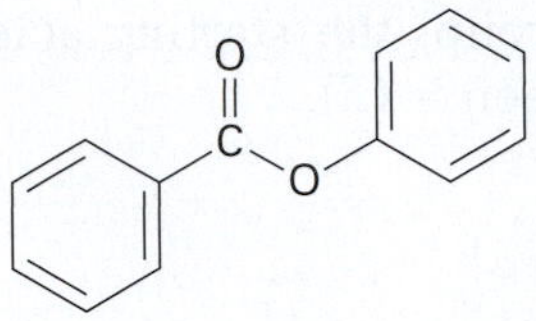

(h) 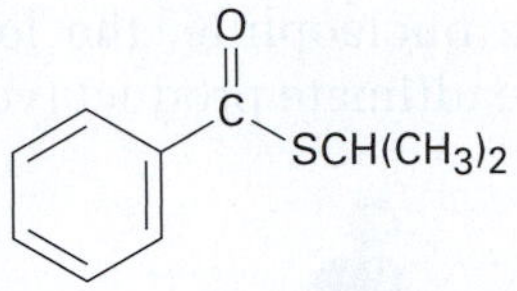 $SCH(CH_3)_2$ with O

16.32 ■ Draw structures corresponding to the following names:

(a) *p*-Bromophenylacetamide

(b) *m*-Benzoylbenzamide

(c) 2,2-Dimethylhexanamide

(d) Cyclohexyl cyclohexanecarboxylate

(e) Ethyl cyclobut-2-enecarboxylate

(f) Succinic anhydride

16.33 Draw and name compounds that meet the following descriptions:

(a) Three thioesters having the formula $C_6H_{10}OS$

(b) Three amides having the formula $C_7H_{11}NO$

16.34 ■ How might you prepare the following compounds from butanoic acid?

(a) Butan-1-ol   (b) Butanal   (c) 1-Bromobutane

(d) Butyl acetate   (e) Pentanenitrile   (f) *N*-Methylpentanamide

16.35 ■ Predict the product(s) of the following reactions:

(a) 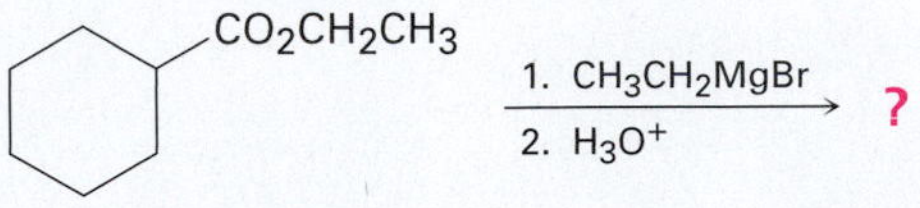 $CO_2CH_2CH_3$ (cyclohexane ring) → 1. CH_3CH_2MgBr / 2. H_3O^+ → **?**

(b) 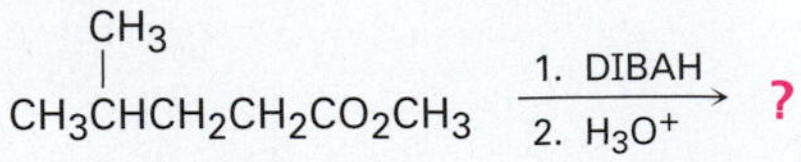 $CH_3CHCH_2CH_2CO_2CH_3$ with CH_3 → 1. DIBAH / 2. H_3O^+ → **?**

(c) 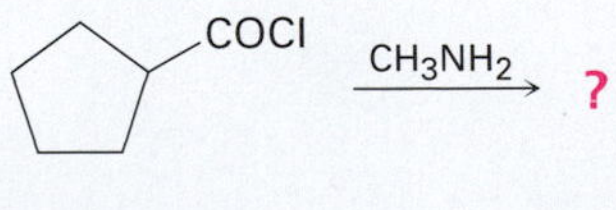 $COCl$ (cyclopentane ring) → CH_3NH_2 → **?**

(d) 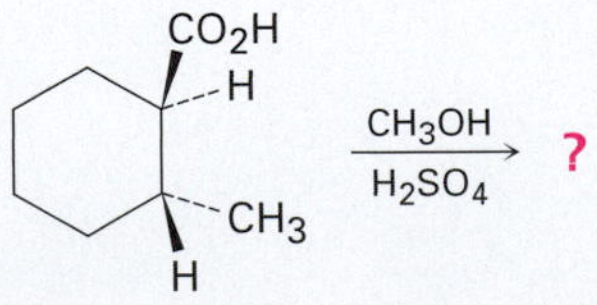 cyclohexane with CO_2H, H, CH_3, H → CH_3OH / H_2SO_4 → **?**

(e)

$$H_2C{=}CHCHCH_2CO_2CH_3 \quad \xrightarrow[\text{2. } H_3O^+]{\text{1. LiAlH}_4} \quad ?$$

with CH₃ substituent on the CH position.

(f)

cyclohexanol with OH $\xrightarrow[\text{Pyridine}]{\text{CH}_3\text{CO}_2\text{COCH}_3}$?

(g)

benzene ring with CONH₂ and CH₃ substituents $\xrightarrow[\text{2. H}_2\text{O}]{\text{1. LiAlH}_4}$?

(h)

benzene ring with CH₂CO₂H and Br substituents $\xrightarrow{\text{SOCl}_2}$?

16.36 ■ Predict the product, if any, of reaction between methyl acetate and the following reagents:

(a) $LiAlH_4$, then H_3O^+ (b) CH_3MgBr, then H_3O^+

(c) $NaOH, H_2O$ (d) Aniline

16.37 ■ Answer Problem 16.36 for reaction of the listed reagents with propanamide.

16.38 Treatment of 5-aminopentanoic acid with DCC (dicyclohexylcarbodiimide) yields a lactam. Show the structure of the product and the mechanism of the reaction.

16.39 ■ The following reactivity order has been found for the saponification of *p*-substituted methyl benzoates:

$$Y = NO_2 > Br > H > CH_3 > OCH_3$$

How can you explain this reactivity order? Where would you expect Y = CHO and Y = NH₂ to be in the reactivity list?

benzene ring with CO₂CH₃ and Y substituents $\xrightarrow[\text{H}_2\text{O}]{^-\text{OH}}$ benzene ring with CO₂⁻ and Y substituents + CH_3OH

16.40 ■ The following reactivity order has been found for the saponification of alkyl acetates by aqueous NaOH. Explain.

$$CH_3CO_2CH_3 > CH_3CO_2CH_2CH_3 > CH_3CO_2CH(CH_3)_2 > CH_3CO_2C(CH_3)_3$$

16.41 Explain the observation that attempted Fischer esterification of 2,4,6-trimethylbenzoic acid with methanol and HCl is unsuccessful. No ester is obtained, and the acid is recovered unchanged. What alternative method of esterification might be successful?

16.42 Draw the structure of the polymer you would expect to obtain from reaction of dimethyl terephthalate with a triol such as glycerol. What structural feature would this new polymer have that was not present in Dacron? How do you think this new feature might affect the properties of the polymer?

16.43 ■ Fats are biosynthesized from glycerol 3-phosphate and fatty-acyl CoA's by a reaction sequence that begins with the following step. Show the mechanism of the reaction.

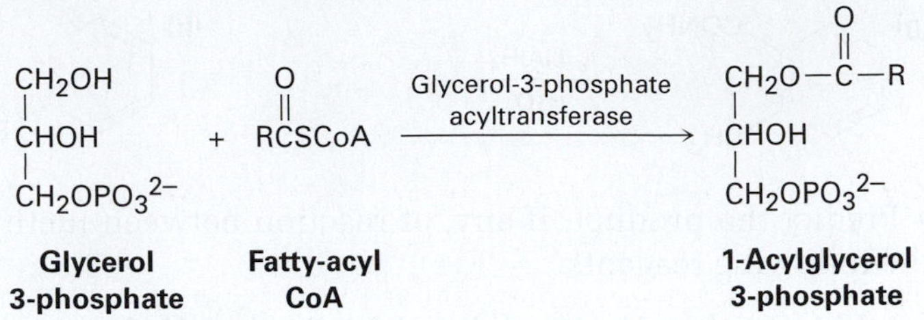

| Glycerol 3-phosphate | Fatty-acyl CoA | | 1-Acylglycerol 3-phosphate |

16.44 When a carboxylic acid is dissolved in isotopically labeled water, the label rapidly becomes incorporated into *both* oxygen atoms of the carboxylic acid. Explain.

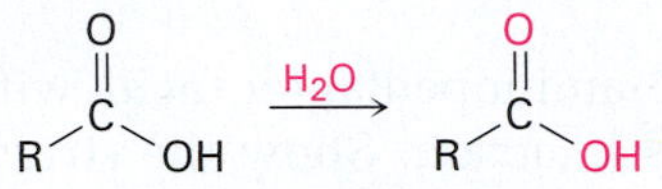

16.45 ■ When *ethyl* benzoate is heated in methanol containing a small amount of HCl, *methyl* benzoate is formed. Propose a mechanism for the reaction.

16.46 ■ *tert*-Butoxycarbonyl azide, a reagent used in protein synthesis, is prepared by treating *tert*-butoxycarbonyl chloride with sodium azide. Propose a mechanism for this reaction.

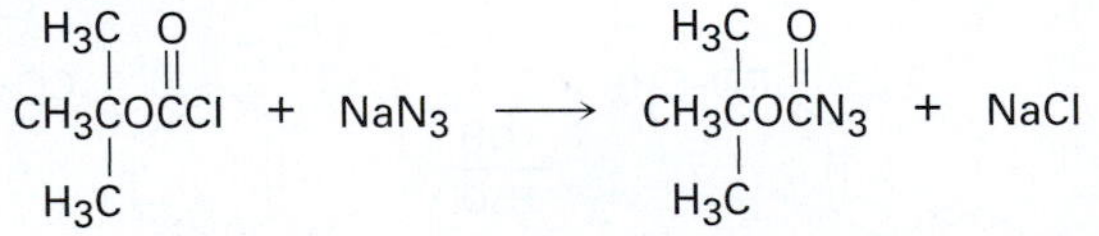

16.47 We said in Section 16.6 that mechanistic studies on ester hydrolysis have been carried out using ethyl propanoate labeled with ^{18}O in the ether-like oxygen. Assuming that ^{18}O-labeled acetic acid is your only source of isotopic oxygen, propose a synthesis of the labeled ethyl propanoate.

16.48 Treatment of an α-amino acid with DCC yields a 2,5-diketopiperazine. Propose a mechanism.

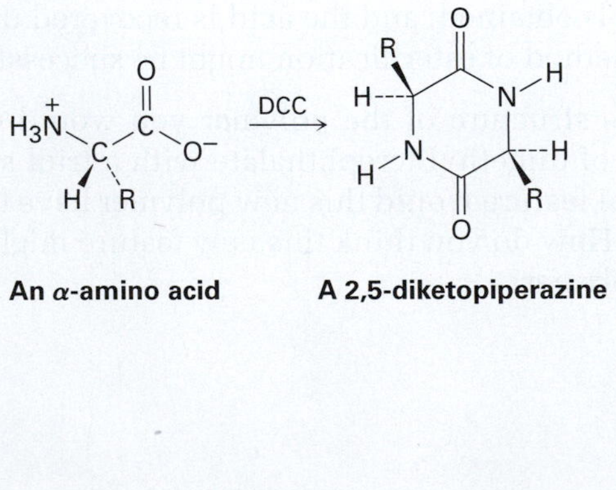

An α-amino acid   **A 2,5-diketopiperazine**

16.49 ▪ Treatment of a carboxylic acid with trifluoroacetic anhydride leads to an unsymmetrical anhydride that rapidly reacts with alcohol to give an ester:

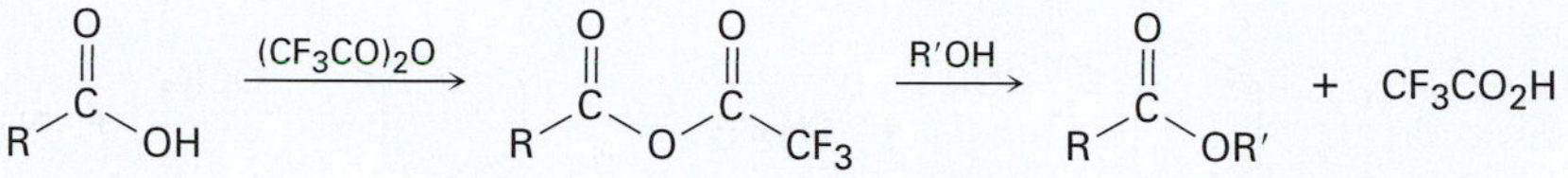

(a) Propose a mechanism for formation of the unsymmetrical anhydride.

(b) Why is the unsymmetrical anhydride unusually reactive?

(c) Why does the unsymmetrical anhydride react as indicated rather than giving a trifluoroacetate ester plus carboxylic acid?

16.50 ▪ Succinic anhydride yields succinimide when heated with ammonium chloride at 200 °C. Propose a mechanism for this reaction. Why do you suppose such a high reaction temperature is required?

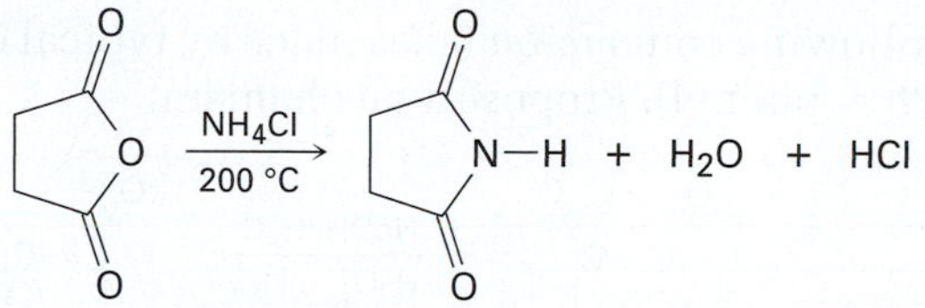

16.51 ▪ Phenyl 4-aminosalicylate is a drug used in the treatment of tuberculosis. Propose a synthesis of this compound starting from 4-nitrosalicylic acid.

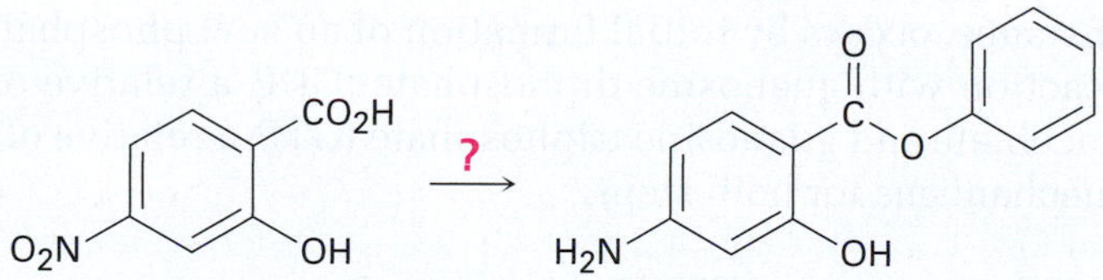

4-Nitrosalicylic acid **Phenyl 4-aminosalicylate**

16.52 ▪ *N,N*-Diethyl-*m*-toluamide (DEET) is the active ingredient in many insect-repellent preparations. How might you synthesize DEET from *m*-bromotoluene?

***N,N*-Diethyl-*m*-toluamide**

▪ Problems assignable in Organic OWL.

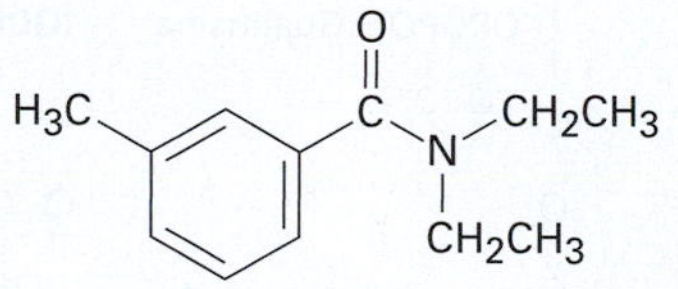

16.53 One frequently used method for preparing methyl esters is by reaction of carboxylic acids with diazomethane, CH_2N_2:

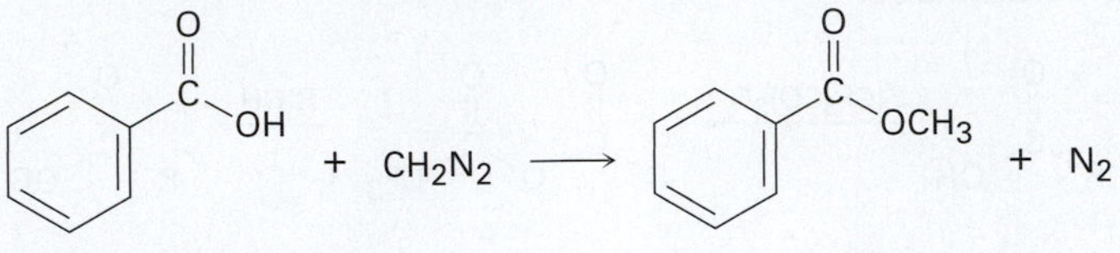

Benzoic acid   **Diazomethane**   **Methyl benzoate (100%)**

The reaction occurs in two steps: (1) protonation of diazomethane by the carboxylic acid to yield methyldiazonium ion, $CH_3N_2^+$, plus a carboxylate ion, and (2) reaction of the carboxylate ion with $CH_3N_2^+$.

(a) Draw two resonance structures of diazomethane, and account for step 1.

(b) What kind of reaction occurs in step 2?

16.54 ▪ The following conversion takes place by typical carbonyl-group reactions (Ph = phenyl). Propose a mechanism.

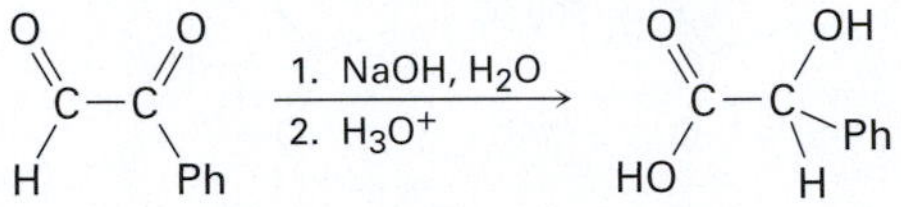

16.55 ▪ The hydrolysis of a biological thioester to the corresponding carboxylate is often more complex than the overall result might suggest. The conversion of succinyl CoA to succinate in the citric acid cycle, for instance, occurs by initial formation of an acyl phosphate, followed by reaction with guanosine diphosphate (GDP, a relative of ADP) to give succinate and guanosine triphosphate (GTP, a relative of ATP). Suggest mechanisms for both steps.

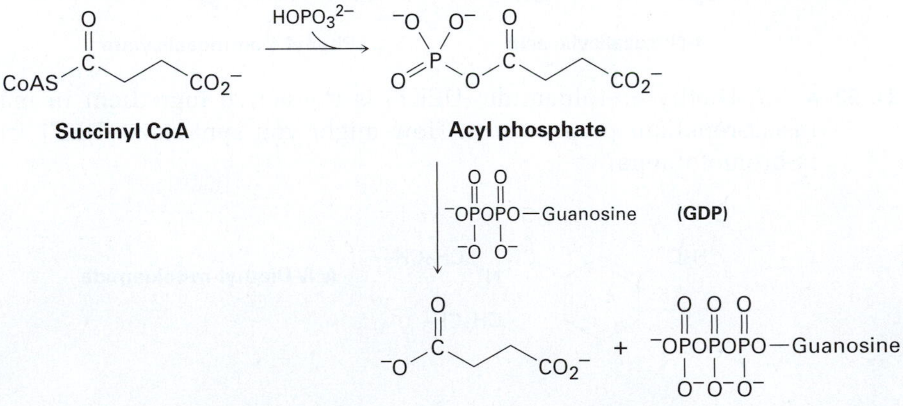

Succinyl CoA   **Acyl phosphate**

(GDP)

Succinate   **GTP**

16.56 One step in the *gluconeogenesis* pathway for the biosynthesis of glucose is the partial reduction of 3-phosphoglycerate to give glyceraldehyde 3-phosphate. The process occurs by phosphorylation with ATP to give 1,3-bisphosphoglycerate, reaction with a thiol group on the enzyme to give an enzyme-bound thioester, and reduction with NADH. Suggest mechanisms for all three reactions.

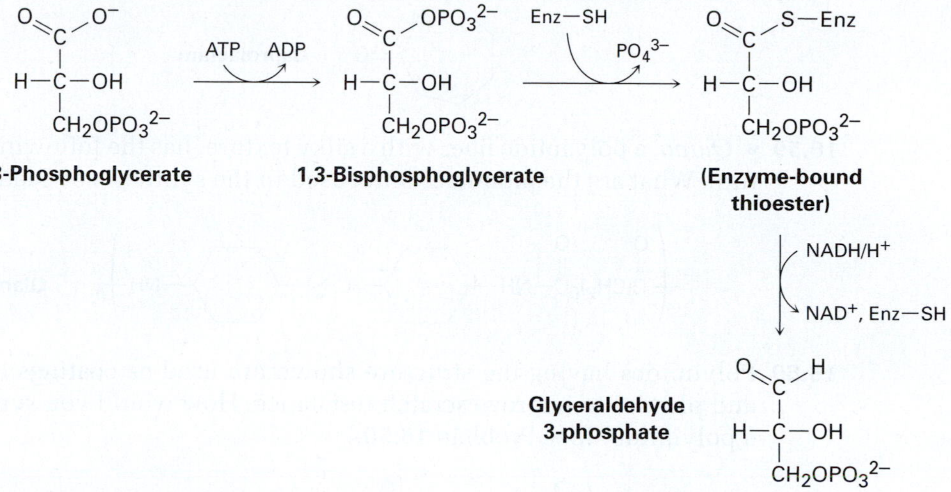

3-Phosphoglycerate   **1,3-Bisphosphoglycerate**   **(Enzyme-bound thioester)**

Glyceraldehyde 3-phosphate

16.57 Penicillins and other β-lactam antibiotics (see the *Lagniappe* in this chapter) typically develop a resistance to bacteria due to bacterial synthesis of β-lactamase enzymes. Tazobactam, however, is able to inhibit the activity of the β-lactamase by trapping it, thereby preventing resistance from developing.

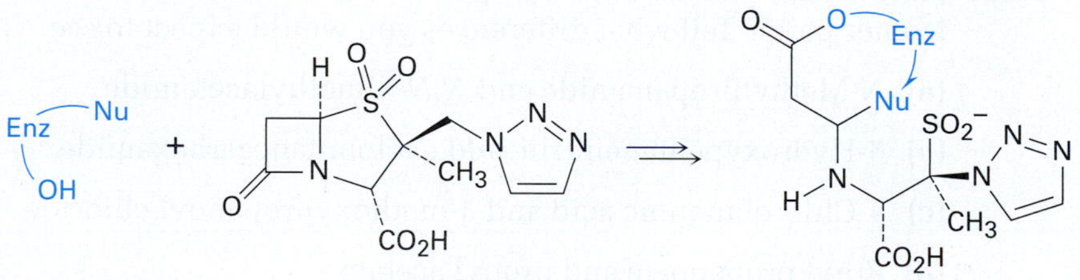

β-Lactamase   **Tazobactam**   **Trapped β-lactamase**

(a) The first step in trapping is reaction of a hydroxyl group on the β-lactamase to open the β-lactam ring of tazobactam. Show the mechanism.

(b) The second step is opening of the sulfur-containing ring in tazobactam to give an acyclic iminium ion intermediate. Show the mechanism.

(c) Cyclization of the iminium ion intermediate gives the trapped β-lactamase product. Show the mechanism.

16.58 ▪ The step-growth polymer nylon 6 is prepared from caprolactam. The reaction involves initial reaction of caprolactam with water to give an intermediate open-chain amino acid, followed by heating to form the polymer. Propose mechanisms for both steps, and show the structure of nylon 6.

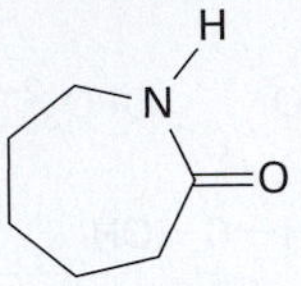

Caprolactam

16.59 ▪ *Qiana*, a polyamide fiber with a silky texture, has the following structure. What are the monomer units used in the synthesis of Qiana?

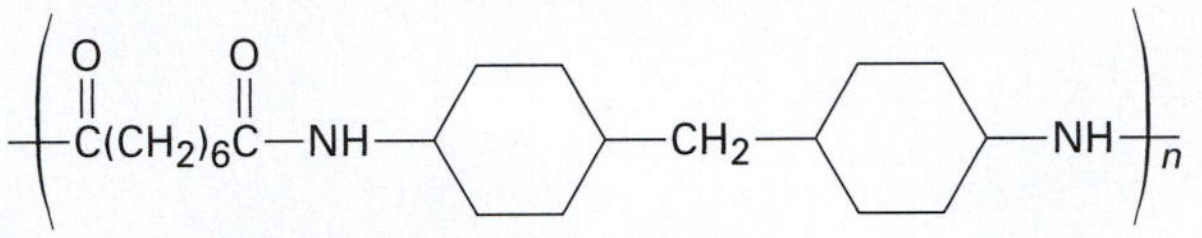

Qiana

16.60 Polyimides having the structure shown are used as coatings on glass and plastics to improve scratch resistance. How would you synthesize a polyimide? (See Problem 16.50.)

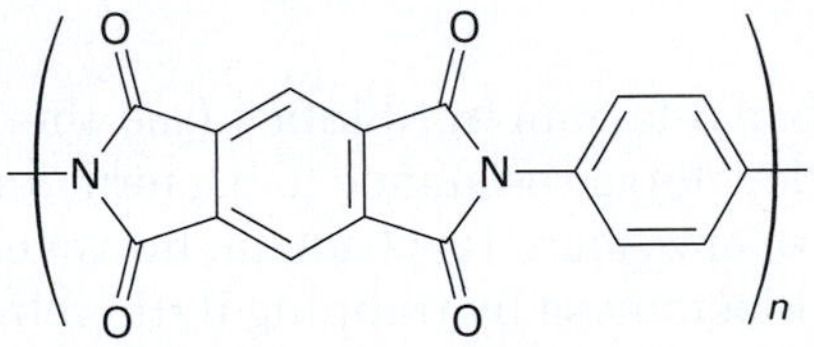

A polyimide

16.61 How would you distinguish spectroscopically between the following isomer pairs? Tell what differences you would expect to see.

(a) *N*-Methylpropanamide and *N,N*-dimethylacetamide

(b) 5-Hydroxypentanenitrile and cyclobutanecarboxamide

(c) 4-Chlorobutanoic acid and 3-methoxypropanoyl chloride

(d) Ethyl propanoate and propyl acetate

16.62 ▪ Propose a structure for a compound, $C_4H_7ClO_2$, that has the following IR and ^1H NMR spectra:

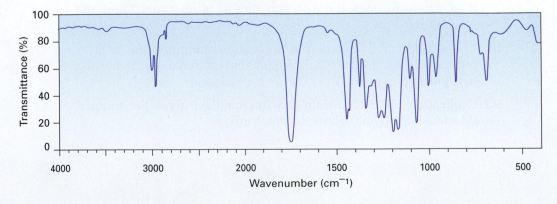

▪ Problems assignable in Organic OWL.

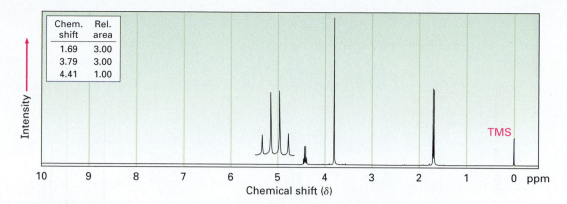

16.63 ■ Assign structures to compounds with the following ^1H NMR spectra:

(a) C_4H_7ClO

IR: 1810 cm^{-1}

(b) $C_5H_7NO_2$

IR: 2250, 1735 cm^{-1}

■ Problems assignable in Organic OWL.

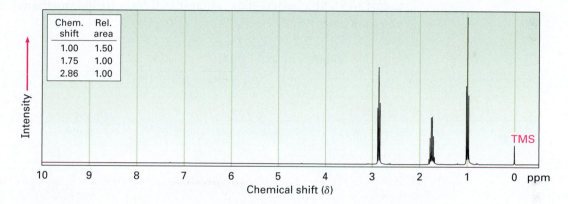

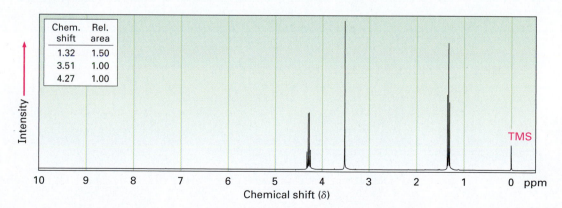

(c) $C_5H_{10}O_2$

IR: 1735 cm^{-1}

Chem. shift
Rel. area
1.22 6.00
2.01 3.00
4.99 1.00
Intensity
TMS
10 9 8 7 6 5 4 3 2 1 0 ppm
Chemical shift (δ)

16.64 In the *iodoform reaction,* a triiodomethyl ketone reacts with aqueous NaOH to yield a carboxylate ion and iodoform (triiodomethane). Propose a mechanism.

OH⁻
H2O
+ HCI3

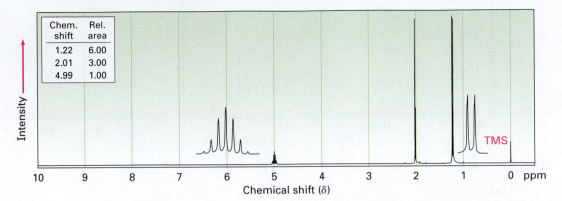

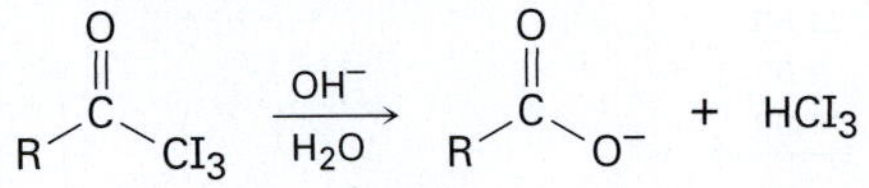

17 Carbonyl Alpha-Substitution and Condensation Reactions

β-Ketoacyl-CoA thiolase catalyzes the cleavage of a β-ketoacyl-CoA to give acetyl CoA, the final step in the β-oxidation cycle of fatty-acid metabolism.

We said in the *Preview of Carbonyl Chemistry* that much of the chemistry of carbonyl compounds can be explained by just four fundamental reaction types: nucleophilic additions, nucleophilic acyl substitutions, α substitutions, and carbonyl condensations. Having now studied the first two of these reactions, we'll look in this chapter at the remaining two major carbonyl-group processes—the **α-substitution reaction** and the **carbonyl condensation reaction**.

Alpha-substitution reactions occur at the position *next to* the carbonyl group—the *α position*—and involve the substitution of an α hydrogen atom by an electrophile, E, through either an *enol* or *enolate ion* intermediate.

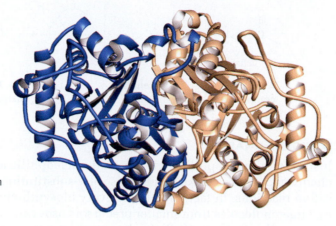

A carbonyl compound

An enolate ion

An enol

An alpha-substituted carbonyl compound

Carbonyl condensation reactions take place between two carbonyl partners and involve a *combination* of α-substitution and nucleophilic addition steps. One partner is converted into its enolate ion and undergoes an α-substitution

OWL Online homework for this chapter can be assigned in Organic OWL.

695

reaction when it carries out a nucleophilic addition to the second partner, giving a β-hydroxy carbonyl compound as product.

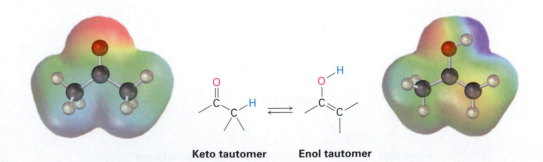

| A carbonyl compound | An enolate ion | A β-hydroxy carbonyl compound |

WHY THIS CHAPTER?

Just as with nucleophilic additions and nucleophilic acyl substitutions, biochemical pathways make frequent use of α-substitution and carbonyl condensation reactions. In fact, practically every biosynthetic pathway for building up larger molecules from smaller precursors uses carbonyl condensation reactions for the purpose. We'll see how and why these reactions occur in this chapter.

17.1 Keto–Enol Tautomerism

A carbonyl compound with a hydrogen atom on its α carbon rapidly equilibrates with its corresponding **enol** isomer (*ene* + alcoh*ol*). This spontaneous interconversion between two isomers, usually with the change in position of a hydrogen, is called *tautomerism*, from the Greek *tauto*, meaning "the same," and *meros*, meaning "part." The individual isomers are called **tautomers**.

Keto tautomer   **Enol tautomer**

Note the difference between tautomers and resonance forms, discussed in Section 2.5: tautomers are constitutional isomers—different compounds with different structures—while resonance forms are different representations of a single structure. Tautomers have their *atoms* arranged differently, while resonance forms differ only in the position of their *electrons*.

Most carbonyl compounds exist almost exclusively in the keto form at equilibrium, and it's usually difficult to isolate the pure enol. For example, cyclohexanone contains only about 0.0001% of its enol tautomer at room temperature, and acetone contains only about 0.000 000 1% enol. The percentage of enol tautomer is even less for carboxylic acids, esters, and amides. Even though enols are difficult to isolate and are present only to a small extent at equilibrium, they are nevertheless responsible for much of the chemistry of carbonyl compounds because they are so reactive.

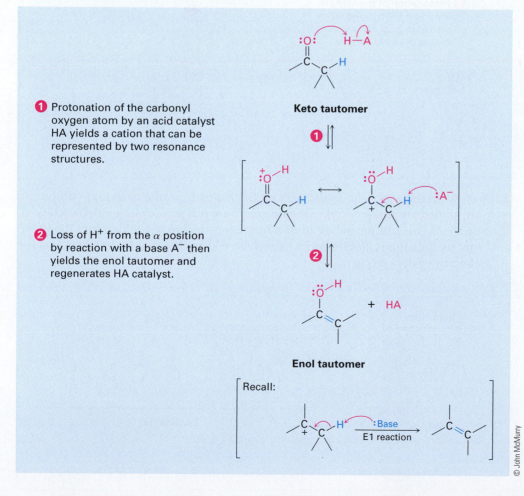

99.999 9%	0.000 1%		99.999 999 9%	0.000 000 1%
Cyclohexanone			**Acetone**	

Keto–enol tautomerism of carbonyl compounds is catalyzed by both acids and bases. Acid catalysis occurs by protonation of the carbonyl oxygen atom to give an intermediate cation that loses H^+ from its α carbon to yield a neutral enol (Figure 17.1). This proton loss from the cation intermediate is similar to what occurs during an E1 reaction when a carbocation loses H^+ to form an alkene (Section 12.13).

❶ Protonation of the carbonyl oxygen atom by an acid catalyst HA yields a cation that can be represented by two resonance structures.

❷ Loss of H^+ from the α position by reaction with a base A^- then yields the enol tautomer and regenerates HA catalyst.

Keto tautomer

Enol tautomer

Recall:

E1 reaction

FIGURE 17.1 MECHANISM: Mechanism of acid-catalyzed enol formation. The protonated intermediate can lose H^+, either from the oxygen atom to regenerate the keto tautomer or from the α carbon atom to yield an enol.

© John McMurry

Base-catalyzed enol formation occurs because the carbonyl group makes the hydrogens on the α carbon weakly acidic. Thus, a carbonyl compound can donate one of its α hydrogens to the base, giving an **enolate ion** that is then protonated. Because the enolate ion is a resonance hybrid of two forms, it can be protonated either on the α carbon to regenerate the keto tautomer or on oxygen to give the enol tautomer (Figure 17.2).

FIGURE 17.2 MECHANISM: Mechanism of base-catalyzed enol formation. The intermediate enolate ion, a resonance hybrid of two forms, can be protonated either on carbon to regenerate the starting keto tautomer or on oxygen to give an enol.

1 Base removes an acidic hydrogen from the α position of the carbonyl compound, yielding an enolate anion that has two resonance structures.

2 Protonation of the enolate anion on the oxygen atom yields an enol and regenerates the base catalyst.

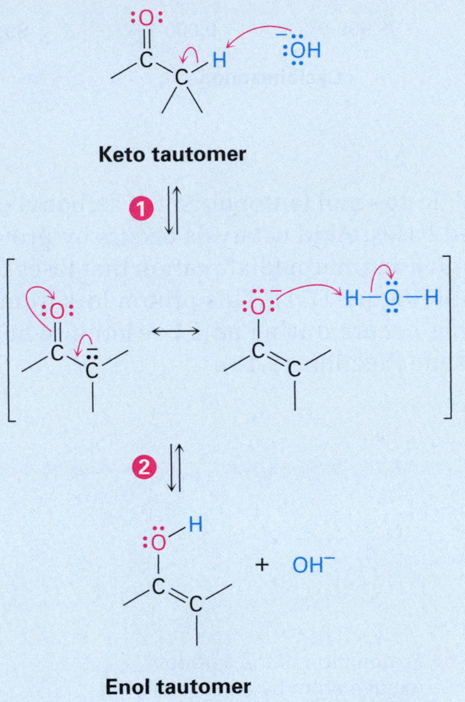

© John McMurry

Note that only the hydrogens on the α position of a carbonyl compound are acidic. Hydrogens at β, γ, δ, and so on, are not acidic and can't be removed by base. This unique behavior of α hydrogens is due to the fact that the resultant enolate ion is stabilized by a resonance form that places the charge on the electronegative oxygen.

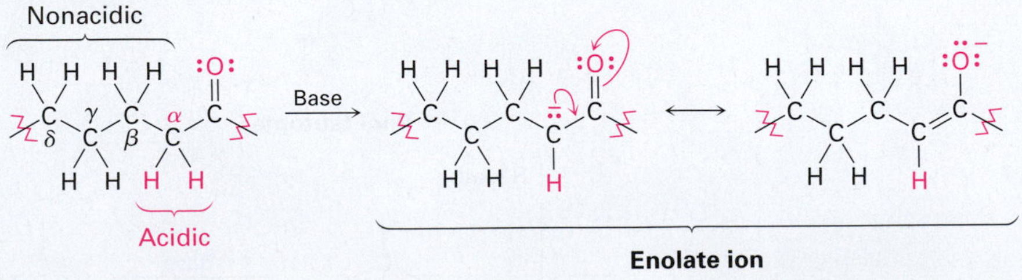

Problem 17.1

Draw structures for the enol tautomers of the following compounds, and identify all acidic hydrogens in each:

(a) Cyclopentanone   **(b)** Methyl thioacetate   **(c)** Ethyl acetate
(d) Propanal   **(e)** Acetic acid   **(f)** Phenylacetone

Problem 17.2

Draw structures for all monoenol forms of the following molecule. Which would you expect to be most stable? Explain.

17.2 Reactivity of Enols: α-Substitution Reactions

What kind of chemistry do enols have? Because their double bonds are electron-rich, enols behave as nucleophiles and react with electrophiles in much the same way that alkenes do. But because of resonance electron donation of a lone-pair of electrons on the neighboring oxygen, enols are more electron-rich and correspondingly more reactive than alkenes. Notice in the following electrostatic potential map of ethenol ($H_2C=CHOH$) how there is a substantial amount of electron density (yellow/red) on the α carbon.

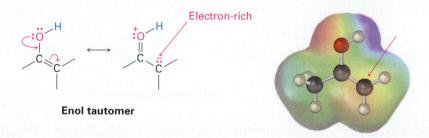

Enol tautomer

When an *alkene* reacts with an electrophile, such as HCl, initial addition of H^+ gives an intermediate cation and subsequent reaction with Cl^- yields an addition product (Section 7.6). When an *enol* reacts with an electrophile, however, only the initial addition step is the same. Instead of reacting with Cl^- to give an addition product, the intermediate cation loses the —OH proton to give an α-substituted carbonyl compound. The general mechanism is shown in Figure 17.3.

FIGURE 17.3 MECHANISM:
General mechanism of a carbonyl
α-substitution reaction on an
enol. The initially formed cation
loses H⁺ to regenerate a
carbonyl compound.

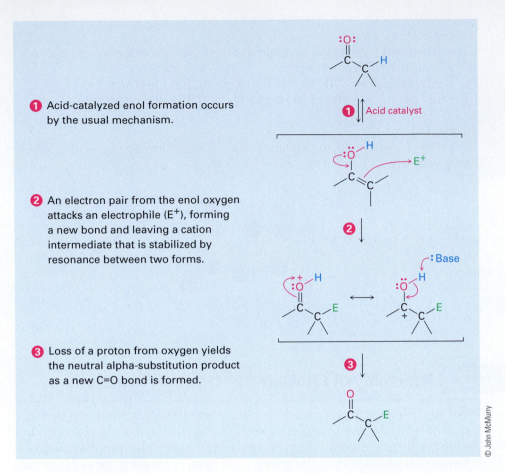

1 Acid-catalyzed enol formation occurs
by the usual mechanism.

1 Acid catalyst

2 An electron pair from the enol oxygen
attacks an electrophile (E⁺), forming
a new bond and leaving a cation
intermediate that is stabilized by
resonance between two forms.

3 Loss of a proton from oxygen yields
the neutral alpha-substitution product
as a new C=O bond is formed.

© John McMurry

One particularly common α-substitution reaction in the laboratory is the
halogenation of aldehydes and ketones at their α positions by reaction with
Cl₂, Br₂, or I₂ in acidic solution. Bromine in acetic acid solvent is often used.

Acetophenone   **α-Bromoacetophenone (72%)**

Remarkably, ketone halogenation also occurs in biological systems, particularly in marine alga, where dibromoacetaldehyde, bromoacetone, 1,1,1-tri-
bromoacetone, and other related compounds have been found.

From the Hawaiian alga *Asparagopsis taxiformis*

The reaction proceeds by acid-catalyzed formation of an enol intermediate, as shown in Figure 17.4.

FIGURE 17.4 MECHANISM: Mechanism of the acid-catalyzed bromination of acetone.

1 The carbonyl oxygen atom is protonated by acid catalyst.

2 Loss of an acidic proton from the alpha carbon takes place in the normal way to yield an enol intermediate.

3 An electron pair from the enol attacks bromine, giving an intermediate cation that is stabilized by resonance between two forms.

4 Loss of the –OH proton then gives the alpha-halogenated product and generates more acid catalyst.

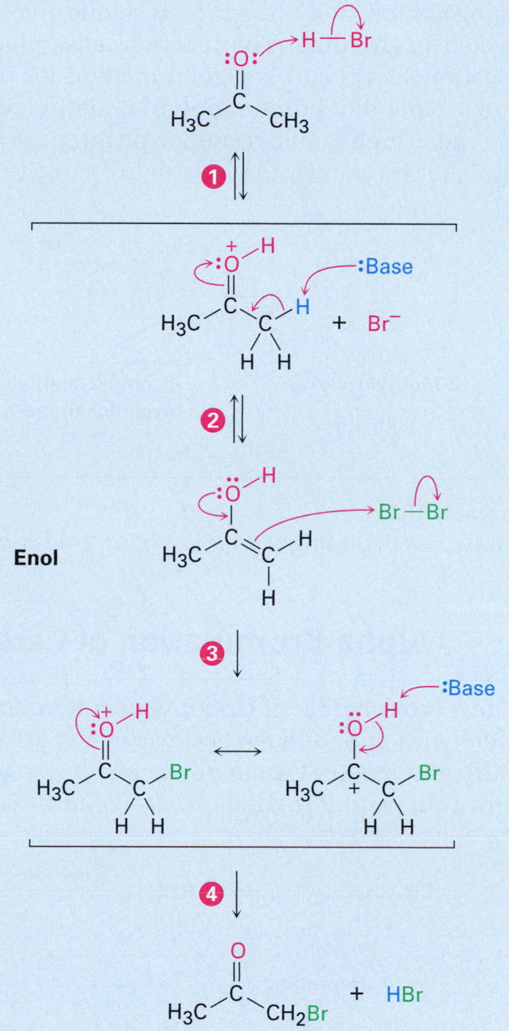

© John McMurry

Evidence for the mechanism shown in Figure 17.4 comes from deuterium-exchange experiments. If an aldehyde or ketone is treated with D_3O^+, the acidic α hydrogens are replaced by deuterium. For a given ketone, the rate of deuterium exchange is identical to the rate of halogenation, implying that the same intermediate—presumably the enol—is involved in both processes.

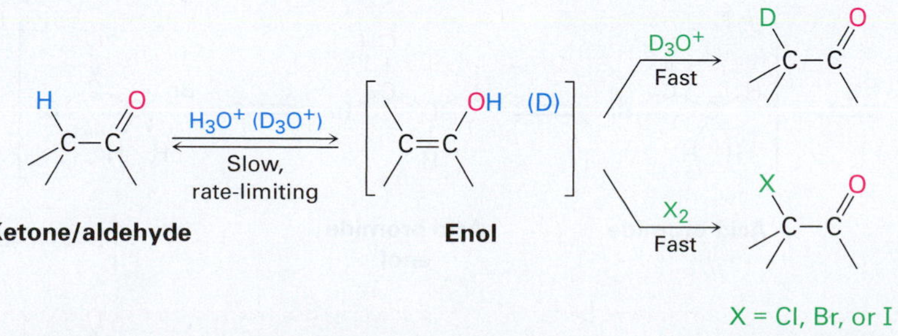

α-Bromo ketones are useful in laboratory synthesis because they can be dehydrobrominated by base treatment to yield α,β-unsaturated ketones. For example, 2-methylcyclohexanone gives 2-bromo-2-methylcyclohexanone on halogenation, and the α-bromo ketone gives 2-methylcyclohex-2-enone when heated in pyridine. The reaction takes place by an E2 elimination pathway (Section 12.12) and is a good method for introducing C=C bonds into molecules. Note that bromination of 2-methylcyclohexanone occurs primarily on the more highly substituted α position because the more highly substituted enol is favored over the less highly substituted one (Section 7.5).

2-Methylcyclo-hexanone   **2-Bromo-2-methyl-cyclohexanone**   **2-Methylcyclo-hex-2-enone (63%)**

Problem 17.3
Show how you might prepare pent-1-en-3-one from pentan-3-one.

17.3 Alpha Bromination of Carboxylic Acids

The α bromination of carbonyl compounds by Br_2 in acetic acid is limited to aldehydes and ketones because acids, esters, and amides don't enolize to a sufficient extent. Carboxylic acids, however, can be α brominated by a mixture of Br_2 and PBr_3 in the *Hell–Volhard–Zelinskii (HVZ) reaction.*

Heptanoic acid   **2-Bromoheptanoic acid (90%)**

The Hell–Volhard–Zelinskii reaction is a bit more complex than it looks and actually involves α substitution of an *acid bromide enol* rather than a carboxylic acid enol. The process begins with reaction of the carboxylic acid with PBr_3 to form an acid bromide plus HBr (Section 16.4). The HBr then catalyzes enolization of the acid bromide, and the resultant enol reacts with Br_2 in an α-substitution reaction to give an α-bromo acid bromide. Addition of water hydrolyzes the acid bromide in a nucleophilic acyl substitution reaction and yields the α-bromo carboxylic acid product.

Carboxylic acid   **Acid bromide**   **Acid bromide enol**

Problem 17.4

If methanol rather than water is added at the end of a Hell–Volhard–Zelinskii reaction, an ester rather than an acid is produced. Show how you could carry out the following transformation, and propose a mechanism for the ester-forming step.

$$\underset{CH_3CH_2CHCH_2COH}{\overset{\overset{CH_3}{|}\quad\overset{O}{\|}}{}} \quad \overset{?}{\longrightarrow} \quad \underset{CH_3CH_2CHCHCOCH_3}{\overset{\overset{CH_3}{|}\quad\overset{O}{\|}}{\underset{\overset{|}{Br}}{}}}$$

17.4 Acidity of α Hydrogen Atoms: Enolate Ion Formation

As noted in Section 17.1, a hydrogen on the α position of a carbonyl compound is weakly acidic and can be removed by a strong base to yield an enolate ion. In comparing acetone ($pK_a = 19.3$) with ethane ($pK_a \approx 60$), for instance, the presence of a neighboring carbonyl group increases the acidity of the ketone over the alkane by a factor of 10^{40}.

Acetone
($pK_a = 19.3$)

Ethane
($pK_a \approx 60$)

Proton abstraction from a carbonyl compound occurs when the α C–H bond is oriented roughly parallel to the p orbitals of the carbonyl group. The α carbon atom of the enolate ion is sp^2-hybridized and has a p orbital that overlaps the neighboring carbonyl p orbitals. Thus, the negative charge is shared by the electronegative oxygen atom, and the enolate ion is stabilized by resonance (Figure 17.5).

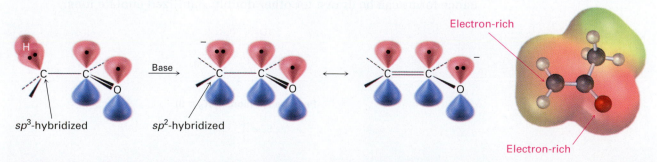

Electron-rich

sp^3-hybridized sp^2-hybridized

Electron-rich

FIGURE 17.5 Mechanism of enolate ion formation by abstraction of an α proton from a carbonyl compound. The enolate ion is stabilized by resonance, and the negative charge (red) is shared by the oxygen and the α carbon atom, as indicated by the electrostatic potential map.

Because carbonyl compounds are only weakly acidic, a strong base is needed for enolate ion formation. If an alkoxide such as sodium ethoxide is used as base, deprotonation takes place only to the extent of about 0.1% because acetone is a weaker acid than ethanol ($pK_a = 16$). If, however, a more powerful base such as sodium hydride (NaH) or lithium diisopropylamide [$LiN(i-C_3H_7)_2$] is used, a carbonyl compound can be completely converted into its enolate ion. Lithium diisopropylamide (LDA), which is easily prepared by reaction of the strong base butyllithium with diisopropylamine, is widely used in the laboratory as a base for preparing enolate ions from carbonyl compounds at low temperature.

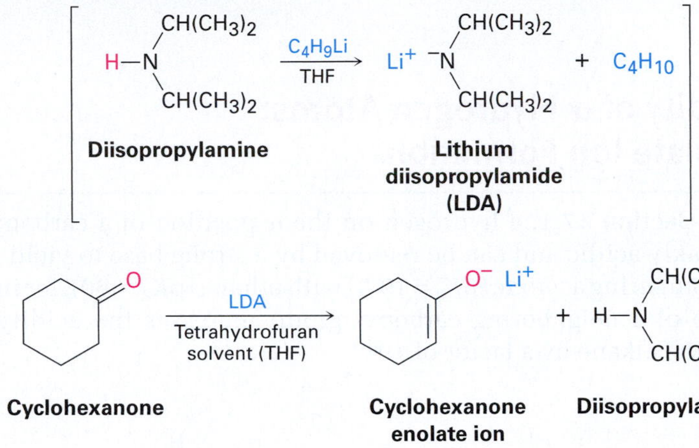

Many types of carbonyl compounds, including aldehydes, ketones, esters, thioesters, acids, and amides, can be converted into enolate ions by reaction with LDA. Table 17.1 lists the approximate pK_a values of different types of carbonyl compounds and shows how these values compare to other acidic substances we've seen. Note that nitriles are also acidic and can be converted into enolate-like anions.

When a hydrogen atom is flanked by two carbonyl groups, its acidity is enhanced even more. Table 17.1 thus shows that compounds such as 1,3-diketones (β-diketones), 3-keto esters (β-keto esters), and 1,3-diesters are more acidic than water. This enhanced acidity of β-dicarbonyl compounds is due to the stabilization of the resultant enolate ions by delocalization of the negative charge over both carbonyl groups. The enolate ion of pentane-2,4-dione, for instance, has three resonance forms. Similar resonance forms can be drawn for other doubly stabilized enolate ions.

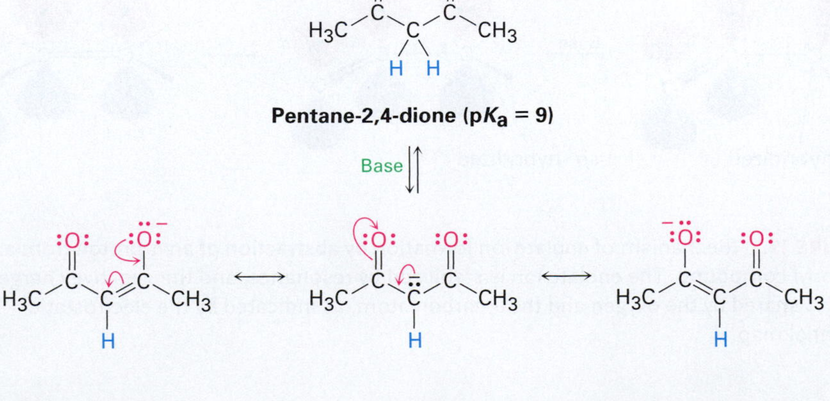

Pentane-2,4-dione ($pK_a = 9$)

TABLE 17.1
Acidity Constants for Some Organic Compounds

Functional group	Example	pK_a
Carboxylic acid	O‖CH$_3$COH	5
1,3-Diketone	O O‖ ‖CH$_3$CCH$_2$CCH$_3$	9
3-Keto ester	O O‖ ‖CH$_3$CCH$_2$COCH$_3$	11
1,3-Diester	O O‖ ‖CH$_3$OCCH$_2$COCH$_3$	13
Alcohol	CH$_3$OH	16
Acid chloride	O‖CH$_3$CCl	16
Aldehyde	O‖CH$_3$CH	17
Ketone	O‖CH$_3$CCH$_3$	19
Thioester	O‖CH$_3$CSCH$_3$	21
Ester	O‖CH$_3$COCH$_3$	25
Nitrile	CH$_3$C≡N	25
N,N-Dialkylamide	O‖CH$_3$CN(CH$_3$)$_2$	30
Dialkylamine	HN(i-C$_3$H$_7$)$_2$	40

WORKED EXAMPLE 17.1 Identifying Acidic Hydrogens in a Compound

Identify the most acidic hydrogens in each of the following compounds, and rank the compounds in order of increasing acidity:

(a)

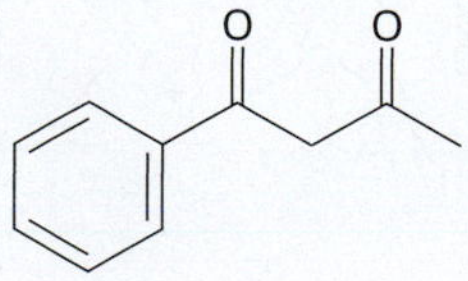

(b)   CH$_3$CHCOCH$_3$ with O above and CH$_3$ below

(c)

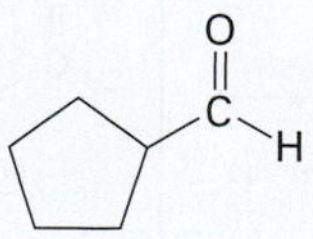

Strategy

Hydrogens on carbon next to a carbonyl group are acidic. In general, a β-dicarbonyl compound is most acidic, a ketone or aldehyde is next most acidic, and a carboxylic acid derivative is least acidic. Remember that alcohols, phenols, and carboxylic acids are also acidic because of their –OH hydrogens.

Solution

The acidity order is (a) > (c) > (b). Acidic hydrogens are shown in red:

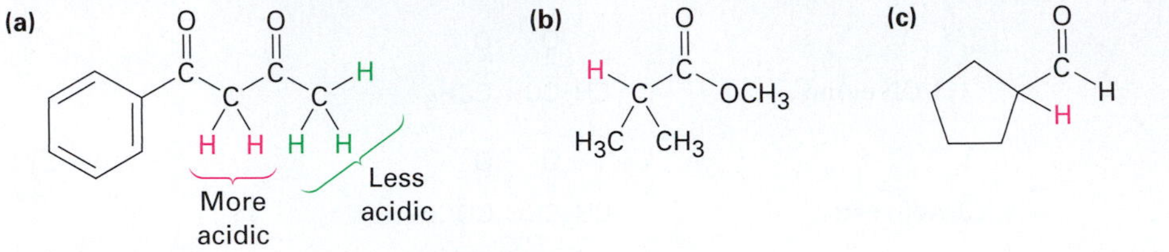

Problem 17.5

Identify the most acidic hydrogens in each of the following molecules:
(a) CH_3CH_2CHO **(b)** $(CH_3)_3CCOCH_3$ **(c)** CH_3CO_2H
(d) Benzamide   **(e)** $CH_3CH_2CH_2CN$ **(f)** $CH_3CON(CH_3)_2$

Problem 17.6

Draw a resonance structure of the acetonitrile anion, $^-:CH_2C\equiv N$, and account for the acidity of nitriles.

17.5 Alkylation of Enolate Ions

Enolate ions are more useful than enols for two reasons. First, pure enols can't normally be isolated but are instead generated only as short-lived intermediates in low concentration. By contrast, stable solutions of pure enolate ions are easily prepared from most carbonyl compounds by reaction with a strong base. Second, enolate ions are more reactive than enols and undergo many reactions that enols don't. Whereas enols are neutral, enolate ions are negatively charged, making them much better nucleophiles. As a result, enolate ions are more common than enols in both laboratory and biological chemistry.

Because they are resonance hybrids of two nonequivalent forms, enolate ions can be looked at either as vinylic alkoxides (C=C–O$^-$) or as α-keto carbanions ($^-$C–C=O). Thus, enolate ions can react with electrophiles either on oxygen or on carbon. Reaction on oxygen yields an enol derivative, while reaction on carbon yields an α-substituted carbonyl compound (Figure 17.6). Both kinds of reactivity are known, but reaction on carbon is more common.

Perhaps the single most important reaction of enolate ions is their *alkylation* by treatment with an alkyl halide or tosylate, thereby forming a new C–C bond and joining two smaller pieces into one larger molecule. Alkylation occurs when the nucleophilic enolate ion reacts with the electrophilic alkyl halide in an S_N2 reaction and displaces the leaving group by backside attack.

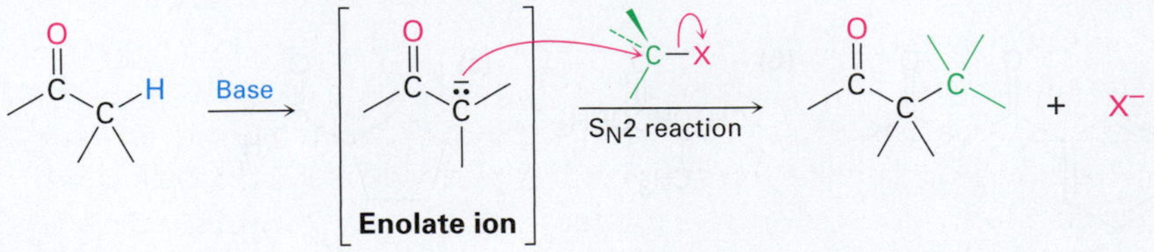

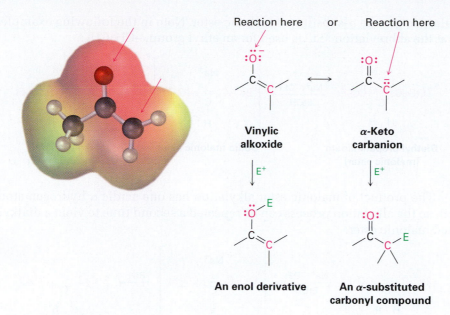

Reaction here   or   Reaction here

Vinylic alkoxide   **α-Keto carbanion**

An enol derivative   **An α-substituted carbonyl compound**

FIGURE 17.6 The electrostatic potential map of acetone enolate ion shows how the negative charge is delocalized over both the oxygen and the α carbon. As a result, two modes of reaction of an enolate ion with an electrophile E$^+$ are possible. Reaction on carbon to yield an α-substituted carbonyl product is more common.

Alkylation reactions are subject to the same constraints that affect all S_N2 reactions (Section 12.7). Thus, the leaving group X in the alkylating agent R–X can be chloride, bromide, iodide, or tosylate. The alkyl group R should be primary or methyl and preferably should be allylic or benzylic. Secondary halides react poorly, and tertiary halides don't react at all because a competing E2 elimination of HX occurs instead. Vinylic and aryl halides are also unreactive because backside approach is sterically prevented.

$$R\!-\!X \begin{cases} -X\text{: Tosylate} > -I > -Br > -Cl \\ R-\text{: Allylic} \approx \text{Benzylic} > H_3C- > RCH_2- \end{cases}$$

The Malonic Ester Synthesis

One of the oldest and best known carbonyl alkylation reactions in the laboratory is the **malonic ester synthesis**, a method for preparing a carboxylic acid from an alkyl halide while lengthening the carbon chain by two atoms.

$$R\!-\!X \xrightarrow[\text{synthesis}]{\text{Malonic ester}} \underset{H\quad H}{R\diagdown \underset{\displaystyle C}{}\diagup CO_2H}$$

Diethyl propanedioate, commonly called diethyl malonate or *malonic ester,* is more acidic than monocarbonyl compounds ($pK_a = 13$) because its α hydrogens are flanked by two carbonyl groups. Thus, malonic ester is easily converted into its enolate ion by reaction with sodium ethoxide in ethanol. The enolate ion, in turn, is a good nucleophile that reacts rapidly with an alkyl halide to give an α-substituted malonic ester. Note in the following examples that the abbreviation "Et" is used for an ethyl group, –CH$_2$CH$_3$:

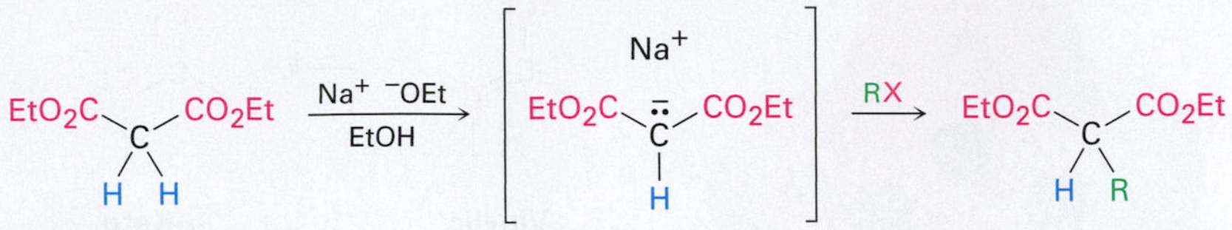

Diethyl propanedioate (malonic ester) ... **Sodio malonic ester** ... **An alkylated malonic ester**

The product of malonic ester alkylation has one acidic α hydrogen atom left, so the alkylation process can be repeated a second time to yield a dialkylated malonic ester:

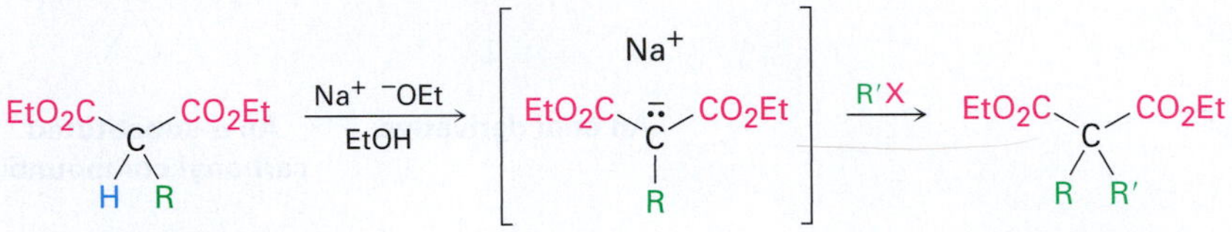

An alkylated malonic ester ... **A dialkylated malonic ester**

On heating with aqueous hydrochloric acid, the alkylated (or dialkylated) malonic ester undergoes hydrolysis of its two ester groups followed by *decarboxylation* (loss of CO$_2$) to yield a substituted monoacid:

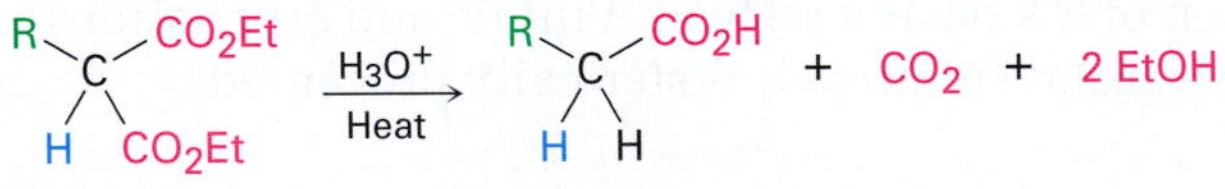

An alkylated malonic ester ... **A carboxylic acid**

Decarboxylation is not a general reaction of carboxylic acids. Rather, it is unique to compounds that have a *second* carbonyl group two atoms away from the –CO$_2$H. That is, only substituted malonic acids and β-keto acids undergo loss of CO$_2$ on heating. The decarboxylation reaction occurs by a cyclic mechanism and involves initial formation of an enol, thereby accounting for the need to have a second carbonyl group appropriately positioned.

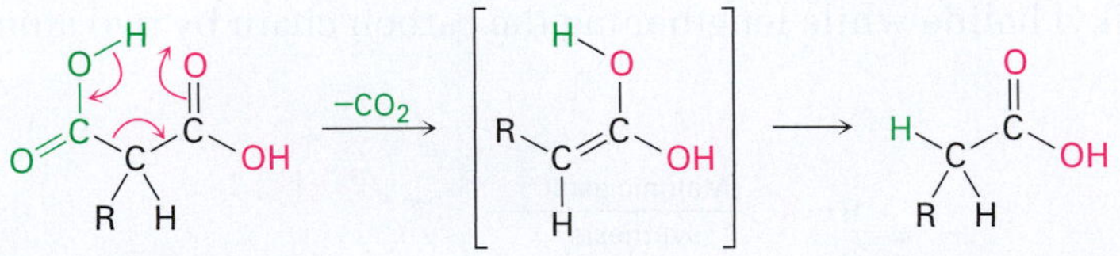

A diacid ... **An acid enol** ... **A carboxylic acid**

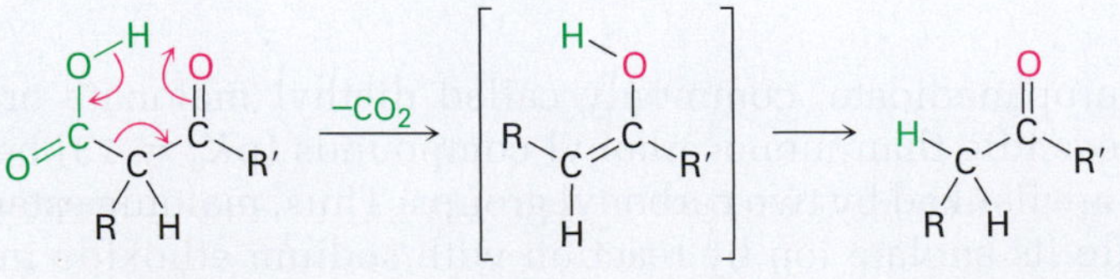

A β-keto acid ... **An enol** ... **A ketone**

As noted previously, the overall effect of the malonic ester synthesis is to convert an alkyl halide into a carboxylic acid while lengthening the carbon chain by two atoms.

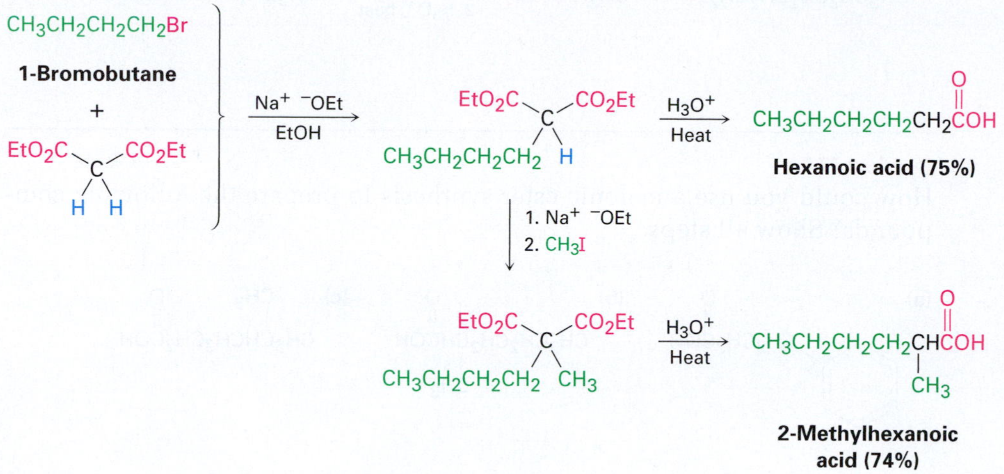

The malonic ester synthesis can also be used to prepare *cyclo*alkane-carboxylic acids. For example, when 1,4-dibromobutane is treated with diethyl malonate in the presence of 2 equivalents of sodium ethoxide base, the second alkylation step occurs *intramolecularly* to yield a cyclic product. Hydrolysis and decarboxylation then give cyclopentanecarboxylic acid. Three-, four-, five-, and six-membered rings can all be prepared in this way.

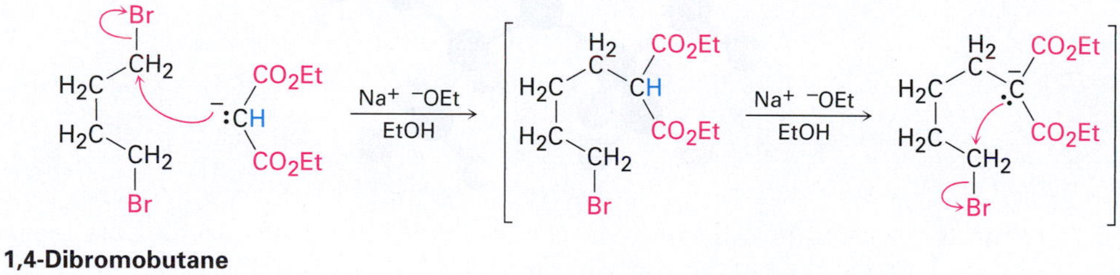

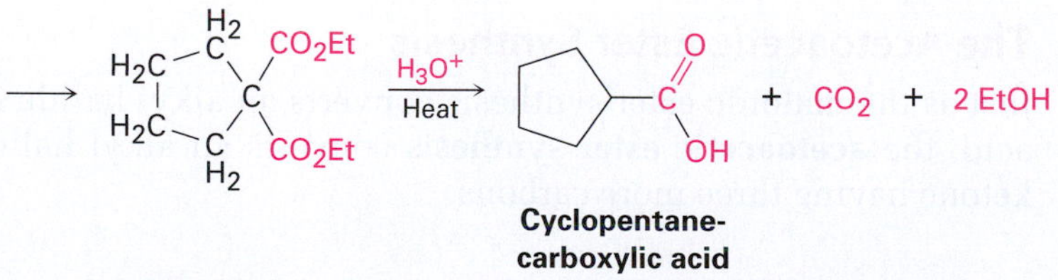

1,4-Dibromobutane

Cyclopentane-carboxylic acid

WORKED EXAMPLE 17.2 Using the Malonic Ester Synthesis to Prepare a Carboxylic Acid

How would you prepare heptanoic acid using a malonic ester synthesis?

Strategy

The malonic ester synthesis converts an alkyl halide into a carboxylic acid having two more carbons. Thus, a *seven*-carbon acid chain must be derived from the *five*-carbon alkyl halide 1-bromopentane.

Solution

$$CH_3CH_2CH_2CH_2CH_2Br + CH_2(CO_2Et)_2 \xrightarrow[\text{2. } H_3O^+\text{, heat}]{\text{1. } Na^+\ {}^-OEt} CH_3CH_2CH_2CH_2CH_2CH_2COH$$

Problem 17.7

How could you use a malonic ester synthesis to prepare the following compounds? Show all steps.

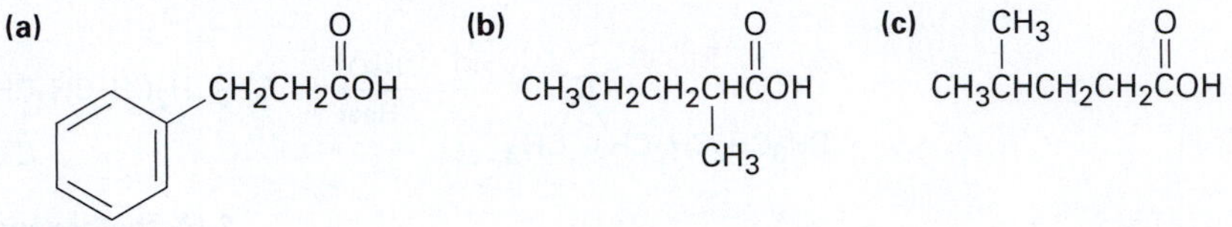

Problem 17.8

How could you use a malonic ester synthesis to prepare the following compound?

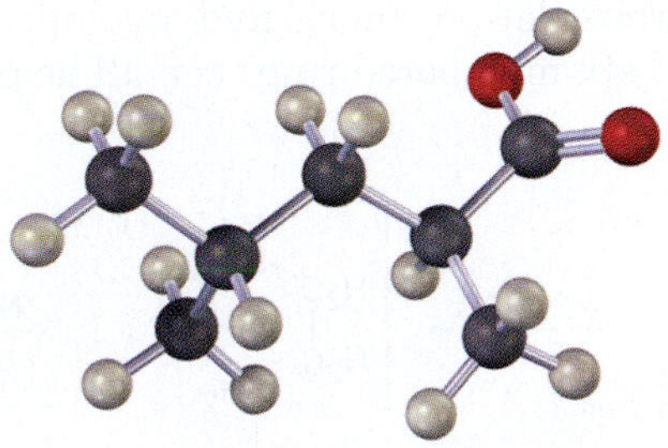

The Acetoacetic Ester Synthesis

Just as the malonic ester synthesis converts an alkyl halide into a carboxylic acid, the **acetoacetic ester synthesis** converts an alkyl halide into a methyl ketone having three more carbons.

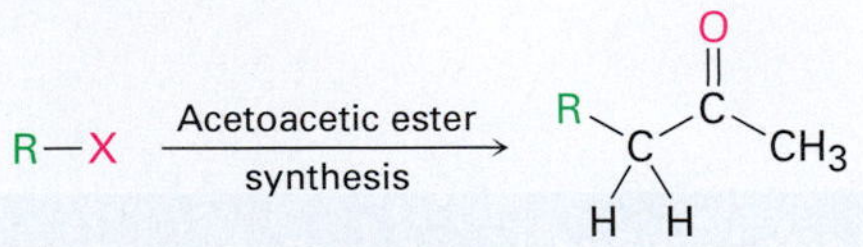

Ethyl 3-oxobutanoate, commonly called ethyl acetoacetate or *acetoacetic ester,* is much like malonic ester in that its α hydrogens are flanked by two carbonyl groups. It is therefore readily converted into its enolate ion, which can be alkylated by reaction with an alkyl halide. A second alkylation can also be carried out if desired, since acetoacetic ester has two acidic α hydrogens.

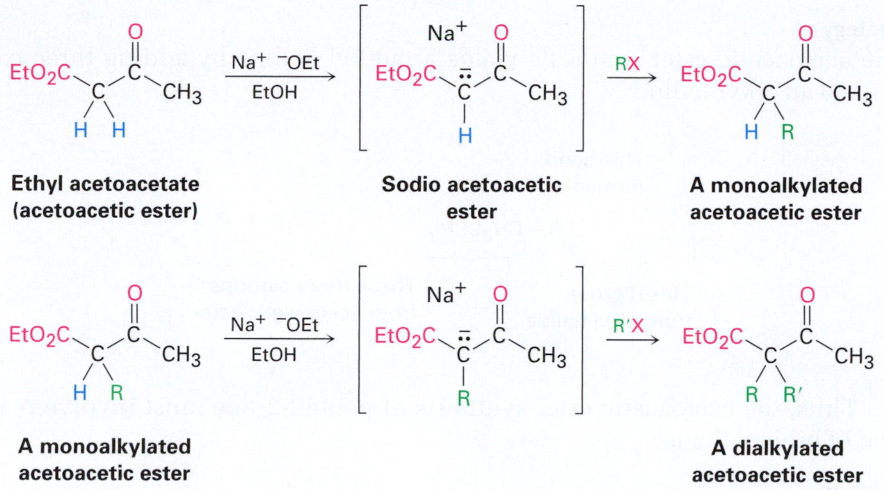

Ethyl acetoacetate (acetoacetic ester)   **Sodio acetoacetic ester**   **A monoalkylated acetoacetic ester**

A monoalkylated acetoacetic ester   **A dialkylated acetoacetic ester**

On heating with aqueous HCl, the alkylated (or dialkylated) acetoacetic ester is hydrolyzed to a β-keto acid, which then undergoes decarboxylation to yield a ketone product. The decarboxylation occurs in the same way as in the malonic ester synthesis and involves a ketone enol as the initial product.

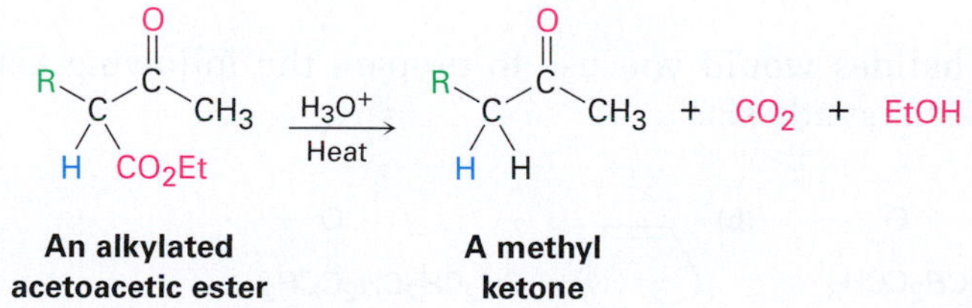

An alkylated acetoacetic ester   **A methyl ketone**

The three-step sequence of (1) enolate ion formation, (2) alkylation, and (3) hydrolysis/decarboxylation is applicable to all β-keto esters with acidic α hydrogens, not just to acetoacetic ester itself. For example, *cyclic* β-keto esters such as ethyl 2-oxocyclohexanecarboxylate can be alkylated and decarboxylated to give 2-substituted cyclohexanones.

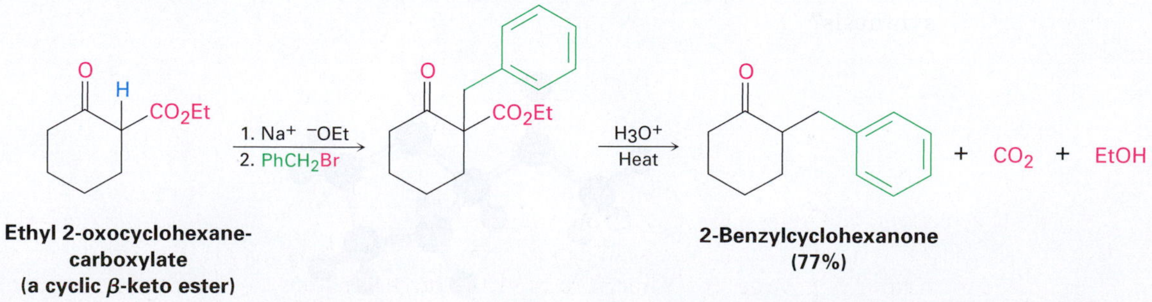

Ethyl 2-oxocyclohexane-carboxylate (a cyclic β-keto ester)   **2-Benzylcyclohexanone** (77%)

WORKED EXAMPLE 17.3 Using the Acetoacetic Ester Synthesis to Prepare a Ketone

How would you prepare pentan-2-one by an acetoacetic ester synthesis?

Strategy

The acetoacetic ester synthesis yields a methyl ketone by adding three carbons to an alkyl halide:

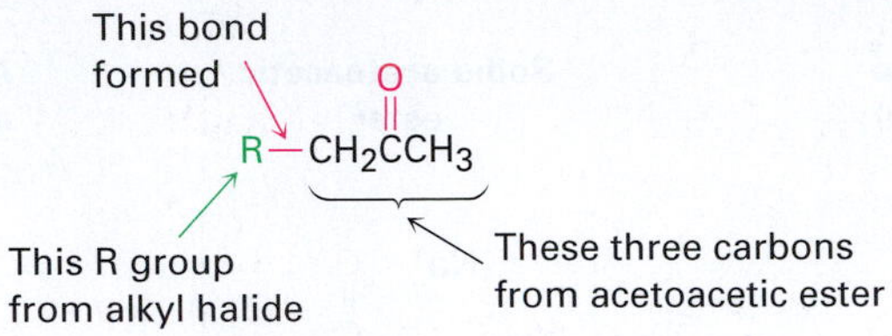

$$
\overset{\text{This bond}}{\underset{\text{This R group}}{\text{formed}}} \quad R-CH_2CCH_3
$$

This bond formed

R—CH₂CCH₃

This R group from alkyl halide

These three carbons from acetoacetic ester

Thus, the acetoacetic ester synthesis of pentan-2-one must involve reaction of bromoethane.

Solution

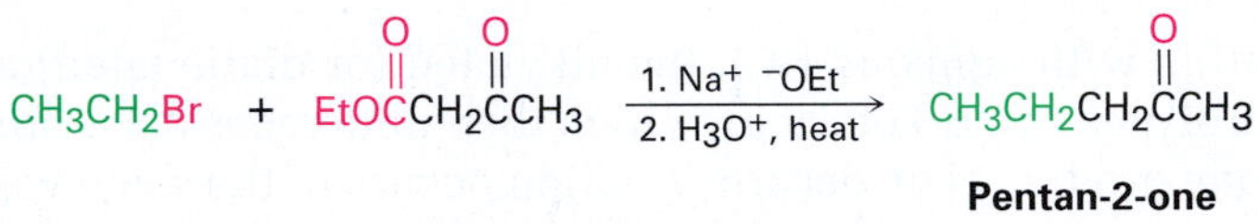

$$
CH_3CH_2Br \;+\; EtOCCH_2CCH_3 \xrightarrow[\text{2. H}_3O^+,\text{ heat}]{\text{1. Na}^+ \; {}^-OEt} CH_3CH_2CH_2CCH_3
$$

Pentan-2-one

Problem 17.9

What alkyl halides would you use to prepare the following ketones by an acetoacetic ester synthesis?

(a) CH₃
CH₃CHCH₂CH₂CCH₃

(b) ⬡—CH₂CH₂CH₂CCH₃

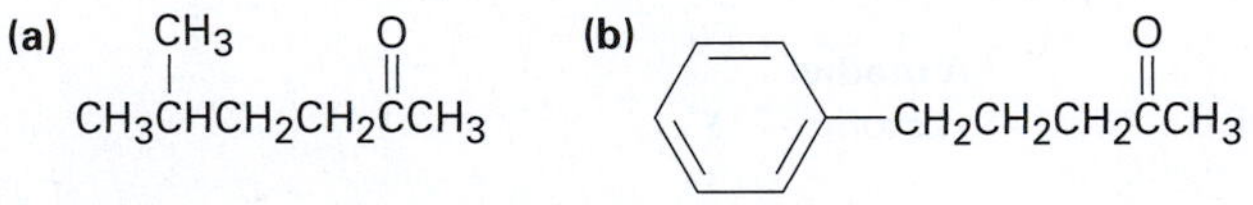

Problem 17.10

Which of the following compounds *cannot* be prepared by an acetoacetic ester synthesis? Explain.
(a) Phenylacetone **(b)** Acetophenone **(c)** 3,3-Dimethylbutan-2-one

Problem 17.11

How would you prepare the following compound using an acetoacetic ester synthesis?

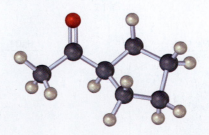

Direct Alkylation of Ketones, Esters, and Nitriles

Both the malonic ester synthesis and the acetoacetic ester synthesis are easy to carry out because they involve unusually acidic dicarbonyl compounds. As a result, relatively mild bases such as sodium ethoxide in ethanol as solvent can be used to prepare the necessary enolate ions. Alternatively, however, it's also possible in many cases to directly alkylate the α position of *mono*carbonyl compounds. A strong, sterically hindered base such as LDA is needed so that complete conversion to the enolate ion takes place rather than a nucleophilic addition, and a nonprotic solvent must be used.

Ketones, esters, and nitriles can all be alkylated using LDA or related dialkylamide bases in THF. Aldehydes, however, rarely give high yields of pure products because their enolate ions undergo carbonyl condensation reactions instead of alkylation. Some specific examples of alkylation reactions are shown:

Lactone

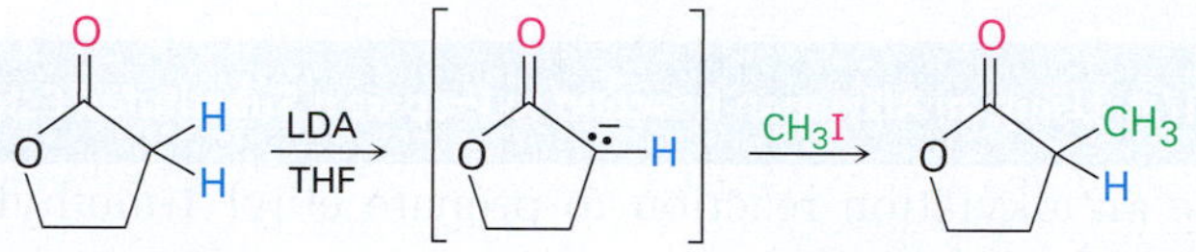

Butyrolactone   **2-Methylbutyrolactone (88%)**

Ester

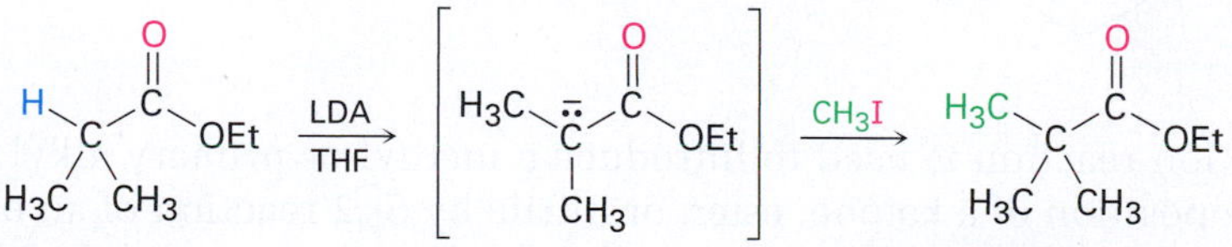

Ethyl 2-methylpropanoate   **Ethyl 2,2-dimethylpropanoate (87%)**

Ketone

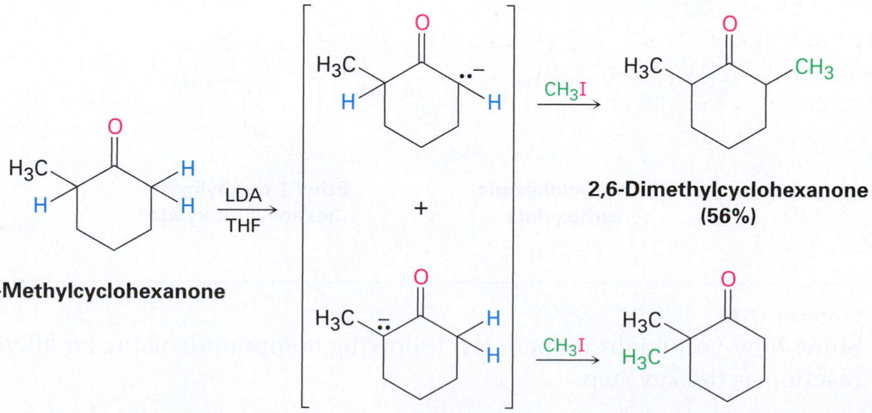

2-Methylcyclohexanone

2,6-Dimethylcyclohexanone (56%)

2,2-Dimethylcyclohexanone (6%)

Nitrile

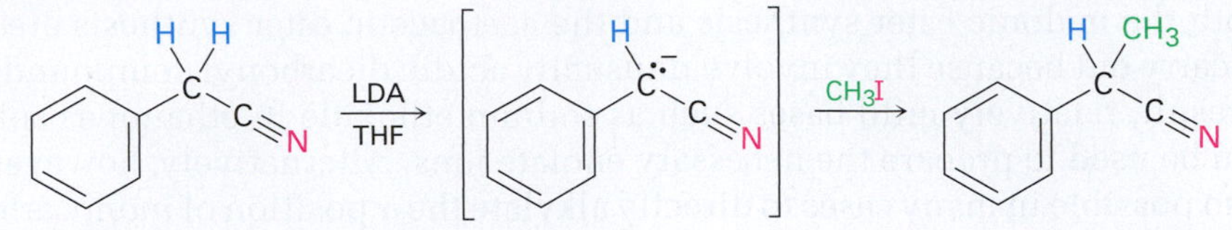

Phenylacetonitrile

2-Phenylpropane-nitrile (71%)

Note in the ketone example that alkylation of 2-methylcyclohexanone leads to a mixture of products because both possible enolate ions are formed. In general, the major product in such cases occurs by alkylation at the less hindered, more accessible position. Thus, alkylation of 2-methylcyclohexanone occurs primarily at C6 (secondary) rather than at C2 (tertiary).

WORKED EXAMPLE 17.4 Using an Alkylation Reaction to Prepare a Substituted Ester

How might you use an alkylation reaction to prepare ethyl 1-methylcyclohexanecarboxylate?

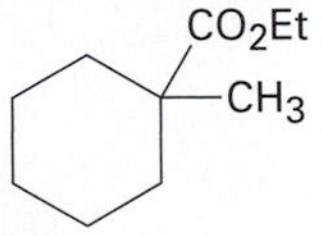

Ethyl 1-methylcyclohexanecarboxylate

Strategy

An alkylation reaction is used to introduce a methyl or primary alkyl group onto the α position of a ketone, ester, or nitrile by S_N2 reaction of an enolate ion with an alkyl halide. Thus, we need to look at the target molecule and identify any methyl or primary alkyl groups attached to an α carbon. In the present instance, the target has an α methyl group, which might be introduced by alkylation of an ester enolate ion with iodomethane.

Solution

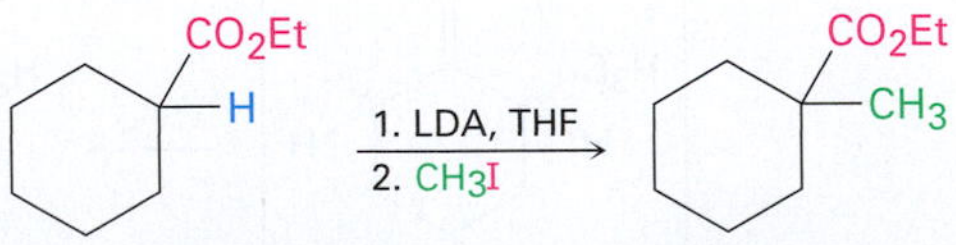

Ethyl cyclohexane-carboxylate

Ethyl 1-methylcyclo-hexanecarboxylate

Problem 17.12

Show how you might prepare the following compounds using an alkylation reaction as the key step:

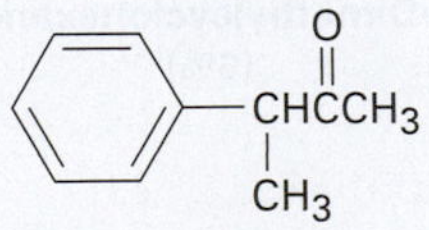

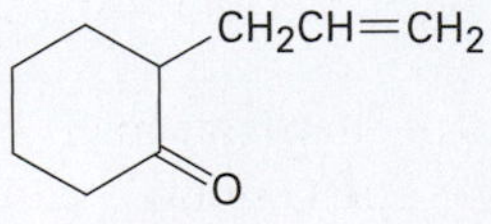

(d)

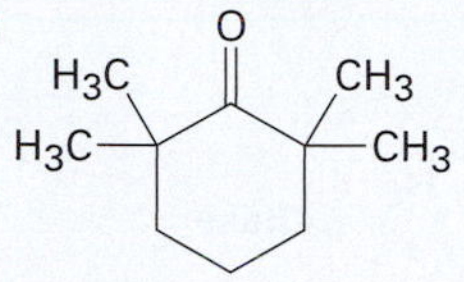

H₃C, CH₃ / H₃C, CH₃ (cyclohexanone derivative with O)

(e)

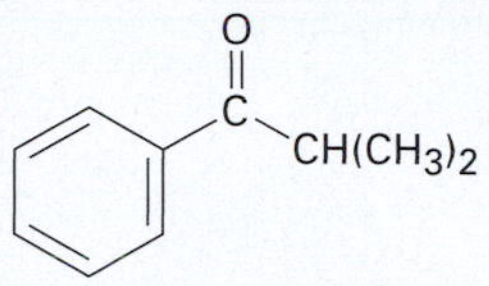

phenyl—C(=O)—CH(CH₃)₂

(f)

CH₃CHCHCOCH₃ with CH₃ and CH₂CH₃ substituents

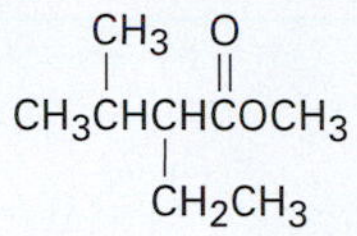

Biological Alkylations

Alkylations are rare but not unknown in biological chemistry. One example occurs during biosynthesis of the antibiotic indolmycin from indolylpyruvate when a base abstracts an acidic hydrogen from an α position and the resultant enolate ion carries out an S$_N$2 alkylation reaction on the methyl group of *S*-adenosylmethionine (SAM; Section 12.10). Although it's convenient to speak of "enolate ion" intermediates in biological pathways, it's unlikely that they exist for long in an aqueous cellular environment. Rather, proton removal and alkylation probably occur at essentially the same time (Figure 17.7).

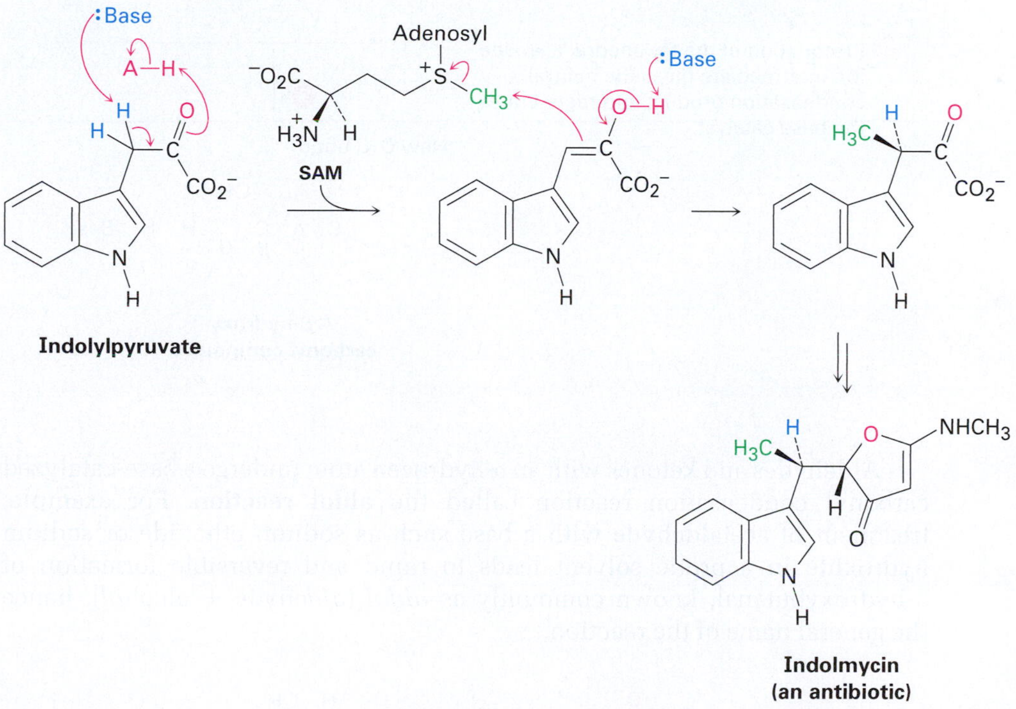

Indolylpyruvate

Indolmycin (an antibiotic)

FIGURE 17.7 The biosynthesis of indolmycin from indolylpyruvate occurs through a pathway that includes an alkylation reaction of a short-lived enolate ion intermediate.

17.6 Carbonyl Condensations: The Aldol Reaction

As noted in the chapter introduction, carbonyl condensation reactions take place between two carbonyl partners and involve a combination of α-substitution and nucleophilic addition steps. One partner is converted into an enolate-ion nucleophile and adds to the electrophilic carbonyl group of the second partner. In so doing, the nucleophilic partner undergoes an α-substitution reaction and the electrophilic partner undergoes a nucleophilic addition. The general mechanism of the process is shown in Figure 17.8.

FIGURE 17.8 MECHANISM: General mechanism of a carbonyl condensation reaction. One partner becomes a nucleophilic donor and adds to the second partner as an electrophilic acceptor. The product is a β-hydroxy carbonyl compound.

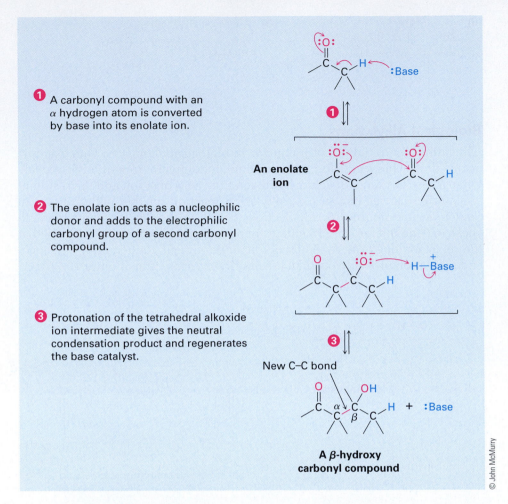

1 A carbonyl compound with an α hydrogen atom is converted by base into its enolate ion.

An enolate ion

2 The enolate ion acts as a nucleophilic donor and adds to the electrophilic carbonyl group of a second carbonyl compound.

3 Protonation of the tetrahedral alkoxide ion intermediate gives the neutral condensation product and regenerates the base catalyst.

New C–C bond

A β-hydroxy carbonyl compound

© John McMurry

Aldehydes and ketones with an α hydrogen atom undergo a base-catalyzed carbonyl condensation reaction called the **aldol reaction**. For example, treatment of acetaldehyde with a base such as sodium ethoxide or sodium hydroxide in a protic solvent leads to rapid and reversible formation of 3-hydroxybutanal, known commonly as *aldol* (*ald*ehyde + alcoh*ol*), hence the general name of the reaction.

Acetaldehyde   **3-Hydroxybutanal (aldol)**

The position of the aldol equilibrium depends both on reaction conditions and on substrate structure. The equilibrium generally favors condensation product in the case of aldehydes with no α substituent (RCH_2CHO) but favors reactant for disubstituted aldehydes (R_2CHCHO) and for most ketones. Steric factors are probably responsible for these trends, since increased substitution near the reaction site increases steric congestion in the aldol product.

Aldehydes

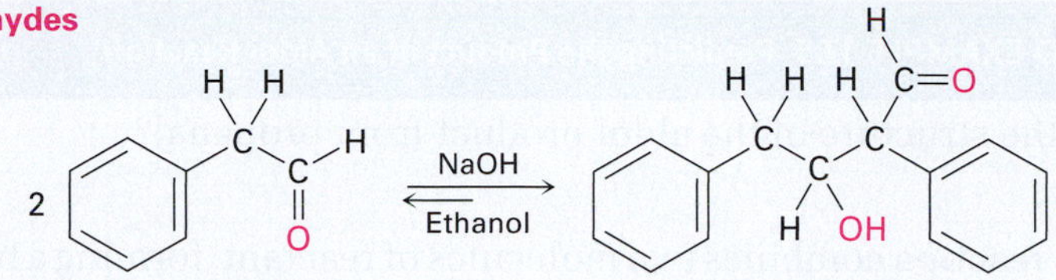

**Phenylacetaldehyde
(10%)**

NaOH
Ethanol

(90%)

Ketones

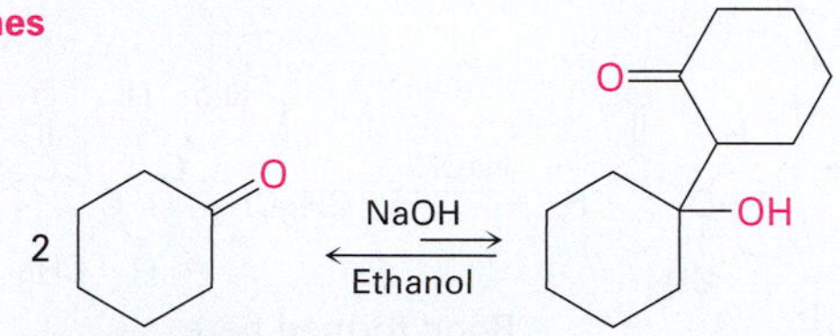

**Cyclohexanone
(78%)**

NaOH
Ethanol

(22%)

Aldol reactions, like all carbonyl condensations, occur by nucleophilic addition of the enolate ion of the donor molecule to the carbonyl group of the acceptor molecule. The resultant tetrahedral intermediate is then protonated to give an alcohol product (Figure 17.9). The reverse process occurs in exactly the opposite manner: base abstracts the –OH hydrogen from the aldol to yield a β-keto alkoxide ion, which cleaves to give one molecule of enolate ion and one molecule of neutral carbonyl compound.

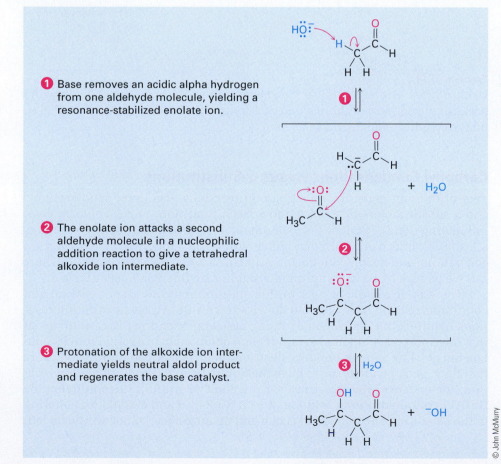

① Base removes an acidic alpha hydrogen from one aldehyde molecule, yielding a resonance-stabilized enolate ion.

② The enolate ion attacks a second aldehyde molecule in a nucleophilic addition reaction to give a tetrahedral alkoxide ion intermediate.

③ Protonation of the alkoxide ion intermediate yields neutral aldol product and regenerates the base catalyst.

ACTIVE FIGURE 17.9

MECHANISM: Mechanism of the aldol reaction, a typical carbonyl condensation. **Go to this book's student companion site at** www.cengage.com/chemistry/mcmurry **to explore an interactive version of this figure.**

© John McMurry

WORKED EXAMPLE 17.5 Predicting the Product of an Aldol Reaction

What is the structure of the aldol product from propanal?

Strategy

An aldol reaction combines two molecules of reactant, forming a bond between the α carbon of one partner and the carbonyl carbon of the second partner.

Solution

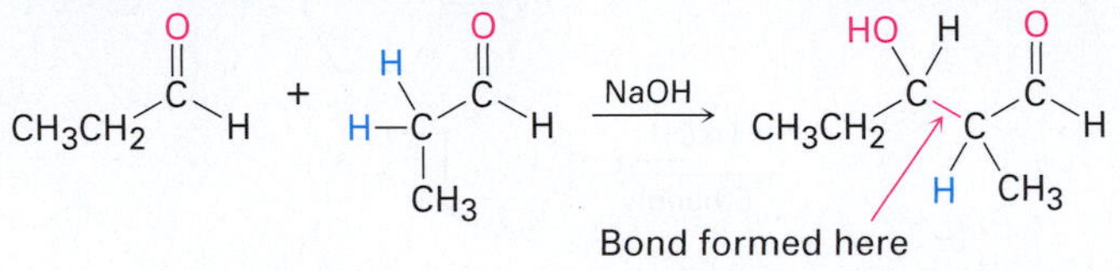

Bond formed here

Problem 17.13

Predict the aldol reaction product of the following compounds:

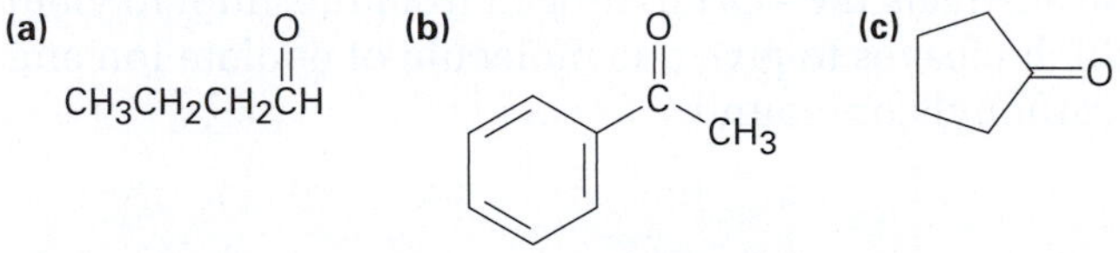

(a)

$$CH_3CH_2CH_2CH$$ (with O double bonded)

(b)

(c)

Problem 17.14

Using curved arrows to indicate the electron flow in each step, show how the base-catalyzed retro-aldol reaction of 4-hydroxy-4-methylpentan-2-one takes place to yield 2 equivalents of acetone.

Carbonyl Condensations versus α-Substitutions

Two of the four general carbonyl-group reactions—carbonyl condensations and α substitutions—take place under basic conditions and involve enolate-ion intermediates. Because the experimental conditions for the two reactions are similar, how can we predict which will occur in a given case? When we generate an enolate ion with the intention of carrying out an α alkylation, how can we be sure that a carbonyl condensation reaction won't occur instead?

There is no simple answer to this question, but the exact experimental conditions usually have much to do with the result. Alpha-substitution reactions require a full equivalent of strong base and are normally carried out so that the carbonyl compound is rapidly and completely converted into its enolate ion at a low temperature. An electrophile is then added rapidly to ensure that the reactive enolate ion is quenched quickly. In a ketone alkylation reaction, for instance, we might use 1 equivalent of lithium diisopropylamide (LDA) in tetrahydrofuran solution at $-78\ °C$. Rapid and complete generation of the ketone enolate ion would occur, and no unreacted ketone would be left, so no condensation reaction could take place. We would then immediately add an alkyl halide to complete the alkylation reaction.

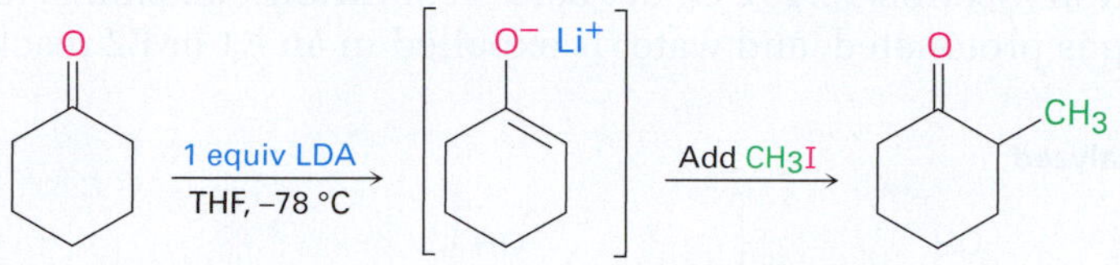

On the other hand, carbonyl condensation reactions require only a *catalytic* amount of a relatively weak base rather than a full equivalent so that a small amount of enolate ion is generated *in the presence* of unreacted carbonyl compound. Once a condensation has occurred, the basic catalyst is regenerated. To carry out an aldol reaction on propanal, for instance, we might dissolve the aldehyde in methanol, add 0.05 equivalent of sodium methoxide, and then warm the mixture to give the aldol product.

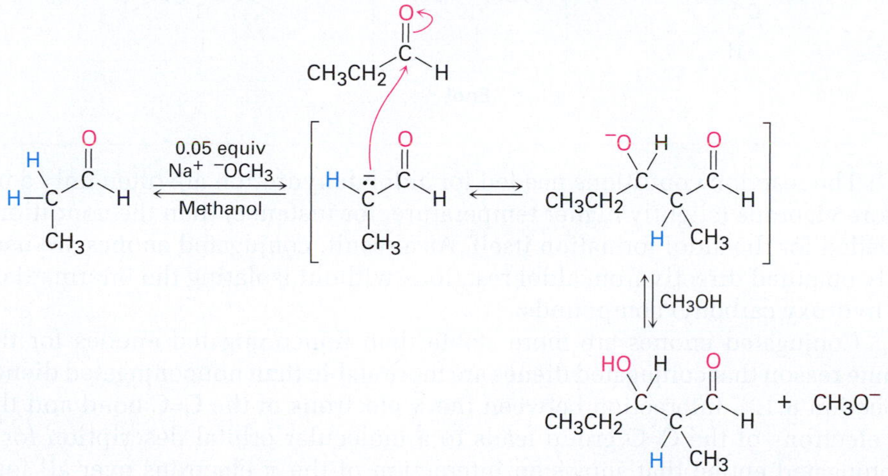

17.7 Dehydration of Aldol Products

The β-hydroxy aldehydes or ketones formed in aldol reactions can be easily dehydrated to yield α,β-unsaturated products, or *conjugated enones*. In fact, it's this loss of water that gives the carbonyl *condensation* reaction its name, because water condenses out of the reaction when the enone product forms.

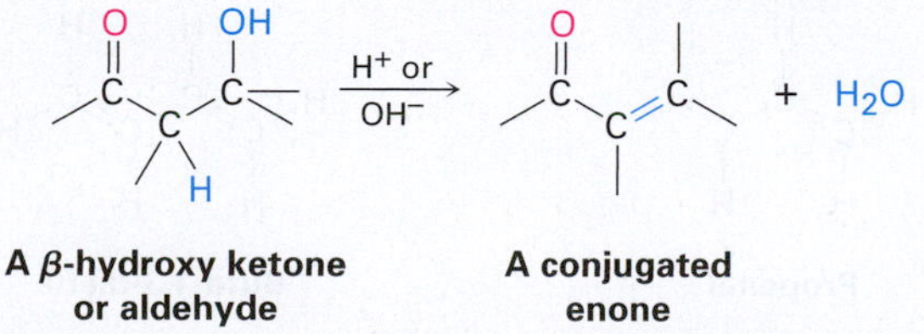

A β-hydroxy ketone or aldehyde   **A conjugated enone**

Most alcohols are resistant to dehydration by base (Section 13.4) because hydroxide ion is a poor leaving group, but aldol products dehydrate easily because of the carbonyl group. Under *basic* conditions, an acidic α hydrogen is removed, yielding an enolate ion that expels the $^-$OH leaving group in an E1cB reaction (Section 12.13). Under *acidic* conditions, an enol is formed, the —OH group is protonated, and water is expelled in an E1 or E2 reaction.

Base-catalyzed

Enolate ion

Acid-catalyzed

Enol

The reaction conditions needed for aldol dehydration are often only a bit more vigorous (slightly higher temperature, for instance) than the conditions needed for the aldol formation itself. As a result, conjugated enones are usually obtained directly from aldol reactions without isolating the intermediate β-hydroxy carbonyl compounds.

Conjugated enones are more stable than nonconjugated enones for the same reason that conjugated dienes are more stable than nonconjugated dienes (Section 8.12). Interaction between the π electrons of the C=C bond and the π electrons of the C=O group leads to a molecular orbital description for a conjugated enone that shows an interaction of the π electrons over all four atomic centers (Figure 17.10).

FIGURE 17.10 The π bonding molecular orbitals of a conjugated enone (propenal) and a conjugated diene (buta-1,3-diene) are similar in shape and are spread over the entire π system.

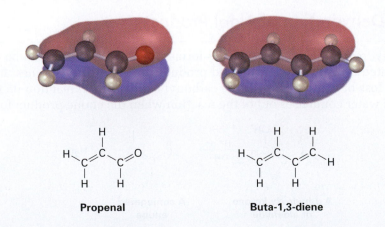

Propenal   **Buta-1,3-diene**

The real value of aldol dehydration is that removal of water from the reaction mixture can be used to drive the aldol equilibrium toward product. Even though the initial aldol step itself may be unfavorable (as it usually is for ketones), the subsequent dehydration step nevertheless allows many aldol condensations to be carried out in good yield. Cyclohexanone, for example, gives cyclohexylidenecyclohexanone in 92% yield even though the initial equilibrium is unfavorable.

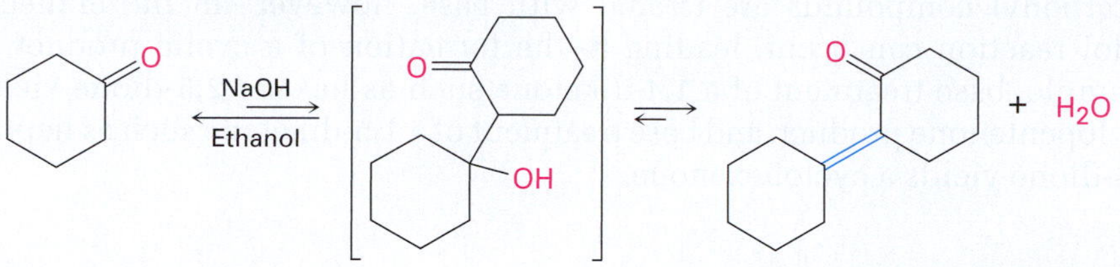

Cyclohexanone

Cyclohexylidenecyclohexanone (92%)

WORKED EXAMPLE 17.6 Predicting the Product of an Aldol Reaction

What is the structure of the enone obtained from aldol condensation of acetaldehyde?

Strategy

In the aldol reaction, H_2O is eliminated and a double bond is formed by removing two hydrogens from the acidic α position of one partner and the carbonyl oxygen from the second partner.

Solution

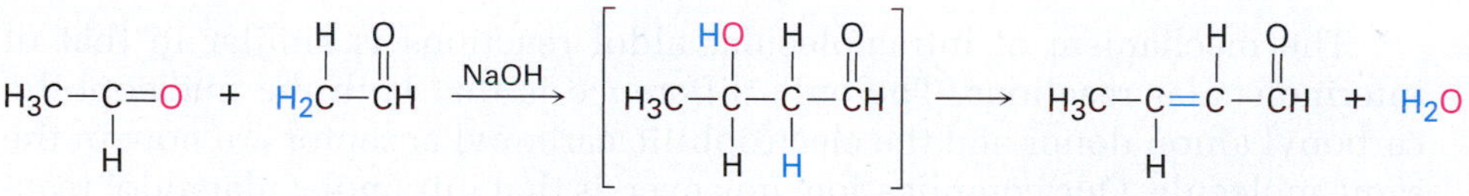

But-2-enal

Problem 17.15

What enone product would you expect from aldol condensation of each of the following compounds?

(a)

(b)

(c)

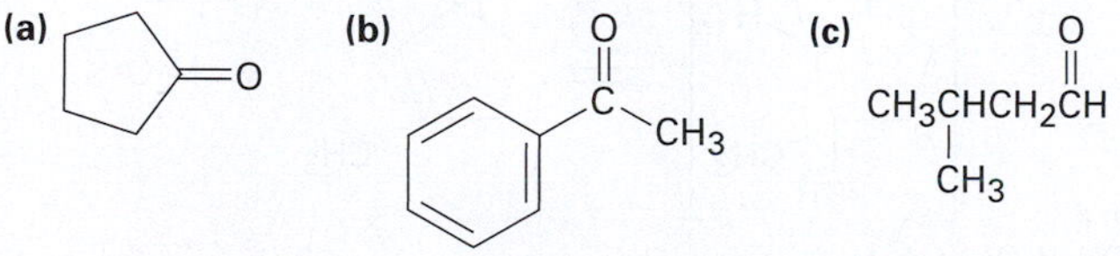

Problem 17.16

Aldol condensation of 3-methylcyclohexanone leads to a mixture of two enone products, not counting double-bond isomers. Draw them.

Problem 17.17

Which of the following compounds is an aldol condensation product? What is the aldehyde or ketone precursor of each?

(a) 2-Hydroxy-2-methylpentanal (b) 5-Ethyl-4-methylhept-4-en-3-one

17.8 Intramolecular Aldol Reactions

The aldol reactions we've seen thus far have all been intermolecular, meaning that they have taken place between two different molecules. When certain *di*carbonyl compounds are treated with base, however, an *intra*molecular aldol reaction can occur, leading to the formation of a cyclic product. For example, base treatment of a 1,4-diketone such as hexane-2,5-dione yields a cyclopentenone product, and base treatment of a 1,5-diketone such as heptane-2,6-dione yields a cyclohexenone.

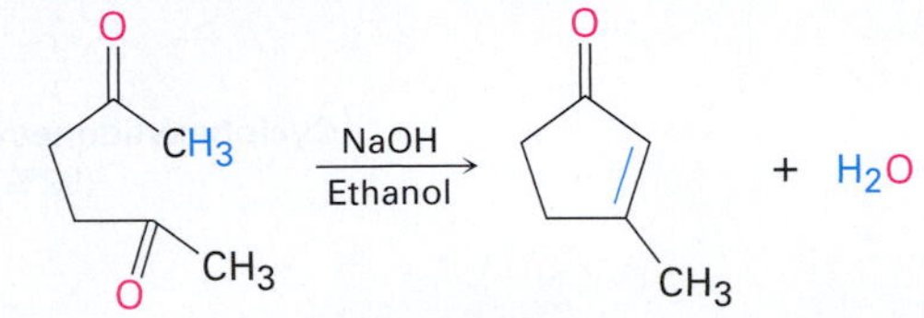

Hexane-2,5-dione   **3-Methylcyclopent-2-enone**
(a 1,4-diketone)

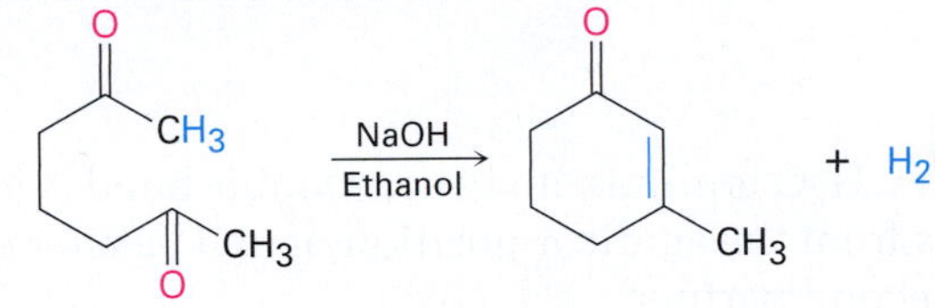

Heptane-2,6-dione   **3-Methylcyclohex-2-enone**
(a 1,5-diketone)

The mechanism of intramolecular aldol reactions is similar to that of intermolecular reactions. The only difference is that both the nucleophilic carbonyl anion donor and the electrophilic carbonyl acceptor are now in the same molecule. One complication, however, is that intramolecular aldol reactions might lead to a mixture of products, depending on which enolate ion is formed. For example, hexane-2,5-dione might yield either the five-membered-ring product 3-methylcyclopent-2-enone or the three-membered-ring product (2-methylcyclopropenyl)ethanone (Figure 17.11). In practice, though, only the cyclopentenone is formed.

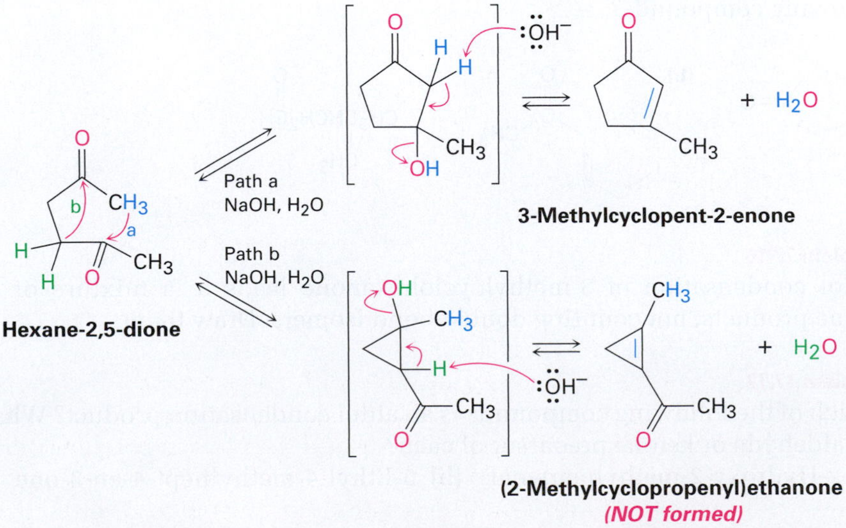

FIGURE 17.11 Intramolecular aldol reaction of hexane-2,5-dione yields 3-methyl-cyclopent-2-enone rather than the alternative cyclopropene.

Path a
NaOH, H₂O

3-Methylcyclopent-2-enone

Hexane-2,5-dione

Path b
NaOH, H₂O

(2-Methylcyclopropenyl)ethanone
(NOT formed)

The selectivity observed in the intramolecular aldol reaction of hexane-2,5-dione is due to the fact that all steps in the mechanism are readily reversible, so an equilibrium is reached. Thus, the relatively strain-free cyclopentenone product is considerably more stable than the highly strained cyclopropene alternative. For similar reasons, intramolecular aldol reactions of 1,5-diketones lead only to cyclohexenone products rather than to acyl cyclobutenes.

Problem 17.18

Treatment of a 1,3-diketone such as pentane-2,4-dione with base does not give an aldol condensation product. Explain.

Problem 17.19

What product would you expect to obtain from base treatment of cyclodecane-1,6-dione?

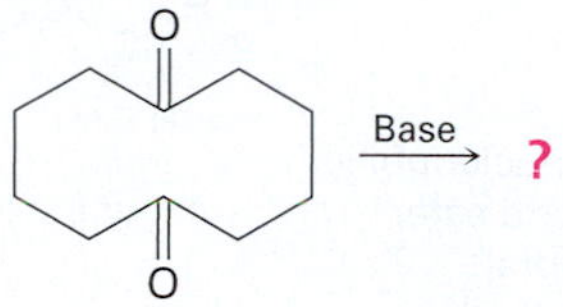

17.9 The Claisen Condensation Reaction

Esters, like aldehydes and ketones, are weakly acidic. When an ester with an α hydrogen is treated with 1 equivalent of a base such as sodium ethoxide, a reversible carbonyl condensation reaction occurs to yield a β-keto ester. For example, ethyl acetate yields ethyl acetoacetate on base treatment. This reaction between two ester molecules is known as the **Claisen condensation reaction**. (We'll use ethyl esters, abbreviated "Et," for consistency, but other esters will also work.)

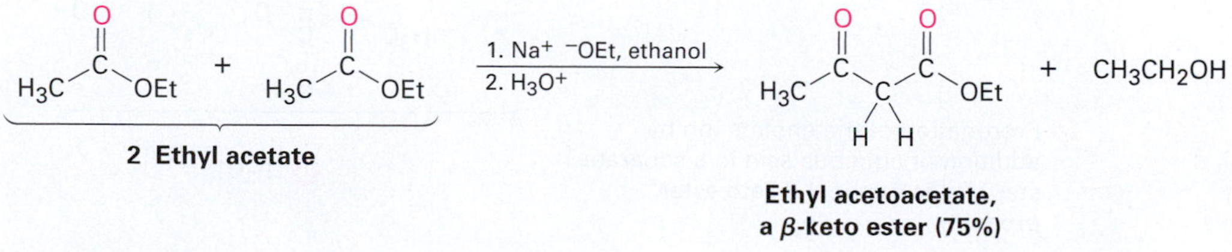

The mechanism of the Claisen condensation is similar to that of the aldol condensation and involves the nucleophilic addition of an ester enolate ion to the carbonyl group of a second ester molecule. The only difference between the aldol condensation of an aldehyde or ketone and the Claisen condensation of an ester involves the fate of the initially formed tetrahedral intermediate. The tetrahedral intermediate in the aldol reaction is protonated to give an alcohol product—exactly the behavior previously seen for aldehydes and ketones (Section 14.4). The tetrahedral intermediate in the Claisen reaction, however, expels an alkoxide leaving group to yield an acyl substitution product—exactly the behavior previously seen for esters (Section 16.6). The mechanism of the Claisen condensation reaction is shown in Figure 17.12.

FIGURE 17.12 MECHANISM:
Mechanism of the Claisen condensation reaction.

① Base abstracts an acidic alpha hydrogen atom from an ester molecule, yielding an ester enolate ion.

② The enolate ion adds in a nucleophilic addition reaction to a second ester molecule, giving a tetrahedral alkoxide intermediate.

③ The tetrahedral intermediate expels ethoxide ion to yield a new carbonyl compound, ethyl acetoacetate.

④ But ethoxide ion is a strong enough base to deprotonate ethyl acetoacetate, shifting the equilibrium and driving the overall reaction to completion.

⑤ Protonation of the enolate ion by addition of aqueous acid in a separate step yields the final β-keto ester product.

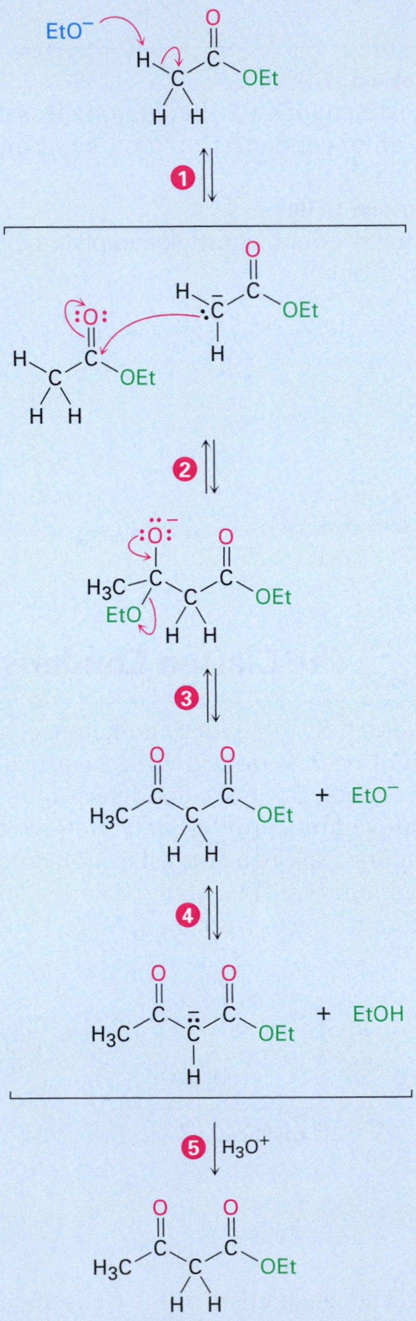

© John McMurry

If the starting ester has more than one acidic α hydrogen, the product β-keto ester has a highly acidic, doubly activated hydrogen atom that can be abstracted by base. This deprotonation of the product requires that a full equivalent of base rather than a catalytic amount be used in the reaction. Furthermore, the deprotonation serves to drive the equilibrium completely to the product side so that high yields are usually obtained in Claisen condensations.

WORKED EXAMPLE 17.7 Predicting the Product of a Claisen Condensation Reaction

What product would you obtain from Claisen condensation of ethyl propanoate?

Strategy

The Claisen condensation of an ester results in loss of one molecule of alcohol and formation of a product in which an acyl group of one reactant bonds to the α carbon of the second reactant. The product is a β-keto ester.

Solution

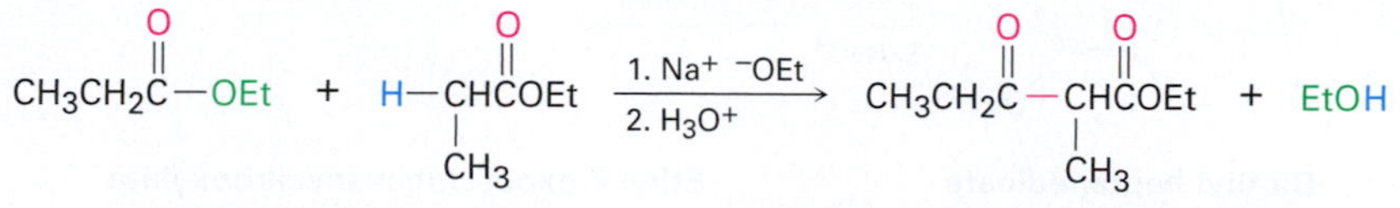

2 Ethyl propanoate   Ethyl 2-methyl-3-oxopentanoate

Problem 17.20

Show the products you would expect to obtain by Claisen condensation of the following esters:

(a) $(CH_3)_2CHCH_2CO_2Et$
(b) Ethyl phenylacetate
(c) Ethyl cyclohexylacetate

Problem 17.21

As shown in Figure 17.12, the Claisen reaction is reversible. That is, a β-keto ester can be cleaved by base into two fragments. Using curved arrows to indicate electron flow, show the mechanism by which this cleavage occurs.

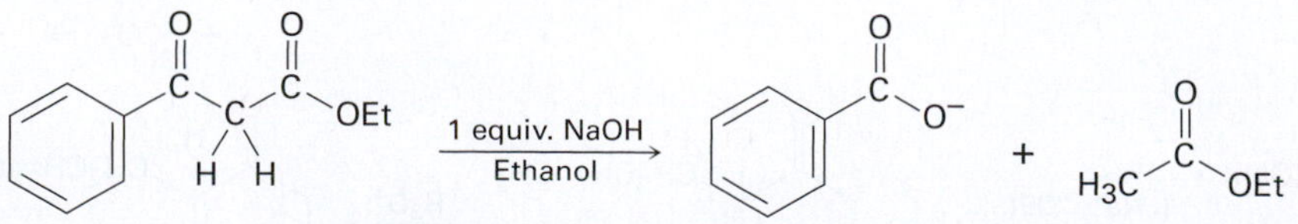

17.10 Intramolecular Claisen Condensations

Intramolecular Claisen condensations can be carried out with diesters, just as intramolecular aldol condensations can be carried out with diketones (Section 17.8). Called the *Dieckmann cyclization,* the reaction works best on 1,6-diesters and 1,7-diesters. Intramolecular Claisen cyclization of a 1,6-diester gives a five-membered cyclic β-keto ester, and cyclization of a 1,7-diester gives a six-membered cyclic β-keto ester.

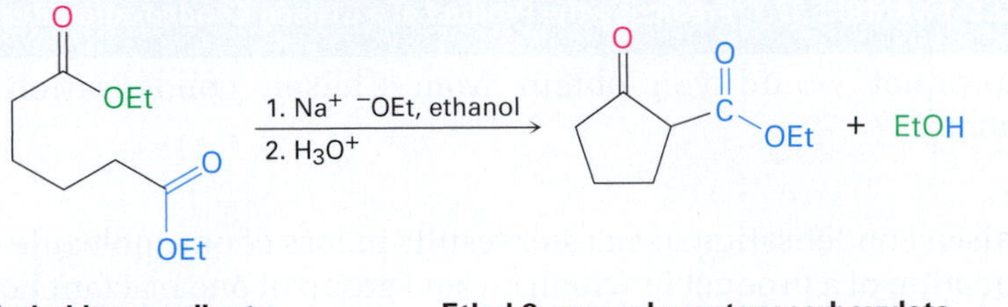

Diethyl hexanedioate
(a 1,6-diester)

Ethyl 2-oxocyclopentanecarboxylate
(82%)

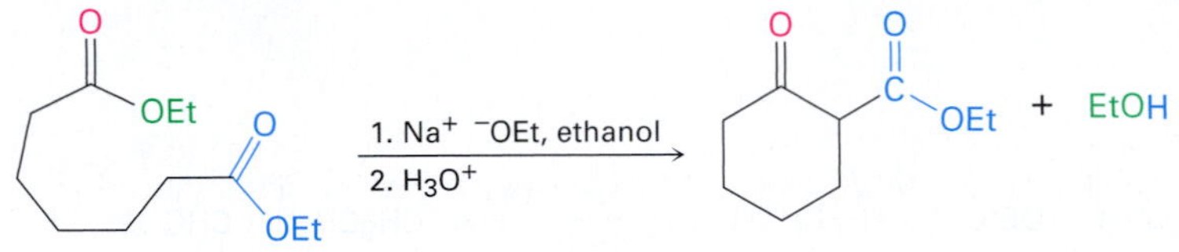

Diethyl heptanedioate
(a 1,7-diester)

Ethyl 2-oxocyclohexanecarboxylate

The mechanism of the intramolecular Claisen cyclization, shown in Figure 17.13, is the same as that of the intermolecular Claisen condensation. One of the two ester groups is converted into an enolate ion, which then carries out a nucleophilic acyl substitution on the second ester group at the other end of the molecule. A cyclic β-keto ester product results.

The cyclic β-keto ester produced in an intramolecular Claisen cyclization can be further alkylated and decarboxylated by a series of reactions analogous to those used in the acetoacetic ester synthesis (Section 17.5). Alkylation and subsequent decarboxylation of ethyl 2-oxocyclohexanecarboxylate, for instance, yields a 2-alkylcyclohexanone. The overall sequence of (1) intramolecular Claisen cyclization, (2) β-keto ester alkylation, and (3) decarboxylation is a powerful method for preparing 2-substituted cyclohexanones and cyclopentanones.

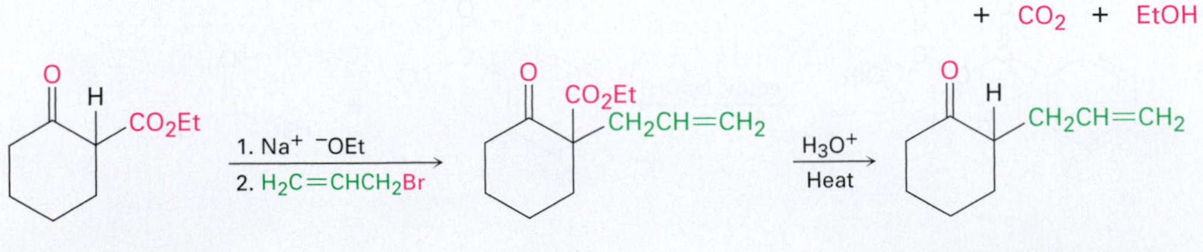

Ethyl 2-oxocyclo-
hexanecarboxylate

2-Allylcyclohexanone
(83%)

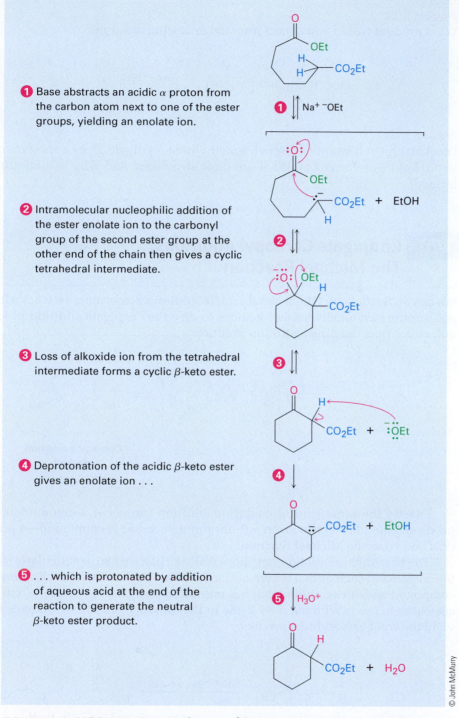

① Base abstracts an acidic α proton from the carbon atom next to one of the ester groups, yielding an enolate ion.

② Intramolecular nucleophilic addition of the ester enolate ion to the carbonyl group of the second ester group at the other end of the chain then gives a cyclic tetrahedral intermediate.

③ Loss of alkoxide ion from the tetrahedral intermediate forms a cyclic β-keto ester.

④ Deprotonation of the acidic β-keto ester gives an enolate ion . . .

⑤ . . . which is protonated by addition of aqueous acid at the end of the reaction to generate the neutral β-keto ester product.

© John McMurry

FIGURE 17.13 MECHANISM: Mechanism of the intramolecular Claisen cyclization of a 1,7-diester to yield a cyclic β-keto ester product.

Problem 17.22

What product would you expect from the following reaction?

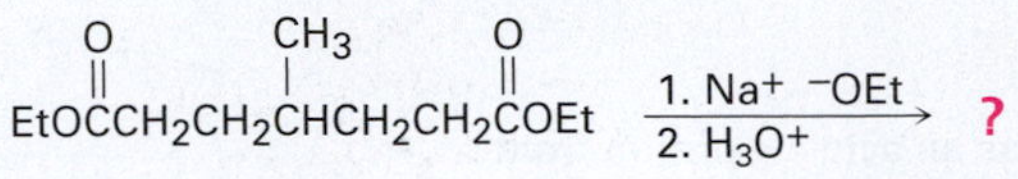

$$\underset{\text{EtOCCH}_2\text{CH}_2\text{CHCH}_2\text{CH}_2\text{COEt}}{\overset{\overset{\text{O}}{\|}\qquad\overset{\text{CH}_3}{|}\qquad\overset{\text{O}}{\|}}{}} \quad \xrightarrow[\text{2. H}_3\text{O}^+]{\text{1. Na}^+ \ ^-\text{OEt}} \quad \textbf{?}$$

Problem 17.23

Dieckmann cyclization of diethyl 3-methylheptanedioate gives a mixture of two β-keto ester products. What are their structures, and why is a mixture formed?

17.11 Conjugate Carbonyl Additions: The Michael Reaction

We saw in Section 14.11 that certain nucleophiles, such as amines, react with α,β-unsaturated aldehydes and ketones to give the conjugate addition product, rather than the direct addition product:

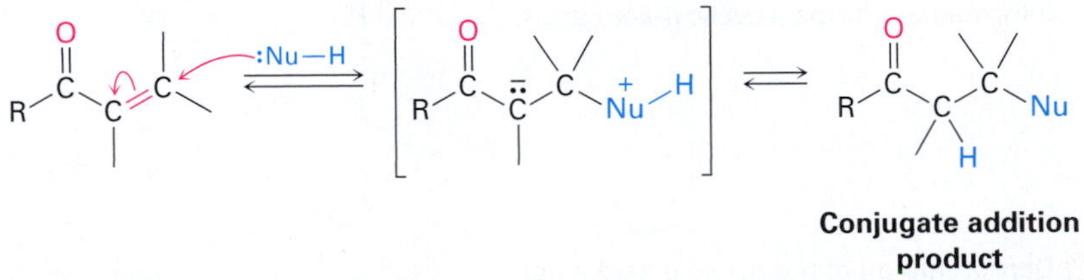

Conjugate addition product

Exactly the same kind of conjugate addition can occur when a nucleophilic enolate ion reacts with an α,β-unsaturated carbonyl compound—a process known as the **Michael reaction**.

The best Michael reactions are those that take place when a particularly stable enolate ion such as that derived from a β-keto ester or other 1,3-dicarbonyl compound adds to an unhindered α,β-unsaturated ketone. For example, ethyl acetoacetate reacts with but-3-en-2-one in the presence of sodium ethoxide to yield the conjugate addition product.

Ethyl acetoacetate   **But-3-en-2-one**

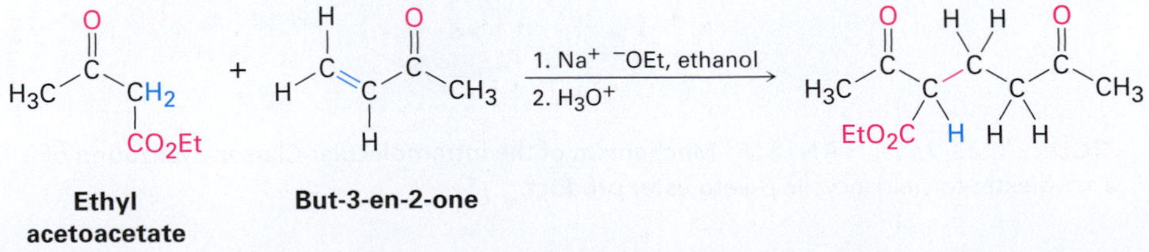

Michael reactions take place by addition of a nucleophilic enolate ion donor to the β carbon of an α,β-unsaturated carbonyl acceptor, according to the mechanism shown in Figure 17.14.

FIGURE 17.14 MECHANISM: Mechanism of the Michael reaction between a β-keto ester and an α,β-unsaturated ketone.

① The base catalyst removes an acidic alpha proton from the starting β-keto ester to generate a stabilized enolate ion nucleophile.

② The nucleophile adds to the α,β-unsaturated ketone electrophile in a Michael reaction to generate a new enolate as product.

③ The enolate product abstracts an acidic proton, either from solvent or from starting keto ester, to yield the final addition product.

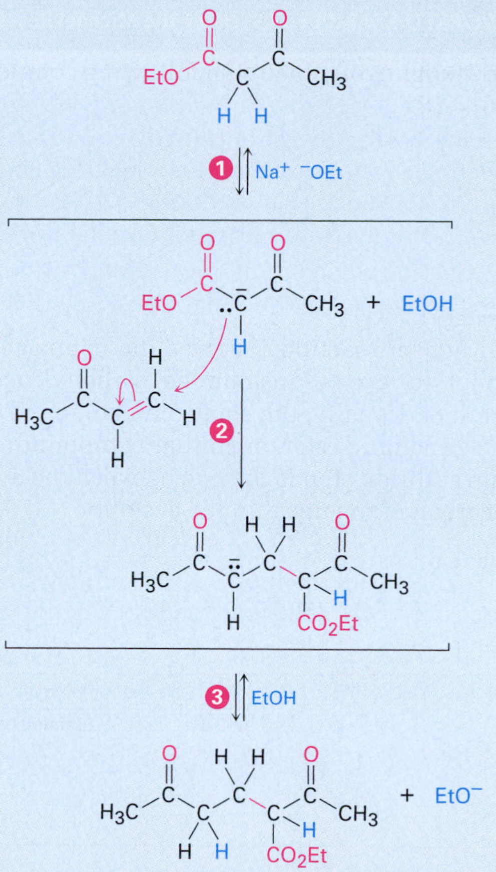

The Michael reaction occurs with a variety of α,β-unsaturated carbonyl compounds, not just conjugated ketones. Unsaturated aldehydes, esters, thioesters, nitriles, and amides can all act as the electrophilic acceptor component in Michael reactions (Table 17.2). Similarly, a variety of different donors can be used, including β-diketones, β-keto esters, malonic esters, and β-keto nitriles.

TABLE 17.2
Some Michael Acceptors and Michael Donors

Michael acceptors		Michael donors	
$H_2C=CHCH$ (O)	Propenal	$RCCH_2CR'$ (O, O)	β-Diketone
$H_2C=CHCCH_3$ (O)	But-3-en-2-one	$RCCH_2COEt$ (O, O)	β-Keto ester
$H_2C=CHCOEt$ (O)	Ethyl propenoate	$EtOCCH_2COEt$ (O, O)	Diethyl malonate
$H_2C=CHCNH_2$ (O)	Propenamide	$RCCH_2C\equiv N$ (O)	β-Keto nitrile
$H_2C=CHC\equiv N$	Propenenitrile		

© John McMurry

WORKED EXAMPLE 17.8 Using the Michael Reaction

How might you obtain the following compound using a Michael reaction?

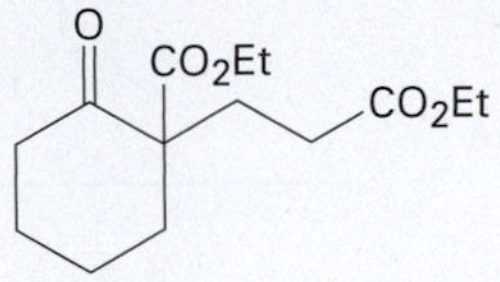

Strategy

A Michael reaction involves the conjugate addition of a stable enolate ion donor to an α,β-unsaturated carbonyl acceptor, yielding a 1,5-dicarbonyl product. Usually, the stable enolate ion is derived from a β-diketone, β-keto ester, malonic ester, or similar compound. The C–C bond made in the conjugate addition step is the one between the α carbon of the acidic donor and the β carbon of the unsaturated acceptor.

Solution

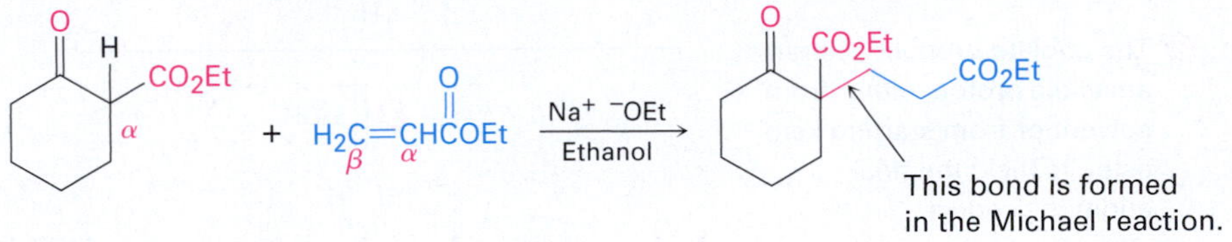

Problem 17.24

What product would you obtain from a base-catalyzed Michael reaction of pentane-2,4-dione with each of the following α,β-unsaturated acceptors?
(a) Cyclohex-2-enone **(b)** Propenenitrile **(c)** Ethyl but-2-enoate

Problem 17.25

What product would you obtain from a base-catalyzed Michael reaction of but-3-en-2-one with each of the following nucleophilic donors?

(a)

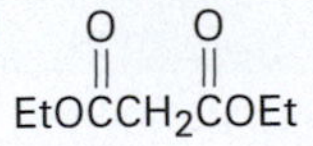

$$EtOCCH_2COEt$$

(b)

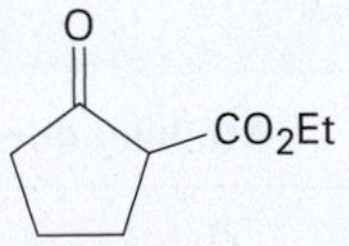

17.12 Carbonyl Condensations with Enamines: The Stork Reaction

In addition to enolate ions, other kinds of carbon nucleophiles also add to α,β-unsaturated acceptors in Michael-like reactions. Among the most important such nucleophiles, particularly in biological chemistry, are *enamines,* which are readily prepared by reaction between a ketone and a secondary amine, as we saw in Section 14.7. For example:

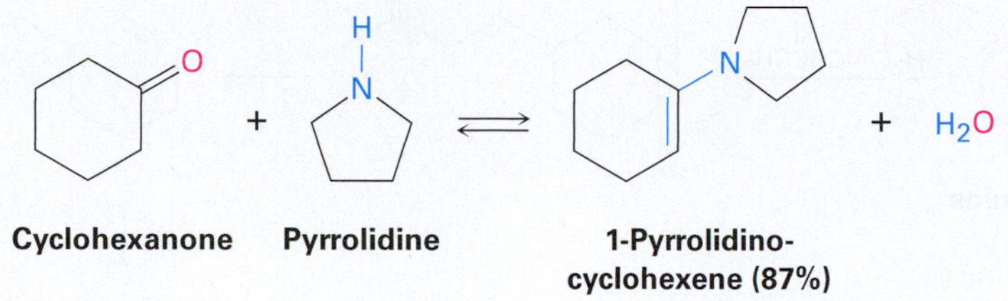

Cyclohexanone   **Pyrrolidine**   **1-Pyrrolidino-**
                                          cyclohexene (87%)

As the following resonance structures indicate, enamines are electronically similar to enolate ions. Overlap of the nitrogen lone-pair orbital with the double-bond p orbitals leads to an increase in electron density on the α carbon atom, making that carbon nucleophilic. An electrostatic potential map of N,N-dimethylaminoethylene shows this shift of electron density (red) toward the α position.

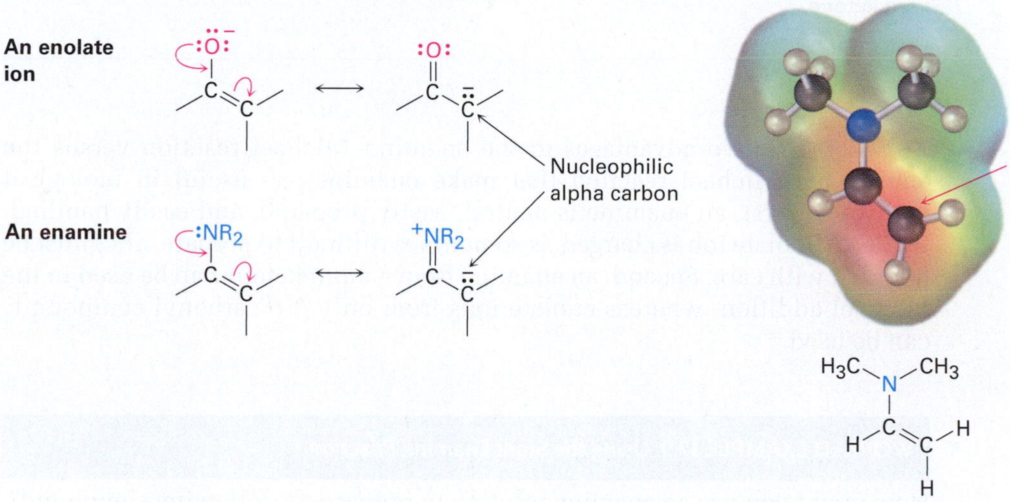

An enolate ion

An enamine

Nucleophilic alpha carbon

Enamines behave in much the same way as enolate ions and enter into many of the same kinds of reactions. In the *Stork reaction,* for example, an enamine adds to an α,β-unsaturated carbonyl acceptor in a Michael-like process. The initial product is then hydrolyzed by aqueous acid (Section 14.7) to yield a 1,5-dicarbonyl compound. The overall reaction is thus a three-step sequence of (1) enamine formation from a ketone, (2) Michael addition to an α,β-unsaturated carbonyl compound, and (3) enamine hydrolysis back to a ketone.

The net effect of the Stork reaction is a Michael addition of a ketone to an α,β-unsaturated carbonyl compound. For example, cyclohexanone reacts with the cyclic amine pyrrolidine to yield an enamine; further reaction with an enone such as but-3-en-2-one yields a Michael adduct; and aqueous hydrolysis completes the sequence to give a 1,5-diketone (Figure 17.15).

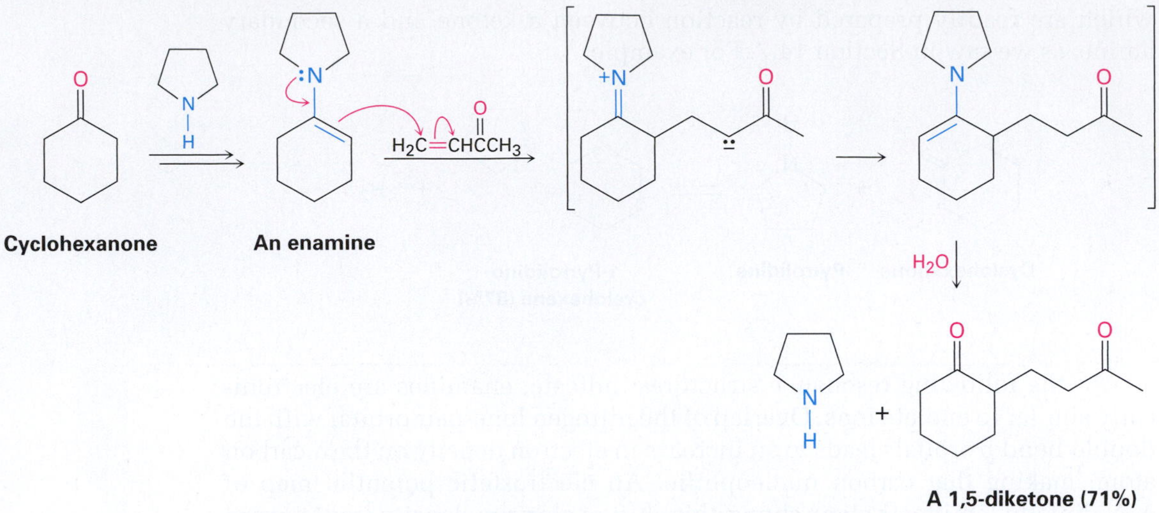

Cyclohexanone   **An enamine**

A 1,5-diketone (71%)

FIGURE 17.15 The Stork reaction between cyclohexanone and but-3-en-2-one. Cyclohexanone is first converted into an enamine, the enamine adds to the α,β-unsaturated ketone in a Michael reaction, and the conjugate addition product is hydrolyzed to yield a 1,5-diketone.

There are two advantages to the enamine–Michael reaction versus the enolate-ion–Michael reaction that make enamines so useful in biological pathways. First, an enamine is neutral, easily prepared, and easily handled, while an enolate ion is charged, is sometimes difficult to prepare, and must be handled with care. Second, an enamine from a *mono*ketone can be used in the Michael addition, whereas enolate ions from only β-*di*carbonyl compounds can be used.

WORKED EXAMPLE 17.9 Using the Stork Enamine Reaction

How might you use an enamine reaction to prepare the following compound?

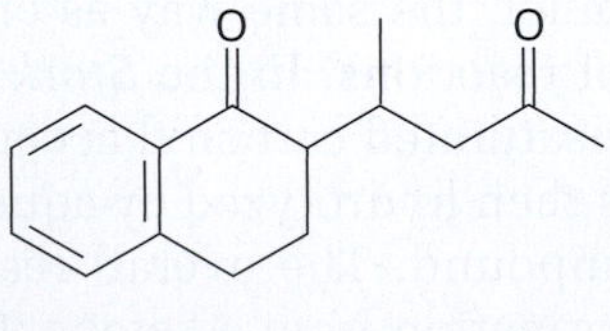

Strategy

The overall result of an enamine reaction is the Michael addition of a ketone as donor to an α,β-unsaturated carbonyl compound as acceptor, yielding a 1,5-dicarbonyl product. The C–C bond made in the Michael addition step is the one between the α carbon of the ketone donor and the β carbon of the unsaturated acceptor.

Solution

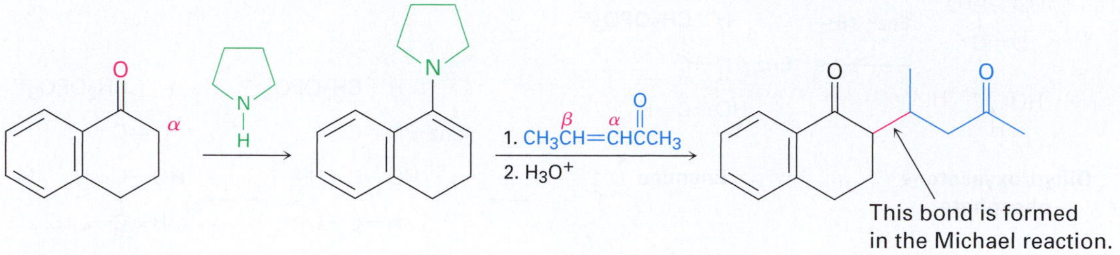

This bond is formed in the Michael reaction.

Problem 17.26

What products would result after hydrolysis from reaction of the enamine prepared from cyclopentanone and pyrrolidine with the following α,β-unsaturated acceptors?

(a) $H_2C{=}CHCO_2Et$ **(b)** $H_2C{=}CHCHO$ **(c)** $CH_3CH{=}CHCOCH_3$

Problem 17.27

Show how you might use an enamine reaction to prepare each of the following compounds:

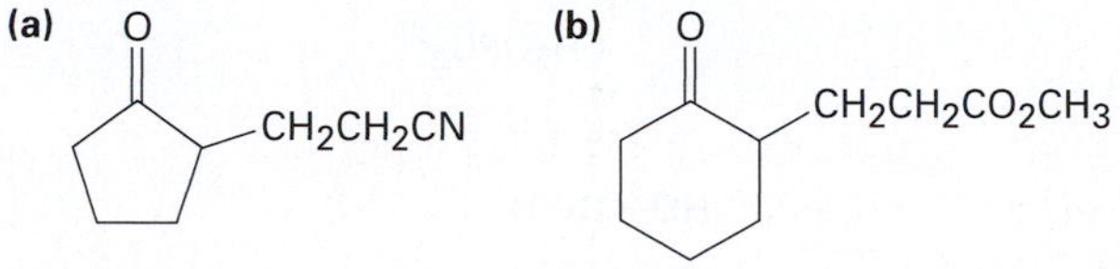

(a)

CH_2CH_2CN

(b)

$CH_2CH_2CO_2CH_3$

17.13 Biological Carbonyl Condensation Reactions

Biological Aldol Reactions

Aldol reactions occur in many biological pathways but are particularly important in carbohydrate metabolism, where enzymes called *aldolases* catalyze the addition of a ketone enolate ion to an aldehyde. Aldolases occur in all organisms and are of two types. Type I aldolases occur primarily in animals and higher plants; type II aldolases occur primarily in fungi and bacteria. Both types catalyze the same kind of reaction, but type I aldolases operate through an enamine, while type II aldolases require a metal ion (usually Zn^{2+}) as Lewis acid and operate through an enolate ion.

An example of an aldolase-catalyzed reaction occurs in glucose biosynthesis when dihydroxyacetone phosphate reacts with glyceraldehyde 3-phosphate to give fructose 1,6-bisphosphate. In animals and higher plants, dihydroxyacetone phosphate is first converted into an enamine by reaction with the $-NH_2$ group on a lysine amino acid in the enzyme. The enamine then adds to glyceraldehyde 3-phosphate, and the iminium ion that results is hydrolyzed. In bacteria and fungi, the aldol reaction occurs directly, with the ketone carbonyl group of glyceraldehyde 3-phosphate complexed to a Zn^{2+} ion to make it a better acceptor (Figure 17.16).

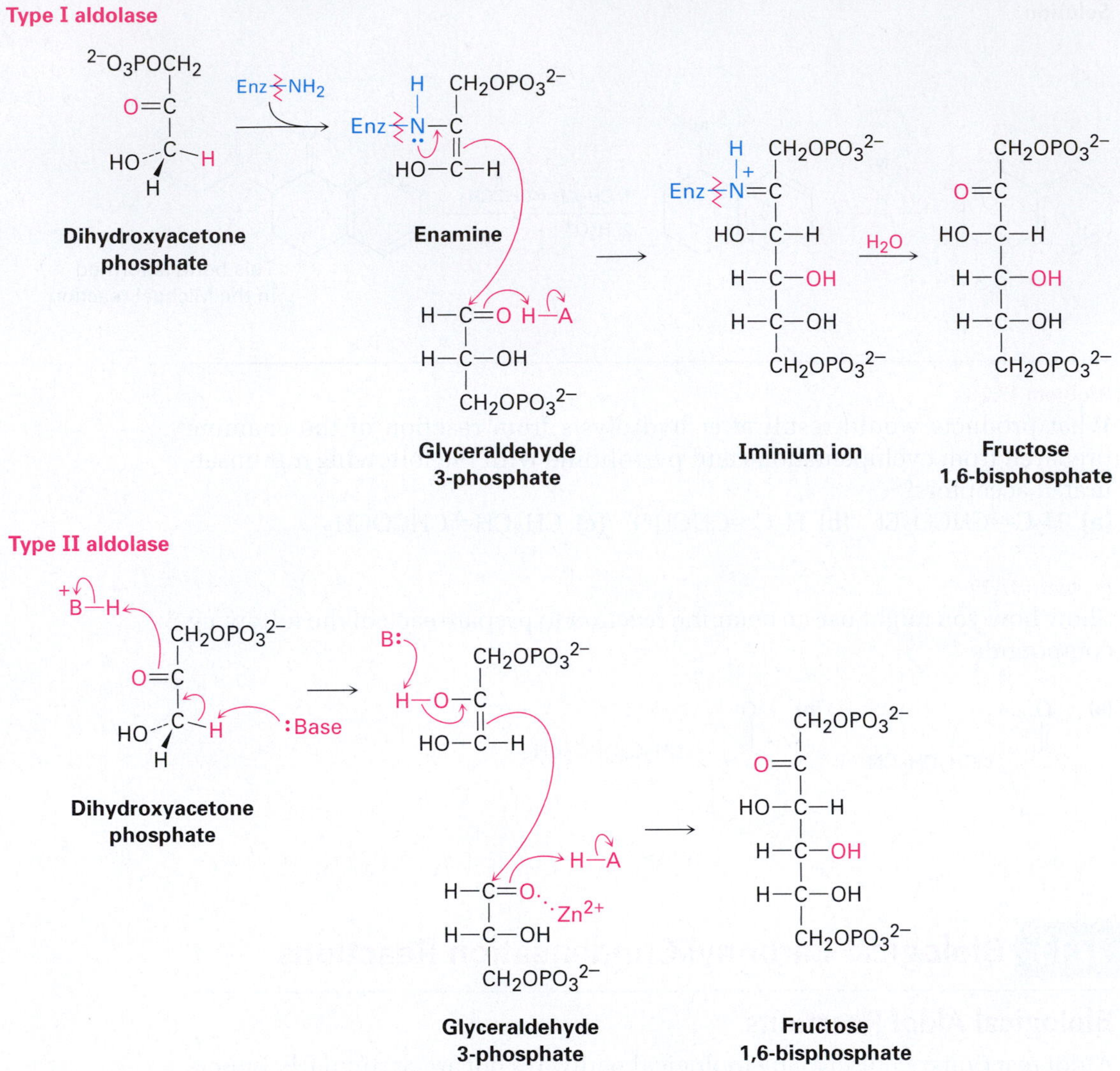

FIGURE 17.16 Mechanisms of type I and type II aldolase reactions in glucose biosynthesis.

Note that the reactions shown in Figure 17.16 are *mixed* aldol reactions, which take place between two different partners, as opposed to the symmetrical aldol reactions between identical partners usually carried out in the laboratory. Mixed aldol reactions between different partners often give mixtures of products in the laboratory but are successful in living systems because of the selectivity of the enzyme catalysts.

Biological Claisen Condensations

Claisen condensations, like aldol reactions, also occur in a large number of biological pathways. In fatty-acid biosynthesis, for instance, an enolate ion generated by decarboxylation (Section 17.5) of malonyl ACP adds to the carbonyl group of another acyl group bonded through a thioester linkage to a synthase enzyme. The tetrahedral intermediate that results then expels the synthase, giving acetoacetyl ACP (Figure 17.17).

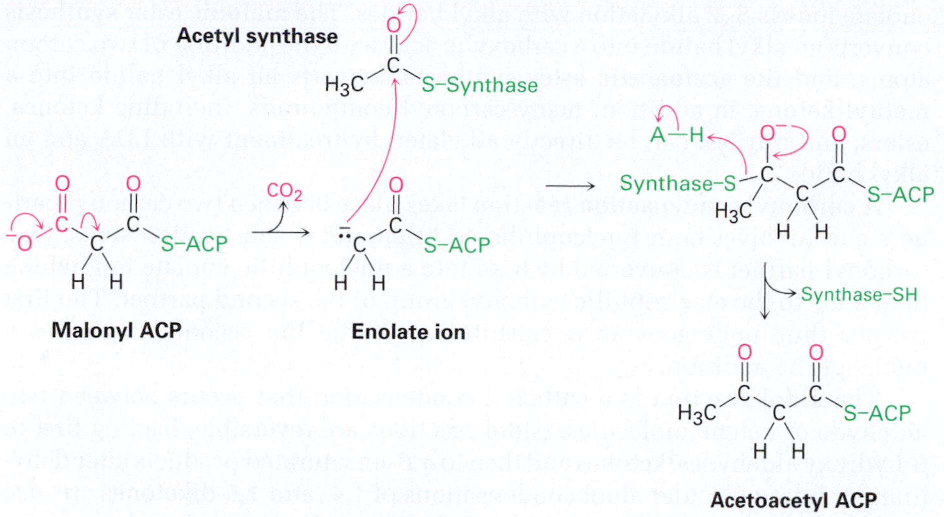

FIGURE 17.17 A Claisen condensation between two thioesters occurs as the first step in fatty-acid biosynthesis.

Like the mixed aldol reaction between different partners shown in Figure 17.16, mixed Claisen condensations also occur frequently in living organisms, particularly in the pathway for fatty-acid biosynthesis that we'll discuss in Section 23.6. Butyryl synthase, for instance, reacts with malonyl ACP in a mixed Claisen condensation to give 3-ketohexanoyl ACP.

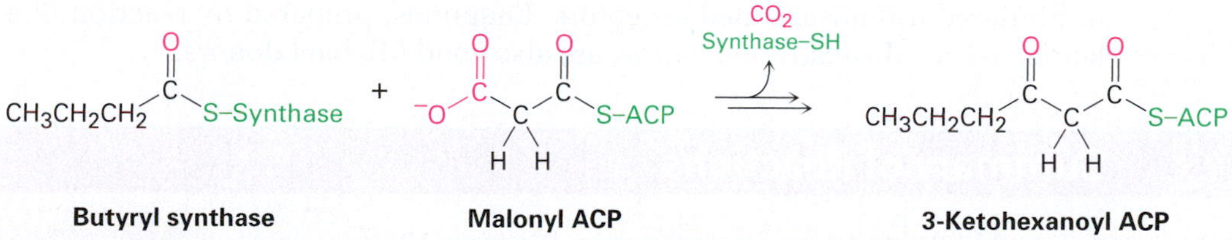

Summary

Biochemical pathways make frequent use of α-substitution and carbonyl condensation reactions. In fact, practically every biosynthetic pathway for building up larger molecules from smaller precursors uses carbonyl condensation reactions for the purpose. In this chapter, we saw how and why these reactions occur.

Carbonyl compounds are in a rapid equilibrium with their **enols**, a process called keto–enol tautomerism. Although enol **tautomers** are normally present to only a small extent at equilibrium and usually can't be isolated in pure form, they nevertheless contain a highly nucleophilic double bond and react with electrophiles in an **α-substitution reaction**. An example is the α halogenation of ketones on treatment with Cl_2 or Br_2 in acid solution. Alpha bromination of carboxylic acids can be similarly accomplished by the Hell–Volhard–Zelinskii (HVZ) reaction, in which an acid is treated with Br_2 and PBr_3.

Alpha hydrogen atoms of carbonyl compounds are weakly acidic and can be removed by strong bases, such as lithium diisopropylamide (LDA), to yield strongly nucleophilic **enolate ions**. Among the most useful reactions of

Key Words

acetoacetic ester synthesis, 710

aldol reaction, 716

α-substitution reaction, 695

carbonyl condensation
   reaction, 695

Claisen condensation
   reaction, 723

enol, 696

enolate ion, 698

malonic ester synthesis, 707

Michael reaction, 728

tautomer, 696

enolate ions is S_N2 alkylation with alkyl halides. The **malonic ester synthesis** converts an alkyl halide into a carboxylic acid with the addition of two carbon atoms, and the **acetoacetic ester synthesis** converts an alkyl halide into a methyl ketone. In addition, many carbonyl compounds, including ketones, esters, and nitriles, can be directly alkylated by treatment with LDA and an alkyl halide.

A **carbonyl condensation reaction** takes place between two carbonyl partners and involves both nucleophilic addition and α-substitution steps. One carbonyl partner is converted by base into a nucleophilic enolate ion, which then adds to the electrophilic carbonyl group of the second partner. The first partner thus undergoes an α substitution, while the second undergoes a nucleophilic addition.

The **aldol reaction** is a carbonyl condensation that occurs between two aldehyde or ketone molecules. Aldol reactions are reversible, leading first to β-hydroxy aldehydes/ketones and then to α,β-unsaturated products after dehydration. Intramolecular aldol condensations of 1,4- and 1,5-diketones are also successful and provide a good way to make five- and six-membered rings.

The **Claisen condensation reaction** is a carbonyl condensation that occurs between two ester components and gives a β-keto ester product. Intramolecular Claisen condensations, called *Dieckmann cyclizations,* yield five- and six-membered cyclic β-keto esters starting from 1,6- and 1,7-diesters.

The conjugate addition of a carbon nucleophile to an α,β-unsaturated acceptor is known as the **Michael reaction**. The best Michael reactions take place between unusually acidic donors (β-keto esters or β-diketones) and unhindered α,β-unsaturated acceptors. Enamines, prepared by reaction of a ketone with a disubstituted amine, are also good Michael donors.

Summary of Reactions

1. Aldehyde/ketone halogenation (Section 17.2)

2. Hell–Volhard–Zelinskii bromination of acids (Section 17.3)

3. Alkylation of enolate ions (Section 17.5)

   (a) Malonic ester synthesis

(b) Acetoacetic ester synthesis

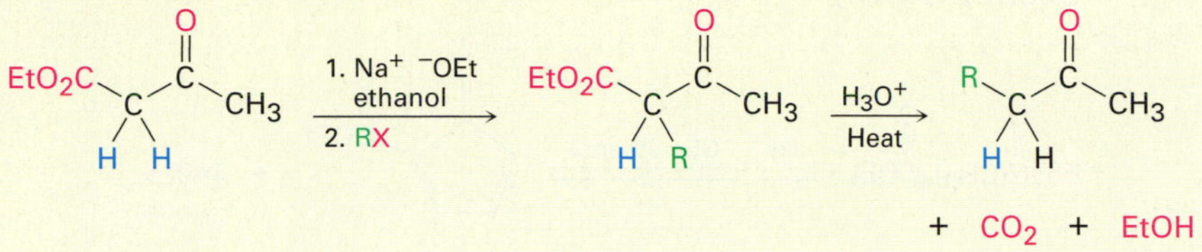

(c) Direct alkylation of ketones, esters, and nitriles

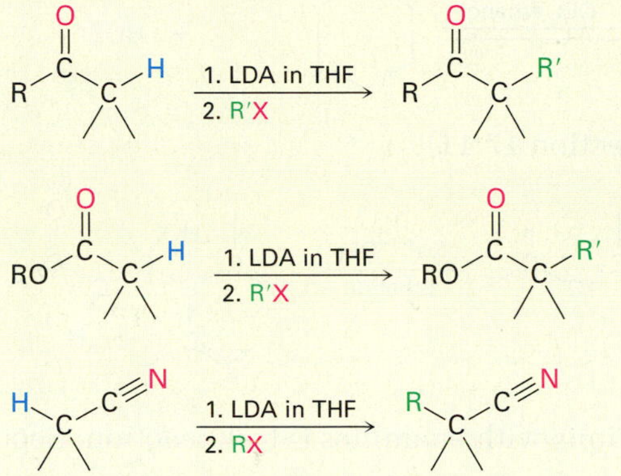

4. Aldol reaction (Section 17.6)

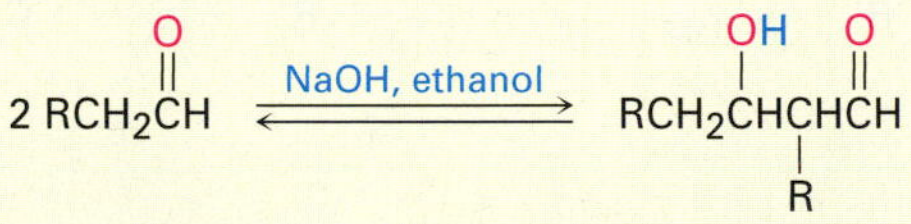

5. Intramolecular aldol reaction (Section 17.8)

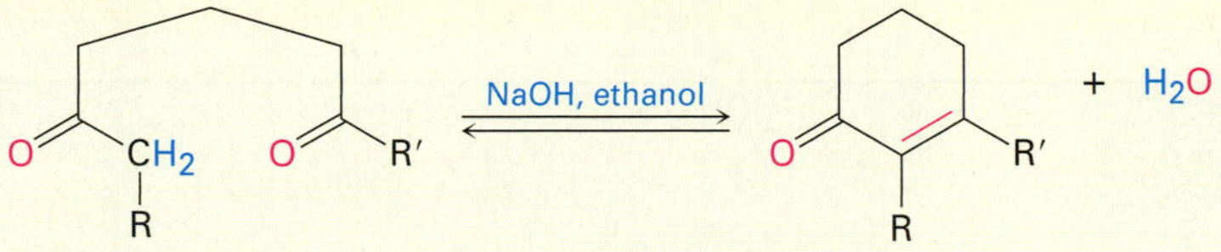

6. Dehydration of aldol products (Section 17.7)

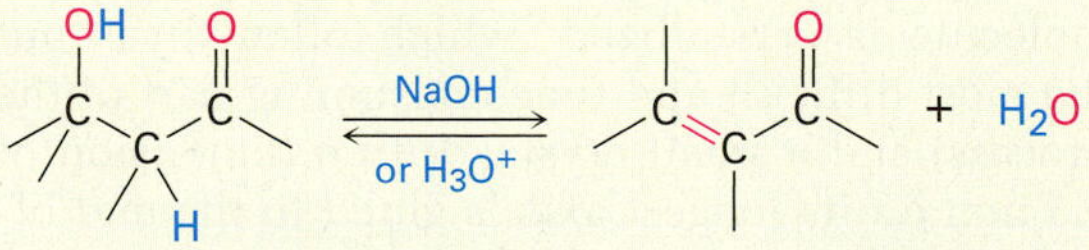

7. Claisen condensation reaction (Section 17.9)

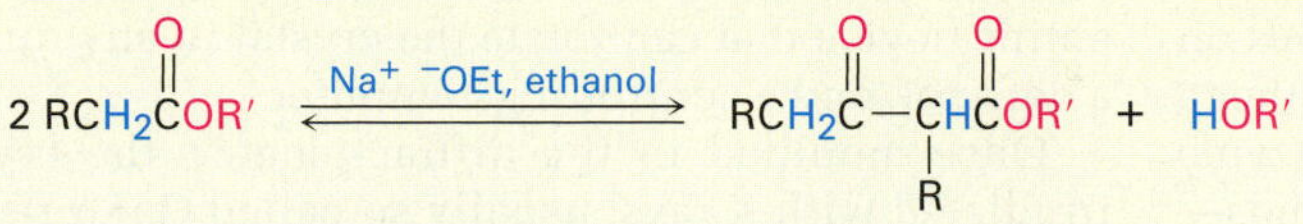

8. Intramolecular Claisen condensation (Dieckmann cyclization; Section 17.10)

$$EtOC(CH_2)_4COEt \rightleftharpoons[\text{Na}^+ \ ^-\text{OEt, ethanol}]{} \quad + \quad HOEt$$

$$EtOC(CH_2)_5COEt \rightleftharpoons[\text{Na}^+ \ ^-\text{OEt, ethanol}]{} \quad + \quad HOEt$$

9. Michael reaction (Section 17.11)

$$\xrightarrow[\text{Ethanol}]{\text{Na}^+ \ ^-\text{OEt}}$$

10. Carbonyl condensations with enamines (Stork reaction; Section 17.12)

$$\xrightarrow[\text{2. H}_3\text{O}^+]{\text{1. Mix in THF solvent}}$$

Lagniappe

X-Ray Crystallography

Determining the three-dimensional shape of an object around you is easy—you just look at it, let your eyes focus the light rays reflected from the object, and let your brain assemble the data into a recognizable image. If the object is small, you use a microscope and let the microscope lens focus the visible light. Unfortunately, there is a limit to what you can see, even with the best optical microscope. Called the *diffraction limit,* you can't see anything smaller than the wavelength of light you are using for the observation. Visible light has wavelengths of several hundred nanometers, but atoms in molecules have dimensions on the order of 0.1 nm. Thus, to "see" a molecule—whether a small one in the laboratory or a large, complex enzyme with a molecular weight in the hundreds of thousands— you need wavelengths in the 0.1 nm range, which corresponds to X rays.

Let's say that we want to determine the structure and shape of an enzyme or other biological molecule. The technique used is called *X-ray crystallography*. First, the molecule is crystallized (which often turns out to be the most difficult and time-consuming part of the entire process) and a small crystal with a dimension of 0.4 to 0.5 mm on its longest axis is glued to the end of a glass fiber. The fiber and attached crystal are then mounted in an instrument called an *X-ray diffractometer,* which consists of a radiation source, a sample positioning and orienting device that can rotate the crystal in any direction, a detector, and a controlling computer.

Once mounted in the diffractometer, the crystal is irradiated with X rays, usually so-called $\text{Cu}K\alpha$ radiation with a wavelength of 0.154 nm. When the X rays strike the enzyme crystal, they interact with electrons in the

continued

Lagniappe *continued*

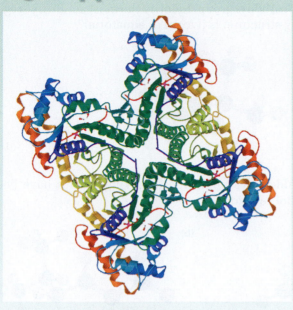

The structure of human muscle fructose-1,6-bisphosphate aldolase, as determined by X-ray crystallography and downloaded from the Protein Data Bank.

molecule and are scattered into a diffraction pattern, which, when detected and visualized, appears as a series of intense spots against a null background.

Manipulation of the diffraction pattern to extract three-dimensional molecular data is a complex process, but the final result is that an electron-density map of the molecule is produced. Because electrons are largely localized around atoms, any two centers of electron density located within bonding distance of each other are assumed to represent bonded atoms, leading to a recognizable chemical structure. So important is this structural information for biochemistry that an online database of more than 60,000 biological structures has been created. Operated by Rutgers University and funded by the U.S. National Science Foundation, the Protein Data Bank (PDB) is a worldwide repository for processing and distributing three-dimensional structural data for biological macromolecules. All the enzyme models used at the beginning of chapters in this book are deposited in and can be downloaded from the PDB. We'll see how to access the PDB in the Chapter 19 *Lagniappe*.

Exercises

VISUALIZING CHEMISTRY

(Problems 17.1–17.27 appear within the chapter.)

17.28 ■ Show the steps in preparing each of the following substances, using either a malonic ester synthesis or an acetoacetic ester synthesis:

■ indicates problems that are assignable in Organic OWL.

Go to this book's companion website at **www.cengage.com/ chemistry/mcmurry** to explore interactive versions of the Active Figures from this text.

(a)

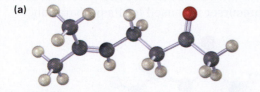

(b)

■ Problems assignable in Organic OWL.

17.29 ■ For a given α hydrogen atom to be acidic, the C–H bond must be parallel to the *p* orbitals of the C=O bond, that is, perpendicular to the plane of the adjacent carbonyl group. Identify the most acidic hydrogen atom in the following structure. Is it axial or equatorial?

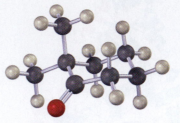

17.30 ■ What ketones or aldehydes might the following enones have been prepared from by aldol reaction?

(a)   (b)

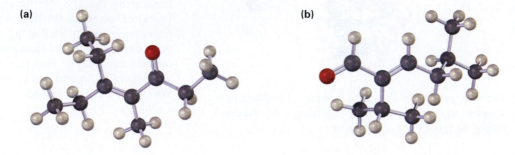

17.31 ■ The following structure represents an intermediate formed by addition of an ester enolate ion to a second ester molecule. Identify the reactant, the leaving group, and the product.

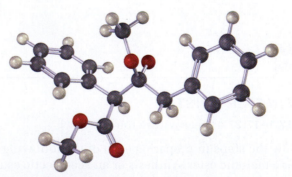

17.32 ■ The following molecule was formed by an intramolecular aldol reaction. What dicarbonyl precursor was used for its preparation?

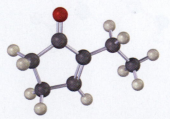

■ Problems assignable in Organic OWL.

ADDITIONAL PROBLEMS

17.33 ■ Identify all the acidic hydrogens (pK_a < 25) in the following molecules:

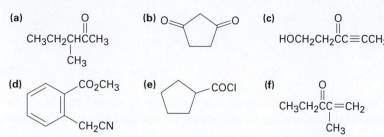

17.34 ■ Write resonance structures for the following anions:

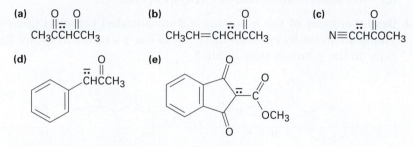

17.35 ■ Which, if any, of the following compounds can be prepared by a malonic ester synthesis? Show the alkyl halide you would use in each case.

(a) Ethyl pentanoate   (b) Ethyl 3-methylbutanoate

(c) Ethyl 2-methylbutanoate   (d) Ethyl 2,2-dimethylpropanoate

17.36 ■ Which, if any, of the following compounds can be prepared by an acetoacetic ester synthesis? Explain.

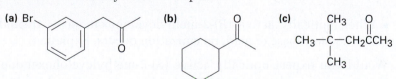

17.37 ■ How would you prepare the following ketones using an acetoacetic ester synthesis?

(a)

CH₃CH₂CHCCH₃
        |
        CH₂CH₃

(b)

CH₃CH₂CH₂CHCCH₃
          |
          CH₃

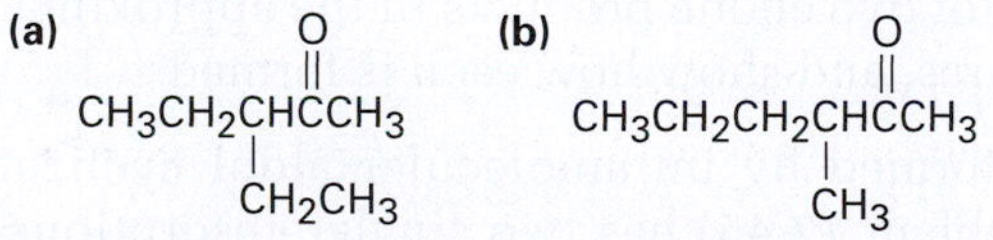

17.38 ■ How would you prepare the following compounds using either an acetoacetic ester synthesis or a malonic ester synthesis?

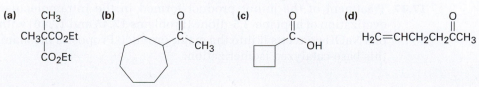

■ Problems assignable in Organic OWL.

17.39 ■ How might you synthesize each of the following compounds using an aldol reaction? In each case, show the structure of the starting aldehyde(s) or ketone(s) you would use.

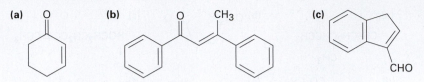

(a)   **(b)**   **(c)**

17.40 ■ What condensation products would you expect to obtain by treatment of the following substances with sodium ethoxide in ethanol?

(a) Ethyl butanoate   **(b)** Cycloheptanone

(c) Nonane-3,7-dione   **(d)** 3-Phenylpropanal

17.41 Base treatment of the following α,β-unsaturated carbonyl compound yields an anion by removal of H^+ from the γ carbon. Why are hydrogens on the γ carbon atom acidic?

17.42 Treatment of 1-phenylprop-2-enone with a strong base such as LDA does *not* yield an anion, even though it contains a hydrogen on the carbon atom next to the carbonyl group. Explain.

1-Phenylprop-2-enone

17.43 ■ When optically active (*R*)-2-methylcyclohexanone is treated with either aqueous base or acid, racemization occurs. Explain.

17.44 Would you expect optically active (*S*)-3-methylcyclohexanone to be racemized on acid or base treatment in the same way as 2-methylcyclohexanone (Problem 17.43)? Explain.

17.45 Intramolecular aldol cyclization of heptane-2,5-dione with aqueous NaOH yields a mixture of two enone products in the approximate ratio 9:1. Write their structures, and show how each is formed.

17.46 ■ The major product formed by intramolecular aldol cyclization of heptane-2,5-dione (Problem 17.45) has two singlet absorptions in the ^1H NMR spectrum at 1.65 δ and 1.90 δ and has no absorptions in the range 3 to 10 δ. What is its structure?

17.47 Treatment of the minor product formed in the intramolecular aldol cyclization of heptane-2,5-dione (Problems 17.45 and 17.46) with aqueous NaOH converts it into the major product. Propose a mechanism for this base-catalyzed isomerization.

■ Problems assignable in Organic OWL.

17.48 ▪ The aldol reaction is catalyzed by acid as well as by base. What is the reactive nucleophile in the acid-catalyzed aldol reaction? Propose a mechanism.

17.49 Butan-1-ol is prepared commercially by a route that begins with an aldol reaction. What steps are likely to be involved?

17.50 ▪ Leucine, 1 of the 20 amino acids found in proteins, is metabolized by a pathway that includes the following step. Propose a mechanism.

3-Hydroxy-3-methyl-
glutaryl CoA   **Acetyl CoA**   **Acetoacetate**

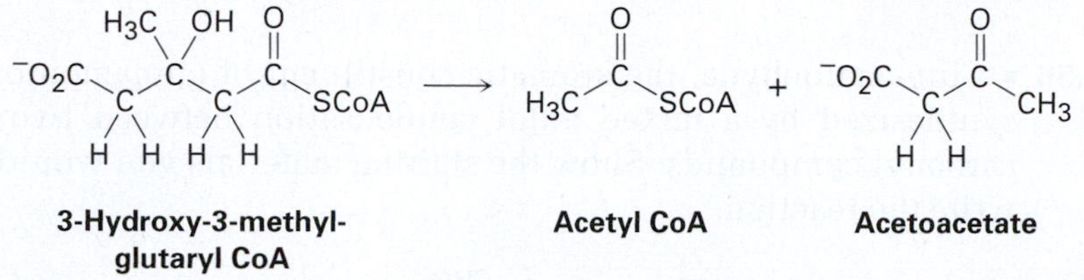

17.51 ▪ Isoleucine, another of the 20 amino acids found in proteins, is metabolized by a pathway that includes the following step. Propose a mechanism.

2-Methyl-3-keto-
butyryl CoA   **Acetyl CoA**   **Propionyl CoA**
                                                      (propanoyl CoA)

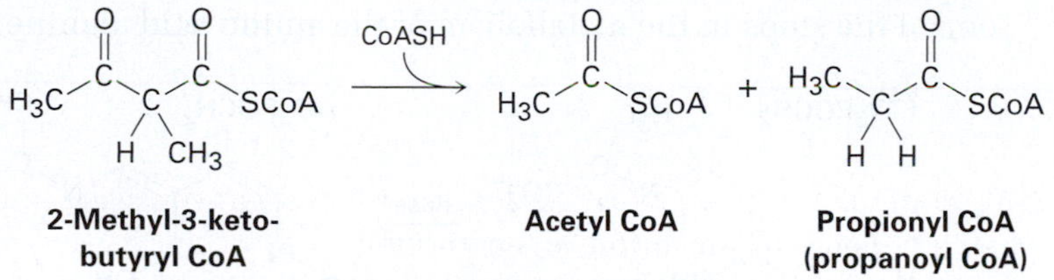

17.52 ▪ Fill in the reagents **a**–**c** that are missing from the following scheme:

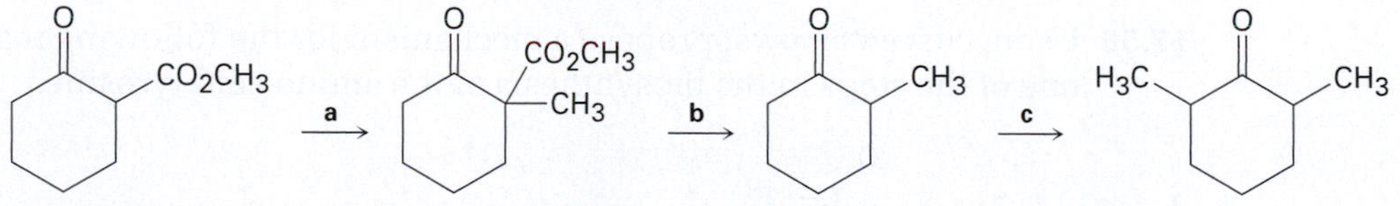

17.53 ▪ Nonconjugated β,γ-unsaturated ketones, such as cyclohex-3-enone, are in both an acid- and a base-catalyzed equilibrium with their conjugated α,β-unsaturated isomers. Propose a mechanism.

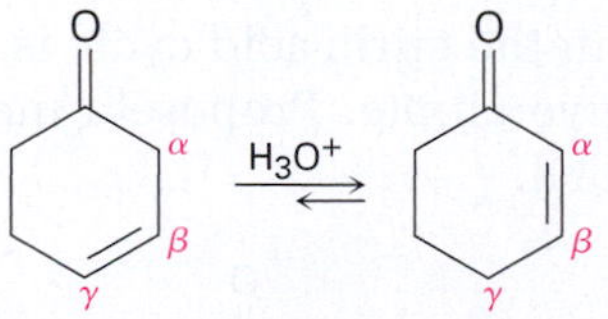

17.54 ▪ A consequence of the base-catalyzed isomerization of unsaturated ketones described in Problem 17.53 is that 2-substituted cyclopent-2-enones can be interconverted with 5-substituted cyclopent-2-enones. Propose a mechanism.

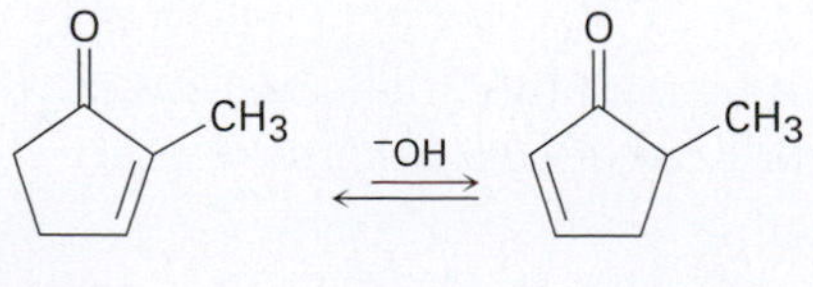

▪ Problems assignable in Organic OWL.

17.55 Although 2-substituted cyclopent-2-enones are in a base-catalyzed equilibrium with their 5-substituted cyclopent-2-enone isomers (Problem 17.54), the analogous isomerization is not observed for 2-substituted cyclohex-2-enones. Explain.

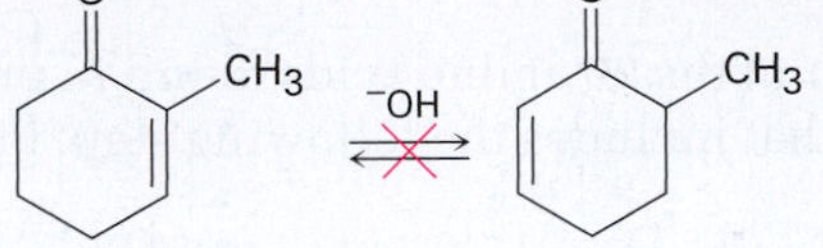

17.56 ▪ Cinnamaldehyde, the aromatic constituent of cinnamon oil, can be synthesized by a mixed aldol condensation between two different carbonyl compounds. Show the starting materials you would use, and write the reaction.

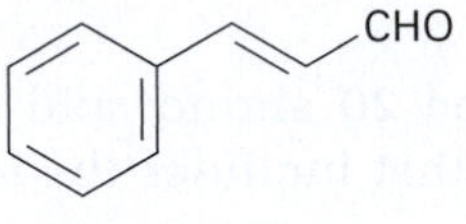

Cinnamaldehyde

17.57 Using curved arrows, propose a mechanism for the following reaction, one of the steps in the metabolism of the amino acid alanine:

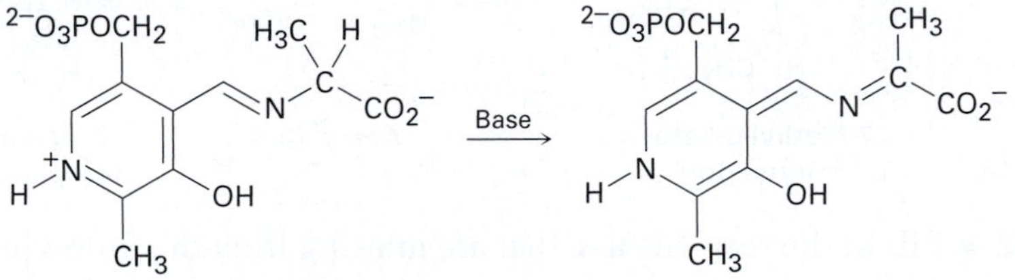

17.58 Using curved arrows, propose a mechanism for the following reaction, one of the steps in the biosynthesis of the amino acid tyrosine.

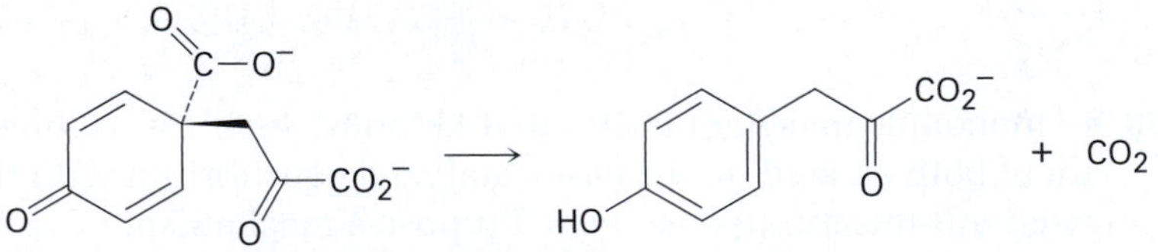

17.59 ▪ The first step in the citric acid cycle is reaction of oxaloacetate with acetyl CoA to give citrate. Propose a mechanism, using acid or base catalysis as needed.

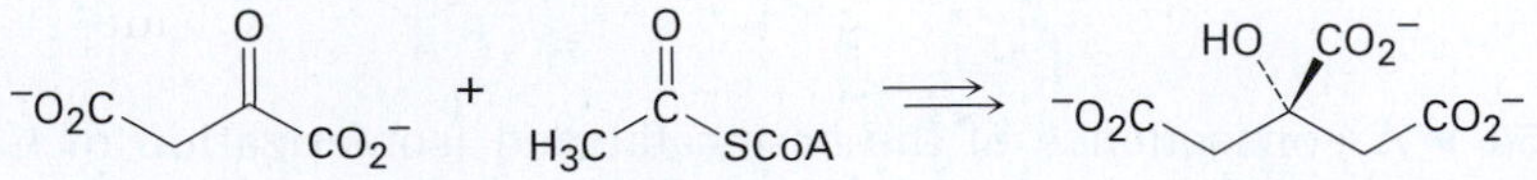

| Oxaloacetate | Acetyl CoA | Citrate |

17.60 One of the later steps in glucose biosynthesis is the isomerization of fructose 6-phosphate to glucose 6-phosphate. Propose a mechanism, using acid or base catalysis as needed.

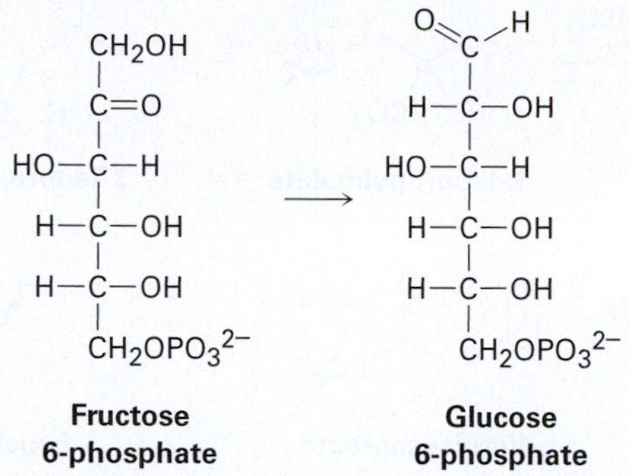

CH2OH
|
C=O
|
HO—C—H
|
H—C—OH
|
H—C—OH
|
CH2OPO3^{2-}

**Fructose
6-phosphate**

O   H
\\ //
C
|
H—C—OH
|
HO—C—H
|
H—C—OH
|
H—C—OH
|
CH2OPO3^{2-}

**Glucose
6-phosphate**

17.61 ■ How might the following compounds be prepared using Michael reactions? Show the nucleophilic donor and the electrophilic acceptor in each case.

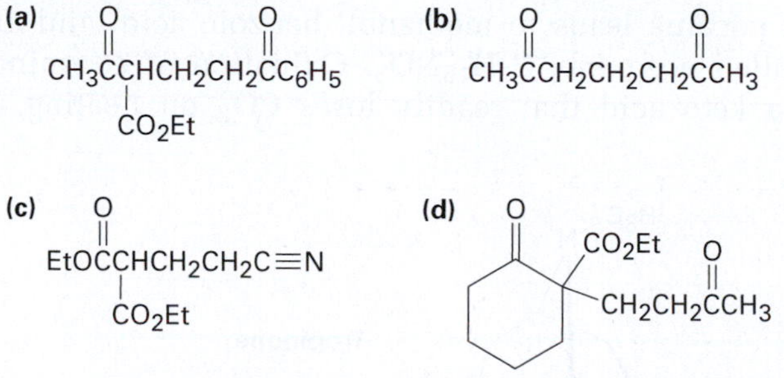

(a)

$$CH_3\overset{O}{\underset{}{C}}CHCH_2CH_2\overset{O}{\underset{}{C}}C_6H_5$$
|
CO2Et

(b)

$$CH_3\overset{O}{\underset{}{C}}CH_2CH_2CH_2\overset{O}{\underset{}{C}}CH_3$$

(c)

$$EtO\overset{O}{\underset{}{C}}CHCH_2CH_2C{\equiv}N$$
|
CO2Et

(d)

A cyclohexanone ring bearing a CO2Et group and a CH2CH2CCH3 (with C=O) group at the same carbon.

17.62 ■ Fill in the missing reagents **a–d** in the following scheme:

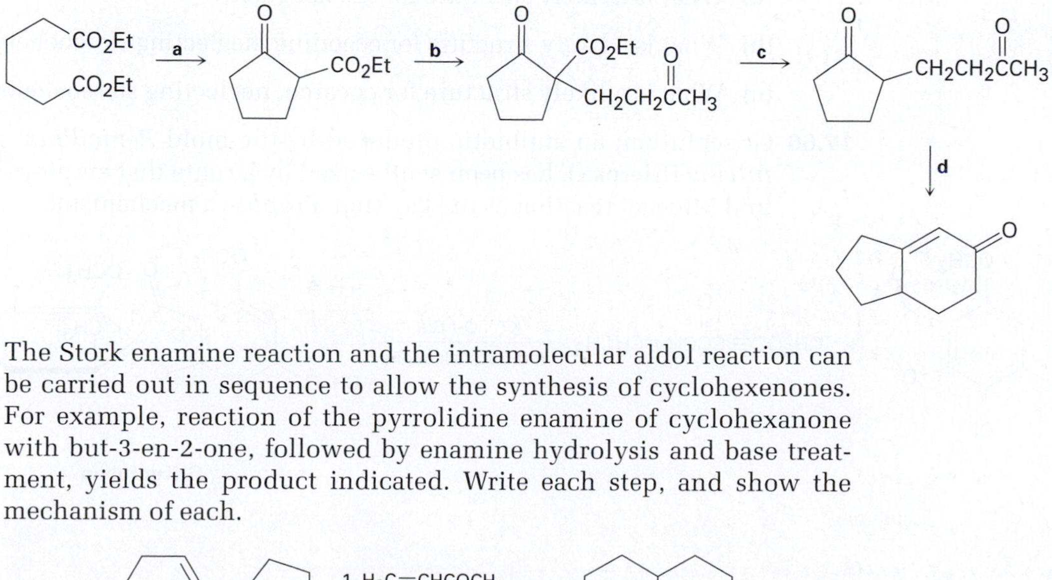

A cyclohexane ring with two CO2Et groups →ᵃ cyclopentanone with CO2Et substituent →ᵇ cyclopentanone with CO2Et and CH2CH2CCH3 substituents →ᶜ cyclopentanone with CH2CH2CCH3 substituent →ᵈ bicyclic enone.

17.63 The Stork enamine reaction and the intramolecular aldol reaction can be carried out in sequence to allow the synthesis of cyclohexenones. For example, reaction of the pyrrolidine enamine of cyclohexanone with but-3-en-2-one, followed by enamine hydrolysis and base treatment, yields the product indicated. Write each step, and show the mechanism of each.

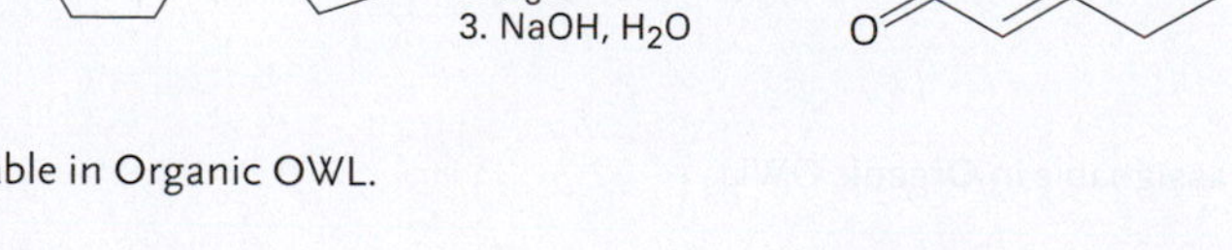

Pyrrolidine enamine of cyclohexanone

1. H2C=CHCOCH3
2. H3O+
3. NaOH, H2O

→ octahydronaphthalenone product

■ Problems assignable in Organic OWL.

17.64 The amino acid leucine is biosynthesized from α-ketoisovalerate by the following sequence of steps. Show the mechanism of each.

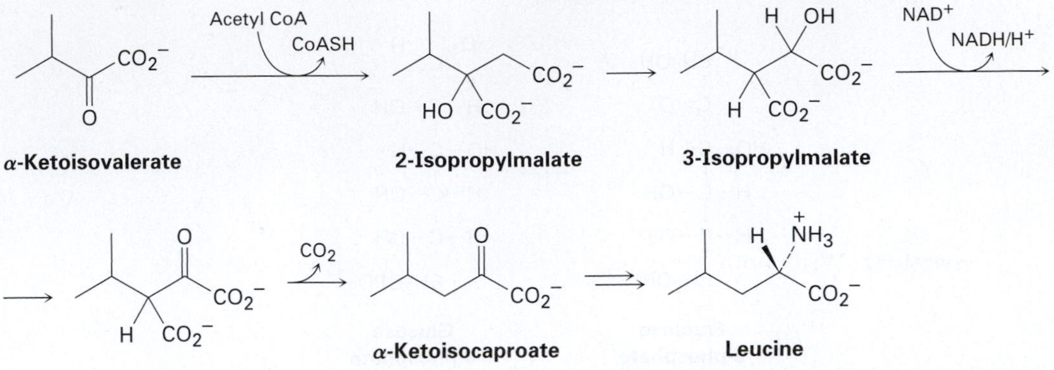

α-Ketoisovalerate   2-Isopropylmalate   3-Isopropylmalate

α-Ketoisocaproate   Leucine

17.65 As far back as the 16th century, South American Incas chewed the leaves of the coca bush, *Erythroxylon coca*, to combat fatigue. Chemical studies of *Erythroxylon coca* by Friedrich Wöhler in 1862 resulted in the discovery of *cocaine*, $C_{17}H_{21}NO_4$, as the active component. Basic hydrolysis of cocaine leads to methanol, benzoic acid, and another compound called *ecgonine*, $C_9H_{15}NO_3$. Oxidation of ecgonine with CrO_3 yields a keto acid that readily loses CO_2 on heating, giving tropinone.

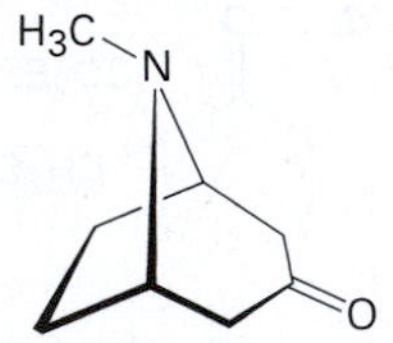

Tropinone

(a) What is a likely structure for the keto acid?

(b) What is a likely structure for ecgonine, neglecting stereochemistry?

(c) What is a likely structure for cocaine, neglecting stereochemistry?

17.66 Griseofulvin, an antibiotic produced by the mold *Penicillium griseo-fulvum* (Dierckx), has been synthesized by a route that employs a two-fold Michael reaction as the key step. Propose a mechanism.

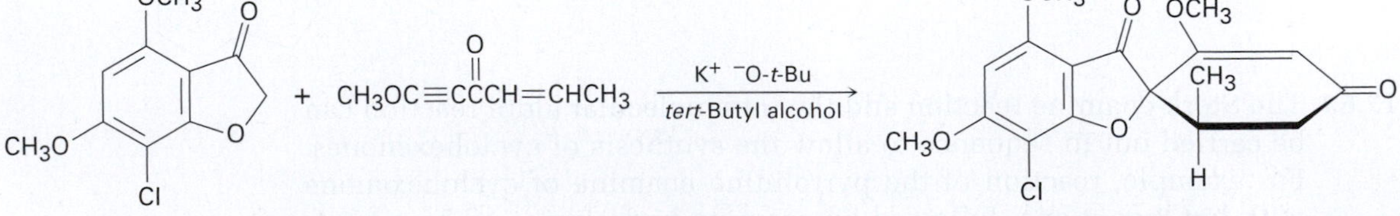

Griseofulvin

17.67 The following reaction involves an intramolecular Michael reaction followed by an intramolecular aldol reaction. Write both steps, and show their mechanisms.

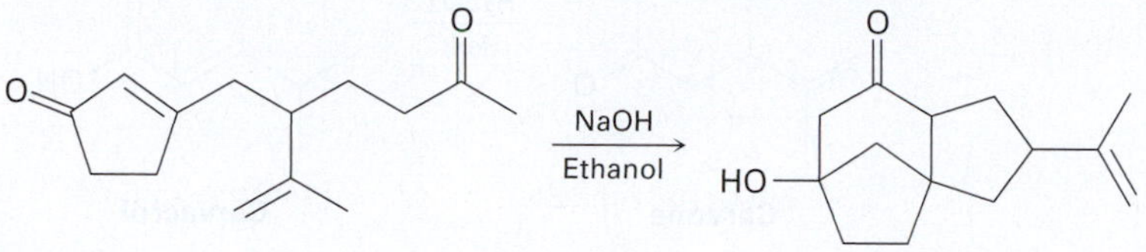

17.68 The following reaction involves two successive intramolecular Michael reactions. Write both steps, and show their mechanisms.

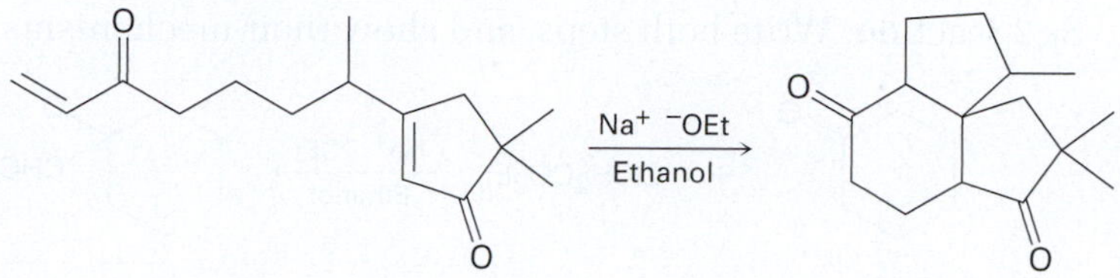

17.69 The following reaction involves an intramolecular aldol reaction followed by a *retro* aldol-like reaction. Write both steps, and show their mechanisms.

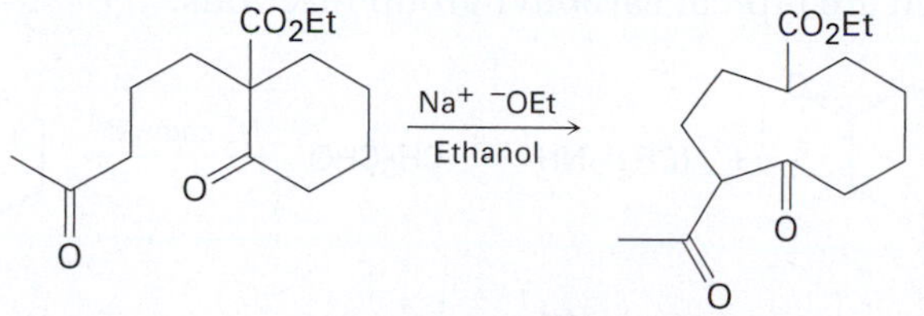

17.70 Amino acids can be prepared by reaction of alkyl halides with diethyl acetamidomalonate, followed by heating the initial alkylation product with aqueous HCl. Show how you would prepare alanine, $CH_3CH(NH_2)CO_2H$, 1 of the 20 amino acids found in proteins, and propose a mechanism for acid-catalyzed conversion of the initial alkylation product to the amino acid.

$$CH_3\overset{O}{\overset{\|}{C}}NH\underset{CO_2Et}{\underset{|}{C}}H\overset{O}{\overset{\|}{C}}OEt$$ **Diethyl acetamidomalonate**

17.71 Amino acids can also be prepared by a two-step sequence that involves Hell–Volhard–Zelinskii reaction of a carboxylic acid followed by treatment with ammonia. Show how you would prepare leucine, $(CH_3)_2CHCH_2CH(NH_2)CO_2H$, and identify the mechanism of the second step.

■ Problems assignable in Organic OWL.

17.72 Heating the terpene carvone with aqueous sulfuric acid converts it into carvacrol. Propose a mechanism for the isomerization.

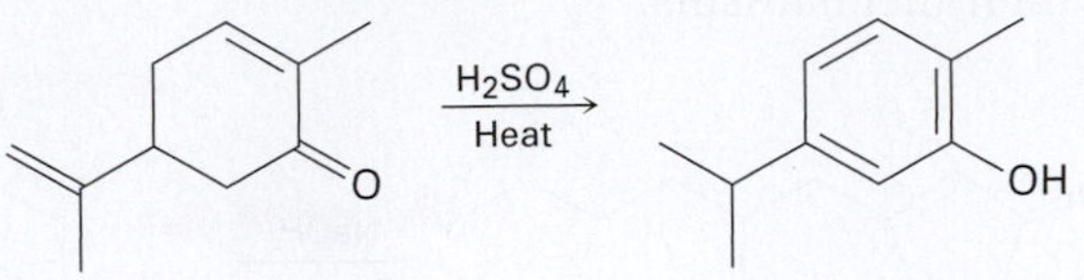

Carvone   **Carvacrol**

17.73 The *Darzens reaction* involves a two-step, base-catalyzed condensation of ethyl chloroacetate with a ketone to yield an epoxy ester. The first step is a carbonyl condensation reaction, and the second step is an S_N2 reaction. Write both steps, and show their mechanisms.

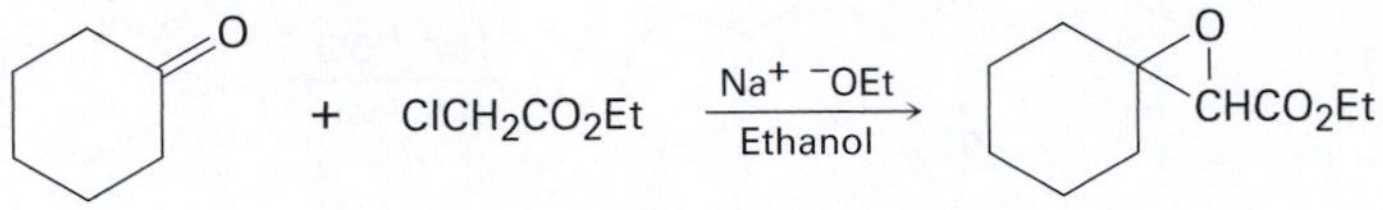

17.74 The *Mannich reaction* of a ketone, an amine, and an aldehyde is one of the few three-component reactions in organic chemistry. Cyclohexanone, for example, reacts with dimethylamine and acetaldehyde to yield an amino ketone. The reaction takes place in two steps, both of which are typical carbonyl-group reactions.

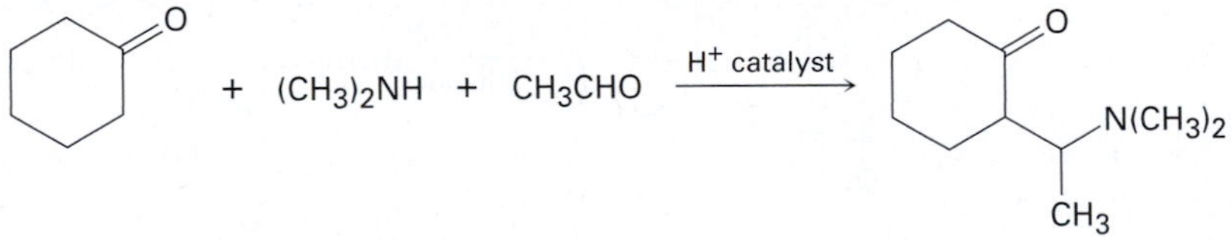

(a) The first step is reaction between the aldehyde and the amine to yield an intermediate iminium ion ($R_2C=NR_2^+$) plus water. Propose a mechanism, and show the structure of the intermediate iminium ion.

(b) The second step is reaction between the iminium ion intermediate and the ketone to yield the final product. Propose a mechanism.

17.75 Cocaine has been prepared by a sequence beginning with a Mannich reaction (Problem 17.74) between dimethyl acetonedicarboxylate, an amine, and a dialdehyde. Show the structures of the amine and dialdehyde.

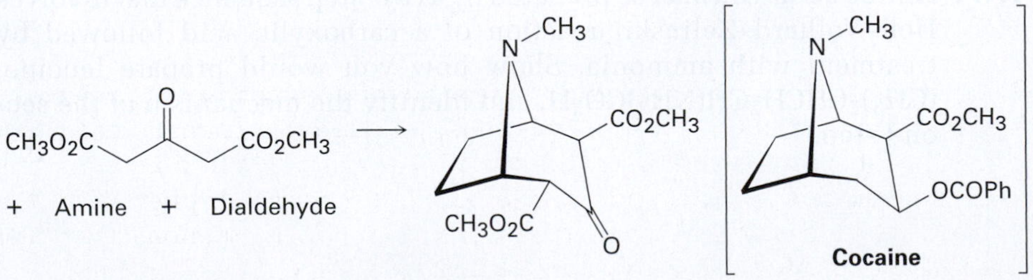

+   Amine   +   Dialdehyde   **Cocaine**

18 Amines and Heterocycles

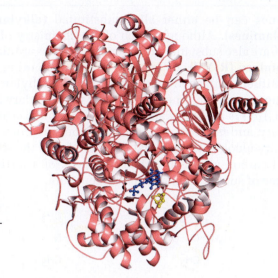

Glutamine synthase catalyzes the reductive amination of α-ketoglutarate to give glutamate, a step in amino acid metabolism.

Amines are organic derivatives of ammonia in the same way that alcohols and ethers are organic derivatives of water. Like ammonia, amines contain a nitrogen atom with a lone pair of electrons, making amines both basic and nucleophilic. We'll soon see, in fact, that most of the chemistry of amines depends on the presence of this lone pair of electrons.

Amines occur widely in all living organisms. Trimethylamine, for instance, occurs in animal tissues and is partially responsible for the distinctive odor of fish, nicotine is found in tobacco, and cocaine is a stimulant found in the leaves of the South American coca bush. In addition, amino acids are the building blocks from which all proteins are made, and cyclic amine bases are constituents of nucleic acids.

Trimethylamine   **Nicotine**   **Cocaine**

WHY THIS CHAPTER?

By the end of this chapter, we will have seen all the common functional groups that occur in biomolecules. Of those groups, amines and carbonyl compounds are the most abundant and have the richest chemistry. In addition to the proteins and nucleic acids already mentioned, the majority of pharmaceutical agents contain amine functional groups, and many of the common coenzymes necessary for biological catalysis are amines.

CONTENTS

OWL Online homework for this chapter can be assigned in Organic OWL.

18.1 Naming Amines

Amines can be either alkyl-substituted (**alkylamines**) or aryl-substituted (**arylamines**). Although much of the chemistry of the two classes is similar, there are also substantial differences. Amines are classified as **primary (RNH_2)**, **secondary (R_2NH)**, or **tertiary (R_3N)**, depending on the number of organic substituents attached to nitrogen. Thus, methylamine (CH_3NH_2) is a primary amine, dimethylamine [$(CH_3)_2NH$] is a secondary amine, and trimethylamine [$(CH_3)_3N$] is a tertiary amine. Note that this usage of the terms *primary, secondary,* and *tertiary* is different from our previous usage. When we speak of a tertiary alcohol or alkyl halide, we refer to the degree of substitution at the alkyl carbon atom, but when we speak of a tertiary amine, we refer to the degree of substitution at the nitrogen atom.

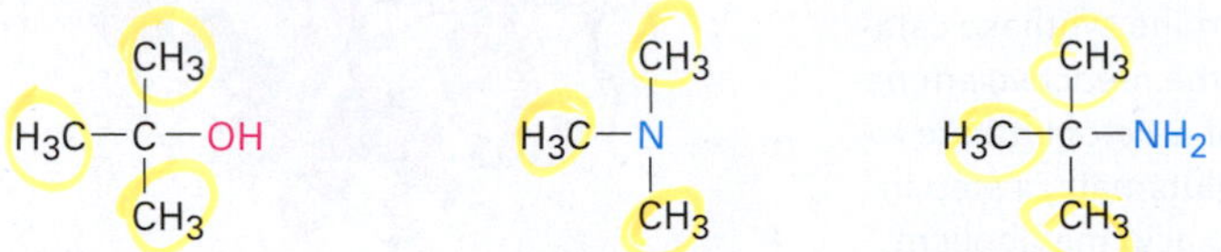

tert-Butyl alcohol
(a tertiary alcohol)

Trimethylamine
(a tertiary amine)

tert-Butylamine
(a primary amine)

Compounds containing a nitrogen atom with four attached groups also exist, but the nitrogen atom must carry a formal positive charge. Such compounds are called **quaternary ammonium salts**.

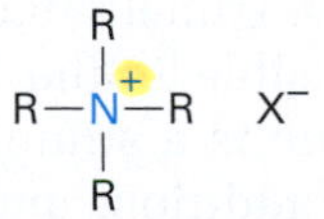

A quaternary ammonium salt

Primary amines are named in the IUPAC system in several ways. For simple amines, the suffix *-amine* is added to the name of the alkyl substituent. You might also recall from Chapter 9 that phenylamine, $C_6H_5NH_2$, has the common name *aniline*.

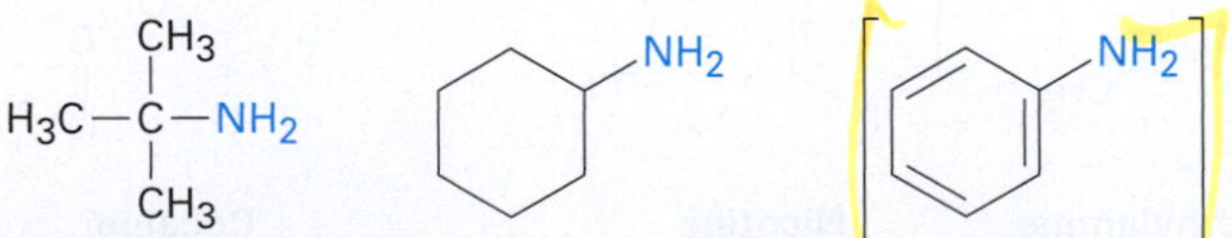

tert-Butylamine   **Cyclohexylamine**   **Aniline**

Alternatively, the suffix *-amine* can be used in place of the final *-e* in the name of the parent compound:

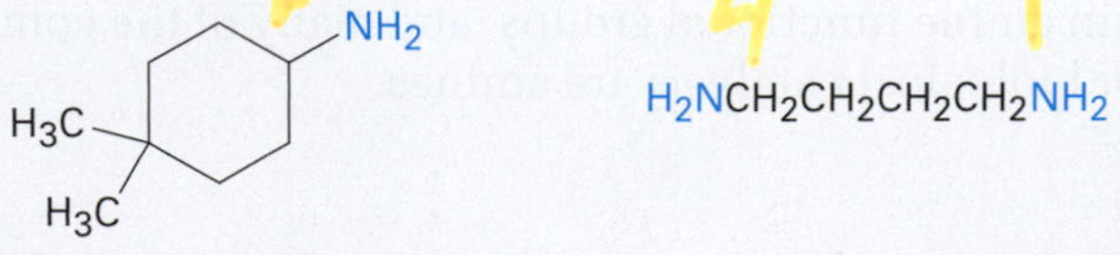

4,4-Dimethylcyclohexanamine   **Butane-1,4-diamine**

Amines with <mark>more than one functional group</mark> are named by considering the $-NH_2$ as an *amino* substituent on the parent molecule:

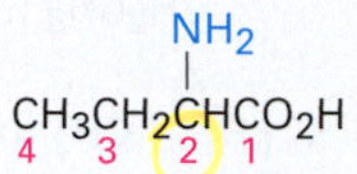

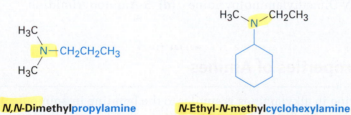

2-Aminobutanoic acid 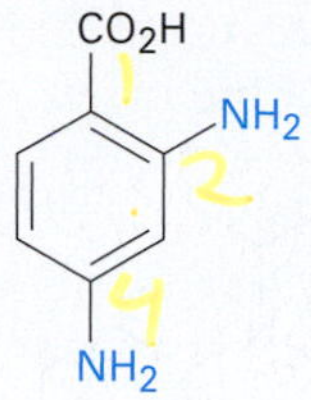   **2,4-Diamino**benzoic acid   **4-Amino**butan-2-one

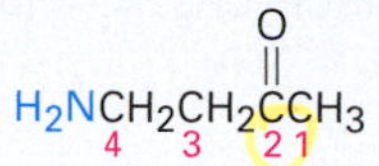

Symmetrical secondary and tertiary amines are named by adding the prefix *di-* or *tri-* to the alkyl group:

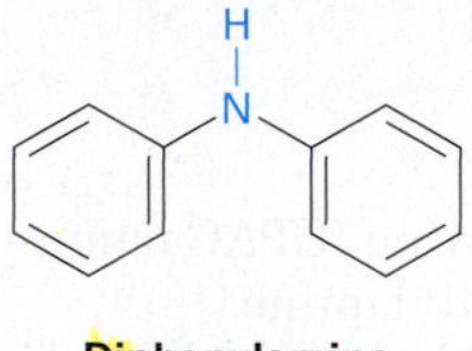

Diphenylamine 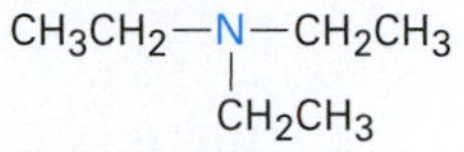   **Triethylamine**

Unsymmetrically substituted secondary and tertiary amines are named as *N*-substituted primary amines. The largest alkyl group is chosen as the parent name, and the other alkyl groups are considered *N*-substituents on the parent (*N* because they're attached to nitrogen).

N,N-Dimethylpropylamine   **N-Ethyl-N-methyl**cyclohexylamine

<mark>Heterocyclic amines</mark>—compounds in which the nitrogen atom occurs as part of a ring—are also common, and each different heterocyclic ring system has its own parent name. The <mark>heterocyclic nitrogen atom is always numbered as position 1.</mark>

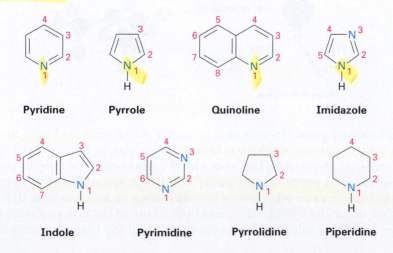

Pyridine   Pyrrole   Quinoline   Imidazole

Indole   Pyrimidine   Pyrrolidine   Piperidine

Problem 18.1

Name the following compounds:

(a) $CH_3NHCH_2CH_3$ (b)   (c)

(d)   (e)   (f) $H_2NCH_2CH_2\underset{\underset{\displaystyle CH_3}{|}}{C}HNH_2$

Problem 18.2

Draw structures corresponding to the following IUPAC names:

(a) Triisopropylamine   (b) Diallylamine
(c) *N*-Methylaniline   (d) *N*-Ethyl-*N*-methylcyclopentylamine
(e) *N*-Isopropylcyclohexylamine   (f) *N*-Ethylpyrrole

Problem 18.3

Draw structures for the following heterocyclic amines:

(a) 5-Methoxyindole   (b) 1,3-Dimethylpyrrole
(c) 4-(*N*,*N*-Dimethylamino)pyridine   (d) 5-Aminopyrimidine

18.2 Properties of Amines

The bonding in alkylamines is similar to the bonding in ammonia. The nitrogen atom is sp^3-hybridized, with the three substituents occupying three corners of a tetrahedron and the lone pair of electrons occupying the fourth corner. As you might expect, the C–N–C bond angles are close to the 109° tetrahedral value—108° in trimethylamine, for example.

Trimethylamine

One consequence of tetrahedral geometry is that an amine with three different substituents on nitrogen is chiral, as we saw in Section 5.10. Unlike chiral carbon compounds, however, chiral amines can't usually be resolved because the two enantiomeric forms rapidly interconvert by a *pyramidal inversion*, much as an alkyl halide inverts during an S_N2 reaction. Pyramidal inversion occurs by a momentary rehybridization of the nitrogen atom to planar, sp^2 geometry, followed by rehybridization of the planar intermediate to tetrahedral, sp^3 geometry (Figure 18.1). The barrier to inversion is about 25 kJ/mol (6 kcal/mol), an amount only twice as large as the barrier to rotation about a C–C single bond.

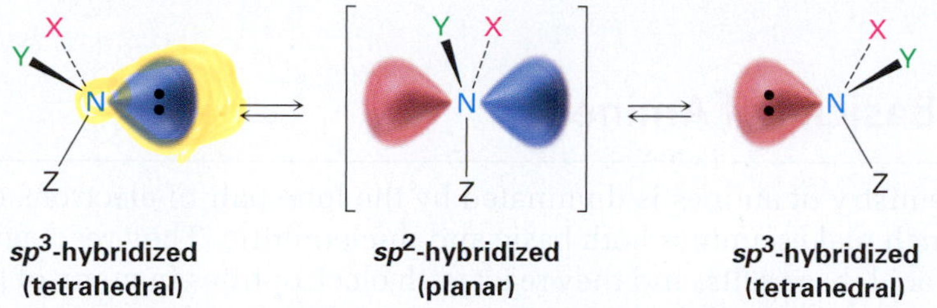

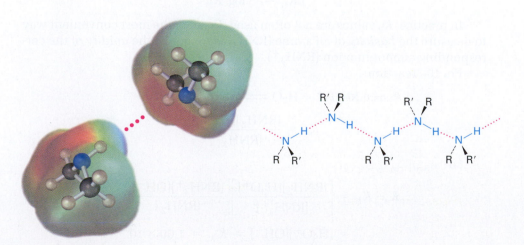

sp³-hybridized (tetrahedral)   **sp²-hybridized (planar)**   **sp³-hybridized (tetrahedral)**

ACTIVE FIGURE 18.1 Pyramidal inversion rapidly interconverts the two mirror-image (enantiomeric) forms of an amine. **Go to this book's student companion site at** www.cengage.com/chemistry/mcmurry **to explore an interactive version of this figure.**

Alkylamines have a variety of applications in the chemical industry as starting materials for the preparation of insecticides and pharmaceuticals. Labetalol, for instance, a so-called β-blocker used for the treatment of high blood pressure, is prepared by S_N2 reaction of an epoxide with a primary amine. The substance marketed for drug use is a mixture of all four possible stereoisomers, but the biological activity derives primarily from the (R,R) isomer.

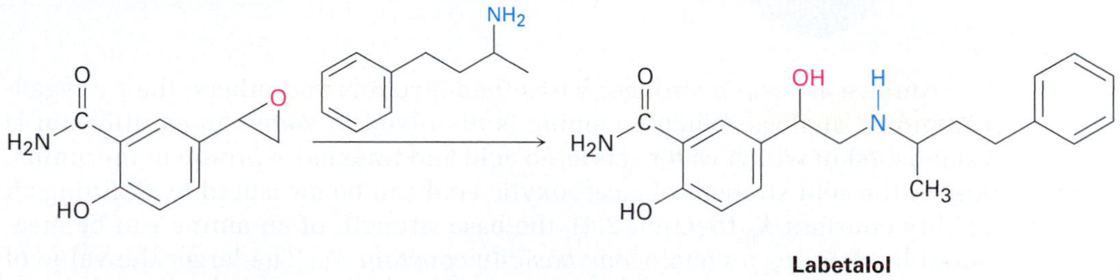

Labetalol

Like alcohols, amines with fewer than five carbon atoms are generally water-soluble. Also like alcohols, primary and secondary amines form hydrogen bonds and are highly associated. As a result, amines have higher boiling points than alkanes of similar molecular weight. Diethylamine (MW = 73 amu) boils at 56.3 °C, for instance, while pentane (MW = 72 amu) boils at 36.1 °C.

One other characteristic of amines is their *odor*. Low-molecular-weight amines such as trimethylamine have a distinctive fishlike aroma, while diamines such as pentane-1,5-diamine, commonly called cadaverine, have the appalling odors you might expect from their common names.

18.3 Basicity of Amines

The chemistry of amines is dominated by the lone pair of electrons on nitrogen, which makes amines both basic and nucleophilic. They react with acids to form acid–base salts, and they react with electrophiles in many of the polar reactions seen in past chapters. Note in the following electrostatic potential map of trimethylamine how the negative (red) region corresponds to the lone pair of electrons on nitrogen.

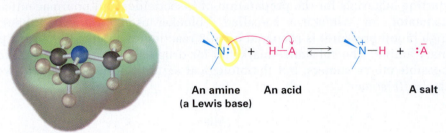

An amine An acid A salt
(a Lewis base)

Amines are much stronger bases than alcohols and ethers, their oxygen-containing analogs. When an amine is dissolved in water, an equilibrium is established in which water acts as an acid and transfers a proton to the amine. Just as the acid strength of a carboxylic acid can be measured by defining an acidity constant K_a (Section 2.8), the base strength of an amine can be measured by defining an analogous *basicity constant* K_b. The larger the value of K_b, and the smaller the value of pK_b, the more favorable the proton-transfer equilibrium and the stronger the base.

For the reaction:

$$RNH_2 + H_2O \rightleftharpoons RNH_3^+ + OH^-$$

$$K_b = \frac{[RNH_3^+][OH^-]}{[RNH_2]}$$

$$pK_b = -\log K_b$$

In practice, K_b values are not often used. Instead, the most convenient way to measure the *basicity* of an amine (RNH_2) is to look at the *acidity* of the corresponding ammonium ion (RNH_3^+).

For the reaction:

$$RNH_3^+ + H_2O \rightleftharpoons RNH_2 + H_3O^+$$

$$K_a = \frac{[RNH_2][H_3O^+]}{[RNH_3^+]}$$

so

$$K_a \cdot K_b = \left[\frac{[RNH_2][H_3O^+]}{[RNH_3^+]}\right]\left[\frac{[RNH_3^+][OH^-]}{[RNH_2]}\right]$$

$$= [H_3O^+][OH^-] = K_w = 1.00 \times 10^{-14}$$

Thus,

$$K_a = \frac{K_w}{K_b} \quad \text{and} \quad K_b = \frac{K_w}{K_a}$$

so

$$pK_a + pK_b = 14$$

These equations say that the K_b of an amine multiplied by the K_a of the corresponding ammonium ion is equal to K_w, the ion-product constant for water (1.00×10^{-14}). Thus, if we know K_a for an ammonium ion, we also know K_b for the corresponding amine base because $K_b = K_w/K_a$. The more acidic the ammonium ion, the less tightly the proton is held and the weaker the corresponding base. That is, a weaker base has an ammonium ion with a smaller pK_a, and a stronger base has an ammonium ion with a larger pK_a.

Weaker base: Smaller pK_a for ammonium ion

Stronger base: Larger pK_a for ammonium ion

Table 18.1 lists pK_a values of some ammonium ions and indicates that there is a substantial range of amine basicities. Most simple alkylamines are

TABLE 18.1
Basicity of Some Common Amines

Name	Structure	pK_a of ammonium ion
Ammonia	NH_3	9.26
Primary alkylamine		
Methylamine	CH_3NH_2	10.64
Ethylamine	$CH_3CH_2NH_2$	10.75
Secondary alkylamine		
Diethylamine	$(CH_3CH_2)_2NH$	10.98
Pyrrolidine	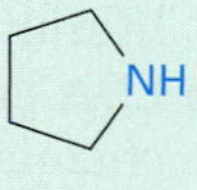	11.27
Tertiary alkylamine		
Triethylamine	$(CH_3CH_2)_3N$	10.76
Arylamine		
Aniline	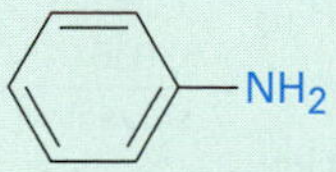	4.63
Heterocyclic amine		
Pyridine	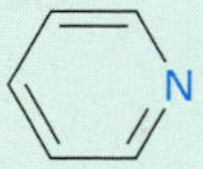	5.25
Pyrimidine	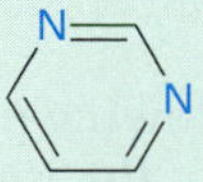	1.3
Pyrrole	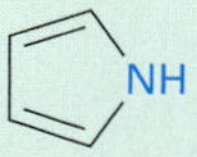	0.4
Imidazole	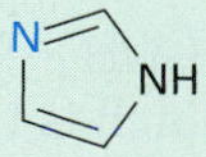	6.95

similar in their base strength, with pK_a's for their ammonium ions in the narrow range 10 to 11. Arylamines, however, are considerably less basic than alkylamines, as are the heterocyclic amines pyridine and pyrrole.

In contrast with amines, *amides* ($RCONH_2$) are nonbasic. Amides don't undergo substantial protonation by aqueous acids, and they are poor nucleophiles. The main reason for this difference in basicity between amines and amides is that an amide is stabilized by delocalization of the nitrogen lone-pair electrons through orbital overlap with the carbonyl group. In resonance terms, amides are more stable and less reactive than amines because they are hybrids of two resonance forms. This amide resonance stabilization is lost when the nitrogen atom is protonated, so protonation is disfavored. Electrostatic potential maps show clearly the decreased electron density on the amide nitrogen.

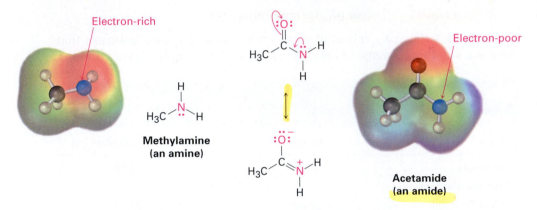

Methylamine
(an amine)

Acetamide
(an amide)

In addition to their behavior as bases, primary and secondary amines can also act as very weak acids because an N–H proton can be removed by a sufficiently strong base. We've seen, for example, how diisopropylamine (p$K_a \approx 40$) reacts with butyllithium to yield lithium diisopropylamide (LDA; Section 17.4). Dialkylamine anions like LDA are extremely powerful bases that are often used in laboratory organic chemistry for the generation of enolate ions from carbonyl compounds (Section 17.5). They are not, however, encountered in biological chemistry.

$$C_4H_9Li \ + \ H-N\substack{CH(CH_3)_2 \\ CH(CH_3)_2} \ \xrightarrow[\text{solvent}]{\text{THF}} \ Li^+ \ ^-\!\!:N\substack{CH(CH_3)_2 \\ CH(CH_3)_2} \ + \ C_4H_{10}$$

Butyllithium   **Diisopropylamine**   **Lithium diisopropylamide**
                                                              (LDA)

Problem 18.4

Which compound in each of the following pairs is more basic?
(a) $CH_3CH_2NH_2$ or $CH_3CH_2CONH_2$ **(b)** NaOH or CH_3NH_2
(c) CH_3NHCH_3 or pyridine

Problem 18.5

The benzylammonium ion ($C_6H_5CH_2NH_3{}^+$) has pK_a = 9.33, and the propylammonium ion has pK_a = 10.71. Which is the stronger base, benzylamine or propylamine? What are the pK_b's of benzylamine and propylamine?

18.4 Basicity of Arylamines

As noted previously, arylamines are generally less basic than alkylamines. Anilinium ion has $pK_a = 4.63$, for instance, whereas methylammonium ion has $pK_a = 10.64$. Arylamines are less basic than alkylamines because the nitrogen lone-pair electrons are delocalized by interaction with the aromatic ring π electron system and are less available for bonding to H^+. In resonance terms, arylamines are stabilized relative to alkylamines because of their five resonance forms:

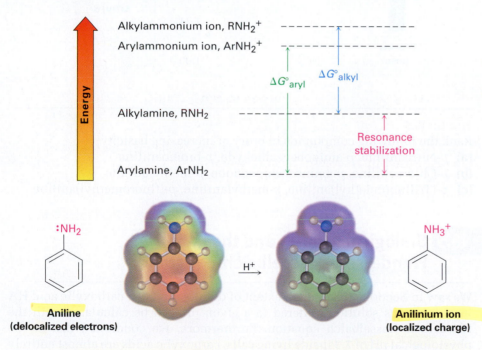

Much of the resonance stabilization is lost on protonation, however, so the energy difference between protonated and nonprotonated forms is higher for arylamines than it is for alkylamines, making arylamines less basic. Figure 18.2 illustrates the difference.

FIGURE 18.2 Arylamines have a larger positive $\Delta G°$ for protonation and are therefore less basic than alkylamines, primarily because of resonance stabilization of the ground state. Electrostatic potential maps show that lone-pair electron density is delocalized in the amine but the charge is localized in the corresponding ammonium ion.

Substituted arylamines can be either more basic or less basic than aniline, depending on the substituent. Electron-donating substituents, such as $-CH_3$ and $-OCH_3$, which increase the reactivity of an aromatic ring toward electrophilic substitution (Section 9.8), also increase the basicity of the corresponding arylamine. Electron-withdrawing substituents, such as $-Cl$, $-NO_2$, and $-CN$, which decrease ring reactivity toward electrophilic substitution, also decrease arylamine basicity. Table 18.2 considers only *p*-substituted anilines, but similar trends are observed for ortho and meta derivatives.

TABLE 18.2
Base Strength of Some _p_-Substituted Anilines

$$Y-\!\!\!\!\bigcirc\!\!\!\!-\ddot{N}H_2 \;+\; H_2O \;\rightleftharpoons\; Y-\!\!\!\!\bigcirc\!\!\!\!-\overset{+}{N}H_3 \;+\; {}^-OH$$

Substituent, Y	pK_a	
$-NH_2$	6.15	
$-OCH_3$	5.34	Activating groups
$-CH_3$	5.08	
$-H$	4.63	
$-Cl$	3.98	
$-Br$	3.86	Deactivating groups
$-CN$	1.74	
$-NO_2$	1.00	

Stronger base ↑ (top) → Weaker base (bottom)

Problem 18.6

Rank the following compounds in order of increasing basicity:
(a) _p_-Nitroaniline, _p_-aminobenzaldehyde, _p_-bromoaniline
(b) _p_-Chloroaniline, _p_-aminoacetophenone, _p_-methylaniline
(c) _p_-(Trifluoromethyl)aniline, _p_-methylaniline, _p_-(fluoromethyl)aniline

18.5 Biological Amines and the Henderson–Hasselbalch Equation

We saw in Section 15.3 that the extent of dissociation of a carboxylic acid HA in an aqueous solution buffered to a given pH can be calculated with the Henderson–Hasselbalch equation. Furthermore, we concluded that at the physiological pH of 7.3 inside living cells, carboxylic acids are almost entirely dissociated into their carboxylate anions, RCO_2^-.

Henderson–Hasselbalch equation:
$$pH = pK_a + \log\frac{[A^-]}{[HA]}$$

$$\log\frac{[A^-]}{[HA]} = pH - pK_a$$

What about amine bases? In what form do they exist at the physiological pH inside cells—as the amine ($A^- = RNH_2$) or as the ammonium ion ($HA = RNH_3^+$)? Let's take a 0.0010 M solution of methylamine at pH = 7.3, for example. According to Table 18.1, the pK_a of methylammonium ion is 10.64, so from the Henderson–Hasselbalch equation, we have:

$$\log\frac{[RNH_2]}{[RNH_3{}^+]} = pH - pK_a = 7.3 - 10.64 = -3.34$$

$$\frac{[RNH_2]}{[RNH_3{}^+]} = \text{antilog}(-3.34) = 4.6 \times 10^{-4}$$

so $\qquad [RNH_2] = (4.6 \times 10^{-4})[RNH_3{}^+]$

In addition, we know that

$$[RNH_2] + [RNH_3{}^+] = 0.0010\ M$$

Solving the two simultaneous equations gives $[RNH_3{}^+] = 0.0010\ M$ and $[RNH_2] = 5 \times 10^{-7}\ M$. In other words, at a physiological pH of 7.3, essentially 100% of the methylamine in a 0.0010 M solution exists in its protonated form as methylammonium ion. The same is true of other amine bases, so we always write cellular amines in their protonated form and amino acids in their ammonium carboxylate form to reflect their structures at physiological pH.

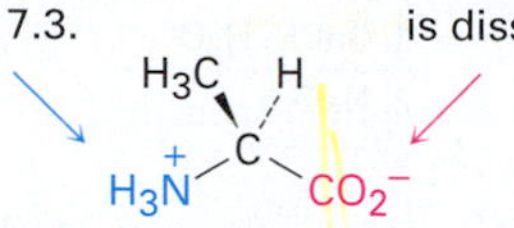

The amino group is protonated at pH = 7.3.

The carboxylic acid group is dissociated at pH = 7.3.

Alanine (an amino acid)

Problem 18.7

Calculate the percentages of neutral and protonated forms present in a solution of 0.0010 M pyrimidine at pH = 7.3. The pK_a of pyrimidinium ion is 1.3.

18.6 Synthesis of Amines

Reduction of Nitriles, Amides, and Nitro Compounds

We've already seen in Sections 15.7 and 16.7 how amines can be prepared by reduction of nitriles and amides with LiAlH$_4$. The two-step sequence of S$_N$2 displacement with CN$^-$ followed by reduction thus converts an alkyl halide into a primary alkylamine having one more carbon atom. Amide reduction converts carboxylic acids and their derivatives into amines with the same number of carbon atoms.

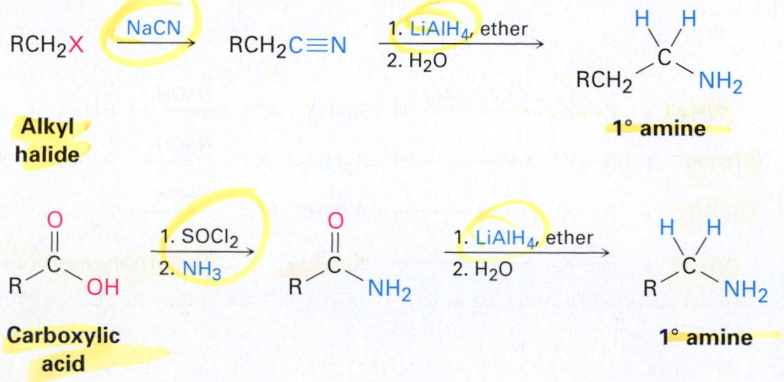

Arylamines are usually prepared by nitration of an aromatic starting material, followed by reduction of the nitro group (Section 9.6). The reduction step can be carried out in many different ways, depending on the circumstances. Catalytic hydrogenation over platinum works well but is often incompatible with the presence elsewhere in the molecule of other reducible groups, such as C=C bonds or carbonyl groups. Iron, zinc, tin, and tin(II) chloride (SnCl$_2$) are also effective when used in acidic aqueous solution. Tin(II) chloride is particularly mild and is often used when other reducible functional groups are present.

p-tert-Butylnitrobenzene

$\xrightarrow[\text{ethanol}]{\text{H}_2 \atop \text{Pt catalyst,}}$

p-tert-Butylaniline (100%)

m-Nitrobenzaldehyde

$\xrightarrow[\text{2. NaOH}]{\text{1. SnCl}_2, \text{H}_3\text{O}^+}$

m-Aminobenzaldehyde (90%)

Problem 18.8

Propose structures for either a nitrile or an amide that might be a precursor of each of the following amines:

(a) CH$_3$CH$_2$CH$_2$NH$_2$ **(b)** (CH$_3$CH$_2$CH$_2$)$_2$NH

(c) Benzylamine, C$_6$H$_5$CH$_2$NH$_2$ **(d)** N-Ethylaniline

S$_N$2 Reactions of Alkyl Halides

Ammonia and other amines are good nucleophiles in S$_N$2 reactions. As a result, the simplest method of alkylamine synthesis is by S$_N$2 alkylation of ammonia or an alkylamine with an alkyl halide. If ammonia is used, a primary amine results; if a primary amine is used, a secondary amine results; and so on. Even tertiary amines react rapidly with alkyl halides to yield quaternary ammonium salts, R$_4$N$^+$ X$^-$.

Ammonia	$\overset{..}{\text{N}}\text{H}_3$ + R—X	$\xrightarrow{\text{S}_N\text{2 reaction}}$	RNH$_3^+$ X$^-$	$\xrightarrow{\text{NaOH}}$ RNH$_2$	Primary
Primary	$\overset{..}{\text{R}}\text{NH}_2$ + R—X	$\longrightarrow$	R$_2$NH$_2^+$ X$^-$	$\xrightarrow{\text{NaOH}}$ R$_2$NH	Secondary
Secondary	$\overset{..}{\text{R}_2\text{NH}}$ + R—X	$\longrightarrow$	R$_3$NH$^+$ X$^-$	$\xrightarrow{\text{NaOH}}$ R$_3$N	Tertiary
Tertiary	$\overset{..}{\text{R}_3\text{N}}$ + R—X	$\longrightarrow$	R$_4$N$^+$ X$^-$	Quaternary ammonium salt	

Unfortunately, these reactions don't stop cleanly after a single alkylation has occurred. Because ammonia and primary amines have similar reactivity, the initially formed monoalkylated substance often undergoes further reaction to yield a mixture of mono-, di-, and trialkylated products. A better method for preparing primary amines from alkyl halides is to use azide ion, N_3^-, as the nucleophile rather than ammonia. The product is an alkyl azide, which is not nucleophilic, so overalkylation can't occur. Subsequent reduction of the alkyl azide with $LiAlH_4$ then leads to the desired primary amine.

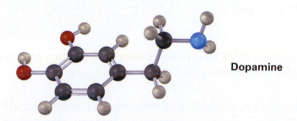

1-Bromo-2-phenylethane   **2-Phenylethyl azide**   **2-Phenylethylamine (89%)**

Problem 18.9

Show two methods for synthesizing dopamine, a neurotransmitter involved in regulation of the central nervous system.

Dopamine

Reductive Amination of Aldehydes and Ketones

Amines can be synthesized in a single step by treatment of an aldehyde or ketone with ammonia or an amine in the presence of a reducing agent, a process called **reductive amination**. For example, amphetamine, a central nervous system stimulant, is prepared commercially by reductive amination of phenylpropan-2-one with ammonia, using hydrogen gas over a nickel catalyst as the reducing agent. In the laboratory, $NaBH_4$ is often used as the reducing agent rather than H_2 and nickel.

Phenylpropan-2-one   **Amphetamine**

Reductive amination takes place by the pathway shown in Figure 18.3. An imine intermediate is first formed by a nucleophilic addition reaction (Section 14.7), and the C=N bond of the imine is then reduced.

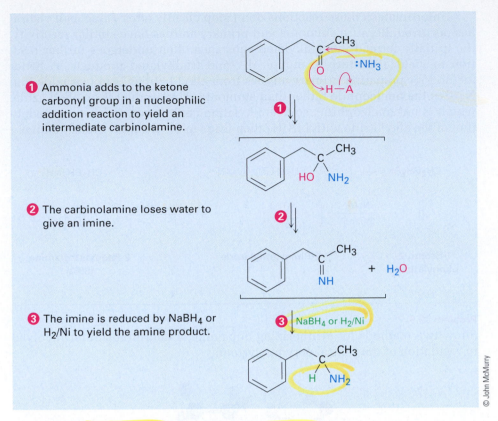

1. Ammonia adds to the ketone carbonyl group in a nucleophilic addition reaction to yield an intermediate carbinolamine.

2. The carbinolamine loses water to give an imine.

3. The imine is reduced by $NaBH_4$ or H_2/Ni to yield the amine product.

© John McMurry

FIGURE 18.3 MECHANISM: Mechanism of reductive amination of a ketone to yield an amine. Details of the imine-forming step were shown in Figure 14.8.

Ammonia, primary amines, and secondary amines can all be used in the reductive amination reaction, yielding primary, secondary, and tertiary amines, respectively.

Primary amine   **Secondary amine**   **Tertiary amine**

Reductive aminations also occur in various biological pathways. In the biosynthesis of the amino acid proline, for instance, glutamate 5-semialdehyde undergoes internal imine formation to give 1-pyrrolinium 5-carboxylate, which is then reduced by nucleophilic addition of hydride ion to the C=N bond. Reduced nicotinamide adenine dinucleotide, NADH, acts as the biological reducing agent.

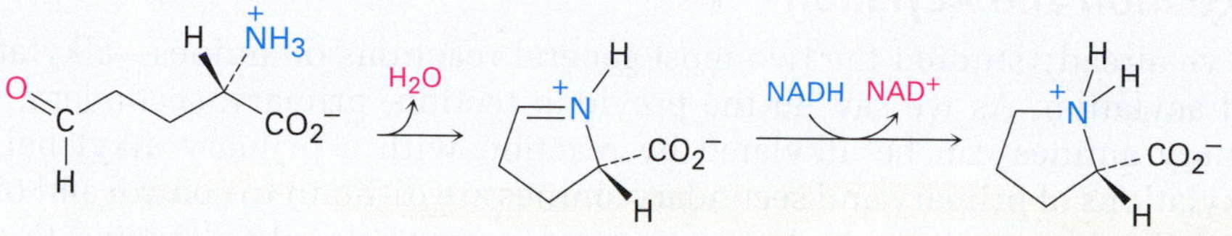

H
NH_3^+
O
C
H
CO_2^-
H_2O
H
N
CO_2^-
H
NADH
NAD^+
H
H
N
CO_2^-
H
Glutamate
5-semialdehyde
1-Pyrrolinium
5-carboxylate
Proline

WORKED EXAMPLE 18.1 Using a Reductive Amination Reaction

How might you prepare *N*-methyl-2-phenylethylamine using a reductive amination reaction?

$NHCH_3$

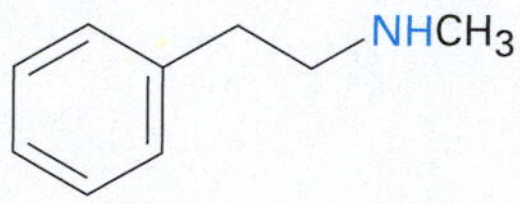

***N*-Methyl-2-phenylethylamine**

Strategy

Look at the target molecule, and identify the groups attached to nitrogen. One of the groups must be derived from the aldehyde or ketone component, and the other must be derived from the amine component. In the case of *N*-methyl-2-phenylethylamine, there are two combinations that can lead to the product: phenylacetaldehyde plus methylamine or formaldehyde plus 2-phenylethylamine. In general, it's usually better to choose the combination with the simpler amine component—methylamine in this case—and to use an excess of that amine as reactant.

Solution

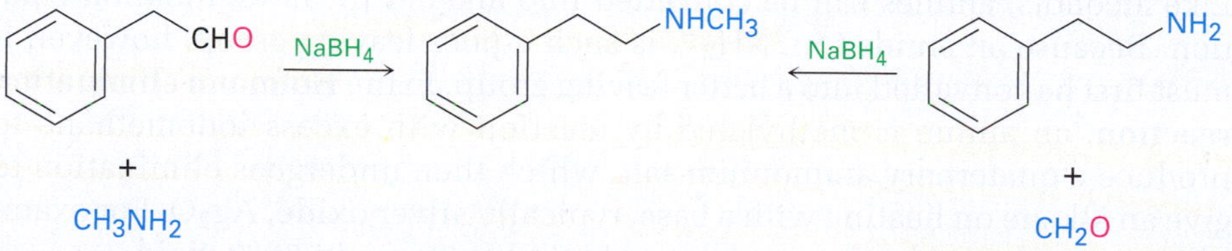

CHO
$NaBH_4$
$NHCH_3$
$NaBH_4$
NH_2
+
CH_3NH_2
+
CH_2O

Problem 18.10

How could you prepare the following amine using a reductive amination reaction?

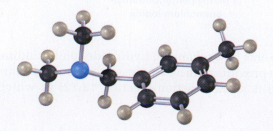

18.7 Reactions of Amines

Alkylation and Acylation

We've already studied the two most general reactions of amines—alkylation and acylation. As we saw in the previous section, primary, secondary, and tertiary amines can be alkylated by reaction with a primary alkyl halide. Alkylations of primary and secondary amines are difficult to control and often give mixtures of products, but tertiary amines are cleanly alkylated to give quaternary ammonium salts. Primary and secondary (but not tertiary) amines can also be acylated by nucleophilic acyl substitution reaction with an acid chloride or an acid anhydride to yield amides (Sections 16.4 and 16.5). Note that overacylation of the nitrogen does not occur because the amide product is much less nucleophilic and less reactive than the starting amine.

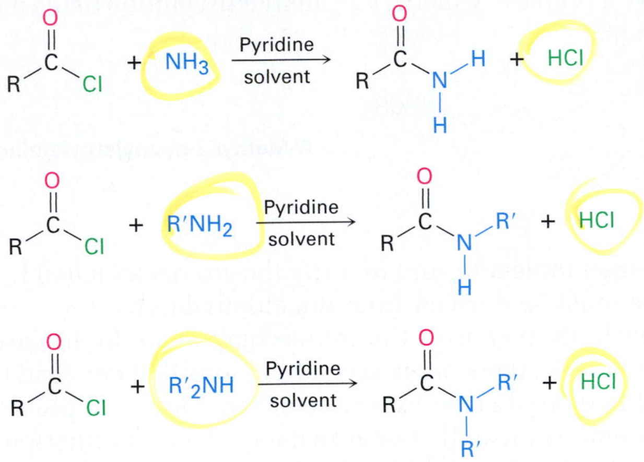

Hofmann Elimination

Like alcohols, amines can be converted into alkenes by an elimination reaction. Because an amide ion, NH_2^-, is such a poor leaving group, however, it must first be converted into a better leaving group. In the **Hofmann elimination reaction**, an amine is methylated by reaction with excess iodomethane to produce a quaternary ammonium salt, which then undergoes elimination to give an alkene on heating with a base, typically silver oxide, Ag_2O. For example, 1-methylpentylamine is converted into hex-1-ene in 60% yield.

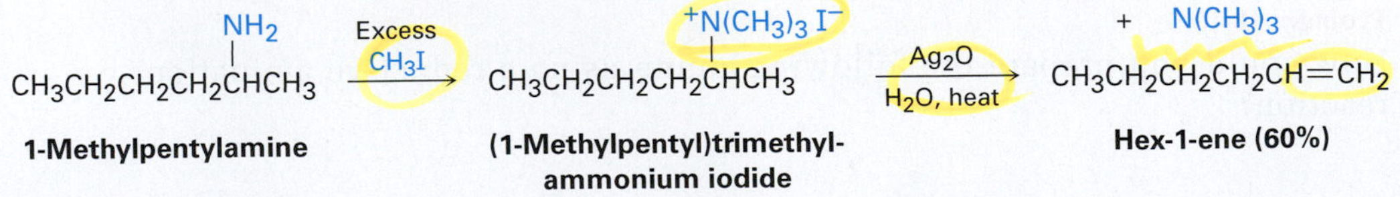

1-Methylpentylamine (1-Methylpentyl)trimethyl-ammonium iodide **Hex-1-ene (60%)**

Silver oxide acts by exchanging hydroxide ion for iodide ion in the quaternary salt, thus providing the base necessary to cause elimination. The actual elimination step is an E2 reaction (Section 12.12) in which hydroxide ion removes a proton at the same time that the positively charged nitrogen atom leaves.

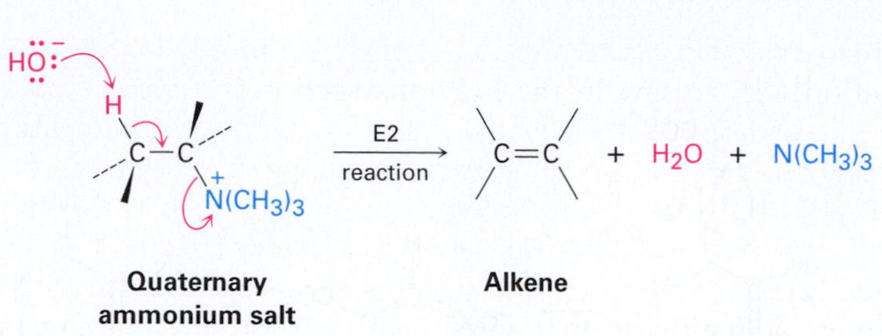

Quaternary ammonium salt → **Alkene**

Unlike what happens in other E2 reactions, the major product of the Hofmann elimination is the *less* highly substituted alkene rather than the more highly substituted one, as shown by the reaction of (1-methylbutyl)-trimethylammonium hydroxide to give pent-1-ene rather than the alternative pent-2-ene. The reason for this non-Zaitsev stereochemistry is probably steric. Because of the large size of the trialkylamine leaving group, the base must abstract a hydrogen from the most accessible, least hindered position.

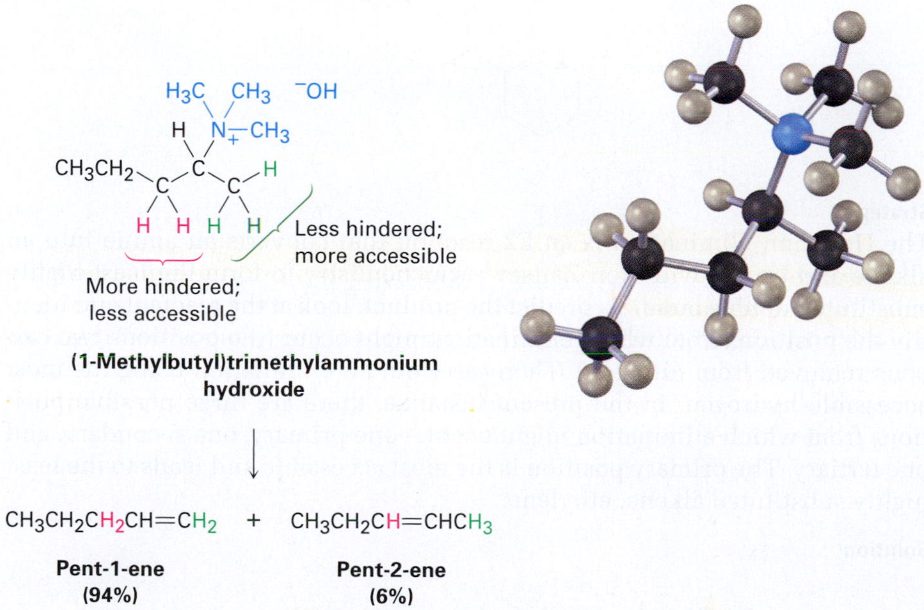

More hindered; less accessible

Less hindered; more accessible

(1-Methylbutyl)trimethylammonium hydroxide

$CH_3CH_2CH_2CH=CH_2$ + $CH_3CH_2CH=CHCH_3$

Pent-1-ene (94%)   **Pent-2-ene (6%)**

The Hofmann elimination reaction is not often used today in the laboratory, but analogous biological eliminations occur frequently, although usually with protonated ammonium ions rather than quaternary ammonium salts. In the biosynthesis of nucleic acids, for instance, a substance called adenylosuccinate undergoes an elimination of a positively charged nitrogen to give fumarate plus adenosine monophosphate.

Adenylosuccinate   **Fumarate**   **Adenosine monophosphate**

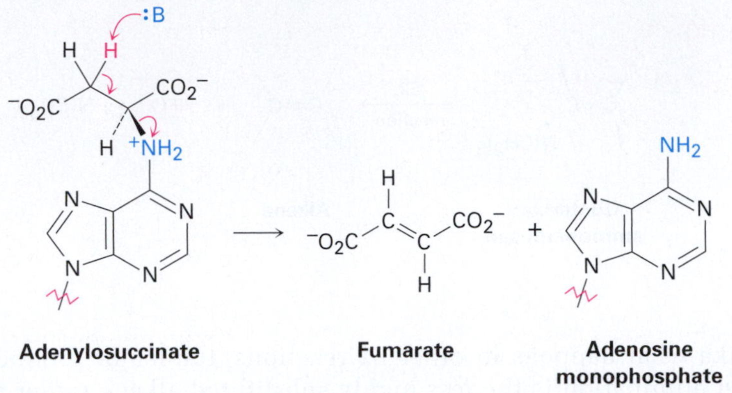

WORKED EXAMPLE 18.2 Predicting the Product of a Hofmann Elimination

What product would you expect from Hofmann elimination of the following amine?

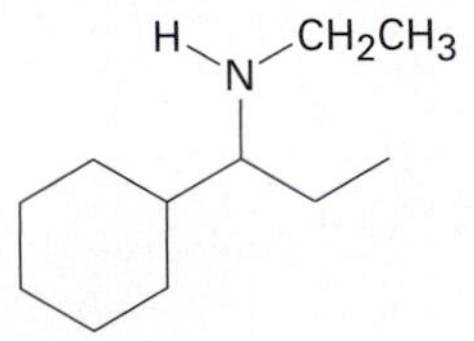

Strategy

The Hofmann elimination is an E2 reaction that converts an amine into an alkene and occurs with non-Zaitsev regiochemistry to form the least highly substituted double bond. To predict the product, look at the reactant and identify the positions from which elimination might occur (the positions two carbons removed from nitrogen). Then carry out an elimination using the most accessible hydrogen. In the present instance, there are three possible positions from which elimination might occur—one primary, one secondary, and one tertiary. The primary position is the most accessible and leads to the least highly substituted alkene, ethylene.

Solution

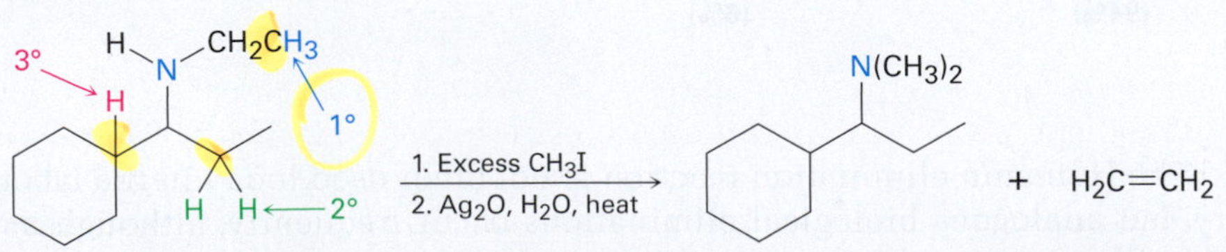

Problem 18.11

What products would you expect from Hofmann elimination of the following amines? If more than one product is formed, tell which is major.

(a) 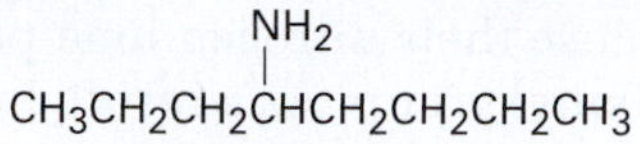
$$CH_3CH_2CH_2CHCH_2CH_2CH_2CH_3$$
with NH$_2$

(b) 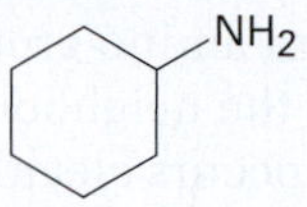 cyclohexane with NH$_2$

(c) 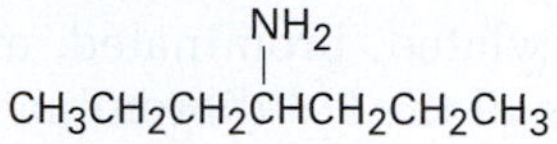
$$CH_3CH_2CH_2CHCH_2CH_2CH_3$$
with NH$_2$

(d) 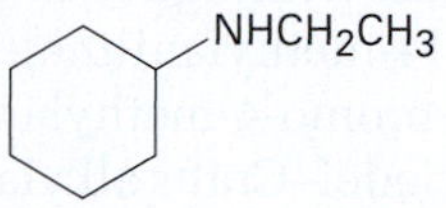 cyclohexane with NHCH$_2$CH$_3$

Problem 18.12

What product would you expect from Hofmann elimination of a heterocyclic amine such as piperidine? Write all the steps.

Piperidine

Electrophilic Aromatic Substitution

An amino group is strongly activating and ortho- and para-directing in electrophilic aromatic substitution reactions (Section 9.8). This high reactivity of amino-substituted benzenes can be a drawback at times because it's often difficult to prevent polysubstitution. For example, reaction of aniline with Br$_2$ takes place rapidly and yields the 2,4,6-tribrominated product. The amino group is so strongly activating that it's not possible to stop at the monobromo stage.

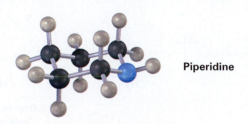

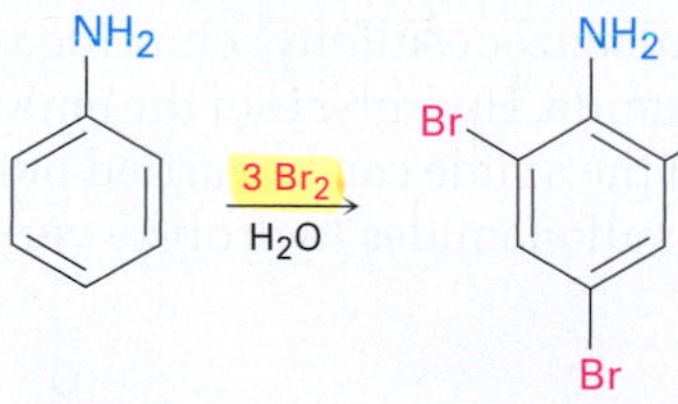

o, p-

Aniline

2,4,6-Tribromoaniline
(100%)

Another drawback to the use of amino-substituted benzenes in electrophilic aromatic substitution reactions is that Friedel–Crafts reactions are not successful (Section 9.7). The amino group forms an acid–base complex with the AlCl$_3$ catalyst, which prevents further reaction from occurring. Both drawbacks can be overcome, however, by carrying out electrophilic aromatic substitution reactions on the corresponding *amide* rather than on the free amine.

As we saw in Section 16.5, treatment of an amine with acetic anhydride yields the corresponding acetyl amide, or acetamide. Although still activating and ortho-, para-directing, amido substituents (–NHCOR) are less strongly activating and less basic than amino groups because their nitrogen lone-pair electrons are delocalized by the neighboring carbonyl group. As a result, bromination of an *N*-arylamide occurs cleanly to give a monobromo product, and hydrolysis of the amide with aqueous base then gives the free amine. For example, *p*-toluidine (4-methylaniline) can be acetylated, brominated, and hydrolyzed to yield 2-bromo-4-methylaniline. None of the 2,6-dibrominated product is obtained. Friedel–Crafts alkylations and acylations of *N*-arylamides also proceed normally.

p-Toluidine

2-Bromo-4-methyl-
aniline (79%)

Modulating the reactivity of an amino-substituted benzene by forming an amide is a useful trick that allows many kinds of electrophilic aromatic substitutions to be carried out that would otherwise be impossible. An example is the preparation of the sulfa drugs, such as sulfanilamide. Sulfa drugs were among the first pharmaceutical agents to be used clinically against bacterial infection. Although they have largely been replaced today by safer and more powerful antibiotics, sulfa drugs were credited with saving the lives of thousands of wounded during World War II, and they are still prescribed for urinary-tract infections. They are prepared by chlorosulfonation of acetanilide, followed by reaction of *p*-(*N*-acetylamino)benzenesulfonyl chloride with ammonia or some other amine to give a sulfonamide. Hydrolysis of the amide then yields the sulfa drug. Note that hydrolysis of the amide can be carried out in the presence of the sulfonamide group because sulfonamides hydrolyze very slowly.

Acetanilide

Sulfanilamide
(a sulfa drug)

Problem 18.13

Propose a synthesis of the drug sulfathiazole from benzene and any necessary amine.

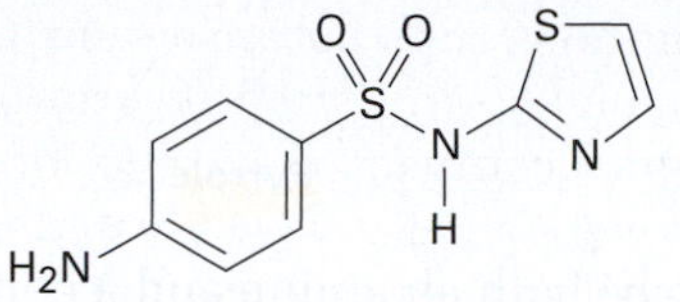

Sulfathiazole

Problem 18.14

Propose syntheses of the following compounds from benzene:
(a) *N,N*-Dimethylaniline **(b)** *p*-Chloroaniline **(c)** *m*-Chloroaniline

18.8 Heterocyclic Amines

As noted in Section 9.4 in connection with a discussion of aromaticity, a cyclic organic compound that contains atoms of two or more elements in its ring is called a *heterocycle*. Heterocyclic amines are particularly common, and many have important biological properties. Pyridoxal phosphate, a coenzyme; sildenafil (Viagra), a well-known pharmaceutical; and heme, the oxygen carrier in blood, are examples.

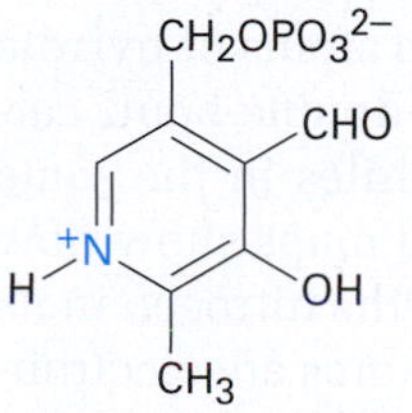

Pyridoxal phosphate
(a coenzyme)

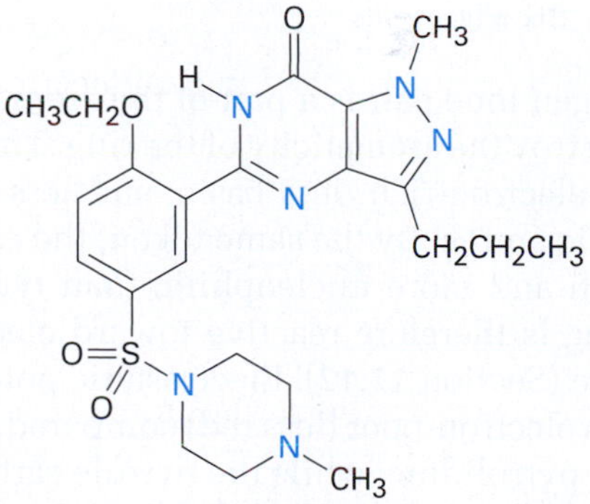

Sildenafil
(Viagra)

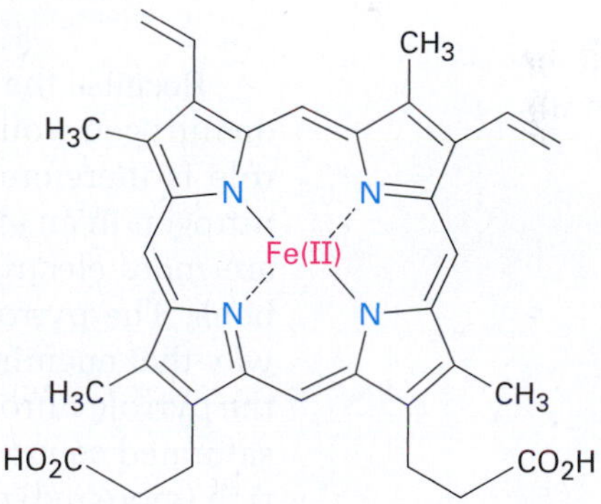

Heme

Most heterocycles have the same chemistry as their open-chain counterparts. Lactones and acyclic esters behave similarly, lactams and acyclic amides behave similarly, and cyclic and acyclic ethers behave similarly. In certain cases, however, particularly when the ring is unsaturated, heterocycles have unique and interesting properties.

Pyrrole and Imidazole

Pyrrole, the simplest five-membered unsaturated heterocyclic amine, is obtained commercially by treatment of furan with ammonia over an alumina catalyst at 400 °C. Furan, the oxygen-containing analog of pyrrole, is obtained by acid-catalyzed dehydration of the five-carbon sugars found in oat hulls and corncobs.

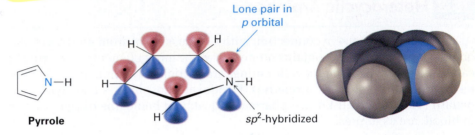

Furan   **Pyrrole**

Although pyrrole appears to be both an amine and a conjugated diene, its chemical properties are not consistent with either of these structural features. Unlike most other amines, pyrrole is not basic—the pK_a of the pyrrolinium ion is 0.4; unlike most other conjugated dienes, pyrrole undergoes electrophilic substitution reactions rather than additions. The reason for both these properties, as noted previously in Section 9.4, Figure 9.6, is that pyrrole has six π electrons and is aromatic. Each of the four carbons contributes one π electron, and the sp^2-hybridized nitrogen contributes two more from its lone pair.

Lone pair in
p orbital

Pyrrole

sp^2-hybridized

Six π electrons

Because the nitrogen lone pair is a part of the aromatic sextet, protonation on nitrogen would destroy the aromaticity of the ring. The nitrogen atom in pyrrole is therefore less electron-rich, less basic, and less nucleophilic than the nitrogen in an aliphatic amine. By the same token, the *carbon* atoms of pyrrole are *more* electron-rich and more nucleophilic than typical double-bond carbons. The pyrrole ring is therefore reactive toward electrophiles in the same way that enamines are (Section 17.12). Electrostatic potential maps show how the pyrrole nitrogen is electron-poor (less red) compared with the nitrogen in its saturated counterpart pyrrolidine, while the pyrrole carbon atoms are electron-rich (more red) compared with the carbons in cyclopenta-1,3-diene.

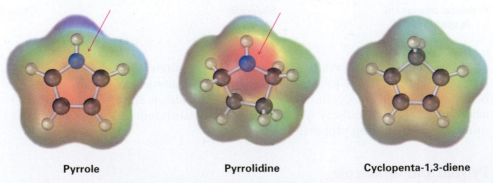

Pyrrole   **Pyrrolidine**   **Cyclopenta-1,3-diene**

The chemistry of pyrrole is similar to that of activated benzene rings. In general, however, the heterocycles are more reactive toward electrophiles than benzene rings are, and low temperatures are often necessary to control the reactions. Halogenation, nitration, sulfonation, and Friedel–Crafts acylation can all be accomplished. For example:

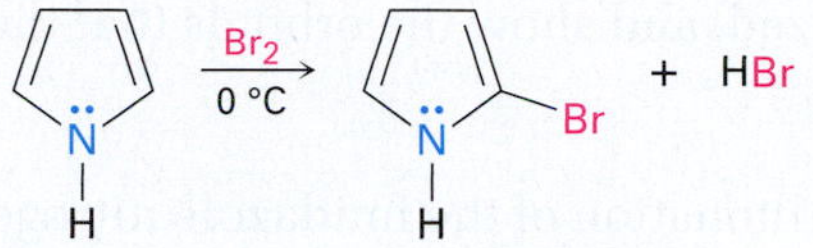

Pyrrole   **2-Bromopyrrole**
(92%)

Electrophilic substitutions normally occur at C2, the position next to the nitrogen, because reaction at this position leads to a more stable intermediate cation having three resonance forms, whereas reaction at C3 gives a less stable cation with only two resonance forms (Figure 18.4).

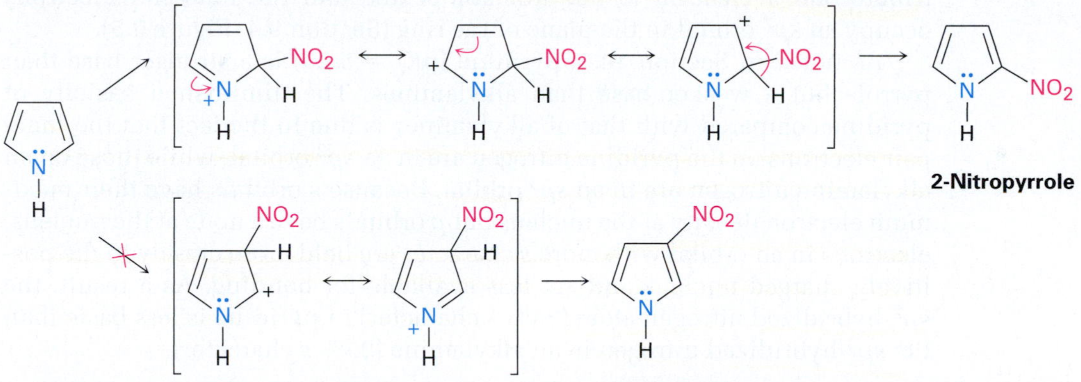

2-Nitropyrrole

3-Nitropyrrole
(NOT formed)

FIGURE 18.4
Electrophilic nitration of pyrrole. The intermediate produced by reaction at C2 is more stable than that produced by reaction at C3.

Other common five-membered heterocyclic amines include imidazole and thiazole. Imidazole, a constituent of the amino acid histidine, has two nitrogens, only one of which is basic. Thiazole, the five-membered ring system on which the structure of thiamin (vitamin B_1) is based, also contains a basic nitrogen that is alkylated in thiamin to form a quaternary ammonium ion.

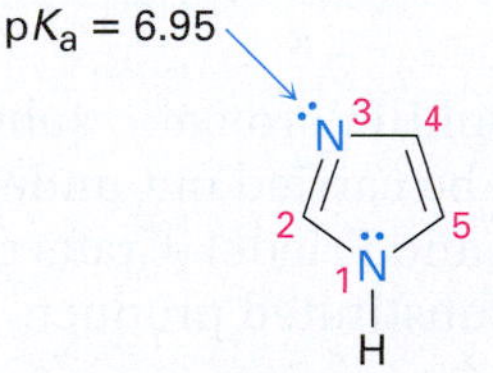

Imidazole 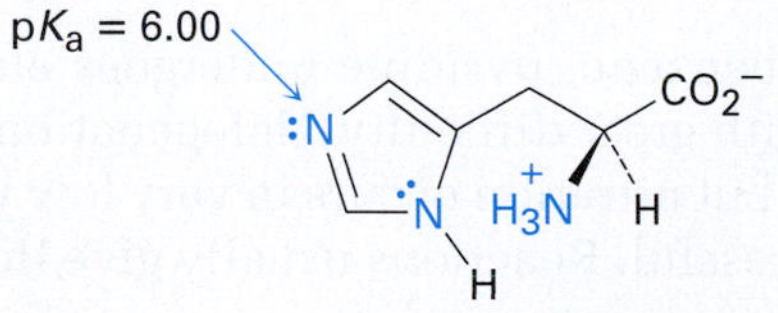   **Histidine**

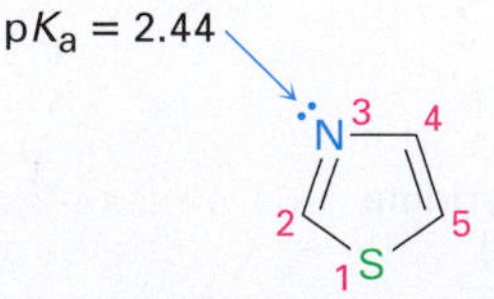

Thiazole 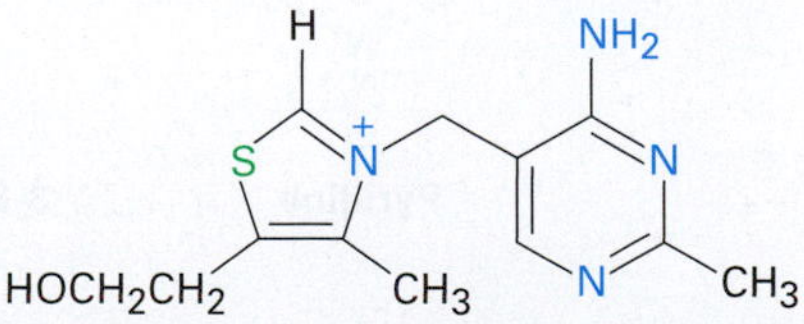   **Thiamin**
(vitamin B_1)

Problem 18.15

Draw an orbital picture of thiazole. Assume that both the nitrogen and sulfur atoms are sp^2-hybridized, and show the orbitals that the lone pairs occupy.

Problem 18.16

What is the percent protonation of the imidazole nitrogen atom in histidine at a physiological pH of 7.3? (See Section 18.5.)

Pyridine and Pyrimidine

Pyridine is the nitrogen-containing heterocyclic analog of benzene. Like benzene, pyridine is a flat, aromatic molecule, with bond angles of 120° and C–C bond lengths of 139 pm, intermediate between typical single and double bonds. The five carbon atoms and the sp^2-hybridized nitrogen atom each contribute one π electron to the aromatic sextet, and the lone-pair electrons occupy an sp^2 orbital in the plane of the ring (Section 9.4, Figure 9.5).

As noted in Section 18.3, pyridine (pK_a = 5.25) is a stronger base than pyrrole but a weaker base than alkylamines. The diminished basicity of pyridine compared with that of alkylamines is due to the fact that the lone-pair electrons on the pyridine nitrogen are in an sp^2 orbital, while those on an alkylamine nitrogen are in an sp^3 orbital. Because s orbitals have their maximum electron density at the nucleus but p orbitals have a node at the nucleus, electrons in an orbital with more s character are held more closely to the positively charged nucleus and are less available for bonding. As a result, the sp^2-hybridized nitrogen atom (33% s character) in pyridine is less basic than the sp^3-hybridized nitrogen in an alkylamine (25% s character).

Pyridine

Unlike benzene, pyridine undergoes electrophilic aromatic substitution reactions with great difficulty. Halogenation can be carried out under drastic conditions, but nitration occurs in very low yield and Friedel–Crafts reactions are not successful. Reactions usually give the 3-substituted product.

Pyridine   **3-Bromopyridine**
                              (30%)

The low reactivity of pyridine toward electrophilic aromatic substitution is caused by a combination of factors. One is that acid–base complexation between the basic ring nitrogen atom and the incoming electrophile places a positive charge on the ring, thereby deactivating it. Equally important is that the electron density of the ring is decreased by the electron-withdrawing inductive effect of the electronegative nitrogen atom. Thus, pyridine has a substantial dipole moment ($\mu = 2.26$ D), with the ring carbons acting as the positive end of the dipole. Reaction of an electrophile with the positively polarized carbon atoms is therefore difficult.

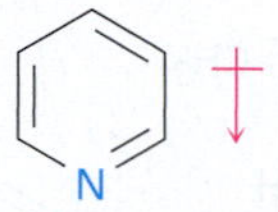

$\mu = 2.26$ D

In addition to pyridine, the six-membered diamine pyrimidine is also found commonly in biological molecules, particularly as a constituent of nucleic acids. With a pK_a of 1.3, pyrimidine is substantially less basic than pyridine because of the inductive effect of the second nitrogen.

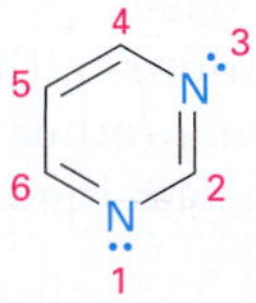

Pyrimidine
$pK_a = 1.3$

Problem 18.17

Electrophilic aromatic substitution reactions of pyridine normally occur at C3. Draw the carbocation intermediates resulting from reaction of an electrophile at C2, C3, and C4, and explain the observed result.

18.9 Fused-Ring Heterocycles

Quinoline, isoquinoline, indole, and purine are common *fused-ring heterocycles*. The first three contain both a benzene ring and a heterocyclic aromatic ring, while purine contains two heterocyclic rings fused together. All four ring systems occur commonly in nature, and many compounds with these rings have pronounced physiological activity. The quinoline alkaloid quinine, for instance, is widely used as an antimalarial drug;

tryptophan is a common amino acid; and the purine adenine is a constituent of nucleic acids.

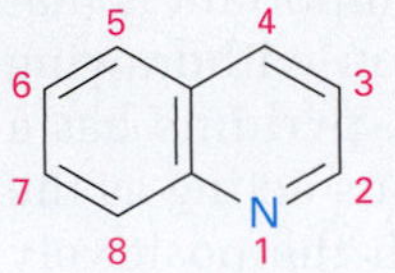

Quinoline 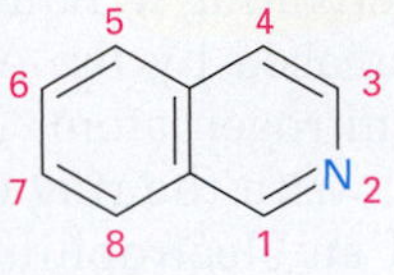 **Isoquinoline** 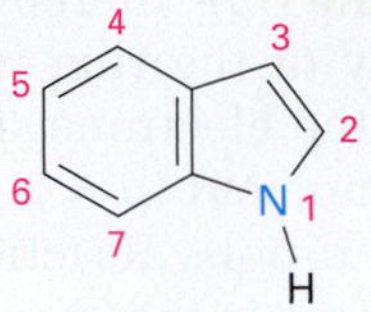 **Indole** 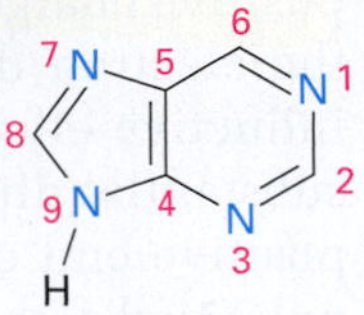 **Purine**

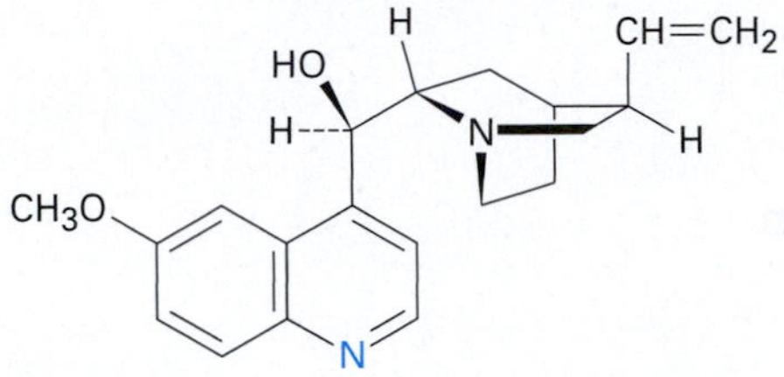

Quinine
(antimalarial)

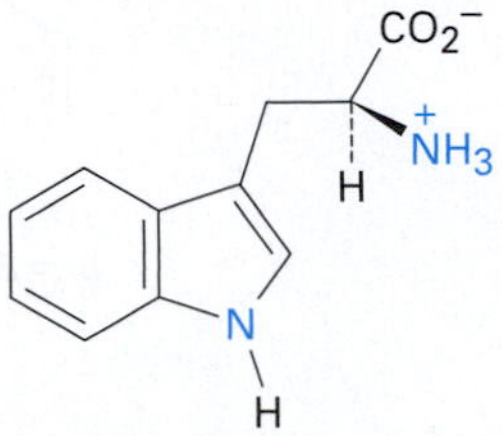

Tryptophan
(amino acid)

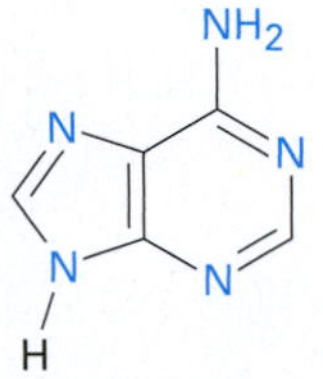

Adenine
(DNA constituent)

The chemistry of these fused-ring heterocycles is just what you might expect based on knowledge of the simpler heterocycles pyridine and pyrrole. Quinoline and isoquinoline both have basic, pyridine-like nitrogen atoms, and both undergo electrophilic substitutions, although less easily than benzene. Reaction occurs on the benzene ring rather than on the pyridine ring, and a mixture of substitution products is obtained.

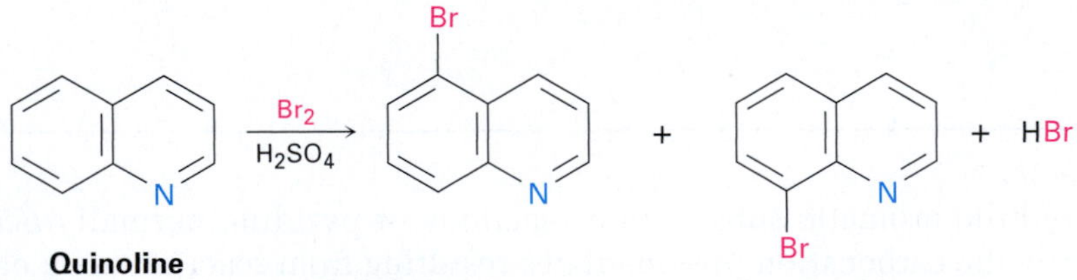

Quinoline

5-Bromoquinoline   **8-Bromoquinoline**

A 51 : 49 ratio

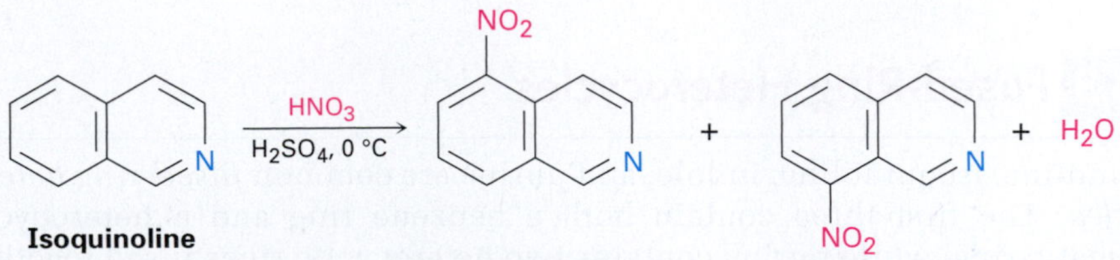

Isoquinoline

5-Nitroisoquinoline   **8-Nitroisoquinoline**

A 90 : 10 ratio

Indole has a nonbasic, pyrrole-like nitrogen and undergoes electrophilic substitution more easily than benzene. Substitution occurs at C3 of the electron-rich pyrrole ring, rather than on the benzene ring.

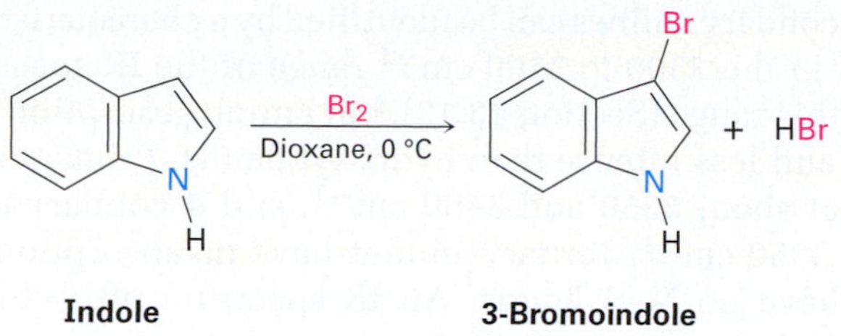

Indole   3-Bromoindole

Purine has three basic, pyridine-like nitrogens with lone-pair electrons in sp^2 orbitals in the plane of the ring. The remaining purine nitrogen is non-basic and pyrrole-like, with its lone-pair electrons as part of the aromatic π electron system.

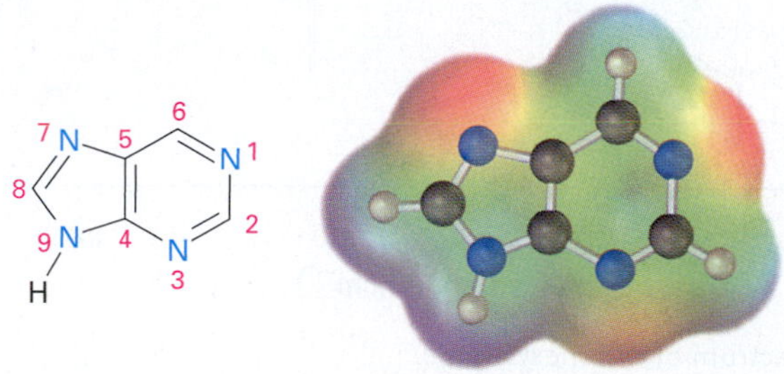

Purine

Problem 18.18
Which nitrogen atom in the hallucinogenic indole alkaloid *N,N*-dimethyl-tryptamine is more basic? Explain.

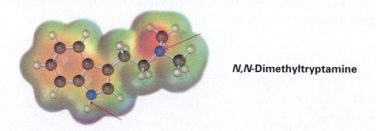

N,N-Dimethyltryptamine

Problem 18.19
Indole reacts with electrophiles at C3 rather than at C2. Draw resonance forms of the intermediate cations resulting from reaction at C2 and C3, and explain the observed results.

18.10 Spectroscopy of Amines

Infrared Spectroscopy

Primary and secondary amines can be identified by a characteristic N–H stretching absorption in the 3300 to 3500 cm^{-1} range of the IR spectrum. Alcohols also absorb in this range (Section 13.12), but amine absorption bands are generally sharper and less intense than hydroxyl bands. Primary amines show a pair of bands at about 3350 and 3450 cm^{-1}, and secondary amines show a single band at 3350 cm^{-1}. Tertiary amines have no absorption in this region because they have no N–H bonds. An IR spectrum of cyclohexylamine is shown in Figure 18.5.

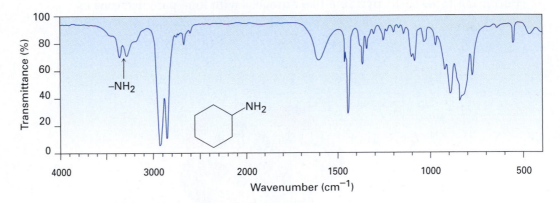

FIGURE 18.5 IR spectrum of cyclohexylamine.

Nuclear Magnetic Resonance Spectroscopy

Amines are difficult to identify solely by ^{1}H NMR spectroscopy because N–H hydrogens tend to appear as broad signals without well-defined coupling to neighboring C–H hydrogens. As with O–H absorptions, amine N–H absorptions can appear over a wide range and are best identified by adding a small amount of D_2O to the sample tube. Exchange of N–D for N–H occurs, and the N–H signal disappears from the NMR spectrum.

$$\ce{>N-H ->[D2O] >N-D + HDO}$$

Hydrogens on the carbon next to nitrogen are deshielded because of the electron-withdrawing effect of the nitrogen, and they therefore absorb at lower field than alkane hydrogens. N-Methyl groups are particularly distinctive because they absorb as a sharp three-proton singlet at 2.2 to 2.6 δ. The N-methyl resonance at 2.42 δ is easily seen in the ^{1}H NMR spectrum of N-methylcyclohexylamine (Figure 18.6).

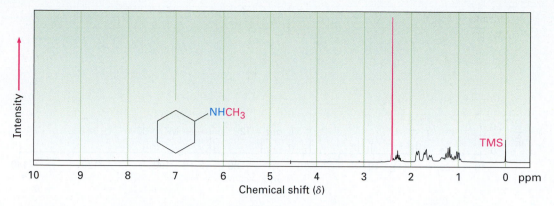

FIGURE 18.6 Proton NMR spectrum of *N*-methylcyclohexylamine.

Carbons next to amine nitrogens are slightly deshielded in the ^{13}C NMR spectrum and absorb about 20 ppm downfield from where they would absorb in an alkane of similar structure. In *N*-methylcyclohexylamine, for example, the ring carbon to which nitrogen is attached absorbs at a position 24 ppm lower than that of any other ring carbon.

Mass Spectrometry

The *nitrogen rule* of mass spectrometry says that a compound with an odd number of nitrogen atoms has an odd-numbered molecular weight. Thus, the presence of nitrogen in a molecule is detected simply by observing its mass spectrum. An odd-numbered molecular ion usually means that the unknown compound has one or three nitrogen atoms, and an even-numbered molecular ion usually means that a compound has either zero or two nitrogen atoms. The logic behind the rule derives from the fact that nitrogen is trivalent, thus requiring an odd number of hydrogen atoms. For example, morphine has the formula $C_{17}H_{19}NO_3$ and a molecular weight of 285 amu.

Alkylamines undergo a characteristic α cleavage in the mass spectrometer, similar to the cleavage observed for alcohols (Section 13.12). A C–C bond nearest the nitrogen atom is broken, yielding an alkyl radical and a nitrogen-containing cation:

$$\left[RCH_2 \overset{}{\underset{}{C}} NR_2 \right]^{+\cdot} \xrightarrow[\text{cleavage}]{\text{Alpha}} RCH_2 \cdot \; + \; \left[\overset{:NR_2}{\underset{}{C^+}} \longleftrightarrow \overset{^+NR_2}{\underset{}{C}} \right]$$

As an example, the mass spectrum of *N*-ethylpropylamine shown in Figure 18.7 has peaks at $m/z = 58$ and $m/z = 72$, corresponding to the two possible modes of α cleavage.

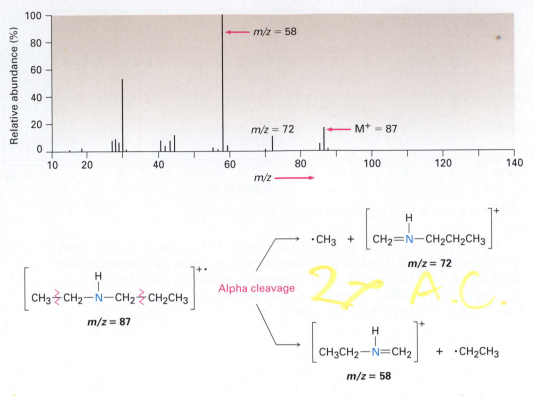

FIGURE 18.7 Mass spectrum of *N*-ethylpropylamine. The two possible modes of α cleavage lead to the observed fragment ions at $m/z = 58$ and $m/z = 72$.

Summary

Key Words

alkylamine, 750

amine, 749

arylamine, 750

heterocyclic amine, 751

Hofmann elimination
   reaction, 764

primary amine (RNH$_2$), 750

quaternary ammonium salt, 750

reductive amination, 761

secondary amine (R$_2$NH), 750

tertiary amine (R$_3$N), 750

We've now seen all the common functional groups that occur in biomolecules. Of those groups, amines are among the most abundant and have among the richest chemistry. In addition to proteins and nucleic acids, the majority of pharmaceutical agents contain amine functional groups and many of the common coenzymes necessary for biological catalysis are amines.

Amines are organic derivatives of ammonia. They are named in the IUPAC system either by adding the suffix -*amine* to the name of the alkyl substituent or by considering the amino group as a substituent on a more complex parent molecule.

The chemistry of amines is dominated by the lone-pair electrons on nitrogen, which make amines both basic and nucleophilic. The base strength of **arylamines** is generally lower than that of **alkylamines** because the nitrogen lone-pair electrons are delocalized by interaction with the aromatic π system. Electron-withdrawing substituents on the aromatic ring further weaken the basicity of a substituted aniline, while electron-donating substituents increase basicity. Alkylamines are sufficiently basic that they exist almost entirely in their protonated form at the physiological pH of 7.3 inside cells.

Heterocyclic amines are compounds that contain one or more nitrogen atoms as part of a ring. Saturated heterocyclic amines usually have the same chemistry as their open-chain analogs, but unsaturated heterocycles such as pyrrole, imidazole, pyridine, and pyrimidine are aromatic. All four are unusually stable, and all undergo aromatic substitution on reaction with electrophiles. Pyrrole is nonbasic because its nitrogen lone-pair electrons are part of the aromatic π system. Fused-ring heterocycles such as quinoline, isoquinoline, indole, and purine are also commonly found in biological molecules.

Arylamines are prepared by nitration of an aromatic ring followed by reduction. Alkylamines are prepared by S_N2 reaction of ammonia or an amine with an alkyl halide as well as by a number of reductive methods, including LiAlH$_4$ reduction of amides and nitriles. Also important is the **reductive amination** reaction in which an aldehyde or ketone is treated with an amine in the presence of a reducing agent.

Many of the reactions of amines are familiar from past chapters. Thus, amines react with alkyl halides in S_N2 reactions and with acid chlorides in nucleophilic acyl substitution reactions. Amines also undergo E2 elimination to yield alkenes if they are first quaternized by treatment with iodomethane and then heated with silver oxide, a process called the **Hofmann elimination**.

Summary of Reactions

1. Synthesis of amines (Section 18.6)

(a) Reduction of nitriles

$$RCH_2X \xrightarrow{NaCN} RCH_2C{\equiv}N \xrightarrow[\text{2. H}_2\text{O}]{\text{1. LiAlH}_4\text{, ether}} RCH_2\overset{\overset{\text{H}}{|}}{\underset{\underset{\text{NH}_2}{}}{C}}\text{H}$$

(b) Reduction of amides

$$\underset{R}{\overset{O}{\underset{}{\overset{\|}{C}}}}NH_2 \xrightarrow[\text{2. H}_2\text{O}]{\text{1. LiAlH}_4\text{, ether}} R\overset{\overset{\text{H}}{|}}{\underset{\underset{\text{NH}_2}{}}{C}}\text{H}$$

(c) Reduction of nitrobenzenes

$$\text{C}_6\text{H}_5{-}NO_2 \xrightarrow[\text{or Fe, H}_3\text{O}^+]{\text{H}_2\text{, Pt}} \text{C}_6\text{H}_5{-}NH_2$$

(d) S_N2 Alkylation of alkyl halides

Ammonia	$:NH_3$ + RX $\longrightarrow$ RNH_3^+ X$^-$	$\xrightarrow{NaOH}$	RNH_2	Primary
Primary	$:NH_2R$ + RX $\longrightarrow$ $R_2NH_2^+$ X$^-$	$\xrightarrow{NaOH}$	R_2NH	Secondary
Secondary	$:NHR_2$ + RX $\longrightarrow$ R_3NH^+ X$^-$	$\xrightarrow{NaOH}$	R_3N	Tertiary
Tertiary	$:NR_3$ + RX $\longrightarrow$ R_4N^+ X$^-$		Quaternary ammonium salt	

(e) Reductive amination of aldehydes/ketones

$$\underset{R}{\overset{O}{\underset{R'}{\overset{\|}{C}}}} \xrightarrow[\text{NaBH}_4]{NH_3} \underset{R}{\overset{\text{H}\quad NH_2}{\underset{R'}{C}}}$$

2.   Reactions of amines (Section 18.7)

(a) Alkylation with alkyl halides; see reaction 1(d)

(b) Acylation with acid chlorides

Ammonia

$$R-C(=O)-Cl + NH_3 \xrightarrow[\text{solvent}]{\text{Pyridine}} R-C(=O)-N(H)(H) + HCl$$

Primary

$$R-C(=O)-Cl + R'NH_2 \xrightarrow[\text{solvent}]{\text{Pyridine}} R-C(=O)-N(R')(H) + HCl$$

Secondary

$$R-C(=O)-Cl + R'_2NH \xrightarrow[\text{solvent}]{\text{Pyridine}} R-C(=O)-N(R')(H) + HCl$$

(c) Hofmann elimination

$$\overset{H}{\underset{NR_2}{C-C}} \xrightarrow[\text{2. Ag}_2\text{O, heat}]{\text{1. CH}_3\text{I}} C=C$$

Lagniappe

Green Chemistry II: Ionic Liquids

Liquids made of ions? Usually when we think of ionic compounds, we think of high-melting solids: sodium chloride, magnesium sulfate, lithium carbonate, and so forth. But yes, there are also ionic compounds that are liquid at room temperature, and they are gaining importance as reaction solvents, particularly for use in green chemistry processes (see the Chapter 12 *Lagniappe*).

Ionic liquids have been known for nearly a century; the first to be discovered was ethylammonium nitrate, $CH_3CH_2NH_3^+\ NO_3^-$, with a melting point of 12 °C. More generally, however, the ionic liquids in use today are salts in which the cation is unsymmetrical and in which one or both of the ions are bulky so that the charges are dispersed over a large volume. Both factors minimize the crystal lattice energy and disfavor formation of the solid. Typical cations are quaternary ammonium ions from heterocyclic amines, either 1,3-dialkylimidazolium ions, *N*-alkylpyridinium ions, or ring-substituted *N*-alkylpyridinium ions.

$$\left[\begin{array}{l} R = -CH_3, -CH_2CH_3, -CH_2CH_2CH_2CH_3, \\ -CH_2CH_2CH_2CH_2CH_2CH_2CH_2CH_3 \end{array} \right]$$

1,3-Dialkylimidazolium ions

$$\left[\begin{array}{l} R = -CH_2CH_3, -CH_2CH_2CH_2CH_3, \\ -CH_2CH_2CH_2CH_2CH_2CH_3 \end{array} \right]$$

N-Alkylpyridinium ions

continued

Lagniappe *continued*

Image provided by Peg Williams, USAFA, Dept. of Chemistry Research Center

Anions are just as varied as the cations, and more than 250 different ionic liquids with different anion/cation combinations are commercially available. Hexafluorophosphate, tetrafluoroborate, alkyl sulfates, trifluoromethanesulfonates (triflates), and halides are some anion possibilities.

Hexafluoro-phosphate   **Tetrafluoro-borate**   **Methyl sulfate**   **Trifluoromethane-sulfonate**   **Halide**

Ionic liquids have several features that make them attractive for use as solvents, particularly in green chemistry:

- They dissolve both polar and nonpolar organic compounds, giving high solute concentrations and thereby minimizing the amount of solvent needed.
- They can be optimized for specific reactions by varying cation and anion structures.
- They are nonflammable.
- They are thermally stable.
- They have negligible vapor pressures and do not evaporate.
- They are generally recoverable and can be reused many times.

...s, these liquids really do ...nsist of ionic rather than ...olecular substances.

As an example of their use in organic chemistry, the analgesic drug pravadoline has been synthesized in two steps using 1-butyl-3-methylimidazolium hexafluorophosphate, abbreviated [bmim][PF$_6$], as the solvent for both steps. The first step is a base-induced S$_N$2 reaction of 2-methylindole with a primary alkyl halide, and the second is a Friedel–Crafts acylation. Both steps take place in 95% yield, and the ionic solvent is recovered simply by washing the reaction mixture, first with toluene and then with water. We'll be hearing a lot more about ionic solvents in coming years.

$$\xrightarrow[\text{[bmim][PF}_6\text{]}]{\text{KOH}}$$

$$\xrightarrow[\text{[bmim][PF}_6\text{]}]{}$$

Pravadoline

Exercises

■ indicates problems that are assignable in Organic OWL.

Go to this book's companion website at www.cengage.com/chemistry/mcmurry to explore interactive versions of the Active Figures from this text.

VISUALIZING CHEMISTRY

(Problems 18.1–18.19 appear within the chapter.)

18.20 ■ Name the following amines, and identify each as primary, secondary, or tertiary:

(a)

(b)

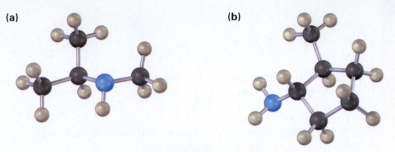

(c)

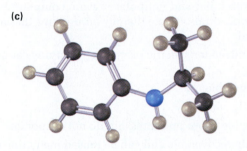

18.21 ■ The following compound contains three nitrogen atoms. Rank them in order of increasing basicity.

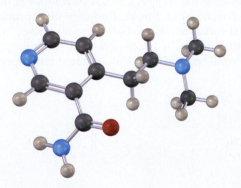

18.22 Name the following amine, including *R,S* stereochemistry, and draw the product of its reaction with excess iodomethane followed by heating with Ag$_2$O (Hofmann elimination). Is the stereochemistry of the alkene product *Z* or *E*? Explain.

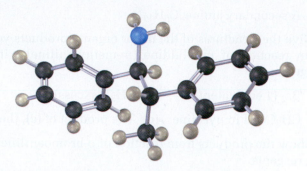

18.23 The following molecule has three nitrogen atoms. List them in order of increasing basicity, and explain your ordering.

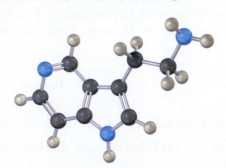

ADDITIONAL PROBLEMS

18.24 ■ Draw structures corresponding to the following IUPAC names:

(a) *N,N*-Dimethylaniline

(b) (Cyclohexylmethyl)amine

(c) *N*-Methylcyclohexylamine

(d) (2-Methylcyclohexyl)amine

(e) 3-(*N,N*-Dimethylamino)propanoic acid

18.25 ■ Name the following compounds:

(a)

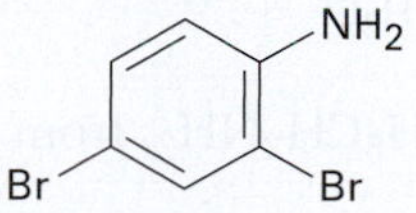

Br and Br substituents, NH$_2$

(b) 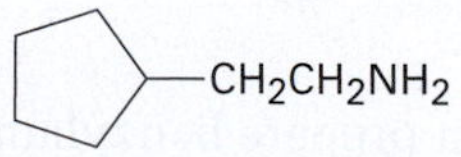 cyclopentane—CH$_2$CH$_2$NH$_2$

(c) 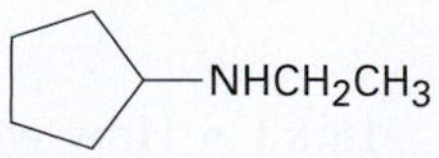 cyclopentane—NHCH$_2$CH$_3$

(d) 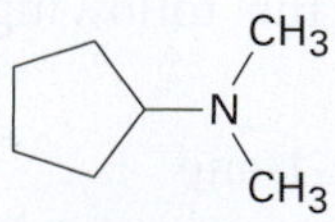 cyclopentane—N with CH$_3$ and CH$_3$

(e) 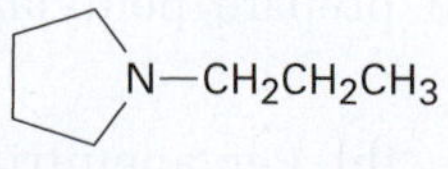 N—CH$_2$CH$_2$CH$_3$ (pyrrolidine ring)

(f) H$_2$NCH$_2$CH$_2$CH$_2$CN

■ Problems assignable in Organic OWL.

18.26 Propose structures for substances that fit the following descriptions:

(a) A chiral quaternary ammonium salt

(b) A six-membered heterocyclic diamine

(c) A secondary amine, $C_6H_{11}N$

18.27 ■ Give the structures of the major organic products you would expect from reaction of *m*-toluidine (*m*-methylaniline) with the following reagents:

(a) Br_2 (1 equivalent)　　**(b)** CH_3I (excess)

(c) CH_3COCl in pyridine　　**(d)** The product of **(c)**, then HSO_3Cl

18.28 ■ Show the products from reaction of *p*-bromoaniline with the following reagents:

(a) CH_3I (excess)　**(b)** HCl　**(c)** CH_3COCl　**(d)** CH_3MgBr

18.29 Oxazole is a five-membered aromatic heterocycle. Draw an orbital picture of oxazole, showing all *p* orbitals and all lone-pair orbitals. Would you expect oxazole to be more basic or less basic than pyrrole? Explain.

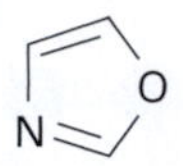

Oxazole

18.30 ■ How would you prepare the following substances from butan-1-ol?

(a) Butylamine　**(b)** Dibutylamine　**(c)** Pentylamine　**(d)** Butanamide

18.31 Substituted pyrroles are often prepared by treatment of a 1,4-diketone with ammonia. Suggest a mechanism.

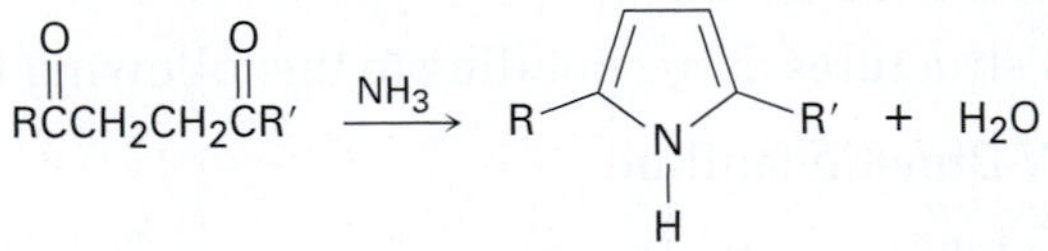

18.32 3,5-Dimethylisoxazole is prepared by reaction of pentane-2,4-dione with hydroxylamine. Propose a mechanism.

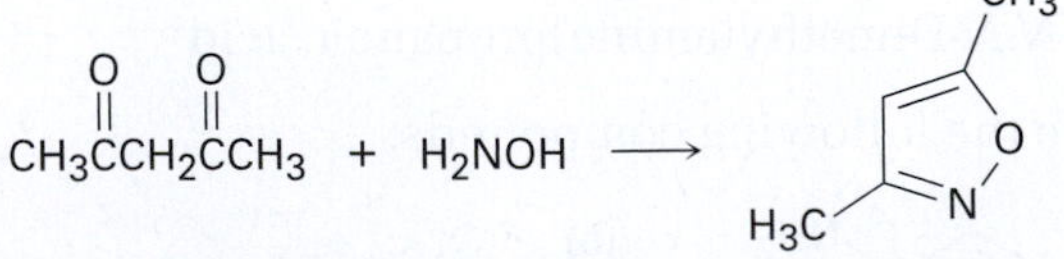

18.33 ■ How would you prepare benzylamine, $C_6H_5CH_2NH_2$, from benzene? More than one step is needed.

18.34 ■ How might you prepare pentylamine from the following starting materials?

(a) Pentanamide　**(b)** Pentanenitrile　**(c)** But-1-ene

(d) Butan-1-ol　　**(e)** Pentanoic acid

18.35 ▪ What are the major products you would expect from Hofmann elimination of the following amines?

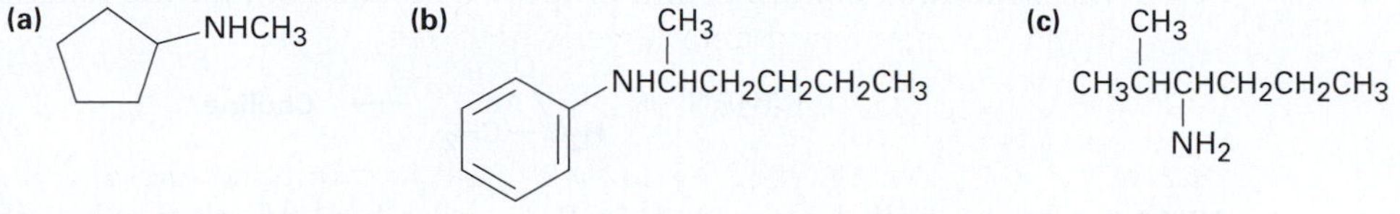

(a)

NHCH₃

(b)

CH₃
|
NHCHCH₂CH₂CH₂CH₃

(c)

CH₃
|
CH₃CHCHCH₂CH₂CH₃
|
NH₂

18.36 ▪ Fill in the missing reagents **a–e** in the following scheme:

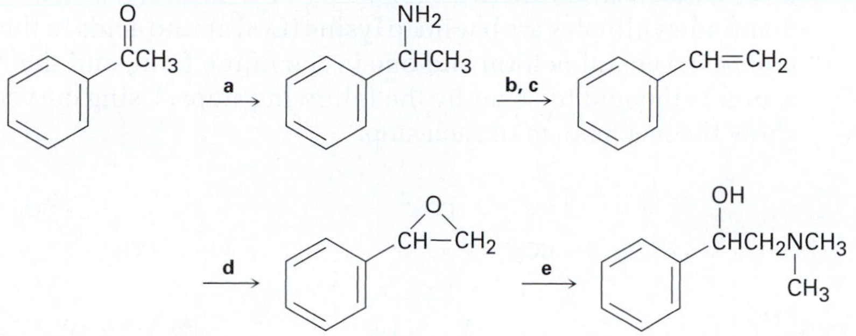

18.37 Protonation of an amide using strong acid occurs on oxygen rather than on nitrogen. Suggest a reason for this behavior, taking resonance into account.

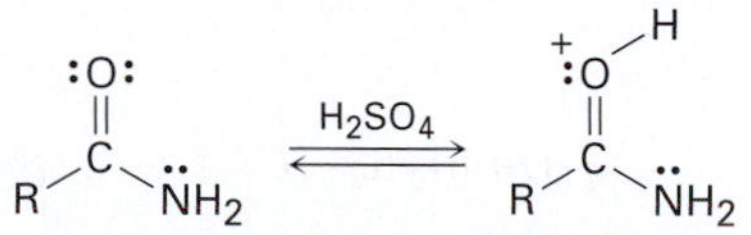

18.38 *p*-Nitroaniline (pK_a = 1.0) is less basic than *m*-nitroaniline (pK_a = 2.5) by a factor of 30. Explain, using resonance structures. (The pK_a values refer to the corresponding ammonium ions.)

18.39 Pyrrole has a dipole moment μ = 1.8 D, with the *nitrogen* atom at the positive end of the dipole. Explain.

18.40 Show the mechanism of reductive amination of cyclohexanone and dimethylamine with NaBH₄.

18.41 ▪ Fill in the missing reagents **a–d** in the following synthesis of racemic methamphetamine from benzene:

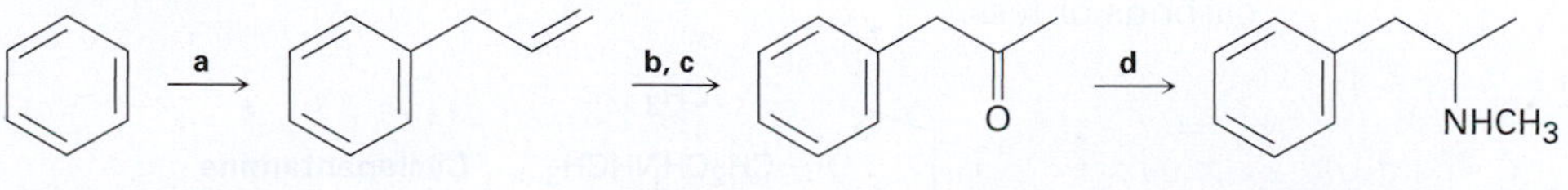

(*R,S*)-Methamphetamine

18.42 One problem with reductive amination as a method of amine synthesis is that by-products are sometimes obtained. For example, reductive amination of benzaldehyde with methylamine leads to a mixture of *N*-methylbenzylamine and *N*-methyldibenzylamine. How do you suppose the tertiary amine by-product is formed? Propose a mechanism.

▪ Problems assignable in Organic OWL.

18.43 ■ Choline, a component of the phospholipids in cell membranes, can be prepared by S_N2 reaction of trimethylamine with ethylene oxide. Show the structure of choline, and propose a mechanism for the reaction.

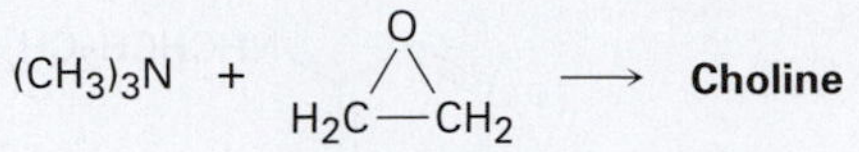

$$(CH_3)_3N \; + \; \underset{H_2C-CH_2}{\overset{O}{\triangle}} \; \longrightarrow \; \textbf{Choline}$$

18.44 ■ Chlorophyll, heme, vitamin B_{12}, and a host of other substances are biosynthesized from porphobilinogen (PBG), which is itself formed from condensation of two molecules of 5-aminolevulinate. The two 5-aminolevulinates are bound to lysine (Lys) amino acids in the enzyme, one in the enamine form and one in the imine form, and their condensation is thought to occur by the following steps. Using curved arrows, show the mechanism of each step.

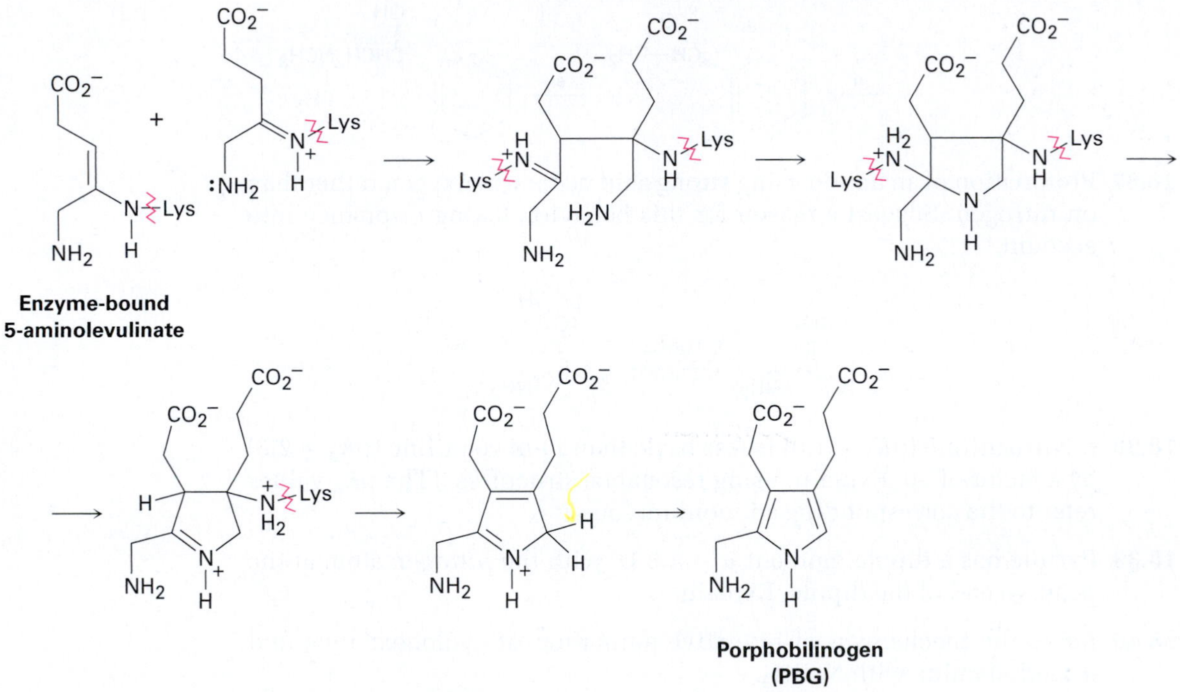

**Enzyme-bound
5-aminolevulinate**

**Porphobilinogen
(PBG)**

18.45 ■ Cyclopentamine is an amphetamine-like central nervous system stimulant. Propose a synthesis of cyclopentamine from materials of five carbons or less.

$$\text{CH}_2\text{CHNHCH}_3 \quad \textbf{Cyclopentamine}$$
$$\overset{\text{CH}_3}{|}$$

18.46 Tetracaine is a substance used as a spinal anesthetic. How would you prepare tetracaine from *p*-nitrobenzoic acid?

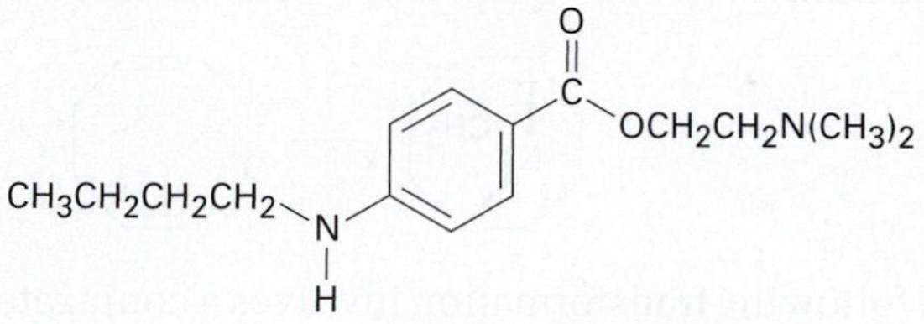

Tetracaine

18.47 Atropine, $C_{17}H_{23}NO_3$, is a poisonous alkaloid isolated from the leaves and roots of *Atropa belladonna,* the deadly nightshade. In small doses, atropine acts as a muscle relaxant; 0.5 ng (nanogram, 10^{-9} g) is sufficient to cause pupil dilation. On basic hydrolysis, atropine yields tropic acid, $C_6H_5CH(CH_2OH)CO_2H$, and tropine, $C_8H_{15}NO$. Tropine is an optically inactive alcohol that yields tropidene on dehydration with H_2SO_4. Propose a structure for atropine.

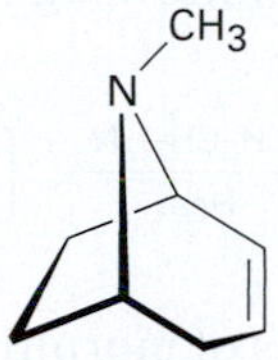

Tropidene

18.48 Propose a structure for the product with formula $C_9H_{17}N$ that results when 2-(2-cyanoethyl)cyclohexanone is reduced catalytically.

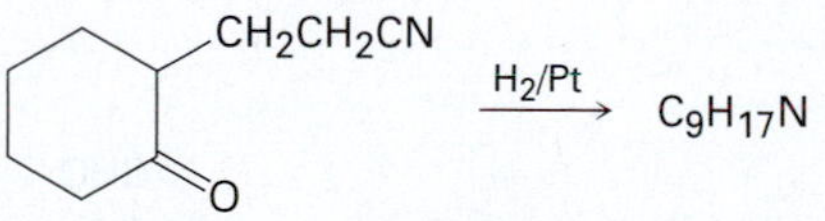

18.49 ■ Coniine, $C_8H_{17}N$, is the toxic principle of the poison hemlock drunk by Socrates. When subjected to Hofmann elimination, coniine yields 5-(*N,N*-dimethylamino)oct-1-ene. If coniine is a secondary amine, what is its structure?

18.50 How would you synthesize coniine (Problem 18.49) from acrylonitrile ($H_2C=CHCN$) and ethyl 3-oxohexanoate ($CH_3CH_2CH_2COCH_2CO_2Et$)? (See Problem 18.48.)

■ Problems assignable in Organic OWL.

18.51 Cycloocta-1,3,5,7-tetraene was first synthesized in 1911 by a route that involved the following transformation. How would you accomplish this reaction?

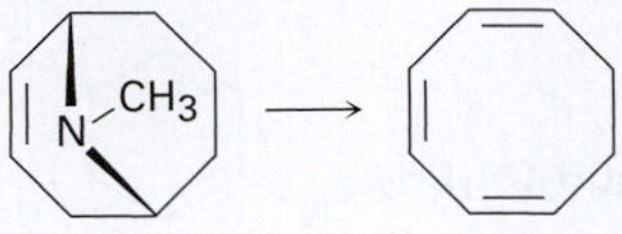

18.52 ▪ The following transformation involves a conjugate nucleophilic addition reaction (Section 14.11) followed by an intramolecular nucleophilic acyl substitution reaction (Section 16.2). Show the mechanism.

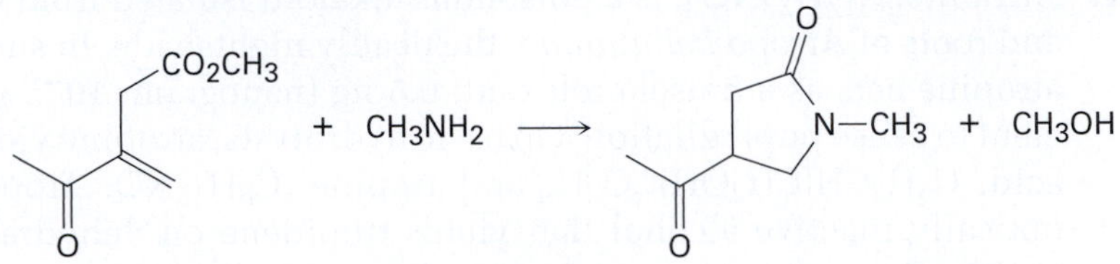

18.53 Propose a mechanism for the following reaction:

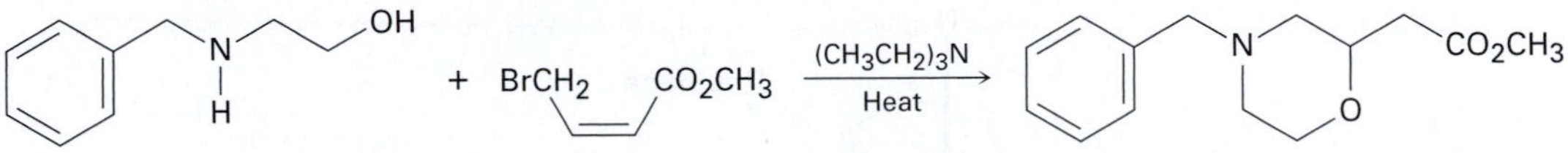

18.54 ▪ One step in the biosynthesis of morphine is the reaction of dopamine with *p*-hydroxyphenylacetaldehyde to give (*S*)-norcoclaurine. Assuming that the reaction is acid-catalyzed, propose a mechanism.

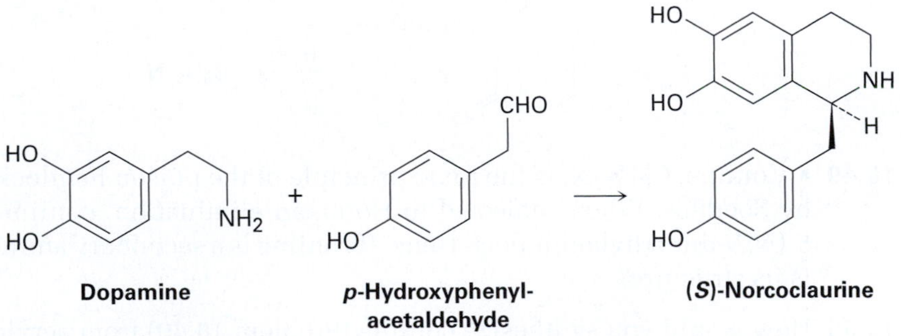

| Dopamine | *p*-Hydroxyphenyl-acetaldehyde | (*S*)-Norcoclaurine |

▪ Problems assignable in Organic OWL.

18.55 The antitumor antibiotic mitomycin C functions by forming cross-links in DNA chains.

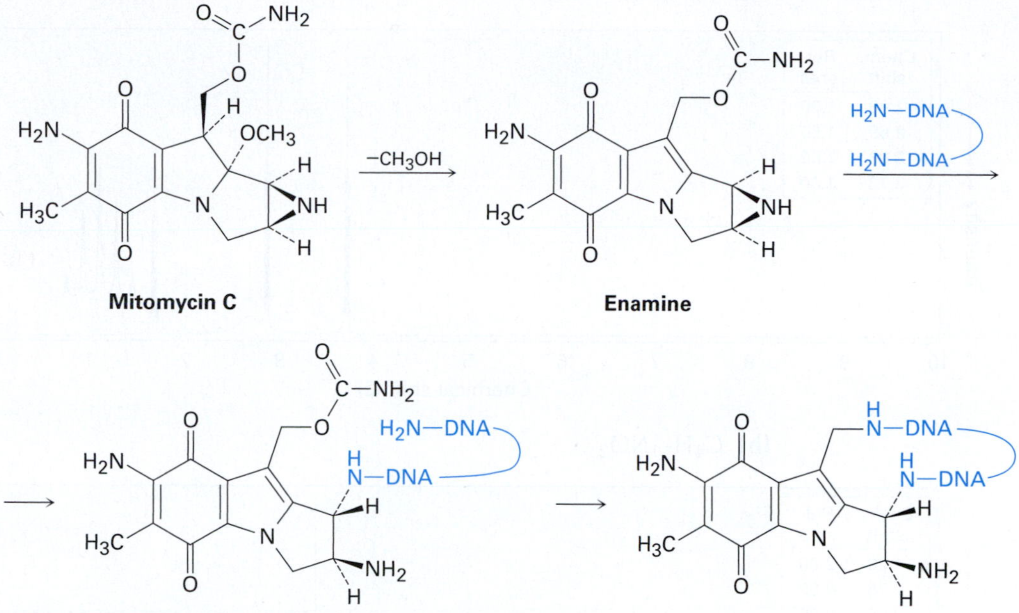

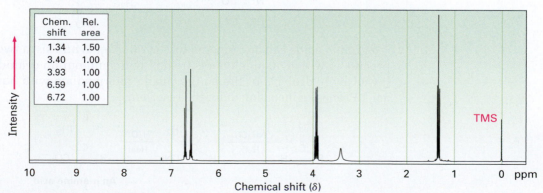

Mitomycin C   **Enamine**

(a) The first step is loss of methoxide and formation of an iminium ion intermediate that is deprotonated to give an enamine. Show the mechanism.

(b) The second step is reaction of the enamine with DNA to open the three-membered, nitrogen-containing (aziridine) ring. Show the mechanism.

(c) The third step is loss of carbamate ($NH_2CO_2^-$) and formation of an unsaturated iminium ion, followed by a conjugate addition of another part of the DNA chain. Show the mechanism.

18.56 ■ Phenacetin, a substance formerly used in over-the-counter headache remedies, has the formula $C_{10}H_{13}NO_2$. Phenacetin is neutral and does not dissolve in either acid or base. When warmed with aqueous NaOH, phenacetin yields an amine, $C_8H_{11}NO$, whose ^1H NMR spectrum is shown. When heated with HI, the amine is cleaved to an aminophenol, C_6H_7NO. What is the structure of phenacetin, and what are the structures of the amine and the aminophenol?

Chem. shift	Rel. area
1.34	1.50
3.40	1.00
3.93	1.00
6.59	1.00
6.72	1.00

TMS

Intensity

10   9   8   7   6   5   4   3   2   1   0 ppm
Chemical shift (δ)

■ Problems assignable in Organic OWL.

18.57 ■ Propose structures for amines with the following ^{1}H NMR spectra:

(a) C_3H_9NO

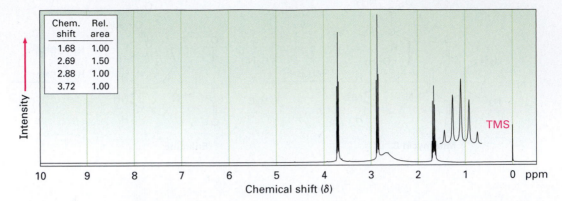

Chem. shift	Rel. area
1.68	1.00
2.69	1.50
2.88	1.00
3.72	1.00

(b) $C_4H_{11}NO_2$

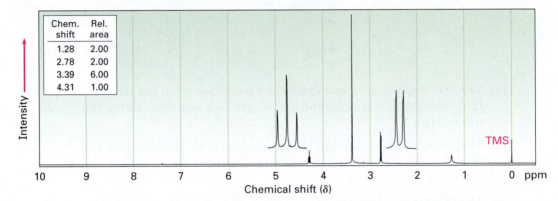

Chem. shift	Rel. area
1.28	2.00
2.78	2.00
3.39	6.00
4.31	1.00

18.58 One of the reactions used in determining the sequence of nucleotides in a strand of DNA is reaction with hydrazine. Propose a mechanism for the following reaction, which occurs by an initial conjugate addition followed by internal amide formation:

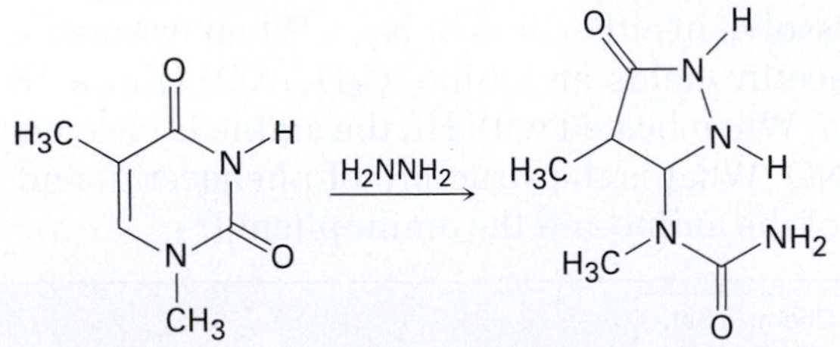

18.59 ■ α-Amino acids can be prepared by the *Strecker synthesis,* a two-step process in which an aldehyde is treated with ammonium cyanide followed by hydrolysis of the amino nitrile intermediate with aqueous acid. Propose a mechanism for the reaction.

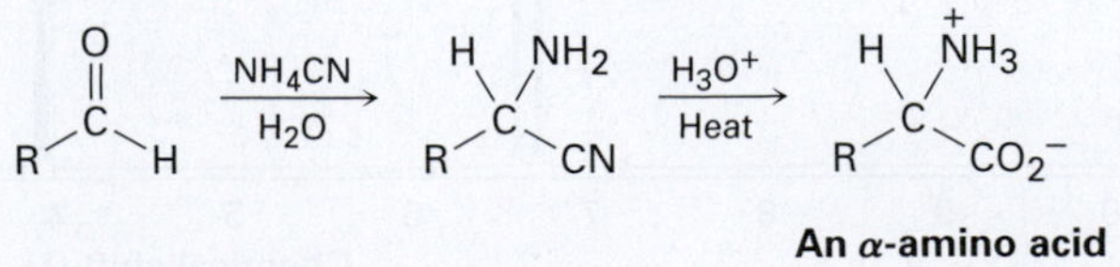

An α-amino acid

■ Problems assignable in Organic OWL.

19 Biomolecules: Amino Acids, Peptides, and Proteins

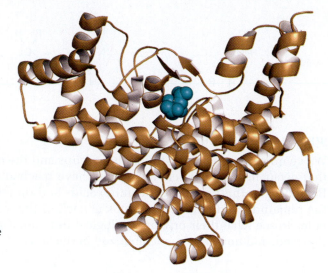

Citrate synthase catalyzes the reaction of acetyl CoA with oxaloacetate to give citrate, the first step in the citric acid cycle of food metabolism.

Proteins occur in every living organism, are of many different types, and have many different biological functions. The keratin of skin and fingernails, the fibroin of silk and spider webs, and the estimated 50,000 or so enzymes that catalyze the biological reactions in our bodies are all proteins. Regardless of their function, all proteins are made up of many *amino acids* linked together in a long chain.

Amino acids, as their name implies, are difunctional. They contain both a basic amino group and an acidic carboxyl group:

Alanine, an amino acid

Their value as building blocks to make proteins stems from the fact that amino acids can join together into long chains by forming amide bonds between the $-NH_2$ of one amino acid and the $-CO_2H$ of another. For classification

OWL Online homework for this chapter can be assigned in Organic OWL.

purposes, chains with fewer than 50 amino acids are often called **peptides**, while the term **protein** is generally used for larger chains.

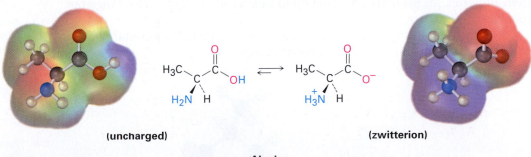

WHY THIS CHAPTER?

We've now seen the major functional groups and the common reaction types that occur in biological chemistry and have reached the heart of this book. Beginning in this chapter with amino acids and proteins, and continuing for the remainder of the text, we'll look at each of the main classes of biomolecules to see what their primary biological functions are, how they're biosynthesized, and how they're metabolized in the body.

19.1 Structures of Amino Acids

We saw in Sections 15.3 and 18.5 that a carboxyl group is deprotonated and exists as the carboxylate anion at a pH of 7.3 in the body (the *physiological* pH), while an amino group is protonated and exists as the ammonium cation. Thus, amino acids exist in aqueous solution primarily in the form of a dipolar ion, or **zwitterion** (German *zwitter,* meaning "hybrid").

(uncharged)   **(zwitterion)**

Alanine

Amino acid zwitterions are internal salts and therefore have many of the physical properties associated with salts. They have large dipole moments, are relatively soluble in water but insoluble in hydrocarbons, and are crystalline. In addition, amino acids are *amphiprotic:* they can react either as acids or as bases, depending on the circumstances. In aqueous acid solution, an amino acid zwitterion is a base that *accepts* a proton to yield a cation; in aqueous base solution, the zwitterion is an acid that *loses* a proton to form an anion. Note that it is the carboxylate anion, $-CO_2^-$, that acts as the basic site and accepts a proton in acid solution, and it is the ammonium cation, $-NH_3^+$, that acts as the acidic site and donates a proton in base solution.

In acid solution

$$R-\underset{\underset{H_3N}{+}}{\overset{\overset{O}{\parallel}}{C}}-\underset{H}{\overset{\overset{O}{\parallel}}{C}}-O^- \;+\; H_3O^+ \;\rightleftharpoons\; R-\underset{\underset{H_3N}{+}}{\overset{\overset{O}{\parallel}}{C}}-\underset{H}{\overset{\overset{O}{\parallel}}{C}}-OH \;+\; H_2O$$

In base solution

$$R-\underset{\underset{H_3N}{+}}{\overset{\overset{O}{\parallel}}{C}}-\underset{H}{\overset{\overset{O}{\parallel}}{C}}-O^- \;+\; OH^- \;\rightleftharpoons\; R-\underset{\underset{H_2N}{}}{\overset{\overset{O}{\parallel}}{C}}-\underset{H}{\overset{\overset{O}{\parallel}}{C}}-O^- \;+\; H_2O$$

The structures, abbreviations (both three- and one-letter), and pK_a values of the 20 amino acids commonly found in proteins are shown in Table 19.1. All are **α-amino acids**, meaning that the amino group in each is a substituent on the α carbon atom—the one next to the carbonyl group. Nineteen of the twenty amino acids are primary amines, RNH_2, and differ only in the nature of the substituent attached to the α carbon, called the **side chain**. Proline is a secondary amine and the only amino acid whose nitrogen and α carbon atoms are part of a ring.

A primary α-amino acid　　　**Proline, a secondary α-amino acid**

In addition to the 20 amino acids commonly found in proteins, 2 others—selenocysteine and pyrrolysine—are found in some organisms, and more than 700 nonprotein amino acids are also found in nature. γ-Aminobutyric acid (GABA), for instance, is found in the brain and acts as a neurotransmitter; homocysteine is found in blood and is linked to coronary heart disease; and thyroxine is found in the thyroid gland, where it acts as a hormone.

Selenocysteine　　　**Pyrrolysine**

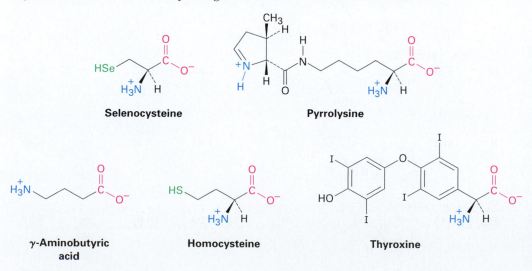

γ-Aminobutyric acid　　　**Homocysteine**　　　**Thyroxine**

TABLE 19.1
The 20 Common Amino Acids in Proteins

Name	Abbreviations		MW	Stucture	pK_a α-CO$_2$H	pK_a α-NH$_3^+$	pK_a side chain	pI
Neutral Amino Acids								
Alanine	Ala	A	89	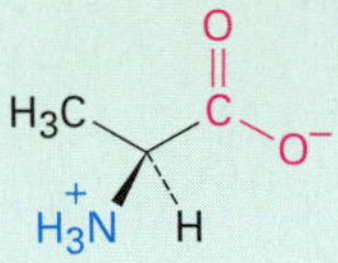	2.34	9.69	—	6.01
Asparagine	Asn	N	132	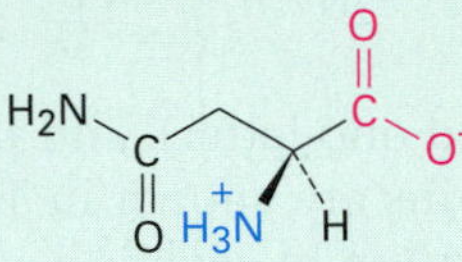	2.02	8.80	—	5.41
Cysteine	Cys	C	121	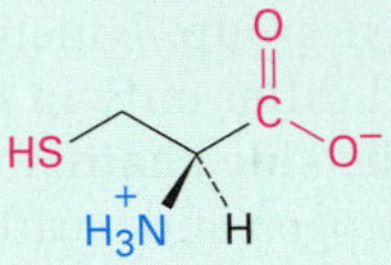	1.96	10.28	8.18	5.07
Glutamine	Gln	Q	146	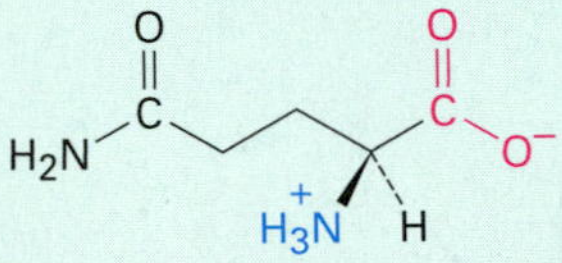	2.17	9.13	—	5.65
Glycine	Gly	G	75	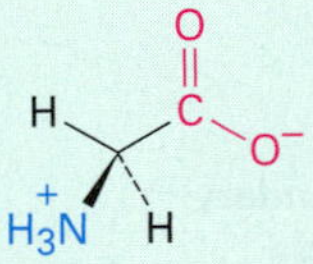	2.34	9.60	—	5.97
Isoleucine	Ile	I	131	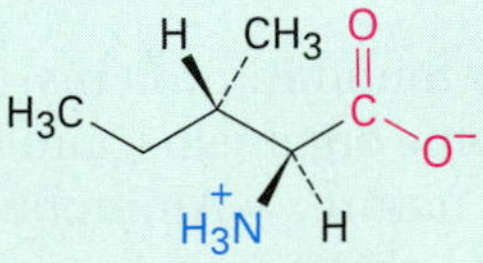	2.36	9.60	—	6.02
Leucine	Leu	L	131	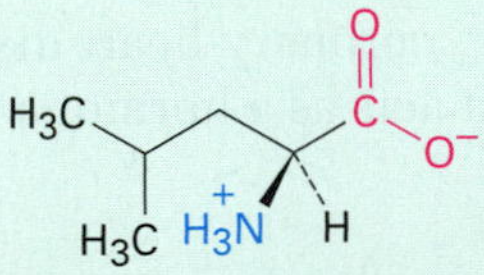	2.36	9.60	—	5.98
Methionine	Met	M	149	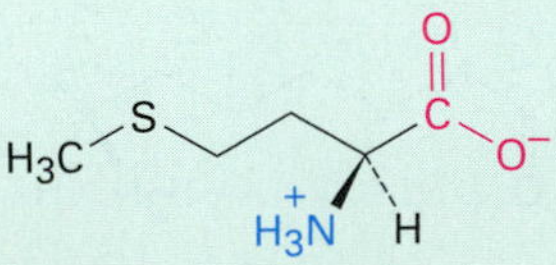	2.28	9.21	—	5.74
Phenylalanine	Phe	F	165	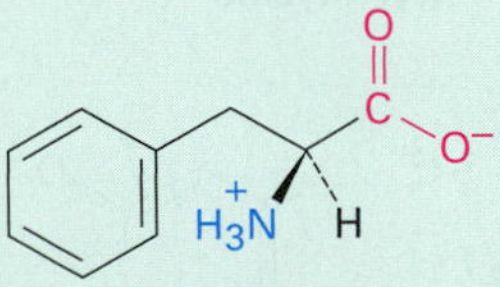	1.83	9.13	—	5.48
Proline	Pro	P	115	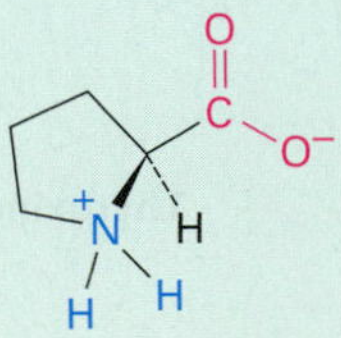	1.99	10.60	—	6.30

continued

TABLE 19.1
The 20 Common Amino Acids in Proteins *continued*

Name	Abbreviations		MW	Stucture	pK_a α-CO$_2$H	pK_a α-NH$_3^+$	pK_a side chain	pI
Neutral Amino Acids *continued*								
Serine	Ser	S	105	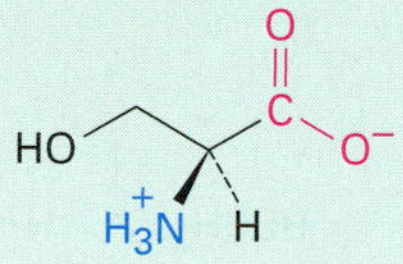	2.21	9.15	—	5.68
Threonine	Thr	T	119	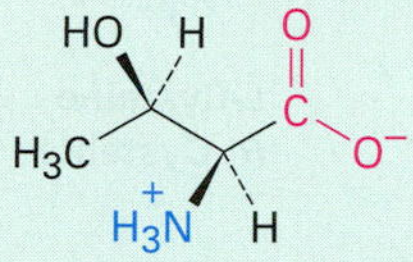	2.09	9.10	—	5.60
Tryptophan	Trp	W	204	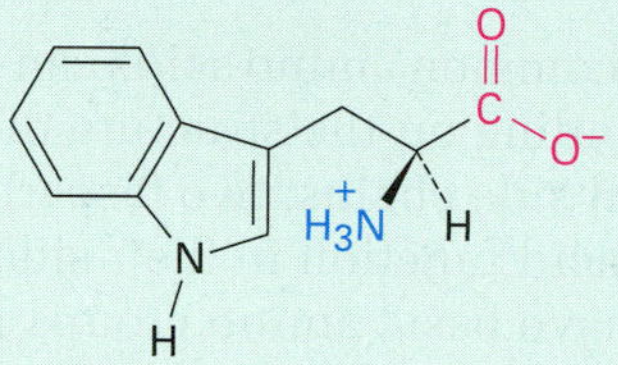	2.83	9.39	—	5.89
Tyrosine	Tyr	Y	181	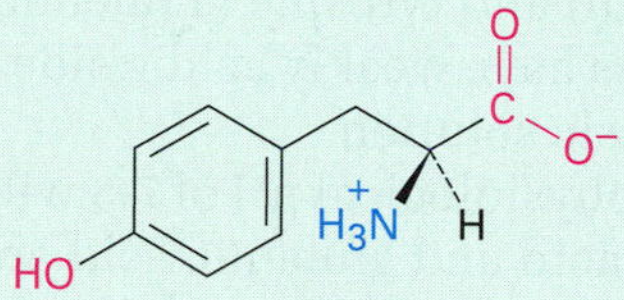	2.20	9.11	10.07	5.66
Valine	Val	V	117	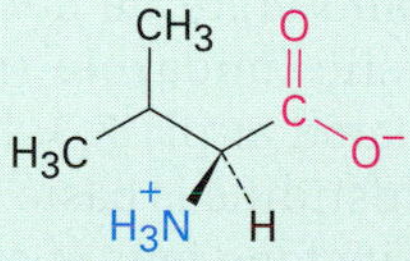	2.32	9.62	—	5.96
Acidic Amino Acids								
Aspartic acid	Asp	D	133	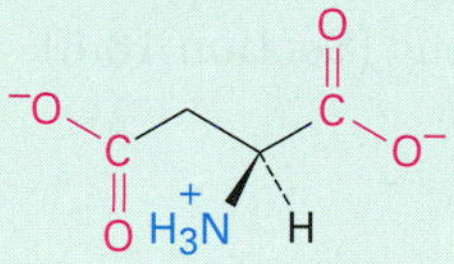	1.88	9.60	3.65	2.77
Glutamic acid	Glu	E	147	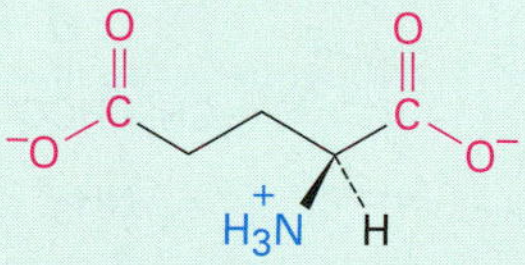	2.19	9.67	4.25	3.22
Basic Amino Acids								
Arginine	Arg	R	174	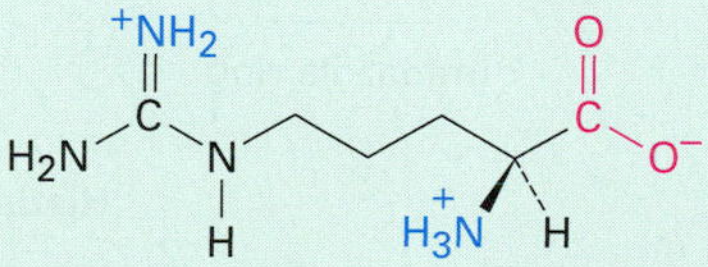	2.17	9.04	12.48	10.76
Histidine	His	H	155	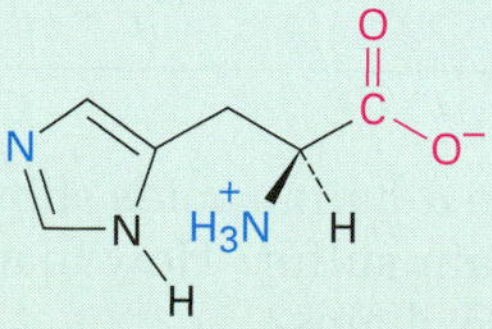	1.82	9.17	6.00	7.59
Lysine	Lys	K	146	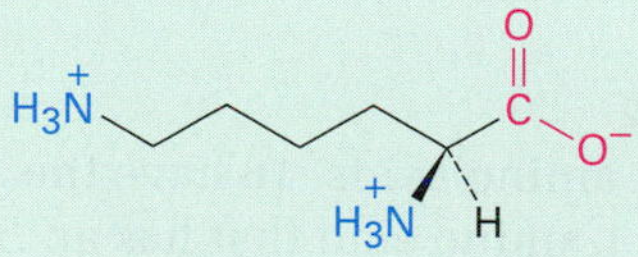	2.18	8.95	10.53	9.74

Except for glycine, $^+H_3NCH_2CO_2^-$, the α carbons of amino acids are chirality centers. Two enantiomers of each are therefore possible, but nature uses only one to build proteins. Because of their stereochemical similarity to L sugars, which we'll look at in Section 21.3, the naturally occurring α-amino acids are often referred to as L amino acids. The nonnaturally occurring enantiomers are called D amino acids.

L-Serine
(S)-Serine

L-Cysteine
(R)-Cysteine

L-Alanine
(S)-Alanine

D-Alanine
(R)-Alanine

The 20 common amino acids can be further classified as neutral, acidic, or basic, depending on the structure of their side chains. Fifteen of the twenty have neutral side chains, two (aspartic acid and glutamic acid) have an extra carboxylic acid function in their side chains, and three (lysine, arginine, and histidine) have basic amino groups in their side chains. Note that both cysteine (a thiol) and tyrosine (a phenol), although usually classified as neutral, nevertheless have weakly acidic side chains that can be deprotonated in a sufficiently basic solution.

At the physiological pH of 7.3 within cells, the side-chain carboxyl groups of aspartic acid and glutamic acid are deprotonated and the basic side-chain nitrogens of lysine and arginine are protonated. Histidine, however, which contains a heterocyclic imidazole ring in its side chain, is not quite basic enough to be protonated at pH 7.3. Note that only the pyridine-like, doubly bonded nitrogen in histidine is basic. The pyrrole-like singly bonded nitrogen is nonbasic because its lone pair of electrons is part of the six-π-electron aromatic imidazole ring (Section 18.8).

Histidine

Problem 19.1
How many of the α-amino acids shown in Table 19.1 contain aromatic rings? How many contain sulfur? How many contain alcohols? How many contain hydrocarbon side chains?

Problem 19.2
Of the 19 L amino acids, 18 have the S configuration at the α carbon. Cysteine is the only L amino acid that has an R configuration. Explain.

Problem 19.3

The amino acid threonine, $(2S,3R)$-2-amino-3-hydroxybutanoic acid, has two chirality centers.

(a) Draw threonine, using normal, wedged, and dashed lines to show dimensionality.

(b) Draw a diastereomer of threonine, and label its chirality centers as R or S.

19.2 Amino Acids and the Henderson–Hasselbalch Equation: Isoelectric Points

According to the Henderson–Hasselbalch equation (Sections 15.3 and 18.5), if we know both the pH of a solution and the pK_a of an acid HA, we can calculate the ratio of $[A^-]$ to $[HA]$ in the solution. Furthermore, when $pH = pK_a$, the two forms A^- and HA are present in equal amounts because $\log 1 = 0$.

$$pH = pK_a = \log\frac{[A^-]}{[HA]} \quad \text{or} \quad \log\frac{[A^-]}{[HA]} = pH - pK_a$$

To apply the Henderson–Hasselbalch equation to an amino acid, let's find out what species are present in a 1.00 M solution of alanine at $pH = 9.00$. According to Table 19.1, protonated alanine $[^+H_3NCH(CH_3)CO_2H]$ has $pK_{a1} = 2.34$ and neutral zwitterionic alanine $[^+H_3NCH(CH_3)CO_2^-]$ has $pK_{a2} = 9.69$:

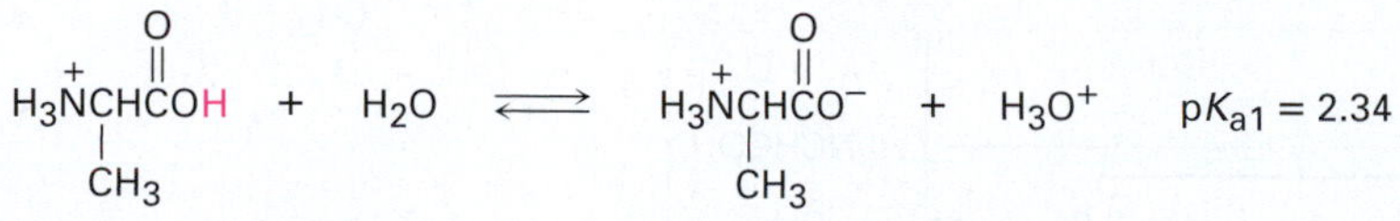

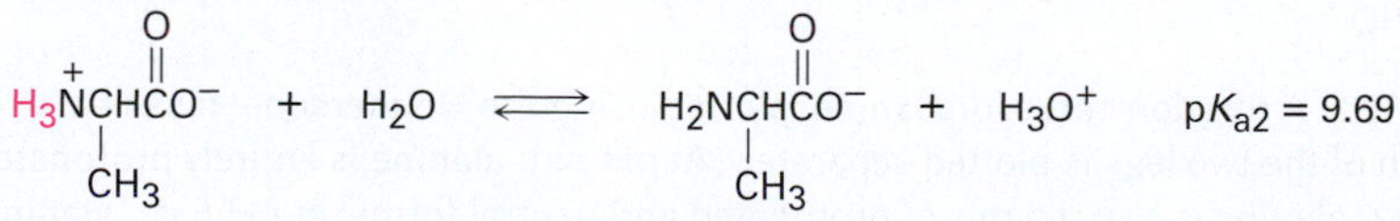

Since the pH of the solution is much closer to pK_{a2} than to pK_{a1}, we need to use pK_{a2} for the calculation. From the Henderson–Hasselbalch equation, we have:

$$\log\frac{[A^-]}{[HA]} = pH - pK_a = 9.00 - 9.69 = -0.69$$

so

$$\frac{[A^-]}{[HA]} = \text{antilog}(-0.69) = 0.20 \quad \text{and} \quad [A^-] = 0.20\,[HA]$$

In addition, we know that

$$[A^-] + [HA] = 1.00 \text{ M}$$

Solving the two simultaneous equations gives $[HA] = 0.83$ and $[A^-] = 0.17$. In other words, at $pH = 9.00$, 83% of alanine molecules in a 1.00 M solution are neutral (zwitterionic) and 17% are deprotonated. Similar calculations can be done at any other pH and the results plotted to give the *titration curve* shown in Figure 19.1.

Each leg of the titration curve is calculated separately. The first leg, from pH 1 to 6, corresponds to the dissociation of protonated alanine, H_2A^+. The second leg, from pH 6 to 11, corresponds to the dissociation of zwitterionic alanine, HA. It's as if we started with H_2A^+ at low pH and then titrated with NaOH. When 0.5 equivalent of NaOH is added, the deprotonation of H_2A^+ is 50% done; when 1.0 equivalent of NaOH is added, the deprotonation of H_2A^+ is complete and HA predominates; when 1.5 equivalent of NaOH is added, the deprotonation of HA is 50% done; and when 2.0 equivalents of NaOH is added, the deprotonation of HA is complete.

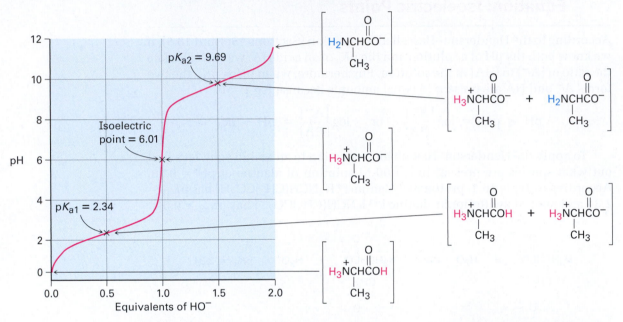

FIGURE 19.1 A titration curve for alanine, plotted using the Henderson–Hasselbalch equation. Each of the two legs is plotted separately. At pH < 1, alanine is entirely protonated; at pH = 2.34, alanine is a 50:50 mix of protonated and neutral forms; at pH 6.01, alanine is entirely neutral; at pH = 9.69, alanine is a 50:50 mix of neutral and deprotonated forms; at pH > 11.5, alanine is entirely deprotonated.

Look carefully at the titration curve in Figure 19.1. In acid solution, an amino acid is protonated and exists primarily as a cation. In basic solution, an amino acid is deprotonated and exists primarily as an anion. In between the two is an intermediate pH at which the amino acid is exactly balanced between anionic and cationic forms and exists primarily as the neutral zwitterion. This pH is called the amino acid's **isoelectric point, pI**.

The isoelectric point of an amino acid depends on its structure, with values for the 20 common amino acids given in Table 19.1. The 15 neutral amino acids have isoelectric points near neutrality, in the pH range 5.0 to 6.5. The two acidic amino acids have isoelectric points at lower pH so that deprotonation of the side-chain $-CO_2H$ does not occur at their pI, and the three basic amino acids have isoelectric points at higher pH so that protonation of the side-chain amino group does not occur at their pI.

More specifically, the pI of any amino acid is the average of the two acid-dissociation constants that involve the neutral zwitterion. For the 13 amino acids with a neutral side chain, pI is the average of pK_{a1} and pK_{a2}. For the four amino acids with either a strongly or weakly acidic side chain, pI is the average of the two *lowest* pK_a values. For the three amino acids with a basic side chain, pI is the average of the two *highest* pK_a values.

pK_a = 3.65   pK_a = 1.88   pK_a = 2.34   pK_a = 10.53   pK_a = 2.18

HOCCH_2CHCOH   CH_3CHCOH $^+H_3NCH_2CH_2CH_2CH_2CHCOH$
         NH_3^+   NH_3^+   NH_3^+

pK_a = 9.60   pK_a = 9.69   pK_a = 8.95

$$pI = \frac{1.88 + 3.65}{2} = 2.77 \qquad pI = \frac{2.34 + 9.69}{2} = 6.01 \qquad pI = \frac{8.95 + 10.53}{2} = 9.74$$

Acidic amino acid   **Neutral amino acid**   **Basic amino acid**
Aspartic acid   **Alanine**   **Lysine**

Just as individual amino acids have isoelectric points, entire proteins have an overall pI because of the acidic or basic amino acids they may contain. The enzyme lysozyme, for instance, has a preponderance of basic amino acids and thus has a high isoelectric point (pI = 11.0). Pepsin, however, has a preponderance of acidic amino acids and a low isoelectric point (p$I \sim 1.0$). Not surprisingly, the solubilities and properties of proteins with different pI's are strongly affected by the pH of the medium. Solubility in water is usually lowest at the isoelectric point, where the protein has no net charge, and is higher both above and below the pI, where the protein is charged.

We can take advantage of the differences in isoelectric points to separate a mixture of proteins into its pure constituents. Using a technique known as *electrophoresis*, a mixture of proteins is placed near the center of a strip of paper or gel. The paper or gel is moistened with an aqueous buffer of a given pH, and electrodes are connected to the ends of the strip. When an electric potential is applied, those proteins with negative charges (those that are deprotonated because the pH of the buffer is above their isoelectric point) migrate slowly toward the positive electrode. At the same time, those amino acids with positive charges (those that are protonated because the pH of the buffer is below their isoelectric point) migrate toward the negative electrode.

Different proteins migrate at different rates, depending on their isoelectric points and on the pH of the aqueous buffer, thereby effecting a separation of the mixture into its pure components. Figure 19.2 illustrates this separation for a mixture containing basic, neutral, and acidic components.

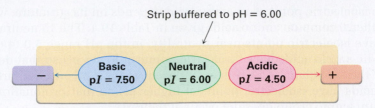

FIGURE 19.2 Separation of a protein mixture by electrophoresis. At pH = 6.00, a neutral protein does not migrate, a basic protein is protonated and migrates toward the negative electrode, and an acidic protein is deprotonated and migrates toward the positive electrode.

Problem 19.4
Hemoglobin has pI = 6.8. Does hemoglobin have a net negative charge or net positive charge at pH = 5.3? At pH = 7.3?

19.3 | Synthesis of Amino Acids

The Amidomalonate Synthesis

α-Amino acids can be synthesized in the laboratory using some of the reactions discussed in previous chapters. For instance, the *amidomalonate synthesis* of amino acids is a straightforward extension of the malonic ester synthesis (Section 17.5). The reaction begins with conversion of diethyl acetamidomalonate into an enolate ion by treatment with base, followed by S_N2 alkylation with a primary alkyl halide. Hydrolysis of both the amide group and the esters occurs when the alkylated product is warmed with aqueous acid, and decarboxylation then takes place to yield an α-amino acid. For example, aspartic acid can be prepared from ethyl bromoacetate, $BrCH_2CO_2Et$:

Diethyl acetamidomalonate
$\xrightarrow[\text{2. } BrCH_2CO_2Et]{\text{1. } Na^+ \ ^-OEt}$
$\xrightarrow[\text{Heat}]{H_3O^+}$
$^-O_2CCH_2CHCO_2^-$
(R,S)-Aspartic acid (55%)

Problem 19.5
What alkyl halides would you use to prepare the following α-amino acids by the amidomalonate method?
(a) Leucine   **(b)** Histidine   **(c)** Tryptophan   **(d)** Methionine

Reductive Amination of α-Keto Acids

Another method for the synthesis of α-amino acids is by reductive amination of an α-keto acid with ammonia and a reducing agent. Alanine, for instance, is prepared by treatment of pyruvic acid with ammonia in the presence of $NaBH_4$.

As described in Section 18.6, the reaction proceeds through formation of an intermediate imine that is then reduced.

Pyruvic acid   **Imine intermediate**   **(R,S)-Alanine**

Enantioselective Synthesis

The synthesis of an α-amino acid from an achiral precursor by either of the methods just described yields a racemate, with equal amounts of S and R enantiomers. To use an amino acid in the laboratory synthesis of a naturally occurring protein, however, the pure S enantiomer must be obtained.

   Two methods are used in practice to obtain enantiomerically pure amino acids. One way is to resolve the racemate into its pure enantiomers (Section 5.8). A more efficient approach, however, is to use an *enantioselective synthesis* to prepare only the desired S enantiomer directly. As discussed in the Chapter 14 *Lagniappe,* the idea behind enantioselective synthesis is to find a chiral reaction catalyst that will temporarily hold a substrate molecule in an unsymmetrical environment. While in that unsymmetrical environment, the substrate may be more open to reaction on one side than on another, leading to an excess of one enantiomeric product over another.

   William Knowles at the Monsanto Company discovered some years ago that α-amino acids can be prepared enantioselectively by hydrogenation of a Z enamido acid with a chiral hydrogenation catalyst. (S)-Phenylalanine, for instance, is prepared in 98.7% purity contaminated by only 1.3% of the (R) enantiomer when a chiral rhodium catalyst is used. For this discovery, Knowles shared the 2001 Nobel Prize in Chemistry.

A (Z) enamido acid   **(S)-Phenylalanine**

The most effective catalysts for enantioselective amino acid synthesis are coordination complexes of rhodium(I) with cycloocta-1,5-diene (COD) and a chiral diphosphine such as (R,R)-1,2-bis(o-anisylphenylphosphino)ethane, the so-called DiPAMP ligand. The complex owes its chirality to the presence of the trisubstituted phosphorus atoms (Section 5.12).

[Rh(R, R-DiPAMP)(COD)]+ BF4−

Problem 19.6
Show how you could prepare the following amino acid enantioselectively:

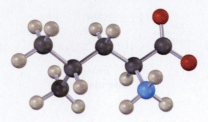

19.4 Peptides and Proteins

Peptides and proteins are amino acid polymers in which the individual amino acids, called **residues**, are joined together by amide bonds, or *peptide bonds*. An amino group from one residue forms an amide bond with the carboxyl of a second residue, the amino group of the second forms an amide bond with the carboxyl of a third, and so on. For example, alanylserine is the dipeptide that results when an amide bond forms between the alanine carboxyl and the serine amino group:

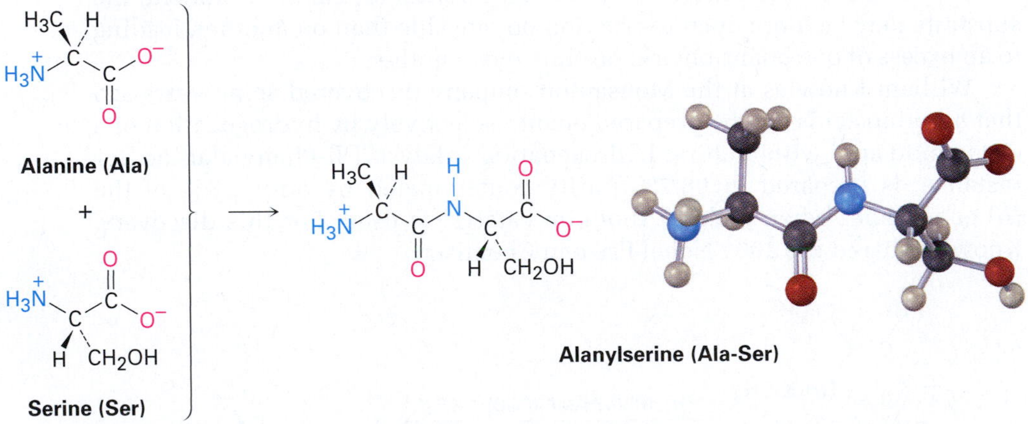

Alanine (Ala)

+

Serine (Ser)

→

Alanylserine (Ala-Ser)

Note that two dipeptides can result from reaction between alanine and serine, depending on which carboxyl group reacts with which amino group. If the alanine amino group reacts with the serine carboxyl, serylalanine results:

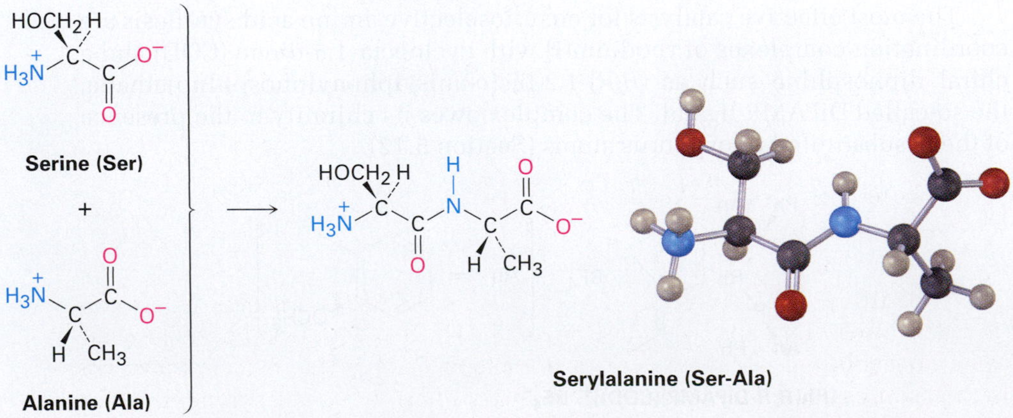

Serine (Ser)

+

Alanine (Ala)

→

Serylalanine (Ser-Ala)

The long, repetitive sequence of –N–CH–CO– atoms that makes up a continuous chain is called the protein's **backbone**. By convention, peptides are written with the **N-terminal amino acid** (the one with the free –NH$_3^+$ group) on the left and the **C-terminal amino acid** (the one with the free –CO$_2^-$ group) on the right. The name of the peptide is indicated by using the abbreviations listed in Table 19.1 for each amino acid. Thus, alanylserine is abbreviated Ala-Ser or A-S, and serylalanine is abbreviated Ser-Ala or S-A. Needless to say, the one-letter abbreviations are more convenient than the older three-letter abbreviations.

The amide bond that links different amino acids together in peptides is no different from any other amide bond (Sections 16.7 and 18.3). An amide nitrogen is nonbasic because its unshared electron pair is delocalized by interaction with the carbonyl group. This overlap of the nitrogen p orbital with the p orbitals of the carbonyl group imparts a certain amount of double-bond character to the C–N bond and restricts rotation around it. The amide bond is therefore planar, and the N–H is oriented 180° to the C=O.

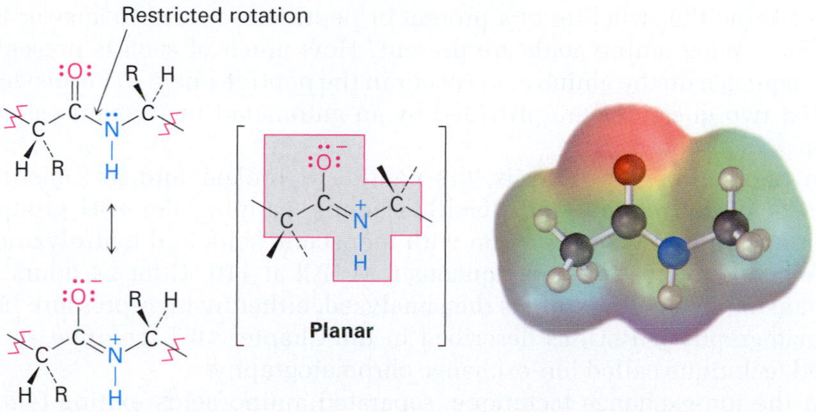

A second kind of covalent bonding in peptides occurs when a disulfide linkage, RS–SR, is formed between two cysteine residues. As we saw in Section 13.7, a disulfide is formed by mild oxidation of a thiol, RSH, and is cleaved by mild reduction.

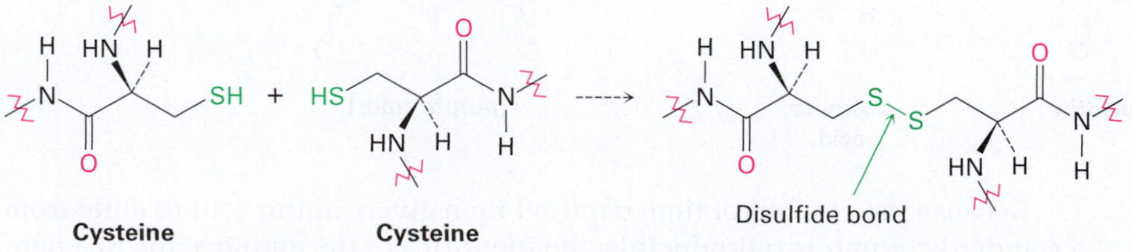

A disulfide bond between cysteine residues in different peptide chains links the otherwise separate chains together, while a disulfide bond between cysteine residues in the same chain forms a loop. Such is the case, for instance, with vasopressin, an antidiuretic hormone found in the pituitary gland. Note that the C-terminal end of vasopressin occurs as the primary amide, $-CONH_2$, rather than as the free acid.

Disulfide bridge

Cys-Tyr-Phe-Glu-Asn-Cys-Pro-Arg-Gly-NH_2

Vasopressin

Problem 19.7
There are six isomeric tripeptides containing valine, tyrosine, and glycine. Name them using both three- and one-letter abbreviations.

Problem 19.8
Draw the structure of M-P-V-G, and indicate the amide bonds.

19.5 Amino Acid Analysis of Peptides

To determine the structure of a protein or peptide, we need to answer three questions: What amino acids are present? How much of each is present? In what sequence do the amino acids occur in the peptide chain? The answers to the first two questions are provided by an automated instrument called an *amino acid analyzer.*

In preparation for analysis, the peptide is broken into its constituent amino acids by reducing all the disulfide bonds, capping the –SH groups of cysteine residues by S_N2 reaction with iodoacetic acid, and hydrolyzing the amide bonds by heating with aqueous 6 M HCl at 110 °C for 24 hours. The resultant amino acid mixture is then analyzed, either by high-pressure liquid chromatography (HPLC) as described in the Chapter 10 *Lagniappe,* or by a related technique called ion-exchange chromatography.

In the ion-exchange technique, separated amino acids exiting *(eluting)* from the end of the chromatography column mix with a solution of a substance called *ninhydrin* and undergo a rapid reaction that produces an intense purple color. The color is detected by a spectrometer, and a plot of elution time versus spectrometer absorbance is obtained.

Ninhydrin   *α*-Amino   **(purple color)**
                         acid

Because the amount of time required for a given amino acid to elute from a standard column is reproducible, the identities of the amino acids in a peptide can be determined. The amount of each amino acid in the sample is determined by measuring the intensity of the purple color resulting from its reaction with ninhydrin. Figure 19.3 shows the results of amino acid analysis of a standard equimolar mixture of 17 *α*-amino acids. Typically, amino acid analysis requires about 100 picomoles (2–3 *μ*g) of sample for a protein containing about 200 residues.

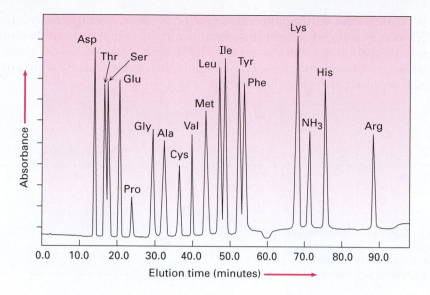

FIGURE 19.3 Amino acid analysis of an equimolar mixture of 17 amino acids.

Problem 19.9

Show the structure of the product you would expect to obtain by S_N2 reaction of a cysteine residue with iodoacetic acid.

Problem 19.10

Show the structures of the products obtained on reaction of valine with ninhydrin.

19.6 Peptide Sequencing: The Edman Degradation

With the identities and relative amounts of amino acids known, the peptide is then *sequenced* to find out in what order the amino acids are linked together. Much peptide sequencing is now done by mass spectrometry, using either electrospray ionization (ESI) or matrix-assisted laser desorption ionization (MALDI) linked to a time-of-flight (TOF) mass analyzer, as described in Section 10.4. Also in common use is a chemical method of peptide sequencing called the *Edman degradation*.

The general idea of peptide sequencing by Edman degradation is to cleave one amino acid at a time from an end of the peptide chain. That terminal amino acid is then separated and identified, and the cleavage reactions are repeated on the chain-shortened peptide until the entire peptide sequence is known. Automated protein sequencers are available that allow as many as 50 repetitive sequencing cycles to be carried out before a buildup of unwanted by-products interferes with the results. So efficient are these instruments that sequence information can be obtained from as little as 1 to 5 picomoles of sample—less than 0.1 μg.

Edman degradation involves treatment of a peptide with phenyl isothiocyanate (PITC), C_6H_5—N=C=S, followed by treatment with trifluoroacetic acid, as shown in Figure 19.4. The first step attaches the PITC to the —NH_2 group of the N-terminal amino acid, and the second step splits the N-terminal residue from the peptide chain, yielding an anilinothiazolinone (ATZ) derivative plus the chain-shortened peptide. Further acid-catalyzed rearrangement of the ATZ derivative with aqueous acid converts it into a phenylthiohydantoin (PTH), which is identified chromatographically by comparison of its elution time with the known elution times of PTH derivatives of the 20 common amino acids. The chain-shortened peptide is then automatically resubmitted to another round of Edman degradation.

FIGURE 19.4

MECHANISM:

Mechanism of the Edman degradation for N-terminal analysis of peptides.

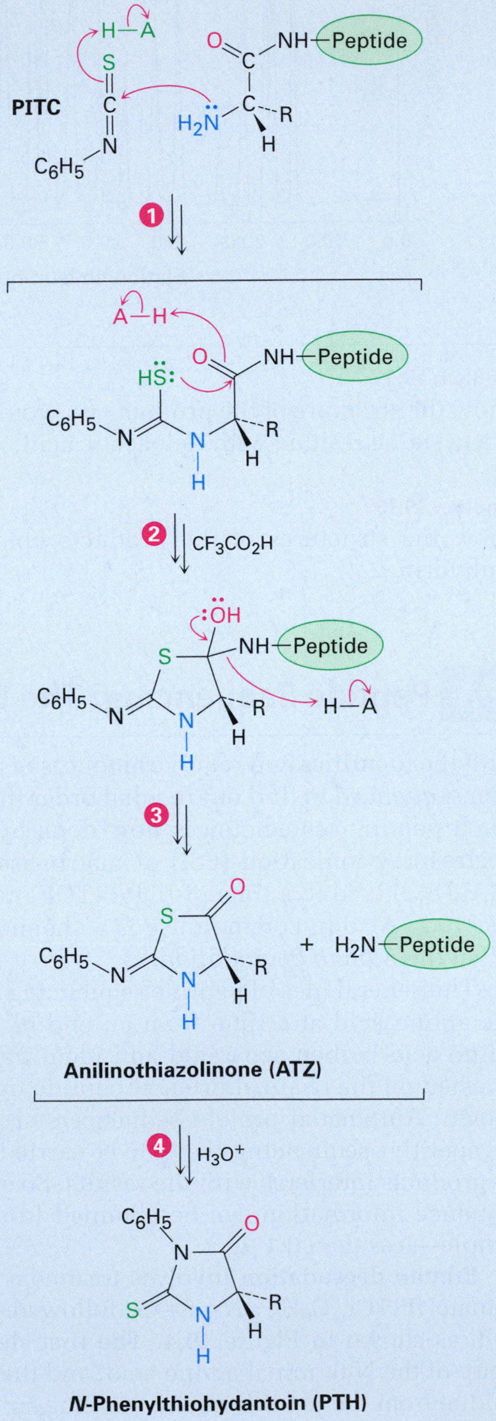

1 Nucleophilic addition of the peptide terminal amino group to phenyl isothiocyanate (PITC) gives an *N*-phenylthiourea derivative.

2 Acid-catalyzed cyclization of the phenylthiourea yields a tetrahedral intermediate . . .

3 . . . which expels the chain-shortened peptide and forms an anilino-thiazolinone (ATZ) derivative.

Anilinothiazolinone (ATZ)

4 The ATZ rearranges in the presence of aqueous acid to an isomeric *N*-phenylthiohydantoin (PTH) as the final product.

N-Phenylthiohydantoin (PTH)

© John McMurry

Complete sequencing of large proteins by Edman degradation is impractical because of the buildup of unwanted by-products. To get around the problem, a large peptide chain is first cleaved by partial hydrolysis into a number of smaller fragments, the sequence of each fragment is determined, and the individual fragments are fitted together by matching the overlapping ends. In this way, protein chains with more than 400 amino acids have been sequenced.

Partial hydrolysis of a peptide can be carried out either chemically, using aqueous acid, or enzymatically. Acidic hydrolysis is unselective and leads to a more or less random mixture of small fragments, but enzymatic hydrolysis is quite specific. The enzyme trypsin, for instance, catalyzes hydrolysis of peptides only at the carboxyl side of the basic amino acids arginine and lysine; chymotrypsin cleaves only at the carboxyl side of the aryl-substituted amino acids phenylalanine, tyrosine, and tryptophan.

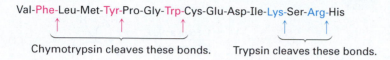

Val-Phe-Leu-Met-Tyr-Pro-Gly-Trp-Cys-Glu-Asp-Ile-Lys-Ser-Arg-His

Chymotrypsin cleaves these bonds.   Trypsin cleaves these bonds.

Problem 19.11

The octapeptide angiotensin II has the sequence Asp-Arg-Val-Tyr-Ile-His-Pro-Phe. What fragments would result if angiotensin II were cleaved with trypsin? With chymotrypsin?

Problem 19.12

What N-terminal residue on a peptide gives the following PTH derivative on Edman degradation?

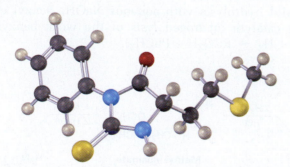

Problem 19.13

Draw the structure of the PTH derivative that would be formed on Edman degradation of angiotensin II (Problem 19.11).

19.7 Peptide Synthesis

Once its structure is known, the synthesis of a peptide can then be undertaken—perhaps to obtain a larger amount for biological evaluation. A simple amide might be formed by treating an amine and a carboxylic acid with dicyclohexylcarbodiimide (DCC; Section 16.3), but peptide synthesis is a more difficult problem because many different amide bonds must be formed in a specific order rather than at random.

The solution to the specificity problem is to protect those functional groups we want to render unreactive while leaving exposed only those functional groups we want to react. If, for example, we wanted to couple alanine with leucine to synthesize Ala-Leu, we could protect the $-NH_2$ group of alanine and the $-CO_2H$ group of leucine to render them unreactive, then form the desired amide bond, and then remove the protecting groups. (You might recall that the general idea of protecting functional groups to render them temporarily unreactive was discussed in Section 13.6 for alcohols and Section 14.8 for aldehydes and ketones.)

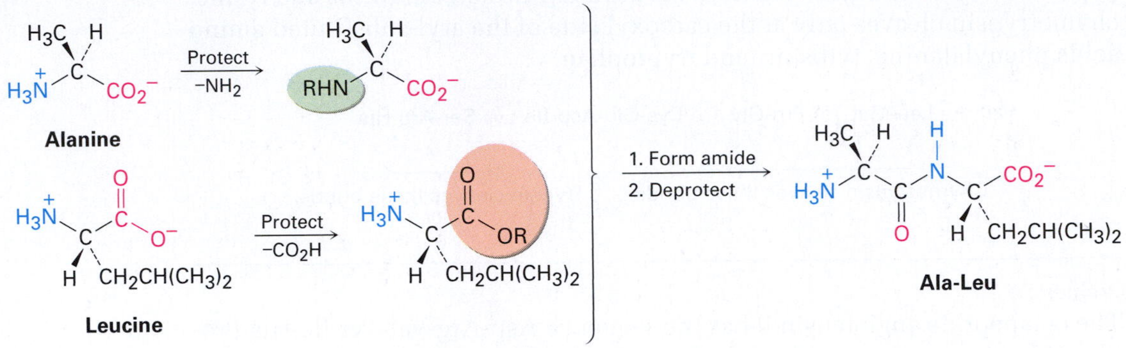

Many different amino- and carboxyl-protecting groups have been devised, but only a few are widely used. Carboxyl groups are often protected simply by converting them into methyl or benzyl esters. Both groups are easily introduced by standard methods of ester formation (Section 16.6) and are easily removed by mild hydrolysis with aqueous NaOH. Benzyl esters can also be cleaved by catalytic *hydrogenolysis* of the weak benzylic C–O bond (RCO_2–CH_2Ph + H_2 → RCO_2H + $PhCH_3$).

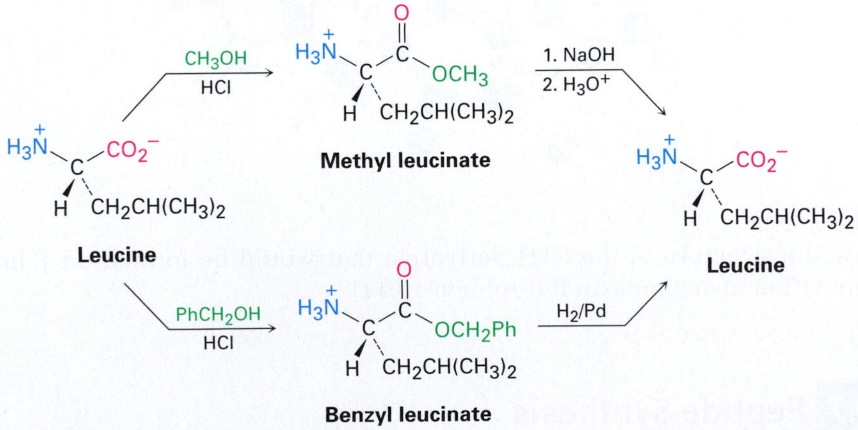

Amino groups are often protected as their *tert*-butyloxycarbonyl amide, or Boc, derivatives. The Boc protecting group is introduced by reaction of the amino acid with di-*tert*-butyl dicarbonate in a nucleophilic acyl substitution reaction and is removed by brief treatment with a strong organic acid such as trifluoroacetic acid, CF_3CO_2H.

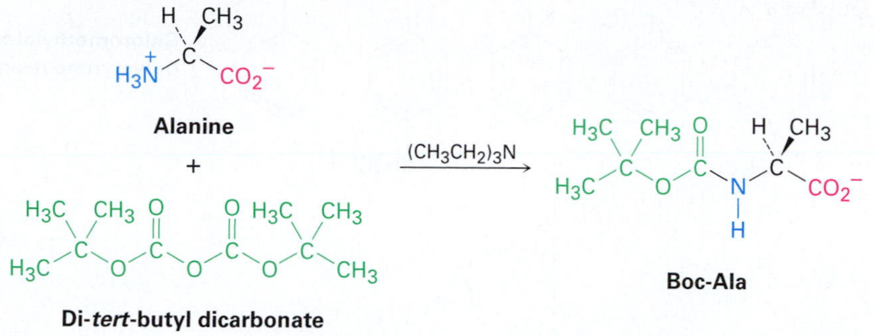

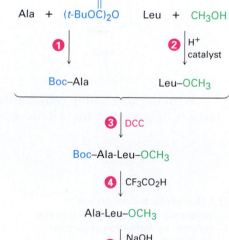

Alanine

+

Di-*tert*-butyl dicarbonate

$(CH_3CH_2)_3N$

Boc-Ala

Thus, five steps are needed to synthesize a dipeptide such as Ala-Leu:

1 The amino group of alanine is protected as the Boc derivative, and

2 the carboxyl group of leucine is protected as the methyl ester.

$$Ala + (t\text{-}BuOC)_2O \qquad Leu + CH_3OH$$

1 ↓   **2** ↓ H^+ catalyst

Boc–Ala   Leu–OCH$_3$

3 The two protected amino acids are coupled using DCC.

3 ↓ DCC

Boc–Ala-Leu–OCH$_3$

4 The Boc protecting group is removed by acid treatment.

4 ↓ CF_3CO_2H

Ala-Leu–OCH$_3$

5 The methyl ester is removed by basic hydrolysis.

5 ↓ NaOH H_2O

Ala-Leu

Although the steps just shown can be repeated to add one amino acid at a time to a growing chain, the synthesis of a large peptide by this sequential addition is long and arduous. An immense simplification is possible, however, using the *Merrifield solid-phase* method. In the Merrifield method, peptide synthesis is carried out with the growing amino acid chain covalently bonded to small beads of a polymer resin rather than in solution. In the original Merrifield procedure, polystyrene resin was used, prepared so that 1 of every 100 or so benzene rings contained a chloromethyl (–CH$_2$Cl) group, and a Boc-protected C-terminal amino acid was then bonded to the resin through an ester bond formed by S$_N$2 reaction.

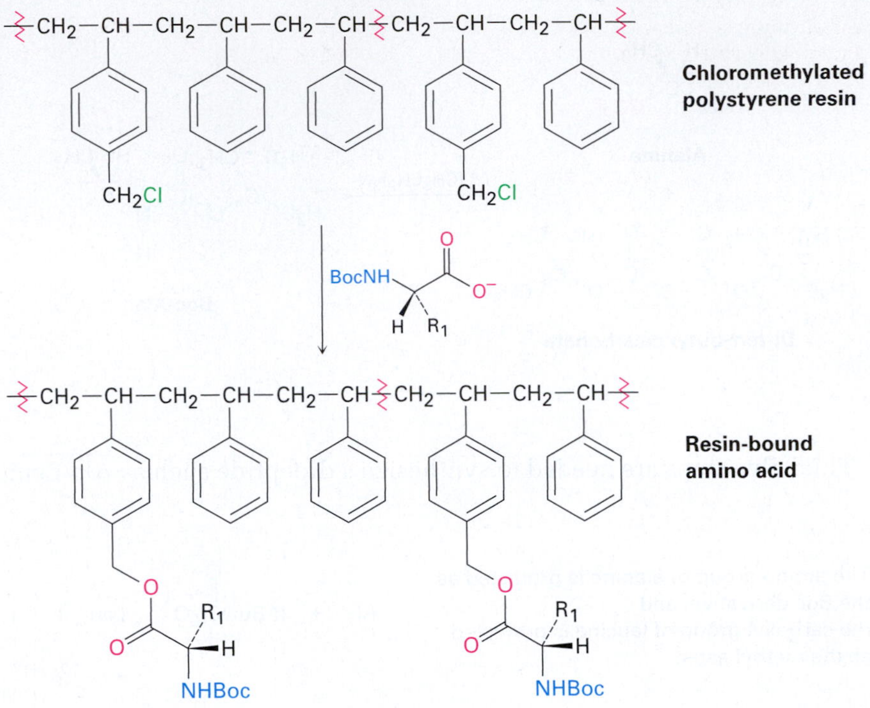

Chloromethylated polystyrene resin

Resin-bound amino acid

With the first amino acid bonded to the resin, a repeating series of four steps is then carried out to build a peptide:

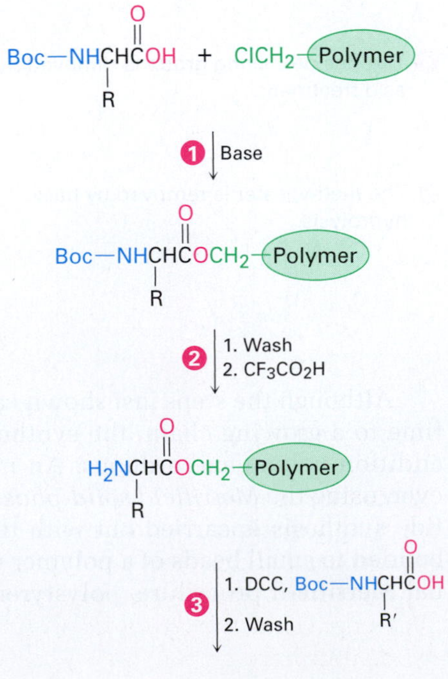

❶ A Boc-protected amino acid is covalently linked to the polystyrene polymer by formation of an ester bond (S$_N$2 reaction).

❷ The polymer-bonded amino acid is washed free of excess reagent and then treated with trifluoroacetic acid to remove the Boc group.

❸ A second Boc-protected amino acid is coupled to the first by reaction with DCC. Excess reagents are removed by washing them from the insoluble polymer.

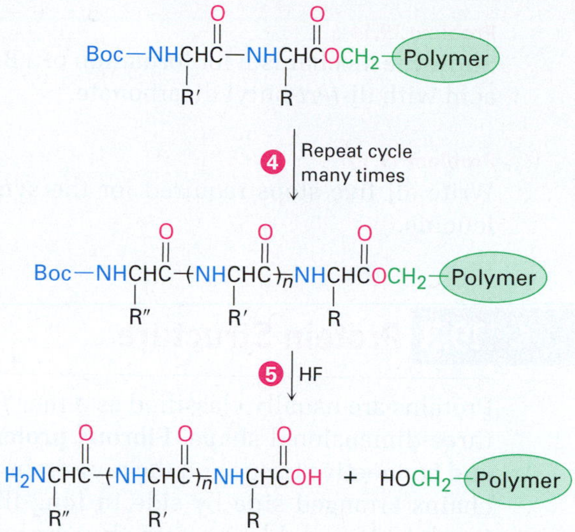

4 The cycle of deprotection, coupling, and washing is repeated as many times as desired to add amino acid units to the growing chain.

5 After the desired peptide has been made, treatment with anhydrous HF removes the final Boc group and cleaves the ester bond to the polymer, yielding the free peptide.

The solid-phase technique has been improved substantially over the years, but the fundamental idea remains the same. The most commonly used resins at present are either the Wang resin or the PAM (phenylacetamido-methyl) resin, and the most commonly used N-protecting group is the fluorenyl-methyloxycarbonyl, or Fmoc group, rather than Boc.

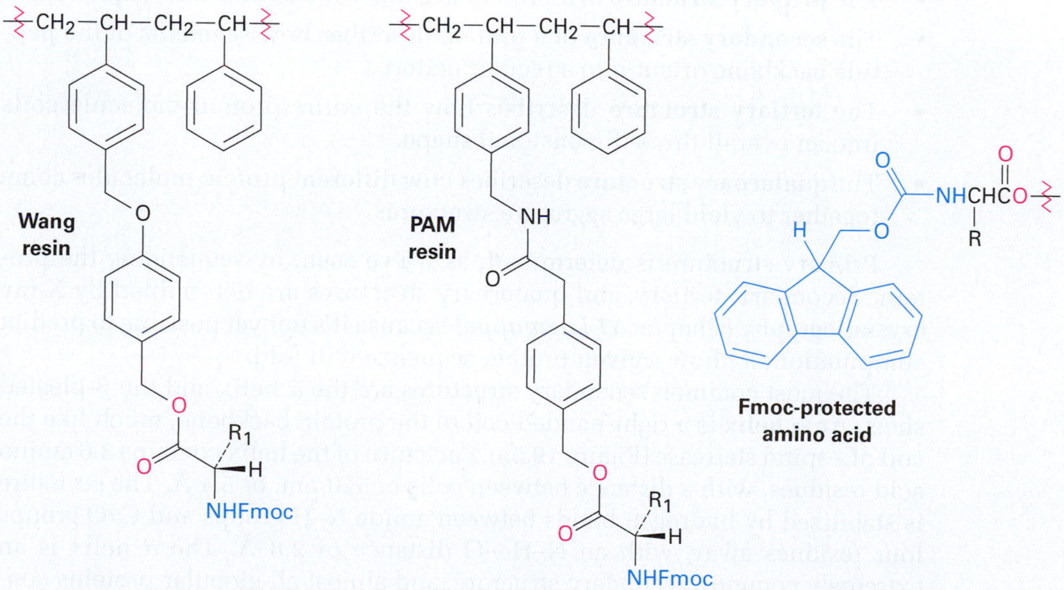

Wang resin

PAM resin

Fmoc-protected amino acid

Robotic peptide synthesizers are now used to automatically repeat the coupling, washing, and deprotection steps with different amino acids. Each step occurs in high yield, and mechanical losses are minimized because the peptide intermediates are never removed from the insoluble polymer until the final step. Using this procedure, up to 30 mg of a peptide with 20 amino acids can be routinely prepared.

Problem 19.14

Show the mechanism for formation of a Boc derivative by reaction of an amino acid with di-*tert*-butyl dicarbonate.

Problem 19.15

Write all five steps required for the synthesis of Leu-Ala from alanine and leucine.

19.8 Protein Structure

Proteins are usually classified as either *fibrous* or *globular,* according to their three-dimensional shape. **Fibrous proteins**, such as the collagen in tendons and connective tissue and the myosin in muscle tissue, consist of polypeptide chains arranged side by side in long filaments. Because these proteins are tough and insoluble in water, they are used in nature for structural materials. **Globular proteins**, by contrast, are usually coiled into compact, roughly spherical shapes. These proteins are generally soluble in water and are mobile within cells. Most of the 3000 or so enzymes that have been characterized to date are globular proteins.

Proteins are so large that the word *structure* takes on a broader meaning than it does with simpler organic compounds. In fact, chemists speak of four different levels of structure when describing proteins:

- The **primary structure** of a protein is simply the amino acid sequence.
- The **secondary structure** of a protein describes how *segments* of the peptide backbone orient into a regular pattern.
- The **tertiary structure** describes how the *entire* protein molecule coils into an overall three-dimensional shape.
- The **quaternary structure** describes how different protein molecules come together to yield large aggregate structures.

Primary structure is determined, as we've seen, by sequencing the protein. Secondary, tertiary, and quaternary structures are determined by X-ray crystallography (Chapter 17 *Lagniappe*) because it's not yet possible to predict computationally how a given protein sequence will fold.

The most common secondary structures are the α helix and the β-pleated sheet. An **α helix** is a right-handed coil of the protein backbone, much like the coil of a spiral staircase (Figure 19.5a). Each turn of the helix contains 3.6 amino acid residues, with a distance between coils of 540 pm, or 5.4 Å. The structure is stabilized by hydrogen bonds between amide N–H groups and C=O groups four residues away, with an N–H····O distance of 2.8 Å. The α helix is an extremely common secondary structure, and almost all globular proteins contain many helical segments. Myoglobin, a small globular protein containing 153 amino acid residues in a single chain, is an example (Figure 19.5b).

A **β-pleated sheet** differs from an α helix in that the peptide chain is fully extended rather than coiled and the hydrogen bonds occur between residues in adjacent chains (Figure 19.6a). The neighboring chains can run either in the same direction (parallel) or in opposite directions (antiparallel), although the antiparallel arrangement is more common and energetically somewhat more favorable. Concanavalin A, for instance, consists of two identical chains of 237 residues with extensive regions of antiparallel β sheets (Figure 19.6b).

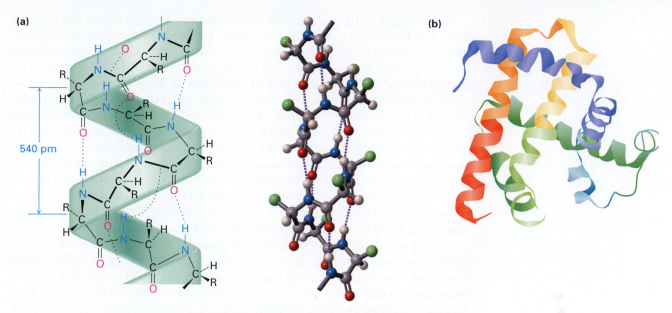

ACTIVE FIGURE 19.5 **(a)** The α-helical secondary structure of proteins is stabilized by hydrogen bonds between the N–H group of one residue and the C=O group four residues away. **(b)** The structure of myoglobin, a globular protein with extensive helical regions that are shown as ribbons in this representation. **Go to this book's student companion site at www.cengage.com/chemistry/mcmurry to explore an interactive version of this figure.**

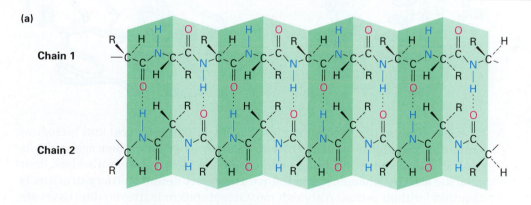

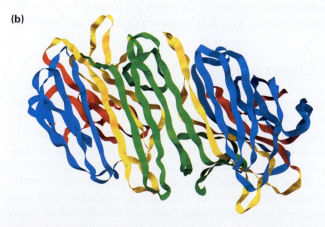

FIGURE 19.6 **(a)** The β-pleated sheet secondary structure of proteins is stabilized by hydrogen bonds between parallel or antiparallel chains. **(b)** The structure of concanavalin A, a protein with extensive regions of antiparallel β sheets, shown as ribbons.

What about tertiary structure? Why does any protein adopt the shape it does? The forces that determine the tertiary structure of a protein are the same forces that act on all molecules, regardless of size, to provide maximum stability. Particularly important are the hydrophilic (water-loving; Section 2.12) interactions of the polar side chains on acidic or basic amino acids. Those acidic or basic amino acids with charged side chains tend to congregate on the exterior of the protein, where they can be solvated by water. Those amino acids with neutral, nonpolar side chains tend to congregate on the hydrocarbon-like interior of a protein molecule, away from the aqueous medium.

Also important for stabilizing a protein's tertiary structure are the formation of disulfide bridges between cysteine residues, the formation of hydrogen bonds between nearby amino acid residues, and the presence of ionic attractions, called *salt bridges*, between positively and negatively charged sites on various amino acid side chains within the protein.

Because the tertiary structure of a globular protein is delicately held together by weak intramolecular attractions, a modest change in temperature or pH is often enough to disrupt that structure and cause the protein to become **denatured**. Denaturation occurs under such mild conditions that the primary structure remains intact but the tertiary structure unfolds from a specific globular shape to a randomly looped chain (Figure 19.7).

FIGURE 19.7 A representation of protein denaturation. A globular protein loses its specific three-dimensional shape and becomes randomly looped.

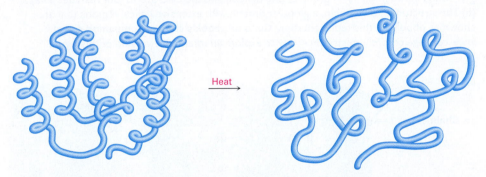

Heat

Denaturation is accompanied by changes in both physical and biological properties. Solubility is drastically decreased, as occurs when egg white is cooked and the albumins unfold and coagulate. Most enzymes also lose their catalytic activity when denatured, since a precisely defined tertiary structure is required for their action. Although most denaturation is irreversible, cases are also known where spontaneous *renaturation* of an unfolded protein to its stable tertiary structure occurs, accompanied by a full recovery of biological activity.

19.9 Enzymes and Coenzymes

An **enzyme**—usually a large protein—is a substance that acts as a catalyst for a biological reaction. Like all catalysts, an enzyme doesn't affect the equilibrium constant of a reaction and can't bring about a chemical change that is otherwise unfavorable. An enzyme acts only to lower the activation energy for a reaction, thereby making the reaction take place more rapidly. Sometimes, in fact, the rate acceleration brought about by enzymes is extraordinary. Millionfold rate increases are common, and the glycosidase enzymes that hydrolyze polysaccharides increase the reaction rate by a factor of more than 10^{17}, changing the time required for the reaction from millions of years to milliseconds.

Unlike many of the catalysts that chemists use in the laboratory, enzymes are usually specific in their action. Often, in fact, an enzyme will catalyze only a single reaction of a single compound, called the enzyme's *substrate.* For example, the enzyme amylase, found in the human digestive tract, catalyzes only the hydrolysis of starch to yield glucose. Cellulose and other polysaccharides are untouched by amylase.

Different enzymes have different specificities. Some, such as amylase, are specific for a single substrate, but others operate on a range of substrates. Papain, for instance, a globular protein of 212 amino acids isolated from papaya fruit, catalyzes the hydrolysis of many kinds of peptide bonds. In fact, it's this ability to hydrolyze peptide bonds that makes papain useful as a meat tenderizer and a cleaner for contact lenses.

$$\text{+NHCHC} \overset{O}{\underset{R}{||}} \text{—NHCHC} \overset{O}{\underset{R'}{||}} \text{—NHCHC} \overset{O}{\underset{R''}{||}} \text{+} \xrightarrow[\text{H}_2\text{O}]{\text{Papain}} \text{+NHCHCOH} \overset{O}{\underset{R}{||}} \text{ + } \text{H}_2\text{NCHC} \overset{O}{\underset{R'}{||}} \text{—NHCHC} \overset{O}{\underset{R''}{||}} \text{+}$$

Enzymes function through a pathway that involves initial formation of an enzyme–substrate complex E · S, followed by a multistep chemical conversion of the enzyme-bound substrate into enzyme-bound product E · P and final release of product from the complex.

$$\text{E + S} \rightleftharpoons \text{E · S} \rightleftharpoons \text{E · P} \rightleftharpoons \text{E + P}$$

The overall rate constant for conversion of the E · S complex to products E + P is called the *turnover number* because it represents the number of substrate molecules a single enzyme molecule turns over into product per unit time. A value of about 10^3 per second is typical.

The rate acceleration achieved by enzymes is due to several factors. Particularly important is the ability of the enzyme to stabilize and thus lower the energy of the transition state(s). That is, it's not the ability of the enzyme to bind the *substrate* that matters but rather its ability to bind and thereby stabilize the *transition state.* Often, in fact, the enzyme binds the transition structure as much as 10^{12} times more tightly than it binds the substrate or products. An energy diagram for an enzyme-catalyzed process might look like that in Figure 19.8.

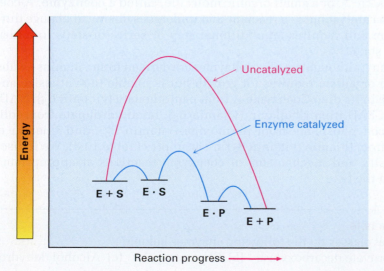

FIGURE 19.8 Energy diagrams for uncatalyzed (red) and enzyme-catalyzed (blue) processes. The enzyme makes available an alternative, lower-energy pathway. Rate enhancement is due to the ability of the enzyme to bind to the transition state for product formation, thereby lowering its energy.

Enzymes are classified into six categories depending on the kind of reaction they catalyze, as shown in Table 19.2. *Oxidoreductases* catalyze oxidations and reductions; *transferases* catalyze the transfer of a group from one substrate to another; *hydrolases* catalyze hydrolysis reactions of esters, amides, and related substrates; *lyases* catalyze the elimination or addition of a small molecule such as H_2O from or to a substrate; *isomerases* catalyze isomerizations; and *ligases* catalyze the bonding together of two molecules, often coupled with the hydrolysis of ATP. The systematic name of an enzyme has two parts, ending with *-ase.* The first part identifies the enzyme's substrate, and the second part identifies its class. For example, hexose kinase is a transferase that catalyzes the transfer of a phosphate group from ATP to a hexose sugar.

TABLE 19.2
Classification of Enzymes

Class	Some subclasses	Function
Oxidoreductases	Dehydrogenases	Introduction of double bond
	Oxidases	Oxidation
	Reductases	Reduction
Transferases	Kinases	Transfer of phosphate group
	Transaminases	Transfer of amino group
Hydrolases	Lipases	Hydrolysis of ester
	Nucleases	Hydrolysis of phosphate
	Proteases	Hydrolysis of amide
Lyases	Decarboxylases	Loss of CO_2
	Dehydrases	Loss of H_2O
Isomerases	Epimerases	Isomerization of chirality center
Ligases	Carboxylases	Addition of CO_2
	Synthetases	Formation of new bond

In addition to their protein part, most enzymes also contain a small nonprotein part called a *cofactor.* A **cofactor** can be either an inorganic ion, such as Zn^{2+}, or a small organic molecule, called a **coenzyme**. A coenzyme is not a catalyst but is a reactant that undergoes chemical change during the reaction and requires an additional step or series of steps to return to its initial state.

Many, although not all, coenzymes are derived from vitamins—substances that an organism requires for growth but is unable to synthesize and must receive in its diet. Coenzyme A from pantothenate (vitamin B_3), NAD^+ from niacin, FAD from riboflavin (vitamin B_2), tetrahydrofolate from folic acid, pyridoxal phosphate from pyridoxine (vitamin B_6), and thiamin diphosphate from thiamin (vitamin B_1) are examples (Table 19.3). We'll discuss the chemistry and mechanisms of coenzyme reactions at appropriate points later in the text.

Problem 19.16
To what classes do the following enzymes belong?
(a) Pyruvate decarboxylase   **(b)** Chymotrypsin   **(c)** Alcohol dehydrogenase

TABLE 19.3
Structures of Some Common Coenzymes

Adenosine triphosphate—ATP (phosphorylation)

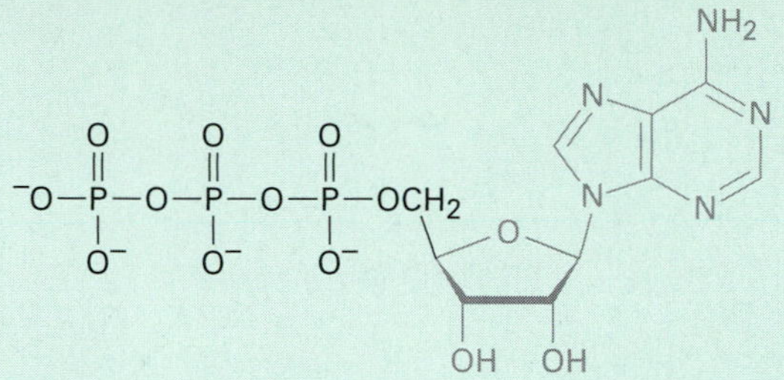

Coenzyme A (acyl transfer)

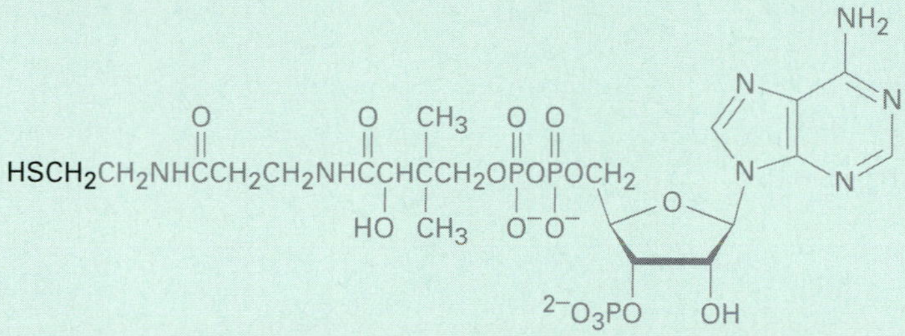

Nicotinamide adenine dinucleotide—NAD⁺ (oxidation/reduction)
(NADP⁺)

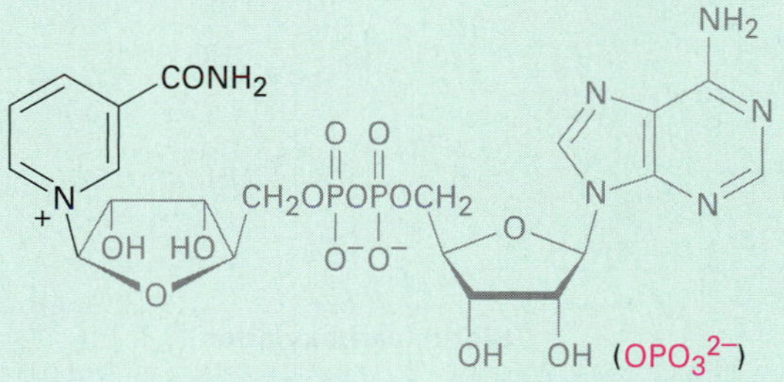

Flavin adenine dinucleotide—FAD (oxidation/reduction)

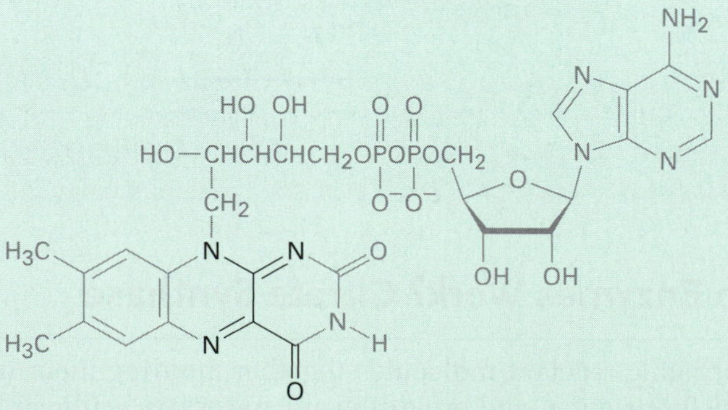

continued

TABLE 19.3
Structures of Some Common Coenzymes *continued*

Tetrahydrofolate (transfer of C_1 units)

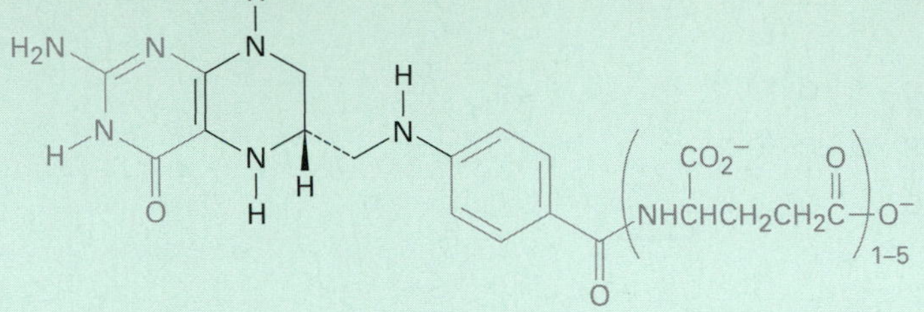

S-Adenosylmethionine (methyl transfer)

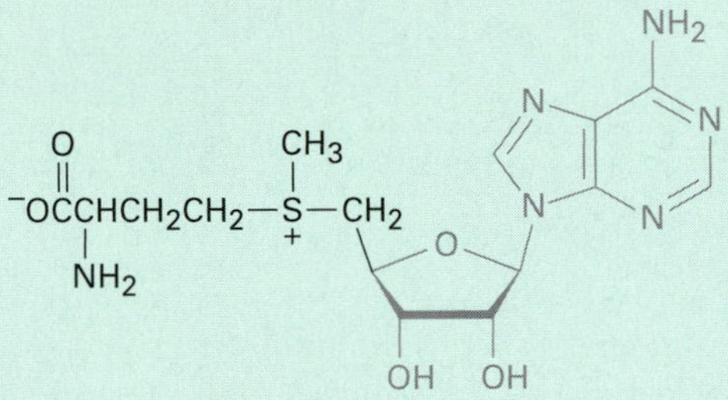

Lipoic acid (acyl transfer)

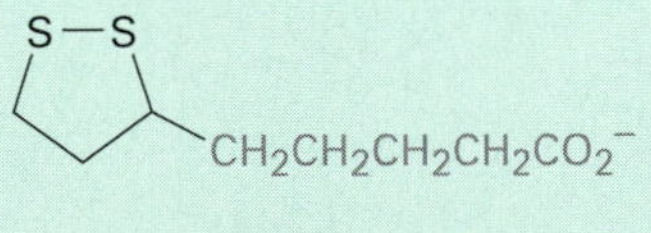

Pyridoxal phosphate (amino acid metabolism)

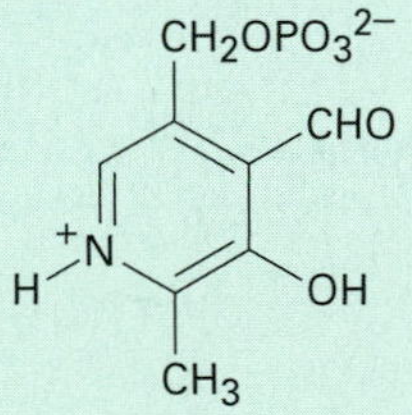

Thiamin diphosphate (decarboxylation)

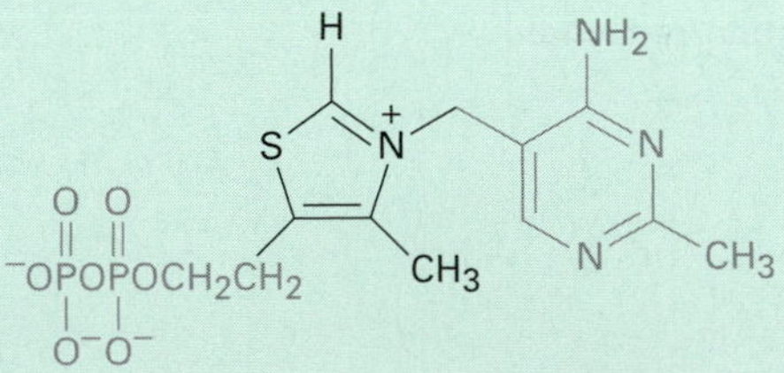

Biotin (carboxylation)

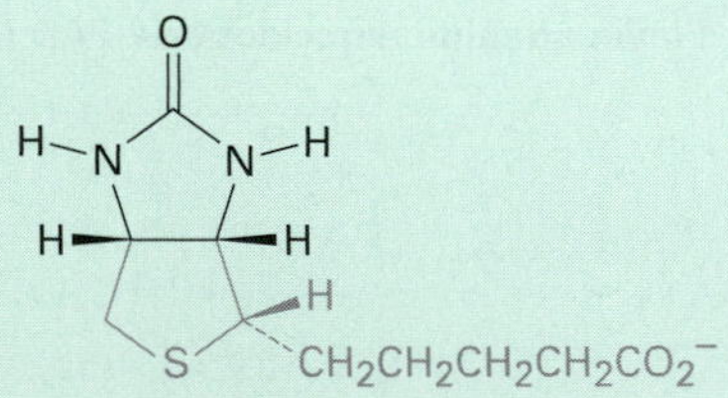

19.10 How Do Enzymes Work? Citrate Synthase

Enzymes work by bringing reactant molecules together, holding them in the orientation necessary for reaction, and providing any necessary acidic or basic sites to catalyze specific steps. As an example, let's look at citrate synthase, an enzyme that catalyzes the aldol-like addition of acetyl CoA to oxaloacetate to give citrate. The reaction is the first step in the citric acid cycle, in which acetyl groups produced by degradation of food molecules are metabolized to yield CO_2 and H_2O. We'll look at the details of the citric acid cycle in Section 22.4.

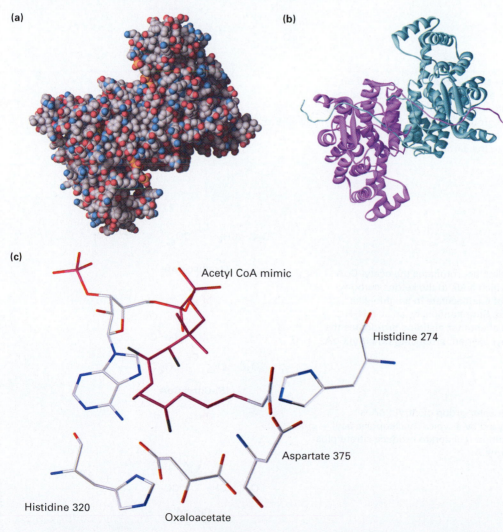

Oxaloacetate   **Acetyl CoA**   **Citrate**

Citrate synthase is a globular protein of 433 amino acids with a deep cleft lined by an array of functional groups that can bind to the substrate oxaloacetate. On binding oxaloacetate, the original cleft closes and another opens up nearby to bind acetyl CoA. This second cleft is also lined by appropriate functional groups, including a histidine at position 274 and an aspartic acid at position 375. The two reactants are now held by the enzyme in close proximity and with a suitable orientation for reaction. Figure 19.9 shows the structure of citrate synthase as determined by X-ray crystallography, along with a close-up of the active site.

(a)

(b)

(c)

Acetyl CoA mimic

Histidine 274

Aspartate 375

Histidine 320

Oxaloacetate

FIGURE 19.9 X-ray crystal structure of citrate synthase. Part **(a)** is a space-filling model and part **(b)** is a ribbon model, which emphasizes the α-helical segments of the protein chain and indicates that the enzyme is dimeric; that is, it consists of two identical chains held together by hydrogen bonds and other intermolecular attractions. Part **(c)** is a close-up of the active site, in which oxaloacetate and an unreactive acetyl CoA mimic are bound.

As shown in Figure 19.10, the first step in the aldol reaction is generation of the enol of acetyl CoA. The side-chain carboxyl of an aspartate residue acts as base to abstract an acidic α proton, while at the same time the side-chain imidazole ring of a histidine donates H⁺ to the carbonyl oxygen. The enol thus produced then does a nucleophilic addition to the ketone carbonyl group of oxaloacetate. The first histidine acts as a base to remove the –OH hydrogen from the enol, while a second histidine residue simultaneously donates a proton to the oxaloacetate carbonyl group, giving citryl CoA. Water then hydrolyzes the thiol ester group in citryl CoA in a nucleophilic acyl substitution reaction, releasing citrate and coenzyme A as the final products. We'll look in similar detail at other enzyme mechanisms as the need arises.

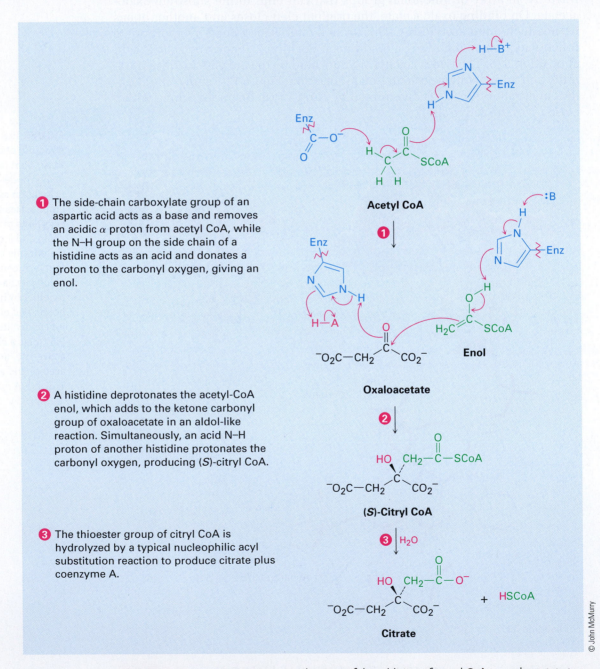

1. The side-chain carboxylate group of an aspartic acid acts as a base and removes an acidic α proton from acetyl CoA, while the N–H group on the side chain of a histidine acts as an acid and donates a proton to the carbonyl oxygen, giving an enol.

2. A histidine deprotonates the acetyl-CoA enol, which adds to the ketone carbonyl group of oxaloacetate in an aldol-like reaction. Simultaneously, an acid N–H proton of another histidine protonates the carbonyl oxygen, producing (S)-citryl CoA.

3. The thioester group of citryl CoA is hydrolyzed by a typical nucleophilic acyl substitution reaction to produce citrate plus coenzyme A.

FIGURE 19.10 **MECHANISM:** Mechanism of the addition of acetyl CoA to oxaloacetate to give (S)-citryl CoA, catalyzed by citrate synthase.

Summary

We've now reached the heart of this book. Beginning in this chapter with a look at amino acids and proteins, and continuing for the remainder of the text, we'll examine each of the main classes of biomolecules to see what their primary biological functions are, how they're biosynthesized, and how they're metabolized in the body.

Proteins are large biomolecules made up of **α-amino acid residues** linked together by amide, or *peptide,* bonds. Chains with fewer than 50 amino acids are often called **peptides**, while the term **protein** is reserved for larger chains. Twenty amino acids are commonly found in proteins; all are α-amino acids, and all except glycine have stereochemistry similar to that of L sugars. In neutral solution, amino acids exist as dipolar **zwitterions**.

Amino acids can be synthesized in racemic form by several methods, including alkylation of diethyl acetamidomalonate and reductive amination of an α-keto acid. Alternatively, an enantioselective synthesis of amino acids can be carried out using a chiral hydrogenation catalyst.

To determine the structure of a peptide or protein, the identity and amount of each amino acid present is first found by amino acid analysis. The peptide is hydrolyzed to its constituent α-amino acids, which are separated and identified. Next, the peptide is sequenced. **Edman degradation** by treatment with phenyl isothiocyanate (PITC) cleaves one residue from the N terminus of the peptide and forms an easily identifiable phenylthiohydantoin (PTH) derivative of the **N-terminal amino acid**. A series of sequential Edman degradations allows the sequencing of a peptide chain up to 50 residues in length.

Peptide synthesis requires the use of selective protecting groups. An N-protected amino acid with a free carboxyl group is coupled to an O-protected amino acid with a free amino group in the presence of dicyclohexylcarbodiimide (DCC). Amide formation occurs, the protecting groups are removed, and the sequence is repeated. Amines are often protected as their *tert*-butyloxycarbonyl (Boc) derivatives, and acids are protected as esters. This synthetic sequence is often carried out by the Merrifield solid-phase method, in which the peptide is esterified to an insoluble polymeric support.

Proteins have four levels of structure. **Primary structure** describes a protein's amino acid sequence, **secondary structure** describes how segments of the protein chain orient into regular patterns—either **α-helix** or **β-pleated sheet**, **tertiary structure** describes how the entire protein molecule coils into an overall three-dimensional shape, and **quaternary structure** describes how individual protein molecules aggregate into larger structures.

Proteins are classified as either globular or fibrous. **Fibrous proteins** such as α-keratin are tough, rigid, and water-insoluble; **globular proteins** such as myoglobin are water-soluble and roughly spherical in shape. Many globular proteins are **enzymes**—substances that act as catalysts for biological reactions. Enzymes are grouped into six classes according to the kind of reaction they catalyze. They function by bringing reactant molecules together, holding them in the orientation necessary for reaction, and providing any necessary acidic or basic sites to catalyze specific steps.

Key Words

α-amino acid, 793
α helix, 812
backbone, 803
β-pleated sheet, 812
C-terminal amino acid, 803
coenzyme, 816
cofactor, 816
denatured, 814
Edman degradation, 805
enzyme, 814
fibrous protein, 812
globular protein, 812
isoelectric point (p*I*), 798
N-terminal amino acid, 803
peptide, 792
primary structure, 812
protein, 792
quaternary structure, 812
residue, 802
secondary structure, 812
side chain, 793
tertiary structure, 812
zwitterion, 792

Summary of Reactions

1.   Amino acid synthesis (Section 19.3)

   (a) Diethyl acetamidomalonate synthesis

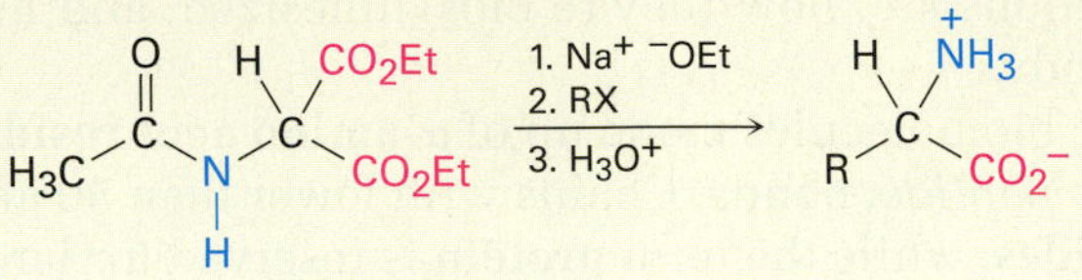

   (b) Reductive amination of an α-keto acid

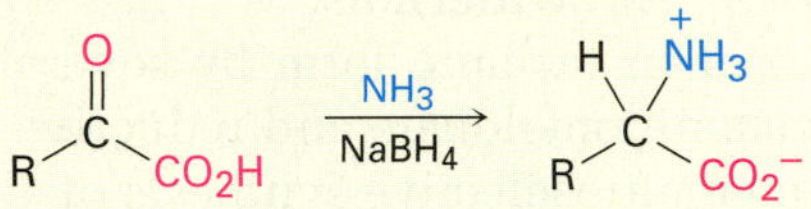

   (c) Enantioselective synthesis

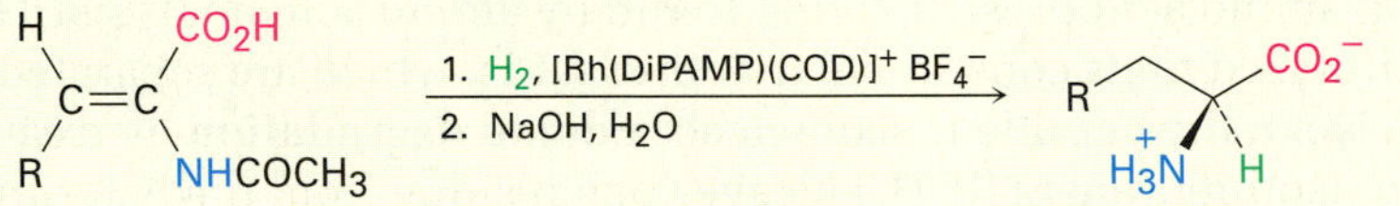

   A (Z) enamido acid   **An (S)-amino acid**

2.   Peptide sequencing by Edman degradation (Section 19.6)

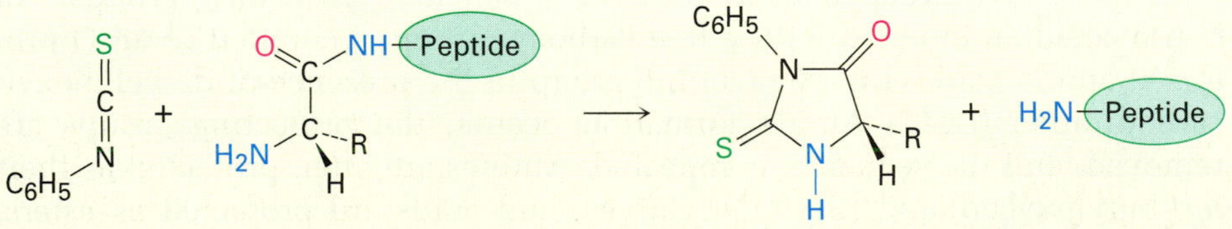

3.   Peptide synthesis (Section 19.7)

   (a) Amine protection

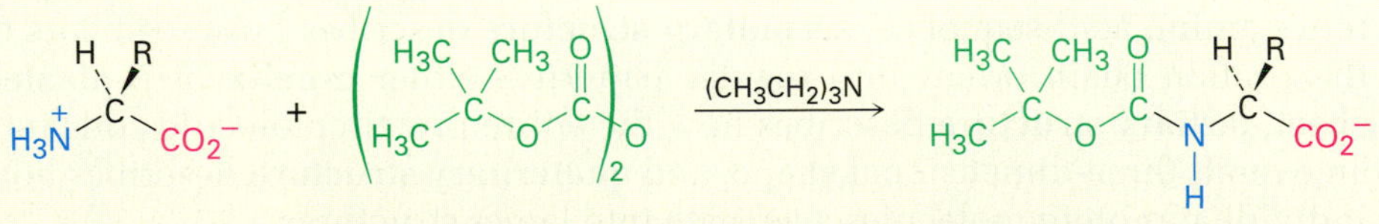

   Boc-protected amino acid

   (b) Carboxyl protection

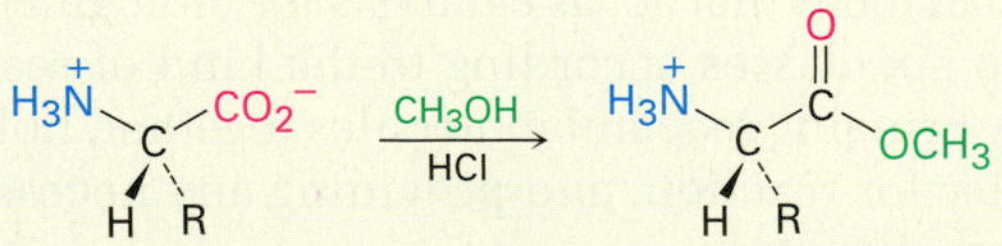

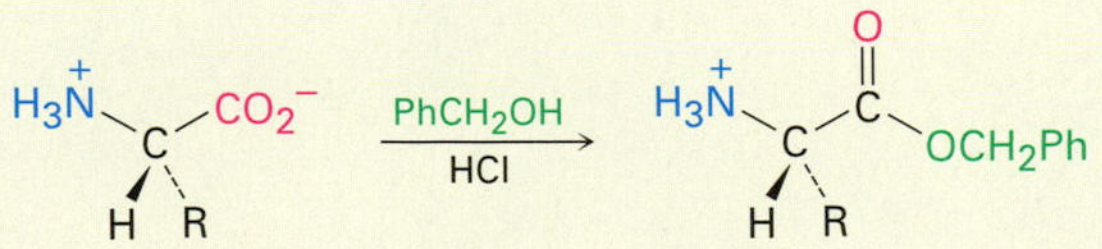

Lagniappe

The Protein Data Bank

Enzymes are so large, so structurally complex, and so numerous that the use of computer databases and molecular visualization programs has become an essential tool for studying biological chemistry. Of the various databases available online, the Kyoto Encyclopedia of Genes and Genomes (KEGG) database (http://www.genome.ad.jp/kegg), maintained by the Kanehisa Laboratory of Kyoto University Bioinformatics Center, is useful for obtaining information on biosynthetic pathways of the sort we'll be describing in the next few chapters. For obtaining information on a specific enzyme, the BRENDA database (http://www.brenda.uni-koeln.de), maintained by the Institute of Biochemistry at the University of Cologne, Germany, is particularly valuable.

Perhaps the most useful of all biological databases is the Protein Data Bank (PDB) operated by the Research Collaboratory for Structural Bioinformatics (RCSB). The PDB is a worldwide repository of X-ray and NMR structural data for biological macromolecules. In early 2009, data for more than 58,000 structures were available, and more than 6000 new ones were being added yearly. To access the Protein Data Bank, go to http://www.rcsb.org/pdb/ and a home page like that shown in Figure 19.11 will appear. As with much that is available online, however, the PDB site is changing rapidly, so you may not see quite the same thing.

To learn how to use the PDB, begin by running the short tutorial listed under Getting Started at the top of the blue sidebar on the left of the screen. After that introduction, start exploring. Let's say you want to view citrate synthase, the enzyme shown previously in Figure 19.9 that catalyzes the addition of acetyl CoA to oxaloacetate to give citrate. Type "citrate synthase" into the small search window on the top line, click on "Site Search," and a list of 30 or so structures will appear. Scroll down near the end of the list until you find the entry with a PDB code of 5CTS and the title "Proposed Mechanism for the Condensation Reaction of Citrate Synthase: 1.9-Angstroms Structure of the Ternary Complex with Oxaloacetate and Carboxymethyl Coenzyme A." Alternatively, if you know the code of the enzyme you want, you can enter it directly into the search window. Click on the PDB code of entry 5CTS, and a new page containing information about the enzyme will open.

If you choose, you can download the structure file to your computer and open it with any of numerous molecular graphics programs to see an image like that in Figure 19.12. The biologically active molecule is a dimer of two identical subunits consisting primarily of α-helical regions displayed as coiled ribbons. For now, just click on "Display Molecule," followed by "Image Gallery," to see some of the tools for visualizing and further exploring the enzyme. We'll look at some of these visualization tools in the Chapter 20 *Lagniappe*.

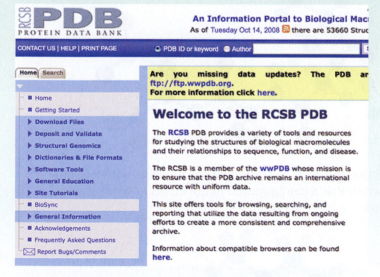

FIGURE 19.11 The Protein Data Bank home page.

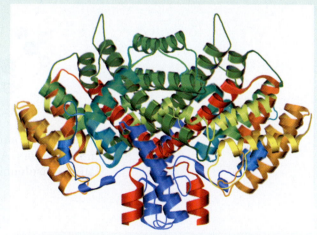

FIGURE 19.12 An image of citrate synthase, downloaded from the Protein Data Bank.

Exercises

■ indicates problems that are assignable in Organic OWL.

Go to this book's companion website at **www.cengage.com/ chemistry/mcmurry** to explore interactive versions of the Active Figures from this text.

VISUALIZING CHEMISTRY

(Problems 19.1–19.16 appear within the chapter.)

19.17 ■ Identify the following amino acids:

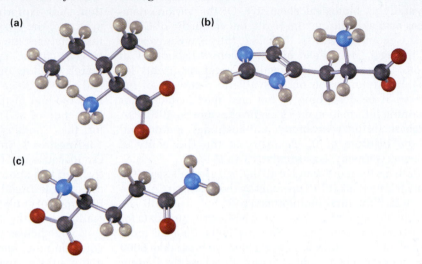

(a)   (b)

(c)

19.18 ■ Give the sequence of the following tetrapeptide (yellow = S):

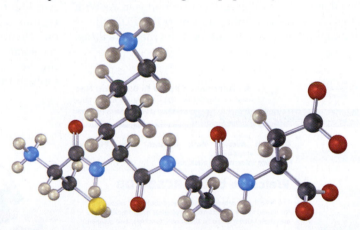

19.19 ■ Isoleucine and threonine are the only two amino acids with two chirality centers. Assign *R* or *S* configuration to the methyl-bearing carbon atom of isoleucine:

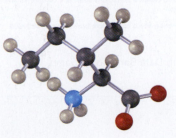

19.20 ▪ Is the following structure a D amino acid or an L amino acid? Identify it.

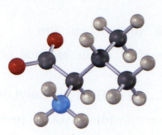

19.21 ▪ Give the sequence of the following tetrapeptide:

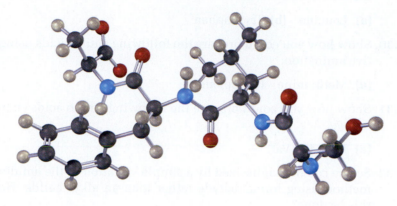

ADDITIONAL PROBLEMS

19.22 Except for cysteine, only *S* amino acids occur in proteins. Several *R* amino acids are also found in nature, however. (*R*)-Serine is found in earthworms, and (*R*)-alanine is found in insect larvae. Draw the structures of (*R*)-serine and (*R*)-alanine. Are these D or L amino acids?

19.23 Cysteine is the only amino acid that has L stereochemistry but an *R* configuration. Make up a structure for another L amino acid of your own creation that also has an *R* configuration.

19.24 Draw the structure of (*S*)-proline.

19.25 ▪ Show the structures of the following amino acids in their zwitterionic forms:

(a) Trp **(b)** Ile **(c)** Cys **(d)** His

19.26 ▪ Proline has $pK_{a1} = 1.99$ and $pK_{a2} = 10.60$. Use the Henderson–Hasselbalch equation to calculate the ratio of protonated and neutral forms at pH = 2.50. Calculate the ratio of neutral and deprotonated forms at pH = 9.70.

▪ Problems assignable in Organic OWL.

19.27 ■ Using both three- and one-letter codes for amino acids, write the structures of all possible peptides containing the following amino acids:

(a) Val, Ser, Leu   (b) Ser, Leu$_2$, Pro

19.28 ■ Predict the product of the reaction of valine with the following reagents:

(a) CH_3CH_2OH, acid   (b) Di-*tert*-butyl dicarbonate

(c) KOH, H_2O (d) CH_3COCl, pyridine; then H_2O

19.29 ■ Show how you could use the acetamidomalonate method to prepare the following amino acids:

(a) Leucine   (b) Tryptophan

19.30 Show how you could prepare the following amino acids using a reductive amination:

(a) Methionine   (b) Isoleucine

19.31 Show how you could prepare the following amino acids enantioselectively:

(a) Pro   (b) Val

19.32 Serine can be synthesized by a simple variation of the amidomalonate method using formaldehyde rather than an alkyl halide. How might this be done?

19.33 ■ Write full structures for the following peptides:

(a) C-H-E-M   (b) E-A-S-Y   (c) P-E-P-T-I-D-E

19.34 ■ Propose two structures for a tripeptide that gives Leu, Ala, and Phe on hydrolysis but does not react with phenyl isothiocyanate.

19.35 Show the steps involved in a synthesis of Phe-Ala-Val using the Merrifield procedure.

19.36 ■ Draw the structure of the PTH derivative product you would obtain by Edman degradation of the following peptides:

(a) I-L-P-F   (b) D-T-S-G-A

19.37 Look at the side chains of the 20 amino acids in Table 19.1, and then think about what is *not* present. None of the 20 contain either an aldehyde or a ketone carbonyl group, for instance. Is this just one of nature's oversights, or is there a likely chemical reason? What complications might an aldehyde or ketone carbonyl group cause?

19.38 The α-helical parts of myoglobin and other proteins stop whenever a proline residue is encountered in the chain. Why do you think a proline is never present in a protein α helix?

19.39 ▪ Which amide bonds in the following polypeptide are cleaved by trypsin? By chymotrypsin?

Phe-Leu-Met-Lys-Tyr-Asp-Gly-Gly-Arg-Val-Ile-Pro-Tyr

19.40 What kinds of reactions do the following classes of enzymes catalyze?

(a) Hydrolases **(b)** Lyases **(c)** Transferases

19.41 ▪ Which of the following amino acids are more likely to be found on the outside of a globular protein, and which on the inside? Explain.

(a) Valine **(b)** Aspartic acid **(c)** Phenylalanine **(d)** Lysine

19.42 The chloromethylated polystyrene resin used for Merrifield solid-phase peptide synthesis is prepared by treatment of polystyrene with chloromethyl methyl ether and a Lewis acid catalyst. Propose a mechanism for the reaction.

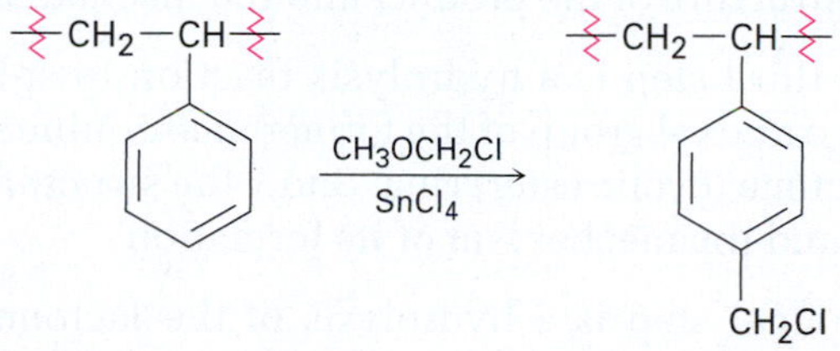

19.43 ▪ An Fmoc amine protecting group is removed by treatment with aqueous base. The mechanism involves initial removal of the relatively acidic hydrogen on the five-membered ring, followed by elimination of the adjacent leaving group and loss of CO_2. Write the mechanism, and explain why the Fmoc group is acidic. (See Section 9.4.)

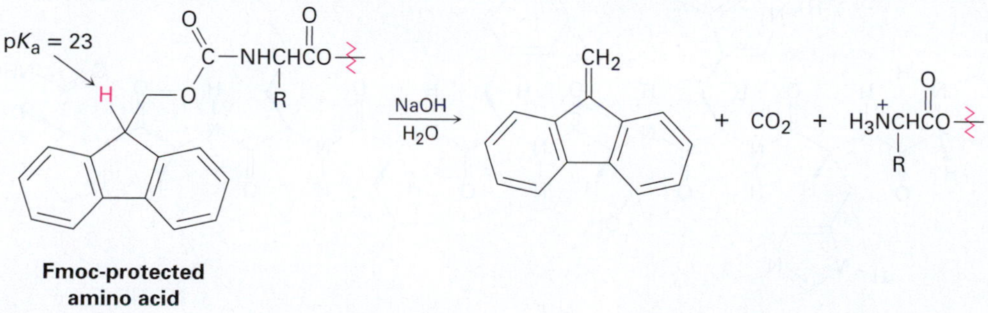

Fmoc-protected amino acid

19.44 ■ Proteins can be cleaved specifically at the amide bond on the carboxyl side of methionine residues by reaction with cyanogen bromide, $BrC\equiv N$:

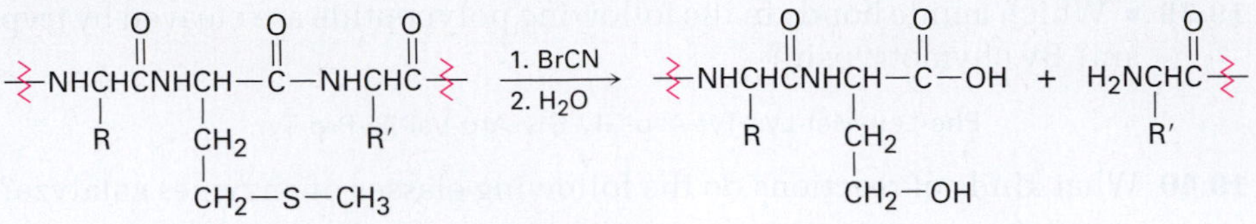

The reaction occurs in several steps:

(a) The first step is a nucleophilic substitution reaction of the sulfur on the methionine side chain with BrCN to give a cyanosulfonium ion, R_2SCN^+. Show the structure of the product, and propose a mechanism for the reaction.

(b) The second step is an internal S_N2 reaction, with the carbonyl oxygen of the methionine residue displacing the positively charged sulfur leaving group and forming a five-membered ring product. Show the structure of the product and the mechanism of its formation.

(c) The third step is a hydrolysis reaction to split the peptide chain. The carboxyl group of the former methionine residue is now part of a lactone (cyclic ester) ring. Show the structure of the lactone product and the mechanism of its formation.

(d) The final step is a hydrolysis of the lactone to give the product shown. Write the mechanism of the reaction.

19.45 ■ Leuprolide is a synthetic nonapeptide used to treat both endometriosis in women and prostate cancer in men.

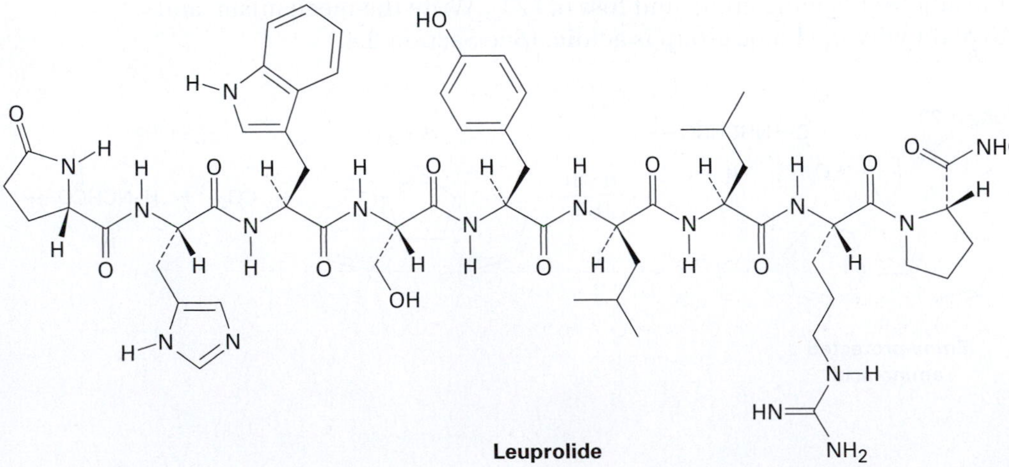

Leuprolide

(a) Both C-terminal and N-terminal amino acids in leuprolide have been structurally modified. Identify the modifications.

(b) One of the nine amino acids in leuprolide has D stereochemistry rather than the usual L. Which one?

■ Problems assignable in Organic OWL.

(c) Write the structure of leuprolide using both one- and three-letter abbreviations.

(d) What charge would you expect leuprolide to have at neutral pH?

19.46 ■ A clever new method of peptide synthesis involves formation of an amide bond by reaction of an α-keto acid with an N-alkylhydroxyl-amine:

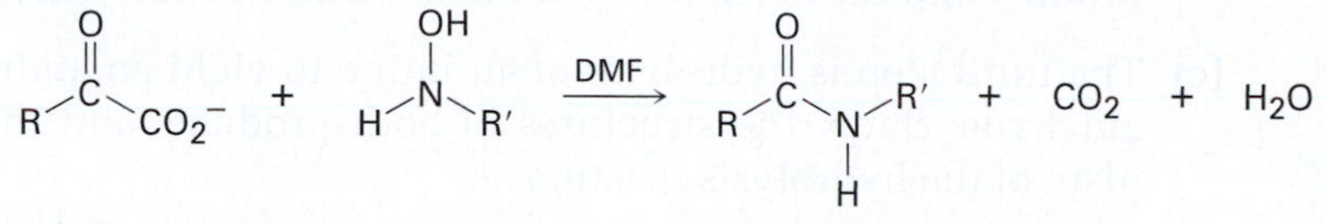

An α-keto acid   **A hydroxylamine**   **An amide**

The reaction is thought to occur by nucleophilic addition of the N-alkylhydroxylamine to the keto acid as if forming an imine (see Section 14.7), followed by decarboxylation and elimination of water. Show the mechanism.

19.47 Arginine, the most basic of the 20 common amino acids, contains a *guanidino* functional group in its side chain. Explain, using resonance structures to show how the protonated guanidino group is stabilized.

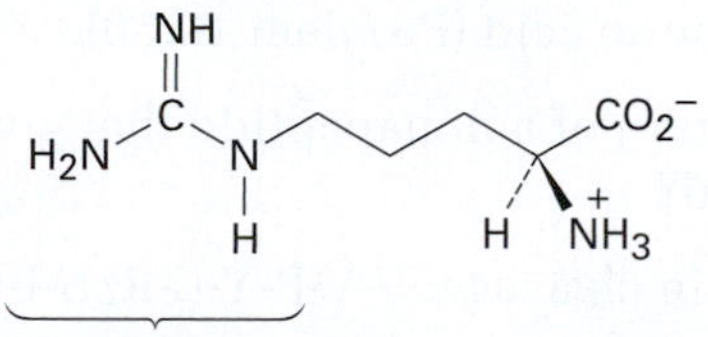

Guanidino
group

19.48 Cytochrome c is an enzyme found in the cells of all aerobic organisms. Elemental analysis of cytochrome c shows that it contains 0.43% iron. What is the minimum molecular weight of this enzyme?

19.49 Evidence for restricted rotation around amide CO–N bonds comes from NMR studies. At room temperature, the ^{1}H NMR spectrum of N,N-dimethylformamide shows three peaks: 2.9 δ (singlet, 3 H), 3.0 δ (singlet, 3 H), 8.0 δ (singlet, 1 H). As the temperature is raised, however, the two singlets at 2.9 δ and 3.0 δ slowly merge. At 180 °C, the ^{1}H NMR spectrum shows only two peaks: 2.95 δ (singlet, 6 H) and 8.0 δ (singlet, 1 H). Explain this temperature-dependent behavior.

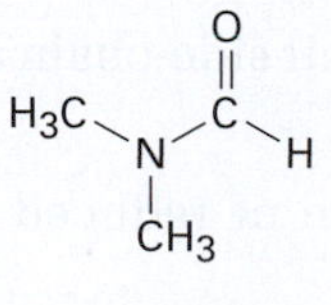

N,N-Dimethylformamide

19.50 ■ The reaction of ninhydrin with an α-amino acid occurs in several steps:

(a) The first step is formation of an imine by reaction of the amino acid with ninhydrin. Show its structure and the mechanism of its formation.

(b) The second step is a decarboxylation. Show the structure of the product and the mechanism of the decarboxylation reaction.

(c) The third step is hydrolysis of an imine to yield an amine and an aldehyde. Show the structures of both products and the mechanism of the hydrolysis reaction.

(d) The final step is formation of the purple anion. Show the mechanism of the reaction.

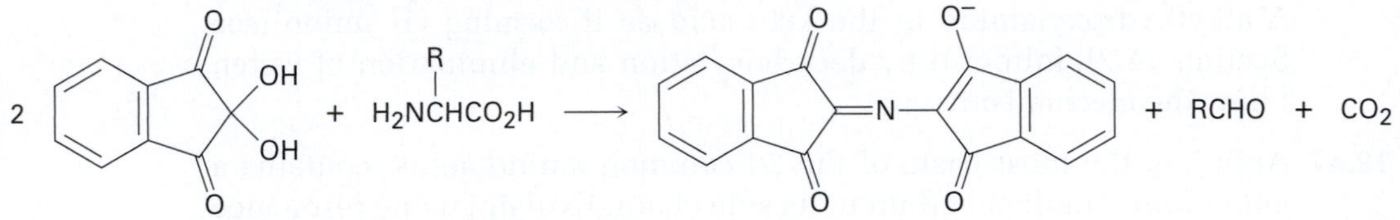

19.51 Draw resonance forms for the purple anion obtained by reaction of ninhydrin with an α-amino acid (Problem 19.50).

19.52 ■ What is the structure of a nonapeptide that gives the following fragments when cleaved?

Trypsin cleavage: V-V-P-Y-L-R, S-I-R

Chymotrypsin cleavage: L-R, S-I-R-V-V-P-Y

19.53 Oxytocin, a nonapeptide hormone secreted by the pituitary gland, functions by stimulating uterine contraction and lactation during childbirth. Its sequence was determined from the following evidence:

1. Oxytocin is a cyclic compound containing a disulfide bridge between two cysteine residues.

2. When the disulfide bridge is reduced, oxytocin has the constitution N, C_2, Q, G, I, L, P, Y.

3. Partial hydrolysis of reduced oxytocin yields seven fragments: D-C, I-E, C-Y, L-G, Y-I-E, E-D-C, C-P-L.

4. Gly is the C-terminal group.

5. Both E and D are present as their side-chain amides (Q and N) rather than as free side-chain acids.

What is the amino acid sequence of reduced oxytocin? What is the structure of oxytocin itself?

19.54 *Aspartame,* a nonnutritive sweetener marketed under the trade name NutraSweet, is the methyl ester of a simple dipeptide, Asp-Phe-OCH_3.

(a) Draw the structure of aspartame.

(b) The isoelectric point of aspartame is 5.9. Draw the principal structure present in aqueous solution at this pH.

(c) Draw the principal form of aspartame present at physiological pH = 7.3.

19.55 Refer to Figure 19.3 and propose a mechanism for the final step in the Edman degradation—the acid-catalyzed rearrangement of the ATZ derivative to the PTH derivative.

19.56 ■ Amino acids are metabolized by a transamination reaction in which the $-NH_2$ group of the amino acid changes places with the keto group of an α-keto acid. The products are a new amino acid and a new α-keto acid. Show the product from transamination of isoleucine.

■ Problems assignable in Organic OWL.

20 Amino Acid Metabolism

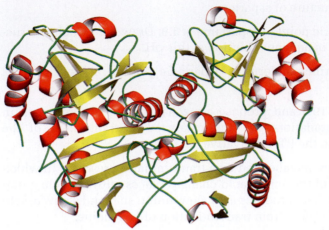

D-Amino-acid aminotransferase catalyzes the deamination of amino acids, the first step in amino acid metabolism.

Anyone who wants to understand or contribute to the revolution now taking place in the biological sciences must first understand life processes at the molecular level. This understanding, in turn, must be based on a detailed knowledge of the chemical reactions and paths used by living organisms. Just knowing *what* occurs is not enough; it's also necessary to understand *how* and *why* organisms use the chemistry they do.

Biochemical reactions are not mysterious. It's true that many of the biological reactions occurring in even the simplest living organism are more complex than those carried out in any laboratory, yet they follow the same rules of reactivity as laboratory reactions and they take place by the same mechanisms. In past chapters, we've seen many biological reactions used as examples, but it's now time to focus specifically on biological reactions, paying particular attention to the metabolic pathways that organisms use to synthesize and degrade biomolecules.

A word of caution: some of the molecules we'll be encountering are substantially larger and more complex than any we've been dealing with up to this point. As always, *keep your focus on the functional groups in those parts of the molecules where changes occur.* The reactions themselves are the same sorts of additions, eliminations, substitutions, carbonyl condensations, and so forth, that we've been dealing with all along.

WHY THIS CHAPTER?

We'll begin a study of biological reactions with a general overview of metabolism and then focus specifically on amino acids, the fundamental building blocks from which the estimated 500,000 or so proteins in our bodies are made. We'll see in this chapter both how amino acids are biosynthesized for incorporation into proteins and how they are ultimately degraded when proteins are broken down.

CONTENTS

 Online homework for this chapter can be assigned in Organic OWL.

20.1 An Overview of Metabolism and Biochemical Energy

The many reactions that go on in the cells of living organisms are collectively called **metabolism**. The pathways that break down larger molecules into smaller ones are called **catabolism**, and the pathways that synthesize larger biomolecules from smaller ones are known as **anabolism**. Catabolic reaction pathways are usually exergonic and release energy, while anabolic pathways are often endergonic and absorb energy. Catabolism can be divided into the four stages shown in Figure 20.1.

Stage 1   Bulk food is hydrolyzed in the stomach and small intestine to give small molecules.

Stage 2   Fatty acids, monosaccharides, and amino acids are degraded in cells to yield acetyl CoA.

Stage 3   Acetyl CoA is oxidized in the citric acid cycle to give CO_2.

Stage 4   The energy released in the citric acid cycle is used by the electron-transport chain to oxidatively phosphorylate ADP and produce ATP.

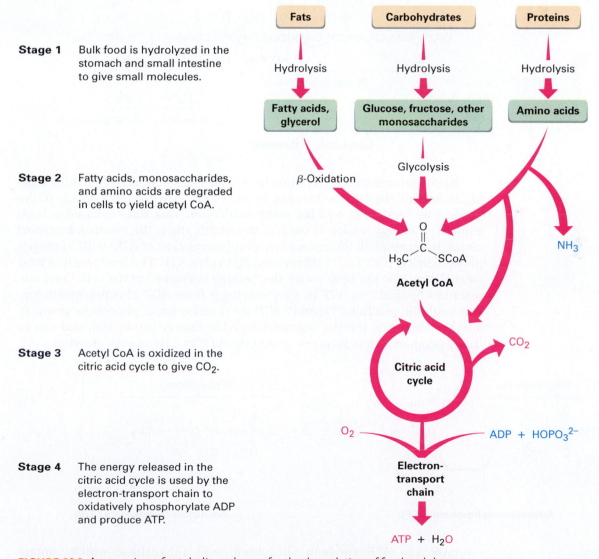

FIGURE 20.1 An overview of catabolic pathways for the degradation of food and the production of biochemical energy. The ultimate products of food catabolism are CO_2 and H_2O, with the energy released in the citric acid cycle used to drive the endergonic synthesis of adenosine triphosphate (ATP) from adenosine diphosphate (ADP) plus phosphate ion, $HOPO_3^{2-}$.

In the first catabolic stage, commonly called digestion, food is broken down in the mouth, stomach, and small intestine by hydrolysis of ester, glycoside (acetal), and peptide (amide) bonds to yield primarily fatty acids plus glycerol, simple sugars, and amino acids. These smaller molecules are further degraded in the cytoplasm of cells in the second stage of catabolism to yield acetyl groups attached by a thioester bond to the large carrier molecule coenzyme A. The resultant compound, acetyl coenzyme A (acetyl CoA), is a key substance both in the metabolism of food molecules and in numerous other biological pathways. As noted in Section 16.8, the acetyl group in acetyl CoA is linked to the sulfur atom of phosphopantetheine, which is itself linked to adenosine 3′,5′-bisphosphate.

Acetyl CoA—a thioester

Acetyl groups are oxidized inside cellular mitochondria in the third stage of catabolism, the *citric acid cycle,* to yield CO_2. (We'll see the details of the process in Section 22.4.) Like many oxidations, this stage releases a large amount of energy, which is used in the fourth stage, the *electron-transport chain,* to accomplish the endergonic phosphorylation of ADP with hydrogen phosphate ion ($HOPO_3^{2-}$, abbreviated P_i) to give ATP. The final result of food catabolism, ATP has been called the "energy currency" of the cell. Catabolic reactions "pay off" in ATP by synthesizing it from ADP plus phosphate ion, and anabolic reactions "spend" ATP by transferring a phosphate group to another molecule, thereby regenerating ADP. Energy production and use in living organisms thus revolves around the ATP ⇌ ADP interconversion.

Adenosine diphosphate (ADP)   **Adenosine triphosphate (ATP)**

ADP and ATP are both phosphoric acid anhydrides, which contain $-\overset{O}{\overset{\|}{P}}-O-\overset{O}{\overset{\|}{P}}-$ linkages analogous to the $-\overset{O}{\overset{\|}{C}}-O-\overset{O}{\overset{\|}{C}}-$ linkage in carboxylic acid anhydrides. Just as carboxylic acid anhydrides react with alcohols by breaking a C–O bond and forming a carboxylic ester, ROCOR′ (Section 16.5), phosphoric acid anhydrides react with alcohols by breaking a P–O bond and forming a phosphate ester, $ROPO_3{}^{2-}$. The reaction is, in effect, a nucleophilic acyl substitution at phosphorus. Note that phosphorylation reactions with ATP generally require the presence of a divalent metal cation in the enzyme, usually Mg^{2+}, to form a Lewis acid–base complex with the phosphate oxygen atoms and neutralize some negative charge.

A phosphate ester   **ADP**

How does the body use ATP? Recall from Section 6.7 that the free-energy change ΔG must be negative and energy must be released for a reaction to be favorable and occur spontaneously. If ΔG is positive, the reaction is energetically unfavorable and the process can't occur spontaneously.

For an energetically unfavorable reaction to occur, it must be "coupled" to an energetically favorable reaction so that the overall free-energy change for the two reactions together is favorable. To understand what it means for reactions to be coupled, imagine that reaction 1 does not occur to any reasonable extent because it has a small equilibrium constant and is energetically unfavorable; that is, the reaction has $\Delta G > 0$.

$$(1) \quad \textbf{A} + m \; \xrightleftharpoons{} \; \textbf{B} + n \qquad \Delta G > 0$$

where **A** and **B** are the biochemically "interesting" substances undergoing transformation, while m and n are enzyme cofactors, H_2O, or other substances.

Imagine also that product n can react with substance o to yield p and q in a second, strongly favorable reaction that has a large equilibrium constant and $\Delta G \ll 0$:

$$(2) \quad n + o \; \xrightleftharpoons{} \; p + q \qquad \Delta G \ll 0$$

Taking the two reactions together, they share, or are coupled through, the common intermediate n, which is a product in the first reaction and a reactant in the second. When even a tiny amount of n is formed in reaction 1, it undergoes essentially complete conversion in reaction 2, thereby removing it from the first equilibrium and forcing reaction 1 to continually replenish n until the reactant **A** is gone. That is, the two reactions added together have a favorable $\Delta G < 0$, and we say that the favorable reaction 2 "drives" the unfavorable reaction 1. Because the two reactions are coupled through n, the transformation of **A** to **B** becomes favorable.

(1) $\text{A} + m \rightleftharpoons \text{B} + \cancel{n}$ $\Delta G > 0$

(2) $\cancel{n} + o \rightleftharpoons p + q$ $\Delta G \ll 0$

Net: $\text{A} + m + o \rightleftharpoons \text{B} + p + q$ $\Delta G < 0$

As an example of two reactions that are coupled, look at the phosphorylation reaction of glucose to yield glucose 6-phosphate plus water, the first step in the breakdown of dietary carbohydrates. The reaction of glucose with $HOPO_3^{2-}$ does not occur spontaneously because it is energetically unfavorable, with $\Delta G^{\circ\prime} = +13.8$ kJ/mol. (As noted in Section 6.7, the standard free-energy change for a biological reaction is denoted $\Delta G^{\circ\prime}$ and refers to a process in which reactants and products have a concentration of 1.0 M in a solution with pH = 7.0.)

$$\underset{\text{Glucose}}{\text{HOCH}_2\text{CHCHCHCHCH}} \rightleftharpoons \underset{\text{Glucose 6-phosphate}}{{}^-\text{OPOCH}_2\text{CHCHCHCHCH}} + \text{H}_2\text{O} \quad \Delta G^{\circ\prime} = +13.8 \text{ kJ}$$

With ATP, however, glucose undergoes an energetically favorable reaction to yield glucose 6-phosphate plus ADP. The overall effect is as if $HOPO_3^{2-}$ reacted with glucose and ATP then reacted with the water by-product, making the coupled process favorable by about 16.7 kJ/mol (4.0 kcal/mol). That is, ATP drives the phosphorylation reaction of glucose:

Glucose + $HOPO_3^{2-}$ $\longrightarrow$ Glucose 6-phosphate + H_2O $\Delta G^{\circ\prime} = +13.8$ kJ/mol

ATP + H_2O $\longrightarrow$ ADP + $HOPO_3^{2-}$ + H^+ $\Delta G^{\circ\prime} = -30.5$ kJ/mol

Net: Glucose + ATP $\longrightarrow$ Glucose 6-phosphate + ADP + H^+ $\Delta G^{\circ\prime} = -16.7$ kJ/mol

It's this ability to drive otherwise unfavorable phosphorylation reactions that makes ATP so useful. The resultant phosphates are much more reactive as leaving groups in nucleophilic substitutions and eliminations than the corresponding alcohols they're derived from and are therefore more chemically useful.

Problem 20.1
One of the early steps in the urea cycle by which ammonia is excreted from the body is the reaction of bicarbonate ion (HCO_3^-) with ATP to yield carboxy phosphate. Write the reaction, and draw the structure of carboxy phosphate. You can check your answer in Figure 20.4.

20.2 Catabolism of Amino Acids: Deamination

Let's now begin a study of some common metabolic pathways by starting with amino acid catabolism. Although the subject is complicated by the fact that each of the 20 α-amino acids in proteins is biologically degraded by its own unique pathway, there are some common themes that tie the different pathways together.

Amino acid catabolism occurs in three stages: (1) removal of the α amino group as ammonia, (2) conversion of the ammonia into urea, and (3) conversion of the remaining amino acid carbon skeleton (usually an α-keto acid) into an intermediate that can enter the citric acid cycle.

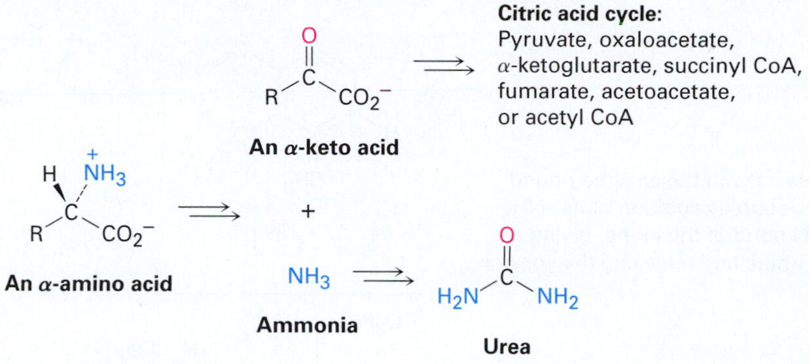

Transamination

The first stage in the metabolic degradation of most α-amino acids is **deamination**, the removal of the α amino group. Deamination is usually accomplished by a **transamination** reaction, in which the $-NH_2$ group of the amino acid is exchanged with the keto group of α-ketoglutarate, forming a new α-keto acid plus glutamate. The overall process occurs in two parts, is catalyzed by aminotransferase enzymes, and involves participation of the coenzyme pyridoxal phosphate, abbreviated PLP, a derivative of pyridoxine (vitamin B_6). Different aminotransferases differ in their specificity for amino acids, but the mechanism remains the same.

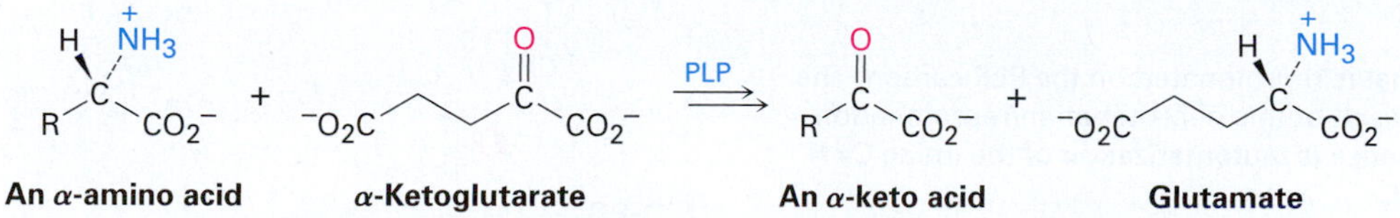

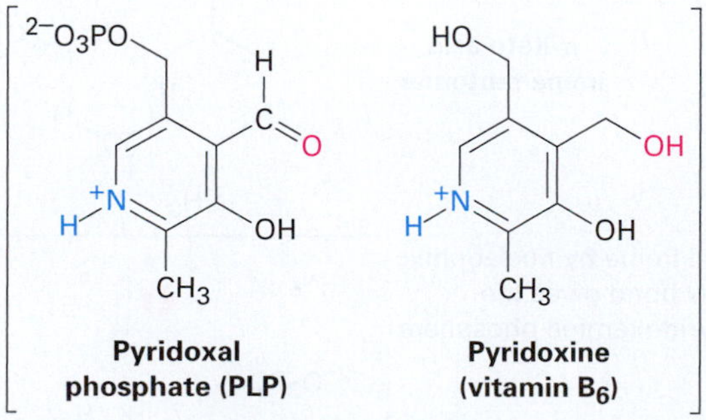

The mechanism of the first part of transamination is shown in Figure 20.2 and occurs through steps that use familiar reactions we've seen in previous chapters. The process begins with reaction between the α-amino acid and pyridoxal phosphate, which is covalently bonded to the aminotransferase by an imine linkage (Section 14.7) between the side-chain $-NH_2$ group of a lysine residue in the enzyme and the PLP aldehyde group. Deprotonation/reprotonation of the PLP–amino acid imine in steps 2 and 3 effects tautomerization of the imine C=N bond, and hydrolysis of the tautomerized imine in step 4 gives an α-keto acid plus pyridoxamine phosphate (PMP).

FIGURE 20.2

MECHANISM: Overall mechanism of the enzyme-catalyzed, PLP-dependent transamination of an α-amino acid to give an α-keto acid. Individual steps are explained in the text.

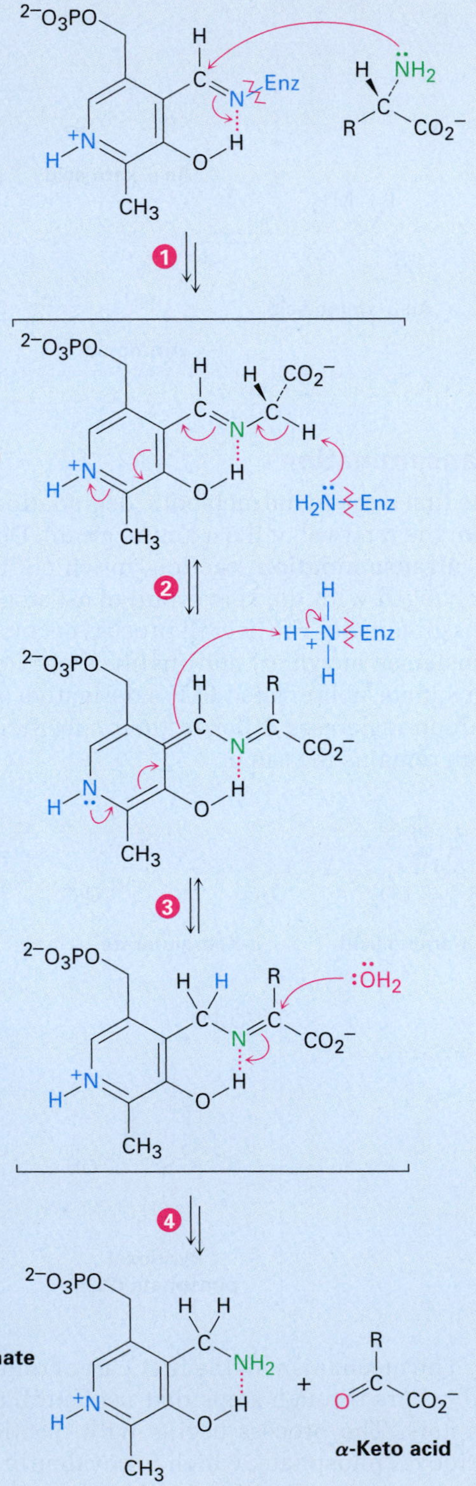

❶ An amino acid reacts with the enzyme-bound PLP imine by nucleophilic addition of its –NH₂ group to the C=N bond of the imine, giving a PLP–amino acid imine and releasing the enzyme amino group.

PLP–amino acid imine (Schiff base)

❷ Deprotonation of the acidic α carbon of the amino acid gives an intermediate α-keto acid imine . . .

α-Keto acid imine

❸ . . . that is reprotonated on the PLP carbon. The net result of this deprotonation/reprotonation sequence is tautomerization of the imine C=N bond.

α-Keto acid imine tautomer

❹ Hydrolysis of the α-keto acid imine by nucleophilic addition of water to the C=N bond gives the transamination products pyridoxamine phosphate (PMP) and α-keto acid.

Pyridoxamine phosphate (PMP)

α-Keto acid

© John McMurry

STEP 1 OF FIGURE 20.2: TRANSIMINATION The first step in trans*amination* is trans*imination*—the reaction of the PLP–enzyme imine with an α-amino acid to give a PLP–amino acid imine plus expelled enzyme as the leaving group. The reaction occurs by nucleophilic addition of the amino acid –NH$_2$ group to the C=N bond of the PLP imine, much as an amine adds to the C=O bond of a ketone or aldehyde in a nucleophilic addition reaction as shown previously in Figure 14.8 on page 577. The protonated diamine intermediate undergoes a proton transfer and then expels the lysine amino group in the enzyme to complete the step (Figure 20.3).

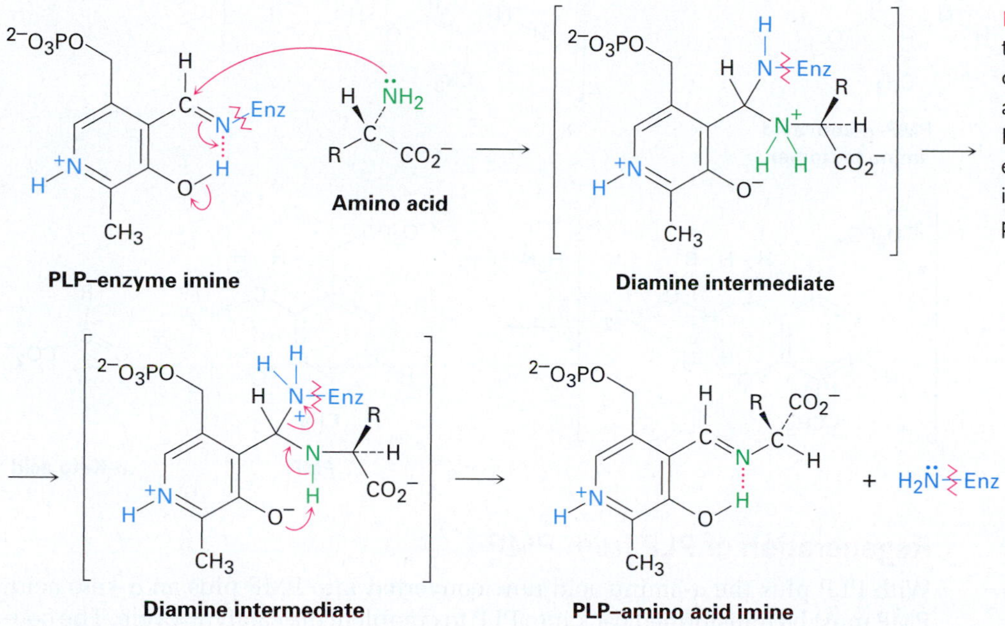

FIGURE 20.3 Mechanism of the transimination reaction of a PLP–enzyme imine with an α-amino acid to give a PLP–amino acid imine plus expelled enzyme. The reaction is analogous to that shown previously in Figure 14.8.

STEPS 2–4 OF FIGURE 20.2: TAUTOMERIZATION AND HYDROLYSIS Following formation of the PLP–amino acid imine in step 1, a tautomerization of the C=N bond occurs in step 2. The basic lysine residue in the enzyme that was expelled as a leaving group during transimination deprotonates the acidic α position of the amino acid, with the protonated pyridine ring of PLP acting as the electron acceptor as shown in step 2 of Figure 20.2. Reprotonation occurs on the carbon atom next to the ring (step 3), generating a tautomeric product that is the imine of an α-keto acid with pyridoxamine phosphate, abbreviated PMP.

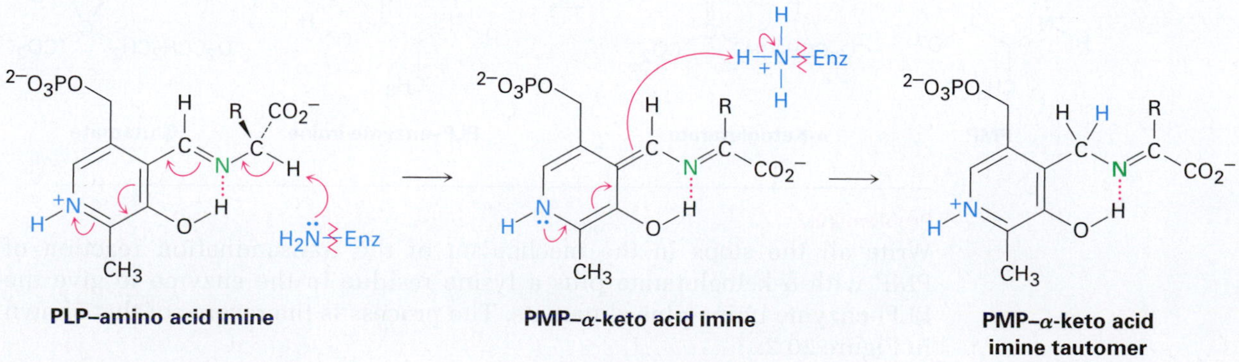

Hydrolysis of this PMP–α-keto acid imine in step 4 then completes the first part of the deamination reaction. The hydrolysis is the exact mechanistic reverse of imine formation (see Figure 14.8) and occurs by nucleophilic addition of water to the imine, followed by proton transfer and expulsion of PMP as leaving group.

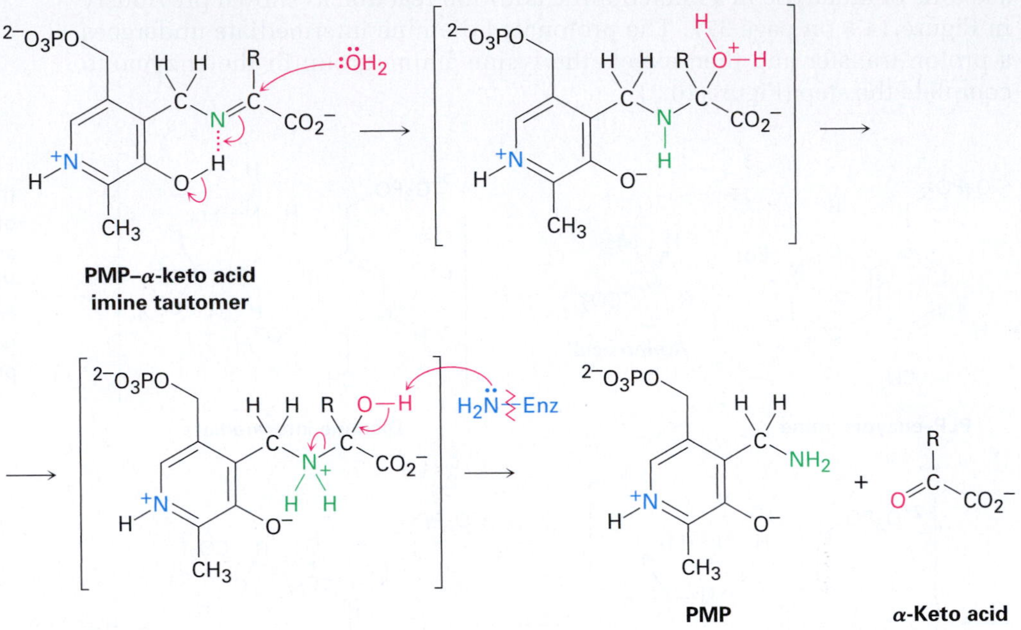

PMP–α-keto acid imine tautomer

PMP   **α-Keto acid**

Regeneration of PLP from PMP

With PLP plus the α-amino acid now converted into PMP plus an α-keto acid, PMP must be transformed back into PLP to complete the catalytic cycle. The conversion occurs by another transamination reaction, this one between PMP and an α-keto acid, usually α-ketoglutarate. The products are PLP plus glutamate, and the mechanism of the process is the exact reverse of that shown in Figure 20.2. That is, PMP and α-ketoglutarate give an imine; the PMP–α-ketoglutarate imine undergoes tautomerization of the C=N bond to give a PLP–glutamate imine; and the PLP–glutamate imine reacts with a lysine residue on the enzyme in a transimination process to yield PLP–enzyme imine plus glutamate.

PMP   **α-Ketoglutarate**   **PLP–enzyme imine**   **Glutamate**

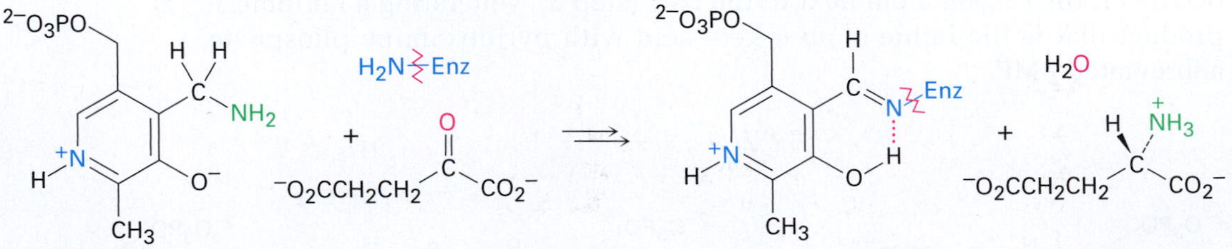

Problem 20.2

Write all the steps in the mechanism of the transamination reaction of PMP with α-ketoglutarate plus a lysine residue in the enzyme to give the PLP–enzyme imine plus glutamate. The process is the reverse of that shown in Figure 20.2.

Oxidative Deamination of Glutamate

Following transferal of the $-NH_2$ group from an amino acid to PLP and then to α-ketoglutarate, the glutamate product undergoes **oxidative deamination** by glutamate dehydrogenase to give ammonia plus regenerated α-ketoglutarate. The reaction occurs by oxidation of the primary amine to an imine, followed by hydrolysis. The mechanism of this biological amine oxidation is analogous to that of biological alcohol oxidation, as shown previously in Figure 13.6 on page 522. That is, a basic histidine residue in the enzyme removes a proton from nitrogen at the same time that the adjacent hydrogen on the α carbon of glutamate is transferred as hydride ion to an oxidizing coenzyme. Either NAD^+ or $NADP^+$ can function as the oxidizing coenzyme, depending on the organism.

Glutamate α-Iminoglutarate α-Ketoglutarate

Problem 20.3

In the oxidative deamination of glutamate, is the hydride ion transferred to the *Re* face or the *Si* face of NAD^+? (Review Section 5.11.)

20.3 The Urea Cycle

The ammonia resulting from amino acid deamination is eliminated in one of three ways depending on the organism. Fish and other aquatic animals simply excrete the ammonia to their aqueous surroundings, but terrestrial organisms must first convert the ammonia into a nontoxic substance—either urea for mammals or uric acid for birds and reptiles.

Urea   Uric acid

The conversion of ammonia into urea begins with its reaction with bicarbonate ion and ATP to give carbamoyl phosphate. The reaction is catalyzed by carbamoyl phosphate synthetase I and occurs by initial activation of HCO_3^- by ATP to give carboxy phosphate, followed by nucleophilic acyl substitution with ammonia to produce carbamate plus phosphate ion (P_i) as the leaving group. Subsequent phosphorylation of carbamate by a second equivalent of ATP then gives carbamoyl phosphate (Figure 20.4). As noted in Section 20.1, the reaction requires Mg^{2+} to form a Lewis acid–base complex with the phosphate oxygen atoms.

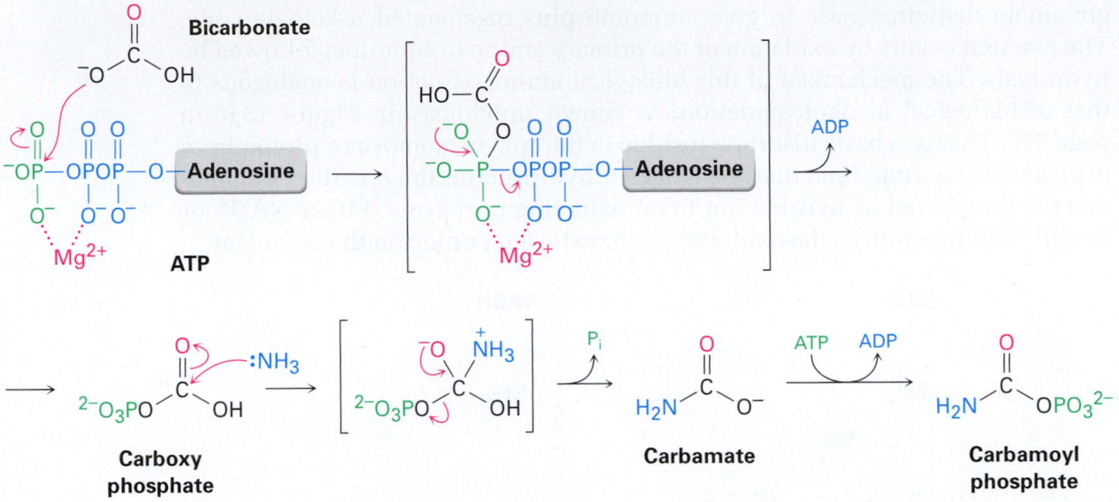

FIGURE 20.4 Mechanism of the formation of carbamoyl phosphate from bicarbonate. Bicarbonate ion is first activated by phosphorylation with ATP, and a nucleophilic acyl substitution with ammonia then occurs.

Carbamoyl phosphate next enters the four-step **urea cycle**, whose overall result can be summarized as

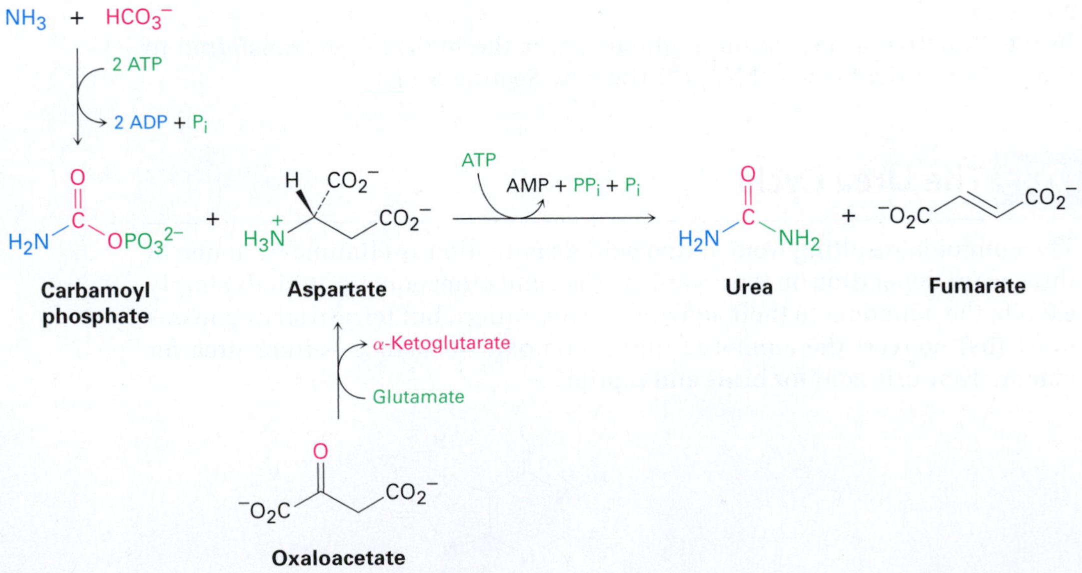

Note that only one of the two nitrogen atoms in urea comes from ammonia; the other nitrogen comes from aspartate, which is itself produced from glutamate by transamination with oxaloacetate, $^-O_2CCOCH_2CO_2^-$. The reactions of the urea cycle are shown in Figure 20.5.

STEPS 1–2 OF FIGURE 20.5: ARGININOSUCCINATE SYNTHESIS The urea cycle begins with a nucleophilic acyl substitution reaction of the nonprotein amino acid ornithine with carbamoyl phosphate to produce citrulline. The side-chain $-NH_2$ group of ornithine is the nucleophile, phosphate ion is the

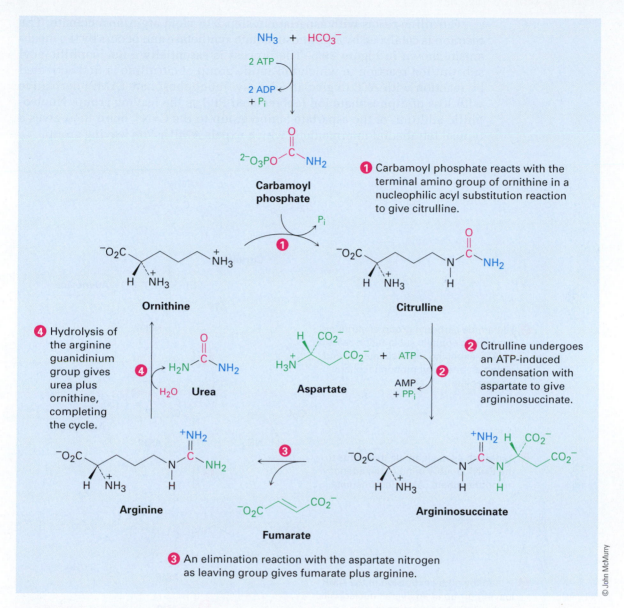

FIGURE 20.5 MECHANISM: The urea cycle is a four-step series of reactions that converts ammonia into urea. Individual steps are explained in the text.

leaving group, and the reaction is catalyzed by ornithine transcarbamoylase. Note that ornithine, although not one of the 20 amino acids in proteins, is similar to lysine but contains one less carbon in its side chain.

Citrulline reacts with aspartate in step 2 to yield argininosuccinate. The reaction is catalyzed by argininosuccinate synthetase and occurs by the mechanism shown in Figure 20.6. The process is essentially a nucleophilic acyl substitution reaction in which the amide group of citrulline is first activated by reaction with ATP to give an adenosyl monophosphate (AMP) derivative with loss of diphosphate ion (abbreviated PP$_i$) as the leaving group. Nucleophilic addition of the aspartate amino group to the C=N$^+$ bond then gives a typical tetrahedral intermediate, which expels AMP as the leaving group.

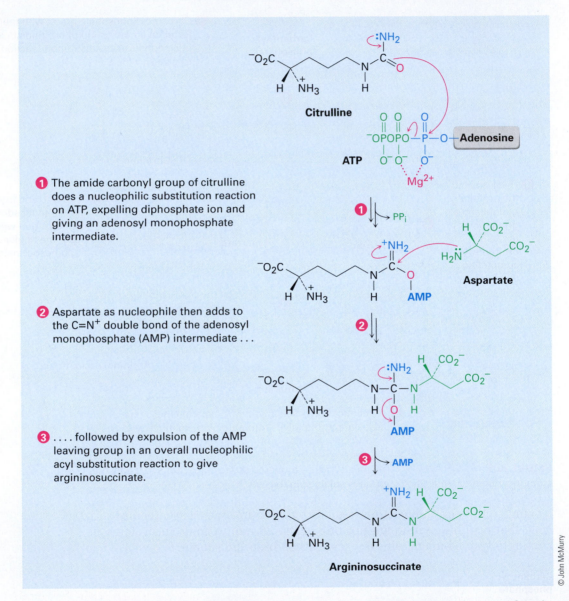

① The amide carbonyl group of citrulline does a nucleophilic substitution reaction on ATP, expelling diphosphate ion and giving an adenosyl monophosphate intermediate.

② Aspartate as nucleophile then adds to the C=N$^+$ double bond of the adenosyl monophosphate (AMP) intermediate . . .

③ . . . . followed by expulsion of the AMP leaving group in an overall nucleophilic acyl substitution reaction to give argininosuccinate.

© John McMurry

FIGURE 20.6 MECHANISM: Mechanism of step 2 in the urea cycle, the reaction of citrulline with aspartate to give argininosuccinate.

STEP 3 OF FIGURE 20.5: FUMARATE ELIMINATION The third step in the urea cycle, conversion of argininosuccinate to arginine plus fumarate, is an elimination reaction catalyzed by argininosuccinate lyase. The process occurs by an E1cB mechanism (Section 17.7), with a histidine residue on the enzyme acting as the base to carry out the deprotonation and form the anion inter-mediate. Note that the *pro-R* hydrogen on argininosuccinate is specifically abstracted in the elimination. Such specificity is typical for enzyme-catalyzed reactions, although it's not usually possible to predict the stereochemistry nor is it something for you to be concerned with at this point.

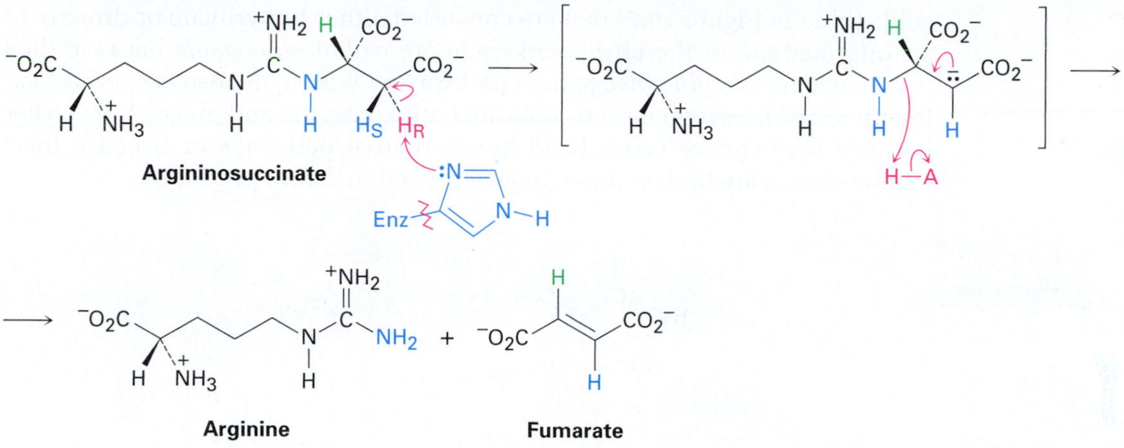

STEP 4 OF FIGURE 20.5: ARGININE HYDROLYSIS The final step to complete the urea cycle is hydrolysis of arginine to give ornithine and urea. The reaction is catalyzed by the Mn^{2+}-containing enzyme arginase and occurs by addition of H_2O to the $C=N^+$ bond, followed by proton transfer and elimination of orni-thine from the tetrahedral intermediate.

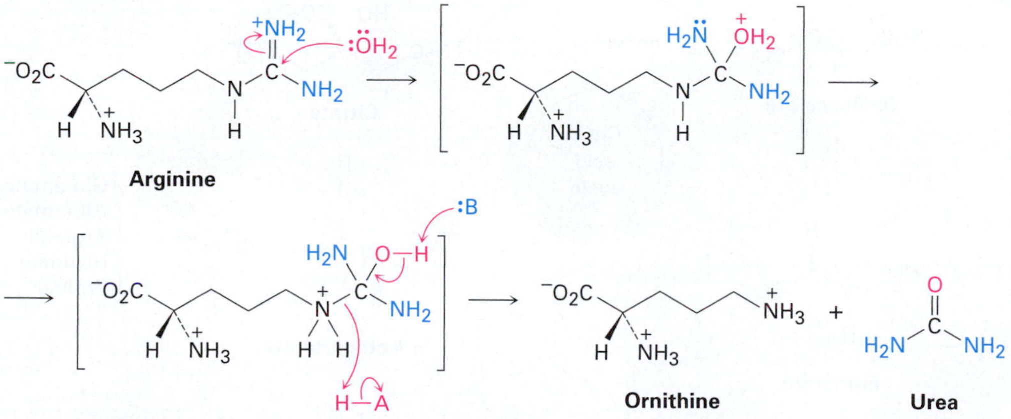

Problem 20.4

Draw the full structure of the adenosyl monophosphate intermediate formed by reaction of citrulline with ATP (Figure 20.6).

20.4 Catabolism of Amino Acids: The Carbon Chains

With the amino group removed by transamination and the resultant ammonia converted into urea, the third and final stage of amino acid catabolism is the degradation of the carbon chains. As indicated in Figure 20.7, the carbon chains of α-amino acids are commonly converted into one of seven intermediates that enter the citric acid cycle for final degradation (Section 22.4).

Those amino acids (red in Figure 20.7) that are converted to either aceto-acetate or acetyl CoA are called *ketogenic* because they can also enter the fatty-acid biosynthesis pathway (Section 23.6) or be converted into so-called ketone bodies—acetoacetate, β-hydroxybutyrate, and acetone. Those amino acids (blue in Figure 20.7) that are converted either to pyruvate or directly to an intermediate in the citric acid cycle are called *glucogenic* because they can also enter the *gluconeogenesis* pathway by which glucose is synthesized (Section 22.5). Several amino acids are both ketogenic and glucogenic, either because they can be catabolized by alternative pathways or because their carbon chains are broken down to give several different products.

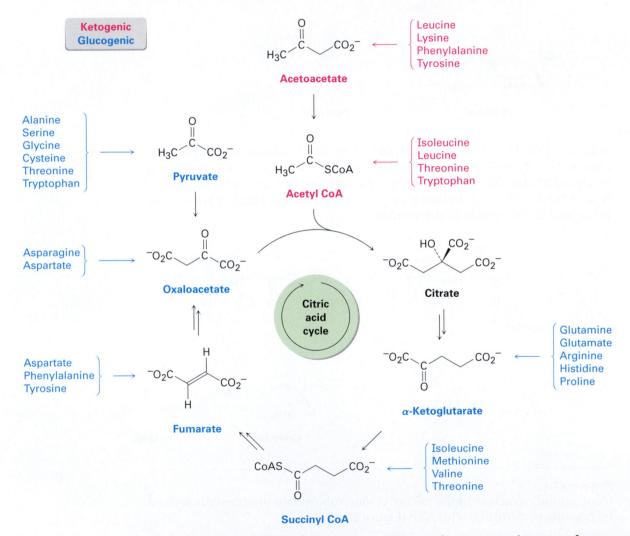

FIGURE 20.7 Carbon chains of the 20 common amino acids are converted into one of seven intermediates for further breakdown in the citric acid cycle. Ketogenic amino acids (red) can also enter the pathway for fatty-acid biosynthesis; glucogenic amino acids (blue) can also enter the gluconeogenesis pathway for glucose biosynthesis.

A detailed coverage of the catabolic pathways for all 20 protein amino acids would take far too much space—tryptophan catabolism alone requires

14 steps—and would be far too complex for this book. The mechanisms of these pathways *are* understandable, but we won't attempt to cover them. Instead, we'll just look at several fairly straightforward schemes to see the kinds of chemistry involved in amino acid catabolism.

Alanine Catabolism

Alanine is one of the six amino acids that are catabolized to give pyruvate. The pathway is a straightforward PLP-dependent transamination reaction, as discussed in Section 20.2, with the PMP intermediate then converted back to PLP by reaction with α-ketoglutarate.

Alanine   **Pyruvate**

α-**Ketoglutarate**   **Glutamate**

Problem 20.5

Review Section 20.2, and write all the steps in the PLP-dependent transamination reaction of alanine plus α-ketoglutarate to give pyruvate plus glutamate.

Serine Catabolism

Serine, like alanine, is converted into pyruvate by a PLP-dependent pathway, but the two reaction sequences are not the same. Whereas alanine catabolism involves a PLP-dependent *transamination,* serine catabolism involves a PLP-dependent *dehydration* to form an intermediate enamine that is then hydrolyzed.

Serine α-**Amino acrylate**   **Pyruvate**
                    (an enamine)

Serine catabolism begins with formation of a PLP–serine imine by reaction of the amino acid with the PLP–enzyme imine, as described in Section 20.2. This PLP–serine imine is then deprotonated by a lysine residue in the serine dehydratase enzyme, just as occurs in a typical deamination reaction (Figure 20.2, step 1). But because serine has a leaving group (the –OH) at its β position, an E1cB elimination takes place and gives an unsaturated imine. Transimination to a lysine residue in the enzyme regenerates the enzyme–PLP imine and releases α-amino acrylate, which tautomerizes to the corresponding imine and is hydrolyzed to pyruvate (Figure 20.8).

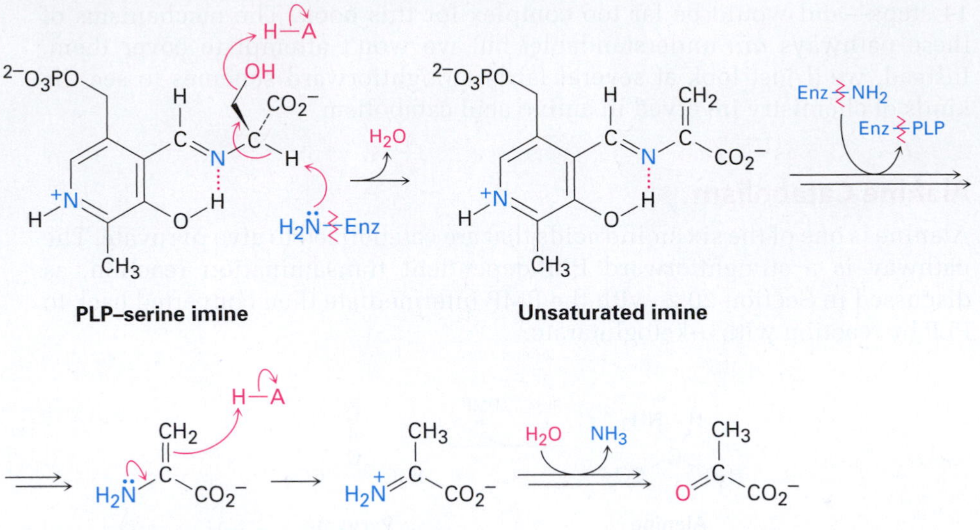

PLP–serine imine   **Unsaturated imine**

α-Amino acrylate   **Imine**   **Pyruvate**

FIGURE 20.8 Mechanism of the PLP-dependent conversion of serine to yield pyruvate. The key step is dehydration by an E1cB process.

Problem 20.6
Show the mechanisms of the final step in serine catabolism, hydrolysis of the imine to give pyruvate.

Asparagine and Aspartate Catabolism

Depending on the organism, asparagine and aspartate are converted into either oxaloacetate or fumarate, both of which are intermediates in the citric acid cycle. The amide bond of asparagine is first hydrolyzed by a nucleophilic acyl substitution reaction to yield aspartate, and aspartate then undergoes either a PLP-dependent transamination to give oxaloacetate or an E1cB elimination of ammonium ion to give fumarate (Figure 20.9).

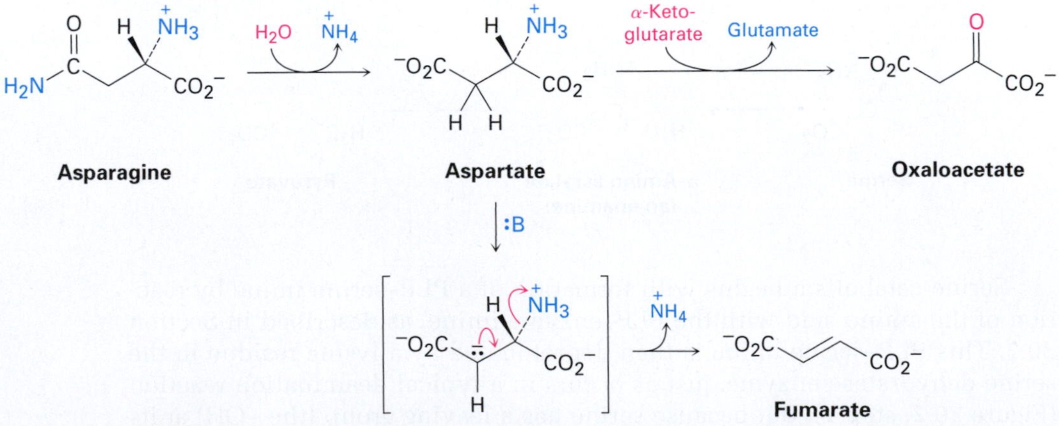

Asparagine   **Aspartate**   **Oxaloacetate**

Fumarate

FIGURE 20.9 Mechanism of the conversion of asparagine and aspartate to oxaloacetate and fumarate. Initial PLP-dependent transamination is followed by an E1cB elimination of ammonium ion.

Threonine Catabolism

Threonine is catabolized by several different pathways. Most commonly, it is oxidized to 2-amino-3-ketobutyrate and then converted by a PLP-dependent retro-Claisenlike reaction into acetyl CoA and glycine (Figure 20.10). The initial oxidation is carried out by NAD$^+$ and takes place by the mechanism described previously in Section 13.5. Formation of a PLP–imine in the usual way is then followed by nucleophilic addition of coenzyme A to the ketone carbonyl group to give a tetrahedral intermediate. Subsequent cleavage in a retro-Claisen reaction yields acetyl CoA plus the PLP–imine of glycine, which is hydrolyzed. The mechanism of the retro-Claisen is the reverse of the forward Claisen reaction, shown previously in Figure 17.12 on page 724, with the pyridinium ring of PLP acting as the electron acceptor.

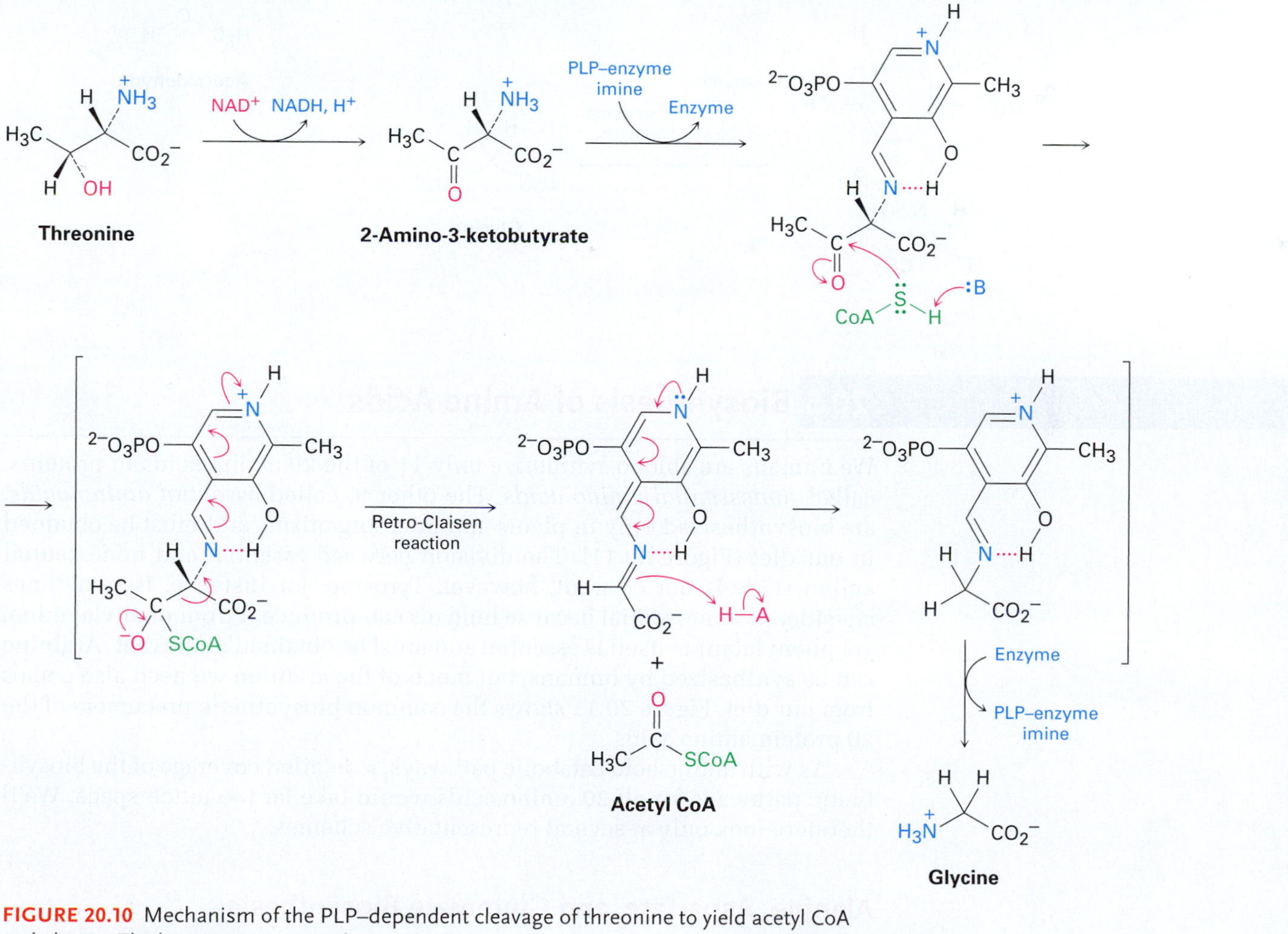

FIGURE 20.10 Mechanism of the PLP–dependent cleavage of threonine to yield acetyl CoA and glycine. The key step is a retro-Claisen reaction.

Alternatively, threonine can also be catabolized by cleavage through a PLP-dependent retro-aldol reaction (Section 17.6) to yield acetaldehyde plus glycine. The acetaldehyde is then oxidized to acetate and converted into acetyl CoA.

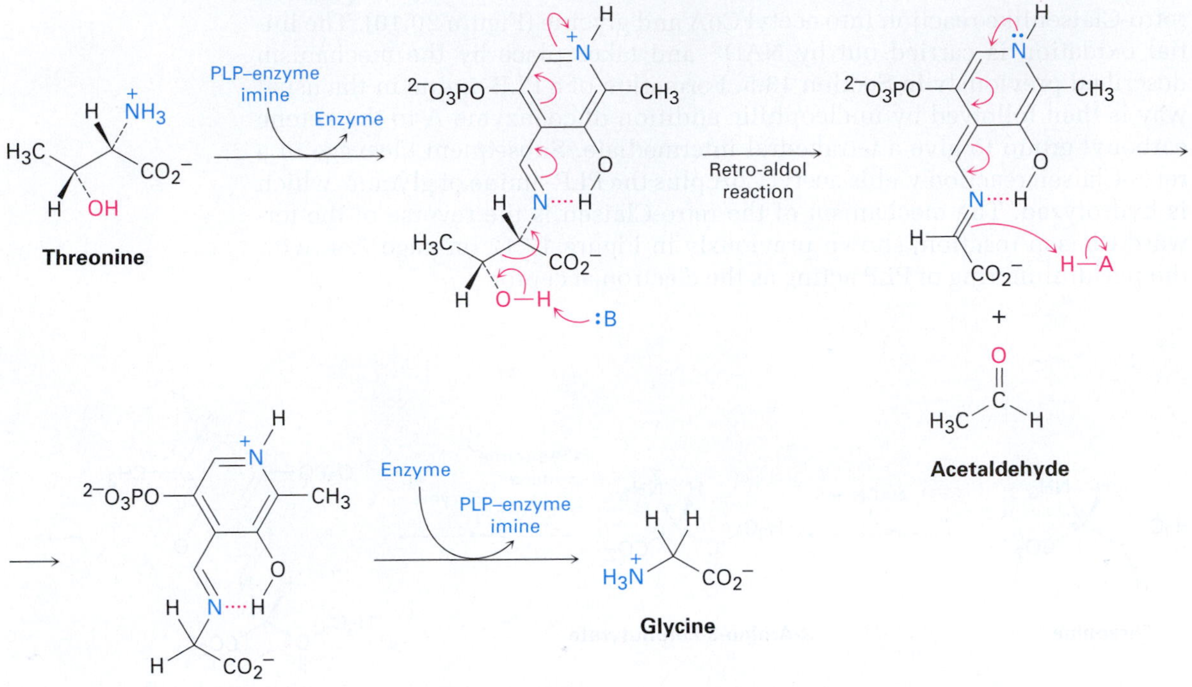

20.5 Biosynthesis of Amino Acids

We humans are able to synthesize only 11 of the 20 amino acids in proteins, called *nonessential amino acids.* The other 9, called *essential amino acids,* are biosynthesized only in plants and microorganisms and must be obtained in our diet (Figure 20.11). The division between essential and nonessential amino acids is not clearcut, however. Tyrosine, for instance, is sometimes considered nonessential because humans can produce it from phenylalanine, yet phenylalanine itself is essential and must be obtained in the diet. Arginine can be synthesized by humans, but much of the arginine we need also comes from our diet. Figure 20.11 shows the common biosynthetic precursors of the 20 protein amino acids.

As with amino acid catabolic pathways, a detailed coverage of the biosynthetic pathways for all 20 amino acids would take far too much space. We'll therefore look only at several representative schemes.

Alanine, Aspartate, and Glutamate Biosynthesis

Seven of the eleven nonessential amino acids are synthesized either from pyruvate or from the citric acid cycle intermediates oxaloacetate and α-ketoglutarate. Alanine is biosynthesized by transamination of pyruvate, aspartate from oxaloacetate, and glutamate from α-ketoglutarate. The mechanisms of

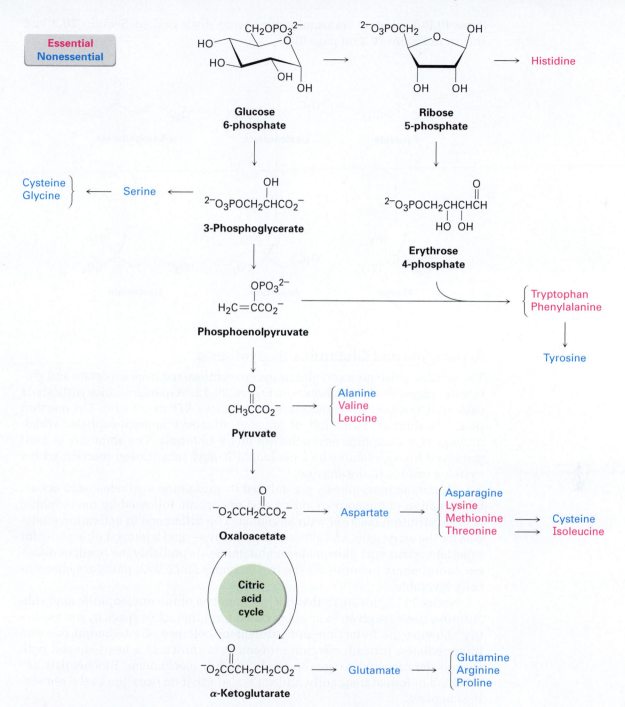

FIGURE 20.11 Biosynthesis of the 20 protein amino acids. Essential amino acids (red) are synthesized in plants and bacteria and must be obtained in our diet. Humans can synthesize only the nonessential amino acids (blue).

these PLP-dependent transaminations were discussed in Section 20.2 and shown in Figure 20.2 on page 838.

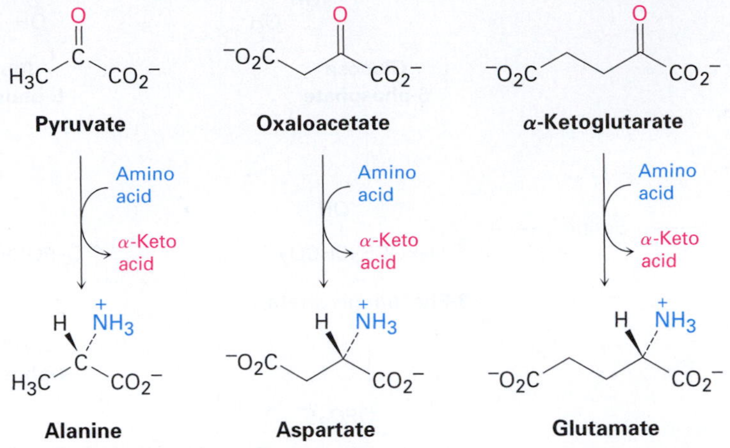

Asparagine and Glutamine Biosynthesis

The amides asparagine and glutamine are synthesized from aspartate and glutamate, respectively, as shown in Figure 20.12. Asparagine biosynthesis is catalyzed by asparagine synthetase and requires ATP as cofactor. The reaction proceeds through formation of an acyl adenosyl monophosphate, which undergoes nucleophilic acyl substitution by ammonia. The ammonia is itself produced from glutamine by a nucleophilic acyl substitution reaction with a cysteine residue in the enzyme.

Glutamine biosynthesis is catalyzed by glutamine synthetase and occurs by formation of the corresponding acyl phosphate followed by nucleophilic acyl substitution reaction with ammonia. The difference in activation strategies for the asparagine and glutamine pathways—acyl adenosyl phosphate for aspartate versus acyl phosphate for glutamate—is probably the result of different evolutionary histories for the two enzymes since both paths are energetically favorable.

Notice in Figure 20.12 that the mechanisms of the nucleophilic acyl substitution steps are given in an abbreviated form that saves space by not explicitly showing the formation and subsequent collapse of tetrahedral reaction intermediates. Instead, electron movement is shown as a heart-shaped path around the carbonyl oxygen to imply the full mechanism. Biochemists use this kind of format frequently, and we'll also use it on occasion in the remaining chapters.

Problem 20.7
Show the full mechanism for the formation of glutamine by reaction of glutamate 5-phosphate with ammonia, and compare that full mechanism to the abbreviated mechanism shown in Figure 20.12 to see the difference.

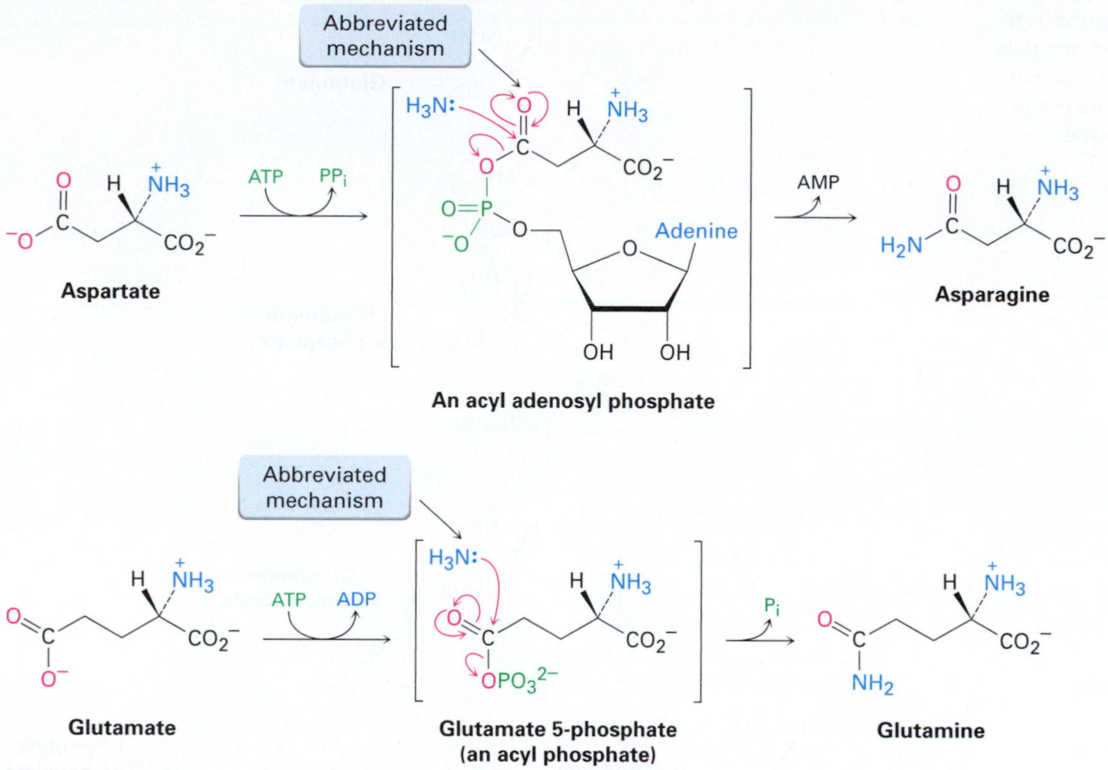

FIGURE 20.12 Biosynthetic pathways for asparagine and glutamine by amide formation from aspartate and glutamate, respectively. As explained in the text, the mechanisms of the acyl substitution reactions are given in abbreviated form without showing the tetrahedral intermediates.

Arginine and Proline Biosynthesis

In humans, arginine is synthesized from glutamate by the pathway shown in Figure 20.13. Reaction of glutamate with ATP gives the same acyl phosphate intermediate as in glutamine biosynthesis (Figure 20.12), which is reduced by NADH in a nucleophilic acyl substitution reaction with hydride ion to yield the corresponding aldehyde, glutamate 5-semialdehyde. You might recall that a similar partial reduction of esters to aldehydes can be carried out in the laboratory by reaction of the ester with diisobutylaluminum hydride (DIBAH; Section 16.6). A related example of a partial reduction of a thioester to an aldehyde by reaction with NADPH was discussed in Section 16.8.

PLP-mediated transamination of the glutamate 5-semialdehyde carbonyl group by reaction with glutamate then gives ornithine, which is converted to arginine in the urea cycle, as discussed previously in Section 20.3 (Figure 20.5).

Proline also is synthesized from glutamate 5-semialdehyde by nonenzymatic formation of a cyclic imine followed by enzymatic reduction of the C=N bond with NADH in a nucleophilic addition reaction.

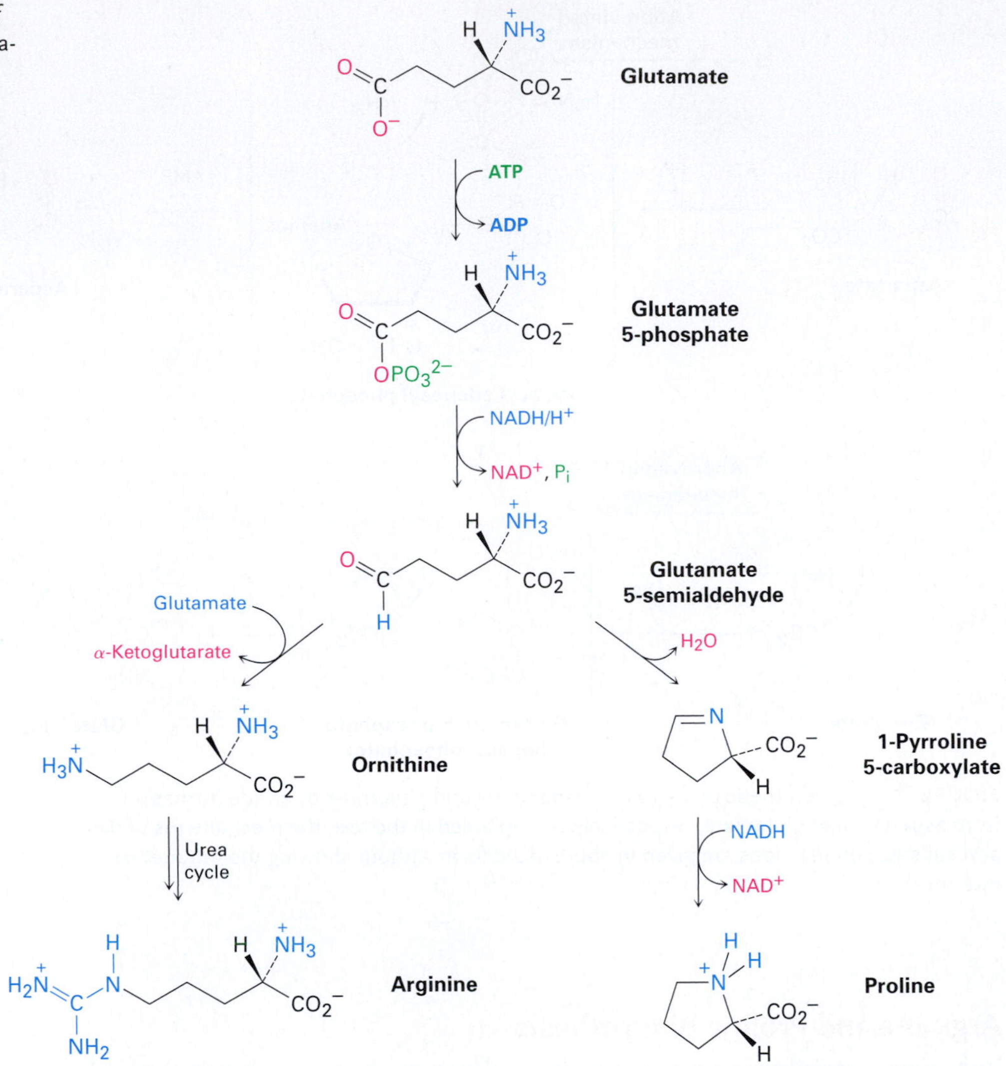

FIGURE 20.13 Biosynthesis of arginine and proline from glutamate. The key step is a partial reduction of glutamate to the corresponding aldehyde.

Problem 20.8

Review Section 16.8, and show the mechanism of the partial reduction of glutamate 5-phosphate with NADH to give glutamate 5-semialdehyde.

Problem 20.9

Show the mechanisms of both the nonenzymatic cyclization of glutamate 5-semialdehyde to give 1-pyrroline 5-carboxylate and the subsequent enzymatic reduction with NADH to yield proline.

Summary

In this chapter, we began a study of biological reactions, focusing specifically on amino acids, the fundamental building blocks from which the estimated 500,000 or so proteins in our bodies are made. We looked both at how amino acids are biosynthesized for incorporation into proteins and how they are ultimately degraded when proteins are broken down.

The many reactions that go on in the cells of living organisms are collectively called **metabolism**. The pathways that break down larger molecules into smaller ones are called **catabolism**, and the pathways that synthesize larger biomolecules from smaller ones are known as **anabolism**. Catabolic reaction pathways are usually exergonic, while anabolic reaction pathways are often endergonic. Catabolism is carried out in four stages: (1) digestion, in which food is hydrolyzed to fatty acids, simple sugars, and amino acids; (2) degradation of small molecules to give acetyl CoA; (3) oxidation of acetyl CoA in the *citric acid cycle* to give CO_2 and release energy; and (4) energy utilization by the *electron-transport chain* to phosphorylate ADP and give ATP. The ATP then drives many other biological reactions.

Amino acid catabolism occurs in three stages: (1) removal of the α amino group as ammonia, (2) conversion of the ammonia into urea, and (3) conversion of the remaining amino acid carbon skeleton, usually an α-keto acid, into an intermediate that can enter the citric acid cycle.

Deamination of an α-amino acid is accomplished by a pyridoxal phosphate (PLP)-dependent **transamination** reaction in which the $-NH_2$ group of the amino acid is exchanged with the keto group of α-ketoglutarate, forming a new α-keto acid plus glutamate. The glutamate is then oxidatively deaminated to give ammonia plus regenerated α-ketoglutarate, and the ammonia is converted into urea in the four-step **urea cycle**. Once the amino acids have been deaminated, their carbon chains are converted into one of seven intermediates that are further degraded in the citric acid cycle. Each amino acid has its own unique degradation pathway.

Humans synthesize only 11 of the 20 amino acids in proteins, called *nonessential amino acids.* The other 9, called *essential amino acids,* are biosynthesized only in plants and microorganisms and must be obtained in the diet. Each amino acid is biosynthesized by a unique pathway.

Key Words

anabolism, 833

catabolism, 833

deamination, 837

metabolism, 833

oxidative deamination, 841

transamination, 837

urea cycle, 842

Lagniappe

Visualizing Enzyme Structures

In the Chapter 19 *Lagniappe,* we discussed how to access enzyme structural data from the Protein Data Bank. Once the data for a specific enzyme have been located, it's then possible to visualize, manipulate, and study the structure. You can do this either by downloading the data file to your own computer and opening it with a free visualization program, such as DeepView (Swiss PDB Viewer) available at http://us.expasy.org/spdbv/, or you can use one of the display options built in to the PDB site.

Let's say that you want to study one of the more complex and interesting amino acid catabolic pathways and that you need to view urocanase, a key enzyme in histidine catabolism that catalyzes the addition of water to *trans*-urocanate.

Histidine   *trans*-Urocanate   Imidazolone 5-propionate

continued

Lagniappe *continued*

Go to the PDB site at http://rcsb.org/pdb/, type "urocanase" into the search window, and choose the structure with a PDB code of 1UWK. After clicking on the code, a screen with information on the selected structure appears, and several display options are presented in the "Images and Visualization" box on the right of the screen. An image of urocanase downloaded from the PDB is shown in Figure 20.14, with helical regions of the dimeric protein represented as coiled ribbons and pleated-sheet regions as flat ribbons.

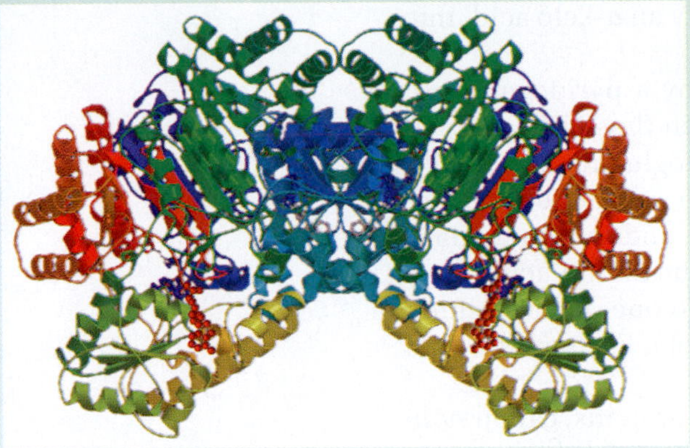

the enzyme. You can then examine the various interactions between the ligands and amino acid residues in the enzyme. Figure 20.15 shows the urocanate substrate and the pyridinium ring of NAD$^+$ cofactor bound at the active site. The –OH group of tyrosine-52 is hydrogen bonded to the urocanate nitrogen, while arginine-362 and threonine-133 are hydrogen bonded to the urocanate carboxylate.

Explore on your own; there is an immense amount of detailed information you can learn.

FIGURE 20.14 An image of urocanase, downloaded from the Protein Data Bank. Urocanase is a dimer composed of two identical subunits.

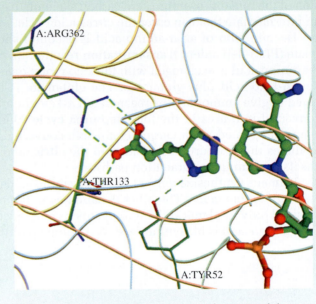

Among the display options available on the PDB site, one of the more useful lets you obtain detailed information on an enzyme's active site. Scroll down the 1UWK information screen until you come to a gray box titled "Ligand Chemical Component." Clicking on the [View] link under "Ligand Interaction" opens an applet called Ligand Explorer, which gives you a close-up view of the urocanate and NAD$^+$ ligands bound at the active site of

FIGURE 20.15 A view of the urocanate substrate and the pyridinium ring of NAD$^+$ cofactor bound at the active site of urocanase. The –OH group of tyrosine-52 is hydrogen bonded to the urocanate nitrogen, while arginine-362 and threonine-133 are hydrogen bonded to the urocanate carboxylate. Notice how urocanate is poised directly over the pyridinium ring of NAD$^+$.

Exercises

VISUALIZING CHEMISTRY

(Problems 20.1–20.9 appear within the chapter.)

20.10 ■ What amino acid is the following α-keto acid derived from?

20.11 The following compound is an intermediate in the biosynthesis of 1 of the 20 common α-amino acids. Which one is it likely to be, and what kind of chemical change must take place to complete the biosynthesis?

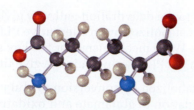

ADDITIONAL PROBLEMS

20.12 ■ What general kind of reaction does ATP carry out?

20.13 ■ Draw the structure of adenosine 5′-*mono*phosphate (AMP), an intermediate in numerous biochemical pathways.

20.14 ■ Cyclic adenosine monophosphate (cyclic AMP), a modulator of hormone action, is related to AMP (Problem 20.13) but has its phosphate group linked to *two* hydroxyl groups, at C3′ and C5′ of the sugar. Draw the structure of cyclic AMP.

■ indicates problems that are assignable in Organic OWL.

Go to this book's companion website at www.cengage.com/chemistry/mcmurry to explore interactive versions of the Active Figures from this text.

20.15 In addition to the two pathways for threonine catabolism discussed in Section 20.4, a third pathway converts threonine into α-ketobutyrate through a multistep mechanism that involves a PLP-dependent dehydration reaction analogous to what occurs in serine catabolism (Figure 20.8).

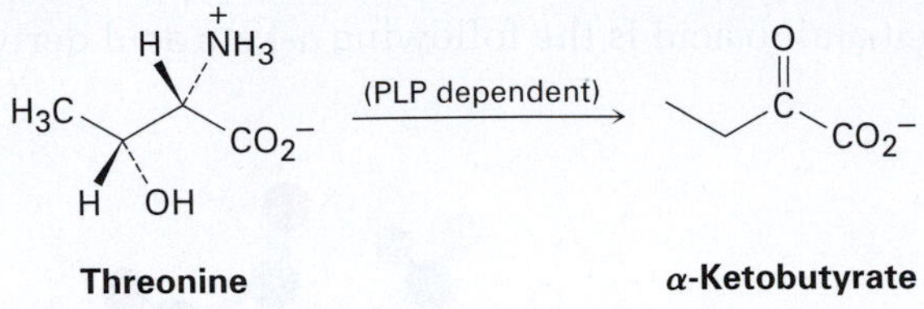

Threonine   **α-Ketobutyrate**

(a) The PLP–threonine imine formed by reaction of threonine with a PLP–enzyme imine undergoes an E1cB dehydration reaction to give an unsaturated PLP–imine. Propose a mechanism, and show the product.

(b) The unsaturated PLP–imine reacts with enzyme to give an enamine plus regenerated PLP–enzyme imine. Propose a mechanism, and show the product.

(c) The enamine is hydrolyzed to give α-ketobutyrate. Show the mechanism of the reaction.

20.16 ■ In addition to the dehydration pathway giving pyruvate (Figure 20.8), serine is also catabolized by an alternative PLP-dependent pathway that gives glycine. Write mechanisms for the key step: a base-catalyzed loss of CH_2O from the PLP–serine imine to give the PLP–glycine imine.

20.17 Proline is catabolized by conversion to glutamate 5-semialdehyde, followed by oxidation to glutamate and oxidative deamination to α-ketoglutarate.

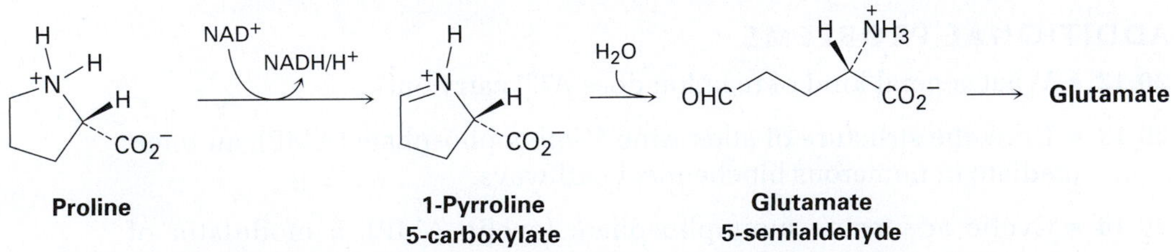

Proline   **1-Pyrroline 5-carboxylate**   **Glutamate 5-semialdehyde**   **Glutamate**

(a) The oxidation of proline to 1-pyrroline 5-carboxylate is analogous to what occurs in the oxidative deamination of glutamate to α-ketoglutarate (Section 20.2). Propose a mechanism.

(b) Show the mechanism of the hydrolysis of 1-pyrroline 5-carboxylate to give glutamate 5-semialdehyde.

(c) What coenzyme is probably required for the oxidation of glutamate 5-semialdehyde to glutamate?

■ Problems assignable in Organic OWL.

20.18 ■ Tyrosine is catabolized by a series of steps that include the following transformations:

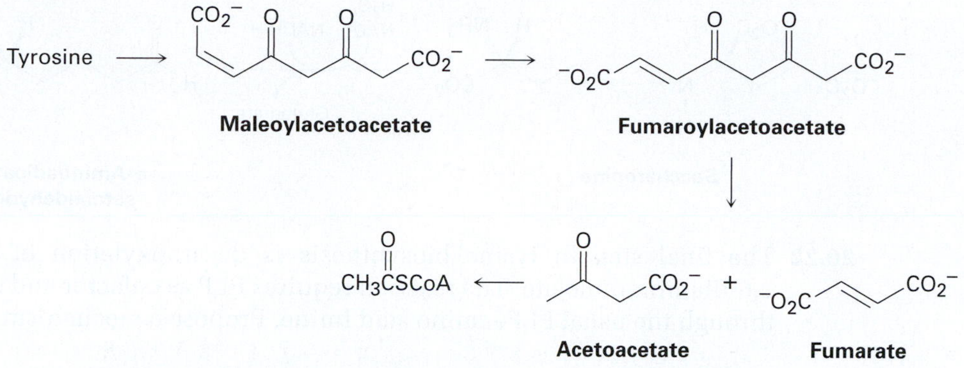

(a) The double-bond isomerization of maleoylacetoacetate to fumaroylacetoacetate is catalyzed by practically any nucleophile, :Nu⁻. Review Section 14.11, and then propose a mechanism.

(b) Propose a mechanism for the biological conversion of fumaroylacetoacetate to fumarate plus acetoacetate.

20.19 ■ Cysteine, $C_3H_7NO_2S$, is biosynthesized from a substance called cystathionine by a multistep pathway:

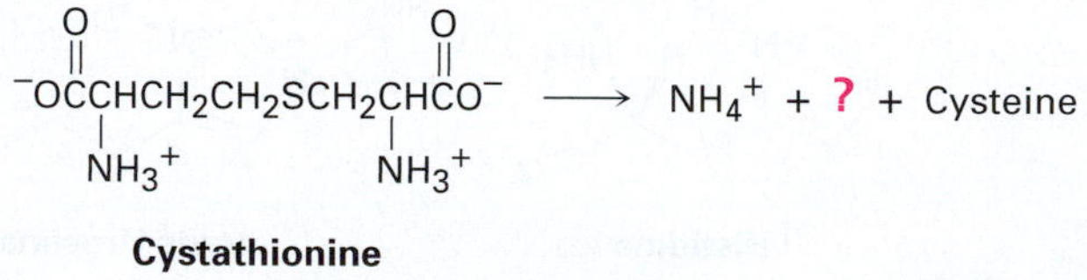

(a) The first step is a transamination. What is the product?

(b) The second step is an E1cB reaction. Show the products and the mechanism of the reaction.

(c) The final step is a double-bond reduction. What product is represented by the question mark in the equation?

20.20 ■ Lysine catabolism begins with reductive amination of α-ketoglutarate to give saccharopine. Show the mechanism.

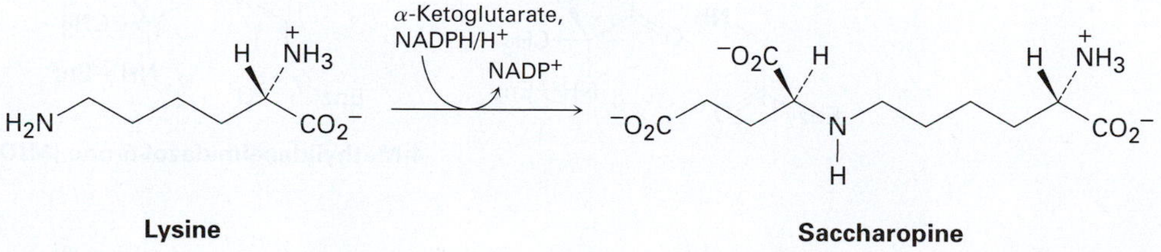

20.21 ■ The second step in lysine catabolism is oxidative deamination of saccharopine to give α-aminoadipate semialdehyde. Show the mechanism.

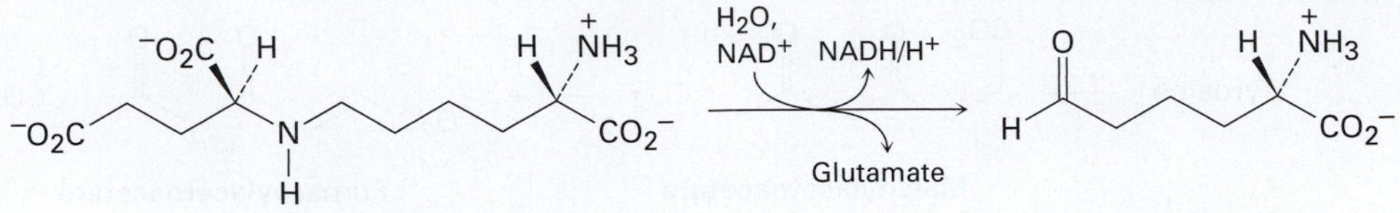

Saccharopine **α-Aminoadipate semialdehyde**

20.22 The final step in lysine biosynthesis is decarboxylation of *meso*-2,6-diaminopimelate. The reaction requires PLP as cofactor and occurs through the usual PLP–amino acid imine. Propose a mechanism.

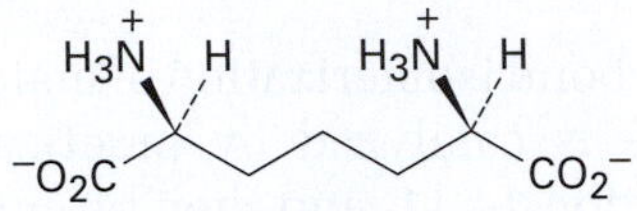

***meso*-2,6-Diaminopimelate**

20.23 Histidine catabolism begins with elimination of ammonia to give *trans*-urocanate in a step catalyzed by histidine ammonia lyase.

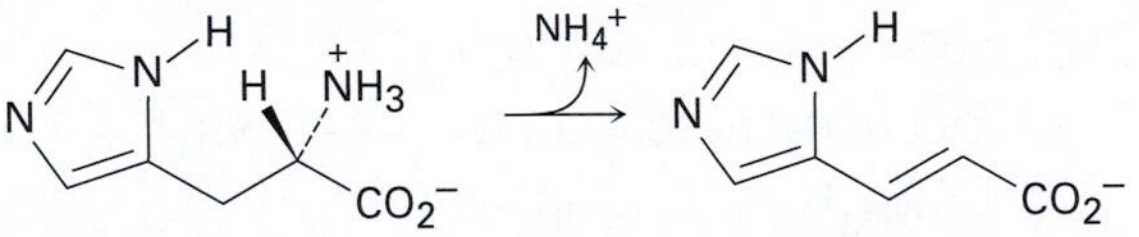

Histidine ***trans*-Urocanate**

The process is more complex than it appears and involves initial formation of a 4-methylideneimidazol-5-one (MIO) ring that arises by cyclization and dehydration of an –Ala–Ser–Gly– segment within the histidine ammonia lyase enzyme. Propose a mechanism for the formation of the MIO ring.

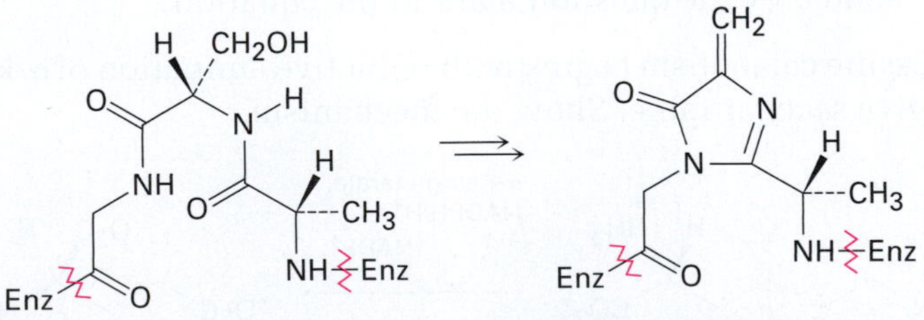

4-Methylideneimidazol-5-one (MIO)

■ Problems assignable in Organic OWL.

20.24 After the MIO ring is formed (Problem 20.23), histidine adds to the MIO in a conjugate nucleophilic addition reaction, producing an iminium ion on histidine that makes the neighboring –CH₂– hydrogens acidic and allows a subsequent E1cB reaction. Show the mechanism.

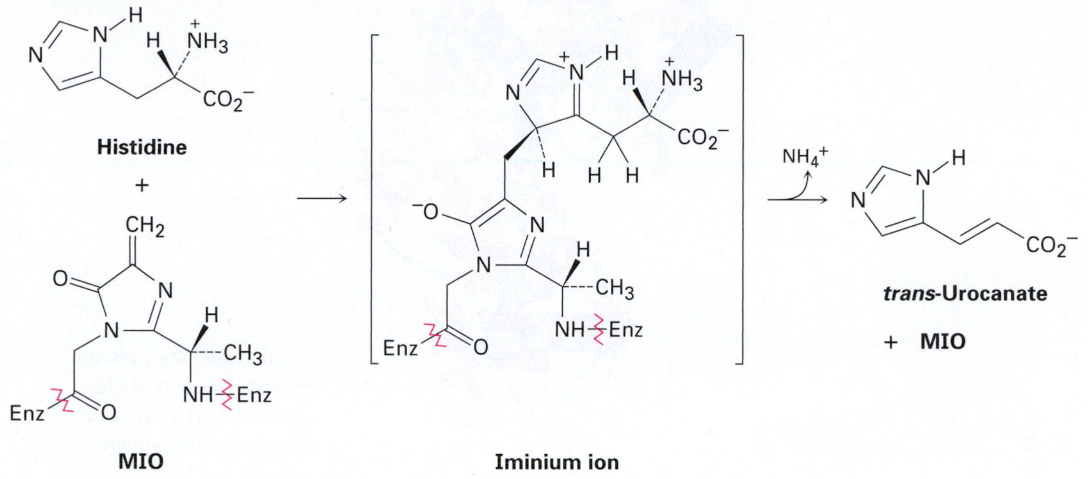

Histidine

MIO

Iminium ion

***trans*-Urocanate**

+ MIO

Following the E1cB reaction, expulsion of *trans*-urocanate by a mechanism that is the opposite of the conjugate addition step regenerates the MIO. Show the mechanism.

20.25 ■ Leucine is biosynthesized from α-ketoisocaproate, which is itself formed from α-ketoisovalerate by a multistep route that involves:

(1) reaction of α-ketoisocaproate with acetyl CoA in an aldol-like reaction

(2) hydrolysis of the thioester

(3) dehydration by an E1cB mechanism

(4) hydration by a conjugate addition reaction

(5) oxidation of an alcohol to a ketone

(6) decarboxylation of a β-keto acid

Show the steps in the biosynthesis, and propose a mechanism for each.

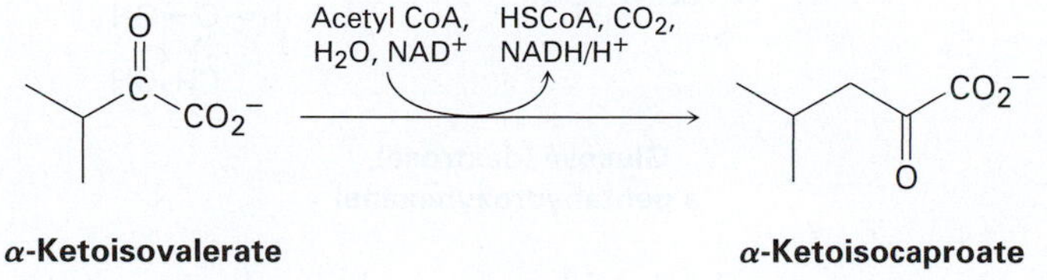

α-Ketoisovalerate

α-Ketoisocaproate

21 Biomolecules: Carbohydrates

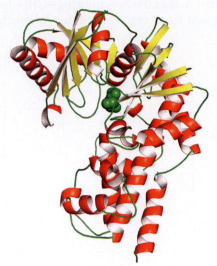

Hexokinase catalyzes the phosphorylation of glucose, the first step in carbohydrate metabolism.

Carbohydrates occur in every living organism. The sugar and starch in food and the cellulose in wood, paper, and cotton are nearly pure carbohydrates. Modified carbohydrates form part of the coating around living cells, other carbohydrates are part of the nucleic acids that carry our genetic information, and still others are used as medicines.

The word **carbohydrate** derives historically from the fact that glucose, the first simple carbohydrate to be obtained in pure form, has the molecular formula $C_6H_{12}O_6$ and was originally thought to be a "hydrate of carbon, $C_6(H_2O)_6$." This view was soon abandoned, but the name survived. Today, the term *carbohydrate* is used to refer loosely to the broad class of polyhydroxylated aldehydes and ketones commonly called *sugars*. Glucose, also known as *dextrose* in medical work, is the most familiar example.

Glucose (dextrose),
a pentahydroxyhexanal

Carbohydrates are synthesized by green plants during photosynthesis, a complex process in which sunlight provides the energy to convert CO_2 and H_2O into glucose plus oxygen. Many molecules of glucose are then chemically linked for storage by the plant in the form of either cellulose or starch. It has

OWL Online homework for this chapter can be assigned in Organic OWL.

been estimated that more than 50% of the dry weight of the earth's biomass—all plants and animals—consists of glucose polymers. When eaten and metabolized, carbohydrates then provide animals with a source of readily available energy. Thus, carbohydrates act as the chemical intermediaries by which solar energy is stored and used to support life.

$$6\ CO_2\ +\ 6\ H_2O\ \xrightarrow{\text{Sunlight}}\ 6\ O_2\ +\ C_6H_{12}O_6\ \longrightarrow\ \text{Cellulose, starch}$$

Glucose

Because humans and most other mammals lack the enzymes needed for digestion of cellulose, they require starch as their dietary source of carbohydrates. Grazing animals such as cows, however, have microorganisms in their first stomach that are able to digest cellulose. The energy stored in cellulose is thus moved along the biological food chain when these ruminant animals eat grass and are themselves used for food.

WHY THIS CHAPTER?

Carbohydrates are the second major class of biomolecules to be discussed. We'll see in this chapter what the structures and primary biological functions of carbohydrates are, and we'll look in the following chapter at how carbohydrates are biosynthesized and degraded in organisms.

21.1 Classification of Carbohydrates

Carbohydrates are generally classed as either *simple* or *complex*. **Simple sugars**, or **monosaccharides**, are carbohydrates like glucose and fructose that can't be converted into smaller sugars by hydrolysis. **Complex carbohydrates** are made of two or more simple sugars linked together by acetal bonds (Section 14.8). Sucrose (table sugar), for example, is a *disaccharide* made up of one glucose linked to one fructose. Similarly, cellulose is a *polysaccharide* made up of several thousand glucose units linked together. Enzyme-catalyzed hydrolysis of a polysaccharide breaks it down into its constituent monosaccharides.

Sucrose
(a disaccharide)

$\xrightarrow{H_3O^+}$ **1 Glucose + 1 Fructose**

Cellulose
(a polysaccharide)

$\xrightarrow{H_3O^+}$ **~3000 Glucose**

Monosaccharides are further classified as either **aldoses** or **ketoses**. The *-ose* suffix designates a carbohydrate, and the *aldo-* and *keto-* prefixes identify the kind of carbonyl group present in the molecule, whether aldehyde or ketone. The number of carbon atoms in the monosaccharide is indicated by the appropriate numerical prefix *tri-*, *tetr-*, *pent-*, *hex-*, and so forth, in the name. Putting it all together, glucose is an *aldohexose,* a six-carbon aldehydo sugar; fructose is a *ketohexose,* a six-carbon keto sugar; ribose is an *aldopentose,* a five-carbon aldehydo sugar; and sedoheptulose is a *ketoheptose,* a seven-carbon keto sugar. Most of the common simple sugars are either pentoses or hexoses.

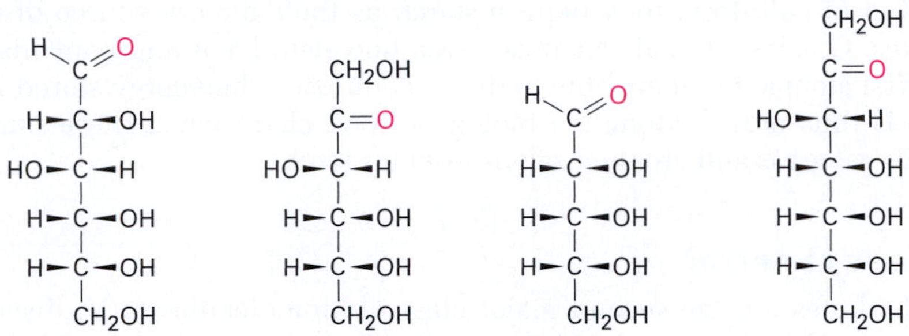

Glucose
(an aldohexose)

Fructose
(a ketohexose)

Ribose
(an aldopentose)

Sedoheptulose
(a ketoheptose)

Problem 21.1

Classify each of the following monosaccharides:

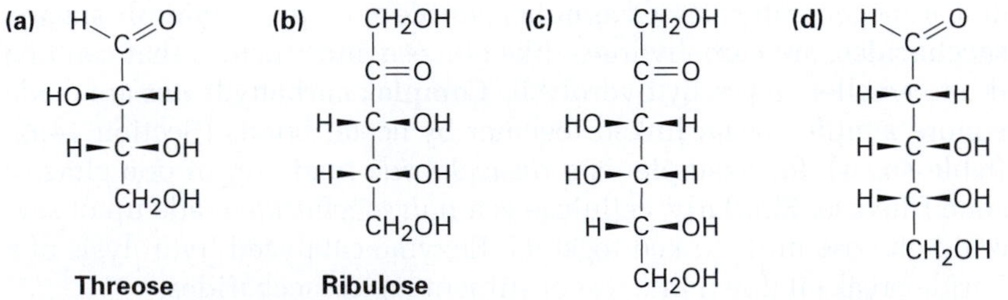

(a) Threose

(b) Ribulose

(c) Tagatose

(d) 2-Deoxyribose

21.2 Depicting Carbohydrate Stereochemistry: Fischer Projections

---

Because carbohydrates usually have numerous chirality centers, it was recognized long ago that a quick method for representing stereochemistry is needed. In 1891, Emil Fischer suggested a method based on the projection of a tetrahedral carbon atom onto a flat surface. These **Fischer projections** were soon adopted and are now a common means of representing stereochemistry at chirality centers, particularly in carbohydrate chemistry.

A tetrahedral carbon atom is represented in a Fischer projection by two crossed lines. The horizontal lines represent bonds coming out of the page, and the vertical lines represent bonds going into the page:

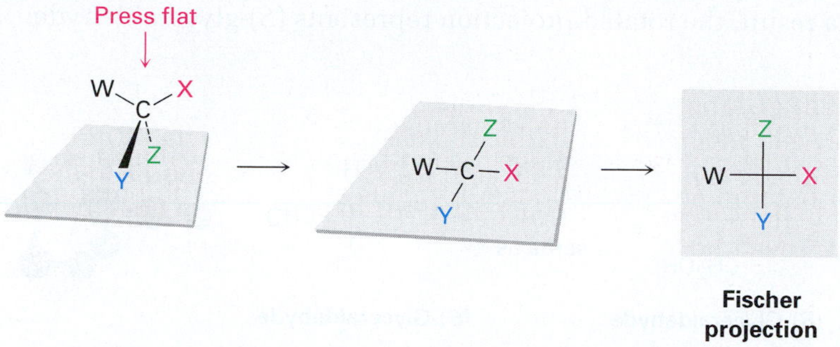

Press flat

Fischer
projection

For example, (*R*)-glyceraldehyde, the simplest monosaccharide, can be drawn as in Figure 21.1.

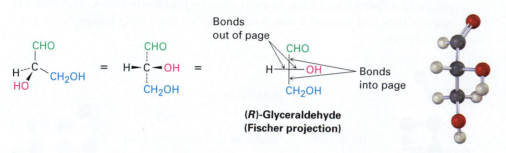

Bonds
out of page

Bonds
into page

(*R*)-Glyceraldehyde
(Fischer projection)

ACTIVE FIGURE 21.1
A Fischer projection of
(*R*)-glyceraldehyde. **Go to this
book's student companion
site at** www.cengage.com/
chemistry/mcmurry **to explore
an interactive version of this
figure.**

Because a given molecule can be drawn in many ways, it's sometimes necessary to compare two projections to see if they represent the same or different enantiomers. To test for identity, Fischer projections can be moved around on the paper, but only two kinds of motions are allowed; moving a Fischer projection in any other way inverts its meaning.

• A Fischer projection can be rotated on the page by 180°, but *not by 90° or 270°*. Only a 180° rotation maintains the Fischer convention by keeping the same substituent groups going into and coming out of the plane. In the following Fischer projection of (*R*)-glyceraldehyde, for example, the –H and –OH groups come out of the plane both before and after a 180° rotation:

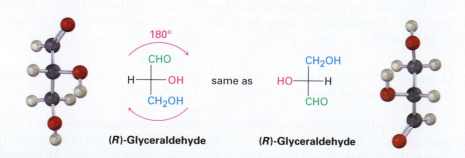

180°

same as

(*R*)-Glyceraldehyde   (*R*)-Glyceraldehyde

A 90° rotation breaks the Fischer convention by exchanging the groups that go into the plane and those that come out. In the following Fischer projections of (*R*)-glyceraldehyde, the –H and –OH groups come out of the plane before rotation but go into the plane after a 90° rotation. As a result, the rotated projection represents (*S*)-glyceraldehyde:

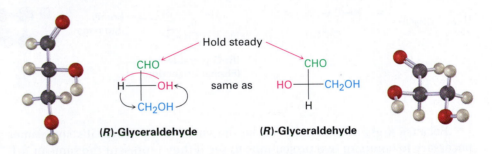

- A Fischer projection can have one group held steady while the other three rotate in either a clockwise or a counterclockwise direction. The effect is simply to rotate around a single bond, which does not change the stereochemistry.

R,S stereochemical designations (Section 5.5) can be assigned to the chirality center in a Fischer projection by following three steps, as shown in Worked Example 21.1.

Step 1
Assign priorities to the four substituents in the usual way (Section 5.5).

Step 2
Place the group of lowest priority, usually H, at the top of the Fischer projection by using one of the allowed motions. This means that the lowest-priority group is oriented back, away from the viewer, as required for assigning configuration.

Step 3
Determine the direction of rotation 1 → 2 → 3 of the remaining three groups, and assign *R* or *S* configuration.

Carbohydrates with more than one chirality center are shown in Fischer projections by stacking the centers on top of one another. By convention, the carbonyl carbon is always placed either at or near the top. Glucose, for example, has four chirality centers stacked on top of one another in a Fischer projection. Such representations don't, however, give an accurate picture of the true three-dimensional conformation of a molecule, which is curled around on itself like a bracelet.

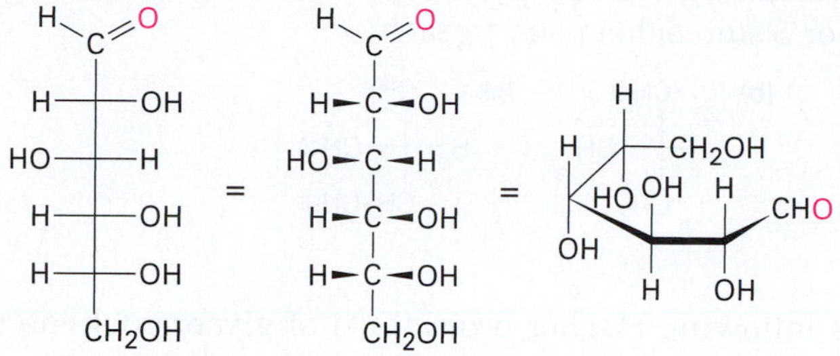

Glucose
(carbonyl group at top)

WORKED EXAMPLE 21.1 Assigning _R_ or _S_ Configuration to a Fischer Projection

Assign _R_ or _S_ configuration to the following Fischer projection of alanine:

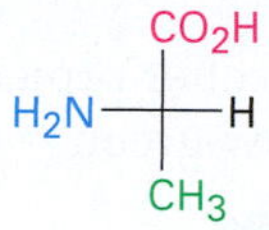

Alanine

Strategy

Follow the steps listed in the text. (1) Assign priorities to the four substituents on the chiral carbon. (2) Manipulate the Fischer projection to place the group of lowest priority at the top by carrying out one of the allowed motions. (3) Determine the direction $1 \rightarrow 2 \rightarrow 3$ of the remaining three groups.

Solution

The priorities of the groups are (1) $-NH_2$, (2) $-CO_2H$, (3) $-CH_3$, and (4) $-H$. To bring the group of lowest priority ($-H$) to the top, we might want to hold the $-CH_3$ group steady while rotating the other three groups counterclockwise:

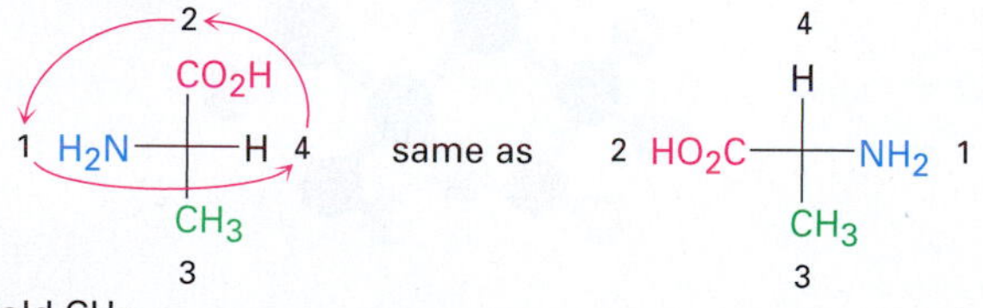

Going from first- to second- to third-highest priority requires a counterclockwise turn, corresponding to _S_ stereochemistry.

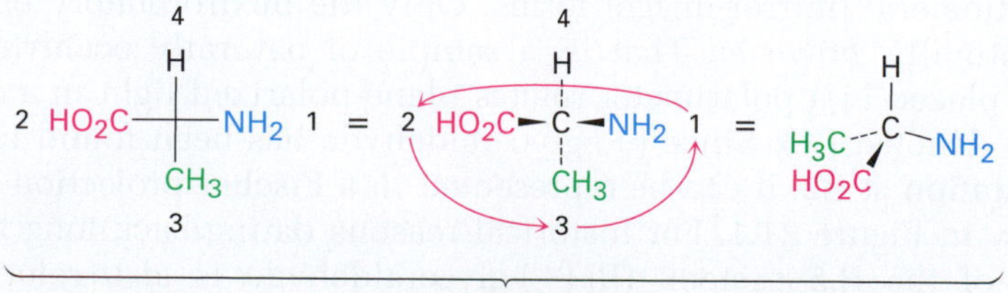

_S_ configuration

Problem 21.2

Convert the following Fischer projections into tetrahedral representations, and assign R or S stereochemistry to each:

(a)
$$\begin{array}{c} CO_2H \\ H_2N \text{---} \!\!\!\!\!\underset{|}{\overset{|}{}}\!\!\!\!\! \text{---} H \\ CH_3 \end{array}$$

(b)
$$\begin{array}{c} CHO \\ H \text{---} \!\!\!\!\!\underset{|}{\overset{|}{}}\!\!\!\!\! \text{---} OH \\ CH_3 \end{array}$$

(c)
$$\begin{array}{c} CH_3 \\ H \text{---} \!\!\!\!\!\underset{|}{\overset{|}{}}\!\!\!\!\! \text{---} CHO \\ CH_2CH_3 \end{array}$$

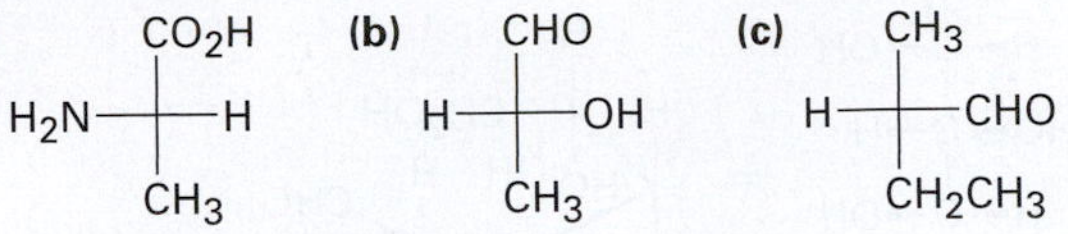

Problem 21.3

Which of the following Fischer projections of glyceraldehyde represent the same enantiomer?

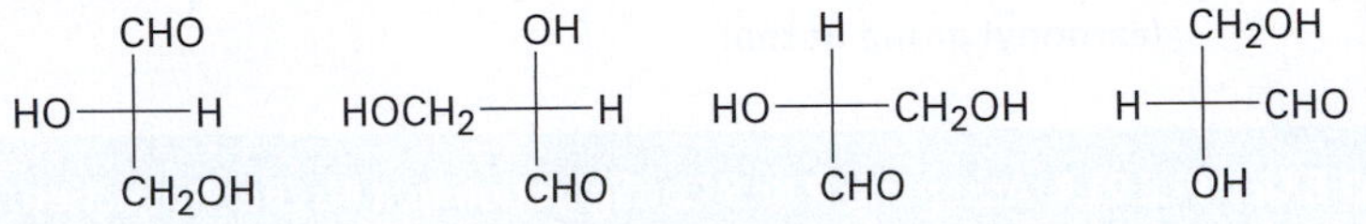

$$\begin{array}{c} CHO \\ HO \text{---} \!\!\!\!\!\underset{|}{\overset{|}{}}\!\!\!\!\! \text{---} H \\ CH_2OH \end{array}$$
A

$$\begin{array}{c} OH \\ HOCH_2 \text{---} \!\!\!\!\!\underset{|}{\overset{|}{}}\!\!\!\!\! \text{---} H \\ CHO \end{array}$$
B

$$\begin{array}{c} H \\ HO \text{---} \!\!\!\!\!\underset{|}{\overset{|}{}}\!\!\!\!\! \text{---} CH_2OH \\ CHO \end{array}$$
C

$$\begin{array}{c} CH_2OH \\ H \text{---} \!\!\!\!\!\underset{|}{\overset{|}{}}\!\!\!\!\! \text{---} CHO \\ OH \end{array}$$
D

Problem 21.4

Redraw the following molecule as a Fischer projection, and assign R or S configuration to the chirality center (yellow-green = Cl):

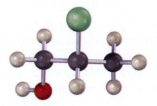

Problem 21.5

Redraw the following aldotetrose as a Fischer projection, and assign R or S configuration to each chirality center:

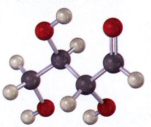

21.3 D,L Sugars

Glyceraldehyde, the simplest aldose, has only one chirality center and thus has two enantiomeric (mirror-image) forms. Only the dextrorotatory enantiomer occurs naturally, however. That is, a sample of naturally occurring glyceraldehyde placed in a polarimeter rotates plane-polarized light in a clockwise direction, denoted (+). Since (+)-glyceraldehyde has been found to have an R configuration at C2, it can be represented in a Fischer projection as shown previously in Figure 21.1. For historical reasons dating back long before the adoption of the R,S system, (R)-(+)-glyceraldehyde is also referred to as D-glyceraldehyde (D for dextrorotatory). The other enantiomer, (S)-(−)-glyceraldehyde, is known as L-glyceraldehyde (L for levorotatory).

Because of the way monosaccharides are biosynthesized in nature, glucose, fructose, and most (although not all) other naturally occurring monosaccharides have the same *R* stereochemical configuration as D-glyceraldehyde at the chirality center farthest from the carbonyl group. In Fischer projections, therefore, most naturally occurring sugars have the hydroxyl group at the bottom chirality center pointing to the right (Figure 21.2). Such compounds are referred to as **D sugars**.

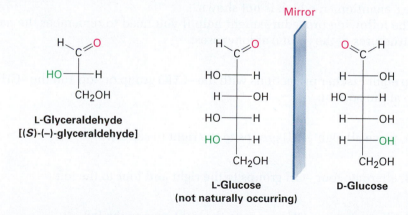

D-Glyceraldehyde
[(R)-(+)-glyceraldehyde]

D-Ribose

D-Glucose

D-Fructose

FIGURE 21.2 Some naturally occurring D sugars. The –OH group at the chirality center farthest from the carbonyl group has the same configuration as (R)-(+)-glyceraldehyde and points toward the right in Fischer projections.

In contrast with D sugars, **L sugars** have an *S* configuration at the lowest chirality center, with the bottom –OH group pointing to the *left* in Fischer projections. Thus, an L sugar is the mirror image (enantiomer) of the corresponding D sugar and has the opposite configuration from the D sugar at all chirality centers. Note that the D and L notations have no relation to the direction in which a given sugar rotates plane-polarized light; a D sugar can be either dextrorotatory or levorotatory. The prefix D indicates only that the –OH group at the lowest chirality center has *R* stereochemistry and points to the right when the molecule is drawn in a Fischer projection. Note also that the D,L system of carbohydrate nomenclature describes the configuration at only one chirality center and says nothing about the configuration of other chirality centers that may be present.

Mirror

L-Glyceraldehyde
[(S)-(–)-glyceraldehyde]

L-Glucose
(not naturally occurring)

D-Glucose

Problem 21.6

Assign *R* or *S* configuration to each chirality center in the following monosaccharides, and tell whether each is a D sugar or an L sugar:

(a)

(b)

(c)

Problem 21.7

(+)-Arabinose, an aldopentose that is widely distributed in plants, is systematically named (2R,3S,4S)-2,3,4,5-tetrahydroxypentanal. Draw a Fischer projection of (+)-arabinose, and identify it as a D sugar or an L sugar.

21.4 Configurations of the Aldoses

Aldotetroses are four-carbon sugars with two chirality centers and an aldehyde carbonyl group. Thus, there are $2^2 = 4$ possible stereoisomeric aldotetroses, or two D,L pairs of enantiomers, named *erythrose* and *threose.*

Aldopentoses have three chirality centers and a total of $2^3 = 8$ possible stereoisomers, or four D,L pairs of enantiomers. These four pairs are called *ribose, arabinose, xylose,* and *lyxose.* All except lyxose occur widely. D-Ribose is an important constituent of RNA (ribonucleic acid), L-arabinose is found in many plants, and D-xylose is found in both plants and animals.

Aldohexoses have four chirality centers and a total of $2^4 = 16$ possible stereoisomers, or eight D,L pairs of enantiomers. The names of the eight are *allose, altrose, glucose, mannose, gulose, idose, galactose,* and *talose.* Only D-glucose, from starch and cellulose, and D-galactose, from gums and fruit pectins, are found widely in nature. D-Mannose and D-talose also occur naturally but in lesser abundance.

Fischer projections of the four-, five-, and six-carbon D aldoses are shown in Figure 21.3. Starting with D-glyceraldehyde, we can imagine constructing the two D aldotetroses by inserting a new chirality center just below the aldehyde carbon. Each of the two D aldotetroses then leads to two D aldopentoses (four total), and each of the four D aldopentoses leads to two D aldohexoses (eight total). In addition, each of the D aldoses in Figure 21.3 has a mirror-image L enantiomer, which is not shown.

The following procedure might help if you need to remember the names and structures of the eight D aldohexoses:

Step 1

Set up eight Fischer projections with the –CHO group on top and the –CH$_2$OH group at the bottom.

Step 2

At C5, place all eight –OH groups to the right (D series).

Step 3

At C4, alternate four –OH groups to the right and four to the left.

Step 4

At C3, alternate two –OH groups to the right, two to the left.

Step 5

At C2, alternate –OH groups right, left, right, left.

Step 6

Name the eight isomers using the mnemonic "**All alt**ruists **gl**adly **ma**ke **gu**m in **gal**lon **ta**nks."

The structures of the four D aldopentoses can be generated in a similar way and named by the mnemonic suggested by a Cornell University undergraduate: "**Rib**s **ar**e **ex**tra **l**ean."

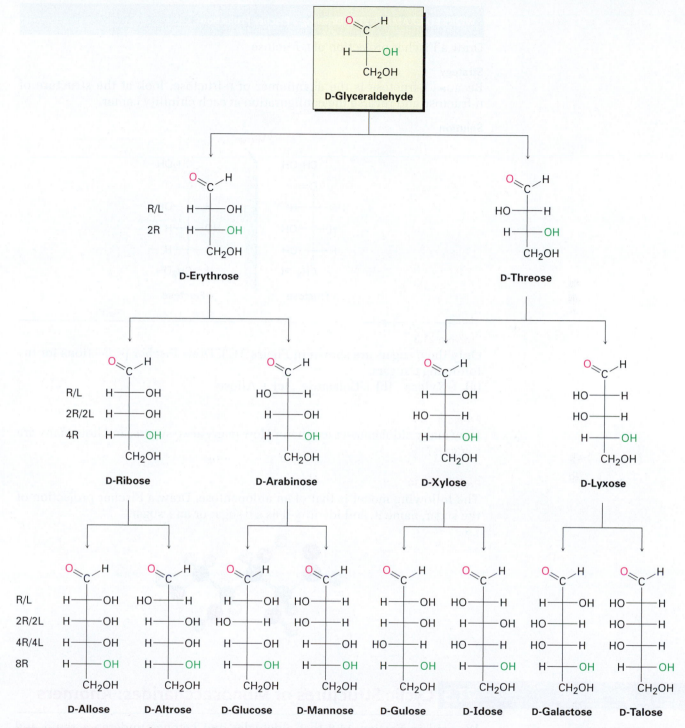

ACTIVE FIGURE 21.3 Configurations of D aldoses. The structures are arranged from left to right so that the –OH groups on C2 alternate right/left (R/L) in going across a series. Similarly, the –OH groups at C3 alternate two right/two left (2R/2L), the –OH groups at C4 alternate 4R/4L, and the –OH groups at C5 are to the right in all eight (8R). Each D aldose has a mirror-image L enantiomer, which is not shown. **Go to this book's student companion site at www.cengage.com/chemistry/mcmurry to explore an interactive version of this figure.**

WORKED EXAMPLE 21.2 Drawing a Fischer Projection

Draw a Fischer projection of L-fructose.

Strategy

Because L-fructose is the enantiomer of D-fructose, look at the structure of D-fructose and reverse the configuration at each chirality center.

Solution

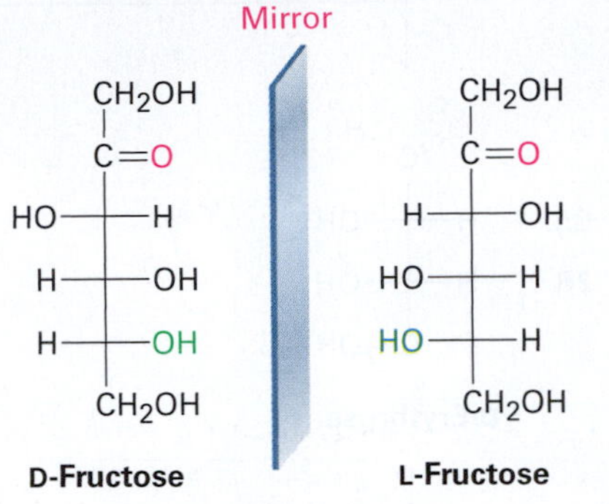

Mirror

```
      CH₂OH              CH₂OH
       |                  |
       C=O                C=O
  HO ──┼── H         H ──┼── OH
   H ──┼── OH        HO ──┼── H
   H ──┼── OH        HO ──┼── H
      CH₂OH              CH₂OH
```

D-Fructose   **L-Fructose**

Problem 21.8

Only the D sugars are shown in Figure 21.3. Draw Fischer projections for the following L sugars:
(a) L-Xylose **(b)** L-Galactose **(c)** L-Allose

Problem 21.9

How many aldoheptoses are there? How many are D sugars, and how many are L sugars?

Problem 21.10

The following model is that of an aldopentose. Draw a Fischer projection of the sugar, name it, and identify it as a D sugar or an L sugar.

21.5 Cyclic Structures of Monosaccharides: Anomers

We said in Section 14.8 that aldehydes and ketones undergo a rapid and reversible nucleophilic addition reaction with alcohols to form hemiacetals:

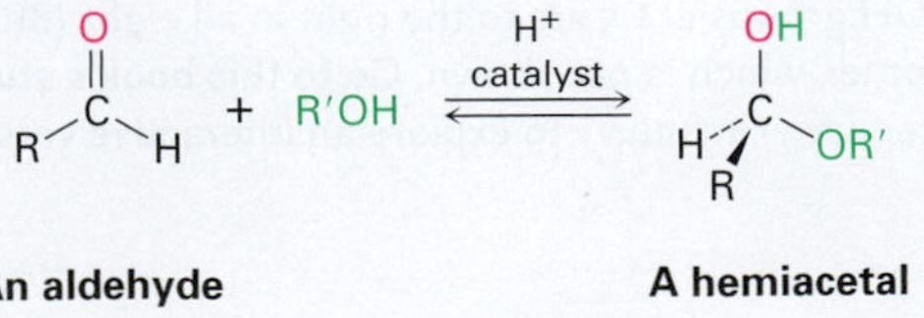

$$
\underset{\textbf{An aldehyde}}{R-\overset{\overset{\displaystyle O}{\|}}{C}-H} \; + \; R'OH \; \underset{}{\overset{H^+ \text{ catalyst}}{\rightleftharpoons}} \; \underset{\textbf{A hemiacetal}}{H-\overset{\overset{\displaystyle OH}{|}}{\underset{R}{C}}-OR'}
$$

If the carbonyl and the hydroxyl group are in the same molecule, an intra-molecular nucleophilic addition can take place, leading to the formation of a cyclic hemiacetal. Five- and six-membered cyclic hemiacetals are relatively strain-free and particularly stable, and many carbohydrates therefore exist in an equilibrium between open-chain and cyclic forms. Glucose, for instance, exists in aqueous solution primarily in the six-membered, **pyranose** form resulting from intramolecular nucleophilic addition of the –OH group at C5 to the C1 carbonyl group (Figure 21.4). The name *pyranose* is derived from *pyran*, the name of the unsaturated six-membered cyclic ether.

Like cyclohexane rings (Section 4.6), pyranose rings have a chairlike geometry with axial and equatorial substituents. By convention, the rings are usually drawn by placing the hemiacetal oxygen atom at the right rear, as shown in Figure 21.4. Note that an –OH group on the *right* in a Fischer projection is on the *bottom* face of the pyranose ring, and an –OH group on the *left* in a Fischer projection is on the *top* face of the ring. For D sugars, the terminal –CH₂OH group is on the top of the ring, whereas for L sugars, the –CH₂OH group is on the bottom.

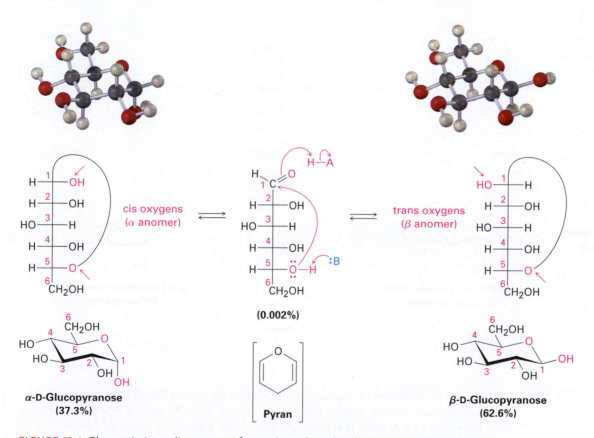

FIGURE 21.4 Glucose in its cyclic pyranose forms. As explained in the text, two anomers are formed by cyclization of glucose. The molecule whose newly formed –OH group at C1 is cis to the oxygen atom on the lowest chirality center (C5) in a Fischer projection is the α anomer. The molecule whose newly formed –OH group is trans to the oxygen atom on the lowest chirality center in a Fischer projection is the β anomer.

When an open-chain monosaccharide cyclizes to a pyranose form, a new chirality center is generated at the former carbonyl carbon and two diastereomers, called **anomers**, are produced. The hemiacetal carbon atom is referred to as the **anomeric center**. For example, glucose cyclizes reversibly in aqueous solution to a 37:63 mixture of two anomers (Figure 21.4). The compound with its newly generated –OH group at C1 *cis* to the –OH at the lowest chirality center in a Fischer projection is called the **α anomer** and has the full name α-D-glucopyranose. The compound with its newly generated –OH group *trans* to the –OH at the lowest chirality center in a Fischer projection is called the **β anomer** and has the full name β-D-glucopyranose. Note that in β-D-gluco-pyranose, all the substituents on the ring are equatorial. Thus, β-D-gluco-pyranose is the least sterically crowded and most stable of the eight D aldohexoses.

Some monosaccharides also exist in a five-membered cyclic hemiacetal form called a **furanose**. D-Fructose, for instance, exists in water solution as 70% β-pyranose, 2% α-pyranose, 0.7% open-chain, 23% β-furanose, and 5% α-furanose. The pyranose form results from addition of the –OH at C6 to the carbonyl group, while the furanose form results from addition of the –OH at C5 to the carbonyl group (Figure 21.5).

ACTIVE FIGURE 21.5
Pyranose and furanose forms of fructose in aqueous solution. The two pyranose anomers result from addition of the C6 –OH group to the C2 carbonyl, while the two furanose anomers result from addition of the C5 –OH group to the C2 carbonyl. **Go to this book's student companion site at** www.cengage.com/chemistry/mcmurry **to explore an interactive version of this figure.**

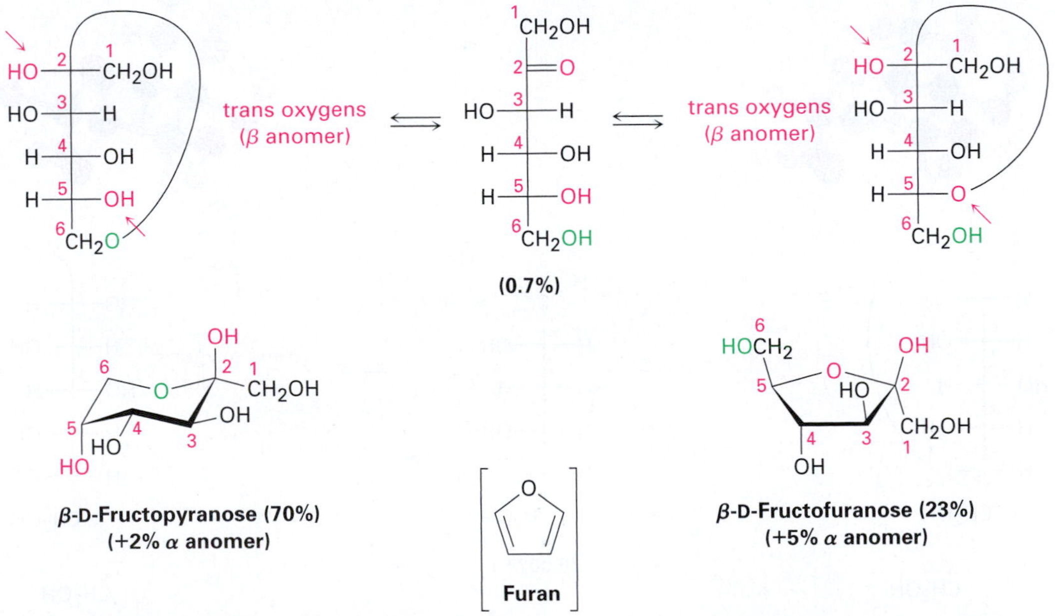

Both anomers of D-glucopyranose can be crystallized and purified. Pure α-D-glucopyranose has a melting point of 146 °C and a specific rotation $[\alpha]_D = +112.2$; pure β-D-glucopyranose has a melting point of 148 to 155 °C and a specific rotation $[\alpha]_D = +18.7$. When a sample of either pure anomer is dissolved in water, however, its optical rotation slowly changes until it reaches a constant value of +52.6. That is, the specific rotation of the α-anomer solution decreases from +112.2 to +52.6, and the specific rotation of the β-anomer solution increases from +18.7 to +52.6. Called **mutarotation**, this change in optical rotation is due to the slow conversion of the pure anomers into a 37:63 equilibrium mixture.

Mutarotation occurs by a reversible ring-opening of each anomer to the open-chain aldehyde, followed by reclosure (Figure 21.4). Although the equilibration is slow at neutral pH, it is catalyzed by both acid and base.

WORKED EXAMPLE 21.3 Drawing the Chair Conformation of an Aldohexose

D-Mannose differs from D-glucose in its stereochemistry at C2. Draw D-mannose in its chairlike pyranose form.

Strategy

First draw a Fischer projection of D-mannose. Then, lay it on its side and curl it around so that the –CHO group (C1) is toward the right front and the –CH$_2$OH group (C6) is toward the left rear. Now, connect the –OH at C5 to the C1 carbonyl group to form the pyranose ring. In drawing the chair form, raise the leftmost carbon (C4) up and drop the rightmost carbon (C1) down.

Solution

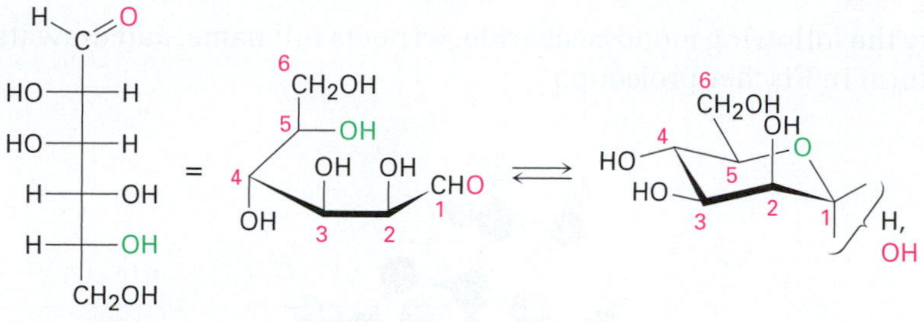

D-Mannose   **(Pyranose form)**

WORKED EXAMPLE 21.4 Drawing the Chair Conformation of a Pyranose

Draw β-L-glucopyranose in its more stable chair conformation.

Strategy

It's probably easiest to begin by drawing the chair conformation of β-D-glucopyranose. Then draw its mirror-image L enantiomer by changing the stereochemistry at every position on the ring, and carry out a ring-flip to give the more stable chair conformation. Note that the –CH$_2$OH group is on the bottom face of the ring in the L enantiomer as is the anomeric –OH.

Solution

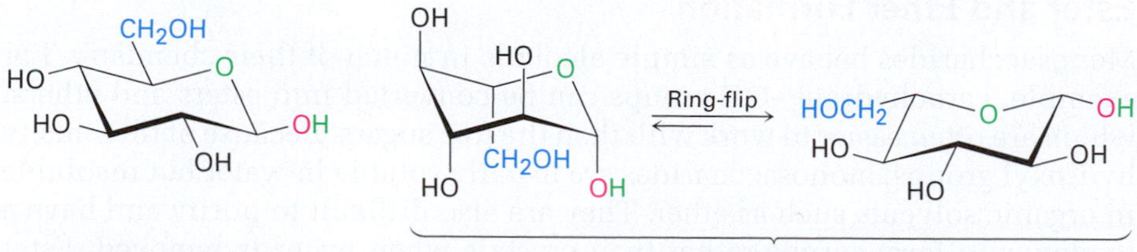

β-D-Glucopyranose   **β-L-Glucopyranose**

Problem 21.11

Ribose exists largely in a furanose form, produced by addition of the C4 –OH group to the C1 aldehyde. Draw D-ribose in its furanose form.

Problem 21.12

Figure 21.5 shows only the β-pyranose and β-furanose anomers of D-fructose. Draw the α-pyranose and α-furanose anomers.

Problem 21.13

Draw β-D-galactopyranose and β-D-mannopyranose in their more stable chair conformations. Label each ring substituent as either axial or equatorial. Which would you expect to be more stable, galactose or mannose?

Problem 21.14

Draw β-L-galactopyranose in its more stable chair conformation, and label the substituents as either axial or equatorial.

Problem 21.15

Identify the following monosaccharide, write its full name, and draw its open-chain form in Fischer projection:

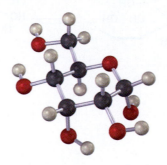

21.6 Reactions of Monosaccharides

Because monosaccharides contain only two kinds of functional groups, hydroxyls and carbonyls, most of the chemistry of monosaccharides is the familiar chemistry of these two groups. As we've seen, alcohols can be converted to esters and ethers and can be oxidized; carbonyl compounds can react with nucleophiles and can be reduced.

Ester and Ether Formation

Monosaccharides behave as simple alcohols in much of their chemistry. For example, carbohydrate —OH groups can be converted into esters and ethers, which are often easier to work with than the free sugars. Because of their many hydroxyl groups, monosaccharides are usually soluble in water but insoluble in organic solvents such as ether. They are also difficult to purify and have a tendency to form syrups rather than crystals when water is removed. Ester and ether derivatives, however, are soluble in organic solvents and are easily purified and crystallized.

Esterification is normally carried out by treating the carbohydrate with an acid chloride or acid anhydride in the presence of a base (Sections 16.4 and 16.5). All the —OH groups react, including the anomeric one. For example,

β-D-glucopyranose is converted into its pentaacetate by treatment with acetic anhydride in pyridine solution.

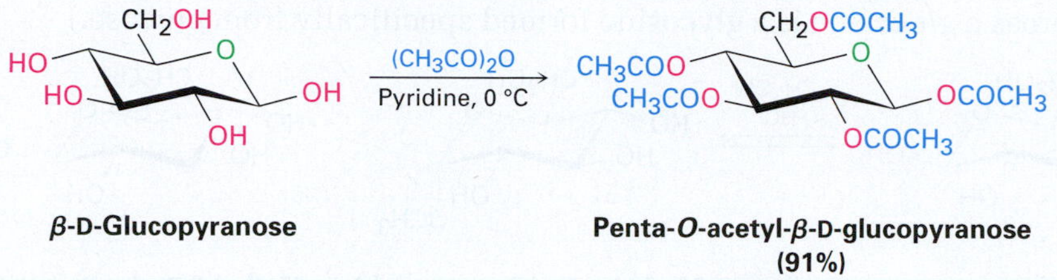

β-D-Glucopyranose   **Penta-O-acetyl-β-D-glucopyranose**
                                              (91%)

Carbohydrates are converted into ethers by treatment with an alkyl halide in the presence of base—the Williamson ether synthesis (Section 13.9). Standard Williamson conditions using a strong base tend to degrade sensitive sugar molecules, but silver oxide works well as a mild base and gives high yields of ethers. For example, α-D-glucopyranose is converted into its pentamethyl ether in 85% yield on reaction with iodomethane and Ag_2O.

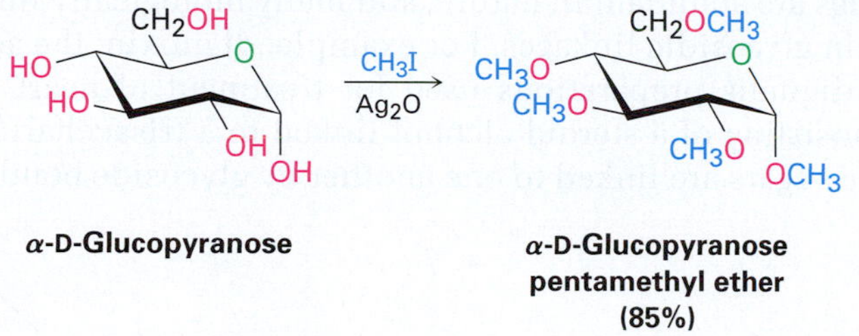

α-D-Glucopyranose   **α-D-Glucopyranose
                                       pentamethyl ether
                                       (85%)**

Problem 21.16

Draw the products you would obtain by reaction of β-D-ribofuranose with:
(a) CH_3I, Ag_2O **(b)** $(CH_3CO)_2O$, pyridine

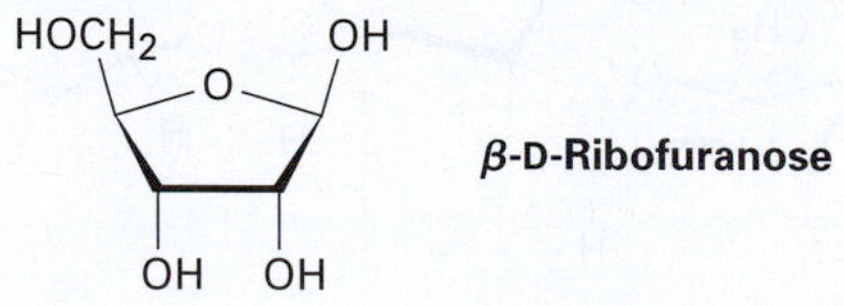

β-D-Ribofuranose

Glycoside Formation

We saw in Section 14.8 that treatment of a hemiacetal with an alcohol and an acid catalyst yields an acetal:

A hemiacetal   **An acetal**

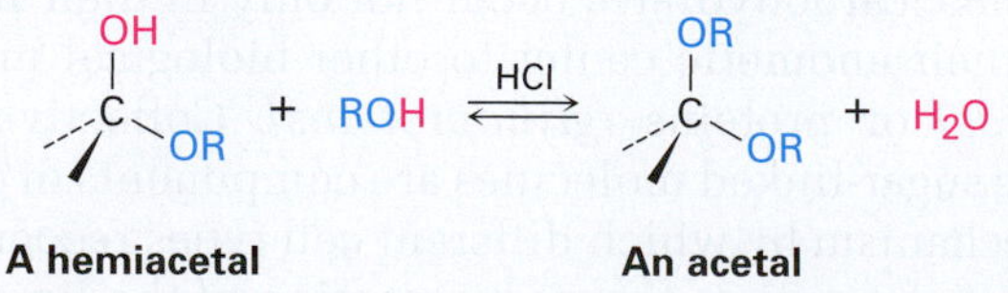

In the same way, treatment of a monosaccharide hemiacetal with an alcohol and an acid catalyst yields an acetal called a **glycoside**, in which the anomeric —OH has been replaced by an —OR group. For example, reaction of β-D-glucopyranose with methanol gives a mixture of α and β methyl D-glucopyranosides. (Note that a *glyc*oside is the functional-group name for any sugar, whereas a *gluc*oside is a glycoside formed specifically from glucose.)

β-D-Glucopyranose (a cyclic hemiacetal)　　**Methyl α-D-glucopyranoside (66%)**　　**Methyl β-D-glucopyranoside (33%)**

Glycosides are named by first citing the alkyl group and then replacing the *-ose* ending of the sugar with *-oside*. Like all acetals, glycosides are stable to neutral water. They aren't in equilibrium with an open-chain form, and they don't show mutarotation. They can, however, be hydrolyzed to give back the free monosaccharide plus alcohol on treatment with aqueous acid (Section 14.8).

Glycosides are abundant in nature, and many biologically important molecules contain glycosidic linkages. For example, digitoxin, the active component of the digitalis preparations used for treatment of heart disease, is a glycoside consisting of a steroid alcohol linked to a trisaccharide. Note also that the three sugars are linked to one another by glycoside bonds.

Digitoxigenin, a glycoside

Biological Ester Formation: Phosphorylation

In living organisms, carbohydrates occur not only in their free form but also linked through their anomeric center to other biological molecules such as lipids *(glycolipids)* or proteins *(glycoproteins).* Collectively called *glycoconjugates,* these sugar-linked molecules are components of cell walls and are crucial to the mechanism by which different cell types recognize one another.

Glycoconjugate formation occurs by reaction of the lipid or protein with a glycosyl nucleoside diphosphate, itself formed by initial phosphorylation of a monosaccharide with ATP to give a glycosyl phosphate. The glycosyl phosphate then reacts with a second nucleoside triphosphate, usually uridine triphosphate (UTP), to give a glycosyl uridine diphosphate. The purpose of the phosphorylation is to activate the anomeric –OH group of the sugar and make it a better leaving group in a nucleophilic substitution reaction with a protein or lipid (Figure 21.6).

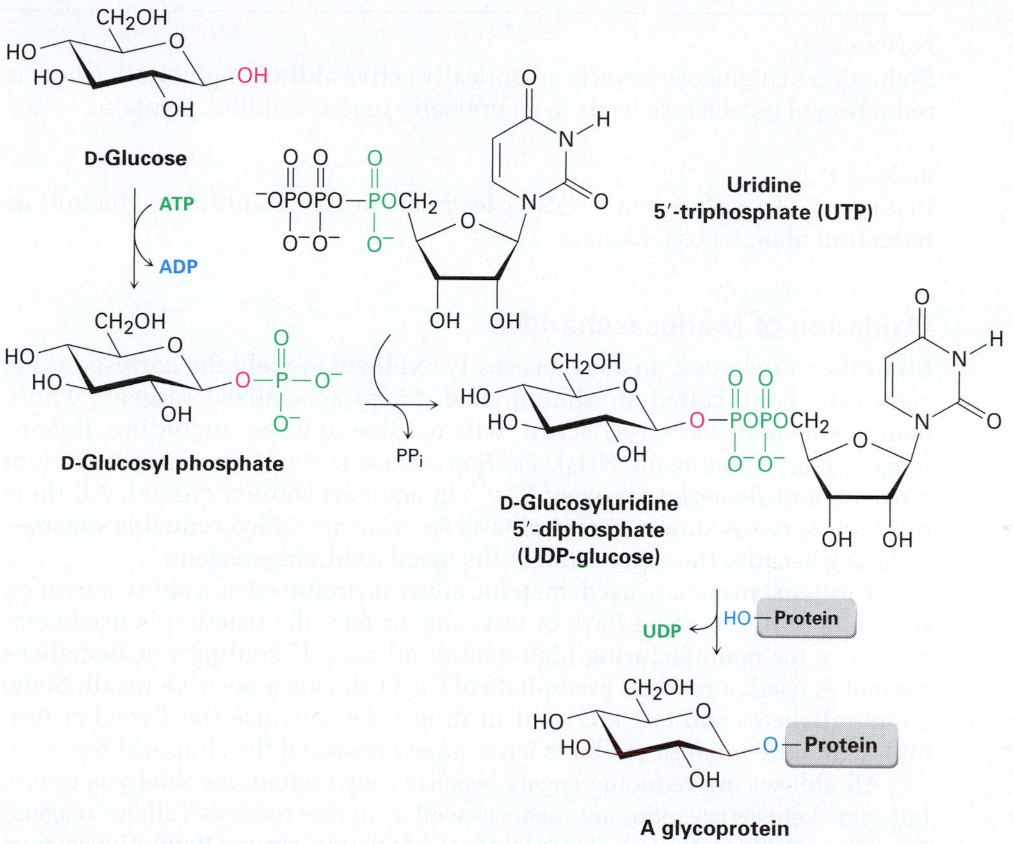

FIGURE 21.6 Glycoprotein formation occurs by initial phosphorylation of the starting carbohydrate to a glycosyl phosphate, followed by reaction with UTP to form a glycosyl uridine 5′-diphosphate. Nucleophilic substitution by an –OH (or –NH$_2$) group on a protein then gives the glycoprotein.

Reduction of Monosaccharides

Treatment of an aldose or ketose with NaBH$_4$ (Section 14.6) reduces it to a polyalcohol called an **alditol**. The reduction occurs by reaction of the open-chain form present in the aldehyde/ketone ⇌ hemiacetal equilibrium. Although only a small amount of the open-chain form is present at any given time, that small amount is reduced, more is produced by opening of the pyranose form, that additional amount is reduced, and so on, until the entire sample has undergone reaction.

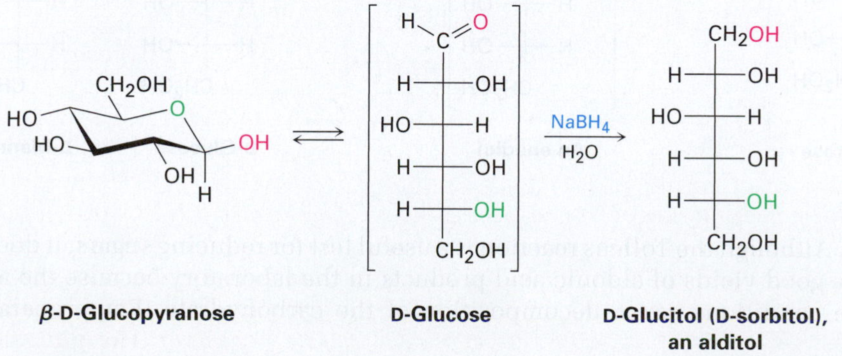

D-Glucitol, the alditol produced by reduction of D-glucose, is itself a naturally occurring substance present in many fruits and berries. It is used under its alternative name D-sorbitol as a sweetener and sugar substitute in foods.

Problem 21.17

Reduction of D-glucose leads to an optically active alditol (D-glucitol), whereas reduction of D-galactose leads to an optically inactive alditol. Explain.

Problem 21.18

Reduction of L-gulose with $NaBH_4$ leads to the same alditol (D-glucitol) as reduction of D-glucose. Explain.

Oxidation of Monosaccharides

Like other aldehydes, an aldose is easily oxidized to yield the corresponding carboxylic acid, called an **aldonic acid**. Many specialized reagents whose names you may have run across will oxidize aldoses, including *Tollens' reagent* (Ag^+ in aqueous NH_3), *Fehling's reagent* (Cu^{2+} in aqueous sodium tartrate), and *Benedict's reagent* (Cu^{2+} in aqueous sodium citrate). All three reactions serve as simple chemical tests for what are called **reducing sugars**—*reducing* because the sugar reduces the metal oxidizing reagent.

If Tollens' reagent is used, metallic silver is produced as a shiny mirror on the walls of the reaction flask or test tube. In fact, the reaction is used commercially for manufacturing high-quality mirrors. If Fehling's or Benedict's reagent is used, a reddish precipitate of Cu_2O signals a positive result. Some simple diabetes self-test kits sold in drugstores still use the Benedict test, although more modern methods have largely replaced the chemical test.

All aldoses are reducing sugars because they contain an aldehyde group, but some ketoses are reducing sugars as well. Fructose reduces Tollens' reagent, for example, even though it contains no aldehyde group. Reduction occurs because fructose is readily isomerized to a mixture of aldoses (glucose and mannose) in basic solution by a series of keto–enol tautomeric shifts (Section 17.1), as shown in Figure 21.7. Glycosides, however, are nonreducing because the acetal group is not hydrolyzed to an aldehyde under basic conditions.

FIGURE 21.7 Fructose, a ketose, is a reducing sugar because it undergoes two base-catalyzed keto–enol tautomerizations that result in conversion to a mixture of aldoses.

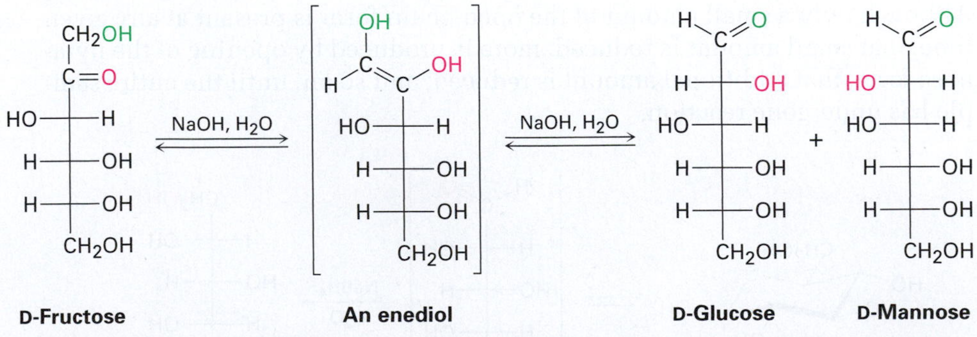

Although the Tollens reaction is a useful test for reducing sugars, it doesn't give good yields of aldonic acid products in the laboratory because the alkaline conditions cause decomposition of the carbohydrate. For preparative purposes, a buffered solution of aqueous Br_2 is a better oxidant. The reaction is specific for aldoses; ketoses are not oxidized by aqueous Br_2.

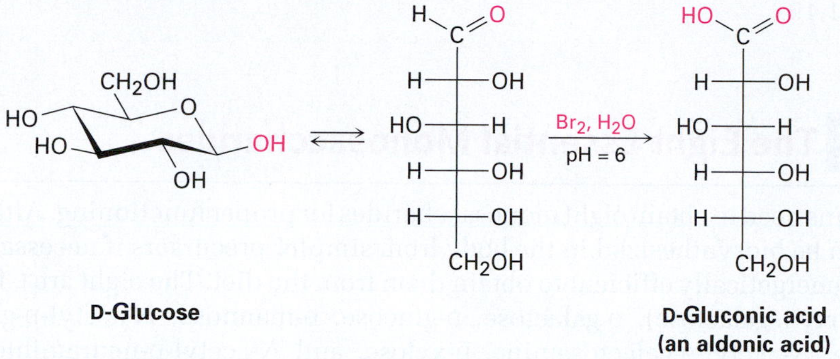

D-Glucose   **D-Gluconic acid (an aldonic acid)**

If a more powerful oxidizing agent such as warm dilute HNO_3 is used, an aldose is oxidized to a dicarboxylic acid, called an **aldaric acid**. Both the –CHO group at C1 and the terminal –CH_2OH group are oxidized in this reaction.

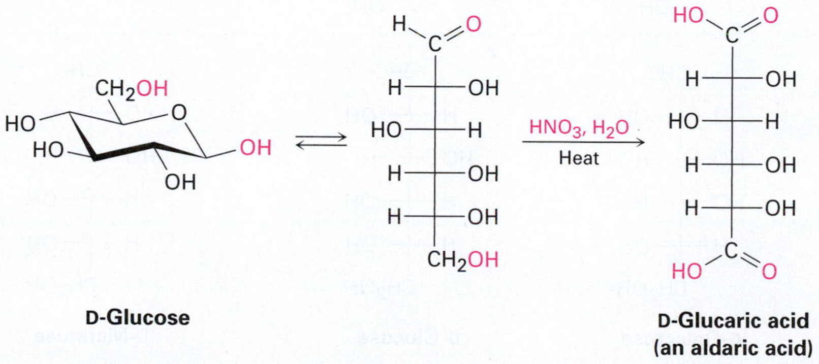

D-Glucose   **D-Glucaric acid (an aldaric acid)**

Finally, if only the –CH_2OH end of the aldose is oxidized without affecting the –CHO group, the product is a monocarboxylic acid called a **uronic acid**. The reaction must be done enzymatically; no chemical reagent is known that can accomplish this selective oxidation in the laboratory.

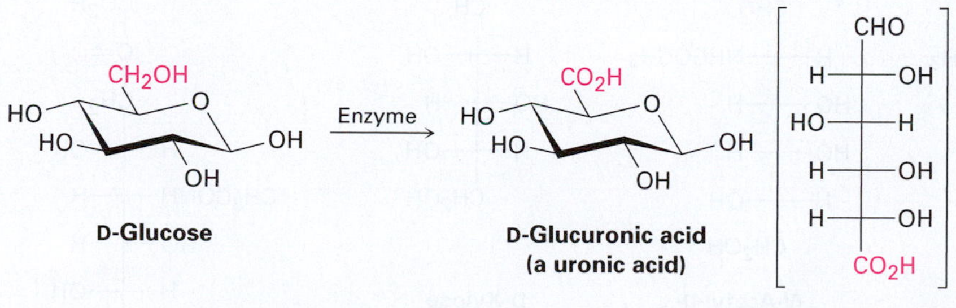

D-Glucose   **D-Glucuronic acid (a uronic acid)**

Problem 21.19

D-Glucose yields an optically active aldaric acid on treatment with HNO_3, but D-allose yields an optically inactive aldaric acid. Explain.

Problem 21.20

Which of the other six D aldohexoses yield optically active aldaric acids on oxidation, and which yield optically inactive (meso) aldaric acids? (See Problem 21.19.)

21.7 The Eight Essential Monosaccharides

Humans need to obtain eight monosaccharides for proper functioning. Although all can be biosynthesized in the body from simpler precursors if necessary, it's more energetically efficient to obtain them from the diet. The eight are L-fucose (6-deoxy-L-galactose), D-galactose, D-glucose, D-mannose, N-acetyl-D-glucosamine, N-acetyl-D-galactosamine, D-xylose, and N-acetyl-D-neuraminic acid (Figure 21.8). All are used for the synthesis of the glycoconjugate components of cell walls, and glucose is also the body's primary source of energy.

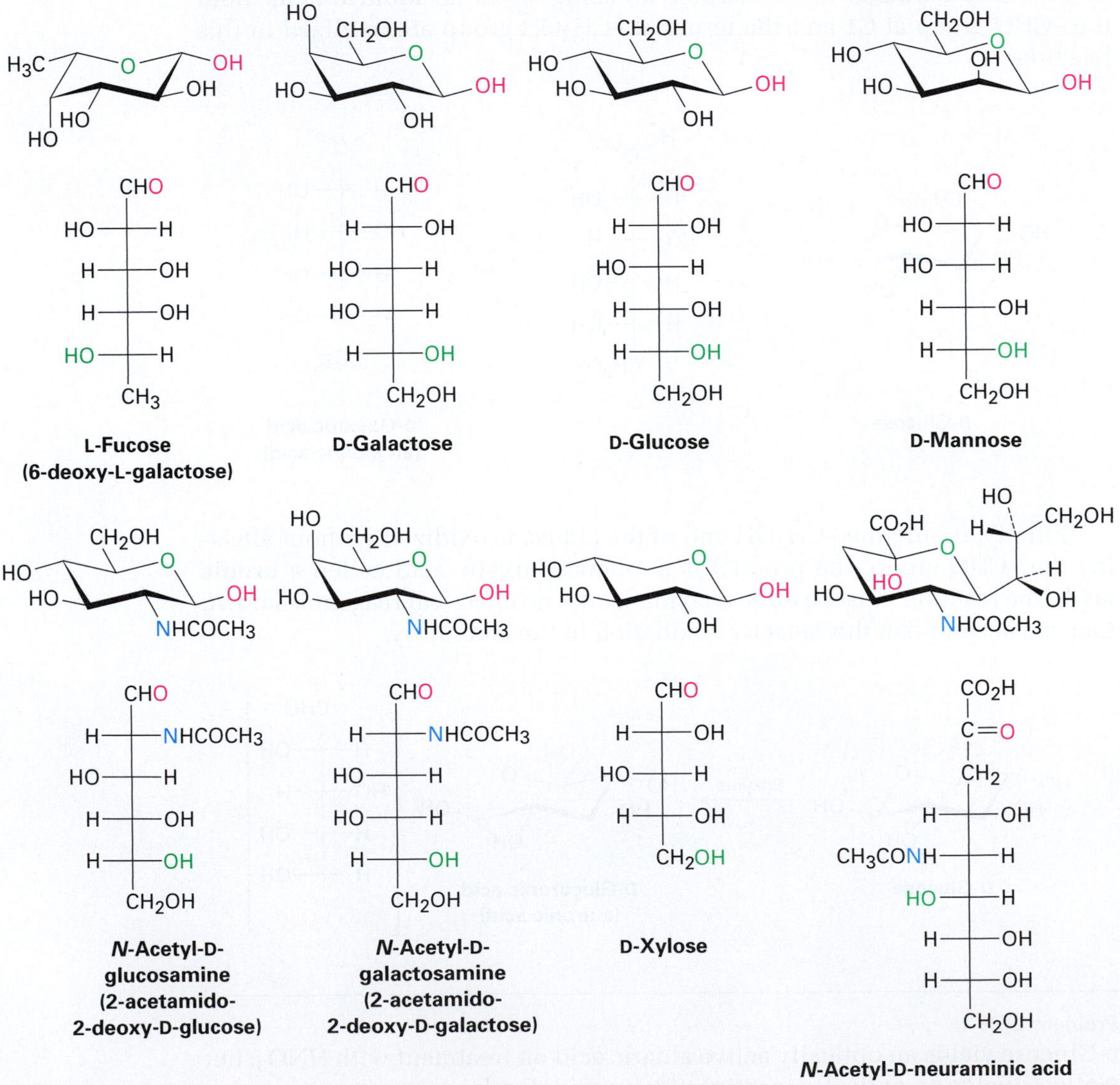

FIGURE 21.8 Structures of the eight monosaccharides essential to humans.

Of the eight essential monosaccharides, galactose, glucose, and mannose are simple aldohexoses, while xylose is an aldopentose. Fucose is a **deoxy sugar**, meaning that it has an oxygen atom "missing." That is, an –OH group (the one at C6) is replaced by an –H. *N*-Acetylglucosamine and *N*-acetylgalactosamine are amide derivatives of **amino sugars** in which an –OH (the one at C2) is replaced by an –NH₂ group. *N*-Acetylneuraminic acid is the parent compound of the *sialic acids,* a group of more than 30 substances with different modifications, including various oxidations, acetylations, sulfations, and methylations. Note that neuraminic acid has nine carbons and is an aldol reaction product of *N*-acetylmannosamine with pyruvate ($CH_3COCO_2^-$). We'll see in the Chapter 22 *Lagniappe* that neuraminic acid is crucially important to the mechanism by which an influenza virus spreads.

All the essential monosaccharides arise from glucose, by the conversions summarized in Figure 21.9. We'll not look specifically at these conversions, but might note that Problems 22.19, 22.20, 22.21, and 22.25 at the end of the next chapter lead you through several of the biosynthetic pathways.

FIGURE 21.9 An overview of biosynthetic pathways for the eight essential monosaccharides.

```
Galactose  ←  Glucose  →  Fructose  →  Mannose
   ↓            ↓            ↓
Xylose      Glucosamine    Fucose
   ↓            ↓
Galactosamine  Mannosamine
               ↓
           Neuraminic acid
```

Problem 21.21

Show how *N*-acetylneuraminic acid can arise by an aldol reaction of *N*-acetylmannosamine with pyruvate, $CH_3COCO_2^-$.

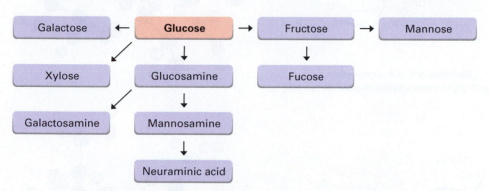

N-Acetylmannosamine

21.8 Disaccharides

We saw in Section 21.6 that reaction of a monosaccharide with an alcohol yields a glycoside, in which the anomeric –OH is replaced by an –OR group. If the alcohol is itself a sugar, the glycosidic product is a **disaccharide**.

Cellobiose and Maltose

Disaccharides contain a glycosidic acetal bond between the anomeric carbon of one sugar and an –OH group at any position on the other sugar. A glycosidic bond between C1 of the first sugar and the –OH at C4 of the second sugar is particularly common. Such a bond is called a *1→4 link*.

The glycosidic bond to an anomeric carbon can be either α or β. Maltose, the disaccharide obtained by enzyme-catalyzed hydrolysis of starch, consists of two α-D-glucopyranose units joined by a 1→4-α-glycoside bond. Cellobiose, the disaccharide obtained by partial hydrolysis of cellulose, consists of two β-D-glucopyranose units joined by a 1→4-β-glycoside bond.

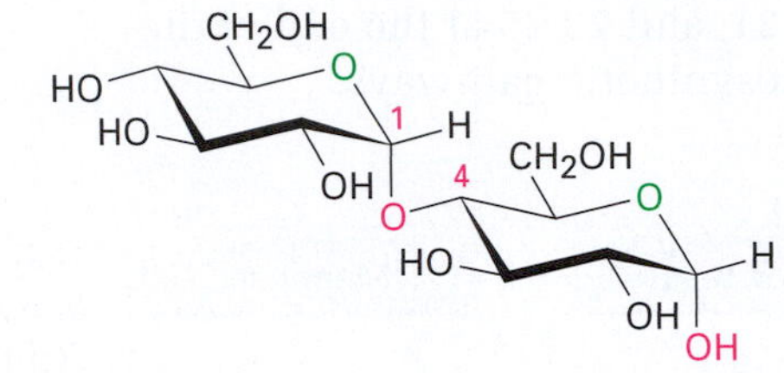

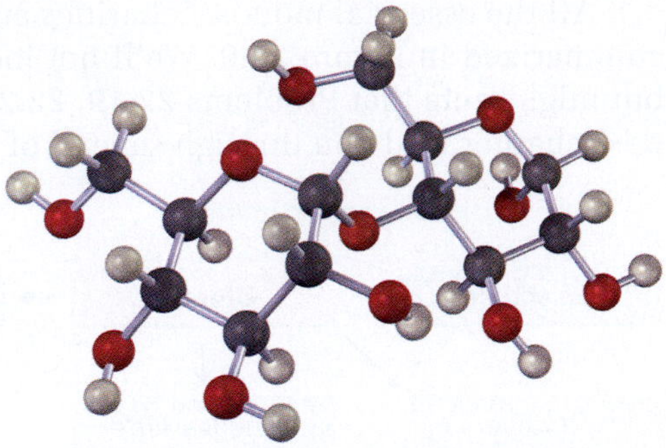

Maltose, a 1→ 4-α-glycoside
[4-*O*-(α-D-glucopyranosyl)-α-D-glucopyranose]

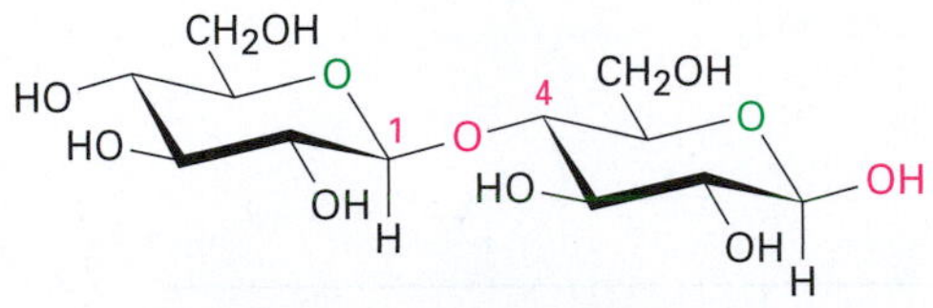

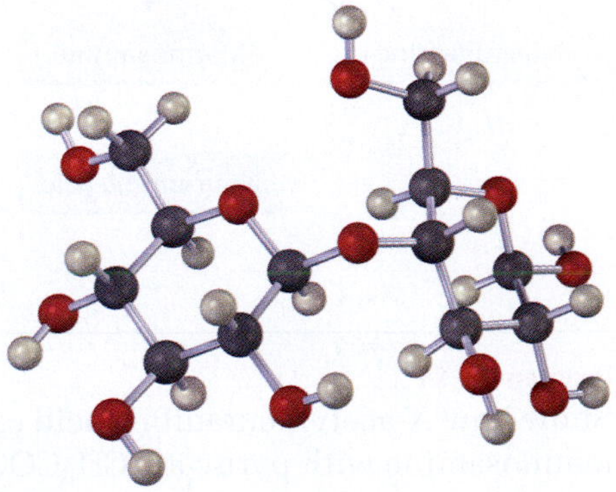

Cellobiose, a 1→ 4-β-glycoside
[4-*O*-(β-D-glucopyranosyl)-β-D-glucopyranose]

Maltose and cellobiose are both reducing sugars because the anomeric carbons on the right-hand glucopyranose units have hemiacetal groups and are in equilibrium with aldehyde forms. For a similar reason, both maltose and cellobiose exhibit mutarotation of α and β anomers of the glucopyranose unit on the right.

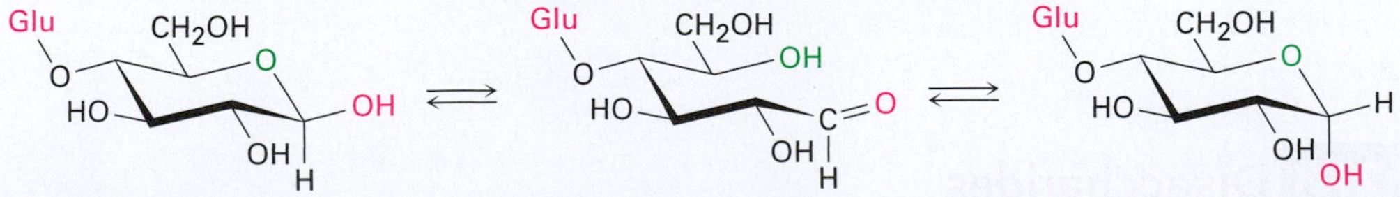

Maltose or cellobiose   **Maltose or cellobiose**   **Maltose or cellobiose**
(β anomers)   **(aldehydes)**   **(α anomers)**

Despite the similarities of their structures, cellobiose and maltose have dramatically different biological properties. Cellobiose can't be digested by humans and can't be fermented by yeast. Maltose, however, is digested without difficulty and is fermented readily.

Problem 21.22
Show the product you would obtain from the reaction of cellobiose with the following reagents:
(a) $NaBH_4$ **(b)** Br_2, H_2O **(c)** CH_3COCl, pyridine

Lactose

Lactose is a disaccharide that occurs naturally in both human and cow's milk. It is widely used in baking and in commercial milk formulas for infants. Like cellobiose and maltose, lactose is a reducing sugar. It exhibits mutarotation and is a 1→4-β-linked glycoside. Unlike cellobiose and maltose, however, lactose contains two different monosaccharides—D-glucose and D-galactose— joined by a β-glycosidic bond between C1 of galactose and C4 of glucose.

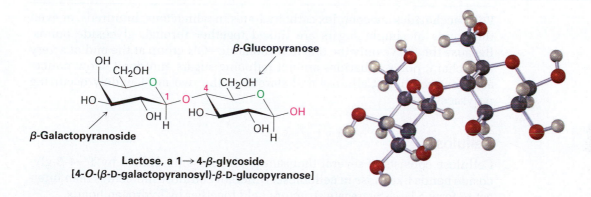

Lactose, a 1→4-β-glycoside
[4-*O*-(β-D-galactopyranosyl)-β-D-glucopyranose]

Sucrose

Sucrose, or ordinary table sugar, is among the most abundant pure organic chemicals in the world and is the one most widely known to nonchemists. Whether from sugar cane (20% sucrose by weight) or sugar beets (15% by weight), and whether raw or refined, all table sugar is sucrose.

Sucrose is a disaccharide that yields 1 equivalent of glucose and 1 equivalent of fructose on hydrolysis. This 1:1 mixture of glucose and fructose is often referred to as *invert sugar* because the sign of optical rotation changes, or inverts, during the hydrolysis from sucrose ($[\alpha]_D$ = +66.5) to a glucose/fructose mixture ($[\alpha]_D$ = −22.0). Insects such as honeybees have enzymes called invertases that catalyze the hydrolysis of sucrose to a glucose + fructose mixture. Honey, in fact, is primarily a mixture of glucose, fructose, and sucrose.

Unlike most other disaccharides, sucrose is not a reducing sugar and does not undergo mutarotation. These observations imply that sucrose is not a hemiacetal and suggest that glucose and fructose must both be glycosides. This can happen only if the two sugars are joined by a glycoside link between the anomeric carbons of both sugars: C1 of glucose and C2 of fructose.

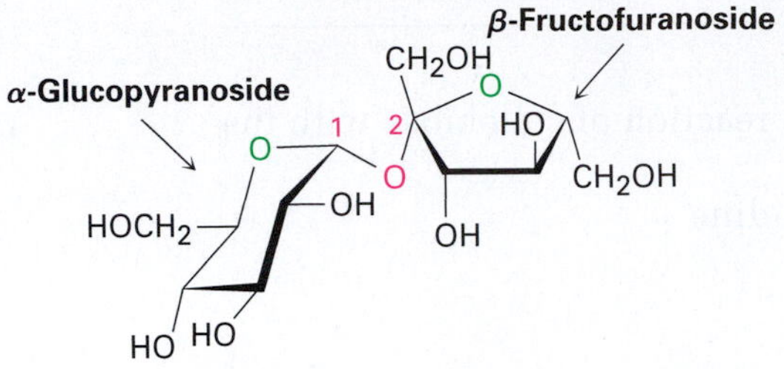

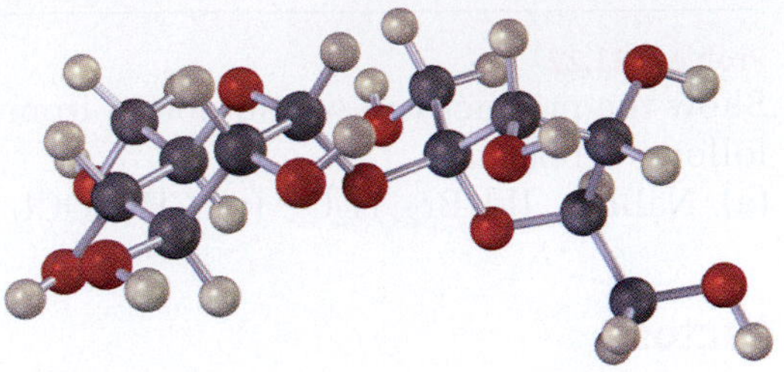

Sucrose, a 1→2-glycoside
[2-*O*-(α-D-glucopyranosyl)-β-D-fructofuranoside]

21.9 Polysaccharides and Their Synthesis

Polysaccharides are complex carbohydrates in which tens, hundreds, or even thousands of simple sugars are linked together through glycoside bonds. Because they have only the one free anomeric –OH group at the end of a very long chain, polysaccharides are not reducing sugars and don't show noticeable mutarotation. Cellulose and starch are the two most widely occurring polysaccharides.

Cellulose

Cellulose consists of several thousand D-glucose units linked by 1→4-β-glycoside bonds like those in cellobiose. Different cellulose molecules then interact to form a large aggregate structure held together by hydrogen bonds.

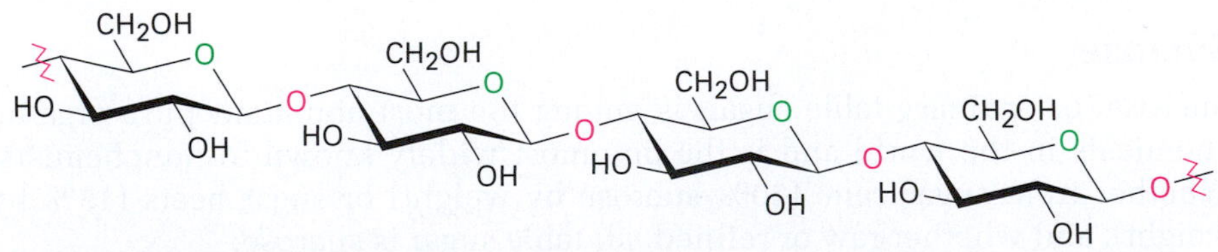

Cellulose, a 1→ 4-*O*-(β-D-glucopyranoside) polymer

Nature uses cellulose primarily as a structural material to impart strength and rigidity to plants. Leaves, grasses, and cotton, for instance, are primarily cellulose. Cellulose also serves as raw material for the manufacture of cellulose acetate, known commercially as acetate rayon, and cellulose nitrate, known as guncotton. Guncotton is the major ingredient in smokeless powder, the explosive propellant used in artillery shells and ammunition for firearms.

Starch and Glycogen

Potatoes, corn, and cereal grains contain large amounts of *starch,* a polymer of glucose in which the monosaccharide units are linked by 1→4-α-glycoside bonds like those in maltose. Starch can be separated into two fractions: amylose, which is insoluble in cold water, and amylopectin, which is soluble in cold water. Amylose accounts for about 20% by weight of starch and consists of several hundred glucose molecules linked together by 1→4-α-glycoside bonds.

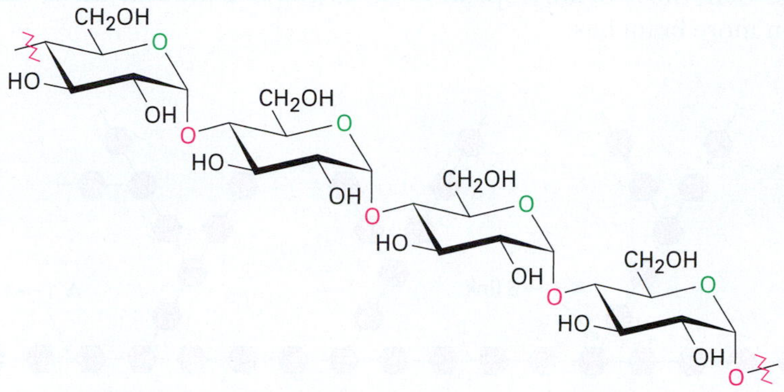

Amylose, a 1→4-*O*-(α-D-glucopyranoside) polymer

Amylopectin accounts for the remaining 80% of starch and is more complex in structure than amylose. Unlike cellulose and amylose, which are linear polymers, amylopectin contains 1→6-α-glycoside branches approximately every 25 glucose units.

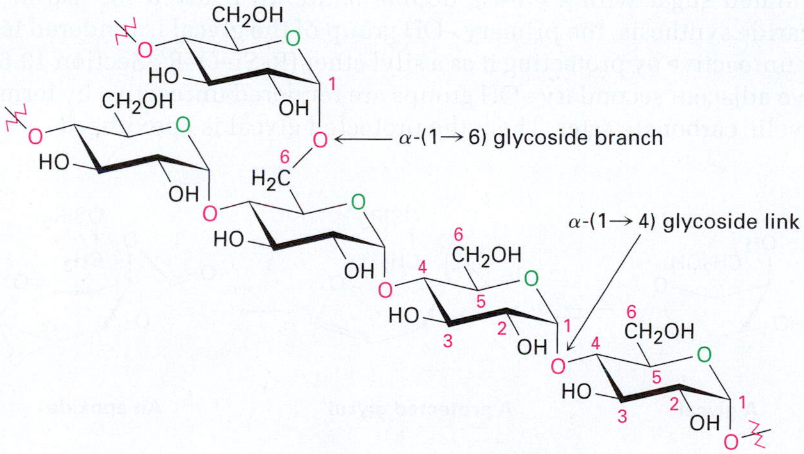

**Amylopectin: α-(1→4) links
with α-(1→6) branches**

Starch is digested in the mouth and stomach by α-glycosidases, which catalyze the hydrolysis of glycoside bonds and release individual molecules of glucose. Like most enzymes, α-glycosidases are highly selective in their action. They hydrolyze only the α-glycoside links in starch and leave the β-glycoside links in cellulose untouched. Thus, humans can digest potatoes and grains but not grass and leaves.

Glycogen is a polysaccharide that serves the same energy storage function in animals that starch serves in plants. Dietary carbohydrates not needed for immediate energy are converted by the body to glycogen for long-term storage. Like the amylopectin found in starch, glycogen contains a complex branching structure with both 1→4 and 1→6 links (Figure 21.10). Glycogen molecules are larger than those of amylopectin—up to 100,000 glucose units—and contain even more branches.

FIGURE 21.10 A representation of the structure of glycogen. The hexagons represent glucose units linked by 1→4 and 1→6 glycoside bonds.

Polysaccharide Synthesis

With numerous –OH groups of similar reactivity, polysaccharides are so structurally complex that their laboratory synthesis has been a particularly difficult problem. Several methods have been devised, however, that have greatly simplified the problem. Among these approaches is the *glycal assembly method*.

Easily prepared from the appropriate monosaccharide, a *glycal* is an unsaturated sugar with a C1–C2 double bond. To ready it for use in polysaccharide synthesis, the primary –OH group of the glycal is rendered temporarily unreactive by protecting it as a silyl ether (R_3Si–O–R'; Section 13.6) and the two adjacent secondary –OH groups are rendered unreactive by formation of a cyclic carbonate ester. Then, the protected glycal is epoxidized.

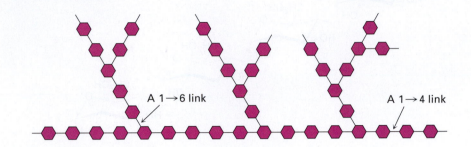

A glycal   A protected glycal   An epoxide

Treatment of the protected glycal epoxide in the presence of $ZnCl_2$ as a Lewis acid with a *second* glycal having a free –OH group causes acid-catalyzed opening of the epoxide ring by S_N2 backside attack and yields a disaccharide. The disaccharide is itself a glycal, so it can be epoxidized and coupled again to yield a trisaccharide, and so on. Using the appropriate sugars at each step, a great variety of polysaccharides can be prepared. After the appropriate sugars are linked, the silyl ethers and cyclic carbonate protecting groups are removed by hydrolysis.

A disaccharide glycal

Among the numerous complex polysaccharides that have been synthesized in the laboratory is the Lewis Y hexasaccharide, a tumor marker that is currently being explored as a potential cancer vaccine.

Lewis Y hexasaccharide

21.10 Cell-Surface Carbohydrates and Carbohydrate Vaccines

It was once thought that carbohydrates were useful in nature only as structural materials and energy sources. Although carbohydrates do indeed serve these purposes, they have many other important biochemical functions as well. As noted in Section 21.6, for instance, glycoconjugates are centrally involved in cell–cell recognition, the critical process by which one type of cell distinguishes another. Small polysaccharide chains, covalently bound by glycosidic links to –OH or –NH$_2$ groups on proteins, act as biochemical markers on cell surfaces, as illustrated by the human blood-group antigens.

It has been known for more than a century that human blood can be classified into four blood-group types (A, B, AB, and O) and that blood from a donor of one type can't be transfused into a recipient with another type unless the two types are compatible (Table 21.1). Should an incompatible mix be made, the red blood cells clump together, or *agglutinate*.

The agglutination of incompatible red blood cells, which indicates that the body's immune system has recognized the presence of foreign cells in the body and has formed antibodies against them, results from the presence of

TABLE 21.1
Human Blood-Group Compatibilities

Donor blood type	Acceptor blood type			
	A	B	AB	O
A	O	X	O	X
B	X	O	O	X
AB	X	X	O	X
O	O	O	O	O

polysaccharide markers on the surface of the cells. Types A, B, and O red blood cells each have their own unique markers, or *antigenic determinants;* type AB cells have both type A and type B markers. The structures of all three blood-group determinants are shown in Figure 21.11. Note that the monosaccharide constituents of each marker are among the eight essential sugars shown previously in Figure 21.8.

FIGURE 21.11 Structures of the A, B, and O blood-group antigenic determinants.

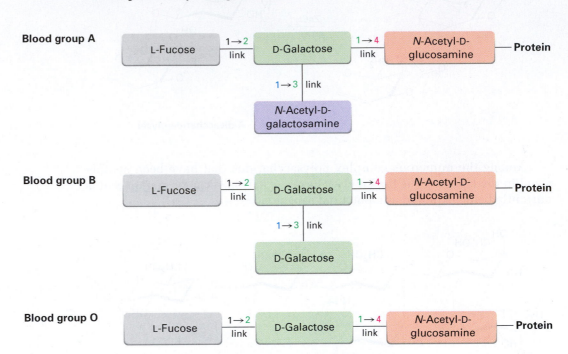

Elucidation of the role of carbohydrates in cell recognition is a vigorous area of current research that offers hope of breakthroughs in treating a wide range of diseases from bacterial infections to cancer. Particularly exciting is the possibility of developing carbohydrate-based vaccines to help mobilize the body's immune system. Diseases currently being studied for vaccine development include pneumonia, malaria, several cancers, and AIDS.

Summary

Key Words

Carbohydrates are polyhydroxy aldehydes and ketones. They are classified according to the number of carbon atoms and the kind of carbonyl group they contain. Glucose, for example, is an aldohexose, a six-carbon aldehydo sugar. **Monosaccharides** are further classified as either **D sugars** or **L sugars**, depending on the stereochemistry of the chirality center farthest from the carbonyl group. Carbohydrate stereochemistry is frequently depicted using **Fischer projections**, which represent a chirality center as the intersection of two crossed lines.

Monosaccharides normally exist as cyclic hemiacetals rather than as open-chain aldehydes or ketones. The hemiacetal linkage results from reaction of the carbonyl group with an –OH group three or four carbon atoms away. A five-membered cyclic hemiacetal is called a **furanose**, and a six-membered cyclic hemiacetal is called a **pyranose**. Cyclization leads to the formation of a new chirality center and production of two diastereomeric hemiacetals, called **α** and **β anomers**.

Much of the chemistry of monosaccharides is the familiar chemistry of alcohols and aldehydes/ketones. Thus, the hydroxyl groups of carbohydrates form esters and ethers. The carbonyl group of a monosaccharide can be reduced with $NaBH_4$ to form an **alditol**, oxidized with aqueous Br_2 to form an **aldonic acid**, oxidized with HNO_3 to form an **aldaric acid**, oxidized enzymatically to form a **uronic acid**, or treated with an alcohol in the presence of acid to form a **glycoside**.

Disaccharides are complex carbohydrates in which simple sugars are linked by a glycoside bond between the **anomeric center** of one unit and a hydroxyl of the second unit. The sugars can be the same, as in maltose and cellobiose, or different, as in lactose and sucrose. The glycosidic bond can be either α (maltose) or β (cellobiose, lactose) and can involve any hydroxyl of the second sugar. A 1→4 link is most common (cellobiose, maltose), but others such as 1→2 (sucrose) are also known. **Polysaccharides**, such as cellulose, starch, and glycogen, are used in nature as structural materials, as a means of long-term energy storage, and as cell-surface markers.

Summary of Reactions

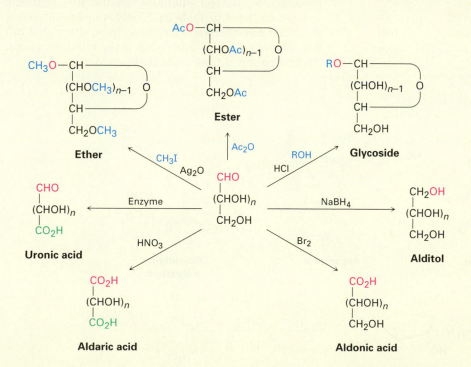

Lagniappe

Sweetness

Say the word *sugar* and most people immediately think of sweet-tasting candies, desserts, and such. In fact, most simple carbohydrates *do* taste sweet, but the degree of sweetness varies greatly from one sugar to another. With sucrose (table sugar) as a reference point, fructose is nearly twice as sweet, but lactose is only about one-sixth as sweet. Comparisons are difficult, though, because perceived sweetness varies depending on the concentration of the solution being tasted. Nevertheless, the ordering in Table 21.2 is generally accepted.

TABLE 21.2
Sweetness of Some Sugars and Sugar Substitutes

Name	Type	Sweetness
Lactose	Disaccharide	0.16
Glucose	Monosaccharide	0.75
Sucrose	**Disaccharide**	**1.00**
Fructose	Monosaccharide	1.75
Aspartame	Synthetic	180
Acesulfame-K	Synthetic	200
Saccharin	Synthetic	350
Sucralose	Semisynthetic	600
Alitame	Semisynthetic	2000

The desire of many people to cut their caloric intake has led to the development of synthetic sweeteners such as saccharin, aspartame, acesulfame, and sucralose. All are far sweeter than natural sugars, so the choice of one or another depends on personal taste, government regulations, and (for baked goods) heat stability. Saccharin, the oldest synthetic sweetener, has been used for more than a century, although it has a somewhat metallic aftertaste. Doubts about its safety and potential carcinogenicity were raised in the early 1970s, but it has now been cleared of suspicion.

The real thing comes from cane fields like this one.

Acesulfame potassium, one of the most recently approved sweeteners, is proving to be extremely popular in soft drinks because it has little aftertaste. Sucralose, another recently approved sweetener, is particularly useful in baked goods because of its stability at high temperatures. Alitame, marketed in some countries under the name Aclame, is not approved for sale in the United States. It is some 2000 times as sweet as sucrose and, like acesufame-K, has no aftertaste. Of the five synthetic sweeteners listed in Table 21.2, only sucralose has clear structural resemblance to a carbohydrate, although it differs dramatically in containing three chlorine atoms. Aspartame and alitame are both dipeptides.

Saccharin

Aspartame

Acesulfame potassium

Sucralose

Alitame

Exercises

■ indicates problems that are assignable in Organic OWL.

Go to this book's companion website at **www.cengage.com/ chemistry/mcmurry** to explore interactive versions of the Active Figures from this text.

VISUALIZING CHEMISTRY

(Problems 21.1–21.22 appear within the chapter.)

21.23 ■ Identify the following aldoses, and tell whether each is a D or L sugar:

(a)   **(b)**

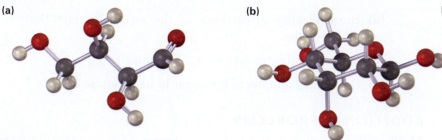

21.24 Draw Fischer projections of the following molecules, placing the carbonyl group at the top in the usual way, and identify each as a D or L sugar:

(a)   **(b)**

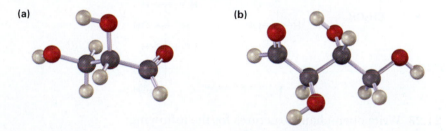

21.25 ■ The following structure is that of an L aldohexose in its pyranose form. Identify it, and tell whether it is an α or β anomer.

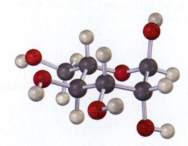

21.26 ■ The following model is that of an aldohexose:

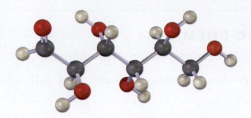

(a) Draw Fischer projections of the sugar, its enantiomer, and a diastereomer.

(b) Is this a D sugar or an L sugar? Explain.

(c) Draw the β anomer of the sugar in its furanose form.

ADDITIONAL PROBLEMS

21.27 ■ Classify each of the following sugars. (For example, glucose is an aldohexose.)

(a)
```
CH₂OH
|
C=O
|
CH₂OH
```

(b)
```
    CH₂OH
H ──┼── OH
    C=O
H ──┼── OH
    CH₂OH
```

(c)
```
       CHO
 H  ──┼── OH
HO  ──┼── H
 H  ──┼── OH
HO  ──┼── H
 H  ──┼── OH
      CH₂OH
```

21.28 Write open-chain structures for the following:

(a) A ketotetrose   (b) A ketopentose

(c) A deoxyaldohexose   (d) A five-carbon amino sugar

21.29 Does ascorbic acid (vitamin C) have a D or L configuration?

```
            OH
            |
      HO    C
        \\ //
         C   C=O
         |   |
     H ──┼── O      Ascorbic acid
    HO ──┼── H
         CH₂OH
```

21.30 Draw the three-dimensional furanose form of ascorbic acid (Problem 21.29), and assign *R* or *S* stereochemistry to each chirality center.

21.31 ■ Assign *R* or *S* configuration to each chirality center in the following molecules:

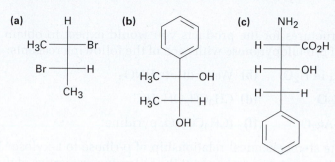

(a)

H₃C——Br
Br——H
CH₃

(b)

H₃C——OH
H₃C——H
OH

(c)

H——CO₂H
H——OH
H——H

21.32 Draw Fischer projections for the two D aldoheptoses whose stereochemistry at C3, C4, C5, and C6 is the same as that of D-glucose at C2, C3, C4, and C5.

21.33 ■ The following cyclic structure is that of allose. Is this a furanose or pyranose form? Is it an *α* or *β* anomer? Is it a D or L sugar?

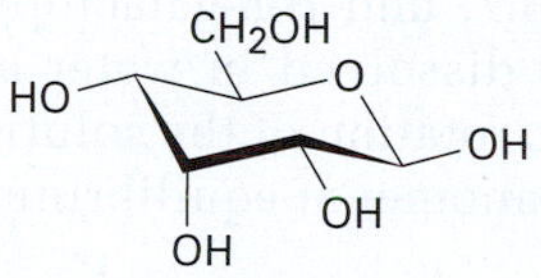

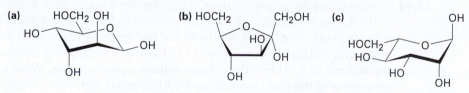

21.34 ■ What is the full name of the following sugar?

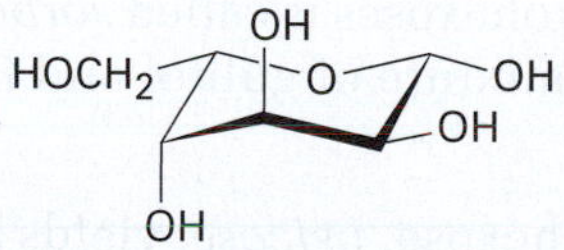

21.35 ■ Write the following sugars in their open-chain forms:

(a)

(b)

(c)

21.36 ■ Draw D-ribulose in its five-membered cyclic *β*-hemiacetal form.

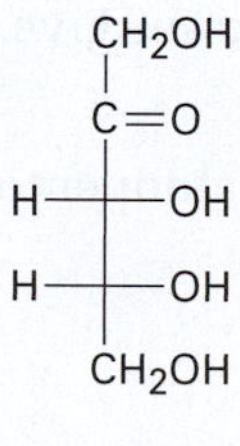

CH₂OH
C=O
H——OH   **Ribulose**
H——OH
CH₂OH

21.37 ▪ Look up the structure of D-talose in Figure 21.3, and draw the β anomer in its pyranose form. Identify the ring substituents as axial or equatorial.

21.38 ▪ Draw structures for the products you would expect to obtain from reaction of β-D-talopyranose with each of the following reagents:

(a) $NaBH_4$ in H_2O (b) Warm dilute HNO_3

(c) Br_2, H_2O (d) CH_3CH_2OH, HCl

(e) CH_3I, Ag_2O (f) $(CH_3CO)_2O$, pyridine

21.39 What is the stereochemical relationship of D-ribose to L-xylose? What generalizations can you make about the following properties of the two sugars?

(a) Melting point   (b) Solubility in water

(c) Specific rotation   (d) Density

21.40 All aldoses exhibit mutarotation. For example, α-D-galactopyranose has $[\alpha]_D$ = +150.7, and β-D-galactopyranose has $[\alpha]_D$ = +52.8. If either anomer is dissolved in water and allowed to reach equilibrium, the specific rotation of the solution is +80.2. What are the percentages of each anomer at equilibrium? Draw the pyranose forms of both anomers.

21.41 How many D-2-ketohexoses are possible? Draw them.

21.42 One of the D-2-ketohexoses is called *sorbose*. On treatment with $NaBH_4$, sorbose yields a mixture of gulitol and iditol. What is the structure of sorbose?

21.43 Another D-2-ketohexose, *psicose,* yields a mixture of allitol and altritol when reduced with $NaBH_4$. What is the structure of psicose?

21.44 L-Gulose can be prepared from D-glucose by a route that begins with oxidation to D-glucaric acid, which cyclizes to form two six-membered-ring lactones. Separating the lactones and treating them with sodium amalgam, Na(Hg), reduces the –CO_2H group to a primary alcohol and the lactone to an aldehyde, giving D-glucose and L-gulose. What are the structures of the two lactones, and which one is reduced to L-gulose?

21.45 ▪ What other D aldohexose gives the same alditol as D-talose?

21.46 Which of the eight D aldohexoses give the same aldaric acids as their L enantiomers?

21.47 ▪ Which of the other three D aldopentoses gives the same aldaric acid as D-lyxose?

21.48 Draw the structure of L-galactose, and then answer the following questions:

    (a) Which other aldohexose gives the same aldaric acid as L-galactose on oxidation with warm HNO_3?

    (b) Is this other aldohexose a D sugar or an L sugar?

    (c) Draw this other aldohexose in its most stable pyranose conformation.

21.49 ▪ Gentiobiose, a rare disaccharide found in saffron and gentian, is a reducing sugar and forms only D-glucose on hydrolysis with aqueous acid. Reaction of gentiobiose with iodomethane and Ag_2O yields an octamethyl derivative, which can be hydrolyzed with aqueous acid to give 1 equivalent of 2,3,4,6-tetra-*O*-methyl-D-glucopyranose and 1 equivalent of 2,3,4-tri-*O*-methyl-D-glucopyranose. If gentiobiose contains a β-glycoside link, what is its structure?

21.50 Amygdalin, or laetrile, is a cyanogenic glycoside isolated in 1830 from almond and apricot seeds. Acidic hydrolysis of amygdalin liberates HCN, along with benzaldehyde and 2 equivalents of D-glucose. If amygdalin is a β-glycoside of benzaldehyde cyanohydrin with gentiobiose (Problem 21.49), what is its structure? [A cyanohydrin has the structure $R_2C(OH)CN$ and is formed by reversible nucleophilic addition of HCN to an aldehyde or ketone.]

21.51 Trehalose is a nonreducing disaccharide that is hydrolyzed by aqueous acid to yield 2 equivalents of D-glucose. Methylation followed by hydrolysis yields 2 equivalents of 2,3,4,6-tetra-*O*-methylglucose. How many structures are possible for trehalose?

21.52 Trehalose (Problem 21.51) is cleaved by enzymes that hydrolyze α-glycosides but not by enzymes that hydrolyze β-glycosides. What is the structure and systematic name of trehalose?

21.53 Isotrehalose and neotrehalose are chemically similar to trehalose (Problems 21.51 and 21.52) except that neotrehalose is hydrolyzed only by β-glycosidases, whereas isotrehalose is hydrolyzed by both α- and β-glycosidases. What are the structures of isotrehalose and neotrehalose?

21.54 D-Glucose reacts with acetone in the presence of acid to yield the nonreducing 1,2:5,6-diisopropylidene-D-glucofuranose. Propose a mechanism.

1,2:5,6-Diisopropylidene-D-glucofuranose

21.55 D-Mannose reacts with acetone to give a diisopropylidene derivative (Problem 21.54) that is still a reducing sugar. Propose a likely structure for this derivative.

21.56 ■ Glucose and mannose can be interconverted (in low yield) by treatment with dilute aqueous NaOH. Propose a mechanism.

21.57 Propose a mechanism to account for the fact that D-gluconic acid and D-mannonic acid are interconverted when either is heated in pyridine solvent.

21.58 The *cyclitols* are a group of carbocyclic sugar derivatives having the general formulation cyclohexane-1,2,3,4,5,6-hexol. How many stereoisomeric cyclitols are possible? Draw them in their chair forms.

21.59 ■ The *Kiliani–Fischer chain extension* is a method for lengthening an aldose chain by one carbon, giving two new aldoses that differ in stereochemistry at C2. D-Erythrose, for instance, yields a mixture of D-ribose and D-arabinose on Kiliani–Fischer chain extension.

D-Erythrose   **D-Ribose**   **D-Arabinose**

(a) What product(s) would you expect from Kiliani–Fischer reaction of D-ribose?

(b) What aldopentose would give a mixture of L-gulose and L-idose on Kiliani–Fischer chain extension?

21.60 ■ The *Wohl degradation* is the opposite of the Kiliani–Fischer chain extension (Problem 21.59) in that it shortens an aldose chain by one carbon, converting C2 of the reactant into C1 of the product. Which two of the four D aldopentoses yield D-threose on Wohl degradation?

21.61 ■ Compound A is a D aldopentose that can be oxidized to an optically inactive aldaric acid B. On Kiliani–Fischer chain extension (Problem 21.59), A is converted into C and D; C can be oxidized to an optically active aldaric acid E, but D is oxidized to an optically inactive aldaric acid F. What are the structures of A–F?

21.62 Simple sugars undergo reaction with phenylhydrazine, $PhNHNH_2$, to yield crystalline derivatives called *osazones*. The reaction is a bit complex, however, as shown by the fact that glucose and fructose yield the same osazone.

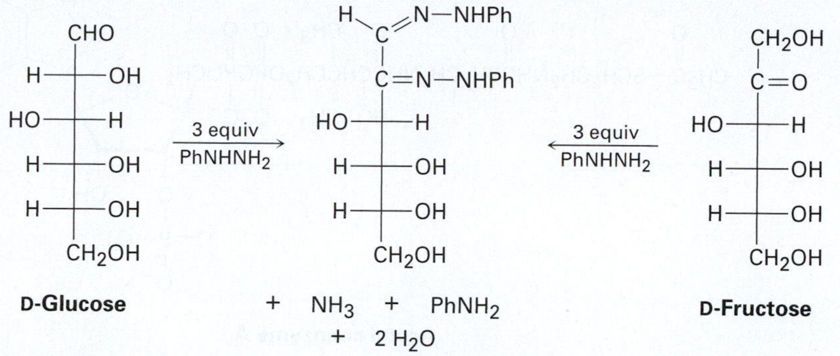

D-Glucose   + NH_3 + $PhNH_2$ **D-Fructose**
                + $2\ H_2O$

(a) Draw the structure of a third sugar that yields the same osazone as glucose and fructose.

(b) Using glucose as the example, the first step in osazone formation is reaction of the sugar with phenylhydrazine to yield an imine called a *phenylhydrazone*. Draw the structure of the product.

(c) The second and third steps in osazone formation are tautomerization of the phenylhydrazone to give an enol, followed by elimination of aniline to give a keto imine. Draw the structures of both the enol tautomer and the keto imine.

(d) The final step is reaction of the keto imine with 2 equivalents of phenylhydrazine to yield the osazone plus ammonia. Propose a mechanism for this step.

21.63 When heated to 100 °C, D-idose undergoes a reversible loss of water and exists primarily as 1,6-anhydro-D-idopyranose.

D-Idose   **1,6-Anhydro-D-idopyranose**

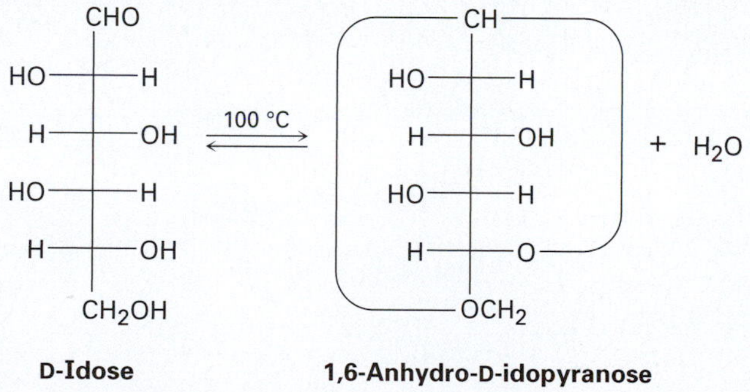

(a) Draw D-idose in its pyranose form, showing the more stable chair conformation of the ring.

(b) Which is more stable, α-D-idopyranose or β-D-idopyranose? Explain.

(c) Draw 1,6-anhydro-D-idopyranose in its most stable conformation.

(d) When heated to 100 °C under the same conditions as those used for D-idose, D-glucose does not lose water and does not exist in a 1,6-anhydro form. Explain.

■ Problems assignable in Organic OWL.

21.64 Acetyl coenzyme A (acetyl CoA) is the key intermediate in food metabolism. What sugar is present in acetyl CoA?

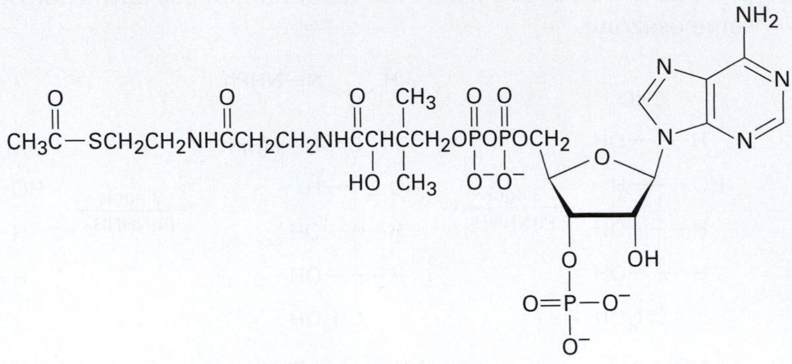

Acetyl coenzyme A

21.65 ■ One of the steps in the biological pathway for carbohydrate metabolism is the conversion of fructose 1,6-bisphosphate into dihydroxyacetone phosphate and glyceraldehyde 3-phosphate. Propose a mechanism for the transformation.

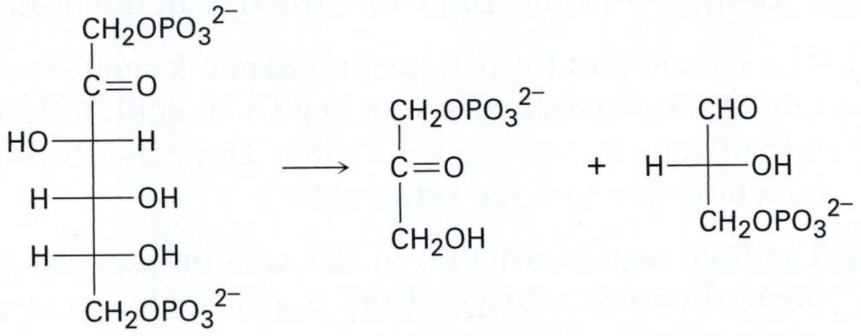

Fructose 1,6-bisphosphate	**Dihydroxyacetone phosphate**	**Glyceraldehyde 3-phosphate**

22 Carbohydrate Metabolism

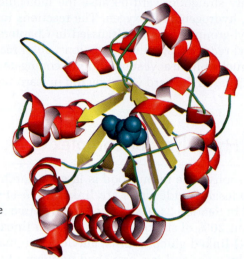

Triose-phosphate isomerase catalyzes the interconversion of dihydroxyacetone phosphate and glyceraldehyde 3-phosphate during glycolysis.

Carbohydrates are the chemical intermediaries by which carbon atoms from CO_2 are incorporated into growing organisms and by which solar energy is stored and used to support life. Thus, carbohydrates are the biological starting point, and their metabolism is intimately interconnected with the metabolism of all other biomolecules. The catabolism of glucose, in fact, has been called the backbone of all metabolic pathways.

We'll look at that metabolic backbone in this chapter, beginning with the hydrolysis of dietary starch to give glucose, which is then catabolized to pyruvate in the *glycolysis* pathway. We'll then see how pyruvate is decarboxylated to yield acetyl CoA and how acetyl CoA is degraded to CO_2 in the *citric acid cycle*. Following this look at catabolism, we'll finish the chapter by seeing how glucose is biosynthesized from pyruvate in the *gluconeogenesis* pathway.

Starch

H_2O

α-Glucose → **Glycolysis** → 2 Pyruvate → Acetyl CoA → **Citric acid cycle** → 2 CO_2

HSCoA CO_2

Online homework for this chapter can be assigned in Organic OWL.

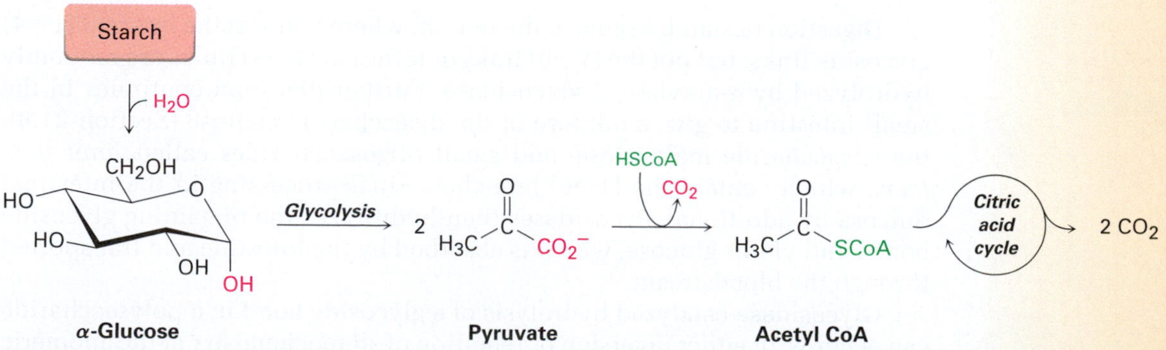

WHY THIS CHAPTER?

Glucose metabolism is at the center of biological chemistry. Fortunately, it's also relatively straightforward because the molecules are small and contain only carbon, hydrogen, and oxygen. The reactions involved are almost entirely the carbonyl-group processes discussed in Chapters 14 to 17: alcohol oxidations, carbonyl reductions, imine formations, aldol reactions, keto–enol tautomerizations, nucleophilic acyl substitutions, conjugate nucleophilic additions, and so forth. Thus, all the hard work you did in learning that material now pays off.

22.1 Hydrolysis of Complex Carbohydrates

The carbohydrate in our food is largely starch, a glucose polymer in which the monosaccharide units are linked by α-(1→4) glycoside bonds. As discussed in Section 21.9, starch consists of two main fractions: amylose makes up about 20% of starch by mass and is a linear polymer of several hundred α-(1→4)-linked glucose units; amylopectin makes up the remaining 80% of starch by mass and is a branched polymer of up to 5000 glucose units with α-(1→6) branches every 25 or so units (Figure 22.1).

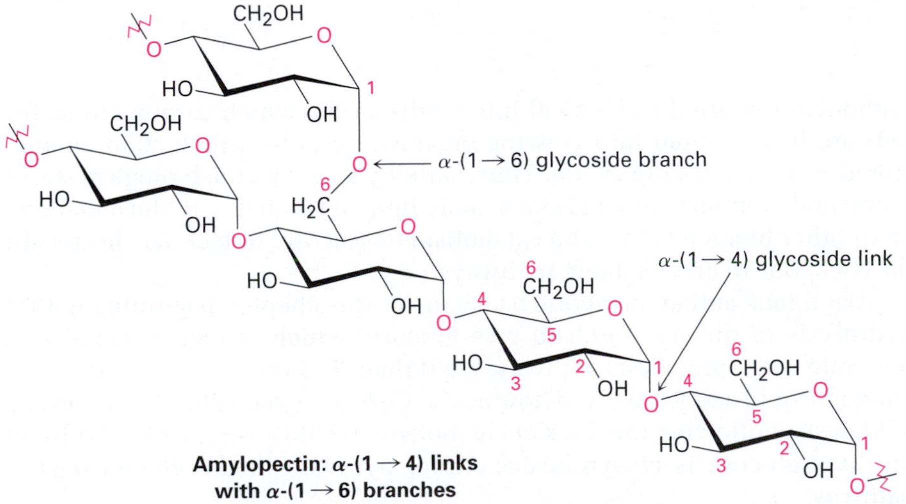

FIGURE 22.1 The structure amylopectin, the major carbohydrate found in starch.

Amylopectin: α-(1 → 4) links
with α-(1 → 6) branches

Digestion of starch begins in the mouth, where many of the internal (1→4) glycoside links, but not the (1→6) links or terminal (1→4) links, are randomly hydrolyzed by α-amylase, a glycosidase. Further digestion continues in the small intestine to give a mixture of the disaccharide maltose (Section 21.8), the trisaccharide maltotriose, and small oligosaccharides called *limit dextrins,* which contain the (1→6) branches. Final processing in the intestinal mucosa by additional glycosidases then hydrolyzes the remaining glycoside bonds and yields glucose, which is absorbed by the intestine and transported through the bloodstream.

Glycosidase-catalyzed hydrolysis of a glycoside bond in a polysaccharide can occur with either inversion or retention of stereochemistry at the anomeric center. Both mechanisms probably proceed through a short-lived oxonium ion (R_3O^+) intermediate in which one face of the carbohydrate is effectively shielded by the leaving group, leaving the other face open to nucleophilic attack (Figure 22.2a). Inverting glycosidases operate through a single S_N2-like inversion in which a carboxylate residue in the enzyme, either aspartate or glutamate, acts as a base to deprotonate water, which then adds to the oxonium ion from the side opposite the leaving group (Figure 22.2b). Retaining glycosidases operate through two inversions. In the first, a carboxylate group in the enzyme adds to the oxonium ion from the side opposite the leaving group, giving a covalently bonded, glycosylated enzyme. In the second, water displaces the carboxylate through another oxonium ion (Figure 22.2c).

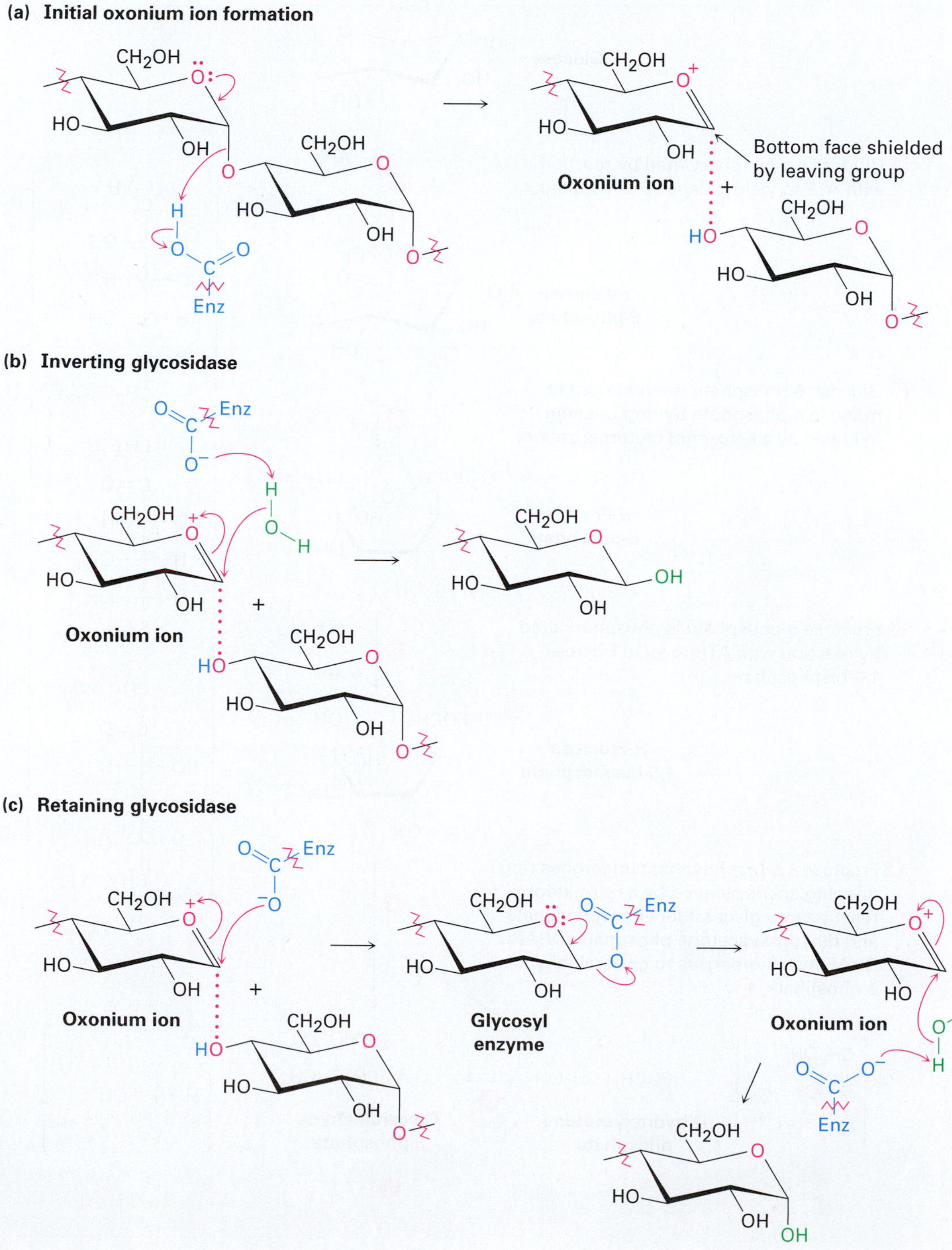

(a) Initial oxonium ion formation

(b) Inverting glycosidase

(c) Retaining glycosidase

FIGURE 22.2 General mechanism of polysaccharide hydrolysis by glycosidases. **(a)** Initial formation of an oxonium ion is followed by either one inversion or two. **(b)** Inverting glycosidases use a single inversion by nucleophilic attack of water. **(c)** Retaining glycosidases use two inversions, the first by a carboxylate ion to give a glycosylated enzyme intermediate and the second by nucleophilic attack of water.

22.2 Catabolism of Glucose: Glycolysis

Glucose is the body's primary source of short-term energy. Its catabolism begins with **glycolysis**, a series of ten enzyme-catalyzed reactions that break down glucose into 2 equivalents of pyruvate, $CH_3COCO_2^-$. The steps of glycolysis, also called the *Embden–Meyerhoff pathway* after its discoverers, are summarized in Figure 22.3.

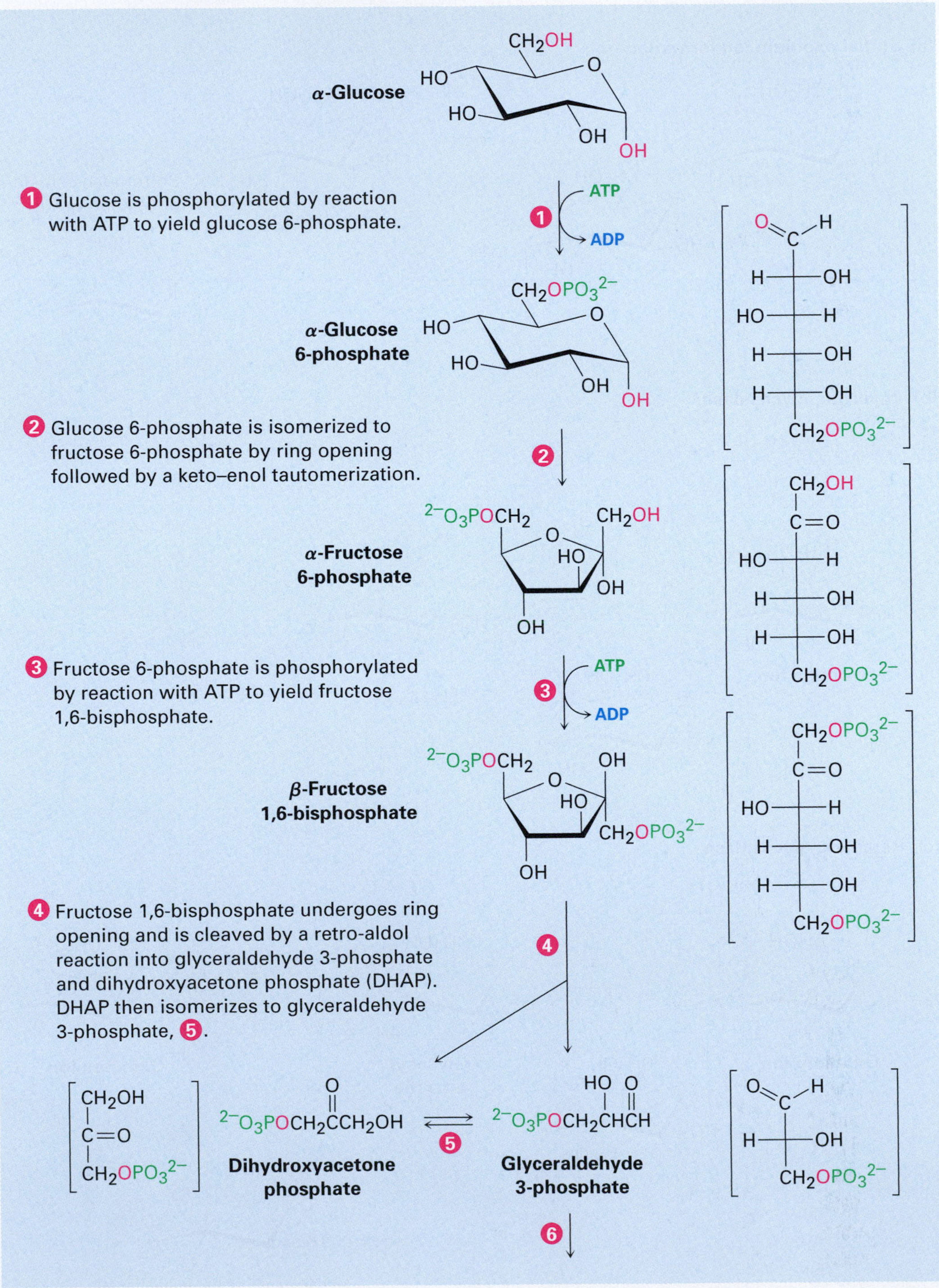

ACTIVE FIGURE 22.3
MECHANISM:
The ten-step glycolysis pathway for catabolizing glucose to two molecules of pyruvate. The individual steps are described in the text. **Go to this book's student companion site at www.cengage .com/chemistry/ mcmurry to explore an interactive version of this figure.**

1 Glucose is phosphorylated by reaction with ATP to yield glucose 6-phosphate.

2 Glucose 6-phosphate is isomerized to fructose 6-phosphate by ring opening followed by a keto–enol tautomerization.

3 Fructose 6-phosphate is phosphorylated by reaction with ATP to yield fructose 1,6-bisphosphate.

4 Fructose 1,6-bisphosphate undergoes ring opening and is cleaved by a retro-aldol reaction into glyceraldehyde 3-phosphate and dihydroxyacetone phosphate (DHAP). DHAP then isomerizes to glyceraldehyde 3-phosphate, 5.

© John McMurry

ACTIVE FIGURE 22.3
(continued)

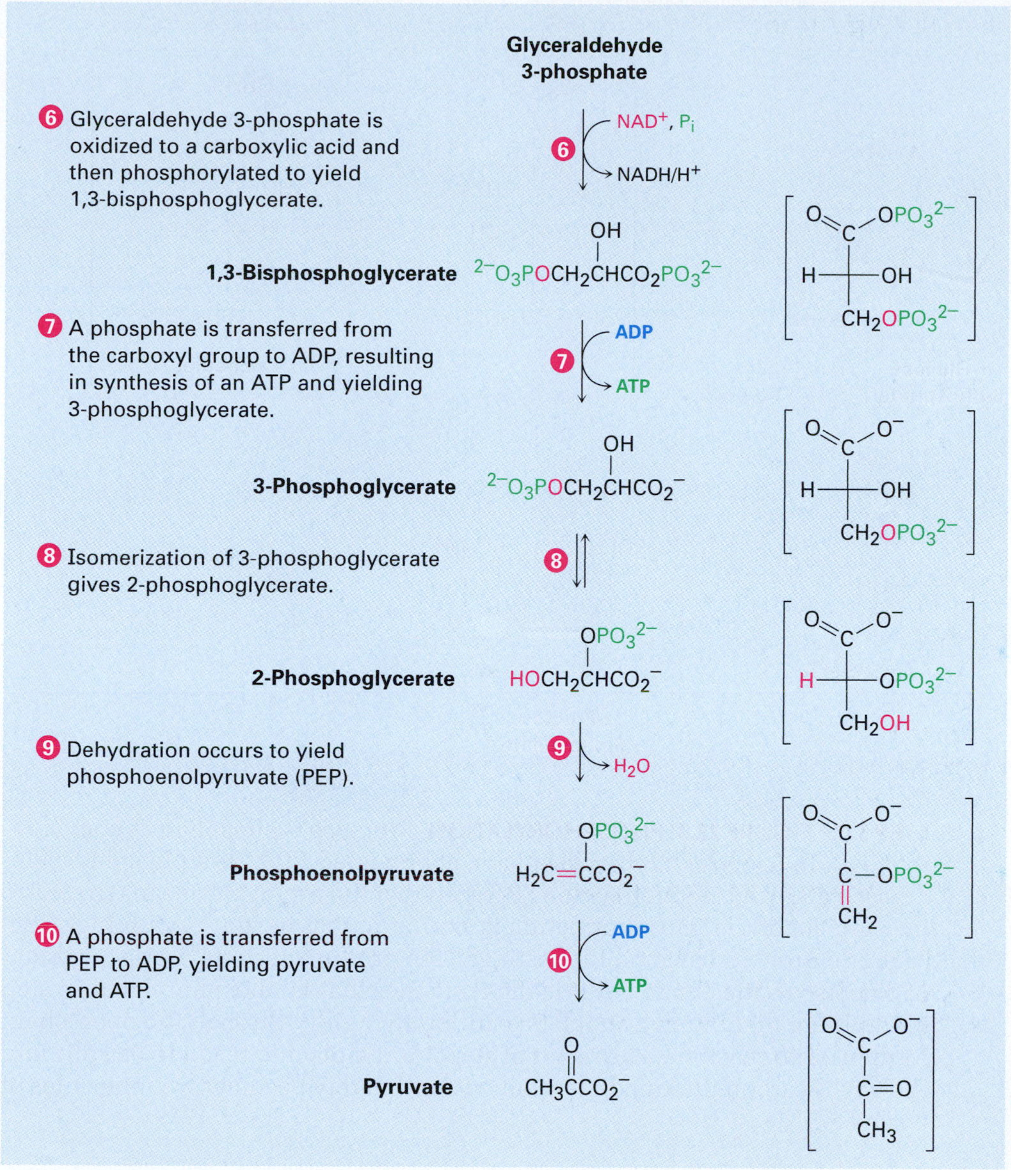

**Glyceraldehyde
3-phosphate**

6 Glyceraldehyde 3-phosphate is oxidized to a carboxylic acid and then phosphorylated to yield 1,3-bisphosphoglycerate.

6 $\mid$ NAD$^+$, P$_i$
     $\rightsquigarrow$ NADH/H$^+$

OH
$\mid$
1,3-Bisphosphoglycerate $^{2-}$O$_3$POCH$_2$CHCO$_2$PO$_3$$^{2-}$

$$\left[\begin{array}{c} O=C-OPO_3{}^{2-} \\ H-\!\!\!-OH \\ CH_2OPO_3{}^{2-} \end{array} \right]$$

7 A phosphate is transferred from the carboxyl group to ADP, resulting in synthesis of an ATP and yielding 3-phosphoglycerate.

7 $\mid$ ADP
     $\rightsquigarrow$ ATP

OH
$\mid$
3-Phosphoglycerate $^{2-}$O$_3$POCH$_2$CHCO$_2$$^-$

$$\left[\begin{array}{c} O=C-O^- \\ H-\!\!\!-OH \\ CH_2OPO_3{}^{2-} \end{array} \right]$$

8 Isomerization of 3-phosphoglycerate gives 2-phosphoglycerate.

8 $\updownarrow$

OPO$_3$$^{2-}$
$\mid$
2-Phosphoglycerate HOCH$_2$CHCO$_2$$^-$

$$\left[\begin{array}{c} O=C-O^- \\ H-\!\!\!-OPO_3{}^{2-} \\ CH_2OH \end{array} \right]$$

9 Dehydration occurs to yield phosphoenolpyruvate (PEP).

9 $\rightsquigarrow$ H$_2$O

OPO$_3$$^{2-}$
$\mid$
Phosphoenolpyruvate H$_2$C=CCO$_2$$^-$

$$\left[\begin{array}{c} O=C-O^- \\ C-OPO_3{}^{2-} \\ \parallel \\ CH_2 \end{array} \right]$$

10 A phosphate is transferred from PEP to ADP, yielding pyruvate and ATP.

10 $\mid$ ADP
      $\rightsquigarrow$ ATP

O
$\parallel$
Pyruvate CH$_3$CCO$_2$$^-$

$$\left[\begin{array}{c} O=C-O^- \\ C=O \\ CH_3 \end{array} \right]$$

© John McMurry

STEPS 1–2 OF FIGURE 22.3: PHOSPHORYLATION AND ISOMERIZATION Glucose is first phosphorylated at the C6 hydroxyl group by reaction with ATP in a process catalyzed by hexokinase. As noted in Section 20.1, the reaction requires Mg^{2+} as a cofactor to complex with the negatively charged phosphate oxygens.

The glucose 6-phosphate that results is isomerized in step 2 by glucose-6-phosphate isomerase to give fructose 6-phosphate. The isomerization takes place by initial opening of the glucose hemiacetal ring to the open-chain form, followed by keto–enol tautomerization (Section 17.1) to a cis enediol, HO–C=C–OH. But because glucose and fructose share a common enediol, further tautomerization to a different keto form produces open-chain fructose. Cyclization to a hemiacetal completes the process (Figure 22.4).

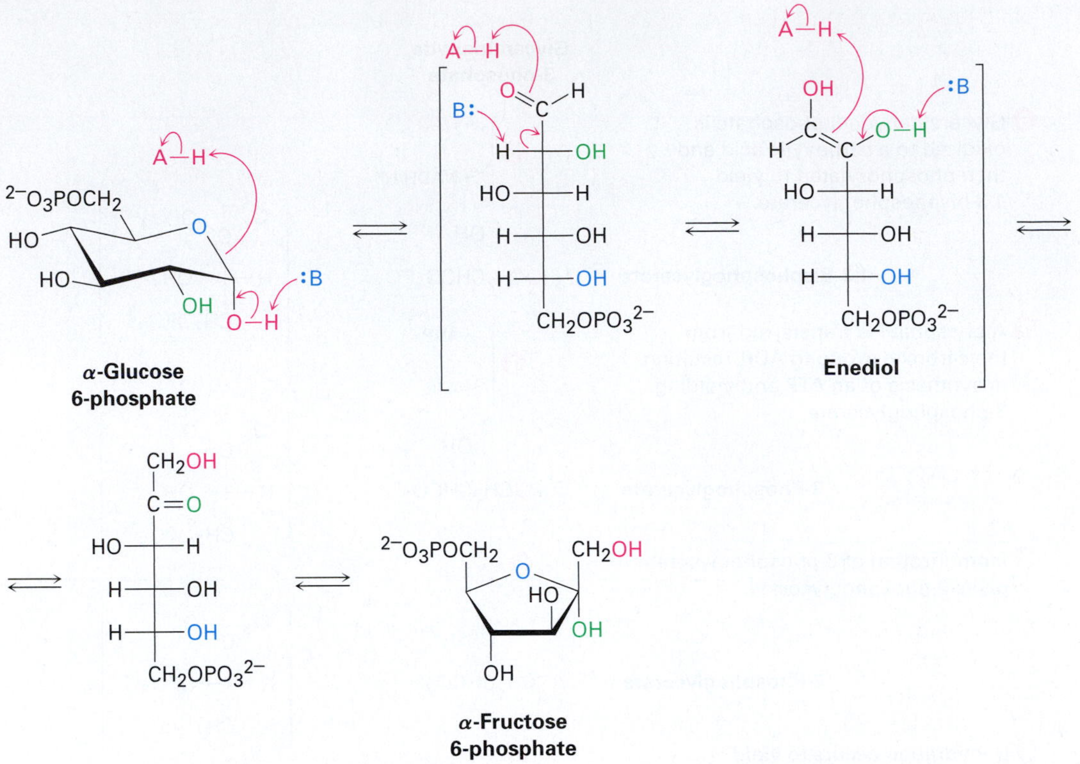

FIGURE 22.4
Mechanism of step 2 in glycolysis, the isomerization of glucose 6-phosphate to fructose 6-phosphate. The reaction occurs by keto–enol tautomerization.

STEP 3 OF FIGURE 22.3: PHOSPHORYLATION Fructose 6-phosphate is converted in step 3 to fructose 1,6-bisphosphate, abbreviated FBP, by a phosphofructo-kinase-catalyzed reaction with ATP (recall that the prefix *bis-* means two). The mechanism of the phosphorylation is similar to that in step 1, with Mg^{2+} ion again required as cofactor. Interestingly, the product of step 2 is the α anomer of fructose 6-phosphate but it is the β anomer that is phosphorylated in step 3, implying that the two anomers equilibrate rapidly through the open-chain form prior to reaction. The result of step 3 is a molecule ready to be split into the two 3-carbon intermediates that will ultimately become two molecules of pyruvate.

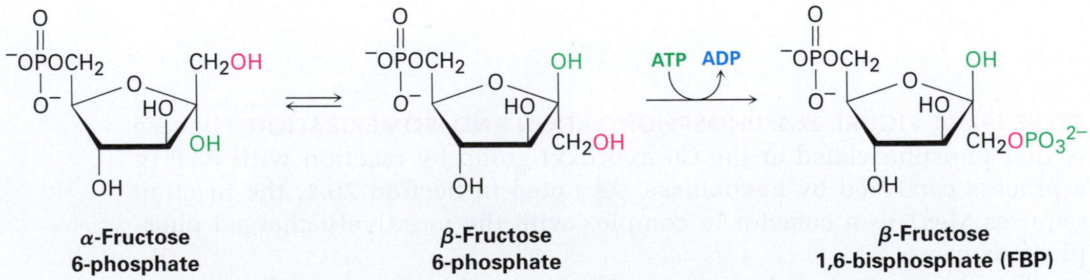

STEP 4 OF FIGURE 22.3: CLEAVAGE Fructose 1,6-bisphosphate is cleaved in step 4 into two 3-carbon pieces, dihydroxyacetone phosphate (DHAP) and glyceraldehyde 3-phosphate (GAP). The bond between C3 and C4 of fructose 1,6-bisphosphate breaks, and a C=O group is formed at C4. Mechanistically, the cleavage is the reverse of an aldol reaction (Section 17.6) and is catalyzed by an aldolase. A *forward* aldol reaction joins two aldehydes or ketones to

give a β-hydroxy carbonyl compound, while a *retro* aldol reaction such as that occurring here cleaves a β-hydroxy carbonyl compound into two aldehydes or ketones.

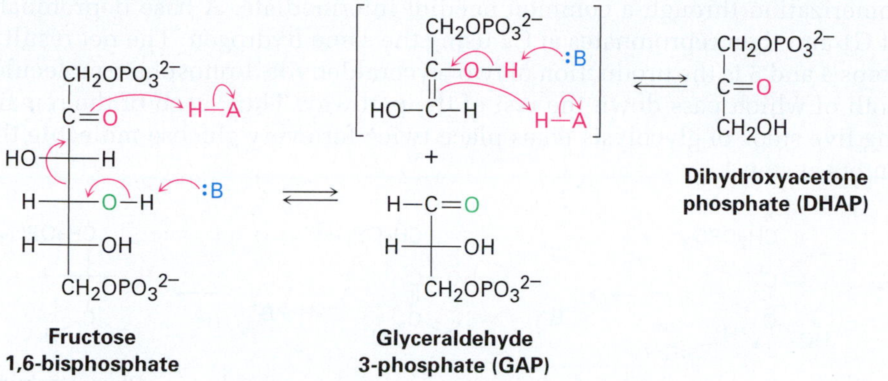

Two classes of aldolases are used by organisms to catalyze the retro-aldol reaction. In fungi, algae, and some bacteria, the retro-aldol reaction is catalyzed by class II aldolases, which function by coordination of the fructose carbonyl group with Zn^{2+} as Lewis acid. In plants and animals, the reaction is catalyzed by class I aldolases and does not take place on the free ketone. Instead, fructose 1,6-bisphosphate undergoes reaction with the side-chain $-NH_2$ group of a lysine residue on the aldolase to yield a protonated, enzyme-bound imine (Section 14.7), often called a **Schiff base** in biochemistry. Because of its positive charge, the iminium ion is a better electron acceptor than a ketone carbonyl group. Retro-aldol reaction ensues, giving glyceraldehyde 3-phosphate and an enamine, which is protonated to give another iminium ion that is hydrolyzed to yield dihydroxyacetone phosphate (Figure 22.5).

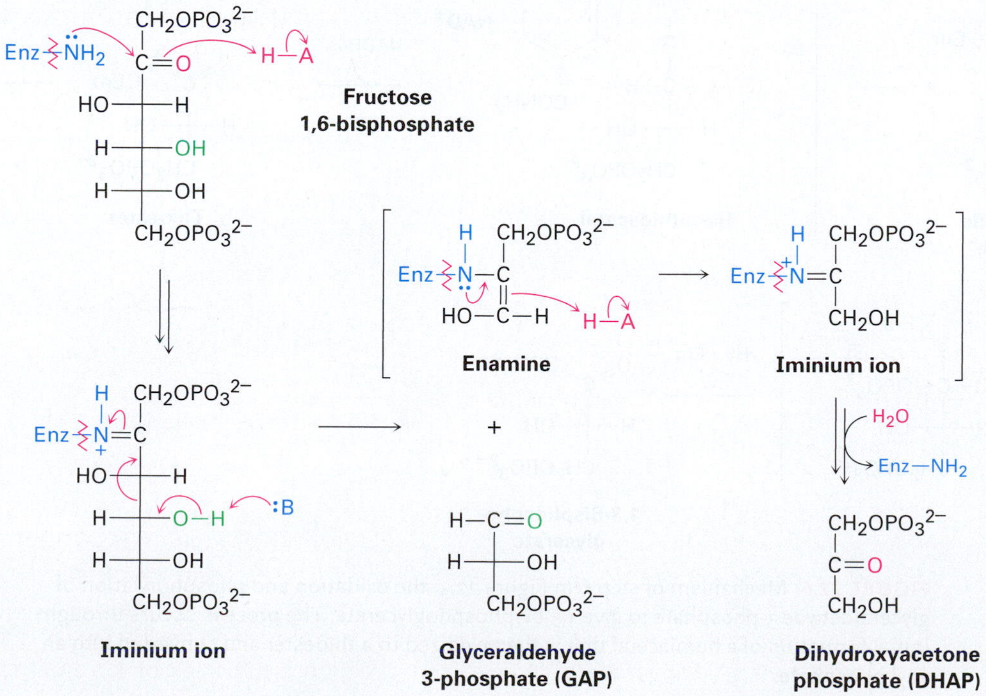

FIGURE 22.5 Mechanism of step 4 in Figure 22.3, the cleavage of fructose 1,6-bisphosphate to yield glyceraldehyde 3-phosphate and dihydroxyacetone phosphate. The reaction occurs through an iminium ion formed by reaction with a lysine residue in the enzyme.

STEP 5 OF FIGURE 22.3: ISOMERIZATION Dihydroxyacetone phosphate is isomerized in step 5 by triose phosphate isomerase to form a second equivalent of glyceraldehyde 3-phosphate. As in the conversion of glucose 6-phosphate to fructose 6-phosphate in step 2, the isomerization takes place by keto–enol tautomerization through a common enediol intermediate. A base deprotonates at C1 and then reprotonates at C2 using the same hydrogen. The net result of steps 4 and 5 is the production of two glyceraldehyde 3-phosphate molecules, both of which pass down the rest of the pathway. Thus, each of the remaining five steps of glycolysis takes place twice for every glucose molecule that enters at step 1.

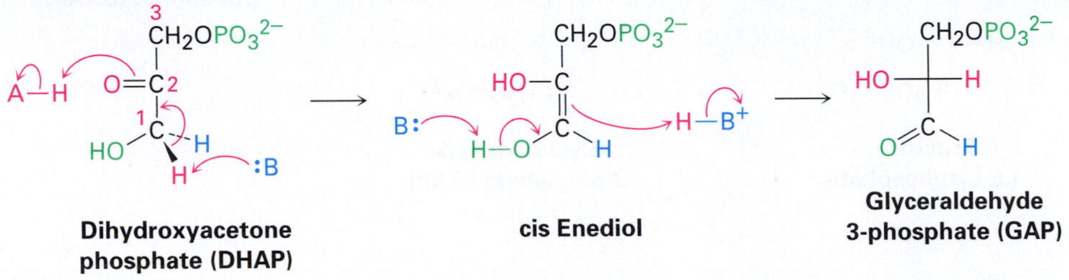

STEPS 6–7 OF FIGURE 22.3: OXIDATION, PHOSPHORYLATION, AND DEPHOSPHORYLATION Glyceraldehyde 3-phosphate is oxidized and phosphorylated in step 6 to give 1,3-bisphosphoglycerate (Figure 22.6). The reaction is catalyzed by glyceraldehyde 3-phosphate dehydrogenase and begins by nucleophilic addition of the –SH group of a cysteine residue in the enzyme to the aldehyde carbonyl group to yield a *hemithioacetal* (RS—C—OH), the sulfur analog of a hemiacetal. Oxidation of the hemithioacetal –OH group by NAD^+ then yields a thioester, which reacts with phosphate ion in a nucleophilic acyl substitution step to give the acyl phosphate 1,3-bisphosphoglycerate, a mixed anhydride between a carboxylic acid and phosphoric acid.

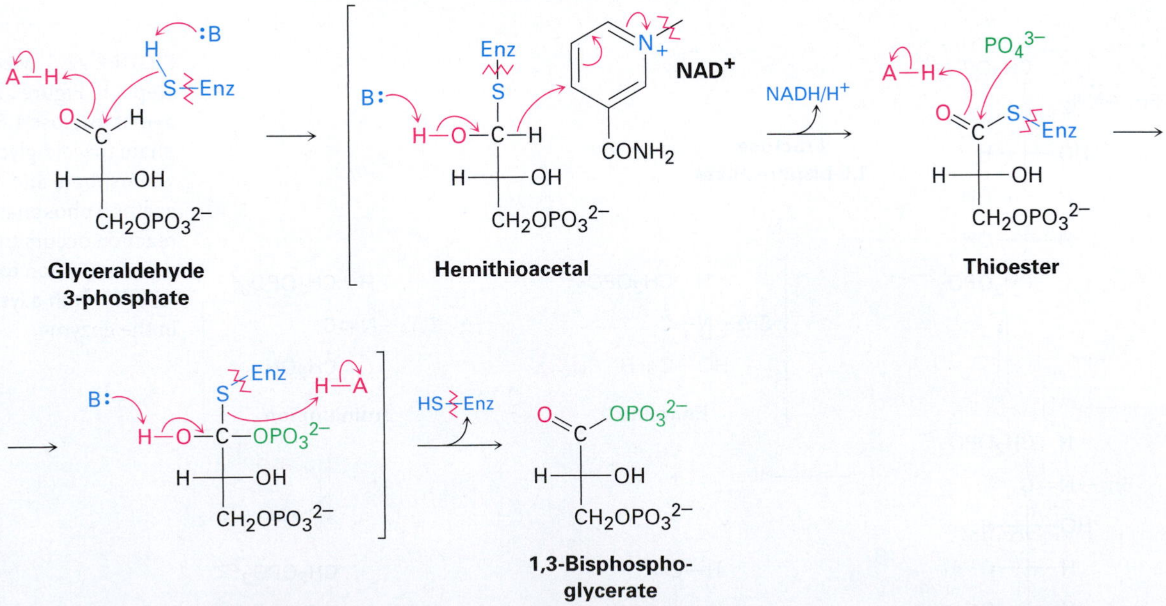

FIGURE 22.6 Mechanism of step 6 in Figure 22.3, the oxidation and phosphorylation of glyceraldehyde 3-phosphate to give 1,3-bisphosphoglycerate. The process occurs through initial formation of a hemiacetal that is then oxidized to a thioester and converted into an acyl phosphate.

Like all anhydrides (Section 16.5), the mixed carboxylic–phosphoric anhydride is a reactive substrate in nucleophilic acyl (or phosphoryl) substitution reactions. Reaction of 1,3-bisphosphoglycerate with ADP occurs in step 7 by substitution on phosphorus, resulting in transfer of a phosphate group to ADP and giving ATP plus 3-phosphoglycerate. The process is catalyzed by phosphoglycerate kinase and requires Mg^{2+} as cofactor. Note that steps 6 and 7 together accomplish the oxidation of an aldehyde to a carboxylic acid.

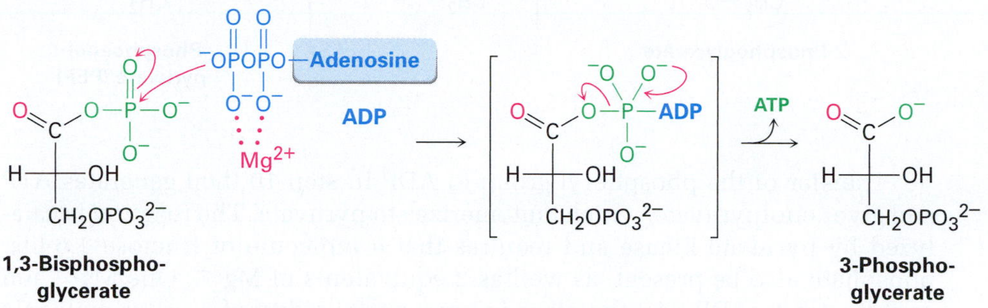

1,3-Bisphospho-glycerate

3-Phospho-glycerate

STEP 8 OF FIGURE 22.3: ISOMERIZATION 3-Phosphoglycerate isomerizes to 2-phosphoglycerate in a step catalyzed by phosphoglycerate mutase. In plants, 3-phosphoglycerate transfers its phosphoryl group from its C3 oxygen to a histidine residue on the enzyme in one step and then accepts the same phosphoryl group back onto the C2 oxygen in a second step. In animals and yeast, however, the enzyme contains a phosphorylated histidine, which transfers its phosphoryl group to the C2 oxygen of 3-phosphoglycerate and forms 2,3-bisphosphoglycerate as intermediate. The same histidine then accepts a phosphoryl group from the C3 oxygen to yield the isomerized product plus regenerated enzyme. As explained in Section 20.5, we'll occasionally use an abbreviated mechanism for nucleophilic acyl substitution reactions to save space.

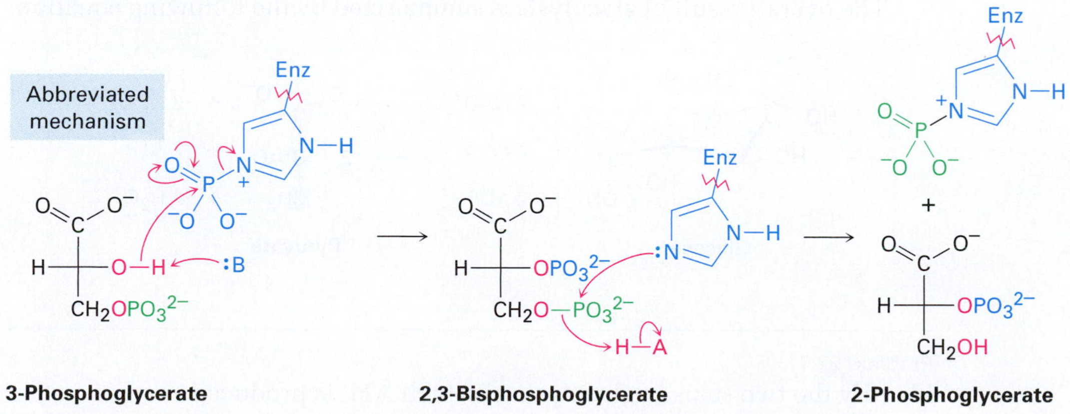

3-Phosphoglycerate

2,3-Bisphosphoglycerate

2-Phosphoglycerate

STEPS 9–10 OF FIGURE 22.3: DEHYDRATION AND DEPHOSPHORYLATION Like most β-hydroxy carbonyl compounds, 2-phosphoglycerate undergoes a ready dehydration in step 9 by an E1cB mechanism (Section 17.7). The process is catalyzed by enolase, and the product is phosphoenolpyruvate, abbreviated

PEP. Two Mg^{2+} ions are associated with the 2-phosphoglycerate to neutralize the negative charges.

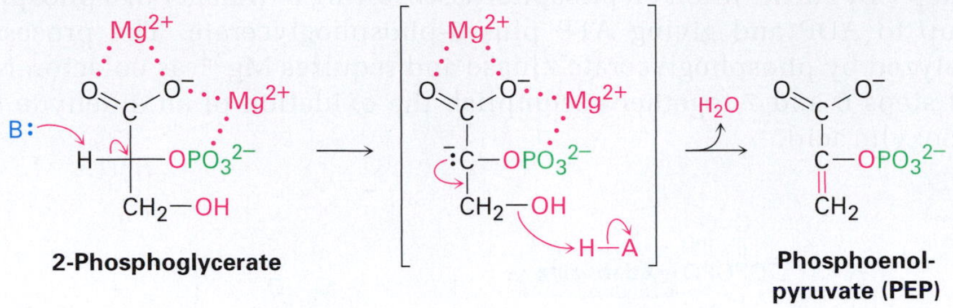

2-Phosphoglycerate   **Phosphoenol-pyruvate (PEP)**

Transfer of the phosphoryl group to ADP in step 10 then generates ATP and gives enolpyruvate, which tautomerizes to pyruvate. The reaction is catalyzed by pyruvate kinase and requires that a molecule of fructose 1,6-bisphosphate also be present, as well as 2 equivalents of Mg^{2+}. One Mg^{2+} ion coordinates to ADP, and the other increases the acidity of a water molecule necessary for protonation of the enolate ion. The requirement for a molecule of fructose 1,6-bisphosphate is not fully understood.

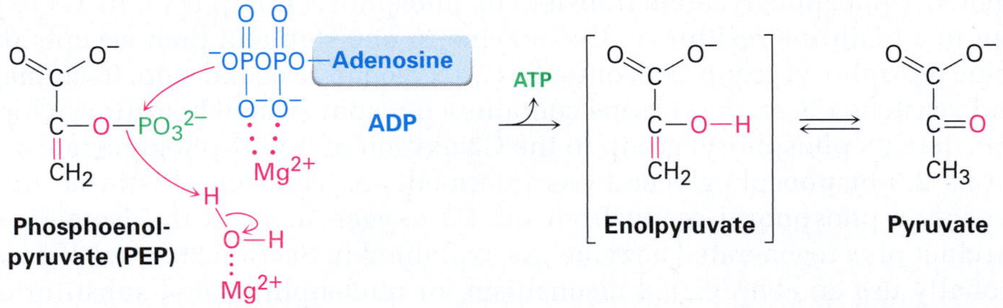

Phosphoenol-pyruvate (PEP)   **Enolpyruvate**   **Pyruvate**

The overall result of glycolysis is summarized by the following equation:

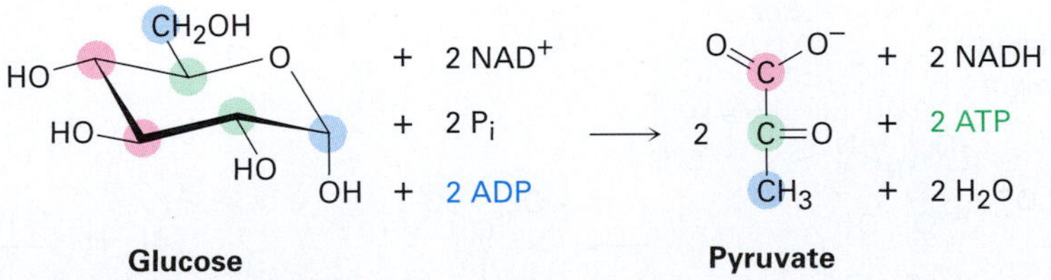

Glucose   **Pyruvate**

Problem 22.1

Identify the two steps in glycolysis in which ATP is produced.

Problem 22.2

Look at the entire glycolysis pathway and make a list of the kinds of organic reactions that take place—nucleophilic acyl substitutions, aldol reactions, imine formations, E1cB reactions, and so forth.

Problem 22.3

Is it the *pro-R* or *pro-S* hydrogen that is removed in step 5 of glycolysis, the isomerization of dihydroxyacetone phosphate to glyceraldehyde 3-phosphate? You might want to review Section 5.11 on prochirality.

Dihydroxyacetone   cis Enediol
phosphate (DHAP)

Problem 22.4

In step 6 of glycolysis (Figure 22.6) the hydride ion from glyceraldehyde 3-phosphate adds to the *Si* face of NAD^+. Draw the structure of the NADH that results, and indicate which hydrogen in NADH is the one added.

22.3 Conversion of Pyruvate to Acetyl CoA

Pyruvate, which is produced both by catabolism of glucose and by degradation of several amino acids (Section 20.4), can undergo several further transformations depending on the conditions and on the organism. In the absence of oxygen, pyruvate can be either reduced by NADH to yield lactate $[CH_3CH(OH)CO_2^-]$ or, in yeast, fermented to give ethanol. Under typical aerobic conditions in mammals, however, pyruvate is converted by a process called *oxidative decarboxylation* to give acetyl CoA plus CO_2. (*Oxidative* because the oxidation state of the carbonyl carbon rises from that of a ketone to that of a thioester.)

The conversion occurs through a multistep sequence of reactions catalyzed by a complex of enzymes and cofactors called the *pyruvate dehydrogenase complex.* The process occurs in three stages, each catalyzed by one of the enzymes in the complex, as outlined in Figure 22.7 on the next page. Acetyl CoA, the ultimate product, then acts as fuel for the final stage of catabolism, the citric acid cycle.

STEP 1 OF FIGURE 22.7: ADDITION OF THIAMIN DIPHOSPHATE The conversion of pyruvate to acetyl CoA begins by reaction of pyruvate with thiamin diphosphate, a derivative of vitamin B_1. Because it was originally called thiamin *pyro*phosphate, thiamin diphosphate is usually abbreviated as TPP. The spelling *thiamine* is also correct and frequently used.

The key structural element in thiamin diphosphate is the presence of a thiazolium ring—a five-membered, unsaturated heterocycle containing a sulfur atom and a positively charged nitrogen atom. The thiazolium ring is weakly acidic, with a pK_a of approximately 18 for the C–H ring hydrogen between N and S. Bases can therefore deprotonate thiamin diphosphate, leading to formation of an *ylide*—a neutral species with adjacent + and − charges such as the phosphonium ylides used in the Wittig reaction (Section 14.9). As in the

1 Nucleophilic addition of thiamin diphosphate (TPP) ylide to pyruvate gives an alcohol addition product.

2 Decarboxylation occurs in a step analgous to the loss of CO_2 from a β-keto acid, yielding the enamine hydroxyethylthiamin diphosphate (HETPP).

3 The enamine double bond attacks a sulfur atom of lipoamide and carries out an S_N2-like displacement of the second sulfur to yield a hemithioacetal.

4 Elimination of thiamin diphosphate ylide from the hemithioacetal intermediate yields acetyl dihydrolipoamide . . .

5 . . . which reacts with coenzyme A in a nucleophilic acyl substitution reaction to exchange one thioester for another and give acetyl CoA plus dihydrolipoamide.

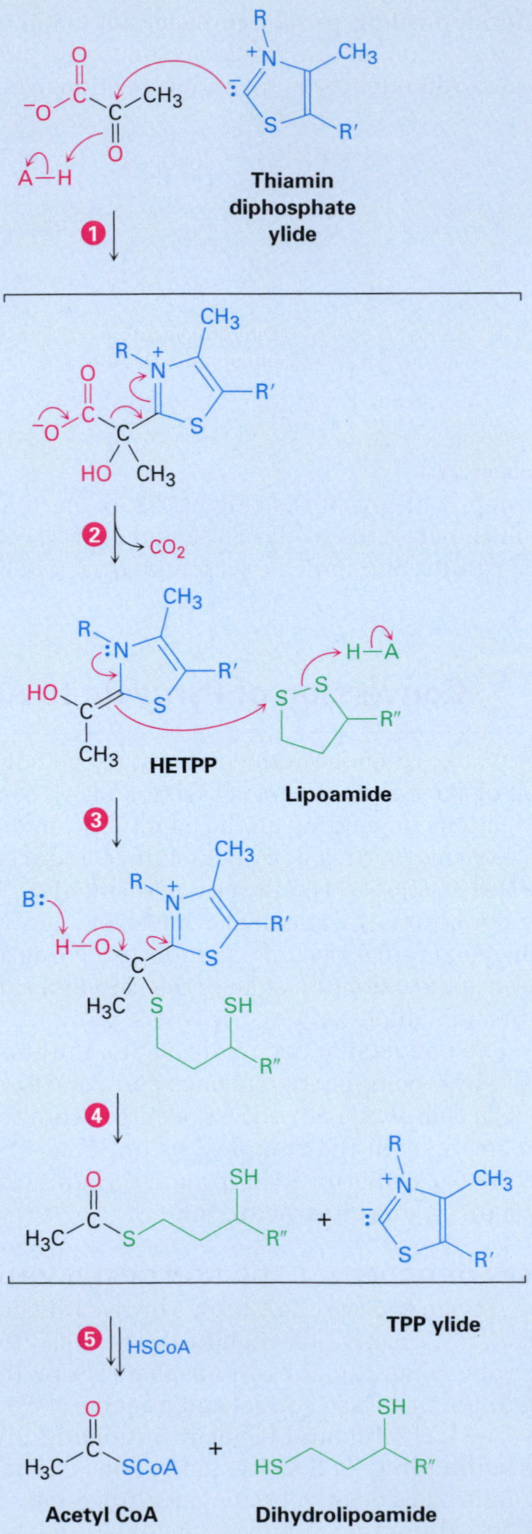

© John McMurry

FIGURE 22.7 MECHANISM: Mechanism of the conversion of pyruvate to acetyl CoA through a multistep sequence of reactions that requires three different enzymes and five different coenzymes. The individual steps are explained in the text.

Wittig reaction, the TPP ylide is a nucleophile and adds to the ketone carbonyl group of pyruvate to yield an alcohol addition product.

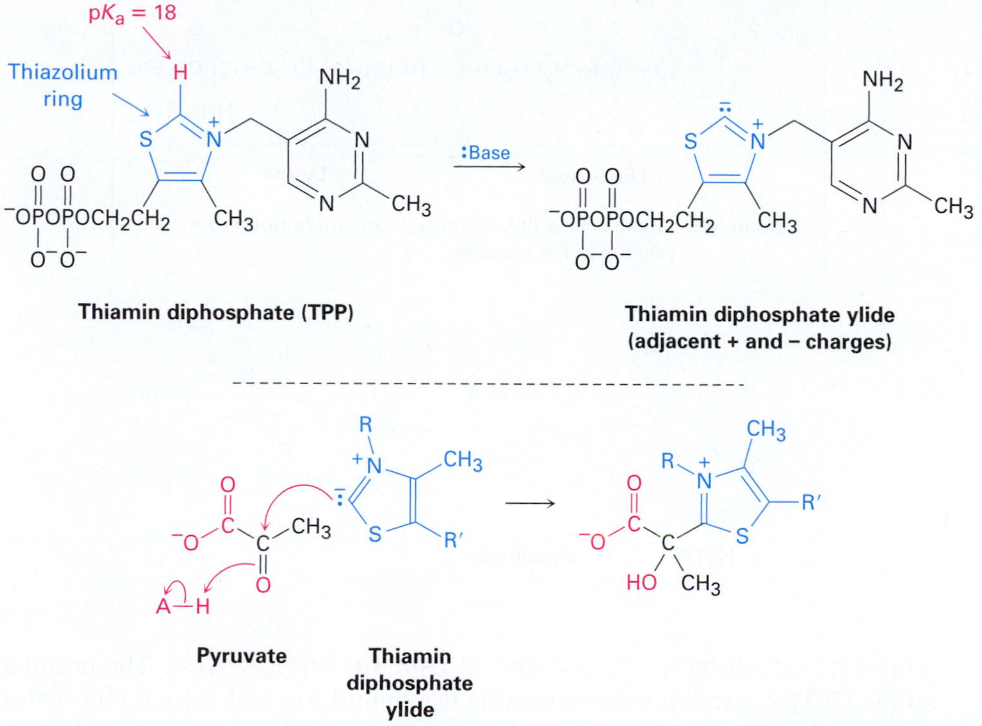

Thiamin diphosphate (TPP)

Thiamin diphosphate ylide (adjacent + and − charges)

Pyruvate   **Thiamin diphosphate ylide**

STEP 2 OF FIGURE 22.7: DECARBOXYLATION The TPP addition product, which contains an iminium ion β to a carboxylate anion, undergoes decarboxylation in much the same way that a β-keto acid decarboxylates in the acetoacetic ester synthesis (Section 17.5). The C=N$^+$ bond of the pyruvate addition product acts like the C=O bond of a β-keto acid to accept electrons as CO_2 leaves, giving hydroxyethylthiamin diphosphate (HETPP).

Thiamin addition product

Hydroxyethylthiamin diphosphate (HETPP) $+$ CO_2

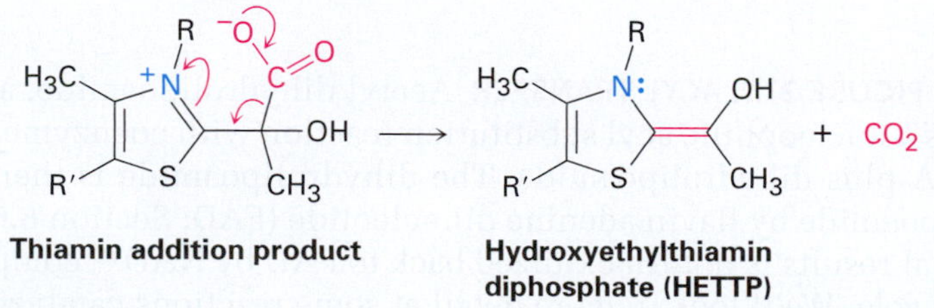

STEP 3 OF FIGURE 22.7: REACTION WITH LIPOAMIDE Hydroxyethylthiamin diphosphate is an enamine (R_2N—C=C), which, like all enamines, is nucleophilic (Section 17.12). It therefore reacts with the enzyme-bound disulfide

lipoamide by nucleophilic attack on a sulfur atom, displacing the second sulfur in an S_N2-like substitution process.

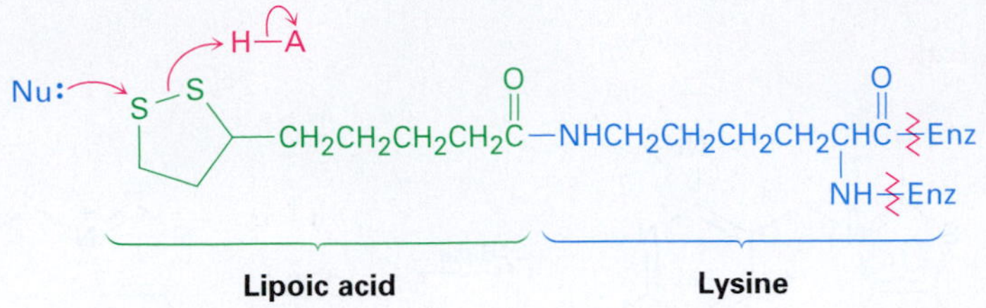

Lipoamide: Lipoic acid is linked through an amide bond to a lysine residue in the enzyme

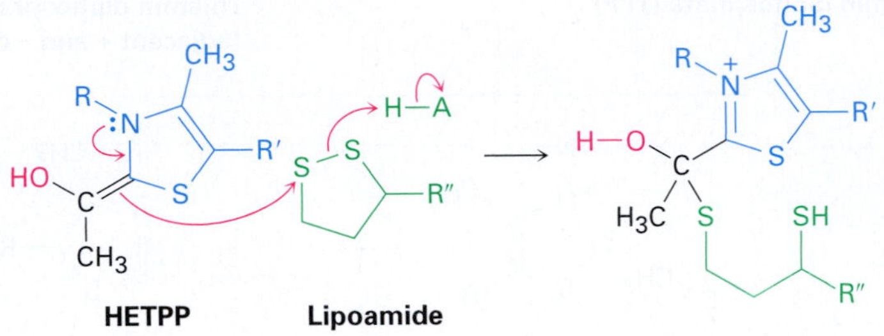

STEP 4 OF FIGURE 22.7: ELIMINATION OF THIAMIN DIPHOSPHATE The product of the HETPP reaction with lipoamide is a hemithioacetal, which eliminates thiamin diphosphate ylide. This elimination is just the reverse of the ketone addition in step 1 and generates acetyl dihydrolipoamide.

Acetyl dihydrolipoamide   **TPP ylide**

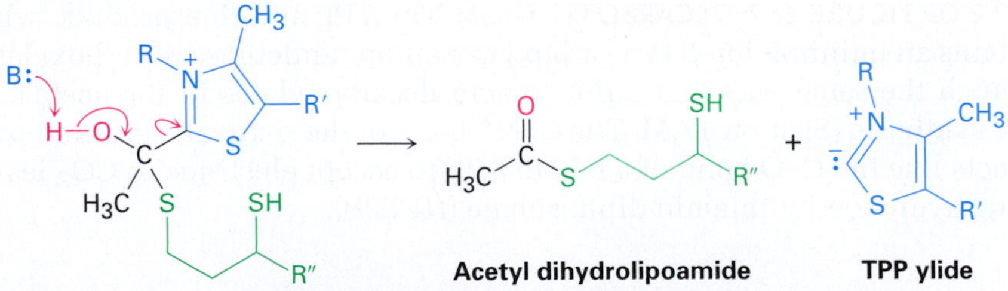

STEP 5 OF FIGURE 22.7: ACYL TRANSFER Acetyl dihydrolipoamide, a thioester, undergoes a nucleophilic acyl substitution reaction with coenzyme A to yield acetyl CoA plus dihydrolipoamide. The dihydrolipoamide is then oxidized back to lipoamide by flavin adenine dinucleotide (FAD; Section 8.6), and the $FADH_2$ that results is in turn oxidized back to FAD by NAD^+, completing the catalytic cycle. We'll look in more detail at some reactions catalyzed by FAD in Section 23.5.

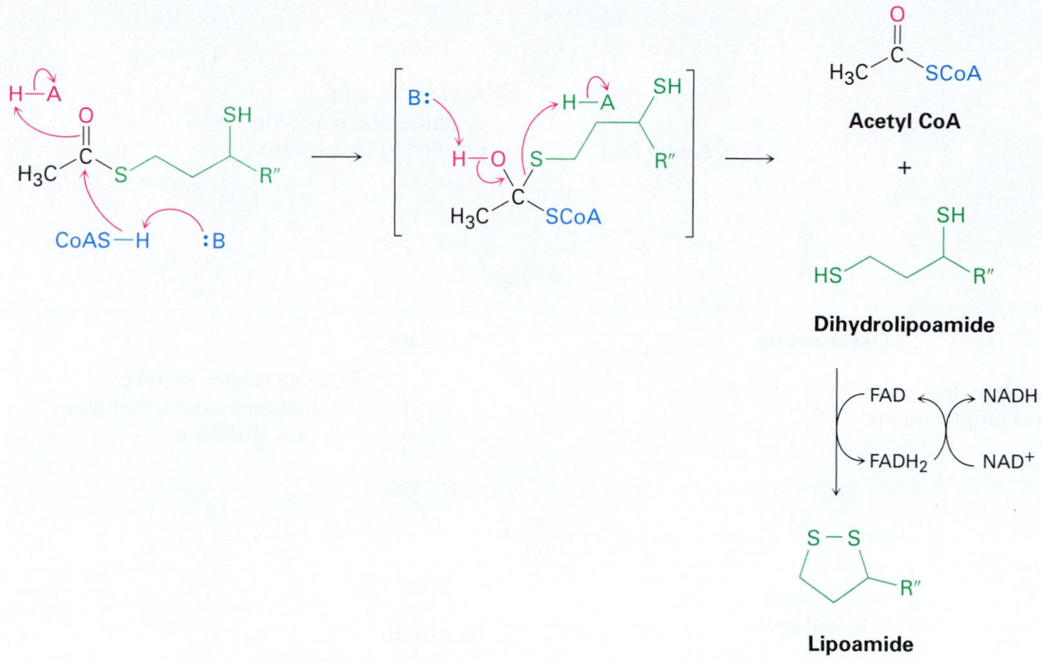

Acetyl CoA

Dihydrolipoamide

Lipoamide

Problem 22.5

Which carbon atoms in glucose end up as $-CH_3$ carbons in acetyl CoA, and which carbons end up as CO_2?

22.4 The Citric Acid Cycle

The initial stages of catabolism result in the conversion of carbohydrates into acetyl groups that are bonded through a thioester link to coenzyme A. Acetyl CoA then enters the next stage of catabolism—the **citric acid cycle**, also called the *tricarboxylic acid (TCA) cycle,* or *Krebs cycle,* after Hans Krebs, who unraveled its complexities in 1937. The overall result of the cycle is the conversion of an acetyl group into two molecules of CO_2 plus reduced coenzymes by the eight-step sequence of reactions shown in Figure 22.8.

As its name implies, the citric acid *cycle* is a closed loop of reactions in which the product of the final step (oxaloacetate) is a reactant in the first step. The intermediates are constantly regenerated and flow continuously through the cycle, which operates as long as the oxidizing coenzymes NAD^+ and FAD are available. To meet this condition, the reduced coenzymes NADH and $FADH_2$ must be reoxidized via the electron-transport chain, which in turn relies on oxygen as the ultimate electron acceptor. Thus, the cycle is dependent on the availability of oxygen and on the operation of the electron-transport chain.

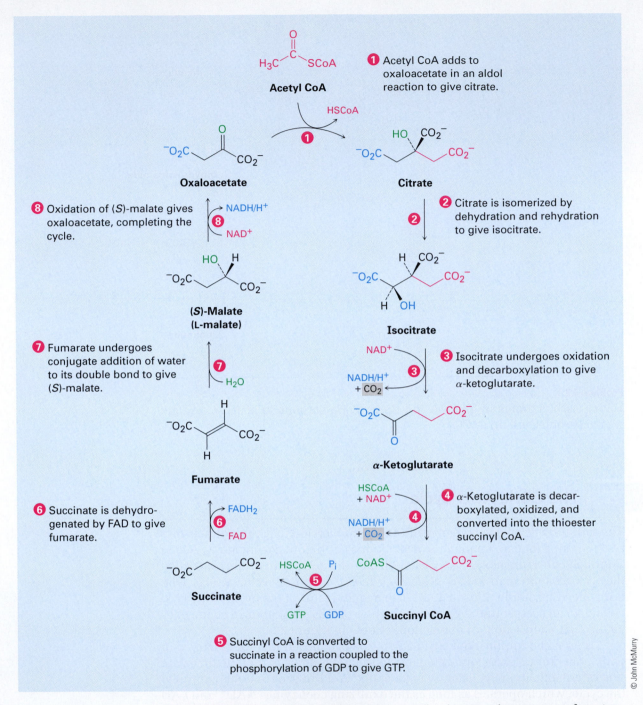

ACTIVE FIGURE 22.8 MECHANISM: The citric acid cycle is an eight-step series of reactions that results in the conversion of an acetyl group into two molecules of CO_2 plus reduced coenzymes. Individual steps are explained in the text. **Go to this book's student companion site at** www.cengage.com/chemistry/mcmurry **to explore an interactive version of this figure.**

STEP 1 OF FIGURE 22.8: ADDITION TO OXALOACETATE Acetyl CoA enters the citric acid cycle in step 1 by nucleophilic addition to the oxaloacetate carbonyl group to give (S)-citryl CoA. The addition is an aldol reaction and is catalyzed by citrate synthase, as discussed in Section 19.10. (S)-Citryl CoA is then hydrolyzed to citrate by a typical nucleophilic acyl substitution reaction with water, catalyzed by the same citrate synthase enzyme.

Note that the hydroxyl-bearing carbon of citrate is a prochirality center that contains two identical "arms." Because the initial aldol reaction of acetyl CoA to oxaloacetate occurs specifically from the *Si* face of the ketone carbonyl group, the *pro-S* arm of citrate is derived from acetyl CoA and the *pro-R* arm is derived from oxaloacetate. You might want to review Section 5.11 on prochirality to brush up on the meanings of the various terms *pro-R, pro-S, Si,* and *Re.*

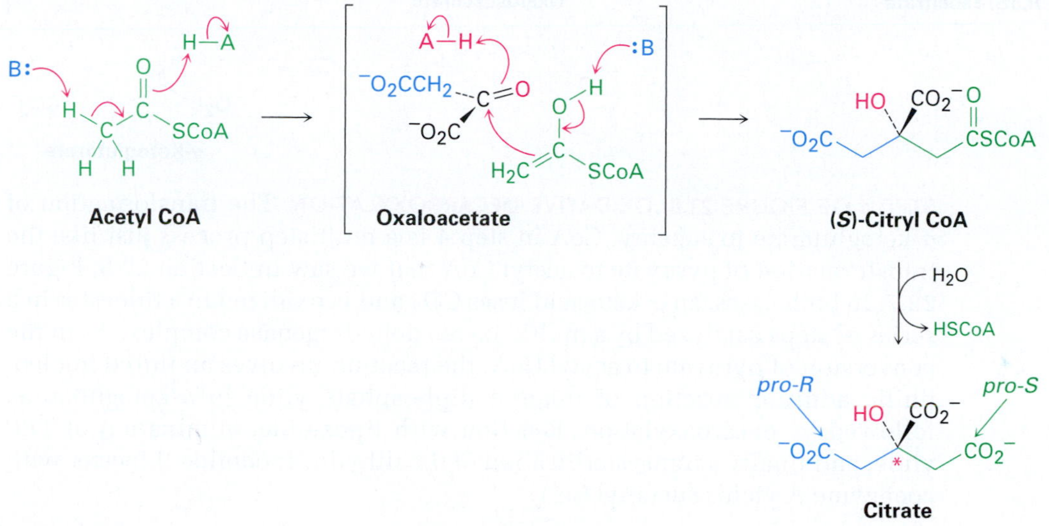

Acetyl CoA   **Oxaloacetate**   **(S)-Citryl CoA**

Citrate

STEP 2 OF FIGURE 22.8: ISOMERIZATION

Citrate, a prochiral tertiary alcohol, is next converted into its isomer, (2R,3S)-isocitrate, a chiral secondary alcohol. The isomerization occurs in two steps, both of which are catalyzed by the same aconitase enzyme. The initial step is an E1cB dehydration of a β-hydroxy acid to give *cis*-aconitate, the same sort of reaction that occurs in step 9 of glycolysis (Figure 22.3). The second step is a conjugate nucleophilic addition of water to the C=C bond (Section 14.11). The dehydration of citrate takes place specifically on the *pro-R* arm—the one derived from oxaloacetate—rather than on the *pro-S* arm derived from acetyl CoA.

Citrate   ***cis*-Aconitate**   **(2R,3S)-Isocitrate**

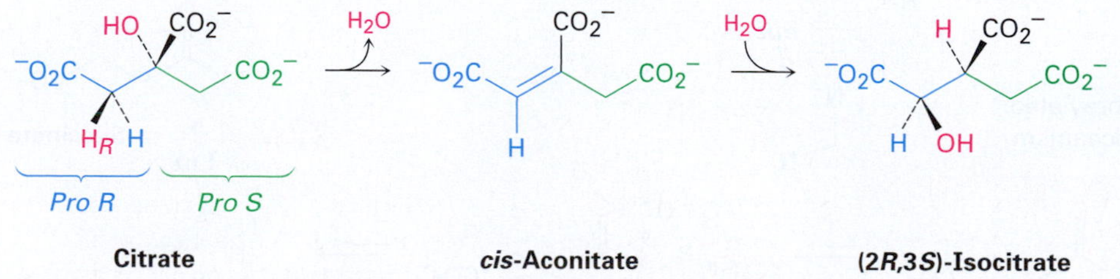

STEP 3 OF FIGURE 22.8: OXIDATION AND DECARBOXYLATION

(2R,3S)-Isocitrate, a secondary alcohol, is oxidized by NAD^+ in step 3 to give the ketone oxalosuccinate, which loses CO_2 to give α-ketoglutarate. Catalyzed by isocitrate dehydrogenase, the decarboxylation is a typical reaction of a β-keto acid, just like that in the acetoacetic ester synthesis (Section 17.5). The enzyme requires a divalent cation as cofactor to polarize the ketone carbonyl group and make it a better electron acceptor.

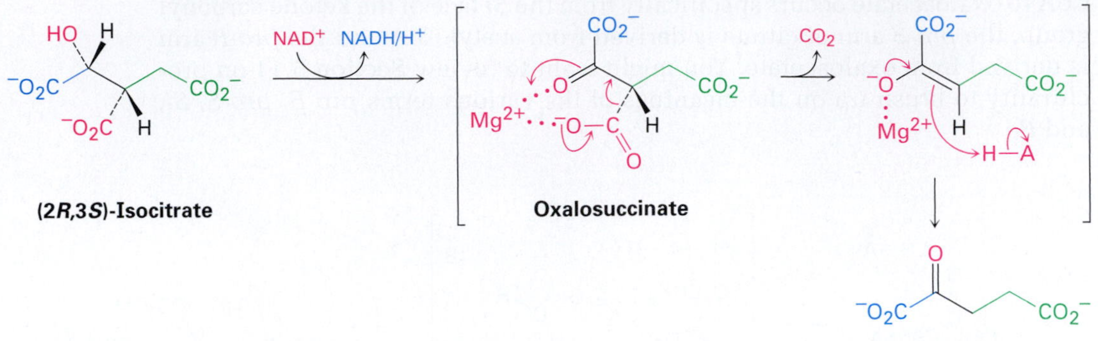

(2R,3S)-Isocitrate

Oxalosuccinate

α-Ketoglutarate

STEP 4 OF FIGURE 22.8: OXIDATIVE DECARBOXYLATION The transformation of α-ketoglutarate to succinyl CoA in step 4 is a multistep process just like the transformation of pyruvate to acetyl CoA that we saw in Section 22.3, Figure 22.7. In both cases, an α-keto acid loses CO_2 and is oxidized to a thioester in a series of steps catalyzed by a multienzyme dehydrogenase complex. As in the conversion of pyruvate to acetyl CoA, the reaction involves an initial nucleophilic addition reaction of thiamin diphosphate ylide to α-ketoglutarate, followed by decarboxylation. Reaction with lipoamide, elimination of TPP ylide, and finally a transesterification of the dihydrolipoamide thioester with coenzyme A yields succinyl CoA.

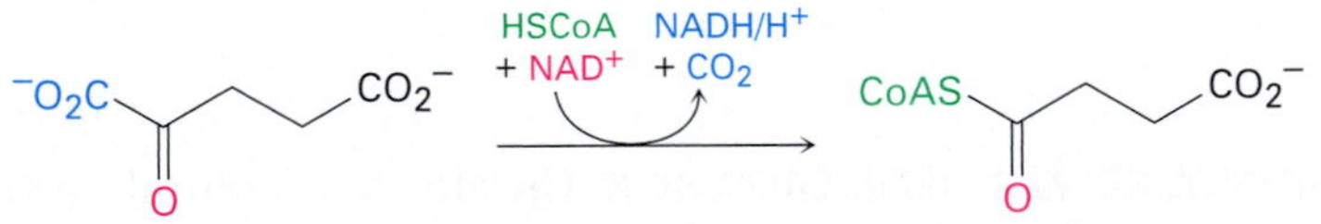

α-Ketoglutarate   **Succinyl CoA**

STEP 5 OF FIGURE 22.8: ACYL COA CLEAVAGE Succinyl CoA is converted to succinate in step 5. The reaction is catalyzed by succinyl CoA synthetase and is coupled with phosphorylation of guanosine diphosphate (GDP) to give guanosine triphosphate (GTP). The overall transformation is similar to that of steps 6 to 8 in glycolysis (Figure 22.3), in which a thioester is converted into an acyl phosphate and a phosphate is then transferred to ADP. The overall result is a "hydrolysis" of the thioester group, but without involving water.

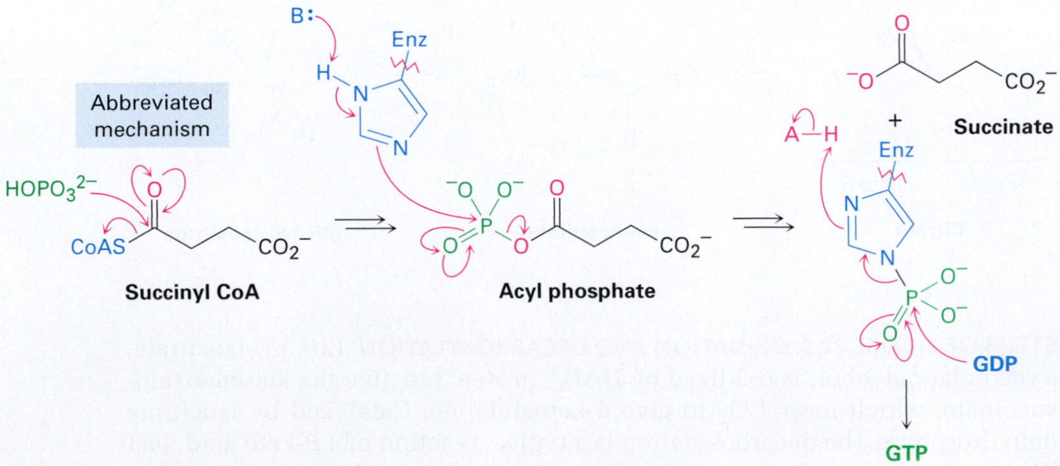

Succinyl CoA   **Acyl phosphate**

STEP 6 OF FIGURE 22.8: DEHYDROGENATION Succinate is dehydrogenated in step 6 by the FAD-dependent succinate dehydrogenase to give fumarate. The process is analogous to what occurs in fatty-acid catabolism, but the mechanism is a bit complex so we'll defer comments about it until Section 23.5. The reaction is stereospecific, removing the *pro-S* hydrogen from one carbon and the *pro-R* hydrogen from the other.

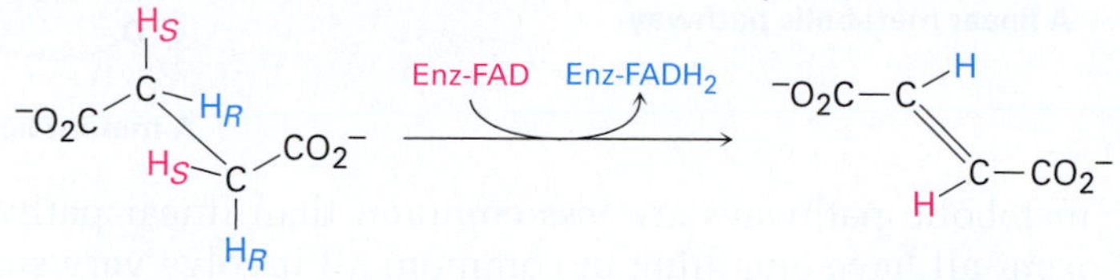

Succinate   **Fumarate**

STEPS 7–8 OF FIGURE 22.8: HYDRATION AND OXIDATION The final two steps in the citric acid cycle are the conjugate nucleophilic addition of water to fumarate to yield (S)-malate and the oxidation of (S)-malate by NAD^+ to give oxaloacetate. The addition is catalyzed by fumarase and is mechanistically similar to the addition of water to *cis*-aconitate in step 2. The reaction occurs through an enolate-ion intermediate, which is protonated on the side opposite the OH, leading to a net anti addition.

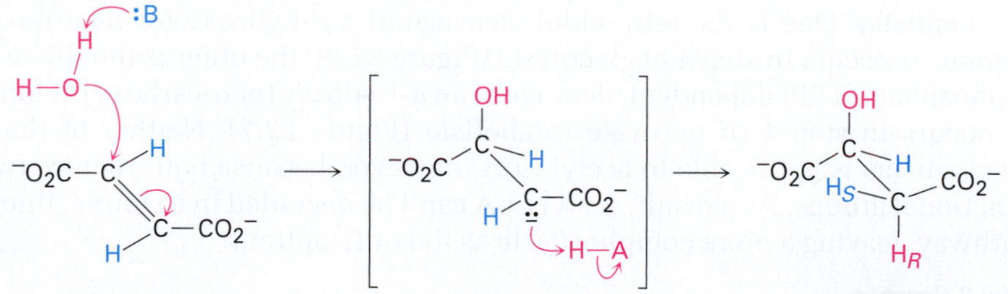

Fumarate   **(S)-Malate**

The final step is the oxidation of (S)-malate by NAD^+ to give oxaloacetate, a reaction catalyzed by malate dehydrogenase. The citric acid cycle has now returned to its starting point, ready to revolve again. The overall result of the cycle is

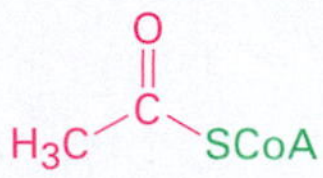

Acetyl CoA $+ \; 3\,NAD^+ \; + \; FAD \; + \; GDP \; + \; P_i \; + \; 2\,H_2O$

$\longrightarrow \quad 2\,CO_2 \; + \; HSCoA \; + \; 3\,NADH \; + \; 2\,H^+ \; + \; FADH_2 \; + \; GTP$

One further point before leaving this discussion of the citric acid cycle: most metabolic pathways—glycolysis, for example—are *linear,* starting with one substance and ending some number of steps later with the final product. The citric acid cycle, however, is a *cycle*—a closed loop of reactions that starts and ends with the same substance, while throwing off the product at some intermediate point. Why is a cyclic pathway needed for acetyl CoA metabolism?

$$A \longrightarrow B \longrightarrow C \longrightarrow D \longrightarrow E$$

A linear metabolic pathway

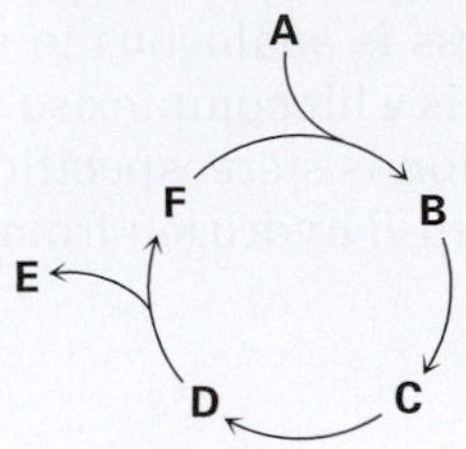

A metabolic cycle

Cyclic metabolic pathways are less common than linear pathways, and those that occur all have one thing in common: all involve very small molecules with few functional groups. The urea cycle, for instance, begins with NH_3 (Section 20.3); the citric acid cycle begins with acetyl CoA; the photosynthetic Calvin cycle used by green plants to synthesize carbohydrates begins with CO_2; and so forth. When starting with a relatively large multifunctional molecule like glucose, the number of potential reaction choices is also large, so an efficient linear pathway is energetically feasible, but when starting with a small monofunctional molecule like NH_3, CO_2, or acetyl CoA, limited reaction choices are available and a linear pathway may not be possible.

Take the citric acid cycle, for example. The metabolic purpose of the cycle is to convert the two-carbon molecule acetyl CoA into two molecules of CO_2, which means that a C–C bond must be broken. But there are very few organic reaction mechanisms for breaking C–C bonds, and only two are common in biochemistry. One is the retro-aldol cleavage of a β-hydroxy (or β-carboxy) ketone, as occurs in step 4 of glycolysis (Figure 22.3); the other is the thiamin diphosphate (TPP) dependent cleavage of an α-hydroxy (or α-carboxy) ketone, as occurs in step 1 of pyruvate catabolism (Figure 22.7). Neither of these mechanisms is applicable to acetyl CoA, however, because both require two functional groups. As a result, acetyl CoA can't be degraded in a simple, linear pathway, leaving a more complex cycle as the only option.

Retro-aldol β cleavage

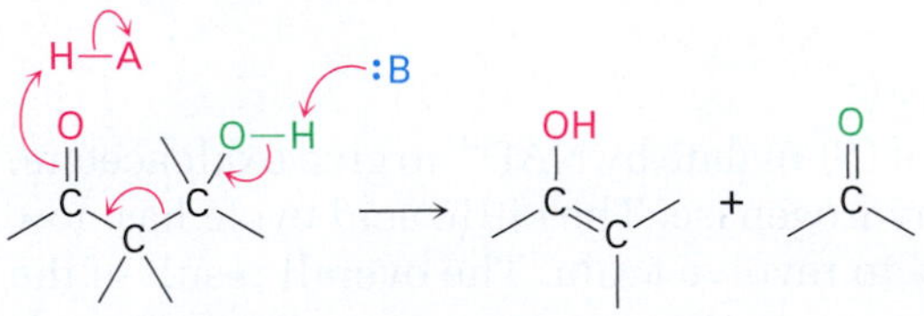

TPP-dependent α cleavage

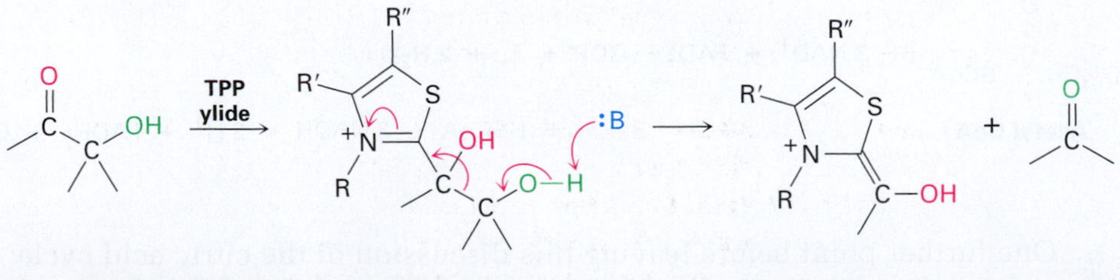

Problem 22.6
Which of the substances in the citric acid cycle are tricarboxylic acids, thus giving the cycle its alternative name?

Problem 22.7

Write mechanisms for step 2 of the citric acid cycle, the dehydration of citrate and the addition of water to *cis*-aconitate.

Problem 22.8

Is the *pro-R* or *pro-S* hydrogen removed from citrate during the dehydration in step 2 of the citric acid cycle? Does the elimination reaction occur with syn or anti geometry?

Citrate   ***cis*-Aconitate**

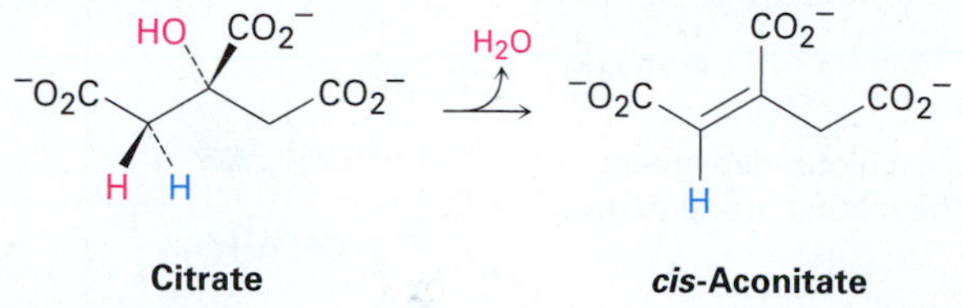

Problem 22.9

Does –OH add to the *Si* face or the *Re* face of *cis*-aconitate in step 2 of the citric acid cycle? Does the addition of water occur with syn or anti geometry?

***cis*-Aconitate**   **(2*R*,3*S*)-Isocitrate**

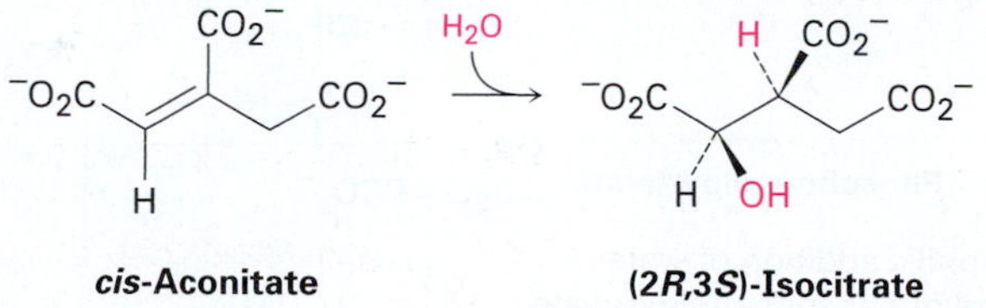

22.5 Biosynthesis of Glucose: Gluconeogenesis

Glucose is the body's primary fuel when food is plentiful, but in times of fasting or prolonged exercise, glucose stores can become depleted. Most tissues then begin metabolizing fats as their source of acetyl CoA, but the brain is different. The brain relies almost entirely on glucose for fuel and is dependent on receiving a continuous supply in the blood. When the supply of glucose fails for even a brief time, irreversible damage can occur. Thus, a pathway for synthesizing glucose from simple precursors is crucial.

Higher organisms are not able to synthesize glucose from acetyl CoA but must instead use one of the three-carbon precursors (*S*)-lactate, alanine, or glycerol, all of which are readily converted into pyruvate.

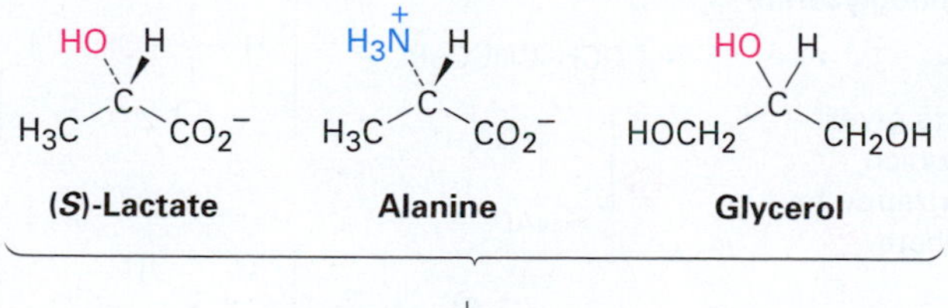

(*S*)-Lactate   **Alanine**   **Glycerol**

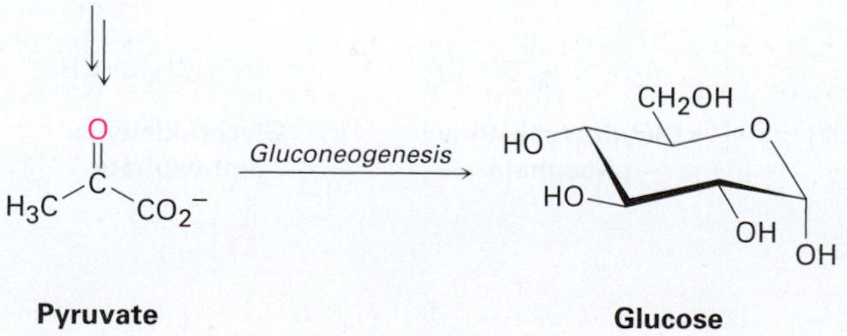

Pyruvate   **Glucose**

Pyruvate then becomes the starting point for **gluconeogenesis**, the 11-step biosynthetic pathway by which organisms make glucose (Figure 22.9). The gluconeogenesis pathway by which glucose is *made*, however, is not the exact reverse of the glycolysis pathway by which it is degraded. The catabolic and anabolic pathways *must* differ in at least some details for both to be energetically favorable and for independent regulatory mechanisms to operate.

FIGURE 22.9

MECHANISM: The gluconeogenesis pathway for the biosynthesis of glucose from pyruvate. Individual steps are explained in the text.

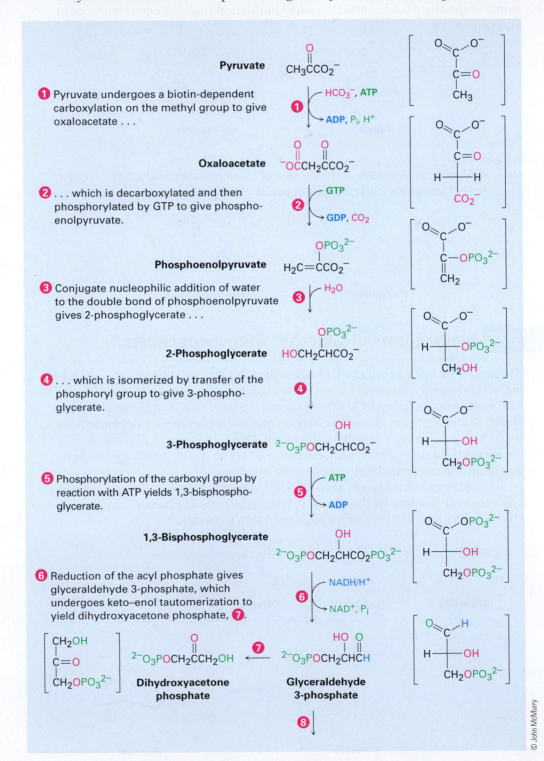

1 Pyruvate undergoes a biotin-dependent carboxylation on the methyl group to give oxaloacetate . . .

2 . . . which is decarboxylated and then phosphorylated by GTP to give phospho-enolpyruvate.

3 Conjugate nucleophilic addition of water to the double bond of phosphoenolpyruvate gives 2-phosphoglycerate . . .

4 . . . which is isomerized by transfer of the phosphoryl group to give 3-phospho-glycerate.

5 Phosphorylation of the carboxyl group by reaction with ATP yields 1,3-bisphospho-glycerate.

6 Reduction of the acyl phosphate gives glyceraldehyde 3-phosphate, which undergoes keto–enol tautomerization to yield dihydroxyacetone phosphate, **7**.

© John McMurry

FIGURE 22.9 *(continued)*

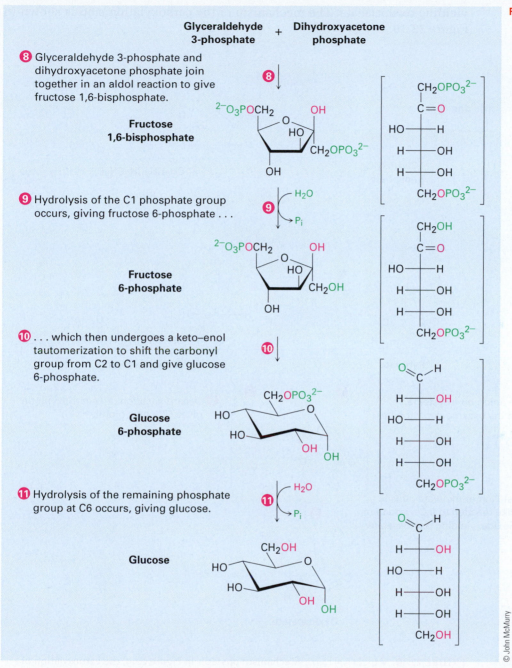

8 Glyceraldehyde 3-phosphate and dihydroxyacetone phosphate join together in an aldol reaction to give fructose 1,6-bisphosphate.

9 Hydrolysis of the C1 phosphate group occurs, giving fructose 6-phosphate . . .

10 . . . which then undergoes a keto–enol tautomerization to shift the carbonyl group from C2 to C1 and give glucose 6-phosphate.

11 Hydrolysis of the remaining phosphate group at C6 occurs, giving glucose.

© John McMurry

STEP 1 OF FIGURE 22.9: CARBOXYLATION Gluconeogenesis begins with the carboxylation of pyruvate to yield oxaloacetate. The reaction is catalyzed by pyruvate carboxylase and requires ATP, bicarbonate ion, and the coenzyme biotin, which acts as a carrier to transport CO_2 to the enzyme active site. ATP first reacts with bicarbonate to give carboxy phosphate, as in the urea cycle (Figure 20.4, page 842), and carboxy phosphate then undergoes decarboxylation. Biotin reacts with the released CO_2 to give *N*-carboxybiotin, which itself then decarboxylates. Simultaneously, and in close proximity on the enzyme, pyruvate is deprotonated to give an anion that immediately adds to the CO_2, yielding oxaloacetate. The mechanism of the carboxylation step is shown in Figure 22.10.

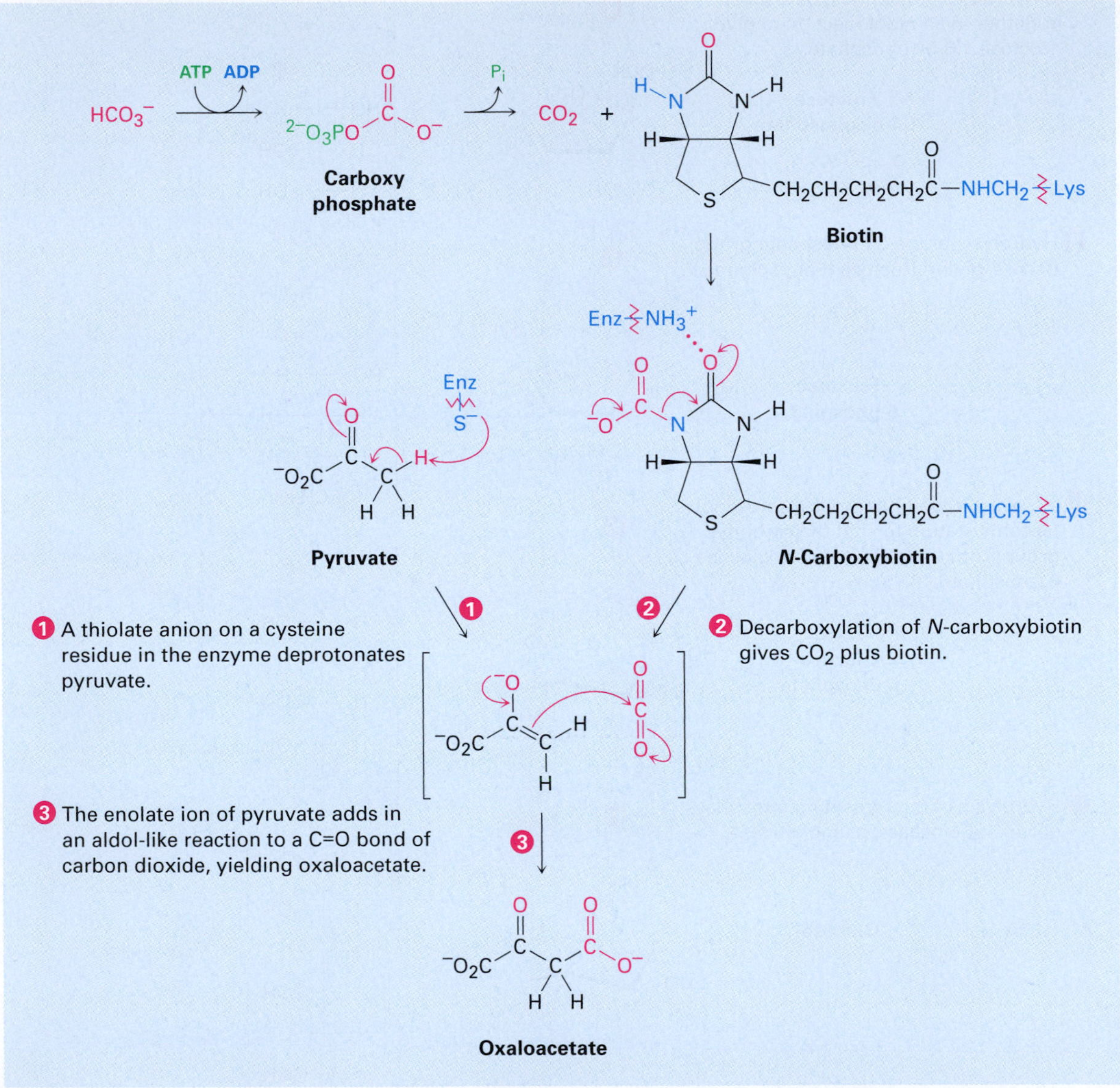

FIGURE 22.10 MECHANISM: Mechanism of step 1 in Figure 22.9, the carboxylation of pyruvate to give oxaloacetate. Biotin acts as a carrier of CO_2, moving it to the appropriate position in the enzyme so that it can react with pyruvate.

STEP 2 OF FIGURE 22.9. DECARBOXYLATION AND PHOSPHORYLATION Decarboxylation of oxaloacetate, a β-keto acid, occurs by the typical retro-aldol mechanism like that in step 3 in the citric acid cycle (Figure 22.8), and phosphorylation of the resultant pyruvate enolate ion by GTP occurs concurrently to give phosphoenolpyruvate. The reaction is catalyzed by phosphoenol-pyruvate carboxykinase.

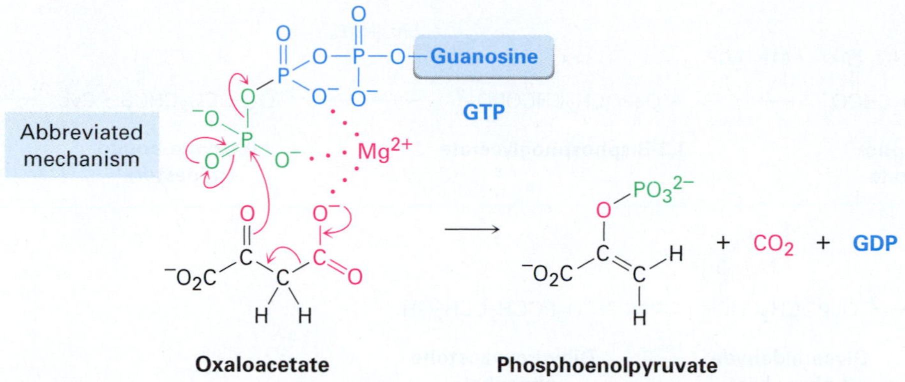

Oxaloacetate   **Phosphoenolpyruvate**

Why is carbon dioxide added in step 1 and then immediately removed in step 2? Why doesn't the conversion of pyruvate to phosphoenolpyruvate proceed directly in a single step by reaction of pyruvate enolate ion with GTP rather than indirectly in two steps? The likely answer lies in the energetics of the process. Phosphoenolpyruvate is sufficiently high in energy that its direct synthesis from pyruvate is unfavorable even though it consumes a molecule of GTP. In the two-step process involving oxaloacetate, however, *two* molecules of nucleoside triphosphate are consumed (one ATP and one GTP), releasing sufficient energy to make the overall process favorable.

STEPS 3–4 OF FIGURE 22.9: HYDRATION AND ISOMERIZATION Conjugate nucleophilic addition of water to the double bond of phosphoenolpyruvate gives 2-phosphoglycerate by a process similar to that of step 7 in the citric acid cycle (Figure 22.8). Phosphorylation of C3 and dephosphorylation of C2 then yields 3-phosphoglycerate. Mechanistically, these steps are the reverse of steps 9 and 8 in glycolysis (Figure 22.3), both of which have equilibrium constants near 1.

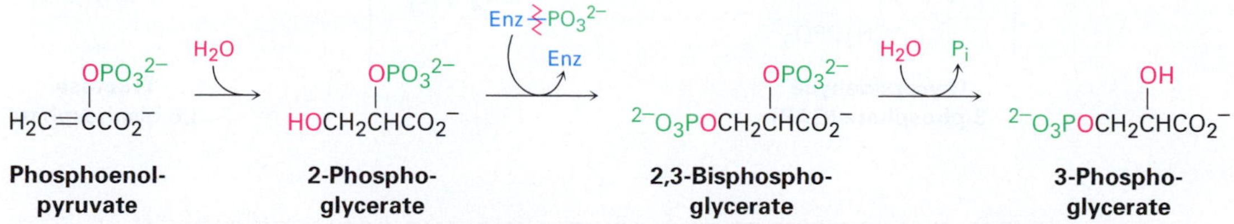

Phosphoenol-pyruvate   **2-Phospho-glycerate**   **2,3-Bisphospho-glycerate**   **3-Phospho-glycerate**

STEPS 5–7 OF FIGURE 22.9: PHOSPHORYLATION, REDUCTION, AND TAUTOMERIZATION Reaction of 3-phosphoglycerate with ATP generates the corresponding acyl phosphate, 1,3-bisphosphoglycerate, which binds to the glyceraldehyde 3-phosphate dehydrogenase by a thioester bond to a cysteine residue. Reduction of the thioester by NADH/H$^+$ then yields the corresponding aldehyde, and keto–enol tautomerization of the aldehyde gives dihydroxyacetone phosphate. All three steps are mechanistically the reverse of the corresponding steps 7, 6, and 5 of glycolysis and have equilibrium constants near 1.

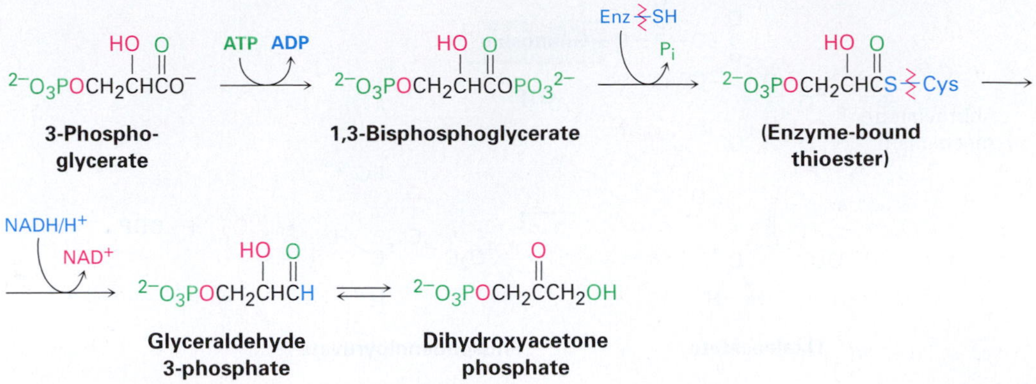

3-Phospho-glycerate   **1,3-Bisphosphoglycerate**   **(Enzyme-bound thioester)**

Glyceraldehyde 3-phosphate   **Dihydroxyacetone phosphate**

STEP 8 OF FIGURE 22.9: ALDOL REACTION Dihydroxyacetone phosphate and glyceraldehyde 3-phosphate, the two 3-carbon units produced in step 7, join by an aldol reaction to give fructose 1,6-bisphosphate, the reverse of step 4 in glycolysis. As in glycolysis (Figure 22.3), the reaction is catalyzed in plants and animals by a class I aldolase and takes place on an iminium ion formed by reaction of dihydroxyacetone phosphate with a side-chain lysine $-NH_2$ group on the enzyme. Loss of a proton from the neighboring carbon then generates an enamine, an aldol-like reaction ensues, and the product is hydrolyzed.

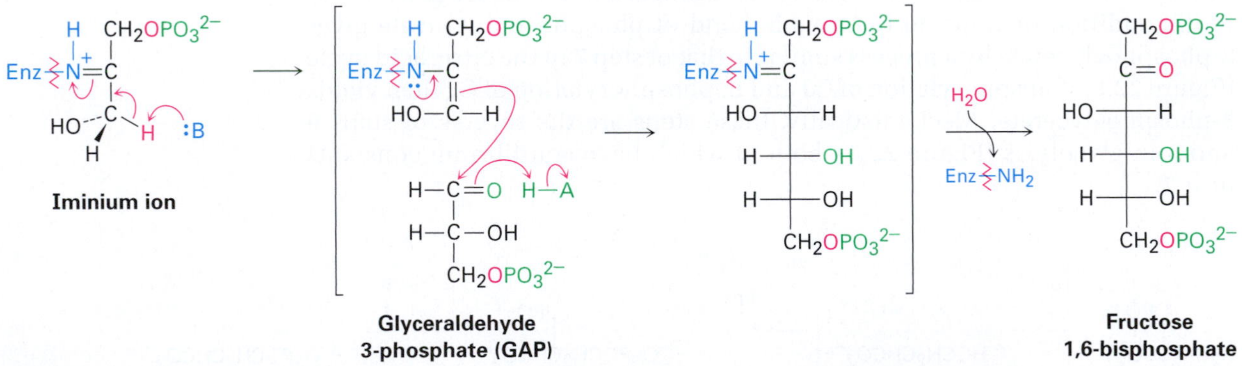

Iminium ion   **Glyceraldehyde 3-phosphate (GAP)**   **Fructose 1,6-bisphosphate**

STEPS 9–10 OF FIGURE 22.9: HYDROLYSIS AND ISOMERIZATION Hydrolysis of the phosphate group at C1 of fructose 1,6-bisphosphate gives fructose 6-phosphate. Although the *result* of the reaction is the opposite of step 3 in glycolysis, the mechanism is not. In glycolysis, the phosphorylation is accomplished by reaction of fructose with ATP, with formation of ADP as by-product. The reverse of that process, however—the reaction of fructose 1,6-bisphosphate with ADP to give fructose 6-phosphate and ATP—is energetically unfavorable because ATP is too high in energy. Thus, an alternative pathway is used in which the C1 phosphate group is removed by a direct hydrolysis reaction, catalyzed by fructose 1,6-bisphosphatase.

Following hydrolysis, keto–enol tautomerization of the carbonyl group from C2 to C1 gives glucose 6-phosphate. The isomerization is the reverse of step 2 in glycolysis (Figure 22.3).

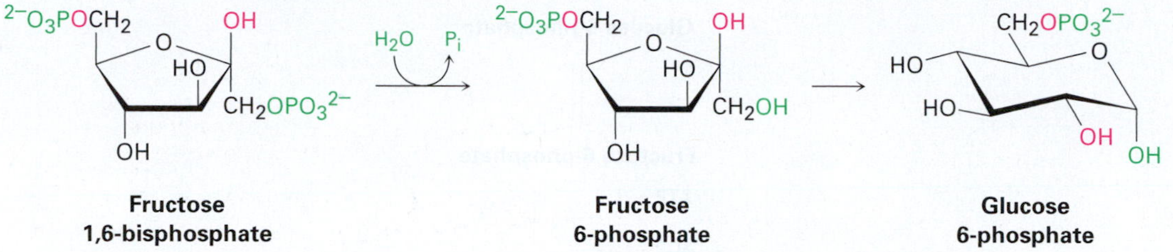

Fructose 1,6-bisphosphate → **Fructose 6-phosphate** → **Glucose 6-phosphate**

STEP 11 OF FIGURE 22.9: HYDROLYSIS The final step in gluconeogenesis is the conversion of glucose 6-phosphate to glucose by a second phosphatase-catalyzed hydrolysis reaction. As just discussed for the hydrolysis of fructose 1,6-bisphosphate in step 9, and for the same energetic reasons, the mechanism of the glucose 6-phosphate hydrolysis is not the reverse of the corresponding step 1 in glycolysis.

Interestingly, however, the mechanisms of the two phosphate hydrolysis reactions in steps 9 and 11 are not the same. In step 9, water is the nucleophile, but in the glucose 6-phosphate reaction of step 11, a histidine residue on the enzyme attacks phosphorus, giving a phosphoryl enzyme intermediate that subsequently reacts with water.

Abbreviated mechanism

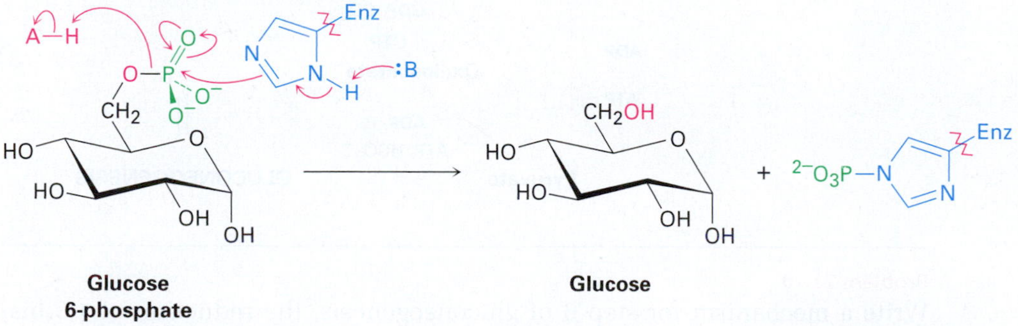

Glucose 6-phosphate → **Glucose**

The overall result of gluconeogenesis is summarized by the following equation, and a comparison of the gluconeogenesis and glycolysis pathways is given in Figure 22.11. The pathways differ at the three steps indicated by red reaction arrows.

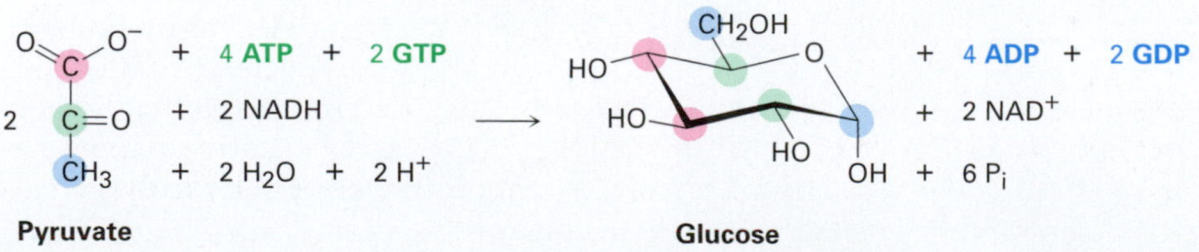

2 **Pyruvate** $+$ 4 ATP $+$ 2 GTP $+$ 2 NADH $+$ 2 H_2O $+$ 2 H^+ $\longrightarrow$ **Glucose** $+$ 4 ADP $+$ 2 GDP $+$ 2 NAD^+ $+$ 6 P_i

FIGURE 22.11 A comparison of glycolysis and gluconeogenesis pathways. The pathways differ at the three steps indicated by red reaction arrows.

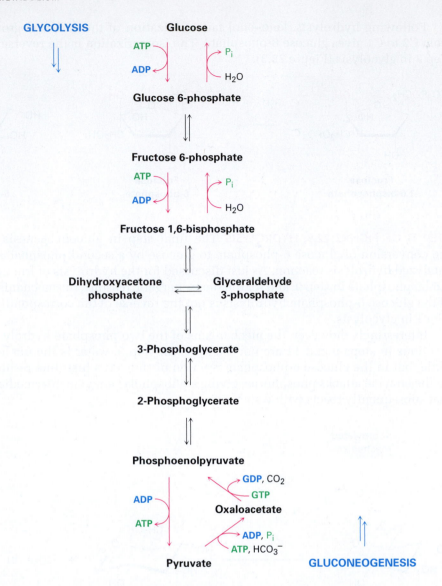

Problem 22.10

Write a mechanism for step 6 of gluconeogenesis, the reduction of 1,3-bis-phosphoglycerate with NADH/H$^+$ to yield glyceraldehyde 3-phosphate.

Summary

Glucose metabolism is at the center of biological chemistry. Fortunately, it's also relatively straightforward because the molecules are small and the reactions involved are almost entirely the carbonyl-group processes discussed previously in Chapters 14 to 17.

Carbohydrates are the chemical intermediaries by which carbon atoms from CO_2 are incorporated into growing organisms. Thus, carbohydrates are the biological starting point. In humans, carbohydrate metabolism begins with glycosidase-catalyzed digestion of starch to give glucose. Glucose catabolism then begins with **glycolysis**, a series of ten enzyme-catalyzed reactions that break down glucose into 2 equivalents of pyruvate, $CH_3COCO_2^-$. Depending on the organism, pyruvate is then converted either into lactate, ethanol, or (in mammals) acetyl CoA. The key step in the formation of acetyl CoA is the decarboxylation of pyruvate catalyzed by thiamin diphosphate, TPP.

Acetyl CoA next enters the **citric acid cycle**, also called the *tricarboxylic acid (TCA) cycle,* or *Krebs cycle.* The overall result of the cycle is the conversion of an acetyl group into two molecules of CO_2 plus reduced coenzymes by an eight-step sequence. The cycle is a closed loop of reactions in which the product of the final step (oxaloacetate) is a reactant in the first step. The intermediates are constantly regenerated and flow continuously through the cycle, which operates as long as the oxidizing coenzymes NAD^+ and FAD are available.

Higher organisms are not able to synthesize glucose from acetyl CoA but must instead use one of the three-carbon precursors lactate, alanine, or glycerol, all of which are readily converted into pyruvate. Pyruvate then becomes the starting point for **gluconeogenesis**, the 11-step biosynthetic pathway by which organisms make glucose.

Key Words

citric acid cycle, 915
gluconeogenesis, 922
glycolysis, 904
Schiff base, 907

Lagniappe

Influenza Pandemics

Each year, seasonal outbreaks of influenza occur throughout the world, usually without particular notice. These outbreaks are caused by subtypes of known flu viruses that are already present in the population, and they can usually be controlled or prevented by vaccination. Every 10 to 40 years, however, a new and virulent subtype never before seen in humans appears. The result can be a worldwide pandemic capable of causing great disruption and killing millions.

Three such pandemics struck in the 20th century, the most serious of which was the 1918–1919 "Spanish flu" that killed an estimated 50 million people worldwide, including many healthy young adults. It has now been about 40 years since the last pandemic, an outbreak of "Hong Kong flu" in 1968–1969, and many public heath officials fear that another may occur soon. The Hong Kong flu was relatively mild compared to the Spanish flu—worldwide casualties were only 750,000—but there is no way of knowing how deadly the next outbreak will be.

Current worries center on two recent influenza outbreaks. The first, discovered in 1997, is commonly called "bird flu"; the second, found in early 2009, is "swine flu." Bird flu is caused by the transfer to humans of an avian H5N1 virus that has killed tens of millions of birds, primarily in Southeast Asia. Human infection by this virus was first noted in Hong Kong in 1997, and by mid 2009, 413 cases with 256 deaths had been confirmed in 16 countries. The virus is transmitted primarily from poultry to humans rather than between humans as of 2009, but the H5N1 strain is highly pathogenic, mutates rapidly, and is able to acquire genes from viruses that infect other animal

continued

Lagniappe *continued*

species. Thus, there is a fear that the capability for human-to-human transmission may increase rapidly.

Swine flu is caused by an H1N1 virus related to those found in pigs, although the exact origin of the virus is not yet known. The virus appears to spread rapidly in humans—more than 3000 cases were found in the first 2 months since it was identified—but its effects appear relatively mild.

The classifications H5N1 and H1N1 are based on the antigenic behavior of two kinds of glycoproteins that coat the viral surface—hemagglutinin (H, type 5 or type 1) and neuraminidase (N, type 1), which, as its name implies, is an enzyme. Infection occurs when a viral particle, or *virion,* binds to the sialic acid part (Section 21.7) of a receptor glycoprotein on the target cell and is then engulfed by the cell. New viral particles are produced inside the infected cell, pass back out, and are again held by sialic acid bonded to glycoproteins in cell-surface receptors. Finally, the neuraminidase present on the viral

surface cleaves the bond between receptor glycoprotein and sialic acid, thereby releasing the virion and allowing it to invade a new cell (Figure 22.12).

So what can be done to limit the severity of an influenza pandemic? Development of a vaccine is the only means to limit the spread of the virus, but work can't begin until the contagious strain of virus has appeared. Once a person has been infected, there is some hope that the recently developed antiviral drug oseltamivir might limit the severity of infection. Oseltamivir phosphate, sold under the name Tamiflu, is one of only a handful of known substances able to inhibit the neuraminidase enzyme. With the enzyme blocked, newly formed virions are not released, and spread of the infection within the body is thus limited. You might notice in Figure 22.12 the similarity in shape between *N*-acetylneuraminic acid and oseltamivir that allows the drug to block the active site of neuraminidase. Unfortunately, the virus is rapidly acquiring resistance to Tamiflu.

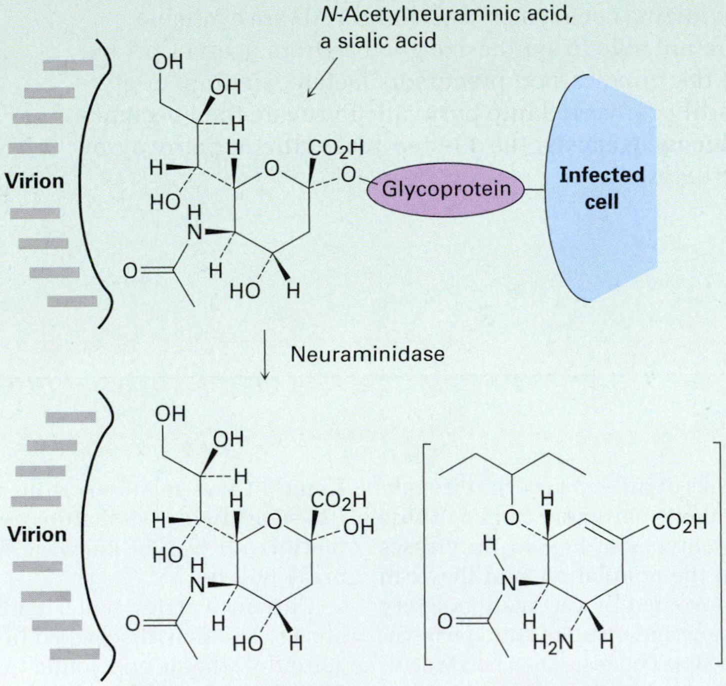

FIGURE 22.12 Release of a newly formed virion from an infected cell occurs when neuraminidase, present on the surface of the virion, cleaves the bond holding the virion to a sialic acid molecule in a glycoprotein receptor on the infected cell. Oseltamivir, sold under the trade name Tamiflu, inhibits the neuraminidase enzyme by binding to its active site, thus preventing release of the virion.

Exercises

VISUALIZING CHEMISTRY

(Problems 22.1–22.10 appear within the chapter.)

22.11 ▪ Identify the following intermediate in the citric acid cycle, and tell whether it has *R* or *S* stereochemistry:

▪ indicates problems that are assignable in Organic OWL.

Go to this book's companion website at **www.cengage.com/ chemistry/mcmurry** to explore interactive versions of the Active Figures from this text.

22.12 The following compound is an intermediate in the pentose phosphate pathway, an alternative route for glucose metabolism. Identify the sugar it is derived from.

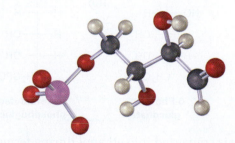

ADDITIONAL PROBLEMS

22.13 ▪ What coenzyme is typically associated with each of the following transformations?

(a) The phosphorylation of an alcohol to give a phosphate

(b) The oxidative decarboxylation of an α-keto acid to give a thioester

(c) The carboxylation of a ketone to give a β-keto acid

22.14 ▪ Lactate, a product of glucose catabolism in oxygen-starved muscles, can be converted into pyruvate by oxidation. What coenzyme do you think is needed? Write the equation in the usual biochemical format using a curved arrow.

$$\underset{\text{Lactate}}{CH_3\overset{\overset{\displaystyle OH}{|}}{C}HCO_2^-}$$

22.15 ▪ Write a mechanism for the conversion of α-ketoglutarate to succinyl CoA in step 4 of the citric acid cycle (Figure 22.8).

22.16 Plants, but not animals, are able to synthesize glucose from acetyl CoA by a pathway that begins with the *glyoxalate cycle.* One of the steps in the cycle is the conversion of isocitrate to glyoxalate plus succinate, a process catalyzed by isocitrate lyase. Propose a mechanism for the reaction.

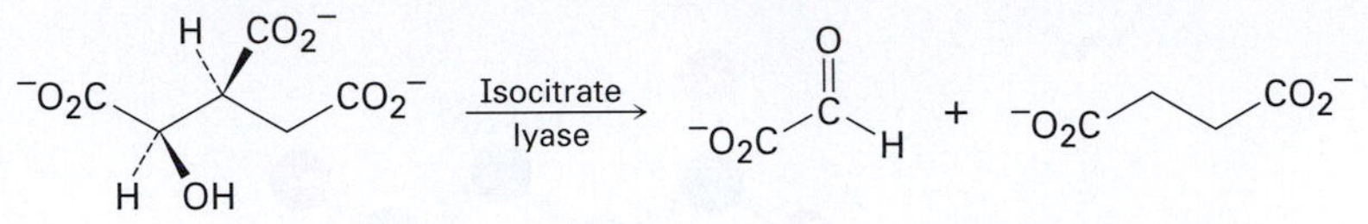

Isocitrate **Glyoxalate** **Succinate**

22.17 Propose a mechanism for the conversion of 6-phosphogluconate to 2-keto-3-deoxy-6-phosphogluconate, a step in the Entner–Douderoff bacterial pathway for glucose catabolism.

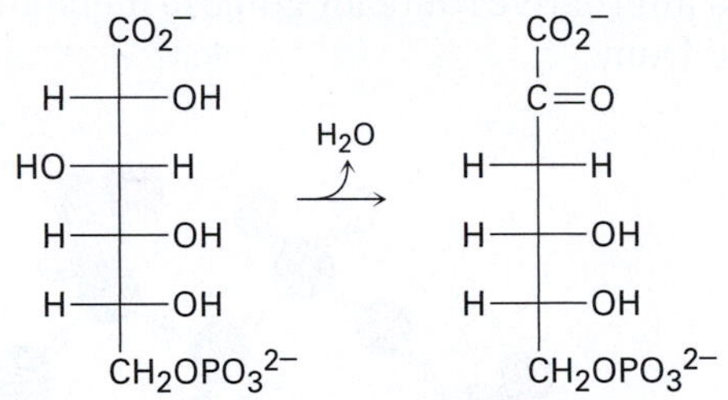

6-Phospho-gluconate **2-Keto-3-deoxy-6-phosphogluconate**

22.18 ▪ Pyruvate is converted into ethanol during fermentation in yeast.

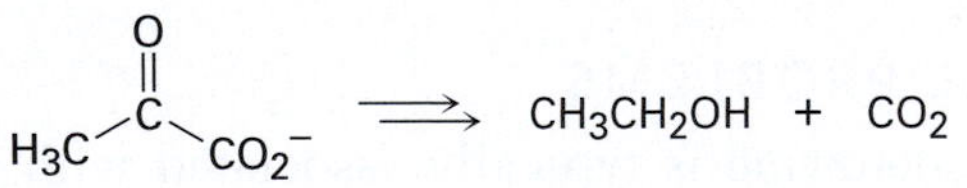

Pyruvate **Ethanol**

(a) The first step is a TPP-dependent decarboxylation of pyruvate to give HETPP. Show the mechanism.

(b) The second step is a protonation followed by elimination of TPP ylide to give acetaldehyde. Show the mechanism.

(c) The final step is a reduction with NADH. Show the mechanism.

▪ Problems assignable in Organic OWL.

22.19 ▪ Galactose, one of the eight essential monosaccharides (Section 21.7), is biosynthesized from UDP-glucose by galactose 4-epimerase, where UDP = uridylyl diphosphate. The enzyme requires NAD^+ for activity, but it is not a stoichiometric reactant and NADH is not a final reaction product. Propose a mechanism.

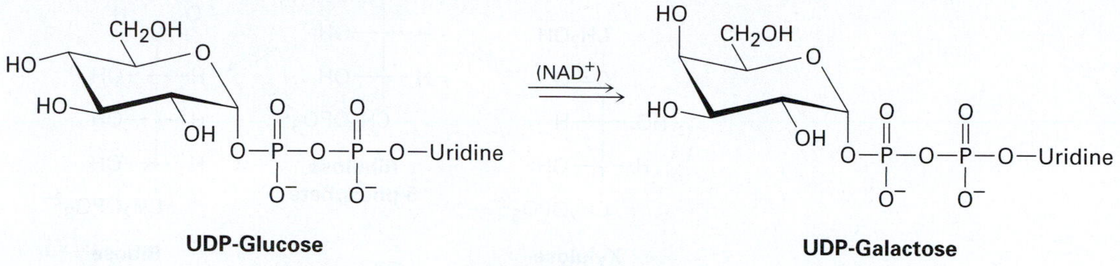

UDP-Glucose   **UDP-Galactose**

22.20 ▪ Mannose, one of the eight essential monosaccharides (Section 21.7), is biosynthesized as its 6-phosphate derivative from fructose 6-phosphate. No cofactor is required. Propose a mechanism.

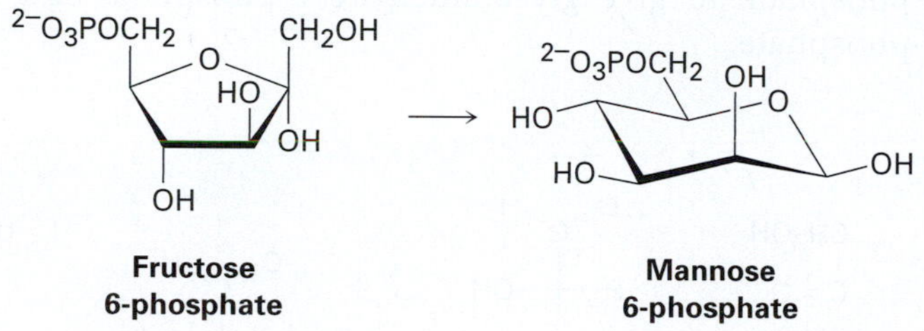

Fructose   **Mannose**
6-phosphate   **6-phosphate**

22.21 ▪ Glucosamine, one of the eight essential monosaccharides (Section 21.7), is biosynthesized as its 6-phosphate derivative from fructose 6-phosphate by reaction with ammonia. Propose a mechanism.

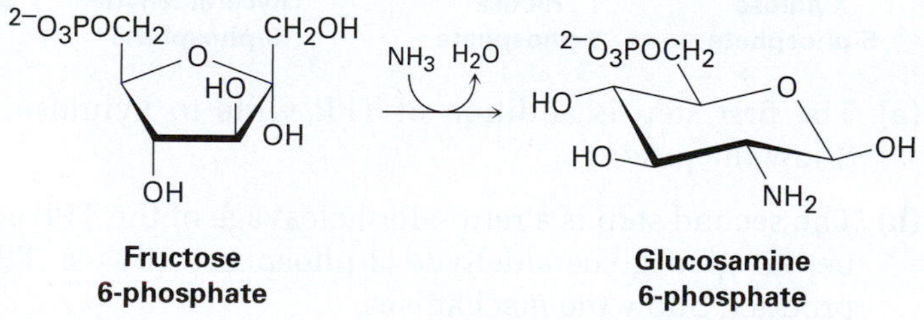

Fructose   **Glucosamine**
6-phosphate   **6-phosphate**

▪ Problems assignable in Organic OWL.

22.22 ■ In the pentose phosphate pathway for glucose metabolism, ribulose 5-phosphate undergoes reversible isomerizations to both ribose 5-phosphate and xylulose 5-phosphate. Show mechanisms for both.

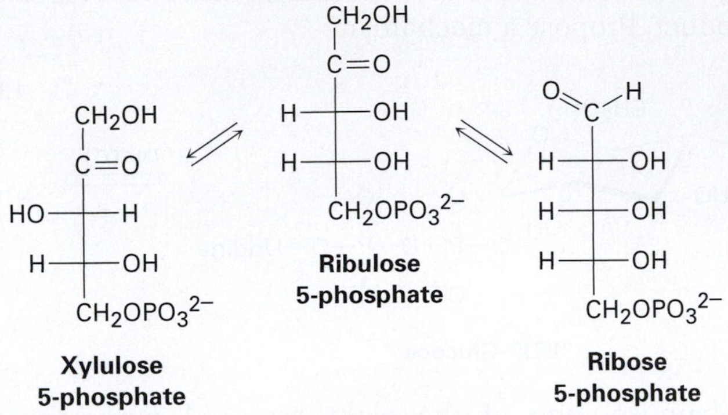

Xylulose 5-phosphate   **Ribulose 5-phosphate**   **Ribose 5-phosphate**

22.23 ■ One of the steps in the pentose phosphate pathway for glucose metabolism is the TPP-dependent reaction of xylulose 5-phosphate with ribose 5-phosphate to give glyceraldehyde 3-phosphate and sedoheptulose 7-phosphate.

Xylulose 5-phosphate   **Ribose 5-phosphate**   **Glyceraldehyde 3-phosphate**   **Sedoheptulose 7-phosphate**

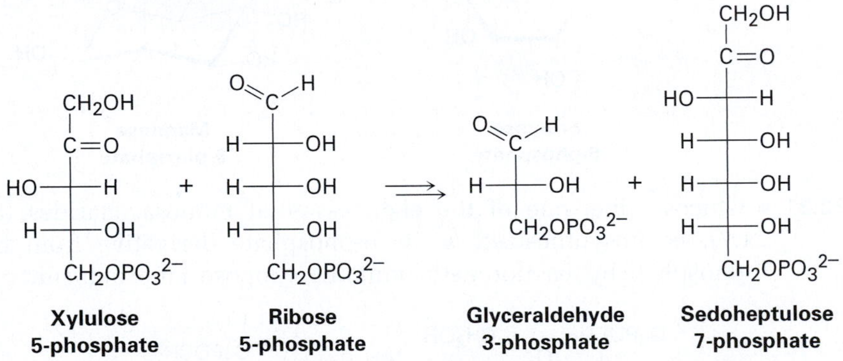

(a) The first step is addition of TPP ylide to xylulose 5-phosphate. Show the product.

(b) The second step is a retro-aldol cleavage of the TPP addition product to give glyceraldehyde 3-phosphate plus a TPP-containing product. Show the mechanism.

(c) The third step is an aldol addition of the TPP-containing product from step 2 to ribose 5-phosphate. Show the product and the mechanism.

(d) The final step is an elimination of TPP ylide to give sedoheptulose 7-phosphate. Show the mechanism.

22.24 ■ One of the steps in the photosynthesis cycle is conversion of ribulose 1,5-bisphosphate to 3-phosphoglycerate.

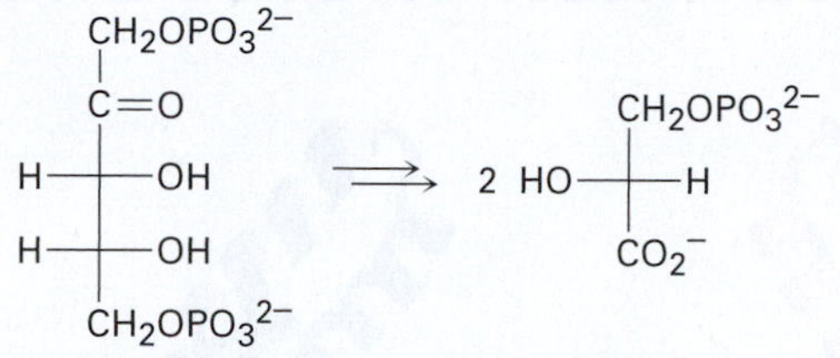

Ribulose 1,5-bisphosphate

3-Phospho-glycerate

(a) The first step is tautomerization of the carbonyl group. Show the product and the mechanism.

(b) The second step is a biotin-dependent carboxylation to give a β-keto acid. Show the product and the mechanism.

(c) The final step is a retro-aldol–like reaction to yield two molecules of 3-phosphoglycerate. Show the mechanism.

22.25 ■ L-Fucose, one of the eight essential monosaccharides (Section 21.7), is biosynthesized from GDP-D-mannose by the following three-step reaction sequence:

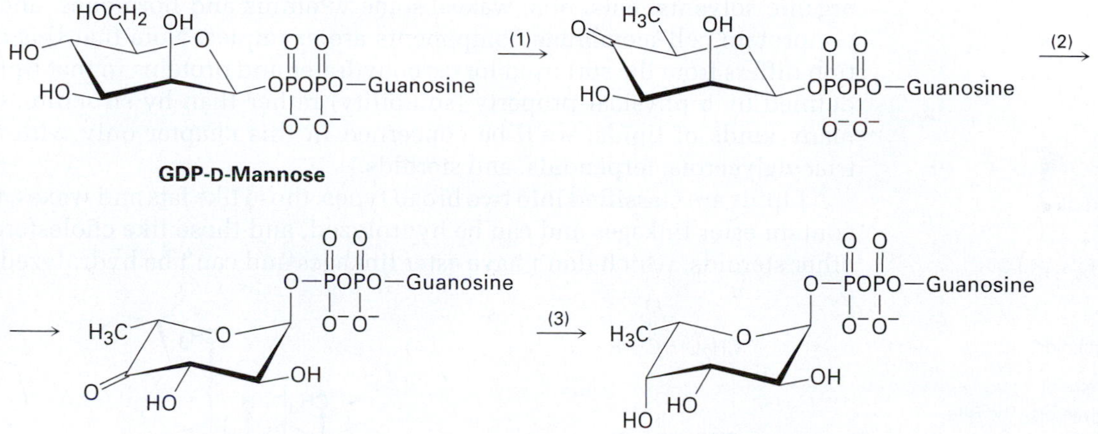

GDP-D-Mannose

GDP-L-Fucose

(a) Step 1 involves an oxidation, a dehydration, and a reduction. The step requires NADP$^+$, but no NADPH is formed as a final reaction product. Propose a mechanism.

(b) Step 2 accomplishes two epimerizations and utilizes acidic and basic sites in the enzyme but does not require a coenzyme. Propose a mechanism.

(c) Step 3 requires NADPH as coenzyme. Show the mechanism.

■ Problems assignable in Organic OWL.

23 Biomolecules: Lipids and Their Metabolism

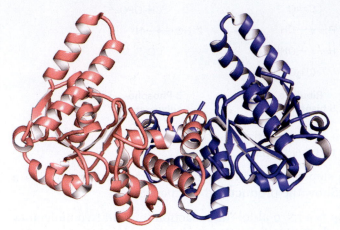

Hydroxyacyl-CoA dehydrogenase catalyzes the oxidation of a β-hydroxyacyl CoA to a β-ketoacyl CoA during fatty-acid metabolism.

Lipids are naturally occurring organic molecules that have limited solubility in water and can be isolated from organisms by extraction with nonpolar organic solvents. Fats, oils, waxes, some vitamins and hormones, and most nonprotein cell-membrane components are examples. Note that this definition differs from the sort used for carbohydrates and proteins in that lipids are defined by a physical property (solubility) rather than by structure. Of the many kinds of lipids, we'll be concerned in this chapter only with a few: triacylglycerols, terpenoids, and steroids.

Lipids are classified into two broad types: those like fats and waxes, which contain ester linkages and can be hydrolyzed, and those like cholesterol and other steroids, which don't have ester linkages and can't be hydrolyzed.

Animal fat—a triester
(R, R′, R″ = C₁₁–C₁₉ chains)

Cholesterol

WHY THIS CHAPTER?

We've now covered two of the four major classes of biomolecules—proteins and carbohydrates—and have two remaining. We'll cover lipids, the largest and most diverse class of biomolecules, in this chapter, looking both at their structure and function and at their metabolism.

OWL Online homework for this chapter can be assigned in Organic OWL.

23.1 Waxes, Fats, and Oils

Waxes are mixtures of esters of long-chain carboxylic acids with long-chain alcohols. The carboxylic acid usually has an even number of carbons from 16 through 36, while the alcohol has an even number of carbons from 24 through 36. One of the major components of beeswax, for instance, is triacontyl hexadecanoate, the ester of the C_{30} alcohol triacontan-1-ol and the C_{16} acid hexadecanoic acid. The waxy protective coatings on most fruits, berries, leaves, and animal furs have similar structures.

$$CH_3(CH_2)_{14}\overset{\overset{\displaystyle O}{\|}}{C}O(CH_2)_{29}CH_3$$

Triacontyl hexadecanoate (from beeswax)

Animal **fats** and **vegetable oils** are the most widely occurring lipids. Although they appear different—animal fats like butter and lard are solids, whereas vegetable oils like corn and peanut oil are liquid—their structures are closely related. Chemically, fats and oils are *triglycerides,* or **triacylglycerols**—triesters of glycerol with three long-chain carboxylic acids called **fatty acids**. Animals use fats for long-term energy storage because they are much less highly oxidized than carbohydrates and provide about six times as much energy as an equal weight of stored, hydrated glycogen.

A triacylglycerol

Hydrolysis of a fat or oil with aqueous NaOH yields glycerol and three fatty acids. The fatty acids are generally unbranched and contain an even number of carbon atoms between 12 and 20. If double bonds are present, they have largely, although not entirely, *Z*, or cis, geometry. The three fatty acids of a specific triacylglycerol molecule need not be the same, and the fat or oil from a given source is likely to be a complex mixture of many different triacylglycerols. Table 23.1 lists some of the commonly occurring fatty acids, and Table 23.2 lists the approximate composition of some fats and oils from different sources.

More than 100 different fatty acids are known, and about 40 occur widely. Palmitic acid (C_{16}) and stearic acid (C_{18}) are the most abundant saturated fatty acids; oleic and linoleic acids (both C_{18}) are the most abundant unsaturated ones. Oleic acid is *monounsaturated* because it has only one double bond, whereas linoleic, linolenic, and arachidonic acids are **polyunsaturated fatty acids** because they have more than one double bond. Linoleic and linolenic

TABLE 23.1
Structures of Some Common Fatty Acids

Name	No. of carbons	Melting point (°C)	Structure
Saturated			
Lauric	12	43.2	$CH_3(CH_2)_{10}CO_2H$
Myristic	14	53.9	$CH_3(CH_2)_{12}CO_2H$
Palmitic	16	63.1	$CH_3(CH_2)_{14}CO_2H$
Stearic	18	68.8	$CH_3(CH_2)_{16}CO_2H$
Arachidic	20	76.5	$CH_3(CH_2)_{18}CO_2H$
Unsaturated			
Palmitoleic	16	−0.1	(Z)-$CH_3(CH_2)_5CH{=}CH(CH_2)_7CO_2H$
Oleic	18	13.4	(Z)-$CH_3(CH_2)_7CH{=}CH(CH_2)_7CO_2H$
Linoleic	18	−12	(Z,Z)-$CH_3(CH_2)_4(CH{=}CHCH_2)_2(CH_2)_6CO_2H$
Linolenic	18	−11	$(all\ Z)$-$CH_3CH_2(CH{=}CHCH_2)_3(CH_2)_6CO_2H$
Arachidonic	20	−49.5	$(all\ Z)$-$CH_3(CH_2)_4(CH{=}CHCH_2)_4CH_2CH_2CO_2H$

TABLE 23.2
Approximate Composition of Some Fats and Oils

Source	Saturated fatty acids (%)				Unsaturated fatty acids (%)	
	C_{12} lauric	C_{14} myristic	C_{16} palmitic	C_{18} stearic	C_{18} oleic	C_{18} linoleic
Animal fat						
Lard	—	1	25	15	50	6
Butter	2	10	25	10	25	5
Human fat	1	3	25	8	46	10
Whale blubber	—	8	12	3	35	10
Vegetable oil						
Coconut	50	18	8	2	6	1
Corn	—	1	10	4	35	45
Olive	—	1	5	5	80	7
Peanut	—	—	7	5	60	20

acids occur in cream and are essential in the human diet; infants grow poorly and develop skin lesions if fed a diet of nonfat milk for prolonged periods.

$$CH_3CH_2CH_2CH_2CH_2CH_2CH_2CH_2CH_2CH_2CH_2CH_2CH_2CH_2CH_2CH_2CH_2\overset{\overset{\displaystyle O}{\|}}{C}OH$$

Stearic acid

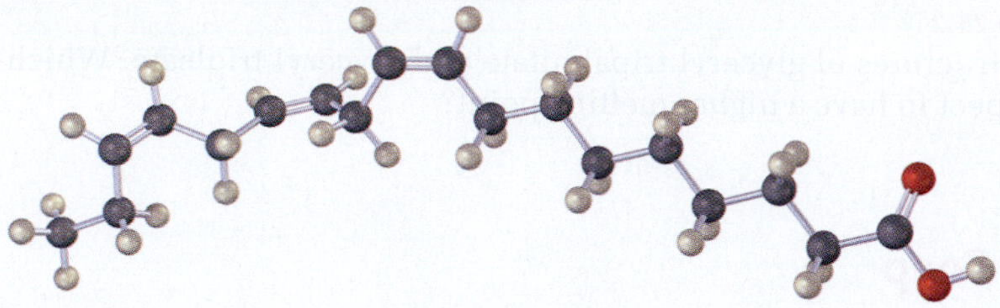

CH₃CH₂CH=CHCH₂CH=CHCH₂CH=CHCH₂CH₂CH₂CH₂CH₂CH₂CH₂COH

Linolenic acid, a polyunsaturated fatty acid

The data in Table 23.1 show that unsaturated fatty acids generally have lower melting points than their saturated counterparts, a trend that is also true for triacylglycerols. Since vegetable oils generally have a higher proportion of unsaturated to saturated fatty acids than animal fats (Table 23.2), they have lower melting points. The difference is a consequence of structure. Saturated fats have a uniform shape that allows them to pack together efficiently in a crystal lattice. In unsaturated vegetable oils, however, the C=C bonds introduce bends and kinks into the hydrocarbon chains, making crystal formation more difficult and lowering the melting point of the oil.

The C=C bonds in vegetable oils can be reduced by catalytic hydrogenation, typically carried out at high temperature using a nickel catalyst, to produce saturated solid or semisolid fats. Margarine and shortening are produced by hydrogenating soybean, peanut, or cottonseed oil until the proper consistency is obtained. Unfortunately, the hydrogenation reaction is accompanied by some cis–trans isomerization of the double bonds that remain, producing fats with about 10% to 15% trans unsaturated fatty acids. Dietary intake of trans fatty acids increases cholesterol levels in the blood, thereby increasing the risk of heart problems. The conversion of linoleic acid into elaidic acid is an example.

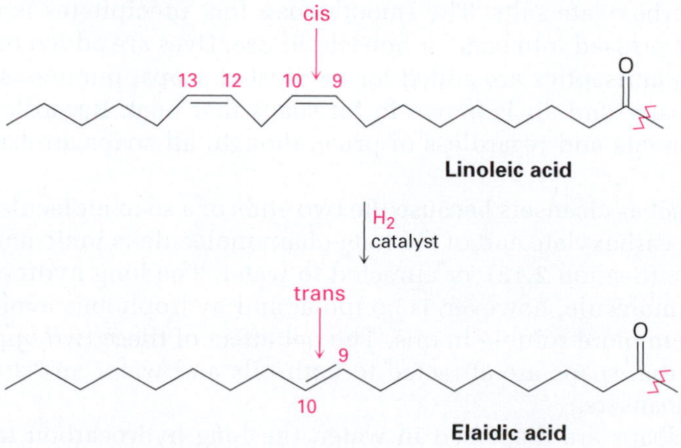

Linoleic acid

Elaidic acid

Problem 23.1

Carnauba wax, used in floor and furniture polishes, contains an ester of a C_{32} straight-chain alcohol with a C_{20} straight-chain carboxylic acid. Draw its structure.

Problem 23.2
Draw structures of glyceryl tripalmitate and glyceryl trioleate. Which would you expect to have a higher melting point?

23.2 Soap

Soap has been known since at least 600 BC, when the Phoenicians prepared a curdy material by boiling goat fat with extracts of wood ash. The cleansing properties of soap weren't generally recognized, however, and the use of soap did not become widespread until the 18th century. Chemically, soap is a mixture of the sodium or potassium salts of the long-chain fatty acids produced by hydrolysis *(saponification)* of animal fat with alkali. Wood ash was used as a source of alkali until the early 1800s, when the development of the LeBlanc process for making Na_2CO_3 by heating sodium sulfate with limestone (CaO) became available.

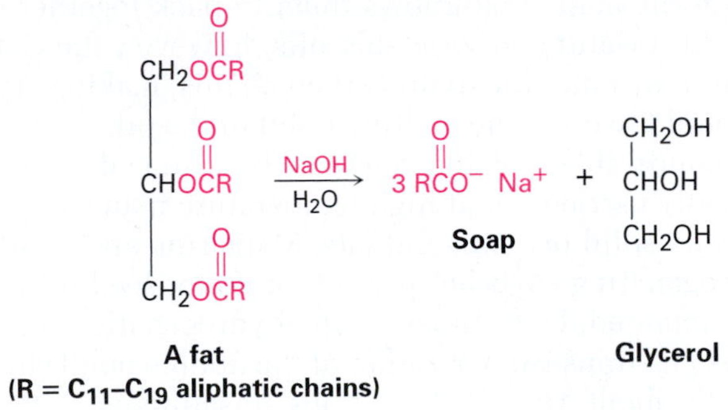

A fat
(R = C_{11}–C_{19} aliphatic chains)
Soap
Glycerol

Crude soap curds contain glycerol and excess alkali as well as soap but can be purified by boiling with water and adding NaCl or KCl to precipitate the pure carboxylate salts. The smooth soap that precipitates is dried, perfumed, and pressed into bars for household use. Dyes are added to make colored soaps, antiseptics are added for medicated soaps, pumice is added for scouring soaps, and air is blown in for soaps that float. Regardless of these extra treatments and regardless of price, though, all soaps are basically the same.

Soaps act as cleansers because the two ends of a soap molecule are so different. The carboxylate end of the long-chain molecule is ionic and therefore hydrophilic (Section 2.12), or attracted to water. The long hydrocarbon portion of the molecule, however, is nonpolar and hydrophobic, avoiding water and therefore more soluble in oils. The net effect of these two opposing tendencies is that soaps are attracted to both oils and water and are therefore useful as cleansers.

When soaps are dispersed in water, the long hydrocarbon tails cluster together on the inside of tangled, hydrophobic balls, while the ionic heads on the surface of the clusters stick out into the water layer. These spherical clusters, called **micelles**, are shown schematically in Figure 23.1. Grease and oil droplets are solubilized in water when they are coated by the nonpolar tails of soap molecules in the center of micelles. Once solubilized, the grease and dirt can be rinsed away.

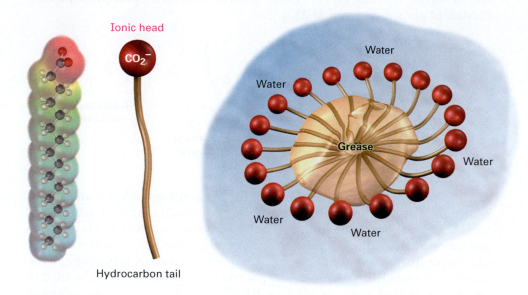

Ionic head

CO_2^-

Grease

Water

Water

Water

Water

Water

Water

Hydrocarbon tail

ACTIVE FIGURE 23.1
A soap micelle solubilizing a grease particle in water. An electrostatic potential map of a fatty-acid carboxylate shows how the negative charge is located in the head group. **Go to this book's student companion site at** www.cengage.com/ chemistry/mcmurry **to explore an interactive version of this figure.**

As useful as they are, soaps also have some drawbacks. In hard water, which contains metal ions such as Mg^{2+}, Ca^{2+}, and Fe^{3+}, soluble sodium carboxylates are converted into insoluble metal salts, leaving the familiar ring of scum around bathtubs and the gray tinge on white clothes. Chemists have circumvented these problems by synthesizing a class of synthetic detergents based on salts of long-chain alkylbenzenesulfonic acids. The principle of synthetic detergents is the same as that of soaps: the alkylbenzene end of the molecule is attracted to grease, while the anionic sulfonate end is attracted to water. Unlike soaps, though, sulfonate detergents don't form insoluble metal salts in hard water and don't leave an unpleasant scum.

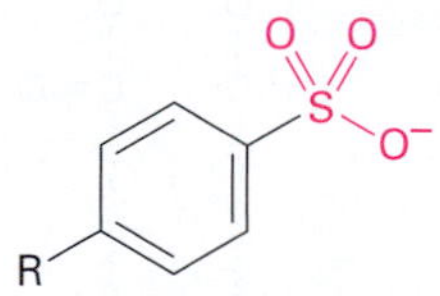

A synthetic detergent
(R = a mixture of C_{12} chains)

Problem 23.3
Draw the structure of magnesium oleate, a component of bathtub scum.

Problem 23.4
Write the saponification reaction of glyceryl dioleate monopalmitate with aqueous NaOH.

23.3 Phospholipids

Just as waxes, fats, and oils are esters of carboxylic acids, **phospholipids** are diesters of phosphoric acid, H_3PO_4:

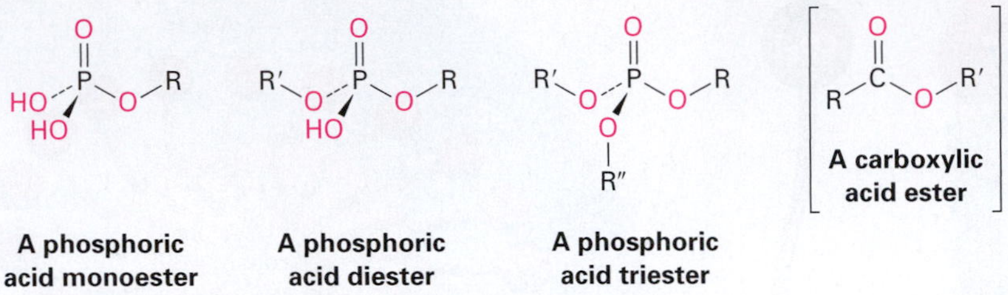

A phosphoric   A phosphoric   A phosphoric   A carboxylic
acid monoester   acid diester   acid triester   acid ester

Phospholipids are of two general kinds: *glycerophospholipids* and *sphingomyelins*. Glycerophospholipids are based on phosphatidic acid, which contains a glycerol backbone linked by ester bonds to two fatty acids and one phosphoric acid. Although the fatty-acid residues can be any of the C_{12}–C_{20} units typically present in fats, the acyl group at C1 is usually saturated and the one at C2 is usually unsaturated. The phosphate group at C3 is also bonded to an amino alcohol such as choline $[HOCH_2CH_2N(CH_3)_3]^+$, ethanolamine $(HOCH_2CH_2NH_2)$, or serine $[HOCH_2CH(NH_2)CO_2H]$. The compounds are chiral and have an L, or *R*, configuration at C2.

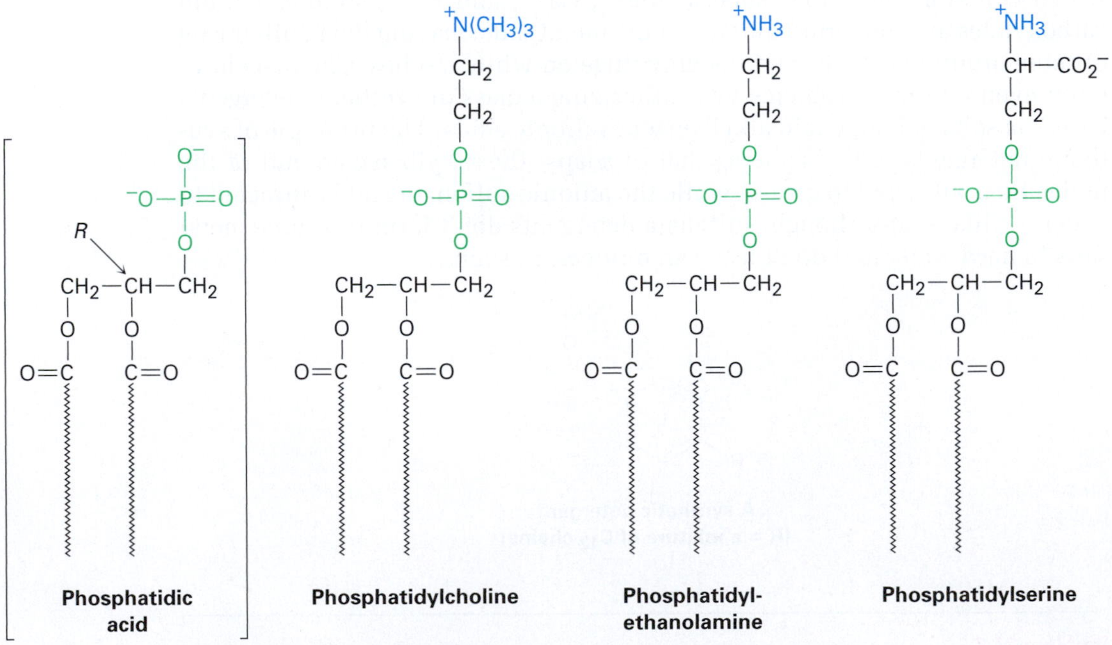

Phosphatidic   Phosphatidylcholine   Phosphatidyl-   Phosphatidylserine
acid   ethanolamine

Sphingomyelins are the second major group of phospholipids. These compounds have sphingosine or a related dihydroxyamine as their backbone and are particularly abundant in brain and nerve tissue, where they are a major constituent of the coating around nerve fibers.

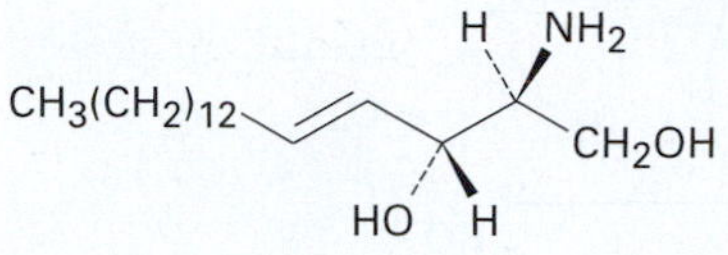

Sphingosine

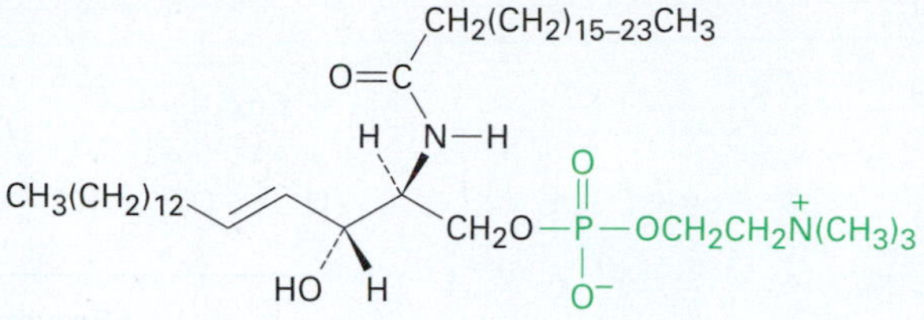

A sphingomyelin

Phospholipids are found widely in both plant and animal tissues and make up approximately 50% to 60% of cell membranes. Because they are like soaps in having a long, nonpolar hydrocarbon tail bound to a polar ionic head, phospholipids in the cell membrane organize into a **lipid bilayer** about 5.0 nm (50 Å) thick. As shown in Figure 23.2, the nonpolar tails aggregate in the center of the bilayer in much the same way that soap tails aggregate in the center of a micelle. This bilayer serves as an effective barrier to the passage of water, ions, and other components into and out of cells.

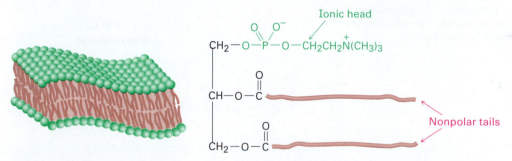

ACTIVE FIGURE 23.2 Aggregation of glycerophospholipids into the lipid bilayer that composes cell membranes. **Go to this book's student companion site at www.cengage.com/chemistry/mcmurry to explore an interactive version of this figure.**

23.4 Catabolism of Triacylglycerols: The Fate of Glycerol

Triacylglycerol catabolism begins with hydrolysis in the stomach and small intestine to yield glycerol plus fatty acids. The reaction is catalyzed by a lipase, whose mechanism of action is shown in Figure 23.3. The active site of the enzyme contains a catalytic triad of aspartic acid, histidine, and serine residues, which act cooperatively to provide the acidic and basic catalysis for the individual steps. Hydrolysis is accomplished by two sequential nucleophilic acyl substitution reactions, one that covalently binds an acyl group to the side-chain –OH of a serine residue on the enzyme and a second that frees the fatty acid from the enzyme.

STEPS 1–2 OF FIGURE 23.3: ACYL ENZYME FORMATION The first nucleophilic acyl substitution step—reaction of the triacylglycerol with the active-site serine to give an acyl enzyme—begins with deprotonation of the serine alcohol by histidine to form the more strongly nucleophilic alkoxide ion. This proton transfer is facilitated by a nearby side-chain carboxylate anion of aspartic

1 The enzyme active site contains an aspartic acid, a histidine, and a serine. First, histidine acts as a base to deprotonate the –OH group of serine, with the negatively charged carboxylate of aspartic acid stabilizing the nearby histidine cation that results. Serine then adds to the carbonyl group of the triacylglycerol, yielding a tetrahedral intermediate.

2 This intermediate expels a diacylglycerol as leaving group in a nucleophilic acyl substitution reaction, giving an acyl enzyme. The diacylglycerol is protonated by the histidine cation.

3 Histidine deprotonates a water molecule, which adds to the acyl group. A tetrahedral intermediate is again formed, and the histidine cation is again stabilized by the nearby carboxylate.

4 The tetrahedral intermediate expels the serine as leaving group in a second nucleophilic acyl substitution reaction, yielding a free fatty acid. The serine accepts a proton from histidine, and the enzyme has now returned to its starting structure.

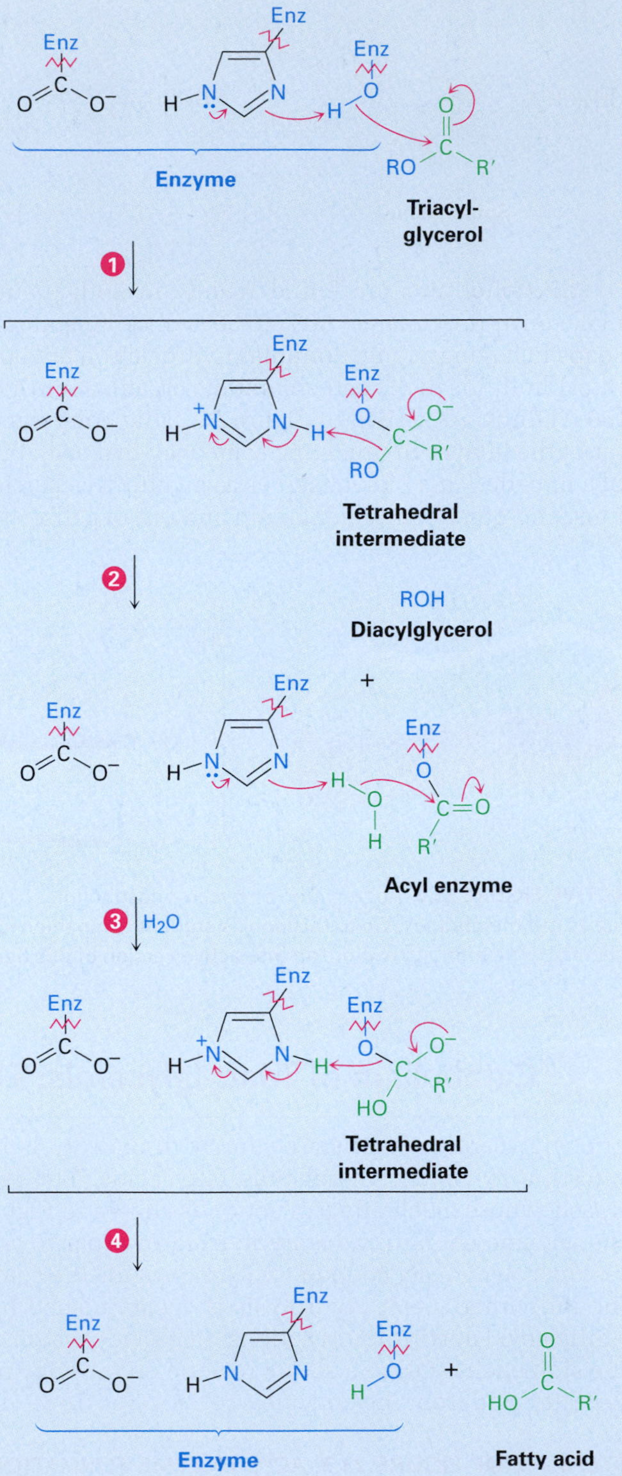

FIGURE 23.3 MECHANISM: Mechanism of action of lipase. The active site of the enzyme contains a catalytic triad of aspartic acid, histidine, and serine, which react cooperatively to carry out two nucleophilic acyl substitution reactions. Individual steps are explained in the text.

acid, which makes the histidine more basic and stabilizes the resultant histidine cation by electrostatic interactions. The deprotonated serine adds to a carbonyl group of a triacylglycerol to give a tetrahedral intermediate.

The tetrahedral intermediate expels a diacylglycerol as the leaving group and produces an acyl enzyme. The step is catalyzed by a proton transfer from histidine to make the leaving group a neutral alcohol.

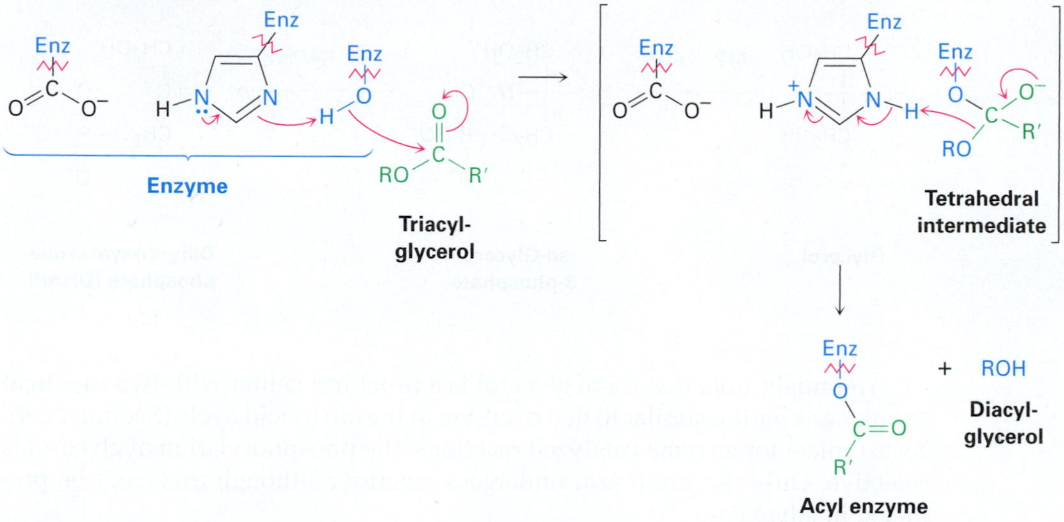

STEPS 3–4 OF FIGURE 23.3: HYDROLYSIS The second nucleophilic acyl substitution step hydrolyzes the acyl enzyme and gives the free fatty acid by a mechanism analogous to that of the first two steps. Water is deprotonated by histidine and then adds to the enzyme-bound acyl group. The tetrahedral intermediate then expels the neutral serine residue as the leaving group, freeing the fatty acid and returning the enzyme to its active form.

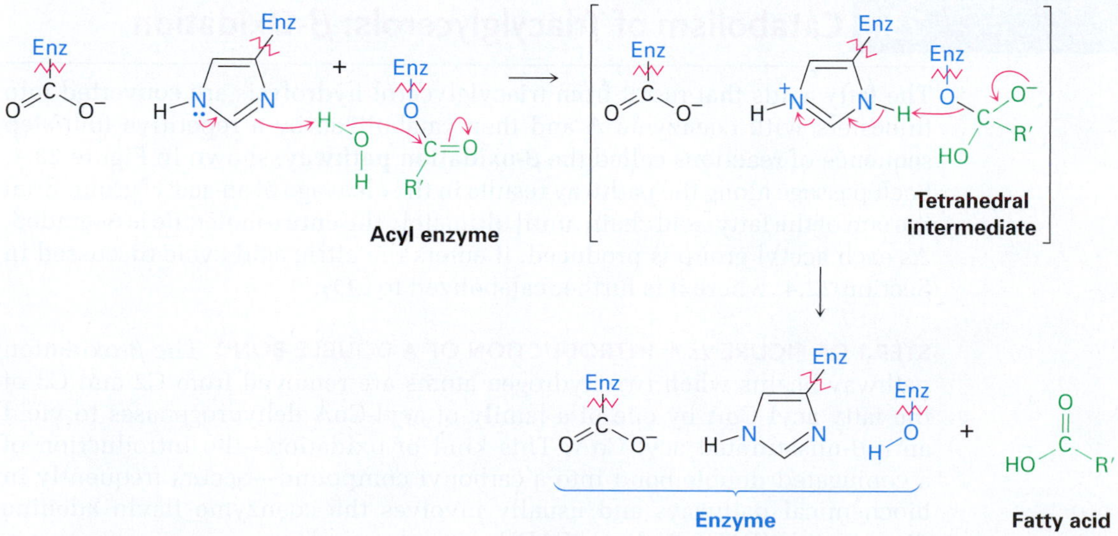

The fatty acids released on triacylglycerol hydrolysis are transported to mitochondria and degraded to acetyl CoA, while the glycerol is carried to the liver for further metabolism. In the liver, glycerol is first phosphorylated by reaction with ATP and then oxidized by NAD^+. The dihydroxyacetone phosphate (DHAP) that results enters the carbohydrate glycolysis pathway that we saw in Section 22.2.

Glycerol   ***sn*-Glycerol 3-phosphate**   **Dihydroxyacetone phosphate (DHAP)**

You might note that C2 of glycerol is a prochiral center with two identical "arms," a situation similar to that of citrate in the citric acid cycle (Section 22.4). As is typical for enzyme-catalyzed reactions, the phosphorylation of glycerol is selective. Only the *pro-R* arm undergoes reaction, although this can't be predicted in advance.

Note also that the phosphorylation product is named *sn*-glycerol 3-phosphate, where the *sn*- prefix means "stereospecific numbering." In this convention, the molecule is drawn in Fischer projection with the –OH group at C2 pointing to the left and the glycerol carbon atoms are numbered beginning at the top.

23.5 Catabolism of Triacylglycerols: β-Oxidation

The fatty acids that result from triacylglycerol hydrolysis are converted into thioesters with coenzyme A and then catabolized by a repetitive four-step sequence of reactions called the **β-oxidation pathway**, shown in Figure 23.4. Each passage along the pathway results in the cleavage of an acetyl group from the end of the fatty-acid chain, until ultimately the entire molecule is degraded. As each acetyl group is produced, it enters the citric acid cycle discussed in Section 22.4, where it is further catabolized to CO_2.

STEP 1 OF FIGURE 23.4: INTRODUCTION OF A DOUBLE BOND The β-oxidation pathway begins when two hydrogen atoms are removed from C2 and C3 of the fatty acyl CoA by one of a family of acyl-CoA dehydrogenases to yield an *α,β*-unsaturated acyl CoA. This kind of oxidation—the introduction of a conjugated double bond into a carbonyl compound—occurs frequently in biochemical pathways and usually involves the coenzyme flavin adenine dinucleotide (FAD). Reduced $FADH_2$ is the by-product.

FAD (Flavin adenine dinucleotide)   **FADH₂**

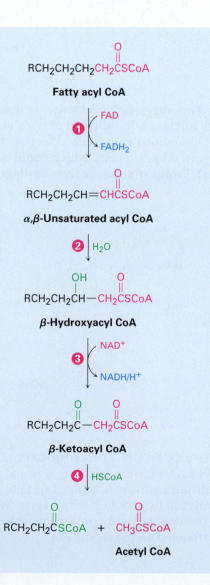

Fatty acyl CoA

$RCH_2CH_2CH_2CH_2CSCoA$

**① ** A conjugated double bond is introduced by removal of hydrogens from C2 and C3 by the coenzyme flavin adenine dinucleotide (FAD).

FAD → FADH₂

$RCH_2CH_2CH{=}CHCSCoA$

α,β-Unsaturated acyl CoA

**② ** Conjugate nucleophilic addition of water to the double bond gives a β-hydroxyacyl CoA.

H₂O

$RCH_2CH_2CH{-}CH_2CSCoA$ (OH)

β-Hydroxyacyl CoA

**③ ** The alcohol is oxidized by NAD⁺ to give a β-keto thioester.

NAD⁺ → NADH/H⁺

$RCH_2CH_2C{-}CH_2CSCoA$

β-Ketoacyl CoA

**④ ** Nucleophilic addition of coenzyme A to the keto group occurs, followed by a retro-Claisen condensation reaction. The products are acetyl CoA and a chain-shortened fatty acyl CoA.

HSCoA

RCH_2CH_2CSCoA + CH_3CSCoA

Acetyl CoA

FIGURE 23.4 MECHANISM: The four steps of the β-oxidation pathway, resulting in the cleavage of an acetyl group from the end of the fatty-acid chain. The key chain-shortening step is a retro-Claisen reaction of a β-keto thioester. Individual steps are explained in the text.

© John McMurry

The mechanisms of FAD-catalyzed reactions are often difficult to establish because flavin coenzymes can operate by both two-electron (polar) and one-electron (radical) pathways. As a result, extensive studies of the family of acyl-CoA dehydrogenases have not yet provided a clear picture of how these enzymes function. What is known is that: (1) The first step is abstraction of the *pro-R* hydrogen from the acidic α position of the acyl CoA to give a thioester enolate ion. Hydrogen-bonding between the acyl carbonyl group and the ribitol hydroxyls of FAD increases the acidity of the acyl group. (2) The *pro-R* hydrogen at the β position is transferred to FAD. (3) The α,β-unsaturated acyl CoA that results has a trans double bond.

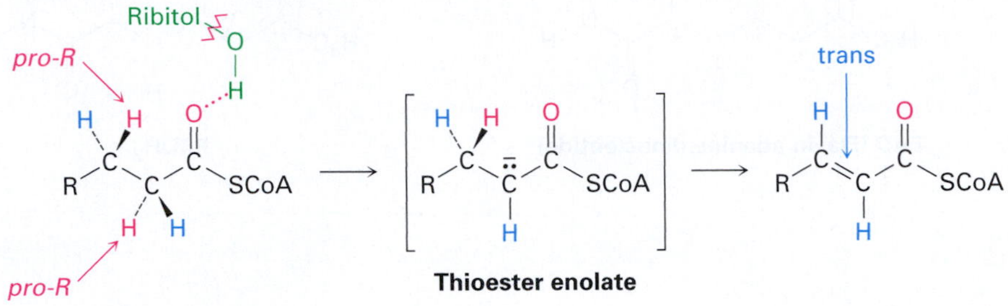

One suggested mechanism is that the reaction may take place by a conjugate nucleophilic addition of hydride, analogous to what occurs during alcohol oxidations with NAD$^+$ (Section 13.5). Electrons on the enolate ion might expel a β hydride ion, which could add to the doubly bonded N5 nitrogen on FAD. Protonation of the intermediate at N1 would give the product:

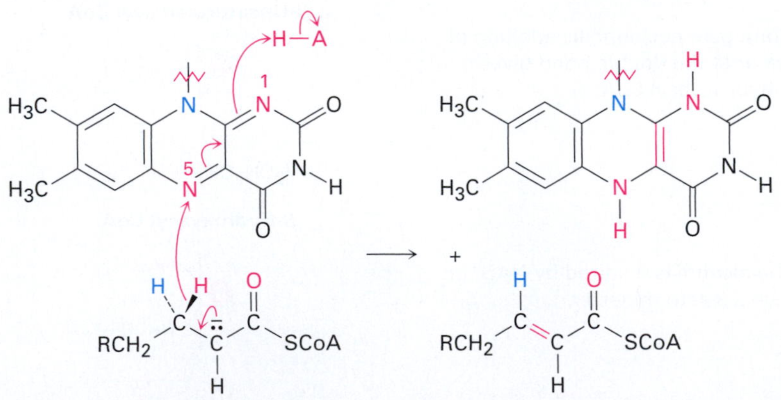

STEP 2 OF FIGURE 23.4: CONJUGATE ADDITION OF WATER The α,β-unsaturated acyl CoA produced in step 1 reacts with water by a conjugate nucleophilic addition pathway (Section 14.11) to yield a β-hydroxyacyl CoA in a process catalyzed by enoyl-CoA hydratase. Water as nucleophile adds to the β carbon of the double bond, yielding an intermediate thioester enolate ion, which is protonated on the α position.

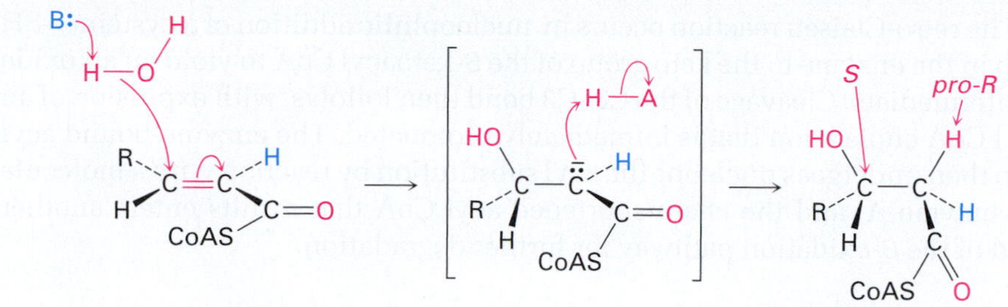

(3S)-Hydroxyacyl CoA

STEP 3 OF FIGURE 23.4: ALCOHOL OXIDATION

The β-hydroxyacyl CoA from step 2 is oxidized to a β-ketoacyl CoA in a reaction catalyzed by one of a family of L-3-hydroxyacyl-CoA dehydrogenases, which differ in substrate specificity according to the chain length of the acyl group. As in the oxidation of *sn*-glycerol 3-phosphate to dihydroxyacetone phosphate mentioned in Section 23.4, this alcohol oxidation requires NAD$^+$ as a coenzyme and yields reduced NADH/H$^+$ as by-product. The reaction is facilitated by deprotonation of the hydroxyl group by a histidine residue at the active site.

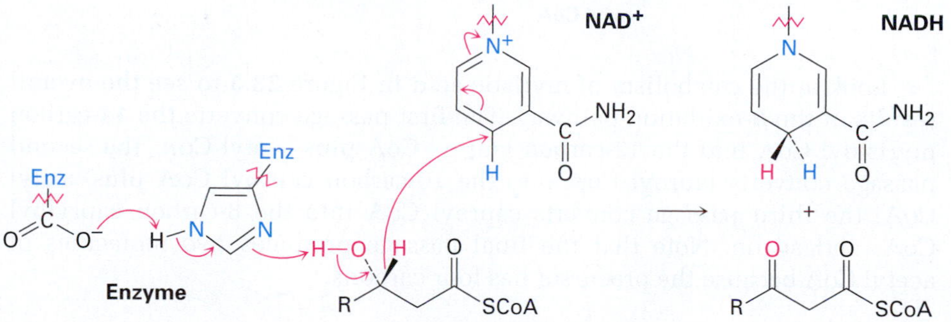

β-Hydroxyacyl CoA   β-Ketoacyl CoA

STEP 4 OF FIGURE 23.4: CHAIN CLEAVAGE

Acetyl CoA is split off from the chain in the final step of β-oxidation, leaving an acyl CoA that is two carbon atoms shorter than the original. The reaction is catalyzed by β-ketoacyl-CoA thiolase and is mechanistically the reverse of a Claisen condensation reaction (Section 17.9). In the *forward* direction, a Claisen condensation joins two esters together to form a β-keto ester product. In the *reverse* direction, a retro-Claisen reaction splits apart a β-keto ester (or β-keto thioester in this case) to form two esters (or two thioesters).

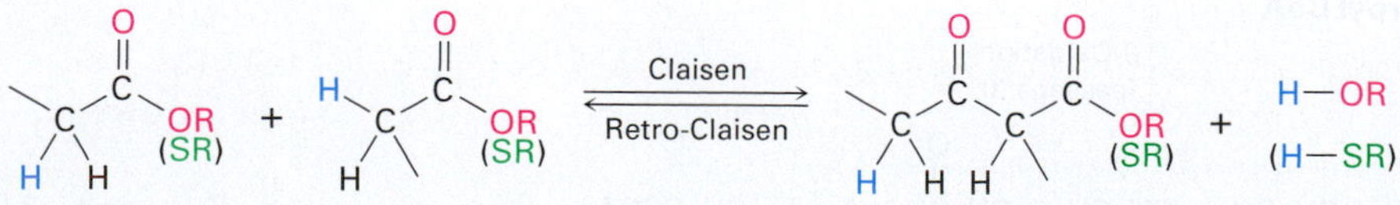

The retro-Claisen reaction occurs by nucleophilic addition of a cysteine –SH group in the enzyme to the keto group of the β-ketoacyl CoA to yield an alkoxide ion intermediate. Cleavage of the C2–C3 bond then follows, with expulsion of an acetyl CoA enolate ion that is immediately protonated. The enzyme-bound acyl group then undergoes nucleophilic acyl substitution by reaction with a molecule of coenzyme A, and the chain-shortened acyl CoA that results enters another round of the β-oxidation pathway for further degradation.

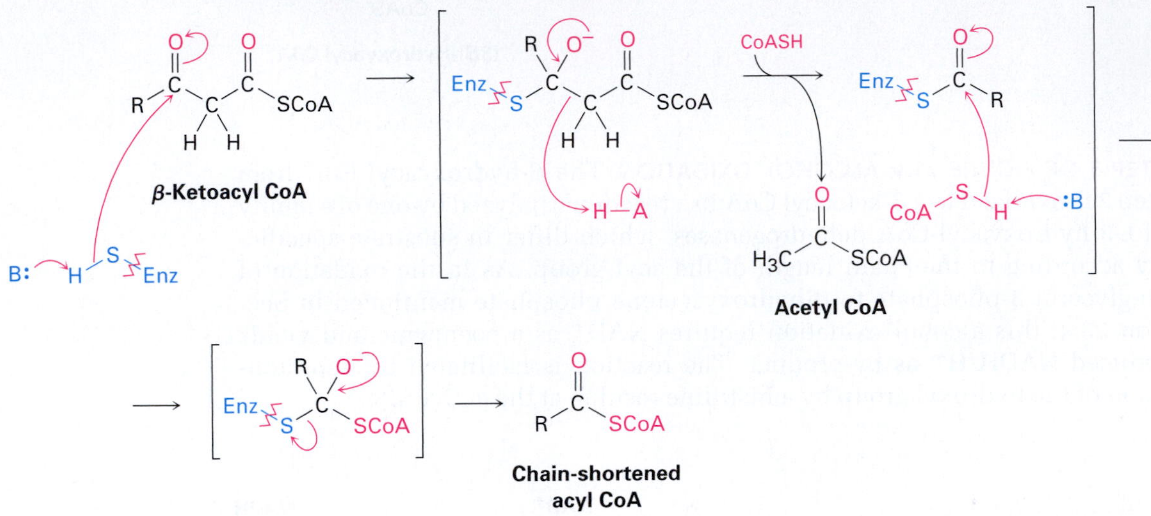

Look at the catabolism of myristic acid in Figure 23.5 to see the overall results of the β-oxidation pathway. The first passage converts the 14-carbon myristoyl CoA into the 12-carbon lauroyl CoA plus acetyl CoA, the second passage converts lauroyl CoA into the 10-carbon caproyl CoA plus acetyl CoA, the third passage converts caproyl CoA into the 8-carbon capryloyl CoA, and so on. Note that the final passage produces *two* molecules of acetyl CoA because the precursor has four carbons.

FIGURE 23.5 Catabolism of the 14-carbon myristic acid by the β-oxidation pathway yields seven molecules of acetyl CoA after six passages.

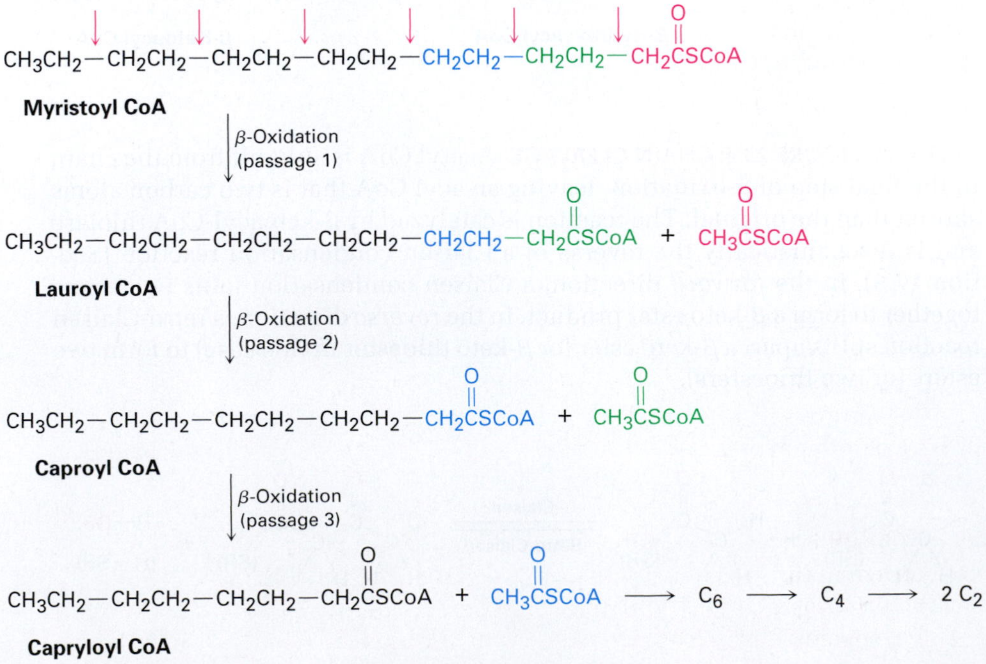

Most fatty acids have an even number of carbon atoms, so none are left over after β-oxidation. Those fatty acids with an odd number of carbon atoms yield the three-carbon propionyl CoA in the final β-oxidation. Propionyl CoA is then converted to succinate by a multistep radical pathway, and succinate enters the citric acid cycle (Section 22.4). Note that the three-carbon propionyl group should properly be called *propanoyl,* but biochemists generally use the nonsystematic name.

Problem 23.5

Write the equations for the remaining passages of the β-oxidation pathway following those shown in Figure 23.5.

Problem 23.6

How many molecules of acetyl CoA are produced by catabolism of the following fatty acids, and how many passages of the β-oxidation pathway are needed?

(a) Palmitic acid, $CH_3(CH_2)_{14}CO_2H$
(b) Arachidic acid, $CH_3(CH_2)_{18}CO_2H$

23.6 Biosynthesis of Fatty Acids

One of the most striking features of the common fatty acids is that they have an even number of carbon atoms (Table 23.1). This even number results because all fatty acids are derived biosynthetically from acetyl CoA by sequential addition of two-carbon units to a growing chain. The acetyl CoA, in turn, arises primarily from the metabolic breakdown of carbohydrates in the glycolysis pathway (Section 22.2). Thus, dietary carbohydrates consumed in excess of immediate energy needs are turned into fats for storage.

As noted previously in Section 22.5 when discussing carbohydrate biosynthesis, the anabolic pathway by which a substance is made is not the reverse of the catabolic pathway by which it is degraded. Thus, the β-oxidation pathway that converts fatty acids *into* acetyl CoA and the biosynthesis pathway that prepares fatty acids *from* acetyl CoA are related but are not exact opposites. Differences include the identity of the acyl-group carrier, the stereochemistry of the β-hydroxyacyl reaction intermediate, and the identity of the redox coenzyme. FAD is used to introduce a double bond in β-oxidation, while NADPH is used to reduce the double bond in fatty-acid biosynthesis.

In bacteria, each step in fatty-acid synthesis is catalyzed by separate enzymes. In vertebrates, however, fatty-acid synthesis is catalyzed by a large, multienzyme complex called a *synthase* that contains two identical subunits of 2505 amino acids each and catalyzes all steps in the pathway. An overview of fatty-acid biosynthesis is shown in Figure 23.6.

STEPS 1–2 OF FIGURE 23.6: ACYL TRANSFERS The starting material for fatty-acid biosynthesis is the thioester acetyl CoA, the final product of carbohydrate breakdown in the glycolysis pathway (Section 22.2). The pathway begins with several *priming reactions,* which transport acetyl CoA and convert it into more reactive species. The first priming reaction is a nucleophilic acyl substitution

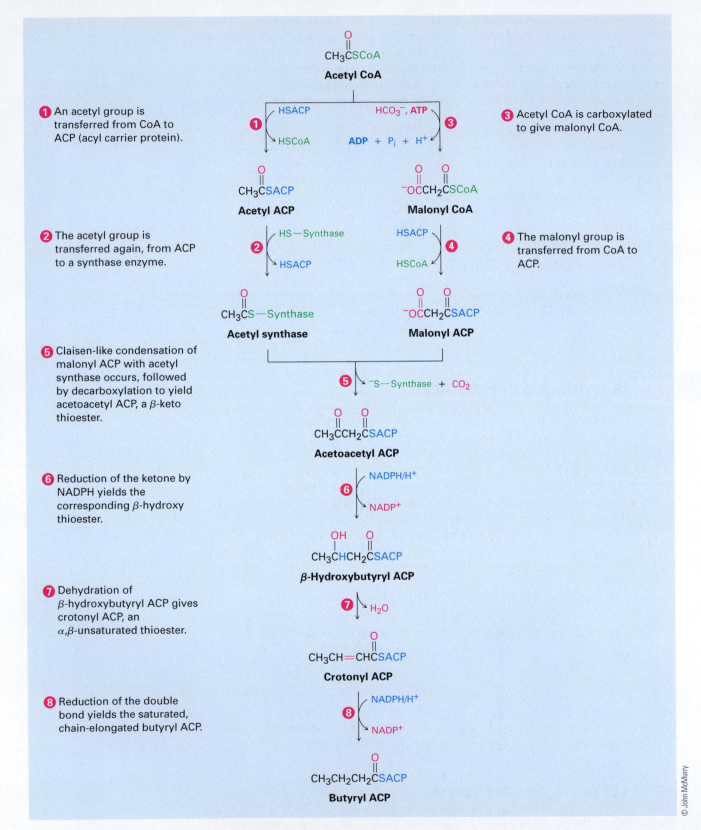

FIGURE 23.6 MECHANISM: The pathway for fatty-acid biosynthesis from the two-carbon precursor, acetyl CoA. Individual steps are explained in the text.

that converts acetyl CoA into acetyl ACP (acyl carrier protein). The reaction is catalyzed by ACP transacylase.

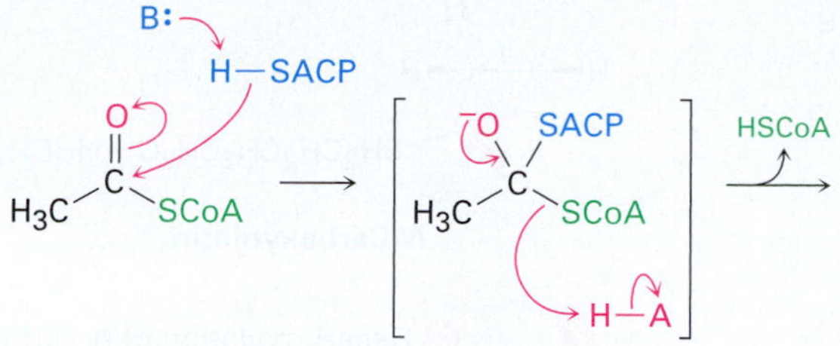

Tetrahedral intermediate

In bacteria, ACP is a small protein of 77 residues that transports an acyl group from one enzyme to another. In vertebrates, however, ACP appears to be a long arm on a multienzyme synthase complex, whose apparent function is to shepherd an acyl group from site to site within the complex. As in acetyl CoA, the acyl group in acetyl ACP is linked by a thioester bond to the sulfur atom of phosphopantetheine. The phosphopantetheine is in turn linked to ACP through the side-chain –OH group of a serine residue in the enzyme.

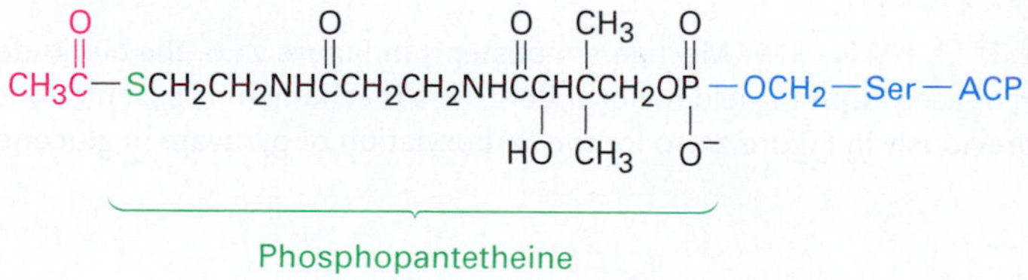

Phosphopantetheine

Acetyl ACP

Step 2, another priming reaction, involves a further exchange of thioester linkages by another nucleophilic acyl substitution and results in covalent bonding of the acetyl group to a cysteine residue in the synthase complex that will catalyze the upcoming condensation step.

STEPS 3–4 OF FIGURE 23.6: CARBOXYLATION AND ACYL TRANSFER The third step is a *loading* reaction in which acetyl CoA is carboxylated by reaction with HCO_3^- and ATP to yield malonyl CoA plus ADP. As in the first step of gluconeogenesis, in which pyruvate is carboxylated to yield oxaloacetate, the coenzyme biotin acts as a carrier of CO_2 by forming *N*-carboxybiotin. The mechanism of the reaction, shown in Figure 23.7, is essentially identical to that of the pyruvate carboxylation given previously in Figure 22.10 on page 924.

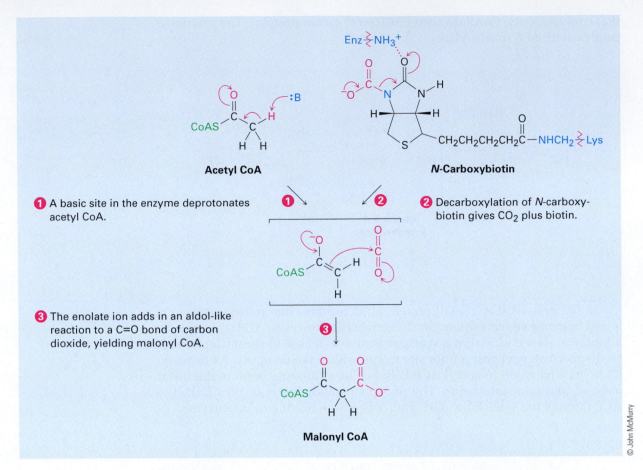

1 A basic site in the enzyme deprotonates acetyl CoA.

2 Decarboxylation of *N*-carboxy-biotin gives CO_2 plus biotin.

3 The enolate ion adds in an aldol-like reaction to a C=O bond of carbon dioxide, yielding malonyl CoA.

Acetyl CoA

***N*-Carboxybiotin**

Malonyl CoA

© John McMurry

FIGURE 23.7 MECHANISM: Mechanism of step 3 in Figure 23.6, the biotin-dependent carboxylation of acetyl CoA to yield malonyl CoA. This mechanism is essentially identical to that shown previously in Figure 22.10 for the carboxylation of pyruvate in gluconeogenesis.

Following the formation of malonyl CoA, another nucleophilic acyl substitution reaction occurs in step 4 to form the more reactive malonyl ACP, thereby binding the malonyl group to an ACP arm of the multienzyme synthase. At this point, both acetyl and malonyl groups are bound to the enzyme and the stage is set for their condensation.

STEP 5 OF FIGURE 23.6: CONDENSATION The key carbon–carbon bond-forming reaction that builds the fatty-acid chain occurs in step 5. This step is simply a Claisen condensation between acetyl synthase as the electrophilic acceptor and malonyl ACP as the nucleophilic donor. The mechanism of the condensation is thought to involve decarboxylation of malonyl ACP to give an enolate ion, followed by immediate nucleophilic addition of the enolate ion to the carbonyl group of acetyl synthase. Breakdown of the tetrahedral intermediate then gives the four-carbon condensation product acetoacetyl ACP and frees the synthase binding site for attachment of the chain-elongated acyl group at the end of the sequence.

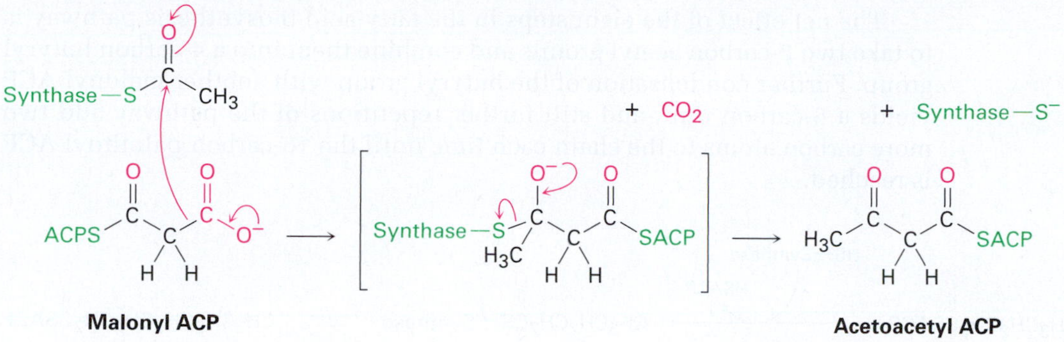

Malonyl ACP   + CO_2 + Synthase—S⁻   **Acetoacetyl ACP**

STEPS 6–8 OF FIGURE 23.6: REDUCTION AND DEHYDRATION The ketone carbonyl group in acetoacetyl ACP is next reduced to the alcohol β-hydroxybutyryl ACP by β-keto thioester reductase and NADPH. *R* Stereochemistry results at the newly formed chirality center in the β-hydroxy thioester product. (Note that the systematic name of a butyryl group is *butanoyl*.)

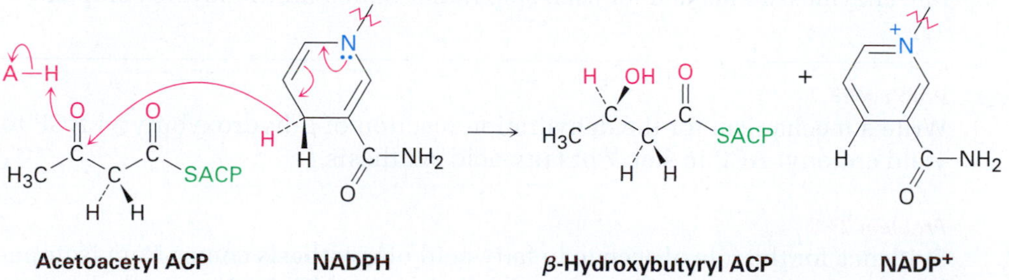

Acetoacetyl ACP   **NADPH**   **β-Hydroxybutyryl ACP**   **NADP⁺**

Subsequent dehydration of β-hydroxybutyryl ACP by an E1cB reaction in step 7 yields *trans*-crotonyl ACP, and the carbon–carbon double bond of crotonyl ACP is reduced by NADPH in step 8 to yield butyryl ACP. The double-bond reduction occurs by conjugate nucleophilic addition of a hydride ion from NADPH to the β carbon of *trans*-crotonyl ACP. In vertebrates, the reduction occurs by an overall *syn* addition, but other organisms carry out a similar transformation with different stereochemistry.

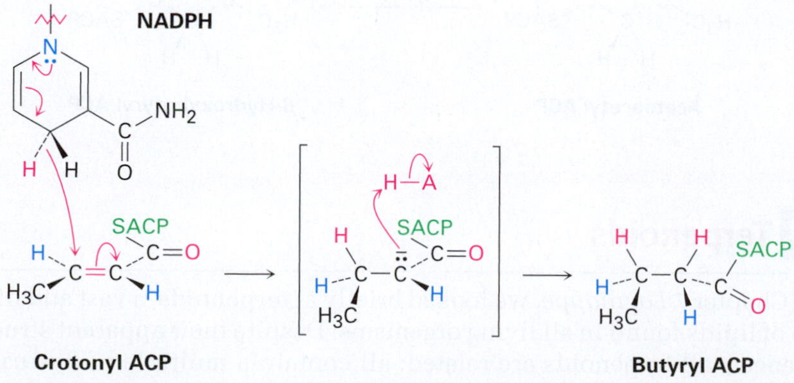

Crotonyl ACP   **Butyryl ACP**

The net effect of the eight steps in the fatty-acid biosynthesis pathway is to take two 2-carbon acetyl groups and combine them into a 4-carbon butyryl group. Further condensation of the butyryl group with another malonyl ACP yields a 6-carbon unit, and still further repetitions of the pathway add two more carbon atoms to the chain each time until the 16-carbon palmitoyl ACP is reached.

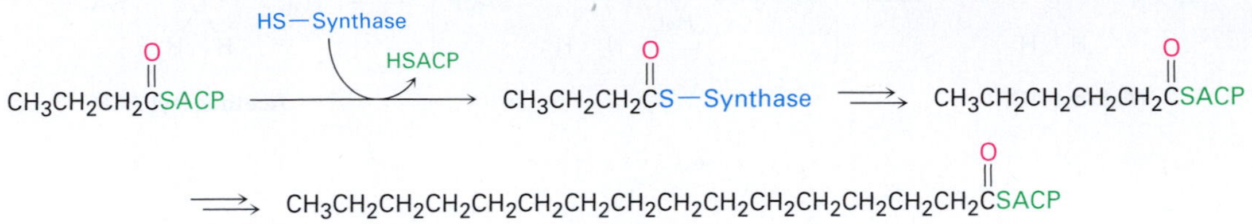

Palmitoyl ACP

Further chain elongation of palmitic acid occurs by reactions similar to those just described, but CoA rather than ACP is the carrier group and separate enzymes are needed for each step rather than a multienzyme complex.

Problem 23.7

Write a mechanism for the dehydration reaction of β-hydroxybutyryl ACP to yield crotonyl ACP in step 7 of fatty-acid synthesis.

Problem 23.8

Evidence for the role of acetate in fatty-acid biosynthesis comes from isotope-labeling experiments. If acetate labeled with ^{13}C in the methyl group ($^{13}CH_3CO_2H$) were incorporated into fatty acids, at what positions in the fatty-acid chain would you expect the ^{13}C label to appear?

Problem 23.9

Does the reduction of acetoacetyl ACP in step 6 occur on the *Re* face or the *Si* face of the carbonyl group?

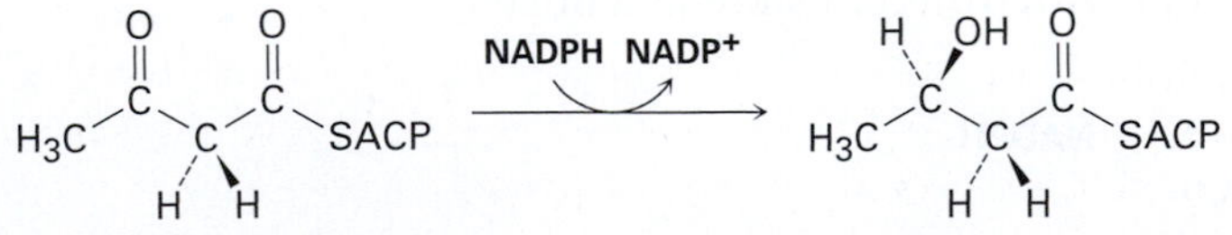

Acetoacetyl ACP   **β-Hydroxybutyryl ACP**

23.7 Terpenoids

In the Chapter 7 *Lagniappe*, we looked briefly at **terpenoids**, a vast and diverse group of lipids found in all living organisms. Despite their apparent structural differences, all terpenoids are related: all contain a multiple of five carbons and are derived biosynthetically from the five-carbon precursor isopentenyl diphosphate (Figure 23.8). Although formally a *terpenoid* contains oxygen, while a *terpene* is a hydrocarbon, we'll use the term terpenoid to refer to both for simplicity.

FIGURE 23.8 Structures of some representative terpenoids.

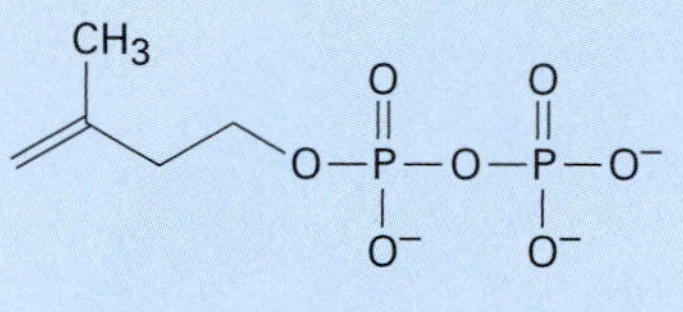

Isopentenyl diphosphate

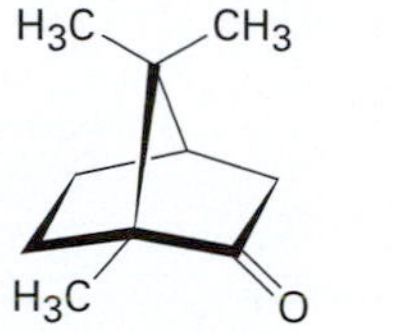

Camphor
(a monoterpene—C_{10})

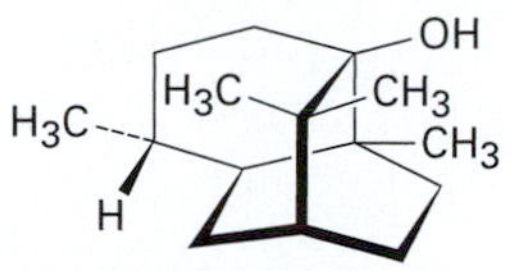

Patchouli alcohol
(a sesquiterpene—C_{15})

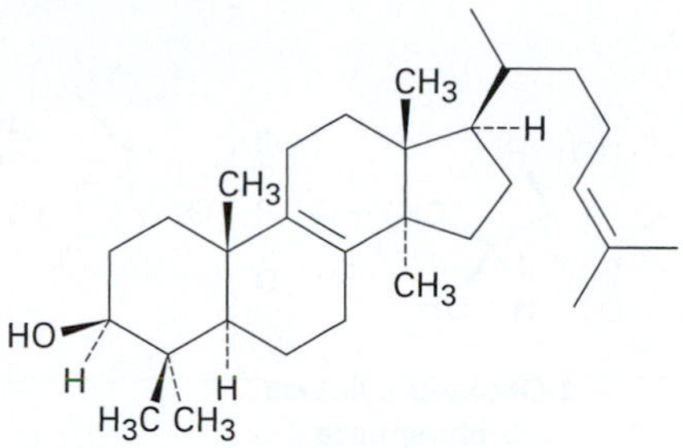

Lanosterol
(a triterpene—C_{30})

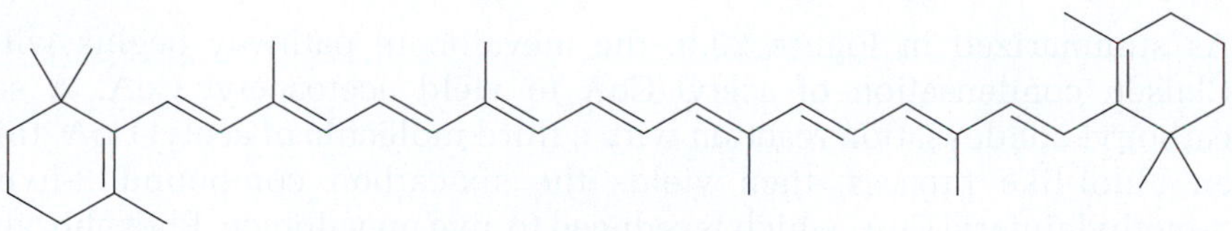

β-Carotene
(a tetraterpene—C_{40})

Terpenoids are classified according to the number of five-carbon multiples they contain. *Monoterpenoids* contain 10 carbons and are derived from two isopentenyl diphosphates, *sesquiterpenoids* contain 15 carbons and are derived from three isopentenyl diphosphates, *diterpenoids* contain 20 carbons and are derived from four isopentenyl diphosphates, and so on, up to triterpenoids (C_{30}) and tetraterpenoids (C_{40}). Monoterpenoids and sesquiterpenoids are found primarily in plants, bacteria, and fungi, but the higher terpenoids occur in both plants and animals. The triterpenoid lanosterol, for example, is the precursor from which steroid hormones are made, and the tetraterpenoid β-carotene is a dietary source of vitamin A (Figure 23.8).

The terpenoid precursor isopentenyl diphosphate, formerly called isopentenyl pyrophosphate and thus abbreviated IPP, is biosynthesized by two different pathways depending on the organism and the structure of the final product. In animals and higher plants, sesquiterpenoids and triterpenoids arise primarily from the *mevalonate* pathway, whereas monoterpenoids, diterpenoids, and tetraterpenoids are biosynthesized by the *1-deoxyxylulose 5-phosphate (DXP)* pathway. In bacteria, both pathways are used. We'll look only at the mevalonate pathway, which is more common and better understood at present.

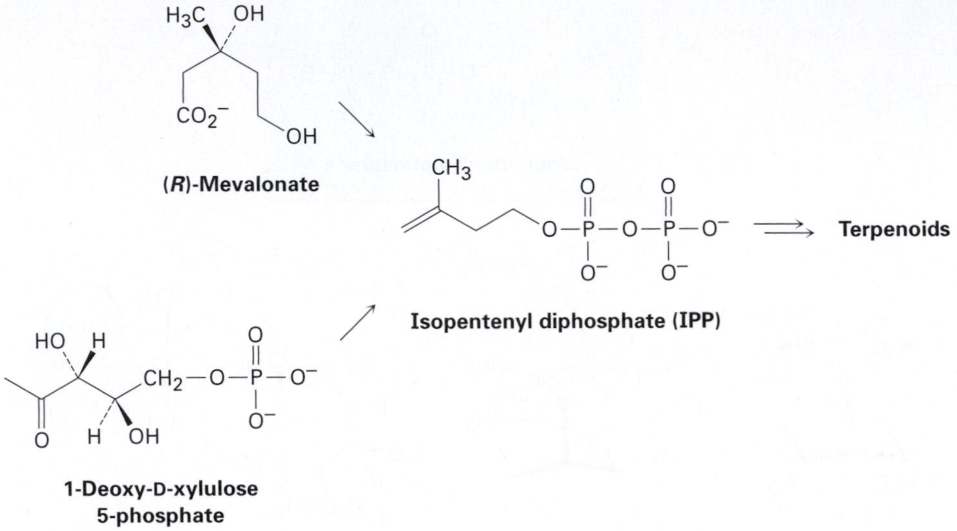

(R)-Mevalonate

Isopentenyl diphosphate (IPP)

1-Deoxy-D-xylulose 5-phosphate

→ **Terpenoids**

The Mevalonate Pathway to Isopentenyl Diphosphate

As summarized in Figure 23.9, the mevalonate pathway begins with the Claisen condensation of acetyl CoA to yield acetoacetyl CoA. A second carbonyl condensation reaction with a third molecule of acetyl CoA, this one an aldol-like process, then yields the six-carbon compound 3-hydroxy-3-methylglutaryl CoA, which is reduced to give mevalonate. Phosphorylation, followed by loss of CO_2 and phosphate ion, completes the process.

STEP 1 OF FIGURE 23.9: CLAISEN CONDENSATION The first step in mevalonate biosynthesis is a Claisen condensation to yield acetoacetyl CoA, a reaction catalyzed by acetoacetyl-CoA acetyltransferase. An acetyl group is first bound to the enzyme by a nucleophilic acyl substitution reaction with a cysteine –SH group. Formation of an enolate ion from a second molecule of acetyl CoA, followed by Claisen condensation, then yields the product.

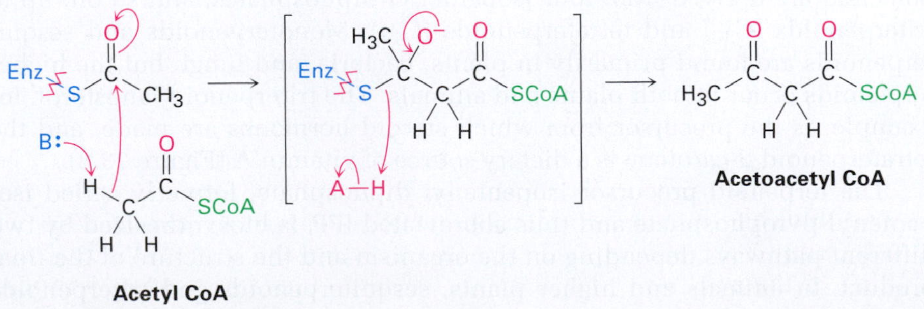

Acetyl CoA

Acetoacetyl CoA

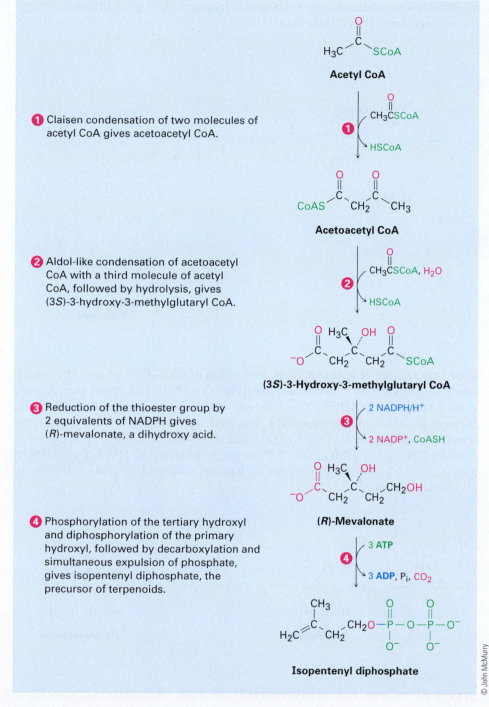

FIGURE 23.9 MECHANISM:
The mevalonate pathway for the biosynthesis of isopentenyl diphosphate from three molecules of acetyl CoA. Individual steps are explained in the text.

① Claisen condensation of two molecules of acetyl CoA gives acetoacetyl CoA.

② Aldol-like condensation of acetoacetyl CoA with a third molecule of acetyl CoA, followed by hydrolysis, gives (3S)-3-hydroxy-3-methylglutaryl CoA.

③ Reduction of the thioester group by 2 equivalents of NADPH gives (R)-mevalonate, a dihydroxy acid.

④ Phosphorylation of the tertiary hydroxyl and diphosphorylation of the primary hydroxyl, followed by decarboxylation and simultaneous expulsion of phosphate, gives isopentenyl diphosphate, the precursor of terpenoids.

© John McMurry

STEP 2 OF FIGURE 23.9: ALDOL CONDENSATION Acetoacetyl CoA next undergoes an aldol-like addition of an acetyl CoA enolate ion in a reaction catalyzed by 3-hydroxy-3-methylglutaryl-CoA synthase. The reaction occurs by initial binding of the substrate to a cysteine —SH group in the enzyme, followed by enolate-ion addition and subsequent hydrolysis to give (3*S*)-3-hydroxy-3-methylglutaryl CoA (HMG-CoA).

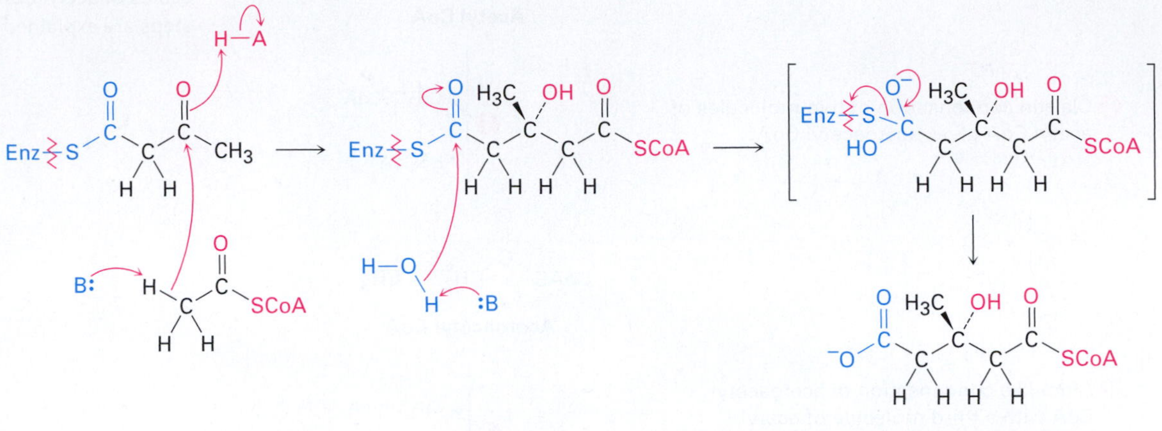

(3*S*)-3-Hydroxy-3-methylglutaryl CoA (HMG-CoA)

STEP 3 OF FIGURE 23.9: REDUCTION Reduction of HMG-CoA to give (*R*)-mevalonate is catalyzed by 3-hydroxy-3-methylglutaryl-CoA reductase and requires 2 equivalents of NADPH. The reaction occurs in two steps and proceeds through an aldehyde intermediate. The first step is a nucleophilic acyl substitution reaction involving hydride transfer from NADPH to the thioester carbonyl group of HMG-CoA. Following expulsion of HSCoA as leaving group, the aldehyde intermediate undergoes a second hydride addition to give mevalonate.

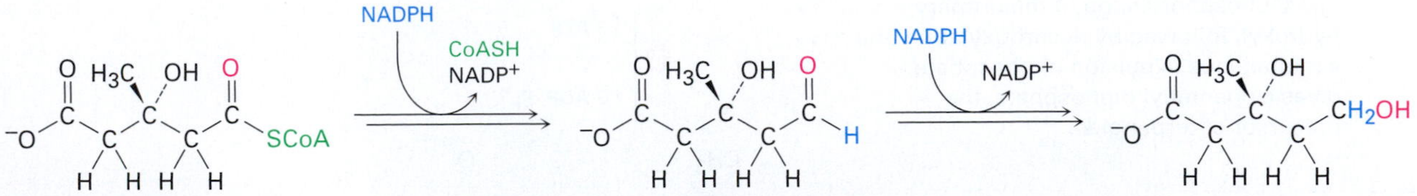

HMG-CoA **Mevaldehyde** **(*R*)-Mevalonate**

STEP 4 OF FIGURE 23.9: PHOSPHORYLATION AND DECARBOXYLATION Three additional reactions are needed to convert mevalonate to isopentenyl diphosphate. The first two are straightforward phosphorylations by ATP that occur through nucleophilic substitution reactions on the terminal phosphorus. Mevalonate is first converted to mevalonate 5-phosphate (phosphomevalonate) by reaction with ATP, and mevalonate 5-phosphate then reacts with a second ATP to give mevalonate 5-diphosphate (diphosphomevalonate). The third reaction results in phosphorylation of the tertiary hydroxyl group, followed by decarboxylation and loss of phosphate ion.

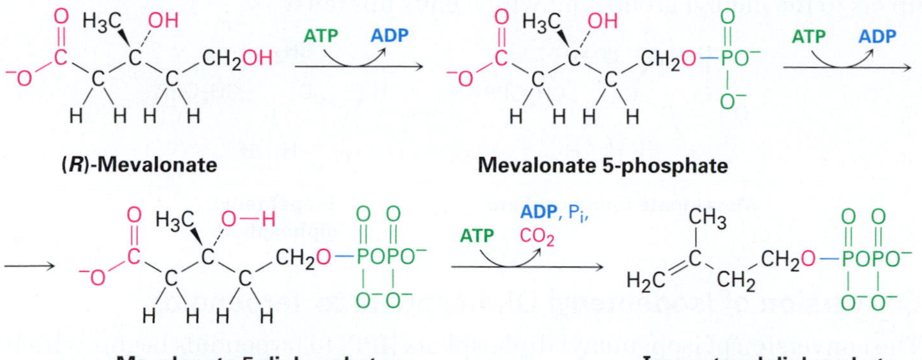

(R)-Mevalonate   **Mevalonate 5-phosphate**

Mevalonate 5-diphosphate   **Isopentenyl diphosphate**

The final decarboxylation of mevalonate 5-diphosphate appears unusual because decarboxylations of acids do not typically occur except in β-keto acids and malonic acids, in which the carboxylate group is two atoms away from an additional carbonyl group. As discussed in Section 17.5, the function of this second carbonyl group is to act as an electron acceptor and stabilize the charge resulting from loss of CO_2. In fact, though, the decarboxylation of a β-keto acid and the decarboxylation of mevalonate 5-diphosphate are closely related.

Catalyzed by mevalonate-5-diphosphate decarboxylase, the substrate is first phosphorylated on the tertiary –OH group by reaction with ATP to give a tertiary phosphate, which undergoes spontaneous dissociation to give a tertiary carbocation. The positive charge then acts as an electron acceptor to facilitate decarboxylation in exactly the same way a β carbonyl group does, giving isopentenyl diphosphate. (In the following structures, the diphosphate group is abbreviated OPP.)

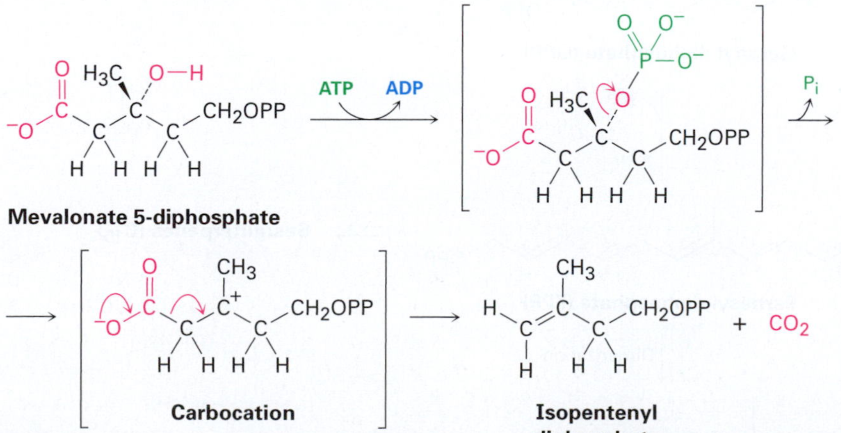

Mevalonate 5-diphosphate

Carbocation   **Isopentenyl diphosphate** $+$ CO_2

Problem 23.10

Studies of the conversion of mevalonate 5-phosphate to isopentenyl diphosphate have shown the following result. Which hydrogen, *pro-R* or *pro-S,* ends up cis to the methyl group, and which ends up trans?

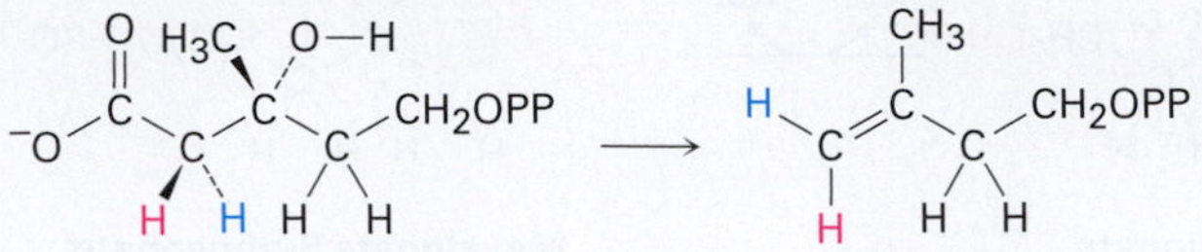

Mevalonate 5-diphosphate   **Isopentenyl diphosphate**

Conversion of Isopentenyl Diphosphate to Terpenoids

The conversion of isopentenyl diphosphate (IPP) to terpenoids begins with its isomerization to dimethylallyl diphosphate, abbreviated DMAPP and formerly called dimethylallyl pyrophosphate. These two C_5 building blocks then combine to give the C_{10} unit geranyl diphosphate (GPP). The corresponding alcohol, geraniol, is itself a fragrant terpenoid that occurs in rose oil.

Further combination of GPP with another IPP gives the C_{15} unit farnesyl diphosphate (FPP), and so on, up to C_{25}. Terpenoids with more than 25 carbons—that is, triterpenoids (C_{30}) and tetraterpenoids (C_{40})—are synthesized by dimerization of C_{15} and C_{20} units, respectively. Triterpenoids and steroids, in particular, arise from dimerization of farnesyl diphosphate to give squalene (Figure 23.10).

FIGURE 23.10 An overview of terpenoid biosynthesis from isopentenyl diphosphate.

Isopentenyl diphosphate (IPP)   **Dimethylallyl diphosphate (DMAPP)**

Geranyl diphosphate (GPP) ⟹ **Monoterpenes (C_{10})**

Farnesyl diphosphate (FPP) ⟹ **Sesquiterpenes (C_{15})**

Squalene ⟹ **Triterpenes (C_{30})**

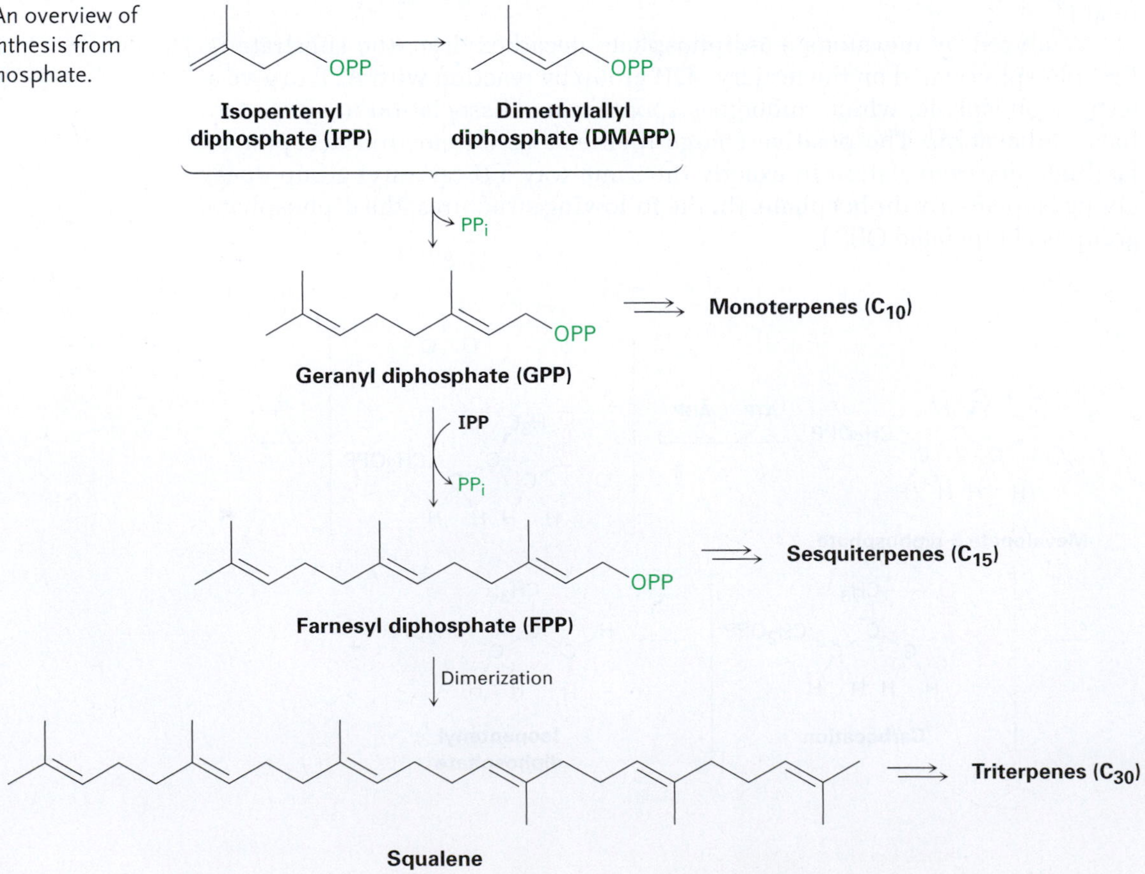

The isomerization of isopentenyl diphosphate to dimethylallyl diphosphate is catalyzed by IPP isomerase and occurs through a carbocation pathway. Protonation of the IPP double bond by a hydrogen-bonded cysteine residue in the enzyme gives a tertiary carbocation intermediate, which is deprotonated by a glutamate residue as base to yield DMAPP. X-ray structural studies on the enzyme show that it holds the substrate in an unusually deep, well-protected pocket to shield the highly reactive carbocation from reaction with solvent or other external substances.

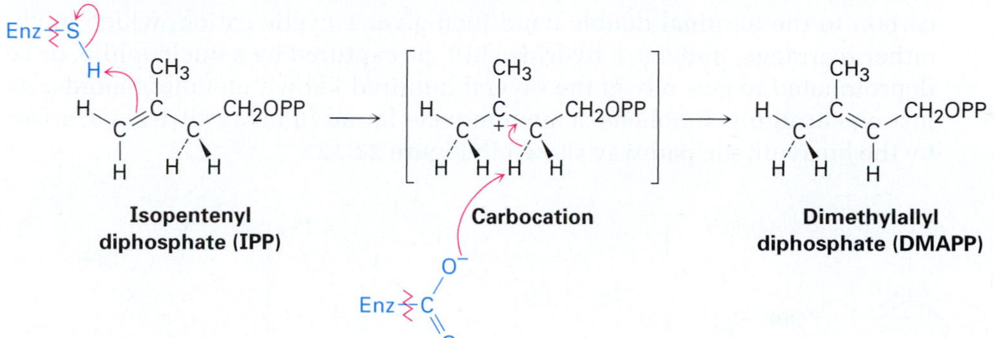

Isopentenyl diphosphate (IPP) → **Carbocation** → **Dimethylallyl diphosphate (DMAPP)**

Both the initial coupling of DMAPP with IPP to give geranyl diphosphate and the subsequent coupling of GPP with a second molecule of IPP to give farnesyl diphosphate are catalyzed by farnesyl diphosphate synthase. The process requires Mg^{2+} ion, and the key step is a nucleophilic substitution reaction in which the double bond of IPP behaves as a nucleophile in displacing a diphosphate ion leaving group (PP_i) on DMAPP. Evidence suggests that the DMAPP develops considerable cationic character and that the reaction probably proceeds by spontaneous dissociation of the allylic diphosphate ion in an S_N1-like pathway (Figure 23.11).

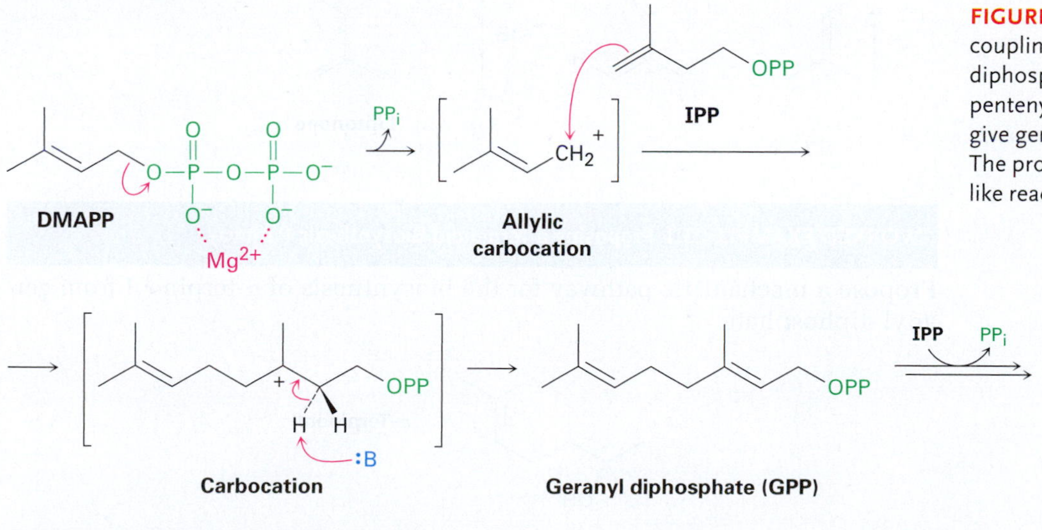

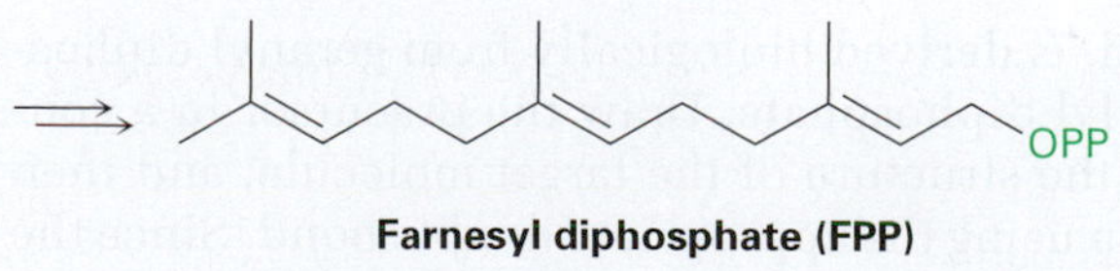

FIGURE 23.11 Mechanism of the coupling reaction of dimethylallyl diphosphate (DMAPP) and isopentenyl diphosphate (IPP), to give geranyl diphosphate (GPP). The process is probably an S_N1-like reaction.

DMAPP — Mg^{2+} — **Allylic carbocation** — **IPP**

Carbocation — **Geranyl diphosphate (GPP)**

Farnesyl diphosphate (FPP)

The further conversion of geranyl diphosphate into monoterpenoids typically involves carbocation intermediates and multistep reaction pathways that are catalyzed by terpene cyclases. Monoterpene cyclases function by first isomerizing geranyl diphosphate to its allylic isomer linalyl diphosphate (LPP), a process that occurs by spontaneous S_N1-like dissociation to an allylic carbocation, followed by recombination. The effect of this isomerization is to convert the C2–C3 double bond of GPP into a single bond, thereby making cyclization possible and allowing E/Z isomerization of the double bond. Further dissociation and cyclization by electrophilic addition of the cationic carbon to the terminal double bond then gives a cyclic cation, which might either rearrange, undergo a hydride shift, be captured by a nucleophile, or be deprotonated to give any of the several hundred known monoterpenoids. As just one example, limonene, a monoterpene found in many citrus oils, arises by the biosynthetic pathway shown in Figure 23.12.

FIGURE 23.12 Mechanism of the formation of the monoterpene limonene from geranyl diphosphate.

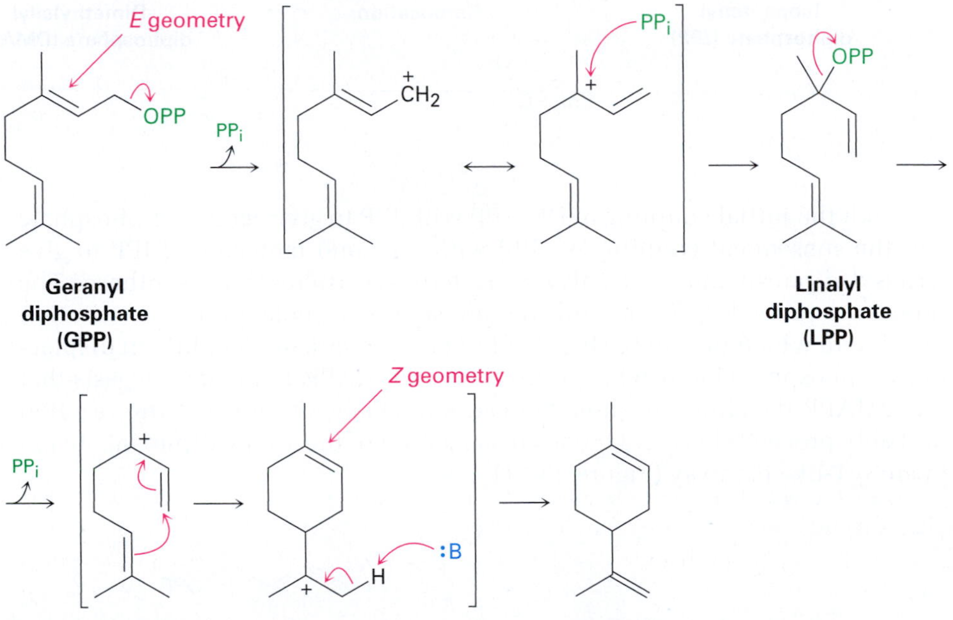

WORKED EXAMPLE 23.1 Proposing a Terpenoid Biosynthesis Pathway

Propose a mechanistic pathway for the biosynthesis of α-terpineol from geranyl diphosphate.

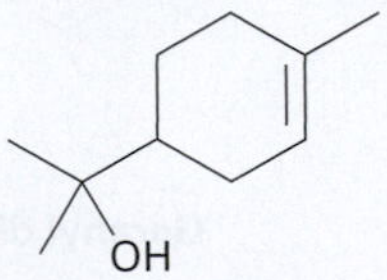

α-**Terpineol**

Strategy

α-Terpineol, a monoterpenoid, is derived biologically from geranyl diphosphate through its isomer linalyl diphosphate. Draw the precursor in a conformation that approximates the structure of the target molecule, and then carry out a cationic cyclization using the appropriate double bond. Since the target is an alcohol, the carbocation resulting from cyclization evidently reacts with water.

Solution

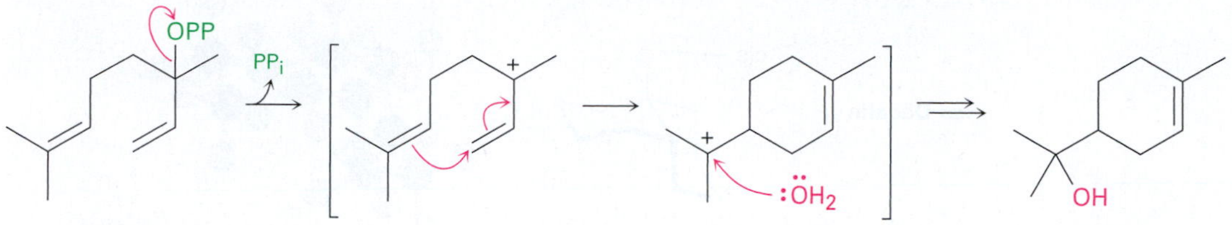

Linalyl diphosphate PP_i **α-Terpineol**

Problem 23.11

Propose mechanistic pathways for the biosynthesis of the following terpenoids:

(a)

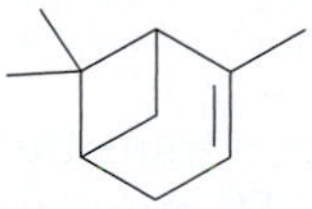

α-Pinene

(b)

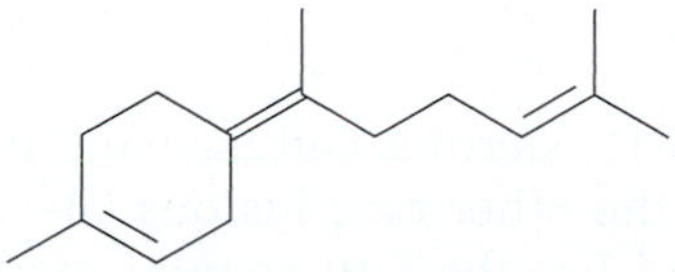

γ-Bisabolene

23.8 Steroids

In addition to fats, phospholipids, and terpenoids, the lipid extracts of plants and animals also contain **steroids**, molecules that are derived from the triterpenoid lanosterol (Figure 23.8) and whose structures are based on a tetracyclic ring system. The four rings are designated A, B, C, and D, beginning at the lower left, and the carbon atoms are numbered beginning in the A ring. The three 6-membered rings (A, B, and C) adopt chair conformations but are prevented by their rigid geometry from undergoing the usual cyclohexane ring-flips (Section 4.6).

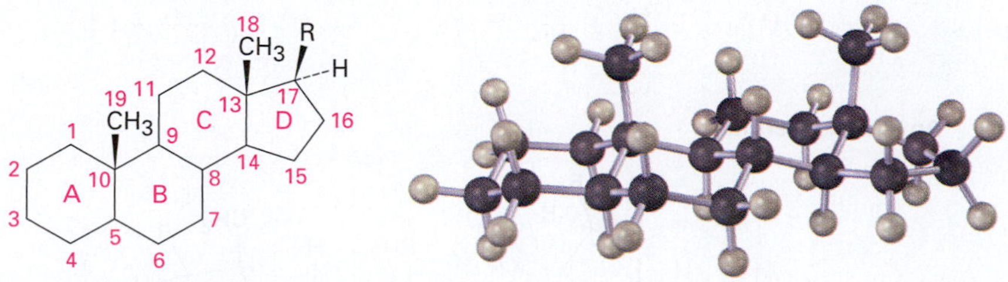

A steroid
(R = various side chains)

Two cyclohexane rings can be joined in either a cis or a trans manner. With cis fusion to give *cis*-decalin, both groups at the ring-junction positions (the *angular* groups) are on the same side of the two rings. With trans fusion to give *trans*-decalin, the groups at the ring junctions are on opposite sides.

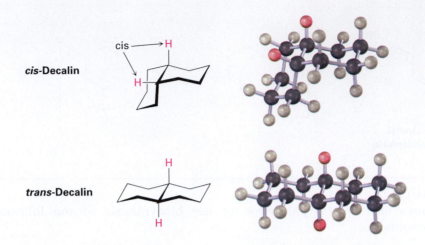

cis-Decalin

cis → H

H

trans-Decalin

H

H

As shown in Figure 23.13, steroids can have either a cis or a trans fusion of the A and B rings, but the other ring fusions (B–C and C–D) are usually trans. An A–B trans steroid has the C19 angular methyl group up, denoted β, and the hydrogen atom at C5 down, denoted α, on opposite sides of the molecule. An A–B cis steroid, by contrast, has both the C19 angular methyl group and the C5 hydrogen atom on the same side *(β)* of the molecule. Both kinds of steroids are relatively long, flat molecules that have their two methyl groups protruding axially above the ring system. The A–B trans steroids are the more common, although A–B cis steroids are found in liver bile.

FIGURE 23.13 Steroid conformations. The three 6-membered rings have chair conformations but are unable to undergo ring-flips. The A and B rings can be either cis-fused or trans-fused.

An A–B trans steroid

CH₃ R
     H
CH₃ H
   H   H
H

CH₃ R
     H
CH₃ H
   H   H
H

An A–B cis steroid

CH₃ R
     H
CH₃ H
   H   H
H

CH₃ R
     H
CH₃ H
H   H   H

Substituent groups on the steroid ring system can be either axial or equatorial. As with simple cyclohexanes (Section 4.7), equatorial substitution is generally more favorable than axial substitution for steric reasons. The hydroxyl group at C3 of cholesterol, for example, has the more stable equatorial orientation. Unlike what happens with simple cyclohexanes, however, steroids are rigid molecules whose geometry prevents cyclohexane ring-flips.

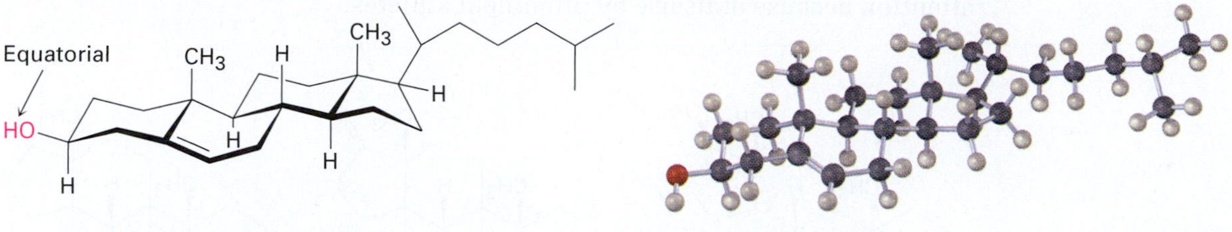

Cholesterol

Problem 23.12

Draw the following molecules in chair conformations, and tell whether the ring substituents are axial or equatorial:

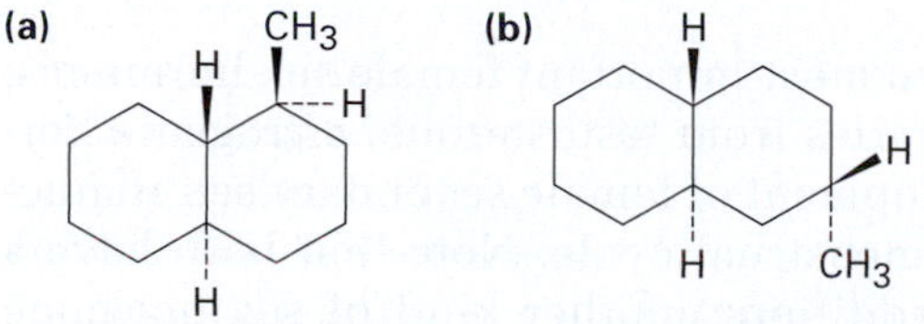

Problem 23.13

Lithocholic acid is an A–B cis steroid found in human bile. Draw lithocholic acid showing chair conformations as in Figure 23.13, and tell whether the hydroxyl group at C3 is axial or equatorial.

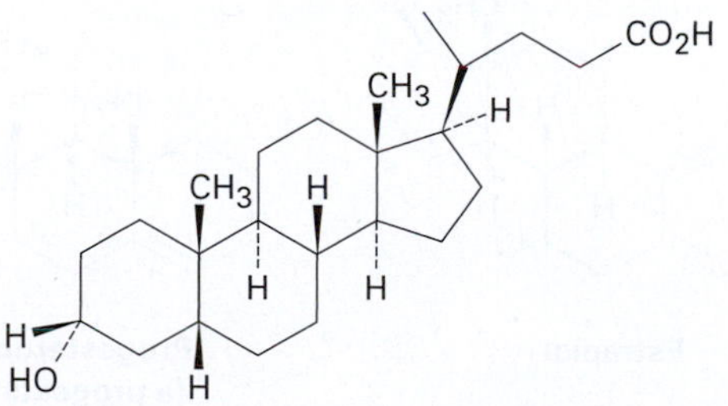

Lithocholic acid

Steroid Hormones

In humans, most steroids function as *hormones,* chemical messengers that are secreted by endocrine glands and carried through the bloodstream to target tissues. There are two main classes of steroid hormones: the *sex hormones,* which control maturation, tissue growth, and reproduction, and the *adreno-cortical hormones,* which regulate a variety of metabolic processes.

SEX HORMONES Testosterone and androsterone are the two most important male sex hormones, or *androgens.* Androgens are responsible for the development of male secondary sex characteristics during puberty and for promoting tissue and muscle growth. Both are synthesized in the testes from cholesterol. Androstenedione is another minor hormone that has received particular attention because of its use by prominent athletes.

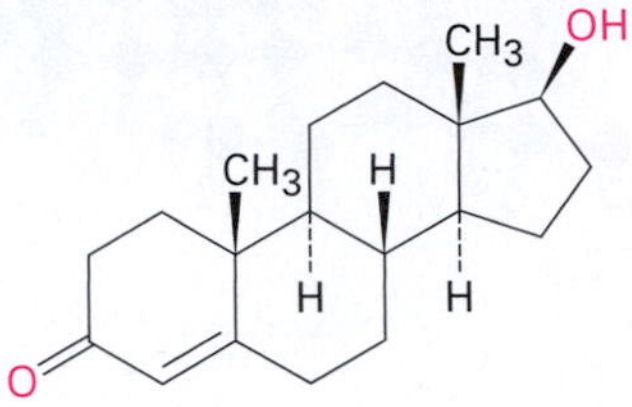

Testosterone 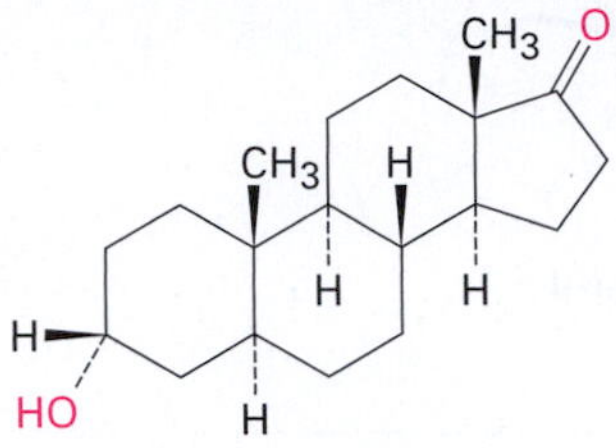 **Androsterone** 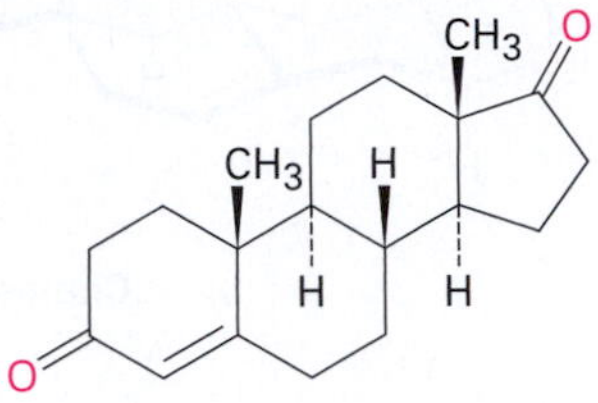 **Androstenedione**

(Androgens)

Estrone and estradiol are the two most important female sex hormones, or *estrogens.* Synthesized in the ovaries from testosterone, estrogenic hormones are responsible for the development of female secondary sex characteristics and for regulation of the menstrual cycle. Note that both have a benzene-like aromatic A ring. In addition, another kind of sex hormone called a *progestin* is essential for preparing the uterus for implantation of a fertilized ovum during pregnancy. Progesterone is the most important progestin.

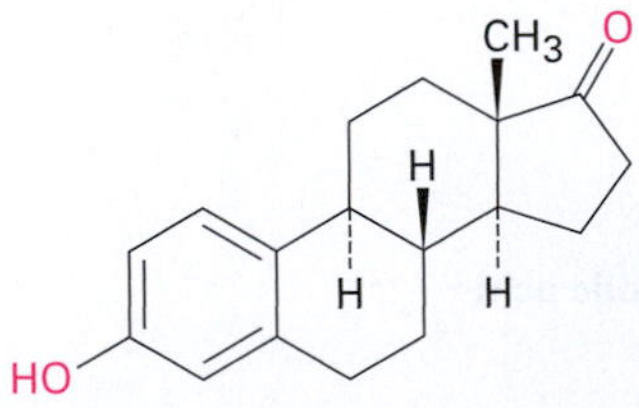

Estrone 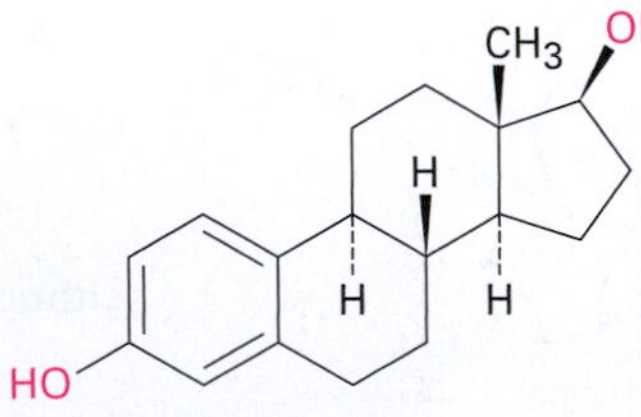 **Estradiol** 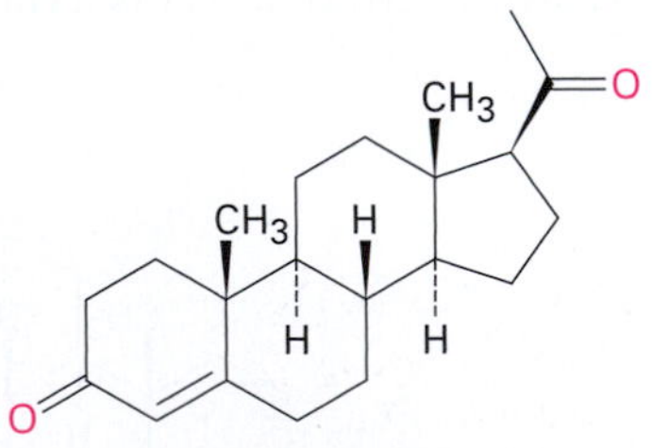 **Progesterone (a progestin)**

(Estrogens)

ADRENOCORTICAL HORMONES Adrenocortical steroids are secreted by the adrenal glands, small organs located near the upper end of each kidney. There are two types of adrenocortical steroids, called *mineralocorticoids* and *glucocorticoids.* Mineralocorticoids, such as aldosterone, control tissue swelling by regulating cellular salt balance between Na$^+$ and K$^+$. Gluco-corticoids, such as hydrocortisone, are involved in the regulation of glucose metabolism and in the control of inflammation. Glucocorticoid ointments are widely used to bring down the swelling from exposure to poison oak or poison ivy.

Aldosterone
(a mineralocorticoid)

Hydrocortisone
(a glucocorticoid)

SYNTHETIC STEROIDS In addition to the many hundreds of steroids isolated from plants and animals, thousands more have been synthesized in pharmaceutical laboratories in a search for new drugs. Among the best-known synthetic steroids are oral contraceptives and anabolic agents. Most birth-control pills are a mixture of two compounds, a synthetic estrogen, such as ethynyl-estradiol, and a synthetic progestin, such as norethindrone. Anabolic steroids, such as methandrostenolone (Dianabol), are synthetic androgens that mimic the tissue-building effects of natural testosterone.

Ethynylestradiol
(a synthetic estrogen)

Norethindrone
(a synthetic progestin)

Methandrostenolone
(Dianabol)

23.9 Biosynthesis of Steroids

Steroids are heavily modified triterpenoids that are biosynthesized in living organisms from farnesyl diphosphate (C_{15}). A reductive dimerization first converts farnesyl diphosphate to the acyclic hydrocarbon squalene (C_{30}), which is converted into lanosterol (Figure 23.14). Further rearrangements and degradations then take place to yield various steroids. The conversion of squalene to lanosterol is among the most intensively studied of all biosynthetic transformations. Starting from an achiral, open-chain polyene, the entire process requires only two enzymes and results in the formation of six carbon–carbon bonds, four rings, and seven chirality centers.

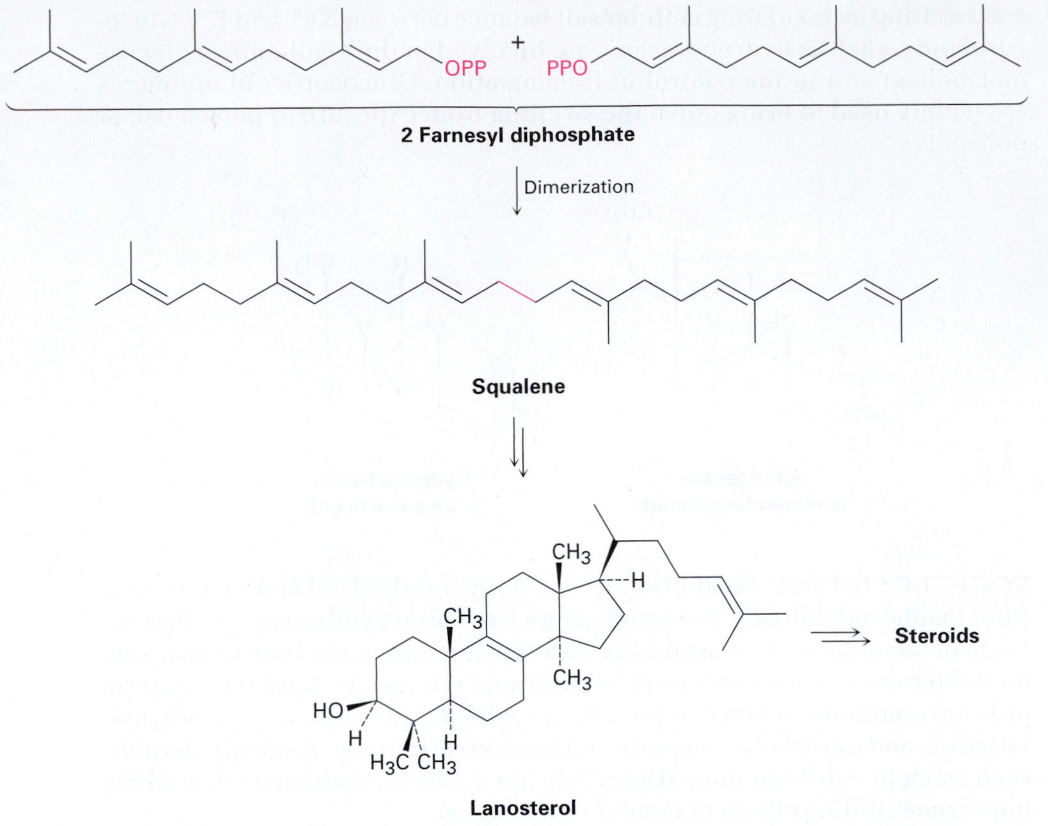

FIGURE 23.14 An overview of steroid biosynthesis from farnesyl diphosphate.

FIGURE 23.14 An overview of steroid biosynthesis from farnesyl diphosphate.

Lanosterol biosynthesis begins with the selective conversion of squalene to its epoxide, (3S)-2,3-oxidosqualene, catalyzed by squalene epoxidase. Molecular O_2 provides the source of the epoxide oxygen atom, and NADPH is required, along with a flavin coenzyme. The proposed mechanism involves reaction of $FADH_2$ with O_2 to produce a flavin hydroperoxide intermediate (ROOH), which transfers an oxygen to squalene in a pathway initiated by nucleophilic attack of the squalene double bond on the terminal hydroperoxide oxygen (Figure 23.15). The flavin alcohol formed as a by-product loses H_2O to give FAD, which is reduced back to $FADH_2$ by NADPH. As noted in Section 8.6, this biological epoxidation mechanism is closely analogous to the mechanism by which peroxyacids (RCO_3H) react with alkenes to give epoxides in the laboratory.

The second part of lanosterol biosynthesis is catalyzed by oxidosqualene:lanosterol cyclase and occurs as shown in Figure 23.16. Squalene is folded by the enzyme into a conformation that aligns the various double bonds for undergoing a cascade of successive intramolecular electrophilic additions, followed by a series of hydride and methyl migrations. Except for the initial epoxide protonation/cyclization, the process is probably stepwise and appears to involve discrete carbocation intermediates that are stabilized by electrostatic interactions with electron-rich aromatic amino acids in the enzyme.

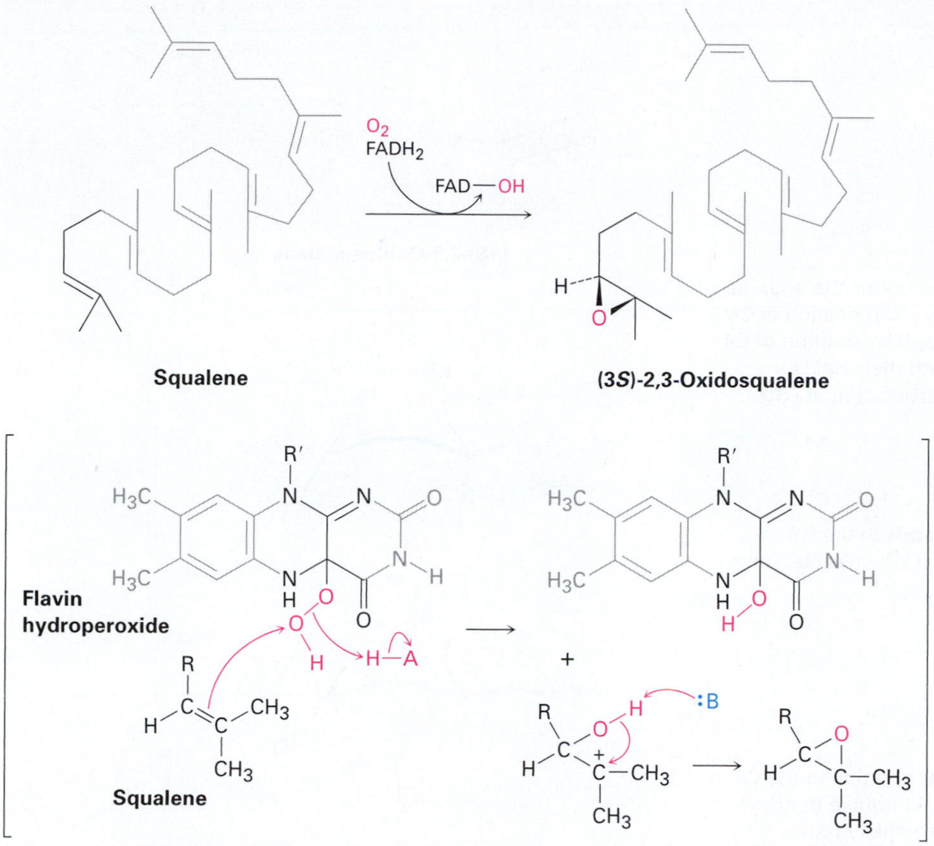

FIGURE 23.15 Proposed mechanism of the oxidation of squalene by flavin hydroperoxide.

STEPS 1–2 OF FIGURE 23.16: EPOXIDE OPENING AND INITIAL CYCLIZATIONS Cyclization begins in step 1 with protonation of the epoxide ring by an aspartic acid residue in the enzyme. Nucleophilic opening of the protonated epoxide by the nearby 5,10 double bond (steroid numbering; Section 23.8) then yields a tertiary carbocation at C10. Further addition of C10 to the 8,9 double bond in step 2 next gives a bicyclic tertiary cation at C8.

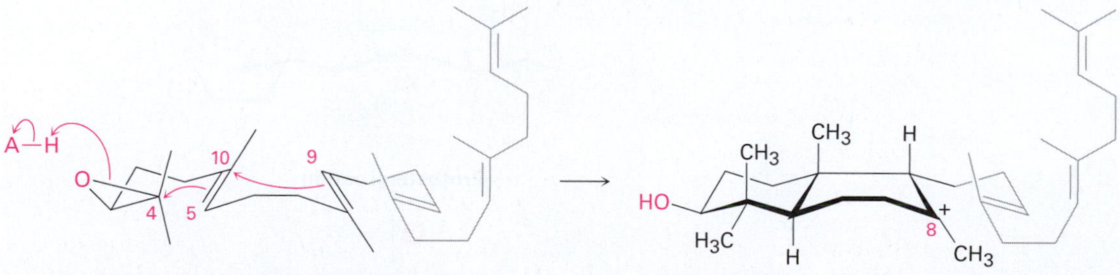

(3*S*)-2,3-Oxidosqualene

STEP 3 OF FIGURE 23.16: THIRD CYCLIZATION The third cationic cyclization is unusual because it occurs with non-Markovnikov regiochemistry and gives a secondary cation at C13 rather than the alternative tertiary cation at C14.

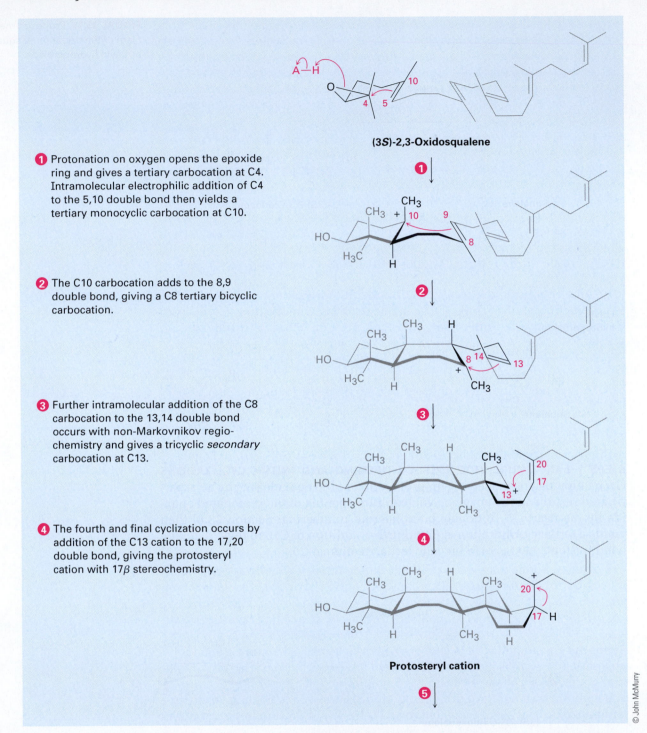

1 Protonation on oxygen opens the epoxide ring and gives a tertiary carbocation at C4. Intramolecular electrophilic addition of C4 to the 5,10 double bond then yields a tertiary monocyclic carbocation at C10.

2 The C10 carbocation adds to the 8,9 double bond, giving a C8 tertiary bicyclic carbocation.

3 Further intramolecular addition of the C8 carbocation to the 13,14 double bond occurs with non-Markovnikov regiochemistry and gives a tricyclic *secondary* carbocation at C13.

4 The fourth and final cyclization occurs by addition of the C13 cation to the 17,20 double bond, giving the protosteryl cation with 17β stereochemistry.

(3*S*)-2,3-Oxidosqualene

Protosteryl cation

© John McMurry

FIGURE 23.16 MECHANISM: Mechanism of the conversion of 2,3-oxidosqualene to lanosterol. Four cationic cyclizations are followed by four rearrangements and a final loss of H⁺ from C9. The steroid numbering system is used for referring to specific positions in the intermediates (Section 23.8). Individual steps are explained in the text.

Protosteryl cation

5 Hydride migration from C17 to C20 occurs, establishing *R* stereochemistry at C20.

6 A second hydride migration takes place, from C13 to C17, establishing the final 17β stereochemistry of the side chain.

7 Methyl migration from C14 to C13 occurs.

8 A second methyl migration occurs, from C8 to C14.

9 Loss of a proton from C9 forms an 8,9 double bond and gives lanosterol.

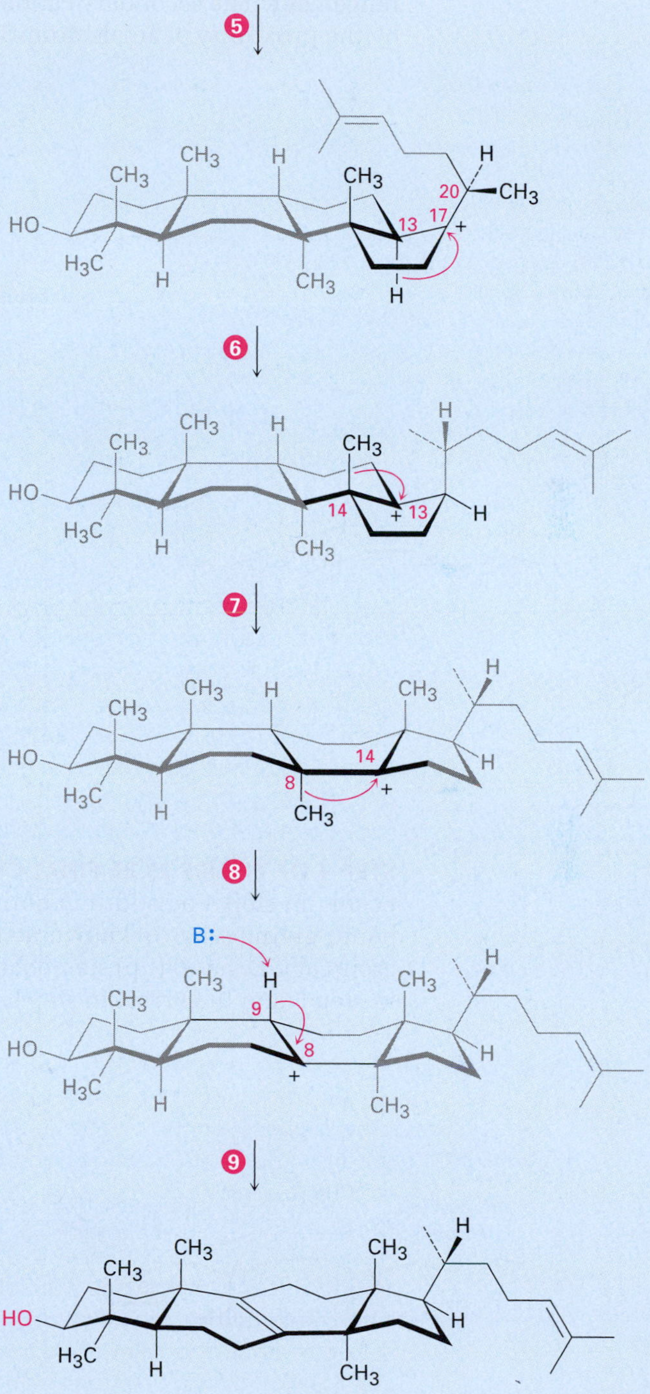

Lanosterol

© John McMurry

FIGURE 23.16 *(continued)*

There is growing evidence, however, that the tertiary carbocation may in fact be formed initially and that the secondary cation arises by subsequent rearrangement. The secondary cation is probably stabilized in the enzyme pocket by the proximity of an electron-rich aromatic ring.

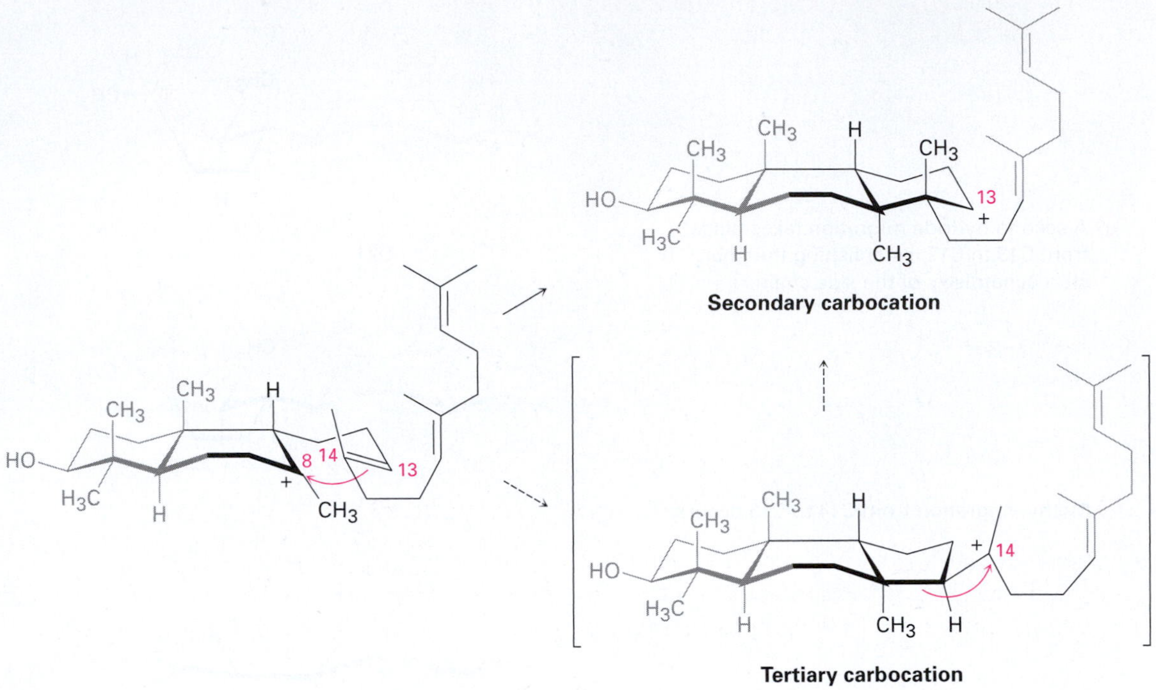

Secondary carbocation

Tertiary carbocation

STEP 4 OF FIGURE 23.16: FINAL CYCLIZATION The fourth and last cyclization occurs in step 4 by addition of the cationic center at C13 to the 17,20 double bond, giving what is known as the *protosteryl* cation. The side-chain alkyl group at C17 has β (up) stereochemistry, although this stereochemistry is lost in step 5 and then reset in step 6.

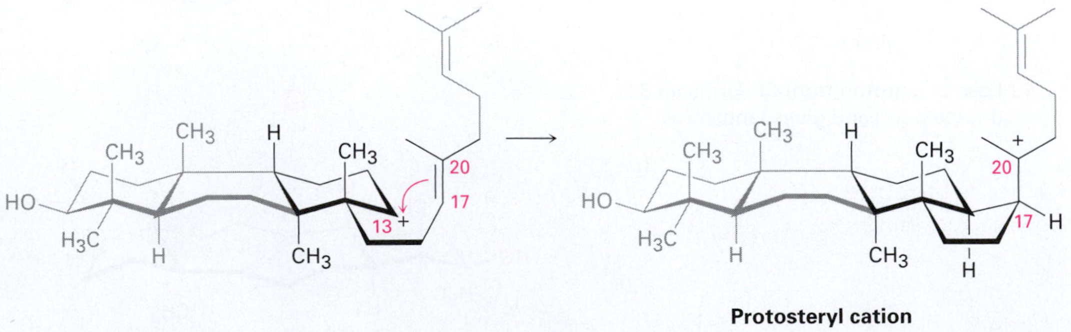

Protosteryl cation

STEPS 5–9 OF FIGURE 23.16: CARBOCATION REARRANGEMENTS Once the tetracyclic carbon skeleton of lanosterol has been formed, a series of carbocation rearrangements occur (Section 7.10). The first rearrangement, hydride migration from C17 to C20, occurs in step 5 and results in establishment of *R* stereochemistry at C20 in the side chain. A second hydride migration then occurs from C13 to C17 on the bottom (α) face of the ring in step 6 and reestablishes the 17β orientation of the side chain. Finally, two methyl group migrations, the first from C14 to C13 on the top (β) face and the second from C8 to C14 on the bottom (α) face, place the positive charge at C8. A basic histidine residue in the enzyme then removes the neighboring β proton from C9 to give lanosterol.

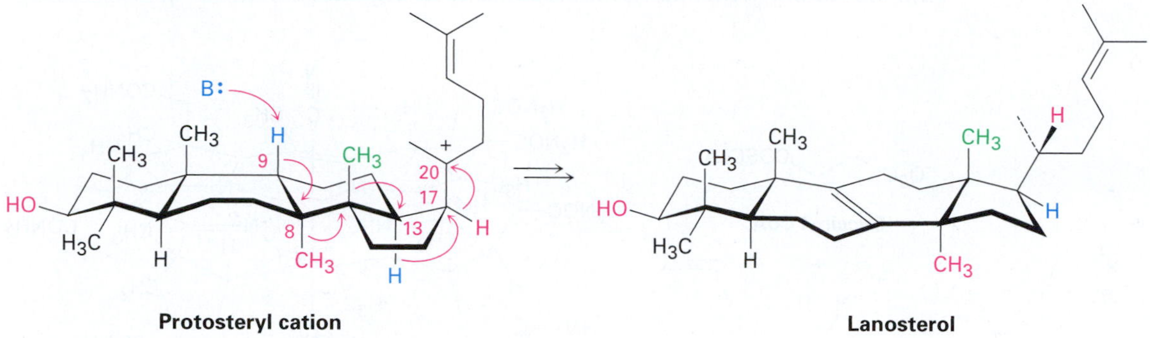

Protosteryl cation   **Lanosterol**

From lanosterol, the pathway for steroid biosynthesis continues on to yield cholesterol. Cholesterol then becomes a branch point, serving as the common precursor from which all other steroids are derived.

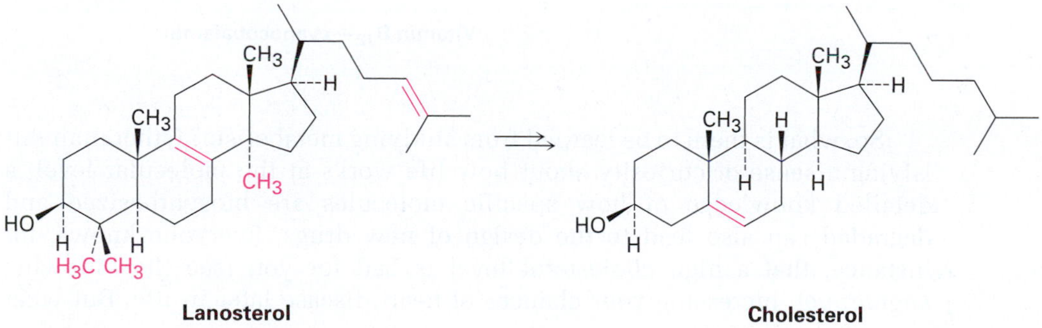

Lanosterol   **Cholesterol**

Problem 23.14
Compare the structures of lanosterol and cholesterol, and catalog the changes needed for the transformation.

23.10 Some Final Comments on Metabolism

In the last several chapters, we've had a brief introduction to metabolism. There are, however, many more pathways we've not mentioned and thousands of biological molecules whose biosynthetic schemes have been explored and elucidated. Take vitamin B_{12} (cyanocobalamin), for instance. This extraordinarily complex molecule is biosynthesized from the simple precursors glycine and succinyl CoA by a route of 60 or so steps (it depends on how you count), all of which have been worked out.

If you were to look at the steps of vitamin B_{12} biosynthesis, you would see exactly the same kinds of reactions we've been seeing throughout the text—nucleophilic substitutions, eliminations, aldol reactions, nucleophilic acyl substitutions, and so forth. There are, of course, some complexities, but the fundamental mechanisms of organic chemistry remain the same, whether in the laboratory with smaller molecules or in organisms with larger molecules.

Vitamin B_{12}—cyanocobalamin

So, what is there to be learned from studying metabolism? Other than satisfying a sense of curiosity about how life works at the molecular level, a detailed knowledge of how specific molecules are biosynthesized and degraded can also lead to the design of new drugs. Everyone knows, for instance, that a high cholesterol level is bad for you (see the following *Lagniappe*), increasing your chances of heart disease later in life. But what can you do about it?

The cholesterol in your body comes from two sources: from your diet and from what you synthesize in your liver. You can change your diet to limit cholesterol intake, but what can you do to limit your own cholesterol synthesis? That's where a detailed chemical knowledge of cholesterol metabolism comes in. We saw in Sections 23.7 and 23.9 that all steroids, including cholesterol, are biosynthesized from the triterpene lanosterol, which in turn comes from acetyl CoA through isopentenyl diphosphate. If you knew all the enzyme-catalyzed mechanisms for all the chemical steps in cholesterol biosynthesis, you might be able to devise a drug that would block one of those steps, thereby short-circuiting the biosynthetic process and controlling the amount of cholesterol produced.

But we do know those mechanisms!

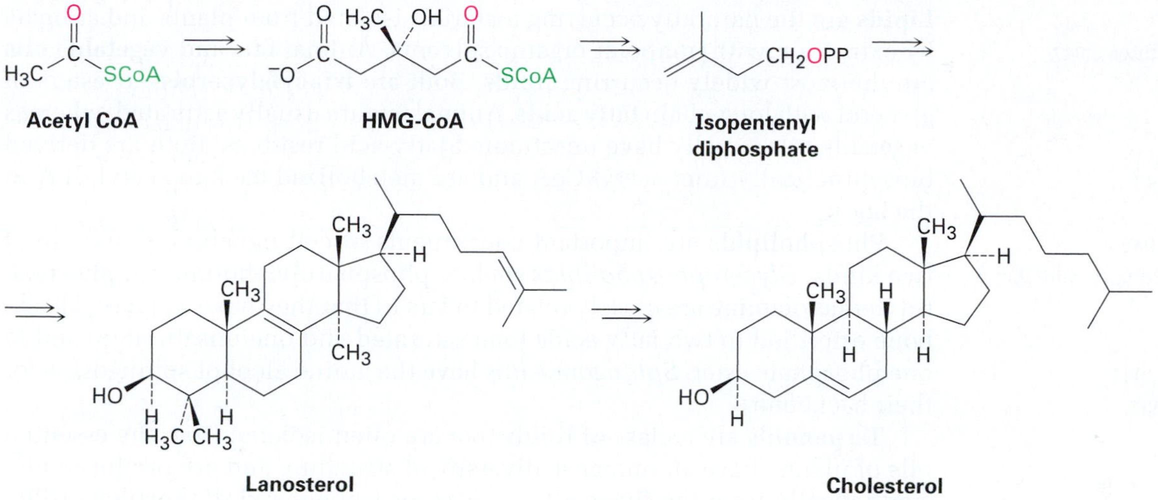

Acetyl CoA   **HMG-CoA**   **Isopentenyl diphosphate**

Lanosterol   **Cholesterol**

Look back in Figure 23.9 on page 959 at the pathway for the biosynthesis of isopentenyl diphosphate from acetyl CoA. The rate-limiting step in that pathway is the third one, the reduction of 3-hydroxy-3-methylglutaryl-CoA to give (*R*)-mevalonate, catalyzed by 3-hydroxy-3-methylglutaryl-CoA reductase (HMG-CoA reductase). By finding a molecule that binds to the active site of the enzyme, the enzyme can be blocked and cholesterol synthesis inhibited. And, as described on page 1 of this text, that is exactly what the so-called *statin* drugs, such as simvastatin, marketed as Zocor, and atorvastatin, marketed as Lipitor, do. These drugs are now among the most widely prescribed pharmaceuticals in the world and are credited with saving the lives of many millions of people. In the 10-year period from 1994 to 2004, the death rate from coronary heart disease in the United States decreased by 33%. Knowing the chemistry of metabolic pathways has certainly paid off in this instance.

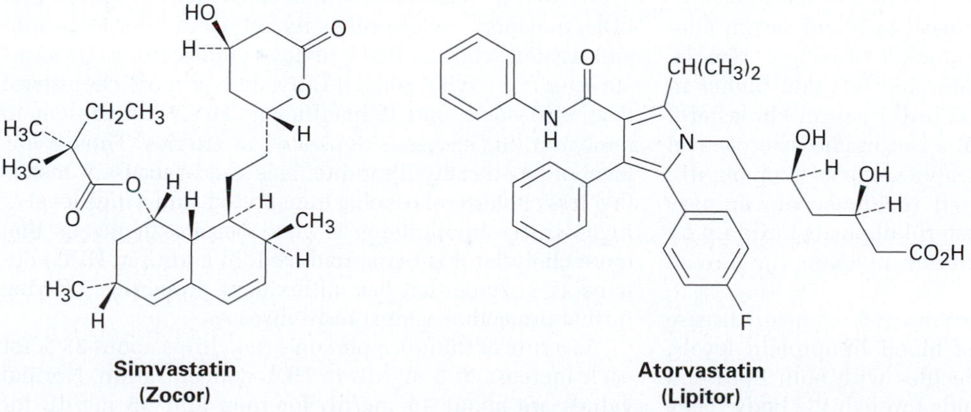

Simvastatin (Zocor)   **Atorvastatin (Lipitor)**

Summary

Key Words

β-oxidation pathway, 947

fat, 937

fatty acid, 937

lipid, 936

lipid bilayer, 943

micelle, 940

phospholipid, 942

polyunsaturated fatty acid, 937

steroid, 965

terpenoid, 956

triacylglycerol, 937

vegetable oil, 937

wax, 937

Lipids are the naturally occurring materials isolated from plants and animals by extraction with nonpolar organic solvents. Animal **fats** and **vegetable oils** are the most widely occurring lipids. Both are **triacylglycerols**—triesters of glycerol with long-chain **fatty acids**. Animal fats are usually saturated, whereas vegetable oils usually have unsaturated fatty-acid residues. Both are derived biosynthetically from acetyl CoA and are metabolized back to acetyl CoA in the body.

Phospholipids are important constituents of cell membranes and are of two kinds. *Glycerophospholipids* such as phosphatidylcholine and phosphatidylethanolamine are closely related to fats in that they have a glycerol backbone esterified to two fatty acids (one saturated and one unsaturated) and to one phosphate ester. *Sphingomyelins* have the amino alcohol sphingosine for their backbone.

Terpenoids are a class of lipids that are often isolated from the essential oils of plants, have an immense diversity of structure, and are produced biosynthetically from the five-carbon precursor isopentenyl diphosphate (IPP). Isopentenyl diphosphate is itself biosynthesized from 3 equivalents of acetate in the mevalonate pathway.

Steroids are plant and animal lipids with a characteristic tetracyclic carbon skeleton. Steroids occur widely in body tissues and have a large variety of physiological activities. They are closely related to terpenoids and arise biosynthetically from the triterpene lanosterol, which itself arises from cationic cyclization of the acyclic hydrocarbon squalene.

Lagniappe

Saturated Fats, Cholesterol, and Heart Disease

We hear a lot these days about the relationships between saturated fats, cholesterol, and heart disease. What are the facts? It's well established that a diet rich in saturated animal fats often leads to an increase in blood serum cholesterol, particularly in sedentary, overweight people. Conversely, a diet lower in saturated fats and higher in polyunsaturated fats leads to a lower serum cholesterol level. Studies have shown that a serum cholesterol level greater than 240 mg/dL (a desirable value is <200 mg/dL) is correlated with an increased incidence of coronary artery disease, in which cholesterol deposits build up on the inner walls of coronary arteries, blocking the flow of blood to the heart muscles.

A better indication of a person's risk of heart disease comes from a measurement of blood lipoprotein levels. *Lipoproteins* are complex molecules with both lipid and protein parts that transport lipids through the body. They can be divided into three types according to density, as shown in Table 23.3. Very-low-density lipoproteins (VLDLs) act primarily as carriers of triglycerides from the intestines to peripheral tissues, whereas low-density lipoproteins (LDLs) and high-density lipoproteins (HDLs) act as carriers of cholesterol to and from the liver. Evidence suggests that LDLs transport cholesterol as its fatty-acid ester *to* peripheral tissues, whereas HDLs remove cholesterol as its stearate ester *from* dying cells. If LDLs deliver more cholesterol than is needed, and if insufficient HDLs are present to remove it, the excess is deposited in arteries. Thus, a *low* level of *low*-density lipoproteins is good because it means that less cholesterol is being transported, and a *high* level of *high*-density lipoproteins is good because it means that more cholesterol is being removed. In addition, HDL contains an enzyme that has antioxidant properties, offering further protection against heart disease.

As a rule of thumb, a person's risk drops about 25% for each increase of 5 mg/dL in HDL concentration. Normal values are about 45 mg/dL for men and 55 mg/dL for women, perhaps explaining why premenopausal women appear to be somewhat less susceptible than men to heart disease.

continued

Lagniappe *continued*

It's hard to resist, but a high intake of saturated animal fat doesn't do much for your cholesterol level. This marbled Kobe beef costs $100/lb.

© Michael Freeman/CORBIS

TABLE 23.3
Serum Lipoproteins

Name	Density (g/mL)	% Lipid	% Protein	Optimal (mg/dL)	Poor (mg/dL)
VLDL	0.940–1.006	90	10	—	—
LDL	1.006–1.063	75	25	<100	>130
HDL	1.063–1.210	60	40	>60	<40

Not surprisingly, the most important factor in gaining high HDL levels is a generally healthful lifestyle. Obesity, smoking, and lack of exercise lead to low HDL levels, whereas regular exercise and a sensible diet lead to high HDL levels. Distance runners and other endurance athletes have HDL levels nearly 50% higher than the general population. Failing that—not everyone wants to run 50 miles per week—diet is also important. Diets high in cold-water fish, like salmon and whitefish, raise HDL and lower blood cholesterol because these fish contain almost entirely polyunsaturated fat. Animal fat from red meat and cooking fats should be minimized because saturated fats and monounsaturated trans fats raise blood cholesterol.

Exercises

VISUALIZING CHEMISTRY

(Problems 23.1–23.14 appear within the chapter.)

23.15 ■ Identify the following fatty acid, and tell whether it is more likely to be found in peanut oil or in red meat:

■ indicates problems that are assignable in Organic OWL.

Go to this book's companion website at **www.cengage.com/chemistry/mcmurry** to explore interactive versions of the Active Figures from this text.

23.16 ■ The following model is that of cholic acid, a constituent of human bile. Locate the three hydroxyl groups, and identify each as axial or equatorial. Is cholic acid an A–B trans steroid or an A–B cis steroid?

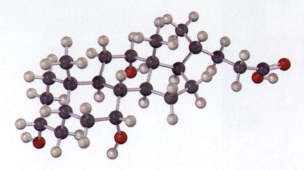

23.17 Propose a biosynthetic pathway for the sesquiterpene helmintho-germacrene from farnesyl diphosphate:

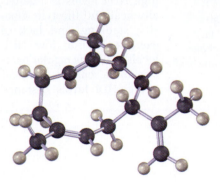

ADDITIONAL PROBLEMS

23.18 Fats can be either optically active or optically inactive, depending on their structure. Draw the structure of an optically active fat that yields 2 equivalents of stearic acid and 1 equivalent of oleic acid on hydrolysis. Draw the structure of an optically inactive fat that yields the same products.

23.19 Spermaceti, a fragrant substance from sperm whales, was much used in cosmetics until it was banned in 1976 to protect the whales from extinction. Chemically, spermaceti is cetyl palmitate, the ester of cetyl alcohol (n-$C_{16}H_{33}OH$) with palmitic acid. Draw its structure.

23.20 The *plasmalogens* are a group of lipids found in nerve and muscle cells. How do plasmalogens differ from fats?

$$CH_2OCH{=}CHR$$
$$|$$
$$\overset{\displaystyle O}{\overset{\displaystyle \|}{CHOCR'}}$$
$$|$$
$$\overset{\displaystyle O}{\overset{\displaystyle \|}{CH_2OCR''}}$$

A plasmalogen

23.21 What products would you obtain from hydrolysis of a plasmalogen (Problem 23.20) with aqueous NaOH? With H_3O^+?

23.22 *Cardiolipins* are a group of lipids found in heart muscles. What products would be formed if all ester bonds, including phosphates, were saponified by treatment with aqueous NaOH?

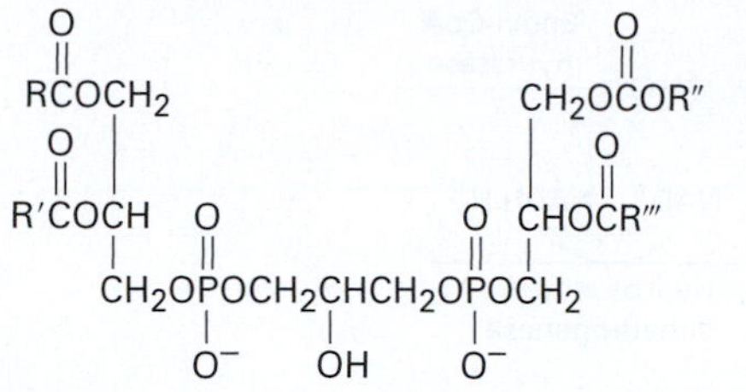

A cardiolipin

23.23 ▪ Show the products you would expect to obtain from reaction of glyceryl trioleate with the following reagents:

(a) Excess Br_2 in CH_2Cl_2

(b) H_2/Pd

(c) $NaOH/H_2O$

(d) $LiAlH_4$, then H_3O^+

(e) CH_3MgBr, then H_3O^+

23.24 ▪ How would you convert oleic acid into the following substances?

(a) Methyl oleate

(b) Methyl stearate

(c) Pentatriacontan-18-one, $CH_3(CH_2)_{16}CO(CH_2)_{16}CH_3$

23.25 ▪ Cold-water fish like salmon are rich in *omega-3* fatty acids, which have a double bond three carbons in from the noncarboxyl end of the chain and have been shown to lower blood cholesterol levels. Draw the structure of eicosa-5,8,11,14,17-pentaenoic acid, a common example. (Eicosane = $C_{20}H_{42}$)

23.26 ▪ *Prostaglandins* are a group of C_{20} lipids found in all human tissues and fluids. They are derived biosynthetically from all-*cis* eicosa-5,8,11,14-tetraenoic acid, known commonly as arachidonic acid. Prostaglandin E_1 is an example. Draw arachidonic acid in such a way that it is structurally similar to PGE_1 in geometry, and identify the two carbons that must form a bond.

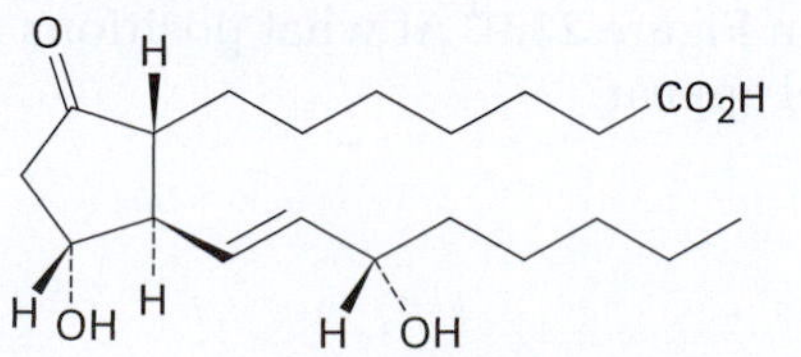

Prostaglandin E_1 (PGE$_1$)

▪ Problems assignable in Organic OWL.

23.27 Show the product of each of the following reactions:

(a)

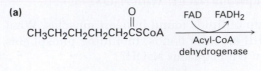

$$CH_3CH_2CH_2CH_2CH_2\overset{O}{\overset{\|}{C}}SCoA \xrightarrow[\substack{\text{Acyl-CoA}\\\text{dehydrogenase}}]{\text{FAD} \quad \text{FADH}_2}$$

(b) Product of **(a)**   + H_2O $\xrightarrow{\substack{\text{Enoyl-CoA}\\\text{hydratase}}}$

(c) Product of **(b)** $\xrightarrow[\substack{\beta\text{-Hydroxyacyl-CoA}\\\text{dehydrogenase}}]{\text{NAD}^+ \quad \text{NADH/H}^+}$

23.28 Draw a Fischer projection of *sn*-glycerol 1-phosphate, and assign *R* or *S* configuration to the chirality center. Do the same for *sn*-glycerol 2,3-diacetate.

23.29 Without proposing an entire biosynthetic pathway, draw the appropriate precursor, either geranyl diphosphate or farnesyl diphosphate, in a conformation that shows a likeness to each of the following terpenoids:

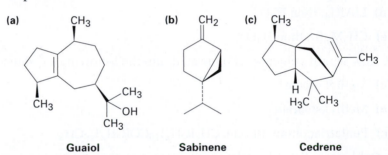

(a)   **(b)**   **(c)**

Guaiol   Sabinene   Cedrene

23.30 ▪ Indicate by asterisks the chirality centers present in each of the terpenoids shown in Problem 23.29. What is the maximum possible number of stereoisomers for each?

23.31 ▪ Assume that the three terpenoids in Problem 23.29 are derived biosynthetically from isopentenyl diphosphate and dimethylallyl diphosphate, each of which was isotopically labeled at the diphosphate-bearing carbon atom (C1). At what positions would the terpenoids be isotopically labeled?

23.32 ▪ Assume that acetyl CoA containing a ^{14}C isotopic label in the carboxyl carbon atom is used as starting material for the biosynthesis of mevalonate, as shown in Figure 23.9. At what positions in mevalonate would the isotopic label appear?

▪ Problems assignable in Organic OWL.

23.33 ■ Assume that acetyl CoA containing a ^{14}C isotopic label in the carboxyl carbon atom is used as starting material and that the mevalonate pathway is followed. Identify the positions in α-cadinol where the label would appear.

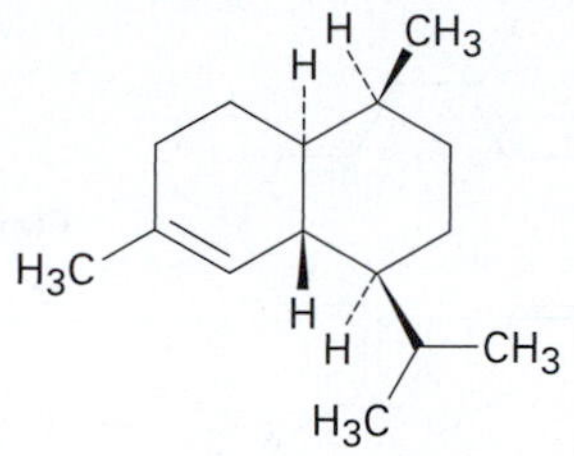

α-Cadinol

23.34 Assume that acetyl CoA containing a ^{14}C isotopic label in the carboxyl carbon atom is used as starting material and that the mevalonate pathway is followed. Identify the positions in squalene where the label would appear.

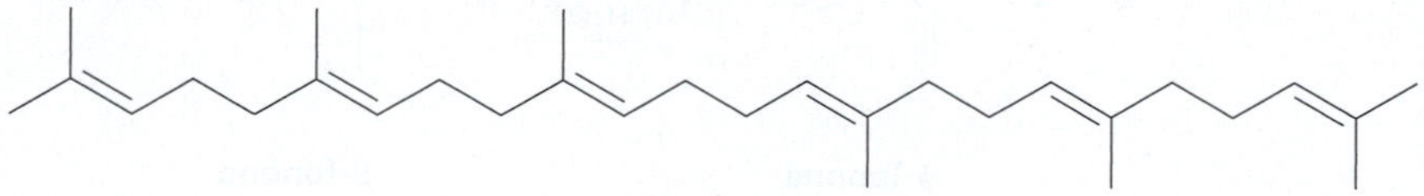

Squalene

23.35 Assume that acetyl CoA containing a ^{14}C isotopic label in the carboxyl carbon atom is used as starting material and that the mevalonate pathway is followed. Identify the positions in lanosterol where the label would appear.

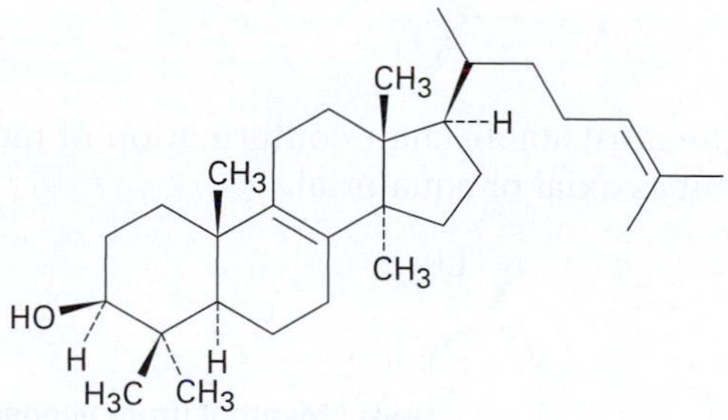

Lanosterol

23.36 Propose a mechanistic pathway for the biosynthesis of caryophyllene, a substance found in clove oil.

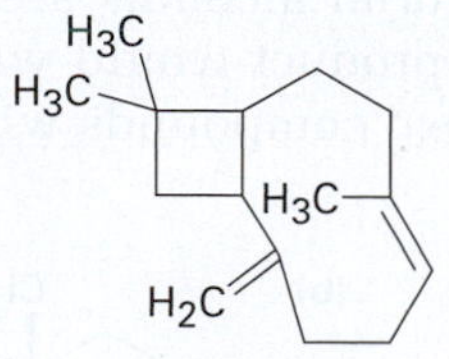

Caryophyllene

■ Problems assignable in Organic OWL.

23.37 Flexibilene, a compound isolated from marine coral, is the only known terpenoid to contain a 15-membered ring. What is the structure of the acyclic biosynthetic precursor of flexibilene? Show the mechanistic pathway for the biosynthesis.

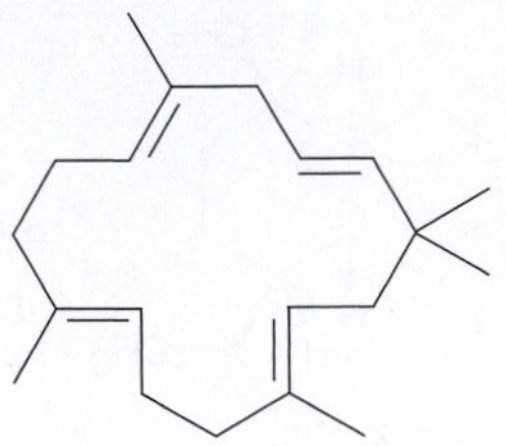

Flexibilene

23.38 ▪ Suggest a mechanism by which ψ-ionone is transformed into β-ionone on treatment with acid.

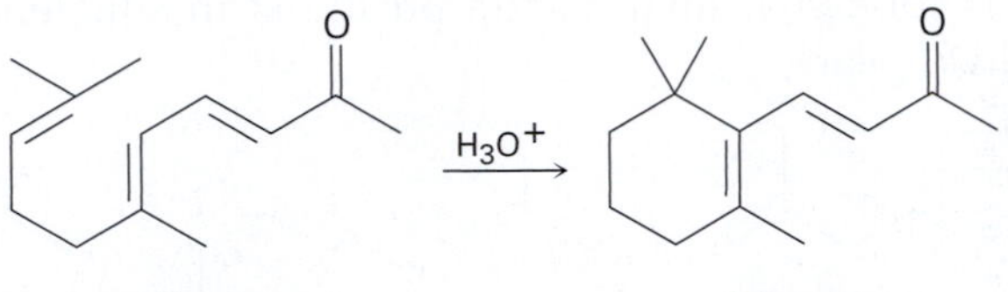

H_3O^+

ψ-Ionone   β-Ionone

23.39 ▪ Draw the most stable chair conformation of dihydrocarvone.

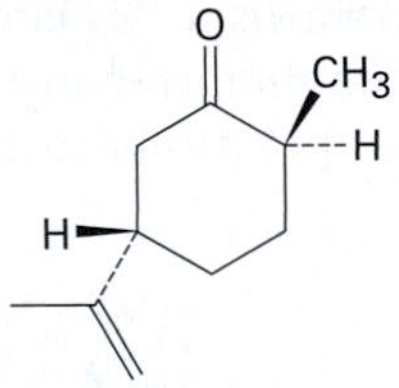

Dihydrocarvone

23.40 ▪ Draw the most stable chair conformation of menthol, and label each substituent as axial or equatorial.

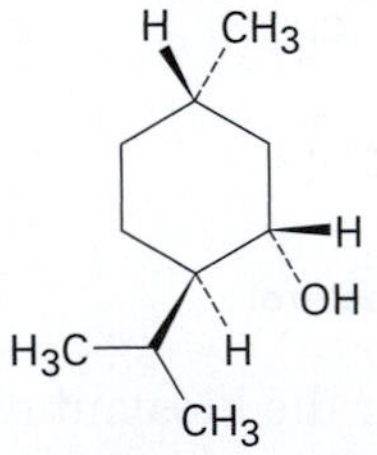

Menthol (from peppermint oil)

23.41 ▪ As a general rule, equatorial alcohols are esterified more readily than axial alcohols. What product would you expect to obtain from reaction of the following two compounds with 1 equivalent of acetic anhydride?

(a)

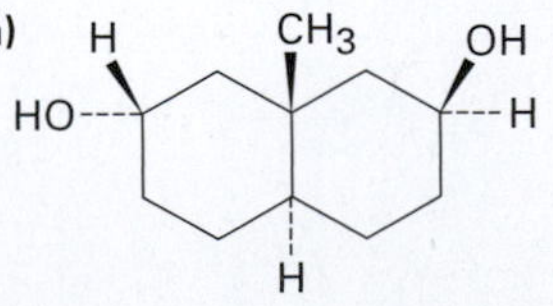

(b)

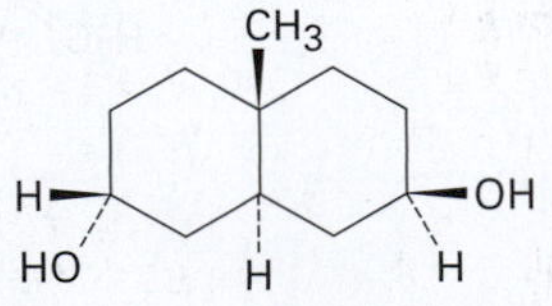

▪ Problems assignable in Organic OWL.

23.42 ▪ Propose a mechanistic pathway for the biosynthesis of isoborneol. A carbocation rearrangement is needed at one point in the scheme.

Isoborneol

23.43 ▪ Isoborneol (Problem 23.42) is converted into camphene on treatment with dilute sulfuric acid. Propose a mechanism for the reaction, which involves a carbocation rearrangement.

Isoborneol   **Camphene**

23.44 Eleostearic acid, $C_{18}H_{30}O_2$, is a rare fatty acid found in the tung oil used for finishing furniture. On ozonolysis followed by treatment with zinc, eleostearic acid furnishes one part pentanal, two parts glyoxal (OHC—CHO), and one part 9-oxononanoic acid [OHC(CH$_2$)$_7$CO$_2$H]. What is the structure of eleostearic acid?

(Note that alkenes undergo ozonolysis followed by treatment with zinc to give carbonyl compounds in which each of the former C=C carbons becomes a C=O carbon.)

23.45 Diterpenoids are derived biosynthetically from geranylgeranyl diphosphate (GGPP), which is itself biosynthesized by reaction of farnesyl diphosphate with isopentenyl diphosphate. Show the structure of GGPP, and propose a mechanism for its biosynthesis from FPP and IPP.

23.46 Cembrene, $C_{20}H_{32}$, is a diterpene hydrocarbon isolated from pine resin. Cembrene has a UV absorption at 245 nm, but dihydrocembrene ($C_{20}H_{34}$), the product of hydrogenation with 1 equivalent H$_2$, has no UV absorption. On exhaustive hydrogenation, 4 equivalents H$_2$ react and octahydrocembrene, $C_{20}H_{40}$, is produced. On ozonolysis of cembrene, followed by treatment of the ozonide with zinc, four carbonyl-containing products are obtained:

$$CH_3CCH_2CH_2CH \; + \; CH_3CCHO \; + \; HCCH_2CH \; + \; CH_3CCH_2CH_2CHCHCH_3$$

Propose a structure for cembrene that is consistent with its formation from geranylgeranyl diphosphate (Problem 23.45).

▪ Problems assignable in Organic OWL.

23.47 α-Fenchone is a pleasant-smelling terpenoid isolated from oil of lavender. Propose a pathway for the formation of α-fenchone from geranyl diphosphate. A carbocation rearrangement is required.

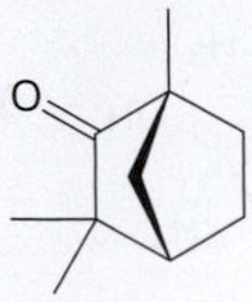

α-Fenchone

23.48 Propose a mechanism for the biosynthesis of the sesquiterpene trichodiene from farnesyl diphosphate. The process involves cyclization to give an intermediate secondary carbocation, followed by several carbocation rearrangements.

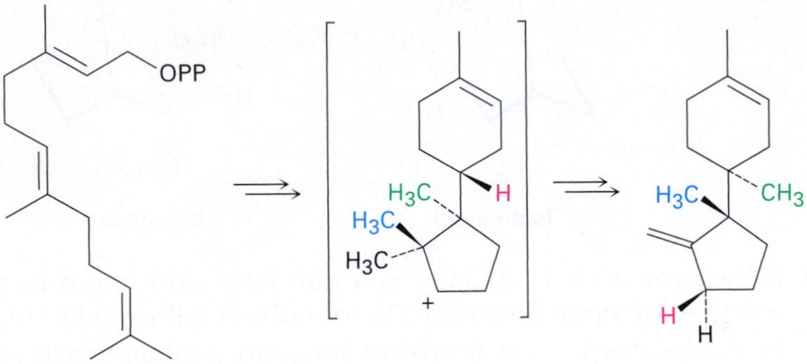

Farnesyl
diphosphate (FPP)

Trichodiene

24 Biomolecules: Nucleic Acids and Their Metabolism

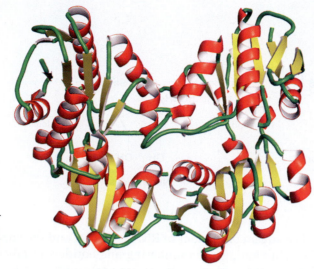

Phosphoribosyl-diphosphate synthetase catalyzes the phosphorylation of ribose 5-phosphate during the biosynthesis of pyrimidine nucleotides.

The nucleic acids, **deoxyribonucleic acid (DNA)** and **ribonucleic acid (RNA)**, are the chemical carriers of a cell's genetic information. Coded in a cell's DNA is all the information that determines the nature of the cell, controls the cell's growth and division, and directs biosynthesis of the enzymes and other proteins required for cellular functions.

In addition to nucleic acids themselves, nucleic acid derivatives such as ATP are involved as phosphorylating agents in many biochemical pathways, and several important coenzymes, including NAD^+, FAD, and coenzyme A, have nucleic acid components.

WHY THIS CHAPTER?

Nucleic acids are the last of the four major classes of biomolecules we'll consider. So much has been written and spoken about DNA in the media that the basics of DNA replication and transcription are probably known to you. Thus, we'll move fairly quickly though the fundamentals and then focus more closely on the chemical details of DNA sequencing, synthesis, and metabolism.

24.1 Nucleotides and Nucleic Acids

Just as proteins are biopolymers made of amino acids, nucleic acids are biopolymers made of **nucleotides** joined together to form a long chain. Each nucleotide is composed of a **nucleoside** bonded to a phosphate group, and each nucleoside is composed of an aldopentose sugar linked through its

CONTENTS

Online homework for this chapter can be assigned in Organic OWL.

anomeric carbon to the nitrogen atom of a heterocyclic purine or pyrimidine base.

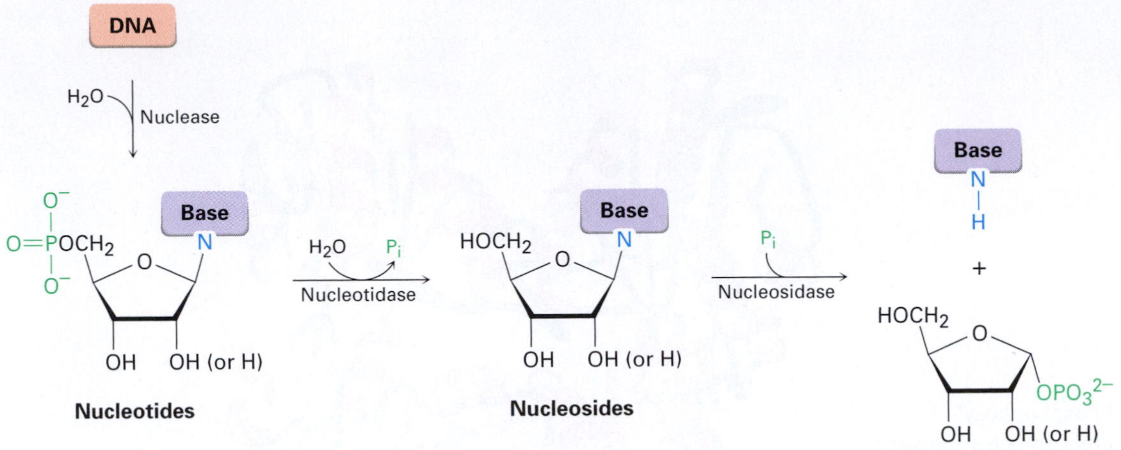

The sugar component in RNA is ribose, and the sugar in DNA is 2'-deoxyribose. (In naming and numbering nucleotides, numbers with a prime superscript refer to positions on the sugar, and numbers without a prime superscript refer to positions on the heterocyclic base. Thus, the prefix 2'-deoxy indicates that oxygen is missing from C2' of ribose.) DNA contains four different amine bases, two substituted purines (adenine and guanine) and two substituted pyrimidines (cytosine and thymine). Adenine, guanine, and cytosine also occur in RNA, but thymine is replaced in RNA by a closely related pyrimidine base called uracil.

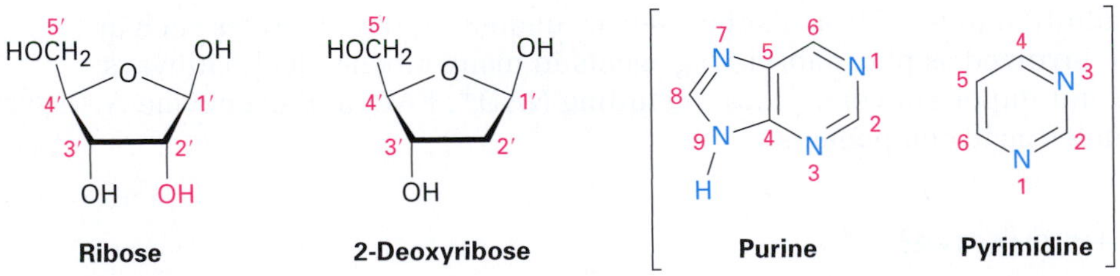

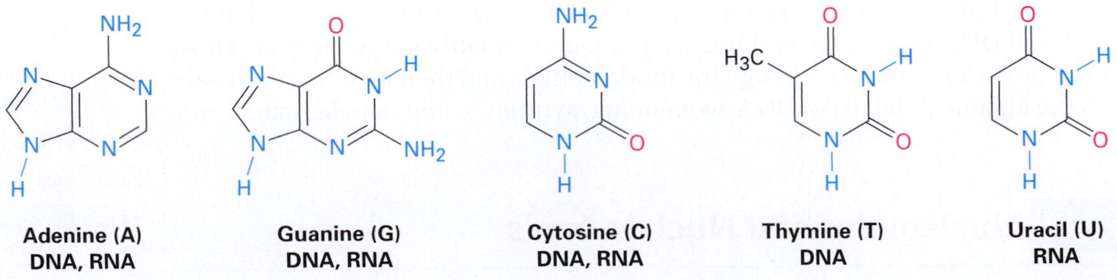

The structures of the four deoxyribonucleotides and the four ribonucleotides are shown in Figure 24.1. Although similar chemically, DNA and RNA differ dramatically in size. Molecules of DNA are enormous, containing as many as 245 million nucleotides and having molecular weights as high as 75 billion. Molecules of RNA, by contrast, are much smaller, containing as few as 21 nucleotides and having molecular weights as low as 7000.

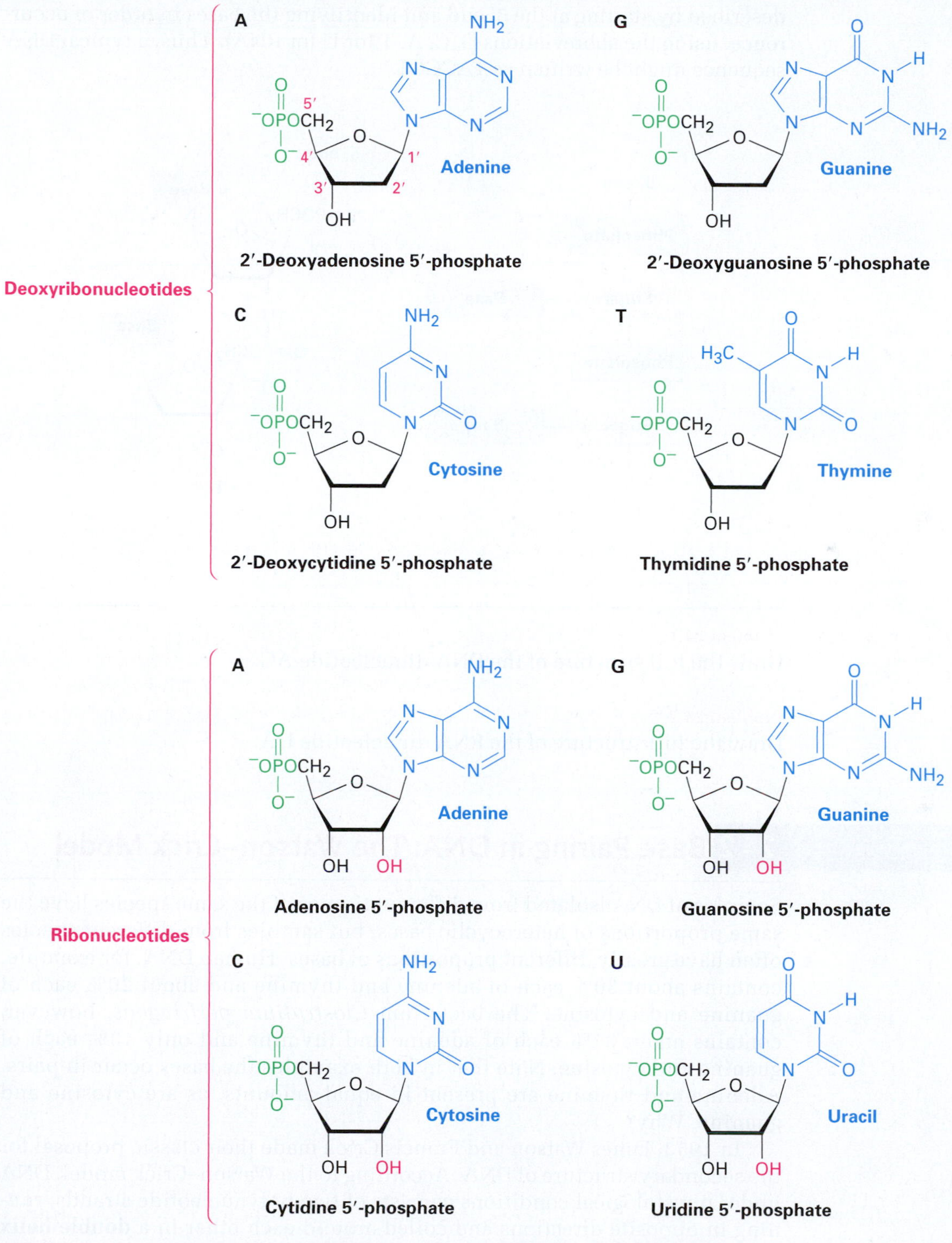

FIGURE 24.1 Structures of the four deoxyribonucleotides and the four ribonucleotides.

Nucleotides are linked together in DNA and RNA by *phosphodiester* bonds [RO—(PO$_2$$^-$)—OR′] between phosphate, the 5′ hydroxyl group on one nucleoside, and the 3′-hydroxyl group on another nucleoside. One end of the nucleic acid polymer has a free hydroxyl at C3′ (the **3′ end**), and the other end has a phosphate at C5′ (the **5′ end**). The sequence of nucleotides in a chain is described by starting at the 5′ end and identifying the bases in order of occurrence, using the abbreviations G, C, A, T (or U for RNA). Thus, a typical DNA sequence might be written as TAGGCT.

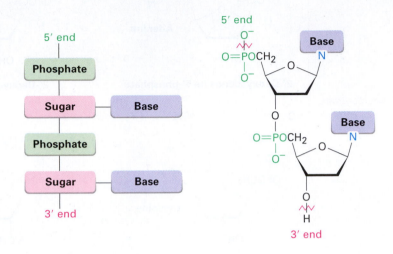

Problem 24.1

Draw the full structure of the DNA dinucleotide AG.

Problem 24.2

Draw the full structure of the RNA dinucleotide UA.

24.2 Base Pairing in DNA: The Watson–Crick Model

Samples of DNA isolated from different tissues of the same species have the same proportions of heterocyclic bases, but samples from different species often have greatly different proportions of bases. Human DNA, for example, contains about 30% each of adenine and thymine and about 20% each of guanine and cytosine. The bacterium *Clostridium perfringens*, however, contains about 37% each of adenine and thymine and only 13% each of guanine and cytosine. Note that in both examples the bases occur in pairs. Adenine and thymine are present in equal amounts, as are cytosine and guanine. Why?

In 1953, James Watson and Francis Crick made their classic proposal for the secondary structure of DNA. According to the Watson–Crick model, DNA under physiological conditions consists of two polynucleotide strands, running in opposite directions and coiled around each other in a **double helix** like the handrails on a spiral staircase. The two strands are complementary rather than identical and are held together by hydrogen bonds between specific pairs of bases, A with T and C with G. That is, whenever an A base occurs in one strand, a T base occurs opposite it in the other strand; when a C base occurs in one, a G occurs in the other (Figure 24.2). This complementary base pairing thus explains why A and T are always found in equal amounts, as are G and C.

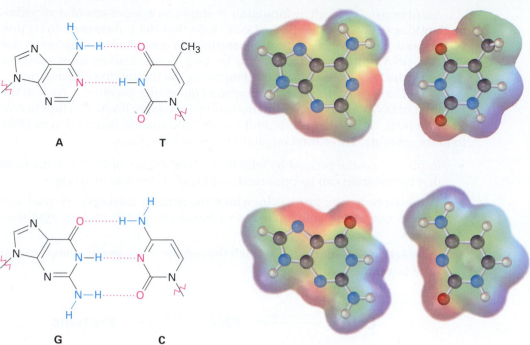

ACTIVE FIGURE 24.2
Hydrogen-bonding between base pairs in the DNA double helix. Electrostatic potential maps show that the faces of the bases are relatively neutral (green), while the edges have positive (blue) and negative (red) regions. Pairing G with C and A with T brings together oppositely charged regions. **Go to this book's student companion site at** www.cengage.com/ chemistry/mcmurry **to explore an interactive version of this figure.**

A full turn of the DNA double helix is shown in Figure 24.3. The helix is 20 Å wide, there are 10 base pairs per turn, and each turn is 34 Å in length. Notice in Figure 24.3 that the two strands of the double helix coil in such a way that two kinds of "grooves" result, a *major groove* 12 Å wide and a *minor groove* 6 Å wide. The major groove is slightly deeper than the minor groove, and both are lined by hydrogen bond donors and acceptors. As a result, a variety of flat, polycyclic aromatic molecules are able to slip sideways, or *intercalate,* between the stacked bases. Many cancer-causing and cancer-preventing agents function by interacting with DNA in this way.

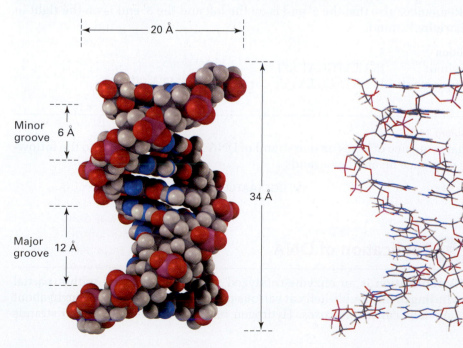

ACTIVE FIGURE 24.3 A turn of the DNA double helix in both space-filling and wire-frame formats. The sugar–phosphate backbone runs along the outside of the helix, and the amine bases hydrogen bond to one another on the inside. Both major and minor grooves are visible. **Go to this book's student companion site at** www.cengage.com/chemistry/ mcmurry **to explore an interactive version of this figure.**

An organism's genetic information is stored as a sequence of deoxyribonucleotides strung together in the DNA chain. For the information to be preserved and passed on to future generations, a mechanism must exist for copying DNA. For the information to be used, a mechanism must exist for decoding the DNA message and implementing the instructions it contains.

What Crick called the "central dogma of molecular genetics" says that the function of DNA is to store information and pass it on to RNA. The function of RNA, in turn, is to read, decode, and use the information received from DNA to make proteins. Three fundamental processes take place:

- **Replication**—the process by which identical copies of DNA are made so that information can be preserved and handed down to offspring.

- **Transcription**—the process by which the genetic messages are read and carried out of the cell nucleus to ribosomes, where protein synthesis occurs.

- **Translation**—the process by which the genetic messages are decoded and used to synthesize proteins.

| WORKED EXAMPLE 24.1 | **Predicting the Complementary Base Sequence in Double-Stranded DNA** |

What sequence of bases on one strand of DNA is complementary to the sequence TATGCAT on another strand?

Strategy

Remember that A and G form complementary pairs with T and C, respectively, and then go through the sequence replacing A by T, G by C, T by A, and C by G. Remember also that the 5′ end is on the left and the 3′ end is on the right in the original strand.

Solution

Original: (5′) TATGCAT (3′)
Complement: (3′) ATACGTA (5′)   or   (5′) ATGCATA (3′)

Problem 24.3

What sequence of bases on one strand of DNA is complementary to the following sequence on another strand?

(5′) GGCTAATCCGT (3′)

24.3 Replication of DNA

DNA **replication** is an enzyme-catalyzed process that begins with a partial unwinding of the double helix at various points along the chain, brought about by enzymes called *helicases*. Hydrogen bonds are broken, the two strands separate to form a "bubble," and bases are exposed. New nucleotides then line up on each strand in a complementary manner, A to T and G to C, and two new strands begin to grow from the ends of the bubble, called the *replication forks*. Each new strand is complementary to its old template strand, so two identical DNA double helices are produced (Figure 24.4). Because each of the new DNA molecules contains one old strand and one new strand, the process is described as *semiconservative replication*.

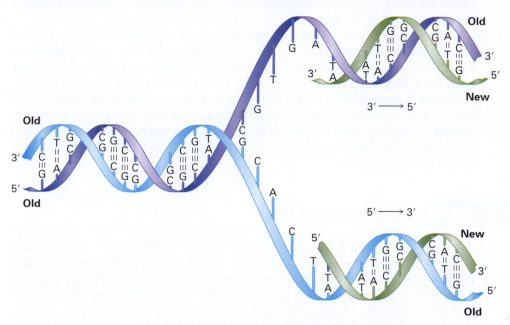

FIGURE 24.4 A representation of semiconservative DNA replication. The original double-stranded DNA partially unwinds, bases are exposed, nucleotides line up on each strand in a complementary manner, and two new strands begin to grow. Both strands are synthesized in the same 5′ → 3′ direction, one continuously and one in fragments.

Addition of nucleotides to a growing chain takes place in the 5′ → 3′ direction and is catalyzed by DNA polymerase. The key step is the addition of a nucleoside 5′-triphosphate to the free 3′-hydroxyl group of the growing chain, with loss of a diphosphate leaving group.

Because both new DNA strands are synthesized in the $5' \rightarrow 3'$ direction, they can't be made in exactly the same way. One new strand must have its 3' end nearer a replication fork, while the other new strand has its 5' end nearer the replication fork. What happens is that the complement of the original $5' \rightarrow 3'$ strand is synthesized continuously in a single piece to give a newly synthesized copy called the *leading strand,* while the complement of the original $3' \rightarrow 5'$ strand is synthesized discontinuously in small pieces called *Okazaki fragments* that are subsequently linked by DNA ligases to form the *lagging strand.*

The magnitude of the replication process is staggering. The nucleus of every human cell contains 2 copies of 22 chromosomes plus an additional 2 sex chromosomes, for a total of 46. Each chromosome consists of one very large DNA molecule, and the sum of the DNA in each of the two sets of chromosomes is estimated to be 3.0 billion base pairs, or 6.0 billion nucleotides. Despite the size of these enormous molecules, their base sequence is faithfully copied during replication. The entire copying process takes only a few hours and, after proofreading and repair, an error gets through only about once each 10 to 100 billion bases.

24.4 Transcription of DNA

As noted previously, RNA is structurally similar to DNA but contains ribose rather than deoxyribose and uracil rather than thymine. There are three major kinds of RNA, each of which serves a specific purpose. In addition, there are a number of small RNAs that appear to control a wide variety of important cellular functions. All RNA molecules are much smaller than DNA, and all remain single-stranded rather than double-stranded.

- **Messenger RNA (mRNA)** carries genetic messages from DNA to ribosomes, small granular particles in the cytoplasm of a cell where protein synthesis takes place.

- **Ribosomal RNA (rRNA)** complexed with protein provides the physical makeup of the ribosomes.

- **Transfer RNA (tRNA)** transports amino acids to the ribosomes, where they are joined together to make proteins.

- **Small RNAs**, also called *functional RNAs,* have a variety of functions within the cell, including silencing transcription and catalyzing chemical modifications of other RNA molecules.

The genetic information in DNA is contained in segments called *genes,* each of which consists of a specific nucleotide sequence that encodes a specific protein. The conversion of that information from DNA into proteins begins in the nucleus of cells with the synthesis of mRNA by **transcription** of DNA. In bacteria, the process begins when RNA polymerase recognizes and binds to a *promoter sequence* on DNA, typically consisting of around 40 base pairs located upstream (5') of the transcription start site. Within the promoter are two hexameric *consensus sequences,* one located 10 base pairs upstream of the start and the second located 35 base pairs upstream.

Following formation of the polymerase–promoter complex, several turns of the DNA double helix unwind, forming a "bubble" and exposing 14 or so base pairs of the two strands. Appropriate ribonucleotides then line up by hydrogen-bonding to their complementary bases on DNA, bond formation occurs in the 5′ → 3′ direction, the RNA polymerase moves along the DNA chain, and the growing RNA molecule unwinds from DNA (Figure 24.5). At any one time, about 12 base pairs of the growing RNA remain hydrogen-bonded to the DNA template.

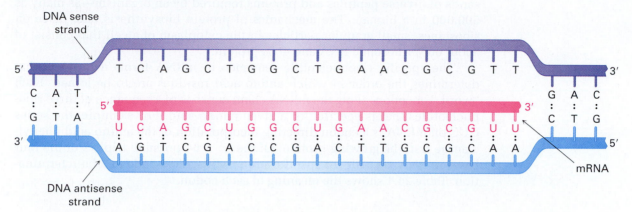

FIGURE 24.5 Biosynthesis of RNA using a DNA base segment as template.

Unlike what happens in DNA replication, where both strands are copied, only one of the two DNA strands is transcribed into mRNA. The DNA strand that contains the gene is often called the **sense strand**, or *coding strand,* and the DNA strand that gets transcribed to give RNA is called the **antisense strand**, or *noncoding strand.* Because the sense strand and the antisense strand in DNA are complementary, and because the DNA antisense strand and the newly formed RNA strand are also complementary, *the RNA molecule produced during transcription is a copy of the DNA sense strand.* That is, the complement of the complement is the same as the original. The only difference is that the RNA molecule has a U everywhere the DNA sense strand has a T.

Another part of the picture in vertebrates and flowering plants is that genes are often not continuous segments of the DNA chain. Instead, a gene will begin in one small section of DNA called an *exon,* then be interrupted by a noncoding section called an *intron,* and then take up again farther down the chain in another exon. The final mRNA molecule results only after the non-coded sections are cut out and the remaining pieces are joined together by spliceosomes. The gene for triose phosphate isomerase in maize, for instance, contains nine exons accounting for approximately 30% of the DNA base pairs and eight introns accounting for 70% of the base pairs.

Problem 24.4
Show how uracil can form strong hydrogen bonds to adenine.

Problem 24.5
What mRNA base sequence is complementary to the following DNA base sequence?

(5′) GATTACCGTA (3′)

Problem 24.6
From what DNA base sequence was the following mRNA sequence transcribed?

(5′) UUCGCAGAGU (3′)

24.5 Translation of RNA: Protein Biosynthesis

The primary cellular function of mRNA is to direct biosynthesis of the thousands of diverse peptides and proteins required by an organism—as many as 500,000 in a human. The mechanics of protein biosynthesis take place on ribosomes, small granular particles in the cytoplasm of a cell that consist of about 60% ribosomal RNA and 40% protein.

The specific ribonucleotide sequence in mRNA forms a message that determines the order in which amino acid residues are to be joined. Each "word," or **codon**, along the mRNA chain consists of a sequence of three ribonucleotides that is specific for a given amino acid. For example, the series UUC on mRNA is a codon directing incorporation of the amino acid phenylalanine into the growing protein. Of the $4^3 = 64$ possible triplets of the four bases in RNA, 61 code for specific amino acids and 3 code for chain termination. Table 24.1 shows the meaning of each codon.

TABLE 24.1
Codon Assignments of Base Triplets

First base (5′ end)	Second base	Third base (3′ end)			
		U	C	A	G
U	U	Phe	Phe	Leu	Leu
	C	Ser	Ser	Ser	Ser
	A	Tyr	Tyr	Stop	Stop
	G	Cys	Cys	Stop	Trp
C	U	Leu	Leu	Leu	Leu
	C	Pro	Pro	Pro	Pro
	A	His	His	Gln	Gln
	G	Arg	Arg	Arg	Arg
A	U	Ile	Ile	Ile	Met
	C	Thr	Thr	Thr	Thr
	A	Asn	Asn	Lys	Lys
	G	Ser	Ser	Arg	Arg
G	U	Val	Val	Val	Val
	C	Ala	Ala	Ala	Ala
	A	Asp	Asp	Glu	Glu
	G	Gly	Gly	Gly	Gly

The message embedded in mRNA is read by transfer RNA (tRNA) in a process called **translation**. There are 61 different tRNAs, one for each of the 61 codons that specifies an amino acid. A typical tRNA is single-stranded and has roughly the shape of a cloverleaf, as shown in Figure 24.6. It consists of about 70 to 100 ribonucleotides and is bonded to a specific amino acid by an ester linkage through the 3′ hydroxyl on ribose at the 3′ end of the tRNA. Each tRNA also contains on its middle leaf a segment called an **anticodon**, a sequence of three ribonucleotides complementary to the codon sequence. For example, the codon sequence UUC present on mRNA is read by a phenylalanine-bearing tRNA having the complementary anticodon base sequence GAA. [Remember that nucleotide sequences are written in the 5′ → 3′ direction, so the sequence in an anticodon must be reversed. That is, the complement to (5′)-UUC-(3′) is (3′)-AAG-(5′), which is written as (5′)-GAA-(3′).]

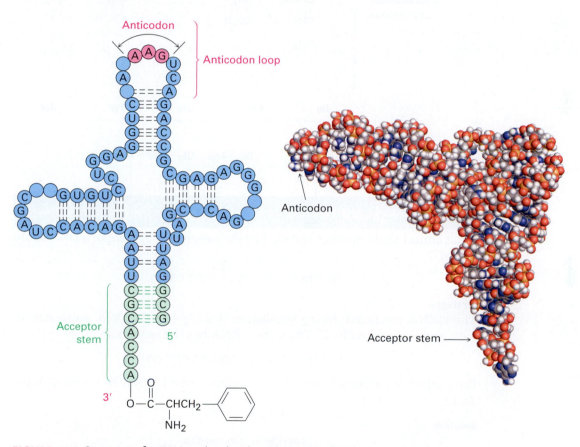

FIGURE 24.6 Structure of a tRNA molecule. The tRNA molecule is roughly cloverleaf-shaped and contains an anticodon triplet on one "leaf" and an amino acid unit attached covalently at its 3′ end. The example shown is a yeast tRNA that codes for phenylalanine. The nucleotides not specifically identified are chemically modified analogs of the four common ribonucleotides.

As each successive codon on mRNA is read, different tRNAs bring the correct amino acids into position for enzyme-mediated transfer to the growing peptide. When synthesis of the proper protein is completed, a "stop" codon signals the end and the protein is released from the ribosome. The process is illustrated in Figure 24.7.

FIGURE 24.7 A representation of protein biosynthesis. The codon base sequences on mRNA are read by tRNAs containing complementary anticodon base sequences. Transfer RNAs assemble the proper amino acids into position for incorporation into the growing peptide.

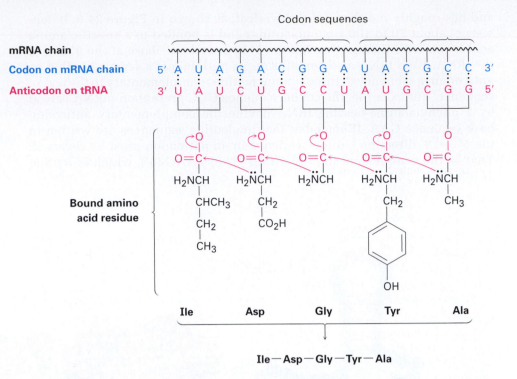

WORKED EXAMPLE 24.2 Predicting the Amino Acid Sequence Transcribed from DNA

What amino acid sequence is coded by the following segment of a DNA sense strand?

(5') CTA-ACT-AGC-GGG-TCG-CCG (3')

Strategy

The mRNA produced during translation is a copy of the DNA sense strand, with each T replaced by U. Thus, the mRNA has the sequence

(5') CUA-ACU-AGC-GGG-UCG-CCG (3')

Each set of three bases forms a codon, whose meaning can be found in Table 24.1.

Solution

Leu-Thr-Ser-Gly-Ser-Pro

Problem 24.7
List anticodon sequences on the tRNAs carrying the following amino acids:
(a) Ala **(b)** Phe **(c)** Leu **(d)** Tyr

Problem 24.8
What amino acid sequence is coded by the following mRNA base sequence?

CUU-AUG-GCU-UGG-CCC-UAA

Problem 24.9
What is the base sequence in the original DNA strand on which the mRNA sequence in Problem 24.8 was made?

24.6 DNA Sequencing

One of the greatest scientific revolutions in history is now underway in molecular biology, as scientists are learning how to manipulate and harness the genetic machinery of organisms. None of the extraordinary advances of the past two decades would have been possible, however, were it not for the discovery in 1977 of methods for sequencing immense DNA chains.

The first step in DNA sequencing is to cleave the enormous chain at known points to produce smaller, more manageable pieces, a task accomplished by the use of *restriction endonucleases*. Each different restriction enzyme, of which more than 3500 are known and approximately 200 are commercially available, cleaves a DNA molecule at a point in the chain where a specific base sequence occurs. For example, the restriction enzyme *Alu*I cleaves between G and C in the four-base sequence AG-CT. Note that the sequence is a *palindrome*, meaning that the *sequence* (5′)-AGCT-(3′) is the same as its *complement* (3′)-TCGA-(5′) when both are read in the same 5′ → 3′ direction. The same is true for other restriction endonucleases.

If the original DNA molecule is cut with another restriction enzyme that has a different specificity for cleavage, still other segments are produced whose sequences partially overlap those produced by the first enzyme. Sequencing of all the segments, followed by identification of the overlapping regions, allows complete DNA sequencing.

Two methods of DNA sequencing are available. The *Maxam–Gilbert method* uses chemical techniques, while the **Sanger dideoxy method** uses enzymatic reactions. The Sanger method is the more commonly used of the two and is the method responsible for sequencing the entire human genome of 3.0 billion base pairs. In commercial sequencing instruments, the dideoxy method begins with a mixture of the following:

- The restriction fragment to be sequenced
- A small piece of DNA called a *primer,* whose sequence is complementary to that on the 3′ end of the restriction fragment
- The four 2′-deoxyribonucleoside triphosphates (dNTPs)
- Very small amounts of the four 2′,3′-*dideoxy*ribonucleoside triphosphates (ddNTPs), each of which is labeled with a fluorescent dye of a different color. (A 2′,3′-dideoxyribonucleoside triphosphate is one in which both 2′ and 3′ –OH groups are missing from ribose.)

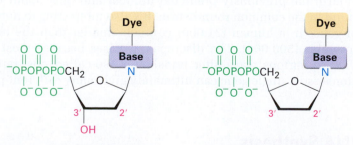

A 2′-deoxyribonucleoside
triphosphate (dNTP)

A 2′,3′-dideoxyribonucleoside
triphosphate (ddNTP)

DNA polymerase is added to the mixture, and a strand of DNA complementary to the restriction fragment begins to grow from the end of the primer.

Most of the time, only normal deoxyribonucleotides are incorporated into the growing chain because of their much higher concentration in the mixture, but every so often, a dideoxyribonucleotide is incorporated. When that happens, DNA synthesis stops because the chain end no longer has a 3′-hydroxyl group for adding further nucleotides.

When reaction is complete, the product consists of a mixture of DNA fragments of all possible lengths, each terminated by one of the four dye-labeled dideoxyribonucleotides. This product mixture is then separated according to the size of the pieces by gel electrophoresis (Section 19.2), and the identity of the terminal dideoxyribonucleotide in each piece—and thus the sequence of the restriction fragment—is identified by noting the color with which it fluoresces. Figure 24.8 shows a typical result.

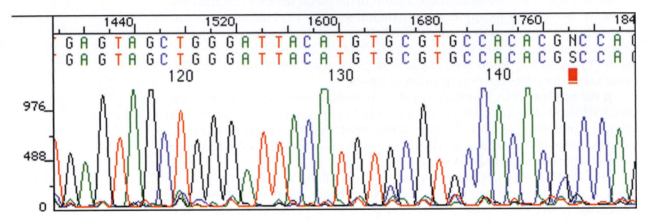

FIGURE 24.8 The sequence of a restriction fragment determined by the Sanger dideoxy method can be read by noting the colors of the dye attached to each of the various terminal nucleotides.

So efficient is the automated dideoxy method that sequences up to 1100 nucleotides in length, with a throughput of up to 19,000 bases per hour, can be sequenced with 98% accuracy. After a decade of work, preliminary sequence information for the entire human genome of 3.0 billion base pairs was announced early in 2001 and complete information was released in 2003. More recently, the genome sequencing of specific individuals, including that of James Watson, discoverer of the double helix, has been accomplished.

Remarkably, our genome appears to contain only about 21,000 genes, less than one-fourth the previously predicted number and only about twice the number found in the common roundworm. It's also interesting to note that the number of genes in a human (21,000) is much smaller than the number of kinds of proteins (500,000). The discrepancy arises because most proteins are modified in various ways after translation—so-called posttranslational modifications—so a single gene can ultimately give many different proteins.

24.7 DNA Synthesis

The ongoing revolution in molecular biology has brought with it an increased demand for the efficient chemical synthesis of short DNA segments, called *oligonucleotides,* or simply *oligos.* The problems of DNA synthesis are similar to those of protein synthesis (Section 19.7) but are more difficult because of the complexity of the nucleotide monomers. Each nucleotide has multiple reactive sites that must be selectively protected and deprotected at the proper times, and coupling of the four nucleotides must be carried out in the proper sequence. Automated DNA synthesizers are available, however, that allow the fast and reliable synthesis of DNA segments up to 200 nucleotides in length.

DNA synthesizers operate on a principle similar to that of the Merrifield solid-phase peptide synthesizer (Section 19.7). In essence, a protected nucleotide is covalently bonded to a solid support, and one nucleotide at a time is added to the growing chain by the use of a coupling reagent. After the final nucleotide has been added, all the protecting groups are removed and the synthetic DNA is cleaved from the solid support. Five steps are needed:

Step 1

The first step in DNA synthesis is to attach a protected deoxynucleoside to a silica (SiO_2) support by an ester linkage to the 3′ –OH group of the deoxynucleoside. Both the 5′ –OH group on the sugar and free –NH_2 groups on the heterocyclic bases must be protected. Adenine and cytosine bases are protected by benzoyl groups, guanine is protected by an isobutyryl group, and thymine requires no protection. The deoxyribose 5′ –OH is protected as its *p*-dimethoxytrityl (DMT) ether.

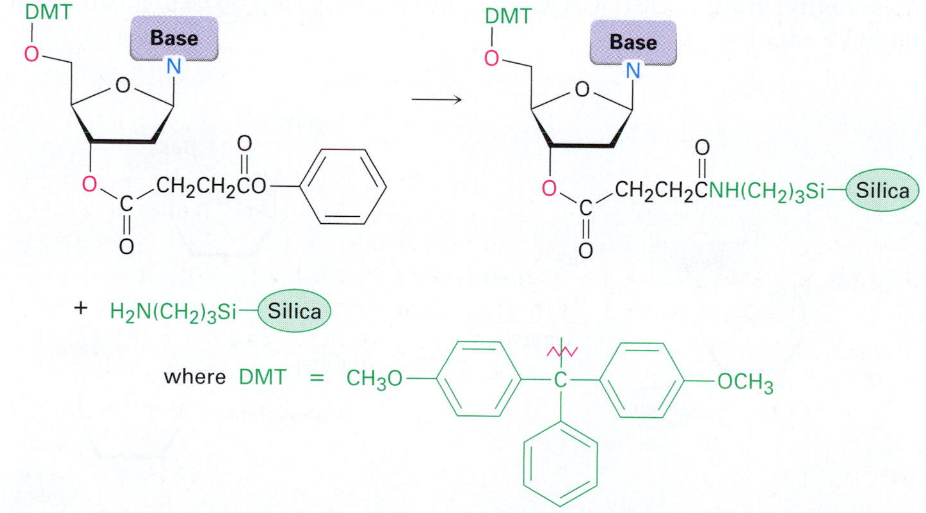

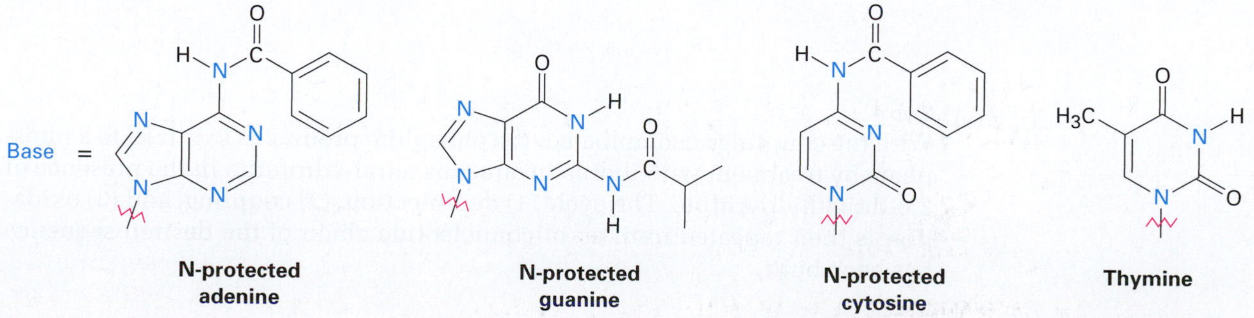

Step 2

The second step is removal of the DMT protecting group by treatment with dichloroacetic acid in CH_2Cl_2. The reaction occurs by an S_N1 mechanism and proceeds rapidly because of the stability of the tertiary, benzylic dimethoxytrityl cation.

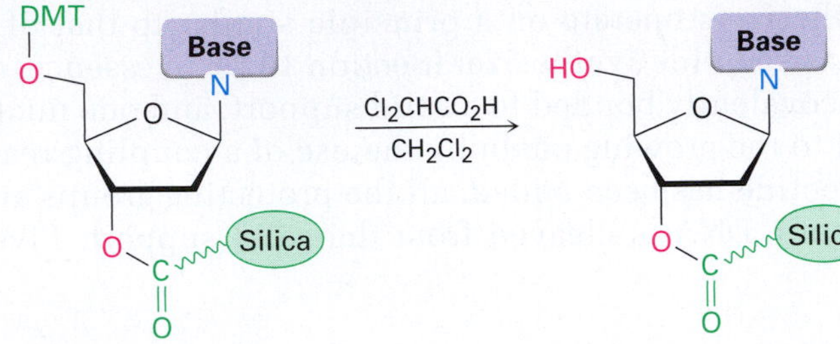

Step 3

The third step is the coupling of the polymer-bonded deoxynucleoside with a protected deoxynucleoside containing a *phosphoramidite* group at its 3′ position. [A phosphoramidite has the structure $R_2NP(OR)_2$.] The coupling reaction takes place in the polar aprotic solvent acetonitrile, requires catalysis by the heterocyclic amine tetrazole, and yields a *phosphite*, $P(OR)_3$, as product. Note that one of the phosphorus oxygen atoms is protected by a β-cyanoethyl group, $-OCH_2CH_2C\equiv N$. The coupling step takes place in better than 99% yield.

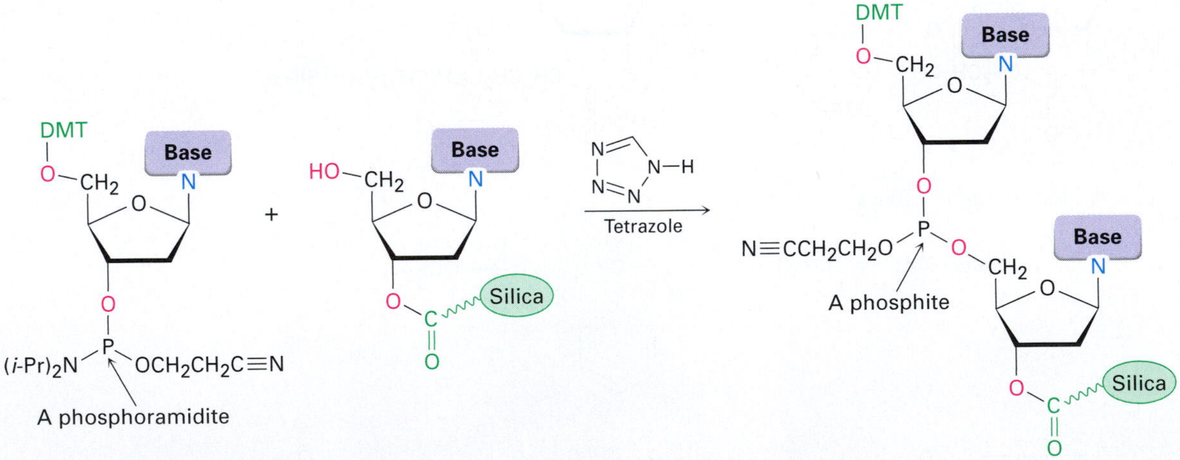

Step 4

With the coupling accomplished, the phosphite product is oxidized to a phosphate by treatment with iodine in aqueous tetrahydrofuran in the presence of 2,6-dimethylpyridine. The cycle (1) deprotection, (2) coupling, and (3) oxidation is then repeated until an oligonucleotide chain of the desired sequence has been built.

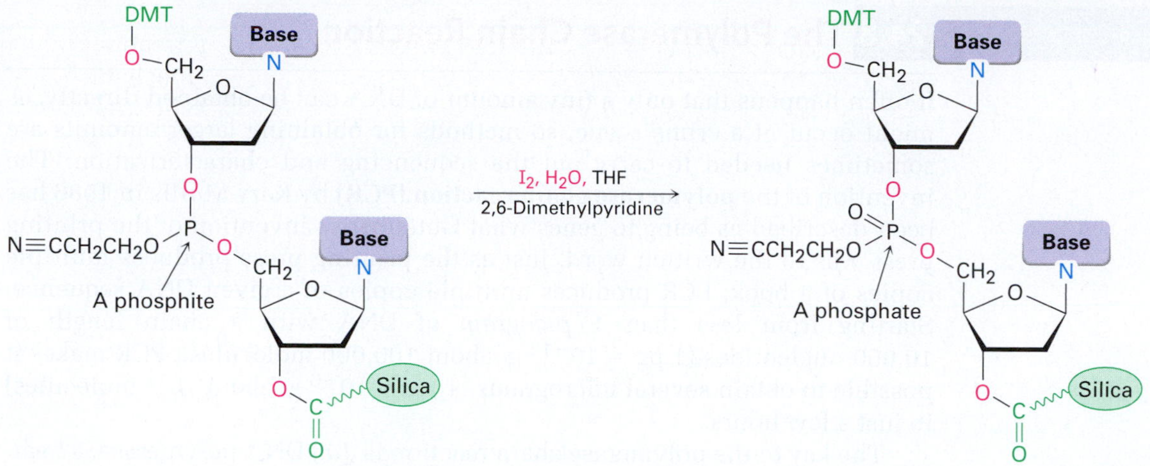

Step 5

The final step is removal of all protecting groups and cleavage of the ester bond holding the DNA to the silica. All these reactions are done at the same time by treatment with aqueous NH_3. Purification by electrophoresis then yields the synthetic DNA.

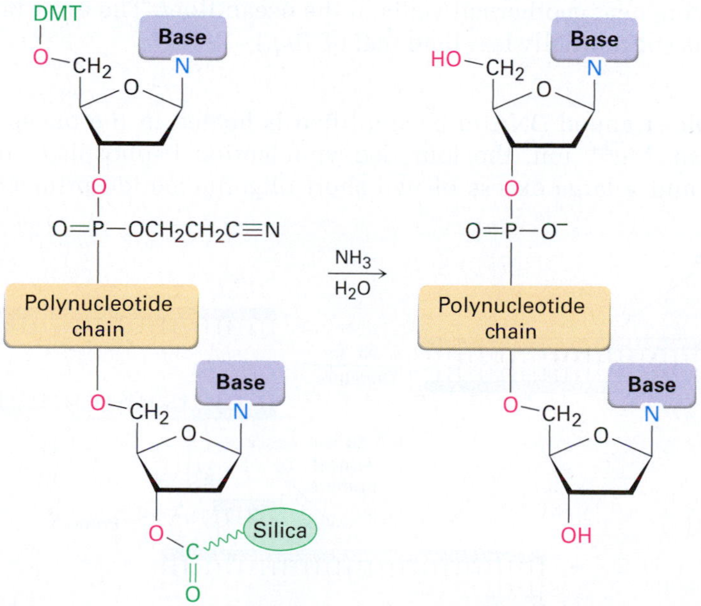

Problem 24.10

p-Dimethoxytrityl (DMT) ethers are easily cleaved by mild acid treatment. Show the mechanism of the cleavage reaction.

Problem 24.11

Propose a mechanism to account for cleavage of the β-cyanoethyl protecting group from the phosphate groups on treatment with aqueous ammonia. (Acrylonitrile, $H_2C{=}CHCN$, is a by-product.) What kind of reaction is occurring?

24.8 The Polymerase Chain Reaction

It often happens that only a tiny amount of DNA can be obtained directly, as might occur at a crime scene, so methods for obtaining larger amounts are sometimes needed to carry out the sequencing and characterization. The invention of the **polymerase chain reaction (PCR)** by Kary Mullis in 1986 has been described as being to genes what Gutenberg's invention of the printing press was to the written word. Just as the printing press produces multiple copies of a book, PCR produces multiple copies of a given DNA sequence. Starting from less than 1 *picogram* of DNA with a chain length of 10,000 nucleotides (1 pg = 10^{-12} g; about 100,000 molecules), PCR makes it possible to obtain several micrograms (1 μg = 10^{-6} g; about 10^{11} molecules) in just a few hours.

The key to the polymerase chain reaction is *Taq* DNA polymerase, a heat-stable enzyme isolated from the thermophilic bacterium *Thermus aquaticus* found in a hot spring in Yellowstone National Park. *Taq* polymerase is able to take a single strand of DNA that has a short, primer segment of complementary chain at one end and then finish constructing the entire complementary strand. The overall process takes three steps, as shown in Figure 24.9. (More recently, improved heat-stable DNA polymerase enzymes have become available, including Vent polymerase and *Pfu* polymerase, both isolated from bacteria growing near geothermal vents in the ocean floor. The error rate of both enzymes is substantially less than that of *Taq*.)

Step 1
The double-stranded DNA to be amplified is heated in the presence of *Taq* polymerase, Mg^{2+} ion, the four deoxynucleotide triphosphate monomers (dNTPs), and a large excess of two short oligonucleotide primers of about

FIGURE 24.9 The polymerase chain reaction. Details are explained in the text.

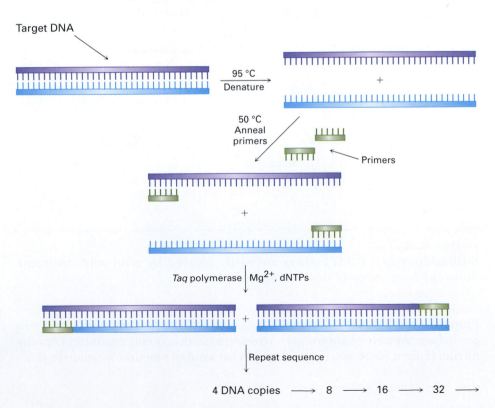

20 bases each. Each primer is complementary to the sequence at the end of one of the target DNA segments. At a temperature of 95 °C, double-stranded DNA denatures, spontaneously breaking apart into two single strands.

Step 2

The temperature is lowered to between 37 and 50 °C, allowing the primers, because of their relatively high concentration, to anneal by hydrogen-bonding to their complementary sequence at the end of each target strand.

Step 3

The temperature is then raised to 72 °C, and *Taq* polymerase catalyzes the addition of further nucleotides to the two primed DNA strands. When replication of each strand is finished, *two* copies of the original DNA now exist. Repeating the denature–anneal–synthesize cycle a second time yields four DNA copies, repeating a third time yields eight copies, and so on, in an exponential series.

PCR has been automated, and 30 or so cycles can be carried out in an hour, resulting in a theoretical amplification factor of 2^{30} ($\sim 10^9$). In practice, however, the efficiency of each cycle is less than 100%, and an experimental amplification of about 10^6 to 10^8 is routinely achieved for 30 cycles.

24.9 Catabolism of Nucleotides

The catabolism of nucleotides is generally more complex than that of amino acids, carbohydrates, or fatty acids because the structures of the nucleotides themselves are more complex. As a result, we'll treat the subject lightly and look only at one example.

Dietary nucleic acids first pass through the stomach to the intestines, where they are hydrolyzed to their constituent nucleotides by a variety of different nucleases. Dephosphorylation by various nucleotidases next gives nucleosides, and cleavage by nucleosidases then gives the constituent bases, which are catabolized to produce intermediates that enter other metabolic processes or are excreted.

As an example of nucleoside catabolism, let's look at guanosine, which is degraded by a three-step pathway that begins with cleavage to give guanine. Hydrolysis of guanine then yields xanthine, and oxidation of xanthine gives uric acid, which is excreted in the urine (Figure 24.10).

FIGURE 24.10
Pathway for the
catabolism of
guanosine to
uric acid. Indi-
vidual steps
are explained
in the text.

FIGURE 24.10
Pathway for the
catabolism of
guanosine to
uric acid. Indi-
vidual steps
are explained
in the text.

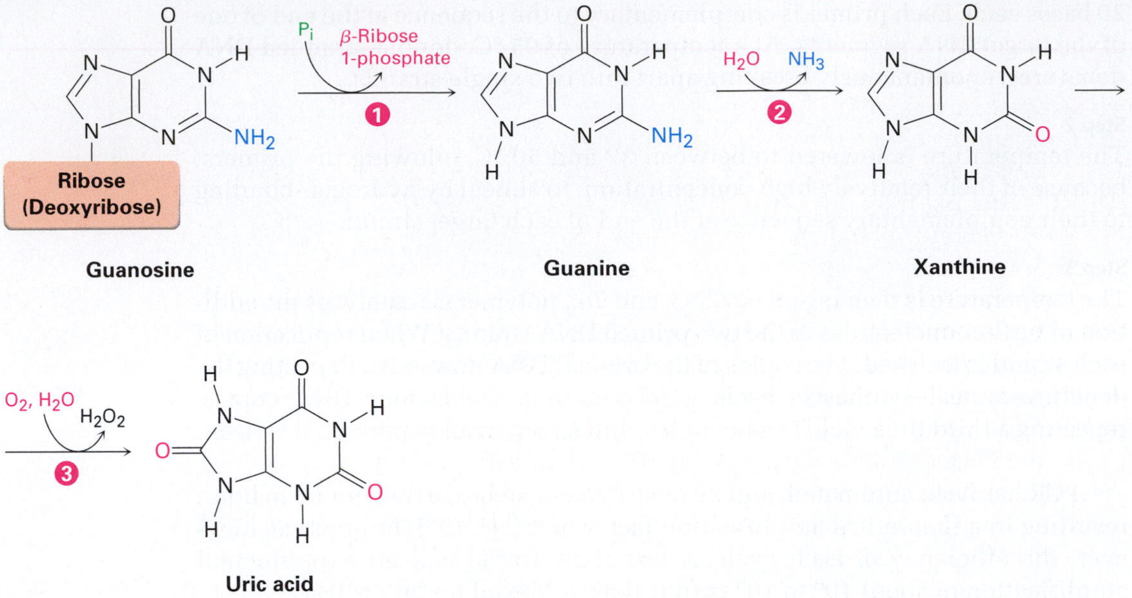

STEP 1 OF FIGURE 24.10: PHOSPHOROLYSIS The phosphorolysis of guanosine is catalyzed by purine nucleoside phosphorylase and gives β-ribose
1-phosphate plus guanine. The reaction probably occurs by an S_N1-like
replacement of guanine by phosphate ion through an oxonium-ion intermediate, analogous to what occurs during the hydrolysis of a glycoside with an
inverting glycosidase (Figure 22.2 on page 903).

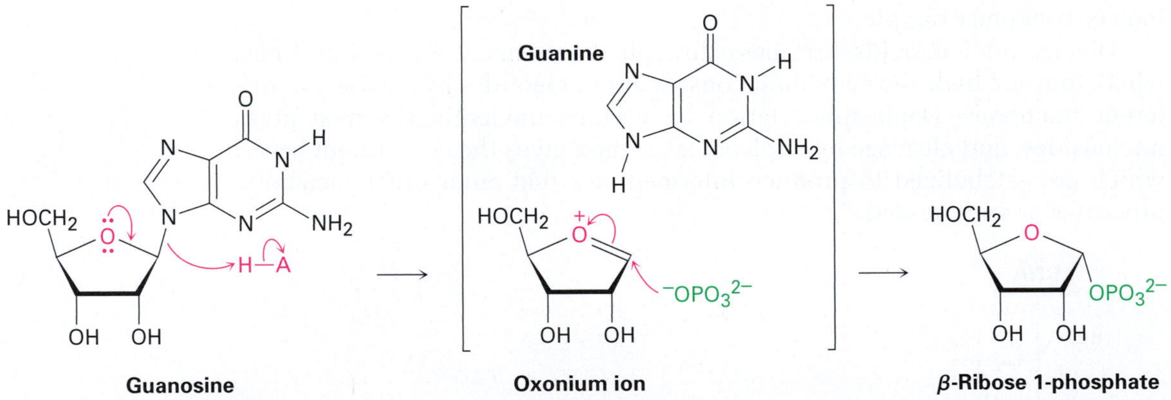

STEP 2 OF FIGURE 24.10: HYDROLYSIS The hydrolysis of guanine to give xanthine is catalyzed by guanine deaminase and occurs by nucleophilic addition
of water to the C=N bond, followed by expulsion of ammonium ion—essentially a nucleophilic acyl substitution reaction.

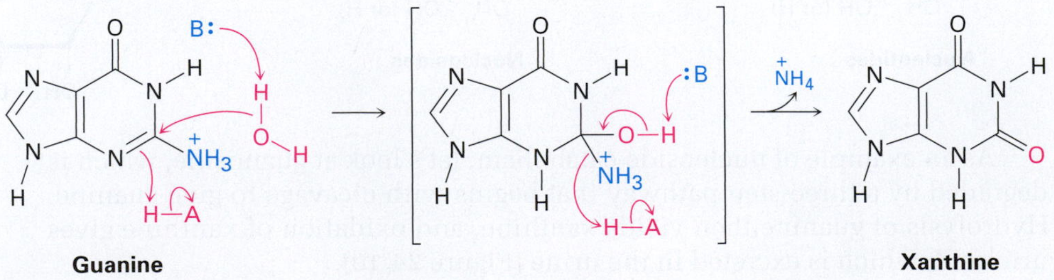

STEP 3 OF FIGURE 24.10: OXIDATION The only unusual step in guanosine catabolism is the oxidation of xanthine by xanthine oxidase, a complex enzyme that contains FAD and an oxo–molybdenum(VI) cofactor. Current evidence suggests the mechanism in Figure 24.11, in which a base deprotonates the Mo–OH group and the resulting anion does a nucleophilic addition to a C=N bond in xanthine. The nitrogen anion then expels hydride ion, which adds to an Mo=S bond, thereby reducing the molybdenum center from Mo(VI) to Mo(IV). Hydrolysis of the Mo–O bond gives an enol that tautomerizes to uric acid, and the reduced molybdenum is reoxidized by O_2 in a complex redox pathway.

The transformation may look complicated because you're probably unfamiliar with molybdenum. Note, though, that the reactions taking place on xanthine *are* familiar and we've seen them numerous times. Thus, the initial nucleophilic addition of an oxygen anion to C=N is similar to what occurs in step 2 when water adds to guanine, and the subsequent expulsion of hydride ion by the adjacent nitrogen atom is analogous to what occurs during NADH reductions (Section 14.10).

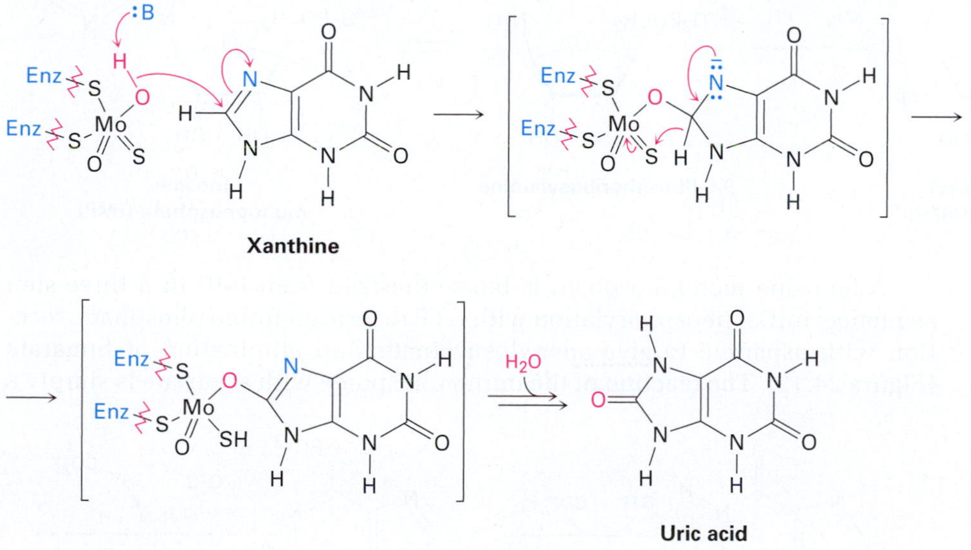

Xanthine

Uric acid

FIGURE 24.11 The mechanism of step 3 in Figure 24.10, oxidation of xanthine to yield uric acid.

Adenosine, the other purine nucleotide, is degraded by a strategy similar to that used for guanosine, but the order of steps is different. Rather than having the base first cleaved off and then degraded, the base in adenosine is first degraded and then removed.

Problem 24.12

Write a likely mechanism for the first step in adenosine catabolism, the hydrolysis of adenosine to yield inosine.

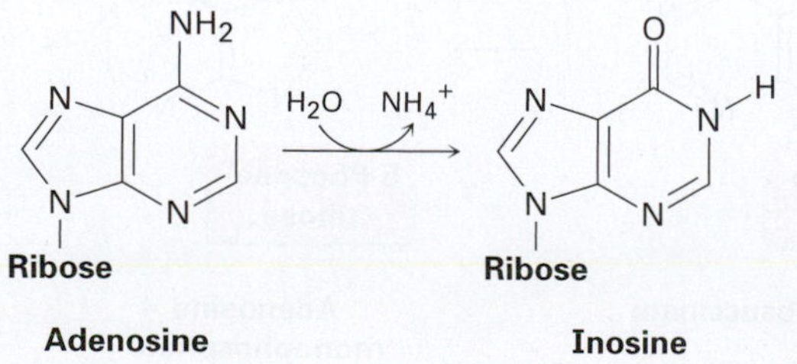

Ribose

Adenosine

Ribose

Inosine

24.10 Biosynthesis of Nucleotides

Nucleotide biosynthesis, like nucleotide catabolism, is relatively complex. Thus, we'll again look at only one example, adenosine monophosphate. Purine nucleotides are formed by initial attachment of an $-NH_2$ group to ribose, followed by multistep buildup of the heterocyclic base. The attachment of $-NH_2$ takes place by a nucleophilic substitution reaction of ammonia with 5-phosphoribosyl α-diphosphate to give β-5-phosphoribosylamine and probably involves an S_N1-like loss of diphosphate ion with formation of an oxonium-ion intermediate. Although we'll not cover the details of its formation, inosine monophosphate (IMP) is the first fully formed purine ribonucleotide, with adenosine monophosphate (AMP) derived from it.

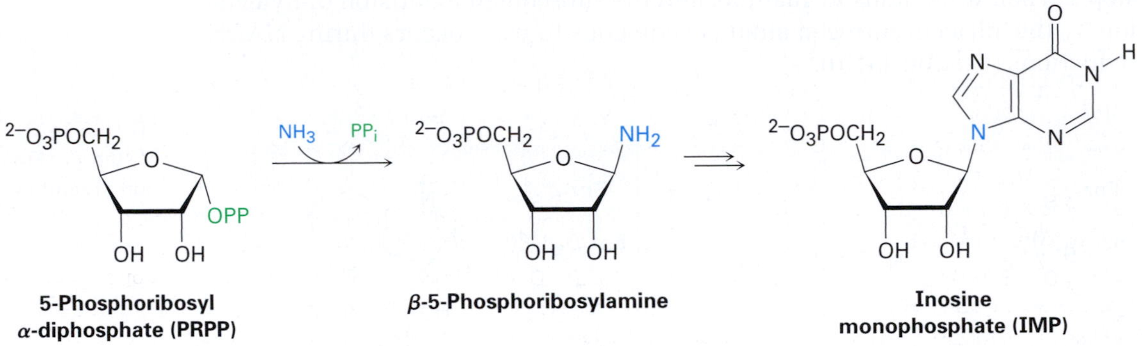

5-Phosphoribosyl α-diphosphate (PRPP)   **β-5-Phosphoribosylamine**   **Inosine monophosphate (IMP)**

Adenosine monophosphate is biosynthesized from IMP in a three-step sequence: initial phosphorylation with GTP to form an imino phosphate, reaction with aspartate to give adenylosuccinate, and elimination of fumarate (Figure 24.12). The reaction of the imino phosphate with aspartate is simply a

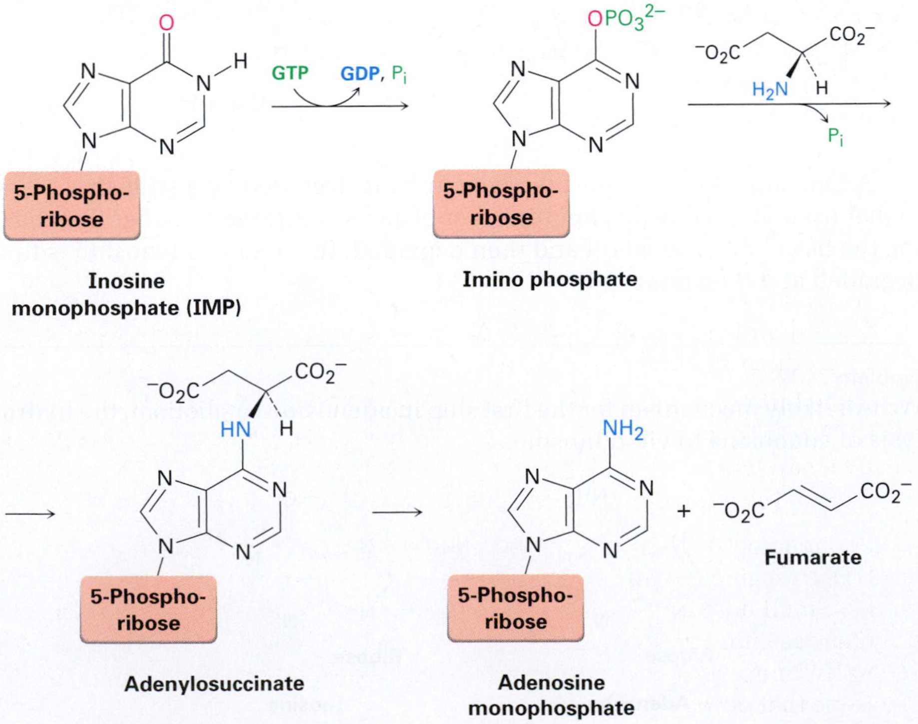

FIGURE 24.12 Pathway for the conversion of inosine monophosphate to adenosine monophosphate.

Inosine monophosphate (IMP)   **Imino phosphate**

Adenylosuccinate   **Adenosine monophosphate**   **Fumarate**

nucleophilic acyl substitution reaction, and the elimination of fumarate is an E1cB reaction, analogous to the third step in the urea cycle in which arginino-succinate is converted to arginine (Figure 20.5 on page 843).

Problem 24.13
Write the mechanism of the formation of adenylosuccinate from inosine monophosphate, the second step in adenosine biosynthesis (Figure 24.12).

Problem 24.14
Show the mechanism of the formation of adenosine monophosphate from adenylosuccinate, the third step in adenosine biosynthesis (Figure 24.12).

Summary

We've now covered the last of the four major classes of biomolecules—the nucleic acids, **DNA (deoxyribonucleic acid)** and **RNA (ribonucleic acid)**. So much has been written and spoken about DNA in the media that our focus has been on the chemical details of DNA sequencing, synthesis, and metabolism rather than on simpler fundamentals.

DNA and RNA are biological polymers that act as chemical carriers of an organism's genetic information. Enzyme-catalyzed hydrolysis of nucleic acids yields **nucleotides**, the monomer units from which RNA and DNA are constructed. Further enzyme-catalyzed hydrolysis of the nucleotides yields **nucleosides** plus phosphate. Nucleosides, in turn, consist of a purine or pyrimidine base linked to C1 of an aldopentose sugar—ribose in RNA and 2-deoxyribose in DNA. The nucleotides are joined by phosphate links between the 5′ phosphate of one nucleotide and the 3′ hydroxyl on the sugar of another nucleotide.

Molecules of DNA consist of two complementary polynucleotide strands held together by hydrogen bonds between heterocyclic bases on the different strands and coiled into a **double helix**. Adenine and thymine form hydrogen bonds to each other, as do cytosine and guanine.

Three processes take place in deciphering the genetic information of DNA:

- **Replication** of DNA is the process by which identical DNA copies are made. The DNA double helix unwinds, complementary deoxyribonucleotides line up in order, and two new DNA molecules are produced.

- **Transcription** is the process by which RNA is produced to carry genetic information from the nucleus to the ribosomes. A short segment of the DNA double helix unwinds, and complementary ribonucleotides line up to produce **messenger RNA (mRNA)**.

- **Translation** is the process by which mRNA directs protein synthesis. Each mRNA is divided into **codons**, ribonucleotide triplets that are recognized by small amino acid–carrying molecules of **transfer RNA (tRNA)**, which deliver the appropriate amino acids needed for protein synthesis.

Sequencing of DNA is carried out by the **Sanger dideoxy method**, and small DNA segments can be synthesized in the laboratory by automated instruments. Small amounts of DNA can be amplified by a factor of 10^6 using the **polymerase chain reaction (PCR)**. Nucleotide catabolism and biosynthesis is generally more complex than that of other classes of biomolecules, but the reactions that occur are similar.

Key Words
anticodon, 997
antisense strand, 995
codon, 996
deoxyribonucleic acid (DNA), 987
double helix, 990
3′ end, 989
5′ end, 989
messenger RNA (mRNA), 994
nucleoside, 987
nucleotide, 987
polymerase chain reaction (PCR), 1004
replication, 992
ribonucleic acid (RNA), 987
ribosomal RNA (rRNA), 994
Sanger dideoxy method, 999
sense strand, 995
small RNAs, 994
transcription, 994
transfer RNA (tRNA), 994
translation, 996

Lagniappe

DNA Fingerprinting

The invention of DNA sequencing has affected society in many ways, few more dramatic than those stemming from the development of *DNA fingerprinting.* DNA fingerprinting arose from the discovery in 1984 that human genes contain short, repeating sequences of noncoding DNA, called *short tandem repeat* (STR) loci. Furthermore, the STR loci are slightly different for every individual, except identical twins. By sequencing these loci, a pattern unique to each person can be obtained.

Perhaps the most common and well-publicized use of DNA fingerprinting is that carried out by crime laboratories to link suspects to biological evidence—blood, hair follicles, skin, or semen—found at a crime scene. Thousands of court cases have now been decided based on DNA evidence.

For use in criminal cases, forensic laboratories in the United States have agreed on 13 core STR loci that are most accurate for identification of an individual. Based on these 13 loci, a Combined DNA Index System (CODIS) has been established to serve as a registry of convicted offenders. When a DNA sample is obtained from a crime scene, the sample is subjected to cleavage with restriction endonucleases to cut out fragments containing the STR loci, the fragments are amplified using the polymerase chain reaction, and the sequences of the fragments are determined.

If the profile of sequences from a known individual and the profile from DNA obtained at a crime scene match, the probability is approximately 82 billion to 1 that the DNA is from the same individual. In paternity cases, where the DNA of father and offspring are related but not fully identical, the identity of the father can be established with a probability of around 100,000 to 1. Even after several generations have passed, paternity can still be inferred from DNA analysis of the Y chromosome of direct male-line descendants. The most well-known such case is that of Thomas Jefferson, who likely fathered a child by his slave Sally Hemings. Although Jefferson himself has no male-line descendants, DNA analysis of the male-line descendants of Jefferson's paternal uncle contained the same Y chromosome as a male-line descendant of Eston Hemings, the youngest son of Sally Hemings. Thus, a mixing of the two genomes is clear, although the male individual responsible for that mixing can't be conclusively identified.

Among its many other applications, DNA fingerprinting is widely used for the diagnosis of genetic disorders, both prenatally and in newborns. Cystic fibrosis, hemophilia, Huntington's disease, Tay–Sachs disease, sickle cell anemia, and thalassemia are among the many diseases that can be detected, enabling early treatment of an affected child. Furthermore, by studying the DNA fingerprints of relatives with a history of a particular disorder, it's possible to identify DNA patterns associated with the disease and perhaps obtain clues for eventual cure. In addition, the U.S. Department of Defense now requires blood and saliva samples from all military personnel. The samples are stored, and DNA is extracted should the need for identification of a casualty arise.

© Bettmann/CORBIS

Historians have wondered for many years whether Thomas Jefferson fathered a child by his slave, Sally Hemings. DNA fingerprinting evidence strongly suggests that he did.

Exercises

VISUALIZING CHEMISTRY

(Problems 24.1–24.14 appear within the chapter.)

24.15 ■ Identify the following bases, and tell whether each is found in DNA, RNA, or both:

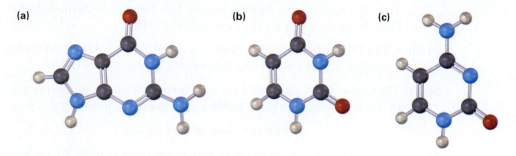

(a)   (b)   (c)

24.16 ■ Identify the following nucleotide, and tell how it is used:

24.17 Amine bases in nucleic acids can react with alkylating agents in typical S_N2 reactions. Look at the following electrostatic potential maps, and tell which is the better nucleophile, guanine or adenine. The reactive positions in each are indicated.

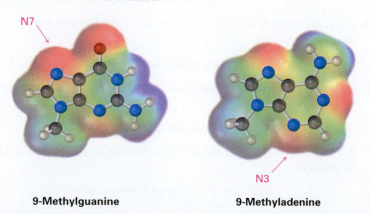

N7

N3

9-Methylguanine   9-Methyladenine

■ indicates problems that are assignable in Organic OWL.

Go to this book's companion website at **www.cengage.com/ chemistry/mcmurry** to explore interactive versions of the Active Figures from this text.

ADDITIONAL PROBLEMS

24.18 Human brain natriuretic peptide (BNP) is a small peptide of 32 amino acids used in the treatment of congestive heat failure. How many nitrogen bases are present in the DNA that codes for BNP?

24.19 Human and horse insulin both have two polypeptide chains, with one chain containing 21 amino acids and the other containing 30 amino acids. They differ in primary structure at two places. At position 9 in one chain, human insulin has Ser and horse insulin has Gly; at position 30 in the other chain, human insulin has Thr and horse insulin has Ala. How must the DNA for the two insulins differ?

24.20 ■ The DNA of sea urchins contains about 32% A. What percentages of the other three bases would you expect in sea urchin DNA?

24.21 The codon UAA stops protein synthesis. Why does the sequence UAA in the following stretch of mRNA not cause any problems?

-GCA-UUC-GAG-GUA-ACG-CCC-

24.22 ■ Which of the following base sequences would most likely be recognized by a restriction endonuclease? Explain.

(a) GAATTC **(b)** GATTACA **(c)** CTCGAG

24.23 ■ For what amino acids do the following ribonucleotide triplets code?

(a) AAU **(b)** GAG **(c)** UCC **(d)** CAU

24.24 ■ From what DNA sequences were each of the mRNA codons in Problem 24.23 transcribed?

24.25 ■ What anticodon sequences of tRNAs are coded for by the codons in Problem 24.23?

24.26 Draw the complete structure of the ribonucleotide codon UAC. For what amino acid does this sequence code?

24.27 Draw the complete structure of the deoxyribonucleotide sequence from which the mRNA codon in Problem 24.26 was transcribed.

24.28 Give an mRNA sequence that will code for synthesis of metenkephalin:

Tyr-Gly-Gly-Phe-Met

24.29 Give an mRNA sequence that will code for the synthesis of angiotensin II:

Asp-Arg-Val-Tyr-Ile-His-Pro-Phe

24.30 ■ What amino acid sequence is coded for by the following DNA sense strand?

(5′) CTT-CGA-CCA-GAC-AGC-TTT (3′)

24.31 ■ What amino acid sequence is coded for by the following mRNA base sequence?

(5′) CUA-GAC-CGU-UCC-AAG-UGA (3′)

■ Problems assignable in Organic OWL.

24.32 If the DNA sense sequence -CAA-CCG-GAT- were miscopied during replication and became -CGA-CCG-GAT-, what effect would there be on the sequence of the protein produced?

24.33 ▪ Show the steps involved in a laboratory synthesis of the DNA fragment with the sequence CTAG.

24.34 Write a mechanism for the oxidation of malonic semialdehyde to give malonyl CoA, one of the steps in uracil catabolism. The process is similar to what occurs in step 6 of glycolysis.

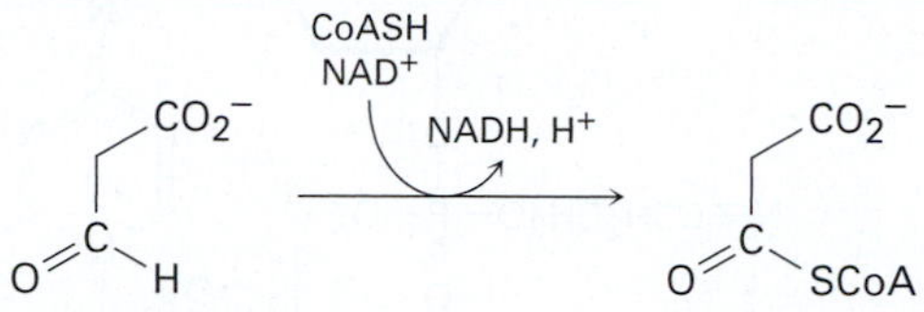

Malonic semialdehyde   **Malonyl CoA**

24.35 ▪ One of the steps in the biosynthesis of inosine monophosphate is the formation of aminoimidazole ribonucleotide from formylglycinamidine ribonucleotide. Propose a mechanism.

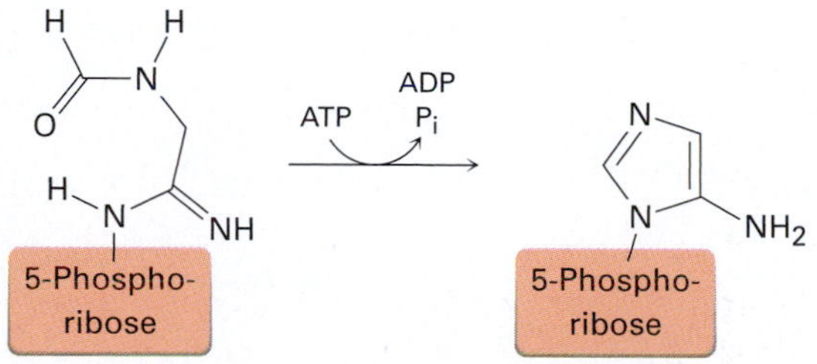

Formylglycinamidine ribonucleotide   **Aminoimidazole ribonucleotide**

24.36 ▪ One of the steps in the biosynthesis of uridine monophosphate is the reaction of aspartate with carbamoyl phosphate to give carbamoyl aspartate followed by cyclization to form dihydroorotate. Zn^{2+} ion is required as a Lewis acid to catalyze the cyclization. Propose mechanisms for both steps.

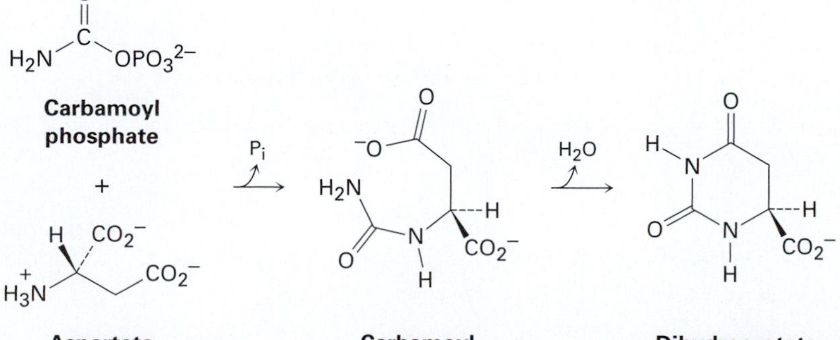

Carbamoyl phosphate

+

Aspartate   **Carbamoyl aspartate**   **Dihydroorotate**

▪ Problems assignable in Organic OWL.

24.37 ■ The final step in DNA synthesis is deprotection by treatment with aqueous ammonia. Show the mechanisms by which deprotection occurs at the points indicated in the following structure:

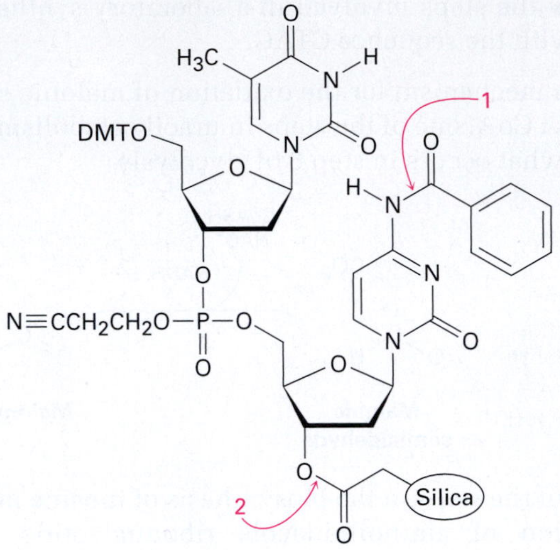

25 Secondary Metabolites: An Introduction to Natural Products Chemistry

Norcoclaurine synthase catalyzes the coupling of dopamine with *p*-hydroxyphenylacetaldehyde, a step in morphine biosynthesis.

In the past six chapters, we've looked at the chemistry and metabolism of the four major classes of biomolecules—proteins, carbohydrates, lipids, and nucleic acids. But there is far more to do, for all living organisms also contain a vast diversity of substances usually grouped under the heading *natural products*. The term **natural product** really refers to *any* naturally occurring substance but is generally taken to mean a **secondary metabolite**—a small molecule that is not essential to the growth and development of the producing organism and is not classified by structure.

It has been estimated that well over 300,000 secondary metabolites exist, and it's thought that their primary function is to increase the likelihood of an organism's survival by repelling or attracting other organisms. Alkaloids, such as morphine; eicosanoids, such as prostaglandin E_1; and antibiotics, such as erythromycin and the penicillins, are examples.

Morphine

Prostaglandin E_1

OWL Online homework for this chapter can be assigned in Organic OWL.

1015

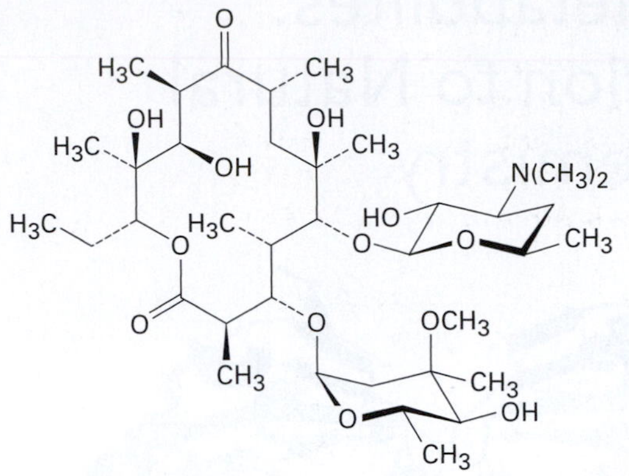

Erythromycin A

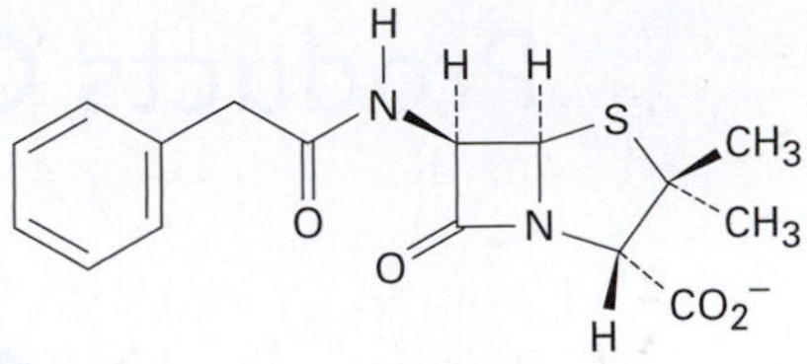

Benzylpenicillin

WHY THIS CHAPTER?

This brief chapter merely tickles the surface of natural-products chemistry, for hundreds, if not thousands, of books have been written on the subject. Rather than pretending to be comprehensive, this chapter is meant only to provide a brief introduction to a large and immensely important area of modern biochemistry, perhaps tempting you to learn more on your own. To provide that introduction, we'll look at the pathways by which several well-known natural products are synthesized in living organisms: pyridoxal phosphate (PLP), morphine, and erythromycin A. The molecules may appear complex (erythromycin A, in particular), but the individual chemical steps by which they are made should be familiar to you at this point.

25.1 Classification of Natural Products

There is no rigid scheme for classifying natural products—their immense diversity in structure, function, and biosynthesis is too great to allow them to fit neatly into a few simple categories. In practice, however, workers in the field often speak of five main classes of natural products: terpenoids and steroids, fatty acid–derived substances and polyketides, alkaloids, nonribosomal polypeptides, and enzyme cofactors.

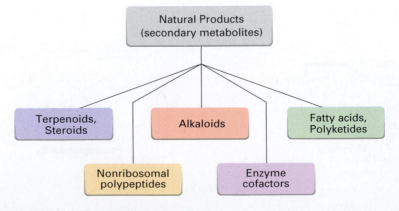

- **Terpenoids** and **steroids**, as discussed previously in Chapter 23, are a vast group of substances—more than 35,000 are known—derived biosynthetically from isopentenyl diphosphate. Terpenoids have an immense variety of apparently unrelated structures, while steroids have a common tetracyclic carbon skeleton and are modified terpenoids that are biosynthesized from the triterpene lanosterol. We looked at terpenoid and steroid biosynthesis in Sections 23.7 through 23.9.

- **Alkaloids**, like terpenoids, are a large and diverse class of compounds, with more than 12,000 examples known at present. They contain a basic amine group in their structure and are derived biosynthetically from amino acids. We'll look at morphine biosynthesis as an example in Section 25.3.

- **Fatty acid–derived substances** and **polyketides**, of which more than 10,000 are known, are biosynthesized from simple acyl precursors such as acetyl CoA, propionyl CoA, and methylmalonyl CoA. Natural products derived from fatty acids generally have most of the oxygen atoms removed, but polyketides, such as the antibiotic erythromycin A, often have many oxygen substituents remaining. We'll look at erythromycin biosynthesis in Section 25.4.

- **Nonribosomal polypeptides** are peptide-like compounds that are biosynthesized from amino acids by a multifunctional enzyme complex without direct RNA transcription. The penicillins are good examples, but their chemistry is a bit complicated and we'll not discuss their biosynthesis.

- **Enzyme cofactors** don't fit one of the other general categories of natural products and are usually classed separately. We've seen numerous examples of coenzymes in past chapters (see the list in Table 19.3) and will look at the biosynthesis of pyridoxal phosphate (PLP) in Section 25.2.

As you might imagine, unraveling the biosynthetic pathways by which specific natural products are made is difficult and time-consuming work. Small precursor molecules have to be identified, guesses about likely routes made, and individual enzymes that catalyze each step isolated, characterized, and mechanistically studied. The payoff for all this painstaking work is a fundamental understanding of how organisms function at the molecular level, an understanding that can be used to design new pharmaceutical agents.

25.2 Biosynthesis of Pyridoxal Phosphate

Let's begin this quick tour of natural-products chemistry by looking at the biosynthesis of pyridoxal 5′-phosphate (PLP), a relatively simple enzyme cofactor we've encountered several times in different metabolic pathways. An overview of PLP biosynthesis is shown in Figure 25.1.

STEPS 1–2 OF FIGURE 25.1: OXIDATION Pyridoxal phosphate biosynthesis begins with oxidation of the aldehyde group in D-erythrose 4-phosphate to give the corresponding carboxylic acid, D-erythronate 4-phosphate. The oxidation requires NAD^+ as cofactor and occurs by a mechanism similar to that of step 6 in glycolysis, in which glyceraldehyde 3-phosphate is oxidized to the corresponding acid (Figure 22.6 on page 908). A cysteine –SH group in the enzyme adds to the aldehyde carbonyl group of D-erythrose 4-phosphate

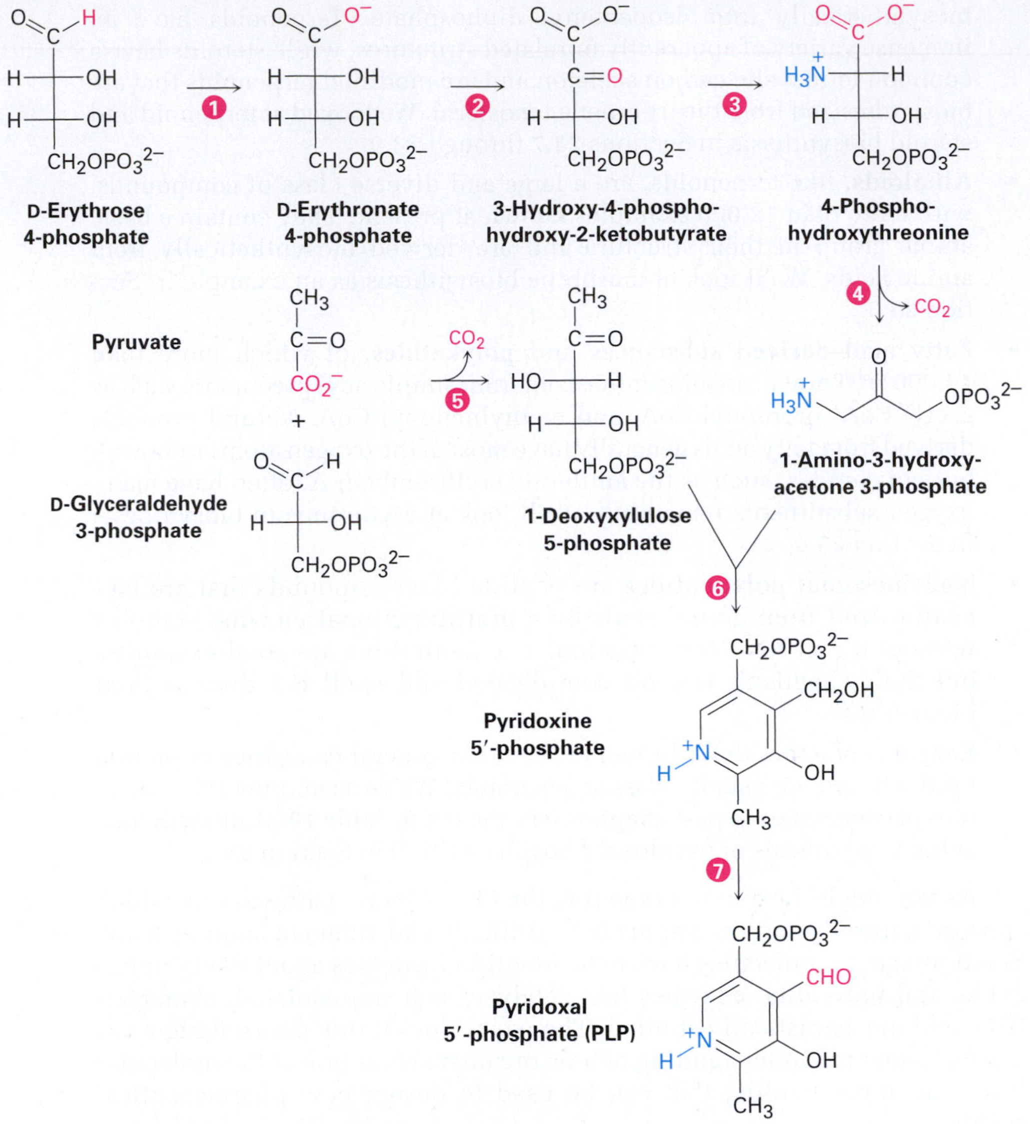

FIGURE 25.1 An overview of the pathway for pyridoxal 5′-phosphate biosynthesis. Individual steps are explained in the text.

to give an intermediate hemithioacetal, which is then oxidized by NAD^+ to a thioester. Hydrolysis of the thioester yields erythronate 4-phosphate, and a further oxidation of the –OH group at C2 by NAD^+ gives 3-hydroxy-4-phosphohydroxy-2-ketobutyrate (Figure 25.2).

STEPS 3–4 OF FIGURE 25.1: TRANSAMINATION AND OXIDATION/DECARBOXYLATION 3-Hydroxy-4-phosphohydroxy-2-ketobutyrate undergoes a transamination in step 3 on reaction with α-ketoglutarate by the usual PLP-dependent mechanism, shown previously in Figure 20.2 on page 838. The product, 4-phosphohydroxythreonine, is then oxidized by NAD^+ to give an intermediate β-keto ester, which undergoes concurrent decarboxylation and yields 1-amino-3-hydroxyacetone 3-phosphate. The reactions are shown in Figure 25.3.

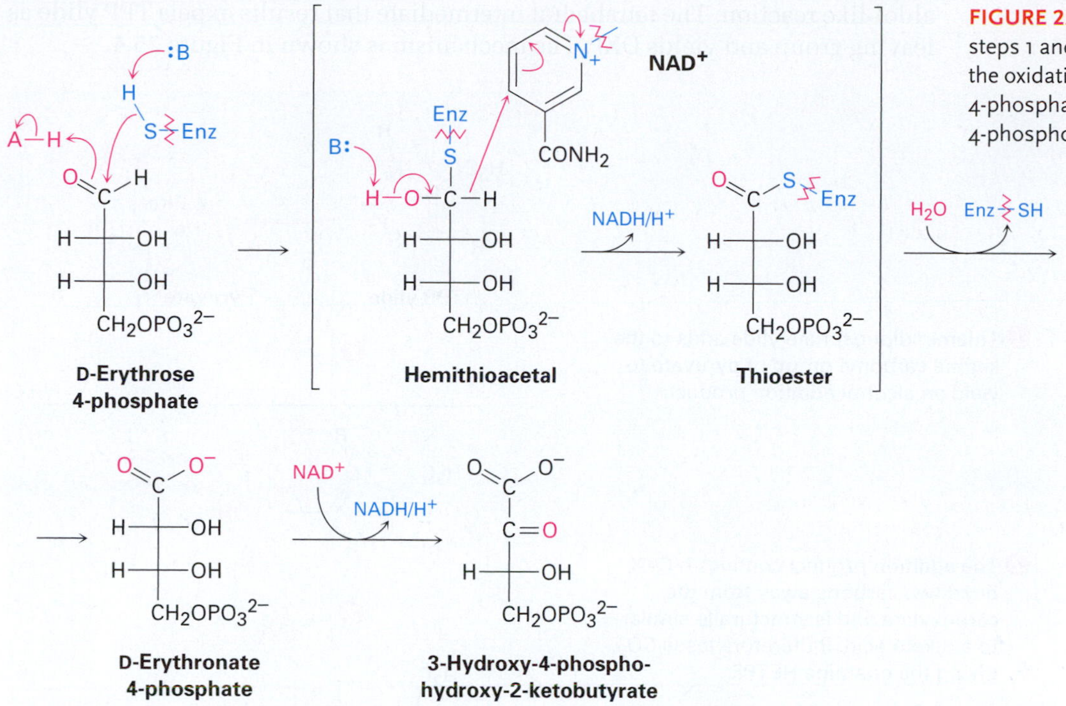

FIGURE 25.2 Mechanism of steps 1 and 2 in PLP biosynthesis, the oxidation of D-erythrose 4-phosphate to give 3-hydroxy-4-phosphohydroxy-2-ketobutyrate.

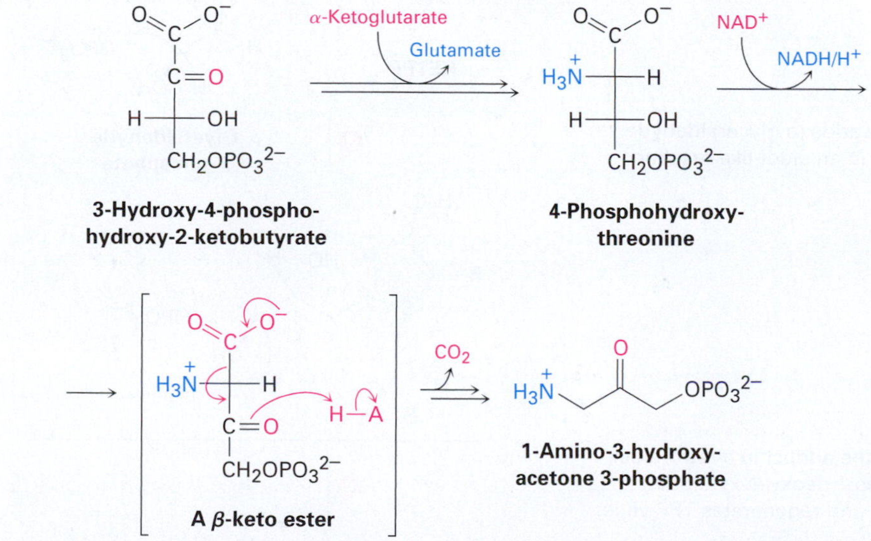

FIGURE 25.3 Mechanism of steps 3 and 4 in PLP biosynthesis.

STEP 5 OF FIGURE 25.1: FORMATION OF 1-DEOXYXYLULOSE 5-PHOSPHATE The
1-amino-3-hydroxyacetone 3-phosphate formed in step 4 of PLP biosynthesis reacts in step 6 with 1-deoxyxylulose 5-phosphate (DXP). DXP arises in step 5 by an aldol-like condensation of D-glyceraldehyde 3-phosphate with pyruvate in a thiamin-dependent reaction catalyzed by DXP synthase.

You might recall from Figure 22.7 on page 912 that pyruvate is converted to acetyl CoA by a process that begins with addition of thiamin diphosphate (TPP) ylide to the ketone carbonyl group, followed by decarboxylation to give hydroxyethylthiamin diphosphate (HETPP). Exactly the same reaction occurs in DXP biosynthesis, but instead of reacting with lipoamide to give a thioester, as in the formation of acetyl CoA, HETPP adds to glyceraldehyde 3-phosphate in an aldol-like reaction. The tetrahedral intermediate that results expels TPP ylide as leaving group and yields DXP. The mechanism is shown in Figure 25.4.

FIGURE 25.4
MECHANISM:
Mechanism of step 5 in pyridoxal phosphate biosynthesis, the thiamin-dependent aldol reaction of D-glyceraldehyde 3-phosphate with pyruvate to give 1-deoxyxylulose 5-phosphate.

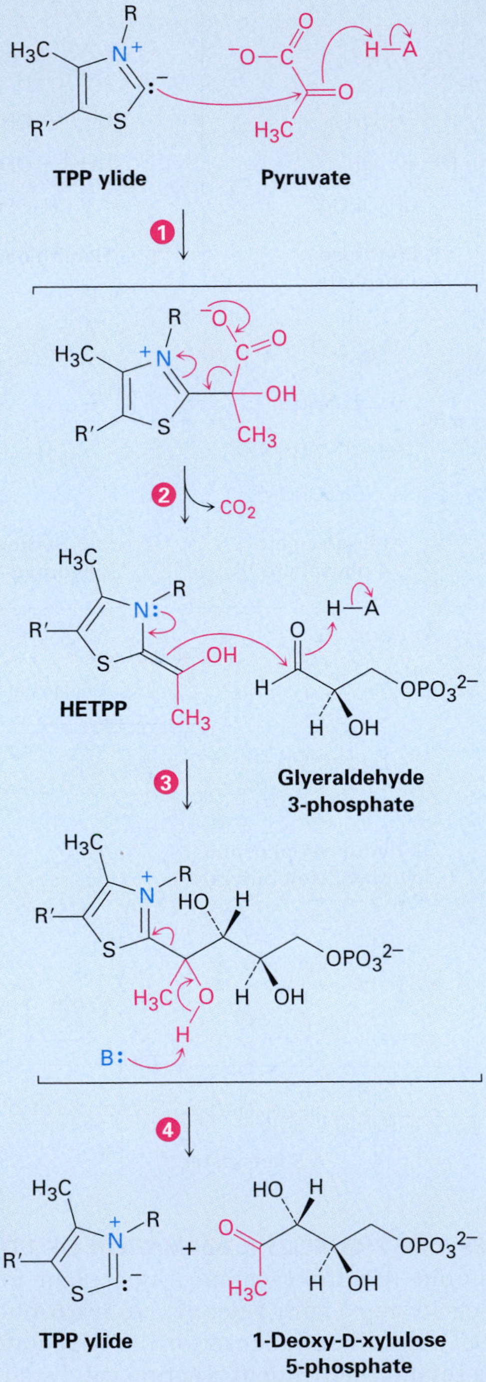

1 Thiamin diphosphate ylide adds to the ketone carbonyl group of pyruvate to yield an alcohol addition product.

2 The addition product contains a C=N bond two carbons away from the carboxylate and is structurally similar to a β-keto acid. It therefore loses CO_2, giving the enamine HETPP.

3 The enamine adds to glyceraldehyde 3-phosphate in an aldol-like reaction.

4 Cleavage of the adduct in a retro-aldol reaction gives 1-deoxy-D-xylulose 5-phosphate and regenerates TPP ylide.

TPP ylide   Pyruvate

HETPP   Glyeraldehyde 3-phosphate

TPP ylide   1-Deoxy-D-xylulose 5-phosphate

© John McMurry

STEP 6 OF FIGURE 25.1: CONDENSATION AND CYCLIZATION 1-Deoxy-D-xylulose 5-phosphate is dephosphorylated and then condenses with 1-amino-3-hydroxyacetone 3-phosphate in step 6 to give pyridoxine 5′-phosphate. The reaction begins with formation of an enamine, followed by loss of water to form an enol that also contains a ketone group six atoms away. The enol adds to the ketone in an intramolecular aldol reaction (Section 17.8) to form a six-membered ring, which then loses water. Tautomerization of the resultant unsaturated ketone gives an aromatic pyridine ring. Note that a loss of phosphate ion occurs at some point in the process, although the exact point at which this happens is not known. The mechanism is shown in Figure 25.5.

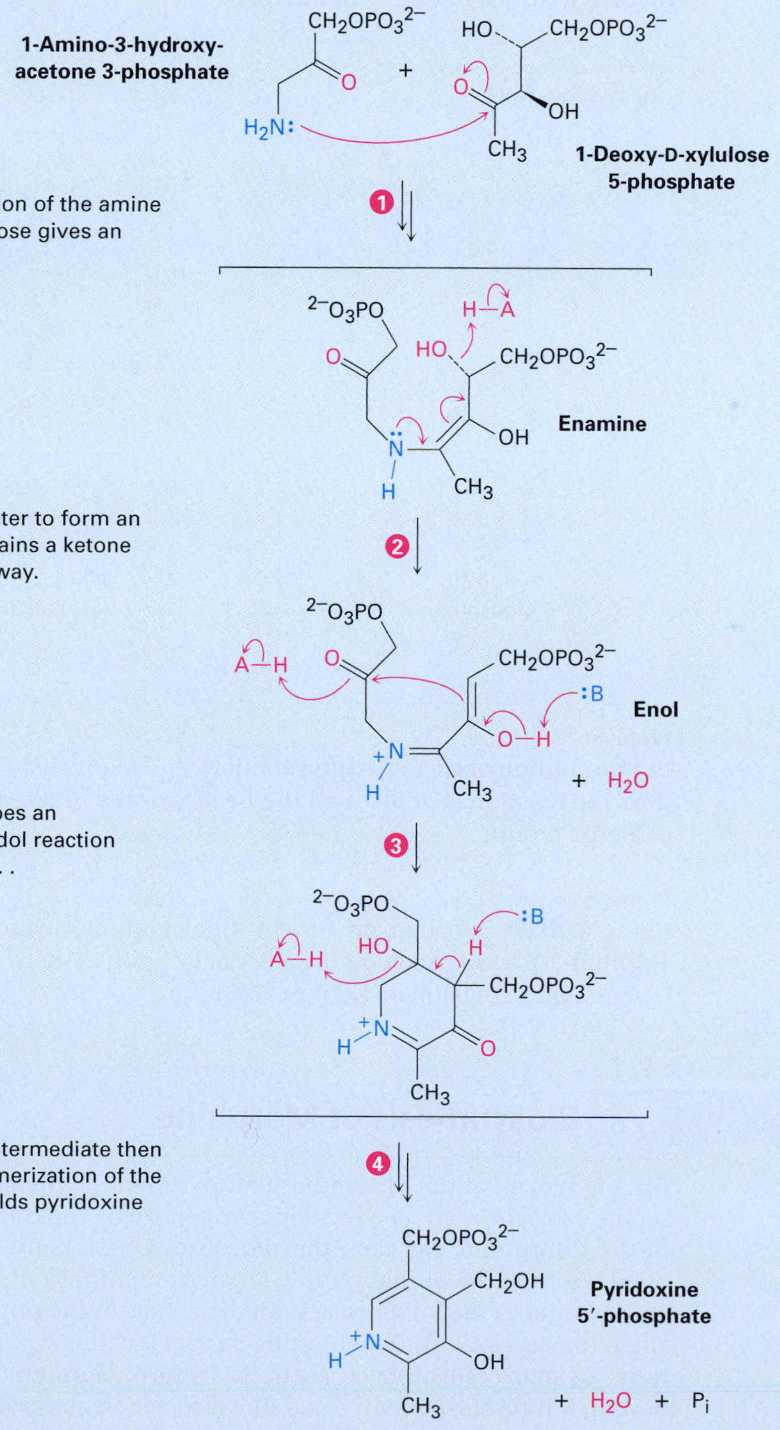

1 Amino-3-hydroxy-acetone 3-phosphate

1-Deoxy-D-xylulose 5-phosphate

❶ Nucleophilic addition of the amine to 1-deoxy-D-xylulose gives an enamine . . .

Enamine

❷ . . . which loses water to form an enol that also contains a ketone group six atoms away.

Enol

❸ The enol undergoes an intramolecular aldol reaction with the ketone . . .

❹ . . . and the aldol intermediate then loses water. Tautomerization of the carbonyl group yields pyridoxine 5′-phosphate.

Pyridoxine 5′-phosphate

FIGURE 25.5

MECHANISM:
Mechanism of step 6 in PLP biosynthesis, the reaction of 1-amino-3-hydroxy-acetone 3-phosphate with 1-deoxy-D-xylu-lose 5-phosphate to give pyridoxine 5′-phosphate.

© John McMurry

STEP 7 OF FIGURE 25.1: OXIDATION The final step in PLP biosynthesis is oxidation of the primary alcohol group in pyridoxine 5'-phosphate to the corresponding aldehyde. Typically, as we've seen on numerous occasions, alcohol oxidations are carried out by either NAD$^+$ or NADP$^+$. In this instance, however, flavin mononucleotide (FMN) is involved as the oxidizing coenzyme and reduced flavin mononucleotide (FMNH$_2$) is the by-product. The details of the reaction are not clear, but evidence suggests that a hydride transfer is involved, just as in NAD$^+$ oxidations.

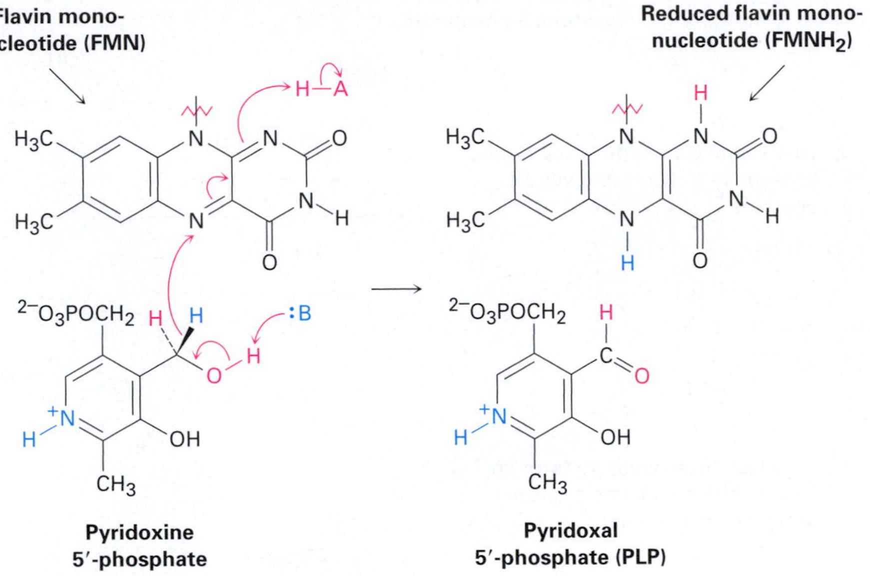

Pyridoxine 5'-phosphate

Pyridoxal 5'-phosphate (PLP)

Problem 25.1

In the addition of HETPP to glyceraldehyde 3-phosphate shown in Figure 25.4, does the reaction take place on the *Re* face or the *Si* face of the glyceraldehyde carbonyl group?

Problem 25.2

Show a likely mechanism for the final tautomerization in the reaction of 1-amino-3-hydroxyacetone 3-phosphate with 1-deoxy-D-xylulose to give pyridoxine 5'-phosphate (Figure 25.5).

25.3 Biosynthesis of Morphine

Having looked at the biosynthesis of pyridoxal 5'-phosphate in the previous section, let's now go up a level in complexity by looking at morphine biosynthesis. Morphine, perhaps the oldest and best known of all alkaloids, is obtained from the opium poppy, *Papaver somniferum,* which has been cultivated for more than 6000 years. Medical uses of the poppy have been known since the early 1500s, when crude extracts, called *opium,* were used for the relief of pain. Morphine was the first pure compound to be isolated from opium, but its close relative codeine also occurs naturally. Codeine, which is simply the methyl ether of morphine and is converted to morphine in the body, is used in prescription cough medicines and as an analgesic. Heroin, another close relative of morphine, does not occur naturally but is synthesized in the laboratory by diacetylation of morphine.

Morphine   **Codeine**   **Heroin**

Chemical investigations into the structure of morphine occupied some of the finest chemical minds of the 19th and early 20th centuries, and it was not until 1924 that the puzzle was finally solved by Robert Robinson, who received the 1947 Nobel Prize in Chemistry for this and other work with alkaloids.

Morphine and its relatives are extremely useful pharmaceutical agents, yet they also pose an enormous social problem because of their addictive properties. Much effort has therefore gone into understanding how morphine works and into developing modified morphine analogs that retain the analgesic activity but don't cause physical dependence. Our present understanding is that morphine functions by binding to so-called mu opioid receptor sites in both the spinal cord, where it interferes with the transmission of pain signals, and brain neurons, where it changes the brain's reception of the signal.

Hundreds of morphine-like molecules have been synthesized and tested for their analgesic properties. Research has shown that not all the complex framework of morphine is necessary for biological activity. According to the "morphine rule," biological activity requires (1) an aromatic ring attached to (2) a quaternary carbon atom, followed by (3) two more carbon atoms and (4) a tertiary amine. Meperidine (Demerol), a widely used analgesic, and methadone, a substance used in the treatment of heroin addiction, are two compounds that fit the morphine rule.

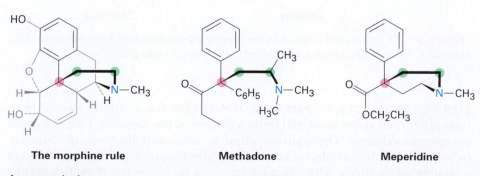

The morphine rule   **Methadone**   **Meperidine**

An aromatic ring
attached to a quaternary carbon (●)
followed by two more carbons (●)
and a tertiary amine (N)

Morphine is biosynthesized from two molecules of the amino acid tyrosine. One tyrosine is converted into dopamine, the second is converted into *p*-hydroxyphenylacetaldehyde, and the two are coupled to give morphine. The entire pathway is a bit complex at several points, but an abbreviated scheme is given in Figure 25.6.

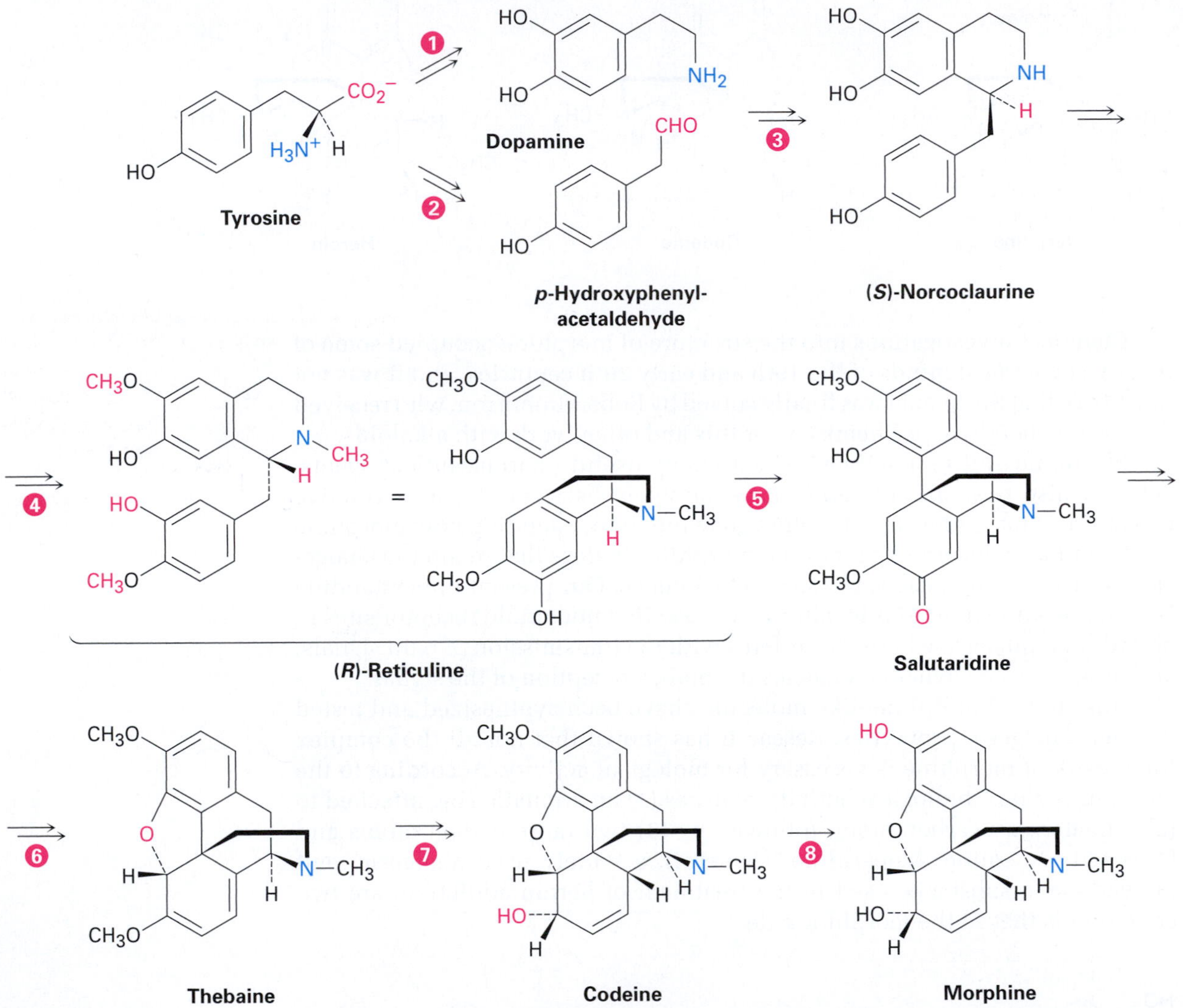

FIGURE 25.6 An abbreviated pathway for the biosynthesis of morphine from two molecules of tyrosine. The individual steps are explained in more detail in the text.

STEP 1 OF FIGURE 25.6: DOPAMINE BIOSYNTHESIS Dopamine is formed from tyrosine in two steps: an initial hydroxylation of the aromatic ring, followed by decarboxylation. The hydroxylation is catalyzed by tyrosine 3-monooxygenase, requires a cofactor called tetrahydrobiopterin, and occurs through a somewhat complex pathway that involves an iron–oxo (Fe=O) complex analogous to that involved in prostaglandin biosynthesis (Figure 8.11). The decarboxylation is catalyzed by the PLP-dependent enzyme aromatic L-amino acid decarboxylase.

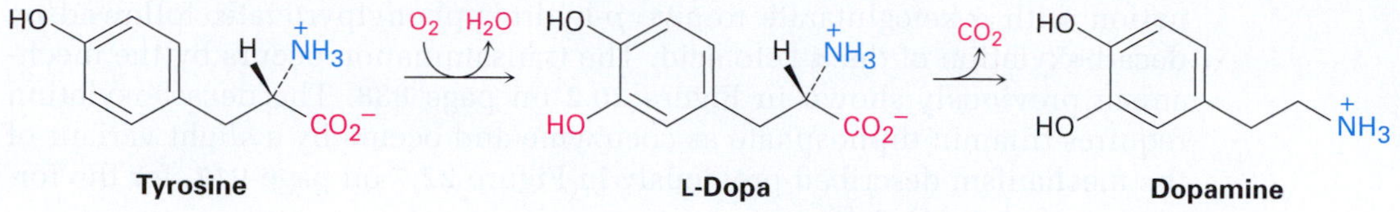

Recall from Section 20.2 that pyridoxal 5'-phosphate reacts with the α amino group of an α-amino acid to form an imine, or Schiff base. When L-dopa reacts with PLP, the resultant imine undergoes decarboxylation, with the pyridinium ion of PLP acting as the electron acceptor. Hydrolysis then gives dopamine and regenerated PLP. The mechanism is shown in Figure 25.7.

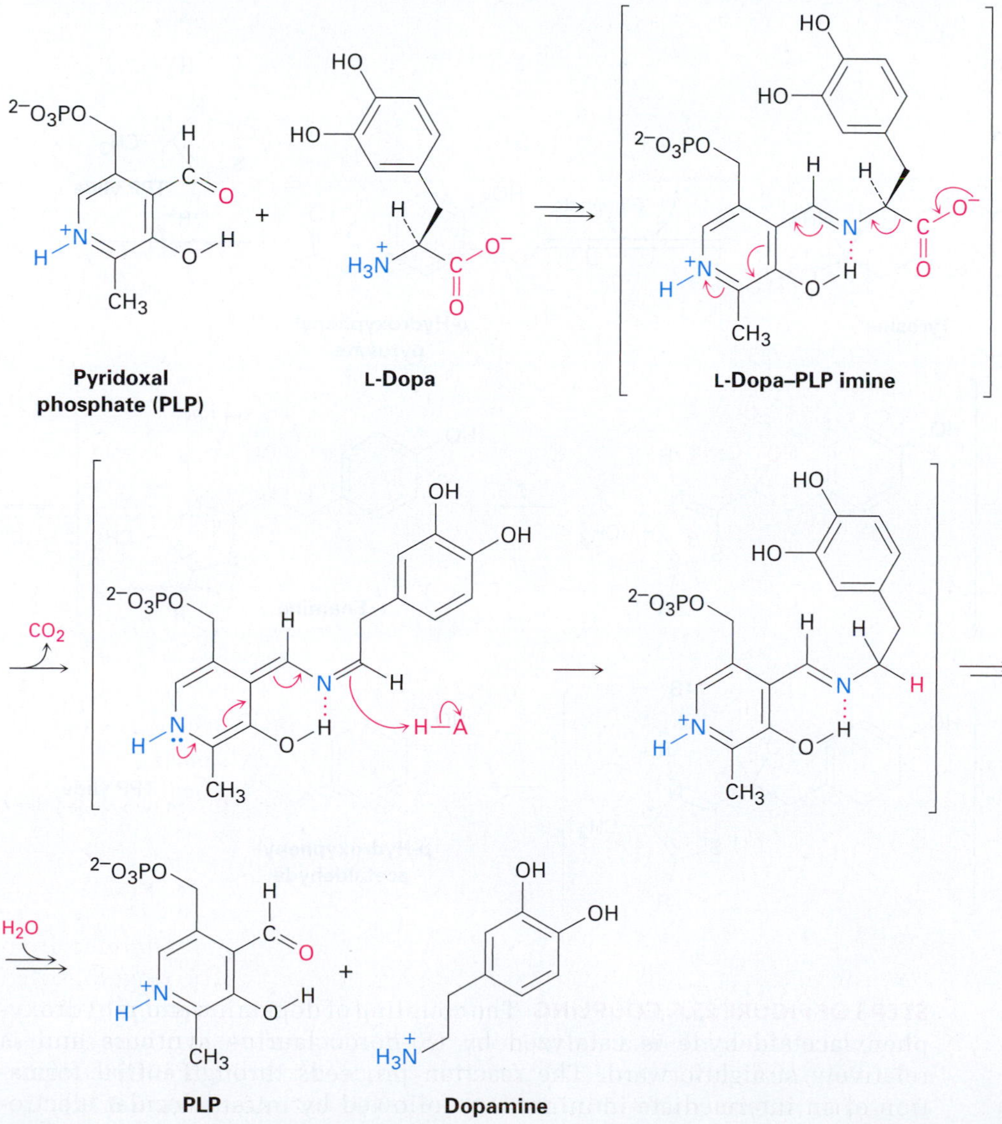

FIGURE 25.7 Mechanism of step 1 in morphine biosynthesis, the PLP-dependent decarboxylation of L-dopa to give dopamine.

STEP 2 OF FIGURE 25.6: *p*-HYDROXYPHENYLACETALDEHYDE BIOSYNTHESIS

p-Hydroxyphenylacetaldehyde, the second tyrosine-derived precursor of morphine, is also formed in two steps: an initial PLP-dependent transamination with α-ketoglutarate to give *p*-hydroxyphenylpyruvate, followed by decarboxylation of the α keto acid. The transamination occurs by the mechanism previously shown in Figure 20.2 on page 838. The decarboxylation requires thiamin diphosphate as coenzyme and occurs by a slight variant of the mechanism described previously in Figure 22.7 on page 912, for the formation of acetyl CoA from pyruvate.

Decarboxylation of *p*-hydroxyphenylpyruvate begins with nucleophilic addition of TPP ylide to the ketone carbonyl group, followed by loss of CO_2 to give an enamine in the usual way. But whereas the enamine formed from pyruvate decarboxylation reacts with lipoamide to give a thioester and regenerated TPP ylide, the enamine from *p*-hydroxyphenylpyruvate decarboxylation is simply protonated to give an aldehyde plus TPP ylide. The mechanism is shown in Figure 25.8.

FIGURE 25.8 Mechanism of step 2 in morphine biosynthesis, the TPP-dependent decarboxylation of *p*-hydroxyphenylpyruvate to give *p*-hydroxyphenylacetaldehyde.

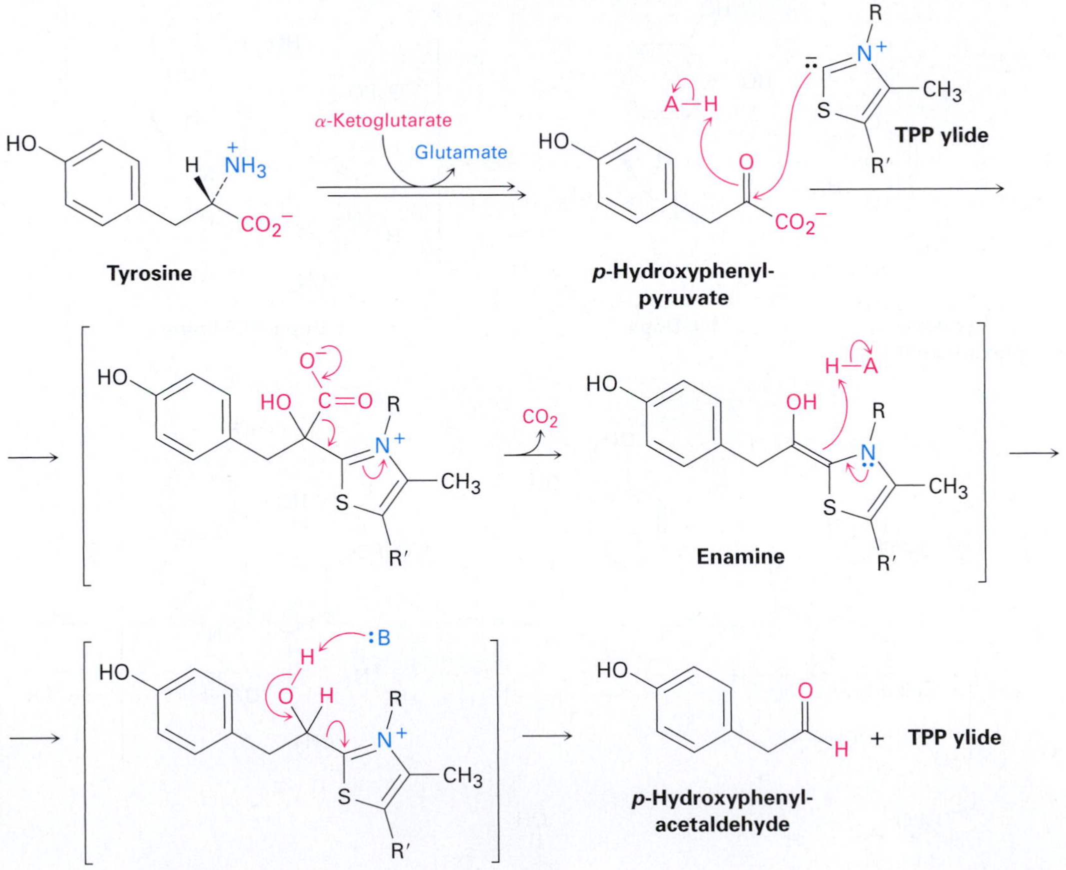

STEP 3 OF FIGURE 25.6: COUPLING

The coupling of dopamine and *p*-hydroxyphenylacetaldehyde is catalyzed by (*S*)-norcoclaurine synthase and is relatively straightforward. The reaction proceeds through initial formation of an intermediate iminium ion, followed by intramolecular electrophilic aromatic substitution at a position *para* to one of the hydroxyl groups (Figure 25.9).

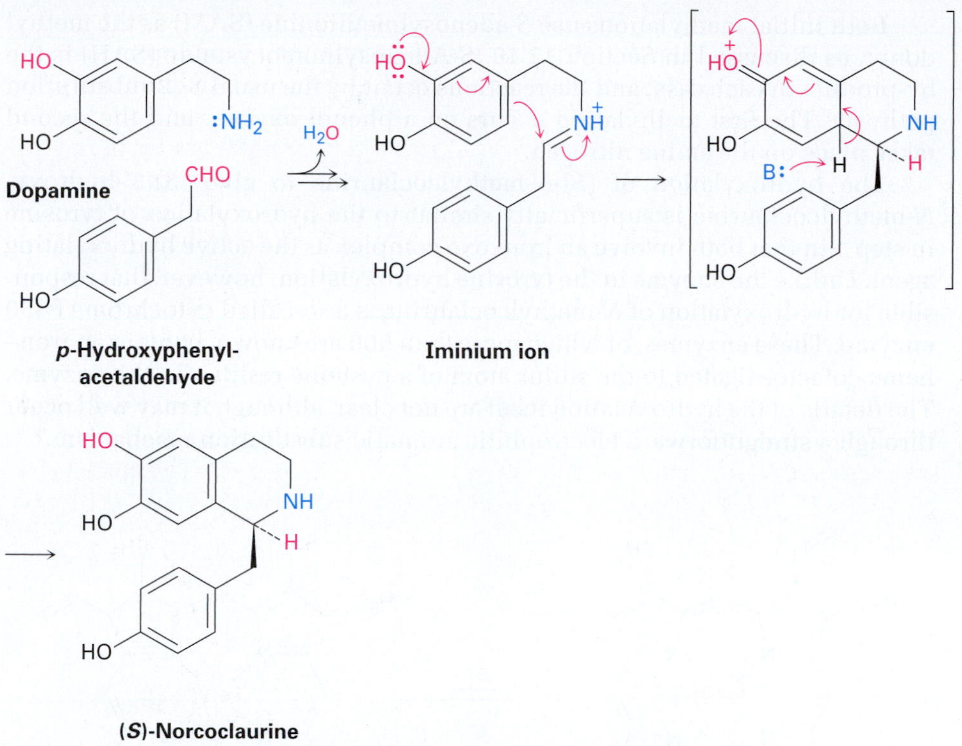

FIGURE 25.9
Mechanism of step 3 in morphine biosynthesis, the coupling of dopamine and *p*-hydroxyphenylacetaldehyde to give (*S*)-norcoclaurine.

STEP 4 OF FIGURE 25.6: METHYLATION, HYDROXYLATION, AND EPIMERIZATION

(*S*)-Norcoclaurine next undergoes two methylations and a hydroxylation to give (*S*)-3′-hydroxy-*N*-methylcoclaurine, which is methylated a third time to produce (*S*)-reticuline. Epimerization of (*S*)-reticuline then yields (*R*)-reticuline (Figure 25.10).

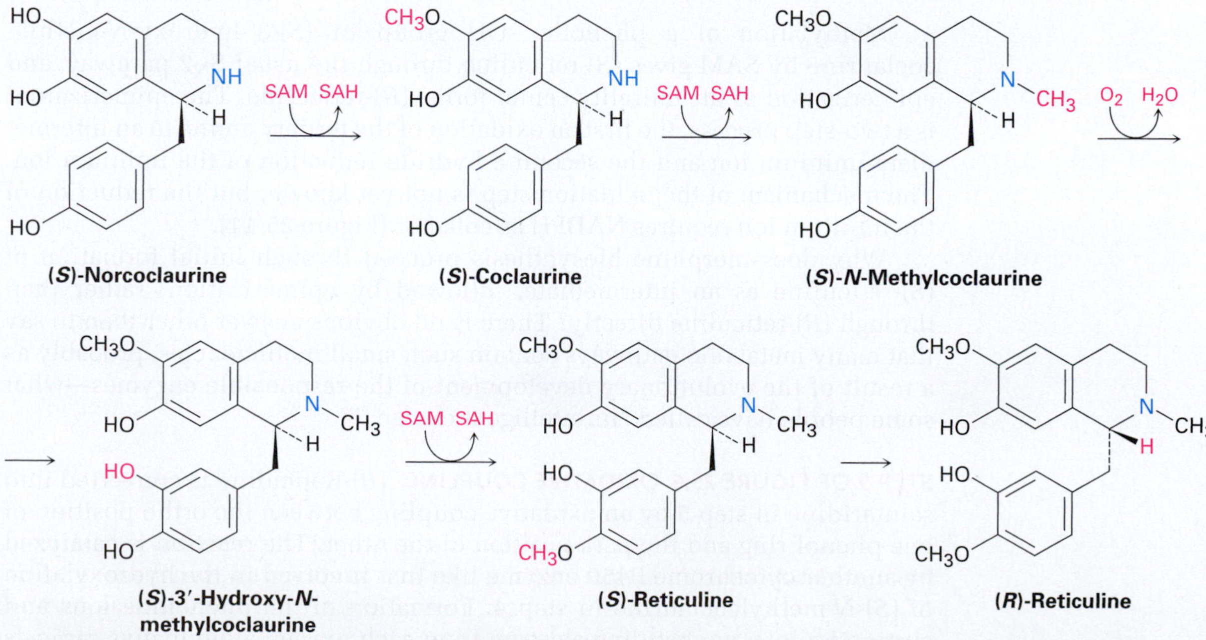

FIGURE 25.10 An overview of the reactions in step 4 of morphine biosynthesis, the conversion of (*S*)-norcoclaurine to (*R*)-reticuline.

Both initial methylations use *S*-adenosylmethionine (SAM) as the methyl donor, as discussed in Section 12.10. *S*-Adenosylhomocysteine (SAH) is the by-product in each case, and the reactions occur by the usual S_N2 substitution pathway. The first methylation occurs on a phenol oxygen, and the second takes place on the amine nitrogen.

The hydroxylation of (*S*)-*N*-methylcoclaurine to give (*S*)-3′-hydroxy-*N*-methylcoclaurine is superficially similar to the hydroxylation of tyrosine in step 1 in that both involve an iron–oxo complex as the active hydroxylating agent. Unlike the enzyme in the tyrosine hydroxylation, however, that responsible for hydroxylation of *N*-methylcoclaurine is a so-called cytochrome P450 enzyme. These enzymes, of which more than 500 are known, contain an iron–heme cofactor ligated to the sulfur atom of a cysteine residue in the enzyme. The details of the hydroxylation itself are not clear, although it may well occur through a straightforward electrophilic aromatic substitution mechanism.

Heme   **Heme iron–oxo complex**

Methylation of a phenolic –OH group in (*S*)-3′-hydroxy-*N*-methyl-coclaurine by SAM gives (*S*)-reticuline through the usual S_N2 pathway, and epimerization of the chirality center forms (*R*)-reticuline. The epimerization is a two-step process, the first an oxidation of the tertiary amine to an intermediate iminium ion and the second a hydride reduction of the iminium ion. The mechanism of the oxidation step is not yet known, but the reduction of the iminium ion requires NADPH as cofactor (Figure 25.11).

Why does morphine biosynthesis proceed through initial formation of (*S*)-reticuline as an intermediate, followed by epimerization, rather than through (*R*)-reticuline directly? There is no obvious answer other than to say that many metabolic pathways contain such small inefficiencies, probably as a result of the evolutionary development of the responsible enzymes—what some people have called "unintelligent design."

STEP 5 OF FIGURE 25.6: OXIDATIVE COUPLING (*R*)-Reticuline is converted into salutaridine in step 5 by an oxidative coupling between the ortho position of one phenol ring and the para position of the other. The reaction is catalyzed by another cytochrome P450 enzyme like that involved in the hydroxylation of (*S*)-*N*-methylcoclaurine in step 4. Formation of the phenoxide ions and abstraction of a nonbonding electron from each oxygen atom to give radicals occurs, followed by radical coupling and a keto–enol tautomerization to yield salutaridine (Figure 25.12).

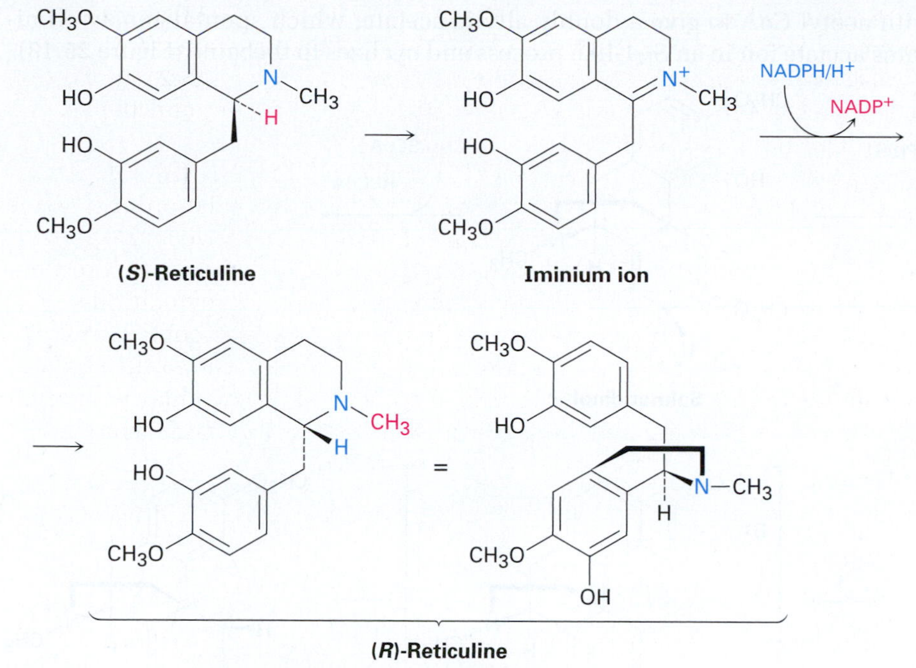

FIGURE 25.11 Mechanism of the epimerization of (S)-reticuline to (R)-reticuline in step 4 of morphine biosynthesis.

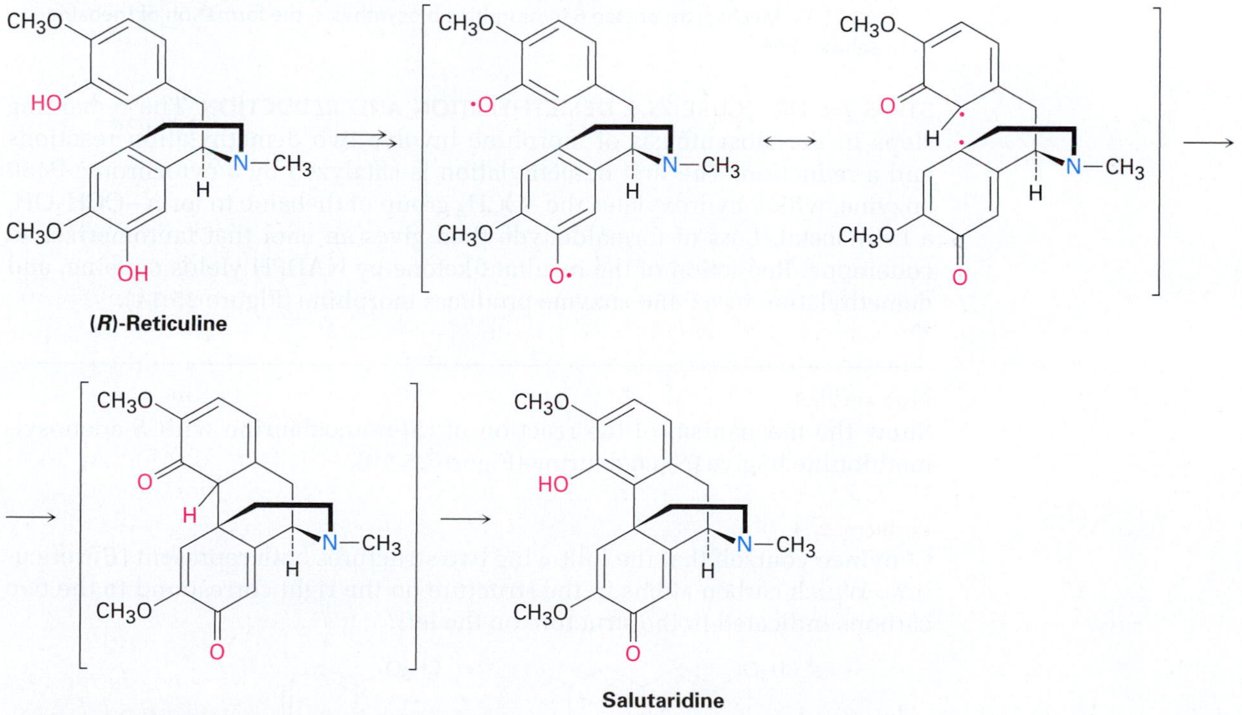

FIGURE 25.12 Mechanism of step 5 in morphine biosynthesis, the oxidative phenol coupling of (R)-reticuline to salutaridine.

STEP 6 OF FIGURE 25.6: REDUCTION AND CYCLIZATION

Reduction of salutaridine to salutaridinol is catalyzed by salutaridine reductase, with NADPH as cofactor. This alcohol then undergoes a nucleophilic acyl substitution reaction with acetyl CoA to give a doubly allylic acetate, which spontaneously eliminates acetate ion in an S_N1-like process and cyclizes to thebaine (Figure 25.13).

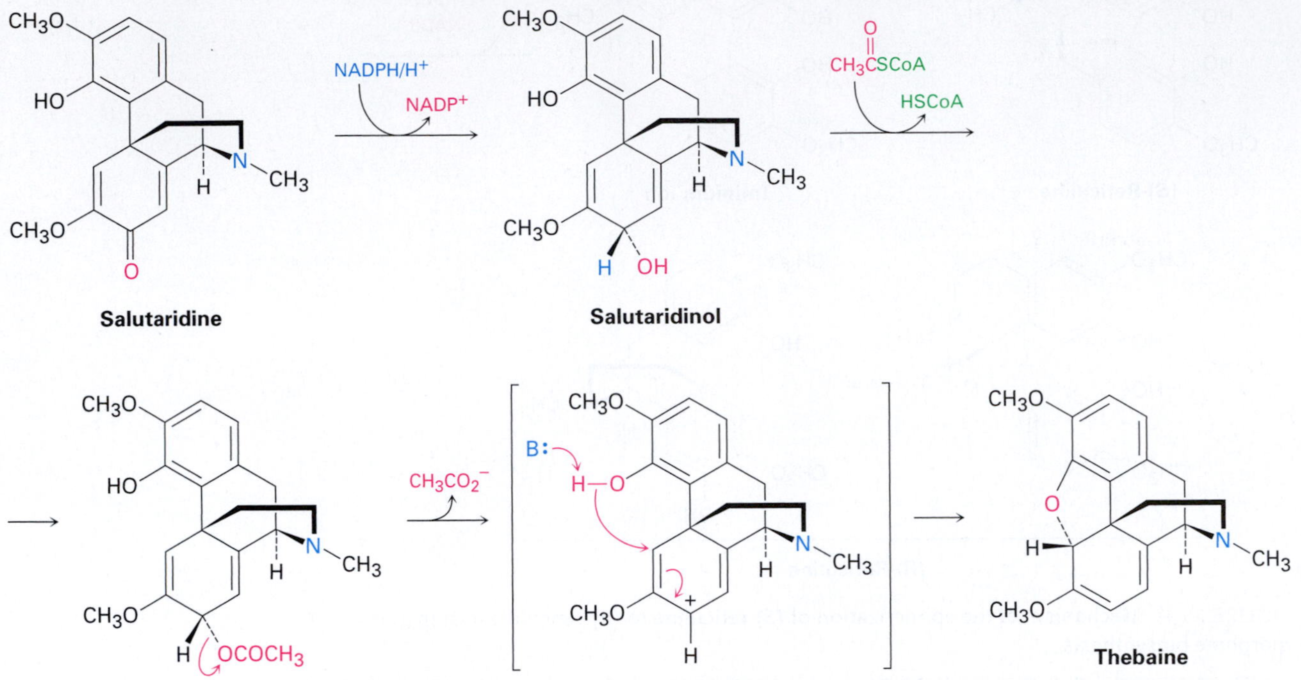

Salutaridine

Salutaridinol

Thebaine

FIGURE 25.13 Mechanism of step 6 in morphine biosynthesis, the formation of thebaine from salutaridine.

STEPS 7–8 OF FIGURE 25.6: DEMETHYLATION AND REDUCTION The remaining steps in the biosynthesis of morphine involve two demethylation reactions and a reduction. The first demethylation is catalyzed by a cytochrome P450 enzyme, which hydroxylates the $-OCH_3$ group of thebaine to form $-OCH_2OH$, a hemiacetal. Loss of formaldehyde then gives an enol that tautomerizes to codeinone. Reduction of the resultant ketone by NADPH yields codeine, and demethylation by a P450 enzyme produces morphine (Figure 25.14).

Problem 25.3

Show the mechanism of the reaction of (*S*)-norcoclaurine with *S*-adenosyl-methionine to give (*S*)-coclaurine (Figure 25.10).

Problem 25.4

Convince yourself that the following two structures both represent (*R*)-reticuline. Which carbon atoms in the structure on the right correspond to the two carbons indicated in the structure on the left?

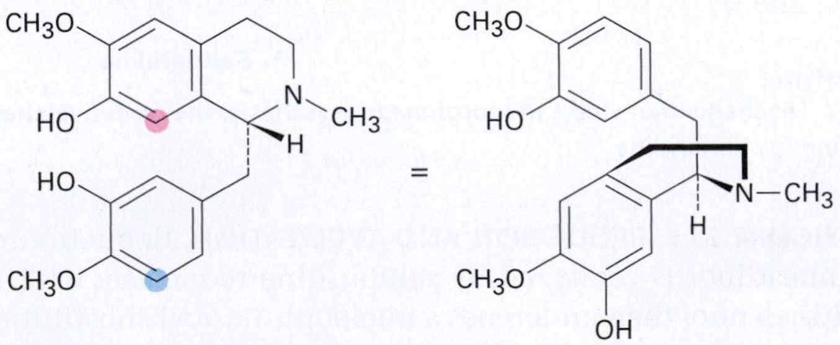

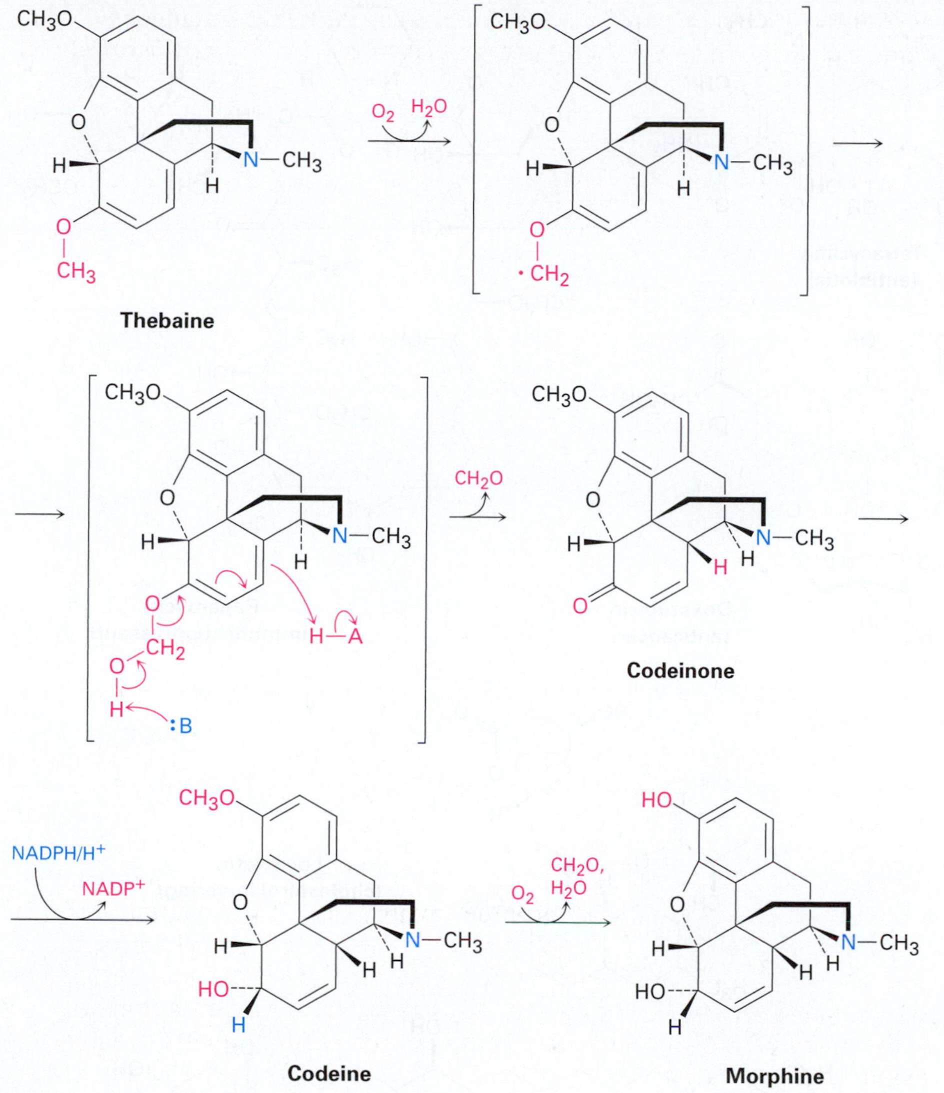

FIGURE 25.14 Mechanism of step 7 in morphine biosynthesis, the demethylation of thebaine to give codeinone, catalyzed by a P450 enzyme. Reduction of codeinone with NADPH then yields codeine, and a final demethylation produces morphine.

25.4 Biosynthesis of Erythromycin

Having discussed the biosynthesis of pyridoxal phosphate and morphine in the preceding two sections, we'll end this chapter on natural-products chemistry by going up yet one more level in complexity and looking at polyketide biosynthesis. Unlike what happens in many metabolic pathways, where each separate step is catalyzed by a separate, relatively small enzyme, erythromycin and other polyketides are assembled by a single massive enzyme called a *synthase*. The synthase contains many enzyme domains linked together, with each domain catalyzing a specific biosynthetic step in sequence.

Polyketides are an extraordinarily valuable class of natural products, numbering over 10,000 compounds. Commercially important polyketides include antibiotics (erythromycin A, tetracycline) and immunosuppressants (rapamycin), as well as anticancer (doxorubicin), antifungal (amphotericin B), and cholesterol-lowering (lovastatin) agents (Figure 25.15). It has been estimated that the sales of these and other polyketide pharmaceuticals total more than $15 billion per year.

FIGURE 25.15 Structures of some polyketides used as pharmaceutical agents.

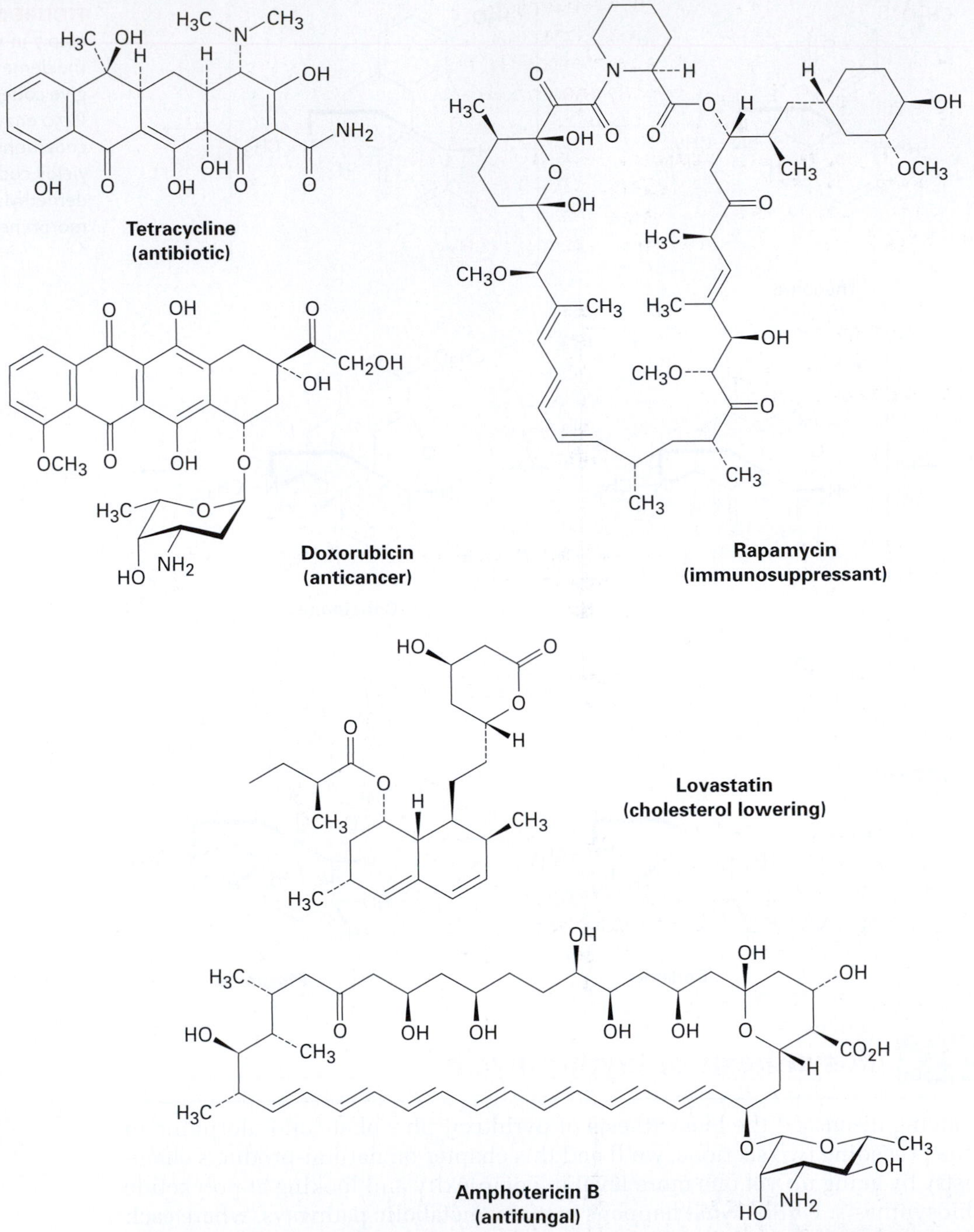

Tetracycline
(antibiotic)

Doxorubicin
(anticancer)

Rapamycin
(immunosuppressant)

Lovastatin
(cholesterol lowering)

Amphotericin B
(antifungal)

Polyketides are biosynthesized by the joining together of the simple acyl CoA's acetyl CoA, propionyl CoA, methylmalonyl CoA, and (less frequently) butyryl CoA. The key carbon–carbon bond-forming step in each joining is a Claisen condensation (Section 17.9). Once the carbon chain is assembled and released from the enzyme, further transformations take place to give the final product. Erythromycin A, for instance, is prepared from one propionate and six methylmalonate units by the pathway outlined in Figure 25.16. Following initial assembly of the acyl units into the macrocyclic lactone 6-deoxyerythronolide B, two hydroxylations, two glycosylations, and a final methylation complete the biosynthesis.

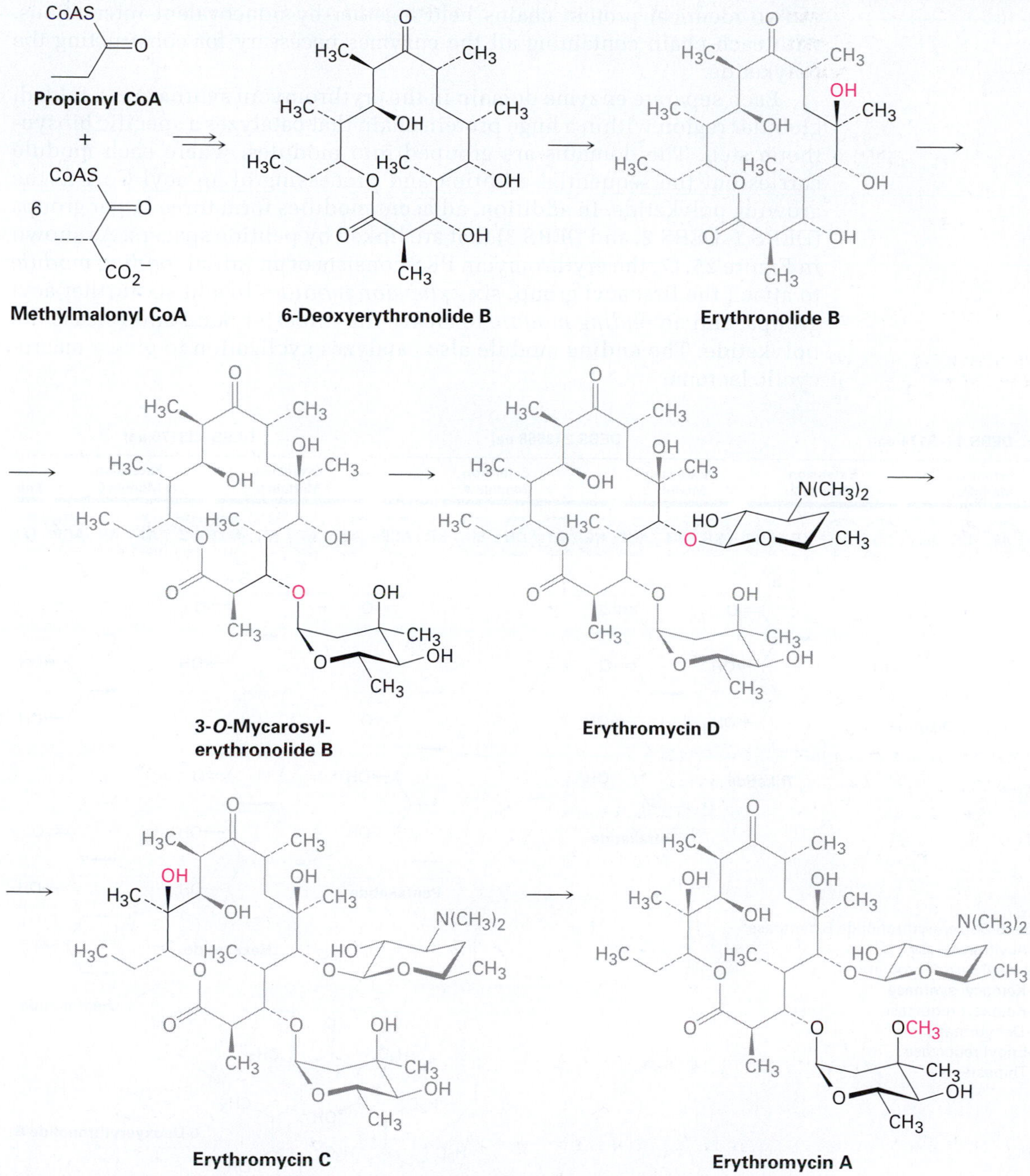

FIGURE 25.16 An outline of the pathway for the biosynthesis of erythromycin A. One propionate and six methylmalonate units are first assembled into the macrocyclic lactone 6-deoxyerythronolide B, which is then hydroxylated, glycosylated by two different sugars, hydroxylated again, and finally methylated.

The initial assembly of seven acyl CoA precursors to build a polyketide carbon chain is carried out by a multienzyme complex called a *polyketide synthase*, or PKS. The 6-deoxyerythronolide B synthase (DEBS) is a massive structure of greater than 2 million molecular weight and containing more than 20,000 amino acids. Furthermore, it is a *homodimer*, meaning that it consists of two identical protein chains held together by noncovalent interactions, with each chain containing all the enzymes necessary for constructing the polyketide.

Each separate enzyme domain in the erythromycin synthase is a folded, globular region within a huge protein chain that catalyzes a specific biosynthetic step. The domains are grouped into modules, where each module carries out the sequential addition and processing of an acyl CoA to the growing polyketide. In addition, adjacent modules form three larger groups (DEBS 1, DEBS 2, and DEBS 3) that are linked by peptide spacers. As shown in Figure 25.17, the erythromycin PKS consists of an initial *loading module* to attach the first acyl group, six *extension modules* to add six further acyl groups, and an *ending module* to cleave the thioester bond and release the polyketide. The ending module also catalyzes cyclization to give a macrocyclic lactone.

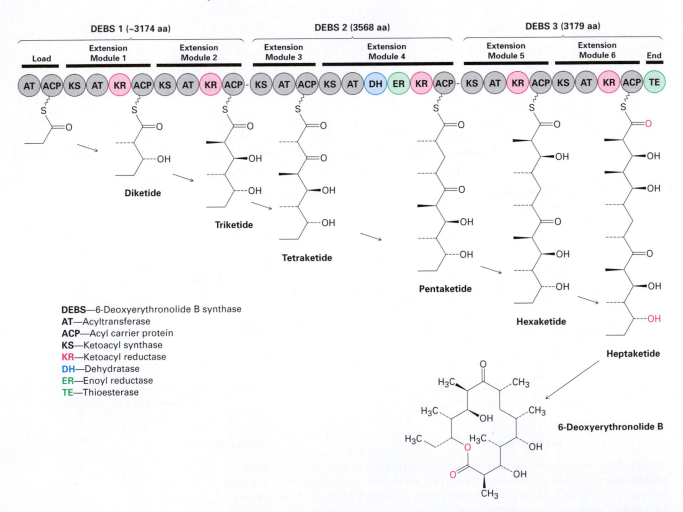

FIGURE 25.17 A schematic view of the 6-deoxyerythronolide B synthase (DEBS), showing the locations of the enzyme domains within the loading module and the six extension modules. The figure is explained in detail in the text.

The loading module has two domains: an acyl transfer (AT) domain and an acyl carrier protein (ACP) domain. The AT selects the first acyl CoA (propionyl CoA in the case of erythromycin) and transfers it to the adjacent ACP, which binds it through a thioester linkage and holds it for further reaction. Each extension module has a minimum of three domains: an AT, an ACP, and a ketosynthase (KS), which catalyzes the Claisen condensation reaction that builds the polyketide chain. In addition to the three minimum domains, some extension modules also contain a ketoreductase (KR) to reduce a ketone carbonyl group and produce an alcohol, a dehydratase (DH) to dehydrate the alcohol and produce a C=C bond, and an enoyl reductase (ER) to reduce the C=C bond. Finally, the ending domain is a thioesterase (TE), which releases the product by catalyzing a lactonization.

Polyketide chain extension occurs when an extension module AT selects a new acyl CoA, transfers it to the ACP, and the KS then catalyzes a Claisen condensation reaction between the newly bonded acyl group and the acyl group of the previous module. Figure 25.18 shows the steps occurring in the first extension cycle; other extension cycles take place similarly.

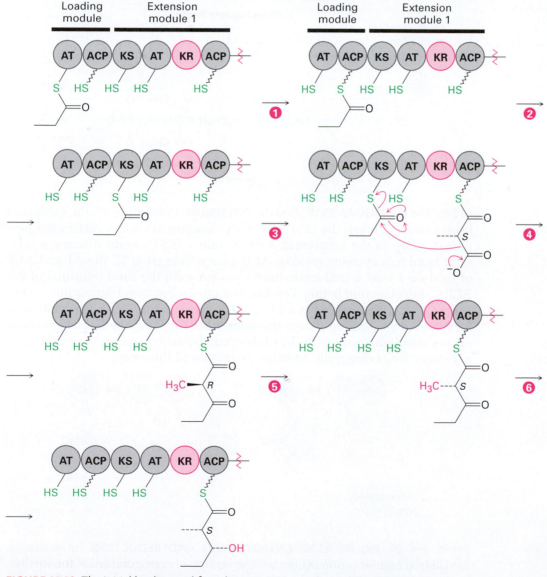

FIGURE 25.18 The initial loading and first chain-extension cycle catalyzed by the erythromycin PKS. Individual steps are explained in the text.

STEP 1 OF FIGURE 25.18: LOADING The loading AT domain begins the erythromycin biosynthesis by binding a propionyl CoA through a thioester bond to the –SH of a cysteine residue. The AT then transfers the propionyl group to the adjacent ACP. Each ACP in the synthase contains a phosphopantetheine bonded to the hydroxyl of a serine residue, and bonding of the acyl group to the enzyme occurs by thioester formation with the phosphopantetheine –SH (Figure 25.19). The phosphopantetheine effectively acts as a long, flexible arm to allow movement of the acyl group from one catalytic domain to another.

FIGURE 25.19 *Formation of an acyl ACP during polyketide biosynthesis. Phosphopantetheine, symbolized by a zigzag line between S and ACP, acts as a long, flexible arm to allow the acyl group to move from one catalytic domain to another.*

STEPS 2–4 OF FIGURE 25.18: CHAIN EXTENSION Polyketide chain extension begins (step 2) when the acyl ACP of the loading module transfers the propionyl group to the ketosynthase of module 1 (KS1), again forming a thioester bond to a cysteine residue. At the same time (step 3), the AT and ACP of module 1 load a (2S)-methylmalonyl CoA onto the thiol terminus of the ACP1 phosphopantetheine. The key carbon–carbon bond formation occurs (step 4) when KS1 catalyzes a Claisen condensation and decarboxylation to form an enzyme-bound β-keto thioester. It's likely that the decarboxylation occurs simultaneously with the Claisen condensation, giving the enolate ion necessary for nucleophilic addition to the second thioester.

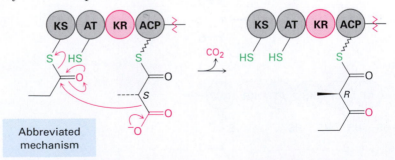

STEPS 5–6 OF FIGURE 25.18: EPIMERIZATION AND REDUCTION Interestingly, the Claisen condensation occurs with inversion of configuration at the methyl-bearing chirality center so that the initially formed diketide has (R) stereochemistry. Base catalyzed epimerization of the (R) product, an acidic β-diketone,

occurs in step 5, however, so the product that goes on to the next step regains the (S) configuration. Finally, KR1 reduces the ketone to a β-hydroxy thioester in step 6 by transfer of the *pro-S* hydrogen from NADPH as cofactor. Module 1 is now finished, so the diketide is transferred to KS2 for another chain extension.

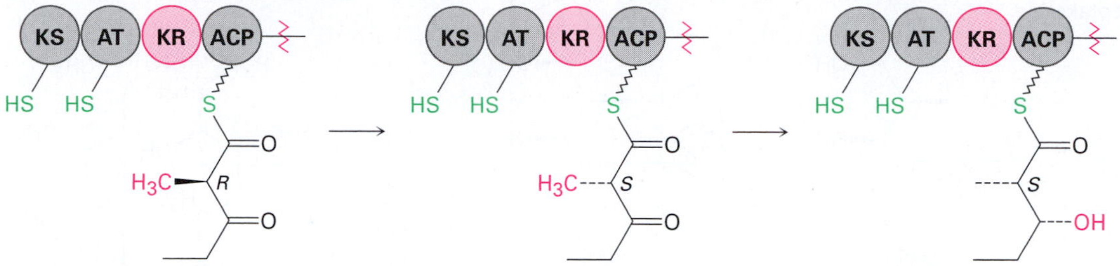

The reactions catalyzed by extension modules 2, 5, and 6 are similar to those of module 1, although the stereochemistries of the Claisen condensation and reduction steps may differ. The reactions in modules 3 and 4, however, are different. Module 3 lacks a KR domain, so no reduction occurs and the tetraketide product contains a ketone carbonyl group (Figure 25.17). Module 4 contains a KR and two additional enzyme domains, so it catalyzes a ketone reduction plus two additional reactions. Following the reduction by KR4 of the pentaketide, a dehydratase (DH) dehydrates the pentaketide alcohol to an α,β-unsaturated thioester and the double bond is then reduced by an enoyl reductase (ER) domain (Figure 25.20).

Note that the complete sequence of reactions carried out by module 4—Claisen condensation, ketone reduction, dehydration, and double-bond reduction—is identical to the series of reactions found in fatty-acid biosynthesis (Figure 23.6 on page 952). In fact, all fatty-acid synthases have the same set of AT, ACP, KS, KR, DH, and ER domains as the polyketide synthases.

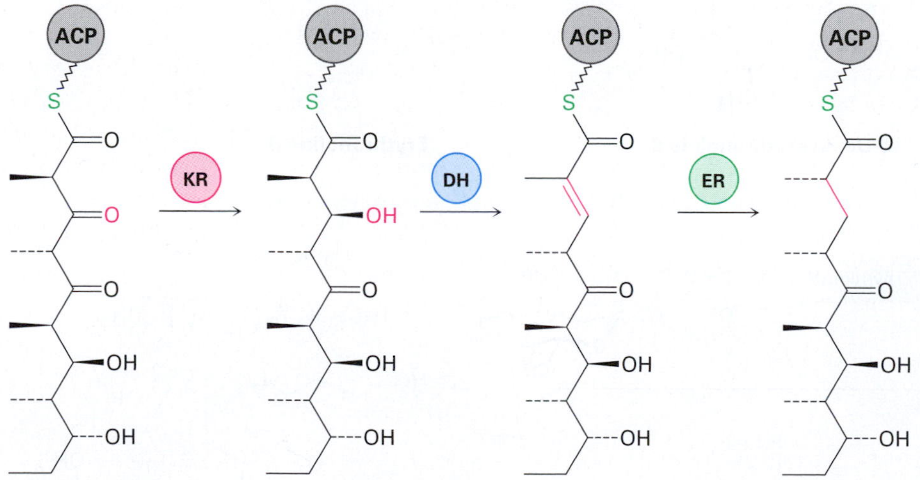

FIGURE 25.20 Additional processing of the pentaketide intermediate in module 4 removes a carbonyl group by a reduction–dehydration–reduction sequence.

A pentaketide

Release of 6-deoxyerythronolide B from the PKS is catalyzed by the ending thioesterase module. A serine residue on the TE module first carries out a nucleophilic acyl substitution on the ACP-bound heptaketide, and the acyl enzyme that results undergoes lactonization. A histidine residue in the TE acts as base to catalyze nucleophilic acyl substitution of the serine ester by the terminal –OH group in the heptaketide (Figure 25.21).

FIGURE 25.21 Release of 6-deoxyerythronolide from the PKS occurs by lactonization of an acyl enzyme, formed by reaction of a serine residue in the TE module with the heptaketide.

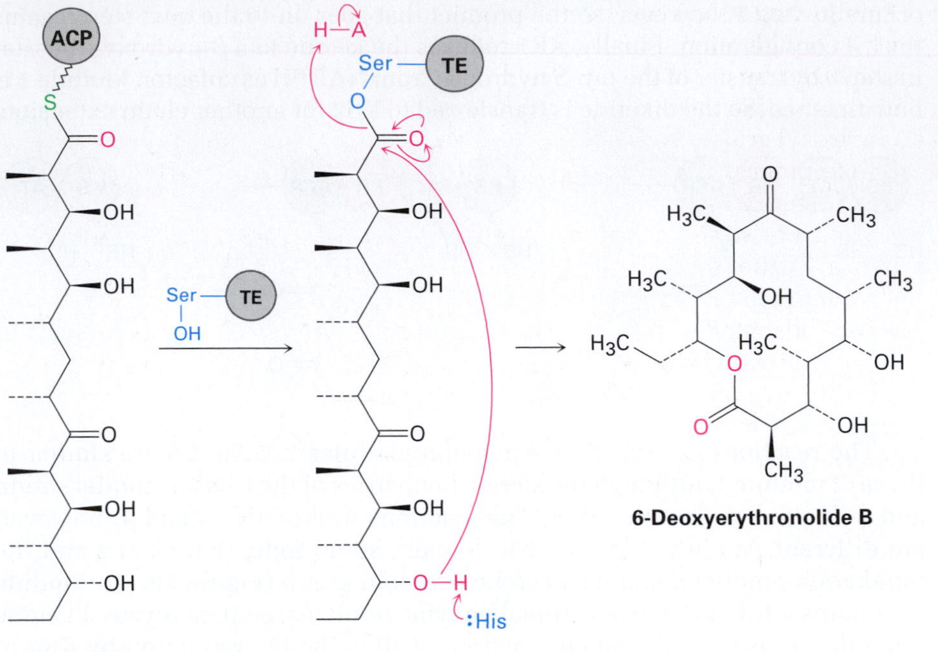

Heptaketide

6-Deoxyerythronolide B

FIGURE 25.22 Hydroxylation and glycosylation of 6-deoxyerythronolide B to give 3-O-mycarosyl-erythronolide B.

6-Deoxyerythronolide B

Erythronolide B

3-O-Mycarosyl-erythronolide B

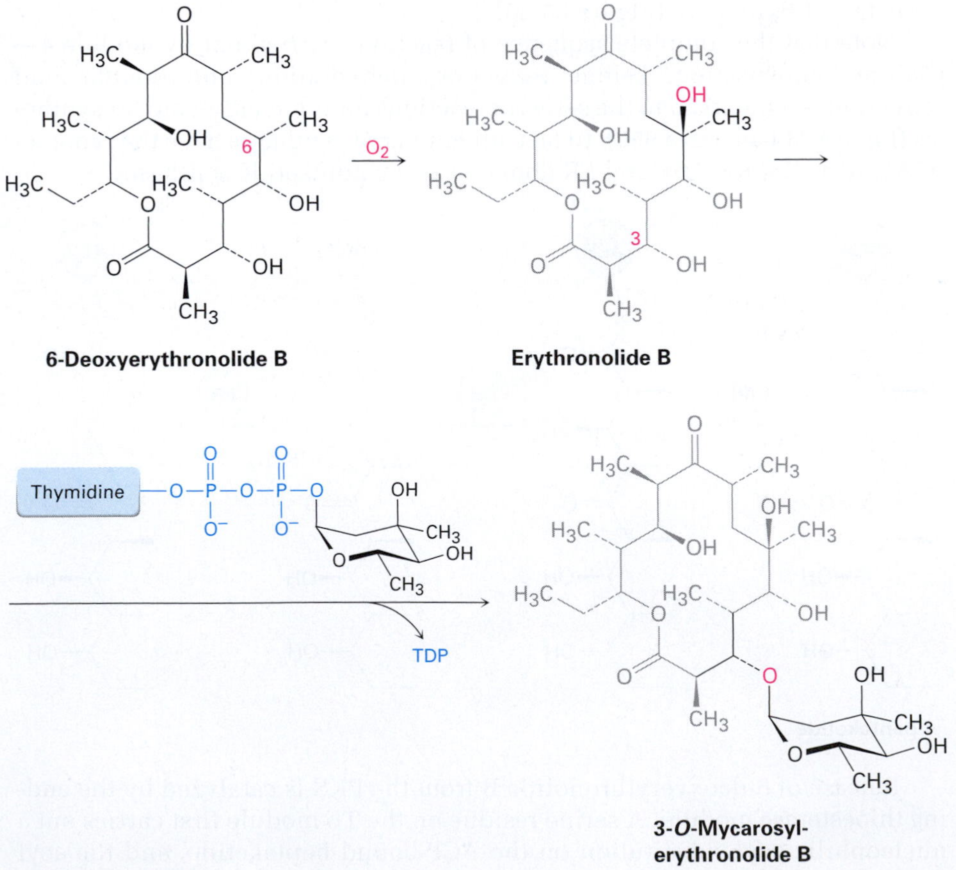

Following its release from the PKS, 6-deoxyerythronolide B is hydroxylated at C6 with retention of configuration to give erythronolide B. The reaction is catalyzed by a P450 hydroxylase analogous to that involved in morphine biosynthesis (Section 25.3, Figure 25.14). L-Mycarose is then attached to the C3 hydroxyl group by reaction with thymidyl diphosphomycarose through an S_N1-like process that proceeds by initial formation of the mycarosyl carbocation (Figure 25.22).

The final steps in erythromycin A biosynthesis are a further glycosylation, a further hydroxylation, and a methylation (Figure 25.23). As in the attachment of mycarose, the attachment of the amino sugar D-desosamine also takes place by transfer from a thymidyl diphosphosugar. C12 hydroxylation by another P450 enzyme occurs with retention of configuration to give erythromycin C, and methylation of the C3′ hydroxyl group of the mycarose unit by reaction with *S*-adenosylmethionine gives erythromycin A.

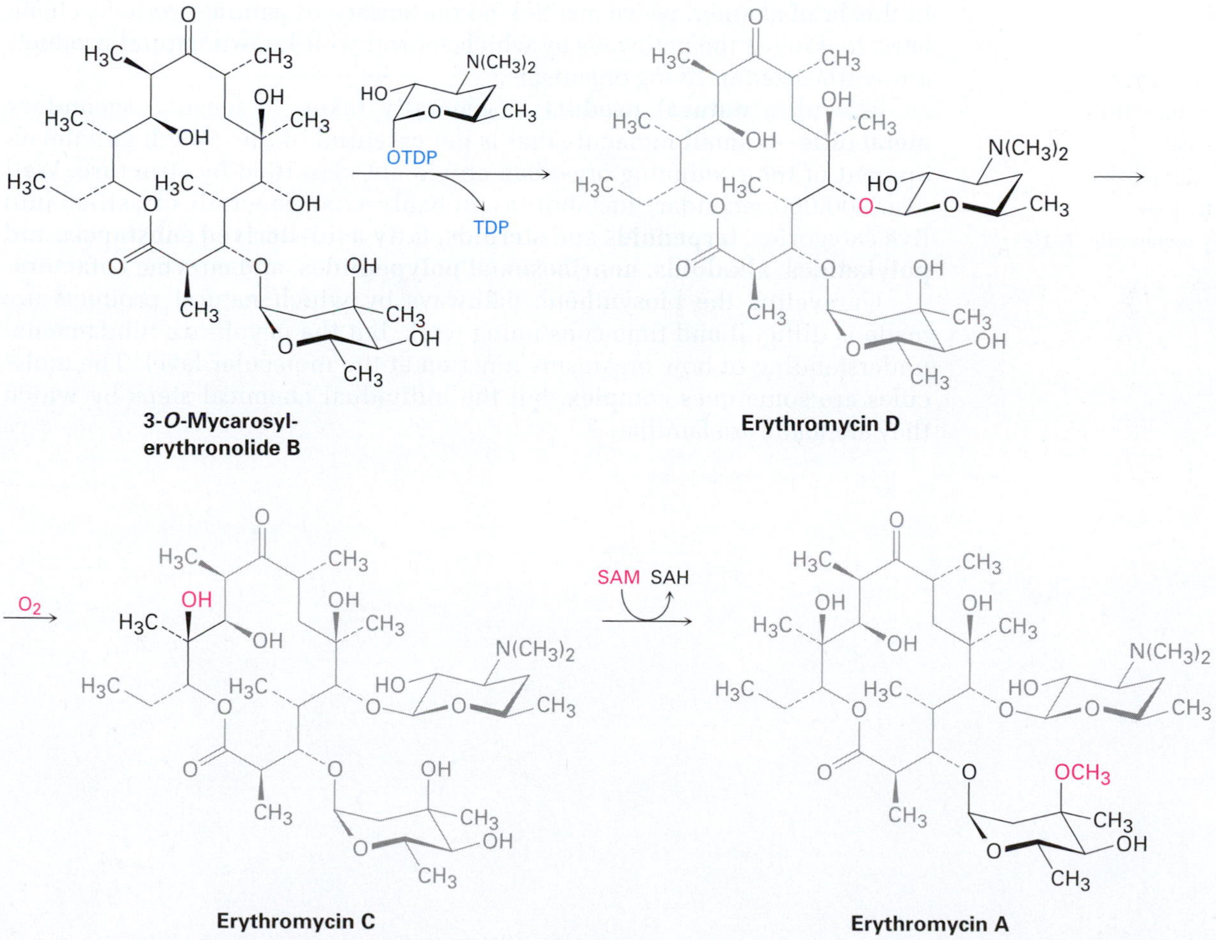

FIGURE 25.23 Final steps in the biosynthesis of erythromycin A.

Problem 25.5

Show a likely mechanism for the epimerization that occurs in step 5 of Figure 25.18.

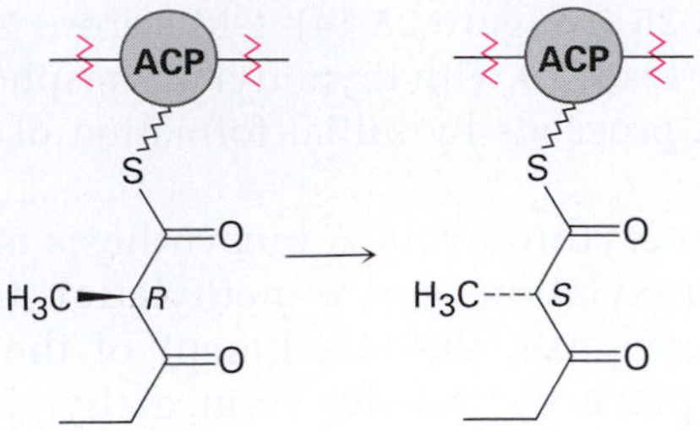

Problem 25.6

Propose a mechanism for the reaction of erythronolide B with thymidyl diphosphomycarose to give 3-*O*-mycarosylerythronolide B (Figure 25.22).

Summary

In this brief chapter, we've just tickled the surface of natural-products chemistry, looking at the pathways by which several well-known natural products are synthesized in living organisms.

The term **natural product** is generally taken to mean a **secondary metabolite**—a small molecule that is not essential to the growth and development of the producing organism and is not classified by structure. Well over 300,000 secondary metabolites probably exist, generally classified into five categories: **terpenoids** and **steroids**, **fatty acid–derived substances** and **polyketides**, **alkaloids**, **nonribosomal polypeptides**, and **enzyme cofactors**.

Unraveling the biosynthetic pathways by which natural products are made is difficult and time-consuming work, but the payoff is a fundamental understanding of how organisms function at the molecular level. The molecules are sometimes complex, but the individual chemical steps by which they are made are familiar.

Key Words

fatty-acid derived
  substance, 1017
natural product, 1015
nonribosomal
  polypeptide, 1017
polyketide, 1017
secondary metabolite, 1015

Lagniappe

Bioprospecting: Hunting for Natural Products

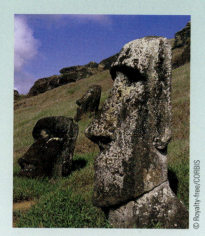

Rapamycin, an immunosuppressant natural product used during organ transplants, was originally isolated from a soil sample found on Easter Island, or Rapa Nui, an island 2200 miles off the coast of Chile known for its giant Moai statues.

Most chemists and biologists spend the majority of their time in the laboratory. A few, however, spend their days scuba diving off South Pacific islands or trekking through the rainforests of South America and Southeast Asia. They aren't on vacation, though; they're at work as bioprospectors, and their job is to hunt for new and unusual natural products that might be useful as drugs.

As noted in the Chapter 5 *Lagniappe,* more than half of all new drug candidates come either directly or indirectly from natural products. All four natural products shown in the introduction to this chapter, for instance, are used as drugs: morphine from the opium poppy, prostaglandin E$_1$ from sheep prostate glands, erythromycin A from a *Streptomyces erythreus* bacterium cultured from a Philippine soil sample, and benzylpenicillin from *Penicillium notatum.* Still other examples include rapamycin (Figure 25.15), an immunosuppressant isolated from a *Streptomyces hygroscopicus* bacterium first found in a soil sample from Easter Island (Rapa Nui), and paclitaxel (Taxol),

an anticancer drug isolated from the bark of the Pacific yew tree found in the American Northwest.

Paclitaxel (Taxol)

With less than 1% of living organisms yet investigated, bioprospectors have a lot of work to do. But there is a race going on. Rainforests throughout the world are being destroyed at an alarming rate, causing many species of both plants and animals to become extinct before they can even be examined. Fortunately, the governments in many countries seem aware of the problem, but there is as yet no international treaty on biodiversity that could help preserve vanishing species.

Exercises

ADDITIONAL PROBLEMS

(Problems 25.1–25.6 appear within the chapter.)

25.7 Which hydrogen, *pro-R* or *pro-S*, is removed from pyridoxine 5'-phosphate in the final step of PLP biosynthesis?

Pyridoxine 5'-phosphate → **Pyridoxal 5'-phosphate (PLP)**

■ Problems assignable in Organic OWL.

■ indicates problems that are assignable in Organic OWL.

Go to this book's companion website at **www.cengage.com/chemistry/mcmurry** to explore interactive versions of the Active Figures from this text.

25.8   Does the ketone reduction step catalyzed by KR1 in erythromycin biosynthesis occur on the *Re* or the *Si* face of the substrate carbonyl group? (See Figure 25.18.)

25.9   When the enoyl reductase domain (ER4) in the erythromycin PKS is deactivated by gene mutation, all further steps still occur normally. What is the structure of the lactone that results?

25.10 ▪ One of the steps in the biosynthesis of the alkaloid berbamunine is an epimerization of (*S*)-*N*-methylcoclaurine. Review the morphine biosynthesis in Figure 25.6, and propose a mechanism for the epimerization.

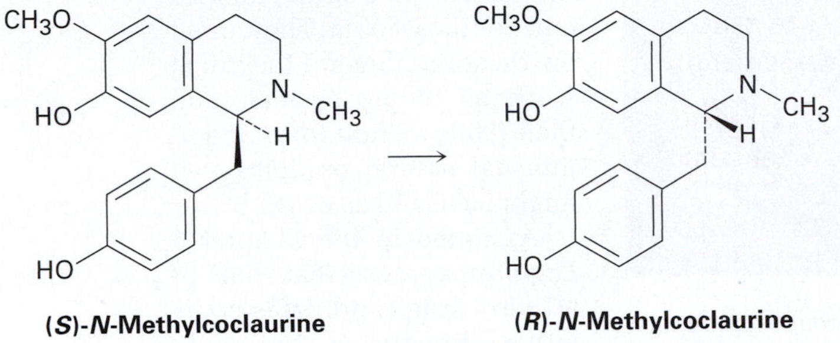

(S)-N-Methylcoclaurine   **(R)-N-Methylcoclaurine**

25.11 ▪ The final step in the biosynthesis of berbamunine is a coupling reaction of (*S*)-*N*-methylcoclaurine with (*R*)-*N*-methylcoclaurine (Problem 25.10). Propose a mechanism.

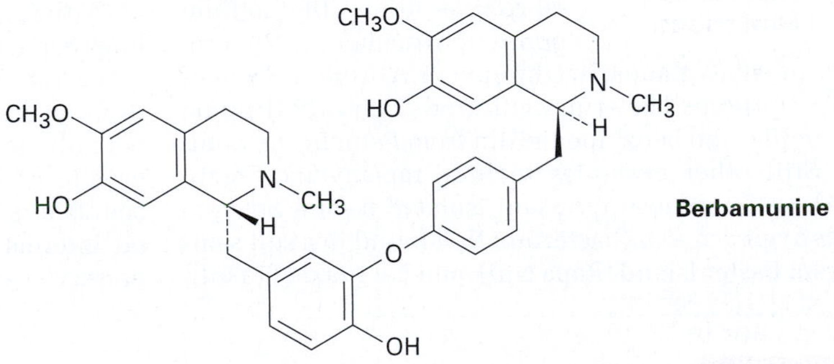

Berbamunine

25.12 ▪ 5-Aminolevulinate is the precursor from which the large class of alkaloids called *tetrapyrroles* are biosynthesized. It arises by a PLP-dependent reaction of glycine and succinyl CoA. Review the mechanism of the formation of dopamine from L-dopa in Figure 25.7, and propose a mechanism for 5-aminolevulinate biosynthesis.

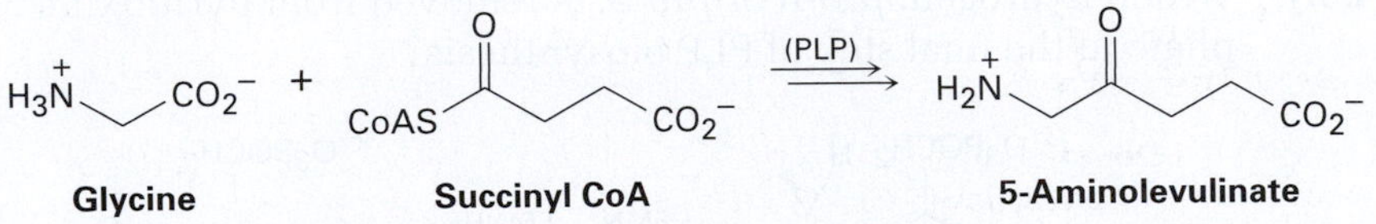

Glycine   **Succinyl CoA**   **5-Aminolevulinate**

25.13 ▪ One of the steps in the biosynthesis of penicillins is a PLP-dependent epimerization of isopenicillin N to penicillin N.

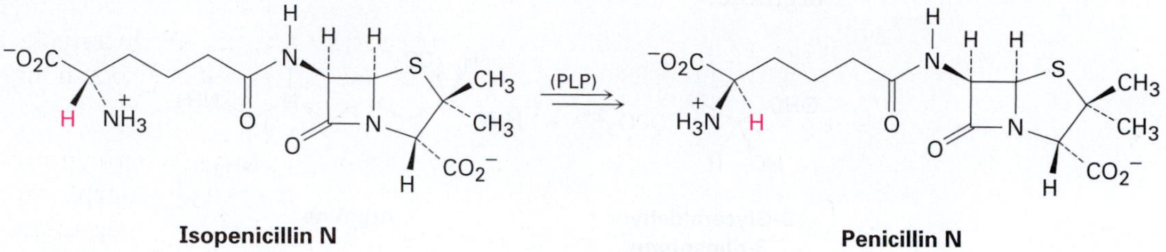

Isopenicillin N   **Penicillin N**

The reaction occurs by initial formation of an imine, followed by a base-catalyzed isomerization. Propose a mechanism.

25.14 ▪ Propose a mechanism for the following biosynthetic conversion. What cofactors are likely to be involved?

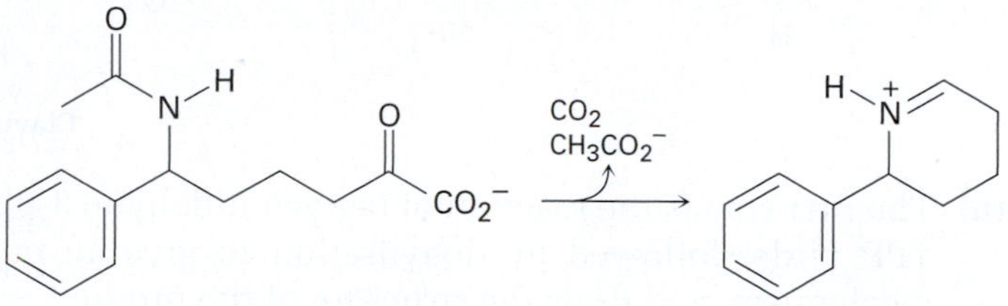

25.15 ▪ The enzyme acetolactate synthase catalyzes the thiamin diphosphate-dependent conversion of two molecules of pyruvate to acetolactate. Propose a mechanism.

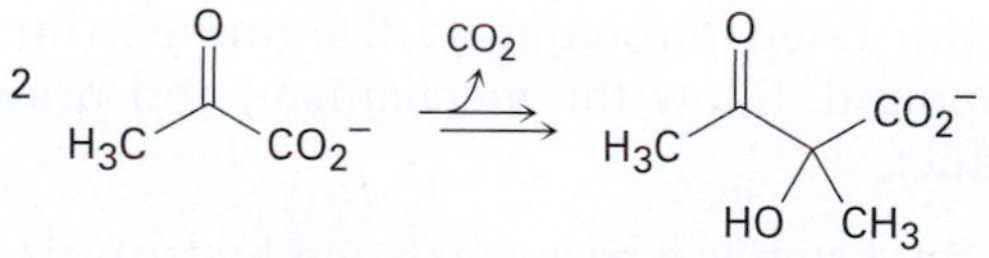

25.16 ▪ 1-Deoxy-D-xylulose 5-phosphate (DXP), in addition to being a precursor to PLP, is also a precursor to isopentenyl diphosphate in terpenoid biosynthesis. The initial step in the pathway is a base-catalyzed rearrangement, followed by reduction with NADPH to give 2C-methyl-D-erythritol 4-phosphate. Show the structure of the rearranged intermediate, and propose a mechanism for its formation.

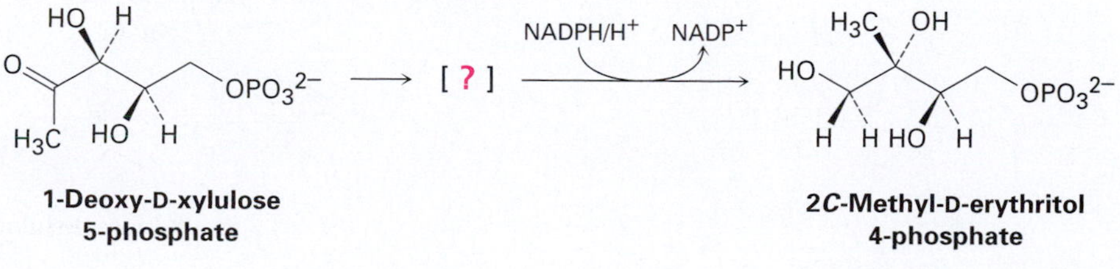

1-Deoxy-D-xylulose   **2C-Methyl-D-erythritol**
5-phosphate   **4-phosphate**

25.17 ■ Biosynthesis of the β-lactam antibiotic clavulanic acid begins with a TPP-dependent reaction between D-glyceraldehyde 3-phosphate and arginine.

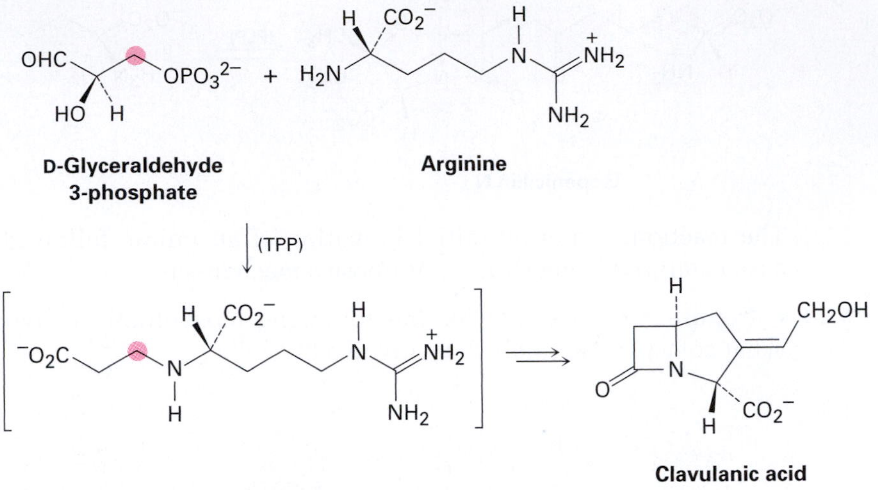

D-Glyceraldehyde
3-phosphate

Arginine

(TPP)

Clavulanic acid

(a) The first step is the reaction of D-glyceraldehyde 3-phosphate with TPP ylide, followed by dehydration to give an enol. Show the mechanism, and draw the structure of the product.

(b) The second step is loss of hydrogen phosphate from the enol to give an unsaturated carbonyl compound. Show the mechanism, and draw the structure of the product.

(c) The third step is a conjugate of arginine to the unsaturated carbonyl compound. Show the mechanism, and draw the structure of the product.

(d) The final step is a base-catalyzed hydrolysis to give the final product and regenerate TPP ylide. Show the mechanism.

A Nomenclature of Polyfunctional Organic Compounds

With more than 37 million organic compounds now known and several thousand more being created daily, naming them all is a real problem. Part of the problem is due to the sheer complexity of organic structures, but part is also due to the fact that chemical names have more than one purpose. For Chemical Abstracts Service (CAS), which catalogs and indexes the worldwide chemical literature, each compound must have only one correct name. It would be chaos if half the entries for CH_3Br were indexed under "M" for methyl bromide and half under "B" for bromomethane. Furthermore, a CAS name must be strictly systematic so that it can be assigned and interpreted by computers; common names are not allowed.

People, however, have different requirements than computers. For people—which is to say chemists in their spoken and written communications—it's best that a chemical name be pronounceable and that it be as easy as possible to assign and interpret. Furthermore, it's convenient if names follow historical precedents, even if that means a particularly well-known compound might have more than one name. People can readily understand that bromomethane and methyl bromide both refer to CH_3Br.

As noted in the text, chemists overwhelmingly use the nomenclature system devised and maintained by the International Union of Pure and Applied Chemistry, or IUPAC. Rules for naming monofunctional compounds were given throughout the text as each new functional group was introduced, and a list of where these rules can be found is given in Table A.1.

Naming a monofunctional compound is reasonably straightforward, but even experienced chemists often encounter problems when faced with naming a complex polyfunctional compound. Take the following compound, for instance. It has three functional groups, ester, ketone, and C=C, but how should it be named? As an ester with an *-oate* ending, a ketone with an *-one* ending, or an alkene with an *-ene* ending? It's actually named methyl 3-(2-oxo-cyclohex-6-enyl)propanoate.

Methyl 3-(2-oxocylohex-6-enyl)propanoate

TABLE A.1
Nomenclature Rules for Functional Groups

Functional group	Text section	Functional group	Text section
Acid anhydrides	16.1	Aromatic compounds	9.1
Acid halides	16.1	Carboxylic acids	15.1
Acyl phosphates	16.1	Cycloalkanes	4.1
Alcohols	13.1	Esters	16.1
Aldehydes	14.1	Ethers	13.8
Alkanes	3.4	Ketones	14.1
Alkenes	7.2	Nitriles	15.1
Alkyl halides	12.1	Phenols	13.1
Alkynes	7.2	Sulfides	13.8
Amides	16.1	Thiols	13.1
Amines	18.1	Thioesters	16.1

The name of a polyfunctional organic molecule has four parts—suffix, parent, prefixes, and locants—which must be identified and expressed in the proper order and format. Let's look at each of the four.

Name Part 1: The Suffix—Functional-Group Precedence

Although a polyfunctional organic molecule might contain several different functional groups, we must choose just one suffix for nomenclature purposes. It's not correct to use two suffixes. Thus, keto ester **1** must be named either as a ketone with an -*one* suffix or as an ester with an -*oate* suffix, but it can't be named as an -*onoate*. Similarly, amino alcohol **2** must be named either as an alcohol (-*ol*) or as an amine (-*amine*), but it can't be named as an -*olamine* or -*aminol*.

1.
$$\underset{CH_3CCH_2CH_2COCH_3}{\overset{O \quad\quad O}{\parallel \quad\quad \parallel}}$$

2.
$$\underset{CH_3CHCH_2CH_2CH_2NH_2}{\overset{OH}{|}}$$

The only exception to the rule requiring a single suffix is when naming compounds that have double or triple bonds. Thus, the unsaturated acid $H_2C=CHCH_2CO_2H$ is but-3-enoic acid, and the acetylenic alcohol $HC\equiv CCH_2CH_2CH_2OH$ is pent-5-yn-1-ol.

How do we choose which suffix to use? Functional groups are divided into two classes, **principal groups** and **subordinate groups**, as shown in Table A.2. Principal groups can be cited either as prefixes or as suffixes, while subordinate groups are cited only as prefixes. Within the principal groups, an order of priority has been established, with the proper suffix for a given compound determined by choosing the principal group of highest priority. For example, Table A.2 indicates that keto ester **1** should be named as an ester rather than as a ketone because an ester functional group is higher in priority than a ketone.

TABLE A.2
Classification of Functional Groups[a]

Functional group	Name as suffix	Name as prefix
Principal groups		
Carboxylic acids	-oic acid -carboxylic acid	carboxy
Acid anhydrides	-oic anhydride -carboxylic anhydride	—
Esters	-oate -carboxylate	alkoxycarbonyl
Thioesters	-thioate -carbothioate	alkylthiocarbonyl
Acid halides	-oyl halide -carbonyl halide	halocarbonyl
Amides	-amide -carboxamide	carbamoyl
Nitriles	-nitrile -carbonitrile	cyano
Aldehydes	-al -carbaldehyde	oxo
Ketones	-one	oxo
Alcohols	-ol	hydroxy
Phenols	-ol	hydroxy
Thiols	-thiol	mercapto
Amines	-amine	amino
Imines	-imine	imino
Ethers	ether	alkoxy
Sulfides	sulfide	alkylthio
Disulfides	disulfide	—
Alkenes	-ene	—
Alkynes	-yne	—
Alkanes	-ane	—
Subordinate groups		
Azides	—	azido
Halides	—	halo
Nitro compounds	—	nitro

[a]Principal groups are listed in order of decreasing priority; subordinate groups have no priority order.

Similarly, amino alcohol **2** should be named as an alcohol rather than as an amine. Thus, the name of **1** is methyl 4-oxopentanoate, and the name of **2** is 5-aminopentan-2-ol. Further examples are shown:

$$\text{CH}_3\overset{\overset{\displaystyle O}{\|}}{C}\text{CH}_2\text{CH}_2\overset{\overset{\displaystyle O}{\|}}{C}\text{OCH}_3$$

1. Methyl 4-oxopentanoate
(an ester with a ketone group)

$$\text{CH}_3\overset{\overset{\displaystyle OH}{|}}{C}\text{HCH}_2\text{CH}_2\text{CH}_2\text{NH}_2$$

2. 5-Aminopentan-2-ol
(an alcohol with an amine group)

$$\text{CH}_3\overset{\overset{\displaystyle CHO}{|}}{C}\text{HCH}_2\text{CH}_2\text{CH}_2\overset{\overset{\displaystyle O}{\|}}{C}\text{OCH}_3$$

3. Methyl 5-methyl-6-oxohexanoate
(an ester with an aldehyde group)

$$\text{H}_2\text{N}\overset{\overset{\displaystyle O}{\|}}{C}\text{CH}_2\overset{\overset{\displaystyle OH}{|}}{C}\text{HCH}_2\text{CH}_2\overset{\overset{\displaystyle O}{\|}}{C}\text{OH}$$

4. 5-Carbamoyl-4-hydroxypentanoic acid
(a carboxylic acid with amide and alcohol groups)

5. 3-Oxocyclohexanecarbaldehyde
(an aldehyde with a ketone group)

Name Part 2: The Parent—Selecting the Main Chain or Ring

The parent, or base, name of a polyfunctional organic compound is usually easy to identify. If the principal group of highest priority is part of an open chain, the parent name is that of the longest chain containing the largest number of principal groups. For example, compounds **6** and **7** are isomeric aldehydo amides, which must be named as amides rather than as aldehydes according to Table A.2. The longest chain in compound **6** has six carbons, and the substance is therefore named 5-methyl-6-oxohexanamide. Compound **7** also has a chain of six carbons, but the longest chain that contains both principal functional groups has only four carbons. The correct name of **7** is 4-oxo-3-propylbutanamide.

$$\text{H}\overset{\overset{\displaystyle O}{\|}}{C}\text{CHCH}_2\text{CH}_2\text{CH}_2\overset{\overset{\displaystyle O}{\|}}{C}\text{NH}_2$$
$$\underset{\text{CH}_3}{|}$$

6. 5-Methyl-6-oxohexanamide

$$\text{CH}_3\text{CH}_2\text{CH}_2\overset{\overset{\displaystyle CHO}{|}}{C}\text{HCH}_2\overset{\overset{\displaystyle O}{\|}}{C}\text{NH}_2$$

7. 4-Oxo-3-propylbutanamide

If the highest-priority principal group is attached to a ring, the parent name is that of the ring system. Compounds **8** and **9**, for instance, are isomeric keto nitriles and must both be named as nitriles according to Table A.2. Substance **8** is named as a benzonitrile because the —CN functional group is a substituent on the aromatic ring, but substance **9** is named as an acetonitrile because the —CN functional group is on an open chain. The correct names are 2-acetyl-(4-bromomethyl)benzonitrile **(8)** and (2-acetyl-4-bromophenyl)-acetonitrile **(9)**. As further examples, compounds **10** and **11** are both keto acids and must be named as acids, but the parent name in **(10)** is that of a ring system (cyclohexanecarboxylic acid) and the parent name in **(11)** is that of an open chain (propanoic acid). The full names are *trans*-2-(3-oxopropyl)cyclo-hexanecarboxylic acid (**10**) and 3-(2-oxocyclohexyl)propanoic acid (**11**).

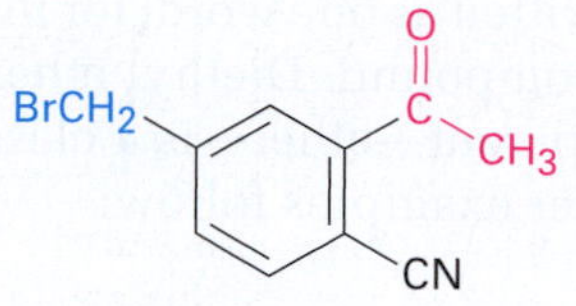

8. 2-Acetyl-(4-bromomethyl)benzonitrile

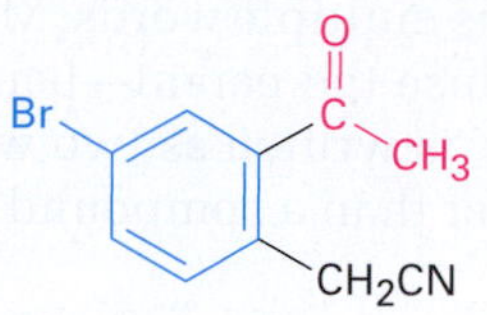

9. (2-Acetyl-4-bromophenyl)acetonitrile

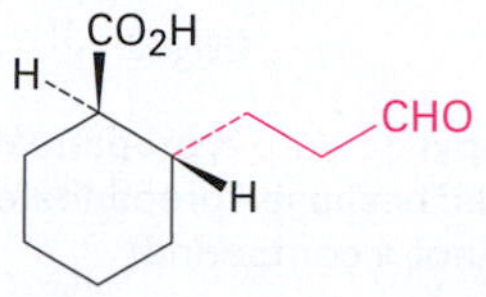

10. *trans*-**2-(3-oxopropyl)cyclo-hexanecarboxylic acid**

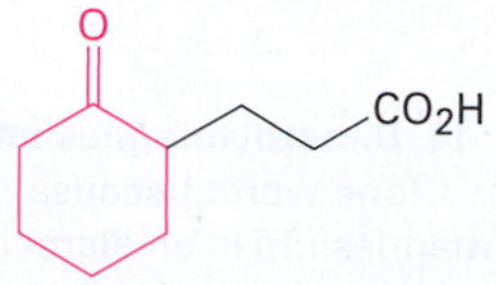

11. 3-(2-Oxocyclohexyl)propanoic acid

Name Parts 3 and 4: The Prefixes and Locants

With the parent name and the suffix established, the next step is to identify and give numbers, or *locants,* to all substituents on the parent chain or ring. These substituents include all alkyl groups and all functional groups other than the one cited in the suffix. For example, compound **12** contains three different functional groups (carboxyl, keto, and double bond). Because the carboxyl group is highest in priority and because the longest chain containing the functional groups has seven carbons, **12** is a heptenoic acid. In addition, the main chain has a keto (oxo) substituent and three methyl groups. Numbering from the end nearer the highest-priority functional group, **12** is named (2*E*)-2,5,5-trimethyl-4-oxohept-2-enoic acid. Look back at some of the other compounds we've named to see other examples of how prefixes and locants are assigned.

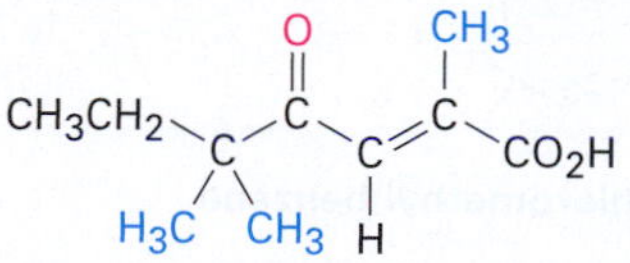

12. (2*E*)-2,5,5-Trimethyl-4-oxohept-2-enoic acid

Writing the Name

Once the name parts have been established, the entire name is written out. Several additional rules apply:

1. **Order of prefixes**   When the substituents have been identified, the main chain has been numbered, and the proper multipliers such as *di-* and *tri-* have been assigned, the name is written with the substituents listed in alphabetical, rather than numerical, order. Multipliers such as *di-* and *tri-* are not used for alphabetization purposes, but the prefix *iso-* is used.

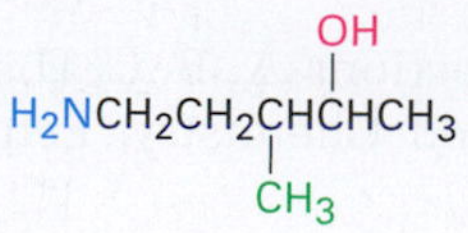

$H_2NCH_2CH_2CHCHCH_3$ **13. 5-Amino-3-methylpentan-2-ol**

2. **Use of hyphens; single- and multiple-word names**   The general rule is to determine whether the parent is itself an element or compound. If it is, then the name is written as a single word; if it isn't, then the name is written as multiple words. Methylbenzene is written as one word, for instance, because the parent—benzene—is itself a compound. Diethyl ether, however, is written as two words because the parent—ether—is a class name rather than a compound name. Some further examples follow:

H₃C—Mg—CH₃

14. Dimethylmagnesium
(one word, because
magnesium is an element)

$$HOCH_2CH_2COCHCH_3$$
$$\overset{O}{\overset{\|}{}}$$
CH₃

15. Isopropyl 3-hydroxypropanoate
(two words, because "propanoate"
is not a compound)

16. 4-(Dimethylamino)pyridine
(one word, because pyridine
is a compound)

17. Methyl cyclopentanecarbothioate
(two words, because "cyclopentanecarbothioate" is not a compound)

3. **Parentheses**   Parentheses are used to denote complex substituents when ambiguity would otherwise arise. For example, chloromethylbenzene has two substituents on a benzene ring, but (chloromethyl)benzene has only one complex substituent. Note that the expression in parentheses is not set off by hyphens from the rest of the name.

18. *p*-Chloromethylbenzene

19. (Chloromethyl)benzene

$$HOCCHCH_2CH_2COH$$
CH₃CHCH₂CH₃

20. 2-(1-Methylpropyl)pentanedioic acid

ADDITIONAL READING
Further explanations of the rules of organic nomenclature can be found online at http://www.acdlabs.com/iupac/nomenclature/ and in the following references:

1. "A Guide to IUPAC Nomenclature of Organic Compounds," CRC Press, Boca Raton, FL, 1993.

2. "Nomenclature of Organic Chemistry, Sections A, B, C, D, E, F, and H," International Union of Pure and Applied Chemistry, Pergamon Press, Oxford, 1979.

B Acidity Constants for Some Organic Compounds

Compound	pK_a	Compound	pK_a	Compound	pK_a
CH_3SO_3H	−1.8	CH_2ClCO_2H	2.8	[3-chlorobenzoic acid: Cl-substituted benzene with CO_2H]	3.8
$CH(NO_2)_3$	0.1	$HO_2CCH_2CO_2H$	2.8; 5.6	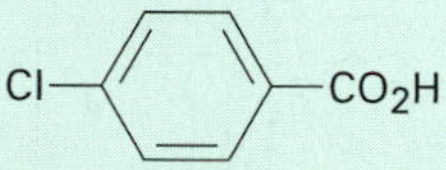	
[2,4,6-trinitrophenol: benzene with O_2N, NO_2, NO_2 and OH]	0.3	CH_2BrCO_2H	2.9	[4-chlorobenzoic acid: Cl-substituted benzene with CO_2H]	4.0
		[2-chlorobenzoic acid: benzene with CO_2H and Cl]	3.0	$CH_3BrCH_2CO_2H$	4.0
CCl_3CO_2H	0.5	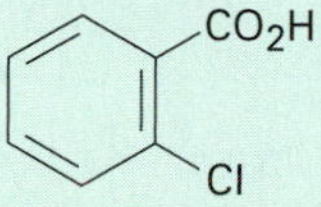 [salicylic acid: benzene with CO_2H and OH]	3.0	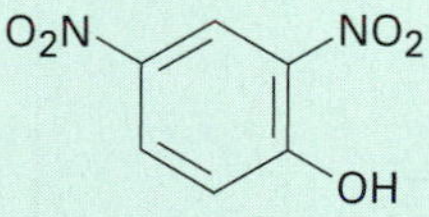 [2,6-dinitrophenol: benzene with O_2N, NO_2 and OH]	4.1
CF_3CO_2H	0.5	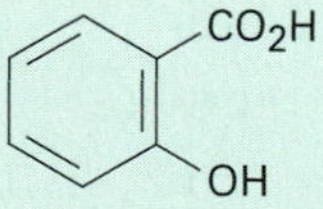		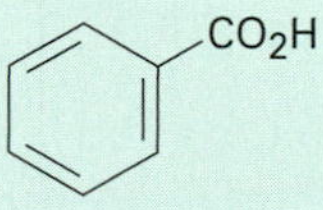	
CBr_3CO_2H	0.7	CH_2ICO_2H	3.2	[benzoic acid: benzene with CO_2H]	4.2
$HO_2CC{\equiv}CCO_2H$	1.2; 2.5	$CHOCO_2H$	3.2		
HO_2CCO_2H	1.2; 3.7	[4-nitrobenzoic acid: O_2N-substituted benzene with CO_2H]	3.4	$H_2C{=}CHCO_2H$	4.2
$CHCl_2CO_2H$	1.3			$HO_2CCH_2CH_2CO_2H$	4.2; 5.7
$CH_2(NO_2)CO_2H$	1.3	[2,4-dinitrobenzene: O_2N, O_2N-substituted benzene with CO_2H]	3.5	$HO_2CCH_2CH_2CH_2CO_2H$	4.3; 5.4
$HC{\equiv}CCO_2H$	1.9				
Z $HO_2CCH{=}CHCO_2H$	1.9; 6.3	$HSCH_2CO_2H$	3.5; 10.2	[pentachlorophenol: benzene with five Cl and OH]	4.5
[2-nitrobenzoic acid: benzene with CO_2H and NO_2]	2.4	$CH_2(NO_2)_2$	3.6		
		$CH_3OCH_2CO_2H$	3.6		
CH_3COCO_2H	2.4	$CH_3COCH_2CO_2H$	3.6	$H_2C{=}C(CH_3)CO_2H$	4.7
$NCCH_2CO_2H$	2.5	$HOCH_2CO_2H$	3.7	CH_3CO_2H	4.8
$CH_3C{\equiv}CCO_2H$	2.6	HCO_2H	3.7		
CH_2FCO_2H	2.7				

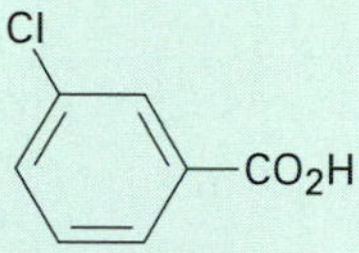

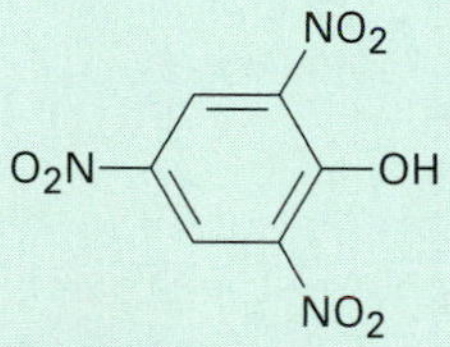

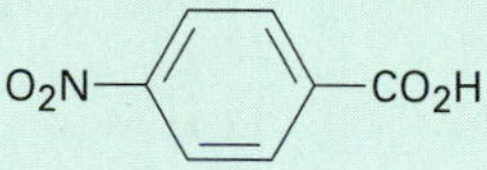

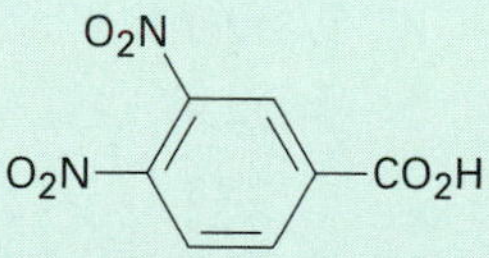

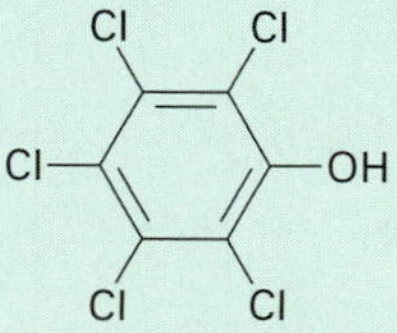

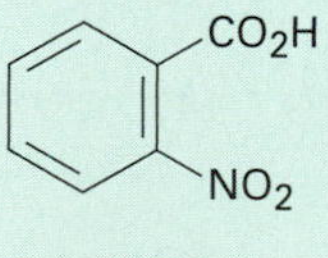

continued

Compound	pK_a	Compound	pK_a	Compound	pK_a
$CH_3CH_2CO_2H$	4.8	$CH_3COCH_2COCH_3$	9.0	[benzyl alcohol, –CH_2OH]	15.4
$(CH_3)_3CCO_2H$	5.0	[1,3-benzenediol (resorcinol), HO–C$_6$H$_4$–OH]	9.3; 11.1	CH_3OH	15.5
$CH_3COCH_2NO_2$	5.1			$H_2C{=}CHCH_2OH$	15.5
[1,3-cyclohexanedione]	5.3	[1,2-benzenediol (catechol)]	9.3; 12.6	CH_3CH_2OH	16.0
$O_2NCH_2CO_2CH_3$	5.8			$CH_3CH_2CH_2OH$	16.1
[2-oxocyclopentanecarbaldehyde]	5.8	[benzyl mercaptan, –CH_2SH]	9.4	CH_3COCH_2Br	16.1
[2,4,6-trichlorophenol]	6.2	[1,4-benzenediol (hydroquinone)]	9.9; 11.5	[cyclohexanone]	16.7
[benzenethiol, –SH]	6.6	[phenol, –OH]	9.9	CH_3CHO	17
HCO_3H	7.1	$CH_3COCH_2SOCH_3$	10.0	$(CH_3)_2CHCHO$	17
[2-nitrophenol]	7.2	[2-methylphenol (o-cresol)]	10.3	$(CH_3)_2CHOH$	17.1
$(CH_3)_2CHNO_2$	7.7	CH_3NO_2	10.3	$(CH_3)_3COH$	18.0
[2,4-dichlorophenol]	7.8	CH_3SH	10.3	CH_3COCH_3	19.3
CH_3CO_3H	8.2	$CH_3COCH_2CO_2CH_3$	10.6	[fluorene]	23
[2-chlorophenol]	8.5	CH_3COCHO	11.0	$CH_3CO_2CH_2CH_3$	25
$CH_3CH_2NO_2$	8.5	$CH_2(CN)_2$	11.2	$HC{\equiv}CH$	25
[F_3C–C$_6$H$_4$–OH, 4-(trifluoromethyl)phenol]	8.7	CCl_3CH_2OH	12.2	CH_3CN	25
		Glucose	12.3	$CH_3SO_2CH_3$	28
		$(CH_3)_2C{=}NOH$	12.4	$(C_6H_5)_3CH$	32
		$CH_2(CO_2CH_3)_2$	12.9	$(C_6H_5)_2CH_2$	34
		$CHCl_2CH_2OH$	12.9	CH_3SOCH_3	35
		$CH_2(OH)_2$	13.3	NH_3	36
		$HOCH_2CH(OH)CH_2OH$	14.1	$CH_3CH_2NH_2$	36
		CH_2ClCH_2OH	14.3	$(CH_3CH_2)_2NH$	40
		[cyclopentadiene]	15.0	[toluene, –CH_3]	41
				[benzene]	43
				$H_2C{=}CH_2$	44
				CH_4	~60

An acidity list covering more than 5000 organic compounds has been published: E.P. Serjeant and B. Dempsey (eds.), "Ionization Constants of Organic Acids in Aqueous Solution," IUPAC Chemical Data Series No. 23, Pergamon Press, Oxford, 1979.

C Glossary

Absolute configuration (Section 5.5): The exact three-dimensional structure of a chiral molecule. Absolute configurations are specified verbally by the Cahn–Ingold–Prelog R,S convention.

Absorbance (Section 10.9): In optical spectroscopy, the logarithm of the intensity of the incident light divided by the intensity of the light transmitted through a sample; $A = \log I_0/I$.

Absorption spectrum (Section 10.5): A plot of wavelength of incident light versus amount of light absorbed. Organic molecules show absorption spectra in both the infrared and the ultraviolet regions of the electromagnetic spectrum.

Acetal (Section 14.8): A functional group consisting of two –OR groups bonded to the same carbon, $R_2C(OR')_2$. Acetals are often used as protecting groups for ketones and aldehydes.

Acetoacetic ester synthesis (Section 17.5): The synthesis of a methyl ketone by alkylation of an alkyl halide, followed by hydrolysis and decarboxylation.

Acetyl group (Section 14.1): The CH_3CO- group.

Acetylide anion (Section 8.15): The anion formed by removal of a proton from a terminal alkyne.

Achiral (Section 5.2): Having a lack of handedness. A molecule is achiral if it has a plane of symmetry and is thus superimposable on its mirror image.

Acid anhydride (Chapter 16 Introduction): A functional group with two acyl groups bonded to a common oxygen atom, RCO_2COR'.

Acid halide (Chapter 16 Introduction): A functional group with an acyl group bonded to a halogen atom, RCOX.

Acidity constant, K_a (Section 2.8): A measure of acid strength in water. For any acid HA, the acidity constant is given by the expression

$$K_a = \frac{[H_3O^+][A^-]}{[HA]}$$

Activating group (Section 9.8): An electron-donating group such as hydroxyl (–OH) or amino (–NH$_2$) that increases the reactivity of an aromatic ring toward electrophilic aromatic substitution.

Activation energy, $\Delta G^{\ddagger}$ (Section 6.9): The difference in energy between ground state and transition state in a reaction. The amount of activation energy determines the rate at which the reaction proceeds. Most organic reactions have activation energies of 40–100 kJ/mol.

Active site (Sections 6.11, 19.10): The pocket in an enzyme where a substrate is bound and undergoes reaction.

Acyl group (Sections 9.7, 14.1): A –COR group.

Acyl phosphate (Chapter 16 Introduction): A functional group with an acyl group bonded to a phosphate, $RCO_2PO_3{}^{2-}$.

Acylation (Section 9.7): The introduction of an acyl group, –COR, onto a molecule. For example, acylation of an alcohol yields an ester, acylation of an amine yields an amide, and acylation of an aromatic ring yields an alkyl aryl ketone.

Acylium ion (Section 9.7): A resonance-stabilized carbocation in which the positive charge is located at a carbonyl-group carbon, $R-C^+=O \leftrightarrow R-C\equiv O^+$. Acylium ions are strongly electrophilic and are involved as intermediates in Friedel–Crafts acylation reactions.

Adams catalyst (Section 8.5): The PtO_2 catalyst used for alkene hydrogenations.

1,2-Addition (Sections 8.13, 14.11): The addition of a reactant to the two ends of a double bond.

1,4-Addition (Sections 8.13, 14.11): Addition of a reactant to the ends of a conjugated π system. Conjugated dienes yield 1,4-adducts when treated with electrophiles such as HCl. Conjugated enones yield 1,4-adducts when treated with nucleophiles such as amines.

Addition reaction (Section 6.1): The reaction that occurs when two reactants add together to form a single new product with no atoms "left over."

Adrenocortical hormone (Section 23.8): A steroid hormone secreted by the adrenal glands. There are two types of adrenocortical hormones: mineralocorticoids and glucocorticoids.

Alcohol (Chapter 13 Introduction): A compound with an –OH group bonded to a saturated, alkane-like carbon, ROH.

Aldaric acid (Section 21.6): The dicarboxylic acid resulting from oxidation of an aldose.

Aldehyde (Chapter 14 Introduction): A compound containing the –CHO functional group.

Alditol (Section 21.6): The polyalcohol resulting from reduction of the carbonyl group of a sugar.

Aldol reaction (Section 17.6): The carbonyl condensation reaction of an aldehyde or ketone to give a β-hydroxy carbonyl compound.

Aldonic acid (Section 21.6): The monocarboxylic acid resulting from oxidation of the aldehyde group of an aldose.

Aldose (Section 21.1): A carbohydrate with an aldehyde functional group.

Alicyclic (Section 4.1): An aliphatic cyclic hydrocarbon such as a cycloalkane or cycloalkene.

Aliphatic (Section 3.2): A nonaromatic hydrocarbon such as a simple alkane, alkene, or alkyne.

Alkaloid (Chapter 2 *Lagniappe,* Section 25.1): A naturally occurring organic base, such as morphine.

Alkane (Section 3.2): A compound of carbon and hydrogen that contains only single bonds.

Alkene (Chapter 7 Introduction): A hydrocarbon that contains a carbon–carbon double bond, $R_2C=CR_2$.

Alkoxide ion (Section 13.2): The anion RO⁻ formed by deprotonation of an alcohol.

Alkyl group (Section 3.3): The partial structure that remains when a hydrogen atom is removed from an alkane.

Alkyl halide (Chapter 12 Introduction): A compound with a halogen atom bonded to a saturated, sp^3-hybridized carbon atom.

Alkylamine (Section 18.1): An amino-substituted alkane, RNH_2, R_2NH, or R_3N.

Alkylation (Sections 9.7, 17.5): Introduction of an alkyl group onto a molecule. For example, aromatic rings can be alkylated to yield arenes, and enolate anions can be alkylated to yield α-substituted carbonyl compounds.

Alkyne (Chapter 7 Introduction): A hydrocarbon that contains a carbon–carbon triple bond, RC≡CR.

Allyl group (Section 7.2): An $H_2C=CHCH_2$– substituent.

Allylic (Sections 8.13, 12.2): The position next to a double bond. For example, $H_2C=CHCH_2Br$ is an allylic bromide.

α-Amino acid (Section 19.1): A difunctional compound with an amino group on the carbon atom next to a carboxyl group, $RCH(NH_2)CO_2H$.

α Anomer (Section 21.5): The cyclic hemiacetal form of a sugar that has the hemiacetal –OH group on the side of the ring opposite the terminal –CH_2OH.

α Helix (Section 19.8): A coiled secondary structure of a protein.

α Position (Chapter 17 Introduction): The position next to a carbonyl group.

α-Substitution reaction (Section 17.2): The substitution of the α hydrogen atom of a carbonyl compound by reaction with an electrophile.

Amide (Chapter 16 Introduction): A compound containing the –$CONR_2$ functional group.

Amidomalonate synthesis (Section 19.3): A method for preparing an α-amino acid by alkylation of diethyl amidomalonate with an alkyl halide.

Amine (Chapter 18 Introduction): A compound containing one or more organic substituents bonded to a nitrogen atom, RNH_2, R_2NH, or R_3N.

Amino acid (Section 19.1): *See* α-Amino acid.

Amino sugar (Section 21.7): A sugar with one of its –OH groups replaced by –NH_2.

Amphiprotic (Section 19.1): Capable of acting either as an acid or as a base. Amino acids are amphiprotic.

Amplitude (Section 10.5): The height of a wave measured from the midpoint to the maximum. The intensity of radiant energy is proportional to the square of the wave's amplitude.

Anabolism (Section 20.1): The group of metabolic pathways that build up larger molecules from smaller ones.

Androgen (Section 23.8): A male steroid sex hormone.

Angle strain (Section 4.3): The strain introduced into a molecule when a bond angle is deformed from its ideal value. Angle strain is particularly important in small-ring cycloalkanes, where it results from compression of bond angles to less than their ideal tetrahedral values.

Anomeric center (Section 21.5): The hemiacetal carbon atom in the cyclic pyranose or furanose form of a sugar.

Anomers (Section 21.5): Cyclic stereoisomers of sugars that differ only in their configuration at the hemiacetal (anomeric) carbon.

Anti conformation (Section 3.7): The geometric arrangement around a carbon–carbon single bond in which the two largest substituents are 180° apart as viewed in a Newman projection.

Anti periplanar (Section 12.12): Describing a stereochemical relationship whereby two bonds on adjacent carbons lie in the same plane at an angle of 180°.

Anti stereochemistry (Section 8.2): The opposite of syn. An anti addition reaction is one in which the two ends of the double bond are attacked from different sides. An anti elimination reaction is one in which the two groups leave from opposite sides of the molecule.

Antiaromatic (Section 9.3): Describing a planar, apparently conjugated molecule with $4n$ π electrons. Delocalization of the π electrons leads to an increase in energy.

Antibonding MO (Section 1.11): A molecular orbital that is higher in energy than the atomic orbitals from which it is formed.

Anticodon (Section 24.5): A sequence of three bases on tRNA that reads the codons on mRNA and brings the correct amino acids into position for protein synthesis.

Antisense strand (Section 24.4): The template, noncoding strand of double-helical DNA that does not contain the gene.

Arene (Section 9.1): An alkyl-substituted benzene.

Aromaticity (Chapter 9 Introduction): The special characteristics of cyclic conjugated molecules, including unusual stability and a tendency to undergo substitution reactions rather than addition reactions on treatment with electrophiles. Aromatic molecules are planar, cyclic, conjugated species that have $4n + 2$ π electrons.

Arylamine (Section 18.1): An amino-substituted aromatic compound, $ArNH_2$.

Atomic mass (Section 1.1): The average mass number of the atoms of an element.

Atomic number, Z (Section 1.1): The number of protons in the nucleus of an atom.

ATZ derivative (Section 19.6): An anilinothiazolinone, formed from an amino acid during Edman degradation.

Axial position (Section 4.6): A bond to chair cyclohexane that lies along the ring axis perpendicular to the rough plane of the ring.

Backbone (Section 19.4): The continuous chain of atoms running the length of a protein or other polymer.

Base peak (Section 10.1): The most intense peak in a mass spectrum.

Basicity constant, K_b (Section 18.3): A measure of base strength in water. For any base B, the basicity constant is given by the expression

$$B + H_2O \rightleftharpoons BH^+ + OH^-$$

$$K_b = \frac{[BH^+][OH^-]}{[B]}$$

Bent bonds (Section 4.4): The bonds in small rings such as cyclopropane that bend away from the internuclear line and overlap at a slight angle, rather than head-on. Bent bonds are highly strained and highly reactive.

Benzoyl group (Section 14.1): The C_6H_5CO- group.

Benzyl group (Section 9.1): The $C_6H_5CH_2-$ group.

Benzylic (Section 9.10): The position next to an aromatic ring.

β Anomer (Section 21.5): The cyclic hemiacetal form of a sugar that has the hemiacetal –OH group on the same side of the ring as the terminal $-CH_2OH$.

β Diketone (Section 17.4): A 1,3-diketone.

β-Keto ester (Section 17.4): A 3-keto ester.

β-Oxidation pathway (Section 23.5): The metabolic pathway for degrading fatty acids.

β-Pleated sheet (Section 19.8): A type of secondary structure of a protein.

Bimolecular reaction (Section 12.6): A reaction whose rate-limiting step occurs between two reactants.

Boat cyclohexane (Section 4.5): A conformation of cyclohexane that bears a slight resemblance to a boat. Boat cyclohexane has no angle strain but has a large number of eclipsing interactions that make it less stable than chair cyclohexane.

Boc derivative (Section 19.7): A butyloxycarbonyl N-protected amino acid.

Bond angle (Section 1.6): The angle formed between two adjacent bonds.

Bond dissociation energy, D (Section 6.8): The amount of energy needed to break a bond and produce two radical fragments.

Bond length (Section 1.5): The equilibrium distance between the nuclei of two atoms that are bonded to each other.

Bond strength (Section 1.5): An alternative name for bond dissociation energy.

Bonding MO (Section 1.11): A molecular orbital that is lower in energy than the atomic orbitals from which it is formed.

Branched-chain alkane (Section 3.2): An alkane that contains a branching connection of carbons as opposed to a straight-chain alkane.

Bridgehead atom (Section 4.9): An atom that is shared by more than one ring in a polycyclic molecule.

Bromohydrin (Section 8.3): A 1,2-disubstituted bromo-alcohol; obtained by addition of HOBr to an alkene.

Bromonium ion (Section 8.2): A species with a divalent, positively charged bromine, R_2Br^+.

Brønsted–Lowry acid (Section 2.7): A substance that donates a hydrogen ion (proton; H^+) to a base.

Brønsted–Lowry base (Section 2.7): A substance that accepts H^+ from an acid.

C-terminal amino acid (Section 19.4): The amino acid with a free $-CO_2H$ group at the end of a protein chain.

Cahn–Ingold–Prelog sequence rules (Sections 5.5, 7.4): A series of rules for assigning relative priorities to substituent groups on a double-bond carbon atom or on a chirality center.

Cannizzaro reaction (Section 14.10): The disproportionation reaction of an aldehyde to yield an alcohol and a carboxylic acid on treatment with base.

Carbanion (Sections 12.4, 14.6): A carbon anion, or substance that contains a trivalent, negatively charged carbon atom ($R_3C:^-$). Carbanions are sp^3-hybridized and have eight electrons in the outer shell of the negatively charged carbon.

Carbene (Section 8.9): A neutral substance that contains a divalent carbon atom having only six electrons in its outer shell ($R_2C:$).

Carbinolamine (Section 14.7): A molecule that contains the $R_2C(OH)NH_2$ functional group. Carbinolamines are produced as intermediates during the nucleophilic addition of amines to carbonyl compounds.

Carbocation (Sections 6.5, 7.8): A carbon cation, or substance that contains a trivalent, positively charged carbon atom having six electrons in its outer shell (R_3C^+).

Carbohydrate (Chapter 21 Introduction): A polyhydroxy aldehyde or ketone. Carbohydrates can be either simple sugars, such as glucose, or complex sugars, such as cellulose.

Carbonyl condensation reaction (Section 17.6): A reaction that joins two carbonyl compounds together by a combination of α-substitution and nucleophilic addition reactions.

Carbonyl group (*Preview of Carbonyl Chemistry*): The C=O functional group.

Carboxyl group (Section 15.1): The $-CO_2H$ functional group.

Carboxylation (Section 15.5): The addition of CO_2 to a molecule.

Carboxylic acid (Chapter 15 Introduction): A compound containing the $-CO_2H$ functional group.

Carboxylic acid derivative (Chapter 16 Introduction): A compound in which an acyl group is bonded to an electronegative atom or substituent that can act as a leaving group in a substitution reaction. Esters, amides, and acid halides are examples.

Catabolism (Section 20.1): The group of metabolic pathways that break down larger molecules into smaller ones.

Cation radical (Section 10.1): A species typically formed in a mass spectrometer, having both a positive charge and an odd number of electrons.

Chain reaction (Section 6.3): A reaction that, once initiated, sustains itself in an endlessly repeating cycle of propagation steps. The radical chlorination of alkanes is an example of a chain reaction that is initiated by irradiation with light and then continues in a series of propagation steps.

Chain-growth polymer (Sections 8.10, 16.9): A polymer whose bonds are produced by chain reactions. Polyethylene and other alkene polymers are examples.

Chair conformation (Section 4.5): A three-dimensional conformation of cyclohexane that resembles the rough shape of a chair. The chair form of cyclohexane is the lowest-energy conformation of the molecule.

Chemical shift (Section 11.3): The position on the NMR chart where a nucleus absorbs. By convention, the chemical shift of tetramethylsilane (TMS) is set at zero, and all other absorptions usually occur downfield (to the left on the chart). Chemical shifts are expressed in delta units, δ, where 1 δ equals 1 ppm of the spectrometer operating frequency.

Chiral (Section 5.2): Having handedness. Chiral molecules are those that do not have a plane of symmetry and are therefore not superimposable on their mirror image. A chiral molecule thus exists in two forms, one right-handed and one left-handed. The most common cause of chirality in a molecule is the presence of a carbon atom that is bonded to four different substituents.

Chiral environment (Section 5.12): Chiral surroundings or conditions in which a molecule resides.

Chirality center (Section 5.2): An atom (usually carbon) that is bonded to four different groups.

Chromatography (Chapter 10 *Lagniappe,* Section 19.5): A technique for separating a mixture of compounds into pure components. Different compounds adsorb to a stationary support phase and are then carried along it at different rates by a mobile phase.

Cis–trans isomers (Sections 4.2, 7.3): Stereoisomers that differ in their stereochemistry about a double bond or ring.

Citric acid cycle (Section 22.4): The metabolic pathway by which acetyl CoA is degraded to CO_2.

Claisen condensation reaction (Section 17.9): The carbonyl condensation reaction of an ester to give a β-keto ester product.

Claisen rearrangement reaction (Section 13.10): The conversion of an allyl phenyl ether to an *o*-allylphenol by heating.

Coding strand (Section 24.4): The sense strand of double-helical DNA that contains the gene.

Codon (Section 24.5): A three-base sequence on a messenger RNA chain that encodes the genetic information necessary to cause a specific amino acid to be incorporated into a protein. Codons on mRNA are read by complementary anticodons on tRNA.

Coenzyme (Section 19.9): A small organic molecule that acts as a cofactor.

Cofactor (Section 19.9): A small nonprotein part of an enzyme that is necessary for biological activity.

Complex carbohydrate (Section 21.1): A carbohydrate that is made of two or more simple sugars linked together.

Condensed structure (Sections 1.12, 3.2): A shorthand way of writing structures in which C–H and C–C bonds are understood rather than shown explicitly. Propane, for example, has the condensed structure $CH_3CH_2CH_3$.

Configuration (Section 5.5): The three-dimensional arrangement of atoms bonded to a chirality center.

Conformation (Section 3.6): The three-dimensional shape of a molecule at any given instant, assuming that rotation around single bonds is frozen.

Conformational analysis (Section 4.8): A means of assessing the energy of a substituted cycloalkane by totaling the steric interactions present in the molecule.

Conformer (Section 3.6): A conformational isomer.

Conjugate acid (Section 2.7): The product that results from protonation of a Brønsted–Lowry base.

Conjugate addition (Section 14.11): Addition of a nucleophile to the β carbon atom of an α,β-unsaturated carbonyl compound.

Conjugate base (Section 2.7): The anion that results from deprotonation of a Brønsted–Lowry acid.

Conjugation (8.12): A series of overlapping *p* orbitals, usually in alternating single and multiple bonds. For example, buta-1,3-diene is a conjugated diene, but-3-en-2-one is a conjugated enone, and benzene is a cyclic conjugated triene.

Constitutional isomers (Sections 3.2, 5.9): Isomers that have their atoms connected in a different order. For example, butane and 2-methylpropane are constitutional isomers.

Coupled reactions (Section 20.1): Two reactions that share a common intermediate so that the energy released in the favorable step allows the unfavorable step to occur.

Coupling constant, *J* (Section 11.11): The magnitude (expressed in hertz) of the interaction between nuclei whose spins are coupled.

Covalent bond (Section 1.4): A bond formed by sharing electrons between atoms.

Cyanohydrin (Section 15.7): A compound with an –OH group and a –CN group bonded to the same carbon atom; formed by addition of HCN to an aldehyde or ketone.

Cycloalkane (Section 4.1): An alkane that contains a ring of carbons.

D Sugar (Section 21.3): A sugar whose hydroxyl group at the chirality center farthest from the carbonyl group points to the right when drawn in Fischer projection.

***d,l* form** (Section 5.8): The racemic mixture of a chiral compound.

Deactivating group (Section 9.8): An electron-withdrawing substituent that decreases the reactivity of an aromatic ring toward electrophilic aromatic substitution.

Deamination (Section 20.2): The removal of an amino group from a molecule, as occurs with amino acids during metabolic degradation.

Debye, D (Section 2.2): The unit for measuring dipole moments; $1 D = 3.336 \times 10^{-30}$ coulomb meter (C · m).

Decarboxylation (Section 17.5): The loss of carbon dioxide from a molecule. β-Keto acids decarboxylate readily on heating.

Degree of unsaturation (Section 7.1): The number of rings and/or multiple bonds in a molecule.

Dehydration (Sections 8.1, 13.4): The loss of water from an alcohol to yield an alkene.

Dehydrohalogenation (Sections 8.1, 12.11): The loss of HX from an alkyl halide. Alkyl halides undergo dehydrohalogenation to yield alkenes on treatment with strong base.

Delocalization (Section 8.12): A spreading out of electron density over a conjugated π electron system. For example, allylic cations and allylic anions are delocalized because their charges are spread out over the entire π electron system.

Delta scale (Section 11.3): An arbitrary scale used to calibrate NMR charts. One delta unit (δ) is equal to 1 part per million (ppm) of the spectrometer operating frequency.

Denaturation (Section 19.8): The physical changes that occur in a protein when secondary and tertiary structures are disrupted.

Deoxy sugar (Section 21.7): A sugar with one of its –OH groups replaced by an –H.

Deoxyribonucleic acid, DNA (Section 24.1): The biopolymer consisting of deoxyribonucleotide units linked together through phosphate–sugar bonds. Found in the nucleus of cells, DNA contains an organism's genetic information.

DEPT-NMR (Section 11.6): An NMR method for distinguishing among signals due to CH_3, CH_2, CH, and quaternary carbons. That is, the number of hydrogens attached to each carbon can be determined.

Deshielding (Section 11.2): An effect observed in NMR that causes a nucleus to absorb downfield (to the left) of tetramethylsilane (TMS) standard. Deshielding is caused by a withdrawal of electron density from the nucleus.

Deuterium isotope effect (Section 12.12): A tool used in mechanistic investigations to establish whether a C–H bond is broken in the rate-limiting step of a reaction.

Dextrorotatory (Section 5.3): A word used to describe an optically active substance that rotates the plane of polarization of plane-polarized light in a right-handed (clockwise) direction.

Diastereomers (Section 5.6): Non–mirror-image stereoisomers; diastereomers have the same configuration at one or more chirality centers but differ at other chirality centers.

Diastereotopic (Section 11.8): Two hydrogens in a molecule whose replacement by some other group leads to different diastereomers.

1,3-Diaxial interaction (Section 4.7): The strain energy caused by a steric interaction between axial groups three carbon atoms apart in chair cyclohexane.

Dideoxy DNA sequencing (Section 24.6): A biochemical method for sequencing DNA strands.

Dieckmann cyclization reaction (Section 17.10): An intramolecular Claisen condensation reaction to give a cyclic β-keto ester.

Diels–Alder cycloaddition reaction (Section 8.14): The cycloaddition reaction of a diene with a dienophile to yield a cyclohexene.

Dienophile (Section 8.14): A compound containing a double bond that can take part in the Diels–Alder cycloaddition reaction. The most reactive dienophiles are those that have electron-withdrawing groups on the double bond.

Digestion (Section 20.1): The first stage of catabolism, in which food is broken down by hydrolysis of ester, glycoside (acetal), and peptide (amide) bonds to yield fatty acids, simple sugars, and amino acids.

Dihedral angle (Section 3.6): The angle between two bonds on adjacent carbons as viewed along the C–C bond.

Dipole moment, μ (Section 2.2): A measure of the net polarity of a molecule. A dipole moment arises when the centers of mass of positive and negative charges within a molecule do not coincide.

Dipole–dipole force (Section 2.12): A noncovalent electrostatic interaction between polar molecules.

Disaccharide (Section 21.8): A carbohydrate formed by linking two simple sugars through an acetal bond.

Dispersion force (Section 2.12): A noncovalent interaction between molecules that arises because of constantly changing electron distributions within the molecules.

Disulfide (Section 13.7): A compound of the general structure RSSR′.

DNA (Section 24.1): *See* Deoxyribonucleic acid.

Double helix (Section 24.2): The structure of DNA in which two polynucleotide strands coil around each other.

Doublet (Section 11.11): A two-line NMR absorption caused by spin–spin splitting when the spin of the nucleus under observation couples with the spin of a neighboring magnetic nucleus.

Downfield (Section 11.3): Referring to the left-hand portion of the NMR chart.

E **geometry** (Section 7.4): A term used to describe the stereochemistry of a carbon–carbon double bond. The two groups on each carbon are assigned priorities according to the Cahn–Ingold–Prelog sequence rules, and the two carbons are compared. If the high-priority groups on each carbon are on opposite sides of the double bond, the bond has *E* geometry.

E1 reaction (Section 12.13): A unimolecular elimination reaction in which the substrate spontaneously dissociates to give a carbocation intermediate, which loses a proton in a separate step.

E1cB reaction (Section 12.13): A unimolecular elimination reaction in which a proton is first removed to give a carbanion intermediate, which then expels the leaving group in a separate step.

E2 reaction (Section 12.12): A bimolecular elimination reaction in which C–H and C–X bond cleavage are simultaneous.

Eclipsed conformation (Section 3.6): The geometric arrangement around a carbon–carbon single bond in which the bonds to substituents on one carbon are parallel to the bonds to substituents on the neighboring carbon as viewed in a Newman projection.

Eclipsing strain (Section 3.6): The strain energy in a molecule caused by electron repulsions between eclipsed bonds. Eclipsing strain is also called torsional strain.

Edman degradation (Section 19.6): A method for N-terminal sequencing of peptide chains.

Electromagnetic spectrum (Section 10.5): The range of electromagnetic energy, including infrared, ultraviolet, and visible radiation.

Electron configuration (Section 1.3): A list of the orbitals occupied by electrons in an atom.

Electron-dot structure (Section 1.4): A representation of a molecule showing valence electrons as dots.

Electron-transport chain (Section 20.1): The final stage of catabolism in which ATP is produced.

Electronegativity (Section 2.1): The ability of an atom to attract electrons in a covalent bond. Electronegativity increases across the periodic table from right to left and from bottom to top.

Electrophile (Section 6.4): An "electron-lover," or substance that accepts an electron pair from a nucleophile in a polar bond-forming reaction.

Electrophilic addition reaction (Section 7.6): The addition of an electrophile to an alkene to yield a saturated product.

Electrophilic aromatic substitution reaction (Section 9.6): A reaction in which an electrophile (E^+) reacts with an aromatic ring and substitutes for one of the ring hydrogens.

Electrophoresis (Sections 19.2, 24.6): A technique used for separating charged organic molecules, particularly proteins and DNA fragments. The mixture to be separated is placed on a buffered gel or paper, and an electric potential is applied across the ends of the apparatus. Negatively charged molecules migrate toward the positive electrode, and positively charged molecules migrate toward the negative electrode.

Electrostatic potential map (Section 2.1): A molecular representation that uses color to indicate the charge distribution in the molecule as derived from quantum-mechanical calculations.

Elimination reaction (Section 6.1): What occurs when a single reactant splits into two products.

Elution (Chapter 10 *Lagniappe*): The removal of a substance from a chromatography column.

Embden–Meyerhof pathway (Section 22.2): An alternative name for glycolysis.

Enamine (Section 14.7): A compound with the $R_2N—CR=CR_2$ functional group.

Enantiomers (Section 5.1): Stereoisomers of a chiral substance that have a mirror-image relationship. Enantiomers have opposite configurations at all chirality centers.

Enantioselective synthesis (Chapter 14 *Lagniappe,* Section 19.3): A method of synthesis from an achiral precursor that yields only a single enantiomer of a chiral product.

Enantiotopic (Section 11.8): Two hydrogens in a molecule whose replacement by some other group leads to different enantiomers.

3′ End (Section 24.1): The end of a nucleic acid chain with a free hydroxyl group at C3′.

5′ End (Section 24.1): The end of a nucleic acid chain with a free hydroxyl group at C5′.

Endergonic (Section 6.7): A reaction that has a positive free-energy change and is therefore nonspontaneous. In a reaction energy diagram, the product of an endergonic reaction has a higher energy level than the reactants.

Endothermic (Section 6.7): A reaction that absorbs heat and therefore has a positive enthalpy change.

Enol (Section 17.1): A vinylic alcohol that is in equilibrium with a carbonyl compound, C=C–OH.

Enolate ion (Section 17.1): The anion of an enol, $C=C–O^-$.

Enthalpy change, ΔH (Section 6.7): The heat of reaction. The enthalpy change that occurs during a reaction is a measure of the difference in total bond energy between reactants and products.

Entropy change, ΔS (Section 6.7): The change in amount of molecular randomness. The entropy change that occurs during a reaction is a measure of the difference in randomness between reactants and products.

Enzyme (Sections 6.11, 19.9): A biological catalyst. Enzymes are large proteins that catalyze specific biochemical reactions.

Epimers (Section 5.6): Diastereomers that differ in configuration at only one chirality center but are the same at all others.

Epoxide (Section 8.6): A three-membered-ring ether functional group.

Equatorial bond (Section 4.6): A bond to cyclohexane that lies along the rough equator of the ring.

ESI (Section 10.4): Electrospray ionization, a "soft" ionization method used for mass spectrometry of biological samples of very high molecular weight.

Essential amino acid (Section 20.5): One of nine amino acids that are biosynthesized only in plants and microorganisms and must be obtained by humans in the diet.

Essential oil (Chapter 7 *Lagniappe*): The volatile oil obtained by steam distillation of a plant extract.

Ester (Chapter 16 Introduction): A compound containing the $-CO_2R$ functional group.

Estrogen (Section 23.8): A female steroid sex hormone.

Ether (Chapter 13 Introduction): A compound that has two organic substituents bonded to the same oxygen atom, ROR'.

Exergonic (Section 6.7): A reaction that has a negative free-energy change and is therefore spontaneous. On a reaction energy diagram, the product of an exergonic reaction has a lower energy level than that of the reactants.

Exon (Section 24.4): A section of DNA that contains genetic information.

Exothermic (Section 6.7): A reaction that releases heat and therefore has a negative enthalpy change.

Fat (Section 23.1): A solid triacylglycerol derived from an animal source.

Fatty acid (Section 23.1): A long, straight-chain carboxylic acid found in fats and oils.

Fatty acid–derived substance (Section 25.1): A natural product biosynthesized from simple acyl precursors such as acetyl CoA and propionyl CoA.

Fibrous protein (Section 19.8): A protein that consists of polypeptide chains arranged side by side in long threads. Such proteins are tough, insoluble in water, and used in nature for structural materials such as hair, hooves, and fingernails.

Fingerprint region (Section 10.7): The complex region of the infrared spectrum from 1500 to 400 cm^{-1}.

First-order reaction (Section 12.8): A reaction whose rate-limiting step is unimolecular and whose kinetics therefore depend on the concentration of only one reactant.

Fischer esterification reaction (Section 16.3): The acid-catalyzed reaction of an alcohol with a carboxylic acid to yield an ester.

Fischer projection (Section 21.2): A means of depicting the absolute configuration of a chiral molecule on a flat page. A Fischer projection uses a cross to represent the chirality center. The horizontal arms of the cross represent bonds coming out of the plane of the page, and the vertical arms of the cross represent bonds going back into the plane of the page.

Fmoc derivative (Section 19.7): A fluorenylmethyloxy-carbonyl N-protected amino acid.

Formal charge (Section 2.3): The difference in the number of electrons owned by an atom in a molecule and by the same atom in its elemental state.

Formyl group (Section 14.1): A $-CHO$ group.

Frequency, ν (Section 10.5): The number of electromagnetic wave cycles that travel past a fixed point in a given unit of time. Frequencies are expressed in units of cycles per second, or hertz.

Friedel–Crafts reaction (Section 9.7): An electrophilic aromatic substitution reaction to alkylate or acylate an aromatic ring.

FT-NMR (Section 11.4): Fourier-transform NMR; a rapid technique for recording NMR spectra in which all magnetic nuclei absorb at the same time.

Functional group (Section 3.1): An atom or group of atoms that is part of a larger molecule and that has a characteristic chemical reactivity.

Functional RNA (Section 24.4): An alternative name for small RNAs.

Furanose (Section 21.5): The five-membered-ring form of a simple sugar.

Gauche conformation (Section 3.7): The conformation of butane in which the two methyl groups lie 60° apart as viewed in a Newman projection. This conformation has 3.8 kJ/mol steric strain.

Geminal (Section 14.5): Referring to two groups attached to the same carbon atom. For example, a 1,1-diol is a geminal diol.

Gibbs free-energy change, ΔG (Section 6.7): The free-energy change that occurs during a reaction, given by the equation $\Delta G = \Delta H - T\Delta S$. A reaction with a negative free-energy change is spontaneous, and a reaction with a positive free-energy change is nonspontaneous.

Globular protein (Section 19.8): A protein that is coiled into a compact, nearly spherical shape. These proteins, which are generally water-soluble and mobile within the cell, are the structural class to which enzymes belong.

Glucogenic amino acid (Section 20.4): An amino acid that is metabolized either to pyruvate or to an intermediate of the citric acid cycle.

Gluconeogenesis (Section 22.5): The anabolic pathway by which organisms make glucose from simple three-carbon precursors.

Glycal (Section 21.9): An unsaturated sugar with a C1–C2 double bond.

Glycal assembly method (Section 21.9): A method for linking monosaccharides together to synthesis polysaccharides.

Glycerophospholipid (Section 23.3): A lipid that contains a glycerol backbone linked to two fatty acids and a phosphoric acid.

Glycoconjugate (Section 21.6): A molecule in which a carbohydrate is linked through its anomeric center to another biological molecule such as a lipid or protein.

Glycol (Section 8.7): A diol, such as ethylene glycol, $HOCH_2CH_2OH$.

Glycolipid (Section 21.6): A biological molecule in which a carbohydrate is linked through a glycoside bond to a lipid.

Glycolysis (Section 22.2): A series of ten enzyme-catalyzed reactions that break down glucose into 2 equivalents of pyruvate, $CH_3COCO_2^-$.

Glycoprotein (Section 21.6): A biological molecule in which a carbohydrate is linked through a glycoside bond to a protein.

Glycoside (Section 21.6): A cyclic acetal formed by reaction of a sugar with another alcohol.

Green chemistry (Chapter 12 *Lagniappe,* Chapter 18 *Lagniappe*): The design and implementation of chemical products and processes that reduce waste and attempt to eliminate the generation of hazardous substances.

Grignard reagent (Section 12.4): An organomagnesium halide, RMgX.

Ground state (Section 1.3): The most stable, lowest-energy electron configuration of a molecule or atom.

Halogenation (Sections 8.2, 9.6): The reaction of halogen with an alkene to yield a 1,2-dihalide addition product or with an aromatic compound to yield a substitution product.

Halohydrin (Section 8.3): A 1,2-disubstituted haloalcohol, such as that obtained on addition of HOBr to an alkene.

Hammond postulate (Section 7.9): A postulate stating that we can get a picture of what a given transition state looks like by looking at the structure of the nearest stable species. Exergonic reactions have transition states that resemble reactant; endergonic reactions have transition states that resemble product.

Heat of hydrogenation (Section 7.5): The amount of heat released when a carbon–carbon double bond is hydrogenated.

Heat of reaction (Section 6.7): An alternative name for the enthalpy change in a reaction, ΔH.

Hell–Volhard–Zelinskii (HVZ) reaction (Section 17.3): The reaction of a carboxylic acid with Br_2 and phosphorus to give an α-bromo carboxylic acid.

Hemiacetal (Section 14.8): A functional group consisting of one –OR and one –OH group bonded to the same carbon.

Hemithioacetal (Section 22.2): The sulfur analog of an acetal, resulting from nucleophilic addition of a thiol to a ketone or aldehyde.

Henderson–Hasselbalch equation (Sections 15.3, 18.5): An equation for determining the extent of deprotonation of a weak acid at various pH values.

Hertz, Hz (Section 10.5): A measure of electromagnetic frequency, the number of waves that pass by a fixed point per second.

Heterocycle (Sections 9.4, 18.8): A cyclic molecule whose ring contains more than one kind of atom. For example, pyridine is a heterocycle that contains five carbon atoms and one nitrogen atom in its ring.

High-energy compound (Section 6.8): A term used in biochemistry to describe substances such as ATP that undergo highly exothermic reactions.

Hofmann elimination reaction (Section 18.7): The elimination reaction of an amine to yield an alkene by reaction with iodomethane, followed by heating with Ag_2O.

HOMO (Section 10.9): The highest occupied molecular orbital.

Homotopic (Section 11.8): Hydrogens that give the identical structure on replacement by X and thus show identical NMR absorptions.

Hormone (Section 23.8): A chemical messenger that is secreted by an endocrine gland and carried through the bloodstream to a target tissue.

HPLC (Chapter 10 *Lagniappe*): High-pressure liquid chromatography; a variant of column chromatography using high pressure to force solvent through very small absorbent particles.

Hückel's rule (Section 9.3): A rule stating that monocyclic conjugated molecules having $4n + 2$ π electrons ($n =$ an integer) are aromatic.

Hund's rule (Section 1.3): If two or more empty orbitals of equal energy are available, one electron occupies each, with their spins parallel, until all are half-full.

Hybrid orbital (Section 1.6): An orbital derived from a combination of atomic orbitals. Hybrid orbitals, such as the sp^3, sp^2, and sp hybrids of carbon, are strongly directed and form stronger bonds than atomic orbitals do.

Hydration (Section 8.4): Addition of water to a molecule, such as occurs when alkenes are treated with aqueous sulfuric acid to give alcohols.

Hydride shift (Section 7.10): The shift of a hydrogen atom and its electron pair to a nearby cationic center.

Hydroboration (Section 8.4): Addition of borane (BH_3) or an alkylborane to an alkene. The resultant trialkylborane products can be oxidized to yield alcohols.

Hydrocarbon (Section 3.2): A compound that contains only carbon and hydrogen.

Hydrogen bond (Sections 2.12, 13.2): A weak attraction between a hydrogen atom bonded to an electronegative atom and an electron lone pair on another electronegative atom.

Hydrogenation (Section 8.5): Addition of hydrogen to a double or triple bond to yield a saturated product.

Hydrogenolysis (Section 19.7): Cleavage of a bond by reaction with hydrogen. Benzylic ethers and esters, for instance, are cleaved by hydrogenolysis.

Hydrophilic (Sections 2.12, 23.2): Water-loving; attracted to water.

Hydrophobic (Section 2.12, 23.2): Water-fearing; not attracted to water.

Hydroquinone (Section 13.5): A 1,4-dihydroxybenzene.

Hydroxylation (Section 8.7): Addition of two –OH groups to a double bond.

Hyperconjugation (Section 7.5): An interaction that results from overlap of a vacant p orbital on one atom with a neighboring C–H σ bond. Hyperconjugation is important in stabilizing carbocations and in stabilizing substituted alkenes.

Imine (Section 14.7): A compound with the $R_2C{=}NR$ functional group; also called a Schiff base.

Inductive effect (Sections 2.1, 7.8, 9.8): The electron-attracting or electron-withdrawing effect transmitted through σ bonds. Electronegative elements have an electron-withdrawing inductive effect.

Infrared (IR) spectroscopy (Section 10.6): A kind of optical spectroscopy that uses infrared energy. IR spectroscopy is particularly useful in organic chemistry for determining the kinds of functional groups present in molecules.

Initiator (Section 6.3): A substance with an easily broken bond that is used to initiate a radical chain reaction. For example, radical chlorination of alkanes is initiated when light energy breaks the weak Cl–Cl bond to form Cl· radicals.

Integration (Section 11.10): A technique for measuring the area under an NMR peak to determine the relative number of each kind of proton in a molecule. Integrated peak areas are superimposed over the spectrum as a "stair-step" line, with the height of each step proportional to the area underneath the peak.

Intermediate (Section 6.10): A species that is formed during the course of a multistep reaction but is not the final product. Intermediates are more stable than transition states but may or may not be stable enough to isolate.

Intramolecular, intermolecular (Section 17.8): A reaction that occurs within the same molecule is intramolecular; a reaction that occurs between two molecules is intermolecular.

Intron (Section 24.4): A section of DNA that does not contain genetic information.

Ion pair (Section 12.8): A loose complex between two ions in solution. Ion pairs are implicated as intermediates in S_N1 reactions to account for the partial retention of stereochemistry that is often observed.

Isoelectric point, p*I* (Section 19.2): The pH at which the number of positive charges and the number of negative charges on a protein or an amino acid are equal.

Isomers (Sections 3.2, 5.9): Compounds that have the same molecular formula but different structures.

Isoprene rule (Chapter 7 *Lagniappe*): An observation to the effect that terpenoids appear to be made up of isoprene (2-methylbuta-1,3-diene) units connected head-to-tail.

Isotopes (Section 1.1): Atoms of the same element that have different mass numbers.

IUPAC system of nomenclature (Section 3.4): Rules for naming compounds, devised by the International Union of Pure and Applied Chemistry.

Kekulé structure (Section 1.4): A method of representing molecules in which a line between atoms indicates a bond.

Ketal (Section 14.8): An alternative name for an acetal that is derived from a ketone rather than an aldehyde.

Keto–enol tautomerism (Section 17.1): The rapid equilibration between a carbonyl form and vinylic alcohol form of a molecule.

Ketogenic amino acid (Section 20.4): An amino acid that is metabolized into an intermediate that can enter fatty-acid biosynthesis.

Ketone (Chapter 14 Introduction): A compound with two organic substituents bonded to a carbonyl group, $R_2C{=}O$.

Ketone body (Section 20.4): One of the substances acetoacetate, β-hydroxybutyrate, or acetone resulting from amino acid catabolism.

Ketose (Section 21.1): A carbohydrate with a ketone functional group.

Kinetics (Section 12.6): Referring to reaction rates. Kinetic measurements are useful for helping to determine reaction mechanisms.

Krebs cycle (Section 22.4): An alternative name for the citric acid cycle, by which acetyl CoA is degraded to CO_2.

ʟ **Sugar** (Section 21.3): A sugar whose hydroxyl group at the chirality center farthest from the carbonyl group points to the left when drawn in Fischer projection.

Lactam (Section 16.7): A cyclic amide.

Lactone (Section 16.6): A cyclic ester.

Lagging strand (Section 24.3): The complement of the original 3′ → 5′ DNA strand that is synthesized discontinuously in small pieces that are subsequently linked by DNA ligases.

Lagniappe: A word in the Creole dialect of southern Louisiana meaning an extra benefit, or a little something extra. See the *Lagniappes* at the end of each chapter.

LD₅₀ (Chapter 1 *Lagniappe*): The amount of a substance per kilogram body weight that is lethal to 50% of test animals.

Leading strand (Section 24.3): The complement of the original 5′ → 3′ DNA strand that is synthesized continuously in a single piece.

Leaving group (Section 12.5): The group that is replaced in a substitution reaction.

Levorotatory (Section 5.3): An optically active substance that rotates the plane of polarization of plane-polarized light in a left-handed (counterclockwise) direction.

Lewis acid (Section 2.11): A substance with a vacant low-energy orbital that can accept an electron pair from a base. All electrophiles are Lewis acids.

Lewis base (Section 2.11): A substance that donates an electron lone pair to an acid. All nucleophiles are Lewis bases.

Lewis structure (Section 1.4): A representation of a molecule showing valence electrons as dots.

Lindlar catalyst (Section 8.15): A hydrogenation catalyst used to convert alkynes to cis alkenes.

Line-bond structure (Section 1.4): A representation of a molecule showing covalent bonds as lines between atoms.

1→4 Link (Section 21.8): An acetal link between the C1 –OH group of one sugar and the C4 –OH group of another sugar.

Lipid (Chapter 23 Introduction): A naturally occurring substance isolated from cells and tissues by extraction with a nonpolar solvent. Lipids belong to many different structural classes, including fats, terpenes, prostaglandins, and steroids.

Lipid bilayer (Section 23.3): The ordered lipid structure that forms a cell membrane.

Lipoprotein (Chapter 23 *Lagniappe*): A complex molecule with both lipid and protein parts that transports lipids through the body.

Lone-pair electrons (Section 1.4): Nonbonding valence-shell electron pairs. Lone-pair electrons are used by nucleophiles in their reactions with electrophiles.

LUMO (Section 10.9): The lowest unoccupied molecular orbital.

Magnetic resonance imaging, MRI (Chapter 11 *Lagniappe*): A medical diagnostic technique based on nuclear magnetic resonance.

Major groove (Section 24.2): The larger of two grooves in the DNA double helix.

MALDI (Section 10.4): Matrix-assisted laser desorption ionization, a "soft" ionization method used for mass spectrometry of biological samples of very high molecular weight.

Malonic ester synthesis (Section 17.5): The synthesis of a carboxylic acid by alkylation of an alkyl halide, followed by hydrolysis and decarboxylation.

Markovnikov's rule (Section 7.7): A guide for determining the regiochemistry (orientation) of electrophilic addition reactions. In the addition of HX to an alkene, the hydrogen atom bonds to the alkene carbon that has fewer alkyl substituents.

Mass number, A (Section 1.1): The total of protons plus neutrons in an atom.

Mass spectrometry (Section 10.1): A technique for measuring the mass, and therefore the molecular weight (MW), of ions.

McLafferty rearrangement (Section 10.3): A mass-spectral fragmentation pathway for carbonyl compounds.

Mechanism (Section 6.2): A complete description of how a reaction occurs. A mechanism must account for all starting materials and all products and must describe the details of each individual step in the overall reaction process.

Meisenheimer complex (Section 9.9): The intermediate in a nucleophilic aromatic substitution reaction, formed by addition of a nucleophile to a halo-substituted aromatic ring.

Mercapto group (Section 13.1): An alternative name for the thiol group, –SH.

Meso compound (Section 5.7): A compound that contains chirality centers but is nevertheless achiral because it contains a symmetry plane.

Messenger RNA (Section 24.4): A kind of RNA formed by transcription of DNA and used to carry genetic messages from DNA to ribosomes.

Meta, m- (Section 9.1): A naming prefix used for 1,3-disubstituted benzenes.

Metabolism (Section 20.1): A collective name for the many reactions that go on in the cells of living organisms.

Methylene group (Section 7.2): A $-CH_2-$ or $=CH_2$ group.

Micelle (Section 23.2): A spherical cluster of soaplike molecules that aggregate in aqueous solution. The ionic heads of the molecules lie on the outside, where they are solvated by water, and the organic tails bunch together on the inside of the micelle.

Michael reaction (Section 17.11): The conjugate addition reaction of an enolate ion to an unsaturated carbonyl compound.

Minor groove (Section 24.2): The smaller of two grooves in the DNA double helix.

Molar absorptivity (Section 10.9): A quantitative measure of the amount of UV light absorbed by a sample.

Molecular ion (Section 10.1): The cation produced in the mass spectrometer by loss of an electron from the parent molecule. The mass of the molecular ion corresponds to the molecular weight of the sample.

Molecular mechanics (Chapter 4 *Lagniappe*): A computer-based method for calculating the minimum-energy conformation of a molecule.

Molecular orbital (MO) theory (Section 1.11): A description of covalent bond formation as resulting from a mathematical combination of atomic orbitals (wave functions) to form molecular orbitals.

Molecule (Section 1.4): A neutral collection of atoms held together by covalent bonds.

Molozonide (Section 8.8): The initial addition product of ozone with an alkene.

Monomer (Section 8.10): The simple starting unit from which a polymer is made.

Monosaccharide (Section 21.1): A simple sugar.

Monoterpene (Chapter 7 *Lagniappe,* Section 23.7): A ten-carbon lipid.

Multiplet (Section 11.11): A pattern of peaks in an NMR spectrum that arises by spin–spin splitting of a single absorption because of coupling between neighboring magnetic nuclei.

Mutarotation (Section 21.5): The change in optical rotation observed when a pure anomer of a sugar is dissolved in water. Mutarotation is caused by the reversible opening and closing of the acetal linkage, which yields an equilibrium mixture of anomers.

***n* + 1 rule** (Section 11.11): A hydrogen with *n* other hydrogens on neighboring carbons shows *n* + 1 peaks in its ^{1}H NMR spectrum.

N-terminal amino acid (Section 19.4): The amino acid with a free $-NH_2$ group at the end of a protein chain.

Natural gas (Chapter 3 *Lagniappe*): A naturally occurring hydrocarbon mixture consisting chiefly of methane, along with smaller amounts of ethane, propane, and butane.

Natural product (Chapter 6 *Lagniappe,* Chapter 25): A catchall term generally taken to mean a secondary metabolite found in bacteria, plants, and other living organisms.

Neopentyl group (Section 3.4): The 2,2-dimethylpropyl group, $(CH_3)_3CCH_2-$.

Neuraminidase (Chapter 22 *Lagniappe*): An enzyme present on the surface of viral particles that cleaves the bond holding the newly formed viral particles to host cells.

New molecular entity, NME (Chapter 6 *Lagniappe*): A new biologically active chemical substance approved for sale as a drug by the U.S. Food and Drug Administration.

Newman projection (Section 3.6): A means of indicating stereochemical relationships between substituent groups on neighboring carbons. The carbon–carbon bond is viewed end-on, and the carbons are indicated by a circle. Bonds radiating from the center of the circle are attached to the front carbon, and bonds radiating from the edge of the circle are attached to the rear carbon.

Nitration (Section 9.6): The substitution of a nitro group onto an aromatic ring.

Nitrile (Section 15.1): A compound containing the $C{\equiv}N$ functional group.

Nitrogen rule (Section 18.10): A compound with an odd number of nitrogen atoms has an odd-numbered molecular weight.

Node (Section 1.2): A surface of zero electron density within an orbital. For example, a *p* orbital has a nodal plane passing through the center of the nucleus, perpendicular to the axis of the orbital.

Nonbonding electrons (Section 1.4): Valence electrons that are not used in forming covalent bonds.

Noncovalent interaction (Section 2.12): One of a variety of nonbonding interactions between molecules, such as dipole–dipole forces, dispersion forces, and hydrogen bonds.

Nonessential amino acid (Section 20.5): One of the eleven amino acids that are biosynthesized by humans.

Nonribosomal polypeptide (Section 25.1): A peptidelike compound biosynthesized from an amino acid without direct RNA transcription. The penicillins are examples.

Normal alkane (Section 3.2): A straight-chain alkane, as opposed to a branched alkane. Normal alkanes are denoted by the suffix *n*, as in *n*-C_4H_{10} (*n*-butane).

NSAID (Chapter 9 *Lagniappe*): A nonsteroidal anti-inflammatory drug, such as aspirin or ibuprofen.

Nuclear magnetic resonance, NMR (Chapter 11): A spectroscopic technique that provides information about the carbon–hydrogen framework of a molecule. NMR works by detecting the energy absorption accompanying the transitions between nuclear spin states that occur when a molecule is placed in a strong magnetic field and irradiated with radiofrequency waves.

Nucleophile (Section 6.4): A "nucleus-lover," or species that donates an electron pair to an electrophile in a polar bond-forming reaction. Nucleophiles are also Lewis bases.

Nucleophilic acyl substitution reaction (Section 16.2): A reaction in which a nucleophile attacks a carbonyl compound and substitutes for a leaving group bonded to the carbonyl carbon.

Nucleophilic addition reaction (Section 14.4): A reaction in which a nucleophile adds to the electrophilic carbonyl group of a ketone or aldehyde to give an alcohol.

Nucleophilic aromatic substitution reaction (Section 9.9): The substitution reaction of an aryl halide by a nucleophile.

Nucleophilic substitution reaction (Section 12.5): A reaction in which one nucleophile replaces another attached to a saturated carbon atom.

Nucleophilicity (Section 12.7): The ability of a substance to act as a nucleophile in an S_N2 reaction.

Nucleoside (Section 24.1): A nucleic acid constituent, consisting of a sugar residue bonded to a heterocyclic purine or pyrimidine base.

Nucleotide (Section 24.1): A nucleic acid constituent, consisting of a sugar residue bonded both to a heterocyclic purine or pyrimidine base and to a phosphoric acid. Nucleotides are the monomer units from which DNA and RNA are constructed.

Nylon (Section 16.9): A synthetic polyamide step-growth polymer.

Olefin (Chapter 7 Introduction): An alternative name for an alkene.

Optical isomers (Section 5.4): An older, alternative name for enantiomers. Optical isomers are isomers that have a mirror-image relationship.

Optically active (Section 5.3): A substance that rotates the plane of polarization of plane-polarized light.

Orbital (Section 1.2): A wave function, which describes the volume of space around a nucleus in which an electron is most likely to be found.

Organic chemistry (Chapter 1 Introduction): The study of carbon compounds.

Organohalide (Chapter 12 Introduction): A compound that contains one or more halogen atoms bonded to carbon.

Organometallic compound (Section 12.4): A compound that contains a carbon–metal bond. Grignard reagents, RMgX, are examples.

Organophosphate (Section 1.10): A compound that contains a phosphorus atom bonded to four oxygens, with one of the oxygens also bonded to carbon.

Ortho, *o-* (Section 9.1): A naming prefix used for 1,2-disubstituted benzenes.

Oxidation (Section 8.6): A reaction that causes a decrease in electron ownership by carbon, either by bond formation between carbon and a more electronegative atom (usually oxygen, nitrogen, or a halogen) or by bond-breaking between carbon and a less electronegative atom (usually hydrogen).

Oxidative deamination (Section 20.2): The conversion of a primary amine into a ketone by oxidation to an imine followed by hydrolysis.

Oxidative decarboxylation (Section 22.3): A decarboxylation reaction, usually of an α-keto acid, that is accompanied by a change in oxidation state of the carbonyl carbon from that of a ketone to that of a carboxylic acid or ester.

Oxirane (Section 8.6): An alternative name for an epoxide.

Oxymercuration (Section 8.4): A method for double-bond hydration using aqueous mercuric acetate as the reagent.

Ozonide (Section 8.8): The product formed by addition of ozone to a carbon–carbon double bond. Ozonides are usually treated with a reducing agent, such as zinc in acetic acid, to produce carbonyl compounds.

Para, *p-* (Section 9.1): A naming prefix used for 1,4-disubstituted benzenes.

Paraffin (Section 3.5): A common name for alkanes.

Parent peak (Section 10.1): The peak in a mass spectrum corresponding to the molecular ion. The mass of the parent peak therefore represents the molecular weight of the compound.

Pauli exclusion principle (Section 1.3): No more than two electrons can occupy the same orbital, and those two must have spins of opposite sign.

Peptide (Chapter 19 Introduction): A short amino acid polymer in which the individual amino acid residues are linked by amide bonds.

Peptide bond (Section 19.4): An amide bond in a peptide chain.

Pericyclic reaction (Section 8.14): A reaction that takes place in a single step without intermediates by a cyclic redistribution of bonding electrons.

Periplanar (Section 12.12): A conformation in which bonds to neighboring atoms have a parallel arrangement. In an eclipsed conformation, the neighboring bonds are syn periplanar; in a staggered conformation, the bonds are anti periplanar.

Peroxyacid (Section 8.6): A compound with the $-CO_3H$ functional group.

Petroleum (Chapter 3 *Lagniappe*): A complex mixture of naturally occurring hydrocarbons derived from the decomposition of plant and animal matter.

Phenol (Chapter 13 Introduction): A compound with an $-OH$ group directly bonded to an aromatic ring, ArOH.

Phenoxide ion (Section 13.2): The anion of a phenol, ArO^-.

Phenyl group (Section 9.1): The name for the $-C_6H_5$ unit when the benzene ring is considered as a substituent. A phenyl group is abbreviated as $-Ph$.

Phosphite (Section 24.7): A compound with the structure $P(OR)_3$.

Phospholipid (Section 23.3): A lipid that contains a phosphate residue. For example, glycerophospholipids contain a glycerol backbone linked to two fatty acids and a phosphoric acid.

Phosphoramidite (Section 24.7): A compound with the structure $R_2NP(OR)_2$.

Phosphoric acid anhydride (Section 20.1): A substance that contains PO_2PO link, analogous to the CO_2CO link in carboxylic acid anhydrides.

Physiological pH (Section 15.3): The pH of 7.3 that exists inside cells.

Pi (π) bond (Section 1.8): The covalent bond formed by sideways overlap of atomic orbitals. For example, carbon–carbon double bonds contain a π bond formed by sideways overlap of two p orbitals.

PITC (Section 19.6): Phenylisothiocyanate, used in the Edman degradation of proteins.

pK_a (Section 2.8): The negative common logarithm of the K_a; used to express acid strength.

Plane of symmetry (Section 5.2): A plane that bisects a molecule such that one half of the molecule is the mirror image of the other half. Molecules containing a plane of symmetry are achiral.

Plane-polarized light (Section 5.3): Ordinary light that has its electromagnetic waves oscillating in a single plane rather than in random planes. The plane of polarization is rotated when the light is passed through a solution of a chiral substance.

Plasticizer (Section 16.6): A small organic molecule added to polymers to act as a lubricant between polymer chains.

Polar aprotic solvent (Section 12.7): A polar solvent that can't function as a hydrogen ion donor. Polar aprotic solvents such as dimethyl sulfoxide (DMSO) and dimethylformamide (DMF) are particularly useful in S_N2 reactions because of their ability to solvate cations.

Polar covalent bond (Section 2.1): A covalent bond in which the electron distribution between atoms is unsymmetrical.

Polar reaction (Section 6.4): A reaction in which bonds are made when a nucleophile donates two electrons to an electrophile and in which bonds are broken when one fragment leaves with both electrons from the bond.

Polarity (Section 2.1): The unsymmetrical distribution of electrons in a molecule that results when one atom attracts electrons more strongly than another.

Polarizability (Section 6.4): The measure of the change in a molecule's electron distribution in response to changing electric interactions with solvents or ionic reagents.

Polycyclic aromatic compound (Section 9.5): A compound with two or more benzene-like aromatic rings fused together.

Polycyclic compound (Section 4.9): A compound that contains more than one ring.

Polyketide (Section 25.1): A natural product biosynthesized from simple acyl precursors such as acetyl CoA, propionyl CoA, and methylmalonyl CoA by a large multifunctional enzyme complex.

Polymer (Sections 8.10, 16.9): A large molecule made up of repeating smaller units. For example, polyethylene is a synthetic polymer made from repeating ethylene units, and DNA is a biopolymer made of repeating deoxyribonucleotide units.

Polymerase chain reaction, PCR (Section 24.8): A method for amplifying small amounts of DNA to produce larger amounts.

Polysaccharide (Section 21.9): A carbohydrate that is made of many simple sugars linked together by acetal bonds.

Polyunsaturated fatty acid (Section 23.1): A fatty acid containing two or more double bonds.

Primary, secondary, tertiary, quaternary (Section 3.3): Terms used to describe the substitution pattern at a specific site. A primary site has one organic substituent attached to it, a secondary site has two organic substituents, a tertiary site has three, and a quaternary site has four.

	Carbon	Carbo-cation	Hydrogen	Alcohol	Amine
Primary	RCH_3	RCH_2^+	RCH_3	RCH_2OH	RNH_2
Secondary	R_2CH_2	R_2CH^+	R_2CH_2	R_2CHOH	R_2NH
Tertiary	R_3CH	R_3C^+	R_3CH	R_3COH	R_3N
Quaternary	R_4C				

Primary structure (Section 19.8): The amino acid sequence in a protein.

Prochiral (Section 5.11): A molecule that can be converted from achiral to chiral in a single chemical step.

Prochirality center (Section 5.11): An atom in a compound that can be converted into a chirality center by changing one of its attached substituents.

Propagation step (Section 6.3): The step or series of steps in a radical chain reaction that carry on the chain. The propagation steps must yield both product and a reactive intermediate.

***pro-R* configuration** (Section 5.11): One of two identical atoms in a compound whose replacement leads to an *R* chirality center.

***pro-S* configuration** (Section 5.11): One of two identical atoms in a compound whose replacement leads to an *S* chirality center.

Protecting group (Sections 13.6, 14.8, 19.7, 21.9): A group that is introduced to protect a sensitive functional group toward reaction elsewhere in the molecule. After serving its protective function, the group is removed.

Protein (Chapter 19 Introduction): A large peptide containing 50 or more amino acid residues. Proteins serve both as structural materials and as enzymes that control an organism's chemistry.

Protein Data Bank (Chapter 19 *Lagniappe*): A worldwide online repository of X-ray and NMR structural data for biological macromolecules. To access the Protein Data Bank, go to http://www.rcsb.org/pdb/.

Protic solvent (Section 12.7): A solvent such as water or alcohol that can act as a proton donor.

Pyramidal inversion (Section 18.2): The rapid inversion of configuration of an amine.

Pyranose (Section 21.5): The six-membered-ring form of a simple sugar.

Quartet (Section 11.11): A set of four peaks in an NMR spectrum, caused by spin–spin splitting of a signal by three adjacent nuclear spins.

Quaternary: *See* Primary.

Quaternary ammonium salt (Section 18.1): An ionic compound containing a positively charged nitrogen atom with four attached groups, $R_4N^+ X^-$.

Quaternary structure (Section 19.8): The highest level of protein structure, involving a specific aggregation of individual proteins into a larger cluster.

Quinone (Section 13.5): A cyclohexa-2,5-diene-1,4-dione.

R configuration (Section 5.5): The configuration at a chirality center as specified using the Cahn–Ingold–Prelog sequence rules.

R group (Section 3.3): A generalized abbreviation for an organic partial structure.

Racemate (Section 5.8): A mixture consisting of equal parts (+) and (−) enantiomers of a chiral substance.

Radical (Section 6.2): A species that has an odd number of electrons, such as the chlorine radical, Cl·.

Radical reaction (Section 6.3): A reaction in which bonds are made by donation of one electron from each of two reactants and in which bonds are broken when each fragment leaves with one electron.

Rate constant (Section 12.6): The constant k in a rate equation.

Rate equation (Section 12.6): An equation that expresses the dependence of a reaction's rate on the concentration of reactants.

Rate-limiting step (Section 12.7): The slowest step in a multistep reaction sequence. The rate-limiting step acts as a kind of bottleneck in multistep reactions.

***Re* face** (Section 5.11): One of two faces of a planar, sp^2-hybridized atom.

Reaction energy diagram (Section 6.9): A representation of the course of a reaction in which free energy is plotted as a function of reaction progress. Reactants, transition states, intermediates, and products are represented, and their appropriate energy levels are indicated.

Reaction intermediate (Section 6.10): *See* Intermediate.

Reaction mechanism (Section 6.2): *See* Mechanism.

Rearrangement reaction (Section 6.1): What occurs when a single reactant undergoes a reorganization of bonds and atoms to yield an isomeric product.

Reducing sugar (Section 21.6): A sugar that reduces silver ion in the Tollens test or cupric ion in the Fehling or Benedict tests.

Reduction (Section 8.5): A reaction that causes an increase of electron ownership by carbon, either by bond-breaking between carbon and a more electronegative atom or by bond formation between carbon and a less electronegative atom.

Reductive amination (Sections 18.6, 19.3): A method for preparing an amine by reaction of an aldehyde or ketone with ammonia and a reducing agent.

Refining (Chapter 3 *Lagniappe*): The process by which petroleum is converted into gasoline and other useful products.

Regiospecific (Section 7.7): A term describing a reaction that occurs with a specific regiochemistry to give a single product rather than a mixture of products.

Replication (Section 24.3): The process by which double-stranded DNA uncoils and is replicated to produce two new copies.

Replication fork (Section 24.3): The point of unraveling in a DNA chain where replication occurs.

Residue (Section 19.4): An amino acid in a protein chain.

Resolution (Section 5.8): The process by which a racemic mixture is separated into its two pure enantiomers.

Resonance effect (Section 9.8): The donation or withdrawal of electrons through orbital overlap with neighboring π bonds. For example, an oxygen or nitrogen substituent donates electrons to an aromatic ring by overlap of the O or N orbital with the aromatic ring p orbitals.

Resonance form (Section 2.4): An individual Lewis structure of a resonance hybrid.

Resonance hybrid (Section 2.4): A molecule, such as benzene, that can't be represented adequately by a single Kekulé structure but must instead be considered as an average of two or more resonance structures. The resonance structures themselves differ only in the positions of their electrons, not their nuclei.

Restriction endonuclease (Section 24.6): An enzyme that is able to cleave a DNA molecule at points in the chain where a specific base sequence occurs.

Retrosynthetic (Section 9.11): Planning an organic synthesis by working backward from product to starting material.

Ribonucleic acid, RNA (Section 24.1): The biopolymer found in cells that serves to transcribe the genetic information found in DNA and uses that information to direct the synthesis of proteins.

Ribosomal RNA (Section 24.4): A kind of RNA used in the physical makeup of ribosomes.

Ring-flip (Section 4.6): A molecular motion that converts one chair conformation of cyclohexane into another chair conformation. The effect of a ring-flip is to convert an axial substituent into an equatorial substituent.

RNA (Section 24.1): *See* Ribonucleic acid.

S configuration (Section 5.5): The configuration at a chirality center as specified using the Cahn–Ingold–Prelog sequence rules.

s-Cis conformation (Section 8.14): Describing a conformation that is "cis-like" about a single bond.

Saccharide (Section 21.1): A sugar.

Salt bridge (Section 19.8): The ionic attraction between two oppositely charged groups in a protein chain.

Sanger dideoxy method (Section 24.6): A biochemical method for sequencing DNA strands.

Saponification (Section 16.6): An old term for the base-induced hydrolysis of an ester to yield a carboxylic acid salt.

Saturated (Section 3.2): A molecule that has only single bonds and thus can't undergo addition reactions. Alkanes are saturated, but alkenes are unsaturated.

Sawhorse structure (Section 3.6): A manner of representing stereochemistry that uses a stick drawing and gives a perspective view of the conformation around a single bond.

Schiff base (Sections 14.7, 22.2): An alternative name for an imine, $R_2C{=}NR'$, used primarily in biochemistry.

Secondary: *See* Primary.

Secondary metabolite (Chapter 25 Introduction): A small naturally occurring molecule that is not essential to the growth and development of the producing organism and is not classified by structure.

Secondary structure (Section 19.8): The level of protein substructure that involves organization of chain sections into ordered arrangements such as β-pleated sheets or α helices.

Second-order reaction (Section 12.6): A reaction whose rate-limiting step is bimolecular and whose kinetics are therefore dependent on the concentration of two reactants.

Semiconservative replication (Section 24.3): The process by which DNA molecules are made containing one strand of old DNA and one strand of new DNA.

Sense strand (Section 24.4): The coding strand of double-helical DNA that contains the gene.

Sequence rules (Sections 5.5, 7.4): A series of rules for assigning relative priorities to substituent groups on a double-bond carbon atom or on a chirality center.

Sesquiterpene (Chapter 7 *Lagniappe*, Section 23.7): A 15-carbon lipid.

Sharpless epoxidation (Chapter 14 *Lagniappe*): A method for enantioselective synthesis of a chiral epoxide by treatment of an allylic alcohol with *tert*-butyl hydroperoxide, $(CH_3)_3C{-}OOH$, in the presence of titanium tetraisopropoxide and diethyl tartrate.

Shell (electron) (Section 1.2): A group of an atom's electrons with the same principal quantum number.

Shielding (Section 11.2): An effect observed in NMR that causes a nucleus to absorb toward the right (upfield) side of the chart. Shielding is caused by donation of electron density to the nucleus.

***Si* face** (Section 5.11): One of two faces of a planar, sp^2-hybridized atom.

Sialic acid (Section 21.7): One of a group of more than 300 carbohydrates based on acetylneuramic acid.

Side chain (Section 19.1): The substituent attached to the α carbon of an amino acid.

Sigma (σ) bond (Section 1.5): A covalent bond formed by head-on overlap of atomic orbitals.

Silyl ether (Section 13.6): A substance with the structure R_3Si—O—R. The silyl ether acts as a protecting group for alcohols.

Simple sugar (Section 21.1): A carbohydrate that cannot be broken down into smaller sugars by hydrolysis.

Skeletal structure (Section 1.12): A shorthand way of writing structures in which carbon atoms are assumed to be at each intersection of two lines (bonds) and at the end of each line.

Small RNAs (Section 24.4): A type of RNA that has a variety of functions within the cell, including silencing transcription and catalyzing chemical modifications of other RNA molecules.

S_N1 reaction (Section 12.8): A unimolecular nucleophilic substitution reaction.

S_N2 reaction (Section 12.6): A bimolecular nucleophilic substitution reaction.

Solid-phase synthesis (Section 19.7): A technique of synthesis whereby the starting material is covalently bound to a solid polymer bead and reactions are carried out on the bound substrate. After the desired transformations have been effected, the product is cleaved from the polymer.

Solvation (Section 12.7): The clustering of solvent molecules around a solute particle to stabilize it.

***sp* Hybrid orbital** (Section 1.9): A hybrid orbital derived from the combination of an *s* and a *p* atomic orbital. The two *sp* orbitals that result from hybridization are oriented at an angle of 180° to each other.

***sp²* Hybrid orbital** (Section 1.8): A hybrid orbital derived by combination of an *s* atomic orbital with two *p* atomic orbitals. The three sp^2 hybrid orbitals that result lie in a plane at angles of 120° to each other.

***sp³* Hybrid orbital** (Section 1.6): A hybrid orbital derived by combination of an *s* atomic orbital with three *p* atomic orbitals. The four sp^3 hybrid orbitals that result are directed toward the corners of a regular tetrahedron at angles of 109° to each other.

Specific rotation, $[\alpha]_D$ (Section 5.3): The optical rotation of a chiral compound under standard conditions.

Sphingomyelin (Section 23.3): A phospholipid that has sphingosine as its backbone.

Spin–spin splitting (Section 11.11): The splitting of an NMR signal into a multiplet because of an interaction between nearby magnetic nuclei whose spins are coupled. The magnitude of spin–spin splitting is given by the coupling constant, J.

Staggered conformation (Section 3.6): The three-dimensional arrangement of atoms around a carbon–carbon single bond in which the bonds on one carbon bisect the bond angles on the second carbon as viewed end-on.

Step-growth polymer (Section 16.9): A polymer in which each bond is formed independently of the others. Polyesters and polyamides (nylons) are examples.

Stereochemistry (Section 3.6; Chapters 3, 4, and 5): The branch of chemistry concerned with the three-dimensional arrangement of atoms in molecules.

Stereoisomers (Section 4.2): Isomers that have their atoms connected in the same order but have different three-dimensional arrangements. The term *stereoisomer* includes both enantiomers and diastereomers.

Stereospecific (Section 8.9): Describing a reaction in which a single product stereoisomer is formed.

Steric strain (Sections 3.7, 4.7): The strain imposed on a molecule when two groups are too close together and try to occupy the same space. Steric strain is responsible both for the greater stability of trans versus cis alkenes and for the greater stability of equatorially substituted versus axially substituted cyclohexanes.

Steroid (Section 23.8): A lipid whose structure is based on a tetracyclic carbon skeleton with three 6-membered and one 5-membered ring. Steroids occur in both plants and animals and have a variety of important hormonal functions.

Stork enamine reaction (Section 17.12): The conjugate addition of an enamine to an α,β-unsaturated carbonyl compound, followed by hydrolysis to yield a 1,5-dicarbonyl product.

STR loci (Chapter 24 *Lagniappe*): Short tandem repeat sequences of noncoding DNA that are unique to every individual and allow DNA fingerprinting.

Straight-chain alkane (Section 3.2): An alkane whose carbon atoms are connected without branching.

Substitution reaction (Section 6.1): What occurs when two reactants exchange parts to give two new products. S_N1 and S_N2 reactions are examples.

Sulfide (Chapter 13 Introduction): A compound that has two organic substituents bonded to the same sulfur atom, RSR′.

Sulfonation (Section 9.6): The substitution of a sulfonic acid group onto an aromatic ring.

Sulfone (Section 13.11): A compound of the general structure $RSO_2R′$.

Sulfoxide (Section 13.11): A compound of the general structure RSOR′.

Symmetry plane (Section 5.2): A plane that bisects a molecule such that one half of the molecule is the mirror image of the other half. Molecules containing a plane of symmetry are achiral.

Syn periplanar (Section 12.12): Describing a stereochemical relationship in which two bonds on adjacent carbons lie in the same plane and are eclipsed.

Syn stereochemistry (Section 8.4): The opposite of anti. A syn addition reaction is one in which the two ends of the double bond react from the same side. A syn elimination is one in which the two groups leave from the same side of the molecule.

Tautomers (Section 17.1): Isomers that interconvert spontaneously, usually with the change in position of a hydrogen.

Template strand (Section 24.4): The strand of double-helical DNA that does not contain the gene.

Terpenoid (Chapter 7 *Lagniappe*, Section 23.7): A lipid that is formally derived by head-to-tail polymerization of isoprene units.

Tertiary: *See* Primary.

Tertiary structure (Section 19.8): The level of protein structure that involves the manner in which the entire protein chain is folded into a specific three-dimensional arrangement.

Thioester (Chapter 16 Introduction): A compound with the RCOSR′ functional group.

Thiol (Chapter 13 Introduction): A compound containing the –SH functional group.

TMS (Section 11.3): Tetramethylsilane, used as an NMR calibration standard.

TOF (Section 10.4): A time-of-flight mass spectrometer.

Tollens' reagent (Section 21.6): A solution of Ag_2O in aqueous ammonia; used to oxidize aldehydes to carboxylic acids.

Torsional strain (Section 3.6): The strain in a molecule caused by electron repulsion between eclipsed bonds. Torsional strain is also called eclipsing strain.

Tosylate (Section 12.5): A *p*-toluenesulfonate ester.

Transamination (Section 20.2): The exchange of an amino group and a keto group between reactants.

Transcription (Section 24.4): The process by which the genetic information encoded in DNA is read and used to synthesize RNA in the nucleus of the cell. A small portion of double-stranded DNA uncoils, and complementary ribonucleotides line up in the correct sequence for RNA synthesis.

Transfer RNA (Section 24.4): A kind of RNA that transports amino acids to the ribosomes, where they are joined together to make proteins.

Transimination (Section 20.2): The exchange of an amino group and an imine group between reactants.

Transition state (Section 6.9): An activated complex between reactants, representing the highest energy point on a reaction curve. Transition states are unstable complexes that can't be isolated.

Translation (Section 24.5): The process by which the genetic information transcribed from DNA onto mRNA is read by tRNA and used to direct protein synthesis.

Tree diagram (Section 11.12): A diagram used in NMR to sort out the complicated splitting patterns that can arise from multiple couplings.

Triacylglycerol (Section 23.1): A lipid, such as that found in animal fat and vegetable oil, that is a triester of glycerol with long-chain fatty acids.

Tricarboxylic acid cycle (Section 22.4): An alternative name for the citric acid cycle by which acetyl CoA is degraded to CO_2.

Triplet (Section 11.11): A symmetrical three-line splitting pattern observed in the ^1H NMR spectrum when a proton has two equivalent neighbor protons.

Turnover number (Section 19.9): The number of substrate molecules acted on by an enzyme per unit time.

Twist-boat conformation (Section 4.5): A conformation of cyclohexane that is somewhat more stable than a pure boat conformation.

Ultraviolet (UV) spectroscopy (Section 10.9): An optical spectroscopy employing ultraviolet irradiation. UV spectroscopy provides structural information about the extent of π electron conjugation in organic molecules.

Unimolecular reaction (Section 12.8): A reaction that occurs by spontaneous transformation of the starting material without the intervention of other reactants. For example, the dissociation of a tertiary alkyl halide in the S_N1 reaction is a unimolecular process.

Unsaturated (Section 7.1): A molecule that has one or more multiple bonds.

Upfield (Section 11.3): The right-hand portion of the NMR chart.

Urea cycle (Section 20.3): The metabolic pathway for converting ammonia into urea.

Uronic acid (Section 21.6): The monocarboxylic acid formed by oxidizing the –CH_2OH end of a sugar without affecting the –CHO end.

Valence bond theory (Section 1.5): A bonding theory that describes a covalent bond as resulting from the overlap of two atomic orbitals.

Valence shell (Section 1.4): The outermost electron shell of an atom.

Van der Waals forces (Section 2.12): Intermolecular forces that are responsible for holding molecules together in the liquid and solid states.

Vegetable oil (Section 23.1): A liquid triacylglycerol derived from a plant source.

Vinyl group (Section 7.2): An $H_2C{=}CH-$ substituent.

Vinyl monomer (Section 8.10): A substituted alkene monomer used to make chain-growth polymers.

Vinylic (Section 9.7): A term that refers to a substituent at a double-bond carbon atom. For example, chloroethylene is a vinylic chloride.

Virion (Chapter 22 *Lagniappe*): A viral particle.

Vitamin (Section 19.9): A small organic molecule that must be obtained in the diet and is required in trace amounts for proper growth and function.

Vulcanization (Chapter 8 *Lagniappe*): A technique for cross-linking and hardening a diene polymer by heating with a few percent by weight of sulfur.

Walden inversion (Section 12.5): The inversion of configuration at a chirality center that accompanies an S_N2 reaction.

Wave equation (Section 1.2): A mathematical expression that defines the behavior of an electron in an atom.

Wave function (Section 1.2): A solution to the wave equation for defining the behavior of an electron in an atom. The square of the wave function defines the shape of an orbital.

Wavelength, λ (Section 10.5): The length of a wave from peak to peak. The wavelength of electromagnetic radiation is inversely proportional to frequency and inversely proportional to energy.

Wavenumber, $\tilde{\nu}$ (Section 10.6): The reciprocal of the wavelength in centimeters.

Wax (Section 23.1): A mixture of esters of long-chain carboxylic acids with long-chain alcohols.

Williamson ether synthesis (Section 13.9): A method for synthesizing ethers by S_N2 reaction of an alkyl halide with an alkoxide ion.

Wittig reaction (Section 14.9): The reaction of a phosphorus ylide with an aldehyde or ketone to yield an alkene.

X-ray crystallography (Chapter 17 *Lagniappe*): A technique using X rays to determine the structure of molecules.

Ylide (Sections 14.9, 22.3): A neutral species with adjacent + and − charges, such as the phosphoranes used in Wittig reactions.

Z geometry (Section 7.4): A term used to describe the stereochemistry of a carbon–carbon double bond. The two groups on each carbon are assigned priorities according to the Cahn–Ingold–Prelog sequence rules, and the two carbons are compared. If the high-priority groups on each carbon are on the same side of the double bond, the bond has *Z* geometry.

Zaitsev's rule (Section 12.11): A rule stating that E2 elimination reactions normally yield the more highly substituted alkene as major product.

Zwitterion (Section 19.1): A neutral dipolar molecule in which the positive and negative charges are not adjacent. For example, amino acids exist as zwitterions, $H_3N^+{-}CHR{-}CO_2^-$.

D   Answers to In-Text Problems

The following answers are meant only as a quick check while you study. Full answers for all problems are provided in the accompanying *Study Guide and Solutions Manual*.

CHAPTER 1

1.1 (a) $1s^2\, 2s^2\, 2p^4$

(b) $1s^2\, 2s^2\, 2p^6\, 3s^2\, 3p^3$

(c) $1s^2\, 2s^2\, 2p^6\, 3s^2\, 3p^4$

1.2 (a) 2   (b) 2 (+ 7)   (c) 6

1.3

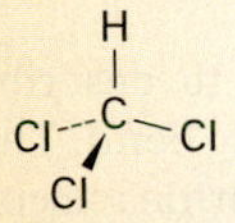

$$Cl\!-\!\overset{\textstyle H}{\underset{\textstyle Cl}{C}}\!-\!Cl$$

1.4

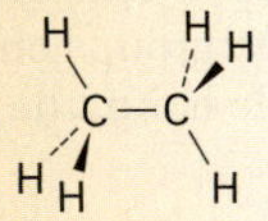

1.5 (a) CH_2Cl_2 (b) CH_3SH (c) CH_3NH_2

1.6 (a)

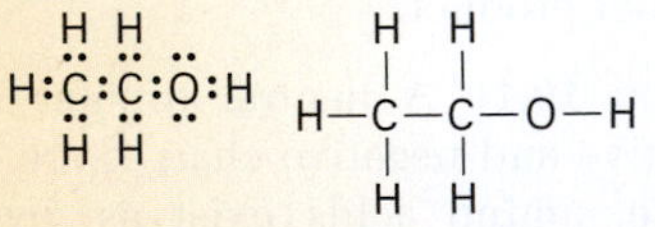

(b)

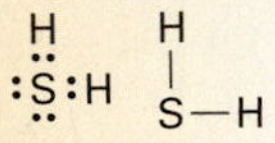

(c)

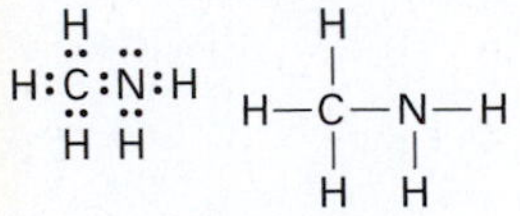

(d)

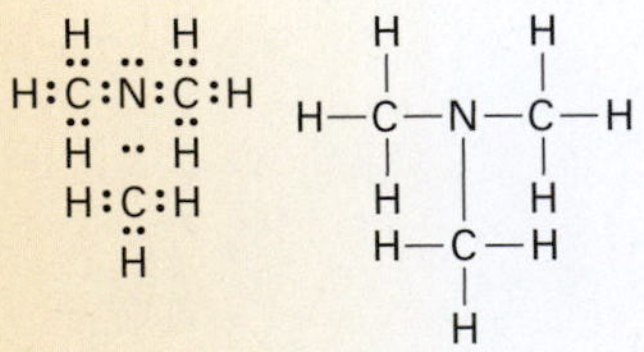

1.7 C_2H_7 has too many hydrogens for a compound with 2 carbons.

1.8

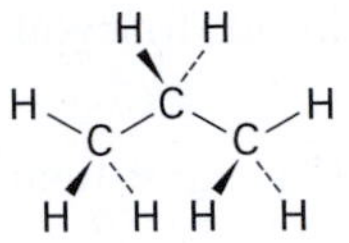

All bond angles are near 109°.

1.9

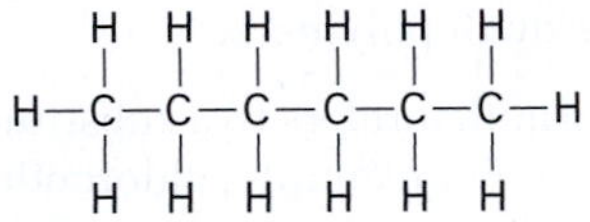

1.10 The CH_3 carbon is sp^3; the double-bond carbons are sp^2; the C=C–C and C=C–H bond angles are approximately 120°; other bond angles are near 109°.

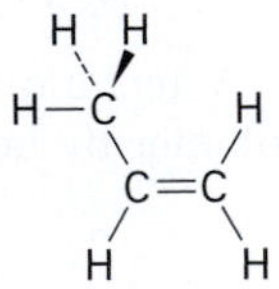

1.11 All carbons are sp^2, and all bond angles are near 120°.

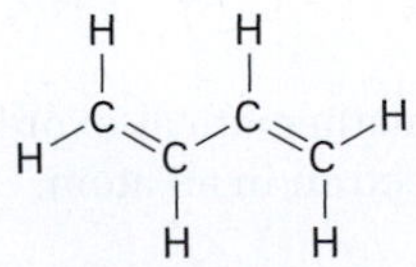

1.12 All carbons except CH_3 are sp^2.

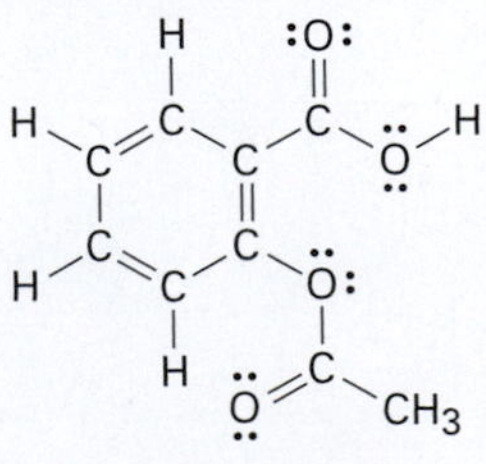

1.13 The CH_3 carbon is sp^3; the triple-bond carbons are sp; the C≡C–C and H–C≡C bond angles are approximately 180°.

$$H-C\equiv C-C\overset{\displaystyle H}{\underset{\displaystyle H}{\big|}}H$$

1.14

(a)

$$H\overset{\displaystyle H}{\underset{\displaystyle H}{C}}\ddot{O}\overset{\displaystyle H}{\underset{\displaystyle H}{C}}H$$

sp^3—tetrahedral

(b)

$$H_3C-\overset{\displaystyle \cdot\cdot}{N}-CH_3 \\ H_3C$$

sp^3—tetrahedral

(c)

$$H-\overset{\displaystyle \cdot\cdot}{\underset{\displaystyle H}{P}}-H$$

sp^3—tetrahedral

(d)

$$H_3C\diagdown\overset{\displaystyle \cdot\cdot}{\underset{\displaystyle \cdot\cdot}{S}}\diagup CH_2CH_2\overset{O}{\overset{\|}{C}}\underset{NH_2}{CHCOH}$$

sp^3—tetrahedral

1.15 (a)

Adrenaline—$C_9H_{13}NO_3$

(b)

Estrone—$C_{18}H_{22}O_2$

1.16 There are numerous possibilities, such as:

(a) C_5H_{12} $CH_3CH_2CH_2CH_2CH_3$ $CH_3CH_2\underset{\underset{CH_3}{|}}{C}HCH_3$ $CH_3\underset{\underset{CH_3}{|}}{\overset{\overset{CH_3}{|}}{C}}CH_3$

(b) C_2H_7N $CH_3CH_2NH_2$ CH_3NHCH_3

(c) C_3H_6O $CH_3\overset{O}{\overset{\|}{C}}H$ $H_2C=CHCH_2OH$ $H_2C=CHOCH_3$

(d) C_4H_9Cl $CH_3CH_2CH_2CH_2Cl$ $CH_3CH_2\underset{\underset{Cl}{|}}{C}HCH_3$ $CH_3\underset{\underset{CH_3}{|}}{C}HCH_2Cl$

1.17

$$\underset{H_2N}{\diagdown}\text{—}\overset{O}{\overset{\|}{C}}\diagdown OH$$

CHAPTER 2

2.1 (a) H (b) Br (c) Cl (d) C

2.2 (a) $\overset{\delta+\ \ \delta-}{H_3C-Cl}$ (b) $\overset{\delta+\ \ \ \delta-}{H_3C-NH_2}$ (c) $\overset{\delta-\ \ \ \delta+}{H_2N-H}$

(d) H_3C-SH (e) $\overset{\delta-\ \ \ \delta+}{H_3C-MgBr}$ (f) $\overset{\delta+\ \ \delta-}{H_3C-F}$

Carbon and sulfur have identical electronegativities.

2.3 $H_3C-OH < H_3C-MgBr < H_3C-Li = H_3C-F < H_3C-K$

2.4 The nitrogen is electron-rich, and the carbon is electron-poor.

$$\overset{\delta-}{:}NH_2 \\ | \\ H-\overset{C}{\underset{H}{|}}-H \\ H\ \delta+$$

2.5 The two C–O dipoles cancel because of the symmetry of the molecule:

$$HO\overset{\overset{H\ \ H}{\diagup C\diagdown}}{\underset{\underset{H\ \ H}{\diagdown C\diagup}}{}}OH$$

2.6 (a) $H\diagdown_{\text{—}}C=C\diagup_{\text{—}}H$ (b)

No dipole moment

(c)

(d)

2.7 (a) For carbon: FC = 4 − 8/2 − 0 = 0;
for the middle nitrogen: FC = 5 − 8/2 − 0 = +1;
for the end nitrogen: FC = 5 − 4/2 − 4 = −1

(b) For nitrogen: FC = 5 − 8/2 − 0 = +1;
for oxygen: FC = 6 − 2/2 − 6 = −1

(c) For nitrogen: FC = 5 − 8/2 − 0 = +1;
for the end carbon: FC = 4 − 6/2 − 2 = −1

2.8

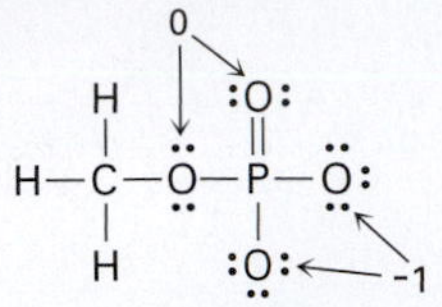

2.9

(a)

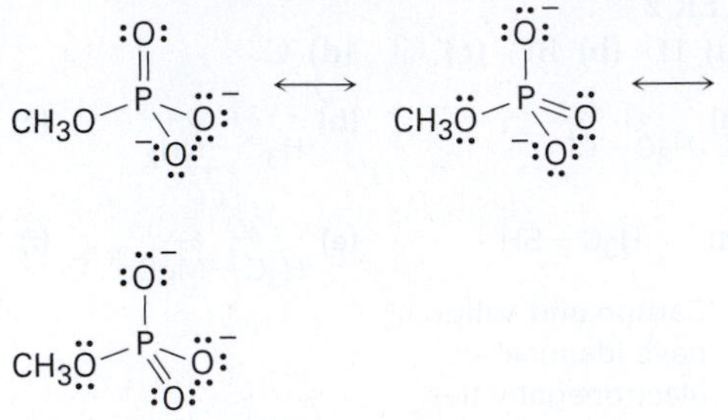

(b)

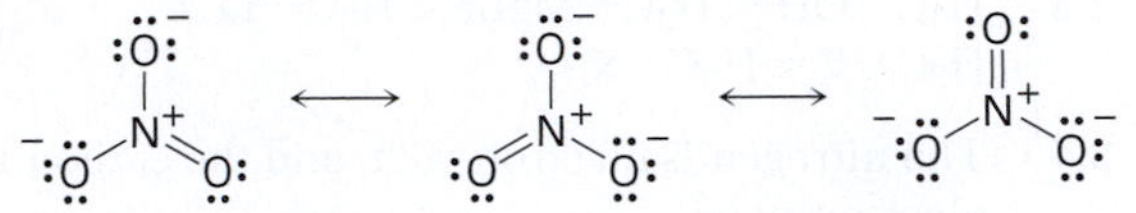

(c) $H_2C=CH-CH_2^+ \longleftrightarrow H_2\overset{+}{C}-CH=CH_2$

(d)

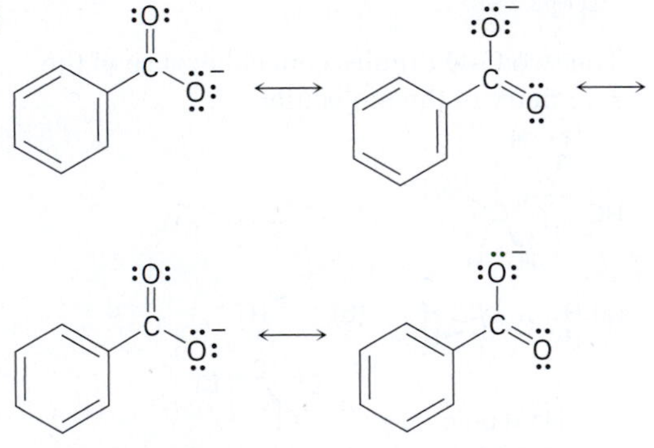

2.10 $H-NO_3 + :NH_3 \rightleftharpoons NO_3^- + NH_4^+$

       Acid   Base   Conjugate Conjugate
                          base   acid

2.11 Phenylalanine is stronger.

2.12 Water is a stronger acid.

2.13 Neither reaction will take place.

2.14 Reaction will take place.

2.15 $K_a = 4.9 \times 10^{-10}$

2.16 (a)

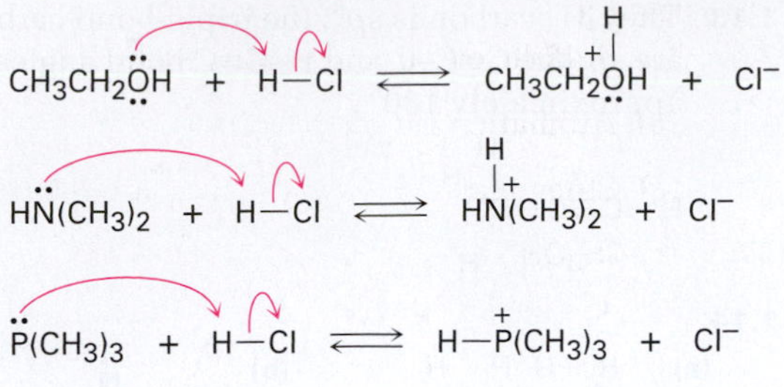

(b)

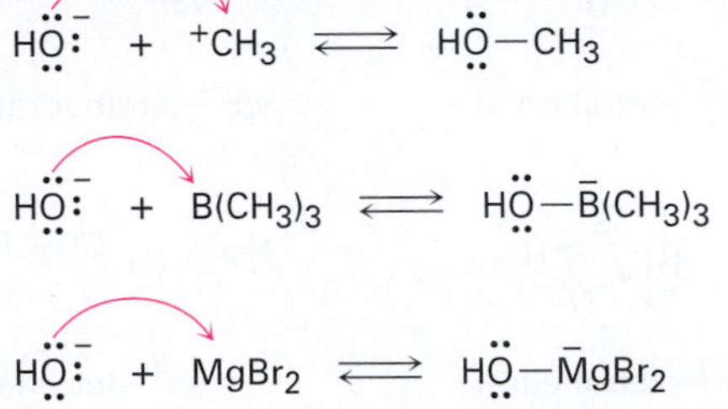

2.17 (a) More basic (red)   Most acidic (blue)

Imidazole

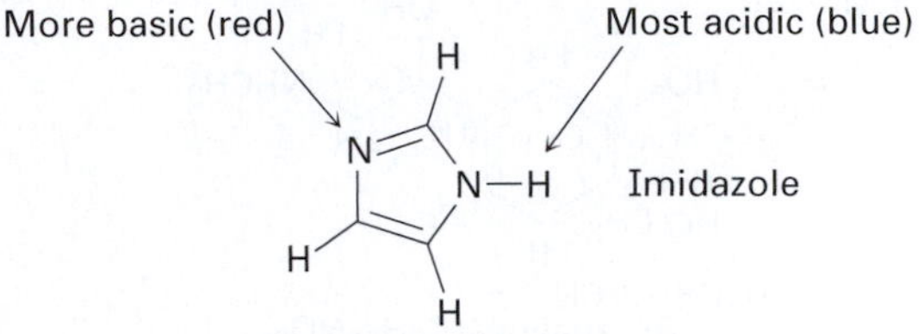

(b)

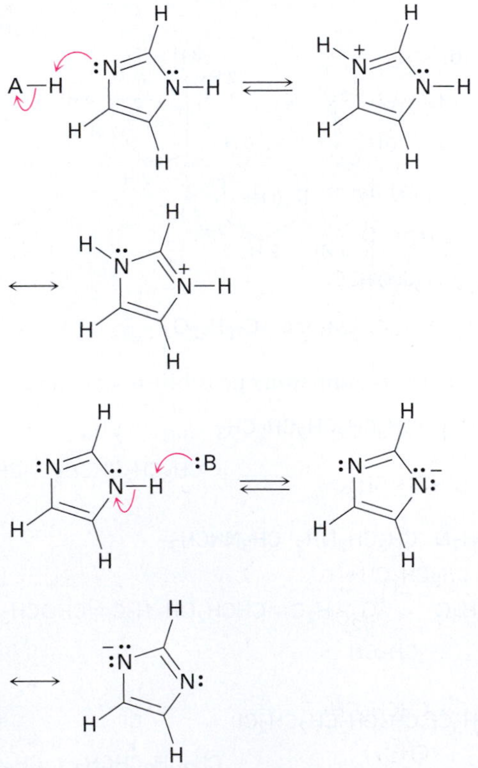

2.18 Vitamin C is water-soluble (hydrophilic); vitamin A is fat-soluble (hydrophobic).

CHAPTER 3

3.1 **(a)** Sulfide, carboxylic acid, amine

(b) Aromatic ring, carboxylic acid

(c) Ether, alcohol, aromatic ring, amide, C=C bond

3.2 **(a)** CH_3OH **(b)**

(c)

(d) CH_3NH_2 **(e)**

(f)

3.3

$C_8H_{13}NO_2$

3.4

$CH_3CH_2CH_2CH_2CH_2CH_3$

$CH_3CHCH_2CH_2CH_3$ with CH_3

$CH_3CH_2CHCH_2CH_3$ with CH_3

$CH_3CCH_2CH_3$ with CH_3 above and below

$CH_3CHCHCH_3$ with CH_3 above and CH_3 below

3.5 Part **(a)** has nine possible answers.

(a)

$CH_3CH_2CH_2COCH_3$ $CH_3CH_2COCH_2CH_3$ $CH_3COCHCH_3$ with O and CH_3

(b) $CH_3CH_2SSCH_2CH_3$ $CH_3SSCH_2CH_2CH_3$ $CH_3SSCHCH_3$ with CH_3

3.6 **(a)** Two **(b)** Four **(c)** Four

3.7

$CH_3CH_2CH_2CH_2CH_2-$ $CH_3CH_2CH_2CH-$ with CH_3

CH_3CH_2CH- with CH_2CH_3 $CH_3CH_2CHCH_2-$ with CH_3

$CH_3CHCH_2CH_2-$ with CH_3 CH_3CH_2C- with CH_3 above and CH_3 below

CH_3CHCH- with CH_3 above and CH_3 below CH_3CCH_2- with CH_3 above and CH_3 below

3.8 **(a)**

(b)

(c)

3.9 Primary carbons have primary hydrogens, secondary carbons have secondary hydrogens, and tertiary carbons have tertiary hydrogens.

3.10 **(a)**

$CH_3CHCHCH_3$ with CH_3 above and CH_3 below

(b) CH_3CHCH_3 / $CH_3CH_2CHCH_2CH_3$

(c)

$CH_3CCH_2CH_3$ with CH_3 above and CH_3 below

3.11 **(a)** Pentane, 2-methylbutane, 2,2-dimethylpropane

(b) 2,3-Dimethylpentane

(c) 2,4-Dimethylpentane

(d) 2,2,5-Trimethylhexane

3.12 **(a)**

$CH_3CH_2CH_2CH_2CH_2CHCHCH_2CH_3$ with CH_3 above and CH_3 below

(b)

$CH_3CH_2CH_2C-CHCH_2CH_3$ with CH_3 above, CH_3 and CH_2CH_3 below

(c)

$CH_3CH_2CH_2CH_2CHCH_2C(CH_3)_3$ with $CH_2CH_2CH_3$

(d)

$CH_3CHCH_2CCH_3$ with CH_3, CH_3 above and CH_3 below

3.13 Pentyl, 1-methylbutyl, 1-ethylpropyl, 2-methylbutyl, 3-methylbutyl, 1,1-dimethylpropyl, 1,2-dimethylpropyl, 2,2-dimethylpropyl

3.14

3,3,4,5-Tetramethylheptane

3.15

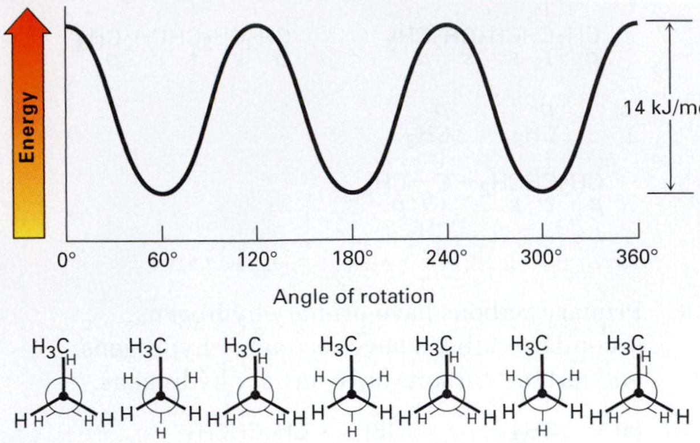

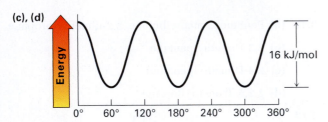

3.16

(a)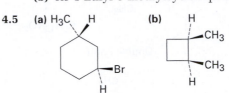

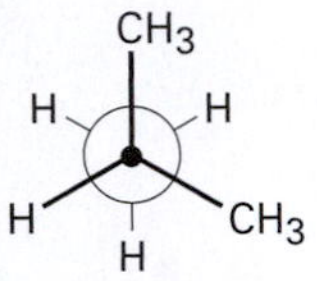

(b) 4.0 kJ/mol   6.0 kJ/mol

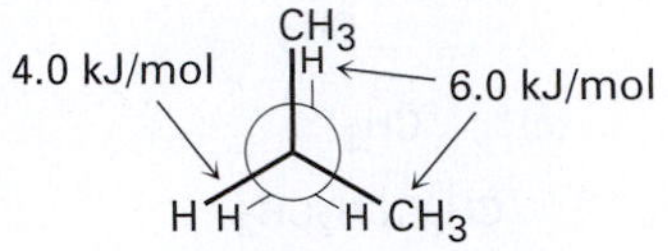

(c), (d)

16 kJ/mol

3.17

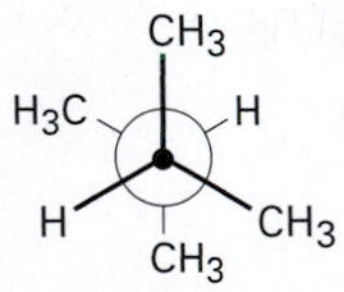

CH₃
H₃C   H
H   CH₃
CH₃

3.18

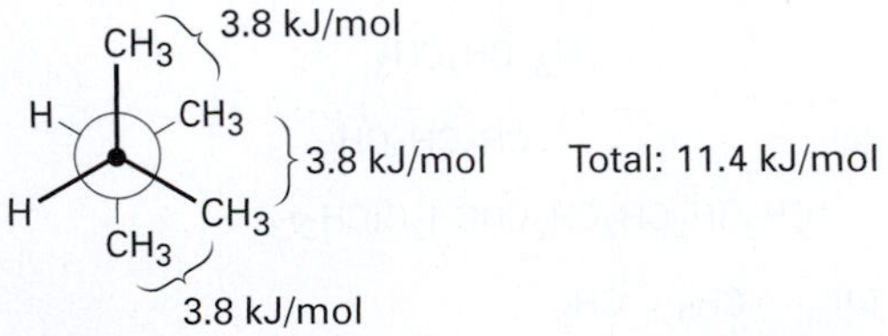

CH₃   3.8 kJ/mol
H   CH₃
    }3.8 kJ/mol   Total: 11.4 kJ/mol
H   CH₃
CH₃
    3.8 kJ/mol

CHAPTER 4

4.1 **(a)** 1,4-Dimethylcyclohexane

   (b) 1-Methyl-3-propylcyclopentane

   (c) 3-Cyclobutylpentane

   (d) 1-Bromo-4-ethylcyclodecane

   (e) 1-Isopropyl-2-methylcyclohexane

   (f) 4-Bromo-1-*tert*-butyl-2-methylcycloheptane

4.2 **(a)** 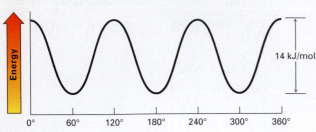 **(b)**

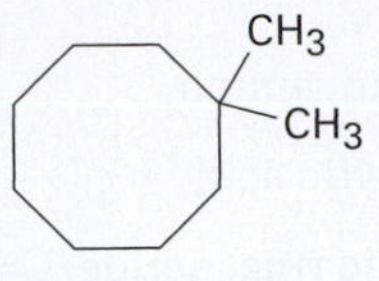

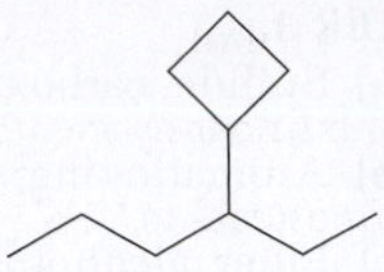

(c) 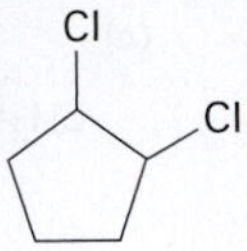 Cl, Cl **(d)** CH₃, Br, Br

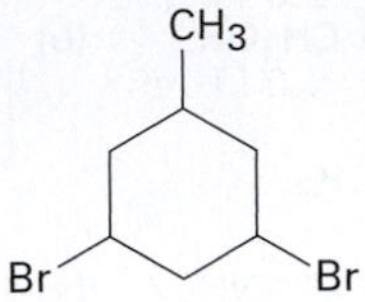

4.3 3-Ethyl-1,1-dimethylcyclopentane

4.4 **(a)** *trans*-1-Chloro-4-methylcyclohexane

   (b) *cis*-1-Ethyl-3-methylcycloheptane

4.5 **(a)** H₃C, H, Br, H **(b)** H, CH₃, CH₃, H

   (c) CH₂CH₃, H, C(CH₃)₃, H

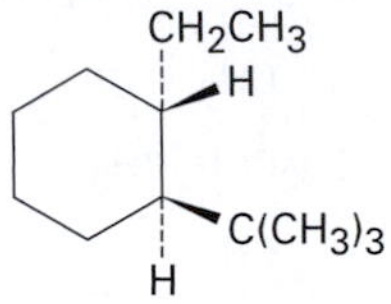

4.6 The two hydroxyl groups are cis. The two side chains are trans.

4.7 **(a)** *cis*-1,2-Dimethylcyclopentane

   (b) *cis*-1-Bromo-3-methylcyclobutane

4.8 Six interactions; 21% of strain

4.9 The cis isomer is less stable because the methyl groups eclipse each other.

4.10 Ten eclipsing interactions; 40 kJ/mol; 35% is relieved.

4.11 Conformation **(a)** is more stable because the methyl groups are farther apart.

4.12

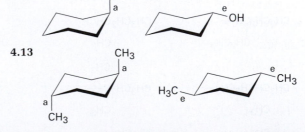

OH
a
e
OH

4.13

CH₃
a
H₃C
e
CH₃
a
CH₃

4.14 Before ring-flip, red and blue are equatorial and green is axial. After ring-flip, red and blue are axial and green is equatorial.

4.15 4.2 kJ/mol

4.16 The linear cyano group points straight up.

4.17 Equatorial is 70%; axial is 30%

4.18 **(a)** 2.0 kJ/mol **(b)** 11.4 kJ/mol
(c) 2.0 kJ/mol **(d)** 8.0 kJ/mol

4.19

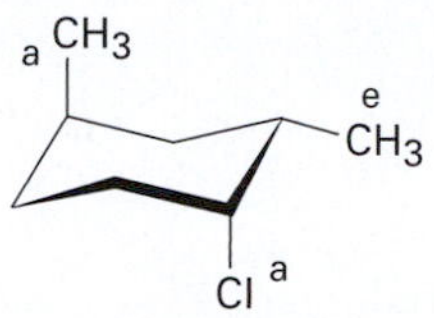

Less stable chair form

4.20 *trans*-Decalin is more stable because it has no 1,3-diaxial interactions.

CHAPTER 5

5.1 Chiral: screw, beanstalk, shoe

5.2 **(a)**

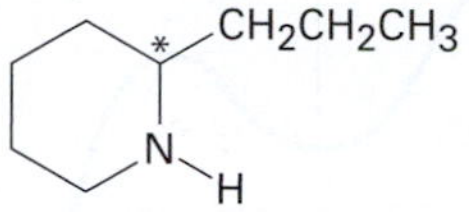

(b)

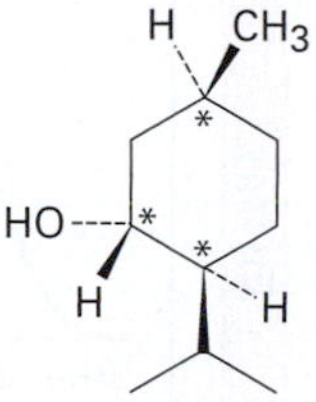

(c)

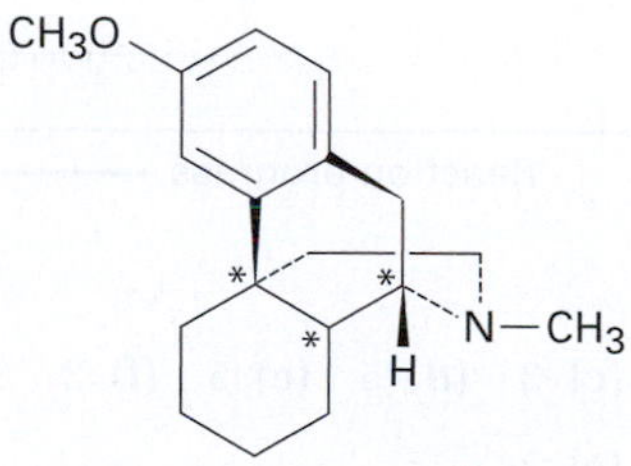

5.3

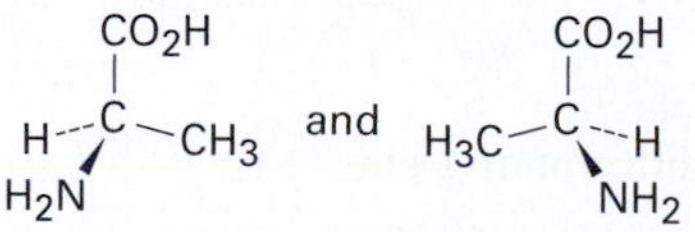

5.4 **(a)**

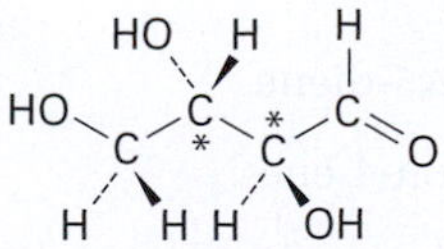

(b)

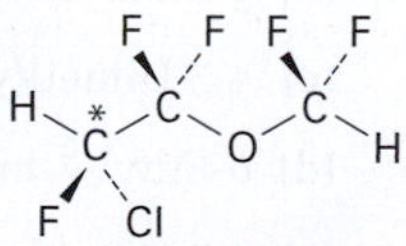

5.5 Levorotatory

5.6 +16.1

5.7 **(a)** –Br **(b)** –Br **(c)** –CH$_2$CH$_3$
(d) –OH **(e)** –CH$_2$OH **(f)** –CH=O

5.8 **(a)** –OH, –CH$_2$CH$_2$OH, –CH$_2$CH$_3$, –H
(b) –OH, –CO$_2$CH$_3$, –CO$_2$H, –CH$_2$OH
(c) –NH$_2$, –CN, –CH$_2$NHCH$_3$, –CH$_2$NH$_2$
(d) –SSCH$_3$, –SH, –CH$_2$SCH$_3$, –CH$_3$

5.9 **(a)** *S* **(b)** *R* **(c)** *S*

5.10 **(a)** *S* **(b)** *S* **(c)** *R*

5.11

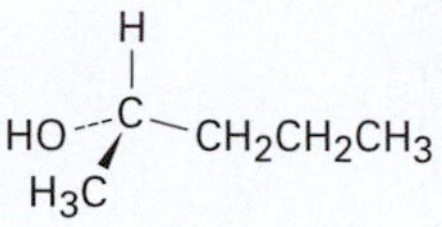

5.12 *S*

5.13 **(a)** *R,R* **(b)** *S,R* **(c)** *R,S* **(d)** *S,S*

Compounds **(a)** and **(d)** are enantiomers and are diastereomeric with **(b)** and **(c)**.

5.14 *S,S*

5.15 Five chirality centers; 32 stereoisomers

5.16 Compounds **(a)** and **(d)** are meso.

5.17 Compounds **(a)** and **(c)** have meso forms.

5.18

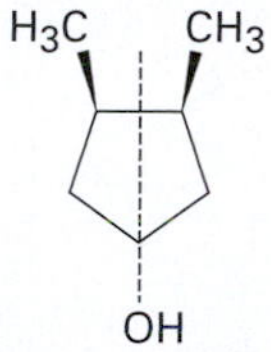

Meso

5.19 The product retains its *S* stereochemistry.

5.20 Two diastereomeric salts are formed: (*R*)-lactic acid plus (*S*)-1-phenylethylamine and (*S*)-lactic acid plus (*S*)-1-phenylethylamine.

5.21 **(a)** Constitutional isomers **(b)** Diastereomers

5.22 **(a)**

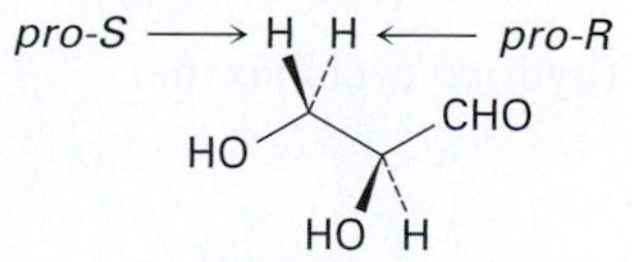

(b)

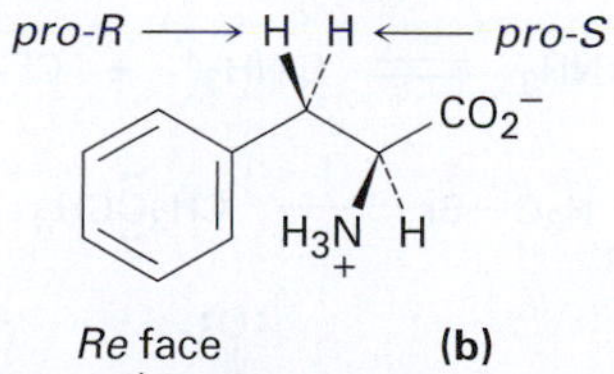

5.23 **(a)**

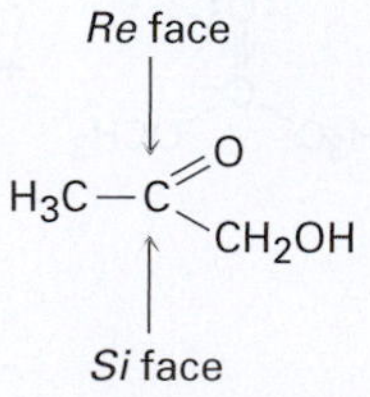

(b)

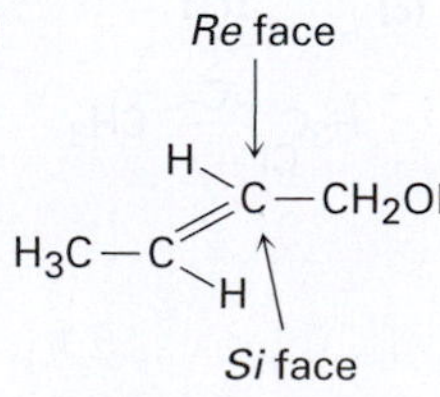

5.24 (*S*)-Lactate

5.25 The –OH adds to the *Re* face of C2, and –H adds to the *Re* face of C3. The overall addition has anti stereochemistry.

CHAPTER 6

6.1 **(a)** Substitution **(b)** Elimination **(c)** Addition

6.2 1-Chloro-2-methylpentane,
2-chloro-2-methylpentane,
3-chloro-2-methylpentane,
2-chloro-4-methylpentane,
1-chloro-4-methylpentane

6.3 Radical addition reaction

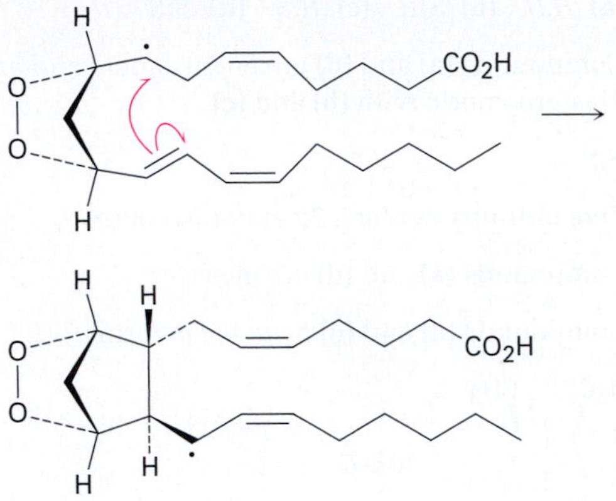

6.4 **(a)** Carbon is electrophilic.

  (b) Sulfur is nucleophilic.

  (c) Nitrogens are nucleophilic.

  (d) Oxygen is nucleophilic; carbon is electrophilic.

6.5 Electrophilic; vacant *p* orbital

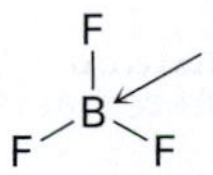

6.6 Cyclohexanol (hydroxycyclohexane)

6.7

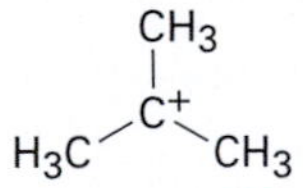

6.8 **(a)** Cl—Cl + :NH₃ ⇌ ClNH₃⁺ + Cl⁻

  (b) CH₃Ö:⁻ + H₃C—Br ⟶ CH₃ÖCH₃ + Br⁻

  (c)

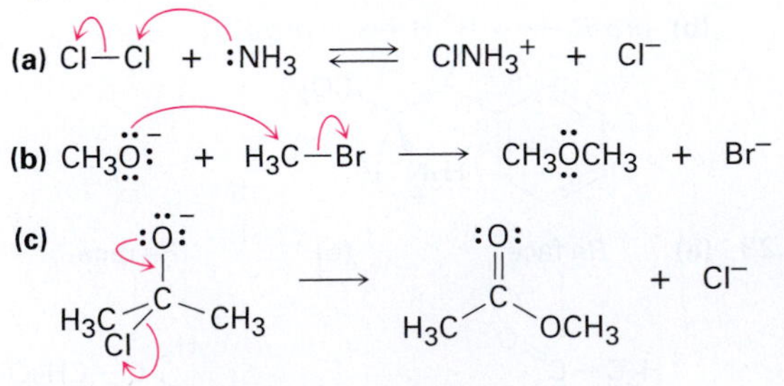

6.9

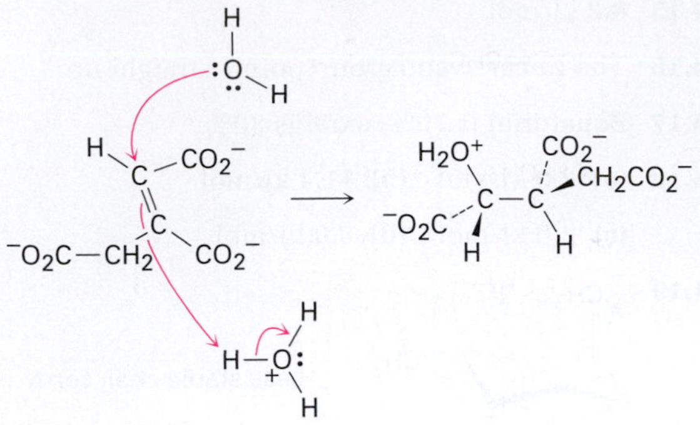

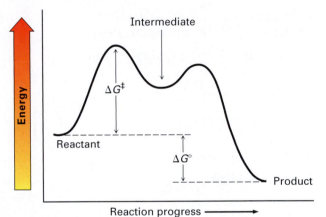

6.10 Negative $\Delta G°$ is more favored.

6.11 Larger K_{eq} is more exergonic.

6.12 Lower $\Delta G^{‡}$ is faster.

6.13

CHAPTER 7

7.1 **(a)** 2 **(b)** 3 **(c)** 3 **(d)** 5 **(e)** 5 **(f)** 3

7.2 **(a)** 1 **(b)** 2 **(c)** 2

7.3 $C_{16}H_{13}ClN_2O$

7.4 **(a)** 3,4,4-Trimethylpent-1-ene

  (b) 3-Methylhex-3-ene

  (c) 4,7-Dimethylocta-2,5-diene

  (d) 6-Ethyl-7-methylnon-4-ene

  (e) 1,2-Dimethylcyclohexene

  (f) 4,4-Dimethylcycloheptene

  (g) 3-Isopropylcyclopentene

7.5 **(a)**

$$CH_3$$
$$H_2C=CHCH_2CH_2C=CH_2$$

(b)

$$CH_2CH_3$$
$$CH_3CH_2CH_2CH=CC(CH_3)_3$$

(c)

$$CH_3 \quad CH_3$$
$$CH_3CH=CHCH=CHC-C=CH_2$$
$$CH_3$$

(d)

$$CH_3 \qquad CH_3$$
$$CH_3CH \qquad CHCH_3$$
$$C=C$$
$$CH_3CH \qquad CHCH_3$$
$$CH_3 \qquad CH_3$$

7.6 **(a)** 2,5-Dimethylhex-3-yne

(b) 3,3-Dimethylbut-1-yne

(c) 3,3-Dimethyloct-4-yne

(d) 2,5,5-Trimethylhept-3-yne

(e) 6-Isopropylcyclodecyne

7.7

(a) 2,5,5-Trimethylhex-2-ene   **(b)** 2,2-Dimethylhex-3-yne

$$CH_3 \qquad CH_3 \qquad\qquad CH_3$$
$$CH_3C=CHCH_2CCH_3 \qquad CH_3CC\equiv CCH_2CH_3$$
$$CH_3 \qquad\qquad CH_3$$

7.8

7.9 Compounds **(c)**, **(e)**, and **(f)** have cis–trans isomers.

7.10 **(a)** *cis*-4,5-Dimethylhex-2-ene

(b) *trans*-6-Methylhept-3-ene

7.11 **(a)** –CH$_3$ **(b)** –Cl

(c) –CH=CH$_2$ **(d)** –OCH$_3$

(e) –CH=O   **(f)** –CH=O

7.12 **(a)** –Cl, –OH, –CH$_3$, –H

(b) –CH$_2$OH, –CH=CH$_2$, –CH$_2$CH$_3$, –CH$_3$

(c) –CO$_2$H, –CH$_2$OH, –C≡N, –CH$_2$NH$_2$

(d) –CH$_2$OCH$_3$, –C≡N, –C≡CH, –CH$_2$CH$_3$

7.13 **(a)** Z **(b)** E **(c)** Z **(d)** E

7.14

7.15 **(a)** 2-Methylpropene is more stable than but-1-ene.

(b) *trans*-Hex-2-ene is more stable than *cis*-hex-2-ene.

(c) 1-Methylcyclohexene is more stable than 3-methylcyclohexene.

7.16 **(a)** Chlorocyclohexane

(b) 2-Bromo-2-methylpentane

(c) 2-Hydroxy-4-methylpentane

(d) 1-Bromo-1-methylcyclohexane

7.17 **(a)** Cyclopentene

(b) 1-Ethylcyclohexene or ethylidenecyclohexane

(c) Hex-3-ene   **(d)** Cyclohexylethylene

7.18 **(a)**

$$CH_3 \quad CH_3$$
$$CH_3CH_2\overset{+}{C}CH_2\overset{}{C}HCH_3$$

(b)

7.19 In the conformation shown, only the methyl-group C–H that is parallel to the carbocation *p* orbital can show hyperconjugation.

7.20 The second step is exergonic; the transition state resembles the carbocation.

7.21

CHAPTER 8

8.1 2-Methylbut-2-ene and 2-methylbut-1-ene

8.2 Five

8.3 *trans*-1,2-Dichloro-1,2-dimethylcyclohexane

8.4

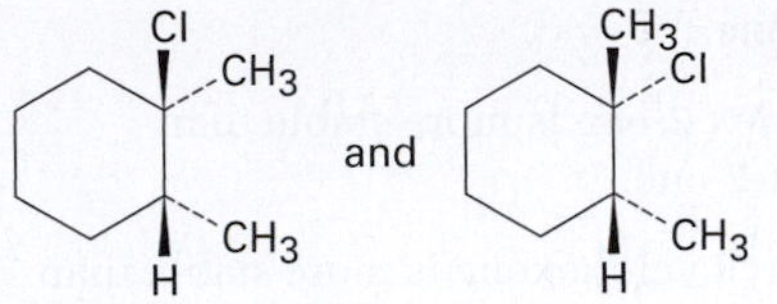

and

8.5 Markovnikov orientation

8.6 **(a)** Oxymercuration: 2-methylpentan-2-ol; hydroboration: 2-methylpentan-3-ol

(b) Oxymercuration: 1-ethylcyclohexanol; hydroboration: 1-cyclohexylethanol

8.7 **(a)** From 3-methylbut-1-ene by hydroboration

(b) From 2-methylbut-2-ene by hydroboration or from 3-methylbut-1-ene by oxymercuration

(c) From methylenecyclohexane by hydroboration

8.8

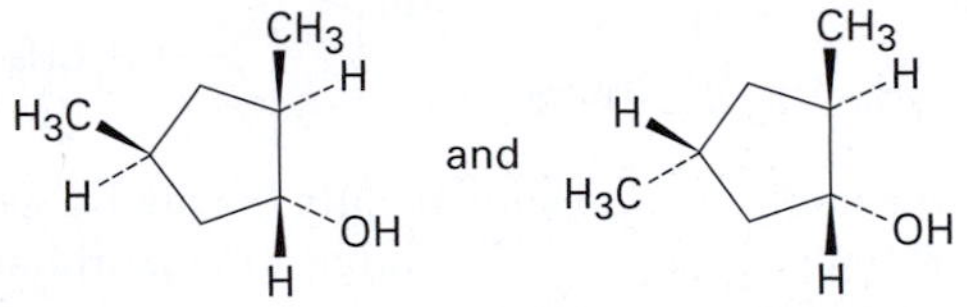

and

8.9 **(a)** 2-Methylpentane

(b) 1,1-Dimethylcyclopentane

8.10

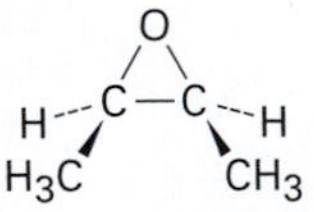

cis-2,3-Epoxybutane

8.11 **(a)** 1-Methylcyclohexene

(b) 2-Methylpent-2-ene

(c) Buta-1,3-diene

8.12 **(a)** $CH_3COCH_2CH_2CH_2CH_2CO_2H$

(b) $CH_3COCH_2CH_2CH_2CH_2CHO$

8.13 **(a)** 2-Methylpropene **(b)** Hex-3-ene

8.14

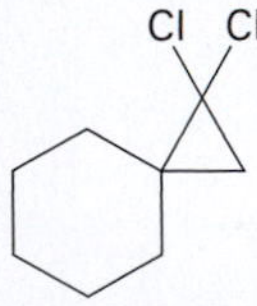

8.15 **(a)** H_2C=$CHOCH_3$ **(b)** $ClCH$=$CHCl$

8.16

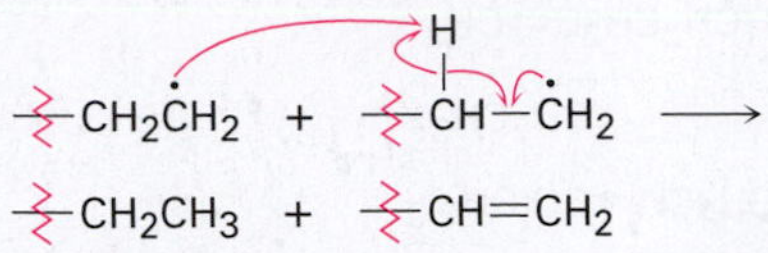

8.17 1,2-Addition: 4-chloropent-2-ene, 3-chloropent-1-ene
1,4-Addition: 4-chloropent-2-ene, 1-chloropent-1-ene

8.18 1,2-Addition: 6-bromo-1,6-dimethylcyclohexene
1,4-Addition: 3-bromo-1,2-dimethylcyclohexene

8.19

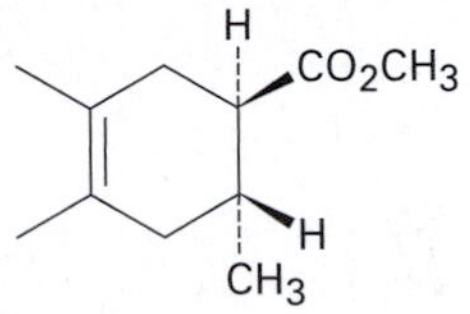

8.20 Good dienophiles: **(a)**, **(d)**

8.21 Compound **(a)** is *s*-cis. Compound **(c)** can rotate to *s*-cis.

8.22 **(a)** 1,1,2,2-Tetrachloropentane

(b) 1-Bromo-1-cyclopentylethylene

(c) 2-Bromohept-2-ene and 3-bromohept-2-ene

CHAPTER 9

9.1 **(a)** Meta **(b)** Para **(c)** Ortho

9.2 **(a)** *m*-Bromochlorobenzene

(b) (3-Methylbutyl)benzene

(c) *p*-Bromoaniline

(d) 2,5-Dichlorotoluene

(e) 1-Ethyl-2,4-dinitrobenzene

(f) 1,2,3,5-Tetramethylbenzene

9.3 **(a)**

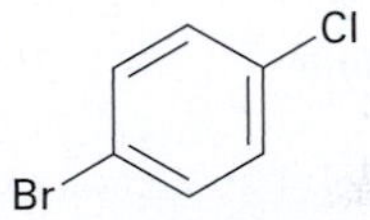

(b)

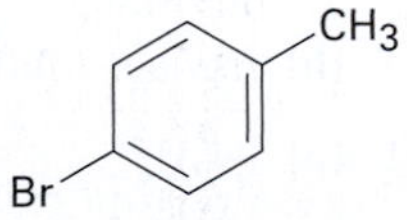

(c) 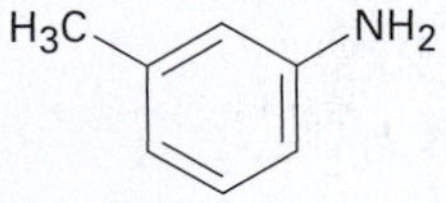 **(d)**

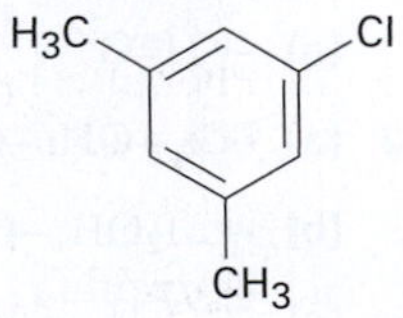

9.4 Pyridine has an aromatic sextet of electrons.

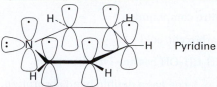

Pyridine

9.5 Cyclodecapentaene is not flat because of steric interactions.

9.6 The cyclooctatetraenyl dianion is aromatic (ten π electrons) and flat.

9.7

Furan

9.8 The thiazolium ring has six π electrons.

9.9

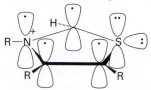

9.10 The three nitrogens in double bonds each contribute one; the remaining nitrogen contributes two.

9.11 *o*-, *m*-, and *p*-Bromotoluene

9.12 *o*-Xylene: 2; *m*-xylene: 3; *p*-xylene: 1

9.13 D$^+$ does electrophilic substitutions on the ring.

9.14 No rearrangement: **(a)**, **(b)**, **(e)**

9.15 *tert*-Butylbenzene

9.16 **(a)** $(CH_3)_2CHCOCl$ **(b)** $PhCOCl$

9.17 **(a)** Phenol > Toluene > Benzene > Nitrobenzene

(b) Phenol > Benzene > Chlorobenzene > Benzoic acid

(c) Aniline > Benzene > Bromobenzene > Benzaldehyde

9.18 The initial alkylation product is more reactive than the starting material, but the initial acylation product is less reactive.

9.19 Toluene is more reactive; the trifluoromethyl group is electron-withdrawing.

9.20 **(a)** Methyl *m*-nitrobenzoate

(b) *m*-Bromonitrobenzene

(c) *o*- and *p*-Chlorophenol

(d) *o*- and *p*-Bromoaniline

9.21 Ortho intermediate:

Para intermediate:

Meta intermediate:

9.22 **(a)** *m*-Chlorobenzonitrile

(b) *o*- and *p*-Bromochlorobenzene

9.23

[chemical structures showing the reaction mechanism with F₃C, Cl, F, OCH₂CH₃, N⁺, O⁻ groups]

F_3C ... Cl ... OCH_2CH_3 ... N^+ ... O^-

[structure of Oxyfluorfen]

$+ \ F^-$

Oxyfluorfen

9.24 (a) *m*-Nitrobenzoic acid

   (b) *p-tert*-Butylbenzoic acid

9.25 1. PhCOCl, AlCl₃; 2. H₂/Pd

9.26 (a) 1. HNO₃, H₂SO₄; 2. Cl₂, FeCl₃

   (b) 1. CH₃COCl, AlCl₃; 2. Cl₂, FeCl₃; 3. H₂, Pd

   (c) 1. CH₃CH₂COCl, AlCl₃; 2. H₂, Pd; 3. Cl₂, FeCl₃

   (d) 1. CH₃Cl, AlCl₃; 2. SO₃, H₂SO₄; 3. Br₂, FeBr₃

9.27 (a) The Friedel–Crafts reaction in step 1 will not take place on a cyano-substituted benzene.

   (b) The Friedel–Crafts reaction will occur with a carbocation rearrangement, and the wrong isomer will be obtained on chlorination.

CHAPTER 10

10.1 $C_{19}H_{28}O_2$

10.2 (a) 2-Methylpent-2-ene   **(b)** Hex-2-ene

10.3 (a) $43, 71$ **(b)** 82 **(c)** 58 **(d)** 86

10.4 102 (M⁺), 84 (dehydration), 87 (alpha cleavage), 59 (alpha cleavage)

10.5 X-ray energy is higher.
$\lambda = 9.0 \times 10^{-6}$ m is higher in energy.

10.6 (a) 2.4×10^6 kJ/mol   **(b)** 4.0×10^4 kJ/mol

   (c) 2.4×10^3 kJ/mol   **(d)** 2.8×10^2 kJ/mol

   (e) 6.0 kJ/mol   **(f)** 4.0×10^{-2} kJ/mol

10.7 (a) Ketone or aldehyde

   (b) Nitro compound

   (c) Carboxylic acid

10.8 (a) CH₃CH₂OH has an –OH absorption.

   (b) Hex-1-ene has a double-bond absorption.

   (c) CH₃CH₂CO₂H has a very broad –OH absorption.

10.9 (a) 1715 cm⁻¹   **(b)** $1730, 2100, 3300$ cm⁻¹

   (c) $1720, 2500$–3100 cm⁻¹, 3400–3650 cm⁻¹

10.10 $1690, 1650, 2230$ cm⁻¹

10.11 300–600 kJ/mol; UV energy is greater than IR energy.

10.12 1.46×10^{-5} M

10.13 All except **(a)** have UV absorptions.

CHAPTER 11

11.1 7.5×10^{-5} kJ/mol for ¹⁹F; 8.0×10^{-5} kJ/mol for ¹H

11.2 The vinylic C–H protons are nonequivalent.

b H   CH₃ a
   C=C
c H   Cl

11.3 (a) $7.27 \, \delta$ **(b)** $3.05 \, \delta$ **(c)** $3.46 \, \delta$ **(d)** $5.30 \, \delta$

11.4 (a) 420 Hz   **(b)** $2.1 \, \delta$ **(c)** 1050 Hz

11.5 (a) 4 **(b)** 7 **(c)** 4 **(d)** 5 **(e)** 5 **(f)** 7

11.6 (a) 1,3-Dimethylcyclopentene

   (b) 2-Methylpentane

   (c) 1-Chloro-2-methylpropane

11.7 –CH₃, $9.3 \, \delta$; –CH₂–, $27.6 \, \delta$; C=O, $174.6 \, \delta$; –OCH₃, $51.4 \, \delta$

11.8 $23, 26 \, \delta$

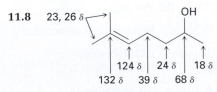

11.9

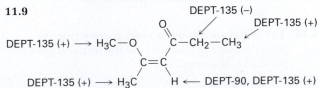

11.10

11.11 A DEPT-90 spectrum would show two absorptions for the non-Markovnikov product (RCH=CHBr) but no absorptions for the Markovnikov product (RBrC=CH$_2$).

11.12 (a) Enantiotopic   (b) Diastereotopic

   (c) Diastereotopic   (d) Diastereotopic

   (e) Diastereotopic   (f) Homotopic

11.13 (a) 2 (b) 4 (c) 3 (d) 4 (e) 5 (f) 3

11.14 4

11.15 (a) 1.43 δ (b) 2.17 δ (c) 7.37 δ

   (d) 5.30 δ (e) 9.70 δ (f) 2.12 δ

11.16 Seven kinds of protons

11.17 Two peaks; 3:2 ratio

11.18 (a) –CHBr$_2$, quartet; –CH$_3$, doublet

   (b) CH$_3$O–, singlet; –OCH$_2$–, triplet; –CH$_2$Br, triplet

   (c) ClCH$_2$–, triplet; –CH$_2$–, quintet

   (d) CH$_3$–, triplet; –CH$_2$–, quartet; –CH–, septet; (CH$_3$)$_2$, doublet

   (e) CH$_3$–, triplet; –CH$_2$–, quartet; –CH–, septet; (CH$_3$)$_2$, doublet

   (f) =CH, triplet; –CH$_2$–, doublet; aromatic C–H, two multiplets

11.19 (a) CH$_3$OCH$_3$ (b) CH$_3$CH(Cl)CH$_3$

   (c) ClCH$_2$CH$_2$OCH$_2$CH$_2$Cl

   (d) CH$_3$CH$_2$CO$_2$CH$_3$ or CH$_3$CO$_2$CH$_2$CH$_3$

11.20 CH$_3$CH$_2$OCH$_2$CH$_3$

11.21 J_{1-2} = 16 Hz; J_{2-3} = 8 Hz

11.22 1-Chloro-1-methylcyclohexane has a singlet methyl absorption; 1-chloro-2-methylcyclohexane has a doublet.

CHAPTER 12

12.1 (a) 1-Iodobutane

   (b) 1-Chloro-3-methylbutane

   (c) 1,5-Dibromo-2,2-dimethylpentane

   (d) 1,3-Dichloro-3-methylbutane

   (e) 1-Chloro-3-ethyl-4-iodopentane

   (f) 2-Bromo-5-chlorohexane

12.2 (a) CH$_3$CH$_2$CH$_2$C(CH$_3$)$_2$CH(Cl)CH$_3$

   (b) CH$_3$CH$_2$CH$_2$C(Cl)$_2$CH(CH$_3$)$_2$

   (c) CH$_3$CH$_2$C(Br)(CH$_2$CH$_3$)$_2$

   (d)

   (e)

   (f)

12.3

12.4 The intermediate allylic radical reacts at the more accessible site and gives the more highly substituted double bond.

12.5 (a) 3-Bromo-5-methylcycloheptene and 3-bromo-6-methylcycloheptene

   (b) Four products

12.6 (a) 2-Methylpropan-2-ol + HCl

   (b) 4-Methylpentan-2-ol + PBr$_3$

   (c) 5-Methylpentan-1-ol + PBr$_3$

   (d) 2,4-Dimethylhexan-2-ol + HCl

12.7 Both reactions occur.

12.8 React Grignard reagent with D$_2$O.

12.9 (R)-1-Methylpentyl acetate, CH$_3$CO$_2$CH(CH$_3$)CH$_2$CH$_2$CH$_2$CH$_3$

12.10 (S)-Butan-2-ol

12.11

(S)-2-Bromo-4-methylpentane $\longrightarrow$ (R) $\underset{\underset{\text{CH}_3}{|}}{\text{CH}_3\text{CHCH}_2}\underset{\underset{\text{SCH}_3}{|}}{\text{CHCH}_3}$

12.12 **(a)** 1-Iodobutane **(b)** Butan-1-ol

(c) Hex-1-yne **(d)** Butylammonium bromide

12.13 **(a)** $(\text{CH}_3)_2\text{N}^-$ **(b)** $(\text{CH}_3)_3\text{N}$ **(c)** H_2S

12.14 $\text{CH}_3\text{OTos} > \text{CH}_3\text{Cl} > (\text{CH}_3)_2\text{CHCl} > \text{CH}_3\text{NH}_2$

12.15 Similar rate to protic solvents because the transition state is not stabilized

12.16 Racemic 1-ethyl-1-methylhexyl acetate

12.17 90.1% racemization and 9.9% inversion

12.18 Racemic 2-phenylbutan-2-ol

12.19 $\text{H}_2\text{C}=\text{CHCH(Br)CH}_3 > \text{CH}_3\text{CH(Br)CH}_3 >$
$\text{CH}_3\text{CH}_2\text{Br} > \text{H}_2\text{C}=\text{CHBr}$

12.20 The same allylic carbocation intermediate is formed in both reactions.

12.21 **(a)** $\text{S}_\text{N}1$ **(b)** $\text{S}_\text{N}2$

12.22

Linalyl diphosphate

Limonene

12.23 **(a)** Major: 2-methylpent-2-ene;
minor: 4-methylpent-2-ene

(b) Major: 2,3,5-trimethylhex-2-ene;
minor: 2,3,5-trimethylhex-3-ene
and 2-isopropyl-4-methylpent-1-ene

(c) Major: ethylidenecyclohexane;
minor: cyclohexylethylene

12.24 **(a)** 1-Bromo-3,6-dimethylheptane

(b) 4-Bromo-1,2-dimethylcyclopentane

12.25 (Z)-1-Bromo-1,2-diphenylethylene

12.26 (Z)-3-Methylpent-2-ene

12.27 The cis isomer reacts faster because the bromine is axial.

12.28 **(a)** $\text{S}_\text{N}2$ **(b)** E2 **(c)** $\text{S}_\text{N}1$ **(d)** E1cB

CHAPTER 13

13.1 **(a)** 5-Methylhexane-2,4-diol

(b) 2-Methyl-4-phenylbutan-2-ol

(c) 4,4-Dimethylcyclohexanol

(d) trans-2-Bromocyclopentanol

(e) 2-Methylheptane-4-thiol

(f) Cyclopent-2-ene-1-thiol

13.2 **(a)**

(b)

(c)

(d) $\underset{\underset{\text{SH}}{|}}{\text{CH}_3\text{CH}}\text{CH}_2\text{CH}_2\text{CH}_2\text{SH}$

(e)

(f)

13.3 **(a)** p-Methylphenol < Phenol <
p-(Trifluoromethyl)phenol

(b) Benzyl alcohol < Phenol <
p-Hydroxybenzoic acid

13.4 The electron-withdrawing nitro group stabilizes an alkoxide ion, but the electron-donating methoxyl group destabilizes the anion.

13.5 Thiophenol is more acidic because the anion is resonance-stabilized.

13.6 **(a)** Benzaldehyde or benzoic acid (or ester)

(b) Acetophenone **(c)** Cyclohexanone

(d) 2-Methylpropanal or 2-methylpropanoic acid (or ester)

13.7 (a) 1-Methylcyclopentanol

(b) 3-Methylhexan-3-ol

13.8 (a) Acetone + CH_3MgBr,
or ethyl acetate + 2 CH_3MgBr

(b) Cyclohexanone + CH_3MgBr

(c) Butan-2-one + PhMgBr,
or ethyl phenyl ketone + CH_3MgBr,
or acetophenone + CH_3CH_2MgBr

(e) Formaldehyde + PhMgBr

13.9 Cyclohexanone + CH_3CH_2MgBr

13.10 (a) 2-Methylpent-2-ene

(b) 3-Methylcyclohexene

(c) 1-Methylcyclohexene

13.11 (a) 1-Phenylethanol

(b) 2-Methylpropan-1-ol

(c) Cyclopentanol

13.12 (a) Hexanoic acid, hexanal

(b) Hexan-2-one

(c) Hexanoic acid, no reaction

13.13 S_N2 displacement of alkoxide ion by F^- ion

13.14 1. $LiAlH_4$; 2. PBr_3; 3. $(H_2N)_2C=S$; 4. H_2O, NaOH

13.15 (a) Diisopropyl ether

(b) Cyclopentyl propyl ether

(c) p-Bromoanisole or
4-bromo-1-methoxybenzene

(d) 1-Methoxycyclohexene

(e) Benzyl methyl sulfide

(f) Allyl methyl sulfide

13.16 A mixture of diethyl ether, dipropyl ether, and
ethyl propyl ether is formed in a 1:1:2 ratio.

13.17 (a) $CH_3CH_2CH_2O^- + CH_3Br$

(b) $PhO^- + CH_3Br$ (c) $(CH_3)_2CHO^- + PhCH_2Br$

(d) $(CH_3)_3CCH_2O^- + CH_3CH_2Br$

13.18 (a) Bromoethane > 2-Bromopropane >>
Bromobenzene

(b) Bromoethane > Chloroethane >>
1-Iodopropene

13.19 (a)

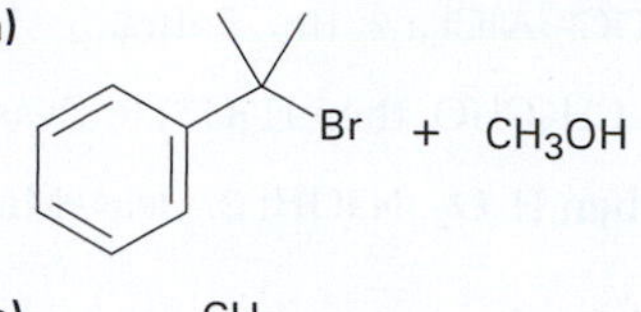

(b)

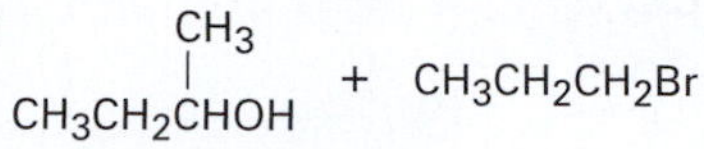

13.20 Protonation of the oxygen atom, followed by
E1 reaction

13.21 o-(1-Methylallyl)phenol

PREVIEW OF CARBONYL CHEMISTRY

1. Acetyl chloride is more electrophilic than
acetone.

2.

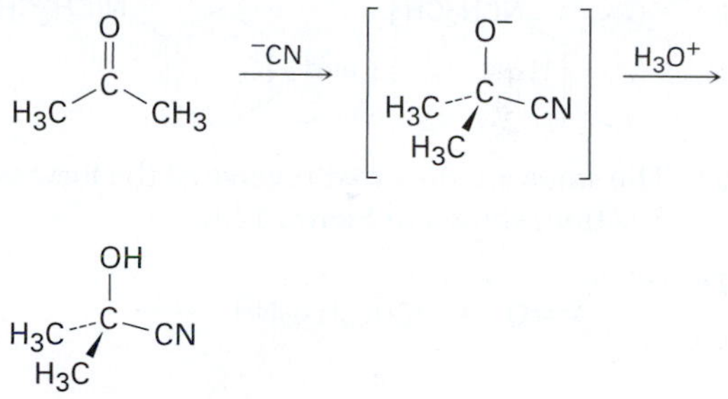

3. (a) Nucleophilic acyl substitution

(b) Nucleophilic addition

(c) Carbonyl condensation

CHAPTER 14

14.1 (a) 2-Methylpentan-3-one

(b) 3-Phenylpropanal

(c) Octane-2,6-dione

(d) trans-2-Methylcyclohexanecarbaldehyde

(e) Hex-4-enal

(f) cis-2,5-Dimethylcyclohexanone

14.2 (a)

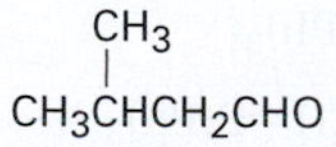

(b)

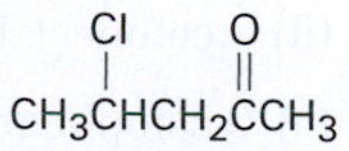

(c)

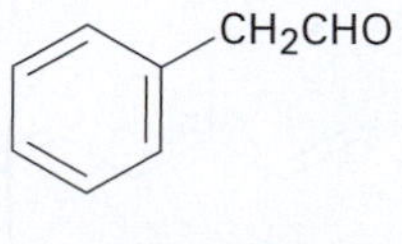

(d)

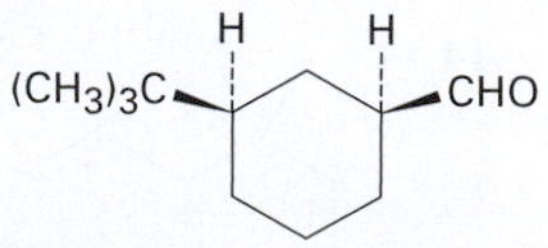

(e)

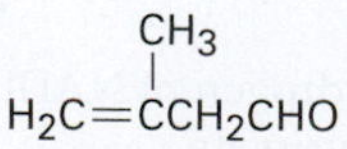

(f)

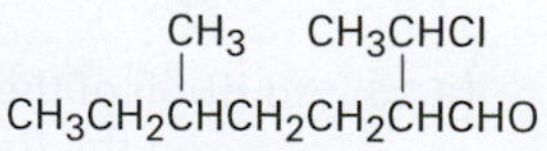

14.3 (a) 1. CH_3COCl, $AlCl_3$; 2. Br_2, $FeBr_3$

(b) 1. Mg; 2. CH_3CHO, then H_3O^+; 3. Dess–Martin

(c) 1. BH_3, then H_2O_2, NaOH; 2. Dess–Martin

14.4

14.5 The electron-withdrawing nitro group in *p*-nitrobenzaldehyde polarizes the carbonyl group.

14.6 $CCl_3CH(OH)_2$

14.7 Labeled water adds reversibly to the carbonyl group.

14.8

and

14.9 The steps are the exact reverse of the forward reaction, shown in Figure 14.8.

14.10

14.11 The mechanism is identical to that between a ketone and 2 equivalents of a monoalcohol, shown in Figure 14.10.

14.12

14.13 (a) Cyclohexanone + $CH_3CH{=}P(Ph)_3$

(b) Cyclohexanecarbaldehyde + $H_2C{=}P(Ph)_3$

(c) Acetone + $CH_3CH_2CH_2CH{=}P(Ph)_3$

(d) Acetone + $PhCH{=}P(Ph)_3$

(e) Acetophenone + $PhCH{=}P(Ph)_3$

(f) Cyclohex-2-enone + $H_2C{=}P(Ph)_3$

14.14

14.15 Addition of the *pro-R* hydrogen of NADH takes place on the *Re* face of pyruvate.

14.16 The –OH group adds to the *Re* face at C2, and –H adds to the *Re* face at C3, to yield (2*R*,3*S*)-isocitrate.

14.17

14.18 (a) But-3-en-2-one + $(CH_3CH_2CH_2)_2CuLi$

(b) 3-Methylcyclohex-2-enone + $(CH_3)_2CuLi$

(c) 4-*tert*-Butylcyclohex-2-enone + $(CH_3CH_2)_2CuLi$

(d) Unsaturated ketone + $(H_2C{=}CH)_2CuLi$

14.19 Look for the presence or absence of a saturated ketone absorption in the product.

14.20 (a) 1715 cm^{-1} (b) 1685 cm^{-1} (c) 1750 cm^{-1}

(d) 1705 cm^{-1} (e) 1715 cm^{-1} (f) 1705 cm^{-1}

14.21 (a) Different peaks due to McLafferty rearrangement

(b) Different peaks due to *a* cleavage and McLafferty rearrangement

(c) Different peaks due to McLafferty rearrangement

14.22 IR: 1750 cm^{-1}; MS: 140, 84

CHAPTER 15

15.1 (a) 3-Methylbutanoic acid

(b) 4-Bromopentanoic acid

(c) 2-Ethylpentanoic acid

(d) *cis*-Hex-4-enoic acid

(e) 2,4-Dimethylpentanenitrile

(f) *cis*-Cyclopentane-1,3-dicarboxylic acid

15.2

(a)

$$H_3C\ CH_3$$
$$CH_3CH_2CH_2CHCHCO_2H$$

(b)

$$CH_3$$
$$CH_3CHCH_2CH_2CO_2H$$

(c)

(d)

(e)

(f) $CH_3CH_2CH{=}CHCN$

15.3   Dissolve the mixture in ether, extract with aqueous NaOH, separate and acidify the aqueous layer, and extract with ether.

15.4   43%

15.5   **(a)** 82% dissociation   **(b)** 73% dissociation

15.6   Lactic acid is stronger because of the inductive effect of the –OH group.

15.7   The dianion is destabilized by repulsion between charges.

15.8   More reactive

15.9   **(a)** 1. Mg; 2. CO_2, then H_3O^+

   (b) 1. Mg; 2. CO_2, then H_3O^+,
   or 1. NaCN; 2. H_3O^+ with heat

15.10   1. NaCN; 2. H_3O^+; 3. $LiAlH_4$,
   or Grignard carboxylation, then $LiAlH_4$

15.11   **(a)** Propanenitrile + CH_3CH_2MgBr, then H_3O^+

   (b) *p*-Nitrobenzonitrile + CH_3MgBr, then H_3O^+

15.12   1. Heat 4-methylpentanenitrile with H_3O^+;
   2. $LiAlH_4$

15.13   Cyclopentanecarboxylic acid has a very broad –OH absorption at 2500–3300 cm^{-1}.

15.14   4-Hydroxycyclohexanone: **H**–C–O absorption near 4 δ in ^{1}H spectrum and **C**=O absorption near 210 δ in ^{13}C spectrum.

   Cyclopentanecarboxylic acid: –CO$_2$**H** absorption near 12 δ in ^{1}H spectrum and –CO$_2$H absorption near 170 δ in ^{13}C spectrum.

CHAPTER 16
16.1   **(a)** 4-Methylpentanoyl chloride

   (b) Cyclohexylacetamide

   (c) Isopropyl 2-methylpropanoate

   (d) Benzoic anhydride

   (e) Isopropyl cyclopentanecarboxylate

   (f) Cyclopentyl 2-methylpropanoate

   (g) *N*-Methylpent-4-enamide

   (h) (*R*)-2-Hydroxypropanoyl phosphate

   (i) Ethyl 2,3-dimethylbut-2-enethioate

16.2

(a) $C_6H_5CO_2C_6H_5$ **(b)** $CH_3CH_2CH_2CON(CH_3)CH_2CH_3$

(c) $(CH_3)_2CHCH_2CH(CH_3)COCl$ **(d)**

(e)

$$CH_3CH_2\overset{O}{\underset{\|}{C}}CH_2\overset{O}{\underset{\|}{C}}OCH_2CH_3$$

(f)

(g)

$$H\overset{O}{\underset{\|}{C}}O\overset{O}{\underset{\|}{C}}CH_2CH_3$$

(h)

16.3

16.4   The electron-withdrawing trifluoromethyl group polarizes the carbonyl carbon.

16.5   **(a)** $CH_3CO_2^- Na^+$ **(b)** CH_3CONH_2

   (c) $CH_3CO_2CH_3 + CH_3CO_2^- Na^+$

   (d) $CH_3CONHCH_3$

16.6

16.7   **(a)** Acetic acid + butan-1-ol

   (b) Butanoic acid + methanol

16.8

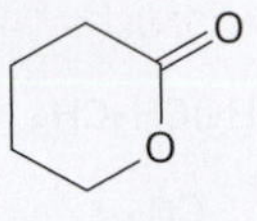

16.9 **(a)** Propanoyl chloride + methanol

(b) Acetyl chloride + ethanol

(c) Benzoyl chloride + ethanol

16.10 Benzoyl chloride + cyclohexanol

16.11 This is a typical nucleophilic acyl substitution reaction, with morpholine as the nucleophile and chloride as the leaving group.

16.12 A nucleophilic acyl substitution of H^- ion for Cl^- ion gives benzaldehyde, which undergoes a nucleophilic addition reaction to give benzyl alcohol.

16.13 **(a)** Propanoyl chloride + methylamine

(b) Benzoyl chloride + diethylamine

(c) Propanoyl chloride + ammonia

16.14 **(a)** Benzoyl chloride + $[(CH_3)_2CH]_2CuLi$, or 2-methylpropanoyl chloride + Ph_2CuLi

(b) Prop-2-enoyl chloride + $(CH_3CH_2CH_2)_2CuLi$, or butanoyl chloride + $(H_2C=CH)_2CuLi$

16.15 This is a typical nucleophilic acyl substitution reaction, with *p*-hydroxyaniline as the nucleophile and acetate ion as the leaving group.

16.16 Monomethyl ester of benzene-1,2-dicarboxylic acid

16.17 Reaction of a carboxylic acid with an alkoxide ion gives the carboxylate ion.

16.18 $HOCH_2CH_2CH_2CHO$

16.19 **(a)** $CH_3CH_2CH_2CH(CH_3)CH_2OH$

(b) $PhOH + PhCH_2OH$

16.20 **(a)** H_2O, NaOH

(b) Product of **(a)**, then $LiAlH_4$

(c) $LiAlH_4$

16.21 1. Mg; 2. CO_2, then H_3O^+; 3. $SOCl_2$; 4. $(CH_3)_2NH$; 5. $LiAlH_4$

16.22

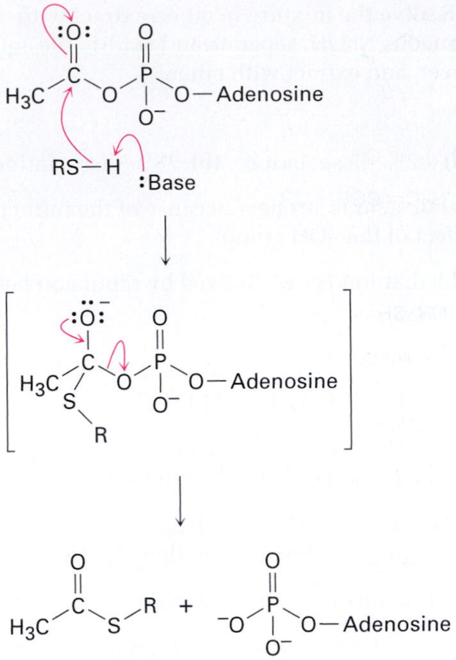

Acetyl CoA

16.23 **(a)**

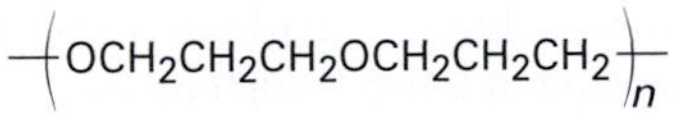

(b)

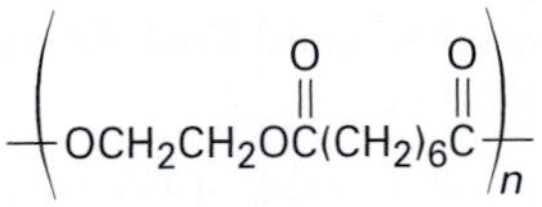

(c)

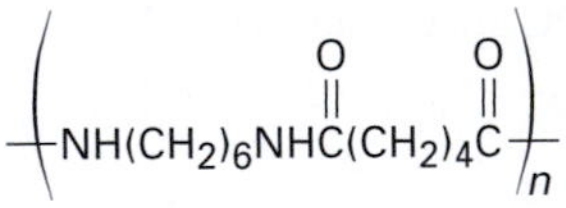

16.24

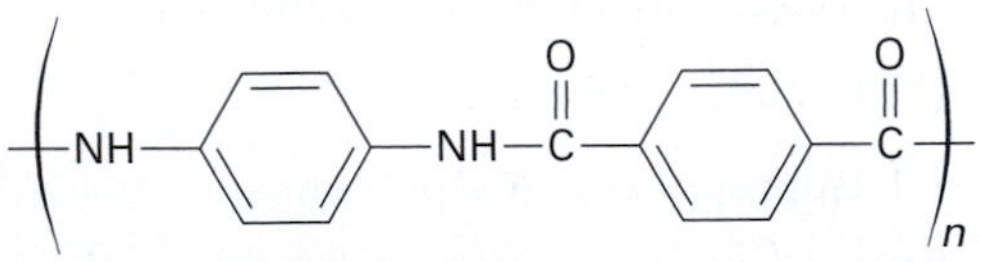

16.25 **(a)** Ester   **(b)** Acid chloride   **(c)** Carboxylic acid

(d) Aliphatic ketone or cyclohexanone

16.26 **(a)** $CH_3CH_2CH_2CO_2CH_2CH_3$ and other possibilities

(b) $CH_3CON(CH_3)_2$

(c) $CH_3CH=CHCOCl$ or $H_2C=C(CH_3)COCl$

CHAPTER 17

17.1 **(a)** (cyclopentene with OH)

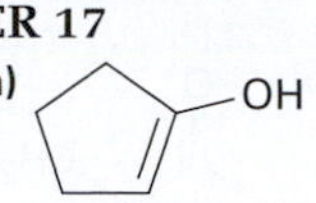

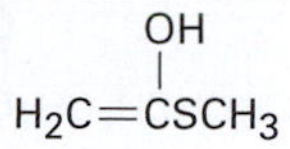

(b) $H_2C=\overset{\overset{OH}{|}}{C}SCH_3$

(c) $H_2C=\overset{\overset{OH}{|}}{C}OCH_2CH_3$

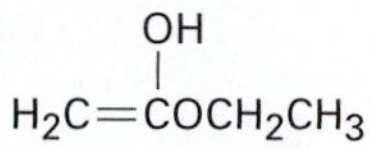

(d) $CH_3CH=CHOH$

(e) $CH_3CH=\overset{\overset{OH}{|}}{C}OH$

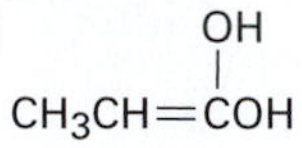

(f) $PhCH=\overset{\overset{OH}{|}}{C}CH_3$ or $PhCH_2\overset{\overset{OH}{|}}{C}=CH_2$

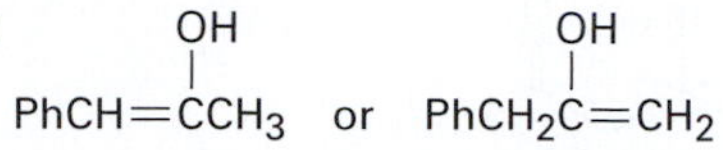

17.2

(structures showing tautomeric equilibria)

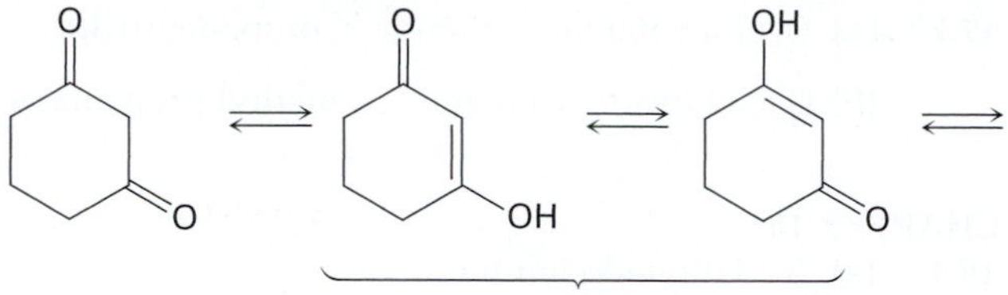

Equivalent;
more stable

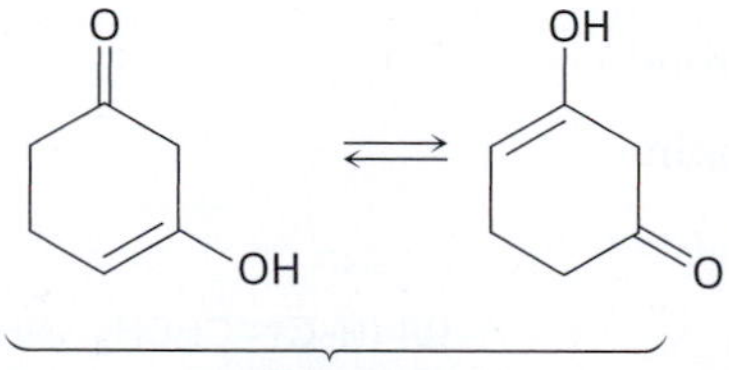

Equivalent;
less stable

17.3 1. Br$_2$; 2. Pyridine, heat

17.4 The intermediate *a*-bromo acid bromide undergoes a nucleophilic acyl substitution reaction with methanol to give an *a*-bromo ester.

17.5 **(a)** $CH_3C\textbf{H}_2CHO$ **(b)** $(CH_3)_3CCOC\textbf{H}_3$

(c) $CH_3CO_2\textbf{H}$ **(d)** $PhCON\textbf{H}_2$

(e) $CH_3CH_2C\textbf{H}_2CN$ **(f)** $\textbf{CH}_3CON(CH_3)_2$

17.6 $^-\!:CH_2C\equiv N: \longleftrightarrow H_2C=C=\ddot{N}:^-$

17.7 **(a)** 1. Na$^+$ $^-$OEt; 2. PhCH$_2$Br; 3. H$_3$O$^+$

(b) 1. Na$^+$ $^-$OEt; 2. CH$_3$CH$_2$CH$_2$Br; 3. Na$^+$ $^-$OEt; 4. CH$_3$Br; 5. H$_3$O$^+$

(c) 1. Na$^+$ $^-$OEt; 2. (CH$_3$)$_2$CHCH$_2$Br; 3. H$_3$O$^+$

17.8 1. Na$^+$ $^-$OEt; 2. (CH$_3$)$_2$CHCH$_2$Br; 3. Na$^+$ $^-$OEt; 4. CH$_3$Br; 5. H$_3$O$^+$

17.9 **(a)** (CH$_3$)$_2$CHCH$_2$Br **(b)** PhCH$_2$CH$_2$Br

17.10 None can be prepared.

17.11 1. 2 Na$^+$ $^-$OEt; 2. BrCH$_2$CH$_2$CH$_2$CH$_2$Br; 3. H$_3$O$^+$

17.12 **(a)** Alkylate phenylacetone with CH$_3$I

(b) Alkylate pentanenitrile with CH$_3$CH$_2$I

(c) Alkylate cyclohexanone with H$_2$C=CHCH$_2$Br

(d) Alkylate cyclohexanone with excess CH$_3$I

(e) Alkylate C$_6$H$_5$COCH$_2$CH$_3$ with CH$_3$I

(f) Alkylate methyl 3-methylbutanoate with CH$_3$CH$_2$I

17.13 **(a)**

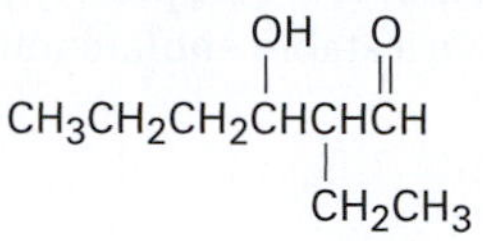

$CH_3CH_2CH_2\overset{\overset{OH}{|}}{C}HCH\overset{\overset{O}{\|}}{C}H$
    $\overset{|}{C}H_2CH_3$

(b) (1,3-diphenyl structure with O, HO, CH$_3$)

(c) (bicyclic structure with OH and O)

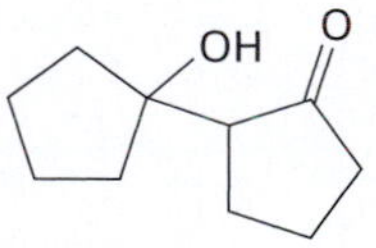

17.14 The reverse reaction is the exact opposite of the forward reaction, shown in Figure 17.9

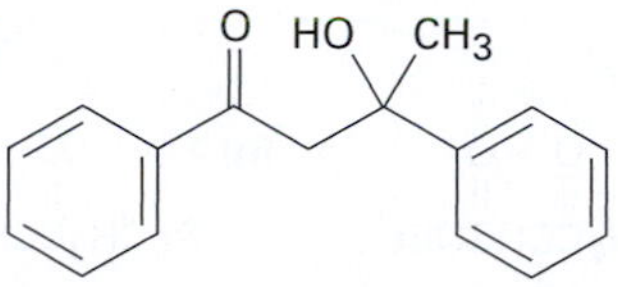

17.15 **(a)** (cyclopentanone fused structure)

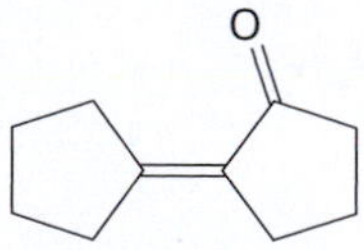

(b) (structure with CH$_3$, O, H and two phenyl groups)

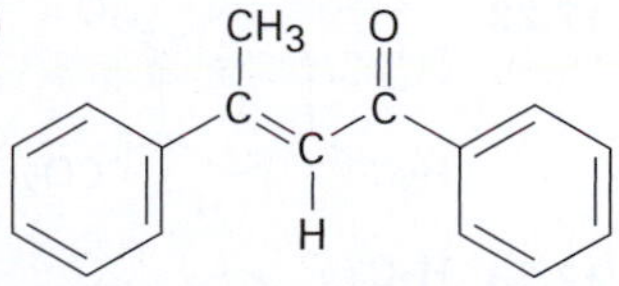

(c) $(CH_3)_2CHCH_2CH=\overset{\overset{O}{\|}}{C}CH$
          $\overset{|}{C}H(CH_3)_2$

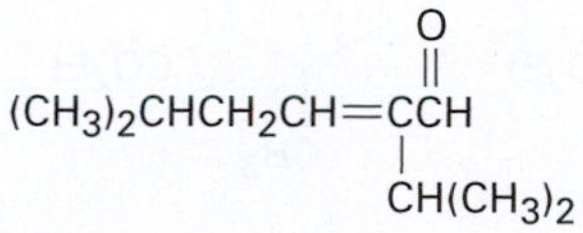

17.16

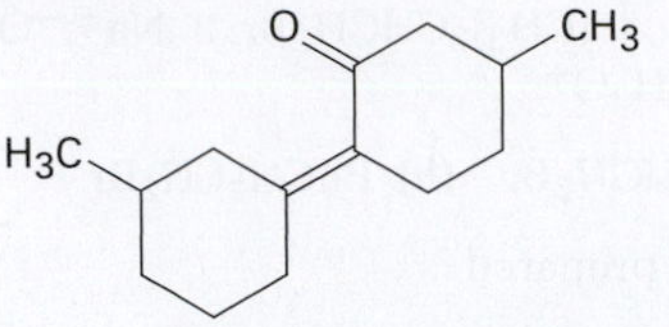

and

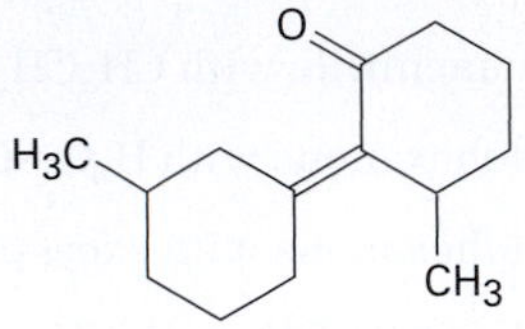

17.17 **(a)** Not an aldol product   **(b)** Pentan-3-one

17.18 The CH_2 position between the two carbonyl groups is so acidic that it is completely deprotonated to give a stable enolate ion.

17.19

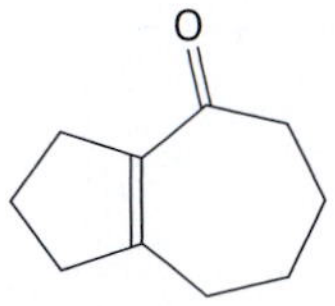

17.20 **(a)**

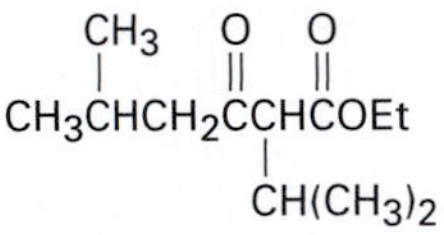

$$CH_3CHCH_2CCHCOEt$$
with CH_3 and $CH(CH_3)_2$ substituents

(b)

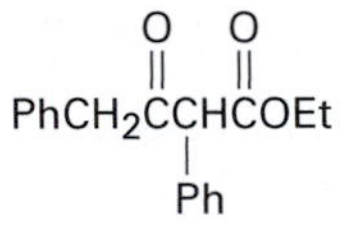

$$PhCH_2CCHCOEt$$
with Ph substituent

(c)

$$C_6H_{11}CH_2CCHCOEt$$
with C_6H_{11} substituent

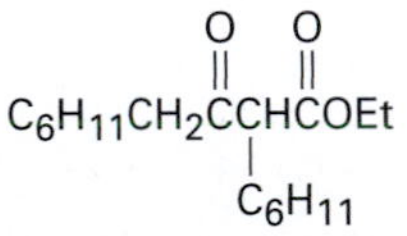

17.21 The cleavage reaction is the exact reverse of the forward reaction, shown in Figure 17.12.

17.22

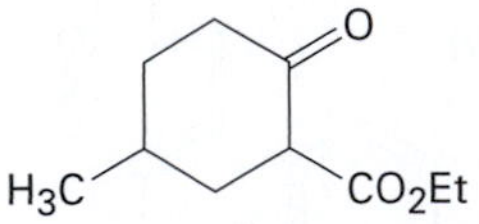

17.23

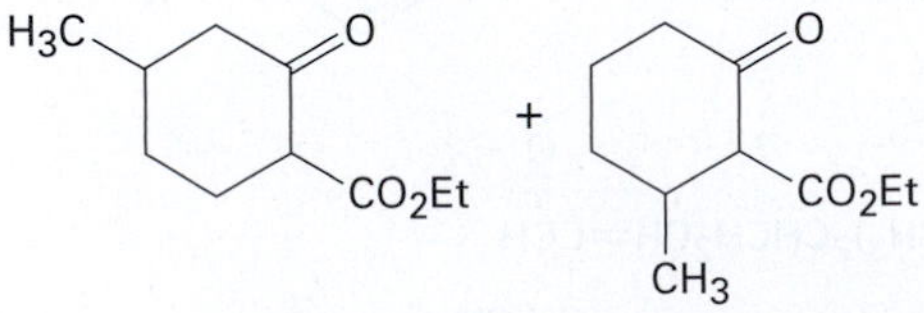

17.24 **(a)**

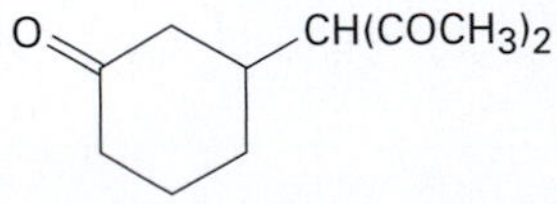

(b) $(CH_3CO)_2CHCH_2CH_2CN$

(c)

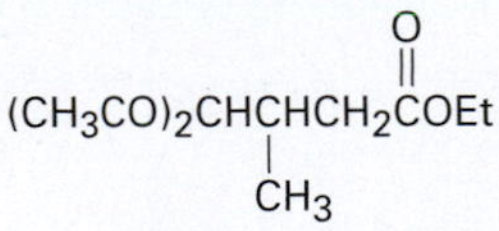

$$(CH_3CO)_2CHCHCH_2COEt$$
with CH_3 substituent

17.25
(a) 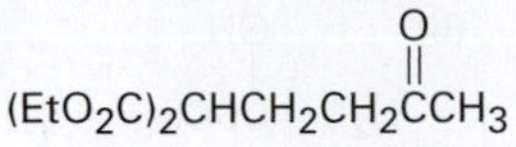 $(EtO_2C)_2CHCH_2CH_2CCH_3$ **(b)**

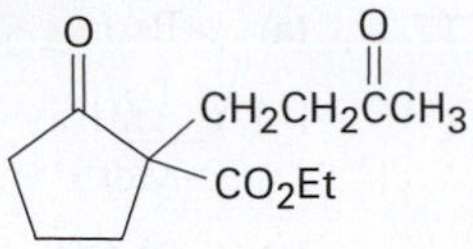

17.26
(a)

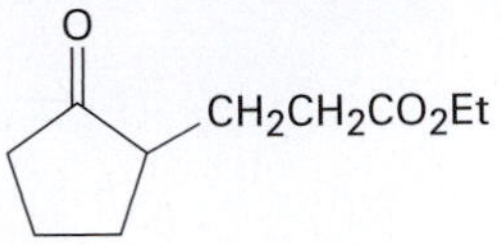

(b)

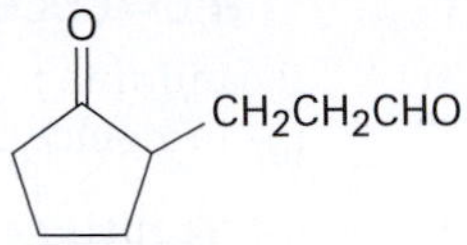

(c)

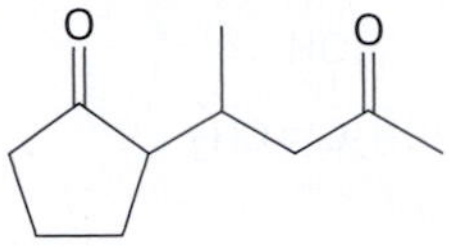

17.27 **(a)** Cyclopentanone enamine + propenenitrile

(b) Cyclohexanone enamine + methyl propenoate

CHAPTER 18

18.1 **(a)** N-Methylethylamine

(b) Tricyclohexylamine

(c) N-Ethyl-N-methylcyclohexylamine

(d) N-Methylpyrrolidine

(e) Diisopropylamine

(f) Butane-1,3-diamine

18.2 **(a)** $[(CH_3)_2CH]_3N$ **(b)** $(H_2C=CHCH_2)_2NH$

(c)

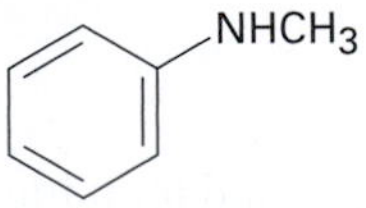

(d)

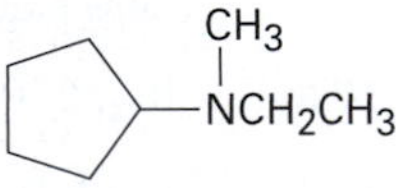

(e)

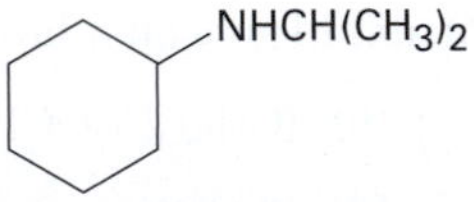

(f)

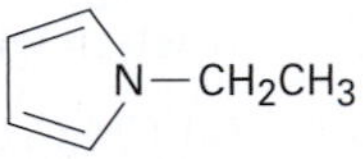

18.3 **(a)**

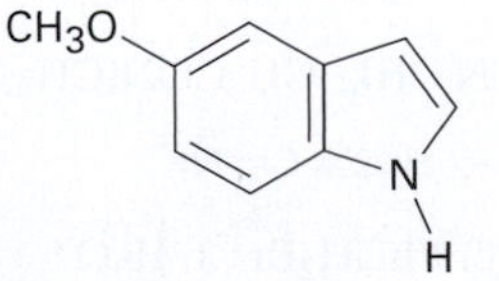

(b)

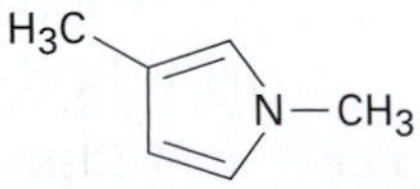

(c)

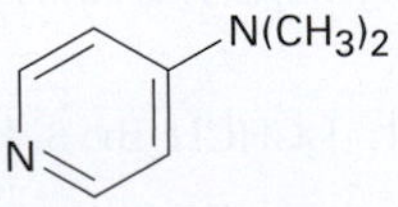

(d)

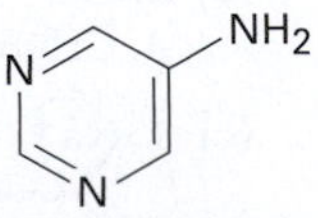

18.4 **(a)** $CH_3CH_2NH_2$ **(b)** NaOH   **(c)** CH_3NHCH_3

18.5 Propylamine is stronger; benzylamine $pK_b = 4.67$; propylamine $pK_b = 3.29$

18.6 **(a)** *p*-Nitroaniline < *p*-Aminobenzaldehyde < *p*-Bromoaniline

(b) *p*-Aminoacetophenone < *p*-Chloroaniline < *p*-Methylaniline

(c) *p*-(Trifluoromethyl)aniline < *p*-(Fluoromethyl)aniline < *p*-Methylaniline

18.7 Pyrimidine is essentially 100% neutral (unprotonated).

18.8 **(a)** Propanenitrile or propanamide

(b) *N*-Propylpropanamide

(c) Benzonitrile or benzamide

(d) *N*-Phenylacetamide

18.9

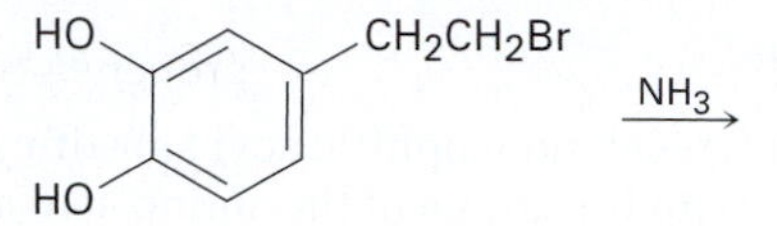

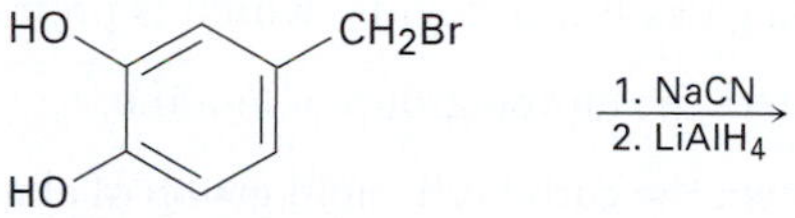

18.10

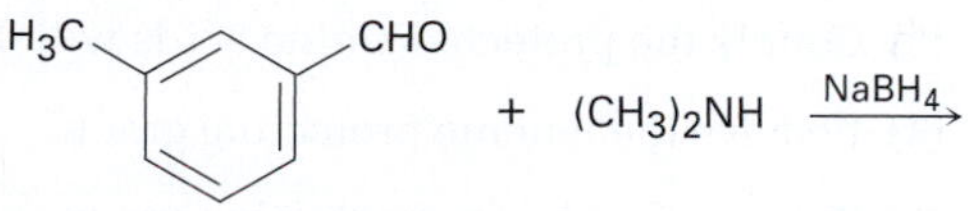

18.11 **(a)** Oct-3-ene and oct-4-ene

(b) Cyclohexene

(c) Hept-3-ene

(d) Ethylene and cyclohexene

18.12 $H_2C=CHCH_2CH_2CH_2N(CH_3)_2$

18.13 1. HNO_3, H_2SO_4; 2. H_2/PtO_2; 3. $(CH_3CO)_2O$; 4. $HOSO_2Cl$; 5. aminothiazole; 6. H_2O, NaOH

18.14 **(a)** 1. HNO_3, H_2SO_4; 2. H_2/PtO_2; 3. 2 CH_3Br

(b) 1. HNO_3, H_2SO_4; 2. H_2/PtO_2; 3. $(CH_3CO)_2O$; 4. Cl_2; 5. H_2O, NaOH

(c) 1. HNO_3, H_2SO_4; 2. Cl_2, $FeCl_3$; 3. Sn

18.15

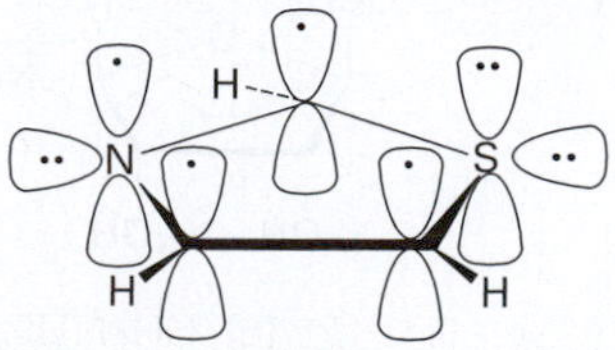

18.16 4.1% protonated

18.17
Attack at C2:

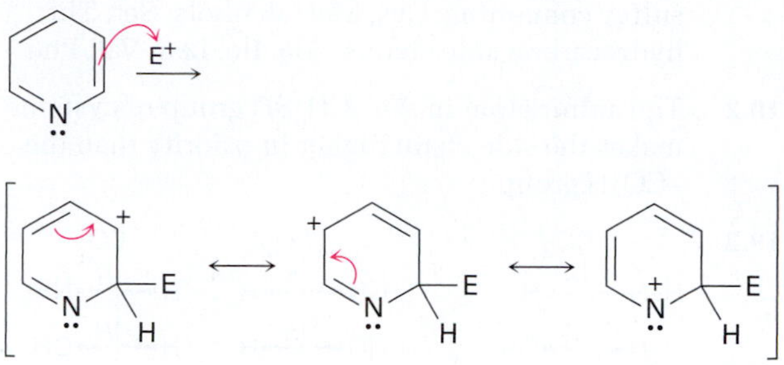

Unfavorable

Attack at C3:

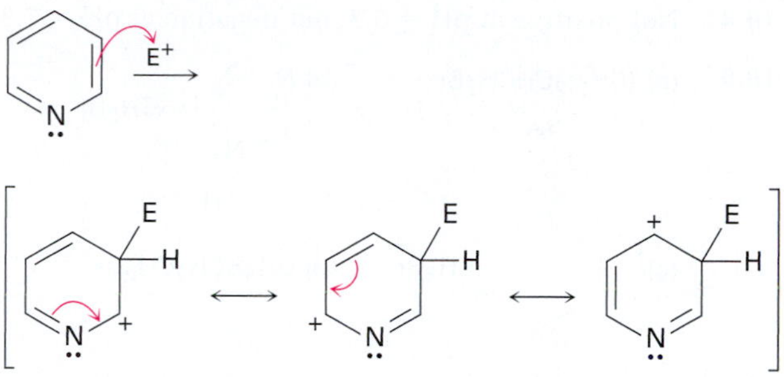

Attack at C4:

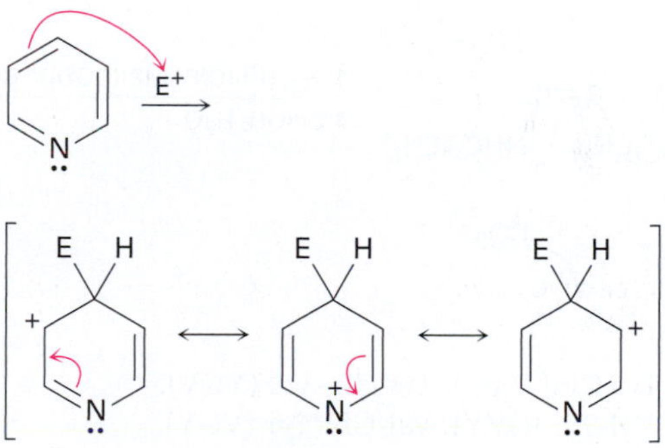

Unfavorable

18.18 The side-chain nitrogen is more basic than the ring nitrogen.

18.19 Reaction at C2 is disfavored because the aromaticity of the benzene ring is lost.

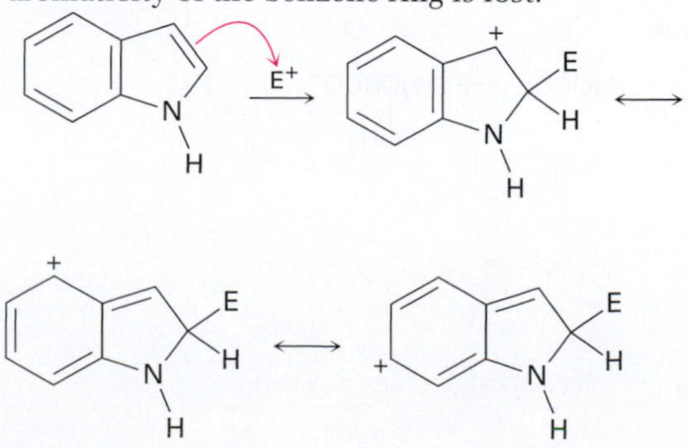

CHAPTER 19

19.1 Aromatic: Phe, Tyr, Trp, His; sulfur-containing: Cys, Met; alcohols: Ser, Thr; hydrocarbon side chains: Ala, Ile, Leu, Val, Phe

19.2 The sulfur atom in the $-CH_2SH$ group of cysteine makes the side chain higher in priority than the $-CO_2H$ group.

19.3

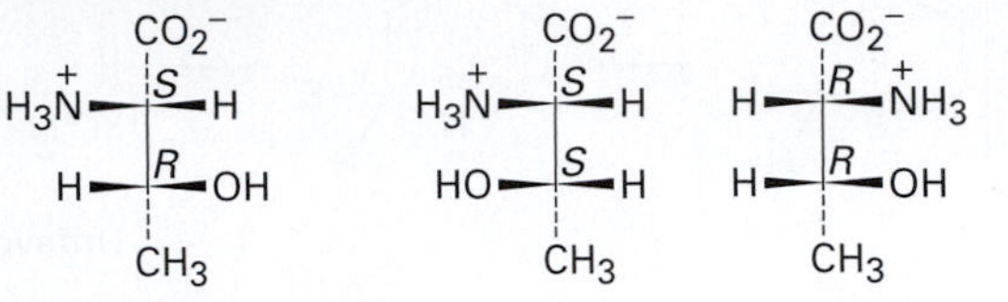

L-Threonine   Diastereomers of L-threonine

19.4 Net positive at pH = 5.3; net negative at pH = 7.3

19.5 (a) $(CH_3)_2CHCH_2Br$ (b)

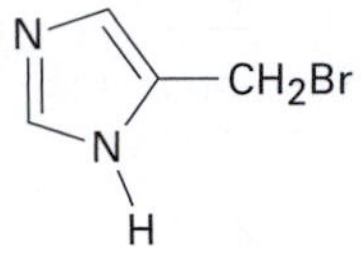

(c)

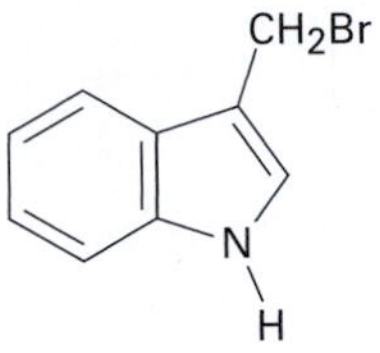

(d) $CH_3SCH_2CH_2Br$

19.6

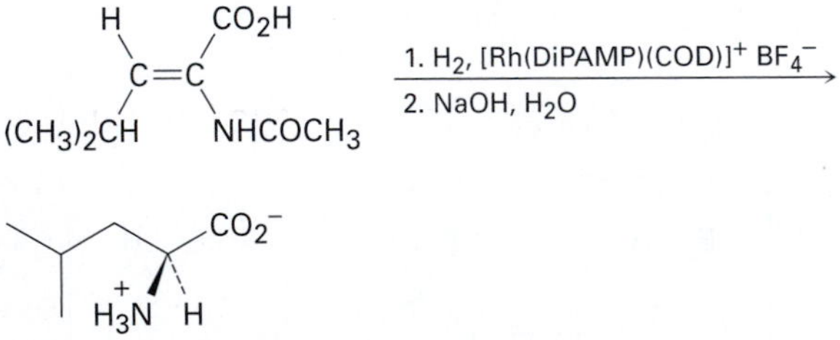

19.7 Val-Tyr-Gly (VYG), Tyr-Gly-Val (YGV), Gly-Val-Tyr (GVY), Val-Gly-Tyr (VGY), Tyr-Val-Gly (YVG), Gly-Tyr-Val (GYV)

19.8

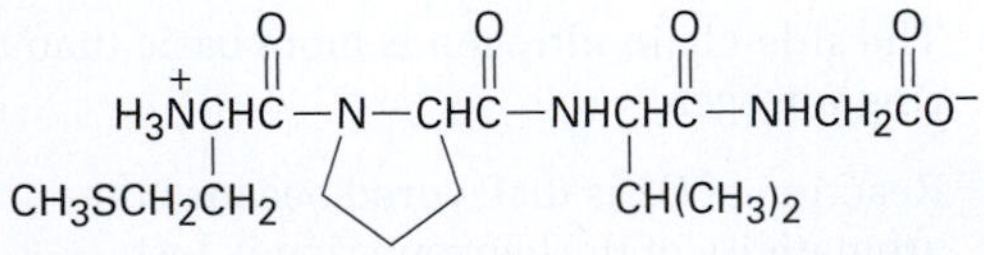

19.9

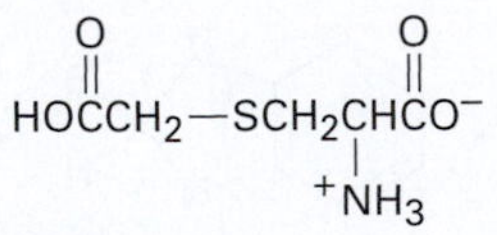

19.10

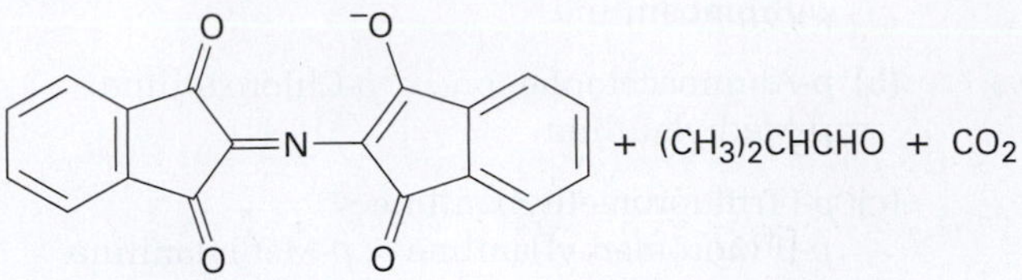

$+ (CH_3)_2CHCHO + CO_2$

19.11 Trypsin: Asp-Arg + Val-Tyr-Ile-His-Pro-Phe
Chymotrypsin: Asp-Arg-Val-Tyr + Ile-His-Pro-Phe

19.12 Methionine

19.13 C_6H_5

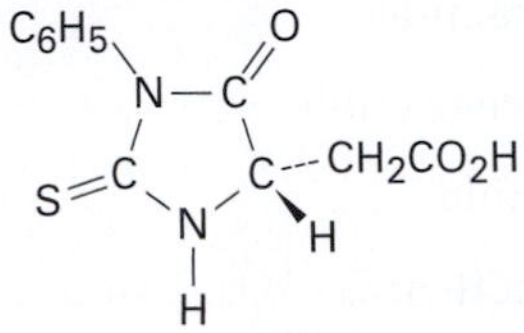

19.14 This is a typical nucleophilic acyl substitution reaction, with the amine of the amino acid as the nucleophile and *tert*-butyl carbonate as the leaving group. The *tert*-butyl carbonate then loses CO_2 and gives *tert*-butoxide, which is protonated.

19.15 (1) Protect the amino group of leucine.

(2) Protect the carboxylic acid group of alanine.

(3) Couple the protected amino acids with DCC.

(4) Remove the leucine protecting group.

(5) Remove the alanine protecting group.

19.16 (a) Lyase   (b) Hydrolase   (c) Oxidoreductase

CHAPTER 20

20.1

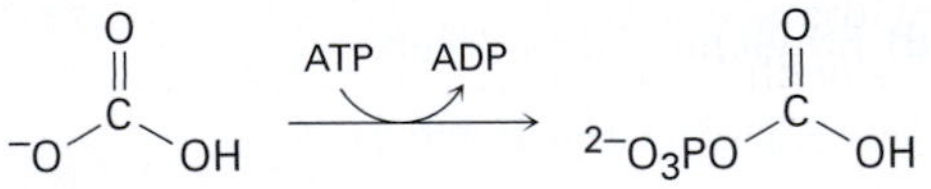

20.2 The mechanism is the reverse of that shown in Figure 20.2.

20.3 The *Re* face

20.4

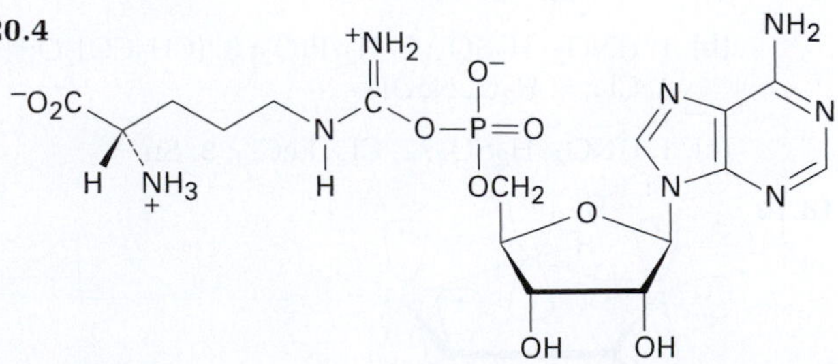

20.5 The mechanism is the same as that shown in Figure 20.2.

20.6

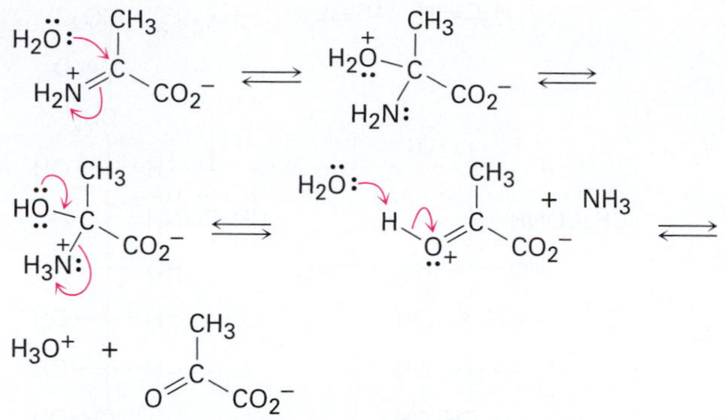

20.7 The full mechanism involves formation of a tetrahedral intermediate, followed by expulsion of ammonia.

20.8

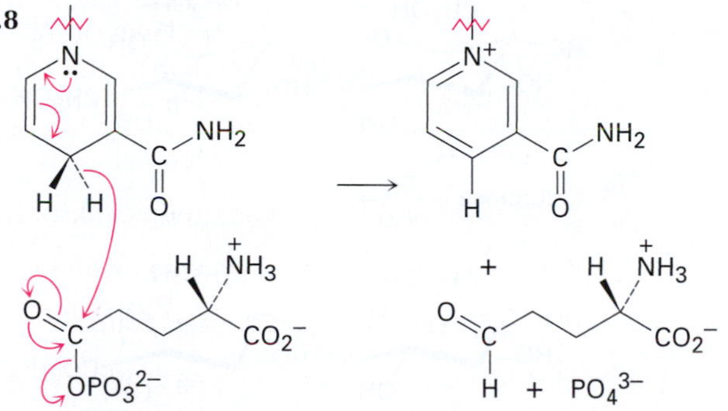

20.9 The nonenzymatic cyclization is an internal imine formation that occurs by nucleophilic addition of the amine to the carbonyl group followed by loss of water. The enzymatic reduction is a nucleophilic addition to the iminium ion:

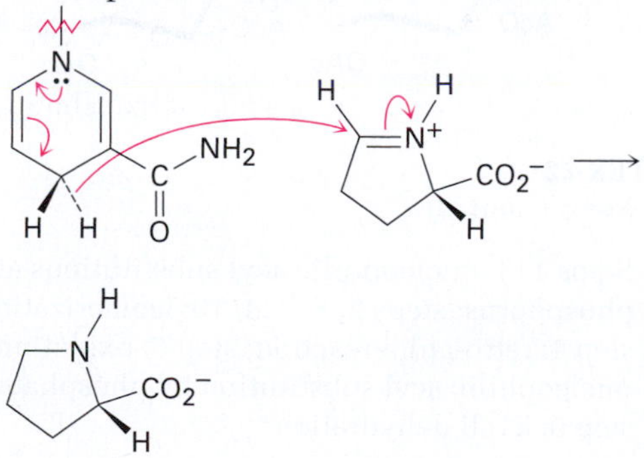

CHAPTER 21

21.1 **(a)** Aldotetrose **(b)** Ketopentose **(c)** Ketohexose **(d)** Aldopentose

21.2 **(a)** *S* **(b)** *R* **(c)** *S*

21.3 A, B, and C are the same.

21.4

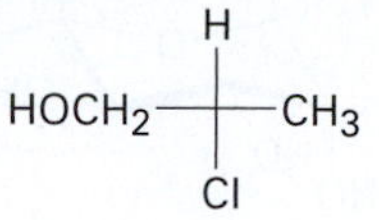

R

21.5

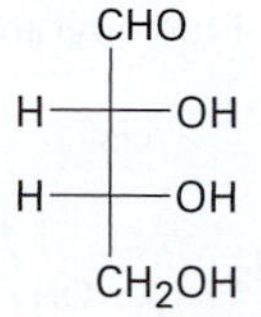

R
R

21.6 **(a)** L-Erythrose; 2*S*,3*S*

(b) D-Xylose; 2*R*,3*S*,4*R*

(c) D-Xylulose; 3*S*,4*R*

21.7

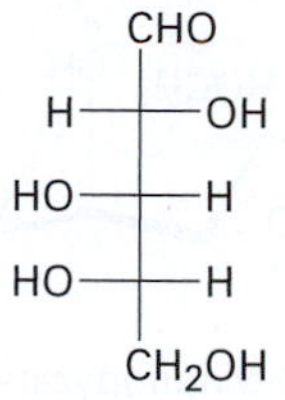

L-(+)-Arabinose

21.8

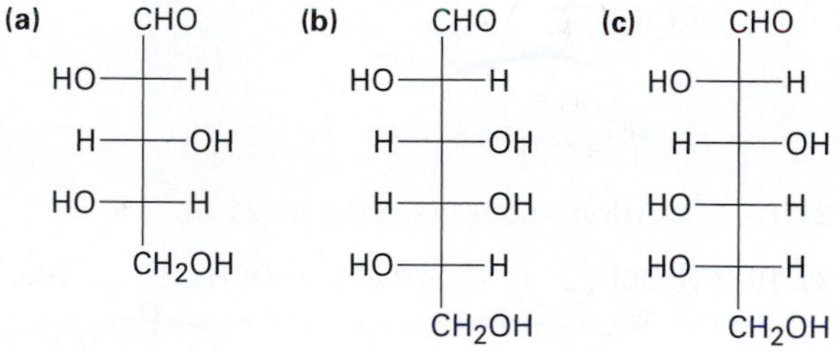

21.9 16 D and 16 L aldoheptoses

21.10

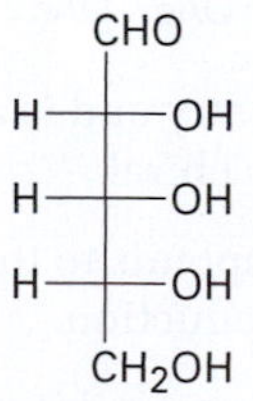

D-Ribose

21.11

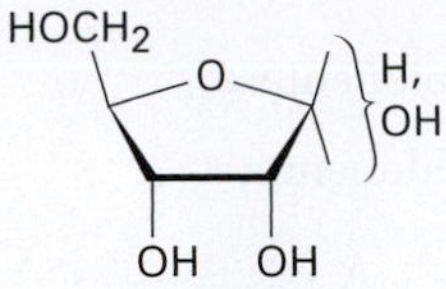

21.12

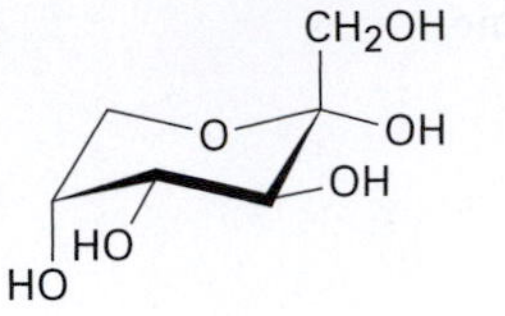

α-D-Fructopyranose   β-D-Fructopyranose

trans   cis

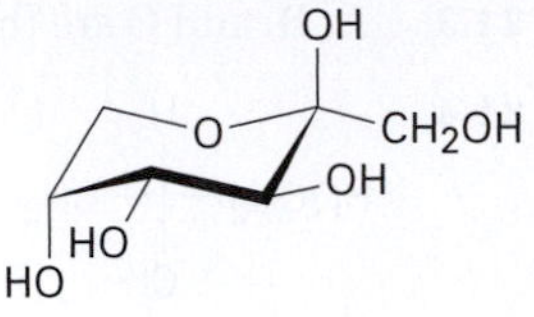

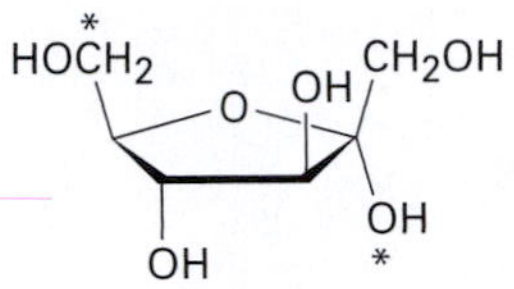

α-D-Fructofuranose 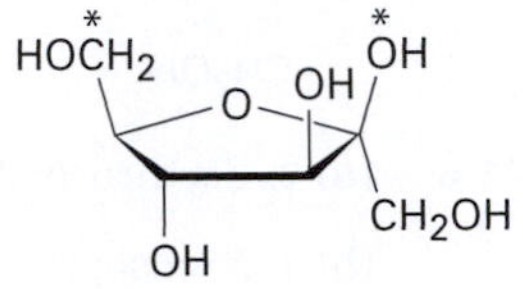   β-D-Fructofuranose

21.13

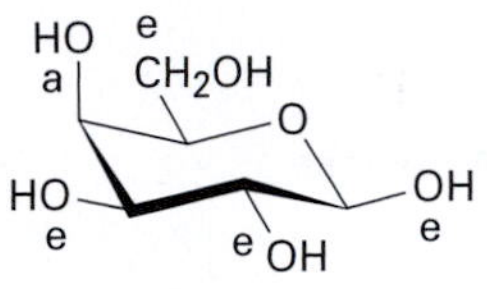

β-D-Galactopyranose 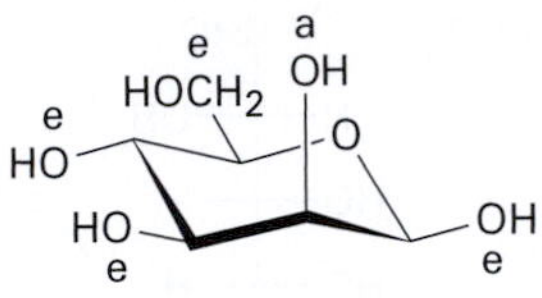   β-D-Mannopyranose

21.14

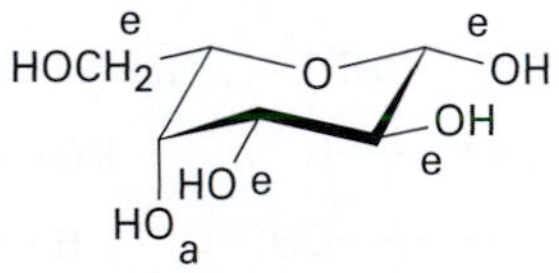

21.15 α-D-Allopyranose (see Figure 21.3)

21.16

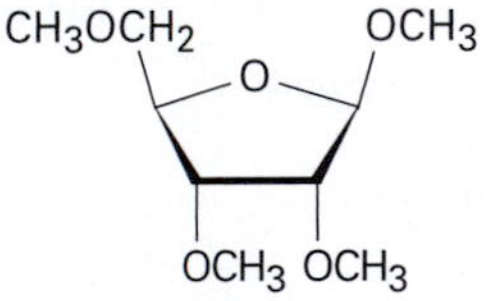

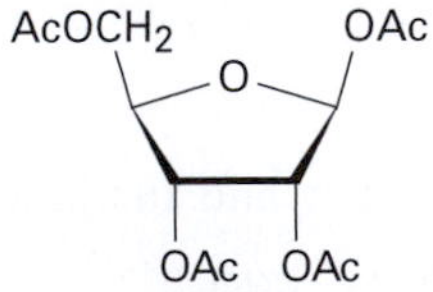

21.17 D-Galactitol has a plane of symmetry and is a meso compound, whereas D-glucitol is chiral.

21.18 The –CHO end of L-gulose corresponds to the –CH₂OH end of D-glucose after reduction.

21.19 D-Allaric acid has a symmetry plane and is a meso compound, but D-glucaric acid is chiral.

21.20 D-Allose and D-galactose yield meso aldaric acids; the others yield optically active aldaric acids.

21.21

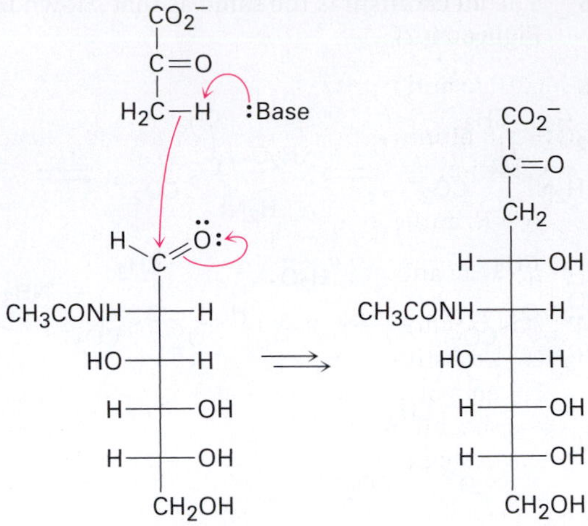

21.22 **(a)** Cellobiose $\xrightarrow[\text{2. H}_2\text{O}]{\text{1. NaBH}_4}$

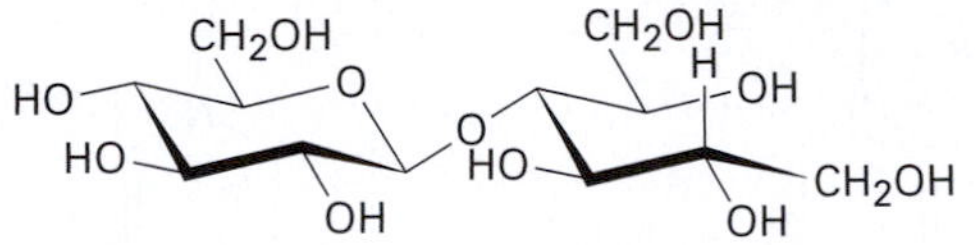

(b) Cellobiose $\xrightarrow[\text{H}_2\text{O}]{\text{Br}_2}$

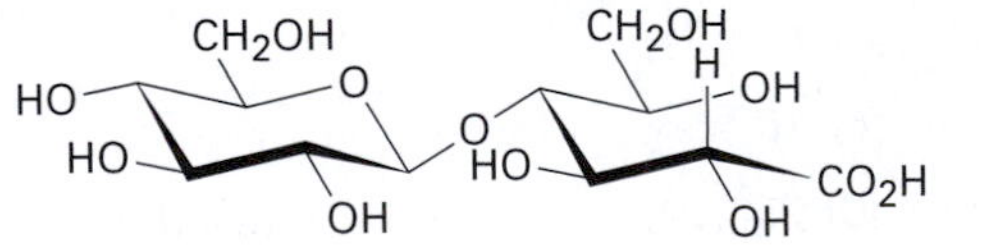

(c) Cellobiose $\xrightarrow[\text{pyridine}]{\text{CH}_3\text{COCl}}$

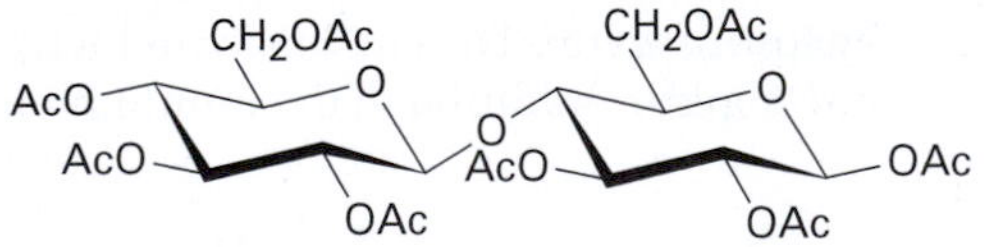

CHAPTER 22

22.1 Steps 7 and 10

22.2 Steps 1, 3: nucleophilic acyl substitutions at phosphorus; steps 2, 5, 7, 8, 10: isomerizations; step 4: retro-aldol reaction; step 6: oxidation and nucleophilic acyl substitution by phosphate; step 9: E1cB dehydration

22.3 *pro-R*

22.4

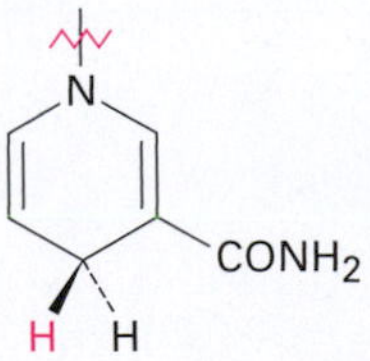

22.5   C1 and C6 of glucose become $-CH_3$ groups; C3 and C4 become CO_2.

22.6   Citrate and isocitrate

22.7   E1cB elimination of water, followed by conjugate addition

22.8   *pro-R;* anti geometry

22.9   *Re* face; anti geometry

22.10 The reaction occurs by two sequential nucleophilic acyl substitutions, the first by a cysteine residue in the enzyme, with phosphate as leaving group, and the second by hydride donation from NADH, with the cysteine residue as leaving group.

CHAPTER 23

23.1 $CH_3(CH_2)_{18}CO_2CH_2(CH_2)_{30}CH_3$

23.2   Glyceryl tripalmitate is higher melting.

23.3 $[CH_3(CH_2)_7CH{=}CH(CH_2)_7CO_2^-]_2 \ Mg^{2+}$

23.4   Glyceryl dioleate monopalmitate $\rightarrow$ Glycerol + 2 Sodium oleate + Sodium palmitate

23.5   Capryloyl CoA $\rightarrow$ Hexanoyl CoA $\rightarrow$ Butyroyl CoA $\rightarrow$ 2 Acetyl CoA

23.6   **(a)** 8 acetyl CoA; 7 passages

      (b) 10 acetyl CoA; 9 passages

23.7   The dehydration is an E1cB reaction.

23.8   At C2, C4, C6, C8, and so forth

23.9   The *Si* face

23.10 The *pro-S* hydrogen is cis to the $-CH_3$ group; the *pro-R* hydrogen is trans.

23.11
(a)

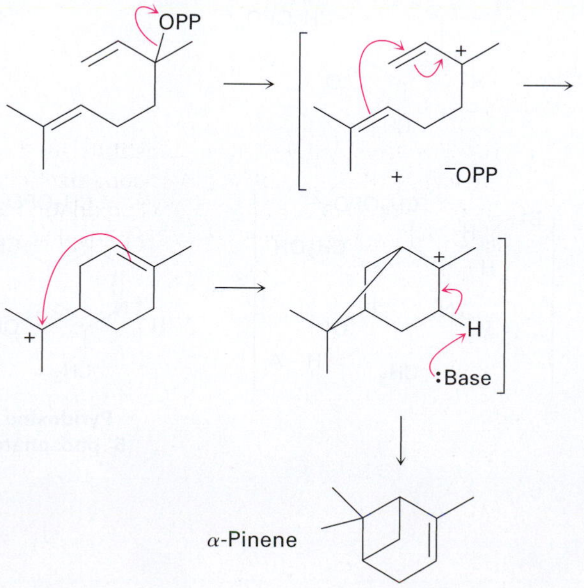

α-Pinene

(b)

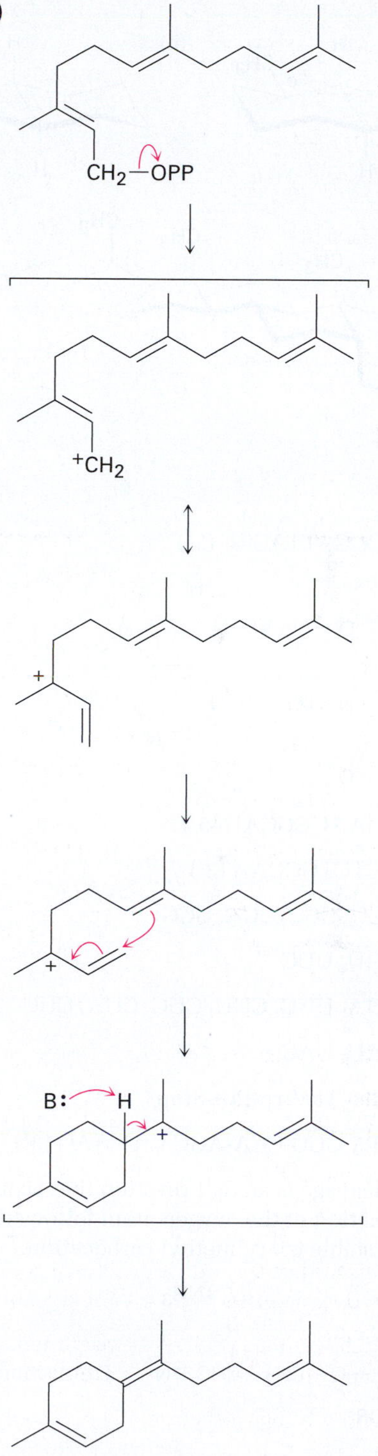

γ-Bisabolene

23.12
(a)   **(b)**

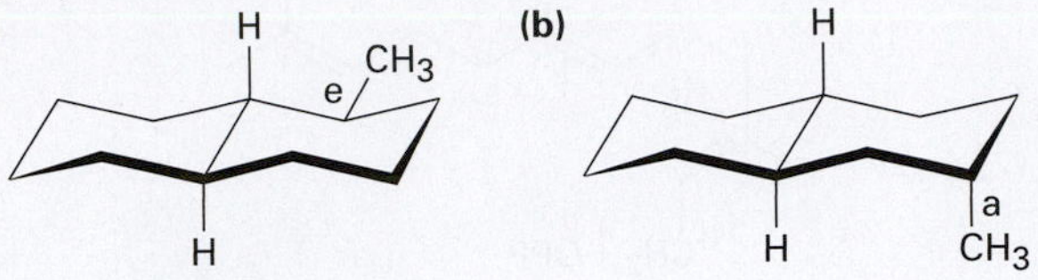

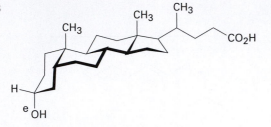

23.13

CHAPTER 24

24.3   (5′) ACGGATTAGCC (3′)

24.4

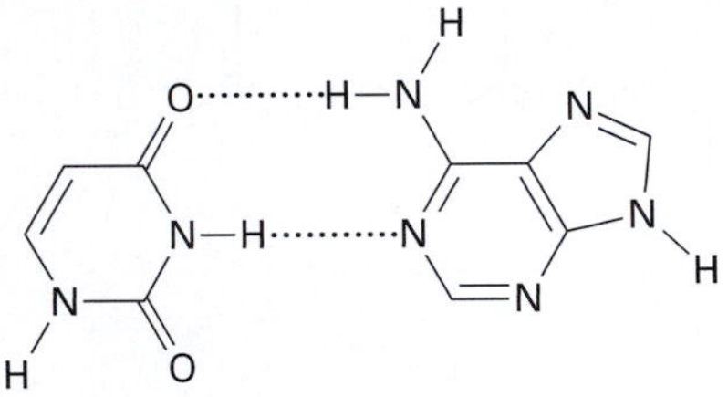

24.5   (3′) CUAAUGGCAU (5′)

24.6   (5′) ACTCTGCGAA (3′)

24.7   **(a)** GCU, GCC, GCA, GCG

(b) UUU, UUC

(c) UUA, UUG, CUU, CUC, CUA, CUG

(d) UAU, UAC

24.8   Leu-Met-Ala-Trp-Pro-Stop

24.9   (5′) TTA-GGG-CCA-AGC-CAT-AAG (3′)

24.10 The cleavage is an S_N1 reaction that occurs by protonation of the oxygen atom followed by loss of the stable triarylmethyl carbocation.

24.11

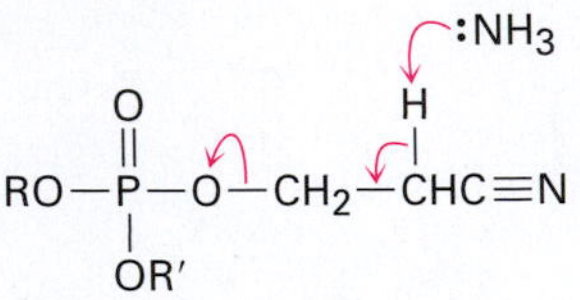

RO—P—O—CH$_2$—CHC≡N   E2 reaction

24.12

Adenosine

Inosine

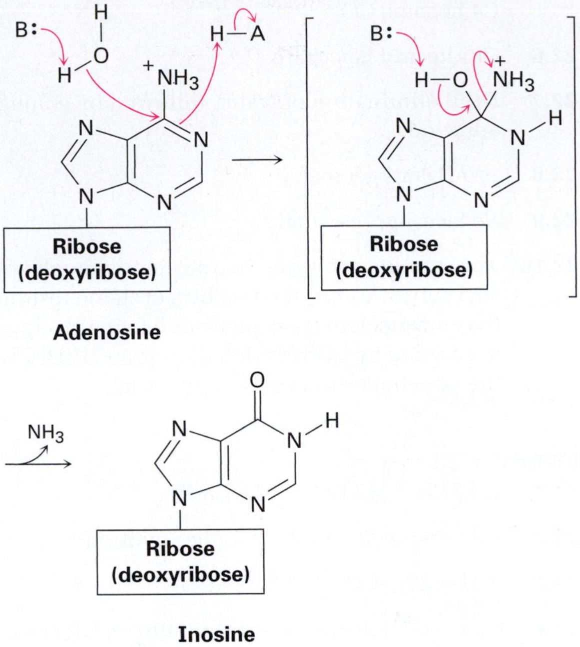

24.13 The mechanism occurs by (1) phosphorylation of inosine monophosphate by reaction with GTP, (2) acid-catalyzed nucleophilic addition of aspartate to an imine, and (3) loss of phosphate by an E1cB reaction.

24.14 The reaction is an E1cB elimination.

CHAPTER 25

25.1   *Si* face

25.2

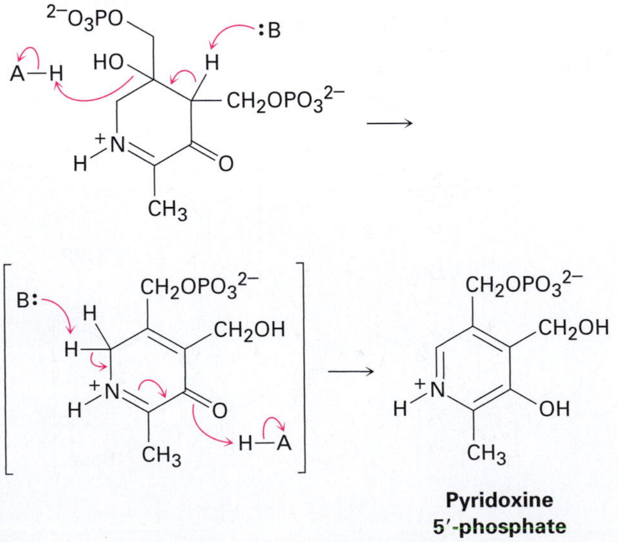

**Pyridoxine
5′-phosphate**

25.3

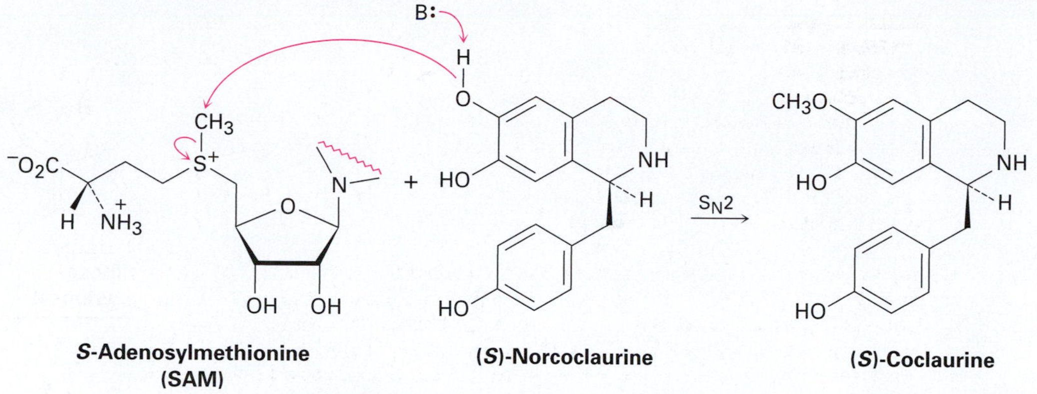
B:
S_N2
S-Adenosylmethionine (SAM)
(*S*)-Norcoclaurine
(*S*)-Coclaurine

25.4

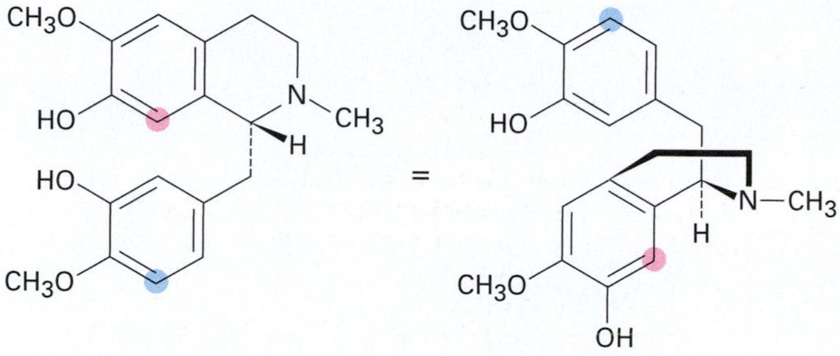

25.5

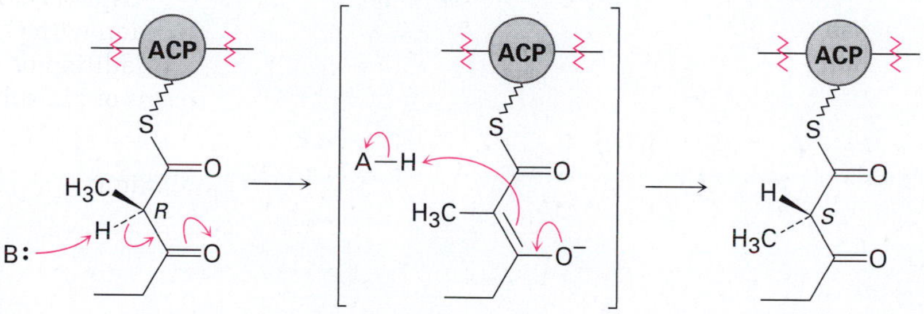
ACP

25.6

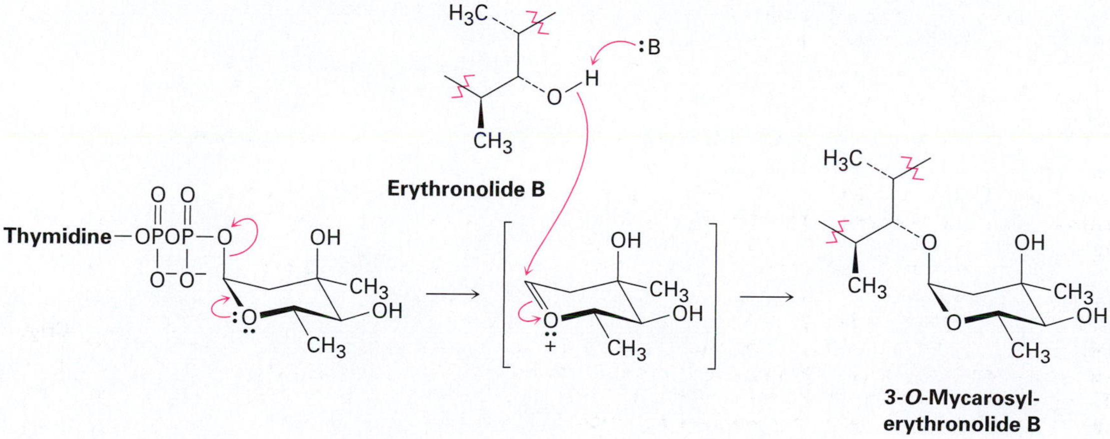
Erythronolide B
Thymidine
3-*O*-Mycarosyl-erythronolide B

Index

The boldfaced references refer to pages where terms are defined.

Periodic Table of the Elements

Key

79
Au — Symbol
Gold — Name
196.9665 — Atomic mass

An element

- 79 — Atomic number

Legend:
- ☐ Metals
- ☐ Semimetals
- ☐ Nonmetals

Group number, U.S. system / IUPAC system

Period number

Numbers in parentheses are mass numbers of radioactive isotopes.

Period	1A (1)	2A (2)	3B (3)	4B (4)	5B (5)	6B (6)	7B (7)	8B (8)	8B (9)	8B (10)	1B (11)	2B (12)	3A (13)	4A (14)	5A (15)	6A (16)	7A (17)	8A (18)
1	1 **H** Hydrogen 1.0079																	2 **He** Helium 4.0026
2	3 **Li** Lithium 6.941	4 **Be** Beryllium 9.0122											5 **B** Boron 10.811	6 **C** Carbon 12.011	7 **N** Nitrogen 14.0067	8 **O** Oxygen 15.9994	9 **F** Fluorine 18.9984	10 **Ne** Neon 20.1797
3	11 **Na** Sodium 22.9898	12 **Mg** Magnesium 24.3050											13 **Al** Aluminum 26.9815	14 **Si** Silicon 28.0855	15 **P** Phosphorus 30.9738	16 **S** Sulfur 32.066	17 **Cl** Chlorine 35.4527	18 **Ar** Argon 39.948
4	19 **K** Potassium 39.0983	20 **Ca** Calcium 40.078	21 **Sc** Scandium 44.9559	22 **Ti** Titanium 47.88	23 **V** Vanadium 50.9415	24 **Cr** Chromium 51.9961	25 **Mn** Manganese 54.9380	26 **Fe** Iron 55.847	27 **Co** Cobalt 58.9332	28 **Ni** Nickel 58.693	29 **Cu** Copper 63.546	30 **Zn** Zinc 65.39	31 **Ga** Gallium 69.723	32 **Ge** Germanium 72.61	33 **As** Arsenic 74.9216	34 **Se** Selenium 78.96	35 **Br** Bromine 79.904	36 **Kr** Kryrpton 83.80
5	37 **Rb** Rubidium 85.4678	38 **Sr** Strontium 87.62	39 **Y** Yttrium 88.9059	40 **Zr** Zirconium 91.224	41 **Nb** Niobium 92.9064	42 **Mo** Molybdenum 95.94	43 **Tc** Technetium (98)	44 **Ru** Ruthenium 101.07	45 **Rh** Rhodium 102.9055	46 **Pd** Palladium 106.42	47 **Ag** Silver 107.8682	48 **Cd** Cadmium 112.411	49 **In** Indium 114.82	50 **Sn** Tin 118.710	51 **Sb** Antimony 121.757	52 **Te** Tellurium 127.60	53 **I** Iodine 126.9045	54 **Xe** Xenon 131.29
6	55 **Cs** Cesium 132.9054	56 **Ba** Barium 137.327	57 **La** Lanthanum 138.9055	72 **Hf** Hafnium 178.49	73 **Ta** Tantalum 180.9479	74 **W** Tungsten 183.85	75 **Re** Rhenium 186.207	76 **Os** Osmium 190.2	77 **Ir** Iridium 192.22	78 **Pt** Platinum 195.08	79 **Au** Gold 196.9665	80 **Hg** Mercury 200.59	81 **Tl** Thallium 204.3833	82 **Pb** Lead 207.2	83 **Bi** Bismuth 208.9804	84 **Po** Polonium (209)	85 **At** Astatine (210)	86 **Rn** Radon (222)
7	87 **Fr** Francium (223)	88 **Ra** Radium 227.0278	89 **Ac** Actinium (227)	104 **Rf** Rutherfordium (261)	105 **Db** Dubnium (262)	106 **Sg** Seaborgium (263)	107 **Bh** Bohrium (262)	108 **Hs** Hassium (265)	109 **Mt** Meitnerium (266)	110 **Ds** Darmstadtium (269)	111 **Rg** Roentgenium (272)							

Lanthanides (6)

58 **Ce** Cerium 140.115	59 **Pr** Praseodymium 140.9076	60 **Nd** Neodymium 144.24	61 **Pm** Promethium (145)	62 **Sm** Samarium 150.36	63 **Eu** Europium 151.965	64 **Gd** Gadolinium 157.25	65 **Tb** Terbium 158.9253	66 **Dy** Dysprosium 162.50	67 **Ho** Holmium 164.9303	68 **Er** Erbium 167.26	69 **Tm** Thulium 168.9342	70 **Yb** Ytterbium 173.04	71 **Lu** Lutetium 174.967

Actinides (7)

90 **Th** Thorium 232.0381	91 **Pa** Protactinium 231.0359	92 **U** Uranium 238.00289	93 **Np** Neptunium (237)	94 **Pu** Plutonium (244)	95 **Am** Americium (243)	96 **Cm** Curium (247)	97 **Bk** Berkelium (247)	98 **Cf** Californium (251)	99 **Es** Einsteinium (252)	100 **Fm** Fermium (257)	101 **Md** Mendelevium (258)	102 **No** Nobelium (259)	103 **Lr** Lawrencium (260)